PLEASE STAMP DATE DUE, BOTH BELOW AND ON CARD

DATE DUE	DATE DUE	DATE DUE	DATE DUE
9-27-10			

GL-15

The Origin and Evolution of the Caribbean Plate

Geological Society books refereeing procedures

The Society makes every effort to ensure that the scientific and production quality of its books matches that of its journals. Since 1997, all book proposals have been refereed by specialist reviewers as well as by the Society's Books Editorial Committee. If the referees identify weaknesses in the proposal, these must be addressed before the proposal is accepted.

Once the book is accepted, the Society Book Editors ensure that the volume editors follow strict guidelines on refereeing and quality control. We insist that individual papers can only be accepted after satisfactory review by two independent referees. The questions on the review forms are similar to those for *Journal of the Geological Society*. The referees' forms and comments must be available to the Society's Book Editors on request.

Although many of the books result from meetings, the editors are expected to commission papers that were not presented at the meeting to ensure that the book provides a balanced coverage of the subject. Being accepted for presentation at the meeting does not guarantee inclusion in the book.

More information about submitting a proposal and producing a book for the Society can be found on its web site: www.geolsoc.org.uk.

It is recommended that reference to all or part of this book should be made in one of the following ways:

JAMES, K. H., LORENTE, M. A. & PINDELL, J. L. (eds) 2009. *The Origin and Evolution of the Caribbean Plate*. Geological Society, London, Special Publications, **328**.

STANEK, K. P., MARESCH, W. V. & PINDELL, J. L. 2009. The geotectonic story of the northwestern branch of the Caribbean Arc: implications from structural and geochronological data of Cuba. *In*: JAMES, K. H., LORENTE, M. A. & PINDELL, J. L. (eds) *The Origin and Evolution of the Caribbean Plate*. Geological Society, London, Special Publications, **328**, 361–398.

GEOLOGICAL SOCIETY SPECIAL PUBLICATION NO. 328

The Origin and Evolution of the Caribbean Plate

EDITED BY

K. H. JAMES
Aberystwyth University, Wales, UK

M. A. LORENTE
Central University of Venezuela, Venezuela

and

J. L. PINDELL
Tectonic Analysis Ltd, West Sussex, UK
Rice University, Texas, USA

2009
Published by
The Geological Society
London

THE GEOLOGICAL SOCIETY

The Geological Society of London (GSL) was founded in 1807. It is the oldest national geological society in the world and the largest in Europe. It was incorporated under Royal Charter in 1825 and is Registered Charity 210161.

The Society is the UK national learned and professional society for geology with a worldwide Fellowship (FGS) of over 9000. The Society has the power to confer Chartered status on suitably qualified Fellows, and about 2000 of the Fellowship carry the title (CGeol). Chartered Geologists may also obtain the equivalent European title, European Geologist (EurGeol). One fifth of the Society's fellowship resides outside the UK. To find out more about the Society, log on to www.geolsoc.org.uk.

The Geological Society Publishing House (Bath, UK) produces the Society's international journals and books, and acts as European distributor for selected publications of the American Association of Petroleum Geologists (AAPG), the Indonesian Petroleum Association (IPA), the Geological Society of America (GSA), the Society for Sedimentary Geology (SEPM) and the Geologists' Association (GA). Joint marketing agreements ensure that GSL Fellows may purchase these societies' publications at a discount. The Society's online bookshop (accessible from www.geolsoc.org.uk) offers secure book purchasing with your credit or debit card.

To find out about joining the Society and benefiting from substantial discounts on publications of GSL and other societies worldwide, consult www.geolsoc.org.uk, or contact the Fellowship Department at: The Geological Society, Burlington House, Piccadilly, London W1J 0BG: Tel. +44 (0)20 7434 9944; Fax +44 (0)20 7439 8975; E-mail: enquiries@geolsoc.org.uk.

For information about the Society's meetings, consult *Events* on www.geolsoc.org.uk. To find out more about the Society's Corporate Affiliates Scheme, write to enquiries@geolsoc.org.uk.

Published by The Geological Society from:
The Geological Society Publishing House, Unit 7, Brassmill Enterprise Centre, Brassmill Lane, Bath BA1 3JN, UK

(*Orders*:   Tel. +44 (0)1225 445046, Fax +44 (0)1225 442836)
Online bookshop: www.geolsoc.org.uk/bookshop

British Library Cataloguing in Publication Data

A catalogue record for this book is available from the British Library.
ISBN 978-1-86239-288-5

Typeset by Techset Composition Ltd., Salisbury, UK
Printed by Antony Rowe Ltd, Chippenham, UK

Distributors

North America
For trade and institutional orders:
The Geological Society, c/o AIDC, 82 Winter Sport Lane, Williston, VT 05495, USA
Orders: Tel. +1 800-972-9892
           Fax +1 802-864-7626
           E-mail: gsl.orders@aidcvt.com

For individual and corporate orders:
AAPG Bookstore, PO Box 979, Tulsa, OK 74101-0979, USA
Orders: Tel. +1 918-584-2555
           Fax +1 918-560-2652
           E-mail: bookstore@aapg.org
           Website: http://bookstore.aapg.org

India
Affiliated East-West Press Private Ltd, Marketing Division, G-1/16 Ansari Road, Darya Ganj, New Delhi 110 002, India
Orders: Tel. +91 11 2327-9113/2326-4180
           Fax +91 11 2326-0538
           E-mail: affiliat@vsnl.com

Contents

vi CONTENTS

Preface

This book records the proceedings of a conference, held in Spain, 2006, that addressed aspects of the geology of the area between North and South America, with focus upon the origin of the Caribbean Plate. Additional papers provide new data. The book follows the structure of the conference, beginning with papers that describe different understandings of the plate origins, continuing with a geological tour of the Caribbean and ending in the plate interior.

The papers express a wide variety of understanding of Caribbean geology. They cannot all be correct. We, and the Geological Society, hope that a book that brings them together will facilitate debate and help readers to make their own analyses. The first four papers in particular describe considerably different visions of Caribbean Plate origins and make-up.

Caribbean geology is complicated by wide dispersal over many geographic elements, some with poor accessibility, tropical weathering and/or young volcanic cover and some just poorly studied. Much of the plate interior lies below deep water and is directly sampled only by a few cores. Literature is spread over many sources, some international but many local, in English, French, German and Spanish. Given its location between the Gulf Mexico and northern South America, both well-explored places where surprises continue to surface, it is obvious that a lot remains to be learned about the Caribbean.

Prevailing understanding is that the Caribbean Plate formed in the Pacific Ocean and migrated between the Americas. It is built, therefore, of oceanic and intra-oceanic volcanic arc rocks. Alternatively, it could have formed in place by extension between separating North and South America. It could include continental crust. Some models are hybrids of these possibilities.

Pindell & Kennan provide an update on Caribbean evolution in which the Americas engulf a swath of Pacific-origin oceanic lithosphere during their westward drift from Africa, from Early Cretaceous to Present, after a period of Jurassic passive margin formation around the Gulf of Mexico and Proto-Caribbean Sea. References, frames, origins of arcs, arc-continent interactions, mantle plume exrusion, back arc spreading events and neotectonic settings are integrated to form an intrinsically consistent evolutionary model. **Giunta & Oliveri** also suggest oceanic spreading and formation of an oceanic plateau but between the Americas. **James** synthesizes data that indicate an *in situ* origin of the plate by extreme continental extension and serpentinization. **James** describes a plate evolution interpretation of these in a separate paper, along with implications and outstanding problems. Existing data/samples could provide many answers if revisited by studies designed to distinguish between different ideas.

Milsom compares the Caribbean with basin/orocline pairs in the Mediterranean area and the Banda Sea. **Giner-Robles** et al. describe the structure of the Atlantic Plate as it descends below the Scotia Plate, using earthquake focal mechanism solutions. This offers an analogue for the Lesser Antilles subduction zone of the eastern Caribbean.

Keppie *et al.* discuss Oligocene–Miocene events in southwestern Mexico. Instead of being influenced by movements of the Chortís Block, the geology results from forearc subduction resulting from collision of an oceanic plateau with the trench and change of plate relative motion associated with the breakup of the Farallon Plate. **Cerca** *et al.* describe analogue models that reproduce the tectonic evolution of SW Mexico and the Xolapa complex related to oblique sinistral transpressional movement of the Chortís Block. **Guzmán-Speziale** uses earthquake data to suggest that the Chortís Block is extruding to the SE in response to interaction with the North American and Cocos Plates and **Valls Alvarez** details geological history recorded in Guatemala in similar terms. **Ratschbacher** *et al.* provide new data relevant to plate history in Guatemala and Honduras, while **Solari** *et al.* focus upon zircon data to unravel the Palaeozoic history of the area.

Cobiella-Reguera details the record of volcanic arc activity and Late Cretaceous–Middle Eocene emplacement of ophiolites in Cuba. Volcanic activity ceased from the Late Campanian to the Danian when a new arc with north-dipping subduction formed. This is not consistent with the concept of a coeval and continuous north-facing Great Arc of the Caribbean. In western and central Cuba, the Northern Ophiolite Belt and Cretaceous volcanic arc rocks were thrust several tens of kilometres to the north by the Middle Eocene. In contrast, **Stanek** *et al.* suggest that volcanic activity moved southwards to the Cayman Ridge with south-dipping subduction. This, plus history of HP metamorphic rocks on Cuba, derived from the southern Yucatán, supports the development of the Caribbean Great Arc in the Pacific.

Hastie *et al.* describe the geochemistry and significance of Cretaceous island arc tholeiites and calc-alkaline rocks on Jamaica. Both occur before and after the Aptian/Albian, thus commonly seen volcanic arc subduction polarity reversal at this time is not supported. In a separate paper, **Hastie** shows that Caribbean primitive island arc (PIA) rocks are chemically indistinguishable from modern island arc tholeiites (IAT) and recommends abandonment of PIA usage.

Buchs et al. report on detailed field studies of the Osa and Burica Peninsulas (Costa Rica) that recorded a long-term history of accretion alternating with subduction erosion. An igneous complex includes a Coniacian–Santonian oceanic plateau and Coniacian–Santonian to Middle Eocene seamounts that accreted along the Caribbean Plate between the Paleocene and the Late Eocene. A mélange formed along the Igneous Complex in the Late Eocene by accretion of mass wasting deposits reworked from the igneous complex and the Central American Arc. The study suggests that some sequences of Late Cretaceous oceanic plateau(s) in Central America are unrelated to the Caribbean Large Igneous Province (CLIP) and formed in the Pacific before being accreted along the Caribbean Plate.

Montgomery & Kerr reinterpret red cherts on La Désirade, Lesser Antilles, relating them to hydrothermal sedimentation at an Upper Jurassic, Pacific spreading ridge.

Jaillard et al. relate upper Cretaceous ocean plateau rocks of western Ecuador to the Caribbean Oceanic Plateau by their similar evolutions. If correct, the latter formed on the Farallon Plate and not above the Galapagos hot spot. Similarly, **Kennan & Pindell** relate the history of western rocks from northern Peru to Colombia to migration of the Caribbean Plate and arcs at its leading and trailing margins.

Grande & Urbani report the first findings of Grenvillian basement or derived rocks in northwestern Venezuela. Their data will help to complete ancient palaeogeography of Pangaea. **Weber et al.** describe new data on serpentinites, gabbros and andesites from the Guajira Peninsula, relating these to a surfacing supra-subduction zone. This paper also supports migration of the Caribbean Plate.

Audemard details the progressive collision of a migrating Caribbean volcanic arc along northern Venezuela and proposes that the Falcón, Bonaire, Blanquilla and Grenada Basins of northern Venezuela and the eastern Caribbean began life as a single, back-arc basin. **Baquero et al.** describe Late Cretaceous–Middle Eocene nappe emplacement in the Falcón area of Venezuela, followed by Late Eocene–Early Miocene graben formation and then Middle Miocene inversion. The classic Upper Cretaceous petroleum source overmatured during graben formation and oils of this region are sourced by Cenozoic rocks.

Higgs proposes a radical change of timing of Caribbean evolution, with rifting ending in the Coniacian instead of the Jurassic, and with Santonian–Campanian spreading lasting only 10–15 Ma instead of 140 Ma, followed by subduction of the Caribbean below northern South America. Entry of the Caribbean arc occurred in the Oligocene, not in the Palaeocene, and Caribbean relative motion became eastward in the Pliocene, not in the Middle Miocene. **Higgs** also presents indirect evidence of a former Berriasian–Valanginian Carib Halite Formation, formed in a graben extending from Colombia to Trinidad and seen today only in Colombia. Indications include solution subsidence, saline springs, heat-flow and gravity anomalies, and thrust belt structural style.

Cooney & Lorente present data that show an unrecognized Campanian structural event and resultant unconformity in the Maracaibo Basin that could be related to changes in Caribbean Plate tectonics.

Maresch et al. discuss HP/LT metamorphic rocks of Margarita, overlain by greenschists. HP/LT protoliths were both oceanic and continental (Palaeozoic and Mesozoic) and metamorphism is seen to have occurred to the west, in accord with Pacific models for the Caribbean.

Pindell et al. discuss heavy mineral content of rocks from eastern Venezuela, Trinidad and Barbados, and outline a history of Cretaceous passive margin, Palaeogene subduction below South America, Eocene–Miocene foredeep development related to passage of the Caribbean Plate and Miocene–Recent sedimentation of the Orinoco River.

Diebold presents seismic data over the Caribbean Plateau of the western Venezuela Basin. Two separate sequences, possibly coeval lower intrusions and upper extrusions, were folded during emplacement of the upper sequence, prior to uplift of the Beata Ridge. This could have occurred as the Caribbean Plate was entering the Caribbean area in the Late Cretaceous. **Kerr et al.** discuss chemistry of basalts drilled at DSDP Site 1001 and conclude that they are most likely derived from depleted mantle plume, possibly mixed with depleted upper mantle source.

Drs A. W. Bally, M. A. Lorente and D. Roberts chaired the 2006 meeting. The conference convenors, Keith James and Maria Antonieta Lorente, gratefully acknowledge financial support for the meeting from BP, Repsol, SEPM, Shell and Statoil and the support of the Spanish Asociación de Geólogos y Geofísicos Españoles del Petróleo, especially President Wenceslao Martínez, Vice-President Susanna Torrescusa Villaverde and Secretary Aurelio Jiménez Fernández. We also wish to thank

Cristina Rzepka de Lombas, Lorenzo Villalobos, Natalia Villalobos Vencelá and Amparo Donderis for their enthusiastic energy in the organization and running of the conference. The Town Hall of Sigüenza and the Sigüenza Parador Hotel generously provided transport, tours and meeting facilities. Lorenzo Villalobos and Ramón Mas led a field trip to excellent local outcrops and places of historic interest.

K. H. JAMES
M. A. LORENTE
J. L. PINDELL

Tectonic evolution of the Gulf of Mexico, Caribbean and northern South America in the mantle reference frame: an update

JAMES L. PINDELL[1,2]* & LORCAN KENNAN[1]

[1]*Tectonic Analysis Ltd, Chestnut House, Duncton, West Sussex GU28 0LH, UK*

[2]*Department of Earth Science, Rice University, Houston, TX 77002, USA*

Corresponding author (e-mail: jim@tectonicanalysis.com)

Abstract: We present an updated synthesis of the widely accepted 'single-arc Pacific-origin' and 'Yucatán-rotation' models for Caribbean and Gulf of Mexico evolution, respectively. Fourteen palaeogeographic maps through time integrate new concepts and alterations to earlier models. Pre-Aptian maps are presented in a North American reference frame. Aptian and younger maps are presented in an Indo-Atlantic hot spot reference frame which demonstrates the surprising simplicity of Caribbean–American interaction. We use the Müller *et al.* (*Geology* 21: 275–278, 1993) reference frame because the motions of the Americas are smoothest in this reference frame, and because it does not differ significantly, at least since *c.* 90 Ma, from more recent 'moving hot spot' reference frames. The Caribbean oceanic lithosphere has moved little relative to the hot spots in the Cenozoic, but moved north at *c.* 50 km/Ma during the Cretaceous, while the American plates have drifted west much further and faster and thus are responsible for most Caribbean–American relative motion history. New or revised features of this model, generally driven by new data sets, include: (1) refined reconstruction of western Pangaea; (2) refined rotational motions of the Yucatán Block during the evolution of the Gulf of Mexico; (3) an origin for the Caribbean Arc that invokes Aptian conversion to a SW-dipping subduction zone of a trans-American plate boundary from Chortís to Ecuador that was part sinistral transform (northern Caribbean) and part pre-existing arc (eastern, southern Caribbean); (4) acknowledgement that the Caribbean basalt plateau may pertain to the palaeo-Galapagos hot spot, the occurrence of which was partly controlled by a Proto-Caribbean slab gap beneath the Caribbean Plate; (5) Campanian initiation of subduction at the Panama–Costa Rica Arc, although a sinistral transform boundary probably pre-dated subduction initiation here; (6) inception of a north-vergent crustal inversion zone along northern South America to account for Cenozoic convergence between the Americas ahead of the Caribbean Plate; (7) a fan-like, asymmetric rift opening model for the Grenada Basin, where the Margarita and Tobago footwall crustal slivers were exhumed from beneath the southeast Aves Ridge hanging wall; (8) an origin for the Early Cretaceous HP/LT metamorphism in the El Tambor units along the Motagua Fault Zone that relates to subduction of Farallon crust along western Mexico (and then translated along the trans-American plate boundary prior to onset of SW-dipping subduction beneath the Caribbean Arc) rather than to collision of Chortis with Southern Mexico; (9) Middle Miocene tectonic escape of Panamanian crustal slivers, followed by Late Miocene and Recent eastward movement of the 'Panama Block' that is faster than that of the Caribbean Plate, allowed by the inception of east–west trans-Costa Rica shear zones. The updated model integrates new concepts and global plate motion models in an internally consistent way, and can be used to test and guide more local research across the Gulf of Mexico, the Caribbean and northern South America. Using examples from the regional evolution, the processes of slab break off and flat slab subduction are assessed in relation to plate interactions in the hot spot reference frame.

The realization that the Bullard *et al.* (1965) reconstruction of the Equatorial Atlantic margins was dramatically in error due to the inclusion of post-rift sediment build-up along the Amazon margin (Pindell & Dewey 1982; Pindell 1985*a*; Klitgord & Schouten 1986) led to major advances in the understanding of the evolution of the Gulf of Mexico and Caribbean regions. By backstripping the margin and tightening the crustal fit between northern Brazil and western Africa, Pindell & Dewey (1982) and Pindell (1985*a*) showed that the gap between Texas and Venezuela upon Atlantic closure was far smaller than that shown by Bullard *et al.* and that a satisfactory Alleghanian reconstruction could only be achieved with Yucatán inserted into the Gulf, in an orientation that was rotated some 45–60° clockwise relative to its present orientation. In addition, this adjustment to the Atlantic closure greatly simplified the Cretaceous relative motion history between the Americas over earlier kinematic models (e.g. Ladd 1976; Sclater *et al.* 1977), leading to the conclusion that the Americas have moved little with respect to each other since the Campanian while the relative

From: JAMES, K. H., LORENTE, M. A. & PINDELL, J. L. (eds) *The Origin and Evolution of the Caribbean Plate.*
Geological Society, London, Special Publications, **328**, 1–55.
DOI: 10.1144/SP328.1   0305-8719/09/$15.00 © The Geological Society of London 2009.

eastward migration of the Pacific-derived Caribbean Plate has been the dominant story (Pindell 1985b; Pindell et al. 1988; Burke 1988). It was also evident that this relative migration history was due mainly to the westward drift of the Americas past a Caribbean Plate that was nearly stationary in the hot spot reference frame (Pindell & Dewey, 1982; Duncan & Hargraves 1984; Pindell et al. 1988; Pindell 1993). Since these realizations, most recently corroborated by Müller et al. (1999), both the rotation of Yucatán during the opening of the Gulf of Mexico and the Pacific origin of the Caribbean oceanic lithosphere have gained increasing favour as the concepts and implications have been digested and tested by expanding data sets (Stephan et al. 1990; Schouten & Klitgord 1994; Stöckhert et al. 1995; Diebold et al. 1999; Driscoll & Diebold 1999; Kerr et al. 1999, 2003; Mann 1999; Dickinson & Lawton 2001; Miranda et al. 2003; Jacques et al. 2004; Bird et al. 2005; Imbert 2005; Imbert & Philippe 2005; Pindell et al. 2005).

In this paper, we update the 'Yucatán-rotation' model for the Gulf of Mexico (Pindell & Dewey 1982; Fig. 1) and the 'single-arc Pacific-origin' model for the Caribbean region (Pindell 1985b; Pindell et al. 1988; Fig. 2) by integrating into the original models a number of concepts and the implications of key data sets developed in recent years. We believe the collected arguments for a Pacific origin of the Caribbean oceanic lithosphere are overwhelmingly clear (Pindell 1990, 1993; Pindell et al. 2005, 2006, 2009) so we will not repeat them here. However, we will take the opportunity to highlight key pro-Pacific factors when expedient, as well as to point out why various objections to the Pacific model put forth in recent years are invalid.

Plate reconstructions and reference frames

Our circum-Atlantic assembly uses the Central Atlantic reconstruction of Le Pichon & Fox (1971) which, despite being an early paper, best superposes the East Coast and West African magnetic anomalies, and the Equatorial Atlantic reconstruction of Pindell et al. (2006). For spreading history, we use the marine magnetic anomaly reconstructions of Müller et al. (1999), Pindell et al. (1988) and Roest et al. (1992) for various anomaly pairs in the Equatorial and Central Atlantic, the integration of which was checked for internal consistency. Our palaeogeographic maps are drawn in the North American reference frame prior to the Aptian, and in the Indo-Atlantic hot spot reference frame of Müller et al. (1993) for times since the Aptian, when such a reference frame is more likely to be meaningful. Torsvik et al. (2008) has compared different hot spot reference frames,

including fixed Indo-Atlantic (or African) hot spots, moving Indo-Atlantic hot spots and moving global hot spots, and has found that all are similar within error back to 84 Ma, and agree well with palaeomagnetic data. Thus, the choice of a particular Indo-Atlantic reference frame for Late Cretaceous–Recent reconstructions is not critical. Prior to 84 Ma, the positions of major continents calculated from hot spot tracks drift south and rotate with respect to their positions calculated from palaeomagnetic data, perhaps indicating significant hot spot motion or true polar wander.

Both the relative and the absolute positions of the major continents on our maps since anomaly 34 (84 Ma) are quite reliable. Our 100 Ma reconstruction (interpolation) within the Cretaceous magnetic quiet period (124.61–84 Ma) is subject to greater uncertainty (but still less than c. 100 km) because there are no magnetic anomaly determinations for this period, although satellite depictions of fracture zones do define the flow lines, if not the rates of motion, between Africa and the Americas for that interval. The M0 (124.61 Ma, Early Aptian, in the recent Gradstein et al. 2004 timescale) and older Mesozoic anomalies are reliably identified and we have a high degree of confidence in the Aptian and older Equatorial Atlantic closure fit; thus, the 125 Ma and older reconstructions reliably show the relative positions of the major continents. Their absolute positions are less certain because of the Albian and older differences between various hot spot and palaeomagnetic reference frames. Early Cretaceous palaeo-longitudes of the continents are consistent to less than 5° between different models, but there is significant latitudinal variation and some rotation. The Müller et al. (1993) fixed Indo-Atlantic hot spot model used here places the Americas approximately 10–15° to the south of moving Indo-Atlantic hot spot or palaeomagnetic–hot spot hybrid models (Torsvik et al. 2008). However, regardless of choice of reference frame, or even if alternative models for the origin of hot spot tracks are chosen (e.g. the propagating crack and mantle counterflow model of Anderson 2007), the maps serve well to illustrate the westward flight of the Americas from a slowly drifting and rotating Africa at the core of the former Pangaea. The relatively slow motion of Africa reflects its being surrounded by oceanic spreading ridges rather than convergent plate boundaries. We find that the Caribbean oceanic lithosphere has moved little to the east or west in the hot spot reference frame (Pindell 1993; Pindell & Tabbutt 1995) and evolutionary maps drawn in this reference frame convey the surprising simplicity of the Pacific origin model for the Caribbean lithosphere.

Cretaceous motions of plates in the Pacific with respect to the Americas are harder to constrain than

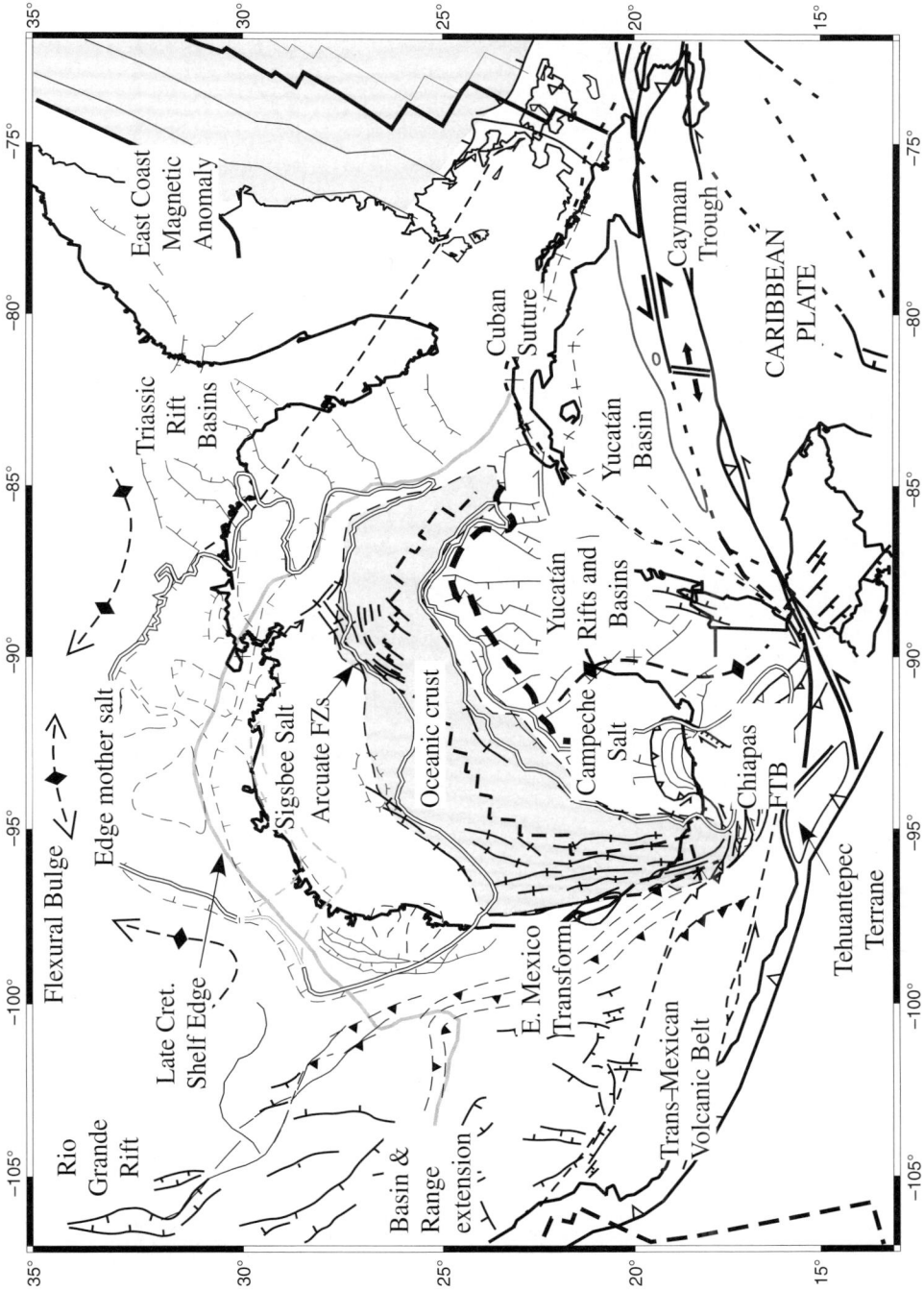

Fig. 1. Present day tectonic map of the Gulf of Mexico region.

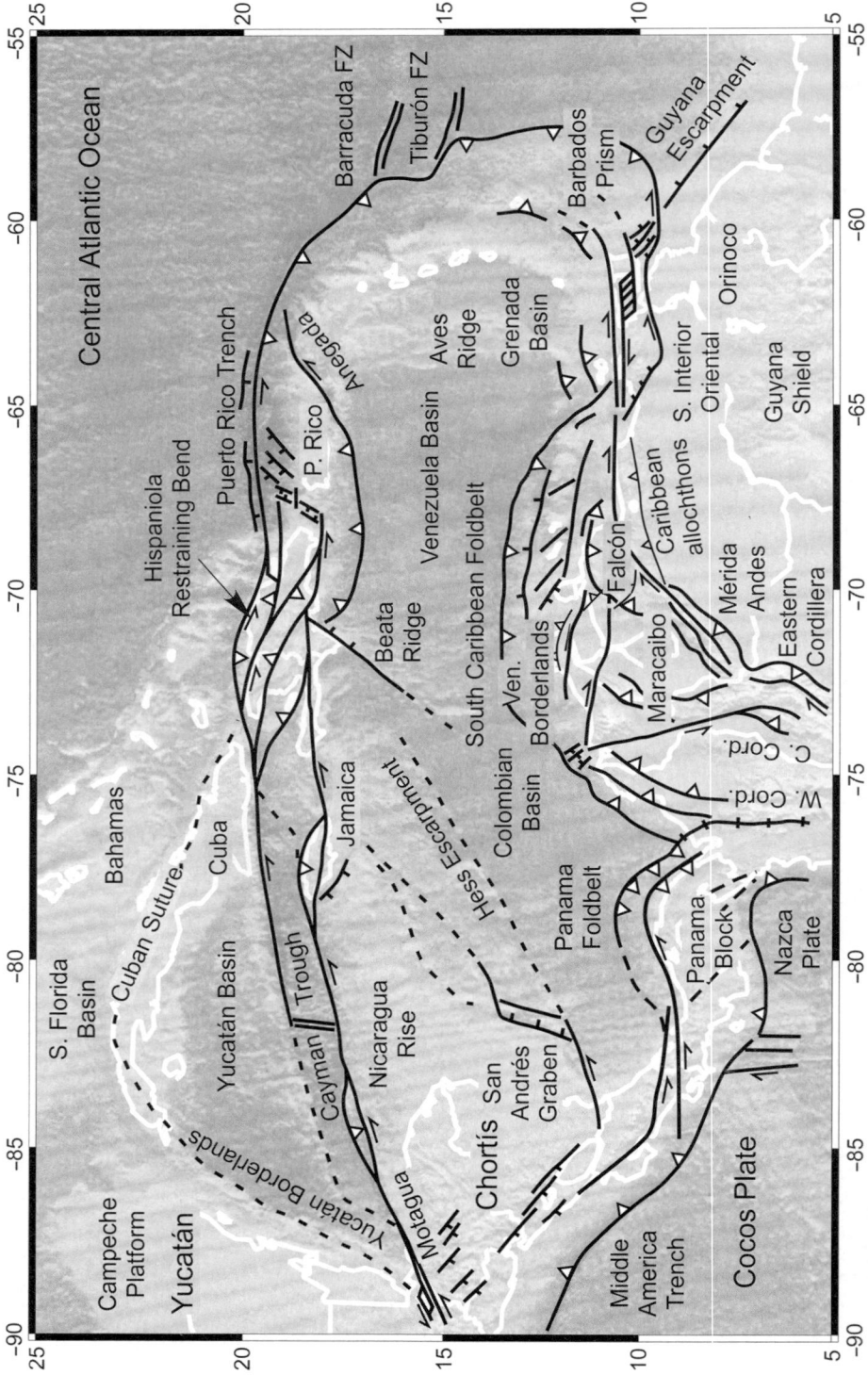

Fig. 2. Present day tectonic map of the Caribbean region.

circum-Atlantic motions. Models that assume no relative motion between Pacific and Indo-Atlantic hot spots (such as Engebretson *et al.* 1985) fit progressively worse with both hot spot track and palaeomagnetic data back into the Late Cretaceous (e.g. Tarduno & Gee 1995) and it is clear that Pacific hot spots were moving NW with respect to Indo-Atlantic hot spots at *c.* 30–50 km/Ma (see Steinberger 2000; Steinberger *et al.* 2004; Torsvik *et al.* 2008 for discussion of moving hot spot models). However, quantifying such relative motion prior to 84 Ma remains elusive, and here we employ a hybrid model that allows for only a moderate amount of westward drift of the Pacific hot spots relative to the African hot spots, preferring to base the approximate palaeopositions of the Caribbean Plate relative to the Americas mostly on geological criteria from the circum-Caribbean and the American Cordilleran regions. Geometric constraints (e.g. avoiding 'eduction', or pulling subducted slabs back out of their subduction channel) allow for slow (perhaps 0.5°/Ma) counterclockwise rotation and northwestward drift of the Pacific hot spot reference frame (as seen from the Caribbean region) relative to the Indo-Atlantic hot spots.

We begin our discussion with the Early Jurassic reconstruction of western Pangaea and the opening of the Central Atlantic and Gulf of Mexico, and of the early development of Mexico, the northern Andes, and the Proto-Caribbean passive margins. We then progress to the evolution of the Caribbean lithosphere and its interactions with the Americas, working forward in time, ending with an assessment of the 'Neo-Caribbean Phase' of deformation over the last 10 Ma.

Western Pangaea, the Gulf of Mexico, and the Early Proto-Caribbean Seaway

The circum-Atlantic closure reconstruction (Fig. 3) shows the fault zones and plate boundaries responsible for Early and Middle Jurassic (190–158 Ma) dispersion of the continental blocks of the time. Seafloor spreading proceeded in the Central Atlantic for this interval, following Appalachian and Central Atlantic margin rifting, but more diffuse continental rifting continued in the margins of the Gulf of Mexico and Proto-Caribbean regions until probably the Early Oxfordian (158 Ma). This syn-rift phase in the Gulf of Mexico margins appears to have been of a low-angle, asymmetric nature, with Yucatán detaching from the US and northeast Mexican Gulf margins in a relative southeastward direction with probable minor counter-clockwise rotation (Pindell & Kennan 2007a). The Tamaulipas Arch, Balcones trend and the southern flanks of the Sabine and Wiggins 'arches' are probable asymmetric rift

footwalls that were tectonically unroofed by extension along a low-angle detachment. Thus, Eagle Mills red beds often appear to be in depositional contact with, rather than faulted against, basement on their northern and western depositional limits. The Chiapas Massif also appears to us as a low-angle footwall detachment where the bulk of Yucatán detached to the east (present-day coordinates) to form the salt-bearing Chiapas Foldbelt Basin in the Middle Jurassic; our reconstruction positions the Massif as a southerly projection of the Tamaulipas Arch prior to rotational seafloor spreading in the Gulf, such that the two granitic trends have a common rift history (footwalls) in addition to similar lithologies and geochronologies. The Yucatán Block has been reduced by about 20% north–south (Fig. 3; Pindell & Dewey 1982), or roughly NW–SE in today's coordinates, accounting for probable rift structures interpreted from gravity maps (Fig. 1).

We maintain that stretched continental crust underlies the Great Bank of the Bahamas (where Jurassic salt is present) and the South Florida Basin, but not the southeastern Bahamas (east of Acklin Island), which is probably underlain by a hot spot track. This continental crust must be restored to normal thickness as well as retracted back into the eastern Gulf to avoid overlap with the Demerara–Guinea Plateau of Gondwana (Pindell 1985a; Pindell & Kennan 2001). In Mexico, sinistral transform motions of blocks whose geometries remain debated persisted into the Late Jurassic, the effect of which was to postpone significant divergence between southern Mexico and Colombia until long after the Atlantic had begun to open. Subduction at the Cordilleran margin was probably strongly left-lateral, which helped to drive the continental crust of southern and western Mexico into the position formerly occupied by Colombia: that is these blocks were sinistrally sheared along the southwestern flank of the North American Plate as the latter took flight from Gondwana. We show the western limit of North America's continental crust along the Arcelia–Guanajuato trend, because continental terrane is either absent or poorly presented in the arc terranes to the west, despite the ubiquitous presence of Precambrian and Palaeozoic zircons in those terranes (Talavera-Mendoza *et al.* 2007). We believe that these zircon populations argue against a distal intra-Pacific origin for the Guerrero arc or arcs (as was proposed by Dickinson & Lawton 2001), and prefer to migrate the terranes southward along the Mexican Cordillera outboard of relatively narrow intra-arc basins capable of receiving old zircons from cratonic areas to the east and north. Based on the geometrical requirements of Pangaea assembly, we place the pre-Jurassic Central Mexican, Southern Mexican, Chortís, Tahami–Antioquia

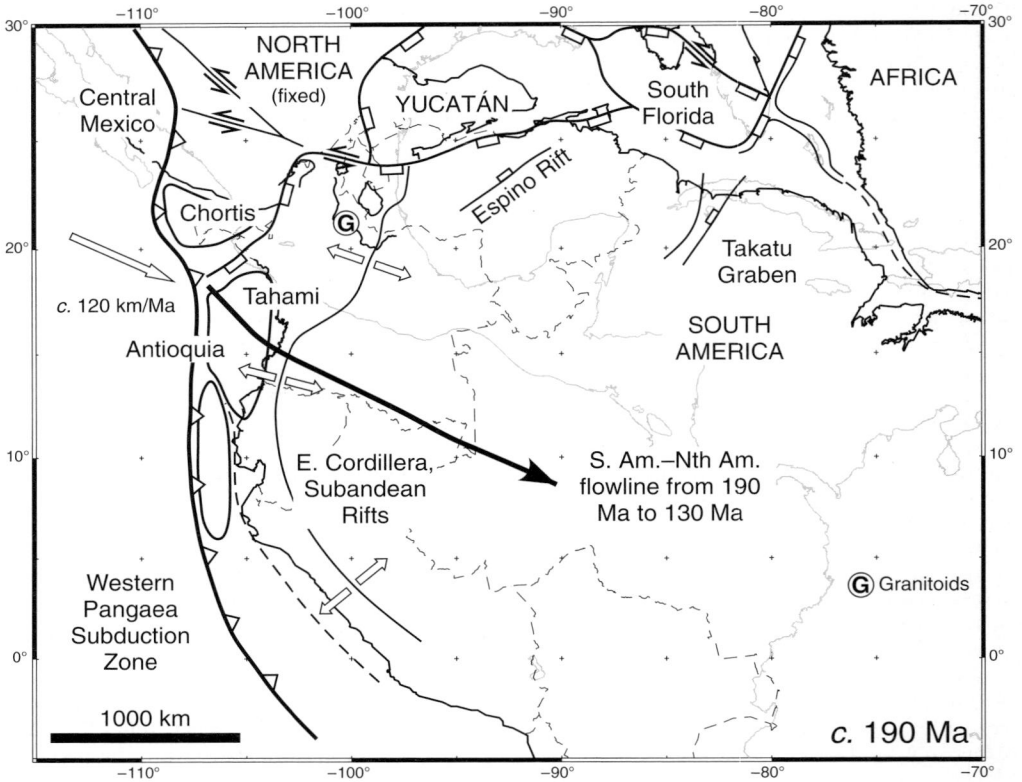

Fig. 3. 190 Ma reconstruction of the circum-Gulf of Mexico region, employing the Central Atlantic closure fit of Le Pichon & Fox (1971) and the Equatorial Atlantic fit of Pindell *et al.* (2006), in fixed North America reference frame (also Figs 4–7). Plate motions modified from Engebretson *et al.* (1985), Pindell *et al.* (1988) and Roest *et al.* (1992). Positions of circum-Atlantic continents are well-defined, but motions of Pacific plates relative to the Americas require the assumption of fixity between Pacific and Indo-Atlantic hot spots. Position of Yucatán in this syn-rift stage is constrained by closure geometry and subsequent Late Jurassic rotational ocean crustal fabric in the central and eastern Gulf of Mexico. Position of southern Mexico is constrained by the need to avoid overlap with the northwestern Andes. One or more transtensional NW–SE trending fault systems were active in Mexico during Early–Middle Jurassic, allowing Mexican terranes to move SE relative to the rest of North America. We show an 'Antioquia–Tahami' terrane as the conjugate margin to Chortís, and crudely restore the effects of subsequent northward translation and cross-strike shortening. The position shown is consistent with restoring estimated dextral strike–slip and shortening in the Colombian Andes and suggests that the Medellín dunites may have analogues in the Baja California forearc. The continental blocks that are found within the Arquia and possibly Chaucha Terranes are inferred to originate SW of Antioquia, opposite present-day Ecuador.

and Chaucha–Arquia terranes outboard of the more stable cratonic areas of northeast Mexico and the Guayana Shield.

By Late Callovian time (158 Ma, Fig. 4), the majority of intra-continental extension in the Gulf region and Cordilleran terrane migration in Mexico had occurred, and was followed by initial seafloor spreading in the Gulf. This is the first reconstruction in which there is space between the Americas to accommodate the area of highly stretched continental crust, US and Mexican salt basins, and possible zones of serpentinized mantle

flanking today's central Gulf oceanic crust (Fig. 1). It is difficult to determine the time of initial salt deposition, but this reconstruction is near to its end (Pindell & Kennan 2007a). When seafloor spreading began, the pole of rotation was situated nearby in the deep southeastern Gulf, and thus fracture zone trends in the Gulf of Mexico are highly curvilinear (Figs 1 & 5; Imbert 2005; Imbert & Philippe 2005), recording the strong counterclockwise rotation of the Yucatán Block during the seafloor spreading stage first predicted by Pindell & Dewey (1982). The trends of Triassic

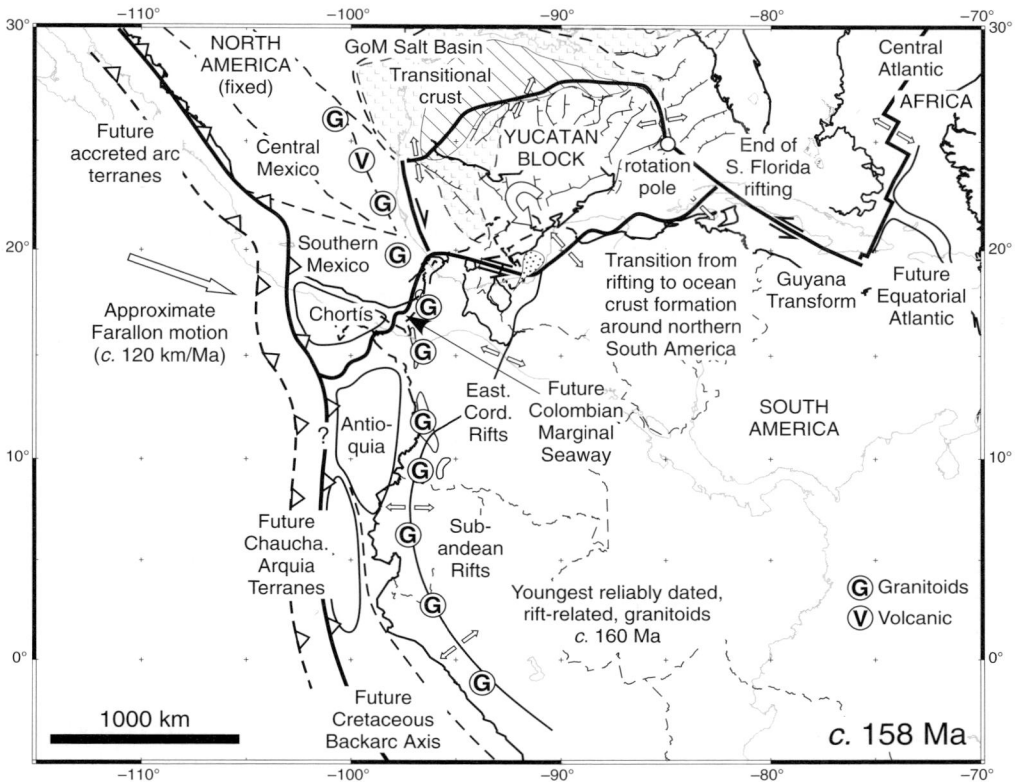

Fig. 4. A 158 Ma reconstruction of the circum-Gulf of Mexico region; Atlantic palaeopositions are interpolated between the Blake Spur Magnetic Anomaly fit of Pindell *et al.* (1988) and M25 of Roest *et al.* (1992). Pangaea breakup has reached incipient oceanic crust formation in the Gulf of Mexico, the Proto-Caribbean Seaway between Yucatán and Venezuela, and possibly between Colombia and Chortís. Rifting is active and there is a continuous belt of granitoids approximately 500 km from the trans-American trench, some of which are associated with rifting or extensional arc tectonics. Chortís and Antioquia are inferred to be in a forearc position relative to these granitoids and associated Jurassic volcanic rocks.

and Jurassic rifts in Georgia, Florida, Yucatán and central and eastern Venezuela and Trinidad are most parallel (orientated toward 070°) when Yucatán is rotated 30° to 40° clockwise relative to the present (Fig. 1, Pindell *et al.* 2006), a situation which had been achieved by the end of rifting but before the onset of seafloor spreading in the Gulf (Fig. 4). This period also marked the initial stages of spreading in the Proto-Caribbean and Colombian Marginal seaways, including probable hot spot activity along the Bahamas trend (Pindell & Kennan 2001).

The nature of the continent–ocean boundary in the Gulf of Mexico is not well defined. The flat basement in the deep central Gulf (Fig. 1) is normal oceanic crust as suggested by backstripping and the fact that basal sediment reflectors onlap toward a central, magnetically positive strip of crust in the

central Gulf continuing from the southeast Gulf to Veracruz Basin, which we believe is the position of the former spreading axis (Pindell & Kennan 2007*a*), including the area of 'buried hills' in the northeastern deep Gulf. The buried hills (Fig. 1), which form curvilinear trends nearly concentric around the Late Jurassic–earliest Cretaceous spreading pole, are not rift shoulders resulting from NW–SE extension (e.g. Stephens 2001) but leaky transforms, formed entirely in deep water as Yucatán rotated away from Florida. Flanking the northern, eastern and southern limits of flat oceanic basement in the deep Gulf is a downward step in basement closely matching the edge of mother salt. The nature of basement at the base of this downward step is not yet clear, but options are: (1) highly thinned continental crust that initially had a syn-rift halite section far thicker than the

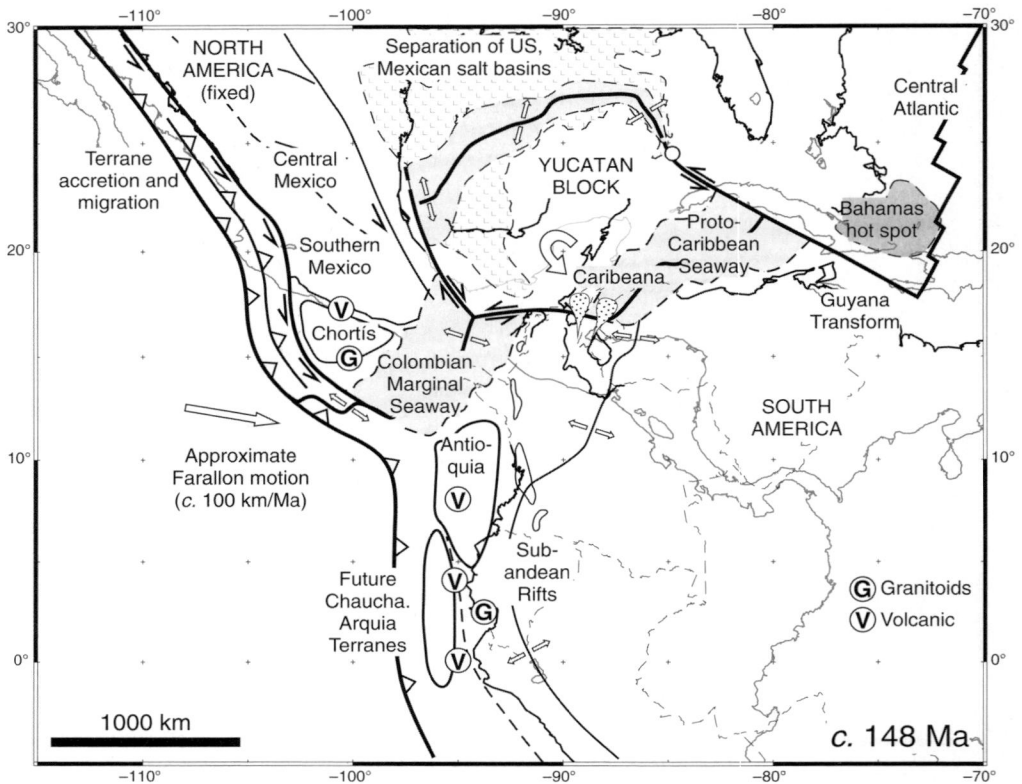

Fig. 5. A 148 Ma (Anomaly M21) reconstruction of the circum-Gulf of Mexico region. Relative palaeopositions of North and South America after either Müller *et al.* (1999) or Roest *et al.* (1992). At this time, towards the end of extension in the Chihuahua Trough, southern Mexico is close to its final position, and a *c.* 1000 km seaway, not yet fully connected to the Proto-Caribbean, is inferred to separate Colombia from Chortís. A discontinuous volcanic arc is present, and back-arc extension-related volcanism continued locally in Colombia, Ecuador and Peru. Off Mexico, the trench may have advanced westward relative to North America through southward forearc migration and terrane accretion. The trans-American trench is interpreted to have connected western Chortís and the southern Colombian portion of the Andean margin. The youngest granitoids in Ecuador and central Colombia (Ibagué) may be associated with subduction at this trench. It is kinematically impossible for the Andean subduction zone to have continued north of Ibague, where the margin was more or less passive and the conjugate of Chortís. Note that separation of North and South America resulted in a halving of the rate of Farallon subduction beneath South America compared to Mexico.

c. 2.7 km water depth (below sea level) at which the oceanic crust of the central Gulf was later emplaced; (2) landward-dipping footwall extrusions of serpentinized mantle peridotite from beneath detached, more landward continental crustal limits, thus implying a non-volcanic style of rifting and transition from continental to oceanic crust; (3) a mafic, quasi-oceanic crust that was not able to acquire the layered structure of normal oceanic crust (i.e. layered gabbro, dykes, pillows, sediments) due to being accreted beneath thick salt (5–6 km) rather than open seawater. This last option was explored by Pindell & Kennan (2007*a*): the basement step up could be explained by basinward spilling and thinning of salt after salt deposition stopped while opening of the Gulf of Mexico continued, thereby allowing progressively shallower accretion of oceanic crust until the salt pinched out (stopped spilling basinward), thus defining the line where the oceanic crust proceeded to form thereafter at 2.7 km depth (open seawater). However, all options remain viable until further data are released or collected. Our reconstructions (Figs 3–5) show that the eastern Gulf underwent a sharp (roughly 90°) change in extensional direction when seafloor spreading began in about Early Oxfordian time.

The kinematics of the creation of the basement step up in the first two of the above three options will adhere to the NW–SE extensional stage, whereas the third option will adhere to the spreading stage. Along the western Gulf margin, the continent–ocean boundary is a fracture zone rather than a rift (Pindell 1985a; Pindell et al. 2006), with high continental basement rather than a deep rift to its west such that the downward step noted above is not seen. There, the reconstructed Tamaulipas Arch–Chiapas Massif formed the footwall to the low-angle Yucatán detachment, whose hanging wall cut-off now lies below the Campeche salt basin, but at the onset of seafloor spreading the new spreading system cut into this former footwall and carried the Chiapas Massif portion of it southward with Yucatán.

Along the Cordillera, a fairly continuous belt of granitoids and extrusive volcanic rocks, generally with subduction-related calc-alkaline arc geochemistry (e.g. Bartolini et al. 2003), lies some 300–500 km inboard from the proposed site of the trench axis, when plotted palinspastically (Fig. 4). The relatively inboard position of this arc (compared with typically 150–200 km) with respect to the trench suggests flat-slab subduction, which may have pertained to the rate of plate convergence (fast), the age of the downgoing plate (young), and/or to the motion of the Americas over the mantle (westward drifting). Note that the Chortís, Tahami–Antioquia and Chaucha–Arquia terranes are interpreted to lie in a continental forearc position in this reconstruction, the along-strike position of which remains unclear.

The rotational phase of seafloor spreading continued in the Gulf of Mexico until the Late Jurassic or Early Cretaceous. Yucatán cannot have overlapped with the northern Andes, but palinspastic reconstructions of the northern Andes vary enough that a given reconstruction is only a soft constraint on the period of Gulf spreading. Marton & Buffler (1999) showed that extensional faulting ceased in the southeastern Gulf in earliest Cretaceous time, perhaps at about 135 Ma, which we agree should mark the end of significant movement of Yucatán with respect to North America, of which Florida was a part by this time. Along the eastern Mexico shear zone along which Yucatán had migrated, the Tuxpan portion of the margin was a fracture zone with little or no Jurassic faulting upward into the sedimentary section, whereas the Veracruz-Tehuantepec portion became a dead transform when Yucatán's migration stopped (Pindell 1985a). Along the latter portion, the Miocene-Recent invasion of igneous activity associated with the Middle American Arc now masks possible Jurassic deformations. Also, the fracture zone/palaeo-transform

margin along eastern Mexico has undergone subtle fault inversion with probably greater vertical displacements (west side up–east side down) due to flexure during the Eocene and Neogene tectonic phases in Cordilleran Mexico, as shown by the uplift history of the Mexican margin east of the Sierra Madre thrustfront and seismic data interpretation (Gray et al. 2003; Horbury et al. 2003; Le Roy et al. 2008). This development can be viewed as backthrusting with respect to compressional subduction at the Middle American Trench, with analogy to the Limón Basin of Costa Rica but on a grander crustal scale. Taken significantly further, this presently active process could develop in future to bonafide subduction, but at present appears to be responsible for extremely deep oceanic basement depths in the SW Gulf of Mexico. The young volcanism in the eastern Trans-Mexican Volcanic Belt continuing southward into the Chiapas Foldbelt gives the Mexican margin a high degree of buoyancy that probably increases the vertical shear along the deforming margin, as well as thermally softening the crust, both facilitating the onset of backthrusting at the Gulf of Mexico's Jurassic ocean–continent transition zone.

The Neocomian marked the final separation and continued seafloor spreading between NW South America from the Yucatán and Chortís Blocks in the early Proto-Caribbean Seaway and the Colombian Marginal Basin (Figs 6 & 7; Pindell & Erikson 1994; Pindell & Kennan 2001). Spanning the gap between the Americas, a lengthening plate boundary of debated nature and complexity must have connected east-dipping subduction zones to the west of the North and South American cordilleras, because it is kinematically impossible for the Proto-Caribbean spreading centre to project into the Pacific (in contrast to the maps of Jaillard et al. 1990, their fig. 9). That is, a plate boundary separating North and South America cannot also separate oceanic plate or plates of the Pacific that are subducting beneath the Americas. Thus, a 'trans-American' plate boundary most likely projected southeastward from the southwest flank of the Chortís Block, much like the Shackleton Fracture Zone at the southern tail of Chile today (Fig. 8), which may be a good analogue. We take the view that a highly sinistral-oblique trench, possibly with local transform segments (Pindell 1985a; Pindell et al. 2005, their fig. 7c), connected the Early Cretaceous Guerrero Arc of southern Mexico (Talavera-Mendoza 2000) and the Manto Arc of Chortís (Rogers et al. 2007a) with the Peru Trench of the Andes. To a first approximation, the position of this trans-American plate boundary can be estimated by projecting the Late Jurassic and Early Cretaceous North America–South America flowline

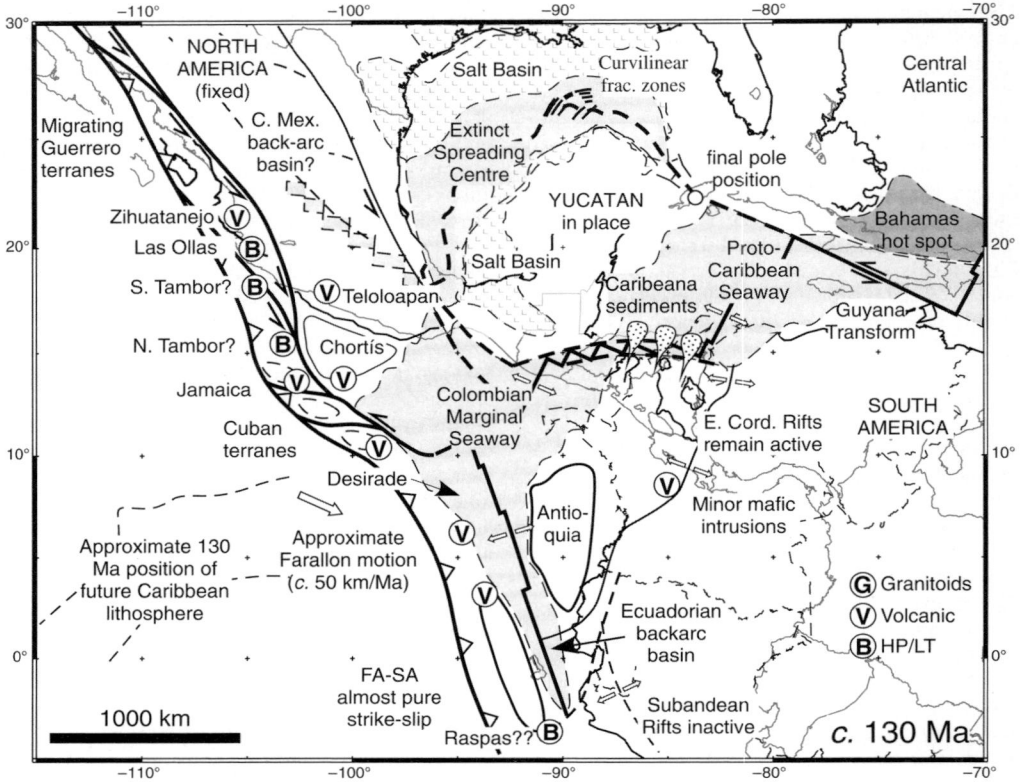

Fig. 6. A 130 Ma reconstruction of the circum-Gulf of Mexico and Caribbean region. Rotational oceanic crust formation is completed in the Gulf of Mexico, and Yucatán has stopped migrating with respect to North America. An oceanic back-arc basin is inferred to separate the trans-American arc from southern Colombia and Ecuador and to be the source of many of the 140–130 Ma ultramafic and mafic rocks that separate the Arquia and Quebradagrande terranes in Colombia from the rest of the Central Cordillera. The southern end of the arc joins South America in the vicinity of the Celica Arc near the present-day Peru–Ecuador border. In Colombia east of this back-arc basin, there is no subduction-related arc activity and no evidence for a subduction zone trending NE along the Colombian margin. Some of the separation of the Americas was accommodated by ongoing rifting in the Eastern Cordillera, with associated minor mafic magmatism. The trans-American plate boundary had lengthened by both internal extension and southward migration of arc and forearc terranes along Mexico/Chortís, assisted by oblique subduction of the Farallon Plate beneath North America. The positions shown for the ancestral Nicaragua Rise and Cuban terranes outside southern Mexico are compatable with the likely rates of subduction, strike–slip and separation of the Americas. Note that Farallon subduction beneath South America may have been slow (c. 25 mm/annum) west of Ecuador swinging towards trench-parallel strike–slip further south. The indicated palaeoposition of future Caribbean crust (assuming an Early Cretaceous basement) is consistent with calculated rates of Farallon motion with respect to the Americas (Engebretson *et al.* 1985), but subject to considerable error because the relative motions of Pacific and Indo-Atlantic hot spots cannot be constrained prior to c. 84 Ma. Possible palaeo-positions of the El Tambor c. 130 Ma HP/LT rocks are shown between the future Nicaragua Rise and Siuna terranes, south of the Las Ollas blueschists of southern Mexico. Very low geothermal gradients inferred for the southern El Tambor HP/LT rocks may suggest an origin in a cold, relatively rapid and long-lived subduction zone such as Farallon–North America rather than a narrow, transient subduction zone between Chortís and southern Mexico (e.g. Mann *et al.* 2007). Strike–slip displacement of these terranes from southern Mexico may play a role in their exhumation prior to emplacement against the Yucatán Block later in the Cretaceous. The Raspas blueschist of southern Ecuador (Arculus *et al.* 1999; Bosch *et al.* 2002) may also originate at a west-facing trench.

from southernmost Chortís (i.e. Chortís defined the SW extent of the Proto-Caribbean Basin). North of the intersection of this flowline with South America (close to the Ecuador–Colombia border),

Pacific plates would not have been present and thus could not have been subducted beneath South America. Thus, we show a rifted margin rather than a NE-trending subduction zone along

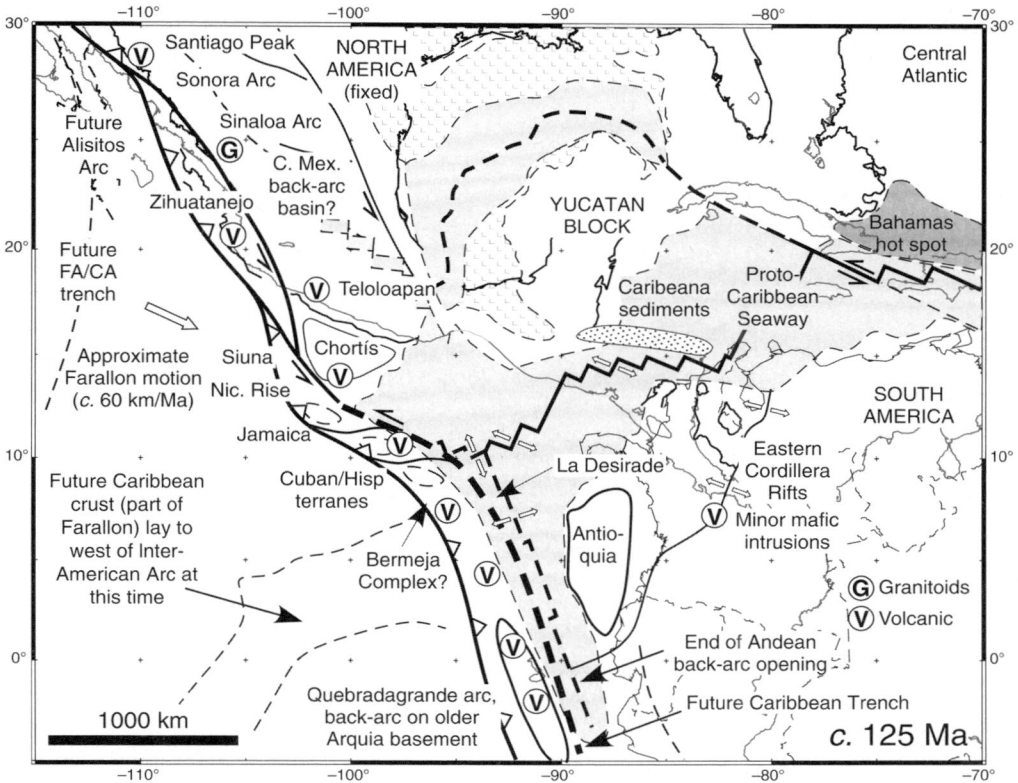

Fig. 7. A 125 Ma reconstruction of the circum-Gulf of Mexico and Caribbean region, showing the trans-American Arc immediately before the initiation of west-dipping subduction and onset of Caribbean Arc volcanism, and prior to development of the Alisitos arc of Baja Mexico. The Sonora, Sinaloa, Zihuatanejo and Teloloapan arcs in Mexico are shown 200–500 km inboard of a single Farallon–Mexico subduction zone, possibly on a basement of previously accreted oceanic crust and continental sediment without continental basement (hence their oceanic island arc character). Southward migration of Zihuatanejo terrane during Aptian–Albian time later results in an apparent double arc in SW Mexico. The Americas are still separating and transform faults continued to draw the Siuna, Nicaragua Rise/Jamaica and Cuban terranes SE of Chortís. The position of the future Caribbean trench is shown at this northern transform margin and within the Andean back-arc basin in the south (dashed). The width of the Andean back-arc is not constrained. In this relatively autochthonous interpretation of the Guerrero Arc, the 'Arperos Ocean' is interpreted as one or more narrow intra-arc or back-arc basins that may link to Proto-Caribbean Seaway via the Cuicateco Terrane, rather than being a broad oceanic basin separating an east-facing Guerrero Arc from Chortís and central Mexico (e.g. Freydier *et al.* 1996, 2000).

Colombia. Large regions of the Colombian Marginal Seaway between Colombia and Chortís formed in a supra-subduction zone environment with respect to the Pacific, and we expect associated rocks now preserved within the Caribbean orogen to have a backarc geochemical character even though it was an Atlantic type ocean basin with respect to the Americas. The end-Jurassic cherts and basalts of La Desirade Island (Montgomery & Kerr 2009) were probably deposited on the eastern flank of the trans-American plate boundary.

Complicating this simple scenario, several lines of evidence suggest that this plate boundary may have had an Andean intra-arc basin toward its eastern end before merging onshore with the Celica Arc (Jaillard *et al.* 1999) of northwestern Peru and southern Ecuador (Pindell *et al.* 2005, 2006, their figs 7c and 8, respectively, see also Kennan & Pindell 2009). First, an autochthonous Early Cretaceous continental arc was never developed in Ecuador and Colombia, in contrast to Peru. Second, the Arquia and Quebradagrande Complexes in Colombia are separated from the Antioquia–Tahami terrane and most of the Central Cordillera by a discontinuous belt of sheared mafic and ultramafic rocks that may mark the axis of the intra-arc (evolving to back-arc) basin (Kennan & Pindell 2009). The Quebradagrande volcanic rocks

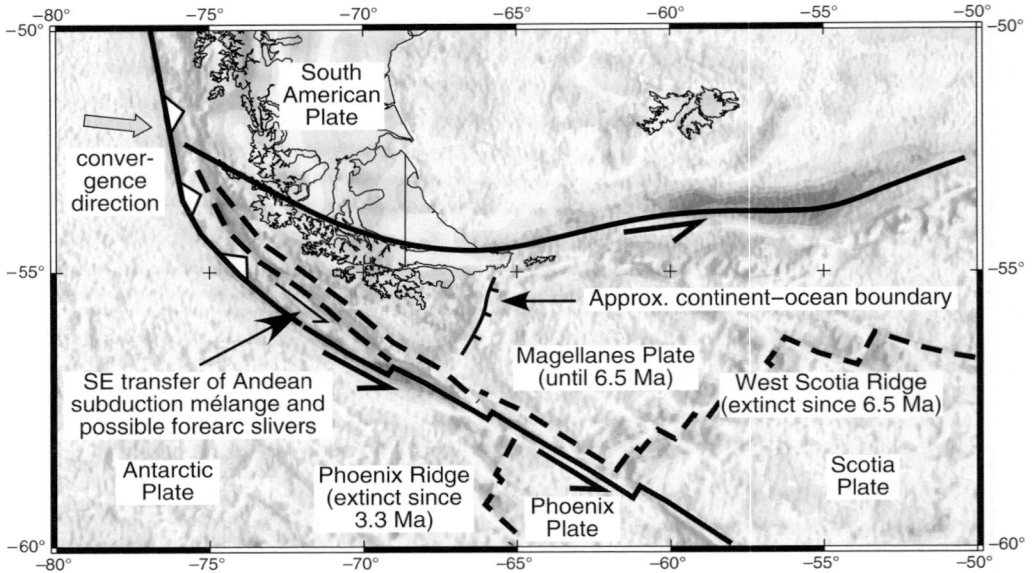

Fig. 8. Free-air gravity map, from Sandwell & Smith (1997), of the Tierra del Fuego and Shackleton Fracture Zone area, southern Chile. Major plates and plate boundaries active during the Late Cenozoic are simplified after Thomson (1985) and Eagles *et al.* (2004). Note how the Andes Trench veers eastwards into the transform/fracture zone, allowing sinistral secondary faults in the forearc to carry arc, forearc and subduction channel terranes some distance along the greater transform zone. In addition, the transform ties into the western Scotia spreading centre in a geometry that would satisfy the perceived plate kinematics of the Late Jurassic southern tip of the North American Cordillera and Proto-Caribbean spreading centre, providing a possible analogue.

(Nivia *et al.* 2006) in Colombia, which lie between the older Arquia metamorphic belt to the west and the Central Cordillera to the east, are interpreted here as a southern continuation of the Aptian–Albian Caribbean Arc that was accreted to the Colombian margin rather than migrating north with the rest of the arc (oblique collision). Third, a number of continental fragments occur in the allochthonous Caribbean Arc along northern South America that appear to have affinity with rocks along the western flank of the Central Cordillera. These include the Juan Griego basement rocks of Margarita (Stöckhert *et al.* 1995; Maresch *et al.* 2009), the Tinaco–Caucagua terrane of central Venezuela (Stephan *et al.* 1980; Bellizzia 1985; Beck 1986), the Grenvillian granulites and marbles of Falcón (Grande & Urbani 2009), continental knockers of the Cordillera de la Costa terrane, central Venezuela (Smith *et al.* 1999; Sisson *et al.* 2005), and the Dragon Gneiss of Paria Peninsula (Speed *et al.* 1997). Of these, at least the Juan Griego unit of Margarita appears to represent the eastern flank of an intra-arc basin, while at least the Tinaco–Caucagua Terrane, with its Albian unconformity and basal conglomerates followed by arc volcanic rocks (Bellizzia 1985), appears to represent the active arc side of the intra-arc basin.

Aptian–Maastrichtian (125–71 Ma) closure of the Colombian Marginal Seaway

Conversion of the Trans-American Plate Boundary to the NE-facing Caribbean Arc System

The trans-American plate boundary linking Chortís and Peru (Fig. 7) underwent a major transformation in the Early Aptian as it was converted to a SW-dipping subduction zone beneath the future Caribbean Arc (Fig. 9). Subduction of oceanic crust of the Colombian Marginal Seaway is responsible for the Late Aptian to Maastrichtian, generally unmetamorphosed, parts of the Caribbean Arc. Some formations such as the Los Ranchos and Water Island formations of Hispaniola and Virgin Islands were once thought to pre-date the onset of SW-dipping subduction, but new dating and geochemical characterization support the view that these formations post-date the polarity reversal (Kesler *et al.* 2005; Lidiak *et al.* 2008; Jolly *et al.* 2008). As North America took flight from Gondwana, the Chortís–Peru (trans-American) plate boundary lengthened and became more transcurrent. Where the trans-American plate boundary had remained an east-dipping subduction zone, arc polarity reversal resulted with the potential for the pre- and post-Aptian arc axes to be superposed; however, any transform portions of the boundary would have been the site of subduction initiation only (Pindell et al. 2005, 2006; Pindell 2008). The palaeo-geometry of the margin suggests that subduction polarity reversal more probably occurred in the southeastern part of the arc (to become the Aves Ridge?), while subduction initiation in the NW part (now the Greater Antilles) more probably occurred at more of a transform boundary. In both settings, initiation of SW-dipping subduction can be constrained, in general, by the oldest ages of HP/LT metamorphism in circum-Caribbean subduction complexes/sutures and the onset of arc magmatism related to that subduction. Both aspects point to the Aptian, or 125–114 Ma (Pindell 1993; Stöckhert et al. 1995; Smith et al. 1999; Snoke et al. 2001; Harlow et al. 2004; Pindell et al. 2005; García Casco et al. 2006; Maresch et al. 2009; Stanek 2009).

However, there may well have been the added complexity at the western end of the trans-American arc that Late Jurassic and/or Early Cretaceous trench, forearc and arc materials lying originally west of Mexico and Chortís (e.g. Las Ollas Complex, Talavera-Mendoza 2000; and west-central Baja California, Baldwin & Harrison 1989) were dragged by the sinistral component of oblique subduction some distance southeast along the trans-American boundary. Some insight on this process comes from considering similar tectonic settings such as the southern tip of Chile today (Fig. 8), where slivers of Andean forearc rocks, or mélange containing continental blocks, may be moving SE along the sinistral Shackleton Fracture Zone. If so, such terranes would become amalgamated within the roots of the western parts of the Caribbean Arc upon the onset of SW-dipping subduction. We offer this as an explanation for why two western Caribbean HP/LT localities are significantly older than (1) other circum-Caribbean HP/LT rocks and (2) the Late Aptian/Albian–Eocene Antillean magmatic cycle: the 139 Ma age for HP/LT metamorphism in the Siuna terrane of Nicaragua (Flores et al. 2007; Baumgartner et al. 2008) and the 132 Ma ages for HP/LT rocks in the El Tambor unit of central Guatemala (Brueckner et al. 2005). It may also be the mechanism by which Grenvillian aged blocks found their way into the allochthonous subduction mélange of central Cuba (Renne et al. 1989); such basement rock types are not known in the autochthonous margins of the Proto-Caribbean, but do occur in SW Mexico.

Arc volcanism became more prevalent in the Caribbean Arc during Late Aptian–Albian time, including sections in Jamaica, Cuba, Hispaniola,

Puerto Rico, the Villa de Cura Group of Venezuela, Tobago, and elsewhere. However, there is a general lack of arc-derived tuffs in the Proto-Caribbean passive margins until the Maastrichtian–Cenozoic (initial contamination of these margins by arc-derived tuffs youngs eastward), a primary argument by Pindell (1990) for the Pacific origin of the Caribbean arcs. Significant spatial separation between the volcanic Caribbean arcs and the non-volcanic Proto-Caribbean passive margins is clearly indicated. However, there are a few examples of Early Cretaceous volcanic rocks in these margins. First, an 11 cm bentonite is known from a well in the Albian level of the La Luna Group (La Grita unit, see Villamil & Pindell 1998) in the Maracaibo Basin (PDVSA pers. comm. 1994), but the mineralogy (and any possible arc relationship) is unknown to us. Second, there are low volumes of mafic rocks mostly associated with extensional faults in the Eastern Cordillera of Colombia (Vasquez & Altenberger 2005) and in the Oriente Basin of Ecuador (Barragán et al. 2005). Where geochemical data are available, an alkaline, extensional or possibly plume-related character is indicated, rather than a supra-subduction zone or arc character. Most of these rocks post-date the onset of southwest dipping subduction in the Caribbean Arc, but a few are as old as 136–132 Ma, approximately of the same age as many of the mafic rocks inferred to define the trace of the former Andean back-arc basin. These data, deriving from the only known magmatic rocks of the time, reinforce our view that there was no arc and hence no subduction zone along the Colombian margin during Early Cretaceous time (see also Kennan & Pindell 2009). Third, Early Cretaceous 'bentonites' have been identified in the Punta Gorda borehole in southern Belize (Punta Gorda Formation; Ramanathan & García 1991), cuttings of which have recently been obtained by us courtesy of Brian Holland (Belize Minerals). Analyses for mineralogy are pending to determine magmatic affinity. Should these prove to be arc-related, we would judge that the nearest known coeval arc volcanism, in the Chortís Block (Ratschbacher et al. 2009), was able to reach Belize, 800 km away (a distance that is commonly covered by airfall tuffs today) and at a palaeolatitude of about 3°N (Fig. 9). However, another possibility is that they pertain to the transtensional plate boundary separating Yucatán and Guajíra during Neocomian time.

The NW South America–Caribbean Plate boundary zone in the Cretaceous

Following the probable Aptian onset of SW-dipping subduction beneath the Caribbean Arc, motion of

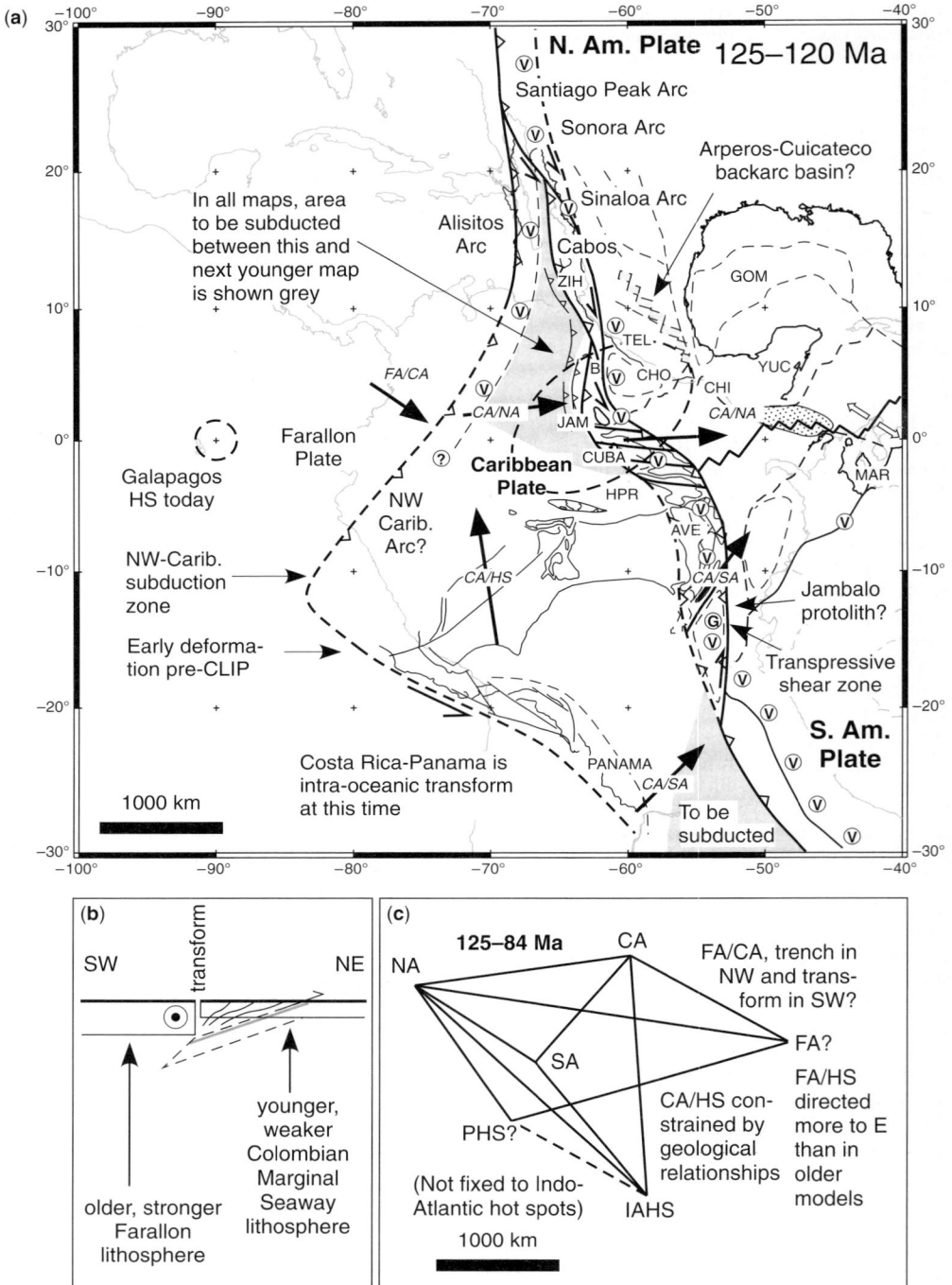

(a)

N. Am. Plate 125–120 Ma

Santiago Peak Arc
Sonora Arc
Arperos-Cuicateco backarc basin?
Sinaloa Arc
Alisitos Arc   Cabos,
In all maps, area to be subducted between this and next younger map is shown grey
ZIH
GOM
TEL
YUC
FA/CA
B   CHO   CHI
CA/NA   CA/NA
JAM
Farallon Plate
?   Caribbean Plate
CUBA
MAR
Galapagos HS today
HPR
NW Carib. Arc?
AVE
NW-Carib. subduction zone
CA/HS
CA/SA
Jambalo protolith?
Early deforma-tion pre-CLIP
Transpressive shear zone
S. Am. Plate
1000 km
Costa Rica-Panama is intra-oceanic transform at this time
PANAMA
CA/SA
To be subducted

(b)

transform
SW   NE
younger, weaker Colombian Marginal Seaway lithosphere
older, stronger Farallon lithosphere

(c)

125–84 Ma   CA
NA   FA/CA, trench in NW and trans-form in SW?
FA?
SA
FA/HS directed more to E than in older models
PHS?   CA/HS con-strained by geological relationships
(Not fixed to Indo-Atlantic hot spots)   IAHS
1000 km

Fig. 9. (a) A 125–120 Ma reconstruction of the circum-Caribbean region, shown in the Indo-Atlantic hot spot reference frame of Müller *et al.* (1993) as are all younger reconstructions. The map shows proposed plate boundary relationships immediately after initiation of SW-dipping subduction beneath the Caribbean Arc. Heavy black arrows show relative plate motions. The age, setting and reconstruction of western Mexican terranes are speculative and still debated.

the arc and future Caribbean lithosphere behind it relative to South America was almost parallel to the overall NNE trend of the Ecuador–Colombia margin, particularly after about 100 Ma. Associated structures are dextral strike–slip to dextral transpressive throughout the Ecuadorian Cordillera Real and Colombian Central Cordilleran terranes and initial cooling ages in these areas range from 120–85 Ma, consistent with the plate boundaries shown (Figs 9–11, see Kennan & Pindell 2009 for more detail). Dextral shearing started the slow migration of Antioquia north towards its present position. Deformation was initially ductile, becoming brittle towards the end of the Cretaceous, when we suspect the Huancabamba–Palestina Fault Zone became active. Further, we consider that a STEP fault ('subduction–transform edge propagator', Govers & Wortel 2005) may have defined the termination of the Caribbean trench at the South American continent–ocean boundary for this transcurrent stage; the tear was propagated along the boundary by the loading effect of the advancing Caribbean Arc.

We identify the former existence of a mainly tonalite–trondhjemite belt of intrusive rocks along the Albian–Early Eocene Andes–Caribbean Plate boundary that becomes apparent when Caribbean–South American Plate motions are restored for that time. Candidates for this belt include Tobago (Tobago Plutonic Series, Snoke et al. 2001), at least some parts of the Leeward Antilles Islands (e.g. Aruba Batholith, Wright et al. 2008), the

Guayacán trondhjemite of Margarita (Maresch et al. 2009), several intrusives in Guajíra and Santa Marta (Cardona et al. 2008), and the Antioquia, Buga, and several other nearby plutons (Kennan & Pindell 2009). The interesting aspect about all these intrusions is that they lie within 100 km, and on both sides of or within, our reconstructed Caribbean–South America Plate boundary zone (Figs 11 & 12), which is too close for these to be normal arc-related intrusions. Instead, we propose a model of tonalite/trondhjemite production by the re-melting of mafic crust of the 'slab nose' upon subduction initiation (e.g. Nikolaeva et al. 2008; García-Casco et al. 2008a), where basaltic crust of the downgoing plate was juxtaposed with lower lithosphere of an adjacent plate that was still hot because the cooling effect from subduction had been minimal by the time of melting. Hence, the basalts underwent anatexis and intruded other subducted components (e.g. Guayacán metatrondhjemite of Margarita; Maresch et al. 2009) and stocks and plutons along the plate boundary at shallower levels. Figure 13 offers settings where subduction initiation could occur along the northern Andes, which should have been diachronous northwards. However, this new hypothesis for the origin of these magmas needs to be tested and refined as there are large uncertainties concerning the location of various plutons relative to the plate boundary in this model. For example, the Aruba Batholith (89 Ma gabbrotonalite; Wright et al. 2008) has a very similar

Fig. 9. (*Continued*) Here, the Guerrero Arc is interpreted to reflect subducton of Caribbean crust under Mexico, building an arc on migrating former forearc terranes comprising accreted oceanic crust and continent-derived sediments. Outboard of the Guerrero Arc we show the inception of a new Farallon–Caribbean Plate boundary. To the SE along South America, oblique south or west-dipping subduction led to closure of the Andean back-arc basin. Abbreviations; TEL, Teloloapan; CHO, Chortís; CHI, Chiapas; CLIP, Caribbean large igneous province; YUC, Yucatán; GOM, Gulf of Mexico; MAR, Maracaibo; HPR, Hispaniola–Puerto Rico; JAM, Jamaica. The initial location of the El Tambor blueschists is shown as B, immediately to the west of Chortís. Circled V indicates approximate location of arc volcanism at this time; circled G approximate location of granitoid intrusion. (**b**) Model for subduction initiation at a pre-existing transform boundary along the northwestern part of the Caribbean Arc. Upon Aptian onset of convergence at the transform, subduction polarity became SW-dipping as the weaker side buckled and imbricated. Material in the new subduction melange comprises MORB basalts, transform metamorphic rocks, supra-subduction basalts, HP/LT metamorphic rocks from western Mexico/Chortís and arc fragments. As a result, the Caribbean Arc began to wrap transpressively around Chortís (future Siuna Terrane). Concurrently, Caribbean crust underthrust Chortís from the west and south while accreting the Mesquito Terrane. (**c**) A semi-schematic vector nest for 125–84 Ma suggests that the Farallon Plate moved east in a Pacific hot spot reference frame while geological constraints suggest that the Caribbean Plate was migrating north with respect to an Indo-Atlantic reference frame. Thus a Farallon–Caribbean Plate boundary is required unless the Pacific hot spots were migrating to the NW relative to the Indo-Atlantic hot spots faster than 75–100 km/Ma, which is unlikely. In the NW, this boundary was probably the site of south-dipping subduction, possibly explaining the Aptian–Albian onset of arc magmatism in the Alisitos Arc in Baja California (Sedlock 2003), which we show outboard of the Sonora–Sinaloa Arc (Henry et al. 2003) and the Zihuatanejo Arc where Farallon-cum-Caribbean crust continued to be subducted. Southwards along this new boundary in Costa Rica to Panama, Farallon–Caribbean motion could have been accommodated along an oceanic transform that would become the site of east-dipping subduction only after a dramatic turn in Farallon–Americas motion at c. 84 Ma. The rate of subduction and transform motion is estimated at c. 25–50 km/Ma. Accretion of the arc portion (i.e. Alisitos Arc) of this boundary along Mexico, due to subduction of Caribbean crust beneath Zihuatanejo Arc, began at c. 110 Ma and younged to the south.

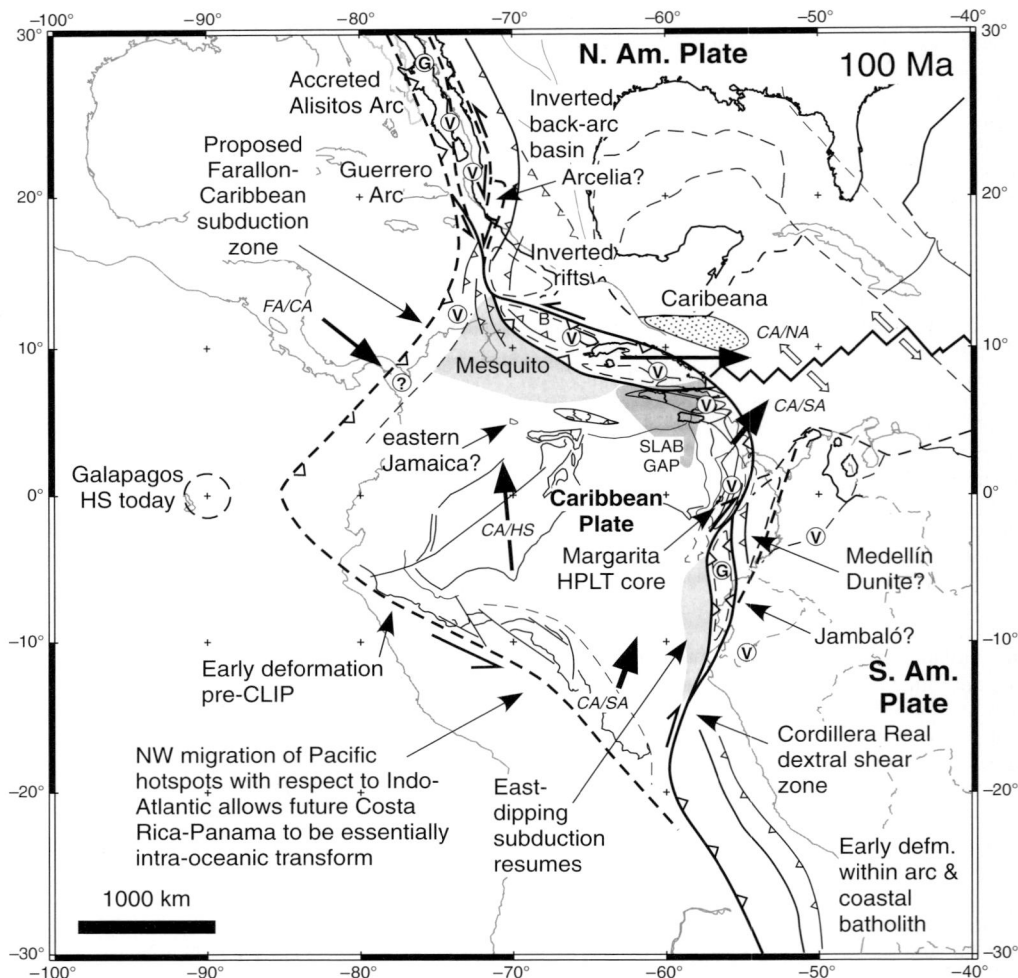

Fig. 10. A 100 Ma reconstruction of the circum-Caribbean region. Motion of the Caribbean Plate relative to the hot spots is towards the north, and toward the east relative to North America. Continued Proto-Caribbean spreading results in almost pure dextral motion between the Northern Andes and the Caribbean Plate. By this time, the Andean back arc has closed, most circum-Caribbean HP/LT metamorphic complexes have formed, and eastward transpressive terrane migration is occurring on the north and south flanks of the Caribbean. Along the South American margin, STEP-fault and subduction initiation processes result in tonalitic/trondhjemitic magmatism within the lengthening dextral Cordillera Real–Central Cordillera Plate boundary zone (including possibly Pujili, Altavista, Antioquia, Aruba, Salado granitoids). The oceanic basin between the inner and outer arcs in Mexico has been closed as far south as Chortís. Eastward motion of the Caribbean with respect to North America has drawn the Nicaragua Rise/Jamaica and Cuban terranes SE of Chortís. The extent of the slab gap in the Proto-Caribbean slab beneath the eastern Caribbean is shown (shaded mid-grey). Note that approximately two-thirds of the Caribbean Plate will be subducted beneath North or South America, much of which is seen in seismic tomographic data. The Alisitos and Zihuatanejo arc terranes in Mexico are shown more or less in place. We speculate that the Albian submarine pillow basalts in the Arcelia area may have been deposited in a small pull-apart basin along the faults that linked the southward migrating Zihuatanejo and Siuna–Nicaragua Rise terranes off Chortís.

geochemistry to the Turonian–Coniacian (*c.* 94–90 Ma) Aruba lava formation in which it sits (White *et al.* 1999; Wright *et al.* 2008); the pluton may simply be a late equivalent of the extrusive

lavas, all of which relate to the Caribbean large igneous province (LIP) (see below), initially situated on the Caribbean Plate some distance SW (prior to accretion) of the new east-dipping

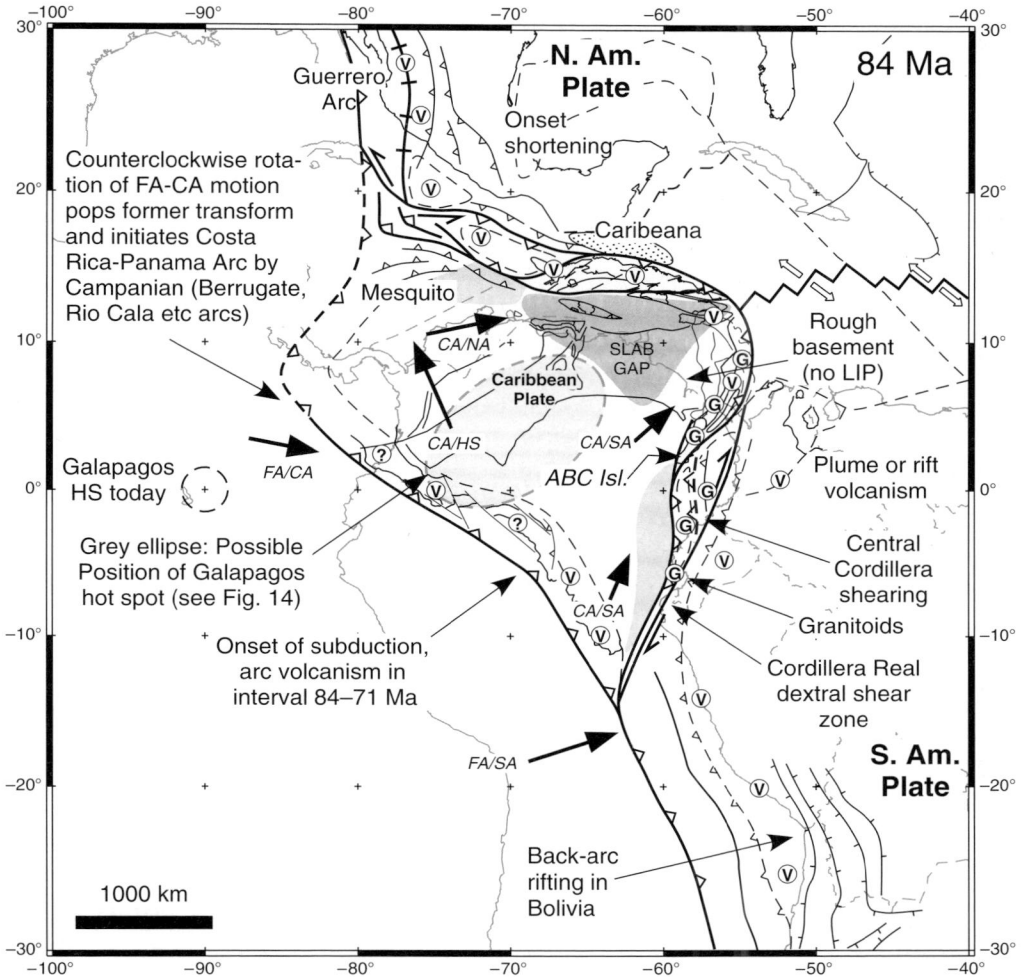

Fig. 11. A 84 Ma reconstruction of the circum-Caribbean region. Relative motion between the Farallon and Caribbean has rotated resulting in onset of oblique subduction at the former Costa Rica–Panama transform. The gap in the Proto-Caribbean slab extended to approximately Beata Ridge and may have played a role in allowing deep plume-derived melts into the NE Caribbean region. The Caribbean Plate has migrated far enough to the north to underthrust and entrain the Antioquia terrane in the North Andean Plate boundary zone. Relative motions of the Farallon Plate to the Caribbean and South America are based on our own calculations (modified from Doubrovine & Tarduno 2008).

accretionary plate boundary, rather than being due to the hypothetical mechanism outlined above (Fig. 13). The onset of subduction here pertains to the Late Cretaceous slowing/cessation of spreading between the Americas (Pindell *et al.* 1988; Müller *et al.* 1999), such that Caribbean–South American relative plate motion evolved from dextral strike–slip to dextral convergence (Fig. 12). However, no magmatic arc has developed above this Benioff Zone at typical distances from the trench, due mainly to the flat geometry and slow rate of subduction of the buoyant Caribbean slab.

The North America–Caribbean Plate boundary zone in the Cretaceous

In the western part of the Caribbean Arc, the onset of SW-dipping subduction (possibly at a transform boundary) produced an east–west-trending transpressive shear zone that lengthened with time by sinistral shear along cross faults, and by axis parallel extension. Continued oblique convergence of the arc, and any pre-Aptian rocks within it, with the southern and eastern margins of the Chortís Block would have led to north-vergent

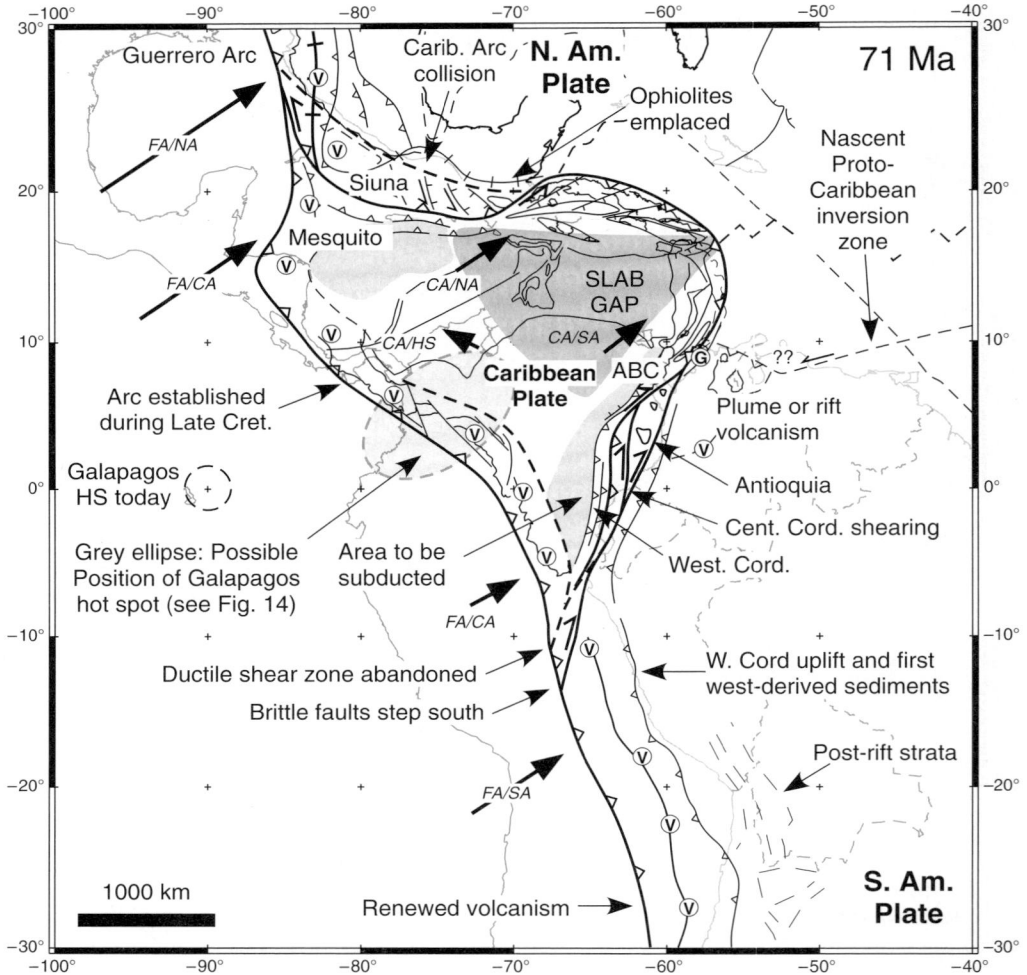

Fig. 12. A 71 Ma reconstruction of the circum-Caribbean region. North and South America cease diverging, resulting in more head-on subduction of the Caribbean beneath the northern Andes and northward zippering of Panama against the Andes. Suturing of the Caribbean Arc along the Chortís–Yucatán margin is nearly complete, resulting in backthrusting and further convergence being taken up at the Lower Nicaragua Rise. Chortís was dislodged from North America at this time, and began to move as an independent terrane eastward along Mexico due to partial coupling with the underlying Caribbean crust, much like Maracaibo Block moves today between the Caribbean and stable South America. Note that Farallon motions with respect to the Americas suggest a trebling of the rate of subduction under the Costa Rica–Panama Arc from SE to NW.

emplacement of the Siuna Terrane (Figs 10 & 11). We generally follow the syntheses of Pindell *et al.* (2005) and Rogers *et al.* (2007*b, c*) but further propose that the Siuna Belt of Nicaragua and Honduras continues on our palinspastic reconstruction to the ENE into the Chontal arc remnants in southeasternmost Mexico (Carfantan 1986), and then into the 'Tehuantepec Terrane' in the Gulf of Tehuantepec (see below, and Fig. 18), and on to the east into the Nicaragua Rise and Jamaica and into

Cuba in the Caribbean Arc. This belt comprises arc and HP/LT subduction channel rocks that appear to be thrust northward onto the former North American margin. The emplacement was diachronous to the NE, culminating in the Maastrichtian with the overthrusting of the southern Yucatán margin and Caribeana sediment pile, and creating the Sepur foredeep section of northern Guatemala (Pindell & Dewey 1982; Rosenfeld 1993; García-Casco *et al.* 2008*b*). The occurrence

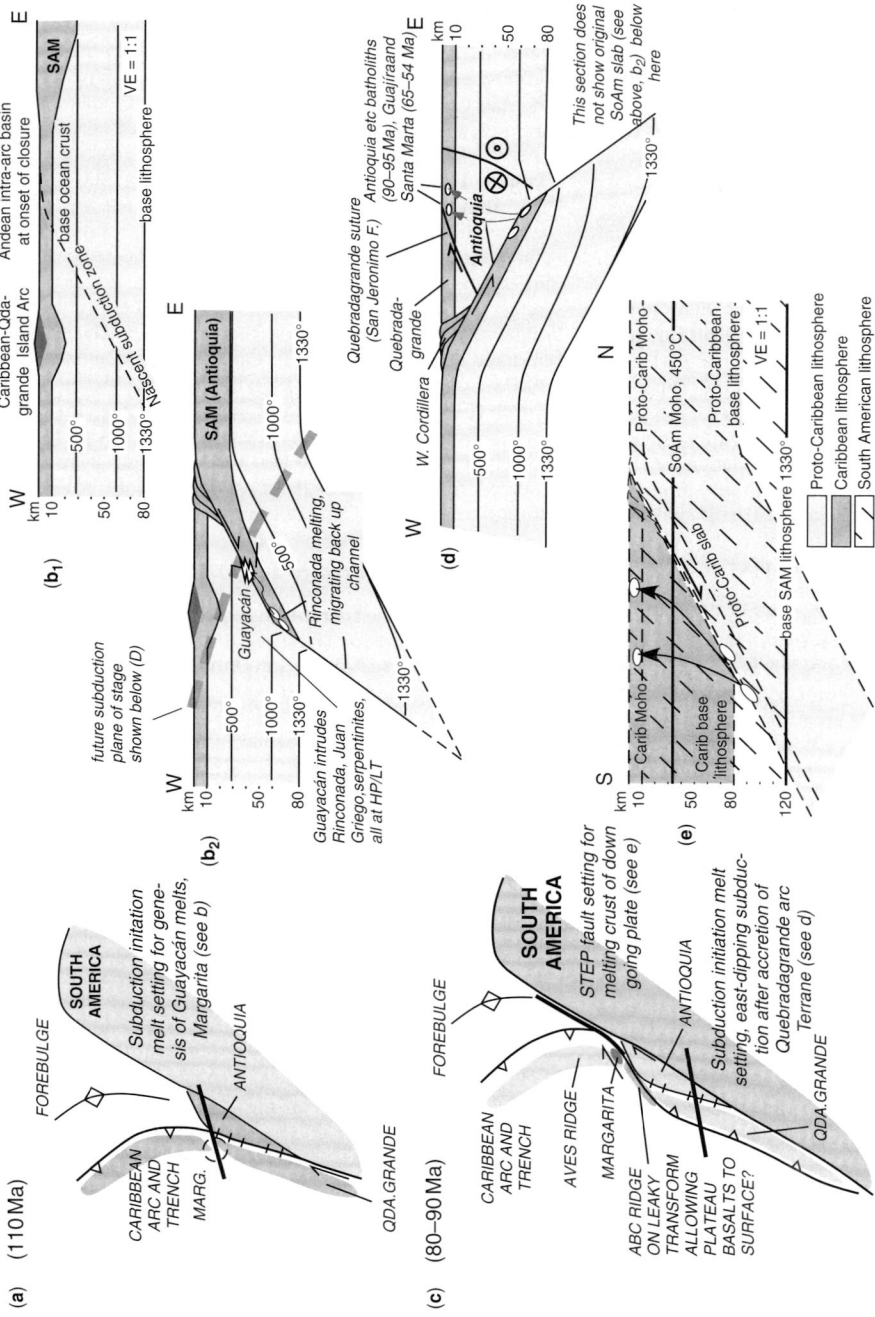

Fig. 13. Tectonic settings and proposed mechanisms for production of 'subduction-initiation' (cross-section on map **a**, and the southern of the two cross-sections on map **c**) and 'STEP fault' (northern cross-section in map **c**) melts, which seem to form plutons very close to the plate boundaries (<100 km). Upon subduction initiation (i.e. about 140 km of convergence, achievable in <3 Ma for a plate convergence of 50 mm/annum), the basaltic upper crust of the new downgoing lithosphere (cross-sections **b₁**, **b₂**, **d**) must pass along the lower lithosphere of the hanging wall, which is hot (>750 °C) because it has not yet lost heat into the downgoing slab (i.e. subduction zone isotherms have not yet equilibrated to steady state). Thus, heat transfer can melt the hydrous, often sodic (due to metasomatism) basaltic oceanic crust and any subducted sediments, producing melts of gabbro–tonalitic and/or trondhjemitic compositions which can (1) intrude other, firmer lithologies in the subduction channel, or (2) move up the subduction channel some distance depending on volume and apparently intrude the hanging wall very near to the trench, possibly along active faults. In the side-on viewpoint of cross-section **e**, a potential melt setting adjacent to STEP faults is shown. The South American (SAM) lithosphere is shown in dashed pattern, with the oceanic Caribbean lithosphere shown behind in grey. Setting where hydrous basalts and sediments contact hot SoAm lower lithosphere is indicated as the deep ovals. Examples of subduction initiation melts may include the Albian Guayacán unit of Margarita (Maresch *et al.* 2009) while an example of a STEP fault melt may be the Antioquia Batholith of the Antioquia Terrane.

of 132 and 139 Ma HP/LT rocks in this belt (Brueckner *et al.* 2005; Flores *et al.* 2007; Baumgartner *et al.* 2008) indicates to us that such Early Cretaceous material in this belt was dragged by transcurrent shear along the trans-American plate boundary from the western flank of Chortís (Figs 7–9).

In response to the collision of the Caribbean Arc with eastern Chortís and southern Yucatán, northward subduction beneath the accreted terranes (Siuna, Tehuantepec, Nicaragua Rise, Jamaica) was established or renewed by backthrusting along a trend which may have been the site of pre-120 Ma eastward dipping subduction, with arc development continuing therein through the Early Eocene. Underthrusting of Caribbean lithosphere beneath the Chortís continental block was instrumental in the eventual acquisition of Chortís as part of the Caribbean Plate: we suspect that the subduction angle was low such that Chortís was effectively obducted onto the Caribbean Plate, although short-ening continued, much like the Maracaibo Block has been obducted onto the Caribbean Plate since the Oligocene (also flat slab, and still undergoing minor relative motion), such that the Maracaibo 'block' is loosely being carried upon the Caribbean Plate as well. From a seismological perspective, the Mérida Andes today define the primary present Caribbean–South America Plate boundary, where-as the South Caribbean foldbelt is the petrological (and longer term evolutionary) plate boundary. Like Maracaibo today, upon the underthrusting of Caribbean crust beneath Chortís in a flat slab geo-metry, basal coupling was probably strong enough by the Campanian–Maastrichtian to tear the Chortís hanging wall promontory from North America as the latter continued to drift to the west in the hot spot reference frame, thereby gradually transferring Chortís to the Caribbean lithosphere, a process completed by Eocene time.

Initiation of the western Caribbean Plate boundary

The age of initiation of the western Caribbean Plate boundary, defined today and during the Ceno-zoic by the Panama–Costa Rica Arc, remains a critical issue for two reasons. First, it defines when the Caribbean and Farallon Plates became kinemati-cally independent. Provided there are no additional plates in the eastern Pacific, Farallon Plate motions should define the motion and development of the Caribbean Arc until the western Caribbean boundary was formed. Second, if the inception of the western Caribbean subduction zone post-dated the general 88–92 Ma age of most Caribbean LIP extrusion, then the 'Caribbean' LIP would actually have been a 'Farallon' LIP in the absence of a boundary to differentiate the two plates.

Discrepancies for the age of inception range from the Aptian (Pindell & Kennan 2001), through Campanian (e.g. Pindell & Barrett 1990) to Palaeo-gene (Ross & Scotese 1988). The Aptian age proposed by Pindell & Kennan (2001) was based on the Calvo & Bolz (1994) claim that island arc volcaniclastic sandstones in the accretionary Nicoya Complex of Costa Rica are as old as Albian. However, Flores *et al.* (2003a, b, 2004; also Bandini *et al.* 2008) have since dated this section, called the Berrugate Formation, as Conia-cian to lowest Campanian (88–83 Ma), and hence the stratigraphic inferrence for an Albian arc no longer exists. Arc magmatism was more certainly underway by 75 Ma based on geochemical analysis of dated exposed outcrops in Panama (Buchs *et al.* 2007; Buchs 2008). However, if the 'arc' desig-nation (Flores *et al.* 2004; Calvo & Bolz 1994) for the Berrugate Formation volcaniclastic rocks is correct, then it is possible that the sediments were sourced from unidentified arc rocks possibly now buried beneath the Cenozoic arc. In either case, a reasonable age for subduction initiation might be 80–88 Ma, considering that a slab needs several million years to reach depths where melt can be gen-erated. Such an age is at the young end of the period of most LIP extrusion (Kerr *et al.* 2003).

From the above, subduction at the SW Caribbean Plate boundary appears to have begun just after the period of LIP extrusion. Thus the following Mid-Cretaceous setting can be proposed for the western Caribbean. In the absence of a western Caribbean Benioff Zone, there would be no necessary south-western limit to the area that might have been intruded by plume-type magmatism rising in or near the Proto-Caribbean slab gap, and the field of LIP magmatism might have extended further SW within the Farallon Plate than the future Panama–Costa Rica Trench. It is thus possible that the trench formed within the LIP field with perhaps some LIP extrusive rocks situated or still forming to the SW of the impending plate boundary. Subsequent subduction at the trench would have led quickly to the accretion of LIP seamounts and plateau material at the Panama–Costa Rica accre-tionary complexes (e.g. Osa and Nicoya penin-sulas, Hoernle *et al.* 2002; Buchs *et al.* 2009; Baumgartner *et al.* 2008). These accreted rocks would be potentially genetically and temporally correlative to the LIP rocks on the internal Carib-bean Plate, such as those in Southern Hispaniola, Aruba, Curacaõ, eastern Jamaica, the lower Nicara-gua Rise and the basinal DSDP holes, because there was no subducting plate boundary to separate them when they formed. Such a site for subduction initiation adheres to the mechanical modelling of

Niu *et al.* (2003), in which lateral buoyancy contrast between the thick/depleted oceanic plateau lithosphere and normal oceanic lithosphere plays a key role in initiating subduction beneath the more buoyant feature, which in this case would have been the core of the recently extruded Caribbean LIP. Also in this case, the initiation of NE-dipping subduction agrees with a first-order change in motion of the Farallon Plate with respect to the Caribbean. Preliminary calculations suggest that, in the few million years prior to 84 Ma, Farallon motion was to the SE with respect to the Caribbean, at some 85–120 km/Ma more or less parallel to the proposed Costa Rica–Panama transform margin (Figs 9–11). After 84 Ma, Farallon motion with respect to the Caribbean turned towards the east at about 55 km/Ma, which would substantially add to the horizontal stress at the margin of the LIP.

The idea of initiating the Costa Rica–Panama subduction zone within an active LIP field has another potential implication for the northwestern Nicoya Complex. There, highly deformed Jurassic radiolarites are encased with intrusive contact in younger (Mid-Cretaceous) LIP type basalts (Denyer & Baumgartner 2006; Baumgartner *et al.* 2008). These authors offer two mechanisms for how this may have been achieved: (1) Mid-Cretaceous LIP intrusion incorporated the original sedimentary strata on older crust as it formed a plateau; and (2) the deformed radiolarite slumped from the terrane at or north of the Santa Elena Peninsula (Costa Rica) and onto the LIP surface as it was extruded. Here, we offer a third option, which is that the nascent Costa Rica–Panama subduction zone continued to be the site of local LIP magmatism while initial shortening was beginning. The radiolarite may have been deformed by Coniacian–Santonian accretionary tectonism, concurrent with or followed by Santonian/younger basaltic melt flowing up the juvenile lithospheric scale fault zone that would become the subduction channel. The Nicoya Complex, then, could have formed in exactly the same setting where it occurs today, in the hanging wall of the Costa Rica–Panama Trench, with no need of further tectonic complexity, accretion or translation. A fourth option will be suggested in the following section.

We accept that subduction at the Panama–Costa Rica Arc was initiated by Campanian time, defining a southwestern trailing edge of a 'Caribbean Plate' (Fig. 11), with the boundary probably continuing SW towards northern Peru. As with most of Costa Rica and Panama (except the Berrugate Formation), new age data for primitive island arc rocks from the southern end of this plate boundary (present-day Ecuadorian forearc) also suggest a post-Santonian, most likely Campanian, age for subduction initiation (Luzieux 2007; Vallejo 2007). In addition,

the position and orientation of the Caribbean lithosphere shown (Fig. 11) leaves a large oceanic gap between northern Costa Rica and Chortís–Nicaragua Rise–Jamaica. As pointed out by Pindell & Barrett (1990), such a swath of crust between these arcs allows for contraction between them in the form of northward-dipping subduction. This allowed (1) Chortís–Jamaica to move east along Mexico while the Caribbean lithosphere moves NE; (2) provides an explanation for continuous arc magmatism in Nicaragua Rise–Jamaica through the Early Eocene that otherwise is difficult to conceive of; and (3) predicts that the area of rough bathymetry of the lower Nicaragua Rise (below San Pedro Escarpment) was the site of subduction accretion of Caribbean upper crustal elements. However, apart from the Santa Rosa south-vergent accretionary episode near Santa Elena Peninsula (Baumgartner *et al.* 2008), such accretion remains unproved for the Nicaragua Rise and this is one of larger outstanding questions regarding Caribbean evolution.

Beyond the above considerations for the time of subduction initiation at the western Caribbean boundary, there remains a larger issue associated with this boundary that involves the relative motions of the Pacific and Indo-Atlantic hot spots. As noted earlier, plate circuit determinations of Farallon Plate motion with respect to the Americas back to 84 Ma differ substantially from those based on motions with respect to Pacific hot spots assumed to be fixed to Indo-Atlantic hot spots. The assumption of global hot spot fixity is invalid but there do appear to be two independent hot spot reference frames, Indo-Atlantic and Pacific, within which the member hot spots have remained more or less fixed. Relative motion of these two reference frames can be calculated back to 84 Ma using plate circuits (see above) and older relative motion can only be crudely estimated. Thus, while the motion of the Caribbean Plate with respect to the Americas shown here (Figs 9–11) is very similar to motions of the Farallon Plate with respect to the Americas in Engebretson *et al.* (1985), and would appear to suggest that the Farallon Plate and Caribbean Plate may not have become differentiated until about 84 Ma, we view this as coincidental.

Recent models for Farallon Plate motion with respect to the Pacific Plate (Müller *et al.* 2008) combined with either fixed Pacific hot spots (Wessel & Kroenke 2008) or models in which Pacific and Indo-Atlantic hot spots have moved with respect to one another after 84 Ma (Torsvik *et al.* 2008) give quite different results to Engebretson *et al.* (1985). Whether we assume hot spot fixity prior to 84 Ma, or estimate motion between Pacific and Indo-Atlantic hot spots, is not particularly important; Farallon motion between 120 and 84 Ma in both cases is directed to the SE, parallel to the proposed Costa Rica–Panama transform (pre-trench) boundary. Hot spot drift largely controls the rate (85–120 km/Ma) but not gross direction of relative motion. In order for Caribbean and Farallon motion to have been the same (one plate), the Pacific hot spot reference frame would have to migrate NW with respect to the Indo Atlantic hot spot reference frame at a rate of at least 50–60 km/Ma from 125 to 100 Ma and 50–100 km/Ma from 100 to 84 Ma. These rates are equal to or exceed the northwestward motion of North America in the Indo-Atlantic hot spot reference frame, which we consider implausible.

A hybrid solution for prior to 84 Ma (semi-schematic vector nest inset on Fig. 9) allows for northwestward migration of the Pacific hot spots with respect to the Indo-Atlantic hot spots, but more slowly than the motion of North America. Our solution's rate of motion between the reference frames for this time is broadly comparable to measurable rates after 84 Ma; unfortunately, there is at present no unique solution to this problem, as there is no available plate circuit, palaeomagnetic or other data that can be brought to bear. We consider it most likely that the Farallon and Pacific Plates differentiated from each other prior to 84 Ma, probably at the same time as the onset of westward-dipping subduction beneath the eastern Caribbean at 125 Ma. The suggested SE-directed Farallon–Caribbean motion of 25–50 km/Ma (125–100 Ma), rising to 85 km/Ma (100–84 Ma) requires the development of a subduction zone (probably SE-dipping) in the NW Caribbean that terminates against a sinistral transform fault approximately parallel to the future Panama–Costa Rica Arc (Figs 9–11). Accepting this proposition, strain associated with the transform is a fourth possible mechanism for deforming Jurassic oceanic sediments in the Nicoya Complex of Costa Rica prior to the extrusion of Caribbean LIP basalts into them. The existence of an arc-to-transform transition in this boundary also provides a possible solution to the appearance of volcaniclastic sandstone of the Berrugate Formation in Costa Rica earlier than the Campanian volcanic rocks dated elsewhere. They may derive from the SW end of the SW–NE-trending arc connecting Costa Rica to Mexico, have been deposited within the transform fault zone on the Farallon Plate, and transported perhaps 300 km towards the southeast from their origin in as little as c. 3 Ma.

The model suggests that a new intra-oceanic arc may have developed in the NW Caribbean that would link to Mexico at a trench–trench–trench triple junction in the vicinity of the US–Mexico border. A good candidate is the intra-oceanic Alisitos Arc of Baja California (Sedlock 2003), which

initiated at about 125 Ma probably not far from the continent (explaining the presence of older detrital zircons in associated volcaniclastic sediments) and accreted to the Mexican margin by 105 Ma. Between 125 Ma and eruption of the Caribbean LIP at c. 90 Ma the subduction of 750–1500 km of Farallon crust beneath the NW Caribbean would not prevent the eruption of plume-derived plateau basalts further south.

Accretion of the Alisitos Arc and southward triple junction migration is a necessary consequence of the proposed plate configuration (Fig. 9). Intra-oceanic arc fragments accreted further south than Baja may include the forearc of Central America (Geldmacher et al. 2008). The Caribbean–Chortís relative motions shown in our maps suggest that the trench–trench–trench triple junction migrated south until about 100 Ma, and thereafter the NE-trending plate boundary was subducted beneath Chortís (Fig. 11). The rate of this plate boundary subduction would have increased markedly at about 84 Ma, when Farallon–Caribbean relative motion direction rotated towards the east. Associated burial, imbrication and uplift may be the origin of the c. 80 Ma thermal event that affected Guatemalan forearc rocks (Geldmacher et al. 2008). Much of the Mesquito Composite Oceanic Terrane (Baumgartner et al. 2008) between the continental Chortís Block and the Central American trench may be the result of the accretion–subduction of the trailing edge Caribbean Arc, while the Siuna Terrane southeast of Chortís may comprise leading edge Caribbean Arc and HP/LT rocks accreted to Chortís prior to the Albian, immediately before Mesquito accretion started. The 84 Ma change in Farallon–Caribbean relative motion initiated NE-dipping subduction at the site of the proposed transform fault southwest of Costa Rica–Panama, leading to the onset of arc volcanism in those areas (Fig. 11). At the same time, slower and more oblique subduction on the proposed NE-trending trench may have led to reduced arc volcanism between Costa Rica and Central America.

The Caribbean LIP

Between the North and South American zones of Caribbean Plate boundary deformation, the Caribbean large igneous province (LIP), or plateau, was extruded across much of the pre-existing Caribbean oceanic lithosphere, in which coeval NE–SW extensional faulting was occurring (Driscoll & Diebold 1999; Diebold 2009). Pindell (2004) and Pindell et al. (2006) pointed out that the concurrence of seafloor spreading between North and South America and the consumption of the Colombian Marginal Seaway beneath the Caribbean lithosphere leads to the nearly inescapable conclusion that subduction of the Proto-Caribbean spreading ridge produced a slab gap beneath the Caribbean lithosphere from 125 Ma (onset of SW-dipping subduction) through about 71 Ma (termination of Proto-Caribbean seafloor spreading). These authors loosely suggested that the Caribbean LIP might relate to mantle convection (i.e. to the Proto-Caribbean spreading cell) associated with this slab gap, as this age range effectively brackets the age of most Caribbean LIP extrusion (Kerr et al. 2003). Indeed, our plate reconstructions herein (Figs 10–12) place the slab gap directly beneath much of, but certainly not all, the Caribbean LIP's known occurrence at the appropriate time. This includes our interpretation for the original area of the Bath–Dunrobin Formation of eastern Jamaica, recently classified as plume-related (Hastie et al. 2008), although the Bath–Dunrobin Formation may not have merged with the rest of Jamaica until Early Eocene time, after a history of end-Cretaceous accretion into the Lower Nicaragua Rise and subsequent NE-trending sinistral shear along with the Blue Mountain HP/LT suite. However, it is difficult to model the slab gap as having reached the SW Caribbean region: areas such as Costa Rica, Panama and the Pacific coastal zone down to Ecuador probably did not overlie the Proto-Caribbean slab gap, so the slab gap concept is probably not a sole explanation for the Caribbean LIP. Having said that, it remains difficult to judge whether exposed 'plateau-related rocks' along the Pacific forearc such as at the Nicoya and Azueros Peninsulas and Gorgona Island represent the Caribbean Plate's hanging wall, with direct implications for the Caribbean LIP, or Farallon Plate seamounts/plateaus that were accreted into the Caribbean Plate's forearc during subduction, with little implication for the Caribbean. Nonetheless, other areas of Mid-Cretaceous 'LIP-like' magmatism include the Oriente Basin of Ecuador (Barragán et al. 2005), Texas (Byerly 1991) and the Eastern Cordillera of Colombia (Vasquez & Altenberger 2005), which of course cannot pertain to a Proto-Caribbean slab gap model. In addition, geochemical arguments seem to require a deep mantle plume source for many of the Caribbean LIP magmas (Kerr et al. 2003), at odds with the idea of a convective spreading cell source in a slab gap. Thus, the Mid-Cretaceous was a time of widespread igneous activity in the region with a probable deep mantle source, and only some of this activity occurred above the Proto-Caribbean slab gap. For these various reasons, we presently consider that the Caribbean LIP was largely fed by deep mantle plume(s), but that the Proto-Caribbean slab gap allowed plume magmatism to reach the central and northeastern parts of the Caribbean lithosphere, perhaps focused by rising along the site of the subducted Proto-Caribbean convective spreading cell (a subducted Icelandic-type setting). Once the plume(s) reached the base of the Caribbean lithosphere, plume magma may have spread laterally over a larger area (possibly beyond the strict limits of the slab gap), from which it was locally able to propagate toward the surface along extensional faults at crustal (brittle) levels.

The slab gap concept appears to reconcile how large areas of the Mid-Cretaceous Caribbean LIP show no sign of a supra-subduction signature, despite the strong probability that the LIP was extuded while SW-dipping subduction of Proto-Caribbean lithosphere beneath the Caribbean Arc had occurred since the Aptian (Pindell 2004). We might also expect the LIP magmas above the Proto-Caribbean slabs flanking the slab gap to show some slab contamination, although no such contamination has yet been recognized. However, areas where this might have occurred have not necessarily been analysed. One such area that is predicted by our reconstructions to have overlain a subducted Proto-Caribbean slab flank, and that might show such contamination with further study, is the southwestern portion of Hispaniola (Sierras des Neiba and Bahoruco).

Accepting a mantle plume role in the Caribbean LIP, a point of ongoing debate is whether the palaeo-Galapagos hotspot was involved (Duncan & Hargraves 1984), if it indeed existed in the Mid-Cretaceous (Hoernle et al. 2004). In view of the discussion above, integration of plate circuit data back to 84 Ma (Doubrovine & Tarduno 2008) and Pacific Plate motion with respect to Pacific hot spots (Pilger 2003, after Raymond et al. 2000; Wessel et al. 2006; Wessel & Kroenke 2008) allow us to identify a significant westward drift of the Pacific hot spot reference frame relative to the Müller et al. (1993) Indo-Atlantic hot spot reference frame (Fig. 14). In addition, we extrapolate the curves back to 92 Ma, the approximate onset of most Caribbean Plateau basalt magmatism. By placing the end points of these curves on the Galapagos Islands in the Indo-Atlantic hot spot projection of our maps, the curves denote the migration of the Galapagos hot spot by some 2200 km relative to the Indo-Atlantic reference frame. In addition, Steinberger (2002) proposed that the Easter Island hot spot drifts west relative to other Pacific hot spots at 10–20 mm/annum due to return mantle flow from the Andes Trench. We suggest that the Galapagos hot spot may have behaved similarly with respect to Panama–Costa Rica Trench since its inception at about 75 Ma. If so, then this drift may add perhaps 800 km to the movement of the hot spot relative to the Indo-Atlantic hot spots compared with the plate circuit calculations (heavy grey arrow on Fig. 14, deviating from the Pacific drift curves

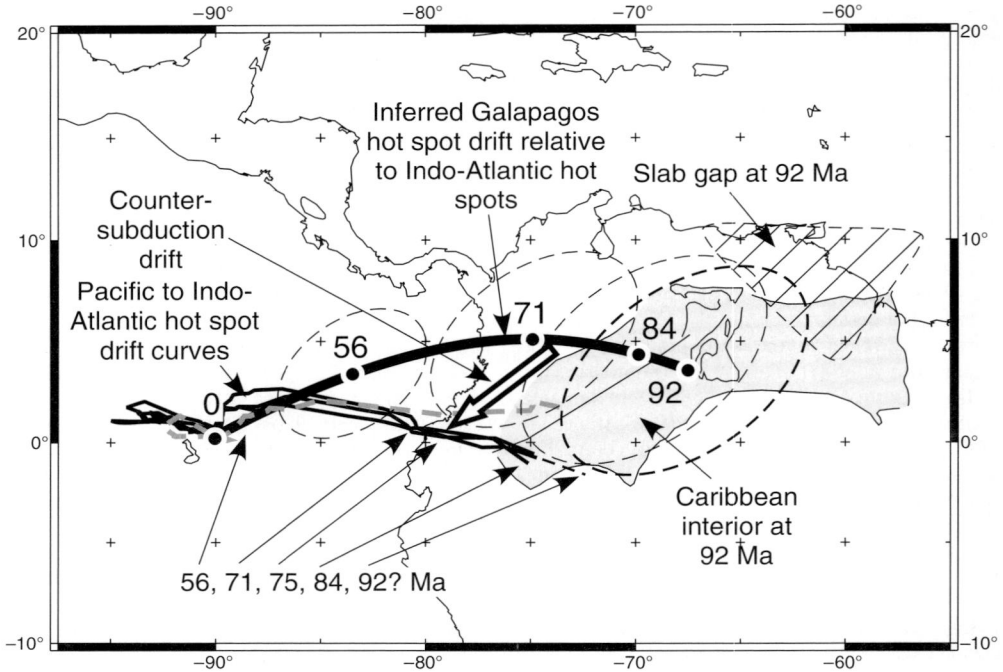

Fig. 14. Possible migration path since 92 Ma of the Galapagos hot spot relative to the Indo-Atlantic reference frame of our map set (heavy black line with ages shown). Ellipses show generously estimated errors. Finer weight curves emanating from the Galapagos Islands: calculated motion histories of the Pacific hot spot reference frame relative to the Indo-Atlantic frame, determined for the Galapagos hot spot ($0°/90°W$); grey line, Pilger (2003); black lines, Wessel (Wessel *et al.* 2006; Wessel & Kroenke 2008—models 08A and 08G). Heavy arrow is a subjective correction to the above curves following concepts of Steinberger (2002; see text). Slant-ruled area is the estimated position of the Proto-Caribbean slab gap at 92 Ma, interpolated from Figs 10 & 11; note the proposed position of the hot spot lies entirely in line with the slab gap, which we perceive allowed the deep mantle plume to reach the base of the overriding Caribbean Plate. The Caribbean interior basin at 92 Ma is shown in grey. The palaeopositions of the Caribbean lithosphere and the Galapagos hot spot become superposed at 92 Ma, the age of most of the Caribbean Basalt Plateau. Also, the deep hot spot probably passed beneath the Panama–Costa Rica Trench in the Maastrichtian–Paleocene, just after arc inception, but following most plateau magmatism. Palaeogene plateau-type basalts at Azuero Peninsula (Hoernle *et al.* 2002) were probably accreted from the subducting plate after the passage of the hot spot beneath the arc, but some Palaeogene basalts along Central America may pertain directly to the passage of the hot spot beneath the arc itself.

at about 75 Ma). The solid flow line shown with ages is our estimate of the sum of these two processes. The two curves are drawn parallel from 92 to 75 Ma, after which time subduction may have driven the hot spot westwards relative to the central Pacific hotspots. We have crudely estimated the possible area (subject to large error) in which Galapagos hot spot magmatism may have occurred at the times shown. We suggest that there is a plausible match between the 92 Ma position of the predicted area of Galapagos hot spot magmatism and the 92 Ma position of the Caribbean Basalt Plateau (interpolated between the position on our 100 Ma and 84 Ma maps, Figs 10 & 11). Larger or smaller values for the subduction-related drift than

800 km would produce a less satisfactory fit. We chose this value because it is reasonable and provides a good match, but there is no independent way of refining the estimate, defining errors, or proving the Galapagos hot spot–Caribbean Plateau relationship. Models that track the position of the Galapagos hot spot in the Indo-Atlantic reference frame or assume Cretaceous to Present fixity of Pacific and Indo-Atlantic hot spots fail to place the Galapagos hot spot beneath the Caribbean Plate (Pindell *et al.* 2006), but accounting for relative motion between these reference frames since 84 Ma (the oldest possible plate circuit) and possible westward or southwestward additional drift of the Galapagos hot spot due to deep return flow in

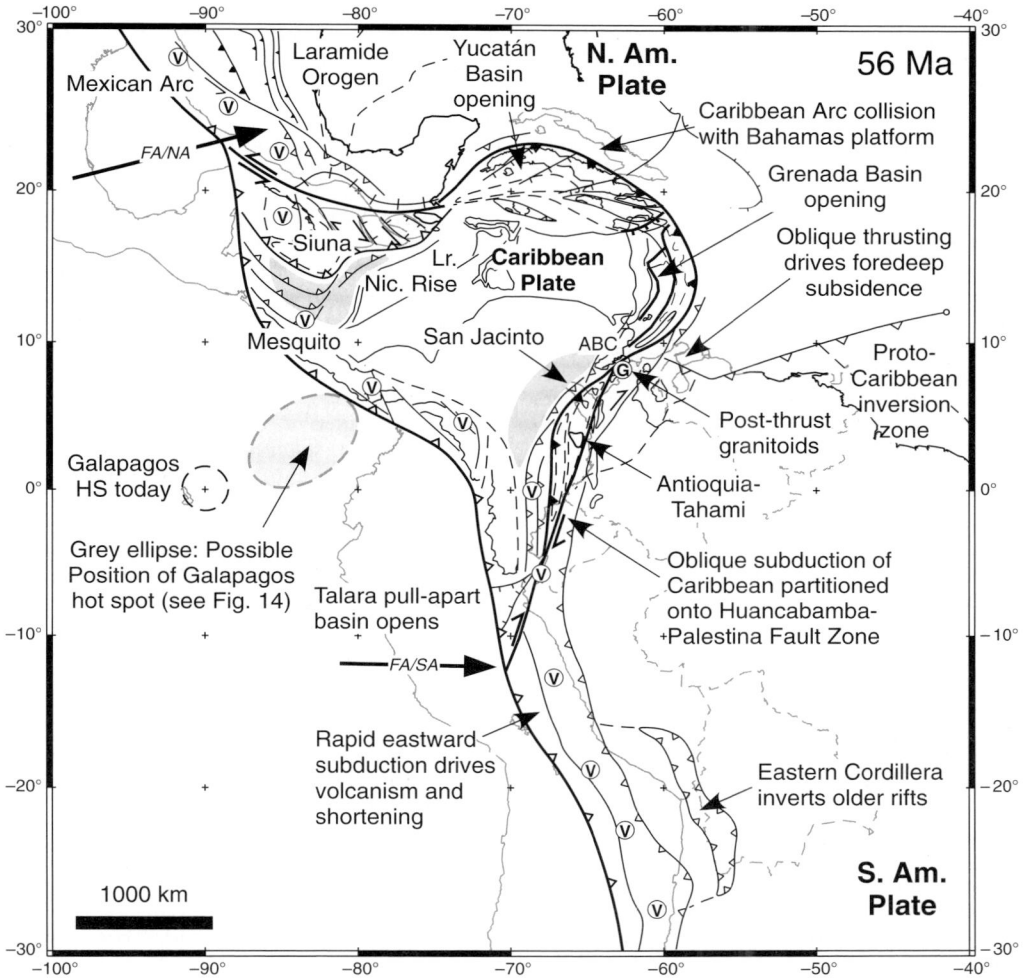

Fig. 15. A 56 Ma reconstruction of the circum-Caribbean region, shown in the Indo-Atlantic hot spot reference frame. By this time, oblique intra-arc basins were opening as the Caribbean spreads into the wider Proto-Caribbean seaway towards the Florida–Bahamas platform (Yucatán intra-arc Basin) and South America (Grenada intra-arc Basin). Subduction of Caribbean crust beneath Chortís–Jamaica arc trend is almost complete, which continues to accrete the composite Mesquite accretionary terrane. The northward zippering of the Panama Arc outside the Western Cordillera in Colombia continues. Subduction of Caribbean crust beneath Colombia was becoming more head-on. Slow convergence was underway by this time between North and South America: this shortening was probably accommodated at the Proto-Caribbean subduction zone along northern South America, although it is not clear if this structure had formed along its entire length, or if it propagated east ahead of the Caribbean Plate with time.

the mantle driven by subduction suggests that a palaeo-Galapagos hot spot may well have been the source of the Caribbean Plateau, with the added factor that the Proto-Caribbean slab gap helped to focus the basalts very near to the Antilles volcanic arc. We show the position of the possible palaeo-Galapagos hot spot in relation to the Caribbean Plate, as reconstructed in Figure 14, on Figures 11, 12 & 15.

Comparison with alternative scenarios for Aptian–Maastrichtian evolution of the Caribbean

In addition to the above Pacific-origin Caribbean model, there are two other types of Pacific-origin model for Cretaceous time: the 'far-travelled Farallon–Guerrero Arc model' (e.g. Dickinson &

Lawton 2001; Mann *et al.* 2007), and the 'delayed polarity reversal model' (e.g. Burke 1988; Kerr *et al.* 2003). Both these models have features that appear to be incompatible with geological observations in the circum-Caribbean region.

Far-travelled Farallon–Guerrero Arc models place a subduction zone at the leading edge of the Farallon Plate (that is, future Caribbean lithosphere) far to the west of the Americas at *c.* 125 Ma, migrating east and consuming 'Mescalera' or 'Arperos' oceanic lithosphere which itself is presumed to concurrently subduct eastward beneath the Cordilleran and Trans-American Plate boundary. Apparent geological contradictions include: (1) the lack of an explanation for continental crustal fragments or continent-derived sediment in the Caribbean and Guerrero arcs; (2) the timing of interaction of the Caribbean Plate with the Trans-American Arc (90 Ma) that we believe is 30 Ma too late (in the Mann *et al.* 2007 version at least); and (3) lack of evidence for amalgamation of two discrete arcs and subduction complexes in the Caribbean Arc.

Delayed polarity-reversal models call for east-dipping subduction of Farallon lithosphere beneath the Trans-American Arc until *c.* 80–88 Ma, when it is proposed that the Trans-American trench was choked by the newly erupted Caribbean Basalt Plateau, thus forcing subduction polarity to reverse. A key argument for this model is that the basalt plateau occurs very near to the Caribbean Arc (especially in Hispaniola), and that this magmatic incompatibility can be resolved by allowing for subduction of some amount of intervening crust (between the plateau and arc) prior to juxtaposition. However, accepting that the line of juxtaposition (lower Nicaragua Rise, Los Pozos Fault zone in central Hispaniola and the Muertos Trough) is the site of post-plateau shortening and dislocation (Pindell & Barrett 1990; Dolan *et al.* 1991; maps herein), this argument becomes less compelling. Further, delayed reversal models do not explain: (1) the history of HP/LT metamorphism in the northern Caribbean beginning at about 118 Ma; (2) the lack a disruptive event at 80–90 Ma in the *P–T–t* paths of such rocks; or (3) lack of evidence for major arc-wide uplift, erosion and cooling at 80–90 Ma. Further still, we wish to emphasize that the slab gap aspect of our model herein does allow for plume rocks to be emplaced quite near to, but not within, the active Caribbean Arc (Fig. 11).

Maastrichtian–Palaeogene expansion of the Caribbean Plate into the Proto-Caribbean Seaway

The Maastrichtian–Palaeogene evolutionary interval (Figs 12 and 15–17) involves: (1) the cessation of Proto-Caribbean seafloor spreading by 71 Ma (Pindell *et al.* 1988; Müller *et al.* 1999); (2) north-vergent inversion (potentially developing into south-dipping subduction) along the foot of the northern South American rifted margin (Pindell *et al.* 1991, 2006; Pindell & Kennan 2007*b*); (3) the migration of the Caribbean Arc from the Yucatán–Guajíra 'bottleneck' to the Bahamas and western Venezuelan collision zones (Pindell *et al.* 1988, 2005); (4) the opening of the Yucatán (Pindell *et al.* 2005) and Grenada (revised model proposed here) intra-arc basins as a means of the arc expanding into the Proto-Caribbean Seaway, which was wider than the Yucatán–Guajíra bottleneck, and maintaining collisional continuity with the American margins (Pindell & Barrett 1990); (5) the migration of Chortís–Nicaragua Rise–Jamaica along southwestern Mexico/Yucatán (Pindell *et al.* 1988); (6) polarity reversal/onset of northward dipping subduction of Caribbean lithosphere at the Lower Nicaragua Rise, which we believe was the cause of arc magmatism in the eastward migrating Nicaragua Rise–Jamaica and took up the convergence between that terrane and the Caribbean Plate while the latter migrated northeast into the Proto-Caribbean Seaway (Pindell & Barrett 1990); and (7) the poorly-dated Eocene amalgamation of the Chortís and Panama–Costa Rica arcs into a single Middle American arc. Here, we will focus new considerations on the Motagua Fault Zone of Guatemala and the opening history of the Grenada and the Tobago basins.

Figure 18 shows the relationship of the Motagua Fault Zone of Guatemala to the broad and diffuse Cocos–North America–Caribbean triple junction in southern Mexico, Guatemala and El Salvador. A smooth and continuous eastward migration of Chortís from a position off SW Mexico is commonly portrayed using the eastward-younging onset of arc magmatism in southern Mexico as a yardstick (Pindell & Barrett 1990; Ferrari *et al.* 1999). However, as shown here (Fig. 18) and as pointed out by Keppie & Morán-Zenteno (2005) and Guzman-Speziale & Meneses-Rocha (2000), this is not necessarily a simple case of triple junction migration. A precise definition of the plate boundaries in the region is not yet to hand, and thus it is not clear how to restore the crustal elements in the region back in time. This complexity, along with an apparent lack of disruption in gravity data and seismic lines in the Gulf of Tehuantepec, led Keppie & Morán-Zenteno (2005) to question the commonly inferred westward trace of the Motagua Fault toward the Middle America Trench, and hence to doubt whether Chortís and Mexico had been adjacent in the Cretaceous, despite the lithological similarities between Chortís and the Oaxaca and Mixteca terranes of Mexico (e.g. Rogers *et al.* 2007*b*). These authors

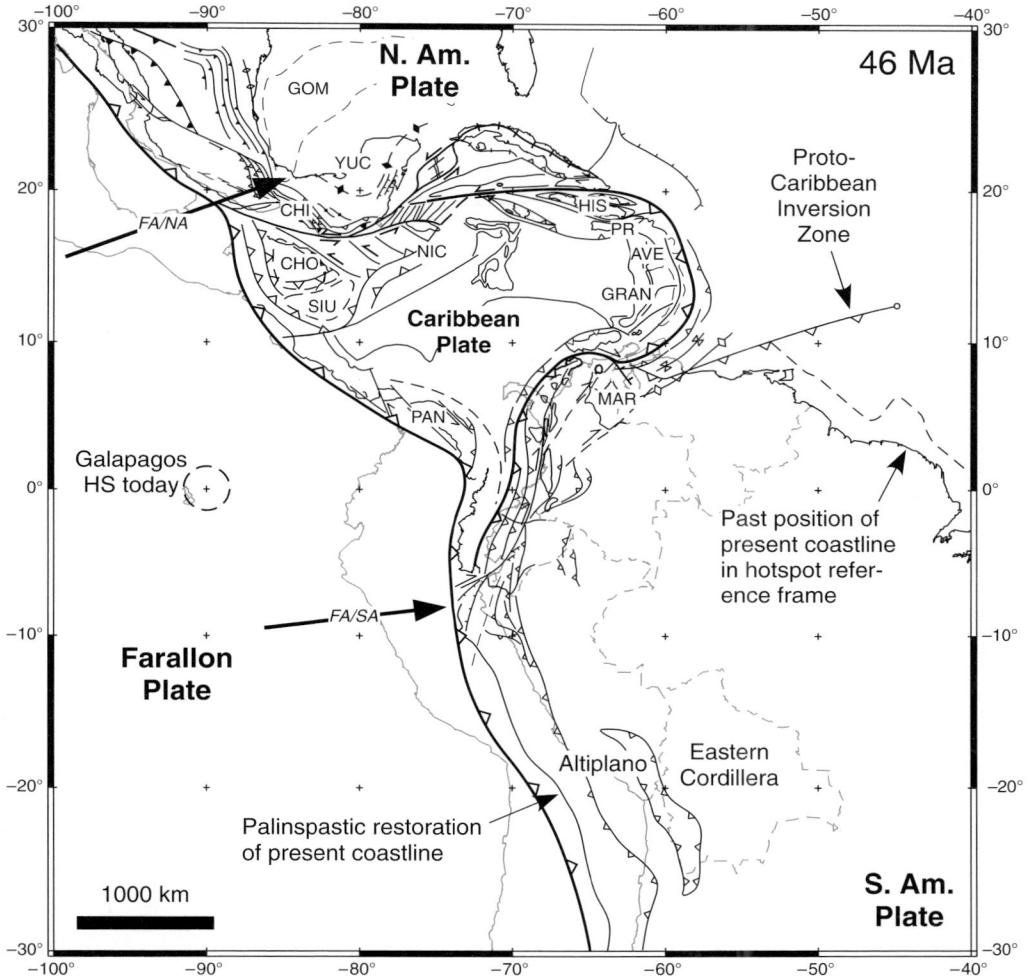

Fig. 16. A 46 Ma reconstruction of the circum-Caribbean region, shown in the Indo-Atlantic hot spot reference frame. Northward drift of the Caribbean (in the hot spot frame) has stopped. Collision of Cuba with the Bahamas Platform terminated the opening of the Yucatán Basin and resulted in continued Caribbean–North America relative motion occurring on the Cayman Trough. The end of subduction beneath Chortís and Nicaragua Rise resulted in their being incorporated into (but actually onto) the Caribbean Plate. The southeastern Caribbean Plate advanced SE toward the central Venezuelan margin along the Lara transfer zone northeast of Lake Maracaibo. The southern part of the Panama Arc was accreting into the Ecuadorian forearc. Caribbean–South America motion rotates almost orthogonal to the Huancabamba–Palestina Fault Zone slowing the rate of northward terrane migration in the Northern Andes.

position Chortís out in the Pacific away from Mexico, employing faults visible on seismic at about 14.3°N in the offshore forearc as a means of deriving Chortís from the WSW. These concerns caused us to question the nature of the crust in the Gulf of Tehuantepec, which we refer to here as the Tehuantepec Terrane, and which Keppie & Morán portray as a fairly stable block bounded by the Middle America Trench to the SW and by the Chiapas Massif to the NE (Fig. 18). First, we ruled out that this crust belongs to Yucatán, as we are

confident that the transform that carried Yucatán and Chiapas Massif to their present positions crosses the Isthmus of Tehuantepec from the Vera-cruz Basin and runs parallel to and along the south-east flank of the Chiapas Massif (the 'Tonalá Fault' *sensu* Geological Survey of Mexico), and not further west (see Figs 5–7). Second, we could not accept that the Tehuantepec Terrane once belonged to Chortís, because the magnitude of shortening in the Chiapas Foldbelt (less than *c.* 70 km) is too small for the terrane to restore south of the eastward

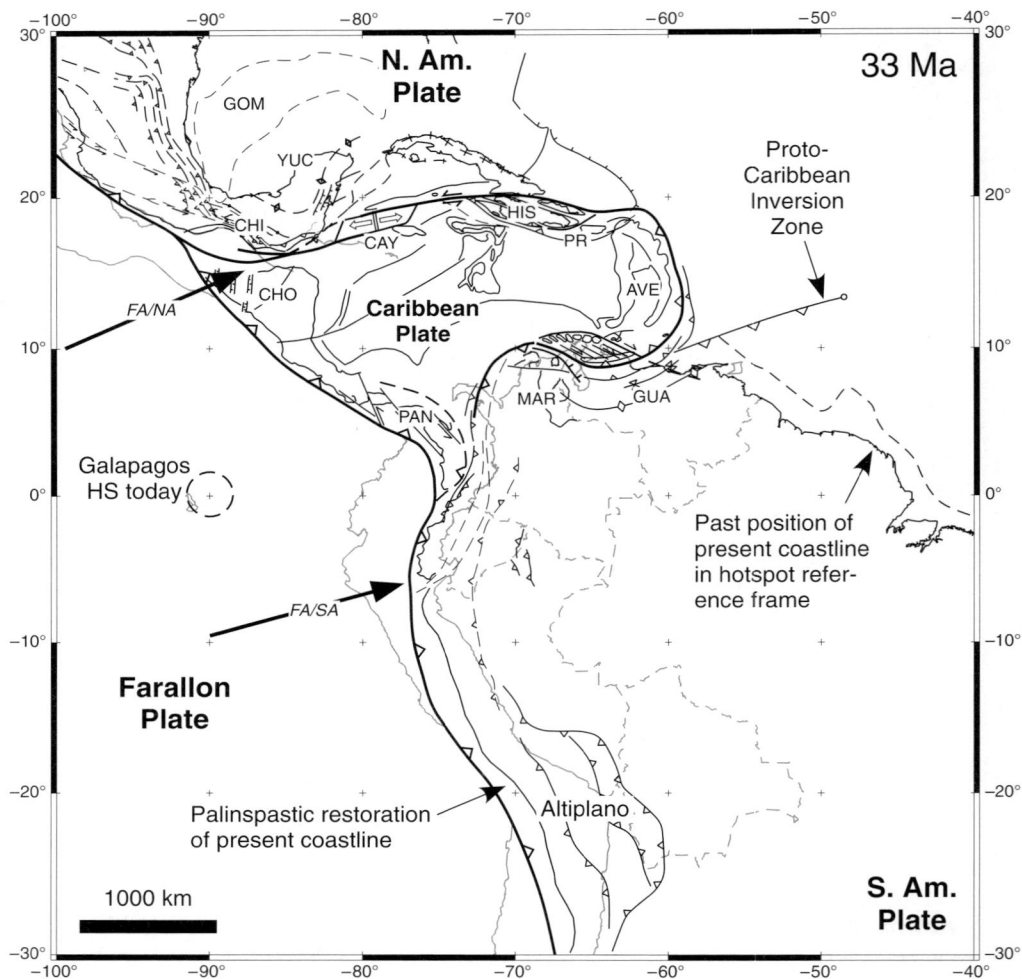

Fig. 17. A 33 Ma reconstruction of the circum-Caribbean region, shown in the Indo-Atlantic hot spot reference frame. North America–Caribbean Plate boundary is taking on the form of today's boundary system. South America–Caribbean motion is ESE-directed, resulting in overthrusting of Caribbean terranes onto central and eastern Venezuela. Southeast dipping subduction beneath the northern Andes at the western South Caribbean Foldbelt was propagating eastward to the north of Maracaibo Block. As the oblique collision progressed along Venezuela, continued convergence would necessarily transfer to this eastward-propagating, south-dipping South Caribbean Foldbelt.

projection of the southern Mexican trench, which we would expect had the terrane originated from the southwestern margin of Mexico. Third, the terrane could be an ESE extension of the Sierra Madre del Sur of Mexico, but the presence of Upper Cretaceous volcanic rocks ('Turonian–Santonian basalt, dacite, and tonalitic agglomerate'; Sanchez-Barreda 1981; Keppie & Morán-Zenteno 2005) in well SC-1 (Fig. 18) from this terrane is atypical of southern Mexico. Thus, we consider that it may be a remnant fragment of the Caribbean Arc which, rather than Chortís, was judged to have

collided with Mexico here by Pindell & Dewey (1982). The allochthonous Tehuantepec terrane would logically connect the allochthonous Siuna Belt with the Nicaragua Rise and Jamaican portions of the Caribbean Arc, collectively forming the arc's western end (Figs 11 & 12). In addition, a small area of poorly dated Cretaceous volcanic rocks onshore Mexico (the Chontal Arc volcanic rocks, Carfantan 1986; Fig. 18) could be equivalent and also a part of this allochthonous trend. The Tehuantepec Terrane would thus have been isolated and acquired by North America upon the

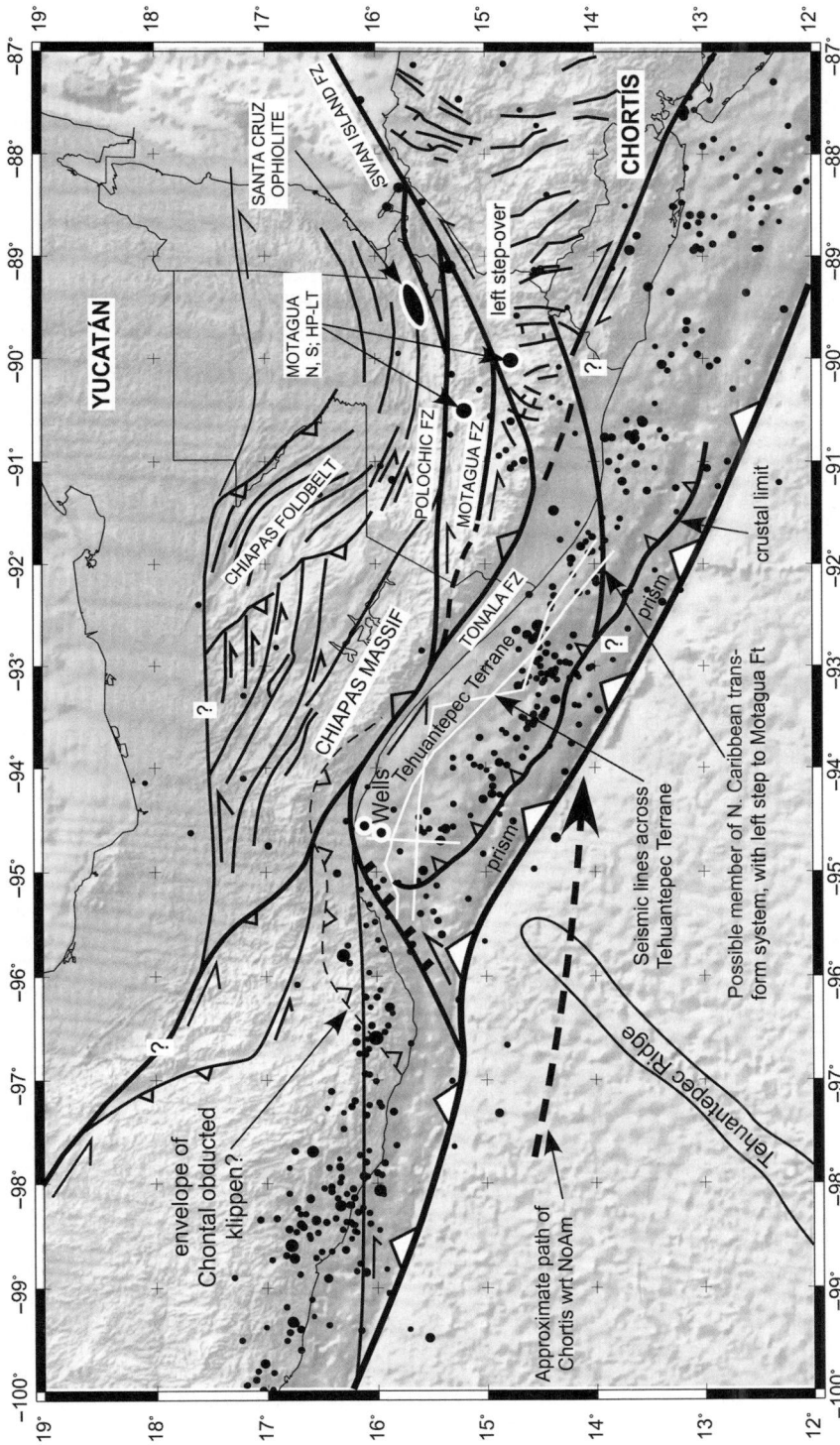

Fig. 18. North America–Cocos–Caribbean diffuse triple junction, showing seismicity, gravity, major and lesser faults, and our proposed Caribbean arc fragments (Chontal klippen and parts of Tehuantepec terrane). We interpret the primary Chortís-North America plate boundary to lie outboard of Tehuantepec Terrane (trajectory of dashed line, prior to Chiapas shortening). The kink in the trench in SW Gulf of Tehuantepec, associated with a break in the forearc basement, may result from Mid-Miocene shortening in Chiapas Massif/Foldbelt, and movement along Tonalá Fault. Numerous north–south grabens that reflect west–east stretching in the tail of Chortís may form a step allowing transfer of sinistral shear to the western (thrusted) flank of the Tehuantepec Terrane.

Maastrichtian onset of transcurrent motion along the North Chortís–Motagua Fault Zone after the collision of the arc with Yucatán/Chiapas Massif.

Considering the Tehuantepec Terrane may have once been part of the Caribbean Arc, it is possible that most of the Caribbean–North American relative plate motion has passed south of the Tehuantepec terrane in the zone of intense seismicity at about 14.3°N and 93°W (Fig. 18). This also satisfies Keppie & Morán-Zenteno's concerns about the apparent paucity of faulting further north where most authors have drawn the westward extension of the Motagua Fault. However, broad strain is also evident in Chiapas, where perhaps 100–200 km of dextral transpressive movement has occurred on the Polochic Fault in Neogene–Quaternary times. Displacement of the Tehuantepec Terrane seems to be a part of this story as is implied by the strong NE–SW trending negative gravity anomaly in the northwesternmost Gulf of Tehuantepec, which may be a break-away detachment between the Tehuantepec and Sierra Madre del Sur. Restoration of about 70 km shortening in the Chiapas Foldbelt (Tectonic Analysis Inc., unpublished data) would appear to realign the SW flank of the Tehuantepec terrane with a smooth east–southeastward projection of the Mexican trench. If the bulk dextral strain can be shown to be larger than 100–200 km in the Chiapas–Tehuantepec region, then we would expect transfer of motion into the area from as far inland as the Trans-Mexican Volcanic Belt (Fig. 18). In conclusion, the Chortís Block appears to have passed the Gulf of Tehuantepec to the south of the Tehuantepec terrane, using faults acknowledged on seismic by Keppie & Morán-Zenteno (2005) that lie in a zone of strong seismicity. The deformation in southeastern Mexico is only secondary by this reasoning, and may be very much the result of this area becoming the hanging wall to a subduction zone only since the Miocene by the eastward movement of Chortís, especially one where a buoyant ridge (Tehuantepec Ridge) is entering the trench.

Acknowledging the possible existence of a swath of Caribbean Arc forearc in the Gulf of Tehuantepec, which should possess HP/LT metamorphic rocks like all the other circum-Caribbean forearc terranes, is potentially significant with regard to assessing the history of the Motagua Fault Zone. Donnelly et al. (1990) built a case for a Chortís–Yucatán collision, and argued that the nearby occurrences of the El Tambor HP/LT rocks on the northern and southern flanks of the Motagua Valley disproved a large strike–slip displacement along the Motagua Fault. This view requires the Cayman Trough to be seen as something other than a Cenozoic pull-apart basin, which in turn makes it difficult to reconcile the Eocene to Recent history of

subduction related magmatism in the Lesser Antilles, which requires significant (c. 1000 km) Caribbean–North America displacement. Since then, ^{40}Ar–^{39}Ar cooling ages on the northern and southern El Tambor HP/LT rocks have been shown to be different, that is, c. 120 and 70 Ma, respectively, and this discovery, in conjunction with an acceptance of the overwhelming evidence for large displacements on the Motagua Fault Zone, led to the proposal of the former existence of two entirely distinct subduction zones with opposing polarities and different times of collisional uplift (Harlow et al. 2004). The 120 Ma cooling event was interpreted as an emplacement of the southern El Tambor rocks onto Chortís (north-dipping subduction) which occurred between Chortís and SW Mexico, while the 70 Ma collision, emplacing the northern El Tambor rocks onto Yucatán (south-dipping subduction), was interpreted as Pindell & Dewey (1982) did as marking the collision between the Caribbean Arc and Yucatán. The different collisional settings were proposed in order to allow the acknowledged large strike–slip offset to bring the southern and northern Tambor units together today. This complex model survives, despite the more recent acquisition of Nm–Nd ages on both the northern and southern Tambor units of about 132 Ma (Brueckner et al. 2005; Ratschbacher et al. 2009), which suggests instead to us that they may both have formed in the same subduction zone, though not necessarily in the same place. In addition, a hypothetical Late Jurassic rifting event between Chortís and Mexico is proposed as part of this model (Mann 2007) in order to create an oceanic basin that might have started to close by 130 Ma and been sutured by 120 Ma.

We do not accept that the proposed rift event led to the opening of a seaway with oceanic crust basement between Chortís and southern Mexico; we see no evidence for a rifted margin on northern Chortís on a scale compatible with creation of an oceanic basin, and neither is there any sign of a Late Jurassic–Early Cretaceous north-facing sedimentary margin or syn-collisional foredeep basin in northern Chortís onto which the southern El Tambor was supposedly emplaced during the Aptian, which appears to have been a time of extension in central Chortís (Rogers et al. 2007a). Instead, we stick to the original Chortís–Mexico relationship of Pindell & Dewey (1982), Pindell et al. (1988) and Rosenfeld (1993) in which the Caribbean Arc, rather than Chortís, collided with southern Chiapas Massif and southern Yucatán to create the Motagua ophiolitic Suture with its HP/LT rocks, and in which Chortís later migrated eastward to create the Motagua shear zone. At issue is the mode and timing of juxtaposition of the El Tambor South unit with the Las Ovejas metamorphic rocks and San Diego Phyllite of the Chortís Block. Pindell & Barrett (1990) stated in their note added in proof, that 'emplacement of the [southern] El Tambor onto Chortís could be a Cenozoic extrusion (flower structure) during strike slip [on Motagua Fault], prior to most Neogene motion through Guatemala on the Polochic Fault (Burkart 1983). In cross section only, the resulting orogen appears as a collision between Chortís and Yucatán'. Similarly, the appearance of a collision between Chortís and southern Mexico may be misleading. If, during southeastward transpressive migration of Chortís towards its present position, strain were strongly partitioned between sinistral slip and orthogonal thrusting, it would be possible to superimpose Cenozoic sinistral shear on slightly older thrust structures while Chortís lay south of Tehuantepec.

Pindell et al. (2005) compiled data to show that HP/LT metamorphic ages in the Caribbean Arc span the period of active Caribbean Arc subduction, from the onset of SW-dipping subduction to collision. In Cuba and in Hispaniola, cooling ages on HP/LT mineral suites begin at about 118 Ma and continue up to about 70 Ma in Cuba (García-Casco 2008b; Stanek 2009) and younger in Hispaniola (Krebs et al. 2008), defining the period of subduction from initiation to collision with 'Caribeana', a sediment pile deposited along southern Yucatán, and the Bahamas. The Cuban ages in particular are within the errors of the $^{40}Ar-^{39}Ar$ cooling ages for the two groups of El Tambor HP/LT rocks (northern and southern), a fact that we expect is significant, possibly placing both El Tambor HP/LT units along the same Caribbean Arc trench, although originally separated by many hundreds of kilometres along strike. However, 132 Ma Sm–Nd ages from the Guatemalan rocks have not yet been found in Cuba or Hispaniola.

We propose a history for the southern and northern El Tambor suites that is intimately related to that of the Caribbean Arc, and has nothing to do with a hypothetical Chortís–Mexico collision. In the Late Jurassic and Early Cretaceous, eastward-dipping subduction beneath the Americas is indicated by continental volcanic arc belts east of coeval subduction complexes at the coast. We infer that Chortís was part of this continental belt, and that the dense rocks beneath the Sandino Basin of western Chortís pertain to a primary east-dipping Benioff Zone that remains active today, although it has stepped westward somewhat since the Early Cretaceous. As the gap between the Americas grew between 140 and 125 Ma, a largely transcurrent boundary spanned the gap between Chortís and Ecuador, along which arc, forearc and subduction complex terranes of western Mexico and Chortís, mostly oceanic but partly continental,

should have migrated southeast due to the obliquity of convergence, taking up a position south of Chortís along the transcurrent plate boundary. We consider the El Tambor North and South as well as the Siuna terrane with its 139 Ma $^{40}Ar-^{39}Ar$ age (Baumgartner et al. 2008), as well as Grenville age continental blocks in the subduction mélange of central Cuba (Renne et al. 1989), were carried along in this manner. We suggest that these HP/LT terranes lay SW of Acapulco at 130 Ma (Fig. 6). By the early Aptian, Farallon–North American relative motion became much more NE–SW (Engebretson et al. 1985), triggering convergence at the previously transcurrent boundary which we argue was manifested as the onset of SW-dipping subduction of Proto-Caribbean lithosphere beneath the band of terranes that would go on to form the underpinnings of the Caribbean Arc (Pindell et al. 2005). The inception of SW-dipping subduction began the cooling and uplift, possibly by subduction zone counterflow, of HP/LT metamorphic rocks in the new Caribbean Arc hanging wall. The southern El Tambor eclogites were uplifted early on (c. 125–118 Ma), and subsequently remained above the $^{40}Ar-^{39}Ar$ blocking temperature in the hanging wall. Continued SW-dipping subduction into the Late Cretaceous produced younger HP/LT rocks as well, and the initial collision of the Tehuantepec–Nicaragua Rise–Jamaica terrane with the southern Chiapas Massif and Yucatán margin caused HP/LT metamorphism in some of the passive margin sediment wedge strata (Caribeana, García-Casco et al. 2008b) and some Yucatán marginal basement slices (e.g. Chuacús Formation, Martens et al. 2008). This belt of marginal basement and overlying sediments was uplifted and cooled in the Maastrichtian by obduction onto Yucatán, imbricating Proto-Caribbean oceanic crust (Santa Cruz Ophiolite) and Yucatán shelf strata (Cobán, Campur), and producing the Sepur foredeep basin (eastward younging) of northern Guatemala and Belize (Pindell & Dewey 1982). The obduction set the Maastrichtian $^{40}Ar-^{39}Ar$ cooling ages for northern El Tambor and Chuacús HP/LT rocks (Harlow et al. 2004). In this model, the fact that the northern El Tambor unit carries a 132 Ma Sm–Nd age, like the southern El Tambor, could imply that (1) the northern El Tambor is equivalent to but along strike of the southern El Tambor, but that it was subducted deeper again after 120 Ma to reset the $^{40}Ar-^{39}Ar$ age in the subduction channel prior to uplift at 70 Ma, or (2) some of the original 132 Ma HP/LT material remained continuously above the $^{40}Ar-^{39}Ar$ blocking temperature (i.e. presumably deeper) until the Maastrichtian. Either possibility is interesting: the former suggests that rock can flow upwards and downwards in a subduction channel before final exhumation, the latter that rock may reside at depth in subduction channels for long periods of time (60 Ma).

From the above, we appear to have a simple means of introducing HP/LT complexes with cooling ages ranging from Early Cretaceous (and potentially older) through Maastrichtian along the southern flank of Chiapas Massif–southern Yucatán, including in the Gulf of Tehuantepec, without invoking collision of Chortís with Mexico or Yucatán, for which evidence is lacking. This is the also the case in Cuba and Hispaniola where Chortís–Yucatán collision obviously never occurred. The question is, then, can the southern El Tambor rocks be emplaced onto Chortís during a non-collisional strike–slip migration of Chortís along the zone of arc accretion against Yucatán? Sisson et al. (2008) reported fission track ages for the northern basement rocks of Chortís of 35–15 Ma, demonstrating that these rocks cooled through 200–100 °C and were situated at considerable depth prior to this time. The uplift presumably pertains to compressional extrusion of rock adjacent to the Motagua Fault Zone. Nearby occurrences of Upper Cretaceous and Palaeogene strata (Valle del Angeles and Subinal) attest to this vertical uplift being only local, adjacent to the fault. In addition, Ratschbacher et al. (2009) shows that the metamorphism and migmatisation of the Las Ovejas unit associated with the southern El Tambor pertains to a Mid-Cenozoic deformation and cooling event. Thus, it seems likely that all these rocks were uplifted significantly during the Cenozoic transcurrent phase. A collection of tectonic flakes, caught between southward-vergent thrusts to the south and the Motagua shear zone on the north, would allow these rocks to be juxtaposed and to shallow and cool, without any stratigraphic record of the juxtaposition. As for the place of origin of the southern El Tambor HP/LT rocks, today's near juxtaposition with the northern Tambor (only 80 km displacement) is probably coincidence only. It cannot be ruled out that the same relationship does not extend east and west, if only outcrop permitted it to be seen; the strike–slip offset cannot be measured because it is potentially larger than the exposed area over which total displacement markers might be found. This brings us back to the Tehuantepec terrane, which may well have been the original pre-transcurrent site of southern El Tambor HP/LT rocks. If so, they have been uplifted by some 8–10 km while migrating along the flank of a transcurrent fault zone some 400–700 km. Such as history of uplift should not be surprising.

Another key aspect of this evolutionary stage is the opening of the Grenada–Tobago Basin, one of the two Caribbean intra-arc spreading basins of Palaeogene age, the other being the Yucatán Basin (Pindell & Barrett 1990; Rosencrantz 1990). Pindell et al. (2005) addressed the opening kinematics of the Yucatán Basin so we will focus on the Grenada Basin here, generally regarded as an intra-arc basin that opened as arc magmatism died at the Aves Ridge (remnant arc) and either began or continued at the Lesser Antilles frontal arc in the Palaeogene (Pindell & Barrett 1990; Bird et al. 1999). Although the basement of the Grenada Basin remains unsampled, seismic stratigraphy and heat flow measurements also suggest a Palaeogene age (Speed et al. 1984). Speed & Walker (1991) go so far as to suggest that Eocene MORB-type basalts on Mayreau are uplifted Grenada Basin oceanic crust.

An important clue to the opening kinematics of the Grenada Basin is that the 'Caribbean Arc' collided obliquely with both the Bahamas and western Venezuela concurrently, in the Palaeogene. The progressive oblique collision in the south is recorded by foredeep loading of the western Venezuelan margin (Fig. 19), and Caribbean volcanic terranes were clearly providing the tectonic load as shown by sandstone compositions of the foredeep fill (Zambrano et al. 1971; Pindell et al. 2009). Although slow convergence between the Americas was underway, the combined Palaeogene north–south shortening in the northern and southern Caribbean was far greater and faster, such that a single Caribbean 'Plate' could not have driven both collisions. In the absence of any evidence for internal expansion at this time of the Caribbean Plate itself, Pindell et al. (1988) and Pindell & Barrett (1990) therefore proposed that the Grenada Basin had a north–south component of opening great enough for a frontal arc terrane east of the forming basin to have driven the southern oblique collision, and suggested that rollback of the South American margin was responsible. The basin was drawn by these authors as a dextral pull-apart type intra-arc basin with north–south extension which accorded with possible magnetic anomaly lineations (Speed et al. 1984) in the presumed oceanic crust (based on refraction; Officer et al. 1957, 1959) of the deep basin, although Bird et al. (1999) refute the idea that the magnetic lineations are spreading related.

Here, we consider a modified opening model with a north–south kinematic component. Both gravity trends and regional structure contours to basement for the greater Grenada and Tobago basins define a fan-like shape (Speed et al. 1984; Speed & Walker 1991) whose apex is in the direction of the Bonaire Basin to the west. In the northern part of the Grenada basin, thought to comprise foundered arc basement, linear basement features trend ENE, which we interpret as shoulders of normal

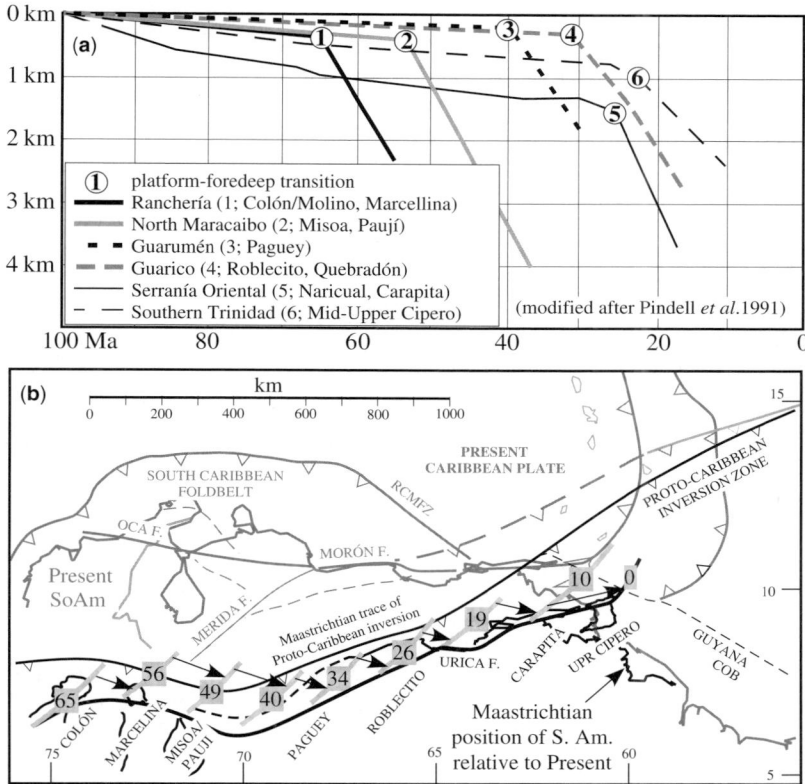

Fig. 19. Foredeep subsidence method of tracking Caribbean–South America displacement history, revised after Pindell *et al.* (1991). (**a**) Sediment accumulation curves for six autochthonous or parautochthonous locations along the margin from west to east. Typical passive margin subsidence histories persist until the times of Caribbean arrival, thereby loading the margin and initiating foredeep subsidence whose basal formations in each sub-basin are indicated in (**b**). Foredeep onset clearly youngs eastward. However, the distance of foredeep advance along the margin is larger (*c.* 1500 km) than the true relative plate displacement (*c.* 1200 km) due to obliquity of convergence (indicated by the arrows in b). (b) Map of Caribbean advance relative to a palinspastically restored South America that is also rotated back to its Maastrichtian position relative to North America when convergence began, showing the times of forearc collision in Ma and the positions and names of formations recording foredeep subsidence. Note that motion since 10 Ma has been essentially transcurrent in Eastern Venezuela and Trinidad (Pindell *et al.* 1998).

faults having SSE motion on them. In the deeper southern, presumably oceanic, part of the basin, and in the Tobago Trough as well, the basement structural grain strikes ENE, again hinting at an SSE extensional direction. East of the Lesser Antilles islands, the Caribbean crystalline forearc appears to be rifted into an array of basement blocks (Tobago Terrane, St Lucia Ridge, La Desirade High) with intervening gaps (Tobago Trough and the basement lows east of Martinique, north of La Desirade, and east of Barbuda). Superimposed upon this composite Eocene basement fabric, the Eocene and younger Lesser Antilles Arc volcanic pile has loaded (and deepened) the crust in the flanking basins. To the south, the Margarita–Los

Testigos Ridge (which may continue northeastward as a basement horst coring the Lesser Antilles Arc) is flanked by two linear basinal trends, the La Blanquilla and the Caracolito basins (Ysaccis 1997; Clark *et al.* 2008), that have been inverted by perhaps 40 km and 20 km during the Middle Miocene collision between the Caribbean crust and Eastern Venezuela, respectively. These two basins deepen to the northeast into the oceanic domains of the Grenada Basin and Tobago Trough.

We have reconstructed the Grenada and related basins to a pre-rift configuration, relative to the Caribbean Plate, in a model of Eocene NNW–SSE radial intra-arc rifting and seafloor spreading that employs the above noted structures and fabrics

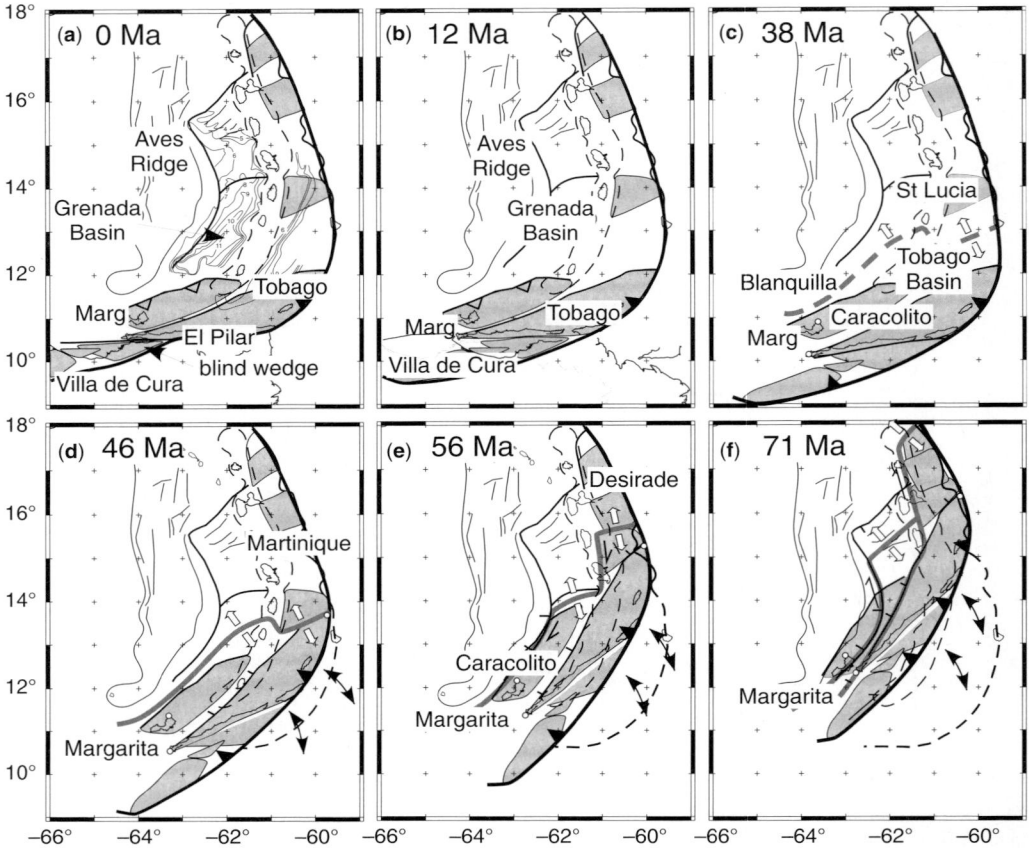

Fig. 20. Fan-like opening/closure model for the Palaeogene intra-arc Grenada and Tobago basins, and the migration of the Tobago, Margarita and Villa de Cura terranes from the Aves Ridge, driven by subduction zone roll back after the Caribbean Arc had rounded the Guajíra corner, and before collisional choking of the trench by the central Venezuelan margin. The north–south component of opening allowed the leading Caribbean terranes to move transpressionally southeastward along the western Venezuelan margin, while the rest of the Caribbean Plate moved more easterly relative to South America. The resulting Maastrichtian palinspastic reconstruction portrays the SE edge of the Caribbean crust as a fairly straight margin, the result of having migrated along the NW flank of the Guajíra salient in the Maastrichtian, with possible shedding of additional Caribbean forearc material along the western flank of Colombia.

(Fig. 20). We presume that the opening was driven by gravitational collapse of the Caribbean Arc in the direction allowed by roll back of Jurassic Proto-Caribbean oceanic lithosphere (Pindell 1993), as the arc rounded the Guajíra salient of Colombia. Extensional opening involved the southeastward expulsion (NW-dipping asymmetric rift) of the Villa de Cura, Margarita and Tobago forearc terranes (effectively comprising the subduction channel) from beneath the Aves Ridge remnant arc hanging wall (hence little *apparent* extension in the eastern Aves Ridge but note the 10 km depth to the Aves Ridge hanging wall cut-off), but ceased when the forearc terrane collided with the Venezuelan margin, by the Oligocene. Beginning

with a simplified basement terrane map (Fig. 20a), we then restore 200 km of post-10 Ma dextral movement on the El Pilar Fault (Fig. 20b). We then restore Early and Middle Miocene NW–SE compressional basement inversion structures in the Blanquilla and Caracolito sub-basins (Fig. 20c), keeping the Gulf of Barcelona primitive arc volcanic zone (Ysaccis 1997) and the correlative three main pieces of the Villa de Cura Klippe as part of the southeastern Caribbean forearc. This produces, for the purposes of this paper, an end Middle Eocene, pre-collisional, post-Grenada Basin opening, shape for the southeastern Caribbean forearc. Next, we progressively close the eastern basins by rotating the southeast Caribbean forearc composite terrane northwards, roughly orthogonal to structural trends. Figure 20d closes the Tobago Basin, restoring the eastern Tobago Terrane against the St Lucia Ridge. Figure 20e then closes most of the oceanic part of the Grenada Basin. Finally, Figure 20f closes both the Caracolito and La Blanquilla basins, whose early faults appear to have been oblique low-angle detachment normal faults that dipped to the NW, as well as the gravitational low east of Barbuda. In this model, the Bonaire intra-arc basin is viewed as having a genetic association with the Grenada basin system, although with far less extension (nearer to the gross pole of rotation) and hence little to no Palaeogene oceanic crust. The regional Caribbean reconstructions herein employ the reconstructed shape for the southeastern Caribbean (Fig. 20f) for Maastrichtian and older times.

This model for the Grenada Basin seems viable enough and explains most structures, but still remains to be tested. Another option that would also allow for continuous convergence between Caribbean terranes and western Venezuela is for the Northern Andes to have migrated 100–200 km northeastward along precursor faults (e.g. Caparo Fault) in the Mérida Andes during the Paleocene–Eocene. Aymard *et al.* (1990) showed that local fault controlled basins and local uplifts with truncation to the Jurassic level are overlain by Cenozoic molasse in the Apure Basin, but the age of the molasse is debated and thus it is not clear if this deformation is Palaeogene or Neogene. Nonetheless, the idea of a limited amount of Palaeogene dextral shear, during Lara Nappe emplacement, accords with the Maastrichtian onset of Caribbean subduction beneath Colombia.

Eocene–Middle Miocene transcurrence and oblique collision along northern South America

During this period, the American plates further engulfed the Caribbean lithosphere and arc between them (Figs 17, 21 & 22). Caribbean–American relative motion was recorded by (1) opening of the Cayman Trough pull-apart basin [note: Cayman Trough magnetic anomalies (Rosencrantz *et al.* 1988; Leroy *et al.* 2000; ten Brink *et al.* 2002) may record the basement fault fabric rather than/in addition to seafloor spreading anomalies, but nevertheless they strike north–south over some 900–1000 km, making a north–south opening direction highly unlikely]; (2) Eocene and younger Lesser Antillean arc magmatism (Briden *et al.* 1979); (3) the eastward migration of arc magmatism in SW Mexico as the motion of Chortís exposed that margin to subduction (Pindell

et al. 1988; Schaaf *et al.* 1995; Ferrari *et al.* 1999); (4) the migrating Caribbean foredeep along northern South America (Dewey & Pindell 1986; Pindell *et al.* 1991; Fig. 19); (5) the progressive collision and closure between the trailing edge of Caribbean lithosphere (Panama) and Colombia (Keigwin 1978; Pindell *et al.* 1998; Kennan & Pindell 2009); and (6) the transcurrent separation of northern Hispaniola from Cuba, and the transpressional assembly of the Hispaniolan terranes, along eastward strands of the Cayman Trough transform system (Pindell *et al.* 1988, 1998; Iturralde-Vinent & McPhee 1999). In addition, the east–west compression resulting from progressive flat-slab overthrusting of South America onto the relatively buoyant Caribbean lithosphere (Pindell *et al.* 1998, 2009) undoubtedly played a major role in the Late Oligocene and younger northeastward tectonic escape of the Northern Andes terranes along the Mérida Andes.

Relative to North America, Caribbean motion during this period was roughly parallel to the Cayman Trough. However, because the American plates were slowly converging, accumulating 200–360 km of shortening increasing westward from Trinidad to Colombia (Pindell *et al.* 1988; Müller *et al.* 1999), the southern Caribbean boundary was much more convergent. Where the Caribbean–South America Plate boundary was developing (i.e. west of the Lesser Antilles Trench), collision proceeded obliquely. To the east of the Lesser Antilles, shortening was probably initiated before Caribbean arrival by inversion or possibly even minor subduction at the Proto-Caribbean inversion zone or trench (Pindell *et al.* 1991, 2006). In the Caribbean–South America oblique collision zone, Caribbean forearc rocks such as the Villa de Cura complex, Carúpano Basin and Tobago Terrane basements as well as outer parts of the former continental margin ahead of them were thrust SE onto the inner margin and underwent axis-parallel extension (Fig. 23). We reiterate Pindell & Barrett (1990) that the majority of the total Caribbean–South America displacement is situated at the sole of the Caribbean allochthonous belt and thus is not measurable with offset markers along strike–slip faults at the surface. The high-angle strike–slip faults (e.g. Oca, Boconó, Morón and El Pilar Faults) that cut the thrust-soled allochthons have developed well after allochthon emplacement, and mostly in relation to the Neo-Caribbean Phase (see below; Dewey & Pindell 1986; Pindell & Barrett 1990). These faults certainly should have displacements far less than the total predicted relative Caribbean–South America displacement; they post-date and have little or nothing to do with the Caribbean–South American collision which emplaced the

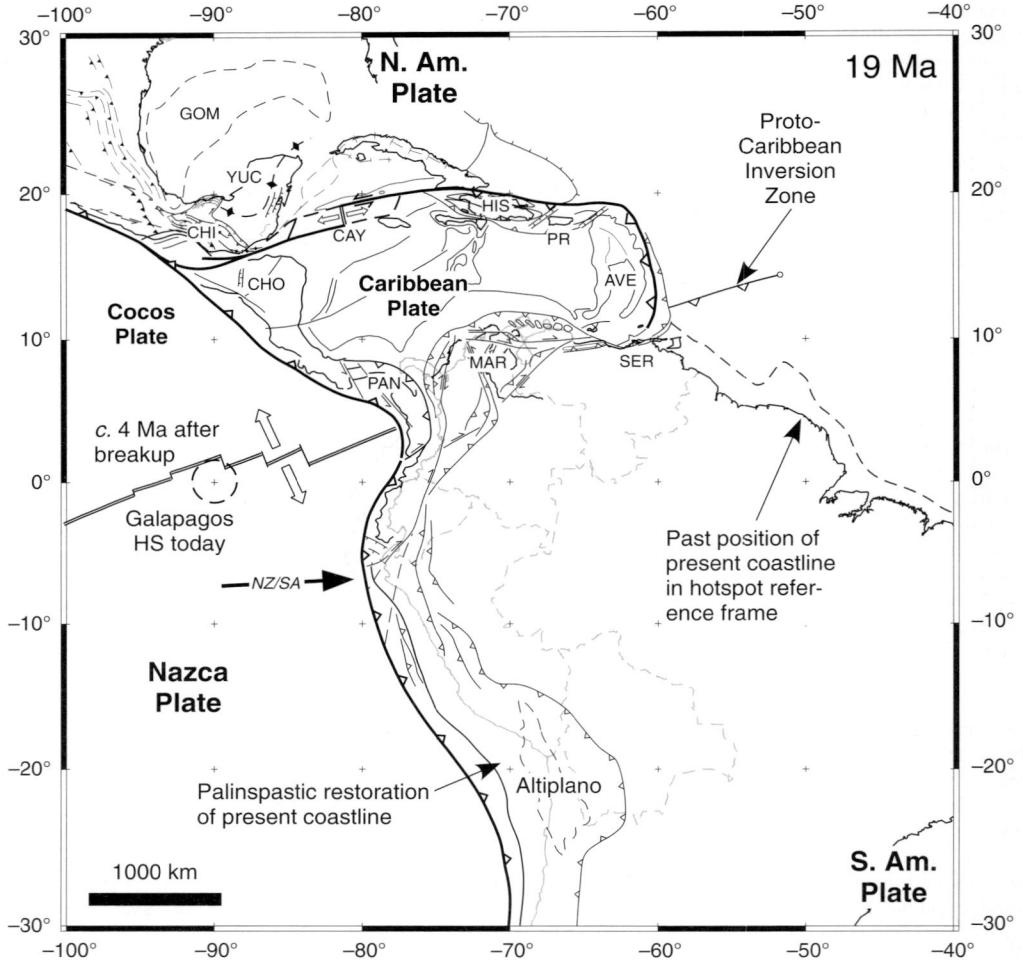

Fig. 21. A 19 Ma reconstruction of the circum-Caribbean region, shown in the Indo-Atlantic hot spot reference frame. At this time, the tail of Chortís has moved far enough to the east that any north–south sinistral shear is not required, but west–east extension continues. Oblique collision along South America has started to encompass the Serranía Oriental of Venezuela and Trinidad, and the South Caribbean foldbelt is now taking up most of the continued convergence to the west. The Margarita (or Roques Canyon) transfer fault is feeding into the Urica transfer, thus allowing shortening to proceed in the Serranía Oriental. The Panama (PAN) Arc is choking the Western Cordillera–Sinú Trench and starts to escape to the NW, relative to the Caribbean, bounded by NW-trending sinistral faults and driving NW-directed thrusting in the western North Panama Fold Belt. Shortening at Colombia's Eastern Cordillera and northeastward migration of the Maracaibo Block is underway, adding in turn to the shortening at the South Caribbean Foldbelt. The Galapagos Ridge was subducting somewhere along the Panama or Colombian margin.

allochthonous belt of Caribbean rocks along northern South America.

Concerning the progressive collision of Panama with Colombia, the tectonic escape model employed by Wadge & Burke (1983), Mann & Corrigan (1990) and Pindell (1993) probably occurred during the Early and Middle Miocene rather than being active today. The NW-trending faults in Panama that those authors employed as escape structures are apparent on radar topography images (e.g. Farr *et al.* 2007), but it is not clear that these remain active or significant today. For reasons given in the next section, we favour the view that these faults and their associated folds were active *until* about 9 Ma rather than continuing to younger times, and thus that the tectonic escape

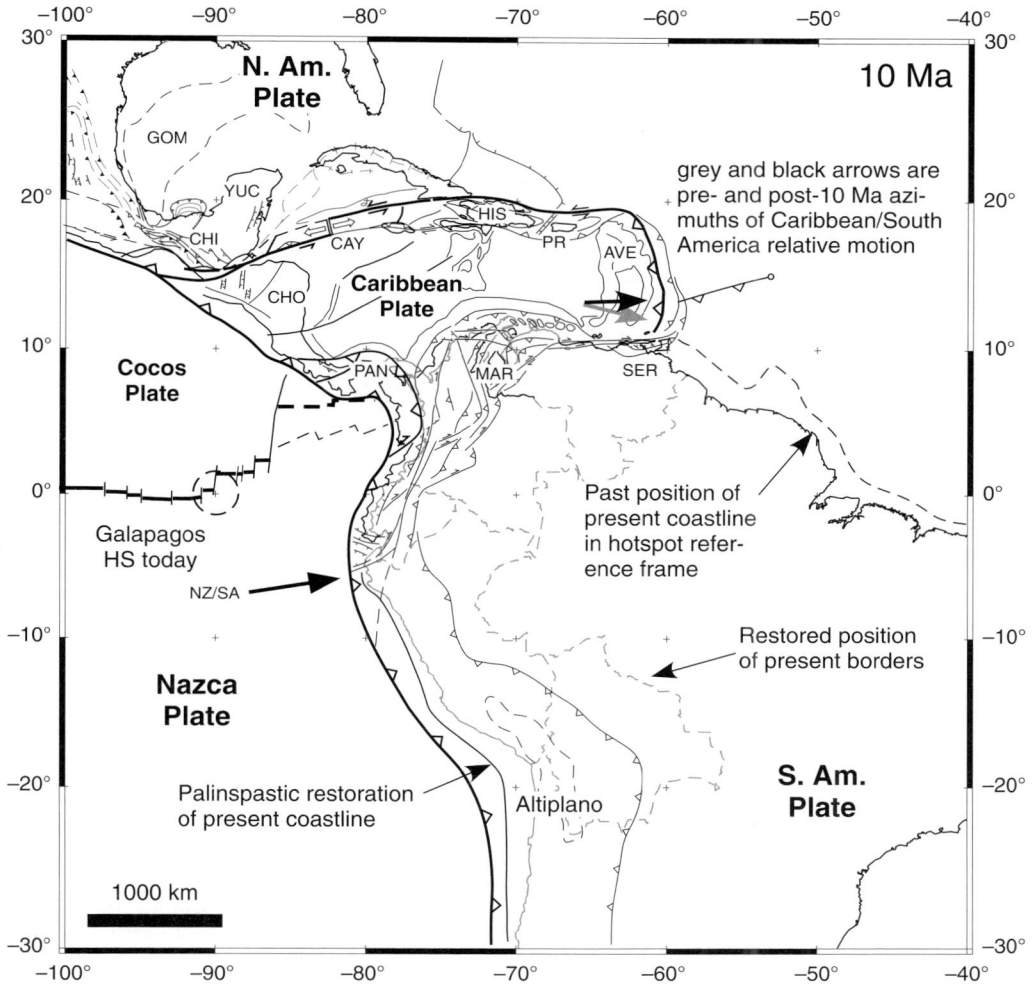

Fig. 22. A 10 Ma reconstruction of the circum-Caribbean region, shown in the Indo-Atlantic hot spot reference frame. At 10 Ma, a fundamental shift in Caribbean motion with respect to the Americas, resulting in 085°-directed dextral shear dominating the SE Caribbean, and 070°-directed transpression dominating the northern Caribbean. The Cocos–Nazca Plate boundary jumped at this time to the Panama Fracture Zone. The Panama Block has become partly coupled to the Nazca Plate, resulting in a Panama–Colombia collision that is presently occurring nearly twice as fast as Caribbean–South America relative motion; thus, the NW escape of the Panama slivers has ceased.

mechanism ended at that time, to be replaced by a setting in which the Choco Block is driven eastward relative to the Caribbean Plate by the underriding Nazca Plate (see Neo-Caribbean Phase, below).

The 'Neo-Caribbean Phase' of Caribbean evolution: 10 Ma to present

Dewey & Pindell (1986) showed that the eastward-diachronous Caribbean foredeep basin along northern South America (Fig. 19) advanced at an average rate of 20 mm/annum over the Cenozoic. Concerning the azimuth of motion, an essentially east–west azimuth for the southeast Caribbean was employed by Robertson & Burke (1989) in the north Trinidad offshore. Algar & Pindell (1993) confirmed that Trinidad had a younger structural style which accords with east–west transcurrence (085°), but that this was superposed onto an older style (pre-10 Ma) that was more compressive. The 085° azimuth in the southeast Caribbean was

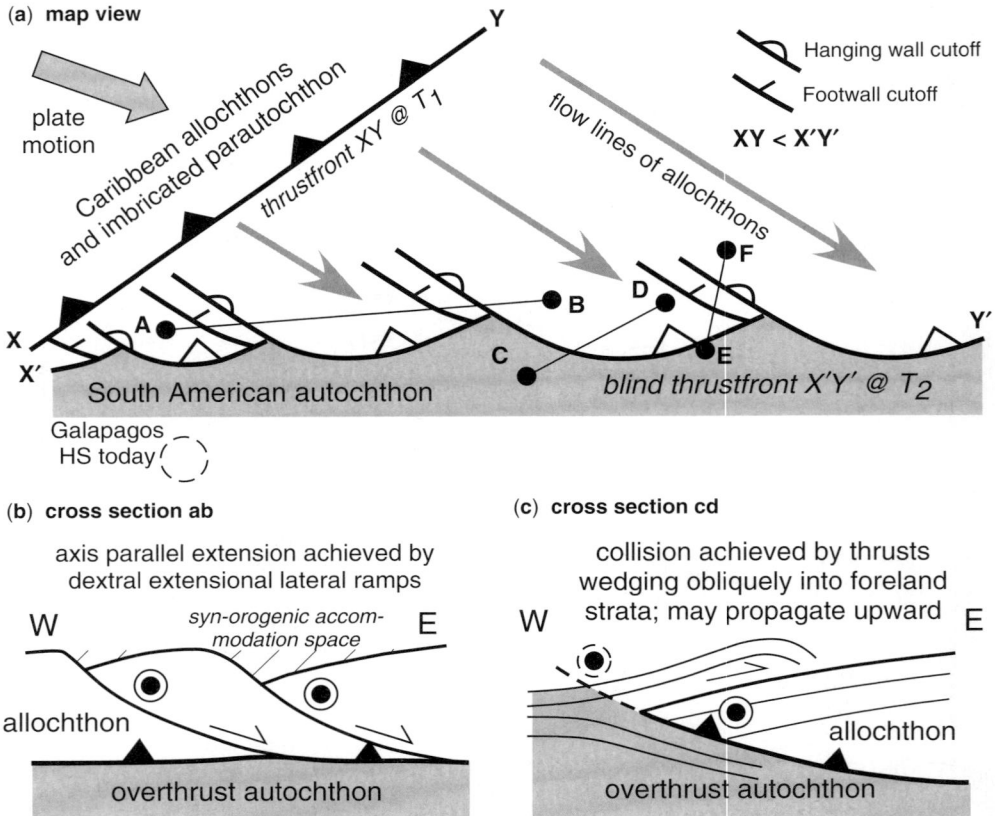

Fig. 23. Tectonic style in the allochthonous thrust belt during Eocene–Middle Miocene dextral oblique collision along northern South America. Thrustfront X–Y migrates to thrustfront X′–Y′, and necessarily becomes longer by axis parallel extension. At shallow levels (<6–8 km) near the thrustfront, length increase is achieved by low angle extensional detachment along lateral ramps and tear faults. These rarely propagate down into the autochthon, however piano key faults in the autochthon allow for differential amounts of load-induced subsidence; these seem to nucleate on former Jurassic marginal offsets in the margin (e.g. Urica and Bohordal faults). Section A–B (in **b**) shows extensional style in the allochthon; accommodation space on the allochthon (piggy-back basins) can be created if the east–west thinning in the allochthon exceeds the uplift due to north–south shortening. Where the strike–slip component of lateral ramps allows for advance of the thrustfront into the foreland basin (**c**), the structural style is commonly that of a convergent blind wedge where north-vergent backthrusting of foreland strata occurs above the advancing blind wedge. If motion is sufficient, tear faulting may propagate up into the foreland strata along the lateral ramp, while foreland folding occurs ahead of the thrusts. In seismic sections in positions such as section E–F, extension and compression would be coeval. We suggest that when interpreting seismic data in oblique collision zones, axis parallel extension along lateral ramps should be the first working hypothesis to be considered.

corroborated by an updated assessment of circum-Caribbean Plate boundary seismicity, which also indicated an 070° azimuth of relative motion in the NE Caribbean (Deng & Sykes 1995). GPS positioning results (Dolan *et al.* 1998; Perez *et al.* 2001; Weber *et al.* 2001) have confirmed the 20 mm/annum rate and the 070° and 085° azimuths for the NE and SE Caribbean, respectively (Fig. 24).

These azimuths and rate afford a fairly good understanding of Caribbean neotectonics and structural development in most areas back to about 10 Ma (compare Figs 22 & 24). However, the present is not a very satisfactory key to the past in the Caribbean because extending the current azimuths of motion prior to about 10 Ma produces various unacceptable crustal overlaps between the Caribbean and South America. The last 10 Ma, called the 'Neo-Caribbean Phase' by Pindell & Barrett (1990), has seen a range of sub-regional tectonic developments whose differences with pre-Late

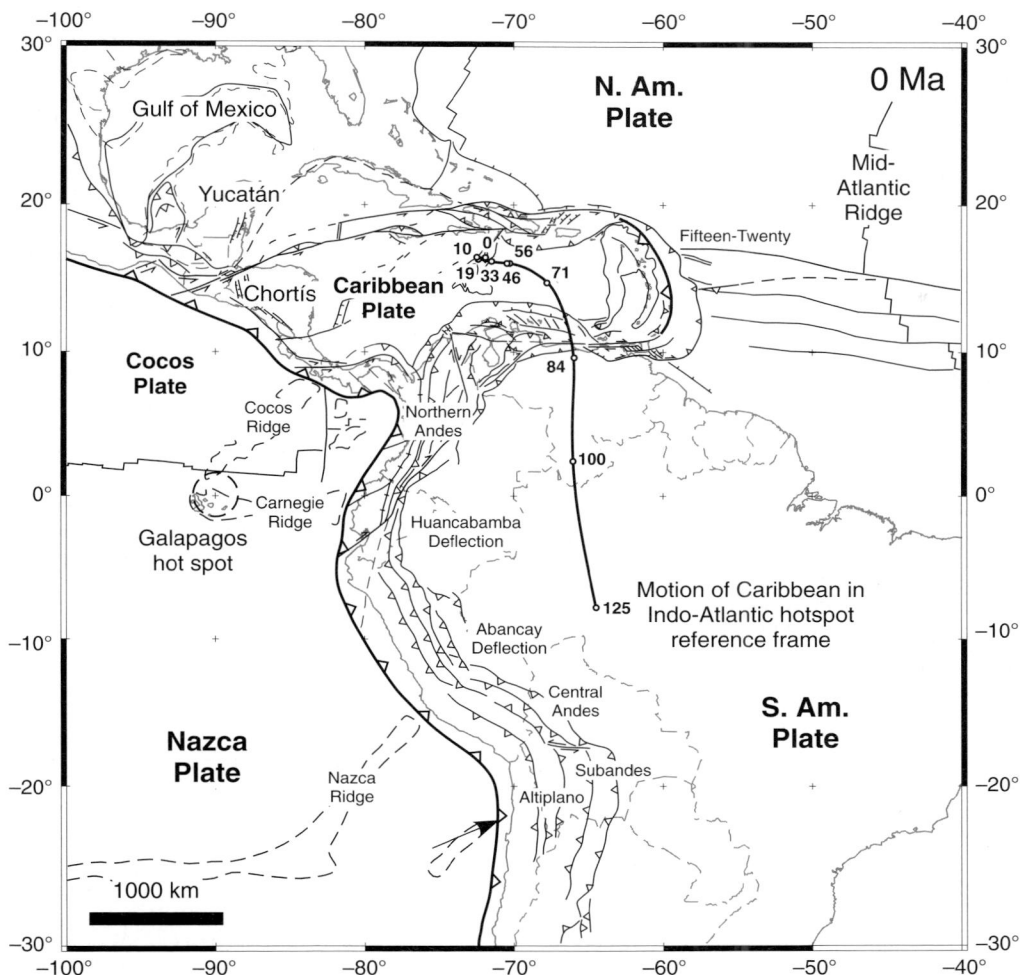

Fig. 24. Present day plate boundary map of the Caribbean region, continuing the format used in the preceding evolutionary figures, and showing the overall migration history of the central Caribbean oceanic lithosphere in the Indo-Atlantic hot spot reference frame.

Miocene tectonic patterns have been, in our view, under-appreciated. As a result, there has been a tendency for workers to either project the significance of young features such as the El Pilar Fault of Eastern Venezuela (Fig. 2) too far back in time, and to neglect the full significance of older features such as the basal thrusts of Caribbean allochthons along northern South America (Fig. 2), where most Caribbean–South American displacement has occurred (Pindell *et al.* 1988). The cause of the Neo-Caribbean Phase appears to have been, in the eastern Caribbean at least, a late Middle Miocene change of about 15° in the Caribbean Plate's azimuth of motion relative to the American

plates, from 085° to 070° for North America and from 100° to 085° for South America (Algar & Pindell 1993; Pindell *et al.* 1998), and possibly to the hot spot reference frame as well (Müller *et al.* 1993; see below). It is difficult to reconcile pre-Late Miocene evolution of the Caribbean margins with the Present azimuth of relative motion.

Figure 22 shows the Caribbean region at 10 Ma, in which the Caribbean interior has been retracted 200 km westward along the present azimuths noted above, and the bulk of Neo-Caribbean Phase structures have been palinspastically restored. Prior to 10 Ma, Caribbean flow lines were probably convex northward, mimicking the bounding faults of the Cayman Trough, at potentially varying curvatures (Pindell *et al.* 1998). We now expand upon the following five sub-regional developments, in addition to the Gulf of Tehuantepec area already discussed, where the Neo-Caribbean Phase has most strongly obscured earlier tectonic patterns.

(1) *Southeast Caribbean*: the linear Morón–Cariaco Basin–El Pilar–Gulf of Paria Basin fault system and additional splays through Trinidad crosscut the Middle Miocene fold-thrust structures of Eastern Venezuela and Trinidad that had resulted from Caribbean collision of that age (Pindell & Kennan 2007*b*). The new (post-10 Ma) plate boundary configuration is also associated with a primary change in Late Miocene and younger deposition (Algar & Pindell 1993; Ysaccis 1997; Pindell *et al.* 1998, 2005). An important result of this reconstruction is that the Orchila Basin–Margarita Fault aligns with the Urica Fault. These two faults are lateral ramps to the South Caribbean and Serranía Oriental fold-thrust belts, respectively. For at least the Early and Middle Miocene and possibly older, these presently displaced faults served as a primary transfer fault crossing the orogenic float between the South Caribbean Foldbelt and the Serranía del Interior Oriental of Venezuela.

(2) *Northern Andes*: the northeastward extrusion of the Maracaibo Block (Mann & Burke 1984) is suspected of having begun in the Late Oligocene (Pindell *et al.* 1998) or Early Miocene (Bermúdez-Cella *et al.* 2008), but the coarsening of flanking orogenic molasse, increase in foreland subsidence history and the ratio of fission track ages on basement rocks younger and older than 10 Ma in the Mérida Andes indicates that uplift and, probably, tectonic escape have accelerated at that time. This in turn has the effect of strengthening the rate of shortening along the South Caribbean Foldbelt, which is the free face that takes up much of the northerly component of Andean/Maracaibo extrusion. The effect of this development is to amplify the appearance that the Caribbean Plate is subducting beneath South America, which is true, but this detracts from the fact that in the Eocene–Oligocene the Caribbean Plate's leading fringe was obducted southeastwards onto the South American margin in a west-to-east diachronous history of oblique collision. As that collision culminated, the polarity of shortening was reversed, earlier in the west, and the site of continued shortening stepped out to the South Caribbean foldbelt.

It is important to recognize that many hundreds of kilometres of relative plate displacement between South America and the Caribbean had occurred prior to the onset of this back-thrusting at the South Caribbean Foldbelt.

(3) *The 'Panama Block'*: GPS data (e.g. Trenkamp *et al.* 2002) show that Panama and the Sierra Baudó are converging with South America faster (40 mm/annum) than the Caribbean Plate is converging with South America (20 mm/annum). Thus the tectonic escape model invoked by Wadge & Burke (1983), Mann & Corrigan (1990) and Pindell (1993), wherein slices of Panama are being backthrust to the NW onto the Caribbean Plate, is not currently operating, although it probably did so earlier in the collision (Middle to Late Miocene). In our view, Panama is probably moving east faster than the Caribbean Plate because the former is partially coupled at its crustal base to the north-dipping Nazca Plate which moves east toward South America at >60 mm/annum. Panama is now overthrusting Caribbean crust on Panama's northeastern flank, and not its NW flank (Fig. 2). Thus, we deduce that there should be east–west shear zones crossing Costa Rica that account for this late eastward displacement. Inspection of radar imagery shows that indeed there are strong topographic lineaments precisely where differences in GPS motions predict them to be, although seismicity along these zones rarely exceeds magnitude 4 events. Here we employ the term 'Panama Block' to denote the general area that moves east faster than the Caribbean Plate, subject to refinement. We consider that the onset of coupling with the Nazca Plate was coeval with the *c.* 9 Ma jump in plate boundary position from the Malpelo Ridge (now extinct) to the Panama Fracture Zone; thus, if the Panama tectonic escape model is valid, it probably was a Middle Miocene to earliest Late Miocene phenomenon. The folds recording motion along the escape faults (Mann & Corrigan 1990) appear to be onlapped by flanking strata, rather than the youngest strata being folded, suggesting that Late Miocene termination of folding might be supported by the geology. Careful dating of these sediments may better demonstrate when the folds were active.

(4) *Hispaniola*: as the eastern tip of the Bahamas has progressively approached the terranes of Hispaniola over the last 10 Ma, several previously strike–slip or moderately transpressive structures in Hispaniola have become greatly tightened, the result of which has

been an increase in shortening relative to transcurrence leading to 3000 m topography in the Central Cordillera and the creation of the Hispaniola restraining bend of the North Caribbean Plate boundary (Pindell & Draper 1991). However, the geology of Hispaniola is very diverse, and prior to this stage numerous terranes oriented generally WNW–ESE had been amalgamated by large values of sinistral strike–slip offset (Pindell & Barrett 1990; Lewis & Draper 1990; Mann *et al.* 1991). The entire southwestern half of the island, probably everywhere south of the San Juan Valley, comprises elevated Caribbean seafloor rather than island arc material, whose clean micritic siliceous and chalky Mid-Cenozoic limestones received no arc-derived clastic detritus until well into the Miocene as a result of transcurrent motions on the Los Pozos Fault Zone (McLaughlin & Sen Gupta 1991; Pindell & Barrett 1990).

(5) *Jamaica*: like Hispaniola, Jamaica occupies a transpressional bend and hence is being uplifted by transpression onto the southeast flank of the Cayman Trough (Case & Holcombe 1980; Pindell *et al.* 1988). Sykes *et al.* (1982) showed that the southeastern Cayman Trough is seismically active, allowing for an uncertain amount of east–west transcurrent slip through the Jamaica area. Although Late Neogene faulting is known onshore Jamaica (Burke *et al.* 1980), radar and other topographic imagery appears to discount the probability of primary onshore through-going faults that may define the main locus of slip. Pindell *et al.* (1988) inferred that the primary site of such slip lies instead at the foot of the northern Jamaican slope. The Late Miocene–Recent uplift of Jamaica (by transpression) probably records the onset of transcurrent motion along this flank of the Cayman Trough; up to 20 km of transpressional movement may have occurred here in that time, judging from offset markers along the zone.

Figure 22, which accounts for the above aspects of the Neo-Caribbean Phase, may be used as a template to better understand Middle Miocene Caribbean processes and developments. For example, it can be used to assess the southeastern Caribbean collision zone without the complication of having been dissected and offset 200 km by the east–west El Pilar transcurrent fault (e.g. Pindell & Kennan 2007*b*). Also, by retraction of transpression in the Chiapas Foldbelt of southern Mexico, the southern flank of the Tehuantepec terrane aligns with the SW Mexican Trench, restoring the smooth curvilinear transform trend along which the Chortís

Block migrated since the Maastrichtian. Palinspastic reconstructions such as this afford more accurate interpretations of progressive history through time: for a region like the Caribbean, assessing tectonic evolution is best done palinspastically, so that the effects of younger events are removed from the period in question.

Discussion

Caribbean motion in the hot spot reference frame

We can readily reconstruct the motion history of a point (southern Hispaniola) in the centre of the stable Caribbean oceanic lithosphere (i.e. not including the accreted Chortís Block) relative to the Indo-Atlantic reference frame (Fig. 24). This history can be broken into two main stages. The Cretaceous stage involves northward translation of about 25° palaeo-latitude with little vertical-axis rotation. During the Cenozoic stage, the Caribbean Plate has been nearly stationary in the hot spot reference frame. It seems remarkable that the absolute plate migration of the Caribbean lithosphere is so minimal given the regional geological complexity of the plate boundaries: the Americas have moved much further over the hot spot reference frame in the same period, and most of the geological complexity of the Caribbean region results from the plate boundary interactions that result from those larger scale motions. As the American margins were wrapped around the Caribbean Plate, mélanges, blocks and slivers of crust from the former North American and South American Cordillera have been left behind on the edges of the Caribbean Plate and are now found mixed with Caribbean rocks along the mobile North and South American–Caribbean Plate boundary zones as fault and subduction mélange, olistostromes and remnant klippes of former thrust sheets. Caribbean evolution has influenced the geology and evolution of the American Cordillera from Baja California to northern Peru, and assessments of Cordilleran history between these widespread localities will need to consider the former interactions with the Caribbean lithosphere.

To summarize the plate motions, Figure 25 shows the motion of North and South America in the Müller *et al.* (1993) Indo-Atlantic hot spot reference frame (grey lines younging westward, net Cenozoic convergence is shown in the inset at upper right). Note how closely the North America/hot spot line mimics the Cayman Trough (grey shape) in length and average trend, suggesting that not only does the Trough record Caribbean/North America motion history back to 50 Ma, but also North America/hot spot motion history. In addition,

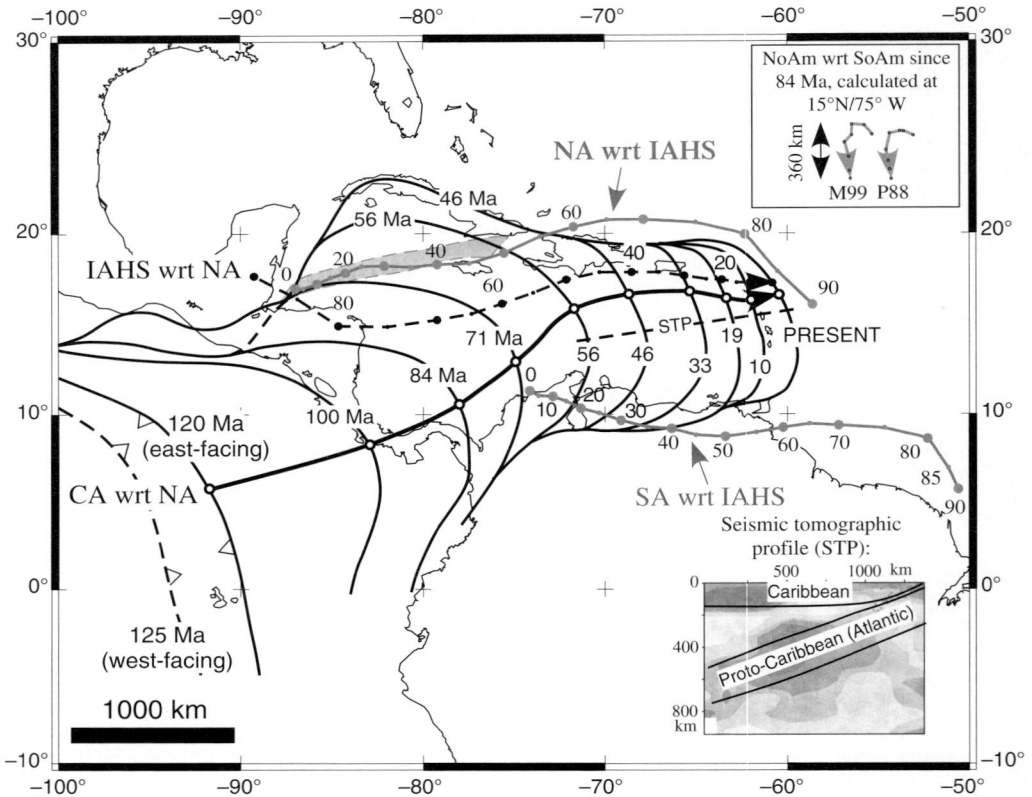

Fig. 25. Motion histories of: North (NA) and South America (SA) relative to Indo-Atlantic hot spot (IAHS) Müller *et al.* (1993) reference frame (grey lines; NA wrt IAHS and SA wrt IAHS); hot spots relative to North America (dashed black line; IAHS wrt NA); Caribbean relative to North America (heaviest black line; Car wrt NA), as summarized from former relative positions of the Caribbean Trench (lighter black lines). Also shown: Cayman Trough (grey outline); Cenozoic convergence between the Americas (inset upper right; P88 = Pindell *et al.* 1988; M99 = Müller *et al.* 1999); seismic tomographic profile of van der Hilst (1990) (inset, lower right).

we have inverted the North America motion history to show the motion of the hot spot reference frame relative to North America measured at a point in the eastern Caribbean (dashed black line younging eastward). We also show the progressive advance of the Caribbean lithosphere relative to the Americas (lighter black lines), summarized by the heaviest black line younging eastward. Comparison between the dashed and the heavy black lines provides a measure of how closely the Caribbean has remained in the Indo-Atlantic reference frame through time; for the Cenozoic, the two lines are equivalent within probable error, but in the Cretaceous the Caribbean begins to drift southward back in time, in accord with the curve in Figure 24. Finally, the seismic tomographic profile (line STP and inset at lower right; van der Hilst 1990) shows at least 1500 km of subducted Atlantic slab beneath the

Caribbean, providing a direct visual measure of Caribbean–American migration.

Implications of Caribbean evolution for slab break off and flat slab subduction

Since at least the Campanian, the Caribbean Plate has been anchored in the Indo-Atlantic mantle reference frame by its two bounding Benioff Zones (Pindell *et al.* 1988; Pindell 1993). The above evolutionary model comprises a number of tectonic settings and events that can be assessed for tectonic processes. Here we address two such settings for their implications for slab break off and flat slab subduction. The first setting is where North and South America serve as the downgoing (choking) plate during collision with an arc that is stationary in the mantle reference frame; examples include

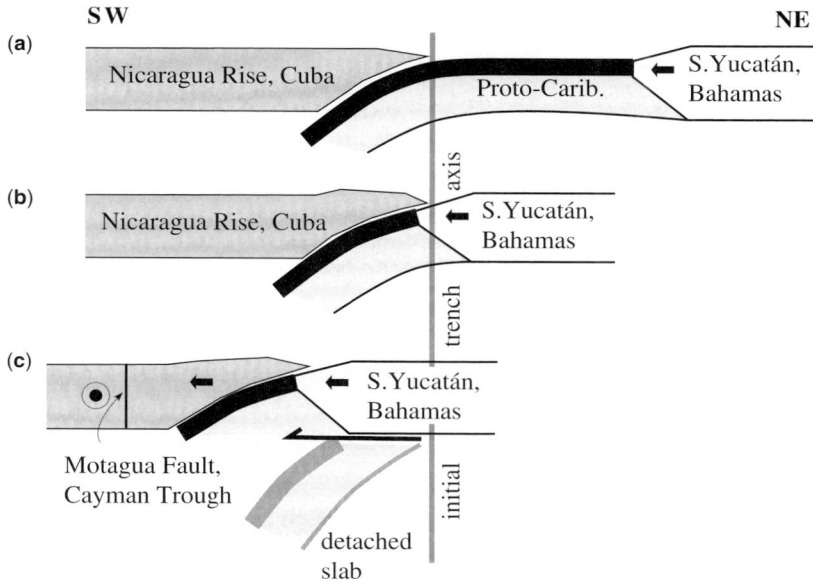

Fig. 26. Schematic interpreted histories of southwest Yucatán and Bahamian collisions with the stationary Caribbean Arc (Nicaragua Rise–Jamaica and Cuban portions, respectively). North American slab is subducted beneath stationary (in mantle reference frame) arc (**a**) until buoyant continental crust arrives at and chokes the trench (**b**). Then, continued westward drift of American crust across the mantle can only be accommodated if the slab detaches and founders in place (**c**). Accreted arc and lithosphere behind it must then move either with North America (Cuban example), or take up independent motion that allows the continenental crust to continue moving (Guatemalan example; i.e. Motagua–Cayman Trough system).

the Maastrichtian South Yucatán–Caribbean Arc collision and the Eocene Bahamas–Caribbean Arc collision. The second is where these westward drifting plates serve as the overriding plate during east-dipping subduction of oceanic crust beneath them; examples include the Cenozoic history of Caribbean subduction beneath Colombia, and the Neogene history of Cocos subduction beneath southwestern Mexico, the latter of which has been a consequence of the Cenozoic eastward translation of the Chortís Block from along the Mexican margin (Pindell *et al.* 1988).

Cross sections representing the SW Yucatán and SW Bahamas collisional events, each of which involved west-dipping subduction of oceanic slab attached to westward-migrating American continental crust, are shown in Figure 26. Prior to arc–continent collision at each, convergence occurred by American (Proto-Caribbean) oceanic crust entering sub-Caribbean mantle, such as is occurring at the Lesser Antilles today. Upon collision, however, buoyant continental crust choked the subduction zone such that continued westward drift of the American continental lithospheres could only be accommodated by the continents detaching from and overthrusting their former oceanic slabs. This

is because the dipping slabs cannot move laterally through the mantle as fast as the plate at the surface can move; we conclude that the slabs must be left behind to founder in the mantle near the point of collision. Thus, there may be a horizontal shear parameter in addition to negative buoyancy (e.g. Davies & von Blanckenburg 1995) involved in severing the lithosphere during slab break-off. Slab break-off from the Yucatán margin has been suspected previously due to the apparent post-collisional uplift (Pindell *et al.* 2005) as well as some late collisional igneous activity (Ratschbacher *et al.* 2009). It is not clear from any existing mantle tomography where these slabs currently lie. However, the palaeo-sites of the Yucatán and Bahamian collisions relative to today's geography are the eastern Colombian Basin and the Silver Plain off the NE flank of the Bahamas (23°N, 70°W), respectively; relative to the mantle, those are the positions where southern Yucatán was situated in the Campanian, and where the SW Bahamian margin was situated in the Eocene. It may be that a tear in the slab was initiated along the Yucatán margin, which then progressively migrated eastward with continuing collision along the foot of the Belize margin (Pindell *et al.* 2005), and eventually along the foot of the Bahamas Platform. Pindell & Kennan (2007b) interpreted the tomography of van der Hilst (1990) to suggest that the tear has reached Hispaniola at present. If this progressive tear model is correct, then there may be a very large remnant of the Proto-Caribbean slab accumulating in the mantle in a zone between the Colombian Basin and the Silver Plain. Unfortunately, little tomographic data is available for this area to test this idea. Finally, we wish to point out that such a progressive tear would provide an elegant

explanation for the Late Maastrichtian–Middle Eocene opening of the Yucatán intra-arc basin. This opening is normally attributed to rollback (Pindell et al. 1988) that is often taken as a passive gravitational process, but the model outlined here is dynamic in that the apparent rollback is driven not by gravitational subsidence of the slab but rather by the locking of the slab in the mantle while the North American Plate drifted west, thereby actively tearing the slab northward along the Belize margin (implying in turn that the foot

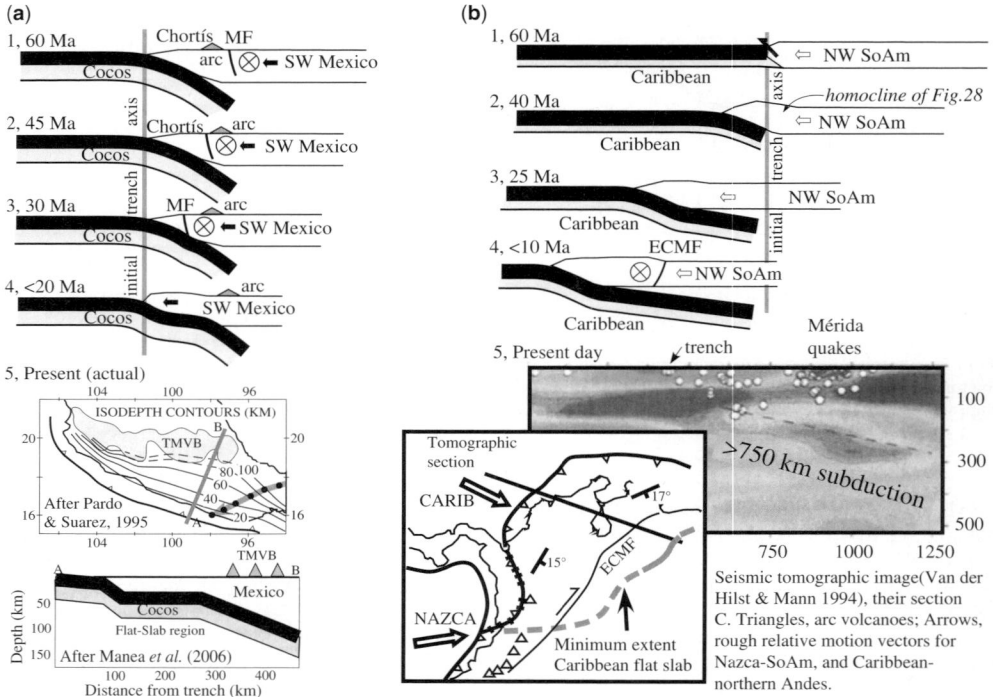

Fig. 27. (a) Cross sections of the development of SW Mexico, showing the conversion of SW Mexico from the north flank of an intra-continental transform to the hanging wall of the Middle American Trench as a result of the progressive transcurrent removal of the westwardly tapering Chortís Block from the cross section (section drawn in North American reference frame). At about 18–20 Ma, Mexican basement impinged on Cocos Plate, and continued westward migration of Mexico occurred by telescoping of Mexico onto the northeast-dipping slab, which probably involved subduction erosion of lower hanging wall crust and also produced large uplift (>5 km). Present day geometry (result of sections) is also shown according to Pardo & Suarez (1995) and Manea et al. (2006). Movement of SW Mexico across the mantle for the last 25 million years is shown by 5 Ma increments (dots on heavy grey line). The flat-slab geometry of some 300 km from the trench corresponds well to the 300 km of North America–mantle motion since the time Chortís left the cross section (c. 18 Ma). (b) Cross sections of the development of northwest Colombia. showing initiation and continuation of Cenozoic overriding by NW South America onto the stationary Caribbean Plate. Uplift was large in the Eocene (>5 km, section 2), as South American crust was initially ramped onto buoyant Caribbean lithosphere, roughly doubling the crustal thickness. After a Middle Cenozoic relaxation due to slower westward drift by South America then (Pindell et al. 1998), convergence has intensified again since the Late Oligocene, and in addition to orthogonal orogenic shortening this convergence has also been relieved by tectonic escape along the Eastern Cordillera–Mérida transpressive fault system (ECMF). Note that the Caribbean slab is not subducting into the mantle, but South America is overthrusting it. No arc has developed, as the rate of aqueous addition to the mantle wedge is presumably too slow in the absence of a subduction component of motion by the Caribbean. Subduction-related arc volcanoes (triangles) are restricted to the hanging wall above the Nazca Plate only.

of the Belize passive margin was weaker than the oceanic lithosphere of the Proto-Caribbean Sea). The NW flank of the Caribbean Arc (i.e. central Cuba) accordingly collapsed into the site of dynamic subsidence (trench) caused by the westard motion of North American lithosphere, hence opening the Yucatán Basin.

Moving now to cases of flat slab subduction, Benioff Zone seismicity (Pardo & Suarez 2005; Manea *et al.* 2006) and seismic tomography (van der Hilst & Mann 1994) show that both SW Mexico and NW Colombia are sites of flat-slab subduction, where east-dipping subduction of oceanic slabs occurs beneath the hanging walls of westward-migrating American continental lithospheres (see cross-sections in Fig. 27). In these localities, the continental hanging walls either continuously advance across the trench axis (Colombia) or progressively approach the trench axis as the intervening Chortís Block escapes east, and then advance across it (SW Mexico). Both are specifically because of westward drift of the Americas across

the mantle. The effect is to superpose the footwall and hanging wall crusts to create areas of roughly double crustal thickness where only a single crustal thickness had existed previously. This is because, for these examples at least, slab roll back is slower than the westward drift of the Americas. Only some 750 km of Caribbean lithosphere remains visible beneath Colombia in seismic tomography (Fig. 27), but some 1150–1200 km of total plate displacement has occurred since the Maastrichtian onset of subduction (Fig. 25). If we consider that Andean shortening accounts for perhaps 150 km of that, we judge that the eastern 250–300 km of the subducted slab has become thermally equilibrated and can no longer be seen in the tomography.

The two processes (slab break-off and flat slab suduction) produce different effects at the surface, namely in heat flow history/igneous activity and development of unconformities. Concerning unconformities, the examples of slab break (Guatemala and Cuba) show modest post-orogenic uplift and

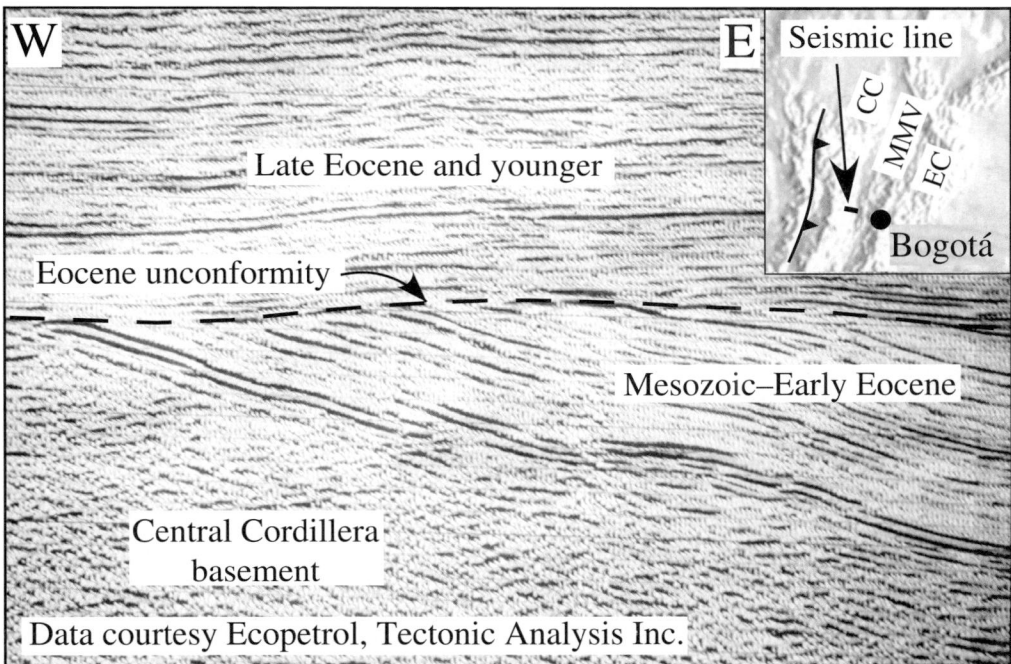

Fig. 28. East–west seismic section (about 20 km long), Middle Magdalena Basin (MMV), Colombia, location roughly shown in inset which also shows approximate position of Maastrichtian trench initiation for subduction of Caribbean lithosphere, the Romeral Fault Zone. The section shows the Eocene subaerial unconformity at several km depth subsurface, which had cut down to basement level in the Central Cordillera (CC) and which had exposed most of Colombia at that time. Homoclinal dip of Jurassic–Early Eocene stratal section records westward increasing uplift of basement as South American crust was first thrust (ramped) onto the Caribbean lithosphere at the onset of subduction. The unconformity is buried by Late Eocene–Oligocene and younger section, most of which is foredeep fill related to Eastern Cordillera (EC) uplift. Data are courtesy of Ecopetrol and Tectonic Analysis Inc.

erosion on the order of 1–2 km, judging from the fact that foredeep basin sections as young as the collisions remain locally preserved (Rosenfeld 1993; Iturralde-Vinent *et al.* 2008). Such unconformities are presumably produced or enhanced by isostatic rebound upon detachment of the slab (loss of negative buoyancy). In contrast, the examples of flat slab subduction (western Colombia, SW Mexico) show far more extreme uplift and erosion, perhaps as much as 5–7 km. In Colombia, the Maastrichtian onset of subduction generated the regional 'Middle Eocene' unconformity which, judging from seismic reflection records in the western Middle Magdalena Valley and denudation of the Central Cordillera, cut downward by more than 5 km through the entire Jurassic–Cretaceous section over the Cordillera Central (Fig. 28). Basement was broadly exposed and deeply eroded at this time in the Central Cordillera, supplying clastic detritus from Jurassic, Palaeozoic and Precambrian source terranes to the Paleocene–Early Eocene Matatere and other flysch units of western and central (but not eastern) Venezuela. Such drastic uplift is the effect of nearly doubling the crustal thickness by crustal scale ramping. Much of Colombia's basaltic and deep water clastic accretionary Western Cordillera, San Jacinto and Sinú belts were scraped from the Caribbean Plate during this Cenozoic history of plate convergence. In SW Mexico, subduction did not begin until the transcurrent removal of the Chortís Block during the Miocene. Since then, large amounts (>5 km) of hanging wall uplift can be demonstrated by eastwardly younging Cenozoic $^{40}Ar-^{39}Ar$ cooling ages in Precambrian rocks of the Xolapa Terrane along the Mexican coast (Morán-Zenteno *et al.* 1996; Ducea *et al.* 2004). The reason for large, homoclinal uplift is, again, the effective doubling of the crustal thickness by ramping of the continental hanging wall onto the crust of the downgoing plate (Fig. 27).

Concerning heat flow and igneous activity, slab break away should cause increased heat flow for a time (perhaps for 10 Ma. after the event; no longer easily detectable) due to the former cold slab being replaced by hot asthenosphere. In addition, igneous activity may result from decompression melting of this asthenosphere as it replaces the slab, or by melting of the remaining crust in the suture zone by heat transfer from the rising asthenosphere. Ratschbacher *et al.* (2009) considers that Maastrichtian pegmatites in the southern Maya Block may pertain to slab break-off, but we are aware of no Eocene magmatism in Cuba.

Note added in proof

The Pacific origin of Caribbean oceanic lithosphere requires either an arc polarity reversal from east- to west-dipping subduction at the Greater Antillean arc, or the inception of west-dipping subduction at perhaps an oceanic transform from the North- to South American Cordillera, in order to allow Pacific-derived Caribbean lithosphere to be engulfed between the Americas during their westward drift from Africa (Pindell & Dewey 1982). This event is commonly thought to have occurred between distinct periods of primitive v. calc-alkaline magmatism in the Caribbean arcs, but evidence for such a distinct boundary is waning as new geochronological and stratigraphic data are developed. The present paper acknowledges the Los Ranchos and Water Island formations of Hispaniola and the Virgin Islands, respectively (Kesler *et al.* 2005; Jolly *et al.* 2008) as postdating the onset of west-dipping subduction, largely because the new Aptian–Albian ages for these units post-date the commonly perceived *c.* 120–125 Ma onset of HP-LT metamorphism/cooling in rocks of the circum-Caribbean suture zone (e.g. Pindell *et al.* 2005; Garcia-Casco *et al.* 2006; Krebs *et al.* 2008). However, two additional arc units, the Lower Devil's Racecourse of Jamaica and the Los Pasos of Cuba, date to the Hauterivian (130–135 Ma) and have been highlighted as integral elements of the Caribbean arc (Hastie *et al.* 2009; Stanek *et al.* 2009). These older ages for arc activity accord with the ages on HP-LT rocks from the Siuna (139 Ma) and Motagua (*c.* 132 Ma) parts of the circum-Caribbean suture (Brueckner *et al.* 2005; Baumgartner *et al.* 2008), which are herein suggested to have migrated along the Trans-American transform from the western flank of Chortis. However, given that both the arc and HP-LT ages extend back to the 130s, we now consider that west-dipping subduction beneath the Greater Antilles arc likely dates to 135 Ma or even older. Referring to fig. 6 in Pindell *et al.* (2005), such an age would place the inception of west-dipping subduction prior to most/all arc magmatism in the Caribbean arcs (note: the Mt. Charles unit of Jamaica is now known to be Late Cretaceous; A. Hastie, pers. comm., 2009). This in turn suggests that an arc polarity reversal did not necessarily occur, and that subduction initiation occurred instead at a pre-existing transform or fracture zone, possibly the 'Trans-America Transform' of this paper, such that the arc developed directly on Jurassic or earliest Cretaceous ocean crust when the transform/fracture zone became convergent, by about 135 Ma.

We dedicate this paper to our academic supervisor, Professor John F. Dewey FRS, whose love for geology and its relationship to plate kinematics have inspired and defined a standard for the kind of work presented in this paper. We have benefited from and are grateful for exposure to data while collaborating with Pemex,

Ecopetrol, Petrotrin and PDVSA on research programmes that are far more detailed than the regional story told here. Without that basic input of information, many of the principles and evolutionary events outlined herein would need to be presented with less confidence. J. L. Pindell has also benefited from collaboration with NSF BOLIVAR Program (EAR-0003572 to Rice University) co-researchers A. Levander, J. Wright, P. Mann, B. Magnani, G. Christeson, M. Schmitz, S. Clark and H. Avé Lallemant. J. L. Pindell. also thanks M. Iturralde, A. García-Casco, Y. Rojas, K. Stanek and W. Maresch for collaboration on the Cuban sub-region; W. Maresch for joint development of working hypotheses concerning Margarita; J. Sisson, H. Avé Lallemant and L. Ratschbacher for discussions regarding Guatemala and the Chortís Block; P. Baumgartner, D. Buchs, K. Flores and A. Bandini of the University of Lausanne for workshops on Costa Rica and Panama; A. Kerr, I. Neill and A. Hastie for sharing viewpoints about Tobago, Jamaica and the Caribbean LIP; G. Draper, E. Lidiak and J. Lewis for discussions on the Greater Antilles; and A. Cardona of the Smithsonian Tropical Institute in Panama for discussions on the age and occurrence of intrusive rocks in Colombia. L. Kennan acknowledges R. Spikings and E. Jaillard for many helpful suggestions that improved our understanding of the Northern Andes. We are grateful to K. H. James and M. A. Lorente for organizing the June 2006 Sigüenza Caribbean meeting, at which many of the concepts presented herein were outlined. Keith's persistent questioning of long-held interpretations keeps us working harder.

References

ALGAR, S. T. & PINDELL, J. L. 1993. Structure and deformation history of the northern range of Trinidad and adjacent areas. *Tectonics*, **12**, 814–829.

ANDERSON, D. L. 2007. *The New Theory of the Earth.* Cambridge University Press.

ARCULUS, R. J., LAPIERRE, H. & JAILLARD, E. 1999. A geochemical window into subduction–accretion processes: the Raspas Metamorphic Complex, Ecuador. *Geology*, **27**, 547–550.

AYMARD, R., PIMENTEL, L. *ET AL.* 1990. Geological integration and evaluation of northern Monagas, Eastern Venezuelan Basin. *In*: BROOKS, J. (ed.) *Classic Petroleum Provinces.* Geological Society of London, Special Publications, **50**, 37–53.

BALDWIN, S. L. & HARRISON, T. M. 1989. Geochronology of blueschists from west-central Baja California and the timing of uplift in subduction complexes. *Journal of Geology*, **97**, 149–163.

BANDINI, A. N., FLORES, K., BAUMGARTNER, P. O., JACKETT, S.-J. & DENYER, P. 2008. Late Cretaceous and Paleogene Radiolaria from the Nicoya Peninsula, Costa Rica: a tectonostratigraphic application. *Stratigraphy*, **5**, 3–21.

BARRAGÁN, R., BABY, P. & DUNCAN, R. 2005. Cretaceous alkaline intra-plate magmatism in the Ecuadorian Oriente Basin: geochemical, geochronological and tectonic evidence. *Earth and Planetary Science Letters*, **236**, 670–690.

BARTOLINI, C., LANG, H. & SPELL, T. 2003. Geochronology, geochemistry, and tectonic setting of the Mesozoic Nazas Arc in north-central Mexico, and its continuation to northern South America. *In*: BARTOLINI, C., BUFFLER, R. T. & BLICKWEDE, J. F. (eds) *The Circum-Gulf of Mexico and the Caribbean; Hydrocarbon Habitats, Basin Formation, and Plate Tectonics.* American Association of Petroleum Geologists, Memoirs, **79**, 427–461.

BAUMGARTNER, P. O., FLORES, K., BANDINI, A. N., GIRAULT, F. & CRUZ, D. 2008. Upper Triassic to Cretaceous radiolaria from Nicaragua and northern Costa Rica–the Mesquito composite oceanic terrane. *Ofioliti*, **33**, 1–19.

BECK, C. 1986. Collision caraibe, derive andine, et evolution geodynamique mesozoique–cenozoique des Caraibes [Caribbean collision, Andean drift, and Mesozoic–Cenozoic geodynamic evolution of the Caribbean islands]. *Revue de Geologie Dynamique et de Geographie Physique*, **27**, 163–182.

BELLIZZIA, A. 1985. Sistema montanosa del Caribe – una Cordillera aloctona en la parte norte de America del Sur. *Memorias del VI Congreso Geologico Venezolano.* Sociedad Venezolana de Geologia, Caracas, **10**, 6657–6835.

BERMÚDEZ-CELLA, M., VAN DER BEEK, P. & BERNET, M. 2008. Fission-track thermochronological evidence for km-scale vertical offsets across the Bocono strike–slip fault, central Venezuelan Andes. *Geophysical Research Abstracts*, **10**, EGU2008-A-07173.

BIRD, D. E., HALL, S. A., CASEY, J. F. & MILLEGAN, P. S. 1999. Tectonic evolution of the Grenada Basin. *In*: MANN, P. (ed.) *Caribbean Basins.* Sedimentary Basins of the World, **4**. Elsevier Science, Amsterdam, 389–416.

BIRD, D. E., BURKE, K., HALL, S. A. & CASEY, J. F. 2005. Gulf of Mexico tectonic history: hot spot tracks, crustal boundaries, and early salt distribution. *American Association of Petroleum Geologists Bulletin*, **89**, 311–328.

BOSCH, D., GABRIELE, P., LAPIERRE, H., MALFERE, J. L. & JAILLARD, E. 2002. Geodynamic significance of the Raspas metamorphic complex (SW Ecuador): geochemical and isotopic contraints. *Tectonophysics*, **345**, 83–102.

BRIDEN, J., REX, D. C., FALLER, A. M. & TOMBLIN, J. F. 1979. K–Ar geochronology and palaeomagnetism of volcanic rocks in the Lesser-Antilles island arc. *Philosophical Transactions Royal Society of London, Series A*, **291**, 485–528.

BRUECKNER, H. K., HEMMING, S., SORENSEN, S. S. & HARLOW, G. E. 2005. Synchronous Sm–Nd mineral ages from HP terranes on both sides of the Motagua Fault of Guatemala: convergent suture and strike–slip fault? *Eos Transactions, AGU Fall Meeting Supplement*, **86**, 52.

BUCHS, D. 2008. *Late Cretaceous to Eocene geology of the south Central American forearc area (southern Costa Rica and western Panama): initiation and evolution of an intra-oceanic convergent margin.* PhD Thesis, Université de Lausanne, Switzerland.

BUCHS, D., BAUMGARTNER, P. & ARCULUS, R. 2007. Late Cretaceous arc initiation on the edge of an oceanic plateau (southern Central America). *Eos Transactions AGU 88(52), Fall Meeting Supplement*, Abstract T13C-1468.

BUCHS, D. M., BAUMGARTNER, P. O. *ET AL.* 2009. Late Cretaceous to Miocene seamount accretion and mélange formation in the Osa and Burica peninsulas (southern Costa Rica): episodic growth of a convergent margin. *In*: JAMES, K. H., LORENTE, M. A. & PINDELL, J. L. (eds) *The Origin and Evolution of the Caribbean Plate*. Geological Society, London, Special Publications, **328**, 411–456.

BULLARD, E., EVERETT, J. E. & SMITH, A. G. 1965. The fit of the continents around the Atlantic. *In*: BLACKETT, P. M. S., BULLARD, E. & RUNCORN, S. K. (eds) *A Symposium on Continental Drift*, Philosophical Transactions of the Royal Society, **258**, 41–51.

BURKART, B. 1983. Neogene North American–Caribbean Plate boundary across northern central America: offset along the Polochic Fault. *Tectonophysics*, **99**, 251–270.

BURKE, K. 1988. Tectonic evolution of the Caribbean. *Annual Review of Earth and Planetary Sciences*, **16**, 210–230.

BURKE, K., GRIPPI, J. & SENGOR, A. M. C. 1980. Neogene structures in Jamaica and the tectonic style of the northern Caribbean Plate boundary zone. *Journal of Geology*, **88**, 375–386.

BYERLY, G. R. 1991. Nature of igneous activity. *In*: SALVADOR, A. (ed.) *The Gulf of Mexico Basin*. The Geology of North America, **J**. Geological Society of America, Boulder, CO, 91–108.

CALVO, C. & BOLZ, A. 1994. Der älteste kalkalkaline Inselbogen-Vulkanismus in Costa Rica. Marine Pyroklastika der Formation Loma Chumico (Alb bis Campan) [The oldest calcalkaline island arc volcanism in Costa Rica. Marine tephra deposits from the Loma Chumico Formation (Albian to Campanian)]. *Profil*, **7**, 235–264.

CARDONA, A., DUQUE, J. F. *ET AL.* 2008. Geochronology and tectonic implications of granitoids rocks from the northwestern Sierra Nevada de Santa Marta and surrounding basins, northeastern Colombia: Late Cretaceous to Paleogene convergence, accretion and subduction interactions between the Caribbean and South American plates. *Abstract Volume of the 18th Caribbean Geological Conference*, 24–28 March 2008, Santo Domingo, Dominican Republic. World Wide Web Address: http://www.ugr.es/~agcasco/igcp546/DomRep08/Abstracts_CaribConf_DR_2008.pdf.

CARFANTAN, J.-C. 1986. *Du systeme Cordillerain Nord-Americain au domain Caraibe – Etude Geologique du Mexique Meridional*. PhD Thesis, Universite de Savoie, France.

CASE, J. E. & HOLCOMBE, T. L. 1980. *Geologic–Tectonic Map of the Caribbean (Scale: 1:2 500 000)*. United States Geological Survey Miscellaneous Investigations Series Map, **I-1100**.

CLARK, S. A., ZELT, C. A., MAGNANI, M. B. & LEVANDER, A. 2008. Characterizing the Caribbean–South American Plate boundary at 64°W using wide-angle seismic data. *Journal of Geophysical Research*, **113**, B07401.

DAVIES, J. H. & VON BLANCKENBURG, F. 1995. Slab breakoff: a model of lithosphere detachment and its test in the magmatism and deformation of collisional orogens. *Earth and Planetary Science Letters*, **129**, 85–102.

DENG, J. S. & SYKES, L. R. 1995. Determination of Euler pole for contemporary relative motion of Caribbean and North American plates using slip vectors of interplate earthquakes. *Tectonics*, **14**, 39–53.

DENYER, P. & BAUMGARTNER, P. O. 2006. Emplacement of Jurassic–Lower Cretaceous radiolarites of the Nicoya Complex (Costa Rica). *Geologica Acta*, **4**, 203–218.

DEWEY, J. F. & PINDELL, J. L. 1985. Neogene block tectonics of eastern Turkey and northern South America: continental applications of the finite difference method. *Tectonics*, **4**, 71–83.

DEWEY, J. F. & PINDELL, J. L. 1986. Neogene block tectonics of eastern Turkey and northern South America: continental applications of the finite difference method: Reply. *Tectonics*, **5**, 703–705.

DICKINSON, W. R. & LAWTON, T. F. 2001. Carboniferous to Cretaceous assembly and fragmentation of Mexico. *Geological Society of America Bulletin*, **113**, 1142–1160.

DIEBOLD, J. 2009. Submarine volcanic stratigraphy and the Caribbean LIP's formational environment. *In*: JAMES, K. H., LORENTE, M. A. & PINDELL, J. L. (eds) *The Origin and Evolution of the Caribbean Plate*. Geological Society, London, Special Publications, **328**, 799–808.

DIEBOLD, J., DRISCOLL, N. & EW-9501-SCIENCE TEAM. 1999. New insights on the formation of the Caribbean basalt province revealed by multichannel seismic images of volcanic structures in the Venezuelan Basin. *In*: MANN, P. (ed.) *Caribbean Basins*. Sedimentary Basins of the World, **4**. Elsevier Science, Amsterdam, 561–589.

DOLAN, J., MANN, P., DE ZOETEN, R., HEUBECK, C. & SHIROMA, J. 1991. Sedimentologic, stratigraphic, and tectonic synthesis of Eocene-Miocene sedimentary basins, Hispaniola and Puerto Rico. *In*: MANN, P., DRAPER, G. & LEWIS, J. F. (eds) *Geologic and Tectonic Development of the North America–Caribbean Plate Boundary in Hispaniola*. Geological Society of America, Special Papers, **262**, 17–264.

DOLAN, J. F., MULLINS, H. T. & WALD, D. J. 1998. Active tectonics of the north-central Caribbean: oblique collision, strain partitioning, and opposing subducted slabs. *In*: DOLAN, J. F. & MANN, P. (eds) *Active Strike-Slip and Collisional Tectonics of the Northern Caribbean Plate Boundary Zone*. Geological Society of America Special Papers, **326**, 1–61.

DONNELLY, T. W., HORNE, G. S., FINCH, R. C. & LOPEZ-RAMOS, E. 1990. Northern Central America: The Maya and Chortís blocks. *In*: DENGO, G. & CASE, J. E. (eds) *The Caribbean Region*. The Geology of North America, **H**. Geological Society of America, 371–396.

DOUBROVINE, P. V. & TARDUNO, J. A. 2008. A revised kinematic model for the relative motion between Pacific Oceanic plates and North America since the Late Cretaceous. *Journal of Geophysical Research*, **113**, B12101.

DRISCOLL, N. W. & DIEBOLD, J. B. 1999. Tectonic and stratigraphic development of the eastern Caribbean: new constraints from multichannel seismic data. *In*: MANN, P. (ed.) *Caribbean Basins*. Sedimentary Basins of the World, **4**. Elsevier Science, Amsterdam, 591–626.

DUCEA, M. N., GEHRELS, G. E., SHOEMAKER, S., RUIZ, J. & VALENCIA, V. A. 2004. Geologic evolution of the Xolapa Complex, southern Mexico: evidence from U–Pb zircon geochronology. *Geological Society of America Bulletin*, **116**, 1016–1025.

DUNCAN, R. A. & HARGRAVES, R. B. 1984. Plate tectonic evolution of the Caribbean region in the mantle reference frame. *In*: BONINI, W. E., HARGRAVES, R. B. & SHAGAM, R. (eds) *The Caribbean–South American Plate Boundary and Regional Tectonics*. Geological Society of America, Memoirs, **162**, 81–93.

EAGLES, G., LIVERMORE, R. A., FAIRHEAD, J. D. & MORRIS, P. 2004. Tectonic evolution of the west Scotia Sea. *Journal of Geophysical Research*, **110**, B02401.

ENGEBRETSON, D. C., COX, A. & GORDON, R. G. 1985. *Relative Motions Between Oceanic and Continental Plates in the Pacific Basin*. Geological Society of America, Special Papers, **206**.

FARR, T. G., ROSEN, P. A. ET AL. 2007. The Shuttle Radar Topography Mission. *Reviews of Geophysics*, **45**, RG2004. World Wide Web Address: Image of Central America available at http://photojournal.jpl.nasa.gov/catalog/PIA03364.

FERRARI, L., LOPEZ-MARTINEZ, M., AGUIRRE-DIAZ, G. & CARRASCO-NUNEZ, G. 1999. Space-time patterns of Cenozoic arc volcanism in central Mexico: from the Sierra Madre Occidental to the Mexican Volcanic Belt. *Geology*, **27**, 303–306.

FLORES, K., DENYER, P. & AGUILAR, T. 2003a. Nueva propuesta estratigráfica: geología de las hojas Matambú y Talolinga, Guanacaste, Costa Rica. *Revista Geologica de America Central*, **28**, 131–138.

FLORES, K., DENYER, P. & AGUILAR, T. 2003b. Nueva propuesta estratigráfica: geología de la hoja Abangares, Guanacaste, Costa Rica. *Revista Geologica de America Central*, **29**, 127–136.

FLORES, K., BAUMGARTNER, P. O., DENYER, P., BANDINI, A. N. & BAUMGARTNER-MORA, C. 2004. Pre-Campanian Terranes in Nicoya (Costa Rica, Middle America). *2nd Swiss Geoscience Meeting*, Lausanne, 19–20 November, 2004. World Wide Web Address: http://geoscience-meeting.scnat-web.ch/sgm2004/abstracts_2004/Flores_Baumgart-ner_et_al.pdf.

FLORES, K., BAUMGARTNER, P. O., SKORA, S., BAUMGARTNER, L., MÜNTENER, O., COSCA, M. & CRUZ, D. 2007. The Siuna Serpentinite Mélange: an Early Cretaceous subduction/accretion of a Jurassic Arc. *Eos Transactions AGU 88(52), Fall Meeting Supplement*, Abstract T-11D-03.

FREYDIER, C., MARTINEZ, J., LAPIERRE, H., TARDY, M. & COULON, C. 1996. The Early Cretaceous Arperos oceanic basin (western Mexico). Geochemical evidence for an aseismic ridge formed near a spreading center. *Tectonophysics*, **259**, 343–367.

FREYDIER, C., LAPIERRE, H., RUIZ, J., TARDY, M., MARTINEZ, J. & COULON, C. 2000. The Early Cretaceous Arperos basin: an oceanic domain dividing the Guerrero arc from nuclear Mexico evidenced by the geochemistry of the lavas and sediments. *Journal of South American Earth Sciences*, **13**, 325–336.

GARCÍA-CASCO, A., TORRES-ROLDÁN, R. L. ET AL. 2006. High pressure metamorphism of ophiolites in Cuba. *Geologica Acta*, **4**, 63–88.

GARCÍA-CASCO, A., LAZARO, C. ET AL. 2008a. Partial melting and counterclockwise *P T* path of subducted oceanic crust (Sierra del Convento Melange, Cuba). *Journal of Petrology*, **49**, 129–161.

GARCÍA-CASCO, A., ITURRALDE-VINENT, M. A. & PINDELL, J. L. 2008b. Latest Cretaceous Collision/accretion between the Caribbean Plate and caribeana: origin of metamorphic terranes in the Greater Antilles. *International Geology Review*, **50**, 781–809.

GELDMACHER, J., HOERNLE, K., VAN DEN BOGAARD, P., HAUFF, F. & KLÜGEL, A. 2008. Age and geochemistry of the Central American Forearc Basement (DSDP Leg 67 and 84): insights into Mesozoic Arc volcanism and seamount accretion on the fringe of the Caribbean LIP. *Journal of Petrology*, **49**, 1781–1815.

GOVERS, R. & WORTEL, M. J. R. 2005. Lithosphere tearing at STEP faults: response to edges of subduction zones. *Earth and Planetary Science Letters*, **236**, 505–523.

GRADSTEIN, F. M., OGG, J. G. & SMITH, A. G. 2004. *A Geologic Timescale 2004*. Cambridge University Press, Cambridge.

GRANDE, S. & URBANI, F. 2009. Presence of high-grade rocks in NW Venezuela of possible Grenvillian affinity. *In*: JAMES, K. H., LORENTE, M. A. & PINDELL, J. L. (eds) *The Origin and Evolution of the Caribbean Plate*. Geological Society, London, Special Publications, **328**, 533–548.

GRAY, G., POTTORF, R. J., YUREWICZ, D. A., MAHON, K. I., PEVEAR, D. R. & CHUCHLA, R. J. 2003. Thermal and chronological record of syn- to post-Laramide burial and exhumation, Sierra Madre Oriental, Mexico. *In*: BARTOLINI, C., BUFFLER, R. T. & CANTU-CHAPA, A. (eds) *The Western Gulf of Mexico Basin: Tectonics, Sedimentary Basins, and Petroleum Systems*. American Association of Petroleum Geologists, Memoirs, **75**, 159–181.

GUZMAN-SPEZIALE, M. & MENESES-ROCHA, J. J. 2000. The North America–Caribbean Plate boundary west of the Motagua–Polochic fault system: a fault jog in Southeastern Mexico. *Journal of South American Earth Sciences*, **13**, 459–468.

HARLOW, G. E., HEMMING, S. R., AVE-LALLEMANT, H. G., SISSON, V. B. & SORENSEN, S. S. 2004. Two high-pressure–low-temperature serpentinite-matrix mélange belts, Motagua fault zone, Guatemala: a record of Aptian and Maastrichtian collisions. *Geology*, **32**, 17–20.

HASTIE, A. R., KERR, A. C., MITCHELL, S. F. & MILLER, I. L. 2008. Geochemistry and petrogenesis of Cretaceous oceanic plateau lavas in eastern Jamaica. *Lithos*, **101**, 323–343.

HENRY, C. D., MCDOWELL, F. W. & SILVER, L. T. 2003. Geology and geochronology of granitic batholithic complex, Sinaloa, México: implications for Cordilleran magmatism and tectonics. *In*: JOHNSON, S. E., PATERSON, S. R., FLETCHER, J. M., GIRTY, G. H., KIMBROUGH, D. L. & MARTÍN-BARAJAS, A. (eds) *Tectonic Evolution of Northwestern México and the Southwestern USA*. Geological Society of America, Special Papers, **374**, 237–273.

HOERNLE, K., VAN DEN BOGAARD, P. ET AL. 2002. Missing history (16–71 Ma) of the Galpapagos hot spot: implications for the tectonic and biological evolution of the Americas. *Geology*, **30**, 795–798.

HOERNLE, K., HAUFF, F. & VAN DEN BOGAARD, P. 2004. A 70 m.y. history (139–69 Ma) for the Caribbean large igneous province. *Geology*, **32**, 697–700.

HORBURY, A., HALL, S. *ET AL.* 2003. Tectonic sequence stratigraphy of the western margin of the Gulf of Mexico in the Late Mesozoic and Cenozoic: less passive than previously imagined. *In*: BARTOLINI, C., BUFFLER, R. T. & BLICKWEDE, J. F. (eds) *The Circum-Gulf of Mexico and the Caribbean; Hydrocarbon Habitats, Basin Formation, and Plate Tectonics*. American Association of Petroleum Geologists, Memoirs, **79**, 184–245.

IMBERT, P. 2005. The Mesozoic opening of the Gulf of Mexico: Part 1. Evidence for oceanic accretion during and after salt deposition. *In*: POST, P. J., ROSEN, N. C., OLSON, D. L., PALMES, S. L., LYONS, K. T. & NEWTON, G. B. (eds) *Transactions of the 25th Annual GCSSEPM Research Conference: Petroleum Systems of Divergent Continental Margins*, 1119–1150.

IMBERT, P. & PHILIPPE, Y. 2005. The Mesozoic opening of the Gulf of Mexico: Part 2. Integrating seismic and magnetic data into a general opening model. *In*: POST, P. J., ROSEN, N. C., OLSON, D. L., PALMES, S. L., LYONS, K. T. & NEWTON, G. B. (eds) *Transactions of the 25th Annual GCSSEPM Research Conference: Petroleum Systems of Divergent Continental Margins*, 1151–1189.

ITURRALDE-VINENT, M. A. & MACPHEE, R. D. E. 1999. Paleogeography of the Caribbean region: implications for Cenozoic biogeography. *Bulletin of the American Museum of Natural History*, **238**, 1–95.

ITURRALDE-VINENT, M. A., DÍAZ OTERO, C., GARCÍA-CASCO, A. & VAN HINSBERGEN, D. J. J. 2008. Paleogene foredeep basin deposits of North-Central Cuba: a record of arc–continent collision between the Caribbean and North American Plates. *International Geology Review*, **50**, 863–884.

JACQUES, J. M., PRICE, A. D. & BAIN, J. E. 2004. Digital integration of potential fields and geologic data sets for Plate tectonic and basin dynamic modeling – the first step toward identifying new play concepts in the Gulf of Mexico Basin. *The Leading Edge*, **23**, 384–389.

JAILLARD, E., SOLER, P., CARLIER, G. & MOURIER, T. 1990. Geodynamic evolution of the northern and central Andes during Early to Middle Mesozoic times: a Tethyan model. *Journal of the Geological Society of London*, **147**, 1009–1022.

JAILLARD, E., LAUBACHER, G., BENGTSON, P., DHONDT, A. V. & BULOT, L. G. 1999. Stratigraphy and evolution of the Cretaceous forearc Celica – Lancones basin of southwestern Ecuador. *Journal of South American Earth Sciences*, **12**, 51–68.

JOLLY, W. T., LIDIAK, E. G. & DICKIN, A. P. 2008. The case for persistent southwest-dipping Cretaceous convergence in the northeast Antilles: geochemistry, melting models, and tectonic implications. *Geological Society of America Bulletin*, **120**, 1036–1052.

KEIGWIN, L. D. JR. 1978. Late Cenozoic paleoceanography of the Panama, Colombia and Venezuela basins. *In*: HARRIS, B. J. (ed.) *CICAR II; Symposium on Progress in Marine Research in the Caribbean and Adjacent Regions*. FAO Fisheries Report, **2**, 387–392.

KENNAN, L. & PINDELL, J. 2009. Dextral shear, terrane accretion and basin formation in the Northern Andes: best explained by interaction with a Pacific-derived Caribbean Plate. *In*: JAMES, K. H., LORENTE, M. A. & PINDELL, J. L. (eds) *The Origin and Evolution of the Caribbean Plate*. Geological Society, London, Special Publications, **328**, 487–531.

KEPPIE, J. D. & MORÁN-ZENTENO, D. J. 2005. Tectonic implications of alternative Cenozoic reconstructions for Southern Mexico and the Chortís Block. *International Geology Review*, **47**, 476–491.

KERR, A. C., ITURRALDE VINENT, M. A., SAUNDERS, A. D., BABBS, T. L. & TARNEY, J. 1999. A new plate tectonic model of the Caribbean: implications from a geochemical reconnaissance of Cuban Mesozoic volcanic rocks. *Geological Society of America Bulletin*, **111**, 1581–1599.

KERR, A. C., WHITE, R. V., THOMPSON, P. M. E., TARNEY, J. & SAUNDERS, A. D. 2003. No oceanic plateau – no Caribbean Plate? The seminal role of an oceanic plateau in Caribbean Plate evolution. *In*: BARTOLINI, C., BUFFLER, R. T. & BLICKWEDE, J. F. (eds) *The Circum-Gulf of Mexico and the Caribbean; Hydrocarbon Habitats, Basin Formation, and Plate Tectonics*. American Association of Petroleum Geologists, Memoirs, **79**, 126–168.

KESLER, S. E., CAMPBELL, I. H. & ALLEN, C. M. 2005. Age of the Los Ranchos Formation, Dominican Republic: timing and tectonic setting of primitive island arc volcanism in the Caribbean region. *Geological Society of America Bulletin*, **117**, 987–995.

KLITGORD, K. & SCHOUTEN, H. 1986. Plate kinematics of the central Atlantic. *In*: VOGT, P. R. & TUCHOLKE, B. E. (eds) *The Western North Atlantic Region*. The Geology of North America, **M**. Geological Society of America, 351–378.

KREBS, M., MARESCH, W. V. *ET AL.* 2008. The dynamics of intra-oceanic subduction zones: a direct comparison between fossil petrological evidence (Rio San Juan Complex, Dominican Republic) and numerical simulation. *Lithos*, **103**, 106–137.

LADD, J. W. 1976. Relative motion of South America with respect to North America and Caribbean tectonics. *Geological Society of America Bulletin*, **87**, 969–976.

LE PICHON, X. & FOX, P. J. 1971. Marginal offsets, fracture zones, and the early opening of the North Atlantic. *Journal of Geophysical Research*, **76**, 6294–6308.

LE ROY, C., RANGIN, C., LE PICHON, X., NGUYIN, H., NGOC, T., ANDREANI, L. & ARANDA-GARCÍA, M. 2008. Neogene crustal shear zone along the western Gulf of Mexico margin and its implications for gravity sliding processes. Evidences from 2D and 3D multichannel seismic data. *Bulletin de la Societe Geologique de France*, **179**, 175–193.

LEROY, S., MAUFFRET, A., PATRIAT, P. & DE LEPINAY, B. M. 2000. An alternative interpretation of the Cayman Trough evolution from a re-identification of magnetic anomalies. *Geophysical Journal International*, **141**, 539–557.

LEWIS, J. F. & DRAPER, G. 1990. Geological and tectonic evolution of the northern Caribbean margin. *In*:

DENGO, G. & CASE, J. E. (eds) *The Caribbean Region.* The Geology of North America, **H**. Geological Society of America, 77–140.

LIDIAK, E., JOLLY, W. & DICKIN, A. 2008, *Geochemical and Tectonic Evolution of Albian to Eocene Volcanic Strata in the Virgin Islands and Eastern and Central Puerto Rico.* Geological Society of America, Abstracts with Programs, **40**, 105.

LUZIEUX, L. 2007. *Origin and Late Cretaceous–Tertiary evolution of the Ecuadorian forearc.* PhD Thesis, ETH, Zurich.

MANEA, V. C., MANEA, M., KOSTOGLODOV, V. & SEWELL, G. 2006. Intraslab seismicity and thermal stress in the subducted Cocos Plate beneath central Mexico. *Tectonophysics*, **420**, 389–408.

MANN, P. 1999. Caribbean sedimentary basins: classification and tectonic setting from Jurassic to Present. *In*: MANN, P. (ed.) *Caribbean Basins.* Sedimentary Basins of the World, **4**, Elsevier Science, Amsterdam, 3–31.

MANN, P. 2007. Overview of the tectonic history of northern Central America. *In*: MANN, P. (ed.) *Geologic and Tectonic Development of the Caribbean Plate Boundary in Northern Central America.* Geological Society of America, Special Papers, **428**, 1–19.

MANN, P. & BURKE, K. 1984. Neotectonics of the Caribbean. *Review of Geophysics and Space Physics*, **22**, 309–362.

MANN, P. & CORRIGAN, J. C. 1990. Model for Late Neogene deformation in Panama. *Geology*, **18**, 558–562.

MANN, P., DRAPER, G. & LEWIS, J. F. 1991. An overview of the geologic and tectonic development of Hispaniola. *In*: MANN, P., DRAPER, G. & LEWIS, J. F. (eds) *Geologic and Tectonic Development of the North America–Caribbean Plate Boundary in Hispaniola.* Geological Society of America, Special Papers, **262**, 1–28.

MANN, P., ROGERS, R. & GAHAGAN, L. 2007. Overview of Plate tectonic history and its unresolved tectonic problems. *In*: BUNDSCHUH, J. & ALVARADO, G. (eds) *Central America: Geology, Resources and Hazards.* Taylor & Francis, London, **1**, 201–237.

MARESCH, W. V., KLUGE, R. *ET AL.* 2009. The occurrence and timing of high-pressure metamorphism on Margarita Island, Venezuela: a constraint on Caribbean-South America interaction. *In*: JAMES, K. H., LORENTE, M. A. & PINDELL, J. L. (eds) *The Origin and Evolution of the Caribbean Plate.* Geological Society, London, Special Publications, **328**, 705–741.

MARTENS, U., SOLARI, L., MATTINSON, C. G., WOODEN, J. & LIOU, J. G. 2008. Polymetamorphism at the southern boundary of the North American Plate: the El Chol unit of the Chuacús Complex, Central Guatemala. *Abstract Volume of the 18th Caribbean Geological Conference*, 24–28 March 2008, Santo Domingo, Dominican Republic. World Wide Web Address: http://www.ugr.es/~agcasco/igcp546/DomRep08/Abstracts_CaribConf_DR_2008.pdf.

MARTON, G. L. & BUFFLER, R. T. 1999. Jurassic–Early Cretaceous tectono-paleogeographic evolution of the southeastern Gulf of Mexico Basin. *In*:

MANN, P. (ed.) *Caribbean Basins.* Sedimentary Basins of the World, Elsevier Science, Amsterdam, **4**, 63–91.

MCLAUGHLIN, P. & SEN GUPTA, B. K. 1991. Migration of Neogene marine environments, southwestern Dominican Republic. *Geology*, **19**, 222–225.

MIRANDA-C., E., PINDELL, J. L. *ET AL.* 2003. Mesozoic tectonic evolution of Mexico and Southern Gulf of Mexico: framework for basin evaluation in Mexico. *American Association of Petroleum Geologists International Conference & Exhibition*, 21–24 September, Barcelona, Spain. World Wide Web Address: http://aapg.confex.com/aapg/barcelona/techprogram/paper_83820.htm.

MONTGOMERY, H. & KERR, A. C. 2009. Rethinking the origins of the red chert at La Désirade, French West Indies. *In*: JAMES, K. H., LORENTE, M. A. & PINDELL, J. L. (eds) *The Origin and Evolution of the Caribbean Plate.* Geological Society, London, Special Publications, **328**, 457–467.

MORAN-ZENTENO, D. J., CORONA-CHAVEZ, P. & TOLSON, G. 1996. Uplift and subduction erosion in southwestern Mexico since the Oligocene: pluton geobarometry constraints. *Earth and Planetary Science Letters*, **141**, 51–65.

MÜLLER, R. D., ROYER, J.-Y. & LAWVER, L. A. 1993. Revised plate motions relative to the hot spots from combined Atlantic and Indian Ocean hot spot tracks. *Geology*, **21**, 275–278.

MÜLLER, R. D., ROYER, J.-Y., CANDE, S. C., ROEST, W. R. & MASCHENKOV, S. 1999. New constraints on the Late Cretaceous/Tertiary plate tectonic evolution of the Caribbean. *In*: MANN, P. (ed.) *Caribbean Basins.* Sedimentary Basins of the World, **4**. Elsevier Science, Amsterdam, 33–59.

MÜLLER, R. D., SDROLIAS, M., GAINA, C. & ROEST, W. R. 2008. Age, spreading rates, and spreading asymmetry of the world's ocean crust. *Geochemistry Geophysics Geosystems*, **9**, Q04006.

NIKOLAEVA, K., GERYA, T. V. & CONNOLLY, J. A. D. 2008. Numerical modelling of crustal growth in intraoceanic volcanic arcs. *Physics of the Earth and Planetary Interiors*, **171**, 336–356; doi: 10.1016/j.pepi.2008.06.026.

NIU, Y., O'HARA, M. J. & PEARCE, J. A. 2003. Initiation of subduction zones as a consequence of lateral compositional buoyancy contrast within the lithosphere: a petrological perspective. *Journal of Petrology*, **44**, 851–866.

NIVIA, A., MARRINER, G. F., KERR, A. C. & TARNEY, J. 2006. The Quebradagrande Complex: a Lower Cretaceous ensialic marginal basin in the Central Cordillera of the Colombian Andes. *Journal of South American Earth Sciences*, **21**, 423–436.

OFFICER, C. B. J., EWING, J. I., EDWARDS, R. S. & JOHNSON, H. R. 1957. Geophysical investigations in the eastern Caribbean: Venezuelan Basin, Antilles island arc, and Puerto Rico Trench. *Geological Society of America Bulletin*, **68**, 359–378.

OFFICER, C., EWING, J., HENNION, J., HARKINDER, D. & MILLER, D. 1959. Geophysical investigations in the eastern Caribbean – summary of the 1955 and 1956 cruises. *Physics and Chemistry of the Earth*, **3**, 17–109.

PARDO, M. & SUÁREZ, G. 1995. Shape of the subducted Rivera and Cocos plates in southern Mexico: seismic and tectonic implications. *Journal of Geophysical Research*, **100**, B7, 12,357–12,373.

PEREZ, O. J., BILHAM, R. ET AL. 2001. Velocity field across the southern Caribbean Plate boundary and estimates of Caribbean/South American Plate motion using GPS geodesy 1994–2000. *Geophysical Research Letters*, **28**, 2987–2990.

PILGER, R. H. 2003. *Geokinematics: Prelude to Dynamics.* Springer, Berlin.

PINDELL, J. L. 1985a. Alleghanian reconstruction and subsequent evolution of the Gulf of Mexico, Bahamas, and proto-Caribbean. *Tectonics*, **4**, 1–39.

PINDELL, J. L. 1985b. *Plate tectonic evolution of the Gulf of Mexico and Caribbean region.* PhD Thesis, University of Durham, Durham.

PINDELL, J. L. 1990. Geological arguments suggesting a Pacific origin for the Caribbean Plate. *In*: LARUE, D. K. & DRAPER, G. (eds) *Transactions of the 12th Caribbean Geologic Conference, St Croix, 7–11 August 1989.* Miami Geological Society, Miami, FL, 1–4.

PINDELL, J. L. 1993. Regional synopsis of Gulf of Mexico and Caribbean evolution. *In*: PINDELL, J. L. & PERKINS, R. F. (eds) *Transactions of the 13th Annual GCSSEPM Research Conference: Mesozoic and Early Cenozoic Development of the Gulf of Mexico and Caribbean Region*, 251–274.

PINDELL, J. L. 2004. Origin of Caribbean Plateau basalts, the arc–arc Caribbean–South America collision, and upper level axis parallel extension in the Southern Caribbean Plate Boundary Zone. *EOS Transactions*, American Geophysical Union, **85** (47), Fall Meeting Supplement, Abstract T33B-1365.

PINDELL, J. L. 2008. Early Cretaceous Caribbean tectonics: models for genesis of the Great Caribbean Arc. *Abstract Volume of the 18th Caribbean Geological Conference*, 24–28 March 2008, Santo Domingo, Dominican Republic. World Wide Web Address: http://www.ugr.es/~agcasco/igcp546/DomRep08/Abstracts_CaribConf_DR_2008.pdf.

PINDELL, J. L. & BARRETT, S. F. 1990. Geological evolution of the Caribbean region: a Plate tectonic perspective. *In*: DENGO, G. & CASE, J. E. (eds) *The Caribbean Region.* The Geology of North America, **H**. Geological Society of America, 405–432.

PINDELL, J. L. & DEWEY, J. F. 1982. Permo-Triassic reconstruction of western Pangaea and the evolution of the Gulf of Mexico/Caribbean region. *Tectonics*, **1**, 179–211.

PINDELL, J. L. & DRAPER, G. 1991. Stratigraphy and geological history of the Puerto Plata area, northern Dominican Republic. *In*: MANN, P., DRAPER, G. & LEWIS, J. F. (eds) *Geologic and Tectonic Development of the North America–Caribbean Plate Boundary in Hispaniola.* Geological Society of America, Special Papers, **262**, 97–114.

PINDELL, J. L. & ERIKSON, J. P. 1994. Mesozoic passive margin of northern South America. *In*: SALFITY, J. A. (ed.), *Cretaceous Tectonics in the Andes.* International Monograph Series. Vieweg Publishing, Earth Evolution Sciences, Wiesbaden, 1–60.

PINDELL, J. L. & KENNAN, L. 2001. Processes and events in the terrane assembly of Trinidad and eastern Venezuela. *In*: FILLON, R. H., ROSEN, N. C. ET AL. (eds) *Transactions of the 21st GCSSEPM Annual Bob F. Perkins Research Conference: Petroleum Systems of Deep-Water Basins*, 159–192.

PINDELL, J. L. & KENNAN, L. 2007a. Rift models and the salt-cored marginal wedge in the northern Gulf of Mexico: implications for deep water Paleogene Wilcox deposition and basinwide maturation. *In*: KENNAN, L., PINDELL, J. L. & ROSEN, N. C. (eds) *Transactions of the 27th GCSSEPM Annual Bob F. Perkins Research Conference: The Paleogene of the Gulf of Mexico and Caribbean Basins: Processes, Events and Petroleum Systems*, 146–186.

PINDELL, J. L. & KENNAN, L. 2007b. Cenozoic kinematics and dynamics of oblique collision between two convergent plate margins: the Caribbean–South America collision in eastern Venezuela, Trinidad, and Barbados. *In*: KENNAN, L., PINDELL, J. L. & ROSEN, N. C. (eds) *Transactions of the 27th GCSSEPM Annual Bob F. Perkins Research Conference: The Paleogene of the Gulf of Mexico and Caribbean Basins: Processes, Events and Petroleum Systems*, 458–553.

PINDELL, J. L. & TABBUTT, K. D. 1995. Mesozoic-Cenozoic Andean paleogeography and regional controls on hydrocarbon systems. *In*: TANKARD, A. J., SUÁREZ, S. R. & WELSINK, H. J. (eds) *Petroleum Basins of South America.* American Association of Petroleum Geologists, Memoir, **62**, 101–128.

PINDELL, J. L., CANDE, S. C. ET AL. 1988. A plate-kinematic framework for models of Caribbean evolution. *Tectonophysics*, **155**, 121–138.

PINDELL, J. L., ERIKSON, J. P. & ALGAR, S. T. 1991. The relationship between Plate motions and the sedimentary basin development in northern South America: from a Mesozoic passive margin to a Cenozoic eastwardly-progressive transpressional orogen. *In*: GILLEZEAU, K. A. (ed.) *Transactions of the Second Geological Conference of the Geological Society of Trinidad and Tobago*, 3–8 April, 1991, Port of Spain, 191–202.

PINDELL, J. L., HIGGS, R. & DEWEY, J. F. 1998. Cenozoic palinspastic reconstruction, paleogeographic evolution, and hydrocarbon setting of the northern margin of South America. *In*: PINDELL, J. L. & DRAKE, C. L. (eds) *Paleogeographic Evolution and Non-Glacial Eustasy, Northern South America.* SEPM (Society for Sedimentary Geology), Special Publications, **58**, 45–86.

PINDELL, J. L., KENNAN, L., MARESCH, W. V., STANEK, K. P., DRAPER, G. & HIGGS, R. 2005. Plate-kinematics and crustal dynamics of circum-Caribbean arc-continent interactions, and tectonic controls on basin development in proto-Caribbean margins. *In*: AVÉ-LALLEMANT, H. G. & SISSON, V. B. (eds) *Caribbean–South American Plate Interactions, Venezuela.* Geological Society of America, Special Papers, **394**, 7–52.

PINDELL, J. L., KENNAN, L., STANEK, K. P., MARESCH, W. V. & DRAPER, G. 2006. Foundations of Gulf of Mexico and Caribbean evolution: eight controversies resolved. *Geologica Acta*, **4**, 89–128.

PINDELL, J., KENNAN, L., WRIGHT, D. & ERIKSON, J. 2009. Clastic domains of sandstones in central/eastern

Venezuela, Trinidad and Barbados: heavy mineral and tectonic constraints on provenance and palaeogeography. *In*: JAMES, K. H., LORENTE, M. A. & PINDELL, J. L. (eds) *The Origin and Evolution of the Caribbean Plate*. Geological Society, London, Special Publications, **328**, 743–797.

RAMANATHAN, R. & GARCÍA, E. 1991. Cretaceous paleography, foraminiferal biostratigraphy and paleoecology of the Belize Basin, Belize. *In*: GILLEZEAU, K. A. (ed.) *Transactions of the Second Geological Conference of the Geological Society of Trinidad & Tobago*, 3–8 April, 1991, Port of Spain, 203–211.

RATSCHBACHER, L., FRANZ, L. *ET AL*. 2009. The North American–Caribbean plate boundary in Mexico–Guatemala–Honduras. *In*: JAMES, K. H., LORENTE, M. A. & PINDELL, J. L. (eds) *The Origin and Evolution of the Caribbean Plate*. Geological Society, London, Special Publications, **328**, 219–293.

RAYMOND, C. A., STOCK, J. M. & CANDE, S. C. 2000. Fast Paleogene motion of the Pacific hotspots from revised global plate circuit constraints. *In*: RICHARDS, M. A., GORDON, R. G. & VAN DER HILST, R. D. (eds) *The History and Dynamics of Global Plate Motions*. American Geophysical Union, Monographs, **121**, 359–375.

RENNE, P. R., MATTINSON, J. M. *ET AL*. 1989. ^{40}Ar/^{39}Ar and U–Pb evidence for Late Proterozoic (Grenville-age) continental crust in north-central Cuba and regional tectonic implications. *Precambrian Research*, **42**, 325–341.

ROBERTSON, P. M. & BURKE, K. 1989. Evolution of the southern Caribbean plate boundary in the vicinity of Trinidad and Tobago. *American Association of Petroleum Geologists Bulletin*, **73**, 490–509.

ROEST, W. R., VERHOEF, J. & PILKINGTON, M. 1992. Magnetic interpretation using 3-D analytic signal. *Geophysics*, **57**, 116–125.

ROGERS, R. D., MANN, P., SCOTT, R. W. & PATINO, L. 2007a. Cretaceous intra-arc rifting, sedimentation, and basin inversion in east-central Honduras. *In*: MANN, P. (ed.) *Geologic and Tectonic Development of the Caribbean Plate Boundary in Northern Central America*. Geological Society of America Special Papers, **428**, 129–149.

ROGERS, R. D., MANN, P. & EMMET, P. A. 2007b. Tectonic terranes of the Chortís block based on integration of regional aeromagnetic and geologic data. *In*: MANN, P. (ed.) *Geologic and Tectonic Development of the Caribbean Plate Boundary in Northern Central America*. Geological Society of America Special Papers, **428**, 65–88.

ROGERS, R. D., MANN, P., EMMET, P. A. & VENABLE, M. E. 2007c. Colon fold belt of Honduras: evidence for Late Cretaceous collision between the continental Chortís block and intra-oceanic Caribbean arc. *In*: MANN, P. (ed.) *Geologic and Tectonic Development of the Caribbean Plate Boundary in Northern Central America*, Geological Society of America Special Papers, **428**, 129–149.

ROSENCRANTZ, E. 1990. Structure and tectonics of the Yucatán Basin, Caribbean Sea, as determined from seismic reflection studies. *Tectonics*, **9**, 1037–1059.

ROSENCRANTZ, E., ROSS, M. I. & SCLATER, J. G. 1988. Age and spreading history of the Cayman Trough as determined from depth, heat flow, and magnetic anomalies. *Journal of Geophysical Research*, **93**, 2141–2157.

ROSENFELD, J. H. 1993. Sedimentary rocks of the Santa Cruz Ophiolite, Guatemala – a proto-Caribbean history. *In*: PINDELL, J. L. & PERKINS, R. F. (eds) *Transactions of the 13th Annual GCSSEPM Research Conference: Mesozoic and Early Cenozoic Development of the Gulf of Mexico and Caribbean Region*, 173–180.

ROSS, M. I. & SCOTESE, C. R. 1988. A hierarchical tectonic model of the Gulf of Mexico and Caribbean region. *Tectonophysics*, **155**, 139–168.

SANCHEZ-BARREDA, L. A. 1981. *Geologic evolution of the continental margin of the Gulf of Tehuantepec in Southwestern Mexico*. PhD Thesis, Department of Geological Sciences, The University of Texas at Austin.

SANDWELL, D. T. & SMITH, W. H. F. 1997. Marine gravity anomaly from Geosat and ERS-1 satellite altimetry. *Journal of Geophysical Research*, **102**, 10039–10054.

SCHAAF, P., MORAN-ZENTENO, D. J., DEL SOL HERNANDEZ-BERNAL, M., SOLIS-PICHARDO, G. N., TOLSON, G. & KOEHLER, H. 1995. Paleogene continental margin truncation in southwestern Mexico; geochronological evidence. *Tectonics*, **14**, 1339–1350.

SCHOUTEN, H. & KLITGORD, K. D. 1994. Mechanistic solutions to the opening of the Gulf of Mexico. *Geology*, **22**, 507–510.

SCLATER, J. G., HELLINGER, S. & TAPSCOTT, C. 1977. The paleobathymetry of the Atlantic Ocean from the Jurassic to the present. *Journal of Geology*, **85**, 509–522.

SEDLOCK, R. L. 2003. Geology and tectonics of the Baja California peninsula and adjacent areas. *In*: JOHNSON, S. E., PATERSON, S. R., FLETCHER, J. M., GIRTY, G. H., KIMBROUGH, D. L. & MARTÍN-BARAJAS, A. (eds) *Tectonic Evolution of Northwestern México and the Southwestern USA*. Geological Society of America, **374**, 1–42.

SISSON, V. B., AVÉ LALLEMANT, H. G. *ET AL*. 2005. Overview of radiometric ages in three allochthonous belts of northern Venezuela: old ones, new ones, and their impact on regional geology. *In*: AVÉ LALLEMANT, H. G. & SISSON, V. B. (eds) *Caribbean–South American Plate Interactions, Venezuela*. Geological Society of America, Special Paper, **394**, 91–118.

SISSON, V. B., AVÉ-LALLEMANT, H. G. *ET AL*. 2008. New U/Pb and fission track geochronologic constraints on the subduction history of central Guatemala. *Abstract Volume of the 18th Caribbean Geological Conference*, 24–28 March 2008, Santo Domingo, Dominican Republic. World Wide Web Address: http://www.ugr.es/~agcasco/igcp546/DomRep08/Abstracts_CaribConf_DR_2008.pdf.

SMITH, C. A., SISSON, V. B., AVE LALLEMANT, H. G. & COPELAND, P. 1999. Two contrasting pressure–temperature–time paths in the Villa de Cura blueschist belt, Venezuela; possible evidence for Late Cretaceous initiation of subduction in the Caribbean. *Geological Society of America Bulletin*, **111**, 831–848.

SNOKE, A. W., ROWE, D. W., YULE, J. D. & WADGE, G. 2001. *Petrologic and Structural History of Tobago,*

West Indies: a Fragment of the Accreted Mesozoic Oceanic-arc of the Southern Caribbean. Geological Society of America, Special Papers, **354**.

SPEED, R. & WALKER, J. A. 1991. Oceanic crust of the Grenada Basin in the southern Lesser-Antilles arc platform. *Journal of Geophysical Research*, **96**, 3835–3852.

SPEED, R. C., WESTBROOK, G. *ET AL*. 1984. *Lesser-Antilles Arc and Adjacent Terranes*. Marine Science International, Woods Hole, Massachusetts, Ocean Margin Drilling Program, Regional Atlas Series, **10**.

SPEED, R. C., SHARP, W. D. & FOLAND, K. A. 1997. Late Paleozoic granitoid gneisses of northeastern Venezuela and the North America–Gondwana collision zone. *The Journal of Geology*, **105**, 457–470.

STANEK, K. P., MARESCH, W. V. & PINDELL, J. 2009. The geotectonic story of the northwestern branch of the Caribbean arc: implications from structural and geochronological data of Cuba. *In*: JAMES, K. H., LORENTE, M. A. & PINDELL, J. L. (eds) *The Origin and Evolution of the Caribbean Plate*. Geological Society, London, Special Publications, **328**, 361–398.

STEINBERGER, B. 2000. Plumes in a convecting mantle: Models and observations for individual hot spots. *Journal of Geophysical Research*, **105**, 11127–11152.

STEINBERGER, B. 2002. Motion of the Easter hot spot relative to Hawaii and Louisville hot spots. *Geochemistry Geophysics Geosystems*, **3**, 8503.

STEINBERGER, B., SUTHERLAND, R. & O'CONNELL, R. J. 2004. Prediction of Emperor-Hawaii seamount locations from a revised model of global plate motion and mantle. *Nature*, **430**, 167–173.

STEPHAN, J. F., BECK, C., BELLIZZIA, A. & BLANCHET, R. 1980. La chaine Caraibe du Pacifique a l'Atlantique [The Caribbean chain from Pacific to Atlantic]. *Geologique de Chaines Alpines Issues de la Tethys: 216th Congres Geologique Internationalle Colloque [Geology of the alpine chains born of the Tethys: 216th Geologic International Congress]*. Editions du Bureau de Recherche Geologique et Minière, Paris, France, **15**, 38–59.

STEPHAN, J. F., MERCIER DE LEPINAY, B. *ET AL*. 1990. Paleogeodynamic maps of the Caribbean: 14 steps from Lias to Present. *Bulletin de la Societe Geologique de France*, **8**, 915–919.

STEPHENS, B. 2001. Basement controls on hydrocarbon systems, depositional pathways, exploration plays beyond the Sigsbee escarpment in the central Gulf of Mexico. *In*: FILLON, R. H., ROSEN, N. C. *ET AL*. (eds) *Transactions of the 21st Annual GCSSEPM Foundation Bob F. Perkins Research Conference: Petroleum Systems of Deep-Water Basins: Global and Gulf of Mexico Experience*, 129–157.

STÖCKHERT, B., MARESCH, W. V. *ET AL*. 1995. Crustal history of Margarita Island (Venezuela) in detail: constraint on the Caribbean Plate-tectonic scenario. *Geology*, **23**, 787–790.

SYKES, L. R., MCCANN, W. R. & KAFKA, A. L. 1982. Motion of the Caribbean Plate during last 7 million years and implications for earlier Cenozoic movements. *Journal of Geophysical Research*, **87**, 10656–10676.

TALAVERA-MENDOZA, O. 2000. Mélange in southern Mexico: geochemistry and metamorphism of the Las Ollas complex (Guerrero Terrane). *Canadian Journal of Earth Sciences*, **37**, 309–1320.

TALAVERA-MENDOZA, O., RUIZ, J., GEHRELS, G. E., VALENCIA, V. A. & CENTENO-GARCÍA, E. 2007. Detrital zircon U/Pb geochronology of southern Guerrero and western Mixteca arc successions (southern Mexico): new insights for the tectonic evolution of southwestern North America during the Late Mesozoic. *Geological Society of America Bulletin*, **119**, 1052–1065.

TARDUNO, J. A. & GEE, J. 1995. Large-scale motion between Pacific and Atlantic hot spots. *Nature*, **378**, 477–480.

TEN BRINK, U. S., COLEMAN, D. F. & DILLON, W. P. 2002. The nature of the crust under Cayman Trough from gravity. *Marine and Petroleum Geology*, **19**, 971–987.

THOMSON, J. W. 1985. *Tectonic Map of the Scotia Arc, 1:3000000*. Miscellaneous Publications, British Antarctic Survey, Cambridge.

TORSVIK, T. H., MÜLLER, R. D., VOO, R. V. D., STEINBERGER, B. & GAINA, C. 2008. Global plate motion frames: toward a unified model. *Reviews of Geophysics*, **46**, RG3004.

TRENKAMP, R., KELLOGG, J. N., FREYMUELLER, J. T. & MORA, H. P. 2002. Wide Plate margin deformation, southern Central America and northwestern South America, CASA GPS observations. *Journal of South American Earth Sciences*, **15**, 157–171.

VALLEJO, C. 2007. *Evolution of the Western Cordillera in the Andes of Ecuador (Late Cretaceous–Paleogene)*. PhD Thesis, ETH, Zurich.

VAN DER HILST, R. 1990. *Tomography with P, PP, pP delay-time data and the three dimensional mantle structure below the Caribbean region*. PhD thesis, University of Utrecht.

VAN DER HILST, R. & MANN, P. 1994. Tectonic implications of tomographic images of subducted lithosphere beneath northwestern South America. *Geology*, **22**, 451–454.

VASQUEZ, M. & ALTENBERGER, U. 2005. Mid-Cretaceous extension-related magmatism in the eastern Colombian Andes. *Journal of South American Earth Sciences*, **20**, 193–210.

VILLAMIL, T. & PINDELL, J. L. 1998. Mesozoic paleogeographic evolution of northern South America: Foundations for sequence stratigraphic studies in passive margin strata deposited during non-glacial times. *In*: PINDELL, J. L. & DRAKE, C. (eds) *Paleogeographic Evolution and Non-Glacial Eustacy: Northern South America*. SEPM (Society for Sedimentary Geology), Special Publications, **58**, 283–318.

WADGE, G. & BURKE, K. 1983. Neogene Caribbean Plate rotation and associated Central American tectonic evolution. *Tectonics*, **2**, 633–643.

WEBER, J. C., DIXON, T. H. *ET AL*. 2001. GPS estimate of relative motion between the Caribbean and South American plates, and geologic implications for Trinidad and Venezuela. *Geology*, **29**, 75–78.

WESSEL, P. & KROENKE, L. W. 2008. Pacific absolute Plate motion since 145 Ma: an assessment of the fixed hot spot hypothesis. *Journal of Geophysical Research*, **113**, B06101.

WESSEL, P., HARADA, Y. & KROENKE, L. W. 2006. Toward a self-consistent, high-resolution absolute plate motion model for the Pacific. *Geochemistry Geophysics Geosystems*, **7**, Q03L12.

WHITE, R. V., TARNEY, J. *ET AL.* 1999. Modification of an oceanic plate, Aruba, Dutch Caribbean: implications for the generation of continental crust. *Lithos*, **46**, 43–68.

WRIGHT, J. E., WYLD, S. J. & URBANI, F. 2008. *Late Cretaceous Subduction Initiation Leeward Antilles/Aves Ridge: Implications for Caribbean Geodynamic Models.* Geological Society of America, Abstracts with Programs, **40**, 103.

YSACCIS, R. 1997. *Tertiary evolution of the northeastern Venezuela offshore.* PhD Thesis, Rice University, Houston, TX.

ZAMBRANO, E., VASQUEZ, E., DUVAL, B., LATRIELLE, M. & COFFINIERES, B. 1971. Sintesis paleogeografica y petrolera del occidente de Venezuela [Paleogeographic reconstruction and petroleum in western Venezuela]. *Memorias del IV Congreso Geologico Venezolano*, **1**, 481–552.

Some remarks on the Caribbean Plate kinematics: facts and remaining problems

G. GIUNTA* & E. OLIVERI

Dipartimento di Geologia, Università di Palermo, Via Archirafi 20, 90100 Palermo, Italy

Corresponding author (e-mail: giuntape@unipa.it)

Abstract: Caribbean Plate margins are assemblages of terranes located, since the Mid-Cretaceous, along transform boundaries between the Caribbean, North and South America and the Pacific and Atlantic oceans. Litho-stratigraphic, petrological and metamorphic features of the main units and their regional correlations allow definition of the main geotectonic elements (continental margins, oceanic basins, subduction zones, magmatic arcs) involved in the evolution of Caribbean Plate margins. They provide valuable constraints on plate evolution since the Jurassic. This involved proto-Caribbean ocean opening, thickening into an oceanic plateau, beginning of convergence in the Early Cretaceous, atypical evolution of a supra-subduction system during the Mid-Cretaceous, subduction of rifted continental margins, Late Cretaceous convergence related to eastward migration of two opposite triple-junctions and strike–slip tectonics. Using these data, we compare different models and suggest improvements.

The Caribbean Plate (Fig. 1) is an independent lithospheric element of more than 4000 km^2, consisting of undeformed or little deformed Cretaceous oceanic plateau crust (Colombia and Venezuela Basins; almost 1000 km^2) and the Palaeozoic–Mesozoic Chortís continental block (about 700 000 km^2). These are bounded by deformed marginal belts (about 2300 km^2) resulting from Mesozoic to Present interactions with the adjacent Nazca, Cocos and Americas plates. The northern (Guatemala and Greater Antilles) and southern (northern Venezuela) plate margins consist mainly of ophiolite decorated shear zones that represent collisional sutures between the plate and North and South America. Magmatic arcs of the Central American Isthmus and the Lesser Antilles characterize the western and eastern convergent boundaries with the Pacific and Atlantic Plates. Jurassic–Cretaceous ophiolitic terranes comprise about 40–50% of these belts. These assemblages record several compressional episodes, beginning in the Cretaceous, followed by tensional and/or strike–slip tectonics. More internal Caribbean marginal areas in Venezuela, Colombia, Panama and Hispaniola were involved in accretionary prisms (Stephan *et al.* 1986). Fragments of original Caribbean lithosphere are now accreted on adjacent plate margins.

Bivergent 'flower structures' exist along the northern and southern Caribbean margins where diachronous oblique movements have occurred along the Cenozoic.

Structure of the Caribbean Plate margins

Caribbean marginal belts are the product of complex interaction between several first-order geotectonic elements, characterized by different tectono-magmatic features and formed in different areas, that now lie fragmented and dispersed along the margins. They include continental margins, rifted continental margins, oceanic crust, oceanic plateau, intra-oceanic and sub-continental subduction zones and foredeep basins. The most important geological features are reported below.

Costa Rica

The western margin of the Caribbean Plate involves tectonic juxtaposition of three main composite blocks, Chortís, Chorotega and Chocó, of the Central American Isthmus. Accreted terranes are overlain by recent volcanic (Kuijpers 1980; Azema *et al.* 1984; Dengo 1985; Sinton *et al.* 1998; Beccaluva *et al.* 1999; Giunta *et al.* 2002a; Baumgartner & Denyer 2006; Denyer *et al.* 2006; Gazel *et al.* 2006). They form part of the Chortís continental block and most of Chorotega and Chocó.

The Santa Elena and Nicoya complexes of Costa Rica (Fig. 2) are two of the most important ophiolitic occurrences in the western margin of the Caribbean Plate. The first consists mainly of a peridotitic body, cut by a number of doleritic dykes and with subordinate breccias, thrusted onto basaltic rocks in the southwesternmost part of the

From: JAMES, K. H., LORENTE, M. A. & PINDELL, J. L. (eds) *The Origin and Evolution of the Caribbean Plate.*
Geological Society, London, Special Publications, **328**, 57–75.
DOI: 10.1144/SP328.2   0305-8719/09/$15.00 © The Geological Society of London 2009.

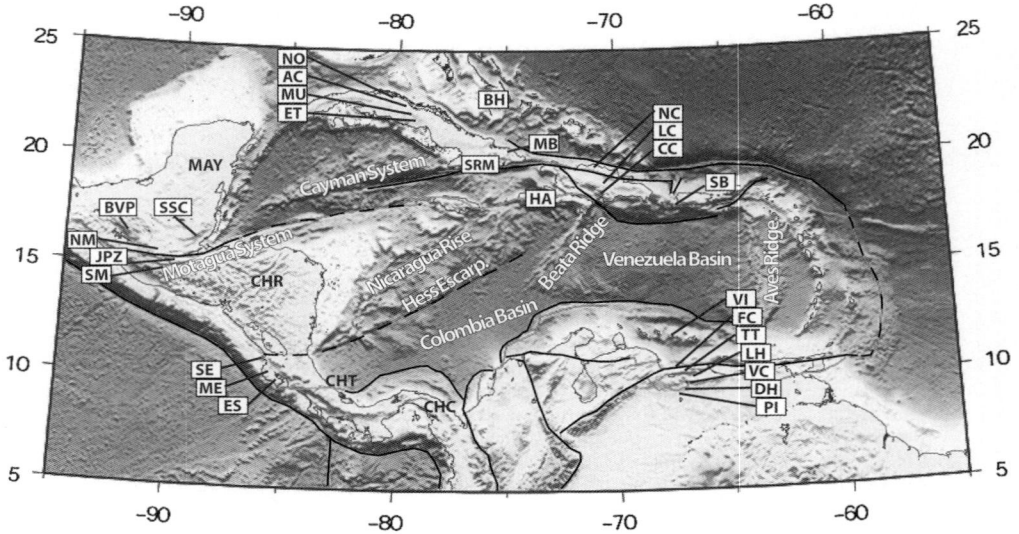

Fig. 1. Location map of the studied peri-Caribbean tectonic units. Northern margin units. BVP, Baja Verapaz; SSC, Sierra Santa Cruz; JPZ, Juan de Paz; NM, North Motagua; SM, South Motagua (Guatemala); BH, Bahamas; NO, Northern Ophiolite; AC, Cretaceous Arc; MU, Mabujna; ET, Escambray; MB, Mayarì–Baracoa; SRM, Sierra Maestra (Cuba); NC Northern Cordillera; LC, Loma Caribe; CC, Central Cordillera; HA, Dumisseau and Massif de la Hotte (Hispaniola); SB, Sierra Bermeja (Puerto Rico). Southern margin units: VI, Dutch and Venezuelan Islands; FC, Franja Costera; TT, Caucagua–El Tinaco; LH, Loma de Hierro; VC, Villa de Cura; DH, Dos Hermanas; PI, Piemontine (Venezuela). Western margin units: SE, Santa Elena; ME, Metapalo; ES, Esperanza (Costa Rica). Minor blocks. MAY, Maya; CHR, Chortís; CHT, Chorotega; CHC, Chocó.

peninsula. The Nicoya complex consists of an intrusive suite (gabbros, Fe-gabbros, Fe-diorites and plagiogranites) and basaltic rocks (basalts and dolerites), discontinuously covered by radiolarites. This complex was originally divided into two main units, Metapalo (ME) and Esperanza (ES). The older ME consists of basalts with scarce gabbros (Potrero intrusives) and sills, overlain by radiolarites of Late Jurassic (?) to Early Cretaceous age (Punta Conchal Fm). The ES, dated Mid- to

Late Cretaceous, consists of basalts and diabases with widespread gabbroic (Potrero) and plagiogranitic intrusions, with scattered radiolaritic cover. Several décollements affect the complexes though locally the contact between the units can be interpreted as high-angle faults. These units are unconformably overlain by the Campanian to Cenozoic turbiditic sandstones, andesite and carbonate rocks of the Sabana Grande, El Viejo, Rivas, Las Palmas, Samara and Barraonda Fms.

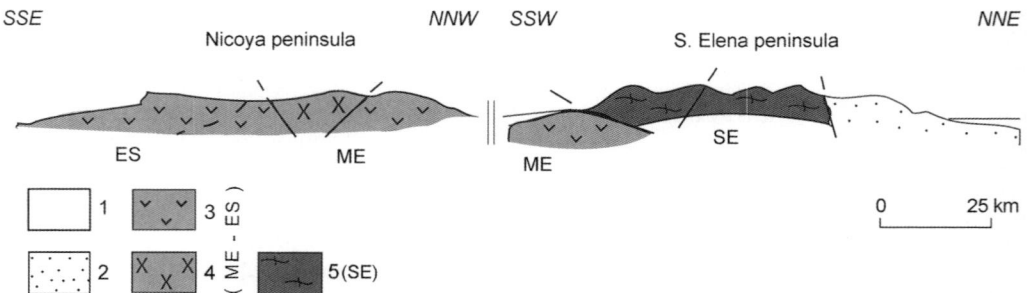

Fig. 2. Cross section of Guanacaste province of Costa Rica (modified from Beccaluva *et al.* 1999). Main units: SE, Santa Elena; ME, Metapalo; ES, Esperanza. Legend: 1, Recent deposits; 2, terrigenous and carbonatic sequences (Late Cretaceous–Cenozoic); 3, basalts and diabases with MOR affinity (Late Jurassic–Late Cretaceous); 4, gabbroic and scattered plagiogranitic intrusions with MOR affinity; 5, serpentnized mantle peridotites with doleritic dykes.

Guatemala

The present northwestern margin of the Caribbean Plate crops out along the Motagua Suture Zone (Fig. 3) in Guatemala, which links the Middle-American trench with the Cayman Trough extensional system (Muller 1980; Rosenfeld 1981; Finch & Dengo 1990; Beccaluva et al. 1995; Giunta et al. 2002a; Valls Alvarez 2006). The suture is a sinistral shear-zone between the Maya and Chortís continental blocks, consisting of east–west and ENE–WSW strike–slip fault systems (Polochic, Motagua, Cabañas, Yucatán). Remarkable west–east trending uplift structures (Sierra Chuacus, Sierra de Las Minas, Montañas del Mico), pull-apart basins (Izabal Lake, Bananeras) and grabens elongated in a prevalent north–south direction (Guatemala, Chiquimula), occur in this zone, which is a typical transpressional

'flower structure', with north and south vergence. The following are the main components:

(1) the Sierra Santa Cruz (SSC) and Baja Verapaz (BVP) units overthrust northward onto the Maya Block, the former onto the Late Cretaceous–Eocene carbonatic–terrigenous sequences of the Sepur formation, the latter onto Palaeozoic metamorphites of the Chuacus Group or the Mesozoic evaporitic–terrigenous–carbonatic deposits of the Todos Santos, Coban and Campur formations;

(2) the Juan de Paz unit (JPZ) overthrusts the Palaeozoic metamorphic basement of the Sierra de Las Minas and Montañas del Mico;

(3) the South Motagua (SM) and North Motagua (NM) units overthrust both the Palaeozoic continental basement (Las Ovejas and San Diego formations) of the Chortís Block to

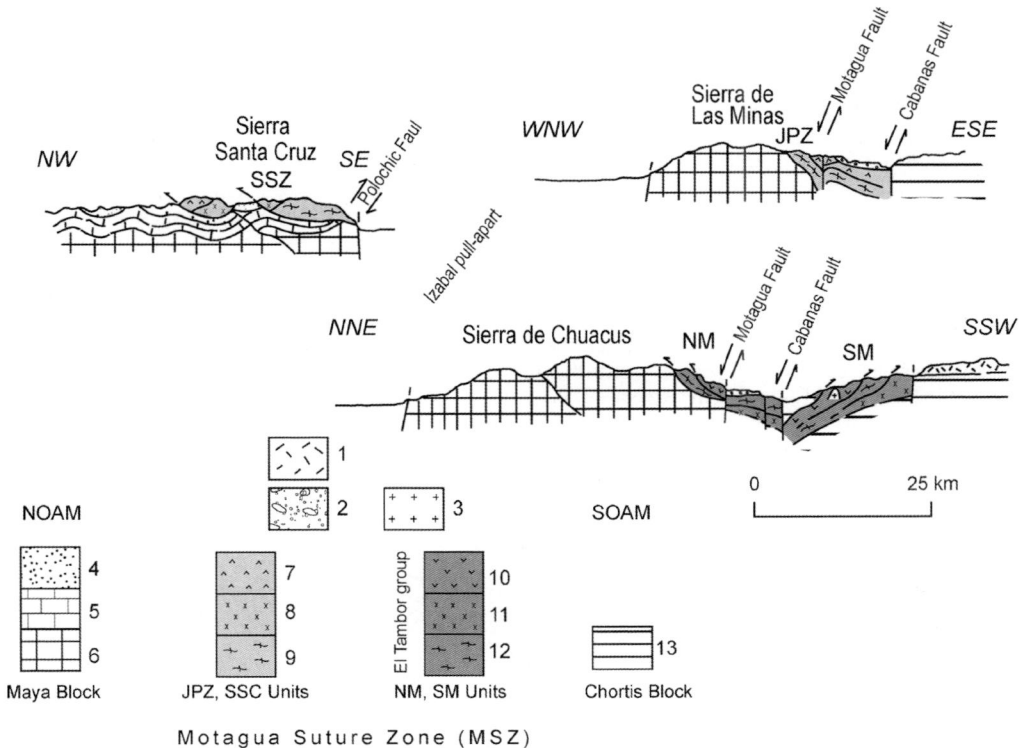

Fig. 3. Cross section of the Motagua suture Zone (MSZ), in Guatemala (modified from Beccaluva et al. 1995). Main units: MAY, Maya Cont. Block; BVP, Baja Verapaz; SSC, Sierra Santa Cruz; JPZ, Juan de Paz; NM, North Motagua; SM, South Motagua; GR, Zacapa granitoids; CHR, Chortís Cont. Block. Legend: 1, Cenozoic–Quaternary volcanics; 2, flysch and molassic deposits, Late Cretaceous–Eocene (Subinal Fm); 3, arc tonalitic magmatism, Cretaceous (Zacapa granitoids, GR); 4, Late Cretaceous–Paleocene pre-flysch and flysch (Sepur Fm); 5, Late Jurassic–Cretaceous terrigenous and carbonatic covers; 6, Palaeozoic continental basement; 7, Mid- to Late Cretaceous basalts, limestones and andesitic basalts; 8, gabbros and dolerites; 9, peridotites; 10, Late Jurassic–Early Cretaceous basalts, radiolarites, phyllites and metalimestones; 11, gabbros; 12, peridotites; 13, Palaeozoic continental basement and Mesozoic sedimentary covers.

the south (SM), and the Palaeozoic metamorphic terranes of the Sierras de Chuacus and Las Minas of the Maya Block (?) to the north (NM). These units are imbricated with 'out of sequence' basement slices.

The SSC, BVP and JPZ units consist of generally serpentinized mantle harzburgites, layered gabbros, dolerites and andesitic basalts. The SSC is locally covered by small outcrops of terrigenous and volcanoclastic sequences including andesitic and dacitic fragments (Cretaceous Tzumuy Fm), while the JPZ is covered by basic volcanoclastic and andesitic breccias passing up to carbonatic breccias and calcarenites, with sandstone and microconglomerates containing acid volcanic fragments (Late Cretaceous Cerro Tipon Fm).

The SM and NM consist of the so-called El Tambor Group, made-up of serpentinized peridotites and foliated gabbros, followed by a thick basaltic pillow lava sequence, radiolarian cherts, metasiltites and metarenites with intercalations of basaltic flows, in places metamorphosed to blueschist and eclogite facies. The top of the sequence is formed by phyllitic metasiltites alternating with marbles and metacalcarenites (Mid- to Late Cretaceous Cerro de La Virgen limestones). Along the Motagua Valley the JPZ, SM and NM units are unconformably overlain by the Eocene continental molasse of the Subinal Fm.

Cuba

The northernmost portion of the original Caribbean Plate margin crops out in Cuba, overthrust onto the southern edge of the North American Plate (Pardo 1975; Iturralde-Vinent 1989, 1999, 2000; Beccaluva *et al.* 1996; Cobiella-Reguera 2005;

Marchesi *et al.* 2006; Meschede *et al.* 2006; Stanek *et al.* 2006). It is separated from the rest of the Greater Antilles by the Bartlett sinistral strike–slip structure, which represents the present-day northern boundary of the Caribbean Plate. The plate boundary shifted from the Cuba area to the Cayman Trough in the Eocene (e.g. Pindell 2006).

The Cuban thrust belt (Fig. 4) includes two continental elements: the northern Bahamas Platform and the southern Guaniguanico–Piños–Escambray (ET). They are overthrusted by oceanic elements of the Northern Ophiolite (NO), a Cretaceous Arc (AC) and a Palaeogene Arc. The deformation front extends onto a Palaeogene foredeep as a series of north-vergent slices with associated flyschoid sequences and olistostromes.

(1) The Bahamas units (BH) are structurally the lowest in the folded system and consist of at least four tectonic sheets, Cayo Coco, Remedios, Camajunì and Placetas, which represent the original edge of the North American margin. In the Placetas Unit, Late Jurassic tholeiitic lavas are also present. A Paleocene–Eocene foredeep basin overlying the Bahamas margin dates the collision of the Northern Ophiolite and Cretaceous Arc against NOAM.

(2) The NO unit is a metamorphic mélange which includes blocks of peridotites and cumulate gabbros cut by dykes of diabases and overlain by basaltic lavas, hyaloclastites, radiolarites and volcanoclastites. In eastern Cuba the NO consists of peridotitic and gabbroic massifs, such as Holguin and Mayarì-Baracoa, and big bodies of metamorphic rocks (Purial, Asuncion, Sierra del Convento).

(3) The AC units overlie the ophiolitic mélange to the north and the Escambray Terranes to the

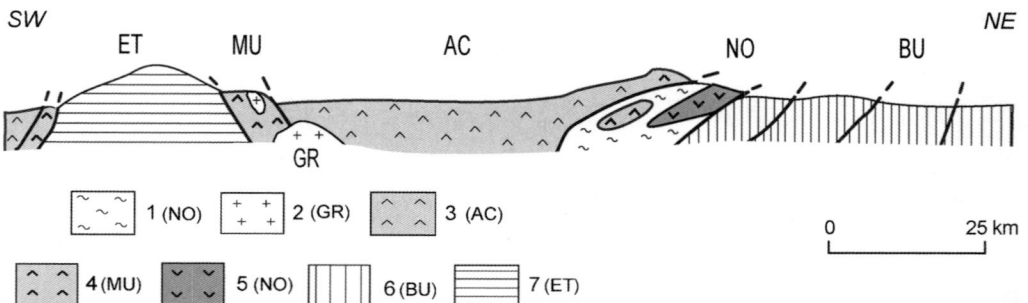

Fig. 4. Cross sections of central Cuba (modified from Iturralde-Vinent 1994). Main units: BH, Bahamas; NO, Northern Ophiolite; AC, Cretaceous Arc; MU, Mabujna; ET, Escambray. Legend: 1, melanges with variably terrigenous matrix (Late Cretaceous–Palaeogene), including 4; 2, tonalitic intrusive (Late Cretaceous–Eocene); 3, arc volcanics (CA affinity) with scattered reefal limestones (Cretaceous–Paleocene); 4, metamorphosed vulcano–plutonic arc sequences with CA affinity of supra-subduction complex (Cretaceous); 5, peridotites, cumulitic gabbros, basalts (MOR magmatism) and radiolarites (Late Jurassic–Early Cretaceous), involved in mélanges; 6, sedimentary sequences (Jurassic–Cretaceous) of Bahamas continental margin; 7, metamorphic continental basement of Escambray.

south. They consist of lava flows, pyroclastites and volcanoclastic rocks, sometimes unconformably overlain by carbonate and terrigenous sequences. Rudist-bearing limestone horizons of Late Albian, Santonian and Early Campanian ages are present in this unit.

(4) The Mabujina subduction complex (MU) consists of metavolcanics and metaplutonics with calc-alkaline magmatic affinity; it underlies the Cretaceous Arc units and overthrusts the Escambray continental terranes.

(5) Different plutonic complexes intrude the AC and the MU units, with variably thick cornubianitic aureolae, made up of locally foliated granitoid and tonalitic bodies, from Aptian to Campanian age. Palaeogene volcanic and intrusive rocks occur in the Sierra Maestra.

(6) The ET, considered to be equivalent to the Guaniguanico and Los Piños terranes, tectonically underlies the Mabujina unit and crops out as a complex thrust system in a large dome-shaped structure. The sequence generally consists of metamorphosed Mesozoic terrigenous and carbonate deposits of continental margin origin. Slices of metavolcanites, metagabbros and serpentinites have been reported in some localities, together with a metamorphic mélange.

Hispaniola

Hispaniola represents a transpressional shear zone (Lewis & Draper 1990; Draper & Nagle 1991;

Draper *et al.* 1994; Escuder-Viruete *et al.* 2002) with terranes of the Central Cordillera juxtaposed against Northern Cordillera ophiolites to the NE and a portion of emerged oceanic plateau to the SW (Massif de la Hotte-Bahoruco, Southern Peninsula of Haiti). Large, sinistral strike–slip faults (Septentrional, Enriquillo, Hispaniola) separate these sectors (Fig. 5).

(1) the Northern Cordillera (NC) is a heterogeneous ophiolitic terrane where the following tectonic units have been recognized:

  (a) the Puerto Plata complex of variably sized bodies of serpentinized peridotites, layered metagabbros and pillowed metabasalts, with scattered Early Cretaceous radiolarites;

  (b) the Rio San Juan complex, composed of
    (i) the Gaspar Hernandez serpentinites,
    (ii) the Hicotea and Puerca Gorda schists,
    (iii) the Jagua Clara Mélange, with several high pressure metamorphic blocks in an ultramafic matrix,
    (iv) the Cuaba amphibolites and the Rio Boba gabbroic layered sequences.

  (c) the Samanà-Punta Balandra complex, consisting of a continuous sequence of foliated marbles with intercalations of micaschists, including boudins of metagabbros and metadolerites metamorphosed into blueschist and eclogite facies.

(2) the Central Cordillera (CC) terrane, consisting mainly of the Duarte and Tireo complexes,

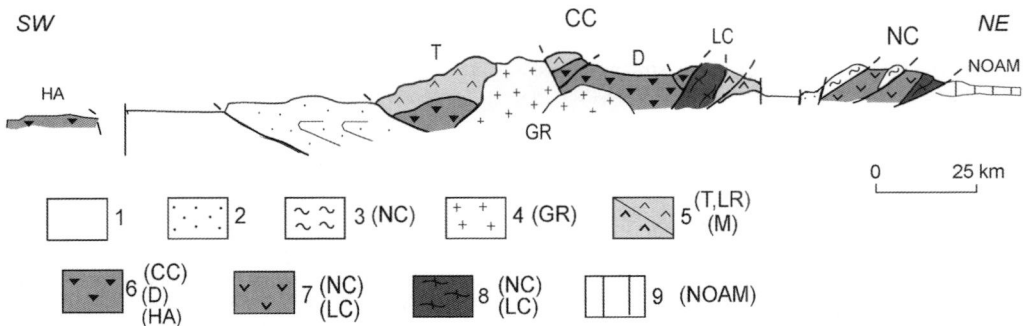

Fig. 5. Cross section of Hispaniola (modified from Lewis & Draper 1990). Main units: NC,Northern Cordillera; CC, Central Cordillera; LC, Loma Caribe–Ortega; D, Duarte complex, T, Tireo complex; GR, tonalitic arc magmatism; SC, Southern Cordillera; HA, Massif de la Hotte–Salle–Bahoruco; NOAM, North American continental margin. Legend: 1, Cenozoic–Quaternary deposits; 2, terrigenous sequences (Late Cretaceous–Palaeogene); 3, mélanges and olistostromes with variably terrigenous matrix (Late Cretaceous–Palaeogene), including blocks of 6 and 7; 4, gabbroid to granitoid intrusives (tonalitic arc magmatism, Late Cretaceous); 5, metabasalts and metadolerites (D), and basic to acidic metavolcanics (T) (MOR to OIB affinity), with intercalations of radiolarites (Late Jurassic–Early Cretaceous, and Late Cretaceous); 6, MOR ophiolites (mantle peridotites, metagabbros, metadolerites and metabasalts) with scattered radiolarites (Late Jurassic–Early Cretaceous), also included in 3; 7, serpentinized mantle peridotites belonging to MOR ophiolites, also included in 3; 8, foliated marbles and micaschists, with boudins of metagabbros and metadolerites with MOR affinity (Late Jurassic–Cretaceous); 9, North American Continental Plate.

overthrusts northeastward above the Loma Caribe–Ortega (LC) unit, which in turn overthrusts the Maimon–Amina and Los Ranchos units, which are separated by the Hatillo Thrust. Southwestward the CC overthrusts the Late Cretaceous–Palaeogene terrigenous sequences of the Trois Rivieres belt. The generalized stratigraphical sequence of the CC, reconstructed from the dismembered components, consists of basal serpentinized harzburgites (LC), covered by metadolerites and frequently pillowed metabasalts with intercalation of Late Jurassic–Early Cretaceous radiolarites (Duarte), and basalts, tuffs and volcanoclastites locally associated with Late Cretaceous radiolarites and siltstones (Tireo). The Amina–Maimon and Los Ranchos units are respectively composed of dacitic metatuffs intercalated with metasediments and of pillowed basalts and breccias with dacitic and rhyolitic composition, the latter underlying reefal limestones of the Middle Cretaceous. Several gabbroic and tonalitic plutons intrude into Duarte–Tireo complex of the Central Cordillera.

(3)   The Massif de la Hotte–Bahoruco terrane (HA) of the Southern Peninsula in Haiti is mainly a monotonous sequence of Middle–Late Cretaceous basalts and dolerites, with pelagic intercalations of limestones, cherts and siltstones, the most representative of which is the Dumisseau Fm.

Puerto Rico

Most of Puerto Rico is made up of volcanic arc 'strata' (lavas, tuffs and volcanoclastic products) of Aptian to Eocene age (Jolly et al. 1998; Mitchell 2006). This complex, intruded by granitoid plutonites since the Late Cretaceous, may be subdivided into three main districts separated by sinistral NW–SE strike–slip faults.

The southwesternmost portion of the island consists of the Sierra Bermeja complex, a tectonic assemblage of serpentinized harzburgites, metabasalts and amphibolites (Las Palmas unit) and scattered pelagic sediments with Cretaceous radiolarites (Mariquita Fm).

Much of western Puerto Rico is characterized by tectonic slices with variable vergence, separated by at least three NW–SE elongated peridotitic bodies (Monte del Estado, Rio Guanajbo, Sierra Bermeja).

The central and eastern district of the island is mainly made by Late Cretaceous–Early Cenozoic volcanic sequences and plutonic bodies.

Venezuela

The Dutch–Venezuelan islands, the Coastal Cordillera of Venezuela and the Northern Range of Trinidad form the southern margin of the Caribbean Plate, extending from the Barquisimeto depression, in the west, to Trinidad and Tobago in the east (Bellizzia 1986; Stephan et al. 1986; Beccaluva et al. 1996; Giunta et al. 1997, 2002c). The northern uplifts are bounded to the south by the Interior and Central Ranges of Venezuela and Trinidad, which overthrust the Guayana continental foreland of the South American Plate. Further west the north-vergent South Caribbean Deformed Belt 'accretionary prism' of Colombia and Venezuela forms the Curaçao Ridge, while to the south the Gulf of Venezuela, Maracaibo and Falcón pull-apart/intermontane basins lie between uplifts of the Guajira and Paraguaná peninsulas, the Sierra de Perijá and the Mérida Andes (James 2006a).

The whole belt (Fig. 6) consists of imbricated tectonic units, highly dismembered and affected by severe brittle and ductile/brittle deformation, related to a west–east dextral shear zone with strike–slip faults (Oca, San Sebastian, El Pilar) and associated synthetic (Tacata, Charallave, Urica, San Juan) and subordinate antithetic fault systems.

The CC consists of pre-Mesozoic continental basement covered by Late Jurassic–Cretaceous metamorphosed carbonate–terrigenous sediments (with local volcanic intercalations). The CC is overthrust by the ophiolitic mélange of the Franja Costera (FC) unit and part of the Caucagua–El Tinaco units (TT) to the north, and by the Caucagua–El Tinaco, Loma de Hierro (LH), Villa de Cura, (VC) and Dos Hermanas (DH) units to the south:

(1)   The Franja Costera (FC) consists of a Cretaceous metamorphic volcano-sedimentary and carbonate-terrigenous sequence with boudins of serpentinized peridotites, metagabbros and eclogites.

(2)   The TT units, south of the CC terrane, consist of pre-Mesozoic basement (El Tinaco complex) covered by a Cretaceous metavolcanosedimentary sequence (Tucutunemo Fm), including Los Naranjos basalts and Sabana Larga dolerites and gabbros. They are overlain by the Tinaquillo thrust sheets, which consist of serpentinized mantle lherzolites and metagabbros. On Margarita Island, north of CC, Mesozoic continental basement (Juan Griego Complex) is intruded by layered metagabbros (La Rinconada Fm). These overthrust the metacarbonatic–terrigenous sequence of the Cretaceous Los Robles Fm, overlain by the Cerro Matasiete peridotites. Late

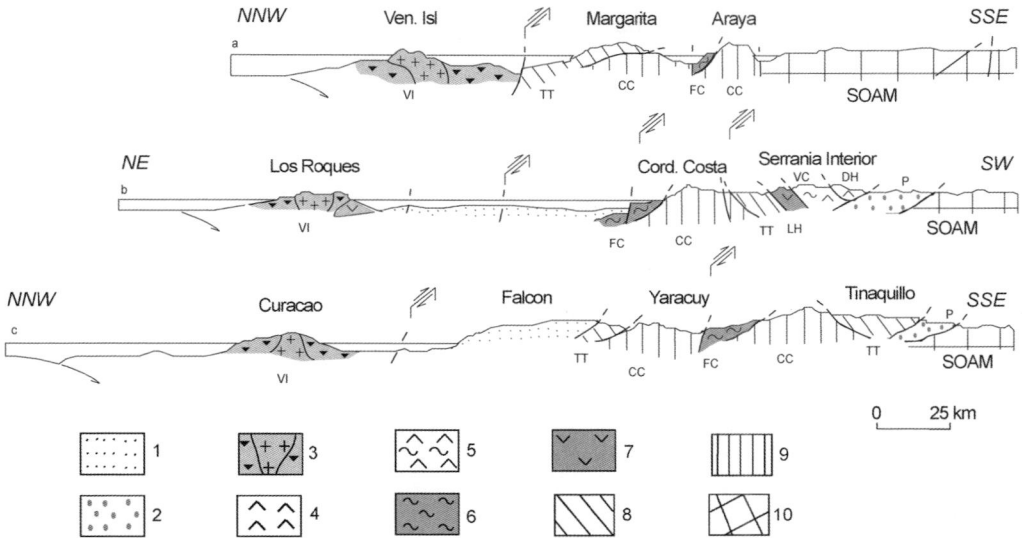

Fig. 6. Cross sections of the Sistema Montanoso del Caribe, in Venezuela (modified from Giunta *et al.* 1977). Main units: P, Piemontine foredeep; VI, Venezuelan Islands; DH, Dos Hermanas; VC, Villa de Cura; FC, Franja Costera; LH, Loma de Hierro; TT, Caucagua–El Tinaco; CC, Cordillera de la Costa; SOAM, South America continent. Legend: 1, Cenozoic terrigenous deposits; 2, Terrigenous flysch-like sequences (Late Cretaceous–Palaeogene); 3, Oceanic Plateau intruded by Late Cretaceous 'Tonalitic' arc magmatism (quarzo-diorites and granitoids) (Aruba, White *et al.* 1999; Los Roques, Giunta *et al.* 1997); Cretaceous Arc volcanics at Bonaire and Tobago (Donnelly *et al.* 1990). (VI, Dutch and Venezuelan Islands Unit); 4, Arc volcanics (basaltic–andesitic lava breccias) with IAT affinity, Mid- to Late Cretaceous (DH, Dos Hermanas Unit); 5, Metamorphosed volcano-plutonic arc sequences (serpentinized peridotites, metabasalts, metatuffs and metavolcano-sedimentary sequence with rhyolites and cherts) with IAT affinity of supra-subduction complex, Early to Mid-Cretaceous (VC, Villa de Cura Units); 6, Volcano-sedimentary mélanges including boudins of peridotites, metagabbros and metabasalts with MOR affinity, Early to Mid-Cretaceous (FC, Franja Costera Unit); 7, MOR ophiolites (serpentinized peridotites, basalts and dolerites) with scattered radiolarites, metalimestones and siltites, Late Jurassic–Cretaceous (LH, Loma de Hierro Unit); 8, Continental crystalline basement (Palaeozoic) overlain by serpentinized mantle lherzolites and metagabbros (Tinaquillo), volcano–sedimentary sequence with basalts and dolerites (WPTh affinity), Late Jurassic–Early Cretaceous (TT, Caucagua–El Tinaco Units); 9, Continental crystalline basement, pre-Cambrian–Palaeozoic, covered by metacarbonate–terrigenous sequences, Late Jurassic–Cretaceous (CC, Cordillera de la Costa Unit); 10, South American Continental Plate (SOAM).

Cretaceous granitoid bodies (El Salado) intrude the previous formations.

(3) The Loma de Hierro (LH) unit consists of serpentinized mantle peridotites, layered gabbroic cumulates (Rio Mesia), and basaltic lavas and dolerites (Tiara Fm), discontinuously covered by Late Jurassic–Early Cretaceous radiolarites (Capas Rio Guare), and Cretaceous silicified metalimestones and siltites (Paracotos Fm).

(4) The Villa de Cura (VC) units consist of serpentinized mantle peridotites and wehrlite–clinopyroxenite cumulates (Chacao complex), massive metabasalts (El Carmen Fm), metatuffs and subordinate metalavas (El Chino–El Cano Fm) and an Early–Mid-Cretaceous metavolcano–sedimentary sequence, prevalently comprising rhyolites, siltstones and

cherts (S. Isabel Fm). High pressure/low temperature (HP/LT) metamorphism (blueschist facies) characterizes several tectonic slices.

(5) The Dos Hermanas (DH) unit is represented by less metamorphic basaltic–andesitic lava breccias and volcanoclastites. The LH, VC and DH units were thrusted southwards onto the Piemontine foredeep-terrigenous units in the Late Cretaceous–Early Cenozoic.

(6) The Dutch–Venezuelan Islands (VI) unit, off northern Venezuela, include a basement made up of basaltic and picritic lavas, dolerites and gabbros, intruded by Late Cretaceous tonalitic plutons and rhyolitic dykes. The tectonic relationship between the VI and the rest of the orogen is poorly known. It probably involves dextral strike–slip high angle faults.

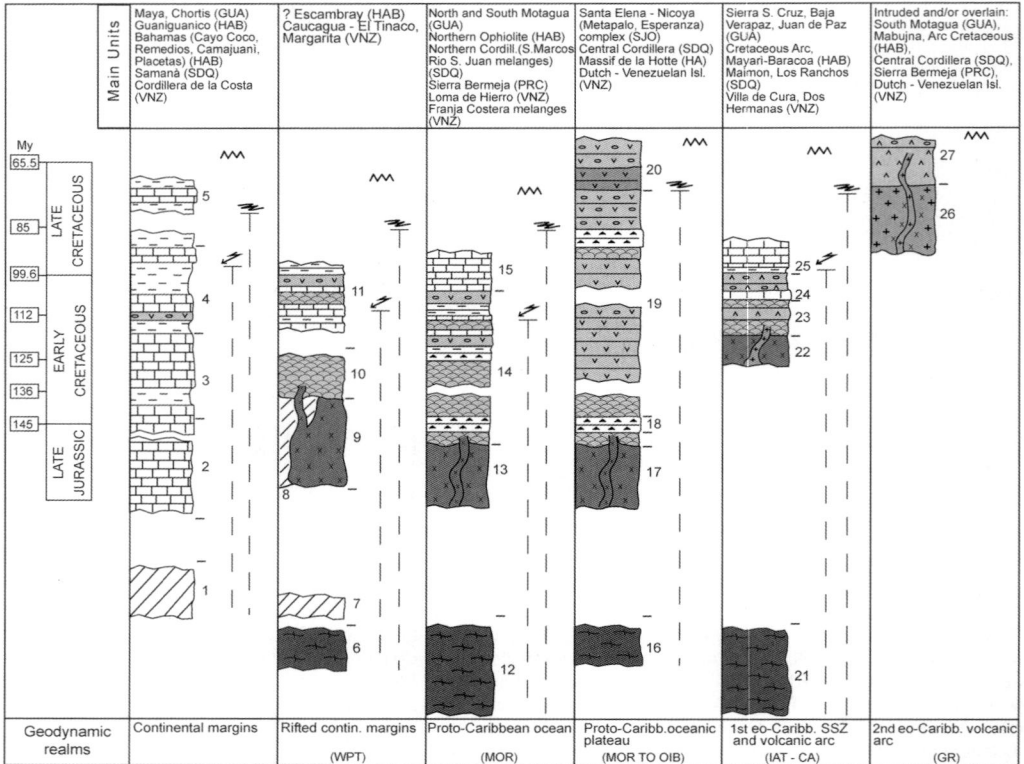

Fig. 7. Regional correlations between the main sedimentary, magmatic, metamorphic, and deformational events recorded in the peri-Caribbean terranes. Abbreviations: SJO, Costa Rica; GUA, Guatemala; HAB, Cuba; SDQ, Hispaniola; HA, Haiti; PRC, Puerto Rico; VNZ, Venezuela; WPT, within-plate tholeiite; MOR, mid-ocean ridge; OIB, ocean islands basalt; IAT, island arc tholeiite; CA, island arc calc-alkaline; GR, tonatitic arc magmatism (gabbroid to granitoid). 1–27: most representative formations. Continental margins: 1, Sebastopol complex (VNZ): continental metamorphic basement (mainly gneiss), with granitic intrusions; 2, Las Brisas (VNZ): phyllite, metasiltites and meta-arenites, with intercalations of metacalcarenites; Todos Santos (GUA): reddish arenites and siltites, with intercalations of limestones and poligenic conglomerates; 3, Antimano (VNZ): metacalcarenites and marbles, with intercalations of amphibolites and metabasites; 4, Las Mercedes (VNZ): graphitic phyllites, metacalcarenites, with scarce intercalations of metabasites; Chuspita (VNZ): meta-arenites, metasiltites and metaconglomerates, with intercalations of marbles; Coban (GUA): evaporites, dolomites and limestones; 5, Campur (GUA): limestones, with intercalations of siltites. Rifted continental margins: 6, Tinaquillo, Cerro Matasiete (VNZ): serpentinized lherzolites; 7, El Tinaco complex (VNZ): continental basement made by gneiss with few amphibolites, phillites, metaconglomerates and meta-arenites; 8, Juan Griego (VNZ): continental crystalline (quartz, feldspar, rich schist and ortho or paragneisses) basement; 9, Rinconada, Tinaquillo p.p. (VNZ): metagabbro cumulates intruded and overlying the 8; 10, Sabana Larga (VNZ): basalts (mainly pillow lavas), dolerites and gabbro breccias; 11, Tucutunemo, Los Naranjos, Los Robles (VNZ): metalimestones and metasiltites with basaltic pillow lavas. Proto-Caribbean ocean: 12, Gaspar Hernandez (SDQ), Monte del Estado (PRC): serpentinized peridotites; 13, Cuaba (SDQ), Rio Mesia (VNZ): layered metagabbros; 14, El Tambor group (GUA), Punta Balandra (SDQ), Sierra Bermeja (PRC), Tiara (VNZ): metasiltites, meta-arenites, metacalcarenites, with intercalations of basaltic (flows or pillow lavas) and radiolarian cherts; 15, Cerro de la Virgen (GUA): metacalcarenites and metacalcasiltites with breccia and siltitic intercalations. Proto-Caribbean oceanic plateau: 16, Loma Caribe (SDQ): serpentinized peridotites; 17, Potrero (SJO): gabbro cumulates, Fe-gabbros, Fe-diorites and subordinate plagiogranites; 18, Punta Conchal (SJO): radiolarites intercalated in pillow and massive basalts; 19, Duarte complex (SDQ), Lava (CUR): pillow and massive basalts with intercalations of radiolarites and volcanoclastites; 20, Siete Cabezas (SDQ), Knip group (CUR): volcanoclastites, radiolarites; pillow and massive basalts; 19, 20, Dumissieu (HA): pillow and massive basalts, volcanoclastites, with intercalations of radiolarites. First eo-Caribbean SSZ and volcanic arc: 21, serpentinized peridotites; 22, layered gabbros with scattered intrusions of granites; Chacao complex 21 + 22 (VNZ), Mayarì Baracoa (HAB); 23, Mabujina, Purial (HAB), El Carmen, El Chino, El Cano, S. Isabel (VNZ): metabasalts (mainly pillow lavas), metadolerites and meta-andesites; 24, Hatillo (SDQ): reefal limestones; 25, Tzumuy, Cerro Tipon (GUA): calcarenites and carbonatic breccias

After Giunta *et al.* (1997), the VI is an independent unit, in disagreement with some authors, who consider it a westward continuation of the Villa de Cura unit, following Stephan *et al.* (1986).

Regional correlations

The peri-Caribbean tectonic units and their numerous formations described above can be grouped into at least six litho-stratigraphic sections with similar lithological characteristics but very different tectonic origins, mainly based on the petrological data characterizing the geochemical affinities of the magmatic lithotypes. As shown in Fig. 7, each is the litho-stratigraphic product of different tectono-magmatic environments (Giunta *et al.* 2002*b*; Lewis *et al.* 2002; Lewis *et al.* 2006*a*, *b*; Cobiella-Reguera 2005):

(1) Mesozoic continental margins (Bahamas Platform, northern South America, the Maya and Chortís Blocks of Central America, Guaniguanico in Cuba, Cordillera de la Costa in Venezuela), including a pre-Mesozoic basement;

(2) rifted continental margins (Escambray in Cuba, Caucagua-El Tinaco and Tinaquillo in Venezuela) related to Jurassic-Early Cretaceous tensional episodes which continued to affect the continental margins, characterized by within-plate tholeiitic (WPT) magmatism;

(3) Jurassic–Early Cretaceous oceanic crust, with mid-ocean ridge (MOR) affinity (North and South Motagua in Guatemala, Northern Ophiolite in Cuba, Northern Cordillera in Hispaniola, Sierra Bermeja in Puerto Rico, Loma de Hierro and Franja Costera in Venezuela);

(4) 'thin' oceanic crust, in places evolved into an oceanic plateau structure with related ocean island basalts (OIB) during the Cretaceous (Santa Elena, Matapalo and Esperanza in Costa Rica, Loma Caribe, Central Cordillera and Massif de la Hotte in Hispaniola, Dutch–Venezuelan Islands);

(5a) Mid-Cretaceous intra-oceanic supra subduction zones (SSZ) and related volcanic arc magmatism, with island arc tholeiitic (IAT) and/or calc-alkaline (CA) affinities (Sierra Santa Cruz, Baja Verapaz and Juan de Paz in Guatemala, Mabujina, Cretaceous Arc, Mayarì-Baracoa and Purial in Cuba, Maimon–Los

Ranchos in Hispaniola, Villa de Cura and Dos Hermanas in Venezuela), in places affected by HP/LT metamorphism;

(5b) Mid-Cretaceous, sub-continental subduction, HP/LT metamorphosed ophiolitic melanges, including mafic blocks with MOR affinity (Franja Costera in Venezuela);

(6) Late Cretaceous tonalitic arc magmatism with CA affinity (intruding: South Motagua in Guatemala, Mabujna and Cretaceous Arc in Cuba, Cordillera Central in Hispaniola, Dutch–Venezuelan Islands and part of Caucagua–El Tinaco in Venezuela);

(7) Late Cretaceous mélanges (Northern Ophiolite in Cuba, Northern Cordillera in Hispaniola), followed by Palaeogene olistostromes, involving blocks of different origin (MOR, SSZ) in the deformation fronts, colliding against the NOAM and SOAM continental plates, through the progressive activation of foredeep basins (Sepur in Guatemala, Piemontine in Venezuela).

Constraints

The time–space distribution and regional correlation of the main peri-Caribbean units provide important geological constraints for reconstruction of the history of the Caribbean Plate. This is summarized the following different stages.

- *Continental margin rifting* – Pangaean breakup and separation of South and North America, Maya and Chortís continental blocks favoured rifting and tholeiitic magmatism (WPT), from Triassic–Early Jurassic.

- *Oceanization* – MOR oceanic crust formed at multiple spreading centres during the Jurassic and Early Cretaceous, forming the 'proto-Caribbean' ocean.

- *Oceanic plateau* – during the Cretaceous, parts of the oceanic crustal domain thickened into an oceanic plateau of MOR to OIB affinity.

- *Subduction* – several lines of geological evidence, such as relict HP/LT distribution, record different subduction complexes (eo-Caribbean stage), both intra-oceanic (IAT and CA arc magmatism associated with SSZ) and/or sub-continental (mélanges). They also indicate trapped or back-arc oceanic crust.

Fig. 7. (*Continued*) (with andesitic fragments) and andesitic volcanoclastites. Second eo-Caribbean volcanic arc: 26, Yautia (SDQ): tonalites, quarzo-diorites and granites, with gabbroic and mafic differentiates; 27, Tireo (SDQ), Washikemba (BON): volcanoclastites and sedimentary layers, basalts, basalt–andesites and rhyolites. ✦: HP/LT (blueschist to eclogite facies) subduction related metamorphism and ductile deformation; ➤: greenschist to amphibolite facies metamorphism and ductile deformation; ᴧᴧ: ductile deformation.

- *Early convergence* – onset of the accretionary stage seems to have been diachronous, between 115 and 95 Ma (Smith *et al.* 1999; Escuder-Viruete *et al.* 2006; Giunta *et al.* 2003*b*), according to HP/LT assemblages related to ocean–ocean subduction and to ocean–continent subduction that locally reached eclogite facies.
- *Double-arc magmatism* – Mid-Cretaceous (100–95 Ma) and Late Cretaceous (85–75 Ma) peaks of IAT and CA arc magmatism associated with the SSZ occurred in the eo-Caribbean accretionary stage, which has been separated in first and second eo-Caribbean phases of subduction.
- *HP/LT assemblages* – widespread occurrences of blueschist and eclogite assemblages (Garcia-Casco *et al.* 2006*a*, *b*) in both oceanic and continental terranes of the Caribbean Plate margins requires a geodynamic model where portions of both oceanic and continental lithospheres were simultaneously subducted. Development of HP/LT conditions in MOR-type units is commonly related to subduction of oceanic lithosphere in either sub-continental or intra-oceanic settings. Metamorphism of continental margin elements requires more complex subduction mechanisms (e.g. continental collision, tectonic erosion) that allow underthrusting of thinned continental crust. There is also the possibility that minor continental terranes, formed during breakup and proto-Caribbean formation, were involved in subduction, as suggested by Avé Lallemant & Sisson (2005). Moreover, if large rivers delivered continental sediments to the ocean, these also could be subducted (Cardona, pers. comm.).
- *Atypical evolution of the older volcanic arc* – the Mid-Cretaceous island-arc system was frequently deeply subducted and metamorphosed to blueschist facies. In general, supra-subduction zone units obduct onto continental margins and only locally become involved in the deep subduction. In these cases an unusual geodynamic evolution must be imagined, such as subduction polarity reversal following subduction blocking or 'tectonic erosion' of the overriding plate. It has been also proposed that intra-oceanic arcs can be subducted, mainly controlled by lithosphere thickness (Boutelier *et al.* 2003).
- *Diachronous tonalitic magmatism* – the second calc-alkaline magmatic arc, mainly tonalitic, seems to have been diachronously connected to the Aves–Lesser Antilles arc system since 85 Ma. In the north the tonalitic magmatic arc generally rests on both older arc systems and oceanic plateau. In the south it is tectonically decoupled from the older arc and is intruded into both undeformed and deformed oceanic plateau and into rifted continental margin units.
- *Strike–slip tectonic regime* – contrasts in the *P–T* paths of various HP/LT metamorphic units (Sisson *et al.* 1997; Garcia-Casco *et al.* 2006*a*) probably indicate that converging zones were subdivided into different tectonic sectors during the Mid-Cretaceous. According to Giunta *et al.* (2003*a* and references therein), some eclogites and blueschists followed an 'Alpine-type' retrograde trajectory, suggesting that shut-off of subduction occurred before the beginning of exhumation. In contrast, blueschists locally show a 'Franciscan-type' path, which commonly characterizes decompression in intra-oceanic settings during still active subduction processes.

On the Venezuelan Caribbean margin deformation occurred under retrograde *P–T* conditions during exhumation from the Campanian onwards. The beginning of exhumation is often characterized by well-developed foliation showing mineral or stretching lineations and transected folds. The present more or less east–west trend of the second lineations suggests that displacement during the first stage of exhumation was nearly orthogonal to the north–south axis of subduction, as demonstrated by the trend of L1 lineations sub-parallel to the Caribbean–NOAM–SOAM plate boundaries.

According to Avé Lallement (1997), these structural features can be explained by an exhumation dominated by strike–slip tectonics.

Remarks and models comparison

The above observations allow reconstruction of Caribbean Plate evolution (Beccaluva *et al.* 1999; Giunta *et al.* 2002*a*–*c*; 2003*a*, *b*, 2006) that differs in some aspects from models proposed by others (Iturralde-Vinent 1989, 1994, 2003; James 2003, 2006*a*, *b*; Pindell 2003; Draper & Pindell 2006).

Major Caribbean events (Fig. 8) occurred in different tectonic regimes (rifting, spreading, plume, accretion, collision), from the early proto-Caribbean stage of oceanization through two eo-Caribbean accretionary stages of subduction to the collisional stage leading to the present Caribbean Plate.

During the Jurassic, tensional and transtensional stress related to Central Atlantic opening and separation of North (NOAM) and South (SOAM) American plates produced several spreading centres, offset by transform faults, forming a proto-Caribbean oceanic realm, between the Atlantic and Pacific. During the Cretaceous, proto-Caribbean oceanic crust thickened at its westernmost end, becoming structurally and petrologically

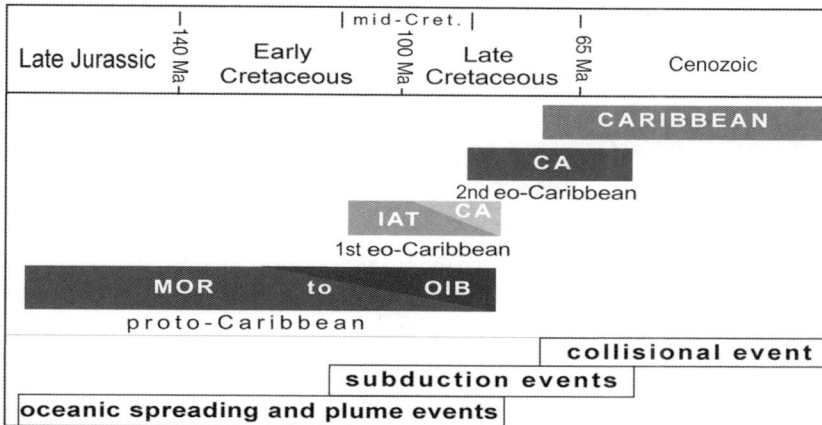

Fig. 8. Timeline of major Caribbean events. Abbreviations: MOR, mid-ocean ridge; OIB, ocean islands basalt; IAT, island arc tholeiite; CA, island arc calc-alkaline rocks.

comparable to typical oceanic plateaus (e.g. Ontong–Java). The resulting rocks are tholeiitic basalts with flat REE patterns, sometimes associated with picrites, such as those recorded at Curaçao (Kerr *et al.* 1996; Giunta *et al.* 2002*c*), Aruba (White *et al.* 1999), Los Roques (Giunta *et al.* 1997, 2002*c*; Kerr & Hastie 2006), the Tortugal komatiitic suite in Costa Rica (Alvarado *et al.* 1997), Central Cordillera of Dominican Republic (Lapierre *et al.* 2000; Lewis *et al.* 2002; Giunta *et al.* 2002*a*) and the western Colombian ophiolites (Kerr *et al.* 1997). Crustal accretion by vertical overthickening (intrusive and extrusive) was probably predominant, with production of large amounts of basaltic magmas and picrites. Some basaltic rocks, occurring as seamounts, further contributed to thickening of the Caribbean oceanic crust (Kerr & Hastie 2006).

Since Late Jurassic–Early Cretaceous nucleation this oceanic domain separated the Bahamas, Maya and Chortís continental margins to the north from the Guayana shield to the south, developing through rifted continental margins with WPT magmatism, which created space for the proto-Caribbean oceanic domain. This scenario supports the proposed 'near American' location of the proto-Caribbean ocean (Giunta 1993; Meschede & Frisch 1998; Beccaluva *et al.* 1999; Giunta *et al.* 2002*a*, *c*, *d*, 2003*a*, *b*, 2006; Iturralde-Vinent 2003; James 2003, 2006*a*, 2009) and is favoured over the classic idea that the Caribbean Plate is a 'Pacific promontory' (Duncan & Hargraves 1984; Burke 1988; Pindell & Barrett 1990; Pindell 2003; Draper & Pindell 2006; Maresch 2009; Stanek & Maresch 2006, 2009).

The original location of the proto-Caribbean is still debated. It could be thought of as a westward projection of the Tethyan ocean, forming an

Atlantic–Pacific link. It would be 'inter-American' because it formed during Pangaean separation of the Americas. The controversy concerns the location of oceanic crust thickening – inter- or extra-American.

The proto-Caribbean realm constituted a Large Igneous Province (LIP) that has partly been subducted, so its original extension is unknown. The relationship between thin and thickened oceanic crust, with part of the first evolving to the latter, depends on the original distance between them. Close proximity suggests a near mid-American location. Distance suggests the Pacific model.

Ophiolites on Caribbean Plate margins, relics of Mid- to Late Cretaceous tectonic phases, indicate that the original LIP experienced two main stages of intraoceanic subduction and subordinate continental interaction with mélange formation and HP/LT metamorphism, involving both the proto-Caribbean oceanic lithosphere and/or suprasubduction complexes. These two stages, called eo-Caribbean phases (Giunta 1993), are respectively recorded by volcano–plutonic sequences with IAT or CA affinities and tonalitic intrusions of CA affinity. The two phases are recorded in several places along the plate margins. The first occurred in the Mid-Cretaceous (before 96 Ma); the second in the Late Cretaceous (from 86 Ma). One of the main problems of Caribbean geology is whether these record a single system continuously evolving in the Mid- and Late Cretaceous (Great Volcanic Arc model), or separate arcs (first and second eo-Caribbean).

Beginning in the Early Cretaceous, South Atlantic opening and westward–northwestward motion of the American plates led to ocean–ocean and ocean–continent plate convergence (first eo-Caribbean phase), producing several Cordilleran-like ophiolites

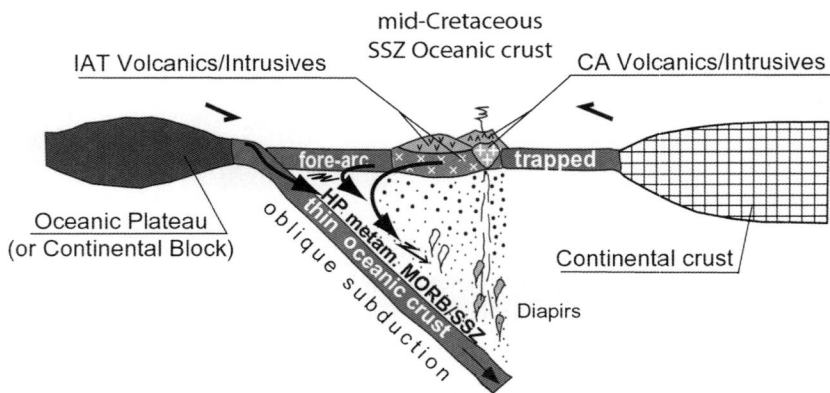

Fig. 9. First eo-Caribbean phase (Mid-Cretaceous): generalized cross-section across the intra-oceanic subduction zones. Metamorphism and deformation realized in a strike–slip tectonic regime. ➤: HP/LT (blueschist to eclogite facies) subduction related metamorphism and ductile deformation; ➤: greenschist to amphibolite facies metamorphism and ductile deformation.

(Beccaluva *et al.* 2004) related to SSZ and magmatic arcs (Fig. 9). Evidence of involvement of proto-Caribbean oceanic lithosphere in subduction zones comes from the above-mentioned HP/LT metamorphosed units of the plate margins. They formed during ocean–ocean subduction or, subordinately, an ocean–continent subduction. Parts of the previously rifted continental margins were also subducted, reaching eclogite facies in places. Subduction models have been proposed by Garcia-Casco *et al.* (2006*a*, *b*), Escuder-Viruete *et al.* (2006) and Maresch (2006, 2009). According to Giunta *et al.* (2003*b*, and references therein), the oldest available radiometric age for the HP/LT metamorphic climax is 96.3 ± 0.4 Ma. Since some units (e.g. Villa de Cura, Venezuela) formed in a supra-subduction setting, subduction must have commenced earlier than this HP/LT event. Peak HP/LT metamorphism of some rifted continental margin units (e.g. La Rinconada Fm of the Caucagua–El Tinaco complex in Venezuela) was probably younger than 114–105 Ma. A time span of about 11–16 Ma, during which the subduction continued, is suggested by the younger radiometric ages of climax conditions (79.8 ± 0.4 Ma), as well as the oldest radiometric ages of retrogradation (84.5 ± 0.2 Ma).

The Mid-Cretaceous accretionary stage is interpreted in different ways and is the most controversial of Caribbean evolution. The main features and problems are:

(1) *Locations of either ocean–ocean or ocean–continent convergence.* The oldest intra-oceanic convergence probably involved the eastern proto-Caribbean, where thinner oceanic lithosphere favoured subduction.

Simultaneously the western part of the plate was thickening to an oceanic plateau (Late Cretaceous). At the same time subduction occurred below the continental crust of the main plates and minor blocks, recorded by metamorphic mélanges and continental-derived units. Various models propose several magmatic arcs in a near mid-American area. Others propose a single great arc located in the Pacific, far from the present Central America Isthmus.

(2) *Both volcanic arc complexes and thinned continental crust were involved in subduction.* Geology indicates existance of two coeval subduction settings: intra-oceanic and sub-continental. HP/LT assemblages (blueschist and eclogitic) in both oceanic and continental lithosphere require deep subduction. As mentioned earlier, HP/LT metamorphism of arc complexes (e.g. Villa de Cura), and in particular continental margin units (e.g. Caucagua–El Tinaco, Venezuela) require more complex subduction than sinking of dense oceanic lithosphere. It could have involved tectonic erosion (likely) or underthrusting of thinned continental crust or flakes during continental collision.

(3) *Ocean floor or back-arc origin of MOR-type ophiolitic units.* This problem is very difficult to resolve with petrological data alone. Depending on the location of the SSZ in different models, data suggest that some units better fit an intra- or back-arc supra-subduction origin or originated as trapped oceanic fragments. How many ophiolitic units can be referred to the un-thickened oceanic proto-Caribbean crust is difficult to establish.

(4) *Sinking direction of oceanic slabs in ocean–ocean or ocean–continent convergences – eastward or westward.* In an attempt to accommodate all recognized constraints, at least three models have been proposed (Giunta *et al.* 2003*a*, *b*, 2006), with strike–slip boundaries separating accretionary–subduction environments (Fig. 10). Palaeomagnetic data may distinguish between these models.

The models differ in strike–slip fault locations. In model A, more or less east–west trending transform faults separate subduction zones with different dipping directions (westward in intra-oceanic convergence, eastward in sub-continental convergence). These are seen to have been located inside the oceanic domain, with micro-continents present. In model B, all the oceanic domains could have been located between the main transform boundaries. This reconstruction requires both west-dipping subduction and very complicated continental margin morphology of continental promontories. Both these models can explain the early evolution of the so-called Great Volcanic Arc of Draper & Pindell (2006) and Pindell (2006). They differ in the Pacific or near mid-American origin of the arc system and the geometrical relationships of intra-oceanic and sub-continental subduction. Model C proposes widespread east-dipping subduction. Transform faults allowed coeval intra-oceanic and sub-continental subduction. Back-arc origin of some MOR-type units is also easily explained. Since few, dismembered arc portions have been referred to the first eo-Caribbean accretionary stage, doubts remain that they formed one original single arc. The first eo-Caribbean accretionary phase ended in the Late Cretaceous when the un-thickened oceanic realm was involved in westward subduction below the oceanic plateau (e.g. Burke 1988; Lewis & Draper 1990; Lewis *et al.* 2002; Meschede & Frisch 1998; Beccaluva *et al.* 1999; Giunta *et al.* 2002*b*, *c*, 2003*b*). This implies a flip of intraoceanic subduction polarity in model C (Kerr *et al.* 1996, 1999; White *et al.* 1999) or continuous westward subduction in models A and B.

Late Cretaceous kinematics (Fig. 11; second eo-Caribbean) involved eastward drift of the proto-Caribbean oceanic plateau, resulting in diachronous tonalitic magmatism (from 85 Ma) associated with westward-dipping oblique subduction of proto-Caribbean-Atlantic ocean floor and eastward youngling obduction of dismembered subduction complexes along northern and southern, east–west trending continental margins. This seems to be the consequence of eastward migration of northern and southern triple-junctions, resulting in bending of the Aves–Lesser Antilles arc (Giunta *et al.* 2003*a*, 2006). Tonalitic arc magmatism along the northern margin intruded both deformed older arc complexes and a new accretionary wedge migrating from the west (Motagua Suture Zone of Guatemala) to east (Cuba and Hispaniola). The Paleocene–Eocene volcanic arc of the Sierra Maestra in eastern Cuba (Iturralde-Vinent 2000; Kerr *et al.* 1999; Rojas-Agramonte *et al.* 2006*b*) may be connected with northward subduction of 'residual' oceanic crust below the older volcanic arc, as a second-order segment of the triple-point (Giunta *et al.* 2003*b*). At the same time the Late Cretaceous magmatic arc–accretionary wedge couple (western and central Cuba) collided against the NOAM, becoming progressively inactive.

Along the southern margin, tonalitic magmatism intruded both undeformed (85–82 Ma, Aruba; White *et al.* 1999) and deformed (Venezuelan islands) oceanic plateau, as well as the northernmost metamorphic complexes of rifted continental margin terrane (Caucagua–El Tinaco unit, Margarita island). No younger tonalitic magmatism is recorded in deformed units of the first eo-Caribbean volcanic arc. This implies that the older portion of the deformed belt was disconnected from Late Cretaceous active subduction. During Late Cretaceous–Paleocene collision the previously decoupled portions of the deformed belt were juxtaposed along the west–east southern margin of the Caribbean Plate (Giunta *et al.* 2002*c*, 2003*a*, *b*).

The proposed models for Late Cretaceous evolution differ in important aspects, as follows:

(1) *Flip or no flip of the intra-oceanic subduction direction between first and second eo-Caribbean stages.* In Pacific models early subduction direction of the leading edge arc is supposed to have dipped westward. This changed to eastward dip after collision of the Caribbean plateau with the subduction zone. The change would have to have occurred in a very short time interval.

(2) *Timing of plateau insertion into the mid-American area, and onset of two triple-junctions.* The Coniacian (85 Ma) seems a probable age of eastward drifting of the plateau and related triple-junctions, followed by collisional and suture zones from the west. The triple-junctions model can explain the evolution of the plate margins since at least the Late Cretaceous. The problem is to measure the distance travelled by the plateau. In some Pacific models this was more than 1000 km. 'In-place' models do not invoke plateau insertion and movement of the Caribbean Plate relative to its North and

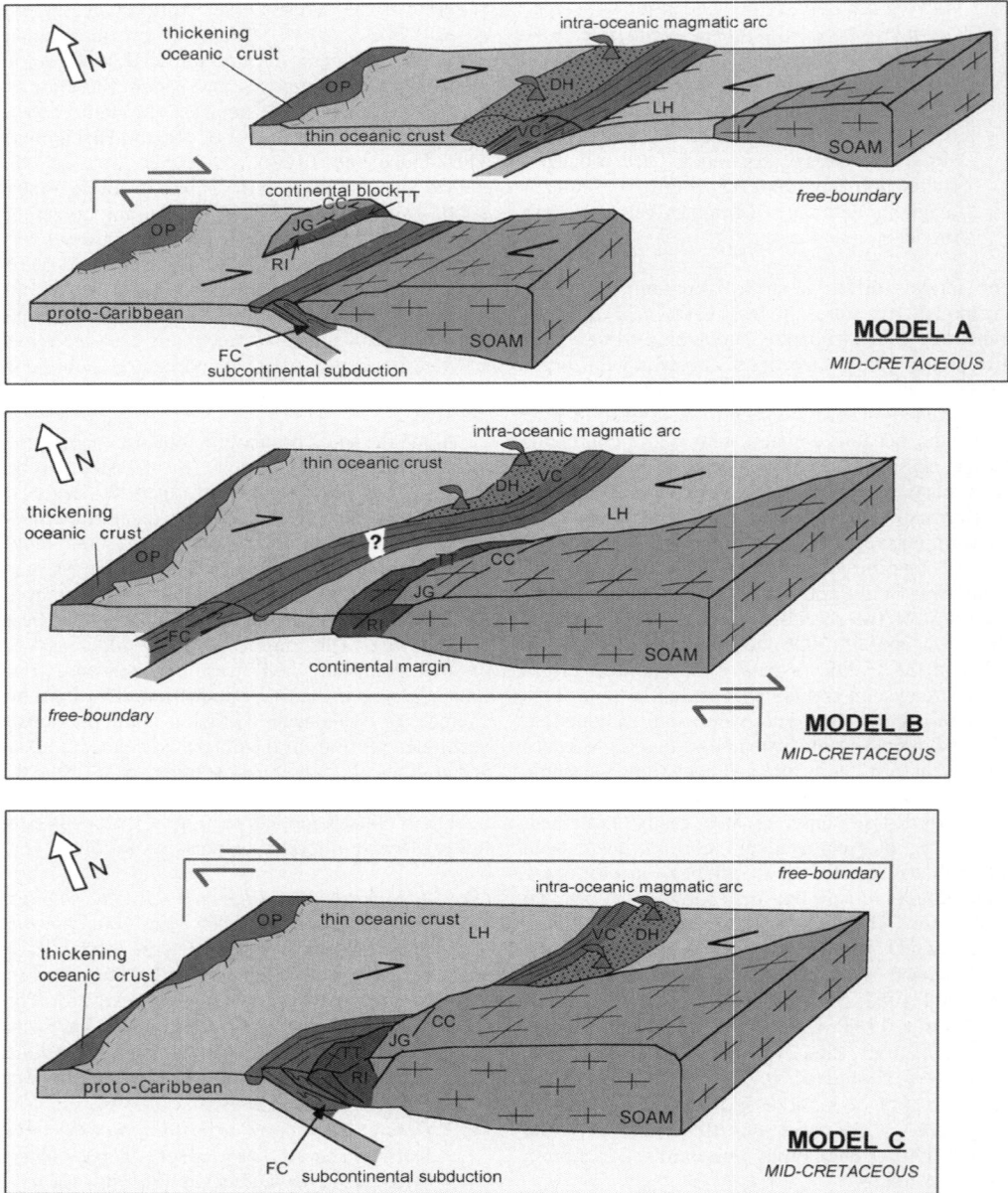

Fig. 10. Possible evolutionary models during the first (Mid-Cretaceous) eo-Caribbean phase related to the southern Caribbean Plate margin of Venezuela, depending on different locations of first-order free-boundaries (modified from Giunta *et al.* 2003*b*). These models could be proposed also for northern Caribbean margin, in a left-lateral convergence. Abbreviations: SOAM, South American Plate; JG and CC, continental margin of the Juan Griego and Cordillera de la Costa groups; RI and TT, sub-continental mantle and crust of rifted margins of La Rinconada and Caucagua-El Tinaco units characterized by WPT magmatism; FC and LH, MOR oceanic lithosphere of the Franja Costera group and Loma de Hierro unit; VC and DH, island arc showing IAT magmatism of the Villa de Cura and Dos Hermanas units; OP, thickening oceanic lithosphere (future Caribbean Plateau). See text for details and more explanations of models A, B, and C.

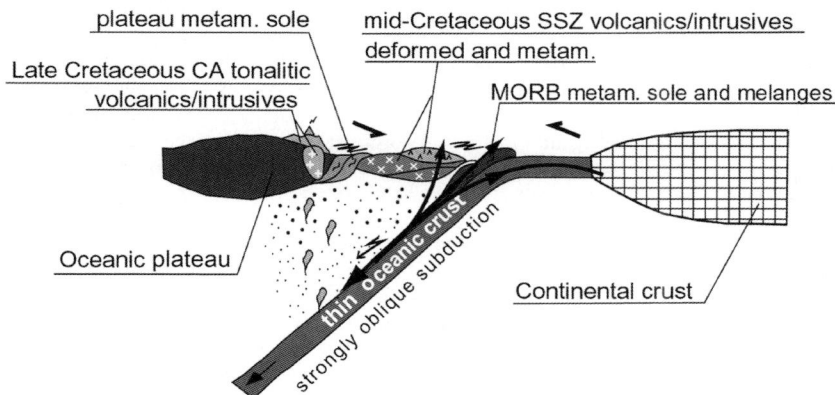

Fig. 11. Second eo-Caribbean phase (Late Cretaceous): generalized cross-section across the northern Caribbean accretionary system. Metamorphism and deformation realized in a strike–slip tectonic regime. ✐: HP/LT (blueschist to eclogite facies) subduction related metamorphism and ductile deformation; ✐: greenschist to amphibolite facies metamorphism and ductile deformation.

South American neighbours might have been as little as 300 km (James 2006*a*, *b*). If the western Colombian terranes are considered as remnants comparable to the Caribbean Plate, there is some palaeomagnetic data that suggests significant latitudinal displacement.

(3) *Driving forces induced by the plateau.* From the Late Cretaceous, the three models of Figure 10 converge, with eastward motion of the Caribbean Plateau driving evolution of the Caribbean margins (Escuder-Viruete & Pérez-Estaún 2006). Exhumation of HP/LT units and retrograde tectono-metamorphism occurred from the Campanian onward, with uplift related to displacement in shear zones sub-parallel to the plate boundaries. Strike–slip led to progressive dismembering of the orogenic system and development of separate accreted terranes.

(4) *Timing of granitoid and tonalitic magmatism.* In the model of Beccaluva *et al.* (1999), and Giunta *et al.* (2002*a*, *c*, *d*, 2003*a*, *b*) arc intrusions are divided in two main peaks of activity, connected with Mid-Cretaceous (first eo-Caribbean) and Late Cretaceous (second eo-Caribbean) intra-oceanic subduction. In contrast, Stanek & Maresch (2009) reconstruct continuous (110–73 Ma) arc activity in Cuba, supporting the existence of a single Great Volcanic Arc. Other arc regions (e.g. Andes or North American Cordilleras) may show the fact that, although magmatism can be continuous, there are some peaks where the magmatic volume is more significant, related to major tectonic changes. Diachronism of tonalitic magmatism, younging from the west since the Late Cretaceous, supports both the shifting of triple

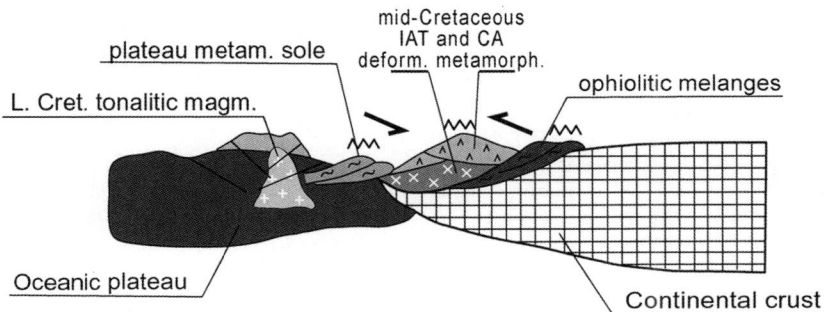

Fig. 12. Generalized cross-section of the peri-Caribbean collisional belts (Late Cretaceous–Palaeogene). ⋀⋀: ductile deformation, realized in a strike–slip tectonic regime.

junctions and bending of the Aves–Lesser
Antilles arc.

(5) *Construction of Central America isthmus,
along the Pacific subduction zone.* From the
Late Cretaceous, the Chortís continental
block moved eastward, rotating anticlockwise
with respect to Maya and building the
Motagua Suture Zone in Guatemala. At the
same time, intra-oceanic convergence resulted
in construction of the Chorotega block (Costa
Rica), in an undefined position west of the
Caribbean Plateau. The southern part of this
arc, the Chocó block, corresponds to eastern
Panama and the northwesternmost Andes.

Eastward drift of the Caribbean Plateau continued
from Late Cretaceous to Present, enhancing the
Lesser Antilles arc by eastward shift of the two
triple-junctions to their present-day positions,
north of Puerto Rico and east of Tobago. The end
of the second eo-Caribbean accretionary phase was
marked by Late Cretaceous–Palaeogene collision
and/or obduction of the proto- and eo-Caribbean
complexes against or onto NOAM and SOAM continental margins. Suture zones formed along east–
west strike–slip zones (Fig. 12). As a result, both
the northern and southern boundaries of the
Caribbean Plate are collisional belts that followed
the eastward migrating triple-junctions. They are
broad zones where large-scale tear faulting
(still seismologically active) facilitates eastward
dispersion and uplift of tectonic units, juxtaposing
them within deformed terranes (Giunta *et al.*
2003*a*, *b*, 2006; Guzman-Speziale *et al.* 2006).

The Caribbean oceanic plateau became defined
to the east by the Aves–Lesser Antilles arc-back
arc system and to the west by the Chortís, Chorotega
and Chocí blocks (Beccaluva *et al.* 1999). Fore-arc,
back-arc and piggy-back basins on the deforming
plate borders were filled with clastic and volcaniclastic sediments. Foredeep systems (Sepur Basin
in Mexico-Guatemala; Foreland Basin in Cuba;
Piemontine Basin in Venezuela) developed at
thrust fronts (Giunta *et al.* 2002*b*, 2003*a*).

Models of Cenozoic–Present Caribbean kinematics are generally similar. Minor differences at
the large scale concern the role of the strike–slip
tectonics in plate margin evolution since at least
the Mid-Cretaceous exhumation of HP/LT units.
The Motagua zone of Guatemala is perhaps the
best example of collisional belt evolution (Giunta
et al. 2002*b*) in a strike–slip tectonic regime.

Conclusions

In order to compare different geodynamic models
of the Caribbean Plate, facts and unresolved
problems are distinguished and considered. Major

disagreements currently concern: (a) the original
location of the 'proto-Caribbean' oceanic realm or
realms; (b) the Early Cretaceous palaeogeography
and morphology of North and South American continental margins and minor blocks, in particular
original locations of the latter; (c) polarity of the
Cretaceous subduction zones; (d) the locations of
and relationships between coeval intra-oceanic
and sub-continental subduction zones; (e) the progressive insertion of rifted continent and supra-
subduction complexes in subduction; (f) the possible subduction polarity reversal; and (g) the
number of magmatic arcs and their peaks of activity.

From the Jurassic–Early Cretaceous until the
Present plate evolution involved ocean spreading
and plume activity, accretionary and collisional tectonics, dominated by a strongly oblique tectonic
regime, constraining seafloor spreading, subduction,
crustal exhumation, emplacement and dismembering processes, the evidence of which has been
recorded in the oceanic remnants of a lost large
igneous province.

A recent conference, held in Sigüenza, Spain,
discussed in detail current regional problems.
Although there are sufficient data to unravel an evolution outline, many different order problems remain
open or insufficiently explained, so that current
models seem to be far too speculative. The detailed
new data that are being obtained by the Caribbean
geological community certainly will show a more
realistic picture of both plate convergence and
ocean dynamics.

The authors are very grateful to A. Cardona and K. James
for their careful revision of the manuscript, and for
the suggestions improving it. Many thanks also go to the
colleagues attending the Sigüenza conference for the
interesting discussion on the main problems of the
Caribbean tectonics.

References

ALVARADO, G. E., DENYER, P. & SINTON, C. W. 1997.
The 89 Ma Tortugal komatiitic suite, Costa Rica:
implications for a common geological origin of the
Caribbean and Eastern Pacific region from a mantle
plume. *Geology*, **25**, 439–442.
AVÉ LALLEMANT, H. G. 1997. Transpression, displace-
ment partioning, and exhumation in the eastern Carib-
bean/South America plate boundary zone. *Tectonics*,
16, 272–289.
AVÉ LALLEMANT, H. G. & SISSON, V. B. 2005.
Exhumation of eclogites and blueschists in northern
Venezuela: constraints from kinematic analysis of
deformation structures. *In*: AVÉ LALLEMANT, H. G.
& SISSON, V. B. (eds) *Caribbean/South American
plate interactions, Venezuela.* Geological Society of
America, Special Papers, **394**, 193–206.
AZEMA, J., BOURGOIS, J., BAUMGARTNER, P. O.,
TOURNON, J., DESMET, A. & AUBOIN, J. 1984.

A Tectonic Cross-section of the Costa Rica Pacific Littoral as Key to the Structure of the Landward Slope of Middle America Trench of Guatemala. Initial Reports of the Deep Sea Drilling Program.

BAUMGARTNER, P. O. & DENYER, P. 2006. Evidence for Middle Cretaceous accretion at Santa Elena Peninsula (Santa Rosa Accretionary Complex), Costa Rica. Geologia Acta, 4, 179–191.

BECCALUVA, L., BELLIA, S. ET AL. 1995. The northwestern border of the Caribbean Plate in Guatemala: new geological and petrological data on the Motagua ophiolitic belt. Ofioliti, 20, 1–15.

BECCALUVA, L., COLTORTI, M. ET AL. 1996. Cross sections through the ophiolitic units of the Southern and Northern Margins of the Caribbean Plate, in Venezuela (Northern Cordilleras) and Central Cuba. Ofioliti, 21, 85–103.

BECCALUVA, L., CHINCHILLA CHAVES, A. L., COLTORTI, M., GIUNTA, G., SIENA, F. & VACCARO, C. 1999. The, St. Helena–Nicoya Ophiolitic Complex in Costa Rica and its geodynamic implications for the Caribbean Plate Evolution. European Journal of Mineralogy, 11, 1091–1107.

BECCALUVA, L., COLTORTI, M., GIUNTA, G. & SIENA, F. 2004. Tethyan vs Cordilleran Ophiolites: a reappraisal of distinctive tectono-magmatic features of supra-subduction complexes in relation to the subduction mode. Tectonophysics, 393, 163–174.

BELLIZZIA, A. 1986. Sistema Montañoso del Caribe, una cordillera alóctona en la parte norte de América del Sur. 6th Congreso Géologico Venezolano, Caracas, Sociedad Venezolana de Geòlogos, Memoria, 10, 6657–6836.

BOUTELIER, D., CHEMENDA, A. & BURG, J.-P. 2003. Subduction versus accretion of intra-oceanic volcanic arcs: insight from thermo-mechanical analogue experiments Earth and Planetary Science Letters, 212, 31–45.

BURKE, K. 1988. Tectonic evolution of the Caribbean. Annual Review Earth and Planetary Science Letters, 16, 201–230.

COBIELLA-REGUERA, J. L. 2005. Emplacement of Cuban Ophiolites. Geologica Acta, 3, 273–294.

DENGO, G. 1985. Mid America: tectonic setting for the Pacific margin from Southern Mexico to North Western Colombia. In: NAIRN, A. E. M., STEHLI, F. G. & UYEDA, S. (eds) The Ocean Basin and Margins, Plenum Press, New York, 7, 15–37.

DENYER, P., BAUMGARTNER, P. O. & GAZEL, E. 2006. Characterization and tectonic implications of Mesozoic–Cenozoic oceanic assemblages of Costa Rica and Western Panama. Geologica Acta, 4, 219–235.

DONNELLY, T. W., BEETS, D. ET AL. 1990. History and tectonic setting of Caribbean magmatism. In: DENGO, G. & CASE, J. E. (eds) The Geology of North America, H, The Caribbean Region. Geological Society of America, 339–374.

DRAPER, G. & NAGLE, F. 1991. Geology, Structure, and Tectonic Development of the Rio San Juan Complex, Northern Dominican Republic. Geological Society of America, Special Papers, 262, 77–95.

DRAPER, G. & PINDELL, J. L. 2006. Plate tectonic view of Greater Antillean geology and evolution. Geology of the area between North and South America, with focus on the origin of the Caribbean Plate. International Research Conference, Sigüenza, Spain, 2006.

DRAPER, G., MANN, P. & LEWIS, J. F. 1994. Hispaniola. In: DONOVAN, S. K. & JACKSON, T. A. (eds) Caribbean Geology: An Introduction. University of the West Indies Publication Association, Kingston, 129–150.

DUNCAN, R. A. & HARGRAVES, R. B. 1984. Plate tectonic evolution of the Caribbean region in the mantle reference frame. In: BONINI, W. E., HARGRAVES, R. B. & SHAGAM, R. (eds) The Caribbean–South America Plate Boundary and Regional Tectonics. Geological Society of America, Boulder, CO, Memoirs, 162, 81–93.

ESCUDER VIRUETE, J., HERNÁIZ HUERTA, P. P., DRAPER, G., GUTIÉRREZ, G., LEWIS, J. F. & PÉREZ-ESTAÙN, A. 2002. Metamorfismo y estructura de la Formación Maimón y los Complejos Duarte y Río Verde, Cordillera Central Dominicana: implicaciones en la estructura y la evolucion del primitivo Arco Isla Caribeño. Geologica Acta Hispanica, 37, 123–162.

ESCUDER VIRUETE, J., DÍAZ DE NEIRA, A. ET AL. 2006a. Island-arc tholeiites, boninites and related felsic rocks in Hispaniola: magmatic and age constraints on initiation of intra-oceanic subduction and development of the Caribbean island arc. Geophysical Research Abstracts, 8, 01492.

ESCUDER VIRUETE, J. & PÉREZ-ESTAÚN, A. 2006b. P–T path of high-P metabasic rocks from the Samaná Peninsula complex, Hispaniola: record of subduction and collision processes in the northern edge of the Caribbean Plate. Geophysical Research Abstracts, 8, 01494.

FINCH, R. C. & DENGO, G. 1990. NOAM–CARIB Plate boundary in Guatemala: a Cretaceous suture reactivated as a Neogene transform fault. Field trip 17 guidebook, November 2–7 1990.

GARCÍA-CASCO, A., TORRES-ROLDÀN, R. L. ET AL. 2006a. High pressure metamorphism of ophiolites in Cuba. Geologica Acta, 4, 63–88.

GARCÍA-CASCO, A., ITURRALDE-VINENT, M. A. ET AL. 2006b. Birth and demise of subduction in the northern margin of the Caribbean Plate (Cuba). Geophysical Research Abstracts, 8, 05003.

GAZEL, E., DENYER, P. & BAUMGARTNER, P. O. 2006. Magmatic and geotectonic significance of Santa Elena Peninsula, Costa Rica. Geologica Acta, 4, 193–202.

GIUNTA, G. 1993. Los margenes mesozoicos de la Placa Caribe: Problematicas sobre nucleacion y evolucion. 6th Congreso Colombiano de Geologia, Memoria, III, 729–747.

GIUNTA, G., BECCALUVA, L., COLTORTI, M. & SIENA, F. 1997. Ophiolitic units of the southern margin of Caribbean Plate in Venezuela: a reappraisal of their petrogenesis and original tectonic setting. Memorias del VIII Congreso Geologico Venezolano, Porlamar, 1, 331–337.

GIUNTA, G., BECCALUVA, L., COLTORTI, M., MORTELLARO, D., SIENA, F. & CUTRUPIA, D. 2002a. The peri-Caribbean ohiolites: structure, tectono-magmatic significance and geodynamic implications. Caribbean Journal of Earth Science, 36, 1–20.

GIUNTA, G., BECCALUVA, L. *ET AL.* 2002*b*. The Motagua suture zone in Guatemala. Field-trip guide book of the IGCP. 433 Workshop and 2nd Italian–Latin American Geological Meeting 'In Memory of Gabriel Dengo'. *Ofioliti*, **27**, 47–72.

GIUNTA, G., BECCALUVA, L., COLTORTI, M., SIENA, F. & VACCARO, C. 2002*c*. The southern margin of the Caribbean Plate in Venezuela: tectono-magmatic setting of the ophiolitic units and kinematic evolution. *Lithos*, **63**, 19–40.

GIUNTA, G., BECCALUVA, L., COLTORTI, M. & SIENA, F. 2002*d*. Tectono-magmatic significance of the peri-Caribbean ophiolitic units and geodynamic implications. Proceedings of 15th CGC, IGCP Project 364. *In*: JACKSON, T. A. (ed.) *Caribbean Geology into the Third Millennium*. University of the West Indies Press, Jamaica, 15–34.

GIUNTA, G., MARRONI, M., PADOA, E. & PANDOLFI, L. 2003*a*. Geological constraints for the Geodynamic evolution of the southern margin of the Caribbean Plate. *In*: BARTOLINI, C., BUFFLER, R. T. & BLICKWEDE, J. (eds) *The Circum-Gulf of Mexico and the Caribbean: Hydrocarbon Habitats, Basin Formation, and Plate Tectonics*. American Association of Petroleum Geologists, Memoirs, **79**, 104–125.

GIUNTA, G., BECCALUVA, L., COLTORTI, M. & SIENA, F. 2003*b*. *The Peri-Caribbean Ophiolites and Implications for the Caribbean Plate Evolution*, International Conference, Barcelona. American Association of Petroleum Geologists, 1–6.

GIUNTA, G., BECCALUVA, L. & SIENA, F. 2006. Caribbean Plate margin evolution: constraints and current problems. *Geologia Acta*, **4**, 265–277.

GUZMÁN-SPEZIALE, M., KOSTOGLODOV, V., MANEA, M., MANEA, V. & FRANCO, S. 2006. GPS measurements in the northwestern Caribbean: implications for the North America–Caribbean Plate boundary zone. *Geophysical Research Abstracts*, **8**, 09844.

ITURRALDE-VINENT, M. 1989. Role of ophiolites in the geological constitution of Cuba. *Geotectonics*, **4**, 63–74.

ITURRALDE-VINENT, M. 1994. Cuban geology: a new plate tectonic synthesis. *Journal of Petroleum Geology*, **17**, 38–70.

ITURRALDE-VINENT, M. 2000. Sinopsis de la constitución geológica de Cuba. *Acta Geológica Hispánica*, **33**, 9–56.

ITURRALDE-VINENT, M. 2003. The conflicting paleontologic versus stratigraphic record of the formation of the Caribbean Seaway. *In*: BARTOLINI, C., BUFFLER, R. T. & BLICKWEDE, J. (eds) *The Circum-Gulf of Mexico and the Caribbean: Hydrocarbon Habitats, Basin Formation, and Plate Tectonics*. American Association Petroleum Geology, Memoirs, **79**, 75–88.

JAMES, K. H. 2003. *Caribbean Plate Origin: Discussion of Arguments Claiming to Support a Pacific Origin; Arguments for an In-Situ Origin*. International Conference, Barcelona. American Association of Petroleum Geologists, 8–9.

JAMES, K. H. 2006*a*. Arguments for and against the Pacific origin of the Caribbean Plate: discussion, finding for an inter-American origin. *Geologica Acta*, **4**, 279–302.

JAMES, K. H. 2006*b*. Caribbean geology simplified. *In*: *Geology of the Area Between North and South America, with Focus on the Origin of the Caribbean Plate*. International Research Conference, Sigüenza, Spain, 2006.

JAMES, K. H. 2009. *In-situ* origin of the Caribbean: discussion of data. *In*: JAMES, K. H., LORENTE, M. A. & PINDELL, J. L. (eds) *The Origin and Evolution of the Caribbean Plate*. Geological Society, London, Special Publications, **328**, 77–126.

JOLLY, W. T., LIDIAK, E. G., SCHELLEKENS, J. H. & SANTOS, H. 1998. *Volcanism, Tectonics, and Stratigraphic Correlations in Puerto Rico*. Geological Society of America, Special Papers, **322**, 1–34.

KERR, A. C. & HASTIE, A. R. 2006. An unequivocal hot mantle plume origin for the thickened oceanic crust of the Caribbean Plate and its margins. *In*: *Geology of the Area Between North and South America, with Focus on the Origin of the Caribbean Plate*. International Research Conference, Sigüenza, Spain, 2006.

KERR, A. C., TARNEY, J., MARRINER, G. F., KLAVER, G. T., SOUNDERS, A. D. & THIRLWALL, M. F. 1996. The geochemistry and petrogenesis of the Late Cretaceous picrites and basalts of Curaçao, Netherlands Antilles: a remnant of an oceanic plateau. *Contributions to Mineralogy and Petrology*, **124**, 29–43.

KERR, A. C., MARRINER, G. F. *ET AL.* 1997. Cretaceous basaltic terranes in Western Colombia: elemental, chronological and Sr-Nd isotopic constraints on petrogenesis. *Journal of Petrology*, **38**, 677–702.

KERR, A. C., ITURRALDE-VINENT, M., SAUNDERS, A. D., BABBS, T. L. & TARNEY, J. 1999. A new plate tectonic model of the Caribbean: implications from a geochemical reconnaissance of Cuban Mesozoic volcanic rocks. *Geological Society of America Bulletin*, **111**, 1581–1599.

KUIJPERS, J. 1980. The geological history of the Nicoya ophiolite complex, Costa Rica and its geotectonic significance. *Tectonophysics*, **68**, 233–255.

LAPIERRE, H., BOSCH, D. *ET AL.* 2000. Multiple plume events in the genesis of the peri-Caribbean Cretaceous oceanic plateau province. *Journal of Geophysical Research – Solid Earth*, **105**, 8403–8421.

LEWIS, J. F. & DRAPER, G. 1990. Geology and tectonic evolution of the Northern Caribbean, margin. *In*: DENGO, G. & CASE, J. E. (eds) *The Caribbean Region*. The Geology of North America, **H**. Geological Society of America, 77–110.

LEWIS, J. F., ESCUDER-VIRUETE, J., HERNAIZ-HUERTA, P. P., GUTIERREZ, G., DRAPER, G. & PÉREZ-ESTAÚN, A. 2002. Geochemical subdivision of the Circum-Caribbean Island Arc, Dominican Cordillera Central: implications for crustal formation, accretion and growth within an intra-oceanic setting. *Acta Geologica*, **37**, 81–122.

LEWIS, J. F., DRAPER, G., PROENZA, J. A., ESPAILLAT, J. & JIMENÈZ, J. 2006*a*. Ophiolite-related ultramafic rocks (serpentinites) in the Caribbean region: a review of their occurrence, composition, origin, emplacement and Ni-laterite soil formation. *Geologica Acta*, **4**, 237–263.

LEWIS, J. F., PROENZA, J. A., JOLLY, W. T. & LIDIAK, E. G. 2006*b*. Monte del Estado (Puerto Rico) and Loma Caribe (Dominican Republic) peridotites: a look at two different Mesozoic mantle sections within northern Caribbean region. *Geophysical Research Abstracts*, **8**, 08798.

MARCHESI, C., BOSCH, D. *ET AL.* 2006. New geochemical and geochronological constraints on the Cretaceous volcanism from northeastern Cuba. *Geophysical Research Abstracts*, **8**, 08595.

MARESCH, W. V. 2006. Dynamics and timing of collisional events recorded in metamorphic rocks exposed along the northern and southern margins of the Caribbean: no leeway for 'in situ' models. *In*: *Geology of the Area Between North and South America, with Focus on the Origin of the Caribbean Plate*. International Research Conference, Sigüenza, Spain, 2006.

MARESCH, W. V., KLUGE, R., BAUMANN, A., PINDELL, J. L., KRÜCKHANS-LUEDER, G. & STANEK, K. 2009. The occurrence and timing of high-pressure metamorphism on Margarita Island, Venezuela: a constraint on Caribbean–South America interaction. *In*: JAMES, K. H., LORENTE, M. A. & PINDELL, J. L. (eds) *The Origin and Evolution of the Caribbean Plate*. Geological Society, London, Special Publications, **328**, 705–741.

MESCHEDE, M. & FRISCH, W. 1998. A plate tectonic model for the Mesozoic and Early Cenozoic history of the Caribbean Plate. *Tectonophysics*, **296**, 269–291.

MESCHEDE, M., SOMMER, M., HUENEKE, H. & COBIELLA-REGUERA, J. 2006. Transfer of the northwestern Caribbean Plate to the North American continental margin: geodynamic evolution of the Cuban archipelago. *Geophysical Research Abstracts*, **8**, 01964.

MITCHELL, S. F. 2006. Timing and implications of Late Cretaceous tectonic and sedimentary events in Jamaica. *Geologica Acta*, **4**, 171–178.

MULLER, P. D. 1980. *Geology of the Los Amates quadrangle and vicinity, Guatemala, Central America*. Unpublished PhD dissertation, NY State University.

PARDO, G. 1975. Geology of Cuba. *In*: NAIRN, A. E. M. & STEHI, F. G. (eds) *The Ocean Basins and Margins*. Plenum Press, New York, 553–613.

PINDELL, J. L. 2003. *Pacific Origin of Caribbean Oceanic Lithosphere and Circum-Caribbean Hydrocarbon Systems*. International Conference, Barcelona, American Association of Petroleum Geologists, 11–12.

PINDELL, J. L. 2006. Plate tectonics and basin development in the Gulf of Mexico, Caribbean, and northern South America. *In*: *Geology of the Area Between North and South America, with Focus on the Origin of the Caribbean Plate*. International Research Conference, Sigüenza, Spain, 2006.

PINDELL, J. L. & BARRETT, S. F. 1990. *Geological Evolution of the Caribbean Region: a Plate-tectonic Perspective*. *In*: DENGO, G. & CASE, J. E. (eds) *The Caribbean Region. The Geology of North America*, **H**. Geological Society of America, 339–374.

ROJAS-AGRAMONTE, Y., KRÖNER, A., GARCÍA-CASCO, A., ITURRALDE-VINENT, M. A., WINGATE, M. T. D. & LIU, D. Y. 2006a. Geodynamic implications of zircon ages from Cuba. *Geophysical Research Abstracts*, **8**, 04943.

ROJAS-AGRAMONTE, Y., NEUBAUER, F., BOJAR, A. V., HEJL, E., HANDLER, R. & GARCIA-DELGADO, D. E. 2006b. Geology, age and tectonic evolution of the Sierra Maestra Mountains, southeastern Cuba. *Geologica Acta*, **4**, 123–150.

ROSENFELD, L. 1981. *Geology of the western Sierra de Santa Cruz, Guatemala*. Unpublished PhD thesis, NY State University.

SINTON, C. W., DUNCAN, R. A., STOREY, M., LEWIS, J. & ESTRADA, J. J. 1998. An oceanic flood basalt province within the Caribbean Plate. *Earth and Planetary Science Letters*, **155**, 221–235.

SISSON, V. B., EVREN ERTAN, I. & AVÉ LALLEMANT, H. G. 1997. High-pressure (-2000 MPa) kyanite- and glaucophane-bearing pelitic schist and eclogite from Cordillera de la Costa Belt, Venezuela. *Journal of Petrology*, **38**, 65–83.

SMITH, C. A., SISSON, V. B., AVÉ LALLEMANT, H. G. & COPELAND, P. 1999. Two contrasting pressure–temperature paths in the Villa de Cura blueschist belt, Venezuela: possible evidence for Late Cretaceous initiation of subduction in the Caribbean. *Geological Society of America Bulletin*, **111**, 831–848.

STANEK, K. P., MARESCH, W. V., GRAFE, F., GRAVEL, C. H. & BAUMANN, A. 2006. Structure, tectonics and metamorphic development of the Sancti Spiritus Dome (eastern Escambray Massif, Central Cuba). *Geologica Acta*, **4**, 151–170.

STANEK, K. P. & MARESCH, W. V. 2006. The geotectonic story of the Great Antillean Arc – implications of structural and geochronological data from Central Cuba. *In*: *Geology of the Area Between North and South America, with Focus on the Origin of the Caribbean Plate*. International Research Conference, Sigüenza, Spain, 2006.

STANEK, K. P., MARESCH, W. V. & PINDELL, J. L. 2009. The geotectonic story of the northwestern branch of the Caribbean Arc: implications from structural and geochronological data of Cuba. *In*: JAMES, K. H., LORENTE, M. A. & PINDELL, J. L. (eds) *The Origin and Evolution of the Caribbean Plate*. Geological Society, London, Special Publications, **328**, 361–398.

STEPHAN, J. F., BLANCHET, R. & MERCIER DE LEPINAY, B. 1986. Northern and southern Caribbean festoons (Panama, Colombia, Venezuela, Hispaniola, Puerto Rico) interpreted as subductions induced by the east west shortening of the Pericaribbean continental frame. *In*: WEZEL, F. C. (ed.) *The Origin of Arcs*. Development in Geotectonics, **21**. Elsevier, New York, 35–51.

VALLS ALVAREZ, R. A. 2006. Geology and geochemical evolution of the ophiolitic belts in Guatemala. A field guide to nickel-bearing, laterites. *Geophysical Research Abstracts*, **8**, 10820.

WHITE, R. V., TARNEY, J. *ET AL.* 1999. Modification of an oceanic plateau, Aruba, Dutch Caribbean: implication for the generation of continental crust. *Lithos*, **46**, 43–68.

In situ origin of the Caribbean: discussion of data

KEITH H. JAMES

*Institute of Geography and Earth Sciences, Aberystwyth, Wales, UK and
Consultant Geologist, Plaza de la Cebada, 3, 09346 Covarrubias, Burgos, Spain
(e-mail: khj@aber.ac.uk)*

Abstract: Compiled and synthesized geological data suggest that the Caribbean Plate consists of dispersed continental basement blocks, wedges of ?Triassic–Jurassic clastic rocks, Jurassic–Late Cretaceous carbonate rocks, volcanic arc rocks, widespread, probably subaerial basalts and serpentinized upper mantle. This points to an *in situ* origin of the Caribbean Plate as part of Middle America, continuing the geology of the eastern North America margin in a more extensional tectonic setting. Extension increases from the Gulf of Mexico through the Yucatán Basin to the Caribbean.

The Caribbean Plate was formerly seen to be an *in situ* part of Middle America. Stainforth (1969), for example, interpreted the area as the product of extension between North and South America, with much dispersed continental material. In 1966, Wilson, who had observed ice rafts from the air, suggested that the Caribbean and Scotia plates resembled tongues of lithosphere intruding between North and South America and South America and Antarctica. There are now numerous models for the area, admirably summarized by Morris *et al.* (1990) and Rueda-Gaxiola (2003). Most support the idea that the Caribbean Plate migrated from the Pacific.

Understanding of Caribbean evolution is complicated by:

(1) Dispersal of geology over a large number of geographic elements (Fig. 1), ranging from the well-documented, onshore South and North America (thorough exploration for abundant hydrocarbons, but with new learnings still coming to light) to the less well known Central America and the discontinuous Caribbean islands. Large submarine extensions of Florida–Bahamas, Yucatán, Nicaragua, Panamá, the Greater Antilles and northern South America are not known in detail. The Lower Nicaragua Rise and the Aves and Barbados ridges (Barbados island excepted) are poorly sampled. Young volcanic rocks cover large parts of Central America. Geological synthesis requires familiarity with literature ranging from local to regional focus and from academic to industrial (mainly hydrocarbon) interest. As always, the whole is greater than the sum of the parts.

(2) Absence (unexplained) of oceanic spreading anomalies and fractures from the whole of Middle America, apart from the centre of the Cayman Trough.

(3) Interpretation of data premised on an oceanic, Pacific origin of the Caribbean Plate.

This paper respects the call of Meyerhoff & Meyerhoff (1973) to honour data. It discusses data synthesized from more than 5000 articles. Together they point to an *in situ* origin of the Caribbean Plate by severe extension of continental crust and serpentinization of mantle, between separating North and South America. An accompanying paper (James 2009) presents an interpretation of *in situ* Caribbean Plate evolution and considers implications, predictions, outstanding questions and tests of the model. This volume also includes other interpretations, especially those of Giunta & Oliveri and Pindell & Kennan, which should be compared and contrasted with these papers.

Regional setting

The Caribbean Plate (Fig. 1), roughly 3000 km east–west and 800 km north–south (2.64 million km^2), forms part of Middle America, between North and South America. Middle America comprises four marine areas, the Gulf of Mexico, the Yucatán Basin, the Cayman Trough and the Caribbean Sea, dispersed between NW sinistrally offset North and South America. The Gulf of Mexico is intracontinental, surrounded by southern North America and the Florida and Campeche platforms. Cuba, with basement and Mesozoic carbonate cover intimately related to the Florida–Bahamas platform, bounds the Yucatán Basin to the north, while the Cayman Ridge separates the basin from the Cayman Trough. South of the Trough lies the Nicaragua Rise, the marine extension of the Chortís Block – the only place where continental

From: JAMES, K. H., LORENTE, M. A. & PINDELL, J. L. (eds) *The Origin and Evolution of the Caribbean Plate.*
Geological Society, London, Special Publications, **328**, 77–125.
DOI: 10.1144/SP328.3   0305-8719/09/$15.00 © The Geological Society of London 2009.

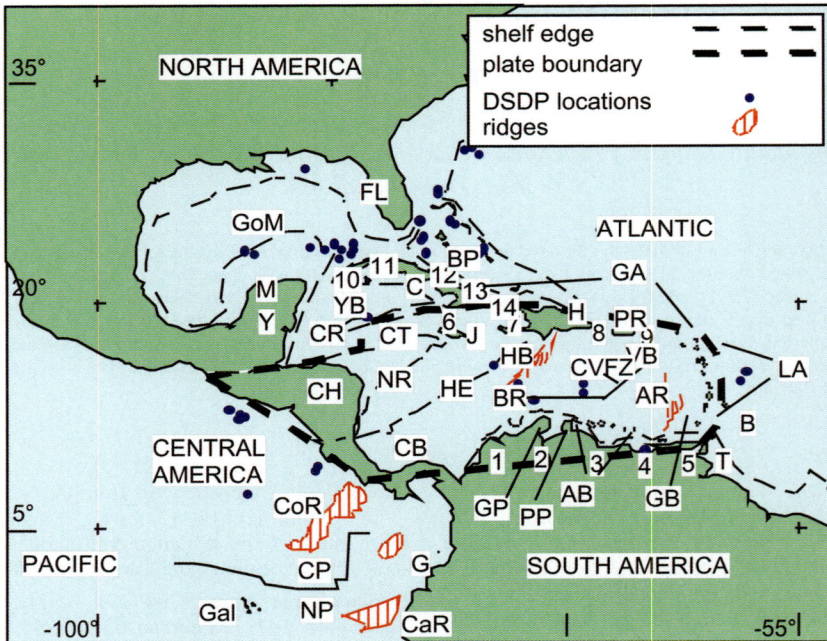

Fig. 1. Middle America, Caribbean Plate outlined by heavy dashed line. AB, Aruba–Blanquilla; AR, Aves Ridge; B, Barbados; BP, Bahamas Platform; BR, Beata Ridge; C, Cuba; CaR, Carnegie Ridge; CB, Colombia Basin; CH, Chortís; CoR, Cocos Ridge; CP, Cocos Plate; CR, Cayman Ridge; CT, Cayman Trough; CVFZ, Central Venezuela fault zone; FL, Florida; G, Gorgona; GA, Greater Antilles; Gal, Galapagos Islands; GB, Grenada Basin; GoM, Gulf of Mexico; GP, Guajira Peninsula; H, Hispaniola; HB, Haiti Basin; J, Jamaica; HE, Hess Escarpment; LA, Lesser Antilles; M, Maya; SCDB, South Caribbean Deformed Belt; NP, Nazca Plate; PFB, Panamá Fold Belt; NR, Nicaragua Rise; PP, Paraguaná Peninsula; T, Trinidad; VB, Venezuela Basin; Y, Yucatán; YB, Yucatán Basin. Numbers refer to basins named in text.

crust is currently recognized on the Caribbean Plate. Together they form about a third of the plate. The Chorotega and Chocó blocks link Chortís to South America. The Greater Antillean islands of Hispaniola, Puerto Rico and the northern Virgin Islands lie on the northern Caribbean Plate boundary, diminishing in size from west to east. A series of small islands, Aruba-Blanquilla, Margarita, Los Frailes and Los Testigos, lie offshore from northern South America, along the southern plate boundary. The east-convex volcanic arc of the active Lesser Antilles bounds the Caribbean Plate to the east. Another active volcanic arc runs along the western margin of Central America.

Water depths range up to 9 km. The greatest depths lie in the Puerto Rico and Muertos troughs, supposedly subduction trenches north and south of Puerto Rico, and along the margins of the Cayman Trough. Depths in the Gulf of Mexico range to more than 3000 m, in the Yucatán Basin to more than 4000 m and in the Caribbean to more than 5000 m – a trend that points towards increasing extension. Stratigraphy shows that parts of the

Caribbean and the Gulf of Mexico have suffered severe subsidence. Basement blocks in the deep Gulf subsided in the Late Cretaceous–Early Cenozoic (Roberts *et al.* 2005). The Cayman Ridge was a shallow carbonate bank until subsidence began in the Miocene (Perfit & Heezen 1978). The Nicaraguan plateau sank to its present depth by the Pliocene. Late Cretaceous–Paleocene granodiorites that may have been exposed in the Late Paleocene–Early Eocene are now at 4000 m on the Cayman Ridge and Eocene and Oligocene shallow water carbonates occur at 3000 m of both sides of the Cayman Trench. The Aves and Beata Ridges have similar subsidence histories (Perfit & Heezen 1978). Cretaceous–Cenozoic intertidal–subtidal limestones lie as deep as 6500 m on the south wall of the Puerto Rico Trench (Schneidermann *et al.* 1972). Turonian basalts of Caribbean seismic Horizon B'' that probably formed subaerially now lie below thousands of metres of water.

The Caribbean Plate interacts with the western Atlantic Plate, which carries North and South America, to the north, south and east and with the

Nazca and Cocos plates to the west. Atlantic elements are moving westward relative to the Caribbean Plate. Broad zones (>250 km wide) of east–west sinistral and dextral strike–slip on the northern and southern Caribbean Plate boundaries and subduction below the Lesser Antilles volcanic arc accommodate these movements. The Pacific Cocos Plate converges NE relative to Central America, where subduction and volcanism also occur. The east–west Nazca/Caribbean boundary is strike–slip and little volcanic activity occurs in Panamá. However, spreading along the east–west Carnegie Ridge provides a northerly component of convergence that drives NW South America and the Panamá block northward. Ahead of these lie the north-verging Southern Caribbean Deformed Belt and the Panamá Fold Belt, respectively. The Beata Ridge separates the Colombian Basin and Venezuelan Basins (contiguous south of the Ridge) while the Aves Ridge separates the Venezuelan and Grenada basins.

Prevailing understanding

Consensus is that ocean spreading in Middle America first occurred in the Jurassic (Gulf of Mexico–Callovian age, peripheral data from the NE Gulf of Mexico; Marton & Buffler 1999, or Toarcian–Aalenian, palaeogeographic reconstruction; Rueda-Gaxiola 2003). This produced 'Proto-Caribbean' oceanic crust. Spreading occurred again in the Palaeogene when the Yucatán and Grenada basins and the Cayman Trough formed (heat-flow and depth-to-basement estimates; Bouysse 1988; Rosencrantz *et al.* 1988).

The Caribbean Plate formed in the Pacific during the Jurassic and thickened in the Cretaceous above a mantle plume/hotspot or above a 'slab gap' in subducting 'Proto-Caribbean' crust. While migrating eastwards the resulting plateau collided with a linear volcanic arc at the eastern, east-facing subduction margin of the plate, blocking and reversing subduction polarity. On entering between North and South America, the leading edge arc collided with Yucatán and Colombia, subducting pieces of continent to great depths (70–80 km), where they suffered HP/LT metamorphism. Volcanic activity ceased during Eocene to Oligocene oblique and diachronous arc collision with the Florida–Bahamas platform and northern South America. Slab roll-back, in two different directions, opened the Yucatán Basin south of the Cuban segment of the arc. The subducted and metamorphosed fragments of Yucatán and Colombia resurfaced in Cuba and along northern Venezuela. Cuba and the Yucatán Basin detached from the Caribbean Plate and joined North America as the Caribbean Plate boundary jumped south to the Cayman Trough. Spreading in the trough accompanied some 1100 km of Caribbean Plate eastward movement. The remaining plate continued eastwards, generating northern uplifts/foreland basins along northern South America and fragmenting and dispersing the Greater Antilles. Cenozoic Grenada Basin inter- or back-arc spreading separated the Aves Ridge from the Lesser Antilles, the active remains of the arc (Bouysse 1988).

This history invokes rotation of the large continental blocks of Yucatán (up to 135° counter clockwise or 100° clockwise) from the Gulf of Mexico, and Chortís (up to 180 counter clockwise or 80° clockwise; e.g. Freeland & Dietz 1972; Pindell *et al.* 2001; Rogers *et al.* 2007*a*). The latter followed the Caribbean Plate into place and accreted to its north-western corner.

According to these models the Caribbean Plate comprises oceanic crust surrounded by volcanic arc rocks, with just one continental component, Chortís. Other models also see the Caribbean Plate as largely oceanic, but formed between the Americas (Meschede 1998; Giunta & Oliveri 2009). Iturralde-Vinent & García-Casco (2007) recognize the element 'Caribeana', 'a thick sedimentary prism represented by a submarine promontory extended eastward into the Proto-Caribbean realm from the Maya Block, somehow as a southern counterpart of the Bahamas'.

Absence of oceanic magnetic anomalies

Middle America carries no identified ocean fractures or magnetic anomalies apart from the centre of the Cayman Trough, a large pull-apart within the sinistral, northern Caribbean Plate boundary. Here, there are two main sets of magnetic anomalies. Central, slow-spreading anomalies record 300 km of extension since the Early Miocene. More distal anomalies vary greatly in shape and have low amplitudes. They are 'hardly recognizable'; identification is problematic and based on models (Leroy *et al.* 2000).

On the Caribbean Plate itself, Donnelly (1973*a*) identified linear magnetic anomalies over thick crust (the Caribbean 'oceanic plateau') in the western Venezuela Basin where Edgar *et al.* (1973) described a corresponding structural grain of buried scarps and seismic isopachs. Diebold *et al.* (1981) combined these in the same map. Ghosh *et al.* (1984) attributed the magnetic anomalies to Early Cretaceous spreading but Donnelly (1989) and Diebold *et al.* (1999) emphasized that they correspond to structure. Age estimations based upon supposed spreading magnetic anomalies have also been proposed for the Colombian

Basin (Late Cretaceous, Christofferson 1976), the western Yucatán Basin (Maastrichtian–Paleocene or Late Paleocene–Middle Eocene, Rosencrantz 1990) and the Grenada Basin (Early Cenozoic, Bird *et al.* 1999). Again, these are areas of thick crust and magnetic grain probably reflects structure rather than ocean spreading.

Tectonic setting

Several interpretations of magnetic anomalies and fracture patterns in the Atlantic show NW–SE separation of North and South America until the Eocene, followed by 250–700 km of NE–SW or north–south convergence (Pindell & Barrett 1990, Pindell *et al.* 1998, 2006; Müller *et al.* 1999). They imply significant and abrupt changes in the drift behaviour of North and/or South America. However, there is no indication of convergence

where the plates meet in the Atlantic at 15°N, east of the Lesser Antilles, and post Middle Eocene subsidence, with Middle Eocene carbonates now lying kilometres deep, argues for continued extension.

Ball *et al.* (1969) pointed out that, since the west-convex spreading ridge off NW Africa is moving away from the continent, it has to be increasing in length. Satellite-derived bathymetry details an east–west zone of north–south extension across the Atlantic between West Africa and the Caribbean (Figs 2 & 3, the Vema Wedge; James *et al.* 1998; see also Funnell & Smith 1968). The wedge began to form in the Albian when South America commenced westward drift. North America was drifting N60°W at that time, and gradually assumed N10°W movement (James 2003*a*, fig. 4; James 2003*b*, figs 3a & b). Fractures within the wedge show that the Central Atlantic Ridge continues to extend over Caribbean latitudes.

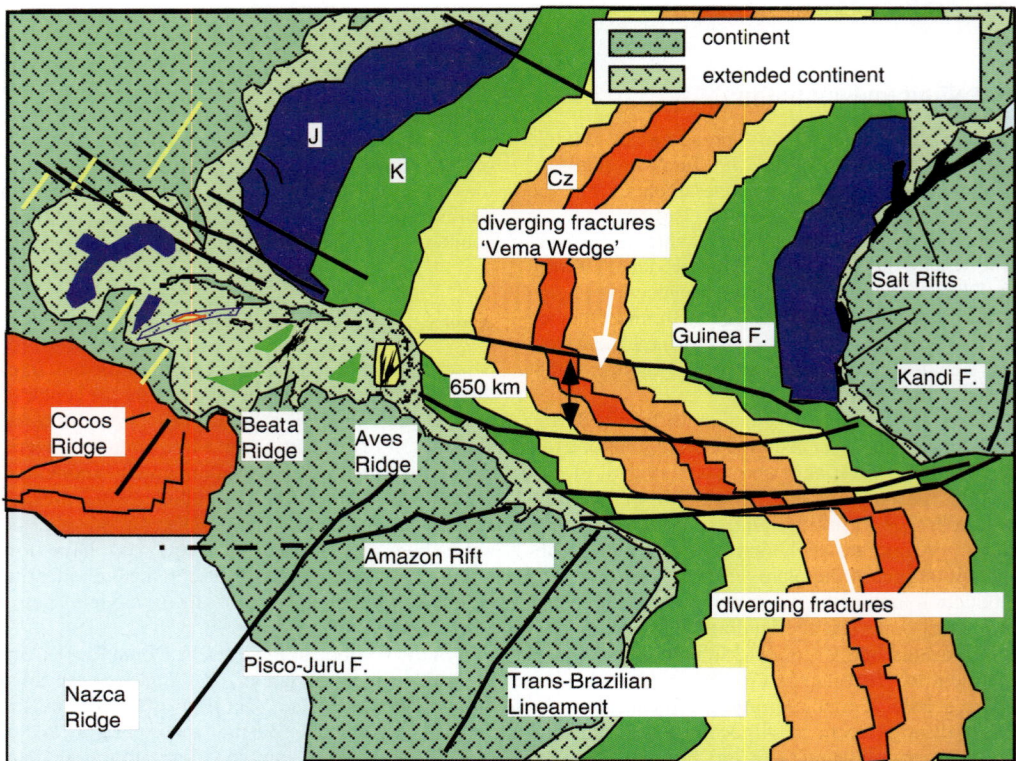

Fig. 2. The Caribbean lies west of a zone of diverging ocean fractures (the Vema Wedge), just as the Amazon Rift lies west of diverging fractures further south. These, and highlighted ocean fractures and intracontinental faults, show continued (Cretaceous–Cenozoic) divergence between North and South Americas, focused on Caribbean latitudes (James 2003*a*, fig. 4). This figure also highlights regional NE-trending faults crossing continents and extending along ocean ridges. In North America, displacements revealed by offsets (of yellow lines) in the Appalachian–Ouachita Palaeozoic suture as it enters the highly extended Gulf of Mexico–Caribbean region can be used to restore Middle America, the Caribbean included, which this map suggests is extended continental crust. This study after Szatmari (1983), Müller *et al.* (1997), Fairhead & Wilson (2005), Davison (2005).

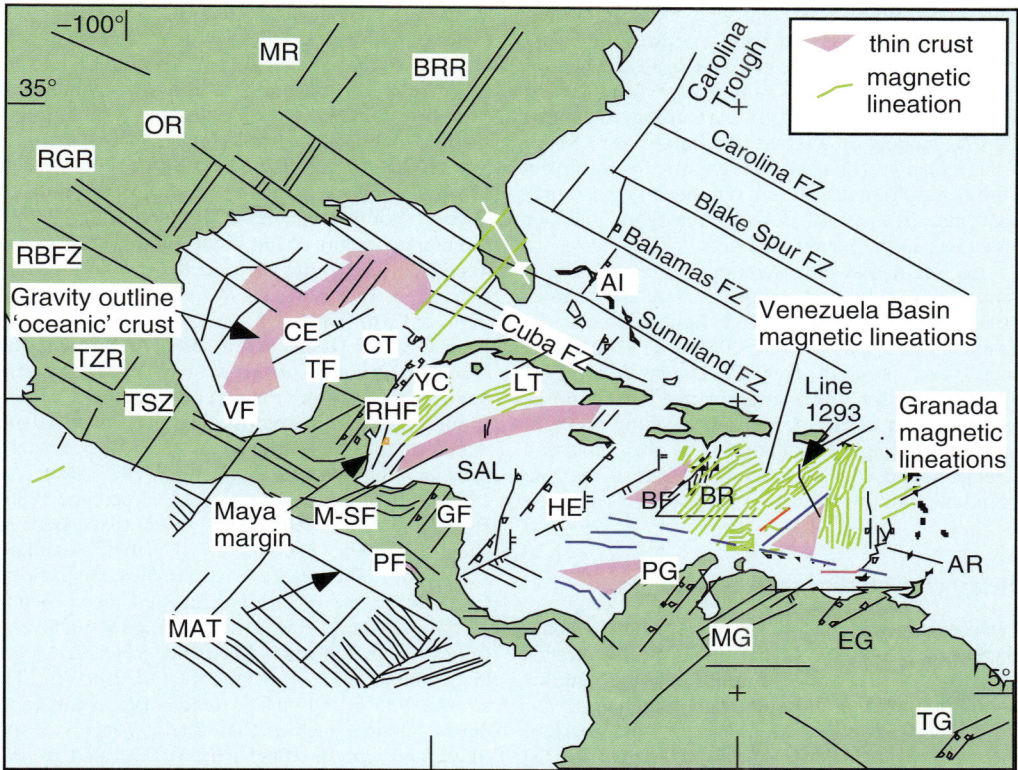

Fig. 3. Tectonic fabric of Middle America. AI, Aklins–Inagua–Caicos; BF, Beata F; BR, Beata Ridge; BRR, Blue Ridge Rift; CE, Campeche Escarpment; CT, Catoche Tongue; EG, Espino Graben; GF, Guayape FM; HE, Hess Escarpment; LT, La Trocha FM; MG, Mérida Graben; MiG, Mississippi Graben; M-SF, Motagua–Swan FM; OR, Oachita Rift; PF, Patuca FM; PG, Perijá Graben–Urdaneta; RBFZ, Río Bravo fault zone; RGR, Río Grande Rift; RHF, Río Hondo FM; SAL, San Andrés Lineament; TF, Ticul FM; TG, Takutu Graben; TT, Texas Transform; TSZ, Tenochtitlan Shear Zone; TZR, Tepic–Zacoalco Rift; VF, Veracruz FM; YC, Yucatán Channel. Red line, SE limit of Caribbean Plateau, the Central Venezuela FZ. Green lines, magnetic anomalies (Gough 1967; Hall & Yeung 1980; Ghosh *et al.* 1984). Compiled from many sources including the Exxon world geological map, IFP map of Caribbean Geology (Mascle *et al.* 1990) and GSA Geological Map of North America (Reed *et al.* 2005).

Models that derive the Caribbean Plate from the Pacific propose that it overrode 'Proto-Caribbean' crust between North and South America. Tomographic data are quoted in support. Tomography indicates an anomaly descending to 600 km below the southern Lesser Antilles (Van der Hilst 1990, who qualified his results as preliminary, tentative and a working hypothesis). Since the Wadati–Benioff zone here descends at 60° (vertical south of Grenada), only around 700 km of crust would have been subducted below some 340 km of eastern Caribbean crust. At the present rate of relative eastward Caribbean movement of 2 cm/annum that corresponds to the 35 Ma Oligocene commencement of Central Cayman Trough opening and the approximate 300 km of subsequent strike–slip along northern and southern plate boundaries.

The data support neither tomographic suggestion of >1500 km of subducted crust below the Caribbean nor overriding of more than 2500 km of 'Proto-Caribbean' crust by a migrating plate (Pindell *et al.* 2006, fig. 6).

If spreading symmetry is preserved in Caribbean latitudes, absence of Jurassic crust along West Africa south of the Guinea Fault indicates little if any Jurassic crust below the Caribbean. However, some do see Jurassic crust SE of a boundary that passes NW below St Vincent (Maury *et al.* 1990).

Magnetic anomaly data in the equatorial Atlantic remain poorly known (a blank area on the magnetic anomaly map of Cande *et al.* 1989), especially in the area of the Vema Wedge. Rift/drift transition in the Central Atlantic is poorly constrained (limited/contested magnetic data, lack of drilling calibration; Withjack & Schlische 2005). Published flow paths for North and South America are based upon fracture and magnetic data from the central North and Southern Atlantic regions (e.g. Pindell 1991; Pindell *et al.* 2006). According to Eagles (2007), models of relative motion between South America and Africa fail to recognize large offsets within South America and significantly misrepresent the azimuth of Lower Cretaceous seafloor spreading in the South Atlantic.

An alternative interpretation of fracture orientations and published magnetic anomalies shows continued but declining separation of North and South America (James 2002, 2003*b*). The separation locus was focused over Caribbean and Vema Wedge latitudes, where considerable but diminishing extension occurred. Volcanism along northern and southern plate boundaries in the Jurassic–Cretaceous and its decline in the Early Cenozoic correlate with this history.

Regional tectonic fabric

Middle America manifests a regional tectonic pattern of N60°W, N35°E and N60°E trends (Fig. 3). They occur on Caribbean Plate margins and interior and on its continental neighbours.

N60°W trending lineaments of the western Atlantic and the Middle American Trench (Fig. 3) bracket Middle America. Major faults ('mega-lineaments') continue the trend into southern North America where they offset the NE-trending Palaeozoic Appalachian suture of eastern North America (Harry *et al.* 2003, fig. 1). The trend appears briefly on the Yucatán Peninsula as the Ticul F. Projected NW this perhaps continues as the Río Bravo Fault Zone of SW Texas (Flotté *et al.* 2008). Thick crustal blocks in the Colombian Basin follow this trend (Bowland 1993). Arches in Venezuela (El Baúl, Arauca, Mérida and Vaupes) and the Marañon Basin of Colombia continue the NW trend into NW South America. They parallel Proterozoic structures of the craton (Krüger *et al.* 2002, fig. 11).

N60°E trending faults are seen as the Hess Escarpment, basement structure of the eastern Yucatán Basin, the La Trocha Fault, the Swan–Eastern Motagua Fault, the NW Campeche Escarpment and the east central Gulf of Mexico. The eastern side of the Espino Graben, eastern Venezuela and parts of the Takutu Graben, magnetic lineations in the Grenada and Venezuela Basin, also exhibit this trend.

The N35°E structural trends are followed by many elements in Middle America. They appear as normal faults in the northwestern Gulf of Mexico and the Tenochtitlan Shear Zone of Mexico.

Magnetic trends and the structural grain of the southeastern Gulf of Mexico link the Catoche Tongue, a Jurassic (?Triassic) graben, of the Campeche Platform to Florida, where drilling encountered extensional rhyolites and basalts of a possible Triassic rift (Gough 1967; Heatherington & Mueller 1991; Marton & Buffler 1999; Phair & Buffler 1983; Shaub 1983). Basement structure of the deep Gulf of Mexico exhibits the trend (Roberts *et al.* 2005). Structure contours of salt in the Isthmian Salt Basin trend NE towards the NE-oriented Comalcalco and Macuspana Basins of the SW Gulf of Mexico (Contreras & Castillón 1968).

The Maya Mountains of southern Yucatán, the basement structure of the western Yucatán Basin, fault blocks at the distal ends of the Cayman Trough and faulting across Grand Cayman Island manifest the same trend, as does the western margin of the Beata Ridge (Bateson 1972; Donnelly 1973*b*; Edgar *et al.* 1973; Case & Holcombe 1980; Holcombe *et al.* 1990; Leroy *et al.* 1996; Bain & Hamilton 1999; Diebold *et al.* 1999; Andreani *et al.* 2008). The trend crosses Jamaica, Hispaniola, Puerto Rico, the Aklins and Inagua-Caicos islands of the Bahamas, Barbados and La Désirade and follows the Grenada–Mustique volcanic ridge (Trechmann 1937; Westercamp *et al.* 1985, fig. 1).

The N35°E and N60°E trends often combine to define features such as the eastern margins of the Maya and Chortís Blocks (best illustrated by the Exxon World Geological Map) and the Espino and Takuto grabens in northern South America.

The regional integrity of these structures in Middle America indicates a tectonic fabric and history shared with its continental neighbours. It does not support rotation of major blocks such as Maya and Chortís (see below).

Structural periodicity along northern and southern margins of the Caribbean Plate

Along northern South America there is a large-scale (*c.* 350 km), periodic repetition of extensional basins (see numbers on Fig. 1): (1) Lower Magdalena Valley–Río Magdalena mouth; (2) Maracaibo Basin–Gulf of Venezuela; (3) Guarumen Basin (overthrust and hidden but revealed by seismic and drilling)–Golfo Triste; (4) El Hatillo–Cariaco; and (5) Gulf of Paria–Carupano. The periodicity is repeated at the same scale along the Greater Antilles from west to east: (6) Windward Passage–Gonave Basin; (7) Asua Basin; (8) Puerto Rico–Hispaniola Mona Passage; (9) Anegada Passage. It appears also in Cuba, from west to east: (10) Pinar fault–Los Palacios Basin; (11) Llabre Lineament; (12) La Trocha Fault–Central Basin; (13) Camagüey Lineament–Vertientes

Basin; (14) Oriente Fault–San Luis–Guantánamo Basin. The periodicity suggests a shared geology of a common, reworked basement pattern (Stainforth 1969; Mora *et al.* 1993; James 2000).

Origin of tectonic fabric

Rifting progressed southward in North America and Europe/North Africa in the Early Triassic and affected Middle America by the Late Triassic (Ager 1986; Davison 2005). At 200 ± 4 Ma the Central Atlantic Magmatic Province (CAMP) formed, heralding Pliensbachian–Toarcian (190–180 Ma) Pangaean break-up (Marzoli *et al.* 1999; McHone 2002, fig. 1 or http://www.mantle-plumes.org/CAMP.html, fig. 2). Large volumes of tholeiitic magma intruded Triassic rocks as sills and later extruded as continental flood basalt lavas (Marzoli *et al.* 1999; McHone 2008). The CAMP event extended along eastern North America/western Africa and as far south as northern Brazil (Mohriak *et al.* 2008), but its focus on Middle American latitudes indicates greatest extension there. That this remained the case until the Early Cenozoic.

Rifting followed Caledonian/Hercynian tectonic fabric in North America and West Africa (Davison 2005; Tommasi & Vauchez 2001, fig. 1). Late Triassic–Early Jurassic rifts accommodated

2000 m or more of non-marine beds in NE-trending basins between continental and thick transitional crust in the Gulf of Mexico and along the Eastern Continental Margin of Mexico (Antoine & Bryant 1969; Horbury *et al.* 2003).

The Central Atlantic began to open in the Jurassic, as North America drifted NW from Gondwana. Rift/drift transition commenced in the south in the Late Triassic and progressed northwards to the NE by the Early Jurassic (Withjack & Schlische 2005). Opening of the South Atlantic began with Early Cretaceous rifting in the south, propagating northwards so that westward drift of South America at Caribbean latitudes occurred around 100 Ma (Albian; Eagles 2007). Consequently, Jurassic crust is present in the Central Atlantic, but not in the Equatorial Atlantic, and Central Atlantic Cretaceous crust is wider (Fig. 2). This map (Fig. 2) is simplified: ridge jumps occurred in the Jurassic along eastern North America, stranding the earliest crust instead of sharing it with Africa (Davison 2005; Bird *et al.* 2008).

Fractures in the western Atlantic show that the drift path of North America from South America at this time was N60°W (Figs 2–4). Regional N60°E trends fit extensional strain in this sinistral offset. East–west trends, such as the northern Caribbean Plate boundary, are synthetic to the system (James 2009, fig. 2). The N35°E grain reflects Palaeozoic sutures reactivated during

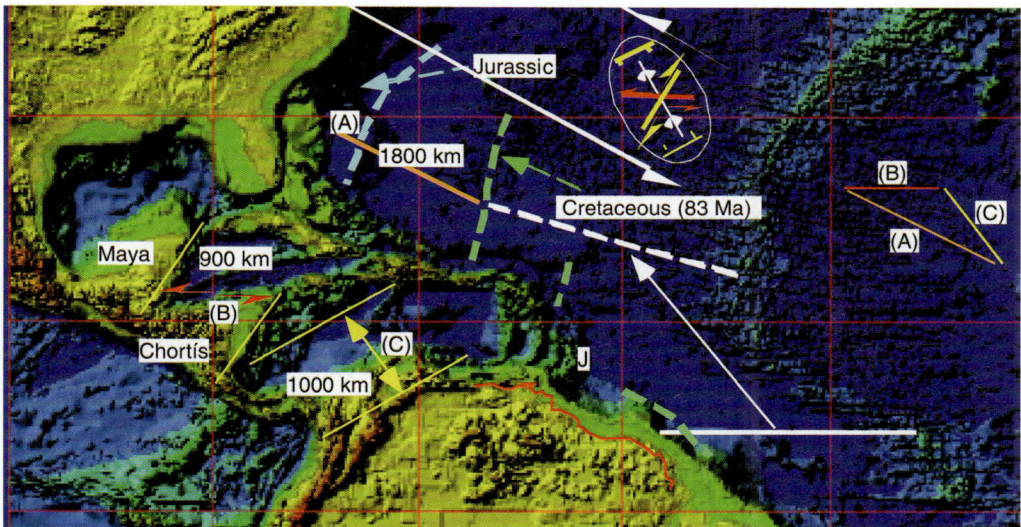

Fig. 4. Continent margin–Mid Atlantic Ridge distance is *c.* 1800 km greater in the Central Atlantic than in the Equatorial Atlantic (broken white line reproduces the solid white line; red line indicates additional distance). The difference relates to Jurassic crust, not present in the Equatorial Atlantic, and wider Early Cretaceous crust. The distance (**a**) equates to the vector sum of offsets between Maya and Chortís (**b**, early Cayman offset), and Chortís (Hess Escarpment)–South America (**c**). The stress–strain ellipse indicates that N35°E and N60°E faults were dextral antithetic faults and extensional strain generated by sinistral slip along N60°W fractures.

Triassic–Jurassic rifting and Jurassic–Cretaceous drifting as dextral and then, normal faults. Cenozoic convergence of the Cocos Plate with the western Caribbean reactivated the trend as sinistral faults (James 2007*a*).

Rotation of Maya and Chortís, origin of the 'Motagua Orocline'

Most plate reconstructions of Middle America rotate Maya and Chortís by as 80° or more anti-clockwise from the Gulf of Mexico and SW Mexico respectively (e.g. Ross & Scotese 1988; Pindell *et al.* 2005). Others rotate both blocks from the Gulf (Freeland & Dietz 1972).

Restoration of Chortís against SW Mexico is denied by geology. Two features of the Xolapa Complex of southern Mexico – southward thickening Jurassic–Cretaceous sedimentary rocks and southward increase in Early Cretaceous (*c.* 132 Ma) HT/LP metamorphism – are not seen in Chortís (Ortega-Gutiérrez & Elias-Herrera 2003; Keppie & Morán-Zenteno 2005). Instead the non-metamorphosed El Plan Fm and Valle de Angeles Gp. record the Cretaceous (Horne *et al.* 1990). None of the north–south terrane boundaries of southern Mexico, truncated at the coast, is seen in the Chortís Block.

Jurassic sections on Maya and Chortis develop eastwards from continental to marginal to marine facies and the crust thins across the Río Hondo and Guayape faults (Mills *et al.* 1967; López-Ramos 1975; Mascle *et al.* 1990; Rogers & Mann 2007). These N35°E trending Jurassic faults remain parallel to the regional trend of Triassic–Jurassic rifts and show that the blocks have not rotated relative to their large continental neighbours. The N60°W Ticul Fault of Maya, parallel to bounding fractures in Middle America, provides a second, fixing coordinate for this block. An unnamed N60°E fault crossing Maya (Mascle *et al.* 1990, map) and the parallel NW Campeche Escarpment provide a third. When the Caribbean faulted margins of Maya (Yucatán Basin margin; Baie 1970) and Chortís (Providencia–San Andres trough; Holcombe *et al.* 1990) are aligned, the Río Hondo and Guayape faults also line up, suggesting a formerly continuous, major Jurassic graben (James 2007*a*). This restoration indicates *c.* 900 km of sinistral offset along the Motagua Fault Zone (MFZ).

Maya and Chortís meet along the MFZ, bounded north and south by the Polochic and Jocotán faults. Ortega-Guttiérez *et al.* (2007) emphasize that this is an important zone of fault-bounded, often rootless, crystalline Jurassic–Early Cretaceous terranes and Palaeozoic rocks. The Jurassic cover common on Chortís and Maya is absent.

The Maastrichtian El Tambor Gp of the Motagua Valley contains rocks similar to those dredged from walls of the Cayman Trough (and denies Cenozoic trough opening). There are large masses (up to 80 km long) of serpentinites, wacke, phyllites and schists, and Valanginian–Aptian and Cenomanian fossils occur in basalt interbeds (Donnelly *et al.* 1990). Serpentinites mica dates (Ar–Ar) are 77–65 Ma (Campanian–Maastrichtian) to the north and 125–113 Ma (Barremian–Aptian) to the south of the Motagua F (Harlow *et al.* 2004).

The MFZ curves eastwards from NW to NE. Mountains of southern Maya also follow this tend, forming a 'Motagua orocline'. Major sinistral movement along NE trending faults such as the Guayape, Río Hondo and eastern boundary faults of Maya began in the Palaeogene, causing inversion and oroclinal bending of NW trending Jurassic and Cretaceous depocentres (Rogers *et al.* 2007*a*; James 2007*a*). Sinistral movement transforms into contraction, curving west in the south and east in the north (James 2007*a*, figs 11 & 12). The strike–slip/orocline systems are 'closed' – deformation does not affect areas further north or south. Northward convex western Cuba mirrors the southward convex Motagua zone. Both restore to linear features when some 350 km of sinistral movement is removed. The oroclines and faults affect Middle Eocene rocks. Movement began in the Late Eocene, coeval with the beginning of transpressional uplift along the Eastern Cordillera–Mérida Andes of NW South America, where similar strike–slip transformation to compression is common (James 2000, figs 4 & 7). Movement along these systems continues today.

Displacement along northern and southern plate boundaries; two-phase opening of the Cayman Trough

Faster westward movement of the Atlantic Plate relative to the Caribbean results today in sinistral and dextral strike–slip along the northern and southern plate boundaries at around 2 cm/year (GPS data, DeMets *et al.* 2000; Weber *et al.* 2001). Estimates of maximum offset along the major faults of northern South America sum to 565 km (James 2000). Measured offset along faults of the Motagua fault zone (Motagua, Polochic, Jocotán–Chamelecon) sum to a maximum of 655 km (Gordon & Ave Lallemant 1995). There is no measured fault record of the >2500 km translation required by Pacific models of a migrating Caribbean Plate.

Northern plate boundary offset is generally assumed to be calibrated by the 1200 km long Cayman Trough. This is pivotal to Pacific models of Caribbean Plate history: 'If smaller estimates of offset are assumed, an inter-American formation of the Caribbean Plate is required' (Pindell & Barrett 1990).

The Cayman Trough is a pull-apart within the sinistral, northern Caribbean Plate boundary. The only recorded Middle American spreading centre, with Early Miocene to Recent, north–south ridges and magnetic anomalies, lies in its central 300 km (Leroy *et al.* 2000). Gradual increase in apparent crustal thickness and shallowing of basement indicate that the (undated) distal parts of Cayman Trough are underlain serpentinized upper mantle/ highly attenuated crust (Ten Brink *et al.* 2003). 'Ocean'/continent transition is seen on seismic data from the distal eastern part of the trough (Leroy *et al.* 1996), where NE structural trends indicate Jurassic or reworked Jurassic structures.

North–south trending extension parallels grabens in Chortís and offsets in the Hess Escarpment. It fits the extensional strain expected of major movement along N35°E trending faults, sinistrally active since the Late Eocene, that dominate the western Caribbean (James 2007*a*). Coeval pull-apart extension dispersing the Greater Antilles and the Aruba-Blanquilla islands along the northern and southern strike–slip plate boundaries also sums to around 300 km. The N60°E trend of the eastern part of the Motagua Fault and Swan Fault as far as Swan Island (Fig. 3; Pinet 1971, fig. 2) parallels the extensional strain expected in regional N60°W sinistral slip (James 2009, fig. 2).

Literature commonly quotes a 45 Ma commencement of Cayman opening (depth-to-basement and heat flow studies, Rosencrantz *et al.* 1988). However, the continent margin–Mid Atlantic Ridge distance is *c.* 1800 km greater in the Central Atlantic than in the Equatorial Atlantic. The difference relates to presence of Jurassic crust, absent from the Equatorial Atlantic, and wider Cretaceous crust in the Central Atlantic, north of the Caribbean northern boundary (Fig. 2). The difference is matched by the vector sum of offsets between the Maya and Chortís Blocks and of the Hess Escarpment from the Mérida Andes trend (Fig. 4). These offsets respectfully fit the combined extension/ synthetic and extension within N60°W sinistral offset of North from South America.

These data indicate that *c.* 900 km of Cayman Trough offset occurred during Late Jurassic–Early Cretaceous (James 2006). Lack of disturbance in Upper Cretaceous–Recent sediments in the Gulf of Tehuantepec, west of the Motagua zone (Sanchez-Barreda 1981) supports this early offset, which relates only to the northern plate boundary. The Miocene–Recent 300 km central Cayman spreading and strike–slip along the northern and southern Caribbean boundaries carries total Cayman offset to 1200 km. East–west sinistral faulting and folding and thrusting in southern Maya offset absorb the later offset (Guzman-Speziale & Meneses-Rocha 2000; Guzman-Speziale 2008).

The Cayman Trough boundary continues onshore in Guatemala as the Motagua Fault Zone, regarded as a Late Cretaceous suture between Maya and Chortís (Burke *et al.* 1984; Donnelly *et al.* 1990, Draper 1993). Here, two major fault zones are separated by a horst of Palaeozoic and older metamorphic and granitic rocks. The northern zone, Polochic, contains areas of cataclastic rocks overlain by undeformed Permian and Mesozoic sediments. Thus the structural trend was active in pre-Permian time (Bonis 1969). The Sierra de Santa Cruz of eastern Guatemala consists of serpentinized peridotite, gabbro, sheeted dolerite, pillow basalt and minor pelagic sediment and abundant metamorphosed hemipelagic sediment with overlying volcanic wackes (Donnelly 1985). It was emplaced northwards during Late Cretaceous 'suturing' of Chortís and Maya. However, a well on Turneffe Cay, offshore Belize, drilled mafic volcanic rock overlain by flysch, with the same lithological section as the Sierra de Santa Cruz. Since the two sections lie on opposite sides of the Late Palaeozoic Maya Mountains an alternative to the suture explanation is needed.

Global analogues

Since the Caribbean Plate lacks oceanic fractures and spreading magnetic anomalies, I turn to the Scotia and Banda plates for highly relevant information (James 2005*a*, fig. 10). Both carry spreading ridges and dated magnetic anomalies and are known to have formed in place by eastward-migrating back-arc spreading (Honthaas *et al.* 1998; Barker 2001).

The plates lie between sinistrally offset, major continental blocks to the north and south. Each is around 3000 km long and 700–800 km wide. Each has a curved volcanic arc in the east. Northwest trending volcanic arcs follow continental Sumatra/Java of Banda and Chortís of the Caribbean in the west. The Scotia Plate is bounded in the west by the shallow (700 m), NW-trending Shackleton Fracture Zone, built of continental slivers (Livermore *et al.* 2004) and there are active volcanoes in the Drake Passage (Dalziel 1972). Fractures in the ocean east and west of the plates show divergent spreading directions – the plates are located in zones of extension.

Syn-subduction, backarc crust lies behind east-facing western Pacific arcs, while much older crust occurs in the lower plates (Garzanti *et al.* 2007). Jurassic or Cretaceous oceanic crust is being subducted below the Pacific arcs, the Lesser Antilles and the Scotia Arc. Palaeozoic–Mesozoic continental crust subducts below the Banda Arc (Charlton 2004).

The Scotia Plate formed by backarc extension that distributed continental blocks originally a continuous continental connection between South America and Antarctica (Barker 2001, fig. 3). Magnetic anomalies indicate that spreading was faster than the motion of the South American and Antarctic plates and occurred in different directions, beginning in the south in the Protector and Dove basins (Eagles *et al.* 2005; Ghiglione *et al.* 2008). It then expanded the plate eastward above Atlantic crust roll-back.

Similarities between the Caribbean and Scotia plates are remarkable, not only in the general tectonic framework but also in Mesozoic volcanism, orogenic deformation and batholith emplacement (Dalziel 1972). East–west strike–slip zones, dextral in the north and sinistral in the south bound Scotia and the Caribbean. The West and East Scotia ridges coincide with the Caribbean Beata and Aves ridges. Gravity lows along the northern margins connect with lows along the trenches east of the South Sandwich and Lesser Antilles volcanic arcs. The North Scotia Ridge has transgressed northwards over oceanic crust; the Puerto Rico–Virgin Islands segment of the Greater Antilles thrusts northwards over Atlantic ocean crust. In both the South Sandwich and the Lesser Antilles arcs older oceanic crust is being subducted in the north and suffers an east–west tear.

The North and South Scotia ridges carry continental rocks that correlate with South America and Antarctica. Parts of the Scotia Sea more elevated than normal ocean floor may be continental crust thinned by extension (Barker 2001). The Arctic Peninsula contains Palaeozoic basement and evidence of subduction since the Early Mesozoic or earlier, similar to NW South America (Colombia–Ecuador).

The Neogene Banda Sea also opened in an extensional setting (Hall 1997). Dredged sedimentary and metamorphic rocks show that internal ridges are continental slivers from New Guinea. Oceanic crust forms only a small part of the Flores Sea.

Implications of these plate similarities are:

(1) The Caribbean Plate formed in place by north–south extension in the centre, NW–SE extension in the west (Beata Ridge = West Scotia Ridge?) and then east–west back-arc spreading in the east (Aves Ridge = East Scotia Ridge?).
(2) The eastern Greater Antilles have migrated northwards over Atlantic oceanic crust.
(3) The Lesser Antilles are migrating eastwards over Atlantic crust.
(4) The Caribbean Plate is rimmed by continental fragments and carries internal, extended continental crust.

Continental crust in the Caribbean

Is there continental crust on the Caribbean Plate? Several, independent lines of data suggest there is.

Crustal thicknesses

Crustal thicknesses (Fig. 5) decline progressively from continental (45–50 km) on cratonal North and South America to anomalously thin (*c.* 3 km) in the SE Venezuela Basin, the centre of the Cayman Trough and the deep Gulf of Mexico. Thicknesses of around 20–30 km characterize areas of known extended continental crust such as the Yucatán Peninsula, Cuba, the Chortís Block and the proximal Bahamas Plateau. Similar thicknesses are common on Caribbean Plate margins. Relatively thick crust (8–20 km) occurs in the Colombian, Venezuelan and Yucatán and the Tobago Trough. The data suggest increasing extension towards the deep Gulf of Mexico and the Caribbean area.

Gravity data

The gravity map of Westbrook (1990) shows steep gradient positive anomalies over the active Caribbean Plate margins of the Greater, Lesser and Leeward Antilles–Paraguaná–Guajira–Santa Marta, along the eastern margin of the Maya Block, the SW and NE margins of the Cayman Ridge, over northern Panamá and part of western Central America. On the Caribbean Plate interior positive anomalies occur over the Aves Ridge, the western/southern parts of the Beata Ridge and the Lower Nicaragua Rise. Positive anomalies also occur over the central part of the Cayman Trough. These anomalies correspond to uplifted 'oceanic' and volcanic arc rocks, major fault margins and areas of high extension.

Steep gradient negative anomalies characterize the trenches of the Lesser Antilles, the Puerto Rico and Muertos troughs north and south of Puerto Rico, Central America and foreland basins and inverted rifts of northern South America. Elsewhere, rather featureless, neutral gravity anomalies characterize continental South America, the Maya and Chortís Blocks, the Upper Nicaragua Rise and the interiors of the Colombian, Grenada, Venezuelan and Yucatán basins.

Gravity data witness the dynamic nature of Caribbean Plate boundaries (data in Bowin 1976). The world's largest negative sea-level Bouguer anomaly (−200 mgal) corresponds to the Eastern Venezuela Maturín Basin, which lies south of the southward-moving Interior Ranges (a root whose mountain is on the way). A large negative anomaly (−150 mgal) in the southeastern Maracaibo Basin

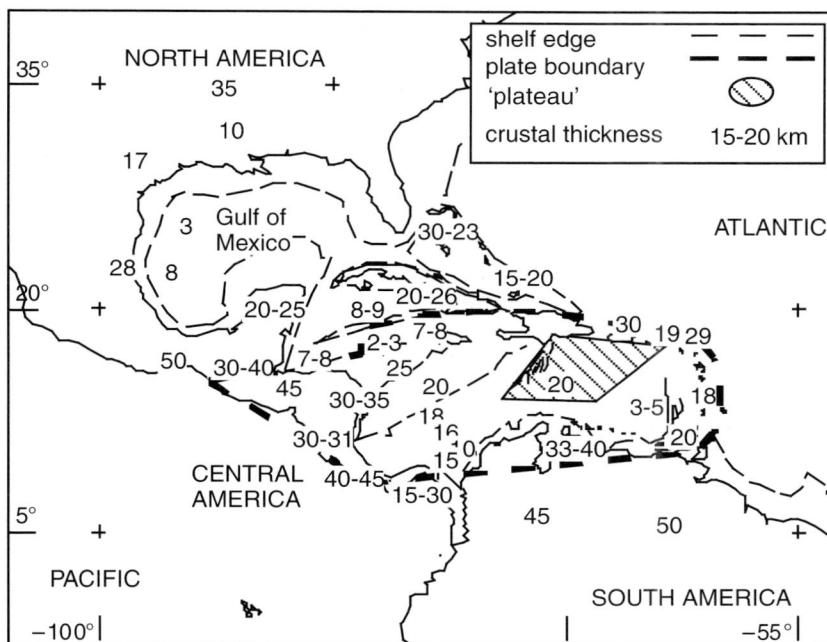

Fig. 5. Crustal thicknesses in Middle America. Compiled from many sources.

is half overthrust by the Mérida Andes (a root receiving its mountain). A large positive anomaly (+210 mgal) characterizes the 5800 m high Sierra Nevada de Santa Marta, NW Colombia (a mountain without a root). Positive anomalies (up to +222 mgal in the Blue Mountains and over the Cretaceous central inlier) show that most of Jamaica is under-compensated. Cayman Trough gravity lows lie on the northern margin in the east (extension of the Puerto Rico Trough) and on the southern margin in the west, close to the active Oriente and Swan strike–slip faults.

There is little indication of vigorous igneous/ tectonic activity in the basins of Middle America, no indication of large igneous provinces and no indication of shallow Moho. Maximum Bouguer anomaly values are lower over the Caribbean than over the neighbouring Atlantic. The Caribbean Sea is not as deep and the combined mass of the crust and the uppermost mantle is less, indicating greater depths to mantle and/or lower densities (Bowin 1976). These data are consistent with the possibility that the area is built of extended continental crust.

Northern Central America–Chortís

Chortís is the only recognized continental part of the Caribbean Plate (Schuchert 1935; Dengo 1975). Grenvillian and Palaeozoic crust along with thin Jurassic red beds and Cretaceous carbonates and clastic rocks crop out in the north (Manton 1996). Jurassic–Upper Cretaceous rocks occur in the east. Jurassic red beds thicken from tens of metres to more than 2000 m and crustal thicknesses decrease SE of N35°E-trending faults on Chortís (Maya also), recording continental margin extension.

Chortís has been derived from various locations in the Gulf of Mexico or alongside Colombia or SW Mexico. The most popular model relates Chortís to SW Mexico but, despite attempts to seek geological continuity between the two areas (e.g. Pindell 1993; Rogers et al. 2007b), it does not exist (Keppie & Morán-Zenteno 2005). More-over, this understanding moves Chortís south-eastward to join the rear of the NE-moving Caribbean Plate in the Eocene by an unexplained process (e.g. Mann 2007).

The restoration of Chortís suggested by this paper brings the older crustal rocks into contact with the southern Mexico Oaxaca, Acatlán and Xolapa complexes. All have Grenvillian or inherited Grenvillian components (U–Pb, Sm–Nd data, Nelson et al. 1997). The Precambrian granulite–facies Oaxacan Complex of southern Mexico are reported from Chortís as c. 1 Ga amphibolite gneis-ses and a Permo-Carboniferous tectonomagmatic event seen in northern Honduras may correlate with a similar event in the Acatlán Complex of southern Mexico (Manton 1996; Keppie *et al.* 2003).

The 20–25 km thick Siuna complex of eastern Chortís is a mélange of blocks of gabbros, peridotites, greenstones, greenschists, metamafics, schists, metacherts, detrital quartzites, radiolarian cherts, black shales and radiolarites bearing Middle and Late Jurassic radiolaria in a serpentinite matrix (Flores *et al.* 2006). Conglomerates with abundant quartz and fragments of schists and quartzite indicate a nearby continental source (Venable 1993). The mélange is followed unconformably by thin-bedded calcareous hemipelagites containing Aptian/Albian planktonic foraminifera (Flores *et al.* 2006). This continues up section into thick-bedded, shallow-water limestones that correlate with the Aptian/Albian Atima Formation of Chortís. These data are incompatible with formation of the complex in ?Jurassic–Early Cretaceous time in a Pacific location and accretion in the Late Cretaceous at the leading edge of a migrating plate (e.g. Rogers *et al.* 2007a, fig. 9A). The Siuna complex is related to Chortís as its eastern rift/drift margin.

Southern Central America–Chorotega and Chocó

Southern Central America exposes only troughs of marine Cenozoic deposits with volcanic and plutonic igneous rocks above Mesozoic oceanic rocks and is thought to have intra-oceanic origins (Dengo 1985; Escalante 1990).

Crust thickens from 30–31 km below Nicaragua, on the Chortís Block, to 40–45 km below Costa Rica (Chorotega Block) where gravity data indicate continental crust (Case 1974; Case *et al.* 1990; Auger *et al.* 2004). Seismic velocities over the Cocos Plate and Costa Rica show transition from oceanic to continental crust, with continental Moho at around 40 km depth (Sallares *et al.* 2001).

The Costa Rican volcano Arenal produces granulite xenoliths and micaschists and amphibolites occur in the Talamanca Range and on the Osa and Azuero peninsulas (Tournon *et al.* 1989; Krawinkel & Seyfried 1994; Sachs & Alvaredo 1996). Geochemistry of high silica ignimbrite and granitoid rocks of Costa Rica is very similar to continental rocks and to ignimbrites in Guatemala where Palaeozoic crust occurs (Deering *et al.* 2004; Vogel *et al.* 2004, 2007).

Albian (Iturralde-Vinent 2004) and Miocene (Escalante 1990) quartz sands are present in Costa Rica and the ?Albian–Santonian volcanic arc section of Santa Elena probably contains continentally derived sediments (Iturralde-Vinent 2004). The southern extension of southern Central America (Choco Block) is accreted to western Colombia as the Serranía de Baudo. Westward-coarsening quartz sands in Late Cretaceous turbidites of the inboard Colombian Cordillera Occidental indicate continental basement for the Serranía (Bourgois *et al.* 1987).

The data indicate that the area is underlain by continental crust.

Nicaragua Rise, Cayman Ridge

The submarine Nicaragua Rise extends Chortís eastwards to Jamaica. The crustal thickness of Jamaica is around 20 km. Palaeozoic rocks are unknown but drilled platform carbonate rocks suggest continental foundations. Arden (1975) suggested that the Rise rifted away from the Beata Ridge. Parallel to the Rise, on the opposite side of the Cayman Trough, the Cayman Ridge extends to southeast Cuba.

For many authors the Rise and Ridge are built of volcanic arc rocks (e.g. Holcombe *et al.* 1990). Others note that Northern Nicaragua Rise crust is up to 25 km thick and think it is continental, possibly with Palaeozoic basement (Holcombe *et al.* 1990; Muñoz *et al.* 1997). Late Precambrian to Early Palaeozoic metaigneous rocks (equivalent to the Grenville or younger gneisses of the Oaxaca or Acatlán complex of southern Mexico) form a ridge along the southern margin of the western Cayman Trough (Donnelly *et al.* 1990).

Exploration wells on the Nicaragua Rise encountered andesite, granodiorite and metasedimentary rocks as far east as Rosalind Bank (Dengo 1975). However, Upper Cretaceous–Paleocene granitoid rocks of the northern Rise, chemically similar to granitoids from Jamaica, Haiti and the Sierra Maestra, southern Cuba, show affinity with mature oceanic arc rocks uncontaminated by continental material (Lewis *et al.* 2008).

According to Pacific models Chortís and the Upper Nicaragua Rise were obducted southeastwards from SW Mexico onto the trailing edge of the Caribbean Plate as it entered between the Americas. However, seismic data show northwest vergence of folds and thrusts on Chortís and the adjacent Rise, where continental Jurassic–Cretaceous geology of the N35°E Colón Mountains, Honduras, continues into the offshore, but trending N60°E (Rogers *et al.* 2007b, fig. 2).

The Miskito Basin, on the Nicaragua Rise east of Nicaragua, is bounded to the north and south by the parallel, N60°E Pedro Fracture and Hess Escarpment. It comprises intermediate, stretched crust and southward decreasing thermal gradient indicates transition to oceanic crust (Muñoz *et al.* 1997). This, Lower Nicaragua Rise, is shown as a distinct and unexplained element lying north of the Caribbean Plateau in some models, suturing to the Upper Rise (along the Pedro F.) in the Campanian (e.g. Mann 2007, fig. 4; Pindell & Kennan 2009, fig. 10). Undisturbed, upper Cretaceous sediments lie next to the Hess Escarpment, which therefore is older (Edgar et al. 1973). Both faults represent extensional strain of N60°W sinistral movement of North America (James 2009, fig. 2).

The Cayman Ridge extends to southeast Cuba. Built of tilted fault blocks, it has near-continental crustal thickness and low magnetic susceptibility in the west, similar to rift blocks of the margin of British Honduras (Dillon & Vedder 1973; Rosencrantz 1990). Acoustic basement is interpreted to be continental (Malin & Dillon 1973; Dillon & Vedder 1973). Paleocene continental granitoids (U–Pb and Ar/Ar 62–64 Ma) occur on the walls of the Cayman Trough (Lewis et al. 2005) along with red beds, greywackes and arkose (Perfit & Heezen 1978).

Cuba

The well-exposed Cuban geology of oceanic and volcanic arc rocks emplaced upon continental basement by northward thrusting mirrors geology of northern South America. Together, these areas are key to understanding the segmented northern Caribbean boundary east of Cuba.

Late Proterozoic–Early Palaeozoic continental basement of the Florida Platform extends through the eastern Bahamas and northern Cuba (Meyerhoff & Hatten 1974; Pardo 1975). DSDP sites 537 and 538, off northwestern Cuba, penetrated Cambrian–Ordovician phyllites, gneisses, amphibolites, intruded by Early–Middle Jurassic diabase dykes, correlative with rift volcanism in the North Atlantic, covered by lower Cretaceous sediments (Schlager et al. 1984). Northwest–southeast gravity and magnetic trends of western and southwest Florida continue uninterrupted across the Bahamas to Cuba (Meyerhoff & Hatten 1974) and to Yucatán.

Rigassi-Studer (1961) and Hatten (1967), quoting radiogenic ages determined by the Cuban Academy of Sciences, considered that metamorphic and igneous rocks on Cuba are Variscan metamorphosed Palaeozoic rocks, though Khudoley (1967) claimed they are Jurassic. Details of these older rocks appear in Stanek et al. (2009).

While no Triassic rocks are known, Triassic gneiss and granite pebbles in a Palaeogene conglomerate in the west indicate Triassic magmatism (zircon dates, Somin et al. 2006). Early Jurassic–?Callovian/Early Oxfordian clastic, syn-rift sediments occur in western Cuba (Guaniguanico). They indicate quartzose continental basement of Precambrian–Palaeozoic age (zircons, Rojas-Agramonte et al. 2006), suggesting Grenvillian

rocks related to SW Mexico and Chortís. Similar rocks occur as exotics on diapirs of Jurassic salt in north-central Cuba. Thick marbles overlain by pelitic and quartzo-pelitic schist occur in the Sierra de Trinidad and rocks on the Isla de Pinos and in eastern Oriente show stratigraphic and lithological similarities (Pszczolkowski 1999). Thicknesses are probably more than 3000 m (Hatten 1967).

Pre-Tithonian arkosic palaeosols overly basement in west-central Cuba while Early–Middle Tithonian extrusive tholeiites occur in east-central Cuba (Iturralde-Vinent 1994). In western Cuba the northwestern, coastal-shallow-water/neritic, Early–Middle Jurassic San Cayetano (Pszczolkowski 1999), Jagua and Francisco Fms change to the deep marine Sábalo Fm in the southeast (Pszczolkowski 1999). The Oxfordian–Early Kimmeridgian El Sabalo Fm and part of the Encrucijada Fm (Aptian–Albian) have been interpreted as oceanic crust contaminated with continental crust. Allibon et al. (2008) relate dolerite dykes of the El Sabalo to continental margin rift basalts.

Tithonian and Berriasian ammonite assemblages of north-central (Bahama platform) and western Cuba show biogeographic/palaeogeographic coupling (Pszczolkowski & Myczynski 2003). Radiolaria in cherts of central and western Cuba in both deep-water continental-margin rocks and ophiolites suggest common palaeogeography (Aiello et al. 2004). Arc rocks of the Mabujina Unit of central Cuba, dated by spores and pollen as Upper Jurassic–Lower Cretaceous, show sedimentary contamination (Stanek et al. 2006).

In the Albian–Cenomanian carbonate platforms and banks, hundreds to thousands of metres thick, surrounded much of the Gulf of Mexico (Sheridan et al. 1981; Salvador 1991; Carrasco-V 2003). At least 11 km of horizontal carbonate rocks overlie magnetic basement on the southern margin of the Bahama Platform (Lewis 1990, summarizing Mossakovsky & Albear 1978; Pszczolkowski 1976; see also Iturralde-Vinent 1994). In northern Cuba (Cayo Coco) the section is at least 5 km thick and ranges from Upper Jurassic dolomites and anhydrites through Neocomian–Aptian shallow water limestones, Albian–Coniacian pelagic limestones and marls, Maastrichtian limestones and Paleocene to Middle Eocene marls and limestones (Pardo 1975). The deep-water equivalent, one-fifth as thick, is seen further south through windows in allochthonous ophiolites. Tithonian to Maastrichtian rocks include pelagic limestones, calcareous turbidites, radiolarian cherts, sandstones and marls. Tuffs and radiolarian cherts record nearby volcanism (Cabaigan Belt of Pardo 1975).

According to Pacific models, sedimentary Jurassic continental rocks in western Cuba (Guaniguanico) and their metamorphosed equivalents in south Central Cuba (Isle of Pines, Escambray) were picked up from southern Yucatán and underplated to the forearc of a migrating volcanic arc in the Campanian (Pszczolkowski 1999; Draper & Pindell 2004). However, palaeomagnetic studies indicate no significant latitudinal differences between Jurassic–Cretaceous rocks of western Cuba relative to North America (Alva-Valdivia et al. 2001). Rocks in this area are different from and do not suggest affinity with Chortís and Mexico (Pszczolkowski & Myczynski 2003).

Cuban continental basement clearly is the autochthonous continuation of the Florida–Bahamas Platform/Maya Block, with eastward-increasing extension. The next section shows that it continues below the island blocks further east.

The Greater Antilles–northern Lesser Antilles, southern Lesser Antilles

The Greater Antilles extend some 2500 km from western Cuba to the Virgin Islands and, for this paper, through the northern Lesser Antilles (Limestone Caribbees) to their abrupt termination at Marie Galante (Lewis & Draper 1990, saw them continuing as far as Guadeloupe). Their limited subaerial appearance is deceptive. They are massive areas, 200 km wide but two-thirds below sea level, contiguous but for Oligocene/younger pull-apart lows of the narrow Windward, Mona and Anegada Passages (Fig. 1).

The Greater Antilles are generally attributed to an extinct Cretaceous volcanic arc (Mattson 1966; Mattson & Schwartz 1971; Donnelly 1989). This began life as the 'Caribbean Great Arc' in the Pacific and entered the Caribbean at the leading edge of the migrating Caribbean Plate (Burke 1988). Diachronous collision with the Bahamas and northern South America supposedly caused arc volcanism to cease progressively from west to east (Pindell et al. 2006, fig. 7b; Levander et al. 2004). However, arc activity ceased synchronously from Cuba to Puerto Rico/Virgin Islands, a distance of some 2000 km, in the Middle Eocene (Iturralde-Vinent 1995; Lidiak 2008).

There are several indications that continental crust underpins the Greater Antilles east of Cuba. The idea is not new – Stainforth (1969), Nagle et al. (1982), Joyce & Aronson (1983), Alonso et al. (1987) and Lidz (1988) all observed relevant data.

Schists and marbles on the Samaná Peninsula of Hispaniola and on Tortuga Island north of Hispaniola (Nagle 1970; Khudoley & Meyerhoff 1971; Fox & Heezen 1975; Nagle et al. 1982; Lewis et al. 1990; Goncalves et al. 2002; Draper et al. 2008) continue the trend of continental rocks on Cuba (Guaniguanico–Isle of Pines–Escambray). Samaná rocks are 'strikingly similar' (Nagle 1974)

to the eclogite, eclogite-amphibolite and garnet amphibolite tabular masses, boudins and blocks in graphitic schist and marbles of the Jurassic–Cretaceous (pre-Albian), continental margin Las Mercedes Fm of northern Venezuela.

Marble occurs in continuous outcrop from 2600 to 3600 m off the Samaná Peninsula (Heezen et al. 1976, quoted by Nagle et al. 1978). Two dredges from the south wall of the Puerto Rico Trench north of the Mona Passage retrieved black marble (Fox & Heezen 1975). Two others, east of Cape Samaná, Hispaniola, recovered siltstone and recrystallized carbonate with deformed bands of white mica. Dredges from between 6000 and 3500 m on the south wall of the Trench north of Puerto Rico and the Virgin Islands indicate a stratigraphy of upper Cretaceous to lower Miocene, shallow marine platform carbonates (Fox & Heezen 1975). Albian to at least Early Miocene inter-tidal to sub-tidal deposits on the Silver–Navidad Banks show palaeogeographic continuity of these areas (Schneidermann et al. 1972). These are not the foundations of an intra-oceanic volcanic arc.

Amphibolitic gneiss crops out at the western end of the Sierra Bermeja of Hispaniola (Mattson 1960; Renz & Verspyk 1962, quoted by Donnelly 1964). Garnet peridotite, normally associated with subducted continental rock, occurs in the Cuaba amphibolite of the Dominican Republic (Draper et al. 2002). Geophysical data indicate continental crustal thickness of about 30 km in the centre of Puerto Rico and 29 km below the Anegada Platform (Shurbet et al. 1956; Lidz 1988). Rocks known from outcrop on Puerto Rico account for only 6 km of the total thickness – some 24 km of section are unknown (Mattson 1966). The Moho lies at around 29 km below St Thomas and St Croix and 22 km below the Anegada trough (Shurbet et al. 1956).

These data indicate that Hispaniola–North Virgin Islands continue the Cuban geology of basalt and volcanic arc rocks thrust over thick Mesozoic limestones lying on Triassic–Jurassic rift section in Palaeozoic basement.

The Grenada Basin and Aves Ridge terminate against the shallow marine platform of Saba Bank in the north. Drilling on the Bank encountered 2858 m of Cenozoic sedimentary cover and terminated after penetrating 119 m of porphyritic andesite. Seismic data indicate an extensive sedimentary section (upper Cretaceous?) below the andesite to as deep as 4100 mbsl (Bouysse 1984).

Northern South America

As in North America, Pangaean rifting began in the Triassic in northern South America. The upper Triassic to lower Jurassic section consists of rift red beds and volcanic rocks in the Sierra de Perijá,

Mérida Andes, on the Guajira Peninsula and in the subsurface San Fernando-Matecal, Espino, Anibal and Takutu grabens (Bellizia 1972; Feo-Codecido *et al.* 1984; Maze 1984; Crawford *et al.* 1985; Gonzalez & Lander 1990). Similar rocks occur in the Arauquita-1 well and Guafita areas of Venezuela's Llanos Basin (McCollough & Carver 1992). Triassic-Jurassic volcanic rocks occur on Venezuela's El Baul Uplift (Kiser 1987). The Tinaco–Tinaquillo belt of northern Venezuela is a 3 km thick, Jurassic peridotite complex (Ostos 2002; Ostos & Sisson 2002). It formed in NE–SW rifted Cambro–Ordovician basement in the Cordillera de la Costa and could be related to Appalachian–Caledonian events and the Acatlán complex, also affected by Jurassic rifting, of southern Mexico (Ostos & Sisson 2002, fig. 9; Nance *et al.* 2006).

Upper Triassic continental sediments, sills and flows in the Guajira Peninsula region are followed by several thousands of metres of shallow marine, Jurassic to Late Cretaceous limestones (Middle Jurassic molluscs, Late Jurassic ammonites, Cretaceous molluscs, ammonites, foraminifera; Rollins 1965; Lockwood 1971). Upper Jurassic shales occur on the Paraguaná Peninsula (ammonites; Bartok *et al.* 1985). While this section is metamorphosed along most of northern South America, these areas carry sedimentary rocks. They escaped metamorphism because they lay at least 300 km SW of the plate boundary until the Eocene.

Thick sections of Jurassic metasediments occur in the Coastal and Northern Ranges of Venezuela and Trinidad (Late Jurassic pelecypods, tintinnids; Kugler 1953; Feo-Codecido 1962; Bellizia 1972; Bellizia & Dengo 1990) and similar rocks crop out on Margarita Island (Gonzalez de Juana *et al.* 1980). Protoliths were shallow (?Callovian gypsum; Kugler, 1953) to deep marine, passive margin sandstones, organic shales and limestones (Stainforth, 1969; Gonzalez de Juana *et al.* 1980). In the Coastal Range of Venezuela the section rests upon Precambrian–Palaeozoic basement (Urbani 2004). Northward increasing silica content and lenses of volcanic ash in the Jurassic–Cretaceous (pre-Albian) Las Mercedes Fm record nearby volcanic activity. Metasediments on the Araya Paria and Trinidad's Northern Range record a vast thickness of ?Triassic–Jurassic–Lower Cretaceous marine beds above the Late Permian–Early Triassic Dragon Gneiss of Paria and Sebastopol Gneiss of central Venezuela (Stainforth 1969; Kugler 1972).

Most of Venezuela's eastern offshore platform is underlain by metamorphic gneisses and passive margin metasedimentary rocks, meta-ophiolites and MORB volcanic rocks of Jurassic–Early Cretaceous age (Ysaccis 1997). Exposed geology of Margarita shows basement of metamorphosed (100–90 Ma) MORB rocks and schists and gneisses

(Jurassic–Cretaceous protoliths, Pennsylvanian zircons; Stockhert *et al.* 1995). East of Margarita, wells have penetrated section ranging from ?Jurassic to Early Cretaceous (shallow water Bocas metabasalts and low grade metasediments), Early to Late Cretaceous deep marine, euxinic shales, limestones, cherts, volcaniclastic rocks and lavas (Mejillones Gp.) and Late Eocene–Lower Early Oligocene arc volcanic (Los Testigos) igneous–metamorphic complexes (Castro & Mederos 1985). The Upper Jurassic–Lower Cretaceous (ammonites, foraminifera) North Coast Schist (greenschist) of Tobago contains metatuffs, graphitic siliceous schist, graphitic quartzose phyllite. The chemically related Aptian–Albian Tobago Volcanic Group includes volcaniclastic breccias and lavas. The volcanic–intrusive complex continues south of the island to the Jurassic (Tithonian ammonites, aptychi) metasediments black shales and Albian Sans Souci tuffs, tuff breccias, conglomerates and andesitic lava flows of Trinidad's Northern Range (Ramroop 1985).

For Stockhert *et al.* (1995) the geology of Margarita is representative of the 70 000 km^2 area of (eastern) coastal Venezuela, Trinidad and Tobago. They derive it from NW South America by arc–continent collision and subduction, followed by resurrection along the transcurrent boundary following passage of the arc complex, in a manner similar to the proposed history of the Greater Antilles. Kerr *et al.* (2003) attribute the Sans Souci of Trinidad and parts of the North Coast Schist of Tobago to the Cretaceous Caribbean Plateau. However, they manifest Albian continental input and lie SE (outboard) of the volcanic arc (Lesser Antilles) the plateau is supposed to have followed into place. According to Wadge & Macdonald (1985), Sans Souci tholeiites erupted onto the passive margin of South America in the Aptian–Santonian.

The above units mirror those of southern North America–Cuba. Together with rocks on Maya and Chortís they show a regionally coherent Jurassic–Cretaceous palaeogeography of intra-continental rifts to deep passive margin flanked by volcanic activity along northern South America, southern North America and along the eastern margins of continental fragments in the west.

Igneous rocks

Abundant high-silica rocks indicate the widespread presence of continental crust on Caribbean margins. Tonalites, a chemical signal of continental crust, are common. Many date radiometrically in the region of 86–80 Ma, close to a major pulse of Caribbean Plate thickening at 90–88 Ma (Donnelly 1989; Kerr *et al.* 2003). Four tonalites on Hispaniola are Albian in age (U–Pb, Ar/Ar 109–106 Ma, Escuder Viruete et al. 2006), a tonalite–gabbro batholith in the Netherlands Antilles is dated Middle Albian to Coniacian (Beets et al. 1984) and cooling of a Mid-Cretaceous diorite and tonalite on Tobago occurred at 103 Ma (zircon fission track; Cerveny & Snoke 1993). Tonalites have been dredged from the Cayman Trough (Perfit & Heezen 1978). The geochemistry of six of these is typical of continental arc granitoids (Lewis et al. 2005). Their radiometric age, 62–64 Ma, is similar to that of granodiorite intrusions in Jamaica (K–Ar, Ru–Sr, 65 ± 5 Ma, Chubb & Burke 1963).

The silica content of plutonic rocks in the northeast Caribbean (Puerto Rico–northern Lesser Antilles) ranges from 45 to 78% weight (Smith et al. 1998). The Virgin Gorda batholith on the Virgin Islands Tortola and Virgin Gorda consists of diorites, tonalites and granodiorites. Basement of La Désirade consists of Tithonian trondhjemite and rhyolite and coarse-grained, quartz-rich granite (Fink 1970; Bouysse et al. 1983; Westercamp 1988).

Lesser Antilles basalts are directly comparable in chemical composition and mineralogy with basalts from calc-alkaline suites of circum-oceanic islands and continental margin orogenic belts (Lewis 1971). Lewis & Gunn (1972) regarded basalt–andesites of the Lesser Antilles as products of mantle fusion and differentiation at shallow depths; Benioff zone fusion did not appear to be involved. Siliceous metasedimentary xenoliths occur in basalts and rounded quartz grains in andesites (Nicholls et al. 1971).

Silicic magmas in Costa Rica are chemically similar to those in Guatemala where Palaeozoic crust occurs (Vogel et al. 2007). Crustal thickness (refraction data) and gravity values for Costa Rica are continental, Albian and Miocene quartz sands are present and the Arenal volcano produces granulite xenoliths (Case 1974; Sachs & Alvarado 1996; Iturralde-Vinent 2004; Auger et al. 2004).

The Aves Ridge has a relief of 2–3 km and locally reaches sea level. Dredges from steep pedestals recovered glassy and brecciated basaltic rocks, suggesting volcanic origin. Seismic indicates 5 km capping of sediments and/or volcanics above a crustal layer with velocity of 6.2–6.3 km s^{-1} (Fox et al. 1971). Similar layering is seen in the Grenada and Venezuela basins. Dredging on Aves Ridge also recovered granodiorites, close to Venezuela. Three gave K–Ar ages of 78–89, 65–67 and 57–58 Ma (Lower and Upper Senonian, Upper Paleocene) (Fox et al. 1971; Fox & Heezen 1975). The in situ confining pressure of the 6.0 km s^{-1} under the Aves Ridge is 1–2 kbar. Calculated velocity of the granitic samples at this pressure is 5.8–6.1 km s^{-1}, similar to velocities of continental granodiorites.

Andesites, dacites, diorites and granodiorites (ages not known to the author), also continental signals, occur in Panamá, on the Nicaragua Rise, Jamaica, north Yucatán, Cuba, Hispaniola, the Saba and Mariner Banks, the Virgin Islands and the Lesser Antilles (Butterlin 1956; Donnelly 1966; Lidiak 1970; Case 1974; Arden 1975; Banks 1975; Tomblin 1975; Bouysse et al. 1985; Despretz et al. 1985; Jackson et al. 1987; Lewis 1990; Lewis & Draper 1990; Blein et al. 2003). The Testigos-1 and -2 wells offshore NE Venezuela bottomed in andesitic basalts of calc-alkaline type, dated 40–35 Ma (Ysaccis 1997).

The Caribbean 'oceanic plateau'; continental foundations?

Donnelly et al. (1990) wrote comprehensive early discussions of the Caribbean 'oceanic plateau' while Kerr et al. (1999, 2003) provided later data and an update on analytical/dating techniques. A detailed discussion of the Caribbean plateau is given by James (2007b).

Thick crust in the Venezuela Basin was the 'original' Caribbean Plateau of Donnelly et al. (1973). Later works have implicated accreted oceanic rocks in Cuba, western Colombia, Curaçao, Aruba, Venezuela, Trinidad, Jamaica, Hispaniola and Central America and even Ecuador. Kerr et al. (1997) described the combined Caribbean–Colombian large igneous province as one of the world's best-exposed example of a plume-derived oceanic plateau. However, 'plateau' rocks on Cuba lie on the North American plate and those of western Colombia and Ecuador lie on the South American Plate. The rocks on Costa Rica lie west of the arc that bounds the Caribbean Plate/plateau and those on Tobago lie outboard of the volcanic arcs they are supposed to have followed during plate migration. Thick parts of the Caribbean Plate are bounded by older or coeval thin (3 km) crust in the Colombia and Venezuela basins. Such crust in the SE Venezuela Basin would not be capable of driving the Aves/Lesser Antilles volcanic arc over Atlantic crust of normal thickness.

Kerr et al. (2003, table 2) summarized occurrences and ages of plateau rocks in the Pacific and the Caribbean. Ages cluster at 124–112 Ma (Barremian–Aptian), 91–88 Ma (Turonian) and 78–59 Ma (Campanian–Danian). Activity also occurred at 124–112 Ma on Ontong–Java and Manihiki and at 90 Ma on Ontong–Java (Larson 1997; Birkhold et al. 1999; Larson et al. 2002). Oceanic rocks accreted to Ecuador also fall into ages c. 120 and c. 90 Ma (Jaillard et al. 2009). The data suggest widespread pulses of igneous activity at these times rather than spatial relation of the localities.

The data compiled by Kerr *et al.* (2003) include rocks with a wide range of age and chemistry. The Duarte complex of Hispaniola ranges in Ar/Ar age from 86 to 69 Ma but Jurassic radiolaria occur locally. The Curaçao Lava Fm yields Ar/Ar dates of 88–90 Ma but Middle Albian ammonites are present (Wiedmann 1978). Basalts at Site 1001 on the Lower Nicaragua Rise/Hess Escarpment are overlain by Campanian limestones (nannofossils with minimum age 77 Ma, Sigurdsson *et al.* 1996) and give a mean Ar/Ar age of 80.9 ± 0.9 Ma (Sinton *et al.* 2000). It is important to note that drilling, dredging and submersible sampling of the Venezuela Plateau have tested only the uppermost components of the 'plateau' and its faulted (rifted–intruded?) margins. The bulk of the plateau remains uncalibrated and should not be assumed to consist of a huge volcanic pile. Rocks accreted to the plate margins are assumed, not known, to represent the plateau of the plate interior. Older rocks accreted to Ecuador in the Campanian, younger rocks in the Late Maastrichtian and Late Paleocene (Jaillard *et al.* 2009). This suggests separate origins.

Kerr *et al.* (2003, see also Kerr *et al.* 2000) discuss means of identifying oceanic plateaus. Indicators are basalts and high Mg lavas, La–Nb ratios around 1 (less than arc rocks), flat rare earth element patterns and narrow radiogenic isotope ranges. While these parameters individually do not identify a plateau, in conjunction they provide 'a powerful set of discriminants'. However, Kerr *et al.* (2003) also note that basalts of marginal basins of island and continental arcs are potentially the most difficult to distinguish from oceanic plateaus.

The Caribbean 'oceanic plateau' is thought by many to have been generated by a mantle plume (e.g. Kerr *et al.* 2009). On the other hand, since Caribbean 'plateau' magmas formed at three different times show similar petrological and chemical compositions, Révillon *et al.* (2000) questioned mantle plume genesis, noting instead that lithospheric thinning could produce the same melting conditions. There is no indication anywhere in the Caribbean region of the radial strain expected over mantle plume (Glen & Ponce 2002). Instead, the whole of Middle America manifests NE structural grain that parallels Triassic–Jurassic rifts in neighbouring continental masses (Fig. 4).

Oceanic plateaus are defined as anomalous rises above the seafloor that are not parts of known continents, volcanic arcs or spreading ridges (Nur & Ben-Avraham 1983). Neither the Colombia Basin nor the Venezuela basin 'plateau' is elevated today. They mostly lie below more than 3000 m of water (the Beata Ridge, which limits the Venezuela basin 'plateau' to the west, rises locally to 1000 m). It is correct to note, however, that Cretaceous basalts

of smooth Horizon B″ could have formed under shallow marine/subaerial conditions (below). Reflection data show a basal onlapping stratigraphic section over acoustic basement in the Venezuelan Basin (Driscoll *et al.* 1995). Correlation from DSDP holes shows this to be older than Middle Eocene (50 ma) and younger than Senonian (88 ma). The section is absent from the Caribbean 'plateau'. Together with the presence of Middle Eocene shallow marine carbonates sampled from the Beata Ridge, this shows that the plateau subsided to its present depths since the Middle Eocene.

The term 'plateau' in the Caribbean today therefore refers more to its crustal thickness than its elevation. It is not known whether it is part of a continent or spreading ridge. It is presumed to be oceanic because the Caribbean Plate came from the Pacific.

Many oceanic plateaus have crustal thicknesses of 20–40 km and an upper crustal velocity of 6.0–6.3 km s^{-1}, typical of granitic rocks in continental crust. The Caribbean 'plateau' is up to 20 km thick. Parts of the Caribbean are floored by thick crust with velocity 6.1–6.5 km s^{-1}.

Rosendahl *et al.* (1992) suggested that continental crust might occur up to hundreds of kilometres from continental margins. The Kerguelen, Ontong Java Iceland and Rockall plateaus are known from dredge samples or ancient zircons to carry continental rocks (Roberts 1975; Doucet *et al.* 2002; Frey *et al.* 2002; Klingelhöfer *et al.* 2005; Paquette *et al.* 2006; Ishikawa *et al.* 2007). Granitic basement is exposed on the Seychelles, the Parcel Islands, Kerguelen and Agulhas and possibly underlies the Iceland Plateau (Foulger *et al.* 2005).

Diebold *et al.* (1999) noted: 'The concept that the Colombia and Venezuela basins are capped uniformly by a Cretaceous igneous body persists' (see illustrations in Mann 2007; Rogers *et al.* 2007a, b). In reality, crustal thickness of the Caribbean Plate varies from normal, 6–8 km, in the Haiti Basin, west of the Beata Ridge, to thickened, up to 20 km, between the Venezuela Basin Fault Zone, and the western Beata Ridge boundary, to abnormally thin, 3–5 km, in the southeastern Venezuela Basin (Diebold *et al.* 1999). Crust is also thick in the Yucatán (8–9 km, Hall 1995), Colombian and Grenada Basins (10–22 km; 18 km; Case *et al.* 1990). These thick areas all display the NE tectonic fabric of continental neighbours (James 2006).

Seismic data shown by Diebold *et al.* (1999) and Diebold (2009) over the Caribbean 'plateau' in the Venezuela Basin show large (35 km wide) NE-trending highs flanked by dipping wedges of reflections and structures that pierce the sea floor. Similar data are seen in the Colombia Basin where thick crust continues uninterrupted to Panamá and Costa Rica (Bowland & Rosencrantz 1988). Diebold *et al.* (1999, fig. 15) interpret the architecture as blocks of vertical dykes flanked by wedges of igneous flows and seamounts. These authors outline a history of oceanic crust formation by spreading, followed by extension and thinning with widespread eruption of basaltic flows. The top of these forms smooth seismic Horizon B″, sampled by DSDP and ODP drilling.

Smooth Horizon B″ is reminiscent of a continental flood basalt and reduced velocities in the upper sub-B″ sequence could indicate vesicles, brecciation, weathering or interbedded sediments (Diebold *et al.* 1999). This would explain the great lateral extent of smooth B″. Vesicularity of the uppermost basalts and shallow water fauna in overlying sediments drilled by the DSDP supports this subaerial/shallow origin (Sigurdsson *et al.* 1996). Other sites of presumed plateau rocks also indicate shallow/subaerial conditions. Aruba and Curaçao became emergent in the Late Cretaceous (Beets 1972; Wright 2004). The Curaçao Fm has a presumed palaeosol at its top and a palaeosol occurs above weathered basalt on the correlative Aruba Lava Fm (Beets 1977; Snoke 1990). Spheroidal weathering of gabbros and dolerites from the Beata Ridge indicates subaerial weathering (Révillon *et al.* 2000).

Diebold *et al.* (1999) recognized upper and lower volcanic sequences on the plateau. Diebold (2009) suggests that the 5 km thick Albian–Cenomanian Curaçao Lava Fm (Klaver 1987) is analogous to the upper 5 km of the upper volcanic sequence of the Caribbean Plateau. If they are age equivalent, then the lower part of Diebold's upper volcanic sequence must be pre-Albian. The lower sequence and its structure must be older still, at least Early Cretaceous, possibly Triassic–Jurassic. This was the time of rift/drift, when the regional inter-American N35°E structural grain developed (below).

Seaward dipping wedges (SDRs) are typical of extended continental crust at the continent–ocean transition (Hinz 1983; Rosendahl *et al.* 1992). They are common in the North and South Atlantic (Jackson *et al.* 2000). Characteristic features are oceanward-directed, upwardly convex and diverging dips and seismic velocities increasing from 2.6–4 to 6.4 km s^{-1} (Mutter *et al.* 1982; Hinz 1983). They could consist of shallow marine to subaerial volcanic layers (Hinz 1983), volcanic layers on ocean crust (Mutter *et al.* 1982) or sediments in half grabens with continentward-dipping listric faults (Bally, pers. comm. 2008).

Rifting between North America and Europe/North Africa followed the Newfoundland–Honduras Palaeozoic tectonic belt in the Early Triassic–Hettangian and affected Middle America by the Late Triassic (Ager 1986; Helwig 1975; Manspiezer 1988; Davison *et al.* 2003; Withjack &

Schlische 2005). Basins formed along low angle detachments and sinistral faults, reactivating Palaeozoic thrusts or dextral faults, and filled with red beds and volcanic rocks. Tectonism along eastern North America then abandoned inboard rifts and moved to the Atlantic, where seafloor spreading followed. There, basin fill comprises a syn-rift section separated by a break-up unconformity from post-rift/drift section. Seismic data indicate up to 5 km of Triassic (?) syn-rift deposits in the Baltimore Canyon, Carolina Trough and Blake Plateau basins (Manspiezer 1988). Postrift, wedge strata, 8–13 km thick, are cut by salt diapirs. COST well G-2, on Georges Bank, bottomed in upper Triassic salt (Manspeizer 1988). Figure 6 illustrates a section over these basins.

Attenuated continental crust and SDRs are present below the Gulf of Mexico, where there may be only little true oceanic crust (Johnson *et al.* 2005; Post 2005). Violet siltstone, with Carboniferous K–Ar age, dredged from a Sigsbee salt diapir records Palaeozoic basement (Pequegnat *et al.* 1971) below the deep Gulf of Mexico. Seismic data and DSDP/ODP data reveal large continental horsts flanked by deep wedges of dipping reflections (Jurassic sediments, flows, volcaniclastics, and possibly salt) in the SE Gulf of Mexico (Phair & Buffler 1983), where NE tectonic grain links the Yucatán Peninsula to Florida. Continental signal in oceanic basalts of the Oxfordian–Early Kimmeridgian El Sabalo and Encrucijada Fms (Aptian–Albian) of nearby western Cuba suggests eruption through continental crust like North Atlantic seaward-dipping sequences (Kerr *et al.* 1999, 2003). Eastern North America geology continues into the Gulf of Mexico.

This paper proposes that Caribbean 'plateau' architecture continues the same geology into a more extensional location. In this scenario seismic line 1293 (Diebold *et al.* 1999) shows blocks of extended continental crust flanked by wedges of dipping Jurassic sediments and flows below the smooth basalt flows of the western Venezuela Basin (fig. 7, James 2007*b*). The unexplained 'ski-jump' (Diebold *et al.* 1999) at the edge the plateau may be a marginal reef/carbonate mound. Rough Horizon B″ is the equivalent 'oceanic', serpentinized mantle crust formed during extreme crustal attenuation. 'Oceanic' crust derived in this manner does not have organized magnetic anomalies. Downlap of Turonian smooth B″ onto rough B″ by 20–30 km beyond the plateau boundary (Diebold 2009) for this paper indicates a similar age. Similar geometry appears in the Colombian Basin (Bowland & Rosencrantz 1988).

If all this is true, then (a) the Caribbean plateau formed over a long interval of time, not by a Late Cretaceous 'large igneous plateau' event, (b) it is

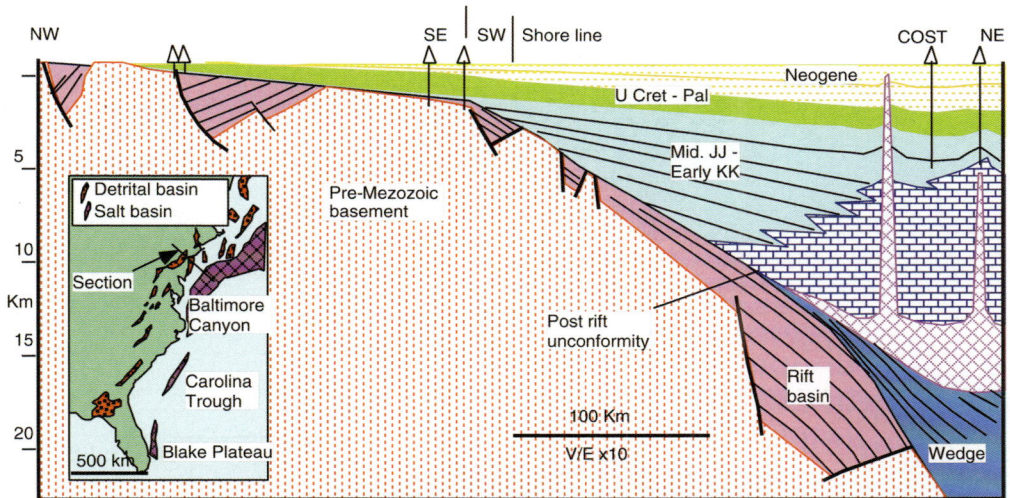

Fig. 6. Cross section, eastern North America continental margin. After Benson & Doyle (1988) and Manspeizer (1988) (inset).

underlain by extended continental crust, (c) there is little 'oceanic' crust in Middle America and (d) it formed by serpentinization of upper mantle during extreme extension of continental crust and thus shows no spreading signature. Thick crust in the Yucatán, northern Grenada (both with NE tectonic grain) and Colombia basins are likely to have similar origins. Seismic over the Colombian Basin shows the character and the same architecture of horsts flanked by wedges of dipping reflections (Bowland & Rosencrantz 1988, figs 7 & 9).

If the 'plateau' formerly were shallow, Middle Eocene emplacement (see below) of its upper rocks onto neighbouring continent would be more easily explained than uplift from oceanic depths. Seaways in the area would have been restricted, offering an explanation for the prolific upper Cretaceous hydrocarbon source rocks of northern South America. Reduced velocities in the upper sub-B″ sequence could indicate presence of source rocks on the plateau.

Salt diapirs

The 'seamount' shown on line 1293 near CDP 2000 (Fig. 7; Diebold *et al.* 1999, fig. 2) looks like a piercing diapir and so could consist of serpentinite or salt. The diapir rises at least 700 m above the sea floor and resembles Sigsbee Knoll diapirs that rise 200–400 m above the 3600 m deep seafloor of the Gulf of Mexico (Fig. 8). Seafloor sediment push-up indicates that the feature on Line 1293 is active, but there is no reported volcanism in the area. Onlap (arrow of flat onto upturned reflections adjacent to

the diapir on its northern side indicates pre-B″ growth. It does not push up Horizon B″, despite being some 14 km wide at this level, which suggests considerable normal fault extension, consistent with the *c.* 150 m seafloor drop south to north across the feature and an apparently very thick (at least 1.7 secs twt; 4 km?) pre-B″ section on the down thrown side. Dip of Horizon B″ and deeper reflections towards both sides of a similar feature near CDP 6000 suggests a withdrawal rim syncline. Both diapirs appear to root deep in the pre-B″ section. If they are built of salt, this is older Mesozoic in age. Analogy with eastern North America and the Gulf of Mexico suggests Triassic and/or Jurassic age. N35°E projection from the diapir leads to the SE coast of Puerto Rico (Fig. 8). The coastal village of Salinas shows 5 salt knolls (apparently not dated) on its flag.

Diebold *et al.* (1999) note that positive magnetic anomalies correspond to their seamounts. However, diapirism is often triggered by faults. If the NE trends of the Caribbean are reactivated older structures, they could have been the locus of older intrusions.

Beata Ridge and Salt Mountain, Hispaniola

The Beata Ridge is about 23 km thick and its crustal velocity structure is similar to the Nicaragua Rise (Edgar *et al.* 1971). Gravity anomalies indicate that the ridge is not oceanic crust uplifted 3 km (anomalies would be larger) but that topography is compensated at depth by downflexing of lithosphere (Bowin 1976).

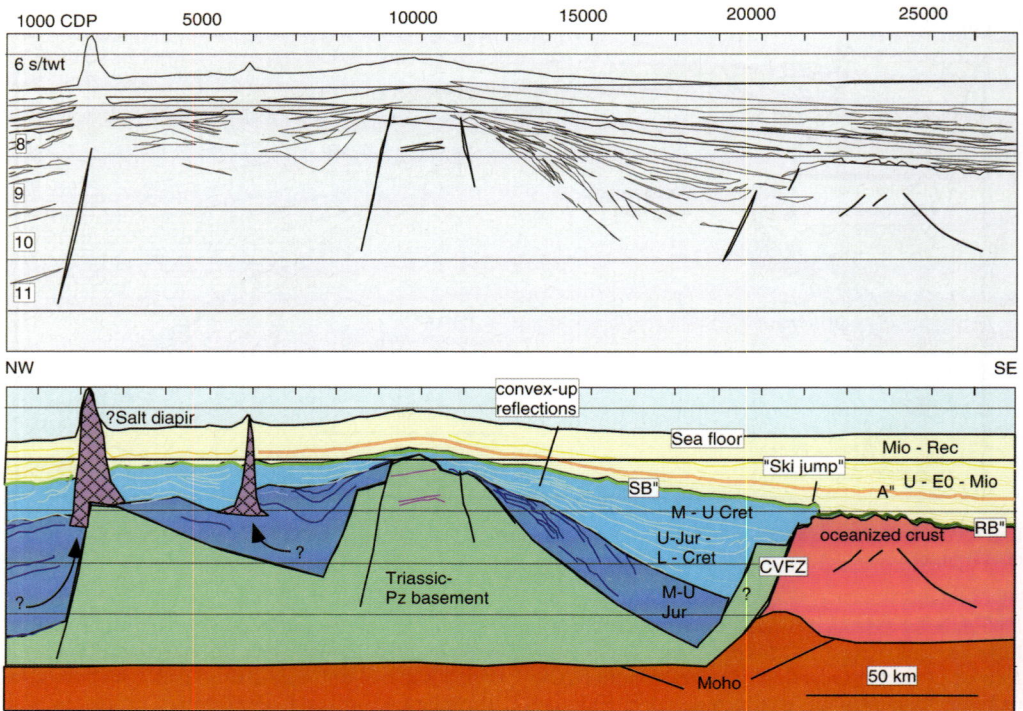

Fig. 7. Line drawing and geology, this study, of seismic line 1293 (located on Fig. 3) from Diebold *et al.* (1999, fig. 2). The figure suggests that Caribbean geology continues that of eastern offshore North America (Fig. 6). Triassic–Jurassic rifting accommodated continental–shallow marine sediments, volcanic flows and salt. Drifting introduced open marine Jurassic–Cretaceous sediments, probably very thick carbonate sections. Late Cretaceous extension resulted in subaerial basalt flows over this architecture, forming smooth Horizon B″ (SB″), and serpentinization of adjacent mantle, forming rough Horizon B″ (RB″). Horizon A″, the Middle Eocene contact between chert and overlying unconsolidated sediment correlates with cessation of volcanism along the north and south plate boundaries. CVFZ, Central Venezuelan Fault Zone.

The ridge has been seen as a trench–trench transform hinge fault, a ridge–trench transform, as thrusting of the Colombian Plate over the Venezuelan Plate and as part of the Nicaragua Rise (Malfait & Dinkleman 1972; Arden 1975; Anderson & Schmidt 1983; Mauffret & Leroy 1997).

Submarine sampling indicates that the ridge includes hypabyssal intrusive rocks (gabbros and dolerites) alternating with sedimentary rocks, probably in tectonic contacts (Révillon *et al.* 2000). There are also rare pillow basalts. Diebold (2009) relates the ridge to thick volcanic flows and presents seismic evidence of compressional structure.

Constant seismic interval B″–A″ thickness in the south indicates uplift after the Middle Eocene (Moore & Flaquist 1976). Coring recovered shallow-water Mid-Eocene carbonates from the crest of the ridge (Fox & Heezen 1975) and DSDP Site 151 recorded an Early Oligocene–Middle Eocene hiatus. Deformation increases northwards towards Hispaniola where reverse faults cut upper

Miocene strata onshore (Biju-Duval *et al.* 1982). The ridge is thus seen to have formed since the Miocene by strong transpression, with reverse faults, pop-up structures and strike–slip faults observed in the east (Mauffret & Leroy 1997).

The Beata ridge carries scattered 'volcanic seamounts' that rise several hundred metres above the sea floor immediately south of the Bahoruco Peninsula, Hispaniola (Biju-Duval *et al.* 1982). The western margin of the Ridge, the Beata Fault, follows the regional N35°E tectonic trend and runs east of the Bahoruco Peninsula of Hispaniola to meet the coast in the Azua area (Ramirez *et al.* 1995). 'Basement highs' on seismic offshore Azua are strikingly similar to salt diapirs on the Sigsbee Scarp (Ladd *et al.* 1981, fig. 1; Buffler, 1983, fig. 4).

Offshore Azua geology is exposed onshore to the west on the Bahoruco Peninsula. Here, the upper part of the Cretaceous section in the central Massif de la Selle includes chaotic blocks of upper Albian to upper Coniacian–Santonian radiolarian cherts,

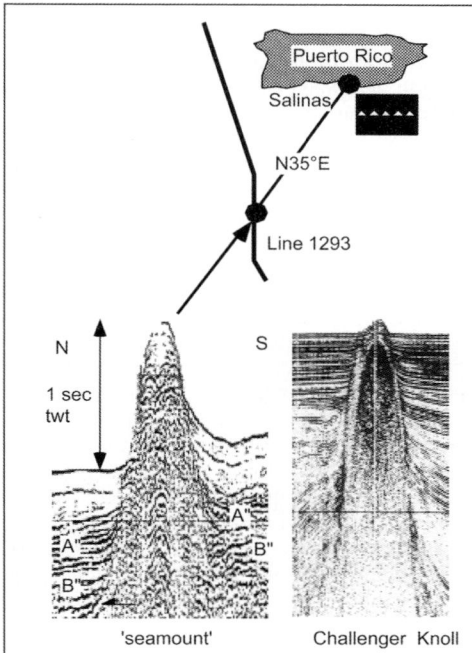

Fig. 8. Similarity of Caribbean 'seamount' (after Diebold *et al.* 1999, seismic line 1293, fig. 2) to a drilled salt dome in the Gulf of Mexico (after Burk *et al.* 1969). Rim syncline adjacent to diapir in deeper reflections indicates withdrawal. Projection from the diapir along regional structural trend N35°E leads to SE Puerto Rico and the coastal village of Salinas, whose flag illustrates five salt knolls. Seismic line is located on Fig. 4. A″, B″, seismic horizons.

siliceous limestones and dolerites in a volcanic sedimentary matrix of Campanian–Late Maastrichtian age, overlain by Coniacian cherts (Lewis *et al.* 1990). These deposits can be seen as uplifted samples of the Muertos Trough accretionary prism further east. Cerro del Sal (Salt Mountain), on the north flank of the Bahoruco Range is a 21 km^2, vertically dipping outcrop of salt and gypsum, one of the world's largest salt deposits. It seems to be part of the accreted Massif de la Selle geology, structured in the Miocene. Horizontal salt deposits in the Enriquillo Trough, to the north, between Middle and Late Miocene beds (Largo-1 well) are likely to be redeposited. Features on seismic over the Beata Ridge further south (Holcombe *et al.* 1990, plate 8, fig. U) are very similar to the diapir of Figure 8. They also root below Upper Cretaceous seismic Horizon B″.

Other knolls, diapirs, volcanoes

Probable salt piercement structures appear on seismic east of Honduras (Pinet 1972). When Maya–

Chortís offset is removed, they lie next to the Chiapas salt basin of southern Maya. 'Volcanoes' (or diapirs?) occur throughout the area of the Lower Nicaragua Rise, with some showing Neogene–Recent activity (Holcombe *et al.* 1990). A series of regularly spaced knolls, 800 m high, occur adjacent to the Muertos Trough and one occurs near the Aruba Gap (Holcombe *et al.* 1990). A 'seamount' domain near the centre of the Yucatán Basin trends NE (Rosencrantz 1990). To the NE, salt diapirs crop out on the north coast of Cuba, along the trend of the major N60°E trending La Trocha Fault that crosses the basin (Fig. 3). A pre-Jurassic ridge in central Cuba, extending to the longitude of the eastern Bahamas, played a major role in localizing salt deposits of the Early and Middle Jurassic Punta Alegre Fm (Kirkland & Gerhard 1971; Meyerhoff & Hatten 1974). Recent exploration has encountered salt diapirs in the Florida Straits further north (Cobiella, pers. comm. 2008).

The Yucatán Basin

This roughly triangular basin between Cuba and the Cayman Ridge has not been drilled and basement is not dated. It is attributed to spreading behind the Cuban volcanic arc (Holcombe *et al.* 1990) or part of the Caribbean abandoned during a plate boundary jump from north of Cuba to the Cayman Trough (e.g. Pindell *et al.* 2006).

A characteristic seismic reflection resembles Horizon B″ of the Caribbean Basin (Rosencrantz 1990) and the basin exhibits structural grain of N35°E west of the La Trocha Fault and N60°E to the east. Tilted fault blocks occur in the west (Holcombe *et al.* 1990). A rise south of the Cuban Isle of Pines could be an extension of continental crust (Rosencrantz 1990). Anomalously thick crust (10–15 km) occurs in the southeast adjacent to the Cayman Ridge (Holcombe *et al.* 1990; Leroy *et al.* 1996). Rosencrantz (1990) noted that this could be Aptian–Albian or possibly Late Jurassic in age. Its thickness suggests that extended continental crust is present (James 2007*a*).

Continental rocks on the Cayman Ridge show that Cretaceous volcanic arc rocks accreted to Cuba cannot have come from the Pacific. Since Jurassic 'oceanic' rocks have been thrust onto Cuba from the south (Cobiella-Reguera 2009), the Yucatán Basin has to be at least in part Jurassic in age, not Paleocene as proposed by Pacific models.

Thin Caribbean 'oceanic' crust, serpentinite

Thin Caribbean crust, seen on seismic SE of thick crust in the Colombia and Venezuela basins, has not been sampled in place. Obducted Jurassic rocks on Cuba, Hispaniola, Puerto Rico, La Désirade, Costa Rica and northern Venezuela suggest that oldest Mesozoic Caribbean crust is of that age. However, some thin crust could have formed much later (James 2007b).

Hess (1938) observed a circum-Caribbean belt of serpentinized peridotite across Guatemala, the whole length of Cuba, through northern Hispaniola and across Puerto Rico. In the south it runs across northern Venezuela from Margarita through Orchila and El Roque to Cabo Vela on Guajira and southward into the serpentinite belt of the Cordillera Central of Colombia. The most spectacular and widespread occurrences are on Cuba, where more than 6500 sq. km of serpentinite are exposed (Khudoley & Meyerhoff 1971). Yet more lies beneath Late Eocene and younger strata, so the total area is at least 15 000 sq. km (Kozary 1968). According to Donnelly et al. (1990) some (onshore) 'plateau' occurrences are dominantly serpentinite, others basalt. Many consist of highly deformed and scattered mafic lithologies in a serpentinite matrix. Gabbros and peridotite with cumulate texture occur in a serpentinite matrix on NE Nicaragua (Flores et al. 2006). Dredging recovered serpentinite from the centre of the Cayman Trough (Eggler et al. 1973).

Hess (1966) suggested that serpentinized peridotite in the Caribbean region was hydrated upper mantle. The idea is supported by studies of the Galicia Bank and Iberian margins that indicate exhumation and serpentinization of upper mantle near the base of extremely thinned continental crust to form a layer with crust-like seismic velocities (Hopper et al. 2004). Diebold et al. (1999) noted low velocities in the mantle beneath thin crust southeast of the Caribbean plateau. Possible listric faults extend from the seafloor to the Moho. Serpentinite in the Cuban Jarahueca oilfield passing abruptly to almost unaltered peridotite (Rigassi-Studer 1961) supports this origin. Hydration results in density decrease and many serpentinites are diapiric or intruded along fault planes. Hence the typical sheared texture. Intrusion of the circum-Caribbean belt probably occurred around end Middle Eocene, with serpentinite occurring in and perhaps lubricating zones of greatest deformation (Hess 1938).

Volcanic arc rocks

Active, andesitic volcanic arcs on the southwestern and eastern Caribbean Plate margins in Central America and the Lesser Antilles relate to subduction of the Cocos and Atlantic plates, respectively, marked by the Middle America Trench, the Barbados Accretionary Prism in the south and a trench further north.

Volcanism in the Lesser Antilles terminates in the north near Saba, at the southern margin of the Greater Antilles, and in the south at Grenada, near the northern limit of the South American shelf. Tear faults in Atlantic crust in the north and south are envisaged to accommodate eastward movement of the Caribbean relative to the Americas. Central American volcanism terminates abruptly at the limits of convergence between the Cocos and Caribbean plates, at the Mexico–Guatemala border in the north (northern Caribbean Plate boundary) and at the Cocos Ridge in the south.

Ages of arc rocks are important for Caribbean understanding. Donnelly (1989) pointed out that if the Caribbean Plate migrated from the Pacific there should be no subduction zone on its western margin until eastward migration became impeded by Palaeogene collision at its leading edge. Thus it is important to note that Liassic–Lower Dogger volcaniclastic rocks occur in Costa Rica (De Wever et al. 1985) and there is a Middle–Upper Jurassic subduction mélange in NE Nicaragua (Flores et al. 2006).

Upper Jurassic volcanic rocks also occur on Cuba, Hispaniola and Puerto Rico, La Désirade and perhaps Tobago (Fink 1972; Bouysse et al. 1983; Snoke et al. 2001) and volcaniclastic rocks at least as old as Albian are known from the northeastern Lesser Antilles (Donnelly et al. 1990). They suggest that volcanism around the Caribbean area began during Jurassic extension.

Stratigraphic and magmatic similarity of the upper parts of Cuban proximal and distal Jurassic sections shows that this volcanic-arc sequence was related to the continental margin (Cobiella-Reguera 2000) and ancient zircons in Cretaceous arc rocks (Rojas-Agramonte et al. 2006) confirm this. Basalts of the Oxfordian–Early Kimmeridgian El Sabalo Fm and part of the Encrucijada Fm (Aptian–Albian) appear to be oceanic crust contaminated by/erupted through continental crust (Kerr et al. 2003). Tobago arc rocks show continental input from the Albian and suffered tonalitic intrusion at that time (Snoke 1990; Snoke et al. 1990).

Cretaceous volcanic rocks and associated sediments are present on Cuba, Hispaniola, Puerto Rico, Aruba, Curaçao, Venezuela, Trinidad and Tobago. Together with the Aves Ridge, assumed to be an abandoned volcanic arc, and the active Lesser Antilles, assumed to be a Cenozoic volcanic arc, these are regarded as a 'Great Caribbean Arc' (Burke 1988), formed in the Pacific and driven between the Americas ahead of a migrating Caribbean Plate. The total length of this arc, consisting of the Greater Antilles, Lesser Antilles, Netherlands–Venezuelan Antilles is around 4000 km (300 km of

Oligocene–Recent, pull-apart extension removed). Pacific models illustrate the 4000 km Great Arc as originally nearly straight, trending NW or SE, then entering a gap of around 700 km wide and becoming highly curved, extinct and accreted to northern and southern plate margins and extant along the Lesser Antilles. This is, at best, difficult to imagine.

Cretaceous–Eocene Caribbean and Cuban arc rocks were seen to subdivide into Late Jurassic–Early Cretaceous primitive island arc (PIA) and Late Cretaceous–Oligocene calc-alkaline (CA) suites (Donnelly *et al.* 1990). The chemical change occurred abruptly in the Albian (rocks on Hispaniola, Puerto Rico and Tobago, Lebron & Perfit 1993; Frost & Snoke 1989). However, calc-alkaline arc-like rocks occur in Jurassic meta-sedimentary rocks in the Escambray Massif of Cuba and PIA rocks range up to the Turonian in Cuba and the northern Virgin Islands (Kerr *et al.* 2003; Lardeaux *et al.* 2004; Proenza *et al.* 2006; Jolly *et al.* 2006).

The change from PIA to CA chemistry has been attributed to a reversal of subduction direction (Mattson 1984, 110 Ma; Pindell & Barrett 1990, 84 Ma; Lewis & Draper 1990, 95–90 Ma; Lebron & Perfit 1993, Albian). Frost & Snoke (1989) noted the influence of continent-derived sediments in the (oceanic arc) North Coast Schist of Tobago by the Albian. PIA Albian–Lower Cenomanian (planktonic foraminifera; K–Ar 91–102 Ma, Castro & Mederos 1985) volcanic rocks penetrated by the well Patao-1 offshore eastern Venezuela formed above metamorphosed continental crust. If the change of chemistry implies continental input, as suggested by Lebron & Perfit (1993), the arc could not have been in the Pacific when it occurred.

Reversal seems equivocal, with Maresch & Gerya (2005) even questioning its possibility. Kerr *et al.* (2003) conclude that there was no sudden change of chemistry and no Albian subduction polarity reversal. Samples of arc-related basalts of the same stratigraphic unit and age and just a few metres apart on Cuba have different geochemical signatures, reflecting different parental magmas (Kerr *et al.* 1999). Both island arc–tholeiite and calc-alkaline rocks occur in the Colombia basalt of east-central Cuba. Kerr *et al.* (2003) and Jolly *et al.* (2006) suggest that chemical change resulted from increasing sediment input. The latest ideas are that chemistry changed diachronously eastwards and that it resulted from variation in subducted material (Lidiak 2008; Mitchell 2008).

Somoza (2008) suggests that early evolution of the Caribbean area was likely associated with opening of the central Atlantic, with NW–SE extension accommodated in corridors of thinned continental and oceanic lithosphere, separated by strike–slip/transform faults. Tholeiitic lavas emanated along NE trending faults. These faults changed to oblique sinistral transpression and even subduction when motion between the Americas changed to left lateral around 125 Ma. The NE faults became the locus of calc-alkaline activity and HP/LT metamorphism, producing a record similar to a subduction zone.

Volcanic arc activity ceased along 1000 km in Cuba in the Campanian and a new arc, with different orientation, formed in southeastern Cuba, Hispaniola and Puerto Rico in the Palaeogene (Iturralde-Vinent 1995; Mattietti-Kysar 1999; Rojas-Agramonte *et al.* 2006). It formed at 45° to the Cretaceous arc after a 15 Ma gap and faced south or SE, contrary to Pacific models (Iturralde-Vinent 1995). Blein *et al.* (2003) also concluded that two different island arcs were tectonically juxtaposed in central Cuba: the classical Lower and Upper Cretaceous suites of the Greater Antilles arc and a Jurassic to Lower Cretaceous island-arc suite with a Pacific provenance. Contrast in structural grain and lithology between southern Haiti and the rest of Hispaniola suggests that two island arcs existed in the Late Cretaceous, one in southern Haiti, Jamaica and the Nicaragua Rise and the other in northern Hispaniola, Puerto Rico and Cuba (Maurrasse 1981).

Activity in the younger Cuban arc died in the Middle Eocene. Volcanism also largely ceased in Greater Antilles (36 Ma in the Virgin Islands) and along northern South America in the Middle Eocene. Volcanism switched off synchronously from Cuba to Puerto Rico, a distance of around 2000 km.

These multiple arcs, ages and chemistries do not reconcile with a single, Pacific-derived, Great Arc (Burke 1988) that collided diachronously with Cuba–Hispaniola–Puerto Rico.

The relation between subduction and volcanism seems problematic. Most of the Grenadines archipelago has not experienced calc-alkaline activity since the end of the Early Pliocene (*c.* 3.5 Ma), during which time a slab of 90×70 km has been subducted without producing surface volcanic activity (Westercamp 1988). Lesser Antilles basalt–andesites are products of mantle fusion and differentiation at shallow depths; fusion down a Benioff zone does not appear to be involved (Lewis & Gunn 1972). While Caribbean crust supposedly dips to around 300 km below the Maracaibo Basin (tomography, Hilst & Mann 1994), there is no volcanism and little seismicity in the area. No volcanism occurs in NW Colombia above a supposed Caribbean oceanic slab at 158 km (seismicity, Pennington 1981). No volcanism occurs in the Greater Antilles where subduction is supposed to be recorded by both the Muertos and Puerto Rico troughs.

Ancient zircons in arc rocks

Precambrian and Palaeozoic zircons occur in Cretaceous calc-alkaline volcanic arc rocks in central and eastern Cuba and earliest Triassic or older zircons occur in gneiss in western Cuba (Rojas-Agramonte *et al.* 2006). Zircons of the same age occur in metasediments of the continental Escambray Massif. Detrital zircons in Cretaceous metavolcano-sedimentary rocks accreted to NW South America show proximity to continental margin (Cardona *et al.* 2008).

Upper Cretaceous turbidites on Curaçao contain Mesozoic, Paleozoic and Precambrian zircons along with Barremian–Albian and Santonian–Campanian arc-derived grains (Wright 2004). Orthogneiss on the Macanao Peninsula, Margarita Island, contains zircons with Carboniferous crystallization ages (U–Pb 315 +35/−24 Ma, Stockhert *et al.* 1995).

Pacific models maintain that the zircons were picked up by a migrating volcanic arc. The alternative is that they reflect autochthonous basement.

Metamorphic rocks

Metamorphic rocks, commonly HP/LT and attributed to subduction, are widespread on Caribbean Plate margins. Pacific models explain HP/LT rocks by capture of continental crust from Yucatán and Colombia by the entering Caribbean Plate in the Late Cretaceous. After burial to great depths (40–80 km) but at low temperatures, HP/LT metamorphic rocks surfaced 'by arc-parallel stretching resulting from displacement partitioning along an oblique plate margin' (Sisson & Lallemant 2002).

Along both the northern and southern plate boundaries uplift/elevation, size of islands/width of mountains, age of exposed basement, metamorphic grade and age of metamorphism decline from west to east, reflecting eastward-migrating transpression.

Greater Antilles island size diminishes from Cuba through Hispaniola and Puerto Rico to the Virgin Islands. Lower Cretaceous arc rocks are metamorphosed to blueschists in formerly contiguous SE Cuba–NW Hispaniola (Draper 1988) and in Jamaica (Abbott *et al.* 2003), greenschists–amphibolites occur on Puerto Rico while unmetamorphosed, Late Aptian–Early Albian Water Island volcanic rocks occur on the Virgin Islands (Jolly & Lidiak 2005). Jurassic metabasalts occur on La Désirade (Westercamp 1988). Jamaican blueschist formed at 5.1–6.2 kbar (Abbott *et al.* 2003). Eclogite and garnet glaucophanite from the Samaná complex of Hispaniola formed at 22–24 kbar and part of the Cuban Escambray Massif at 16–25 kbar (Stanek & Maresch 2006; *c.* 80 km lithostatic depth).

Along northern S America uplift diminishes from the Sierra Nevada de Santa Marta (>5000 m) through the Coastal Range of Venezuela (2400 m) to the Northern Range of Trinidad (1000 m), finally disappearing below sea level further east. Blueschists and eclogites derived from Late Jurassic–Cretaceous continental slope deposits occur between Puerto Cabello and Choroní in western Venezuela and in the Villa de Cura rocks of north central Venezuela (Menéndez 1966; Sisson & Ave Lallemant 1992). On the Paria Peninsula metamorphism is of extremely low grade (greenschist–chlorite); regional metamorphism grades from quartz–chlorite and quartz–mica schist to phyllites and slates and then sediments (Rodriguez 1968). Volcanic rocks of the islands Frailes and Testigos show, at most, low level metamorphism (Pereira 1985). Metamorphism of Palaeozoic rocks on the Paria Peninsula, Venezuela and in the Northern Range, Trinidad, occurred at 20–30 Ma (Speed *et al.* 1991). However, arc rocks on Tobago were metamorphosed to lower greenschists in the Albian (Snoke *et al.* 2001).

Greenschist to blueschists facies occur in the Purial area of eastern Cuba but in some sections the rocks are only slightly recrystallized (Boiteau *et al.* 1972). Interlayering of carbonate, pelitic and mafic metamorphic rocks is common on the Samaná Peninsula of NE Hispaniola (Joyce 1991). Cretaceous pelagic mudstones and limestone interlayered with basalt sills and flows occur on the Atlantic seafloor north of the Puerto Rico Trench and are exposed on the Presqu'ile de Sud of Haiti. The Samaná rocks may be metamorphosed equivalents (Joyce 1991).

Metamorphic grade in the Caracas Group of northern Venezuela increases northward towards major transcurrent faults of the plate boundary but metamorphism remains locally low (Oxburgh 1966; Wehrmann 1972). HP metamorphic rocks of the western Coastal Range of Venezuela had Jurassic–Cretaceous continental margin protoliths and include reworked Palaeozoic rocks (Sisson *et al.* 2005). Further east greenschists of the Caribbean Series on the north of the Araya–Paria Peninsula are separated by the El Pilar Fault from Late Jurassic–Mid-Cretaceous sedimentary equivalents (Christensen 1961; Cruz *et al.* 2004).

According to Pacific models sedimentary Jurassic continental rocks in western Cuba (Guaniguanico) and their metamorphosed equivalents in south Central Cuba (Isle of Pines, Escambray) were picked up from southern Yucatán and underplated to the forearc of a migrating volcanic arc in the Campanian (Pszczolkowski 1999; Draper & Pindell 2004). However, similar rocks continue through northern Hispaniola to the walls of the Puerto Rico Trough and as far as the Lesser Antilles (e.g. Boiteau *et al.* 1972), a distance of some 2500 km. It is unlikely that they all came from Yucatán. It is also unlikely that some of rocks were buried as much as 80 km to become metamorphosed while others remained unmetamorphosed. Similarly, metamorphic rocks along northern South America occur along with sedimentary rocks, a complication highlighted by Giunta & Oliveri (2009).

Metamorphic rocks systematically occur close to strike–slip faults along Caribbean margins (Goncalves *et al.* 2002). El Tambor HP/LT rocks of Guatemala occur in fault slices associated with the Motagua Fault Zone and preserve primary lithological features (Chiari *et al.* 2004; Harlow *et al.* 2004). Generally slight metamorphism in the Organos Belt of Cuba becomes extreme along the Pinar Fault (Pardo 1975). Pardo (1975) emphasized that Jurassic or older metasediments, marbles, dolomites, schists, graphitic schists, quartzite, gneiss and serpentinite of his Trinidad Belt are surrounded to the west, north and east by rocks of the Cabaigan belt (volcanic sediments, basalts, oolitic limestones tuffs, sandtones). Completely unaltered and unmetamorphosed Upper Jurassic oolitic limestones and Neocomian limestones occur along faults between the belts.

The oldest rock fabrics in eclogite- and blueschist-facies rocks of the Upper Jurassic–Lower Cretaceous Gavilanes unit of south-central Cuba indicate arc-parallel extension and the NW-trending mineral lineations and sheath folds parallel present-day structure (Stanek *et al.* 2006). Fabric in eclogite and blueschist of western Venezuela manifests dextral shear (Avé Lallemant & Sisson 1992), parallel to major faults. Arc-parallel shear affected the North Coast Schist of Tobago during Albian, producing low-grade regional metamorphism (Snoke 2003). Strain must have occurred in the present location/orientation and not along a north–south trending arc.

Paired metamorphic belts are observed on northern Jamaica and Hispaniola (Abbott *et al.* 2003; Nagle 1974) alone and both indicate continent to the south during the Cretaceous. On Cuba thrusting of Bathonian (170 Ma, zircon data) ecologite and metasediments metamorphosed in the Middle to Late Cretaceous (U–Pb, Rb–Sr, Ar/Ar dates) occurred to the north (Stanek *et al.* 2005).

For this paper, it is not reasonable to attribute Mesozoic HP/LT metamorphism in the Caribbean area to ancient subduction zones when no such rocks are seen along active zones. There no evidence of return-path (HP/LT) material in the Lesser Antilles or Central America, where volcanic arc activity has been underway since at least the Albian and probably since the Jurassic. Some

blueschists obducted in the Caribbean show no evidence of coeval volcanism (Maresch & Gerya 2006). Andesites are common in subduction volcanic arcs (Leat & Larter 2002), but there is little mention of andesites along northern South America (Eocene andesite on Los Testigos). This at odds with Mid-Cretaceous subduction invoked to explain eclogites and blueschists from Late Jurassic–Cretaceous continental slope protoliths (e.g. Avé Lallemant & Guth 1990; Avé Lallemant & Sisson 1992; Sisson *et al.* 2005). Bellizia (1972) also noted the absence of paired metamorphism along this area.

The above data suggest development of metamorphism close to plate boundary faults where great burial depths occur in foredeeps associated with strike–slip. Since transpression is migrating eastward along northern and southern plate boundaries, we can look to the east for a modern scenario where HP/LT metamorphism might occur.

The Columbus Basin south of Trinidad today contains up to 25 km of Cenozoic sediments above Mesozoic section perhaps equally thick (e.g. Heppard *et al.* 1998, fig. 5). Heat flow in the contiguous Maturín Basin is low to moderate ($1.3F°/100'$, $1-1.5$ hfu, $35-40$ mW/m^2, George & Socas 1994; Summa *et al.* 2003), as is common in foreland basins. These are conditions capable of generating blueschists.

Maturation of organic matter, clay diagenesis, mineral transformation, osmosis and fluid volume increase with rising temperature can all add to lithostatic pressure. Such mechanisms, coupled with clay seals of zero permeability, are responsible for pressures up to twice lithostatic in the Columbus Basin where Late Cenozoic subsidence (up to 9 km Plio–Pliocene sediments) has been dramatic (Heppard *et al.* 1998).

Transpression west of Trinidad led to thrusting and exhumation of deeply buried rocks. The horizontal stress required to form mountains is estimated to be as high as 6.0 kbar (Price 2001). Transpression that emplaced nappes of oceanic crust several kilometres thick and hundreds of kilometres long onto Caribbean continental margins and rapidly lifted metamorphosed Mesozoic section to form mountain ranges must play an important role.

Palaeontology

Central American land bridge/barrier

Just as Central America forms a land bridge between North and South America today, so it did at times in the past. Faunal studies indicate times of connection and thus separation of Pacific from the Caribbean.

Morris *et al.* (1990) summarized palaeontological evidence of connection between North and South America from the Early Carboniferous, through the Triassic, all of the Jurassic and parts of the Cretaceous (Kauffman 1973; Abouin *et al.* 1982; Rémane 1980). Widespread Jurassic calpionellids in the Caribbean, absent from coeval strata in the Pacific (Rémane 1980), record a barrier between the areas. Dinosaur and marsupial fossils show that North and South America were connected in the Cretaceous (Keast 1972; Estes & Báez (1985). Mammalian faunas show North–South America connection in the Late Cretaceous but little evidence of interchange in the Paleocene and Eocene (Gingerich 1985). Disappearance of hermatypic corals from the Caribbean in the Late Oligocene resulted from isolation from the Pacific by Central America (Frost 1972). An Oligo–Miocene turtle from Costa Rica and Middle–Late Miocene terrestrial mammalian fossils from Panama, Honduras and El Salvador show North American affinities (Lucas *et al.* 2006). The Middle Miocene–Upper Pliocene saw the steady disappearance of genera still living in the Indo-Pacific. Early–Middle Miocene formations in Panamá show sub-aerial environments (Morón & Jaramillo 2008).

In summary, there is evidence of a long-lived Central America land bridge.

Radiolaria

In the Caribbean region ribbon-bedded radiolarites occur on the Nicoya Complex (NW Costa Rica), the Siuna Oceanic Complex (NE Nicaragua), El Tambor Group (Montagua Suture Zone, Guatemala), Duarte Complex (Hispaniola), Mariquita Chert (SW Puerto Rico) and in the Phare Unit (La Désirade) (Baumgartner *et al.* 2005). The only clear occurrence as cover to MORB-type ocean floor seems to be in the Duarte Complex, Hispaniola. On La Désirade they are interbedded with back-arc pillow basalts (Gauchat 2004).

Since no DSDP/ODP site in the Atlantic has penetrated red radiolarian chert, Caribbean bedded cherts are seen to indicate a Pacific origin (e.g. Montgomery & Kerr 2009). However, Jurassic radiolaria occur in sediments just above basalt at Site 534, on the Blake Plateau, close by the Caribbean islands (Bartolini & Larson 2001) and it has to be possible that Atlantic red cherts exist but remain unsampled. Jurassic red cherts are present throughout the Mediterranean and Middle East.

Late Tithonian radiolaria occur in the volcano-sedimentary component of the Northern Ophiolite Belt of Cuba (Cobiella-Reguera 2009). The rocks were obducted from the south. Since continental rocks occur in the walls of the Cayman Trough these oceanic rocks must have come from the Yucatán Basin. Jurassic ribbon-bedded radiolarites in the lower Cretaceous Siuna mélange must also have formed in a Caribbean location.

Some bedded cherts are not associated with ophiolites and mélanges and some formed in deep, elongate basins like the Gulf of California or Red Sea (Blatt *et al.* 2006). The continental margin rift origin for the Caribbean discussed in this paper fits this scenario and might explain some of the radiolarites in the area.

Rudists

Rojas (2004) noted rudistid correlations between the Caribbean margin and Cretaceous volcanic arc rocks. Early Aptian rudistids (*Amphitriscoelus waringi* association) extended from Mexico and Trinidad to France and were linked to the Tethyan continental margin. Albian, Santonian and Campanian *Tepeyacia corrugata, Durania curasavica* and *Barrettia monilifera* associations (Antillean region and Mexico) appear to have characterized the Cretaceous volcanic island arc while the Maastrichtian *Tepeyacia* giganteus association characterized carbonate platforms of the largest Antilles, the Bahamas (Cuba), Central America and Mexico (Chiapas).

Cretaceous rudists from Cuba, Hispaniola and the continental margins and Aptian–Cenomanian gastropods of Mexico, the Gulf of Mexico and the Caribbean show strong affinity with Tethyan faunas (Buitrón-Sánchez & Gómez-Espinosa 2003; Rojas 2004; Mycxynski & Iturralde-Vinent 2005).

Organic material

Upper Cretaceous organic-rich sediments occur at several widely separated sites in the Caribbean and Atlantic. Absence of carbonaceous material in Pacific cores of the same age shows that a barrier separated bottom waters of the Pacific and Caribbean (Saunders *et al.* 1973).

Oil offshore Belize is typed to Jurassic source rocks with chemistry similar to the Smackover of the Gulf of Mexico. If the volcanic arc of Pacific models had passed by, removing continental basement, Jurassic source rocks would not be preserved at this location. Similar oil on Jamaica shows that the island did not come from the Pacific.

Lower Cretaceous rocks on Hispaniola resemble the La Luna of Venezuela (black, chert nodules, Bowin 1975). They indicate similar palaeogeography of restriction close to continent, not open ocean conditions.

Palaeomagnetic data

Palaeomagnetic data show that ophiolite complexes of the Pacific coast of Costa Rica and western Panama formed in an equatorial position and moved approximately 10° since, conforming with the movement of South America (Frisch *et al.* 1992, Calvo & Bolz 1994). Data show that Chorotega has not rotated relative to South America (Di Marco *et al.* 1995). Palaeomagnetism of rocks of the Guaniguanico Terrane, western Cuba indicates a palaeopole not significantly different from North American Jurassic–Cretaceous poles (Alva-Valdivia *et al.* 2001).

Drilling at DSDP Sites 146, 150, 151, 152, 153 penetrated diabase sills below upper Cretaceous sediments. Palaeomagnetic measurements on 23 igneous specimens and a further 11 from adjacent sediments at Sites 146 and 152 conform to those from the surrounding area including Jamaica, Puerto Rico, northern Colombia and Venezuela (Lowrie & Opdyke 1975). Two holes at Site 1001 penetrated Mid-Campanian volcanic rocks. Magnetic directions recorded by the flows indicate that the plateau was near the palaeoequator in the Mid-Campanian (Sigurdsson *et al.* 1996).

In contrast to these '*in situ*' indications, other studies claim to document rotation of elements during plate or block migrations. Cretaceous and Cenozoic intrusive rocks of the Villa de Cura and Tinaco belts of Venezuela are seen to have rotated clockwise by 90° during accretion onto the northern margin of South America as a north–south trending arc collided with the continent (Skerlec & Hargraves 1980). Similar data from the northern plate boundary, where anticlockwise rotation mirror imaging of this model would be expected, are lacking. While data from the Chiapas Massif, southern Mexico, are consistent with a pre-rift location of the Maya Block rotated to the coast of Texas and Louisiana (Molina-Garza *et al.* 1992), rotation of the block is denied by parallelism of its structural trends with regional tectonic fabric. These magnetic data are all from locations of oroclinal bending and strike–slip.

Gose & Swartz (1977) show Honduras migrating since the Aptian–Albian from 10°N to 30°N, back to 20°N then to 15°N to 20°N to 12°N and finally to its present position at 15°N. These unlikely wanderings are the result of poorly controlled stratigraphy (Wilson & Meyerhoff 1978). Steiner (2005) shows a similarly unlikely history for the Maya Block (Yucatán), migrating from a Permian position off NW South America to SW Mexico in the Triassic, then into the northern Gulf of Mexico (Middle–Late Jurassic) and finally south to its present position in the Late Jurassic.

Stratigraphy

Cretaceous

There is geographic continuity from Mesozoic sedimentary to metamorphic rocks and indications of nearby volcanoes along northern South America. The Cenomanian–Turonian La Luna Fm, source of the rich resources of the Maracaibo Basin, contains tuffs and shows silica content increasing to the north. Here, graphitic schists of Las Mercedes Fm also show northward-increasing silica content and lenses of volcanic ash occur in the northernmost outcrops (Wehrmann 1972). Some units are only lightly metamorphosed and contain concretions identical to those in the La Luna Fm (Wehrmann 1972). Further north the Las Mercedes interdigitates with the metamorphic Tacagua Fm, originally tuff or volcanic ash (Dengo 1953; Feo Codecido 1962). The northern sections thicken to several thousand metres and possibly extend down to the Jurassic.

The eastern Venezuela equivalent of the La Luna, the Querecual Fm, also exhibits northward-increasing chert content. Lower Cretaceous (?) serpentinites and low grade metagabbro, metatuffs and metapillow lavas on the Araya Peninsula and Margarita are possible stratigraphic equivalents to the Querecual (Chevalier 1987). Offshore the low-grade metamorphic rocks of the deep-marine, Lower–Upper Mejillones Complex include lavas, volcaniclastics, radiolarites, basalts, massive limestones and organic brown cherts. The well Patao-1 found tholeiitic basalts (radiometric ages 91–102 Ma, Albian–Cenomanian planktonic forams) associated with metamorphosed continental crust (Pereira 1985).

Yet further east the bituminous, cherty argillites of the Albian–Campanian Naparima Hill and Gautier limestone Fms source hydrocarbons in Trinidad. In the Northern Range tuffs occur in the Barremian Toco Fm. The overlying Sans Souci Volcanic Fm is a series of volcanic tuffs, tuff breccias, agglomerates and andesitic lavas erupted onto the passive margin of South America during the Aptian–Santonian (Kugler 1953). Associated sediments are limestones, black, carbonaceous shale and quartz sandstones and conglomerates with continental provenance (Wadge & Macdonald 1985). Albian tuff on Tobago shows continental chemical signal (Sharp 1988; Frost & Snoke 1989).

On the northern Caribbean margin the clay content of the Florida Cretaceous shows montmorillonite in Upper Cretaceous and illite in Lower Cretaceous rocks increasing from north to south (Weaver & Stevenson, 1971, quoted by Meyerhoff & Hatten 1974). Illite comes from a mica–schist terrane such as south Cuba; the montmorillonite comes from a volcanic terrane, such as Cuba. Tuffs and radiolarian cherts in autochthonous deep water Tithonian to Maastrichtian rocks on Cuba (Lewis 1990) record nearby volcanism.

Late Cretaceous–Paleocene

At Site 146/149 aphanitic limestones and claystones of Campanian age are followed by siliceous limestones and black cherts with Maastrichtian claystones and Paleocene laminated claystones interbedded with siliceous limestones (Saunders et al. 1973). The same sequence occurs in the Río Chávez Fm and in the Mucaria Fm of the Piemontine Nappe in northern Venezuela (Vivas & Macsotay 2002). The abyssal sediments remained in their original position, while the Mucaria and Río Chávez formations were imbricated and thrust above the South American passive margin during a Middle Eocene event.

Eocene

Middle Eocene seismic horizons and cherts show Caribbean–Atlantic affinities. Regional Caribbean seismic Horizon A″ is identified at ODP sites 146/ 149 and 153 as the onset of lithification at the Early–Middle Eocene level (Saunders et al. 1973). Chert (cristobalite) and interbedded compact chalks and limestones underlie radiolarian–foraminiferal oozes and chalks. Chert formation involves redistribution of silica from radiolarian tests and no additional silica is involved according to Saunders et al. (1973). However, dispersed volcanic ash forms 10–20% of the deeper sedimentary column at Site 1001 and ash alteration may have provided silica for the chert (Sigurdsson et al. 1996). Coincidence of Horizon A″ with cessation of volcanism along the northern and southern margins of the Caribbean Plate supports this idea. Cristobalite occurs in ash from the Soufriere Hills Volcano of Montserrat (Baxter et al. 1999).

Eocene chert occurs also in cores from the Straits of Florida where rapid deepening (below the CCD?) occurred from the Late Cretaceous through Early Cenozoic (Schlager & Buffler 1984; Roberts et al. 2005). Radiolaria-rich sediments overlying earliest Middle Eocene siliceous limestones and cherts correlated with horizon A″ are also found in the Middle Eocene at Sites 27 and 28 east of the Lesser Antilles and north of the Greater Antilles (Edgar et al. 1973). Similar radiolarian rich sediments appeared suddenly in the Middle Eocene of Barbados, in the Atlantic. Paleocene and Eocene green cherty tuffs and pelagic sediments in Puerto Rico and the Dominican Republic correlate with oceanic horizons Layer A in the Atlantic Ocean and Layer A″ in the Caribbean (Mattson et al. 1972).

Early Middle Eocene Atlantic Layer Ac (Mountain & Tucholke 1985), calibrated by Joides 6 and 7, also comprises cherty turbidites, lutites and siliceous mudstones, rich in radiolaria, diatoms and sponge spicules, overlain by red clays.

There is no record of Middle Eocene cherts in the Pacific.

The Scotland Group of Barbados; the Barbados Ridge and thick sediments in the SW Caribbean. The Barbados accretionary prism includes the Middle Eocene Scotland Group of Barbados, an almost pure quartzite with blue quartz derived from the Guayana Shield. It resembles the Lower– Middle Eocene Mirador and Misoa Fms of Colombia and Venezuela (Senn 1940; Meyerhoff & Meyerhoff 1972; Dickey 1980). The Scotland sandstones are interbedded with hemipelagic units containing Middle and Late Eocene radiolaria (Cuevas & Maurasse 1995).

Models of Caribbean Plate migration along northern South America suggest that Scotland sediments were shunted from a location north of the Maracaibo Basin (Dickey 1980; Burke et al. 1984; Pindell 1993). However, zircon fission track ages for this area range from 50 to 126 Ma. Zircons from Scotland sandstones give fission track ages of 20–80, 200–350 and >500 Ma (Baldwin et al. 1986). Other models have shown the sands arriving north of the Araya Peninsula in the Middle Eocene (Kasper & Larue 1986) or Oligocene (Pindell et al. 1998). The latest model suggests that the ridge comprises two prisms merged in the Miocene (Pindell & Kennan 2009).

As long ago as 1908 Suess noted that older formations of Barbados are genetically related to correlative units on Trinidad. The sandstones lie outboard of the Lesser Antilles arc but contain no volcanic quartz. Instead, polycrystalline quartz, feldspars and heavy minerals reflect orogenic and cratonal metamorphic/plutonic crystalline sources (Kasper & Larue 1986). Exotic rocks like the Naparima Hill and blocks of Albian limestone bearing fauna identical to rocks of Trinidad and eastern Venezuela show northeastern South America provenance (Vaughan & Wells 1945; Kugler 1953; Douglass 1961; Tomblin 1970; Meyerhoff & Meyerhoff 1972). Poor sorting and angular to sub-rounded grains (Senn 1944) indicate abrupt or short distance rather than prolonged transport though Trechmann (1925) described rounded quartz pebbles and sand grains.

Basement ridges on the Atlantic floor cause ponding of modern Orinoco fan sediments (Peter & Westbrook 1976). The Barbados Ridge dies out at the Tiburón Rise (Dolan et al. 1990) where

ODP Leg 110 found Middle Eocene–Oligocene (Middle Eocene peak) 'coarse' sand. Mineralogy of these sands also suggests a South American source. Such sediments are absent from Site 543, just 19 km to the north but DSDP Site 672, on the Atlantic Plate to the east, encountered correlative Middle and Upper Eocene sands (Mascle *et al.* 1986). Northward transport of Barbados Ridge sediments was stopped by the Tiburón Rise, which lies on the Atlantic Plate and is moving west.

Scotland deposition is thought by most to be a deepwater deposit (e.g. Pudsey & Reading 1982), though earlier investigators recognized shallow indicators (Hess 1938 – shallow water, perhaps terrestrial; Baadsgaard 1960 – polygonal mudcracks; Barker & McFarlane 1980 – supratidal flat). Fresh and unbroken fresh-water genera (*Unio, Cyrena, Ampullaria*) occur in 'beds full of large fresh-water mollusca' and show affinity with Soldado Rock, Trinidad (Trechmann 1925). Widespread Middle Eocene shallow water carbonates in the Caribbean region support a shallow origin for the Middle Eocene Scotland.

Barbados Ridge basement is not calibrated by the drill. Seismic data show what appears to be rifted basement to the west and east of the Barbados Accretionary Prism. It could be transitional crust, extended in the Jurassic. This would explain the NE trend of structures on Barbados (Trechmann 1937; Meyerhoff & Meyerhoff 1972), which conform to the regional tectonic grain of Middle America.

The eastern Caribbean was the site of a thick Eocene deposition at sites west of the Aves Ridge, in the Grenada Basin and in front of the Lesser Antilles. A fan of probably Cretaceous–Eocene turbidites lies in the southern part of the Venezuela Basin and the Grenada Basin contains up to 9 km of sediments (Driscoll *et al.* 1995). They do not support opening of the Grenada Basin by Oligocene back-arc spreading (e.g. Pindell *et al.* 2005, fig. 5G) and such a thickness does not support Pacific origins of the Caribbean Plate.

Major Caribbean 'events'

Mattson (1984) noted unconformities and hiatuses around the Caribbean in the Aptian–Albian, Santonian, Paleocene, Middle–Late Eocene and Oligocene. To these we can add Campanian and Miocene unconformities. The latter is seen along northern South America, in Panamá, the Dominican Republic and on the Caribbean Plate interior (Biju-Duval *et al.* 1982; Okaya & Ben-Avraham 1987; Mauffret & Leroy 1997; Jacques & Otto 2003; García-Senz & Pérez-Estaún 2008). The Panamá arc collided with the SW Caribbean in the Middle Miocene and a structural event is observed

in the Serranía del Interior of eastern Venezuela (Benkovics *et al.* 2006). The Campanian event is recognized in western Venezuela (Cooney & Lorente 2009).

Data relevant to three circum-Caribbean unconformities are discussed here because they underscore regional, contiguous geology and they raise questions about responsible processes.

Aptian–Albian

Albian shallow marine limestones, commonly associated with unconformities and often karstified, are regionally distributed around margins of Middle America. Albian plate margin collision, intrusion and change of arc chemistry characterize Caribbean margins (James 2006, fig. 5).

Donnelly (1989) observed that shallow-water, Albian limestones overlie an unconformity around the Caribbean. Shallow marine Albian Atima limestones overlie metasedimentary basement in Honduras (Rogers & Mann 2007). In Nicaragua, thick-bedded limestones that correlate with the Atima Fm contain well-rounded/sorted and imbricated volcanic pebble conglomerates recording a high energy/beach environment (Flores *et al.* 2006). Albian–Campanian platform limestones occur above thin continental Jurassic or Palaeozoic basement rocks on Maya (Cobán/Ixcoy, Campur Fms; Donnelly 1989). Oceanic and continental basement of western Mexico (Guerrero) is capped by Albian limestones (Tardy *et al.* 1994).

On Cuba, Aptian–Albian limestones occur both in the North American passive margin section of the north (Palenque and Guajaibon Fms) and in the volcanic arc terrane of the south (Guaos Fm, Cobiella-Reguera 2001, pers. comm.). On Hispaniola and Puerto Rico the limestones are the Hatillo Fm and the Barrancas and Río Matón limestones (Lewis 2002; Mycxynski & Iturralde-Vinent 2005). The Andros-1 well (Bahamas) bottomed in Albian backreef carbonates (Meyerhoff & Hatten 1974). Albian intertidal–supratidal carbonates form the easternmost Bahamas Platform (Silver and Navidad Banks) and on the southern wall of the Puerto Rico Trench (Schneidermann *et al.* 1972). A shallow-water carbonate platform-complex extended from the West Florida Shelf across what is now the Straits of Florida and northern Cuba to the Bahamas (Iturralde-Vinent 1996).

Albian limestones occur across northern South America (e.g. James 2000). The Late Aptian Machiques Member of the Maracaibo Basin consists of bituminous shales and limestones. Limestones of the overlying Lisure and Maraca Formations range to Late Albian and record open marine conditions. In Eastern Venezuela Albian bituminous limestones and shales occur in the Cutacual Formation of the

Serranía del Interior. They interfinger with open marine El Cantil Formation limestones.

Lewis (2002) listed Albian unconformities on Hispaniola, Cuba, Puerto Rico, Jamaica and the Virgin Islands. Aptian–Albian (110 Ma) metamorphism, deformation and intrusion occurred northeastern Nicaragua, Cuba, Hispaniola and Puerto Rico and a coeval break is present along the southern plate boundary in the Caribbean Mountains, Santa Marta Massif and on the Colombia–Caribbean coast (Mattson 1984). Obduction of peridotites onto Hispaniola occurred in the Aptian–Albian (Draper *et al.* 1996). Metamorphism of the Venezuelan Caracas Group occurred in the Early Albian. There is a probable karstification surface within Aptian carbonates in the Maracaibo Basin (Castillo & Mann 2006) and cavernous limestones at the Lower–Upper Cretaceous boundary indicate subaerial exposure on the Bahamas Platform (Cay Sal well, Paulus 1972). The Late Albian–Early Cenomanian El Abra Fm of Mexico is karstified and there is a karsted unconformity at the top of the Albian Edwards and Orizaba limestones of the San Marcos and Cordoba platforms of Texas and Mexico (Carrasco 2003).

Volcanic arc rock chemistry changed from primitive to calc-alkaline in many areas, reflecting continental input. Aptian U–Pb ages of zircons and equivalent trace-element geochemistry link volcanic rocks and tonalites on Hispaniola, while Albian Ar/Ar ages of tonalite hornblende record final cooling after emplacement (Escuder Viruete *et al.* 2006). Zircon fission track cooling of diorite and tonalite occurred at 103 Ma on Tobago (Cerveny & Snoke 1993), where volcaniclastic rocks of the North Coast Schist show continental input from the Albian (Frost & Snoke 1989).

Foundering of carbonate platforms occurred during the Mid-Cretaceous. DSDP sites along the top of the Campeche Escarpment found Albian shallow-water sediments at depths of 1500–1800 m water depth unconformably overlain by Late Cretaceous–Paleocene deep-water deposits (Worzel *et al.* 1973; Antoine *et al.* 1974). Drowning and breakup of the west Florida–Cuba–Bahamas platform began during the Late Albian (?) to Middle Cenomanian when the southern Straits of Florida became a deep water trough (Iturralde-Vinent 1996).

These data point to regional convergence, uplift, shallow-water carbonate deposition, exposure and karstification and accretion. Perhaps they are related to the Albian separation of South America from Africa (Ladd 1976).

Cenomanian

An angular unconformity developed on Chortís between the marine shales of the Late Albian to Early Cenomanian Krausirpe Fm and the overlying Late Cretaceous Valle de Angeles Fm clastics and there is a palaeokarst on top of the Albian–Cenomanian Atima limestone (Rogers & Mann 2007). An angular unconformity separates Cenomanian or Turonian and the overlying Campanian or Maastrichtian throughout the Greater Antilles (Khudoley & Meyerhoff, 1971). Late Senonian reef/platform carbonates lie above a Cenomanian unconformity on the Nicoya Peninsula, Costa Rica (Calvo & Bolz 2003, pers. comm.). Shallow-water limestones formed in Cuba (Pardo 1975) Puerto Rico (Santos 2002; Laó-Dávila *et al.* 2004) and Jamaica (Mitchell 2005).

A Cenomanian hiatus is recorded on the Guajira Peninsula and Aruba and Curaçao became emergent in the Late Cretaceous (Alvarez 1968; Beets 1972; Wright 2004). Aruba suffered tonalite intrusion at 89 Ma.

Metamorphism affected Guatemala, the Isthmus of Tehuantepec (Dengo 1985) and Aruba (Wright 2004). Cretaceous volcanism ceased along 1000 km of the Cuban arc (Iturralde-Vinent 1995; Cobiella-Reguera 2009), in Jamaica (Draper 1986), the Netherlands Antilles (Beets *et al.* 1984) and Costa Rica (Calvo 2003).

Regional Caribbean seismic Horizon B″ (Case 1974; Donnelly *et al.* 1990; Sigurdsson *et al.* 1996), calibrated by DSDP Leg 15, formed at 90–88 Ma.

Middle Eocene

Kugler (1953) introduced the term wildflysch to South America to describe slump masses of Barremian, Aptian, Albian and Turonian rocks in Paleocene shales in Trinidad and eastern Venezuela. The concept was rapidly applied to units containing large (several to many kilometres, formerly mapped as distinct formations) olistoliths/olistostromes in western Venezuela. The American Geological Institute defines wildflysch as a mappable stratigraphic unit of flysch containing large and irregularly sorted blocks and boulders resulting from tectonic fragmentation, and twisted, contorted and confused beds resulting from slumping or sliding under the influence of gravity (Jackson 1997). In many cases the wildflysch lies beneath nappes that supplied the detritus. The deposits in this sense record approach and emplacement of displaced rocks that supplied and came to overlie the wildflysch. In many Caribbean cases the wildflysch formed and the nappes, some of very great thickness and area, were emplaced rapidly in the Middle Eocene, seemingly recording a regional, abrupt and violent event (Stainforth 1969; James 1997, 2005*a*, 2006, fig. 6).

Flysch and wildflysch deposits occur in Central America (Rivas, Las Palmas and Brito Fms), between the Maya and Chortís Blocks (Sepur Fm), in the Guaniguanico Cordillera (Manacas Fm) and in north central Cuba Vegas and Vega Alta Fms (Cobiella-Reguera 2001, pers. comm.), on Jamaica (Richmond Fm, Wagwater) and Puerto Rico (San German Fm), in NW Colombia (Luruaco Fm; Maco Conglomerate, Aleman 1997, pers. comm.), in western and central Venezuela (Matatere, Río Guache, Guárico, Paracotos Fms, in Trinidad (Pointe-a-Pierre, Chaudiere and Lizard Springs Fms), offshore Venezuela on Bonaire (Rincón Fm) and Margarita (Punta Mosquito/Carnero Fms) and in Golfo Triste wells, on Grenada (Tufton Hall Fm) and on Barbados (Scotland Group) (see James 2005*a*, for comprehensive discussion of these units).

They also occur in SE Mexico (Ocozocuautla Fm, Dengo 1968), in the Parras and Chicontepec basins of NNE Mexico (Tardy *et al.* 1994) and the Veracruz Basin of eastern Mexico (Mossman & Viniegra 1976).

The main phase of arc and ophiolite emplacement in Cuba occurred in the Middle Eocene (Pardo 1975). In Bahía Honda, west Cuba, an apparently overturned sequence includes ultramafic rocks, mafic lavas and gabbros upon the distal margin assemblage mafic tuffs, lavas, slates and laminated limestones (Lewis 1990), recording unroofing of the source area (Pszczolkowski 1999). The Pinar-1 well of western Cuba penetrated thrust sheets with lower Cretaceous, Eocene and Jurassic rocks and bottomed in a thick Jurassic section. Overlying Jurassic sections are thinner upsection, suggesting that the allochthons came from more distal sites. By late Middle or Early Late Eocene time, faulting and subsidence were so intense that great 'chaos breccias' formed in northern Pinar del Río, northern Las Villas Province, and northern Camaguey (Khudoley & Meyerhoff 1971). Blocks of crushed serpentinites are as much as $50-100 \text{ km}^2$ in area.

Metamorphosed mid-oceanic ridge basalt and peridotite were thrusted northward onto early arc rocks of Hispaniola and Jamaica in the Middle Eocene (Draper & Lewis 1991; Draper *et al.* 1996). Jamaican Eocene deposits (the Wagwater Belt) contain flysch derived from the north and NE. Conglomerates contain Cretaceous rudist-bearing limestones, gneiss, schist, quartzite, slate, marble and various igneous rocks indicate a continental provenance with metamorphic basement and Cretaceous carbonate platform (Cuba? Trechmann 1925).

This geology is mirrored in northern Venezuela where metamorphic grade increases upwards (again, unroofing, Stainforth 1966, quoting Menéndez 1975 and Seiders 1965) in the Villa de Cura Group, a $4-5 \text{ km}$ thick section of lower Cretaceous gabbros, diorites, volcanic and volcaniclastic rocks, emplaced from the north in the Middle Eocene above

Paleocene flysch. The western continuation of the Villa de Cura allochthons today lies offset to the north and dispersed along the Aruba–Blanquilla islands (James 2005*a*).

This was a violent and unexplained event, clearly distinct from Oligocene–Recent, diachronous transpression along the northern and southern Caribbean Plate boundaries. Its energy is recorded by the size of emplaced bodies. Cuban ophiolites extend for some 1000 km, are up to 5 km thick and suffered up to 140 km of transport (Cobiella-Reguera 2008, 2009). In Venezuela the Villa de Cura nappe is 250 km long and 5 km thick. Up to 18 thrust slices in Mexico's Veracruz Basin stack 6 km high and moved at least 30 km (Mossman & Viniegra 1976, fig. 2).

The deposits record regional, shared, coeval history between the Caribbean Plate and its continental neighbours. They do not support diachronous passage of a migrating plate or localized 'Proto-Caribbean' subduction along northern South America (Pindell & Barrett 1990, Pindell *et al.* 2006; Pindell & Kennan 2009).

Middle Eocene unconformity overlain by shallow marine limestones

The Middle Eocene event resulted in uplift to wave-base where a regional unconformity formed. Overlying shallow marine, Middle Eocene limestones developed in the photic zone (James 2005*a*).

Several authors refer to this synchronous tectonic event in northern South America. Bell (1972) recognized a Middle Eocene orogeny, involving crustal shortening, overthrusting, uplift and strike-slip faulting in Venezuela and Trinidad. Guedez (1985) noted late Middle Eocene uplift of the Monay–Carora area. Uplift of the Mérida Andes and the Sierra de Perijá began in the Eocene (Chigne & Hernandez 1990; Audemard 1991). Fission-track data indicate Middle Eocene uplift of the Cordillera de la Costa, Serrania del Interior and the Maracaibo Basin (Perez de Armas 1999; Sisson *et al.* 2005) and apatite annealing at 45 Ma shows uplift was occurring on Tobago (Cerveny & Snoke 1993). Maresch *et al.* (1993) and Kluge *et al.* (1995) reported (K–Ar) radiometric data indicating uplift of Margarita Island at 50–55 Ma. A regional unconformity ('Post-Eocene Unconformity') truncates the Eocene section in the Maracaibo Basin. The greatest erosional vacuity (3000 m) is in the north of the basin but yet further north an estimated 4000 m of Eocene were removed from the Guajira area. An Eocene unconformity separates the primary reservoirs above from the secondary reservoir and source rocks below in Colombia's Middle Magdalena Valley (Mora *et al.* 1996).

In the Golfo Triste, western Venezuela, sedimentary upper Eocene lies above lightly metamorphosed Eocene-Paleocene (Ysaccis 1997). The Middle Eocene section of the northern Tuy–Cariaco Basin, east-central Venezuela, comprises lower, deep-water shales and upper, shallow, algal, platform limestones.

In the Greater Antilles tectonic unconformities are observed on land and at sea in the Upper Eocene of Cuba and Hispaniola (Calais & Mercier de Lepinay 1995). In western and central Cuba the Upper Eocene overlies the Lower and Middle Eocene with prominent angular unconformity (Khudoley & Meyerhoff 1971). Arc volcanism in Cuba that had resumed in the Mid or Late Danian along the east–west trending, south-facing Turquino–Cayman Ridge–Hispaniola(?) Arc died in the Middle Eocene (Iturralde-Vinent 1995; Cobiella-Reguera 2009). Arc volcanism also died along 2000 km of the Greater Antilles at this time (Gestel et al. 1999). Isotopic data indicate exhumation of metamorphic rocks on Roatan Island, Honduras, in the Late Eocene–Early Oligocene (Avé Lalleman & Gordon 1999).

A regional unconformity separates the upper Middle Eocene through the lower Oligocene in northern Puerto Rico and Late Eocene–Oliogocene uplift from basinal to shallow marine conditions occurred in the south (collision), with erosion of 1–2 km of section (Montgomery 1998; Larue et al. 1998). Middle Eocene emergence of the Greater Antilles allowed terrestrial flora and fauna to colonize the area (Graham 2003). A prominent Middle to Late Eocene unconformity in the deep western approaches to the Straits of Florida, calibrated by DSDP site 540 and by seismic data, reflects slope instability and mass wasting (Angstadt et al. 1983).

In Central America a Middle Eocene tectonic uplift and erosion terminated activity of the Santa Elena subduction zone (Lew 1985). A hiatus covers most of the Late Eocene–Early Oligocene interval in eastern Panamá, with deep-water sedimentation occurring again in the Middle Oligocene (Bandy & Casey 1973). Continental areas formed in the arc in the Middle–Late Eocene in Costa Rica (Barbosa et al. 1997).

Hunter (1995) described a line of late Middle Eocene algal/foraminiferal limestones from the Central American Isthmus, Colombia (Ciénega de Oro or Tolu) to Venezuela (Tinajitas), capping highly deformed flysch and other deep-water sediments. They are best developed on the frontal thrusts. They occur on Aruba (Helmers & Beets 1977) and Curaçao (Beets 1977). On Bonaire a ?Paleocene/Eocene fluvial conglomerate (Soebi Blanco Fm) is followed by upper Eocene limestones (Lagaay 1969).

Middle Eocene limestones occur on Jamaica (Fonthill limestone, White Limestone, Chapleton Fm), Cuba, Haiti (Plaisance or Hidalgo limestones), St Barts (St Bartholomew Fm), Tortola and Virgin Gorda (Tortola, Necker Fms; Christman 1953; Butterlin 1956; Burke and Robinson 1965; Robinson 1967; Tomblin 1975; Lewis & Draper 1990; Iturralde-Vinent 1994; Pubellier et al. 2000; James & Mitchell 2005). They are known also from the southern wall of the Puerto Rico Trench and from the Mona Canyon (Fox & Heezen 1975; Perfit et al. 1980). Jordan Knoll may have been uplifted in Middle Eocene time (erosion; Bryant et al. 1969).

Shallow marine limestones occur in Costa Rica at Parritilla (southern Valle Central), Damas (Parrita), Punta Catedral (Quepos promontory), Penón de Arío (Nicoya Peninsula), and Quebrada Piedra Azul (Burica Peninsula) (Bolz & Calvo 2002). Carbonate platforms developed on late Middle Eocene folds and thrusts in the South Limón Basin (Fernández et al. 1994). A widespread limestone at the top of the turbiditic Brito Fm in the Nicoya Complex continues through the Nicoya Peninsula (Junquillal and Punta Cuevas Limestones), the Central Pacific provinces (Damas Limestone) and the Térraba Basin (El Cajón and Fila de Cal Limestones) to the Chiriquí (David Limestone) and Tuira-Chucunaque (Corcona Limestone) basins of Panamá (Escalante 1990). Middle Eocene reef limestone (Río Tonosí Fm) overlies upper Cretaceous basalt in SW Panamá (Kolarsky et al. 1995).

In the Lesser Antilles the northeastern branch of low-lying Lesser Antillean islands is called the Limestone Caribees because of extensive Middle Eocene–Pleistocene calcareous cover (Maury et al. 1990). Siliceous limestones with Middle–Early Eocene foraminifera occur on Mayreau and of Upper Eocene, reefal limestones on Carriacou (Tomblin 1970).

Middle Eocene limestone also occurs within the Caribbean Plate on the Aves Ridge, Saba Bank and the Beata Ridge (Fox et al. 1970, 1971; Nagle 1972; Pinet et al. 1985; Bouysse et al. 1990). The Middle Eocene Punta Gorda limestone occurs above deformed Late Cretaceous Valle de Angeles beds on the Nicaragua Rise (Alivia et al. 1984). A Middle to Late Eocene hiatus is recorded in DSDP Site 540 (44–38.5 Ma).

Conclusions

Many aspects of geology between North and South America (Middle America) show regional harmony and a shared history among the many geographic components. Crustal thicknesses, gravity data and high silica content of igneous rocks indicate presence of continental fragments dispersed around and within the Caribbean Plate. They underlie the whole of Central America and the Greater and Lesser Antilles at least the northern and southern Lesser Antilles. They underpin the thick 'plateau' areas of the Venezuelan Basin and probably of the Colombian, Yucatán and Grenada basins as well.

Following Triassic/Jurassic extrusion of the Central Atlantic Magmatic Province North America separated from South America/Africa. Triassic–Jurassic rifting reactivated N35°E trending basement lineaments along eastern North America south to the Gulf of Mexico and on into northern South America. Jurassic–Cretaceous N60°W drift and sinistral offset of North from South America added N60°E extensional strain, strikingly illustrated by the Hess Escarpment, and synthetic, east–west, sinistral offset along the northern Caribbean boundary. The N670°W, N35°E and N60°E structural trends are regionally preserved today in Middle America. They occur on the Caribbean Plate interior and on Maya and Chortís. They prove that the latter blocks have not rotated.

Subsidence of proximal areas (Bahamas and Yucatán–Campeche platforms, Nicaragua Rise) accommodated kilometres-thick carbonate sections. Horsts of continental crust flanked by wedges of Jurassic–Cretaceous sediments, flows and salt formed in more distal areas along the eastern margin of North America and within Middle America (Yucatán, Colombian and Venezuelan Basins). Shallow/subaerial flows of smooth seismic Horizon B″ capped the Caribbean areas in the Late Cretaceous. Areas of extreme extension suffered serpentinization of upper mantle, forming rough Horizon B″. Areas of both occur in the Venezuela, Colombia and Yucatán basins.

While the Gulf of Mexico remained largely intra-continental, the Caribbean area, west of diverging fractures in the Central Atlantic and a lengthening Mid Atlantic Ridge, suffered greater extension and volcanism. Convergence in the Middle and Late Cretaceous and Middle Eocene led to pause or cessation of activity, uplift to wavebase or subaerial erosion and development of unconformities and shallow marine carbonates. Change of chemical composition from primitive to calc-alkaline recorded continental input. The Middle and Late Cretaceous events in the Colombia and Venezuela basins are recorded by abundant extrusion (subaerial flood basalts) and organic-rich sediments of restricted marine conditions. The Middle Eocene regional, convergent event terminated most volcanic activity along the northern and southern Caribbean Plate boundaries, where Oligocene–Recent strike–slip followed. The only oceanic spreading

in Middle America, with recognizable spreading ridges and magnetic anomalies, occurred in the central 300 km of the Cayman Trough during this latest tectonic phase. The Caribbean Plate has extended eastwards over Atlantic crust. Analogy with the Scotia Plate, which occupies a similar tectonic setting, suggests that this occurred by back-arc spreading along the Aves Ridge.

I thank B. Bally for many detailed, perceptive comments on this paper and M. A. Lorente, Y. Chevalier, G. Giunta and C. Rangin for several general suggestions. H. Krause and A. Aleman provide continuing encouragement in my efforts to change the 'Pacific Paradigm'. Thanks to M. A. Lorente, C. Repzka de Lombas, Lorenzo and N. Villalobos for their great enthusiasm and energy in the organization of the Sigüenza 2006 conference, which led to this publication. Thanks to those who attended, including chairpersons B. Bally, T. Lorente and D. Roberts, and made that meeting an enjoyable success. Thanks to all who contributed to this volume.

References

ABBOTT, R. N. JR, BANDY, B. R., JACKSON, T. A. & SCOTT, P. W. 2003. Blueschist–Greenschist Transition in the Mt. Hibernia Schist, Union Hill, Parish of St. Thomas, Jamaica. *International Geology Review*, **45**, 1–15.

ABOUIN, J., AZEMA, J. *ET AL.* 1982. The Middle America Trench in the geological framework of Central America. *Initial Reports of the Deep Sea Drilling Project*, **67**, 747–775.

AGER, D. V. 1986. Migrating fossils, moving plates and an expanding Earth. *Modern Geology*, **10**, 377–390.

AIELLO, I. W., CHIARI, M., PRINCIPI, G. & GARCÍA, D. 2004. Stratigraphy of Cretaceous radiolarian cherts of Cuba (absent). *32nd IGC*, Florence, 2004, IGCP Project 433 Caribbean Plate Tectonics, Abstracts.

ALIVIA, F., TAPPMEYER, D., AVES, H., GILLET, M. & KLENK, C. 1984. Recent studies of basins are encouraging for future exploration of Honduras. *Oil and Gas Journal*, **17**, 139–149.

ALLIBON, J., LAPIERRE, H., BUSSY, F., TARDY, M., CRUZ GAMEZ, E. M. & SENEBIER, F. 2008. Late Jurassic continental flood basalt doleritic dykes in northwestern Cuba: remnants of the Gulf of Mexico opening. *Bulletin Societe Géologie Français*, **179**, 445–452.

ALONSO, R. M., KRIEG, E. A. & MEYERHOFF, A. A. 1987. Age and origin of the Puerto Rico Trench. *Transactions, X Caribbean Geological Conference*, Cartagena, 1983, 82–103.

ALVAREZ, W. 1968. *Geology of the Simarua and Carpintero areas, Guajira Peninsula, Colombia*. PhD thesis, Princeton University.

ALVA-VALDIVIA, L. M., GOGUITCHAIVHVILI, A. *ET AL.* 2001. Palaeomagnetism of the Guaniganico Cordillera, Western Cuba: a pilot study. *Cretaeous Research*, **22**, 705–718.

ANDERSON, T. H. & SCHMIDT, V. A. 1983. The evolution of Middle America and the Gulf of Mexico-Caribbean

Sea region during Mesozoic time. *Geological Society of America Bulletin*, **94**, 941–966.

ANDREANI, L., RANGIN, C. *ET AL.* 2008. The Neogene Veracruz Fault: evidences for left-lateral slip along the southern Mexico block. *Bulletin de la Société Géologique de France*, **179**, 195–208.

ANGSTADT, D. M., AUSTIN, J. A. JR & BUFFLER, R. T. 1983. Deep-sea erosional unconformity in the southeastern Gulf of Mexico. *Geology*, **11**, 215–218.

ANTOINE, J. W. & BRYANT, W. R. 1969. Distribution of salt and salt structures in Gulf of Mexico. *American Association of Petroleum Geologists Bulletin*, **53**, 2543–255.

ANTOINE, J. W., MARTIN, R. G., PYLE, T. G. & BRYANT, W. R. 1974. Continental margins of the Gulf of Mexico. *In*: BURK, C. A. & DRAKE, C. L. (eds) *The Geology of Continental Margins*. Springer, Berlin, 683–693.

ARDEN, D. D. 1975. The geology of Jamaica and the Nicaragua rise. *In*: NAIRN, A. E. M. & STEHLI, F. G. (eds) *The Ocean Basins and Margins*, **3**, *The Gulf of Mexico and the Caribbean*. Plenum Press, New York, 617–661.

AUDEMARD, F. M. 1991. *Tectonics of Western Venezuela*. PhD thesis, Rice University, Houston, Texas.

AUGER, L. S., ABERS, G. *ET AL.* 2004. Crustal thickness along the Central American Volcanic Arc. *EOS Transactions, American Geophysical Union*, **85**, Fall Meeting Supplement, Abstract.

AVÉ LALLEMENT, H. G. 1997. Transpression, displacement partitioning, and exhumation in the eastern Caribbean/South American plate boundary. *Tectonics*, **16**, 272–289.

AVÉ LALLEMENT, H. G. & GUTH, L. R. 1990. Role of extensional tectonics in exhumation of eclogites and blueschists in an oblique subduction setting: Northeastern Venezuela. *Geology*, **18**, 950–953.

AVÉ LALLEMENT, H. G. & SISSON, V. B. 1992. Burial and ascent of blueschists and eclogites, Venezuela, part 1: structural constraints. *Geological Society of America Annual Meeting*, Abstracts, A149.

AVÉ LALLEMENT, H. G. & GORDON, M. B. 1999. Deformation history of Roatan Island: implications for the origin of the Tela Basin (Honduras). *In*: MANN, P. (ed.) *Caribbean Sedimentary Basins, Sedimentary Basins of the World*. Elsevier, 197–218.

BAADSGAARD, P. H. 1960. Barbados, W. I. Exploration Results 1950–1958. *In*: KALE, A. & METZGER, A. (eds) *Structure of the Earth's Crust and Deformation of Rocks: 21st International Geological Congress*, Copenhagen, 18.

BAIE, L. F. 1970. Possible structural link between Yucatán and Cuba. *American Association of Petroleum Geologists Bulletin*, **54**, 2204–2207.

BAIN, J. E. & HAMILTON, K. J. 1999. Finding new structural features that focus sediment flows. *Offshore*, 78–80.

BALDWIN, S. L., HARRISON, T. M. & BURKE, K. 1986. Fission track evidence for the source of accreted sandstones, Barbados. *Tectonics*, **5**, 457–468.

BALL, M. M., HARRISON, C. G. A. & SUPKO, P. R. 1969. Atlantic opening and the origin of the Caribbean. *Nature*, **223**, 167–168.

BANDY, O. L. & CASEY, R. E. 1973. Reflector Horizons and Paleobathymetric History, Eastern Panama.

Geological Society of America Bulletin, **84**, 3081–3086.

BANKS, P. O. 1975. Basement rocks bordering the Gulf of Mexico and the Caribbean Sea. *In*: NAIRN, A. E. M. & STEHLI, F. G. (eds) *The Ocean Basins and Margins*, **3**, *The Gulf of Mexico and the Caribbean*. Plenum Press, New York, 181–199.

BARBOSA, G., FERNÁNDEZ, J. A., BARRIENTOS, J. & BOTTAZI, G. 1997. Costa Rica: petroleum geology of the Caribbean margin. *The Leading Edge*, 1787–1794.

BARKER, P. F. 2001. Scotia Sea tectonic evolution: implications for mantle flow and palaeocirculation. *Earth-Science Reviews*, **55**, 1–39.

BARKER, L. & MCFARLANE, N. 1980. Notes on some sedimentological evidence for shallow water origin of parts of the Scotland Formation of Barbados. *Journal of the Geological Society of Jamaica*, **xix**, 46.

BARTOK, P. E., RENZ, O. & WESTERMAN, G. E. G. 1985. The sisquisique ophiolites, Northern Lara State, Venezuela: a discussion on their Middle Jurassic ammonites and tectonic implications. *Geological Society of America Bulletin*, **96**, 1050–1055.

BARTOLINI, A. & LARSON, R. L. 2001. Pacific microplate and the Pangea supercontinent in the Early to Middle Jurassic. *Geology*, **29**, 735–738.

BATESON, J. H. 1972. New interpretation of geology of Maya Mountains, British Honduras. *American Association of Petroleum Geologists Bulletin*, **56**, 956–963.

BAUMGARTNER, P. O., BANDINI, A. N. & DENYER, P. 2005. Remnants of Pacific ocean floor: Jurassic-lower Cretaceous radiolarites in central America and the Caribbean. *17th Caribbean Geological Conference*, Puerto Rico, Abstracts, 5.

BAXTER, P. J., BONADONNA, C. *ET AL.* 1999. Cristobalite in volcanic ash of the Soufriere Hills Volcano, Montserrat, British West Indies. *Science*, **283**, 1142–1145.

BEETS, D. J. 1972. *Lithology and stratigraphy of the Cretaceous and Danian succession of Curaçao*. Uitgaven Natuurweternschappelijke Studierkring voor Suriname en Nederlandse Antillen, Utrecht, **70**, 153.

BEETS, D. J. 1977. Cretaceous and Early Tertiary of Curaçao. *In*: *8th Caribbean Geological Conference Guide to Field Excursions*. Geophysical Union of America, Papers of Geology, Amsterdam, **10**, 18–28.

BEETS, D. J., MARESCH, W. V. *ET AL.* 1984. Magmatic rock series and high-pressure metamorphism as constraints on the tectonic history of the southern Caribbean. *In*: BONINI, W. E. *ET AL.* (eds) *The Caribbean–South American Plate Boundary and Regional Tectonics*, Geological Society of America Memoirs, **162**, 95–130.

BELL, J. S. 1972. Geotectonic evolution of the southern Caribbean area. *In*: *Studies in Earth and Space Sciences*. Geological Society of America Memoirs, **132**, 367–386.

BELLIZIA, A. G. 1972. Sistema Montañoso del Caribe, Borde Sur de la Placa Caribe ¿Es Una Cordillera Aloctona? *VI Conferencia Geológica del Caribe*, Margarita, Venezuela, Memorias, 247–257.

BELLIZIA, A. & DENGO, C. 1990. The Caribbean mountain system, northern South America; a summary. *In*: DENGO, G. & CASE, J. E. (eds) *The Geology of*

North America, Vol. H, The Caribbean Region, Geological Society of America, 167–175.

BENKOVICS, L., FRANCO, A., ETEMADI, M., FINTINA, C. & NOVOA, E. 2006. Analogies and differences between the Eastern Venezuela fold and thrust belt and the Trinidad Nariva fold and thrust belt. *In*: JAMES, K. H. & LORENTE, M. A. (eds) *Geology of the Area between North and South America, with Focus on the Origin of the Caribbean Plate*. An International Research Conference, Sigüenza, Spain, Abstracts.

BENSON, R. N. & DOYLE, R. G. 1988. Early Mesozoic rift basins and the development of the United States middle Atlantic continental margin. *In*: MANZPEIZER, W. (ed.) *Triassic–Jurassic Rifting*, Part A. Elsevier, Amsterdam, 99–127.

BIJU-DUVAL, B. G., BIZON, A., MASCLE & MULLER, C. 1982. Active margin processes: field observations in Southern Hispaniola. *In*: WATKINS, J. S. & DRAKE, C. L. (eds) *Studies in Continental Margin Geology*. American Association of Petroleum Geologists Memoirs, **34**, 325–344.

BIRD, D. E., HALL, S. A. & CASEY, J. F. 1999. *Tectonic Evolution of the Grenada Basin: Caribbean Basins*. Sedimentary Basins of the World Series, **4**. Elsevier, Amsterdam, 389–416.

BIRD, D., HALL, S., BURKE, K., CASEY, J. & SAWYER, D. 2008. *Mesozoic seafloor spreading history of the Central Atlantic Ocean*. World Wide Web Address: http://www.conjugatemargins.com/abstracts/by_session/1

BIRKHOLD, A. B., NEAL, C. R., MAHONEY, J. J. & DUNCAN, R. A. 1999. The Ontong Java Plateau: episodic growth along the SE margin. *American Geophysical Union Fall Meeting*, San Francisco, EOS, **80**, F1103.

BLATT, H., TRACEY, R. J. & OWENS, B. E. 2006. *Petrology, Igneous, Sedimentary and Metamorphic*. Freeman, New York.

BLEIN, O., GUILLOT, S. H. *ET AL*. 2003. Geochemistry of the Mabujina Complex, Central Cuba: Implications on the Cuban Cretaceous Arc Rocks. *Journal of Geology*, 2003, **111**, 89–101.

BOITEAU, A., MICHARD, A. & SALIOT, P. 1972. High pressure metamorphism within the Purial Ophiolite Complex (Oriente, Cuba). *Comptes Rendus Académie de Science (Paris), Series D*, **274**, 2137–2140.

BOLZ, A. & CALVO, C. 2002. Calizas Lutetianas del arco interno Paleógeno de Costa Rica. *Revista Geológica de América Central*, **26**, 7–24.

BONIS, S. B. 1969. *Evidence for a Paleozoic Cayman Trough*. Geological Society of America Special Papers, **121**, Abstracts, 32.

BOURGOIS, J., TOUSSAINT, J.-F. *ET AL*. 1987. Geological history of the Cretaceous ophiolitic complexes of northwestern South America (Colombian Andes). *Tectonophysics*, **143**, 307–327.

BOUYSSE, P. 1984. The Lesser Antilles Island Arc: structure and geodynamic evolution. *Initial Reports, DSDP*, **LXXVIIA**, 83–103.

BOUYSSE, P. 1988. Opening of the Granada back-arc basin and evolution of the Caribbean Plate during the Mesozoic and Early Paleogene. *Tectonophysics*, **149**, 121–143.

BOUYSSE, P., SCHMIDT-EFFING, R. & WESTERCAMP, D. 1983. La Desirade Island (Lesser Antilles) revisited: Lower Cretaceous radiolarian cherts and arguments against an ophiolitic origin for the basal complex. *Geology*, **11**, 244–247.

BOUYSSE, P., ANDREIEFF, P. *ET AL*. 1985. Aves Swell and northern Lesser Antilles Ridge: rock dredging results from Arcante 3 Cruise. *Géodynamique des Caraibes*, Symposium Paris, Editions Technip, **27**, Rue Ginoux, 65–76.

BOUYSSE, P., WESTERCAMP, D. & ANDREIEFF, P. 1990. *The Lesser Antilles Island Arc: Proceedings of the Ocean Drilling Program, Scientific Results*. College Station, TX, **110**, 29–44.

BOWIN, C. 1975. The geology of Hispaniola. *In*: NAIRN, A. E. M. & STEHLI, F. G. (eds) *The Ocean Basins and Margins*, **3**. Plenum Press, New York, 501–552.

BOWIN, C. 1976. Caribbean gravity field and plate tectonics. *Geological Society of America Special Papers*, **169**.

BOWLAND, C. L. 1993. Depositional history of the western Colombian Basin, Caribbean Sea, revealed by seismic stratigraphy. *Geological Society of America Bulletin*, **105**, 1321–1345.

BOWLAND, C. L. & ROSENCRANTZ, E. 1988. Upper crustal study of the western Colombian Basin, Caribbean Sea. *Geological Society of America Bulletin*, **100**, 534–546.

BUITRÓN-SÁNCHEZ, B. E. & GÓMEZ-ESPINOSA, C. 2003. Cretaceous (Aptian-Cenomanian) gastropods of Mexico and their biogeographic implications. *In*: BARTOLINI, C., BUFFLER, R. T. & BLICKWEDE, I. F. (eds) *The Circum-Gulf of Mexico and the Caribbean: Hydrocarbon Habitats, Basin Formation, and Plate Tectonics*. American Association of Petroleum Geologists, Memoir **79**, p. 403–418.

BUFFLER, R. T. 1983. Structure of the Sigsbee Scarp, Gulf of Mexico. *In*: BALLY, A. W. (ed.) *Seismic Expression of Structural Styles*. American Association of Petroleum Geologists Studies in Structural Geology, Series 15, **2**.

BURK, C. A., EWING, M. *ET AL*. 1969. Deep-sea drilling into the Challenger Knoll, central Gulf of Mexico. *American Association of Petroleum Geologists Bulletin*, **53**, 1338–1347.

BURKE, K. 1988. Tectonic evolution of the Caribbean. *Annual Reviews in Earth and Planetary Science*, **16**, 201–230.

BURKE, K. & ROBINSON, E. 1965. Sedimentary structures in the Wagwater Belt, Eastern Jamaica. *The Journal of the Geological Society of Jamaica*, **VII**, 1–10.

BURKE, K., COOPER, C., DEWEY, J. F., MANN, P. & PINDELL, J. L. 1984. Caribbean tectonics and relative plate motions. *In*: BONINI, W. E. *ET AL*. (eds) *The Caribbean-South American Plate Boundary and Regional Tectonics*. Geological Society of America Memoirs, **162**, 31–63.

BUTTERLIN, J. 1956. *La constitucion géologique et la structure des Antilles*. Centre National de la Recherche Scientifique, Paris.

BRYANT, W. R., MEYERHOFF, A. A., BROWN, N. K., FURRER, M. A., PYLE, T. E. & ANTOINE, J. W. 1969. Escarpments, reef trends, and diapiric structures,

Eastern Gulf of Mexico. *American Association of Petroleum Geologists Bulletin*, **53**, 2506–2542.

CALAIS, E. & MERCIER DE LEPINAY, B. 1995. Strike–slip tectonic processes in the northern Caribbean between Cuba and Hispaniola (Windward Passage). *Marine Geophysical Research*, **17**, 63–95.

CALVO, C. 2003. Provenance of plutonic detritus in cover sandstones of Nicoya Complex, Costa Rica: Cretaceous unroofing history of a Mesozoic ophiolite sequence. *Geological Society of America Bulletin*, **115**, 832–844.

CALVO, C. & BOLZ, A. 1994. The oldest calcalkaline island arc volcanism in Costa Rica. Marine tephra deposits from the Loma Chumico Formation (Albian to Campanian). *In*: SEYFRIED, H. & HELLMANN, W. (eds) *Geology of an Evolving Island Arc*, **7**. Institute of Geology and Paleontology, University of Stuttgart, 235–264.

CANDE, S. C., LABREQUE, J. L., LARSON, R. L., PITMAN III, W. C., GOLOVCHENKO, X. & HAXBY, W. F. 1989. *Magnetic Lineations of the World's Ocean Basins*. American Association of Petroeleum Geologists Bookstore, Tulsa, OK. Map.

CARDONA, A., VALENCIA, V. *ET AL.* 2008. Cenozoic exhumation of the Sierra Nevada De Santa Marta, Colombia: implications on the interactions between the Carribean and South American Plate. *2008 Joint Meeting of the Geological Society of America, American Society of Agronomy, Crop Science Society of America, Gulf Coast Association of Geological Societies with the Gulf Coast Section of the Society for Sedimentary Petrology*, Abstracts, paper no. 128-5.

CARRASCO-V, B. 2003. Paleokarst in the marginal Cretaceous rocks, Gulf of Mexico. *In*: BARTOLINI, C. *ET AL.* (eds) *The Circum-Gulf of Mexico and the Caribbean: Hydrocarbon habitats, basin formation, and plate tectonics. American Association of Petroleum Geologists Memoirs*, **79**, 169–183.

CASE, J. E. 1974. Oceanic crust forms basement of Eastern Panamá. *Geological Society of America, Bulletin*, **85**, 645–652.

CASE, J. E. & HOLCOMBE, T. L. (compilers). 1980. Geologic–tectonic map of the Caribbean region. *U.S. Geological Survey Miscellaneous Investigation Series Map* **I-1100**, scale 1 : 2 500 000.

CASE, J. E., MACDONALD, W. D. & FOX, P. J. 1990. Caribbean crustal provinces; Seismic and gravity evidence. *In*: DENGO, G. & CASE, J. E. (eds) *The Geology of North America*, **H**, *The Caribbean Region*. Geological Society of America, 15–36.

CASTRO, M. & MEDEROS, A. 1985. Litoestratigrafia de la Cuenca de Carúpano. *VI Congreso Geologico Venezolano*, Caracas, Memoria, **I**, 201–225.

CASTILLO, M. V. & MANN, P. 2006. Deeply buried, Early Cretaceous paleokarst terrane, southern Maracaibo Basin, Venezuela. *American Association of Petroleum Geologists Bulletins*, **90**, 529–565.

CERVENY, P. F. & SNOKE, A. W. 1993. Thermochronologic data from Tobago, West Indies: constraints on the cooling and accretion history of Mesozoic oceanic-arc rocks in the southern Caribbean. *Tectonics*, **12**, 433–440.

CHARLTON, T. R. 2004. The petroleum potential of inversion anticlines in the Banda Arc. *American*

Association of Petroleum Geologists Bulletins, **88**, 565–585.

CHEVALIER, Y. 1987. *Les zones internes de la chaîne sud-caraibe sur la transect: Ile de Margarita–Pénsula d'Araya (Venezuela)*. PhD thesis, Université de Bretagne Occidentale, Brest.

CHIARI, M., DUMITRICA, P., MARRONI, M., PANDOLFI, L. & GIANFRANCO, G. 2004. Paleontological evidences for a Late Jurassic age of the Guatemala ophiolites. *32nd IGC*, Florence, 2004, IGCP Project 433 Caribbean Plate Tectonics, Abstracts.

CHIGNE, N. & HERNANDEZ, L. 1990. Main aspects of petroleum exploration in the Apure area of southwestern Venezuela, 1985–1987. *In*: BROOKS, J. (ed.) *Classic Petroleum Provinces*. Geological Society, London, Special Publications, **50**, 55–75.

CHRISTENSEN, R. M. 1961. *Geology of the Paria-Araya Peninsula, Northeastern Venezuela*. PhD University of Nebraska.

CHRISTMAN, R. A. 1953. Geology of St. Bartholomew, St. Martin, and Anguilla, Lesser Antilles. *Geological Society of America Bulletin*, **64**, 65–96.

CHRISTOFFERSON, E. 1976. Colombian Basin magmatism and Caribbean Plate tectonics. *Geological Society of America Bulletin*, **87**, 1255–1258.

CHUBB, L. J. & BURKE, K. 1963. Age of the Jamaican granodiorite. *Geology Magazine*, **100**, 524–532.

COBIELLA-REGUERA, J. L. 2000. Jurassic and Cretaceous geological history of Cuba. *International Geology Review*, **42**, 594–616.

COBIELLA-REGUERA, J. L. 2008. Reconstrucción palin-spástica del paleomargen de América del Norte en Cuba occidental y el sudeste del Golfo de México. Implicaciones para la evolución del SE del Golfo de México. *Revista Mexicana de Ciencias Geológicos*, **25**, 382–401.

COBIELLA-REGUERA, J. L. 2009. Emplacement of the northern ophiolites of Cuba and the Campanian–Eocene geological history of the northwestern Caribbean–SE Gulf of Mexico region. *In*: JAMES, K. H., LORENTE, M. A. & PINDELL, J. L. (eds) *Origin and Evolution of the Caribbean Plate*. Geological Society, London, Special Publications, **328**, 315–338.

CONTRERAS, V. H. & CASTILLÓN, M. B. 1968. Morphology and origin of salt domes of Isthmus of Tehuantepec. *In*: BRAUNSTEIN, J. & O'BRIAN, G. D. (eds) *Diapirism and Diapirs*. American Association of Petroleum Geologists Memoirs, **8**, 244–260.

COONEY, P. M. & LORENTE, M. A. 2009. A structuring event of Campanian age in western Venezuela, interpreted from seismic and palaeontological data. *In*: JAMES, K. H., LORENTE, M. A. & PINDELL, J. L. (eds) *Origin and Evolution of the Caribbean Plate*. Geological Society, London, Special Publications, **328**, 687–703.

CRAWFORD, F. D., SZELEWSKI, C. E. & ALVEY, G. D. 1985. Geology and exploration in the Takutu Graben of Guyana and Brazil. *Journal of Petroleum Geology*, **8**, 5–36.

CRUZ, L., FAYON, A., TESSIER, C. & WEBER, J. 2004. Exhumation and deformation processes in transpressional orogens: the Venezuelan Paria Peninsula, SE Caribbean–South American plate boundary. *In*: TILL, A. B. *ET AL.* (eds) *Exhumation Associated with*

Continental Strike–Slip Fault Systems. Geological Society of America Special Papers, **434**, 149–165.

CUEVAS, E. D. & MAURASSE, F. J-M. R. 1995. Radiolarian biostratigraphy of the Conset Bay Series, Barbados, West Indies. *3rd Geological Conference of the Geological Society of Trinidad and Tobago and 14th Caribbean Geological Conference*, Abstracts, 20.

DALZIEL, I. W. D. 1972. The tectonic framework of the southern Antilles (Scotia Arc). Its possible bearing on the evolution of the Antilles. *VI Conferencia Geológica del Caribe–Margarita, Venezuela, Memorias*, 300–301.

DAVISON, I. 2005. Central Atlantic margin basins of North West Africa: Geology and hydrocarbon potential (Morocco to Guinea). *Journal of African Earth Sciences*, **43**, 254–274.

DAVISON, I., TAYLOR, M. & LONGACRE, M. 2003. Central Atlantic Salt Basins Correlation. *AAPG International Meeting*, Barcelona, Extended Abstracts.

DEERING, C. D., VOGEL, T. A., PATINO, L. C. & ALVARADO, G. E. 2004. Implications for the petrogenesis of distinct silicic magma types from the Lower Pleistocene Guachipelin Caldera, NW Costa Rica. *EOS Transaction of the American Geophysical Union*, 85, Fall Meeting Supplement, Abstract.

DEMETS, C., JANSMA, P. E. *ET AL.* 2000. GPS geodetic constraints on Caribbean–North America plate motion: implications for plate rigidity and oblique plate boundary convergence. *Geophysical Research Letters*, **27**, 437–440.

DENGO, G. 1953. Geology of the Caracas region in Venezuela. *Bulletin of the Geological Society of America*, **64**, 7–40.

DENGO, G. 1968. *Estructura geológico, historia tectónica y morfología de América Central*. Guatemala Instituto Centroamericano de Investigación y Tecnología Industrial, Centro Regional de Ayuda Técnica, Agencia para el Desarrollo Internacional.

DENGO, G. 1975. Paleozoic and Mesozoic tectonic belts in Mexico and Central America. *In*: NAIRN, A. E. M. & STEHLI, F. G. (eds) *The Ocean Basins and Margins*, **3**. *The Gulf of Mexico and the Caribbean*. Plenum Press, New York, 283–323.

DENGO, G. 1985. Mid America: tectonic setting for the Pacific margin from southern Mexico to northwestern Colombia. *In*: NAIRN, A. E. M., STEHLI, F. G. & UYEDA, S. (eds) *The Ocean Basins and Margins*, **7A**. *The Gulf of Mexico and the Caribbean*. Plenum Press, New York, 123–180.

DESPRETZ, J. M., DALY, T. E. & ROBINSON, E. 1985. Geology and petroleum potential of Saba Bank area, northeastern Caribbean. *American Association of Petroleum Geologists Bulletin*, **69**, 249, Abstract.

DE WEVER, P., AZEMA, J., TOURNON, J. & DESMET, A. 1985. Discovery of Lias–Lower Dogger Ocean material on Santa Elena Peninsula (Costa Rica, Central America). *Pierre & Marie Curie Université, Nancy Université Compte Renduez*, **300**, 759–764.

DICKEY, P. A. 1980. Barbados as a fragment of South America ripped off by continental drift. *Transactions Caribbean Geological Conference*, 51–52.

DIEBOLD, J. 2009. Submarine volcanic stratigraphy and the Caribbean LIP's formational environment. *In*:

JAMES, K. H., LORENTE, M. A. & PINDELL, J. L. (eds) *Origin and Evolution of the Caribbean Plate*. Geological Society, London, Special Publications, **328**, 799–808.

DIEBOLD, J., DRISCOLL, N. & THE EW-9501 SCIENCE TEAM. 1999. New insights on the formation of the Caribbean Basalt Province revealed by multichannel seismic images of volcanic structures in the Venezuela Basin. *In*: MANN, P. (ed.) *Caribbean Sedimentary Basins, Sedimentary Basins of the World*. Elsevier, Amsterdam, 561–589.

DIEBOLD, J. B., STOFFA, P. L., BUHL, P. & TRUCHAN, M. 1981. Venezuela Basin crustal structure. *Journal of Geophysical Research*, **86**, 7901–7923.

DILLON, W. P. & VEDDER, J. G. 1973. Structure and development of the continental margin of British Honduras. *Geological Society of America Bulletin*, **84**, 2713–2732.

DI MARCO, G., BAUMGARTNER, P. O. & CHANNELL, J. E. T. 1995. Late Cretaceous–Early Tertiary paleomagnetic data and a revised tectonostratigraphic subdivision of Costa Rica and western Panama. *In*: MANN, P. (ed.) *Geologic and Tectonic Development of the Caribbean Plate Boundary in Southern Central America*. Geological Society of America Special Papers, **295**, 1–28.

DOLAN, J. F., BECK, C., OGAWA, Y. & KLAUS, A. 1990. Eocene–Oligocene sedimentation in the Tiburón Rise/ODP Leg 110 area: an example of significant upslope flow of distal turbidity currents. *Proceedings of the Ocean Drilling Program, Scientific Results*, **110**, 47–63.

DONNELLY, T. W. 1964. Evolution of eastern Greater Antilles island arc. *American Association of Petroleum Geologists Bulletin*, **48**, 680–696.

DONNELLY, T. W. 1966. Geology of St. Thomas and St. John, Virgin Islands. *In*: HESS, H. H. (ed.) *Caribbean Geological Investigations*. Geological Society of America Memoirs, **98**, 85–176.

DONNELLY, T. W. 1973a. *Magnetic Anomaly Observations in the Eastern Caribbean Sea*. Initial Reports of the Deep Sea Drilling Project. US Government Printing Office, Washington, DC, **15**, 1023–1025.

DONNELLY, T. W. 1973b. Late Cretaceous basalts from the Caribbean, a possible flood-basalt province of vast size. *EOS American Geophysical Union Transactions*, **54**, 1004.

DONNELLY, T. W. 1985. Mesozoic and Cenozoic plate evolution of the Caribbean region. *In*: STEHLI, F. G. & WEBB, S. D. (eds) *The Great American Biotic Interchange*. Plenum Press, London, 89–121.

DONNELLY, T. W. 1989. Geologic history of the Caribbean and Central America. *In*: BALLY, A. W. & PALMER, A. R. (eds) *The Geology of North America*, **A**, *An Overview*. Geological Society of America, 299–321.

DONNELLY, T. W., MELSON, W., KAY, R. & ROGERS, J. J. W. 1973. Basalts and dolerites of Late Cretaceous age from the central Caribbean. *Initial Reports of the Deep Sea Drilling Project*, **XV**, 989–1011.

DONNELLY, T. W., BEETS, D. *ET AL.* 1990. History and tectonic setting of Caribbean magmatism. *In*: DENGO, G. (ed.) *The Geology of North America*, **H**, *The Caribbean Region*, chapter 13, 339–374.

DOUCET, S., WEIS, D., SCOATES, J. S., NICOLAYSEN, K., FREY, F. A. & GIRET, A. 2002. The depleted mantle component in Kerguelen Archipelago Basalts: petrogenesis of tholeiitic–transitional basalts from the Loranchet Peninsula. *Journal of Petrology*, **43**, 1341–1366.

DOUGLASS, R. C. 1961. Orbitolinas from Caribbean Islands. *Journal of Paleontology*, **35**, 475–479.

DRAPER, G. 1986. Blueschists and associated rocks in eastern Jamaica and their significance for Cretaceous plate-margin development in the northern Caribbean. *Geological Society of America Bulletin*, **97**, 48–60.

DRAPER, G. 1988. Tectonic reconstruction of N. Hispaniola and S. E. Cuba: dissection of a Cretaceous island arc. *Annual Geological Society of America Centennial Celebration Meeting Abstracts*, No. 9991.

DRAPER, G. 1993. Metamorphic map of the Caribbean region, including Central America and Northern South America. *Geological Society of America, Abstracts*, **25**, A-285.

DRAPER, G. & LEWIS, J. F. 1991. Metamorphic belts of central Hispaniola. *In*: MANN, P., DRAPER, G. & LEWIS, J. F. (eds) *Geologic and Tectonic Development of the North America–Caribbean Plate Boundary in Hispaniola*. Geological Society of America Special Papers, **262**, 29–45.

DRAPER, G. & PINDELL, J. L. 2004. Arc-continent collision on the southern margin of North America: Cuba and Hispaniola. *EOS Transactions American Geophysical Union*, **85**, Fall Meeting Supplement, Abstract.

DRAPER, G., GUTIERREZ, G. & LEWIS, J. F. 1996. Thrust emplacement of the Hispaniola peridotite belt: orogenic expression of the Mid-Cretaceous Caribbean arc polarity reversal? *Geology*, **24**, 1143–1146.

DRAPER, G., ABBOTT, R. A. ET AL. 2008. Tectonic overview of the Late Cretaceous–Paleogene subduction zone rocks of Hispaniola. *18th Caribbean Geological Conference*, Santo Domingo, Abstracts.

DRISCOLL, N. W., DIEBOLD, J. B. & LAINE, E. P. 1995. New seismic evidence for Late Cretaceous to Eocene turbidite deposition in the Venezuela Basin. *EOS, Transactions, American Geophysical Union*, **76**, supplement, F615.

EAGLES, G. 2007. New angles on South Atlantic opening. *Geophysical Journal International*, **168**, 353–361.

EAGLES, G., LIVERMORE, R. & MORRIS, P. 2005. Small basins in the Scotia Sea: the Eocene Drake Passage gateway. *Earth and Planetary Science Letters*, **242**, 343–353.

EDGAR, N. T., EWING, J. I. & HENNION, J. 1971. Seismic refraction and reflection in Caribbean Sea. *American Association of Petroleum Geologists Bulletin*, **55**, 833–870.

EDGAR, N. T., SAUNDERS, J. B. ET AL. 1973. *Initial Reports of the Deep Sea Drilling Project*. US Government Printing Office, Washington, DC, **15**.

EGGLER, D. H., FAHLQUIST, D. A., PEQUEGNAT, W. E. & HERNDON, J. M. 1973. Ultrabasic rocks from the Cayman Trough, Caribbean Sea. *Geological Society of America Bulletin*, **84**, 2133–2138.

ESCALANTE, G. 1990. The geology of southern Central America and western Colombia. *In*: DENGO, G. &

CASE, J. E. (eds) *The Caribbean Region, The Geology of North America*. Geological Society of America, **H**, 201–223.

ESCUDER VIRUETE, J., DÍAZ DE NEIRA, A. ET AL. 2006. Island-arc tholeiites, boninites and related felsic rocks in Hispaniola: magmatic and age constraints on initiation of intra-oceanic subduction and development of the Caribbean island arc. *Geophysical Research Abstracts*, **8**.

ESTES, R. & BÁEZ, A. 1985. Herpetofaunas of North and South America during the Late Cretaceous and Cenozoic: evidence for interchange? *In*: STEHLI, F. G. & WEBB, S. D. (eds) *The Great American Biotic Interchange*. Plenum Press, London, 139–197.

FAIRHEAD, J. D. & WILSON, M. 2005. Plate tectonic processes in the South Atlantic Ocean: Do we need deep mantle plumes? *In*: FOULGER, G. R., NATLAND, J. H., PRESNALL, D. C. & ANDERSON, D. L. (eds) *Plates, Plumes, and Paradigms*. Geological Society of America Special Papers, **388**, 537–553.

FEO-CODECIDO, G. 1962. Contribution to the geology of north-central Venezuela: Boletín Informativo. *Asociacion Venezolana de Geología, Minería y Petroleo*, **5**, 119–142.

FEO-CODECIDO, G., SMITH, F. D., ABOUD, N. & DI GIACOMO, E. 1984. Basement and Paleozoic rocks of the Venezuelan Llanos basins. *In*: *The Caribbean–South America Plate Boundary and Regional Tectonics*. Geological Society of America Memoirs, **162**, 175–187.

FERNÁNDEZ, J. A., BOTTAZZI, G., BARBOZA, G. & ASTORGA, A. 1994. Tectónica y estratigrafía de la Cuenca Limón Sur. *Revista Geológica de América Central*, volumen especial, Terremoto de Limón, 15–28.

FINK, L. K. 1970. Field guide to the island of La Desirade with notes on the regional history and development of the Lesser Antilles island arc. *International Field Institute Guidebook to the Caribbean Island–Arc system*. American Geological Institute/National Science Foundation.

FINK, L. K. 1972. Bathymetric and geologic studies of the Guadeloupe Region, Lesser Antilles Island Arc. *Marine Geology*, **12**, 267–288.

FLORES, K., BAUMGARTNER, P. O., SKORA, S., BAUMGARTNER, L., BAUMGARTNER-MORA, C. & RODRIGUEZ, D. 2006. Jurassic oceanic remnants. *In*: The Siuna area (NE-Nicaragua)—Tracing the Chortís–Caribbean paleo-plate boundary. *EOS Transactions of the American Geophysical Union*, **87**, Fall Meeting Supplement, Abstract.

FLOTTÉ, N., RANGIN, C., MARTINEZ-REYES, J., LE PICHON, X., HUSSON, L. & TARDY, M. 2008. The Rio Bravo Fault, a major Late Oligocene left-lateral shear zone. *Bulletin de la Société Géologique de France*, **179**, 147–160.

FOULGER, G. R., NATLAND, J. H. & ANDERSON, D. L. 2005. Genesis of the Iceland melt anomaly by plate tectonic processes. *In*: FOULGER, G. R., NATLAND, J. H., PRESNALL, D. C. & ANDERSON, D. L. (eds) *Plates, Plumes, and Paradigms*. Geological Society of America Special Papers, **388**, 595–625.

FOX, P. J. & HEEZEN, B. C. 1975. The geology of the Caribbean Crust. *In*: NAIRN, A. E. M. & STEHLI,

F. G. (eds) *The Ocean Basins and Margins*. Plenum Press, New York, **3**, 421–466.

FOX, P. J., RUDDIMAN, W. F., RYAN, W. B. F. & HEEZEN, B. C. 1970. The geology of the Caribbean crust, I: Beata Ridge. *Tectonophysics*, **10**, 495–513.

FOX, P. J., SCHREIBER, E. & HEEZEN, B. C. 1971. The geology of the Caribbean crust, Tertiary sediments, granitic and basic rocks from the Aves Ridge. *Tectonophysics*, **12**, 89–109.

FREELAND, G. L. & DIETZ, R. S. 1972. Plate tectonic evolution of the Caribbean–Gulf of Mexico region. *6th Caribbean Geological Conference, Venezuela, Transactions*, 259–264.

FREY, F. A., WEIS, D., BORISOVA, A. Yu. & XU, G. 2002. Involvement of continental crust in the formation of the Cretaceous Kerguelen Plateau: new perspectives from ODP Leg 120 Sites. *Journal of Petrology*, **43**, 1207–1239.

FRISCH, W., MESCHEDE, M. & SICK, M. 1992. Origin of the Central American ophiolites: Evidence from paleomagnetic results. *Geological Society of America Bulletin*, **104**, 1301–1314.

FROST, C. D. & SNOKE, A. W. 1989. Tobago, West Indies, a fragment of a Mesozoic oceanic island arc: petrochemical evidence. *Journal of the Geological Society, London*, **146**, 953–964.

FROST, S. H. 1972. Evolution of Cenozoic Caribbean coral faunas. *VI Conferencia Geologica del Caribe–Margarita*, Venezuela, Memorias, 461–464.

FUNNELL, B. M. & SMITH, A. G. 1968. Opening of the Atlantic Ocean. *Nature*, **219**, 1328.

GARCÍA-SENZ, J. & PÉREZ-ESTAÚN, A. 2008. Miocene to Recent tectonic elevation in Eastern Dominican Republic. *18th Caribbean Geological Conference*, Santo Domingo, Abstracts.

GARZANTI, E., DOGLIONI, C., VEZZOLI, G. & ANDO, S. 2007. Orogenic belts and orogenic sediment provenance. *Journal of Geology*, **115**, 315–334.

GAUCHAT, K. 2004. *Geochemistry of Desirade Islands rocks (Guadeloupe, French Antilles)*. Unpublished Diploma, University of Lausanne.

GEORGE, R. P. JR & SOCAS, M. B. 1994. Historia de maduracion termal de rocas madre del Cretacico superior y Mioceno en la subcuenca de Maturin. *V Simposio Bolivariano, Puerto La Cruz, Memoria*, 405–407.

GHIGLIONE, M. C., YAGUPSKY, D., GHIDELLA, M. & RAMOS, V. A. 2008. Continental stretching preceding the opening of the Drake Passage: evidence from Tierra del Fuego. *Geology*, **36**, 643–646.

GHOSH, N., HALL, S. A. & CASEY, J. F. 1984. Sea floor spreading magnetic anomalies in the Venezuelan Basin. *In: The Caribbean–South American Plate Boundary*. Geological Society of America Memoirs, **162**, 65–80.

GINGERICH, P. D. 1985. South American mammals in the Palaeocene of North America. *In*: STEHLI, F. G. & WEBB, S. D. (eds) *The Great American Biotic Interchange*. Plenum Press, London, 123–137.

GIUNTA, G. & OLIVERI, E. 2009. Some remarks on the Caribbean Plate kinematics: facts and remaining problems. *In*: JAMES, K. H., LORENTE, M. A. & PINDELL, J. (eds) *Origin and Evolution of the Caribbean Plate*. Geological Society, London, Special Publications, **328**, 57–75.

GLEN, J. M. G. & PONCE, D. A. 2002. Large-scale fractures related to inception of the Yellowstone hotspot. *Geology*, **30**, 647–650.

GONCALVES, P., GUILLOTE, S., LARDEAUX, J.-M., NICOLLETA, C. & MERCIER DE LEPINAY, B. 2002. Thrusting and wrenching in a pre-Eocene HP-LT Caribbean accretionary wedge (Samaná Peninsula, Dominican Republic). *Geodinamica Acta*, **13**, 119–132.

GONZALEZ, A. & LANDER, R. 1990. Regimes tectonicos desde el Triasico hasta el Neogeno en el area occidental de la Cuenca Oriental de Venezuela. *V Congreso Venezolano de Geofisica, Caracas, Memorias*, 134–141.

GONZALEZ DE JUANA, C., ITURRALDE DE AROZENA, J. M. & CADILLAT, X. P. 1980. *Geología de Venezuela y de sus Cuencas Petrolíferas*. Ediciones Foninves, Caracas.

GORDON, M. B. & AVE LALLEMENT, H. G. 1995. Cryptic strike–slip faults of the Chortís block. *GSA Abstracts with Programs*, **27**, 227–228.

GOSE, W. A. & SWARTZ, D. K. 1977. Paleomagnetic results from Cretaceous sediments in Honduras: tectonic implications. *Geology*, **5**, 505–508.

GOUGH, D. I. 1967. Magnetic anomalies and crustal structure in eastern Gulf of Mexico. *American Association of Petroleum Geologists Bulletin*, **51**, 200–211.

GRAHAM, A. 2003. Geohistory models and Cenozoic paleoenvironments of the Caribbean region. *Journal of Systematic Botany*, **28**, 378–386.

GUEDEZ, V. 1985. Revision geologíca y evaluación exploratoria del sector de Monay–Carora. *VI Congreso Geológico Venezolano*, 3103–3143.

GUZMÁN-SPEZIALE, M. 2008. Beyond the Motagua and Polochic faults: active transform faults in the North America–Caribbean Plate boundary zone. *18th Caribbean Geological Conference*, Santo Domingo, Abstracts.

GUZMAN-SPEZIALE, M. & MENESES-ROCHA, J. J. 2000. The North America-Caribbean Plate boundary west of the Motagua–Polochic fault system, a jog in southeastern Mexico. *Journal of South American Earth Sciences*, **13**, 459–468.

HALL, R. 1997. Cenozoic plate tectonic reconstruction of SE Asia. *In*: FRASER, A. J. *ET AL.* (eds) *Petroleum Geology of Southeast Asia*. Geological Society, London, Special Publications, **126**, 11–23.

HALL, S. A. 1995. Oceanic basement of the Caribbean Basins. *Annual Geological Society of America meeting*, New Orleans, poster, abstract in proceedings, **27**, A-153.

HALL, S. A. & YEUNG, T. 1980. A study of magnetic anomalies in the Yucatán Basin: Transactions, *9th Caribbean Geological Conference*, 519–526.

HARLOW, G. E., HEMMING, S. R., AVÉ LALLEMENT, H. G., SISSON, V. B. & SORENSEN, S. S. 2004. Two high-pressure–low-temperature serpentinite-matrix mélange belts, Motagua fault zone, Guatemala: a record of Aptian and Maastrictian collisions. *Geology*, **32**, 17–20.

HARRY, D. L., LODONO, J. & HUERTA, A. 2003. Early Paleozoic transform-margin structure beneath the Mississippi coastal plain, southeast United States. *Geology*, **31**, 969–972.

HATTEN, C. W. 1967. Principal features of Cuban geology: discussion. *American Association of Petroleum Geologists Bulletin*, **51**, 780–789.

HEATHERINGTON, A. L. & MUELLER, P. A. 1991. Geochemical evidence for Triassic rifting in southwest Florida. *Tectonophysics*, **188**, 291–302.

HELMERS, H. & BEETS, D. J. 1977. Cretaceous of Aruba. In: *Guide to Geological Excursions on Curaçao, Bonaire and Aruba. 8th Caribbean Geological Conference*, 29–35.

HELWIG, J. 1975. Tectonic evolution of the Southern Continental margin of North America from a Palaeozoic perspective. In: NAIRN, A. E. M. & STEHLI, F. G. (eds) *The Ocean Basins and Margins*, **3**, *The Gulf of Mexico and the Caribbean*, 243–256.

HEPPARD, P. D., CANDER, H. S. & EGGERTSON, E. B. 1998. Abnormal pressure and the occurrence of hydrocarbons in offshore Eastern Trinidad, West Indies. In: LAW, B. E., ULMISHEK, G. F. & SLAVIN, V. I. (eds) *Abnormal Pressures in Hydrocarbon Environments*. American Association of Petroleum Geologists Memoirs, **70**, 215–246.

HESS, H. H. 1938. Gravity anomalies and island arc structure with particular reference to the West Indies. *American Philosophical Society Proceedings*, **79**, 71–96.

HESS, H. H. 1966. *Caribbean Geological Investigations*. Geological Society of America Memoirs, **98**.

HINZ, K. 1983. Line BFB (24-fold stack) from the Norwegian continental margin/outer Voring Plateau. In: BALLY, A. W. (ed.) *Seismic Expression of Structural Styles – a Picture and Work Atlas*. American Association of Petroleum Geologists Studies in Geology, **15**, 2.2.3–39.

HOLCOMBE, T. L., LADD, J. W., WESTBROOK, G., EDGAR, N. T. & BOWLAND, C. L. 1990. Caribbean marine geology; ridges and basins of the plate interior. In: DENGO, G. & CASE, J. E. (eds) *The Caribbean Region. The Geology of North America*, **H**. Geological Society of America, 231–260.

HONTHAAS, C., R'HAULT, J.-P. *ET AL*. 1998. A Neogene back-arc origin for the Banda Sea basins: geochemical and geochronological constraints from the Banda ridges (East Indonesia). *Tectonophysics*, **298**, 297–317.

HOPPER, J. R., FUNCK, T. & TUCHOLKE, B. E. 2004. Continental break up and the onset of ultraslow seafloor spreading off Flemish Cap on the Newfoundland margin. *Geology*, **32**, 93–96.

HORBURY, A. D., HALL, S. *ET AL*. 2003. Tectonic sequence stratigraphy of the western margin of the Gulf of Mexico in the Late Mesozoic and Cenozoic: less passive than previously imagined. In: BARTOLINI, C. *ET AL*. (eds) *The Circum-Gulf of Mexico and the Caribbean: Hydrocarbon Habitats, Basin Formation, and Plate Tectonics*. American Association of Petroleum Geologists Memoirs, **79**, 184–245.

HORNE, G. S., FINCH, R. C. & DONNELLY, T. W. 1990. The Chortís Block. In: DENGO, G. & CASE, J. E. (eds) *Geology of North America*, **H**. Geological Society of America, 55–76.

HUNTER, V. F. 1995. Limestone buildups as ecotectonic indicators within the Tertiary of the southern Caribbean region. *3rd Geological Conference of the Geological Society of Trinidad and Tobago and 14th Caribbean Geological Conference*, Abstracts, 35.

ISHIKAWA, A., KURITANI, T., MAKISHIMA, A. & NAKAMURA, E. 2007. Recycling of the Rodinia supercontinent in the Cretaceous Ontong Java Plateau? *Earth and Planetary Science Letters*, **259**, 134–148.

ITURRALDE-VINENT, M. A. 1994. Cuban Geology: a new plate-tectonic synthesis. *Journal of Petroleum Geology*, **17**, 39–70.

ITURRALDE-VINENT, M. A. 1995. Late Paleocene to early Middle Eocene Cuban Island Arc. *3rd Geological Conference of the Geological Society of Trinidad and Tobago*, Program and Abstracts, 37.

ITURRALDE-VINENT, M. 1996. *IGCP Project 364. Correlation of Episodes, Caribbean Ophiolites and Volcanic Arcs*. **19**, 1–2.

ITURRALDE-VINENT, M. 2004. IGCP Project 433 Caribbean Plate Tectonics: State of the debate and future developments. *32nd IGC, Florence, 2004, IGCP Project 433. Caribbean Plate Tectonics*, Abstracts.

ITURRALDE-VINENT, M. A. & GARCÍA-CASCO, A. 2007. Caribeana, a possible solution to a long standing puzzle: the Caribbean latest Cretaceous tectonic events. *Field-Workshop of Caribbean Geology, 2da Convención Cubana de Ciencias de la Tierra, Havana, Cuba*, UNESCO/IUGS – IGCP Project 546, Abstracts GEO10-02.

JACKSON, J. 1997. *Glossary of Geology*. American Geological Institute.

JACKSON, T. A., SMITH, T. E. & HUANG, C. H. 1987. The geochemistry and geochemical variations of Cretaceous andesites and dacites from Jamaica West Indies. In: DUQUE-CARO, H. (ed.) *10th Caribbean Geological Conference*, Cartegena, Colombia, 473–479.

JACKSON, M. P. A., CRAMEZ, C. & FONCK, J. M. 2000. Role of subaerial volcanic rocks and mantle plumes in creation of South Atlantic margins: implications for salt tectonics and source rocks. *Marine and Petroleum Geology*, **17**, 477–498.

JACQUES, J. M. & OTTO, S. 2003. Two major tectonic events expressed in the tectonostratigraphic evolution of the Caribbean, Gulf of Mexico and Sub-Andean Basins: 25 and 12 Ma. *AAPG Bulletin*, **87**, 13 (supplement).

JAILLARD, E., LAPIERRE, H., ORDOÑEZ, M., ÁLAVA, J. T., AMÓRTEGUI, A. & VANMELLE, J. 2009. Accreted oceanic terranes in Ecuador: southern edge of the Caribbean Plate? In: JAMES, K. H., LORENTE, M. A. & PINDELL, J. L. (eds) *Origin and Evolution of the Caribbean Plate*. Geological Society, London, Special Publications, **328**, 469–485.

JAMES, E. & MITCHELL, S. F. 2005. Comparison of the stratigraphy of the Tertiary limestone across central Jamaica. *17th Caribbean Geological Conference*, Puerto Rico, Abstracts, 39.

JAMES, K. H. 1997. Distribution and tectonic significance of Cretaceous–Eocene flysch/wildflysch deposits of Venezuela and Trinidad. Sociedad Venezolana de Geologos, *VIII Venezuelan Geological Congress*, Margarita, 415–421.

JAMES, K. H. 2000. The Venezuelan hydrocarbon habitat. Part 1: Tectonics, structure, palaeogeography and source rocks. *Journal of Petroleum Geology*, **23**, 5–53.

JAMES, K. H. 2002. A discussion of arguments for and against the far-field origin of the Caribbean Plate, finding for an *in situ* origin. *16th Caribbean Geological Conference*, Barbados, Abstracts, 89.

JAMES, K. H. 2003*a*. Caribbean Plate origin: discussion of arguments claiming to support a Pacific origin; arguments for an *in situ* Origin. *American Association of Petroleum Geologists Bulletin*, **87**, 13 (supplement).

JAMES, K. H. 2003*b*. A simple synthesis of Caribbean geology. *American Association of Petroleum Bulletin*, **87** (supplement).

JAMES, K. H. 2005*a*. Arguments for and against the Pacific origin of the Caribbean Plate and arguments for an *in situ* origin: transactions, 16th Caribbean Geological Conference, Barbados. *Caribbean Journal of Earth Sciences*, **39**, 47–67.

JAMES, K. H. 2005*b*. A simple synthesis of Caribbean geology: transactions, 16th Caribbean Geological Conference, Barbados. *Caribbean Journal of Earth Sciences*, **39**, 71–84.

JAMES, K. H. 2005*c*. Palaeocene to Middle Eocene flysch–wildflysch deposits of the Caribbean area: a chronologicla compilation of literature reports, implications for tectonic history and recommendations for further investigation: transactions, 16th Caribbean Geological Conference, Barbados. *Caribbean Journal of Earth Sciences*, **39**, 29–46.

JAMES, K. H. 2006. Arguments for and against the Pacific origin of the Caribbean Plate: discussion, finding for an inter-American origin. *In*: ITURRALDE-VINENT, M. A. & LIDIAK, E. G. (eds) *Caribbean Plate Tectonics. Geologica Acta*, **4**, 279–302.

JAMES, K. H. 2007*a*. Structural geology: from local elements to regional synthesis. *In*: BUNDSCHUH, J. & ALVARADO, G. E. (eds) *Central America: Geology, Resources and Hazards*. Balkema, Rotterdam, 277–321.

JAMES, K. H. 2007*b*. The Caribbean Plateau. World Wide Web Address: http://www.mantleplumes.org/Caribbean.html.

JAMES, K. H. 2009. Evolution of Middle America and the *in situ* Caribbean Plate model. *In*: JAMES, K. H., LORENTE, M. A. & PINDELL, J. L. (eds) *Origin and Evolution of the Caribbean Plate*. Geological Society, London, Special Publications, **328**, 127–138.

JAMES, K. H., SOOFI, K. A. & WEINZAPFEL, A. C. 1998. Ocean floor mapped from space. *American Association of Petroleum Geologists Explorer*, 20–23.

JOHNSON, E. A. E., KACEWICZ, M., BLICKWEDE, J. F. & HUSTON, H. H. 2005. An interpretation of the crustal framework and continent–oceanic boundary in U.S. OCS of the Gulf of Mexico, based on gravity and refraction data analysis. *25th Annual Gulf Coast Section SEPM Foundation Bob F. Parker Research Conference*, Abstracts, 48.

JOLLY, W. T. & LIDIAK, E. G. 2005. Role of crustal melting in petrogenesis of the Cretaceous Water Island Formation, Virgin Islands northeast Antilles Island Arc. *17th Caribbean Geological Conference*, Puerto Rico, Abstracts, 41.

JOLLY, W. T., LIDIAK, E. G. & DICKIN, A. P. 2006. Cretaceous to Mid-Eocene pelagic sediment budget in Puerto Rico and the Virgin Islands (northeast Antilles Island arc). *In*: ITURRALDE-VINENT, M. A. & LIDIAK, E. G. (eds) *Caribbean Plate Tectonics. Geologica Acta*, **4**, 35–62.

JOYCE, J. 1991. Blueschist metamorphism and deformation on the Samana Peninsula – a record of subduction and collision in the Greater Antilles. *In*: MANN, P., DRAPER, G. & LEWIS, J. F. (eds) *Geologic and Tectonic Development of the North America–Caribbean Plate Boundary in Hispaniola*. Geological Society of America Special Papers, **262**, 29–46.

JOYCE, J. & ARONSON, J. 1983. K–Ar age dates for blueschist metamorphism on the Samaná Peninsula, Dominican Republic. *10th, Caribbean Geological Conference*, Cartagena.

KASPER, D. C. & LARUE, D. K. 1986. Paleogeographic and tectonic implications of quartzose sandstones of Barbados. *Tectonics*, **5**, 837–854.

KAUFFMAN, E. G. 1973. Cretaceous bivalvia. *In*: HALLAM, A. (ed.) *Atlas of Paleogeography*. Elsevier, New York, 353–383.

KEAST, A. 1972. Continental drift and evolution of biota on southern continents. *In*: KEAST, A. *ET AL*. (eds) *Evolution, Mammals, and Southern Continents*. State University of NY Press, Albany, NY, 23–88.

KEPPIE, J. D. & MORÁN-ZENTENO, D. J. 2005. Tectonic implications of alternative Cenozoic reconstructions for Southern Mexico and the Chortís Block. *International Geology Review*, **47**, 473–491.

KEPPIE, J. D., DOSTAL, J., CAMERON, K. L., SOLARI, L. A., ORTEGA-GUTIÉRREZ, F. & LOPEZ, R. 2003. Geochronology and geochemistry of Grenvillian igneous suites in the northern Oaxacan Complex, southern Mexico: tectonic implications. *Precambrian Research*, **120**, 365–389.

KERR, A. C., TARNEY, J., MARRINER, G. F., NIVIA, A. & SAUNDERS, A. D. 1997. The Caribbean–Colombian Cretaceous Igneous Province: the internal anatomy of an oceanic plateau. *In*: MAHONEY, J. J. & COFFIN, M. F. (eds) *Large Igneous Provinces, Continental, Oceanic, and Planetary Flood Volcanism*. American Geophysical Union Geophysical Monographs, **100**, 123–144.

KERR, A. C., ITURRALDE-VINENT, M. A., SAUNDERS, A. D., BABBS, T. L. & TARNEY, J. 1999. A new plate tectonic model of the Caribbean: implications from a geochemical reconnaissance of Cuban Mesozoic volcanic rocks. *Geological Society of America Bulletin*, **111**, 1581–1599.

KERR, A. C., WHITE, R. V. & SAUNDERS, A. D. 2000. LIP reading: Recognizing oceanic plateaux in the geological record. *Journal of Petrology*, **41**, 1041–1056.

KERR, A. C., WHITE, R. V., THOMPSON, P. M. E., TARNEY, J. & SAUNDERS, A. D. 2003. No Oceanic Plateau – no Caribbean Plate? The seminal role of oceanic plateau(s) in Caribbean Plate evolution. *In*: BARTOLINI, C., BUFFLER, R. T. & BLICKWEDE, J. (eds) *The Gulf of Mexico and Caribbean Region: Hydrocarbon Habitats, Basin Formation and Plate Tectonics*. American Association of Petroleum Geologists Memoirs, **79**, 126–268.

KERR, A. C., PEARSON, D. G. & NOWELL, G. M. 2009. Magma source evolution beneath the Caribbean

oceanic plateau: new insights from elemental and Sr–
Nd–Pb–Hf isotopic studies of ODP Leg 165 Site 1001
basalts. *In*: JAMES, K. H., LORENTE, M. A. &
PINDELL, J. L. (eds) *Origin and Evolution of the Car-
ibbean Plate*. Geological Society, London, Special
Publications, **328**, 809–827.

KHUDOLEY, K. M. 1967. Principal features of Cuban
geology: discussion. *American Association of Petroleum Geologists Bulletin*, **51**, 789–791.

KHUDOLEY, K. M. & MEYERHOFF, A. 1971. *Paleogeography and Geological History of the Greater Antilles*.
Geological Society of America Memoirs, **129**.

KIRKLAND, D. W. & GERHARD, J. E. 1971. Jurassic salt,
central Gulf of Mexico, and its temporal relation to
circum-Gulf evaporites. *American Association of
Petroleum Geologist Bulletin*, **55**, 680–686.

KISER, G. D. 1987. Exploration results, Machete area,
Orinoco Oil Belt, Venezuela. *Journal of Petroleum
Geology*, **10**, 149–162.

KLAVER, G. Th. 1987. *The Curaçao Lava Formation,
an ophiolitic analogue of the anomalous thick layer
2B of the Mid-Cretaceous oceanic plateaus in the
western Pacific and central Caribbean*. PhD thesis,
Amsterdam.

KLINGELHÖFER, F., EDWARDS, R. A., HOBBS, R. W. &
ENGLAND, R. W. 2005. Crustal structure of the NE
Rockall Trough from wide-angle seismic data modeling. *Journal of Geophysical Research*, **110**, B11105;
doi:10.1029/2005JB003763.

KLUGE, R., BAUMANN, A., TOETZ, A., MARESCH, W.
V., STOCKHERT, B. & TROESCH, M. 1995. New
geochronological constraints on the crustal history of
Margarita Island. *3rd Geological Conference of the
Geological Society of Trinidad and Tobago and 14th
Caribbean Geological Conference*, Abstracts, 41.

KOLARSKY, R. A., MANN, P. & MONECHI, S. 1995.
Stratigraphic development of southwestern Panama
as determined from integration of marine seismic
data and onshore geology. *In*: MANN, P. (ed.) *Geologic
and Tectonic Development of the Caribbean Plate
Boundary in Southern Central America*. Geological
Society of America Special Papers, **295**, 159–200.

KOZARY, M. T. 1968. Ultramafic rocks in thrust zones of
northwestern Oriente Province, Cuba. *American
Association of Petroleum Geologists Bulletin*, **52**,
2298–2317.

KRAWINKEL, J. & SEYFRIED, H. 1994. A review of plate-
tectonic processes involved in the formation of the
southwestern edge of the Caribbean Plate. *Profil*, **7**,
47–61.

KRÜGER, F., SCHERBAUM, F., ROSA, J. W. C., KIND, R.,
ZETSCHE, F. & HOHNE, J. 2002. Crustal and upper
mantle structure in the Amazon region (Brazil)
determined with broadband mobile stations. *Journal
of Geophysical Research*, **107**, 2265.

KUGLER, H. G. 1953. Jurassic to Recent sedimentary
environments in Trinidad. *Bulletin Asociacion Suisse
des Géologues et Ingénieurs du Petrole*, **20**, 27–60.

KUGLER, H. G. 1972. The Dragon Gneiss of Paria
Peninsula (Eastern Venezuela). *VI Conferencia Geológica del Caribe*, Margarita, Venezuela, Memorias,
113–116.

LADD, J. W. 1976. Relative motion of South America
with respect to North America and Caribbean

tectonics. *Geological Society of America Bulletin*, **87**,
969–976.

LADD, J. W., SHIH, T-Ch. & TSAI, C. J. 1981. Cenozoic
tectonics of Central Hispaniola and adjacent Caribbean
Sea. *American Association of Petroleum Geologists
Bulletin*, **65**, 466–489.

LAGAAY, R. A. 1969. *Geophysical investigations of
the Netherlands Leeward Antilles*. Verhandelingen
der Koningkliike Nederlandse Akademie van
Wetenschappen, Afd. Natuurkunde, V. **25**, 2.

LAÓ-DÁVILA, C. A., ANDERSON, T. H. & LLERANDI-
ROMÁN, P. A. 2004. Olistostromes and allochthonous
serpentinite: major tectonostratigraphic elements in
southwest Puerto Rico. *32nd IGC*, Florence, 2004,
IGCP Project 433 Caribbean Plate Tectonics,
Abstracts.

LARDEAUX, J. M., TORRES-ROLDÁN, R. L. & MILLÁN
TRUJILLO, G. 2004. Origin and evolution of the
Escambray Massif (Central Cuba): an example of
HP/LT rocks exhumed during intraoceanic subduc-
tion. *Journal of Metamorphic Geology*, **22**, 227.

LARSON, R. L. 1997. Superplumes and ridge interactions
between Ontong Java and Manihiki Plateaus and the
Nova-Canton Trough. *Geology*, **25**, 779–782.

LARSON, R. L., POCKALNY, R. A. *ET AL.* 2002. Mid-
Cretaceous tectonic evolution of the Tongareva triple
junction in the southwestern Pacific Basin. *Geology*,
30, 67–70.

LARUE, D. K., TORRINI, R. JR, SMITH, A. L. & JOYCE, J.
1998. North Coast Tertiary Basin of Puerto Rico: From
arc basin to carbonate platform to arc-massif slope. *In*:
LIDIAK, E. G. & LARUE, D. K. (eds) *Tectonics and
Geochemistry of the Northeastern Caribbean*. Geological Society of America, 155–176.

LEAT, P. T. & LARTER, R. D. 2003. Intra-oceanic subduction systems: introduction. *In*: LARTER, R. D. & LEAT,
P. T. (eds) *Intra-oceanic Subduction Systems: Tectonic
and Magmatic Processes*. Geological Society, London,
Special Publications, **219**, 1–17.

LEBRON, M. C. & PERFIT, M. R. 1993. Stratigraphic and
petrochemical data support subduction polarity reversal of the Cretaceous Caribbean Arc. *Journal of
Geology*, **101**, 389–396.

LEROY, S., MERCIER DE LEPINAY, B., MAUFFRET, A. &
PUBELLIER, M. 1996. Structural and tectonic evolution of the eastern Cayman Trough (Caribbean Sea)
from seismic reflection data. *American Association of
Petroleum Geologists Bulletin*, **80**, 222–247.

LEROY, S., MAUFFRET, A., PATRIAT, P. & MERCIER DE
LÉPINAY, B. 2000. An alternative interpretation of the
Cayman Trough evolution from a reidentification of
magnetic anomalies. *Geophysical Journal International*, **141**, 539–557.

LEVANDER, A., MANN, P., AVE LALLEMENT, H. G. &
SCHMITZ, M. 2004. Bolivar: the SE Caribbean Conti-
nental Dynamics Project. *EOS Transactions American
Geophysical Union*, **85**, Fall Meeting Supplement,
Abstract.

LEW, L. R. 1985. *The geology of the Santa Elena Peninsula, Costa Rica, and its implications for the tectonic
evolution of the Central America–Caribbean region*.
PhD thesis, Pennsylvania State University.

LEWIS, J. F. 1971. Composition, origin, and differentiation of basalt magma in the Lesser Antilles.

In: DONNELLEY, T. W. (ed.) *Caribbean Geophysical, Tectonic and Petrologic Studies.* Geological Society of America Memoirs, **130**, 159–179.

LEWIS, J. F. 1990. Cuba. *In*: DENGO, G. & CASE, J. E. (eds) *The Caribbean Region.* The Geology of North America, **H**. Geological Society of America, 78–93.

LEWIS, J. 2002. Albian unconformities. *16th Caribbean Geological Conference*, Barbados.

LEWIS, J. F. & GUNN, B. M. 1972. Aspects of island arc evolution and magmatism in the Caribbean: geochemistry of some West Indian plutonic and volcanic rocks. *VI Conferencia Geológica del Caribe–Margarita*, Venezuela, Memorias, 171–177.

LEWIS, J. F. & DRAPER, G. 1990. Geology and tectonic evolution of the northern Caribbean margin. *In*: DENGO, G. & CASE, J. E. (eds) *The Caribbean Region.* The Geology of North America, **H**. Geological Society of America, 77–140.

LEWIS, J. F., DRAPER, G., BOWIN, C., BOURDON, L., MAURRASSE, F. & NAGLE, F. 1990. Hispaniola. *In*: LEWIS, J. F. & DRAPER, G. Geology and tectonic evolution of the northern Caribbean margin. *In*: DENGO, G. & CASE, J. E. (eds) *The Caribbean Region.* The Geology of North America, Geological Society of America, 94–112.

LEWIS, J. F., PERFIT, M. R. ET AL. 2005. Anomalous granitoid compositions from the northwestern Cayman Trench: implications for the composition and evolution of the Cayman Ridge. *17th Caribbean Geological Conference*, Puerto Rico, Abstracts, 49.

LEWIS, J. G., MATTIETTI, K., PERFIT, M. & KAMENOV, G. 2008. Geochemistry and petrology of three granitoid rock cores from the Nicaragua Rise, Caribbean Sea: implication for its composition, structure and tectonic evolution. *18th Caribbean Geological Conference*, Santo Domingo, Abstracts.

LIDIAK, E. G. 1970. *Volcanic rocks in the Puerto Rican Orogen: international field institute guidebook to the Caribbean Island Arc system.* American Geological Institute (article individually paginated).

LIDIAK, E. G. 2008. Geochemical and tectonic evolution of Albian to Eocene volcanic strata in the Virgin Islands and eastern and central Puerto Rico. *2008 Joint Meeting of the Geological Society of America, American Society of Agronomy, Crop Science Society of America, Gulf Coast Association of Geological Societies with the Gulf Coast Section of the Society for Sedimentary Geology*, Abstracts, paper **128-12**.

LIDZ, B. H. 1988. Upper Cretaceous (Campanian) and Cenozoic stratigraphic sequence, Northeast Caribbean (St. Croix, U. S. Virgin Islands). *Geological Society of America Bulletin*, **100**, 282–298.

LIVERMORE, R., EAGLES, G., MORRIS, P. & MALDONADO, A. 2004. Shackleton Fracture Zone: no barrier to early circumpolar ocean circulation. *Geology*, **32**, 797–800.

LOCKWOOD, J. P. 1971. Detrital serpentinite from the Guajira Peninsula, Colombia. *In*: DONNELLY, T. W. (ed.) *Caribbean Geophysical, Tectonic and Petrologic Studies.* Geological Society of America Memoirs, **130**, 55–75.

LÓPEZ-RAMOS, E. 1975. Geological summary of the Yucatán peninsula. *In*: NAIRN, A. E. M. & STEHLI,

F. G. (eds) *The Ocean Basins and Margins*, **3**. Plenum Press, New York, 257–282.

LOWRIE, W. & OPDYKE, N. D. 1975. Paleomagnetism of igneous and sedimentary samples. *In*: EDGAR, N. T. ET AL. (eds) *Initial Reports of the Deep Sea Drilling Project*, **15**. US Government Printing Office, Washington, DC, 1017–1021.

LUCAS, S. G., ALVARADO, G. E., GARCÍA, R., ESPINOSA, E., CISNEROS, J. C. & MARTENS, U. 2006. Vertebrate paleontology. *In*: BUNDSCHUH, J. & ALVARADO, G. E. (eds) *Central America, Geology, Resources and Hazards.* Taylor & Francis, Balkema, The Netherlands, Chapter 16, 443–451.

MALIN, P. E. & DILLON, W. P. 1973. Geophysical reconnaissance of the western Cayman Ridge. *Journal of Geophysical Research*, **78**, 7769–7775.

MALFAIT, B. T. & DINKLEMAN, M. G. 1972. Circum-Caribbean tectonic and igneous activity and the evolution of the Caribbean Plate. *Geological Society of America Bulletin*, **83**, 251–272.

MANN, P. 2007. Overview of the tectonic history of northern Central America. *In*: MANN, P. (ed.) *Geologic and Tectonic Development of the Caribbean Plate Boundary in Northern Central America.* Geological Society of America Special Paper, **428**, 1–19.

MANSPEIZER, W. 1988. Triassic–Jurassic rifting and opening of the Atlantic: an overview. *In*: MANZPEIZER, W. (ed.) *Triassic–Jurassic Rifting*, Part A. Elsevier, Amsterdam, 41–79.

MANTON, W. 1996. The Grenville of Honduras, Abstracts. *Geological Society of America*, **28**, 493.

MARESCH, W. V. & GERYA, T. V. 2006. Blueschists and blue amphiboles: how much subduction do they need? *In*: LIOU, J. G. & CLOOS, M. (eds) *Phase Relations, High Pressure Terranes, P–T-Ometry, and Plate Pushing: a Tribute to W. G. Ernst.* Geological Society of America, International Book Series, **9**, 23–37.

MARESCH, W. V., STOCKHERT, B., KRUCKHANS-LUDER, G., TOETZ,A. & BRIX, M. 1993. Structural geology, fabric analysis, petrology and geochronology in concert: how does the Margarita crustal block fit the Caribbean tectonic puzzle? *Geological Society of America Annual Meeting*, Boston, Abstract A-71.

MARTON, G. & BUFFLER, R. T. 1993. Application of simple-shear model to the evolution of passive continental margins of the Gulf of Mexico basin. *Geology*, **21**, 495–498.

MARZOLI, A., RENNE, P. R., PICCIRILLO, E. M., ERNESTO, M., BELLIENI, G. & DE MIN, A. 1999. Extensive 200-million-year-old continental flood basalts of the central Atlantic magmatic province. *Science*, **284**, 616–618.

MASCLE, A., MOORE, J. C. ET AL. 1988. Synthesis of shipboard results: Leg 110 transect of the Northern Barbados Ridge. *In*: MOORE, C., MASCLE, A. J. ET AL. (eds) *Proceedings of the Ocean Drilling Program, Part A, Initial Reports*, 577–591.

MASCLE, D., LETOUZEY, P. ET AL. 1990. *Geological Map of the Caribbean.* IFP, Beicip, Editions Technip, Paris.

MATTIETTI-KYSAR, G. 1999. The role of Paleogene volcanism in the evolution of the northern Caribbean margin. *Penrose Conference: Subduction to Strike–Slip Transitions on Plate Boundaries*, Abstracts, 63–64.

MATTSON, P. H. 1966. Geological characteristics of Puerto Rico. *In*: POOLE, W. H. (ed.) *Continental Margins and Island Arcs*. Geological Survey of Canada Papers, 66–15.

MATTSON, P. H. 1984. *Caribbean Structural Breaks and Plate Movements*. Geological Society of America Memoirs, **162**, 131–152.

MATTSON, P. H. & SCHWARTZ, D. P. 1971. Control of intensity of deformation in Puerto Rico by mobile peridotite basement. *In*: DONNELLEY, T. W. (ed.) *Caribbean Geophysical, Tectonic and Petrologic Studies*. Geological Society of America Memoirs, **130**, 97–106.

MATTSON, P., PESSAGNO, E. A. & HELSEY, C. E. 1972. *Outcropping Later A and A″ Correlatives in the Greater Antilles*. Geological Society of America Memoirs, **132**, 57–66.

MAUFFRET, A. & LEROY, S. 1997. Seismic stratigraphy and structure of the Caribbean igneous province. *Tectonophysics*, **283**, 61–104.

MAURASSE, F. 1981. Relations between the geologic setting of Hispaniola and the origin and evolution of the Caribbean. *In*: MAURASSE, F. *ET AL*. (eds) *Presentations Transactions du ler Colloque sur la Geologie d'Haiti*, Port-au-Prince, 1980, 246–264.

MAURY, R. C., WESTBROOK, G. K., BAKER, P. E., BOUYSSE, Ph. & WESTERCAMP, D. 1990. Geology of the Lesser Antilles. *In*: DENGO, G. & CASE, J. E. (eds) *The Caribbean Region*. The Geology of North America, **H**, 141–166.

MAZE, W. B. 1984. Jurassic La Quinta Formation in the Sierra de Perija, northwestern Venezuela: geology and tectonic environment of red beds and volcanic rocks. *In*: *The Caribbean–South American Plate Boundary and Regional Tectonics*. Geological Society of America Memoirs, **162**, 263–282.

MCCOLLOUGH, C. N. & CARVER, J. A. 1992. The Giant Caño Limon Field, Llanos Basin, Colombia. *In*: HALBOUTY, M. T. (ed.) *Giant Oil and Gas Fields of the Decade 1978–1988*. American Association of Petroleum Geologists Memoirs, **54**, 175–195.

MCHONE, J. G. 2002. Volatile emissions from Central Atlantic Magmatic Province basalts: mass assumptions and environmental consequences. *In*: *The Central Atlantic Magmatic Province*. AGU Monograph, 1–13.

MENÉNDEZ, A. 1966. Tectónica de la parte central de las montañas occidentales del Caribe, Venezuela. *Boletín de Geología*, **8**, 116–139.

MESCHEDE, M. 1998. The impossible Galapagos connection: geometric constraints for a near-American origin of the Caribbean Plate. *Geologische Rundschau*, **87**, 200–205.

MEYERHOFF, A. A. & HATTEN, C. W. 1974. Bahamas salient of North America: tectonic framework, stratigraphy, and petroleum potential. *In*: BURK, C. A. & DRAKE, C. L. (eds) *The Geology of Continental Margins*. Springer, New York, 429–446.

MEYERHOFF, A. A. & MEYERHOFF, H. A. 1972. Continental drift, IV: the Caribbean "Plate". *Journal of Geology*, **80**, 34–60.

MILLS, R. A., HUGH, K. E., FERAY, F. E. & SWOLFS, H. C. 1967. Mesozoic stratigraphy of Honduras. *American Association of Petroleum Geologists Bulletin*, **51**, 1711–1786.

MITCHELL, S. 2005. Cretaceous and Cainozoic evolution of Jamaica. How does it test the current models? *17th Caribbean Geological Conference*, Puerto Rico, Abstracts, 60.

MITCHELL, S. F. 2008. Biostratigraphy and its implications for the early tectonic evolution of the Great Arc of the Caribbean. *2008 Joint Meeting of the Geological Society of America, American Society of Agronomy, Crop Science Society of America, Gulf Coast Association of Geological Societies with the Gulf Coast Section of the Society for Sedimentary Geology*, Abstracts, **128-10**.

MOHRIAK, W. U., BROWN, D. E. & TARI, G. 2008. Sedimentary basins in the central and south Atlantic conjugate margins: deep structures and salt tectonics. World Wide Web Address: http://www.conjugatemargins.com/abstracts/by_session/1.

MOLINA-GARZA, R. S., VAN DER VOO, R. & URRUTIA-FUCUGAUCHI, J. 1992. Paleomagnetism of the Chiapas Massif, southern Mexico: evidence for rotation of the Maya Block and implications for the opening of the Gulf of Mexico. *Geological Society of America Bulletin*, **104**, 1156–1168.

MONTGOMERY, H. 1998. Paleogene stratigraphy and sedimentology, North Coast, Puerto Rico. *In*: LIDIAK, E. G. & LARUE, D. K. (eds) *Tectonics and Geochemistry of the Northeastern Caribbean*. Geological Society of America, 177–192.

MONTGOMERY, H. & KERR, A. C. 2009. Rethinking the origins of the red chert at La Désirade, French West Indies. *In*: JAMES, K. H., LORENTE, M. A. & PINDELL, J. L. (eds) *Origin and Evolution of the Caribbean Plate*. Geological Society, London, Special Publications, **328**, 457–467.

MOORE, G. T. & FAHLQUIST, D. A. 1976. Seismic profiling tying Caribbean DSDP sites 153, 151, and 152. *Geological Society of America Bulletin*, **87**, 1609–1614.

MORA, J., LOUREIRO, D. & OSTOS, M. 1993. Pre-Mesozoic rectangular network of crustal discontinuities: one of the main controlling factors of the tectonic evolution of northern South America. *American Association of Petroleum Geologists/Sociedad Venezolana de Geología International Congress and Exhibition*, Caracas, Abstracts, 58.

MORA, C., CORDOBA, F. *ET AL*. 1996. Petroleum systems of the Middle Magdalena Valley, Colombia. *American Association of Petroleum Geologists/Sociedad Geológica Venezolana International Congress and Exhibition*, Caracas, Abstracts A 32.

MORÓN, S. & JARAMILLO, C. 2008. Miocene Culebra-Cucaracha Formation boundary in Panama. Climatic or tectonic change. *18th Caribbean Geological Conference*, Santo Domingo, Abstracts.

MORRIS, A. E. L., TANER, I., MEYERHOFF, H. A. & MEYERHOFF, A. A. 1990. Tectonic evolution of the Caribbean region; Alternative hypothesis. *In*: DENGO, G. & CASE, J. E. (eds) *The Caribbean Region*. The Geology of North America, **H**, 433–457.

MOSSMAN, W. & VINIEGRA-O, F. 1976. Complex structures in Veracruz province of Mexico. *American Association of Petroleum Geologists Bulletin*, **60**, 379–388.

MOUNTAIN, G. S. & TUCHOLKE, B. E. 1985. Mesozoic and Cenozoic geology of the U.S. Atlantic continental slope and rise. *In*: POAG, C. W. (ed.) *Geologic Evolution of the U. S. Atlantic Margin*. Van Nostrand Reinhold, Stroudsburg, PA, 293–341.

MÜLLER, D., ROEST, W. R., ROYER, J. Y., GALAGHAN, L. M. & SCLATER, J. G. 1997. Digital isochrons of the world's ocean floor. *Journal of Geophysical Research*, **102**, 3211–3214.

MULLER, R. D., ROYER, J.-Y., CANDE, S. C., ROEST, W. R. & MASCHENKOV, S. 1999. New constraints on the Late Cretaceous/Tertiary Plate tectonic evolution of the Caribbean. *In*: MANN, P. (ed.) *Caribbean Sedimentary Basins, Sedimentary Basins of the World*. Elsevier, Amsterdam, 33–59.

MUÑOZ, A., BACA, D., ARTILES, V. & DUARTE, M. 1997. Nicaragua: Petroleum geology of the Caribbean margin. *The Leading Edge*, 1799–1805.

MUTTER, J. C., TALWANI, M. & STOFFA, P. 1982. Origin of eastward-dipping reflectors in oceanic crust off the Norwegian margin by subaerial seafloor spreading. *Geology*, **10**, 353–357.

MYCXYNSKI, R. & ITURRALDE-VINENT, M. 2005. The Late Lower Albian invertebrate fauna of the Río Hatillo Formation of Pueblo Viejo, Dominican Republic. *Caribbean Journal of Earth Science*, **41**, 782–796.

NAGLE, F. 1970. Caribbean geology. *Bulletin of Marine Science*, **21**, 375–439.

NAGLE, F. 1972. Rocks from seamounts and escarpments on the Aves Ridge. *VI Conferencia Geologica Del Caribe*, Margarita, Venezuela, Memorias, 409–413.

NAGLE, F. 1974. Blueschist, eclogite paired metamorphic belts and the early tectonic history of Hispaniola. *Geological Society of America Bulletin*, **85**, 1461–1466.

NAGLE, F., ERLICH, R. N. & CANOVI, C. J. 1978. Caribbean dredge haul compilation: summary and implications. *Geologie en Mijnbouw*, **57**, 267–270.

NAGLE, F., WASSAL, H., TARASIEWICZ, G. & TARASIEWICZ, E. 1982. Metamorphic rocks and stratigraphy of central Tortué Island, Haiti: Transactions. *IX Caribbean Geological Conference*, Santo Domingo, **2**, 409–415.

NANCE, R. D., MILLER, B. V., KEPPIE, J. D., MURPHY, J. B. & DOSTAL, J. 2006. Acatlán Complex, southern Mexico: Record spanning the assemblage and breakup of Pangea. *Geology*, **34**, 857–860.

NELSON, B. K., HERMANN, U. R., GORDON, M. B. & RATSBACHER, L. 1997. Sm–Nd and U–Pb evidence for Proterozoic crust in the Chortís Block, Central America: Comparison with the crustal history of southern Mexico. *Terra Nova*, **9**, Abstract Supplement No 1, 496.

NICHOLLS, I. A., CARMICHAEL, I. S. E. & STORMER, J. C. 1971. Silica activity and Ptotal in igneous rocks. *Contributions to Mineralogy and Petrology*, **33**, 1–20.

NUR, A. & BEN-AVRAHAM, Z. 1983. Displaced terranes and mountain building. *In*: HSU, K. J. (ed.) *Mountain Building Processes*, Academic Press, 73–84.

OKAYA, D. A. & BEN-AVRAHAM, Z. 1987. Structure of the continental margin of southwestern Panama. *Geological Society of America Bulletin*, **99**, 792–802.

ORTEGA-GUTIERREZ, F. & ELIAS-HERRERA, M. 2003. *Wholesale Melting of the Southern Mixteco Terrane and Origin of the Xolapa Complex*. Geological Society of America, abstracts with program, **35**, 66.

ORTEGA-GUTIÉRREZ, F., SOLARI, L. A. *ET AL*. 2007. The Maya–Chortís boundary: a tectonostratigraphic approach. *International Geology Review*, **49**, 996–1024.

OSTOS, M. 2002. The alpine-type Tinaquillo peridotite complex, Venezuela: Fragment of a Jurassic rift zone? *In*: AVÉ LALLEMENT, H. G. & SISSON, V. B. (eds) *Caribbean/South American Plate Interactions, Venezuela*. Geological Society of America, Special Papers, **394**, 207–222.

OSTOS, M. & SISSON, V. B. 2002. Geochemistry and tectonic setting of igneous and metaigneous rocks of northern Venezuela. *In*: AVÉ LALLEMENT, H. G. & SISSON, V. B. (eds) *Caribbean/South American Plate Interactions, Venezuela*. Geological Society of America, Special Papers, **394**, 119–156.

OXBURGH, E. R. 1966. Geology and metamorphism of Cretaceous rocks in eastern Carabobo State, Venezuelan coast ranges. *In*: HESS, H. H. (ed.) *Caribbean Geological Investigations*, Geological Society of America Memoirs, **98**, 241–310.

PAQUETTE, J., SIGMARSSON, O. & TIEPOLO, M. 2006. Continental basement under Iceland revealed by old zircons. *EOS Trans. AGU*, **87**, Fall Meeting Supplement, Abstract V33A-064.

PARDO, G. 1975. Geology of Cuba. *In*: NAIRN, A. E. M. & STEHLI, F. G. (eds) *The Ocean Basins and Margins. The Gulf of Mexico and the Caribbean*. Plenum Press, New York, 553–615.

PAULUS, F. J. 1972. The geology of site 98 and the Bahama Platform. *In*: HOLLISTER *ET AL*. (eds) *Initial Reports of the Deep Sea Drilling Program*, **XI**, 877–897.

PENNINGTON, W. D. 1981. Subduction of the Eastern Panama Basin and seismotectonics of northwestern South America. *Journal of Geophysical Research*, **86**, 10753–10770.

PEQUEGNAT, W. E., BRYANT, W. R. & HARRIS, J. E. 1971. Carboniferous sediments from Sigsbee Knolls, Gulf of Mexico. *American Association of Petroleum Geologists Bulletin*, **55**, 116–123.

PEREIRA, J. G. 1985. Evolucion tectonica de la Cuenca de Carupano durante el Terciario. *VI Congreso Geologica Venezolano*, Caracas, Memoria, **IV**, 2618–2648.

PEREZ DE ARMAS, J. 1999. Transpressional history of the Serrania del Interior Foreland Fold and Thrust Belt, north-central Venezuela: New evidence from apatite fission-track modeling and 2D seismic reflection analysis. *Penrose Conference: Subduction to Strike–Slip Transitions on Plate Boundaries*, Abstracts, 83–84.

PERFIT, M. R. & HEEZEN, B. C. 1978. The geology and evolution of the Cayman Trench. *Geological Society of America Bulletin*, **89**, 1155–1174.

PERFIT, M. R., HEEZEN, B. C., RAWSON, M. & DONNELLY, T. W. 1980. Chemistry, origin and tectonic significance of metamorphic rocks from the Puerto Rico Trench. *Marine Geology*, **34**, 125–156.

PETER, G. & WESTBROOK, G. K. 1976. Tectonics of southwestern North Atlantic and Barbados Ridge Complex. *American Association of Petroleum Geologists Bulletin*, **60**, 1078–1106.

PHAIR, R. L. & BUFFLER, R. T. 1983. Pre-Middle Cretaceous geologic history of the deep southeastern Gulf of Mexico. *In*: BALLY, A. W. (ed.) *Seismic Expression of Structural Styles – a Picture and Work Atlas*. American Association of Petroleum Geologists, Studies in Geology, **15**, 141–147.

PINDELL, J. L. 1991. Geological arguments suggesting a Pacific origin for the Caribbean Plate. *Transactions 12th Caribbean Geological Conference*, St Croix.

PINDELL, J. L. 1993. Regional synopsis of the Gulf of Mexico and Caribbean evolution. *GCSSEOM Foundation 13th Annual Research Conference*, 251–274.

PINDELL, J. L. & BARRETT, S. F. 1990. Geological evolution of the Caribbean region; a plate-tectonic perspective. *In*: DENGO, G. & CASE, J. E. (eds) *The Caribbean Region*. The Geology of North America, **H**. Geological Society of America, 405–432.

PINDELL, J. L. & KENNAN, L. 2009. Tectonic evolution of the Gulf of Mexico, Caribbean and northern South America in the mantle reference frame: an update. *In*: JAMES, K. H., LORENTE, M. A. & PINDELL, J. L. (eds) *Origin and Evolution of the Caribbean Plate*. Geological Society, London, Special Publications, **328**, 1–55.

PINDELL, J. L., HIGGS, R. & DEWEY, J. F. 1998. Cenozoic palinspastic reconstruction, palegeographic evolution, and hydrocarbon setting of the northern margin of South America. *In*: PINDELL, J. L. & DRAKE, C. L. (eds) *Paleogeographic Evolution and Non-Glacial Eustasy, Northern South America*. SEPM Special Publications, **58**, 45–86.

PINDELL, J. L., DRAPER, G., KENNAN, L., STANEK, K. P. & MARESCH, W. V. 2001. Evolution of the northern portion of the Caribbean Plate: Pacific origin to Bahamian collision. *SA Abstracts from the IGCP Project 433 – Caribbean Plate Tectonics*.

PINDELL, J. L., KENNAN, L., MARESCH, W. V., STANEK, K. P., DRAPER, G. & HIGGS, R. 2005. Plate-kinematics and crustal dynamics of circum-Caribbean arc-continent interactions: Tectonics controls of basin development in Proto-Caribbean margins. *In*: AVÉ LALLEMENT, H. & SISSON, V. B. (eds) *Caribbean–South American Plate Interactions, Venezuela*. Geological Society of America Special Papers, **394**, 7–52.

PINDELL, J. L., KENNAN, L., STANEK, K.-P, MARESCH, W. V. & DRAPER, G. 2006. Foundations of Gulf of Mexico and Caribbean evolution: eight controversies resolved. *In*: ITURRALDE-VINENT, M. A. & LIDIAK, E. G. (eds) *Caribbean Plate Tectonics. Geologica Acta*, **4**, 303–341.

PINET, P. R. 1971. Structural configuration of the northwestern Caribbean Plate boundary. *Geological Society of America Bulletin*, **82**, 2027–2032.

PINET, P. R. 1972. Diapirlike features offshore Honduras: implications regarding tectonic evolution of Cayman Trough and Central America. *Geological Society of America Bulletin*, **83**, 1911–1922.

PINET, B., LAJAT, D., LE QUELLEC, P. & BOUYSSE, Ph. 1985. Structure of Aves Ridge and Grenada Basin from Multichannel Seismic Data. *Géodynamique des Caraibes, Symposium*. Editions Technip, Paris, 53–64.

POST, P. 2005. Constraints on the interpretation of the origin and early development of the Gulf of Mexico Basin. *25th Annual Gulf Coast Section SEPM Foundation Bob F. Parker Research Conference, Abstracts*, 46.

PRICE, N. J. 2001. *Major Impacts and Plate Tectonics: A Model for Phanerozoic Evolution of the Earth's Lithosphere*. Routledge, Abingdon.

PROENZA, J. A., DÍAZ-MARTÍNEZ, R. *ET AL*. 2006. Primitive Cretaceous island-arc volcanic rocks in eastern Cuba: the Téneme Formation. *In*: ITURRALDE-VINENT, M. A. & LIDIAK, E. G. (eds) *Caribbean Plate Tectonics. Geologica Acta*, **4**, 103–121.

PSZCZOLKOWSKI, A. 1976. Stratigraphic facies sequences in the Sierra de Rosario, Cuba. *Academie Polonaise des Sciences, Serie des Sciences de la Terre*, **24**, 193–203.

PSZCZOLKOWSKI, A. 1999. The exposed passive margin of North America in Western Cuba. *In*: MANN, P. (ed.) *Caribbean Sedimentary Basins*. Sedimentary Basins of the World. Elsevier, Amsterdam, 93–121.

PSZCZOLKOWSKI, A. & MYCZYNSKI, R. 2003. Stratigraphic constraints on the Late Jurassic–Cretaceous paleotectonic interpretations of the Placetas Belt in Cuba. *In*: BARTOLINI, C. *ET AL*. (eds) *The Circum-Gulf of Mexico and the Caribbean: Hydrocarbon Habitats, Basin Formation, and Plate Tectonics*. American Association of Petroleum Geologists Memoirs, **79**, 545–581.

PUBELLIER, M., MAUFFRET, A., LEROY, S., VILA, J.-M. & AMILCAR, H. 2000. Plate boundary readjustment in oblique convergence: Example of the Neogene of Hispaniola, Greater Antilles. *Tectonics*, **19**, 630–648.

PUDSEY, C. J. & READING, H. G. 1972. Sedimentology and structure of the Scotland Group, Barbados. *In*: *Trench-Forearc Geology; Sedimentation and Tectonics on Modern and Ancient Active Plate Margins*. Geological Society, London Special Publications, **10**, 291–308.

RAMIREZ, Mi. I., DRAPER, G., SMITH, L., HORNAFIUS, J. S., MUNTHE, J. & GORDON, M. 1995. Neotectonic structures and paleostress in south-central Hispaniola. *3rd Geological Conference Geological Society of Trinidad and Tobago, 14th Caribbean Geological Conference, Transactions*, 214.

RAMROOP, C. 1985. Structure and composition of the "acoustic" basement of the North Coast Complex, Trinidad. *First Geological Conference of the Geological Society of Trinidad and Tobago*, 53–66.

REED, J. C., WHEELER, J. O. & TUCHOLKE, B. E. 2005. *Geologic Map of North America*. Decade of North American Geology. Geological Society of America.

RÉMANE, R. 1980. Calpionellids. *In*: HAQ, B. V. & BOERSMA, A. (eds) *Introduction to Micropalaeontology*. Elsevier, New York, 161–170.

RÉVILLON, S., HALLOT, E., ARNDT, N. T., CHAUVEL, C. & DUNCAN, R. A. 2000. A complex history for the

Caribbean Plateau: petrology, geochemistry, and geochronology of the Beata Ridge, South Hispaniola. *Journal of Geology*, **108**, 641–661.

RIGASSI-STUDER, D. 1961. Quelques vues nouvelles sur la géologie cubaine. *Chronique des Mines et de la Recherche Miniere*, **29**, 3–7.

ROBERTS, D. G. 1975. Marine geology of the Rockall Plateau and Trough. *Philosophical Transactions for the Royal Society of London. Series A, Mathematical and Physical Sciences*, **278**, 447–509.

ROBERTS, M., HOLLISTER, C., YARGERAND, H. & WELCH, R. 2005. Regional geologic and geophysical observations basinward of the Sigsbee Escarpment and Mississippi Fan Fold Belt, Central Deep-Water Gulf of Mexico. *25th Annual Gulf Coast Section SEPM Foundation Bob F. Parker Research Conference*, Extended Abstracts, 1190–1199.

ROBINSON, E. 1967. Submarine slides in white limestone group, Jamaica. *American Association of Petroleum Geologists Bulletin*, **51**, 569–578.

RODRIGUEZ, S. E. 1968. Estudio sobre el metamorfismo regional en la Peninsula de Paria, Estado Sucre. *Boletin Informativo, Asociación Venezolana de Geología, Minería y Petróleo*, **11**, 61–88.

ROGERS, R. & MANN, P. 2007. Transtensional deformation of the western Caribbean–North America plate boundary zone. *In*: MANN, P. (ed.) *Geologic and Tectonic Development of the Caribbean Plate Boundary in Northern Central America*. Geological Society of America Special Papers, 37–64.

ROGERS, R. D., MANN, P. & EMMET, P. A. 2007*a*. Tectonic terranes of the Chortís Block based on integration of regional aeromagnetic and geologic data. *In*: MANN, P. (ed.) *Geologic and Tectonic Development of the Caribbean Plate Boundary in Northern Central America*. Geological Society of America, Special Papers, **428**, 65–88.

ROGERS, R. D., MANN, P., EMMET, P. A. & VENABLE, M. M. 2007*b*. Colon fold belt of Honduras: Evidence for Late Cretaceous collision between the continental Chortís Block and intra-oceanic Caribbean Arc. *In*: MANN, P. (ed.) *Geologic and Tectonic Development of the Caribbean Plate Boundary in Northern Central America*. Geological Society of America Special Papers, **428**, 129–149.

ROJAS, C. R. 2004. The Cuban rudist mollusks, an important tool in the biostratigraphical correlation of the Caribbean Cretaceous. *32nd International Geological Conference*, Florence, 2004, IGCP Project 433 Caribbean Plate Tectonics, Abstracts.

ROJAS-AGRAMONTE, Y., KRÖNER, A., GARCÍA-CASCO, A., ITURRALDE-VINENT, M. A., WINGATE, M. T. D. & LIU, D. Y. 2006. Geodynamic implications of zircon ages from Cuba. *Geophysical Research*, Abstracts, **8**.

ROLLINS, J. F. 1965. *Stratigraphy and Structure of the Goajira Peninsula, Northwestern Venezuela and Northeastern Colombia*. University of Nebraska Studies: New Series, **30**.

ROSENCRANTZ, E. 1990. Structure and tectonics of the Yucatán Basin, Caribbean Sea, as determined from seismic reflection studies. *Tectonics*, **9**, 1037–1059.

ROSENCRANTZ, E., ROSS, M. I. & SCLATER, J. G. 1988. Age and spreading history of the Cayman Trough as determined from depth, heat flow, and magnetic anomalies. *Journal of Geophysical Research*, **93**, 2141–2157.

ROSENDAHL, B. R., MEYERS, J. & GROSCHEL, H. 1992. Nature of the transition from continental to oceanic crust and the meaning of reflection Moho. *Geology*, **20**, 721–724.

ROSS, M. I. & SCOTESE, C. R. 1988. A hierarchical tectonic model of the Gulf of Mexico and Caribbean region. *Tectonophysics*, **155**, 139–168.

RUEDA-GAXIOLA, J. 2003. The Origin of the Gulf of Mexico Basin and its petroleum subbasins in Mexico, based on Red Bed and salt palynostratigraphy. *In*: BARTOLINI *ET AL*. (eds) *The Circum-Gulf of Mexico and the Caribbean: Hydrocarbon Habitats, Basin Formation and Plate Tectonics*. American Association of Petroleum Geologists Memoirs, **79**, 246–282.

SACHS, P. M. & ALVARADO, G. E. 1996. Mafic metaigneous lower crust beneath Arenal Volcano (Costa Rica): evidence from xenoliths. *Boletin Observatorio Vulcano Arenal, Año 6*, **11–12**, 71–78.

SALLARES, V., DANOBEITIA, J. J. & FLUEH, E. R. 2001. Lithospheric structure of the Costa Rica Isthmus: effects of subduction zone magmatism on a oceanic plateau. *Journal of Geophysical Research*, **106**, 621–643.

SALVADOR, A. 1991. Triassic–Jurassic. *In*: SALVADOR, A. (ed.) *The Gulf of Mexico Basin*. The Geology of North America. Geological Society of America, 131–180.

SANCHEZ-BARREDA, L. A. 1981. *Geologic evolution of the continental margin of the Gulf of Tehuantepec in southern Mexico*. PhD thesis, University of Texas, Austin, TX.

SANTOS, H. 2002. Stratigraphy and depositional history of the upper Cretaceous strata in the Cabo Rojo-San German structural block, southwestern Puerto Rico. *16th Caribbean Geological Conference*, Barbados, Abstracts, 67.

SAUNDERS, J. B., EDGAR, N. T., DONNELLY, T. W. & HAY, W. W. 1973. Cruise synthesis. *In*: EDGAR, N. T. *ET AL*. (eds) *Initial Reports of the Deep Sea Drilling Project*. US Government Printing Office, Washington, DC, **15**, 1077–1111.

SCHLAGER, W., BUFFLER, R. T. *ET AL*. 1984. Deep Sea Drilling Project, Leg 77, southeastern Gulf of Mexico. *Geological Society of America Bulletin*, **95**, 226–236.

SCHNEIDERMANN, N., BECKMANN, J. P. & HEEZEN, B. C. 1972. Shallow water carbonates from the Puerto Rico Trench region. *VI Conferencia Geología del Caribe*, Margarita, Venezuela, Memorias, 423–425.

SCHUCHERT, C. 1935. *Historical Geology, Antillean – Caribbean Region*. Wiley, Chichester.

SEIDERS, V. M. 1965. Geología de Miranda Central, Venezuela. *Ministerio de Minas e Hidrocarburos Boletin Geologico*, **6**, 289–416.

SENN, A. 1940. Paleogene of Barbados and its bearing on history and structure of Antillean–Caribbean region. *Bulletin American Association of Petroleum Geologists*, **24**, 1548–1610.

SENN, A. 1944. *Inventory of the Barbados Rocks and Their Possible Utilization*. Department of Science and Agriculture, Barbados, Bulletin **1**, 1–40.

SHARP, W. D. 1988. Tobago, West Indies, geochronological study of a fragment of a composite mesozoic oceanic island arc. *Annual Geological Society of America Centennial Celebration Meeting*, Abstracts, **11964**.

SHAUB, F. J. 1983. Origin of the Catoche Tongue. *In*: BALLY, A. W. (eds) *Seismic Expression of Structural Styles – A Picture and Work Atlas*. American Association of Petroleum Geologists Studies in Geology, **15**, 129.

SHERIDAN, R. E., CROSBY, J. T., BRYAN, G. M. & STOFFA, L. 1981. Stratigraphy and structure of southern Blake Plateau, northern Florida Straits, and northern Bahamas Platform from multichannel seismic reflection data. *American Association of Petroleum Geologists Bulletin*, **65**, 2571–2593.

SHURBET, G. L., WORZEL, J. L. & EWING, M. 1956. Gravity reconnaissance survey of Puerto Rico. *Geological Society of America Bulletin*, **67**, 511–534.

SIGURDSSON, H., LECKIE, R. M. & ACTON, G. D. 1996. *Caribbean Ocean History and the Cretaceous/Tertiary Boundary Event*. Ocean Drilling Program, Leg 165 Preliminary Report.

SINTON, C. W., SIGURDSSUN, H. & DUNCAN, R. A. 2000. Geochronology and petrology of the igneous basement at the lower Nicaraguan Rise, Site 1001. *Proceedings of the Ocean Drilling Program*, Scientific Results, Leg 165, 233–236.

SISSON, V. B. & AVÉ LALLEMENT, H. G. 1992. Burial and ascent of blueschists and eclogites, Venezuela, part 1: petrologic constraints. *Geological Society of America Annual Meeting*, Abstracts, A149.

SISSON, V. B., AVÉ LALLEMENT, H. G. *ET AL*. 2005. Overview of radiometric ages in three allochthonous belts of northern Venezuela: old ones, new ones, and their impact on regional geology. *In*: AVÉ LALLEMENT, H. G. & SISSON, V. B. (eds) *Caribbean–South America Plate Interactions, Venezuela*. Geological Society of America Special Papers, **394**, 91–117.

SKERLEC, G. M. & HARGRAVES, R. B. 1980. Tectonic significance of paleomagnetic data from northern Venezuela. *Journal of Geophysical Research*, **85**, 5303–5315.

SMITH, A. L., SCHELLEKENS, J. H. & DÍAZ, A-L. M. 1998. Batholiths as markers of tectonic change in the northeastern Caribbean. *In*: LIDIAK, E. G. & LARUE, D. K. (eds) *Tectonics and Geochemistry of the Northeastern Caribbean*. Geological Society of America, 99–122.

SNOKE, A. W. 1990. An evaluation of the petrogenesis of the accreted Mesozoic Island Arc of the southern Caribbean. *Second Geological Conference of the Geological Society of Trinidad and Tobago, Transactions*, 222–243.

SNOKE, A. W. 2003. *Tobago, West Indies, as a Guide for the Mesozoic Evolution of the Great Arc of the Caribbean*. Geological Society of America Abstracts with Program, **35**, paper no. 33-1.

SNOKE, A. W., YULE, J. D., ROWE, D. W., WADGE, G. & SHARP, W. D. 1990. Stratigraphic and structural relationships on Tobago and some tectonic implications. *In*: LARUE, D. K. & DRAPER, G. (eds) *Transactions of the 12th Caribbean Geological Conference*, St Croix, 7–11 August 1989, 393–403.

SNOKE, A. W., YULE, J. D., ROWE, D. W., WADGE, G. & SHARP, W. D. 2001. *Geologic History of Tobago, West Indies*. Geological Society of America Special Papers, **354**.

SOMIN, M. L., LEPEKHINA, E. N. & TOLMACHEVA, E. V. 2006. El Guaybo gneiss sialic basement boulder in western Cuba. *Geophysical Research Abstracts*, **8**, 03377.

SOMOZA, R. 2008. Paleomagnetic data from the Americas admit the possibility of an extensional origin for the Caribbean LIP. World Wide Web Address: http://www.mantleplumes.org/CaribbeanPMag.html.

SPEED, R., RUSSO, R., WEBER, J. & ROWLEY, K. C. 1991. Evolution of southern Caribbean Plate boundary, vicinity of Trinidad and Tobago: discussion. *American Association of Petroleum Geologists Bulletin*, **75**, 1789–1794.

STAINFORTH, R. M. 1966. Gravitational deposits in Venezuela [Depositos gravitationales en Venezuela]. *Boletin Informativo, Asociacion Venezolana de Geologia, Mineria y Petroleo*, 277–287.

STAINFORTH, R. M. 1969. The concept of sea-floor spreading applied to Venezuela. *Associacion Venezolana Geología, Minería y Petróleo, Boletin Informativo*, **12**, 257–274.

STANEK, K. P. & MARESCH, W. V. 2006. The geotectonic story of the Great Antillean Arc – implications of structural and geochronological data from Central Cuba. *In*: JAMES, K. H. & LORENTE, M. A. (convenors), *Geology of the Area between North and South America, with Focus on the Origin of the Caribbean Plate: an International Research Conference*, Sigüenza, Spain, Abstracts.

STANEK, K. P., MARESCH, W. V. *ET AL*. 2005. Contrasting high-pressure and low-pressure P-T-t-d paths in a single nappe pile: a case study from the Cuban collisional suture. *17th Caribbean Geological Conference*, Puerto Rico, abstracts, 84.

STANEK, K. P., MARESCH, W. V., GRAFE, F., GREVEL, CH. & BAUMANN, A. 2006. Structure, tectonics and metamorphic development of the Sancti Spiritus Dome (eastern Escambray Massif, Central Cuba). *In*: ITURRALDE-VINENT, M. A. & LIDIAK, E. G. (eds) *Caribbean Plate Tectonics. Geologica Acta*, **4**, 151–170.

STANEK, K. P., MARESCH, W. V. & PINDELL, J. L. 2009. The geotectonic story of the northwestern branch of the Caribbean Arc: implications from structural and geochronological data of Cuba. *In*: JAMES, K. H., LORENTE, M. A. & PINDELL, J. L. (eds) *Origin and Evolution of the Caribbean Plate*. Geological Society, London, Special Publications, **328**, 361–398.

STEINER, M. B. 2005. Pangean reconstruction of the Yucatán Block: its Permian, Triassic, and Jurassic geologic and tectonic history. *In*: ANDERSON, T. H. *ET AL*. (eds) *The Mojave–Sonora Megashear Hypothesis*. Geological Society of America Special Papers, **393**, 457–480.

STOCKHERT, B., MARESCH, W. V. *ET AL*. 1995. Crustal history of Margarita Island (Venezuela) in detail: constraint on the Caribbean Plate-tectonic scenario. *Geology*, **23**, 787–790.

SUMMA, L. L., GOODMAN, E. D., RICHARDSON, M., NORTON, L. O. & GREEN, A. R. 2003. Hydrocarbon systems of northeastern Venezuela: plate through molecular scale-analysis of the genesis and evolution of the eastern Venezuela Basin. *Marine and Petroleum Geology*, **20**, 323–349.

SZATMARI, P. 1983. Amazon rift and the Pisco-Juruá Fault: their relation to the separation of North America from Gondwana. *Geology*, **11**, 300–304.

TARDY, M., LAPIERRE, H. ET AL. 1994. The Guerrero suspect terrane (western Mexico) and coeval arc terranes (the Greater Antilles and the Western Cordillera of Colombia): a Late Mesozoic intra-oceanic arc accreted to cratonal America during the Cretaceous. *Tectonophysics*, **230**, 49–73.

TEN BRINK, U. S., COLEMAN, D. F. & DILLON, W. P. 2003. The nature of the crust under Cayman Trough from gravity. *Marine and Petroleum Geology*, **19**, 971–987.

TOMBLIN, J. F. 1970. *Field Guide to the Grenadines, Lesser Antilles: International Field Institute Guidebook to the Caribbean Island-Arc System*. American Geological Institute (article individually paginated).

TOMBLIN, J. F. 1975. The Lesser Antilles and Aves Ridge. *In*: NAIRN, A. E. M. & STEHLI, F. G. (eds) *The Ocean Basins and Margins*, **3**. Plenum Press, New York, 467–500.

TOMMASI, A. & VAUCHEZ, A. 2001. Continental rifting parallel to ancient collisional belts: an effect of the mechanical anisotropy of the lithospheric mantle. *Earth and Planetary Science Letters*, **185**, 199–210.

TOURNON, J., TRIBOULET, C. & AZEMA, J. 1989. Amphibolites from Panama: anticlockwise $P–T$ paths from a pre-Upper Cretaceous metamorphic basement in Isthmian Central America. *Journal of Metamorphic Geology*, **7**, 539–546.

TRECHMANN, C. T. 1925. The Scotland beds of Barbados. *Geological Magazine*, **62**, 481–504.

TRECHMANN, C. T. 1937. The base and top of the coral-rock in Barbados. *Geological Magazine*, **LXXIV**, 337–359.

URBANI, F. 2004. New geological map of the Cordillera de la Costa, Northern Venezuela. *32nd IGC*, Florence, 2004. IGCP Project 433 Caribbean Plate Tectonics, abstracts.

VAN DER HILST, R. 1990. Tomography with P, PP and pP delay-time data and the three dimensional mantle structure below the Caribbean region *Geologica Ultraiectina*, Mededelingen van de Faculteit Aardwetenschahppen der Rijsuniversiteit te Utrecht, **67**, 250p.

VAN DER HILST, R. & MANN, P. 1994. Tectonic implications of tomographic images of subducted lithosphere beneath northwestern South America. *Geology*, **22**, 451–454.

VAN GESTEL, J. P., MANN, P., GRINDLAY, N. R. & DOLAN, J. F. 1999. Three-phase tectonic evolution of the northern margin of Puerto Rico as inferred from an integration of seismic reflection, well, and outcrop data. *Marine Geology*, **16**, 257–286.

VAUGHAN, T. W. & WELLS, J. W. 1945. *American Old and Middle Tertiary Larger Foraminifera and Corals*. Geological Society of America Memoirs, **9**.

VENABLE, M. E. 1993. An interpretation of the tectonic development of N. E. Nicaragua. *Geological Society of America Abstracts*, **25**, A-286.

VIVAS, V. & MACSOTAY, O. 2002. Cretaceous (Senonian)–Paleocene paleoceanographic continuity

between the Caribbean Plate's abyssal facies (Venezuelan Basin) and the South American Plate's slope and shelf facies (northern Venezuela). *16th Caribbean Geological Conference*, Barbados, Abstracts, 70.

VOGEL, T. A., PATINO, L. C., ALVARADO, G. E. & GANS, P. B. 2004. Silicic ignimbrites within the Costa Rican volcanic front: evidence for the formation of continental crust. *Earth and Planetary Science Letters*, **226**, 149–159.

VOGEL, T. A., PATINO, L. C., ALVARADO, G. E. & ROSE, W. I. 2007. Petrogenesis of ignimbrites. *In*: BUNDSCHUH, J. & ALVARADO, G. E. (eds) *Central America: Geology, Resources and Hazards*. Taylor & Francis/Balkema, The Netherlands, 591–618.

WADGE, G. & MACDONALD, R. 1985. Cretaceous tholeiites of the northern margin of South America: the Sans Souci Formation of Trinidad. *Journal of the Geological Society, London*, **142**, 297–308.

WEBER, J. C., DIXON, T. H. ET AL. 2001. GPS estimate of relative motion between the Caribbean and South American plates, and geologic implications for Trinidad and Venezuela. *Geology*, **29**, 75–78.

WEHRMANN, M. 1972. Geologia de la region de Guatire-Colonia Tovar. *IV Congreso Geologico Venezolano, Caracas, Memoria, Boletín Geología*, Publicación Especial, **5**, 2093–2219.

WESTBROOK, G. K. 1990. Gravity anomaly map of the Caribbean. *In*: DENGO, C. & CASE, J. E. (eds) *The Caribbean Region*. The Geology of North America, **H**. Geological Society of America, plate 7.

WESTERCAMP, D. 1988. Magma generation in the Lesser Antilles: geological constraints. *Tectonophysics*, **149**, 145–163.

WESTERCAMP, D., ANDREIEFF, P., BOUYSE, Ph., MASCLE, A. & BAUBRON, J. C. 1985. The Grenadines, Southern Lesser Antilles. Part 1. Stratigraphy and volcano-structural evolution. *Géodynamique des Caraïbes*, Symposium, Édition Technip, 109–118.

WIEDMANN, J. 1978. Ammonites from the Curaçao Lava Formation, Curaçao, Caribbean. *Geologie en Mijnbouw*, **57**, 361–364.

WILSON, H. H. & MEYERHOFF, A. A. 1978. Paleomagnetic results from Cretaceous sediments in Honduras: Tectonic implications: Comments and reply. *Geology*, **6**, 440–442.

WILSON, J. T. 1966. Did the Atlantic close and then re-open? *Nature*, **211**, 676–681.

WITHJACK, M. O. & SCHLISCHE, R. W. 2005. A review of tectonic events on the passive margin of eastern North America. *25th Annual Gulf Coast Section SEPM Foundation Bob F. Parker Research Conference*, Extended Abstracts, 203–235.

WORZEL, J. L., BRYANT, W. ET AL. 1973. *Initial Report of the Deep Sea Drilling Project*, **S**. US Government Printing Office, Washington, DC.

WRIGHT, J. E. 2004. Aruba and Curaçao: remnants of a collided Pacific oceanic plateau? Initial geologic results from the BOLIVAR project. *EOS Transactions of the American Geophysical Union*, **85**, Fall Meeting Supplement, Abstract.

YSACCIS, R. 1997. *Tertiary Evolution of the Venezuelan Northeastern Offshore*. PhD thesis, Rice University, Houston.

Evolution of Middle America and the *in situ* Caribbean Plate model

KEITH H. JAMES

Institute of Geography and Earth Sciences, Aberystwyth, Wales, UK and
Consultant Geologist, Plaza de la Cebada 3, 09346 Covarrubias, Burgos, Spain
(e-mail: khj@aber.ac.uk)

Abstract: Regional geological data and global analogues suggest Caribbean Plate geology continues that seen along the margin of eastern North America in a more extensional setting, between the diverging Americas. From west to east there are continental masses with Triassic rifts, proximal continental blocks with kilometres-thick Mesozoic carbonates, more distal areas of Palaeozoic horsts flanked by Triassic–Jurassic dipping wedges of sediments, including salt and overlain by Cretaceous basalts, and most distal areas of serpentinized upper mantle. Plate history began along with the Late Triassic formation of the Central Atlantic Magmatic Province and involved Triassic–Jurassic rifting, Jurassic–Early Cenozoic extension and Oligocene–Recent strike–slip. Great extension promoted volcanism, foundering, eastward growth of the plate by backarc spreading and distribution of continental fragments on the plate interior and along its margins. Hydrocarbons probably are present. Caribbean geology has important implications for understanding of oceanic plateaus, intra-oceanic volcanic arcs, the 'andesite problem' and genesis of 'subduction' HP/LT metamorphic rocks. The model can be tested by re-examination of existing samples and seismic data and by deep sea drilling.

Middle America – crustal makeup

Based upon data discussed in a sister article this paper suggests an *in situ* evolution of the Caribbean Plate between North and South America. The data indicate that Middle America is built mainly of extended/distributed continental crust and smaller areas of pseudo-oceanic crust (serpentinized mantle) (the only recognized spreading crust, with magnetic anomalies, lies in the centre of the Cayman Trough; Fig. 1). The geology continues that along the eastern margin of North America but in a more extensional setting between the diverging Americas. Rifted and extended continent with thick Mesozoic carbonate cover recording subsidence surrounds the Gulf of Mexico and Yucatán basins and underpins the Florida–Bahamas platform and the submerged extension, as far as Jamaica, of continental Chortís along the Nicaragua Rise. The internal Gulf of Mexico, Yucatán, Colombian, Venezuelan and Grenada basins carry extended continental blocks flanked by half-grabens with wedge-shape fill, bounded by areas of thin (*c.* 3 km) pseudo oceanic crust of serpentinized mantle.

Tectonic fabric

Middle America manifests a regional tectonic fabric (Fig. 2) that demonstrates regional geological coherence and shows that no major block rotations or plate migration occurred. The stress–strain ellipse (inset) for sinistral movement of North America away from South America along N60°W fractures (a) reactivated N35°E Palaeozoic sutures as Triassic–Jurassic rifts with dextral component of movement; (b) generated N60°E extensional faults such as the Hess and NW Campeche escarpments; and (c) generated the Florida Arch. Curved faults in northern Central America reflect oroclinal bending and shortening during Cenozoic sinistral reactivation of N35°E trends (James 2007).

The following sketches suggest a Pangaean reconstruction and the evolution of Middle America. They do not pretend to be quantitative.

Pangaean reconstruction

Pre-drift restoration of Middle America (Fig. 3a) requires removal of sinistral offset and extension between North and South America and removal of volcanic-arc crust and serpentinized mantle (Fig. 1). The Gulf of Mexico (NAm) is closed up around the Maya Block. Maya and Chortís are united (removal of *c.* 900 km of Jurassic–Cretaceous, early Cayman offset and 300 km of Oligocene–Recent Cayman offset) by aligning their faulted eastern margins and Jurassic rifts associated with the Río Hondo and Guayapé faults (Fig. 2). This also restores Cuba north of Hispaniola and Puerto Rico, themselves closed up by removal of Cenozoic pull-apart. Thick crust of the Caribbean 'Plateau', western Venezuela Basin, lies close to South America. Similar crust seems to be present in the Yucatán and Colombian basins. Cenozoic oroclinal bending of the Motagua and Agalta areas and western Cuba (James 2007) is removed. The Hess Escarpment/southern limit of the Nicaragua Rise restores against the Mérida trend of NW Venezuela.

From: JAMES, K. H., LORENTE, M. A. & PINDELL, J. L. (eds) *The Origin and Evolution of the Caribbean Plate*.
Geological Society, London, Special Publications, **328**, 127–138.
DOI: 10.1144/SP328.4   0305-8719/09/$15.00 © The Geological Society of London 2009.

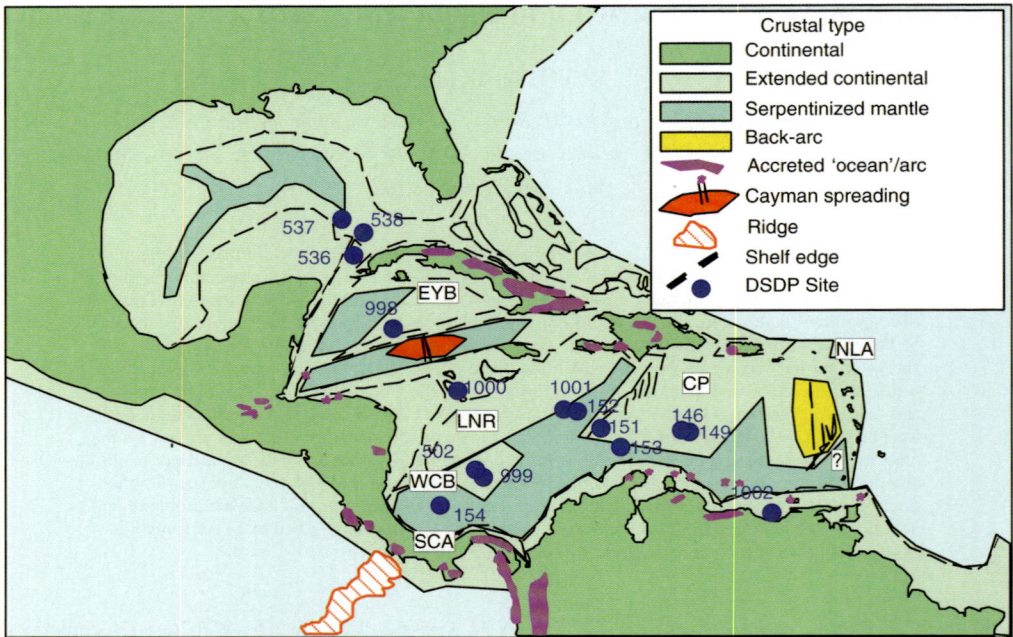

Fig. 1. Middle America crustal types/distribution. Continental blocks, indicated by crustal thickness (gravity, seismic), high silica rocks and dredge samples, beneath southern Central America (SCA) and the Greater Antilles–northern Lesser Antilles (NLA) are hidden beneath obducted volcanic arc/oceanic crust. Thick crustal areas on the Lower Nicaragua Rise (LNR), eastern Yucatán Basin (EYB), Caribbean 'Plateau' (CP) and west Colombia Basin (WCB) are underpinned by extended continent and locally overlain by Upper Cretaceous basalts. Serpentinized mantle possibly includes extremely attenuated continental crust. The Oligocene–Recent area in the Cayman Trough (red) is the only area with spreading magnetic anomalies.

Northwestern South America – the 'Bolivar Block' is restored several hundred kilometres to the SW along the Mérida Andes – Eastern Cordillera of Venezuela–Colombia. The Aruba–Blanquilla island chain does not exist at this time. Southern Central America (Chorotega–Chocó) restores to the SW of Chortís (present day Gulf of Tehuantepec, James 2007).

This schematic reconstruction suggests continuity of major faults crossing Maya and Chortís with southern Florida and the eastern seaboard of North America and its continuation into NW South America, possibly through the Sierra Nevada de Santa Marta in Colombia.

Geological evolution

Triassic–Early Jurassic rifting (Fig. 3b) reactivated Palaeozoic continental sutures as rifts, accommodating red beds and basalts (Manspeizer 1988; for a more regional vision of the Central Atlantic Magmatic Province and continental margin wedges see McHone *et al.* 2005, Fig. 1). Inboard basins were then abandoned as extension moved to the

continental margin. Here, asymmetric basins accommodated red beds, carbonates and evaporites as seaward-dipping wedges along the eastern seaboard of North America to Caribbean latitudes. Major N60°E trending features such as the Hess and NW Campeche escarpments and the La Trocha F. formed as the extensional strain within the regional N60°W sinistral system of offset between North and South America (Fig. 2). The resultant new basins accommodated salt deposition. Since salt diapirs and ultramafic rocks are seen along northern Honduras (Pinet 1971, 1972) and since serpentinites occur in the Motagua fault zone in the Cretaceous (Harlow *et al.* 2004), this paper suggests that the early Cayman Trough extended between Maya and Chortís as a salt basin at this time (Figs 4 & 5).

Late Jurassic–Early Cretaceous spreading in the Central Atlantic resulted in WNW drift of North America from Gondwana (South America–Africa) and great extension in Middle America (Fig. 4a, James 2009, fig. 4). Slip along major NW transfer faults within southern North America continued, offsetting the formerly linear Appalachian–Ouachita

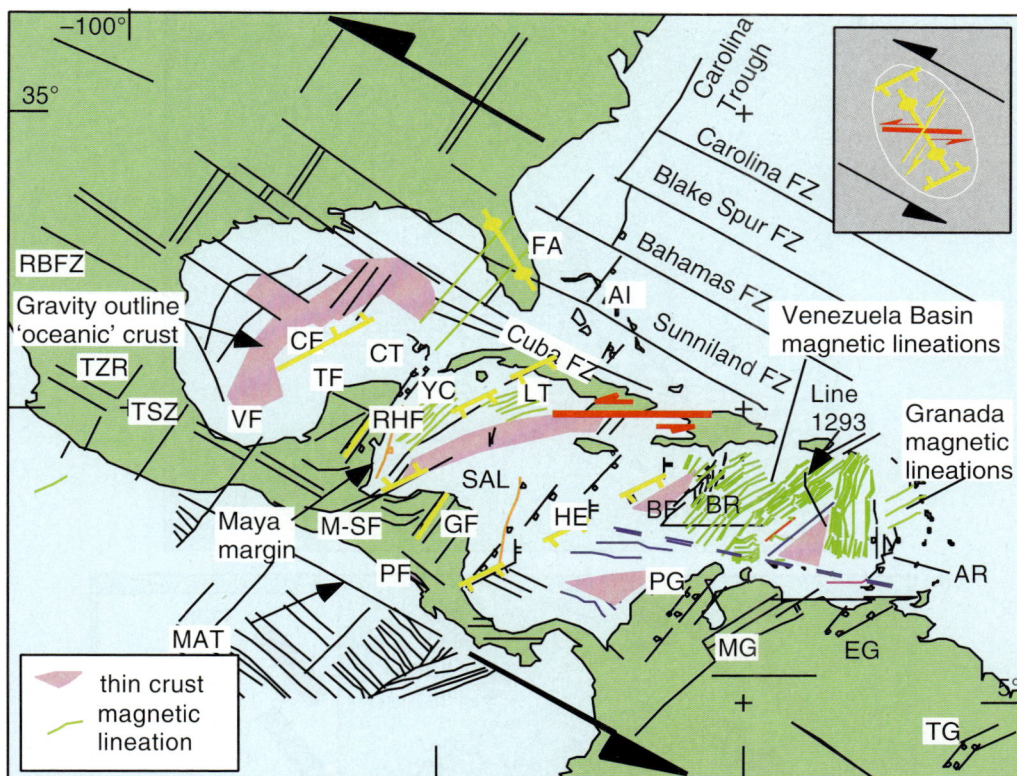

Fig. 2. Interpretation of Middle America tectonic fabric as the result of (1) reactivation of ancient lineaments (N35°E), (2) extension (N60°E) and strike–slip (east–west) strain within the sinistral offset (N60°W) of North from South America. AI, Aklins–Inagua–Caicos; BF, Beata F; BR, Beata Ridge; BRR, Blue Ridge Rift; CE, Campeche Escarpment; CT, Catoche Tongue; EG, Espino Graben; GF, Guayape FM; HE, Hess Escarpment; LT, La Trocha FM; MG, Mérida Graben; MiG, Mississippi Graben; M-SF, Motagua–Swan FM; OR, Oachita Rift; PF, Patuca FM; PG, Perijá Graben – Urdaneta; RBFZ, Río Bravo fault zone; RGR, Río Grande Rift; RHF, Río Hondo FM; SAL, San Andrés Lineament; TF, Ticul FM; TG, Takutu Graben; TT, Texas Transform; TSZ, Tenochtitlan Shear Zone; TZR, Tepic–Zacoalco Rift; VF, Veracruz FM; YC, Yucatán Channel. Red line, SE limit of Caribbean Plateau, the Central Venezuela FZ. Green lines, magnetic anomalies. Compiled from many sources.

trend and opening the Gulf of Mexico as the continent interior pulled away from the Maya Block and the NW Campeche Escarpment.

Maya moved west relative to Chortís by some 900 km (early Cayman displacement). Chortís moved around 900 km NW relative to South America and 400 km NW relative to Chorotega/ Chocó. Marginal areas subsided to accommodate platform carbonates several kilometres thick (Florida–Bahamas platform–Greater Antilles, Limestone Caribbees, Campeche–Yucatán– Nicaragua Rise). Further outboard, NE-trending normal faults (Triassic–Jurassic rifts) became listric as continental crust thinned into horsts flanked by wedges of sediments.

Albian commencement of westward drift of South America from Africa at Caribbean latitudes introduced strike–slip along the southern Caribbean Plate boundary, possibly provoking HP/LT metamorphism in rapidly subsiding/filling deeps, enhanced Caribbean extension and resulted in a first phase of basalt extrusion (120 Ma). Atlantic– Caribbean convergence resulted in subduction below dispersed continental fragments along the Lesser Antilles with resultant volcanism. Uplift to wave base, formation of an unconformity capped by shallow marine limestones and change of arc chemistry from primitive to calc-alkaline (continental input) occurred plate-wide, possibly reflecting intra-plate expansion due to decompression melting. A further plate-wide unconformity formed in the Cenomanian. Extrusion occurred again in the Turonian (90–88 Ma), forming the probably subaerial smooth seismic Horizon B″,

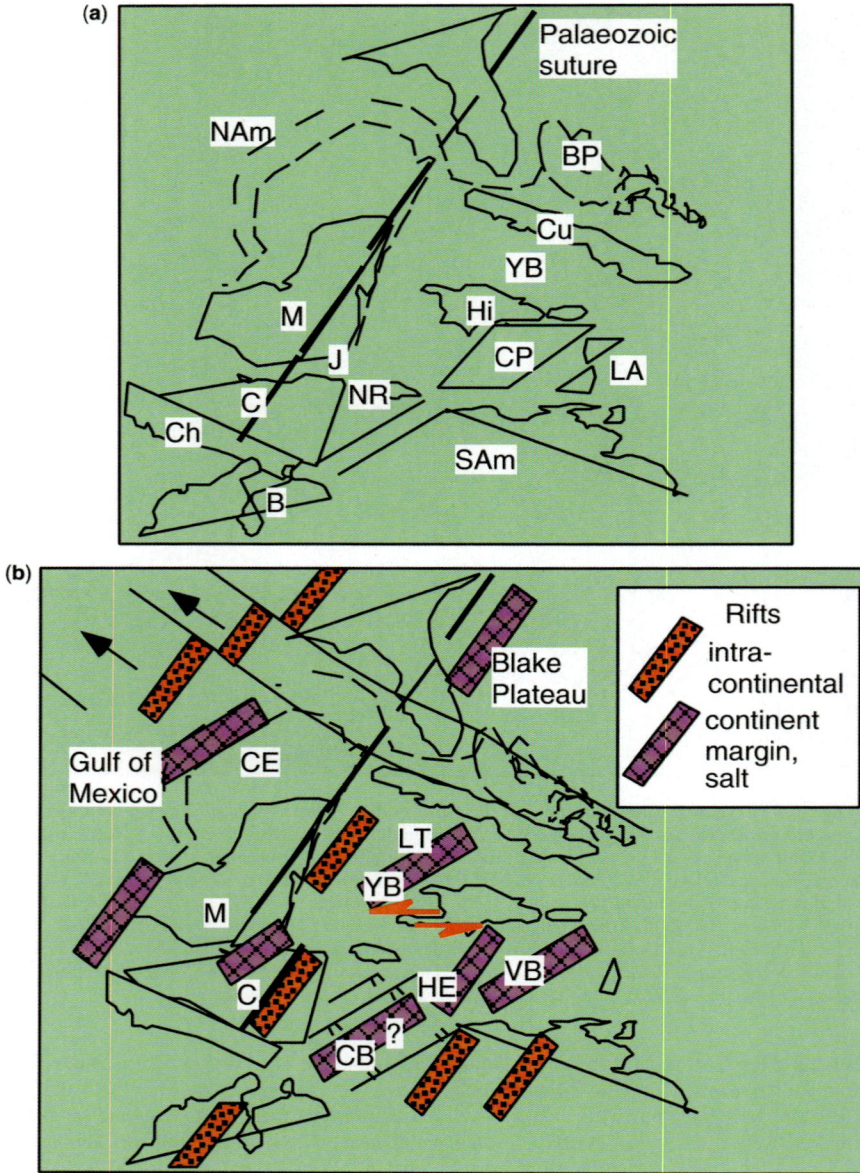

Fig. 3. (a) Schematic Pangaean restoration of middle America, modern coastlines shown for reference. Cayman offset between Maya and Chortís removed. Chorotega/Chocó restored to SW Chortís. NW South America (Bolivar Block, B) restored S60°W along Mérida Andes–Eastern Cordillera. Lesser Antilles basement represented by a single block, later dispersed by radial spreading. Extension of Bahamas Plateau, Nicaragua Rise and oroclinal bending of western Cuba removed. B, Bolivar Block; BP, Bahamas Platform; C, Chortís; Ch, Chorotega/Chocó; Cu, Cuba; Hi, Hispaniola; J, Jamaica; LA, Lesser Antilles; M, Maya; NAm, North America; NR, Nicaragua Rise; SAm, South America; YB, Yucatán Basin. (b) Triassic–Early Jurassic rifting accommodated intracontinental Triassic red beds (Jurassic in South America) and continent margin red beds and salt (Callovian? in Gulf of Mexico). Rifts reactivated N35°E Palaeozoic sutures. Offset along major transforms in southern North America pulled the Maya Block from the Gulf of Mexico and the Nicaragua Rise away from NW South America – the parallel Campeche and Hess Escarpments are the extensional strain of N60°E sinistral movement of North relative to South America (inset Fig. 2). Early Cayman offset (east–west sinistral synthetic to N60°W movement) began to separate Maya from Chortís along the northern Caribbean. C, Chortís Block; CB, Colombia Basin; CE, Campeche Escarpment; HE, Hess Escarpment; LT, La Trocha F; M, Maya Block; VB, Venezuela Basin; YB, Yucatán Basin.

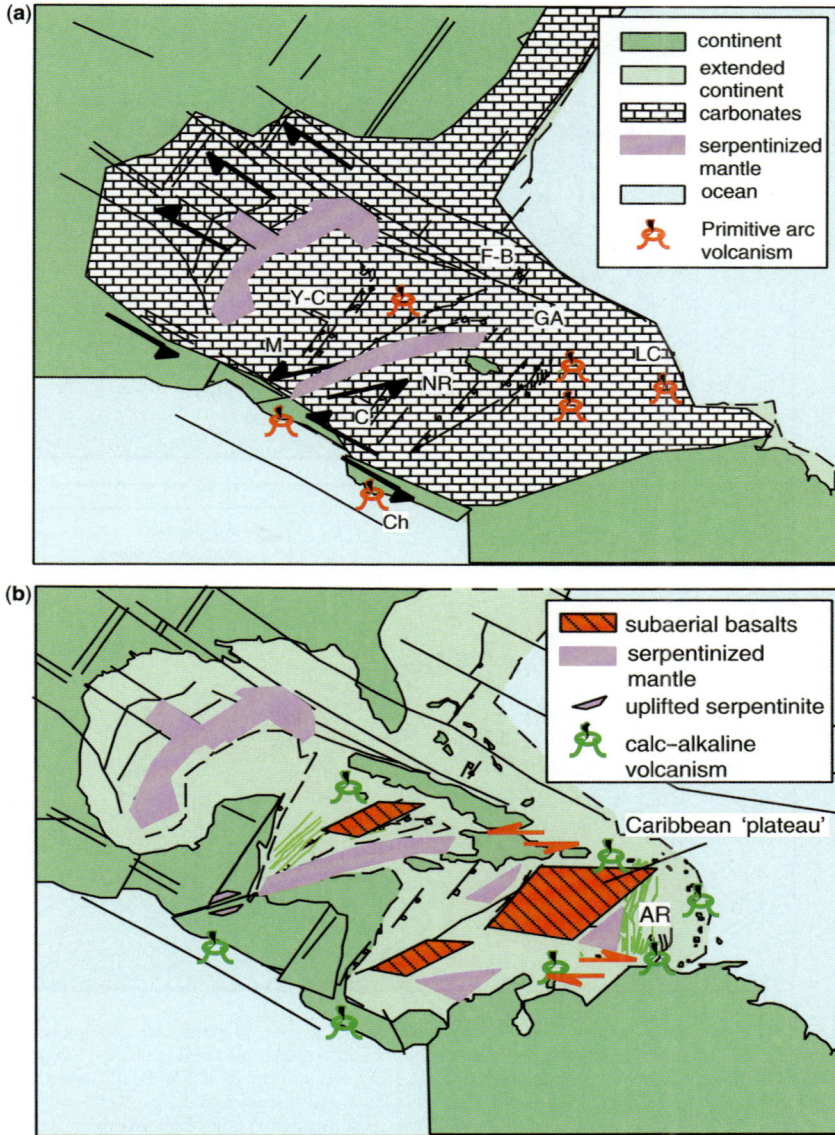

Fig. 4. (a) Late Jurassic–Early Cretaceous. Severe extension and subsidence of continental margin crust accommodated kilometres-thick carbonate platforms (Florida–Bahamas–Greater Antilles–Limestone Caribees, Yucatán–Campeche, Nicaragua Rise). Early 'Cayman' offset of Maya and Chortís continued. Chortís was offset NW from Chorotega/Chocó. NW trending grabens formed in SW Mexico–West Central America. Subduction volcanism began in Central America and the Lesser Antilles. C, Chortís; Ch, Chorotega/Chocó; F-B, Florida–Bahamas Platform; GA, Greater Antilles; LC, Limestone Caribbees; NR, Nicaragua Rise; Y-C, Yucatán Campeche Platform. (b) Middle–Late Cretaceous. Basaltic intrusion/extrusion occurred over highly extended continental crust, forming the 'Caribbean Plateau' in the Venezuela Basin. Similar seismic signature is seen in Colombia, north Grenada and Yucatán basins. Severe extension resulted in serpentinization of upper mantle (pink areas). Back-arc spreading at the Aves Ridge (AR) drove the Lesser Antilles eastwards over Atlantic crust. Peripheral volcanic arc rock chemistry shows continental input since the Albian and rocks contain ancient zircons.

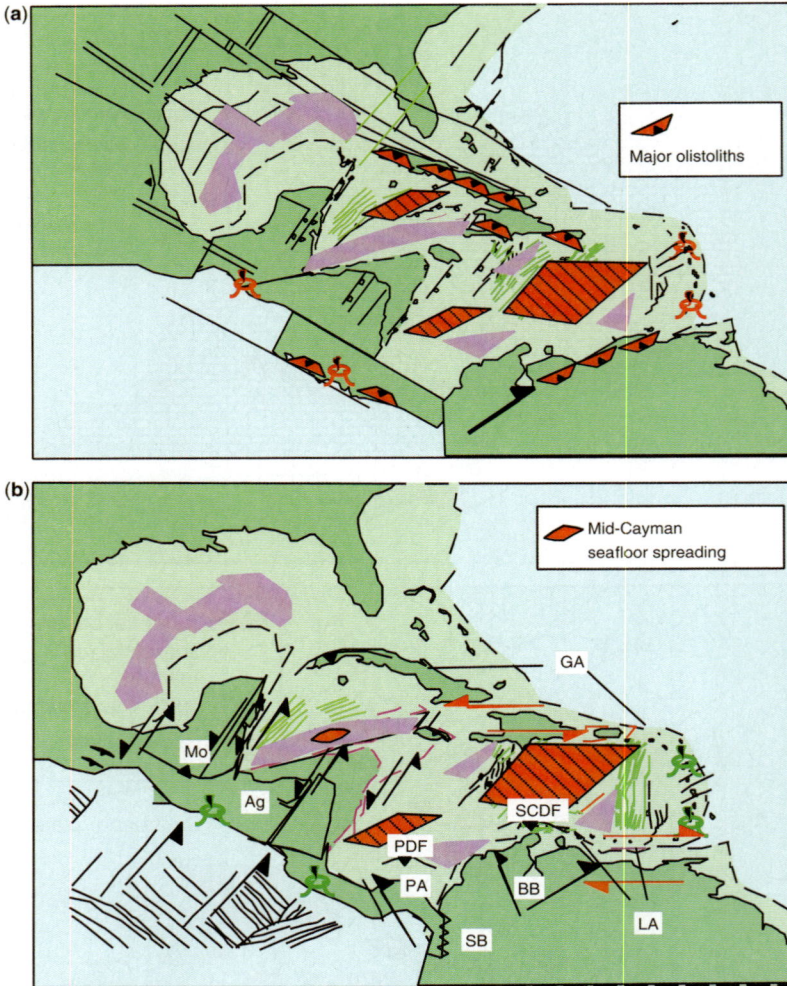

Fig. 5. (**a**) Middle Eocene. Allochthons of 'oceanic', volcanic arc and continental margin rocks emplaced around the Caribbean, covered by erosional unconformity and shallow marine limestones (uplift to wave base). Volcanism ceased along northern and southern Caribbean Plate boundaries. Beginning NE movement of NW South America along the Mérida Andes/Eastern Cordillera. (**b**) Oligocene–Recent. East Pacific Rise spreading drives the Cocos Plate orthogonally against Central America, reactivating Jurassic rifts as sinistral faults, resulting in oroclinal bending of Motagua (Mo), Agalta (Ag) and western Cuba (James 2007, figs 11 & 12). It also drives the Caribbean Plate eastwards, triggering central Cayman Trough spreading and strike–slip (300 km) along the northern and southern boundaries. Transtensional extension disperses the Greater (GA) and Leeward (LA) Antilles. Southern Chocó sutures to NW South America (SB, Serranía de Baudo); northward extrusion of the Panamá arc (PA) and the Bolivar Block (BB) drives the Panamá (PDF) and South Caribbean (SCDB) Deformed Belts.

capping the horsts and half grabens of extended continental crust in the Caribbean Plate interior. The presence of basalt flows dated 120 and 90 Ma in areas such as Ontong Java suggests episodes of global activity. This paper suggests that the thick areas of crust in the Venezuelan, Yucatán and Colombian basins include thick carbonate sections sandwiched between the 120 and 90 Ma basalts.

Middle Eocene

Flysch deposition became common in the Late Cretaceous. It culminated violently with Middle

Eocene emplacement of extremely large olistoliths of 'oceanic', volcanic arc and continental margin rocks in the Greater Antilles and along northern South America, where arc volcanism ceased (Fig. 5a, James 2005*a*). Uplift raised allochthons through wave base where a regional Middle Eocene unconformity developed, overlain by regional Middle Eocene shallow-water carbonates (James 2005*a*). Allochthons included the 250 km long Villa de Cura Nappe (Venezuela), with its former western extension, the Aruba–Blanquilla island chain, the 1000 km long ophiolite belt of Cuba and the Duarte and Bermeja complexes of Hispaniola and Puerto Rico. Horizontal movement was as much as 140 km in Cuba (Cobiella-Reguera 2009). At the same time thrust slices in Mexico's Veracruz Basin stacked 6 km high and moved eastward at least 30 km (Mossman & Viniegra-O 1976, fig. 2).

Rapid (204 ± 80 mm/annum, Daly 1989) Late Eocene convergence between the Nazca Plate and northwestern South America began to drive the Bolivar Block (Fig. 5a) of northwestern South America (James 2000) NE.

Sometime in the Late Cretaceous, possibly Early Cenozoic, back arc spreading in the eastern Caribbean extended the plate eastwards over Atlantic crust and dispersed continental blocks below the Lesser Antilles.

The most recent phase of Caribbean history (Fig. 5b) involved some 300 km of Oligocene–Recent strike–slip, sinistral and dextral respectively, along the northern and southern plate boundaries, driven by convergence of the Cocos Plate with the western Caribbean. Along both margins uplift occurred earlier and is greater in the west. The size of Greater Antillean islands and the height and depth of erosion in the Coastal Range–Northern Range of Venezuela and Trinidad diminish eastwards. Erosion has exposed the strata lying below allochthons in the west (continental basement in Cuba and the Venezuelan Coastal Range as far as the Gulf of Paria). In the east these rocks remain occulted. From Hispaniola to the Virgin Islands only uplifted oceanic and volcanic arc rocks crop out; however, gravity and chemical data indicate thick and siliceous basement.

The episode began with a pulse of block faulting and pull-apart extension, possibly triggered by the Middle Eocene event. Once again, highly extended areas subsided. Middle Eocene, shallow marine carbonates on the Beata and Aves Ridges now lie in thousands of metres water depths. It was followed by eastward-migrating thrusting over complementary foredeep basins in Venezuela–Trinidad (Guárico Basin–Oligocene; Maturín Basin–Miocene; Gulf of Paria–Pliocene; Columbus Basin–Pleistocene–Recent). Earth's largest negative gravity anomaly at sea level over the Maturín

Basin and great thickness of young Cenozoic sediments in the Columbus Basin testify to the highly dynamic nature of the strike–slip system, capable of driving rocks quickly to depths where HP metamorphism can occur at low temperatures.

Along the northern plate boundary a second phase of Cayman offset was accompanied by spreading in the Trough and separation of the eastern Greater Antilles along the Mona, Windward and Anegada Passages (Fig. 5b). North–south faults divided the Hess Escarpment into three *c*. 200 km long elements and the Nicaragua Rise became segmented (Muttia *et al.* 2005).

Convergence of the Farallon/Nazca plate with NW South America, low (44 ± 26 mm/annum) during the Oligocene–Early Miocene and high (125 ± 33 mm/annum) from Middle Miocene–Recent (Daly 1989), continued to drive the Bonaire Block NE. The northern part of the block transgressed the South America–Caribbean dextral boundary, driving the South Caribbean Deformed Belt ahead of it. Extrusion occurred along the NNW Santa Marta–Bucarramanga Fault and NE faults in the Colombian Eastern Cordillera–Venezuelan Mérida Andes. A large positive gravity anomaly associated with the 5800 m high Sierra Nevada de Santa Marta, Colombia, indicates the absence of isostatic equilibrium and witnesses the dynamism of this system.

Pull-apart extended the northern Bonaire Block, formerly the western continuation of the Villa de Cura nappe, as it crossed the plate boundary. The Aruba–Blanquilla islands, highs of 5.4 km s^{-1} material separated by basins with thick Oligocene–Recent sediments, exhibit the same structural periodicity as the nappe (Edgar *et al.* 1971; Curet 1992; James 2005*b, c,* fig. 14).

Northern Chocó separated from Chorotega along the Panamá F., moving northwards to form the Panamá arc and driving the Panama Deformed Belt.

NE convergence of the Cocos Plate along Central America drives a regional system of NE trending sinistral faults in the western Caribbean (James 2007). They transform in the north and south into oroclines and fold and thrust-belts and accommodate shortening between the bounding fracture zones of North Cuba and Central America. The most impressive system runs along eastern Maya, linking the Motagua 'orocline' with western Cuba. An estimated 350 km of shortening (Rosencrantz 1990; James 2007) here closed the western extension of the Cayman Trough by NE-moving Chortís.

Implications

Data (James 2005*c*, 2006, 2009) and the interpretation of this paper suggest learnings for aspects of global geology. Indications of continental fragments below the Caribbean 'Plateau' and the Greater and Lesser Antillean and Central American volcanic arcs imply that it is perilous to assume purely intra-oceanic origins for oceanic plateaus and volcanic arcs (Leat & Larter 2003; Kerr *et al.* 2009). Accepted discriminatory chemical/isotope data for such areas need to be statistically qualified and examined independently of presumed origins to see what messages they carry. Geology of the Caribbean volcanic arcs suggests answers to the 'andesite problem' – it is not a case of understanding how subducting basalt gives rise to such silica-rich rocks (Takahashi *et al.* 2007) but rather of recognizing that arc roots, related by seismic velocities to continental rocks (Tatsumi & Kosigo 2003), involve original continental fragments.

HP/LT metamorphic rocks are not necessarily signals of subduction. Transcurrent faulting is capable of generating rapid, deep burial and exhumation of such rocks.

The Caribbean Plate lies between the giant oil provinces of the Gulf of Mexico and northern South America. Oil is seen on Puerto Rico, Hispaniola, Jamaica, Belize, Guatemala, Honduras, Nicaragua, Costa Rica, Panamá, Colombia, Venezuela, Trinidad and Barbados, circumscribing practically the whole of the plate. This signals overlooked potential on the Caribbean Plate. Some oil is thought to come from Cenozoic sources but oils of Guatemala, Belize, Jamaica, Costa Rica and Barbados have Jurassic or Cretaceous chemical signatures (Larue & Warner 1991; Babaie *et al.* 1992; Burggraf *et al.* 2002; Cameron 2004; Emmet 2002; Lawrence *et al.* 2002).

Source rocks are likely in Jurassic and Cretaceous sections on the Caribbean Plate interior. The former developed in narrow marine basins; the latter in seas restricted by subaerial basalt extrusion over extended continental crust, since subsided below thousands of metres of water. A risk is that they were overmatured by igneous activity. Gas might be expected.

Questions

Juxtaposition of continental, volcanic arc and ophiolitic rocks in locations such as the supposed suture between Maya and Chortís and along northern Venezuela is problematic (Giunta & Oliveri 2009). Ophiolites, once attributed to mid-ocean spreading, today are seen to form in several different environments, including pull-apart intra-arc basins, back-arc basins and extensional zones in forearcs during development of island arcs (Moores 2003). All these settings can be envisaged in Middle America as extension occurred during the Jurassic–Cenozoic. Ophiolite thicknesses up to

5 km, common around the Caribbean, testify to significant extension, presumably close to thinned continental crust.

What caused repeated episodes of convergence, metamorphism, uplift, erosion and unconformities covered by shallow marine limestones? Transpression? Vigorous plate expansion, driven by decompression melting? Pulses of Pacific Plate convergence?

The Middle Eocene event was remarkable for its suddenness and violence (Stainforth 1969). It abruptly emplaced sections of serpentinite up to 5 km thick and up to 1000 km long by as much as 140 km onto continental margins in Cuba and Venezuela – outward moving in response to a plate interior compressional stress or sudden convergence of North and South America. Coeval deposits in Peru and Ecuador and an eastward verging stack of thrusts 6 km high in the Veracruz Basin of Mexico rule out the latter. Conventional wisdom suggests convergence of such material should lead to subduction, not obduction. Did salt and/or serpentinite assist décollement? Did hydro-pressure due to hydrocarbon generation play a role? Whatever the cause, the regional and coeval occurrence of these deposits correlates continental margin with Caribbean Plate boundary and internal geology.

What is the age and nature of thin crust in the Haiti Basin, the SW Colombian and Venezuelan basins? Jurassic, Cretaceous, Cenozoic?

Did high standing blocks in the Caribbean and/or shallow/subaerial basalt outpouring cause restriction and supply nutrients for Albian and Turonian organic rocks?

Why are there no return path metamorphic rocks associated with the Lesser Antilles or Central American subduction arcs?

Why is there no major difference in chemistry between the northern (descent of oceanic crust close to the arc) and southern (descent of ocean crust east of the Barbados Accretionary Prism and Tobago Trough) parts of this arc?

Are there hydrocarbons on the Caribbean Plate interior?

Predictions

The model described by this paper suggests the following.

El Tambor rocks of the Motagua 'suture' are the exhumed western continuation of the Cayman Trough.

Thin, 'oceanic' crust in the southeastern Colombian and Venezuela Basins and in the Grenada Basin is serpentinized mantle, Cretaceous or Cenozoic in age.

Dispersed continental blocks underpin the Cayman Ridge, the Upper Nicaragua Rise, Costa

Rica, Hispaniola, Puerto Rico, the Virgin Islands, the northern Lesser Antilles, Chorotega/Chocó.

Thick crust of the Venezuela, Colombia, Yucatán and Grenada basins is underlain by blocks of continental rocks flanked by wedges of Triassic–Jurassic–Cretaceous rift sediments, evaporites and carbonates, overlain by subaerial basalt flows.

The Lower Nicaragua Rise is underpinned by highly extended/transitional crust – there are local continental blocks.

The N60°E Hess Escarpment is the outer marginal limit of the Nicaragua Rise/Chortís continental block. It formed as a major extension in the Jurassic–Cretaceous sinistral N60°W offset of North from South America. It became segmented by north–south trending faults during Cenozoic extension.

Pull-apart extension along the Nicaragua Rise/Hess Escarpment, the Greater Antilles and Aruba–Blanquilla sums to around 300 km, equal to Central Cayman spreading in the north and extension of the Falcón Basin and separation of the Leeward Antilles in the south.

The Greater Antilles mirror-image northern South America. They are built of north verging thrust sheets, some inverting stratigraphy, and mixed blocks of allocthonous, mainly Jurassic–Cretaceous, oceanic and volcanic arc rocks deposited as Middle Eocene wildflysch above continental blocks.

The Beata and Aves Ridges were the focus of back-arc spreading in the Mid–Late Cretaceous and Late Cretaceous–Early Cenozoic, respectively.

There are salt diapirs in the Caribbean, predicted salt age Jurassic.

Hydrocarbons are present on the Caribbean interior.

Tests

The understanding of this paper suggests the following tests.

Many data are easily obtainable. Dredge samples of the Cayman Trough walls, walls of the Puerto Rico Trough and the Aves Ridge and core samples of the deeper Venezuelan and Colombian basins could be re-investigated for lithology, age and zircon content. Existing seismic data could be revisited and new data gathered that would distinguish between *in situ* and allochthonous models. It should be possible to distinguish definitively between seamounts and salt diapirs, to see if thickened parts of the Colombian, Grenada and Yucatán basins have the same architecture as the western Venezuelan Basin and to see if combined velocity/magnetic/gravity data can distinguish between igneous/volcanic and extended continental origins. Much could be learned by systematic reprocessing/reinterpretation of existing seismic data.

- Age date zircon in volcanic/oceanic rocks in Central America, Hispaniola, Puerto Rico, Lesser Antilles and Aves Ridge.
- Age date zircon in volcanic rocks from DSDP samples of the Caribbean Plateau.
- Age date zircon in dredge samples from the Caribbean.
- Age date clasts in rocks dredged from the Cayman Trough.
- Palynological study of salt/cap rock at Salinas, Puerto Rico, Cerro Sal, Dominican Republic and Salt Pond Pen, Jamaica.
- Gravity and heat flow investigation of seamounts v. diapirs.

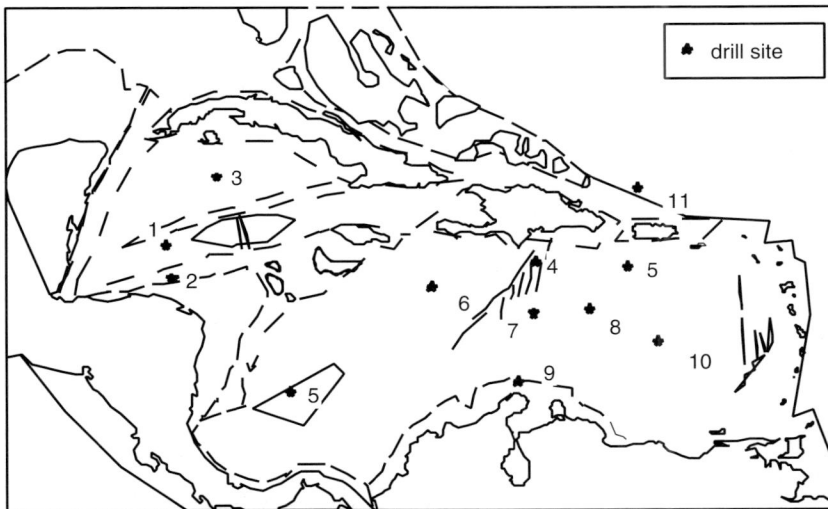

Fig. 6. Suggested drill sites to test ideas of this paper.

- Examination of Middle Eocene sections for evidence of impact (tektites, iridium).
- IODP drilling of Cayman Trough basement, predicted to be Early Cretaceous serpentinite (Site 1, Fig. 6).
- IODP drilling of 'seamounts' – predicted to be salt diapirs, north of Honduras, Yucatán Basin, Beata Ridge, seismic line 1293 (Sites 2–5).
- IODP drilling with greater penetration of the Caribbean Plateau (Sites 5 and 8).
- IODP drilling to 'oceanic crust' Haiti and Venezuela basins (Sites 5 and 10).
- IODP drilling to test a circular bathymetric low in the Venezuela Basin (Site 8), a possible impact crater (James 1997).
- IODP drilling to date early sediments exposed in the Guajira Canyon (Site 9).
- IODP drilling to test deep section on the southern wall of the Puerto Rico Trough, predicted continental crust (Site 11).

Conclusions

The geology of the area between North and South America shows regional harmony and a shared history among the various geographic components, all of which are autochthonous. Following Triassic/Jurassic extrusion of the Central Atlantic Magmatic Province North America separated from South America/Africa. Triassic–Jurassic rifting reactivated ancient NE-trending basement lineaments. This structural grain is regionally preserved today in Middle America, including the Caribbean Plate interior. No major block rotations have occurred.

Volcanism accompanied extension but compressional events in the Middle and Late Cretaceous and Middle Eocene led to pause or cessation of activity, uplift to wavebase and subaerial erosion, development of unconformities and shallow marine carbonates, karstification. Abundant extrusion accompanied the Middle and Late Cretaceous events in the Colombia and Venezuela basins. Coeval high organic productivity resulted from restriction and formed hydrocarbon source rocks.

Subsidence of proximal areas (Bahamas and Yucatán–Campeche platforms, Nicaragua Rise) accommodated kilometres-thick carbonate sections. Horsts of continental crust flanked by wedges of Jurassic–Cretaceous sediments, flows and salt formed in more distal areas along the eastern margin of North America and within Middle America (Yucatán, Colombian and Venezuelan Basins). Shallow/subaerial flows of smooth seismic Horizon B″ capped the Caribbean areas in the Late Cretaceous. Areas of extreme extension

suffered serpentinization of upper mantle, forming rough Horizon B″.

The Middle Eocene regional, convergent event terminated most volcanic activity along the northern and southern Caribbean Plate boundaries, where Oligocene–Recent strike–slip followed. The only spreading crust in Middle America, with recognizable spreading ridges and magnetic anomalies, formed in the central 300 km of the Cayman Trough during this latest tectonic phase.

Continued divergence of North and South America, shown by diverging fractures in the Atlantic ocean east of North and South America and by a wedge of fractures east of Caribbean latitudes recording 650 km of north–south growth of the Mid-Atlantic Ridge, created space for back-arc growth of the Caribbean Plate. Analogy with the Scotia Plate, which occupies as similar tectonic setting, suggests that this occurred firstly along the Beata Ridge and later along the Aves Ridge.

Many unrecognized fragments of continental crust are dispersed around and on the Caribbean Plate. They underlie the whole of Central America and the Greater and Lesser Antilles and at least the northern and southern Lesser Antilles.

The strength of the *in situ* model is that it incorporates data in a regionally coherent, simple evolution that conforms to the wider geology of eastern North America and the Gulf of Mexico. It compares with the well-calibrated geology of the analogous Scotia Plate. It is a model that can be tested. In contrast, models deriving the Caribbean Plate from the Pacific models are complicated. Required processes such as slab-rollback beneath an overriding plate and below the Yucatán Basin are not recorded by data and are not testable.

I pay tribute to those geologists of the past who, with far less data than I have, also saw Caribbean geology in simple terms.

References
BABAIE, H. A., SPEED, R. C., LARUE, D. K. & CLAYPOOL, G. E. 1992. Source rock and maturation evaluation of the Barbados accretionary prism. *Marine and Petroleum Geology*, **9**, 623–632.

BURGGRAF, D., LUNG-CHUAN, K., WEINZAPFEL, A. & SENNESETH, O. 2002. A new assessment of Barbados onshore oil characteristics and implications for regional petroleum exploration. *16th Caribbean Geological Conference*, Barbados, Abstracts, 3.

CAMERON, N. 2004. Jamaican oil biomarkers require a re-examination of the petroleum geology of the northern Caribbean. *American Association of Petroleum Geologists International Conference*, Cancún, Mexico, Abstracts, A11.

COBIELLA-REGUERA, J. L. 2009. Emplacement of the northern ophiolites of Cuba and the Companian–Eocere geological history of the northwest Caribbean–SE Gulf of Mexico region. *In*: JAMES, K. H., LORENTE, M. A. & PINDELL, J. L. (eds) *The Origin and Evolution of the Caribbean Plate.* Geological Society, London, Special Publications, **328**, 315–338.

CURET, E. A. 1992. Stratigraphy and evolution of the Tertiary Aruba Basin. *Journal of Petroleum Geology*, **15**, 283–304.

DALY, M. C. 1989. Correlations between Nazca/Farallon Plate kinematics and forearc basin evolution in Ecuador. *Tectonics*, **8**, 769–790.

EDGAR, N. T., EWING, J. I. & HENNION, J. 1971. Seismic refraction and reflection in Caribbean Sea. *American Association of Petroleum Geologists Bulletin*, **55**, 833–870.

EMMET, P. 2002. Structure and stratigraphy of the Gracias A Dios platform and the Mosquitia basin, offshore eastern Honduras, and implications for the tectonic history of the Chortís block. *Caribbean Workshop, University of Texas Institute for Geophysics*, Austin, TX. World Wide Web Address: http://www.ig.utexas.edu/CaribPlate/forum/emmet/emmet_abstract.htm.

GIUNTA, G. & OLIVERI, E. 2009. Some remarks on the Caribbean Plate kinematics: facts and remaining problems. *In*: JAMES, K. H., LORENTE, M. A. & PINDELL, J. L. (eds) *The Origin and Evolution of the Caribbean Plate.* Geological Society, London, Special Publications, **328**, 57–75.

HARLOW, G. E., HEMMING, S. R., AVÉ LALLEMENT, H. G., SISSON, V. B. & SORENSEN, S. S. 2004. Two high-pressure–low-temperature serpentinite-matrix mélange belts, Motagua fault zone, Guatemala: A record of Aptian and Maastrictian collisions. *Geology*, **32**, 17–20.

JAMES, K. H. 1997. A possible impact crater in the Venezuelan Basin, Caribbean. *In*: AVÉ LALLEMANT, H. G. (ed.) *Final Report: Evolution of the Eastern Caribbean–South American Plate Boundary Zone: Transpression Volcanic Arc Accretion, and Orogenic Float.* Workshop, Rice University, Houston, poster.

JAMES, K. H. 2000. The Venezuelan hydrocarbon habitat. Part 1: Tectonics, structure, palaeogeography and source rocks. *Journal of Petroleum Geology*, **23**, 5–53.

JAMES, K. H. 2005a. Palaeocene to Middle Eocene flysch-wildflysch deposits of the Caribbean area: a chronological compilation of literature reports, implications for tectonic history and recommendations for further investigation: Transactions, 16th Caribbean Geological Conference, Barbados. *Caribbean Journal of Earth Sciences*, **39**, 29–46.

JAMES, K. H. 2005b. A simple synthesis of Caribbean geology: Transactions, 16th Caribbean Geological Conference, Barbados. *Caribbean Journal of Earth Sciences*, **39**, 71–84.

JAMES, K. H. 2005c. Arguments for and against the Pacific origin of the Caribbean Plate and arguments for an in situ origin: Transactions, 16th Caribbean Geological Conference, Barbados. *Caribbean Jounral of Earth Sciences*, **39**, 47–67.

JAMES, K. H. 2006. Arguments for and against the Pacific origin of the Caribbean Plate: discussion, finding for an inter-American origin. *Geologica Acta*, **4**, 279–302.

JAMES, K. H. 2007. Structural geology: from local elements to regional synthesis. *In*: BUNDSCHUH, J. & ALVARADO, G. E. (eds) *Central America: Geology, Resources and Hazards.* Taylor & Francis/Balkema, The Netherlands, 277–321.

JAMES, K. H. 2009. *In situ* origin of the Caribbean: discussion of data. *In*: JAMES, K. H., LORENTE, M. A. & PINDELL, J. L. (eds) *The Origin and Evolution of the Caribbean Plate.* Geological Society, London, Special Publications, **328**, 77–126.

KERR, A. C., PEARSON, G. & NOWELL, G. M. 2009. Magma source evolution beneath the Caribbean oceanic plateau: new insights from elemental and Sr–Nd–Pb–Hf isotopic studies of ODP Leg 165 site 1001 basalts. *In*: JAMES, K. H., LORENTE, M. A. & PINDELL, J. L. (eds) *The Origin and Evolution of the Caribbean Plate.* Geological Society, London, Special Publications, **328**, 809–827.

LARUE, D. K. & WARNER, A. J. 1991. Sedimentary basins of the NE Caribbean Plate boundary zone and their petroleum potential. *Journal of Petroleum Geology*, **14**, 275–290.

LAWRENCE, S. R., CORNFORD, C., KELLY, R., MATHEWS, A. & LEAHY, K. 2002. Kitchens on a conveyor belt-petroleum systems in accretionary prisms. *16th Caribbean Geological Conference*, Barbados, Abstracts, 41.

LEAT, P. T. & LARTER, R. D. 2003. Intra-oceanic subduction systems: introduction. *In*: LARTER, R. D. & LEAT, P. T. (eds) *Intra-oceanic Subduction Systems: Tectonic and Magmatic Processes.* Geological Society, London, Special Publications, **219**, 1–17.

MANSPEIZER, W. 1988. Triassic–Jurassic rifting and opening of the Atlantic: an overview. *In*: MANZPEIZER, W. (ed.) *Triassic–Jurassic Rifting, Part A*, Elsevier, Amsterdam, 41–79.

MCHONE, J. G., ANDERSON, D. L., BEUTEL, E. K. & FIALKO, Y. A. 2005. Giant dikes, flood basalts, and plate tectonics: A contention of mantle models. *In*: FOULGER, G. R., NATLAND, J. H., PRESNALL, D. C. & ANDERSON, D. L. (eds) *Plates, Plumes, and Paradigms.* Geological Society of America, Special Papers, **388**, 401–420.

MOORES, E. M. 2003. A personal history of the ophiolite concept. *In*: DILEK, Y & NEWCOMBE, S. (eds) *Ophiolite Concept and the Evolution of Geological Thought.* Geological Society of America, Special Papers, **373**, 17–29.

MOSSMAN, W. & VINIEGRA-O, F. 1976. Complex structures in Veracruz province of Mexico. *American Association of Petroleum Geologists Bulletin*, **60**, 379–388.

MUTTIA, M., DROXLER, A. W. & CUNNINGHAM, A. D. 2005. Evolution of the northern Nicaragua Rise during the Oligocene–Miocene: drowning by environmental factors. *Sedimentary Geology*, **175**, 237–258.

PINET, P. R. 1971. Structural configuration of the northwestern Caribbean Plate boundary. *Geological Society of America Bulletin*, **82**, 2027–2032.

PINET, P. R. 1972. Diapirlike features offshore Honduras: implications regarding tectonic evolution of Cayman

Trough and Central America. *Geological Society of America Bulletin*, **83**, 1911–1922.

ROSENCRANTZ, E. 1990. Structure and tectonics of the Yucatan Basin, Caribbean Sea, as determined from seismic reflection studies. *Tectonics*, **9**, 1037–1059.

STAINFORTH, R. M. 1969. The concept of sea-floor spreading applied to Venezuela. *Asociación Venezolana Geología, Minería y Petróleo, Boletin Informativo*, **12**, 257–274.

TAKAHASHI, N., KODAIRA, S., KLEMPERER, S. L., TATSUMI, Y., KANEDA, Y. & SUYEHIRO, K. 2007. Crustal evolution of the Mariana intra-oceanic island arc. *Geology*, **35**, 203–206.

TATSUMI, Y. & KOSIGO, T. 2003. The subduction factory: its role in the evolution of the Earth's crust and mantle. *In*: LARTER, R. D. & LEAT, P. T. (eds) *Intra-oceanic Subduction Systems: Tectonic and Magmatic Processes*. Geological Society, London, Special Publications, **219**, 55–80.

The Caribbean: an oroclinal basin?

JOHN MILSOM*

Department of Earth Sciences, University of Hong Kong, Hong Kong, China

*Address for correspondence: The Camp, Gladestry, Kington HR5 3NY, UK
(e-mail: kivibatu-5@yahoo.co.uk)

Abstract: Tightly curved mountain belts are prominent features of global topography. Typically, these 'oroclines' occur in areas of regional compression but enclose basins where extension has been contemporaneous with outward directed thrusting in the orogens. Examples of such basin–orogen pairs include the Alboran Sea–Gibraltar Arc, Tyrrhenian Sea–Aeolian Arc, Aegean Sea–Hellenic Arc and Pannonian Basin–Carpathian Arc, all in the western Tethys but matched in the eastern Tethys by the Banda Sea and Outer Banda Arc. The development of the basins has been variously explained by gravitational collapse of rapidly elevated mountain blocks and by extrusion prompted by asthenospheric flows, but it is not even universally agreed that similar processes have operated in all cases. Critics have cited gross differences in volcanic activity (absent from the Gibraltar Arc, modest in the Carpathians but intense in other examples) and in the presence or absence of recognizable Wadati–Benioff Zones. The superficial similarities between the Caribbean Sea–Antilles Arcs and typical oroclinal basin–orocline pairs have recently been invoked in support of an *in situ* Caribbean evolutionary model, even though the disputed origins of oroclines limit their reliability as analogues. The Caribbean's considerably greater area further emphasizes the need for caution, while the most obvious objection to identifying it as a member of the oroclinal group is its very long history. Oroclinal basins typically pass from initiation to effective stabilization in a few tens of millions of years, whereas the original Caribbean oceanic crust, which is now bounded to the east and west by active subduction zones, was probably formed in the Jurassic. Rather than invoking an overall common origin for the Caribbean and the Tethyan basins, it is more useful to look for shared causes of specific individual similarities. The impact of a rigid block might be as effective in imposing curvature on a mountain belt as rapid expansion in an adjacent area. However, it does seem that the case for the crust of the Caribbean being typical of oceanic large igneous provinces (LIPs) may have been overstated and, in the light of oroclinal analogues, that some features of the still poorly understood Beata Ridge and Lower Nicaragua Rise may be most easily explained by east–west extension promoted by the convergence between North and South America.

The Caribbean comprises a deep, complex and mainly oceanic basin that is partly enclosed by an intensely curved and fragmented mountain belt. Comparisons with the sub-Antarctic Scotia Sea and the Indonesian Banda Sea have been so frequently made as to be almost commonplace. Similarities can, in turn, be recognized between the Banda Sea and the Alboran, Tyrrhenian, Ligurian and Aegean Seas. Also, and despite its location within the European land mass north of the Mediterranean (Fig. 1), the Pannonian Basin is widely regarded as conforming to the same pattern of extension within the zone of Tethyan collision (e.g. Horvath & Tari 1999). The mountain ranges encircling such basins have come to be known as oroclines, and the internal depressions are consequently referred to as oroclinal basins. The term 'orocline' seems to have been first used by Carey (1955), who applied it to mountain chains exhibiting at least 90° of curvature. Many are curved through angles approaching 180°, and the Western Alps,

which partly enclose the Po Basin, actually exceed this value (Fig. 1). In every one of these cases, development of the internal basin was accompanied by outwards-directed thrusting in the orogen.

Despite obvious similarities, it is not clear that the Caribbean could ever be properly regarded as a member, even if an end member, of the group of oroclinal basins. The point is important, since the fact that the basins listed above all developed in very much their present locations with respect to the surrounding major plates has been cited in support of an *in situ* origin for the Caribbean and as evidence against the alternative, Pacific, origin (James 2006). The question of the extent to which comparisons with other basins can provide useful insights into the processes involved in Caribbean origin and evolution is therefore worth considering. In this paper, significant features of the western Tethyan basins are mentioned where appropriate, but the discussion focuses on the Banda Sea, as

From: JAMES, K. H., LORENTE, M. A. & PINDELL, J. L. (eds) *The Origin and Evolution of the Caribbean Plate*.
Geological Society, London, Special Publications, **328**, 139–154.
DOI: 10.1144/SP328.5 0305-8719/09/$15.00 © The Geological Society of London 2009.

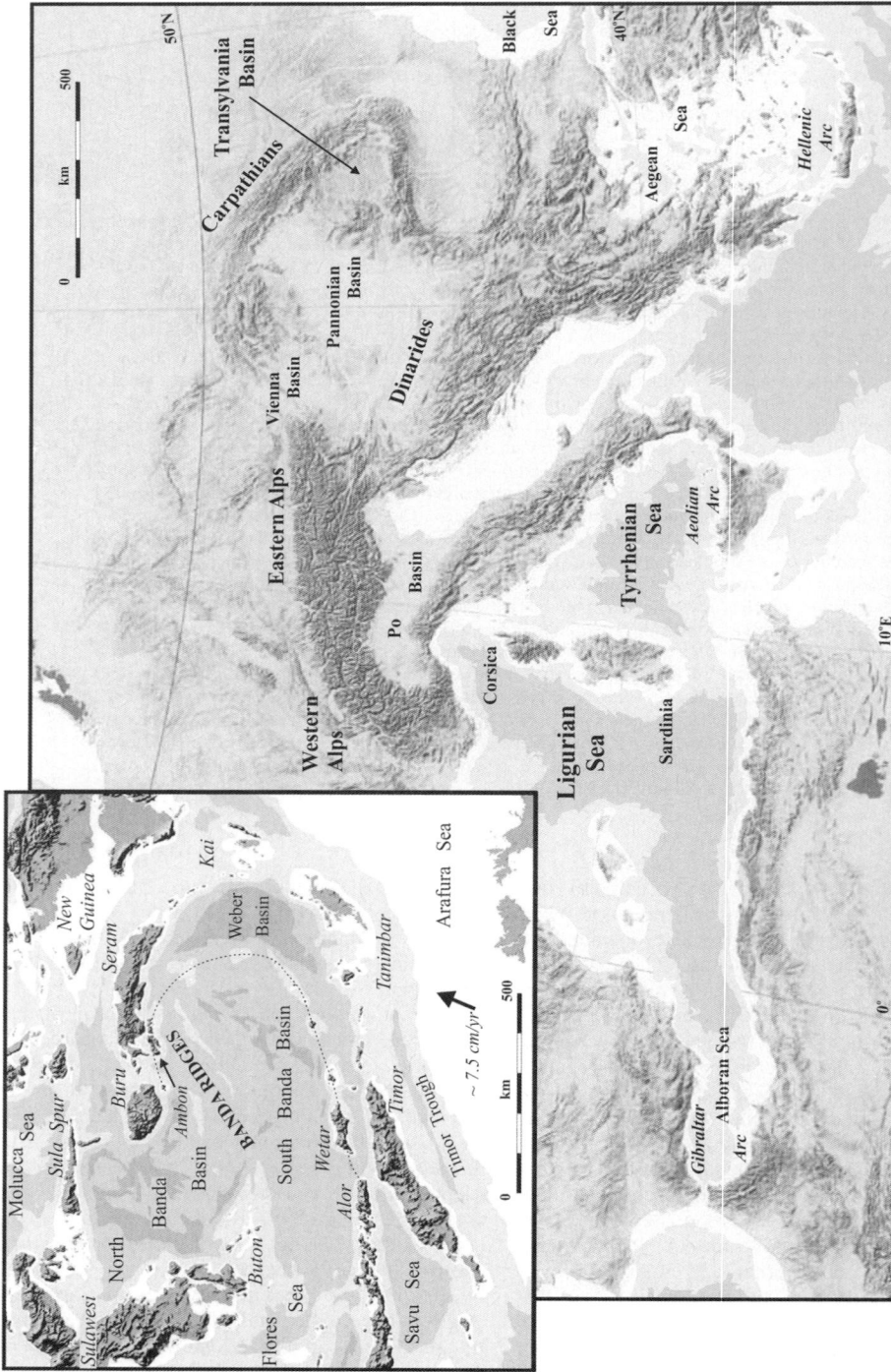

Fig. 1. Oroclines and topography in Europe and the Mediterranean. Inset: the Banda Sea, to the same scale. Topographic bases prepared in Global Mapper using the GTOPO30 global topographic grid. Bathymetries from the GEBCO digital atlas. See Figure 4 for depth scale. The dotted line defines the axis of the volcanic Inner Banda Arc. The essentially arbitrary boundary between this and the Sunda arc is conventionally placed immediately west of Alor. The arrow in the Arafura Sea indicates the direction of motion of Australia relative to Southeast Asia, after Genrich *et al.* (1996).

being one of the two or three oroclinal basins most commonly cited as a Caribbean analogue, and also the basin with which the author is most familiar.

Mediterranean oroclines

The Eurasia–Africa collision zone in the Mediterranean (Fig. 1) includes a number of basins that have been formed by rapid extension within the regional compressional setting (Jolivet et al. 1999; Jolivet & Faccena 2000). In each case extension in the basin appears to have been accompanied by folding and thrusting in the enclosing orocline, but there are wide individual variations. Only in the Tyrrhenian and Ligurian seas has extension led to the creation of new oceanic crust (see Carminati et al. 1998; Séranne 1999). The Alboran and Aegean basins are floored by highly extended continental crust (Platt & Vissers 1989; Hatzfield 1999) and this also seems to be true of the Pannonian basin (Horvath & Tari 1999), which is now infilled with thick sediments. The relation of the Carpathian orocline to extension appears particularly dramatic. The eastern Alps were disrupted and ultimately separated into the now distinct Dinarides and Carpathians by Pannonian expansion, in the course of which the Carpathian part of the orogen seems to have been not only bent but detached from the Alpine chain by the opening of the Vienna Basin (Fig. 1).

Volcanic activity in the European oroclines varies widely, and links to subduction are in some cases controversial. The Tyrrhenian and Aegean seas are clearly back-arc to, respectively, the Aeolian and Hellenic arcs, but conflicting hypotheses have been presented for the Alboran Sea by Lonergan & White (1997) and by Platt & Vissers (1989). Only the Lonergan & White (1997) interpretation involves subduction. Volcanism may now have ceased in the Carpathians and, although it was widespread during the Cenozoic and continued into the Holocene (Seghedi et al. 2005), the past presence of a subduction zone is still disputed (e.g. Knapp et al. 2005). The situation as regards the Ligurian Sea is uncertain, because of overprinting by more recent events (Séranne 1999).

It is a common assumption that oroclinal curvatures were imposed on formerly much straighter mountain belts. The necessary stresses could have been produced by rapid expansion in small areas adjacent to the mountains, and this seems to have been the case in all the European examples cited above, as well as in the Banda and Scotia seas. For bending to occur, there must be directions in which movement is subject to only minimal constraints and also directions in which it is very strictly constrained. It is therefore essential that there are, or were, areas of oceanic crust in some, but not all,

places on the far side of the orogens from the sources of stress. In the Carpathians at least, the ocean has now been entirely eliminated and arc–continent collision has followed.

The Banda Sea

Eastern Indonesia is the site of the oblique collision between the Asian and Australasian (Australia and New Guinea) continental masses, which is a consequence of the northward movement of the Indo-Australian Plate towards an Asia that is almost stationary in a global reference frame. The Banda Sea (Fig. 1, inset) is the most important geographic element of this region, interposing as it does a number of blocks of oceanic crust between the continental masses of New Guinea to the east and Borneo–Sulawesi to the west. The sea is enclosed on three sides by a double arc. The inner arc consists of a chain of small to medium sized volcanic islands, while the outer arc is dominated by the large islands of Timor, Tanimbar, Kai, Seram and Buru. Igneous rocks in the outer arc are pre-Neogene, and mainly pre-Cenozoic.

There is no general consensus concerning the origin of the Banda Sea, but early hypotheses that invoked the trapping of old Indian or other ocean floor (e.g. Lee & McCabe 1986) have now been generally discarded. Such ideas were largely prompted by the depths of the Banda oceanic basins, which are more appropriate to Mesozoic than Cenozoic features. However, Cenozoic and, in fact, Late Cenozoic, ages are suggested by the results of dredging and, in the case of the southern basin, by analysis of linear magnetic anomalies (Hinschberger et al. 2001).

If the Banda Sea did not exist prior to the Late Cenozoic, it must have relatively recently displaced other crustal elements. The identification beneath it of a scoop-shaped Wadati–Benioff Zone (Hamilton 1979; Milsom 2001) indicates expansion over pre-existing oceanic crust, while the presence of Australasian continental fragments in the Outer Banda Arc to the north as well as to the south and east (Milsom 2000 and references therein) strongly suggests that this subducted crust must have been part of the Indo-Australian Plate. Virtually all published palaeogeographies (e.g. Charlton 1986; Daly et al. 1991; Hall 2002) therefore show an embayment of the Indian Ocean lying between westernmost New Guinea and the mainland of northwestern Australia in the Early Cenozoic. It is also commonly accepted that there was collision in the Late Oligocene or Early Miocene between the margin of Southeast Asia and forerunners of the Australian continent (Smith & Silver 1991).

Despite these areas of agreement, there are important differences in Banda Sea reconstructions that have implications for comparisons with the Caribbean. Figure 2 is based on the 20 Ma reconstruction of Hall (2002). It shows the outer arc islands of Buru, Seram, Tanimbar and Timor, together with terranes that now make up eastern and southeastern Sulawesi, at the margins of an Indian Ocean embayment but firmly attached to Australasia. However, the material subducted beneath the Banda Sea, which represents the oceanic crust of the former embayment, now sub-crops the seafloor between the outer arc and Australia/New Guinea, that is, it has somehow reached the position shown by the dotted line in Figure 2. There is thus a problem with this reconstruction that can only be solved by complicated transfers across the subduction zone.

Other hypotheses avoid this problem by supposing that the Miocene collision was between a southeast Asian marginal arc (represented by modern West Sulawesi) and a microcontinental fragment that was rifted away from the Australian margin in the Early Jurassic and then drifted north in advance of the main continental mass (e.g. Smith & Silver 1991). The emplacement of the extensive ophiolite sheets of eastern Sulawesi above Australian-derived basement occurred in the

course of this collision. One specific example of such a hypothesis supposes that, following collision, the orogen developed to a level of gravitational instability that caused collapse and rapid extension, in a manner similar to that suggested by Platt & Vissers (1989) for the Alboran Sea. Dispersal (Fig. 3) would have been facilitated by subduction of the Indian Ocean, with the Banda Sea developing as a back-arc basin. Banda Sea expansion was limited to the south by the advance of Australia and was eventually terminated in the east by collision in the Kai region. The continuing convergence of Australia on Southeast Asia is now creating a new orogen in eastern Indonesia. The stratigraphic and palaeogeographic arguments for this scenario have been discussed by Milsom (2000) and by Milsom et al. (2000) and will not be repeated here.

Comparisons

In comparing the Caribbean with the Banda Sea, or any other oroclinal basin, note must first be taken of its much longer geological history. This history has been discussed by many authors, and is

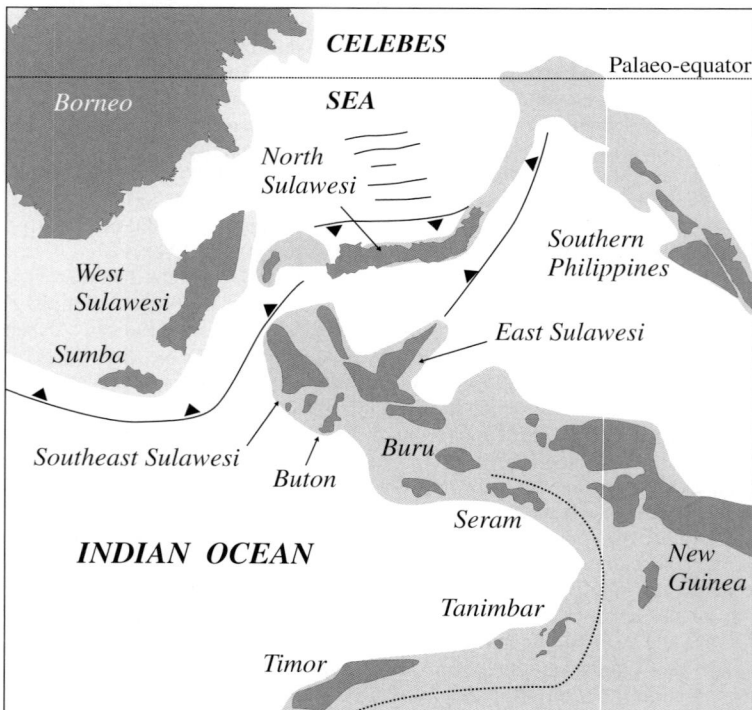

Fig. 2. Reconstruction at 20 Ma of the Banda Sea area, after Hall (2002). The dotted line, which is not part of Hall's illustration, shows the present-day surface subcrop of subducted lithosphere, as indicated by earthquake hypocentres.

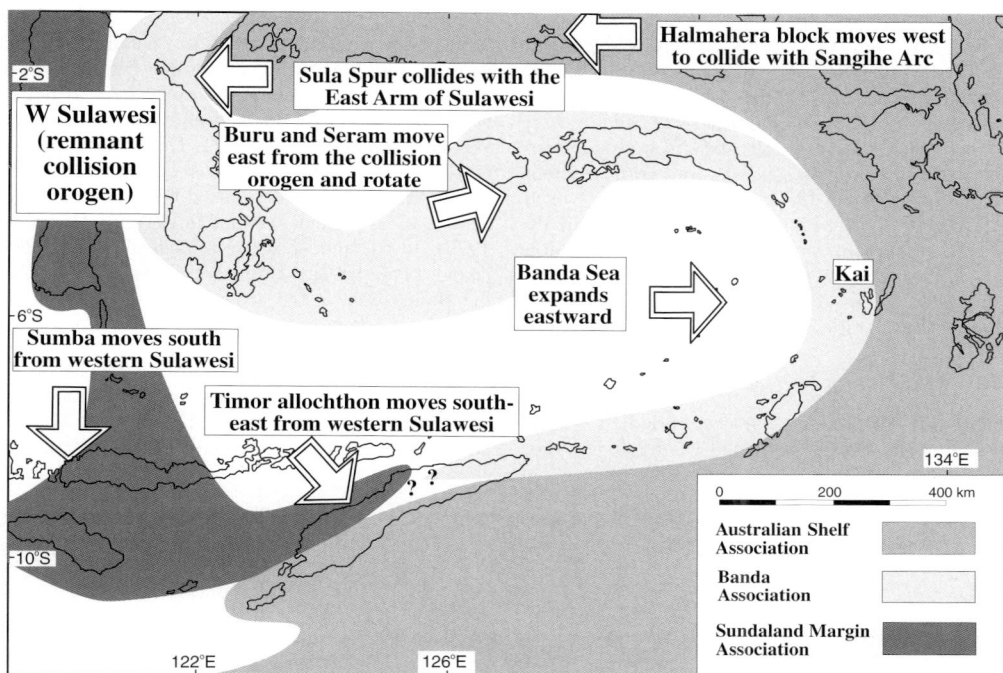

Fig. 3. Neogene movements in the Banda Sea region, as suggested by Milsom (2000). See Figure 4 for the locations of the east-facing Sangihe Arc and the west-facing Halmahera Arc.

the theme of the present volume. The summary by Driscoll & Diebold (1999) is typical. They regarded the original Proto-Caribbean crust as having been formed by conventional seafloor spreading in the Late Jurassic and Early Cretaceous and as having been modified by one or more large igneous provinces (LIPs)-style magmatic episodes that led to the formation of an oceanic plateau in the Late Cretaceous. Subsequently, subduction zones developed to both east and west, and north–south compression was produced by the convergence of North and South America (Müller *et al.* 1999). There was also significant extension at times but the amount of crustal thinning is poorly constrained. Although many aspects of this history are contested in detail, the major disagreements are concerned more with where certain events took place than with when. In particular, James (2005, 2006) has argued strongly against the widely accepted development of the LIP in the Pacific (see, for instance, Pindell & Barrett 1990) and for an origin directly between the Americas.

There is a significant difference in area as well as in age between the Banda Sea and the Caribbean, and this suggests that it may be more appropriate to use the whole of Indonesia in comparisons rather than the Banda Sea alone. Doing so also introduces further geological similarities, because there are then active continental margins underlain by well-defined subduction zones at the western boundaries of both areas (Fig. 4). Inevitably, the general appearance of similarity is the product of a number of quite separate details that vary in importance. These are considered individually below. Aspects of Caribbean geology that are not relevant to these comparisons are not discussed.

Thrust Belts

Indonesia. Outward-directed thrusting is characteristic of oroclinal mountain belts, and the thrust structures of Timor, the largest island in the Outer Banda Arc, have been described by numerous authors (e.g. Barber *et al.* 1977; Barber 1978; Karig *et al.* 1987; Charlton *et al.* 1991*a*; Sawyer *et al.* 1993; Reed *et al.* 1996). Expeditions to Seram have examined mirror-image structures on the opposite side of the Banda Sea (e.g. Audley-Charles *et al.* 1979). Seismic reflection sections from around the arc, from Timor to Seram, have imaged foreland-directed thrusts surfacing in the peripheral troughs (Schlüter & Fritsch 1985; Jongsma *et al.* 1989; Hughes *et al.* 1996). Folding occurs but is relatively minor.

Caribbean. The well documented thrust structures of northern Venezuela and the Greater Antilles islands of Cuba and Hispaniola (Iturralde-Vinent 1998; Giunta *et al.* 2003, 2006) effectively replicate the patterns of thrusting around the Banda Arc. Almost equally striking are the discontinuities in both the thrust orogens. Each is formed of discrete blocks represented by the major islands, in a pattern consistent with stretching of an originally coherent mountain range. The north coast region of South America may be previewing the Sunda–Banda arc system at some future stage in its collision with Australia.

Active continental margins

Indonesia. The large NW–SE oriented island of Sumatra at the western end of the Indonesian archipelago (Fig. 4a) is the Andean margin of the Sundaland continental block, which includes Borneo and the Malay Peninsula (Barber *et al.* 2005). Sumatra is dominated by a chain of active volcanoes which are located above a well defined Wadati–Benioff Zone (WBZ). There is a transition eastwards from continental crust beneath Sundaland proper via Java, where exposed basement has been described as subduction mélange (e.g. Hamilton 1979), to the intra-oceanic arcs further east. The island of Sulawesi, immediately to the west of the Banda Sea, is the easternmost element of the Sundaland complex.

Caribbean. Middle America west of the Caribbean is noted for its volcanic activity, which is associated with the WBZ that dips NE beneath the Chortís Block from the Middle America Trench (Sykes &

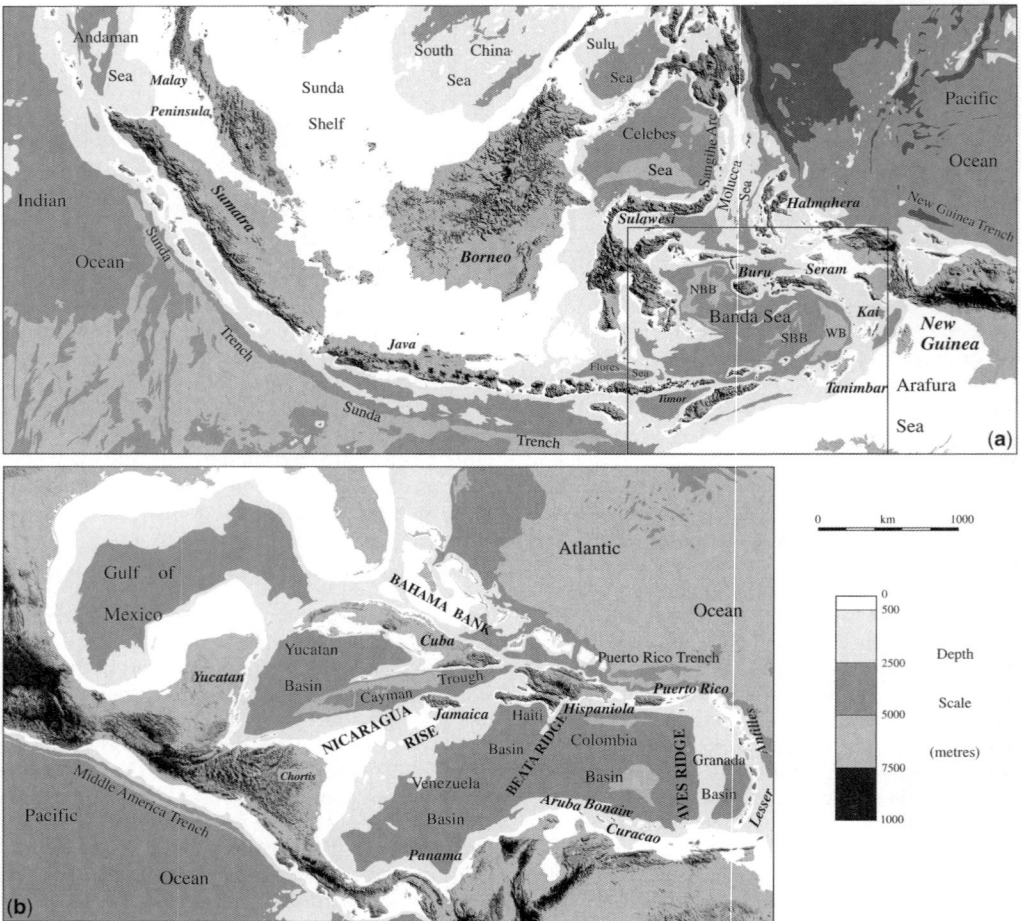

Fig. 4. (**a**) Indonesia. (**b**) the Caribbean. The two illustrations are to the same scale. Topography in both cases is based on the GTOPO30 grids and bathymetry on the GEBCO digital atlas. The rectangle indicates the Banda Sea area, shown in more detail in the inset to Figure 1.

Ewing 1965). The block is believed to be underlain, at least in part, by continental crust (e.g. Pindell & Barrett 1990). There are transitions from continental crust beneath Chortís eastwards via the quasi-continental Nicaragua Rise, which extends out towards Jamaica, to the fully oceanic Caribbean basins, and southeastwards to the Panama intra-oceanic arc. The Middle America and Sumatra trenches can be validly compared, but equating the Gulf of California to the Andaman Sea might be taking the analogy too far.

Volcanic arcs

Indonesia. The Banda Sea is bounded to the east by an active volcanic arc that consists of relatively small islands (Fig. 1, inset). There are also subsidiary, east–west trending lines of islands young enough for volcanic landforms to be preserved but with no recorded activity. One such line is formed by the relatively large islands of Alor and Wetar and smaller centres, some of which are entirely submerged, further to the east. A second, shorter, line runs through the islands south of Seram, including Ambon, westwards to a small inactive volcano south of Buru. The cessation of activity can be explained as a consequence of the eastward migration of the site of active subduction and, additionally or alternatively, by the severance of the magma conduits by backthrusts.

There has been a similar migration of volcanic activity in the Carpathians. Volcanic centres in the easternmost Carpathians have been active during the last few tens of thousands of years and it would be unwise to completely rule out the possibility of further eruptions (e.g. Seghedi *et al.* 2005). This is consistent with steady eastward migration of the Carpathian chain to eventual collision with the Russian Platform.

Caribbean. The volcanic line of the Lesser Antilles provides one of the most obvious of the resemblances between the Caribbean and the Banda Sea. The volcanoes are relatively young in Caribbean terms, although still much older than those in the Banda Arc. The oldest dated igneous rocks in the northern Antilles, which merges with the Aves Ridge, are Late Eocene, but in the southern Antilles, which are separated from the Aves Ridge by the Granada Basin, the oldest known igneous rocks are Early Miocene (see tabulation in Pindell & Barrett 1990).

There does not seem to be a subduction-related volcanic province in the northern Caribbean that can be compared with that in the northern Banda Sea, despite the presence of the Puerto Rico Trench, but in the south the Netherlands Antilles (Aruba, Curaçao and Bonaire; Fig. 4) can be seen as analogues of the Alor–Wetar line. Recent studies by White *et al.* (1999) for Aruba & Kerr *et al.* (1996) for Curaçao have suggested that igneous rocks on these two islands may not be subduction-related, but the island-arc character of Bonaire seems clear (Thompson *et al.* 2004). More work is needed. Beardsley & Avé Lallemant (2007) compiled age data on the youngest volcanic rocks of each of the islands in the chain (including the small Venezuelan islands east of Bonaire) that suggested west-to-east diachronous termination of volcanic activity during the period from 73 to 44 Ma, and reported palaeomagnetic and structural evidence for similarly diachronous clockwise rotations through angles in excess of 90°.

Seismicity and seismic tomography

Indonesia. The WBZ associated with the Sumatran Andean margin is generally unremarkable, but seismic travel-time tomography has imaged subducted lithosphere extending hundreds of kilometres deeper than the deepest earthquakes (Widiyantoro & van der Hilst 1996). It is possible that the lower part of the slab has detached at a depth of about 200 km and is now sinking independently into the mantle.

The pattern in eastern Indonesia is very different. In east–west cross-sectional views the seismic activity beneath the Banda Sea is bounded by a gently west-dipping line (Fig. 5a), but when narrow north–south swathes are viewed, the activity resolves into U- or V-shaped zones defining a scoop or shoehorn-shaped surface which reaches a maximum depth of almost 500 km beneath the central part of the sea and which subcrops along the axes of the Timor, Tanimbar and Seram troughs (Hamilton 1979; Milsom 2001). Concentrated seismic activity beneath the region of maximum curvature east of Timor may indicate incipent detachment of parts of the subducted slab, a process thought to have been already completed in the Carpathians (Milsom 2005).

Caribbean. Seismic activity in the Caribbean region is associated principally with WBZs that dip northeast beneath the Central American Andean margin in the west and west beneath the Lesser Antilles intra-oceanic arc in the east (Molnar & Sykes 1969). As in Indonesia, the western WBZ is unremarkable. In the east, travel-time tomography has been used to trace the southern end of the Lesser Antilles slab around its pronounced bend to the west, and to define what appears to be a detachment at crustal levels (VanDecar *et al.* 2003). Because of the greater overall dimensions of the Caribbean, the contortions required of the downgoing slab are much less extreme than are those in eastern Indonesia. Moreover, tomography suggests that, far from flattening out into a shoehorn, the Antilles slab

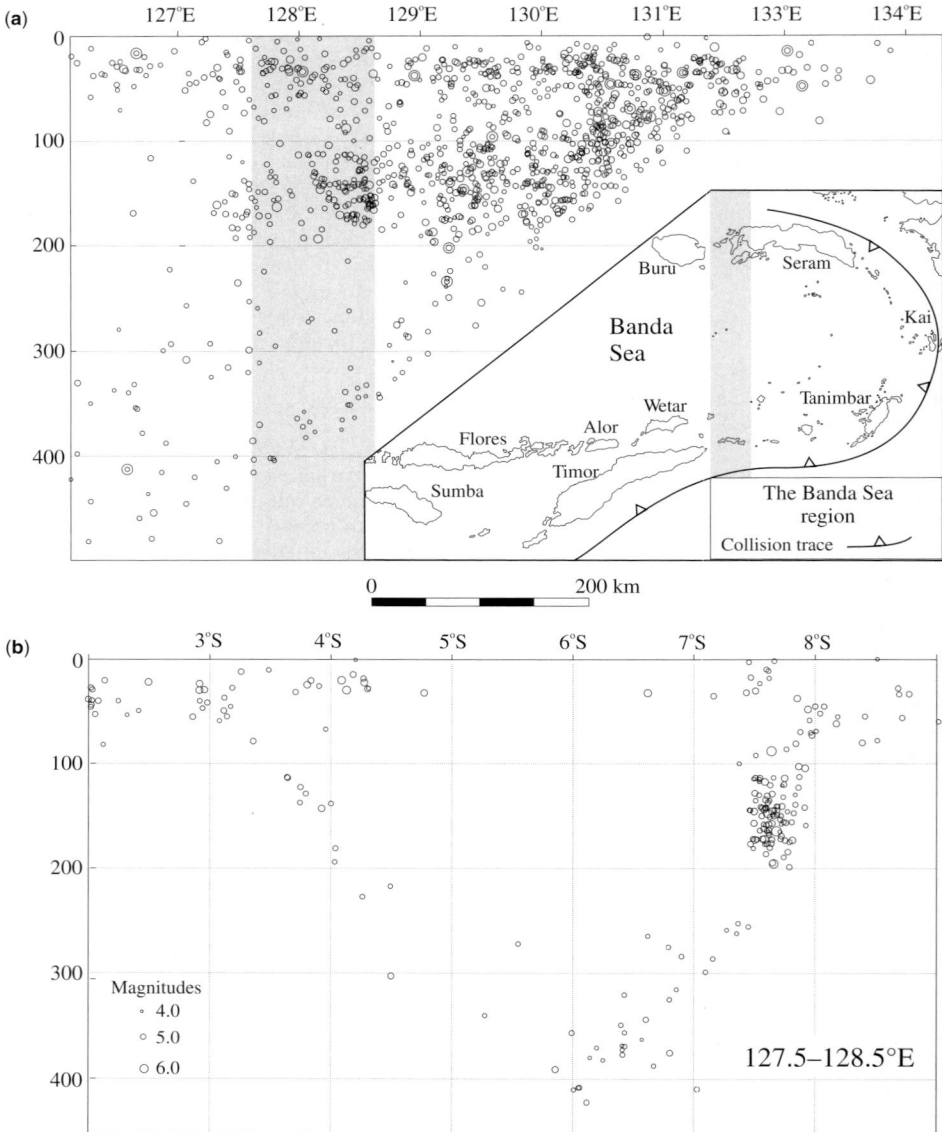

Fig. 5. Seismicity of the eastern Indonesia area. (**a**) Distribution of earthquake hypocentres on a cross-section viewed from the south, after Milsom (2001). Shading, on main figure and inset, indicates location of the swathe viewed in (**b**). (**b**) Cross-section along a typical north–south swathe, after Milsom (2001). The cluster of intermediate-depth shocks between 7 and 8°S may be indicating incipient delamination or detachment of the subducting slab. Hypocentre locations from supporting material to Engdahl *et al.* (1998).

dips steeply into the asthenosphere (VanDecar *et al.* 2003).

Strike–slip provinces

Indonesia. Although the eastward roll-back of the Banda slab seems to require transcurrent motion

along both the northern and southern borders of the Banda Sea, strike–slip features are not obvious in the south. To the north, however, the Sula Spur (Fig. 1, inset) has some claim to be considered one of the most remarkable features on the Earth's surface. Essentially, the spur consists of a long, narrow strip of Australasian continental crust

(Garrard *et al.* 1988) that moved west (Fig. 3) and collided with the East Arm of Sulawesi in the Late Miocene (Davies 1990). Such motion would have required major left-lateral faulting along the spur's southern margin and probably also right-lateral faulting along its northern margin. The total displacements are measured in hundreds of kilometres, yet the spur is in places only about a hundred kilometres across. Tectonic processes in this region are still only poorly understood, but are evidently extremely complex.

Caribbean. As is the case with the southern margin of the Banda Sea, the southern margin of the Caribbean is more notable for north–south compressional structures than east–west transcurrent ones, although the latter undoubtedly exist in the El Pilar fault system (Avé Lallemant & Sisson 2005). The El Pilar has a northern counterpart in the Guatemalan Motagua–Polochic fault system (Giunta *et al.* 2006) but, as in the Banda Sea, transcurrent displacement at the northern margin of the Caribbean has been accommodated in part by a feature (the Cayman Trough) that has no counterpart anywhere else on the globe. Although the north–south oriented spreading centre in the trough is little more than 100 km long, it has produced some 1100 km of oceanic crust in an extensional episode that has lasted for much of the Cenozoic (Leroy *et al.* 1996). The North America–South America relative motions shown diagrammatically in Figure 6 indicate only about 200 km of relative

dextral displacement between these two major blocks between 83 and 38 Ma and none thereafter. The much greater (and largely later) displacement at the northern (Cayman Trough) margin of the Caribbean therefore requires there to have been also large-scale shear at the opposite (southern) margin.

The Cayman Trough clearly differs from the Sula Spur in almost every respect apart from its approximate dimensions and its similar positioning with respect to the ocean basin to the south, but both trough and spur seem lie at the extreme limits of the structures that might be anticipated within the plate-tectonic paradigm.

Arc-associated basins

Indonesia. The extreme east of the Banda Sea is occupied by the Weber Basin (Fig. 7a) which, with a maximum depth of more than 7000 m, contains the deepest seafloor known anywhere outside the subduction trenches. There is no consensus explanation for these great depths in what is actually a forearc basin; the classic roll-back forearc basins in the Marianas and Bonin arcs are only about half this depth, despite having far greater along-strike extent, and therefore more room for subsidence. Moreover, roll-back cannot plausibly still be operating in the Weber area, because the Banda forearc has collided with, and now rests upon, Australian continental crust (Charlton *et al.* 1991*b*). Seismic reflection surveys have clearly imaged the former

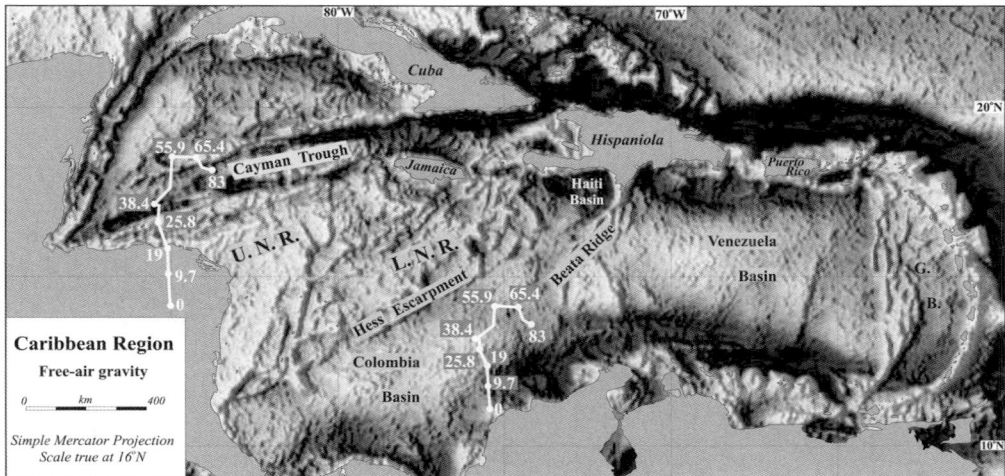

Fig. 6. Free-air gravity of the Caribbean, based on satellite derived grids (Sandwell & Smith 1995). U.N.R., Upper Nicaragua Rise; L.N.R., Lower Nicaragua Rise. The white traces record the motion of North America relative to South America at longitudes 75 and 85°W, after Müller *et al.* (1999). The white numbers are ages, in Ma. The latitudinal positions of the traces have been selected so as to minimize conflicts with other information, and have no other significance. GB, Granada Basin.

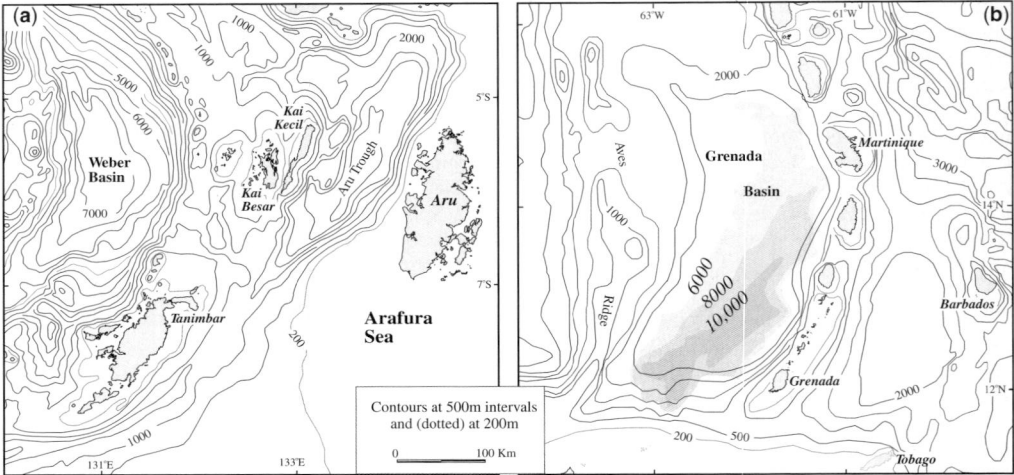

Fig. 7. (a) Bathymetry of the Weber Basin. (b) Bathymetry of the Grenada Basin, to the same scale. Shading shows depth to basement at 6000, 8000 and 10 000 m within the basin, after Holcombe *et al.* (1990). Bathymetric contours from the GEBCO digital bathymetric atlas in both cases.

subduction fault reaching the seafloor in the narrow strait between Kai Besar and Kai Kecil (Milsom *et al.* 1996).

The Weber Basin depth problem is often simply ignored in regional discussions, but astheno-spheric flow prompted by convergent tectonics could provide an explanation, and an additional drive for Banda Sea expansion (Fig. 8). To the north, and beyond the narrow strip of continental crust represented by the Sula Spur, the Molucca Sea provides a rare example of arc–arc collision (McCaffrey *et al.* 1980). Asthenosphere escaping south from the collision zone between the

converging Halmahera and Sangihe arcs would inevitably tend to force down the scoop-shaped slab of subducted Indian Ocean lithosphere beneath the Banda Sea, taking the overlying crust with it. The Kai collision, far from being a problem for this explanation, is a probable necessary condition for the development of the anomalous depths, because it prevents the stress being relieved by simple subduction zone retreat.

A wider applicability for the mantle-flow hypothesis is suggested by the occurrence of rather similar features within other oroclines. The Transyl-vania Basin in the eastern loop of the Carpathians

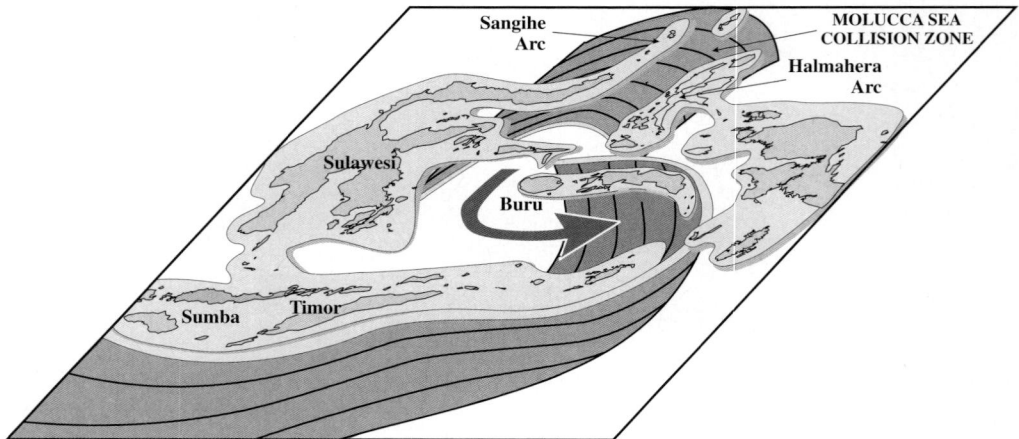

Fig. 8. Asthenospheric flow in eastern Indonesia, as suggested by Milsom (2001).

is separated from the main Pannonian Basin by the old volcanic high of the Apuseni mountains (Horvath & Tari 1999), while the palaeogeographic maps presented by Séranne (1999) suggest that Corsica and Sardinia similarly separated the younger Tyrrhenian from the older Ligurian basin (Fig. 1). There may also be an analogous drive producing the region of ductile extension recognized by Jolivet & Patriat (1999) in the southern Aegean Sea.

Caribbean. Flows in the sub-lithospheric mantle could well have been important in the post-Paleocene Caribbean also, despite the absence of arc–arc collision. Convergence between the deep lithospheric keels of North and South America would inevitably pressurize the asthenosphere trapped behind the subducted slabs beneath Middle America and the Lesser Antilles. It is therefore not surprising that the Grenada Basin at the eastern end of the Caribbean (Fig. 7b) appears at first sight to be a direct analogue of the Weber Basin.

Closer inspection reveals important differences. Not only is the Grenada Basin only about half as deep, but it occupies a back-arc rather than a forearc position. It is also almost certainly much older. Although the actual ages are not definitely known in either case, the most commonly cited dating for the Grenada Basin places its origin in the Paleocene–Eocene (e.g. Pindell & Barrettt 1990), whereas the Weber Basin is probably Late Neogene (Honthaas *et al.* 1997). This may have a bearing on the other differences. For example, the contrast in depth is much less significant if the sediment fill, which is negligible in the Weber Basin but considerable in the Grenada Basin, is taken into account (Fig. 7b). This sediment cover renders the seafloor of the Grenada Basin much the less rugged of the two, but the maximum depth to basement in its southern part, close to the prolific sediment sources of the South American continent, is more than 10 km (Holcombe *et al.* 1990).

Normal age-depth relationships applied to the Grenada Basin suggest a stabilization age of about 40 Ma (Holcombe *et al.* 1990), but the Weber Basin demonstrates the danger of using this approach in such an environment. A younger age is possible, and would be consistent with the apparent youth of the southern Lesser Antilles. It is quite possible that the Grenada Basin originally developed as the forearc to a volcanically active Aves Ridge and only became back-arc when continued roll-back forced volcanic activity to transfer further east. This differs from the way in which active arc–remnant arc pairs usually evolve by arc splitting (e.g. Karig 1972), but is rendered plausible by the situation in eastern Indonesia. Were the Banda Sea able to continue its expansion eastwards,

volcanic activity would surely eventually have to move eastwards also, but the Weber Basin is probably already too deep for this transfer to be accomplished by rifting of the present volcanic arc.

Whatever the precise history of the Lesser Antilles, it is very difficult to see the Aves Ridge as anything but an abandoned 'remnant arc' This interpretation is supported by the maximum ages of the igneous rocks, which are greater in the northern part of the Lesser Antilles ridge, where it merges with the Aves Ridge, than in the south (Pindell & Barrett 1990). The suggestion (James 2005) that the Aves was itself a constructive plate margin seems incompatible with its morphology and what is known of its geology (Holcombe *et al.* 1990). Because of the very low magnetic latitude, magnetic anomalies in the Grenada Basin are poorly defined (Bird *et al.* 1999), and certainly not dateable, but comparison with other basins developed between active and remnant arcs (Karig 1972) suggests that there may have been a series of short spreading axes, offset by transforms and oblique to a dominantly east–west direction of extension.

Internal basins

Indonesia. In addition to the Weber forearc basin, the Banda Sea includes two sub-basins that are more typically oceanic. The South Banda Basin is clearly back-arc with respect to the present-day Banda Arc, but the tectonic position of the North Banda Basin is less clear. Its nearest analogue would appear to be the Vienna Basin, which separates the Carpathians from the Eastern Alps in much the same way as the North Banda Basin separates the Outer Banda Arc from Sulawesi (Fig. 1).

Because of the absence of deep drilling in the Banda Sea, none of its sea-floor can be considered securely dated. Magnetic lineations in the South Basin were interpreted by Hinschberger *et al.* (2001) as Late Miocene to Pliocene (6–3 Ma), but identifications of relatively short magnetic anomalies in restricted basins are notoriously uncertain. Dredge hauls from the North Basin led Réhault *et al.* (1994) to suggest that this had a slightly earlier, Late Miocene (c. 10 Ma), origin. The extremely rugged seafloor in both basins, and the thinness of the sedimentary cover, would be hard to reconcile with greater ages, even given the absence of major sediment sources.

Caribbean. The Caribbean is often described as consisting of two main sub-basins (the Colombia and Venezuela Basins; Fig. 4), each floored by oceanic crust modified by massive Late Cretaceous outpourings of basaltic lavas. The upper surface of these flows is identified with the widespread B″

seismic reflector. However, Driscoll & Diebold (1999) have maintained that this is a very considerable oversimplification and that in some areas, notably the south-eastern part of the Venezuela Basin, the crust is actually thinner than normal oceanic. They ascribed a major role to extension in controlling basin structure but argued convincingly that, in the Venezuela Basin at least, this took place either during or soon after the B'' volcanic episode.

The Colombia Basin is 3000–4000 m deep, which is shallow for an oceanic basin but unsurprising given the basin's location and history. Not only was the crust supposedly thickened by the Cretaceous basaltic events, but the basin is infilled by thick sediments derived in the main from erosion of the Colombian Andes. It is actually quite hard to believe that the term 'oceanic plateau', so widely used in describing the Caribbean, could ever have been appropriate, since modern oceanic plateaus form very prominent bathymetric highs and in many cases rise locally above sea level. Bowland & Rosencrantz (1988) considered that, although features in the western Colombia Basin were of plateau type, its centre was underlain by relatively unaltered oceanic crust.

The Haiti Basin is often regarded merely as a northeastern extension of the Colombia Basin but is linked to it only through a narrow gap between the Beata Ridge and the Lower Nicaragua Rise (Fig. 6). The colinearity of the southeastern margin of the rise (the Hess Escarpment) with the western margin of the ridge is very striking and it is geometrically attractive to suppose that the Haiti Basin opened when transcurrent faulting developed along this line. The date at which this might have occurred is completely unconstrained but is likely to have been contemporaneous with the extensional faulting of the Nicaragua Rise discussed below.

Internal ridges

Indonesia. The Banda Sea sub-basins are separated by high-standing ridges of complex origin (Fig. 1, inset). The South Banda Basin is separated from the partly oceanic Flores Sea by a region of rough and shallow sea floor capped by small islands. Similarly, the Banda Ridges province between the North and South Banda basins rises above sea level only in a few low islands on which only Recent coral limestones are exposed, but dredge hauls on the ridge flanks have recovered a wide variety of rocks, including Triassic platform carbonates (Villeneuve *et al.* 1994), Palaeogene gabbros and basalts and Neogene calc-alkaline volcanics (Silver *et al.* 1985). The carbonates can be correlated with Australasian crust exposed on Seram,

Buru, Buton and East Sulawesi, the gabbros and basalts may be fragments of the East Sulawesi ophiolite and the calc-alkaline rocks could be parts of a remnant arc related to the present-day Inner Banda Arc. All of these elements are consistent with the fragmentation of a collision orogen in the Sulawesi region in the Mid-Cenozoic (Fig. 3). An impressive sub-sea scarp suggestive of large-scale extensional faulting forms the boundary between the Banda Ridges province and the North Banda Basin (R. V. Charles Darwin cruise CD30, unpublished data, 1990).

Caribbean. In the Caribbean, the Nicaragua Rise may match the Indonesian Banda Ridges in fragmentation and heterogeneity. This is well illustrated by the free-air gravity patterns (Fig. 6), which show basement structures more clearly than do bathymetric maps because they are less affected by the masking effect of the sedimentary cover. The picture, particularly along the lower rise, is one of pervasive extensional faulting. Much of this faulting and accompanying alkaline volcanism may be Miocene to Recent (see Fig. 5 in Holcombe *et al.* 1990). The entire lower rise appears to be broken up into blocks that are only a few tens of kilometres across but, despite this, the Hess Escarpment that separates it from the Colombia Basin is remarkably linear.

The Colombia and Venezuela basins are separated by the Beata Ridge, which is underlain by crust that is up to 12 km thick (Holcombe *et al.* 1990), and broadens out towards the SW into a complex of diverging ridges. The reason for the relatively thick crust is uncertain. Mauffret & Leroy (1997) interpreted seismic reflection data as indicating overthrusting of the Colombia Basin on the Venezuela Basin, but Driscoll & Diebold (1999), using similar but possibly better quality data, saw only extension in the same area. They presented a seismic section that showed the ridge as a highly asymmetric feature, with a very steep, fault-defined slope into the Haiti Basin to the west but a much gentler slope into the Venezuela Basin to the east.

Analogy with the Banda Ridges suggests that the Beata Ridge may be a remnant of the original, unextended, Caribbean LIP, positioned adjacent to areas in which later extension has produced markedly thinner crust. Holcombe *et al.* (1990) suggested that the changes in crustal thickness between the Lower Nicaragua Rise, the Colombia Basin and the Beata Ridge indicated by seismic refraction surveys could have been one of the factors responsible for the faulting that defines the various bathymetric features. It seems just as reasonable to suggest that it was the extensional faulting that produced the changes in crustal thickness.

North–south convergence

Indonesia. The Asian land mass has been almost stationary in the mantle reference frame throughout the Cenozoic, but during that period Australia has been moving steadily northwards as part of the Indian Ocean plate. Its present-day convergence on Southeast Asia in the vicinity of Timor is estimated to be about 7.5 cm/annum at about N15°E (Genrich *et al.* 1996). Given their present relative positions, it appears that Australia will strike Asia only a glancing blow, but at relatively high speed.

Caribbean. The Late Mesozoic and Cenozoic history of relative motion between North and South America, as estimated by Müller *et al.* (1999), is summarized in Figure 6. Beginning in the latest Paleocene (*c.* 56 Ma), South America has converged on North America at a rate that increases from east to west but which is currently about 8 mm/annum in the central region. This is only about a tenth of the rate of advance of Australia on Asia but not very different from the Africa–Europe convergence within which the oroclinal basins of western Tethys have developed. Regardless of whether the Caribbean developed between the Americas as suggested by James (2006), or in the Pacific to the west as argued by Pindell *et al.* (2006), it would have been located between North and South America as they approached each other. The observed structures of the Tethyan oroclines suggest that east–west extension of the Caribbean would be expected during this compressional phase.

Backthrusting

Indonesia. Continental crust has now entered the subduction zone south of Timor, and the continuing northward movement of Australia must be accommodated elsewhere. North-directed thrusts in the southern Banda Sea have been imaged on a number of research cruises (e.g. Silver & Reed 1988) and are consistent with GPS campaigns that have shown Timor moving with the Australian continent and converging on Seram (e.g. Genrich *et al.* 1996). The locus of convergence appears to be a thrust expressed as a narrow trough just north of the inner-arc island of Wetar (Breen *et al.* 1989). The existence of a mirror-image feature south of Seram has been the subject of much informal speculation, but no conclusive evidence has yet been presented.

Caribbean. Although the rates are much slower than in Indonesia, continental convergence in the Caribbean has a history spanning tens of millions of years and it is unsurprising to find backthrusts analogous to those produced in the Banda Sea. In the Caribbean there is good evidence for thrusting from the north, at the Muertos Trough south of Puerto Rico, as well as from the south, where an accretionary prism has developed north of the Netherlands Antilles (e.g. Giunta *et al.* 2006).

Conclusions

In the preceding section, similarities between the Caribbean region and Indonesia have been identified and discussed. The list is impressive, but the generally-accepted two-phase origin of most of the Caribbean basement, first as normal oceanic crust and then as a volcanic oceanic plateau, suggests very great differences between the two areas. Clearly, arguments for specific evolutionary hypotheses that are based on supposed analogies between the Caribbean and the Banda Sea (or any other oroclinal basin) must be treated with extreme caution. Although the stress needed to create an orocline from a rectilinear mountain chain could be produced by rapid expansion in an area adjacent to it, and this process seems to have created the Tethyan oroclines, it seems probable that the impact on a long-established subduction zone of a rigid, but localized, crustal block (a 'rigid indentor' in the terminology of Indo-Asian collision) would have very similar effects. Indentor impact is implicit in a Pacific origin for the main Caribbean Plate, and most of the Caribbean/Indonesia similarities can be explained equally well by either process. There seem to be only a few aspects of Caribbean geology for which orocline analogies can provide fresh insights.

One lesson that can be drawn from the Tethyan basins that might be applicable to the Caribbean is that extension is not only not prohibited by continental convergence but may actually be promoted by it. It may therefore be appropriate to take a second look at the development of the Caribbean during the Cenozoic convergence of North and South America. Is it possible that extension during this period, far from being confined to the Cayman Trough and Granada Basin, was distributed far more widely through the Caribbean than is currently supposed? The free-air patterns on the Upper and, especially, the Lower Nicaragua rises are indicative of pervasive extensional faulting. The eastern end of the lower rise is readily interpreted as conjugate to southeastern Hispaniola and the western flank of the Beata Ridge, suggesting that the Haiti Basin is an extreme consequence of this faulting, that it may be younger than the remainder of the Colombia Basin and that it may even be floored by Neogene oceanic crust. Moreover, just as the Vienna Basin has been opened by the eastward migration of the

Carpathians, the seaways between Jamaica, Hispaniola, Puerto Rico and the Virgin Islands may all have been expanded during the Cenozoic. Hypotheses that relate crustal block movements to asthenospheric flows are unproven but do at least provide process-related explanations for some of the similarities between the Caribbean and Indonesian regions that would not necessarily be produced by a rigid indentor alone.

An additional factor, which as yet lacks any theoretical explanation but which may have been crucial in creating some of the resemblances, is the observed world-wide asymmetry in subduction. All east-facing subduction systems, whether in the Pacific or the Atlantic, have experienced roll-back, producing large scale back-arc extension and generating new oceanic crust. Of the west-facing systems, only the Andaman Islands have an oceanic back-arc, and this exception may be an unavoidable consequence of the partitioning of the oblique convergence west of Sumatra between the Sunda Trench and the Sumatra Fault (Curray 1989). Since the roots of much of the perceived similarity between the Caribbean and Indonesia lie in the eastward migrations of the Lesser Antilles and Banda volcanic arcs, combined with the locking of the Middle Americas and Sumatra subduction zones to their respective continental margins, this global asymmetry cannot be ignored.

References

AUDLEY-CHARLES, M. G., CARTER, D. J., BARBER, A. J., NORVICK, M. S. & TJOKROSAPOETRO, S. 1979. Re-interpretation of the geology of Seram: implications for Banda Arc tectonics. *Journal of the Geological Society, London*, **136**, 547–568.

AVÉ LALLEMANT, H. G. & SISSON, V. 2005. Prologue. *In*: AVÉ LALLEMANT, H. G. & SISSON, V. (eds) *Caribbean–South American Plate Interactions, Venezuela*. Geological Society of America, Special Papers, **394**, 1–5.

BARBER, A. J. 1978. Structural interpretations of the island of Timor. *Proceedings of the South East Asia Petroleum Exploration Society (SEAPEX)*, **9**, 9–21.

BARBER, A. J., AUDLEY-CHARLES, M. G. & CARTER, D. J. 1977. Thrust tectonics in Timor. *Journal of the Geological Society of Australia*, **24**, 51–62.

BARBER, A. J., CROW, M. J. & MILSOM, J. (eds). 2005. *Sumatra: Geology, Resources and Tectonic Evolution*. Geological Society, London Memoirs, **31**.

BEARDSLEY, A. G. & AVÉ LALLEMANT, H. G. 2007. Oblique collision and accretion of the Netherlands Leeward Antilles to South America. *Tectonics*, **26**; doi: 10.1029/2006TC002028.

BIRD, D. E., HALL, S. A. & CASEY, J. F. 1999. Tectonic evolution of the Granada Basin. *In*: MANN, P (ed.) *Caribbean Basins: Sedimentary Basins of the World*, **4**. Elsevier Science, Amsterdam, 389–416.

BOWLAND, C. L. & ROSENCRANTZ, E. 1988. Upper crustal structure of the western Colombian Basin, Caribbean Sea. *Geological Society of America Bulletin*, **100**, 534–546.

BREEN, N. A., SILVER, E. A. & ROOF, S. 1989. The Wetar back-arc thrust belt, eastern Indonesia: the effect of accretion against an irregularly shaped arc. *Tectonics*, **8**, 85–98.

CAREY, S. W. 1955. The orocline concept in geotectonics. *Proceedings of the Royal Society of Tasmania*, **89**, 255–288.

CARMINATI, E., WORTEL, M. J. R., MEIJER, P. T. & SABADINI, R. 1998. The two-stage opening of the western-central Mediterranean basins: a forward modelling test to a new evolutionary model. *Earth and Planetary Science Letters*, **160**, 667–679.

CHARLTON, T. R. 1986. A plate tectonic model of the eastern Indonesia collision zone. *Nature*, **319**, 394–396.

CHARLTON, T. R., BARBER, A. J. & BARKHAM, S. T. 1991a. The structural evolution of the Timor collision complex, eastern Indonesia. *Journal of Structural Geology*, **13**, 489–500.

CHARLTON, T. R., KAYE, S. J., SAMODRA, H. & SARDJONO, 1991b. The geology of the Kai Islands: implications for the evolution of the Aru Trough and the Weber Basin. *Marine and Petroleum Geology*, **8**, 62–69.

CURRAY, J. R. 1989. The Sunda Arc: a model for oblique plate convergence. *Netherlands Journal of Sea Research*, **24**, 131–140.

DALY, M. C., COOPER, M. A., WILSON, I., SMITH, D. G. & HOOPER, B. G. D. 1991. Cainozoic plate tectonics and basin evolution in Indonesia. *Marine and Petroleum Geology*, **8**, 2–21.

DAVIES, I. C. 1990. Geological and exploration review of Tomori PSC, eastern Indonesia. *Proceedings of the Indonesian Petroleum Association*, **19**, 41–67.

DRISCOLL, N. W. & DIEBOLD, J. B. 1999. Tectonic and stratigraphic development of the eastern Caribbean: new constraints from multichannel seismic data. *In*: MANN, P. (ed.) *Caribbean Basins: Sedimentary Basins of the World*, **4**. Elsevier Science, Amsterdam, 35–59.

ENGDAHL, E. R., VAN DER HILST, R. D. & BULAND, R. P. 1998. Global teleseismic earthquake relocation with improved travel times and procedures for depth determination. *Bulletin of the Seismological Society of America*, **88**, 722–743.

GARRARD, R. A., SUPANDJONO, J. B. & SURONO, 1988. The geology of the Banggai–Sula microcontinent, eastern Indonesia. *Proceedings of the Indonesian Petroleum Association*, **17**, 23–52.

GENRICH, J., BOCK, Y., MCCAFFREY, R., CALAIS, E., STEVENS, C. & SUBARYA, C. 1996. Accretion of the southern Banda Arc to the Australian Plate margin determined from Global Positioning System measurements. *Tectonics*, **15**, 288–295.

GIUNTA, G., MARRONI, M., PADOA, E. & PADOLFI, L. 2003. Geological constraints for the geodynamic evolution of the southern margin of the Caribbean Plate. *In*: BARTOLINI, C., BUFFLER, R. T. & BLICKWEDE, J. (eds) *The Circum-Gulf of Mexico and Caribbean: Hydrocarbon Habitats, Basin Formation and Plate Tectonics*. American Association of Petroleum Geologists, 104–125.

GIUNTA, G., BECCALUVA, L. & SIENA, F. 2006. Caribbean Plate margin evolution; constraints and current problems. *Geologica Acta*, **4**, 265–277.

HALL, R. 2002. Cenozoic geological and plate tectonic evolution of SE Asia and the SW Pacific: computer-based reconstructions, models and animations. *Journal of Asian Earth Sciences*, **20**, 353–431.

HAMILTON, W. 1979. *Tectonics of the Indonesian Region*. United States Geological Survey, Professional Papers, **1078**.

HATZFIELD, D. 1999. The present-day tectonics of the Aegean as deduced from seismicity. *In*: DURAND, B., JOLIVET, L., HORVATH, F. & SÉRANNE, M. (eds) *The Mediterranean Basins: Tertiary Extension within the Alpine Orogen*. Geological Society, London, Special Publications, **156**, 415–426.

HINSCHBERGER, F., MALOD, J.-A., DYMENT, J., HONTHAAS, C., REHAULT, J.-P. & BURHANUDDIN, S. 2001. Magnetic lineations constraints for the back-arc opening of the Late Neogene South Banda Basin (eastern Indonesia). *Tectonophysics*, **333**, 47–59.

HOLCOMBE, T. L., LADD, J. W., WESTBROOK, G., EDGAR, N. T. & BOWLAND, C. L. 1990. Caribbean marine geology; ridges and basins of the plate interior. *In*: DENGO, G. & CASE, J. E. (eds) *The Geology of North America; The Caribbean Region (A Decade of North American Geology)*. Geological Society of America, 231–260.

HONTHAAS, C., VILLENEUVE, M. *ET AL.* 1997. L'Ile de Kur: geologie du flanc oriental du bassin de Weber (Indonesie orientale). *Comptes Rendus, Academie des Sciences*, **325**, 883–890.

HORVATH, F. & TARI, G. 1999. *IBS Pannonian Basin project: a Review of the Main Results and their Bearings on Hydrocarbon Exploration*. Geological Society of London, Special Publications, **156**, 195–214.

HUGHES, B. D., BAXTER, K., CLARK, R. A. & SNYDER, D. B. 1996. Detailed processing of seismic reflection data from the frontal part of the Timor trough accretionary wedge, eastern Indonesia. *In*: HALL, R. & BLUNDELL, D. J. (eds) *Tectonic Evolution of Southeast Asia*. Geological Society, London, Special Publications, **106**, 75–83.

ITURRALDE-VINENT, M. 1998. Sinopsis de la constitucion geologica de Cuba. *Acta Geologica Hispanica*, **33**, 9–56.

JAMES, K. H. 2005. A simple synthesis of Caribbean geology. *Caribbean Journal of Earth Science*, **39**, 69–82.

JAMES, K. H. 2006. Arguments for and against the Pacific origin of the Caribbean Plate: discussion, finding for an inter-American origin. *Geologica Acta*, **4**, 279–302.

JOLIVET, L. & FACCENA, C. 2000. Mediterranean extension and the Africa–Eurasia collision. *Tectonics*, **19**, 1095–1106.

JOLIVET, L. & PATRIAT, M. 1999. Ductile extension and the formation of the Aegean Sea. *In*: DURAND, B., JOLIVET, L., HORVATH, F. & SÉRANNE, M. (eds) *The Mediterranean Basins: Tertiary Extension within the Alpine Orogen*. Geological Society, London, Special Publications, **156**, 427–456.

JOLIVET, L., FRIZON DE LAMOTTA, D., MASCLE, A. & SÉRANNE, M. 1999. The Mediterranean: Tertiary extension within the Alpine orogen – an introduction. *In*: DURAND, B., JOLIVET, L., HORVATH, F. & SÉRANNE, M. (eds) *The Mediterranean Basins: Tertiary Extension within the Alpine Orogen*. Geological Society, London, Special Publications, **156**, 1–14.

JONGSMA, D., WOODSIDE, J. M., HUSON, W., SUPARKA, S. & KADARISMAN, D. 1989. Geophysics and tentative Late Cenozoic seismic stratigraphy of the Banda Arc–Australian continent collision zone along three transects. *Netherlands Journal of Sea Research*, **24**, 205–229.

KARIG, D. E. 1972. Remnant arcs. *Geological Society of America Bulletin*, **83**, 1057–1068.

KARIG, D. E., BARBER, A. J., CHARLTON, T. R., KLEMPERER, S. E. & HUSSONG, D. M. 1987. Nature and distribution of deformation across the Banda Arc–Australian collision zone at Timor. *Geological Society of America Bulletin*, **93**, 18–32.

KERR, A. C., TARNEY, J., MARRINER, G. F., KLAVER, G., SANDERS, A. C. & THIRLWALL, M. F. 1996. The geochemistry and petrogenesis of the Late Cretaceous picrites and basalts of Curaçoa, Netherlands Antilles. *Contributions in Mineralogy and Petrology*, **124**, 29–43.

KNAPP, J. H., KNAPP, C. C., RAILEANU, V., MATENCO, L., MOCANU, V. & DINU, C. 2005. Crustal constraints on the origin of mantle seismicity in the Vrancea Zone, Romania: the case for active continental lithosphere delamination. *Tectonophysics*, **410**, 311–323.

LEE, C. S. & MCCABE, R. 1986. Banda–Celebes–Sulu basin: a trapped piece of Cretaceous–Eocene oceanic crust? *Nature*, **322**, 51–54.

LEROY, S., DE LÉPINAY, B. M., MAUFFRET, A. & PUBELLIER, M. 1996. Structural and tectonic history of the eastern Cayman Trough (Caribbean Sea) from seismic reflection data. *Bulletin of the American Association of Petroleum Geologists*, **80**, 222–247.

LONERGAN, L. & WHITE, N. 1997. Origin of the Betic–Rif mountain belt. *Tectonics*, **16**, 504–522.

MAUFFRET, A. & LEROY, S. 1997. Seismic stratigraphy and structure of the Caribbean Sea. *Tectonophysics*, **283**, 61–104.

MCCAFFREY, R., SILVER, E. A. & RAITT, R. W. 1980. Crustal structure of the Molucca Sea collision zone, Indonesia. *In*: HAYES, D. E. (ed.), *The Tectonic and Geologic Evolution of Southeast Asian Seas and Islands*. Monograph 23. American Geophysical Union, Washington, DC.

MILSOM, J. 2000. Stratigraphic constraints on suture models for eastern Indonesia. *Journal of Asian Earth Sciences*, **18**, 761–779.

MILSOM, J. 2001. Subduction in eastern Indonesia: how many slabs? *Tectonophysics*, **338**, 167–178.

MILSOM, J. 2005. The Vrancea seismic zone and its analogue in the Banda Arc, eastern Indonesia. *Tectonophysics*, **410**, 325–336.

MILSOM, J., KAYE, S. & SARDJONO. 1996. Extension, collision and curvature in the eastern Banda Arc. *In*: HALL, R. & BLUNDELL, D. J. (eds) *Tectonic Evolution of Southeast Asia*. Geological Society, London, Special Publications, **106**, 85–94.

MILSOM, J., THUROW, J. & ROQUES, D. 2000. Sulawesi dispersal and the evolution of the northern Banda Arc. *Proceedings of the Indonesian Petroleum Association*, **27**, 495–504.

MOLNAR, P. & SYKES, L. 1969. Tectonics of the Caribbean and Middle America regions from focal mechanisms and seismicity. *Journal of the Geological Society of America*, **80**, 1639–1684.

MÜLLER, R. D., ROYER, J.-Y., CANDE, S. C., ROEST, W. R. & MASCHENKOV, S. 1999. New constraints on the Late Cretaceous/Tertiary plate tectonic evolution of the Caribbean. *In:* MANN, P. (ed.) *Caribbean Basins: Sedimentary Basins of the World*, **4**. Elsevier Science, Amsterdam, 35–59.

PINDELL, J. L. & BARRETT, S. F. 1990. Geological evolution of the Caribbean region; a plate tectonic perspective. *In:* DENGO, G. & CASE, J. E. (eds) *The Geology of North America; The Caribbean Region (A Decade of North American Geology)*. Geological Society of America, 405–432.

PINDELL, J. L., KENNAN, L., STANEK, K. P., MARESCH, W. V. & DRAPER, G. 2006. Foundations of the Gulf of Mexico and Caribbean evolution: eight controversies resolved. *Geologica Acta*, **4**, 303–341.

PLATT, J. & VISSERS, R. L. M. 1989. Extensional collapse of thickened continental lithosphere: a working hypothesis for the Alboran Sea and Gibraltar Arc. *Geology*, **17**, 540–543.

REED, T. A., DE SMET, M. E. M., HARAHAP, B. H. & SJAPAWI, A. 1996. Structural and depositional history of East Timor. *Proceedings of the Indonesian Petroleum Association*, **23**, 297–308.

RÉHAULT, J.-P., MAURY, R. C. *ET AL.* 1994. La Mer de Banda Nord (Indonesie): un bassin arriere-arc du Miocene superieur. *Comptes Rendus, Academie des Sciences. Paris, Series II*, **318**, 969–976.

SANDWELL, D. T. & SMITH, W. H. F. 1995. *Exploring the Ocean Basins with Satellite Altimeter Data*. US Department of Commerce, National Oceanic and Atmospheric Administration, National Geophysical Data Centre & World Data Centre A for Marine Geology and Geophysics.

SAWYER, R. K., SANI, K. & BROWN, S. 1993. The stratigraphy and sedimentology of West Timor, Indonesia. *Proceedings of the Indonesian Petroleum Association*, **22**, 534–574.

SCHLÜTER, H. U. & FRITSCH, J. 1985. Geology and tectonics of the Banda arc between Tanimbar Island and Aru Island (Indonesia). *Geologisches Jahrbuch*, **E30**, 3–41.

SEGHEDI, I., DOWNES, H., HARANGI, S., MASON, P. R. D. & PECKASKY, Z. 2005. Geochemical response of magmas to Neogene-Quaternary collision in the Carpathian–Pannonian region: a review. *Tectonophysics*, **410**, 485–499.

SÉRANNE, M. 1999. *The Gulf of Lions Continental Margin*. Geological Society of London, Special Publications, **156**, 15–36.

SILVER, E. A., GILL, J. B., SCHWARZ, D., PRASETYO, H. & DUNCAN, R. A. 1985. Evidence for a submerged and displaced continental borderland, north Banda Sea, Indonesia. *Geology*, **13**, 687–691.

SILVER, E. A. & REED, D. L. 1988. Backthrusting in accretionary wedges. *Journal of Geophysical Research*, **93**, 3116–3126.

SMITH, R. B. & SILVER, E. A. 1991. Geology of a Miocene collision complex, Buton, eastern Indonesia. *Geological Society of America Bulletin*, **103**, 660–678.

SYKES, L. R. & EWING, M. 1965. The seismicity of the Caribbean region. *Journal of Geophysical Research*, **70**, 5065–5074.

THOMPSON, P. M. E., KEMPTON, P. D. *ET AL.* 2004. Elemental, Hf–Nd isotopic and geochronological constraints on an island arc sequence associated with the Cretaceous Caribbean Plateau: Bonaire, Dutch Antilles. *Lithos*, **74**, 91–116.

VANDECAR, J. C., RUSSO, R., JAMES, D. E., AMBEH, W. B. & FRANKE, M. 2003. Aseismic continuation of the Lesser Antilles slab beneath continental South America. *Journal of Geophysical Research*, **108**; doi: 10.1029/2001JB000884.

VILLENEUVE, M., CORNEE, J.-J. *ET AL.* 1994. Upper Triassic shallow-water limestones in the Sinta Ridge (Banda Sea, Indonesia). *Geo-Marine Letters*, **14**, 29–35.

WHITE, R. V., TARNEY, J. *ET AL.* 1999. The geochemistry and petrogenesis of the Late Cretaceous picrites and basalts of Curaçoa, Netherlands Antilles. *Contributions in Mineralogy and Petrology*, **124**, 29–43.

WIDIYANTORO, S. & VAN DER HILST, R. D. 1996, Structure and evolution of the lithospheric slab beneath the Sunda Arc, Indonesia. *Science*, **271**, 1566–1570.

Present-day strain field on the South American slab underneath the Sandwich Plate (Southern Atlantic Ocean): a kinematic model

J. L. GINER-ROBLES[1]*, R. PÉREZ-LÓPEZ[2], M. A. RODRÍGUEZ-PASCUA[2], J. J. MARTÍNEZ-DÍAZ[3] & J. M. GONZÁLEZ-CASADO[†]

[1]*Departamento de Geología y Geoquímica, Facultad de Ciencias, Universidad Autónoma de Madrid, Madrid 28049, Spain*

[2]*Departamento de Investigación y Prospectiva Geocientífica, Instituto Geológico y Minero de España (IGME), C/Alenza, nº 1, Madrid 28003, Spain*

[3]*Departamento de Geodinámica, Facultad de Ciencias Geológicas, Universidad Complutense de Madrid, C/José A. Novais s/n, Madrid 28040, Spain*

[†]*Deceased*

Corresponding author (e-mail: jorge.giner@uam.es)

Abstract: This work analyses the present-day principal strain orientation on the downgoing slab of the South America Plate (SAM) beneath the Sandwich Plate (SAND). The strain regime was deduced from the study of 331 earthquake focal mechanism solutions examined by fault population analysis methods. In the slab, the maximum horizontal shortening direction (ey) rotates in trend in a clockwise direction from NE in the north, to SE in the south. Based on this rotation, three different areas were defined according to the prevailing focal mechanism type: (1) the North Zone, with ey oriented N058°E and reverse and strike–slip focal mechanisms; (2) the Central Zone, with only reverse focal mechanisms and ey striking N080°E; and (3) the South Zone, with ey oriented N106°E and reverse and strike–slip focal mechanisms. The strain field in the North Zone of the SAND involves decoupling of the slab at approximately 70 km depth. In contrast, the South Zone edge slab exhibits no decoupling and it exhibits different geometry (hook-like shaped) from the North Zone. Finally, we define the dextral strike–slip component acting at the South Sandwich Fracture Zone (SSFZ), according to focal mechanism solutions and the regional tectonic configuration.

The geometry and kinematics of a plate tectonic boundary can be defined using geophysical and tectonic data such as earthquake focal locations and focal mechanism characteristics. Earthquake distribution shows the approximate position of the tectonic boundary if it constitutes a single line (e.g. Bird 2003). The orientations of the principal strain axis can be deduced from the focal mechanism solutions of earthquakes occurring on the plate boundary (e.g. Capote *et al.* 1991; Giner-Robles *et al.* 2003a).

This study analysed 331 earthquake focal mechanism solutions, located within the South Sandwich region (2005 Harvard Catalogue, HCMT). Fault population analysis methods have been applied to these focal mechanisms in order to establish principal strain orientations. Their spatial locations allow definition of the three-dimensional (3-D) variations of the strain field within the slab. The results from the strain field and stress tensors are discussed according to the regional tectonic setting, and with variations of the subduction style along the slab. These results are influenced by the tectonic

context: two triple junctions near the south slab edge (FFT and FFR) and one in the north edge (FRT) (Fig. 1). Processes and features detected at the underthrusting lithosphere are different age of the slab edges at the north and south (Eagles *et al.* 2005), lithosphere warming and thermal anomalies (Cruciani *et al.* 2005). Anomalous mantle inflow in both areas has been also suggested (Livermore *et al.* 1997; Leat *et al.* 2004).

Tectonic setting

The South America (SAM), Antarctica (ANT) and Scotia (SCO) plates surround the Sandwich Plate (SAND) (Fig. 1), which is a microplate basically composed of oceanic lithosphere aged 8–10 Ma (Barker & Hill 1981; Larter *et al.* 2003). The SAND shows an almost perfect semi-elliptical geometry with its semi-major axis orientated north–south. This axis coincides with a divergent tectonic boundary (East Scotia Ridge), whilst the eastern ellipse's perimeter is a convergent boundary at the

From: JAMES, K. H., LORENTE, M. A. & PINDELL, J. L. (eds) *The Origin and Evolution of the Caribbean Plate.* Geological Society, London, Special Publications, **328**, 155–167.
DOI: 10.1144/SP328.6   0305-8719/09/$15.00 © The Geological Society of London 2009.

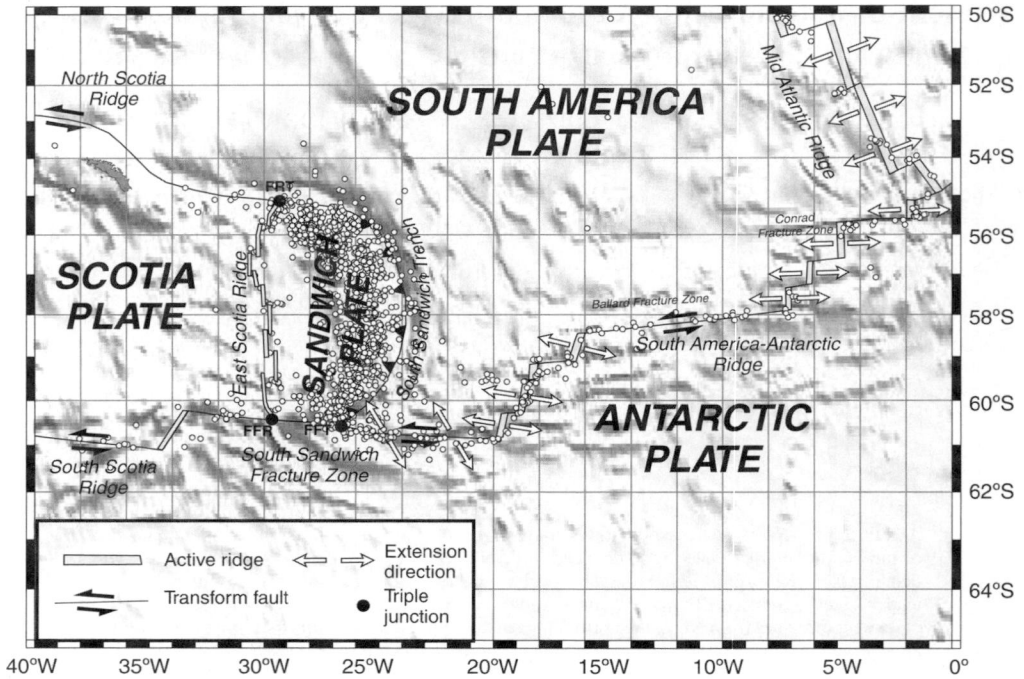

Fig. 1. Geographic location of the Sandwich Plate (SAND) earthquakes recorded within the area (IRIS catalog) and surrounding major plates South America (SAM), Scotia (SCO) and Antarctic (ANT) plates.

South Sandwich Trench. The South Sandwich Fracture Zone (SSFZ), a short east–west trending transform fault, connects the trench to the ridge in the south.

The South Sandwich Islands denote a present-day magmatic arc related to the subduction of SAM oceanic lithosphere beneath the SAND. Subduction began at approximately 83 Ma in the north part and 27 Ma in the south part (Leat *et al.* 2004). Convergence velocity between the SAM and the SAND ranges from 67 mm/annum in the north to 79 mm/annum in the south (Thomas *et al.* 2003).

Bathymetric data and the absence of magnetic anomalies (Bruguier & Livermore 2001) together with the existence of a mid-ocean ridge basalt (MORB-like) oceanic crust (Fretzdorff *et al.* 2002) suggest that back-arc basin (East Scotia Ridge) spreading started recently. The direction of mantle deformation is parallel to the trench (Muller 2001). Consequently, mantle flow has the same orientation. The northern and southern tips of the back-arc spreading centre are inflated as a result of the inflow of mantle into the back-arc region around the edges of both slabs (Livermore *et al.* 1997; Leat *et al.* 2000; Bruguier & Livermore 2001).

Methods and rationale

The methodology applied in this paper to reconstruct the present-day strain field is based on two analyses:

(1) First, we established the spatial and geographical distribution of regional seismicity in the area (epicentral and depth foci distribution) using data from the IRIS catalogue (http://dmc.iris.washington.edu), which includes more than 2000 earthquakes between 1967 and 2005 (magnitude greater than 4). The IRIS catalogue indicates an RMS error of 5 km. However, this error is not relevant in relation to the type of analysis proposed here, because the cluster analysis combines areas of hundred kilometres' width.

(2) Second, in spite of poor seismic station coverage within the studied area, 331 earthquakes from the HCMT were analysed. Earthquakes with focal mechanism solutions from HCMT database (http://www.globalcmt.org/CMTsearch.html) provide a standard error of the centroid latitude of ±5 km. This value is not important at the scale resolution of our study. Transverse cross-sections are *c.* 250 km depth and the longitude of the

South Sandwich trench is 800 km, approximately. The mechanisms of HCMT database agree with a media mechanism correlation coefficient of 0.9 (Helffrich 1997), which is very high quality to perform strain/stress analyses.

Fault population methods have been applied to these focal mechanism solutions; two distinct approaches have been used:

- *Strain analysis* – slip method (Reches 1983) and right dihedral method (Angelier & Mechler 1977).
- *Stress analysis* – stress inversion technique (Reches 1987; Reches *et al.* 1992).

Right dihedral (RD) is a qualitative method based on the stereographic projection of areas with similar first motions, dilatation or contraction (right dihedral diagram). Given a set of focal mechanisms, the right dihedral diagram indicates the common areas for all the focal mechanism solutions analysed. Principal strains must be located inside these common areas; thus the maximum horizontal shortening axis (ey) and minimum horizontal shortening axis (ex) could be established. The principal advantage of RD is that a priori knowledge of which of the two nodal planes corresponds to the auxiliary plane and which to the fault plane is not necessary. Furthermore, the 3-D location of the compressional and extensional areas defines the tectonic regime in the zone associated with the analysed focal mechanism solutions (Capote *et al.* 1991).

The slip model technique (SM) allows us to identify which of the two nodal planes corresponds with the fault plane. This method is based on the Reches fracture model (Reches 1983) and it is explained in detail in Capote *et al.* (1991). With the fault orientation data and using the method described by Capote *et al.* (1991) the ey orientation can be calculated for each analysed focal mechanism.

The stress inversion technique (SI) allows determination of the orientation and shape of the stress ellipsoid that best fits focal mechanism solutions. The method is explained in detail by Herraiz *et al.* (2000) and Giner-Robles *et al.* (2003b). In this analysis we can estimate the orientation of the S_{Hmax} (maximum horizontal stress direction). The SI technique applied to the fault plane deduced from the strain analysis produced high-quality results.

Application of these methods (RD, SM and SI) allowed accurate determination of the position and orientation of the strain and stress axes from focal mechanism solutions. The similarity between the orientation of ey and S_{Hmax} is relevant because it indicates a strong correlation between both fields (coaxial strain and stress) and thus the strain trajectories can be plotted. Consequently, it was possible to determine the 3-D tectonic regime within the area and along the plate boundaries (De Vicente *et al.* 1996; González-Casado *et al.* 2000; Giner-Robles *et al.* 2003a, among others).

Finally, cluster analysis determined areas of homogeneous strain. This shows the 3-D spatial distribution of localized strain field in a regional tectonic context. The strain fields determined from RD and SM can be grouped to indicate the tectonic setting and characteristic features of the area.

Analysis of the earthquake spatial distribution

Earthquakes with magnitude (m_b, M_s or M_w) less than 4 or with a null value and which are also independent of the earthquake agency (NEIC, NEIS, etc.) were removed from the complete IRIS Earthquake Catalogue (2005). The remaining 2000 earthquakes had a maximum magnitude of 7.5 and maximum depth of almost 390 km. Figure 2 shows the spatial hypocentre distribution around the SAND grouped by depth intervals ($\leq$30; 30–70; 70–150; 150–300; and >300 km). Dashed lines (Fig. 2) indicate the upper limit of the subducting slab and the slope angles in the north and south. In the north the subducting slab loses its arcuate form below 70 km (Fig. 2c & d). In the south slab curvature increases with depth (Fig. 2c–e) and the southern limb migrates northwards, giving the slab a hook-like geometry (Barker 1995; Leat *et al.* 2004). The central part shows a slightly shallower angle (Fig. 2f).

Six cross-sections perpendicular to the arc trench (Figs 2f & 3) illustrate the depth variation of earthquake foci. Dips increase at 70–80 km depth (Brett 1977) and show the slab to be curved and consequently extended on its western face. Most earthquake foci plot shallower than 70 km, but a deep cluster between 70 and 150 km appears in the two northern cross-sections, profiles 1–1' and 2–2' (Fig. 3).

At its northern and southern margins, the South Sandwich Trench curves westward to meet the two transform plate boundaries of the North and South Scotia ridges (Fig. 1). The north slab edge corresponds with a FRT triple junction (Fig. 1). Earthquakes with strike–slip focal mechanism solutions characterize these boundaries (Fig. 4). They are triggered by a strain field with maximum horizontal shortening oriented N45°E (Giner-Robles *et al.* 2003b). In the south corner of the SAND there are two triple junctions: (1) FFT, where the South Sandwich Trench ends at the eastern prolongation of the SSFZ (Figs 1 & 4) and focal mechanisms of normal and strike–slip earthquakes occur (Fig. 4); and (2) FFR, defined by the junction between the East

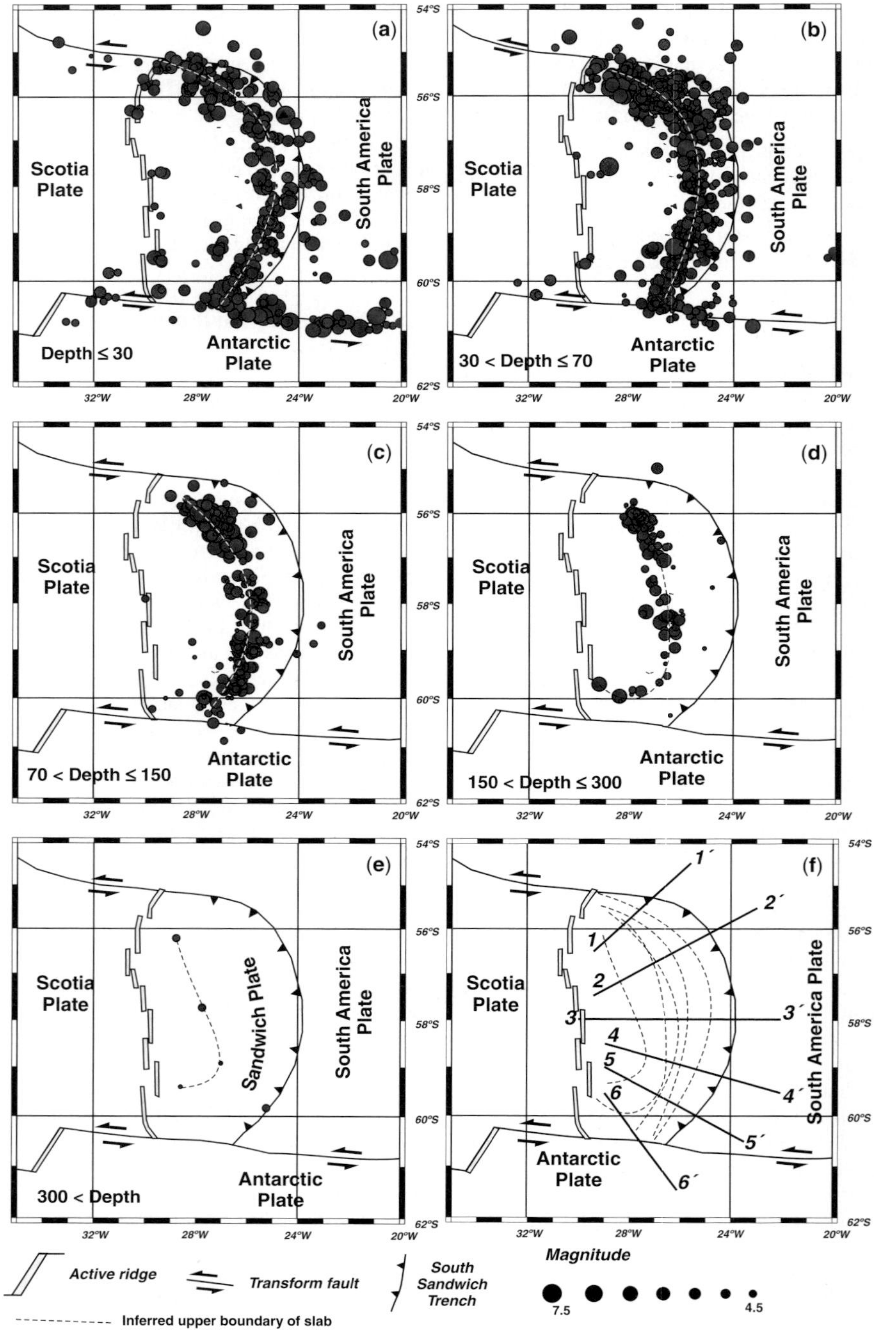

Fig. 2. Spatial distribution of earthquakes by depth: (**a**) 0 ≤ depth ≤ 30 km; (**b**) 30 < depth ≤ 70 km; (**c**) 70 km < depth ≤ 150 km; (**d**) 150 km < depth ≤ 300 km; (**e**) 300 km < depth; (**f**) Location of the cross-section plotted perpendicular to the South Sandwich Trench (see Fig. 3).

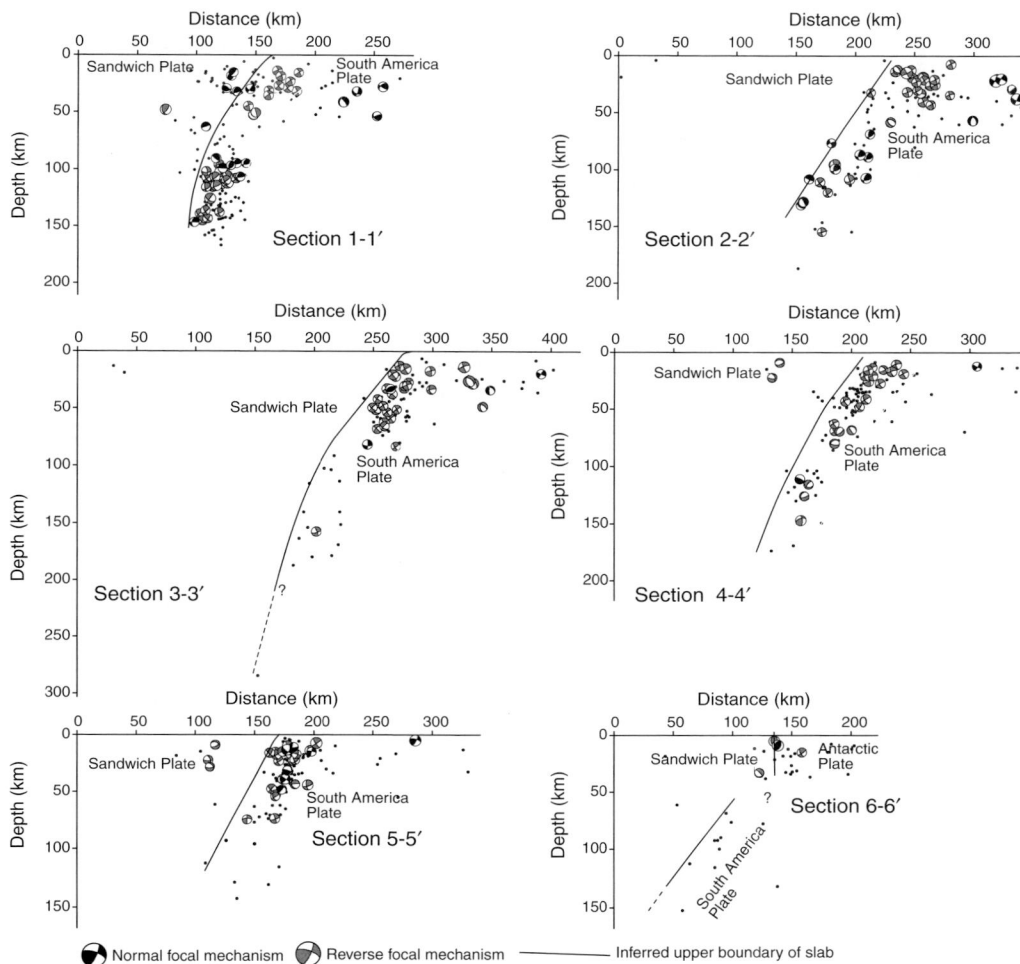

Fig. 3. Cross-sections (located in Fig. 2f) showing the vertical distribution of seismicity with focal mechanism solutions. Black shows normal focal mechanism solutions and grey shows reverse focal mechanisms (Harvard CMT Catalogue). Earthquakes located 50 km each side of the profile were projected onto it. Cross-sections of focal mechanisms were plotted with GMT.

Scotia Ridge, the South Scotia Ridge and the SSFZ (Figs 1 & 4).

Strain and stress analyses

Three-hundred and thirty-one focal mechanism solutions (HCMT) located beneath the SAND were grouped into three sets according to their dynamics (reverse, normal and strike–slip) and their location in the subduction area: north, central and south zones (Fig. 4).

Most earthquakes in the central region exhibit reverse fault focal mechanisms. In the north and south both reverse and strike–slip focal mechanism solutions prevail (Fig. 5). Earthquakes in the north and south reveal similar proportions of reverse and strike–slip focal mechanism solutions and, thus, deduced fault types (Figs 5a & c).

Northern earthquakes cluster into two sets: shallower and deeper than 70 km. The shallow earthquakes are similar to those of the south (Fig. 5), both recording shallow plate subduction (<70 km).

The geometry of the southern border of the southern zone is not clear. Both reverse and strike–slip focal mechanisms occur along the SAM–ANT plate boundary, east of the South Sandwich Trench (Fig. 4). These focal mechanisms are included in the study because their dynamic features (nodal

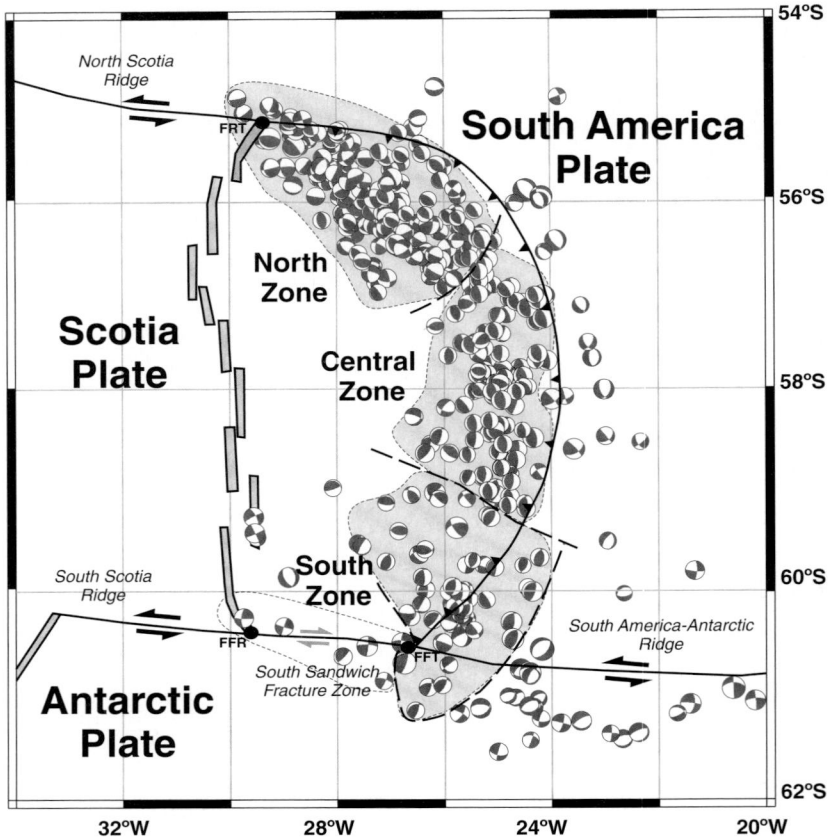

Fig. 4. Spatial location of focal mechanism solutions and proposed zonation: North Zone, Central Zone and South Zone. For details see text.

plane orientations and character) correspond to the faults and mechanisms expected within the area, although they are not related to subduction.

Regional analysis of the seismicity

North Zone analysis

A total of 173 focal mechanism solutions (HCMT) are located within this area. Twenty-one per cent (36) correspond to movements on normal faults; 62% (108) to reverse faults and 17% (29) to strike–slip movements (Fig. 5a).

Strain analysis indicates that 66% of normal faults are oblique. Twenty per cent of the strike–slip faults are dextral and 80% are sinistral. Fifty-two per cent of reverse faults are oblique slip. The inferred orientation of the ey is N58°E.

Stress analysis indicates S_{HMAX} oriented N55°E, with a shape factor $[R = (\sigma_2 - \sigma_3)/(\sigma_1 - \sigma_3)]$ of 0.15 and a stress tensor defined by reverse and oblique reverse faults (Fig. 6).

The depth distribution of earthquakes suggests the following: shallow $(d < 70\ km)$, transitional $(70 \geq d \geq 150\ km)$ and deep $(d > 150\ km)$ zones (Figs 3 & 7, cross-sections 1–1′ and 2–2′). The shallow zone is characterized by focal mechanisms in which the fault plane solution shows low dip (<35°) and oriented N145°E, parallel to South Sandwich Trench. The orientations of S_{HMAX} and ey (Fig. 7a) deduced from these faults are N60°E and N56°E, respectively.

Fault planes related to focal mechanisms in the deep region show north–south trends and steeper dips than in the shallow region. The calculated stress tensor indicates S_{HMAX} oriented N49°E and

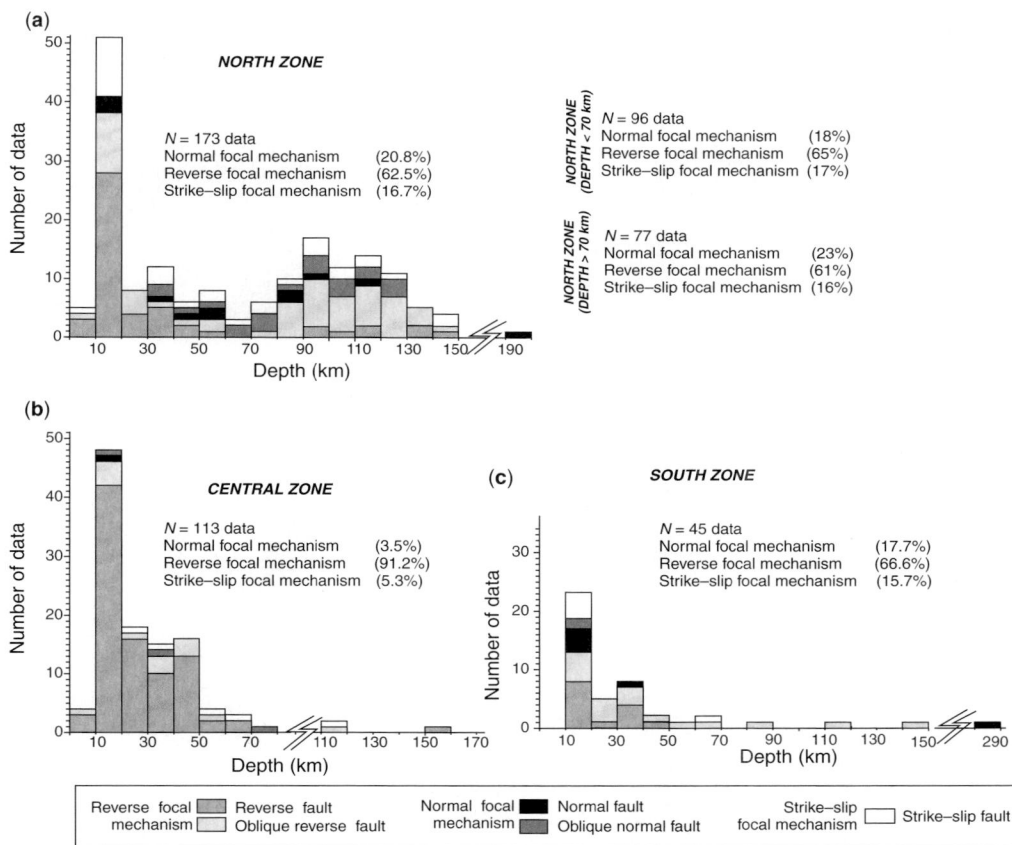

Fig. 5. Histograms of earthquakes showing variations of focal mechanism solutions with depth: (**a**) Northern Zone; (**b**) Central Zone; (**c**) South Zone. For details see text.

shows an important strike–slip component, greater than the shallow data (Figs 5 & 7c) and ey N43°E. This deep area is well defined in the northwestern part of the South Sandwich Trench (Fig. 7, section 1–1′) and disappears towards the southeast (Fig. 7, section 2–2′).

The transition zone defined in this work involves focal mechanisms that indicate east–west trending normal faults with S_{HMAX} oriented N91°E (Fig. 7b) and ey N96°E. This zone exhibits a western cluster of earthquakes at around 80 km depth, representing the inflexion point of the slab (Fig. 7, section 1–1′). This area expands downwards and southeastwards, coinciding with the disappearance of the deep cluster of earthquakes (Fig. 7, section 2–2′). The transition zone could well be in a folded part of the slab, with mechanisms showing extension in the outside of the bend. Reverse focal

mechanisms appear clustered both in the shallow and deep areas of the slab (Fig. 7).

Central Zone analysis

The 113 earthquakes of this zone could be grouped in: reverse focal mechanisms 91% (103), normal 3% and strike–slip 5% (Fig. 5b). Strain analysis indicates that 85% of reverse mechanisms correspond to simple reverse faults while the rest are oblique reverse faults. The trend of S_{HMAX} is N75°E, similar to the orientation of ey, N80°E (Fig. 6).

South Zone analysis

There are only 45 earthquakes with focal mechanism solutions in this region. Sixty-seven per cent are reverse, 18% show normal movement and 15%

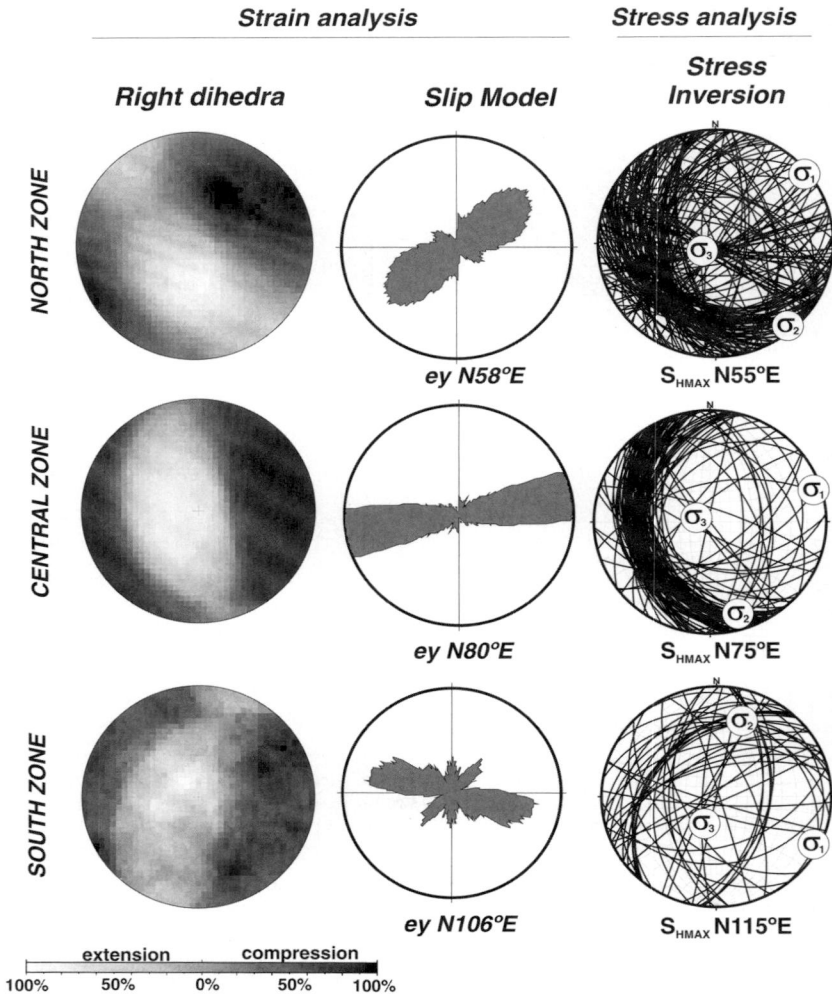

Fig. 6. Comprehensive analyses of focal mechanism solutions by each defined area: North Zone 173 data; Central Zone 113 data, South Zone 45 data. Left, right dihedral diagram; central column, Rose diagram of the horizontal maximum shortening directions (ey) obtained with the slip model analysis; right; stereographic projection of the fault plane solutions and principal axes of the stress tensor ($\sigma_1 > \sigma_2 > \sigma_3$) obtained with the stress inversion method.

strike–slip (Fig. 5c). Strain analysis shows that 53% of the reverse mechanisms are oblique reverse faults, 60% of strike–slip movements are dextral faults and 40% are sinistral.

The results are more scattered than those observed in the other zones (shallow and transition areas). S_{HMAX} is oriented N115°E and is defined by reverse and oblique faults, as in the northern zone (Fig. 6), ey is oriented N106°E. However, the rose diagram of horizontal maximum shortening (ey, Fig. 6), reveals two secondary directions: north–south (normal and oblique normal faults) and

N45°E (oblique reverse and strike–slip faults). The latter can be related to strain generated by the South America–Antarctic Ridge.

Discussion

The South Sandwich Trench exhibits a very complex mixture of normal, reverse and strike–slip focal mechanism solutions. The strain/stress analysis presented here identifies seismotectonically homogeneous areas and considers their significance in the regional tectonic setting.

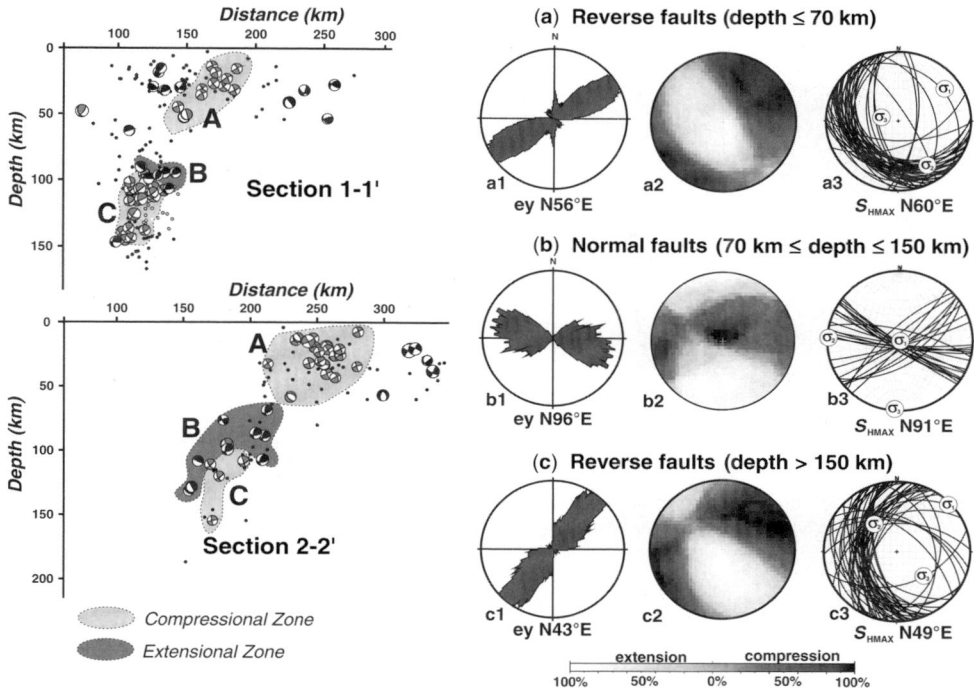

Fig. 7. Analysis of the focal mechanism solutions by depth for the North Zone, cross-sections 1−1′ and 2−2′ (see Figs 2b & 4). Classification of areas by depth: (**a**) shallow area, (**b**) transition zone; and (**c**) deep zone. Results of the fault population analysis are shown in: (1) right dihedral; (2) Rose diagram of maximum horizontal shortening directions (ey) from slip model method; and (3) stereographic projection of the fault plane solutions and principal axis of the stress tensor ($\sigma_1 > \sigma_2 > \sigma_3$), obtained from the stress Inversion method.

The trend of maximum horizontal shortening, ey, deduced principally from reverse faults, changes from N58°E in the north to N106°E in the south (Fig. 8), remaining constantly perpendicular to the subduction arc. Normal faults on the SAM close to the trench indicate bending of the subducting slab (Fig. 8), with normal faults trending parallel to the trench (Figs 5 and 8).

The North and South zones show strain tensors with shear component instead the pure contraction of the Central Zone (Fig. 6). The percentage of normal, reverse and strike–slip faults deduced for the northern and southern zones are similar (Fig. 5). This could be explained by:

(1) Strike–slip structures related to the connection between the South Sandwich Trench with the North Scotia Ridge, South Scotia Ridge and SSFZ.
(2) Suggested mantle inflows (Livermore *et al.* 1997; Leat *et al.* 2004) (Figs 8 & 9).

Nevertheless, the two zones differ in that North Zone earthquakes with magnitude greater than 5

(Mw) form two clusters, one located at *c.* 30 km depth and the other at *c.* 70 km depth (Figs 3 & 7b). In the southern zone there is only one shallow earthquake cluster between 10 and 50 km depth (Figs 3 & 7b). The boundary between shallow and deep seismicity is well-defined in the north (Fig. 7, section 1−1′) but becomes diffuse towards the Central Zone (Fig. 7, section 2−2′), where it disappears.

The deep earthquakes of the North Zone are related to reverse and oblique-reverse faults (zone C, Figs 7c & 9). The ey direction related to these focal mechanisms trends more northerly than the ey direction of shallower (<70 km) mechanisms. This is interpreted to result from a lateral push of the mantle as it flows around the slab edge from the east, as suggested by Livermore *et al.* (1997) and Leat *et al.* (2004) (Fig. 9).

The analyses performed here (Figs 2, 7 & 8) suggest different geometry and kinematics of the South American slab subduction to those suggested by Livermore *et al.* (1997) and Leat *et al.* (2004). Those authors proposed the North and South zones

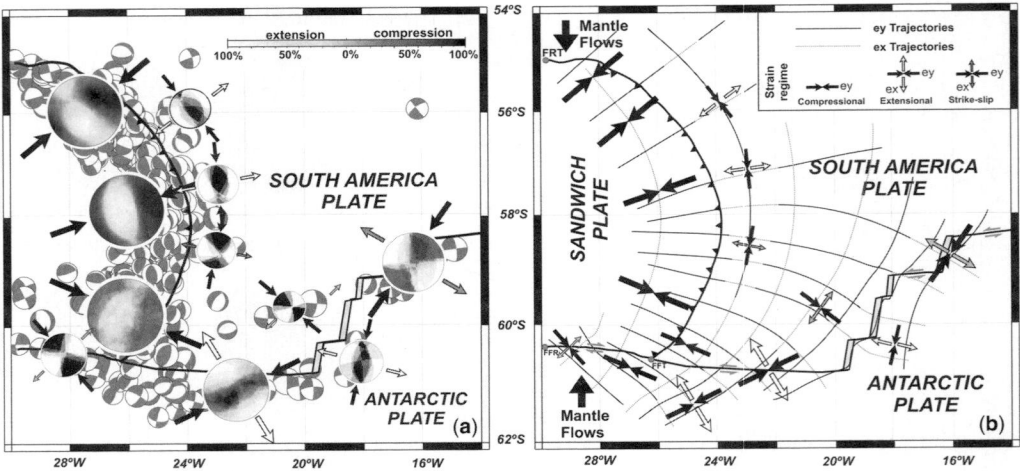

Fig. 8. (a) Kinematics scheme. Individual focal mechanism solutions and right dihedral diagrams, by zones, are shown. (b) Trajectories of principal strain deduced from the analysis of focal mechanisms.

to be quite alike, with a Wadati–Benioff plane whose slope increases with depth. However, the focal mechanisms analysis of this paper suggest that two different processes are at work. We suggest that subduction decoupling occurs at approximately 70 km in the north, in response to the strain regime of a triple junction (FRT) with its transform arms among the North Scotia Ridge, the northern part of the South Sandwich Trench and the northern end of the East Scotia Ridge (Fig. 9 zone B).

Decoupling does not occur in the South Zone. Instead, the western part of the slab dips more to the north as depth increases, with resulting increasing slab curvature (Leat *et al.* 2004) (Figs 2f & 9 zone D). This is seen as a result of interplay between the South Sandwich Trench, the South Sandwich Fracture Zone, the South America–Antarctica Ridge and the East Scotia Ridge at two triple junctions (FFT and FFR, Figs 4, 9 & 10). The SSFZ links the East Scotia Ridge to the South Sandwich Trench via dextral movement driven by spreading at the East Scotia Ridge. This is opposite to spreading at the South America–Antarctic Ridge (Figs 1 & 9). Figure 9 (zone H) suggests that a sinistral transform fault, parallel to the SSFZ, could accommodate the opposed spreading directions.

The proposed structures explain shallow reverse focal mechanism solutions in the South Zone and in neighbouring parts of the Antarctica and South America plates (Figs 5 & 9). Strong N135°E extension along the South America–Antarctic Ridge near the South Sandwich Trench occurs on normal and transtensional sinistral strike–slip faults (Fig. 8).

The proposed strain field configuration (NW–SE maximum shortening direction, ey) at the

Sandwich–South America–Antarctica triple junction (FFT) (Figs 8 & 9, zone H) is constrained by the following processes: (1) subduction at the South Sandwich Trench (Fig. 9, zone D); (2) extension from the South America–Antarctic Ridge (Fig. 9, zone G); and (3) compression derived from the SSFZ (Fig. 9, zone H). This means that the triple junction FFT is deforming as result of the compression from the South America–Antarctic Ridge, from the South Sandwich Trench and from the transpressive SSFZ.

Summation of these processes in shallow areas (<70 km) results in greater convergence between the SAND and SAM in the South Zone (79 mm/annum) than in the North Zone (67 mm/annum), where only one process (NE–SW subduction) is acting (Fig. 10). In the northern part of the trench subducting oceanic crust is 56–84 Ma old (Müller *et al.* 1997) and the convergence velocity is close to *c.* 67 mm/annum. In the south the subducting oceanic lithosphere is 0–20 Ma old (Müller *et al.* 1997) and the velocity of convergence is higher (*c.* 79 mm/annum) (Fig. 10). This might indicate that the brittle thickness of subducting lithosphere is larger in the north than in the south.

Tectonic complexity is higher in the South Zone than in the North Zone. In the south there are four plates (SAM, SCO, SAND and ANT) and two triple junctions (FFR and FFT); in the north there are only three plates (SAM, SCO and SAND) and one triple junction (FRT) (Fig. 10). We propose that this difference explains strong–moderate size deep earthquakes ($d > 150$ km; $M > 5$) with decoupling at 70 km depth of the slab in the north while the south slab shows moderate size deep

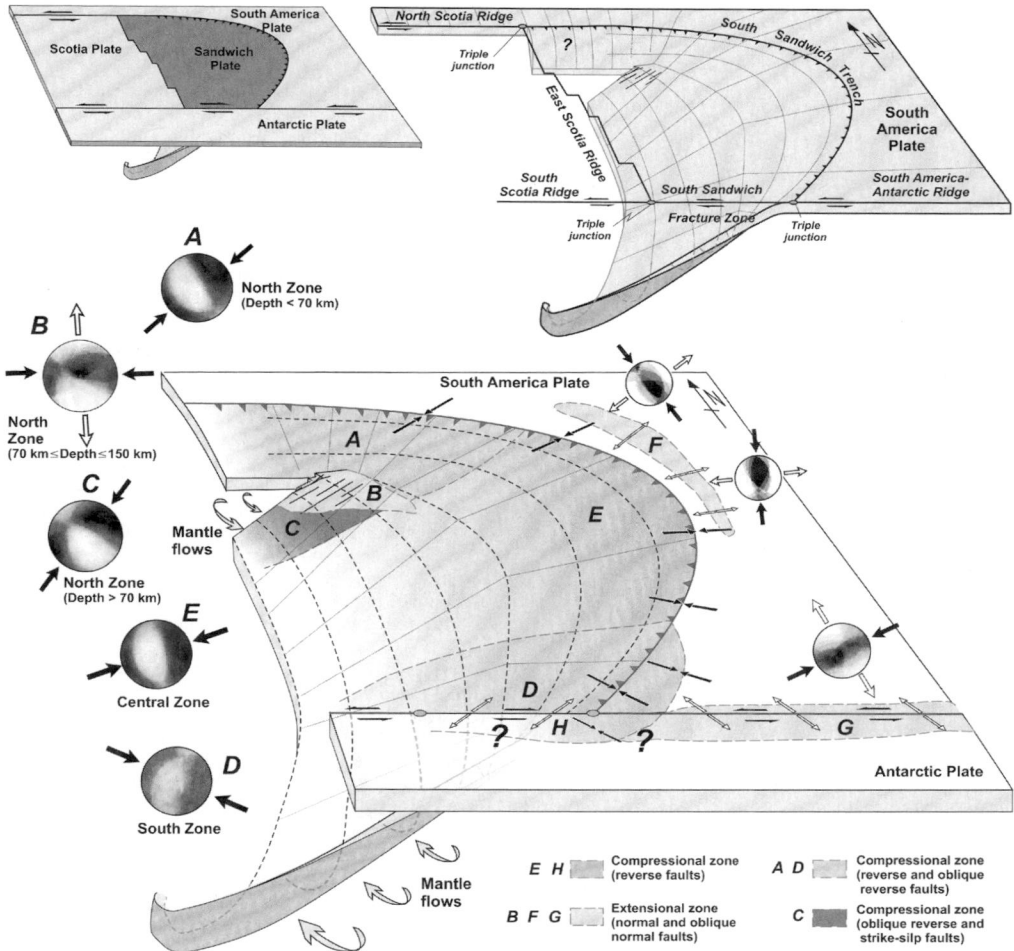

Fig. 9. Schematic sketch of the South Sandwich subduction system. Areas in colour indicate the tectonic process affecting compression, extension and strike–slip. Black arrows indicate the orientation of ey and white arrows indicate the orientation of ex (after Leat *et al.* 2004).

earthquakes ($d > 150$ km; $M < 5$) and slab curvature (hook-shaped).

In both areas strain is affected by lateral mantle inflows. Inflow is greater in the north than in the south (Leat *et al.* 2004) and results in ey trend variation with depth (Fig. 7, zone C).

Conclusions

The orientation of the maximum horizontal shortening direction (ey) along the South American subducting slab rotates from NE in the North Zone to east–west in the Central Zone and SE in the South Zone of the South Sandwich Trench.

In the North and South zones strike–slip and reverse focal mechanism solutions dominate, whereas the Central Zone exhibits reverse focal mechanisms. Strike–slip and reverse focal mechanism solutions are related to: (a) transformation of subduction at the trench into lateral movement along the North Scotia Ridge and along the South Scotia Ridge; and (b) mantle inflows around the ends of the slab.

The North Zone exhibits different geometry from the South Zone. An abrupt change in the slab geometry and increase in slope occurs close to 70 km depth, where slab decoupling occurs. In contrast, the south edge shows a gradual curvature in depth of the downgoing slab. In addition

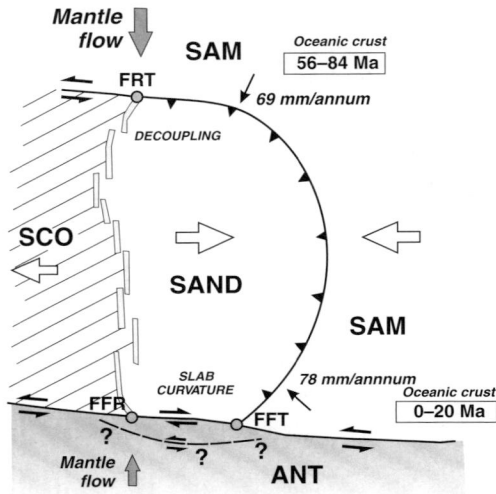

Fig. 10. Tectonic configuration proposed in plan-view from the strain analysis: South America (SAM), Sandwich (SAND), Scotia (SCO) and Antarctic (ANT) plates. White arrows indicate relative movement of plates (ANT is fixed). Size of the grey arrows indicates that in the North Zone is greater the mantle inflows. The age of the underthrusting lithosphere (SAM) is extracted from Müller *et al.* (1997).

subduction turns progressively northward, resulting in a hook-shaped slab.

The geometric slab differences between the North and South could be related to (Fig. 10):

- Different tectonic complexity: the North Zone is characterized by interaction between three plates (South America, Sandwich and Scotia) and the South Zone by interaction between four plates (South America, Sandwich, Scotia and Antarctica).
- Age of the downgoing slab – greater in the North Zone (56–84 Ma) than in the South Zone (0–20 Ma).
- Greater mantle inflow in the North Zone.
- Different convergence rates: 67 mm/annum in the north and 79 mm/annum in the south.

We define a dextral strike–slip regime along the South Sandwich Fracture Zone in agreement with the focal mechanism solutions. This corresponds with the transform segment located between the southern tip of the South Scotia Ridge (FFR junction) and the southern tip of the South Sandwich Trench (FFT junction).

A possible new tectonic feature (a sinistral strike–slip fault) could explain the presence of reverse focal mechanism solutions southward of

the South America–Antarctic boundary near to FFT triple junction.

Thanks are given to K. H. James, P. Baker and G. Eagles for their kind and constructive reviews and good suggestions to improve the final manuscript. The Spanish Interministerial Commission of Science and Education (MEC) provided financial support for this work; research projects FALLADEC (CGL2005-24148-E/ANT) and TECTO2 (CGL2006-28134-E/CLI). We would like to thank Brian Crilly for his linguistic assistance. Sadly, Professor José M. González-Casado died in 2008, and thus this paper represents his last thoughts on the topic. Some of the suggested changes during the review were discussed to preserve Professor Gonzalez-Casado's thoughts.

References

ANGELIER, J. & MECHLER, P. 1977. Sur une méthode graphique de recherche des contraintes principales également utilisables en tectonique et en sismologie: la méthode des dièdres droits. *Bulletin de la Société Géologique de la France*, **9**, 1309–1318.

BARKER, P. F. 1995. Tectonic framework of the east Scotia Sea. *In:* TAYLOR, B. (ed.) *Backarc Basins. Tectonics and Magmatism.* Plenum Press, New York, 281–314.

BARKER, P. F. & HILL, I. A. 1981. Back-arc extension in the Scotia Sea. *Philosophical Transactions of the Royal Society, London Series A*, **300**, 249–262.

BIRD, P. 2003. An updated digital model of plate boundaries, *Geochemistry, Geophysics, Geosystems*, **4**, 1027; doi: 10.1029/2001GC000252.

BRETT, C. P. 1977. Seismicity of the South Sandwich Islands region. *Geophysics Journal of the Royal Astronomical Society*, **5**, 453–464.

BRUGUIER, N. J. & LIVERMORE, R. A. 2001. Enhanced magma supply at the southern East Scotia Ridge: evidence for mantle flow around the subducting slab? *Earth and Planetary Science Letters*, **191**, 129–144.

CAPOTE, R., DE VICENTE, G. & GONZÁLEZ-CASADO, J. M. 1991. An application of the slip model of brittle deformations to focal mechanism analysis in three different plate tectonic situations. *Tectonophysics*, **191**, 339–409.

CRUCIANI, C., CARMINATI, E. & DOGLIONI, C. 2005. Slab dip vs. Lithosphere age: no direct function. *Earth and Planetary Science Letters*, **238**, 298–310.

DE VICENTE, G., GINER-ROBLES, J. L., MUÑOZ-MARTÍN, A., GONZÁLEZ-CASADO, J. M. & VEGAS, R. 1996. Determination of present-day stress tensor and neotectonic interval in the Spanish Central System and Madrid Basin, central Spain. *Tectonophysics*, **266**, 405–424.

EAGLES, G., LIVERMORE, R. A., DEREK-FAIRHEAD, J. & MORRIS, P. 2005. Tectonic evolution of the west Scotia Sea. *Journal of Geophysical Research*, **110**, 1–19.

FRETZDORFF, S., LIVERMORE, R. A., DEVEY, C. W., LEAT, P. T. & STOFFERS, P. 2002. Petrogenesis of the back-arc East Scotia Ridge, South Atlantic Ocean. *Journal of Petrology*, **43**, 1435–1467.

GINER-ROBLES, J. L., GONZÁLEZ-CASADO, J. M., GUMIEL, P. & GARCÍA-CUEVAS, C. 2003a. Strain trajectories in three different plate tectonic margins from earthquake focal mechanism. *Tectonophysics*, **372**, 179–191.

GINER-ROBLES, J. L., GONZÁLEZ-CASADO, J. M., GUMIEL, P., MARTÍN VELÁZQUEZ, S. & GARCÍA-CUEVAS, C. 2003b. Kinematics model for the Scotia Plate (SW Atlantic Ocean). *Journal of South American Earth Sciences*, **16**, 179–191.

GONZÁLEZ-CASADO, J. M., GINER-ROBLES, J. L. & LÓPEZ-MÁRTINEZ, J. 2000. The Bransfield Basin, Antarctic Peninsula: not a 'normal' back-arc basin. *Geology*, **28**, 1043–1046.

HELFFRICH, G. 1997. How good are routinely determined focal mechanisms? Empirical statistics based on a comparison of Harvard, USGS and ERI moment tensors. *Geophysics Journal International*, **131**, 741–750.

HERRAIZ, M., DE VICENTE, G. *ET AL.* 2000. A new perspective about the recent (Upper Miocene to Quaternary) and present tectonic stress distributions in the Iberian Peninsula. *Tectonics*, **19**, 4, 762–786.

LARTER, R. D., VANNESTE, L. E., MORRIS, P. & SMYTH, D. K. 2003. Tectonic evolution and structure of the South Sandwich arc. *In*: LARTER, R. D. & LEAT, P. T. (eds) *Intra-oceanic Subduction Systems. Tectonic and Magmatic Processes*. Geological Society, London, Special Publications, **219**, 255–284.

LEAT, P. T., LIVERMORE, R. A., MILLAR, I. L. & PEARCE, J. A. 2000. Magma supply in back-arc spreading centre segment E2, East Scotia Ridge. *Journal of Petrology*, **41**, 845–866.

LEAT, P. T., PEARCE, J. A., BARKER, P. F., MILLAR, I. L., BARRY, T. L. & LARTER, R. D. 2004. Magma genesis and mantle flow at a subducting slab edge: the South Sandwich arc-basin system. *Earth and Planetary Science Letters*, **227**, 17–35.

LIVERMORE, R. A., CUNNINGHAM, A., VANNESTE, L. & LARTER, R. D. 1997. Subduction influence on magma supply at the East Scotia Ridge. *Earth and Planetary Science Letters*, **150**, 261–275.

MULLER, C. 2001. Upper mantle seismic anisotropy beneath Antarctica and the Scotia Sea region. *Geophysics Journal International*, **147**, 105–122.

MÜLLER, R. D., ROEST, W. R., ROYER, J. Y., GAHAGAN, L. M. & SCLATER, J. G. 1997. Digital isochrons of the world's ocean floor. *Journal of Geophysical Research*, **102**, 3211–3214.

RECHES, Z. 1983. Faulting of rocks in three-dimensional strain fields, II. Theoretical analysis. *Tectonophysics*, **95**, 133–156.

RECHES, Z., 1987. Determination of the tectonic stress tensor from slip along faults that obey the Coulomb yield condition. *Tectonics*, **7**, 849–861.

RECHES, Z., BAER, G. & HATZOR, Y. 1992. Constraints on the strength of the upper crust from stress inversion of fault slip measurements. *Journal of Geophysical Research*, **97**, 12,481–12,493.

THOMAS, C., LIVERMORE, R. A. & POLLITZ, F. F. 2003. Motion of the Scotia Sea plates. *Geophysics Journal International*, **155**, 789–804.

Synchronous 29–19 Ma arc hiatus, exhumation and subduction of forearc in southwestern Mexico

J. DUNCAN KEPPIE*, DANTE J. MORÁN-ZENTENO, BARBARA MARTINY & ENRIQUE GONZÁLEZ-TORRES

Instituto de Geología, Universidad Nacional Autónoma de México, 04510 México D.F., Mexico

Corresponding author (e-mail: duncan@servidor.unam.mx)

Abstract: The geology of southwestern Mexico (102–96°W) records several synchronous events in the Late Oligocene–Early Miocene (29–19 Ma): (1) a hiatus in arc magmatism; (2) removal of a wide (*c.* 210 km) Upper Eocene–Lower Oligocene forearc; (3) exhumation of 13–20 km of Upper Eocene–Lower Oligocene arc along the present day coast; and (4) breakup of the Farallon Plate. Events 2 and 3 have traditionally been related to eastward displacement of the Chortís Block from a position off southwestern Mexico between 105°W and 97°W; however at 30 Ma the Chortís Block would have lain east of 95°W. We suggest that the magmatic hiatus was caused by subduction of the forearc, which replaced the mantle wedge by relatively cool crust. Assuming that the subducted block separated along the forearc–arc boundary, a likely zone of weakness due to magmatism, the subducted forearc is estimated to be wedge-shaped varying from zero to *c.* 90 km in thickness; however such a wedge is not apparent in seismic data across central Mexico. Given the 121 km/Ma convergence rate between 20 and 10 Ma and 67 km/Ma since 10 Ma, it is probable that any forearc has been deeply subducted. Potential causes for subduction of the forearc include collision of an oceanic plateau with the trench, and a change in plate kinematics synchronous with breakup of the Farallon Plate and initiation of the Guadalupe–Nazca spreading ridge.

Based upon the truncated character of the present southwestern margin of Mexico and space and time constraints related to Caribbean Plate reconstructions, interpretation of the Cenozoic history of southern Mexico has been dominated by the hypothesis that the Chortís Block (mainly Honduras and northern Nicaragua) lay adjacent to southwestern Mexico in the Paleocene moving along the Motagua fault zone to its present position between 45 Ma and the present (e.g. Ross & Scotese 1988; Pindell *et al.* 1988; Schaaf *et al.* 1995; Meschede *et al.* 1997). However, Keppie & Morán-Zenteno (2005) proposed that, in the Paleocene, the Chortís Block lay SW of its present position, rotating clockwise about an average pole in the southern hemisphere near Santiago, Chile (Pindell *et al.* 1988), and was bounded on its NW side by transform faults bordering the Cayman Trough. This latter reconstruction is supported by the undeformed Upper Cretaceous–Recent sequence in the Gulf of Tehuantepec that sits astride any westward projection of the Motagua fault zone (Sánchez-Barreda 1981; Keppie & Morán-Zenteno 2005), the lack of significant movement on the Motagua Fault zone (measured displacements vary from zero to <200 km), and the presence of a continuous Upper Eocene–Lower Oligocene arc parallel to the present southwestern margin of Mexico (Fig. 1). One consequence of the latter reconstruction is that

subduction of the Farallon Plate along a WNW-trending trench parallel to the present Acapulco Trench, but located farther to the south, is responsible for the Upper Eocene–Lower Oligocene arc. In this paper we investigate the Upper Eocene–Recent geological record of southwestern Mexico in order to relate geological events to potential plate tectonic mechanisms. Recent reviews by Morán-Zenteno *et al.* (2007), Goméz-Tuena *et al.* (2007) and Nieto-Samaniego *et al.* (2006) allow us to limit the paper to the main points.

Cenozoic geological record of southwestern Mexico

Arc magmatism and reconstructions

Post-80 Ma magmatism in southern Mexico may be divided into two spatial and temporal belts: a *c.* 80–29 Ma belt in the Sierra Madre del Sur parallel to the present coast, and the 19 Ma to Present Trans-Mexican Volcanic Belt running from the Pacific coast near Puerto Vallarta to the Gulf of Mexico (Figs 1 & 2). An exception to this general distribution is the 22–13 Ma arc magmatic rocks located in eastern Oaxaca State, east of what we call the Veracruz-Oaxaca Line. Early geochronological results from the Sierra Madre de Sur suggested that the Late Cretaceous-Oligocene arc

From: JAMES, K. H., LORENTE, M. A. & PINDELL, J. L. (eds) *The Origin and Evolution of the Caribbean Plate.*
Geological Society, London, Special Publications, **328**, 169–179.
DOI: 10.1144/SP328.7   0305-8719/09/$15.00 © The Geological Society of London 2009.

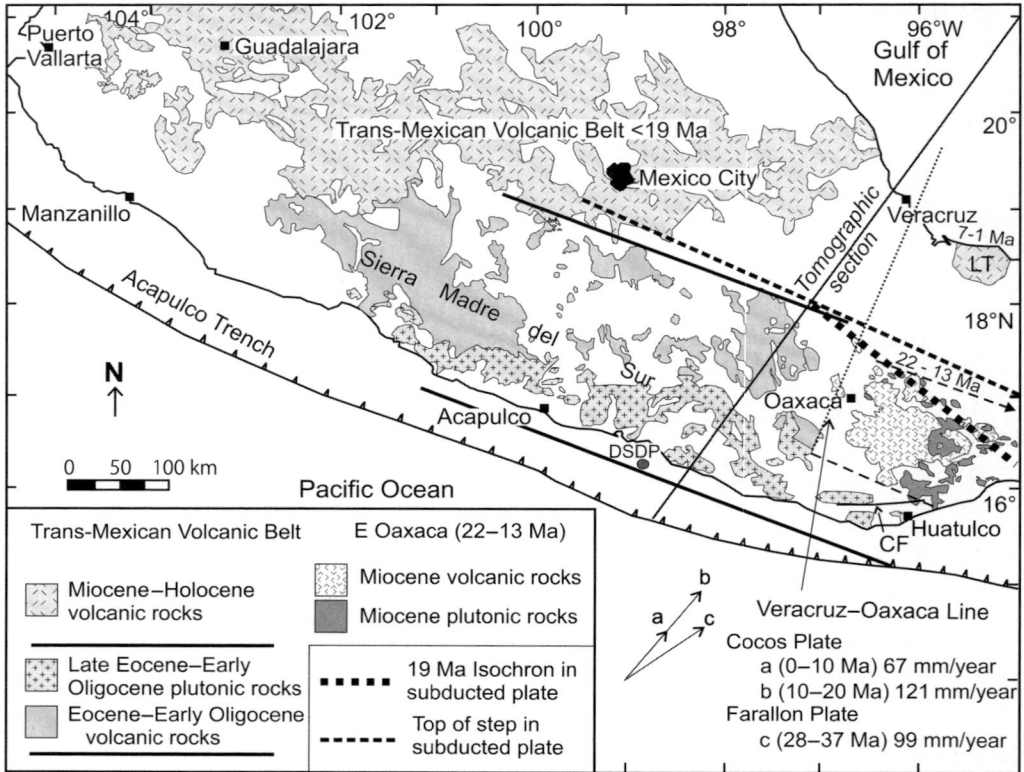

Fig. 1. Geological map of southern Mexico (modified after Morán-Zenteno *et al.* 2007). LT, Los Tuxtlas; CF, Chacalapa Fault.

magmatism migrated from west to east between *c.* 80 and 29 Ma. The part between 45 and 29 Ma was interpreted as result of the passage of the trench–trench–transform triple junction that accompanied the southeastward migration of the Chortís Block (Herrmann *et al.* 1994; Schaaf *et al.* 1995). However, the more comprehensive geochronological database now available (Morán-Zenteno *et al.* 2007 and references therein, Western North America volcanic and intrusive rock database: http://navdat.geongrid.org/) shows a *c.* 220 km wide coast-parallel 38–29 Ma (Upper Eocene–Lower Oligocene) magmatic arc between longitudes 101 and 97°. Between 97 and 96° this arc is represented by a narrow band of coastal plutons; however, undated andesites stratigraphically below Miocene rocks could be the continuation of the Lower Oligocene magmatic belt to the west (Fig. 1). Along-strike to the east, the arc is replaced by an undeformed Upper Cretaceous to Recent sedimentary basin in the Gulf of Tehuantepec with no evidence of Upper Eocene–Lower Oligocene lavas or plutons: the contact in reflection seismic

sections appears to be a series of high-angle faults (Sánchez-Barreda 1981; Keppie and Morán-Zenteno 2005). Between 100°30'W and 96°W, U–Pb granitoid ages migrate from *c.* 35 to 29 Ma, that is a rate of *c.* 75 km/Ma (Fig. 1), much faster than the *c.* 20–30 km/Ma rate deduced in the Cayman Trough (Rosencrantz *et al.* 1988; Ross & Scotese 1988).

Using the empirically derived negative correlation between the angle of subduction and the widths of the arc and forearc derived by Tatsumi & Eggins (1995), the dip of the subduction zone and the forearc width during the Upper Eocene–Lower Oligocene are estimated to have been $11 \pm 9°$ and 280 ± 30 km, respectively (Figs 3 & 4a). These estimates are complicated by factors, such as: (i) variations in the dip of the Benioff zone for example a steep dip near the trench could have changed to a shallow dip under the arc, resulting in a reduction in the width of the forearc; (ii) the original southern margin of the arc may have lain farther south and been removed by subduction erosion; and (iii) neogene extension of the Sierra

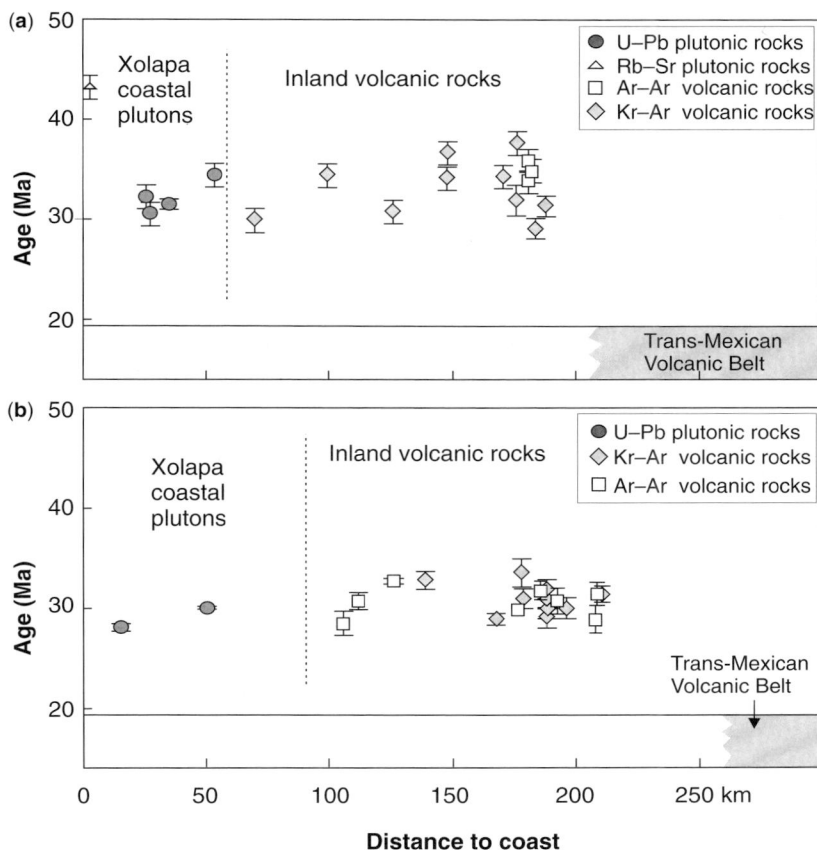

Fig. 2. Geochronology in two NNE-trending transects across southwestern Mexico, from (**a**) Acapulco (age data from Herrmann *et al.* 1994; Ducea *et al.* 2004; Hernández-Pineda 2006) and (**b**) Pinotepa Nacional (age data from Ferrusquía-Villafranca *et al.* 1974; Herrmann *et al.* 1994; Galina-Hidalgo *et al.* 2003; Martiny *et al.* 2000; Cerca *et al.* 2007; Martiny unpublished data).

Madre del Sur may have increased the width of the arc. As these factors cannot presently be quantified, the 280 ± 30 km width for the forearc is used in this paper. In this scenario, the trench would have been located 210 ± 30 km south of the trench in the Late Eocene–Early Oligocene, allowing for 50 km between the southernmost dated arc rocks in borehole DSDP Leg 66, Site 493 (Bellon *et al.* 1982; Fig. 1) and the present trench, and removal of *c*. 20 km width between 19 Ma and the Present; see below (Fig. 4a).

Geochronological data from south-central Mexico (Morán-Zenteno *et al.* 2000, 2007; Gómez-Tuena *et al.* 2007 and references therein) indicates a hiatus in arc magmatism between extinction of the arc magmatism in the central Sierra Madre del Sur at *c*. 29 and the initiation of volcanic activity in the southern Trans-Mexican Volcanic Belt at

c. 19 Ma (Figs 1 & 4c). East of the Veracruz–Oaxaca Line, 22–10 Ma arc magmatism continued closer to the coast in the eastern Oaxaca and Chiapas states, jumping northwards to Los Tuxtlas at *c*. 7 Ma (Figs 1 & 4c). The geochemistry of volcanic rocks from the Trans-Mexican Volcanic Belt includes not only typical calc-alkaline rocks, but also <11 Ma oceanic island basalts, and 12–10 Ma adakites in the eastern Trans-Mexican Volcanic Belt with a few 3.5 Ma adakites near Puerto Vallarta and Mexico City (Gómez-Tuena *et al.* 2007). The Trans-Mexican Volcanic Belt arc is generally *c*. 150 km wide whereas the forearc increases eastwards (*c*. 150–410 km), the latter suggesting a west-to-east variation in the dip of Benioff zone from 50 ± 10° to subhorizontal (Figs 1 & 3; cf. Pardo & Súarez 1995). The reconstruction for the lower Miocene to Recent period

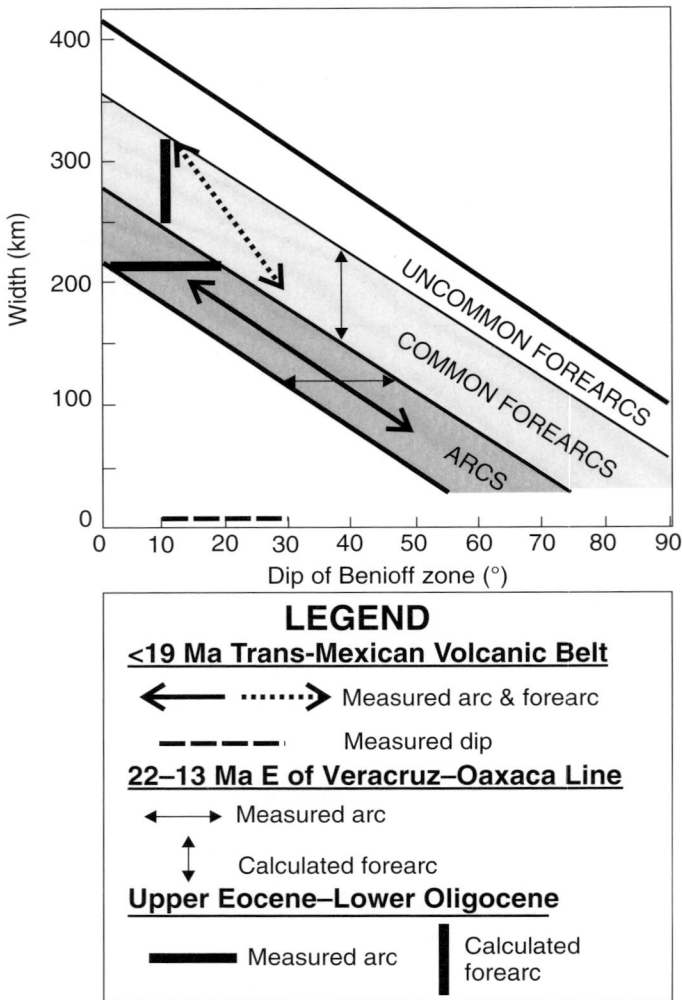

Fig. 3. Width of forearcs and arcs versus dip of Benioff zone (modified from Tatsumi & Eggins 1995) showing data for southwestern Mexico.

(Fig. 4c) has to take into account the *c.* 23 km of subduction erosion that has taken place since 23 Ma (Clift & Vannucchi 2004).

East of the Veracruz–Oaxaca Line, a *c.* 120 km wide belt of Miocene arc magmatism was located along-strike of the Upper Eocene–Lower Oligocene arc and migrated from west to east between 22 and 10 Ma (Fig. 1). Assuming no loss of arc to subduction erosion, the dip of the Benioff zone may have been as high as *c.* 40° with an arc–trench gap of 190 ± 40 km in width (Fig. 3). Adding a distance of *c.* 50 km between the southernmost outcrops of arc rocks at the coast and the present trench to the assumed 23 km of subduction erosion since 23 Ma

(Clift & Vannucchi 2004) places the lower Miocene trench *c.* 117 ± 40 km south of the present Acapulco Trench, that is offset *c.* 93 km south of the position of the lower Miocene Acapulco Trench west of the Veracruz–Oaxaca Line (Fig. 4c). If this offset is real, then the Veracruz–Oaxaca Line may represent a transform fault in the subducting plate. Across the Veracruz–Oaxaca Line, the Benioff zone dip changed from low-angle on the western side to ≤40° on the eastern side. After *c.* 7 Ma the alkalic-calc-alkaline volcanism of the Trans-Mexican Volcanic Belt extended east of the Veracruz–Oaxaca Line to Los Tuxtlas and the Chiapanecan volcanic arc (Figs 1 & 4c; Damon &

(a)

(b)

(c)

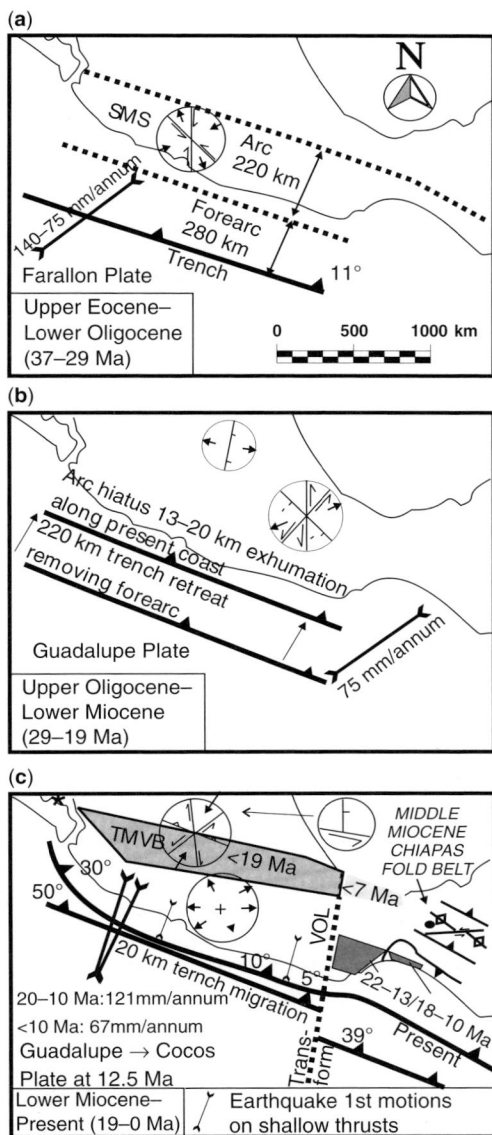

Fig. 4. Cenozoic reconstructions showing the locations of the trench, forearc, and arc and structures at various times. *Cocos Plate pole of rotation from 12.5 Ma to Present. TMVB, Trans-Mexican Volcanic Belt.

Montesinos 1978; Gómez-Tuena *et al.* 2007), suggesting a relative eastward migration of the transform.

Structure

In the Sierra Madre del Sur, the Laramide Orogeny was followed by Upper Eocene–Lower Oligocene

conjugate strike–slip faulting (sinistral WNW–ESE to east–west faults and dextral north–south faults), under an ENE–WSW horizontal shortening direction (Fig. 4a; Alaniz-Álvarez *et al.* 2002; Nieto-Samaniego *et al.* 2006). This was followed by sinistral, 27–23 Ma displacements on east–west faults and river offsets near the coast and along the Chacalapa Fault (Fig. 1; Tolson 2005), whereas farther north horizontal NNE–SSW to east–west extension produced sinistral movements on north–south and NE–SW faults, and normal fault movements on NW–SE faults that post-date the Upper Oligocene (Fig. 4b; Nieto-Samaniego *et al.* 2006): the difference suggests strain partitioning across the Chacalapa Fault.

Exhumation and subsidence

Using the Al–hornblende igneous barometer, Morán-Zenteno *et al.* (1996) have indicated that the uppermost Eocene–lowermost Oligocene coastal plutons were emplaced at depths of 20–13 km, with the amount of exhumation decreasing northwards. (U–Th)/He thermochronology indicates that these plutons were rapidly exhumed before *c.* 25 Ma north of Acapulco and before *c.* 17 Ma north of Puerto Escondido. This was followed by slow exhumation of the coastal zone of the central Sierra Madre del Sur with 85% of the derived sediments being recycled via subduction erosion (Ducea *et al.* 2004). The *c.* 4 km subsidence in the outer forearc indicates *c.* 1 km/Ma retreat of the trench (Clift & Vannucchi 2004) over the last 23 Ma.

Geophysical data

The dip of the Benioff zone south of the Trans-Mexican Volcanic Belt has been shown to vary from 30° in the west to nearly zero in the east (Pardo & Suárez 1995). Preliminary analysis of recent seismic data from Acapulco through Mexico City to Tampico, although generally confirming this geometry, indicates a 15°N dip northwards from the trench to *c.* 100 km, becoming almost subhorizontal between 100 and 275 km at a depth of *c.* 45 km, and dipping at 20°N beyond 275 km to Tampico (Kim *et al.* 2006; Husker & Davis 2006). East of *c.* 95.5°W, the Cocos Plate bends downwards to the east and its dip changes from gentle to moderate (40°NNE) (Bravo *et al.* 2004): the change has been related to the projection of the Tehuantepec Transform beneath the North American Plate (Keppie & Morán-Zenteno 2005). *P*-wave tomography across western Mexico shows a similar geometry with a moderately dipping slab down to *c.* 400 km; between 400 and 600 km the dip is gentler, and below *c.* 600 km there is a considerable gap

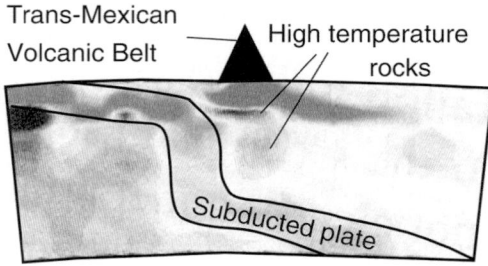

Fig. 5. Tomographic section across southern Mexico showing subducted plate (see Fig. 1 for location). Modified after Gorbatov & Fukao (2005).

before reaching a deeper part of the subducted slab (Gorbatov & Fukao 2005). However, across eastern Mexico, *P*-wave tomography suggests a vertical step down to the north in the Benioff zone that increases eastwards from zero at *c.* 99°W through *c.* 200 km in height at 97°W along the Veracruz section (Figs 1 & 5) to *c.* 350 km in height at 92°W in the section across Chiapas. This step connects a shallow southerly segment from the Acapulco Trench to a deep northerly segment (Figs 1 & 5). Along the Veracruz cross-section, the age of the subducted slab at the top of the step is estimated to be *c.* 19 Ma, increasing gradually towards the east (Fig. 1).

Magnetotelluric data along the Acapulco–Mexico City–Tampico section show bright anomalies near the trench that have been related to serpentinization (Jödicke *et al.* 2006). The lack of anomalies in the flat slab segment suggest that fluids are absent, whereas those beneath and to the Trans-Mexican Volcanic Belt may be related to dehydration of the subducted slab (Jödicke *et al.* 2006). Along the Veracruz–Oaxaca–Puerto Escondido line, the magnetotelluric data indicate several discrete low resistivity anomalies within 150 km of the Pacific coast that have been related to release of water at depths of *c.* 20 and *c.* 40 km (Jödicke *et al.* 2006).

Interpretations and conclusions

The geological record of southwestern Mexico (102–96°W) reveals several major changes during the Upper Oligocene–Lower Miocene (Fig. 6): (i) there was an hiatus in arc magmatism, which separates a WNW-trending, 220 km wide, Upper Eocene–Lower Oligocene arc parallel to the present coast from a more northerly, west-trending Miocene–Present Trans-Mexican Volcanic Belt arc: 22–10 Ma arc magmatism east of the Veracruz–Oaxaca Line lies along-strike of the earlier arc; (ii) 13–20 km of exhumation occurred along the

present coastal part of the Upper Eocene–Lower Oligocene arc; (iii) the strain regime north of the Xolapa Complex associated with conjugate strike–slip faults changed from shortening to extension along a WSW–ENE direction; (iv) the dip of the Benioff changed from a uniform *c.* 11° dip in the Upper Eocene–Lower Oligocene west of the Veracruz–Oaxaca Line to an asymmetric dip in the Miocene from 30 to 50° in the western Trans-Mexican Volcanic Belt to <10° in the eastern Trans-Mexican Volcanic Belt, and ≤40° east of the Veracruz–Oaxaca Line; (v) the Farallon Plate broke into several smaller plates at *c.* 29–25 Ma (Mammerickx & Klitgord 1982); and (vi) *c.* 210 km width of the Upper Eocene–Lower Oligocene forearc was removed.

A 29–19 Ma arc hiatus and subduction erosion

A gradual change in the dip of the Benioff zone through time beneath southwestern Mexico could explain most of the arc rotation. However, the 10 Ma hiatus in arc magmatism requires another explanation, such as: (i) subduction ceased; (ii) the dip of the Benioff zone changed rapidly; and/or (iii) the forearc was subducted. Cessation of subduction is unlikely because subduction of the Farallon and Guadalupe plates appears to have been continuous (Mammerickx & Klitgord 1982), as is confirmed by the tomographic continuity of the subducted plates beneath eastern Mexico. A change in the dip of the Benioff zone should be accompanied by migration of the arc, and this is not observed (Fig. 2). On the other hand, subduction of the forearc may have caused the 10 Ma hiatus by replacing the mantle wedge beneath the arc by relatively cold material. Assuming that the subducted block separated along the boundary between the forearc and the arc, a likely zone of weakness due to magmatism, the subducted forearc is estimated to be wedge-shaped, varying in thickness from zero at the trench to *c.* 90 km at the arc, the depth required to initiate arc magmatism. Such an underthrust wedge-shaped slice of continental margin material varying in thickness from zero to >50 km has been recorded in southern British Columbia (Monger & Price 2002): that it has not been far removed by subduction may be due to the nearly margin-parallel relative motion between the Juan de Fuca and North American plates (Engebretson *et al.* 1985). A similar, but smaller, slice occurs in northern Cascadia where a 20 × 100 km thrust-bounded slice of crustal rocks is being transported downwards by aseismic slow slip (Calvert 2003). The apparent absence of such slices beneath southern Mexico suggests that they have been deeply subducted. If the Veracruz–Oaxaca Line

(Ma) Time Scale	Magmatism South North	Tectonics
0 Pleistocene Pliocene 5.3 Upper 10 11.6 Miocene Middle 16.0 20 Lower 23.0 Oligocene Upper 28.4 30 Lower 33.9 37.2 Upper Eocene Middle 40	U. Crust & sediments in source Adakite (Mitla) (Veracruz–Oaxaca Line) (TMVB) Magmatic hiatus Coastal belt exhumation = 13–20 km	Benioff zone dip = 30°(W)–10°(E) Arc–trench gap = 200–320 km Subduction erosion of 20 km since 20 Ma Slow cooling >90% sediment subducted Change in Pacific–Cocos plates pole of rotation Benioff zone dip = 50°↔subhorizontal to 39° east of Veracruz–Oaxaca Line (VOL) Breakup of Farallon Plate into Guadalupe & Mitla plates & subduction erosion of 260 km width of forearc SIERRA MADRE DEL SUR (SMS) Arc width = 220 km, Benioff zone dip = 11° Arc–trench gap = c. 280 km

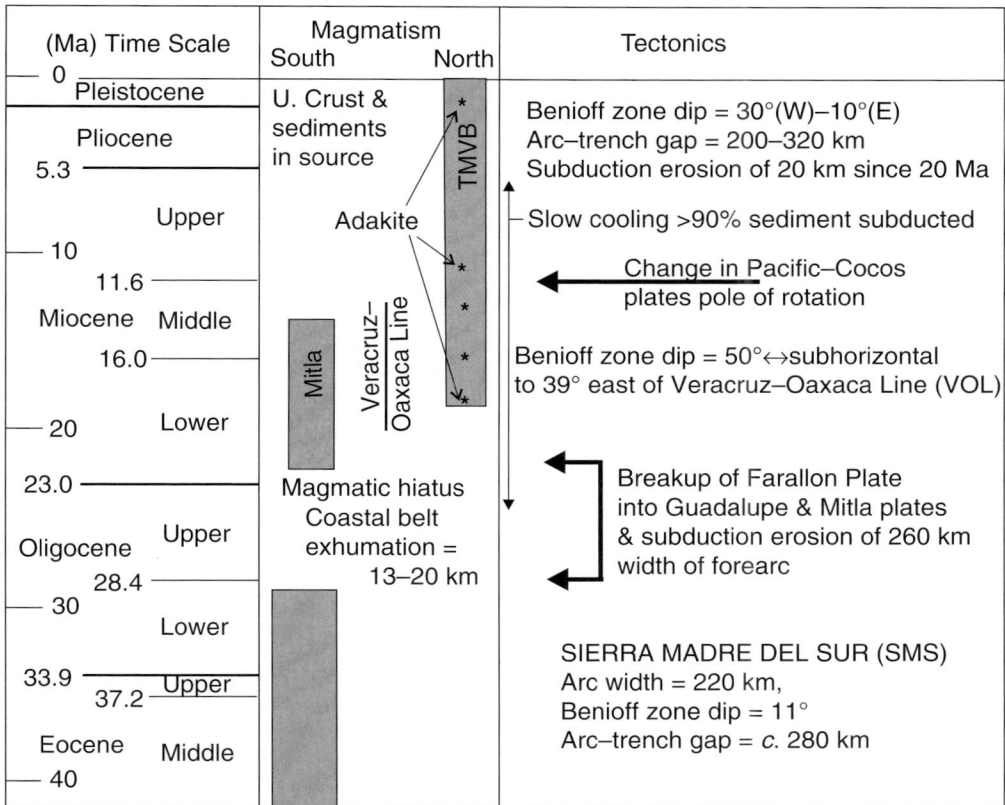

Fig. 6. Time and space diagram showing magmatic and tectonic events during the Cenozoic in southwestern Mexico (41 Ma to Present).

represents a transform fault, across which the dip of the Benioff zone changed, it could explain the age of reinitiation of arc magmatism (at c. 22 Ma east of the Veracruz–Oaxaca Line compared with c. 19 Ma along the Trans-Mexican Volcanic Belt) in terms of the different time of arrival of the trailing edge of the subducted forearc to c. 90 km depth where arc magmatism would generally begin.

Geodynamic modeling of subduction incorporating dehydration of the subducting slab produces two scenarios (Manea & Gurnis 2006): (i) a release of fluids at shallow depths produces serpentinization and a low viscosity forearc mantle wedge, which causes a decrease in the dip of the Benioff zone; and (ii) dehydration at depths up to 400 km causes steepening of the Benioff zone. Thus, subduction of the cold forearc would have initially replaced the corner of the mantle wedge causing a magmatic hiatus. Post-19 Ma gently dipping subduction beneath the eastern part of the Trans-Mexican Volcanic Belt suggests that shallow

dehydration continued behind the subducted forearc. The steeply dipping boundary between cooler, low viscosity serpentinites and the hotter, normal mantle wedge may have coincided with the crustal viscosity contrast at the boundary between the arc and forearc produced by upward magmatic flow. However, continued subduction of the forearc could have increased the depth of dehydration into the mantle wedge causing steepening of the Benioff zone, which perhaps explains the vertical step in the Benioff zone observed in the tomography beneath eastern Mexico (Fig. 5).

Assuming that the 10 my magmatic hiatus records the time of passage of the c. 210 km forearc beneath southern Mexico, the rate of relative motion is c. 21 km/Ma (forearc width/hiatus length). This is much slower than both the pre-30 Ma rate of 99 km/Ma and the 20–10 Ma rate of 121 km/Ma (Engebretson et al. 1985; Pindell et al. 1988; Schaaf et al. 1995). By analogy with Cascadia, this suggests that the forearc was underthrusting by aseismic slip at a slower rate than the subducting slab. Passage of the trailing edge of the subducting forearc would lead to reinitiation of arc magmatism in the Trans-Mexican Volcanic Belt. If the forearc continued subducting at the 21 km/Ma rate after 19 Ma its trailing edge would presently lie 400 km from forearc-arc boundary. In this case, it might have underplated the overriding plate and be revealed in geophysical data as a relatively cool layer. However, such underplating is obscured by recent magmatic activity that produced fluids and melt recorded in both tomographic and magneto-telluric data. On the other hand, once the forearc passed beneath the continental Moho, it may have traveled with the subducting slab at a convergence rate of *c*. 120 km/Ma between 19 and 10 Ma and *c*. 60–70 km/Ma after 10 Ma, which would place its trailing edge at *c*. 1700 km north of the forearc-arc boundary. If the subducted forearc was floored by oceanic lithosphere it would have been converted to eclogite.

Possible causes of 29–19 Ma subduction erosion

What initiated subduction of the forearc is unclear. It does not appear to be related to a change in the rate of convergence, which was 99 mm/annum prior to 28 Ma and increased to 121 mm/annum after 20 Ma (Fig. 1). Nor does it correlate with a discontinuity in the age of the subducting oceanic lithosphere (Mammerickx & Klitgord 1982).

At *c*. 29 Ma, the East Pacific Rise lay *c*. 2000 km from the coast of southwestern Mexico and the subducting Farallon Plate would be *c*. 25 Ma older (= Anomaly 19). Keppie & Morán-Zenteno (2005) proposed that subduction of seamounts (Chumbia seamounts) could have been responsible for the subduction erosion; however, elsewhere in the world seamounts generally only cause an ephemeral rise in the rate of subduction erosion (generally <10 km/Ma: Clift & Vannucchi 2004). Average global rates are <8 km/Ma rising to *c*. 10 km/Ma where mid-oceanic ridges have entered the subduction zone (Clift & Vannucchi 2004). Using estimates calculated earlier in this paper based on the empirical relationship between the dip of the Benioff zone and the width of the arc and forearc suggest that *c*. 210 km forearc width was removed between 29 and 19 Ma, that is an average of *c*. 21 km/Ma. The exceptionally high rate of subduction erosion at 29–19 Ma in southern Mexico may be explained in several ways.

Traditionally, removal of the forearc would be explained by moving the Chortís Block eastwards along the southern coast of Mexico (e.g. Ross & Scotese 1988; Schaaf *et al.* 1995); however, several further problems arise besides those discussed in

Keppie & Morán-Zenteno (2005). Firstly, passage of the Chortís Block through the Gulf of Tehuantepec would have removed and deformed the Upper Cretaceous and Cenozoic basin south of the westward projection of the Motagua Fault zone: this contradicts seismic data that clearly shows a continuous basin astride such a projection (Sánchez-Barreda 1981). Secondly, using the estimated 15 ± 5 mm/annum rate of opening of the Cayman Trough since 30 Ma (Rosencrantz & Sclater 1988) and the *c*. 130 km, post-45 Ma stretching of the Nicaragua Rise (Ross & Scotese 1988), the northwestern tip of the Chortís Block would have lain at 95°W off the Gulf of Tehuantepec at 30 Ma, not further west between 105°W and 97°W off southwestern Mexico (Fig. 7), that is using this model, the Chortís Block would have been removed before removal of the forearc. Thirdly, the western Chortís Block has a triangular shape varying in width from zero to 600 km measured from the present coast of Mexico: thus any Eocene–Oligocene arc would likely have a NNW-strike passing from southern Mexico into the Chortís Block rather than its WNW-trend within southern Mexico (Fig. 1).

Collision of a larger feature, such as an oceanic plateau, is a possibility for producing subduction erosion. Such an oceanic plateau could have lain on the inferred Chumbia seamount chain; however, no mirror-image plateau appears along the Moonless Mountain seamount chain, which Keppie & Morán-Zenteno (2005) deduced to be the mirror-image of the Chumbia seamount chain. On the other hand, an oceanic plateau could have been created just on the east side of the mid-ocean ridge by either plume magmatism or between two temporarily overlapping ridges as has been documented by Mammerickx & Klitgord (1982). The synchroneity of the 29–19 Ma subduction erosion with the 29–25 Ma breakup of the subducting Farallon Plate into the Guadalupe Plate and birth of the Cocos–Nazca spreading centre at 25 Ma suggests a cause and effect relationship. However, it is unclear how such reorganization of the Pacific plates affected the North America Plate.

Exhumation and subduction erosion

Lallemand *et al.* (1994) have shown that whereas overriding of a seamount causes uplift in the outer forearc above the seamount followed by subsidence after its passage, the inner forearc is less affected. By analogy, subduction of a forearc wedge could cause exhumation of the present coastal plutons in southwestern Mexico followed by subsidence of the offshore section. It might also explain the switch from NE–SW contraction to extension farther inland.

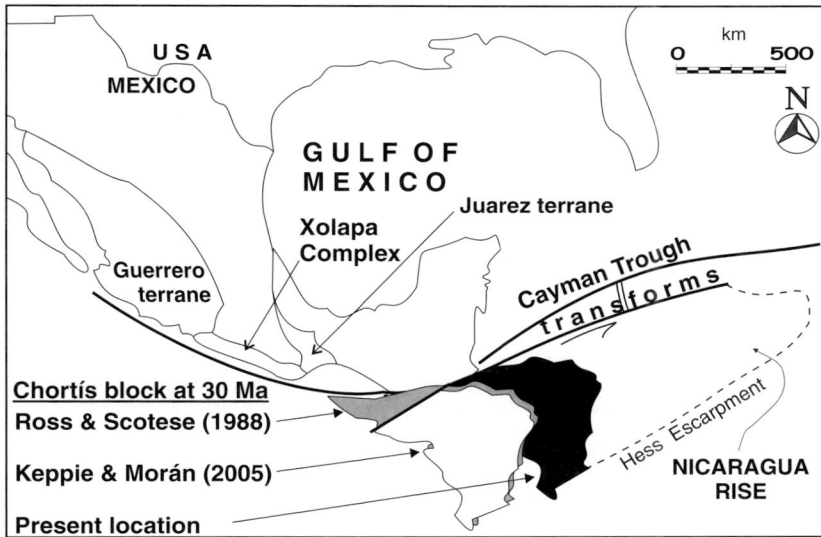

Fig. 7. Reconstructions over 30 Ma showing the position of the Chortís Block according to Keppie & Morán-Zenteno (2005), and Ross & Scotese (1988) including subtraction of the *c*. 130 km stretching in the Nicaragua Rise.

Coupling or decoupling between plates

It has been suggested that coupling between the plates produces parallelism between the direction of convergence and the stress directions in the overriding plate (Tatsumi & Eggins 1995 and references therein). In apparent confirmation of this hypothesis, Meschede *et al.* (1997) related structures in the central Sierra Madre del Sur to stress transmission between subducting and overriding plates, and concluded that stress and strain directions are parallel. However, Nieto-Samaniego *et al.* (2006) show extension axes associated with strike–slip tectonics and convergence directions are slightly oblique (Fig. 4b), which is consistent with the general case where the axis of convergence is not parallel to either strain or stress axes in oblique convergent plate scenarios (Jiang *et al.* 2001). The obliquity could also be due to a divergence between the convective flow pattern in the mantle wedge and the convergence direction between the plates, although an intervening hydrous peridotite/serpentinite layer may severely limit viscous coupling.

Given the high rate of subduction erosion at 29–19 Ma, the absence of a fold-and-thrust belt in southern Mexico appears to be anomalous. Compressional deformation could have been restricted to the removed forearc. However, there is no evidence of such deformation in the present forearc where shallow flat slab of the Cocos Plate is occurring (Kim *et al.* 2006), possibly due to either sediment subduction and/or fluids released from the subducting Cocos Plate that have lubricated the Benioff zone (cf. Bostock *et al.* 2002). On the other hand, Sobolev & Babeyko (2005) have calculated that a 2–3 cm/annum absolute westward motion of the South American Plate combined with the high coefficient of friction (0.05) and crust >40 km are the main factors in producing a fold-and-thrust belt in the Central Andes. Crustal thicknesses and rates of the absolute westward motion of the North and South American plates are similar in the central Andes and Mexico, suggesting that a low friction coefficient may be a significant factor in inhibiting coupling. Furthermore, the Acapulco Trench strikes WNW, highly oblique to the westwards absolute motion of North America, which may also contribute to the lack of a fold-and-thrust belt.

We are grateful to CONACyT grant (0255P-T9506) for funding the project, and thank Drs Luca Ferrari and Albert Bally for their constructive comments on an earlier version of the paper.

References

ALANIZ-ÁLVAREZ, S. A., NIETO-SAMANIEGO, A. F., MORÁN-ZENTENO, D. J. & ALBA-ALDAVE, L. 2002. Rhyolitic volcanism in extension zone associated with strike-slip tectonics in the Taxco region, southern Mexico. *Journal of Volcanology and Geothermal Research*, **118**, 1–14.

BELLON, H., MAURY, R. C. & STEPHAN, J. F. 1982. Dioritic basement, Site 493: petrology, geochemistry, and geodynamics. *Initial Reports of the Deep Sea Drilling Project*, **66**, 723–731.

BOSTOCK, M. G., HYNDMAN, R. D., RONDENAY, S. & PEACOCK, S. M. 2002. An inverted continental Moho and serpentinization of the forearc mantle. *Nature*, **417**, 536–538.

BRAVO, H., REBOLLAR, C. J., URIBE, A. & JIMÉNEZ, O. 2004. Geometry and state of stress of the Wadati-Benioff zone in the Gulf of Tehuatepec, México. *Journal of Geophysical Research*, **109**, B04307; doi: 10.1029/2003JB002854.

CALVERT, A. J. 2003. Seismic reflection imaging of two megathrusts in the northern Cascadia subduction zone. *Nature*, **428**, 163–167.

CERCA, M., FERRARI, L., LÓPEZ-MARTÍNEZ, M., MARTINY, B. & IRIONDO, A. 2007. Late Cretaceous shortening and Early Tertiary shearing in the central Sierra Madre del Sur, southern Mexico: insights into the evolution of the Caribbean–North America Plate interaction. *Tectonics*, **26**, TC3007, doi:10.1029/2006TC001981.

CLIFT, P. & VANNUCCHI, P. 2004. Controls on tectonic accretion versus erosion in subduction zones: implications for the origin and recycling of the continental crust. *Review of Geophysics*, **42**, RG2001.

DAMON, P. E. & MONTESINOS, E. 1978. Late Cenozoic volcanism and metallogenesis over an active Benioff zone in Chiapas, Mexico. *Arizona Geological Society Digest*, **11**, 155–168.

DUCEA, M. N., VALENCIA, V. A. *ET AL.* 2004. Rates of sediment recycling beneath the Acapulco trench: constraints from (U–Th)/He thermochronology. *Journal of Geophysical Research*, **109**, B09404; doi:10.1029/2004JB003112.

ENGEBRETSON, D. C., COX, A. & GORDON, R. G. 1985. *Relative Motions between Oceanic and Continental Plates in the Pacific Basin*. Geological Society of America, Special Papers, **206**.

FERRUSQUÍA-VILLAFRANCA, I., WILSON, J. A., DENISON, R. E., MCDOWELL, F. W. & SOLORIO-MUNGUIA, J. 1974. Tres edades radiométricas oligocénicas y miocénicas de rocas volcánicas de las regiones Mixteca Alta y Valle de Oaxaca, Estado de Oaxaca. *Boletín Asociación Mexicana Geólogos Petroleros*, **26**, 249–262.

GALINA-HIDALGO, S. M., URRUTIA-FUCUGAUCHI, J., RUIZ-CASTELLANOS, M. & TERRELL, D. 2003. K/Ar dating and magnetostratigraphy of Cretaceous and Oligocene igneous rocks from Huajuapan de León-Petlalcingo, Mixteca terrane, Mexico. *99th Annual Meeting, Cordilleran Section*, Puerto Vallarta, Jalisco, Abstracts with Programs, Geological Society of America, 8.

GÓMEZ-TUENA, A., OROZCO-ESQUIVEL, T. & FERRARI, L. 2007. Igneous petrogenesis of the Transmexican Volcanic Belt. *In*: NIETO-SAMANIEGO, Á. F. & ALANIZ-ÁLVAREZ, S. A. (eds) *Geology of México: Celebrating the Centenary of the Geological Society of México*. Geological Society of America Special Papers, **442**, 129–181.

GORBATOV, A. & FUKAO, Y. 2005. Tomographic search for missing link between the ancient Farallon subduction and the present Cocos subduction. *Geophysical Journal International*, **160**, 849–854.

HERNÁNDEZ-PINEDA, G. A. 2006. *Geoquímica y geocronología de granitoides en el área de Tierra Colorada, Guerrero*. Unpublished BSc thesis, Facultad de Ingeniería, Universidad Nacional Autónoma de México.

HERRMANN, U. R., NELSON, B. K. & RATSCHBACHER, L. 1994. The origin of a terrane: U/Pb zircon geochronology and tectonic evolution of the Xolapa complex (southern Mexico). *Tectonics*, **13**, 455–474.

HUSKER, A. L. & DAVIS, P. M. 2006. Seismic tomography of the Cocos Plate. *EOS Transactions of the American Geophysics Union*, **87**, Fall Meeting Supplement, Abstract T13F-06.

JIANG, D., LIN, S. & WILLIAMS, P. F. 2001. Deformation path in high strain zones, with reference to slip partitioning in transpressional plate-boundary regimes. *Journal of Structural Geology*, **23**, 991–1005.

JÖDICKE, H., JORDING, A., FERRARI, L., AZATE, J., MEZGER, K. & RUCKE, L. 2006. Fluid release from the subducted Cocos Plate and partial melting of the crust deduced from magnetotelluric studies in southern Mexico: implications for the generation of volcanism and subduction dynamics. *Journal of Geophysical Research*, **111**, B08102; doi: 10.1029/2005JB003739.

KEPPIE, J. D. & MORÁN-ZENTENO, D. J. 2005. Tectonic implications of alternative Cenozoic reconstructions for southern Mexico and the Chortís Block. *International Geology Review*, **47**, 473–491.

KIM, Y., GREENE, F., ESPEJO, L., PEREZ-CAMPOS, X. & CLAYTON, R. W. 2006. Receiver function analysis of the Middle American subduction zone. *EOS Transactions of the American Geophysics Union*, **87**, Fall Meeting Supplement, Abstract TT53D-1637.

LALLEMAND, S. E., SCHNÜRLE, P. & MALAVIEILLE, J. 1994. Coulomb theory applied to accretionary and non-accretionary wedges: possible causes for tectonic erosion and/or frontal accretion. *Journal of Geophysical Research*, **99**, 12033–12055.

MAMMERICKX, J. & KLITGORD, K. D. 1982. Northern East Pacific Rise: evolution from 25 m.y. B.P. to the Present. *Journal of Geophysical Research*, **87**, 6751–6759.

MANEA, V. C. & GURNIS, M. 2006. Central Mexican subduction zone evolution contolled by a low viscosity mantle wedge. *EOS Transactions of the American Geophysics Union AGU* **87**, Fall Meeting Supplement, Abstract T11B-0437.

MARTINY, B., MARTÍNEZ-SERRANO, R. G., MORÁN-ZENTENO, D. J., MACÍAS-ROMO, C. & AYUSO, R. A. 2000. Stratigraphy, geochemistry and tectonic significance of the Oligocene magmatic rocks of western Oaxaca, southern Mexico. *Tectonophysics*, **318**, 71–98.

MESCHEDE, M., FRISCH, W., HERRMANN, U. R. & RATSCHBACHER, L. 1997. Stress transmission across an active plate boundary: an example from southern Mexico. *Tectonophysics*, **266**, 81–100.

MONGER, J. & PRICE, R. 2002. The Canadian Cordillera: geology and tectonic evolution. *Canadian Society of Economic Geologists Recorder*, February, 1–36.

MORÁN-ZENTENO, D. J., CORONA-CHÁVEZ, P. & TOLSON, G. 1996. Uplift and subduction erosion in southwestern Mexico since the Oligocene: pluton geobarometry constraints. *Earth & Planetary Science Letters*, **141**, 51–65.

MORÁN-ZENTENO, D. J., MARTINY, B. *ET AL.* 2000. Geocronología y características geoquímicas de las rocas magmáticas terciarias de la Sierra Madre del Sur. *Boletín de la Sociedad Geológica Mexicana*, **53**, 27–58.

MORÁN-ZENTENO, D. J., CERCA, M. & KEPPIE, J. D. 2007. The Cenozoic tectonic and magmatic evolution of southwestern México: advances and problems of interpretation. *In*: NIETO-SAMANIEGO, Á. F. & ALANIZ-ÁLVAREZ, S. A. (eds) *Geology of México: Celebrating the Centenary of the Geological Society of México*. Geological Society of America, Special Papers, **422**, 343–357.

NIETO-SAMANIEGO, A. F., ALANIZ-ÁLVAREZ, S. A. *ET AL.* 2006. Latest Cretaceous to Miocene deformation events in the eastern Sierra Madre del Sur, Mexico, inferred from the geometry and age of major structures. *Geological Society of America Bulletin*, **118**, 1868–1882.

PARDO, M. & SUÁREZ, G. 1995. Shape of the subducted Rivera and Cocos Plates in southern Mexico: seismic and tectonic implications. *Journal of Geophysical Research*, **100**, 12357–12373.

PINDELL, J. L., CANDE, S. C. *ET AL.* 1988. A plate kinematic framework for models of Caribbean evolution. *Tectonophysics*, **155**, 121–138.

ROSENCRANTZ, E. & SCLATER, J. G. 1988. Depth and age in the Cayman trough. *Earth and Planetary Science Letters*, **79**, 133–144.

ROSENCRANTZ, E., ROSS, M. I. & SCLATER, J. G. 1988. Age and spreading history of the Cayman Trough as determined from depth, heat flow, and magnetic anomalies. *Journal of Geophysical Research*, **93**, 2141–2157.

ROSS, M. I. & SCOTESE, C. R. 1988. A hierarchical tectonic model of the Gulf of Mexico and Caribbean region. *Tectonophysics*, **155**, 139–168.

SÁNCHEZ-BARREDA, L. A. 1981. *Geologic evolution of the continental margin of the Gulf of Tehuantepec in southern Mexico*. PhD thesis, University of Texas, Austin, TX.

SCHAAF, P., MORÁN-ZENTENO, D., HERNÁNDEZ-BERNAL, M. S., SOLIS-PICHARDO, G., TOLSON, G. & KÖHLER, H. 1995. Paleogene continental margin truncation in southwestern Mexico: geochronological evidence. *Tectonics*, **14**, 1339–1350.

SOBOLEV, S. V. & BABEYKO, A. Y. 2005. What drives orogeny in the Andes? *Geology*, **33**, 617–620.

TATSUMI, Y. & EGGINS, S. 1995. *Subduction Zone Magmatism. Frontiers in Earth Sciences*. Blackwell Science, Massachusetts.

TOLSON, G. 2005. La falla Chacalapa en el sur de Oaxaca. *Boletín de la Sociedad Geológica Mexicana*, **LVII**, 111–122.

Analogue models of an Early Cenozoic transpressive regime in southern Mexico: implications on the evolution of the Xolapa complex and the North American–Caribbean Plate boundary

MARIANO CERCA[1]*, LUCA FERRARI[1], GUSTAVO TOLSON[2], GIACOMO CORTI[3], MARCO BONINI[3] & PIERO MANETTI[3]

[1]Centro de Geociencias, Universidad Nacional Autónoma de México, Campus Juriquilla, Apartado Postal 1-742, Querétaro 76230, Mexico

[2]Instituto de Geología, Universidad Nacional Autónoma de México, Ciudad Universitaria, México, D.F. 04510, Mexico

[3]Consiglio Nazionale delle Ricerche, Istituto di Geoscienze e Georisorse, Unità Operativa di Firenze, via G. La Pira, 4, 50121 Firenze, Italy

*Corresponding author (e-mail: mcerca@geociencias.unam.mx)

Abstract: We present analogue models that illustrate the tectonic evolution of the continental margin of southwestern Mexico and the Early Cenozoic deformation of the Xolapa complex. Together with geological data they suggest that oblique convergence caused distributed deformation and mountain building near the present-day margin of southern Mexico in a general left-lateral transpressional regime. A similar deformation is also observed north of the Xolapa complex in Maastrichtian to Paleocene sedimentary and volcanic rock units. Since post-Oligocene exhumation of middle crust does not significantly affect Late Eocene to Oligocene volcanic rocks, we infer that the evolution of the transform margin led to the formation of discrete boundaries that eventually decoupled exhumed mid-lower crust from the onshore upper-crust sequences since the Late Eocene.

Deciphering the intricate structure of the Xolapa complex represents a major target for understanding the geological evolution of the continental margin of southern Mexico (Fig. 1). The Xolapa complex, also known as Xolapa Terrane (Campa & Coney 1983) or Chatino Terrane (Sedlock et al. 1993) is a trench-parallel belt composed of pre-Cenozoic metamorphic and migmatite rocks (Ortega-Gutiérrez 1981) intruded by plutons of Late Cretaceous to Early Cenozoic age (Herrmann et al. 1994; Morán-Zenteno et al. 1999; Ducea et al. 2004). Several models for its evolution have been proposed but most studies interpret the complex as a middle crust equivalent of the continental crust exposed to the north (e.g. Dickinson & Lawton 2001; Ortega-Gutiérrez & Elías-Herrera 2003; Ducea et al. 2004; Keppie 2004) with two major episodes of arc magmatism occurring in the Paleocene and Late Eocene to Oligocene (Herrmann et al. 1994; Schaaf et al. 1995; Morán-Zenteno et al. 1999, 2004, 2005; Ducea et al. 2004).

Recent studies suggest also that during the Late Cretaceous to Paleocene the area of the Xolapa complex was characterized by north-directed shortening near Tierra Colorada (Torres De León &

Solari 2004) or transpression near Chacalapa (Corona-Chávez et al. 2006; Fig. 1). Similarly, a transpressive regime affecting the sedimentary and volcanic sequence of the Maastrichtian to Early Paleocene in a wide area north of the Xolapa complex from the Guerrero Morelos Platform to western Oaxaca has been recently documented (Cerca et al. 2004; Cerca et al. 2007; Fig. 1). An independent confirmation for this transpressive phase is provided by the palaeomagnetic data of the Mezcala intrusive suite (Molina-Garza & Alva-Aldivia 2006), which shows a decreasing amount of anticlockwise rotation about the vertical axis from Early Cretaceous to Early Cenozoic rocks. The trend of the transpressive structures is nearly perpendicular to older north–south trending shortening structures associated with the Laramide orogeny (Fig. 1). This suggests the existence of a separate phase of deformation between the east-directed shortening that finished in the Late Cretaceous in the Guerrero-Morelos Platform (Fig. 1) to the left-lateral strike-slip regime dominant in the Oligocene (Cerca et al. 2007).

Left-lateral strike–slip associated with extension is widespread in southern Mexico and probably

From: JAMES, K. H., LORENTE, M. A. & PINDELL, J. L. (eds) The Origin and Evolution of the Caribbean Plate. Geological Society, London, Special Publications, **328**, 181–195.
DOI: 10.1144/SP328.8   0305-8719/09/$15.00 © The Geological Society of London 2009.

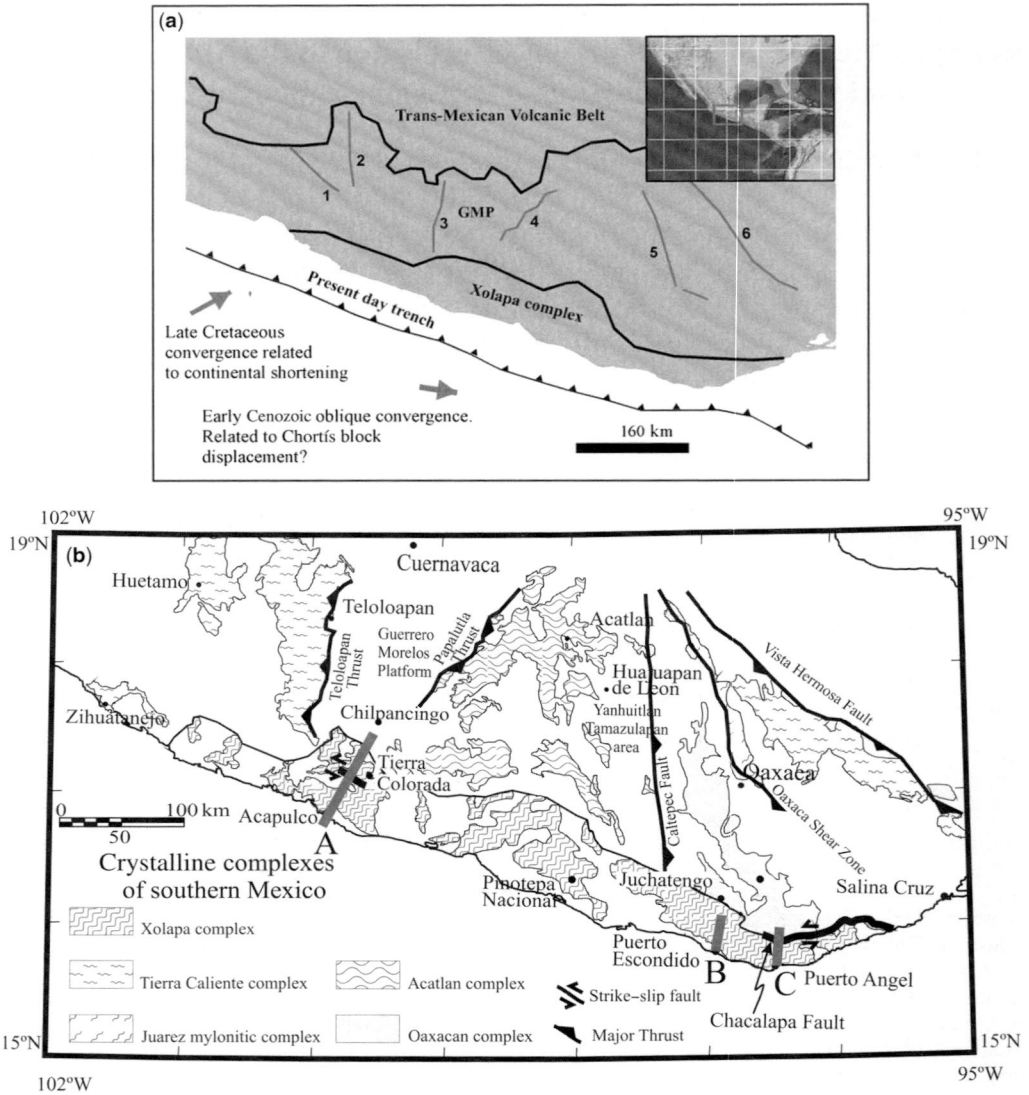

Fig. 1. (a) Schematic view of the features analysed showing the convergence vectors between Farallon and North America plates in the Late Cretaceous and Early Cenozoic for the studied area (after Engebretson *et al.* 1985; Schaaf *et al.* 1995; Meschede *et al.* 1996). The inset shows the location of southern Mexico in North America. Black thick lines in the continental area correspond to the volcanic front of the Trans-Mexican Volcanic Belt and the approximate limits of the middle to lower crust rocks of the Xolapa complex. Note that this complex strikes nearly at right angles to the major crustal structures at the margin of Mexico. Gray lines represent major structures in the continental area include: (1) left-lateral faults affecting Eocene and Oligocene intrusives (Montiel-Escobar *et al.* 2000); (2) Tzitzio anticlinorium formed in the Paleocene to Eocene (Ferrari *et al.* 2004); (3) Late Cretaceous Teloloapan Thrust (Campa *et al.* 1976; Cabral-Cano *et al.* 2000; Rivera-Carranza *et al.* 1997; Campa-Uranga *et al.* 1997); (4) Occidental border of the Acatlán complex and Papalutla Thrust (De Cserna *et al.* 1980; Cerca *et al.* 2007); (5) Oaxaca Fault (see Nieto-Samaniego *et al.* 1995, 2006); (6) Vista Hermosa Thrust (Ham-Wong 1981). A thicker crustal block is inferred between the Papalutla Thrust and the Oaxaca Shear Zone (Cerca *et al.* 2004; Nieto-Samaniego *et al.* 2006). GMP, Guerrero Morelos Platform. (b) Simplified geological map of southern Mexico (modified after Tolson 2005; Sánchez-Zavala 2005); capital letters refer to the geological sections presented in Figure 2: (A) Acapulco–Tierra Colorada; (B) Puerto-Escondido–Juchatengo; (C) Puerto–Angel–Chacalapa.

initiated by the Late Eocene (see Morán-Zenteno *et al.* 2005 for a review of the strike–slip structures). The dominant deformational features near or at the proposed northern boundary of the Xolapa complex have been interpreted as product of a left-lateral regime with associated NE–SW trending extension (Ratschbacher *et al.* 1991; Riller *et al.* 1992; Meschede *et al.* 1996; Tolson 1998, 2005). Deformation is characterized by NW–SE trending mylonitic shear zones localized near the northernmost exposures of mid-lower crustal rocks (Morán-Zenteno *et al.* 1996; Tolson 2005; Corona-Chávez *et al.* 2006). North–south extension associated with left-lateral mylonitic shear zones has been proposed in the Tierra Colorada, Juchatengo (Ratschbacher *et al.* 1991; Riller *et al.* 1992), and Chacalapa zones (Tolson 2005; Corona-Chávez *et al.* 2006), for the Late Eocene–Miocene time span. Exhumation estimates for the Xolapa complex since the Oligocene range from *c.* 13 to 19 km in the Acapulco-Tierra Colorada area (Morán-Zenteno *et al.* 1996) and *c.* 25 km in the Puerto Escondido area (Corona-Chávez *et al.* 2006).

Deformation and magmatism near the continental margin have been linked to displacement of the Chortís Block during the Early Cenozoic. Some authors argue that Chortís detached from its original position on the North American plate along a continental transform boundary, leading to continental margin truncation. In this model, migration of the trench–trench–transform triple junction since the Eocene (*c.* 45 Ma) was associated with the eastward migration of magmatic extinction documented with the ages of the plutons along the coast (Herrmann *et al.* 1994; Schaaf *et al.* 1995; Meschede *et al.* 1996; Morán-Zenteno *et al.* 1996). It also has been suggested that the exhumation of lower crustal rocks is the combined result of truncation and subduction erosion after the Oligocene intrusion of plutons (Morán-Zenteno *et al.* 1996). In contrast, recent tectonic restorations agree that the Caribbean Plate entered the proto-Caribbean seaway in Coniacian to Campanian times, causing deformation along its northern and southern sides (e.g. Kerr *et al.* 1999; Pindell & Kennan 2001; Rogers *et al.* 2003). Collision of the plate and its progressive interaction with the southern margin of Mexico is a probable cause for pre-Eocene transpressive deformation (Cerca *et al.* 2007).

Here, we discuss a possible scenario for the structural evolution of the continental margin of southern Mexico during the transpressive phase of the Early Cenozoic. The geological model is supported by analogue models that explore and focus on the mechanical response of the brittle–ductile crust of southern Mexico to oblique convergence. Thus, we do not intend to explain the causes of the oblique convergence nor all structural details that

led to the present day configuration of the continental margin of southern Mexico.

Shear zones bordering the Xolapa complex and the elusive transpressional phase

Two main deformation phases have been documented in southern Mexico since the Late Cretaceous:

(1) An east-directed 'Laramide' shortening phase that started at approximately 90 Ma and migrated eastwards in time and space. The end of this phase also migrated from west to east since the Late Maastrichtian (Nieto-Samaniego *et al.* 2006; Cerca *et al.* 2007).

(2) A left-lateral strike–slip and extensional phase with a predominant NW–SE trend of structures since the Late Eocene (Alaniz-Álvarez *et al.* 2002; Morán-Zenteno *et al.* 2005; Nieto-Samaniego *et al.* 2006). Associated deformation is more intense within the Xolapa complex.

From Paleocene to Eocene a transitional phase characterized by transpression is documented in a wide area north of the Xolapa complex, the Guerrero Morelos Platform and Yanhuitlan–Tamazulapan areas (Cerca *et al.* 2004, 2007). In the Xolapa complex, geological information is scarce but available data supports the interpretation of a similar transpressional phase during the Early Cenozoic. For example, Tolson (2005) describes a deformational phase D_{n+2} characterized by folds and thrusts with a northward vergence. This deformation episode is posterior to the emplacement of an apparently syntectonic intrusive suite that has yielded contrasting ages of *c.* 130 Ma (Morán-Zenteno 1992) or between 66 and 46 Ma (Herrmann *et al.* 1994) suggesting that these are in fact two magmatic pulses of different age, and predate the main Oligocene to Miocene transtensive activity of the Chacalapa Fault. Moreover, deformation, metamorphism and magmatism of this area suggest that the Xolapa complex evolved as a NE-verging thrust system during the Late Cretaceous–Palaeogene, with syn-kinematic metamorphism (Corona-Chávez *et al.* 2006). Recent geological mapping (Campa-Uranga *et al.* 1997) also shows that northward shortening associated with the La Venta Fault affected Albian to Cenomanian limestone as well as Early Cenozoic red beds *c.* 5 km south of Tierra Colorada (Fig. 1). Preliminary studies of this area (Torres De León & Solari 2004; Solari *et al.* 2006) have also documented north-directed shortening affecting the Late Cretaceous sequences in this area and mylonitization between *c.* 50 and 45 Ma.

Studies of shear zones at or near the northern limits of the Xolapa complex, including the Tierra

Colorada and Juchatengo areas (Ratschbacher et al. 1991; Riller et al. 1992; Morán-Zenteno et al. 1996) and Chacalapa Fault (Tolson 2005; Corona-Chávez et al. 2006) report a mainly left-lateral, oblique, divergent regime, active at least since the Late Eocene constrained by the age of the intrusives in the area (Fig. 1). However, they also give insights on the shortening and transpressive structures formed earlier. The structure of mylonites and cataclastic rocks exposed near Tierra Colorada suggests left-lateral shear (Ratschbacher et al. 1991; Riller et al. 1992; Meschede et al. 1996; Fig. 2a). However, the precise age of mylonite formation is still controversial and ranges between the Late Cretaceous and the Eocene (Ratschbacher et al. 1991; Riller et al. 1992; Herrmann et al. 1994; Morán-Zenteno et al. 1999; Solari et al. 2006).

Furthermore, in the same area, the Tierra Colorada intrusive of c. 34 Ma seems unaffected by mylonitization (Herrmann et al. 1994), setting an upper limit to the age of mylonitic deformation. Subsequently, extensional and strike–slip deformation accompanied the c. 13–20 km rapid exhumation of the Xolapa complex during and after an Oligocene magmatic episode (Morán-Zenteno et al. 1996, 1999, 2005; Ducea et al. 2004). Brittle–ductile and brittle deformation related to strike–slip displacement and associated extension in the Oligocene postdates mylonitic structures as well as earlier strike–slip and low angle thrusts (see Ratschbacher et al. 1991; Riller et al. 1992).

Mylonites and cataclastic rocks in the Chacalapa shear zone record transtensional deformation between 29 and 23 Ma (Fig. 2, Sections b & c).

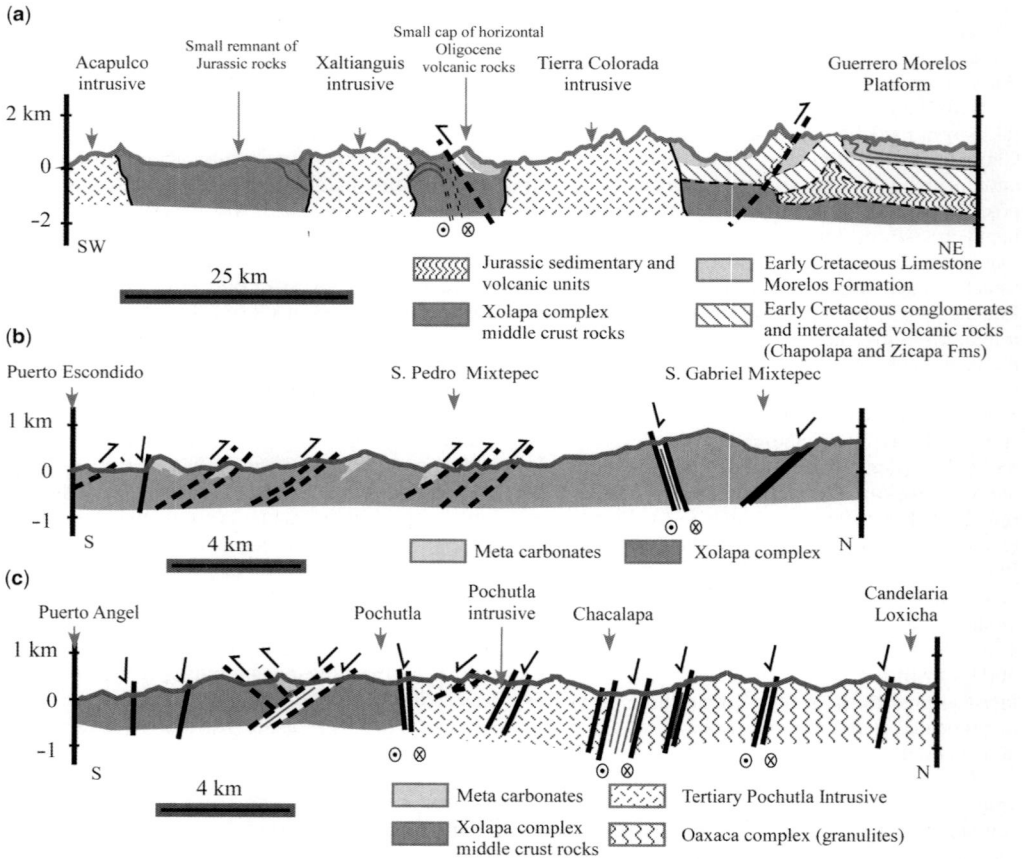

Fig. 2. Schematic and idealized geological sections at: (**a**) Acapulco–Tierra Colorada profile. This section was constructed after Morán-Zenteno et al. (1996, 2003); Campa-Uranga et al. (1997); and our own field data. (**b**) Puerto-Escondido profile redrawn and simplified after Corona-Chávez et al. (2006); (**c**) Chacalapa profile simplified after Corona-Chávez et al. (2006) and Tolson (2005). Note the difference in scale in sections a and c with respect to section a. The structures presented are exclusively related to deformation phases posterior to the Late Cretaceous.

Mylonitization affected the syntectonic Huatulco intrusion, dated at 29 Ma (Tolson 1998, 2005). The same author considers that Chacalapa mylonites formed progressively from a ductile to brittle regime, suggesting a progressive exhumation. The Chacalapa mylonites also overprint older, low-angle ductile shear zones (Tolson 1998, 2005; Corona-Chávez *et al.* 2006). In the same way at Juchatengo, the scarce data available indicate mylonitization in a progressive ductile to brittle deformation during north–south stretching (Ratschbacher *et al.* 1991; Fig. 2, Section b). These observations suggest that the overall deformation of the continental margin evolved from transpressional to transtensional from Paleocene to Eocene.

Our analysis of the available data and the results of analogue models favour the idea that the transpressional phase caused mountain building and probably early uplift of the lower crust rocks of the Xolapa complex. Later, the major uplift of the Xolapa complex was favoured by lateral mechanical decoupling during and after Late Eocene to Oligocene peak magmatism. This can explain why transpressive structures are more evident north of the Xolapa complex. We speculate that the thermal weakening of the lower crust and the ascent and emplacement of large magma bodies favoured mechanical decoupling by decreasing the viscosity of the lower crust and the large-scale exhumation

of the Xolapa complex without affecting substantially the adjacent continental blocks.

Analogue models

Our brittle–ductile two-layer crustal analogue models represent a further development of the series reported by Cerca *et al.* (2004). Two phases of deformation documented in southern Mexico were simulated:

(1) Late Cretaceous (Campanian), Laramide shortening; and
(2) Early Cenozoic left-lateral oblique convergence (Cerca *et al.* 2004).

The models are constrained by the available geological information in southern Mexico (e.g. Cerca *et al.* 2007). In this study, we focus on studying the deformation patterns produced by the proposed transpressional phase (2) in the models. We do not discuss the results of the previous shortening phase that are only given as a reference in the text and are shown in Figure 3. The major transtensional phase of the Late Eocene–Oligocene is not modelled either. During orthogonal shortening models were shortened 42 mm by a moving wall at a velocity of 6 mm h^{-1} (corresponding to *c.* 1 mm/annum in nature). During transpression, an orthogonal moving wall produced left-lateral oblique convergence

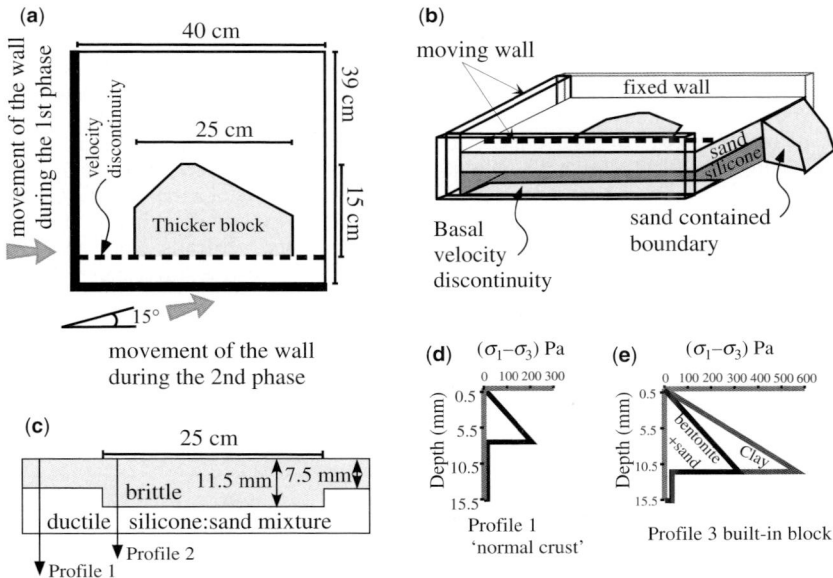

Fig. 3. Experimental setting: (**a**) map view; (**b**) three-dimensional view; (**c**) longitudinal section of the models showing the variation in thickness of the upper brittle layer and the location of the strength profiles of (**d**) the normal brittle–ductile layers and (**e**) the embedded thicker block constructed alternatively with sand and clay. Note that the velocity discontinuity (VD) is only active during the second phase.

resulting in a bulk displacement of 72 mm at 15 mm h^{-1} (corresponding to *c*. 3 mm/annum in nature) with an angle of 75°.

The set-up of previous models was basically preserved, but in the new models described here we added a basal metallic plate parallel to the moving wall during transpression. This acted as a velocity discontinuity (VD) that focused strike–slip in the lower ductile layer, reducing the influence of the vertical moving wall and causing oblique convergence (Fig. 3). The setting implies that deformation in the brittle crust was partially driven by flow of the lower crust as has been proposed for southern Mexico (Tolson 1998). The VD represents the zone in which the deformation is though to be localized. For modelling simplicity we assume that localization of deformation lies south of the stronger block, as could be the case for southern Mexico. Properties of the materials used as analogues in the experiments are presented in Table 1.

Models were scaled in such a way that 1 cm in the model is equivalent to 20 km in nature, according to principles of similarity outlined by Hubbert (1937) and Ramberg (1981). Models were simplified to simulate a two-layer brittle–ductile crust composed of a brittle layer of sand above a ductile layer of a silicon–sand mixture (5:5.5% in weight). A thicker area in the brittle layer representing a more rigid block with a parallelepiped shape was built of fine bentonite sand (Chortís 07), quartz sand (Chortís 08) and humid, high-cohesive, clay (Chortís 09). The built-in block simulates a thicker area of southern Mexico composed by metamorphic terranes that apparently acted as a more stable area during the simulated deformation phases (Cerca *et al.* 2004, 2007; Nieto-Samaniego *et al.* 2006). This setup permitted study of the effects of changes in strength of the block during the oblique convergence phase (Table 1). The model design also implies that crustal strength resides mainly in the upper brittle layer and the deformation of the lithospheric mantle is not considered (Cerca *et al.*

2004). We assumed that greater strength contrast favours mechanical decoupling between the ductile and brittle layers, as in the case of the block constructed with clay (Fig. 3e). We realize that the use of this highly cohesive material is out of scale with respect to natural systems (Table 1), but it has the effect of enhancing the deformation patterns around the stronger block. For further details in scaling and material properties see Cerca *et al.* (2004). Thus deformation patterns in the brittle layer are enhanced.

Limitations of modelling

We use analogue models as an indirect approach to shed light on the poorly known tectonic evolution of the Xolapa complex in the Palaeogene to Oligocene, when left-lateral transpressional deformation occurred in the Sierra Madre del Sur. The analogue models inevitably simplify the geometry and rheology of the natural processes under study. Thus, the inherent limitations of modelling need to be discussed before any attempt at comparison with the natural situation.

From a mechanical point of view, the simple two-layer structure of the model simulates only the upper part of the lithosphere. As in previous models, it was considered that in our case the crustal strength resides mainly in the upper brittle layer (e.g. Jackson 2002). The model considers a uniformly thick continental crust characterized by a region with thicker and stronger brittle crust above a ductile layer that is thinner than elsewhere.

During the experiments compressive stresses causing deformation are transmitted mainly by the underlying velocity discontinuity, although lateral transmission of forces from the rigid moving wall may be also significant. This might represent an appropriate simplification of the natural process where forces causing deformation of the lithosphere are transmitted vertically from below, driven by mantle flow or by the lower crust (e.g. Teyssier &

Table 1. *Model parameters of the materials used in the experiments*

Analogue material	Density (kg cm^{-3})	Cohesion (kPa)	Viscosity η (Pa s)	Coefficient of internal friction μ	Mechanical behaviour analogue
Quartz sand	1450	Negligible	—	0.70	Brittle upper crust
Bentonite	*c*. 1400	—	—		Brittle upper crust
Clay	2500	*c*. 37–59[a]	—		High strength brittle upper crust
Silicone + qz sand (5:5.5)	*c*. 1500		3×10^{5b}		Ductile crust

[a]Estimated at 24% water content with a vane tester.
[b]Determined at *c*. 21 °C using a coni-cylindrical viscometer.

Tikoff 1998) and/or laterally driven by the displacement of a rigid block (see later).

Several other factors not considered by the model may influence structures formed during deformation. Among them are erosion and deposition in basins, the effect of pore pressure in the growth and propagation of structures, thermal evolution, magma migration and emplacement or isostacy effects (e.g. Corti *et al.* 2003). The models did not attempt to reproduce the complex thermal evolution expected in the Xolapa complex (vertical thermal gradient, temperature variations during progressive thickening or uplift), nor chemical or petrological variations occurring during transpression. Thus rheological modifications of the different layers and destabilization of the brittle–ductile transition in the crust, which continuously evolve during lithosphere thickening, were not modelled. In the models the viscosity of the ductile layer is uniform. In nature, viscosity can be strongly temperature dependent changing through time as well as spatially, and lower viscosity regions can also concentrate deformation (e.g. Turcotte & Schubert 1982), as shown by previous analogue models (e.g. Corti *et al.* 2003). In addition, the cohesion of clay used in model Chortís 09 is not properly scaled and clearly exceeds the rigidity of the natural prototype; this might amplify the effect of lateral strength contrasts within the continental crust. Despite all these simplifications, the analogue models give insights into the possible effects of oblique convergence along the southern margin of Mexico.

Results of modelling

The overall structural evolution exhibited by the experiments was similar to previous models of Cerca *et al.* (2004). Our analysis focuses on structures formed during the strike–slip-dominated second phase of deformation with two end-member experiments. The first (A) investigates mechanical coupling between brittle and ductile layers crust during transpression (e.g. Teyssier & Cruz 2004), represented by models Chortís 07 and Chortís 08 (Fig. 3e). Model Chortís 09 represents experiment B, with low mechanical coupling compensated by the high-strength contrast of the thicker block constructed with humid plastic clay (Fig. 3e). The evolution of deformation in the models is portrayed in Figure 4 and is discussed taking the short vertical side of the model as north. Our models assume that strain partitioning is an important process, enhanced by increased obliquity of convergence. Thus, the first structure formed in all the experiments is a band of strike–slip deformation, with a subordinate reverse component of movement, striking parallel to the moving wall and located directly above the velocity discontinuity. After *c.* 25 mm of mobile wall displacement, strain is distributed in a *c.* 4 cm band north of the velocity discontinuity in model A, whereas strain is evidently localized in the southern margin of the rigid block in model B (Fig. 4). Although not displayed in the figure, another feature observed in model B is the effective transmission of strain toward the northeastern edge of the clay block. In both cases, folds formed in

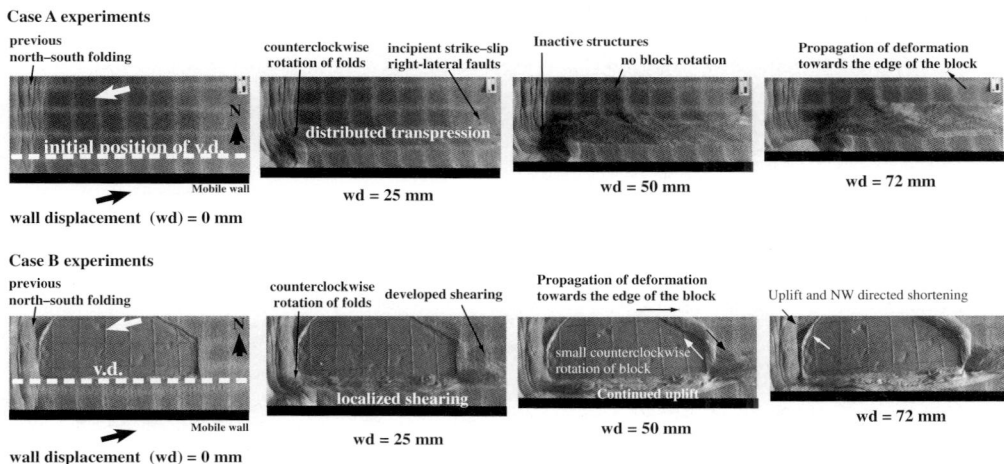

Fig. 4. Evolution of the models deformation in plain view showing the differences in the evolution of deformation of case A and case B. Case A experiments investigate the deformation of brittle–ductile crust during transpression, case B, explores a low mechanical coupling propitiated by the high strength contrast of the thicker block constructed with humid plastic clay.

the first phase rotate counterclockwise in the south-western corner of the model. We show this feature only as a reference since the analysed sections do not cut perpendicularly to the folds.

After *c.* 50 mm of mobile wall displacement in experiment A, the block begins to transmit strain toward the northeastern edge. Effective transmission of mass towards the east causes uplift of this area and, within the deformed area, formation of synthetic structures. In case B, the rigid block begins to rotate and deformation propagates toward the northwest along its boundary. Since there cannot be mass transfer from the cohesive block toward the deformed areas, a deep basin forms adjacent to the block (Fig. 4, end of type B experiments).

The deformation zone at the end of experiment A is characterized by an imbricate system of linked conjugated strike–slip faults and uplift in an area around the velocity discontinuity (Fig. 5). An important characteristic is the northeastward

translation of the uplifted area observed during the evolution of the models Chortís 07 and 08 (Fig. 4). Propagation of strain toward the northeastern edge of the block is not as effective as in experiment B, but at the end of the experiment slight rotation of the block and uplift in the foreland can be still observed.

In experiment B, thickening is concentrated in the boundaries of the block, where a strong rheological contrast exists. Also, concentration of strain does not reflect the final position of the velocity discontinuity, because deformation is concentrated in the edges of the rigid block. This resulted in lateral extrusion of the ductile layer below. The deformation observed at the end of the experiments Chortís 07 and 08 is summarized in Figure 5.

Adding a basal velocity discontinuity allowed separation of the block from the moving wall in the second phase, accommodating lateral displacement that was not previously considered (e.g. Cerca *et al.* 2004). We present model sections that

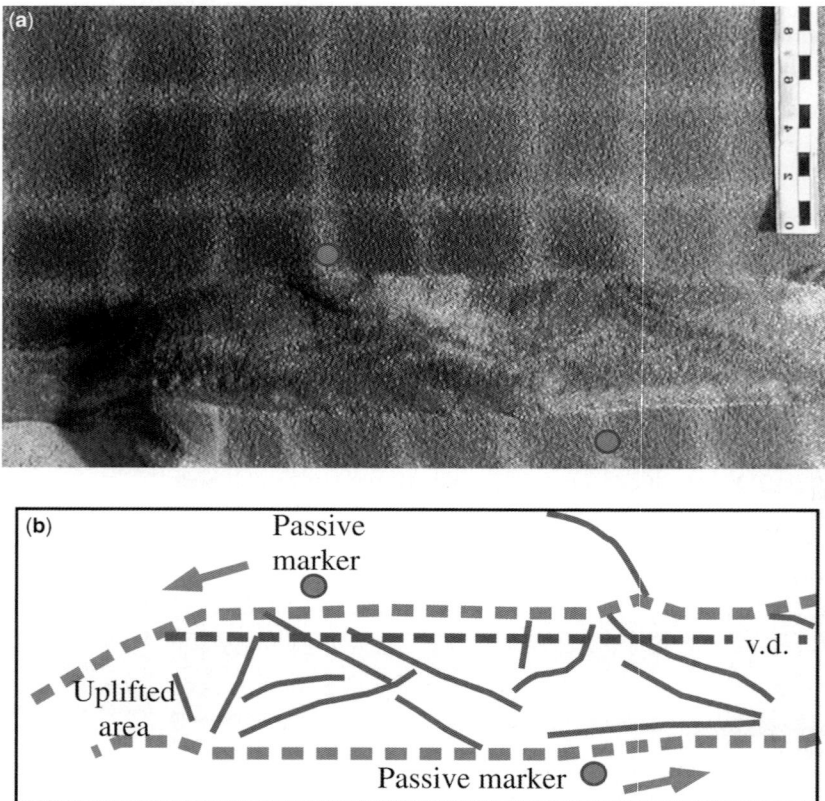

Fig. 5. (a) Photograph of the resulting deformation in the model Chortís 07, representative of the deformational features observed in the experiments. Dots show the left-lateral displacement of two points located originally in the same vertical marker line. **(b)** Line drawing of the resulting structures.

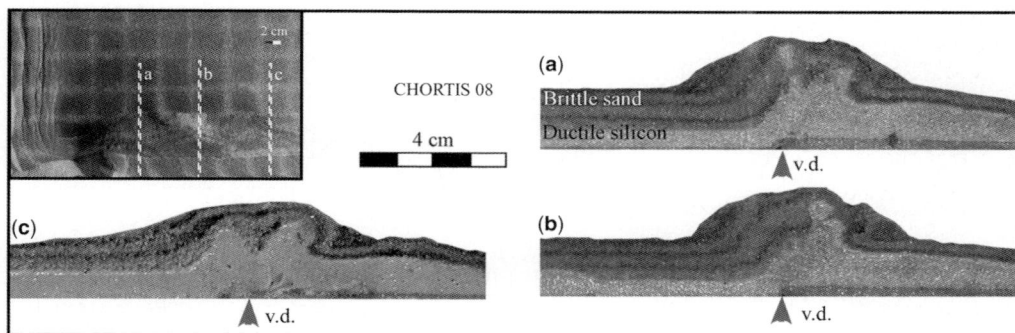

Fig. 6. Plan view of the deformation at the end of the experiment Chortís 08 and cross-sections. The arrows mark the final position of the velocity discontinuity (VD). (**a, b**) Longitudinal sections that cut the stronger block.

illustrate the differences in deformation styles of the two experiments reported in Figures 6–8. In the models, the displacement gradient caused by the velocity discontinuity and the differential mass flow produces a complex strain pattern that cannot be fully described with a single section. Deformation within the ductile layer is characterized by penetrative flow along the velocity discontinuity and thrust of the ductile over the brittle layer is readily observed in the cross sections (Fig. 6). Significant lateral extrusion of the ductile layer occurred during the transpressive phase. In general, deformation associated with the second phase is characterized by thickening of the crust and lateral exhumation of the ductile layer.

Flower structures typical of transpressional regimes do not form in the experiment involving the stronger block (Figs 6a, b, 7a, b, & 8a, b). Instead, lateral extrusion of the ductile layer is enhanced. In the absence of the block (Fig. 6c) or

near the tip of the block (Fig. 7c), a slightly asymmetrical flower structure is formed, with the exception of the experiment B, where lateral mass flow of the block is prevented by its high cohesion.

To sum up, the above results suggest that the structural and topographic evolution of the deformed area depends on the lateral rheological structure of the model (i.e. the composition and presence or absence of the block), and lateral mass transport (e.g. extrusion of ductile layers).

Discussion

Oblique convergence deformation in southern Mexico and Xolapa complex

A large-scale left-lateral shear regime affected the continental lithosphere of southern Mexico during the Cenozoic (Ratschbacher *et al.* 1991; Riller

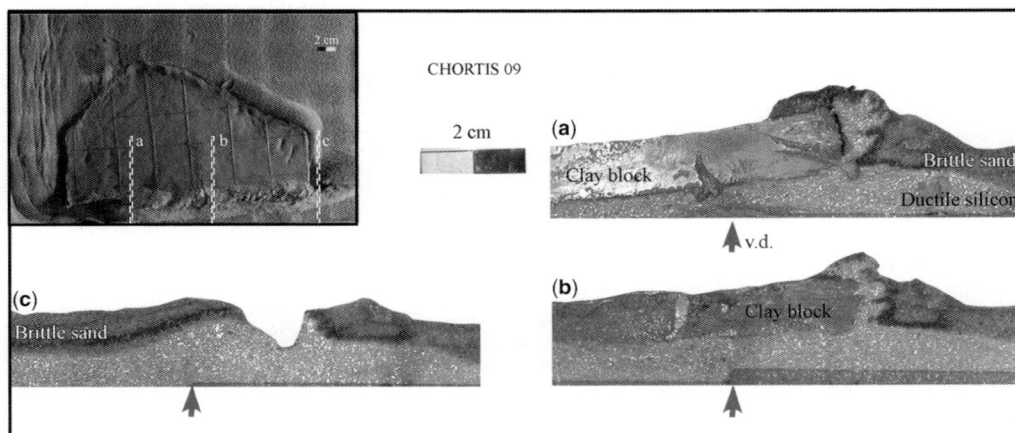

Fig. 7. Plan view of the deformation at the end of the experiment Chortís 07 and cross-sections.

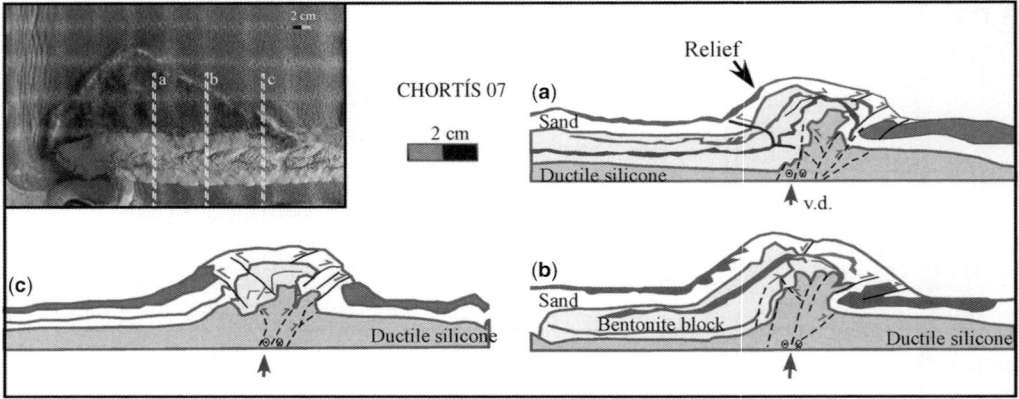

Fig. 8. Cross-sections of model Chortís 09 (**a**–**c**). Because of the high cohesion of the clay block, it remains basically non-deformed and it prevents mass flow toward the adjacent brittle layer, as can be seen in (**c**).

et al. 1992; Ducea *et al.* 2004; Morán-Zenteno *et al.* 2005; Tolson 2005; Corona-Chávez *et al.* 2006; Nieto-Samaniego *et al.* 2006). Deformation associated with this event was more intense near the continental margin and superimposed over previous deformational structure in the middle crustal rocks of the Xolapa complex. Earlier works in this area focus on the progressive ductile–brittle transtensive deformation that began in the Late Eocene (Ratschbacher *et al.* 1991; Riller *et al.* 1992; Ducea *et al.* 2004; Tolson 2005). Our analysis of the available data reveals the existence of a Paleocene–Eocene transpressive phase that is better recorded by upper crustal rocks north of the Xolapa complex (Cerca *et al.* 2007). The results of modelling highlight the importance of the Early Cenozoic transpressive deformation for shaping the lithosphere of southern Mexico and serve as a guide for discussing the major geometrical and mechanical characteristics of this poorly know deformational phase.

Model results indicate that during low angle (15°) oblique convergence, mountain building and strain partitioning are important processes enhanced by rheological contrasts (Fig. 9). If mountain building occurred in southern Mexico, this could have enhanced erosion, and changed the overall hydrographic system. In southern Mexico, the thick continental sequence of sedimentary and intercalated volcanic rocks known as Balsas Group was deposited between Late Cretaceous and Early Cenozoic, thus being mostly contemporaneous with the transpressional regime (Cerca *et al.* 2004, 2007; Morán-Zenteno *et al.* 2005).

Changes in width of the Xolapa complex have been attributed to the presence of major structures nearly perpendicular to the strike of the palaeo plate boundary, such as folds and thrusts of a previous deformation phase (Cerca *et al.* 2004) or the

Palaeozoic–Mesozoic Oaxaca shear zone (Nieto-Samaniego *et al.* 1995; Alaniz-Álvarez *et al.* 1996). The results of modelling suggest that contrasts in strength might have also influenced the changes of direction and width of the Xolapa complex.

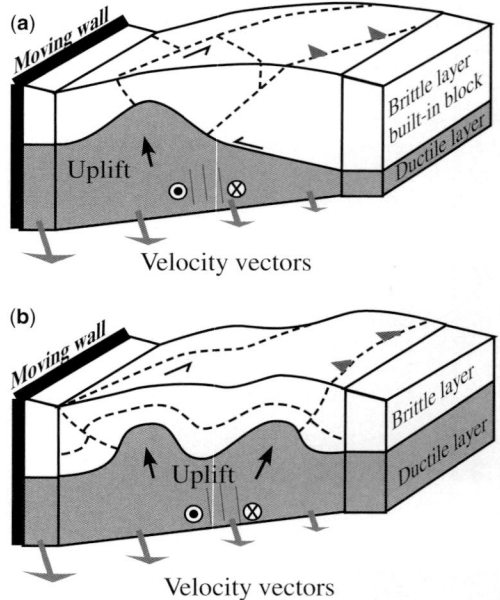

Fig. 9. Schematic block diagrams showing the type of deformation observed in the models: (**a**) an asymmetric transpressive structure develops in the presence of thicker block; (**b**) typical flower structure in the absence of the block.

Our model results show a diversity of structures mainly at three different levels:

(1) Deformation of the brittle layer is characterized by folding parallel to the moving wall and subvertical transpressive structures; the contrast in strength with the thicker block provokes the formation of thrusting in the opposite direction of the convergence.

(2) Mechanical coupling between the brittle–ductile layers forms a transition zone (i.e. the attachment zone of Teyssier & Cruz 2004).

(3) Deformation in the ductile layer is characterized by vertical flattening and stretching along the velocity discontinuity (Fig. 9).

The presence of a thicker block results in an asymmetric structure (Fig. 9a), whereas in the absence of the block deformation produces a typical flower structure (Fig. 9b). Some structural studies observed northward shortening or an oblique convergence regime in the Xolapa complex prior to the Late Eocene (Meschede *et al.* 1996; Campa-Uranga *et al.* 1997; Torres De León & Solari 2004; Cerca *et al.* 2004, 2007; Tolson 2005; Corona-Chávez *et al.* 2006). The upper crust exposed north of the Xolapa complex includes northward-directed thrusting north of the Tierra Colorada area (Campa-Uranga *et al.* 1997; Torres De León & Solari 2004) and counterclockwise rotation of blocks and earlier structures (Cerca *et al.* 2004, 2007). These observations confirm a general left-lateral transpressive regime for the Paleocene. In the lower crust exposed in the Xolapa complex, the evidence includes the northeastward vergent folding and thrusting of the migmatitic foliation and low angle south dipping shear zones (Tolson 2005; Corona-Chávez *et al.* 2006). No kinematic indicators suggesting stretching associated with this event have been reported, and further age constraints for the observed deformation in the lower crust are needed to clarify this issue. In any case, concentration of the deformation in the Xolapa complex suggests that its area corresponds to a weakened lithosphere.

The model results and the analysis of available data permit some inferences on the process of later, post-orogenic exhumation. Ductile–brittle transtensional deformation associated with discrete shear zones in the northern part of the Xolapa complex was active during and immediately after voluminous pluton emplacement (Schaaf *et al.* 1995; Morán-Zenteno *et al.* 1996, 1999, 2005) in the Late Eocene–Oligocene, postdating earlier transpressional deformation (Corona-Chávez *et al.* 2006; Morán-Zenteno *et al.* 1996; Tolson 2005). This setting has been interpreted as a progressive change from ductile to brittle regime during exhumation (Riller *et al.* 1992; Tolson 2005). It can

be argued that the Late Eocene–Oligocene large-scale magmatic episode provoked a decrease in the bulk viscosity of the already weakened Xolapa complex. At the same time, a weakened transpressive orogen could have favoured the emplacement of arc magmas.

Estimates of Oligocene exhumation of the Xolapa complex range from 13 to 19 km east of Acapulco (Morán-Zenteno *et al.* 1996) to *c.* 25 km in the Puerto Escondido area (Corona-Chávez *et al.* 2006). The presence of undeformed Late Eocene and Oligocene volcanic rocks in the Tierra Colorada area, just to the north of the Xolapa complex (Cerca *et al.* 2007), suggests that lateral mechanical decoupling permitted the rapid exhumation of intrusives and host rocks after the Eocene.

The geological evolution of the Xolapa complex in southern Mexico is clearly more complex than presented here. However, with the simple geometry of the models we were able to discuss the possible effects of a transpressional phase poorly documented in southern Mexico. Further fieldwork is mandatory to confirm or discard the model predictions.

Relevance to southern Mexico and Caribbean tectonics

The presence of a left-lateral oblique convergence regime in southern Mexico has important implications for the tectonic evolution of the North American and Caribbean Plates. De Cserna (1967) and Karig *et al.* (1978) noted the truncated nature of the continental margin of southern Mexico. Malfait & Dinkelman (1972) and Ross & Scotese (1988) promoted the idea that the Chortís Block was originally attached to the, now truncated, southern part of the North American plate during most of the Mesozoic. In this model the Chortís Block was transferred to the Caribbean Plate through formation of a transcurrent plate boundary oblique to the NNW-trending Farallon–North America boundary. Since then, many authors have postulated that Chortís was attached to southern Mexico by Late Cretaceous times (Meschede & Frisch 1988; Herrmann *et al.* 1994; Schaaf *et al.* 1995; Morán-Zenteno *et al.* 1996; Meschede *et al.* 1996; Kerr *et al.* 1999; Pindell 1993; Pindell *et al.* 2005; Pindell & Kennan 2001; Rogers *et al.* 2003, 2007; Cerca *et al.* 2004, 2007; Nieto-Samaniego *et al.* 2006). In a recent detailed geological study, Rogers *et al.* (2003, 2007) suggest striking similarities in geological and structural features between the Chortís Block and southern Mexico once the block is restored and rotated back to the inferred Cretaceous position. In an update of their tectonic reconstruction, Pindell *et al.* (2005) suggest that the Chortís Block began to detach from North America in the Maastrichtian and became part of the Caribbean Plate in the Middle Eocene, as the Cayman Trough opened. This process probably resulted from the collision of the leading edge of the Caribbean arc with the south-western margin of North America in the latest Cretaceous (Pindell *et al.* 2005; Rogers *et al.* 2007; Mann *et al.* 2007). Taking into account the convergence vectors suggested for the Early Ceno-zoic for the relative displacement between Farallon and North American plates (Schaaf *et al.* 1995; Meschede *et al.* 1996), transpression is inescapable if Chortís was displaced to the SE from an original position attached to North America. On the other hand, it has been commonly accepted that the Chortís Block moved along a trajectory marked by the present day Middle American Trench (Herrmann *et al.* 1994; Schaaf *et al.* 1995; Tolson 2005; Nieto-Samaniego *et al.* 2006). This idea does not consider that subduction erosion is likely to have occurred after emplacement of the Late Eocene to Oligocene magmatic arc, as suggested by the absence of a well-developed accretionary prism and the presence of arc granites close to the trench (Morán-Zenteno *et al.* 1996, 2005, and references therein). Although no estimates of the amount of subduction erosion are available, it may be argued that at least part of the forearc was

removed during this process. This consideration places the Chortís Block attached to North America in a more southern position. In this scen-ario, we speculate that in an early phase of the Chortís Block motion deformation was distributed over a wide area, mostly located to the south of the present day margin (Fig. 10). During post-Oligocene subduction erosion, a large part of the forearc weakened by the deformation was removed. Placing the Chortís Block in a more southern position is a possibility that needs to be explored in further work since it could explain some of the problems outlined by Keppie & Morán-Zenteno (2005) in linking the Cayman Trough via the Motagua fault zone of Guatemala to the Middle American Trench.

While the discussion presented in this study does not prove that transpressional deformation in southern Mexico was caused by the left-lateral dis-placement of the Chortís Block, it does suggest that it is kinematically possible. If this is the case, the transpressional deformation should have affected a wide area north and south of the evolving plate boundary. Finally, our results show that appar-ent mechanical incompatibilities of structures in the upper and lower crust of southern Mexico can be resolved by the simple model presented here for discussion.

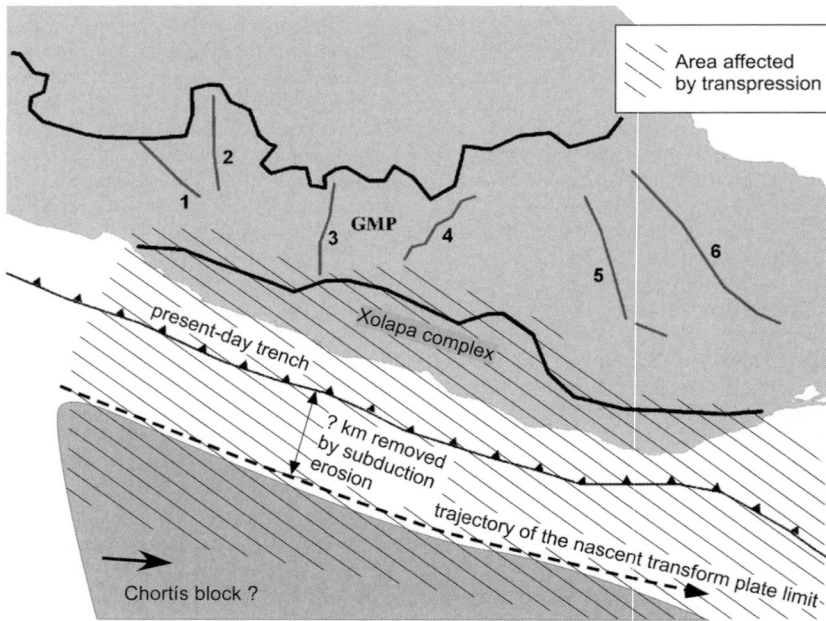

Fig. 10. Cartoon showing the model proposed for the oblique convergence deformation. A triple junction moves along the trajectory of the newly formed plate boundary.

Conclusions

Scaled analogue models provide valuable information that allows discussion of the Early Cenozoic evolution of the southern margin of Mexico. The models illustrate a mechanism capable of explaining the presence of north-directed shortening structures in the Xolapa complex and inland transpressional deformation. Deformation affecting a wide area north and south of the continental margin of southern Mexico may have resulted from movement of the Chortís Block to the SE. With the analysis of the available data and the model results presented here, we propose that the tectonic evolution of the southwestern margin of Mexico may be seen as a spatially heterogeneous, and time transgressive deformation associated with the general North American–Caribbean Plate interaction. The model results provide a plausible explanation for the variety of structures observed in the Xolapa complex.

We acknowledge funds from CONACYT (grant 32509-T to LF), and SRE-MAE bilateral grant. This study was also partially funded by CONACYT grant 46235 and DGAPA-PAPIIT grant IN120305 to MC. Thanks to Dante Morán-Zenteno for the many hours of fruitful discussion about the tectonics of southern Mexico and for providing insights of important geological constrains. Revisions by Bert Bally and Keith James are gratefully acknowledged.

References

ALANIZ-ÁLVAREZ, S. A., VAN DER HEYDEN, P., NIETO-SAMANIEGO, A. F. & ORTEGA-GUTIERREZ, F. 1996. Radiometric and kinematic evidence for Middle Jurassic strike-slip faulting in southern México related to the opening of the Gulf of México. *Geology*, 24, 443–446.

ALANIZ-ÁLVAREZ, S. A., NIETO-SAMANIEGO, A. F., MORÁN-ZENTENO, D. J. & ALVA-ALDAVE, L. 2002. Rhyolitic volcanism in extension zone associated with strike–slip tectonics in the Taxco region, southern Mexico. *Journal of Volcanology and Geothermal Research*, 118, 1–14.

CABRAL-CANO, E., DRAPER, G., LANG, H. R. & HARRISON, C. G. A. 2000. Constraining the Late Mesozoic and Early Cenozoic tectonic evolution of southern Mexico: structure and deformation history of the Tierra Caliente region, southern Mexico. *Journal of Geology*, 108, 427–446.

CAMPA, M. F. & CONEY, P. J. 1983. Tectono-stratigraphic terranes and mineral resource distributions in Mexico. *Canadian Journal of Earth Sciences*, 20, 1040–1051.

CAMPA, M. F., OVIEDO, A. & TARDY, M. 1976. La cabalgadura laramídica del dominio volcánico-sedimentario (Arco de Alisitos–Teloloapán) sobre el miogeosinclinal mexicano en los límites de los estados de Guerrero y México. *In: Resúmenes del III Congreso Latinoamericano de Geología*, México, D. F.,

Universidad Nacional Autónoma de México, Instituto de Geología, 23.

CAMPA-URANGA, M. F., BUSTAMANTE-GARCÍA, J., TORREBLANCA-CASTRO, T., DE, J., GARCÍA-DÍAZ, J. L., AGUILERA-MARTÍNEZ, M. A. & VERGARA-MARTÍNEZ, A. 1997. *Mapa Geológico – Minero Chilpancingo, escala 1:250 000 – hoja INEGI E14-8.* Consejo de Recursos Minerales, Escuela Regional de Ciencias de la Tierra, UAG. 1 map and report.

CERCA, M., FERRARI, L., BONINI, M., CORTI, G. & MANETTI, P. 2004. The role of crustal heterogeneity in controlling vertical coupling during Laramide shortening and the development of the Caribbean–North American transform boundary in southern Mexico: insights from analogue models. *In:* GROCOTT, J., TAYLOR, G. & TIKOFF, B. (eds) *Vertical Coupling and Decoupling in the Lithosphere.* Geological Society, London, Special Publications, 227, 117–140.

CERCA, M., FERRARI, L., LÓPEZ-MARTÍNEZ, M., MARTINY, B. & IRIONDO, A. 2007. Late Cretaceous shortening and Early Cenozoic shearing in the central Sierra Madre del Sur, southern Mexico: insights into the evolution of the Caribbean-North American plate interaction. *Tectonics*, 26, TC3007.

CORONA-CHÁVEZ, P., POLI, S. & BIGIOGGERO, B. 2006. Syndeformational migmatites and magmatic-arc metamorphism in the Xolapa Complex, southern Mexico. *Journal of Metamorphic Geology*, 24, 169–191; doi: 10.111/j.1525-1314.2006.00632.x.

CORTI, G., BONINI, M., CONTICELLI, S., INNOCENTI, F., MANETTI, P. & SOKOUTIS, D. 2003. Analogue modelling of continental extension: a review focused on the relations between the patterns of deformation and the presence of magma. *Earth-Science Reviews*, 63, 169–247.

DE CSERNA, Z. 1967. Tectonic framework of southern Mexico and its bearing on the problem of continental drift. *Boletín de la Sociedad Geológica Mexicana*, 62, 1–77.

DE CSERNA, Z., ORTEGA-GUTIÉRREZ, F. & PALACIOS-NIETO, M. 1980. Reconocimiento geologico de la parte central de la cuenca del Río Balsas, estados de Guerrero y Puebla, en Sociedad Geológica Mexicana. Libro Guía de la excursión geológica a la parte central de la cuenca del alto Río Balsas. *V Convención Geológica Nacional*, 2–33.

DICKINSON, W. R. & LAWTON, T. F. 2001. Carboniferous to Cretaceous assembly and fragmentation of Mexico. *GSA Bulletin*, 113, 1142–1160.

DUCEA, M. N., GEHRELS, G. E., SHOEMAKER, S., RUIZ, J. & VALENCIA, V. A. 2004. Geological Evolution of the Xolapa Complex, Southern Mexico. Evidence from U–Pb zircon geochronology. *Geological Society of America Bulletin*, 116, 1016–1025.

ENGEBRETSON, D. C., COX, A. & GORDON, R. G. 1985. *Relative Plate Motions Between Oceanic and Continental Plates in the Pacific Basin: Boulder, Colorado.* Geological Society of America, Special Papers, 206.

FERRARI, L., CERCA, M, LÓPEZ-MARTÍNEZ, M., SERRANO-DURAN, L. & GONZALEZ-CERVANTES, N. 2004. Age of formation of the Tzitzio antiform and structural control of volcanism in eastern Michoacán and western Guerrero. *Libro de Resúmenes, IV*

Reunión Nacional de Ciencias de la Tierra, Juriquilla, Qro., 68.

HAM-WONG, J. M. 1981. Prospecto: Guelatao, Informe Geológico no. 799 Zona Sur. Petróleos Mexicanos. Inedit.

HERRMANN, U., NELSON, B. K. & RATSCHBACHER, L. 1994. The origin of a terrane: U/Pb zircon geochronology and tectonic evolution of the Xolapa Complex (southern Mexico). *Tectonics*, **13**, 455–474.

HUBBERT, M. K. 1937. Theory of scale models as applied to the study of geologic structures. *Bulletin of the Geological Society of America*, **48**, 1459–1520.

JACKSON, J. 2002. Strength of the continental lithosphere: Time to abandon the jelly sandwich? *GSA Today*, **12**, 4–10; doi: 10.1130/1052-5173(2002)012<0004: SOTCLT>2.0.CO;2.

KARIG, D. E., CARDWELL, R. K., MOORE, G. F. & MOORE, D. G. 1978. Late Cenozoic subduction and continental margin truncation along the northern Middle American Trench. *Geological Society of America Bulletin*, **89**, 265–276.

KEPPIE, J. D. 2004. Terranes of Mexico revisited: a 1.3 billion year odyssey. *International Geology Review*, **46**, 765–794.

KEPPIE, J. D. & MORAN-ZENTENO, D. J. 2005. Tectonic implications of alternative Cenozoic reconstructions for southern Mexico and the Chortis Block. *International Geology Review*, **47**, 473–491.

KERR, A. C., ITURRALDE VINENT, M. A., SAUNDERS, A. D., BABBS, T. L. & TARNEY, J. 1999. A new plate tectonic model of the Caribbean: Implications from a geochemical reconnaissance of Cuban Mesozoic volcanic rocks. *Geological Society of America Bulletin*, **111**, 1581–1599.

MALFAIT, B. T. & DINKELMAN, M. G. 1972. Circum-Caribbean tectonic and igneous activity and the evolution of the Caribbean plate. *Geological Society of America Bulletin*, **83**, 251–272.

MANN, P., ROGERS, R. D. & GAHAGAN, L. 2007. Overview of plate tectonic history and its unresolved tectonic problems. *In*: BUNDSCHUH, J. & ALVARADO, G. E. (eds) *Central America. Geology, Resources and Hazards*. Taylor & Francis, 205–241.

MESCHEDE, M. & FRISCH, W. 1998. A plate-tectonic model for the Mesozoic and Early Cenozoic history of the Caribbean Plate. *Tectonophysics*, **296**, 269–291.

MESCHEDE, M., FRISCH, W., HERRMANN, U. R. & RATSCHBACHER, L. 1996. Stress transmission across an active plate boundary: an example from southern Mexico. *Tectonophysics*, **266**, 81–100.

MOLINA-GARZA, R. & ALVA-ALDIVIA, L. 2006. Paleomagnetism of the latest Cretaceous–Palaeocene intrusive suite of the Mezcala district, southern Mexico. *Earth Planets and Space*, **58**, 1315–1322.

MONTIEL-ESCOBAR, J. E., SEGURA DE LA TEJA, M. A., ESTRADA-RODARTE, G., CRUZ-LÓPEZ, D. & ROSALES-FRANCO, E. 2000. *Carta geológica-minera de la hoja Ciudad Altamirano*. Escala 1 : 250 000 (**E14-4**). Pachuca, Hidalgo, Consejo de Recursos Minerales. 1 map.

MORÁN-ZENTENO, D. J. 1992. *Investigaciones isotópicas de Rb–Sr y Sm–Nd en rocas cristalinas de la región Tierra Colorada–Acapulco–Cruz Grande, estado de Guerrero: México*. PhD Thesis, Universidad Nacional Autónoma de México, Colegio de Ciencias y Humanidades, Unidad Académica de los Ciclos Profesional y de Posgrado, Instituto de Geofísica.

MORÁN-ZENTENO, D. J., CORONA-CHÁVEZ, P. & TOLSON, G. 1996. Uplift and subduction erosion in southwestern Mexico since the Oligocene: pluton geobarometry constraints. *Earth and Planetary Science Letters*, **141**, 51–65.

MORÁN-ZENTENO, D. J., TOLSON, G. *ET AL.* 1999. Cenozoic arc-magmatism of the Sierra Madre del Sur, Mexico, and its transition to the volcanic activity of the Trans-Mexican Volcanic Belt. *Journal of South American Earth Sciences*, **12**, 513–535.

MORÁN-ZENTENO, D. J., MARTINY, B., ALVA-ALDAVE, L., GONZALEZ-TORRES, E., HERNANDEZ-TREVIÑO, T. & ALANIZ-ALVAREZ, S. A. 2003. Tertiary magmatism, tectonic deformation along the Cuernavaca–Acapulco transect. *In*: *Geologic Transects across Cordilleran Mexico*. Guidebook for the fieldtrips of the 99th Geological Society of America Cordilleran Section Annual Meeting, Puerto Vallarta, Jalisco, Mexico, 4–6 April. Publicación Especial, Universidad Nacional Autónoma de México, Fieldtrip 11, 261–280.

MORÁN-ZENTENO, D. J., ALBA-ALDAVE, L. A., SOLÉ, J. & IRIONDO, A. 2004. A major resurgent caldera in southern Mexico: the source of the Late Eocene Tilzapotla ignimbrite. *Journal of Volcanology and Geothermal Research*, **136**, 97–119.

MORÁN-ZENTENO, D. J., CERCA, M. & KEPPIE, J. D. 2005. La evolución tectónica y magmatica Cenozoica del suroeste de México: avances y problemas de interpretación. *In*: NIETO-SAMANIEGO, A. F. & ALANIZ-ÁLVAREZ, S. A. (eds) *Temas selectos de la Geología Mexicana. Boletín de la Sociedad Geológica Mexicana*, **57**, 319–342.

NIETO-SAMANIEGO, Á. F., ALANIZ-ÁLVAREZ, S. A. & ORTEGA-GUTIÉRREZ, F. 1995. Estructura interna de la falla de Oaxaca (México) e influencia de las anisotropías litológicas durante su actividad Cenozoica. *Revista Mexicana de Ciencias Geológicas*, **12**, 1–8.

NIETO-SAMANIEGO, A. F., ALANIZ-ÁLVAREZ, S. A., SILVA-ROMO, G., EGUIZA-CASTRO, M. H. & MENDOZA-ROSALES, C. C. 2006. Latest Cretaceous to Miocene deformation events in the eastern Sierra Madre del Sur, Mexico, inferred from the geometry and age of major structures. *Geological Society of America Bulletin*, **118**, 238–252; doi: 10.1130/ B25730.1.

ORTEGA-GUTIÉRREZ, F. 1981. Metamorphic belts of southern Mexico and their tectonic significance, *Geofisica Internacional*, **20**, 177–202.

ORTEGA-GUTIÉRREZ, F. & ELÍAS-HERRERA, M. 2003. Wholesale melting of the southern Mixteco Terrane and origin of the Xolapa Complex. *Geological Society of America Cordilleran Section Meeting 2003*, Abstract 27-6.

PINDELL, J. L. 1993. Regional synopsis of Gulf of Mexico and Caribbean evolution. *In*: PINDELL, J. L. & PERKINS, R. (eds) *Mesozoic and Early Cenozoic Development of the Gulf of Mexico and Caribbean Region*. GCSSEPM Foundation, 13th Annual Research Conference Proceedings, 251–274.

PINDELL, J. L. & KENNAN, L. 2001. Kinematic evolution of the Gulf of Mexico and Caribbean. *GCSSEPM Foundation 21st Annual Research Conference Transactions, Petroleum Systems of Deep-Water Basins: Global and Gulf of Mexico Experience*, Houston, TX, 193–220.

PINDELL, J. L., KENNAN, L., MARESCH, W. V., STANEK, K.-P., DRAPER, G. & HIGGS, R. 2005. *Plate-Kinematics and Crustal Dynamics of Circum-Caribbean Arc–Continent Interactions: Tectonic Controls on Basin Development in Proto-Caribbean Margins*. Geological Society of America, Special Papers, **394**, 7–52; doi: 10.1130/2005.2394(01).

RAMBERG, H. 1981. *Gravity, Deformation and Earth's Crust*. Academic Press, San Diego California, CA.

RATSCHBACHER, L., RILLER, U., MESCHEDE, M., HERRMANN, U. & FRISCH, W. 1991. Second look at suspect terranes in southern Mexico. *Geology* **19**, 1233–1236.

RILLER, U., RATSCHBACHER, L. & FRISCH, W. 1992. Left-lateral transtension along the Tierra Colorada deformation zone, northern margin of the Xolapa magmatic arc of southern Mexico. *Journal of South American Earth Sciences*, **5**, 237–249.

RIVERA-CARRANZA, E., TEJADA-SEGURA, M. A. *ET AL.* 1998. *Carta geológico-minera de la hoja Cuernavaca. Escala 1:250000 (E14-5)*. Pachuca, Hidalgo, *Consejo de Recursos Minerales*, 1 map.

ROGERS, R. D., PATINO, L. & SCOTT, R. 2003. The Cretaceous margins of the extreme southwest corner of the North American Plate. *Geological Society of America 99th Cordilleran Section Meeting 2003*, Abstract with Programs, Paper 33-6.

ROGERS, R. D., MANN, P. & EMMET, P. 2007. Tectonic terranes of the Chortis block based on integration of regional aeromagnetic and geologic data. *In*: MANN, P. (ed.) *Geologic and Tectonic Development of the Caribbean Plate in Northern Central America*. Geological Society of America Special Paper, **428**, 65–88.

ROSS, M. I. & SCOTESE, C. R. 1988. A hierarchical tectonic model of the Gulf of Mexico and Caribbean region. *Tectonophysics*, **155**, 139–168.

SÁNCHEZ-ZAVALA, J. L. 2005. *Estratigrafía, sedimentología y análisis de procedencia de la Formación Tecomate y su papel en la evolución del Complejo Acatlán, sur de México*. PhD Thesis, Universidad Nacional Autónoma de México, Posgrado en Ciencias de la Tierra.

SCHAAF, P., MORÁN-ZENTENO, D., HERNÁNDEZ-BERNAL, M. S., SOLÍS-PICHARDO, G., TOLSON, G. & KÖHLER, H. 1995. Paleogene continental margin truncation in southwestern México: geochronological evidence. *Tectonics* **14**, 1339–1350.

SEDLOCK, R. L., ORTEGA-GUTIERREZ, F. & SPEED, R. C. 1993. *Tectonostratigraphic Terranes and Tectonic Evolution of Mexico*. Geological Society of America, Special Papers, 278.

SOLARI, L. A., TORRES DE LEÓN, R., HERNÁNDEZ-PINEDA, G., SOLÉ, J., SOLÍS-PICHARDO, G. & HERNÁNDEZ-TREVIÑO, T. 2006. Geochronology, structural evolution of the Tierra Colorada area, and tectonic implications for southern Mexico and Chortís block connections. *EOS Transactions of the American Geophysics Union*, **87**, Joint Assembly Supplement Abstract U44A-04.

TEYSSIER, C. & CRUZ, L. 2004. Strain gradients in transpressional to transtensional attachment zones. *In*: GROCOTT, J., McCAFFREY, K. J. W., TAYLOR, G. & TIKOFF, B. (eds) *Vertical Coupling and Decoupling in the Lithosphere*. Geological Society, London, Special Publications, **227**, 101–115.

TEYSSIER, C. & TIKOFF, B. 1998. Strike–slip partitioned transpression of the San Andreas fault system: a lithospheric-scale approach. *In*: HOLDSWORTH, R. E., STRACHAN, R. A. & DEWEY, J. F. (eds) *Continental Transpressional and Transtensional Tectonics*. Geological Society, London, Special Publications, **135**, 143–158.

TOLSON, G. 1998. *Deformación, exhumación y neotectónica de la margen continental de Oaxaca: Datos estructurales, petrológicos y geotermobarométricos*. PhD Thesis, Universidad Nacional Autónoma de México, UACPyP México, D. F.

TOLSON, G. 2005. La falla Chacalapa en el sur de Oaxaca. *Boletín de la Sociedad Geológica Mexicana*, Volumen conmemorativo del centenario, Grandes fronteras tectónicas de México, **57**, 111–122.

TORRES DE LEÓN, R. & SOLARI, L. 2004. Geología estructural e implicaciones tectónicas de unidades metamórficas del área de La Venta, Estado de Guerrero. *IV Reunión Anual de Ciencias de la Tierra*, GEOS, **24**, 166, GET–27.

TURCOTTE, D. L. & SCHUBERT, G. 1982. *Geodynamics: Applications of Continuum Physics to Geological Problems*. John Wiley & Sons, New York.

A seismotectonic model for the Chortís Block

MARCO GUZMÁN-SPEZIALE

*Centro de Geociencias, Universidad Nacional Autónoma de México, Mexico, Boulevard
Juriquilla 3001, Juriquilla, Querétaro, 76230, México.
(e-mail: marco@geociencias.unam.mx)*

Abstract: I propose a new seismotectonic model for the Chortís Block, at the northwestern
corner of the Caribbean Plate. Shallow seismicity in the area clearly shows three zones of deformation: one along the North America–Caribbean Plate boundary and another along the Central
America volcanic arc, and one in the area of the grabens of northern Central America. Analysis
of Centroid moment–tensor solutions for shallow earthquakes in these three area show that **T** or
tension axes are horizontal and trend away from the corner, and that **P** or compression axes
for the plate boundary and the volcanic arc are also horizontal and trending towards the corner.
Calculation of seismic moment release per unit volume reveals similar values for the volcanic
arc and the plate boundary. The state of stress and similarity in seismic moment release suggest
that the Chortís Block is being extruded towards the ESE. This is probably due to compression
of the large North America and Cocos Plates that surround it.

From the neotectonic point of view the northwestern
corner of the Caribbean Plate is a rather complex
area. To the north, the North America–Caribbean
Plate boundary follows the Swan Islands and
Motagua–Polochic fault zones, a series of active,
left-lateral transform faults. To the west, the Cocos
Plate is subducted below the Caribbean Plate
along the Middle America trench (Fig. 1). For
most authors (e.g. Muehlberger & Ritchie 1975;
Plafker 1976; Burkart 1978, 1983; Guzmán-
Speziale *et al.* 1989) the three plates do not form a
simple trench–trench–transform triple junction, in
the sense of McKenzie & Morgan (1969).

Guzmán-Speziale *et al.* (1989), Guzmán-
Speziale & Meneses-Rocha (2000) proposed that,
north of the Motagua–Polochic fault system, the
structural deformation related to the triple junction
extends into southeastern Mexico. There, a strike–
slip left-lateral system involves more than nine fault
zones to form a complex strike–slip fault province.
These appear to be linked by a fault step or jog, to a
province associated with many reverse faults.

One would expect that, south of the Motagua–
Polochic fault system, deformation associated
with the triple junction would be as widespread
as it is in Mexico. Yet, published works address
only specific subjects and/or areas: there are
papers on block rotation by Plafker (1976) and by
Burkart & Self (1985); the Guayapé fault is the
subject of papers by Finch & Ritchie (1991) and
by Gordon & Muehlberger (1994); Avé Lallemant
& Gordon (1999) studied deformation in the Tela
Basin, in Honduras; Guzmán-Speziale (2001) determined the rate of opening of the grabens of Central
America from seismic strain-rate: Cáceres *et al.*
(2005) analysed crustal seismic deformation in
northern Central America, but focused on seismic

hazard in the region, rather than tectonics; and
seismicity along the Central America volcanic arc
has been addressed in Harlow & White (1985),
White & Harlow (1993), DeMets (2001) and
Guzmán-Speziale *et al.* (2005).

I now present a seismotectonic model for
deformation of the entire northwestern corner
of the Caribbean Plate, that is the Chortís Block.
We base our study on the analysis of Centroid
moment–tensor solutions (CMTs) of crustal earthquakes along the Central America volcanic arc
(CAVA), the North America–Caribbean Plate
boundary zone (NOAM–CARB), and the grabens
of northern Central America. Specifically, we determine directions of principal axes of tension (**T** axes)
and compression (**P** axes), and slip vectors. We also
calculate and compare seismic strain-rate along the
NOAM–CARB and the CAVA.

Tectonic framework

The Chortís Block is located in the northwestern
corner of the Caribbean Plate (Fig. 1). To the SW,
the Cocos Plate subducts the Caribbean Plate along
the Middle America Trench at an approximate rate
of 65–80 mm/annum (DeMets *et al.* 1990). The
northern limit of the block is the Motagua–Polochic
fault system which, together with the Swan
Islands and Oriente fault zones, is the left-lateral,
transform plate boundary between the Caribbean
and North America plates. The relative motion
between these two plates is about 20 mm/annum
(e.g. Dixon *et al.* 1998; DeMets *et al.* 2000), although
several authors have proposed that the Motagua–
Polochic fault system continues to the west of
their known surface trace and meets the Middle
America trench (e.g. Malfait & Dinkelman 1972;

From: JAMES, K. H., LORENTE, M. A. & PINDELL, J. L. (eds) *The Origin and Evolution of the Caribbean Plate.*
Geological Society, London, Special Publications, **328**, 197–204.
DOI: 10.1144/SP328.9   0305-8719/09/$15.00 © The Geological Society of London 2009.

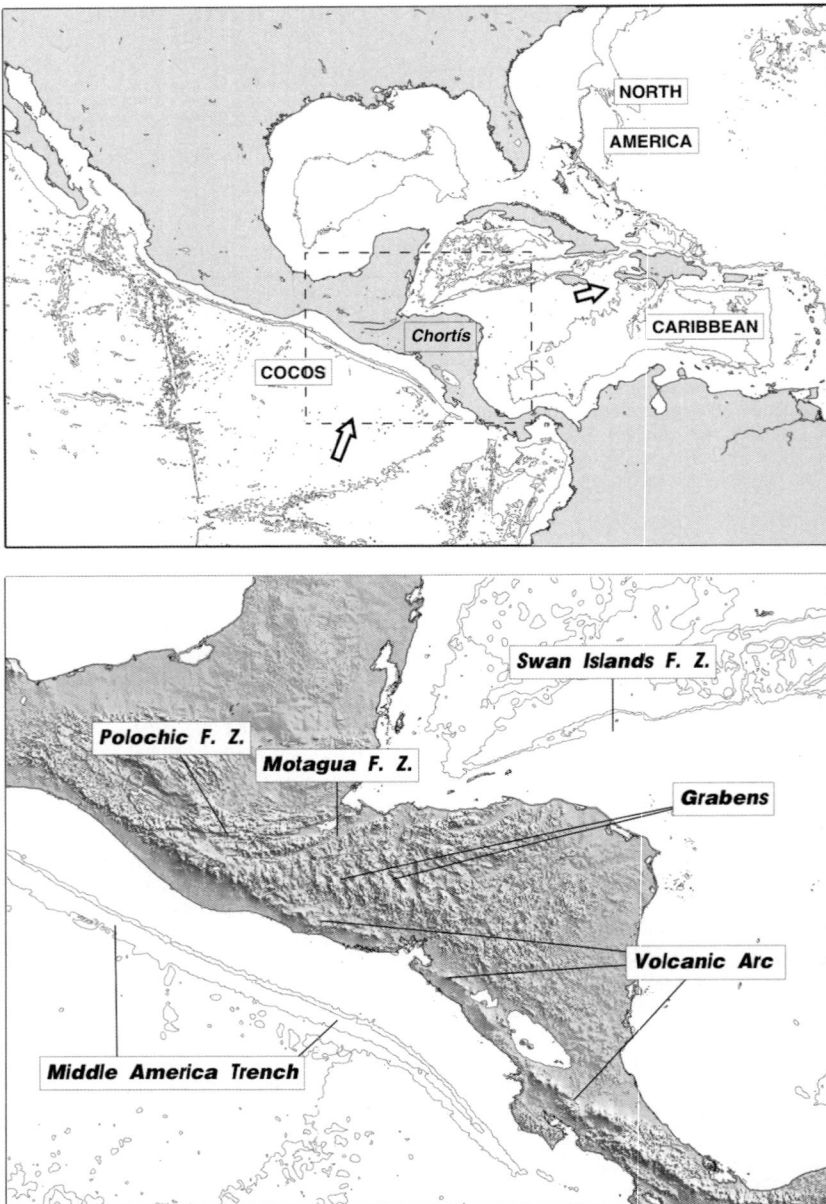

Fig. 1. Tectonic framework of the Chortís Block. **Top**: Plate names are indicated as well as the location of the block. Arrows represent direction of motion of the plate with respect to North America and are proportional to the speed. **Bottom**: Main tectonic elements of the northwestern Caribbean. All figures drawn with the help of the *Generic Mapping Tools* (*GMT*) of Wessel and Smith (1991).

Muehlberger & Ritchie 1975; Plafker 1976; Burkart 1978; Machorro & Mickus 1993). Guzmán-Speziale *et al.* (1989) showed that there is no direct evidence that the Motagua–Polochic system continues to the west to meet the Middle America trench (Fig. 1).

Southeast of the fault system and parallel to the Middle America trench lies the Central America volcanic arc, which consists of about 75 edifices with documented Holocene activity. Thirty-one of these volcanoes have been active in historic times

(Simkin *et al.* 1981, 1994). The arc extends for about 900–1000 km, from the Motagua–Polochic fault sytsem to central Costa Rica, with closely spaced (12–30 km apart) volcanoes, a width of no more than 15 km and elevations ranging from 100 to 4000 m. Volcanic activity is directly related to subdction of the Cocos Plate underneath the Caribbean (e.g. Carr & Stoiber 1990).

Just SSE of the Motagua–Polochic fault system and ENE of the volcanic arc, there is a system of north–south trending grabens, herein referred to collectively as the grabens of northern Central America, which comprise at least 13 structures, the main ones being the Guatemala City, Ipala, and Esquipulas grabens as well as the Honduras depression (Fig. 1) (e.g. Dengo 1968; Bonis *et al.* 1970; Weyl 1980; Burkart 1983; Mann & Burke 1984; Gordon & Muehlberger 1994).

There are various explanations for the formation of these grabens, such as local tensile stresses (Dengo 1968; Mann & Burke 1984), *pinning* of the Caribbean Plate against North America (Malfait & Dinkelman 1972; Plafker 1976), *terminus* of a great strike–slip fault system (Langer & Bollinger 1979), block rotation (Burkart & Self 1985), flexural stresses because of the shape of the Motagua fault (Suter 1991), and response to slip along major strike–slip faults (Gordon & Muehlberger 1994).

The Cocos–Caribbean Plate boundary is seismically very active, with earthquakes sometimes reaching magnitudes close to 8 (Molnar & Sykes 1969; Dean & Drake 1978; Burbach *et al.* 1984; Dewey & Suárez 1991). The North America Caribbean Plate boundary is also seismically active. Worldwide catalogues (e.g. National Earthquake Information Center, International Seismological Center, Harvard University) show many epicentres along the boundary (Fig. 2). The largest recorded event is the Guatemala earthquake of 4 February 1976, which reached a magnitude of 7.5 (Plafker 1976; Kanamori & Stewart 1978; Langer & Bollinger 1979). White (1984) has documented 25 destructive earthquakes in the vicinity of the Motagua–Polochic fault system since 1530.

Along the volcanic arc, shallow-focus, strike–slip earthquakes take place, at an average of one event every 3–5 years, and magnitudes can reach 6.9 (e.g. White 1991; White & Harlow 1993). These earthquakes are clearly of tectonic origin, that is, not related to volcanic activity, and are well documented at least since the sixteenth century (Carr & Stoiber 1977; White & Harlow 1993; Ambraseys & Adams 1996; Peraldo & Montero 1999). In a recent work, Guzmán-Speziale *et al.* (2005) compiled 55 historic (1700–1978) earthquakes along the arc with magnitudes greater than 5 (Fig. 2).

There is also seismic activity in the grabens of northern Central America (Fig. 2). Carr & Stoiber (1977), Osiecki (1981), Montero-Pohly (1989), White & Harlow (1993) and Peraldo & Montero (1999) have published information on historic earthquakes which, as shown by Guzmán-Speziale (2001), are shallow-focus events in the grabens of northern Central America. Our compilation (Guzmán-Speziale 2001; Guzmán-Speziale *et al.* 2005) from these sources yielded 23 events prior to 1978. The first event dates back to 1586.

Data and method

We use earthquake focal mechanisms, regularly published as Centroid Moment tensor solutions (CMT) by the University of Harvard (e.g. Dziewonski & Woodhouse 1983) (see also http://www.globalcmt.org/CMTsearch.html) for the years 1978–2005. Although there are some CMTs in the catalogue for 1976 and 1977, systematic reporting started in 1978. We only include CMTs for crustal earthquakes (shallower than 50 km) and whose focal mechanism is either normal-faulting or strike–slip faulting (Fig. 2).

A centroid moment tensor is just a form of the seismic moment tensor. In turn, the seismic moment tensor is a matrix formulation of an earthquake focal mechanism. This matrix, with elements M_{ij}, contains information on the size of the earthquake (the *scalar seismic moment* or just the *seismic moment Mo*), the orientation and plunge of the faulting plane, and the state of stress that gave rise to the faulting process, that is, the principal axes of compression (**P**), null (**B**) and tension (**T**).

Frohlich & Apperson (1992) showed that the faulting mechanism of an earthquake may be determined by the plunge of one of their principal axes. Inspired by these authors, we define normal-faulting mechanisms as those in which the **P** (compression) axis plunges at least 45°, whereas the **B** (null) axis of a strike–slip mechanism plunges 45° or more. We discard thrust-faulting mechanisms (for which the **T**, or tension, axis plunges ≥45°) because we consider these earthquakes related to the subduction process, although for completeness we show their location in Figure 2.

Earthquake epicentres concentrate in three areas: the North America–Caribbean Plate boundary, the Central America volcanic arc and the grabens of northern Central America. The first two areas are arcuate in shape so, to better visualize deformation along them, we fit small circles to them (Fig. 3). Their respective centres of curvature are given in Table 1.

Earthquakes along the North America–Caribbean Plate boundary show strike–slip left-lateral mechanisms, in accordance with the relative motion between the two plates. There is no fault running along the Central America volcanic

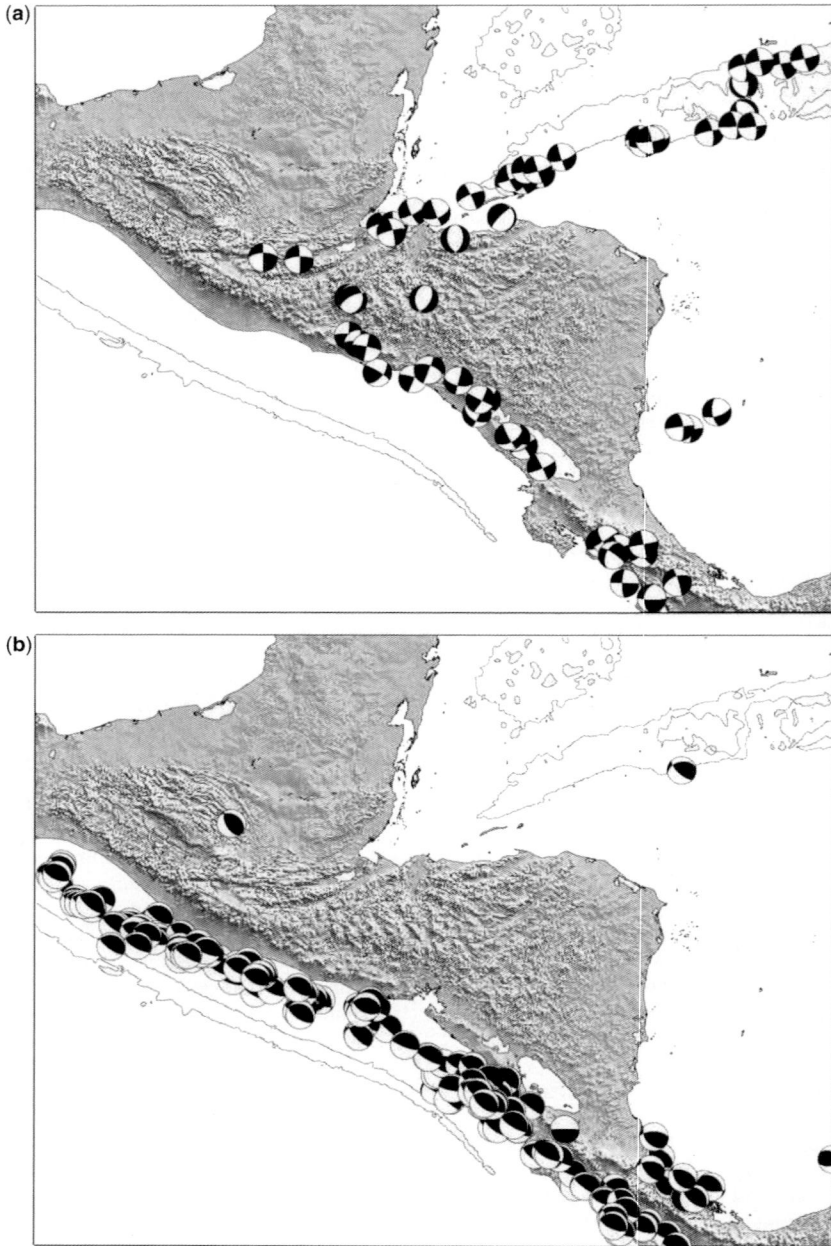

Fig. 2. Focal mechanisms of shallow ($z \leq 50$ km) earthquakes reported as centroid moment tensors in the period 1978–2005. (**a**) Normal-faulting and strike–slip mechanisms. (**b**) Reverse-faulting mechanisms.

arc; focal mechanisms are also strike–slip, either right-lateral parallel to the trend of the arc or left-lateral perpendicular to it. From geological (i.e. local surface faulting) or seismic (distribution of aftershocks) evidence, it is known that faulting

for some of the earthquakes is right-lateral along the arc, and left-lateral perpendicular to it for others (Guzmán-Speziale *et al.* 2005).

Principal stress axes, tension (**T**), null (**B**) and compression (**P**), are parameters routinely

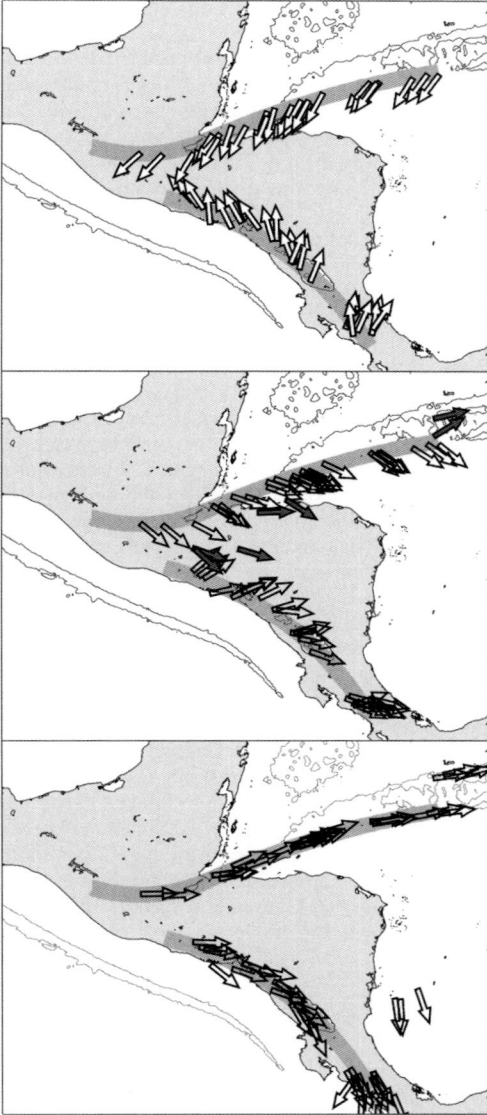

Fig. 3. Results from centroid moment tensors. Shaded areas define the zones of deformation (see text for details). **Top**: Directions of **P** axes from strike–slip earthquakes along the North America–Caribbean Plate boundary and the Central America volcanic arc. **Middle**: Directions of **T** axes for the same earthquakes plus normal faulting earthquakes (solid arrows). **Bottom**: Directions of slip vectors for the strike–slip events.

determined in CMTs. They are the eigenvectors of the moment tensor (e.g. Jost & Herrmann 1989). We directly obtain **T** and **P** axes from published CMTs.

Table 1. *Centres of curvature for deformation areas*

Location	Latitude (N)	Longitude (E)	Distance (deg)
Volcanic arc	4.144	−92.422	10.092
Motagua–Polochic	20.083	−90.770	4.791
Swan Transform	−9.646	−77.036	27.867

Earthquake slip vectors **u**, can be obtained with the vector equation (e.g. Jost & Hermann 1989):

$$\mathbf{u} = \frac{1}{\sqrt{2}}(\mathbf{T} + \mathbf{P}). \tag{1}$$

If the focal mechanism is strike–slip, the principal **T** and **P** axes, as well as the slip vector **u**, are close to horizontal, although **u** is not unique. The slip vector lies on the plane of the fault, but the focal mechanism (or the seismic moment tensor) determines two faulting planes, perpendicular to each other, one of which is the actual physical (geological) fault plane and the other is the auxiliary plane.

We also wish to obtain and compare the total seismic moment release per unit volume:

$$MV = \frac{\sum_{n=1}^{N} M_{ij}^{n}}{V} \tag{2}$$

along the volcanic arc and along the plate boundary. This is a measure of deformation because it is directly proportional to the seismic strain rate of Kostrov (1974), who showed that the average seismic strain-rate due to the N earthquake moment tensors M_{ij} within the volume V and time τ and a modulus of rigidity μ is:

$$\dot{\varepsilon} = \frac{1}{2\mu V \tau} \sum_{n=1}^{N} M_{ij}^{n}. \tag{3}$$

In turn, the sum of seismic moment tensors may be expressed as the product of the total seismic moment release times a *shape* tensor F, the latter of which only yields the directions of the eigenvectors (e.g. Papazachos & Kiratzi 1992):

$$\sum_{n=1}^{N} M_{ij}^{n} = \left(\sum_{n=1}^{N} M_{0}^{n} \right) \bar{F}_{ij}. \tag{4}$$

Finally,

$$\dot{\varepsilon} = MV \cdot \frac{1}{2\mu\tau} \bar{F}. \tag{5}$$

In order to calculate the volume, we take the deformation area to extend to a depth of 15 km (e.g. White & Harlow 1993) and a width of 30 km. Because the areas are arcuate, the volume is that of a circular segment on a sphere and may be calculated as indicated in Guzmán-Speziale *et al.* (2005).

Results and discussion

Slip vectors **u** of strike–slip earthquakes along the plate boundary are aligned along the deformation zone, which is consistent with the relative motion between the North America and Caribbean Plates (Fig. 3). As discussed above, some mechanisms for earthquakes along the volcanic arc are right-lateral along it, and some are are left-lateral perpendicular to it. Consequently, the slip vector is either parallel to the trend of the arc or perpendicular to it.

Compression vectors (**P** axes) are consistently oriented at an angle of about 45° with respect to the deformation zone. They trend toward the NW corner of the Caribbean Plate, that is, those along the North America–Caribbean Plate boundary trend to the SW and those along the volcanic arc trend to the NW (Fig. 3). Tension vectors (**T** axes), on the other hand, also form a 45° angle with the DZ, but they trend outward from the Chortís corner. It is to be noted here that normal-faulting earthquakes in the grabens of northern Central America also have subhorizontal **T** axes and that they trend to the east (Fig. 3).

The deformation volume along the Central America volcanic arc is 3.98×10^{14} m^3, and 5.33×10^{14} m^3 along the North America–Caribbean Plate boundary. Seismic moment release for the 25 earthquakes along the volcanic arc is 3.00×10^{19} N m and for the 23 earthquakes along the plate boundary it is 3.79×10^{19} N m. It could be argued that the great Guatemalan earthquake of 1976 ($Mo = 2.04 \times 10^{20}$ N m) should be included in the calculations. However, according to White (1984) and Peraldo & Montero (1999), the period of recurrence of large ($M > 6.5$) earthquakes along the Motagua–Polochic fault system is about 120–150 years. Therefore the Guatemala earthquake of 1976 is not typical of the 28-year period (1978–2005) considered here. The total seismic moment release per unit volume then becomes 7.54×10^4 N m^{-2} for the volcanic arc and 7.11×10^4 N m^{-2} for the plate boundary.

It has been suggested (e.g. Harlow & White 1985; DeMets 2001) that seismic activity along the Central America volcanic arc is due to oblique subduction of the Cocos Plate but we have presented several lines of evidence that suggest that this activity is not related to oblique subduction (Guzmán-Speziale & Gómez 2002; Guzmán-Speziale *et al.* 2005), although the state of

stress is related to subduction along the Middle America trench.

From the direction of **P** axes determined in this work, it is clear that the Chortís Block is undergoing compression directed towards the northwestern corner of the block. At the same time, tension (**T**) axes trend away from it. **T** axes for normal-faulting earthquakes along the grabens trend almost due east (Fig. 3).

Since the volcanic arc is not connected to the North America–Caribbean Plate boundary, the intervening deformation is resolved as extension along the grabens which are oriented north–south (east–west extension) and whose **T** axes also show east–west extension.

Malfait & Dinkelman (1972) and Plafker (1976) proposed that the northwestern corner of the Caribbean is being *pinned* by compression between the Cocos and North America plates. Furthermore, Plafker (1976) notes that most of the deformation is confined to the area north of the volcanic chain. He suggests that incipient decoupling within the Caribbean Plate may be taking place along the volcanic arc.

The state of stress indicates horizontal **P** axes which trend towards the northwestern corner of the Caribbean Plate, and horizontal **T** axes away from it, along both the North America–Caribbean Plate boundary and the volcanic arc. Horizontal **T** axes along the grabens of Central America trend to the east. This state of stress suggests a model in which, as already proposed by Plafker (1976), the northwestern corner of the Caribbean Plate is being *pinned* by compression of the Cocos Plate underneath the Caribbean, and relative motion of the North America Plate with respect to the Caribbean (Fig. 4). In this model, the volcanic arc plays

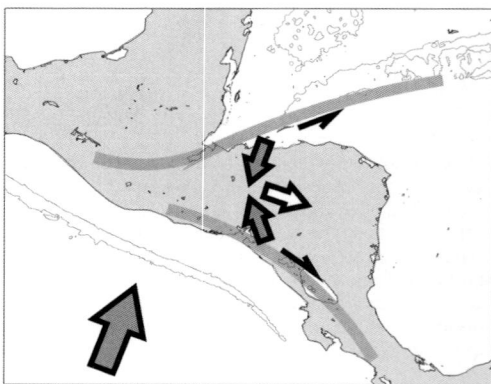

Fig. 4. Model for the state of stress and probable relative motion in the northwestern Caribbean. Dark arrows represent compression, light arrows extension, and half arrows probable displacement.

the role of the zone along which the Chortís Block is being deformed, and even decoupled. Compression along the Middle America trench may be easily transfered to the volcanic arc because it is is a zone of thinned lithosphere and higher thermal gradient, and hence of decreased lithospheric strength.

In a sense, our model proposes that the Chortís Block is being *extruded* to the east, along the North America–Caribbean Plate boundary and the Central America volcanic arc.

Conclusions

The Chortís Block is surounded by zones of active seismic deformation: the North America–Caribbean Plate boundary, the Central America volcanic arc, and the grabens of northern Central America.

P axes of shallow earthquakes are horizontal and tend towards the northwestern corner of the Caribbean Plate. **T** axes are also horizontal and trend away from the corner.

Seismic moment release is almost equal along the North America–Caribbean Plate boundary and the Central America volcanic arc.

These results suggest that the Chortís Block is being extruded towards the east by compression from the large North America and Cocos Plates that surround it.

Very stimulating discussions stemmed from the Conference on the *Geology of the Area between North and South America, with Focus on the Origin of the Caribbean Plate*, held in Sigüenza, Spain, in May 2006. These discussions greatly improved this manuscript. A. W. Bally meticulously read an earlier version of the manuscript and provided many useful comments. Centro de Geociencias UNAM contribution 1107.

References

AMBRASEYS, N. N. & ADAMS, R. D. 1996. Large-magnitude Central America earthquakes, 1898–1994. *Geophysical Journal International*, **127**, 665–692.

AVÉ LALLEMANT, H. G. & GORDON, M. B. 1999. Deformation hisotry of Roatán island: implications for the origin of the Tela basin (Honduras). *In*: MANN, P. (ed.) *Caribbean Basins. Sedimentary Basins of the World*, **4**, Elsevier, Amsterdam, 197–218.

BONIS, S. B., BOHNENBERG, O. H. & DENGO, G. 1970. *Mapa Geológico de la República de Guatemala*, scale 1:500 000.

BURBACH, G. V., FROHLICH, C., PENNINGTON, W. D. & MATUMOTO, T. 1984. Seismicity and tectonics of the subducted Cocos plate. *Journal of Geophysical Research*, **89**, 7719–7735.

BURKART, B. 1978. Offset across the Polochic fault of Guatemala and Chiapas, Mexico. *Geology*, **6**, 328–332.

BURKART, B. 1983. Neogene North American–Caribbean Plate boundary across northern Central America: offset along the Polochic Fault. *Tectonophysics*, **99**, 251–270.

BURKART, B. & SELF, S. 1985. Extension and rotation of crustal blocks in northern Central America and effect on the volcanic arc. *Geology*, **13**, 22–26.

CÁCERES, D., MONTERROSO, D. & TAVAKOLI, B. 2005. Crustal deformation in northern Central America. *Tectonophysics*, **404**, 119–131.

CARR, M. J. & STOIBER, R. E. 1977. Geologic setting of some destructive earthquakes in Central America. *Geological Society of America Bulletin*, **88**, 151–156.

CARR, M. J. & STOIBER, R. E. 1990. Volcanism. *In*: DENGO, G. & CASE, J. E. (eds) *The Caribbean Region*, **H**. Geological Society of America, Boulder, CO, 375–391.

DEAN, B. L. & DRAKE, C. L. 1978. Focal mechanism solutions and tectonics of the Middle America Arc. *Journal of Geology*, **86**, 111–128.

DEMETS, C. 2001. A new estimate for present-day Cocos–Caribbean Plate motion: implications for slip along the Central American volcanic arc. *Geophysical Research Letters*, **28**, 4043–4046.

DEMETS, C., GORDON, R. G., ARGUS, D. F. & STEIN, S. 1990. Current plate motions. *Geophysical Journal International*, **101**, 425–478.

DEMETS, C., JANSMA, P. *ET AL*. 2000. GPS geodetic constraints on Caribbean–North America plate motion. *Geophysical Research Letters*, **27**, 437–440.

DENGO, G. 1968. *Estructura geológica, historia tectónica y morfología de America Central*. Centro Regional de Ayuda Técnica, Agencia para el Desarrollo internacional (AID), México.

DEWEY, J. W. & SUÁREZ, G. 1991. Seismotectonics of Middle America. *In*: SLEMMONS, D. B., ENGDAHL, E. R., ZOBACK, M. D. & BLACKWELL, D. D. (eds) *Neotectonics of North America*. Geological Society of America, Boulder, CO, 309–321.

DIXON, T., FARINA, F., DEMETS, C., JANSMA, P., MANN, P. & CALAIS, E. 1998. Relative motion between the Caribbean and North American plates and related boundary zone deformation from a decade of GPS observations. *Journal of Geophysical Research*, **98**, 15157–15182.

DZIEWONSKI, A. M. & WOODHOUSE, J. H. 1983. An experiment in systematic study of global seismicity: centroid-moment tensor solutions for 201 moderate and large earthquakes of 1981. *Journal of Geophysical Research*, **88**, 3247–3271.

FINCH, R. C. & RITCHIE, A. W. 1991. The Guayapé fault system, Honduras, Central America. *Journal of South American Earth Sciences*, **4**, 43–60.

FROHLICH, C. & APPERSON, K. D. 1992. Earthquake focal mechanisms, moemnt tensors, and the consistency of seismic activity near plate boundaries. *Tectonics*, **11**, 279–296.

GORDON, M. B. & MUEHLBERGER, W. R. 1994. Rotation of the Chortís block causes dextral slip on the Guayapé fault. *Tectonics*, **13**, 858–872.

GUZMÁN-SPEZIALE, M. 2001. Active seismic deformation in the grabens of northern Central America and its relationship to the relative motion of the North America–Caribbean Plate boundary. *Tectonophysics*, **337**, 39–51.

GUZMÁN-SPEZIALE, M. & GÓMEZ, J. M. 2002. Comment on 'A new estimate for present-day Cocos-Caribbean Plate motion: implications for slip along the Central America volcanic arc' by Charles DeMets. *Geophysical Research Letters*, **29**, 1945; doi: 10.1029/2002GL015011.

GUZMÁN-SPEZIALE, M. & MENESES-ROCHA, J. J. 2000. The North America–Caribbean Plate boundary west of the Motagua–Polochic fault system: a fault jog in southeastern México. *Journal of South American Earth Sciences*, **13**, 459–468.

GUZMÁN-SPEZIALE, M., PENNINGTON, W. D. & MATUMOTO, T. 1989. The triple junction of the North America, Cocos, and Caribbean plates: seismicity and tectonics. *Tectonics*, **8**, 981–997.

GUZMÁN-SPEZIALE, M., VALDÉS-GONZÁLEZ, C., MOLINA, E. & GÓMEZ, J. M. 2005. Seismic activity along the Central America volcanic arc: is it related to subduction of the Cocos plate? *Tectonophysics*, **400**, 241–254.

HARLOW, D. H. & WHITE, R. A. 1985. Shallow earthquakes along the volcanic chain in Central America: evidence for oblique subduction (abstract). *Earthquake Notes*, **55**, 28.

JOST, M. L. & HERRMANN, R. B. 1989. A student's guide to and review of moment tensors. *Seismological Research Letters*, **60**, 37–57.

KANAMORI, H. & STEWART, G. S. 1978. Seismological aspects of the Guatemala earthquake of February 4, 1976. *Journal of Geophysical Research*, **83**, 3427–3434.

KOSTROV, V. V. 1974. Seismic moment and energy of earthquakes and seismic flow of rock. *Izvestiya Academy of Sciences of the USSR Earth Physics*, **1**, 23–44.

LANGER, C. J. & BOLLINGER, G. A. 1979. Secondary faulting near the terminus of a seismogenic strike–slip fault; aftershocks of the 1976 Guatemala earthquake. *Bulletin of the Seismological Society of America*, **69**, 427–444.

MACHORRO, R. & MICKUS, K. 1993. Structural continuity of the Polochic Fault into southwest Mexico. *EOS, Transactions of the American Geophysical Union*, **74**, 576.

MALFAIT, B. T. & DINKELMAN, M. G. 1972. Circum-Caribbean tectonic and igneous activity and the evolution of the Caribbean Plate. *Geological Society of America Bulletin*, **83**, 251–272.

MANN, P. & BURKE, K. 1984. Cenozoic rift formation in the northern Caribbean. *Geology*, **12**, 732–736.

MCKENZIE, D. P. & MORGAN, W. J. 1969. Evolution of triple junctions. *Nature*, **224**, 125–133.

MOLNAR, P. & SYKES, L. R. 1969. Tectonics of the Caribbean and Middle America regions from focal mechanisms and seismicity. *Geological Society of America Bulletin*, **80**, 1639–1684.

MONTERO-POHLY, W. 1989. Sismicidad Histórica de Costa Rica. *Geofísica International*, **28**, 521–559.

MUEHLBERGER, W. & RITCHIE, A. W. 1975. Caribbean–Americas plate boundary in Guatemala and southern Mexico as seen on Skylab IV orbital photography. *Geology*, **3**, 232–235.

OSIECKI, P. S. 1981. Estimated intensities and probable tectonic sources of historic (pre 1898) Honduran earthquakes. *Bulletin of the Seismological Society of America*, **71**, 865–881.

PAPAZACHOS, C. & KIRATZI, A. 1992. A formulation for reliable estimation of active crustal deformation and its application to central Greece. *Geophysical Journal International*, **111**, 424–432.

PERALDO, G. & MONTERO, W. 1999. *Sismología Histórica de América Central*. Instituto Panamericano de Geografía e Historia, México.

PLAFKER, G. 1976. Tectonic aspects of the Guatemala earthquake of 4 February 1976. *Science*, **193**, 1201–1208.

SIMKIN, T., SIEBERT, L., MCCLELLAND, L., BRIDGE, D., NEWHALL, C. & LATTER, J. H. 1981. *Volcanoes of the World*. Hutchinson Ross, Stroudsburg, PA.

SIMKIN, T., SIEBERT, L. & FURGANG, K. 1994. *Volcanoes of the World: A Regional Directory, Gazetteer, and Chronology of Volcanism during the Last 10,000 Years*. Geoscience Press, Tucson, AZ.

SUTER, M. 1991. State of stress and active deformation in Mexico and western Central America, *In*: SLEMMONS, D. B., ENGDAHL, E. R., ZOBACK, M. D. & BLACKWELL, D. D. (eds) *Neotectonics of North America*. Geological Society of America, Boulder, CO.

WESSEL, P. & SMITH, W. H. F. 1991. Free software helps map and display data. *EOS Transactions of the American Geophysical Union*, **72**, 445–446.

WEYL, R. 1980. *Geology of Central America*. Gebruder Borntraeger, Berlin.

WHITE, R. A. 1984. *Catalog of Historic Seismicty in the Vicinity of the Chixoy–Polochic and Motagua faults, Guatemala*. US Geological Survey Open-File Report **84–88**, 34.

WHITE, R. A. 1991. Tectonic implications of upper-crustal seismicity in Central America. *In*: SLEMMONS, D. B., ENGDAHL, E. R., ZOBACK, M. D. & BLACKWELL, D. D. (eds) *Neotectonics of North America*. Geological Society of America, Boulder, CO, 323–338.

WHITE, R. A. & HARLOW, D. H. 1993. Destructive upper-crustal earthquakes in Central America since 1900. *Bulletin of the Seismological Society of America*, **83**, 1115–1142.

Geological evolution of the NW corner of the Caribbean Plate

RICARDO A. VALLS ALVAREZ

Nichromet Extraction Inc. 2500-120 Adelaide Street West, Toronto, Ontario, Canada
(e-mail: vallsvg@aol.com)

Abstract: The Caribbean Plate consists of a plateau basalt, formed probably in the Middle Cretaceous, complicated by a continental block, Chortís, several magmatic arcs, strike–slip motions along major fault systems such as the Motagua–Polochic fault zone in Guatemala, the pull-apart basin of the Cayman Trough and subduction zones below Central America and the Lesser Antilles. Five major collisional events have been identified: (i) Late Paleocene–Middle Eocene collision of the Greater Antilles with the Bahamas platform; (ii) Late Cretaceous collision of Chortís with the Maya Block; (iii) emplacement of nappes upon the Venezuelan foreland in the Cenozoic; (iv) collision of the Western Cordillera oceanic complex with the Central Cordillera of Colombia; and (v) Miocene collision of the eastern Costa Rica–Panama arc with the Western Cordillera. All these 'orogenic events' show an eastward movement of the Caribbean Plate relative to the Americas. Migration of the Jamaica Block from the Pacific caused obduction of the oldest ophiolites of Huehuetenango at the western end of the Polochic–Río Negro faults in Guatemala. South-southwest migration of the Chortís Block from west of Mexico and northward towards the Maya Block destroyed a trench associated with the Motagua–Jalomáx fault system and caused the Chuacús Orogeny, emplacing Guatemalan ophiolite complexes and metamorphosing the rocks from the Chuacús Series.

The Caribbean Plate

Roughly 3.2 million square kilometres in area, the Caribbean Plate (Fig. 1) is the result of the pre-Cretaceous to Present interaction of the Nazca, Cocos, North and South American plates. The margins of these plates are large deformed belts resulting from several compressional episodes that started in the Cretaceous, followed by tensional and strike–slip tectonics.

The Caribbean Plate is mostly oceanic, consisting of a plateau basalt formed probably in the Middle Cretaceous, complicated by the presence the continental Chortís Block, several magmatic arcs, a combination of a strike–slip motions along major fault systems, such as the Motagua–Polochic fault zone in Guatemala and the Oriente fault in Cuba, the pull-apart Cayman Trough and sub-duction zones Central America and the Lesser Antilles. Although the Caribbean 'is a single place in the planet, surely with a single history' (Iturralde Vinent 2004) two conflicting theories regard its formation as allochthonous or in-place. This paper contributes data from Guatemala to the current discussions.

Morphotectonic units

Guatemala is physiographically divided into four main morphotectonic units: (1) a narrow Pacific Coastal Plain on the west; (2) a NW-trending Volcanic Province; (3) an east–west trending Central Cordillera centred on the Motagua Suture Zone; and (4) the Petén Lowland to the north.

The Pacific Coastal Plain is about 50 km wide and consists mainly of andesitic and basaltic alluvial pebbles and conglomerates derived from the Volcanic Province. It is the fore-arc basin of the Central American subduction zone.

The Volcanic Province is a chain of active volcanoes to the south and Cenozoic igneous rocks to the north. Magmatic activity results either from subduction or to collision of the Caribbean and North American plates. Volcanism has migrated southwards, becoming more basic. Cretaceous granites are present in the north and Cenozoic granites in the south. Quaternary volcanoes reach up to 4200 m high, and at least seven have been active or in the fumarole stage during the present century. The 80 000 year old Atitlán and Ayarza Lakes occupy calderas.

The Central Cordillera, with the oldest rocks in Guatemala, is composed mainly of Palaeozoic schists, pegmatites, Cretaceous carbonates and pre-Cretaceous to Cenozoic Ophiolite complexes. This area is crossed by the arcuate Motagua–Polochic fault system, considered to be the westward continuation of the Cayman Trench.

The Petén Lowland is the foreland of Palaeozoic and Mesozoic orogenies. This area is characterized by upper Cretaceous to recent carbonates, evaporites, clastics and alluvial deposits, becoming younger and less deformed towards the north. Abundant Cenozoic rhyolitic tuffs occur near the contact with ophiolites of the Polochic Belt (Valls 2006).

From: JAMES, K. H., LORENTE, M. A. & PINDELL, J. L. (eds) *The Origin and Evolution of the Caribbean Plate*.
Geological Society, London, Special Publications, **328**, 205–217.
DOI: 10.1144/SP328.10   0305-8719/09/$15.00 © The Geological Society of London 2009.

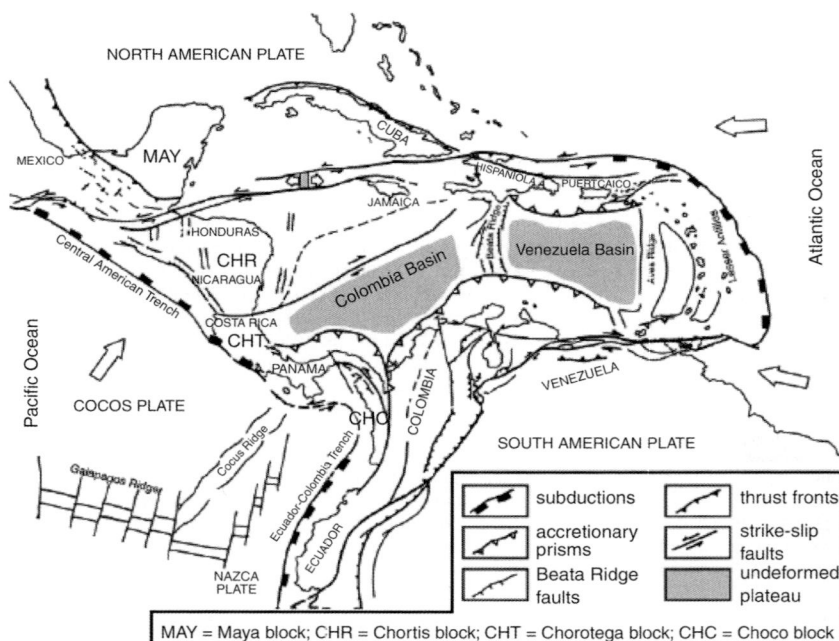

Fig. 1. Structural sketch map of the Caribbean area (from Giunta *et al.* 2002). Arrows show the drifting direction of the main plates. In red is the location of the area of the current study.

Guatemalan stratigraphy

The stratigraphy of Central Guatemala (Fig. 2) is described by Vinson (1962) and the stratigraphic lexicon of SE-central Guatemala (Millan 1985). Following is a brief description of the main stratigraphic units of Central Guatemala.

Chuacús Series (Lower Palaeozoic)

McBirney (1963) named this series of metamorphic rocks in the Central Cordillera, between the Maya and Chortís Blocks. Rocks include schist, gneiss, amphibolites and marbles. Chuacús Series sediments accumulated during Devonian, derived from a Precambrian landmass; U–Pb age dating of zircon, biotite–albite gneiss and biotite–albite–epidote gneiss gives a Proterozoic age of 1075 ± 25 Ma (Gomberg *et al.* 1968). There are three metamorphic zones. A chlorite-sericite zone of sericite schists, metagreywackes, meta-arkose, granitoids quartzites and crystalline limestone is located around the city of Salamá. A biotite zone, composed of biotite–muscovite–hornblende–epidote schist is found in the area of El Chol (El Chol Schist). A garnet–kyanite–muscovite–hornblende zone occurs near Palibatz (Palibatz Schist).

Chiocol Formation (Upper Palaeozoic)

The Chiocol Formation is a sedimentary sequence that crops out on both sides of the Chitxoy–Polochic fault zone, east and SE of San Sebastian Huehuetenango. This formation is a distinctive sequence of interbedded greenish-grey, grey and light blue-grey conglomerate and sandstone, grey-green, grey and maroon tuffs and volcaniclastic beds and less common andesite breccia. Thickness of the formation is in the order of 1000 m and its age is Ordovician–Permian.

Sacapulas Formation (Upper Palaeozoic)

The Sacapulas Formation, along the Chitxoy–Polochic fault zone 35 km to the east of San Sebastian Huehuetenango, consists of 600 m of conglomerates transitional into slate and sandstone with local volcanic and metavolcanic interbeds. The Sacapulas formation is a unit of Santa Rosa Group.

Tactic Formation (Upper Palaeozoic)

The type locality of Tactic Formation lies east of Tactic, Alta Verapaz. The formation is widespread in the Sierra de Los Cuchumatanes. It is also

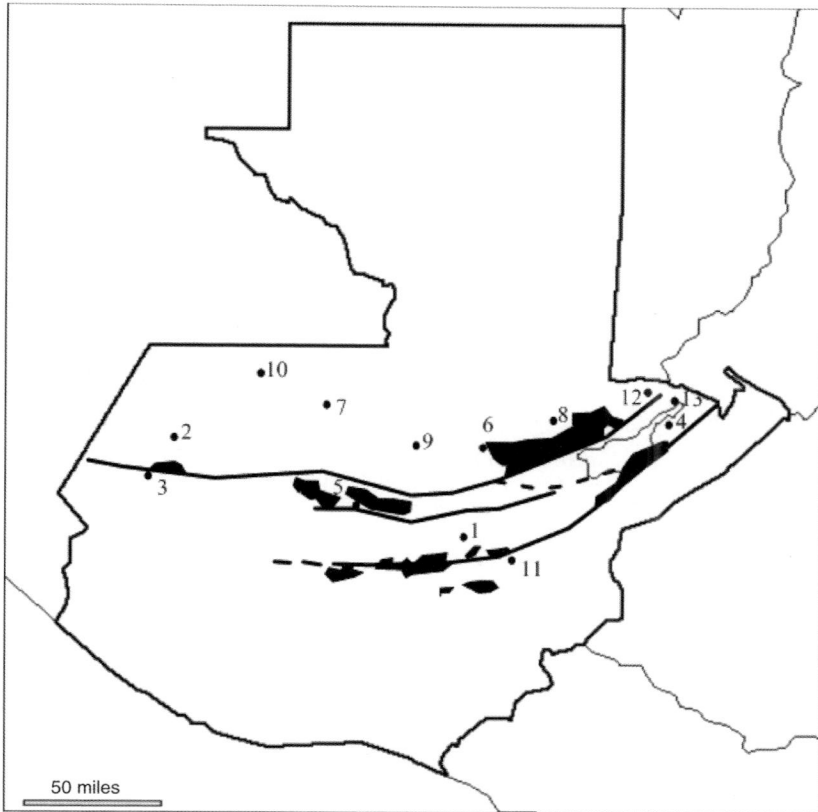

Fig. 2. Typical localities of Guatemalan Formations, also showing the location of the main fault systems and the ophiolite complexes: 1, Chuacús Series, Palaeozoic metamorphic; 2, Chiocol, Sacapulas, Tactic and Esperanza Fms, Carbon–Permian sedimentary rocks; 3, Chochal Fm, Permian sedimentary; 4, Macal Fm, Carbon–Permian sedimentary; 5, Todos Santos Fm, Jurassic–Cretaceous sedimentary rocks; 6, San Lucas Fm, Jurassic volcano-sedimentary rocks; 7, Cobán Fm, Cretaceous sedimentary rocks; 8, Ixcoy Fm, Cretaceous sedimentary rocks; 9, Campur Fm, Cretaceous sedimentary rocks; 10, Verapaz Group, Cretaceous–Cenozoic sedimentary rocks; 11, Subinal Fm, Cretaceous–Cenozoic sedimentary rocks; 12, Desempeño and Lacantun Fms, Cenozoic sedimentary rocks; 13, Caribe, Río Dulce, Herrería and Armas Fms, Cenozoic sedimentary rocks.

recognized in a belt extending across the southern part of the Petén basin from Chiapas, Mexico in the west to the Caribbean Sea in the east. The 800 m thick formation consists of brown to black shale and mudstone with local thin quartzite bed and rare limestone and dolomite. It grades to the overlying Esperanza Formation with gradually increasing limestone. Fossils in the upper part of the formation indicate a Permian age.

Esperanza Formation (Upper Palaeozoic)

The Esperanza Formation occurs in the Altos Cuchumatanes between the Chitxoy–Polochic and Río Ocho fault zones. This unit was first mapped by Blount (1967), Boyd (1966), Davis (1966) and

Anderson (1967) as the Esperanza member of the Santa Rosa Formation. The unit consists of brown to black fossiliferous shale, mudstone and siltstone with limestone and dolomites interbeds. Thickness in the Altos Cuchumatanes is more than 470 m.

Chochal Formation (Upper Palaeozoic)

The Chochal Formation is widespread along the southern Río Ocho and Chixoy–Polochic fault zones. It extends eastward to the Cobán–Purulhá and Senahú area of Alta Verapaz and westwards towards Mexico. The formation consists of massive-bedded, cliff-forming, greyish-black to brownish medium to dark grey dolomite and limestone. The Chochal lithology is similar to the Esperanza Fm.

The unit ranges from 500 m thick to as much as 1000 m along the southern flank of the Cuchumatanes. An angular unconformity separates it from the overlying Todos Santos Formation. The Chochal Formation is part of the Santa Rosa group.

Macal Formation (Upper Palaeozoic)

The Macal Formation extends from the Maya Mountains of Belize to eastern Guatemala. Various authors correlate the Macal Formation with Santa Rosa Group of Guatemala. Fossils indicate an Upper Permian to Pennsylvanian age.

Todos Santos Formation (Triassic–Jurassic)

The Todos Santos Formation lies unconformably on the Chochal and Macal–Santa Rosa formations or on metamorphic basement. The unit comprises a dominant conglomerate member and siltstone-shale member (Richard 1963). The upper part of the Todos Santos Formation is composed of siltstone, sandstone, and dolomite rocks. The unit ranges in thickness from a few metres to at least 1240 m near La Ventosa. Vinson (1962) and Walpe (1960) dated the Todos Santos as the upper Jurassic in northern Guatemala while the type section of central and southeast Guatemala is Middle Jurassic to Middle Cretaceous in age (McBirney 1963).

San Lucas Formation (Upper Jurassic?)

This unit has not been described before though reports from geological companies mention the Jalomáx volcanic unit. The best example is found close to San Lucas. The formation is composed of a thick lower series of conglomerates and siltstone–shales and an upper series of mafic tuffs intensively and pervasively weathered to reddish clay. Fragments of the formation occur as xenoliths in the limestones of the Ixcoy Formation at San Lucas, within the Sierra de Santa Cruz ophiolite complex.

Cobán Formation (Lower to Middle Cretaceous)

Sapper (1937) gave the name Cobán to limestones near Cobán, Alta Verapaz. This thick series of limestones, dolomites and argillaceous to arenaceous clastics, unconformable above the Todos Santos Formation, represents nearly continuous deposition throughout the Cretaceous (Neocomian–Turonian). An evaporitic part of the unit is probably lower Cretaceous in age.

Ixcoy Formation (Middle Cretaceous)

The bituminous, cryptocrystalline Ixcoy Formation, described by Termer (1932) in the Department of Huehuetenango, was thought to be a lower part of the Cobán Formation. This paper observes that it is younger than the Cobán Formation. The unit is very common and spreads to the east central part of Guatemala.

Campur Formation (Upper Cretaceous)

Vinson (1962) proposed this name for Senonian rocks which conformably and gradationally overlie the Cobán Formation in the Alta Verapaz area. The type section occurs along the Cobán-Sebob road approximately 3–6 km south of Finca Campur. The unit consists mainly of gray, gray-brown and dark brown reef limestones and minor dolomites interbedded with thin streaks of shale, siltstone, and limestone breccia or conglomerate.

Verapaz Group (Upper Cretaceous)

According to Vinson (1962) this unit comprises the Chemal, Sepur and Lacandon Formations. Its name comes from Baja and Alta Verapaz, where the group is best developed. The formations consist predominantly of clastic material including shale, sandstone, siltstone, limestone, and conglomerate. Thickness is approximately 600–700 m and rich foraminiferal assemblages give a Campanian–Maastrichtian age.

Chemal Formation

The Chemal Formation is restricted to the Chemal region near the highest point of the Altos Cuchumatanes near Huehuetenango. The upper part of the formation has been eroded. Thickness is 95 m. The unit consists of red and reddish brown shale with minor thin beds of coarse calcarenites and conglomeratic limestones in the lower part and finer calcarenites and dense argillaceous limestone in the upper part. The Chemal is differentiated from the Sepur by its dominant red coloration.

Sepur Formation

The Sepur Formation was named by Sapper (1899) after a place called Sepur near Lanquin Village and Finca Campur in central Alta Verapaz. The formation is composed of brown clays, shales, siltstones, sandstones and marls, interbedded with lenses of limestone. Maximum thickness is about 600 m. The formation lies unconformably on the Upper Cretaceous (Senonian) limestones of the Campur Formation.

Lacandon Formation

The Lacandon Formation occurs in the Lacandon region in northwestern Petén, overlying the Sepur with either gradational contact or local unconformity. This is a thick series of detrital carbonates with local algal beds and microcrystalline limestone of light yellow to light cream color. It is a composite section consisting of three units with thicknesses of 650 and 600–400 m near Lacandon, Petén.

Subinal Formation (Paleocene)

Previously described as a molasse unit, this formation consists of a series of flyschoid, polymictic conglomerates, commonly overlying the ophiolites associated with the Motagua and the Río Negro-Polochic faults. A typical section, at km 77 of the CA-9, close to Guastatoya (UTM E: 816603, N: 1647109), consists of three flyschoid sequences: a fine-grained consolidated polymictic conglomerate, a coarse-grained polymictic conglomerate and a fine-grained unconsolidated sequence.

A possible source for the Subinal Fm is seen south of the Motagua Fault at UTM E: 815968, UTM N: 1636071. This is a polymictic conglomerate, very hard and silicified, which crops out on top of schists of the Chuacús Series and also occurs as big boulders in creek beds. Gold, in many creeks and rivers, may be related to the quartz fragments found on these conglomerates.

Petén Group (Lower Eocene)

Vinson (1962) divided the Petén group into five units, the Cambio, Reforma, Teledo, Santa Amelia and Buena Vista formations, on the basis of tectonic and palaeontological facies. The description of the Petén Group is based on the report compiled by Millan (1985).

Icaiche Formation (Upper Eocene)

This unit was described first by Millan (1985). It is composed mainly of gypsum and marls.

Desempeño Conglomerate Formation (Upper Oligocene)

This formation consists predominantly of massive and hard, grey to black channel conglomerates of quartzitic and siliceous pebbles ranging up to 10 cm in diameter. It occurs as a wedge shaped mass, with a maximum thickness of 200 m, in a low-lying erosional pocket on the south flank of La Libertad arch. The pebbles are very similar to those found in the Caribe and Lacantun Formations and may have the same source.

Lacantun Formation (Upper Oligocene)

The predominantly red bed formation overlying the Petén Group is the Lacantun Formation, with a type locality near the mouth of Río Lacantun in southwestern Petén and eastern Chiapas. Up to 500 m thick, the unit consists of red and brown arkosic and ferruginous sandstones and siltstone, quartz-rich conglomerate, red and brown ferruginous and nodular claystone and mottled, hard to soft nodular shales.

Caribe Formation (Upper Oligocene–Lower Pliocene)

The Caribe Formation is exposed in the Río Salina just south of El Caribe, Petén, where the most complete section, more than 800 m thick, is seen. It also occurs in the adjoining regions of Alta Verapaz and El Quiche. Vinson (1962) suggested it is a time equivalent to the Río Dulce Formation of eastern Guatemala. The formation consists of variegated clays, clay shales, grits, sandstones, siltstones, sandy limestones and quartz-rich conglomerates.

Río Dulce Formation (Lower Miocene)

This formation is known only from the Amatique embayment area of eastern Guatemala. It is characterized by light buff, tan and cream-coloured limestone unconformable on Permian, Cretaceous and Eocene rocks. The type locality of the formation is just above sea level along Río Dulce, upstream from the Town of Livingston, Izabal. The Río Dulce Formation is overlain by the Herrería Formation of Pliocene (?) age, following a low angle unconformity.

Herrería Formation (Pliocene)

The Herrería Formation consists of a long, narrow, north–south strip of clastics overlying the Río Dulce limestone east of Río Dulce gorge. The unit was named by J. P. Gallegher in a private oil company report cited by Vinson (1962) and described by Millan (1985) as poorly consolidated claystones, siltstones, marls and sandstones, which are characteristically conglomeratic. The type section extends from Punta Herrería SW to the contact with the Río Dulce Formation, a distance of about 1 mile. Formation thickness is 240 m.

Armas Formation (age?)

The Armas Formation, composed of red bed and deltaic claystone, siltstone and sandstones, occurs in Motagua River valley in Izabal. The formation is a thick series of young or fresh-appearing strata lying on metamorphic basement and Cretaceous limestones. The unit is divided into two subunits that consist sedimentary rocks derived from the volcanic, metamorphic and sedimentary rocks. Total estimated thickness is 2500–3000 m.

Figure 3 shows a stratigraphic column based on work by Millan (1985) and the author.

PERIOD	EPOCH	COLUMN	UNIT	SYMBOL	
Quaternary	Pleistocene & Holocene		Pumice	Q_p	Thick pumice fills and mantles of diverse origin.
			Undivided volcanic rocks	Q_{uv}	Lava flows, laharic deposits, tuffs, cones and domes.
Tertiary	Miocene-Pliocene		Herrería Fm	PH_{ms}	Conglomerates of claystones, siltstones, marls and sandstones.
	Oligocene		Río Dulce Fm	M_{RDcarb}	Light buff, tab, and cream-coloured limestones.
			Caribe Fm	Oc_{ms}	Variegated clays, clay shales, grits, sandstones, siltstones, sandy limestones and quartz-rich conglomerates.
	Eocene		Icaiche Fm	Em_s	Gypsum and marls.
			Petén Group	EP_{ms}	Marine sediments. Obduction of the Sierra Santa Cruz and Baja Verapaz ophiolitic complexes.
	Paleocene		Subinal Fm	LKE_{rb}	Red beds, mainly Tertiary. Obduction of the Juan de Paz-Los Mariscos ophiolitic complex.
Cretaceous	Late		Verapaz Group Sepur Fm	LKs_{ms}	Brown clays, shales, siltstones, sandstone and marls, interbedded with lenses of limestone. Obduction of the North and South Motagua complexes.
	Medium		Ixcoy Fm	LKl_{carb}	Cryptocrystalline, bituminous-rich limestone, with no fossils remains, containing xenolites of the basaltic series of the San Lucas Formation. Zacapa Island Arc Granitoid.
Jurassic	Late	Hiatus	San Lucas Fm	LJ_{SLrb}	Conglomerates and siltstone-shales and basaltic tuffs intensively and pervasively weathered to form a reddish clay material. Equivalent to San Ricardo Formation.
Permian	Early?	Hiatus	Chochal Fm	EP_{CHcarb}	Massive-bedded, cliff forming dolomite and limestone, ranging from grayish-black to brownish medium to dark grayin colour. Possible time of obduction of the Huehuetenango ophiolitic complex.
Carboniferous	Late		Tactic Fm	LC_{Tphy1}	Dark shales and mudstones. Shales, sandstones, conglomerates and phyllites
Devonian	Late	Hiatus	Chuacús Series. Tambor Fm. La Virgen Fm. Sac apulas Fm.	LDc_{Hum}	Phyllites, chlorite and garnet schists, quartz-mica-feldspar schists and gneisses, marbles and migmatites. Rabinal granite.

Fig. 3. Stratigraphic column for the studied area.

The geological evolution of Central Guatemala

The stratigraphy of Central Guatemala described above suggests the geological evolution shown in Figure 4. The oldest rocks are Palaeozoic schists and other metamorphic rocks of the pre-Permian Chuacús Series, accompanied by granitic and dioritic batholiths (Fig. 4a). Around 300 Ma, during the Carboniferous, shallow marine sediments and conglomerates were followed by deeper sandstones and shales at greater depths (Santa Rosa Group) (Fig. 4b). During the Early Permian (Fig. 4c), limestones and other carbonate rocks accumulated (Chochal Formation).

A hiatus of nearly 51 million years marks the Triassic Period, when the sea retreated and no significant deposition occurred. In this paper, I am suggesting that the pre-Cretaceous passing by of the Jamaica Block from somewhere in the Pacific Ocean on its way to its present location south of Cuba was the responsible for the obduction of the oldest ophiolites of Huehuetenango in the western end of the Polochic–Río Negro faults in Guatemala.

Late Jurassic red beds of the Todos Santos Fm possibly record tropical oxidizing conditions. Volcanic rocks from the San Lucas Fm formed at this time. The Late Cretaceous to Early Cenozoic periods (Fig. 4d) were recorded by carbonate sediments (Cobán Fm, Ixcoy Fm, Petén Fm, Campur Fm and others) and the intrusion of granitic and dioritic bodies of the Zacapa Island Arc. They were followed by marine clastic sediments of the

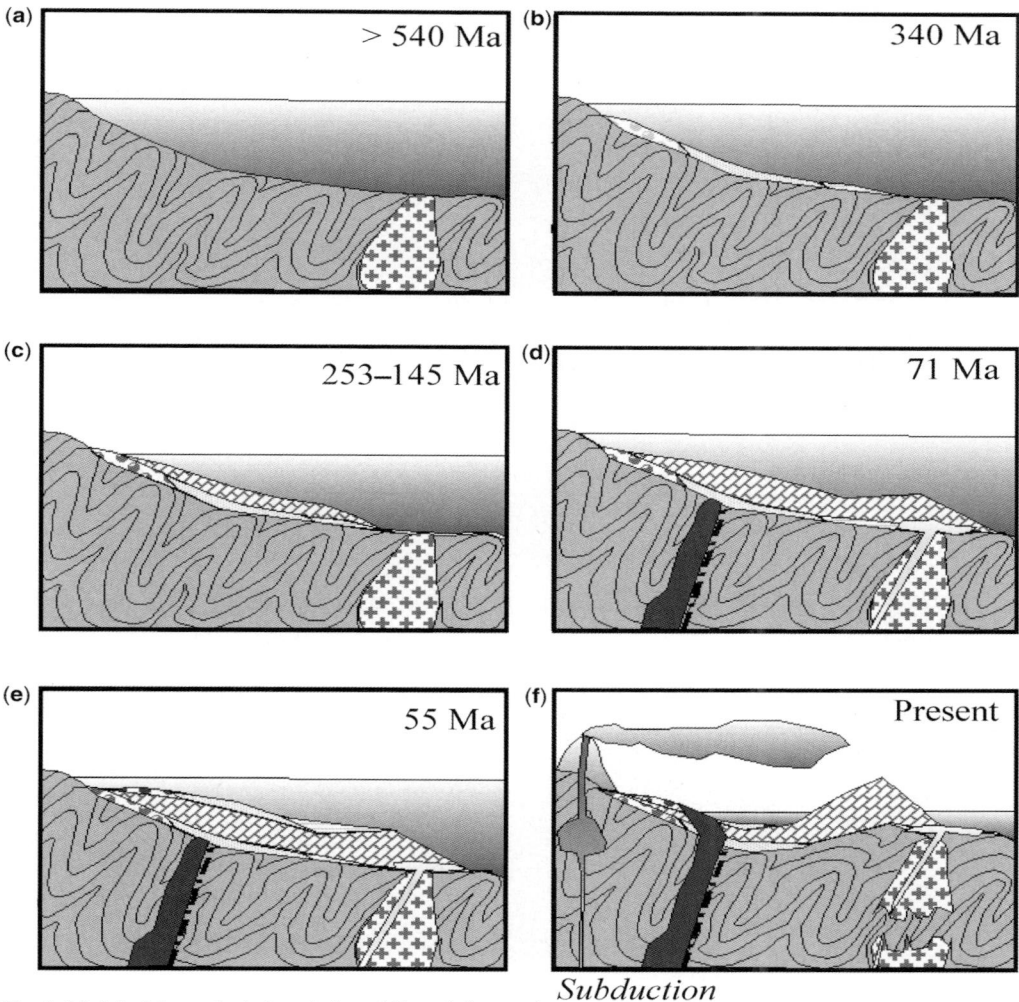

Fig. 4. Model of the geological evolution of Central Guatemala.

Verapaz Group and obduction of the North and South Motagua ophiolite complexes and the Juan de Paz–Los Mariscos ophiolite complex. These were later covered by the red beds of the Subinal Fm, which suggest another period of regression.

The Paleocene witnessed deposition of more marine sediments, mainly shallow marine conglomerates, with deeper sandstones and shales. Obduction of the Sierra de Santa Cruz and the Baja Verapaz ophiolite complexes occurred at this time (Fig. 4e).

Marls and gypsum of the Icaiche Fm formed during the Eocene, while the end of the Cenozoic Period was marked by the deposition of tuffs, lavas and other volcanic rocks accompanied by the intrusion of smaller granitoid bodies.

Finally, Quaternary formations are represented by alluvial and deluvial material as well as by lavas and tuffs from active volcanoes (Guastatoya Fm, Toledo Fm, Desempeño Fm, Río Dulce Fm, and others). Some of the ultramafic intrusives that had undergone serpentinization before were oxidized, resulting in the formation of Ni lateritic zones (Fig. 4f).

Current models for the Caribbean Plate

There are basically two groups of ideas with respect to the formation of the Caribbean Plate. One group defends the idea of the *in situ* formation of the Caribbean Plate (Fig. 5) in contrast to complex Pacific

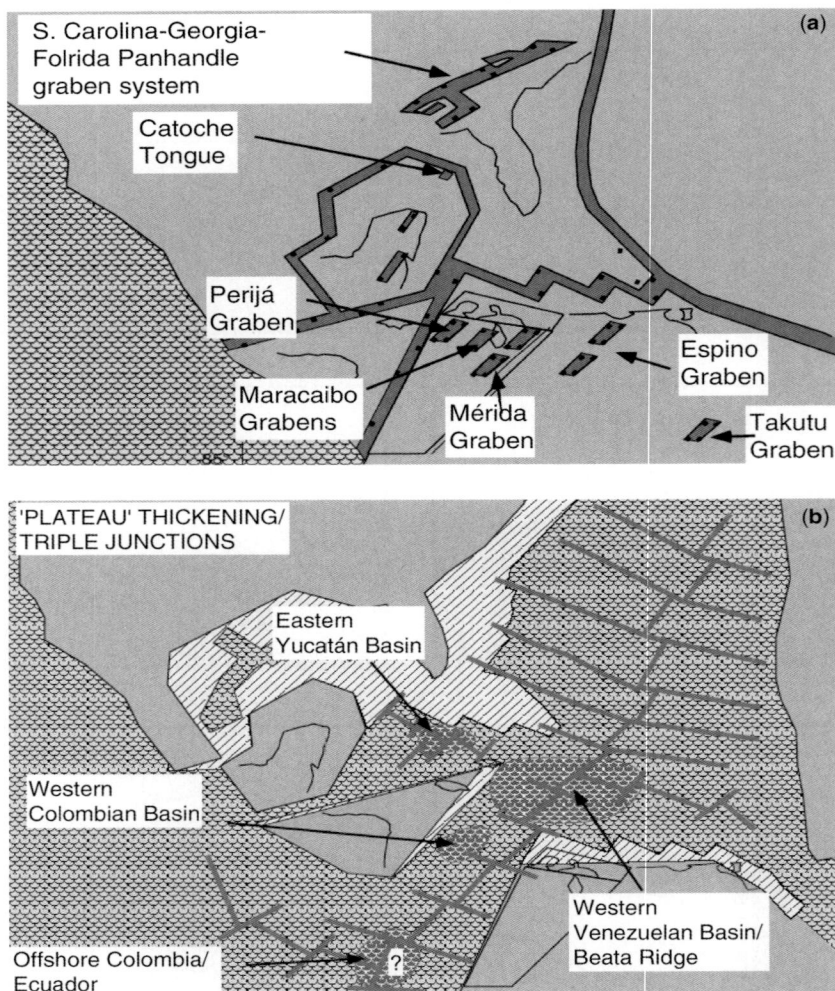

Fig. 5. (a) Triassic–Early Jurassic rifting of Pangaea; (b) Late Jurassic–Early Cretaceous opening of the area between North and South America during drift. Analogy with the Scotia Plate suggests that the Beata Ridge was a spreading centre and that continental fragments lie on the plate margins and interior. From James (2005).

models that require major block rotations, plate migrations, hotspots or plumes (James 2005).

The other group (Pindell 1993; Pindell *et al.* 2000) defends the idea that at least parts of the Caribbean Plate have migrated several thousands of kilometres into its inter-American location from the Pacific (Fig. 6). Let us see now some of the geological features that our current knowledge of the stratigraphy of Guatemala indicates that existed in the area.

The model

Interoceanic channel

The stratigraphy of Central Guatemala calls for the existence of an interoceanic channel since the Early Proterozoic. This is suggested by the presence of sedimentary units along the central part of Guatemala, now constituting the metamorphic

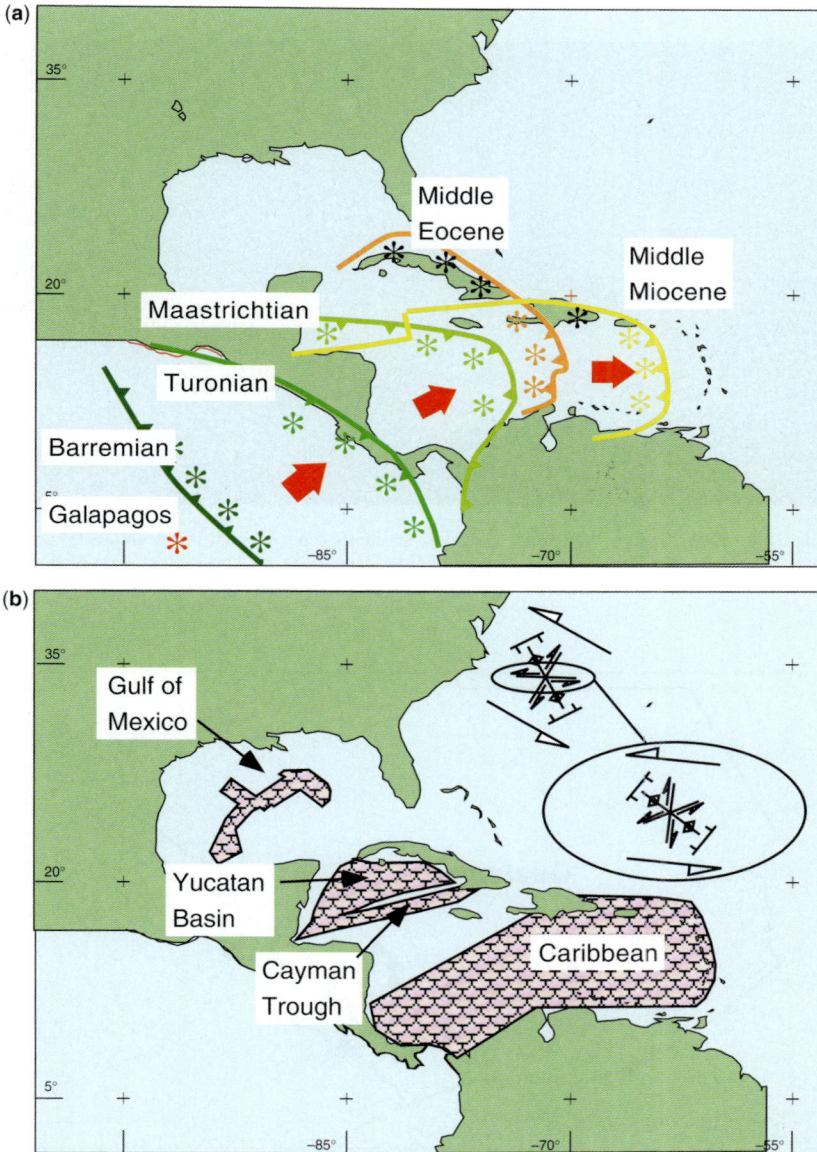

Fig. 6. (**a**) Migration of the Caribbean Plate from the Pacific (red arrows), preceded by a leading edge volcanic arc (colours indicate progress); (**b**) genesis of the Caribbean Plate oceanic crust in place along with the Gulf of Mexico, Yucatán Basin and Cayman Trough during extension as North America drifted NW from South America. From James (2005).

Chuacús Series (Fig. 7). It is generally accepted that both Jamaica and Chortís were located at this time somewhere west of Mexico.

Subduction arcs

Figure 8 shows the location of all volcanoes in Guatemala, Salvador and Honduras. The arcs indicate two subduction zones and the remains of a trench, now collapsed, that can be seen in the ocean floor.

Volcanic activity south of the Maya Block

In 2006, the author discovered vast amounts of rhyolitic tuffs among areas previously believed to be composed only of limestones (Fig. 9) on the northern border of the Polochic Belt ophiolite complex.

Fig. 7. During the Early Proterozoic there must have been an interoceanic channel were continental sediments from the Maya Block were deposited. Note that the position of Jamaica and of the Chortís Block west of Mexico is just a suggestion.

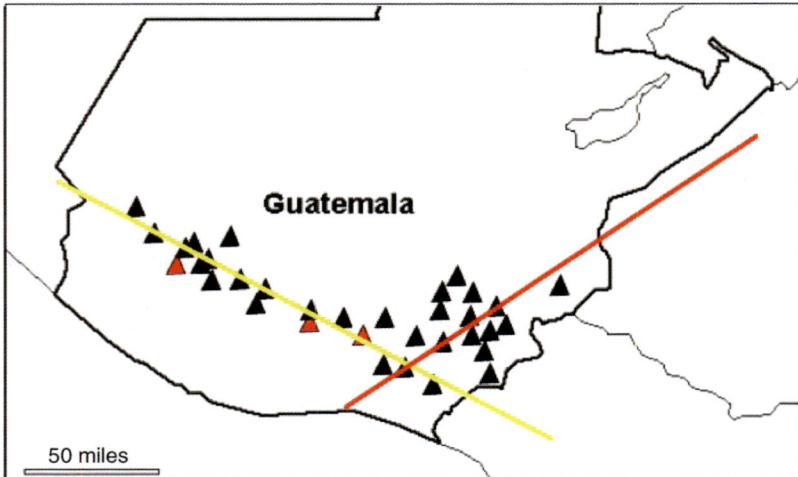

Fig. 8. Position of the two subduction zones, west and south of Guatemala, which are identified by the presence of two different chains of volcanoes.

Fig. 9. Recently discovered massive flows of rhyolite on top of Cretaceous limestones of the Ixcoy and Cobán Formations north of the ophiolites of the Polochic Belt.

These, together with those found above the Baja Verapaz ophiolite complex, record a volcanic arc associated with flat subduction of the Caribbean Plate below the Maya Block.

Interpretation

To explain the geology of the NW corner of the Caribbean Plate, I propose that an orogenic event, the Chuacús Orogeny, coeval with Laramide Orogenesis in North America, occurred as shown in Figures 10–13. If the recently discovered rhyolites on top of the Cretaceous Cobán and Ixcoy limestones north of the Polochic Ophiolite Belt evidence a subduction arc, it is possible that the Caribbean Plate was expanding northwards and southwards. The spreading ridge is represented today by the Lake Izabal graben. The model suggests that the angle of subduction to the north was much shallower than to the south, limiting volcanic activity south of the Maya Block. Figure 10 shows that the Jamaica Block interacted with the Maya Block along the eastern part of the Motagua Suture Zone sometime before the Cretaceous. It was responsible

for the obduction of the Huehuetenango ophiolite complex.

After the migration of Jamaica, the Chortís Block followed in a SE direction. This movement provoked the obduction of the North and South Motagua ophiolite complexes and the fragmentation of the South Motagua ophiolite complex. During the Cenozoic, the Chortís Block was located south of the Maya Block, moving northwards. The subduction zone to the north was more active and the angle of subduction increased, resulting in increasing volcanic activity (Fig. 11).

During the Late Cenozoic northward movement of Chortís intensified, resulting in obduction of the Baja Verapaz complex followed by the Juan de Paz–Los Mariscos and Sierra de Santa Cruz ophiolite complexes. This marked the beginning of the Chuacús Orogeny in the Motagua Suture Zone (Fig. 12).

Finally, collision of the continental blocks activated late volcanic activity over the Maya Block, terminating the Chuacús Orogeny with formation of the different metamorphic rocks of the Chuacús Series (Fig. 13).

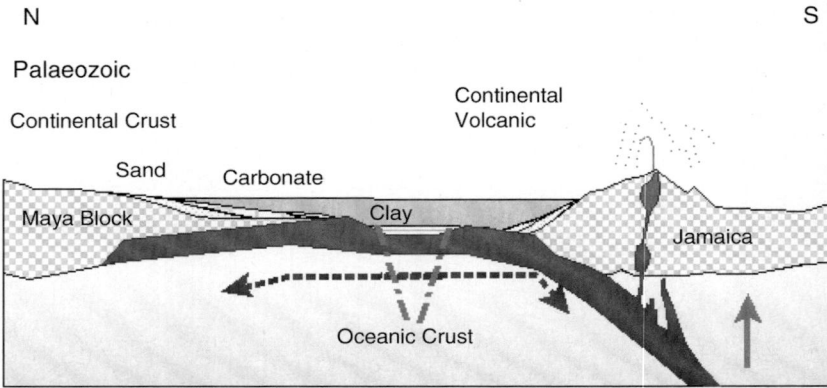

Fig. 10. Initial development of the subduction zones and volcanic arcs during the Palaeozoic.

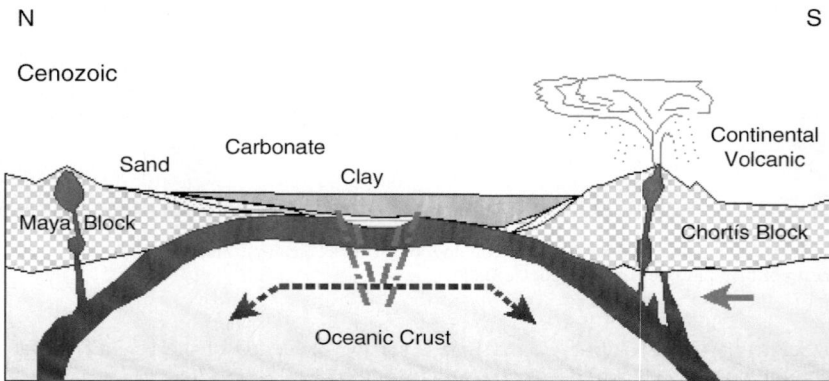

Fig. 11. North–south cross section between the Maya and Chortís Block during Early Cenozoic.

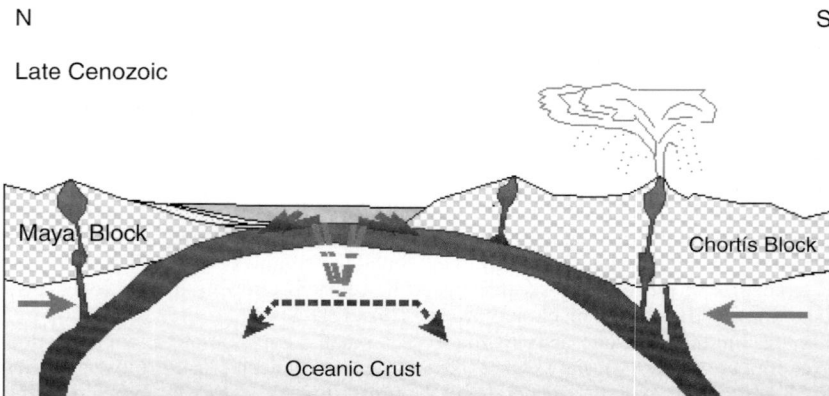

Fig. 12. Continuation of the closure of the interoceanic channel during the Late Cenozoic together with the initiation of the Chuacús Orogeny.

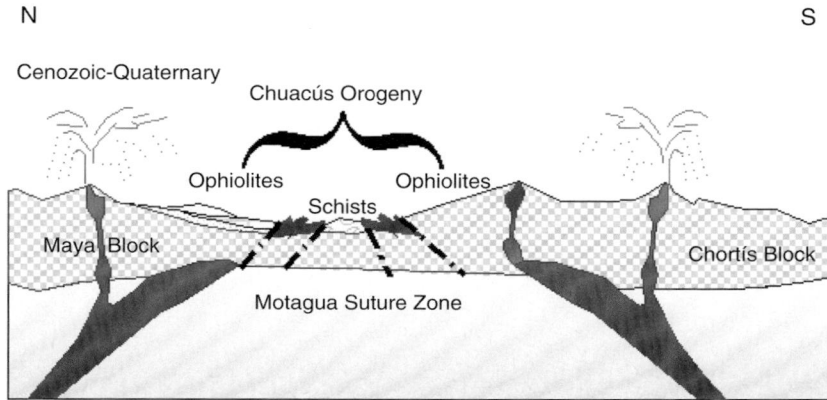

Fig. 13. Collision of the continental blocks and the completion of the Chuacús Orogeny within the Motagua Suture Zone.

Conclusions and recommendations

This paper suggests that the geology of Central Guatemala supports the allochthonous origin of the Caribbean Plate. Further evidence should be sought not only in Guatemala, but also in Honduras and Mexico or even in Jamaica.

This paper was improved by conversations with a large number of people who helped debug it. Particular thanks to K. James, who re-read and revised the manuscript in more than one occasion until it was in compliance with the requirements of the GSL. Also a big thanks goes to Giuseppe Giunta, with whom I had the opportunity to discuss many of the ideas presented in this paper. I am also grateful to my colleagues from Guatemala for their support and their many contributions.

References

ANDERSON, T. H. 1967. *Geology of the Central Third of La Democracia Quadrangle*. MA thesis, Texas University, Austin, TX.

BLOUNT, D. 1967. *Geology of the Chiantla Quadrangle, Huehuetenango, Guatemala*. PhD thesis, Louisiana State University, Baton Rouge, LA.

BOYD, A. 1966. *Geology of the Western Third of La Democracia Quadrangle*. MA thesis, Texas University, Austin, TX.

DAVIS, G. H. 1966. *Geology of the Eastern Third of La Democracia Quadrangle*. MA thesis, Texas University, Austin, TX.

GIUNTA, G., BECCALUVA, L. *ET AL.* 2002. *The Motagua Suture Zone in Guatemala, Field Trip – Guide Book*. MARRONI, M. & PANDOLFI, L. (eds) *Publication Supported by Ofioliti International Journal*.

GOMBERG, D. N., BANKS, P. D. & McBIRNEY, A. R. 1968. Guatemala: preliminary zircon ages from the Central Cordillera. *Science*, **162**, 121–122.

ITURRALDE VINENT, M. 2004. Scientific Report on the Symposium G20.11, *Caribbean Plate Tectonics*, Florence, 20–28 August 2004.

JAMES, K. H. 2005. A simple synthesis of Caribbean geology. Transactions of the 16th Caribbean geological Conference, Barbados, *Carribbean Journal of Earth Sciences*, **39**, 69–82.

McBIRNEY, A. R. 1963. Geology of a part of the Central Guatemalan Coordillera. *Geological Science*, **38**, 177–242.

MILLAN, S. M. 1985. *Preliminary Stratigraphic Lexicon North and Central Guatemala* (a compilation under a contract with the United Nations Development Program).

PINDELL, J. L. 1993. Regional synopsis of the Gulf of Mexico and Caribbean evolution, Gulf Coast Section. *Society of Economic Paleontologists and Mineralogists, 13th Annual Research Conference*, 251–274.

PINDELL, J. L., KENNAN, L. & BARRETT, S. F. 2000, Putting it all together again. *American Association of Petroleum Geologists Explorer*, October, 34–37.

RICHARD, H. G. 1963. Stratigraphy of the earliest Mesozoic sediments in southwestern Mexico and western Guatemala. *American Association of Petroleum Geological Bulletin*, **47**, 1861–1870.

SAPPER, K. 1899. *Über Gebirgsbau und Boden des nördlichen Mittelamerika*. Petermans Mitteilungen, Ergänzungsheft, Gotha J. Perthes, 127.

SAPPER, K. 1937. *Handbuch der Regionalen Geologie*, **8** Mittelame-rika, Heidelberg, 4a.

TERMER, F. 1932. *Geologie von Nordwest-Guatemala*. **7–8** Zeit-Schr. Grssel. f. Erdkunde, Berlin, 241–248.

VALLS, R. A. 2006. *Geology and Geochemical Evolution of the Ophiolitic Belts in Guatemala. A Field Guide to Nickel Bearing Laterites*, 6th edn. Nichromet Extraction Inc., Toronto.

VINSON, G. L. 1962. Upper Cretaceous and Tertiary stratigraphy of Guatemala. *American Association of Petroleum Geologists Bulletin*, **44**, 1273–1315.

WALPE, J. L. 1960. Geology of the Cobán Purulhá Area, Alta Verapaz, Guatemala. *American Association of Petroleum Geologists Bulletin*, **44**, 1273–1315.

The North American–Caribbean Plate boundary in Mexico–Guatemala–Honduras

LOTHAR RATSCHBACHER[1]*, LEANDER FRANZ[1], MYO MIN[1], RAIK BACHMANN[1],
UWE MARTENS[2], KLAUS STANEK[1], KONSTANZE STÜBNER[1], BRUCE K. NELSON[3],
UWE HERRMANN[3], BODO WEBER[4], MARGARITA LÓPEZ-MARTÍNEZ[4],
RAYMOND JONCKHEERE[1], BLANKA SPERNER[1], MARION TICHOMIROWA[1],
MICHAEL O. MCWILLIAMS[2], MARK GORDON[5], MARTIN MESCHEDE[6]
& PETER BOCK[1]

[1]*Geowissenschaften, Technische Universität Bergakademie Freiberg, 09599 Freiberg, Germany*

[2]*Geological and Environmental Sciences, Stanford University, Stanford, CA 94305, USA*

[3]*Earth and Space Sciences, University of Washington, Seattle, WA 98195, USA*

[4]*CICESE, 22860 Ensenada, B.C., Mexico*

[5]*Department of Geology and Geophysics, Rice University, Houston, TX 77251-1892, USA*

[6]*Geowissenschaften, Universität Greifswald, 17487 Greifswald, Germany*

*Corresponding author (e-mail: lothar@geo.tu-freiberg.de)

Abstract: New structural, geochronological, and petrological data highlight which crustal sections of the North American–Caribbean Plate boundary in Guatemala and Honduras accommodated the large-scale sinistral offset. We develop the chronological and kinematic framework for these interactions and test for Palaeozoic to Recent geological correlations among the Maya Block, the Chortís Block, and the terranes of southern Mexico and the northern Caribbean. Our principal findings relate to how the North American–Caribbean Plate boundary partitioned deformation; whereas the southern Maya Block and the southern Chortís Block record the Late Cretaceous–Early Cenozoic collision and eastward sinistral translation of the Greater Antilles arc, the northern Chortís Block preserves evidence for northward stepping of the plate boundary with the translation of this block to its present position since the Late Eocene. Collision and translation are recorded in the ophiolite and subduction–accretion complex (North El Tambor complex), the continental margin (Rabinal and Chuacús complexes), and the Laramide foreland fold–thrust belt of the Maya Block as well as the overriding Greater Antilles arc complex. The Las Ovejas complex of the northern Chortís Block contains a significant part of the history of the eastward migration of the Chortís Block; it constitutes the southern part of the arc that facilitated the breakaway of the Chortís Block from the Xolapa complex of southern Mexico. While the Late Cretaceous collision is spectacularly sinistral transpressional, the Eocene–Recent translation of the Chortís Block is by sinistral wrenching with transtensional and transpressional episodes. Our reconstruction of the Late Mesozoic–Cenozoic evolution of the North American–Caribbean Plate boundary identified Proterozoic to Mesozoic connections among the southern Maya Block, the Chortís Block, and the terranes of southern Mexico: (i) in the Early–Middle Palaeozoic, the Acatlán complex of the southern Mexican Mixteca terrane, the Rabinal complex of the southern Maya Block, the Chuacús complex, and the Chortís Block were part of the Taconic–Acadian orogen along the northern margin of South America; (ii) after final amalgamation of Pangaea, an arc developed along its western margin, causing magmatism and regional amphibolite–facies metamorphism in southern Mexico, the Maya Block (including Rabinal complex), the Chuacús complex and the Chortís Block. The separation of North and South America also rifted the Chortís Block from southern Mexico. Rifting ultimately resulted in the formation of the Late Jurassic–Early Cretaceous oceanic crust of the South El Tambor complex; rifting and spreading terminated before the Hauterivian (*c*. 135 Ma). Remnants of the southwestern Mexican Guerrero complex, which also rifted from southern Mexico, remain in the Chortís Block (Sanarate complex); these complexes share Jurassic metamorphism. The South El Tambor subduction–accretion complex was emplaced onto the Chortís Block probably in the late Early Cretaceous and the Chortís Block collided with

From: JAMES, K. H., LORENTE, M. A. & PINDELL, J. L. (eds) *The Origin and Evolution of the Caribbean Plate.*
Geological Society, London, Special Publications, **328**, 219–293.
DOI: 10.1144/SP328.11   0305-8719/09/$15.00 © The Geological Society of London 2009.

southern Mexico. Related arc magmatism and high-T/low-P metamorphism (Taxco–Viejo–Xolapa arc) of the Mixteca terrane spans all of southern Mexico. The Chortís Block shows continuous Early Cretaceous–Recent arc magmatism.

Supplementary material: Analytical methods and data, and sample description are available at http://www.geolsoc.org.uk/SUP18360.

Structural, geochronological, and petrological data document the Palaeozoic–Cenozoic pressure–temperature–deformation–time (P–T–d–t) history of the southern part of the Maya and northern part of the Chortís Blocks, and the Polochic, Motagua and Jocotán–Chamelecón fault zones along the North American–Caribbean Plate boundary in southern Mexico, Guatemala and northern Honduras. The principal questions addressed are: which crustal sections accommodated the large-scale sinistral offset along the plate boundary, what was the timing of offset and deformation, and in what kinematic framework did it occur? What Palaeozoic to Cenozoic geological correlations exist between the Maya Block, the Chortís Block, the tectono-stratigraphic complexes ('terranes') of southern Mexico, and the arc/subduction complexes of the northern Caribbean.

Northern Caribbean Plate boundary

The Caribbean Plate represents a lithospheric unit between the North and South American plates. Its western and eastern margins consist of subduction zones with magmatic arcs (Central America Isthmus, Lesser Antilles), whereas the northern and southern margins correspond to strike–slip with subsidiary transpression or transtension zones (Polochic–Motagua, Cayman, Greater Antilles, North Andean and South Caribbean, e.g. Pindell *et al.* 2006). The Motagua Suture Zone (MSZ) of southeastern Mexico, Guatemala, and Honduras extends from the Pacific Ocean to the Caribbean Sea for *c.* 400 km east–west and *c.* 80 km north–south (e.g. Beccaluva *et al.* 1995), separating the Maya Block from the Chortís Block (Fig. 1a, Dengo 1969). This belt actually represents a composite structure, encompassing suture zone(s) with relics of Mesozoic ocean floor that characterize the northern Caribbean Plate boundary, and Cenozoic fault zones (mainly the Polochic, Motagua, Jocotán–Chamelecón faults) that link the Cayman trough pull-apart basin in the east with the subduction zone of the Pacific plate in the west (Fig. 1).

The Maya and Chortís Blocks showcase a geological history that encompasses Middle Proterozoic to Quaternary and have long been recognized as key elements in understanding the interaction of Laurentia, Gondwana and the (Proto-)Pacific

domains (e.g. Dickinson & Lawton 2001; Keppie 2004). Most tectonic models suggest that collision of the Maya and Chortís Blocks occurred along south- or north-dipping subduction zones during the Cretaceous, and caused emplacement of Jurassic–Cretaceous Proto-Caribbean oceanic crust onto the blocks; thrusting of metamorphic and sedimentary strata was bivergent north and south of the Motagua fault zone (e.g. Meschede & Frisch 1998; Beccaluva *et al.* 1995 and references therein). Since Late Cretaceous–Early Cenozoic, the MSZ developed into a sinistral wrench zone due to the different velocities of the North American and Caribbean plates (e.g. DeMets 2001). In southern Mexico, Guatemala and Honduras this still active fault system has dismembered the collisional blocks as well as the allochthonous remnants of the former oceanic crust; the latter comprises shaly mélange, containing blocks of serpentinized ophiolite fragments, and metamorphic rocks such as amphibolite, eclogite, albitite and jadeitite (e.g. Tsujimori *et al.* 2005). Inferred Cenozoic displacement accommodated along the MSZ is controversial, but magnetic anomalies within the Cayman trough suggest a minimum of *c.* 1100 km; spreading rates in the trough were *c.* 3 cm/annum from 44 to 30 Ma and *c.* 1.5 cm/annum thereafter (Rosencrantz *et al.* 1988). Active faulting and volcanism related to eastward subduction of the Pacific plate overprinted the ophiolite emplacement and intra-continental translation features within the crust of Mexico, Guatemala and Honduras (e.g. Guzmán-Speziale 2001).

Methods

Methodical and technical aspects of our radiometric dating that includes U–Pb, Ar–Ar, Rb–Sr, and fission-track geochronology are provided as Supplementary material. Also included in the Supplementary materials is our approach to structural analysis of ductile deformation and kinematics, to fault–slip analysis and definition of stress-tensor groups in the brittle crust, and a review of the applied calculation techniques. Methodical and technical aspects of our petrology and geothermobarometry are included in the Supplementary material along with locations of studied outcrops ('stops') and descriptions of analysed samples.

Basement and cover of the Maya and Chortís Blocks

Figure 2 presents lithostratigraphic columns of the southern Maya Block, the MSZ, and the north-central Chortís Block, based mainly on the compilations given in Donnelly *et al.* (1990), Talavera-Mendoza *et al.* (2005, 2007), Vega-Granillo *et al.* (2007) and Rogers (2003). We modified the columns according to our new data and the recent literature (discussed below) and, where available, added stratigraphic range (converted into Ma), thickness, reliable radiometric ages, *PT* estimates and first-order tectono-stratigraphic interpretations. We also show columns of units that we will use in the discussion for correlation and refinement of tectonic models.

Cover sequences

In general, the Maya Block (Fig. 2b) records shelf evolutions during Late Palaeozoic and Cretaceous that were followed by terrestrial–shallow marine rift (Middle–Late Jurassic) and submarine-fan (Late Cretaceous) developments. Along the MSZ, ophiolites are imbricated with probably Upper Jurassic to Upper Cretaceous submarine-fan and ocean-basin deposits (North El Tambor complex and El Pilar Formation, Fig. 2a) and crystalline basement of the Chuacús complex; they are unconformably overlain by continental red beds (Subinal Formation). In the Sierra de Santa Cruz (Fig. 1c), massive ophiolitic rocks that contain basalts of island-arc affinity are imbricated with Maastrichtian to Eocene (*c.* 80–50 Ma) carbonate–arenaceous pre-flysch and flysch (Sepur Formation; Wilson 1974; Beccaluva *et al.* 1995). Aptian and Cenomanian (*c.* 120–95 Ma) sedimentary rocks of the Petén basin of northern Guatemala (Maya Block) record a pelagic facies, reflecting maximum transgression and an open marine connection to the Gulf of Mexico to the north, and the Proto-Caribbean seaway to the south (Fourcade *et al.* 1994, 1999). During the Late Maastrichtian and Danian (*c.* 70–60 Ma), the southern Petén basin (Fig. 1b) was a deep siliciclastic sink, the Sepur foredeep basin. The depocentre of this basin shifted from south during the Late Campanian to north during the Late Maastrichtian and Early Danian; this led to the demise of the pre-existing carbonate platform over more than 100 km from south to north.

The ophiolites south of the Motagua fault (South El Tambor complex, Fig. 2k) are limited to the north by the Cabañas fault and are overlain by Subinal red beds and in the south by the Cenozoic–Recent volcanic arc; they are cut by felsic intrusives. Together with Late Jurassic–Cretaceous 'mélange', they cover basement of the Las Ovejas complex and the San Diego Group of the Chortís Block. Basalts

within these 'southern' ophiolites show mid-ocean ridge basalt (MORB) affinity (Beccaluva *et al.* 1995). In the north-central Chortís Block (Fig. 2j), the clastic, partly turbiditic, rift-related Agua Fría Formation was deformed during Late Jurassic (Viland *et al.* 1996). It is overlain by the shallow-marine, shelf-type Yojoa Group that contains arc and intra-arc rift deposits (Rogers *et al.* 2007a). Possibly two molasse sequences are present: the mainly red beds of the Albian–Cenomanian Valle de Angeles Group and the ?Eocene–Oligocene Subinal Formation; the latter may have been deposited in pull-apart basins along the Motagua fault zone. Thus, molasse-type deposits occur on the Chortís during late Early to Late Cretaceous times (Valle de Angeles Group), whereas marine platform carbonates developed on the Maya Block (Coban and Campur Groups). Top-to-north thrust imbrication affected the Jurassic–Cretaceous sequence of the Chortís Block during Late Cretaceous (Donnelly *et al.* 1990; Colon fold belt of Rogers *et al.* 2007b).

Basement complexes

Clear definition and subdivision of the rock units that comprise the MSZ (Fig. 1c) remain elusive principally because Cenozoic faults delineate parts of these units (see also Ortega-Gutiérrez *et al.* 2007). This section briefly describes the major units of the MSZ from north (Maya Block) to south (Chortís Block).

Maya Mountains and Altos Cuchumatanes (Maya Block). Metamorphic rocks in the Maya Mountains of Belize (Fig. 1c) comprise low-grade metasedimentary rocks with minor rhyolite–dacite intercalations (Steiner & Walker 1996; Martens *et al.* 2010). Lower Palaeozoic beds are intruded by granitoids; Upper Palaeozoic deposits were correlated with the Santa Rosa Group based on lithology and Carboniferous–Permian fossils (Bateson 1972). The Altos Cuchumatanes area (Fig. 1c) exposes low-to high-grade basement, intruded by granitoids, and mantled by Santa Rosa Group metasedimentary rocks (Anderson *et al.* 1973).

Rabinal complex: Salamá schists, San Gabriel unit, and Rabinal granitoids (Maya Block). Van den Boom (1972) called low-grade metasedimentary rocks south of the Polochic fault zone in central Guatemala the Salamá schists. These rocks were grouped into an older clastic and volcanic succession, the San Gabriel unit, which is intruded by the Rabinal granitoids, and a younger clastic and calcareous succession, probably, part of the Santa Rosa Group (Fig. 1c; Donnelly *et al.* 1990; Ortega-Obregón *et al.* 2008). The basal conglomerate beds of the Santa Rosa Group comprise the

Maya block (southern Guatemala)

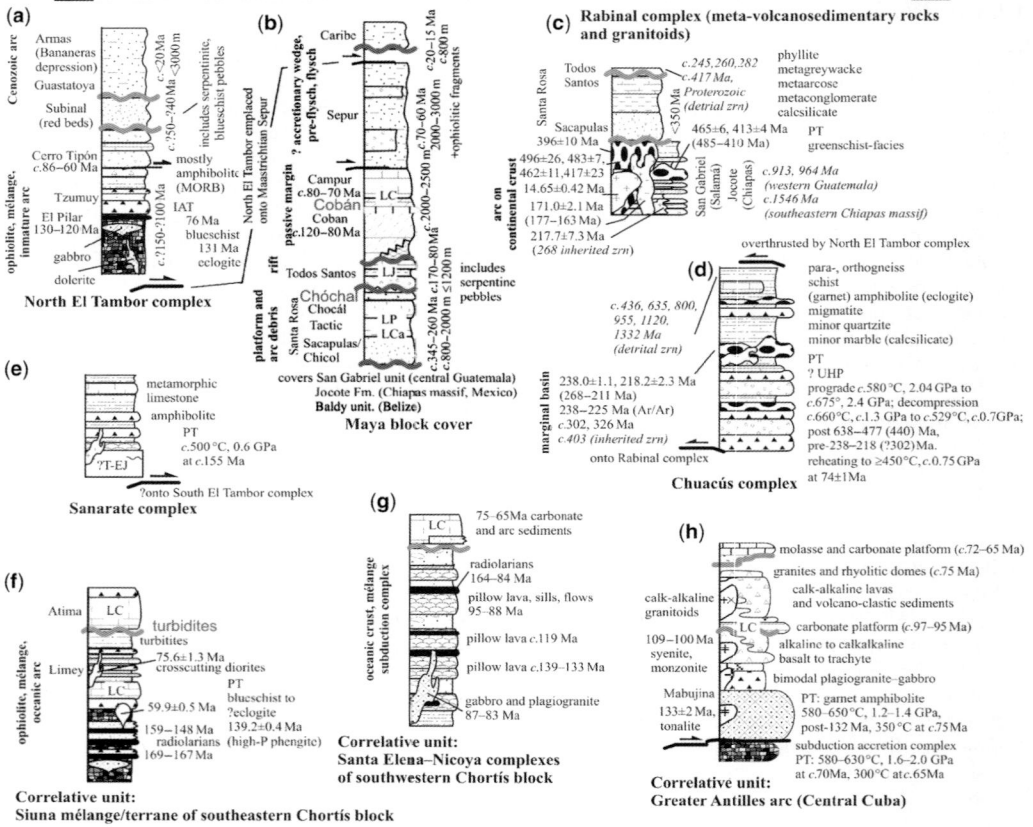

(a)

Cenozoic arc — Armas (Bananeras depression); Guastatoya; Subinal (red beds); Cerro Tipón c.86–60 Ma; Tzumuy; El Pilar 130–120 Ma; gabbro; dolerite

ophiolite, mélange, immature arc

c.20 Ma <3000 m; c.?50–?40 Ma; includes serpentinite, blueschist pebbles; mostly amphibolite (MORB); IAT; 76 Ma blueschist; 131 Ma eclogite; c.?150–?100 Ma

North El Tambor complex

(b)

Caribe; Sepur; Campur c.80–70 Ma; Cobán c.120–80 Ma; Todos Santos; Chóchal; Chocál; Tactic; Sacapulas/Chicol

? accretionary wedge, pre-flysch, flysch; passive margin; rift; platform and arc debris; Santa Rosa

North El Tambor emplaced onto Maastrichtian Sepur

c.20–15 Ma <800 m; c.70–60 Ma; +ophiolitic fragments; 2000–3000 m; 2000–2500 m; c.170–80 Ma; 2000–2000 m ≤1200 m; c.345–260 Ma c.800–2000 m

LC; LJ; LP; LCa

includes serpentine pebbles

covers San Gabriel unit (central Guatemala), Jocote Fm. (Chiapas massif, Mexico), Baldy unit. (Belize)

Maya block cover

(c) Rabinal complex (meta-volcanosedimentary rocks and granitoids)

Todos Santos; Santa Rosa; Sacapulas 396±10 Ma; San Gabriel (Salamá); Jocote (Chiapas)

c.245,260,282; c.417 Ma, Proterozoic (detrital zrn) <350 Ma; 465±6, 413±4 Ma (485–410 Ma); 496±26, 483±7, 462±11,417±23, 14.65±0.42 Ma, 171.0±2.1 Ma (177–163 Ma), 217.7±7.3 Ma (268 inherited zrn); c.913, 964 Ma (western Guatemala); c.1546 Ma (southeastern Chiapas massif)

phyllite; metagreywacke; metaarcose; metaconglomerate; calcsilicate; PT greenschist-facies

arc on continental crust

overthrusted by North El Tambor complex

(d)

c.436, 635, 800, 955, 1120, 1332 Ma (detrital zrn)

238.0±1.1, 218.2±2.3 Ma (268–211 Ma); 238–225 Ma (Ar/Ar); c.302, 326 Ma; c.403 (inherited zrn)

onto Rabinal complex

para-, orthogneiss; schist; (garnet) amphibolite (eclogite); migmatite; minor quartzite; minor marble (calcsilicate); PT; ? UHP; prograde c.580 °C, 2.04 GPa to c.675°, 2.4 GPa; decompression c.660°C, c.1.3 GPa to c.529°C, c.0.7GPa; post 638–477 (440) Ma, pre-238–218 (?302) Ma, reheating to ≥450°C, c.0.75 GPa at 74±1 Ma

marginal basin

Chuacús complex

(e)

metamorphic limestone; amphibolite; PT c.500 °C, 0.6 GPa at c.155 Ma; ?T-EJ; ?onto South El Tambor complex

Sanarate complex

(g)

LC; 75–65Ma carbonate and arc sediments; radiolarians 164–84 Ma; pillow lava, sills, flows 95–88 Ma; pillow lava c.119 Ma; pillow lava c.139–133 Ma; gabbro and plagiogranite 87–83 Ma

oceanic crust, mélange subduction complex

Correlative unit:
Santa Elena–Nicoya complexes of southwestern Chortís block

(f)

Atima; LC; turbidites; turbitites; 75.6±1.3 Ma crosscutting diorites; Limey; LC; PT blueschist to ?eclogite; 59.9±0.5 Ma; 159–148 Ma radiolarians; 139.2±0.4 Ma (high-P phengite); 169–167 Ma

ophiolite, mélange, oceanic arc

Correlative unit:
Siuna mélange/terrane of southeastern Chortís block

(h)

molase and carbonate platform (c.72–65 Ma); granites and rhyolitic domes (c.75 Ma); calk-alkaline lavas and volcano-clastic sediments; carbonate platform (c.97–95 Ma); alkaline to calkalkaline basalt to trachyte; bimodal plagiogranite–gabbro; PT: garnet amphibolite 580–630°C, 1.2–1.4 GPa, post-132 Ma, 350°C at c.75Ma; subduction accretion complex PT: 580–630°C, 1.6–2.0 GPa at c.70Ma, 300°C at c.65Ma

calk-alkaline granitoids; 109–100 Ma syenite, monzonite; Mabujina 133±2 Ma, tonalite; LC

Correlative unit:
Greater Antilles arc (Central Cuba)

Fig. 2. (Litho)stratigraphic and pressure–temperature–time–deformation evolution and tectono-stratigraphic interpretation of the Maya Block, the Motagua suture zone, and the northern Chortís Block and correlative units in southern Mexico and the southern Chortís Block (see text for discussion of correlations). Ages in italics are detrital U–Pb zircon ages or from inherited zircons in granitoids; other ages are from magmatic rocks. All ages are those reported in the captions of Fig. 1c and the text. Pressure–time conditions from this paper and literature are reviewed in the text.

Sacapulas Formation of van den Boom (1972). Ortega-Obregón *et al.* (2004, 2008) proposed an up to 10 km wide, south-dipping shear zone, the Baja Verapaz shear zone, thrusting (with a minor sinistral strike–slip component) Chuacús gneiss onto San Gabriel rocks; deformation is most pronounced within 300 m of the Chuacús–San Gabriel boundary. The lateral continuation of the shear zone is unknown. The San Gabriel unit has a continental depositional environment and is pre-Silurian, given by the age of the Rabinal granitoids (see below). It resembles low-grade metasedimentary rocks in western Guatemala, which are post-Middle Proterozoic based on detrital zircon geochronology (Solari *et al.* 2008). The Santa Rosa Group yielded Early

Carboniferous crinoids (van den Boom 1972) and Mississippian (Tournaisian) conodonts (Ortega-Obregón *et al.* 2008). The Rabinal granitoids are spatially poorly defined, as their sheared margins and the surrounding sediments have similar appearance; they are predominately fault-bounded and widely tectonized. Here we use the term Rabinal complex to describe the in general low-grade rock assemblage that comprises pre-Middle Devonian volcanoclastic rocks and Ordovician–Early Devonian granitoids; it is widely associated with its clastic–calcareous cover, the Santa Rosa Group. We note that there is no obvious difference in Palaeozoic lithostratigraphy and magmatism between the Maya Block (Maya mountains, Alto Cuchumatanes)

(i) Correlative unit: Acatlán complex of southern Mexico

Matzitzi, Olinalá, Patlanoaya, Tecomate

Chazumba

c.297,834,1203 Ma
1546–286 Ma
c.300–250 Ma

Esperanza

372±8 Ma

continental arc

c.719, 917, 977, 1130, 1124, 1448, 1679 Ma

442–440 Ma

c.600–450 Ma

PT
730–830 °C, 1.5–1.7 GPa
*c.*430±5 to 418±18 Ma
640–690 °C, 1.0–1.4 GPa
>372±8 Ma

Xayacatlán

PT
490–610 °C, 1.25 GPa to 600 °C, 0.95 GPa
to 500 °C, 0.67 GPa at *c.*477–490Ma
reheating at 500–525 °C

PT
amphibolite
no high-P

El Rodeo

287±2 Ma

371±34 Ma

461±9 Ma

478–474 Ma
post-high-pressure

c.860–480 Ma

oceanic basin

arc & continental rift

at *c.*0.95 GPa
at *c.*420–380 Ma

*c.*420 Ma

Tecolapa

478–471 Ma
1043±50 Ma
1165±30 Ma

476–471 Ma
?all post-metamorphic

c.988, 1171, 1471 Ma

c.870, 982, c.1135, 1387 Ma

marginal basin

Ixcamilpa

PT
200–580 °C,
0.65–0.9 GPa,
*c.*455–?420 Ma

c.477,603,708,1 946,1128,1821 Ma
(youngest zrn 447 Ma)

PT low-grade **Cosoltepec**

c.410,550,950, 1573 Ma

passive margin

175±3 Ma

pillow basalt
(MORB)

PT
amphibolite

c.275,304,590, 930,1150 Ma
c.<260 Ma
<310Ma

c.410–360 Ma

c.317,525,649,922a

Magdalena

Magdalena migmatite
PT
700–760 °C, 0.55 GPa

stratigraphic columns are not at scale

(j) Chortís Block
(Guatemala, northern and central Honduras)

(l) Correlative unit: Xolapa complex of southern Mexico

Pacific

Pacific arc

Padre Miguel

foreland basin

Matagalpa
Subinal
(red beds)

Wampa/Tabacon
Gualaco
Valle de
Angeles

arc and intra-arc rift

Tayaco
Manto

Tepemechin

Agua

rift

Jacalcapa/
Cacaguapa/
San Diego

Yojoa
andesite
dacite

Aqua Fría

Atima

U.Alima

Las Ovejas

LC

MC

EC

LT,M-LI

Padre Miguel *c.*19–12 Ma
<2000 m
*c.*33–24 Ma <300 m
*c.*50–240 Ma <1000 m

70, 81 Ma

*c.*120–110 Ma *c.*110–95 Ma
<700m

red beds

58, 20 Ma

*c.*120–120 Ma
<2500 m
*c.?*125–120 Ma
<300 m

<1700m

Late Jurassic/
metamorphism/
deformation

c.566,c.873,c.950,c.981, c.1172c.1322 (detrital zrn)

37±3, 130±3, 159±1, 166±1 Ma
245±3Ma (metamorphic), 273±3 Ma
168±1, 396±11, 404±13 Ma
c.88c.19 kc.329c.400, c.480c.728c.965c.1130 Ma (inherited zrn)
c.36c.60c.80c.120c.161c.260 Ma

(k)

?M-LC La Virgen

120 Ma
blueschist
132 Ma
eclogite
MORB

ophiolite, mélange

c.155-?110 Ma

South El Tambor complex

Papagayo
Tilzapotla

Huajuilán
Balsas
Tepetlapa

Morelos

MC

*c.*34 Ma rhyolite
38±1, 35±1 Ma
top-S thrust/fold:*c.*45-34 Ma
44±1Ma

Colorada
La Venta=Tierra Colorado
sinistral transtensive shear zone;
syn*c.*45–35 Ma

*c.*22–24 Ma <2500 m

*c.*110–100 Ma

Colorada Chapolapa
Tierra Colorado 34±1 Ma

Las Piñas 54.2±5.8 Ma
El Salitre 59±1, 55±4 Ma

El Pozuelo 129±0.5 Ma

I-type plutons arc

*c.*126–133 Ma

metapelite, -psammite,
metaandesite. -rhyolite

ortho-, paragneiss, marble, migmatite
pre-129 migmatization

migmatites and gneisses 47,65,132,
160,180,272 Ma
undeformed plutons 27–35 Ma

mud- to siltstones	radiolarian chert	augengneiss
sandstone	basalt	gabbro (amphibolite)
conglomerate	andesite, tuff, breccia	
shale	pillow lavas, sills	IAT, island-arc tholeiite
gypsum	massive lava	MORB, mid-ocean ridge basalt
carbonate (mostly massive limestone)		(serpentinized) ultramafic rock
dolomite		basement: phyllite, schist, quartzite
calcareous shale with sandstones		basement: mica schist, gneiss, migmatite
mudstone, wacke, olistostrome		

L, Late; M, Middle; E, Early; C, Cretaceous; J, Jurassic; T, Triassic, P, Permian; Ca, Carboniferous;
Pz, Palaeozoic; PT, pressure–temperature; UHP, ultra-high pressure

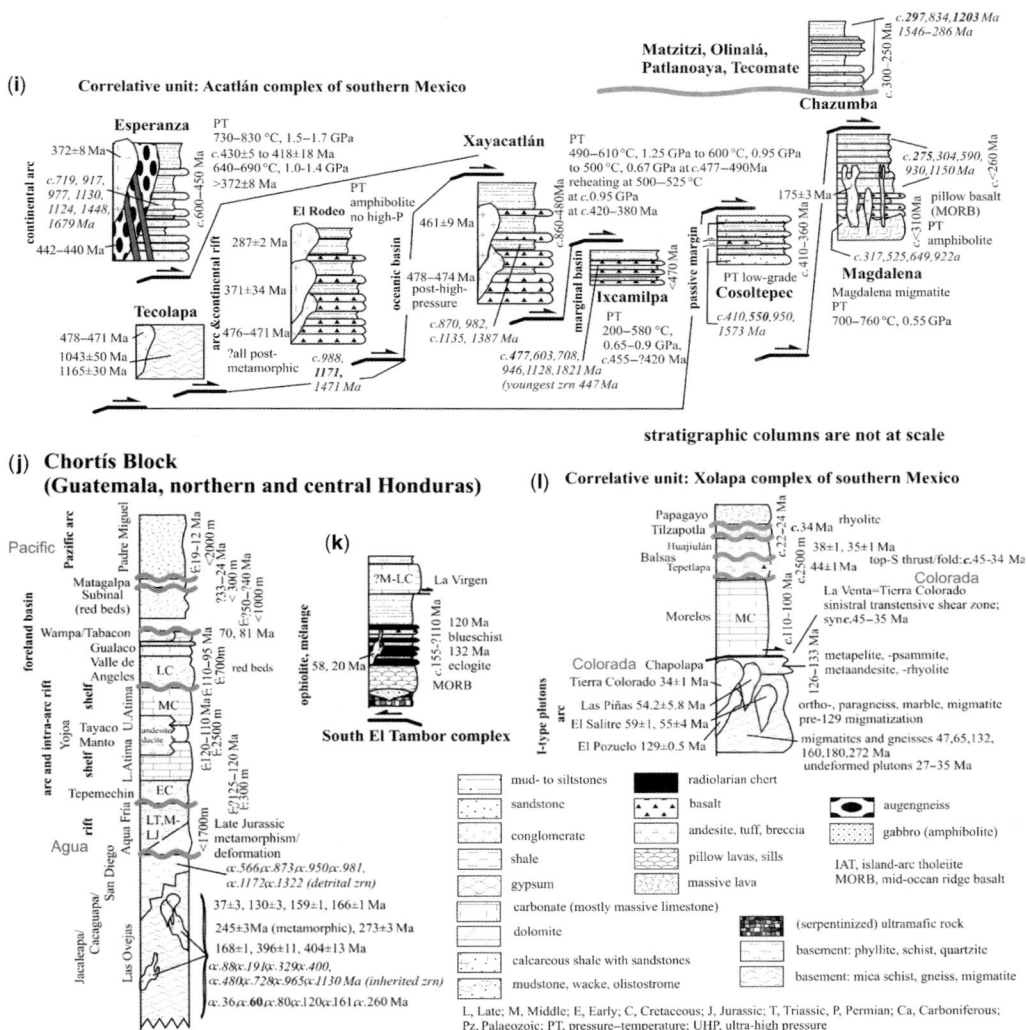

Fig. 2. (*Continued*).

and the Rabinal complex, suggesting that the Polochic fault is not a fundamental plate boundary fault (see Ortega-Gutiérrez *et al.* 2007).

Chuacús complex. In Guatemala, Ortega-Gutiérrez *et al.* (2004) redefined the Chuacús complex as schists and gneisses outcropping between the Motagua fault and the Baja Verapaz shear zone of the Salamá–Rabinal area (Fig. 1c). Chuacús schists and gneisses contain quartz + albite + white mica + epidote/zoisite ± garnet ± biotite ± amphibole ± omphacite ± allanite ± chlorite ± calcite ± titanite ± rutile (Martens *et al.* 2005). Banded migmatite and (garnet) amphibolite (locally relict eclogite),

calcsilicate and marble are uncommon. Local quantitative petrology suggests upper greenschist- to amphibolite-facies conditions up to 650 °C at *c.* 1.1 GPa; retrogression to greenschist facies is poorly documented (van den Boom 1972; Machorro 1993). Both foliation concordant and discordant quartz–albite–white mica pegmatites occur. Eclogitic relicts contain possible ultra-high pressure metamorphism, and pyroxene with up to 45 vol% omphacite and garnet together with phengite + rutile + zoisite + epidote. Thermobarometric analyses yielded *c.* 740 °C at *c.* 2.3 GPa with retrogression at *c.* 590 °C and *c.* 1.4 GPa. Decompression seems associated with local partial melting

(m) Correlative units: Guerrero und Mixteca terranes of southern Mexico

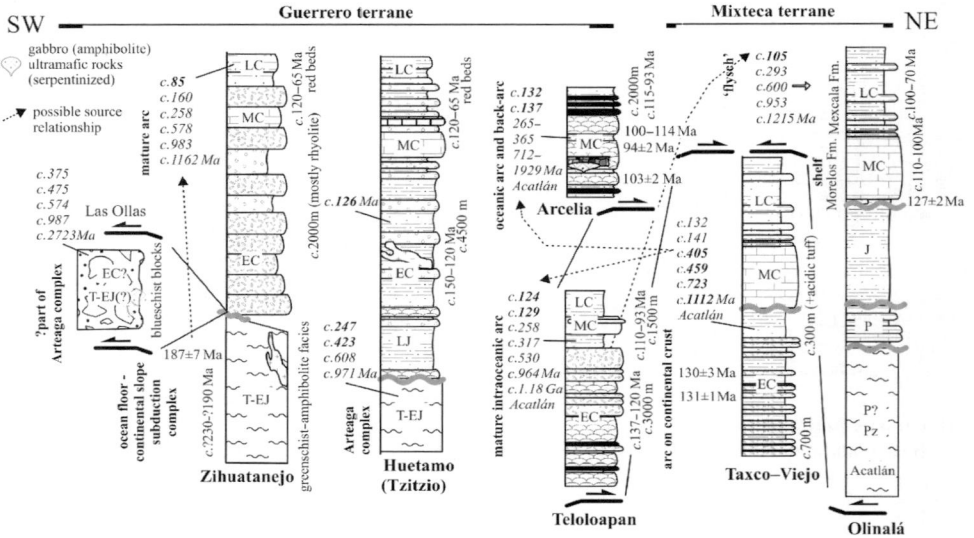

Fig. 2. (*Continued*).

(Ortega-Gutiérrez *et al.* 2004). In central-eastern Guatemala (San Agustín Acasaguastlán), orthogneiss (quartz–monzonite, granodiorite) outcrops together with high-grade, typical Chuacús paragneiss; the latter locally shows sillimanite and abundant K-feldspar. The volcanosedimentary protolith of the Chuacús complex is probably younger than *c.* 440 Ma based on detrital zircon geochronology (Solari *et al.* 2008).

Existing geochronology: Maya Block, Rabinal and Chuacús complexes. U–Pb zircon and monazite ages obtained from two batholiths in the Maya mountains (Steiner & Walker 1996) yielded 404–420 Ma (the best constrained samples are 418 ± 3.6 and 404 ± 3.3 Ma); K–Ar biotite ages of these igneous rocks are 230 ± 9 Ma (weighted mean of seven dates compiled in Steiner & Walker 1996). Martens *et al.* (2010) discriminated a post-520 Ma (detrital zircon geochronology) volcanosedimentary sequence (Baldy unit) that is interbedded with 404 ± 7 Ma rhyolite in its upper section (Bladen Formation). Granitoids of the Altos Cuchumatanes area yielded 269 ± 29 and 391.2 ± 7.4 Ma (Solari *et al.* 2008). Weber *et al.* (2008) reported 482 ± 5 Ma U–Pb zircon and 473 ± 39 Ma Sm–Nd garnet–whole rocks ages from S-type granite from the Chiapas Massif of the southwestern Maya Block. Three zircon fractions of a Rabinal orthogneiss yielded a concordant 396 ± 10 Ma age (recalculated from the data given by Gomberg *et al.* 1968). All zircon fractions of

three Rabinal granites and one pegmatite are discordant (Ortega-Obregón *et al.* 2008); they yielded lower intercepts of 496 ± 26 and 417 ± 23 Ma (granite), 462 ± 11 Ma (pegmatite), and an upper intercept of 483 ± 23 Ma (granite). Muscovite from two pegmatites cutting the San Gabriel unit yielded K–Ar cooling ages of 453 ± 4 and 445 ± 5 Ma. Ortega-Obregón *et al.* (2008) suggested intrusion of the Rabinal granitoids between 462 and 453 Ma. The Matanzas granite of eastern central Guatemala, possibly intruding the Rabinal complex, yielded a two feldspar–whole rock Rb–Sr isochron of *c.* 227 Ma, and Ar/Ar white mica and biotite ages of *c.* 212 and 161 Ma, respectively (Donnelly *et al.* 1990). Four zircon fractions combining two Chuacús gneiss samples south of Salamá in central Guatemala gave a lower intercept U–Pb age of 326 ± 85 Ma (recalculated from Gomberg *et al.* 1968). Three single zircon and two small populations from refolded leucosome interbedded with relict eclogite (El Chol area) yielded intercept U–Pb ages of 302 ± 6 and 1048 ± 10 Ma (Ortega-Gutiérrez *et al.* 2004). K–Ar and Ar/Ar ages from Chuacús-complex rocks are mostly poorly located and documented. Hornblende, white mica and biotite gave Ar/Ar ages of 78–66, 76–67 and 67–64 Ma, respectively (Sutter 1979). Amphibole from garnet amphibolite and white mica from retrograde phengite-bearing gneiss yielded K–Ar ages of *c.* 73 and *c.* 70 Ma, respectively. Coarse- and finer-grained white mica from foliated metapegmatite gave *c.* 73 and *c.* 62 Ma, respectively (weighted mean ages from several fractions given by Ortega-Gutiérrez *et al.* 2004). White mica from Chuacús-complex paragneiss and pegmatite ≥4 km south of the main trace of the Baja Verapaz shear zone gave ages between 74.1 ± 1.3 and 65.3 ± 1 Ma; they were suggested to date shearing along the Baja Verapaz zone (Ortega-Obregón *et al.* 2008). The oldest Ar/Ar age from the Chuacús complex is *c.* 238 Ma (amphibole; Donnelly *et al.* 1990).

El Tambor complexes: metamorphic–metasomatic rocks of the Motagua suture zone. The El Tambor complexes (Fig. 1c) comprise metasedimentary rocks of a subduction–accretionary complex, sheets of serpentinized peridotite, serpentinite mélange and exotic blocks, containing eclogite, jadeitite, mylonitized gabbro, (garnet) amphibolite, pillow lavas and radiolarian cherts; as far as is known, the basal contacts are tectonic (e.g. Bertrand & Vuagnat 1975; Harlow 1994; Beccaluva *et al.* 1995; Tsujimori *et al.* 2005; Chiari *et al.* 2006). Regional prehnite–pumpellyite-facies metamorphism affected the volcanic rocks, forming albite + chlorite, predominantly pumpellyite, and locally prehnite. Centimetre- to metre-scale interlayering of mafic rocks, siliceous marble and quartzite (probably metamorphosed chert) are reported in the La Pita complex of the eastern MSZ (Martens *et al.* 2007); assemblages with barroisite + garnet + epidote + albite + white mica are characteristic. Jadeitite formed sporadically within the serpentinite matrix of the El Tambor mélange; it consists of jadeitic pyroxene with minor paragonite, phengite, phlogopite, albite, titanite, zircon, apatite and garnet and formed at 100–400 °C and 0.5–1.1 GPa (Harlow 1994). Two contrasting high-*P* belts contain lawsonite eclogite and zoisite eclogite south and north of the Motagua valley, respectively. The southern lawsonite eclogites record unusually low-*T*–high-*P* with prograde eclogite facies at *c.* 450 °C and *c.* 1.8–2.4 GPa, followed by retrogression with infiltration of fluids at <300 °C and *c.* 0.7 GPa (Tsujimori *et al.* 2004, 2005, 2006). Amphibolitized, paragonite-bearing zoisite eclogites in antigorite schist north of the Cabañas fault formed at significantly higher *T*, 600–650 °C at 2.0–2.3 GPa (Tsujimori *et al.* 2004).

Existing geochronology: El Tambor complexes. Two Sm–Nd mineral isochrons from eclogites south of the Motagua fault zone gave ages of 131.7 ± 1.7 and 132.0 ± 4.6 Ma, which are identical to the 130.7 ± 6.3 Ma age from eclogite sampled north of the fault (Brueckner *et al.* 2005). The southern eclogites show MORB major and trace element patterns and are isotopically depleted ($^{87}Sr/^{86}Sr$ ratios of 0.70374 and 0.70489, ε_{Nd} of

+8.6 and +9.2). One northern eclogite has a significantly more enriched signature ($^{87}Sr/^{86}Sr =$ 0.70536; $\varepsilon_{Nd} = -2.1$). Harlow *et al.* (2004) obtained Ar/Ar white mica ages from jadeitite, albitite, mica-bearing metamorphic rocks and altered eclogite within the MSZ, ranging between 77–65 Ma for the northern high-*P* belt and 125–113 Ma for the southern belt. In the northern belt, three phengite-only samples are indistinguishable within error at 76 Ma, whereas two phengite–phlogopite–paragonite mixtures are younger at 71 and 65 Ma. The 600–650 °C peak *T* of the northern high-*P* rocks suggests that the phengite (closing *T* *c.* 400 °C, von Blanckenburg *et al.* 1989) dates retrogression under blueschist-facies conditions when Na–Al–Si-rich fluids produced jadeitite and albitite in serpentinite (Harlow 1994). All dated minerals from the southern belt are phengite; four ages are within error at 120 Ma, one is older, two younger. These ages probably date fluid infiltration during blueschist-facies retrograde metamorphism. Bertrand *et al.* (1978) reported K–Ar mineral and whole-rock ages from poorly located samples north of the Motagua fault zone. We divided their samples regionally and recalculated the ages: a metadolerite from the Baja Verapaz ophiolite sheet (Fig. 1c) north of Salamá yielded *c.* 76 Ma, three metadiabase samples from pillow basalt overlying the Chuacús complex gave a weighted mean whole-rock age of *c.* 74 Ma (the $^{40}Ar/^{36}Ar$ v. $^{40}K/^{36}Ar$ isochron gives atmospheric $^{40}Ar/^{36}Ar$), and six amphibole separates from garnet amphibolites of the same unit yielded a weighted mean age of 57.2 ± 5.4 Ma (again, atmospheric $^{40}Ar/^{36}Ar$ with an 'isochron age' of *c.* 56 Ma). All ages probably date metamorphism.

The available petrology (lawsonite and zoisite eclogites south and north of the Motagua fault zone, respectively), geochronology (*c.* 120 and 76 Ma for blueschist-facies retrogression south and north of the Motagua fault zone, respectively) and isotope geochemistry (isotopically depleted v. enriched south and north of the Motagua fault zone, respectively) demonstrate the existence of two distinct subduction–accretion belts along the MSZ, even though ages of crust formation and eclogite–facies metamorphism may appear similar. Herein, we use the terms South and North El Tambor complexes for these two distinct subduction–accretion belts and their ophiolitic sheets, with the South El Tambor complex being located south of the Cabañas strand of the Motagua fault zone, and the North El Tambor complex located to its north.

Las Ovejas complex, San Diego phyllite, and Cacaguapa schist (Chortís Block). The Las Ovejas complex (Fig. 1c) comprises biotite schist, gneiss, migmatite, amphibolite, marble and deformed granitoids. Specifically, amphibolite ($\leq 15\%$ of the Las Ovejas complex in Guatemala), garnetiferous gneiss, staurolite schist, garnet–two-mica schist and marble with calcsilicate layers are exposed together with metaigneous rocks, diorite, tonalite, granodiorite and minor gabbro. Mineral assemblages define amphibolite-facies conditions; metamorphic grade of rocks containing staurolite increases toward the NW where a sillimanite zone is identified (IGN 1978). Both basement and cover of the northern Chortís Block continue as the Bonacca Ridge, forming a series of islands south of the Swan Island Fault (Fig. 1c), the southern boundary fault to the Cayman Trough. Roatán Island, which is the best studied (Avé Lallemant & Gordon 1999), exposes amphibolite-facies basement with (augen)gneiss, schist, amphibolite (partly metagabbro), marble and rare pegmatite.

The San Diego metamorphic rocks consist of phyllite and minor thin quartzite and slate layers (Schwartz 1976). Their stratigraphic age is post-Early Cambrian (youngest detrital zircons are c. 520 Ma; Solari et al. 2008). The Chiquimula batholith thermal overprint produced cordierite and andalusite (Clemons 1966). Greenschist-facies and locally higher-grade metamorphic rocks in Honduras include calcareous phyllite, white mica–chlorite and graphitic schists and slate ('younger sequence' of Horne et al. 1976a). The San Diego phyllite of southern Guatemala and northern Honduras has been correlated with the Cacaguapa schists south and east of the Jocotán–Chamelecón fault system (Fig. 1c; Burkart 1994). The greenschist-facies lithology at Roatán Island comprises (quartz) phyllite, chlorite schist, serpentinite, chert and marble to limestone; locally granite porphyry is present (Avé Lallemant & Gordon 1999).

Sanarate complex. We introduce the Sanarate complex as a distinct unit separate of the South El Tambor and the Las Ovejas complexes due to its distinct lithology and $P–T–d–t$ parameters; however, its boundaries and lithostratigraphy are mostly unknown and we may have exaggerated its extend in the geological maps (e.g. Fig. 1c). The complex constitutes the westernmost basement outcrop south of the MSZ and comprises partly retrograde amphibolite, garnet- and quartz-bearing amphibolite, retrograde leucogabbro, chlorite–epidote–hornblende schist, micaschist, phyllite and chlorite phyllite. Massive amphibolite is characteristic and most widespread. These lithologies probably represent an oceanic environment. The complex is in fault contact with Cenozoic andesite and red beds and is cut by andesite dykes. Along its eastern margin, the Sanarate-complex rocks shows west-dipping foliation and apparently thrusted top-to-east

over South El Tambor complex metaclastic rocks that show distinctly lower metamorphic grade.

Existing geochronology: Las Ovejas complex, San Diego phyllite and Cacaguapa schist, igneous rocks. Most of the available geochronology is, in a modern sense, poorly documented and samples are only approximately located (Fig. 1c). U–Pb zircon ages on igneous rocks are: c. 93 Ma (Rio Julian granite), c. 80 Ma (sheared Rio Cangrejal orthogneiss), 81.0 ± 0.1 (El Carbon granodiorite), c. 29 Ma (Banderos granodiorite; all Manton & Manton 1984), 124 ± 2 Ma (dacite; Drobe & Oliver 1998), and 1017 Ma (orthogneiss; Manton 1996). We recalculated the zircon data of the El Carbon granite of Manton & Manton (1999) and obtained lower and upper intercepts of 30 ± 19 and 126 ± 42 Ma, respectively. Deformed granitic plutons gave a 300 ± 6 Ma Rb–Sr whole-rock isochron for the Quebrada Seca complex (Horne et al. 1976a). Granitoids gave the following Rb–Sr ages: 150 ± 13 Ma (San Marcos granodiorite, whole rock; Horne et al. 1976a), c. 69 Ma (K-feldspar and plagioclase, cooling age of sheared El Carbon granodiorite; Manton & Manton 1984), c. 80 Ma (whole-rock isochron, Trujillo granodiorite; Manton & Manton 1984), 140 ± 15 Ma (Dipilto granite, whole rock; Donnelly et al. 1990), 118 ± 2 Ma (Carrizal diorite, K-feldspar and plagioclase; Manton & Manton 1984), c. 54 Ma (Río Cangrejal granodiorite), 39 Ma (Sula valley (Sula graben) granodiorite), and 35 Ma (Confadia granite; all biotite cooling ages; Manton & Manton 1984). K–Ar cooling ages of igneous rocks comprise 93.3 ± 1.9, 80.5 ± 1.5 and 73.9 ± 1.5 Ma (Tela augite-tonalite, one amphibole and two biotite ages from different samples; Horne et al. 1974), 72.2 ± 1.5 and 56.8 ± 1.1 Ma (Piedras Negras tonalite, amphibole and biotite; Horne et al. 1974), 35.9 ± 0.7 Ma (San Pedro Sula granodiorite, biotite; Horne et al. 1974), 122.7 ± 2.5 and 117.0 ± 1.9 Ma (San Ignacio granodiorite, hornblende and biotite; Emmet 1983), 114.4 ± 1.7 Ma (San Ignacio adamellite, biotite; Horne et al. 1974), 79.6 ± 1.6 Ma (gabbro dike, amphibole; Emmet 1983), 62.2 ± 2.6 Ma (diorite, whole rock; Italian Hydrothermal 1987, reported in Rogers 2003), 60.6 ± 1.3 and 59.3 ± 0.9 Ma (Minas de Oro granodiorite, biotite; Horne et al. 1976b; Emmet 1983), 58.6 ± 0.7 Ma (San Francisco dacite pluton, biotite; Horne et al. 1976b). Ar/Ar ages are from andesite and diorite of the eastern Chortís Block (75.6 ± 1.3 and 59.9 ± 0.5 Ma, amphibole and biotite; Venable 1994). A K–Ar biotite age of 222 ± 8 Ma from micaschist probably reflects a Triassic thermal event (MMAJ 1980). Hornblende from the amphibolite-facies basement of Roatán Island yielded 36.0 ± 1.2 Ma, and a granite porphyry from Barbareta Island, east of Roatán Island, yielded a 39.4 ± 2.8 Ma zircon fission-track age (closing T c. 250 °C; Avé Lallemant & Gordon 1999).

New geochronology

ID-TIMS and ion-probe U–Pb geochronology can be found in the Supplementary materials, and Figs 3 & 4 show the U–Pb data in Concordia diagrams and present representative zircon CL pictures, respectively. Our new Ar/Ar and Rb–Sr mineral ages are summarized in Table 1 and the Supplementary material, respectively, and are plotted in Fig. 5. In this paper, we discuss Phanerozoic U–Pb ages; late Middle Proterozoic ages, which we find in both the Maya and Chortís Blocks, will be reported elsewhere. Our apatite fission-track (AFT) and titanite fission-track (TFT) data are listed in the Supplementary material; confined track-length data appear in the Supplementary material. Figure 6 depicts AFT age v. elevation diagrams for the Rabinal–Chuacús and Las Ovejas complexes of Guatemala/Honduras and the Chiapas Massif of southeastern Mexico (Fig. 1c), and modelled $T-t$ paths for selected samples (Chuacús and Las Ovejas complexes, granitoids). We excluded the samples from western Guatemala from the age v. elevation data (too few readings and widely spaced) and also emphasize that the plotted data do not strictly fulfill the criteria for exhumation estimates from age v. elevation data; for example, the samples cover considerable horizontal distances. We assigned errors to the elevation data, as several samples were collected in the early-1990s when hand-held GPS-receiver precision was still inferior and large-scale topographic maps were locally unavailable. We emphasize that the regression data that we present are first-order estimates. Figure 7 shows the regional distribution of ages, highlights age groups, and depicts $T-t$ fields for selected regions. We employed standard closing temperatures for the $T-t$ diagrams: Ar hornblende, high-T steps, 525 ± 25 °C, low-T steps, 400 and 300 ± 100 °C, depending on grain size; Ar white mica, 350 ± 25 °C; Ar biotite, 300 ± 25 °C; Ar K-feldspar, 400 ± 100 °C for high-T steps and 200 ± 25 °C for low-T steps, depending on the likely presence of single or multiple domains; Rb white mica, 500 ± 25 °C, except for centimetre-sized, pegmatitic white mica, 600 ± 100 °C; Rb biotite, 275 ± 25 °C; TFT, 275 ± 50 °C; AFT, 100 ± 25 °C; exceptionally, errors were set higher, imposed by variable grain size and strong deformation. Zircon crystallization was taken as 800 ± 50 °C, except where zircon growth occurred during peak-T metamorphism and related migmatization, which is constrained by our petrology (see below). We excluded a few samples from the $T-t$

diagrams: ages from sample 5C-36 of the Chuacús complex immediately below North El Tambor complex serpentinite, as these rocks show extreme, late-stage metasomatism; ages from very coarse-grained, pegmatitic white mica. In the following section, we present the data; we relate the ages to the main temperature–deformation $(T-d)$ events in the petrology and structure sections.

U–Pb zircon geochronology

Rabinal complex. Sample G18s (stop G1; Figs 3a & 7a) is a migmatitic gneiss from western Guatemala overprinted by lower greenschist-facies deformation. Four fractions with identical 75–125 μm grains but different magnetic properties were utilized for ID-TIMS analysis. A discordia fit to three fractions yields intercepts of 220.0 ± 3.5 and 1292 ± 39 Ma (MSWD = 0.78); one of these fractions is concordant at 215.92 ± 0.18 Ma. The fourth fraction is concordant at 268.04 ± 0.55 Ma. CL images (Fig. 4) depict well-rounded cores with variable but mostly high U-content, and low-U (on average 63 ppm), oscillatory zoned, idiomorphic rims with Th–U ratios (on average 0.7) that suggest magmatic origin. The weighted average of six grains analysed with the SHRIMP gives a 207-corrected $^{206}Pb–^{238}U$ age of 217.7 ± 7.3 Ma, within error of the concordant TIMS fraction. We interpret this age as the age of leucosome crystallization within the western Rabinal complex; the well-rounded cores suggest a paragneiss protolith and the concordant fraction at c. 268 Ma may indicate an earlier phase of crystallization. Sample G27s (stop G57, western Guatemala, Figs 3a & 7a) is orthogneiss that shows low-T mylonitization along discrete sinistral shear zones. Five fractions varying in grain size or magnetic properties are all discordant but define a chord with lower and upper intercept ages at 284 ± 58 and 958.7 ± 6.8 Ma, respectively (MSWD = 0.63); this rock is probably a Permian intrusion. Sample GM8s (stop G8; Figs 3a & 7a) is granite overprinted by greenschist-facies mylonitization from western Guatemala. Four zircon fractions with variable grain size and magnetic properties define lower and upper intercepts at 165.2 ± 5.4 and c. 990 Ma, respectively (ID-TIMS); one fraction is weakly reverse discordant and two fractions define a concordant age of 163.80 ± 0.38 Ma (Fig. 3a). CL images show oscillatory zoned rims around Proterozoic cores (Fig. 4) and oscillatory zoned coreless grains with marked changes in luminescence from centre to rims; the ages of these grains are identical within error. Consequently, the Th–U ratios obtained from SHRIMP spots are variable and are with one exception >0.2, implying magmatic origin. The 207-corrected ages for a group of six rims and coreless grains yield

Rabinal complex (Maya block)

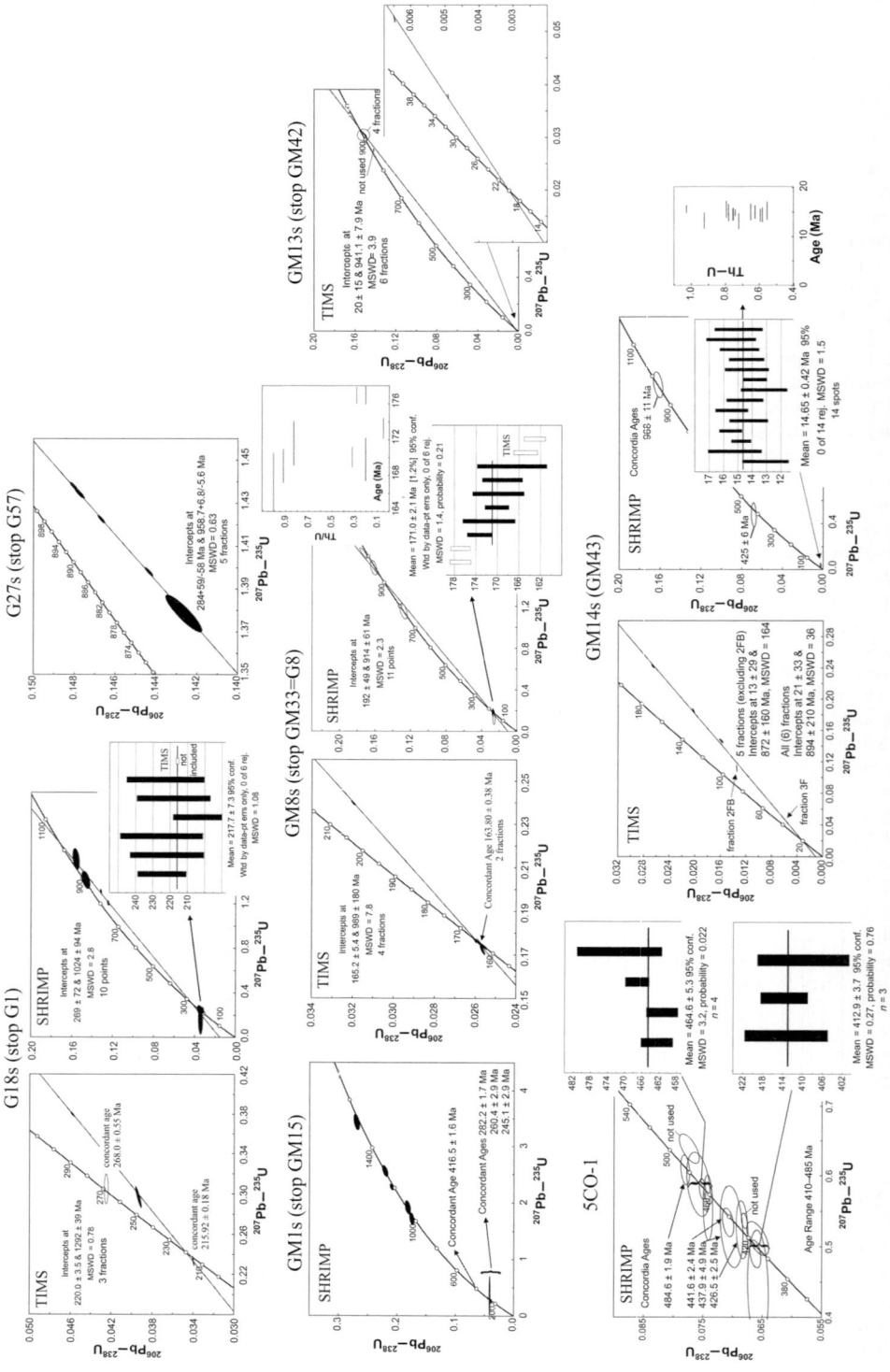

Fig. 3. U–Pb zircon data and age interpretation. See the Supplementary material for sample location and data listings. ID-TIMS results are shown with 2σ error ellipses and discordia are given as lines. SHRIMP results show 2σ error ellipses and are corrected for common Pb. Selected age groups are displayed as weighted average and relative probability diagrams ($^{206}Pb-^{238}U$ ages are 207-corrected). Data evaluation and plotting was supported by the program package of Ludwig (2003). MSWD, mean square weighted deviation. The upper intercept ages and ages older than Early Palaeozoic will be discussed elsewhere. (**a**) Data from the Rabinal complex of the Maya Block.

Chuacús complex

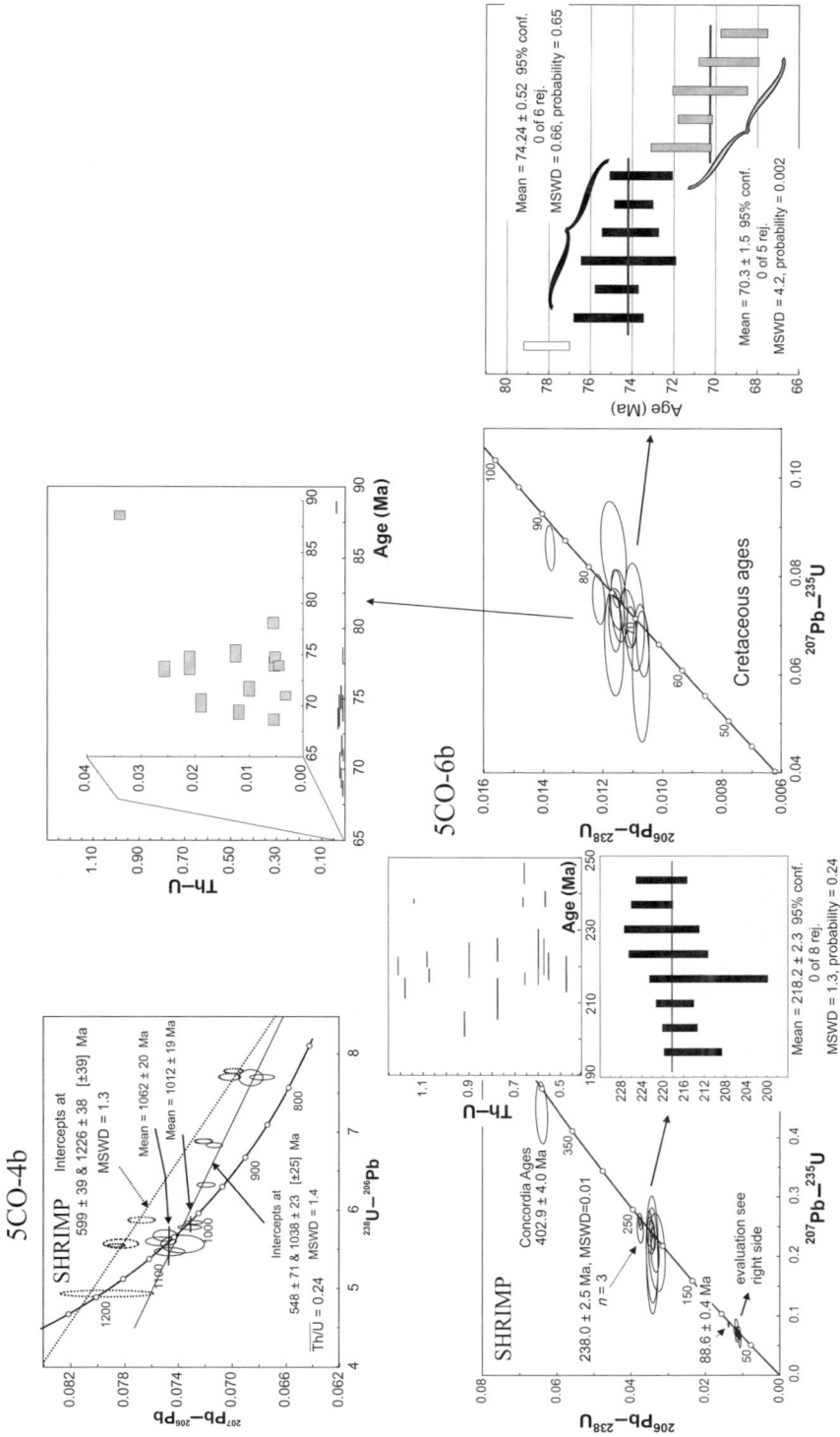

Fig. 3. (*Continued*) (**b**) Data from the Chuacús complex.

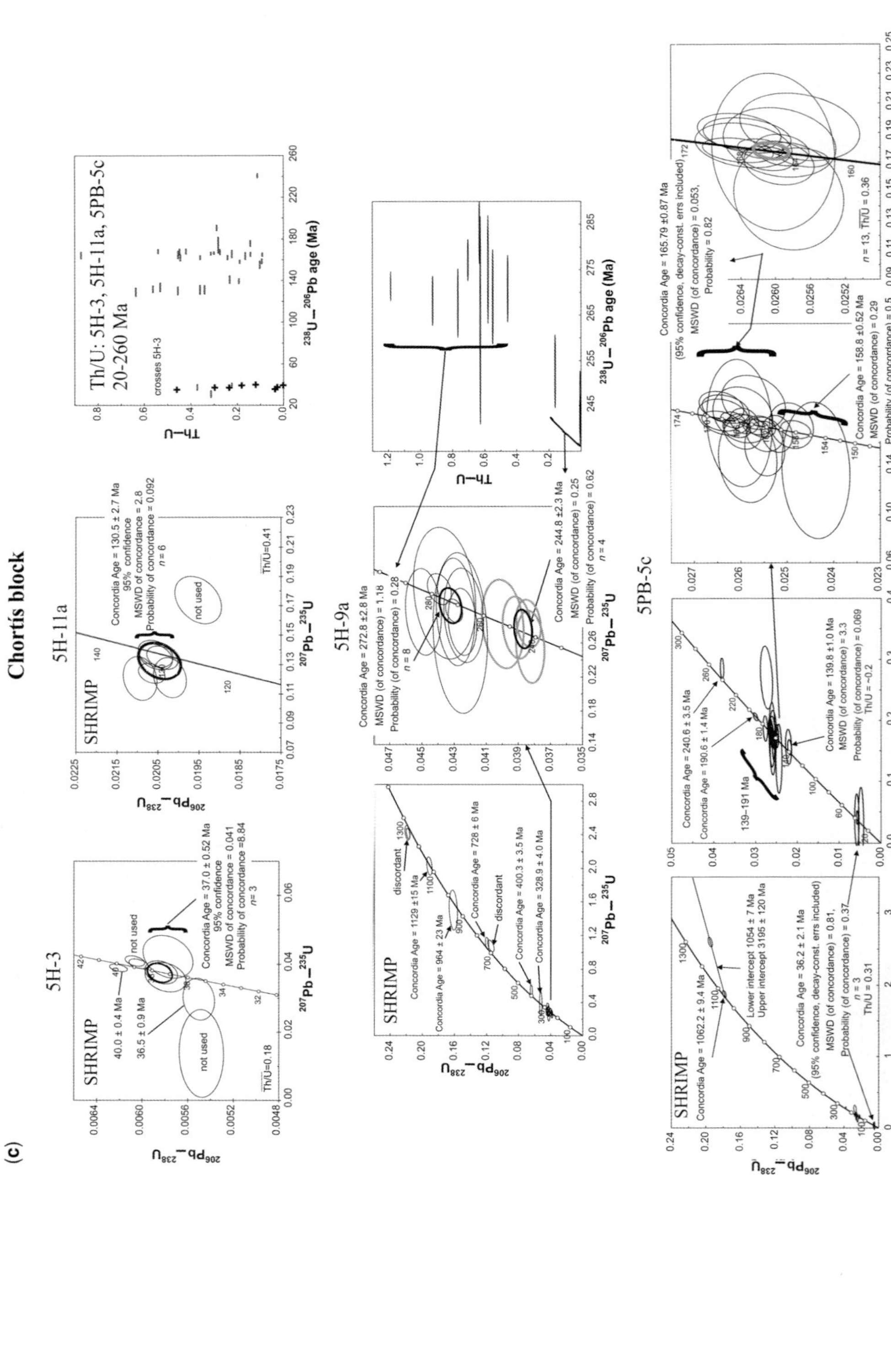

Fig. 3. (*Continued*) (**c**) Data from the Chortís Block; all SHRIMP ages are from the northeastern edge of the Las Ovejas complex and granitoids intruding it. Th–U diagrams aid discrimination of metamorphic and magmatic zircons (or rims).

Chortís block, continuation

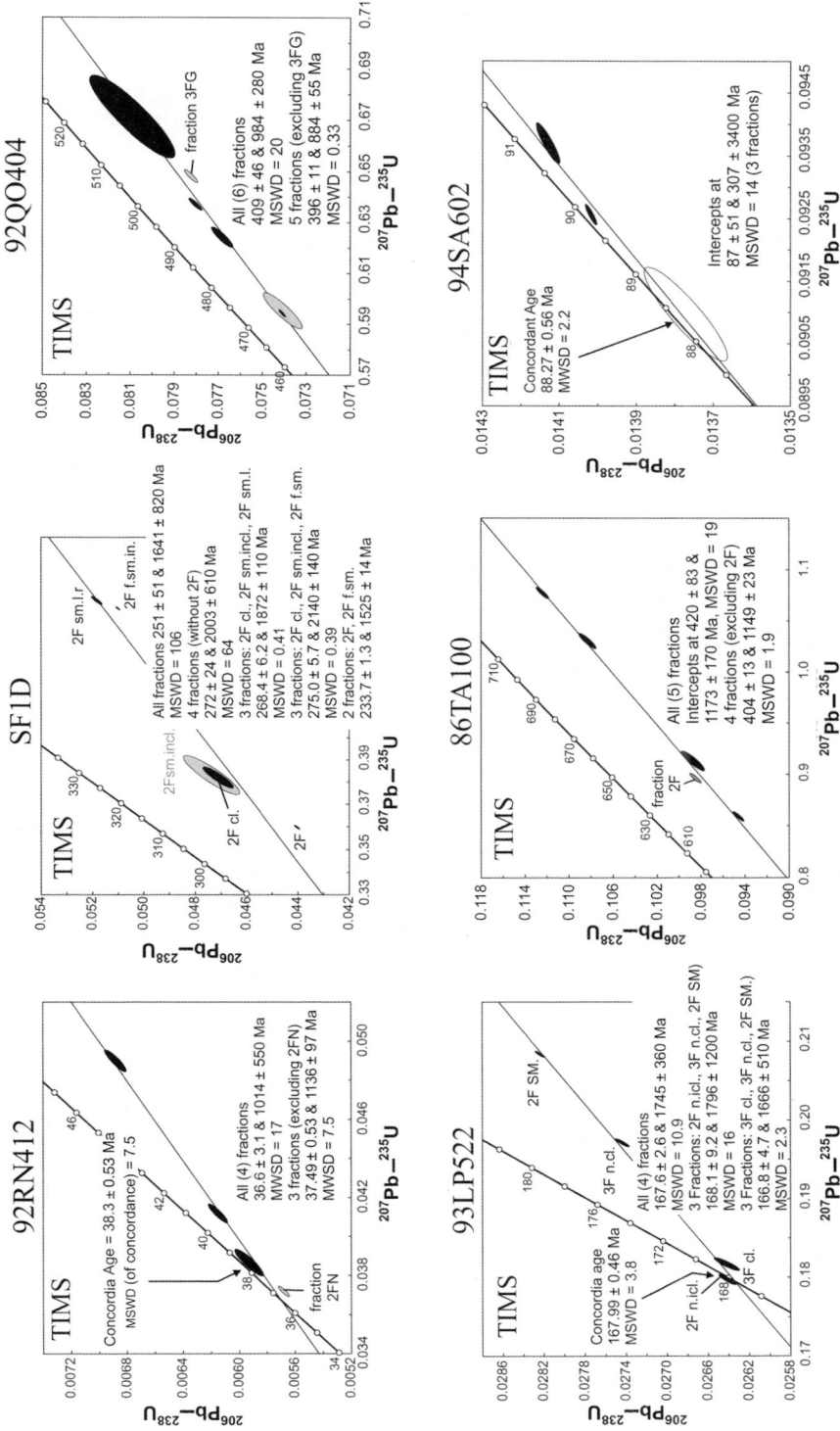

Fig. 3. (*Continued*) (**d**) Data from the Chortís Block, TIMS U–Pb zircon ages.

Rabinal complex (Maya block) and Chuacús complex

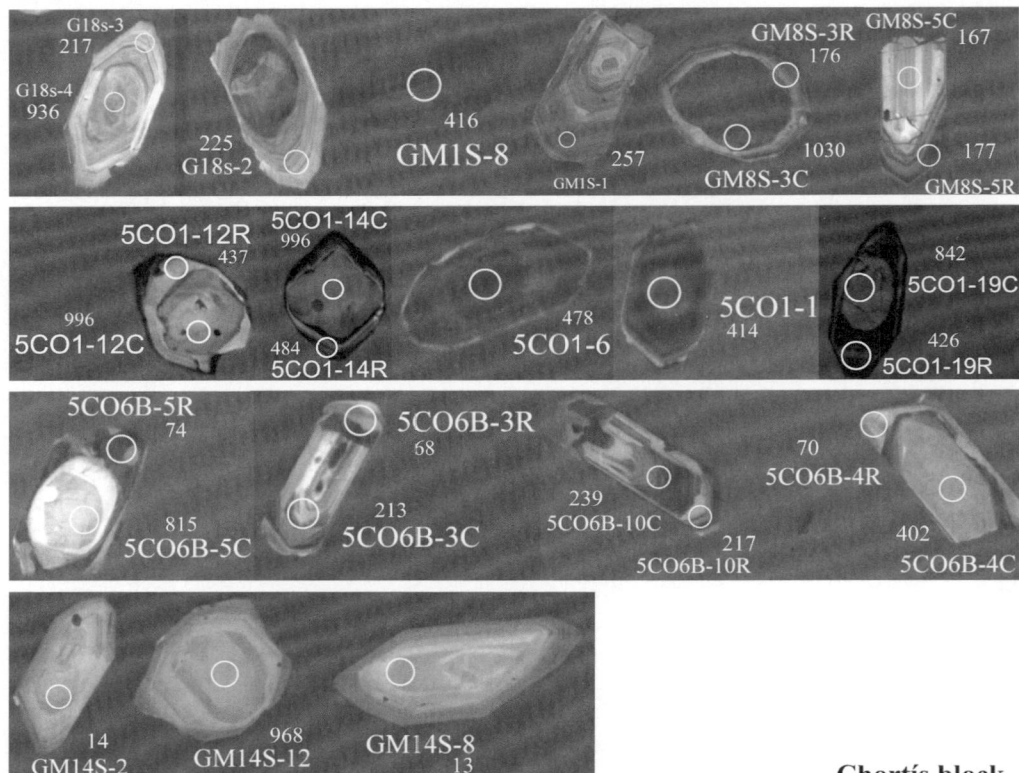

Chortís block

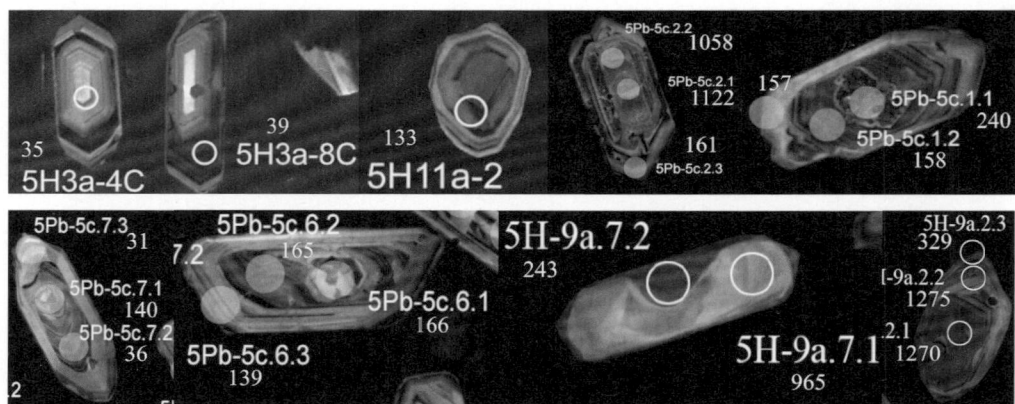

Fig. 4. Cathodoluminescence images of representative zircons. Numbers give sample, spot and spot age (in Ma, below or above the sample number).

a weighted average of 171.0 ± 2.1 Ma. Granite emplacement is either given by this date or is slightly younger (concordant TIMS fractions).

Sample 5CO-1 is two-mica, K-feldspar–porphyroblastic orthogneiss, mylonitized during low-*T* overprint; it defines the 'Rabinal' granitoids in the area of San Miguel Chicaj, central Guatemala (Fig. 7a). Rounded, often mottled, Proterozoic cores are surrounded by in general thin, structure-less, often high-U rims that have variable but

Table 1. $^{40}Ar/^{39}Ar$ data

Sample	Stop	Mineral	J	Weight (mg)	Grain size (μm)	Total fusion age (Ma)	Weighted mean age (Ma)	Isochron age (Ma)	MSWD	$^{40}Ar/^{36}Ar$	$\%^{39}Ar$ used	Steps
Rabinal complex												
G19s	G1	Phe	0.003192	10.1	10–20	66.66 ± 0.33	66.53 ± 0.33	66.50 ± 0.37	4.1/2.0	296 ± 4	61	6–16/21
	G1	Phe	0.003185	10.1	20–40	67.92 ± 0.34	67.38 ± 0.33	67.00 ± 0.37	4.1/1.9	307 ± 5	79	7–19/23
G27s	G57	Bt	0.004555	0.3	250	37.38 ± 1.37	35.06 ± 1.29	37.21 ± 10.06	0.6/2.4	350 ± 74	48	5–10/15
Chuacús complex												
G10s	G29	WM	0.004564	1.2	100–200	66.04 ± 0.65	66.00 ± 0.65	65.96 ± 0.66	14.6/1.7	299 ± 9	93	3–21/26
G12s	G31	WM	0.0007854	n.d.	n.d.	71.80 ± 0.89	72.59 ± 0.65	67.8 ± 5.0	31	297 ± 64	100	1–19/19
5CO-4b	5CO-4	Bt	0.004480	4.3	250	64.31 ± 0.63	64.34 ± 0.63	64.80 ± 0.68	26.6/1.8	264 ± 11	98	2–17/19
		WM	0.004590	2.9	250	67.77 ± 0.67	67.51 ± 0.66	67.56 ± 0.73	174/1.7	286 ± 17	100	1–17/17
5CO-5a	5CO-5	WM	0.004586	2.1	250	234.4 ± 2.2	225.1 ± 2.1	225.2 ± 2.2	0.9/2.6	294 ± 3	5	1–5/23
		Hbl	0.004581	8.9	250	670.61 ± 5.60	—	—	—	—	100	1–20/20
5PB-9b	5PB-9	WM	0.004570	5.1	250	60.27 ± 0.59	—	—	—	—	100	1–27/27
5PB-13b	5PB-13	Hbl	0.0007852	n.d.	n.d.	94 ± 18	69.3 ± 1.4	64.5 ± 5.4	0.31	1283 ± 270	35	5 of 12
5PB-14	5PB-14	WM	0.004568	2.9	250	64.67 ± 0.63	65.01 ± 0.64	64.93 ± 0.64	1.2/2.3	297 ± 1	68	14–20/20
854F	854	Bt	0.014640	28	250	125.3 ± 3.5	—	—	—	—	100	10/10
880C	880	WM	0.014640	16.8	250	102.4 ± 2.7	101.0 ± 2.5	100.4 ± 1.5	0.9/2.6	335 ± 32	88	5–9/10
883B	883	WM	0.014640	17.1	250	75.2 ± 1.8	74.7 ± 1.5	74.2 ± 0.8	0.6/2.4	338 ± 21	93	5–10/10
883B	883	Bt	0.014640	14.9	250	68.2 ± 2.0	68.6 ± 2.0	71.3 ± 1.8	9.0/3.0	146 ± 90	100	1–8/8
887C	887	WM	0.014376	5.4	250	70.5 ± 1.2	69.8 ± 1.2	70.9 ± 3.8	13	321 ± 120	78	1–7/10
Metasomatized southern Chuacús complex												
G6s	G26	WM	0.004565	0.6	c. 63	71.40 ± 0.80	62.01 ± 0.69	59.55 ± 2.37	2.2/2.3	497 ± 367	48	6–12/16
G7s	G27	WR rhyolite	0.004446	20.0	n.a.	68.37 ± 0.94	69.91 ± 1.06	63.66 ± 9.19	9.5/3.0	305 ± 15	59	2–5/7
5C-36a	5C-36	WM	0.004579	1.9	250	48.04 ± 0.47	48.32 ± 0.48	48.20 ± 0.50	16.7/1.8	298 ± 3	86	9–22/22
Sanarate complex												
G3s	G23	WM	0.004534	2.2	c. 63	168.05 ± 1.61	170.56 ± 1.63	151.2 ± 13.22	80/2.4	1339 ± 918	69	3–8/17
G5s	G24	Hbl	0.004519	9.5	250	128.08 ± 1.29	155.85 ± 1.50	155.92 ± 1.91	75/2.7	291 ± 8	36	13–17/17

(Continued)

Table 1. *Continued*

Sample	Stop	Mineral	J	Weight (mg)	Grain size (μm)	Total fusion age (Ma)	Weighted mean age (Ma)	Isochron age (Ma)	MSWD	^{40}Ar/^{36}Ar	%^{39}Ar used	Steps
Western Las Ovejas complex, Guatemala												
5C-2c	5C-2	Hbl	0.000786	n.d.	n.d.	22.1 ± 6.6	19.2 ± 1.5	20.1 ± 6.4	0.32	280 ± 36	73	1–4/6
5C-5	5C-5	Bt	0.004574	0.5	c. 63	29.21 ± 0.29	29.18 ± 0.29	29.90 ± 0.33	3.4/2.6	255 ± 6	88	2–6/7
5C-23d	5C-23	Hbl	0.004609	23.4	250	35.99 ± 0.37	35.43 ± 0.36	35.81 ± 0.40	1.6/2.0	276 ± 9	64	6–15/17
5C-23c	5C-23	Bt	0.004612	0.8	250	25.57 ± 0.26	25.24 ± 0.25	25.34 ± 0.26	0.3/3.0	291 ± 2	60	3–6/9
5C-26c	5C-26	Hbl low-*T*	0.004607	18.9	250	27.84 ± 0.28	17.76 ± 0.18	18.21 ± 0.45	48.3/2.6	272 ± 18	45	1–5/19
		medium-*T*					21.73 ± 0.24	21.72 ± 0.35	2.4/2.4	384 ± 3	13	10–15/19
		high-*T*					38.01 ± 0.38	35.78 ± 2.28	43.8/3.8	314 ± 18	17	15–17/19
5C-26d	5C-26	Bt	0.004605	2.9	250	13.84 ± 0.62	20.49 ± 0.24	21.08 ± 0.84	18.1/2.0	294 ± 2	76	4–9/11
5C-33	5C-33	Kfs	0.004600	12.7	400	30.25 ± 0.59	30.42 ± 0.32	29.94 ± 0.54	6.4/2.3	297 ± 1	89	10–16/17
5C-33		Bt	0.004597	2.5	250	22.42 ± 0.24	23.04 ± 0.23	22.80 ± 0.26	2.2/2.1	297 ± 1	58	6–13/17
		Kfs low-*T*	0.004524	26.8	400	129.9 ± 2.0	7.4 ± 1.1	6.7 ± 3.2	1.7/2.6	298 ± 25	23	6–10/21
		medium-*T*					31.38 ± 2.78	32.98 ± 10.47	0.1/3.8	292 ± 7	7	11–13/21
5C-37b	5C-37	Hbl high-*T*	0.004566	8.6	250	41.06 ± 0.43	39.11 ± 0.55	39.10 ± 3.93	1.2/2.1	512 ± 117	37	8–11/12
		low-*T*					18.23 ± 0.53	18.27 ± 1.63	1.7/2.3	362 ± 20	20	1–3/12
5C-37d	5C-37	Bt	0.004593	2.9	250	17.81 ± 0.18	18.15 ± 0.19	19.49 ± 1.18	60/2.3	289 ± 6	74	4–10/12
5PB-5a	5PB-5	Hbl	0.0007864	n.d.	250	30.4 ± 1.0	31.03 ± 0.70	29.9 ± 1.4	67/–	243 ± 77	96	2–13/15
		Bt	0.004572	1.6	250	28.22 ± 0.28	28.20 ± 0.28	28.29 ± 0.32	18.2/2.1	289 ± 6	100	1–8/8
Eastern Las Ovejas complex, Honduras												
5H-3a	5H-3	Bt (laser)	0.004420	s.g.	250	25.58 ± 0.26	18.98 ± 0.22	19.00 ± 2.78	17.4/2.1	661 ± 412	100	1–10/10
5H-4b	5H-4	Bt (laser)	0.004424	s.g.	250	25.58 ± 0.34	25.70 ± 0.34	23.05 ± 1.86	1.6/3.0	314 ± 13	100	1–4/4
		Hbl	0.004532	5.3	250	36.87 ± 0.40	38.81 ± 0.40	38.86 ± 1.78	93.7/2.6	291 ± 135	92	5–9/10
5H-4c	5H-4	Bt	0.004577	2.3	250	24.77 ± 0.25	25.75 ± 0.26	24.67 ± 0.44	0.6/3.0	311 ± 4	54	5–8/9
		Hbl	0.004527	7.5	250	32.57 ± 0.40	31.53 ± 0.32	31.44 ± 0.65	52.3/2.1	297 ± 6	98	4–11/11
5H-4d	5H-4	Hbl	0.0007881	n.d.	n.d.	28.6 ± 2.0	30.0 ± 1.5	26.1 ± 3.6	6.1/–	334 ± 16	100	1–17/17
Granitoids of the arc complex												
5C-11	5C-11	Bt + Chl	0.004422	s.g.	n.a.	84.51 ± 0.89	89.98 ± 0.89	—	—	297 ± 1	100	1–10/10
5C-21	5C-21	Kfs (laser)	0.004416	s.g.	400	20.03 ± 0.23	19.09 ± 0.25	18.80 ± 0.45	0.3/3.0	291 ± 59	58	5–9/9
5H-5	5H-5	Kfs low-*T*	0.004529	12.3	400	32.98 ± 0.33	28.29 ± 0.28	28.31 ± 0.36	2.4/2.6	265 ± 14	27	10–14/33
5H-13	5H-13	Bt (laser)	0.004425	3.0	250	57.71 ± 0.70	60.89 ± 0.61	61.95 ± 0.86	9.7/2.4		100	1–6/6

Southern Mexico

Sample	Mineral	J	MSWD / steps	n	WMA (Ma)	Total-fusion age (Ma)	Isochron age (Ma)	MSWD	⁴⁰Ar/³⁶Ar	%³⁹Ar	Steps
ML-18	Bt	0.004540	1.1	250	148.82 ± 1.43	162.2 ± 0.8	316.3 ± 0.4	5/3	—	10	19–21/21
ML-37	WM high-T	0.004542	1.7	250	302.76 ± 2.76	315.09 ± 2.87	255.39 ± 22.70	68/3.8	289 ± 2	37	16–20/20
ML-37	WM low-T	0.004542	1.0	250	302.76 ± 2.76	246.70 ± 2.76	246.70 ± 2.76	0.02/3.8	289 ± 14	2	1–3/20
MU-12	Bt	0.0015255	1.0	250	29.69 ± 0.35	28.60 ± 0.29	28.62 ± 0.40	0.5/3.0	295 ± 2	58	13–16/17
MU-13	Bt	0.0015249	1.2	250	35.06 ± 1.24	34.38 ± 0.39	35.12 ± 0.70	1.6/2.6	275 ± 12	56	11–15/16
ML-39	Bt	0.0015307	1.5	250	15.22 ± 0.46	14.57 ± 0.59	13.81 ± 1.90	1.9	300 ± 10	51	1–5/15
PA3-4	Hbl	0.0042502	s.g.	3 exp.	—	8.1 ± 0.6	8.8 ± 1.5	2.1	290 ± 27	100	1–5/5
PA3-4	Bt	0.0042502	203/s.g.	2 exp.	—	—	8.5 ± 0.3	—	Forced	100	1–9/9
CB35	Hbl	0.0036469	198/320	2 exp.	—	—	28.4 ± 1.3	3.0	293 ± 25	—	1–7/9
CB35	Bt	0.0035870	35/203	2 exp.	—	—	8.2 ± 0.1	0.57	297 ± 3	100	1–12/12
CB31	Hbl	0.0049746	207/305	2 exp.	—	—	15.5 ± 1.2	0.89	305 ± 19	—	1–6/8
CB34	Hbl	0.0035870	195/306		—	—	11.0 ± 1.8	0.33	307 ± 6	100	1–9/9
PI28-1	Hbl	0.0042502	s.g.		—	8.1 ± 0.3	—	—	—	100	1–4/4
PI28-1	Bt	0.0042502	191/s.g.	3 exp.	—	—	8.0 ± 0.5	1.3	303 ± 3	100	1–8/8

J is the irradiation parameter; MSWD is the mean square weighted deviation, which expresses the goodness of fit of the isochron; isochron and weighted mean ages are based on fraction of ^{39}Ar and steps listed in the last two columns. WM(P)A is the weighted mean (plateau) age. Abbreviations: sg, single-grains (laser total-fusion ages); Phe, phengite; WR, whole rock; WM, white mica; Hbl, amphibole; Kfs, potassium feldspar; Bt, biotite; Chl, chlorite; low-T, low-temperature steps; medium-T, intermediate-temperature steps; high-T, high-temperature steps; exp., heating or laser degassing experiments on different fractions of minerals from the same sample; n.a., not applicable; n.d., not determined. Errors are 2σ.

Age interpretation – Rabinal complex: G19s fine-grained phengite: unified age for two grain-size fractions: 67 ± 1 Ma; G27s biotite: WMA calculated for ^{40}Ar/^{36}Ar = 350; 37 ± 5 Ma.

Central Chuacús complex: 5CO-4b biotite: 64.5 ± 1.0 Ma; 5CO-4b white mica: 67.5 ± 2.0 Ma. G10s white mica: 66 ± 1 Ma; G12s white mica: 70 ± 3 Ma; 5CO-5a white mica: hump-shaped spectrum indicates ^{40}Ar excess, 219–243 Ma; low-T plateau at 225 ± 5 Ma; 5CO-5a hornblende: loss profile from c. 720 to c. 153 Ma; 5PB-9b sericitic white mica: 52–62 Ma hump-shaped spectrum indicates ^{40}Ar excess; 5PB-13b hornblende: mixing of different age components; minimal age of old component 137 Ma, maximal age of young component 20 Ma; major age component at 65 ± 6 Ma; 5PB-14 white mica: loss profile from c. 65 to c. 50 Ma; mid and high-T 'plateau' at 65.2 ± 0.8 Ma; 854F biotite: age range from 148 to 94 Ma; 880C white mica: age range from 127 to 100 Ma; likely Cretaceous metamorphic overprint on Early Mesozoic or older metamorphic rock; 883B white mica: 75 ± 2 Ma; 883B biotite: 68.6 ± 2.0 Ma; 887C white mica: 70.0 ± 1.5 Ma.

Metasomatized southern Chuacús complex: G6s white mica: U-shaped spectrum indicates ^{40}Ar excess; 60 ± 5 Ma; G7s rhyolite gash filling: 67 ± 5 Ma; 5C-36a white mica: 48 ± 1 Ma.

Sanarate complex: G3s white mica: loss profile from c. 195 to c. 137 Ma; mid-T 'plateau' at 155 ± 10 Ma; G5s hornblende: loss profile from c. 157 to c. 45 Ma; high-T 'plateau' at 156 ± 5 Ma, low-T 'plateau' at 60 ± 20 Ma.

Western Las Ovejas complex, Guatemala: 5C-2c hornblende: 5C-2c biotite: 20 ± 2 Ma; 5C-5 biotite: loss profile from c. 31 to c. 26 Ma; mid-T 'plateau' at 29 ± 2 Ma; 5C-23c biotite: 25.3 ± 0.5 Ma; 5C-23d hornblende: 35.5 ± 1.0 Ma; 5C-26c hornblende: loss profile from c. 64 to 17 Ma; low-T 'plateau' at 18 ± 2 Ma, intermediate-T 'plateau' at 21 ± 2 Ma and high-T 'plateau' at 36 ± 3 Ma; intermediate and low-T 'plateau' and chlorite-biotite ages are identical within error at 20 ± 2 Ma; 5C-26c chlorite-biotite mixture: loss profile from c. 28 to c. 3 Ma; mid-T 'plateau' at 20 ± 2 Ma; 5C-26d potassium-feldspar: 30 ± 1 Ma; 5C-33 biotite: loss profile from c. 26 to c. 15 Ma; mid-T 'plateau' at 23 ± 2 Ma; 5C-33 potassium-feldspar: loss profile from c. 370 to c. 6 Ma, low-T ^{40}Ar excess, fast cooling at 32 ± 4 and 7 ± 2 Ma; 5C-37b hornblende: high-T WMA calculated for ^{40}Ar/^{36}Ar = 553; 39 ± 3 Ma; 5C-37d biotite: loss profile from c. 26 to c. 9 Ma; mid-T 'plateau' at 19 ± 2 Ma; 5PB-5a hornblende: 30 ± 2 Ma; 5PB-5a biotite: 28.2 ± 0.4 Ma.

Eastern Las Ovejas complex, Honduras: 5H-3a biotite: single-grain laser-fusion age (10 grains); WMA calculated from ^{40}Ar/^{36}Ar = 661; 22 ± 4 Ma; 5H-4b biotite: single-grain laser-fusion age (4 grains) at 25 ± 3 Ma; 5H-4c biotite: loss profile from c. 28 to c. 18 Ma; minor ^{40}Ar excess; mid-T 'plateau' at 25 ± 2 Ma; Integrated biotite age of stations 5H-3 and 5H-4 at 24 ± 2 Ma; 5H-4b hornblende: 38 ± 2 Ma; 5H-4c hornblende: 31 ± 5 Ma; 5H-4d hornblende: 29 ± 3 Ma; Integrated hornblende age of stations 5H-4 at 34 ± 4 Ma.

Granitoids: 5C-11 chloritized biotite: single-grain laser-fusion age (10 grains) at 90 ± 10 Ma; 5C-21 potassium-feldspar: single-grain laser-fusion age (8 grains) at 20 ± 1 Ma; 5H-5 K-feldspar: loss profile from c. 37 to c. 28 Ma, low-T 'plateau' indicates fast cooling at 28.3 ± 0.5 Ma; 5H-13 biotite: single-grain laser-fusion age at 59 ± 3 Ma.

Southern Mexico: ML-18 biotite: loss profile from 163 to 88 Ma, high-T 'plateau' at 162 ± 2 Ma, low-T 'steps' at 90 ± 2 Ma; ML-37 white mica: loss profile from 316 to 238 Ma, high-T 'plateau' at 315 ± 3 Ma, low-T 'plateau' at 247 ± 3 Ma; MU-12 biotite: 35 ± 1 Ma; MU-13 biotite: 35 ± 1.5 Ma; PA3–4 biotite: 8.5 ± 1.0 Ma; CB35 hornblende: 28.5 ± 1.5 Ma; CB35 biotite: 8.2 ± 0.5 Ma; CB34 hornblende: 15.5 ± 1.5 Ma; CB31 hornblende: 11.2 ± 0.5 Ma; PA28-1 hornblende: 8.1 ± 0.5 Ma; PI28-1 biotite: 8.0 ± 0.5 Ma.

(a) Rabinal complex

(b) Chuacús complex

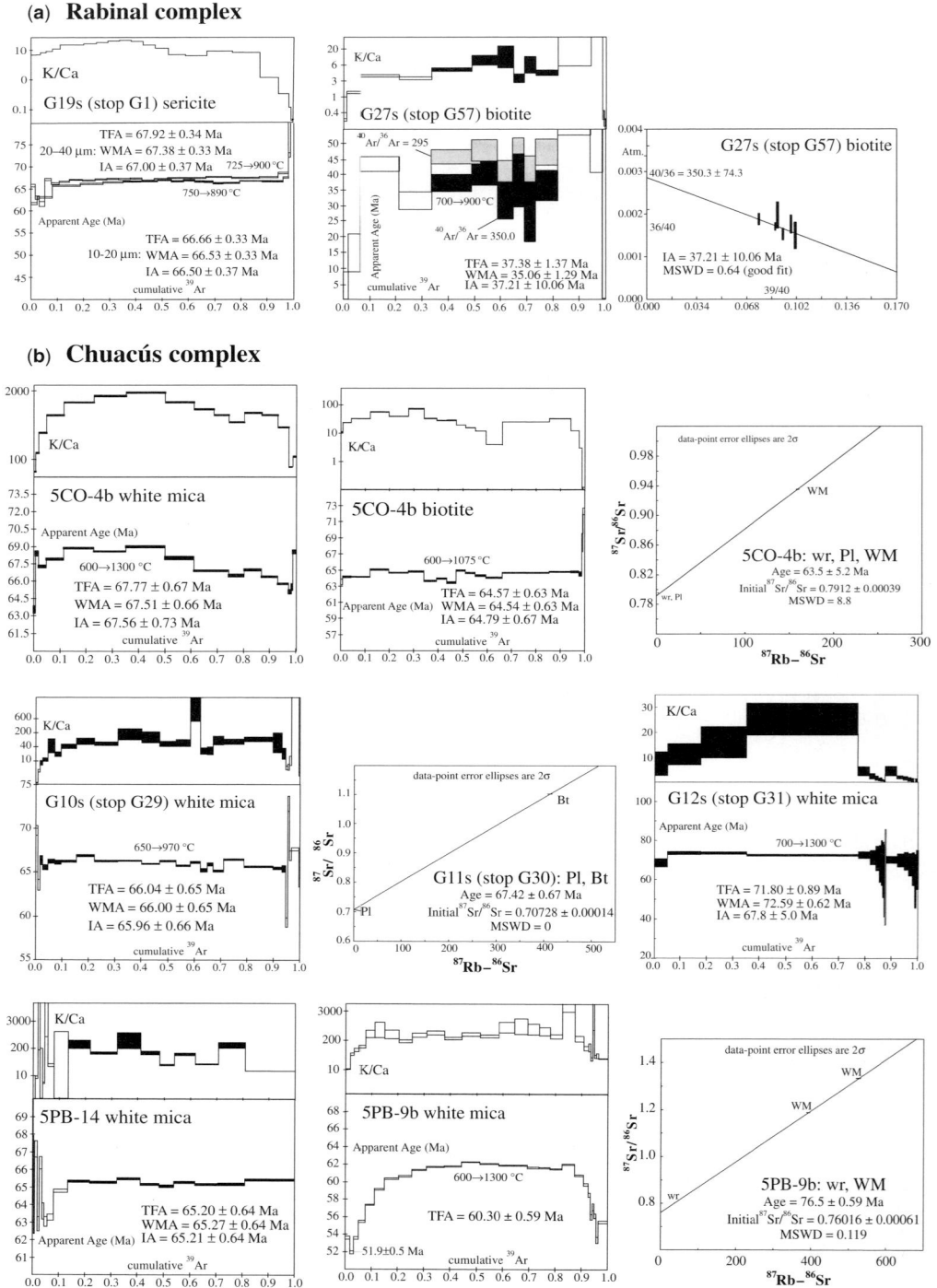

Fig. 5. New ^{40}Ar/^{39}Ar mineral spectra and Rb–Sr whole rock–feldspar–mica 'isochrons'. See the Supplementary material for sample locations, age data and interpretations. Weighted mean ages (WMA) were calculated using black steps of the given temperature range; TFA, total fusion age; IA, isochron age. Uncertainties are ±1σ.

Chuacús complex, continuation

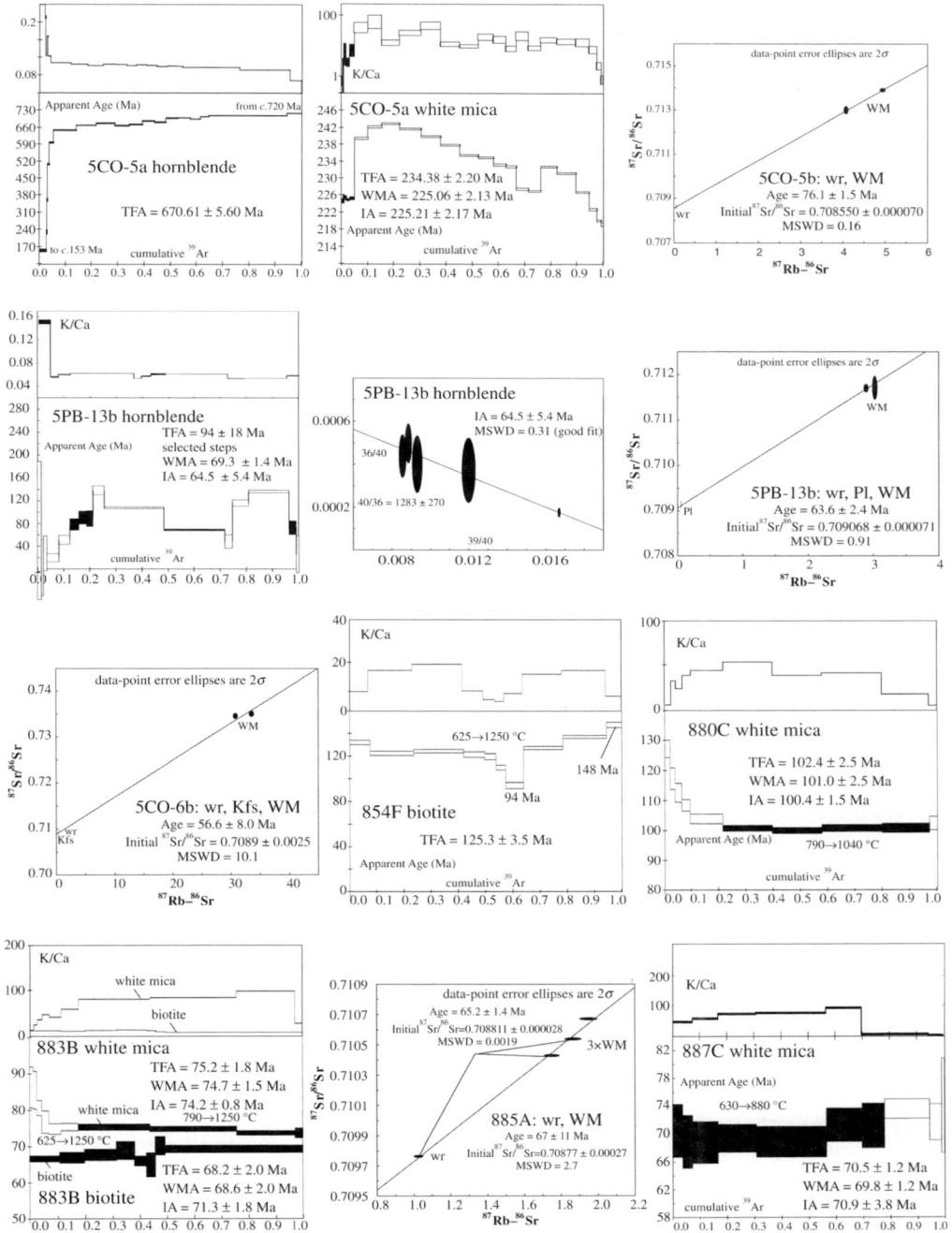

Fig. 5. (*Continued*) 'Atm.' in the isochron diagrams is the $^{40}Ar/^{36}Ar$ ratio of the atmosphere (295.5). MSWD, mean square weighted deviation. Abbreviations: wr, whole rock; Pl, plagioclase; Kfs, K-feldspar; WM, white mica; Bt, biotite.

(c) Metasomatized southern Chuacús complex

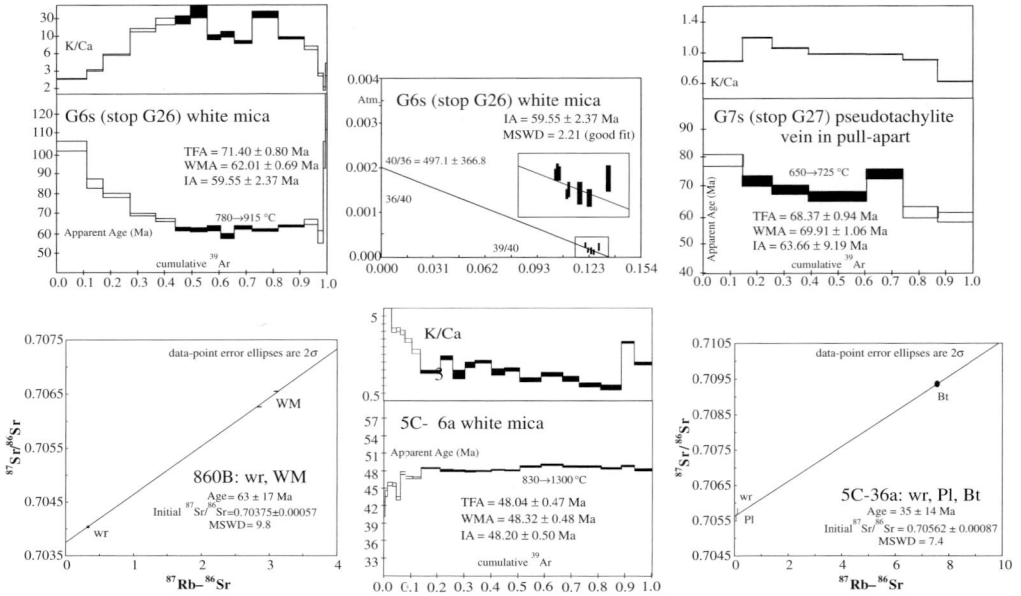

(d) Western Las Ovejas complex (Guatemala)

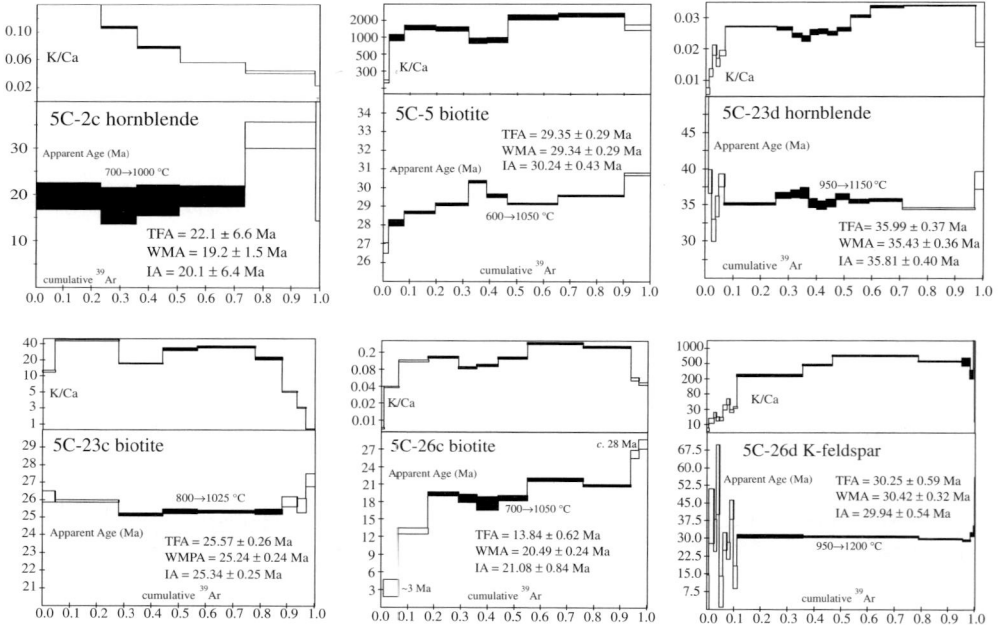

Fig. 5. (*Continued*).

Western Las Ovejas complex (Guatemala), continuation

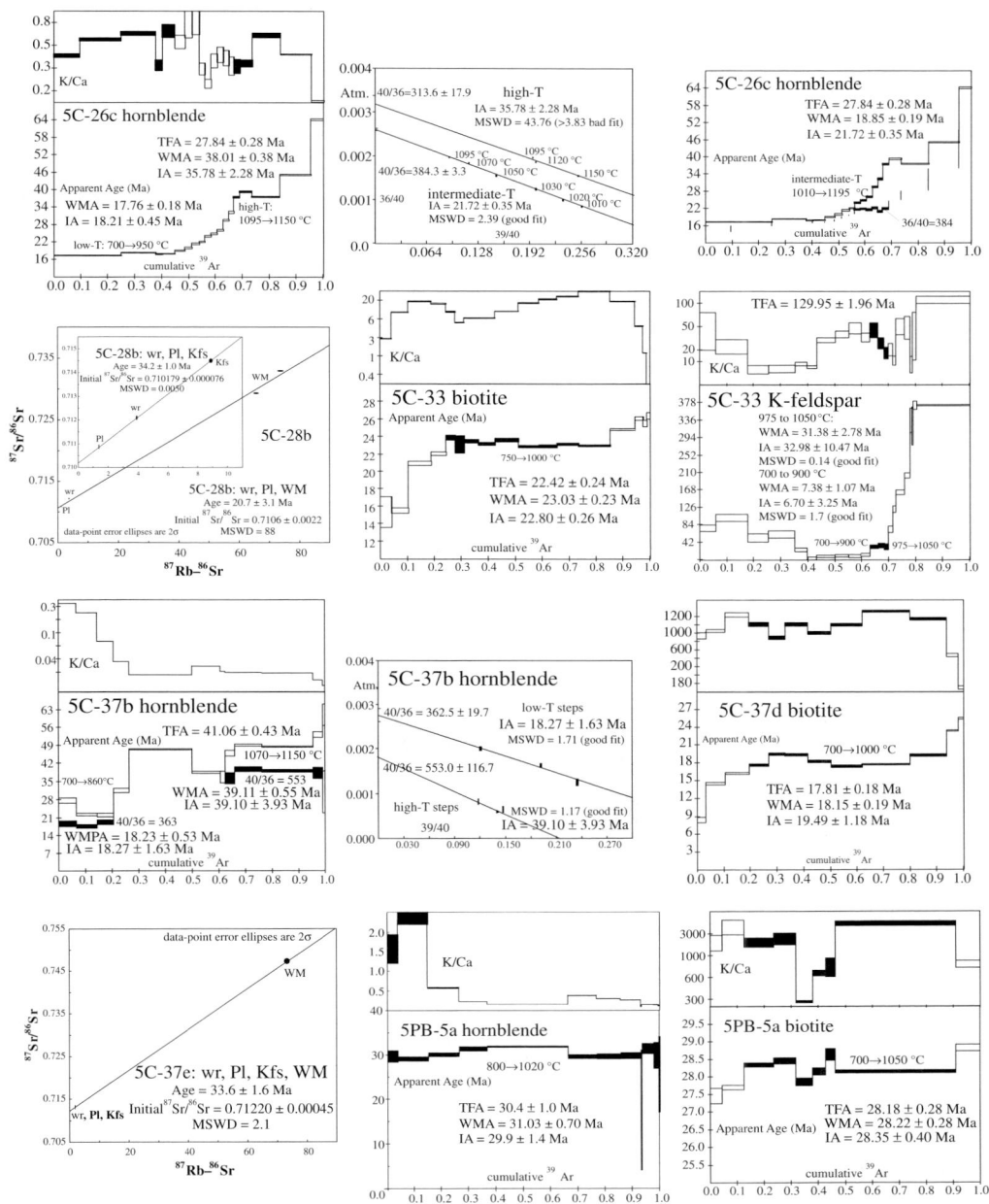

Fig. 5. (*Continued*).

(e) Eastern Las Ovejas complex, Honduras

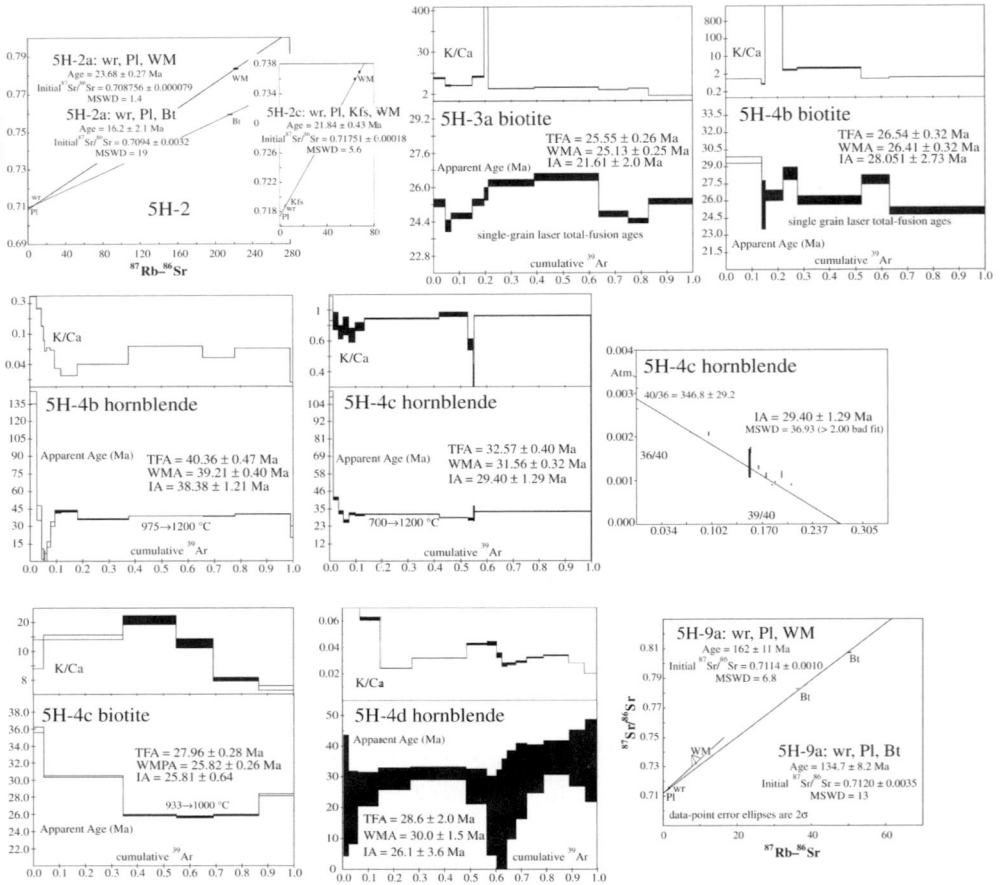

(f) Sanarate complex

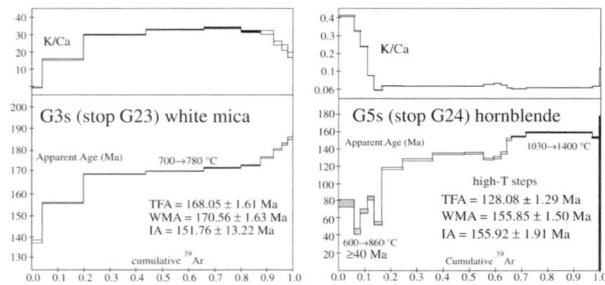

Fig. 5. (*Continued*).

(g) Granitoids

5C-11 biotite-chlorite

single-grain laser total-fusion ages

TFA= 84.51 ± 0.86 Ma
WMA = 89.98 ± 0.89 Ma

5C-21 K-feldspar

TFA = 19.83 ± 0.22 Ma
WMA = 20.06 ± 0.22 Ma
IA = 19.90 ± 0.80 Ma

single-grain laser total-fusion ages

5C-21 K-feldspar
single-grain laser total-fusion ages

40/36 = 296.4 ± 4.6

IA = 19.90 ± 0.80 Ma
MSWD = 5.68 (> 2.15 bad fit)

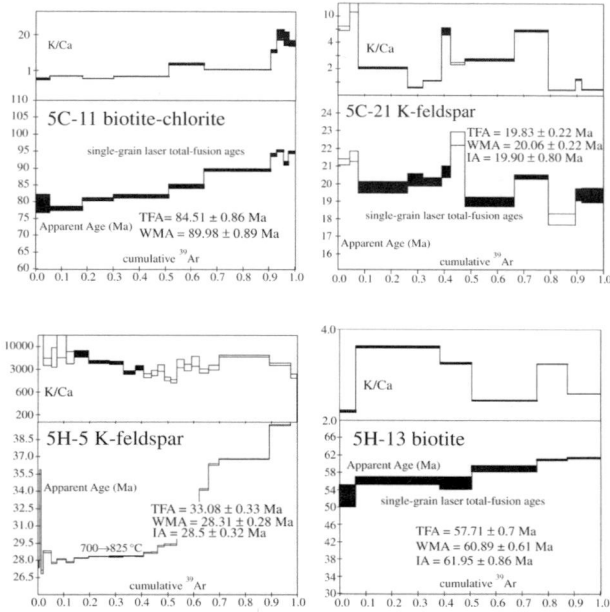

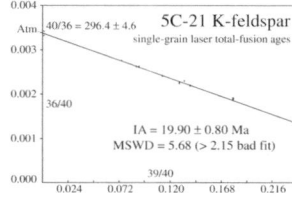

(h) U–Pb, K(Ar)–Ar and Rb–Sr of Chortís Block granitoids and significant age groups

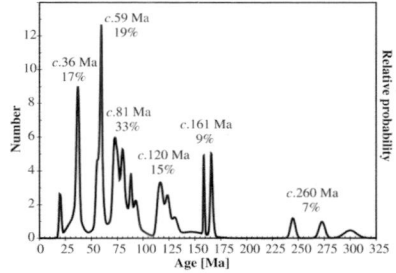

5H-5 K-feldspar

TFA = 33.08 ± 0.33 Ma
WMA = 28.31 ± 0.28 Ma
IA = 28.5 ± 0.32 Ma

700→825 °C

5H-13 biotite

single-grain laser total-fusion ages

TFA = 57.71 ± 0.7 Ma
WMA = 60.89 ± 0.61 Ma
IA = 61.95 ± 0.86 Ma

c.59 Ma 19%
c.36 Ma 17%
c.81 Ma 33%
c.161 Ma 9%
c.120 Ma 15%
c.260 Ma 7%

(i) Southern Mexico

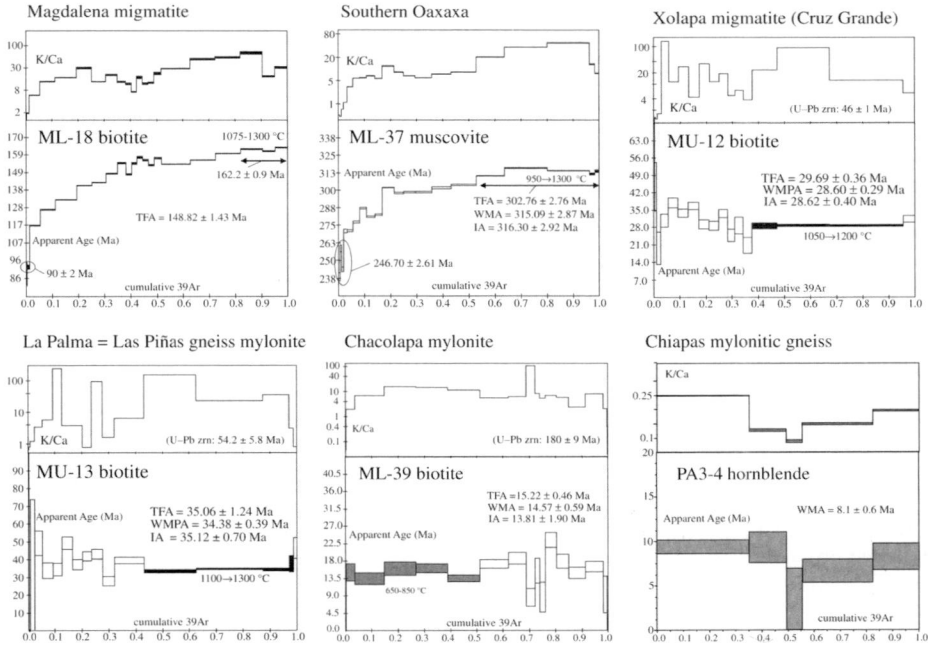

Magdalena migmatite

ML-18 biotite   1075-1300 °C

162.2 ± 0.9 Ma

TFA = 148.82 ± 1.43 Ma

90 ± 2 Ma

Southern Oaxaxa

ML-37 muscovite

950→1300 °C

TFA = 302.76 ± 2.76 Ma
WMA = 315.09 ± 2.87 Ma
IA = 316.30 ± 2.92 Ma

246.70 ± 2.61 Ma

Xolapa migmatite (Cruz Grande)

(U–Pb zrn: 46 ± 1 Ma)

MU-12 biotite

TFA = 29.69 ± 0.36 Ma
WMPA = 28.60 ± 0.29 Ma
IA = 28.62 ± 0.40 Ma

1050→1200 °C

La Palma = Las Piñas gneiss mylonite

(U–Pb zrn: 54.2 ± 5.8 Ma)

MU-13 biotite

TFA = 35.06 ± 1.24 Ma
WMPA = 34.38 ± 0.39 Ma
IA = 35.12 ± 0.70 Ma

1100→1300 °C

Chacolapa mylonite

(U–Pb zrn: 180 ± 9 Ma)

ML-39 biotite

TFA =15.22 ± 0.46 Ma
WMA = 14.57 ± 0.59 Ma
IA = 13.81 ± 1.90 Ma

650-850 °C

Chiapas mylonitic gneiss

PA3-4 hornblende

WMA = 8.1 ± 0.6 Ma

Fig. 5. (*Continued*).

Southern Mexico, continuation

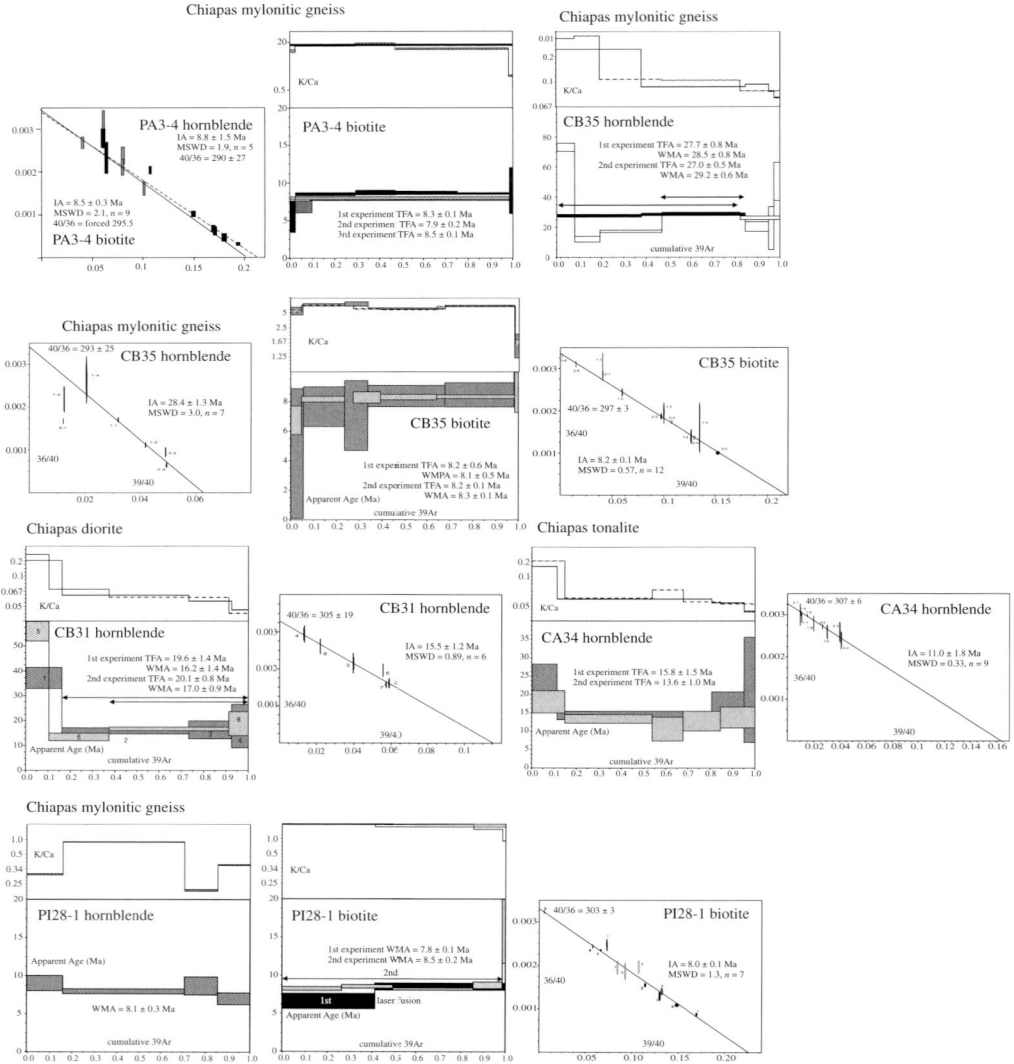

Fig. 5. (*Continued*).

magmatic Th–U ratios (mean 0.4) (Fig. 4). The 207-corrected spot ages, mostly from rims but also from a few coreless grains with relatively flat CL response, range from 410 to 485 Ma (Fig. 3a). There may be two age groups with weighted average ages of 464.6 ± 5.3 and 412.9 ± 3.7 Ma; Th–U ratios are similar. We suggest that these ages indicate prolonged Ordovician–Early Devonian magmatism (see below).

Sample GM1 (stop GM15, central Guatemala; Figs 3a & 7a) is quartz phyllite close to Salamá (central Guatemala); these rocks were mapped as

Santa Rosa Group. Zircons are rounded but a significant portion is sub- to euhedral. Besides ages >1 Ga, we obtained four concordant ages at 416.5 ± 1.6, 282.2 ± 1.7, 260.4 ± 2.9 and 245.1 ± 2.9 Ma. The Devonian spot has low Th–U (0.012), the other zircons (Th–U on average 0.64) probably crystallized from a magma (note difference in luminescence of 416 and 257 Ma grains in Fig. 4). Our ages imply post-Early Triassic deposition for these metasediments.

Samples GM13s (stop GM42) and GM14s (stop GM43) (Figs 3a & 7a) are from western Guatemala.

GM13s is (?ortho)gneiss and GM14s granite intruding gneisses along the Polochic fault. Seven zircon fractions of GM13s with different magnetic properties and, in part, grain size yielded lower and upper intercepts at 20 $\pm$ 15 and 941 $\pm$ 7.9 Ma, respectively. We analysed six fractions of GM14s differing in grain size and magnetic properties. All are discordant and yield lower and upper intercepts at 21 $\pm$ 33 and c. 894 Ma; five fractions yield a lower intercept at 13 $\pm$ 29 Ma. Phanerozoic spot ages of zircons from GM14s gave one concordia age at 425.5 $\pm$ 6.0 Ma and 14 spots with a 14.65 $\pm$ 0.42 Ma weighted average age. These ages demonstrate Middle Miocene Central American arc magmatism.

Chuacús complex. Sample 5CO-4b, high-*P* metamorphic (see below) garnet orthogneiss from the northern Chuacús complex, central Guatemala (Fig. 7a), comprises complex, mostly Proterozoic zircons. Zircon grain shapes include both rounded and prismatic–idiomorphic. In general, low-U cores are surrounded by variable, higher-U rims that both have Proterozoic ages (1180–1017 Ma concordant spots; Fig. 3a). Spots that include the former grain portions and tiny, very luminescent low-U outer rims, irresolvable with the spot size we employed, yielded the youngest ages. The lower intercepts of discordia fitted to two groups of grains indicate crystallization at 638–477 Ma (numbers include errors), predating the high-*P* metamorphic event. Sample 5CO-6b (south-central Chuacús complex; Fig. 7a), a two-mica orthogneiss, is intruded by very coarse-grained pegmatite and interlayered with paragneiss and garnet amphibolite; the latter contains relict eclogite (see petrology for analysis of equivalent eclogite sample 5CO-5a). All rock types are ductilely deformed. The amphibolite forms boudins in the orthogneiss and is more strongly deformed. We interpret the massive orthogneiss as a sill and the eclogite-facies metamorphism pre-dating its emplacement. The zircons of the orthogneiss are complex (Figs 3a & 4): we detected a few well-rounded Proterozoic cores, one core that has flat CL response, low, likely metamorphic Th–U (0.020) and a 207-corrected age of 402.9 $\pm$ 4.0 Ma, and many oscillatory zoned grains with magmatic Th–U (on average 0.82); three of the latter spots give a concordia age of 238.0 $\pm$ 2.5 Ma, and eight spots yield a well-defined group with a weighted average of 218.2 $\pm$ 2.3 Ma. Almost all grains have mostly thin (5–40 μm) overgrowths with flat CL response and low Th–U (on average 0.01) that span 89 to 69 Ma. The oldest grain has the highest Th–U and probably samples different age domains (core and rim). Selected omission of grains from the remaining group yields one statistically significant group of six grains with a weighted average age of

74.24 $\pm$ 0.52 Ma and five younger spots that may constitute a cluster at 70.3 $\pm$ 1.5 Ma; the latter group may be geologically meaningless, if it reflects progressive Pb loss of grains from the former group. Sample 5CO-6b is most straightforwardly interpreted as a Triassic magmatite that inherited Proterozoic and Devonian zircons and experienced high-*T* metamorphism during Late Cretaceous, probably coeval with pegmatite injection; similar very coarse-grained pegmatites also yielded Late Cretaceous K–Ar (Ortega-Gutiérrez et al. 2004) and Rb–Sr white mica cooling ages (see below). The regional significance of magmatism is supported by Triassic hornblende and white-mica cooling ages from this region (238–225 Ma, Donnelly et al. 1990; see below). As we interpret the orthogneiss as a sill, the high-*P* event is probably pre-Middle Triassic.

Chortís Block. Sample 5H-3 is syntectonic, synmigmatitic, plagioclase-rich granite gneiss associated with migmatite, leucogranite and gabbro–diorite; it is typical of the Las Ovejas complex and crops out at the western flank of the Sula graben in northwestern Honduras (Fig. 7a). The 207-corrected spot ages on typically magmatic zircons (Figs 3c & 4) range from 40.0 $\pm$ 0.4 to 36.5 $\pm$ 0.9 Ma with three grains defining a concordia age at 37.0 $\pm$ 0.52 Ma. We interpret this age as the time of magmatism and deformation. Sample 5H-9a comprises orthogneiss and migmatite south of the Jocotán fault zone (Fig. 7a) that are distinctly different in grain-size (coarse) and deformation (weak) from the gneisses and migmatites of the Las Ovejas complex along the MSZ. Besides Proterozoic ages, there are concordant spot ages at 400.3 $\pm$ 3.5 and 328.9 $\pm$ 4.0 Ma. There are two main age groups: 272.8 $\pm$ 2.8 Ma (eight spots) and 244.8 $\pm$ 2.3 Ma (four spots). These groups have distinctly different Th–U ratios and CL-response (Figs 3c & 4). The younger group, probably metamorphic, has mean Th–U at 0.04 and appears dark and unzoned in CL. The older group has mean Th–U of 0.7 and clear oscillatory zoning. We suggest Early Permian protolith crystallization and Early Triassic metamorphism. Sample 5H-11a is low-*T*, mylonitic orthogneiss (phyllonite), probably metarhyolite intercalated with paragneiss and marble; the sample is also from the Jocotán fault zone (Fig. 7a). Six out of seven 207-corrected spot ages yield a 130.5 $\pm$ 2.7 Ma concordia age (Fig. 3c); Th–U ratios are on average 0.41. We interpret this date as reflecting Early Cretaceous magmatism. Sample 5PB-5c is a small orthogneiss pod in massive migmatite that is enclosed in large granite in easternmost Guatemala (Fig. 7a); it is associated with garnet amphibolite (see petrology). Three spots define a 36.2 $\pm$ 2.1 Ma concordia age

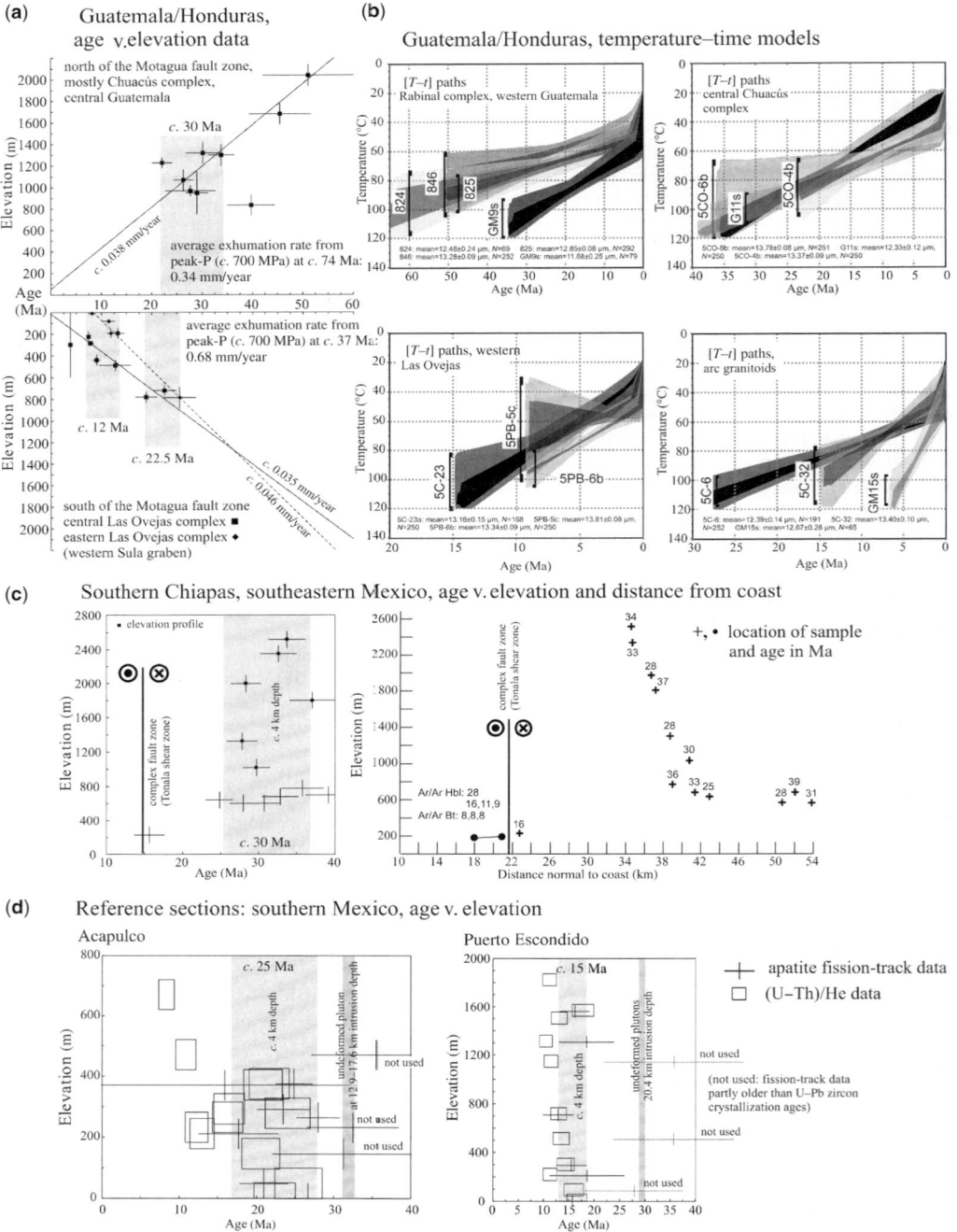

Fig. 6. Low-temperature thermochronology data. (**a**) Apatite fission-track (AFT) age v. elevation plots for the central Chuacús complex (top), the Las Ovejas complex, and undeformed arc granitoids. Locations and fission-track parameters are given in the Supplementary material. (**b**) AFT temperature–time (*T–t*) paths for rocks of the Rabinal (western Guatemala), Chuacús (central Guatemala) and Las Ovejas (western area) complexes, and three arc granitoids; the latter are: 5C-6, Chiquimula batholith in the Las Ovejas complex; 5C-32, small granite east of the Chiquimula batholith; GM15s, augengneiss of the westernmost Rabinal complex overprinted by *c.* 15 Ma granite. Good-fit solutions

(Fig. 3c), probably the age of migmatization. One of the spots is from a relatively high-U, oscillatory zoned central grain portion that is surrounded by a rim poorer in U (Fig. 4); this rim is *c.* 5 Ma (31.2 ± 1.2 Ma) younger. The major group of Phanerozoic ages spans 140–191 Ma with two, probably significant, age groups: 158.8 ± 0.52 Ma ($n = 4$) and 165.79 ± 0.87 Ma ($n = 13$); the consistently lower Th–U ratios (on average 0.12) of the former (the latter has on average 0.36) support the grouping. There are other Phanerozoic concordant spots at 139.8 ± 1.0 Ma ($n = 2$), 190.6 ± 1.4 Ma ($n = 1$) and 240.6 ± 3.5 Ma ($n = 1$). We suggest a Middle Jurassic granite protolith that inherited Proterozoic to Early Jurassic zircons. Sample 92RN412 is granitic gneiss surrounded by amphibolite-facies rocks from the Las Ovejas complex of northwestern Honduras (Fig. 7a); it is grossly the same rock as 5H-3. Rb–Sr and K–Ar biotite cooling ages are *c.* 39 and *c.* 36 Ma, respectively (Horne *et al.* 1974, 1976*a*). Three of four zircon fractions with different grain size and magnetic properties plot close to concordia; one fraction is concordant at 38.3 ± 0.53 Ma. All fractions define intercepts at 36.6 ± 3.1 and *c.* 1014 Ma (Fig. 3c); three fractions are at 37.49 ± 0.53 and *c.* 1136 Ma. The *c.* 38 Ma age is within error identical to sample 5H-3. The age constrains deformation within the eastern Las Ovejas complex to be younger than 39 Ma (see below). Sample 94SA602 is amphibolite within the metamorphic sequence that contains the granite gneiss of sample 92RN412; the protolith was probably gabbro. We obtained three zircon fractions that plot close to and on concordia, defining a lower intercept at 88 ± 51 Ma (Fig. 3c); the concordant fraction is at 88.27 ± 0.56 Ma, probably dating gabbro crystallization. Sample 93LP522 is granite directly south of the Jocotán–Chamelecón fault zone (Fig. 7a). Four zircon fractions define a lower intercept at 167.6 ± 2.6 Ma (MSWD = 10.9), on concordant fraction is at 167.99 ± 0.46 Ma (Fig. 3c); the *c.* 168 Ma age is within error identical to sample 5PB-5c and demonstrates Middle Jurassic magmatism in the Chortís Block. Samples SF1D, 92QO404 and 86TA100 typify orthogneisses embedded in the Cacaguapa schists of the central

Chortís Block in central Honduras (Fig. 1c). SF1D is weakly deformed granite gneiss surrounded by Jurassic metasedimentary rocks of the Agua Fría Formation; the contact relationships are unclear. All five fractions are discordant and lower intercept ages between 275 and 234 Ma can be defined (Fig. 3c). 92QO404 is mega-crystic, K-feldspar metagranite surrounded by phyllite. Five out of six fractions define intercepts at 396 ± 11 and *c.* 884 Ma (MSWD = 0.33; Fig. 3c). 86TA100 is K-feldspar augen schist with white mica, quartz, epidote, microcline, biotite and garnet. Four out of five fractions define 404 ± 13 and 1149 ± 23 Ma lower and upper intercept ages, respectively (MSWD = 1.9; Fig. 3c). Besides Grenville inheritance, these samples indicate Early Devonian magmatism in the Chortís Block.

Ar/Ar and Rb–Sr geochronology

We determined ages of deformation and metamorphic cooling of the Maya and Chortís Block rocks by analysis of syn- to post-tectonic minerals. We also dated cooling in pre- to post-tectonic magmatic rocks, again providing age limits for deformation.

Rabinal complex. Sample G19s (stop G1, Figs 5a & 7a) is weakly migmatitic gneiss from western Guatemala retrogressed to hornblende–epidote schist during localized but high-strain, low-*T* (≥400 °C) sinistral shear. Two sericite fractions of different grain size yielded nearly identical weighted mean ages (WMA). The pooled 67 ± 1 Ma age is interpreted to slightly post-date low-*T* plasticity in the shear zones. Sample G27s (stop G57, western Guatemala, Figs 5a & 7a) is orthogneiss associated with amphibolite and shows low-*T* mylonitization along discrete sinistral shear zones; undeformed, probably Miocene (*c.* 15 Ma, see above) granite occurs in the vicinity. The age spectrum is complex and the *c.* 37 Ma WMA may reflect partially reset Late Cretaceous–Early Cenozoic biotite.

Chuacús complex. Sample 5CO-4b is garnet gneiss associated with K-feldspar–porphyroblastic gneiss and schist from the north-central Chuacús complex

---

Fig. 6. (*Continued*) (all *T*–*t* paths with a merit function value of at least 0.5, Ketcham *et al.* 2000) are dark grey or black; acceptable-fit solutions (all *T*–*t* paths with a merit-function value of at least 0.05) are light grey. (**c**) AFT age v. elevation plot and elevation–age–distance from the Pacific-coast diagram for a vertical profile and regional samples of the south-central Chiapas Massif, southern Mexico. The Tonala shear zone, associated granitoids, metamorphic rocks and mylonites and their ages are indicated (Hbl, hornblende; Bt, biotite); note their influence on one AFT sample. (**d**) AFT and apatite (U–Th)/He data for two coast-normal sections in southern Mexico from Ducea *et al.* (2004*b*) plotted in age v. elevation diagrams; these data are reference sections for our own data and are discussed in the text. Dark grey vertical bars give age range (U–Pb zircon) and intrusion depth of granitoids from about the same area as the thermochronology data. Light grey bars give interpreted closure through 70–100 °C or *c.* 4 km depth.

(Fig. 7a); metamorphism evolved through a high-*P* stage and amphibolite– to epidote–amphibolite-facies retrogression (see below). Low-Si phengite and biotite are part of the recrystallized groundmass ($\leq$510 °C, *c.* 0.7 GPa). The phengite spectrum is weakly hump-shaped but the total-fusion (TFA), WMA, and isochron (IA) ages correspond within error at 67 $\pm$ 2 Ma; the biotite is *c.* 3 Ma younger. A Rb–Sr phengite–plagioclase–whole rock isochron is within error of the Ar/Ar ages (Fig. 5b). Epidote–amphibolite-facies retrogression is thus Late Cretaceous. Sample G10s (stop G29, Fig. 7a) is from the southern Chuacús complex and comprises marble interlayered with garnet micaschist and gneiss just north of El Tambor complex serpentinites. The Ar/Ar age of the coarse white mica in the marble overlaps a Rb–Sr biotite age (G11s; Fig. 7a) of nearby two-mica orthogneiss within error at *c.* 67 Ma (Fig. 5b). Sample G12s (stop G31, Fig. 7a) is fine-grained, mylonitic orthogneiss (*c.* 1 Ga, unpublished U–Pb zircon) north of G10s. The grain size is due to dynamic recrystallization of quartz at the transition between subgrain rotation and grain-boundary migration (*c.* 425 °C, e.g. Stipp *et al.* 2002). White mica is related to the recrystallization fabric, includes albite and is cogenetic with a second generation of garnet and chlorite; the ubiquitous albite may be due to Na-rich fluid exchange between the Chuacús complex and the El Tambor complex rocks in the hanging wall. Mylonitization and fluid exchange occurred at or slightly after 70 $\pm$ 3 Ma (Ar/Ar white mica, Fig. 5b). Samples 5PB-13b and 5PB-14 typify rocks of the east-central Chuacús complex (Fig. 7b): 5PB-13b is coarse-grained garnet amphibolite associated with garnet micaschist metamorphosed at *c.* 540 °C and *c.* 0.7 GPa (see below). 5PB-14 is hornblende–mica–chlorite schist with quartzitic layers formed at identical PT conditions. Quartz recrystallized dynamically by subgrain rotation and grain-boundary migration. The hornblende spectrum of 5PB-13b is complex with an age component older than 137 Ma; several steps are at 65 $\pm$ 6 Ma. The latter are within error of the Ar/Ar (5PB-14) and Rb–Sr (5PB-13b) white mica ages (Fig. 5b). As the closing temperatures for these systems are close to the peak-*T* conditions represented by the synkinematic minerals, deformation is dated at $\geq$65 Ma. Sample 5PB-9b is retrograde biotite–chlorite gneiss from ortho–paragneiss intercalations at the eastern end of the Chuacús-complex outcrop in the Sierra de Las Minas area (below the Juan de Paz serpentinite unit; Fig. 7a). Mylonitization is upper greenschist facies with basal <a> slip and grain-boundary migration recrystallization in quartz and coexisting fracturing and incipient recrystallization in feldspar. The Ar/Ar white mica spectrum is classic hump-shaped (52–62 Ma)

indicating ^{40}Ar excess, the Rb–Sr white mica–whole rock isochron is at 76.5 $\pm$ 0.59 Ma (Fig. 5b). Samples 5CO-5a,b, 5CO-6b, 880C, 883B, 885A and 887C are from the central Chuacús complex and typical for the garnet amphibolite–micaschist–biotite–(?ortho)gneiss–metapegmatite intercalations of this region (Fig. 7a). The pegmatites are variably deformed but massive stocks are weakly strained. All rocks share various stages of post-tectonic annealing. The Ar-release spectrum of phengite of garnet amphibolite 5CO-5a is hump-shaped, indicating ^{40}Ar excess; a low-*T* plateau is at 225 $\pm$ 5 Ma (Fig. 5b). Hornblende shows a classic loss-profile spectrum (*c.* 720–153 Ma). The 70.0 $\pm$ 1.5 Ma Ar/Ar white mica age from intercalated quartzofeldspathic paragneiss sample 887C expresses Late Cretaceous reheating, also highlighted by our U–Pb zircon geochronology (see above). White mica of metapelite sample 880C defines a shear zone related to late stage folding (F$_2$); the individual heating steps cover 127–100 Ma. White mica and biotite from paragneiss sample 883B define a s-c fabric, characteristic of late-stage deformation; sinistral flow is constrained at 74 $\pm$ 2 Ma (white mica) to 68.6 $\pm$ 2.0 Ma (biotite). Sample 854F is migmatitic paragneiss weakly overprinted by late-stage (retrograde) shear zones from the Rio Hondo area of the southeastern Chuacús complex; individual heating steps cover 148–94 Ma, probably reflecting a mixture of age components. Rb–Sr metapegmatite white mica ages are at 76.1 $\pm$ 1.5 (5CO-5b), 57 $\pm$ 8 (5CO-6b) and 67 $\pm$ 11 Ma (885A, extremely coarse grained mica from the Guatemarmol mine west of Granados), demonstrating Late Cretaceous pegmatite emplacement; our ages confirm those from foliated metapegmatite (K–Ar 73–62 Ma) of Ortega-Gutiérrez *et al.* (2004). We interpret our ages as Triassic cooling from granite emplacement and associated metamorphism and Late Cretaceous reheating associated with regional metamorphism and local magmatism. The two samples that suggest cooling in the Early Cretaceous characteristically are from areas with relative weak late-stage tectonic overprint. On one hand, they (880C) support a pre-Late Cretaceous metamorphism in the core of the central Chuacús complex, where sinistral flow is localized and latest Cretaceous (883B). On the other hand, they (854F) demonstrate pre-Cretaceous migmatization in the southern part of the eastern Chuacús.

Samples G6s, G7s, 860B and 5C-36a (Figs 5c & 7a) are Chuacús-complex rocks immediately below the North El Tambor complex rocks. These localities show imbrication of basement (now retrograde phyllonite, chlorite phyllite, epidotized biotite gneiss, carbonate-bearing garnet–actinolite micaschist), marble (San Lorenzo marble), serpentinite and low-grade metasiltstones (Jones Formation). They share a history of extreme fluid flow that resulted in strong retrogression, local carbonate and albite 'metasomatism' and post-tectonic graben formation, and locally intense quartz veining. White mica of G6s (station G26), retrogressed basement, has a U-shaped spectrum indicating ^{40}Ar excess with a flat-portion at 60 ± 5 Ma (Fig. 5c), comprising most of the released Ar. Retrograde biotite gneiss G7s (station G27) exhibits centimetre-sized pull-apart-type tension gashes locally filled with pseudotachylite that spectacularly dates brittle–ductile, sinistral transtension at c. 70 Ma (Fig. 5c). 5C-36a shows coarse white mica from relict gneiss in overall fine-grained, locally biotite-rich phyllonite; its 48 ± 1 Ma Ar/Ar and the 35 ± 14 Ma Rb–Sr biotite ages are the youngest ages encountered in the Chuacús complex rocks and may reflect long-lasting fluid activity.

Chortís Block. The following samples characterize metamorphism and deformation in the Las Ovejas complex and cooling of associated granitoids (Figs 5d & 7a). Sample 5C-2c is partly mylonitic amphibolite intercalated with paragneiss and quartzite and crosscut by diabase dykes. Hornblende is clearly synkinematic to sinistral shear deformation and shows two high-T steps at >30 Ma and a low-T 'plateau' at 20 ± 2 Ma; we attribute limited significance to these ages, as only few steps are available due to a furnace problem. Sample 5C-5 is phyllite faulted against Cenozoic–Quaternary volcanic rocks along the Jocotán fault zone. Synkinematic biotite shows Ar-loss from c. 31 to c. 26 Ma with a mid-T 'plateau' at 29 ± 2 Ma (Fig. 5d). This age indicates Early Oligocene Jocotán fault activity, coeval with mylonitization in the Las Ovejas unit (see below). Samples 5C-23c,d comprise biotite gneiss and amphibolite in a unit of partly migmatitic gneiss, metagabbro and granitic gneiss that is typical Las Ovejas complex lacking retrogression. Synkinematic, fibrous hornblende gives a WMA of 35.5 ± 1.0 Ma; biotite is at 23.3 ± 0.5 Ma and thus >10 Ma younger. Biotite of nearby 5C-33 gneiss depicts a loss profile from c. 26 to c. 15 Ma with a mid-T 'plateau' at 23 ± 2 Ma, identical to biotite of 5C-23c. K-feldspar shows a loss profile from c. 370 to c. 6 Ma; the low-T steps, forming a good-fit IA with atmospheric ^{40}Ar/^{36}Ar at 7 ± 2 Ma, are difficult to interpret geologically. The high-T steps indicate cooling at 32 ± 4 Ma (Fig. 5d). Although from the vicinity of large post-tectonic granite, we interpret our 5C-23 and 5C-33 station ages to date amphibolite-facies metamorphism and associated migmatization, the minerals lack post-tectonic annealing and are in part synkinematic to sinistral strike–slip deformation. The amphibole ages closely approximate the age of deformation

(c. 35 Ma); it remains unclear whether the late cooling of these samples through biotite closure (c. 23 Ma) may indicate long-lasting deformation, reactivation, or slow cooling. Samples 5C-26c,d comprise garnet amphibolite and garnet-bearing biotite gneiss of the lithological unit described above; distinct rocks are migmatitic gneiss and garnet-bearing leucosome. 5C-26c hornblende is aligned in the foliation (low strain) and exhibits a loss profile from c. 64 to 17 Ma. Multiple-step degassing allowed deconvolution into sub-plateaus that were analysed with isochrons (Fig. 5d). A low-T 'plateau' at 18 ± 2 Ma has approximately atmospheric ^{40}Ar/^{36}Ar, an intermediate-T 'plateau' at 21 ± 2 Ma is non-atmospheric but exhibits a good-fit isochron and a high-T 'plateau' at 36 ± 3 Ma is only weakly non-atmospheric. The intermediate and low-T 'plateaus' are within error of the 20 ± 2 Ma WMA for the mid-T steps of a biotite–minor chlorite mixture; this biotite has a loss profile from c. 28 to c. 3 Ma. The high-T steps of K-feldspar give 30 ± 1 Ma. Sample 5C-37b, again typical Las Ovejas complex of eastern Guatemala, is amphibolite associated with migmatitic biotite gneiss and metre-sized bodies of leucogranite. The hornblende spectrum is complex but contains a high-T 'plateau' with a WMA of 39 ± 3 Ma, calculated for ^{40}Ar/^{36}Ar of 553, and a low-T 'plateau' with a WMA of 18 ± 3 Ma, calculated for ^{40}Ar/^{36}Ar of 362; the K–Ca ratios of the latter suggests a contribution of gas released from biotite micro-inclusions. 5C-37d biotite reveals Ar loss c. 26–9 Ma and a mid-T 'plateau' at 19 ± 2 Ma. The leucogranite gives a white mica–K-feldspar–plagioclase–whole rock Rb–Sr isochron of 33.6 ± 1.6 Ma. Located in the far-field of the large granitoids that were mapped crosscutting the Las Ovejas complex, stations 5C-26 and 5C-37 typify cooling from amphibolite-facies metamorphism at c. 37 Ma and, possibly, relatively slow cooling throughout the Oligocene–Miocene. The Rb–Sr leucogranite white mica age supports the field observation that these small melt bodies are cogenetic with Late Eocene metamorphism and migmatization. Sample 5C-28b is pegmatite gneiss crosscutting typical Las Ovejas biotite gneiss; its Rb–Sr whole rock–plagioclase–K-feldspar isochron is at 34.2 ± 1.0 Ma and a whole-rock–plagioclase–white mica isochron at c. 21 Ma. Sample 5PB-5a is weakly deformed garnet amphibolite associated with massive migmatite and small orthogneiss bodies and represents a huge xenolith in granite. Peak-PT estimates using core-sections of minerals such as garnet that lack deformation are c. 630 °C at 0.7 GPa (see petrology); the analysed hornblende (WMA of 30 ± 2 Ma) and biotite (WMA at 28.2 ± 0.4 Ma) are synkinematic matrix minerals that date deformation and rapid cooling during weak, still amphibolite-facies retrograde overprint, also detected in the garnet rims; peak-T conditions are at c. 36 Ma (U–Pb zircon, see above).

The next sample group comprises Las Ovejas complex of northern Honduras and structurally the footwall block of the Sula graben (Figs 5e & 7a); our dating aimed establishing potential along-strike changes in deformation/metamorphism/magmatism and Sula-graben formation. Sample 5H-2a is post-tectonic, weakly deformed garnet–white mica-bearing pegmatite in migmatitic biotite gneiss and rare leucogranite sills, and 5H-2c is a more strongly deformed, finer-grained variety of the pegmatite (augen gneiss); these magmatites are outcrop-scale intrusives. The coarse white mica of 5H-2a is at 23.68 ± 0.27 Ma (Rb–Sr whole rock–plagioclase–white mica; the biotite isochron is c. 7 Ma younger) and provides a younger bound for the intrusion of the pegmatite protolith. The white mica–K-feldspar–plagioclase–whole rock isochron of 5H-2c is c. 2 Ma younger (Fig. 5e), probably reflecting the higher strain. Sample 5H-3a is small, weakly deformed but syntectonic granite in strongly migmatitic biotite gneiss; the 22 ± 4 Ma isochron age of 10 biotite crystals analysed by single-grain laser fusion is >10 Ma younger than the intrusion age of the granite (c. 37 Ma, U–Pb zircon, see above). Nearby station 5H-4 typifies the heterogeneous lithology of the Las Ovejas complex in this area. Sample 5H-4b is biotite and hornblende-bearing pegmatite at the rim of a small, weakly deformed leucogranite body; all magmatites at this locality are deformed and in several cases the strain is concentrated in leucocratic dikes. The 38 ± 2 Ma hornblende is identical to the age of the concordant zircon fraction of nearby granite gneiss 92RN412 (see above); biotite is >10 Ma younger at 25 ± 3 Ma (laser single-grain ages). Sample 5H-4c shows synkinematic hornblende and biotite from a sinistral transtensive, greenschist-facies shear zone in amphibolite: hornblende is at 30 ± 2 Ma, biotite show a 25 ± 2 Ma mid-T 'plateau' (Fig. 5e); these ages best date formation of the Sula graben (see also below). Large hornblende crystals of sample 5H-4d, the weakly deformed host amphibolite to the above shear zone, contain small hornblende crystals and are within error (29 ± 3 Ma) of the synkinematic sample 5H-4c amphiboles. Sample 5H-9a, orthogneiss and migmatite that yielded Permo-Triassic U–Pb zircon ages (see above) and are distinctly different from the Las Ovejas complex rocks, gives Rb–Sr whole rock–plagioclase–white mica and biotite isochrons of c. 162 and c. 135 Ma, respectively. These dates have two possible interpretations: either they reflect very slow cooling from the Permo-Triassic metamorphic/magmatic

event or Middle Jurassic reheating not reflected in the zircon geochronology in this sample but detected in the Las Ovejas complex of Guatemala and in the central Chortís Block of Honduras. We prefer the latter interpretation, as we consider extremely slow cooling unlikely in the active tectonic setting of Early Mesozoic northern Central America.

Sanarate complex. Samples G3s and G5s (stops G23 and G24; Figs 5f & 7a) are phyllite and quartz-bearing amphibolite from the Sanarate complex, respectively. Quartz shows low-T plasticity with grain-growth inhibition and some post-tectonic annealing; together with hornblende and white mica it is synkinematic in discrete mylonite zones. The white mica spectrum of G3s shows Ar-loss from c. 195 to c. 137 Ma with a mid-T 'plateau' at 155 ± 10 Ma, and G5s hornblende from c. 157 to c. 40 Ma; a high-T 'plateau' is at 156 ± 5 Ma and low-T steps at ≥ 40 Ma. We interpret these data as indicating Middle–Late Jurassic deformation and Cenozoic reactivation.

Granitoids of the Chortís Block. We dated cooling in a few of the mostly undeformed granitoids of the northern Chortís Block (Figs 5g & 7a). Sample 5C-11, granodiorite of the Chiquimula batholith, shows biotite partly altered to chlorite; 10 grains dated by laser fusion give a 90 ± 10 Ma WMA. Eight out of 12 K-feldspar grains of sample 5C-21, sub-volcanic granite also from the Chiquimula batholith, have a 20 ± 1 Ma laser-fusion WMA; this demonstrates a composite nature of this batholith. K-feldspar of sample 5H-5, granite from the rift flank of the Sula graben, shows Ar loss from c. 37 to c. 28 Ma; the low-T 'plateau' at 28.3 ± 0.5 Ma indicates rapid cooling (Fig. 5g). In line with the synkinematic hornblende and biotite ages reported from this area (see above), this age suggests initiation of formation of the Sula graben at 28 ± 3 Ma. Six biotite grains of sample 5H-13, granite from south of the Las Ovejas unit, gave a laser-fusion age of 59 ± 3 Ma. The lack of modern U–Pb geochronology still impedes clear age assignments of plutonic events in the Chortís Block. Because of the often sub-volcanic nature of these granitoids, we assume that the available geochronological data reflect, to first-order, these events. Figure 5h pools all granitoid data and gives first-order groupings: there is significant magmatism at c. 36 Ma starting ≥ 40 Ma, at c. 59 Ma, and less significant at c. 81, c. 120, c. 161 and c. 260 Ma.

Southern Mexico. For correlative purposes (see discussion), we analysed samples from southern Mexico with the Ar/Ar method (Figs 5i & 1c).

Sample ML-18 is Magdalena migmatite of the eastern Acatlán complex (Fig. 2g). Biotite shows Ar loss from c. 163 to c. 88 Ma with a high-T 'plateau' at 162 ± 2 Ma; the latter traces cooling from migmatization and San Miguel dyke intrusion (175–171 Ma U–Pb ages; e.g. Keppie et al. 2004; Talavera-Mendoza et al. 2005). White mica from sample ML-37, mylonitic gneiss of the southernmost Oaxaca complex (Fig. 1c), shows Ar loss from a high-T 'plateau' at 315 ± 3 Ma to a low-T 'plateau' at 247 ± 3 Ma; the latter corresponds to ages of Permo-Triassic arc magmatism/metamorphism, exemplified by the Chiapas Massif of southeastern Mexico (e.g. Weber et al. 2006), the former corresponds to a widespread tectonothermal event in the Acatlán complex of southern Mexico (e.g. Vega-Granillo et al. 2007). Biotite from sample MU-12, migmatitic gneiss at Cruz Grande (Xolapa complex, Fig. 1c) yielded 29 ± 1 Ma (WMA), in line with cooling from widespread c. 35–30 Ma Xolapa plutonism (Herrmann et al. 1994; Ducea et al. 2004a). Sample MU-13 is Las Piñas mylonitic orthogneiss, sampled east of La Palma, which forms the structural top of the Xolapa complex. Solari et al. (2007) reported a 54.2 ± 5.8 Ma intrusion age (U–Pb zircon), and 50.5 ± 1.2 and 45.3 ± 1.9 Ma cooling ages (K–Ar and Rb–Sr biotite, respectively) from a moderately deformed variety. Our sample was specifically taken to date low-T, high-strain mylonitic flow (c. 300 °C) along the Tierra Colorada (Riller et al. 1992) or La Venta (Solari et al. 2007) top-to-NW, sinistral-transtensional shear zone; synkinematic biotite is at 35 ± 1 Ma. Sample ML-39, mylonitic gneiss at Pochutla, is from a narrow, late-stage shear zone in the regional Chacalapa shear zone along the northern edge of the eastern Xolapa complex; its major sinistral strike–slip stage is constrained at 29–24 Ma (Tolson 2005; Nieto-Samaniego et al. 2006). Our 15 ± 2 Ma biotite age suggests reactivation, also indicated by faulting in the 29 ± 1 Ma, post-mylonitic Pochutla granite (Herrmann et al. 1994; Meschede et al. 1996).

Samples PA3-4, PI28-1, CB31, CB35 and CA34 are from the Chiapas Massif of southeastern Mexico (Figs 1c & 5i). A belt of small plutons along the southwestern margin of the massif is associated with metamorphic rocks that are partly mylonitized and faulted along a NW-trending, >100 km long zone, named Tonala shear zone by Wawrzyniec et al. (2005). The late ductile and brittle kinematics is sinistral strike–slip, overprinting an older, higher-T fabric; the zone probably connects to similar sinistral strike–slip fabrics and plutons along the Polochic fault (Fig. 1c; see below). Our goal was to obtain ages for the intrusions and sinistral shear. Hornblende of diorite CB31 and tonalite CA34 yielded 15.5 ± 1.5 and 11 ± 2 Ma,

respectively; the former and latter are similar to 14.65 ± 0.42 granite GM14c along the Polochic fault zone of westernmost Guatemala (see above) and a 10.3 ± 0.3 Ma, pervasively sheared pluton (U–Pb zircon, Wawrzyniec et al. 2005) of the Tonala shear zone, respectively. Hornblende and biotite from three mylonitic gneisses (PA3-4, CA35, PI28-1) gave within error identical ages at c. 8.2 Ma, interpreted to date sinistral shear deformation. One hornblende age (CB35) at 28.5 ± 1.5 Ma is speculatively interpreted as approximating the age of the higher-T flow/metamorphism, in line with Oligocene deformation ages along the northern margin of the Xolapa complex (see above).

Fission-track thermochronology

Our TFT and AFT thermochronology is only reconnaissance in nature (Figs 6 & 7a). The two TFT ages support the available Ar/Ar and Rb–Sr geochronology, being close to biotite cooling ages. The AFT ages from Guatemala and Honduras show positive age v. elevation correlations both in the central Chuacús complex and the central and eastern parts of the northern Chortís Block (Fig. 6a); exhumation rates, if significant, are similar and slow for the areas north and south of the Motagua fault zone (c. 0.038 and 0.035 mm/annum, respectively), and possibly slightly more rapid along the western Sula-graben flank of northern Honduras (c. 0.046 mm/annum). The Chuacús-complex ages depict a cluster at c. 30 Ma, the low elevation ages from the Las Ovejas complex at c. 12 Ma (Fig. 6a). Despite high elevation, the youngest ages north of the Motagua fault zone occur in western Guatemala along the Polochic fault in a small, c. 15 Ma granite and the mylonitic augengneiss it intrudes (samples GM13s, GM14s, Fig. 7a). The c. 2 Ma age difference between crystallization and cooling through c. 100 °C of the granite reflects quenching at high crustal levels. The c. 11.5 Ma TFT and the c. 4.8 Ma AFT ages from low-T mylonite approximate the age of ductile–brittle to brittle Polochic faulting. The outcrop conditions prevent assessment of whether the apparently undeformed granite is locally affected by mylonitization but both mylonite and granite show east-trending, sinistral faults. The oldest ages at high elevation in the Chortís Block cluster at c. 22.5 Ma. The trend in the northern Chortís Block for AFT ages to become younger toward major fault zones (Motagua fault, Sula graben), with ages as young as 7.8–3.8 Ma, (Figs 6a & 7a), may be apparent, as these samples also occupy the lowest elevations and thus are certainly also an expression of the age v. elevation trend. The presence of these ages at low elevations along the Motagua fault suggests that its current activity (e.g. M7.5 1976 Guatemala earthquake, incorporating strike–slip and normal components; Plafker 1976) extended at least into the latest Miocene–Pliocene.

The track-length distributions of our samples are unimodal, narrow, and symmetric. Modelled $T–t$ paths give monotonous continuous-cooling type solutions (Fig. 6b). All $T–t$ models (except the $c.$ 15 Ma granite from western Guatemala, see above) show a phase of rapid cooling in the last few Ma that is an artefact caused by annealing at ambient temperatures acting over geological time. Low-T track-length reduction has been described for fossil tracks in age standards (e.g. Donelick *et al.* 1990) and borehole samples (Jonckheere & Wagner 2000). This reduction is not incorporated into the annealing equations derived from laboratory experiments on induced fission tracks, which account for annealing processes that take place within the partial annealing zone (Jonckheere 2003*a*, *b*).

Our samples from the Chiapas Massif, southeastern Mexico, comprise a ≥ 1.5 km elevation profile (Fig. 6c) and two swaths separated by <40 km with little elevation difference; no apparent major fault separates these samples. The youngest sample is at $c.$ 15.7 Ma, close to the Tonala shear zone and the $c.$ 11 Ma tonalite (Fig. 6c); it is most likely thermally influenced, supporting the Miocene magmatism and deformation along this shear zone. The remaining samples range from $c.$ 25 to $c.$ 39 Ma without a clear age v. elevation trend and a weighted average age of 30.4 $\pm$ 2.5 Ma; we suggest that these samples indicate relatively rapid cooling at $c.$ 30 Ma.

Interpretation of new geochronology

Our U–Pb zircon geochronology outlines Early Palaeozoic to Late Cenozoic high-T metamorphism, migmatization, and magmatism (Figs 3 & 7c). In the Rabinal complex of the Maya Block, we resolved prolonged Ordovician–Silurian ($c.$ 410–485 Ma, age groups at 413 and 465 Ma), Permo-Triassic ($c.$ 215–270 Ma, main age group at 218 Ma), and Middle Jurassic ($c.$ 163–177 Ma, main age group at 171 Ma) events (Fig. 3a); the Triassic event includes migmatites. Miocene ($c.$ 15 Ma) granite intruded into shallow crustal levels (nearly identical U–Pb zircon and AFT ages) in western Guatemala. Chuacús-complex rocks inherited Devonian zircons ($c.$ 403 Ma) and experienced Triassic ($c.$ 239–211 Ma, age groups at 238 and 218 Ma) magmatism and Cretaceous high-grade metamorphism ($c.$ 70–78 Ma, main age group at 74 Ma; Th–U < 0.03). These events are partly supported by Ar/Ar and Rb–Sr cooling ages ($c.$ 238–212; $c.$ 161; $c.$ 75–65 Ma). The Late Cretaceous metamorphic event

is about coeval with pegmatite emplacement at $c.$ 74 Ma (Rb–Sr and K–Ar on very coarse-grained white-mica). Local Early Cretaceous cooling ages (two samples) are probably geologically meaningless, possibly reflecting pre-Cretaceous events overprinted by Late Cretaceous deformation. Migmatization in the Río Hondo area is pre-Cretaceous and it remains to be demonstrated that it is Triassic, as in the Rabinal complex of western Guatemala; lithology, structure (see below) and our Ar/Ar data suggest that this region may constitute a distinct unit.

In the Las Ovejas complex of the northern Chortís Block and in the central Chortís Block, we resolved Early Permian ($c.$ 264–283 Ma, main age group at 273 Ma) magmatism followed by Early Triassic ($c.$ 242–253 Ma, main age group at 245 Ma) high-grade metamorphism (including migmatite) that is also reflected by a K–Ar biotite cooling age (222 $\pm$ 8 Ma; MMAJ 1980), and Middle to Late Jurassic ($c.$ 140–191 Ma, age groups at 159 and 166 Ma) magmatism and subsequent cooling (Rb–Sr and Ar/Ar ages at $c.$ 155 Ma). Overwhelmingly, the northern margin of the Las Ovejas complex is dominated by Late Eocene high-grade metamorphism, migmatization, and plutonism at 40–35 Ma (age group at 37 Ma). Cooling ages (Rb–Sr, Ar/Ar) related to this pervasive event cover $c.$ 35–20 Ma and also come from Roatán Island ($c.$ 36 Ma). Our geochronology also supports Silurian ($c.$ 400 Ma) and Jurassic ($c.$ 167 Ma) magmatism, and likely a Permo-Triassic event both in the northern and central Chortís Block. Based on U–Pb zircon geochronology, the Maya and Chortís Blocks share Ordovician to Jurassic events (Fig. 7c; the evident Proterozoic similarity is not addressed here). The northern and central Chortís Blocks are distinctly different from the Rabinal and Chuacús complexes in the Late Cretaceous and Early Cenozoic: Late Cretaceous high-grade metamorphism only occurs in the Chuacús complex and Eocene magmatism and metamorphism only can be found in the northern Chortís Block.

The most striking, first-order result showcased by our Ar/Ar and Rb–Sr geochronology is the clear-cut separation of the Rabinal and Chuacús complexes from northern Chortís Block reflected in all cooling ages (Fig. 7a & b): the Rabinal and Chuacús complexes, rimmed by the Polochic and the Motagua fault zones, cooled through $c.$ 500 °C and through 275–350 °C $c.$ 40 Ma earlier than the Las Ovejas complex south of the Motagua fault zone; cooling through $c.$ 100 °C occurred $c.$ 15 Ma earlier in the north than the south. Obviously, the rock units north and south of the Motagua fault zone experienced distinctly different thermal evolutions during the Late Cretaceous and Cenozoic: the Chuacús complex was heated to high amphibolite-facies metamorphism (see below) and cooled relatively rapidly (c. 28 °C/Ma) to upper crustal temperatures during the Late Cretaceous (Fig. 7d); subsequently, cooling (c. 6 °C/Ma) and exhumation was apparently slow (c. 0.038 mm/ annum; see Figs 6a & 7d). Multi-mineral T–t paths from central and east-central Chuacús samples (Fig. 7e) support this two-stage history. The Las Ovejas complex lacks Cretaceous regional metamorphism; its cooling from ≥40 Ma appears steady-state at c. 16 °C/Ma (Fig. 7d & e). Our geochronology does not prove whether slow cooling actually occurred, thus potentially reflecting long-term wrenching, or represents deformation and cooling at 40–35 Ma and reactivation at 25–18 Ma. As we found clear synkinematic mineral growth defining both age groups (see below), we favour the latter scenario. Along the southern margin of the Chiapas Massif, U–Pb and Ar/Ar geochronology demonstrates Cenozoic plutonism (c. 29, c. 16, c. 11 Ma) and sinistral strike–slip at c. 8 Ma along the Tonala shear zone.

Most AFT ages within the Rabinal and Chuacús complexes group around c. 30 Ma and thus correspond to the c. 30 Ma age cluster in the Chiapas Massif. We suggest that these ages indicate a thermal disturbance at ≥30 Ma induced by Cenozoic arc magmatism and sinistral displacement between the Maya and Chortís Blocks along and across this arc (see below). The young TFT and AFT ages along the Polochic fault zone relate its sinistral strike–slip deformation to the Late Miocene–Recent deformation along the Tonala shear zone, thus supporting their contemporaneous activity (Fig. 1c). These ages also support Burkart's (1983) estimate of major displacement (c. 130 km) along the Polochic fault between 10 and 3 Ma. The presence of very young AFT ages along the Motagua fault zone extends its neotectonic activity into the latest Miocene–Pliocene. The cluster of AFT ages at c. 22.5 Ma in the northern Chortís Block probably reflects continuation of the cooling from Late Eocene–Early Oligocene magmatism and metamorphism; the c. 12 Ma cluster is attributed to a thermal event that is connected with the flare-up of magmatism (ignimbrite event) along Central America (see below).

Taking the peak-P conditions of c. 0.7 GPa (see petrology below) reached in the Chuacús and Las Ovejas complexes during their Cretaceous (Chuacús complex, c. 75 Ma) and Cenozoic (Las Ovejas complex, c. 40 Ma) metamorphism, it appears that (i) early average exhumation rates were at least one order of magnitude more rapid than those derived from the AFT age v. elevation relationship (Fig. 6a); (ii) the presence of AFT ages at high elevations, which approach those of the Ar/Ar and Rb–Sr biotite geochronometers,

suggests that the change from relatively rapid to relatively slow rates occurred relatively early in the exhumation process (c. 60 Ma in the Chuacús complex; c. 20 Ma in the Las Ovejas complex). The fact that the late-stage exhumation rates (AFT age v. elevation data) are grossly similar in the Chuacús and Las Ovejas complexes suggests, given the strongly different cooling rates, much hotter crust in the Las Ovejas complex.

The difference between the southern Maya (here including the Chuacús complex) and northern Chortís Blocks during the Cretaceous–Cenozoic is emphasized by the distinctly different P–T–d–t conditions encountered in the North and South El Tambor complex allochtons (Harlow et al. 2004; Tsujimori et al. 2004): extremely low-T lawsonite eclogites experienced blueschist-facies retrogression at c. 120 Ma south of the Motagua fault, whereas higher-T zoisite eclogites show blueschist-facies retrogression at c. 76 Ma north of the fault. These P–T–d–t conditions illustrate differences in tectonic evolution and time of the emplacement of these subduction–accretion complexes. Significantly however, in the high-P/low-T case, the region south of the Motagua fault zone cooled earlier and cooling in the North El Tambor complex is coeval with the onset of cooling in the Chuacús complex.

New petrology

We selected 12 representative thin sections of magmatic and metamorphic rocks from both sides of the MSZ for electron microprobe analysis. Mineral abbrevations taken from Kretz (1983). Mineral assemblages and summary of the P–T–t estimates can be found in the Supplementry materials, and Figure 8 shows the PT space of our and literature data.

Rabinal complex, western Guatemala

Strongly foliated hornblende–epidote schist G19s contains phengite (Si-content c. 3.3 p.f.u.), green amphibole (edenite to ferro-edenite associated with magnesio- to ferro-hornblende), epidote ($X_{Ps} \geq 0.15$; cf. Dahl & Friberg 1980), and saussuritized plagioclase with an An-content of 0–2 mole%. Biotite forms post-tectonic, topotactic flakes on phengite. Accessory minerals are opaques and titanite. Phengite geobarometry yielded >0.7 GPa at 500 °C (Massonne & Szpurka 1997). The calibrations of Spear (1981) and Plyusnina (1982) gave 0.7–0.8 GPa at 450–500 °C, which are reproduced by Colombi's (1988) Ti-in-hornblende geothermometer. Our 67 ± 1 Ma Ar/Ar phengite age demonstrates latest Cretaceous exhumation through the middle crust with an average rate c. 0.37 mm/annum. Epidote schist

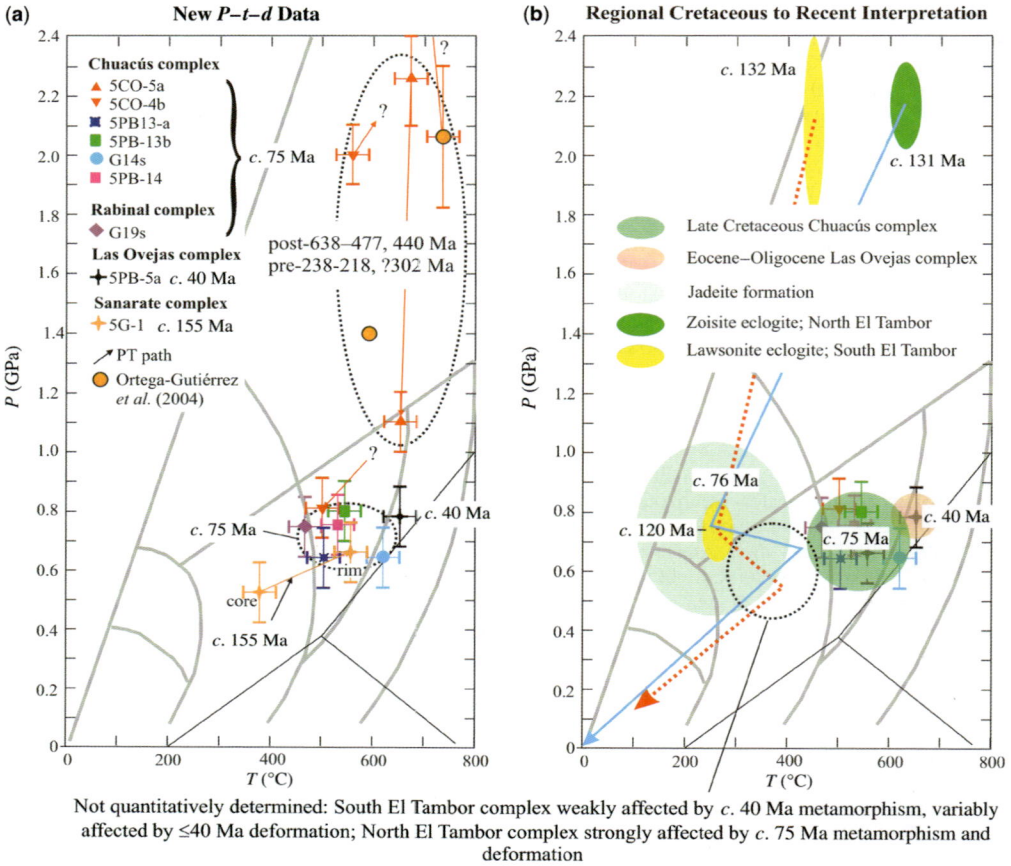

Fig. 8. Pressure–Temperature–time diagrams: (**a**) Pressure–temperature–time space of our new petrology and geochronology, and literature data of Ortega-Gutiérrez *et al.* (2004). A few Chuacús-complex samples plot in the eclogite field. They follow an isothermal decompression path to amphibolite facies. The eclogite-facies metamorphism is at best constrained to between *c.* 440 and 302 Ma (see text). Most of the complex passed through epidote–amphibolite facies metamorphism and subsequent retrogression to greenschist facies; these events are well-dated as Late Cretaceous (≤75 Ma). The Sanarate complex of the northwestern Chortís complex gives Jurassic (*c.* 155 Ma) epidote–amphibolite-facies metamorphism that evolved prograde out of greenschist facies. The Las Ovejas complex (Chortís Block) shows Cenozoic (≤40 Ma) epidote–amphibolite to amphibolite facies. (**b**) Motagua suture zone pressure–temperature–time evolution from this study and Tsujimori *et al.* (2004, 2005, 2006), Harlow (1994), Harlow *et al.* (2004) and Brueckner *et al.* (2005). The Late Cretaceous Chuacús complex and Mid-Cenozoic Las Ovejas complex pressure–temperature fields are from this study. The northern zoisite eclogites (on Maya Block) formed at *c.* 131 Ma and reached mid-crustal conditions at about the same time as the Chuacús rocks experienced reheating at epidote–amphibolite-facies conditions accompanied by local magmatism. We suggest that the two are linked to the same oceanic-to-continental subduction zone (see text). The southern lawsonite eclogites (on Chortís Block) formed at *c.* 132 Ma and were at mid-crustal conditions already by *c.* 120 Ma. Their emplacement is probably not genetically linked to the high-grade metamorphism and magmatism of the Las Ovejas rocks, which is a result of the Caribbean Plate translation and the formation of the Cayman Trough (see text).

G21s is mylonitic with numerous porphyroclasts of epidote (X_{Ps} core = 0.295, X_{Ps} rim = 0.215), albite ($An_{0.3-0.9}$) and microcline ($Kfs_{98}Ab_{1-2}$ $Cel_{1-2}An_{<0.5}$). The feldspars lack recrystallization but quartz recrystallized dynamically by subgrain rotation. Greenish biotite (X_{Mg} *c.* 0.47) and phengite (*c.* 3.3 Si p.f.u.) are common; opaques, titanite, apatite and zircon are accessories. Strong mineral chemical disequilibria between the phases preclude *PT* estimates; the Si-content of phengite suggests *c.* 0.8 GPa (Massonne & Szpurka 1997).

Chuacús complex, central Guatemala

Garnet gneiss 5CO-4b is dominated by orthoclase $(Kfs_{93}Ab_5Cel_1An_{<0.5})$ and minor, partly saussuritized plagioclase $(An_{0.2-1.4})$. Quartz shows subgrain-rotation recrystallization. Chloritized biotite $(X_{Mg} = 0.17)$ and phengite have epidote $(X_{Ps} = 0.23)$ and plagioclase inclusions. Phengite has elevated Si-contents of up to 3.50 p.f.u. in the core, indicating high-P conditions, whereas Si-contents as low as 3.27 p.f.u. occur in the rim sections. Garnet forms porphyroblasts with inclusions of plagioclase, quartz, minor ferro-hornblende/ferro-tschermakite, omphacite (10 mole% jadeite) and rutile. It displays compositional zoning with a bell-shaped spessartine curve (core $Alm_{43}GAU_{44}$-$Sps_{10}Prp_{0.5}$; rim $Alm_{47}GAU_{48}Sps_5Prp_{0.5}$). Carbonate, zircon and titanite form accessory minerals in the matrix. Relict phases highlight the polymetamorphic evolution of 5CO-4b. Assuming equilibrium between the omphacite inclusion in garnet, the adjacent garnet core section and the core of the matrix phengite, PT conditions of c. 583 °C and c. 2.04 GPa are calculated using the garnet–phengite–omphacite geobarometer of Waters & Martin (1993) (phengite activity model of Holland & Powell 1998), clinopyroxene activity model of Holland (1980, 1990), and garnet activity model of Ganguly et al. (1996) in combination with the garnet–clinopyroxene geothermometer of Krogh (2000) (calculation of ferric iron in clinopyroxene according to Ryburn et al. 1976). Similar temperatures are calculated for high-P phengite and garnet core analyses using the thermometer of Green & Hellman (1982). These PT conditions, however, only indicate a point on the prograde PT path as the highest pressure is indicated by the highest $a_{Grs}^2*a_{Prp}$ (Schmid et al. 2000) at an intermediate position between garnet core and rim. The strong retrograde overprint is mirrored by pervasive growth of matrix biotite and low-Si phengite. Outermost garnet rims (highest X_{Prp}) and adjacent biotite yielded c. 510 °C (geothermometer of Hodges & Spear 1982) at c. 0.7 GPa (barometry of Massonne & Szpurka 1997) using paragenetic phengite; this demonstrates re-equilibration under epidote–amphibolite-facies conditions. Because of this overprint, feldspar thermometry (Fuhrman & Lindsley 1988) on coexisting plagioclase $(An_{0.2-1.4})$ and K-feldspar (integrated analysis with a 50 μm beam diameter) indicates cooling temperatures of 400–440 °C. The lower intercept U–Pb zircon age indicates that high-P metamorphism is <638–477 Ma and our Ar/Ar and Rb–Sr phengite and biotite ages date the latest portion of the retrograde path at 68–64 Ma.

Garnet amphibolite (eclogite) 5CO-5a contains minor quartz showing subgrain-rotation recrystallization and foliation-defining, large calcic amphibole (pargasite to edenite). Omphacite with 35 mole% jadeite at the rim to 40 mole% in the core is present as relicts in garnet and hornblende cores and less commonly in the matrix. Well-aligned phengite is the dominant sheet silicate; Si-content decreases from core (3.32 p.f.u.) to rim (3.27 p.f.u.). Garnet displays weak zoning with decreasing grossular and pyrope contents and increasing almandine contents from core $(Alm_{60.5}GAU_{23}$-$Sps_{0.5}Prp_{16})$ to rim $(Alm_{63}GAU_{20.5}Sps_{0.5}Prp_{15.5})$. Plagioclase (An_{2-4}) is secondary, forming along the rims of garnet and hornblende; irregular spots of An_{10-14} are also present. Late-stage biotite $(X_{Mg}$ c. 0.53) grew at the expense of phengite. Apatite, rutile, ilmenite and pyrrhotite are accessories. PT conditions of the eclogite facies event were calculated applying the garnet–phengite–omphacite geobarometer of Waters & Martin (1993) (activity models see sample 5CO-4b) in combination with the garnet–clinopyroxene geothermometer of Krogh (2000). Using the method of Ryburn et al. (1976) to estimate Fe^{3+} in clinopyroxene, PT conditions of c. 675 °C at c. 2.4 GPa were calculated using the core composition of garnet (highest $a_{Grs}^2*a_{Prp}$), the omphacite inclusion in garnet with the highest jadeite content, and the phengite with the highest Si content (Carswell et al. 2000). The application of other methods to estimate the ferric iron content in omphacite (e.g. Carswell et al. 2000; Schmid et al. 2000; Krogh 2000; Krogh & Terry 2004) results in an uncertainty of ±50 °C for the geothermometry. Omphacite in the matrix probably was affected by retrograde processes and consequently shows a wide range of temperatures (480–690 °C). Temperatures for retrograde overprint of the eclogite were estimated at 630–690 °C using the garnet–hornblende thermometer of Graham & Powell (1986) for garnet rims and adjacent hornblende. The GRIPS geobarometer of Bohlen & Liotta (1986) calculates pressures of c. 1.1 GPa for these minerals and associated plagioclase, rutile and ilmenite. Slightly higher pressures of 1.2–1.4 GPa are calculated using the garnet–amphibole–plagioclase geobarometer of Kohn & Spear (1990). These estimates indicate isothermal decompression during exhumation of the eclogite. Ortega-Gutiérrez et al.'s (2004) c. 302 Ma lower intercept zircon age from a leucosome interbedded with relict eclogite from the area of 5CO-5 was interpreted to provide a minimum age for the Chuacús high-P event.

The strong foliation of garnet-bearing micaschist 5PB-13a is formed by alternating mica- and quartz-rich layers; quartz shows subgrain-rotation and grain-boundary-migration recrystallization. Small, almandine-rich and spessartine-poor garnet grains $(Alm_{70}GAU_{10}Sps_1Prp_{18})$ are included in large albite porphyroblasts $(An_{0.2-0.7})$, and larger garnet porphyroblasts with $Alm_{74}GAU_2Sps_{13}Prp_8$ crystallized within the matrix; the garnets are slightly inhomogeneous without regular zonation patterns. Green to brownish biotite (X_{Mg} c. 0.58) is partly altered to oxychlorite. Si-content in phengite decreases from core (3.3 p.f.u.) to rim (<3.2 p.f.u.). This correlates with decreasing Ti-content and indicates a retrograde evolution. Because of the lack of critical mineral assemblages, only a few thermobarometers can be used to evaluate the *PT* conditions of these schists. The Si-content of phengite yields ≥0.6–0.8 GPa; the garnet–biotite geothermometer (Hodges & Spear 1982) and the garnet–phengite geothermometer (Green & Hellman 1982) point to 560–580 °C for the garnet inclusions in albite and 460–475 °C for the matrix garnets. Garnet amphibolite 5PB-13b is well foliated, medium- to coarse-grained, and comprises large, pale-green calcic amphibole (magnesiohornblende to edenite) with numerous garnet and quartz inclusions; the matrix has also phengite and clinozoisite. Phengite Si-content (3.3–3.15 p.f.u.) and X_{Mg} (0.76–0.66) decrease toward the rims. Clinozoisite ($X_{Ps} = 0.08$–0.09) forms large prisms. Quartz shows both dynamic subgrain-rotation and grain-boundary-migration recrystallization. Albite (An_{2-6}) has numerous quartz inclusions. Small, euhedral garnet yields Alm_{55}-$GAU_{35}Sps_5Prp_4$ in the core and $Alm_{54}GAU_{30}Sps_5$-Prp_{11} in the rim and has increasing X_{Mg} values from core to the rim, indicative of prograde growth. One garnet differs compositionally, with high X_{Mg} in the core and abruptly increasing almandine content at the expense of grossular and spessartine towards the rim; this may indicate an earlier, prograde metamorphic evolution. Retrograde chlorite is widespread; apatite, zircon, and titanite are accessories. Pressures of ≥0.7 GPa (Massonne & Szpurka 1997) at 530–550 °C (garnet–hornblende geothermometer of Graham & Powell 1986) are obtained; the Ti-in-hornblende thermometer of Colombi (1988) provides c. 590 °C. Our Ar/Ar amphibole and Rb–Sr phengite ages suggest metamorphism at c. 65 Ma at this locality.

Hornblende–mica–chlorite schist 5PB-14 shows distinct layering with amphibole–mica and quartz that is dynamically recrystallized by subgrain rotation and grain-boundary migration. Pseudomorphs of chlorite (ripidolite after Hey 1954) after relict garnet and lamellae of oligoclase (An_{12}) within albite ($An_{0.5-2}$) demonstrate retrograde overprint. Green amphibole (tschermakite to magnesiohastingsite) and clinozoisite (X_{Ps} c. 0.13) are subordinate. White mica shows increasing Si-content from the core to the rim (3.12–3.20 p.f.u.). Pale brown biotite is rare. Carbonate is Fe-bearing dolomite. Opaque minerals, rutile and titanite form accessory phases. *PT* conditions of

500–550 °C at c. 0.8 GPa are suggested by the geothermobarometer of Plyusnina (1982). This estimate is reproduced by the phengite geobarometry (Massonne & Szpurka 1997) and the amphibole–plagioclase geothermometers of Spear (1981), Blundy & Holland (1990), and Holland & Blundy (1994). A temperature of 600–630 °C is indicated by the Ti-in-hornblende geothermometer of Colombi (1988), probably due to the high bulk Ti-content of the rock. This may, however, also indicate an early stage of medium- to high-grade, amphibolite-facies metamorphism. Our 65 ± 1 Ma Ar/Ar white mica age dates cooling during greenschist-facies retrogression.

Gneiss G14s is inhomogeneous with alternating plagioclase–quartz and mica–hornblende layers. Plagioclase is albite ($An_{0.5-2}$); epidote ($X_{Ps} = 0.2$) and chlorite are present in the mafic layers. Phengite has Si-contents of 3.27–3.34 p.f.u. Biotite yields X_{Mg} of c. 0.51. Green amphibole is magnesio- and ferro-hornblende. Small garnet within the mafic layers and as tiny inclusions in plagioclase has a bell-shaped spessartine curve and rimward decreasing X_{Fe} and X_{Mn}; cores and rims are Alm_{40}-$GAU_{31}Sps_{25}Prp_1$ and $Alm_{48}GAU_{37}Sps_{13}Prp_2$, respectively. Accessories are carbonate, opaques and titanite. Triboulet's (1992) geothermobarometer calculates c. 610 °C at c. 0.62 GPa for the rim composition of the minerals, supported by phengite geobarometry at 0.6–0.7 GPa at 500–600 °C. About 550 °C is obtained by the garnet–hornblende (Graham & Powell 1986) and the garnet–chlorite geothermometer (Ghent et al. 1987). These *PT* estimates indicate peak conditions at the transition from epidote-amphibolite to amphibolite-facies metamorphism.

Intrusive sequence, Juan de Paz ophiolite, North El Tambor complex

Fine- to medium-grained microdiorite 5PB-8 shows agglomeration of hornblende and plagioclase. Plagioclase shows normal zoning with An-content decreasing from 85 to 55 mol% from core to rim. Calcic amphibole (magnesiohornblende) is slightly brownish and short prismatic. A few clinopyroxene relicts yield 95–98 mol% augite (end member calculation after Banno 1959) and are mantled by hornblende. Opaque minerals and apatite are common accessories. Anderson & Smith's (1995) Al-in-hornblende geobarometer indicates c. 140 MPa at 730–860 °C, calculated with the Ti-in-hornblende and the plagioclase–hornblende geothermometers. Bertrand et al. (1978) dated a likely metamorphic overprint in related dolerites from the Baja Verapaz (Fig. 1c) at c. 75 Ma (see above).

Sanarate complex

Chlorite–epidote–amphibole schist 5G-1a contains albite (An$_{1-3.5}$), clinozoisite ($X_{Ps} \geq 0.1$), chlorite (ripidolite), amphibole (edenite) and phengite (3.2–3.3 Si p.f.u.). The application of the geothermobarometer of Triboulet (1992) yields peak *PT* conditions of *c.* 550 °C at *c.* 0.65 GPa for the rims of the minerals. The cores indicate *c.* 380 °C at *c.* 0.55 GPa, outlining a prograde metamorphic loop. These data highlight the transition from lower greenschist to epidote–amphibolite facies and are supported by the geothermometry of Colombi (1988) and Holland & Blundy (1994), which provide *c.* 415 °C for the core and *c.* 560 °C for the rim of the amphibole. Si-contents of syntectonic phengite accordingly yield 0.6–0.65 GPa (Massonne & Szpurka 1997). A likely age of this metamorphism is Middle Jurassic, given by the hornblende and white mica ages from adjacent locations G23, G24, which comprise similar rocks.

Las Ovejas complex

Garnet amphibolite 5PB-5a displays a fine- to medium-grained matrix with synkinematic biotite, amphibole and large, poikiloblastic garnet. Matrix quartz shows numerous subgrains. Biotite is Ti-rich (up to 4.1 wt% TiO$_2$) and has an X_{Mg} of about 0.24. The green amphibole is hastingsite. Large plagioclase grains are chemically homogeneous with An-content of *c.* 35 mol%. Garnet cores show Alm$_{69}$GAU$_{19}$Sps$_8$Prp$_4$; X_{Fe} and X_{Mn} increase towards the rims, indicating retrograde overprint. Accessories are opaques and apatite. Garnet–hornblende geothermometry (Graham & Powell 1986) in combination with garnet–amphibole–plagioclase–quartz geobarometry (Kohn & Spear 1990) yields 627–655 °C at 0.7–0.8 GPa. Comparable temperatures are calculated with garnet–biotite thermometers of Hodges & Spear (1982), Indares & Martignole (1985) and Perchuk & Lavrent'eva (1983). Ar/Ar hornblende and biotite ages out of shear bands date deformation and rapid cooling through the middle crust at 30–28 Ma.

Chiquimula batholith within Las Ovejas complex

Coarse-grained granodiorite 5C-13 is part of the Chiquimula batholith with the mineral assemblage quartz + K-feldspar + plagioclase + biotite + hornblende. Subhedral and anhedral plagioclase crystals are typically altered to sericite and have An-contents of 35–55 mol% with normal zoning patterns. Twinning is in part deformational. Subhedral orthoclase has a K-feldspar component of 80–90 mol% and displays perthitic unmixing.

Prismatic amphibole is magnesiohornblende (classification of Leake *et al.* 1997) with relicts of augite in the cores. Reddish-brown biotite contains up to 4.6 wt% TiO$_2$, has an X_{Mg} of *c.* 0.45 and is partly transformed to chlorite. Accessories are opaque minerals, apatite, and zircon. Anderson & Smith's (1995) Al-in-hornblende geobarometer yields 100–180 MPa at 740–860 °C, calculated with the Ti-in-hornblende and plagioclase–hornblende geothermometers (Colombi 1988; Blundy & Holland 1990; Holland & Blundy 1994). Our Ar/Ar geochronology demonstrates that the Chiquimula granite is a composite batholith (*c.* 90 Ma biotite–chlorite age, 5C-11; *c.* 20 Ma K-feldspar, 5C-13; assuming that the 20 Ma K-feldspar age of 5C-11, adjacent to 5C-13, is close to the intrusion age of this shallow granodiorite (3–5.5 km), the average exhumation rate is 0.15–0.27 mm/annum. This supports our inference (see above) that relatively rapid exhumation in the Las Ovejas complex is pre-20 Ma.

Interpretation of new petrology

Garnet amphibolite and felsic gneiss of the central Chuacús complex in the El Chol area preserve, in accordance with Ortega-Gutiérrez *et al.* (2004), a high-*P* event. Garnet gneiss 5CO-4b probably records part of a prograde path (*c.* 580 °C at *c.* 2 GPa) that culminated at *c.* 675 °C and *c.* 2.4 GPa (5CO-5a). Isothermal decompression to *c.* 660 °C at *c.* 1.3 GPa followed the near ultrahigh pressure conditions and the retrograde path can possibly be traced to *c.* 520 °C at *c.* 0.7 GPa. Only restricted areas of the Chuacús complex escaped the regional overprint at 450–550 °C and 0.7–0.8 GPa. The age of the high-*P* event is pre-238–218 Ma, the age of two-mica orthogneiss intrusion (U–Pb ages) and high-*T* metamorphism (Ar/Ar ages) in the El Chol region. The high-*P* event is probably post-638–477 Ma, the lower intercept U–Pb zircon age (including errors) of high-*P* garnet–K-feldspar gneiss at location 5CO-4. It is possibly older than the *c.* 302 Ma lower intercept U–Pb age of a leucosome interpreted to be associated with eclogite decompression (Ortega-Gutiérrez *et al.* 2004); we have so far not been able to reproduce this age at the same locality using SHRIMP geochronology (unpublished results). The high-*P* event should also be younger than *c.* 440 Ma, the age of the youngest zircons in Chuacús paragneiss (Solari *et al.* 2008). The *c.* 403 Ma age of likely metamorphic zircon incorporated in the Triassic two-mica orthogneiss indicates Devonian metamorphism in the Chuacús complex. The regional ≥450 °C and 0.7–0.8 GPa overprint is Late Cretaceous and the most outstanding feature of most of the Chuacús complex. This regional pressure-dominated metamorphism is associated with local magmatism. The *PT* estimates of the Cretaceous event are grossly similar in the Rabinal complex of western Guatemala and the Chuacús complex of central Guatemala, but somewhat higher temperatures and pressures occur in central Guatemala. The microdiorite from the Juan de Paz ultramafic sheet of the North El Tambor complex probably records seafloor hydrothermal overprint followed by low-grade Cretaceous metamorphism.

Our reconnaissance petrology of the northern Chortís Block identifies Cenozoic amphibolite-facies metamorphism, migmatization, local magmatism and associated deformation for the Las Ovejas complex starting at ≥40 Ma. Peak *PT* conditions are *c.* 650 °C at 0.7–0.8 GPa, veiling earlier high-grade metamorphism associated with magmatism and migmatization that is Permo-Triassic and Jurassic (273–245, 166–158 Ma). Distinctly different, Middle Jurassic (*c.* 155 Ma) prograde, lower greenschist- to epidote–amphibolite-facies metamorphism is preserved in the non-migmatitic Sanarate complex of southern Guatemala, west and outside of the Cenozoic-shaped Las Ovejas complex. The latest Cretaceous–Cenozoic Chiquimula batholith is a composite, shallow crustal level intrusion (3–5.5 km). Sub-volcanic *c.* 20 Ma intrusives in this batholith demonstrate exhumation from *c.* 25 km during 40–20 Ma, supporting a two-stage deformation–exhumation history within the Las Ovejas complex; that is relative rapid pre-20 Ma followed by slower post-20 Ma exhumation (see above).

Figure 8a summarizes the available *P–T–t* data from the Rabinal, Chuacús, Sanarate and Las Ovejas complexes. The intricate metamorphic evolution in the Chuacús complex apparently is related to Silurian–Carboniferous near ultra-high pressure metamorphism and its retrogression into the epidote–amphibolite-facies field; it is overprinted by Triassic magmatism and metamorphism, and regional Cretaceous epidote–amphibolite to amphibolite-facies reheating. The northern Chortís Block contains crust dominated by Jurassic epidote–amphibolite metamorphism, only weakly overprinted in the Cenozoic; this Cenozoic metamorphism defines the Las Ovejas complex, which is characterized by epidote–amphibolite to amphibolite-facies reheating. Figure 8b compares the *P–T–t* evolutions of the mélange complexes (North and South El Tambor) and the continental blocks onto which they were emplaced. Jurassic–Early Cretaceous North El Tambor complex oceanic and immature arc rocks incorporated *c.* 131 Ma eclogite into an accretionary wedge that reached crustal levels in the Late Cretaceous (*c.* 76 Ma) and was incorporated as the roof nappe into the Chuacús complex–southern Maya Block

imbricate stack that initiated contemporaneously (*c.* 75 Ma). Jurassic–Cretaceous South El Tambor complex oceanic rocks incorporated *c.* 132 Ma eclogite into an accretionary wedge, reached crustal levels at *c.* 120 Ma, and were emplaced onto Chortís-Block crust, whose last thermal overprint was during the Middle Jurassic. The South El Tambor complex was variable affected by the Cenozoic deformation and metamorphism that characterizes the Las Ovejas complex (see below). Characteristically, the South El Tambor complex is preserved outside the area of strong Cenozoic tectono-metamorphic overprint, i.e. outside the central and eastern Las Ovejas complex; we may speculate that parts of the latter may actually be South El Tambor complex subjected to Cenozoic high-grade metamorphism and deformation.

Structure and kinematics

Rabinal complex and Late Palaeozoic cover, western Guatemala

Figures 9 and 10 compile structural and kinematic field and laboratory data and Figure 11 shows representative deformation features and dated synkinematic minerals. The Rabinal complex in western Guatemala shows regional, heterogeneously distributed, low-grade deformation (D$_2$) that locally overprints high-grade, partly migmatitic flow structures (D$_1$). Most pronounced is a deflection of the first foliation, s$_1$, into ubiquitous, sub-vertical, sinistral, D$_2$ shear zones. In this study, we concentrated on the Late Cretaceous (see below) greenschist-facies event D$_2$; its low grade is distinct, as the same event reaches amphibolite facies in the central Chuacús complex of southern Guatemala and is coeval with magmatism. Another distinct map-scale feature is a NW structural trend (foliation, stretching lineation, shear planes), oblique to the *c.* east trend of the Polochic fault zone; the latter is clearly younger (see above).

The basement mainly comprises locally migmatitic paragneiss and orthogneiss, amphibolite and (hornblende)–epidote schist, augen gneiss of magmatic protolith, pegmatite and variably strained granitoids. The metagranitoids also intrude possibly Precambrian, low-grade sedimentary rocks (San Gabriel unit of Solari *et al.* 2009). We also studied localities of Late Palaeozoic metasediments (often conglomerate; Chicol and Sacapulas Formations). D$_1$ shows high-*T* plasticity, dominant coaxial flattening (s fabrics), and intrafolial and isoclinal folds, which refold an even earlier, relict foliation. D$_1$ is accompanied by local migmatization, traced by feldspar layers. D$_1$ quartz LPO shows medium- to high-*T* prism <a> glide (G20s of stop G45; Fig. 9, bottom right) that contrasts with D$_2$ low-*T*

Western Guatemala: northern foreland fold-and-thrust belt – Cretaceous deformation

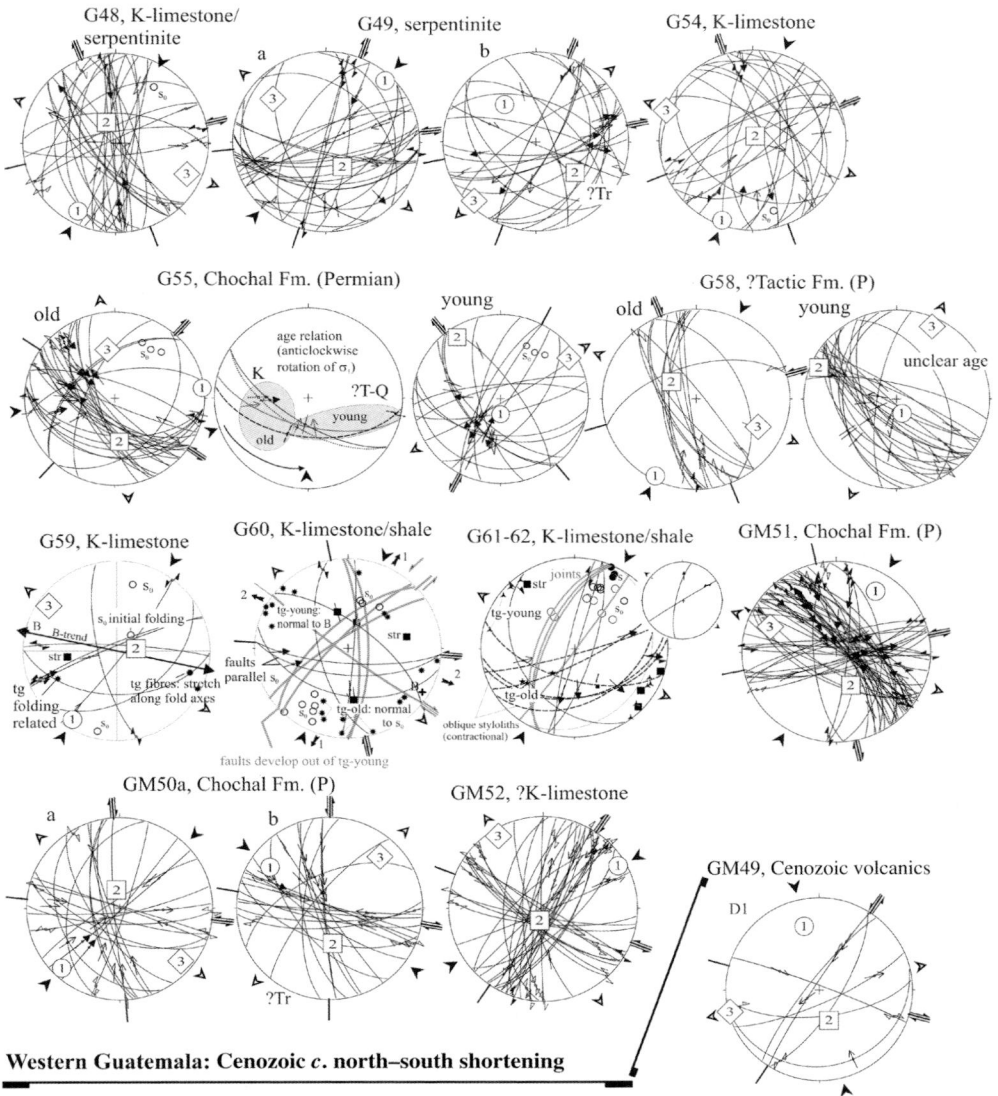

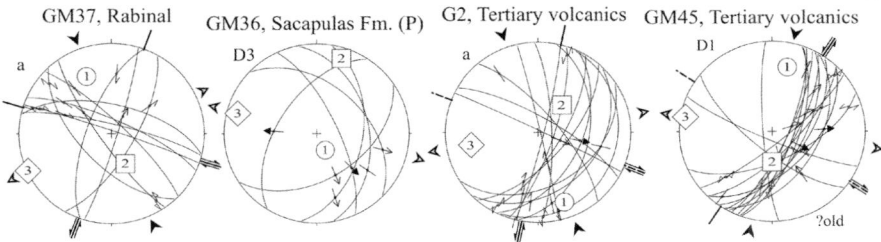

Fig. 10. Additional structural data from the Maya Block of western Guatemala. Legend for structural symbols as in Fig. 9. ccw, counterclockwise.

Western Guatemala: Late Cenozoic sinistral wrenching

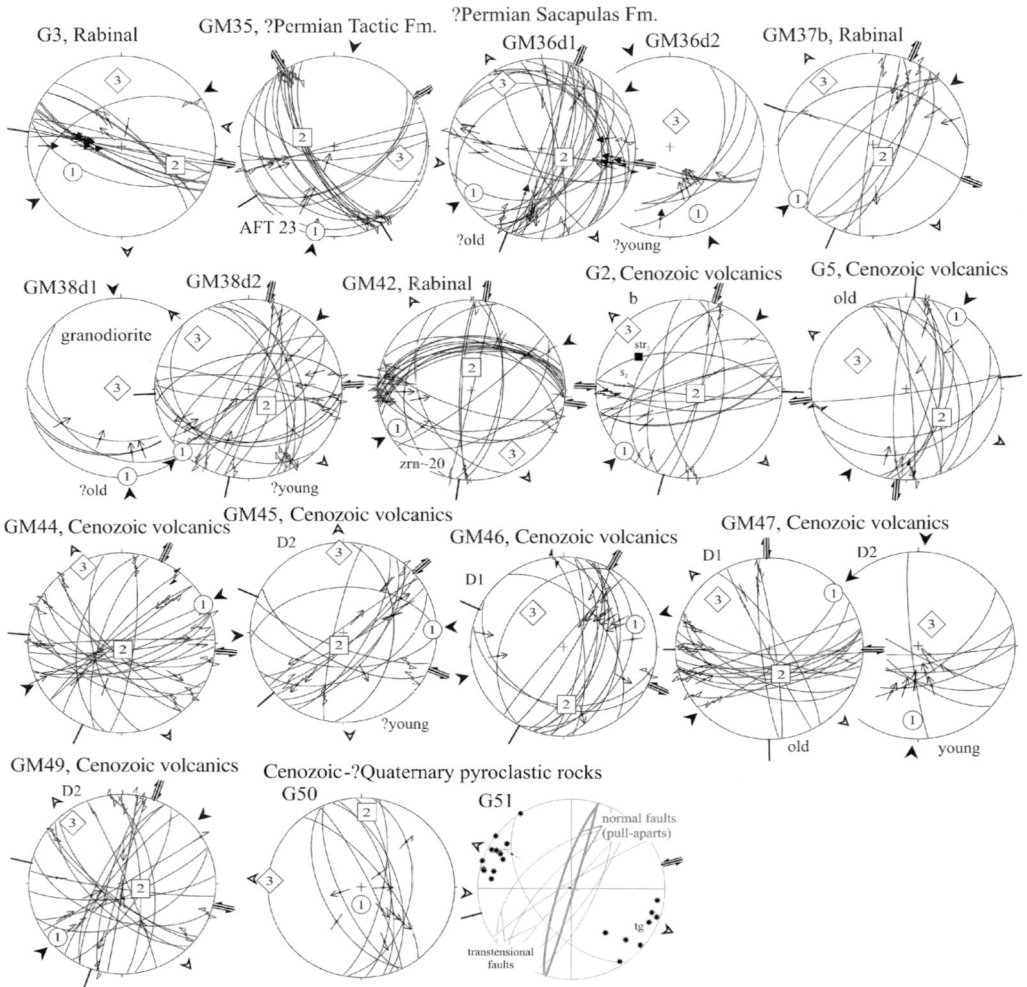

Western Guatemala: ?gravitational normal faulting

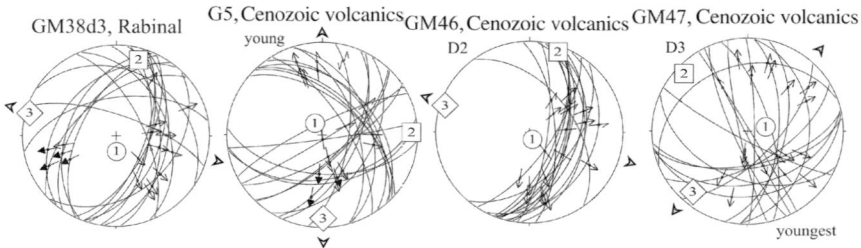

Fig. 10. (*Continued*).

quartz LPO (see below). Quartz is coarse-grained and recrystallized by grain-boundary migration. D_1 is Late Triassic (c. 215 Ma), suggested by our U–Pb zircon age of phyllonitic (due to D_2) migmatitic gneiss (G18s of stop G1). D_2 formed 10–80 cm wide, chlorite- and carbonate-bearing shear zones that boudinaged s_1 (e.g. G1), imposed locally well-developed augen- to ultra-mylonites that evolved during decreasing temperature into ductile–brittle mylonite with clinozoisite-, epidote- and quartz-bearing faults (e.g. G3) and late fracturing (e.g. G45). These discrete shear zones widen into belts of mylonite, transforming the basement gneisses into biotite–chlorite schist and epidote–hornblende phyllite. Strain is locally prolate (dominant l-fabrics; e.g. G56/57). D_2 quartz LPO indicates basal <a> slip (G1) with a strong coaxial flow component (c-axis cross-girdles). Quartz recrystallization mechanisms are variable but typically low-T and dynamic; subgrain rotation is dominant (e.g. G21s of stop G46; Fig. 11a-2) but evolves out of bulging recrystallization in quartz ribbons (e.g. G25s of stop G56, Figs. 11a-2, inset). Hornblende and feldspar are broken. Plagioclase shows rare incipient stages of recrystallization along twin planes; thus, deformation occurred mostly below c. 450 °C, evolving to very low grade (≤300 °C). Sinistral shear criteria are ubiquitous: flow folds, σ clasts, shear bands and s-c- fabrics in layers of quartz recrystallization (Fig. 9). Epidote (±hornblende) schists G19s (stop G1) and G21s (stop G3) provide quantitative PT estimates for the initial, higher-T stages of D_2 at 450–500 °C and 0.7–0.8 GPa. D_2 is precisely dated at 67 ± 1 Ma by synkinematic white mica (G19s of stop G1, Fig. 11a-2). D_2 also developed in the Palaeozoic metasediments and is more pronounced (sinistral) transpressional (e.g. G4, Fig. 9) than in the underlying basement. Location GM36 shows stretched conglomerate (Fig. 9) cut by aplitic veins with the same c. east–west extension that stretched the pebbles; low-T plasticity with dynamic subgrain rotation occurred in quartz clasts. The basement rocks are variably overprinted by faults that are related to the Polochic fault zone (Fig. 10); the principal faults strike WNW (Polochic-parallel) and north. Faulting is ≤20 Ma, given by the c. 20 Ma U–Pb zircon lower intercept of samples GM13s and GM14s (stops GM42, GM43) and the c. 23 Ma AFT age of stop GM35 (sample GM9s); it is best constrained by the TFT and AFT ages along the Polochic fault (11.5–4.8 Ma, see above).

Rabinal complex, Rabinal granitoids and low-grade metasediments, central Guatemala

The Rabinal complex in central Guatemala comprises paragneiss that typically resembles metagreywacke (coarse-grained feldspar and white mica) and K-feldspar porphyroblastic granitoid (U–Pb zircon ages at 485–410 Ma, stop 5CO-1) that were transformed to quartz phyllite and augen gneiss. The distinction between ortho- and paragneiss is problematic at many locations and makes mapped lithological boundaries questionable (e.g. BGR 1971, Fig. 12). Locally marble and amphibolite intercalations occur. The major deformation, attributed to the WNW-trending Baja Verapaz shear zone by Ortega-Gutiérrez et al. (2004), is locally mylonitic and low grade. Foliation dips intermediate to steeply south(west) and the stretching lineation (str_1) plunges intermediate to the SW (Figs 12 & 13). A continuous kinematic evolution from ductile shear bands to ductile–brittle faults and quartz-filled, en-echelon tension gashes (e.g. stops 5CO-1, 5CO-3, 867) indicates sinistral transpression. At station 5CO-1, typical Rabinal augen gneiss, shear bands evolve into ductile–brittle faults (quartz ductile), and faults with large quartz–white mica fibres that are conjugate to kink bands that are partly faulted and indicate north–south shortening and east–west extension along str_1. Quartz shows typical low-T plasticity with subgrain-rotation recrystallization. K-feldspar is brittle. Local ultramylonite (<1 m thick, stop 867) has cogenetic chlorite + sericite and quartz ribbons that show weak bulging- and dominant subgrain-rotation recrystallization. Late kink folds are interpreted as flow heterogeneities at the end of deformation. Quartz fibres in the strain shadow of pyrite are un-recrystallized. The post-245 Ma (see above) phyllite and conglomerate (mostly intermediate volcanic rock pebbles, stops 865–866) records identical, low-T structural geometries and kinematics. Paragneiss (stop 867) preserves pre-Cretaceous deformation with large white mica and pre-existing foliation and felsic veins that are cut by orthogneiss that is typical Rabinal gneiss.

Chuacús complex, central Guatemala

Central Guatemala exposes the deepest part of the Chuacús complex studied by us (Figs 12 & 13) and allows insight into pre-Cretaceous deformation–metamorphism–magmatism. The Cretaceous structural overprint increases both toward south and north, i.e. the North El Tambor complex of the hanging wall. Typical Chuacús-complex rocks are garnet-bearing para- and orthogneiss, garnet amphibolite that is partly retrograde eclogite (see petrology), and marble (e.g. stops 880–884, 886, 5CO-4–6). Felsic and mafic rocks are often concordantly interlayered and resemble a bimodal volcanic sequence dominated by metasediments. The layering is accentuated by felsic sills (dominant) and

(a)

Central Guatemala: Chuacús complex at the footwall of North El Tambor complex, Late Cretaceous deformation

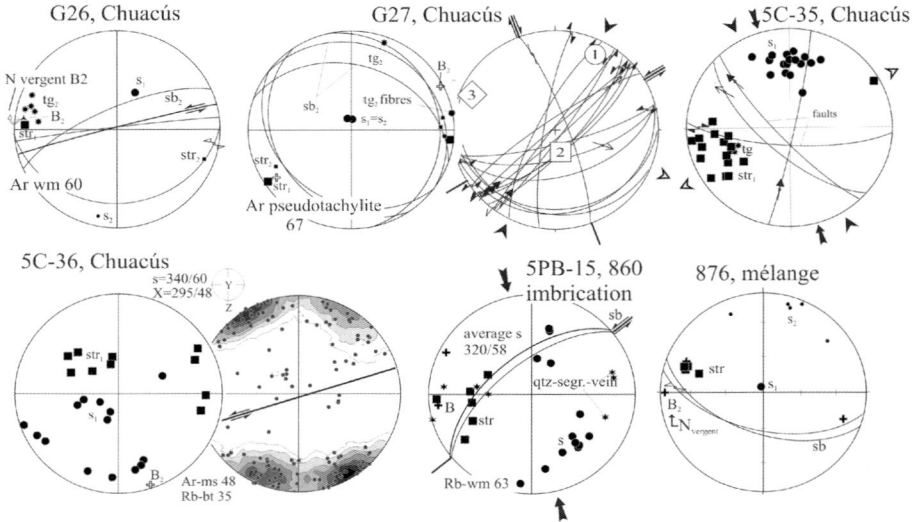

Central Guatemala: Rabinal complex (granitoids), Late Cretaceous deformation

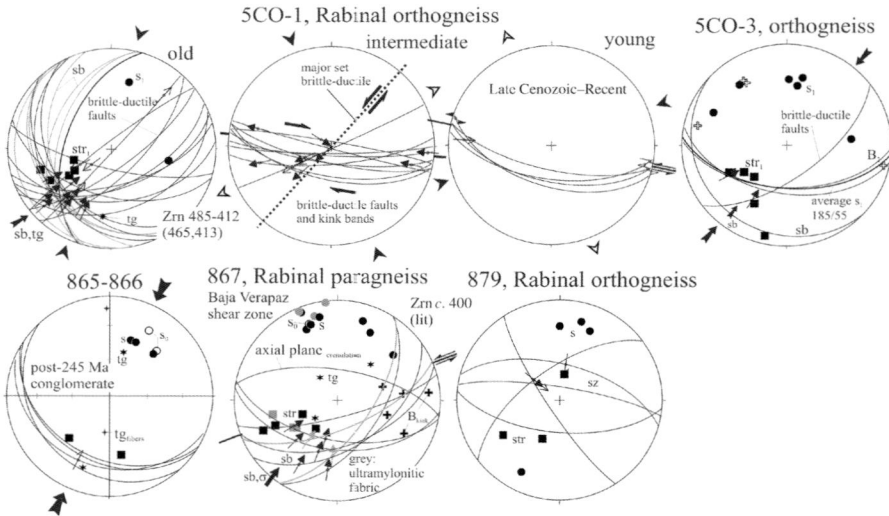

Fig. 13. Additional structural data from the Maya Block and the Early Cenozoic Subinal Formation in central Guatemala. Legend for structural symbols see Figs 9, 10 & 12.

dykes that crop out as partly pegmatitic orthogneiss and locally comprise 50% of the rock volume. Our new geochronology assigns them to the Triassic (see above). This rock association is sheared, folded, and thermally overprinted by syn- to post-tectonic feldspar + hornblende + white mica +

garnet ($\pm$ kyanite in quartz veins) growth. Local, large-scale fluid flow is indicated by up to 1 m thick quartz-segregation veins and ubiquitous albite growth. Locally, but increasingly abundant to the south, a second type of pegmatite is observed (e.g. stop 884). It is muscovite-rich, crosscuts

(b) Central Guatemala: northern foreland fold–thrust belt, Late Cretaceous–Early Cenozoic deformation

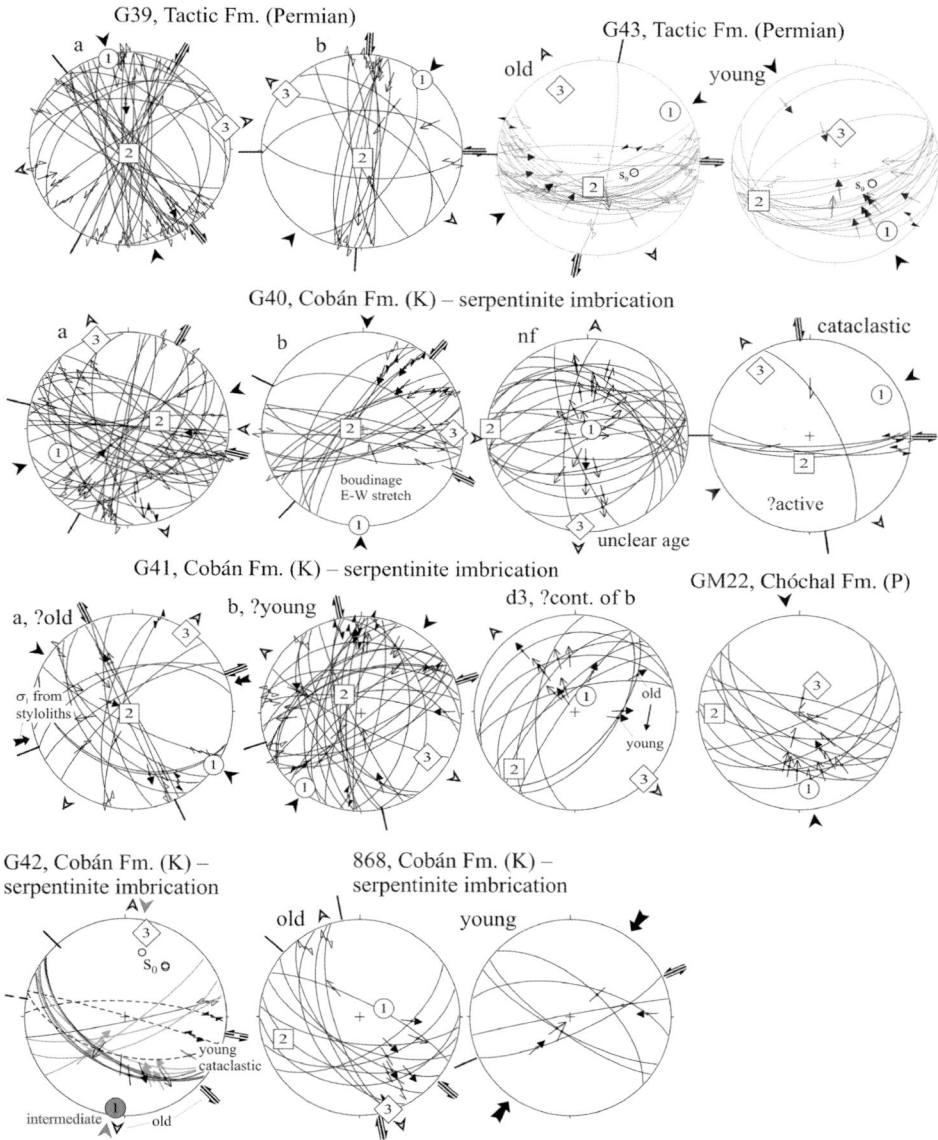

Fig. 13. (*Continued*).

the sills/dykes, and is associated with lower amphibolite- to upper greenschist-facies metamorphism. These pegmatites are undeformed where massive but sheared along quartz-rich layers. They are Late Cretaceous as indicated by *c*. 74 Ma metamorphic overprint in the host rocks

(U–Pb zircon, stop 5CO-6) and their 74–62 Ma white mica ages (see above).

Where preserved, Triassic deformation is distinct. Foliation is sub-vertical and folded tight-isoclinally to ptygmatically with axes parallel to the mineral stretching lineation (F_1 <a> type

**(c)   North El Tambor complex on Chuacús complex: ultramafic unit north of the
Motagua suture zone, Late Cretaceous and Late Cenozoic deformation**

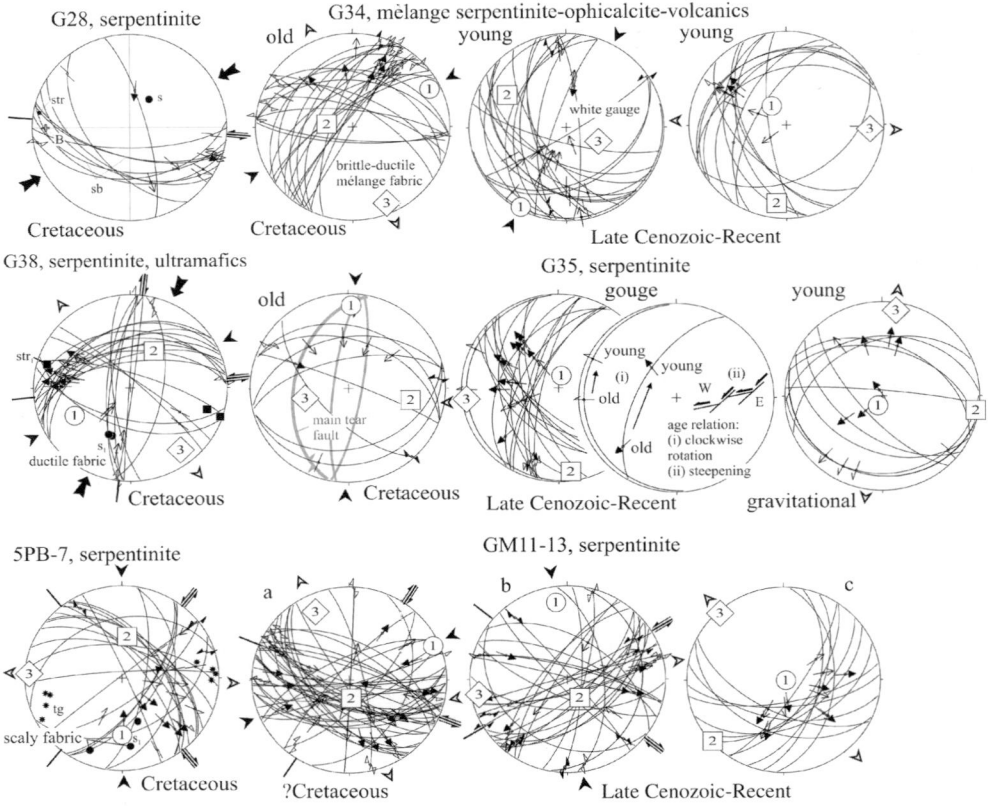

Early Tertiary Formation: Cenozoic and Late Cenozoic deformation

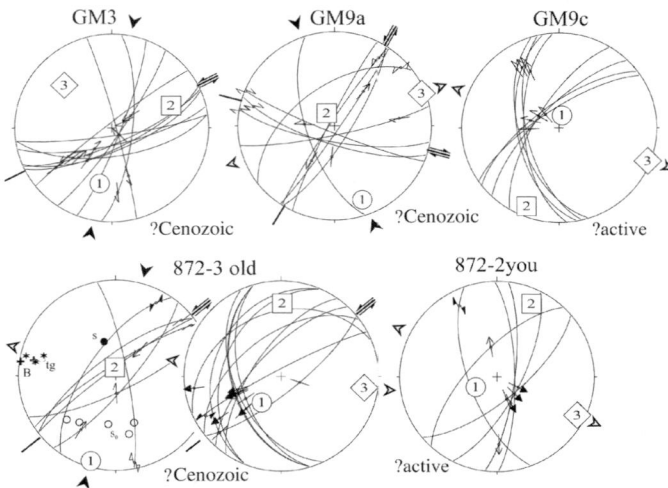

Fig. 13. (*Continued*).

folds) that trends NNW (Fig. 12, stops 5CO-5, 880–886) and thus is clearly different from the overall trend of the Late Cretaceous deformation (c. ENE). F_1 verge mostly (north)east and are locally sheath folds. At these locations, overprinting Cretaceous F_2 folds are steeply plunging, open, northeast-vergent, and associated with dextral and sinistral (conjugate) shear bands (Fig. 12). Garnet–K-feldspar gneiss (5CO-4), resembling Rabinal augen gneiss, has old garnet with an internal foliation that is 90° to the external (Cretaceous) foliation. Overall, Triassic deformation appears coaxial and the structures are annealed (widespread albite overgrowth, see above).

Structural geometry and kinematics of the regionally dominant Cretaceous deformation appear simple (Figs 11a-3, 12 & 13): the stretching lineation consistently trends c. ENE and parallels the ubiquitous B_2 axes; where s_1 is sub-vertical, it is transposed by sinistral shear zones/bands; where s_1 is sub-horizontal, shear zones/bands record sinistral transpressive, top-to-NE flow. Late faults have huge fibres and indicate north–south shortening (e.g. stop G29). This Cretaceous high-strain, NE–SW stretching is suggested by various features. Stop G31, for example, shows relatively idiomorphic, inclusion-free, pre-Cretaceous garnet that is fractured and connected with chlorite; a second generation of garnet and chlorite is co-genetic. Quartz shows transition from subgrain-rotation to grain-boundary-migration recrystallization (e.g. stop G31). At stops 880–886 late, mostly unfoliated/unfolded dykes/veins are exposed; some of these quartz + feldspar + muscovite + biotite pegmatites show high-strain boudinage. At stop 887, a several metres thick, weakly foliated pegmatite crosscuts the major (Triassic) foliation but deflects it sinistrally. F_3 refold F_2 openly. Where F_3 is not developed l-tectonites are common. Our interpretation is that folding is represented by prolate strain (e.g. stop 887). Our geochronology constrains cooling from c. 550 to 275 °C and thus deformation at 75–65 Ma (see above).

Stations 854 and 856 reveal Chuacús complex that is dominated by orthogneiss (Fig. 12). At 856, <0.5 m thick, D_1 mylonite–ultramylonite shear zones cut augen gneiss, biotite granite and aplite dykes. The host rock is weakly deformed at high-T preserving NW trending str_1. Discrete ultramylonite shear zones that differ in degree of localization – the younger are more strongly localized – overprint a first, east striking, dextral, high-T shear zone fabric. D_1 is syn-migmatitic, F_1 are isoclinal. D_2 imprints ENE-trending ultra-mylonitic–mylonitic, Ramsay–Graham-type shear zones with late chlorite–quartz veins, in which quartz is fibrous tracing the NW–SE str_2. Quartz and plagioclase in the

sinistral, often anastomosing shear zones (Fig. 11a-4) show subgrain-rotation recrystallization and reaction softening by transfer to sericite and epidote, respectively; locally feldspar is recrystallized (c. 500 °C). At 854 and 856, undeformed pegmatite and aplite dykes/dykelets cut migmatitic, sillimanite- and white mica-bearing ortho- and para-gneiss and are younger than the D_2 fabric (Fig. 11a-5). Pure quartz layers show prism <a> slip and subgrain-rotation with transitions to grain-boundary migration recrystallization. Common F_2 refold both s_1 and D_2 shear zones, imposing prolate strain. Radiometric age control on these deformations is weak (854F); migmatization and D_1 are pre-149 Ma, however, structural geometry, kinematics and associated metamorphism suggest a Late Cretaceous age for D_2.

We also studied the Chuacús complex along a section at its easternmost outcrop at the base of the Juan de Paz ophiolite unit (Fig. 12). The structural trend changes to ENE, from WNW in western and east in central Guatemala. Deformation is continuous from greenschist facies to brittle (boudinaged quartz rods) and heterogeneous. At station 5PB-9, deformation is high-strain, mostly coaxial with sinistral and dextral shear bands and <5 cm thick mylonite zones, shows coaxial, basal <a> slip in quartz, and open <a> folds. The whole rock–synkinematic white mica (Fig. 11a-6), Rb–Sr isochron at 76.5 ± 0.6 Ma precisely dates deformation. At station 5PB-10 (Fig. 12), ubiquitous vertical, sinistral shear bands, a single- to type-I cross-girdle quartz c-axis orientation and strong preferred orientation of a-axis and m-planes indicate dominant rhomb and prism <a> slip under somewhat higher deformation temperatures than at station 5PB-9 and dominant non-coaxial flow.

Metasomatized southern Chuacús complex rocks below the North El Tambor complex, central Guatemala

The immediate footwall of the North El Tambor serpentinites comprises imbricates that stack Chuacús-complex rocks, Chuacús rocks with low-grade sedimentary rocks (Jones Formation and San Lorenzo marble), and these rocks with serpentinite (Figs 12 & 13). Rocks that are imbricated are phyllite (retrograde biotite gneiss), epidotized biotite gneiss, (± garnet)–actinolite–clinozoisite–chlorite phyllite, marble and low-grade siltstone. These rocks show neocrystallization of calcite, quartz, Mg-chlorite and albite (e.g. stops 5PB-15, 860) depending on the surrounding lithology. The boundary between serpentinite and Chuacús-complex rocks is, where exposed, a shallow-dipping breccia zone with serpentinite and marble clasts (stop

G26, Fig. 13). Nearly all of the stations (Fig. 13) show relict, mostly sub-horizontal, high-grade foliation (s_1) that is folded (e.g. G26, G27, 5C-36). During D_2 sinistral ductile–brittle shear bands developed where s_1 was steeply dipping (e.g. 5PB-15, 860). Stretching is WSW–ENE and exemplified by s_1 boudinage. Quartz-filled tension gashes are ubiquitous (e.g. G26, 5PB-15) and testify to fluid flow. F_2 flexural-glide folds are north(east) vergent and show lengthening along B_2 (e.g. G26, 5C-36). Quartz deformed by subgrain-rotation recrystallization (e.g. G26) and basal <a> glide (5C-36). The LPO indicates dominantly coaxial flow. White mica at stop G26 is sericite and its Ar/Ar age dates deformation at ≥ 60 Ma. This corresponds to the Rb–Sr white mica–whole rock isochron at *c.* 63 Ma from a chlorite phyllite. Faulting started in

the ductile–brittle regime with a dominant northeast striking set. At station G27 pseudotachylite (Figs 11a-1 & 13, sample G7s) fills pull-apart tension fractures. These are again involved in continuous deformation, as east-trending quartz fibres formed as strain shadows. The pseudotachylite dates D_2 at *c.* 67 Ma.

North El Tambor complex on Chuacús complex, central Guatemala

We analysed imbrications of serpentinite, ophicalcite, and felsic volcanic rocks (stop G34) and serpentinites (stops G28, G35, G38, GM11-13, 5PB-7; Figs 13 & 14). All stations show a low-grade, ductile–brittle fabric with shear bands and faults that slipped sinistrally, coeval with

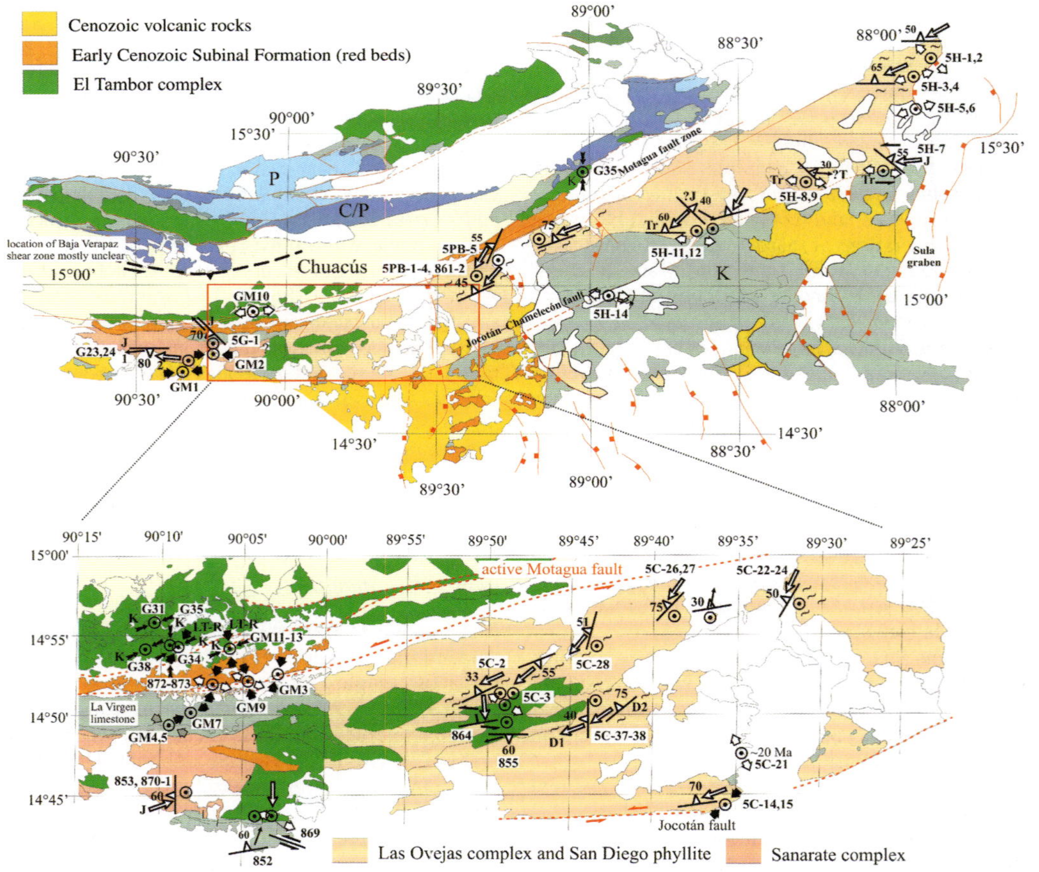

Fig. 14. Geological map (bottom) of the Motagua suture zone with structural data from the northern Chortís Block, the North and South El Tambor complexes, and the Sanarate complex; the outline of the latter is conjectural. In addition structural data of Cretaceous to Late Cenozoic sedimentary and magmatic rocks straddling the suture are shown. Main structural features are plotted both on map (this figure) and in stereonets (lower hemisphere, equal area, Fig. 15). See Figs 9 & 10 for legend. Adapted from Bonis *et al.* (1970).

shortening indicated by folding and oblique-slip faulting. White mica ages (c. 65 Ma) of greenschist-facies rocks (Chuacús complex, stops G26, G29, see above) in the immediate footwall of these El Tambor rocks suggest latest Cretaceous deformation. At several stations, tectonic mélange occurs along major north dipping detachments that contain foliated, moderately serpentinized ultramafic rocks (G34, G38) and pure serpentinite (G35; Figs 13 & 14) with brittle–ductile s–c fabrics. The ultramafic bodies are surrounded by an anastomosing, lenticular, scaly foliation that records overall coaxial deformation. Shortening is NNE–SSW and extension is WNW–ESE. Most stations contain a younger fabric with unclear age relations: sinistral shear and normal faulting both with NW–SE extension. Late-stage, flattening-type extension is possibly caused by topographic collapse.

Northern foreland fold–thrust belt, Guatemala

The 'Laramide' foreland fold–thrust belt north of the Polochic fault zone and the overlying North El Tambor complex allow comparisons of structural geometries and kinematics with those recorded in the Chuacús basement in the south. The stratigraphic range observed at our structural stations ranges from Permian to Upper Cretaceous, that is Permian Chochal and Tactic Formation limestone and shale, and Lower and Upper Cretaceous Cobán and Campur Formation dolomite and limestone (BGR 1971). In the following, we give an overview but emphasize that the stereoplots (Figs 10 & 13) contain a cornucopia of structural detail, for example, geometries, deformation evolution and changes in the palaeostress field. Deformation age is difficult to constrain directly. Structural and kinematic compatibility with the well-dated deformation in the Chuacús complex (see above) and that in the Sierra Madre Oriental–Yucatán fold belt (e.g. Gray et al. 2001) suggest the major deformation, sinistral transpression, is latest Cretaceous–Early Cenozoic.

In western Guatemala, stations G59–G61 within Cretaceous limestone and minor shale record folding with well developed ductile–brittle, NW–SE extension along the regional fold trend (Figs 10 & 11c-1). Deformation starts during layer-parallel shortening before buckling with vertical extension by tension gashes (tg_{old}) and continues with buckling and layer-internal ductile flow with foliation development, foliation boudinage and creation of a fold axis-parallel stretching lineation. Later stages develop conjugate but mostly sinistral faults and oblique stylolites that show strong volume-loss by pressure solution; dissolved material is partly deposited in veins that comprise up to 20 vol% of the rock. Fold axis-parallel extension is accentuated by a second set of tension gashes (tg_{young}) and conjugate faults that represent tilted horst-graben structures; the latter exactly constrain extension at 128° (inset stop G61–G62, Fig. 10). Abundant crinoids in the limestone record up to 50% shortening due to volume reduction that is mostly accommodated by bedding-parallel stylolites (Fig. 11c-1). Other stations, both in limestone and serpentinite, show less ductile flow but more pronounced faulting. Folding is mostly gliding in bedding; the slip direction is oblique to the fold axes (e.g. stop G55). Common to most stations is rotation of σ_1 from east–west to north–south in time and sinistral transpression.

In central Guatemala, Permian limestone and the contact zone between Cretaceous limestone and serpentinite provide a similar albeit less complete deformation history. Sinistral transpression and rotation of σ_1 imply increased shortening and less strike–slip over time (e.g. G43, Fig. 13). The weak structural record of serpentinite emplacement and the overall paucity of shallowly dipping faults are notable. Early fault sets indicate c. northeast–southwest shortening along conjugate strike–slip faults (e.g. G39). At station G40, c. north–south shortening along mostly strike–slip faults formed anastomosing fault zones with boudinage and east–west extension. Normal faults may be topographically induced, and late cataclastites likely record active sinistral wrenching along a strand of the Polochic fault zone. The normal faulting observed at station G41 could also be gravitationally induced or, alternatively, a transtensional continuation of sinistral wrenching.

Cenozoic volcanic rocks and Early Cenozoic Subinal-Formation red beds, western and central Guatemala

In western Guatemala, Cenozoic pyroclastic rocks, welded tuffs and, locally, sedimentary rocks cover the Rabinal complex and are exposed along the Polochic fault zone. Our data (Figs 9 & 10) highlight three strike–slip dominated events. From older to younger these are: (i) c. north–south shortening; (ii) c. NE–SW shortening, mostly along c. east trending sinistral strike–slip faults. This is by far the dominant set and traces the Polochic fault-zone deformation. The strata dip up to 80°, indicating strong local shortening. A similar stress field but dominated by normal faults also occurs in subhorizontal pyroclastic rocks that likely are part of the Cenozoic ignimbrite province of Central America (c. 15 Ma; e.g. Jordan et al. 2007); we thus attribute this major event to the currently active stress field. (iii) Normal faulting with widely dispersed slip directions. This event probably records topographic collapse.

Deformation of the Subinal-Formation red beds started with *c.* north–south shortening that tilted beds (open to tight folds) and produced a fracture cleavage (Figs 13 & 14). At stops 872–873, two sets of older faults have constant ENE–WSW extension (σ_3) but σ_1 and σ_2 permutated. The strike–slip faults have oblique lineations so that faulting probably started during late stages of folding. Faulting recorded at all stations has a clear sinistral-transpressive component. Late normal faulting is interpreted as prolongation of folding-related deformation with extension along the fold axes becoming dominant.

Sanarate complex, central Guatemala

The Sanarate complex, the westernmost basement outcrop south of the MSZ, comprises retrograde amphibolite, chlorite–epidote–hornblende schist, micaschist and chlorite phyllite with peak PT conditions at *c.* 550 °C and *c.* 0.65 GPa. The basement is in fault contact with Cenozoic andesite and red beds (stops G23, G24, 5G-1; Figs 14 & 15). Foliation and relict isoclinal folds constitute D_1. D_2 shows low-*T*, partly mylonitic quartz–white mica fabrics along an east trending, vertical foliation; shear sense is sinistral. Quartz is post-tectonically annealed. D_3 occurs both in the red beds/volcanic rocks and the basement. ENE–WSW shortening thrust the basement in a sinistral-oblique sense onto the red beds. Two generations of hornblende exist in the studied amphibolite. The dominant synkinematic younger generation (Fig. 11b-2) and also synkinematic white mica in the micaschist/phyllite date D_2 as Middle Jurassic (samples G3s, G5s, stations G23, G24, Fig. 5e). The reheating (quartz weakly annealed) is probably Cenozoic (low-*T* steps in G5s). Adjacent stations 853 and 870–871 comprise massive amphibolite and rare garnet amphibolite and leucogabbro close to the boundary with the South El Tambor mélange rocks. Their apparently lower-grade metamorphism distinguishes them from the Las Ovejas complex amphibolites. Asymmetric boudinage of hornblende–quartz veins, σ clasts and late biotite-rich shear bands associated with kink bands record top-to-NE flow with a slight dextral component along west dipping foliations. Tight to isoclinal flow folds are present in high-strain zones (Fig. 15). This deformation remains to be dated.

Las Ovejas complexes and San Diego phyllite, Guatemala and western Honduras

Stations 5C-2, 5C-28, 5C-26, 27 and 5C-22–24 are in basement lacking retrograde metamorphism of the Las Ovejas complex south of the central Motagua valley and north(east) of serpentinite and mélange rocks of the South El Tambor complex (Figs 14 & 15). This area comprises biotite granite with diorite xenoliths, gabbro-amphibolite, migmatitic biotite gneiss and garnet-bearing leucosome. The main deformation is of variable intensity, from weakly foliated to locally mylonitic–ultramylonitic but constant in orientation. Foliation has intermediate dips toward the NW, kinematically related shear bands trend NE and dip, in general, steeper than the foliation, and the stretching lineation is sub-horizontal NE–SW, parallel to axes of <a> type folds. These structures overprint relict earlier deformation fabrics and began syn-migmatic (Fig. 11b-3) with melt locally concentrated along shear bands; shear is overwhelmingly sinistral. Dykes intrude parallel to the foliation (5C-22, subvolcanic granite), mostly normal to the stretching lineation (5C-2, diorite), thus indicating syntectonic magmatism or crosscut irregularly (5C-28, pegmatite). Synkinematic hornblende and white mica date deformation as pre-to syn-35 Ma; a cluster of *c.* 10 Ma younger ages (mostly cooling through 300 °C) indicates later structural reactivation, as suggested by locally well-defined, synkinematic mineral fabrics (e.g. stop 5C-2, Fig. 11b-4).

Paragneiss and micaschist at stop 5C-3 bound South El Tambor serpentinite and show syntectonic, greenschist-facies retrograde metamorphism. Late sinistral faults juxtapose these rocks with the serpentinite. Stations 5C-37 and 38 at the southeastern edge of the same South El Tambor allochton comprise migmatitic biotite gneiss, amphibolite and leucogranite and their retrograde products. D_1 synamphibolite-facies deformation with local migmatites is overprinted by localized but pervasive greenschist-facies D_2 (Fig. 11b-5). The biotite gneiss and the amphibolite are in part extremely deformed and often l-tectonites; s_1 and sinistral-transtensional shear bands dip intermediate to NW and str_1 plunges shallowly. The tourmaline–garnet–white mica leucogranites postdate the migmatitic gneiss, intrude mostly as sills, and are locally deformed synintrusive with high strain. D_2 induced vertical sinistral shear zones, <a> folds, foliation boudinage and syn- and mostly antithetic faulting of leucogranite dykes and quartz segregation veins (Fig. 11b-5). The transition from shallowly dipping D_1 shear bands to vertical, ductile–brittle D_2 ones is abrupt (within *c.* 10 m). Hornblende (amphibolite) and white mica (leucogranite) date the initiation of deformation and syntectonic intrusion at 40–35 Ma. The likely kinematic evolution from ductile ($\leq$40 Ma) to ductile–brittle (*c.* 20 Ma, biotite) could be interpreted in this outcrop as long-lasting progressive deformation instead of structural reactivation (see

(a) Chortís block (Las Ovejas complex), eastern Guatemala: Cenozoic deformation

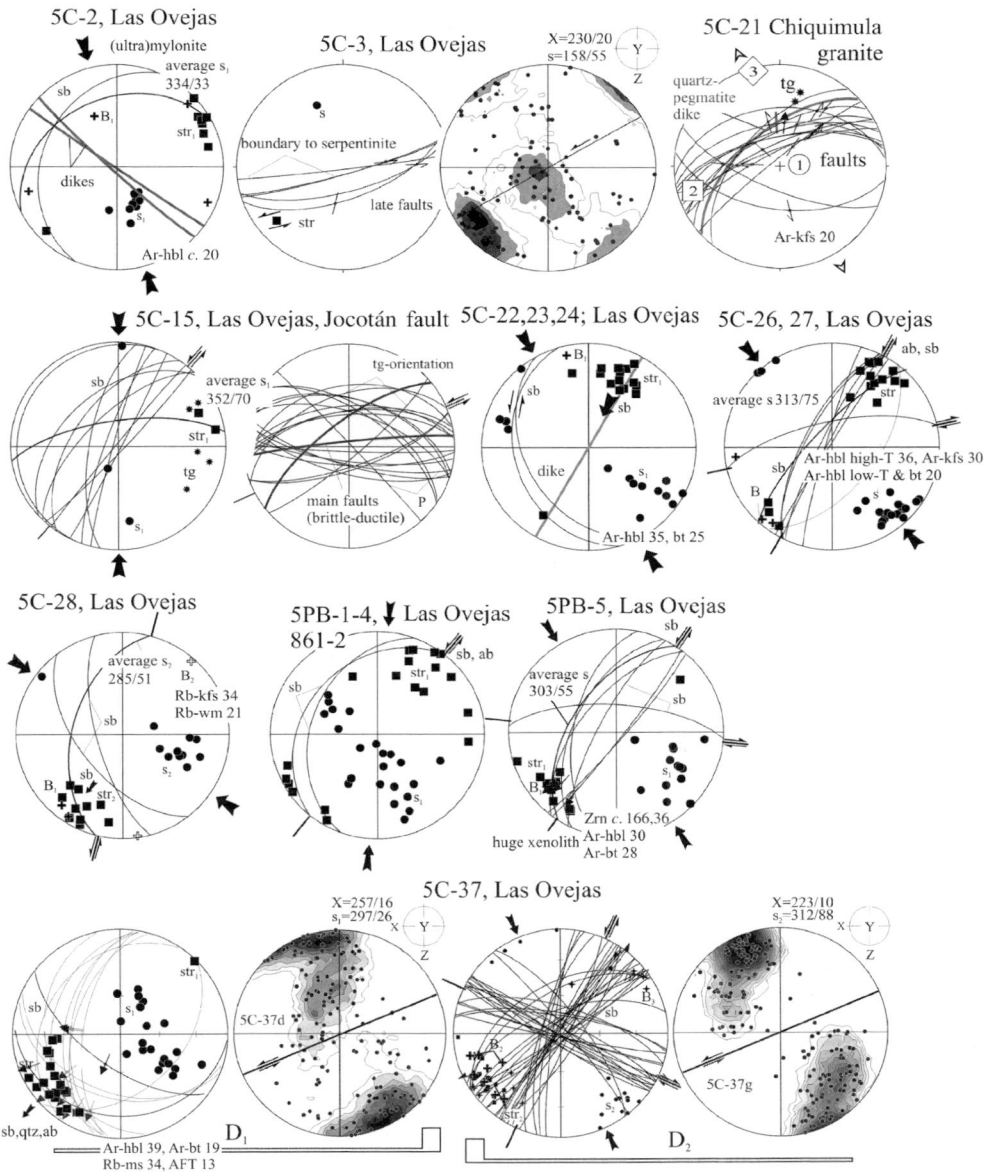

Fig. 15. Structural data from the Chortís Block and Cretaceous to Late Cenozoic sedimentary and magmatic rocks straddling the Motagua suture zone. Legend for structural symbols as in Figs 9, 10 & 12.

above). Stations 5PB-1–5 and 861, 862 expose unretrogressed Las Ovejas complex migmatitic biotite gneiss, intercalations of (garnet) amphibolite, biotite–white mica–garnet–tourmaline pegmatite, marble and calcsilicate rocks. Locally these rocks

are post-tectonically annealed due to their proximity to granitoids (Fig. 14). At location 5PB-5 massive migmatite is enclosed in granite. Garnet amphibolite contained in the migmatite yielded c. 650 °C at 0.7–0.8 GPa peak metamorphic conditions and preserves

(b) Chortís block (Las Ovejas complex), northern Honduras: Cenozoic deformation

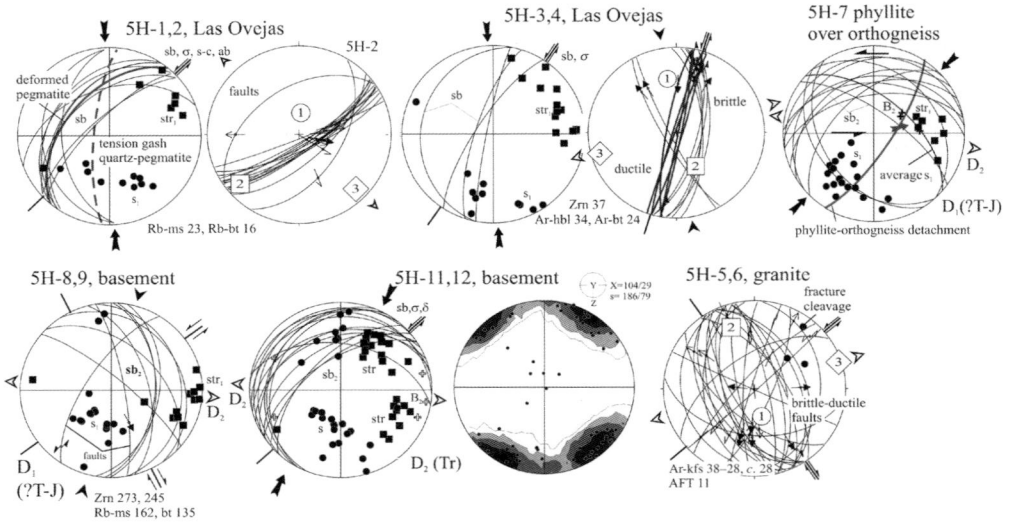

Cretaceous sedimentary rocks south of Motagua Suture Zone: Cenozoic and Late Cenozoic deformation

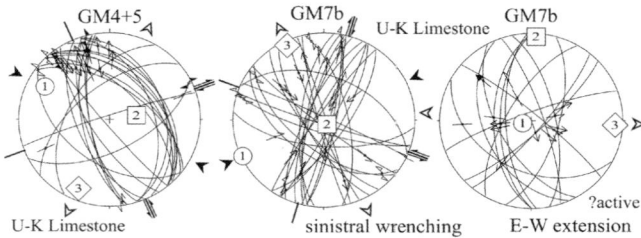

Cenozoic–Quaternary volcanic rocks south of Motagua Suture Zone: Late Cenozoic deformation

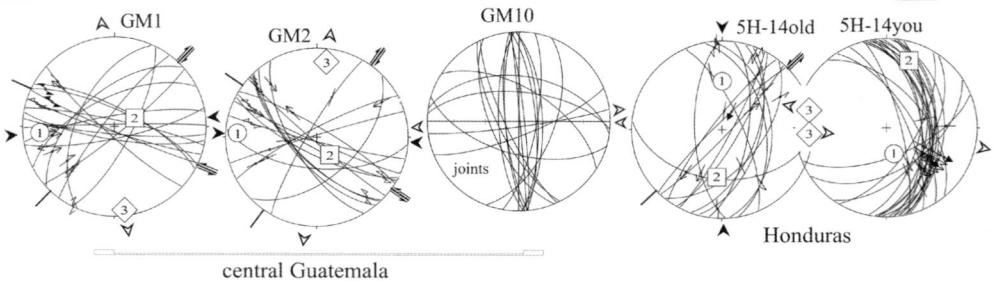

Fig. 15. (*Continued*).

pre-migmatitic garnet relicts. Deformation started prior to migmatization, is locally high-strain with migmatitic biotite–gneiss mylonite, and is sinistral-transpressive with local s–c mylonites. It continued with sinistral ductile–brittle faulting.

Migmatization was at *c.* 36 Ma (U–Pb zircon and Ar/Ar hornblende, see above). Station 5C-15 shows the Jocotán fault as a ductile–brittle sinistral fault zone. North–south ductile shortening with well-developed sinistral shear bands changed to

(c) South El Tambor complex: Cretaceous deformation

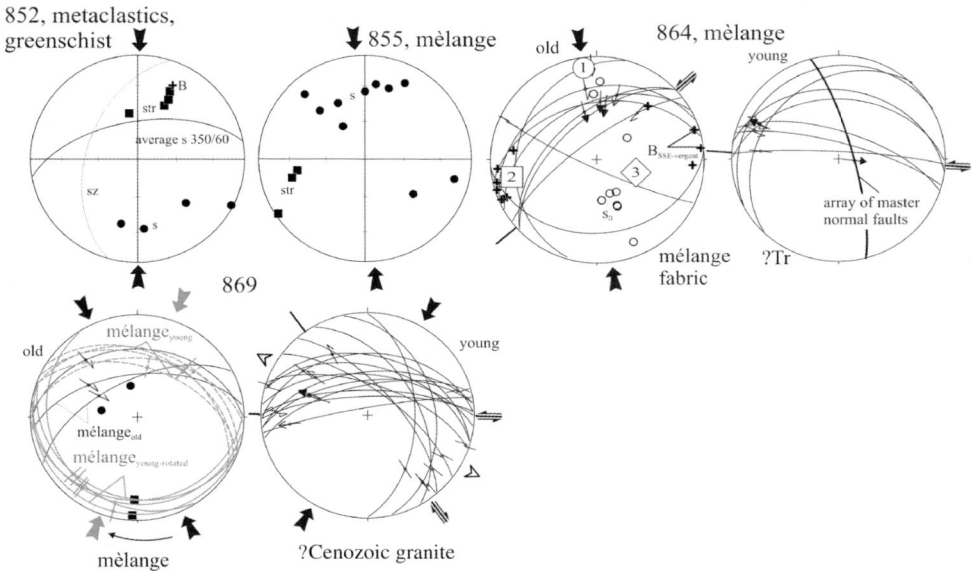

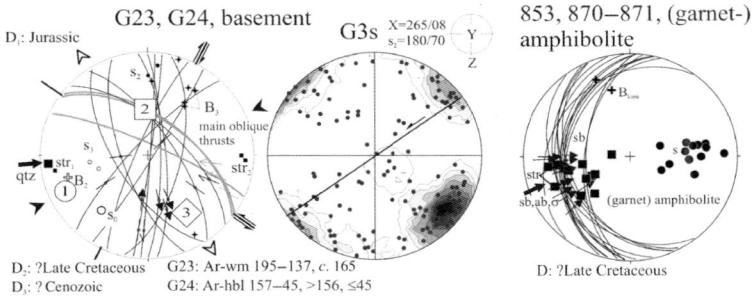

Fig. 15. (*Continued*).

NE–SW compression in the brittle field. Low-*T* plasticity is recorded by bulging to dominant subgrain-rotation recrystallization in quartz.

Las Ovejas complex of northern Honduras

Stations 5H-1–6 characterize deformation of the western footwall of the Sula graben in northern Honduras (Figs 14 & 15). The Las Ovejas complex rocks and their deformation are similar to those in Guatemala. Partly migmatitic biotite gneiss, biotite granite, diorite, micaschist and small white mica- and garnet- or hornblende-bearing pegmatite pods and sills deformed under

sinistral transpression that started during the mig-matitic stage. Deformation continued into the ductile–brittle field with sinistral-transtensive, SE-dipping shear zones/normal faults that we relate to opening of the Sula graben. The pegmatites either show strain concentration with well-developed mylonite and ultramylonite or are variably but mostly weakly deformed, forming irregular pods. In comparison with Guatemala, foliation is rotated, dipping moderately to steeply north with the stretching lineation NE–SW; shear bands dip NW. Our 40–36 Ma U–Pb zircon ages from mig-matitic biotite gneiss and the *c*. 38 Ma Ar/Ar horn-blende age from hornblende pegmatite constrain small-scale magmatism and deformation (see above). Synkinematic hornblende and biotite (Fig. 11b-6) and rapid cooling of K-feldspar in greenschist-facies, ductile–brittle shear zones related to the Sula graben suggest its initiation at *c.* 28 Ma. At stop 5H-7, a top-to-east shear/normal fault zone separates orthogneiss from phyllite and is likely related to the Sula graben. The orthogneiss shows a relict older fabric (D_1) with NE dipping foliation. Greenschist facies with up to mylonitic deformation is present both in phyllite and orthogneiss. Shear bands show sinistral transpression (NE–SW shortening) with transition to brittle–ductile fabrics. Quartz shows subgrain-rotation recrystallization. Stations 5H-8 and 9 comprise orthogneiss and migmatite with mostly Permo-Triassic U–Pb zircon ages and Jurassic–Cretaceous cooling ages. These rocks are distinctly different from the Las Ovejas complex rocks along the MSZ. Migmatization is pervasive and thus different from the *c.* 40 Ma event. D_1 is coeval with migmatization. Again, NE-dipping s_1 is overprinted by discrete ductile–brittle shear bands/zones that probably are Cenozoic. Stations 5H-11 and 12 comprise low-*T*, mylonitic orthogneiss (probably a rhyolitic protolith) interlayered with thin paragneiss and marble (?metavolcanic sequence). A relict high-*T* deformation, again with the characteristic NE-dipping foliation, is cut by greenschist-facies mylonite (locally chlorite out of biotite and sericite on mylonitic shear bands) with <3 cm chlorite-bearing ultramylonite layers. Overall deformation is top-to-SW sinistral transpression and probably related to the Jocotán fault zone.

Late Cretaceous and Cenozoic–Quaternary sedimentary and magmatic rocks south of the Motagua suture zone, central Guatemala

Conjugate strike–slip fault sets that indicate *c.* east–west shortening dominate Upper Cretaceous limestones (stops GM4, 5 and 7; Figs 14 & 15). The quality of the field exposures does not allow us to discriminate among subgroups representing a deformation sequence or to assess block rotations. The lack of a dominant east trending set of sinistral strike–slip faults, which we attribute to the active stress field along the Motagua fault, distinguishes these stations from the Cenozoic–Quaternary rocks (stops GM1, 2 and 10; Figs 14 & 15) but ties them to the deformation recorded in Subinal-Formation red beds. The latest increment of deformation in all these stations is east–west extension by normal faulting. Biotite granite ($\geq$37 Ma) at stations 5H-5 and 6 within the footwall block of the Sula graben shows brittle–ductile normal faults and fracture cleavage related to graben

formation dated by rapid cooling at *c.* 28 Ma (Ar/Ar K-feldspar). Conjugate strike–slip faulting with north–south compression was followed by normal faulting (σ_3 of both events trends east) in red beds and volcanic rocks at station 5H-14 within the Jocotán fault zone.

South El Tambor complex on Las Ovejas complex and San Diego phyllite, central Guatemala

Stations 852, 855, 864 and 869 comprise rare unfoliated blueschist, serpentinite, graphite-bearing metasiltstone to phyllite with quartzite layers, and greenschist of the southern South El Tambor mélange. The metasiltstones show ductile–brittle, heterogeneous fracture cleavage formed by pressure solution (diffusion creep), and quartz veins in metasandstone layers. D_1 shows open–tight, *c.* south vergent folds and thrust imbricates with associated fault-bend folds. We do not have age information on this mélange-forming event. These rocks are cut by sub-volcanic andesite dykes and sub-volcanic granite stocks, which we interpret as apophyses of the Chiquimula batholith. D_2 affects the mélange rocks and these dykes/granitoids by *c.* east-trending (transtensional) strike–slip and normal faults. Their low-*T* nature and the presence of *c.* 20 Ma sub-volcanic granite in the Chiquimula batholith (see above) suggest Cenozoic faulting.

Interpretation of new structural geology

Our work along the MSZ focused on the Cretaceous–Cenozoic deformation. Clearly, however, older events occurred. In the Rabinal and Chuacús complexes, Triassic deformation, studied superficially, shows high-*T* ductile flow and is distinct in structural geometry and kinematics: mostly (north)west striking foliation, NW-trending stretching directions, tight–isoclinal <a> folds. Shortening appears coaxial north(east)–south(west) with strong along-strike lengthening. Associated high-*T* metamorphism, migmatization, and magmatism are Late Triassic (238–215 Ma, major group at *c.* 220 Ma). Foliation trajectories of the Late Cretaceous, Maya-Block (for this discussion it includes the Rabinal and Chuacús complexes and their cover) deformation change from WNW in western, to west in central, and SW in eastern Guatemala. Using the (west-)northwest strike of the Late Cretaceous–Paleocene Mexican thrust belt as a reference frame (e.g. Nieto-Samaniego *et al.* 2006), this implies large-scale sinistral shear. The southern Maya Block in Guatemala apparently comprises a ductile–brittle nappe stack, whose detailed geometry remains to be established. The North El Tambor allochton in the hanging wall, itself an accretionary stack of rocks of variable tectono-stratigraphic origin and subduction depth, is underlain by an up to several hundred metres-thick carapace of imbricated rocks, containing Chuacús basement and cover, altered by fluid infiltration and metasomatism. The central Chuacús complex may either constitute a Late Cretaceous antiformal stack or, our preferred model, an antiform above a blind thrust, exposing pre-Cretaceous metamorphic and magmatic products in its core. The Baja Verapaz shear zone is the sole thrust to this stack. The northern El Tambor sheets (Baja Verapaz, Sierra des Santa Cruz) are cut by out-of-sequence thrusts that are part of the foreland fold–thrust belt. The relatively high-P (0.7–0.8 GPa) and relatively low-T ($\geq 450\ ^\circ$C) metamorphism in the Chuacús complex suggests subduction of oceanic crust transitioning to continental crust (the Chuacús geothermal gradient is $\geq 18\ ^\circ$C/km from our data) and steep continental subduction; this is supported by the nearly coeval blueschist ($c.$ 76 Ma, North El Tambor complex) and epidote–amphibolite- to amphibolite-facies (≤ 75 Ma, Chuacús complex) metamorphism in the accretion stack and the sinistral–transpressive deformation. Penetration of deformation and overburden (PT conditions) appear to increase from western to eastern Guatemala. Although in detail variable, meso-scale Late Cretaceous deformation is simple: north–south shortening by reverse shear zones and up to two generations of folds are accompanied by sinistral strike–slip shearing and major east–west lengthening. Tangential stretching is dominant in most cases, as deduced from the mostly shallowly plunging stretching lineation. In detail, overall sinistral transpression is typically partitioned between oblique stretching within the foliation and strike–slip along steeper dipping shear zones/bands transecting the foliation. Flow has a dominant non-coaxial component, most clearly expressed by quartz LPOs, developed during dynamic, subgrain-rotation dominated recrystallization. There is a clear kinematic continuity from ductile to brittle deformation. Although dominated by diffusion creep, deformation is first-order identical within the northern foreland fold–thrust belt, the sole thrust carapace below the North El Tambor complex sheet, and within it. Emplacement of the El Tambor allochton, poorly constrained by our data, appears to change from tangential to frontal over time (strike–slip before thrusting).

Structural trends in the northern foreland fold–thrust belt and the Rabinal complex of western Guatemala are NW and oblique to the Polochic fault zone, testifying to the younger age of this fault. Polochic mylonitic deformation appears to

be younger than $c.$ 15 Ma (apparent post-tectonic pluton), although our only Cenozoic age ($c.$ 37 Ma) from western Guatemala (G27s, stop G57), from a discrete low-T mylonite zone in gneiss close to the Polochic fault zone, may hint at a Late Eocene slip history. Ductile to brittle slip along the Polochic is dated as Late Neogene–Recent in western Guatemala (TFT and AFT ages) and at $c.$ 8 Ma (Ar/Ar chronology) along the, probably kinematically related, Tonala shear zone in southeastern Mexico.

Cenozoic volcanic rocks record a phase of pronounced north–south shortening along the western Polochic fault zone. Its active deformation is by a kinematically related combination of strike–slip and normal faulting. We interpret the Subinal-Formation red beds as intra-continental molasse related to late-stage, erosional unroofing of mostly the Chuacús complex. Its deformation was developed during north–south shortening by folding and fold-axis parallel extension, accompanied by sinistral shear. The latest stage indicates north trending graben formation.

Our study indicates that the Las Ovejas complex of the northern Chortís Block is less a lithologically defined unit than distinguished from the other basement units by its Cenozoic metamorphism, magmatism and deformation. It forms a westward narrowing belt of overall sinistral transtension along $c.$ NW dipping foliations. Deformation penetration apparently increases toward north, the MSZ, and seems to be more localized and to die out southward and westward. The belt of Cenozoic shear thus appears to trend oblique to the Motagua fault. Deformation interacts with amphibolite-facies metamorphism, widespread migmatization and local magmatism, demonstrated most clearly by small syn- to post-tectonic pegmatites. The close spatial and temporal interaction between high-strain deformation and Cenozoic metamorphism, migmatization and magmatism is confined to the northern Chortís Block. It appears that arc magmatism/metamorphism provided the heat input to localize deformation. Deformation started pre-migmatitic and occurred in a kinematic continuity from ductile to brittle. Batholiths are post-tectonic. Cenozoic sinistral-transtensional wrenching appears, like the Cretaceous deformation of the Chuacús complex, partitioned between oblique stretching along a shallower dipping foliation and strike–slip along steeper dipping shear zones/bands. Deformation began prior to 40 Ma and we suggest that there were two major phases, ≥ 40–35 and 25–18 Ma. This reactivation interpretation still must be tested. The Sula graben started in temporal and kinematic continuity with wrenching along the Las Ovejas complex at $c.$ 28 Ma. The topographically well developed active grabens, however, are a neotectonic feature. Also the Jocotán–Chamelecón fault zone was part of this deformation system, albeit mostly ductile–brittle. The Cenozoic event veils pre-Cenozoic tectono-metamorphic–magmatic events, with distinct structural orientation and kinematics. The latest preserved is Jurassic and has a NW structural trend. The South El Tambor complex was emplaced by south vergent thrusting and associated folding. It is involved into the Cenozoic sinistral wrench deformation.

The Sanarate complex with its distinct lithology, metamorphism and age of metamorphism is affected by the Cenozoic deformation but has a characteristic west dipping foliation and top-to-east flow in its dominant amphibolite lithology. This deformation is Jurassic or younger. Late Cenozoic deformation, preserved in Late Cretaceous and Cenozoic–Quaternary rocks on the Chortís Block, is by distributed sinistral strike–slip. The latest event we recognized in the Chortís Block and the Subinal Formation is east–west extension by normal faulting. The entire MSZ area shows gravitationally induced faulting due to the large topographic relief.

Discussion

Here we return to the principal questions this paper aims to address: (i) which crustal sections of the MSZ accommodated the large-scale sinistral offset along the northern Caribbean Plate boundary, when did it occur, and in which kinematic framework, and (ii) what Palaeozoic–Cenozoic geological correlations exist between the Maya Block, the Chortís Block, the tectono-stratigraphic complexes ('terranes') of southern Mexico, and the arc/subduction complexes of the northern Caribbean. First we focus on geological correlations proceeding from Palaeozoic to Recent, and then we refine Caribbean tectonic models for the Cretaceous and Cenozoic.

Taconic–Acadian (Early–Middle Palaeozoic) orogeny

The Chuacús complex contains a paragneiss-dominated sequence, in which metabasalts prevail among the magmatic rocks, and quartzite and calcsilicate/marble are minor constituents (Fig. 2d). In contrast, the hanging Rabinal complex is orthogneiss-dominated with minor metaclastic and rare metavolcanic rocks (Fig. 2c). The Chuacús–Rabinal boundary is a Cretaceous shear zone but this may represent reactivation of an older structure (e.g. Ortega-Gutiérrez *et al.* 2007). Both complexes have inherited Proterozoic crust (U–Pb zircon).

Ortega-Obregón *et al.* (2008) suggested intrusion of the Rabinal granitoids between 462 and 453 Ma. Combining all available U–Pb zircon ages from the Maya Mountains (Belize), the Altos Cuchumatanes (western Guatemala), the Chiapas Massif (southeastern Mexico) and the Rabinal complex (see above; Figs 1c & 16a), we suggest that the Maya Block experienced prolonged Ordovician–Early Devonian magmatism. We consider the *c.* 396 Ma age of a Rabinal orthogneiss reliable, as 404–391 Ma magmatites (both granite and rhyolite) occur north of the Rabinal complex. Our Rabinal augen–gneiss sample is dominated by 485–400 Ma zircons and shows two age groups at 465 and 413 Ma. Similar ages occur in the Maya Mountains (*c.* 418 Ma; Steiner & Walker 1996) and in other Rabinal granitoids (*c.* 462, *c.* 417 Ma; Ortega-Obregón *et al.* 2008). The Chiapas Massif and the Rabinal area of central Guatemala may also host *c.* 482 Ma granitoids (Weber *et al.* 2008; Ortega-Obregón *et al.* 2008; this study).

The Chuacús-complex rocks apparently recycle these Early Palaeozoic magmatites in their sediments and granitoids: Solari *et al.* (2009) reported, among others, 480–402 Ma (main age group at 436 Ma) detrital zircons in a Chuacús-complex paragneiss, and we found *c.* 403 Ma, probably metamorphic zircon in Triassic orthogneiss. The Chuacús complex shows high-*P*, perhaps ultrahigh-*P* (Ortega-Gutiérrez *et al.* 2004) metamorphism that is, at best, constrained in age to between *c.* 440 Ma, the youngest significant zircon-age group of the Chuacús metasediments and possibly 302 Ma, the age of post-high-*P* migmatite leucosome, and conservatively to between *c.* 440 (see above) and 238–218 Ma, the age of Triassic orthogneiss (Figs 2d & 16a; see above). The Rabinal complex apparently lacks high-*P* metamorphism but this has to be confirmed by quantitative petrology. Lower Carboniferous–Permian Santa Rosa Group sediments cover the Rabinal complex unconformably (Fig. 2c).

Recent studies within and along the southern margin of the Maya Block suggest a subdivision of its lithostratigraphy into pre-Early Ordovician clastic rocks (e.g. Jocote complex, intruded by *c.* 482 Ma granite in the southern Chiapas Massif, Weber *et al.* 2008; post-*c.* 520 Ma Baldy unit of the Maya mountains, Martens *et al.* 2010), and a younger, again mostly clastic sequence that has rhyolite intercalations (*c.* 404 Ma; Bladen Formation of the Maya mountains, Martens *et al.* 2010) and is intruded by granitoids as young as *c.* 404 Ma (Maya mountains, Steiner & Walker 1996). Ortega-Obregón *et al.* (2008) suggested the presence of pre-Middle Ordovician basement in the Rabinal area of central Guatemala (San

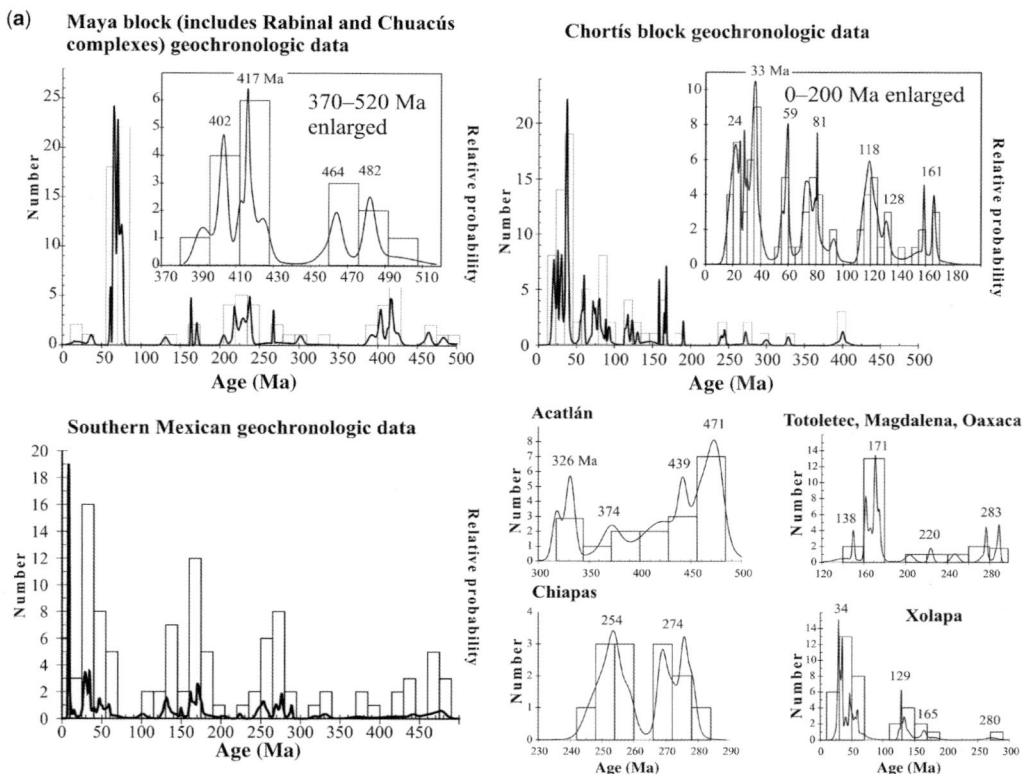

Fig. 16. Tectonic–palaeogeographic evolution models for the North American–Caribbean Plate boundary and Palaeozoic–Cenozoic correlations between southern Mexico and the Chortís Block. (**a**) Overview of our new and published geochronology from the Maya Block, the Chortís Block, and southern Mexico. Data are referenced in the captions of Fig. 1 and in the text.

Gabriel unit intruded by the Rabinal granitoids at *c.* 462 Ma); Solari *et al.* (2009) extended the San Gabriel unit to western Guatemala. Our inference that intrusions in the Rabinal area span 482–396 Ma suggests similar prolonged volcanoclastic deposition and intrusion for the Rabinal complex and firmly ties this complex to the Maya Block. All areas of the Maya Block exposing Ordovician–Early Devonian granitoids are covered by the Santa Rosa Group; characteristically these sediments contain Early Palaeozoic detrital zircons. Tectono-stratigraphically, the Rabinal complex probably represents continental crust overprinted by an arc to back-arc setting (see also geochemical data reviewed in Ortega-Obregón *et al.* 2008). Detrital zircon geochronology ties the Maya Block to the South American side of the Iapetus–Rheic oceans (e.g. Weber *et al.* 2008 and references therein). The Chuacús complex may represent a marginal basin that received detritus from the Ordovician–Silurian arc, and Pan-African (*c.* 0.5–0.7 Ga), and Grenvillian (*c.* 0.9–1.2 Ga)

hinterland and was subducted into the mantle likely in the Devonian.

Next, we discuss possible correlations of the Rabinal and Chuacús complexes with the Acatlán complex of the Mixteca terrane of southern Mexico (Fig. 1a & c). The Acatlán complex comprises a nappe stack containing Middle Proterozoic to Palaeozoic units imbricated during the Palaeozoic. To date, seven nappes containing distinctive stratigraphy have been discriminated (Fig. 2i; Yañez *et al.* 1991; Ortega-Gutiérrez *et al.* 1999; Talavera-Mendoza *et al.* 2005; Keppie *et al.* 2006; Vega-Granillo *et al.* 2007). The stack shares a *c.* 325 Ma regional metamorphism. The hanging Esperanza suite is dominated by Early Silurian (442 ± 5 to 440 ± 14 Ma) mega-crystic, K-feldspar augen gneiss and contains metasediments and metabasalts transformed to eclogite. The entire suite records peak *PT* at 730–830 °C and 1.5–1.7 GPa attained at *c.* 430–420 Ma. Partial overprinting occurred at 640–690 °C and 1.0–1.4 GPa prior to *c.* 375 Ma. The Tecolapa suite consists of

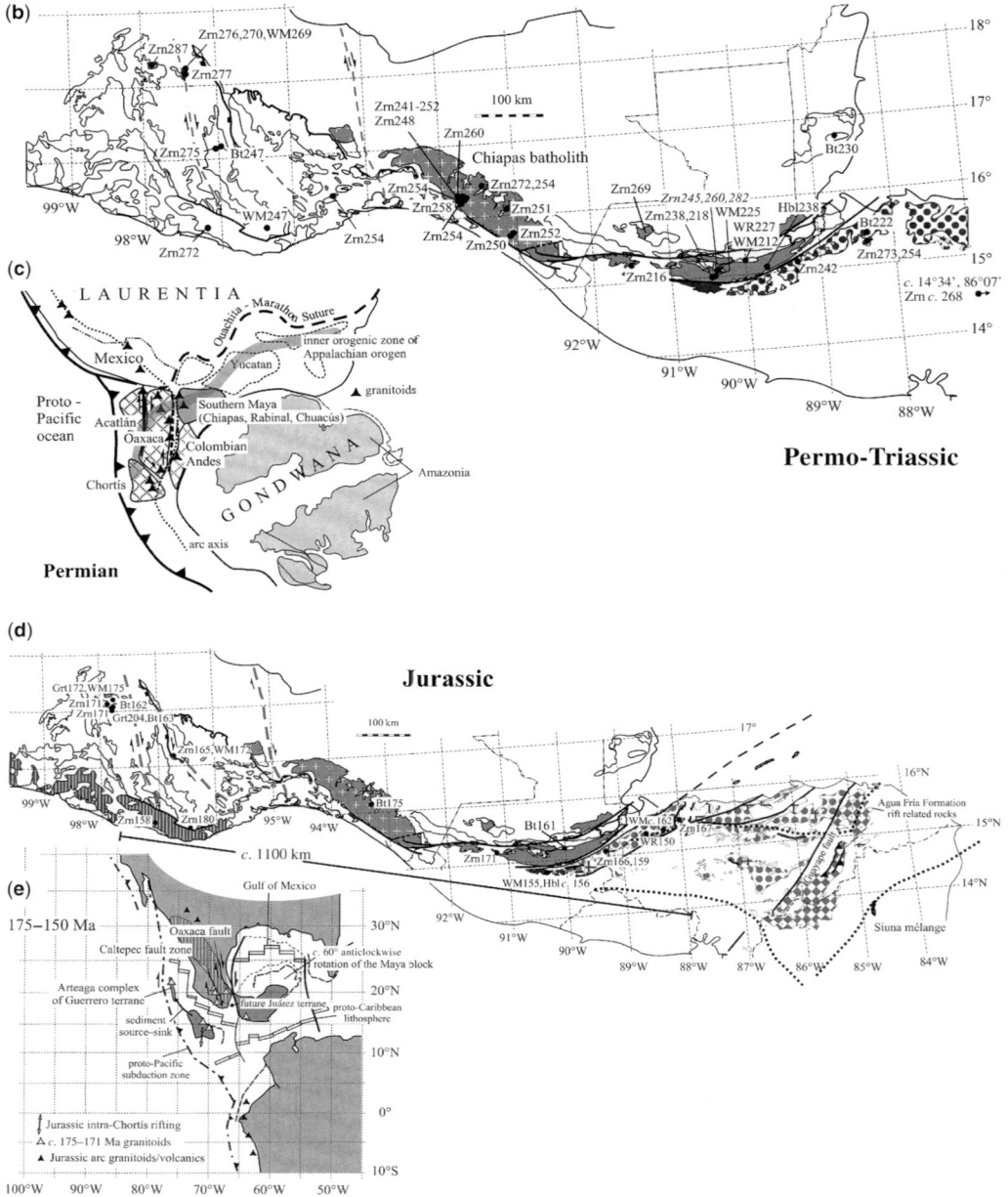

Fig. 16. (*Continued*) (**b**) Permo-Triassic ages of southern Mexico and northern Central America. Ages in italics refer to detrital zircons. (**c**) Palaeogeographic model for the Permian modified from Weber *et al.* (2006) and Vega-Granillo *et al.* (2007) emphasizing arc magmatism along the margins of Laurentia and Gondwana and a common position of the Acatlán complex and the Maya Block in the inner orogenic zone of the Acadian orogen. (**d**) Jurassic ages along the northern Caribbean Plate boundary. Subdivision of the Chortís Block follows Rogers *et al.* (2007c). Mineral abbreviations as in Fig. 1. (**e**) Middle to Late Jurassic tectonics of the Caribbean emphasizing rifting of the Chortís Block from southern Mexico and its connection with the Arteaga complex of the Zihuatanejo terrane of southwestern Mexico.

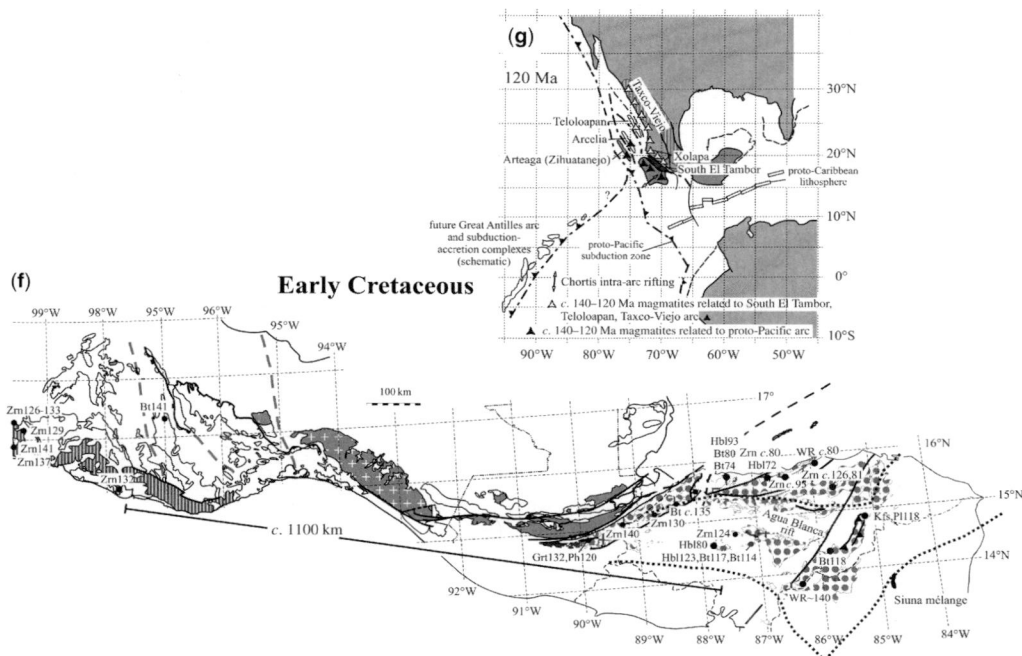

Fig. 16. (*Continued*) (**f**) Early Cretaceous ages of southern Mexico and northern Central America. (**g**) Late Early Cretaceous tectonics of the Caribbean underlining emplacement of the South El Tambor complex rocks onto the Chortís Block and the existence of continuous arcs off southwestern and southern Mexico: Chortís–Xolapa–Taxco–Viejo; South El Tambor complex–Teloloapan; Arteaga (Zihuatanejo)–Sanarate (Chortís).

Middle Proterozoic mega-crystic granitoids and tonalitic gneisses and Early Ordovician leucogranites (478–471 Ma). The greenschist-facies El Rodeo suite contains Cambrian–Early Ordovician clastics and basalts of continental-rift affinity and is intruded by Ordovician (476–471 Ma), Devonian (371 ± 34 Ma), and Early Permian (287 ± 2 Ma) granitoids. The Xayacatlán suite comprises eclogite-facies, mafic and pelitic rocks and is intruded by Ordovician leucogranites (478±5 to 461 ± 9 Ma), post-dating high-P metamorphism. Peak PT conditions were at 490–610 °C and 1.2–1.3 GPa, reached during Early Ordovician (*c.* 490–477 Ma). The suite re-equilibrated in greenschist facies during the Mississippian (*c.* 332–318 Ma). The Ixcamilpa suite consists of greenschists and blueschists (200–580 °C, 0.65–0.9 GPa) intercalated with metaclastic rocks that are younger than *c.* 470 Ma. The high-P event is loosely constrained between 450 and 330 Ma; the absence of Silurian–Devonian detrital zircons and the existence of a major magmatic event of Silurian age within several units of the Acatlán complex, which may be genetically linked to subduction, suggest that high-P metamorphism was Late Ordovician–Early Devonian. The greenschist-facies Cosoltepec suite

is the most widespread unit of the Acatlán complex and consists of monotonous, greenschist-facies clastic rocks that are younger than *c.* 410 Ma and basalts. This passive margin sequence was imbricated with the other nappes before the Early Carboniferous (Figs 2i & 16a, Acatlán).

We see several gross similarities between the Acatlán-complex suites and the Rabinal and Chuacús complexes (Fig. 2c, d & i): (1) the upper nappe stack, from the Xayacatlán suite at the bottom to Esperanza suite at the top, contains exclusively Proterozoic detrital zircons, the lower stack also Ordovician and younger zircons (Magdalena, ≥275 Ma; Cosoltepec, ≥410 Ma; Ixcamilpa, ≥477 Ma). The Maya-Block rocks (Baldy unit, Maya Mountains; Jocote complex, southeastern Chiapas), including the Rabinal complex (San Gabriel unit, Guatemala), contain exclusively Proterozoic detrital zircons as well, the Chuacús complex has a major component of Early Ordovician–Early Devonian detrital zircons (480–402 Ma; *c.* 436 Ma) besides Proterozoic ages. (2) Although elucidating the Proterozoic history of northern Central America is beyond the topic of this paper, we note that the major Proterozoic detrital zircon age clusters in both the upper stack suites

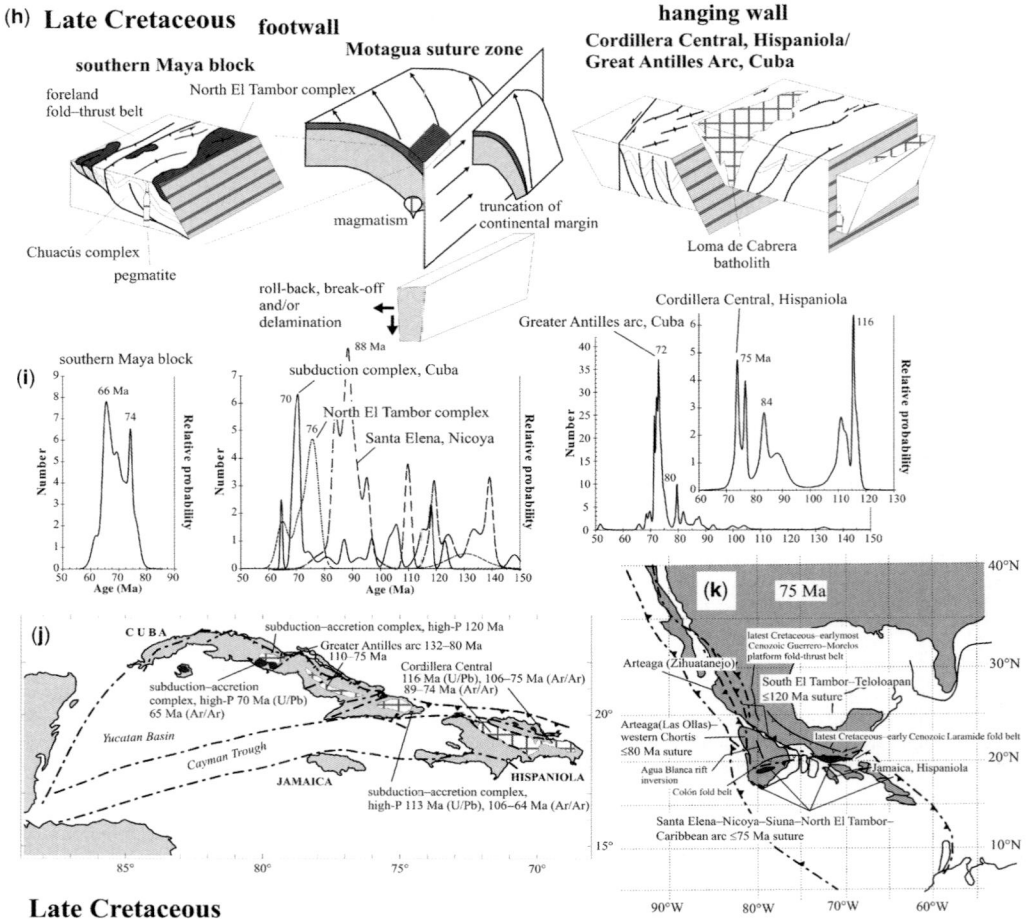

Fig. 16. (*Continued*) (**h**) Block diagrams illustrating north–south swath across the Motagua suture zone at *c*. 70 Ma. Left: the southern Maya Block is imbricated by sinistral transpression. Centre: truncation of the continental margin by sinistral transpression. Roll-back and break-off of the North El Tambor complex slab may have caused magmatism in the southern Maya Block. Right: hanging wall arc magmatism terminated contemporaneously to slightly after (reflecting transfer to the Bahamas continental margin) their collision with the southern Maya and southern Chortís Blocks (see text for further discussion). Hanging wall arc diagram modified from the Cordillera Central, Hispaniola, situation (Escuder-Viruete *et al.* 2006*a*). (**i**) Geochronology of the footwall and hanging wall of the Motagua suture zone. Left: ages of imbrication and associated metamorphism in the southern Maya Block (this paper). Centre: ages in the North El Tambor complex (see references in the text), and the subduction–accretion complexes in Cuba (Garcia-Casco *et al.* 2002, 2007; Schneider *et al.* 2004; Krebs *et al.* 2007). Right: ages from the Greater Antilles arc, Cuba, and Cordillera Central arc, Hispaniola (Grafe *et al.* 2001; Hall *et al.* 2004; Stanek *et al.* 2005; Kesler *et al.* 2005; Rojas-Agramonte *et al.* 2006; Escuder Viruete *et al.* 2006*b*). (**j**) Subduction–accretion and arc complexes of Cuba and Hispaniola and major age groups emphasizing the evolution of the hanging wall of the Motagua suture zone. (**k**) Late Cretaceous tectonics of the northern Caribbean highlighting the two Late Cretaceous suture zones on the Chortís Block.

of the Acatlán complex and the Maya Block/ Rabinal complex are Grenvillian (*c*. 0.95– 1.25 Ga) and South American (*c*. 1.3–1.8 Ga; Rondônia–San Ignacio and Río Negro–Juruena provinces; see Weber *et al.* 2008 for locations; Fig. 2c & i). (3) All upper stack suites (Xayacatlán to Esperanza) have massive Early Ordovician to

Middle Devonian intrusions, whereas the lower stack suites (Cosoltepec to Ixcamilpa) received the erosional detritus of these granitoids. Similarly, the Maya Block/Rabinal complex carries Early Palaeozoic magmatism, whereas the Chuacús complex collected its detritus. Items (1)–(3) link the Maya Block/Rabinal complex to the Xayacatlán

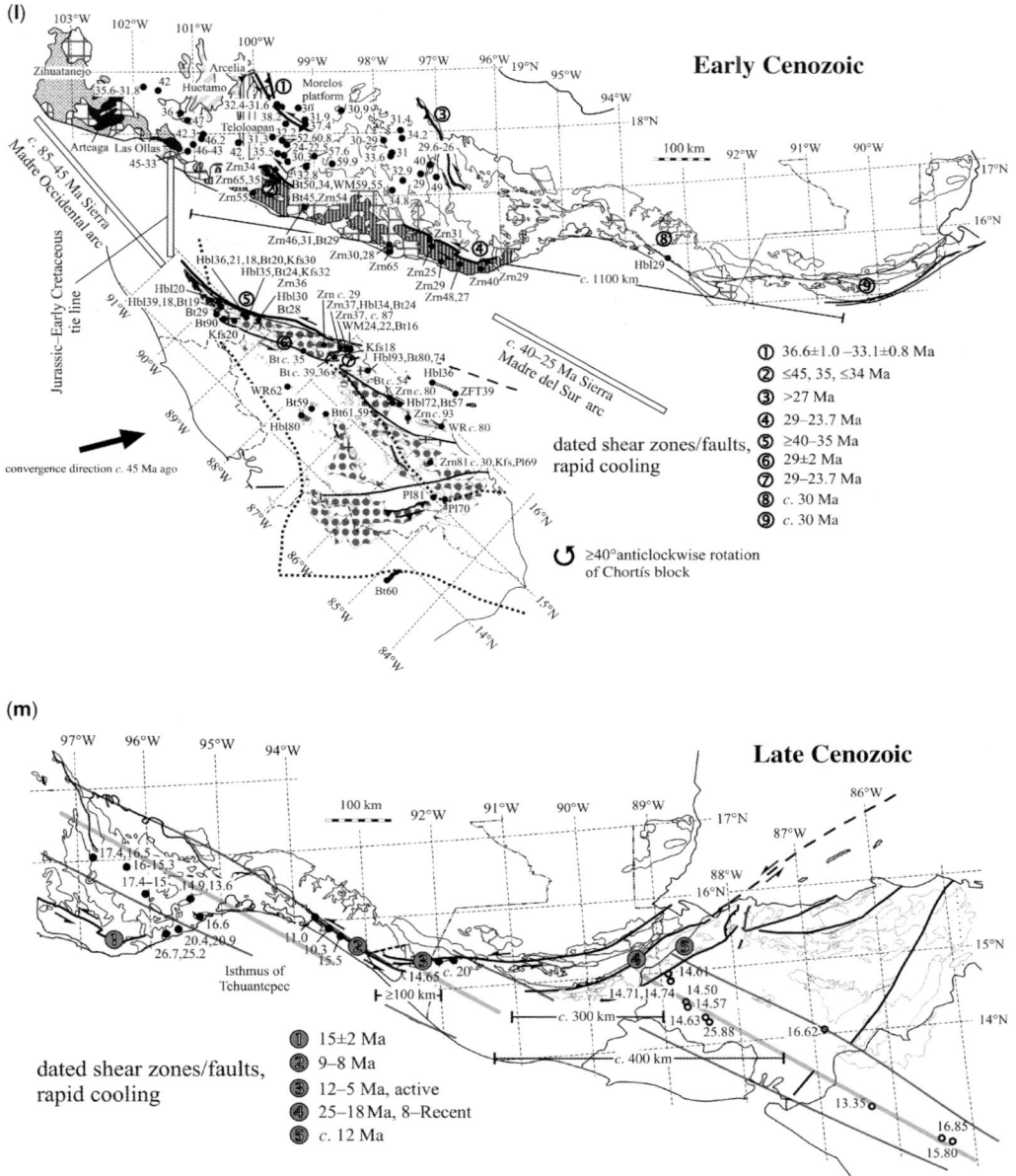

Fig. 16. (*Continued*) (**l**) Late Cretaceous and Early Cenozoic ages along the northern Caribbean Plate boundary bringing the Chortís Block to its pre-*c.* 40 Ma position off southern Mexico. Geochronological data compiled from Herrmann *et al.* (1994), Morán-Zenteno *et al.* (1999), Ducea *et al.* (2004*a*), Cerca *et al.* (2007) and references therein. Deformation zones (numbers 1–9) and their ages are discussed in the text. Convergence direction at *c.* 45 Ma is from Pindell *et al.* (1988). (**m**) Middle Miocene ages along the northern Caribbean Plate boundary. Geochronology is from this paper, Morán-Zenteno *et al.* (1999), and Jordan *et al.* (2007) and references therein. Deformation zones (numbers 1–5) and their ages are discussed in the text.

to Esperanza stack, and the Chuacús complex to the Cosoltepec and Ixcamilpa stack of the Acatlán complex. In both areas, southern Mexico and Guatemala, Permian, Triassic and Jurassic magmatism stitches the various units together (Figs 1c, 2c, d & i). In contrast, eclogite-facies metamorphism is confined to the upper stack in the Acatlán complex, and to the Chuacús complex in Guatemala; the lower stack in Mexico shows blueschist-facies metamorphism only. As outlined above, upper stack high-*P* metamorphism is Ordovician (Xayacatlán suite) and Silurian (Esperanza suite), and the lower stack (blueschist-facies) high-*P* metamorphism is 450–330 Ma (best guess of Vega-Granillo *et al.* 2007 is 458–420 Ma). Our loose constrain for the Chuacús complex eclogite-facies event is 440–302 Ma (see above). These age constrains tie the Chuacús complex high-*P* event either to the Esperanza or the Ixcamilpa suite; disparate lithostratigraphy and age of metaclastic sequence make the Esperanza suite–Chuacús complex correlation unlikely (Fig. 2c, d & i).

Can we go further with the correlations? (1) The most obvious correlation, based on lithostratigraphy and type and age of magmatism, compares the Rabinal complex with the Esperanza suite of the Acatlán complex. In this scenario, at least the pre-Silurian parts of the Rabinal complex should contain high-*P* relicts. However, placing the Rabinal complex (and the Maya Block) further into the hinterland of the Ordovician orogeny than the Esperanza suite would allow variable metamorphic overprint. (2) We predict that the Chuacús complex will see separation into several units. For example, several samples of Chuacús-complex orthogneiss (data not shown in this paper) yielded Proterozoic ages (0.98–1.2 Ga) that are not or are little rejuvenated by Palaeozoic–Mesozoic events; it is possible that these gneisses define a unit that corresponds to the Tecolapa suite of the Acatlán complex. If our correlations are correct, where are the El Rodeo (?arc and continental rift) and Xayacatlán (oceanic basin) suites of southern Mexico in Guatemala? First, these suites may represent relative small-scale, regionally confined successions (as several units of the Cretaceous Mixteca–terrane arc succession; see below). Second, only few parts of the Chuacús complex and the Maya Block/Rabinal complex have been studied in detail. The abundance of mafic rocks in the some parts of Chuacús complex and the strong variability of lithology within it (compare, for example, the amphibolite–paragneiss-dominated El Chol and orthogneiss and migmatite-dominated Río Hondo areas described in the Structure and Kinematics section) suggests that other suites, reminiscent to those of the Acatlán complex, may be discovered. Here we propose that the Chuacús

complex that is rich in mafic rocks best correlates with the Ixcamilpa suite. These units formed a marginal basin that was transferred into a Silurian–Carboniferous subduction–accretion complex; the Chuacús complex may have represented the leading edge of this subduction complex. The clastics-dominated Chuacús complex may correspond to the Cosoltepec suite of the Acatlán complex. (3) Although data from the Chortís Block are still sparse, our U–Pb analysis identified *c*. 400 Ma crystallization ages in the basement rocks, suggesting that this block also experienced Early Devonian magmatism, identical to the youngest ages from the Maya Block/Rabinal complex.

We propose that the southern Maya Block/Rabinal complex and the Chuacús complex were parts of the Taconic–Acadian orogen. The Chuacús complex due to its high-*P* metamorphism and Rabinal complex due to its extensive Ordovician–Silurian magmatism were located together with the Acatlán complex within the orogenic interior along the northern margin of South America. It is likely that the Chortís Block also belonged to this orogen (Fig. 16c). Subduction produced an Early Ordovician arc magmatism on the Tecolapa and El Rodeo suites, which was followed by high-*P* metamorphism and accretion of the Xayacatlán suite to the Tecolapa and El Rodeo suites. Subduction continues beneath the Xayacatlán producing the post-high pressure granitoids within the Xayacatlán suite. Subsequently the Maya Block/Rabinal complex and the Esperanza suite were accreted and the Esperanza suite involved in continental subduction. Subduction again migrated outward, producing the Ixcamilpa and Chuacús complexes and involved them in subduction. In the Carboniferous Laurentia and Gondwana collided producing the nappe stack of the Acatlán complex and Chuacús complex.

Permo-Triassic arc on Maya and Chortís Blocks

Gneiss of the Rabinal complex of western Guatemala was migmatized at *c*. 218 Ma and inherited *c*. 268 Ma zircon; thus the high-grade rocks that accompany the low-grade metasedimentary rocks of the San Gabriel unit are probably a result of local Triassic (and Jurassic magmatism, see below) migmatization. In central Guatemala, Chuacús-complex orthogneiss, intruding high-*P* paragneiss and garnet amphibolite (relict eclogite), records two zircon-crystallization events at 238 and 219 Ma. Regional magmatism and metamorphism is demonstrated by similar hornblende and white mica cooling ages (238–212 Ma; Fig. 16a & b). Phyllite interlayered with conglomerate attributed to the Santa Rosa Group of central Guatemala

(Rabinal–Salamá area) contains detrital zircons at
c. 282, c. 260 and c. 245 Ma. Volcanic pebbles dominate the conglomerate. Triassic deformation in the
Rabinal and Chuacús complex shows high-T flow
with north(east)–south(west) shortening and both
sub-vertical and along-strike, NW–SE extension.
In the Chortís Block, migmatitic orthogneiss crystallized at c. 273 Ma and underwent high-T metamorphism at c. 245 Ma. Inherited c. 241 Ma
zircon in Jurassic orthogneiss and biotite cooling
ages support regional Triassic magmatism and
metamorphism (Fig. 16a & b).

Permo-Triassic granitoids are documented along
the length of Mexico and probably extend into
northwestern South America (e.g. Centeno-García
et al. 1993). Their apparent linear arrangement probably traces an active continental margin that originated after final amalgamation of Pangaea along
the Ouachita–Marathon suture at c. 290 Ma
(Fig. 16c; Torres et al. 1999). Isotopic ages of
these granitoids span 287–210 Ma with reliable
U–Pb zircon ages at 287–270 Ma (Figs 1c & 16a,
Ttoletec/Magdalena/Oaxaca, 16b; Yañez et al.
1991; Solari et al. 2001; Elías-Herrera & Ortega-
Gutiérrez 2002; Ducea et al. 2004a). In contrast to
these mostly undeformed granitoids, the Chiapas
Massif batholith rocks (Fig. 16b) are deformed and
metamorphosed (Weber et al. 2006). Rock types
include orthogneisses that are intruded by calc-
alkaline rocks varying from granite to gabbro,
metasedimentary rocks (metapelite/psammite, calcsilicate; Sepultura unit) and mafic rocks (hornblende
gneiss; Custepec unit). Zircons from orthogneiss
yielded a concordia age at 271.9 ± 2.7 Ma, and
three samples from migmatitic paragneiss and leucosome yielded 254.0 ± 2.3 to 251.8 ± 3.8 Ma ages
(Fig. 16a, Chiapas). The latter ages date the most
prominent event in the Chiapas Massif – partial
anatexis in pelite/psammite and, locally, orthogneiss. High-grade metamorphism is contemporaneous with c. north–south shortening (Weber et al.
2006).

Our new ages indicate that during the Permo-
Triassic the Rabinal and Chuacús complexes and
the northern Chortís Blocks were part of a continuous magmatic arc from northeastern Mexico to
northern South America (e.g. Dickinson & Lawton
2001), which initiated after the formation of
Pangaea along the western continental margin of
Gondwana (Fig. 16c). Sample 5H-9a zircons of migmatitic orthogneiss in the north-central Chortís
Block have within error the same intrusion and high-
grade metamorphic ages as the Chiapas Massif rocks
(Fig. 16a). This identical geological history suggests
a direct correlation between the Chiapas Massif and
the northern Chortís Block along the Permo-Triassic
arc prior to the opening of the Gulf of Mexico
(Fig. 16b & c). The ages in the Rabinal and

Chuacús complexes are c. 30 Ma younger than
those in the Chiapas Massif and also younger than
most Permo-Triassic ages in southern Mexico. We
speculate that the Chuacús complex occupied,
together with the Maya Mountains of Belize (c.
230 Ma reheating), an easterly position along the
main Chiapas–Chortís arc (Fig. 16c). The younger
ages may reflect an episode of flat-slab subduction
that shortened the main arc (Weber et al. 2006)
and caused a younger episode of magmatism and
metamorphism in its hinterland. Shortening
directions are similar in the Rabinal and Chuacús
complexes and the Chiapas Massif. Associated
with the shortening there is probably a dextral-
transpressional component, widespread in southern
Mexico (e.g. Malone et al. 2002), which is consistent with tangential stretching in the Triassic
Chuacús complex. Finally, the conglomerate clasts
that are predominantly sourced from intermediate
volcanic rocks, and c. 245, c. 260 and c. 280 Ma detrital zircons within phyllite demonstrate that parts of
the clastic cover sequence of the Rabinal complex in
central Guatemala is not Lower Carboniferous–
Permian Santa Rosa Group. We suggest that a part
of this Group comprises either a new unit or is
part of the undated basal conglomeratic sequence
of the Jurassic–Cretaceous Todos Santos Formation
(Fig. 2b & d). Detritus and age of detrital zircons
make the Chiapas–Chortís-arc Massif the likely
source of these molasse-type, basal deposits.

Jurassic rifting and arc magmatism

Granite intruded into the Rabinal complex of
western Guatemala at c. 171 Ma (164–177 Ma
zircons; Fig. 16a). Most of the Maya Block was
probably emergent during the Triassic–Jurassic
with the likely Middle Jurassic–Early Cretaceous
Todos Santos Formation reflecting rifting during
Jurassic separation of North and South America
(Fig. 2b; e.g. Donnelly et al. 1990). Granites also
intruded the northern Chortís Block during the
Middle Jurassic (c. 159 and 166 Ma, zircons at
140–191 Ma; 168 Ma; Fig. 16d). The Middle Jurassic, clastic, partly greenschist-facies Agua Fría
Formation (Fig. 2j) thickens from the central to
the eastern Chortís Block and is interpreted to be
related to transtension and rifting during the Jurassic
opening between North and South America; e.g.
the Guayape fault zone may have originated as a
Jurassic normal fault (Fig. 16d; e.g. James 2006;
Gordon 1993). The rocks of the Agua Fría Formation were deformed during the Late Jurassic
(Viland et al. 1996).

Early–Middle Jurassic magmatism–metamorphism affected the Chazumba and Cosoltepec
units of the Acatlán complex and produced the Magdalena migmatites of southern Mexico (Figs 2i &

16a, Totoletec/Magdalena/Oaxaca; e.g. Talavera-Mendoza *et al.* 2005; our new *c.* 162 Ma cooling age); inferred metamorphic conditions were low-*P*/high-*T* (*c.* 730 °C at *c.* 0.55 GPa; Keppie *et al.* 2004). Anatectic granitic dykes of the San Miguel unit (175–172 Ma; e.g. Yañez *et al.* 1991) cut the former units. These events have been interpreted as the result of a plume breaking up Pangaea, opening of the Gulf of Mexico and Triassic–Middle Jurassic magmatic arc activity (e.g. Keppie *et al.* 2004). Approximately at 165 Ma, NNW-trending, dextral transtension shear, kinematically compatible with opening of the Gulf of Mexico, occurred along the Oaxaca fault (Alaniz-Alvarez *et al.* 1996), and NW-trending, Jurassic–earliest Cretaceous dextral transpression took place within the northern Oaxaca complex (Solari *et al.* 2004). The Xolapa complex (Fig. 2l) also shows 165–158 Ma magmatic–metamorphic ages (Fig. 16a, Xolapa; Herrmann *et al.* 1994; Ducea *et al.* 2004*a*). The Zihuatanejo and Huetamo mature arc sections of the Guerrero terrane rest unconformably on the Arteaga–Las Ollas basement, which comprises ocean-floor–continental-slope–subduction complexes that were deposited in the Late Triassic–Early Jurassic, intruded by 187 ± 7 Ma granite, and deformed in the Middle Jurassic (Fig. 2m). Major late Middle–Late Proterozoic–Phanerozoic detrital zircon populations from these complexes are *c.* 247, *c.* 375, *c.* 475, *c.* 575 and *c.* 975 Ma. The overlying Cretaceous arc has major zircon populations at *c.* 160, *c.* 258, *c.* 578 and *c.* 983 Ma (Fig. 2m; Talavera-Mendoza *et al.* 2007).

Our tectonic model (Fig. 16e) integrating the Jurassic evolution of the Maya and Chortís Blocks with the magmatic–tectonic evolution of southern Mexico is modified from Mann (2007). Onset of rifting in the Gulf of Mexico and between North and South America reduced Triassic (see above) to Jurassic volcanic eruption rate in southern Mexico and the Maya and the Chortís Blocks. Late intrusion into active dextral fault zones (Mexico) may explain the concentration of magmatism at 175–166 Ma (Fig. 16d). Rifting of the Chortís Block from southern Mexico probably initiated along the arc and developed either as back-arc extension above the proto-Pacific subduction zone, a failed arm of the Proto-Caribbean rift, or a combination. Additionally, the Arteaga–Las Ollas basement units rifted from southern Mexico, as they likely share the Late Triassic–Jurassic history of southern Mexico (Fig. 2m; e.g. Talavera-Mendoza *et al.* 2007). Our model requires rifting the Chortís Block from southern Mexico and creation of oceanic crust, because MORB oceanic crust and mélange of the South El Tambor complex is present between the Chortís Block and southern Mexico. This must predate 132 Ma, the age of

eclogite formation in the South El Tambor complex and is probably pre-Late Jurassic, the probable age of the oldest radiolarians in the South El Tambor cherts (Chiari *et al.* 2006). Rifting affected also the interior of the Chortís Block (Agua Fría Formation; Fig. 16d). These rift structures trend *c.* NE (present coordinates), sub-parallel to the northwestern and southeastern margins of the Chortís Block and the Guayape fault. This rifting is distinctly different from the WNW-trending Mid-Cretaceous intra-arc rifting (Rogers *et al.* 2007*a*; see below). The Late Jurassic regional metamorphism (165–150 Ma; Fig. 16a) and deformation (Viland *et al.* 1996) may be rift and strike–slip related.

Several likely connections exist between the Chortís Block, the Xolapa complex, and the Guerrero composite terrane. The Sanarate complex (Fig. 2e), the westernmost basement outcrop of the Chortís Block south of the MSZ, was included with the South El Tambor complex because of its metamorphism (lower metamorphic grade than that of the Las Ovejas unit) and rock assemblage (dominated by mafic and clastic rocks; e.g. Tsujimori *et al.* 2004, 2006); however, the Jurassic age for the metamorphism precludes such a correlation. Here we suggest a correlation of this complex with the Arteaga–Las Ollas units (Figs 2m & 16e) based on lithology, likely proximity to the Jurassic continental margin of Mexico and distinct Jurassic metamorphism (Fig. 2e & m). The Sanarate complex probably also includes parts of the Cretaceous Zihuatanejo and Huetamo volcanic and clastic units (Fig. 2m). Both the Arteaga–Las Ollas units and the overlying Cretaceous arc contain zircons that may have originated from the Chortís Block. The *c.* 247, *c.* 257 and 475 Ma clusters, particularly in the ≥85 Ma arc rocks, are difficult to derive from North America because of intervening spreading ridges and subduction zones (Figs 2m & 16e, see below). The pre-migmatitic basement of the Xolapa complex, containing evidence for a Grenville crustal component (Herrmann *et al.* 1994) and Permian magmatism (see above) in an alternation of metagreywacke and metabasic rocks, probably correlate to the Permian and Jurassic section of the Mixteca terrane (Fig. 2m, Olinalá unit). Finally, connecting the Jurassic magmatic–metamorphic rocks of southern Mexico and the northern Chortís Block implies ≥1000 km post-Jurassic offset.

Early Cretaceous subduction, top-to-south emplacement of the South El Tambor complex, and proto-Pacific arc formation

Spreading between the Chortís Block and southern Mexico must have reverted to subduction before 132 Ma (see above). The *c.* 120 Ma phengite ages of the South El Tambor complex blueschists are interpreted as marking fluid infiltration during retrogression from eclogite-facies conditions (see above) and could date exhumation during emplacement of the South El Tambor complex mélange onto the Chortís Block (Fig. 8b). The El Tambor rocks are emplaced top-to-south (present coordinates). Aptian–Albian calc-alkaline, arc volcanic rocks of the Manto Formation and volcanoclastic rocks of the Tayaco Formation, sandwiched between shallow water carbonates, were deposited in a zone of intra-arc extension in the central Chortís Block, the Agua Blanca rift (Rogers *et al.* 2007a; Figs 2j & 16f). The strike of the rift is WNW (present coordinates), oblique to the northern margin of the Chortís Block and the Jurassic extensional structures (see above). Cretaceous (peaks at *c.* 128, *c.* 118 and *c.* 81 Ma; Figs 16a, f & 5h) magmatic rocks are widespread on the Chortís Block (Fig. 16f) and probably mark continuous arc formation above the proto-Pacific subduction zone (Fig. 16g). The Xolapa complex, the largest plutonic and metamorphic mid-crustal basement unit in southern Mexico, preserves abundant migmatites that were produced during a single metamorphic event with peak prograde conditions at 830–900 °C and 0.63–0.95 GPa. This occurred within a continental magmatic arc and overprinted a pre-migmatitic metamorphic assemblage (Corona-Chávez *et al.* 2006). Deformation, likely (north-)northeast–(south-)southwest shortening (Corona-Chávez *et al.* 2006), associated with low-*P* metamorphism and migmatization predates 129 ± 0.5 Ma in the Tierra Colorada area of the western Xolapa complex (Solari *et al.* 2007). The migmatites have been directly dated at 131.8 ± 2.2 Ma (Herrmann *et al.* 1994) and 130–140 Ma (Ducea *et al.* 2004a). The structural top of the migmatized sequences is the 126–133 Ma Chapolapa Formation (Campa & Iriondo 2004; Hernández-Treviño *et al.* 2004) that mostly comprises andesite/rhyolite; its hanging wall, Albian–Cenomanian Morelos Formation shelf limestone, is again in tectonic contact (Figs 2l & 16a, Xolapa; e.g. Riller *et al.* 1992; Solari *et al.* 2007).

Rogers *et al.* (2007a) suggested several features and piercing lines that tie the Chortís Block to southern Mexico; here we elaborate on the tectonic model explaining the correlation (Fig. 16g). We suggest that the 141–126 Ma migmatization–magmatism of the Xolapa complex represents arc magmatism related to subduction of the South El Tambor complex lithosphere. Characteristically, no Cretaceous ages younger than *c.* 126 Ma have been found in Xolapa complex, so the end of arc magmatism in the South El Tambor subduction zone is likely to be contemporaneous with emplacement of the South El Tambor Complex onto the

Chortís Block (≤120 Ma, see above). Laterally, Xolapa-complex magmatism is correlated with that in the Taxco–Viejo unit of the Mixteca terrane, also an arc on continental crust (e.g. Centeno-García *et al.* 1993) and dated at 130.0 ± 2.6 and 131.0 ± 0.85 Ma (Campa & Iridono 2004). This correlation is supported by the, within error, identical ages of the Chapolapa Formation meta-andesite/rhyolite (structural top of the Xolapa complex) and the volcanic flows of the Taxco–Viejo unit (Fig. 2l & m). Magmatism in the Teloloapan intra-oceanic arc ended at *c.* 120 Ma. Magmatism in the oceanic arc and back-arc region of the Arcelia units, outboard (southwesterly) of the Teloloapan unit continued into the Cenomanian (Fig. 2m). The youngest detrital zircon fractions in the clastic rocks above the volcanic succession in the Teloloapan and Taxco–Viejo units, comprising magmatic zircons from local sources, are *c.* 124 and *c.* 132 Ma, respectively (Talavera-Mendoza *et al.* 2007), thus contemporaneous with the Xolapa-complex zircons. Again, these ages indicate that magmatism terminated at that time and suggest a close spatial relationship of the Taxco–Viejo–Xolapa–Teloloapan arcs (Figs 2m & 16g). The NE-dipping subduction zone of the South El Tambor complex may have ended at a transform fault at the eastern end of the Maya Block (Fig. 16g).

The Teloloapan, Taxco–Viejo, and Olinalá units and the Xolapa complex are covered by Albian–Lower Cenomanian Morelos shelf limestones (Fig. 2l & m). These limestones may correspond to the massive La Virgen limestone atop the South El Tambor complex (Fig. 2k). Carbonate passive margin strata (Cobán–Campur Formations; Fig. 2b) were deposited on the Maya Block. The post-suturing deposits on the Chortís Block are probably the red beds and foreland-basin deposits of the Valle de Angeles and Gualaco Formations. The latter is typically intercalated with Upper Cretaceous volcanic rocks indicating establishment of the Sierra Madre Occidental arc in Late Cretaceous, also on the Chortís Block (see below). Characteristically, proto-Pacific subduction continued beneath the Chortís Block and the Zihuatanejo unit far into the Late Cretaceous (Fig. 2j & m). In our interpretation the Agua Blanca rift and the Manto volcanic rock intercalations are related to proto-Pacific subduction and intra-arc extension. We attribute the inversion of the Agua Blanca intra-arc rift to the accretion of the Zihuatanejo–Huetamo units–Sanarate complex to southern Mexico and the western Chortís Block. Top-to-east shear in the greenschist facies, metabasaltic/clastic Sanarate complex of the western Chortís Block reflects this accretion of the Guerrero terrane/Sanarate complex.

Why were so many individual arcs and subduction–accretion complexes developed along the southwestern margin of Mexico (the Guerrero composite terrane) and the Chortís Block (the Sanarate complex) during the Early and Late Cretaceous (e.g. Talavera-Mendoza *et al.* 2007)? If rifting between the Chortís Block–Arteaga unit and southern Mexico was mostly related to spreading between the two Americas and formation of the Gulf of Mexico (Fig. 16e), spreading was probably north–south with *c.* north-trending transform faults. Transformation of this rift to convergence along the *c.* NNW-trending proto-Pacific subduction zone may have allowed activation of some transform faults as subduction zones.

Late Cretaceous North El Tambor complex subduction and Maya Block–Caribbean arc collision

Our structural, petrological, and geochronological data demonstrate sinistral transpressive collision along the southern margin of the Maya Block during Late Cretaceous. Collision transferred the southern Maya Block into a nappe stack with the North El Tambor ophiolitic mélange at top. Characteristically, deformation involved north–south shortening and east–west lengthening. Sinistral transpression was partitioned on all scales between oblique stretching along the foliation and sinistral strike–slip along steeply dipping shear zones (Fig. 16h). To a first order, deformation is identical in the North El Tambor complex, Chuacús complex, and the northern foreland fold–thrust belt; 450–550 °C and 0.7–0.8 GPa regional Cretaceous reheating within the Chuacús complex, associated with local magmatism, is synkinematic. Penetration of deformation and overburden (*PT* conditions) increases from west to east along the collision zone. The relatively high-*P*/low-*T* metamorphism and nearly coeval blueschist- (North El Tambor complex) and amphibolite-facies metamorphism (Chuacús complex) suggest a cold geotherm and steep continental subduction (Fig. 16h). Southern Maya-Block imbrication is dated at 75–65 Ma (Fig. 16i) and cooling to upper crustal levels (*c.* 100 °C) was attained by Early Cenozoic (AFT results at high elevation, see above). The Jurassic–Early Cretaceous North El Tambor oceanic and immature arc rocks incorporated *c.* 131 Ma eclogite into an accretionary wedge that reached crustal levels by Late Cretaceous (*c.* 76 Ma).

What collided with the southern Maya Block in the latest Cretaceous? We follow and broaden the model of Mann (2007) and Rogers *et al.* (2007b) and suggest that the Maya Block–North El Tambor complex suture extends toward the west

into the footwall of the Siuna terrane of the southern Chortís Block and the Nicaragua Rise and to the east into the subduction–accretion and arc complexes of Jamaica, Hispaniola (Dominican Republic) and Cuba (Fig. 16j & k). In comparison with the North El Tambor complex (Fig. 2a), the Siuna mélange (Fig. 2f) contains similar lithology (ultramafic rocks, radiolarite, meta-andesite, volcanic arenites) of proven Late Jurassic age, high-*P* phengites at *c.* 139 Ma, blueschist and eclogite, and Aptian–Albian turbiditic sequence. These rocks are overlain by the Átima Formation and intruded by diorite (*c.* 60 Ma) and andesite dykes (*c.* 76 Ma; Fig. 2f; Flores *et al.* 2007; Venable 1994). The poorly dated (post-80 Ma) Colón fold belt of the eastern Chortís Block and the Nicaragua Rise probably reflects imbrication along the footwall of the Siuna terrane (Rogers *et al.* 2007b). The Siuna terrane probably extends westward into the ocean igneous complexes of the Santa Elena and Nicoya peninsulas of Costa Rica; these complexes contain *c.* 165–83 Ma pillow lavas, sills and flows, radiolarian chert, basalt breccias, and gabbro and plagiogranite. They are unconformably capped by ≤75 Ma reef limestones (Figs 2g & 16i; e.g. Hoernle & Hauff 2005). The Cuban subduction–accretion complex contains *c.* 148–65 Ma high-*P* rocks (major peaks at *c.* 116 and *c.* 70 Ma; Fig. 16i & j; Krebs *et al.* 2007, Schneider *et al.* 2004). The Greater Antilles arc of Cuba was active at 133–66 Ma (Fig. 16i & j; major peaks at *c.* 72 and *c.* 80 Ma; Grafe *et al.* 2001; Hall *et al.* 2004; Rojas-Agramonte *et al.* 2005; Stanek *et al.* 2005), and the Cordillera Central arc (Dominican Republic) spanned *c.* 117–74 Ma (Fig. 16h–j; major peaks at *c.* 75, *c.* 84 and *c.* 116 Ma; Kesler *et al.* 2005; Escuder Viruete *et al.* 2006b). Thus these subduction–accretion complexes and arcs cover time ranges comparable to that of the North El Tambor complex, with subduction and arc formation terminated *c.* 10 Ma after collision along the southern Maya Block. This indicates the time span required to transfer these subduction–accretion and arc complexes to the Bahamas-platform collision zone when the southern Maya Block acted as a sinistral transform fault zone. Accordingly, termination of pronounced magmatism in the Santa Elena–Nicoya subduction-accretion complexes is approximately 10 Ma earlier (Fig. 16i) and reflects their earlier collision with the southwestern Chortís Block. Based on the termination of magmatism at 74 Ma within the Cordillera Central arc and its well-developed sinistral transpressive deformation (Escuder Viruete *et al.* 2006a, b), we speculate this arc was the immediate hanging wall of the collision zone along the southern Maya Block.

What was the origin of the ≤75 Ma magmatism (pegmatites) in the southern Maya Block? Causes might include Pacific-arc magmatism, local anatexis due to crustal thickening, or break-off of the Siuna terrane–North El Tambor complex–Caribbean arc slab. We favour the latter (Fig. 16h), as entry of the Nicoya–Siuna–North El Tambor–Caribbean slab into the Proto-Caribbean sea must have induced its rapid anticlockwise rotation, rapid along-strike length reduction, and thus slab fracturing and roll-back. In contrast, contemporaneous Pacific arc subduction occurred far to the west on the Chortís Block (Fig. 16k) and the thermal overprint due to crustal stacking was probably insufficient and too early in the Chuacús complex to cause anatexis (see above).

Cerca et al. (2007) outlined a mostly east vergent, partly sinistral transpressive, north-trending fold-thrust belt along the western Mixteca terrane, the Guerrero–Morelos platform that originated during the Coniacian ($\leq$90 Ma; Cenomanian, $\leq$100 Ma, after Talavera-Mendoza et al. 2005) and was active into the Cenozoic. They related this east–west shortening to the collision of the Caribbean Plate with the Chortís Block. Rogers et al. (2007b) and our data show, however, that this collision is c. north–south and $\leq$76 Ma on the southern Chortís and Maya Blocks (see above). Probably, the Laramide fold–thrust belt orocline on the Guerrero–Morelos platform, along the Sierra de Juárez (Cerca et al. 2007), and within the southern Maya Block (this study) is a composite structure, related to accretion of the most outboard Guerrero-terrane units, Pacific subduction and collision of the Siuna terrane–North El Tambor complex–Caribbean Plate. We suggest that shortening in the Guerrero–Morelos platform and its continuation into the Chortís Block (Agua Blanca rift inversion) is mainly related to the $\leq$80 Ma Arteaga–western Chortís (Sanarate complex) suture and Pacific subduction and the Sierra de Juárez–southern Maya Block shortening due to the $\leq$76 Ma Caribbean Plate collision (Fig. 16k). Structural superpositions must have occurred mostly in the Mixteca–Oaxaca basement terrane area.

Late Eocene–Recent eastward displacement of the Chortís Block

Most tectonic models place the Chortís Block opposite southern Mexico during Late Cretaceous–Early Cenozoic (e.g. Mann 2007; see above). Rogers et al. (2007a, b) summarized features and piercing lines that support this connection. Here we go over additional evidence and trace the eastward displacement of the Chortís Block to its present position. Approximately 80–45 Ma magmatic rocks of the Sierra Madre Occidental crop out along the coast of southwestern Mexico and inland up to c. 99°W (Fig. 16l; Morán-Zenteno et al. 1999; Schaaf et al.

1995). This magmatic arc continued into the central Chortís Block. Lining up the northwest trend of these rocks in Mexico (the arc continues NW of the area shown in Fig. 16l) and on the Chortís Block yields $\geq$1100 km sinistral displacement and $\leq$40° anticlockwise rotation of the Chortís Block (Figs 1c & 16l). Based on one data point from the central Chortís Block, the palaeomagnetically determined anticlockwise post-60 Ma rotation is c. 32° (Gose 1985). During the Oligocene, eastward migration of Chortís induced a northeastward migration of the trench (Herrmann et al. 1994) and the associated Sierra Madre del Sur arc shifted inland and eastward into southeastern Mexico. The arc trends ESE (Morán-Zenteno et al. 1999), which is particularly evident on the Chortís Block (Fig. 16l).

When did the Chortís Block break away from southern Mexico? In the Chortís Block, high-grade and high-strain deformation began at $\geq$40 Ma and we suggested above that there were major deformation phases at 40–35 Ma and possibly at 25–18 Ma. Sinistral transtension in the La Venta–Tierra Colorada shear zone along the northwestern margin of the Xolapa complex is low-grade at c. 300 °C (low-T quartz ductility; Riller et al. 1992) and started at $\leq$45 Ma (Solari et al. 2007). Local high-strain flow occurred at c. 35 Ma (see above). Post-tectonic intrusions are 34 $\pm$ 2 Ma (Tierra Colorada granite, Herrmann et al. 1994); however, brittle, sinistral transtension continued during cooling of this granite (Riller et al. 1992). The distinctly different exposure level of the continuous Early Cretaceous Taxco–Viejo (west; up to greenschist facies) and Xolapa arc (east; amphibolite facies and migmatization) is ascribed to major exhumation of the latter. Large-scale sinistral transtension along the northwestern Xolapa-complex margin certainly contributed to this difference. Robinson et al. (1990) obtained a c. 40 Ma age for synkinematic tonalite north of Acapulco. Sinistral shear occurred along the northern margin of the eastern Xolapa complex between 29 and 24 Ma (Chacalapa shear zone, point 4 in Fig. 16l; Tolson 2005). Other sinistral transtensive deformation zones in southern Mexico outside the Xolapa complex are dated to between c. 37 and c. 27 Ma (points 1 and 3 in Fig. 16l). Corona-Chávez et al. (2006) estimated that about 70% of the eastern Xolapa complex was affected by discontinuous mylonitic–cataclastic, post-migmatitic deformation. We conclude that the migration of the Chortís Block initiated $\geq$40 Ma and was fully active at 40–35 Ma. This is compatible with the chronology of spreading in the Cayman Trough pull-apart basin (see above).

Using (U–Th)/He thermochronology, Ducea et al. (2004b) estimated average exhumation rates for the Xolapa complex. A western segment at Acapulco yielded ages between 26 and 8.4 Ma and a rate of 0.22 mm/annum; an eastern profile at Puerto Escondido gave 17.7–10.4 Ma ages and 0.18 mm/annum. This slow Neogene exhumation contrasts with petrological data that suggest exhumation from ≥25 km (migmatites; Corona-Chávez *et al.* 2006) and 20–13 km (granitoids; Morán-Zenteno *et al.* 1996) since 34–27 Ma, the age of the granitoids; therefore, most exhumation must have taken place before the Miocene. Pooling the (U–Th)/He and AFT data of Ducea *et al.* (2004*b*) and excluding outliers (e.g. ages older than crystallization and those with very large errors) indicate that most samples of the western profile cooled through 70–100 °C or were exhumed from *c.* 4 km at *c.* 25 Ma. Except for three outliers most samples of the eastern profile have identical ages over 2 km elevation and thus suggest rapid cooling at *c.* 15 Ma (Fig. 6d). Using the three plutons for which U–Pb zircon ages are available (Herrmann *et al.* 1994), we obtained exhumation rates of *c.* 1.5 mm/annum between 31 and 25 Ma (12.9 km depth at 31 Ma and *c.* 4 km at *c.* 25 Ma) and *c.* 1.9 mm/annum between 32 and 25 Ma (17.9 km depth at 32 Ma and *c.* 4 km at *c.* 25 Ma) for the western section, and 1.2 mm/annum (20.4 km depth at 29 Ma and *c.* 4 km at *c.* 15 Ma) for the eastern profile. For the western section, Herrmann *et al.* (1994) obtained two *c.* 47 Ma migmatization ages; this yields *c.* 0.8 mm/annum between 47 and 31 Ma (*c.* 26 km depth at 47 Ma and 12.9 km at 31 Ma). Along the northern rim of the Chortís Block our data indicate an exhumation rate of *c.* 1 mm/annum between 36 and 9 Ma (*c.* 28 km depth at 36 Ma and *c.* 3 km at 9 Ma). These data indicate that the early stages of Chortís-Block migration were associated with rapid exhumation due to sinistral transtension and erosion.

The early stage of distributed sinistral displacement and erosion that affected a broad zone in southern Mexico and the northern Chortís Block also must have affected the southern Maya Block, as its eastern extension, the Nicaragua Rise, was located off the western edge of the Maya Block (Fig. 16k). We interpret our AFT data from the southern Chiapas Massif and the Chuacús complex, both clustering at *c.* 30 Ma (Fig. 6a & c), to record this distributed transtension–transpression–sinistral migration and erosion phase.

Why did the Chortís Block break away at ≥40 Ma? Following Ratschbacher *et al.* (1991) and Riller *et al.* (1992), we propose that the contemporaneous change in several plate tectonic boundary conditions triggered break away. During Eocene, shear stresses across the arc were high due to rapid convergence of relatively young lithosphere under an angle of *c.* 45° to a pre-existing lithospheric weakness, the South El Tambor suture. Most importantly, sinistral break away started in an area that had been an arc at least since Late Cretaceous (see above) and thus was thermally weakened. This is a common process in oblique subduction settings. Consequently, the largest volume (Morán-Zenteno *et al.* 1999) and the concentration in time (*c.* 33 Ma Chortís Block, *c.* 34 Ma Xolapa complex; Fig. 16a) of magmatism and metamorphism seems to have been controlled by the NW–SE shear zones associated with transtensional–transpressional tectonics at the boundary between southern Mexico and the Chortís Block (Herrmann *et al.* 1994; Fig. 16l).

Early and Middle Miocene (*c.* 20–10 Ma) volcanic rocks are distributed around the Isthmus of Tehuantepec (Fig. 16m). In this region, it is evident that volcanism in the inland regions was active after cessation of Oligocene plutonism in the Xolapa complex (Morán-Zenteno *et al.* 1999). The earliest manifestations of magmatic activity along the eastern and central Trans-Mexican volcanic belt are as old as Middle Miocene (*c.* 17 Ma; Ferrari *et al.* 1994) and thus this belt may form a direct continuation of the volcanism around the Isthmus of Tehuantepec. To the east coeval magmatism occurred along the western part of the Polochic fault zone in westernmost Guatemala (Fig. 16m), indicating ≥100 km offset along the Polochic fault since then. This is similar to the *c.* 130 km offset proposed along this fault between 10 and 3 Ma (Burkart 1983; Burkart *et al.* 1987). Still further east, the *c.* 16.9–13.4 Ma ignimbrite province strikes northwest–southeast across the Chortís Block (Fig. 16m; e.g. Jordan *et al.* 2007). Aligning the southeastern Mexican and Chortís Block volcanic provinces yields *c.* 400 km displacement since *c.* 15 Ma and *c.* 300 km offset along the Motagua fault (Fig. 16m). Several elements trace the displacement during this time (Fig. 16m): tectonic and stratigraphic features NW of the Isthmus of Tehuantepec indicate sinistral transtension with *c.* north trending pull-apart basins coeval with volcanism (Morán-Zenteno *et al.* 1999); the easternmost part of the Xolapa complex experienced rapid cooling at *c.* 15 Ma (Fig. 6d); the Chacalapa shear zone was reactivated at *c.* 15 Ma; the Tonala and Polochic shear zones were active at 9–8 and 12–5 Ma, respectively; and most of the AFT ages in the Chortís Block closed at *c.* 12 Ma (all data from this study).

Variations in the Late Cenozoic stress field

North–south rifts that disrupt the ignimbrite province in western Honduras and southeastern Guatemala commenced after *c.* 10.5 Ma (Gordon & Muehlberger 1994). They are related to plateau uplift following slab break-off underneath Central America (Rogers *et al.* 2002). Our 11.2–7.9 Ma AFT data from the western flank of the Sula graben support the Late Miocene onset of rifting, although extension structures formed as early as *c.* 28 Ma, contemporaneous with and kinematically related to sinistral wrenching along the northern margin of the Chortís Block. Numerical modelling suggests that presently the western edge of the Chortís Block is pinned against North America, making the triple junction between the Cocos, North American and Caribbean plates a zone of diffuse deformation (Álvarez-Gómez *et al.* 2008). Our 'Late Cenozoic deformation' structural event (Figs 10, 13 & 15), reflecting sinistral slip along *c.* east-striking strike–slip faults locally interacting with *c.* north-striking normal faults, is distributed across the entire plate boundary from the northern foreland to south of the Jocotán–Chamelecón fault zone. The transition from older dominant strike–slip to younger, prevailing normal faulting, proposed by Gordon & Muehlberger (1994), is also evident from our data. However, normal faulting exemplified by the grabens in southern Guatemala and western Honduras (Fig. 15), extends northward across the Motagua fault zone into the Cenozoic Subinal Formation (Fig. 13), the North El Tambor rocks on the central Chuacús complex (Fig. 13), and the Rabinal complex and Cenozoic–Quaternary volcanic rocks of western Guatemala (Fig. 10). This may indicate that today the Polochic fault constitutes the major plate boundary fault. This is compatible with termination of the active strike–slip and reverse fault province of eastern Chiapas north of the Polochic fault zone (Andreani *et al.* 2008). Furthermore, our data support additional salient features of the Late Cenozoic stress distribution: shortening directions trend *c.* NW–SE in western and west-central Guatemala and in particular in western Guatemala south of the Polochic fault ('Late Cenozoic deformation'; Figs 10, 13 & 15) and outline the dextral, WNW-trending Jalpatagua distributed fault zone along the active volcanic arc (Gordon & Muehlberger 1994; Álvarez-Gómez *et al.* 2008).

Conclusions

New geochronological, petrological, and structural data constrain the *P–T–d–t* history of the northern Caribbean Plate boundary in southern Mexico, central Guatemala and northwestern Honduras. In particular, we investigated which crustal sections of the Motagua suture zone accommodated the large-scale sinistral offset along the plate boundary, and the chronological and kinematic framework of this process. Correlations between the Maya and Chortís Blocks, tectono-stratigraphic complexes of

southern Mexico and arc/subduction complexes of the northwestern Caribbean constrain models for the Early Palaeozoic to Recent evolution of the North American–Caribbean Plate boundary.

During the Early Palaeozoic, southern Mexico (Acatlán complex of the Mixteca terrane), the southern Maya Block (Rabinal complex), the Chuacús complex and also the Chortís Block were part of the Taconic–Acadian orogen along the northern margin of South America. The most obvious correlation, based on lithostratigraphy and type and age of magmatism, compares the Rabinal complex with the Esperanza suite of the Acatlán complex. In this scenario, at least the pre-Silurian parts of the Rabinal complex should contain high-*P* relics, which are not observed. Placing the Rabinal complex (and the Maya Block) further into the hinterland of the Ordovician orogeny than the Esperanza suite allows variable metamorphic overprint. The Chuacús complex best correlates with the Ixcamilpa suite and the Cosoltepec suite of the Acatlán complex. These units formed a marginal basin that was transferred into a Silurian–Carboniferous subduction–accretion complex; the Chuacús complex may have represented the leading edge of this subduction complex.

After the final amalgamation of Pangaea, an arc developed along its western margin, causing magmatism and regional amphibolite-facies metamorphism in southern Mexico (*c.* 290–250 Ma), the Rabinal complex of the southern Maya Block and the Chuacús complex (*c.* 270–215 Ma) and the Chortís Block (*c.* 283–242 Ma). Identical intrusion and metamorphic ages (273 and 254 Ma) in the Chiapas Massif of the southwestern Maya Block and the northern Chortís Block suggest a similar position within the arc. Younger ages in the Rabinal and Chuacús complexes are attributed to their position in the hinterland of the main arc and a speculative episode of flat-slab subduction.

The Jurassic opening of the Gulf of Mexico separated the Maya Block from North America. Granite intruded the Rabinal complex (Maya Block) of western Guatemala at 164–177 Ma. Most of the Maya Block was probably emergent during Triassic–Jurassic with the likely Middle Jurassic–Early Cretaceous Todos Santos Formation reflecting rifting during separation of North and South America. Dextral shear along *c.* NNW trending shear zones within southern Mexico is kinematically related to opening the Gulf of Mexico and may account for the concentration of magmatism at *c.* 175–166 Ma in southern Mexico. Granites also intruded the northern Chortís Block during Middle Jurassic time (*c.* 168–159 Ma). The Middle Jurassic Agua Fría Formation records upper crustal transtension and rifting. Rifting of the Chortís Block from southern Mexico probably initiated within the thermally weakened proto-Pacific arc, possibly as a failed arm of the Proto-Caribbean rift. This rifting separated the northern Chortís Block from the Xolapa complex of southern Mexico and also parts of the Guerrero terrane from parts of southwestern Mexico that have Jurassic-aged basement. The Sanarate complex of the north-western Chortís Block correlates with the Arteaga–Las Ollas units of the Guerrero terrane, sharing a dominantly mafic lithology and Jurassic meta-morphism. Rifting produced the Late Jurassic–Early Cretaceous oceanic crust of the South El Tambor complex and terminated before 132 Ma, the age of eclogite-facies metamorphism in that complex.

Rifting between Chortís Block and southern Mexico reverted to subduction in the Early Cretaceous. Arc magmatism and high-T/low-P metamorphism (141–126 Ma) along the Xolapa complex of southern Mexico is probably related to subduction of South El Tambor complex lithosphere, which may have been emplaced southward onto the Chortís Block at or after *c.* 120 Ma. Laterally, Xolapa-complex magmatism is correlated with that in the Taxco–Viejo unit of the Mixteca terrane. While collision of the Chortís Block with southern Mexico terminated arc magmatism in the Xolapa complex, proto-Pacific subduction continued on the Chortís Block and the Zihuatanejo unit of the Guerrero terrane to its NW. This unit and the Sanarate complex of the Chortís Block amalgamated to southern Mexico during Late Cretaceous.

Spectacular sinistral transpressive collision occurred along the southern margin of the Maya Block during Late Cretaceous ($\leq$76 Ma). Collision transferred the southern Maya Block into a nappe stack with the North El Tambor complex at top. Sinistral transpression was partitioned on all scales between oblique stretching along the foliation and sinistral strike–slip along steeply dipping shear zones. To first order, deformation is identical in the North El Tambor complex, Chuacús complex, and the Laramide, northern foreland fold–thrust belt. The relatively high-P/low-T metamorphism and nearly coeval blueschist- (North El Tambor complex) and lower amphibolite-facies metamorphism (Chuacús complex) suggest a cold geotherm and steep continental subduction. The Maya Block–North El Tambor complex suture extends westward into the footwall of the Santa Elena–Nicoya complexes, the Siuna terrane of the southern Chortís Block, and the Nicaragua Rise and to the east into the subduction–accretion and arc complexes of Cuba and Hispaniola. Postulated break-off of the North El Tambor complex slab may account for $\leq$75 Ma magmatism in the southern Maya Block.

High shear stress across the Caribbean–North American plate boundary during the Eocene triggered breakaway of the Chortís Block from southern Mexico at $\geq$40 Ma, coeval with initiation of the Cayman trough pull-apart basin. Sinistral break away started in an area, the Xolapa complex and the northern Chortís Block, that had been an arc at least since Late Cretaceous and thus was thermally weakened. The Las Ovejas complex of the northern Chortís Block forms a westward narrowing belt of distributed sinistral transtension and represents the southern part of the high-strain displacement zone along which the Chortís Block migrated eastward. Deformation interacted with amphibolite-facies metamorphism, widespread migmatization, and local magmatism, illustrated most clearly by small syn- to post-tectonic pegmatites. The close spatial and temporal interaction between high-strain deformation and Cenozoic metamorphism, migmatization and magmatism is confined to the northern Chortís Block. It appears that arc magmatism provided the heat input to localize deformation. Consequently, the largest volume and the concentration in time (40–20 Ma) of magmatism and metamorphism seems to have been controlled by shear zones associated with transtensional–transpressional tectonics at the boundary between southern Mexico and the Chortís Block. The Sula graben, and possibly other north-trending grabens of southern Guatemala and northern Honduras, started in temporal and kinematic compatibility with wrenching along the Las Ovejas complex. The topographically well expressed active grabens, however, are neotectonic features. Also the Jocotán–Chamelecón fault zone was part of this deformation system. Breakaway and early eastward translation of the Chortís Block induced exhumation rates of 0.8–1.9 mm/annum in the Xolapa complex and the northern Chortís Block. Initial eastward translation is also recorded by rapid cooling within the southern/southwestern Maya Block at *c.* 30 Ma. Approximately 175, *c.* 130 and *c.* 40 Ma magmatic belts both in southern Mexico and on the Chortís Block consistently argue for $\geq$1100 km offset since $\geq$40 Ma along the northern Caribbean Plate boundary. Post-40 Ma structures and a *c.* 15 Ma magmatic belt indicate that *c.* 700 km of that offset occurred prior to *c.* 15 Ma. The Tonala shear zone, bounding the Chiapas Massif to the SW, and Polochic shear zones were major deformation zones at 9–8 and 12–5 Ma, respectively.

North–south rifts in western Honduras and southeastern Guatemala are related to plateau uplift following slab break-off underneath Central America. Apatite–fission track ages of 11.2–7.9 Ma from the western flank of the Sula graben support the Late Miocene onset of rifting. Presently, the triple junction between the Cocos, North American and Caribbean plates is a zone of diffuse deformation. Normal faulting, exemplified by the grabens in southern Guatemala and western Honduras, extends northward across the Motagua fault zone into the Chuacús complex, indicating that today the Polochic fault may constitute the major plate boundary fault. Distributed, dextral, WNW trending fault zones are active along the active volcanic arc.

This work grew out of our research in southern Mexico in the early 1990s and was started by G. Dengo (Guatemala City) in 1993. Stimulating discussions with C. Rangin and his group at College de France, Aix-en-Provence, and the inspiring work of our Mexican colleagues prompted the first author to put together the data that always seemed to be less bright than those from Asia; we hope that this paper proves otherwise. B. Wauschkuhn (Freiberg) compiled Fig. 1b. G. Seward (UCSB) contributed the EBSD LPO measurement of sample 5PB-10. E. Enkelmann (Lehigh) supported the AFT work. W. Frisch (Tübingen), G. Michel (Bermuda) and U. Riller (Toronto) helped in the field. L. Ratschbacher thanks B. Hacker (UCSB) for endless support in the Ar-lab when both were at Stanford in the mid 1990s. M. Kozuch collected sample SF1D, and A. Andrews-Rodbell and J. M. Gutierrez sample 86TA100. Funded by DFG grants RA442/6 and 25 (to L. Ratschbacher), NSF grants EAR 0510325-02 (to U. Martens), EAR 9219284 (to M. Gordon), CONACYT project D41083-F (to B. Weber), and awards from DAAD and Studienstiftung des Deutschen Volkes to K. Stübner. Reviewed by L. Kennan, J. L. Pindell and an anonymous colleague. Finally, we thank the editors to allow a long and data-packed paper into their volume.

References

ALANIZ-ALVAREZ, S., VAN DER HEYDEN, P., NIETO-SAMANIEGO, A. F. & ORTEGA-GUTIÉRREZ, F. 1996. Radiometric and kinematic evidence for Middle Jurassic strike–slip faulting in southern Mexico related to the opening of the Gulf of Mexico. *Geology*, **24**, 443–446.

ÁLVAREZ-GÓMEZ, J. A., MEIJER, P. T., MARTÍNEZ-DIAZ, J. J. & CAPOTE, R. 2008. Constraints from finite element modeling on the active tectonics of northern Central America and the Middle America Trench. *Tectonics*, **27**, TC1008; doi: 10.1029/2007TC002162.

ANDERSON, J. L. & SMITH, D. R. 1995. The effects of temperature and fO_2 on the Al-in-hornblende barometer. *American Mineralogist*, **80**, 549–559.

ANDERSON, T. H., BURKART, B., CLEMONS, R. E., BOHNENBERGER, O. H. & BLOUNT, D. N. 1973. Geology of the western Altos Cuchumatanes, northwestern Guatemala. *Geological Society of America Bulletin*, **84**, 805–826.

ANDREANI, L., LE PICHON, X., RANGIN, C. & MARTÍNEZ-REYES, J. 2008. The southern Mexico Block, main boundaries and new estimation for its Quaternary motion. *Bulletin de la Société Géologique de France*, **179**, 209–223.

AVÉ LALLEMANT, H. G. & GORDON, M. B. 1999. Deformation history of Roatán Island, Implications for the origin of the Tela Basin (Honduras). *In*: MANN, P. (ed.) *Caribbean Basins. Sedimentary Basins of the World*, **4**. Elsevier, Amsterdam, 197–218.

BANNO, S. 1959. Aegirinaugites from crystalline schists in Sikoku. *Journal Geological Society Japan*, **65**, 652–657.

BARGAR, K. E. 1991. Fluid inclusions and preliminary studies of hydrothermal alteration in core hole PLTG-1, Platanares geothermal area, Honduras. *Journal of Volcanology and Geothermal Research*, **45**, 147–160.

BATESON, J. H. 1972. New interpretation of geology of Maya Mountains, British Honduras. *American Association of Petroleum Geologists, Bulletin*, **56**, 956–963.

BECCALUVA, L., BELLIA, S. *ET AL.* 1995. The northwestern border of the Caribbean Plate in Guatemala: new geological and petrological data on the Motagua ophiolitic belt. *Ofioliti*, **20**, 1–15.

BERTRAND, J. & VUAGNAT, M. 1975. Données chimiques diverses sur des ophiolites du Guatémala. *Bulletin Suisse de Minéralogie Pétrographie*, **57**, 466–483.

BERTRAND, J., DELALOYE, M., FONTIGNIE, D. & VUAGNAT, M. 1978. Ages (K–Ar) sur diverses ophiolites et roches associées de la Cordillère centrale du Guatémala. *Bulletin Suisse de Minéralogie Pétrographie*, **58**, 405–412.

BGR, BUNDESANSTALT FÜR BODENFORSCHUNG 1971. *Geologische Übersichtskarte* 1:125 000, Baja Verapaz und Südteil der Alta Verapaz (Guatemala). D-3 Hannover-Buchholz.

BLUNDY, J. D. & HOLLAND, T. 1990. Calcic amphibole equilibria and a new plagioclase-amphibole geothermometer. *Contributions to Mineralogy and Petrology*, **104**, 204–224.

BOHLEN, S. & LIOTTA, J. 1986. A barometer for garnet amphibolites and garnet granulites. *Journal of Petrology*, **27**, 1025–1034.

BONIS, S., BOHNENBERGER, O. H. & DENGO, G. 1970. *Mapa geológio de la República de Guatemala*. Instituto Geográfico Nacional, Guatemala.

BRUECKNER, H. K., HEMMING, S., SORENSEN, S. & HARLOW, G. E. 2005. Synchronous Sm-Nd mineral ages from HP terranes on both sides of the Motagua Fault of Guatemala: convergent suture and strike–slip fault? *AGU Fall Meeting, Supplement*, **86**, 52.

BURKART, B. 1983. Neogene North America–Caribbean Plate boundary across northern Central America: offset along the Polochic fault. *Tectonophysics*, **99**, 251–270.

BURKART, B. 1994. Northern Central America. *In*: DONOVAN, S. & JACKSON, T. (eds) *Caribbean Geology*. UWI Publ. Association, Kingston, 265–283.

BURKART, B., DEATON, B. C., DENGO, C. & MORENO, G. 1987. Tectonic wedges and offset Laramide structures along the Polochic fault of Guatemala and Chiapas, Mexico: reaffirmation and large Neogene displacement. *Tectonics*, **6**, 411–422.

CAMPA, M. F. & IRIONDO, A. 2004. Significado de dataciones cretácicas de los arcos volcánicos de Taxco, Taxco Viejo y Chacalapa, en la evolución de la plataforma Guerrero Morelos. *IV Reunión Nacional de Ciencias de la Tierra*, Juriquilla, Querétaro, 338.

CARSWELL, D., O'BRIAN, P., WILSON, R. & ZHAI, M. 2000. Metamorphic evolution, mineral chemistry and thermobarometry of schists and orthogneisses hosting ultra-high pressure eclogites in the Dabie Shan of Central China. *Lithos*, **52**, 121–155.

CENTENO-GARCÍA, E. 2005. Review of Upper Paleozoic and Lower Mesozoic stratigraphy and depositional environments of central and west Mexico: constraints on terrane analysis and paleogeography. *In*: ANDERSON, T. H., NOURSE, J. A., MCKEE, J. W. & STEINER, M. B. (eds) *The Mojave-Sonora Megashear Hypothesis: Development, Assessment, and Alternatives*. Geological Society of America Special Paper, **393**, 233–258.

CENTENO-GARCÍA, E., RUIZ, J., CONEY, P. J., PATCHETT, P. J. & ORTEGA, G. F. 1993. Guerrero terrane of Mexico: its role in the Southern Cordillera from new geochemical data. *Geology*, **21**, 419–422.

CERCA, M., FERRARI, L., LÓPEZ-MARTÍNEZ, M., MARTINY, B. & IRONDO, A. 2007. Late Cretaceous shortening and Early Cenozoic shearing in the central Sierra Madre des Sur, southern Mexico: insights into the evolution of the Caribbean–North American plate interaction. *Tectonics*, **26**, TC3007; doi: 10.1029/2006TC001981.

CHIARI, M., DUMITRICA, P., MARRONI, M., PANDOLFI, L. & PRINCIPI, G. 2006. Radiolarian biostratigraphic evidence for a Late Jurassic age of the El Tambor Group ophiolites (Guatemala). *Ofioliti*, **31**, 173–182.

CLEMONS, R. E. 1966. *Geology of the Chiquimula quadrangle, Guatemala, Central America*. PhD thesis, University of Texas, Austin, TX.

COLOMBI, A. 1988. *Métamorphisme et géochimie des roches mafiques des Alpes onest centrals (géoprofil Viége–Domodossola–Lacarno*. PhD thesis, University Lausanne, Switzerland.

CORONA-CHÁVEZ, P., POLI, S. & BIGIOGGERO, B. 2006. Syn-deformational migmatites and magmatic-arc metamorphism in the Xolapa Complex, southern Mexico. *Journal of Metamorphic Geology*, **24**, 169–191.

DAHL, P. S. & FRIBERG, L. M. 1980. The occurrence and chemistry of epidote-clinozoisites in mafic gneisses from the Ruby Range, southwestern Montana. *Contributions to Geology, University Wyoming*, **18**, 77–82.

DEMETS, C. 2001. A new estimate for present-day Cocos–Caribbean Plate motion: implications for slip along the Central American volcanic arc. *Geophysical Research Letters*, **28**, 4043–4046.

DENGO, G. 1969. Problems of tectonic relations between Central America and the Caribbean. *Transactions of the Gulf Coast Association of Geological Societies*, **19**, 311–320.

DICKINSON, W. R. & LAWTON, T. F. 2001. Carboniferous to Cretaceous assembly and fragmentation of Mexico. *Geological Society of America Bulletin*, **113**, 1142–1160.

DONELICK, R. A., RODEN, M. K., MOOERS, J. D., CARPENTER, B. S. & MILLER, D. S. 1990. Etchable length reduction of induced fission tracks in apatite at room temperature (=23 °C): crystallographic orientation effects and 'initial' mean lengths. *Nuclear Tracks and Radiation Measurements*, **17**, 261–265.

DONNELLY, T. W., HORNE, G. S., FINCH, R. C. & LÓPEZ-RAMOS, E. 1990. Northern Central America: the Maya and Chortis blocks. *In*: DENGO, G. &

CASE, J. (eds) *The Caribbean Region*. Geology of North America. Geological Society of America, **H**, 37–76.

DROBE, J. & OLIVER, D. 1998. U–Pb age constraints on Early Cretaceous volcanism and stratigraphy in central Honduras, Geological Society of America. *GSA Cordilleran Section Meeting*, **30**, 12.

DUCEA, M., GEHRELS, G. E., SHOEMAKER, S., RUÍZ, J. & VALENCIA, V. A. 2004a. Geologic evolution of the Xolapa Complex, southern Mexico: evidence from U–Pb zircon geochronology. *Geological Society of America Bulletin*, **116**, 1016–1025.

DUCEA, M., VALENCIA, V. A. *ET AL.* 2004b. Rates of sediment recycling beneath the Acapulco trench: constraints from (U-Th)/He thermochronology. *Journal of Geophysical Research*, **109**; doi: 10.1029/2004JB003112.

ELÍAS-HERRERA, M. & ORTEGA-GUTIÉRREZ, F. 2002. Caltepec fault zone: an Early Permian dextral transpressional boundary between the Proterozoic Oaxacan and Palaeozoic Acatlán complexes, southern Mexico and regional tectonic implications. *Tectonics*, **21**, 1–19.

ELÍAS-HERRERA, M., ORTEGA-GUTIÉRREZ, F., SÁNCHEZ-ZAVALA, J. L., MACÍAS-ROMO, C., ORTEGA-RIVERA, A. & IRIONDO, A. 2005. La falla de Caltepec: raíces expuestas de una frontera tectónica de larga vida entre dos terrenos continentales del sur de México. *Boletín de la Sociedad Geológica Mexicana*, **LVII**, 83–109.

EMMET, P. 1983. *Geology of the Agalteca Quadrangle, Honduras, Central America*. MS thesis, The University of Texas, Austin.

ESCUDER VIRUETE, J., CONTRERAS, F. *ET AL.* 2006a. Transpression and strain partitioning in the Caribbean island-arc: fabric development, kinematics and Ar-Ar ages of syntectonic emplacement of the Loma de Cabrera batholith, Dominican Republic. *Journal of Structural Geology*, **28**, 1496–1519.

ESCUDER VIRUETE, J., DÍAZ DE NEIRA, A. *ET AL.* 2006b. Magmatic relationships and ages of Caribbean Island arc tholeiites, boninites and related felsic rocks, Dominican Republic. *Lithos*, **90**, 161–186.

FERRARI, L., GARDUÑO, V. H., PASQUARÉ, G. & TIBALDI, A. 1994. Volcanic and tectonic evolution of central Mexico: oligocene to present. *Geofísica Internacional*, **33**, 91–105.

FLORES, K., BAUMGARTNER, P. O., SKORA, S., BAUMGARTNER, L., COSCA, M. & CRUZ, D. 2007. The Siuna serpentinite mélange: an Early Cretaceous subduction/accretion of a Jurassic arc. *EOS Transactions, AGU Fall Meeting, Supplement*, **88/52**, Abstract T11D-03.

FOURCADE, E., MÉNDEZ, J. *ET AL.* 1994. Dating of the settling and drowning of the carbonate platform, and of the overthrusting of the ophiolites on the Maya Block during the Mesozoic (Guatemala). *Newsletter Stratigraphy*, **30**, 33–43.

FOURCADE, E., PICCIONI, L., ESCRIBA, J. & ROSSELO, E. 1999. Cretaceous stratigraphy and palaeoenvironments of the Southern Petén Basin, Guatemala. *Cretaceous Research*, **20**, 793–811.

FUHRMAN, M. L. & LINDSLEY, D. H. 1988. Ternary feldspar modeling and thermometry. *American Mineralogist*, **73**, 201–215.

GANGULY, J., CHENG, W. & TIRONE, M. 1996. Thermodynamics of aluminosilicate garnet solid solution: new experimental data, an optimized model, and thermometry applications. *Contributions to Mineralogy and Petrology*, **126**, 137–151.

GARCÍA-CASCO, A., TORRES-ROLDÁN, R. L., MILLÁN, G., MONIÉ, P. & SCHNEIDER, J. 2002. Oscillatory zoning of eclogitic garnet and amphibole, Northern Serpentinite melange, Cuba: a record of tectonic instability during subduction. *Journal of Metamorphic Geology*, **20**, 581–598.

GARCÍA-CASCO, A., LAZARO, C., ROJAS AGRAMONTE, Y., KRÖNER, A. & NEUBAUER, F. 2007. From Aptian onset to Danian demise of subduction along the northern margin of the Caribbean Plate (Sierra del Convento Melange, Eastern Cuba). *AGU Fall Meeting, Supplement*, **88/52**, Abstract T11D-06.

GHENT, E. D., STOUT, M. Z., BLACK, P. M. & BROTHERS, R. N. 1987. Chloritoid bearing rocks with blueschists and eclogites, northern New Caledonia. *Journal of Metamorphic Petrology*, **5**, 239–254.

GOMBERG, D. N., BANKS, P. O. & McBIRNEY, A. R. 1968. Guatemala: preliminary zircon ages from the Central Cordillera. *Science*, **162**, 121–122.

GORDON, M. B. 1993. Revised Jurassic and Early Cretaceous (pre-Yojoa Group) stratigraphy of the Chortis block: paleogeographic and tectonic implications. *GCSSEPM Foundation 13th Annual Research Conference, Proceedings*, 143–154.

GORDON, M. B. & MUEHLBERGER, W. R. 1994. Rotation of the Chortis block causes dextral slip of the Guayape fault. *Tectonics*, **13**, 858–872.

GOSE, W. A. 1985. Paleomagnetic results from Honduras and their bearing on Caribbean tectonics. *Tectonics*, **4**, 565–585.

GRAFE, F., STANEK, K. P. *ET AL*. 2001. Rb–Sr and $^{40}Ar–^{39}Ar$ mineral ages of granitoid intrusives in the Mabujina unit, Central Cuba: thermal exhumation history of the Escambray Massif. *Journal of Geology*, **109**, 615–631.

GRAHAM, C. M. & POWELL, R. 1986. A garnet-hornblende geothermometer: calibration, testing, and application to the Pelona Schists, Southern California. *Journal of Metamorphic Geology*, **2**, 13–31.

GRAY, G. G., POTTORF, R. J., YUREWICZ, D. A., MAHON, K. I., PEVEAR, D. R. & CHUCHLA, R. J. 2001. Thermal and chronological record of syn- to post-Laramide burial and exhumation, Sierra Madre Oriental, Mexico. *In*: BARTOLINI, C., BUFFLER, R. T. & CANTÚ-CHAPA, A. (eds) *The Western Gulf of Mexico Basins: Tectonics, Sedimentary Basins, and Petroleum Systems*. American Association of Petroleum Geologists, Memoirs, **75**, 159–181.

GREEN, T. H. & HELLMAN, P. L. 1982. Fe–Mg partitioning between coexisting garnet and phengite at high pressure, and comments on a garnet–phengite geothermometer. *Lithos*, **15**, 253–266.

GUZMÁN-SPEZIALE, M. 2001. Active seismic deformation in the grabens of northern Central America and its relationship to the relative motion of the North America–Caribbean Plate boundary. *Tectonophysics*, **337**, 39–51.

HALL, C. M., KESLER, S. E. *ET AL*. 2004. Age and tectonic setting of the Camagüey volcanic-intrusive arc,

Cuba: evidence for rapid uplift of the western Greater Antilles. *Journal of Geology*, **112**, 521–542.

HARLOW, G. E. 1994. Jadeitites, albitites and related rocks from the Motagua Fault Zone, Guatemala. *Journal of Metamorphic Geology*, **12**, 49–68.

HARLOW, G. E., HEMMING, S. R., AVÉ LALLEMANT, H. G., SISSON, V. B. & SORENSEN, S. S. 2004. Two high-pressure–low-temperature serpentinite-matrix mélange belts, Motagua fault zone, Guatemala: a record of Aptian and Maastrichtian collisions. *Geology*, **32**, 17–20.

HERNÁNDEZ-TREVIÑO, T., TORRES DE LEÓN, R., SOLÍS-PICHARDO, G., SCHAAF, P., HERNÁNDEZ-BERNAL, M. S. & MORALES-CONTRERAS, J. J. 2004. Edad de la Formación Chapolapa en la localidad del Río Cochoapa al este del Ocotito, estado de Guerrero. *IV Reunión Mexicana de Ciencias de la Tierra*, Juriquilla, Querétaro, 338.

HERRMANN, U., NELSON, B. K. & RATSCHBACHER, L. 1994. The origin of a terrane: U–Pb systematics and tectonics of the Xolapa complex (southern Mexico). *Tectonics*, **13**, 455–474.

HEY, M. 1954. A new review of the chlorites, *Mineralogical Magazine*, **30**, 225–234.

HODGES, K. V. & SPEAR, F. S. 1982. Geothermometry, geobarometry and the Al_2SiO_5 triple point at Mt. Moosilauke, New Hampshire. *American Mineralogist*, **67**, 1118–1134.

HOERNLE, K. & HAUFF, F. 2005. Oceanic igneous complexes. *In*: BUNDSCHUH, J. & ALVARADO, G. (eds) *Central America: Geology, Resources, and Natural Hazards*. Taylor and Francis, Abingdom, 523–547.

HOLLAND, T. J. B. 1980. The reaction albite = jadeite + quartz determined experimentally in the range of 600 °C–1200 °C. *American Mineralogist*, **65**, 129–134.

HOLLAND, T. J. B. 1990. Activities of components in omphacitic solid solution. *Contributions Mineralogy Petrology*, **105**, 446–453.

HOLLAND, T. J. B. & BLUNDY, J. D. 1994. Non-ideal interactions in calcic amphiboles and their bearing on amphibole-plagioclase thermometry. *Contributions to Mineralogy and Petrology*, **116**, 208–224.

HOLLAND, T. J. B. & POWELL, R. 1998. An internally consistent thermodynamic data set for phases of petrological interest. *Journal of Metamorphic Geology*, **16**, 309–343.

HORNE, G., ATWOOD, M. & KING, A. 1974. Stratigraphy, sedimentology, and paleoenvironment of Esquias Formation of Honduras. *American Association of Petroleum Geologists Bulletin*, **58**, 176–188.

HORNE, G., CLARK, G. & PUSHKAR, P. 1976*a*. Pre-Cretaceous rocks of northwestern Honduras: basement terrane in Sierra de Omoa. *American Association of Petroleum Geologists Bulletin*, **60**, 566–583.

HORNE, G., PUSHKAR, A. & SHAFIQULLAH, M. 1976*b*. Laramide plutons on the landward continuation of the Bonacca ridge, northern Honduras. *Publicaciones Geologicas del ICAITI (Guatemala)*, **5**, 84–90.

IGN. 1978. Instituto Geográfico Nacional, Mapa Geológico de Guatemala. Escala 1 : 50 000, *Hoja Río Hondo, Guatemala*.

INDARES, A. & MARTIGNOLE, J. 1985. Biotite garnet geothermometry in the granulite facies: the influence of Ti and Al in biotite. *American Mineralogist*, **70**, 272–278.

JAMES, K. H. 2006. Arguments for and against the Pacific origin of the Caribbean Plate: discussion, finding for an inter-American origin. *Geological Acta*, **4**, 279–302.

JONCKHEERE, R. 2003*a*. On the densities of etchable fission tracks in a mineral and co-irradiated external detector with reference to fission-track dating of minerals. *Chemical Geology*, **200**, 41–58.

JONCKHEERE, R. 2003*b*. On the methodical problems in estimating geological temperature and time from measurements of fission tracks in apatite. *Radiation Measurements*, **36**, 43–55.

JONCKHEERE, R. & WAGNER, G. A. 2000. The KTB apatite fission-track profile: the significance of borehole data in fission-track analysis. *Geological Society of Australia Abstracts*, **58**, 193–194.

JORDAN, B. R., SIGURDSSON, H. *ET AL.* 2007. Petrogenesis of Central American Cenozoic ignimbrites and associated Caribbean Sea tephra. *In*: MANN, P. (ed.) *Geologic and Tectonic Development of the Caribbean Plate Boundary in Northern Central America*. Geological Society of America, Special Papers, **428**, 151–179.

KEPPIE, D. J. 2004. Terranes of Mexico revisited: a 1.3 billion year odyssey. *International Geology Review*, **46**, 765–794.

KEPPIE, D. J., NANCE, R. D. *ET AL.* 2004. Mid-Jurassic tectonothermal event superposed on a Palaeozoic geological record in the Acatlán complex of southern Mexico: hotspot activity during the breakup of Pangea. *Gondwana Research*, **7**, 239–260.

KEPPIE, J. D., NANCE, R. D., FERNÁNDEZ-SUÁREZ, J., STOREY, C. D., JEFFRIES, T. E. & MURPHY, J. B. 2006. Detrital zircon data from the eastern Mixteca terrane, southern Mexico: evidence for an Ordovician–Mississippian continental rise and a Permo-Triassic clastic wedge adjacent to Oaxaquia. *International Geology Review*, **48**, 97–111.

KESLER, S. E., CAMPELL, I. H. & ALLEN, CH. M. 2005. Age of the Los Ranchos Formation, Dominican Republic: timing and tectonic setting of primitive island arc volcanism in the Caribbean region. *Geological Society of America Bulletin*, **117**, 987–995.

KETCHAM, R. A., DONELICK, R. A. & DONELICK, M. B. 2000. AFTSolve: a program for multikinetic modeling of apatite fission-track data. *Geological Materials Research*, **2**, 1–32.

KOHN, M. & SPEAR, F. 1990. Two new geobarometers for garnet-amphibolites, with application to southeastern Vermont. *American Mineralogist*, **75**, 89–96.

KREBS, M., STANEK, K. P. *ET AL.* 2007. Age of high pressure metamorphism from the Escambray Massif, Cuba. Abstract, *Goldschmidt Conference*, Cologne, 2007. Cambridge Publications, A522.

KRETZ, R. 1983. Symbols for rock-forming minerals. *American Mineralogist*, **68**, 277–279.

KROGH, E. J. 2000. The garnet-clinopyroxene Fe/Mg geothermometer: an updated calibration. *Journal of Metamorphic Geology*, **18**, 211–219.

KROGH, E. J. & TERRY, M. P. 2004. Geothermobarometry of UHP and HP eclogites and schists: an evaluation

of equilibria among garnet-clinopyroxene-kyanite-phengite-coesite/quartz. *Journal of Metamorphic Geology*, **22**, 579–592.

LEAKE, B. E., WOOLEY, A. R. *ET AL.* 1997. Nomenclature of amphiboles. Report of the subcommittee on amphiboles of the Intern. Mineral. Association Commission on new minerals and mineral names. *European Journal of Mineralogy*, **9**, 623–665.

LUDWIG, K. R. 2003. Isoplot 3.01. *A Geochronological Toolkit for Microsoft Exel*. Berkeley Geochronology Center Special Publications, **4**, 1–70.

MACHORRO, R. 1993. *Geology of the northwestern part of the Granados Quadrangle, Central Guatemala*. MSc thesis, University of Texas, El Paso.

MALONE, J. R., NANCE, R. D., KEPPIE, J. D. & DOSTAL, J. 2002. Deformational history of part of the Acatlán Complex: Late Ordovician Early Permian orogenesis in southern Mexico. *Journal of South American Earth Sciences*, **15**, 511–524.

MANN, P. 2007. Overview of the tectonic history of northern Central America. *In*: MANN, P. (ed.) *Geologic and Tectonic Development of the Caribbean Plate Boundary in Northern Central America*. Geological Society of America, Special Papers, **428**, 1–19.

MANTON, W. I. 1996. The Grenville of Honduras. *Geological Society of America, Annual Meeting Abstracts with Program*, A-493.

MANTON, W. I. & MANTON, R. S. 1984. *Geochronology and Late Cretaceous–Tertiary Tectonism of Honduras*. Direccion General de Minas e Hidrocarburos, Honduras, 1–55.

MANTON, W. I. & MANTON, R. S. 1999. The southern flank of the Tela basin, Republic of Honduras. *In*: MANN, P. (ed.) *Caribbean Basins. Sedimentary Basins of the World*, **4**. Elsevier, Amsterdam, 219–236.

MARTENS, U., ORTEGA-OBREGÓN, C., ESTRADA, J. & VALLE, M. 2007. Metamorphism and metamorphic rocks. *In*: BUNDSCHUH, J. & ALVARADO, G. (eds) *Central America: Geology, Resources, and Natural Hazards*. Taylor and Francis, Abingdom, 485–522.

MARTENS, U., WEBER, B. & VALENCIA, V. 2010. U–Pb geochronology of lower Paleozoic beds in the southwestern Maya block, Central America: its affinity with Avalonian-type Peri-Gondwanan terranes. *Geological Society of America Bulletin*, in press.

MASSONNE, H.-J. & SZPURKA, Z. 1997. Thermodynamic properties of white micas on the basis of high-pressure experiments in the system $K_2O-MgO-Al_2O_3-SiO_2-H_2O$ and $K_2O-FeO-Al_2O_3-SiO_2-H_2O$. *Lithos*, **41**, 229–250.

MESCHEDE, M. & FRISCH, W. 1998. A plate-tectonic model for the Mesozoic and Early Cenozoic history of the Caribbean Plate. *Tectonophysics*, **296**, 269–291.

MESCHEDE, M., FRISCH, W., HERRMANN, U. & RATSCHBACHER, L. 1996. Stress transmission across an active plate boundary: an example from southern Mexico. *Tectonophysics*, **266**, 81–100.

MMAJ, METAL MINING AGENCY OF JAPAN. 1980. *Report on Geology Survey of the Western Area, Olancho*. Japan International Cooperation Agency, Government of Japan, 1–138.

MORÁN-ZENTENO, D. 1992. *Investigaciones isotópicas de Rb–Sr y Sm–Nd en rocas cristalinas de la eregión de Tierra Colorada–Acapulco–Cruz–Grande, Estado de Guerrero.* PhD thesis, Universidad Autonoma de Mexico, Mexico, DF.

MORÁN-ZENTENO, D. J., CORONA-CHÁVEZ, P. & TOLSON, G. 1996. Uplift and subduction erosion in southwestern Mexico since the Oligocene: Pluton geobarometry constraints. *Earth and Planetary Science Letters*, **141**, 51–65.

MORÁN-ZENTENO, D. J., TOLSON, G. *ET AL.* 1999. Tertiary arc-magmatism of the Sierra Madre del Sur, Mexico, and its transition to the volcanic activity of the Trans-Mexican Volcanic Belt. *Journal of South American Earth Sciences*, **12**, 513–535.

NIETO-SAMANIEGO, A. F., ALANIZ-ÁLVAREZ, S. A., SILVA-ROMO, G., EGUIZA-CASTRO, M. H. & MENDOZA-ROSALES, C. C. 2006. Latest Cretaceous to Miocene deformation events in the eastern Sierra Madre del Sur, Mexico, inferred from the geometry and age of major structures. *Geological Society of America Bulletin*, **118**, 238–252.

ORTEGA-GUTIÉRREZ, F., ELÍAS-HERRERA, M., MACÍAS-ROMERO, C. & LÓPEZ, R. 1999. Late Ordovician–Early Silurian continental collisional orogeny in southern Mexico and its bearing on Gondwana–Laurentia connections. *Geology*, **27**, 719–722.

ORTEGA-GUTIÉRREZ, F., SOLARI, L. A., SOLÉ, J., MARTENS, U., GÓMEZ-TUENA, A., MORÁN-ICAL, S., REYES-SALA, M. & ORTEGA-OBREGÓN, C. 2004. Polyphase, high-temperature eclogite-facies metamorphism in the Chuacús complex, Central Guatemala: petrology, geochronology, and tectonic implications. *International Geology Review*, **46**, 445–470.

ORTEGA-GUTIÉRREZ, F., SOLARI, L. A. *ET AL.* 2007. The Maya–Chortís boundary: a tectonostratigraphic approach. *International Geology Review*, **49**, 996–1024.

ORTEGA-OBREGÓN, C., SOLARI, L., ORTEGA-GUTIÉRREZ, F., SOLÉ-VIÑAS, J. & GÓMEZ-TUENA, A. 2004. Caracterización estructural, petrológica y geocronológica de la zona de cizalla "Baja Verapaz", Guatemala. *Libro de Resúmenes, IV Reunión Nacional de Ciencias de la Tierra*, Juriquilla, 2004, 204.

ORTEGA-OBREGÓN, C., SOLARI, L., KEPPIE, J. D., ORTEGA-GUTIÉRREZ, F., SOLÉ, J. & MORÁN, S. 2008. Middle–Late Ordovician magmatism and Late Cretaceous collision in the southern Maya block, Rabinal-Salamá area, central Guatemala: implications for North America–Caribbean Plate tectonics. *Geological Society of America Bulletin*, **120**, 556–570.

PERCHUK, L. L. & LAVRENT'EVA, J. V. 1983. Experimental investigations of exchange equilibria in the system cordierite–garnet–biotite. *In*: SAXENA, S. (ed.) *Kinematics and Equilibria in Mineral Reactions*. Springer, Berlin, 199–239.

PINDELL, J. K., CANDE, S. C. *ET AL.* 1988. A plate-kinematic framework for models of Caribbean evolution. *Tectonophysics*, **155**, 121–138.

PINDELL, J. L., KENNAN, L., STANEK, K. P., MARESCH, W. & DRAPER, G. 2006. Foundations of Gulf of Mexico and Caribbean evolution: eight controversies resolved. *Geologica Acta*, **99**, 303–341.

PLAFKER, G. 1976. Tectonic aspects of the Guatemala earthquake of 4 February 1976. *Science*, **193**, 1201–1208.

PLYUSNINA, L. P. 1982. Geothermometry and geobarometry of plagioclase-hornblende bearing assemblages. *Contributions Mineralogy Petrology*, **80**, 140–146.

RATSCHBACHER, L., RILLER, U., MESCHEDE, M., HERRMANN, U. & FRISCH, W. 1991. Second look at suspect terranes in southern Mexico. *Geology*, **19**, 1233–1236.

RILLER, U., RATSCHBACHER, L. & FRISCH, W. 1992. Left-lateral transtension along the Tierra Colorada deformation zone, northern margin of the Xolapa magmatic arc of southern Mexico. *Journal of South American Earth Sciences*, **5**, 237–249.

ROBINSON, K. L., GASTIL, R. G., CAMPA, M. F. & RAMIREZ-ESPINOSA, J. 1989. Geochronology of basement and metasedimentary rocks in southern Mexico and their relation to metasedimentary rocks in Peninsular California. *Geological Society of America Abstracts with Programs*, **21**, 135.

ROGERS, R. D. 2003. *Jurassic–Recent tectonic and stratigraphic history of the Chortis block of Honduras and Nicaragua (northern Central America).* PhD thesis, University of Texas, Austin, TX.

ROGERS, R. D., KÁRASON, H. & VAN DER HILST, R. D. 2002. Epeirogenic uplift above a detached slab in northern Central America. *Geology*, **30**, 1031–1034.

ROGERS, R. D., MANN, P., SCOTT, R. W. & PATINO, L. 2007a. Cretaceous intra-arc rifting, sedimentation, and basin inversion in east-central Honduras. *In*: MANN, P. (ed.) *Geologic and Tectonic Development of the Caribbean Plate Boundary in Northern Central America*. Geological Society of America, Special Papers, **428**, 89–128.

ROGERS, R. D., MANN, P., EMMET, P. A. & VENABLE, M. E. 2007b. Colon fold belt of Honduras: evidence for Late Cretaceous collision between the continental Chortis block and intra-oceanic Caribbean Arc. *In*: MANN, P. (ed.) *Geologic and Tectonic Development of the Caribbean Plate Boundary in Northern Central America*. Geological Society of America, Special Papers, **428**, 129–149.

ROGERS, R. D., MANN, P. & EMMET, P. A. 2007c. Tectonic terranes of the Chortis block based on integration of regional aeromagnetic and geologic data. *In*: MANN, P. (ed.) *Geologic and Tectonic Development of the Caribbean Plate Boundary in Northern Central America*. Geological Society of America, Special Papers, **428**, 65–88.

ROJAS-AGRAMONTE, Y., KRÖNER, A., GARCÍA-CASCO, A., ITURRALDE-VINENT, M. A., WINGATE, M. T. D. & LIU, D. Y. 2006. Geodynamic implications of zircon ages from Cuba. *Geophysical Research Abstracts*, **8**, 04943; SRef-ID: 1607-7962/gra/EGU06-A-04943.

ROSENCRANTZ, E., ROSS, M. I. & SLATER, J. G. 1988. Age and spreading history of the Cayman Trough as determined from depth, heat flow, and magnetic anomalies. *Journal Geophysical Research*, **93**, 2141–2157.

RYBURN, R. J., RAHEIM, A. & GREEN, D. H. 1976. Determination of the PT paths of natural eclogites during metamorphism – record of a subduction. *Lithos*, **9**, 161–164.

SÁNCHEZ-ZAVALA, J. L., ORTEGA-GUTIÉRREZ, F., KEPPIE, J. D., JENNER, G. A. & BELOUSOVA, E. 2004. Ordovician and Mesoproterozoic zircon from the Tecomate Formation and Esperanza granitoid, Acatlán complex, southern Mexico: local provenance in the Acatlán and Oaxacan complexes. *International Geology Review*, **46**, 1005–1021.

SCHAAF, P., MORÁN-ZENTENO, D., HERNÁNDEZ-BERNAL, M. S., SOLIS-PICHARDO, G., TOLSON, G. & KÖHLER, H. 1995. Paleogene continental margin truncation in southwestern Mexico: geochronological evidence. *Tectonics*, **14**, 1339–1350.

SCHAAF, P., WEBER, B., WEIS, P., GROß, A., ORTEGA-GUTIÉRREZ, F. & KÖHLER, H. 2002. The Chiapas Massif (Mexico) revised: new geologic and isotopic data for basement characteristics. *Neues Jahrbuch Geologie Paläontologie, Abhandlungen*, **225**, 1–23.

SCHMID, R., FRANZ, L., OBERHÄNSLI, R. & DONG, S. 2000. High Si-phengite, mineral chemistry and P-T evolution of ultra-high-pressure eclogites and calc-silicates from the Dabie Shan, eastern PR China. *Geological Journal*, **35**, 185–207.

SCHNEIDER, J., BOSCH, D. *ET AL.* 2004. Origin and evolution of the Escambray Massif (Central Cuba): an example of HP/LT rocks exhumed during intraoceanic subduction. *Journal of Metamorphic Geology*, **22**, 227–247.

SCHWARTZ, D. P. 1976. *Geology of the Zacapa quadrangle and vicinity, Guatemala, Central America*. PhD thesis, State University of New York, Binghamton.

SOLARI, L. A., DOSTAL, J., ORTEGA-GUTIÉRREZ, F. & KEPPIE, J. D. 2001. The 275 Ma arc-related La Carbonara stock in the northern Oaxacan Complex of southern Mexico: U–Pb geochronology and geochemistry. *Revista Mexicana de Ciencias Geológicas*, **18**, 149–161.

SOLARI, L. A., KEPPIE, J. D., ORTEGA-GUTIÉREZ, F., ORTEGA-RIVERA, A., HAMES, W. E. & LEE, J. K. W. 2004. Phanerozoic structures in the Grenville complex, southern Mexico: result of thick-skinned tectonics. *International Geology Review*, **46**, 614–628.

SOLARI, L. A., TORRES DE LEÓN, R., HERNÁNDEZ PINEDA, G., SOLÉ, J., SOLÍS-PICHARDO, G. & HERNÁNDEZ-TREVIÑO, G. 2007. Tectonic significance of Cretaceous–Tertiary magmatic and structural evolution of the northern margin of the Xolapa Complex, Tierra Colorada area, southern Mexico. *Geological Society of America Bulletin*, **119**, 1265–1279.

SOLARI, L. A., ORTEGA-GUTIÉRREZ, F. *ET AL.* 2009. U–Pb zircon geochronology of Palaeozoic units in Western and Central Guatemala: insights into the tectonic evolution of Middle America. *In*: JAMES, K. H., LORENTE, M. A. & PINDELL, J. L. (eds) *The Origins and Evolution of the Caribbean Plate*. Geological Society, London, Special Publications, **328**, 295–313.

SPEAR, F. S. 1981. NaSi–CaAl exchange equilibrium between plagioclase and amphibole. *Contributions to Mineralogy and Petrology*, **72**, 33–41.

STANEK, K. P., RIßE, A., RENNO, A., ROMER, R. & GRAFE, F. 2005. The history of the Great Antillean Island Arc: example from Central Cuba. Abstract, Colloquium on Latin American Geosciences 2005, Potsdam, Terra Nostra, 1.

STEINER, M. B. & WALKER, J. D. 1996. Late Silurian plutons in Yucatan. *Journal Geophysical Research*, **101**, 17727–17735.

STIPP, M., STÜNITZ, H., HEILBRONNER, R. & SCHMID, S. M. 2002. The eastern Tonale fault zone: a natural laboratory for crystal plastic deformation of quartz over a temperature range from 250 to 700°C. *Journal of Structural Geology*, **24**, 1861–1884.

SUTTER, J. F. 1979. Late Cretaceous collisional tectonics along the Motagua fault zone, Guatemala. *Geological Society of America, Abstracts with Programs*, **11**, 525–526.

TALAVERA-MENDOZA, O., RUIZ, J., GEHRELS, G. E., MEZA-FIGUEROA, D., VEGA-GRANILLO, R. & CAMPA-URANGA, M. F. 2005. U–Pb geochronology of the Acatlán complex and implications for the Paleozoic paleogeography and tectonic evolution of southern Mexico. *Earth and Planetary Science Letters*, **235**, 682–699.

TALAVERA-MENDOZA, O., RUIZ, J., GEHRELS, G. E., VALENCIA, V. A. & CENTENO-GARCÍA, E. 2007. Detrital zircon U–Pb geochronology of southern Guerrero and western Mixteca arc successions (southern Mexico): new insights for the tectonic evolution of southwestern North America during the Late Mesozoic. *Geological Society of America Bulletin*, **119**, 1052–1065.

TOLSON, G. 2005. La falla Chacalapa en el sur de Oaxaca. *Boletín de la Sociedad Geológica Mexicana*, **57**, 111–122.

TORRES, R., RUIZ, J., PATCHETT, P. J. & GRAJALES, J. M. 1999. Permo-Triassic continental arc in eastern Mexico: tectonic implications for reconstructions of southern North America. *In*: BARTOLINI, C. W., WILSON, J. L. & LAWTON, T. F. (eds) *Mesozoic Sedimentary and Tectonic History of North-Central Mexico*. Geological Society of America, Special Papers, **340**, 191–196.

TRIBOULET, C. 1992. The (Na–Ca)amphibole–albite–chlorite–epidote–quartz geothermobarometer in the system S–A–F–M–C–N–H_2O. 1. An empirical calibration. *Journal of Metamorphic Geology*, **10**, 545–556.

TSUJIMORI, T., LIOU, J. G. & COLEMAN, R. G. 2004. Comparison of two contrasting eclogites from the Motagua fault zone, Guatemala: southern lawsonite eclogite versus northern zoisite eclogite. *Geological Society of America Abstracts with Programs*, **36**, 136.

TSUJIMORI, T., LIOU, J. G. & COLEMAN, R. G. 2005. Coexisting retrograde jadeite and omphacite in a jadeite-bearing lawsonite eclogite from the Motagua Fault Zone, Guatemala. *American Mineralogist*, **90**, 836–842.

TSUJIMORI, T., SISSON, V. B., LIOU, J. G., HARLOW, G. E. & SORENSEN, S. S. 2006. Petrologic characterization of Guatemalan lawsonite eclogite: Eclogitization of subducted oceanic crust in a cold subduction zone. *In*: HACKER, B. R., MCCLELLAND, W. C. & LIOU, J. G. (eds) *Ultrahigh-pressure Metamorphism: Deep*

Continental Subduction. Geological Society of America, Special Papers, **403**, 147–168.

VAN DEN BOOM, G. 1972. Petrofazielle Gliederung des metamorphen Grundgebirges in der Sierra de Chuacús, Guatemala. *Beihefte Geologisches Jahrbuch*, **122**, 5–49.

VEGA-GRANILLO, R., TALAVERA-MENDOZA, O. *ET AL.* 2007. Pressure–temperature–time evolution of Paleozoic high-pressure rocks of the Acatlán Complex (southern Mexico): implications for the evolution of the Iapetus and Rheic Oceans. *Geological Society of America Bulletin*, **119**, 1249–1264.

VENABLE, M. 1994. *A geological, tectonic, and metallogenetic evaluation of the Siuna terrane (Nicaragua)*. PhD thesis, University of Arizona, Tucson.

VILAND, J., HENRY, B., CALIX, R. & DIAZ, C. 1996. Late Jurassic deformation in Honduras: proposals for a revised regional stratigraphy. *Journal of South American Earth Sciences*, **9**, 153–160.

VON BLANCKENBURG, F. V., VILLA, I., BAUR, H., MORTEANI, G. & STEIGER, R. H. 1989. Time calibration of a *PT*-path in the Western Tauern Window, Eastern Alps: the problem of closure temperatures. *Contributions to Mineralogy and Petrology*, **101**, 1–11.

WATERS, D. J. & MARTIN, H. N. 1993. Geobarometry in phengite-bearing eclogites. *Terra Abstracts*, **5**, 410–411; updated calibration of 1996 at World Wide Web Address: http://www.earth.ox.ac.uk/c.davewa/ecbar.html.

WAWRZYNIEC, T., MOLINA-GARZA, R. S., GEISSMAN, J. & IRIONDO, A. 2005. A newly discovered relic, transcurrent plate boundary – the Tonala shear zone

and paleomagnetic evolution of the western Maya block, SW Mexico. *Geological Society of America Abstract with Programs*, **37**, 68.

WEBER, B., CAMERON, K. L., OSORIO, M. & SCHAAF, P. 2005. A Late Permian tectonothermal event in Grenville crust of the Southern Maya terrane; U–Pb zircon ages from the Chiapas Massif, Southeastern México. *International Geology Review*, **47**, 509–529.

WEBER, B., IRIONDO, A., PREMO, W. R., HECHT, L. & SCHAAF, P. 2006. New insights into the history and origin of the southern Maya Block, SE México: U-Pb-SHRIMP zircon geochronology from metamorphic rocks of the Chiapas Massif. *International Journal of Earth Sciences*; doi: 10.1007/s00531-006-0093-7.

WEBER, B., VALENCIA, V. A., SCHAAF, P., POMPA-MERA, V. & RUIZ, J. 2008. Significance of provenance ages from the Chiapas Massif complex (SE Mexico): redefining the Paleozoic basement of the Maya block and its evolution in a peri-Gondwanan entourage. *Journal of Geology*, **6**, 619–639.

WEILAND, T., SUAYAH, W. & FINCH, R. 1992. Petrologic and tectonic significance of Mesozoic volcanic rocks in the Río Wampú area, eastern Honduras. *Journal of South American Earth Sciences*, **6**, 309–325.

WILSON, H. H. 1974. Cretaceous sedimentation and orogeny in nuclear Central America. *American Association of Petroleum Geologists Bulletin*, **58**, 1348–1396.

YAÑEZ, P., RUIZ, J., PATCHETT, P. J., ORTEGA-GUTIERREZ, F. & GEHRELS, G. E. 1991. Isotopic studies of the Acatlán complex, southern Mexico: implications for Paleozoic North American tectonics. *Geological Society of America Bulletin*, **103**, 817–828.

U–Pb zircon geochronology of Palaeozoic units in Western and Central Guatemala: insights into the tectonic evolution of Middle America

L. A. SOLARI[1]*, F. ORTEGA-GUTIÉRREZ[2], M. ELÍAS-HERRERA[2],
P. SCHAAF[3], M. NORMAN[4], R. TORRES DE LEÓN[2], C. ORTEGA-OBREGÓN[2],
M. CHIQUÍN[5] & S. MORÁN ICAL[5]

[1]Centro de Geociencias, Universidad Nacional Autónoma de México, Campus Juriquilla,
Querétaro 76230, QRO, Mexico

[2]Instituto de Geología, Universidad Nacional Autónoma de México, Ciudad Universitaria,
04510 Del. Coyoacán, México D.F., Mexico

[3]Instituto de Geofísica, Universidad Nacional Autónoma de México, Ciudad Universitaria,
04510 Del. Coyoacán, México D.F., Mexico

[4]Research School of Earth Sciences, Australian National University, Mills Road,
Building 61, Canberra, ACT 0200, Australia

[5]Universidad de San Carlos, Cobán, Guatemala

*Corresponding author (e-mail: solari@servidor.unam.mx)

Abstract: Precambrian and Palaeozoic basements are present in southern Mexico and Central America, where several crustal blocks are recognized by their different geological record, and juxtaposed along lateral faults. Pre-Mesozoic reconstructions must take into account the nature of such crustal blocks, their geological history, age and petrology. Some of those crustal blocks are currently located between southernmost north America (the Maya Block) and Central America (Chortís Block).To better understand the geology of these crustal blocks, and to establish comparisons between their geological history, we performed U–Pb dating of both igneous and metasedimentary key units cropping out in central and western Guatemala. In the Altos Cuchumatanes (Maya Block) granites yield both Permian (269 ± 29 Ma) and Early Devonian (391 ± 7.4 Ma) U–Pb ages. LA-ICPMS detrital zircon ages from rocks of the San Gabriel sequence, interpreted as the oldest metasedimentary unit of the Maya Block, and overlain by the Late Palaeozoic Upper Santa Rosa Group, yield Precambrian detrital zircons bracketed between c. 920 and c. 1000 Ma. The presence of these metasedimentary units, as well as Early Devonian to Silurian granites in the Mayan continental margin, from west (Altos Cuchumatanes), to east (Maya Mountains of Belize) indicates a more or less continuous belt of Lower Palaeozoic igneous activity, also suggesting that the continental margin of the Maya Block can be extended south of the Polochic fault, up to the Baja Verapaz shear zone. A metasedimentary sample belonging to the Chuacús Complex yielded detrital zircons with ages between c. 440 and c. 1325 Ma. The younger ages are similar to the igneous ages reported from the entire southern Maya continental margin, and show proximity of the Complex in the Middle-Late Palaeozoic. The S. Diego Phyllite, which overlies high-grade basement units of the Chortís Block, contains zircons that are Lower Cambrian (c. 538 Ma), Mesoproterozoic (c. 980 to c. 1150 Ma) and even Palaeoproterozoic (c. 1820 Ma). Absence of younger igneous zircons in the San Diego Phyllite indicates that either its sedimentation took place in a close range of time, during the Late Cambrian, or absence of connection between Chortís and Maya Blocks during the Early–Mid-Palaeozoic. The Precambrian zircons could have come from southern Mexico (Oaxaca and Guichicovi Complexes), or from Mesoproterozoic Massifs exposed in Laurentia and Gondwana. Palaeogeographic models for Middle America are limited to post-Jurassic time. The data presented here shed light on Palaeozoic and, possibly, Precambrian relationships. They indicate that Maya and the Chortís did not interact directly until the Mesozoic or Cenozoic, as they approached their current position.

From: JAMES, K. H., LORENTE, M. A. & PINDELL, J. L. (eds) *The Origin and Evolution of the Caribbean Plate*.
Geological Society, London, Special Publications, **328**, 295–313.
DOI: 10.1144/SP328.12   0305-8719/09/$15.00 © The Geological Society of London 2009.

Southern Mexico and Central America are composed of several fault-bounded crustal blocks with different geological records (Fig. 1a). Several authors regard these as suspect terranes (e.g. Campa & Coney 1983; Sedlock *et al.* 1993; Keppie 2004). The geological record ranges back to the Precambrian, as represented by Mesoproterozoic basement of the Zapoteco Terrane (Oaxacan Complex, Keppie *et al.* 2003; Solari *et al.* 2003) of the Maya terrane (Guichicovi Complex, Weber & Kohler 1999) and of the Chortís Block (Manton 1996). The Palaeozoic is recorded in the Acatlán Complex, basement of the Mixteco terrane (e.g. Keppie *et al.* 2004; Ortega-Gutiérrez *et al.* 1999; Talavera-Mendoza *et al.* 2005), the sedimentary cover of the Oaxacan Complex (Gillis *et al.* 2005; Murphy *et al.* 2005), and in the Maya Block as both sedimentary cover (e.g. Ortega-Obregón *et al.* 2008; Weber *et al.* 2006*b*) and igneous rocks associated with metamorphic units (Ortega-Obregón *et al.* 2008; Steiner & Walker 1996; Weber *et al.* 2005, 2006*b*). The Chatino terrane (Xolapa Complex), which defines the southern boundary of the Guerrero, Mixteco and Zapoteco terranes, also contains local Palaeozoic rocks (e.g. Ducea *et al.* 2004). The Maya Block may extend southward of the Polochic Fault as far as the Baja Verapaz shear zone (Ortega-Obregón *et al.* 2008). Yet further south at least three crustal blocks or terranes are recognized by Ortega-Gutiérrez *et al.* (2007), all bounded by left-lateral faults, and lying north of the Chortís. Palaeozoic or older rocks occur in most of these blocks (e.g. Donnelly *et al.* 1990; Horne *et al.* 1976; Ortega-Gutiérrez *et al.* 2004, 2007). According to Dickinson and Lawton (2001) and Elías-Herrera & Ortega-Gutiérrez (2002) the southern Mexico terranes approached their current positions after the Carboniferous–Permian. Some

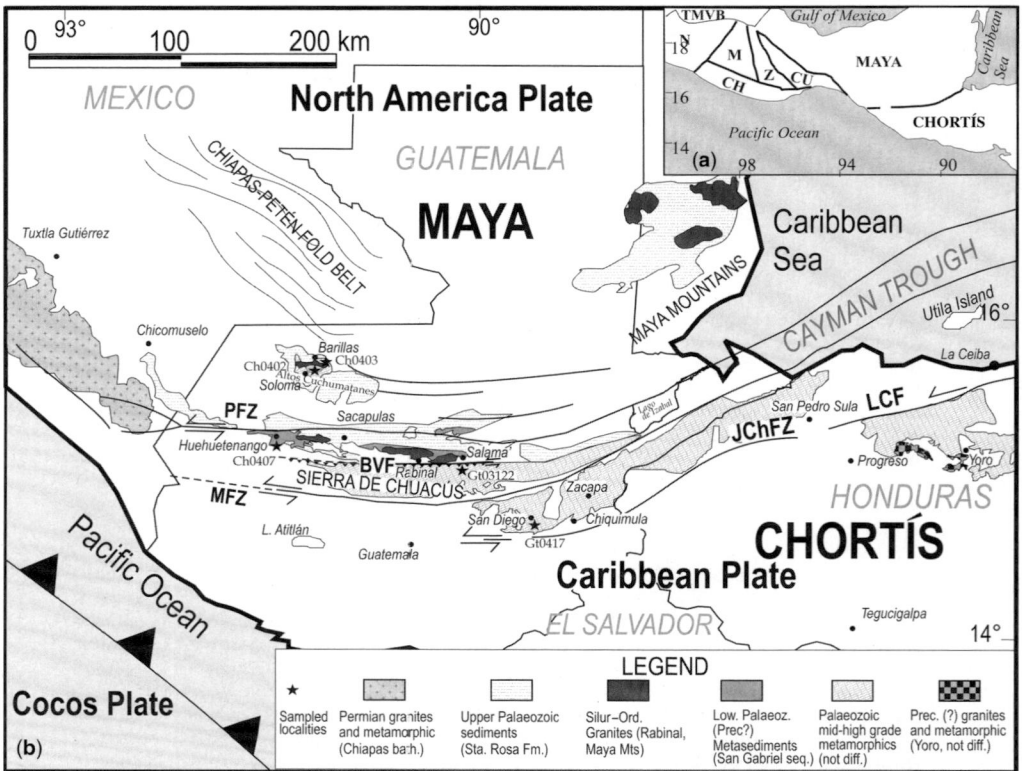

Fig. 1. (**a**) Main subdivision of southern Mexico and Central America in crustal blocks. N, Náhuatl-Guerrero Terrane; TMVB, trans Mexican Volcanic Belt; M, Mixteco terrane; CH, Chatino Terrane; Z, Zapoteco Terrane; CU, Cuicateco Terrane; MAYA, Maya Block; CHORTIS, Chortís Block; PFZ, Polochic Fault Zone; MFZ, Motagua Fault Zone; BVZ, Baja Verapaz Fault Zone; JChFZ, Yucatán–Chamelecón Fault Zone; LCF, La Ceiba Fault. (**b**) Schematic geological map showing basement rocks of central Guatemala, southeasternmost Mexico, and northwestern Honduras. Geology is recompiled from Kesler *et al.* (1970); Anderson *et al.* (1973); Steiner & Walker (1996); Weber *et al.* (2005, 2006*b*); Ortega-Obregón *et al.* (2008) and unpublished data of the authors.

of the bounding faults, such as the Chacalapa–La Venta fault system, are dated as Oligocene NE of Puerto Angel, Oaxaca (Tolson 2005), or as Early Eocene at Tierra Colorada, Guerrero (Solari et al. 2007).

The Polochic–Motagua–Chamelecón–La Ceiba fault system has been active at least since the Miocene (e.g. Burkart 1983). However, data constraining the Palaeozoic palaeogeography of terranes in Central America are scarce. To better understand the palaeogeography of crustal blocks in the southeasternmost Mexican terrane, the Maya Block, and its counterparts in central (Sierra de Chuacús, or Jacalteco terrane of Ortega-Gutiérrez et al. 2007) and southern Guatemala (Chortís Block), and to investigate their possible geological relationships, we performed U–Pb zircon dating by laser ablation inductively coupled plasma (LA-ICPMS) on metasedimentary samples belonging to some of the aforementioned localities, as well as isotopic dilution thermal ionization mass spectrometry (ID-TIMS) U–Pb zircon dating of granites in the Altos Cuchumatanes, western Guatemala.

Regional geology and sample descriptions

Central America basement rocks have been described in several works since the early 1960s. They were previously described by Donnelly et al. (1990), Ortega-Gutiérrez et al. (2007), and Weyl (1980).

In general, the basement of the Maya Block crops out north of the Polochic fault in southeasternmost Mexico (Chiapas Massif, Weber et al. 2005, 2006a) and in western Guatemala (Altos Cuchumatanes, Anderson et al. 1973). It lies between the Polochic and Motagua faults (e.g. McBirney 1963; Kesler et al. 1970; Anderson et al. 1973; Ortega-Gutiérrez et al. 2004; Ortega-Obregón et al. 2008), and south of the Motagua fault, both in Guatemala (Las Ovejas Complex of Lawrence 1975), and Honduras (Sierra de Omoa, Horne et al. 1976).

Altos Cuchumatanes and 'Western Chuacús Group'

The core of the Altos Cuchumatanes anticlinorium (Fig. 1b) exposes low- to high-grade metamorphic rocks such as phyllites, slate, micaschists, subordinate biotite–garnet gneisses and garnet amphibolites, intruded by granites. Phyllites contain andalusite, indicating contact metamorphism and thus an older age than the intrusives. According to Anderson et al. (1973), who performed geological mapping in the area, such rocks correlate with those exposed south of Huehuetenango, south of Polochic fault, described by Kesler et al. (1970) as the Western Chuacús Group.

Unconformably overlying the basement are fusulinid-bearing limestones, and deformed slates and mudstones of the Santa Rosa Group (according to Anderson et al. 1973), cropping out north of Soloma and SE of Barillas, in the core of the Altos Cuchumatanes, suggest that the metamorphic rocks are at least pre-Carboniferous. Higher up in the stratigraphy lie Jurassic red beds of the Todos Santos Formation and Cretaceous limestones of the Ixcoy Formation.

Pink to greenish granitoids intruded Palaeozoic low-grade metasediments and appear at outcrop scale as generally biotite-rich and undeformed. Petrographically they show limited intra-grain deformation and show a pervasive chloritization of biotite, and can be classified as granites to granodiorites.

Granites in the Altos Cuchumatanes were chosen for U–Pb ID-TIMS analyses to define their crystallization ages. Sample CH0402 was collected ENE of Soloma, a small town about 40 km north of Huehuetenango. The sample is made up of quartz, orthoclase, plagioclase, scarce biotite and some accessory minerals such as apatite, pyrite, zircon and magnetite. Secondary minerals, mainly after weathering of plagioclase and biotite, include chlorite, sericite, clay minerals and manganese oxides. Pink colours are evident in fresh outcrops, principally expressed as hues around K-feldspars. Some crustal xenoliths, mainly felsic and banded gneisses, are present in the entire outcrop. Sample CH0403 was collected 5 km ESE of Barillas, near the bed of Amelco River. It consists of a pale grey to green granodiorite made up of quartz, K-feldspar, plagioclase, biotite, rare hornblende, secondary white mica, opaque minerals and accessory phases such as zircon and apatite. Mafic gneisses are also present in the vicinity of the granitic outcrop, including garnet and amphibole-bearing foliated metabasalts. However, we could not define the original contact between granite and gneiss, because of late shearing.

Similar granites, about 5 km south of Huehuetenango, intrude sheared metasediments ranging from very low metamorphic grade up to low greenschist facies. The latter are metagreywackes and metaquartz arenites with local marble lenses and some possible metavolcanic units generally with plagioclase, quartz, chlorite, sericite and some opaque mineral, probably magnetite and ilmenite. Minor units, richer in muscovite, are also present. Abundant granitic and aplitic intrusions are also present east and ESE of Huehuetenango, intruding the low-grade sequence. The only available work on the area (Kesler et al. 1970) referred to the metasediments as the 'Western Chuacús Group', thus correlating them with the rocks cropping out farther to the east, previously described by McBirney (1963). Recent studies on the Sierra de

Chuacús (Ortega-Gutiérrez *et al.* 2004) suggest that typical rocks are medium- to high-grade gneisses, migmatites, paragneisses, mafic rocks and marbles, all affected by high-to very-high-pressure metamorphism. Low-grade metasediments cropping out south of Huehuetenango completely lack high-pressure or partial melting indicators and, although sheared, appear structurally much simpler than rocks of the Sierra de Chuacús. These rocks resemble the sequence exposed just north of the Sierra de Chuacús, from the Baja Verapaz shear zone along the Rabinal-Salamá valley.

San Gabriel sequence and Rabinal suite, Central Guatemala

The San Gabriel sequence (Salamá sequence in Ortega-Obregón *et al.* 2008) crops out between San Gabriel and Rabinal, Baja Verapaz (Fig. 1b). It consists of sheared metasediments, mainly sandstones, minor shales and mudstones, greywackes, and some arkoses. A few metavolcanic members are also present, possibly developed from original ashes of felsic and intermediate composition. Metamorphic grade ranges from very low-grade up to subgreenschist facies. High pressure detrital minerals are absent, precluding a sedimentary provenance from the nearby Chuacús Complex, which overthrusts the San Gabriel sequence showing a top-to the-NE sense of shearing. The Chuacús Complex is characterized by high-pressure minerals, with incipient migmatization apparent in the core of its exposure (Ortega-Gutiérrez *et al.* 2004). It exhibits a lower metamorphic grade (albite-epidote-garnet-biotite) toward the Baja Verapaz thrust zone.

The Rabinal granite suite crops out north of Rabinal town, in Baja Verapaz, and extends along the whole Salamá–Rabinal–San Gabriel–San Miguel area, where it intrudes the San Gabriel sequence and is sheared together with it (Fig. 2b). This is a typical S-type granite, made up of quartz, plagioclase, minor K-feldspar, muscovite, chloritized biotite and minor opaque minerals. Weathering is intense, and generally it is grey-greenish in colour, due to the abundant retrogression. Rabinal granite was previously dated by U–Pb on large zircon fractions (Gomberg *et al.* 1968). They obtained a Grenvillian upper intercept (1075 Ma), and a lower intercept at *c.* 390 Ma, interpreted as a metamorphic event. More recently, Ortega-Obregón *et al.* (2008) reported K–Ar ages on muscovite of 429 and 440 Ma, interpreted as minimum magmatic ages. Martens *et al.* (2005) reported several preliminary U–Pb SHRIMP concordant ages that spread between *c.* 400 and *c.* 500 Ma. The accurate magmatic age of this granite is thus still unresolved.

East and north of Salamá, the San Gabriel sequence is covered by conglomerates with granite

clasts, followed by shales and limestones. The limestones contain Tournaisian (*c.* 345 Ma) conodonts (Ortega-Obregón *et al.* 2008), and thus this last sedimentary package can be considered somewhat equivalent to the Sacapulas Formation (Bohnenberger 1966), the basal member of the Santa Rosa Group, but of Late Pennsylvanian age.

A sheared metasediment sample (CH0407) for detrital zircons analysis was collected from about 6–8 km south of the Polochic fault, near Huehuetenango. It contains quartz, plagioclase, white mica, minor biotite, chlorite, minor epidote and opaque minerals, whereas zircon is a fairly abundant accessory. Undeformed granitic and aplitic dykes cut the metasediment. The area was previously mapped by Kesler *et al.* (1970), who correlated it with the Chuacús Group. Our knowledge of the Sierra de Chuacús lithologies suggests that the metasediments cropping out south of Huehuetenango are similar to those found in the San Gabriel sequence. Dating of detrital zircons in this sample would provide important information on its maximum sedimentary age and would confirm or modify further correlations between the Western Chuacús Group of Kesler *et al.* (1970) and the Chuacús Complex of Ortega-Gutiérrez *et al.* (2004).

Sample Gt03122 was collected from south of Salamá town, along the small paved road that goes towards La Canoa, crossing the Sierra de Chuacús. The sample is a biotite- and muscovite-bearing metasediment, which also contains plagioclase, subordinate quartz, rutile, some garnet, apatite, zircon, magnetite and subordinate pyrite. Because this outcrop lies south of the Baja Verapaz Shear Zone (e.g. Ortega-Obregón *et al.* 2008), we consider this locality as the northernmost outcrop of the Chuacús Complex (Ortega-Gutiérrez *et al.* 2004). Dating of detrital zircons from this sample would provide constraints for depositional age and provenance, as well as the age of accretion between the Chuacús Complex and the metasedimentary San Gabriel sequence.

San Diego Phyllite, southern Guatemala

The San Diego Phyllite constitutes an enigmatic low-grade sedimentary unit, mainly composed of shales, some limestones, local arkoses and sandstones, minor quartzites metamorphosed at low greenschist facies. Lawrence (1975) and Donnelly *et al.* (1990) both observed fault contacts with the underlying El Tambor and Las Ovejas Complex (Schwartz 1977). This latter unit in Guatemala and its counterpart in Honduras (Bañaderos Complex in the Sierra de Omoa, Horne *et al.* 1976) may be considered as the basement of the Chortís Block. It consists of high-grade gneisses and migmatites, characterized by high temperature and mid-low

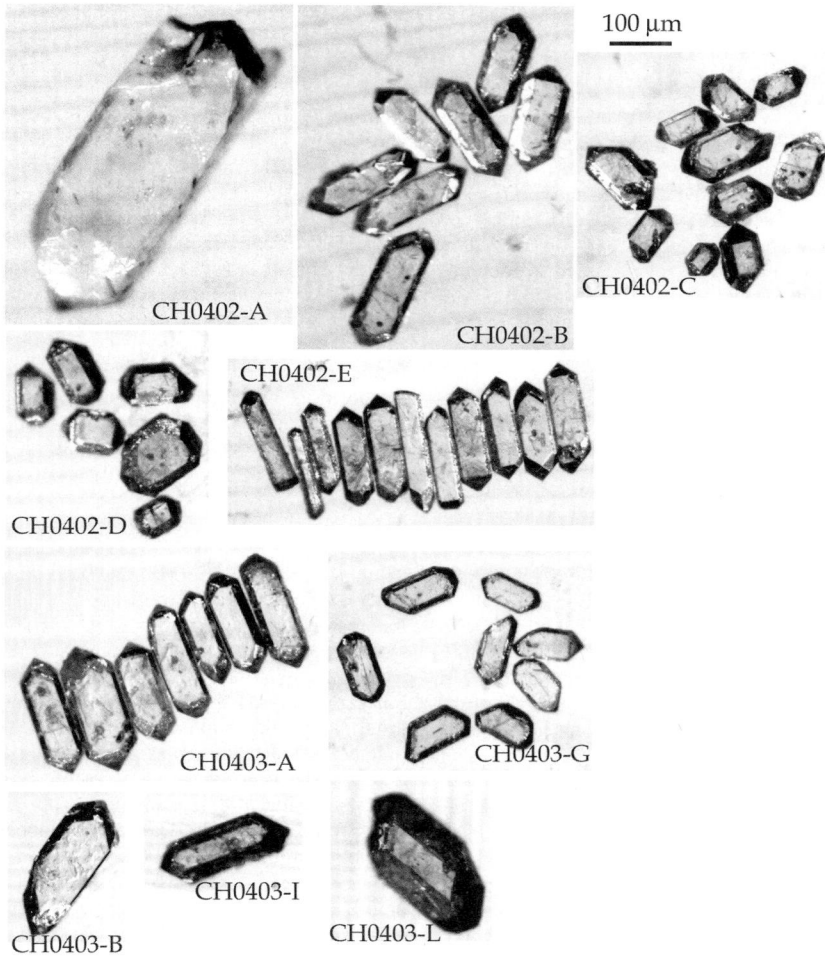

Fig. 2. Photomicrographs of zircons belonging to the samples CH0402 and CH0403, chosen for ID-TIMS dating, and reported in Table 1. Pictures were taken under a Leica S8 APO binocular microscope, under incident light, and prior to abrasion. Scale bar or 100 μm (top right) applies to all zircons.

pressure. No ages are available for either the San Diego Phyllite or the Las Ovejas Complex. In Honduras, Horne *et al.* (1976) dated some gneisses by Rb–Sr, obtaining poorly constrained ages ranging from Carboniferous, to Silurian, to Precambrian. Correlations of the San Diego Phyllite with the Santa Rosa Group (Lawrence 1975), and/or similar units cropping out in Central America (Cacaguapa Group, Fakundiny 1970) suggest that this unit could be pre-Permian. Our field observations suggest that, although bounded by faults, metamorphism in the San Diego Phyllite could have been induced by widespread Mesozoic to Cenozoic magmatism in Central-southern Guatemala (Donnelly *et al.* 1990).

The last sample analysed in this work (Gt0417) belongs to the San Diego Phyllite from its type locality, and consists of a low-grade slate interbedded with thin layers of quartz arenites. It is partially weathered and chloritized, with the main constituents being quartz, plagioclase, chlorite, pyrite cubes, minor opaque minerals, zircon and some apatite. No high-pressure minerals were observed, neither in thin section, nor in the heavy mineral concentrates. Dating of this sample would provide constraints on its age of sedimentation, and shed light on the minimum sedimentation ages on basement rocks in Central America. It would also demonstrate the possible existence of a Palaeozoic cover, which could be comparable with Upper

Palaeozoic Santa Rosa Group sediments (e.g. Weber *et al.* 2006*b*).

Analytical procedures

Sample preparation was carried at Instituto de Geología, UNAM. About 15 kg of each of the selected samples for zircon dating was crushed with a jaw crusher and pulverized in a Bico disk mill. The heavy minerals were concentrated on a Wifley shaking table and were then passed through a Frantz magnetic separator. A current up to 1.5 A is used to generate a magnetic field for samples to concentrate zircons suitable for LA-ICPMS dating, whereas up to 2.0 A is used for samples selected for TIMS dating. Further separation by passing non-magnetic fractions through heavy liquids (methylene iodide), ensured a nearly pure zircon concentration.

ID-TIMS U–Pb

Zircons from those samples chosen for U–Pb ID-TIMS dating were hand-picked under binocular microscope, imaged, and then abraded for 6–8 hours, using pyrite crystals as abrasive (Krogh 1982). They were then washed in warm 4 M HNO_3 for about 20 min and for 10 min in an ultrasonic bath. Further hand-picking under the binocular microscope ensured no surficial contamination by pyrite remnants that could add common Pb to the analyses. Zircons were then weighed on a microbalance, with an error of ± 1 μg, washed again in 8 M HNO_3 on a hot plate for about 20 min, and put into ultra-cleaned Teflon microcapsules, together with concentrated $HF + HNO_3$ acids. Up to nine microcapsules were stored in steel digestion vessels, and samples were digested in an oven during 4 days at 240 °C. Ultrapure acids used throughout the digestion and chemical separation, were prepared by distillation in Savillex Teflon systems (Mattinson 1972). Digested samples were poured into ultraclean PFA Teflon beakers, spiked with a ^{205}Pb–^{235}U solution, evaporated to dryness and redisolved in 0.5 M HBr acid. Chemical separation of U and Pb was performed in 40 μl Teflon microcolumns, filled with EIChrom AG1 X8 100–200 mesh anionic resin. HBr–HCl chemical procedure is a modification of the original method (Krogh 1973), similar to that described by Miller *et al.* (2006). U and Pb collected aliquots were dried with a drop of 0.1 M H_3PO_4, loaded together on previously degassed Re filament, and measured in a Finnigan MAT 262 mass spectrometer equipped with eight Faraday collectors at Laboratorio Universitario de Geoquímica Isotópica (LUGIS), UNAM. Pb was measured in static mode, using Faraday cups for 205, 206, 207 and ^{208}Pb beams, and an

Icm (Ion counter multiplier) for the smaller, ^{204}Pb beam. In general, 25 isotopic ratios were measured for Pb, and 50 for U. Faraday/Icm gain was stable thorough the whole turret, with an error of 0.01%. Routine analyses of NBS 981, 983 and U500 standards were performed to check precision and to correct isotopic ration mass fractioning. Applied corrections are $0.12 \pm 0.04\%$ for Pb ratios, and $0.12 \pm 0.05\%$ for U ratios. Repeated analyses of 91 500 zircon standard allowed calculation of U–Pb errors of $\pm 0.5\%$. Initial Pb ratios determined on feldspar concentrated separated from the same samples were used to correct for initial common Pb into zircons. Reduction of raw data was performed using Pbdat (Ludwig 1991), whereas concordia plots were generated using Isoplot version 3.06 (Ludwig 2004). Common Pb blanks during the analytical work ranged between 15 and 30 pg, whereas U blanks are commonly less than 2 pg.

LA-ICPMS

Zircons from all three samples were hand-picked, mounted in epoxy and polished for age determinations on individual grains by LA-ICPMS. U–Pb isotopic compositions of individual zircon grains were determined at RSES-ANU, Canberra, Australia, using an ArF excimer laser system (193 nm, Lambda Physik) coupled to an Agilent 7500 quadrupole ICPMS (Eggins *et al.* 1998). Ablation was done under a He atmosphere in a custom-built sample chamber using a 40 μm diameter spot and a laser repetition rate of 4 Hz. Data were reduced relative to the reference zircon FC1 (1099 Ma, Paces & Miller 1989), with gas backgrounds collected before each analysis and subtracted from net intensities measured for each isotope. The data were collected in two sessions. Precision on measured ^{207}Pb–^{206}Pb and ^{206}Pb–^{238}U ratios typically was *c.* 0.5% 1σ relative standard deviation. Replicate analyses of the FC1 zircon within each of these sessions indicate an external reproducibility of 0.8% ($n = 22$) and 0.6% ($n = 10$), respectively, on the measured ^{207}Pb–^{206}Pb ratios, 1.8 and 1.9%, respectively, on the measured ^{206}Pb–^{238}U ratios, and 1.4 and 1.6%, respectively, on the ^{208}Pb–^{232}Th ratios (1σ relative standard deviation). These errors are included in the quoted uncertainties for individual analyses of the analysed zircons.

Results

Sample CH0402, Altos Cuchumatanes

Zircons separated from sample CH0402 are generally short prisms. All have sharp facets, and do not show any rounding or irregular shape under binocular microscope. Maximum observed elongation is

Table 1. *U–Pb zircon analyses, TIMS*

Fraction*	Weight† (mg)	U† (ppm)	Total Pb† (ppm)	Composition Pb (pg)	Observed ratios‡			Atomic ratios§			Age (Ma)¶		
					$^{206}Pb/^{204}Pb$	$^{206}Pb/^{207}Pb$	$^{206}Pb/^{208}Pb$	$^{206}Pb–^{238}U$	$^{207}Pb–^{235}U$	$^{207}Pb/^{206}Pb$	$^{206}Pb–^{238}U$	$^{207}Pb–^{235}U$	$^{207}Pb/^{206}Pb$
Sample CH0402, Granite, Altos Cuchumatanes													
N 15°43′37.6″ W 91°23′48.9″													
CH0402 A sng	0.036	114.839	6.79	32	198.22	11.046	4.3053	0.0495027	0.383459	0.0561809	311	330	459 ± 42
CH0402 B 7 xls	0.058	274.242	19.26	120	413.66	11.261	5.5210	0.0619761	0.534709	0.0625737	388	435	694 ± 16
CH0402 C 10 xls	0.045	133.072	8.74	24	375.64	12.896	5.8542	0.0599491	0.519738	0.0628782	375	425	704 ± 22
CH0402 D 6 xls stby	0.03	220.507	13.76	76	196.81	9.3509	3.7662	0.0492862	0.388682	0.0571964	310	333	499 ± 31
CH0402 E 11 xls, long prs	0.043	198.473	11.5	12	936.72	15.313	6.5535	0.0548571	0.449642	0.0594473	344	377	583 ± 12
Sample CH0403, Granite, Altos Cuchumatanes, Rio Amelco													
N 15°44′38.8″ W 91°15′03.8″													
CH0403 A 7 xls	0.064	2099.82	200	1000	691.92	11.893	5.4823	0.0851539	0.752706	0.0641091	527	570	745 ± 9
CH0403 B sng	0.021	170.271	22.9	38	353.06	11.036	4.1251	0.113874	1.10907	0.0706374	695	758	947 ± 19
CH0403 G 8 xls, shrt	0.030	444.23	35.09	45	751.59	14.311	6.0679	0.073318	0.602812	0.0596307	456	479	590 ± 9
CH0403 I sng shrt prsm	0.021	278.778	23.44	55	304.08	11.355	4.9478	0.072284	0.601171	0.0599047	453	478	600 ± 24
CH0403 L sn stby	0.045	411.917	39	71	960.69	13.433	6.6890	0.0887114	0.793045	0.0648362	548	593	769 ± 6

Zircon sample dissolution and ion exchange chemistry modified after Krogh (1973) and Mattinson (1987) in Parrish (1987) type microcapsules.

*All diamagnetic fractions at 2.0 Amp. rnd. round; sh prsm, short prismatic to stubby grains; ov-stby, ovoid to stubby grains; brk xls, broken crystals.

Numbers refer to the micrometric size of the fraction chosen for analysis.

†Concentrations are known at ±30%, due to the weight uncertainty.

‡Observed isotopic ratios are corrected for mass fractionation of 0.12‰ for ^{205}Pb spiked fraction.

Two sigma uncertainties on the $^{207}Pb/^{206}Pb$ and $^{208}Pb/^{206}Pb$ ratios are <0.8‰, generally better than 0.1‰; uncertainties in the $^{206}Pb/^{204}Pb$ ratio vary from 0.1 to 2.4‰.

§Decay constants used: $^{238}U = 1.55125 \times 10^{-10}$, $^{235}U = 9.48485 \times 10^{-10}$, $^{238}U/^{235}U = 137.88$.

Uncertainities on the U–Pb ratio are 0.5%.

¶$^{207}Pb/^{206}Pb$ age uncertainties are 2σ and from the data reduction program PBDAT of Ludwig (1991). Total processing Pb blanks were between 10 and 40 pg.

Initial Pb compositions are from isotopic analyses of feldspar separates.

Isotopic data were measured on a Finnigan MAT 262 mass spectrometer with SEM ion counting at UNAM, Mexico City.

about 5 : 1. They are colourless to light pink or pale amber. Four small fractions, composed of six to 11 crystals, up to 200 μm in size, and a large single grain, were dated from this sample (pictures given in Fig. 2, and data in Table 1). All of the zircons were previously abraded. All analysed zircons yield discordant ages. They define a discordia, with Permian lower intercept of 269 ± 29 Ma, and a Mesoproterozoic upper intercept of 1212 ± 230 Ma (Fig. 3a). We are aware that discordant analyses must be interpreted carefully as geologically meaningful, because they can represent a mixture of more than two components. However, given the absence of metamorphism and tectonic fabric in this granite, and because our zircons were abraded and thus the presence of possible Pb-loss reduced at minimum, we consider the lower intercept as the best approximation of crystallization age, and the upper intercept as indicating a clear Mesoproterozoic component in the zircon cores.

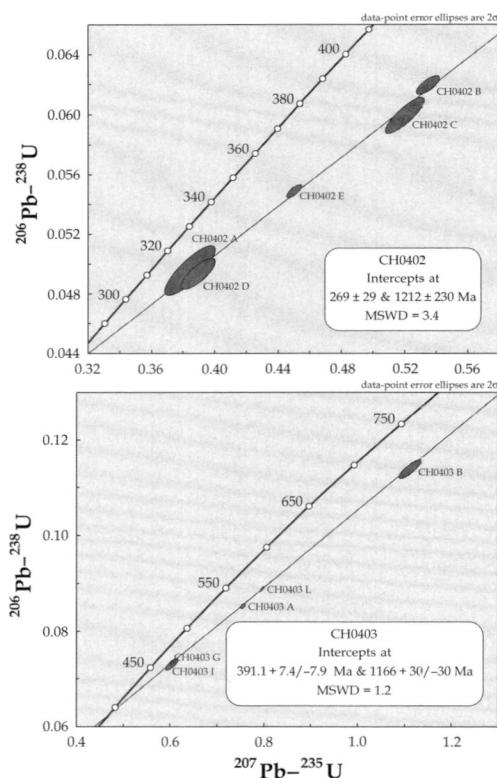

Fig. 3. U–Pb Concordia diagrams for the ID-TIMS dated samples in the Altos Cuchumatanes, western Guatemala. Ellipses in U–Pb Concordia diagrams represent 2σ errors. Given errors in calculated ages are 2σ.

Sample CH0403, Altos Cuchumatanes

Zircons of this sample are generally more intensely coloured than sample CH0402, ranging from yellow-amber to pink or reddish. Colourless crystals are absent. Shapes range from long prismatic, with a maximum ratio of 6 : 1, to stubby and multifaceted, or short prismatic grains (see Fig. 2). Two small fractions, composed of seven and eight crystals, respectively (CH0403-A and CH0403-G, Fig. 2), together with three single grains, were abraded and chosen for analysis. All of them yield discordant results. The data fit a discordia (Fig. 3b) with a well-defined lower intercept of 391.1 ± 7.4 Ma, and an upper of 1166 ± 30 Ma. Also for this sample, due to abrasion and absence of metamorphism or deformation, we can assume that they represent a two-component mixture. The two intercepts are thus interpreted as granite crystallization and Mesoproterozoic inheritance age, respectively.

Sample CH0407, San Gabriel sequence South of Huehuetenango

Zircons separated from this sample were analysed by LA-ICPMS. In general they are darkly coloured or pink to red. Metamictic grains are common, with sizes ranging up to 400 μm. Results are shown in the concordia diagram of Figure 4a together with a histogram plot (Fig. 4b). Of the 60 analyses performed, the oldest two show a slight reversely discordant behaviour, whereas all others are concordant, some within analytical error, or slightly discordant. They span an age range between *c.* 920 and *c.* 1000 Ma (Fig. 4a). All analyses yield, thus, Precambrian, Neoproterozoic to Mesoproterozoic ages. The histogram (Fig. 4b) indicates that the major peak is at *c.* 980 Ma, where some concordant analyses fall. The high Th–U ratio, generally >0.07, is consistent with almost all zircons belonging to this sample being magmatic in origin, or at least did not suffer any high-grade metamorphism (e.g. Rubatto 2002).The restricted range of zircon ages can be interpreted as indicating: (1) a Mesoproterozoic source near the sedimentation basin; (2) a fast sedimentation rate, together with a quick burial of the unit, or isolation of the sedimentary basin from younger sources.

Sample Gt03122, Chuacús Complex metasediment

Zircons separated from this sample are up to 300 μm in size, pale pink to pale amber or colourless. Euhedral forms are prominent, with a prevalence of short prisms (up to 4 : 1 elongation) and stubby grains. Concordia diagram for this sample is presented in Figure 5a, together with a histogram

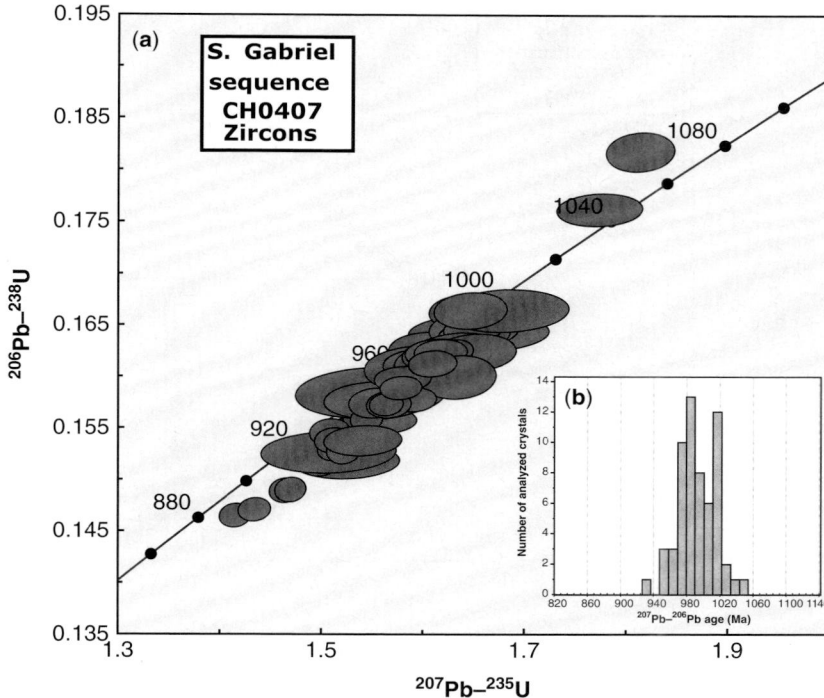

Fig. 4. (**a**) U–Pb concordia diagram for sample CH0407, San Gabriel sequence, south of Huehuetenango, Guatemala, dated by LA-ICPMS. Ellipses in U–Pb Concordia diagrams represent 2σ errors. (**b**) Histogram plot for the same sample, representing all the analysed zircons. See Table 2 for data point analyses.

of ^{206}Pb–^{238}U ages. In general, analysed zircons are concordant within analytical error, to slightly discordant. The age range in this sample is larger than the previous sample (compare Fig. 5a with previous Fig. 4a). The youngest zircons group at about 440 Ma, followed by several analyses slightly discordant up to c. 880 Ma. Beyond that, a group of concordant to slightly discordant grains straddles concordia between c. 920 and c. 990 Ma, with clusters at about 950 Ma, about 980 Ma. Another group of grains cluster at about 1080–1100 Ma (Fig. 5a, b). Only a couple of analyses yielded older ages. The oldest zircon is concordant within analytical error at c. 1325 Ma.

The Th–U ratios are consistent with almost all analyses being igneous zircons.

Sample Gt0417, San Diego Phyllite

Zircons belonging to this sample are clear to pale yellow/amber, a few are pale pink coloured, and they range between 80 and 300 μm. The concordia diagram is shown in Figure 6a, and the histogram of ^{206}Pb–^{238}U ages is plot in Figure 6b. All analysed grains are concordant, some within analytical error,

and just few are slightly discordant. Analysed grains yield ages that range between c. 520 and about 1820 Ma, with main clusters at about 540 Ma, about 980 Ma and about 1130 Ma. Few analyses fall between c. 1150 and c. 1300 Ma. Again, measured Th–U ratios suggest that all but two analyses are likely to be of magmatic origin (Table 2).

Discussion

U–Pb ages of the Altos Cuchumatanes samples are the first geochronologic evidence of Middle Devonian and Permian magmatism in western Guatemala. Early Devonian–Silurian magmatism was previously known from the Maya Mountains of Belize, where granites intruding sediments are overlain by Santa Rosa Group sediments (Steiner & Walker 1996), as well as the Rabinal granite of the Rabinal–Salamá area, north of the Baja Verapaz shear zone, with a minimum K–Ar age of 429–440 Ma (Ortega-Obregón *et al.* 2008). Surprisingly, a similar aged magmatic event is not known from the Chiapas Massif or surrounding areas (e.g. Weber *et al.* 2004, 2006a, b), although Weber *et al.* (2005) recently reported Silurian

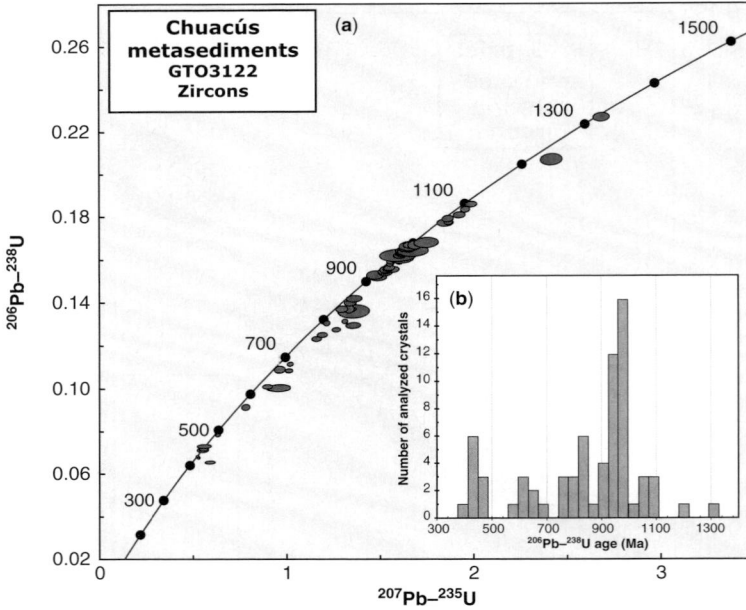

Fig. 5. (**a**) U–Pb concordia diagram for sample Gt03122, metasediment of the Chuacús Complex, south of Salamá, Guatemala, dated by LA-ICPMS. Ellipses in U–Pb Concordia diagrams represent 2σ errors. (**b**) Histogram plot for the same sample, representing all the analysed zircons. See Table 2 for data point analyses.

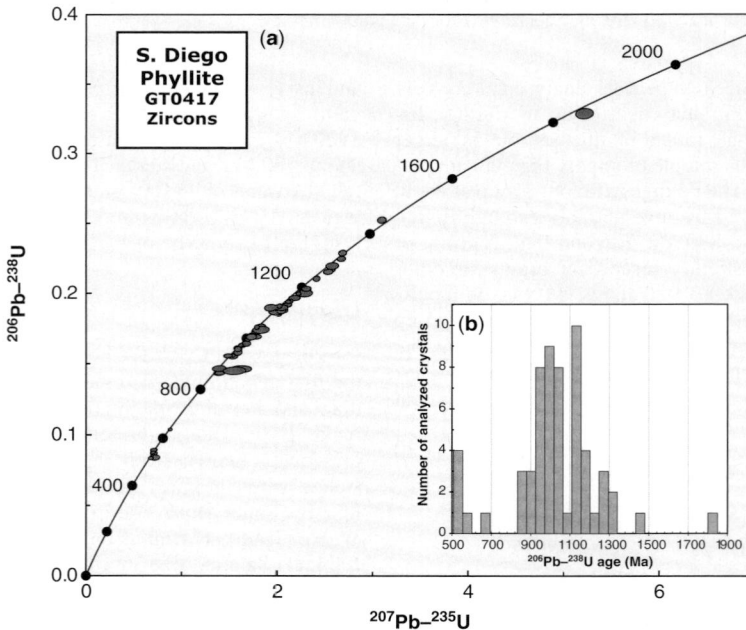

Fig. 6. (**a**) U–Pb concordia diagram for sample Gt0017, S. Diego Phyllite, Guatemala, dated by LA-ICPMS. Ellipses in U–Pb Concordia diagrams represent 2σ errors. (**b**) Histogram plot for the same sample, representing all the analysed zircons. See Table 2 for data point analyses.

Table 2. LA-ICPMS U–Pb analyses, Guatemala

Sample CH0407, San Gabriel sequence, South of Huehuetenango
N 15°17'45.4"
W 91°29'41.4"

	$^{207}Pb/^{235}U$ ratio	$1\sigma\%$	$^{206}Pb/^{238}U$ ratio	$1\sigma\%$	$^{207}Pb/^{206}Pb$ ratio	$1\sigma\%$	$^{208}Pb/^{232}Th$ ratio	$1\sigma\%$	Th–U	$^{206}Pb/^{238}U$ Age (Ma)	$\pm2\sigma$	$^{207}Pb/^{235}U$ Age (Ma)	$\pm2\sigma$	$^{207}Pb/^{206}Pb$ Age (Ma)	$\pm2\sigma$	$^{208}Pb/^{232}Th$ Age (Ma)	$\pm2\sigma$
nv09g16	1.4156	0.46%	0.1463	0.33%	0.07021	0.37%	0.0450	0.66%	0.24	880	5	896	5	934	16	889	11
nv09f07	1.4345	0.47%	0.1468	0.31%	0.07090	0.44%	0.0462	0.82%	0.11	883	5	903	6	954	18	914	15
nv09f10	1.4647	0.41%	0.1487	0.33%	0.07147	0.45%	0.0461	0.49%	0.25	894	6	916	5	970	18	910	9
nv09h06	1.4696	0.42%	0.1488	0.33%	0.07166	0.40%	0.0462	0.71%	0.26	894	6	918	5	976	16	914	13
nv09f13	1.4984	0.47%	0.1512	0.45%	0.07189	0.29%	0.0462	0.44%	0.27	908	8	930	6	982	13	913	8
nv09f05	1.5256	1.40%	0.1515	0.54%	0.07308	1.49%	0.0488	2.91%	0.16	909	9	941	17	1016	60	964	55
nv09h13	1.5074	1.79%	0.1524	0.31%	0.07176	1.86%	0.0463	4.43%	0.07	915	5	933	22	978	76	916	79
nv09f12	1.5128	0.46%	0.1527	0.38%	0.07187	0.49%	0.0466	0.54%	0.22	916	6	936	6	982	20	920	10
nv09f09	1.5218	0.46%	0.1527	0.29%	0.07235	0.46%	0.0467	0.76%	0.23	916	6	939	6	994	18	923	14
nv09g05	1.5337	0.39%	0.1533	0.33%	0.07260	0.35%	0.0477	0.44%	0.22	919	5	944	5	1002	14	942	8
nv09f16	1.5114	0.50%	0.1536	0.33%	0.07142	0.44%	0.0469	0.77%	0.19	921	6	935	6	968	18	926	14
nv09x07	1.5408	1.01%	0.1536	0.40%	0.07282	1.12%	0.0498	3.23%	0.07	921	7	947	12	1008	44	982	62
nv09f14	1.5086	0.54%	0.1544	0.34%	0.07093	0.42%	0.0473	0.73%	0.22	925	6	934	7	954	16	934	13
nv09f09	1.5305	0.61%	0.1548	0.31%	0.07173	0.60%	0.0477	0.94%	0.27	928	5	943	7	978	24	942	17
nv09f07	1.5632	0.82%	0.1555	0.32%	0.07297	0.81%	0.0473	1.25%	0.36	932	6	956	10	1012	32	934	23
nv09g10	1.5349	0.38%	0.1558	0.33%	0.07151	0.34%	0.0478	0.55%	0.26	933	6	944	5	972	14	944	10
nv09h12	1.5448	0.43%	0.1557	0.31%	0.07200	0.36%	0.0482	0.50%	0.21	933	5	948	5	984	14	952	9
nv09j13	1.5663	0.56%	0.1571	0.33%	0.07234	0.56%	0.0481	0.76%	0.23	941	6	957	7	994	22	950	14
nv09j15	1.5641	0.44%	0.1571	0.28%	0.07225	0.42%	0.0476	0.49%	0.26	941	5	956	5	992	16	940	9
nv09h15	1.5522	0.68%	0.1573	0.38%	0.07162	0.68%	0.0471	1.96%	0.07	942	7	951	8	974	28	930	36
nv09h10	1.5551	1.37%	0.1575	0.48%	0.07164	1.24%	0.0484	1.75%	0.24	943	8	953	17	974	50	955	33
nv09j10	1.5910	0.59%	0.1576	0.32%	0.07325	0.59%	0.0489	0.83%	0.28	944	6	967	7	1020	24	965	16
nv09h09	1.5488	1.96%	0.1582	0.65%	0.07105	1.94%	0.0440	4.57%	0.12	947	11	950	24	958	80	870	78
nv09j16	1.5779	0.54%	0.1587	0.30%	0.07215	0.47%	0.0479	0.68%	0.24	950	5	962	7	990	18	946	13
nv09f14	1.5752	0.59%	0.1592	0.33%	0.07181	0.60%	0.0522	1.42%	0.05	952	6	960	7	980	24	1029	28
nv09j08	1.5803	0.73%	0.1596	0.32%	0.07187	0.82%	0.0480	0.94%	0.28	954	6	962	9	982	34	947	17
nv09j07	1.6345	0.94%	0.1598	0.55%	0.07421	0.88%	0.0479	2.87%	0.08	956	10	984	12	1046	36	945	53
nv09g07	1.5946	0.99%	0.1603	0.34%	0.07221	0.95%	0.0501	1.70%	0.07	958	6	968	12	990	38	988	33
nv09f06	1.6124	0.51%	0.1603	0.31%	0.07299	0.52%	0.0488	0.55%	0.33	959	6	975	6	1012	22	962	10
nv09f08	1.6067	1.02%	0.1603	0.33%	0.07272	0.88%	0.0498	0.94%	0.24	959	6	973	13	1006	36	982	18
nv09f16	1.5782	0.86%	0.1604	0.31%	0.07139	0.76%	0.0488	0.99%	0.40	959	6	962	11	968	32	962	19
nv09h11	1.5934	1.30%	0.1606	0.56%	0.07200	1.14%	0.0456	3.98%	0.06	960	10	968	16	984	46	901	70
nv09g13	1.5829	0.68%	0.1608	0.29%	0.07142	0.64%	0.0486	0.76%	0.35	961	5	963	8	968	26	959	14
nv09f05	1.5991	1.00%	0.1610	0.40%	0.07209	1.01%	0.0482	2.77%	0.07	962	7	970	13	988	42	951	52
nv09f06	1.5990	0.69%	0.1610	0.41%	0.07210	0.72%	0.0492	0.85%	0.27	962	7	970	9	988	30	971	16
nv09h05	1.5907	1.11%	0.1610	0.42%	0.07168	1.09%	0.0495	3.12%	0.07	963	8	967	14	976	44	976	60
nv09j12	1.6099	0.60%	0.1611	0.32%	0.07254	0.61%	0.0493	0.78%	0.25	963	6	974	8	1000	24	972	15
nv09f12	1.6318	1.06%	0.1616	0.48%	0.07330	1.01%	0.0470	2.83%	0.07	965	9	983	13	1022	42	929	51
nv09g06	1.6316	0.87%	0.1616	0.40%	0.07325	0.81%	0.0489	2.58%	0.07	966	7	982	11	1020	32	964	49

(Continued)

Table 2. Continued

	$^{207}\text{Pb}\_{235}\text{U}$ ratio	$1\sigma\ \%$	$^{206}\text{Pb}\_{238}\text{U}$ ratio	$1\sigma\ \%$	$^{207}\text{Pb}/^{206}\text{Pb}$ ratio	$1\sigma\ \%$	$^{208}\text{Pb}\_{232}\text{Th}$ ratio	$1\sigma\ \%$	Th–U	$^{206}\text{Pb}\_{238}\text{U}$ Age (Ma)	$\pm2\sigma$	$^{207}\text{Pb}\_{235}\text{U}$ Age (Ma)	$\pm2\sigma$	$^{207}\text{Pb}/^{206}\text{Pb}$ Age (Ma)	$\pm2\sigma$	$^{208}\text{Pb}\_{232}\text{Th}$ Age (Ma)	$\pm2\sigma$
nv09g12	1.6025	0.51%	0.1619	0.32%	0.07183	0.45%	0.0505	0.62%	0.24	967	6	971	6	980	18	995	12
nv09i08	1.6014	0.57%	0.1618	0.33%	0.07184	0.63%	0.0490	1.17%	0.25	967	6	971	7	980	26	967	22
nv09i09	1.6366	1.20%	0.1621	0.51%	0.07328	1.25%	0.0482	1.58%	0.25	968	9	984	15	1020	50	952	29
nv09i10	1.6142	0.82%	0.1621	0.36%	0.07226	0.84%	0.0488	0.77%	0.38	969	6	976	10	992	34	964	14
nv09i11	1.6521	0.99%	0.1622	0.45%	0.07393	1.02%	0.0570	2.68%	0.07	969	8	990	13	1038	40	1119	58
nv09i15	1.6174	0.65%	0.1622	0.31%	0.07235	0.71%	0.0489	0.74%	0.29	969	6	977	8	994	30	965	14
nv09i14	1.6320	0.45%	0.1623	0.29%	0.07297	0.45%	0.0493	0.45%	0.29	970	5	983	6	1012	20	973	9
nv09g09	1.6033	0.93%	0.1625	0.42%	0.07158	0.97%	0.0487	0.90%	0.39	971	8	971	12	974	38	961	17
nv09h07	1.6419	0.68%	0.1637	0.33%	0.07280	0.68%	0.0495	0.69%	0.37	977	6	986	9	1008	28	977	13
nv09f15	1.6550	1.70%	0.1641	0.47%	0.07320	1.76%	0.0511	2.51%	0.19	979	9	991	21	1018	72	1007	49
nv09g14	1.6492	0.92%	0.1640	0.38%	0.07299	1.00%	0.0500	2.46%	0.10	979	7	989	12	1012	40	986	47
nv09g15	1.6397	0.83%	0.1644	0.40%	0.07240	0.92%	0.0502	0.92%	0.42	981	7	986	10	996	36	990	18
nv09h08	1.6481	0.69%	0.1644	0.34%	0.07273	0.64%	0.0492	0.98%	0.38	981	6	989	9	1006	26	971	19
nv09g11	1.6570	0.94%	0.1645	0.43%	0.07308	0.99%	0.0513	1.48%	0.24	982	8	992	12	1016	40	1011	29
nv09f11	1.6845	0.90%	0.1659	0.43%	0.07366	0.80%	0.0503	1.12%	0.28	990	8	1003	11	1032	34	992	22
nv09g08	1.6415	0.88%	0.1662	0.41%	0.07170	0.85%	0.0527	2.63%	0.14	991	8	986	11	976	36	1038	53
nv09j11	1.6820	1.50%	0.1663	0.52%	0.07340	1.59%	0.0528	2.55%	0.19	992	9	1002	19	1024	64	1039	52
nv09j14	1.6474	0.89%	0.1663	0.44%	0.07189	0.93%	0.0505	1.26%	0.22	992	8	989	11	982	38	996	24
nv09j05	1.6706	0.75%	0.1666	0.34%	0.07279	0.69%	0.0527	1.00%	0.29	993	6	997	10	1008	28	1037	20
nv09x05	1.7735	0.97%	0.1761	0.37%	0.07308	1.00%	0.0559	1.32%	0.22	1046	7	1036	13	1016	42	1099	28
nv09i13	1.8133	0.77%	0.1817	0.43%	0.07243	0.79%	0.0601	1.10%	0.17	1076	9	1050	10	996	32	1180	25

Sample Gt 03122, Chuacús Complex metasediment, South of Salamá
N 15°02'01"
W 90°19'2"

	$^{207}\text{Pb}\_{235}\text{U}$ ratio	$1\sigma\ \%$	$^{206}\text{Pb}\_{238}\text{U}$ ratio	$1\sigma\ \%$	$^{207}\text{Pb}/^{206}\text{Pb}$ ratio	$1\sigma\ \%$	$^{208}\text{Pb}\_{232}\text{Th}$ ratio	$1\sigma\ \%$	Th–U	$^{206}\text{Pb}\_{238}\text{U}$ Age (Ma)	$\pm2\sigma$	$^{207}\text{Pb}\_{235}\text{U}$ Age (Ma)	$\pm2\sigma$	$^{207}\text{Pb}/^{206}\text{Pb}$ Age (Ma)	$\pm2\sigma$	$^{208}\text{Pb}\_{232}\text{Th}$ Age (Ma)	$\pm2\sigma$
nv29k05	0.5858	1.91%	0.0643	0.56%	0.0661	2.09%	0.0276	2.38%	0.27	402	4	468	14	808	88	551	26
nv29j14	0.5281	0.48%	0.0666	0.36%	0.0576	0.50%	0.0227	0.97%	0.12	415	3	431	3	512	22	454	9
nv29j16	0.5246	0.55%	0.0673	0.33%	0.0566	0.54%	0.0289	1.86%	0.03	420	3	428	4	474	24	575	21
nv29j13	0.5509	2.27%	0.0702	0.64%	0.0569	2.32%	0.0225	1.32%	0.45	438	5	446	16	486	102	449	12
nv29j06	0.5503	0.76%	0.0705	0.31%	0.0566	0.77%	0.0223	1.00%	0.14	439	3	445	5	474	34	445	9
nv29j12	0.5453	0.98%	0.0705	0.38%	0.0561	0.98%	0.0220	1.21%	0.31	439	3	442	7	456	44	440	11
nv29j10	0.5551	1.75%	0.0711	0.71%	0.0566	1.57%	0.0236	1.87%	0.34	443	6	448	13	476	70	471	17
nv29g07	0.5674	0.85%	0.0720	0.35%	0.0572	0.90%	0.0229	1.39%	0.14	448	3	456	6	498	40	458	13
nv29h07	0.5588	2.75%	0.0721	0.51%	0.0562	2.78%	0.0223	1.77%	1.00	449	4	451	20	460	124	447	16
nv29j10	0.6287	0.97%	0.0774	0.41%	0.0589	0.97%	0.0291	1.22%	0.10	480	4	495	8	564	42	579	14
nv29j13	0.7769	1.17%	0.0905	0.65%	0.0623	0.84%	0.0448	1.35%	0.07	559	7	584	10	682	36	887	23
nv29j05	0.9373	1.79%	0.0991	0.42%	0.0686	1.75%	0.0468	2.36%	0.26	609	5	671	18	886	72	925	43
nv29k09	0.9603	2.50%	0.0996	0.68%	0.0700	2.68%	0.0362	2.46%	0.53	612	8	683	25	926	110	719	35
nv29j16	0.8964	1.34%	0.1002	0.42%	0.0649	1.34%	0.0388	1.42%	0.22	615	5	650	13	770	56	770	22
nv29h08	0.9613	1.37%	0.1082	0.62%	0.0644	1.30%	0.0446	1.52%	0.32	662	8	684	14	754	56	881	26
nv29j14	1.0147	0.81%	0.1077	0.31%	0.0683	0.80%	0.0351	1.21%	0.20	660	4	711	8	878	34	697	17
nv29g05	1.0220	0.68%	0.1108	0.36%	0.0669	0.67%	0.0481	0.79%	0.28	677	5	715	7	834	28	950	15
nv29i10	1.1631	0.94%	0.1226	0.39%	0.0688	0.96%	0.0451	1.32%	0.25	745	6	783	10	892	40	892	23
nv29i13	1.1923	1.06%	0.1244	0.39%	0.0695	1.12%	0.0392	1.13%	0.32	756	6	797	12	912	46	777	17

nv29h15	1.2679	0.72%	0.1268	0.34%	0.0725	0.69%	0.0539	0.81%	0.20	770	5	831	8	998	28	1061	17
nv29g06	1.3592	1.21%	0.1288	0.43%	0.0765	1.26%	0.0489	1.26%	0.43	781	6	872	14	1108	50	965	24
nv29j12	1.2181	0.56%	0.1299	0.37%	0.0680	0.56%	0.0437	0.68%	0.28	787	6	809	6	868	24	864	11
nv29j11	1.3107	0.46%	0.1307	0.33%	0.0727	0.39%	0.0531	0.64%	0.14	792	5	850	5	1006	16	1046	13
nv29g08	1.3606	2.59%	0.1354	0.92%	0.0729	2.74%	0.0419	1.41%	2.42	819	14	872	30	1010	112	830	23
nv29j05	1.3385	1.01%	0.1363	0.34%	0.0713	0.97%	0.0463	1.13%	0.70	823	5	863	12	964	40	915	20
nv29j08	1.3326	0.72%	0.1361	0.31%	0.0710	0.66%	0.0426	0.82%	0.40	823	5	860	8	956	26	842	14
nv29g06	1.2954	0.96%	0.1363	0.47%	0.0689	0.88%	0.0419	1.27%	0.37	824	7	844	11	896	36	829	21
nv29j14	1.3375	0.72%	0.1364	0.56%	0.0711	0.59%	0.0422	0.73%	0.34	824	9	862	8	960	24	836	12
nv29h09	1.3449	0.97%	0.1394	0.38%	0.0700	0.90%	0.0428	1.17%	0.31	841	6	865	11	926	36	846	19
nv29k11	1.3637	1.18%	0.1415	0.44%	0.0699	1.24%	0.0436	1.14%	0.34	853	7	873	14	926	50	863	19
nv29j16	1.4744	1.31%	0.1525	0.53%	0.0701	1.52%	0.0498	0.89%	1.08	915	9	920	16	932	62	983	17
nv29j09	1.4931	1.30%	0.1514	0.42%	0.0715	1.35%	0.0449	1.12%	0.61	909	7	928	16	972	54	888	20
nv29j16	1.4944	0.54%	0.1518	0.27%	0.0714	0.49%	0.0449	0.82%	0.26	911	5	928	7	968	20	888	14
nv29j08	1.5335	1.05%	0.1540	0.48%	0.0723	1.10%	0.0472	1.10%	0.33	923	8	944	13	992	44	932	20
nv29h14	1.5260	0.93%	0.1542	0.35%	0.0718	0.90%	0.0476	0.86%	0.34	925	6	941	11	978	36	941	16
nv29j06	1.5667	1.03%	0.1550	0.38%	0.0733	1.00%	0.0490	1.03%	0.33	929	7	957	13	1022	40	968	19
nv29j15	1.5415	0.53%	0.1549	0.34%	0.0722	0.51%	0.0472	0.68%	0.26	929	6	947	6	990	20	931	12
nv29k13	1.5386	0.95%	0.1551	0.45%	0.0720	1.06%	0.0480	1.35%	0.35	929	8	946	12	984	44	947	25
nv29k14	1.5447	0.80%	0.1554	0.36%	0.0721	0.88%	0.0474	1.04%	0.40	931	6	948	10	988	36	937	19
nv29k10	1.5534	0.78%	0.1559	0.32%	0.0723	0.81%	0.0472	0.86%	0.38	934	6	952	10	992	34	931	16
nv29j08	1.5563	0.51%	0.1576	0.28%	0.0716	0.51%	0.0466	0.72%	0.36	944	5	953	6	974	20	921	13
nv29k16	1.6060	0.57%	0.1588	0.34%	0.0734	0.60%	0.0484	0.80%	0.37	950	6	973	7	1024	24	956	15
nv29g06	1.5663	0.63%	0.1590	0.32%	0.0714	0.68%	0.0480	0.76%	0.19	951	6	957	8	970	26	948	14
nv29g14	1.6161	1.11%	0.1606	0.39%	0.0730	1.16%	0.0495	1.13%	0.65	960	7	976	14	1012	48	977	22
nv29g16	1.6220	0.92%	0.1603	0.36%	0.0734	0.89%	0.0482	0.72%	0.40	959	6	979	12	1024	36	951	13
nv29k07	1.6139	1.70%	0.1605	0.60%	0.0730	1.63%	0.0487	1.93%	0.52	959	11	976	21	1012	66	961	36
nv29j15	1.6001	0.76%	0.1606	0.31%	0.0723	0.72%	0.0487	1.00%	0.41	960	5	970	9	992	28	961	19
nv29j07	1.6223	1.19%	0.1604	0.43%	0.0734	1.28%	0.0480	1.06%	0.76	959	8	979	15	1022	52	947	20
nv29h13	1.5906	0.44%	0.1608	0.30%	0.0718	0.44%	0.0490	0.60%	0.27	961	5	967	5	978	18	966	11
nv29g11	1.6088	0.84%	0.1617	0.33%	0.0722	0.91%	0.0491	0.96%	0.29	966	6	974	11	990	36	969	18
nv29j09	1.5999	2.53%	0.1616	0.73%	0.0718	2.65%	0.0483	2.97%	0.39	966	13	970	32	980	108	953	55
nv29j12	1.6227	0.89%	0.1627	0.39%	0.0723	0.86%	0.0491	0.87%	0.30	972	7	979	11	994	36	968	16
nv29g08	1.6483	0.80%	0.1636	0.40%	0.0731	0.83%	0.0494	1.08%	0.31	977	7	989	10	1014	34	975	21
nv29j11	1.6175	0.73%	0.1633	0.34%	0.0718	0.69%	0.0479	0.84%	0.39	975	6	977	9	980	28	946	15
nv29j07	1.6336	1.15%	0.1639	0.41%	0.0723	1.19%	0.0497	1.15%	0.34	978	7	983	15	994	48	979	22
nv29k12	1.6580	0.77%	0.1645	0.34%	0.0731	0.71%	0.0518	0.86%	0.48	982	6	993	10	1016	30	1021	17
nv29h11	1.6764	1.97%	0.1638	0.64%	0.0743	1.96%	0.0493	1.31%	1.37	978	12	1000	25	1048	78	972	25
nv29g10	1.6337	0.63%	0.1648	0.32%	0.0719	0.60%	0.0497	0.83%	0.41	983	6	983	8	982	24	979	16
nv29j15	1.6770	0.77%	0.1654	0.32%	0.0736	0.73%	0.0492	1.10%	0.31	987	6	1000	10	1028	28	971	21
nv29k05	1.6642	1.43%	0.1660	0.55%	0.0727	1.31%	0.0503	1.52%	0.46	990	10	995	18	1004	52	992	29
nv29j05	1.6614	0.84%	0.1665	0.34%	0.0724	0.81%	0.0506	0.90%	0.31	993	6	994	11	996	34	997	17
nv29g15	1.6914	0.97%	0.1667	0.49%	0.0736	1.01%	0.0512	1.09%	0.40	994	9	1005	12	1030	40	1009	21
nv29j11	1.7480	1.57%	0.1676	0.57%	0.0757	1.58%	0.0491	1.39%	0.50	999	10	1026	20	1086	62	968	26
nv29j07	1.8536	0.98%	0.1770	0.40%	0.0760	0.90%	0.0544	1.08%	0.51	1051	8	1065	13	1094	36	1071	23
nv29j09	1.8703	0.55%	0.1771	0.27%	0.0766	0.58%	0.0554	0.68%	0.33	1051	5	1071	7	1110	24	1090	14
nv29h12	1.8674	0.65%	0.1789	0.35%	0.0757	0.62%	0.0541	0.70%	0.32	1061	7	1070	9	1086	24	1065	15
nv29g13	1.9293	0.66%	0.1805	0.33%	0.0775	0.66%	0.0546	0.63%	0.47	1070	9	1091	9	1134	26	1075	13
nv29j10	1.9604	0.55%	0.1831	0.31%	0.0777	0.55%	0.0557	0.62%	0.42	1084	6	1102	7	1138	22	1096	13

(Continued)

Table 2. *Continued*

	$^{207}Pb-^{235}U$ ratio	$1\sigma\%$	$^{206}Pb-^{238}U$ ratio	$1\sigma\%$	$^{207}Pb/^{206}Pb$ ratio	$1\sigma\%$	$^{208}Pb-^{232}Th$ ratio	$1\sigma\%$	Th–U	$^{206}Pb-^{238}U$ Age (Ma)	$\pm2\sigma$	$^{207}Pb/^{235}U$ Age (Ma)	$\pm2\sigma$	$^{207}Pb/^{206}Pb$ Age (Ma)	$\pm2\sigma$	$^{208}Pb-^{232}Th$ Age (Ma)	$\pm2\sigma$
nv29g09	1.9925	0.65%	0.1857	0.28%	0.0778	0.64%	0.0559	0.75%	0.37	1098	6	1113	9	1142	26	1099	16
nv29k15	2.4218	1.00%	0.2068	0.53%	0.0850	0.98%	0.0576	0.84%	0.60	1212	12	1249	14	1314	38	1131	18
nv29106	2.6928	0.67%	0.2269	0.33%	0.0861	0.66%	0.0668	0.98%	0.31	1318	8	1326	10	1340	26	1307	25

Sample Gt0417, San Diego Phyllite, South of San Diego
N 14°47'14"
W 89°45'36"

	$^{207}Pb-^{235}U$ ratio	$1\sigma\%$	$^{206}Pb-^{238}U$ ratio	$1\sigma\%$	$^{207}Pb/^{206}Pb$ ratio	$1\sigma\%$	$^{208}Pb-^{232}Th$ ratio	$1\sigma\%$	Th–U	$^{206}Pb-^{238}U$ Age (Ma)	$\pm2\sigma$	$^{207}Pb/^{235}U$ Age (Ma)	$\pm2\sigma$	$^{207}Pb/^{206}Pb$ Age (Ma)	$\pm2\sigma$	$^{208}Pb-^{232}Th$ Age (Ma)	$\pm2\sigma$
nv09d12	0.6860	1.79%	0.0842	0.49%	0.0591	1.92%	0.0259	1.80%	0.63	521	5	530	15	572	82	517	18
nv09a06	0.7339	2.98%	0.0846	0.87%	0.0629	2.92%	0.0264	1.63%	1.32	524	9	559	26	706	124	527	17
nv09a15	0.7184	2.09%	0.0875	0.54%	0.0596	2.21%	0.0273	0.67%	0.05	541	6	550	18	586	96	544	7
nv09d07	0.7094	1.32%	0.0881	0.43%	0.0584	1.29%	0.0273	0.95%	0.99	544	4	544	11	546	56	544	10
nv09a11	0.7213	2.00%	0.0898	0.61%	0.0583	2.01%	0.0284	1.83%	0.56	555	6	551	17	538	88	565	20
nv09b09	0.8876	0.89%	0.1045	0.36%	0.0617	0.74%	0.0341	1.08%	0.22	641	4	645	8	662	30	677	14
nv09c08	1.3732	1.03%	0.1440	0.34%	0.0692	0.93%	0.0454	1.05%	0.30	867	6	878	12	904	38	897	19
nv09a09	1.3965	1.68%	0.1442	0.50%	0.0703	1.73%	0.0430	1.40%	0.76	868	8	887	20	936	72	851	23
nv09d10	1.4084	1.51%	0.1445	0.48%	0.0707	1.53%	0.0446	1.67%	0.39	870	8	893	18	948	62	881	29
nv09e16	1.5609	3.13%	0.1459	0.73%	0.0777	3.03%	0.0456	2.72%	0.95	878	12	955	39	1138	120	902	48
nv09b07	1.6260	2.83%	0.1470	0.64%	0.0803	2.74%	0.0541	3.39%	0.34	884	11	980	36	1202	108	1065	70
nv09d13	1.4047	2.20%	0.1472	0.56%	0.0693	2.27%	0.0434	2.06%	0.49	885	9	891	26	906	94	859	35
nv09a10	1.5413	1.35%	0.1561	0.46%	0.0716	1.47%	0.0482	1.56%	0.39	935	8	947	17	974	60	951	29
nv09d15	1.4939	1.18%	0.1564	0.41%	0.0693	1.18%	0.0465	1.37%	0.35	936	7	928	14	908	48	920	25
nv09b15	1.5451	0.57%	0.1564	0.31%	0.0717	0.54%	0.0483	1.64%	0.06	937	5	949	7	976	22	954	31
nv09b05	1.5974	1.05%	0.1585	0.56%	0.0732	1.01%	0.0494	2.05%	0.11	948	10	969	13	1018	42	974	39
nv09b16	1.6189	0.47%	0.1607	0.27%	0.0731	0.38%	0.0487	0.70%	0.37	961	5	978	6	1016	16	961	13
nv09b06	1.6073	0.70%	0.1611	0.31%	0.0724	0.63%	0.0484	0.70%	0.88	963	6	973	9	996	26	955	13
nv09d05	1.5939	0.84%	0.1612	0.42%	0.0718	0.88%	0.0480	1.04%	0.33	963	8	968	11	978	36	947	19
nv09c15	1.5896	0.88%	0.1616	0.37%	0.0714	0.83%	0.0486	0.65%	0.61	966	7	966	11	978	34	959	12
nv09b10	1.6495	0.48%	0.1631	0.30%	0.0734	0.41%	0.0493	1.17%	0.05	974	5	989	6	1024	16	972	22
nv09b08	1.6664	0.36%	0.1637	0.29%	0.0739	0.30%	0.0494	0.60%	0.23	977	5	996	5	1036	12	975	11
nv09d09	1.6273	0.65%	0.1638	0.33%	0.0721	0.66%	0.0498	1.31%	0.13	978	6	981	8	988	28	982	25
nv09c07	1.6770	1.16%	0.1647	0.44%	0.0739	1.18%	0.0500	0.79%	0.91	983	8	1000	15	1038	48	987	15
nv09d14	1.7067	0.83%	0.1678	0.34%	0.0738	0.83%	0.0502	1.02%	0.33	1000	6	1011	11	1036	34	990	20
nv09c13	1.6849	0.53%	0.1682	0.29%	0.0727	0.49%	0.0500	0.62%	0.28	1002	5	1003	7	1004	20	986	12

nv09b14	1.7155	0.70%	0.1690	0.33%	0.0737	0.70%	0.0515	0.88%	0.31	1007	6	1014	9	1032	30	1015	17
nv09e07	1.7397	1.24%	0.1695	0.50%	0.0745	1.17%	0.0514	1.40%	0.39	1009	9	1023	16	1054	46	1013	28
nv09a08	1.7889	1.15%	0.1697	0.43%	0.0765	1.14%	0.0512	1.11%	0.55	1010	8	1041	15	1108	46	1009	22
nv09a13	1.7338	0.63%	0.1712	0.35%	0.0735	0.61%	0.0517	0.79%	0.28	1019	7	1021	8	1028	26	1019	16
nv09e06	1.7994	1.37%	0.1738	0.47%	0.0751	1.28%	0.0551	1.65%	0.34	1033	9	1045	18	1070	52	1084	35
nv09b11	1.8414	1.36%	0.1747	0.48%	0.0765	1.37%	0.0524	1.89%	0.25	1038	9	1060	18	1108	54	1033	38
nv09e10	1.8362	1.95%	0.1748	0.62%	0.0763	2.09%	0.0520	2.09%	0.49	1038	12	1058	26	1100	84	1025	42
nv09e05	1.7958	0.51%	0.1751	0.31%	0.0744	0.45%	0.0534	0.60%	0.46	1040	6	1044	7	1052	18	1051	12
nv09e11	1.7924	0.55%	0.1764	0.37%	0.0738	0.43%	0.0529	0.77%	0.28	1047	7	1043	7	1034	18	1041	16
nv09c06	1.8518	0.88%	0.1768	0.38%	0.0760	0.82%	0.0538	0.65%	0.94	1049	7	1064	12	1094	32	1058	13
nv09e08	1.8334	1.01%	0.1767	0.47%	0.0753	1.08%	0.0549	1.59%	0.35	1049	9	1058	13	1076	42	1081	33
nv09a07	2.0205	0.73%	0.1863	0.36%	0.0787	0.69%	0.0556	1.16%	0.25	1101	7	1122	10	1164	28	1094	25
nv09c10	1.9987	1.62%	0.1879	0.58%	0.0772	1.68%	0.0574	1.98%	0.85	1110	12	1115	22	1126	66	1128	44
nv09c11	2.0469	0.83%	0.1880	0.39%	0.0790	0.83%	0.0569	0.82%	0.65	1111	8	1131	11	1172	34	1118	18
nv09a14	2.0970	0.44%	0.1887	0.28%	0.0807	0.40%	0.0586	0.73%	0.23	1114	6	1148	6	1212	16	1152	16
nv09a16	2.0799	0.82%	0.1900	0.34%	0.0794	0.78%	0.0574	0.97%	0.44	1121	7	1142	11	1182	32	1128	21
nv09d08	1.9347	1.34%	0.1904	0.50%	0.0738	1.26%	0.0526	1.46%	0.28	1123	10	1093	18	1034	50	1036	30
nv09d16	2.0597	1.25%	0.1906	0.42%	0.0784	1.30%	0.0560	1.70%	0.40	1124	9	1136	17	1158	52	1101	36
nv09b12	2.0343	0.52%	0.1916	0.33%	0.0770	0.51%	0.0569	0.86%	0.33	1130	7	1127	7	1122	20	1119	19
nv09c12	2.1121	0.90%	0.1923	0.35%	0.0797	0.93%	0.0574	1.00%	0.41	1134	7	1153	12	1188	38	1128	22
nv09e15	2.1220	0.63%	0.1948	0.27%	0.0791	0.61%	0.0575	0.92%	0.30	1147	6	1156	9	1172	24	1130	20
nv09e13	2.1421	0.64%	0.1952	0.32%	0.0796	0.60%	0.0580	0.62%	0.44	1149	7	1163	9	1188	22	1139	14
nv09b12	2.1890	1.18%	0.1975	0.40%	0.0804	1.14%	0.0584	1.02%	1.13	1162	9	1178	16	1206	44	1147	23
nv09c14	2.2330	0.51%	0.2008	0.27%	0.0807	0.53%	0.0607	0.74%	0.26	1180	6	1191	7	1214	22	1190	17
nv09b13	2.3019	1.37%	0.2011	0.53%	0.0831	1.43%	0.0630	2.46%	0.18	1181	11	1213	19	1270	56	1235	59
nv09d06	2.3148	0.88%	0.2041	0.37%	0.0823	0.78%	0.0592	0.53%	0.82	1197	8	1217	12	1252	32	1161	12
nv09c16	2.4218	0.56%	0.2116	0.40%	0.0831	0.48%	0.0626	0.67%	0.30	1237	9	1249	8	1270	20	1227	16
nv09c09	2.5382	0.79%	0.2167	0.35%	0.0850	0.74%	0.0648	1.04%	0.28	1265	8	1283	12	1314	28	1270	26
nv09c05	2.5740	0.65%	0.2182	0.36%	0.0856	0.62%	0.0655	0.61%	0.61	1272	8	1293	10	1328	24	1283	15
nv09e09	2.5832	1.10%	0.2208	0.42%	0.0849	0.93%	0.0651	0.97%	0.70	1286	10	1296	16	1312	36	1275	24
nv09e14	2.6793	0.60%	0.2258	0.34%	0.0861	0.59%	0.0663	0.61%	0.71	1313	8	1323	9	1340	22	1298	15
nv09d11	2.6949	0.66%	0.2300	0.36%	0.0850	0.62%	0.0697	0.73%	0.55	1335	9	1327	10	1314	24	1362	19
nv09a12	3.1016	0.60%	0.2535	0.36%	0.0888	0.52%	0.0702	0.61%	0.42	1457	9	1433	9	1398	20	1372	16
nv09a05	5.2198	0.72%	0.3295	0.46%	0.1150	0.59%	0.0959	0.67%	0.67	1836	15	1856	12	1878	20	1850	24

(LA-ICPMS) detrital zircons in the Upper Santa Rosa Group sediments. It is unclear what such Devonian granites are intruding, although they are unconformably overlain by Santa Rosa Group.

South of Huehuetenango field relationships are strikingly similar to those observed in the Altos Cuchumatanes. Two granite types, intruding sheared metasediments, are overlain by the Santa Rosa Group which crops out to the east, near Sacapulas (Bohnenberger 1966) and west of the studied area in Chiapas (Weber *et al.* 2006*b*, and references therein). Sample CH0407 indicates its Precambrian nature. Because this is the first sample of such age-span dated by this technique in Guatemala, it is difficult to correlate it. Nevertheless, sediments which are at least older than the Santa Rosa Group have been reported: (1) in Chiapas, where Hernández-García (1973) described a low-grade metamorphic sequence with some schists and abundant phyllites, that he called 'Lower Santa Rosa', and that are overlain by the Upper Santa Rosa Group of Weber *et al.* (2006*b*); (2) in the Rabinal-Salamá area, where Silurian–Ordovician granites intrude low-grade sediments of the San Gabriel sequence (e.g. Ortega-Obregón *et al.* 2008); and (3) in the Maya Mountains of Belize, as described by Steiner & Walker (1996). We believe that such belt of pre-Carboniferous granites intruding low-grade supracrustals could be considered as the southernmost exposure of the Maya Block basement, extending south of the Polochic fault. This fault, in the light of these data, cannot be considered as the southern limit of the Maya Block. Instead our view is that the boundary lies in the Baja Verapaz shear zone (e.g. Ortega-Obregón *et al.* 2008). Shear zone rocks from the type locality, south of Salamá and between Rabinal and Salamá, have yielded ages between 68 and 72 Ma (K–Ar on micas, Ortega-Obregón *et al.* 2008). The ages date the tectonic juxtaposition of the Chuacús Complex on top of the Maya continental margin, with a top-to-the NNE kinematics. Sample Gt03122 of Chuacús metasediments, analysed by LA-ICPMS in this work, and sampled south of Salamá, yielded youngest Late Ordovician zircons. They indicate a possible provenance from either the Silurian–Ordovician granites in the near-Maya Block, which would indicate proximity of Chuacús and Maya, or from Palaeozoic terranes of southern Mexico, where Silurian–Ordovician granites are widespread in the Acatlán Complex (e.g. Ortega-Gutiérrez *et al.* 1999; Talavera-Mendoza *et al.* 2005). Either the Acatlán or the Oaxacan Complex could be the possible source for Mesoproterozoic zircons in both San Gabriel sequence (sample CH0407) and Chuacús Complex (Gt03122). The Zapotecan tectonothermal event at *c.* 990 Ma (Solari *et al.* 2003, 2004) in the northern Oaxacan Complex, in particular, could

source zircons in the range *c.* 975 Ma (post-tectonic pegmatites, Solari *et al.* 2004), up to *c.* 1010 Ma (igneous zircons in anorthosite–mangerite–charnockite–granite suite, Keppie *et al.* 2003).

Zircons from the San Diego Phyllite (sample Gt0417) give some important starting points to interpret the evolution of the Chortís Block. The youngest zircons, of Cambrian age, indicate maximum Early Palaeozoic sedimentation. Absence of Silurian–Ordovician zircons, which are found in the Chuacús metasediments and the Upper Santa Rosa sediments (Weber *et al.* 2006*b*), and the sporadic presence of Pan-African or Braziliano-like zircons in the San Diego Phyllite (e.g. Doblas *et al.* 2002; Veevers 2003; Weber *et al.* 2006*b*) indicate that either the San Diego Phyllite sedimentation took place earlier or that no relationships existed between the Chortís and Maya in the Early–Mid-Palaeozoic. A connection between southern Mexico and the Chortís Block for the Early Palaeozoic–Precambrian, based upon the LA-ICPMS zircon data presented here, is also not straightforward. Unfortunately the only available data for Palaeozoic sequences in southern Mexico are from the Tiñu–Santiago–Ixtaltepec sediments covering the Oaxacan Complex north of Oaxaca (Gillis *et al.* 2005), and from the Acatlán Complex (Keppie *et al.* 2004; Sánchez-Zavala *et al.* 2004; Talavera-Mendoza *et al.* 2005). In both cases Cambrian zircons are absent, although Mesoproterozoic (*sensu lato*) peaks are similar to those observed in the San Diego Phyllite sample. Sources for such Mesoproterozoic zircons are widespread and not limited to the Oaxacan Complex. They could have come from several exposures in both Laurentia and Gondwana (e.g. Gillis *et al.* 2005, and references therein). In the Acatlán Complex, which is a possible alternative source for Mesoproterozoic zircons, samples generally also contain either Permo-Carboniferous zircons, or Archaean to Palaeoproterozoic, which are absent in the samples analysed here. Current models (e.g. Keppie & Morán Zenteno 2005; Nieto-Samaniego *et al.* 2006; Pindell *et al.* 1988, 2006; Pindell & Barret 1990; Ross & Scotese 1988; Schaaf *et al.* 1995) of the tectonic evolution of southern Mexico and Chortís Block only discuss post-Jurassic history because there is a lack of robust data on pre-Jurassic palaeopositions. The preliminary data presented here indicate that Maya, Chuacús Complex and Chortís must be seen as discrete fault-limited Blocks, bordering southern Laurentia (Maya) and/or western Gondwana (Chortís). They did not interact during the assembly of Pangaea (e.g. Dickinson & Lawton 2001; Elías-Herrera & Ortega-Gutiérrez 2002; Steiner 2005), but reached their current position during the Mesozoic or Cenozoic (Keppie & Morán-Zenteno 2005; Pindell *et al.* 2006).

This paper benefited from funds granted to LAS by PAPIIT-DGAPA UNAM project IN100002. Ultraclean laboratory facilities at Instituto de Geología, UNAM, were partially equipped with funds granted to LAS by CONACyT project J-39783.

We would like to thank J. J. Morales and G. Solís-Pichardo from LUGIS, UNAM, for laboratory assistance and their participation during analytical work and data acquisition. U. Martens provided a useful review, as well as important suggestions on Guatemalan geology. K. H. James and B. V. Miller also provided useful reviews, which helped to clarify some points.

References

ANDERSON, T. H., BURKART, B., CLEMONS, R. E., BOHNENBERGER, O. H. & BLOUNT, D. N. 1973. Geology of the western Altos Cuchumatanes, northwestern Guatemala. *Geological Society of America Bulletin*, **84**, 805–826.

BOHNENBERGER, O. H. 1966. Nomenclatura de las Capas Santa Rosa en Guatemala. *Publicaciones Geológicas del ICAITI (Guatemala)*, **1**, 47–51.

BURKART, B. 1983. Neogene North American–Caribbean Plate boundary across northern Central American offset along the Polochic fault. *Tectonophysics*, **99**, 251–270.

CAMPA, M. F. & CONEY, P. J. 1983. Tectono-stratigraphic terranes and mineral resource distributions in Mexico. *Canadian Journal of Earth Sciences*, **20**, 1040–1051.

DICKINSON, W. R. & LAWTON, T. F. 2001. Carboniferous to Cretaceous assembly and fragmentation of Mexico. *Geological Society of America Bulletin*, **113**, 1142–1160.

DOBLAS, M., LÓPEZ-RUÍZ, J., CEBRIS, J. M., YOUBI, N. & DEGROOTE, E. 2002. Mantle insulation beneath the West African craton during the Precambrian–Cambrian transition. *Geology*, **30**, 839–842.

DONNELLY, T. W., HORNE, G. S., FINCH, R. C. & LOPEZ-R, E. 1990. Northern Central America; the Maya and Chortís blocks. *In*: DENGO, G. & CASE, J. E. (eds) *Decade of North American Geology*, **H**. *The Caribbean Region*. Geological Society of America, Boulder, CO, 339–374.

DUCEA, M., GEHRELS, G. E., SHOEMAKER, S., RUÍZ, J. & VALENCIA, V. A. 2004. Geologic evolution of the Xolapa Complex, southern Mexico: evidence from U–Pb zircon geochronology. *Geological Society of America Bulletin*, **116**, 1016–1025.

EGGINS, S. M., KINSLEY, L. P. J. & SHELLEY, J. M. G. 1998. Deposition and element fractionation processes during atmospheric pressure laser sampling for analysis by ICP-MS. *Applied Surface Science*, **127/129**, 278–286.

ELÍAS-HERRERA, M. & ORTEGA-GUTIÉRREZ, F. 2002. The Caltepec Fault Zone: an Early Permian dextral transpressional boundary between the Proterozoic Oaxacan and Palaeozoic Acatlán complexes, southern Mexico, and regional tectonic implications. *Tectonics*, **21**; doi: 10.1029/2000TC001278.

FAKUNDINY, R. H. 1970. *Geology of El Rosario cuadrangle, Honduras, Central America*. PhD Thesis, University of Texas, 234.

GILLIS, R. J., GEHRELS, G. E., RUÍZ, J. & FLORES DE DIOS, A. 2005 Detrital zircon provenance of Cambrian–Ordovician and Carboniferous strata of the Oaxaca terrane, Southern Mexico. *Sedimentary Geology*, **182**, 87–100.

GOMBERG, D. M., BANKS, P. O. & MCBIRNEY, A. R. 1968. Guatemala: preliminary zircon ages from central cordillera. *Science*, **162**, 121–122.

HERNÁNDEZ-GARCÍA, G. R. 1973. Paleogeografía del Palaeozoico de Chiapas, México. *Boletín de la Asociación Mexicana de Geólogos Petroleros*, **25**, 77–134.

HORNE, G. S., CLARK, G. S. & PUSHKAR, P. 1976. Pre-Cretaceous rocks of northwestern Honduras: Basement terrane in Sierra de Omoa. *American Association of Petroleum Geologists Bulletin*, **60**, 566–583.

KEPPIE, J. D. 2004. Terranes of Mexico revisited: a 1.3 billion year odyssey. *International Geology Review*, **46**, 765–794.

KEPPIE, J. D. & MORÁN ZENTENO, D. J. 2005. Tectonic implications of alternative Cenozoic reconstructions for Southern Mexico and the Chortis Block. *International Geology Review*, **47**, 478–491.

KEPPIE, J. D., DOSTAL, J., CAMERON, K. L., SOLARI, L. A., ORTEGA-GUTIÉRREZ, F. & LOPEZ, R. 2003. Geochronology and geochemistry of Grenvillian igneous suites in the northern Oaxacan Complex, southern Mexico: tectonic implications. *Precambrian Research*, **120**, 365–389.

KEPPIE, J. D., SANDBERG, C. A., MILLER, B. V., SÁNCHEZ-ZAVALA, J. L., NANCE, R. D. & POOLE, F. G. 2004. Implications of Latest Pennsylvanian to Middle Permian Paleontological and U–Pb SHRIMP data from the Tecomate Formation to re-dating tectonothermal events in the Acatlán Complex, Southern Mexico. *International Geology Review*, **46**, 745–753.

KESLER, S. E., JOSEY, W. L. & COLLINS, E. M. 1970. Basement rocks of western nuclear Central America: the western Chuacús Group, Guatemala. *Geological Society of America Bulletin*, **81**, 3307–3322.

KROGH, T. E. 1973. A low-contamination method for hydrothermal decomposition of zircon and extraction of U and Pb for isotopic age determinations. *Geochimica et Cosmochimica Acta*, **37**, 485–494.

KROGH, T. E. 1982. Improved accuracy of U–Pb zircon ages by the creation of more concordant systems using an air abrasion technique. *Geochimica et Cosmochimica Acta*, **46**, 637–649.

LAWRENCE, J. 1975. *Petrology and structural geology of the Sanarate-El Progreso area, Guatemala*. PhD thesis, Binghamton State University.

LUDWIG, K. R. 1991. *PbDat: A Computer Program for Processing Pb–U–Th Isotope Data, Version 1.24*. United States Geological Survey, Special Publications, **88–542**.

LUDWIG, K. R. 2004. *Isoplot/Ex, Verion 3, a Geochronological Toolkit for Microsoft Excel*. Berkeley Geochronology Center, Special Publications, **4**.

MANTON, W. I. 1996. The Grenville of Honduras. *Geological Society of America, Abstracts with Programs*, Denver, CO, A-493.

MARTENS, U., RATSCHBACHER, L. & MCWILLIAMS, M. 2005. U–Pb geochronology of the Maya Block, Guatemala. *AGU, San Francisco, CA, USA, Fall Meeting*, T51D-1387.

MATTINSON, J. M. 1972. Preparation of hydrofluoric, hydrochloric, and nitric acids at ultralow lead levels. *Analytical Chemistry*, **44**, 1715–1716.

MATTINSON, J. M. 1987. U–Pb ages of zircons: a basic examination of error propagation. *Chemical Geology*, **66**, 151–162.

MCBIRNEY, A. R. 1963. Geology of a part of the central Guatemalan cordillera. *California University Publications in Geological Sciences*, **38**, 177–242.

MILLER, B. V., FETTER, A. H. & STEWART, K. G.. 2006. Plutonism in three orogenic pulses, Eastern Blue Ridge Province, southern Appalachians. *Geological Society of America Bulletin*, **118**, 171–184.

MURPHY, B., KEPPIE, J. D., BRAID, J. F. & NANCE, R. D. 2005. Geochemistry of the Tremadocian Tinu Formation (southern Mexico); provenance in the underlying approximately 1 Ga Oaxacan Complex on the southern margin of the Rheic Ocean. *International Geology Review*, **47**, 887–900.

NIETO-SAMANIEGO, A. F., ALANIZ-ÁLVAREZ, S. A., SILVA-ROMO, G., EGUIZA-CASTRO, M. H. & MENDOZA-ROSALES, C. C. 2006. Latest Cretaceous to Miocene deformation events in the eastern Sierra Madre del Sur, Mexico, inferred from the geometry and age of major structures. *Geological Society of America Bulletin*, **118**, 238–252.

ORTEGA-GUTIÉRREZ, F., ELIAS-H, M., REYES-SALAS, M., MACIAS-R, C. & LOPEZ, R. 1999. Late Ordovician–Early Silurian continental collision orogeny in southern Mexico and its bearing on Gondwana–Laurentia connections. *Geology*, **27**, 719–722.

ORTEGA-GUTIÉRREZ, F., SOLARI, L. A. *ET AL.* 2007. The Maya–Chortís boundary: a tectonostratigraphic approach. *International Geology Review*, **449**, 996–1024.

ORTEGA-GUTIÉRREZ, F., SOLARI, L. A. *ET AL.* 2004. High Pressure eclogite facies metamorphism in the Chuacús Complex, Sierra de Chuacús, Central Guatemala: petrology, geochronology, and tectonic implications. *International Geology Review*, **46**, 445–470.

ORTEGA-OBREGÓN, C., SOLARI, L. A., KEPPIE, J. D., ORTEGA-GUTIÉRREZ, F., SOLÉ, J. & MORÁN-ICAL, S. 2008. Middle–Late Ordovician magmatism and Late Cretaceous collision in the southern Maya block, Rabinal–Salamá area, central Guatemala: implications for North America–Caribbean Plate tectonics. *Geological Society of America Bulletin*, **120**, 556–570.

PACES, J. B. & MILLER, J. D. 1989. Precise U–Pb ages of Duluth Complex and related mafic intrusions, northeastern Minnesota: geochronological insights to physical, petrogenic, paleomagnetic and tectonomagmatic processes associated with the 1.1 Ga midcontinent rift system. *Journal of Geophysical Research*, **98B**, 13,997–14,013.

PARRISH, R. R. 1987. An improved micro-capsule for zircon dissolution in U–Pb geochronology. *Chemical Geology*, **66**, 99–102.

PINDELL, J. L. & BARRET, S. F. 1990. Geological evolution of the Caribbean region; a plate-tectonic perspective. *In:* DENGO, G. & CASE, J. E. (eds) *The Geology of North America, Vol. II, The Caribbean Region.* Geological Society of America, Boulder, CO, 405–432.

PINDELL, J. L., CANDE, S. C. *ET AL.* 1988. A plate-kinematic framework for models of Caribbean evolution. *Tectonophysics*, **155**, 121–138.

PINDELL, J. L., KENNAN, L., STANEK, K. P., MARESCH, W. V. & DRAPER, G. 2006. Foundations of Gulf of Mexico and Caribbean evolution: eight controversies resolved. *Geologica Acta*, **4**, 303–341.

ROSS, M. I. & SCOTESE, C. E. 1988. A hierarchical model of the Gulf of Mexico and Caribbean region. *Tectonophysics*, **155**, 139–168.

RUBATTO, D. 2002. Zircon trace element geochemistry: partitioning with garnet and the link between U–Pb ages and metamorphism. *Chemical Geology*, **184**, 123–138.

SÁNCHEZ-ZAVALA, J. L., ORTEGA-GUTIÉRREZ, F., KEPPIE, J. D., JENNER, G. A., BELOUSOVA, E. A. & MACÍAS-ROMO, C. 2004. Ordovician and Mesoproterozoic Zircons from the Tecomate Formation and Esperanza Granitoids, Acatlán Complex, Southern Mexico: local provenance in the Acatlán and Oaxacan Complexes. *International Geology Review*, **46**, 1005–1021.

SCHAAF, P., MORÁN-ZENTENO, D. J. *ET AL.* 1995. Paleogene continental margin truncation in Southwestern Mexico: Geochronological evidence. *Tectonics*, **14**, 1339–1350.

SCHWARTZ, D. P. 1977. *Geology of the Zacapa Quadrangle and vicinity, Guatemala, Central America.* PhD thesis, State University New York.

SEDLOCK, R. L., ORTEGA-GUTIÉRREZ, F. & SPEED, R. C. 1993. *Tectonostratigraphic Terranes and Tectonic Evolution of Mexico.* Geological Society of America, Boulder, CO, Special Papers, **278**, 153.

SOLARI, L. A., KEPPIE, J. D., ORTEGA-GUTIÉRREZ, F., CAMERON, K. L., LOPEZ, R. & HAMES, W. E. 2003. Grenvillian tectonothermal events in the northern Oaxacan Complex, southern Mexico: roots of an orogen. *Tectonophysics*, **365**, 257–282.

SOLARI, L. A., KEPPIE, J. D., ORTEGA-GUTIÉRREZ, F., CAMERON, K. L. & LOPEZ, R. 2004. ~990 Ma peak granulitic metamorphism and amalgamation of Oaxaquia, Mexico: U–Pb zircon geochronological and common Pb isotopic data. *Revista Mexicana de Ciencias Geológicas*, **21**, 212–225.

SOLARI, L. A., TORRES DE LEÓN, R., HERNÁNDEZ-PINEDA, G. A., SOLÉ, J., HERNÁNDEZ-TREVIÑO, T. & SOLÍS-PICHARDO, G. 2007. Cretaceous to Tertiary magmatism and evolution of a middle crust shear zone, Tierra Colorada area, southern Mexico: tectonic implications. *Geological Society of America Bulletin*, **119**, 1265–1279.

STEINER, B. & WALKER, J. D. 1996. Late Silurian plutons in Yucatan. *Journal of Geophysical Research*, **101**, 17,727–17,735.

STEINER, M. B. 2005. Pangean reconstruction of the Yucatan Block: its Permian, Triassic, and Jurassic geologic and tectonic history. *In:* ANDERSON, T. H., NOURSE, J. A., MCKEE, J. W. & STEINER, M. B. (eds) *The Mojave–Sonora Megashear Hypothesis: Development, Assessment, and Alternatives.* Geological Society of America, Boulder, CO, Special Papers, **393**, 457–480.

TALAVERA-MENDOZA, O., RUÍZ, J., GEHRELS, G. E., MEZA-FIGUEROA, D., VEGA-GRANILLO, R. &

CAMPA, M. F. 2005. U–Pb geochronology of the Acatlán Complex and implications for the Palaeozoic paleogeography and tectonic evolution of southern Mexico. *Earth and Planetary Science Letters*, **235**, 682–699.

TOLSON, G. 2005. La falla Chacalapa en el sur de Oaxaca. *Boletin Sociedad Geológica Mexicana*, **57**, 111–122.

VEEVERS, J. J. 2003. Pan-African is Pan-Gondwanaland: oblique convergence drives rotation during 650–500 Ma assembly. *Geology*, **31**, 501–504.

WEBER, B. & KOHLER, H. 1999. Sm–Nd, Rb–Sr and U–Pb geochronology of a Grenville terrane in southern Mexico: origin and geologic history of the Guichicovi Complex. *Precambrian Research*, **96**, 245–262.

WEBER, B., CAMERON, K. L., OSORIO, M. & SCHAAF, P. 2005. A Late Permian tectonothermal event in Grenville crust of the Southern Maya Terrane: U–Pb Zircon Ages from the Chiapas Massif, Southeastern Mexico. *International Geology Review*, **47**, 509–529.

WEBER, B., IRIONDO, A., PREMO, W. R., HECHT, L. & SCHAAF, P. 2006a. New insights into the history and origin of the southern Maya block, SE Mexico: U–Pb–SHRIMP zircon geochronology from metamorphic rocks of the Chiapas Massif. *International Journal of Earth Sciences*; doi: 10.1007/s00531–006–0093–7.

WEBER, B., SCHAAF, P., VALENCIA, V. A., IRIONDO, A. & ORTEGA-GUTIÉRREZ, F. 2006b. Provenance ages of Late Paleozoic sandstones (Santa Rosa Formation) from the Maya block, SE Mexico–implications on the tectonic evolution of western Pangea. *Revista Mexicana de Ciencias Geológicas*, **23**, 262–276.

WEYL, R. 1980. *Geology of Central America*. Gebruder Borntraeger, Berlin.

Emplacement of the northern ophiolites of Cuba and the Campanian–Eocene geological history of the northwestern Caribbean–SE Gulf of Mexico region

JORGE L. COBIELLA-REGUERA

Department of Geology, University of Pinar del Rio, Marti #270, Pinar del Rio 20100, Cuba
(e-mail: jcobiella geo.upr.edu.cu)

Abstract: The Mesozoic Proto-Caribbean Plate was consumed in the subduction zone of the Greater Antilles volcanic arc until the Campanian. At this time, volcanic arc magmatism ceased along Cuba. From Late Campanian to Danian, Cuba and its surroundings were a collision zone where the GAC accreted to the North American palaeomargin. In the Danian the almost east–west trending SE Cuba–Cayman Ridge–Hispaniola? volcanic arc was born. The related north dipping subduction zone acted as the SE North American plate boundary. From the Paleocene to Middle Eocene dense Caribbean lithosphere travelled northwards. The location, strike and subduction polarity of the assumed subduction zone are very different from those described by other models. Almost simultaneously the Cuban Orogeny developed in western and central Cuba. During the orogeny the northern ophiolite belt of Cuba and the Cretaceous volcanic rocks were thrust northwards tens of kilometres, onto the Mesozoic North American palaeomargin. In the Middle Eocene subduction stopped. Simultanously(?) a change in the regional stress field originated the near east–west trending sinistral Oriente fault zone, whose position and origin are probably tied to the weakened hot crust to the south of the Palaeogene volcanic arc axis.

At the present time Cuba belongs to the North American plate. Its SE part is located along the Oriente fault zone, one of the major tectonic elements separating the North American and Caribbean plates. It also belongs to the Greater Antilles islands. Despite present positions on different plates, Cuba, Hispaniola and Puerto Rico share some features pointing to a common pre Middle Eocene geological history. Insufficient use has been made of Cuban geological data to test Caribbean tectonic models and the present paper is an attempt to correct this trend.

An outstanding feature of the Cuban geology is the belt of Mesozoic ultramafic and associated rocks that occur as a discontinuous chain of upper Mesozoic ultramafic and mafic rocks, extending for more than 1000 km along the northern half of Cuban mainland. In recent years, several contributions have greatly increased the knowledge on the petrology and geochemistry of the Cuban and other Greater Antilles ophiolites and the spatially related Cretaceous volcanic arc magmatic rocks (Andó *et al.* 1996; Rodríguez *et al.* 1997, 2001; Proenza *et al.* 1999, 2006; Kerr *et al.* 1999; Lewis *et al.* 2006; García Casco *et al.* 2006, among others). The later tectonic emplacement has also been recently studied by Iturralde-Vinent (Iturralde-Vinent & Marphee 1996; Iturralde-Vinent *et al.* 2006), Lewis *et al.* (2006), Cobiella-Reguera (2000, 2002, 2005) and García-Casco *et al.* (2002, 2006), among others. This last event, together with

the development of the Late Cretaceous and Early Palaeogene volcanic arcs, is important in unravelling the North American–Caribbean Plate boundary history.

The first part of the paper presents a brief discussion of the geology of the northern ophiolites. The data and conclusions are used in the second part for a discussion of evolution of the North America–Caribbean Plate boundary as indicated by Cuban data.

Tectonic setting of the northern ophiolites

In order to understand the emplacement of the northen ophiolites some general knowledge of the geology of Cuba, as it relates to the age and origin of these rocks, is neccesary. Two main structural levels can be distinguished in Cuba. The upper level comprises little-deformed Eocene–Quaternary cover. The lower level (sole), is a variably deformed sequence of older rocks (Iturralde-Vinent 1997; Cobiella-Reguera 2000). The sole has two main parts: (1) the pre-Cenozoic basement and (2) a lower Cenozoic folded belt built during the Early Cenozoic Cuban Orogeny. The pre-Cenozoic basement includes three different terranes (Fig. 1): a northern belt of ophiolites (NO), a Cretaceous volcanic arc terrane (KVAT) with its uppermost Cretaceous sedimentary cover and the southern metamorphic terranes (SMT). The SMT protoliths

From: JAMES, K. H., LORENTE, M. A. & PINDELL, J. L. (eds) *The Origin and Evolution of the Caribbean Plate.*
Geological Society, London, Special Publications, **328**, 315–338.
DOI: 10.1144/SP328.13   0305-8719/09/$15.00 © The Geological Society of London 2009.

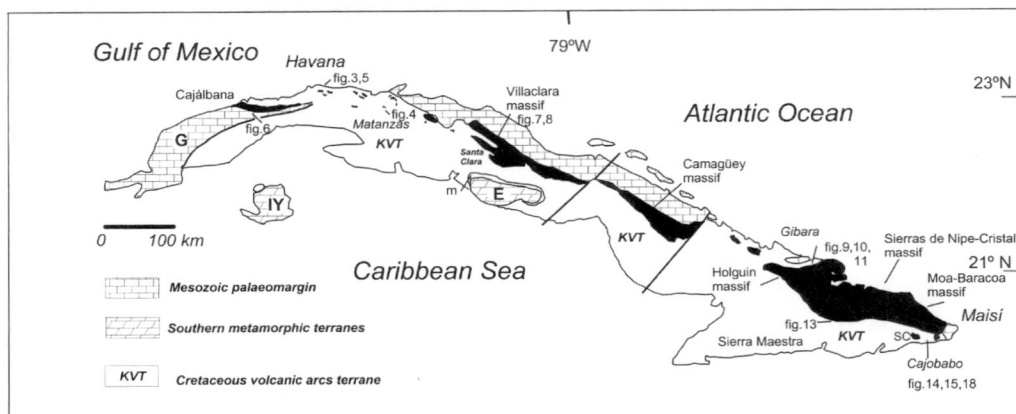

Fig. 1. Mesozoic domains of Cuba after stripping of the Cenozoic cover. G, Guaniguanico Highlands; IY, Isle of Youth; E, Escambray (Guamuhaya) Massif; m, Mabujina complex (metamafites, oceanic basement of the Cretaceous volcanic sequences). In black are ophiolitic massifs. SC, Sierra del Convento ophiolites. Modified after Cobiella-Reguera (2005).

are mainly Mesozoic (Jurassic) continental margin sections, very similar to coeval deposits in Guaniguanico mountains (Somin & Millan 1981; Cobiella-Reguera 2000). However, in central Cuba Escambray (Guamuhaya) mountains, Cretaceous(?) metavolcanic sequences also occur. The different Mesozoic terranes were initially accreted to the North American continental margin during the Cretaceous (Fig. 1; Cobiella-Reguera 1998, 2000, 2005), but the welding process ended in the Middle and Late Eocene (Iturralde-Vinent 1996a; Pszczolkowski 1999). The Northern Ophiolites and the Cretaceous volcanic arc terrane are allochthonous units, resting on the Mesozoic North American passive palaeomargin. The southern metamorphic terranes outcrop in two great tectonic windows below the volcanic terrane. The North American palaeomargin contains Jurassic and Cretaceous mainly marine sections. Precambrian (Grenvillian) rocks are present in small and poor outcrops in its southern fringe (Pszczolkowski & Myczynski 2003). Thin-skinned tectonics is well developed in the palaeomargin, particularly in western Cuba (Guaniguanico mountains) and in the well-stratified deep water sections in central Cuba and Camaguey.

Two different sedimentary, magmatic and tectonic scenarios existed from Danian to Middle Eocene (Fig. 2). West of the Camaguey lineament, the Late Paleocene–Middle (locally Late) Eocene Cuban orogeny developed. During this event the ophiolite massifs and the Cretaceous volcanic terrane are thrusted northward upon the foreland basin built upon the southern fringe of the North American palaeomargin (Pardo 1975).

Olistostrome deposits are preserved between the thrust sheets. Small piggyback basins, filled with turbidites, are locally present on top of the KVAT sections. Eastward of Camaguey lineament orogenic deformations are absent or at least poorly recorded. Marine Danian–Middle Eocene volcanic arc rocks (Turquino arc) attain up to 6000 m in thickness in the Sierra Maestra mountains of SE Cuba and become thinner northward (Lewis & Straczek 1955; Khudoley & Meyerhoff 1971, fig. 2). Intrusive magmatic rocks are abundant in the Sierra Maestra, decreasing towards northern eastern Cuba. Volcanoclastic Middle and Late Eocene deposits attain almost 1000 m in thickness and rest conformably on the underlying lower Palaeogene volcano-sedimentary sequence. Deformations in the entire Paleocene–Eocene section increase southward.

Following (with minor modifications) Iturralde-Vinent's (1996a) classification, three types of ophiolites can be distinguished in Cuba: (1) the northern ophiolites; (2) the metamorphic basement of the Cretaceous volcanic arc terrane (Mabujina complex); and (3) tectonic slices in the Escambray Massif (part of the SMT), in central Cuba. The present paper deals only with the first group (see Iturralde-Vinent 1997; Cobiella-Reguera 2005, for additional information on Cuban ophiolites).

Over 90% of the oceanic lithosphere remains in Cuba occur in the northern ophiolites. It forms an almost continuous, strongly deformed chain of bodies, transported from the south over the continental margin (Kozary 1968; Meyerhoff & Hatten 1968; Iturralde-Vinent 1990; Echevarría-Rodríguez et al. 1991). Most of the NO is a huge mélange,

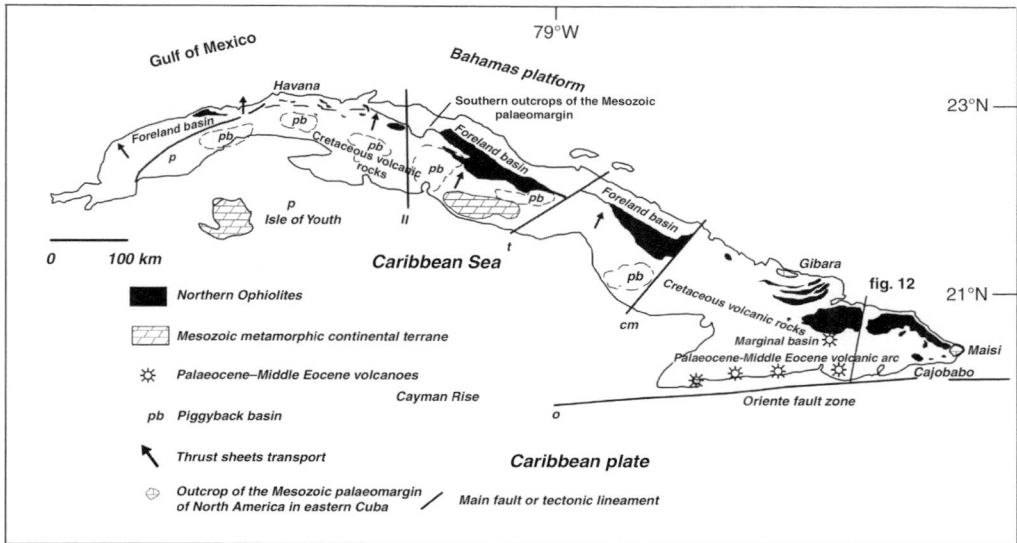

Fig. 2. Danian to Middle Eocene main tectonic elements of Cuba. p, Pinar fault; ll, Llabre lineament; t, La Trocha fault; cm, Camaguey lineament; o, Oriente fault.

extended *c.* 1000 km along the northern half of Cuba (Fig. 1). Blocks of mainly ophiolitic suite components float in a highly deformed serpentinitic matrix. The ophiolites show evidence of pervasive deformation and almost all the contacts are tectonic, with severely crushed bodies cut by tongues of brecciated or foliated serpentinites mm to metres thick (Knipper & Cabrera 1974). Despite internal deformation and mixing, the main members of the suite preserve their identities and in some places the original pretectonic relationships are recorded. Tectonized ultramafic rocks (serpentinites) and the rocks of the cumulative complex (ultramafites + gabbros) are the most common lithologies, while basalts and sedimentary rocks are poorly exposed (Kozary 1968; Knipper & Cabrera 1974; Fonseca *et al.* 1984; Cobiella-Reguera 1984; Iturralde-Vinent 1990, 1996*a*; and others). Biostratigraphic and radiometric data yield Upper Jurassic-Aptian and/or Albian ages (Iturralde-Vinent 1990, 1996*a*; Iturralde-Vinent *et al.* 1996; Andó *et al.* 1996; Millan 1997; Cobiella-Reguera 2005). Some proposed younger ages (Iturralde-Vinent *et al.* 2006; Proenza *et al.* 2006) seem to occur locally and require further research. Recent studies record distinct origins for the NO rocks. Mid Oceanic Ridge (MOR) magmatism has been reported by Beccaluva *et al.* (1996), Giunta *et al.* (2002), Andó *et al.* (1996) and García-Casco *et al.* (2002), while Kerr *et al.* (1999) and Andó *et al.* (1996) found island arc tholeiites and Proenza *et al.* (1999) detected suprasubduction signatures. According to Cobiella-Reguera

(1998, 2000, 2005), Upper Jurassic–Neocomian ophiolites probably have a MOR origin, whereas those of Aptian–Albian age are related to a suprasubduction zone. The NO crops out in several regions, each with distinct characteristics. The areas are: (1) western Cuba; (2) central Cuba and Camaguey; (3) Holguin; (4) eastern Cuba; and (5) Cajobabo. Summary descriptions of these areas have appeared in recent years (Iturralde-Vinent 1996*a*, *b*; Cobiella-Reguera 2005; Lewis *et al.* 2006).

Western Cuba

Western Cuba ophiolitic bodies are represented by the following outcrop areas separated by Cenozoic rocks (from west to east, Fig. 1): Cajalbana–Bahia Honda, NW Havana, NE Havana province and northern Matanzas. These small massifs are located west of the Llabre lineament (defined as a strike–slip fault by Pszczolkowski 1983). In these areas the ophiolites occur as small bodies in tectonic mix with the Cretaceous volcanic arc rocks (volcano-ophiolitic mélange, Figs 3 & 4). Unconformably on the mélange rest strongly deformed upper Campanian, Maastrichtian and K/T boundary deposits (Piotrowska 1986*a*, *b*; Piotrowski 1986; Diaz-Otero *et al.* 2003). The mélange is probably lower Campanian in age (Cobiella-Reguera 2005). The oldest beds in the sedimentary cover containing clasts from the ophiolite suite are upper Campanian conglomerates

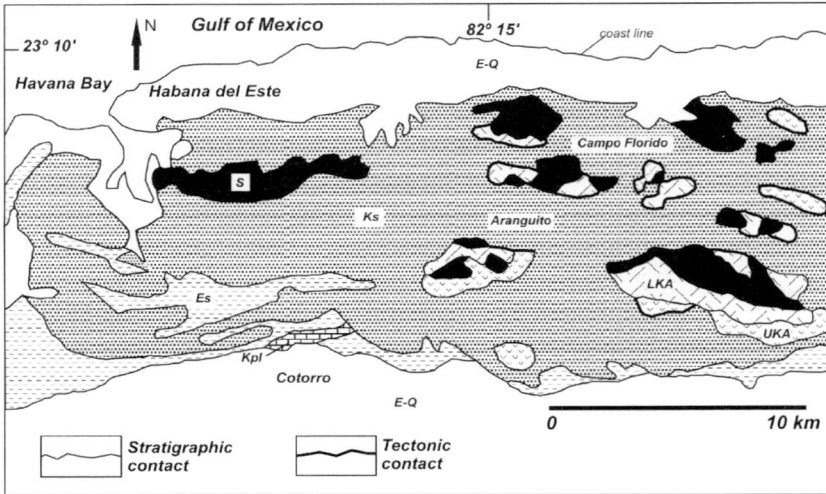

Fig. 3. Geological map of eastern Havana City and its surroundings (simplified and slightly modified after Pushcharovsky, 1988). See location in Figure 1. LKA, Lower Cretaceous volcanic arc rocks (Chirino Fm); UKA, Upper Cretaceous volcanic arc rocks (La Trampa Fm); Ks, upper Campanian, Maastrichtian and K–T boundary sedimentary deposits (Via Blanca and Penalver Formations); Es, Upper Paleocene–Lower Eocene sedimentary beds (Vibora Group and Capdevila Fm); Kpl (?), 'Placetas type' deep water beds; E–Q, upper Lower Eocene–Quaternary deposits. See location in Figure 1. After Cobiella-Reguera (2005).

Fig. 4. Outcrop of the volcano-ophiolitic mélange at Loma Esmeralda, Limonar, Matanzas province. Tectonic serpentinite breccia (**S**) in the background at the summit of the hill. Strongly weathered and deformed thin bedded tuffs (?) (**T**) in the foreground. Contacts are fault planes. See location in Figure 1. Photograph courtesy of Martin Meschede, Greifswald University, Germany. Latitude: 22° 59′ 00″, longitude: 82° 41′ 20″.

(Brönniman & Rigassi 1963; Albear-Fránquiz & Iturralde-Vinent 1985). In some places in northern Havana, upper Paleocene beds lie unconformably upon the uppermost Cretaceous and the K/T boundary beds, recording an Early Paleocene deformation event. A third tectonic event, the Cuban Orogeny, is represented by northward thrusted ophiolite and volcanic arc upon the North American Mesozoic palaeomargin (Meyerhoff & Hatten 1968; Echevarría-Rodríguez et al. 1991). The palaeomargin crops out only to the south of the Cajalbana-Bahia Honda area (Pszczolkowski & Albear 1982; Cobiella-Reguera 2005). In the east (northern Havana and Matanzas provinces) it lies at variable depths below the thrust pile. Thin skinned tectonics characterize the deep water Mesozoic palaeomargin beds and the overlying lower Palaeogene siliciclastic deposits (Bralower & Iturralde-Vinent 1997; Gordon et al. 1997; Fig. 5). Uppermost Paleocene and/or lowermost Lower Eocene chaotic deposits (olistostromes) with clasts from the ophiolites and the Mesozoic palaeomargin accumulated in front of each advancing sheet (Fig. 6, Manacas Fm). Immediatly after deposition these soft water-saturated marine sediments were over-ridden and

mechanically mingled with the sheets, becoming a mélange. The lack of sorting and rounding in the olistostromes suggests very rapid erosion and deposition along submarine scarps. Lower Eocene beds rest unconformably on the thrust sheets (Fig. 5).

In the Cajalbana–Bahia Honda area the Mesozoic palaeomargin and the overlying Manacas Fm form a near east–west trending tectonic fenster in Sierra del Rosario, eastern Guaniguanico Cordillera, containing a north dipping thrust pile (Pushcharovsky 1988) several kilometres thick. To the north of Sierra del Rosario, the basal thrust plane of the overlying volcano-ophiolitic mélange has been found in deep wells near the Gulf of Mexico coast. Recent seismic data shows that the front of the Cuban thrust belt lies several kilometres to the north of the coastline (Moretti et al. 2003). As the ophiolites and the Cretaceous volcanic arc terrane were thrusted northward, the sum of Sierra del Rosario fenster width along a SSE–NNW line (16 km) plus the distance from the northern tectonic boundary of Sierra del Rosario to the coast (10–16 km) gives a minimum value for the tectonic displacement of the volcano-ophiolitic mélange (26–32 km) but more than 100 km seems

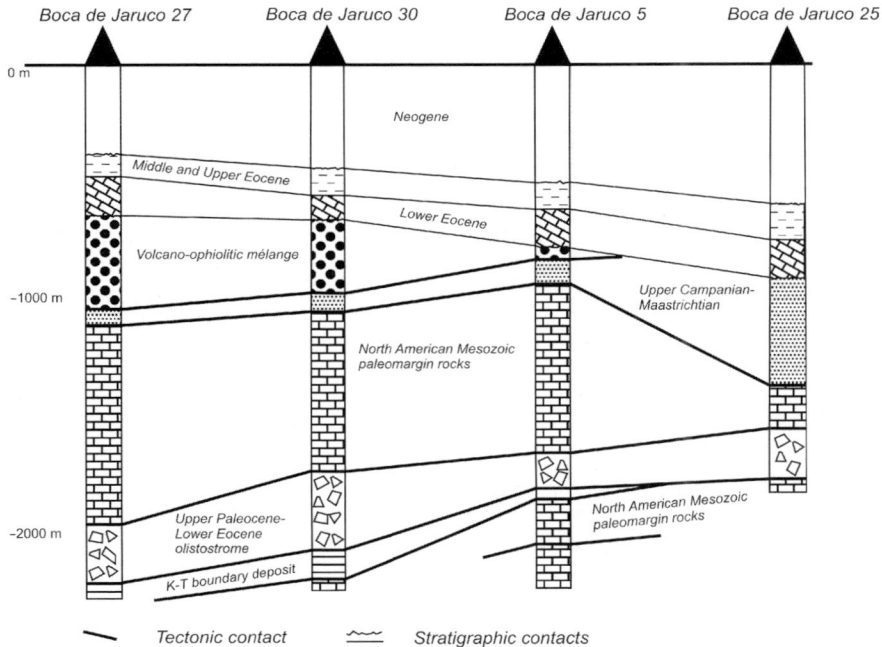

Fig. 5. Sketch correlation in the Boca de Jaruco oil field, located about 20 km to the east of Havana City. A Lower Eocene–Neogene cover rests unconformably on the Cuban orogeny thrust sheets. In this region, the Mesozoic North American palaeomargin rests at least several hundred metres below the Earth's surface and only crops out in small areas (see Fig. 3). The characteristic thin skinned tectonics of the NA palaeomargin is evident (after Furrazola-Bermudez et al. 1979, modified). See location in Figure 1.

Fig. 6. Outcrop of the Manacas Fm at Las Terrazas, Guaniguanico Cordillera (Fig. 1), western Cuba. The unit consists of chaotic submarine deposits (olistostromes), with a foliated, commonly serpentinite rich, silty or sandy matrix, surrounding strongly brecciated blocks of limestones (**bl**, foreground in the photograph), quartzose sandstone, cherts and other lithologies derived from the Mesozoic palaeomargin, and rocks of the ophiolitic suite. Foliation gently dips to the right in the photograph. Similar deposits are present along northern Cuba from the Guaniguanico Highlands to Camaguey (compare with Fig. 8). See location in Figure 1. Latitude: 22° 51′ 20″, longitude: 84° 04′ 06″.

possible. In the same way, in northern Havana and Matanzas province the minimum horizontal movement is between 15 and 25 km.

In deep wells of southern Havana and Matanzas provinces the volcano-ophiolitic mélange is absent and the Cretaceous arc volcanics lie in tectonic contact on top of ophiolitic (?) dolerites (García & Fernandez 1987; Cobiella-Reguera 2005).

Central Cuba and Camaguey

Between the Llabre and Camaguey tectonic lineaments (Fig. 2) the ophiolites form two huge bodies, the Villaclara and Camaguey massifs, resting on the Mesozoic North American palaeomargin. Cretaceous volcanic arc rocks are thrusted from the south upon the ophiolitic massifs (Knipper & Cabrera 1974; Meyerhoff & Hatten 1968; Pardo 1975; Iturralde-Vinent 1997).

In the Villaclara Massif the upper members of the ophiolite suite show the most complete development, particularly in the east. Radiolarites interbedded with basalts and dolerites yield Tithonian ages (Llanes Castro *et al.* 1998). Despite tectonic

mixing along the contact with the Cretaceous volcanic terrane, mélanges similar to those in western Cuba are absent. Upper Maastrichtian–Paleocene beds (Santa Clara Fm) cover both lithologies and the oldest beds in the volcanic arc sedimentary cover are Maastrichtian (Fig. 7). As the youngest volcanics below the unconformity are lower Campanian (Pushcharovsky 1988), a tectonic event more or less coeval with the volcano-ophiolitic mélange formation of western Cuba is indicated. The Early Cenozoic Cuban Orogeny is recorded by deformed olistostromes and turbidites (Fig. 8, Vega Alta Fm, in Pushcharovsky 1988; the 'members' 5 and 6 of Las Villas belt in Pardo 1975). This unit resembles the western Cuba Manacas Fm, occurring as thin tectonic lenses, metres to several hundred of metres thick and hundred of metres to tens of kilometres long, within thin bedded deep water beds (the upper Mesozoic Placetas belt of the North American margin in northern central Cuba). In some places, klippen of the ophiolitic sheet lie upon the southernmost carbonate banks of the Bahamas platform (Remedios zone), showing that in north central Cuba, immediately after the thrusting, the ophiolite

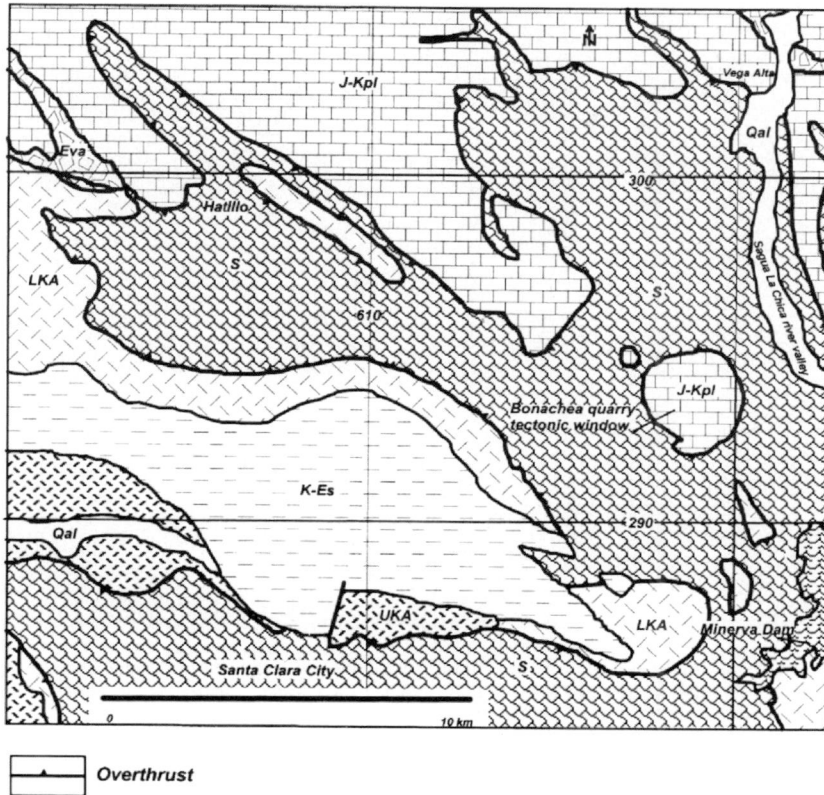

Overthrust

Fig. 7. Simplified geological map of Santa Clara city and its surroundings (local coordinates Cuba North), after Pushcharovsky (1988). The (upper?) Maastrichtian–Eocene gently deformed sedimentary cover rests unconformably on the ophiolitic and volcanic arc thrust sheets. Tectonic mixing between the ophiolites and the Cretaceous volcanic rocks, however present, is less marked than in western Cuba. J–Kpl, Deep water deposits of Placetas zone (Veloz, Santa Teresa, Carmita and Amaro formations); LKA, Lower Cretaceous volcanic arc rocks (Matagua Fm); UKA, Upper Cretaceous volcanic arc rocks (Brujas and Tasajeras formations); K–Es, upper? Maastrichtian–Middle Eocene sedimentary beds (Santa Clara and Ochoa formations); Qal, Quaternary alluvia. See Figure 3 for geological contact symbols. See location in Figure 1.

massifs covered not only the thrust pile belt of the well stratified southern facies of the palaeomargin (Placetas and Camajuani zones, Pszczolkowski 1983), but also reached the southern fringe of the platform. Based on this fact, a minimum of 30 km of horizontal movement can be envisaged for the northern ophiolites of central Cuba. Pszczolkowski (1983) estimated 50–70 km of northward transport for the Placetas belt rocks. Therefore, the horizontal travel of the ophiolite belt rocks, originally located southwads of the Placetas basin, should be much larger. Palaeontological constraints for the Vega Alta Fm are very limited (Kantchev et al. 1978). Together with regional stratigraphic data, they indicate deposition between the Paleocene and the Middle Eocene. However, the chaotic and inmature deposits clearly suggest a much shorter time interval

for deposition. As in western Cuba, the nature of the Vega Alta olistostromes points to rapid erosion and sedimentation along submarine fault scarps.

The Camaguey Massif (Fig. 1) lies between the La Trocha fault (west) and the Camaguey lineament (east). Biostratigraphic data from the volcano-sedimentary member indicate an Aptian–Albian age, whereas a 160 ± 24 Ma K–Ar age was obtained from gabbroids. The massif is a gently south-dipping body resting on the Cretaceous Bahamian carbonate banks (Remedios zone). North of Camaguey city, a Lower Eocene chaotic deposit lies between the ophiolites and the Cretaceous volcanic arc cover (Iturralde-Vinent 1996a). As in central Cuba, the youngest volcanic arc rocks are Campanian in age. Upper Campanian and Maastrichtian sediments rest unconformably on

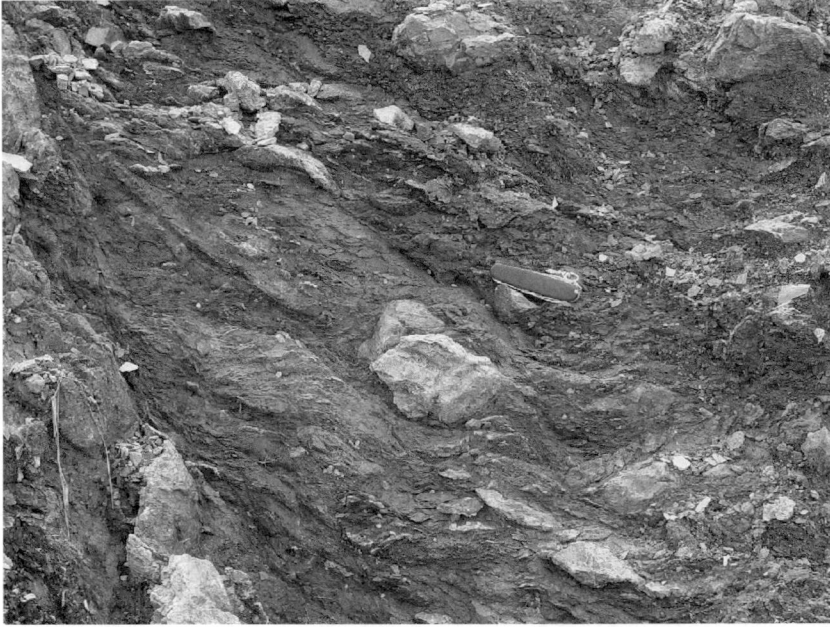

Fig. 8. Vega Alta Fm at Crucero Tarafa, *c*. 20 km NE of Santa Clara city. In this outcrop, small phacoidal blocks of harder rocks (mainly limestones) are surrounded by a fine grained wavy foliated matrix. See location in Figure 1. Photograph courtesy of Martin Meschede, Greifswald University, Germany.

the volcanic arc suite. The >20 km of tectonic overlap of the carbonate banks and their southern deep water facies (Placetas zone) by the Camaguey ophiolites records a second episode of the Cuban Orogeny in this region. The ophiolite emplacement is recorded by olistostromes (Iturralde-Vinent 1997; Quintas 1998; Pszczolkowski & Flores 1986) filling a Middle and/or Upper Eocene foreland basin. In Camaguey, the lowest olistostromes contain only clasts derived from the ophiolitic suite, while sedimentary rock clasts from the Mesozoic palaeomargin are predominant in the upper part (Quintas 1998). The Camaguey lineament represents the eastern limit of the orogeny; there are no Early Palaeogene foreland basin and thrust sheets further east.

Holguin ophiolites

In northern eastern Cuba, ophiolites outcrop in the Maniabon Highlands (NW Holguin province). In the north, the Holguin ophiolites form several gently convex to the SE anastomosing arcuate narrow strips of pervasively deformed serpentinites. South-dipping thrust faults separate the metaultramafic rocks from the intervening strongly deformed sedimentary (mainly greywackes) and volcanic rocks of the Iberia mélange (Fig. 9; Kozary 1968;

Knipper & Cabrera 1974; Pushcharovski 1988). Some narrow, isolated, deep water limestone tectonic lenses (La Morena and Lindero members of 'Iberia Fm' of Jakus 1983), contain an Upper Cretaceous fauna. Further south the ophiolite suite outcrops become wider and continuous (Kozary 1968; Pushcharovsky 1988). This rock complex probably records a forearc accretionary prism (Andó *et al.* 1996). A second mélange in this region is made of serpentinite and gabbro clasts with more or less significant amounts of sedimentary and Cretaceous volcanic rock blocks (Yaguajay Fm). Probably the Yaguajay mélange originally was an olistostrome. The youngest blocks in this last type of mélange are Campanian–Maastrichtian biogenic massive limestones. Similar rocks rest on the serpentinites with tectonic contacts (Fig. 10). Outcrops of the carbonate Bahamas bank (Remedios zone) occur in a small area in Gibara and its western surroundings. The relationships between the North America palaeomargin and the ophiolites at the surface are unclear because of an intervening strip of lower Palaeogene sediments in between (Fig. 9). However, a tectonic contact is almost surely present. In cherts spatially related with basalts Andó *et al.* (1996) found Hauterivian–Barremian radiolaria. K–Ar radiometric ages in basalts and dolerites range from 126.3 ± 8.3 to

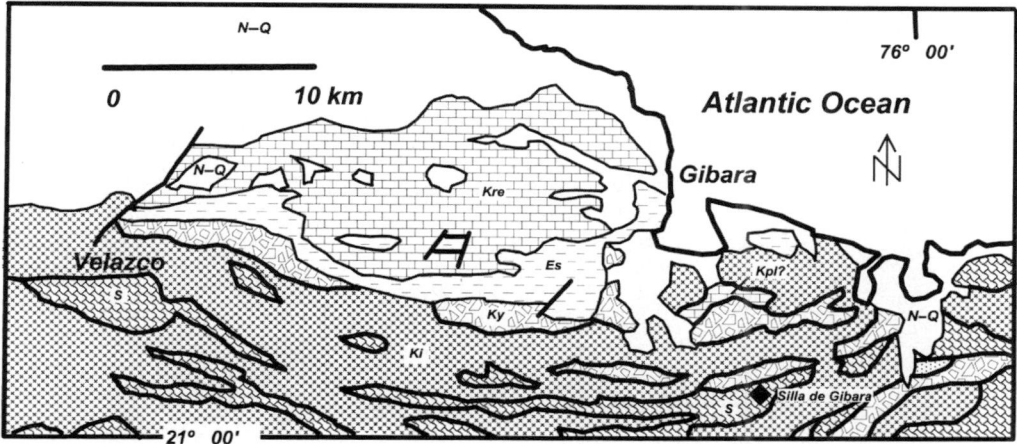

Fig. 9. Geological map of Gibara area, Maniabon Highlands, Holguin (slightly modified after Cobiella-Reguera 2005). Ki, Cretaceous Iberia Fm (mélange); S, ophiolitic rocks (mainly serpentinites); Ky, Maastrichtian olistostrome–mélange (Yaguajay Fm); Kpl, tectonic wedge of Placetas zone-like rocks (Cretaceous deep water sections); Kre, massive shallow water biogenic limestones (Remedios zone–southern fringe of Bahamas platform); Es, Paleocene–Middle Eocene deposits (Embarcadero and Vigia formations); N–Q, Neogene and Quaternary. See location in Fig. 1. See Figure 3 for geological contact symbols.

57.8 ± 5.4 Ma (Iturralde-Vinent *et al.* 1996). The oldest rocks lying with unconformity on the mélanges are upper Maastichtian turbidites (La Jiquima Fm, Gil Gonzalez *et al.* 2003) and Upper Paleocene breccias with some tuffaceous beds (Fig. 11, Haticos Fm) crop out on the southern massif rim. The sum of structural, radiometric and stratigraphic evidence points to Maastrichtian

Fig. 10. Typical landscape at Maniabon Highlands, Holguin. The low areas in the foreground contain mainly outcrops of the Iberia and Yaguajay mélanges. The plateau-like elevation in the cental background is known as Silla de Gibara (Fig. 9). Campanian–Maastrichtian massive south-dipping biogenic limestones (**T**, Tinajita Fm) crops out at its summit and the cliffs below, whereas serpentinites (**S**) are present in the lower, gentler slopes. The contacts between both lithologies are tectonic (Kozary 1968; Pushcharovsky 1988). See location in Figures 1 & 9. Looking from the north. Photograph courtesy of Martin Meschede, Greifswald University, Germany.

Fig. 11. Haticos Fm In the outcrop at Tacajo, Holguin province, beside the typical Haticos Fm breccia, rich in gabbro and dolerite clasts (**H**, left side of the photograph), a second type containing only serpentinite clasts (**SB**) is present. The contact between both units shows some shearing. A crude foliation appears in the serpentinitic breccia, but not in the first type. The breccias seem to be debris flow deposits of very local provenance. Outcrop height: *c.* 2 m. See location in Figure 1. Photograph courtesy of Martin Meschede, Greifswald University, Germany. Latitude: 20° 53′ 02″, longitude: 76° 03′ 30″.

(pre-Late Maastrichtian) emplacement for the Holguin ophiolitic rocks. It may have occurred in two phases, the first related to a Campanian accretionary prism, now probably vanished, and the second recorded by the Maastrichtian Yaguajay mélange.

Regional data suggest north-northwestward tectonic transport (Kozary 1968). Assuming that (1) all along this area the ophiolitic rocks rest with tectonic contact, (2) the tectonic transport was from south to north, at right angle with the the the regional structural trend, and (3) at least the northern border of the ophiolitic mélange probably lies on the carbonate bank facies of the continental palaeomargin (Remedios zone), we can suppose that the northwards transport was no less than the width of the ophiolitic belt (22.5 km) and probably it was much larger.

Several authors claim strong Early Cenozoic deformation (Cuban orogeny) in the Holguin masif (Kozary 1968; Knipper & Cabrera 1974; Andó et al. 1996), but no definitive evidence for such deformation in the lower Cenozoic rocks has been shown. In fact, strong Paleocene–Eocene deformations in the Holguin Massif area were not reported by Nagy (1984), Brezsnyansky & Iturralde-Vinent (1978) and Cobiella-Reguera et al. (1984b). In contrast to massifs west of the Camaguey lineament, the characteristic lower

Cenozoic olistostromes of the Cuban Orogeny, with the mix of ophiolitic and continental palaeo-margin clasts, are absent.

Eastern Cuba ophiolites

At the surface the Holguin Massif is separated from the eastern Cuba massifs by Cenozoic deposits, but in depth the ophiolites probably form a single body (Figs 1 & 12; Knipper & Cabrera 1974; Cobiella-Reguera et al. 1984b). Two great ophiolite outcrops area are present: the Sierra de Nipe–Cristal and Moa–Baracoa massifs, separated by the Sagua de Tanamo River basin. The main feature distinguishing eastern Cuba ophiolites from the other massifs of the NO (except the Holguín area) is their tectonic superposition upon the KVAT (Knipper & Cabrera 1974; Cobiella-Reguera 1978; Iturralde-Vinent 1996a; Iturralde-Vinent et al. 2006). Only its eastern tip lies on the North American palaeomargin, together with KVAT metamorphic rocks (Cobiella-Reguera 2005; Iturralde-Vinent et al. 2006; García-Casco et al. 2006). Most of the eastern Cuba ophiolites are a huge, dismembered, almost flat tectonic prism, about 1 km in maximum thickness (Figs 12 & 13), thrusted several tens of kilometres (Knipper & Cabrera 1974; Cobiella-Reguera 1978) to the north (Nuñez Cambra et al. 2003). In some

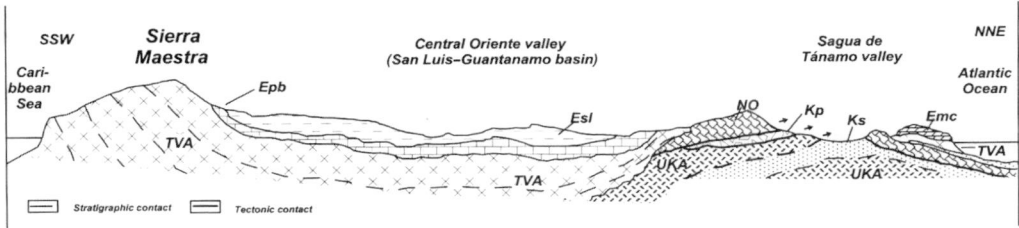

Fig. 12. Schematic geological profile from eastern Sierra Maestra to the northern coast of eastern Cuba, showing the NOB–Palaeogene volcanic arc–Middle–Late Eocene basin relationships. Without scale. Length of the section: 100 km. NO, Northern Ophiolite; UKA, Upper Cretaceous arc rocks (Santo Domingo Fm); Ks, Maastrichtian volcanoclastic deposits; Kp, Maastrichtian olistostromes (La Picota Fm); TVA, Danian–Middle Eocene Turquino volcanic arc rocks (El Cobre and Sabaneta formations); Epb, Middle Eocene carbonate horizon (Puerto Boniato and Charco Redondo formations); Esl, Volcanoclastic Middle and Late Eocene deposits (San Luis and Camarones formations). See Figure 2 for location.

Fig. 13. Sheared massive serpentinite (**S**) resting with almost horizontal contact upon chaotic breccia conglomerates (olistostromes) of La Picota Fm (**P**) at Sabanilla Mayari Arriba, Santiago de Cuba province. All the clasts belong to lithologies of the ophiolite suite. A crude foliation is present in the matrix. A diagenetic hematitic film stains matrix and clasts in many places in this and other outcrops of the unit. East to the right. See location in Figure 1. Photograph courtesy of Martin Meschede, Greifswald University, Germany. Latitude: 20° 24′ 02″, longitude: 76° 25′ 09″.

areas, tectonic deformations and mixing are not as pervasive as in other Cuban ophiolites. A wealth of new data on the primary structure, petrology and geochemistry of several areas belonging to the eastern Cuba ophiolites has been published in the last few years (Cobiella-Reguera 2005 and references therein; see also papers in *Geologica Acta* 1–2, 2006). They suggest that a mantle tectonite more than 5 km thick is exposed in the Sierra de Nipe–Cristal Massif, while the Moa–Baracoa Massif comprises mantle tectonites more than 2.2 km thick, capped by a thin crustal section of lower gabbros (300 m) and discordant volcanics (Quibiján Fm, Proenza *et al.* 2006). Recently, the last volcanic unit was described as part of the volcanic arc section (Iturralde-Vinent *et al.* 2006). Great blocks of high-pressure metamorphic rocks (mainly amphibolites, Somin & Millan 1981; García-Casco *et al.* 2006) are disseminated in both massifs. García-Casco *et al.* (2006) relate this pressure and strong synmetamorphic deformation to collision. K–Ar ages range between 72 ± 3 and 58 ± 4 Ma. The amphibolites belong to the eastern Cuba thrust pile (Cobiella-Reguera *et al.* 1984*a*, Cobiella-Reguera 2005; Iturralde-Vinent 1996*a*; Iturralde-Vinent *et al.* 2006; García-Casco *et al.* 2006) and so are pre-Maastrichtian in age. Maastrichtian shallow water biogenic limestones rest on the ophiolites (Cobiella-Reguera *et al.* 1984*a*; Iturralde-Vinent *et al.* 2006).

In many places, deformed olistostromes (with blocks of serpentinite, dolerite, gabbro, basalt and amphibolite, up to several hundred metres in diameter) of Maastrichtian age (La Picota Fm, Fig. 13; Cobiella-Reguera 1978), lie below the ophiolites. The olistostromes consist almost 100% of serpentinitic and gabbroic clasts. Clasts from the Cretaceous volcanic arc are minor components. Fine grained clastic beds yield Maastrichtian fossils. Some blocks of Maastrichtian shallow limestones with rudists are also present (Iturralde-Vinent *et al.* 2006), but clasts clearly derived from a Mesozoic continental palaeomargin are absent. The Holguin and eastern Cuba Maastrichtian beds are the only chaotic deposits related to the ophiolite emplacement lacking palaeomargin clasts. The oldest beds resting above La Picota Fm are uppermost Maastrichtian or lowermost Paleocene, with abundant serpentinite olistoliths (Iturralde-Vinent *et al.* 2006). Therefore, a Maastrichtian age for the ophiolite emplacement is indicated (Cobiella-Reguera *et al.* 1984*a*; Nuñez Cambra *et al.* 2003; Iturralde-Vinent *et al.* 2006). Regional geology and structural features recorded in the ophiolite massifs and other rocks indicate tectonic transport from south to north (Nuñez Cambra *et al.* 2003; Iturralde-Vinent *et al.* 2006). The massifs travelled no less than 30 km; probably no less than 60 km, if the Sierra del Convento klippe in southern Sierra del Purial (see Fig. 1) is included in the calculation. Emplacement was very rapid and gravity driven (Cobiella-Reguera 1974, 1978; Iturralde-Vinent *et al.* 2006). Subaerial relief on the moving ophiolite thrust is suggested by blocks of weathered serpentinite, derived from lateritic soils, and abundant rounded clasts of mafic rocks found in the olistostromes (Fig. 13).

No record of significant Cenozoic deformation is present in eastern Cuba ophiolitic massifs. Little-deformed Paleocene–Middle Eocene tuffs and sedimentary rocks, belonging to the Early Cenozoic volcanic arc developed in southeastern Cuba, rest upon the older rocks. In easternmost Cuba (Maisí area; Fig. 1) rocks of the Moa–Baracoa Massif lie in tectonic contact upon metamorphic rocks of Jurassic and Cretaceous age, probably a passive continental margin section (Pardo 1975; Somin & Millán 1981; Cobiella-Reguera *et al.* 1984*a*; Iturralde-Vinent 1997; Cobiella-Reguera 2000).

Cajobabo body (easternmost Cuba)

Near the mouth of Cajobabo river, in the Sierra del Purial, easternmost Cuba (Figs 1 & 14), a small serpentinite body rests on Middle–Upper Eocene submarine fan deposits of the San Luis Fm. This includes olistostromes with clasts of serpentine, gabbro, amphibolite and other members of the ophiolite suite, mixed with tuffs and andesite clasts from an Early Cenozoic volcanic arc (Cobiella-Reguera *et al.* 1977). The Early Cenozoic volcanics (El Cobre Fm) are thrust upon the Eocene turbidites and the Cretaceous metavolcanic rocks and lie with unconformity below Upper Miocene–Quaternary strata (Fig. 15, Iturralde-Vinent 1997; Cobiella-Reguera *et al.* 1977, 1984*a*). The volcanic rocks are not recorded in the local pre-Middle Eocene section. Thrusting was probably a late Middle Eocene event, tied to the intense erosion of a nearby mountain chain. Despite its small size, the Cajobabo body is a key piece in northern Caribbean geology, because its emplacement seems to be related to the first recorded horizontal movements along the Oriente fault in the Caribbean–North American plate margin (Cobiella-Reguera *et al.* 1977, 1984*b*).

Some conclusions on the Northern Ophiolites

The NO probably are the obducted remains of NW Proto-Caribbean lithosphere (Pindell & Barrett 1990; Cobiella-Reguera 2005; Pindell 2006). The preceding paragraphs show a complex and varied NO emplacement story. Several facts strongly suggest that the NW Proto-Caribbean oceanic

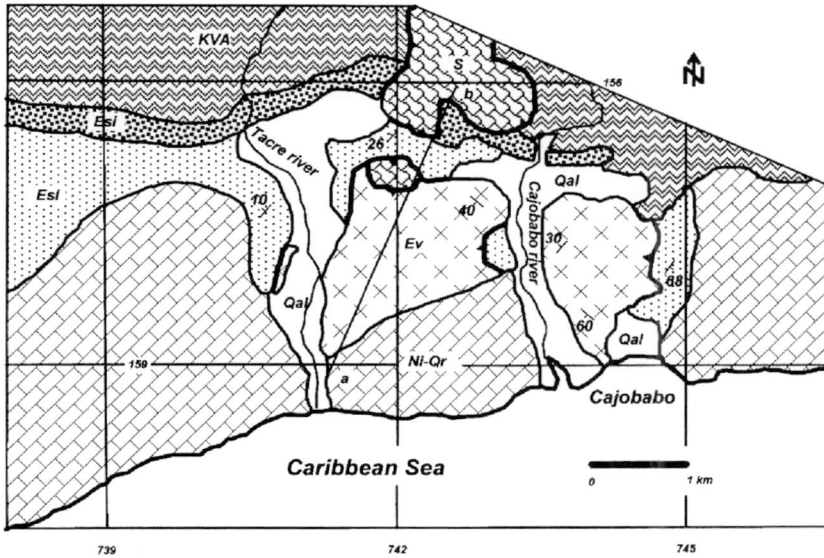

Fig. 14. Geological map of Cajobabo, Guantanamo province (modified after Cobiella-Reguera *et al.* 1977). Local coordinates (Cuba South). See location in Figure 1. KVA, Cretaceous metavolcanic rocks (Sierra del Purial Fm); Ev, Palaeogene volcanic arc rocks (El Cobre Fm Lower Eocene at this locality); Esi, Middle Eocene talus megabreccia; Esl, Middle Eocene (at this locality) volcanomictic deposits (San Luis Fm); Ni–Qr, Upper Miocene–Pliocene (Imias Fm) marine talus deposits and Quaternary fringing reefs limestones; S, serpentinites. Line a–b corresponds to the profile in Figure 15.

basin was closed before the end of the Cretaceous. Among them are:

1. The disappearance of the Late Cretaceous volcanic arc in Cuba (and possibly in Hispaniola) during Campanian time (Iturralde-Vinent 1997; Iturralde-Vinent & Macphee 1999; Draper & Barros 1994). Volcanism ceased for about 15 Ma and was only resumed in SE Cuba in the Early Paleocene (Cobiella-Reguera 1988). The more or less simultaneous break in volcanic activity in a *c.* 1000 km long belt should be tied to the more or less coeval arrival of the North American palaeomargin to the subduction zone.

2. The development of volcanomictic Campanian sediments above the North American Mesozoic palaeomargin in western Cuba (Pszczolkowski 1994, 1999). This could be possible only if the oceanic basin in between the Cretaceous volcanic arc and the NA margin disappeared (Cobiella-Reguera 2000, 2005, fig. 12).

3. The Campanian–Maastrichtian (and K–T boundary) sedimentary rocks unconformably rest on the ophiolites of western Cuba (Piotrowska 1986*a, b*; Pushcharovsky 1988).

4. The upper Maastrichtian–Paleocene sedimentary cover lies on top of the post Early Campanian serpentinite–volcanic tectonic mix in central Cuba (Kantchev *et al.* 1978; Pushcharovsky 1988).

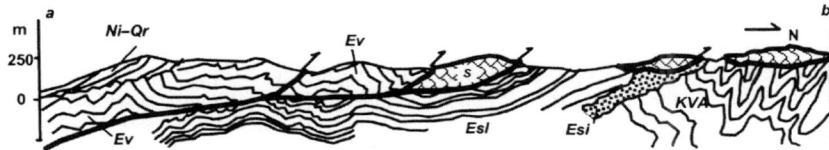

Fig. 15. Sketch structural profile along the a–b line in Figure 14 (slightly modified after Cobiella-Reguera 2005). Same vertical and horizontal scales. KVA, Cretaceous metavolcanic rocks (Sierra del Purial Fm); Ev, Palaeogene volcanic arc rocks (El Cobre Fm, Lower Eocene in this locality); Esi, Middle Eocene talus megabreccia (San Ignacio Fm); Esl, Middle Eocene (in this locality) volcanomictic deposits (San Luis Fm); Ni–Qr, Upper Miocene–Pliocene (Imias Fm) marine talus deposits and Quaternary fringing reefs limestones; S, serpentinites; N, North.

5. The conglomerates and breccias with serpentinite and gabbroid clasts in the upper Campanian beds of western Cuba (Via Blanca Fm, Iturralde-Vinent 1996*a*, *b*), and Maastrichtian (La Picota, Yaguajay, Micara and La Jiquima formations) of eastern Cuba (Jakus 1983; Gil *et al.* 2003). All these units belong to the Cretaceous volcanic terrane cover.

Furthermore, extraordinary K–T boundary deposits with similar characteristics in western and central Cuba rest on the volcanic arc as well as on the continental palaeomargin (Pszczolkowski 1986; Tada *et al.* 2003). This indicates that 65 Ma ago these units were not very separate, leaving very little or no space for the Proto-Caribbean basin at that time.

In several models on the origin of the Caribbean Plate, the NO is considered part of a great ophiolitic suture developed along the northern Caribbean, recording oblique and diachronous collision between the North American and Caribbean plates. However, the data described above show that in most of the cases, the present NO position is due to distinct events of later thrusting, 10–30 Ma after (i.e. Late Maastrichtian to Middle or Early Late Eocene) the Campanian collision. The small Cajobabo body was emplaced in (late?) Middle Eocene, not far away from the Oriente fault zone in SE Cuba. This age is a little younger than the Cayman Trough opening assumed by Leroy *et al.* (2000), supported on very different data.

The association of the ophiolite bodies with olistostromes and the general geological scenario shows that final NOB emplacement events were always violent, and probably, ephemeral events.

Campanian–Late Eocene tectonic history of Cuba and surroundings

The Proto-Caribbean Plate is a vanished oceanic plate developed between the two American continents and connected to the Atlantic realm. Its origin was related to the breakup of western Pangaea, begining in the Late Triassic (Iturralde-Vinent 2003) or Jurassic (Pindell & Barrett 1990; Marton & Buffler 1999). Radiolarites of Upper Jurassic age are preserved in Proto-Caribbean rocks in the Cuban ophiolite belt (Llanes-Castro *et al.* 1998), the Duarte complex in Hispaniola (Lewis *et al.* 2002), and in northern South America (Giunta *et al.* 2002, 2003). Separation of North America and Gondwana plates continued until the Early Cretaceous (Marton & Buffler 1999). The ammonite fauna in Upper Jurassic deposits of the North America palaeomargin in western and central Cuba points to an oceanic seaway open to the Tethys from the Oxfordian

onward (Pszczolkowski & Myczynski 2003; Iturralde-Vinent 2003).

The begining of volcanic arc magmatism during the Early Cretaceous represents a first indication of subduction of Proto-Caribbean lithosphere. In Cuba, the oldest volcanic arc rocks are of Aptian age (Rojas *et al.* 1995). At the same stratigraphic position, rare tuff beds are intercalated within the successions of the North American palaeomargin in western Cuba, which supports the beginning of arc volcanism during Aptian time (Cobiella-Reguera 2000) and its not very distant location from the NA palaeomargin. In Hispaniola, a few fossil finds indicate a Neocomian–Aptian age for the early arc, although robust stratigraphic control on the age of the lower beds of the arc is absent. Kesler *et al.* (2005) reported Aptian and Albian radiometric ages (113.9–110.9 Ma) for PIA rocks in Hispaniola.

The Aptian–Albian volcanic arc was characterized by tholeiitic bimodal magmatism, with some boninitic affinities in Hispaniola and probably in Cuba (Diaz de Villalvilla 1997; Kerr *et al.* 1999; Cobiella-Reguera 2000; Lewis *et al.* 2002). Volcanic activity in Cuba and other parts of the Greater Antilles drastically diminished during the Late Albian–Cenomanian (Kantchev *et al.* 1978; Iturralde-Vinent 1997; Kesler 2005). In the Cenomanian–Campanian (locally Maastrichtian?) volcanic arc magmatism resumed with a prevailing calcalkalic character (Jolly *et al.* 1998; Lebrón & Perfit 1994; Iturralde-Vinent 1996) and with local tholeiitic manifestations in eastern Cuba (Proenza *et al.* 2006).

Campanian and Maastrichtian

For those models of Middle America evolution proposing a Cretaceous–Quaternary 'Great Arc of the Caribbean' (GAC) (Mann 1999) or Great Caribbean Arc (Pindell *et al.* 2006) related to a Pacific origin for the Caribbean Plate (e.g. Pindell & Barrett 1990; Mann 1999; Pindell *et al.* 2006), the Late Cretaceous was a time when the GAC moved to the NE, consuming Upper Jurassic–Lower Cretaceous Proto-Caribbean crust located between the arc and the North American palaeomargin. In these interpretations, the same movement continued in the Early Palaeogene until the arc collided with the North American palaeomargin, first in western Cuba (Paleocene), and thereafter in central and eastern Cuba in different times during the Eocene.

In the Kerr *et al.* (1999) and Iturralde-Vinent (1997) model, the Late Cretaceous volcanic arc travel in the NW Caribbean was in the same direction as in the Cretaceous–Quaternary GAC hypothesis, but the consumed oceanic lithosphere is from the Pacific plate, diving in an east or NE dipping subduction zone. Campanian–Paleocene quiescense in volcanism is considered, but in the accompanying illustrations subduction without volcanism continued during Late Campanian and Maastrichtian (Kerr et al. 1999, fig. 12). The Cretaceous volcanic arc/North American palaeomargin collision was an Early Palaeogene event as in the GAC models. The emplacement from the south of the ophiolites and the volcanic arc terrane upon the North American palaeomargin cannot be explained by the author's proposed mechanism. According to the same authors, the Paleocene–Middle Eocene volcanic arc (Turquino) in SE Cuba is inherited from the Cretaceous subduction zone.

Data from Cuba reviewed here refutes all these models. One of the basic features of the GAC Late Cretaceous travel is its collision with southern Yucatán (Pindell et al. 2006) and the emplacement of the so-called southwestern terranes in Cuba. Iturralde-Vinent (1996a) brought together the Mesozoic sections of Escambray, Isla de la Juventud and Guaniguanico mountains in his 'Southwestern terrains'. In fact, Jurassic sections in the Guaniguanico Cordillera, Isle of Youth and Escambray mountains share many common features. According to Cobiella-Reguera (1996, 2000) they probably were deposited in the same basin, but their present respective location with regard to ophiolites and the Cretaceous volcanic arcs terrane is very different. Regional geology suggests that rocks in Escambray and Isla de la Juventud were always placed south from the Cretaceous volcanic arcs, while Upper Jurassic–Cretaceous sections belonging to the Guaniguanico Cordillera were the western continuation of the deep water Upper Jurassic and Cretaceous sections of the Mesozoic palaeomargin in northern central Cuba (Cobiella-Reguera 1996, 2000; Pszczolkowski & Myczynski 2003). Data from the SE Gulf of Mexico (Marton & Buffler 1994, 1999; Moretti et al. 2003) clearly show close stratigraphic relationships with the Mesozoic sedimentary sections in Guaniguanico Cordillera. Therefore, it seems difficult to consider the last region as a tectono-stratigraphic terrane (e.g. as envisaged by Iturralde-Vinent 1996a; Pszczolkowski 1999; and partially Pindell et al. 2006) since it does not represent crustal blocks exotic to its present tectonic setting.

Recently Pindell et al. (2006) considered the Escambray terrane as a 'tectonically-unroofed, deep level of the Great Arc's forearc where passive margin strata has been subducted and subcreted in the Aptian-Albian'. According to the same authors, the Southern Metamorphic terranes and Guaniguanico were torn from the eastern or southern Yucatán margin much later, during the Campanian and Maastrichtian, when the western leading edge of the Great Caribbean Arc slid

along the Yucatán border. In central Cuba, the Escambray–Cretaceous volcanic arcs terrane tectonic contact is crossed by 88–80 Ma pegmatites (Stanek et al. 2000), showing that the terrane welding in central Cuba probably is a Cretaceous pre-Coniacian event. In that case, and supposing a SE Yucatán origin for the Escambray metamorphic rocks, at least some remains of middle Cretaceous high-pressure rocks or structures pointing to a pre-Coniacian tectonic event should be present in the southeastern Yucatán margin. However, no evidence for such events in SE Yucatán or northern Central America is reported in the geological literature (e.g. Donnelly et al. 1990; Morán-Zenteno 1994). Other authors also pointed out a Yucatán (Maya Block) origin for the Southern Metamorphic terranes and Guaniguanico (Iturralde-Vinent 1996a, b; Kerr et al. 1999; Schaffhausser et al. 2003), but correlation between the Cuban sections and SE Yucatán was not attempted, except by the last mentioned authors. According to Schaffhausser et al. (2003), the Lower Cretaceous 'Guaniguanico like' sections in western Belize were emplaced as thrust sheets between the Maastrichtian and the Eocene, several million years after Pindell et al. (2006) supposed travel and collision of the Great Caribbean Arc with SE Yucatán to have occurred. In the author's opinion, there is no conclusive tectonic or stratigraphic evidence allowing an original position of the Cuban southern metamorphic terranes at the Yucatán SE margin, which accords with Escambray as a Mesozoic poly-genetic unit (terrane) amalgamated into a subduction zone (e.g. García-Casco et al. 2006; Stanek et al. 2000).

Geological information from all of Cuba's territory shows that volcanic arc activity ended during the Campanian and did not resume until late in the Danian (and then only in SE Cuba; Pushcharovsky 1988; Cobiella-Reguera 1988; Iturralde-Vinent 1997). The end of volcanism in Cuba during the Early Campanian suggests that subduction stopped more or less simultaneously along c. 1000 km. Probably, the end of the arc was related to collision and underplating of the North American palaeomargin during the Campanian. Several lines of evidence suggest Campanian emplacement. Indirect evidence of the closing of the oceanic depression between the Late Cretaceous, volcanic arc and the North American Mesozoic palaeomargin occurs in the Campanian sedimentary rocks of western Cuba. Here, the Moreno Fm of the Cordillera de Guaniguanico contains clastic material derived from the KVAT (Pszczolkowski 1994, 1999). This shows that at least in the west there was no Late Campanian oceanic depression between the arc and the palaeomargin (Cobiella-Reguera 2000). A Campanian collision is also known in northern Central America where ophiolite emplacement is recorded by the

siliciclastic Campanian–Lower Cenozoic Sepur Fm
(Donnelly *et al.* 1990; Pindell *et al.* 2006).

In western Cuba, the upper Campanian–
Maastrichtian volcanomictic turbidites (Via Blanca
Fm) lie on the ophiolites and the volcano-ophiolitic
mélange as well as on the KVAT rocks (Pszczolkowski & Albear 1982; Pushcharovsky 1988; Piotrowska 1986*a*, *b*). This points to a Campanian?
(pre-Via Blanca Fm) tectonic event in this area.
Evidence of pre-Late Maastrichtian thrusting and
tectonic mixing is present in central Cuba. After
the initial Campanian collision, the extinguished
Cretaceous volcanic arc was deformed and uplifted
and terrigenous sediments, followed by some carbonates, accumulated in basins during the Late
Campanian until the Late Maastrichtian (Diaz-
Otero *et al.* 2003). Iturralde-Vinent (1997) considers
Campanian–Maastrichtian depressions as piggy
back basins, an interesting proposal that needs
further consideration.

In Hispaniola also, volcanic quiescence occurred
from the Late Campanian or Maastrichtian to the
Early Palaeogene (Lewis & Draper 1990; Draper
& Barros 1994, fig. 7.7). In Puerto Rico, a Late
Maastrichtian–Danian pause in volcanism is
recorded, whereas Maastrichtian volcanic rocks
are limited to the western province of the island
(Jolly *et al.* 1998, fig. 2). In conclusion, at the end
of the Cretaceous–begining of the Palaeogene,

there probably was no GAC in the NW Caribbean.
From Late Maastrichtian to Early Paleocene tectonic unrest occurred in different places. A regional
tectonic event in easternmost Cuba raised ophiolite
massifs to the surface (Kozary 1968; Knipper &
Cabrera 1974), thrusting them to the north during
the (Late?) Maastrichtian (Cobiella-Reguera 2005;
Iturralde-Vinent *et al.* 2006). Overlying Danian
olistostromes and turbidites in the same area point
to continued instability in the earliest Cenozoic.
The unconformity between Maastrichtian (including K–T boundary deposits) and upper Paleocene
sediments in the Havana area, and the poor distribution of upper Maastrichtian and Paleocene
(Pardo 1975) beds in Cuba, suggest a period of
general tectonic deformation at the Cretaceous–
Paleocene transition.

Paleocene–Middle Eocene

In the Early Paleocene (Danian) the tectonic setting
changed (Fig. 16). A new east–west trending
submarine volcanic arc was born in SE Cuba (the
Turquino–Cayman Ridge volcanic arc) on top of
the Cretaceous volcanic terrane and its sedimentary
cover. Several thousand metres of effusive and
pyroclastic rocks, ranging from rhyolites to basalts,
but mainly andesites, with marine sedimentary
intercalations, crop out in the Sierra Maestra

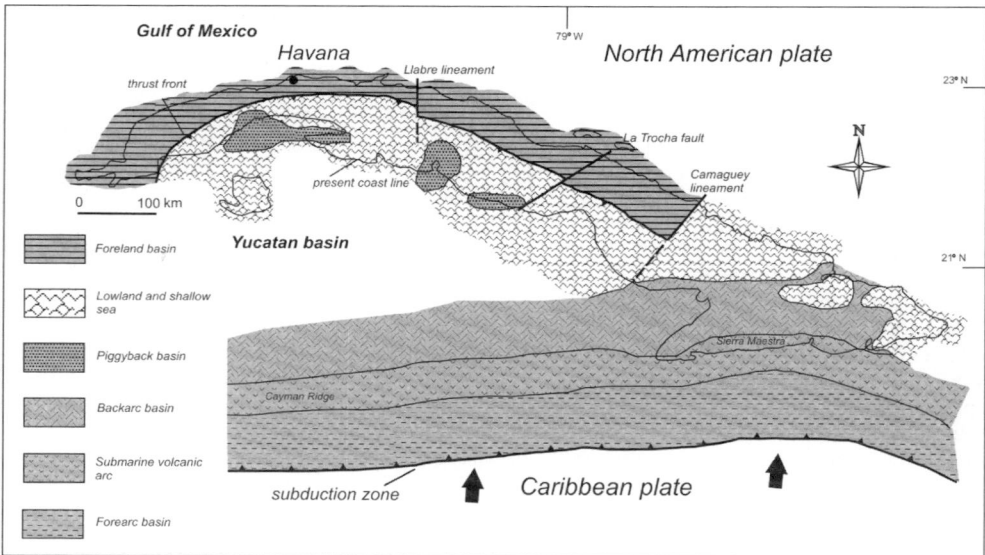

Fig. 16. Late Paleocene tectonic setting of Cuba and its surroundings. West of Camaguey lineament, the Cuban
Orogeny was going on, with northwards moving thrust sheets carrying piggyback basins on top. In SE Cuba the
Turquino–Cayman Ridge volcanic arc was active, with its north-dipping subduction zone acting as the
boundary between the Caribbean and North American plates.

(Lewis & Straczek 1955; Khudoley & Meyerhoff 1971; Cobiella-Reguera 1988; Iturralde-Vinent 1996a, b). Its corresponding back arc basin, filled mainly with pyroclastic rocks, volcanogenic turbidites and sedimentary rocks, lay to the north (Fig. 16, Cobiella-Reguera 1988; Iturralde-Vinent 1996b, 1997). Intruding into the volcanic sequence in the Sierra Maestra are a large number of magmatic bodies with compositions ranging from granites to gabbro. According to Rojas-Agramonte et al. (2005), these are volcanic arc calcalkalic granites of intraoceanic origin. Pb–U analyses from zircons in the granitoids yielded ages from 60 Ma (Paleocene) to 48 Ma (Middle Eocene). This is very similar to the time span for the volcanic rocks of the arc. Westward, Paleocene–Eocene volcanic rocks were recorded in the Cayman Ridge and Nicaragua Plateau by Perfit & Heezen (1978, fig. 9a). According to Siggurdsson et al. (1997), an Early Palaeogene volcanic arc and a northern back arc and basin are present in the Cayman Ridge and Yucatán basin. According to this architecture, a north-dipping suduction zone should flank the Turquino–Cayman Ridge arc to the south (Fig. 17; Cobiella-Reguera 1988; Iturralde-Vinent 1996b; Siggurdsson et al. 1997) and dense oceanic Caribbean crust dived below the arc. Coeval Eocene volcanic arc rocks occur in the Northwestern Peninsula and the Montaignes Noires of Haiti (Butterlin 1960) and are also present in Sierra de Neiba and Seibo of the Dominican Republic (Lewis & Draper 1990; Draper & Barros 1994). If they originally were part of the Turquino–Cayman Ridge

arc, then probably this structure was more than 1500 km long.

This geographic distribution of the main tectonic Early Cenozoic units is an extremely important fact in the Caribbean–North American plate boundary history. The subducted (Caribbean) oceanic crust must have moved with a strong northward component along the 1000–1500 km long subduction zone of the Turquino–Cayman Ridge arc (Fig. 16). This point was suggested by Siggurdsson et al. (1997), but has been largely overlooked. In many of the Caribbean evolution models, at this time (Danian–Middle Eocene) the Caribbean Plate is moving with a strong eastward component (compare Fig. 16 with Pindell et al. 2006, fig. 7). In the present interpretation, the lithosphere to the north of the Turquino–Cayman Ridge subduction zone was already accreted to the North American plate in the Paleocene, and the southeastern boundary of the North American plate was a north-dipping subduction zone from the Paleocene to the Middle Eocene.

From the Danian to the Middle Eocene, in western and central Cuba (west of the Camaguey lineament) a different tectonic environment existed. In these areas volcanic activity was absent and thrust sheets several kilometres thick, containing rocks of the ophiolite belt and the Cretaceous volcanic terrane, with small marine piggyback basins (Hatten et al. 1988; Cobiella-Reguera 1988, 2005; Iturralde-Vinent 1997), moved in a general northward direction toward the North American Mesozoic palaeomargin and the overlying southern

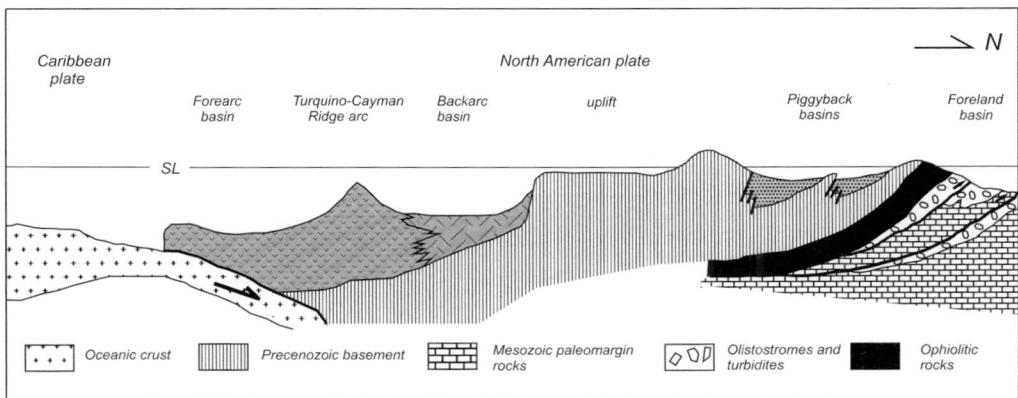

Fig. 17. Paleocene–Early Middle Eocene conceptual palaeotectonic cartoon profile during the Cuban orogeny (modified after Cobiella-Reguera 2005). The profile is located to the west of Camaguey lineament (see Fig. 16) and is about 300 km long. The subduction zone location is assumed from theoretical considerations and regional geology data (Figs 2 & 16, Cobiella-Reguera 1988; Siggurdsson et al. 1997), but no evidence of this structure or the subducted Caribbean lithosphere has been found. The sketch summarizes the proposed relationships between the Turquino–Cayman Ridge arc and the coeval piggyback and foreland basins.

fringe of the foreland basin, that also became involved in deformation (Figs 16 & 17, Cuban Orogeny). The crust north of the thrust belt was depressed below northward-moving nappes and a foreland basin developed from Late Paleocene to Middle (locally Late) Eocene. Subaerial relief was not particularly great, as coarse clastic deposits are rare (except the olistostromes). Rocks of the ophiolite belt were mainly exposed in submarine cliffs. Debris flows and other gravitational deposits moved to the seafloor to form the olistostromes and turbidites. During continued advance the nappes overrode, crushed and partly mingled with the olistostromes, creating mélanges (Cobiella-Reguera 1997). This tectonic event migrated eastwards. In western Cuba, thrusting occurred from Late Paleocene to early Early Eocene. East of the Llabre lineament (Fig. 2) thrusting occurred from Early to Middle Eocene, while east of La Trocha fault it was a late Middle Eocene–?Late Eocene event. Some authors (Brzesnyansky & Iturralde-Vinent 1978) extend this orogeny to Holguin in northern eastern Cuba, but no clear evidence of intense Paleocene or Eocene deformation is present there. In the GAC models, this diachronous orogenic Early Palaeogene event is interpreted as the Late Cretaceous–Early Palaeogene volcanic arc/North American palaeomargin oblique collision (Mann 1999; Pindell 2006). However, as explained earlier, volcanism in Cuba ceased from the Late Campanian to the Danian. The north-dipping subduction zone of the Turquino–Cayman Ridge arc, several hundred kilometres south of the Early Cenozoic deformed belt (Figs 17 & 18),

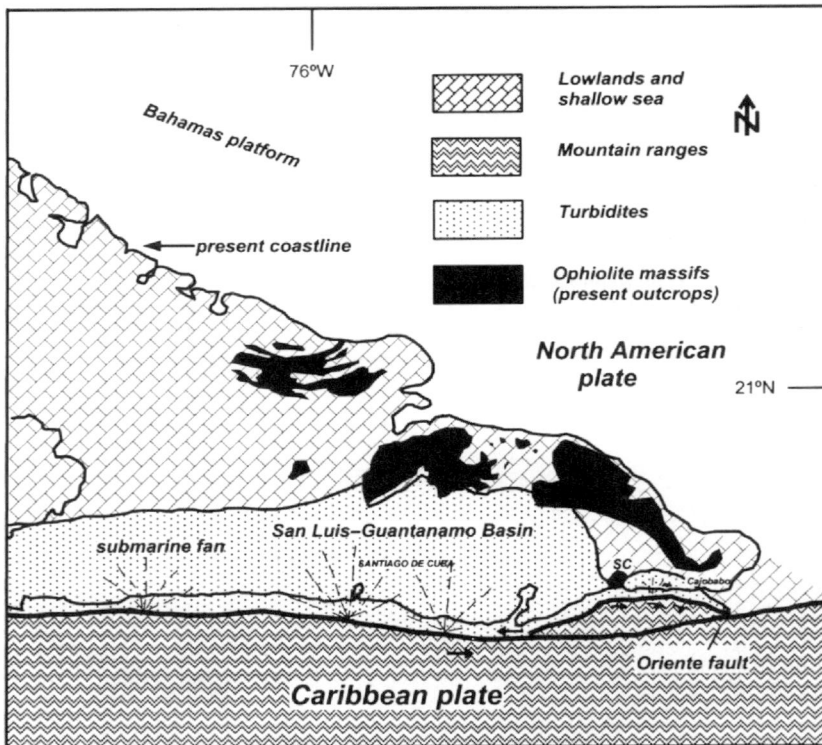

Fig. 18. Palaeogeographic reconstruction of eastern Cuba during late Middle Eocene and Late Eocene made with data from Keijzer (1945), Lewis & Straczek (1955), Bresznyansky & Iturralde-Vinent, M. (1978) and Cobiella-Reguera (1984). The southern highlands were named the Bartlett Land by Taber (1934). Probably their origin is tied to the compressive deformations present along the southern fringe of eastern Cuba, considered Oligocene in age by different authors (Rojas-Agramonte *et al.* 2006; Iturralde-Vinent 1997). However, the thick volcanoclastic Middle–Upper Eocene deposits of the San Luis-Guantanamo basin and the coeval emplacement of the Cajobabo tectonic sheet (Figs 14 & 15), together with the mild deformations in eastern Cuba Oligocene beds, points to a Middle–Late Eocene compressive event. SC, Sierra del Convento ophiolites.

cannot be invoked to explain the deformation of the latter.

Middle–Late Eocene

Early in the Middle Eocene, the Turquino–Cayman Rise volcanic arc became inactive (Cobiella-Reguera 1988; Rojas-Agramonte et al. 2004, 2005, 2006). In SE Cuba, several hundred metres of Middle–Upper Eocene terrigenous volcanomictic deposits (the San Luis, Camarones and related formations) crop out in the San Luis–Guantanamo basin, north of the Sierra Maestra (Keijzer 1945; Lewis & Straczek 1955; Khudoley & Meyerhoff 1971). Highlands with outcrops of the vanished Turquino–Cayman Rise arc (Fig. 18), located in the present Cayman Trough area, not far from the coastline, was the sediment source (Keijzer 1945; Cobiella-Reguera et al. 1984a, b). Erosion unroofed Early Cenozoic granitoid intrusives as their clasts occur in the conglomerates (Lewis & Straczek 1955; Rojas-Agramonte et al. 2004). The sediments conformably lie on top of the underlying Early Cenozoic volcanic arc section (Lewis & Straczek 1955). Compressive deformation also affected the Middle Eocene and older rocks in the Sierra Maestra and its northern foothills. The intensity of deformation increases southward towards the Oriente fault (Iturralde-Vinent 1997; Rojas-Agramonte et al. 2005, 2006). Iturralde-Vinent (1997) assigned an Oligocene age to this event. However, the thick upper Middle and Upper Eocene siliciclastic sections of the San Luis–Guantanamo basin probably are synorogenic deposits related to this tectonic phase. Compression was accompanied by local thrusting in southeasternmost Cuba, where the small Cajobabo allochthon, with Lower Eocene volcanic arc rocks and serpentinite, lies upon Middle Eocene volcanomictic sediments, not far from the Caribbean coast. The sediment source area and the thrust 'roots' were located to the south of the present coastline (Cobiella-Reguera et al. 1977). Because there is no record of autocthonous Turquino volcanic arc rocks in easternmost Cuba, the last two facts could be interpreted as evidence of the first sinistral movement along the Oriente fault, probably in the Middle Eocene. This was an initial short-lived but intense displacement, provoking rapid uplift of the southern block and its relative eastward movement for several tens of kilometres, locating the Turquino volcanic arc sequences in front of the Cajobabo territory (Cobiella-Reguera et al. 1984a, b; Cobiella-Reguera 2005). Local thrusting at Cajobabo should be related to some nearby restraining bend along the fault (Fig. 18). The age of this episode is similar to initiation of Cayman

spreading centre as determined by Leroy et al. (2000): 49 Ma (late Early Eocene, UNESCO & IUGS 2000; Early–Middle Eocene boundary in Witrock et al. 2003).

The position of the Oriente fault zone (Calais & de Lepinay 1991; Rojas-Agramonte et al. 2006) seems to be almost the same as the Turquino–Cayman Ridge volcanic axial zone trace, perhaps a little southward of the latter. However, if the Montaignes Noires of Haiti (Butterlin) and the Sierra de Neiva (Dominican Republic) Eocene sections represent the easternmost area of the Turquino arc, its strike probably turned to ESE (Fig. 16). The volcanic sections in northern Hispaniola, invoked by Pindell (2006) as part of the Palaeogene arc, resemble those of the northern marginal basin in eastern Cuba, including the pre-Cenozoic ophiolite rich basement. Therefore, the Oriente fault zone is, in some way, related to the Turquino–Cayman Ridge volcanic arc. It probably developed as a consequence of changes in the regional tectonic stress field at the begining of the Middle Eocene, focusing sinistral movement along a latitudinal weakened belt between the Turquino–Cayman Rise volcanic arc axis and its subduction zone.

Conclusions

A complicated Campanian–Eocene history is recorded in Cuba. It is fundamental to understanding of Middle America geological development and regional tectonic models.

During the Late Cretaceous the oceanic Proto-Caribbean Plate was consumed in the subduction zone of the Greater Antilles volcanic arc until the Campanian. At this time, volcanic arc magmatism ceased along Cuba (a c. 1000 km long belt) due to collision with the North American passive margin. The strike of the subduction zone must have been more or less orthogonal to descending lithosphere. From Late Campanian to Danian, Cuba and its surroundings were a poorly defined collision zone where the accretion of the Late Cretaceous arc to the North American palaeomargin occurred. The most outstanding example of this tectonic unrest was the obduction of NE Cuban ophiolites and their gravitational Maastrichtian emplacement. Volcanism was absent from the Late Campanian to the Danian in Cuba and, probably, in Hispaniola. Therefore, the presence of a postulated coeval and spatially continuous Great Arc of the Caribbean, critical to Pacific models of Caribbean evolution, is not supported by geological data from Cuba.

In the Mid- or Late Danian the almost east–west trending Turquino–Cayman Ridge–Hispaniola? volcanic arc was born upon the Cretaceous arc.

The related north-dipping subduction zone should act as the SE North American plate boundary and the Yucatán basin was its backarc basin. From the Paleocene to Middle Eocene dense Caribbean lithosphere was subducted. As suggested by Siggurdsson *et al.* (1997), the oceanic lithosphere traveled northwards towards the east–west subduction zone. The location, strike and subduction polarity of the Early Palaeogene subduction zone of this paper are very different from those described by GAC models.

Almost simultaneous with the magmatic event in SE Cuba, the Cuban Orogeny developed in western and central Cuba, west of the Camaguey lineament. The orogeny was diachronous, younging towards the east. During this event the northern ophiolite belt and the Cretaceous volcanic terrane were thrusted northward for at least tens of kilometres, onto the southern fringe of the Mesozoic North American palaeomargin and its overlying Early Cenozoic foreland basin. The preceding facts show that the present NOB position is not the original location of the North American–Proto-Caribbean Plate ophiolitic suture as has been interpreted in several models.

At the beginning of the Middle Eocene subduction stopped, perhaps due to the arrival of an oceanic plateau. Simultanously(?) a change in the regional stress field originated near the east–west trending sinistral Oriente fault zone, whose position and origin are probably tied to the weakened hot crust of the Palaeogene volcanic arc axis.

Recently James (2006) questioned several of the basic arguments of the Pacific origin models of Caribbean origin. The present paper shows many inconsistencies of the GAC (a fundamental element of the Pacific models) with the Cuban geological data and presents an alternative to the GAC hypothesis. The author is conscious that his hypothesis is in a rudimentary state and has several undeveloped questions. Among these are the unexplored relationships between the Palaeogene volcanic arc and neighbouring northern Central America structures (Maya and Chortís Blocks), or with structures bounded by the south the Late Cretaceous volcanic arc. The answers to these questions wait for new research dealing with the poorly known NW Caribbean Sea.

The author is deeply grateful to K. H. James for his invitation to participate in the meeting at Siguenza, Spain (29 May to 2 June 2006) and his support with the English revision of the original manuscript. Some of the ideas presented in this paper were discussed with M. Meschede, H. Hueneke and M. Sommer from Greifswald University, Germany, during the author's stay at this University, in April and May 2005. This paper is a contribution to the project 'Geodynamic Evolution of Western and Central Cuba from Late Jurasssic to Pliocene', developed by the University of Pinar del Rio, Cuba.

References

ALBEAR-FRÁNQUIZ, J. & ITURRALDE-VINENT, M. 1985. Estratigrafia de las provincias de La Habana. *In*: Instituto de Geologia y Paleontologia (ed.) *Contribucion a la Geologia de las provincias de La Habana y Ciudad de La Habana*. Editorial Cientifico-Tecnica, Havana, 12–54.

ANDÓ, J., HARANGI, S., SZAKMANY, L. & DOSZTALY, L. 1996. Petrología de la asociación ofiolítica de Holguin. *In*: ITURRALDE-VINENT, M. (ed.) *Ofiolitas y arcos volcánicos de Cuba*. International Geological Correlation Program, Project 364, Contribution **1**. IUGS/UNESCO, Miami, FL, 154–176.

BECCALUVA, L., COLLORTI, M. *ET AL.* 1996. Cross sections through the ophiolitic units of the southern and northern margins of the Caribbean Plate in Venezuela (northern Cordilleras), and central Cuba. *Ofioliti*, **21**, 85–103.

BRALOWER, T. & ITURRALDE-VINENT, M. 1997. Micropaleontological data of the collision between North American plate and the Greater Antilles arc in western Cuba. *Palaios*, **12**, 133–150.

BRESZNYANSKY, K. & ITURRALDE-VINENT, M. 1978. Paleogeografía del Paleógeno de Cuba oriental. *Geologicae en Mijnbow*, **57**, 123–133.

BRÖNNIMANN, P. & RIGASSI, D. 1963. Contribution to the Geology and Paleontology of the area of the city of La Habana, Cuba and its surroundings. *Eclogae Geologicae Helvetiae*, **56**, 193–430.

BUTTERLIN, J. 1960. *Geologie generale et regionale de la Republique d'Haiti*. University of Paris, Travaux et Memories de l'Institute des Hautes Etudes de l'Amerique Latine, **6**.

CALAIS, E. & DE LEPINAY, B. 1981. From transtension to transpression along the northern Caribbean Plate boundary off Cuba: implications for recent motion of the Caribbean Plate. *Tectonophysics*, **186**, 329–350.

COBIELLA-REGUERA, J. 1974. Los macizos serpentiníticos de Sabanilla, Mayarí Arriba, Oriente. *Revista Tecnológica*, **XII**, 41–50.

COBIELLA-REGUERA, J. L. 1978. Una mélange en Cuba oriental. *La Mineria en Cuba*, **4**, 46–51.

COBIELLA-REGUERA, J. L. 1984. *Sobre el origen del extremo oriental de la fosa de Bartlett*. Editorial Oriente, Santiago de Cuba.

COBIELLA-REGUERA, J. L. 1988. El vulcanismo paleogénico de Cuba. Apuntes para un nuevo enfoque. *Revista Tecnológica*, **XVIII**, 25–32.

COBIELLA-REGUERA, J. 1996. El magmatismo jurásico (caloviano?–oxfordiano) de Cuba occidental: ambiente de formación e implicaciones regionales. *Revista de la Asociación Geológica Argentina*, **51**, 15–28.

COBIELLA-REGUERA, J. L. 1997. Zonación estructuro-facial del corte Paleoceno Eoceno Medio de Cuba. *Minería y Geología*, **XIV**, 3–12

COBIELLA-REGUERA, J. L. 1998. *The Cretaceous System in Cuba – an Overview*. Zentralblatt fur Geologie und Palaontologie, Stüttgart, **1**, 431–440.

COBIELLA-REGUERA, J. L. 2000. Jurassic and Cretaceous Geological History of Cuba. *International Geology Review*, **42**, 594–616.

COBIELLA-REGUERA, J. L. 2002. Remains of oceanic lithosphere in Cuba: types, origins and emplacement ages. *In*: JACKSON, T. (ed.) *Caribbean Geology into the Third Millenium*. Transactions of the Fifteenth Caribbean Geological Conference. University of the West Indies Press, Kingston, 35–46.

COBIELLA-REGUERA, J. L. 2005. Emplacement of Cuban Ophiolites. *Geologica Acta*, **3**, 273–294.

COBIELLA-REGUERA, J. L., BOITEAU, A., CAMPOS, M. & QUINTAS, F. 1977. Geología del flanco sur de la Sierra del Purial. *La Minería en Cuba*, **3**, 54–62.

COBIELLA-REGUERA, J. L., QUINTAS, F., CAMPOS, M. & HERNANDEZ, M. 1984a. *Geología de la región central y suroriental de la provincia de Guantánamo*. Santiago de Cuba, Editorial Oriente.

COBIELLA-REGUERA, J. L., RODRIGUEZ-PEREZ, J. & CAMPOS-DUEÑAS, M. 1984b. Posición de Cuba oriental en la geología del Caribe. *Minería y Geología*, **2**, 65–92.

DIAZ-OTERO, C., ARZ, J., ARENILLAS, I., MOLINA, E. & CORONA, N. 2003. Nuevas consideraciones sobre la edad de la Formación Via Blanca. *Memorias GEOMIN 2003*. La Habana, 24–28 Marzo, 2003, GREG 109–115.

DIAZ DE VILLALVILLA, L. 1997. Caracterizacion geologica de las formaciones volcanicas y volcano-sedimentarias en Cuba central, provincias Cienfuegos, Villaclara y Sancti Spiritus. *In*: FURRAZOLA-BERMUDEZ, G. & NUÑEZ-CAMBRA, (eds) *Estudios sobre Geologia de Cuba*. Centro Nacional de Información Geologica, Havana, 259–270.

DONNELLY, T., HORNE, G., FINCH, R. & LÓPEZ RAMOS, E. 1990. Northern Central America: the Maya and Chortis blocks. *In*: *The Caribbean Region*. The Geology of North America, **H**. The Geological Society of America, Boulder, CO, 37–76.

DRAPER, G. & BARROS, J. 1994. Cuba. *In*: DONOVAN, S. & JACKSON, T. (eds) *Caribbean Geology: an Introduction*. University of West Indies Publishers Association, Kingston, 65–86.

ECHEVARRÍA-RODRÍGUEZ, G., HERNÁNDEZ-PÉREZ, G. ET AL. 1991. Oil and gas exploration in Cuba. *Journal of Petroleum Geologists*, **14**, 259–274.

FONSECA, E., ZELEPUGUIN, V. & HEREDIA, M. 1984. Particularidades de la estructura de la asociación ofiolítica de Cuba. *Ciencias de la Tierra y del Espacio*, **9**, 31–46.

FURRAZOLA-BERMÚDEZ, G., KUZNETSOV, V., GARCÍA-SANCHEZ, R. & BASOV, V. 1979. Estratigrafia de los depositos mesocenozoicos de la costa norte del occidente de Cuba. *La Mineria en Cuba*, **5**, 2–15.

GARCÍA, R. & FERNANDEZ, G. 1987. Estratigrafia del subsuelo de la cuenca Vegas. *Revista Tecnologica*, **XIX**, 3–8.

GARCÍA-CASCO, A., TORRES-ROLDÁN, R., MILLÁN-TRUJILLO, G., MONIE, P. & SCHNEIDER, J. 2002. Oscillatory zoning in eclogitic garnet and amphibole, Northern Serpentinite Melange, Cuba: a record of tectonic unstability during subduction. *Journal of Metamorphic Petrology*, **20**, 581–597.

GARCÍA-CASCO, A., TORRES-ROLDAN, R. ET AL. 2006. High Pressure metamorphism of ophiolites in Cuba. *Geologica Acta*, **4**, 63–88.

GIL GONZALEZ, S., DIAZ OTERO, C. & DELGADO DAMAS, R. 2003. Caracterizacion bioestratigrafica de la Formación La Jiquima. *Memorias V Congreso de*

Geologia y Mineria, Havana, 24–28 March, 2003, GREG 116–123.

GIUNTA, G., BECCALUVA, L., COLTORTI, M. & SIENA, F. 2002. Tectono-magmatic significance of Peri-Caribbean ophiolitic units and geodynamic implications. *In*: JACKSON, T. (ed.) *Caribbean Geology into the Third Millenium. Transactions of the Fifteenth Caribbean Geological Conference*, 15–34.

GIUNTA, G., MARRONI, M., PADOA, E. & PANDOLFI, L. 2003. Geological constrains for the geodynamic evolution of the southern margin of the Caribbean Plate. *In*: BARTOLINI, C., BUFFLER, R. & BLICKWEDE, J. (eds) *The Circum-Gulf of Mexico and the Caribbean: Hydrocarbon Habitats, Basin Formation and Plate Tectonics*. AAPG Memoirs, **79**, 104–125.

GORDON, M., MANN, P., CÁCERES, D. & FLORES, R. 1997. Cenozoic history of the Northamerica–Caribbean Plate boundary in western Cuba. *Journal of Geophysical Research*, **102**, 10055–10082.

HATTEN, C., SOMIN, M., MILLAN, G., RENNE, P., KISTLER, R. & MATTINSON, J. 1988. Tectonostratigraphic units of Central Cuba. *In*: BARKER, L. (ed.) *Transactions of the 11th Caribbean Geological Conference*, Barbados, 1–35.

ITURRALDE-VINENT, M. 1990. Ophiolites in the geological structure of Cuba. *Geotektonika*, **4**, 63–76 (in Russian).

ITURRALDE-VINENT, M. 1996a. Geología de las ofiolitas de Cuba. *In*: ITURRALDE-VINENT, M. (ed.) *Ofiolitas y arcos volcánicos de Cuba*. International Geological Correlation Program, Project 364, Contribution **1**. IUGS/UNESCO, Miami, FL, 83–120.

ITURRALDE-VINENT, M. 1996b. Cuba: el archipielago volcanico Paleoceno-Eoceno Medio. *In*: ITURRALDE-VINENT, M. (ed.) *Ofiolitas y arcos volcánicos de Cuba*. International Geological Correlation Program, Project 364 Contribution **1**, IUGS/UNESCO, Miami, FL, 231–246.

ITURRALDE-VINENT, M. 1997. Introducción a la geología de Cuba. *In*: FURRAZOLA-BERMÚDEZ, G. & NUÑEZ CAMBRA, K. (eds) *Estudios sobre Geología de Cuba*. Centro Nacional de Información Geológica, Havana, 35–68.

ITURRALDE-VINENT, M. 2003. The conflicting paleontologic versus stratigraphic record of the formation of the Caribbean seaway. *In*: BARTOLINI, C., BUFFLER, R. & BLICKWELDE, J. (eds) *The Circum-Gulf of Mexico and the Caribbean: Hydrocarbon Habitats, Basin Formation, and Plate Tectonics*. AAPG Memoir **79**, 75–88.

ITURRALDE-VINENT, M. & MACPHEE, R. 1999. Paleogeography of the Caribbean Region: implications for Cenozoic biogeography. *Bulletin of the American Museum of Natural History*, **238**.

ITURRALDE-VINENT, M., DIAZ-OTERO, C., RODRIGUEZ-VEGA, A. & DIAZ-MARTINEZ, R. 2006. Tectonic implications of paleontologic dating of Cretaceous-Danian sections of Eastern Cuba. *Geologica Acta*, **4**, 89–102.

ITURRALDE-VINENT, M., MILLAN, G., KORPAS, L., NAGY, E. & PAJON, J. 1996. Geological Interpretation of the Cuban K–Ar database. *In*: ITURRALDE-VINENT, M. (ed.) *Ofiolitas y arcos volcánicos de*

Cuba. International Geological Correlation Program, Contribution 1. IUGS/UNESCO, Miami, FL, 48–69.

JAKUS, P. 1983. Formaciones vulcanógeno – sedimentarias y sedimentarias de Cuba oriental. *In*: Instituto de Geologia y Paleontologia (ed.) *Contribucion a la Geologia de Cuba Oriental*. Editorial Cientifico – Tecnica, Havana, 17–89.

JAMES, K. H. 2006. Arguments for and against the Pacific origin of the Caribbean Plate: discussion, finding for an inter-American origin. *Geologica Acta*, **3**, 279–302.

JOLLY, W., LIDIAK, E. & SCHELLEKENS, J. 1998. Volcanism, tectonics and stratigraphic correlations in Puerto Rico. *In*: LIDIAK, E. & LARUE, D. (eds) *Tectonics and Geochemistry of the Eastern Caribbean*. Geological Society of America, Boulder, CO, Special Papers, **322**, 1–34.

KANTCHEV, I., BOYANOV, Y., POPOV, N., CABRERA, R., GORANOV, A., IOLKICEV, N., KANAZIRSKI, M. & STANCHEVA, M. 1978. *Geologia de la provincia de Las Villas. Resultado de las investigaciones geologicas y levantamiento a escala 1 : 250 000 realizado durante el periodo 1969–1975*. Oficina Nacional de Recursos Minerales (unpublished).

KEIJZER, F. 1945. Outline of the geology of eastern part of the province of Oriente, Cuba (E of 76° W.L.). *Utrecht Geographische en Geologische Mededeling Physiographische-Geologie Recks*, **2**, 1–239.

KERR, A., ITURRALDE-VINENT, M., SAUNDERS, A., BABBS, T. & TARNEY, J. 1999. A new plate tectonics model of the Caribbean: implications from a geochemical reconnaissance of Cuban Mesozoic rocks. *GSA Bulletin*, **111**, 1581–1599.

KESLER, S., CAMPPBELL, I. & AEN, C. 2005. Age of Los Ranchos formation of Dominican Republic: time and setting of primitive island arc volcanism in the Caribean region. *GSA Bulletin*, **117**, 987–995.

KHUDOLEY, K. & MEYERHOFF, A. 1971. *Paleogeography and Geological History of Greater Antilles*. Geological Society of America, Boulder, CO, Memoirs, **129**.

KNIPPER, A. & CABRERA, R. 1974. Tectónica y geología histórica de la zona de articulacion entre el mio- y eugeosinclinal de Cuba y del cinturón hiperbasítico de Cuba. *In*: Instituto de Geología y Paleontologia (ed.) *Contribución a la Geología de Cuba*. Academia de Ciencias, Havana, 15–77.

KOZARY, M. 1968. Ultramafic rocks in thrust zones of northwestern Oriente province, Cuba. *AAPG Bulletin*, **52**, 2298–2317.

LEBRÓN, M. & PERFIT, M. 1994. Petrochemistry and tectonic significance of Cretaceous island–arc rocks, Cordillera oriental, Dominican Republic. *Tectonophysics*, **229**, 69–100.

LEROY, S., MAUFFRET, A., PATRIAT, P. & MERCIER DE LEPINAY, B. 2000. An alternative interpretation of Cayman trough evolution from a reidentification of magnetic anomalies. *Geophysical Journal International*, **141**, 539–557.

LEWIS, G. & STRACZEK, J. 1955. *Geology of South Central Oriente, Cuba*. US Geological Survey Bulletin 975-D, Washington DC, 171–336.

LEWIS, J. & DRAPER, G. 1990. Geology and tectonic evolution of the northern Caribbean margin. *In*: DENGO, G.

& CASE, J. H. (eds) *The Caribbean Region*. The Geology of North America, **H**. Geological Society of America, Boulder, CO, 77–140.

LEWIS, J., DRAPER, G., PROENZA, J., ESPAILLAT, J. & JIMENEZ, J. 2006. Ophiolite-related ultramafic rocks (serpentinites) in the Caribbean Region: a review of their occurrence, composition, origin, emplacement and NI-laterite soil formation. *Geologica Acta*, **4**, 237–263.

LEWIS, J., ESCUDER VIRUETE, J., HERNAIZ HUERTA, P., GUTIERREZ, G., DRAPER, G. & PEREZ-ESTAUN, A. 2002. Subdivision geoquimica del Arco Isla Circumcaribeno, Cordillera Central, Republica Dominicana: implicaciones para la formacion, acrecion y crecimiento cortical en un ambiente intraoceanico. *Acta Geologica Hispanica*, **37**, 81–122.

LLANES-CASTRO, A., GARCÍA DELGADO, D. & MEYERHOFF, D. 1998. *Hallazgo de fauna jurásica (Tithoniano) en ofiolitas de Cuba central*. Memorias II Geología y Minería 98. Centro Nacional de Informacion Geológica, Havana, 241–244.

MANN, P. 1999. Caribbean Sedimentary Basins: classification and tectonic setting from Jurassic to present. *In*: MANN, P. (ed., series ed. HSU, K. J.) *Caribbean Basins*. Sedimentary Basins of the World, **4**. Elsevier Science, Amsterdam, 3–31.

MARTON, G. & BUFFLER, R. 1994. Jurassic reconstruction of the Gulf of Mexico Basin. *International Geology Review*, **36**, 545–586.

MARTON, G. & BUFFLER, R. 1999. Jurassic–Early Cretaceous paleogeographic evolution of the Southeastern Gulf of Mexico Basin. *In*: MANN, P. (ed., series ed. HSU, K. J.) *Caribbean Basins*. Sedimentary Basins of the World, **4**. Elsevier Science, Amsterdam, 63–91.

MEYERHOFF, A. & HATTEN, C. 1968. *Diapiric Structures in Central Cuba*. American Association of Petroleum Geologists, Boulder, CO, Memoirs, **8**, 315–357.

MILLÁN TRUJILLO, G. 1997. Posición estratigráfica de las metamorfitas cubanas. *In*: FURRAZOLA BERMÚDEZ, G. & NUÑEZ CAMBRA, K. (Compilators) *Estudios sobre Geología de Cuba*. Instituto de Geología y Paleontología, Centro Nacional de Información Geológica, 251–258.

MORÁN-ZENTENO, D. 1994. *Geology of the Mexican Republic*. American Association of Petroleum Geologists, Cincinnati, OH, Studies in Geology, **39**.

MORETTI, I., TENREYRO, R. *ET AL*. 2003. Petroleum systems of the Cuban northwest offshore zone. *In*: BARTOLINI, C., BUFFLER, R. & BLICKWELDE, J. (eds) *The Circum-Gulf of Mexico and the Caribbean: Hydrocarbon Habitats, Basin Formation, and Plate Tectonics*. AAPG Memoir, **79**, 675–696.

NAGY, E. 1984. Ensayo de las zonas estructuro-faciales de Cuba oriental. *In*: *Contribución a la Geología de Cuba oriental*. Editorial Científico-Técnica, Havana, 9–16.

NUÑEZ CAMBRA, K., CASTELLANOS ABELLA, E., ECHEVARRÍA, B. & LLANES, I. 2003. Estructura del área Merceditas y consideraciones acerca de la procedencia de las ofiolitas del macizo Moa-Baracoa. Abstracts. *5th Cuban Congress on Geology and Mining*, Havana, 291–293.

PARDO, G. 1975. Geology of Cuba. *In*: NAIRN, A. & STEHLI, F. (eds) *The Gulf of Mexico and the*

Caribbean. The Ocean Basins and Margins, **3**, Plenum Press, New York, 553–613.

PERFIT, M. & HEEZEN, B. 1978. The geology and evolution of Cayman Trench. *Geological Society of America Bulletin*, **89**, 1155–1174.

PINDELL, J. L. 1994. Evolution of the Gulf of Mexico and the Caribbean. *In*: DONOVAN, S. & JACKSON, T. (eds) *Caribbean Geology: An Introduction*. The University of the West Indies Publishers' Association, Kingston, 13–40.

PINDELL, J. L. & BARRETT, S. 1990. Geological evolution of the Caribbean Region: a plate tectonics perspective. *In*: DENGO, G. & CASE, J. (eds) *The Caribbean Region*. The Geology of North America, **H**. Geological Society of America, Boulder, CO, 405–432.

PINDELL, J. L., KENNAN, L., STANEK, K., MARESCH, W. & DRAPER, G. 2006. Foundations of Gulf of Mexico and Caribbean evolution: eight controversies resolved. *Geologica Acta*, **4**, 303–341.

PIOTROWSKA, K. 1986*a*. Tectonica de la parte central de la provincia de Matanzas. *Bulletin of the Polish Academy of Sciences. Earth Sciences*, **34**, 3–16.

PIOTROWSKA, K. 1986*b*. Etapas de las deformaciones en la provincia de Matanzas en comparacion con la provincia de Pinar del Rio. *Bulletin of the Polish Academy of Sciences. Earth Sciences*, **34**, 17–27.

PIOTROWSKI, J. 1986. Las unidades de nappes en los valles de Yumuri y de Caunavaco. *Bulletin of the Polish Academy of Sciences. Earth Sciences*, **34**, 29–36.

PROENZA, J., GERVILLA, F. ET AL. 2006. Primitive Cretaceous island-arc volcanic rocks in eastern Cuba: the Teneme Formation. *Geologica Acta*, **4**, 103–121.

PROENZA, J., GERVILLA, F. & MELGAREJO, J. 1999. La Moho Transition Zone en el macizo ofiolítico Moa-Baracoa (Cuba): un ejemplo de interacción magma/peridotita. *Revista Sociedad Geológica de España*, **12**, 309–327.

PSZCZOLKOWSKI, A. 1983. Tectonica del miogeosinclinal cubano en el area limitrofe de las provincias de Matanzas y Villaclara. *Ciencias de La Tierra y del Espacio*, **6**, 53–61.

PSZCZOLKOWSKI, A. 1986. Megacapas del Maastrichtiano de Cuba occidental y central. *Bulletin of the Polish Academy of Sciences*, **34**, 81–87.

PSZCZOLKOWSKI, A. 1994. Lithostratigraphy of Mesozoic and Paleogene rocks of Sierra del Rosario, western Cuba. *Studia Geologica Polonica*, **105**, 39–66.

PSZCZOLKOWSKI, A. 1999. The exposed passive margin of North America in western Cuba. *In*: MANN, P. (ed., series ed. HSU, K. J.). *Caribbean Basins*. Sedimentary Basins of the World, **4**. Elsevier Science, Amsterdam, 93–121.

PSZCZOLKOWSKI, A. & ALBEAR, J. 1982. Subzona estructuro-facial Bahía Honda, Pinar del Río; su tectónica y datos sobre la sedimentación y paleogeografía del Cretácico Superior y Paleógeno. *Ciencias de La Tierra y del Espacio*, **5**, 3–24.

PSZCZOLKOWSKI, A. & FLORES, R. 1986. Fases tectónicas del Paleógeno y Cretácico de Cuba occidental y central. *Bulletin of the Polish Academy of Sciences*, **134**, 95–111.

PSZCZOLKOWSKI, A. & MYCZYNSKI, R. 2003. Stratigraphic constrainsts on the Late Jurassic–Cretaceous Paleotectonic interpretations of the Placetas Belt

in Cuba. *In*: BARTOLINI, C., BUFFLER, R. & BLICKWEDE, J. (eds) *The Circum-Gulf of Mexico and the Caribbean: Hydrocarbon Habitats, Basin Formation and Plate Tectonics*. AAPG Memoirs, **79**, 545–581.

PUSHCHAROVSKY, Y. (ed.) 1988. *Mapa Geológico de la República de Cuba*. Academy of Sciences of Cuba and the Soviet Union, 42 sheets, scale 1 : 250 000.

QUINTAS CABALLERO, F. 1998. Evolución y perspectivas petrolíferas de las cuencas de antepaís en Camagüey. *Minería y Geología*, **XV**, 11–16.

RODRÍGUEZ, R., BLANCO-MORENO, J., PROENZA, J. & OROZCO, G. 2001. Petrología de las rocas plutónicas de afinidad ofiolítica presentes en la zona de Cayo Grande (macizo ofiolítico de Moa-Baracoa) Cuba oriental. *Minería y Geología*, **XVIII**, 31–47.

RODRÍGUEZ, R., SANTA CRUZ PACHECO, M., NAVARRETE, M., FONSECA, E. & ALBEAR, J. 1997. Las cromititas podiformes en las ofiolitas de Cuba. *In*: FURRAZOLA, G. & NUÑEZ CAMBRA, K. (eds) *Estudios sobre Geología de Cuba*. Centro Nacional de Información Geológica, Havana, 301–314.

ROJAS, R., ITURRALDE-VINENT, M. & SKELTON, P. 1995. Stratigraphy, composition and age of Cuban Rudist-bearing deposits. *Revista Mexicana de Ciencias Geologicas*, **12**, 272–291.

ROJAS-AGRAMONTE, Y., NEUBAUER, N. ET AL. 2004. Geochemistry and Early Paleogene SHRIMP zircon ages for island arc granitoids of the Sierra Maestra, southeastern Cuba. *Chemical Geology*, **213** 307–324.

ROJAS-AGRAMONTE, Y., NEUBAUER, N., HANDLER, R., GARCÍA-DELGADO, D., FRIEDL, G. & DELGADO-LAMAS, R. 2005. Variations in paleostress patterns along the Oriente transform wrench corridor, Cuba: significance for Neogene–Quaternary tectonics of the Caribbean realm. *Tectonophysics*, **396**, 161–180.

ROJAS-AGRAMONTE, Y., NEUBAUER, F., BOJAR, A., HEJL, E., HANDLER, R. & GARCÍA-DELGADO, D. 2006. Geology, age and tectonic evolution of the Sierra Maestra, southeastern Cuba. *Geologica Acta*, **4**, 123–150.

SCHAFHAUSER, A., STINNESBECK, W., HOLLAND, B., ADATTE, T. & REMANE, J. 2003. Lower Cretaceous pelagic limestones in southern Belize: protoCaribbean deposits in the southeastern Maya block. *In*: BARTOLINI, C., BUFFLER, R. & BLICKWELDE, J. (eds) *The Circum-Gulf of Mexico and the Caribbean: Hydrocarbon Habitats, Basin Formation, and Plate Tectonics*. AAPG Memoir, **79**, 624–637, 121–130.

SIGGURDSSON, H., LECKIE, R. & ACTON, G. 1997. *Proceedings of the Ocean Drilling Programme, Initial Reports*, **165**. College Station (Ocean Drilling Programme), 49–130.

SOMIN, M. & MILLAN, G. 1981. *Geology of the Metamorphic Complexes of Cuba*. Nauka, Moscow (in Russian).

STANEK, K., COBIELLA-REGUERA, J., MARESCH, W., MILLÁN TRUJILLO, G., GRAFE, F. & GREVEL, C. 2000. Geological Development of Cuba. *Zeitschrift fur Angewandte Geologie*, **SH1**, 259–265.

TABER, S. 1934. Sierra Maestra of Cuba, part of the northern rim of the Bartlett trough. *Geological Society of America Bulletin*, **45**, 567–619.

338   J. L. COBIELLA-REGUERA

TADA, R., ITURRALDE-VINENT, M. *ET AL.* 2003. K/T boundary deposits in the Paleo-western Caribbean Basin. *In*: BARTOLINI, C., BUFFLER, R. & BLICKWEDE, J. (eds) *The Circum-Gulf of Mexico and the Caribbean: Hydrocarbon Habitats, Basin Formation and Plate Tectonics*. AAPG Memoirs, **79**, 582–604.

UNESCO & IUGS 2000. *International Stratigraphic Chart*. Division of Earth Sciences, UNESCO.

WITROCK, R., FRIEDMANN, J., GALLUZO, L., NIXON, P. & ROSS, K. 2003. *Biostratigraphic Chart of the Gulf of Mexico Offshore Region, Jurassic to Quaternary*. United States Interior Department, Mineral Management Service, New Orleans.

Geochemistry and tectonomagmatic significance of Lower Cretaceous island arc lavas from the Devils Racecourse Formation, eastern Jamaica

ALAN R. HASTIE[1]*, ANDREW C. KERR[1], SIMON F. MITCHELL[2] & IAN L. MILLAR[3]

[1]*School of Earth and Ocean Sciences, Cardiff University, Main Building, Park Place, Cardiff CF10 3YE, UK*

[2]*Department of Geography and Geology, University of the West Indies, Mona, Kingston 7, Jamaica*

[3]*NERC Isotope Geoscience Laboratories, Keyworth, Nottingham NG12 5GG, UK*

*Corresponding author (e-mail: hastiear1@cf.ac.uk)

Abstract: The Benbow Inlier in Jamaica contains the Devils Racecourse Formation, which is composed of a Hauterivian to Aptian island arc succession. The lavas can be split into a lower succession of basaltic andesites and dacites/rhyolites, which have an island arc tholeiite (IAT) composition and an upper basaltic and basaltic andesite sequence with a calc-alkaline (CA) chemistry. Trace element and Nd–Hf isotopic evidence reveals that the IAT and CA lavas are derived from two chemically similar mantle wedge source regions predominantly composed of normal mid-ocean ridge-type spinel lherzolite. In addition, Th-light rare earth element/high field strength element–heavy rare earth element ratios, Nd–Hf isotope systematics, $(Ce/Ce^*)_{n-mn}$ and Th/La ratios indicate that the IAT and CA mantle wedge source regions were enriched by chemically distinct slab fluxes, which were derived from both the altered basaltic portion of the slab and its accompanying pelagic and terrigenous sedimentary veneer respectively. The presence of IAT and CA island arc lavas before and after the Aptian–Albian demonstrates that the compositional change in the Great Arc of the Caribbean was the result of the subduction of chemically differing sedimentary material. There is therefore no evidence from the geochemistry of this lava succession to support arc-wide subduction polarity reversal in the Aptian–Albian.

Supplementary material: References for data sources used in figures can be found at: http://www.geolsoc.org.uk/SUP18361.

It is near-universally accepted that the Caribbean Plate was tectonically emplaced between North and South America in the Cretaceous, principally as a result of a subduction polarity reversal from northeast dipping to southwest dipping. This combined with subsequent subduction back-step, and the formation of sinistral and dextral strike–slip faulting on the northern and southern Caribbean Plate margins was the mechanism responsible for isolating the Caribbean as a separate plate between the Americas (Mattson 1979; Burke 1988; Lebron & Perfit 1993, 1994; Schellekens 1998; Müller *et al.* 1999; Kerr *et al.* 1999, 2003; Pindell & Kennan 2001; Kesler *et al.* 2005; Jolly *et al.* 2006; Pindell *et al.* 2006). However, the timing and cause of the subduction polarity reversal is still a significant point of controversy for proponents of the 'Pacific Model' of Caribbean Plate evolution. Many geologists and geochemists (e.g. Lebron & Perfit 1993, 1994; Pindell & Kennan 2001; Kesler *et al.* 2005; Jolly *et al.* 2006; Pindell *et al.* 2006;

Escuder Viruete *et al.* 2007; Marchesi *et al.* 2007) have proposed an Aptian–Albian (125–99.6 Ma) reversal whereas others (e.g. Burke 1988; Schellekens 1998; Sinton *et al.* 1998; Kerr *et al.* 1999, 2003; White *et al.* 1999; Thompson *et al.* 2003; Hastie *et al.* 2008) argue for a Turonian–Santonian (93.5–83.5 Ma) reversal.

Although several lines of evidence for an Aptian–Albian reversal have been put forward by Pindell *et al.* (2006), one of the main arguments consistently used involves an abrupt change in the chemistry of subduction-related lavas in the 'Great Arc of the Caribbean' from an island arc tholeiite (IAT) to a calc-alkaline (CA) signature. It has been suggested that this change in island arc magma type was a result of the reversal of subduction and the derivation of the CA arc magmas from compositionally different source regions associated with subducting proto-Caribbean oceanic crust as opposed to Farallon oceanic crust (Mattson 1979; Donnelly *et al.* 1990; Lebron &

From: JAMES, K. H., LORENTE, M. A. & PINDELL, J. L. (eds) *The Origin and Evolution of the Caribbean Plate.*
Geological Society, London, Special Publications, **328**, 339–360.
DOI: 10.1144/SP328.14   0305-8719/09/$15.00 © The Geological Society of London 2009.

Perfit 1994; Kesler *et al.* 2005; Pindell *et al.* 2006). This change in chemistry is proposed to be accompanied by an Aptian–Albian (125–99.6 Ma) unconformity in Hispaniola and parts of Puerto Rico, which have also been inferred by some (e.g. Lebron & Perfit 1993, 1994; Pindell *et al.* 2006) to represent the Aptian–Albian subduction polarity reversal event. However, Schellekens (1998) and Jolly *et al.* (2006) have noted that the unconformity in Puerto Rico is not as extensive as previously thought and that it is absent in northeastern and parts of central Puerto Rico. In addition, the unconformity was also thought to occur on the Virgin Islands, but recent fieldwork by Rankin (2002) has demonstrated that on St John in the US Virgin islands the volcanic–sedimentary sequence is conformable from the Late Aptian (*c.* 112 Ma) to at least the latest Santonian (*c.* 83.5 Ma).

In contrast to models which invoke an abrupt change in island arc geochemistry in the Aptian–Albian, Kerr *et al.* (2003) argue that the switch from an IAT to a CA composition was more gradual and commenced before, and continued after, the Aptian–Albian unconformity. The compilation of Caribbean geochemical data in Kerr *et al.* (2003) clearly illustrates that both IAT and CA rocks were formed in the Great Arc of the Caribbean both before and after the Aptian–Albian (Kerr *et al.* 2003, fig. 17). This has been supported by more recent geochemical data from volcanic rocks in northeastern Puerto Rico and the Virgin Islands (Jolly *et al.* 2006), the Guamuta, Loma de la Bandera and Cerrajón dykes in Cuba (Marchesi *et al.* 2007) and the Tireo Formation in Hispaniola (Escuder Viruete *et al.* 2007). The rocks in Puerto Rico and the Virgin Islands have IAT compositions which have dates ranging from the Early Albian (112–99.6 Ma) to the Cenomanian (99.6–93.5 Ma). The Cuban dykes and the lower volcanic sequence of the Tireo Formation (composed of tuffs, volcanic breccias and basaltic and andesitic lava flows) have IAT compositions that have been dated from the Aptian (125–112 Ma) to the Turonian (93.5–89.3 Ma). Although Jolly *et al.* (2006), Escuder Viruete *et al.* (2007) and Marchesi *et al.* (2007) still argue for the Aptian–Albian reversal model, we would note that their results, and those of Kerr *et al.* (2003) show that the change in chemistry of the arc magmas occurred gradually due to chemically different material being subducted and incorporated into the island arc magmas above a subduction zone which continued to dip northeast until at least the Turonian.

To further analyse this IAT to CA compositional change in the Cretaceous Greater Antilles island arc lavas this study presents new geochemical data on an Early Cretaceous island arc succession from the Benbow Inlier in eastern Jamaica. The age and composition of the arc lavas is invaluable for determining the Early Cretaceous igneous and tectonic evolution of Jamaica and the Caribbean Plate in the Hauterivian–Aptian (136–112 Ma). Kesler *et al.* (2005) noted that IAT rocks in the western section of the Great Arc of the Caribbean have not been studied in much geochemical detail, thus, this study will provide a full petrological study of an IAT succession from the western part of the Great Arc. Moreover, we will demonstrate that the Early Cretaceous Jamaican island arc sequence evolves from an IAT to a CA composition before the supposed Aptian–Albian reversal and, as such, places further doubt on the argument for a Mid-Cretaceous subduction polarity reversal.

Geological background: the Benbow Inlier

Two-thirds of Jamaica is covered by Cenozoic limestone; however the underlying Cretaceous sedimentary and volcanic rocks are exposed in numerous inliers (Fig. 1) (Robinson *et al.* 1972; Lewis & Draper 1990; Robinson 1994). The Benbow Inlier is small compared with the Blue Mountains, Central and Above Rocks Inliers and contains a thick sequence of lower to early Late Cretaceous volcanic and sedimentary rocks. The stratigraphy of the inlier was originally worked out by Burke *et al.* (1969), but recent geological mapping by Ian Brown and Simon Mitchell (University of the West Indies) indicates that significant revision is needed. This, however, is beyond the scope of this paper, and does not alter the age relationships of the igneous rocks considered herein.

The succession in the Benbow Inlier was divided into three formations by Burke *et al.* (1969), the Devils Racecourse Formation at the base, the succeeding Rio Nuevo Formation and the Tiber Formation at the top. Both the Devils Racecourse and Tiber formations are characterized by the presence of extensive lava flows (Burke *et al.* 1969; Roobol 1972). The ages of the igneous sequences in the Benbow Inlier are well constrained by palaeontology, which allows a robust chronostratigraphic correlation that is much better than in many Early Cretaceous volcanic sequences elsewhere on the Caribbean Plate. In this study, samples were analysed from the Devils Racecourse Formation, and only the geology of the Devils Racecourse and basal Rio Nuevo formations will be reviewed here (Fig. 2).

The Devils Racecourse Formation consists of two major units of lava flows separated by a unit of clastic and carbonate sedimentary rocks, and can be conveniently divided into lower, middle and upper parts for descriptive purposes (Fig. 2). The lower part of the Devils Racecourse Formation

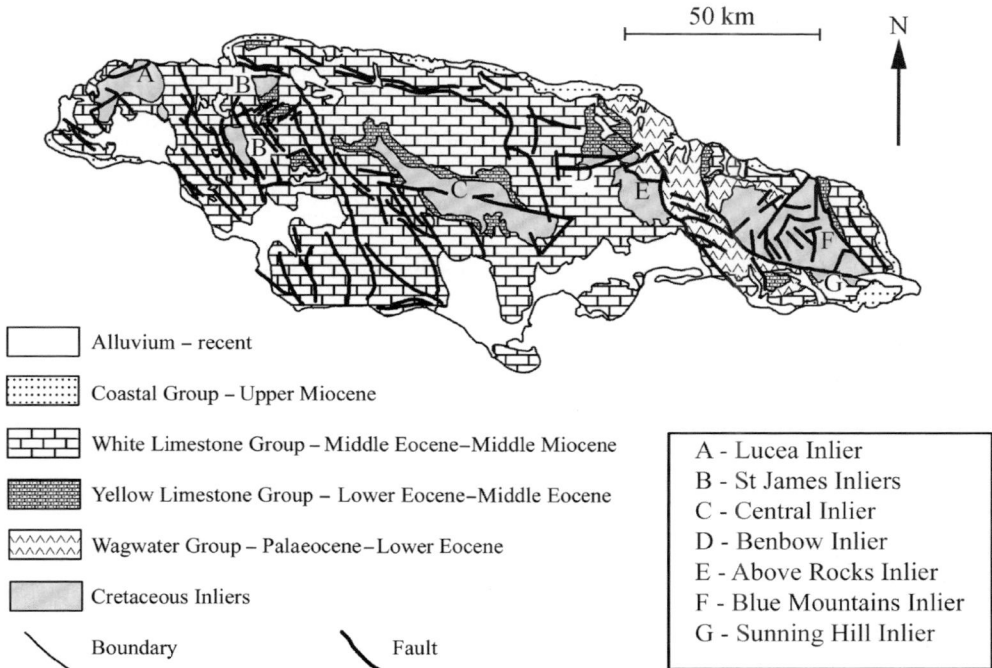

Fig. 1. Geological map of Jamaica (modified from Lewis & Draper 1990).

consists predominantly of a sequence of dark, fine-grained dacitic to basaltic andesite flows with a thickness of 600 m. Flow banding occurs in places, but individual flows cannot be mapped due to extensive weathering and poor exposure. A 5 m succession of thinly bedded lapilli tuffs is present within the lavas at locality 3 (Fig. 3), but cannot be traced elsewhere. Above the lavas is a thin sequence (75 m) of volcaniclastic sandstones and conglomerates. Clasts are angular and have similar compositions to the underlying lavas suggesting reworking in a subaerial environment. The only age constraint on the lavas is that they are older than the overlying limestone (i.e. early or pre-Hauterivian: c. 136 Ma).

The middle part of the Devils Racecourse Formation consists of a succession of rudist-bearing limestones and interbedded clastic sedimentary rocks. The number of limestones present is controversial, with the suggestion that some pinch out laterally (Burke et al. 1969). Our section here is a reinterpretation of the succession given by Burke et al. (1969), but will require modification when mapping is complete (Fig. 2). The Copper limestone (the Bonnett Limestone and Phillipsburg Limestone of Burke et al. 1969) consists of 250 m of micrites, wackestones and packstones. Fossils include rudist bivalves, corals, foraminifers and algae that indicate

a Hauterivian (136–130 Ma) age (Vila 1986; Skelton & Masse 1998). The copper limestone is succeeded by 1300 m of conglomerates containing rounded clasts largely composed of volcanic rocks. The thin Jubilee Limestone also contains rudists indicating a Late Hauterivian or Early Barremian age (Skelton & Masse 1998). The Jubilee limestone is succeeded by 650 m of shales, sandstones and conglomerates that contain a marine fauna. The highest and thickest (400 m) limestone in the middle part of the Devils Racecourse Formation is the Benbow Limestone (Matley & Raw 1942) (Fig. 2). It consists of wackestones and fossils (rudist bivalves, chondrodonts, gastropods, foraminifers and algae), which indicate a lower to upper Barremian age (130–125 Ma) (Sohl 1979; Pisot et al. 1986; Vila 1986; Jiang & Robinson 1987; Skelton & Masse 1998).

The upper part of the Devils Racecourse Formation consists of a lower succession of basic pillow lavas containing a thin limestone unit, and an upper succession of conglomerates. The lower and upper units of pillow lavas (350 and 480 m, respectively) are separated by an 8 m-thick limestone (Fig. 2). The micritic limestone contains rudist bivalves of probable lower Aptian age (c. 125–120 Ma), although more material is needed for more accurate dating. The pillow lavas are

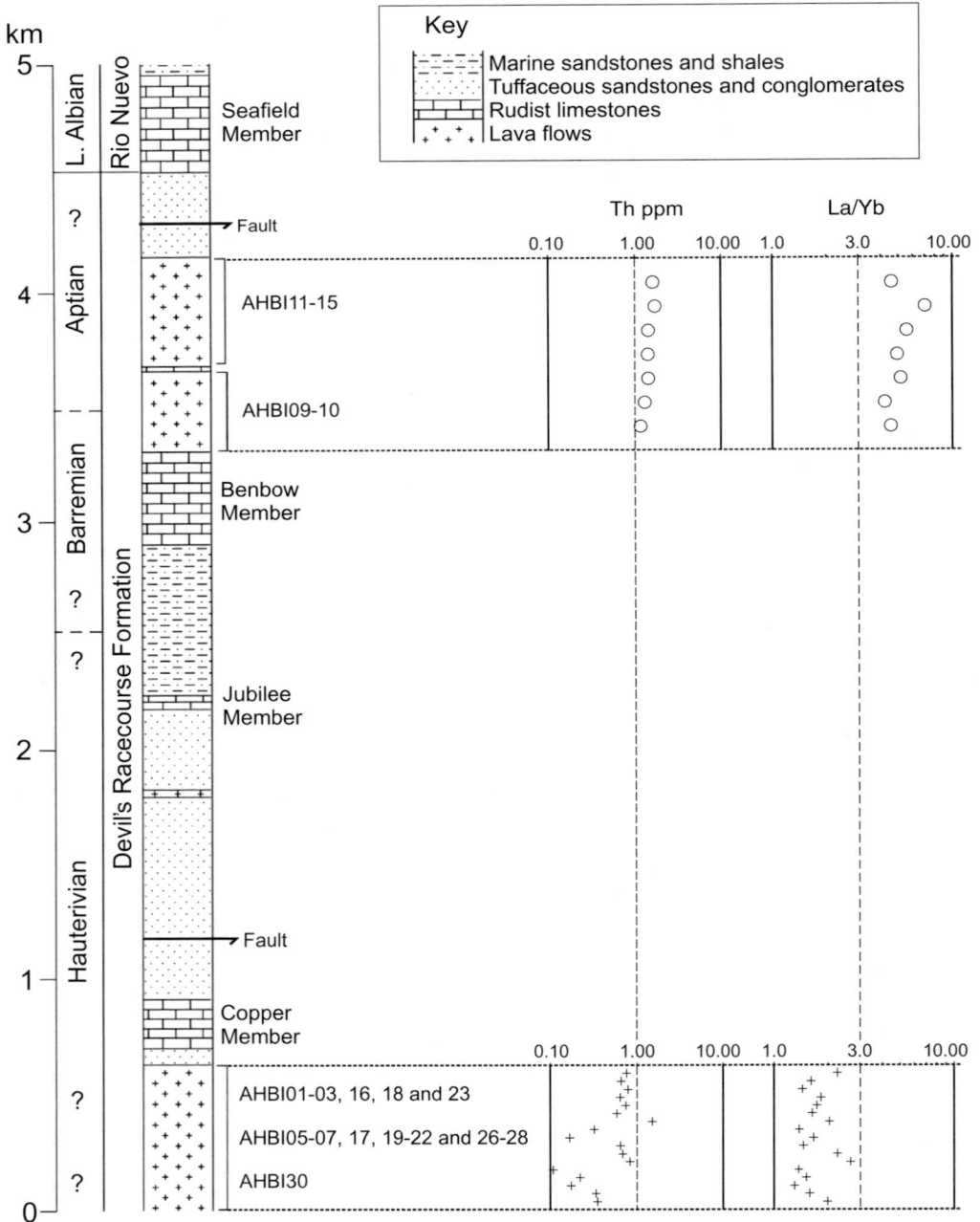

Fig. 2. Stratigraphic column of the Devils Racecourse Formation. Sample localities shown in Figure 3. Th ppm and La/Yb diagrams demonstrate that the upper lavas are more enriched in incompatible trace elements than the lower lavas (see text for more details).

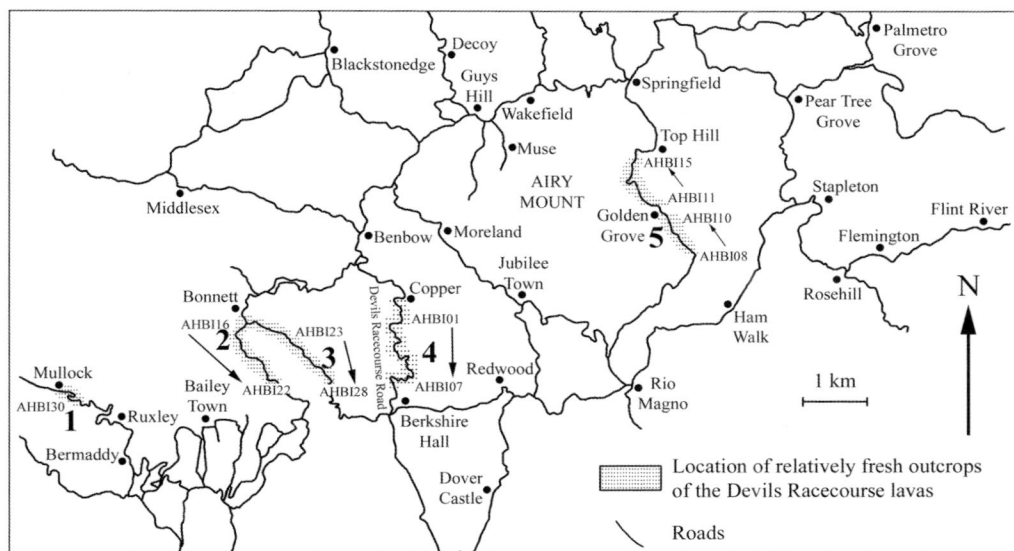

Fig. 3. Location of the Devils Racecourse samples collected in this study. 1–5 refer to the location numbers in Table 1.

succeeded by 360 m of conglomerates containing well-rounded, pebble-sized clasts, with diverse clast compositions that contrast with the monotonous lithologies of the conglomerates lower in the Devils Racecourse Formation.

The Devils Racecourse Formation is succeeded by the Rio Nuevo Formation, which consists of sandstones and dark shales, and contains several units of micritic and bioclastic limestone (Seafield Limestone) in its lower part (Fig. 2). The rudist assemblage in these limestones indicates a lower Albian age.

Sample location and petrography

The location of the Devils Racecourse samples is shown in Figure 3. Samples AHBI01-07 were collected along the Devils Racecourse road (Fig. 3), where the lower Devils Racecourse Formation is represented by 5–8 m tall cliffs of igneous rock. The rocks are relatively fresh, fine-grained, porphyritic and have a darkish grey/blue colour. No volcanic structures were observed during this study, however, Burke *et al.* (1969) described flow banding from these rocks and on the basis of this evidence proposed that they were lava flows and not ash flows; although, flow banding can also develop in sills and dykes.

Samples AHBI16-30 were collected from outcrops that were substantially altered and/or mostly covered with vegetation. It is assumed that they represent laterally equivalent outcrops to the lavas along the Devils Racecourse road (Fig. 3). The rocks are fine-grained, mostly porphyritic and have a darkish blue/grey/brown colour, which occasionally has a green tint because of the presence of chlorite. Samples AHBI09-15 were collected from near the village of Golden Grove where the upper Devils Racecourse Formation consists of vesicular pillow lavas (Fig. 3). The lavas are substantially altered, fine-grained, porphyritic and have a dark grey/brown colour.

Samples AHBI04-07, 17, 19–30 are from the base of the lower section of the Devils Racecourse Formation and have dacitic/rhyolitic compositions (Figs 2 & 3). In thin section they are phyric or aphyric and are composed of plagioclase feldspar, K-feldspar, quartz and Fe–Ti oxide phenocrysts in a groundmass of plagioclase feldspar, K-feldspar, quartz and opaques. The uppermost lavas in the lower section, AHBI01, 03, 16 and 18, have basaltic andesite/andesitic compositions (Fig. 2). Petrographically they are composed of plagioclase feldspar and clinopyroxene phenocrysts in a groundmass of plagioclase feldspar, clinopyroxene and opaques. AHBI09-15 were sampled from the less weathered interior of the pillow lavas that make up the upper Devils Racecourse Formation. These lavas all have basaltic/basaltic andesite compositions, and are composed of plagioclase feldspar and clinopyroxene phenocrysts in a groundmass of plagioclase feldspar, clinopyroxene and opaques.

Table 1. *Selected major and trace element compositions for the Devils Racecourse lavas, Jamaica*

Sample		AHBI01	AHBI03	AHBI05	AHBI06	AHBI07	AHBI09	AHBI10	AHBI11	AHBI12	AHBI13	AHBI14
Section		Lower	Lower	Lower	Lower	Lower	Upper	Upper	Upper	Upper	Upper	Upper
Rock type		BA/A	BA/A	D/R	D/R	D/R	BA	BA	B	BA	B	BA
Rock series		IAT	IAT	IAT	IAT	IAT	CA	CA	CA	CA	CA	CA
Locality number		4	4	4	4	4	5	5	5	5	5	5
ICP-OES analysis	Majors (wt %)											
	TiO_2	0.86	0.86	0.39	0.30	0.64	1.22	1.17	1.10	1.15	1.20	1.17
	K_2O	0.45	0.40	0.16	0.08	0.23	0.61	0.09	0.54	0.31	0.25	1.53
ICP-MS analysis	Traces (ppm)											
	Co	18.7	18.9	3.8	3.7	2.0	23.4	19.9	31.2	24.3	30.4	25.0
	Y	34.5	31.8	25.7	22.2	34.8	38.6	23.9	21.1	23.8	34.9	22.9
	Zr	82.9	82.1	113.2	96.2	123.3	91.6	86.2	67.1	77.7	89.4	84.3
	Nb	1.86	1.85	1.16	0.93	1.58	2.33	2.17	1.72	1.77	2.14	2.00
	Ba	348	219	414	196	183	469	80	301	657	306	433
	La	6.25	5.10	6.80	4.80	5.75	23.00	11.00	8.70	9.10	14.50	11.00
	Ce	14.92	12.91	16.92	11.79	16.53	48.31	27.94	20.20	23.78	35.43	25.90
	Pr	2.06	1.98	2.47	1.76	2.85	5.33	4.12	2.82	3.65	5.30	3.52
	Nd	10.41	9.86	11.38	8.66	14.40	23.90	18.12	12.43	16.59	23.31	15.24
	Sm	3.56	3.41	3.00	2.49	4.27	5.40	4.09	3.24	3.94	5.50	3.76
	Eu	1.15	1.13	0.78	0.60	1.05	1.70	1.17	1.12	1.35	1.74	1.30
	Gd	3.93	3.85	3.44	2.82	4.83	5.52	4.02	3.05	3.99	4.99	3.51
	Tb	0.73	0.72	0.56	0.48	0.81	0.82	0.60	0.52	0.59	0.83	0.62
	Dy	5.02	5.01	3.84	3.29	5.70	5.46	3.95	3.41	3.87	5.32	3.90
	Ho	1.02	1.02	0.75	0.64	1.11	1.06	0.75	0.64	0.72	1.01	0.75
	Er	3.04	3.06	2.41	2.11	3.65	3.37	2.39	1.85	2.27	2.95	2.25
	Tm	0.46	0.45	0.38	0.33	0.60	0.51	0.37	0.28	0.35	0.43	0.34
	Yb	3.15	3.23	2.55	2.14	3.94	3.26	2.39	1.92	2.17	2.81	2.24
	Lu	0.49	0.51	0.46	0.40	0.65	0.54	0.40	0.30	0.42	0.45	0.35
	Hf	2.16	2.22	3.12	2.65	3.38	2.63	2.37	1.52	2.14	2.07	2.03
	Th	0.35	0.34	0.83	0.69	0.65	1.71	1.62	1.17	1.31	1.43	1.41
	U	0.28	0.28	0.50	0.40	0.57	0.48	0.96	0.62	0.59	1.10	0.51

A full list of analysed major and trace elements is presented in Hastie *et al.* (2007). B, basalt; BA, basaltic-andesite; A, andesite; D, dacite; R, rhyolite; IAT, island arc tholeiite; CA, calc-alkaline. The locality numbers refer to Figure 3. The analytical procedure is described in McDonald & Viljoen (2006).

The primary mineralogy of all the rocks has been variably replaced by secondary minerals such as sericite, chlorite, epidote, calcite, clay minerals and iron oxy-hydroxides. This reflects a hydrothermal alteration event post-dated by tropical weathering. The tropical weathering is a particular concern for element mobility, given the high water–rock ratios, high surface temperatures and high concentrations of organic acids (Summerfield 1997).

Geochemistry of the Devils Racecourse lavas

Analytical techniques

Following removal of weathered surfaces and obvious vein material, the samples were crushed in a jaw crusher and powdered using an agate Tema mill. Two grams of powdered sample was then heated in a porcelain crucible to 900 °C for 2 h to determine the loss on ignition and 0.1 ± 0.0005 g of sample fused with 0.4 ± 0.0005 g Li–metaborate flux. This was dissolved in a *c*. 5% HNO_3 solution with a 1 ppm Rh spike and made up to 100 ml using distilled water. Major element abundances were determined using a JY Horiba Ultima 2 inductively coupled plasma optical emission spectrometer (ICP-OES). A 1 ml diquot of the solution was further diluted with 1 ml of In and Tl spike and 8 ml of 2% HNO_3 and analysed by a Thermo X7 series inductively coupled plasma mass spectrometer (ICP-MS) to obtain trace element abundances at Cardiff University, UK (Table 1). A full description of the analytical procedures and equipment at Cardiff University can be found in McDonald & Viljoen (2006).

Multiple analyses of international reference materials JB-1a, BIR-1, W2, JA-2, MRG-1 and JG-3 and four in-house standards ensured the

Table 1. (*Continued*)

	AHBI15	AHBI16	AHBI17	AHBI18	AHBI19	AHBI20	AHBI21	AHBI22	AHBI23	AHBI26	AHBI27	AHBI28	AHBI30
	Upper	Lower	Lower	Lower	Lower	Lower	Lower	Lower	Lower	Lower	Lower	Lower	Lower
	BA	BA	D	BA/A	D	D/R	D/R	D/R	D/R	BA/A	D/R	D/R	D
	CA	IAT	IAT	IAT	IAT	IAT	IAT	IAT	IAT	IAT	IAT	IAT	IAT
	5	2	2	2	2	2	2	2	3	3	3	3	1
	1.11	1.17	0.97	1.27	0.77	0.39	0.40	0.31	1.02	0.43	0.53	0.44	0.50
	0.48	0.18	0.19	0.06	0.05	2.13	0.05	0.29	0.03	0.06	0.06	0.04	0.72
	21.7	31.1	10.2	22.9	9.6	3.5	4.7	2.4	18.9	2.4	4.0	3.4	10.4
	27.1	25.1	21.4	33.7	33.3	42.4	39.7	22.8	22.1	40.4	49.5	34.6	18.5
	69.7	49.9	28.9	59.2	74.1	188.2	103.3	97.2	33.3	108.7	145.7	117.9	74.2
	1.78	1.25	0.96	1.56	1.56	4.57	1.87	1.34	0.97	2.09	2.71	2.08	1.71
	143	191	71	99	58	665	29	211	8	264	43	19	700
	13.25	3.08	3.70	4.24	4.17	9.39	6.54	4.35	2.80	6.73	7.31	5.84	4.55
	26.01	7.86	8.24	10.53	10.33	22.12	15.53	9.25	6.60	16.02	17.48	15.14	10.07
	4.01	1.46	1.10	1.92	1.86	3.58	2.62	1.49	1.17	2.82	3.11	2.63	1.52
	16.98	7.56	5.78	9.58	9.06	16.03	12.64	6.70	6.10	13.78	14.47	12.77	6.59
	4.17	2.60	2.00	3.16	2.99	4.73	3.92	2.09	2.14	4.17	4.62	3.98	1.86
	1.39	1.00	0.80	1.13	0.92	1.01	1.00	0.55	0.91	1.11	1.29	0.86	0.67
	4.20	3.24	2.71	3.95	3.66	5.09	4.55	2.45	2.75	4.88	5.65	4.56	2.08
	0.67	0.59	0.47	0.72	0.67	0.94	0.84	0.48	0.49	0.90	1.09	0.82	0.39
	4.17	3.89	3.16	4.73	4.52	6.35	5.58	3.24	3.29	5.96	7.39	5.47	2.65
	0.84	0.78	0.64	0.95	0.96	1.31	1.18	0.70	0.68	1.23	1.53	1.11	0.55
	2.41	2.29	1.87	2.79	2.91	4.03	3.59	2.17	1.99	3.67	4.64	3.37	1.73
	0.36	0.37	0.30	0.45	0.46	0.67	0.59	0.36	0.31	0.57	0.77	0.55	0.30
	2.38	2.38	1.90	2.81	3.03	4.64	4.01	2.51	2.04	3.69	5.10	3.63	2.02
	0.37	0.35	0.32	0.43	0.47	0.73	0.61	0.41	0.32	0.53	0.74	0.59	0.33
	1.84	1.54	0.91	1.68	2.26	5.42	2.90	2.68	1.07	3.28	4.14	3.43	2.04
	1.44	0.17	0.09	0.22	0.32	1.50	0.59	0.74	0.11	0.64	0.79	0.66	0.76
	0.75	0.13	0.10	0.15	0.24	0.70	0.50	0.37	0.12	0.51	0.78	0.48	0.46

accuracy and precision of the analyses. Most elements did not deviate more than 5% from standard values and have first-order relative standard deviations below 5%.

Sr, Nd and Hf isotope compositions were analysed at the NERC Isotope Geoscience Laboratories, Nottingham, UK. For Hf isotope analysis, samples were fused with Li–metaborate flux, and dissolved in 3 M HCl. Hf was separated using a single-column procedure using LN-Spec resin, following Münker *et al.* (2001), and run on a Nu–Plasma multicollector ICP-MS. Hf blanks are <100 pg. Correction for Lu and Yb was carried out using reverse-mass-bias correction of empirically predetermined ^{176}Yb/^{173}Yb (0.7950) and ^{176}Lu/^{175}Lu (0.02653). Replicate analyses of the JMC475 standard across the period of analysis gave ^{176}Hf/^{177}Hf = 0.282174 ± 0.000010 (2σ, $n = 37$); reported data are normalized to a preferred value of 0.282160 (Nowell & Parrish 2001). Replicate analyses of BCR-1 gave a mean value of 0.282872 ± 0.000009 (2σ, $n = 4$), comparable to previously reported values 0.282879 ± 0.000008 (Blichert-Toft 2001).

Replicate analyses of the in-house standard PK-g-D12 gave 0.283050 ± 0.000005 (2σ, $n = 5$), comparable to previously reported values of 0.283049 ± 0.000018 (2σ, $n = 27$; Kempton *et al.* 2002) and 0.283046 ± 0.000016 (2σ, $n = 9$; Nowell *et al.* 1998).

Determinations of Sr and Nd isotopes followed the procedures of Kempton (1995) and Royse *et al.* (1998). Samples were leached in 6 M HCl prior to analysis. ^{87}Sr/^{86}Sr ratios are normalized to ^{86}Sr/^{88}Sr = 0.1194, and ^{143}Nd/^{144}Nd ratios are normalized to ^{146}Nd/^{144}Nd = 0.7219. Sr was loaded on single Re filaments using a TaO activator, and run using static multicollection on Finnigan MAT262 (NBS987 = 0.710214 ± 0.000028, 2σ, $n = 14$) and Triton (NBS987 = 0.710230 ± 0.000018, 2σ, $n = 40$) mass spectrometers. All Sr isotope data are quoted relative to a value of 0.710240 for the NBS987 standard. Nd was run as the metal species using double Re–Ta filaments on a Finnigan Triton mass spectrometer. Replicate analysis of the in-house J&M standard gave a value of 0.511184 ± 0.000022, 2σ, $n = 24$); Nd isotope data are reported relative to a value of 0.511123 for this standard, equivalent to a value of 0.511864 for La Jolla.

Pb isotopes were analysed on a VG Axiom MC-ICP-MS, with mass fractionation corrected within-run using a Tl-doping method (Thirlwall 2002), using a $^{203}Tl/^{205}Tl$ value of 0.41876, which was determined empirically by cross calibration with NBS 981. On the basis of repeated runs of NBS981, the reproducibility of whole-rock Pb isotope measurements is better than 0.02% (2σ). Blank contribution was <100 pg. Measured values for NBS981 were within error of the preferred values reported by Thirlwall (2002), and so no further correction was made.

Behaviour of elements in the subduction environment

Island arc rocks represent complex melting of a peridotite mantle wedge above a subducting plate, which has been contaminated with slab-related aqueous and/or melt components (e.g. McCulloch & Gamble 1991; Pearce & Peate 1995; Tatsumi & Kogiso 1997; Elliott 2003). They generally have normal mid-ocean ridge basalt (N-MORB)-like abundances of high field strength elements (HFSE), for example, Nb, Ta, Zr, Hf and Y and heavy rare earth elements (HREE) and are relatively enriched in large ion lithophile elements (LILE), for example, K, Rb and Ba, light rare earth elements (LREE), Th and U (e.g. Ewart 1982; McCulloch & Gamble 1991; Pearce 1982; Pearce & Peate 1995; Tatsumi & Kogiso 1997; Elliott 2003).

The enrichment in LILE, LREE, Th and U, relative to the HFSE and HREE, in island arc lavas is derived from the addition of a slab-related component into the mantle wedge. The former elements are defined as 'non-conservative', which means that they are easily removed and transported by aqueous fluids and siliceous melts, thus, concentrating them in any flux from the subducting slab (e.g. Tatsumi *et al.* 1986; Pearce & Peate 1995; Keppler 1996). In contrast, 'conservative elements' include Zr, Hf, Nb, Ta, Y, Ti and the HREE and these remain in the slab due to the low solubility of minerals such as rutile, zircon, apatite, monazite and garnet in the presence of aqueous fluids, that is, they have $D_{slab/fluid}$ values >1. Therefore, unlike the non-conservative elements, they are not concentrated into a slab-related aqueous fluid (Tatsumi *et al.* 1986; Pearce & Peate 1995; Keppler 1996). However, conservative elements can behave non-conservatively in the presence of a siliceous melt. As such, higher temperature melts can have substantial concentrations of certain HFSE and HREE (Tatsumi *et al.* 1986; McCulloch & Gamble 1991; Pearce & Peate 1995).

Consequently, trace element and isotope studies have demonstrated that two different slab components can be responsible for the enriched LILE, LREE, Th and U composition of island arc magmas, these include: (a) aqueous fluid from altered oceanic crust and sediment and (b) silicate melts from the oceanic crust and sediment, which are transported from the subducting slab into the arc mantle wedge source region (e.g. Miller *et al.* 1994; Brenan *et al.* 1995; Pearce & Peate 1995; Regelous *et al.* 1997; Turner & Hawkesworth 1997; Ishikawa & Tera 1999; Class *et al.* 2000; Elliott 2003).

Element mobility and classification

A detailed study of major and trace element mobility in the Devils Racecourse lavas is given in Hastie *et al.* (2007), and will only be briefly summarized here. Figure 4 shows representative variation diagrams whereby an incompatible element is plotted against Nb on the abscissa because it is one of the most immobile trace elements (e.g. Cann 1970; Kurtz *et al.* 2000; Hill *et al.* 2000). Zr, Th, La, Sm and Yb plot as linear trends with high correlation coefficients, indicating that these elements are immobile. However, the Th, La, Sm and Yb diagrams show that the lavas from the lower and upper sections of the Devils Racecourse Formation plot on different, but still coherent trends (Fig. 4). In contrast, the data for U, Ba and K show a much greater degree of scatter with little or no evidence of the expected, pre-alteration linear trend within the lower and upper Devils Racecourse lavas. The lack of a clear trend is interpreted to be the result of sub-solidus alteration of the rocks by hydrothermal and weathering processes (Hastie *et al.* 2007).

The mobility of the elements of low ionic potential limits the use of the standard classification diagrams, such as the total alkali silica and K_2O-SiO_2 diagrams (Peccerillo & Taylor 1976; Le Bas *et al.* 1986, 1992), to study the Jamaican igneous rocks. Hastie *et al.* (2007) addressed this problem by developing the Th–Co diagram, which acts as an immobile element equivalent of the K_2O-SiO_2 diagram (Peccerillo & Taylor 1976). The Th–Co plot (Fig. 5), demonstrates that the lower Devils Racecourse lavas have a tholeiitic (IAT) composition, while the upper lavas are calc-alkaline (CA). Additionally, the lower rocks are predominantly acidic and the upper lavas are basic–intermediate in composition.

Trace element geochemistry

Variation diagrams for Th, La, Sm and Yb show that the upper and lower lavas form two distinct

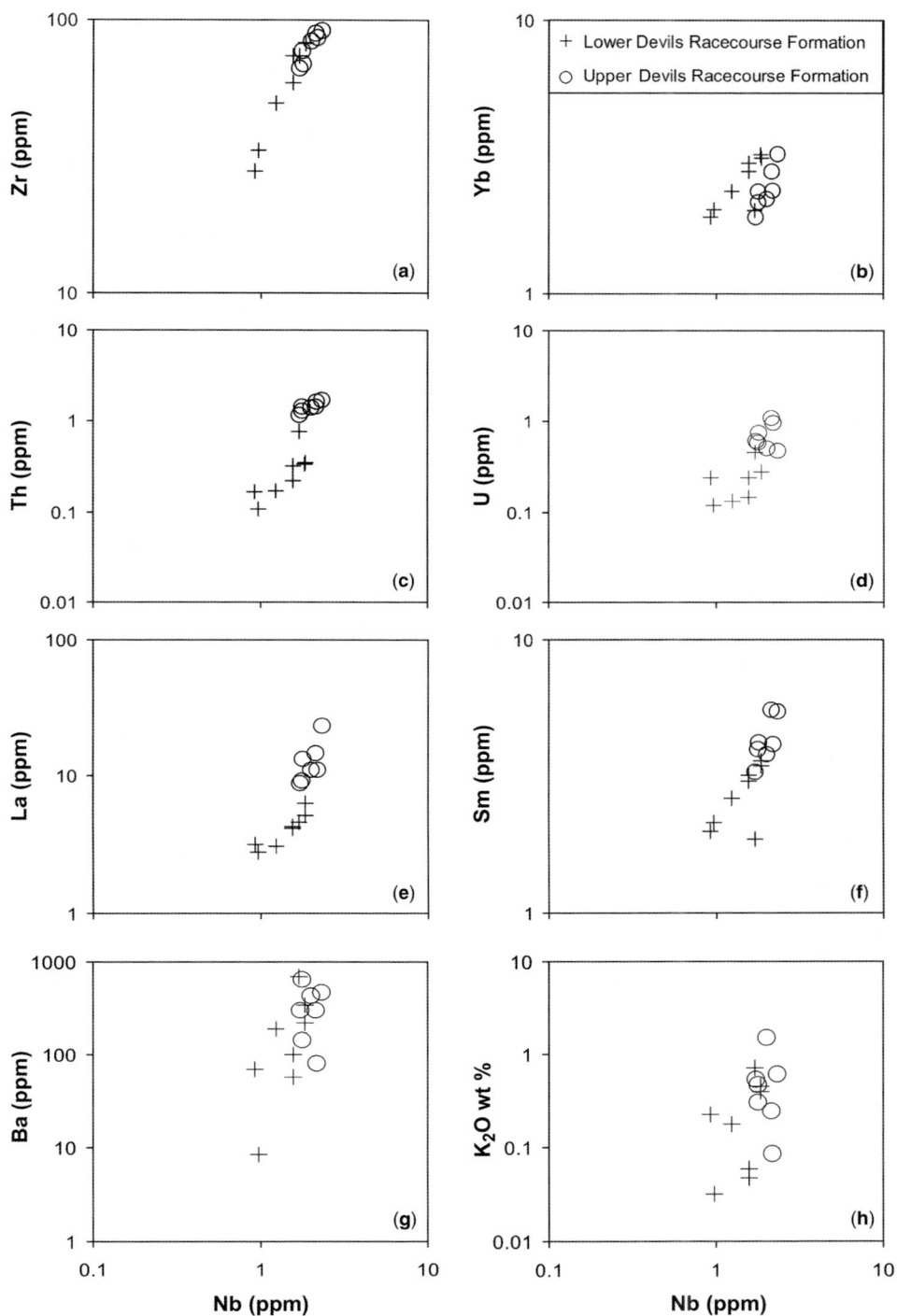

Fig. 4. Variation diagrams for a range of elements plotted against Nb. In all diagrams, variations within the lower and upper Devils Racecourse Formation at basic-intermediate compositions are mainly due to fractional crystallization, which should give near-diagonal (1:1) vectors on log–log plots.

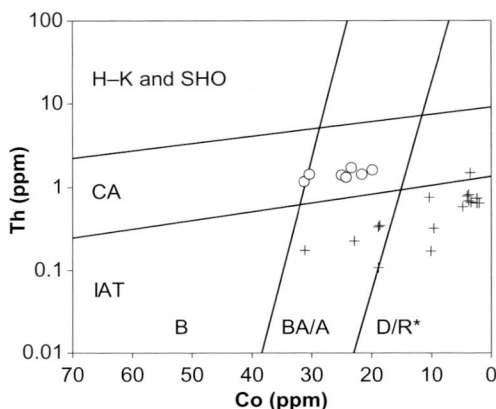

Fig. 5. Th–Co classification of the Devils Racecourse lavas. Key as in Figure 4. IAT, island arc tholeiite; CA, calc-alkaline; H–K and SHO, high-K calc-alkaline and shoshonite; B, basalt; BA/A, basaltic–andesite and andesite; D/R*, dacite and rhyolite; * indicates that latites and trachytes also fall in the D/R fields.

linear trends with the upper lavas having greater abundances of Th, La and Sm and slightly lower concentrations of Yb at a given Nb content (Figs 2 & 4). This enrichment is seen in immobile trace element ratios which reduce the effects of variable fractional crystallization and partial melting. The Devils Racecourse IAT lavas have N-MORB-normalized (*n-mn*) trace element ratios of $(La/Yb)_{n\text{-}mn} = 1.58 - 3.25$ and $(Th/Hf)_{n\text{-}mn} = 1.73 - 6.38$ whereas the upper CA lavas have $(La/Yb)_{n\text{-}mn} = 5.12 - 8.6$ and $(Th/Hf)_{n\text{-}mn} = 10.45 - 13.31$ (Fig. 6a).

The greater LILE and LREE enrichment, in relation to the N-MORB-like HFSE and HREE, in the upper CA lavas, relative to the lower IAT rocks, is seen on an N-MORB-normalized multi-element diagram (Fig. 6b), where both the lower and upper Devils Racecourse lavas have negative Nb and Ta anomalies, which is a typical island arc signature (e.g. McCulloch & Gamble 1991; Pearce & Peate 1995; Turner & Hawkesworth 1997; Elliott 2003). The N-MORB and chondrite-normalized multi-element and rare earth element (REE) plots additionally reveal that some of the upper and lower lavas have negative Ti and Eu anomalies because of the fractional crystallization of Fe–Ti oxides and plagioclase feldspar respectively. Interestingly, most of the lower IAT lavas (e.g. AHBI22; Fig. 6c) and the upper CA lava AHBI15 have small negative Ce anomalies, which are usually associated with the addition of a pelagic sedimentary component into the source region of an island arc rock suite (e.g. Hole *et al.* 1984; Ben Othman *et al.* 1989; Plank & Langmuir 1998; Elliott 2003).

Sr–Pb–Nd–Hf isotope geochemistry

Radiogenic isotope data for samples AHBI01, 03, 13 and 27 are shown in Figure 7 and Table 2. The lavas have been age-corrected to 120 Ma because the Devils Racecourse inter-bedded limestones contain Hauterivian (136.4–130 Ma) to Aptian (125–112 Ma) age rudist bivalves and Nerineid gastropods.

Initial $^{87}Sr/^{86}Sr$ ratios of the Devils Racecourse lavas extend to high radiogenic values (up to 0.70445) at near constant $\varepsilon_{Nd}(i)$ values (Fig. 7a). Thompson *et al.* (2004) also reported that the Cretaceous island arc rocks of the Bonaire Washikemba Formation had similar high $(^{87}Sr/^{86}Sr)i$ ratios (Fig. 7a). The Jamaican samples were acid-leached during preparation and thus the $(^{87}Sr/^{86}Sr)i$ values should be more representative of the primary igneous composition. However, like the Bonaire lavas, the arc lavas have been extensively altered and so it is unclear whether the leaching processes removed all the secondary alteration products. Thompson *et al.* (2004) measured the Sr isotopic compositions of the whole rock and of separated primary igneous apatite grains. They demonstrated that Sr isotope ratios for these apatites were lower than the whole rock values and concluded that these latter values were high because of Sr mobility during sub-solidus seawater alteration processes. A similar study on Caribbean oceanic plateau lavas and separated clinopyroxenes, reached similar conclusions (Révillon *et al.* 2002). Consequently, although not proven, there is a strong possibility that the elevated $(^{87}Sr/^{86}Sr)i$ ratios in the Devils Racecourse lavas are a result of Sr mobility during hydrothermal and chemical weathering processes (Fig. 7a) (e.g. Hastie *et al.* 2007, 2008).

The Pb isotope values for the lavas are: $(^{206}Pb/^{204}Pb)i = 16.51 - 18.75$, $(^{207}Pb/^{204}Pb)i = 15.45 - 15.56$ and $(^{208}Pb/^{204}Pb)i = 37.19 - 38.19$ (Table 2). Pb isotopes are conventionally used to determine the slab contribution in island arc lavas (e.g. Ewart *et al.* 1998; Class *et al.* 2000). However, as with the Sr isotopes, Pb isotopes can not be utilized in determining the petrogenesis of the Jamaican lavas because of the mobility of U and Pb in these, and many other, Caribbean igneous rocks (e.g. Geldmacher *et al.* 2003; Hastie *et al.* 2007, 2008).

In addition to the Sr and Pb isotope systematics the Jamaican lavas possess $\varepsilon_{Nd}(i) = +8.12$ to $+9.03$ and $\varepsilon_{Hf}(i) = +13.22$ to $+14.97$ (Fig. 7b and Table 2). Nd and Hf are considered to be relatively immobile during secondary alteration processes and consequently $\varepsilon_{Nd}(i)$ and $\varepsilon_{Hf}(i)$ values should represent the primary composition of the arc lavas (Fig. 7b) (e.g. White & Patchett 1984). Nd is moderately to highly mobile in a slab-related

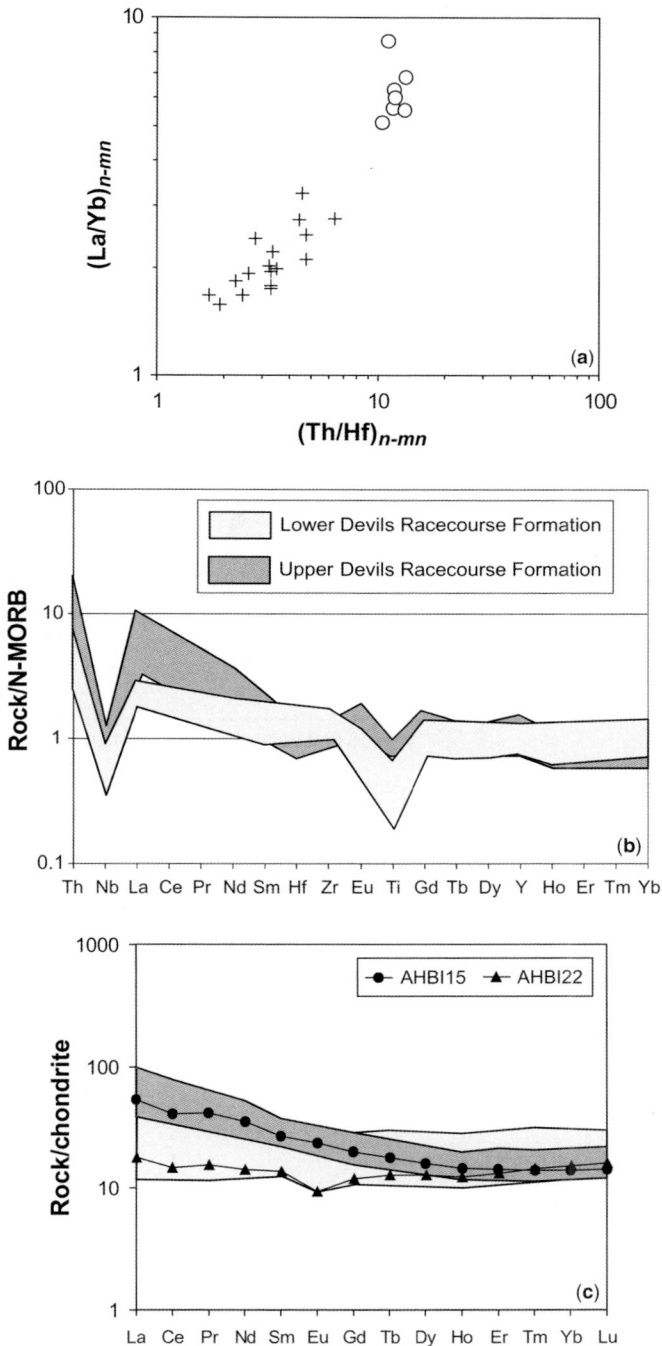

Fig. 6. (**a**) $(La/Yb)_{n-mn}–(Th/Hf)_{n-mn}$ diagram with symbols as in Figure 4. (**b** and **c**) N-MORB-normalized multi-element and chondrite-normalized REE patterns (given as ranges) highlighting the arc-like character (large negative Nb anomalies) of the lavas of the lower and upper Devils Racecourse Formation and the LILE and LREE enrichment in the CA lavas relative to the IAT rocks. Samples AHBI15 and 22 are shown to illustrate the Ce anomalies in the basic to acidic Devils Racecourse lavas. Normalizing values in (b) and (c) are from Sun & McDonough (1989) and McDonough & Sun (1995), respectively.

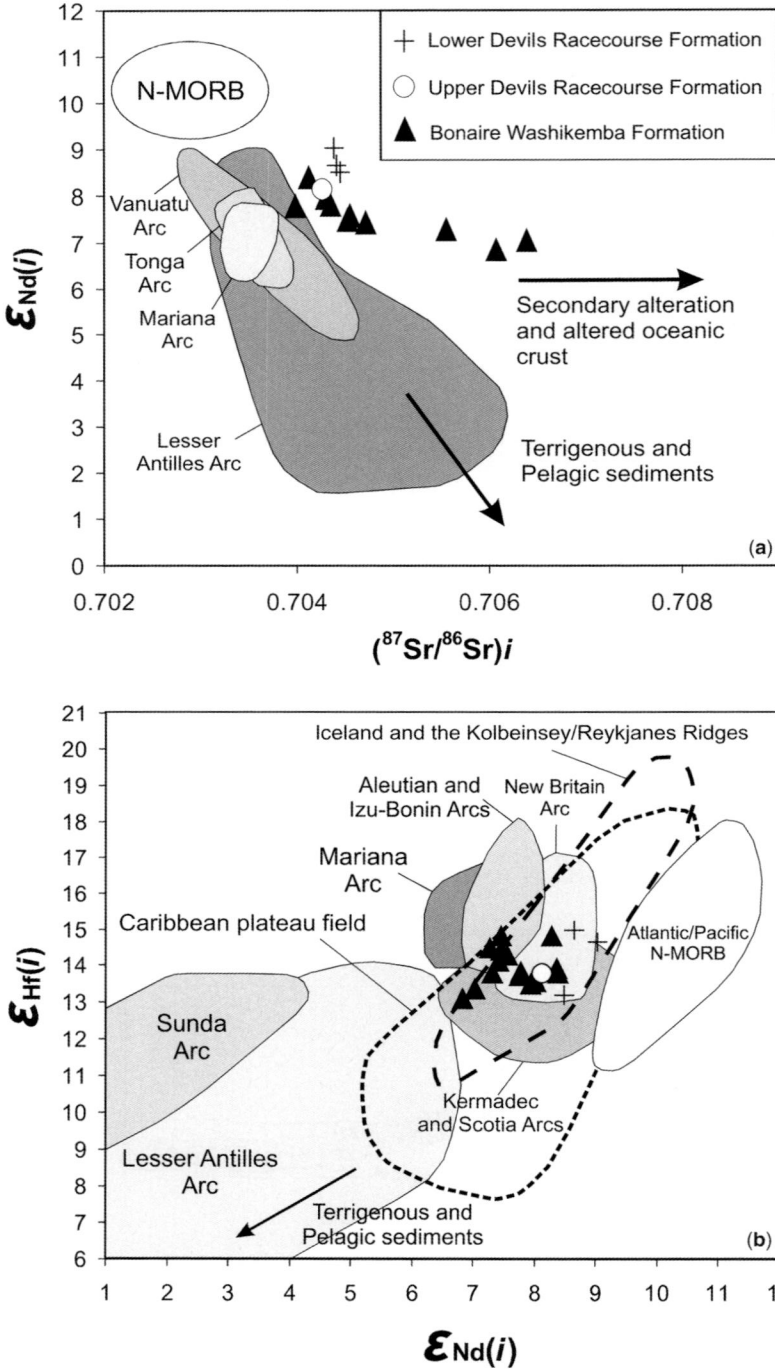

Fig. 7. (a) $(^{87}Sr/^{86}Sr)i-\varepsilon_{Nd}(i)$ and (b) $\varepsilon_{Nd}(i)-\varepsilon_{Hf}(i)$ diagrams for the Devils Racecourse lavas. Data sources used to form the N-MORB and island arc fields can be found in the Supplementary Material. The Caribbean Plateau field was constructed with data from Colombia, Costa Rica, Curaçao, DSDP Leg 15, Galapagos, Gorgona and Jamaica. The references for the plateau data are in the Supplementary Material.

Table 2. *Sr–Pb–Nd–Hf isotope data for Devils Racecourse rocks*

	$^{176}Lu/$ ^{177}Hf	$^{176}Hf/$ ^{177}Hf measured	$^{176}Hf/$ ^{177}Hf initial	$\varepsilon Hf(i)$	$^{147}Sm/$ ^{144}Nd	$^{143}Nd/$ ^{144}Nd measured	$^{143}Nd/$ ^{144}Nd initial	$\varepsilon Nd(i)$	$^{87}Sr/$ ^{86}Sr measured	$^{87}Sr/$ ^{86}Sr initial	$^{206}Pb/$ ^{204}Pb measured	$^{207}Pb/$ ^{204}Pb measured	$^{208}Pb/$ ^{204}Pb measured	$^{206}Pb/$ ^{204}Pb initial	$^{207}Pb/$ ^{204}Pb initial	$^{208}Pb/$ ^{204}Pb initial
AHBI01	0.0324	0.28318	0.28311	14.65	0.207	0.51311	0.51295	9.03	0.70455	0.70439	18.703	15.558	38.158	18.467	15.547	38.061
AHBI03	0.0323	0.28319	0.28312	14.97	0.209	0.51309	0.51293	8.64	0.70452	0.70441	18.704	15.559	38.177	18.528	15.550	38.108
AHBI13	0.0308	0.28316	0.28309	13.80	0.143	0.51301	0.51290	8.12	0.70428	0.70426	18.977	15.573	38.240	16.515	15.454	37.188
AHBI27	0.0254	0.28313	0.28307	13.22	0.193	0.51307	0.51292	8.50	0.70456	0.70445	18.877	15.566	38.227	18.752	15.560	38.186

The chondrite values used to calculate the εHf and Nd values are 0.28273 and 0.51256 respectively (see text for analytical procedures).

fluid and melt respectively (non-conservative) and, as such, Nd abundances are generally enriched in island arc lavas relative to N-MORB (Fig. 6b) (Pearce & Peate 1995). Additionally, the $\varepsilon_{Nd}(i)$ values of the lower and upper Devils Racecourse lavas plot to the left (below) the Atlantic–Pacific MORB field (Fig. 7b). This occurs because the enriched slab component, derived from altered oceanic crust and/or subducted sediments, contaminates the arc magma source region (an overlying mantle wedge) with lower $\varepsilon_{Nd}(i)$ values, thus explaining the position of the IAT Devils Racecourse lavas in Figure 7b (Pearce *et al.* 1999; Woodhead *et al.* 2001).

The CA lava and the Bonaire samples have lower $\varepsilon_{Nd}(i)$ values than the IAT lavas, suggesting that the Jamaican CA samples were derived from a source region with a greater mass fraction of, or a more enriched, slab component. In contrast, the IAT and CA lavas have $\varepsilon_{Hf}(i)$ ratios (+13.22 to +14.97), which are similar to Atlantic/Pacific N-MORB values. This is unsurprising considering that Hf is regarded as a conservative element (i.e. one which does not contribute to the slab component in the subduction environment, cf. Pearce & Peate 1995).

Source regions for the Devils Racecourse Formation

The mantle wedge component

The mantle wedge component can only be studied using 'conservative' immobile trace elements that have $D_{slab/fluid}$ values >1 (e.g. Tatsumi *et al.* 1986; McCulloch & Gamble 1991; Keppler 1996). These include Nb, Ta, Hf, Zr, Y and the HREEs (e.g. McCulloch & Gamble 1991; Pearce & Peate 1995). However, if the subduction component in the Devils Racecourse lavas represents a siliceous melt Nb, Ta, Hf, Zr, Y and the HREEs would be 'non-conservative' and could not be used to study the geochemistry of the mantle wedge component.

A siliceous melt from the subducting oceanic crust would have an adakitic composition [e.g. $SiO_2 > 56\%$, $Al_2O_3 > 15\%$, MgO $< 6\%$, low Y and HREE (Y and Yb < 18 and 1.9 ppm respectively) and high Sr (rarely < 400 ppm)], or would eventually form a high-Mg andesite (e.g. high MgO, Cr and Ni concentrations, Sr up to 3000 ppm, Ba > 1000 ppm and low FeO^*/MgO and high Na/K ratios) depending on the extent to which the melt hybridized with the peridotite in the mantle wedge (e.g. Saunders *et al.* 1987; Drummond & Defant 1990; Defant *et al.* 1992; Yogodzinski *et al.* 1994, 1995; Drummond *et al.* 1996; Calmus *et al.* 2003). In addition, modern Nd–Hf isotope studies (e.g. Pearce *et al.* 1999;

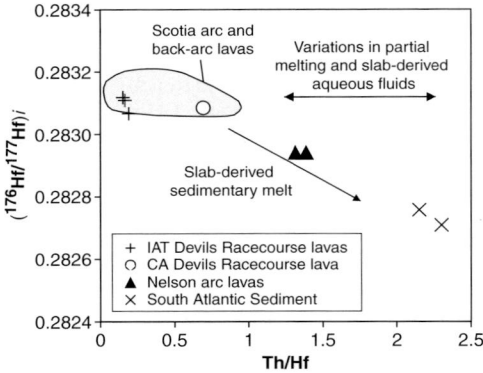

Fig. 8. (^{176}Hf/^{177}Hf)*i* v. Th/Hf diagram modified from Barry *et al.* (2006).

Woodhead *et al.* 2001; Barry *et al.* 2006) on present-day island arc lavas have been used to study the potential mobility of Hf in the subduction environment.

The Devils Racecourse lavas lack the characteristic compositional features of adakites and of high Mg-andesites. Consequently, it is reasonable to assume that the slab component of the Devils Racecourse lavas does not represent a siliceous melt from altered oceanic crust. Furthermore, Barry *et al.* (2006) demonstrate that ^{176}Hf/^{177}Hf and Th/Hf ratios can be used to determine if a subduction component represents a fluid or a melt. The Devils Racecourse lavas in Figure 8 show variable Th/Hf ratios with little variation in (^{176}Hf/^{177}Hf)*i* ratios, similar to the majority of the Scotia arc and back-arc samples. This suggests that the subduction component in the Devils Racecourse lavas is composed of Th, but not Hf, implying that variable Th/Hf ratios at constant (^{176}Hf/^{177}Hf)*i* values are caused by selective addition of Th by aqueous fluids and/or variable partial melting in the arc source region (Fig. 8). In contrast, the Nelson island arc samples (the southernmost volcano in the Scotia Arc) have higher Th/Hf and lower (^{176}Hf/^{177}Hf)*i* ratios, trending towards the pelagic sediments of Barry *et al.* (2006). This scenario requires the addition of Hf and Th via a subduction component which represents a sedimentary melt from the subducting slab (Fig. 8).

Therefore, the slab flux responsible for contaminating the Devils Racecourse source region is likely to have been an aqueous fluid that would not have mobilized the HFSEs and HREEs. Accordingly, the basic and intermediate Devils Racecourse lavas can be plotted on the Nb/Y v. Zr/Y diagram of Fitton *et al.* (1997) to study the composition of the mantle wedge component (Fig. 9a). Fitton

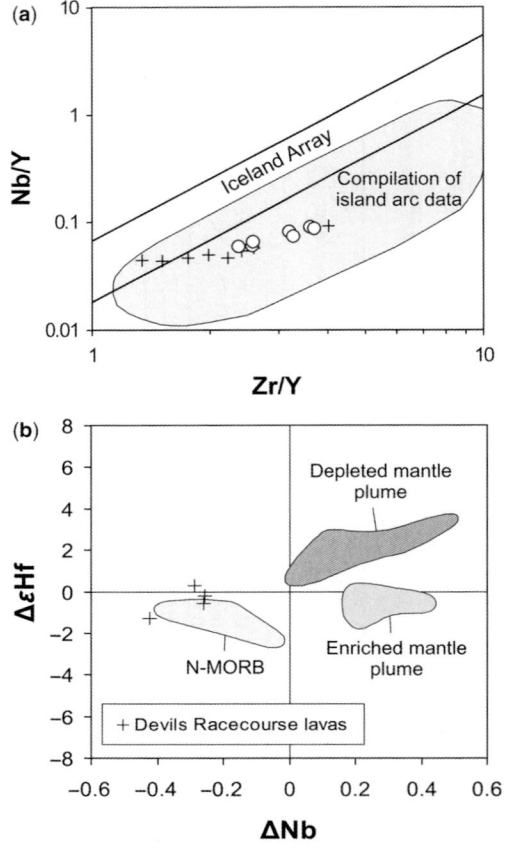

Fig. 9. (**a**) Nb/Y v. Zr/Y diagram of Fitton *et al.* (1997), (**b**) ΔεHf–ΔNb diagram modified from Kempton *et al.* (2000). Compilation of arc data in (a) were taken from Hastie *et al.* (2007).

et al. (1997) showed that these ratios are little affected by low-pressure fractional crystallization and partial melting. The Nb/Y v. Zr/Y diagram was originally constructed to distinguish between plume-related lavas from Iceland which plot between two parallel tramlines and N-MORB which plots below the lower tramline (Fitton *et al.* 1997). Fitton *et al.* (1997) characterized samples that plot in the Icelandic field as having positive ΔNb values whereas samples which lie below the lower tramline have negative ΔNb values. The Nb/Y and Zr/Y ratios should indicate the nature of the mantle wedge component that was involved in the petrogenesis of the Devils Racecourse lavas.

In Figure 9a, most of the IAT and CA Jamaican samples plot below the lower tramline with low to moderate Nb/Y and Zr/Y ratios. This implies that the mantle component involved in the petrogenesis of both the upper and lower basic/intermediate lavas has predominantly an N-MORB-type composition. Nevertheless, two of the IAT samples (AHBI17 and 23) have positive ΔNb values, indicating that they plot above the lower-most tramline. Therefore these samples may have been derived from a mantle source region composed of mantle plume-like material. If so, this would suggest that there was plume material in the mantle wedge of the Great Arc of the Caribbean during the Hauterivian–Aptian. However, Kempton *et al.* (2000) plotted more N-MORB data onto the original Nb/Y v. Zr/Y diagram of Fitton *et al.* (1997) and found that there is more overlap between N-MORB and plume material on the lower tramline than previously thought. Additionally, although isotopic data is unavailable for AHBI17 and 23, the Devils Racecourse lavas are plotted on the ΔεHf–ΔNb diagram of Kempton *et al.* (2000) (Fig. 9b). This diagram uses ΔNb and ΔεHf, which represents the vertical derivation of a given sample from the Hf-Nd mantle regression line of Vervoort *et al.* (1999), to determine an N-MORB or plume affinity for a given sample. Figure 9b shows that the Devils Racecourse lavas have N-MORB-like ΔεHf and ΔNb values.

The Nb/Y–Zr/Y diagram demonstrates that the upper lavas have the highest Nb/Y and Zr/Y ratios. They also have the highest Nb and Zr concentrations and the lowest Yb abundance at a given Nb concentration (Figs 4 and 9a). This is not related to a larger proportion of garnet in the CA mantle source region because the multi-element and REE patterns in Figure 6b, c demonstrate that the HREE patterns are relatively 'flat'. The basic and intermediate samples are also plotted on a Dy–Lu variation diagram which indicates that they lie on the spinel peridotite partial melting and fractional crystalliza-tion vector (Fig. 10a).

In Figure 10b, a Nb/Zr v. Ta/Hf plot, also demonstrates that the conservative elements in the Devils Racecourse lavas have an N-MORB compo-sition. Moreover the similar Nb/Zr and Ta/Hf ratios of the lower and upper lavas suggests that differing degrees of partial melting are not responsible for the high Nb/Y and Zr/Y ratios and low Yb concentrations. Consequently, the high Nb/Y and Zr/Y ratios, together with the higher Nb and Zr con-tents and lower abundance of Yb, is a result of the upper lavas being derived from an N-MORB source region with (1) slightly higher Nb and Zr and lower HREE and Y concentrations than the IAT mantle wedge source region and/or (2) an N-MORB source region with a greater modal pro-portion of clinopyroxene (e.g. Jolly *et al.* 1998*a*, *b*). Furthermore, Figure 10b indicates that the Jamaican IAT and CA lavas have N-MORB-like compositions, more depleted in Nb and Ta than

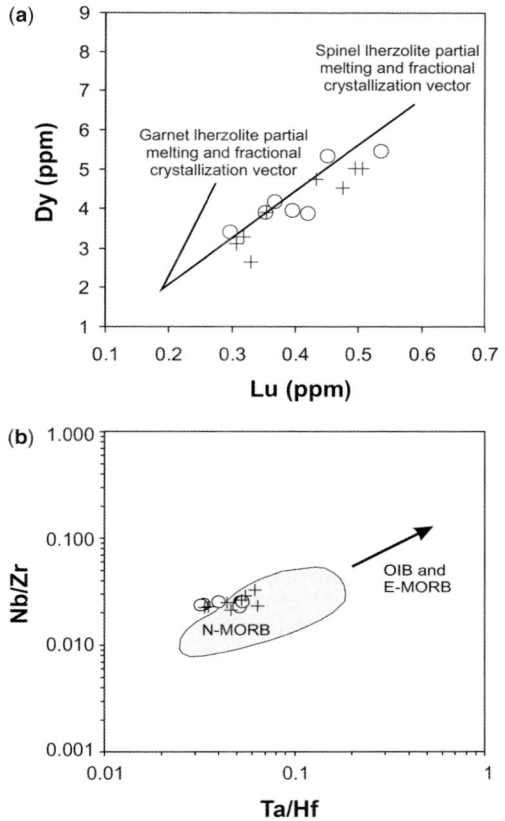

Fig. 10. (**a**) Dy–Lu variation diagram with peridotite vectors modified from Marchesi *et al.* (2007) and (**b**) Nb/Zr v. Ta/Hf diagram. N-MORB data in (b) were taken from references in the Supplementary Material. Symbols as in Figure 4.

E-MORB and ocean island basalt (OIB) material (Fig. 10b).

The slab-related component

Unfortunately, some of the most useful trace elements and isotopes used to identify differing slab components are mobile in the Devils Race-course lavas and cannot be used (e.g. Pb, Sr and Ba) (Hastie *et al.* 2007, 2008). Nevertheless, a study of the immobile, incompatible and 'non-conservative' Th and LREE elements in island arc lavas can be used to distinguish different sedi-mentary signatures within a slab-related component (e.g. Hole *et al.* 1984; Plank 2005).

It was shown in the previous section that the lack of an adakitic or high-Mg andesite composi-tionally signature in the Devils Racecourse lavas implies that the slab component is composed of an aqueous fluid derived from altered oceanic crust and/or sediment. Moreover, the Devils Racecourse lavas plot at the high $\varepsilon_{Nd}(i)$ end of the Bonaire samples and the Mariana, Izu-Bonin, Aleutian and New Britain arcs. These are termed the 'fluid dominated arcs' by Woodhead *et al.* (2001) because, on component discrimination diagrams, their enrichment is considered to be due to the metasomatism of their respective mantle wedges by altered oceanic crust and/or sediment derived aqueous fluids. The Devils Racecourse data do not plot near the Sunda or Lesser Antilles data, which both have enriched (low) $\varepsilon_{Nd}(i)$ and $\varepsilon_{Hf}(i)$ values. It has been interpreted that the Lesser Antilles and Sunda arcs have been enriched by either a sedimentary melt from the descending slab or by assimilation of crustal material as the arc magmas ascend through the island arc crust (e.g. White & Dupre 1986; Davidson 1987; Thirlwall *et al.* 1996).

If in Figure 11a, b a baseline is drawn connecting the 'conservative' elements a qualitative measure of the subduction component can be determined. The abundance of elements up to the baseline represents the elemental contribution from the partial melting of the mantle wedge. The 'non-conservative' elements with abundances which lie above the baseline represent the subduction related component that has been added to the mantle wedge both from slab-derived fluids and from selectively leached elements from the mantle as the fluids propagate through (Table 3) (Pearce 1983; Hawkesworth *et al.* 1991; Pearce & Peate 1995).

Th, La, Ce, Pr and Nd are enriched above the baseline in N-MORB-normalized multi-element plots for the averaged IAT and CA Devils Racecourse lavas (Fig. 11a, b and Table 3). Sm is enriched in the CA arc lavas but appears to be conservative in the IAT lavas; however, this is unsurprising given that Sm is considered to be only slightly non-conservative (e.g. Tatsumi *et al.* 1986; Pearce & Peate 1995). A greater percentage of the Th, LREE and middle rare earth element (MREE) elemental budget in the CA lavas was derived from the subduction-related component compared with the IAT lavas (Fig. 11a, b and Table 3). The CA lavas also have more enriched Th and LREE abundances compared to the N-MORB-like HFSE (including Nb and Ta) and HREE concentrations of both the IAT and CA samples [average $(Th/Nb)_{n-mn}$ is 6.4 and 14.0 for IAT and CA respectively]. This enrichment, together with the more enriched $\varepsilon Nd(i)$ values in the CA lava cannot be generated by variable partial melting or fractional crystallization (Figs 7 & 11a, b). Consequently, the CA lavas are derived from a more enriched source region than the IAT lavas. Therefore, compared with the HFSE and

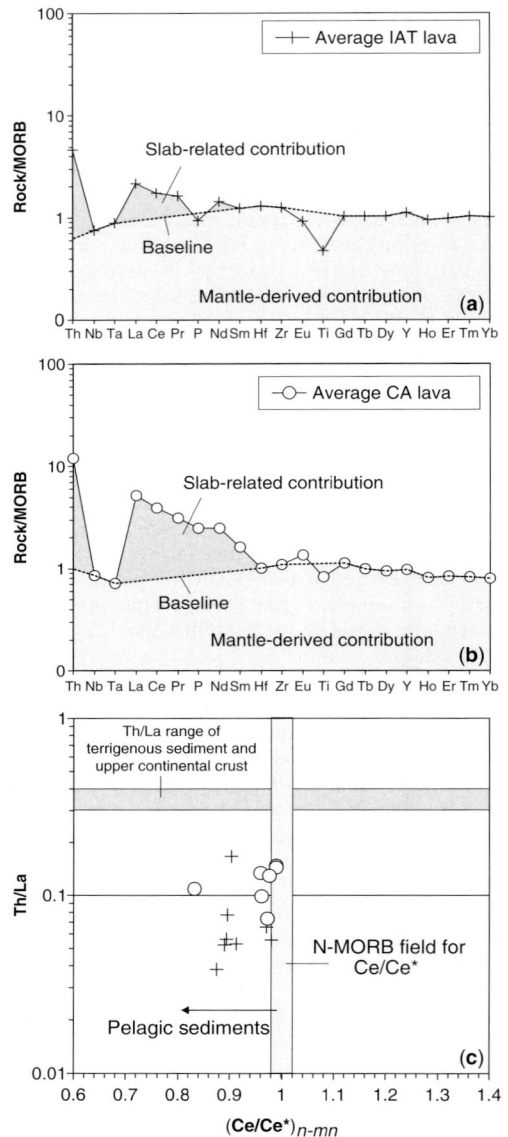

Fig. 11. (**a, b**) N-MORB-normalized multi-element plots of averaged IAT and CA Devils Racecourse lavas and (**c**) Th/La v. $(Ce/Ce^*)_{n-mn}$ diagram. Normalizing values in (a) and (b) are from Sun & McDonough (1989). $(Ce/Ce^*)_{n-mn}$ ratio in (b) is calculated by $Ce/(((La-Pr)/2) + Pr)$. Symbols as in Figure 4.

the HREE, the enrichment in Th, the LREE, the MREE, and $\varepsilon Nd(i)$ values, in the CA lavas relative to the IAT rocks is because of either (i) the input of a more-enriched slab-related fluid into the arc source region or (ii) the input of a larger volume

Table 3. *The estimated percentage contribution of Th, LREE and Sm from the slab-related component for averaged IAT and CA Devils Racecourse lavas (Fig. 11a, b)*

	Th	La	Ce	Pr	Nd	Sm
IAT, Devils Racecourse Formation	85	58	44	38	26	—
CA, Devils Racecourse Formation	94	84	78	71	63	39

of the same slab fluid into the source region of the upper Devils Racecourse lavas.

The presence of a negative Ce anomaly [(Ce/Ce*)$_{n-mn}$ ratios < 1] in most of the lower Devils Racecourse lavas and in AHBI15 suggests that the slab-related component contained a significant proportion of pelagic sedimentary material (Fig. 11c). Negative Ce anomalies are associated with phosphate-rich phases in pelagic sediments and red clay sediments in oxidizing environments where Ce^{3+} is oxidized to Ce^{4+}. The Ce^{4+} in the sediments is more insoluble than La^{3+} and thus the slab-derived component has a negative Ce anomaly which subsequently imparts a negative Ce anomaly to the arc lavas (e.g. Hole *et al.* 1984; Ben Othman *et al.* 1989; McCulloch & Gamble 1991; Plank & Langmuir 1998; Elliott 2003). Thus, the negative Ce anomalies [(Ce/Ce*)$_{n-mn}$ <1] in the basic and intermediate lower Devils Racecourse lavas suggest that at least part of their slab-related component is derived from pelagic sediments (Fig. 11c). In contrast, all of the basic upper CA lavas, apart from AHBI15, have (Ce/Ce*)$_{n-mn}$ ratios of *c.* 1, which indicates that the amount of pelagic sediment in their genesis was negligible (Fig. 11c).

Recent work by Plank (2005) has demonstrated that the Th/La ratios of recent island arc rocks are similar to the Th/La ratio of their locally subducting sediment. It has therefore been suggested that the variations in Th/La in arc lavas are dependent on variations in the composition of the locally subducting sediment beneath the island arc in question, and less so on the composition of the subducting altered slab and the mantle wedge. Terrigenous sediments have Th/La ratios of *c.* 0.3–0.4 (similar to upper continental crust) and metalliferous sediments have low Th/La because of high REE concentrations in fish debris phosphate and hydrogenous Fe–Mn oxides. Volcaniclastic material, which is derived from the island arc itself can also have low Th/La ratios. Figure 11c illustrates that the basic and intermediate lower

IAT lavas have lower Th/La ratios (≪ 0.1) than the basic upper CA lavas that have higher Th/La ratios of > 0.1.

Consequently, the lower and upper Devils Racecourse lavas are derived from N-MORB mantle wedge source regions which have been contaminated with two chemically distinct aqueous subduction components. The source region for the lower IAT lavas was contaminated by a slab-related component derived, in part, from pelagic sediments, whereas the source region for the upper CA lavas was contaminated by a slab-related component derived, in part, from terrigenous sediments. The more enriched terrigenous component imparted larger Th, LREE and MREE abundances relative to N-MORB-like HFSE and HREE contents and lower $\varepsilon_{Nd}(i)$ ratios into the upper CA island arc magmas relative to the IAT magmas.

Discussion

The Aptian–Albian subduction polarity reversal

The identification of post-Aptian IAT rocks by Kerr *et al.* (2003), Jolly *et al.* (2006), Escuder Viruete *et al.* (2007) and Marchesi *et al.* (2007) together with the pre-Aptian CA rocks noted by Kerr *et al.* (2003) and the upper Devils Racecourse Formation demonstrate that an IAT to CA switch in island arc chemistry should no longer be used as an argument for a subduction polarity reversal in the Aptian–Albian. On the contrary, the data suggest that the change in chemistry of the island arc magmas of the 'Great Arc of the Caribbean' was a result of chemically different material (likely to be pelagic v. terrigenous sediment) being subducted and incorporated into the island arc magmas. There is *no* requirement for an Aptian–Albian subduction reversal to explain the *gradual* evolution of the Great Arc from IAT to CA compositions.

As such, the Aptian–Albian unconformity on Hispaniola and parts of Puerto Rico may have been caused by a localized tectonic event and not an arc-wide subduction polarity reversal. This is supported by the review in Rankin (2002) and Jolly *et al.* (2006), who propose that the lack of unconformities in Puerto Rico and the US Virgin Islands suggests that there was no significant tectonic activity in the northeastern section of the Great Arc from the Albian to the Late Cretaceous. It is possible that the Aptian–Albian unconformity could be formed by (1) localized obduction of Jurassic MORB on to the Great Arc to form the Duarte and Bermeja complexes in Hispaniola and Puerto Rico, respectively, and/or (2) the collision of a small Aptian oceanic plateau/seamount chain with the Great Arc (Aptian plateau fragments are found in Cuba; Kerr *et al.* 1999).

Devils Racecourse Formation: related arc rocks on other islands in the Greater Antilles

During the late Early Cretaceous Jamaica was part of the northern section of the 'Great Arc of the Caribbean' and was located north and west of the section of the arc which would eventually form part of present-day Cuba (Pindell *et al.* 2006; Hastie 2007; Hastie *et al.* 2008). The lower IAT Devils Racecourse lavas were erupted over a northeast-dipping subduction zone in the Hauterivian to lower Barremian (*c.* 136.4–122 Ma); in contrast, the upper CA lavas were erupted in the upper Barremian to upper Aptian (*c.* 122–112 Ma) above the same subduction zone. Thus, the IAT to CA transition in Jamaica occurred in the Mid-Barremian (*c.* 122 Ma), and was caused by the subduction of sedimentary material with a greater terrigenous component.

This compositional change in the island arc lavas on Jamaica was not a 'local' phenomenon. IAT and CA island arc rocks of a similar Barremian–Albian age were erupted on other Greater Antilles islands and have IAT lavas postdating the Albian: for example, Aptian to Turonian IAT, CA and boninitic rocks in Cuba (Kerr *et al.* 1999, 2003; Marchesi *et al.* 2007); Barremian–Albian IAT, CA and boninites of the Maimaon Formation, Amina Schists, Los Ranchos Formation, Tireo Formation and the Loma La Vega and Las Guajabas volcanics of Hispaniola (Donnelly *et al.* 1990; Lebron & Perfit 1994; Kerr *et al.* 2003; Kesler *et al.* 2005; Escuder Viruete *et al.* 2006, 2007); upper Aptian to Santonian volcanic phase I (IAT), volcanic phase II (CA) and volcanic phase III (CA, high-K CA and shoshonites) in Puerto Rico (Frost *et al.* 1998; Schellekens 1998; Jolly *et al.* 1998*a*, *b*, 2001, 2006); upper Aptian to Turonian IAT Water Island and Louisenhoj Formations in the Virgin Islands (Donnelly *et al.* 1990; Rankin 2002; Jolly & Lidiak 2006) and upper Albian IAT rocks in Tobago (Kerr *et al.* 2003). In the remaining mid-Late Cretaceous island arc volcanism in the Great Arc progressed, and continually evolved from IAT and CA to high-K calc-alkaline and shoshonitic compositions (e.g. Jolly *et al.* 1998a, b, 2006).

Future work

This paper has principally focused on the basic-intermediate volcanic rocks in the Devils Racecourse Formation to avoid complications with fractional crystallization processes. However, previous studies have focused on the silicic rocks of the Early Cretaceous IAT-series in Puerto Rico and the US Virgin Islands (e.g. Jolly & Lidiak 2006; Jolly *et al.* 2008). These investigations have suggested that the silicic rocks of the bimodal IAT-series have compositions similar to plagiorhyolites and are derived from partially melting a mafic protolith. This petrogenesis could also be true for other silicic IAT rocks in the Greater Antilles arc. Consequently, although it is beyond the scope of this study, it is recommended that a future geochemical analysis should be performed to investigate the potential magma genesis of all of the silicic Early Cretaceous Caribbean IAT-suites.

With regard to this study it should be noted that the silicic lower Devils Racecourse sample AHBI27 has distinctly lower $\varepsilon_{Hf}(i)$ ratios when compared with the intermediate lower Devils Racecourse samples AHBI01 and AHBI03. This may indicate that it is derived from a different source region to that of the basic-intermediate lower Devils Racecourse rocks. As such, more isotope analyses need to be carried out on the lower Devils Racecourse lavas to fully assess the potentially different source components.

Conclusions

1. The lower and upper Devils Racecourse lavas represent a Hauterivian to Aptian island arc succession in eastern Jamaica. The lower lavas are basaltic andesites to dacites/rhyolites and have an IAT composition. In contrast, the upper lavas are CA basalts and basaltic andesites which frequently form pillow lavas.

2. Elemental and isotopic evidence reveal that the IAT and CA Devils Racecourse arc lavas are derived from two chemically similar mantle wedge source regions predominantly composed of N-MORB-type spinel lherzolite. Nevertheless, further isotopic studies are required to investigate the possible occurrence of plume-type material in the mantle wedge of the Great Arc in the Hauterivian to Aptian.

3. Th-LREE/HFSE-HREE ratios, Nd isotope systematics, $(Ce/Ce^*)_{n-mn}$ and Th/La ratios indicate that the IAT and CA mantle wedge source regions were enriched by chemically distinct slab fluxes, which were derived from both the altered basaltic portion of the slab and its accompanying pelagic and terrigenous sedimentary veneer respectively.

4. The presence of IAT and CA island arc lavas before and after the Aptian–Albian suggest that the compositional change in the Great Arc was the result of the subduction of chemically differing sedimentary material with no reason to invoke an arc-wide subduction polarity reversal.

A. Hastie acknowledges NERC PhD Studentship NER/S/A/2003/11215. Thanks to I. McDonald and E. De Vos for assistance with ICP-OES and ICP-MS at Cardiff University, UK. The authors are grateful to Julian Pearce for his geochemical advice and helpful comments. Thanks must also go to T. Jackson, A. Jones, G. Edwards and V. Adams for help with fieldwork and logistics in Jamaica. Constructive reviews by E. Lidiak and J. Schellekens improved the manuscript and are much appreciated.

References

BARRY, T. L., PEARCE, J. A., LEAT, P. T., MILLAR, I. L. & LE ROEX, A. P. 2006. Hf isotope evidence for selective mobility of high-field-strength elements in a subduction setting: South Sandwich Islands. *Earth and Planetary Science Letters*, **252**, 223–244.

BEN OTHMAN, D., WHITE, W. M. & PATCHETT, J. 1989. The geochemistry of marine sediments, island arc magma genesis, and crust-mantle recycling. *Earth and Planetary Science Letters*, **94**, 1–21.

BLICHERT-TOFT, J. 2001. On the Lu-Hf isotope geochemistry of silicate rocks. *Geostandards Newsletter – the Journal of Geostandards and Geoanalysis*, **25**, 41–56.

BRENAN, J. M., SHAW, H. F. & RYERSON, F. J. 1995. Experimental-evidence for the origin of lead enrichment in convergent-margin magmas. *Nature*, **378**, 54–56.

BURKE, K. 1988. Tectonic evolution of the Caribbean. *Annual Review of Earth and Planetary Science*, **16**, 201–230.

BURKE, K., COATES, A. G. & ROBINSON, E. 1969. Geology of the Benbow Inlier and surrounding areas, Jamaica. *In*: SAUNDERS, J. B. (ed.) *Transactions of the Fourth Caribbean Geological Conference, Trinidad, 28 March to 12 April, 1965*, 229–307.

CALMUS, T., AGUILLÓN-ROBLES, A. ET AL. 2003. Spatial and temporal evolution of basalts and magnesian andesites ('bajaites') from Baja California, Mexico: the role of slab melts. *Lithos*, **66**, 77–105.

CANN, J. R. 1970. Rb, Sr, Y, Zr and Nb in some ocean floor basaltic rocks. *Earth and Planetary Science Letters*, **10**, 7–11.

CLASS, C., MILLER, D. M., GOLDSTEIN, S. & LANGMUIR, C. 2000. Distinguishing melt and fluid subduction components in Umnak Volcanics, Aleutian Arc. *Geochemistry Geophysics Geosystems*, **1**, paper 1999GC000010.

DAVIDSON, J. P. 1987. Crustal contamination versus subduction zone enrichment: examples from the Lesser Antilles and implications for the mantle source composition of island arc lavas. *Geochimica et Cosmochimica Acta*, **51**, 2185–2198.

DEFANT, M. J., JACKSON, T. E. ET AL. 1992. The geochemistry of young volcanism throughout western Panama and southeastern Costa Rica: an overview. *Journal of the Geological Society, London*, **149**, 569–579.

DONNELLY, T. W., BEETS, D. ET AL. 1990. History and tectonic setting of Caribbean magmatism. *In*: DENGO, G. & CASE, J. E. (eds) *The Caribbean Region. The Geology of North America*, **H**. Geological Society of America, Boulder, CO, 339–374.

DRUMMOND, M. S. & DEFANT, M. J. 1990. A model for trondhjemite-tonalite-dacite genesis and crustal growth via slab melting: Archean to modern comparisons. *Journal of Geophysical Research*, **95**, 21,503–21,521.

DRUMMOND, M. S., DEFANT, M. J. & KEPEZHINSKAS, P. K. 1996. Petrogenesis of slab-derived trondhjemite-tonalite-dacite/adakite magmas. *Transactions of the Royal Society of Edinburgh: Earth Sciences*, **87**, 205–215.

ELLIOTT, T. 2003. Tracers of the slab. Inside the subduction factory. *Geophysical Monograph*, **138**, 23–45.

ESCUDER VIRUETE, J., DÍAZ DE NEIRA, A. ET AL. 2006. Magmatic relationships and ages of Caribbean Island arc tholeiites, boninites and related felsic rocks, Dominican Republic. *Lithos*, **90**, 161–186.

ESCUDER VIRUETE, J., CONTRERAS, F. ET AL. 2007. Magmatic relationships and ages between adakites, magnesian andesites and Nb-enriched basalt–andesites from Hispaniola: Record of a major change in the Caribbean island arc magma sources. *Lithos*, **99**, 151–177.

EWART, A. 1982. The mineralogy and petrology of Tertiary–Recent orogenic volcanic rocks: with special reference to the andesitic-basaltic compositional range. *In*: THORPE, R. S. (eds) *Andesites*. Wiley, Chichester, 25–83.

EWART, A., COLLERSON, K. D., REGELOUS, M., WENDT, J. I. & NIU, Y. 1998. Geochemical evolution within the Tonga–Kermadec–Lau Arc–Back-arc systems: the role of varying mantle wedge composition in space and time. *Journal of Petrology*, **39**, 331–368.

FITTON, J. G., SAUNDERS, A. D., NORRY, M. J., HARDARSON, B. S. & TAYLOR, R. N. 1997. Thermal and chemical structure of the Iceland plume. *Earth and Planetary Science Letters*, **153**, 197–208.

FROST, C. D., SCHELLEKENS, J. H. & SMITH, A. L. 1998. Nd, Sr, and Pb isotopic characterization of Cretaceous and Paleogene volcanic and plutonic island arc rocks from Puerto Rico. *In*: LIDIAK, E. G. & LARUE, D. K. (eds) *Tectonics and Geochemistry of the Northeast Caribbean*. Geological Society of America, Boulder, CO, Special Papers, **322**, 123–132.

GELDMACHER, J., HANAN, B. B. ET AL. 2003. Hafnium isotopic variations in volcanic rocks from the Caribbean large igneous province and Galapagos hot spot tracks. *Geochemistry Geophysics Geosystems*, **4**, paper number 2002GC000477.

HASTIE, A. R. 2007. *The tectonomagmatic evolution of the Caribbean Plate: insights from igneous rocks on Jamaica*. Cardiff University, unpublished PhD thesis.

HASTIE, A. R., KERR, A. C., PEARCE, J. A. & MITCHELL, S. F. 2007. Classification of altered volcanic island arc rocks using immobile trace elements: Development of the Th-Co discrimination diagram. *Journal of Petrology*, **48**, 2341–2357.

HASTIE, A. R., KERR, A. C., MITCHELL, S. F. & MILLER, I. 2008. Geochemistry and petrogenesis of Cretaceous oceanic plateau lavas in eastern Jamaica. *Lithos*, **101**, 323–343.

HAWKESWORTH, C. J., HERGT, J. M., ELLAM, R. M. &
    MCDERMOTT, F. 1991. Element fluxes associated
    with subduction related magmatism. *Philosophical
    Transactions of the Royal Society of London Series
    A*, **335**, 393–405.
HILL, I. G., WORDEN, R. H. & MEIGHAN, I. G. 2000.
    Yttrium: the immobility-mobility transition during
    basaltic weathering. *Geology*, **28**, 923–926.
HOLE, M. J., SAUNDERS, A. D., MARRINER, G. F. &
    TARNEY, J. 1984. Subduction of pelagic sediments:
    implications for the origin of Ce-anomalous basalts
    from the Mariana Islands. *Journal of the Geological
    Society of London*, **141**, 453–472.
ISHIKAWA, T. & TERA, F. 1999. Two isotopically distinct
    fluid components involved in the Mariana arc:
    Evidence from Nb/B ratios and B, Sr, Nd and Pb
    isotope systematics. *Geology*, **27**, 83–86.
JIANG, M. J. & ROBINSON, E. 1987. Calcareous nanno-
    fossils and larger foraminifera in Jamaican rocks of
    Cretaceous to Early Eocene age. *In*: AHMAD, R. (ed.)
    Proceedings of a Workshop on the Status of Jamaican
    Geology. *Geological Society of Jamaica* (special issue)
    24–51.
JOLLY, W. T. & LIDIAK, E. G. 2006. Role of crustal
    melting in petrogenesis of the Cretaceous water
    island formation (Virgin Islands, northeast Antilles
    Island arc). *Geologica Acta*, **4**, 7–33.
JOLLY, W. T., LIDIAK, E. G., SCHELLEKENS, J. H. &
    SANTOS, H. 1998a. Volcanism, tectonics, and strati-
    graphic correlations in Puerto Rico. *In*: LIDIAK,
    E. G. & LARUE, D. K. (eds) *Tectonics and Geo-
    chemistry of the Northeast Caribbean*. Geological
    Society of America, Boulder, CO, Special Papers,
    322, 1–34.
JOLLY, W. T., LIDIAK, E. G., DICKIN, A. P. & WU, T. W.
    1998b. Geochemical diversity of Mesozoic island arc
    tectonic blocks in eastern Puerto Rico. *In*: LIDIAK,
    E. G. & LARUE, D. K. (eds) *Tectonics and Geo-
    chemistry of the Northeast Caribbean*. Geological
    Society of America, Boulder, CO, Special Papers,
    322, 67–98.
JOLLY, W. T., LIDIAK, E. G., DICKIN, A. P. & WU, T. W.
    2001. Secular geochemistry of central Puerto Rican
    island arc lavas: constraints on Mesozoic tectonism
    in the eastern Greater Antilles. *Journal of Petrology*,
    42, 2197–2214.
JOLLY, W. T., LIDIAK, E. G. & DICKIN, A. P. 2006. Cre-
    taceous to Mid-Eocene pelagic sediment budget in
    Puerto Rico and the Virgin Islands (northeast Antilles
    Island arc). *Geologica Acta*, **4**, 35–62.
JOLLY, W. T., LIDIAK, E. G. & DICKIN, A. P. 2008.
    Bimodal volcanism in northeast Puerto Rico and the
    Virgin Islands (Greater Antilles Island Arc): genetic
    links with Cretaceous subduction of the mid-Atlantic
    ridge Caribbean spur. *Lithos*, **103**, 393–414.
KEMPTON, P. D. 1995. *Common Pb Chemical Procedures
    for Silicate Rocks and Minerals, Methods of Data Cor-
    rection and an Assessment of Data Quality at the
    NERC Isotope Geosciences Laboratory*. NIGL
    Report Series, **78**.
KEMPTON, P. D., FITTON, J. G. *ET AL.* 2000. The Iceland
    plume in space and time: a Sr–Nd–Pb–Hf study of the
    North Atlantic rifted margin. *Earth and Planetary
    Science Letters*, **177**, 255–271.

KEMPTON, P. D., PEARCE, J. A., BARRY, T. L., FITTON,
    J. G., LANGMUIR, C. & CHRISTIE, D. M. 2002. Sr–
    Nd–Pb–Hf isotope results from ODP Leg 187: evi-
    dence for mantle dynamics of the Australian–Antarctic
    discordance and origin of the Indian MORB source.
    Geochemistry Geophysics Geosystems, **3**, paper
    2002GC00320.
KERR, A. C., ITURRALDE-VINENT, M. A., SAUNDERS,
    A. D., BABBS, T. L. & TARNEY, J. 1999. A new
    plate tectonic model of the Caribbean: Implications
    from a geochemical reconnaissance of Cuban Meso-
    zoic volcanic rocks. *Geological Society of America
    Bulletin*, **111**, 1581–1599.
KERR, A. C., WHITE, R. V., THOMPSON, P. M. E.,
    TARNEY, J. & SAUNDERS, A. D. 2003. No Oceanic
    Plateau – no Caribbean Plate? The seminal role of an
    oceanic plateau in Caribbean Plate evolution. *In*:
    BARTOLINI, C., BUFFLER, R. T. & BLICKWEDE, J.
    (eds) *The Circum Gulf of Mexico and Caribbean:
    Hydrocarbon Habitats Basin Formation and Plate
    Tectonics*. American Association of Petroleum
    Geology Memoirs, **79**, 126–268.
KEPPLER, H. 1996. Constraints from partitioning exper-
    iments on the composition of subduction-zone fluids.
    Nature, **380**, 237–240.
KESLER, S. E., CAMPBELL, I. H. & ALLEN, C. M. 2005.
    Age of the Los Ranchos Formation, Dominican Repub-
    lic: timing and tectonic setting of primitive island arc
    volcanism in the Caribbean region. *Geological
    Society of America Bulletin*, **117**, 987–995.
KHUDOLEY, K. M. & MEYERHOFF, A. A. 1971. *Paleo-
    geography and Geological History of Greater Antilles*.
    Geological Society of America Memoirs, **129**, 199.
KURTZ, A. C., DERRY, L. A., CHADWICK, O. A. &
    ALFANO, M. J. 2000. Refractory element mobility in
    volcanic soils. *Geology*, **28**, 683–686.
LEBRON, M. C. & PERFIT, M. R. 1993. Stratigraphic and
    petrochemical support subduction polarity reversal
    of the Cretaceous Caribbean Island arc. *Journal of
    Geology*, **101**, 389–396.
LEBRON, M. C. & PERFIT, M. R. 1994. Petrochemistry
    and tectonic significance of Cretaceous island-arc
    rocks, Cordillera Oriental, Dominican Republic.
    Tectonophysics, **229**, 69–100.
LE BAS, M. J., LE MAITRE, R. W., STRECKEISEN, A. &
    ZANETTIN, B. 1986. A chemical classification of vol-
    canic rocks based on the total alkali-silica diagram.
    Journal of Petrology, **27**, 745–750.
LE BAS, M. J., LE MAITRE, R. W. & WOOLLEY, A. R.
    1992. The constuction of the total alkali-silica chemi-
    cal classification of volcanic rocks. *Mineralogy and
    Petrology*, **46**, 1–22.
LEWIS, J. F. & DRAPER, G. 1990. Geological and tectonic
    evolution of the northern Caribbean Margin. *In*:
    DENGO, G. & CASE, J. E. (eds) *The Caribbean
    Region*. The Geology and North America, **H**. The
    Geological Society of America, Boulder, CO, Special
    Publications, 77–140.
MARCHESI, C., GARRIDO, C. J. *ET AL.* 2007. Geochemis-
    try of Cretaceous magmatism in eastern Cuba:
    recycling of North American continental sediments
    and implications for subduction polarity reversal in
    the Greater Antilles paleo-arc. *Journal of Petrology*,
    48, 1813–1840.

MATLEY, C. A. & RAW, F. 1942. A road section near Guy's Hill, Jamaica. *Geological Magazine*, **79**, 241–252.

MATTSON, P. H. 1979. Subduction, buoyant braking, flipping and strike–slip faulting in the northern Caribbean. *Journal of Geology*, **87**, 293–304.

McCULLOCH, M. T. & GAMBLE, J. A. 1991. Geochemical and geodynamical constraints on subduction zone magmatism. *Earth and Planetary Science Letters*, **102**, 358–374.

McDONALD, I. & VILJOEN, K. S. 2006. Platinum-group element geochemistry of mantle eclogites: a reconnaissance study of xenoliths from the Orapa kimberlite, Botswana. *Applied Earth Science (Transactions of the Institution of Mining and Metallurgy B)*, **115**, 81–93.

McDONOUGH, W. F. & SUN, S.-S. 1995. The composition of the Earth. *Chemical Geology*, **120**, 223–253.

MILLER, D. M., GOLDSTEIN, S. & LANGMUIR, C. 1994. Cerium/lead and lead isotope ratios in arc magmas and the enrichment of lead in the continents. *Nature*, **368**, 514–519.

MÜLLER, R. D., ROYER, J.-Y., CANDE, S. C., ROEST, W. R. & MASCHENKOV, S. 1999. New constraints on the Late Cretaceous/Tertiary plate tectonic evolution of the Caribbean. *In*: MANN, P. (ed.) *Caribbean Basins. Sedimentary Basins of the World* **4**. Elsevier Science, Amsterdam, 33–59.

MÜNKER, C., WEYER, S., SCHERER, E. & MEZGER, K. 2001. Separation of high field strength elements (Nb, Ta, Zr, Hf) and Lu from rock samples for MC-ICPMS measurements. *Geochemistry Geophysics Geosystems*, **2**, paper 2001GC000183.

NOWELL, G. M. & PARRISH, R. R. 2001. Simultaneous acquisition of isotope compositions and parent/daughter ratios by non-isotope dilution solution-mode plasma ionisation multi-collector mass spectrometry (PIMMS). *In*: HOLLAND, G. & TANNER, S. D. (eds) *Plasma Source Mass Spectrometry – The New Millennium*. Royal Society of Chemistry, Cambridge, 298–310.

NOWELL, G. M., KEMPTON, P. D. & NOBLE, S. R. 1998. High precision Hf isotope measurements of MORB and OIB by thermal ionisation mass spectrometry: insights into the depleted mantle. *Chemical Geology*, **149**, 211–233.

PEARCE, J. A. 1982. Trace element characteristics of lavas from destructive plate boundaries. *In*: THORPE, R. S. (ed.) *Andesites*. Wiley, Chichester, 525–547.

PEARCE, J. A. 1983. Role of the sub-continental lithosphere in magma genesis at active continental margins. *In*: HAWKESWORTH, C. J. & NORRY, M. J. (eds) *Continental Basalts and Mantle Xenoliths*. Shiva, Nantwich, 230–249.

PEARCE, J. A. & PEATE, D. W. 1995. Tectonic implications of the composition of volcanic arc magmas. *Annual Reviews Earth and Planetary Science Letters*, **23**, 251–285.

PEARCE, J. A., KEMPTON, P. D., NOWELL, G. M. & NOBLE, S. R. 1999. Hf–Nd element and isotope perspective on the nature and provenance of mantle and subduction components in western Pacific Arc–basin systems. *Journal of Petrology*, **40**, 1579–1611.

PECCERILLO, R. & TAYLOR, S. R. 1976. Geochemistry of eocene calc-alkaline volcanic rocks from the Kastamonu area, northern Turkey. *Contributions to Mineralogy and Petrology*, **58**, 63–81.

PINDELL, J. L. & KENNAN, L. 2001. Kinematic evolution of the Gulf of Mexico and Caribbean. *Transactions, Petroleum Systems of Deep-water Basins: Global and Gulf of Mexico Experience. GCSSEPM 21st Annual Foundation Bob F. Perkins Research Conference*, Houston, TX, 193–220.

PINDELL, J. L., KENNAN, L., STANEK, K. P., MARESH, W. V. & DRAPER, G. 2006. Foundations of Gulf of Mexico and Caribbean evolution: eight controversies resolved. *Geologica Acta*, **4**, 303–341.

PISOT, N., VILA, J.-M., DELOFFRE, R. & MASCLE, A. 1986. Barremian benthic mesogean associations of foraminifera and algae in Benbow Inlier (Jamaica). *11th Caribbean Geological Conference Barbados*, 20–26 July 1986, Abstracts, 79.

PLANK, T. 2005. Constraints from thorium/lanthanum on sediment recycling at subduction zones and the evolution of the continents. *Journal of Petrology*, **46**, 921–944.

PLANK, T. & LANGMUIR, C. 1998. The chemical composition of subducting sediment and its consequences for the crust and mantle. *Chemical Geology*, **145**, 325–394.

RANKIN, D. 2002. *Geology of St. John, U.S. Virgin Islands*. United States Geological Survey Professional Papers, **1631**, 1–36.

REGELOUS, M., COLLERSON, K. D., EWART, A. & WENDT, J. I. 1997. Trace element transport rates in subduction zones: evidence from Th, Sr and Pb isotope data for Tonga–Kermadec arc lavas. *Earth and Planetary Science Letters*, **150**, 291–302.

RÉVILLON, S., CHAUVEL, C., ARNDT, N. T., PIK, R., MARTINEAU, F., FOURCADE, S. & MARTY, B. 2002. Heterogeneity of the Caribbean plateau mantle source: Heterogeneity of the Caribbean plateau mantle source: Sr, O and He isotopic compositions of olivine and clinopyroxene from Gorgona Island. *Earth and Planetary Science Letters*, **205**, 91–106.

ROBINSON, E. 1994. Jamaica. *In*: DONOVAN, S. K. & JACKSON, T. A. (eds) *Caribbean Geology: An Introduction*. University of the West Indies Publisher's Association, Kingston, 111–127.

ROBINSON, E., LEWIS, J. F. & CANT, R. V. 1972. *Field Guide to Aspects of the Geology of Jamaica. International Field Institute Guidebook to the Caribbean Island Arc System 1970*. American Geological Institute, Washington DC, 1–45.

ROOBOL, M. J. 1972. The volcanic geology of Jamaica. *Transactions of the 6th Caribbean Geological Conference, Margarita, Venezuela, 6–14 July, 1971*, 100–107.

ROYSE, K. R., KEMPTON, P. D. & DARBYSHIRE, F. D. 1998. *Procedure for the Analysis of Rubidium-Strontium and Samarium–Neodymium Isotopes at the NERC Isotope Geosciences Laboratory*. NIGL Report Series, **121**.

SAUNDERS, A. D., ROGERS, G., MARRINER, G. F., TERRELL, D. J. & VERMA, S. P. 1987. Geochemistry of Cenozoic volcanic rocks, Baja California,

Mexico: implications for the petrogenesis of post-subduction magmas. *Journal of Volcanology and Geothermal Research*, **32**, 223–245.

SCHELLEKENS, J. H. 1998. Geochemical evolution and tectonic history of Puerto Rico. *In*: LIDIAK, E. G. & LARUE, D. K. (eds) *Tectonics and Geochemistry of the Northeast Caribbean*. Geological Society of America, Special Papers, **322**, 35–66.

SINTON, C. W., DUNCAN, R. A., STOREY, M., LEWIS, J. & ESTRADA, J. J. 1998. An oceanic flood basalt province within the Caribbean Plate. *Earth and Planetary Science Letters*, **155**, 221–235.

SKELTON, P. W. & MASSE, J. P. 1998. Revision of the lower Cretaceous rudist genera *Pachytraga* Paquier and *Retha* Cox (Bivalvia: Hippuritacea) and the origins of the Caprinidae. *Geobios*, **22**, 331–370.

SOHL, N. F. 1979. Notes on middle Cretaceous macrofossils from the Greater Antilles. *Annales de Museum d'Histoire Naturelle de Nice*, **4**, 1–6 (separately numbered).

SUMMERFIELD, M. A. 1997. *Global Geomorphology*. Longman, Singapore, 129–144.

SUN, S.-S. & MCDONOUGH, W. F. 1989. *Chemical and Isotope Systematics of Oceanic Basalts: Implications for Mantle Composition and Processes. Magmatism in the Ocean Basins*. Geological Society, London, Special Publications, **42**, 313–345.

TATSUMI, Y. & KOGISO, T. 1997. Trace element transport during dehydration processes in the subducted oceanic crust: 2. Origin of chemical and physical characteristics in arc magmatism. *Earth and Planetary Science Letters*, **148**, 207–221.

TATSUMI, Y., HAMILTON, D. L. & NESBITT, R. W. 1986. Chemical characteristics of fluid phase released from a subducted lithosphere and origin of arc magmas: evidence from high-pressure experiments and natural rocks. *Journal of Volcanology and Geothermal Research*, **29**, 293–309.

THIRLWALL, M. F. 2002. Multicollector ICP-MS analysis of Pb isotopes using a ^{207}Pb–^{204}Pb double spike demonstrates up to 400 pm/amu systematic errors in Tl-normalization. *Chemical Geology*, **184**, 255–279.

THIRLWALL, M. F., GRAHAM, A. M., ARCULUS, R. J., HARMON, C. G. & MACPHERSON, C. G. 1996. Resolution of the effects of crustal assimilation, sediment subduction and fluid transport in island arc magmas: Pb–Sr–Nd–O isotope geochemistry of Grenada, Lesser Antilles. *Geochimica et Cosmochimica Acta*, **60**, 4785–4810.

THOMPSON, P. M. E., KEMPTON, P. D. *ET AL.* 2003. Hf–Nd isotope constraints on the origin of the Cretaceous Caribbean plateau and its relationship to the Galapagos

plume. *Earth and Planetary Science Letters*, **217**, 59–75.

THOMPSON, P. M. E., KEMPTON, P. D. *ET AL.* 2004. Elemental, Hf-Nd isotopic and geochronological constraints on an island arc sequence associated with the Cretaceous Caribbean plateau: Bonaire, Dutch Antilles. *Lithos*, **74**, 91–116.

TURNER, S. & HAWKESWORTH, C. 1997. Constraints on flux rates and mantle dynamics beneath island arcs from Tonga–Kermadec lava geochemistry. *Nature*, **389**, 568–573.

VERVOORT, J. D., PATCHETT, P. J., BLICHERT-TOFT, J. & ALBARÉDE, F. 1999. Relationships between Lu–Hf and Sm–Nd isotopic systems in the global sedimentary system. *Earth and Planetary Science Letters*, **168**, 79–99.

VILA, J. M. 1986. Evolution sedimentaire et structurale du bassin oligo-miocene de Trois-Rivieres sur la frontiere decrochante nord-caraibe (Massif du Nord d'Haiti, Grandes Antilles) (Sedimentary and structural evolution of the Oligo-Miocene basin of Trois Rivers on the border of the strike–slip fault of the northern Caribbean, North Haiti Massif, Greater Antilles). *Revue de Geologie Dynamique et de Geographie Physique*, **27**, 183–192.

WHITE, W. M. & PATCHETT, P. J. 1984. Hf-Nd-Sr isotopes and incompatible element abundances in island arcs: implications for magma origins and crust-mantle evolution. *Earth and Planetary Science Letters*, **67**, 167–185.

WHITE, W. M. & DUPRE, B. 1986. Sediment subduction and magma genesis in the Lesser Antilles. *Journal of Geophysical Research*, **91B**, 5927–5941.

WHITE, R. V., TARNEY, J. *ET AL.* 1999. Modification of an oceanic plateau, Aruba, Dutch Caribbean: implications for the generation of continental crust. *Lithos*, **46**, 43–68.

WOODHEAD, J. D., HERGT, J. M., DAVIDSON, J. P. & EGGINS, S. M. 2001. Hafnium isotope evidence for 'conservative' element mobility during subduction zone processes. *Earth and Planetary Science Letters*, **192**, 331–346.

YOGODZINSKI, G. M., VOLYNETS, O. N., KOLOSKOV, A. V. & SELIVERSTOV, N. I. 1994. Magnesian andesites and the subduction component in a strongly calcalkaline series at Piip volcano, far western Aleutians. *Journal of Petrology*, **34**, 163–204.

YOGODZINSKI, G. M., KAY, R. W., VOLYNETS, O. N., KOLOSKOV, A. V. & KAY, S. M. 1995. Magnesian andesite in the western Aleutian Komandorsky region: implications for slab melting and processes in the mantle wedge. *Geological Society of America Bulletin*, **107**, 505–519.

The geotectonic story of the northwestern branch of the Caribbean Arc: implications from structural and geochronological data of Cuba

K. P. STANEK[1]*, W. V. MARESCH[2] & J. L. PINDELL[3,4]

[1]TU Bergakademie Freiberg, Institut für Geologie, D-09596 Freiberg, Germany

[2]Ruhr-Universität Bochum, Institut für Geologie, Mineralogie und Geophysik; D-44780 Bochum, Germany

[3]Tectonic Analysis Ltd, Chestnut House, Duncton, West Sussex GU28 0LH, UK

[4]Department of Earth Science, Rice University, Houston, Texas, USA

*Corresponding author (e-mail: stanek@geo.tu-freiberg.de)

Abstract: Within the last decade, modern petrological and geochronological methods in combination with detailed studies of the field geology have allowed the reconstruction of tectonic processes in the northwestern part of the Caribbean Plate. The development of an oceanic Proto-Yucatán Basin can be traced from the Late Jurassic to the Mid-Cretaceous. From the Mid-Cretaceous onward, an interaction of this basin with the Caribbean Arc can be observed. Geochronological data prove continuous magmatic activity and generation of HP mineral suites in the Caribbean Arc from the Aptian to the Campanian/Maastrichtian. Magmatism ceased at least in onshore central Cuba at about 75 Ma, probably as the southern edge of the continental Yucatán Block began to interact with the advancing arc system. Similarly, the youngest recorded ages for peak metamorphism of high-pressure metamorphic rocks in Cuba cluster at 70 Ma; rapid uplift/exhumation of these rocks occurred thereafter. After this latest Cretaceous interaction with the southern Yucatán Block, the northern Caribbean Arc was dismembered as it entered the Proto-Yucatán Basin region. Because of the continued NE-directed movement, Proto-Yucatán Basin sediments were accreted to the arc and now form the North Cuban fold and thrust belt. Parts of the island arc have been thrust onto the southern Bahamas Platform along the Eocene suture zone in Cuba. Between the arc's interaction with Yucatán and the Bahamas (*c.* 70 to *c.* 40 Ma), the Yucatán intra-arc basin opened by extreme extension and local seafloor accretion between the Cayman Ridge (still part of Caribbean Plate) and the Cuban frontal arc terranes, the latter of which were kinematically independent of the Caribbean. Although magmatism ceased in central Cuba by 75 Ma, traces of continuing Early Palaeogene arc magmatism have been identified in the Cayman Ridge, suggesting that magmatism may not have ceased in the arc as a whole, but merely shifted south relative to Cuba. If so, a shallowing of the subduction angle during the opening of the Yucatán Basin would be implied. Further, this short-lived (?) Cayman Ridge arc is on tectonic strike with the Palaeogene arc in the Sierra Maestra of Eastern Cuba, suggesting south-dipping subduction zone continuity between the two during the final stages of Cuba–Bahamas closure. After the Middle Eocene, the east–west opening of the Cayman Trough left the present Yucatán Basin and Cuba as part of the North American Plate. The subduction geometry, *P–T–t* paths of HP rocks in Cuban mélanges, the time of magmatic activity and preliminary palaeomagnetic data support the conclusion that the Great Antillean arc was initiated by intra-oceanic subduction at least 900 km SW of the Yucatán Peninsula in the ancient Pacific. As noted above, the Great Antillean Arc spanned some 70 Ma prior to its Eocene collision with the Bahamas. This is one of the primary arguments for a Pacific origin of the Caribbean lithosphere; there simply was not sufficient space between the Americas, as constrained by Atlantic opening kinematics, to initiate and build the Antillean (and other) arcs in the Caribbean with *in situ* models.

The origin of the Caribbean Plate: contrasting models

Previous interpretations of the geotectonic development of the Caribbean Plate have been based on different data sets and sometimes yielded contradictory results. However, modern analytical methods now allow the geodynamic processes and events involved in plate collisions to be understood on a three-dimensional scale down to deeper crustal and upper mantle levels. This is especially true for the northern and southern margins of the present Caribbean Plate, where petrological, geochronological and geophysical data are providing

From: JAMES, K. H., LORENTE, M. A. & PINDELL, J. L. (eds) *The Origin and Evolution of the Caribbean Plate*. Geological Society, London, Special Publications, **328**, 361–398.
DOI: 10.1144/SP328.15   0305-8719/09/$15.00 © The Geological Society of London 2009.

an increasingly robust basis for objective plate-tectonic interpretations. Most recently, the origin of this largely oceanic plate was discussed extensively in the Sigüenza workshop (Spain) in 2006, and was reduced to two principal models of Caribbean evolution: The 'Pacific-origin' or 'tectonic indenter model' on the one hand, and the 'intra-American' or '*in situ*-model' on the other. The tectonic indenter model invokes east–west displacement of an oceanic piece of Pacific crust into the gap between the westward-drifting (in the mantle reference frame) North and South American continental plates, beginning in the Early Cretaceous (Pindell & Dewey 1982; Pindell 1993; Pindell *et al.* 2005, 2006). Because of the relative motion, an island arc developed at the leading edge of the Pacific crustal fragment (the Caribbean Arc), subducting the 'Proto-Caribbean' (inter-American) oceanic crust between the Late Jurassic–Cretaceous passive margins of the American continents. Multiple plate boundary interactions occurred, such as collision and obduction of parts of the island arc, shifting of the magmatic axis, and lateral translation of fragments of American continental crust. However, the precise time and style of initiation of the subduction zone along the 'bow' of the Pacific/Caribbean Plate remain unclear. The *in situ* model also calls for the opening of space between the two American continents by rifting, but all tectonic features since the Late Jurassic are explained in terms of an inter-American Caribbean oceanic crust which later converged along all the American margins, forming island arc/subduction–zone complexes and collisional suture zones by uncoordinated movements of the Caribbean and American plates (Meschede & Frisch 1998; Giunta *et al.* 2006; James 2002, 2006). In order to acknowledge Caribbean arc histories which started in the Early Cretaceous, this involves active northern and southern Caribbean continental margins early in the Caribbean history rather than the predominantly passive ones acknowledged by the indenter model. The large Cenozoic strike–slip movements in the northern and southern Caribbean as recorded by the Cayman Trough (Rosencrantz *et al.* 1988; Leroy *et al.* 2000), the subduction histories of long-lived island arcs (Pindell & Barrett 1990), and the occurrence of Early Cretaceous island arc related magmatism are not satisfactorily explained by the *in situ* model. Further, the Aptian initiation of circum-Caribbean HP subduction zone metamorphism (Pindell *et al.* 2005) during a time of definite and rapid plate divergence between the Americas (extension) is difficult to reconcile with *in situ* Caribbean models.

For better understanding, the following terms will be used in this paper: all tectonic units that disappeared during the geodynamic evolution by subduction will be characterized by the prefix 'Proto', for example the oceanic crust between the American continental plates after rifting will be called 'Proto-Caribbean', and the subducted oceanic crust in the original triangle between the Yucatán and the Bahamas Blocks will be called 'Proto-Yucatán Basin'. A key area for deciphering the Late Mesozoic to Early Palaeogene geotectonic history of the Caribbean Plate is the triangular region in the northwestern Caribbean between the Yucatán Peninsula, the Cayman Trough and the southern Bahamas Platform (Cuba) (Fig. 1). After the initial opening of the Cayman Trough by an east-trending sinistral strike–slip system in the Eocene (Leroy *et al.* 2000), the Yucatán Basin, the Cayman Ridge and the Cuban terranes of the Caribbean Arc were left behind as part of North America and protected from further significant tectonic activity or magmatism (Fig. 1). The area can be subdivided into several crustal domains, although the data for the southern submerged domains are very poor. The northern shoulder of the Cayman Trough, the Cayman Ridge, shows evidence of Palaeogene arc magmatism (Sigurdson *et al.* 1997; Lewis *et al.* 2005). The Yucatán Basin consists of highly stretched arc (?) and local oceanic crust, as deduced from geophysical data (Rosencrantz 1990). In the western part of the Yucatán Basin, geothermal values have been interpreted as denoting Eocene oceanic crust, suggested to have formed in a pull-apart regime in the final stages of Cuba's migration toward the Bahamas. Along the eastern Yucatán margin, the Hondo Fault Zone separates Yucatán Block continental crust from the stretched arc/oceanic crust in the Yucatán Basin. At this fault zone, obducted bodies of low-grade metasediments (siltstone, shale, quartzite) probably correlate with the San Cayetano Formation of western Cuba (Dillon & Vedder 1973), suggesting a convergent stage between the arc and western Cuba with this continental margin, followed by transcurrence or transtension thereafter to form the current margin. In the NE, the Central Cuban Main Thrust, the western part of the North Caribbean Suture zone (NCSZ), extends along the length of the Cuban archipelago. Along this suture zone, parts of the North American palaeo-continental margin and the Cretaceous Caribbean (Great Antillean) island arc have been amalgamated, and provide a broad exposition of geology for reconstructing the geotectonic history of the northern Caribbean Plate. In the following we will present a short compilation of available stratigraphic, tectonic, geochronological and petrological data for the Yucatán 'triangle', with special emphasis on Cuba, and a geotectonic reconstruction that considers the three-dimensional nature of subduction zones.

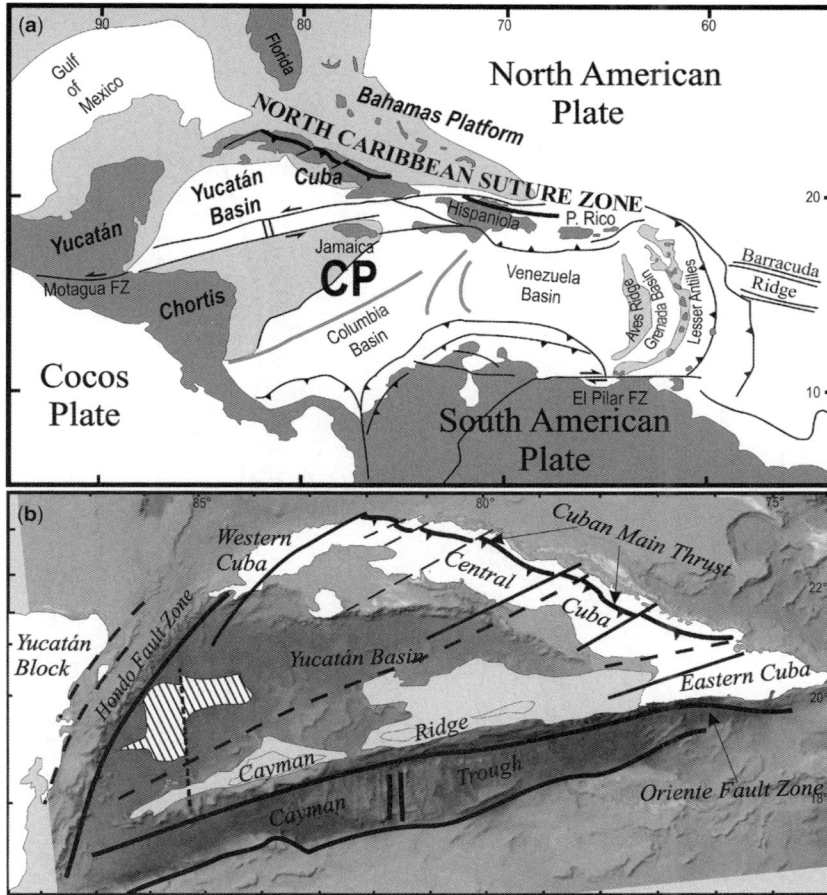

Fig. 1. Sketch map of the geotectonic framework of the Caribbean Plate (CP) (**a**) and the Yucatán Basin (**b**). The Yucatán Basin is bounded by the sinistral Hondo Fault Zone, the Cuban Main Thrust and the sinistral fault system of the northern Cayman Trough. The diagonal patterned area in the western part represents supposed Palaeogene oceanic crust (Rosencrantz 1990).

The geotectonic units of the Cuban part of the North Caribbean Suture Zone: nature and distribution of rocks

The Cuban mainland consists of three major geological regions that differ in structural style and lithological nature (Fig. 2). These areas are separated by regional faults and associated sedimentary basins: in the west the Pinar Fault with the Los Palacios Basin and in the east the Cauto Fault system and the related depression of the same name. Based on this tectonic subdivision of the Cuban mainland, the terms 'western Cuba', 'central Cuba' and 'eastern Cuba' are used for the regions west of the Pinar fault, the area between the Pinar fault and the Cauto Depression, and the area south of the Cauto Depression, respectively. For better

understanding, the stratigraphic names mentioned in this paper are briefly explained in the Summary Lithostratigraphic Chart at the end of the text.

Northwest of the Pinar Fault Zone the West Cuban antiformal stack (the Guaniguanico terrane of Iturralde-Vinent 1994) comprises folded and thrusted sediments and intercalated volcanic rocks of the eastern continental margin of the Yucatán Block. The use of the term 'terrane' in the Caribbean geological literature has been discussed by Iturralde-Vinent & Lidiak (2006). In the present paper the term 'terrane' will be adapted from the cited authors with respect to regional units without semantic changes.

The central part of the Cuban mainland east of the Pinar Fault and west of the Cauto Depression can be subdivided on the basis of geological and

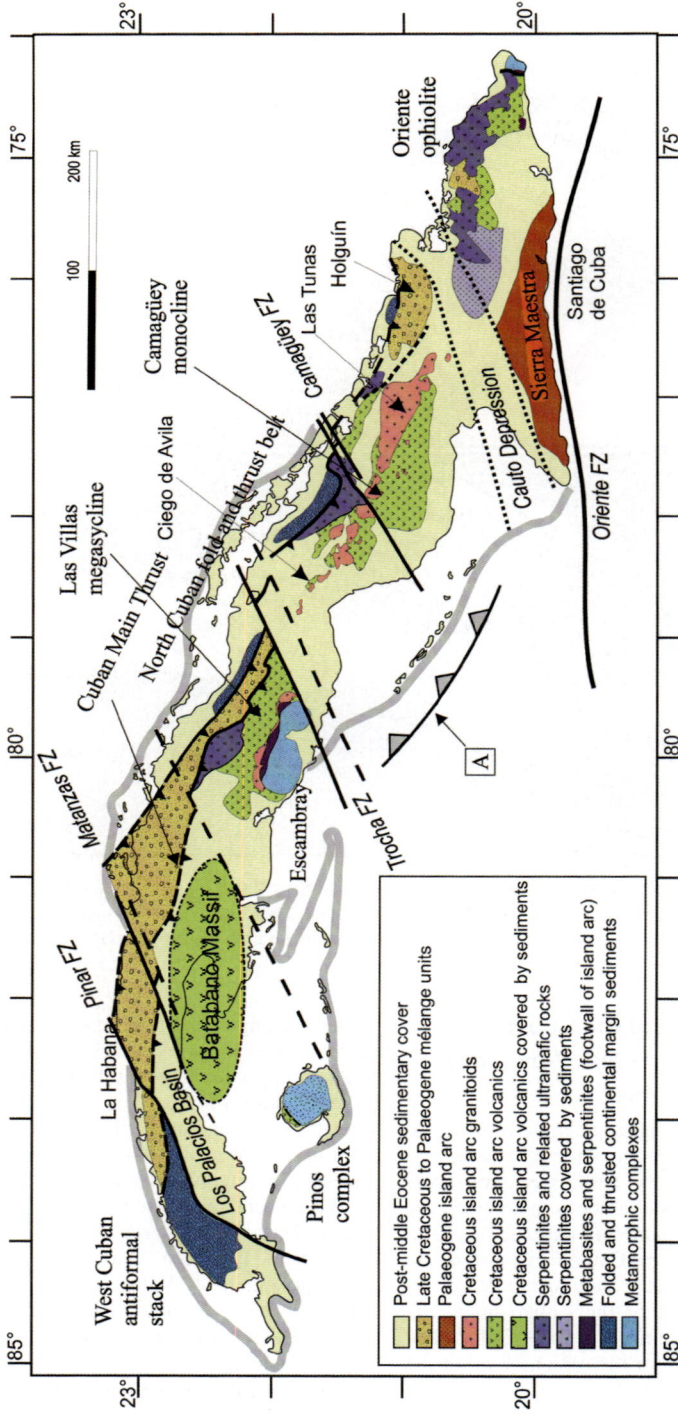

Fig. 2. Generalized geological map of Cuba compiled from the Geological Map of Cuba 1 : 500 000 (Perez Othon & Yarmoliuk 1985) showing the main tectonic units of Cuba. The Batabano Massif has been interpreted on the basis of gravimetric and magnetic data as part of the Cretaceous island arc covered by younger sediments (Pusharovski *et al.* 1989). The broad grey line limits the shelf area deeper than 2000 m. (A) indicates the low angle fault detected by Rosencrantz (1990).

geophysical data into three east–west trending belts. In the northern belt, sediments of the Bahamas carbonate platform, of a continental margin and adjoining deep sea basin have been thrust onto the southern edge of the Bahamas platform and are characterized by typical thin-skin tectonics. This North Cuban fold and thrust belt has been overthrust along the Cuban Main Thrust by a serpentinitic mélange. The serpentinic matrix of the mélange contains in different proportions all rock types of an oceanic crust including ultrabasic lithologies, gabbros as rare basalts and related sedimentary rocks (Kudělásek et al. 1984; Fonseca et al. 1984, 1988). The largest outcrops of serpentinites and gabbros have been mapped in eastern Cuba, but the size of the exposed mélange bodies decreases westward. The serpentinitic mélange has been interpreted as Alpine-type ophiolite suite by Knipper (1975). The occurrences of ultrabasic rocks have been assigned in two types (Kudělásek et al. 1984; Iturralde-Vinent 1994): the northern ophiolites comprise the serpentinitic mélange and the southern ophiolites consist of isolated serpentinitic tectonic slivers in nappe structure of the metamorphic complexes along the southern coast of Cuba. Parts of the Cretaceous Caribbean magmatic arc form the hanging-wall unit of the serpentinitic mélange. In central Cuba both units, mélange and island arc, were subsumed as so-called Zaza zone (Pardo 1975). South of the Cretaceous island-arc sequences, domes of metamorphic rocks like the Pinos complex (Isla de Juventud) and the Escambray complex form isolated, window-like outcrop areas. Together with the serpentinitic mélange, these metamorphic complexes provide a means of studying the subduction–accretionary complex of the Caribbean Arc. Southeast of the Cauto Depression, the Oriente Block consists of metamorphosed parts of the Caribbean Arc, overthrust by ultrabasic and gabbroic rocks of the Nipe–Cristal and Moa–Baracoa massifs and covered by Palaeogene island-arc related volcanic rocks of the Sierra Maestra. Different from central Cuba, the ultrabasic rocks in the easternmost Cuba have been interpreted as supra-subduction related ophiolite (Proenza et al. 1999).

In the following sections, we review these and other aspects of the geology of Cuba. As will be seen, several of the reviewed aspects are important for distinguishing the validity of the in situ and the Pacific origin models for Caribbean evolution.

The continental margin sequences: two types of continental margins

Sediments related to those continental margins formed before and during the opening of the oceanic basin between North and South America are exposed in the Guaniguanico terrane of western Cuba and along the northern edge of central Cuba. Both areas can be distinguished by the range of their stratigraphic profiles and the facies characteristics of their sediments (Fig. 3). The sedimentary sequences of western Cuba have been assigned to the margin off eastern Yucatán (Haczewski 1976; Pindell 1985; Hutson et al. 1998), those of central Cuba to the southern margin of the Bahamas platform (Pardo 1975; Lewis & Draper 1990; Iturralde-Vinent 1994; Draper & Barros 1994; Pszczółkowski 1999).

Western Cuba: the Yucatán type margin

In western Cuba (Guaniguanico terrane), the sediments can be subdivided into four tectonic units on the basis of their facies development and their present tectonic position (Fig. 4). The southern and lowermost tectonic unit comprises the folded sediments of the Sierra de los Organos, (including the Viñales tectonic window) followed to the north by the southern and northern Sierra del Rosario units and the Esperanza unit. At the top of the nappe stack, the island-arc related rocks of the Bahia Honda unit are exotic with respect to the continental margin development. The lowermost part of the stratigraphic record of Guaniguanico comprises the siliciclastic sediments of the San Cayetano Formation, the oldest known sediments of Cuba (see Fig. 3), which form part of the Sierra de los Organos (Pizarras del Norte y del Sur) and the southern Rosario units. The San Cayetano Formation was deposited along the rift-related continental basin between Yucatán and South America during the Lower to Middle Jurassic (Haczewski 1976; Pindell 1985). In the Oxfordian (160–155 Ma), marine carbonates and basic volcanic rocks (Jäger 1972; Pszczółkowski 1994; Cobiella 1996) occur in the stratigraphic succession. The basalts of the El Sabalo Formation show geochemical patterns of a rift-related magmatism (Kerr et al. 1999). This facies change to carbonate-dominated sediments and basaltic magmatism, which can be traced from the southwestern extent of the Esperanza unit to the northern Sierra del Rosario, is the first evidence of seafloor spreading in the region. The upper part of the stratigraphic profile reaches the Middle Cretaceous (Turonian, 90 Ma) and consists of marine carbonates and cherts (Santa Teresa Fm). From the Late Turonian to the Early Campanian (90–80 Ma) there is a hiatus in the stratigraphic profile of Guaniguanico (Iturralde-Vinent 1994). The Upper Cretaceous sediments show the influence of an approaching volcanic arc (Moreno Formation). Detailed descriptions of the biostratigraphy and sedimentology of

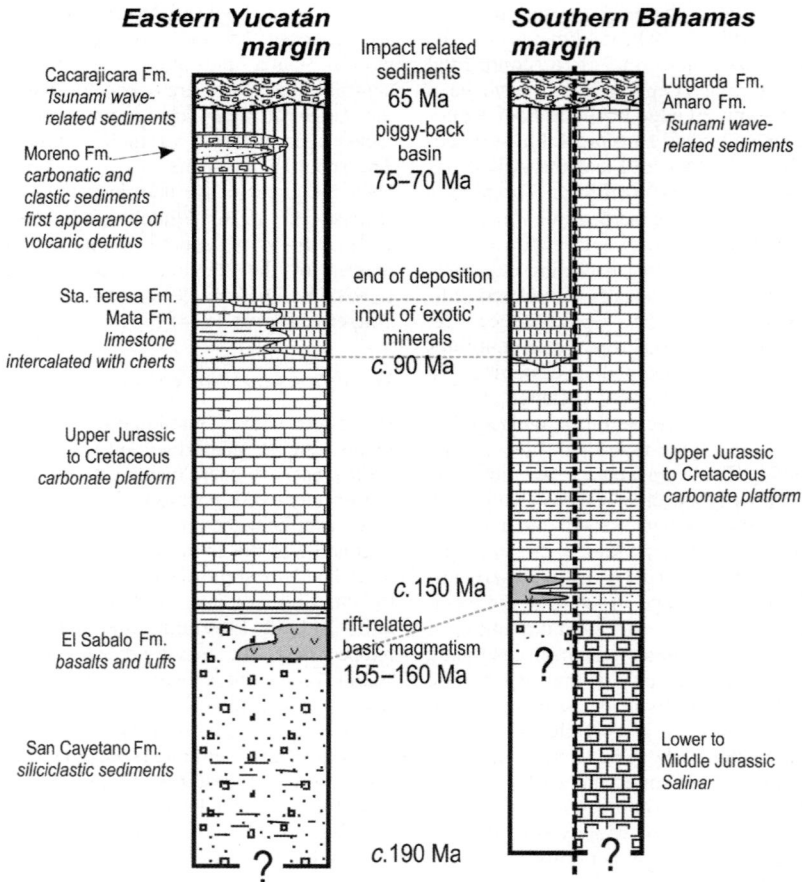

Fig. 3. Generalized stratigraphic profiles of the Yucatán-type continental margin sediments and the Bahamas-type sediments. At the Bahamas column, the right side represents the platform stratigraphic sequence, the left side the sedimentary sequence involved in the North Cuban fold and thrust belt. Note the similarities in both Yucatán-type sequences and the sequences on the left side of the Bahamas column. In both stratigraphic profiles, the Chicxulub impact related sediments cover disconformly the older sedimentary sequences (Pszczółkowski 1986*b*).

the eastern Yucatán continental margin have been summarized by Pszczółkowski (1999). No information is available on the basement and the detachment horizon of the nappe stack in western Cuba.

Central Cuba: the Bahamas-type margin

In central Cuba, the sediments of the North Cuban fold and thrust belt between Matanzas and Holguín (see Fig. 2) have been traditionally subdivided into several sedimentary belts, summarized and interpreted by Ducloz & Vuagnat (1962), Knipper & Cabrera (1974), Pardo (1975), Hatten *et al.* (1988) and Iturralde-Vinent (1994). The three northernmost stratigraphic-facies belts belong to the southern extent of the Bahamas

carbonate platform (Cayo Coco, Canal Viejo de Bahamas and Remedios belts) which comprises Lower Jurassic to Upper Cretaceous evaporitic and carbonate deposits. Stratigraphic hiatuses only occur at the southern edge of the Bahamas platform in the Turonian and Coniacian (*c*. 90–85 Ma). Much more informative in the light of geotectonic reconstructions are the stratigraphic-facies belts which have been interpreted as the slope (Camajuani belt) and basin sediments (Placetas belt) south of the Bahamas platform, thrust onto its southern edge. The known stratigraphic record begins in both zones with Upper Jurassic (Kimmeridgian to Tithonian) deep- and shallow-water carbonate and terrigenous siliciclastic sediments (Pszczółkowski 1987; Iturralde-Vinent 1994) as

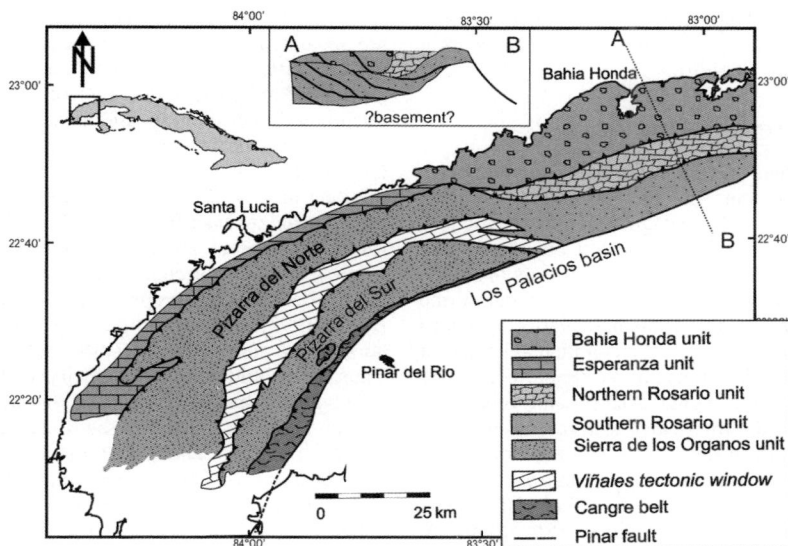

Fig. 4. Tectonic sketch map of the relationship of the thrust sheets in the West Cuban anticlinal stack (Guaniguanico terrane); adapted from Pszczółkowski (1999). The Viñales tectonic window consists of Upper Jurassic to Lower Cretaceous limestones of the Mogote Valley. The bore hole Mariel (Segura Soto *et al.* 1985) is located about 50 km west of Bahia Honda (not shown on the map).

well as rare basic volcanic rocks (Iturralde-Vinent & Morales Marí 1988). Despite the facies differences, the sedimentation in both zones ended in the Turonian (at about 90 Ma). In the Aptian to Cenomanian section of the stratigraphy (Mata Fm and Santa Teresa Fm), an alternating sequence of fine-grained limestones, cherts and bentonitic clays yields detrital minerals such as staurolite, glaucophane, zoisite, white mica and chlorite as well as fragments of K-rich volcanic and ultrabasic rocks (Linares & Smagoulov 1987). This input of 'exotic' minerals in the marine sedimentary rocks has not been investigated in western Cuba. The occurrence of these minerals indicates the erosion of a fore-arc or arc domain. This domain, with its bentonitic clays, could only have been deposited somewhere to the south of today's Cuba, because concurrent deposition of pure carbonate sequences has been described from the Bahamas Platform (Iturralde-Vinent 1998) and basinal DSDP site 540 in the Florida Straits (Schlager & Buffler 1984).

Tectonic environment of the continental margins

The different stratigraphic and facies sections in western and central Cuba (Fig. 3) allow the interpretation that the respective continental margins in which they formed originated in different geotectonic environments. The Sierra Guaniguanico,

representing the east Yucatán margin, formed during the opening of the oceanic basin parallel to the Proto-Caribbean spreading ridge, which steps downward from the Esperanza belt through the Southern Rosario belt to the northern Rosario belt (Pindell 1985; Pszczółkowski 1999). In northern central Cuba, the section represents the southern Bahamas margin, formed under transform control associated with the opening of the Atlantic. Further, in some geotectonic models the south Bahamas margin is shown as developing diachronously from west to east by movement along a palaeo-Bahamas–Guayana transform fault (Pindell 1985; Iturralde-Vinent 1994; Stanek & Voigt 1994; Pindell & Kennan 2001; Pindell *et al.* 2006).

Apart from contrasts in sedimentation on these two palaeo-margins, there are also differences in magmatic activity. In the Guaniguanico Terrane, basic lavas and small subvolcanic intrusions have been observed in the Middle to Late Jurassic (*c.* 155–160 Ma) in all tectonic units (Cobiella 1996; Pszczółkowski 1999). From the southern Bahamas margin, basic volcanic rocks (150 Ma) are only known from the Sierra Camaján north of Camagüey (Iturralde-Vinent & Morales Marí 1988). But the rocks of the Sierra de Camaján have been assigned to the Placetas Belt (Iturralde-Vinent *et al.* 2000). This could suggest that the Sierra de Camaján and thus the whole Placetas do not belong geotectonically to the palaeo-Bahamian margin, but rather to the oceanic basin east of the

368   K. P. STANEK *ET AL.*

Yucatán Block. Until now, no rift-related magmatism has been detected along the southern margin
of the Bahamas Platform.

Is there a continental basement?

Until now, there has been no drilled or otherwise
exposed basement of the Bahamas platform
around the Cuban archipelago. The presence of continental crust beneath central Cuba has been
suggested by the interpretation of gravity data proposing a thinned continental basement below the
southern edge of the Bahamas platform (Otero
et al. 1998). The presence of a continental basement
in this part of the Bahamas is supported by the
occurrence of the Punta Alegre and other salt
diapirs amidst the imbricated shallow-water Bahamian carbonates of the Cayo Coco zone of central
Cuba and by the presence of probable Jurassic red
beds drilled at the Great Isaac borehole, some
300 km north of central Cuba (Meyerhoff &
Hatten 1974). In and near the arc-related parts of
Cuba (Zaza zone of Pardo 1975), there are several
field indications of slivers of a continental basement included in the thrust structure of the Cuban
fold and thrust belt. As for western Cuba, basement
remains unknown, but the lowest unit (San Cayetano
Fm) derives from quartzose continental basement
types with Precambrian and Lower Palaeozoic
ages of detrital mica (Hutson *et al.* 1998) and
zircon grain ages (Rojas-Agramonte *et al.* 2008).

From the above, a fixist view might suggest that
the entire region is underlain by continental crust.
However, it must be kept in mind that the Island
of Cuba represents three distinct domains; two

sedimentary provinces (West and Central Cuba)
are separated from an overthrust arc terrane at an
oceanic suture down the island's axis. Because all
arc-related magmatism lies south of this suture, subduction was SW-dipping and the oceanic basin that
closed originally lay between the Bahamas and the
Cuban part of the Great Antillean arc. Because of
the long history of arc magmatism (at least Aptian
to Campanian/Maastrichtian, some 40 Ma), followed by several hundred kilometres of north–
south tectonic shortening adjacent to the suture
zone after the magmatism ended (Hempton &
Barros 1993), the original palaeogeographic separation was necessarily significant (>1000 km).
Thus, we should not expect any particular relationship between basement rocks of the allochthonous
Cuban arc with those of the Bahamas (Central
Cuba) or of the Yucatán (West Cuba) margins.
This is borne out by the respective geologies as
well: there is no mutual geological affinity
between the Cuban arc and the Bahamas/Florida
Straits DSDP sites until the Maastrichtian or possibly even the Palaeogene (i.e. first arrival of volcanogenic sands/arc-related tuffs in the latter).

In the Cuban arc terranes and adjacent thrust
belt, local areas of metamorphic rocks have been
reported since the 1950s (Fig. 5). Because of their
uncertain nature at the time, a pre-Cretaceous age
was commonly allocated to them. As geochronological work began, Precambrian and Jurassic ages
were reported on metamorphic rocks and spatially
related granites, respectively, from tectonic blocks
involved in the folded and thrusted sequences of
the Placetas belt of the Sierra Morena (Somin &
Millán 1981; Renne *et al.* 1989). An Ar/Ar cooling

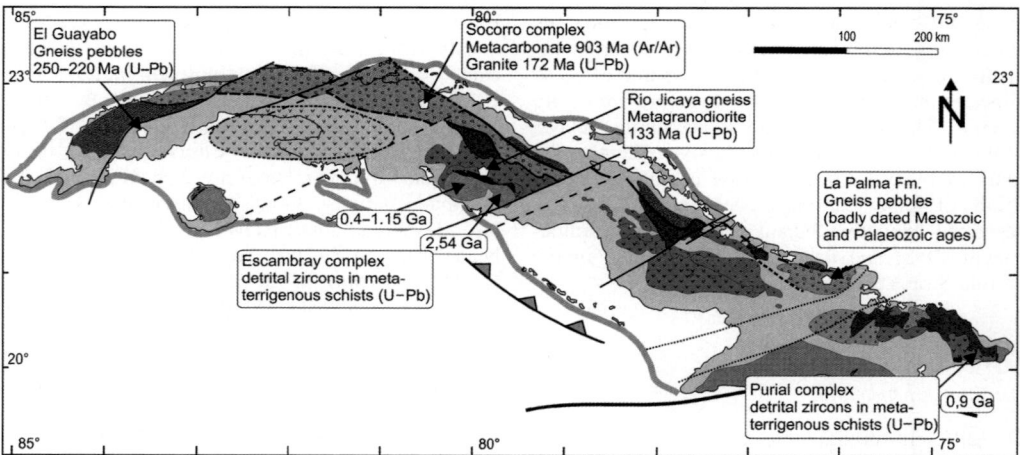

Fig. 5. Occurrences of metamorphic and magmatic rocks interpreted as 'basement' rocks of the continental margin
and the Cretaceous island arc. The zircon data on metaterrigenous schists from the Escambray and the Purial complexes
are given for comparison.

age of 903 Ma for biotite was reported from meta-carbonate rock slivers of the so-called Socorro complex. The intrusive age of pink granites in the Rio Cañas valley (also included in the Placetas belt) was determined by U–Pb on zircon as 172 Ma (Renne et al. 1989). K–Ar data and the observation of weathering crusts above the granites have been interpreted as erosion along the southern Bahamas platform in the Upper Jurassic (140–150 Ma) (Somin & Millán 1981; Pszczółkowski 1986a). However, the tectonic position and the contacts with the hosting units of the Socorro complex and the granites are unknown due to the poor exposure. If the granites and the metamorphic rocks belong to the rock suite of the Placetas belt, then they are far-travelled from the south. If not, both the Socorro complex and the granites are slivers of the Bahamas basement tectonically involved in the fold and thrust belt during the thrusting of the island arc. Additional continental lithologies associated with the Placetas and Camajuaní belts are the felsic gneisses in tectonic mélanges north of Holguín (La Palma Formation); the geochronology of these rocks is not well constrained, but K–Ar determinations on mica suggest Palaeozoic to Early Mesozoic ages (Kosak et al. 1988). Of the above, the 903 Ma age is most interesting, as such ages are not otherwise known from Florida, the Bahamas or Yucatán, and suggest a highly allochthonous origin possibly as far as the Grenvillian terranes of SW Mexico or the Chortís Block.

To the south of the Cuban Main Thrust where the Cuban Cretaceous arc lavas and intrusives occur (Fig. 2), at least two additional areas of metamorphic rocks are found that had originally been thought to represent Cuban basement (see the geological map edited by Perez Othon & Yarmoliuk 1985). These are the Pinos complex (Millán 1975; Somin & Millán 1981; García-Casco et al. 2001) and the Escambray complex (Millán & Myscinski 1978; Somin & Millán 1981). Based on fossils and facies analysis, the protoliths of these metasedimentary and metavolcanic rocks are facies analogous to and coeval with the (Late Jurassic–Early Cretaceous) sedimentary sequences of the West Cuban antiformal nappe stack (Pinar del Rio) (Millán & Myczynski 1978; Piotrowska 1978; Pszczółkowski 1985). The similarities in the sedimentary facies and depositional ages of the metamorphic protoliths with the sedimentary sequences of western Cuba suggest a common palaeo-environment along the NW Proto-Caribbean basin or continental margin. Provenance analysis for the siliciclastic sediments of the Escambray by U–Pb data on detrital zircons gave 0.4–1.15 Ga and 2.45 Ga (Early Palaeozoic to Early Proterozoic) as the age of the sedimentary source (Rojas-Agramonte et al. 2008; Krebs et al. 2007). Ar/Ar data on detrital muscovite grains

provide a Pan-African source area of 700–550 Ma for the sediments of West Cuba (Hutson et al. 1998), but recent SHRIMP U–Pb ages for zircons from the San Cayetano Fm sandstones of West Cuba range from 398–2479 Ma (Rojas-Agramonte et al. 2008), nearly identical to those from Escambray and further verifying the facies and chronological correlation. Of great importance is that the quartzose and zircon-bearing metasediments of the Escambray, and probably also of the Pinos complex (García-Casco, pers. comm. 2006) underwent HP metamorphism. Thus, they represent Jurassic–Early Cretaceous material that was subducted at the Cuban trench (Stanek et al. 2000), meaning that the sedimentary protoliths lay on the Proto-Caribbean seafloor and/or its margins ahead of the advancing arc. These metamorphic complexes do not represent the basement of the original Caribbean Arc, but rather were underplated to it (i.e. accreted to the base of the arc crust by subduction). García-Casco et al. (2008) make a strong case for these sediments having originally been located off the coast of southern Yucatán, forming a palaeogeographic element which they call 'Caribeana', where they were subcreted to the advancing Cuban arc in the Campanian–Maastrichtian.

It is important to realize that the metamorphic rocks noted above in the thrusted subduction zone mélanges of the Cuban suture are continuous beneath the synformal and monoclinal arc belt with the Pinos and Escambray metamorphics to the south. Both regions represent the same subduction channel, although the channel has been tectonically elevated in the south during the opening of the Yucatán Basin to expose a deeper level at Escambray and Pinos (Pindell et al. 2005). Support for this continuity lies in the Cangre Belt along the north flank of the Pinar Fault: there, additional outcrops of the Pinos- and Escambray-type metasediments structurally connect the Cuban Suture with the southern metamorphic rocks in an up-plunge position where the synformal arc terrane projects westward above the surface geology. In addition, pebbles of garnet-bearing gneiss and S-type granite are found in a Palaeogene conglomerate (El Tumbadero unit) in the Los Palacios Basin just south of the Pinar Fault (Somin et al. 2006). SHRIMP dating on zircons from these pebbles reveals a probable Late Triassic protolith age (220–250 Ma) and a metamorphic overprint at about 72 Ma, related to very fast exhumation/cooling determined by K–Ar dating (about 71 Ma). The source of the pebbles is likely to be slivers of Proto-Caribbean continental margin basement that were subducted along with the sediments discussed above. The best fit of geochronological characteristics of the pebbles with regional geology and evolution shows the El Tumbadero unit at the southern edge of the Yucatán platform (Martens *et al.* 2007), similar to the position of Caribeana as proposed by García-Casco *et al.* (2008). These pebbles may provide the first indication of basement involvement in Caribeana.

In summary, the geology of the Cuban 'basement' is known only for a scattered collection of localities. The Great Bank of the Bahamas probably has a stretched continental foundation (Pindell 1985) against which Cuban arc and accretionary prism (Placetas Belt and Maastrichtian–Palaeogene flysch units) collided in the Palaeogene. Unless we accept a 903 Ma metamorphic age for the Bahamian basement, which appears contrary to the regional geology of the rest of the Florida/Bahamas, it is unlikely that any of the Bahamian basement has been caught up in the Cuban thrust belt as the arc approached. It appears that only the sedimentary cover of that basement (carbonates and salt diapirs of Cayo Coco zone) was picked up in the thrusting. As so often is the case, the salt horizon of the southern Bahamas probably served as the décollement for this imbrication. As for the Cuban arc itself, pre-Jurassic continental lithologies occur only as slivers or blocks, mainly in mélange units associated with subduction and subcretion, which could have entered the subduction channel by lateral transport from areas of continental crust along strike in the Early Cretaceous, such as the Chortís Block and southwestern Mexico (Pindell *et al.* 2005). Otherwise, the arc is entirely an intra-oceanic arc thrust onto the Bahamas. Subcreted metasediment serves as 'basement' along the southern flank of Cuba, which is a Maastrichtian–Paleocene rift flank of the Yucatán intra-arc basin whose detachment level cut into the Late Cretaceous subduction channel (Pindell *et al.* 1988, 2006). However, this rifting was situated so close to the Cuban Trench that the central Cuban terrane was thereafter narrower than any typical arc-trench gap, so any latest Cretaceous–Palaeogene arc magmatism probably lies in the southern offshore or Yucatán Basin floor (Pindell & Barrett 1990), and possibly the Cayman Ridge (Sigurdson *et al.* 1997; Lewis *et al.* 2005). This rifting and magmatic shift was apparently associated with the subcretion of the Caribeana sediment pile (García-Casco *et al.* 2008), which itself probably lay above stretched continental crust whose buoyancy caused subduction zone flattening that led to the southward shift in arc magmatism. Finally, the West Cuban thrust pile comprises sediments from the eastern Yucatán margin that were transpressed by the Cuban arc but not subducted as deeply as Pinos and especially Escambray. It is not known if any basement from that margin is incorporated at depth in the sinistral transpressive thrusting. The fact that sediments as old as the Middle Jurassic

San Cayetano occur at the surface today suggests three possibilities: (1) that sediment imbrication is multi-fold, bringing old strata to shallow levels atop other imbrications; (2) that the West Cuban section was thrust onto a nearly autochthonous piece of thick, probably Yucatán, continental crust; or (3) one or more slices of Yucatán marginal basement is/are imbricated in the thrusting with the sediments; if the slice is thick enough, then a single thrust (with simple attendant structure) may suffice to have caused the high structural level.

Timing of thrusting and nappe stacking at the Yucatán and Bahamas platforms

In Central Cuba, northward shifting (migration?) of stratigraphic hiatuses can be observed in the sediments representing the Proto-Yucatán or Proto-Caribbean Basins (i.e. Placetas and Camajuaní Belts). In the southern basin and slope areas of the Proto-Yucatán Basin, deposition stopped at the end of the Turonian (*c.* 90 Ma). Erosional and weathering features have been described in the uppermost part of the basin sediments of the Placetas zone (Pardo 1975). In the Turonian until Campanian a northward-shifting uplift and emergence of the southern Placetas and Camajuaní zones and their involvement in the accretionary prism of the approaching Caribbean arc can be assumed from available stratigraphic data (see Iturralde-Vinent 1994, and citations therein). At this time the Caribbean arc was located near the tip of the southern Yucatán Block. The thrusting of the accretionary prism of the Caribbean arc onto the southern Bahamas platform ended in the Late Eocene, when foreland sediments (Senado Fm) were overthrust by ophiolite nappes (Iturralde-Vinent *et al.* 2000).

In western Cuba (see Fig. 4), the Rosario and the Sierra de los Organos units consist of several nappe sheets which have been thrust from south to north (Pardo 1975; Piotrowska 1987). The tectonically uppermost unit of the stack is represented by the fragments of ultrabasic rocks, gabbro and serpentinite (Sierra de Cajalbana) and the island-arc rocks of the Bahia Honda unit, which were thrust from the south over the continental margin and possibly deeper water sediments (Pszczółkowski & Albear 1982; Pszczółkowski 1994; Iturralde-Vinent 1994). Final thrusting of the island-arc-related rocks and their frontal piggy-back basins took place in the Late Paleocene to Early Eocene between 56 and 50 Ma (Bralower & Iturralde-Vinent 1997). Crucial for the interpretation of the provenance of the tectonic units is the timing of the clastic input of island-arc-derived detritus into the sediments (Pszczółkowski 1994). The first record is the volcanoclastic component of the

Campanian Moreno Formation (northern Rosario unit, c. 75–70 Ma). Fragments of subduction-related igneous rocks are also reported from the latest Maastrichtian Via Blanca Formation (Takayama *et al.* 2000). In the southern Rosario unit the carbonates of the Middle Paleocene Ancon Formation (c. 60 Ma) yield volcanoclastic detritus, whereas the input of such detritus in the Sierra de los Organos unit is delayed until the Upper Paleocene and Lower Eocene Manacas Formation (c. 55 Ma). Comparing the onset of arc-derived sedimentary influence with the indications for northward thrusting and stacking, an inverse position of these tectonic units due to thrusting is apparent (Pszczółkowski 1999; Saura *et al.* 2008). This can be explained by the formation of fault-bend-fold structures and a regional duplex. The duplex has been inverted with northern vergence due to the northward thrusting of the subduction-accretionary complex and the forearc of the Caribbean Arc. Parts of the foreland thrust belt were buried by higher nappes and/or island-arc units. The siliciclastic sediments of the San Cayetano Formation (Sierra de los Organos unit) underwent low-grade metamorphism with estimated pressures of 2 kbar, equivalent to a burial depth of 6–7 km (Hutson *et al.* 1998). Parts of the weak metamorphic overprint can be attributed to the stacking of thrusts. The age of this low-grade overprint is probably post-Maastrichtian–pre-Late Paleocene. The nappe of San Cayetano sediments was tectonically covered by the metamorphic unit of the Cangre belt probably in the Palaeogene. The metabasic rocks of the Cangre belt show indications of a HP metamorphic overprint (Somin & Millán 1981; García-Casco *et al.* 2002; Cruz *et al.* 2007) of uncertain age. On the basis of balanced cross-sections, the shortening by northward thrusting in western Cuba can be estimated from the nappe geometry to be a minimum of 150–200 km (Saura *et al.* 2008).

The Caribbean (Great Antillean) Arc: a record of movement between the Caribbean and North American plates

Island arc-related geological units of Cuba

Outcrops of the Cretaceous Caribbean (island) Arc extend along the Greater Antillean islands from western Cuba to at least Puerto Rico. In Cuba, Cretaceous island-arc-related rocks form a belt from the Pinar Fault through the entire length of the island, and form the largest outcrop of such units in the Caribbean. South of the Eocene suture (the Cuban Main Thrust of Knipper & Cabrera 1974) a 4–5 km thick nappe comprising a serpentinitic mélange with overlying island-arc rocks has overridden the southern edge of the Bahamas Platform (Fig. 2). The serpentinitic mélange contains dislocated slivers of all rock types typical of an ophiolitic profile; this was the reason for calling it the 'Northern Ophiolites' (Kudělásek *et al.* 1984; Iturralde-Vinent 1994). Parts of the thinned island arc are located on the top of the 'ophiolitic' thrust sheets. South of the Cretaceous island-arc units, domes like the Pinos complex (Isla de Juventud) and the Escambray complex form isolated outcrops of relatively high-grade metamorphic rocks (see Fig. 2). These metamorphic complexes and the serpentinitic mélange together represent the subduction–accretionary complex of the Caribbean Arc. In general, thrusting of the island arc onto the southern edge of the Bahamas platform led to incremental tectonic episodes of exhumation which now allow both the stratigraphic succession as well as the lateral extent of the igneous rocks to be studied. On the basis of the tectonic setting and the regional structure, the following key outcrop areas of Caribbean Arc-related rocks can be distinguished (Fig. 2):

• The Bahia Honda unit northwest of the Pinar fault (see Fig. 4), deposited in a forearc position, and now resting as an allochthonous sequence at the top of the West Cuban antiformal nappe stack.

• The mélange unit in the area of La Habana–Matanzas, consisting of completely dislocated island-arc sequences and carbonate rocks with continental margin provenance. South of the mélange zone, the Batabano Massif has been interpreted to be a part of the Cretaceous island arc covered by younger sediments on the base of gravimetric and magnetic data (Bush & Sherbakova 1986).

• The megasyncline of Cretaceous island-arc volcanic rocks in Las Villas, infolded above the metamorphic Escambray complex to the south and the ophiolitic mélange to the north.

• The monoclinal stack of island-arc intrusions and related volcano-sedimentary sequences in the area of Camagüey.

• The mélange-like unit surrounding Holguin.

• The metamorphosed Cretaceous island-arc volcanic sequences in Oriente, forming the footwall of the overthrust Oriente ophiolite.

Subduction related magmatism

In western Cuba, the Bahia Honda unit comprises Early to Mid-Cretaceous island-arc sequences and basic and ultrabasic rocks of oceanic crust provenance (Kerr *et al.* 1999). Tectonically, the unit is an allochthon (Fig. 4), resting on other tectonic units of the Yucatán continental margin (as seen at borehole Mariel 50 km east of the city of Bahia Honda; Segura Soto *et al.* 1985). The inverted stratigraphic profile, the north-vergent tectonic style and the presence of the underlying continental margin (Straits of Florida) were the arguments for the interpretation of the Bahia Honda unit being the uppermost nappe of the West-Cuban antiformal stack (Pardo 1975; Stanek *et al.* 2000; Saura *et al.* 2008). The occurrence of large blocks of ultrabasic rocks (Cajalbana Massif) and tholeiitic basalts of the Encrucijada Formation (Fonseca *et al.* 1984) have been used as arguments for the interpretation of the Bahia Honda unit as a remnant of a back-arc basin (Iturralde-Vinent 1994, 1998). The tholeiitic basalts and the presence of boninitic volcanic rocks of suggested Early Cretaceous age have been tentatively related to an early primitive arc north of the later Caribbean Arc and a subsequent polarity reversal of the subduction (Kerr *et al.* 1999, 2003).

In central Cuba, the largest outcrop areas of the Caribbean Arc are located west and east of the Trocha Fault zone (see Fig. 2). In Las Villas to the west of the fault, the thickness of internally thrusted island arc units including the serpentinitic mélange (arc basement), calculated from seismic data, reaches up to 10 km (Pusharovski *et al.* 1989), which is unusually thin for a long-lived island arc. It seems that the original island arc has been dismembered tectonically and deeply eroded. In Las Villas, the volcanogenic parts of the Cretaceous arc have been folded into a several kilometre-wide megasyncline, comprising the nearly complete stratigraphic section of the Caribbean Arc. In contrast, east of the Trocha fault zone (Ciego de Avila–Camagüey–Las Tunas), Late Cretaceous erosion cut deep into the thrusted island arc sequence exposing the deep-seated and subvolcanic intrusions of the Cretaceous island arc. The alignment of the intrusive bodies clearly defines two magmatic belts, a wider one in the south with dominating alkaline and calc-alkaline granitoids and a northern belt with Na-rich bimodal intrusive bodies (Stanek & Cabrera 1992; Marí Morales 1997; Stanek *et al.* 2005).

In the Las Villas megasyncline, the pre-Albian part of the stratigraphic section starts with a sodium-rich bimodal series of plagio-rhyolites and basalts (Los Pasos Formation) outcropping in the southern fringe of the syncline (Fig. 6). The volcanic rocks show island-arc tholeiitic (IAT) geochemical characteristics. The upper part of the formation includes air fall tuffs, tuffites, marl and stratiform sulphide bodies (Dublan & Alvarez 1986). The age of the Los Pasos volcanic suite is still not constrained. After the extrusion of the Los Pasos lavas, the geochemical behaviour of the magmatism changed from IAT to calc-alkaline lavas (Diaz de Villalvilla 1997). In the southern part of the megasyncline the Matagua and Cabaiguan Formations

consist of basaltic and andesitic lavas, related tuffs and volcanoclastic sediments. Both formations are overlain by the Provincial Formation. The Provincial Formation yields facies transitions from flyschoid to carbonate sediments. The age of the carbonates was constrained on the basis of a rich marine fauna to the Upper Albian and Lower Cenomanian (*c.* 105–98 Ma, Iturralde-Vinent 1996). In the Upper Cretaceous, volcanoclastic rocks and sediments dominate the stratigraphic section. In the lower part of the Late Cretaceous, differentiated calc-alkaline lavas are involved in the stratigraphic sequence (Perera Falcón *et al.* 1998), but subduction related magmatism ceased in Las Villas in the Campanian (Diaz de Villalvilla 1997). The lower part of the volcanic sequences and of related intrusions has been metamorphosed up to the amphibolite facies, forming the Mabujina unit (Somin & Millán 1981). The stratigraphic age of the Mabujina unit has been estimated from poorly preserved sporozoans and pollen as Jurassic–Cretaceous (Dublan & Alvarez 1986). Zircons from two plagiogranitic gneisses dated by the U–Pb method range between 110 Ma (Bibikova *et al.* 1988) and 133 Ma (Rojas-Agramonte *et al.* 2006*a*, *b*). This period probably spans the age of the initiation of the Caribbean Arc, judging from regional geology and Atlantic triggering mechanisms (Pindell 1993). On the basis of isotopic data, Blein *et al.* (2003) suggested that the amphibolites and gneissic granitoids of the Mabujina Unit are correlative with the Mexican Guerrero terrane, which may have been along-strike with the Great Antillean arc prior to the Albian.

A small belt of outcrops of subduction-related granitoids can be mapped along the northern limit of the Escambray Mountains. There are two textural types of granitoids. Undeformed granitoids form the so-called Manicaragua batholith intruding both the volcanic formations in the north as well the amphibolites and gneisses of the Mabujina unit in the south. U–Pb data on zircons (see Fig. 6) have been compiled from Hatten *et al.* (1988), Rojas-Agramonte *et al.* (2006*b*) and Stanek *et al.* (2005). This preliminary data set suggests: (1) an early stage of magmatism from 132 Ma into the 90s Ma; (2) a ductile deformation event at about 90–88 Ma, during which the parts of the subduction-related intrusive rocks have been transformed into gneisses; and (3) the intrusion of the Manicaragua type granitoids between 87 and 80 Ma. The dating of the intrusion of undeformed pegmatites and granitoids into the Mabujina unit (88–80 Ma) constrains the timing of the deformation event (Grafe *et al.* 2001; Rojas-Agramonte *et al.* 2006*b*). Most probably, the ductile deformation of the lower parts of the island arc should be related to the collision of the Caribbean arc with the southern margin of the

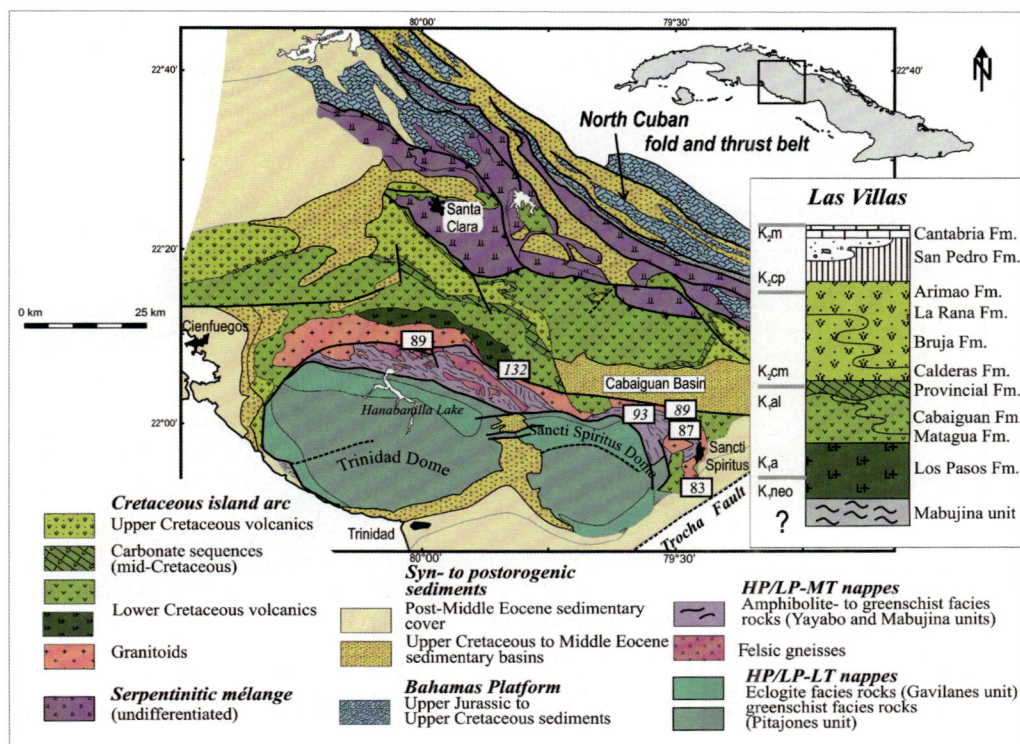

Fig. 6. Tectonic sketch map of western central Cuba (Las Villas), adapted from Kančev (1978) and Belmustakov *et al.* (1981). The North Cuban fold and thrust belt has been overridden by a serpentinitic mélange and Caribbean Arc-related volcanic formations. In the south, the Manicaragua batholith (undeformed granitoids) is in tectonic contact to the Mabujina unit with mostly gneissic granitoids. The numbers in the boxes refer to U–Pb ages on zircon of granitoids; numbers in italics are gneissoid rocks. In the southern backland of the Caribbean Arc, the Escambray metamorphic complex has been exposed since Palaeogene. A generalized stratigraphical section of the Caribbean Arc is given in the inlayer.

Yucatán Block (Pindell *et al.* 2006; Ratschbacher *et al.* 2009).

To the north, the Caribbean Arc-related volcanic sequences of Las Villas are underlain by a serpentinitic mélange including fragments of the entire ophiolite suite. Here the thickness of the mélange has been estimated on the basis of geophysical data as about 1.5 km (Bush & Sherbakova 1986). So far, the age of the ophiolitic fragments in central Cuba is not well constrained. Isolated findings of fossils in cherts intercalated in basaltic lavas span the Upper Jurassic (Llanes-Castro *et al.* 1998) to the Middle Cretaceous (Fonseca *et al.* 1984).

East of the Trocha fault, around Camagüey, erosion has removed the uppermost volcanic structures and revealed a monoclinal stack of igneous massifs. The thrust faults are hidden by Palaeogene basinal sediments (Iturralde-Vinent & Thieke 1987). The intrusive massifs form the magmatic

axis of the Caribbean Arc (Fig. 7) and can be sub-divided into two belts. The northern belt consists of bimodal intrusions of IAT affinity. The geochemical patterns are similar to those of the Los Pasos Formation, giving a reason to correlate the northern intrusive belt of Camagüey with the Los Pasos Formation in Las Villas (Stanek & Cabrera 1992). A single date on zircon (Stanek *et al.* 2005) suggests a Campanian age, similar to small stocks intruding the ophiolitic mélange of Las Villas (Rojas-Agramonte *et al.* 2006b). Thus, IAT type magmatism appears to occur not only in the lowermost part of the Caribbean Arc, but also in the forearc at later times. The southern belt near Camagüey consists of large alkaline (syenites and monzonites) and calc-alkaline differentiated intrusions. The broadly eroded sections of the plutons west of the Camagüey fault (Fig. 7) and the occurrence of hydrothermal alteration zones of the apical part of the intrusions

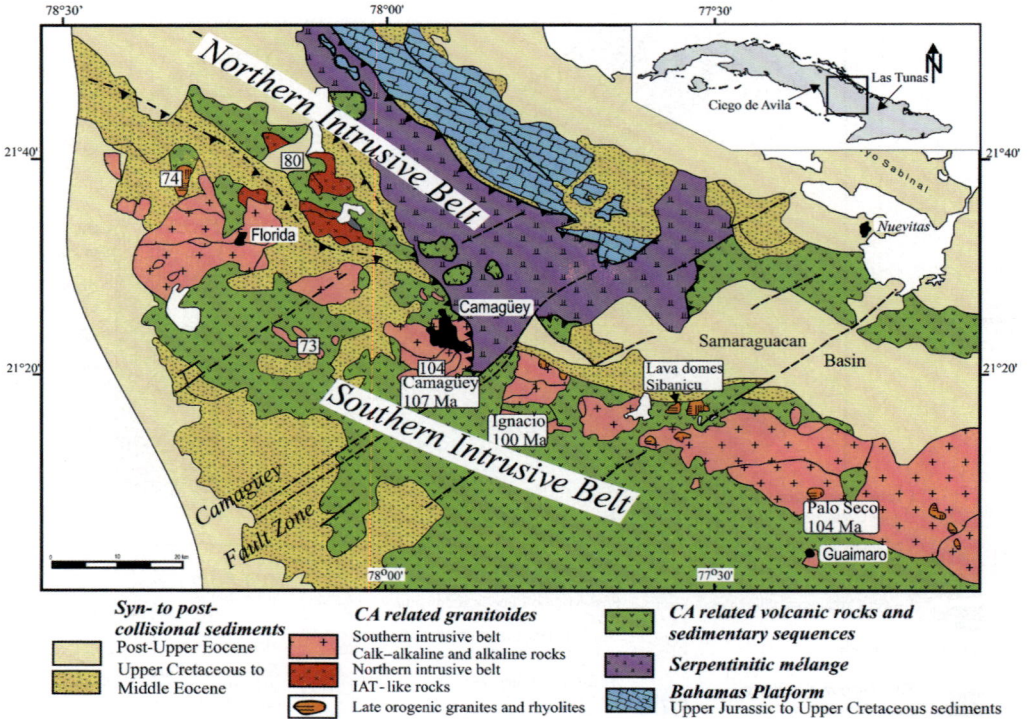

Fig. 7. Tectonic sketch map of eastern Central Cuba (Camagüey), adapted from Iturralde-Vinent & Thieke (1987). The sequences of the Caribbean Arc and the Camagüey batholith form a monoclinal structure; suggested thrust faults have been covered by Palaeogene sedimentary basins. The numbers in boxes refer to U–Pb ages on zircon of granitoids by Stanek *et al.* (2005) and Rojas-Agramonte *et al.* (2007).

east of the fault could be interpreted as east-side-down normal faulting along the Camagüey fault. U–Pb dating on zircon and titanite revealed that the alkaline rocks intruded the arc in the Albian (107–100 Ma), whereas most of the calc-alkaline intrusions yield ages between 95 and 75 Ma (Stanek *et al.* 2005; Rojas-Agramonte *et al.* 2006*b*). The last magmatic pulses in the Caribbean Arc have been reported from felsic lava domes in the area of Sibanicú (show on the map in Fig. 7), east of Camagüey (Hall *et al.* 2004). The Ar/Ar cooling age of about 75 Ma on rhyolites can be considered as the time of extrusion. Similar U–Pb ages on zircons of about 75–78 Ma were obtained from small intrusive stocks west of Camagüey (Stanek *et al.* 2005), but Ar/Ar 'ages' of about 75 Ma have also been obtained from igneous rocks intruded 25–35 Ma earlier (the alkaline rocks of the Camagüey, Ignacio and Palo Seco massifs shown on the map in Fig. 7). The similar Ar/Ar ages in both older intrusive and youngest extrusive rocks give the impression that the island arc was uplifted and cooled down through the 300 °C

isograd in the Late Campanian (at about 75 Ma). The same age range of Ar/Ar cooling ages in the island arc was described in the area north of the Escambray Massif (Grafe *et al.* 2001).

Considering the results of the dating of zircons from the Los Ranchos Formation in Hispaniola between 118 and 110 Ma (Kesler *et al.* 2005; Escuder Viruete *et al.* 2006) and the similar stratigraphic position of the Los Pasos Formation in central Cuba, the initiation of Cretaceous arc magmatism in both countries is at least Early Aptian (>120 Ma). The magmatic activity in the onshore central Cuban part of the Caribbean Arc lasted about 45 Ma (to the Campanian/Maastrichtian). If we acknowledge that the arc axis shifted south at that time, then arc magmatism lasted into at least the Paleocene. This should not be surprising: in Hispaniola the arc persisted into the Eocene, which is also the time of arc collision with the Bahamas (Nagle 1974; Pindell & Draper 1991), recording some 70 million years of south-dipping subduction beneath the arc as it advanced between the Americas from the Pacific.

Over this time the subduction behaviour changed episodically, as reflected in the geochemical characteristics of the intrusive rocks of Camagüey through time (Stanek *et al.* 2005). The geochemical characteristics of the granitoids indicate a subduction origin of the igneous suites (see Fig. 8). The magmatism of the Caribbean Arc starts with island arc tholeiites (the Primitive Island Arc suite in the sense of Donnelly & Rodgers 1980), typical for early subduction of oceanic crust. Development of alkalic magmatism may relate to the onset of subduction of hot (young) oceanic crust, or to subduction of a seafloor spreading ridge. After cooling of the

subduction zone, the magmatism continued in the Late Cretaceous with calc-alkaline intrusive compositions (see also Kerr *et al.* 2003). The final magmatic 'pulse' at about 75 Ma consists of only small felsic magmatic bodies which are timely related to the collision of the Caribbean Arc with the southern edge of the Yucatán block (Ratschbacher *et al.* 2009).

Uplift and erosion of the Caribbean Arc

In southern Central Cuba, a peneplain has formed revealing the emergent island arc formations at the

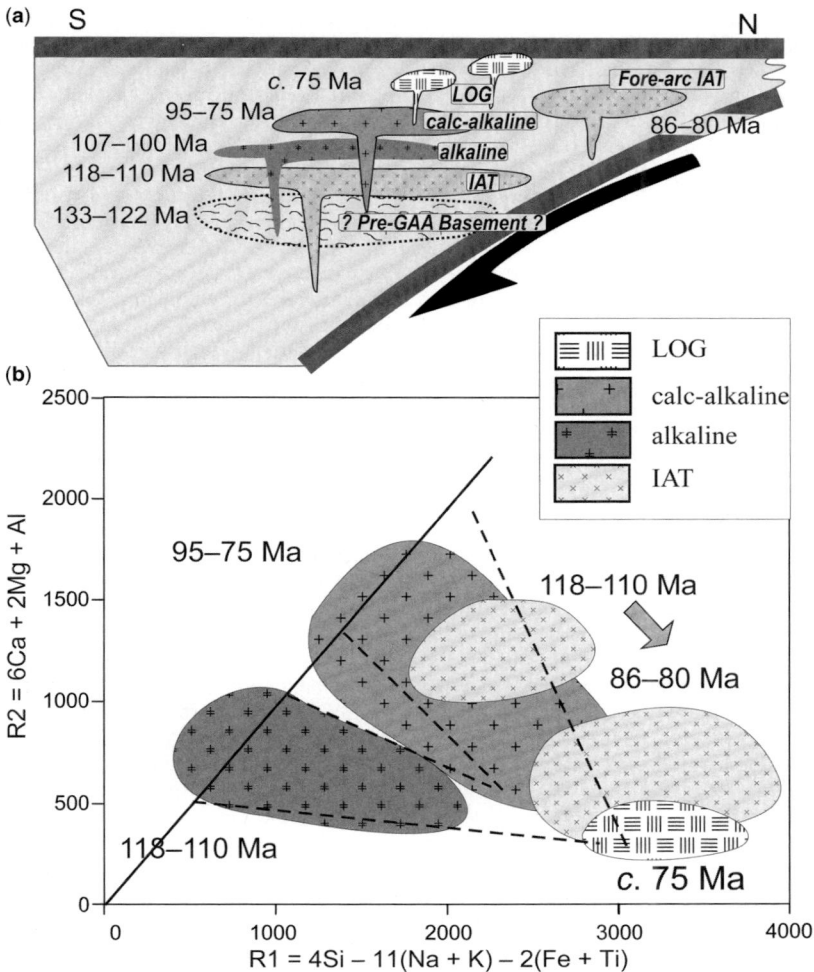

Fig. 8. (**a**) Suggested spatial succesion of the magmatic 'events' in Cuban part of the Caribbean Arc. (**b**) Geochemical evolution in time of the Cuban part of the Caribbean Arc, based on the relations of main elements (Batchelor & Bowden 1985). The magmatism of the Caribbean Arc starts with island arc tholeiits (IAT), shows an episode of alkali magmatism and continues with calc-alkaline composition. The final magmatic 'pulse' is related to small plugs of granitic composition in the field of 'Late Orogenic Granites' (LOG).

Middle Campanian through Early Maastrichtian level. A carbonate platform covering an interval of clastic, molassic sediments on this surface was deposited from the Early Maastrichtian (69.6 Ma) to the Late Maastrichtian (*c.* 67–65.5 Ma) (Pszczółkowski 2002). This platform was designated as the 'Proto-Cuban Maastrichtian Platform' (Iturralde-Vinent 1992). Similar Upper Cretaceous post-arc sediments have been mapped in the Camagüey area, differing only in some facies developments. As the Maastrichtian carbonate platform developed in central Cuba, the basement was intruded by rare basic to intermediate dykes. For example, samples from the Leila mine on the Isla de Juventud have given 68 Ma K–Ar ages; (E. Malinovski, pers. comm. 1988). This particular occurrence may pertain to the southward shift in the magmatic axis noted earlier. However, some young basaltic extrusions (51 Ma Ar/Ar age) are also known from Lavas La Mulata in NE Camagüey (Hall *et al.* 2004), which would not relate to a southward shift. These would instead lie within the deforming forearc (still supra-subduction) as it began collision with the Bahamas.

In central Cuba, the first indication of thrusting in the Caribbean Arc can be related to the onset of piggy-back basins. At the northern side of the Escambray metamorphic complex, the erosional detritus from the exhuming complex accumulated in the Cabaiguan Basin (Fig. 6). The sedimentation started in the Middle Paleocene (62–60 Ma) with olistostrome-like sediments including large blocks of Maastrichtian limestones which covered the Caribbean Arc rocks, as the Proto-Cuban Maastrichtian Platform had earlier. At its southern margin, the sediments of the Cabaiguan Basin were folded and rotated by the rising metamorphic complex in the Late Paleocene and Early Eocene. The first pebbles of HP metamorphic rocks of the Escambray complex appear in the Cabaiguan Basin only in the Eocene (*c.* 45 Ma). North of the Pinos complex to the west, pebbles of marbles, presumably derived from the Pinos complex, have been deposited in the Capdevila Formation (Lower Eocene) near Havana and in the Los Palacios Basin between the Isla de Juventud (isle of Pine) and the Pinar Fault. At the northern flank of the Cuban suture zone, the final thrusting of the ophiolitic units onto the southern Bahamas platform and foreland took place in the Middle Eocene (45–40 Ma), dated by the overriding of ultrabasic rocks onto the foreland olistostromes.

Caribbean Arc related rocks in eastern Cuba

In the easternmost part of eastern Cuba (Oriente) (Fig. 9), the Cretaceous volcano-sedimentary sequences of the Caribbean Arc are referred to as the Santo Domingo and Téneme Formations, as well as the metavolcanic rocks of the Purial Complex (Iturralde-Vinent *et al.* 2006). Here, the arc rocks are metamorphosed to various grades (Purial complex) and occur in a different tectonic position than in central Cuba (see Fig. 10). The blueschists and eclogite-facies rocks of the Sierra del Convento are overthrust by the Cretaceous volcanic arc sequences of the Purial complex, the Santo Domingo Formation and the Oriente ophiolite (Cobiella *et al.* 1984).

Until now, only a few data exist concerning the stratigraphy, structure and geochemical character of the volcanic rock sequences. Calc-alkaline as well as IAT and boninitic compositions have been reported from basic volcanic rocks (Proenza *et al.* 2006). Palaeontological data suggest Middle to Upper Cretaceous age of the volcano-sedimentary sequences (Iturralde-Vinent *et al.* 2006). Upper Cretaceous IAT-like volcanic rocks have been described, supporting the occurrence of a magmatic pulse of forearc related magmatism with IAT-like geochemical characteristics in the Late Cretaceous, which was also described in central Cuba (Rojas-Agramonte *et al.* 2006*b*; Stanek *et al.* 2005).

The volcano-sedimentary sequences have been overthrust by the Moa–Baracoa and the Nipe–Cristal ultrabasic massifs, considered as Oriente ophiolite by Fonseca *et al.* (1984), which thickness no oversteps 2 km (Knipper & Cabrera 1974). The mafic and ultramafic rocks of these massifs show characteristics of a supra-subduction related origin, such as those formed in an intra-arc basin (Proenza *et al.* 1999). The underlying Purial volcanic rock sequences underwent greenschist-facies up to blueschist- and amphibolite-facies metamorphism (Boiteau *et al.* 1972, Somin & Millán 1981; García-Casco *et al.* 2006).

The earliest possible time of metamorphism and thrusting (?) has been estimated as Campanian, based on poorly preserved Turonian–Campanian microfossils in marbles (Somin & Millán 1981; Millán & Somin 1985) and K–Ar ages of about 75 Ma on white mica of granitoid pegmatites (Somin *et al.* 1992). In some locations, the ultrabasic rocks are covered with unconformable stratigraphic contact by the Upper Maastrichtian Yaguaneque limestones (Cobiella *et al.* 1984; Iturralde *et al.* 2006). This stratigraphic relationship constrains the latest timing of thrusting, meaning that the thrusting of the Oriente ophiolite occurred coeval to the collision at the southern Yucatán margin and is probably related to this tectonic event.

The ultrabasic massifs of the Oriente ophiolite are surrounded by exotic breccias and very coarse-grained sediments (La Picota and Mícara

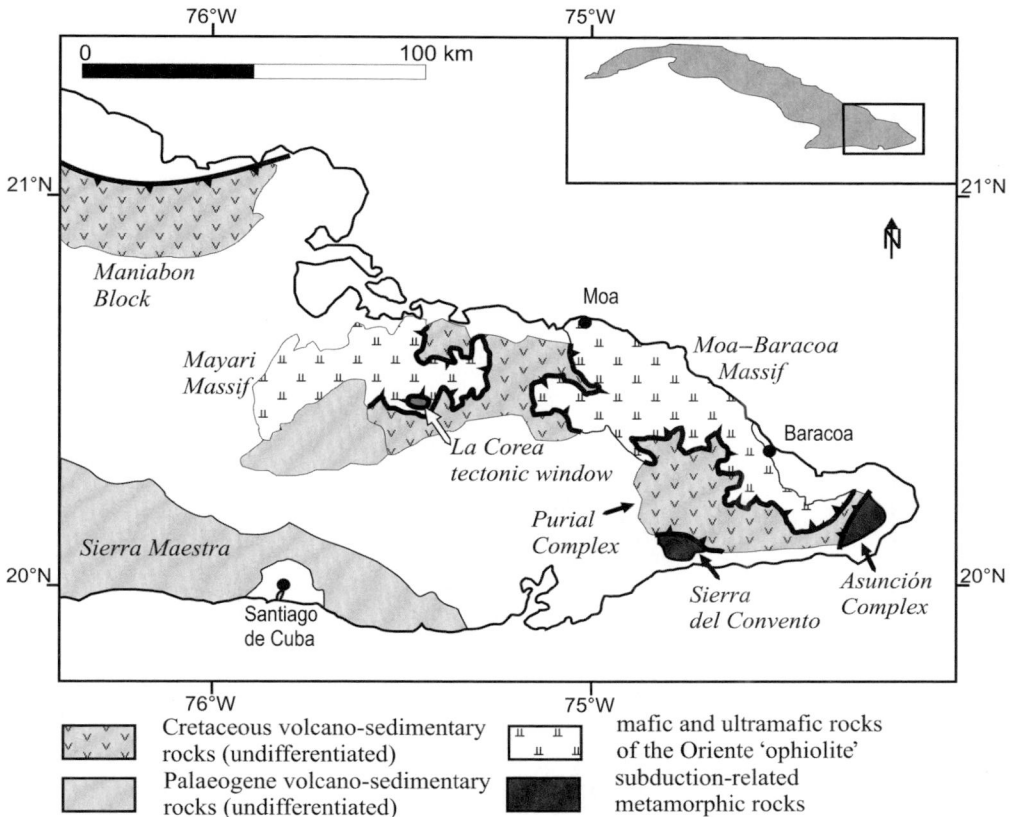

Fig. 9. Sketch of the geological units related to the Cretaceous–Palaeogene subduction in Eastern Cuba (Oriente). The heavy black lines indicate thrust faults. The Cretaceous volcano-sedimentary rocks comprise the Santo Domingo and Téneme Formations, and the metavolcanic rocks of the Purial Complex.

Formations) which contain Maastrichtian fossils (Cobiella *et al.* 1984; Iturralde *et al.* 2006). The La Picota and Mícara Formations are at the stratigraphic level of the Chicxulub impact. The structure of the sedimentary breccias and the large incorporated blocks of serpentinized ultramafic rocks and clastic sediments suggest that the La Picota and Micara Formations may have formed in response to tsunami waves caused by the meteoritic impact and are not the sedimentary fan derived from the thrusting of the ultrabasic massifs. The deposition of ultramafic debris together with littoral fossils suggests that the ultrabasic rocks had been exhumed up to the sea level at the K/T time.

Both the Oriente ophiolite and the underlying Purial meta-arc complex have been thrust onto the Asunción metamorphic complex to the east, which consists of metaterrigenous rocks of continental margin provenance, marbles and amphibolites (see Fig. 12c). Preliminary structural observations (field campaign 2006) in the serpentinites of the Oriente ophiolite and the Purial complex suggest an east- to southeast-directed thrusting of the nappe stack, disregarding possible subsequent larger block rotations.

In the southwestern part of Oriente, Cretaceous volcano-sedimentary sequences of the Caribbean Arc form the basement of Palaeogene volcanics and intrusions of Sierra Maestra. The Sierra Maestra arc rocks represent a short pulse of magmatism (Cazañas *et al.* 1998) that may or may not have been related to the primary SW-dipping subduction zone discussed thus far herein (see discussion in Pindell *et al.* 2006). The intrusions align parallel to the southern coastline of Cuba and have been dated between 60 and 48 Ma (Kysar *et al.* 1998; Rojas-Agramonte *et al.* 2004).

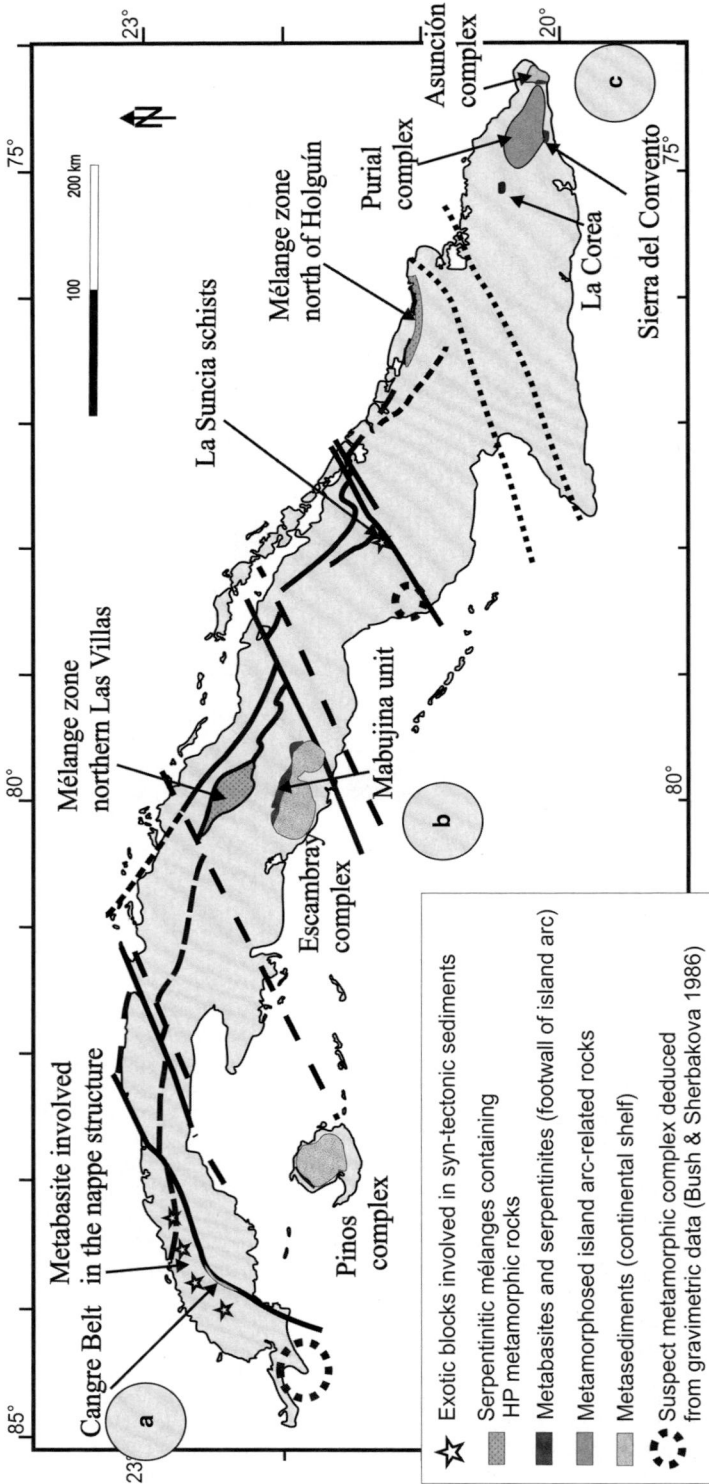

Fig. 10. Sketch map of the outcrops of subduction-related metamorphic rocks in Cuba and suspected complexes hidden by sediments. The letters in the circles indicate the position of the tectonic sections presented in Fig. 12.

Subduction–accretionary complex of the Cretaceous Caribbean arc in Cuba: $P-T$ paths of the metamorphic units and their timing

Geotectonic and petrological background

Today, subduction zones are accepted as an integral part of modern geodynamic concepts. The basic architecture has been clarified, even if details remain to be worked out. The Caribbean Arc had an intra-oceanic character for much of its magmatic life along much of its length, so that this architecture is in fact a relatively simple example of a subduction zone; both of the converging lithospheres had a similar structure. During subduction, the rocks of the downgoing plate will undergo deformation and metamorphism. In the near-surface part of the subduction zone, off-scraping and accretion of sediments and even upper parts of the downgoing oceanic crust may create an accretionary prism. The thickness and extent of this submarine fold and thrust belt will depend on the volume and nature of sediments introduced the angle of subduction, and the velocity of plate convergence. The rocks of the down-going plate face progressive metamorphism and dehydration due to increasing pressure and temperature and dynamic deformation. High-pressure/low-temperature metamorphic suites (such as blueschists and eclogites that require lower than normal temperatures for a given pressure to form) dominate the depth range of 20–80 km along the slab, as isotherms are physically dragged deeper than normal due to the downgoing slab being colder than the surrounding mantle. These high-pressure/low-temperature (often abbreviated simply to HP/LT or HP) suites cannot form in the upper crust, as horizontal tectonic forces are not strong enough to cause the required pressures; sub-horizontal thrusting (and reduction of cumulative stress by tectonic escape) occurs long before diagnostic HP minerals (such as lawsonite, aragonite, jadeite, omphacite, glaucophane, barroisite) can form at such shallow levels. Only in active subduction zones will the normal geotherms be dragged downward so that the pressure–temperature conditions for HP metamorphism can be attained.

The fluids released from the downgoing slab as temperature and pressure increase will serpentinize the peridotites of the overlying mantle wedge (in the corner zone between the two plates), and when critical temperatures are reached, typically at 100–150 km depth, partial melting will trigger the formation of melts that will rise and feed the overlying volcanic arc paralleling the trench of the subduction zone (Gerya & Stöckhert 2006). This depth is independent of slab dip: if the slab dip is low angle ($<30°$), then the arc will lie farther from the trench than an arc above a steeper subduction zone. In examples of newly initiated subduction, there will be a discrete delay between subduction initiation and resultant volcanic arc activity, as time is required for the surface rocks to reach the required depths. A plate entering a $30°$-dipping trench at 30 mm a^{-1} will need 5–8 Ma to initiate an overlying arc after some 160–250 km of subduction. Thus, significant palaeogeographic movements must occur before arc magmatism will be seen. Further, subduction must then continue at rates of 20 mm a^{-1} or more in order to allow enough water to reach the mantle wedge for arc magmatism to be continuous enough to dominate the geology of the arc, as opposed to a few sporadic eruptions here and there. In short, lateral migrations of many hundreds of kilometres are required to build arcs, and the greater the duration of arc magmatism, the greater the amount of required subduction. As noted above, the Great Antillean arc spanned some 70 Ma prior to its Eocene collision with the Bahamas. This is one of the primary arguments for a Pacific origin of the Caribbean lithosphere (Pindell 1990); there simply was not sufficient space between the Americas, as constrained by Atlantic opening kinematics, to initiate and build the Antillean (and other) arcs in the Caribbean with in situ models (Pindell et al. 1988).

Numerical models (e.g. Gerya & Stöckhert 2002; Gerya et al. 2002) have increasingly shown that the 'two-way' transport of material necessary in subduction zones (e.g. Hsu 1971; Cloos 1982; Cloos & Shreve 1988a, b; Shreve & Cloos 1986) is best explained by forced flow in a wedge-shaped subduction channel in which hydrated serpentinized peridotite from the overlying mantle wedge plays a critical role. With regard to the exhumation of HP metamorphic rocks in the Greater Antilles, two scenarios thus present themselves. One is the on-going return flow in the subduction channel towards the accretionary wedge, leading to a subduction–accretionary complex (SAC) in the fore-arc region. During final collision of the oceanic island arc with continental crust of North America, parts of the SAC will be thrust onto the continental margin, marking a suture zone typically characterized by serpentinite mélanges entraining blocks of HP metamorphic rocks (e.g. García-Casco et al. 2002, 2006; Krebs et al. 2008). The pressure–temperature–time ($P-T-t$) paths of these blocks provide critical information on subduction zone dynamics, and therefore also the physical parameters (e.g. convergence rate, subduction angle, lithosphere age) governing the subduction zone itself (e.g. Krebs et al. 2008). Nevertheless, there are also possibilities of 'short-circuiting' this two-way flow in the subduction channel. Rocks from deeper parts of the subduction channel, at or near the thermal stability limits of serpentinite (i.e. where the serpentinite-based subduction channel is thin or non-existent), can be exhumed along 'back-stop thrusts' in the hinterland of the suture zone (e.g. Schwartz *et al.* 2007).

Tectonic settings of subduction-related metamorphic rocks in Cuba

In the northern Caribbean, the SAC of the Caribbean Arc was disrupted during collisional interactions of the arc with the Yucatán Block and later with the Bahamas platform. As a result, subduction-related metamorphic rocks of the SAC were uplifted in the subduction channel very rapidly, and progressively appeared at the surface as meta-morphic nappes and tectonic blocks in serpentinitic mélanges of the accretionary complex as well as exotic blocks in sedimentary foreland basins. The outcrops of subduction-related HP metamorphic rocks in Cuba are schematically summarized in Figure 10.

Two primary settings for HP rocks occur: (1) as coherent slabs or blocks in mélange along the Main Cuban Thrust, which can be called the suture zone; and (2) as extensionally unroofed metamorphic core complexes along the southern flank of Cuba (Draper 2001). The first setting was comprehensively described by Somin & Millán (1981), and is spatially related to the Late Cretaceous sole thrust and the Palaeogene suture. The serpentinitic mél-anges contain tectonic blocks of HP metamorphic rocks, and are reported from northern Las Villas (Somin & Millán 1981; García-Casco *et al.* 2002), from the La Suncia schists east of Camagüey City, and also north of Holguín City (Szakmány *et al.* 1999). Metasedimentary and minor metaigneous rocks such as those in the Cangre belt in western Cuba or the Asunción complex in easternmost Cuba (see Fig. 12a & c) have also been mapped in a footwall position with respect to the suture (Somin & Millán 1981). The second setting includes the Pinos and Escambray complexes, which are dominated by metasedimentary sequences. The pro-toliths were derived from a continental margin setting and were each metamorphosed at different grades. The Pinos and Escambray metamorphic complexes are characterized by typical, slightly positive gravimetric anomalies, two more of which are also observed SW of Camagüey City on the western shoulder of the Camagüey Fault, and SW of the Pinar Fault, suggesting additional unex-posed metamorphic complexes (Fig. 10; Bush & Sherbakowa 1986). Because the Pinos and Escam-bray complexes and the Camagüey anomaly are localized on the upthrown western shoulders of the NE-trending faults which cross-cut and post-date

the thrust structures (Fig. 10), extension is likely to be involved in their exhumation.

In eastern Cuba, the blueschists and eclogite-facies rocks of the Sierra del Convento differ in that they are overthrust by arc sequences of the Purial complex and the Oriente ophiolite. García-Casco *et al.* (2007) and Lázaro & García-Casco (2008) have made a cogent case for suggesting that the amphibolites of the Sierra del Convento rep-resent relatively hot subducted oceanic crust.

Pressure–temperature–time paths of metamorphic rocks

As indicated above, pressure–temperature–time paths of rocks involved in high-pressure meta-morphism in subduction zones can provide valuable information on the petrological and thermal struc-ture as well as on the dynamics of plate convergence and mass movement in such collision zones. In an early summary, Ernst (1988) discussed how the different prograde and mainly retrograde trajec-tories can be logically used to identify specific geodynamic scenarios. Thus $P–T$ trajectories may show clockwise loops denoting essentially isother-mal decompression. These can, for instance, be explained by rapid exhumation associated with ces-sation of active subduction. 'Hair-pin' type $P–T$ paths with exhumation $P–T$ trajectories essentially retracing burial trajectories can logically be related to exhumation during active subduction. However, recent numerical modelling (e.g. Gerya *et al.* 2002; Gerya & Stöckhert 2006; see also Krebs *et al.* 2008) has shown that these two situations are end-member scenarios. As the subduction zone evolves and matures, the serpentinized part of the overlying mantle wedge will widen and a funnel-shaped (downward-narrowing) subduction channel can evolve. Depending on the relationship of mass flow with respect to the local trend of the subduction zone isotherms, isothermal uplift paths are possible in active subduction zones. Counterclockwise paths with isobaric cooling segments are typical of mass trajectories in the early stages of subduction zones, before consistent return flow is established and strong cooling of the system is still underway. As the system matures, the $P–T–t$ paths are con-strained to increasingly higher $P–T$ regimes.

Escambray complex. Along the Cuban suture zone, various different types of $P–T–t$ paths can be distin-guished (Fig. 11). The most comprehensive data set for constructing a complete $P–T–t–d$-path ($d =$ deformation) exists for the metamorphic Escambray complex. Here five tectonic units (see Fig. 12b) have been stacked onto the southern margin of the Bahamas platform by top-to-north thrusting (Stanek *et al.* 2006). The protolith ages of the

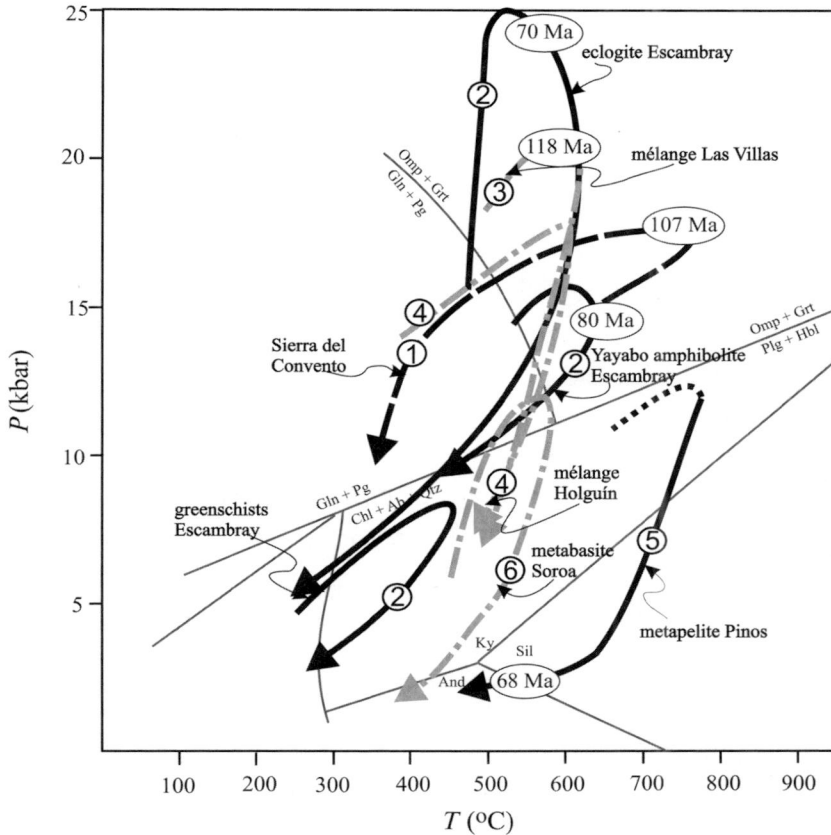

Fig. 11. Examples of metamorphic paths for Cuban SAC related rocks. The broken black line indicates initial subduction conditions, solid black lines metamorphic complexes of the hinterland, grey broken lines HP metamorphic rocks of the footwall mélange. The numbers in circles refer to (1) Sierra del Convento (García-Casco *et al.* 2006, 2007), (2) Escambray complex (Grevel 2000; Schneider *et al.* 2004; Grevel *et al.* 2006), (3) Mélange Las Villas (García-Casco *et al.* 2002), (4) Mélange Holguín (Szakmány *et al.* 1999; García-Casco *et al.* 2006), (5) Pinos complex (García-Casco *et al.* 2001), (6) metamorphic blocks in western Cuba (García-Casco *et al.* 2006). Petrological framework of the diagram from Bucher & Frey (2002).

lower Pitajones and Gavilanes nappes have been estimated as Upper Jurassic to Lower Cretaceous (see above). The protoliths of the uppermost three nappes consist of tholeiitic basalts (Yayabo unit), tholeiitic and IAT-like igneous sequences (Mabujina unit) and the low- to non-metamorphic island arc unit. Gneisses of the Mabujina unit and granitoids of the island arc have been dated between 132 and 80 Ma (Grafe *et al.* 2001; Rojas-Agramonte *et al.* 2006*a*, *b*; Stanek *et al.* 2005). The peak metamorphic conditions in the HP eclogites of the Gavilanes metamorphic unit reached 16–25 kbar at 580–630 °C (Grevel 2000; Schneider *et al.* 2004; Grevel *et al.* 2006; see Fig. 11). The $P-T$ trajectories of these HP metamorphic rocks of the Gavilanes unit describe a near-isothermal

decompression. Different geochronological methods have been applied to determine the timing of the corresponding peak metamorphic conditions, as well as the timing and velocity of exhumation of the metamorphic complex after the beginning of stacking (e.g. closure temperatures of Spear 1993). Earlier dating of zircons from the eclogites by the U–Pb method led to data that spread from about 245 (270–140) Ma (Somin *et al.* 2005), and about 170–106 Ma (Hatten *et al.* 1988; Grafe 2000; Krebs *et al.* 2007). The results near 106 Ma were obtained from rounded zircons thought to be metamorphic in origin, and thus were interpreted as an indicator for peak-metamorphic conditions (Hatten *et al.* 1988). There are no geological arguments to link the older, pre-Late Jurassic ages to subduction

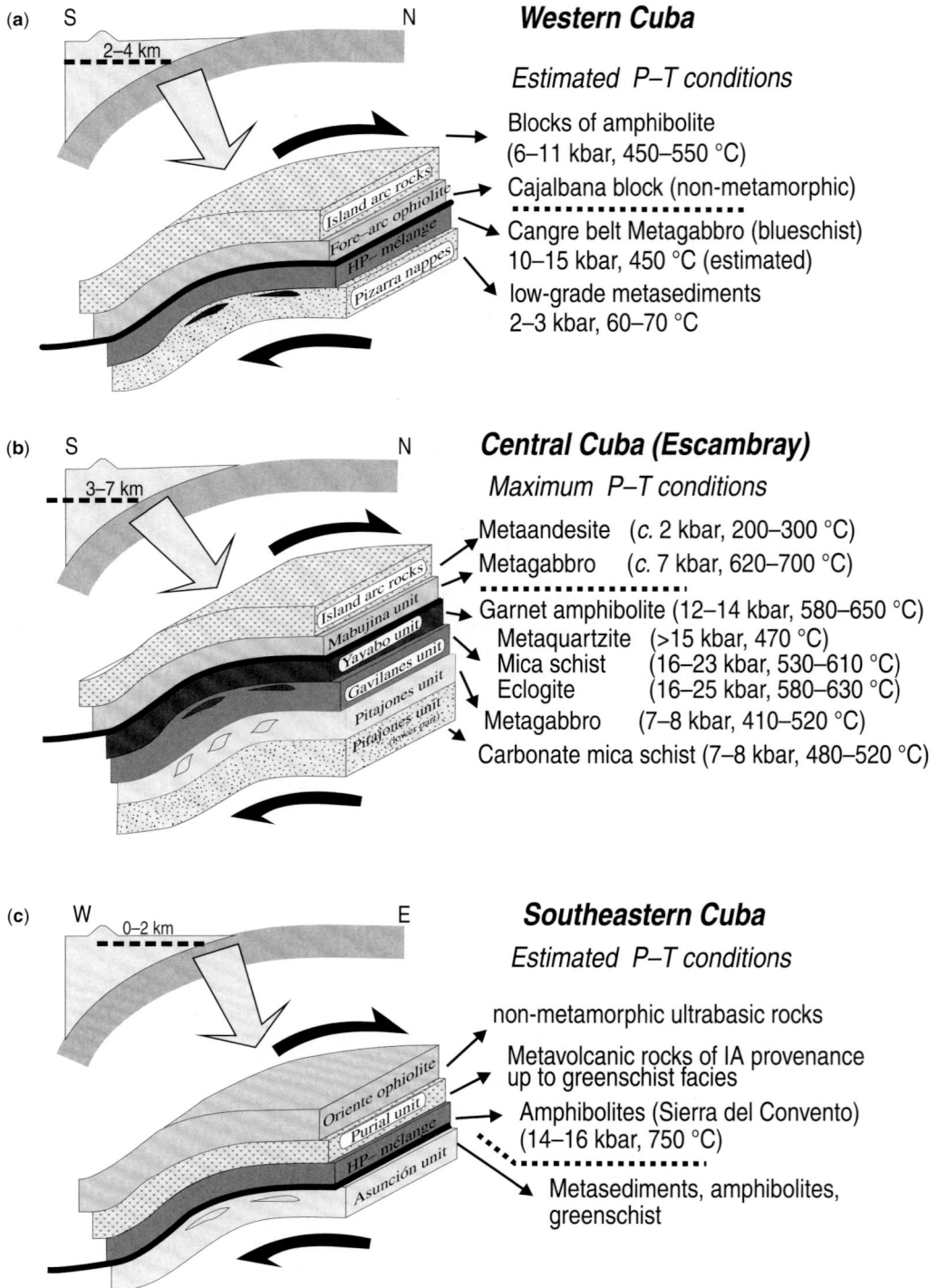

(a) S   N

2–4 km

Western Cuba

Estimated P–T conditions

Blocks of amphibolite
(6–11 kbar, 450–550 °C)

Cajalbana block (non-metamorphic)

Cangre belt Metagabbro (blueschist)
10–15 kbar, 450 °C (estimated)

low-grade metasediments
2–3 kbar, 60–70 °C

Island arc rocks
Fore-arc ophiolite
HP- mélange
Pizarra nappes

(b) S   N

3–7 km

Central Cuba (Escambray)

Maximum P–T conditions

Metaandesite (*c.* 2 kbar, 200–300 °C)

Metagabbro (*c.* 7 kbar, 620–700 °C)

Garnet amphibolite (12–14 kbar, 580–650 °C)

Metaquartzite   (>15 kbar, 470 °C)
Mica schist   (16–23 kbar, 530–610 °C)
Eclogite   (16–25 kbar, 580–630 °C)
Metagabbro   (7–8 kbar, 410–520 °C)

Carbonate mica schist (7–8 kbar, 480–520 °C)

Island arc rocks
Mabujina unit
Yayabo unit
Gavilanes unit
Pitajones unit
Pitajones unit (lower part)

(c) W   E

0–2 km

Southeastern Cuba

Estimated P–T conditions

non-metamorphic ultrabasic rocks

Metavolcanic rocks of IA provenance
up to greenschist facies

Amphibolites (Sierra del Convento)
(14–16 kbar, 750 °C)

Metasediments, amphibolites,
greenschist

Oriente ophiolite
Purial unit
HP- mélange
Asunción unit

Fig. 12. Generalized nappe stacks of (**a**) western Cuba (García-Casco *et al.* 2006; Cruz *et al.* 2007), (**b**) Central Cuba (Escambray; Grevel 2000; Stanek *et al.* 2006) and (**c**) Eastern Cuba (Sierra del Convento; García-Casco *et al.* 2006).

processes of the Caribbean Arc in the area between the spreading American plates. More likely, the U–Pb data on zircons represent protolith ages of basic igneous rocks from the rifting and pre-rifting environment at the Mesozoic continental margin. Rb–Sr isochron and Ar/Ar data on eclogites suggest rapid exhumation after the HP metamorphism at about 70–68 Ma (Schneider *et al.* 2004). New Lu–Hf isochron data (Krebs *et al.* 2007) also constrain peak metamorphism to the Late Cretaceous, that is, of the eclogites of the Gavilanes unit to *c.* 70 Ma, and of the overlying Yayabo unit to *c.* 80 Ma. The 70 Ma Lu–Hf data on the eclogites substantiates the rapid exhumation at this time and corroborates that the 106 Ma U–Pb age represents the protolith. The trend to 'older' cooling ages in the higher tectonic units is also worthy of note in the Ar/Ar data set: *c.* 65 Ma in the lower metamorphic nappes, *c.* 73 Ma for the Mabujina nappe (Grafe *et al.* 2001), and about 75 Ma for the volcanic level of the island arc (Hall *et al.* 2004).

Pinos complex. The *P–T* paths of metapelites and metatrondhjemites of the Pinos complex have still not been completely reconstructed, but are clearly characterized by rapid near-isothermal uplift at unusually high temperatures (*c.* 700–750 °C; Fig. 11). Available Ar/Ar cooling ages indicate uplift of the metamorphic complex in the latest Cretaceous (68 Ma; García-Casco *et al.* 2001).

The *P–T–t* paths of the Escambray complex indicate rapid uplift at 70 Ma under conditions of on-going subduction (or at least non-decay of a subduction thermal regime). The Lu–Hf data on metamorphic garnets show that the prograde subduction path also was rapid (Krebs *et al.* 2007). In other words, we had a sharp 'yo-yo-type' subduction–exhumation path. The Pinos path appears incompatible with the ongoing subduction scenario, but not necessarily so. The numerical simulations of Gerya & Stöckhert (2006) for an active continental margin indicate what could happen, if a 'lid' (e.g. continental crust, thick sedimentary prism, or forearc) existed or was thrust over the subduction–accretionary complex and its underlying subduction zone channel. Exhumation paths with broad clockwise form, with isothermal paths at high *T* down to low *P* can result when exhumation occurs in the 'back-stop' position. This scenario could relate to the Great Antillean Arc colliding with a spur of continental crust (e.g. Caribeana or the southern Yucatán margin; see García-Casco *et al.* 2008), thickening northward.

Southeastern Cuba. In the Purial complex (Sierra del Convento, eastern Cuba), a distinctive counterclockwise path is indicated (Fig. 11). In fact, as suggested by García-Casco *et al.* (2007), the combination of still 'near-normal' geothermal gradients and particularly young, hot subducting crust led to partial melting of the amphibolites, with a subsequent blueschist overprint of these migmatites. This must be considered a rare if not unique example of preserved *in situ* evidence of partial melting of subducted oceanic crust. U–Pb age determination on zircons from the trondhjemitic melting products reveals an age of about 107 Ma (Hatten *et al.* 1988; García-Casco *et al.* 2007) and thus provides critical evidence for subduction initiation in this segment of the Caribbean Arc. A similar U–Pb age has been reported from tonalite-trondhjemites of the Corea mélange exposed in a tectonic window below the Mayarí ophiolite massif (Somin & Millán 1981; Blanco-Quintero *et al.* 2008). The *P–T–t* paths of the HP metamorphic rocks of the Sierra del Convento and the La Corea mélange represent the initial stage of subduction in this segment of the Caribbean Arc, about 10–15 Ma later than in northern Hispaniola (Krebs *et al.* 2007). However, beyond this detailed local data, there is no additional information concerning further ongoing southdipping subduction through the Palaeogene. Preliminary K–Ar and Ar/Ar data on white mica and amphibole of the rocks of the Sierra del Convento Complex and the Corea mélange scatter between 85 and 75 Ma (Somin *et al.* 1992; García-Casco *et al.* 2007). These age data indicate a cooling of the metamorphic rocks in the middle crust which took place somewhat earlier than in central Cuba. The HP metamorphic blocks involved in the serpentinitic mélange of the Rio San Juan complex of northern Hispaniola, located in a tectonic position similar to that in eastern Cuba, demonstrate ongoing subduction from an Early Aptian–Albian stage through to the Eocene (Krebs *et al.* 2008).

Northern and western Cuba. The subduction-related metamorphic rocks of western Cuba, of northern Las Villas and of Holguín are situated in tectonic units near the Palaeogene Main Thust (Fig. 10). In western Cuba, HP metamorphic basic rocks occur in the Cangre Belt (Fig. 12a) and reach blueschist conditions (Millán 1972; Somin & Millán 1981; Cruz *et al.* 2007). Amphibolites occur as exotic blocks in a serpentinitic mélange at the top of the West Cuban anticlinal stack (Felicidades belt in the Bahia Honda unit) and as blocks and olistoliths in syntectonic sediments of the nappe structure (Somin & Millán 1981; García-Casco *et al.* 2006). In northern Las Villas exotic blocks of eclogite are included in the Northern Serpentinite Mélange (García-Casco *et al.* 2002, 2006). The serpentinite mélange forms the footwall mélange of the overthrust Cretaceous island arc. A similar tectonic environment has been mapped near the city of Holguin (Kosak *et al.* 1988; Szakmány *et al.* 1999;

García-Casco *et al.* 2002). The polymictic mélange of the La Palma Formation contains blocks of eclogite, metacarbonate and orthogneiss. The retrograde *P–T* paths of the HP metamorphic rocks included in the mélange of the Cuban Main thrust are quite similar (Fig. 11). The rocks reached different depths and have been uplifted with near-isothermal trajectories. The timing of HP metamorphism and of cooling (exhumation) differs significantly in the various complexes. In northern Las Villas, the data for HP metamorphism and subsequent cooling range from *c.* 118 to 103 Ma (García-Casco *et al.* 2002). In southern Las Villas, in the Escambray complex, the peak of HP metamorphism seems to be simultaneous in the whole complex: the 70 Ma old HP metamorphic overprint was followed by very rapid exhumation and cooling at 65 Ma. In the Pinos complex, Ar/Ar cooling ages indicate latest Cretaceous uplift (68 Ma; García-Casco *et al.* 2001). In western Cuba and in Holguín HP metamorphism and exhumation have not been constrained as yet. The available K–Ar data of the gneisses in Holguín show Middle Mesozoic to Middle Palaeozoic 'ages', which need further detailed geochronological work.

Based on the interpretation by Ernst (1988) of the various possible shapes of subduction-zone *P–T* paths, García-Casco *et al.* (2002) postulated an end of subduction at 120–100 Ma in West and Central Cuba. However, comparing the *P–T–t* paths of West and Central Cuba with the well-documented systematic *P–T–t* paths of the Rio San Juan mélanges in Hispaniola (Krebs *et al.* 2008) as well as the numerical models of Gerya *et al.* (2002) and Gerya & Stoeckhert (2006), the assumption of an end of subduction at this time no longer appears tenable. In any case, magmatism in the 'Cuban' arc requires subduction until at least 75 Ma. The interpretation of the *P–T* path of HP rocks occurring in the Cangre belt still remains problematic. The mélange occurrences of northern Central Cuba (Las Villas) speak for a mature subduction zone at 120–100 Ma (García-Casco *et al.* 2002). The path of the Soroa amphibolites in West Cuba extends to unusually low pressures (although still in the Barrovian regime), and mirrors the Pinos path, although at considerably lower temperatures.

Geotectonic model

To compile an internally consistent geotectonic model of the Cuban part of the North Caribbean suture zone (NCSZ), the structural units of the suture zone must be traced back in space and time to their original positions. This procedure gives a measure of the deficiencies in the available data sets and the reliability of existing models. Using the data of the autochthonous and parautochthonous terranes related to the passive continental margins presented above, and the allochthonous terrane of the Caribbean Arc, three tectonic regions with certain differences in thrust mechanics (western Cuba, central Cuba and eastern Cuba) can be observed along the suture in onshore Cuba. The geotectonic control points in the geological history of the terranes are the onset of sedimentation and magmatism respectively, the time and grade of metamorphism, and the first contact of the terranes and their mutual thrusting. The Caribbean literature is clear about: (1) the formation of passive continental margins between the rifting American continental plates from the Late Jurassic and into the Cretaceous; (2) the subduction-related origin of the Caribbean (Great Antillean) Arc; and (3) interaction of the Caribbean Arc with the continental crust of the Yucatán platform in the Late Cretaceous (80–70 Ma) and final thrusting onto the Bahamas platform in the (Early) Middle to Late Eocene (45–40 Ma). All the data outlined above imply an accumulation of sediments along the continental margins of the Yucatán and the Bahamas platforms and the Proto-Yucatán Basin, depending on the palaeogeographic position, from the Upper Middle Jurassic to the Middle Cretaceous or even to the final collision time in the Eocene (Fig. 13). The subduction-related magmatism at the northwestern arc part of the Caribbean arc reported from Cuban occurrences lasted, with a possible interruption in the Late Campanian–Maastrichtian, over some 70–80 Ma from Early Cretaceous to the Palaeogene. Today, all the terranes discussed above are found in tectonic juxtaposition; however, if we take a minimum of 20 mm a^{-1} convergence rate as a rule of thumb for the continuous generation of an arc, then we see that in excess of 1400 km of subduction and horizontal migration of the Cuban arc terranes toward the Bahamas is indicated for the development of the Cuban arc. This can only be accommodated in Pacific origin-type Caribbean models.

The post-collision structure of the northwestern branch of the North Caribbean suture zone in Cuba

In central Cuba, the WNW-trending Cuban Main thrust suggests a uniform Palaeogene collision belt somewhat offset by NE-trending faults across the thrust belt (Bush & Sherbakova 1986). The crustal thickness (depth to Moho) of the Cuban collision belt has been interpreted from receiver functions (Moreno Toiran 2003) and from gravity and seismic data (Bush & Sherbakova 1986; Otero

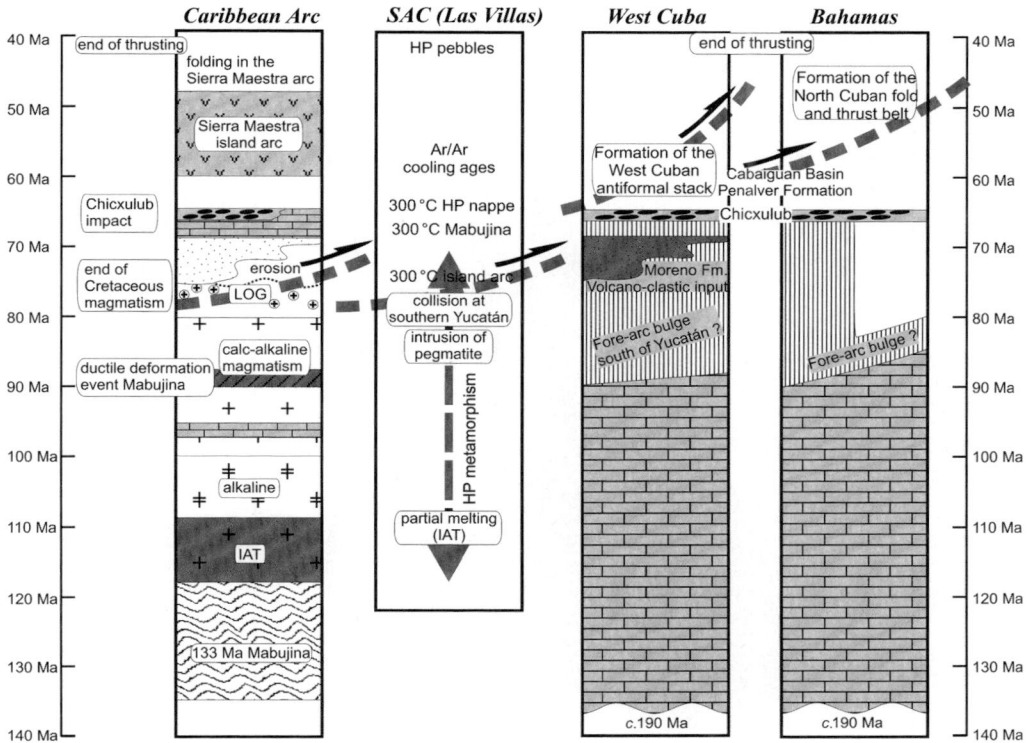

Fig. 13. Comparison of the geological–tectonic events recorded in the geotectonic terranes of the Cuban part of the North Caribbean Suture Zone.

et al. 1998; Fig. 14). According to the existing geophysical data, the crustal thickness gradually steps down from the collision belt (30 km for the Bahamas crust and foreland fold and thrust belt at the southern edge of the Bahamas platform), to the thrust and stacked arc crust (*c.* 20–30 km), the thin crust of the short-lived Sierra Maestra arc (<20 km) to the oceanic crust (<10 km) in the Yucatán Basin (Rosencrantz 1990; Fig. 14). The southeastern margin of the continental crust (Bahamas platform) has been identified east of Holguín by the interpretation of gravity data (Uchupi *et al.* 1971). The vergence of the folded and thrusted Bahamas-type sediments indicates a uniform tectonic transport top to the NE in the Late Paleocene through Middle Eocene. The structures in the overthrust Caribbean Arc sequences follow the northeastern direction of tectonic transport in the eastern part of Central Cuba. The thrust direction of the metamorphic complexes and the fold structures in the western part (Las Villas) show a certain anticlockwise rotation and trend to slightly more northerly directions, possibly due to rotation of the allochthon late in the collision.

In western Cuba, the fold-thrust structures of the sediments of the Guaniguanico terrane, the duplex formation and the overthrusting of the volcano-sedimentary sequences of the Bahia Honda unit indicate tectonic transport and stacking top to NNW (Piotrowska 1987; Saura *et al.* 2008) crosscut by the Pinar Fault (Gordon *et al.* 1997). In eastern Cuba (Oriente), two major tectonic events with divergent directions of compression are observed. In the Late Cretaceous the ultrabasic rocks of the Sierra Cristal and Moa–Baracoa ophiolitic massifs were thrust onto the previously metamorphosed (HP/LT) Purial Caribbean Arc units. In contrast to central Cuba, the lower metamorphosed part of the island arc shows top to the ESE tectonic transport in the Late Cretaceous. During the Palaeogene, the compression changed to north–south, indicated by north-vergent folding of the Sierra Maestra Middle Eocene volcanic rocks.

The tectonic data suggest bending and segmentation of the Cuban suture zone, which is also supported by geophysical studies. Velocity models of seismic waves and the interpretation of data from receiver functions give more detailed information

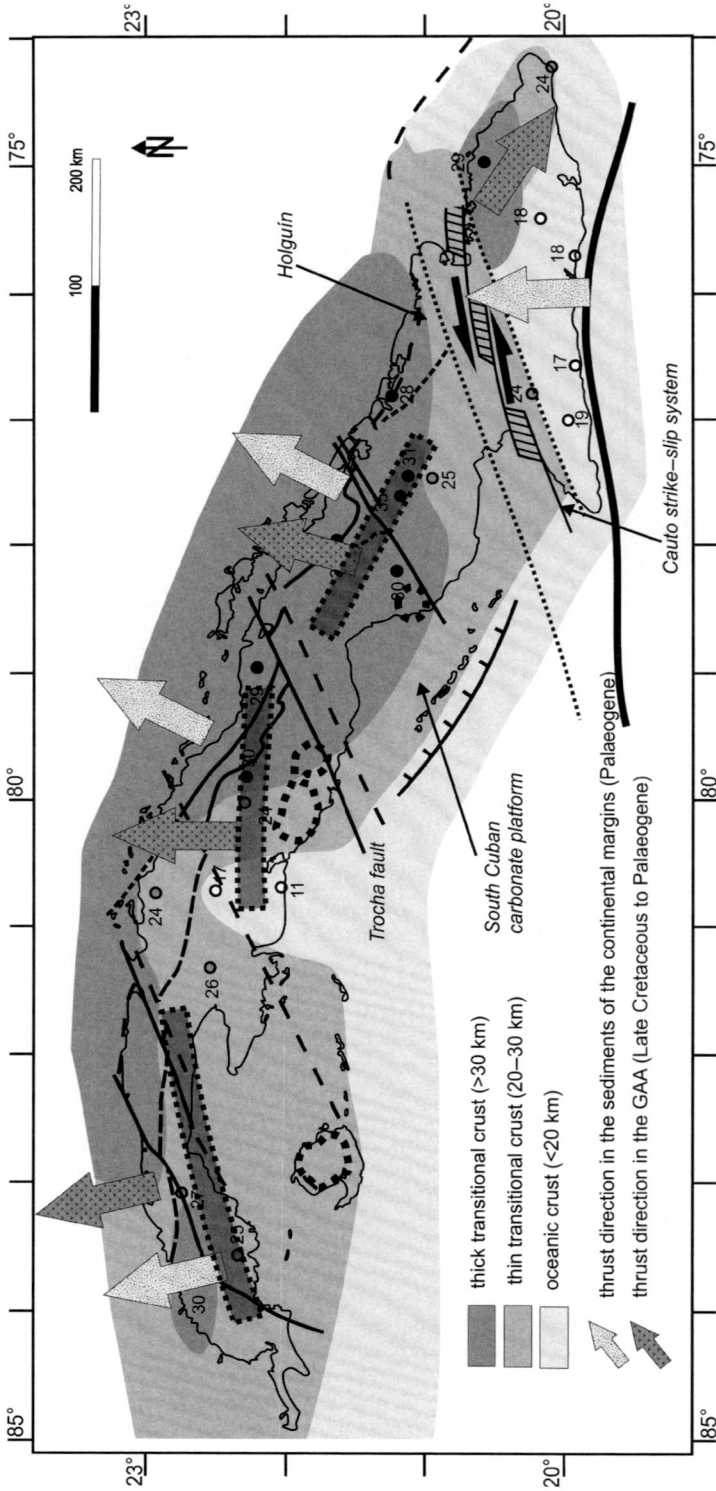

Fig. 14. Sketch map of the crustal thickness (depth to Moho) of the Cuban collision belt. Data compiled from Otero *et al.* (1998), and Moreno Toiran (2003). Grey bars indicate the palaeomagnetic trends (Renne *et al.* 1991; Chauvin *et al.* 1994; Pérez Lazo *et al.* 1994; Bazhenov *et al.* 1996). The Cauto strike–slip system is interpreted from Makarov (1986). The opening of the Cauto Fault Zone is suggested to taken place in the Late Eocene (Leroy *et al.* 2000). The directions of the tectonic transport have been compiled from tectonic field data (1985–2006).

of the crustal thickness (depth to Moho) of the Cuban mainland (Otero *et al.* 1998; Moreno Toiran 2003). The southeastern parts of Cuba (Oriente) consist of thin, probably arc-modified oceanic crust with a juxtaposed arc pile (Sierra Maestra Arc) above (Fig. 14). Only the easternmost part of Oriente shows an area of typically continental crustal thickness, possibly suggesting a piece of continental crust and sedimentary cover involved in subduction in the Late Cretaceous. Considering the amphibolite-facies metamorphic grade of the Asunción complex (Somin & Millán 1981) and its protoliths, this metamorphic suite does not fit logically into the general geology of the Oriente region. The complex may also have been entrained somewhere at the southeastern margin of Yucatán. After the Palaeogene collision, the complex was displaced by the Paleocene–Eocene Cauto fault system (Leroy *et al.* 2000). The Cauto Depression, as modelled from gravity data, could be interpreted as a sinistral strike–slip fault system with three extensional pull-apart basins (Makarov 1986; Fig. 14).

The eastern part of central Cuba has a thick homogenous crust (depth to Moho). In the southeastern offshore area, trending along the recent South Cuban carbonate platform (Fig. 14), a NE-dipping seismic reflector was interpreted as an inactive, NE-dipping (SW-vergent) thrust zone (Rosencrantz 1990). The fault plane is buried by presumed Upper Palaeogene and younger sediments. Relative to the NE-vergent accretionary suture zone of central Cuba, this thrust zone could be interpreted as a backstop thrust in the overall collision (Ellis *et al.* 1999), active during the Palaeogene collision of the Caribbean Arc with the Bahamas platform. The structure of the collision belt west of the Trocha fault appears more complicated, and most of the Caribbean Arc related rocks are hidden below younger sediments (Batabano Massif; Fig. 2). West of the Escambray Mountains a tongue of thin, possibly oceanic crust extends to the north, separating the central Cuban region from western Cuba. This segmentation of the overthrust island arc could explain the different trends and thrust directions in the regions described above. The nature of the thin crust in the collision belt west of the Escambray is still unclear.

The tectonic and geophysical indications of segmentation by faulting are also supported by palaeomagnetic data. Palaeomagnetic determinations from the Bahia Honda unit and the Guaniguanico terrane in western Cuba indicate an anticlockwise rotation of about 90°, and a remagnetization in the Late Cretaceous, which is suggested to be related to the beginning of deformation in this nappe stack (Bazhenov *et al.* 1996). Data from Upper Jurassic sediments gave palaeolatitudes of about 12°N, which place the sediments at the southern margin of the Yucatán Block (Pérez Lazo *et al.* 1994). Reconnaissance palaeomagnetic studies from central Cuba (Las Villas) also reveal a remagnetization in the Campanian and an anticlockwise rotation of about 43–37° (Renne *et al.* 1991; Chauvin *et al.* 1994). The differences in the rotation values are suggested to be related to an oroclinal bending of Cuba (Bazhenov *et al.* 1996). Palaeomagnetic studies of Paleocene and Eocene sediments in Oriente show that the Early Palaeogene position of eastern Cuba was only 1–2° (100–200 km) south of its present location, whereas the position of Upper Eocene sediments shows similar latitude to today's (Pérez Lazo *et al.* 1994).

The continental margin of Yucatán and the Bahamas

The oldest sediments that bear witness to the rifting stage of the Proto-Caribbean ocean are the ?Lower to Middle Jurassic San Cayetano fluvio-deltaic sandstone and shale and the overlying Upper Jurassic marine deposits and intercalated volcanic rocks in the Sierra Guaniguanico, which represents the palaeo-eastern Yucatán margin that rifted from Venezuela (Pindell 1985). As for the Bahamas margin, Atlantic opening kinematics indicate a Late Jurassic–Early Cretaceous diachronous transform fault relationship along the southern Bahamas and the Guyana margin of NE South America (Pindell 1985; Fig. 13). Essentially passive margin conditions are then thought to have prevailed along these margins until the Turonian (*c.* 92 Ma), when arc-derived lithic clasts and minerals such as glaucophane first appear in the Placetas Belt deep marine sediments (Linares & Smagoulov 1987) and Escambray and Pinos complexes. However, contrary to the view (Hempton & Barros 1993) that the Placetas Belt is the basinal facies of the Bahamian margin, the Placetas Belt must have lain farther to the SW, somewhere south of Yucatán, such that it does not constrain Bahamian development at all. As discussed herein and also by Pindell *et al.* (1988, 2005, 2006) and García-Casco *et al.* (2008), the Caribbean Arc did not arrive at southern Yucatán until the Campanian, as recorded by the Sepur foredeep in northern Guatemala. Thus, any Turonian strata with arc debris must have been deposited even farther south or southwest (Pindell *et al.* 2006). The Placetas Belt is best considered as part of the Cuban accretionary prism; it does not physically connect with coherent fold trains in Cuba today, and thus it cannot be used to rigorously reconstruct amounts of total shortening in the Cuban thrustbelt. The 450 km estimated displacement of Hempton & Barros (1993) is, as they openly state, a minimum only.

No Upper Cretaceous pelagic sediments younger than Turonian have been observed in the North Cuban fold and thrust belt. Tada *et al.* (2002) suggested that this may be a hiatus due to erosion from the giant tsunami wave after the Chicxulub impact, but because we are talking about the basin floor of the Proto-Caribbean (*c.* 4 km deep at the time), a more likely explanation is that most Upper Cretaceous pelagic sediment simply was subducted rather than accreted at the arc's accretionary prism, as at Barbados and the south Caribbean Fold Belt today. An Upper Cretaceous hiatus is also known from the immediate area of the Florida Straits (Angstadt *et al.* 1985), probably created by bottom currents flowing through this gap between the Atlantic and the Gulf of Mexico/Western Interior Seaway. It is not known, of course, how far to the SE of the Straits in the Proto-Yucatán Basin this hiatus might have developed. It might be that both explanations are responsible at least in western and central Cuba. The Upper Cretaceous hiatus discussed thus far here is not to be confused with true Bahamian Platform hiati: there, a Maastrichtian hiatus, at which arc-derived tuffs first appear in the Bahamian section, record the advancing Caribbean Plate's peripheral bulge as it arrived some 300 km ahead of the arc at the Bahamas (Pindell *et al.* 1988, 2005). Thereafter, as the arc came nearer, load-induced foreland subsidence and sedimentation rates were strong in the Bahamas (Mullins & Lynts 1977; Pindell 1985).

In the Guaniguanico terrane in western Cuba, the Campanian Moreno Formation contains abundant volcaniclastic detritus (Fig. 13). The facies development and the stratigraphic relationships to the older sediments suggest that the Moreno Fm is a sedimentary piggy-back basin at the top of the fold and thrust belt north of the Cretaceous island arc (Pszczółkowski 1999). The suggested palaeogeographic position of the Upper Cretaceous piggy-back basin is at the eastern margin of Yucatán when the recently dormant island arc started to move into the area east of the Yucatán platform (Saura *et al.* 2008).

Timing of tectonic events in the northwestern Caribbean

Circa 120–80 Ma. Three important evolutionary events in the geotectonic history of the Caribbean Arc are the onset and the cessation of magmatism, and the obduction of the arc onto the Proto-Caribbean continental margins (Fig. 13). The onset of arc magmatism has been dated at least as old as Aptian (*c.* 120 Ma; Kesler *et al.* 2005; Escuder Viruete *et al.* 2006; Rojas-Agramonte *et al.* 2006*a*, *b*; García-Casco *et al.* 2007). Subduction related

magmatism in central Cuba persisted until the Middle Campanian, a period exceeding 45 million years, after which it appears to have shifted south to the offshore. There is no interruption in this interval of Cretaceous island arc activity. Thus, if there was an arc polarity reversal (Pindell & Dewey 1982; Pindell 1993; Lebrón & Perfit 1993; Draper *et al.* 1996), then in Cuba it must have pre-dated 120 Ma, and all subduction that built this arc occurred by southwestward subduction of Proto-Caribbean crust. However, a number of fragments or small terranes of older arc, subduction complexes and possibly continental crust may have migrated SE from the west flank of Chortís along a sinistral transform spanning the Neocomian gap between Chortís and Ecuador, which could now lie beneath the Aptian and younger Great Arc (Pindell 2008).

As argued above, the palaeogeographic position of the arc in the Campanian was along southern Yucatán, such that it still had another 1000 km to travel before reaching the Bahamas. Assuming a minimum convergence rate of 20 mm a^{-1} in order for the Central Cuban arc to develop, the origin of the arc must have been some 900 km farther SW than southern Yucatán. The cessation of magmatism in the Middle Campanian coincides with the time of collision and stacking of the accretionary complex of the Caribbean Arc with the southern continental margin of the Yucatán platform (Fig 15).

80–75 Ma. The blueschist-facies meta-ophiolitic rocks of the El Tambor group, whose metamorphic ages range from about 125 Ma to about 75 Ma (Harlow *et al.* 2004), and amphibolite facies gneisses of the Chuacus complex, were generated during the period of normal (steep) subduction as the Caribbean arc approached southern Yucatán (Ratschbacher *et al.* 2009). These authors further suggest that the slab broke off as the trench was choked by Yucatán, as a means of generating the nearly contemporaneous intrusion of pegmatites in the Maya Block in southern Yucatán, and Pindell & Kennan (2009) argue that slab break-off must have occurred as a means of allowing North American continental crust to continue its westward drift across the mantle. Concurrently to the south, arc magmatism ceased in central Cuba (75 Ma) and HP metamorphism in the Escambray and Pinos sediments was culminating shortly thereafter (70 Ma). The end of arc magmatism would relate to the flattening of the slab as the continental margin entered the trench, while intrusion of the late orogenic granites (LOG) and pegmatites in central Cuba may relate to fluid fronts from the subduction of the continental margin sediments as they began to be metamorphosed. Peak metamorphism of the subducted sediments and igneous rocks of the thinned continental margin would lag (about

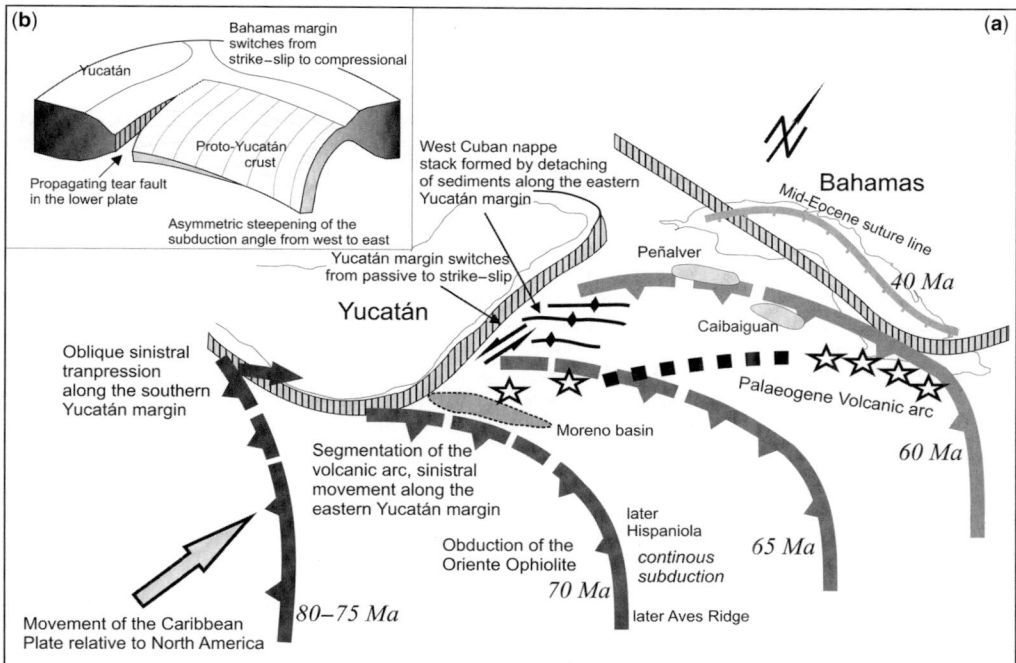

Fig. 15. Model of the 'indenting' northern part of the Caribbean Arc from Maastrichtian to Mid Eocene (adapted from Pindell *et al.* 2005, 2006). (**a**) 80–75 Ma: oblique sinistral collision of the Caribbean Arc with the southern Yucatán margin, subduction of parts of the thinned continental crust, end of subduction-related magmatism in the volcanic arc. 70 Ma: HP metamorphism and rapid uplift of the subducted rocks, movement of the collision complex away from the suture zone into the Proto-Yucatán area. 65 Ma: sinistral movement of the fragmented northwestern part of the Caribbean Arc along the eastern Yucatán margin. The subduction continued in the southern part of the Caribbean Arc. 60 Ma: Formation of a new volcanic arc 'consuming' the Proto-Yucatán oceanic crust by shifting of the magmatic axis. Thrusting started in the North Cuban fold and thrust belt. 40 Ma: final obduction of the remnants of the northern Cretaceous Caribbean Arc onto the Bahamas platform. (**b**) Model of the subducting Proto-Yucatán oceanic crustal slab with asymmetric subduction angles.

70 Ma) while the sediments were heated and subducted by several tens of kilometres. In the area of slab detachment, anomalous heat flow could be expected as rising asthenosphere displaces the slab. This could be a mechanism for explaining the unusually 'warm' $P–T–t$ path of the Pinos complex (Fig. 11), situated over the detaching slab. The Escambray complex, which was situated farther south in the subduction channel, was uplifted in more normal 'colder' geothermal conditions (Fig. 11).

The subducted metasediments then underwent rapid exumation and cooling. This probably began in the subduction channel during flat-slab subduction, but may have been subsequently facilitated by backstop thrust tectonics (see model by Schwartz *et al.* 2007), or, especially at Pinos, by extensional detachment during the initial opening of the Yucatán Basin (Pindell *et al.* 2006). The initial tectonic segmentation of the Central Cuban part

of the Caribbean Arc probably owes its origin to the Campanian–Maastrichtian interaction with southern Yucatán; the Cuban cross faults lie parallel to the sigma one direction of that transpressive interaction, and sediments as old as Maastrichtian lie on the hanging walls of the normal faults (Pindell & Kennan 2009).

70 Ma. By this time, the Cuban part of the Caribbean Arc had been deformed, locally uplifted, deeply eroded as it passed the SE corner of Yucatán, and then covered by the Late Maastrichtian carbonate platform (Iturralde-Vinent 1998) as it began to migrate toward the Bahamas. Arc magmatism probably continued, or was re-established, eastward where slab break-off had not occurred, but south of onshore Cuba due to a flatter slab geometry. With the separation of the Cuban arc terranes from the Cayman Ridge/Caribbean Plate as the Yucatán Basin opened, the northwestern

Caribbean region became the site of a three-plate kinematic problem. The Caribbean Plate continued to move to the NE relative to North America, but the westernmost Cuban arc terranes moved NNW along the Yucatán margin. This difference was manifested by NW–SE extension across the site of the Yucatán Basin (Pindell *et al.* 1988, 2005, 2006). In essence, after the southward shift in arc magmatism, and possible forearc tectonic erosion at 75–70 Ma, the original Cretaceous Cuban arc terranes were now in a forearc tectonic setting, relative to a new and hypothetical arc axis, and within only 20–40 km of the Cuban trench. These forearc slivers were incrementally unroofed as trench rollback sucked them extensionally to the NW relative to the Cayman Ridge/Caribbean Plate, which was concurrently migrating NE relative to North America. Although the initial extensional faulting that helped unroof the southern Cuban HP zones was of Maastrichtian age, the fact that Paleocene arc intrusions occur in the Cayman Ridge (Lewis *et al.* 2005) suggests that much of the extension in the deep Yucatán Basin was Paleocene and Early Eocene, such that any subsequent and final arc magmatism beneath the Yucatan Basin lay north of the Cayman Ridge. The assumed extensional detachments that had cut down through the Cretaceous arc level and into the subduction channel allowed unroofing of the metamorphic complexes on a multi-kilometric scale. The domal structure is due to isostatic rebound of the footwalls as they were tectonically unloaded. Both Pinos and Escambray rapidly cooled between 68 and 45 Ma, as shown by Ar/Ar and zircon/apatite fission track cooling ages. Prior to this crustal level unroofing, the exhumation of deeper parts of the SAC was probably accommodated by backstop thrusts of the accretionary wedge. Following the modelling by Ellis *et al.* (1999) and Goffé *et al.* (2003), mass flow trajectories suggest that parts of the metamorphosed subduction–accretionary complex could rise up to erosional levels along these faults. This could explain the appearance of metamorphosed continental margin sediments to the rear of the overthrust island arc sequences. The Maastrichtian age of initiation of the Cuban cross faults is shown by the basal Maastrichtian sediments in the basins adjacent to them. They are all extensional, and their extension continued into the Eocene. In contrast, at the front of the forearc in the accretionary prism, including the Sierra Guaniguanico, compressional structural development continued as the forearc slivers migrated toward eventual transpressional collision with NE Yucatán and orthogonal collision with the Bahamas.

65–60 Ma. Chicxulub impact-related sediments unconformably cover parts of the Cretaceous sediments of the continental margin, as well as those of the fold and thrust belt. Also, parts of the Caribbean Arc were eroded during the Paleocene, providing a key stratigraphic marker horizon. Most of the white mica in the southern Cuban metamorphic complexes began to pass through the cooling temperature of the K–Ar decay systems at about 65 Ma, implying exhumation to depths of about 10 km by that time. Piggy-back basins developed after the Early Paleocene (*c.* 60 Ma) in a foreland position (Peñalver Formation) on top of the island arc during its folding and thrusting (Caibaiguan Basin), accumulating the clastic debris from the Maastrichtian carbonate platform, the eroding island arc, and the rising metamorphic complexes.

Although there is a 10–15 Ma gap in the record in the latest Cretaceous, subduction related magmatism is again evident along the Cayman Ridge, in southern Cuba (Sierra Maestra Arc, Rojas-Agramonte *et al.* 2004), and extending into Hispaniola, Puerto Rico and at least the Saba Bank of the Aves Ridge. Starting at Oriente, Cuba, this arc axis veers southwards from the Central Cuban trench, indicating a westward-shallowing subduction angle (Pindell *et al.* 2005). Lewis *et al.* (2005) claim a continental influence in the Paleocene Cayman Ridge plutonic rocks, which we suggest was due to subduction of some volume of continental slope and rise sediments off southeastern Yucatán as Cuba moved NNE along the margin. The Sierra Maestra Arc remained active through the Middle Eocene, as it, like the Central Cordillera of Hispaniola, always lay on the arc axis until the time of collision (Pindell *et al.* 1988); the Palaeogene volcanic sequences cover both the Cretaceous volcanic arc and the overthrust Oriente ophiolite. In southern Cuba near Holguín, Paleocene olistostromes are intercalated with felsic tuffs, forming distal deposits of Sierra Maestra arc volcanism.

Accepting this model of southward subduction for all the arc activity noted, the oceanic lithosphere being subducted is that of the Proto-Caribbean Basin, formed between the two American continental plates. A Late Middle to Late Jurassic age for the Proto-Yucatán Basin can be modelled on the basis of plate kinematics (Pindell 1985) and tholeiitic syn–rift intrusions into the passive margin strata of Sierra Guaniguanico. In contrast, a Paleocene to Middle Eocene age for the present Yucatán Basin is predicted as a result of the inferred period of intra-arc rifting and NW–SE spreading, driven by roll-back of the Jurassic Proto-Yucatán slab (Pindell *et al.* 2005); this age is supported by heat-flow data from the basin (Rosencrantz 1990). Note that, although extension in the Yucatán and the Cuban onshore basins was to the NW away from the Cayman Ridge, the direction of collision between the Cuban arc fragments and the

Bahamas Platform (i.e. relative to North America) was mainly toward the NNE. This is why the small pull-apart basin mapped by Rosencrantz (1990) along the western deep Yucatán Basin records NNE–SSW opening: it is a basin along the Cuba–North America boundary, and not the Cuba–Cayman Ridge boundary. Any diachroneity in the Cuba-Bahamas collision, although commonly proposed to be eastward, is actually difficult to prove due to small amounts of Miocene eastward extensional collapse of Cuba above its suture zone.

45–40 Ma. The final thrusting of the Caribbean Arc related units onto the sediments of the eastern Yucatán and southern Bahamas margin took place in the Middle Eocene at about 40–45 Ma (Iturralde-Vinent 1994; Fig. 15). The metamorphic complexes reached the level of erosion; clasts of HP-metamorphic rocks are deposited in Eocene sections of the surrounding basins. In eastern Cuba, the Palaeogene arc was folded with northern vergence. In western Cuba the Bahia Honda ophiolitic unit was thrust to the north on top of the sediment thrust stack. In central Cuba, the ophiolite belt overrode the Eocene foreland basins, and the carbonate sequences of the Bahamas Platform were folded with roughly NNE vergence.

The initial sinistral strike–slip event in the Cauto Depression moved the southeastern part of Cuba (Sierra Maestra) to the east by about 80–100 km, before the fault system jumped to the south to the recent Cayman Trough (Leroy *et al.* 2000). The activation of the strike–slip faults should post-date the magmatism in the Palaeogene arc.

Arguments in favour of an 'indenter' (Pacific provenance) model

1. Considering the period of activity of the Cretaceous Caribbean (Great Antillean) Arc, at least 45 Ma in Cuba, the island arc should have subducted at least 900 km before the Campanian/Maastrichtian collision with southern Yucatán. In Hispaniola, the period of arc activity is 70 Ma, indicating at least 1400 km of SW-dipping subduction before Eocene collision with the Bahamas. Note that these figures are minima, because they assume orthogonal convergence.
2. The oblique northeastward thrusting onto southern Yucatán indicates a south-dipping subduction zone. The subduction polarity and the collision geometry require an origin for the Caribbean Arc somewhere in the Pacific SW of Yucatán.
3. The occurrence of Mid-Cretaceous shallow-water limestones both in the continental margin settings and in the Caribbean Arc was used as an argument in favour of a geographically close position of both units by James (2006). The Albian shallow-water limestones of the Caribbean Arc suggest uplift in the arc, whereas the coeval limestones mentioned by James (2006) belong to carbonate platforms of the Bahamas or Yucatán blocks. If these limestones had been formed in close proximity, there should be a record of the intense magmatism of the Caribbean Arc in the platforms, yet there is little to no evidence of such an influence.
4. The continuous subduction-related magmatism during 45 Ma in Cuba indicates a continuous movement of the Caribbean Arc to the ENE until the Campanian. If a polarity reversal occurred, it must have been pre-120 Ma. In the Late Cretaceous and Palaeogene the northern part of the Caribbean Arc was stretched and split off by intra-arc spreading as the Cuban portion overrode the triangular shaped Proto-Yucatán Basin between the Bahamas and Yucatán Blocks.
5. The data on subduction-related magmatism and the HP metamorphic record reported from Cuba trace the geotectonic history of the Caribbean Arc until the Late Cretaceous collision of the arc with southern Yucatán, and predate the collision with the Bahamas Block. In this framework, the protoliths of the Pinos and Escambray metamorphic complexes were part of the sedimentary margin of the North American Yucatán Block, subcreted to and metamorphosed at the base of the Great Antillean forearc during the Late Cretaceous collision with Yucatán, and progressively exhumed by subduction flow, intra-arc extensional detachment, and eventually uplift and erosion in the Bahamian collision.

Following the above model, the Late Jurassic to Cretaceous oceanic crust of the Proto-Yucatán Basin must have been subducted to allow a north-northeastward movement of the Caribbean Arc. Therefore the oceanic crust bordering the Cayman Trough forms the back arc of the Caribbean Arc and could be Jurassic or Cretaceous in age, but should have a Pacific origin, and is not related to the opening of the Proto-Caribbean ocean with some continental slivers, as suggested by James (2006). The new data on Cuban and other geochronology, petrology, geophysics and structural geology derive largely from work in the last two decades, and indicate continuous subduction over many tens of millions of years. In the plate tectonic concept, we cannot escape the fact that extensive horizontal movements must have been necessary for the generation of such a long-lived magmatic island arc. The NE-, E- and SE-facing island arc at the leading edge of the Caribbean Plate only allowed movement of the Great Arc in these directions, relative to the Americas, supporting a provenance of the Caribbean Arc to the west or SW of its Late Cretaceous to recent position. There are still many gaps to explain in detail, but continued application of modern '4D data' in combination with traditional surface data should continue to bear fruit.

Summary lithostratigraphic chart

Stratigraphic names used in this paper are based on the explanation notes of the Geological Map of Cuba 1985.

- *Punta Alegre Fm*: gypsum breccia with blocks of gypsum, limestone, dolomite, siltstone, sandstone and tuff; Lower to Middle Jurassic, forming diapirs along the north coast of Cuba, Bahamas Platform.
- *San Cayetano Fm*: intercalation of sandstone, siltstone and pelitic schist, rare conglomerates and gravels; Lower to Middle Jurassic, western Cuba, Yucatán margin.
- *El Sabalo Fm*: sequence of basalt and diabase intercalated with limestone, tuff and siltstone; Oxfordian–Lower Kimmeridgian, western Cuba, Yucatán margin (Pszczółkowski 1994).
- *Santa Teresa Fm*: radolarian cherts intercalated with siltstone, bentonite, marl and limestone; Albian to Turonian, western and central Cuba, Proto-Yucatán Basin.
- *Mata Fm*: intercalation of pelitomorphic limestone, marl, siltstone, chert, limy conglomerates, Albian to Turonian, central Cuba, Proto-Yucatán Basin.
- *Moreno Fm*: marly limestone, shale, sandstone and siltstone; Campanian, western Cuba, Proto-Yucatán Basin (Pszczółkowski 1994).
- *Encrucijada-Fm*: aphyric basaltic lavas, lava breccia, chert and rare blocks of massive sulphides; supposed Late Upper Cretaceous, Bahia Honda unit, western Cuba, West Cuba anticlinal stack (Fonseca *et al.* 1984).
- *Via Blanca Fm*: polymictic sandstones, greywacke, conglomerates, tuff, siltstone and marl; Campanian–Lower Maastrichtian, western and western-central Cuba, Proto-Yucatán Basin.
- *Ancon Fm*: limestones with chert lenses, limy conglomerates; Palaeogene, western Cuba, West Cuba anticlinal stack.
- *Capdevila Fm*: polymictic sandstone, siltstone and shale, conglomerates and limestone; Lower Eocene, western Cuba, West Cuba anticlinal stack.

- *Manacas Fm*: olistostrome with blocks of sedimentary and volcanic rocks, the ground mass consists of siltstone with fine-grained limestone layers; Lower–Middle Eocene, western Cuba, West Cuba anticlinal stack.
- *Senado Fm*: polymictic olistostrome containing bolders of serpentinite, volcanic rocks, limestone in conglomeratic groundmass; Lower Eocene, central Cuba, North Cuba fold and thrust belt.
- *La Palma Fm*: polymictic serpentinitic mélange with boulders of metaterrigenous rocks, garnet schist; gneiss and metagranodiorite; supposed Upper Cretaceous, north of Holgúin, North Cuba fold and thrust belt (Kosak *et al.* 1988).
- *Los Pasos Fm*: plagiorhyolitic lavas and basaltic lavas and tuffs, volcanomictic sand- and siltstones; supposed Upper Neocomian, Las Villas, central Cuba, Caribbean Arc.
- *Matagua Fm*: basaltic and andesitic lavas and tuffs, minor limestone; Aptian–Albian, Las Villas, central Cuba, Caribbean Arc.
- *Cabaiguan Fm*: basaltic to dacitic tuffs, sandstone and minor limestone; Aptian–Albian, Las Villas, central Cuba, Caribbean Arc.
- *Provincial Fm*: limestones, marl, polymictic conglomerates and minor tuff; Upper Albian–Cenomanian, Las Villas, central Cuba, Caribbean Arc.
- *Calderas Fm*: volcano-sedimentary sequences with intercalations of basaltic and andesitic lava flows; Cenomanian–Turonian, Las Villas, central Cuba, Caribbean Arc.
- *Arimao Fm*: volcano-sedimentary sequences with intercalations of basaltic and andesitic lava flows; Coniac–Santonian, Las Villas, central Cuba, Caribbean Arc.
- *La Rana Fm*: volcano-sedimentary sequences with intercalations of basaltic and andesitic lava flows; Santonian–Campanian, Las Villas, central Cuba, Caribbean Arc.
- *Bruja Fm*: dacitic and rhyolitic lavas and tuffs, volcanosedimentary sequences; Santonian–Campanian, Las Villas, central Cuba, Caribbean Arc.
- *San Pedro Fm*: sandstones, siltstones, shale, conglomerates, minor limestone; Upper Campanian–Maastrichtian, Las Villas, central Cuba, Proto-Cuban Maastrichtian Platform.
- *Cantabria Fm*: limestone; Maastrichtian, Las Villas, central Cuba, Proto-Cuban Maastrichtian Platform.
- *Santo Domingo Fm*: andesitic to rhyolitic lavas, tuffs, volcanomictic sandstones and shales, minor limestone; Cenomanian–Turonian, Oriente, Caribbean Arc.
- *Téneme Fm*: basaltic lavas and tuffs; supposed Cenomanian–Turonian, Oriente, Caribbean Arc (Proenza *et al.* 2006).

- *La Picota Fm*: sedimentary breccia containing volcanomictic blocks as well as serpentinites, sandstones, siltstone; ?Campanian–Maastrichtian, Oriente, Maastrichtian Platform.
- *Mícara Fm*: gravel, polymictic sandstones, siltstone, marl; ?Campanian–Maastrichtian, Oriente, Maastrichtian Platform.
- *Yaguaneque limestones*: Upper Maastrichtian, Oriente, Maastrichtian Platform (Cobiella *et al.* 1984).

The present paper was funded by grants of the German Science Foundation (grants STA362/3-1, STA362/4-1, STA362/14-1, MA689/9-1, MA689/13-1). We thank Keith James for organizing the meeting on Caribbean Plate tectonics in Siguenza 2006, Antonio García-Casco and Manuel Iturralde-Vinent for many helpful discussions, and Gren Draper and Robert Erlich for the helpful review. This paper is a contribution to the IGCP program 546.

References

ANGSTADT, D. M., AUSTIN, J. A. & BUFFLER, R. T. 1985. Early Late Cretaceous to Holocene seismic stratigraphy and geologic history of southeastern Gulf of Mexico. *American Association of Petroleum Geologists Bulletin*, **69**, 977–995.

BATCHELOR, R. A. & BOWDEN, P. 1985. Petrogenetic interpretation of granitoid rock series using multicationic parameters. *Chemical Geology*, **48**, 43–55.

BAZHENOV, M. L., PSZCZÓIKOWSKI, A. & SHIPUNOV, S. V. 1996. Reconnaissance paleomagnetic results from western Cuba. *Tectonophysics*, **253**, 65–81.

BELMUSTAKOV, I., CABRERA, R., ITURRALDE-VINENT, M. A. & TSHUNEV, D. (eds) 1981. *Informe del levantamiento geologico a escala 1 : 250.000 del teritorio Ciego-Camaguey Las Tunas 1975–1979*. Academia de Ciencias de Cuba, MINBAS, CNFG, La Habana, unpublished internal report.

BIBIKOVA, E. V., SOMIN, M. L., GRACEVA, T. V., MAKAROV, V. A., MILLÁN, G. & CHUKOLJUKOV, J. A. 1988. Pervye rezultaty U–Pb-datirovanija metamorficeskich porod Bolschoi Antillskoi dugi: vozrast kompleksa Mabuchina Kuby. *Doklady Akademii Nauk SSSR, Seria Geologičeskaya*, **301**, 924–928.

BLANCO-QUINTERO, I., GARCÍA-CASCO, A. *ET AL.* 2008. Geochemistry and age of the partially melted subducted slab. La Corea mélange (Eastern Cuba). Abstracts and Program 18th Caribbean Geological Conference, Santo Domingo 2008, 8. World Wide Web Address: http://www.ugr.es/~agcasco/igcp546/DomRep08/Abstracts_CaribConf_DR_2008.pdf.

BLEIN, O., GUILLOT, S. *ET AL.* 2003. Geochemistry of the Mabujina Complex, Central Cuba: Implications on the Cuban Cretaceous Arc Rocks. *Journal of Geology*, **111**, 89–110.

BOITEAU, A., MICHARD, A. & SALIOT, P. 1972. Métamporphisme de haute pression dans le complexe ophiolitique du Purial (Oriente, Cuba). *Comptes Rendus de l'Académie des Sciences Paris*, **274** (série D), 2137–2140.

BRALOWER, T. J. & ITURRALDE-VINENT, M. A. 1997. Micropaleontological dating of the collision between the North American plate and the Greater Antillen arc in western Cuba. *Palaios*, **12**, 133–150.

BUCHER, K. & FREY, M. 2002. *Petrogenesis of Metamorphic Rocks*. Springer, Berlin.

BUSH, V. A. & SHERBAKOVA, I. N. 1986. Novye danny po glubinnoi tektonike Kuby. *Geotektonika*, **3**, 25–41.

CAZAÑAS, X., PROENZA, J. A., MATTIETTI-KYSAR, G., LEWIS, J. & MELGAREJO, J. C. 1998. Rocas volcánicas de las series Inferior y Media del Grupo El Cobre en la Sierra Maestra (Cuba Oriental): volcanismo generado en un arco de islas tholeiítico (Volcanic rocks from the lower and intermediate series of the El Cobre Group, Sierra Maestra, Eastern Cuba: a case of island arc tholeiites). *Acta Geologica Hispanica*, **33**, 57–74.

CHAUVIN, A., BAZHENOV, M. L. & BEAUDOUIN, T. 1994. A reconnaissance paleomagnetic study of Cretaceous tocks from central Cuba. *Geophysical Research Letters*, **12**, 1691–1694.

CLOOS, M. 1982. Flow melanges: Numerical modelling and geologic constraints on their origin in the Franciscan subduction complex, California, *Geological Society of America Bulletin*, **93**, 330–345.

CLOOS, M. & SHREVE, R. L. 1988a. Subduction-channel model of prism accretion, melange formation, sediment subduction, and subduction erosion at convergent plate margins, 1. Background and description. *Pure and Applied Geophysics*, **128**, 455–500.

CLOOS, M. & SHREVE, R. L. 1988b. Subduction-channel model of prism accretion, melange formation, sediment subduction, and subduction erosion at convergent plate margins, 2. Implications and discussion. *Pure and Applied Geophysics*, **128**, 501–545.

COBIELLA, J. L. 1996. El magmatismo Jurásico (Caloviano?–Oxfordiano) de Cuba occidental: Ambiente de formación e implicaciones regionales. *Revista de la Asociación Geologica de Argentina*, **51**, 15–28.

COBIELLA, J. L., QUINTAS, F., CAMPOS, M. & HERNÁNDEZ, M. M. 1984. *Geología de la región central y suroriental de la provincia de Guantanamo*. Editorial Oriente, Santiago de Cuba.

CRUZ, E. M., MARESCH, W. V., CÁCERES, D. & BALCÁZAR, N. 2007. Significado de las paragénesis de anfíboles en metagabros relacionados con secuencias de margen continental en el NW de Cuba. *Revista Mexicana de Ciencias Geológicas*, **24**, 318–327.

DIAZ DE VILLALVILLA, L. 1997. Caracterización geológica de las formaciones volcánicas y volcanosedimentarias en Cuba Central, provincias Cienfuegos–Villa Clara–Sancti Spiritus. *In:* FURRAZOLA BERMÚDEZ, G. F. & NÚÑEZ CAMBRA, K. (eds) *Estudios sobre Geología de Cuba*. Instituto de Geología y Paleotología, La Habana, 325–344.

DILLON, W. P. & VEDDER, J. G. 1973. Structure and development of the continental margin of British Honduras. *Geological Society of America Bulletin*, **84**, 2713–2732.

DONNELLY, T. W. & ROGERS, J. J. W. 1980. Igneous Series in Island Arcs: the Northeastern Caribbean compared with worldwide island arc assemblages. *Bulletin of Volcanology*, **43**, 347–382.

DRAPER, G. 2001. The southern metamorphic terranes of Cuba as metamorphic core complexes exhumed by low angle extensional faulting, *Transactions of the 4th Congreso Cubano de Geologia y Mineria*, La Habana, Cuba, 19–23 March, 2001.

DRAPER, G. & BARROS, J. A. 1994. Cuba. *In*: DONOVAN, S. K. & JACKSON, T. A. (eds) *Caribbean Geology: an Introduction*. University of the West Indies Publishers Association and University of the West Indies Press, Kingston, Jamaica, 65–86.

DRAPER, G., GUTIERREZ, G. & LEWIS, J. F. 1996. Thrust emplacement of the Hispaniola peridotite belt: Orogenic expression of the Mid-Cretaceous Caribbean arc polarity reversal? *Geology*, **24**, 1143–1146.

DUBLAN, L. & ALVAREZ, H. (eds) 1986. *Informe final del levantamiento geologico y evaluacion de minerales utiles, en escala 1:50.000, del poligono CAME I, zona Central*, Academia de Ciencias de Cuba, MINBAS, CNFG, La Habana, unpublished internal report.

DUCLOZ, C. & VUAGNAT, M. 1962. A propose de l'age des serpentinites de Cuba. *Archives de Sciences de Physique et d'Histoire Naturelle, Geneve*, **15**, 309–332.

ELLIS, S., BEAUMONT, C. & PFIFFNER, O. A. 1999. Geodynamic models of subcrustal-scale episodic tectonic accretion and underplating in subduction zones. *Journal of Geophysical Research*, **104**, 15169–15190.

ERNST, W. G. 1988. Tectonic history of subduction zones inferred from retrograde blueschist $P–T$ paths. *Geology*, **16**, 1081–1084.

ESCUDER VIRUETE, J., DÍAZ DE NEIRA, A. *ET AL*. 2006. Magmatic relationships and ages of Caribbean Island arc tholeiites, boninites and related felsic rocks, Dominican Republic. *Lithos*. **90**, 161–186.

FONSECA, E., ZELEPUGUIN, V. & HEREDIA, M. 1984. Particularidades de la estructúra de la asociación ofiolitica de Cuba. *Ciencias de Tierra y Espacio (La Habana)*, **9**, 31–46.

FONSECA, E., DIAZ DE VILLALVILLA, L., MARY, T. & CAPOTE, C. 1988. Nuevos datos acerca de la relación entre las vulcanitas de la asociación ofiolitica y la parte inferior del arco volcanico en Cuba central-oriental. *Serie Geológica (La Habana)*, **2**, 3–12.

GARCÍA-CASCO, A., TORRES-ROLDÁN, R. L., MILLÁN, G., MONIÉ, P. & HAISSEN, F. 2001. High-grade metamorphism and hydrous melting of metapelites in the Pinos terrane (W-Cuba). Evidence for crustal thickening and extension in the northern Caribbean collisional belt. *Journal of Metamorphic Geology*, **19**, 699–715.

GARCÍA-CASCO, A., TORRES-ROLDÁN, R. L., MILLÁN, G., MONIÉ, P. & SCHNEIDER, J. 2002. Oscillatory zoning of eclogitic garnet and amphibole, Northern Serpentinite melange, Cuba: a record of tectonic instability during subduction? *Journal of Metamorphic Geology*, **20**, 581–598.

GARCÍA-CASCO, A., TORRES-ROLDÁN, R. L. *ET AL*. 2006. High pressure metamorphism of ophiolites in Cuba. *Geologica Acta*, **4**, 63–88.

GARCÍA-CASCO, A., LAZARO, C., ROJAS-AGRAMONTE, Y., KRONER, A. & NEUBAUER, F. 2007. From Aptian onset to Danian demise of subduction along the northern margin of the Caribbean Plate (Sierra del Convento Melange, Eastern Cuba). *EOS Transactions AGU*, **88**, Fall Meeting Supplement, Abstract T11D-06.

GARCÍA-CASCO, A., ITURRALDE-VINENT, M. A. & PINDELL, J. L. 2008. Latest Cretaceous collision/accretion between the Caribbean Plate and Caribeana: origin of metamorphic terranes in the Greater Antilles. *International Geology Review*, **50**, 1–29.

GERYA, T. V. & STÖCKHERT, B. 2002. Exhumation rates of high pressure metamorphic rocks in subduction channels: the effect of rheology. *Geophysical Research Letters*, **29**, 1261.

GERYA, T. V. & STÖCKHERT, B. 2006. Two-dimensional numerical modeling of tectonic and metamorphic histories at active continental margins. *International Journal of Earth Sciences*, **94**, 250–274.

GERYA, T. V., STÖCKHERT, B. & PERCHUK, A. L. 2002. Exhumation of high-pressure metamorphic rocks in a subduction channel: a numerical simulation. *Tectonics*, **142**, 6-1–6-19.

GIUNTA, G., BECCALUVA, L. & SIENA, F. 2006. Caribbean Plate margins: constraints and current problems. *Geologica Acta*, **4**, 265–277.

GOFFÉ, B., BOUSQUET, R., HENRY, P. & LE PICHON, X. 2003. Effect of the chemical composition of the crust on the metamorphic evolution of orogenic wedges. *Journal of Metamorphic Geology*, **21**, 123–141.

GORDON, M. B., MANN, P., CÁCERES, D. & FLORES, R. 1997. Cenozoic tectonic history of the North America–Caribbean Plate boundary zone in western Cuba. *Journal of Geophysical Research*, **102**, 10055–10082.

GRAFE, F. 2000. *Geochronologie metamorpher Komplexe am Beispiel der kretazischen Inselbogen-Kontinent-Kollisionszone Zentralkubas*. PhD thesis, Ruhr Universität Bochum.

GRAFE, F., STANEK, K. P. *ET AL*. 2001. Rb-Sr and $^{40}Ar/^{39}Ar$ mineral ages of granitoid intrusives in the Mabujina Unit, Central Cuba: thermal exhumation history of the Escambray Massif. *Journal of Geology*, **109**, 615–631.

GREVEL, C. 2000. *Druck- und Temperaturentwicklung der metamorphen Deckeneinheiten des Escambray Massives, Kuba*. PhD thesis, Ruhr Universität Bochum.

GREVEL, C., MARESCH, W. V., STANEK, K. P., GRAFE, F. & HOERNES, S. 2006. Petrology and geodynamic significance of deerite-bearing metaquartzites from the Escambray Massif, Cuba. *Mineralogical Magazine*, **70**, 527–546.

HACZEWSKI, G. 1976. Sedimentological reconnaissance of the San Cayetano formation: an accumulative continental margin in the Jurassic of western Cuba. *Acta Geologica Polonia*, **26**, 331–353.

HALL, C. M., KESLER, S. E. *ET AL*. 2004. Age and tectonic setting of the Camagüey volcanic-intrusive arc, Cuba: evidence for rapid uplift of the western Greater Antilles. *Journal of Geology*, **112**, 521–542.

HATTEN, C. W., SOMIN, M., MILLÁN, G., RENNE, P., KISTLER, R. W. & MATTINSON, J. M. 1988. Tectonostratigraphic units of Central Cuba. *Transactions of the 11th Caribbean Geological Conference*, Barbados, 35:1–35:13.

HARLOW, G. E., HEMMING, S. R., AVÉ LALLEMANT, H. G., SISSON, V. B. & SORENSEN, S. S. 2004. Two high-pressure–low-temperature serpentinite-matrix mélange belts, Motagua fault zone, Guatemala: a record of Aptian and Maastrichtian collisions. *Geology*, **32**, 17–20.

HEMPTON, M. R. & BARROS, J. A. 1993. Mesozoic stratigraphy of Cuba: deposition architecture of a southeast facing continental margin. *In*: PINDELL, J. L. & PERKINS, R. F. (eds) *Mesozoic and Early Cenozoic Development of the Gulf of Mexico and Caribbean Region – a Context for Hydrocarbon Exploration.* GCSSEPM Foundation Thirteenth Annual Research Conference, 193–209.

HSU, K. J. 1971. Franciscan melange as a model for eugeosynclinal sedimentation and underthrusting tectonics. *Journal of Geophysical Research*, **76**, 1162–1170.

HUTSON, F., MANN, P. & RENNE, P. 1998. ^{40}Ar/^{39}Ar dating of single muscovite grains in Jurassic siliciclastic rocks (San Cayetano Formation). Constraints on the paleoposition of western Cuba. *Geology*, **26**, 83–86.

ITURRALDE-VINENT, M. A. 1992. A short note on the Cuban Late Maastrichtian megaturbifite (an impact-derived deposit?). *Earth and Planetary Science Letters*, **109**, 225–228.

ITURRALDE-VINENT, M. A. 1994. Cuban geology: a new plate-tectonic synthesis. *Journal of Petroleum Geology*, **17**, 39–70.

ITURRALDE-VINENT, M. A. 1996. Estratigrafía del Arco Volcánico en Cuba. *In*: ITURRALDE-VINENT, M. A. (ed.) *Cuban Ophiolites and Volcanic Arcs.* Miami, IGCP Project 364 Special Contributions, **1**, 190–227.

ITURRALDE-VINENT, M. A. 1998. Sinopsis de la Constitución Geológica de Cuba. *In*: MELGAREJO, J. C. & PROENZA, J. A. (eds) *Geología y Metalogénia de Cuba: Una Introducción. Acta Geologica Hispanica*, **33**, 9–56.

ITURRALDE-VINENT, M. A. & THIEKE, U. (eds) 1987. *Informe final sobre los trabajos del levantamiento geológico complejo y busqueda acompañante a escala 1:50.000 en el polígono CAME III 1981–1987.* Academia de Ciencias de Cuba, MINBAS, CNFG, La Habana, unpublished internal report.

ITURRALDE-VINENT, M. A. & MORALES MARÍ, T. 1988. Toleitas del Titoniano medio en la Sierra de Camajan, Camagüey. *Revista Tecnologica, Seria Geologica, La Habana*, **18**, 25–32.

ITURRALDE-VINENT, M. A. & LIDIAK, E. G. 2006. Foreword: Caribbean tectonic, magmatic, metamorphic, and stratigraphic events. Implications for plate tectonics. *Geologica Acta*, **4**, 1–5.

ITURRALDE-VINENT, M. A., STANEK, K. P., WOLF, D., THIEKE, H. U. & MÜLLER, H. 2000. Geology of the Camagüey Region, Central Cuba – evolution of a collisional margin in the northern Caribbean. *Zeitschrift für Angewandte Geologie, Hannover*, **SH 1**, 267–273.

ITURRALDE-VINENT, M. A., DÍAZ-OTERO, C., RODRÍGUEZ-VEGA, A. & DÍAZ-MARTÍNEZ, D. 2006. Tectonic implications of paleontologic dating of Cretaceous–Danian sections of Eastern Cuba. *Geologica Acta*, **4**, 89–102.

JÄGER, W. 1972. Zur Lithlogie, Tektonik und Vererzung im Gebiet vom Matahambre (Provinz Pinar del Rio, Kuba). *Jahrbuch Geologie (Berlin)*, **4** (1968), 347–385.

JAMES, K. H. 2002. A simple synthesis and evolution of the Caribbean region. *In: Abstracts, 16th Caribbean Geological Conference,* Barbados, 16–21, June 2002. World Wide Web Address: http://www.ig.utexas.edu/CaribPlate/forum/james/james_carib_model.pdf.

JAMES, K. H. 2006. Arguments for and against the Pacific origin of the Caribbean Plate: discussion, finding for an inter-American origin. *Geologica Acta*, **4**, 279–302.

KANČEV, I. (ed.). 1978. *Informe geológico de la provincia Las Villas – Resultados de las investigaciones geologicas a escala 1:250.000 durante el periodo 1969–1975.* Academia de Ciencias de Cuba, MINBAS, CNFG, La Habana, unpublished internal report.

KERR, A. C., ITURRALDE-VINENT, M. A., SAUNDERS, A. D., BABBS, T. L. & TARNEY, J. 1999. A new plate tectonic model of the Caribbean: implications from a geochemical reconnaissance of Cuban Mesozoic volcanic rocks. *Geological Society of America Bulletin*, **111**, 1581–1599.

KERR, A. C., WHITE, R. V., THOMPSON, P. M. E., TARNEY, J. & SAUNDERS, A. D. 2003. No oceanic plateau – no Caribbean Plate? The seminal role of an oceanic plateau in Caribbean Plate evolution. *In*: BARTOLINI, C., BUFFLER, R. T. & BLICKWEDE, J. (eds) *The Circum-Gulf of Mexico and the Caribbean: Hydrocarbon Habitats, Basin Formation, and Plate Tectonics.* American Association of Petroleum Geologists, Memoirs, **79**, 126–168.

KESLER, S. E., CAMPELL, I. H. & ALLEN, C. M. 2005. Age of the Los Ranchos Formation, Dominican Republic: timing and tectonic setting of primitive island arc volcanism in the Caribbean region. *Geological Society of America Bulletin*, **117**, 987–995.

KNIPPER, A. L. 1975. *Okeaničeskaja kora v strukturje al'piskoy skladčatoy oblasti. (Jug. Evropy, Zapadnaja čast Azii i Kuba).* Izdatelstvo Nauka, Moskva, 5–205.

KNIPPER, A. L. & CABRERA, R. 1974. *Tectonica y geologia historica de la zona de articulacion entre el mio- y eugeosinclinal y del cinturon hiperbasico de Cuba.* Academia de Ciencias de Cuba Special Publications, **2**, 15–77.

KOSAK, M., ANDO, J., JAKUS, P. & MARTINEZ, Y. R. 1988. Desarollo estructural del arco insular volcanico-Cretacico en la region de Holguin. *Revista Mineria y Geologia, Moa*, **5**, 33–55.

KREBS, M., STANEK, K. P., SCHERER, E., MARESCH, W. V., GRAFE, F., IDLEMAN, B. & RODIONOV, N. 2007. Age of High pressure metamorphism from the Escambray Massif, Cuba. *In: Goldschmidt Conference, Cologne, Germany.* Cambridge Publications, Abstract A522.

KREBS, M., MARESCH, W. V. *ET AL.* 2008. The dynamics of intra-oceanic subduction zones: A direct comparison between fossil petrological evidence (Rio San Juan Complex, Dominican Republic) and numerical simulation, *Lithos*, **103**, 106–137.

KUDĚLÁSEK, V., KUDĚLÁSKOVÁ, M., ZAMARSKÝ, V. & OREL, P. 1984. On the problem of Cuban ophiolites. *Krystalinikum, Prague*, **17**, 159–173.

KYSAR, G., MORTENSEN, J. K. & LEWIS, J. F. 1998. U–Pb zircon age constraints for Paleogene igneous rocks of the Sierra Maestra, Southeastern Cuba; implications for short-lived arc magmatism along the northern Caribbean margin. *Geological Society of America, Abstracts with Programs*, **30**, A185.

LÁZARO, C. & GARCÍA-CASCO, A. 2008. Geochemical and Sr–Nd isotope signatures of pristine slab melts and their residues (Sierra del Convento mélange, eastern Cuba). *Chemical Geology*, **255**, 120–133.

LEBRÓN, M. C. & PERFIT, M. R. 1993. Stratigraphic and petrochemical data support subduction polarity reversal of the Cretaceous Caribbean island arc. *Journal of Geology*, **101**, 389–396.

LEROY, S., MAUFRET, A., PATRIAT, P. & MERCIER DE LÉPINAY, B. 2000. An alternative interpretation of the Cayman trough evolution from a reidentification of the magnetic anomalies. *Geophysical Journal International*, **141**, 539–557.

LEWIS, J. F. & DRAPER, G. 1990. Geological and tectonic evolution of the northern Caribbean margin. *In*: DENGO, G. & CASE, J. E. (eds) *The Caribbean Region*. The Geology of North America, **H**. Geological Society of America, Boulder, CO, 77–140.

LEWIS, J. F., PERFIT, M. R. *ET AL.* 2005. Anomalous granitoid compositions from the northwestern Cayman Trench: Implications for the composition and evolution of the Cayman Ridge. *17th Caribbean Geological Conference*, Puerto Rico, Abstracts, 49–50.

LINARES, E. & SMAGOULOV, R. 1987. Resultados de análisis de minerales pesados de rocas de algunas formaciones lithostratigraficas del miogeosinclinal Cubano. *Revista Tecnologica, Seria Geologica, La Habana*, **4**, 75–99.

LLANES-CASTRO, A., GARCÍA-DELGADO, D. & MEYERHOFF, D. 1998. Hallazgo de fauna jurásica (Tithoniano) en ofiolitas de Cuba central. Memorias II Geología y Minería, **98**, Havana, Centro Nacional de Informacion Geológica, 241–244.

MAKAROV, V. I. 1986. Noveishaja tektonika vostocnoi Kuby, stat'ja pervaja. *Geotektonika*, **6**, 85–96.

MARI MORALES, T. 1997. Particularidades de los granitoides de Ciego-Camagüey-Las Tunas y consideraciones sobre su posición dentro del arco de isla. *In*: FURRAZOLA BERMÚDEZ, G. F. & NÚÑEZ-CAMBRA, K. (eds) *Estudios sobre la geología de Cuba*, Centro Nacional de Información Geológica, La Habana, 399–416.

MARTENS, U., LIOU, J., SOLARI, L., MATTINSON, C. & WOODEN, J. 2007. SHRIMP RG U/Pb age of Chuacús complex zircon: evidence for Cretaceous HP metamorphism in the Maya block. In: MARTENS, U. & GARCÍA-CASCO, A. (eds) *High-Pressure Belts of Central Guatemala: The Motagua Suture and the Chuacús Complex*. IGCP 546 Special Contributions, 1, 7–13.

MESCHEDE, M. & FRISCH, W. 1998. A plate-tectonic model for the Mesozoic and Cenozoic history of the Caribbean Plate. *Tectonophysics*, **296**, 269–291.

MEYERHOFF, A. A. & HATTEN, C. W. 1974. Bahamas salient of North America: tectonic framework, stratigraphy and petroleum potential. *American Association of Petroleum Geologists Bulletin*, **58**, 1201–1239.

MILLÁN, G. 1972. *El metamorfismo y mesodeformaciones de la unidad tectonica mas suroriental de la Sierra de los Organos*. Boletin Actas, Instituto de Geologia, La Habana, **N2**.

MILLÁN, G. 1975. El complejo cristalino Mesozoico de Isla de Pinos. Su metamorfismo. *Serie Geológica, La Habana*, **23**, 3–16.

MILLÁN, G. & MYCZYNSKI, R. 1978. *Fauna Jurasica y consideraciones sobre la edad de las secuencias metamorficas del Escambray*. Informe Cientifico y Tecnico, Academia de Ciencias de Cuba, **80**, 1–14.

MILLÁN, G. & SOMIN, M. L. 1985. *Contribucion al conocimiento geologico de las metamorfitas del Escambray y del Purial*. Academia de Ciencias de Cuba, La Habana, **1985**, 74.

MORENO TOIRAN, B. 2003. The crustal structure of Cuba derived from receiver functions analysis. *Journal of Seismology*, **7**, 359–375.

MULLINS, H. T. & LYNTS, G. W. 1976. Stratigraphy and structure of northeast Providence Channel, Bahamas. *American Association of Petroleum Geologists Bulletin*, **60**, 1037–1053.

NAGLE, F. 1974. Blueschist, eclogite, paired metamorphic belts and the early tectonics of Hispaniola. *Geological Society of America Bulletin*, **85**, 1461–1466.

OTERO, R., PROL, J. L., TENREYRO, R. & ARRIAZA, G. L. 1998. Características de la corteza terrestre de Cuba y su plataforma marina. *Minería y Geología, La Habana*, **15**, 31–35.

PARDO, G. 1975. Geology of Cuba. *In*: NAIRN, A. E. M. & STEHLI, F. G. (eds) *The Ocean Basins and Margins*, **3**. Plenum Press, New York, 553–615.

PERERA-FALCÓN, C. M., BLANCO-BUSTAMANTE, S., SEGURA-SOTO, R., DÍAZ DE VILLALVILLA, L. & PÉREZ-ESTRADA, L. 1998. Estratigrafía del arco Cretácico de Cuenca Central, Cuba. *In*: *Geología y Minería '98*. Centro Nacional de Información Geologica del IGP, **2**, 245–248.

PÉREZ LAZO, J., FUNDORA GRANDA, M., GARCÍA, A., KROPÁČEK, V. & HORÁČEK, J. 1994. Paleomagnetic investigations in Cuba from Late Jurassic to Middle Eocene times and tectonic implications. *Acta Universitatis Carolinae, Geologica, Prague*, **38**, 3–19.

PEREZ-OTHON, J. & YARMOLIUK, V. A. 1985. *Mapa Geologica de Cuba*. MINBAS: CIG, La Habana.

PINDELL, J. L. 1985. Alleghenian reconstruction and subsequent evolution of the Gulf of Mexico, Bahamas, and the Proto-Caribbean. *Tectonics*, **4**, 1–39.

PINDELL, J. L. 1990. Geological arguments suggesting a Pacific origin for the Caribbean Plate. *In*: LARUE, D. K. & DRAPER, G. (eds) *Transactions of the 12th Caribbean Geological Conference*, St Croix. Miami Geological Society, 1–4.

PINDELL, J. L. 1993. Regional synopsis of Gulf of Mexico and Caribbean evolution, *In*: PINDELL, J. L. & PERKINS, R. (eds) *Mesozoic and Early Cenozoic Development of the Gulf of Mexico and Caribbean Region*. GCSSEPM 13th Annual Research Conference Proceedings, 251–274.

PINDELL, J. L. 2008. Early Cretaceous Caribbean tectonics: models for genesis of the Great Caribbean Arc, *Abstract Volume of the 18th Caribbean Geological Conference*, 24–28 March 2008, Santo Domingo, Dominican Republic. World Wide Web Address: http://www.ugr.es/~agcasco/igcp546/DomRep08/Abstracts_CaribConf_DR_2008.pdf.

PINDELL, J. L. & DEWEY, J. F. 1982. Permo-Triassic reconstruction of Western Pangea and the evolution of the Gulf of Mexico/Caribbean region, *Tectonics*, **1**, 179–212.

PINDELL, J. L. & BARRETT, S. F. 1990. Geologic evolution of the Caribbean region: a plate-tectonic perspective. *In*: DENGO, G. & CASE, J. E. (eds) *The Caribbean Region. The Geology of North America*, **H**. Geological Society of America, Boulder, CO, 405–432.

PINDELL, J. L. & DRAPER, G. 1991. Stratigraphy and geological history of the Puerto Plata area, northern Dominican Republic. *In*: MANN, P., DRAPER, G. & LEWIS, J. F. (eds) *Geologic and Tectonic Development if the North American–Caribbean Plate Boundary in Hispaniola*. Geological Society of America, Special Paper, **262**, 97–114.

PINDELL, J. L. & KENNAN, L. 2001. Kinematic evolution of the Gulf of Mexico and the Caribbean. *GCSSEPM Foundation 21st Annual Research Conference, Petroleum Systems of Deep-water Basins*, 193–220.

PINDELL, J. L. & KENNAN, L. 2009. Tectonic evolution of the Gulf of Mexico, Caribbean and northern South America in the mantle reference frame: an update. *In*: JAMES, K. H., LORENTE, M. A. & PINDELL, J. L. (eds) *The Origin and Evolution of the Caribbean Plate*. Geological Society, London, Special Publications, **328**, 1–55.

PINDELL, J. L., CANDE, S. *ET AL*. 1988. A plate-kinematic framework for models of Caribbean evolution, *Tectonophysics*, **155**, 121–138.

PINDELL, J., KENNAN, L., MARESCH, W. V., STANEK, K. P., DRAPER, G. & HIGGS, R. 2005. Plate-kinematics and crustal dynamics of the circum-Caribbean arc-continent interactions: tectonic controls on the basin development in the Proto-Caribbean margins. *In*: AVÉ LALLEMANT, H. G. & SISSON, V. B. (eds) *Caribbean–South American plate interactions, Venezuela*. Geological Society of America, Boulder, CO, Special Papers, **394**, 7–52.

PINDELL, J. L., KENNAN, L., STANEK, K. P., MARESCH, W. V. & DRAPER, G. 2006. Foundations of Gulf of Mexico and Caribbean evolution: eight controversies resolved. *Geologica Acta*, **4**, 303–341.

PIOTROWSKA, K. 1987. Nappe structure in the Sierra de los Organos, western Cuba. *Acta Geologica Polonia*, **28**, 97–170.

PROENZA, J., GERVILLA, F., MELGAREJO, J. C. & BODINIER, J. L. 1999. Al- and Cr-rich chromitites from the Mayarí-Baracoa ophiolitic belt (eastern Cuba) consequence of interaction between volatile-rich melts and peridotites in suprasubduction mantle. *Economic Geology*, **94**, 547–566.

PROENZA, J., DÍAZ-MARTÍNEZ, R. *ET AL*. 2006. Primitive Cretaceous island-arc volcanic rocks in eastern Cuba: the Téneme Formation. *Geologica Acta*, **4**, 103–121.

PSZCZÓŁKOWSKI, A. 1985. Sobre la edad del metamorfismo y la estructura tectónica de la faja Cangre, Provincia de Pinar del Rio, Cuba. *Ciencias de la Tierra y del Espacio*, **10**, 31–36.

PSZCZÓŁKOWSKI, A. 1986a. Secuencia estratigrafica de Placetas en el área limitrofe de las provincias de Matanzas y Villa Clara (Cuba). *Bulletin of Polish Academy of Sciences, Earth Science*, **34**, 67–79.

PSZCZÓŁKOWSKI, A. 1986b. Megacapas del Maestrichtiano en Cuba occidental y central. *Bulletin of Polish Academy of Sciences, Earth Science*, **34**, 81–94.

PSZCZÓŁKOWSKI, A. 1987. Paleogeographic and paleotectonic evolution of Cuba and adjoining areas during the Jurassic–Early Cretaceous. *Annales Societatis Geologorum Poloniae*, **57**, 127–142.

PSZCZÓŁKOWSKI, A. 1994. Geological cross-sections through the Sierra del Rosario thrust belt, western Cuba. *Studia Geologica Polonia*, **105**, 67–90.

PSZCZÓŁKOWSKI, A. 1999. The exposed passive margin of North America in western Cuba. *In*: MANN, P. (ed.) *Caribbean Basins*. Sedimentary Basins of the World, **4**. Elsevier Science, Amsterdam, 93–121.

PSZCZÓŁKOWSKI, A. 2002. Crustacean burrows from the Upper Maastrichtian deposits of south-central Cuba. *Bulletin of Polish Academy of Sciences, Earth Science*, **50**, 147–163.

PSZCZÓŁKOWSKI, A. & ALBEAR, J. F. 1982. Subzona estructuro-facial de Bahia Honda, Pinar del Rio, su tectonica y datos sobre la sedimentación y paleogeografia del Cretacico Superior y del Paleógeno. *Ciencias de la Tierra y del Espacio*, **5**, 3–25.

PUSHAROVSKI, J. M., MOSSAKOVSKI, A. A., NEKRASOV, G. E., ORO, J., FLORES, R. & FORMEL-CORTINA, F. 1989. *Tektonitscheskaja karta Kuby, 1:500 000*. Izdat. Nauka, Moskva.

RATSCHBACHER, L., FRANZ, L. *ET AL*. 2009. The North American–Caribbean Plate boundary in Mexico–Guatemala–Honduras. *In*: JAMES, K. H., LORENTE, M. A. & PINDELL, J. L. (eds) *The Origin and Evolution of the Caribbean Plate*. Geological Society, London, Special Publications, **328**, 219–293.

RENNE, P. R., MATTINSON, J. M. *ET AL*. 1989. ^{40}Ar/^{39}Ar and U–Pb evidence for the Late Proterozoic (Grenville-age) continental crust in north-central Cuba and regional tectonic implications. *Precambrian Research*, **42**, 325–341.

RENNE, P. R., SCOTT, G. R., DOPPELHAMMER, S. K., HARGRAVES, R. B. & LINARES, E. 1991. Discordant Mid-Cretaceous paleomagnetic pole from the Zaza Terrane of central Cuba. *Geophysical Research Letters*, **18**, 455–458.

ROJAS-AGRAMONTE, Y., NEUBAUER, F. *ET AL*. 2004. Geochemistry and age of late orogenic island arc granitoids in the Sierra Maestra, Cuba: evidence for subduction magmatism in the Early Paleogene. *Chemical Geology*, **213**, 307–324.

ROJAS-AGRAMONTE, Y., NEUBAUER, F., BOJAR, A. V., HEJL, E., HANDLER, R. & GARCÍA-DELGADO, D. E. 2006a. Geology, age and tectonic evolution of the Sierra Maestra Mountains, southeastern Cuba. *Geologica Acta*, **4**, 123–150.

ROJAS-AGRAMONTE, Y., KRÖNER, A., GARCÍA-CASCO, A., ITURRALDE-VINENT, M. A., WINGATE, M. T. D. & LIU, D. Y. 2006b. Geodynamic implications of zircon ages from Cuba. *Geophysical Research Abstracts*, **8**, 04943; SRef-ID: 1607- 7962/ gra/EGU06-A-04943.

ROJAS-AGRAMONTE, Y., KRÖNER, A. *ET AL*. 2008. Detrital zircon geochronology of Jurassic sandstones of western Cuba (San Cayetano Formation): implications for the Jurassic paleogeography of the NW Proto-Caribbean. *American Journal of Science*, **308**, 639–656.

ROSENCRANTZ, E. 1990. Structure and tectonics of the Yucatán Basin, Caribbean Sea, as determined from seismic reflection studies. *Tectonics*, **9**, 1037–1059.

ROSENCRANTZ, E., ROSS, M. I. & SCLATER, J. G. 1988. Age and spreading history of the Cayman Trough as determined from depth, heat flow, magnetic anomalies. *Journal of Geophysical Research*, **93**, 2141–2157.

SAURA, E., VERGÉS, J. *ET AL.* 2008. Structural and tectonic evolution of western Cuba fold and thrust belt. *Tectonics*, **27**, TC4002.

SCHLAGER, W. & BUFFLER, R. T. 1984. Deep Sea Drilling Project, Leg 77, southeastern Gulf of Mexico. *Geological Society of America Bulletin*, **95**, 226–236.

SCHNEIDER, J., BOSCH, D. *ET AL.* 2004. Origin and evolution of the Escambray Massif (Central Cuba) an example of HP/LT rocks exhumed during intraoceanic subduction. *Journal of Metamorphic Geology*, **22**, 3, 227–247.

SCHWARTZ, S., LARDEAUX, J. M., TRIEART, P., GUILLOEAND, E. & LABRIN, S. 2007. Diachronous exhumation of HP-LT metamorphic rocks from southwestern Alps: evidence from fission-track analysis. *Terra Nova*, **19**, 133–140.

SEGURA-SOTO, R., MILIAN, E. & FERNANDEZ, J. 1985. Complejos litológicas del extremo noroccidental de Cuba y sus implcaciones estratigráficas de acuerdo con los datos de las perforaciones profundas. *Revista Tecnologica, Seria Geologica, La Habana*, **15**, 32–36.

SHREVE, R. L. & CLOOS, M. 1986. Dynamics of sediment subduction, melange formation, and prism accretion. *Journal of Geophysical Research*, **91**, 10229–10245.

SIGURDSSON, H., LECKIE, M., ACTON, G. & And ODP Leg 165 Scientific Party. 1997. *Proceedings of the Oceanic Drilling Program. Initial Report of Ocean Drilling Project Leg 165*. Ocean Drilling Program, College Station TX.

SOMIN, M. L. & MILLÁN, G. 1981. *Geologija metamorficheskich kompleksov Kuby*. Isdat. Nauka, Moscow, **1981**, 219 p.

SOMIN, M. L., ARAKELJANZ, M. M. & KOLESNIKOV, E. M. 1992. Vozrast i tektoniceskoje znacenije vysokobariceskich metamorficeskich porod Kuby. *Izvestija Akademii Nauk, Rossiiskaja Akademia Nauk, Serija Geologicheskaya*, **N 3**, 91–104.

SOMIN, M. L., MATTINSON, J. M. *ET AL.* 2005. The Arroyo Charcon, an unusual eclogite from the Escambray Massif, Cuba: petrology and zirconology. *Mitteilungen der Österreichischen Mineralogischen*

Gesellschaft, **150**. World Wide Web Address: www. uni-graz.at/IEC-7/PDF-files/Somin.pdf.

SOMIN, M. L., LEPECHINA, E. N. & TOLMACHEVA, E. V. 2006. El Guayabo gneiss sialic basement boulder in Western Cuba. *Geophysical Research Abstracts*, **8**, 03377; Sref-ID: 1607-7962/gra/EGU06-A-03377.

SPEAR, F. S. 1993. *Metamorphic Phase Equilibria and Pressure–Temperature–Time Path*. Mineralogical Society of America Monograph Series, **1**.

STANEK, K. P. & CABRERA, R. 1992. Tectono-magmatic development of Central Cuba. *Zentralblatt für Geologie und Paläontologie*, **6**, 1571–1580.

STANEK, K. P. & VOIGT, S. 1994. Model of Meso-Cenozoic evolution of the NW-Caribbean. *Zentralblatt für Geologie und Paläontologie*, **6**, 1571–1580.

STANEK, K. P., COBIELLA-REGUERA, J. L., MARESCH, W. V., MILLÁN TRUJILLO, G., GRAFE, F. & GREVEL, C. 2000. Geologic development of Cuba. *Zeitschrift für Angewandte Geologie, Hannover*, **SH 1**, 259–265.

STANEK, K. P., RIßE, A., RENNO, A., ROMER, R. & GRAFE, F. 2005. The history of the Great Antillean Island Arc: example from Central Cuba. *Abstract, Colloquium on Latin American Geosciences*, Potsdam, Terra Nostra, **2005/1**, 117.

STANEK, K. P., MARESCH, W. V., GRAFE, F., GREVEL, C. & BAUMANN, A. 2006. Structure, tectonics and metamorphic development of the Sancti Spiritus Dome (eastern Escambray Massif, Central Cuba). *Geologica Acta*, **4**, 151–170.

SZAKMÁNY, G., TÖRÖK, K. & GÁL-SÓLYMOS, K. 1999. Nagynyomású metamorfit blokkok a keletkubai Holguíntól északra húzódó ofiolitos melanzs zónából (High pressure metamorphic blocks from the ophiolitic melange zone, north of Holguín, Eastern Cuba). *Földtani Kötzlöny (Budapest)*, **129**, 541–571.

TADA, R., NAKANO, Y. *ET AL.* 2002. *Complex Tsunami Waves Suggested by the Cretaceous–Tertiary Boundary Deposit at the Moncada Section, Western Cuba*. Geological Society of America, Boulder, CO, Special Papers, **356**, 109–123.

TAKAYAMA, H., TADA, R. *ET AL.* 2000. Origin of the Penalver Formation in northwestern Cuba and its relation to the K/T boundary impact event. *Sedimentary Geology*, **135**, 295–320.

UCHUPI, E., MILLIMAN, J. D., LUYENDYK, B. P., BOWIN, C. O. & EMERY, K. O. 1971. Structure and origin of the southern Bahamas. *American Association of Petroleum Geologists Bulletin*, **55**, 687–704.

Is the Cretaceous primitive island arc series in the circum-Caribbean region geochemically analogous to the modern island arc tholeiite series?

ALAN R. HASTIE

School of Earth and Ocean Sciences, Cardiff University, Main Building, Park Place, Cardiff CF10 3YE, UK (e-mail: hastiear1@cf.ac.uk)

Abstract: The Early Cretaceous island arc lavas in the Caribbean region are frequently assigned to the primitive island arc (PIA) series and not to the island arc tholeiite (IAT) series. However, this review demonstrates that the Caribbean PIA rocks have immobile trace element abundances, trace element ratios and Nd–Hf isotope systematics which are indistinguishable from modern IAT lavas. Thus, it is proposed that the term PIA series be discarded and that the Early Cretaceous island arc rocks in the Caribbean be classified as IAT rocks.

Supplementary material: References for data sources used in figures can be found at: http://www.geolsoc.org.uk/SUP18362.

The Greater Antilles Islands, composed of Cuba, Jamaica, Hispaniola, Puerto Rico and the Virgin islands, are part of an extinct lower-Cretaceous to Early Cenozoic island arc, which was tectonically emplaced between North and South America along with the Caribbean ocean crust in the Cretaceous (Burke 1988; Lebron & Perfit 1993, 1994; Müller *et al.* 1999; Kerr *et al.* 1999, 2003; White *et al.* 1999; Pindell & Kennan 2001; Thompson *et al.* 2003, 2004; Kesler *et al.* 2005; Jolly *et al.* 2006; Pindell *et al.* 2006). This extinct island arc, here named the 'Great Arc of the Caribbean' after Burke (1988), is predominantly composed of two differing island arc-derived volcanic rock suites originally defined by Donnelly *et al.* (1971) and Donnelly & Rogers (1980) as an older primitive island arc (PIA) series, which erupted onto Jurassic–Early Cretaceous oceanic basement (Kesler *et al.* 2005), and a younger calc-alkaline (CA) series that also consists of small volumes of high-K calc-alkaline and shoshonitic rocks.

Figure 1 shows the location of the Caribbean Cretaceous PIA successions in the Greater Antilles islands, which include (1) the lower Devil's Racecourse Formation, Jamaica (Hastie *et al.* 2009); (2) the Sagua la Chica, Los Pasos, Mabujina and parts of the Quivijan Formations and the Guamuta, Loma de la Bandera and Cerrajón dykes in Cuba (Kerr *et al.* 1999; 2003; Marchesi *et al.* 2007); (3) the Maimaon Formation, Amina Schists, Guamira basalt, Los Ranchos Formation and the Tireo Formation in Hispaniola (Donnelly *et al.* 1990; Lebron & Perfit 1994; Kerr *et al.* 2003; Kesler *et al.* 2005; Escuder Viruete *et al.* 2006, 2007);

(4) the Rio Majada Group (Formations A, B and C) in the Central Volcanic Province and the Daguao Formation and Figuera lava in the Northeastern Volcanic Province in Puerto Rico (Frost *et al.* 1998; Schellekens 1998; Jolly *et al.* 1998*a*, *b*, 2001, 2006) and (5) the Water Island and Louisenhoj Formations in the US Virgin Islands (Donnelly *et al.* 1990; Rankin 2002; Jolly & Lidiak 2006). In addition to the island arc successions in the Greater Antilles, the Washikemba Formation on Bonaire has also been assigned to the PIA series, as have arc rocks on Tobago (e.g. Donnelly *et al.* 1990; Kerr *et al.* 2003; Thompson 2002; Thompson *et al.* 2004).

Donnelly *et al.* (1971) and Donnelly & Rogers (1980) demonstrate that the PIA series is composed of volcanic island arc-derived rocks with low Th, U, Rb, Ba and K contents, low Th/U ratios, low rare earth element (REE) abundances and flat chondrite-normalized REE patterns throughout the full range of volcanic rock types (e.g. basalts to rhyolites). Furthermore, they suggested that this chemistry is similar to that of the island arc tholeiite (IAT) series characterized in other island arc volcanic successions worldwide. The younger basic to acidic CA rocks are more enriched in incompatible trace elements and have more enriched radiogenic isotope ratios relative to basic and acidic lavas from the PIA series (Donnelly & Rogers 1980; Donnelly *et al.* 1990; Frost *et al.* 1998; Schellekens 1998; Jolly *et al.* 1998*a*, *b*; Kerr *et al.* 2003).

Donnelly & Rogers (1980) and Donnelly *et al.* (1990) do not, however, use the term 'tholeiite' to

From: JAMES, K. H., LORENTE, M. A. & PINDELL, J. L. (eds) *The Origin and Evolution of the Caribbean Plate*. Geological Society, London, Special Publications, **328**, 399–409.
DOI: 10.1144/SP328.16   0305-8719/09/$15.00 © The Geological Society of London 2009.

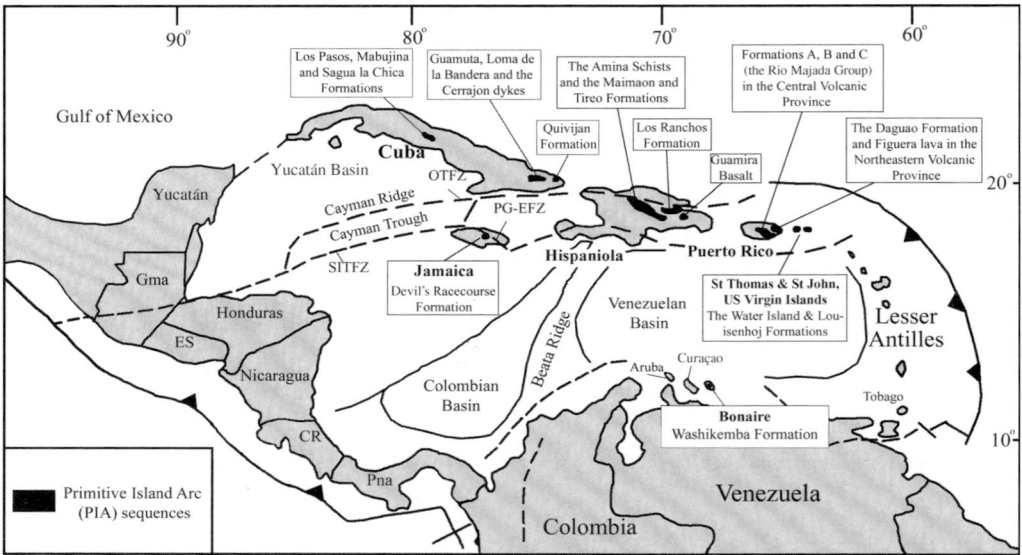

Fig. 1. Map of the Caribbean region showing the location of the Cretaceous PIA successions. Guatemala (Gma), El Salvador (ES), Costa Rica (CR), Panama (Pna), Swan Islands Transform Fault Zone (SITFZ), Oriente Transform Fault Zone (OTFZ), Plantain Garden-Enriquillo Fault zone (PG-EFZ) (modified from Sinton *et al.* 1998).

define the PIA series in the Caribbean because the PIA series contains abundant silicic rock types and they propose that the term 'arc tholeiite' could be confused with other 'non-arc related' tholeiitic rocks in the Caribbean region, for example, the Bermeja Complex in western Puerto Rico, which is predominantly composed of metamorphosed volcanic and sedimentary rocks that have an oceanic crustal affinity (e.g. Donnelly *et al.* 1971; Schellekens 1998). Donnelly *et al.* (1990) also propose that the PIA rocks have higher K_2O and incompatible element contents than IAT lavas. Although they do suggest that future investigations may find that the IAT and PIA series are chemically analogous.

Nevertheless, geochemists currently working in the Caribbean region continue to use both the terms PIA series and IAT series to describe the oldest volcanic arc rocks in the Greater Antilles. For example, Escuder Viruete *et al.* (2006) and Marchesi *et al.* (2007) use the IAT term whereas Kesler *et al.* (2005) and Jolly *et al.* (2006) use the PIA term. Consequently, the purpose of this paper is to compare the geochemistry of the Cretaceous Caribbean PIA rocks with modern IAT lavas to determine if the series are geochemically analogous. If so, it may be wise to abandon the 'PIA series' terminology in favour of the IAT series, which is used for all other modern and ancient volcanic island arcs (e.g. Ewart 1982).

Primitive island arc v. island arc tholeiite compositions

IAT composition and classification

The vast majority of volcanic island arc lavas classify as sub-alkaline on the total alkali–silica (TAS) diagram of Le Bas *et al.* (1986, 1992). These rocks can be further classified into tholeiitic, calc-alkaline, high-K calc-alkaline and shoshonitic series. Originally, the volcanic series were classified on the basis of iron enrichment using diagrams such as $FeO^{tot}/MgO-SiO_2$ (Miyashiro 1974) or $(Na_2O + K_2O)-MgO-FeO^{tot}$ (the AFM plot of Kuno 1968). Commonly, however, the classification of volcanic arc rocks into their rock series is performed using the K_2O-SiO_2 diagram of Peccerillo & Taylor (1976). The advantage of the K_2O-SiO_2 plot over the other diagrams, is that it can assign both a volcanic series (tholeiitic, calc-alkaline, high-K calc-alkaline, shoshonite) based on K_2O enrichment, and a rock type (basalt, basaltic–andesite, andesite, dacite, rhyolite) based on silica content and hence degree of differentiation.

The compositional zonation of an intra-oceanic island arc was recognized five decades ago by workers such as Kuno (1959, 1966). It was shown that the compositional zonation in an island arc can be 'spatial' with IAT rocks occurring closest to the trench with CA and shoshonites (SHO) respectively occurring further towards the rear of the arc. Additionally, Baker (1968); Jakes & White (1969); Gill (1970) and Jakes & Gill (1970) were the first to identify temporal variations in island arcs whereby IAT rocks are erupted prior to the eruption of the CA and SHO. These findings are still supported in recent island arc studies, especially in the Greater Antilles where the PIA series is considered to represent the earliest period of arc development which is subsequently replaced with CA series volcanism (e.g. Jolly et al. 2001; Kesler et al. 2005). Therefore, from a temporal perspective the Caribbean PIA series is similar to the modern day IAT series.

The composition of the IAT series was first defined by Jakes & Gill (1970), who compared it with the CA series and abyssal tholeiites. They demonstrated that, relative to the CA series, IAT lavas have higher Na_2O/K_2O and K/Rb ratios, lower Rb/Sr, Th/U and La/Yb ratios and lower abundances of Rb, Sr, Ba, Cs, Th, U and La at similar silica ranges. A number of processes have been suggested in order to explain the enrichment of the large ion lithophile elements (LILE) and the light rare earth elements (LREE) relative to the high field strength elements (HFSE) and the heavy rare earth elements (HREE) in CA rocks relative to IAT lavas. These include (1) variable crustal contamination of ascending arc magmas, (2) differing depths of fractionation of parental arc magmas, (3) variable fO_2 or PH_2O in parental arc magmas, (4) differing degrees and depths of partial melting and (5) the contamination of the arc magma source region with variable amounts and compositions of material from a subducting oceanic slab (e.g. Gill 1981; Tatsumi et al. 1986; White & Dupre 1986; Davidson 1987; Pearce & Peate 1995; Thirlwall et al. 1996; Tatsumi & Kogiso 1997). The enriched nature of the mid-Late Cretaceous CA lavas, relative to the lower Mid-Cretaceous PIA rocks in the Caribbean region, is commonly attributed to the subduction and incorporation of differing slab-derived sedimentary components into the source region(s) of the Greater Antilles arc magmas (e.g. Jolly et al. 2006; Marchesi et al. 2007; Hastie et al. 2009).

Many geochemists working on altered Cretaceous igneous rocks in the Caribbean, and elsewhere, have demonstrated that SiO_2, MgO and the majority of the LILE, for example, Cs, Na, K, Ba, Rb, Sr, Pb and U, have been variably mobilized by a range of weathering, hydrothermal and metamorphic processes (Pearce 1996; Révillon et al. 2002; Geldmacher et al. 2003; Thompson et al. 2003, 2004; Escuder Viruete et al. 2006, 2007; Jolly et al. 2006; Hastie et al. 2007, 2008). In contrast, elements such as Th, the HFSE, for example, Nb, Ta, Zr and Hf, the rare earth elements (REE) and

the transition elements, for example Co, V, Cr, Ni and Sc, are considered to be relatively immobile during a wide range of alteration conditions and low-temperature metamorphism (up to greenschist-grade; e.g. Floyd & Winchester 1978; Winchester & Floyd 1977; Pearce 1982, 1996; Hastie et al. 2007). This raises an important problem with the majority of the aforementioned discrimination diagrams, major and trace element ratios and major and trace element abundances used to distinguish IAT rocks from CA lavas. They utilize elements which are easily mobilized during alteration processes, for example, the K_2O-SiO_2 and $FeO^{tot}/MgO-SiO_2$ diagrams of Peccerillo & Taylor (1976) and Miyashiro (1974) respectively. However, this problem has been addressed with the construction of the Th/Yb v. Ta/Yb and Th v. Co discrimination diagrams of Pearce (1982) and Hastie et al. (2007), respectively.

The Th/Yb v. Ta/Yb diagram uses the enrichment of Th over Yb to assign a volcanic series to a volcanic arc suite, whereas the enrichment of Ta over Yb illustrates the composition of the mantle source region from which the arc rocks are derived. The Th v. Co diagram similarly uses the abundance of Th to assign an arc rock to a specific volcanic suite, but, additionally uses the Co content as a measure of the degree of differentiation. The Th v. Co diagram is particularly useful for classifying altered arc rocks because it is a direct immobile element equivalent of the K_2O v. SiO_2 diagram. Consequently, immobile element discrimination diagrams coupled with immobile trace element ratios and chondrite normalized REE patterns can be used on altered and low-temperature metamorphosed (up to greenschist-grade) island arc lavas to assign them to a volcanic series and to determine their degree of differentiation.

Major and trace element comparison of the PIA, IAT and CA series

Donnelly & Rogers (1980) have shown that the basic to acidic volcanic island arc rocks of the PIA series have low Th, U, Rb, Ba and K abundances, low Th/U ratios, low REE abundances and flat chondrite-normalized REE patterns. Compositional data for 143 PIA samples were taken from the Earth Reference Data and Models database (http://www.earthref.org; Jolly et al. 1998b; Hastie et al. 2009), which confirms the low Th abundances (Supplementary Material). However, the intense alteration of Cretaceous igneous rocks in the Caribbean region (e.g. Jolly et al. 2001, 2006; Kesler et al. 2005; Escuder Viruete et al. 2006, 2007; Marchesi et al. 2007) has mobilized many of the LILE (see Hastie et al. 2007 for a review). Accordingly, the U, Rb, Ba and K compositions of the PIA rocks cannot be used to either classify the volcanic rocks or study their petrogenesis. The alteration of the PIA rocks also results in artificially increasing the major element oxide abundances because of the removal of mobile elements, indicated by large loss on ignition values and fluid contents (e.g. Hastie *et al.* 2007). Therefore, in order to classify the PIA rocks the Th/Tb v. Ta/Yb and Th v. Co immobile trace element discrimination diagrams of Pearce (1982) and Hastie *et al.* (2007), respectively, have to be used; however, Th, Yb, Ta and Co data is only available for 83 of the 143 PIA samples.

Nevertheless, Figure 2 clearly demonstrates that the majority of the 83 PIA samples can be conclusively classified as having IAT compositions. Interestingly, the bulk of the volcanic Washikemba Formation, which is an upper Cretaceous (*c.* 95 Ma) island arc succession on Bonaire, Dutch Antilles (Fig. 1) (Thompson *et al.* 2004), was originally assigned to the IAT suite by Thompson (2002) using the alkali index [$Na_2O + K_2O/0.17(SiO_2$-43)] v. Al_2O_3 diagram. However, this diagram utilizes many mobile elements and Figure 2c clearly indicates that the rocks of the Washikemba Formation have CA compositions. As such, the Washikemba Formation should no longer be considered to be part of the PIA series (Fig. 2c).

Since the majority of the PIA rocks can be classified as IAT using the Th/Yb v. Ta/Yb and Th v. Co diagrams, it is necessary to compare them with IAT samples from other island arcs to confirm their classification as 'true' IAT rocks. Compositions of relatively young island arc lavas from the Cenozoic–Recent Kermadec, Lesser Antilles, Marianas and Scotia intra-oceanic island arcs were taken from the Earth Reference Data and Models database (http://www.earthref.org; Supplementary Material). These arcs were chosen for comparison because they are intra-oceanic island arcs, much like what the PIA-samples are thought to represent (e.g. Kesler *et al.* 2005). Samples designated in the database as being 'altered' or 'metamorphosed' were removed, as were those with other indications of potential trace element mobility, such as high loss on ignition values. Also, as with the construction of the Th v. Co classification diagram in Hastie *et al.* (2007), the data quality from the Earth Reference Data and Models Database is a slight concern. Although the data has been filtered so that datasets predominantly derived from more recent ICP-MS analyses, older INAA analyses, and (for high Th only) XRF analyses, which usually yield reliable data, are used. It is accepted that full quality control was not available for all data and thus that the dataset is not internally consistent.

The island arc lavas from the Kermadec, Lesser Antilles, Marianas and Scotia island arcs are

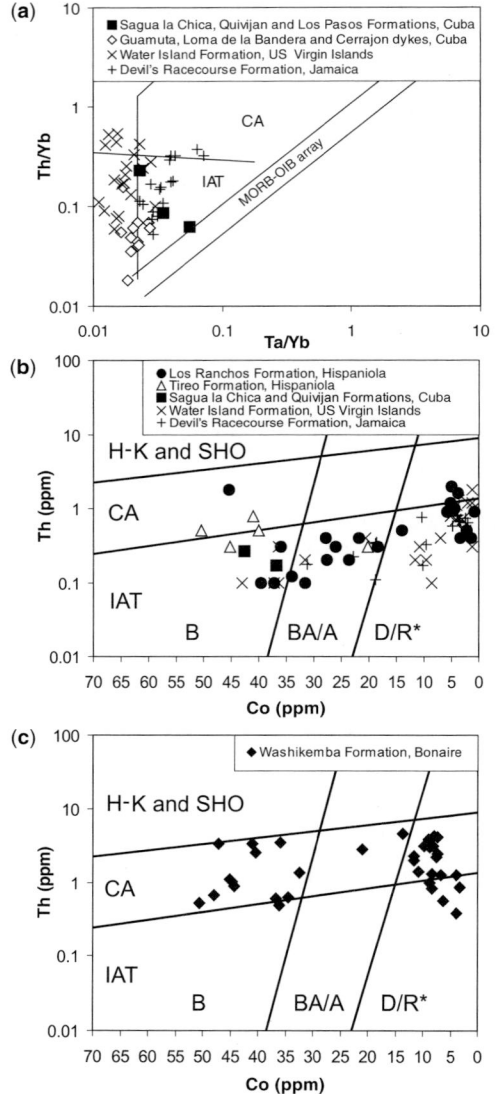

Fig. 2. (**a, b**) Th/Yb v. Ta/Yb (Pearce 1982) and Co v. Th (Hastie *et al.* 2007) discrimination diagrams, respectively, which demonstrate the IAT affinity of the Caribbean PIA rocks. The Co v. Th diagram in (**c**) shows that the arc lavas of the Washikemba Formation in Bonaire have CA compositions. IAT, island arc tholeiite; CA, calc-alkaline; H-K and SHO, high-K calc-alkaline and shoshonite; B, basalt; BA/A, basaltic–andesite and andesite; D/R*, dacite and rhyolite. The asterisk indicates that latites and trachytes also fall in the D/R fields.

classified into their differing volcanic rock suites using the K_2O v. SiO_2, the Th/Yb v. Ta/Yb and the Th v. Co classification diagrams. Table 1 shows the average composition of immobile trace

Table 1. *Selected immobile major and trace element abundances and trace element ratios for PIA rocks in the Caribbean and IAT and CA samples from the Kermadec, Lesser Antilles, Marianas and Scotia intra-oceanic island arcs. The IAT and CA data from the four present-day island arcs are averaged in the last two columns in the table. Caribbean PIA and the Kermadec, Lesser Antilles, Marianas and Scotia IAT and CA data are taken from the references listed in the Supplementary Material*

	PIA samples confirmed as IAT (n = 83)	Kermadec Arc (IAT) (n = 129)	Lesser Antilles Arc (IAT) (n = 404)	Marianas Arc (IAT) (n= 61)	Scotia Arc (IAT) (n= 177)	Kermadec Arc (CA) (n = 61)	Lesser Antilles Arc (CA) (n = 802)	Marianas Arc (CA) (n = 223)	Scotia Arc (CA) (n = 41)	Average IAT (n = 771)	Average CA (n = 1127)
Majors (wt%)											
SiO_2	62.98	58.33	57.33	55.96	53.53	58.58	56.72	55.50	56.83	56.29	56.91
TiO_2	0.62	0.77	0.70	0.69	0.80	0.74	0.67	0.83	1.01	0.74	0.81
Fe_2O_3	6.17	8.47	7.57	10.61	10.74	7.42	7.63	9.52	10.36	9.35	8.73
P_2O_5	0.08	0.12	0.14	0.12	0.08	0.16	0.18	0.27	0.15	0.12	0.19
Traces (ppm)											
V	153	210	167	264	254	176	162	203	252	223.66	198.38
Cr	58	33	41	126	58	26	112	55	33	64.32	56.52
Co	17.3	32.6	18.4	36.6	32.1	35.1	20.2	28.6	28.0	29.92	27.97
Ga	14.4	15.7	17.4	16.8	14.5	14.9	17.2	15.9	17.2	16.09	16.31
Y	26.0	27.7	23.5	23.1	21.3	28.3	20.3	27.7	32.0	23.89	27.08
Zr	69.8	55.0	86.2	64.1	57.1	77.7	88.4	86.2	104.1	65.60	89.11
Nb	1.17	0.81	3.13	0.701	0.57	2.78	4.82	1.81	1.86	1.30	2.82
La	4.18	3.47	6.57	4.05	2.69	13.01	12.97	9.97	5.73	4.20	10.42
Ce	10.26	8.16	16.31	9.64	5.48	12.69	25.70	21.43	18.09	9.90	19.48
Pr	1.69	1.14	2.09	1.54	0.88	2.02	2.66	2.94	2.25	1.41	2.47
Nd	8.79	6.43	10.30	7.87	4.73	9.79	14.05	14.90	11.05	7.33	12.45
Sm	2.74	2.33	2.87	2.52	1.61	3.01	3.13	4.14	3.20	2.33	3.37
Eu	0.87	0.84	0.99	0.91	0.64	0.93	1.02	1.29	1.08	0.84	1.08
Gd	3.49	3.14	3.34	3.10	2.18	3.16	3.43	4.87	3.92	2.94	3.85
Tb	0.64	0.52	0.62	0.62	0.40	0.60	0.53	0.82	0.73	0.54	0.67
Dy	4.15	3.85	3.71	3.35	2.68	3.30	3.91	4.79	4.58	3.40	4.15
Ho	0.89	0.73	0.81	0.78	0.58	0.71	0.67	1.13	1.00	0.72	0.88
Er	2.69	2.30	2.36	2.30	1.69	2.05	1.99	2.94	3.01	2.17	2.50
Tm	0.41	0.33	0.41	0.37	0.27	0.32	0.32	0.41	0.48	0.34	0.38
Yb	2.69	2.52	2.52	2.36	1.76	2.44	2.04	3.14	3.15	2.29	2.69
Lu	0.43	0.40	0.41	0.38	0.27	0.38	0.31	0.49	0.48	0.37	0.42
Hf	2.28	1.25	2.11	1.67	1.05	2.13	2.22	2.64	2.41	1.52	2.35
Ta	0.07	0.03	0.15	0.12	0.04	0.11	0.25	0.22	0.13	0.09	0.18
Th	0.49	0.45	2.13	0.55	0.72	0.67	4.63	1.36	1.27	0.96	1.98

(Continued)

Table 1. *Continued*

	PIA samples confirmed as IAT (n = 83)	Kermadec Arc (IAT) (n = 129)	Lesser Antilles Arc (IAT) (n = 404)	Marianas Arc (IAT) (n= 61)	Scotia Arc (IAT) (n= 177)	Kermadec Arc (CA) (n = 61)	Lesser Antilles Arc (CA) (n = 802)	Marianas Arc (CA) (n = 223)	Scotia Arc (CA) (n = 41)	Average IAT (n = 771)	Average CA (n = 1127)
La/Nd	0.45	0.43	0.58	0.51	0.37	0.53	0.85	0.67	0.53	0.47	0.65
La/Hf	1.89	2.09	2.69	2.44	1.66	3.15	4.25	3.91	2.43	2.22	3.44
La/Zr	0.06	0.07	0.08	0.06	0.07	1.45	0.18	0.11	0.06	0.07	0.45
La/Yb	1.49	1.11	2.34	1.72	0.96	2.52	5.98	3.14	1.89	1.53	3.38
Th/Nd	0.05	0.07	0.14	0.07	0.06	0.07	0.30	0.09	0.12	0.08	0.14
Th/Hf	0.21	0.29	0.53	0.31	0.27	0.39	1.27	0.54	0.56	0.35	0.69
Th/Zr	0.01	0.010	0.025	0.009	0.017	0.011	0.057	0.015	0.015	0.016	0.025
Th/Yb	0.18	0.16	0.50	0.24	0.16	0.31	1.96	0.43	0.44	0.27	0.78
Sm/Yb	1.02	0.94	1.10	1.07	0.89	1.33	1.61	1.32	1.02	1.00	1.32

element abundances and selected Th-LREE/HFSE-HREE ratios in the 83 PIA rocks and present-day IAT and CA lavas from the aforementioned island arcs.

The data for the IAT and CA samples from the four intra-oceanic island arcs are subsequently averaged and compared with the average immobile trace element abundance and Th-LREE/HFSE-HREE ratio composition of the PIA samples from Cuba, Jamaica, Hispaniola and the Virgin Islands. It can be seen in Table 1 and Figure 3 that the PIA rocks have low Th and LREE abundances and low Th-LREE/HFSE-HREE ratios relative to the average CA values. These low Th and LREE abundances and low Th-LREE/HFSE-HREE ratios are similar to the averaged IAT compositions, especially the low La/Yb ratios, which was one of the first immobile element pairs to be routinely used to distinguish IAT from CA (e.g. Jakes & Gill 1970) and has been utilized in other Caribbean geochemical studies (e.g. Kerr *et al.* 1999).

Additionally, Donnelly *et al.* (1990) suggested that the PIA series has higher incompatible element abundances than the IAT series. Elemental abundances can only be compared in similar rock types, preferably with basaltic compositions to limit the effects of fractional crystallization. Therefore, 14 Caribbean basaltic PIA samples are compared with 175 and 282 basaltic IAT and CA lavas from the Kermadec, Lesser Antilles, Marianas and Scotia island arcs in Table 2. IAT and CA samples were chosen with SiO_2 values ranging from 45 to 52 wt%, although basaltic andesites are included

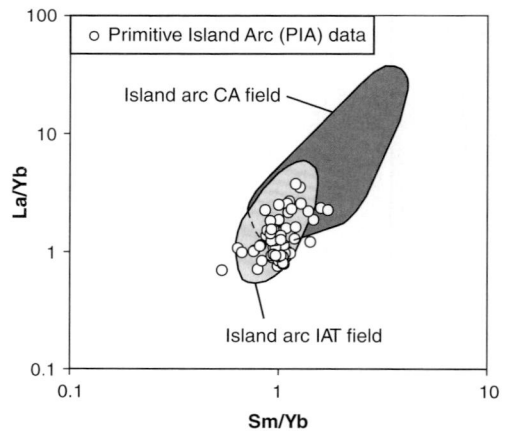

Fig. 3. La/Yb v. Sm/Yb immobile trace element ratio diagram. The 83 PIA samples plot in the IAT field. The island arc data used to define the IAT and CA fields are taken from the Kermadec, Lesser Antilles, Marianas and Scotia references listed in the supplementary material.

Table 2. *Average immobile element compositions of the basaltic PIA rocks from the Caribbean region and the IAT and CA samples from the Kermadec, Lesser Antilles, Marianas and Scotia intra-oceanic island arcs. Data are taken from the references listed in the Supplementary Material*

	Average basaltic PIA samples confirmed as IAT ($n = 14$)	Average basaltic IAT ($n = 175$)	Average basaltic CA ($n = 282$)
Majors (wt%)			
SiO_2	50.06	50.63	50.24
TiO_2	0.76	0.79	0.93
Fe_2O_3	9.58	10.58	11.24
P_2O_5	0.09	0.09	0.14
Traces (ppm)			
V	290	292	309
Cr	65	108	143
Co	40.6	36.5	37.7
Ga	16.5	16.0	16.5
Y	19.9	18.7	21.0
Zr	47.6	47.4	59.2
Nb	1.33	0.96	2.91
La	3.76	2.96	9.18
Ce	9.08	7.35	13.64
Pr	1.41	1.11	1.86
Nd	7.39	5.99	10.00
Sm	2.23	1.90	2.83
Eu	0.89	0.73	0.98
Gd	2.94	2.52	3.35
Tb	0.58	0.47	0.58
Dy	3.24	2.96	3.60
Ho	0.69	0.64	0.76
Er	2.08	1.86	2.24
Tm	0.32	0.29	0.35
Yb	2.10	1.91	2.20
Lu	0.33	0.30	0.34
Hf	1.46	1.26	1.69
Ta	0.07	0.06	0.15
Th	0.39	0.57	1.27
La/Nd	0.45	0.46	0.60
La/Hf	2.53	2.10	3.31
La/Zr	0.09	0.07	0.81
La/Yb	1.49	1.39	3.12
Th/Nd	0.05	0.07	0.12
Th/Hf	0.29	0.29	0.58
Th/Zr	0.010	0.013	0.021
Th/Yb	0.17	0.21	0.66
Sm/Yb	1.03	0.98	1.39

in the Scotia Arc average because of the lack of basaltic samples. For the altered PIA rocks, the basaltic samples were identified using the Th v. Co diagram of Hastie *et al.* (2007). The data shows that the PIA samples have slightly higher LREE contents but lower Th abundances relative to the IAT samples. Consequently, higher incompatible element abundances cannot be relied upon to separate the PIA series from the IAT series rocks. Moreover, Table 2 additionally confirms that the PIA lavas have very similar Th-LREE/HFSE-HREE ratios to the IAT series.

Sr–Pb–Hf–Nd isotope comparison of PIA and IAT series

Several studies have demonstrated that Sr isotopes cannot be used as reliable indicators of magmatic processes in Cretaceous Caribbean igneous rocks due to sub-solidus hydrothermal alteration involving interaction with seawater producing high initial $^{87}Sr/^{86}Sr$ ratios at near constant $\varepsilon_{Nd}(i)$ values (Kerr *et al.* 1996; Hauff *et al.* 2000; Révillon *et al.* 2002; Thompson *et al.* 2003; Hastie *et al.* 2008). Therefore, PIA Sr isotope systematics cannot be compared with those of more recent island arc rocks.

Alternatively, many geochemical studies have utilized Pb radiogenic isotopes (e.g. Cumming & Kesler 1987; Frost *et al.* 1998; Jolly *et al.* 1998*b*, 2001, 2006; Marchesi *et al.* 2007) to identify the source components in Caribbean igneous rocks. However, care should also be taken when interpreting Pb isotope data in Cretaceous Caribbean rocks because of the potential mobility of Pb and U (e.g. Geldmacher *et al.* 2003; Hastie *et al.* 2007). If K, Rb, Sr and Ba are mobilized as a consequence of chemical weathering, hydrothermal and metamorphic processes, U and Pb are also likely to be mobilized. Cumming & Kesler (1987) demonstrate that the U and Pb systematics in the Los Ranchos Formation have been modified by secondary alteration processes since the original formation of the volcanic rocks; however, they continue to use $^{206}Pb/^{204}Pb$, $^{207}Pb/^{204}Pb$ and $^{208}Pb/^{204}Pb$ isotope ratios to study the petrogenesis of the Los Ranchos volcanic rocks.

It is, however, important to note that the mobility of U and Pb in altered Caribbean igneous rocks will only usually substantially affect the U/Pb ratio, not the Pb isotope ratios, as Pb has a much lower abundance in seawater than Sr (Brown *et al.* 1997). Unfortunately, in order to compare samples of different ages, U/Pb ratios need to be used to age-correct the Pb isotope data to 'initial' values, and so this results in anomalous initial Pb isotope ratios (e.g. Hastie *et al.* 2008). Thus initial Pb isotopes cannot be used to quantitatively study the petrogenesis of Caribbean island arc lavas. However, Hastie *et al.* (2008) note that the immobility of ^{232}Th (e.g. Pearce 1982) and its subsequent radioactive decay into ^{208}Pb, suggests that initial $^{208}Pb/^{204}Pb$ ratios can be qualitatively used to identify likely source components in altered igneous rocks.

In contrast, Nd and Hf isotope systems are more resistant to alteration and hence, their values should represent the primary composition of the arc lavas (e.g. White & Patchett 1984). Combined Nd–Hf isotope data has recently been shown to be useful in the petrogenetic interpretation of Caribbean igneous rocks (e.g. Geldmacher *et al.* 2003; Thompson *et al.* 2003, 2004; Hastie *et al.* 2008).

On Figure 4, PIA data from the Devils Racecourse Formation, Jamaica and CA rocks from the Washikemba Formation, Bonaire are plotted along with IAT and CA fields, which are constructed from modern-day island arc data from Woodhead *et al.* (2001) and Barry *et al.* (2006). It can be seen that the PIA samples have positive $\varepsilon_{Nd}(i)$ and $\varepsilon_{Hf}(i)$ values of +8.12 to +9.03 and +13.22 to +14.97, respectively, and lie in the IAT field, indicating that their arc source regions have a long-term

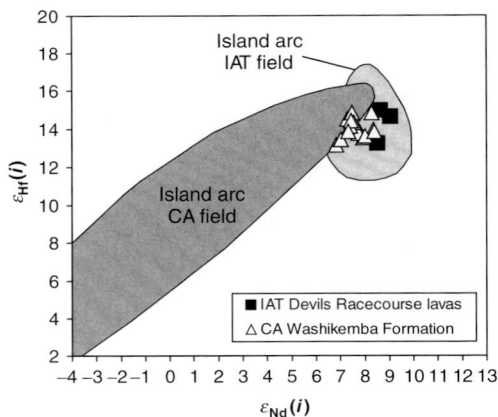

Fig. 4. Plot of $\varepsilon_{Nd}(i)$–$\varepsilon_{Hf}(i)$ for the IAT Devils Racecourse and CA Washikemba samples from Jamaica and Bonaire respectively (Hastie *et al.* 2009; Thompson *et al.* 2004). The IAT and CA fields are constructed from data derived from the Kermadec, Lesser Antilles, Mariana, New Britain, Scotia and Sunda island arcs (Woodhead *et al.* 2001; Barry *et al.* 2006).

history of incompatible element depletion (Hastie *et al.* 2009). This is in contrast to the CA field, which extends from similar positive $\varepsilon_{Nd}(i)$ and $\varepsilon_{Hf}(i)$ values to negative $\varepsilon_{Nd}(i)$ and $\varepsilon_{Hf}(i)$ ratios. In addition, although they did not present Hf isotopic data, Frost *et al.* (1998) also show that PIA rocks have higher initial $^{143}Nd/^{144}Nd$ [higher $\varepsilon_{Nd}(i)$] ratios than Caribbean CA samples. Thus the Nd/Hf isotope systematics demonstrate that the Caribbean PIA and modern-day IAT rocks have similar Nd and Hf isotope ratios, which are derived from more depleted source regions than many other modern-day island arc CA samples. Interestingly, the CA rocks of the Washikemba Formation plot at the depleted end of the CA field, which has similar $\varepsilon_{Nd}(i)$ and $\varepsilon_{Hf}(i)$ values to the PIA and IAT rocks. Therefore, $\varepsilon_{Nd}(i)$ and $\varepsilon_{Hf}(i)$ analyses should be performed on other Cretaceous Caribbean CA suites to determine if Nd–Hf systematics can be used to distinguish between the IAT series and the CA series in the Caribbean.

Conclusions

It has been shown in Figure 2 that when the PIA samples are plotted on the Th/Yb v. Ta/Yb and Th v. Co discrimination diagrams they mostly fall in the IAT fields. In addition, Tables 1 and 2 indicate that immobile, incompatible trace element abundances and ratios are mostly indistinguishable in the PIA and IAT series. Moreover, Figure 4 demonstrates that the Nd–Hf isotope ratios of PIA and IAT rocks are similar and have higher positive $\varepsilon_{Nd}(i)$ and $\varepsilon_{Hf}(i)$ values than many CA rocks.

Gill (1981) discussed the vast, and often contradictory, amount of geochemical and petrographical methods and terminology used in classifying, understanding and modelling the petrogenesis of volcanic rocks. Different geochemical and petrographical methods would lead to igneous rocks being variably classified as different rock types and into different rock series. For example, both the FeO^{tot}/MgO-SiO_2 diagram (Miyashiro 1974; Arculus 2003) and the K_2O v. SiO_2 plot (Peccerillo & Taylor 1976; Rickwood 1989) are used for distinguishing IAT and CA volcanic rock series. However, if a suite of volcanic rocks were plotted, both diagrams would assign differing samples to differing rock series. As such, igneous geochemical methodology and terminology are complex and lead Gill (1981) to comment that 'nomenclature is the Pandora's Box of igneous petrology'. Accordingly, redundant igneous terminology should be discarded in the interests of clarity.

Consequently, there is no geochemical or temporal justification for classifying the early island arc rocks in the Greater Antilles as anything other than IAT. Additionally, the term 'tholeiite' is now utilized for igneous rocks in all tectonic environments and there is no danger in modern geochemists confusing 'island arc tholeiites' with other 'non-arc related tholeiites'. Thus, the conclusion of this review is that the term 'PIA series' should be abandoned and future geochemical investigations in the Caribbean region should classify the appropriate Early Cretaceous island arc lavas in the Greater Antilles as IAT. Furthermore, in order for the altered Early Cretaceous Caribbean IAT rocks to be clearly identified, and their petrogenesis fully investigated, it is recommended that future studies acquire Th, Yb, Ta and Co abundances and Nd–Hf isotope data.

Alan Hastie acknowledges NERC PhD Studentship NER/S/A/2003/11215. The author is grateful to Andrew Kerr, Iain Neill, Julian Pearce and three anonymous reviewers for helpful comments that improved the manuscript.

References

ARCULUS, R. J. 2003. Use and abuse of the terms calc-alkaline and calcalkalic. *Journal of Petrology*, **44**, 929–935.

BAKER, P. E. 1968. Comparative volcanology and petrology of the Atlantic island-arcs. *Bulletin of Volcanology*, **32**, 189–206.

BARRY, T. L., PEARCE, J. A., LEAT, P. T., MILLAR, I. L. & LE ROEX, A. P. 2006. Hf isotope evidence for selective mobility of high-field-strength elements in a subduction setting: South Sandwich Islands. *Earth and Planetary Science Letters*, **252**, 223–244.

BROWN, E., COLLING, A., PARK, D., PHILLIPS, J., ROTHERY, D. & WRIGHT, J. 1997. *Seawater: its Composition, Properties and Behaviour.* Open University Press, Buckingham, 86–87.

BURKE, K. 1988. Tectonic evolution of the Caribbean. *Annual Review of Earth and Planetary Science*, **16**, 201–230.

CUMMING, G. L. & KESLER, S. E. 1987. Lead isotopic composition of the oldest volcanic rocks of the eastern Greater Antilles island arc. *Chemical Geology (Isotope Geoscience Section)*, **65**, 15–23.

DAVIDSON, J. P. 1987. Crustal contamination versus subduction zone enrichment: examples from the Lesser Antilles and implications for the mantle source composition of island arc lavas. *Geochimica et Cosmochimica Acta*, **51**, 2185–2198.

DONNELLY, T. W. & ROGERS, J. J. W. 1980. Igneous series in Island Arcs: the Northeastern Caribbean compared with Worldwide Island Arc assemblages. *Bulletin of Volcanology*, **43**, 347–382.

DONNELLY, T., ROGERS, J., PUSHKAR, P. & ARMSTRONG, R. 1971. Chemical evolution of the igneous rocks of the eastern West Indies: an investigation of thorium, uranium, and potassium distributions and lead and strontium isotopic ratios. *In*: DONNELLY, T. (ed.) *Caribbean Geophysical, Tectonic, and Petrologic Studies.* Geological Society of America, Boulder, CO, Memoirs, 181–224.

DONNELLY, T. W., BEETS, D. *ET AL.* 1990. History and tectonic setting of Caribbean magmatism. *In*: DENGO, G. & CASE, J. E. (eds) *The Caribbean Region.* The Geology of North America, **H**. Geological Society of America, Boulder, CO, 339–374.

ESCUDER VIRUETE, J., DÍAZ DE NEIRA, A. *ET AL.* 2006. Magmatic relationships and ages of Caribbean Island arc tholeiites, boninites and related felsic rocks, Dominican Republic. *Lithos*, **90**, 161–186.

ESCUDER VIRUETE, J., CONTRERAS, F. *ET AL.* 2007. Magmatic relationships and ages between adakites, magnesian andesites and Nb-enriched basalt-andesites from Hispaniola: record of a major change in the Caribbean island arc magma sources. *Lithos*, **99**, 151–177.

EWART, A. 1982. The mineralogy and petrology of Tertiary-Recent orogenic volcanic rocks: with special reference to the andesitic-basaltic compositional range. *In*: THORPE, R. S. (ed.) *Andesites.* John Wiley and Sons, Chichester, 25–83.

FLOYD, P. A. & WINCHESTER, J. A. 1978. Identification and discrimination of altered and metamorphosed volcanic rocks using immobile elements. *Chemical Geology*, **21**, 291–306.

FROST, C. D., SCHELLEKENS, J. H. & SMITH, A. L. 1998. Nd, Sr, and Pb isotopic characterization of Cretaceous and Paleogene volcanic and plutonic island arc rocks from Puerto Rico. *In*: LIDIAK, E. G. & LARNE, D. K. (eds) *Tectonics and Geochemistry of the Northeast Caribbean.* Geological Society of America, Boulder, CO, Special Papers, **322**, 123–132.

GELDMACHER, J., HANAN, B. B. *ET AL.* 2003. Hafnium isotopic variations in volcanic rocks from the Caribbean Large Igneous Province and Galapagos hot spot tracks. *Geochemistry Geophysics Geosystems*, **4**, paper number 2002GC000477.

GILL, J. B. 1970. Geochemistry of Viti Levu, Fiji and its evolution as an island arc. *Contributions to Mineralogy and Petrology*, **27**, 179–203.

GILL, J. B. 1981. *Orogenic Andesites and Plate Tectonics*. Springer, Berlin, 1–12.

HASTIE, A. R., KERR, A. C., PEARCE, J. A. & MITCHELL, S. F. 2007. Classification of altered volcanic island arc rocks using immobile trace elements: development of the Th-Co discrimination diagram. *Journal of Petrology*, **48**, 2341–2357.

HASTIE, A. R., KERR, A. C., MITCHELL, S. F. & MILLER, I. 2008. Geochemistry and petrogenesis of Cretaceous oceanic plateau lavas in eastern Jamaica. *Lithos*, **101**, 323–343.

HASTIE, A. R., KERR, A. C., MITCHELL, S. F. & MILLAR, I. L. 2009. Geochemistry and tectonomagmatic significance of Lower Cretaceous island arc lavas from the Devil's Racecourse Formation, eastern Jamaica. *In*: JAMES, K. H., LORENTE, M. A. & PINDELL, J. L. (eds) *The Origin and Evolution of the Caribbean Plate*. Geological Society, London, Special Publications, **328**, 339–360.

HAUFF, F., HOERNLE, K., TILTON, G., GRAHAM, D. W. & KERR, A. C. 2000. Large volume recycling of oceanic lithosphere over short time scales: geochemical constraints from the Caribbean Large Igneous Province. *Earth and Planetary Science Letters*, **174**, 247–263.

JAKES, P. & WHITE, A. J. R. 1969. Structure of the Melanesian Arcs and correlation with distribution of magma types. *Tectonophysics*, **8**, 223–236.

JAKES, P. & GILL, J. 1970. Rare earth elements and the island arc tholeiitic series. *Earth and Planetary Science Letters*, **9**, 17–28.

JOLLY, W. T. & LIDIAK, E. G. 2006. Role of crustal melting in petrogenesis of the Cretaceous Water Island Formation (Virgin Islands, northeast Antilles Island arc). *Geologica Acta*, **4**, 7–33.

JOLLY, W. T., LIDIAK, E. G., SCHELLECKENS, H. S. & SANTOS, S. 1998a. Volcanism, tectonics, and stratigraphic correlations in Puerto Rico. *In*: LIDIAK, E. G. & LARNE, D. K. (eds) *Tectonics and Geochemistry of the Northeast Caribbean*. Geological Society of America, Boulder, CO, Special Papers, **322**, 1–34.

JOLLY, W. T., LIDIAK, E. G., DICKIN, A. P. & WU, T. W. 1998b. Geochemical diversity of Mesozoic island arc tectonic blocks in eastern Puerto Rico. *In*: LIDIAK, E. G. & LARNE, D. K. (eds) *Tectonics and Geochemistry of the Northeast Caribbean*. Geological Society of America, Boulder, CO, Special Papers, **322**, 67–98.

JOLLY, W. T., LIDIAK, E. G., DICKIN, A. P. & WU, T. W. 2001. Secular geochemistry of central Puerto Rican island arc lavas: constraints on Mesozoic tectonism in the eastern Greater Antilles. *Journal of Petrology*, **42**, 2197–2214.

JOLLY, W. T., LIDIAK, E. G. & DICKIN, A. P. 2006. Cretaceous to Mid-Eocene pelagic sediment budget in Puerto Rico and the Virgin Islands (northeast Antilles Island arc). *Geologica Acta*, **4**, 35–62.

KERR, A. C., TARNEY, J., MARRINER, G. F., KLAVER, G. T., SAUNDERS, A. D. & THIRLWALL, M. F. 1996. The geochemistry and petrogenesis of the Late Cretaceous picrites and basalts of Curacao, Netherlands Antilles: a remnant of an oceanic plateau. *Contributions to Mineralogy and Petrology*, **124**, 29–43.

KERR, A. C., ITURRALDE-VINENT, M. A., SAUNDERS, A. D., BABBS, T. L. & TARNEY, J. 1999. A new plate tectonic model of the Caribbean: implications from a geochemical reconnaissance of Cuban Mesozoic volcanic rocks. *Geological Society of America Bulletin*, **111**, 1581–1599.

KERR, A. C., WHITE, R. V., THOMPSON, P. M. E., TARNEY, J. & SAUNDERS, A. D. 2003. No Oceanic Plateau – no Caribbean Plate? The seminal role of an oceanic plateau in Caribbean Plate evolution. *In*: BARTOLINI, C., BUFFLER, R. T. & BLICKWEDE, J. (eds) *The Circum Gulf of Mexico and Caribbean: Hydrocarbon Habitats Basin Formation and Plate Tectonics*. American Association of Petroleum Geology, Memoirs, **79**, 126–268.

KESLER, S. E., CAMPBELL, I. H. & ALLEN, C. M. 2005. Age of the Los Ranchos Formation, Dominican Republic: timing and tectonic setting of primitive island arc volcanism in the Caribbean region. *Geological Society of America Bulletin*, **117**, 987–995.

KUNO, H. 1959. Origin of Cenozoic petrographic provinces of Japan and surrounding areas. *Bulletin of Volcanology*, **20**, 37–76.

KUNO, H. 1966. Lateral variation of basalt magma across continental margins and island arcs. *Bulletin of Volcanology*, **29**, 195–222.

KUNO, H. 1968. Differentiation of basalt magmas. *In*: HESS, H. H. & POLDERVAART, A. A. (eds) *Basalts: The Poldervaart Treatise on Rocks of Basaltic Composition*, **2**. Interscience, New York, 623–688.

LE BAS, M. J., LE MAITRE, R. W., STECKEISEN, A. & ZANETTIN, B. 1986. A chemical classification of volcanic rocks based on the total alkali–silica diagram. *Journal of Petrology*, **27**, 745–750.

LE BAS, M. J., LE MAITRE, R. W. & WOOLLEY, A. R. 1992. The constuction of the total alkali-silica chemical classification of volcanic rocks. *Mineralogy and Petrology*, **46**, 1–22.

LEBRON, M. C. & PERFIT, M. R. 1993. Stratigraphic and petrochemical support subduction polarity reversal of the Cretaceous Caribbean Island arc. *Journal of Geology*, **101**, 389–396.

LEBRON, M. C. & PERFIT, M. R. 1994. Petrochemistry and tectonic significance of Cretaceous island-arc rocks, Cordillera Oriental, Dominican Republic. *Tectonophysics*, **229**, 69–100.

MARCHESI, C., GARRIDO, C. J. *ET AL.* 2007. Geochemistry of Cretaceous magmatism in eastern Cuba: recycling of North American continental sediments and implications for subduction polarity reversal in the Greater Antilles paleo-arc. *Journal of Petrology*, **48**, 1813–1840.

MIYASHIRO, A. 1974. Volcanic rock series in island arcs and active continental margins. *American Journal of Science*, **274**, 321–355.

MÜLLER, R. D., ROYER, J.-Y., CANDE, S. C., ROEST, W. R. & MASCHENKOV, S. 1999. New constraints on the Late Cretaceous/Tertiary plate tectonic evolution of the Caribbean. *In*: MANN, P. (ed.) *Caribbean Basins. Sedimentary Basins of the World*. Elsevier Science, Amsterdam, 33–59.

MÜNKER, C., WEYER, S., SCHERER, E. & MEZGER, K. 2001. Separation of high field strength elements (Nb, Ta, Zr, Hf) and Lu from rock samples for MC-ICPMS measurements. *Geochemistry Geophysics Geosystems*, **2**, paper no. 2001GC000183.

PEARCE, J. A. 1982. Trace element characteristics of lavas from destructive plate boundaries. *In*: THORPE, R. S. (ed.) *Andesites*. John Wiley and Sons, Chichester, 525–547.

PEARCE, J. A. 1996. A user's guide to basalt discrimination diagrams. *In*: WYMAN, D. A. (ed.) *Trace Element Geochemistry of Volcanic Rocks: Applications for Massive Sulphide Exploration*. Geological Association of Canada, Short Course Notes, **12**, 79–113.

PEARCE, J. A. & PEATE, D. W. 1995. Tectonic implications of the composition of volcanic arc magmas. *Annual Reviews Earth and Planetary Science Letters*, **23**, 251–285.

PECCERILLO, R. & TAYLOR, S. R. 1976. Geochemistry of Eocene calc-alkaline volcanic rocks from the Kastamonu area, northern Turkey. *Contributions to Mineralogy and Petrology*, **58**, 63–81.

PINDELL, J. L. & KENNAN, L. 2001. Kinematic Evolution of the Gulf of Mexico and Caribbean. *Transactions, Petroleum Systems of Deep-water Basins: Global and Gulf of Mexico Experience. GCSSEPM 21st Annual Foundation Bob F. Perkins Research-Conference*, Houston, TX, 193–220.

PINDELL, J. L., KENNAN, L., STANEK, K. P., MARESH, W. V. & DRAPER, G. 2006. Foundations of Gulf of Mexico and Caribbean evolution: eight controversies resolved. *Geologica Acta*, **4**, 303–341.

RANKIN, D. 2002. *Geology of St. John, U.S. Virgin Islands*. United States Geological Survey, Professional Papers, **1631**, 1–36.

RÉVILLON, S., CHAUVEL, C. *ET AL.* 2002. Heterogeneity of the Caribbean plateau mantle source: Sr, O and He isotopic compositions of olivine and clinopyroxene from Gorgona Island. *Earth and Planetary Science Letters*, **205**, 91–106.

RICKWOOD, P. C. 1989. Boundary lines within petrologic diagrams which use oxides of major and minor elements. *Lithos*, **22**, 247–263.

SCHELLECKENS, J. H. 1998. Geochemical evolution and tectonic history of Puerto Rico. *In*: LIDIAK, E. G. & LARNE, D. K. (eds) *Tectonics and Geochemistry of the Northeast Caribbean*. Geological Society of America, Boulder, CO, Special Papers, **322**, 35–66.

SINTON, C. W., DUNCAN, R. A., STOREY, M., LEWIS, J. & ESTRADA, J. J. 1998. An oceanic flood basalt province within the Caribbean Plate. *Earth and Planetary Science Letters*, **155**, 221–235.

TATSUMI, Y. & KOGISO, T. 1997. Trace element transport during dehydration processes in the subducted oceanic crust: 2. Origin of chemical and physical characteristics in arc magmatism. *Earth and Planetary Science Letters*, **148**, 207–221.

TATSUMI, Y., HAMILTON, D. L. & NESBITT, R. W. 1986. Chemical characteristics of fluid phase released from a subducted lithosphere and origin of arc magmas: evidence from high-pressure experiments and natural rocks. *Journal of Volcanology and Geothermal Research*, **29**, 293–309.

THIRLWALL, M. F., GRAHAM, A. M., ARCULUS, R. J., HARMON, C. G. & MACPHERSON, C. G. 1996. Resolution of the effects of crustal assimilation, sediment subduction and fluid transport in island arc magmas: Pb–Sr–Nd–O isotope geochemistry of Grenada, Lesser Antilles. *Geochimica et Cosmochimica Acta*, **60**, 4785–4810.

THOMPSON, P. M. E. 2002. *Petrology and geochronology of an arc sequence, Bonaire, Dutch Antilles, and its relationship to the Caribbean Plateau*. University of Leicester, UK. Unpublished PhD thesis.

THOMPSON, P. M. E., KEMPTON, P. D. *ET AL.* 2003. Hf–Nd isotope constraints on the origin of the Cretaceous Caribbean plateau and its relationship to the Galapagos plume. *Earth and Planetary Science Letters*, **217**, 59–75.

THOMPSON, P. M. E., KEMPTON, P. D. *ET AL.* 2004. Elemental, Hf–Nd isotopic and geochronological constraints on an island arc sequence associated with the Cretaceous Caribbean plateau: Bonaire, Dutch Antilles. *Lithos*, **74**, 91–116.

WINCHESTER, J. A. & FLOYD, P. A. 1977. Geochemical discrimination of different magma series and their differentiation products using immobile elements. *Chemical Geology*, **20**, 325–343.

WHITE, W. M. & DUPRE, B. 1986. Sediment subduction and magma genesis in the Lesser Antilles. *Journal of Geophysical Research*, **91B**, 5927–5941.

WHITE, W. M. & PATCHETT, P. J. 1984. Hf–Nd–Sr isotopes and incompatible element abundances in island arcs: implications for magma origins and crust–mantle evolution. *Earth and Planetary Science Letters*, **67**, 167–185.

WHITE, R. V., TARNEY, J. *ET AL.* 1999. Modification of an oceanic plateau, Aruba, Dutch Caribbean: implications for the generation of continental crust. *Lithos*, **46**, 43–68.

WOODHEAD, J. D., HERGT, J. M., DAVIDSON, J. P. & EGGINS, S. M. 2001. Hafnium isotope evidence for 'conservative' element mobility during subduction zone processes. *Earth and Planetary Science Letters*, **192**, 331–346.

Late Cretaceous to Miocene seamount accretion and mélange formation in the Osa and Burica Peninsulas (Southern Costa Rica): episodic growth of a convergent margin

DAVID MARC BUCHS*, PETER OLIVER BAUMGARTNER,
CLAUDIA BAUMGARTNER-MORA, ALEXANDRE NICOLAS BANDINI,
SARAH-JANE JACKETT, MARC-OLIVIER DISERENS & JÉRÔME STUCKI

*Institut de Géologie et Paléontologie, Université de Lausanne, Bâtiment Anhtropole,
1015 Lausanne, Switzerland*

Corresponding author: (david.buchs@anu.edu.au)

Abstract: Multidisciplinary study of the Osa and Burica peninsulas, Costa Rica, recognizes the Osa Igneous Complex and the Osa Mélange – records of a complex Late Cretaceous–Miocene tectonic–sedimentary history. The Igneous Complex, an accretionary prism (*sensu stricto*) comprises mainly basaltic lava flows, with minor sills, gabbroic intrusives, pelagic limestones and radiolarites. Sediments or igneous rocks derived from the upper plate are absent. Four units delimited on the base of stratigraphy and geochemistry lie in contact along reactivated palaeo-décollement zones. They comprise fragments of a Coniacian–Santonian oceanic plateau (Inner Osa Igneous Complex) and Coniacian–Santonian to Middle Eocene seamounts (Outer Osa Igneous Complex). The units are unrelated to other igneous complexes of Costa Rica and Panama and are exotic with respect to the partly overthickened Caribbean Plate; they formed by multiple accretions between the Late Cretaceous and Middle Eocene, prior to the genesis of the mélange. Events of high-rate accretion alternated with periods of low-rate accretion and tectonic erosion. The NW Osa Mélange in contact with the Osa Igneous Complex has a block-in-matrix texture at various scales, produced by sedimentary processes and later tectonically enhanced. Lithologies are mainly debris flows and hemipelagic deposits. Clastic components (grains to large boulders) indicate Late Eocene mass wasting of the Igneous Complex, forearc deposits and a volcanic arc. Gravitational accumulation of a thick pile of trench sediments culminated with shallow-level accretion. Mass-wasting along the margin was probably triggered by seamount subduction and/or plate reorganization at larger scale. The study provides new geological constraints for seamount subduction and associated accretionary processes, as well as on the erosive/accretionary nature of convergent margins devoid of accreted sediments.

Supplementary material: Sample localities and analytical data can be found at http://www.geolsoc.org.uk/SUP18363.

The Osa and Burica peninsulas lie in southern Costa Rica, on the southwestern edge of the Caribbean Plate, along the convergent margin between Central America and the subducting Cocos Plate (Fig. 1). The peninsulas lie directly above the submarine Cocos ridge, an aseismic volcanic ridge which rises *c.* 2000 m above the adjacent seafloor and constitutes a significant topographic high entering the subduction zone (Walther 2003). The subducting ridge formed above the Galapagos hotspot 13.0–14.5 Ma ago (Werner *et al.* 1999) and is currently moving northeastward at a rate of *c.* 90 mm a^{-1} (Trenkamp *et al.* 2002). It is believed to have started colliding with the Caribbean Plate 8–1 Ma (most probably *c.* 2 Ma) ago (MacMillan *et al.* 2004, and references therein), causing forearc shortening, large-scale block tilting and local cessation of volcanism (Kolarsky *et al.* 1995a; Fisher *et al.* 1998, 2004; Abratis & Wörner 2001; Gräfe *et al.* 2002; Morell *et al.* 2008; Sak *et al.* 2009). The Osa and Burica peninsulas were uplifted and emerged in response to the Cocos Ridge subduction, with an estimated average long-term rate of uplift of 1–2 mm a^{-1} (Corrigan *et al.* 1990; Collins *et al.* 1995; Mann & Kolarsky 1995). Their morphology is controlled by active faults, interpreted to intimately relate to the morphology and roughness of the subducting plate, that define large tilted, variously uplifted blocks (Sak *et al.* 2004, 2009; Vannucchi *et al.* 2006; Fig. 2b).

In southern Central America, subduction began in the Late Campanian along a Coniacian–Early Santonian oceanic plateau (Buchs *et al.* in prep.; see also discussion of ages below). This plateau is

From: JAMES, K. H., LORENTE, M. A. & PINDELL, J. L. (eds) *The Origin and Evolution of the Caribbean Plate.*
Geological Society, London, Special Publications, **328**, 411–456.
DOI: 10.1144/SP328.17   0305-8719/09/$15.00 © The Geological Society of London 2009.

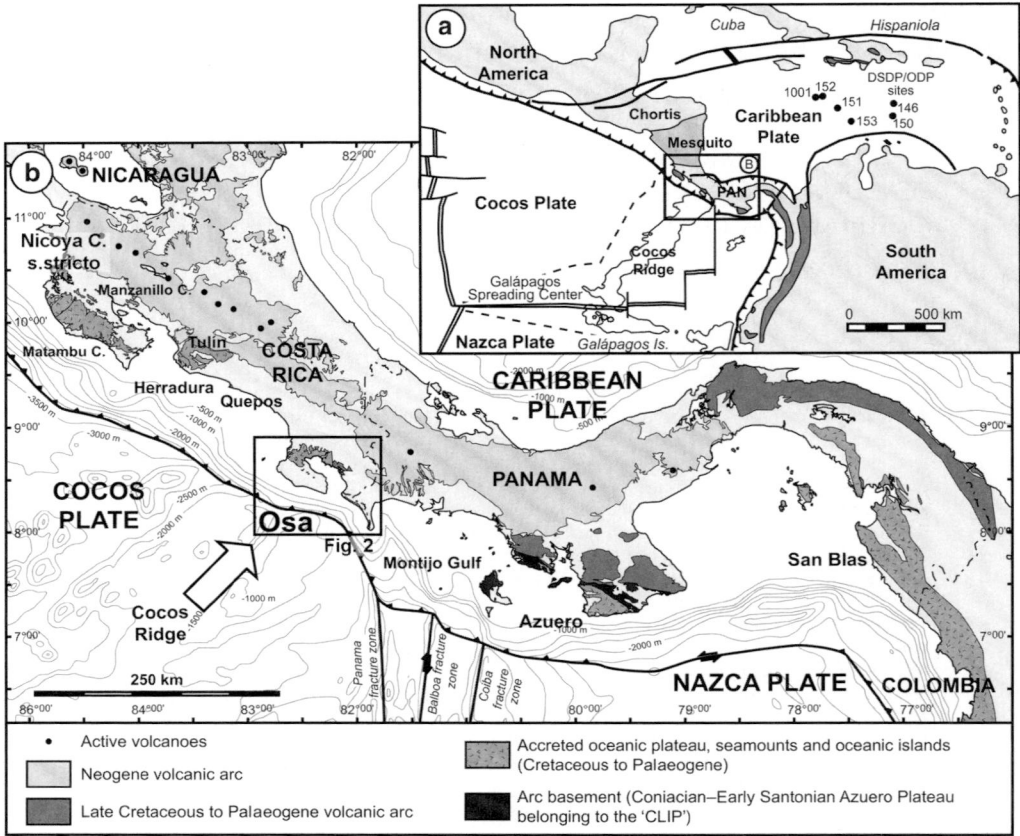

Fig. 1. (**a**) Geological setting of the South Central American Arc. PAN, Panama Microplate; Mesquito, Mesquito Composite Oceanic Terrane (after Baumgartner *et al.* 2008). Dark grey areas represent igneous complexes generally associated to the Caribbean Large Igneous Province (CLIP) or Colombian–Caribbean Oceanic Plateau (CCOP). (**b**) Simplified geological map of southern Central America based on national geological maps of the area and results from this study. Bathymetry based on Smith & Sandwell (1997). Autochtonous and accreted oceanic complexes are defined on the basis of our new results (e.g. Bandini *et al.* 2008; Buchs 2008; Buchs *et al.* in prep.; this study) and stratigraphic data from previous contributions (Bandy & Casey 1973; Baumgartner *et al.* 1984, 2008; Di Marco 1994; Arias 2003; Flores 2003).

exposed in the Azuero Peninsula (Western Panama) and represents the first clear occurrence of arc basement in southern Central America. The Azuero Plateau may further represent the SW edge of a larger oceanic plateau that extends eastward into the Caribbean, generally defined as the Caribbean Large Igneous Province (CLIP) or Caribbean–Colombian Oceanic Plateau (CCOP) (e.g. Kerr *et al.* 2003; Hoernle *et al.* 2004; Fig. 1). Because of (1) location of the South Central American Arc in the western Caribbean and (2) similar geochemistry and ages of the igneous rocks, many igneous complexes exposed along the forearc of Central America have been associated with the CLIP (e.g. Sinton *et al.* 1997, 1998). However these data do not rule out the possibility that complexes belonged

to distinct volcanic edifices in the Pacific, subsequently amalgamated along the margin by accretionary processes. Tectono-stratigraphic constraints are needed to better understand origins of the complexes.

The purpose of this study is to better characterize the age, tectono-stratigraphy and internal arrangement of the exposed rocks from both the inner (isthmus side) and outer southwestern Osa and Burica peninsulas. The area of study lies only 15 km from the Middle American Trench and thus represents one of the outermost parts of the Costa Rican margin. It provides onland access to rocks and structures that usually lie deep under the sea, close to the subduction zone. The rocks exposure on the Osa Peninsula is generally expected to be collected/observed during drilling and dredging

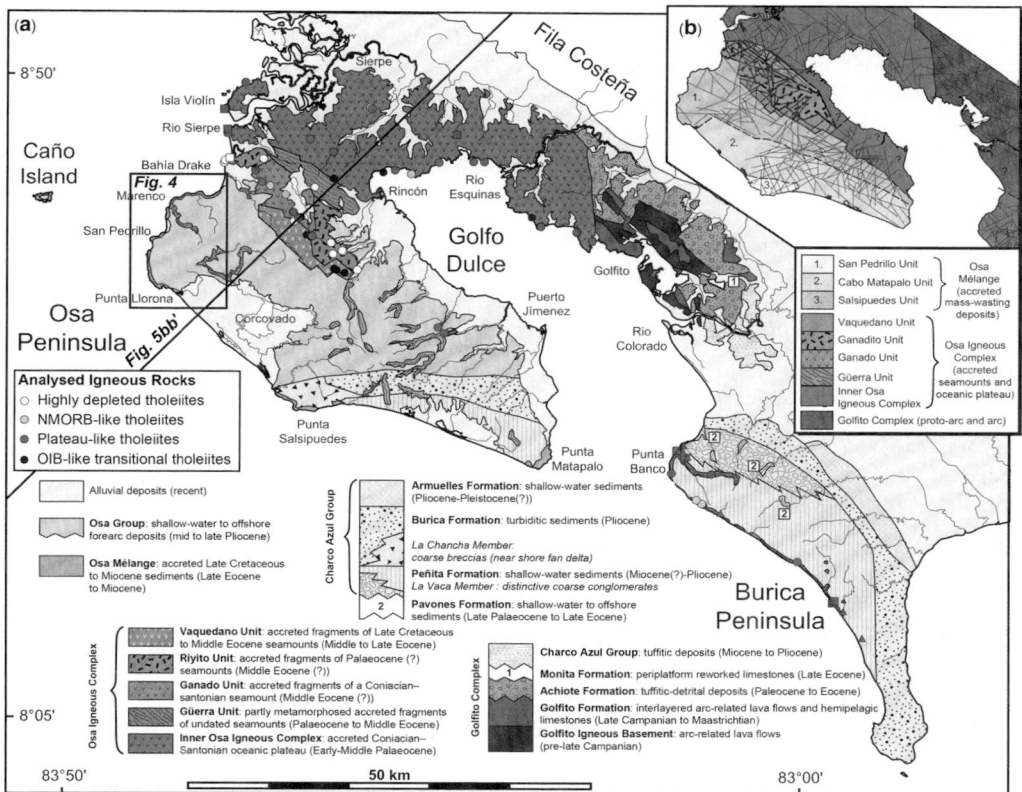

Fig. 2. (**a**) Geological map of the Golfo Dulce area modified after Buchs & Baumgartner (2007) and references therein. Analysed igneous rocks: circles, samples from this study; squares, samples from Hauff *et al.* (2000); triangles, samples from Di Marco (1994), partly re-analysed in this study. (**b**) Geological map of the Golfo Dulce area illustrating basement rocks (i.e. underlying recent alluvial deposits, as well as overlap sediments of the Charco Azul and Osa Groups). Lineaments are based on satellite imagery (pixel size of 28.5 m) and DEM (3 arc sec). Lineaments in the Golfito area are from Mende & Astorga (2007).

only. Their significance in terms of accretionary processes and tectonic erosion is developed here on the basis of: (1) 6 months of field survey, (2) petrological characteristics of 455 samples, (3) geochemical analyses of the igneous rocks and (4) micropalaeontology. We show that igneous complexes exposed on the Osa and Burica peninsulas are exotic with respect to the Caribbean Plate and, as a consequence, are not part of the CLIP.

Geological setting of the Osa and Burica peninsulas

In the literature the Osa Peninsula has been divided into two main units based on dominant lithology: (1) the Osa Igneous Complex (originally defined as the 'Rincon Block' by Di Marco 1994) and (2) the Osa Mélange (originally defined as the 'Osa

Caño Accretionary Complex' by Di Marco 1994) (Fig. 2). We first propose here that the Osa Igneous Complex is not limited to the Osa Peninsula but extends toward the SE in the Burica Peninsula and, possibly, islands in the Montijo Gulf (western Panama). In the Burica Peninsula, the complex comprises a Late Cretaceous igneous complex called the 'Burica Complex' or 'Burica Terrane' by previous authors (e.g. Obando 1986; Di Marco 1994; Hauff *et al.* 2000). Exposures of the Osa Mélange are restricted in the outer part of the Osa Peninsula. Most of the area comprising the Osa Igneous Complex and Osa Mélange is unconformably overlain by Paleocene to Pleistocene sediments called herein 'overlap sequences', in opposition to underlying rocks forming the 'basement' (Fig. 2).

The basement rocks of the Osa Peninsula were seen for a long time to be part of the 'Nicoya Complex', an accreted oceanic plateau cropping out in the NW Nicoya Peninsula (e.g. Denyer & Baumgartner 2006; Bandini *et al.* 2008). However, strong incompatibilities between the Nicoya Complex and the Osa area exist in terms of age, lithology, and tectono-stratigraphy of igneous and sedimentary rocks, as well as geochemistry (e.g. Denyer *et al.* 2006). In this chapter we focus on the geology of the NW Osa Mélange and Osa Igneous Complex. A coloured geological map of the area is given by Buchs & Baumgartner (2007). Figure 2 is an updated greyscale version of this map.

The Osa Igneous Complex

The Osa Igneous Complex is limited on the trench-ward, SW side by the Osa Mélange and on the land-ward, NE side by the Late Cretaceous to Early Cenozoic Golfito Complex. The latter is broadly composed of: (1) a proto-arc that developed on the top of the CLIP in the Late Campanian to Maastrichtian, and (2) younger overlapping sediments (Buchs *et al.* in prep.) (Fig. 2). The lateral extension of the Osa Igneous Complex toward the NW is not known. Toward the SE it extends in the Burica Peninsula and, possibly, in the Montijo Gulf (western Panama, Fig. 2).

The Igneous Complex comprises basalts, micro-gabbros, dolerites and minor gabbros, with rare (<1%) intercalations of radiolarian cherts, black shales, and pelagic limestones (Dengo 1962; Tournon 1984; Obando 1986; Berrangé & Thorpe 1988; Di Marco 1994; Di Marco *et al.* 1995; Buchs 2003). Radiolarian and foraminiferal assemblages from these sediments provide ages ranging from Campanian to Eocene (Di Marco 1994). K–Ar ages of the basalts range from Late Cretaceous to Eocene (Berrangé *et al.* 1989). ^{40}Ar/^{39}Ar datings range from 54.5 ± 1.5 to 62.1 ± 0.6 Ma (Hauff *et al.* 2000; Hoernle *et al.* 2002). Basalt geochemistry indicates compositions ranging across NMORB-like, plateau-like and OIB-like affinities (Berrangé & Thorpe 1988; Di Marco 1994; Meschede & Frisch 1994; Hauff *et al.* 2000). Origins proposed until now for the Complex are: (1) a back-arc basin (Berrangé & Thorpe 1988), (2) a volcanic arc in the Osa Peninsula and an accreted Pacific seamount in the Burica Peninsula (Obando 1986; Di Marco 1994), (3) an uplifted western Caribbean Plateau (Meschede *et al.* 1999; Denyer *et al.* 2006), (4) an accreted aseismic ridge (Hauff *et al.* 2000; Hoernle *et al.* 2002, 2004; Denyer *et al.* 2006; Vannucchi *et al.* 2006), and (5) accreted seamounts (Buchs 2003). This wide range of interpretation is probably due to (1) development of analytical techniques in the last decades, (2) highly heterogeneous geochemical affinities of the igneous rocks, (3) low sedimentary content of the igneous complex and (4) poor

recognition of structures in absence of good stratigraphic markers.

The Osa Mélange

The Osa Mélange was recognized and differentiated from the Nicoya Complex by Baumgartner (1986) and Baumgartner *et al.* (1989), who postulated the presence of an Eocene accretionary prism cropping out on the outer Osa Peninsula and Caño Island. Di Marco (1994) and Di Marco *et al.* (1995) subdivided the Mélange into three tectono-stratigraphic units, from NE to SW: (1) the San Pedrillo Unit, (2) the Cabo Matapalo Unit and (3) the Punta Salsipuedes Unit (Fig. 2).

The San Pedrillo Unit consists of a deformed matrix of detrital and siliceous sediments enclosing variable amounts of blocks of igneous rocks, pelagic sediments and resedimented shallow-water carbonates (Di Marco *et al.* 1995). The siliceous fraction of the matrix was dated as Middle Eocene on the basis of radiolarian assemblages (Azéma *et al.* 1983). The igneous blocks have been interpreted as remnants of accreted seamounts by most authors (e.g. Di Marco *et al.* 1995; Vannucchi *et al.* 2006), but geochemical affinities of the lava blocks are not determined.

The Cabo Matapalo Unit comprises a detrital matrix containing blocks of Late Middle Eocene to Middle Miocene pelagic limestones. Decimetric blocks of basalts occur within the pelagic limestones (Di Marco *et al.* 1995).

The Salsipuedes Unit contains 'large bodies' of limestones in a fine-grained greywacke matrix (Di Marco *et al.* 1995). The ages of the limestones are poorly constrained; a single, Paleocene age was determined by Azéma *et al.* (1983). Interbedded dark shales and greywackes within the limestones (Di Marco *et al.* 1995) indicate that this unit is not a mélange *sensu stricto*, but more likely a dismembered sequence.

Various origins and genetic mechanisms have been proposed for the Osa Mélange. Di Marco *et al.* (1995) suggest that the San Pedrillo Unit represents accreted Eocene trench-fill sediments associated with variable amounts of blocks of igneous and sedimentary rocks originally on the subducting plate. These authors see the most external parts of the Osa Peninsula (the Cabo Matapalo and Salsipuedes units) made by offscraped lenses of Eocene to Miocene pelagic limestones and distal detrital sediments. Meschede *et al.* (1999) postulate the Osa Mélange is a product of underplated material generated by tectonic erosion of an outer arc wedge structure at the interface between the descending and overriding plates. Vannucchi *et al.* (2006, 2007) interpret the Osa Mélange as a tectonically disrupted accreted package of oceanic lithologies that are exotic to the overriding Caribbean Plate. In this interpretation, the mélange suffered pervasive metamorphism and exhibit a 'ghost stratigraphy suggesting dismemberment of a classic sequence of oceanic crust'.

The overlap sequences

The overlap sequences rest unconformably upon the basement rocks of the Osa Igneous Complex and Osa Mélange. They are composed of: (1) the Late Paleocene to Late Eocene Pavones Formation, (2) the Early Pliocene to Early Pleistocene (or younger) Charco Azul Group, and (3) the Mid- to Upper Pleistocene Osa Group (Fig. 2).

The Late Paleocene to Late Eocene Pavones Formation is the oldest overlap sequence in the area (Obando 1986; Di Marco 1994; Di Marco et al. 1995; Fig. 2a). This formation crops out in the Burica Peninsula above deformed igneous rocks of the Osa Igneous Complex (Obando 1986; Di Marco 1994; Fig. 2a). It comprises an association of periplatform, reworked shallow-water limestones and siliceous pelagic limestones (Obando 1986; Di Marco 1994). Although Di Marco (1994) considered this formation to be devoid of arc/continent-derived material and to have deposited on the slope of an intra-oceanic seamount, re-examination of Palaeogene samples collected by Di Marco (1994) in the Rio La Vaca (Di Marco et al. 1995, p. 13, fig. 7) revealed that they contain quartz, zoned plagioclase and green amphibole grains. Similar observations were made by Obando (1986). We propose that these grains indicate depositions close to an arc, presumably a forearc slope. In the Middle Eocene, the formation is characterized in the Quebrada Piedra Azul by the unusual occurrence of thick boulders beds reworking fragments of a carbonate platform (Di Marco et al. 1995).

The Charco Azul Group was comprehensively studied by Corrigan et al. (1990), Coates et al. (1992) and Collins et al. (1995). It has been subdivided into: (1) the Early Pliocene Peñita Formation, which rests unconformably upon the Pavones Formation and basement rocks, (2) the Late Pliocene Burica Formation, conformable on the Peñita Formation and (3) the Early Pleistocene or younger Armuelles Formation, conformable on the Burica Formation (Fig. 2a). The Peñita Formation is composed of as much of 1200 m of clayey, blue-green siltstone and litharenite consistently rich in benthic and planktic foraminifers, deposited in a forearc slope environment (Coates et al. 1992). The basal formation is coarse with locally channelled conglomerates, some of which form a distinctive suite defined as the La Vaca Member by Coates et al. (1992) (Fig. 2a). These coarse deposits record a detrital paralic and fan-delta depositional

environment at the base of the Charco Azul Group. The Burica Formation consists of about 2800 m of mostly fine-grained, volcaniclastic turbidite deposits with local megabreccias formed by large-scale intraformational slumps (Coates et al. 1992). The La Chancha Member is a distinctive coarse facies interpreted to represent canyon fill within the trench slope on which the Burica turbidites were deposited (Coates et al. 1992) (Fig. 2a). Finally, the Armuelles Formation consists of c. 370 m of channelled pebbly conglomerate and unconsolidated greenish-blue litharenite and siltstone in the lower part and predominant grey-blue, clayed siltstone and fine litharenites in the upper part (Coates et al. 1992).

The Osa Group rests on an erosional unconformity cut into the basement rocks in the northern and central Osa Peninsula (Berrangé 1989). This surface is covered by a veneer of palaeosoils and continental deposits hosting the gold deposits of Osa (Berrangé 1989; Berrangé & Thorpe 1988). Mid to Upper Pliocene sediments hosting the placer gold deposits give way upwards to thick turbidite fan deposits representing a deepening sequence, defining a sequence similar to the Charco Azul Group. Whereas Berrangé (1989) defined the Osa Group on the base of dissimilar sedimentary facies in Osa and Burica peninsula, Coates et al. (1992) suggested that the Osa Group may rather represent a lateral extension of the Charco Azul Group. It seems to us initial subdivision between the Osa and Charco Azul groups by Berrangé (1989) should remain until new data from North and Central Osa Peninsula are provided.

In summary of this chapter, we conclude that the overlap sequences exposed in the Osa and Burica peninsulas indicate at least two major tectonic events that occurred prior to the deposition of the sediments. The first preceded unconformable hemipelagic, forearc deposits of the basal Pavones Formation (Late Paleocene) (Di Marco et al. 1995). The second preceded the unconformable, near-shore deposits of the lower Peñita Formation (Early Pliocene) (Coates et al. 1992) and continental deposits of the Osa Group (Mid to Upper Pliocene) (Berrangé 1989). The Charco Azul Group defines a transgressive–regressive cycle between the Early Pliocene and the Pleistocene (Coates et al. 1992; Collins et al. 1995). In the Middle Eocene, thick beds bearing shallow-water limestone boulders are observed in the Pavones Formation (Di Marco et al. 1995), which may also represent a record of a third tectonic event. The nature of the overlap sequences provides constraints on the tectonic evolution of the margin. Their significance is discussed below, along with the accretionary record of the Osa Igneous Complex and Osa Mélange.

Mélanges: nature and origins

In this study special emphasis is given to description and interpretation of the Osa Mélange. Mélanges are generally considered to be very complicated rock associations, frequently described as 'chaotic rock bodies'. We present below a succinct, non-exhaustive review of the mélanges, their origins and commonly associated problems. We also define a nomenclature for our descriptions of the Osa Mélange.

Mélange definition

A recognized definition of a mélange was introduced by Raymond (1984) as: 'a body of rock mappable at a scale of 1:24 000 or smaller and characterized both by the lack internal continuity of contacts and strata and by the inclusion of fragments and blocks of all sizes, both exotic and native, embedded in a matrix of finer-grained material'. The occurrence of fragments and blocks of all size is characteristic of the fractal nature of a mélange – repetition of the same block-in-matrix texture at various scales, or 'scale independence' (Medley 1994).

The use of the term 'mélange' following Raymond (1984) requires an *a priori* knowledge of the origin of the described rock body, because it is necessary to make a distinction between exotic and native blocks. When exotic blocks are not found in a rock body without internal stratal continuity, Raymond (1984) proposes the use of the term 'dismembered unit'. However, identification of genetic characteristics may be problematic in block-in-matrix rocks because it is sometimes difficult to make a distinction between exotic and native blocks. Indeed, Raymond (1984) mentions and discusses this issue. Perhaps for this reason, the term 'mélange' in the literature is widely used with a broader, simpler meaning of 'block-in-matrix rock'. This use tends to be confusing because the term 'mélange' has thus been attributed to rock bodies having very distinctive fabrics, compositions, origins and geological meanings.

The confusion arising from the use of the term 'mélange' led Medley (1994, 2001) to propose the term 'bimrock' for rock bodies that contain competent blocks of varied lithologies, embedded in sheared matrices of weaker rock. Medley (1994, 2001) introduces this term in a geological engineering purpose and defines the bimrock as 'a mixture of rocks, composed of geotechnically significant blocks with a bonded matrix of finer texture'. 'Geotechnically significant blocks' means that there is mechanical contrast between blocks and matrix. Accordingly, Medley's definition is founded on geometric and mechanical parameters. This approach successfully avoids possible confusions related to genetic considerations. On the other hand, it emphasizes deformation that may play a role in the formation of the block-in-matrix rocks by introducing a mechanical contrast between the blocks and the matrix as a fundamental parameter. However, some block-in-matrix rocks (such as the Osa Mélange) may locally lack a significant mechanical contrast between the blocks and the matrix.

In view of mélange definitions by Raymond (1984) and Medley (1994, 2001), it clearly appears that a precise characterization of block-in-matrix rocks is not straightforward and, in absolute, may actually be impossible. Along with complicated fabrics observed in block-in-matrix rocks, this probably explains the significant variations in the meaning attributed to the term mélange in literature. In this chapter, we adopt the geological terminology introduced by Raymond (1984) and use the terms 'mélange' and 'dismembered unit' in the sense presented in the beginning of this chapter. The terms 'block' or 'fragment' are used for geometric bodies regardless of their origins. 'Grain' (size $= 125$ μm to 2 mm), 'granule' (size $= 2-4$ mm), 'pebble' (size $= 4-64$ mm), 'cobble' (size $= 64-256$ mm) and 'boulder' (size > 256 mm) are blocks/fragments of sedimentary origin. We arbitrarily use herein the sedimentary term 'large boulder' as a rock fragment larger than 10 m.

Recognizing the origin of a mélange

A mélange may be produced by (1) purely sedimentary processes, (2) purely tectonic processes or (3) combined sedimentary and tectonic processes (Raymond 1984). Typical examples of mélanges produced by sedimentary processes are olistostromes or debris flow deposits which are commonly produced by erosion or gravitational reworking of older rocks and/or unlithified sediments along steep slopes (e.g. Hampton *et al.* 1996). Tectonic mélanges are often produced at subduction zones along the interface between subducting and overriding plates. They are frequently composed of metamorphosed igneous/sedimentary fragments embedded in a shaley and/or serpentinitic matrix. Literature contains a wide diversity of names for these mélanges, such as flow mélange (Cloos 1984), mélange belt (Doubleday *et al.* 1994), serpentinite mélange (Chang *et al.* 2000) or suture zone ophiolitic mélange (e.g. Dupuis *et al.* 2005). Finally, mélange produced by both sedimentary and tectonic processes ('polygenetic mélanges' after Raymond 1984) are generally a result of sedimentary mélange overprinted by tectonic deformation. Wildflysches are certainly the most illustrative example of a polygenetic mélange. They are produced in orogenic zones by deformation of olistostromic sediments deposited at the front of nappes (e.g. Alonso *et al.* 2006).

All these examples highlight that mélanges are produced by a large spectrum of geological processes encountered along convergent margins. General criteria allowing differentiation between several origins are given by Raymond & Terranova (1984): (1) the nature of contacts (depositional or structural), (2) the relationships to surrounding units, including palaeogeographic considerations, (3) the nature of the matrix, (4) the internal structure of the rock body, (5) the presence or absence of features indicative of soft sediment deformation and (6) the presence or absence of deformation and metamorphism within inclusions. In addition to these criteria, a particular effort was made in our study to characterize the nature and ages of the blocks and matrix. This allowed us to make a precise comparison of the mélange with surrounding units.

Field work in the Osa Mélange

Because of the high complexity generally associated with mélanges, particularly detailed field work was made in the San Pedrillo Unit (inner Osa Mélange). Geological correlations and observations were made at scales ranging from a thin section to the entire unit (i.e. several kilometres). We carried out detailed mapping at 1:5000 scale of the NW edge of the Osa Peninsula and Caño Island (Figs 3 & 4). Comprehensive maps are presented in Buchs & Stucki (2001). Lithologies of variously deformed rocks were characterized on the basis of a systematic sampling over the entire area (Fig. 4). Although our study was restricted in the NW part of the Osa Peninsula, it provides constraints applicable to the entire San Pedrillo Unit. Samples are listed in the supplementary material.

Results

The Osa Igneous Complex

A new tectono-stratigraphy is proposed for the Osa Igneous Complex based on: (1) newly recognized differences in geochemistry of the igneous rocks (Buchs 2008); (2) mapping of distinctive igneous and sedimentary rock formations; and (3) previous and new palaeontological dates from the sediments associated with the igneous rocks.

A comprehensive geochemical study of the Osa Igneous Complex is well beyond the scope of this chapter and is presented in Buchs (2008). However, we choose here to use $(La-Sm)_{CIn}$ and $(Sm-Yb)_{CIn}$ ratios based on the geochemical study exposed in Buchs (2008) as representative parameters of the geochemical variations encountered in the Osa Igneous Complex. In tholeiitic and alkaline rocks of Costa Rica these ratios generally show a good correlation with Sm−Nd and Lu−Hf isotopic

compositions (e.g. Hauff et al. 2000; Buchs 2008). Hence, they are considered here to represent useful discriminative parameters in terms of source composition and origins. In the text, these ratios are called 'highly depleted' when $(La-Sm)_{CIn} = 0.3-0.4$ and $(Sm-Yb)_{CIn} = 0.7-0.9$, 'NMORB-like' when $(La-Sm)_{CIn} = 0.5-0.6$ and $(Sm-Yb)_{CIn} = 0.8-1.1$, 'plateau-like' when $(La-Sm)_{CIn} = 0.7-1.0$ and $(Sm-Yb)_{CIn} = 0.8-1.3$, and 'OIB-like' when $(La-Sm)_{CIn} = 0.9-3.7$ and $(Sm-Yb)_{CIn} = 1.4-4.9$. Plateau-like igneous rocks are distinct from E MORB in terms of Nb and major element contents and resemble typical oceanic plateaus (Buchs 2008).

The Osa Igneous Complex is subdivided into: (1) the Inner Osa Igneous Complex, and an Outer Osa Igneous Complex composed of (2) the Güerra Unit, (3) the Ganado Unit, (4) the Riyito Unit and (5) the Vaquedano Unit (Figs 2 & 5). All these units are bounded by recent (active?) fault zones and are broadly oriented NW−SE, parallel to the Mid American Trench (Fig. 2b).

The Inner Osa Igneous Complex. The Inner Osa Igneous Complex is exposed on the isthmus side of the Osa Peninsula and on the Burica Peninsula. In the NE it is in contact with the arc-related Golfito Complex. In the northern Osa Peninsula it is in contact with the Güerra Unit, whereas in the southern Osa Peninsula it is bordered by the Osa Mélange (Figs 2 & 5b−b′). NW extension of the boundaries of the inner Osa Igneous Complex is poorly or not expressed on reflection profile P 1600 (NW Osa, Fig. 5a−a′). However on-land exposures in close vicinity (c. 15−30 km) tend to indicate the Osa Igneous Complexe and San Pedrillo Unit (inner Osa Mélange) are also present in this part of the margin.

The inner Osa Igneous Complex comprises over 99% of basaltic, massive-columnar lava flows and pillow lavas that constitutes a rock pile several km thick (Fig. 5). The lavas consist of sub-ophitic, intersertal and intergranular basalts and coarse basalts and microgabbros with plagioclase, clinopyroxene, oxides, interstitial glass, minor amounts of sulphides and occasionally altered olivine. Three clear occurrences of intrusive rocks have been observed in the Inner Osa Igneous Complex, which consist of a sill located along the shore close to Rincón (Golfo Dulce), dolerites exposed at Punta Banco (Burica Peninsula) and a gabbro in the Quebrada Sábalo (inner, isthmic side of Osa). The sill is characterized by an atypical microlitic texture with K-feldspar and acicular brown amphibole. The gabbro has a very peculiar poikilitic texture with plagioclase phenocrysts. Occasionally, lenses and layers, 1−3 m thick, of red cherts and black shales are observed between the lava flows. These sediments constitute a very minor amount

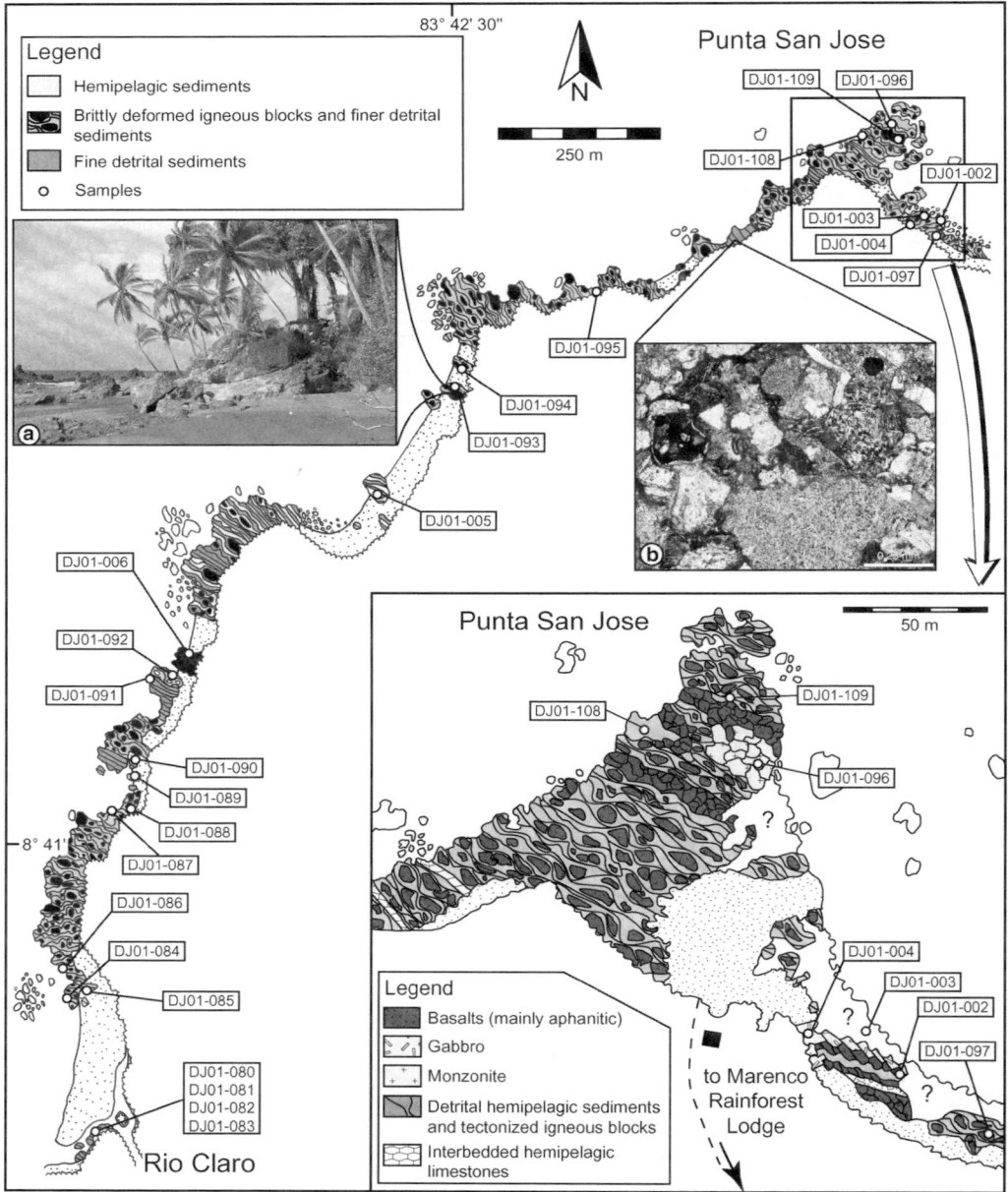

Fig. 3. Detail geological map of the Rio Claro area. (**a**) Basalt olistolith embedded within a cataclasite matrix.
(**b**) Microphotograph of a volcanic microbreccia principally made of basaltic grains. (a–b) highlight the large variability
in the clast size observed in the San Pedrillo Unit. Costal exposures of NW Osa and Caño Island have been integrally
mapped in a similar way (Buchs & Stucki 2001).

of the Inner Osa Igneous Complex. Along the shore
of the Golfo Dulce (Punta Esperanza, 539.9/291.3,
Costa Rican coordinates), a 20 m thick intercalation
of arenite, microbreccia and breccia, named
'Esperanza Formation' by Mende (2001), crops

out within the volcanic sequence. The sediments
contain basaltic material similar to the lavas of the
Inner Osa Igneous Complex. To date, this deposit
constitutes a unique occurrence of detrital material
in the Inner Osa Igneous Complex.

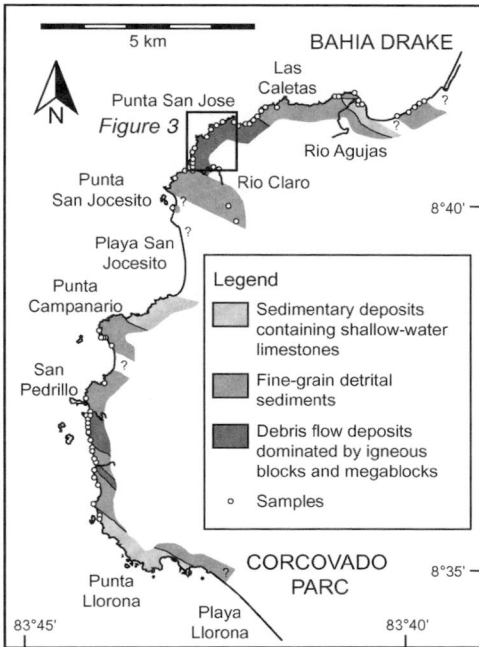

Fig. 4. Simplified geological map of the NW San Pedrillo Unit illustrating recurrent lithological associations.

Lava flows and sediments dip predominantly NE. Orientation of the pillow lavas indicates that the volcanic sequences are in both normal and overturned position. Overturned pillow lavas are found over the entire unit but are much less frequent than rocks in normal position. A low-T submarine alteration in the zeolite facies is shown by low to moderate alteration of the glass and silicate minerals. Samples collected in the Inner Osa Igneous Complex are devoid of greenschist metamorphism. Strong deformation is locally observed and associated with intensive veining of the rocks. Veins are generally filled with calcite and zeolite. Small-scale structures potentially indicating folding of the rocks (e.g. cleavage in sediments) have not been observed. Deformation seems to have predominantly occurred in a brittle mode and resulted in the formation of fault-bounded lenses of relatively undeformed rocks. The lenses are well seen along the shore of the Burica Peninsula and are observed at scales ranging from several metres to possibly several hundreds of metres. In the Inner Osa Igneous Complex the presence of both overturned pillow lavas and lenses of rock bodies is indicative of a tectonic dismemberment of a very large unit or of an imbrication of several 'smaller' rock bodies.

Age of formation of the basalts is defined by radiolarian associations found in intercalated

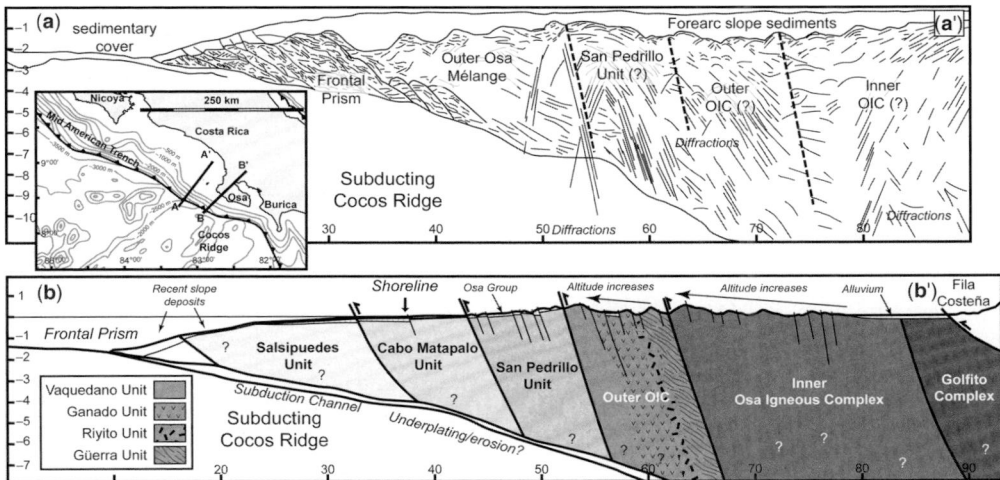

Fig. 5. Comparison of seismic and on-land interpretations for the NW Osa Mélange and Osa Igneous Complex. (**a–a′**) Depth converted interpretation of multichannel reflection profile P 1600 from the upper plate basement (modified after Kolarsky *et al.* 1995*a*; full references and data in Kolarsky *et al.* 1995*a*; vertical exaggeration 2×). The profile is only *c*. 15 and *c*. 30 km distant from the San Pedrillo Unit exposed on Caño Island and other units exposed on the Osa Peninsula, respectively. Dashed lines represent possible extensions of unit boundaries projected from cross-section b–b′. (**b–b′**) SW–NE cross-section through the Osa Peninsula and forearc slope (vertical exaggeration 2×). Topography is from GeoMapApp (online georeferred database). Note particular changes in topography correlating with suture zones between the Inner Osa Igneous Complex, Outer Osa Igneous Complex and San Pedrillo Unit (Inner Osa Mélange).

D. M. BUCHS *ET AL.*

cherts to the Coniacian–Santonian (*c.* 89–84 Ma) (Di Marco 1994; Diserens 2002; this study) (see below for more details). Two tholeiitic (low-K) basalts yield whole rock $^{40}Ar/^{39}Ar$ Ar total fusion isochron ages of 64.2 ± 1.1 and 54.5 ± 1.5 Ma (Hoernle *et al.* 2002). K–Ar ages on similar igneous rocks range from 78.0 ± 2 to 44.8 ± 8 Ma (Berrangé *et al.* 1989). The discrepancy between biochronological and radiometric ages is discussed below.

Geochemistry of the igneous rocks is principally tholeiitic with plateau-like affinities showing consistency with an oceanic, intra-plate origin (Figs 1a & 6). However, igneous rocks with distinct oceanic, intraplate affinities are locally observed, such as NMORB-like and OIB-like tholeiitic lava flows (Figs 2a & 6a). Dolerites from Punta Banco (Burica) have NMORB-like tholeiitic, intraplate affinities and the sill close to Rincón is a basaltic trachyandesite with OIB-like, intraplate oceanic affinities. Local variations of the geochemical affinities point toward distinct origins for the igneous rocks that compose the Inner Osa Igneous Complex (Buchs 2008).

The Güerra Unit. The Güerra Unit is in contact with the Inner Osa Igneous Complex along its landward edge and with the Riyito and Ganado units along its seaward edge (Fig. 2). On a SW–NE cross section the transition from the Inner Osa Igneous

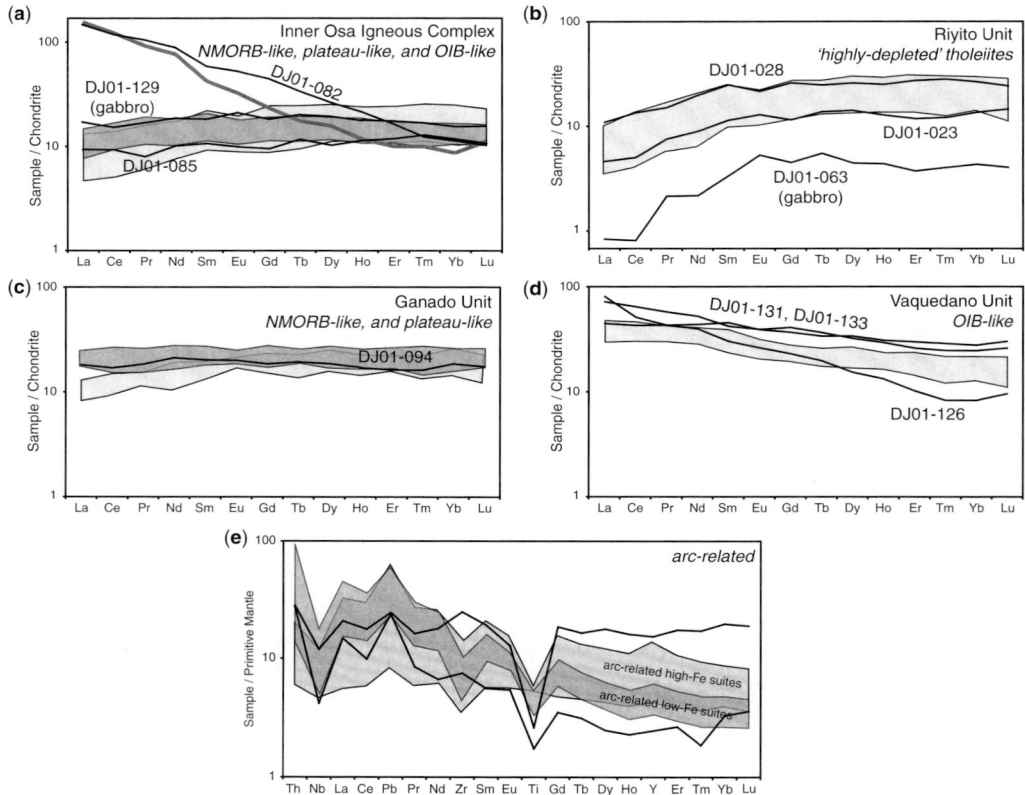

Fig. 6. (**a–d**) REE diagrams for the basalts and gabbros of the Osa Igneous Complex (grey areas, organized per unit) and basaltic olistoliths of the San Pedrillo Unit (black lines with sample references). Samples from the San Pedrillo Unit were individually associated with most similar unit of the Osa Igneous Complex. Normalization values are from McDonough & Sun (1995). 'Highly-depleted', 'NMORB-like', 'plateau-like' and 'OIB-like' denominations are defined in the text. (**e**) Multielement diagram of two dacitic olistoliths from the San Pedrillo Unit. Normalization values are from McDonough & Sun (1995). For comparison, 11–17 Ma high-Fe (tholeiitic) basalts-andesites and 8–12 Ma low-Fe (calc-alkalic) basalts-dacites from Costa Rican forearc volcanics are plotted (grey areas; Abratis 1998). Note that the dacitic olistoliths and igneous rocks from the Osa Igneous Complex have dissimilar affinities in terms of REE, notably.

Complex to the Güerra Unit is marked by changes in the topography (Fig. 5b–b'). In the NE side of the contact, the altitude progressively increases toward the contact and then suddenly drops seaward. This morphology is probably controlled by the presence of a major, possibly active tectonic contact between the Güerra Unit and the Inner Osa Igneous Complex related to overthrusting of the Inner Osa Igneous Complex toward the SW. The contact seems to correlate with Osa shoreline along the Golfo Dulce, indicating that recent tectonic movements occurred along this basement suture (Fig. 2).

In contrast to the Inner Osa Igneous Complex, the Güerra Unit is characterized by stronger deformation and prehnite–pumpellyite to greenschist metamorphic facies. This metamorphism is unique in the area of study. The unit is broadly composed of metamorphosed/altered igneous rocks, marble, recrystallized micritic limestones and metamorphosed volcaniclastic sediments. These lithologies constitute a complicated arrangement of variously deformed lenses extending at scales over 100 m that is interpreted to be a result of a tectonic imbrication of several large rock bodies. The less-deformed lenses exhibit little deformed volcano-sedimentary textures, whereas the most deformed lenses are characterized by pervasive ductile deformation and associated schistosity that erased most pristine textures. There is good correlation between the intensity of the metamorphism and the degree of deformation (i.e. most deformed rocks exhibit highest metamorphic grades), pointing toward metamorphic reactions partly controlled by tectonics. Strike of the rock lenses is NW–SE with a NE moderate/subvertical dip. Similar orientations are marked by the schistosity in the most deformed rocks and by fabrics in the best preserved volcano-sedimentary rocks. This orientation is parallel to the contact of the Güerra Unit with the Inner Osa Igneous Complex.

As a consequence of locally strong deformation, many protoliths of the Güerra Unit are difficult to identify. However it is clear the unit comprises a wide range of rocks and contains a larger proportion of sediments than the Inner Osa Igneous Complex. In the less-deformed areas we notably recognized preserved textures of vesicular pillow lavas interbedded with reddish, recrystallized limestones. These lavas consist of transitional basalts with OIB-like, intraplate affinities. However, due to sediment recrystallization it was not possible to observe preserved fossils in our samples. In other places, spilitized porphyric pillow lavas were found. These lavas have tholeiitic affinities and highly-depleted incompatible element contents. These two examples of 'preserved' volcano-sedimentary sequences are very similar in terms of sedimentary facies and

geochemistry to some sequences of the Vaquedano and Riyito units (see below). Hence, by analogy with other units forming the Outer Osa Igneous Complex and the possible role of the unit in the construction of the Osa Igneous Complex, formation ages of the rocks composing the Güerra Unit are probably encompassed between the Late Cretaceous and the Eocene.

The Ganado Unit. The Ganado Unit is exposed in the SW side of the Güerra Unit. It is in contact with other units of the Outer Osa Igneous Complex and the Osa Mélange (Fig. 2). This unit consists of pillowed and massive basaltic/basaltic-andesitic flows locally intruded by aphyric basaltic dykes, dolerites and gabbros. Intrusive rocks form a minor portion of the sequence (<10%) and are preferentially observed close to the Osa Mélange. Most common textures of the lavas are sub-ophitic, intersertal and intergranular with plagioclases, clinopyroxenes, oxides, interstitial glass and minor amounts of olivines and sulphides. Intrusive rocks have doleritic and ophitic textures with plagioclases, clinopyroxenes, oxides and co-magmatic amphiboles in some samples. In the volcanic rocks low-T submarine alteration in the zeolite facies is common. Strong alteration is also observed in the intrusive rocks and volcanic rocks at the contact. This is notably characterized by extensive argilitization of the plagioclases and epidote formation. It occurred in response to hydrothermal circulation triggered by the emplacement of the intrusive rocks prior to emplacement of the unit along the margin. In some gabbros, curvilinear cleavages are observed. This shows the gabbros were deformed by the time of their cooling, soon after emplacement. A *c.* 2 m thick layer of red radiolarian chert, interbedded with coarse lava flows, was the sole occurrence of sediments encountered in the stratigraphic sequence. Radiolarians provided a Coniacian-Santonian (*c.* 89–84 Ma) age.

The lava flows dip preferentially NE and are found both in inverse and normal positions. The volcanic rocks suffered deformation similar to the Inner Osa Igneous Complex. These observations and occurrences of intrusive bodies (i.e. deep crustal exposures) at the contact with surrounding units indicate the igneous rocks forming the Ganado Unit underwent a significant tectonic dismemberment. The unit may have formed through imbrication of distinct volcanic sequences or dismemberment of a single, very large rock body.

Two geochemically distinct groups of igneous rocks are found in the Ganado Unit. The first group comprises tholeiitic basaltic lava flows and aphyric dykes. It is characterized by plateau-like affinities consistent with an oceanic, intraplate origin (Fig. 6c). Although this group appears to be similar to the plateau-like lavas of the Inner Osa Igneous Complex, discriminative geochemical differences exist between the two, notably in terms of Nb content (Buchs 2008). One exception to this observation comes from a lava flow associated with the Campanian radiolarian chert. Although this lava has plateau-like affinities, it consists of a transitional tholeiite with Nb content similar to the tholeiites from the Inner Osa Igneous Complex. It is interpreted here to represent a record of a late stage of volcanism in the Ganado plateau-like 'series'. The second geochemical group of the Ganado Unit is mainly expressed in the dolerites and gabbros emplaced into the plateau-like group, as well as in a few massive lava flows. This group is composed of NMORB-like tholeiites with oceanic intraplate affinities similar to the NMORB-like tholeiites of the Inner Osa Igneous Complex (Fig. 6c).

The Riyito Unit. The Riyito Unit is a composite (non-contiguous) unit exposed at both the NW and SE edges of the Ganado Unit. It is in contact with the Güerra Unit in the NE and the Vaquedano Unit and Osa Mélange in the SW (Fig. 2).

The unit comprises volcano-sedimentary sequences including small, well-formed pillow lavas, sheet flows and minor occurrences of sediments and hyaloclastites. The lavas are basaltic in composition and locally vesicular. They ubiquitously contain plagioclase phenocrysts embedded in a subophitic–intersertal–microlitic matrix made of plagioclases, clinopyroxenes, oxides, matrix glass and minor amount of sulphides. The porphyric nature of these lavas has been used on the field to constrain the extension of the unit. Vesiculated hyaloclastites are locally observed between pillow lavas, possibly indicating eruption under shallower conditions for some parts of the volcanic rocks. Alteration of the igneous rocks is low to moderate and consistent with low-T oceanic alteration. The sediments represent less than 3% of the total volume of the sequence and are predominantly exposed along the Rio Riyito (central Osa). They comprise turbiditic beds reworking basaltic igneous rock fragments. Plagioclase–phyric dykes intruding lava flows occur in the lower part of the Rio Riyito. At Punta Ganadito (western Osa) crustacean microcoprolites have been encountered in a deep ocean sequence (Buchs *et al.* 2009).

The lava flows dip NE with pillow lavas in both normal and overturned positions. Exposures of the Riyito Unit are clearly non-contiguous and are separated by the Ganado Unit. Hence, similarly to the Inner Osa Igneous Complex, the Riyito Unit shows indications of dismemberment that possibly resulted from an imbrication of distinct rock bodies.

Hauff *et al.* (2000) provided an ^{40}Ar/^{39}Ar age of 62.1 ± 0.6 Ma (Paleocene, total fusion on whole rock matrix) for a basaltic sample of the Punta Ganadito. Fossils allowing a precise dating of the interlayered sediments have not been encountered.

Volcanic rocks are all tholeiitic and exhibit very consistent geochemistry. They are characterized by an unusual high depletion in the most incompatible elements that is consistent with an oceanic origin (Fig. 6b).

The Vaquedano Unit. The Vaquedano Unit is exposed along the Osa Mélange and is in contact with the Ganado and Riyito units (Fig. 2). This unit comprises a complicated imbrication of volcano-sedimentary sequences exhibiting distinct lava morphologies, geochemistry, ages and sedimentary associations. It broadly consists of vesiculated, small-shaped pillow basalts and lava flows, reddish pelagic limestones, detrital sediments and hyaloclastites. Lavas form the bulk of the sequences (>85%) and consist of ophitic, subophitic, intersertal and spherulitic basalts with plagioclase, titanoaugite, olivine pseudomorphs, oxides, glassy matrix and minor amount of sulphides. A few pillow breccias and hyaloclastites are locally interbedded with the lavas. The pelagic limestones locally occur as deformed xenoliths embedded in massive lava flows, indicating an incorporation of unlithified sediments into the lavas by the time of their emplacement. Most commonly, the limestones form interbeds within the pillowed and massive lava flows. Occasional sponge spicules possibly indicate some of the limestones formed in nutrient-rich oceanic currents. A sample of Late Cretaceous micritic limestone contained two badly preserved fragments of red alga. In the Osa Igneous Complex, this is the sole occurrence of material originating from a shallow-water environment. Detrital basaltic sediments that consist of <50 m thick turbidites and pebbly–cobbly breccias are locally intercalated with volcanic rocks.

Ages of the volcano-sedimentary sequences are well constrained by fossil assemblages found in the pelagic limestones. These sediments contain Late Cretaceous Inoceramus fragments and Campanian to Middle Eocene Foraminifera (Di Marco 1994; Buchs 2003). This is the largest range of ages observed for the sequences of the Osa Igneous Complex, which spans *c.* 40 Ma. The igneous rocks yielded K–Ar ages of 55.3 ± 6 to 40.8 ± 4 Ma that are broadly similar to biochronological ages (Berrangé *et al.* 1989).

In both lava flows and sediments the layers predominantly dip NE, with both normal and reverse positions of pillow lavas. Although the lavas are ubiquitously reddish as a response to early, low-T, oceanic oxidization/alteration, no trace of higher temperature alteration or metamorphism has been encountered. The unit is composite and made of large (100 m length/width and bigger) rock bodies identified on the basis of distinct volcano-sedimentary associations, ages and geochemistry. Similar to the Inner Osa Igneous Complex, the Güerra Unit and possibly the Riyito Unit, the Vaquedano Unit resulted from an imbrication of several rock bodies.

Volcanic rocks comprise transitional to alkalic basalts with OIB-like affinities (Fig. 6d). Although these rocks share intraplate oceanic affinities they are characterized by a wide range of immobile incompatible element contents throughout the unit. This is incompatible with a common origin of the volcanic rocks. However, good consistency of the geochemistry is observed in samples collected close to each others, at scales <250 m, which is in agreement with an imbrication of large rock bodies forming the Vaquedano Unit.

The NW Osa Mélange (San Pedrillo Unit)

Sedimentary fabric. Despite the existence of a highly pervasive deformation, the San Pedrillo Unit exhibits a relatively well-defined lithological arrangement (Buchs & Stucki 2001; Buchs & Baumgartner 2003). Compositional variations of the formations and recurrence of a finite number of lithologies are observed, providing the opportunity to map the mélange at 1 : 5000 and larger scales (Figs 3 & 4). In general, the NW Osa Mélange is characterized by alternations of (1) areas dominated by igneous blocks embedded within a deformed sedimentary matrix and (2) areas comprising deformed sedimentary layers lacking a significant amount of large (>1 m) blocks and predominantly made by fine-grained (<5 cm) detrital or hemipelagic sediments (Figs 3 & 8). Alternations of lithologies trend NW–SE, with strong variation in dip (Fig. 8b). Sedimentary structures, such as layering, size grading and laminations are frequently observed throughout the unit and are oriented parallel to lithological alternations. The lithological arrangement and systematic orientation of layered sediments is interpreted to result from imbrication of distinct rock bodies characterized by a lenticular shape (Fig. 8a).

A detrital (sedimentary) fraction is ubiquitously observed in the San Pedrillo Unit. It is indicated by the presence of clasts ranging from grains to large boulders, embedded within sedimentary deposits (Fig. 3). Intense tectonic brecciation of large igneous blocks may lead locally to a possible confusion of the San Pedrillo Unit with a tectonic mélange sensu stricto (e.g. Meschede et al. 1999). However, we observed that (1) the large blocks are systematically found within debris flow layers and are rarely embedded within cataclased rocks and (2) geochemical compositions of the igneous blocks located close to each other and embedded within the same sedimentary deposits preclude a provenance from a unique volcanic series (see below). Thus, the igneous blocks of the San Pedrillo Unit are olistoliths pertaining to the detrital component of the mélange rather than preserved portions of tectonically dismembered igneous sequences. A direct consequence of this observation is that there is a sedimentary fabric throughout the entire San Pedrillo Unit that predates a younger tectonic overprint.

Tectonic fabric. Pervasive tectonic overprint in the San Pedrillo Unit may be divided into (1) active subvertical faulting and (2) earlier deformation. The former has been attributed to uplift and block tilting controlled by the morphology of the subducting Cocos Ridge (Sak et al. 2004). This implies that adjacent coastal exposures may actually represent different structural levels of the Osa Mélange.

Earlier deformation occurred in a low-grade metamorphic environment, below the field stability of greenschists. In a few cases, igneous olistoliths embedded in the sedimentary, low-grade metamorphic matrix show a higher metamorphic facies with the presence of epidote, prehnite and pumpellyite. These facies are probably inherited from their environment of formation prior to their incorporation to the San Pedrillo Unit. Although some deformed rocks are green due to chlorite formation, metamorphism observed in our samples from the Osa Mélange has not exceeded a prehnite–pumpellyite facies. This is an important dissimilarity of the Osa Mélange with other mélanges that commonly contain greenschist, amphibolite or blueschist grade blocks.

Tectonic deformation and related phases that occurred prior to the final exhumation of the San Pedrillo Unit are comprehensively described by Vannuchi et al. (2006). Their results are summarized and discussed conjointly with our observations below.

Lithologies. Large variation in lithological composition occurs in the San Pedrillo Unit. During our 1:5000 mapping, a detailed characterization of the lithologies was performed (Fig. 3). However, the extent of the lithological variations is such that it is fundamentally not possible to constitute a complete report of all the rocks. Hence, we describe here only the most representative associations of lithologies.

The San Pedrillo Unit may be roughly divided into three types of lithologies: (1) sedimentary deposits containing shallow-water limestones

Fig. 7. (*Continued*).

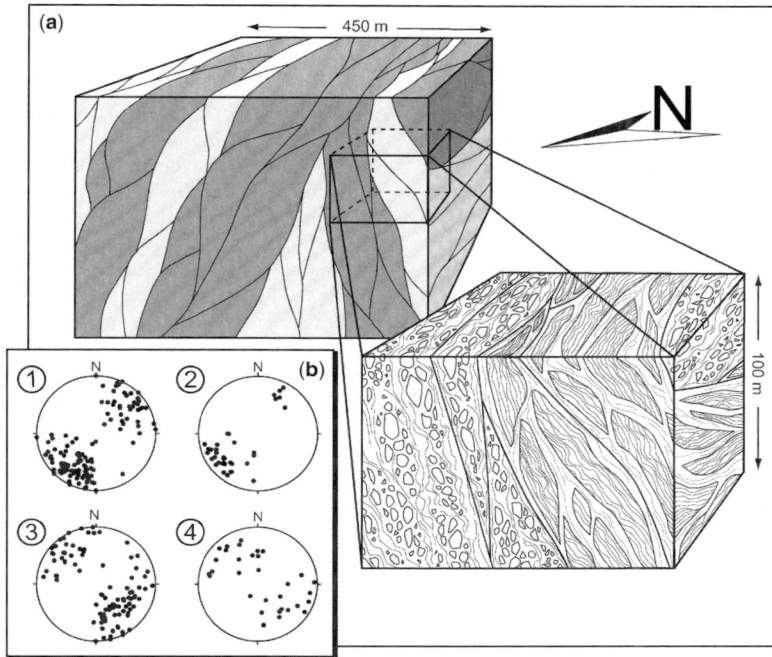

Fig. 8. (**a**) Structural model of the NW Osa Mélange illustrating the block-in-matrix texture and disrupted sedimentary layers at a small scale. At a larger scale, the chaotic arrangement of the mélange is replaced by lithologically coherent lenses. (**b**) Orientations of principal shear planes observed in the mélange. (1) and (2) are shear planes mainly parallel to the orientation of the trench, that crop out on the Osa Peninsula, respectively, on the Caño Island. (3) and (4) represent shear planes predominantly perpendicular to the orientation of the trench, that are exposed on the Osa Peninsula, respectively on the Caño Island. The two families of shear planes show a large variability in their orientation that is consistent with a lenticular arrangement of the lithologies in the San Pedrillo Unit.

(*c*. 20% volume of the San Pedrillo Unit), (2) debris flow deposits dominated by igneous olistoliths (*c*. 25% volume of the San Pedrillo Unit) and (3) fine-grain detrital sediments (*c*. 55% volume of the San Pedrillo Unit).

Sedimentary deposits containing shallow-water limestones. Shallow-water limestones occur mainly

as pebbles and boulders embedded in a dark, tuffitic matrix (Fig. 7a–d). They form matrix-supported breccias that are encountered on Caño Island and between Playa San Josecito and Punta Llorona (Fig. 4). This lithology appears as sedimentary layers that have an apparent thickness of 2–30 m and were probably emplaced by debris flow events. The matrix comprises altered ashy tuffites,

Fig. 7. Outcrops of the principal rock types of the Osa mélange (NW Osa and Caño Island). (**a**) An originally well-bedded, laminated, calciturbidite interbedded with tuffaceous mudstones. The Late Eocene limestone bed lithified before the mudstone became hydrofractured and intruded by the soft mudstone (Agujitas, Bahia Drake, Osa). (**b**) Individual limestone interbeds became boudinaged and faulted Agujitas, Bahia Drake, Osa. (**c**) Limestone–tuffite breccia. This pebbly mudstone is of probable debris flow origin. It is highly deformed as most rocks in the Osa Mélange (Punta Campanario, Osa). (**d**) Boudinaged limestone beds in a siliceous tuffitic mudstone matrix, Caño Island. (**e**) Basalt/radiolarite contact in a Late Cretaceous megablock of probable seamount origin (San Pedrillo, Osa). Scale bar 22 cm. (**f**) Detail of a Late Cretaceous–Paleocene cherty limestone associated with a seamount block. The early lithified chert beds suffered from brittle fracturing, while the limestones responded by pseudo ductile deformation to stress (River mouth of Rio Claro, Marenco, Osa). (**g**) Red and dark green cherty mudstones and cherts that show boudinage in bundles (between Marenco and Agujitas, Osa). Length of hammer 45 cm. (**h**) Mass flow deposit. Base of a basalt block (upper side of the photograph) and boulder breccia including clasts of basalts, dacites and shallow-water limestones within a detrital hemipelagic sediment matrix (Scale bar 22 cm). (**i**) Fine grained, volcaniclastic greywackes and mudstones with well preserved bedding and lamination of turbiditic origin (Punta Campanario, Osa). Scale bar 22 cm. (**j**) Polymict boulder breccia including angular clasts of cherts, rounded clasts of dacites–basalts and tuffites (Punta Campanario, Osa).

intensively deformed during tectonic processes. The limestone clasts are frequently recrystallized and fragmented. Sometimes they comprise over 50% of calcite veins. The matrix was injected within and between the clasts, indicating the tuffites behaved softly and were poorly compacted during the beginning of the tectonic history within the mélange.

Limestone beds, 0.5–3 m thick, are encountered in the areas of Agujitas, Marenco and Caño (Figs 4 & 7a, b). These deposits represent an accumulation of largely dominant carbonate grains without a significant amount of tuffitic matrix. They are interpreted as calci-turbidites interbedded with tuffitic greywackes, volcarenites and micro-breccias that notably contain fragments of mafic to differentiated volcanic sources. During tectonic deformation the beds underwent fracturing, brittle boudinage and injection of under- and overlying muddy sediments, creating networks of sedimentary dykes (Fig. 7a). These limestones allowed us to date the deposition of the sediments forming the bulk of the mélange (see below).

The shallow-water limestone olistoliths and turbidites contain carbonate grains and pebbles formed in a platform and/or peri-platform environment. They include mainly red algae, larger foraminifera, echinoderms and sponge spicules. An associated detrital fraction consists of fragments of pyroxene, zoned plagioclase, quartz, basalts and intermediate to acidic lavas. Detrital quartz and fragments of dacites occasionally occur as nuclei of red algal oncoliths. The larger foraminifera indicate a Late Eocene age (see below). Thus, the depositional age of these shallow-water derived calcareous debris flows and calci-turbidites is Late Eocene or younger.

Similarities of composition, faunal assemblages and age of the shallow-water carbonates of the San Pedrillo Unit probably indicate a common origin. In the NW Osa Mélange they are associated through primarily sedimentary processes with products of a differentiated volcanism that is represented by ashy tuffites, quartz fragments and silica-rich igneous blocks. The geochemical compositions of the silica-rich igneous blocks indicate an arc-related affinity (see below). Thus, it is clear that the carbonates must have formed in the vicinity of a volcanic arc.

Debris flow deposits dominated by igneous olistoliths. Debris flow deposits with large amounts of igneous boulders and large boulders are concentrated in the Marenco and San Pedrillo areas (Figs 3b, 4 & 7h). The olistoliths locally exceed 50 m in size, constraining thickness of the debris flow deposits to an equal or superior size. The olistoliths consist of basalts (*c.* 70%), intermediate-felsic igneous rocks (*c.* 25%), gabbros and hydrothermally altered

gabbros (*c.* 2%), radiolarites (*c.* 1%), limestones (*c.* 1%) and detrital sediments (*c.* 1%). They are embedded within a matrix of detrital sediments comprising mineral fragments, and grains, granules and pebbles of rock fragments. The matrix exhibits a composition analogous to the igneous blocks, indicating a similar provenance of the detrital material. The provenance of the igneous olistoliths is shown by their geochemical composition (see below), which points toward two main sources: (1) a highly heterogeneous basaltic source of oceanic intraplate origin and (2) an arc-related igneous source.

The olistoliths and the matrix are deformed and crosscut by a dense network of zeolite, calcite and quartz veins. An intensive brittle deformation affected the igneous blocks, resulting in strong tectonic brecciation. As a consequence, it may locally be particularly hard to make a distinction between the matrix and the blocks. This may explain the confusion of the Osa Mélange with a tectonically dismembered igneous sequence by some authors. However, association of well-rounded igneous and calcareous pebbles is locally observed in sedimentary layers, clearly indicating a primary sedimentary origin of the igneous blocks (see below). The sedimentary origin of the igneous blocks is further strengthened by the large variability of geochemical affinities observed within the igneous rocks that is clearly incompatible with disruption of a unique magmatic suites or several magmatic suites proceeding from a unique tectonic setting.

Radiolarites and pelagic limestones occur as isolated blocks or in sedimentary association with igneous olistoliths (Fig. 7e, f). Within the igneous olistoliths radiolarites are in stratigraphic contact with lava flows. Pelagic limestones are interbedded within reddish pillow lava sequences. The radiolarians encountered in isolated blocks of sediments or in association with lava flows are Campanian–Maastrichtian in age. Radiolarian assemblages are dissimilar to those of the radiolarian cherts of the Osa Igneous Complex. The pelagic limestones provided ages ranging from the Campanian to the Middle Eocene, similar to the ages of the calcareous sediments found in the Outer Osa Igneous Complex (Vaquedano Unit). Faunal compositions of the pelagic limestones of the San Pedrillo Unit are similar to those of the Vaquedano Unit.

Fine-grain detrital sediments. Fine-grain detrital sediments form the predominant lithology in the San Pedrillo Unit (*c.* 55% of the mapped volume; Fig. 4). They range from mudstone, greywacke, siltite, arenite to microbreccia (Fig. 7g–i). The detrital fraction comprises mineral and lithic fragments (Fig. 3a). A matrix component predominantly of siliceous–calcareous–tuffitic oozes is observed in some samples (up to 95% of the mode). The mineral fragments consist of clinopyroxenes (c. 50%), plagioclases (c. 40%), opaque minerals (c. 10%) and quartz (<1%). Lithoclasts are predominantly fragments of basalts (c. 95%) and, to a lesser extent, intermediate-felsic igneous rocks (c. 5%) and limestones grains (<1%). When observed, the siliceous–calcareous matrix of the detrital sediments contains a few recrystallized radiolarians and planktonic foraminifera, which are not sufficiently preserved to allow age determination.

The occurrence of (1) pelagic faunas and (2) quartz fragments and tuffites within the fine-grain detrital sediments constrains their environment of deposition to a hemipelagic setting. The lithologies in this group range over a wide compositional field that is fully described by a mixing between two end-members: (1) siliceous–calcareous hemipelagic oozes (e.g. mudstones, biomicrites and radiolarian cherts) and (2) a detrital component (e.g. forming litharenites and arkosic arenites).

At Punta Campanario, <2 m-thick debris flow layers are intercalated with the fine-grained detrital sediments (Fig. 7j). Within the debris flow deposits, shallow-water limestone pebbles occur in association with basaltic, andesitic and dacitic pebbles that are both angular and well rounded (Buchs & Baumgartner 2007, p. 4, fig. 2a). The association of mafic, intermediate and differentiated igneous material with grains of shallow-water limestones is in agreement with observations made in other, previously described lithologies of the mélange. In addition, the presence of well-rounded elements of basalts and dacites indicates that this material was transported in a river system or was abraded along a shoreline. As a consequence, the sources of the detrital material encountered in the San Pedrillo Unit were at least partially emerged at the time of their erosion. Large foraminifera in the grains of shallow-water limestones provided Late Eocene ages, similar to the age of the rest of the shallow-water limestones in the San Pedrillo Unit (see below). Late Eocene is thus a limit age of formation of the fine-grained detrital sediments (i.e. they are Late Eocene or younger).

A minor fraction (<2%) of lenses of radiolarian cherts, intercalated within the fine-grained detrital sediments, occurs in the San Pedrillo Unit. These sediments contain <10% of plagioclase, clinopyroxene and glass fragments. Radiolarians are Early Eocene in age, apparently in disagreement with the ages of the calcareous grains. Occurrence of these sediments and the significance of their age are discussed below.

Geochemistry of the igneous olistoliths. Pervasive dense networks of hydrothermal veins within the San Pedrillo Unit indicate that (hydrothermal) fluids circulated throughout the mélange. The

fluids may have altered some of the igneous rocks. However, we observed that, in some cases, the cores of the biggest igneous olistoliths remained surprisingly well preserved (Fig. 3b). We sampled those cores and relatively unaltered igneous blocks to perform geochemical analyses (supplementary material).

Four groups of igneous rocks have been recognized within the Osa Mélange on the basis of immobile trace element contents. We subdivided them into intermediate-felsic igneous rocks and basaltic igneous rocks.

The intermediate-felsic rocks are represented by blocks and megablocks of dacites, rhyodacites, monzonites and granophyres. Three samples of dacites and monzonite were selected in this group for their apparent freshness. Their incompatible element patterns are highly variable, indicating both enrichment and depletion in the LREE and highly incompatible elements (Fig. 6e). We note a systematic $(Nb-La)_{nCI} < 0.6$ and $(Ce-Pb)_{nPM} < 1$ (Fig. 6e). Since La and Nb are known to be immobile during alteration processes (e.g. Verma 1992), the La–Nb ratio is believed to represent a primary feature consistent with an arc-related origin of the samples. A positive Pb anomaly is also observed, which may be due to sediment-derived fluids within the accretionary prism or represent a primary feature. In the light of the high incompatible element contents in the samples, the Pb positive anomaly could hardly be related to secondary processes because it would require a strong alteration that is not observed in the analysed rocks. Accordingly, these rocks have both primary La–Nb and Ce–Pb consistent with a near-arc origin. Indeed, their composition is similar to Miocene high-Fe (tholeiitic) and low-Fe ('calc-alkalic') suites of the Costa Rican forearc (Fig. 6e) (Abratis 1998).

The olistoliths of basaltic rocks span a large domain in terms of geochemical composition (Fig. 6a–d). Groups showing similar affinities on the basis of their immobile element composition were recognized, which point toward oceanic, mostly intraplate origins of the basaltic olistoliths. Three groups of volcanic rocks were recognized with characteristic REE contents: (1) tholeiitic plateau-like basalts with $(La-Sm)_{CIn} \approx 1.0$ and $(Sm-Yb)_{CIn} \approx 1.0$, (2) highly depleted tholeiitic basalts and gabbros with $(La-Sm)_{CIn} = 0.2-0.4$ and $(Sm-Yb)_{CIn} \approx 0.9$ and (3) OIB-like alkali-transitional basalts with $(La-Sm)_{CIn} = 1.0-2.6$ and $(Sm-Yb)_{CIn} = 1.6-4.6$. Igneous samples from the San Pedrillo Unit are compared in terms of REE content to the igneous rocks of the Osa Igneous Complex in Figure 6. Strong geochemical similarities exist between these two groups of rocks. It seems that most of the igneous olistoliths present in the San Pedrillo Unit have an equivalent in the Osa Igneous Complex (Fig. 6), but not in other forearc igneous complexes of South Central America (not shown, see Hauff *et al.* 2000 for a representative dataset of South Central American igneous complexes). Similarities between mafic igneous sequences in the Osa Igneous Complex and mafic igneous blocks in the San Pedrillo Unit are detailed below, conjointly with sedimentary associations.

Similarities between the basaltic olistoliths and the Osa Igneous Complex. Igneous olistoliths occur in debris flow deposits throughout the San Pedrillo Unit. Some of them include a sedimentary cover and interlayered sediments. Hence, both geochemistry and sedimentary associations may be used to point out possible origins for the olistoliths. We make here a detailed description for a representative selection of these olistoliths, with emphasis on the similarities between these rocks and sequences observed in the Osa Igneous Complex.

Basalt DJ01-085, with plateau-like affinities, and OIB-like alkali basalt DJ01-082 were both collected in an olistostromic deposit at the Rio Claro (Fig. 3). An atypical microlitic texture with K-feldspar and acicular brown amphibole is recognized in the alkali basalt that is similar to the texture of the sill exposed close to Rincón in the Inner Osa Igneous Complex. The two samples of the San Pedrillo Unit from the Rio Claro debris flow deposit show strong geochemical similarities with igneous rocks of the Inner Osa Igneous Complex (Fig. 6a).

Plateau-like gabbro DJ01-129 from San Pedrillo area has a very distinctive poikilitic texture and REE content similar to the poikilitic gabbro sampled in the Inner Osa Igneous Complex (Fig. 6a).

Basalt DJ01-094 from Marenco area has geochemical affinities similar to plateau-like igneous rocks of the Ganado Unit (Fig. 6c). These rocks are also characterized by a lower Nb–Y ratio than plateau-like igneous rocks from the Inner Osa Igneous Complex, pointing toward distinct origins (Buchs 2008).

Highly depleted tholeiitic lavas DJ01-023 and DJ01-097 were sampled in olistotromic deposits close to Marenco (Fig. 3). They are characterized by plagioclase-phyric textures and REE contents highly similar to lavas from the Riyito Unit (Fig. 6b). The sample DJ01-023 was embedded in a sedimentary layer conjointly with a cobble of sediment containing *Palaxius osaensis* coprolites. This species was first identified in the sediments of the Riyito Unit (Buchs *et al.* 2009) and thus may be considered as a highly specific marker of this unit.

Reddish alkali/transitional basalts DJ01-131 and DJ01-133 were sampled in olistostromic deposits close to the San Pedrillo ranger station.

These samples have similar geochemical affinities to igneous rocks of the Vaquedano Unit (Fig. 6d). They are associated with Late Cretaceous to Eocene pelagic sediments similar to the limestones of the Vaquedano Unit.

Contact between the Osa Igneous Complex and the Osa Mélange

The Osa Mélange is in contact with the Osa Igneous Complex along a NW–SE fault zone that extents through the Osa Peninsula (Fig. 2). In NW Osa, the Mélange is directly in contact with the Vaquedano Unit, whereas in the SE Osa the Mélange it borders the Inner Osa Igneous Complex. The nature of the contact is similar in both NW and SE Osa. This contact probably extends further toward the NW but has not been imaged on a seismic line *c.* 35 km off the Osa Peninsula (Kolarsky *et al.* 1995*b*) (Fig. 5). Like the fault zone at the Güerra Unit– Inner Osa Igneous Complex interface, the topography is indicative of a major, possibly active tectonic contact between the Osa Igneous Complex and the Osa Mélange (Fig. 5b–b′). In the Outer Osa Igneous Complex, altitude broadly increases toward the contact where it reaches a high and then descends strongly toward the Osa Mélange. This may be caused by thrusting of the Osa Igneous Complex upon the Osa Mélange along an inverse fault.

Deformation of the rocks from the Osa Igneous Complex progressively increases toward the fault zone and marks a transition to the Osa Mélange. The deformation is characterized by the appearance of a pervasive intense brecciation of the igneous rocks associated to a complicated vein network. In an area of <300 m from the mélange, the igneous sequences were brittlely deformed and reduced to a tectonic mélange with 'preserved' igneous blocks embedded within cataclased igneous rocks. In the Vaquedano Unit higher volumes of sediments are intercalated with igneous rocks and preferentially accommodated the deformation, leading to a better preservation of the igneous rocks.

The edge of the San Pedrillo Unit in contact with the Osa Igneous Complex is globally similar in terms of geology and structure to the rest of the San Pedrillo Unit. Tectonic incorporation of brecciated blocks of the Osa Igneous Complex in a *c.* 100 m thick layer may occur at the transition between the two complexes. However, the small thickness of this layer indicates that no significant tectonic incorporation of the Osa Igneous Complex into the Osa Mélange has occurred.

Biostratigraphy

New paleontological determined for the Osa Igneous Complex and San Pedrillo Unit (inner Osa

Mélange) are documented in this chapter. Radiolarian ages from the Osa Igneous Complex given by Diserens (2002) are newly evaluated using updated fossil ranges. Radiolarian ages from the Azuero Plateau (Azuero Complex, western Panama, Buchs *et al.* in prep.) given by Kolarsky *et al.* (1995*b*) are reappraised following the same method so as to provide reliable comparisons of biochronological ages for some Costa Rican–Panamean igneous complexes predominantly composed of plateau-like igneous rocks.

Radiolaria in the Osa Igneous Complex and San Pedrillo Unit. The radiolarian biostratigraphy (Figs 9–11) for the Late Cretaceous is based on work by Riedel & Sanfilippo (1974), Dumitrica (1975), Foreman (1975, 1977), Pessagno (1976), Taketani (1982), Sanfilippo & Riedel (1985), Schaaf (1985), Thurow *et al.* (1988), O'Dogherty (1994), Hollis & Kimura (2001), Vishnevskaya (2001, 2007), while Foreman *et al.* (1973), Sanfilippo & Riedel (1973), Nishimura & Ishiga (1987), Nishimura (1992), Sanfilippo & Nigrini (1998) were used in dating the Palaeogene samples. For more details on the technique used to date samples see Bandini *et al.* (2006, 2008).

Azuero Complex, western Panama. Sample JC-86-A7 (Torio shore, Kolarsky *et al.* 1995*b*): occurrence of *Crucella plana* and *Alievium praegallowayi* gives a Turonian–Early Santonian age. This fauna was illustrated in Kolarsky *et al.* (1995*b*) and is used herein for comparison with the radiolarian assemblages from the Osa Igneous Complex illustrated in this study (see discussion below).

Inner Osa Igneous Complex. Sample GDM 9116 (Punta Banco, Burica Peninsula, Di Marco 1994): the occurrence of *Theocampe urna* and *Praeconocaryomma universa* gives an Early Turonian–Early Maastrichtian age. Sample FBJ 90174 (Punta Esquinas, eastern coast of Golfo Dulce, Diserens 1994): occurrence of *Lithatractus pusillus*, *Acanthocircus yaoi* and *Crucella cachensis* gives a Coniacian–Santonian age.

Ganado Unit. Sample DB02-199 (*c.* 505.6/297.3 Costa Rican coordinates, this study): occurrence of *Archaeospongoprunum bipartitum*, *Eostichomitra* sp. and *Theocampe salillum* gives a Coniacian–Santonian age.

San Pedrillo Unit (inner Osa Mélange). Sample DJ01-114 (this study): occurrence of *Lithatractus pusillus*, *Acanthocircus yaoi* and *Crucella cachensis* gives a Late Campanian–Early Maastrichtian age. Sample DJ01-141 (this study): occurrence of *Archaedictyomitra napaensis* gives a Maastrichtian age.

Sample GDM 9020 (Diserens 2002): occurrence of *Periphaena heliasteriscus* gives a Thanetian–Bartonian age. Sample DJ01-098 (this study): occurrence of *Phormocyrtis striata striata* and *Buryella tetradica* gives a Late Thanetian–Ypresian age. Sample DJ01-140 (this study): occurrence of *Phormocyrtis turgida* and *Buryella tetradica* gives an Ypresian age. Sample DJ01-118 (this study): occurrence of *Phormocyrtis turgida* gives an Ypresian age. Sample DJ01-043 (this study): occurrence of *Theocotylissa alpha, T. auctor, Buryella clinata, B. tetradicta* and *Lychnocanium carinatum*, gives a Late Ypresian age. Sample GDM 90126 (Diserens 2002): occurrence of *Dictyoprora mongolfieri* and *Buryella clinata* gives a Lutetian–Bartonian age.

Discussion. Radiolarian cherts of similar ages (Coniacian–Santonian) have been found in the Azuero Plateau (Kolarsky *et al.* 1995*b*), the Inner Osa Igneous Complex (Di Marco 1994; this study) and the Ganado Unit (this study). All these units are predominantly composed of igneous rocks with plateau affinities (see also Buchs 2008). Despite poor radiolarian preservation and weak abundance in the studied sediments, samples proceeding from these units yielded different faunal assemblages. The fauna of sample JC-86-A7 (Azuero Plateau, Azuero Complex, western Panama) is characterized by the association of *Hemicryptocapsa polyhedra, Praeconocaryomma universa, Pseudoaulophacus lenticulatus, P. venadoensis, Crucella plana* and *Alievium praegallowayi* (Kolarsky *et al.* 1995*b*). The assemblage of sample FBJ 90174 (Inner Osa Igneous Complex) is characterized by the association of *Dictyomitra formosa, Theocampe tina, T. urna, Acanthocircus yaoi, Praeconocaryomma universa, Crucella cachensis, Gongylothorax verbeeki* and *Lithatractus pusillus*. The assemblage of sample DB02-199 is characterized by the association of *Amphipternis stocki, Dictyomitra formosa, Halesium amissum, Praeconocaryomma universa, Theocampe urna, T. salillum, Archaespongoprunum bipartitum* and *Eostichomitra* sp. These three assemblages, from three different units, share only two commonly illustrated species in the published Late Cretaceous radiolarian faunas (*Dictyomitra formosa, Praeconocaryomma universa*). Some precautions are required in the interpretation of faunal assemblages since preservation and methodology may differ between samples and workers. However, we note that in samples of similar ages, associated in the field with similar igneous rocks, faunal assemblages possibly point toward distinct palaeoenvironments. This observation is in good agreement with occurrences of partly distinct sedimentary facies among the units.

Radiolaria occur in a variety of lithologies in the Osa Mélange. Ribbon-bedded red cherts associated

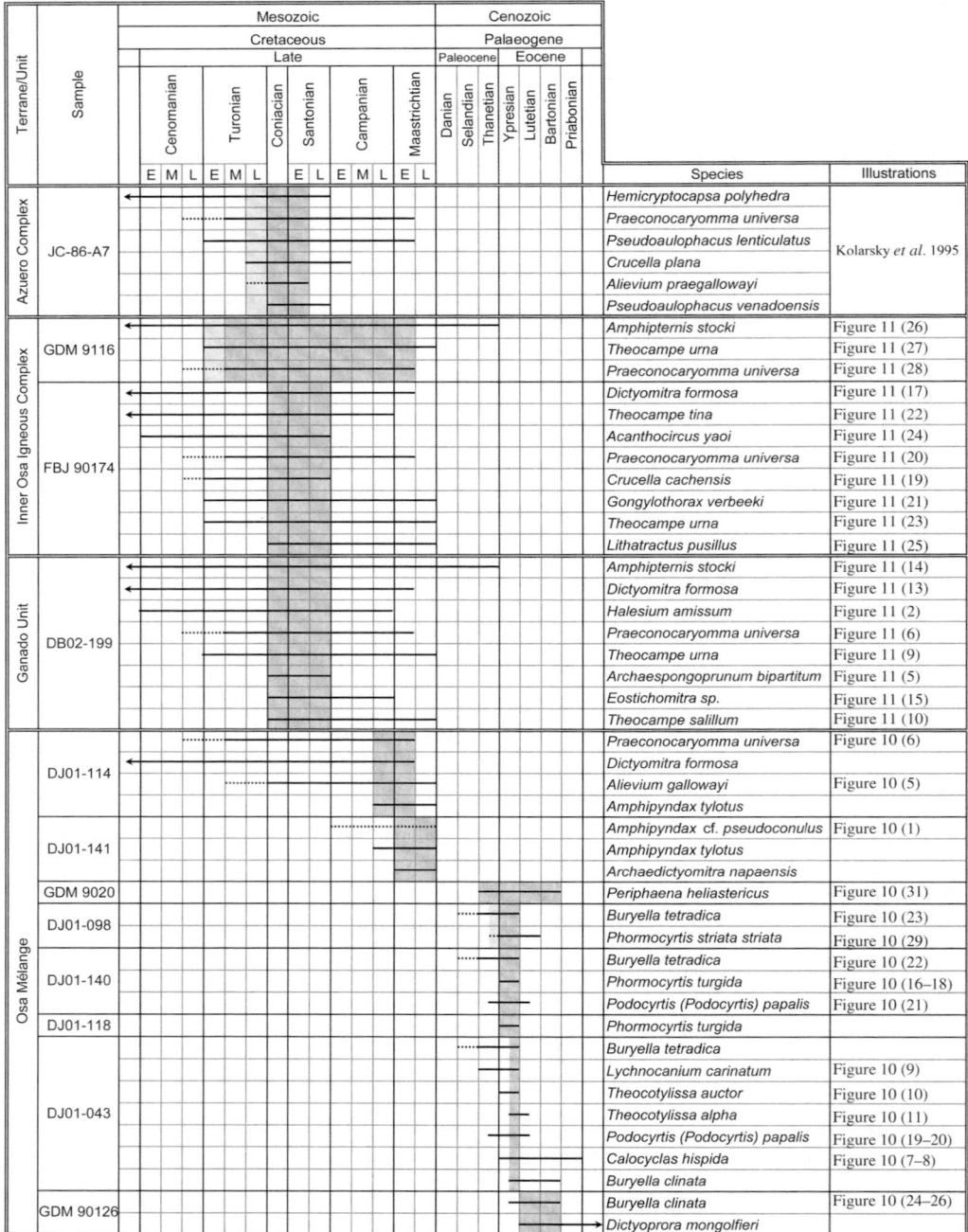

Fig. 9. Ranges and occurrences of Cretaceous and Palaeogene radiolarian species from samples of the Azuero Complex (Azuero Plateau, data from Kolarsky *et al.* 1995*b*), the Osa Igneous Complex (inner Osa Igneous Complex and Ganado Unit, data from Diserens 2002 and this study) and the San Pedrillo Unit (inner Osa Mélange, data from Diserens 2002 and this study). Ages of sample formation are given by grey fields (details in the text).

with basalts in seamount-derived blocks (samples DJ01-114 and DJ01-141) contain Campanian– Maastrichtian radiolarian assemblages. Sample DJ01-141 occurs on the field in association with large blocks of basalt containing interbeds of pelagic limestones of similar ages.

The San Pedrillo Unit contains an important amount of pelitic hemipelagic to pelagic lithologies that range from originally ribbon-bedded muddy chert to siliceous and tuffaceous mudstones that partly form the mélange matrix. Radiolaria and sponge spicules are locally abundant in all these lithologies. Faunal assemblages indicate in general a Late Paleocene–Middle Eocene age, in broad agreement with Middle Eocene ages from Azéma *et al.* (1983). However, the better preserved samples (DJ01-140 and DJ01-043) restrict the age to the Early Eocene.

Larger benthic foraminifera in the San Pedrillo Unit. Larger benthic foraminifera occur together with other shallow-water benthic bioclasts in limestone clasts–olistoliths and calciturbidites of the San Pedrillo Unit in the NW Osa Peninsula and Caño Island. The well-lithified samples only allowed determination in thin sections, studied in light microscopy and cathodoluminescence (Fig. 12). Although there are some compositional variations from one outcrop to another, faunal assemblages are similar and, in general, have a Late Eocene age according to larger foraminifera (see also Azéma *et al.* 1983). On the NE and SE coasts of Caño island, redeposited shallow-water material occurs in isolated blocks of calcarenites and calcirudites showing clasts of lime packstone, grainstone and bindstone and grains of basalt, chert, and feldspar (Mora *et al.* 1989). The limestone clasts contain the following bioclasts: articulate coraline algae (dominant), larger foraminifera, smaller planktonic and benthic foraminifera, echinoderms, siliceous sponge spicules, bryozoans and green algae. The clasts show stylolitized contacts and several phases of fracturing and vein infill by opal-quartz and then calcite. Although the larger foraminifera found at the studied localities are recrystallized, partly silicified and fractured, we could identify the following assemblage (Mora *et al.* 1989): *Amphistegina grimsdalei, Amphistegina lopeztrigoi, Amphistegina parvula, Asterocyclina* sp., *Discocyclina* sp., *Eoconuloides* sp., *Eofabiania cushmani, Eofabiania* sp., *Fabiania cubensis, Lepidocyclina* sp., *Linderina* sp., *Nummulites floridensis, Pararotalia* sp. and *Sphaerogypsina globulus.* This indicates a Late Eocene age with reworking of some Middle Eocene forms.

At Punta Campanario limestone clasts occur in a variety of debris flow breccias with a tuffitic mudstone to greywacke matrix fig. 2 in Buchs &

Baumgartner (2007). Lithoclasts are calcarenites to calcirudites with volcanic content (Fig. 12). Samples DB02-024/026 contain well-preserved, but abraded and sometimes broken larger foraminifera that are clearly size sorted (in general forms of less than 4 mm size). The assemblages include *Operculinoides* sp., small globulose and small flat forms of *Nummulites* sp. such as: *Nummulites* dia, *Nummulites* cf. macquaveri, *Orthophragmina* sp., neolepidine *Lepidocyclina* sp., such as *Lepidocyclina chaperi, Lepidocyclina* cf. antillea, *Lepidocyclina canellei, Lepidocyclina pustulosa, Sphaerogypsina globules, Amphistegina* sp., *Discocyclina* sp. and *Pseudophragmina* sp. Most of these forms indicate Middle to Late Eocene age; some are restricted to Late Eocene.

One sample yielded Late Cretaceous shallow water bioclasts. A spiculite from Agujitas (sample DJ01-019) contains poorly preserved *Pseudorbitoides* sp., *Sulcoperculina* sp. and *Omphalocyclus* sp.; this material is reworked from a probable Late Cretaceous carbonate bank.

Planktonic foraminifera in the San Pedrillo Unit. Planktonic foraminifera occur in pelagic limestones mostly associated with basaltic olistoliths. Rocks were studied in polished thin sections by light and cathodoluminescence microscopy. No isolated forms could be obtained, which leaves the specific determinations somewhat uncertain. The ages range from Campanian–Maastrichtian for a majority of blocks to Middle Eocene. Similar ages and sedimentary facies are observed in the pelagic limestones of the Vaquedano Unit (Di Marco 1994; Buchs 2003; this study).

The Osa Igneous Complex: discussion and interpretation

A well-organized highly composite complex

The Osa Igneous Complex is an imbricate of several rock bodies, recognized on the basis of converging lines of observation: (1) pillow lavas are frequently overturned; (2) strong deformation is locally observed that is very likely related to thrust zones; (3) small-scale (<100 m) tectonic lenses clearly occur in some units; (4) biochronological dating provides Late Cretaceous to Middle Eocene ages of formation for the igneous rocks; and (5) geochemistry points toward a large diversity of mostly intraplate, oceanic origins for the igneous rocks (Fig. 6, Buchs 2008).

Although geochemical and age characteristics are strongly heterogeneous at a scale of the complex (i.e. tens of kilometres), detailed characterization of the volcano-sedimentary sequences allowed us to

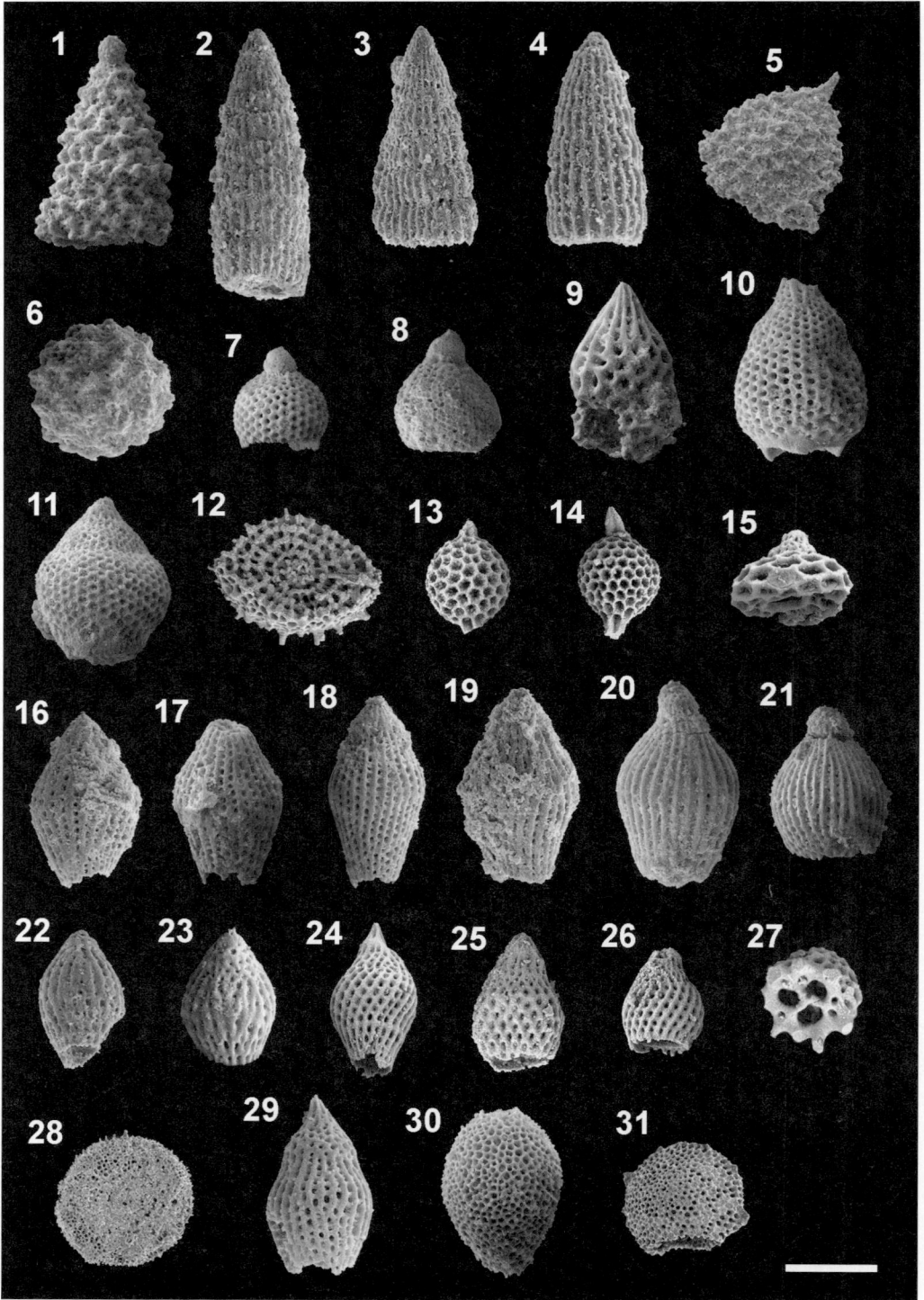

Fig. 10. (*Continued*).

recognize five units that define composite rock belts with similar trench-parallel strikes (Fig. 1b). This arrangement and the presence of various intra-oceanic volcanic rocks are typical of accretionary complexes, or accretionary belts (e.g. Dickinson 2008). This suggests the Osa Igneous Complex is an accretionary complex made up of several pieces of accreted igneous sequences, which may have initially pertained to several oceanic islands, seamounts, and oceanic plateaus. Absence of significant amounts of pelagic sediments and/or geochemical affinities of the igneous rocks (Hauff *et al.* 2000, Buchs 2008) indicates the Osa Igneous Complex does not comprise 'normal', unthickened oceanic crust (i.e. MORB *sensu stricto*).

Size and arrangement of accreted rock bodies

Because of a lack of extended exposures and limited occurrences of stratigraphic markers, size of imbricated rock bodies and imbrication mechanisms are poorly constrained. However, contrary to amalgamated terranes of North America (e.g. Coney 1989), indication of large strike–slip motion has not been observed in the Osa Peninsula. Large fault zones associated with left-lateral strike–slip motion were recognized in western Panama (Kolarsky & Mann 1995), but no precise quantification of possible displacement was provided. The geology of western Panama does not exhibit evidence of large strike–slip displacements (Buchs *et al.* in prep.). Hence, it seems the Osa Igneous Complex provides opportunity to recognize accretion-related structures that normally tend to be altered during later tectonics.

With the exception of the Ganado Unit, large igneous intrusive rocks are lacking in the Osa Igneous Complex. This probably indicates that the imbricated rock bodies represent superficial sequences of volcanic edifices, probably originally restricted in the few uppermost kilometres or hundreds of metres of the crust. Dolerites and gabbroic rocks of the Ganado Unit are devoid of significant crystal accumulations and were possibly originally located in superficial levels of one or several volcanic edifice(s). As a consequence, we propose an estimate of less than 4 km for the maximal thickness of the imbricated, dismembered rock bodies forming the Osa Igneous Complex.

In the Vaquenano and Güerra units, distinct geochemical affinities, volcano-sedimentary sequences and biochronological ages are encountered at scales *c.* 500 m. Thus, these units comprise thin, probably *c.* 250 m thick, accreted sequences. The units are characterized by a high content of accreted sediments (*c.* 20%) compared to other units of the Osa Igneous Complex. Hence, we propose the Vaquedano and Güerra units are made up of several seamount slices accreted by scraping off thin layers of 'strong' igneous rocks along 'weak' sedimentary layers. In absence of sediment interbeds in the volcanic sequences of subducting seamounts several kilometre-thick piles of igneous rocks may detach and accrete to the overriding plate, as notably imaged in Japan (Park *et al.* 1999). Such a mechanism may have been effective for the Ganado Unit, Riyito Unit and Inner Osa Igneous Complex. Hence, several kilometre (<4 km) thick piles of rock probably compose these units. Relatively good preservation of the rocks bodies and frequent occurrences of overturned pillow lavas may be related to an emplacement by duplexing or fault propagation folding at large scale. However, detailed structural observations are still required to better constrain this hypothesis.

Three major sutures bound the Inner and Outer Osa Igneous Complexes and mark the contact with the island arc (Golfito Complex) and the Osa Mélange (Fig. 5b–b'). These sutures correlate to

---

Fig. 10. SEM-illustrations of Late Cretaceous and Palaeogene Radiolaria from the Osa Peninsula, Costa Rica. The numbers of illustrations in this plate correspond to those in Figure 9. Marker = 100 μm. (1) *Amphipyndax sp. Cf. A. pseudoconulus* Pessagno, sample DJ01-141, San Pedrillo. (2–4) *Archaeodictyomitra napaensis* Pessagno, sample DJ01-141, San Pedrillo. (5) *Alievum gallowayi* (White), by sample DJ01-114, Agujitas. (6) *Praeconocaryomma universa* Pessagno, by sample DJ01-114, Agujitas. (7–8) *Calocyclas hispida* (Ehrenberg). sample DJ01-043, between Rio Claro and San Jocesito. (9) *Lychnocanium carinatum* (Ehrenberg), sample DJ01-043, between Rio Claro and San Jocesito. (10) *Theocotyle (Theocotylissa) auctor* Foreman, sample DJ01-043, between Rio Claro and San Jocesito. (11) *Theocotyle (Theocotylissa) alpha* Foreman, sample DJ01-043, between Rio Claro and San Jocesito. (12) *Lithelius* sp., sample GDM 90126, Punta Agujitas. (13–14) *Amphisphaera* spp., sample GDM 90126, Punta Agujitas. (15) *Calocycloma castum* (Haeckel), sample GDM 90126, Punta Agujitas. (16–18) *Phormocyrtis turgida* (Krasheninnikov), sample DJ01-140, San Pedrillo. (19–21) *Podocyrtis (Podocyrtis) papalis* Ehrenberg,19 & 20, sample DJ01-043, between Rio Claro and San Jocesito, 21, sample DJ01-140, San Pedrillo. (22–23) *Buryella tetradica* Foreman, 22, sample DJ01-140, San Pedrillo, 23, sample DJ01-098, Agujitas. (24–26) *Buryella clinata* Foreman, sample GDM 90126, Punta Agujitas. (27) *Tristylospyris* spp., sample GDM 90126, Punta Agujitas. (28) *Spongodiscus* sp., sample GDM 90126, Punta Agujitas. (29) *Phormocyrtis striata striata* Brandt, sample DJ01-098, Agujitas. (30) *Spongurus (?) regularis* (Borisenko), sample DJ01-098, Agujitas. (31) *Periphaena heliastericus* (Clark and Campbell), sample GDM 9020, Rio Cedral.

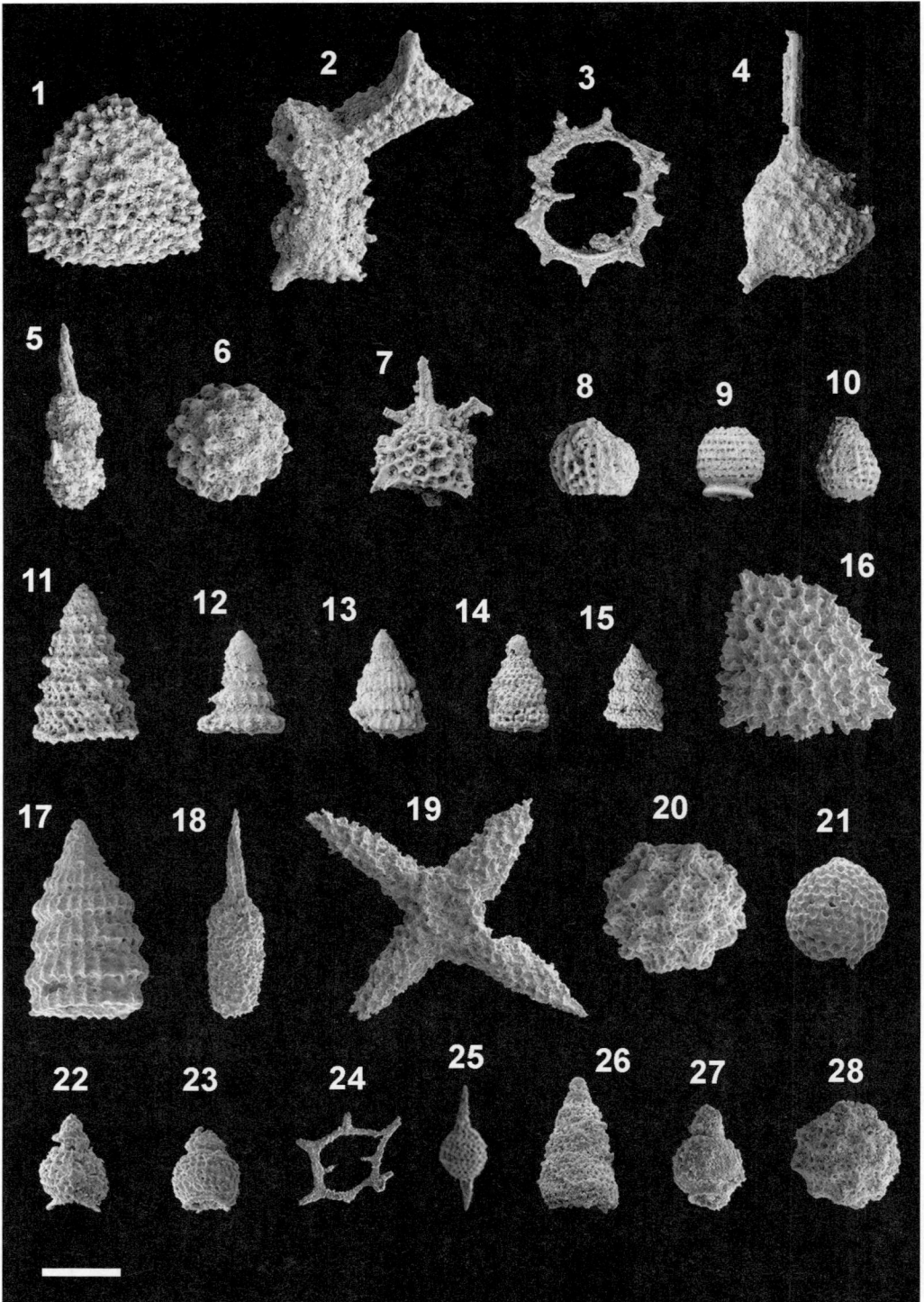

Fig. 11. (*Continued*).

the landscape morphology (incised valleys and slope variations over broad areas). They are interpreted here as ancient tectonic contacts conducting recent and/or active faulting. Although the sutures are well seen in on-land exposures they are not visible on nearby seismic profile P 1600 (Kolarsky et al. 1995a; Fig. 5a–a'). Proximity between on-land exposures and profile P 1600 (c. 15–30 km) certainly indicates allochtonous and accreted igneous rocks forming the outer forearc and the San Pedrillo Unit is also present NW of the Osa Peninsula. Failure of offshore seismic imaging probably results from small velocity and structural contrasts between the units. Location of possibly active fault zones along the edge of the units indicates the compressive regime associated to the incoming of the Cocos Ridge (e.g. Kolarsky et al. 1995a; Morell et al. 2008) or, more generally, topographic irregularities on the top of the oceanic floor, develop faulting preferentially along weaknesses in the crust, such as ancient sutures in the forearc basement. The landscape morphology in the Osa Peninsula is representative of an overthrust of landward units onto seaward units. Although formation of fault-bounded blocks in the outermost southern Costa Rican forearc is triggered by vertical tectonics associated with the topography of the subducting plate (e.g. Sak et al. 2004, 2009; Vannucchi et al. 2006), this mechanism alone does not account for the existence of recent and/or active large inverse fault zones along unit boundaries. Hence, similarly to deformation patterns of the southern Costa Rican innermost forearc (Fila Costeña), we suggest here that block faulting is related to compressive stress in the outermost forearc, in response to the subducting Cocos Ridge. Inverse faulting preferentially develops along weak crustal zones. In the Osa Peninsula these zones correspond to boundaries between accreted units and, hence, to palaeo-décollements exposed in the forearc basement. A direct implication of these observations is that recognition of fault-bounded

blocks by remote imaging is a very useful tool for geological mapping of the South Central American forearc, because faulting in the igneous basement tends to develop along unit boundaries.

Argon loss induced by tectonics?

In the Inner Osa Igneous Complex ages of lava formation are given by sediments and $^{40}Ar/^{39}Ar$ dates. Coniacian–Santonian (c. 89–84 Ma) siliceous pelagic sediments are interbedded with the lavas in several places. Two basalts from the Burica Peninsula and Isla Violín gave dissimilar $^{40}Ar/^{39}Ar$ ages of 54.5 ± 1.5 Ma (Late Paleocene–Early Eocene) and 64.2 ± 1.1 Ma (Early Paleocene), respectively (total fusion on whole rock matrix) (Hoernle et al. 2002).

It may be considered that occurrence of basalts formed at different times is not a big issue in a composite complex, but it is troubling that (1) there is no overlap between the radiometric and biochronological ages and (2) the radiometric method provides the youngest ages. Furthermore, the 54.5 ± 1.5 Ma (Late Paleocene–Early Eocene) intra-oceanic basalt sampled in the Burica Peninsula is stratigraphically covered by the c. 56–59 Ma (Late Paleocene) arc-derived sediments of the Pavones Formation. Older ages of the overlap sediments indicate that the $^{40}Ar/^{39}Ar$ age does not represent an age of formation. As a consequence, the basalt certainly experienced Ar loss after its formation. This loss may be related to syn- or postaccretion tectonics and deformation. Hence, it appears clearly the age of formation of the lava is more efficiently defined by the biochronological dates in the Osa Igneous Complex.

Figure 13 illustrates a comparison between radiometric ages of the igneous rocks and ages of corresponding units based on palaeontological data. Although there is a large variability in the biochronological ages due to heterogeneity of the unit and/or poor age accuracy of the fossil assemblages,

---

Fig. 11. SEM illustrations of Late Cretaceous and Paleogene Radiolaria from the Osa Peninsula, Costa Rica. The numbers of illustrations in this plate correspond to those in Figure 9. Marker = 100 μm, except for (25), where marker equals 50 μm. (1) *Alievium* sp., sample DB02-199. (2) *Halesium amissum* (Squinabol), sample DB02-199. (3) *Acanthocircus* sp. b., sample DB02-199. (4) *Pseudoaulophacus* sp., sample DB02-199. (5) *Archaeospongoprunum bipartitum* Pessagno, sample DB02-199. (6) *Praeconocaryomma universa* Pessagno, sample DB02-199. (7) *Napora* sp., sample DB02-199. (8) *Rhopalosyringium* sp., sample DB02-199. (9) *Theocampe* aff. *urna* (Foreman), sample DB02-199. (10) *Theocampe salillum* (Foreman), sample DB02-199. (11) *Tugurium* sp., sample DB02-199. (12) *Pseudodictyomitra* sp., sample DB02-199. (13) *Dictyomitra formosa* Squinabol, sample DB02-199. (14) *Amphipternis stocki* (Campbell and Clark), sample DB02-199. (15) *Eostichomitra* sp., sample DB02-199. (16) *Alievium* aff. *gallowayi* (White), sample FBJ 90174. (17) *Dictyomitra formosa* Squinabol, sample FBJ 90174. (18) *Archaeospongoprunum* sp., sample FBJ 90174. (19) *Crucella cachensis* Pessagno, sample FBJ 90174. (20) *Praeconocaryomma universa* Pessagno, sample FBJ 90174. (21) *Gongylothorax verbeeki* (Tan Sin Hok), sample FBJ 90174. (22) *Theocampe tina* (Foreman), sample FBJ 90174. (23) *Theocampe urna* (Foreman), sample FBJ 90174. (24) *Acanthocircus yaoi* (Foreman), sample FBJ 90174. (25) *Lithatractus pusillus* (Campbell and Clark), sample FBJ 90174. (26) *Amphipternis stocki* (White), sample GDM 9116. (27) *Theocampe urna* (Foreman), sample GDM 9116. (28) *Praeconocaryomma universa* Pessagno, sample GDM 9116.

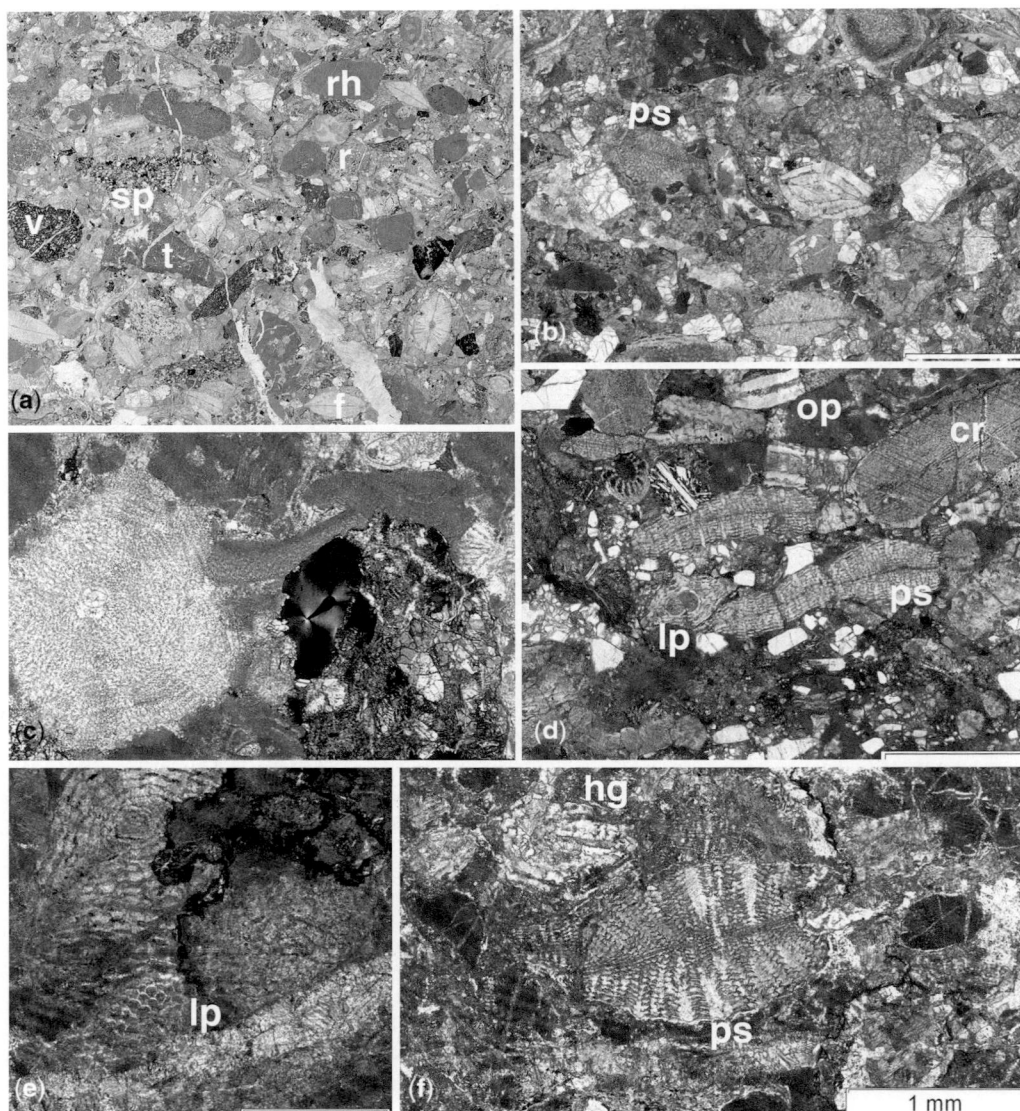

Fig. 12. Transmitted light photomicrographs of thin sections of Late Eocene shallow-water limestone clasts from the San Pedrillo Unit, Osa Peninsula. (**a**) Calcarenite (Punta Campanario, Osa), with abundant larger foraminifera (f), minor red algae rodoliths (rh), green volcanic tuffites (t), dark volcanics (v) and angular feldspars set in a calcareous matrix. In addition there are rare radiolarite (r) and spicultite (sp) clasts. Punta Campanario, Osa. (**b**) Packstone with fragments of larger foraminifera, red algae, crinoids and volcaniclastic material. Ps, oblique and vertical section of *Pseudophragmina*. Nm, small globular Nummulites. sample DB02-026. Punta Campanario, Osa. (**c**) Packstone with *Lithotamnium*, larger foraminifera and volcanic lithoclasts. Ds, equatorial section of *Discocyclina* with a megalospheric embryon of semi-isolepidine type. (**d**) Packstone with fragments of vertical sections of *Pseudophragmina* (ps), *Lepidocyclina* (lp), *Operculinoides* (op) and crinoids (cr). (**e**) Packstone with larger forams. Lp, fragments of equatorial section of *Lepidocyclina* showing the arcuate shape of the equatorial chambers. Sample CM-Caño98F from the Caño Island. (**f**) Bioclastic packstone with larger foraminifera and abundant fragments of echinoderms both very recrystallized. The clasts show mechanic erosion. ps, *Pseudophrágmina* in vertical section. hg, *Helicostegina* in oblique section. Sample: POB 3060 coastal outcrops of Llorona, Corcovado.

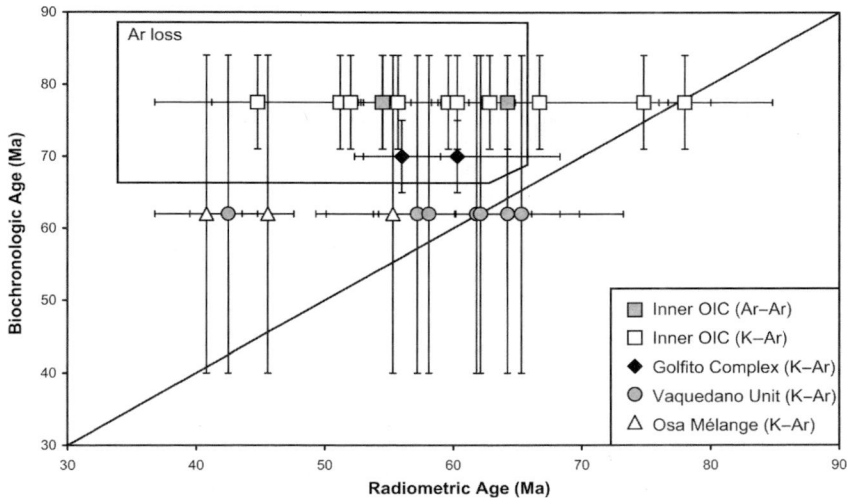

Fig. 13. Radiometric v. biochronological ages for some igneous samples of the Osa Igneous Complex and Osa Mélange, with error bars. Biochronological ages are inferred from the tectono-stratigraphy as defined in this study. Whole rock K–Ar ages are from Berrangé *et al.* (1989) and whole rock $^{40}Ar/^{39}Ar$ total fusion isochron ages from Hauff *et al.* (2000) and Hoernle *et al.* (2002). Large variability in the biochronological ages is due to large range of formation ages for some units and/or poor age accuracy of fossil assemblages. Some samples have inconsistent biochronological and radiometric ages within the incertitude range that is attributed to Ar loss.

many samples from the Inner Osa Igneous Complex and Golfito Complex (considered here solely for the Ar-loss issue) have lower radiometric than biochronological ages. Moreover, we note on Figure 13 that mean ages are globally shifted to the left of a line of similar biochronological and radiometric ages. This indicates that there is a tendency of Ar loss among dated samples. In the view of the geological heterogeneity and tendency of Ar loss in the dated samples, it appears that good agreement between $^{40}Ar/^{39}Ar$ ages and the range of K–Ar ages as pointed out by Hoernle & Hauff (2007) is most likely a result of randomness rather than a convergence of radiometric ages obtained by different methods.

Interestingly, lowest radiometric ages point toward a *c.* 40 Ma (Middle Eocene) age (Fig. 13). This age corresponds broadly to the age of accretion of the Outer Osa Igneous Complex (see below), potentially indicating that massive Ar loss from the igneous rocks may occur in response to increased tectonics along the margin, such as that induced by seamount accretion.

In the light of preceding remarks, it appears that $^{40}Ar/^{39}Ar$ or K–Ar ages have to be interpreted with much caution in the case of the Osa Igneous Complex. Hence, we choose here to consider radiometric dates to represent minimal ages of formation only. Biochronological and radiometric discrepancies pointed out in this study further raise an important question about the interpretation of ages determined by $^{40}Ar/^{39}Ar$ dating for other (low-K)

basaltic samples reported for South Central America (e.g. Sinton *et al.* 1997; Hauff *et al.* 2000; Hoernle *et al.* 2002, 2004). These ages should be used with much caution and controlled by stratigraphic data.

Origin and significance of the units

The units forming the Osa Igneous Complex consist of mostly unmetamorphosed accreted fragments of igneous rocks that range from a few hundred metres to several kilometres in size. In this section, we make a summary of the units and discuss their probable origin and implication in terms of accretionary processes.

One of the most intriguing features of the Osa Igneous Complex is the low occurrence of intercalated arc-derived or continental sediments. Such detrital deposits are very commonly observed in other accretionary complexes (e.g. Isozaki *et al.* 1990; Dickinson 2008). Furthermore, with the exception of the Güerra Unit, the Osa Igneous Complex has apparently not suffered metamorphism (greenschist facies and higher) while metamorphism is very commonly observed in accreted rocks (e.g. Maruyama & Liou 1989; Isozaki *et al.* 1990; Dickinson 2008). Accretionary complexes predominantly composed of poorly metamorphosed igneous rocks are scarce and examples from the Alexander Island in Antarctica (Doubleday *et al.* 1994) and Northern California Coast Ranges

(Shervais *et al.* 2005) show that greenschists may locally represent a significant portion of the accreted material. Indeed, to our knowledge, oceanic terranes of the Solomon Islands (e.g. Petterson *et al.* 1997, 1999) are the sole known occurrence of an accretionary complex similar to the Osa Igneous Complex, that is essentially composed of oceanic sequences devoid of significant metamorphism and trench-fill sediments. Below, we point out other similarities between these two complexes.

The Inner Osa Igneous Complex. The Inner Osa Igneous Complex predominantly comprises Coniacian–Santonian (*c.* 89–84 Ma) tholeiitic basalts with plateau-like, oceanic intraplate affinities. Minor NMORB-like and OIB-like igneous rocks occur as sills, small intrusions and, possibly, interlayered lava flows. The volcano-sedimentary sequences are almost devoid of detrital material and, hence, indicative of submarine volcanism. Rare, thin interbeds of siliceous pelagic sediments point toward high effusive rates for the lavas. Absence of carbonate banks indicates that the lavas probably formed below the CCD. Detrital sediments of the Esperanza Formation are interpreted to be a product of submarine erosion of the plateau-like lavas prior to the accretion. Consistency of rocks forming the Inner Osa Igneous Complex tends to indicate that they initially belonged to a unique volcanic edifice, most probably an oceanic plateau.

The Azuero Plateau (Azuero Complex, western Panama) is made up of tholeiitic, plateau-like basalts and minor intercalations of siliceous pelagic sediments. It represents a Coniacian–Early Santonian (*c.* 89–85 Ma) oceanic plateau that forms the arc basement (Buchs *et al.* in prep.). On this basis, it appears the Inner Osa Igneous Complex and Azuero Plateau are highly similar and could be considered to have the same origin. This interpretation has to be discarded because: (1) dissimilarities exist in terms of sedimentary facies and faunal assemblages between the two units; (2) the Inner Osa Igneous Complex contains NMORB-like and OIB-like igneous rocks not observed yet in the Azuero Plateau; (3) lavas from both units have dissimilar ranges of Mg# (Buchs 2008); and (4) proto-arc igneous rocks crosscut the Azuero Plateau, but have not been observed in the Osa Igneous Complex. Absence of proto-arc dykes in the Inner Osa Igneous Complex which are frequently seen in the Azuero Plateau is a key feature that points toward an allochthonous, Pacific origin for the Inner Osa Igneous Complex. Hence, we propose here that the sequences of the Inner Osa Igneous Complex were part of a Pacific oceanic plateau distinct from the CLIP, which formed prior arc initiation in South Central America and was later accreted to the South Central American Arc and CLIP-related basement.

It is important to note that, despite striking geochemical and age similarities between the Azuero Plateau and Inner Osa Igneous Complex, the former is strictly part of the CLIP whereas the later is exotic and unrelated to the CLIP. Hence, it appears that systematic association of Central American exposures of oceanic plateau with the CLIP on the basis of geochemistry and radiometric dating only (e.g. Sinton *et al.* 1998) is inappropriate. Our results suggest that possible genetic links of plateau-like igneous complexes with the CLIP should be carefully constrained by tectono-stratigraphy and field observations.

The Malaita accretionary prism (Solomon Islands), known to have formed by accretion of the subducting Ontong–Java Plateau (Petterson *et al.* 1999; Mann & Taira 2004; Phinney *et al.* 2004; Taira *et al.* 2004), shares many similarities with the Inner Osa Igneous Complex: (1) accreted rocks are in a very-low metamorphic facies (Petterson *et al.* 1997); (2) occurrences of trench-fill deposits lack (Petterson *et al.* 1997); (3) sequences of the Malaita Volcanic Group (Malaita Island) contain predominantly igneous rocks with plateau affinities and minor interbeds of siliceous sediments (Petterson *et al.* 1997); (4) sequences partly overturned during accretion and development of fault-propagation folds (Petterson *et al.* 1997; Phinney *et al.* 2004); and (5) the Makira Terrane (Makira Island) contains igneous rocks having NMORB signatures interbedded with plateau igneous rocks (Petterson *et al.* 1999). These similarities are another argument for an accreted oceanic plateau origin for the Inner Osa Igneous Complex. Thick (900–1770 m) pelagic limestones observed on the top of the Malaita Volcanic Group (e.g. Petterson *et al.* 1997) have no apparent equivalent in the Inner Osa Igneous Complex. We interpret this dissimilarity as a possible consequence of: (1) distinct travel times of the two oceanic plateaus on the oceanic floor before their accretion (i.e. <25 Ma for the Inner Osa Igneous Complex and *c.* 94–105 Ma for the Ontong–Java Plateau (Petterson *et al.* 1999, and references therein), which resulted in few occurrences and preservation of calcareous sedimentary covers in the Inner Osa Igneous Complex; and/or (2) summital areas of the Inner Osa Igneous oceanic plateau were below the CCD, pointing toward a lower crustal thickness for the Inner Osa Igneous Plateau or distinct palaeo-oceanographic conditions.

The Güerra Unit. The Güerra Unit is characterized by the highest metamorphic and deformation grades observed in the Osa Igneous Complex and Osa Mélange. It marks the contact between the Inner and Outer Osa Igneous Complexes and comprises various rock bodies, similar to the Riyito and Vaquedano units in terms of igneous geochemistry and sediment facies. The Güerra Unit is exposed along the Outer Osa Igneous Complex but is apparently lacking along the contact between the Outer Osa Igneous Complex and the San Pedrillo Unit (Fig. 2b). This geometrical arrangement and compositional similarities with some units of the Outer Osa Igneous Complex indicate that the Güerra Unit formed after the formation of the Inner Osa Igneous Complex and prior to the emplacement of the Outer Osa Igneous Complex. Along-strike disappearance of the unit may be an artefact due to poor exposures in southern and northern Osa or linked to tectonic erosion at some places along the margin, prior to the accretion of the San Pedrillo Unit. Compositional heterogeneity of the Güerra Unit indicates it was likely triggered by low rates of accretion and several seamount subductions before the emplacement of the Vaquedano, Ganado and Riyito Units. In this interpretation, the Güerra Unit remained close to the décollement zone for a relatively long time, potentially recording some of the processes that occur at the interface between the overriding and subduction plates. As suggested in the case of sediment underplating (Moore 1989), long duration of residence close to the décollement may result in strong deformation of accreted material, similar to that of the Güerra Unit. Deformation and/or fluid flows through the unit may have been a source of metamorphic catalysis, in a similar way than formation of metabasalts in the Franciscan complex (e.g. Nelson 1995).

The Ganado Unit. The Ganado Unit is composed of Coniacian–Santonian plateau-like and NMORB-like tholeiites emplaced in an oceanic intraplate setting. The plateau-like lavas have lower Nb–Y ratios than similar rocks of the Inner Osa Igneous Complex, indicating that these two units have dissimilar origins (Buchs 2008). The Ganado Unit may represent a large fragment of accreted seamount that originally developed in submarine conditions on the Pacific Plate. The occurrence of large intrusive rocks in the unit likely indicates that the seamount underwent a severe dismemberment in the vicinity of the trench as a response to slab flexuration (e.g. Kobayashi *et al.* 1987) and/ or during the accretion due to coupling between the overriding and subducting plates. Dismemberment of the seamount in the trench environment may have caused a removal of pelagic sediments initially capping the edifice, presently lacking in the unit.

The Riyito Unit. The Riyito Unit is mostly composed of Paleocene or older (?) highly depleted tholeiites. It represents an assemblage of various seamount fragments. Peculiar geochemical affinities indicate that these seamounts were generated very close to a mid-ocean ridge (Buchs 2008). Although the top of the seamounts may have reach relatively shallow water, allowing formation of highly vesiculated hyaloclastite, all accreted lavas and detrital material issued from the erosion of the seamounts were emplaced in submarine conditions. Similarly to the Ganado Unit, pelagic sediments deposited on the top of the seamounts before their arrival at the subduction zone may have been removed close to the trench through gravitational collapsing.

The Vaquedano Unit. The Vaquedano Unit principally comprises Campanian to Eocene alkali-transitional basalts with OIB-like signatures. These igneous rocks are an example of igneous rocks generally regarded as typical 'Oceanic Island Basalts'. Geochemical and age variations indicate the unit is composed of an imbricate of relatively thin (<250 m) volcano-sedimentary sequences with distinct origins. The unit is devoid of shallow-water limestones, showing the accreted rock bodies initially formed in submarine environments.

Interbeds of pelagic limestones are common among the lavas of the Vaquedano Unit. Alkaline igneous rocks and frequent sedimentary intercalations indicate low eruption rates, generally associated with the latest stage of eruption of large oceanic intraplate volcanoes (e.g. Clague & Dalrymple 1987). Alternatively some seamounts of the Vaquedano Unit were possibly small edifices, which never experienced intense volcanism. In both cases, the Vaquedano Unit comprises superficial layers of seamounts peeled off at shallow-level during subduction.

Important differences exist between the Vaquedano Unit and the Riyito Unit, Ganado Unit and Inner Osa Igneous Complex: (1) accreted seamount fragments are much smaller in the Vaquedano Unit; (2) detachment of rock bodies was easier in the Vaquedano Unit, due to the occurrence of weak layers in the superficial parts of the subducting seamounts; (3) the Vaquedano Unit has more heterogeneous geochemical affinities; and (4) ages of the accreted material spans a longer period of time (*c.* 40 Ma). External position of the Vaquedano Unit (i.e. in contact with the Osa Mélange) may be attributed to an emplacement during the latest stage of growth of the Osa Igneous Complex. These observations tend to indicate that the Vaquedano Unit has a particular significance in terms of accretionary processes. We propose this unit marks a transition between two subduction regimes: (1) a regime characterized by strong coupling between overriding and underlying plates and accretion of thick/large rock bodies proceeding from subducting seamounts (Ganado and Riyito units); and (2) a regime characterized by a lower coupling and/or an absence of large seamount subduction leading to little accretion or tectonic erosion. Similarities in the size of accreted rock bodies and composition between the Vaquedano and Güerra Units arise very likely from the transitional character of these two units in terms of subduction regime. We suggest the Güerra Unit marks a transition from little seamount accretion to seamount accretion and the Vaquedano Unit from seamount accretion to little seamount accretion (see also below). The Vaquedano Unit has remained mostly undeformed and poorly metamorphosed in comparison to the Güerra Unit. This may be due to a rapid emplacement of the Osa Mélange on the

edge of the Vaquedano Unit and to a shorter time of residence of the Vaquedano Unit in the vicinity of the décollement relatively to the Güerra Unit.

Construction of the Osa Igneous Complex

Our model for the construction of the Osa Igneous Complex is principally constrained by the tectono-stratigraphy of the area (Fig. 14), assuming that the complex did not suffer from post-accretion reorganization due to along-strike displacements. There are several regional, stratigraphic and structural aspects that have been taken into account to constrain the model: (1) accretion may have occurred since the Late Campanian only (*c.* 73–70 Ma), after the

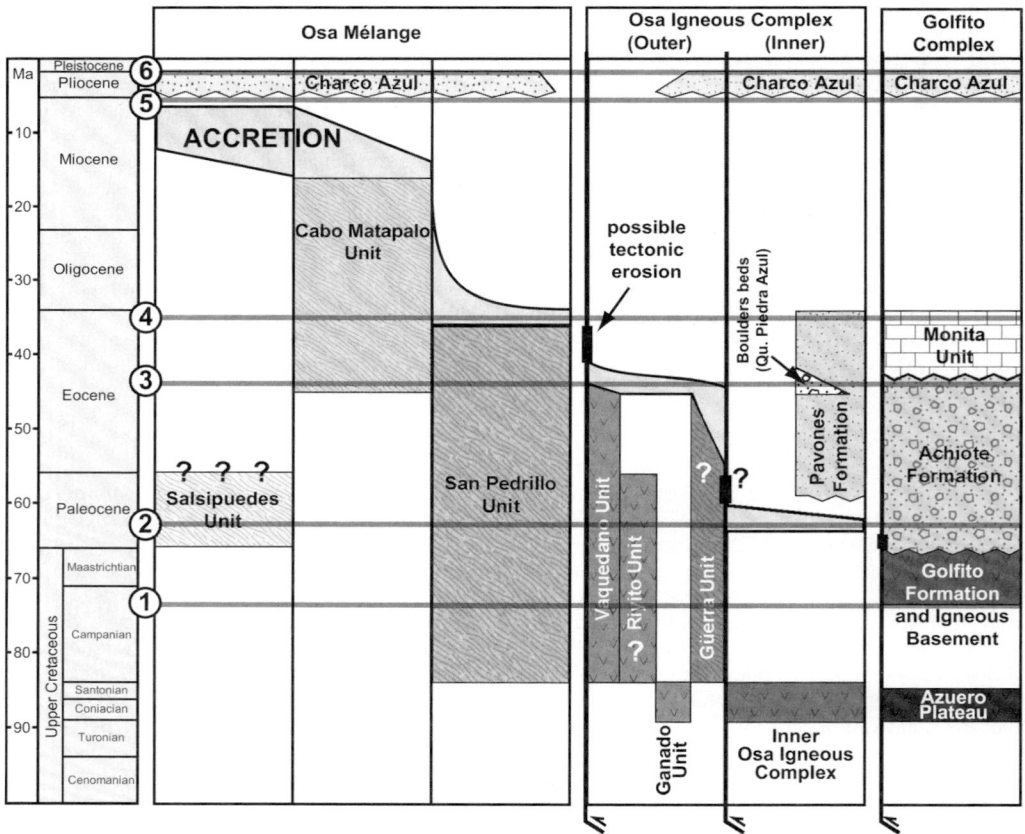

Fig. 14. Tectono-stratigraphy of the Osa and Burica Peninsulas (partly based on data from Berrangé & Thorpe, 1988; Coates *et al.* 1992; Di Marco *et al.* 1995). Horizontal lines represent major tectonic events along the South Costa Rican margin: (1) subduction initiation (Buchs *et al.* in prep.); (2) accretion of the Inner Osa Igneous Complex; (3) accretion of the Ganado and Riyito Units; (4) possible seamount subduction leading to formation of the San Pedrillo Unit; (5) subsidence of the margin (unidentified origin, tectonic erosion?); and (6) possible incoming of the Cocos Ridge. Grey areas represent age ranges of possible accretion for corresponding units. Note that ages of the accreted units roughly define a development of the margin toward the trench through time. Dark bars represent age ranges of possible tectonic erosion. Succinct definition of the formations and units is provided in Figure 2.

initiation of the subduction in southern Central America (Buchs *et al.* in prep.); (2) the oldest possible age of accretion of a volcano-sedimentary sequence is defined by the youngest age of formation of the sequence; (3) units in external (seaward) position accreted after units in internal (landward) position; (4) overlap sequences (i.e. arc-derived sediments unconformably resting on accreted rocks) define a minimal age of accretion of underlying accreted rocks; (5) tectonic erosion may possibly occur between periods of accretion; and (6) deformed, metamorphosed volcano-sedimentary sequences remained longer in the vicinity of the décollement than undeformed, unmetamorphosed volcano-sedimentary sequences. Growth of the margin by successive stacks of rock bodies is well documented in the Osa area by a chiefly progressive decline of ages toward the trench (Fig. 14).

Construction of the Osa Igneous Complex may be summarized as follows. Subsequent to subduction initiation along the southwestern Caribbean Plate in the Late Campanian (event 1 on Fig. 14), portions of arc basement and proto-arc were removed from the margin by subduction erosion. This resulted in progressive migration of the trench toward the Caribbean. Because of this erosion the Golfito Complex was located close to the trench in the Early Paleocene and sediments proceeding from the arc were entirely subducted (Fig. 15a). In the Early–Middle Paleocene the Inner Osa Igneous Complex formed by partial accretion of a subducting oceanic plateau (Fig. 15b, event 2 on Fig. 14). Subsequent to emplacement of the Inner Osa Igneous Complex, possible tectonic erosion and/or subduction of the oceanic plateau caused the outer margin to subside, allowing deposition of arc-derived hemipelagic sediments of the Late Paleocene Pavones Formation (Fig. 15b). From the Late Paleocene to the Early Eocene, the subduction regime was characterized by low rates of superficial accretion of small seamounts that led to the formation of the Güerra Unit and continuous deposition of the Pavones Formation on the forearc slope (Fig. 15c). In the Middle Eocene, a group of Late Cretaceous–Paleocene seamounts entered the subduction zone and accreted, resulting in the formation of the Ganado and Riyito units (Fig. 15d, event 3 on Fig. 14). Seamount accretion squeezed the previously formed Güerra Unit between the Inner Osa Igneous Complex and newly accreted Ganado and Riyito units, preserving it from subsequent tectonic erosion. This event may have been recorded in the Pavones Formation by deposition of the limestone boulders found in the Quebrada Piedra Azul (Di Marco *et al.* 1995; Fig. 14). The area comprising the Golfito Complex was possibly uplifted, partly eroded and subsided, leading to subsequent deposition of peri-platform deposits of the Monita Unit (Fig. 10, Buchs 2008). After this event, the subduction regime was again characterized by low rates of accretion with possible tectonic erosion that led to the construction of the Vaquedano Unit and removal of some portions of the margin (Fig. 15e).

In summary, we observe the construction of the Osa Igneous Complex by multiple events of high-rate accretion alternating with low-rate accretion and/or tectonic erosion. Although this is broadly similar to other well-studied convergent margins (e.g. North America: Byrne & Fisher 1987; Dickinson 2008; Japan: Isozaki *et al.* 1990) the Osa Igneous Complex is characterized by pulses of high-rate accretion, or episodic accretionary events, that are associated with the emplacement of fragments of an oceanic plateau and seamounts. Such events may be facilitated by incoming of large topographic highs into the subduction zone and, possibly, regional tectonic changes in the Middle Eocene that resulted in a stronger coupling between the overriding and subducting plates (Buchs 2008). Arc-derived sediments lack in the Osa Igneous Complex. This indicates the sedimentary supply in the trench remained low or the convergence rate between the overriding and subducting plates was sufficiently high to allow entire subduction of the trench-fill deposits (see discussion on the San Pedrillo Unit below). Whether these characteristics are symptomatic of an erosive or accretionary margin, they are discussed below along with constraints from the Osa Mélange.

Construction of the Osa Mélange

The San Pedrillo Unit: an unusual polygenetic mélange

The San Pedrillo Unit is a complex of lenticular, dismembered sequences composed of variously clastic sediments (Fig. 8). Although the sediments may contain blocks very similar to portions of the Osa Igneous Complex, some components such as shallow-water limestones, differentiated igneous rocks, ash deposits or finer hemipelagic sediments are clearly lacking in the Osa Igneous Complex. The unit appears to be derived from a volcanic arc. In a thin, *c.* 100 m thick layer along the Osa Igneous Complex, the San Pedrillo Unit tectonically incorporates some material of the igneous complex. Extension of this tectonic incorporation is, however, very limited in comparison to the size of the San Pedrillo Unit. Hence, in regard to ubiquitous occurrence of sediments forming the San Pedrillo Unit and poor tectonic incorporation of the Osa Igneous Complex at the contact with the mélange, it appears the Osa Mélange was not simply produced

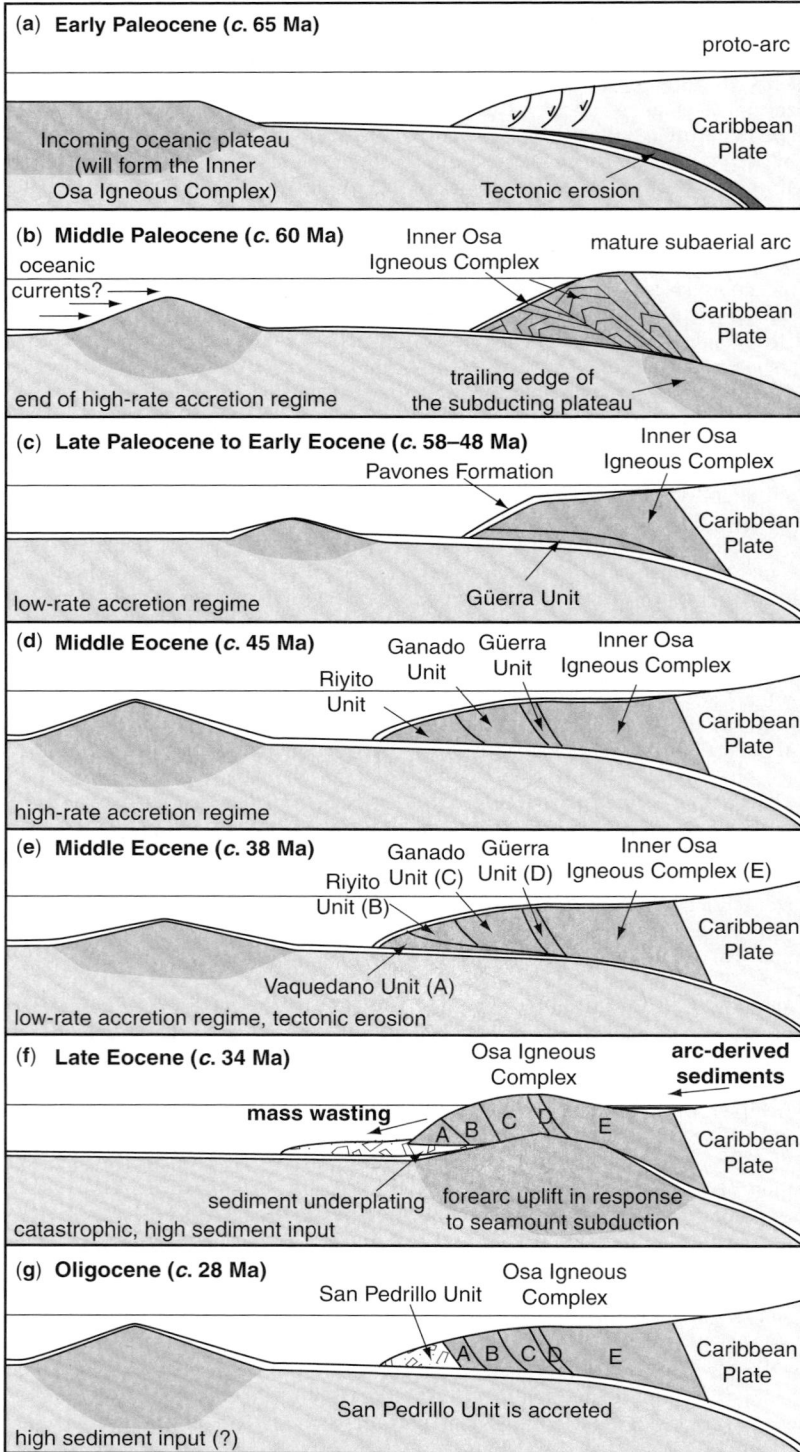

(a) Early Paleocene (*c.* 65 Ma)

proto-arc

Incoming oceanic plateau
(will form the Inner
Osa Igneous Complex)

Tectonic erosion

Caribbean
Plate

(b) Middle Paleocene (*c.* 60 Ma)

Inner Osa
Igneous Complex

mature subaerial arc

oceanic
currents?

Caribbean
Plate

end of high-rate accretion regime

trailing edge of
the subducting plateau

(c) Late Paleocene to Early Eocene (*c.* 58–48 Ma)

Inner Osa
Igneous Complex

Pavones Formation

Caribbean
Plate

low-rate accretion regime

Güerra Unit

(d) Middle Eocene (*c.* 45 Ma)

Ganado
Unit

Güerra
Unit

Inner Osa
Igneous Complex

Riyito
Unit

Caribbean
Plate

high-rate accretion regime

(e) Middle Eocene (*c.* 38 Ma)

Ganado
Unit (C)

Güerra
Unit (D)

Inner Osa
Igneous Complex (E)

Riyito
Unit (B)

Caribbean
Plate

Vaquedano Unit (A)

low-rate accretion regime, tectonic erosion

(f) Late Eocene (*c.* 34 Ma)

Osa Igneous
Complex

**arc-derived
sediments**

mass wasting

A B C D E

Caribbean
Plate

sediment underplating

forearc uplift in response
to seamount subduction

catastrophic, high sediment input

(g) Oligocene (*c.* 28 Ma)

Osa Igneous
Complex

San Pedrillo Unit

A B C D E

Caribbean
Plate

San Pedrillo Unit is accreted

high sediment input (?)

Fig. 15. (*Continued*).

by tectonic, or basal erosion of the Osa Igneous Complex as suggested by Meschede *et al.* (1999).

Vannucchi *et al.* (2006, 2007) proposed that the rocks forming the San Pedrillo Unit are dominated by basalt, chert and shallow-water limestone resulting from accretion of seamounts. In this interpretation, the Osa Mélange is a dismembered unit characterized by a pervasive metamorphism (partly in greenschist facies) that formed by accretion at depth of a 'classic sequence of oceanic crust'. According to the same interpretation the material forming the San Pedrillo Unit consists of uppermost crustal sequences initially part of a 'seamount system' that accreted to form the Osa Igneous Complex (Vannucchi *et al.* 2006, p. 16, fig. 14). This interpretation is inconsistent with our results for several reasons. Our mapping shows the San Pedrillo Unit comprises detrital sediment that incorporates various amounts of clasts ranging from grains to large boulders. This is atypical in classic sequences of Pacific oceanic crust. Furthermore, rocks forming the San Pedrillo Unit are not dominated by basalt, chert and shallow-water limestones but by fine-grain detrital sediments and olistostromic deposits (Fig. 4). A significant portion of the material forming the San Pedrillo Unit proceeded from a volcanic arc. Arc-related and seamount-related olistoliths sedimentarily embedded in the same debris flow deposits are observed throughout the San Pedrillo Unit, indicating that the mélange is not a product of the dismemberment/erosion of the Osa Igneous Complex only. The Osa Igneous Complex is not a 'seamount system' accreted at the same time of the San Pedrillo Unit but is a composite accretionary complex that developed between the Early–Middle Paleocene and Middle Eocene. Hence, accretion of the bulk of the Osa Igneous Complex cannot have triggered the formation of the San Pedrillo Unit. Although some deformed rocks are green due the presence secondary chlorite, we have not observed metamorphosed rocks in the greenschist facies. Hence, it seems the San Pedrillo Unit remained at relatively shallow depth during its formation.

We propose here the occurrence of highly deformed sedimentary deposits throughout the San Pedrillo Unit is representative of a polygenetic mélange, with a block-in-matrix texture primarily controlled by clastic sedimentation and subsequently deformed during accretion. Numerous subduction mélanges around the world are characterized by hard metamorphic blocks embedded in a weaker metamorphosed matrix generally made up of shales or serpentinite (e.g. in the Franciscan Complex, Dickinson 2008, and references therein; Kodiak Convergent Margin, Byrne & Fisher 1987; Japanese convergent margin, Isosaki *et al.* 1990; Alexander Island of Antartica, Doubleday *et al.* 1994; Lichi Mélange of Taiwan, Chang *et al.* 2000 and Burma-Java Subduction Complex, Pal *et al.* 2003). Unlike these mélanges, the San Pedrillo Unit is predominantly composed of unmetamorphosed hard blocks embedded in an unmetamorphosed hard matrix. This specificity of the San Pedrillo Unit with respect to other mélanges worldwide is further discussed in the following sections.

The sedimentary record in the San Pedrillo Unit

Origins of the sediments. We have shown the San Pedrillo Unit is composed of three main types of lithologies that form the bulk of the inner Osa Mélange: (1) sedimentary deposits containing shallow-water limestones; (2) debris flow deposits dominated by igneous olistoliths; and (3) fine-grain detrital sediments (Fig. 4). All these lithologies are characterized by a detritic, clastic component that ranges from grains to large boulders embedded within a finer sedimentary matrix. On the basis of composition of the matrix and clasts, three main, distinct sources are recognized for the material composing the San Pedrillo Unit: (1) a palaeo-Osa Igneous Complex, (2) a subaerial volcanic arc and (3) pelagic–hemipelagic sediments.

Striking similarities exist between the basaltic olistoliths embedded in the sedimentary matrix of the San Pedrillo Unit and the volcano-sedimentary sequences of the Osa Igneous Complex. These similarities are well defined in terms of geochemistry, age (Late Cretaceous to Middle Eocene), faunal

Fig. 15. Model of accretionary processes in the Osa transect of the mid-American Trench (Middle Campanian to Oligocene, not to scale). (**a**) Margin just after the subduction initiation, undergoing overall tectonic erosion. (**b**) Accretion by large-scale duplexing of pieces of a Coniacian–Santonian oceanic plateau; initiation of the construction of the Osa Igneous Complex. (**c**) Period of low-rate accretion; formation of the metamorphic, highly composite Güerra Unit. (**d**) Various fragments of large seamounts accrete and squeeze the Güerra Unit onto the Inner Osa Igneous Complex; Ganado and Riyito Units are emplaced. (**e**) Period of low-rate accretion and intermittent tectonic erosion; formation of the highly composite Vaquedano Unit. (**f**) a seamount subduction causes a strong uplift of the forearc, leading to the gravitational collapse of the Palaeo–Osa Igneous Complex and large production of sediments. Large volumes of sediment proceeding form the forearc quickly deposit into the trench and lead to the formation of the San Pedrillo Unit. (**g**) Possible growth of the San Pedrillo Unit by successive accretion of sediments derived from the forearc and arc areas.

assemblages encountered in the sediments associated with the lava flows and facies of the sediments. Although the Osa Igneous Complex is a composite unit made up of a wide range of Late Cretaceous to Middle Eocene igneous and sedimentary rocks, some of these rocks have very unusual geochemical affinities or contain very particular fossils such as Crustacean microcoprolites. Indeed, similarities between the San Pedrillo basaltic olistoliths and the Osa Igneous Complex are very specific in some cases. Hence, the bulk of the basaltic olistoliths encountered in the San Pedrillo Unit were very likely originally part of the Palaeo-Osa Igneous Complex. The sedimentary mode of emplacement of the olistoliths is well constrained by the sedimentary fabric recognized throughout the San Pedrillo Unit and points toward mass wasting along the margin by the time of formation of the Osa Mélange. Parts of the finer detrital material of the San Pedrillo Unit (e.g. unidentifiable rock fragments and crystals) may also be sedimentarily related to the Palaeo-Osa Igneous Complex.

A subaerial volcanic arc is believed to have provided significant parts of the material forming the San Pedrillo Unit. Arc-derived material consists of fragments (grains to large boulders) of felsic volcanic rocks and reworked and interlayered Late Eocene shallow-water limestones, as well as dark ashy tuffites in the mélange matrix. The association of dark ashy tuffites with fragments of shallow-water limestones form one of the three principal lithological associations recognized in the San Pedrillo Unit, described here as the 'sedimentary deposits containing shallow-water limestones' (Fig. 4). This recurrent rock association further constrain the origin of the shallow-water limestones to an arc environment. The felsic and basaltic olistoliths are locally associated within the same debris flow deposits, indicating that these rocks proceeded from a similar palaeo-environment. This is another argument pointing toward a Palaeo-Osa Igneous Complex for the bulk of the basaltic olistoliths.

The pelagic–hemipelagic sediments consist of undated siliceous–calcareous hemipelagic oozes intercalated with fine-grained detrital sediments and deformed lenses of Early Eocene radiolarian cherts. We relate the hemipelagic oozes to background sedimentation commonly observed in near-trench environments (Carter 1979). The detritic fraction variously expressed in these sediments has a composition pointing toward a predominant basaltic source and a minor differentiated source that may represent an erosion of the Palaeo–Osa Igneous Complex and a subaerial volcanic arc, respectively. The Early Eocene radiolarian cherts are devoid of arc-derived material and, thus, may have originated in a pelagic environment, presumably on the slope of a seamount or on the ocean floor. Scarce igneous olistoliths associated to Campanian–Maastrichtian radiolarites (not found yet in the Osa Igneous Complex), may have been tectonically incorporated from subducting seamounts into the San Pedrillo Unit in a process similar to that proposed by Okamura (1991) for some mélanges in the Mino Terrane (Japan).

Emplacement of the sediments into the trench. The sedimentary record preserved in the San Pedrillo Unit is a highly valuable source of information, allowing a better understanding of the southern Costa Rican margin at the time of the mélange formation (Fig. 16).

Sediments of the San Pedrillo Unit are indicative of a significant detrital input variously expressed among the lithologies, which culminates in the debris flow deposits and the sedimentary deposits containing shallow-water limestones. Large boulders of igneous rocks derived from both the Palaeo–Osa Igneous Complex and a subaerial volcanic arc were embedded together in thick debris flow layers prior to accretion of the mélange. Hence, the San Pedrillo Unit was partly produced by removal of large portions of the forearc area through mass wasting. At the time of deposition of the sediments, the Palaeo–Osa Igneous Complex was presumably undergoing a strong uplift that led to disaggregation of unstable slopes by gravity. The material was gravitationally driven to the trench, partly incorporating arc-derived sediments transiting through by-passes, and ultimately formed thick pile of sediments close to the subduction zone. The carbonate material was displaced or reworked either grain by grain or as limestone clasts and gravitationally transported together with ashy tuffites toward the trench. Grain by grain displacement/reworking, possibly triggered by storm waves or earthquakes, resulted in calci-turbidites relatively poor in ashy material owing to hydrodynamic winnowing. Limestone clast reworking occurred together with the mobilization of abundant arc-derived volcanic material, possibly triggered by earthquakes and/or slope instabilities and resulted in debris flows with a variable amount of carbonates.

Fine-grain detrital sediments and hemipelagic siliceous limestones deposited on the forearc slope and in the vicinity of the trench during periods of lower detrital input. In this view, the hemipelagic limestones represent the background sedimentation in the forearc environment that was sporadically affected by catastrophic events leading to deposition of significant amounts of detrital material (e.g. Underwood & Bachman 1982). Hence, the sedimentary fabric of the San Pedrillo Unit, characterized by layers of various clast size (i.e. grains to large boulders; Fig. 3), may be regarded as a consequence

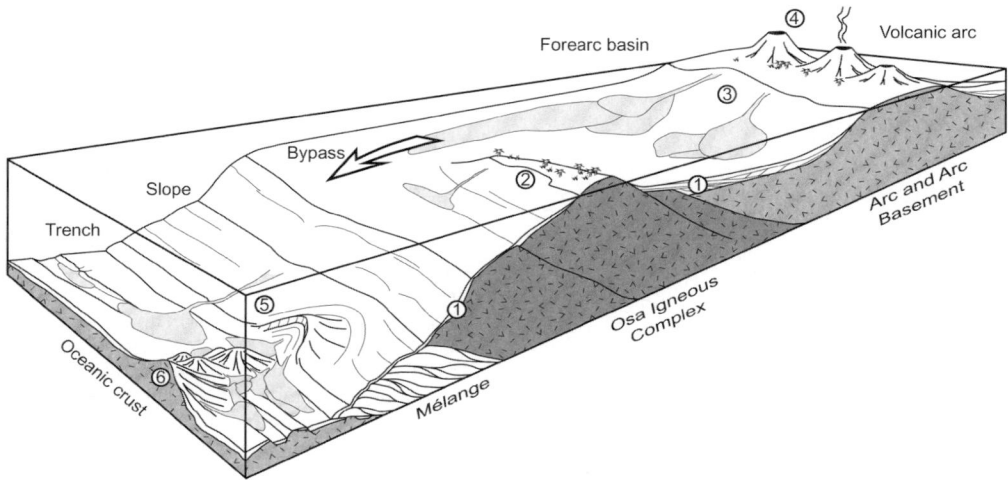

Fig. 16. Palinspastic reconstruction of the southern Costa Rican margin in the Late Eocene, based on the sedimentary record observed in the San Pedrillo Unit. The forearc area undergoes an important uplift possibly in response to seamount subduction. Large amount of sediments are deposited in the trench and quickly accreted to form the San Pedrillo Unit. The sediments are composed of volcanic fragments of the arc (1); Late Eocene shallow-water carbonates originally deposited/produced along the arc (2); Late Cretaceous–Middle Eocene clasts (grains to large boulders) of the Palaeo–Osa Igneous Complex (3); and Early–Middle Eocene hemipelagic and tuffaceous sediments of the forearc slope-basin (4). Minor reworking of previously accreted portions of the San Pedrillo Unit is possible (5). Minor volumes of sediments and basalt derived from collapsing seamounts may have been deposited into the trench or may be tectonically incorporated into the mélange during the subduction of the seamounts (6), but remain very limited in comparison to the volume of detrital material coming from the upper plate.

of variations in degree of clastic input in the trench environment.

The age of deposition of the sediments of the San Pedrillo Unit in the trench is constrained to the Late Eocene by the limestone beds containing shallow-water fauna. This age is in agreement with the age of the youngest rocks of the Cabo Matapalo Unit (Middle Miocene), which were accreted subsequently to the San Pedrillo Unit. This age is also in relatively good agreement with the stratigraphy of the forearc area inboard of the Osa Peninsula that contains >1000 m-thick arc-derived turbidites deposited during the Oligocene–Early Miocene (Henningsen 1963). Occurrence of these deposits points toward high rates of clastic sediment production close to the arc at this time, which has remained preserved from erosion.

Deformation

Vannucchi *et al.* (2006) provided a detailed structural analysis of the San Pedrillo Unit and outlined four tectonic phases that occurred prior to the exhumation. According to their results and our observations, the tectonic record indicates the clastic sediments were not lithified at the beginning of subduction. During the first stages of underthrusting,

the sediments progressively lost their fluid content and deformation changed from an underthrust regime to underplating and shortening within the upper plate wedge. After the constitution of the San Pedrillo Unit and prior to its uplift, new tectonic events occurred, possibly in response to repeated collisions of seamounts with the margin.

Shear planes, presumably related to the SP-D$_3$ and SP-D$_4$ phases (Vannucchi *et al.* 2006), which may correspond to an underplating regime, are defined by the precipitation of calcite, zeolites and quartz minerals. They crosscut the stratification and trend NE–SW with steep dips and, in conjunction with the orientations of stratification, they delimit lenses of consistent sedimentary facies (Fig. 8). Calcite, zeolites and quartz veins are ubiquitously encountered within the San Pedrillo Unit. This indicates pervasive fluid circulation took place during the first stages of the tectonic history of the mélange (SP-D$_1$), as previously suggested by Vannucchi *et al.* (2006). The density of the veins locally increases and may exceed 50% in the formations. It is thought to be associated with hydrofracturation. Preferential fluid drainage is indicated by high density of vein networks and occurred within areas dominated by igneous and calcareous olistoliths. These observations probably indicate fluids associated to poorly lithified sediments plaid a predominant role during accretion of the San Pedrillo Unit.

Accretion of the San Pedrillo Unit

An accretion primarily triggered by catastrophic sediment supply. The presence of debris flow deposits containing large boulders of igneous rocks in the San Pedrillo Unit has to be considered as a record of highly catastrophic events that affected the trench environment in the Late Eocene. We propose that a rapid increase of the sediment supply in the trench triggered the formation of the San Pedrillo Unit (Fig. 15f, g).

Cloos & Shreve (1988) first proposed the existence of an inlet, or 'subduction channel', at the interface between the subducting and overriding plates. In their model, the geometry of the inlet is temporally linked to the tectonic environment and tends to accommodate the subducting plate morphology, the sedimentary supply and the fluid content of the subducting sediments. A rapid increase of the sediment supply may fill topographic depressions and dramatically increase the sediment pile in the vicinity of the subduction zone. With sufficiently low rates of convergence and/or sufficient sediment supply, trench deposits may outgrow the subducting capacity of the subduction channel and ultimately lead to growth of an accretionary prism (e.g. Moore 1989; Le Pichon *et al.* 1993; Lallemand *et al.* 1994). Recent geological evidence from Italy shows the subduction channel at shallow depth may be exposed in uplifted ancient convergent margins (Vannucchi *et al.* 2008).

It is generally considered that convergence between the Farallon and Caribbean Plate was nearly orthogonal in the Eocene (Meschede & Barckhausen 2001; Pindell *et al.* 2005). The rate of convergence may thus have been similar or faster to present day (*c.* 10 mm a^{-1}, DeMets 2001) so sediments in the trench were driven quickly (most probably in less than 500 ka) into the subduction zone. This is corroborated by the presence of low sediment induration in the San Pedrillo Unit at the onset of deformation (Vannucchi *et al.* 2006). In this context, deposition of large volumes of sediments over a short period of time in the trench provided an excess of material in the subduction zone, which over-filled the subduction channel. The excess sediments accumulated at shallow level progressively lost fluid content and accreted to form the San Pedrillo Unit. Whereas this kind of process is generally associated to frontal accretion of the sediment (Le Pichon *et al.* 1993; Lallemand *et al.* 1994), structures in the San Pedrillo Unit may indicate the sediments accreted by underplating (Vannucchi *et al.* 2006).

The structural fabric of the NW Osa Mélange indicates massive sediment fluid loss at the beginning of underthrusting, provoking changes in the regime of deformation (Vannucchi *et al.* 2006). Along erosive margins, fluid content within subducting sediments plays an important role in the control of subduction erosion. Fluids released from the subducting plate may increase hydrofracturing of the upper plate and trigger the removal and subduction of lenses of several kilometres in width and length (Ranero & von Huene 2000). High fluid content of subducting sediments tends to prevent accretion and enhance subduction (Cloos & Shreve 1988; Moore 1989; Seno 2007, 2008). Thus rapid fluid expulsion likely occurred during underthrusting of the San Pedrillo Unit. Large amounts of clastic sediments and low argileous component in the unit probably enhanced the ability of the subducted material to quickly dewater. Fluid expulsion caused a migration of the décollement toward the subducting plate and, consequently, accretion of subducting material. Because of strongly disturbed internal arrangement of the San Pedrillo Unit, it is unclear if this process repeatedly occurred or not to form the unit. For a similar reason it is not possible to know if significant portions of accreted material were removed between possible multiple events of accretion. Nonetheless, the presence of the San Pedrillo Unit indicates overall net accretion in southern Costa Rica during the Late Eocene continuing until at least the Middle Miocene with the subsequent emplacements of the Cabo Matapalo and Salsipuedes units.

The San Pedrillo Unit: a record of a margin collapse in response to seamount subduction. Large volumes of sediment deposited in the trench and produced by mass wasting along the Costa Rican margin during the Late Eocene indicate that significant changes in subduction dynamics took place in the area at that time. This event was likely caused by seamount subduction along the Palaeo–Osa Igneous Complex.

Seamount subduction is a common process along many active margins. It has been notably observed along the Nankai Trough (Park *et al.* 1999; Kodaira *et al.* 2000), the New Hebrides Trench (Collot & Fisher 1989), the New Zealand margin (Collot *et al.* 2001) and the Mid American Trench (von Huene *et al.* 1995). Subduction of a seamount is accompanied by a zone of tectonic deformation in the upper plate and a re-entrant in the margin on the steep slope behind the subducting seamount (von Huene *et al.* 1995). This process creates a zone of slope failure that migrates upward and generates the deposition of reworked sediments in the trench area (Dominguez *et al.* 2000; Hühnerbach *et al.* 2005). The amount of material disrupted in the margin by seamount subduction is about four to five times the volume of the subducting seamount, of which about three quarters seems to be recycled downslope, backfilling the scar (Hühnerbach *et al.* 2005). The volume of reworked material of the margin may be even larger, depending on the brittleness of the margin (e.g. Collot *et al.* 2001). For example, along the Hikurangi margin (New Zealand), Collot *et al.* (2001) reported the presence of (1) a giant debris avalanche constituted by up to 2 km thick blocky deposits covering an area of *c.* 3400 km^2 and (2) a 65–170 m thick debris flow covering an area of *c.* 8000 km^2 that extends over 100 km from the trench fill. According to these authors, the deposits were generated in response to oblique seamount subduction that removed several thousands of cubic kilometre of material from the margin and created an indentation of the slope 65 km landward from the trench. Such present-day observations indicate a seamount subduction may generate large volumes of clastic material redepositing into the trench and forming deposits similar to the sediments of the San Pedrillo Unit. Therefore, we propose the sediments of the San Pedrillo Unit were triggered by the subduction of a seamount (or several seamounts) during the Late Eocene. The presence of subducting seamounts along the Central American margin during this period is in good accordance with the presence of accreted seamounts in the Azuero Peninsula (western Panama) (Buchs 2008).

Subducting seamount along the Palaeo–Osa Igneous Complex in the Late Eocene may have possibly enhanced the ability of the San Pedrillo Unit to accrete by underplating at shallow-level, rather than in the greenschist metamorphic facies or by frontal off-scrapping. Sandbox experiments have shown subducting seamounts may trigger temporary migration of the décollement upward (Dominguez *et al.* 2000), thus temporarily and locally increasing the thickness of the subduction channel. Such mechanism may allow partial subduction of sediments in the wake of downing seamounts and underplating at shallow level (Bangs *et al.* 2006; Fig. 15f). If this mechanism led to the accretion of the San Pedrillo Unit it may indicate that the unit formed very rapidly, possibly through a unique event of accretion.

The bulk of the reworked material in the San Pedrillo Unit proceeded from the partial collapse of the upper plate. It remains possible, however, that some igneous and sedimentary blocks were tectonically incorporated into the mélange matrix directly from disaggregating subducting seamounts. If a similar process is involved in the genesis of some mélanges (e.g. Okamura 1991), this was nonetheless an insignificant process in the formation of the San Pedrillo Unit because (1) the fabric is

primarily controlled by sedimentary processes and (2) the bulk of the basaltic blocks and associated pelagic sediments proceeded from an older accretionary prism (the Palaeo–Osa Igneous Complex). Early Eocene radiolarian cherts sporadically encountered in the mélange as deformed lenses and minor proportion of basaltic olistoliths associated to Campanian–Maastrichtian radiolarites may represent a record of partial tectonic incorporation of subducting seamounts into the San Pedrillo Unit.

Accretion of the Cabo Matapalo and Salsipuedes units

The Cabo Matapalo and Punta Salsipuedes units constitute the outermost extension of the Osa Mélange, presumably emplaced onto the San Pedrillo Unit by accretion processes (Di Marco 1994; Di Marco *et al.* 1995; Fig. 2). Review and interpretation of these units is not the aim of this study, but their presence indicates that accretionary processes may have continued until the Miocene, prior to the exhumation of the Osa basement and subsequent deposition of the Charco Azul Group (Fig. 14).

According to Di Marco (1994), the Cabo Matapalo Unit may have formed similarly to the San Pedrillo Unit, whereas the Salsipuedes Unit may represent sediments detached from the subducting plate that accreted onto the Cabo Matapalo Unit. Further stratigraphic, structural and geochemical data are nonetheless required to better constrain these interpretations.

The southern Central American margin: erosive or accretionary?

Erosive and accretionary margins

Since early work by von Huene & Lallemand (1990) it has been widely recognized that convergent margins may be subdivided into two classes, one dominated by accretion and net growth of accretionary prisms and the second characterized by tectonic erosion, absence of accreted material and trench retreat (von Huene & Scholl 1991; Clift & Vannucchi 2001). Each type composes approximatively half of the convergent margins observed at the present time around the world. Distinction between these two types of margins has been initially based on long-term forearc subsidence and associated landward retreat of the trench (von Huene & Lallemand 1990). New contributions have shown erosive or nonaccretionary margins are notably characterized by higher forearc slope angle, higher rate of orthogonal convergence between the descending and overriding plates, and lower rate of sediment delivery into the trench than accretionary margins (Clift & Vannucchi 2004). The higher forearc slope angle in accretionary margins has been attributed to material removal at the base of the overriding plate, causing subsidence and normal faulting of the forearc taper (Ranero & von Huene 2000). Lower rate of trench sediment supply and higher degree of orthogonal convergence between the plates is intrinsically related to the ability of the subduction zone to accommodate and subduct the sedimentary input. Basal erosion of the overriding plate has been attributed to hydro-fracturation induced by percolating slab-related fluids (Le Pichon *et al.* 1993; von Huene *et al.* 2004) or abrasion/tunnelling by subducting sea-mounts and ridges (Clift & Vannucchi 2004; Kukowski & Oncken 2006). It is important to note here that the long term distinction between erosive and accretionary margins has been principally based on the recognition on seismic profiles of the occurrence or non-occurrence of an accretionary prism composed of sedimentary material (von Huene & Scholl 1991, fig. 3; Clift & Vannucchi 2004, fig. 1).

The Central American margin is generally considered to be currently nonaccretionary (Ranero & von Huene 2000; Meschede 2003; Ranero *et al.* 2007). In northern Costa Rica, recent active subduction erosion has been demonstrated on the basis of geophysical studies and drilling that shown that the outer forearc underwent a recent, rapid subsidence associated with normal faulting (Vannucchi *et al.* 2001, 2003). On the other hand, Late Cretaceous–Cenozoic accretionary complexes are commonly observed along the margin (for review see Denyer *et al.* 2006). This apparent contradiction between present-day observations and the geological record is commonly attributed to a hypothetical recent change of the subduction regime from accretion to erosion (Vannucchi *et al.* 2006). We show below that mechanisms presently observed along the margin are on the contrary in good agreement with the geological record and are actually extremely difficult to interpret in terms of subduction accretion/erosion over long periods of time.

Hidden accretionary complexes along intra-oceanic subduction systems

We have shown exposed forearc in southern Costa Rica records several accretion events, indicating that the margin developed seaward through episodic growth. The Osa Igneous Complex records two events of significant growth by seamount accretion in the Maastrichtian/Early Paleocene with the emplacement of the Inner Osa Igneous Complex and, probably, in the Middle Eocene with the

emplacement of the Ganado and Riyito units (Figs 14 & 15). The San Pedrillo Unit was emplaced onto the Osa Igneous Complex in the Late Eocene subsequently to rapid, large sedimentary supply into the trench possibly in response to seamount subduction. To our knowledge, development of convergent margins over long periods of time (Late Cretaceous to Middle Eocene) by accretion of mostly igneous rocks has not been observed elsewhere. Whether this is due to observation bias or to a peculiarity of the Costa Rican margin is difficult to assess. It is possible the unusual nature of the Costa Rican margin (an arc built on an oceanic plateau, Buchs *et al.* in prep.) has led to unusual accretionary processes between the Late Cretaceous and Middle Eocene. Alternatively, igneous accretionary complexes may be very common along intra-oceanic convergent margins and only poorly exposed.

While sediment accretion along convergent margins is easily demonstrated by seismic imaging of accreted sediments, seamount accretion as recorded in the Osa Igneous Complex may be difficult to recognize in intra-oceanic subduction systems. Density and velocity contrasts between basaltic terranes and arc-related igneous rocks inboard of accretionary complexes may be very low and reflectors poorly developed in the igneous rocks. Density and velocity contrasts may be particularly low along the Costa Rican margin because the arc basement comprises an oceanic plateau highly similar to accreted volcanic rocks (Buchs *et al.* in prep.). Seismic profiles over the seaward extension of the Nicoya Peninsula (Sallarès *et al.* 1999, 2001) failed to image the composite, partly accretionary nature of the area, well documented by on-land, tectono-stratigraphic reconstructions (Flores 2003; Bandini *et al.* 2008). Magnetic signatures of upper plate structures investigated between Nicoya and Osa peninsulas point toward a composition of the margin wedge similar to the igneous complex found onshore on the Nicoya Peninsula (Barckhausen *et al.* 1998). However, accreted material exposed in southern Costa Rica on the Osa Peninsula has not been recognized on seismic profiles in close proximity (Kolarsky *et al.* 1995*a*; Ranero & von Huene 2000; Fig. 5), although these profiles very likely comprise accreted igneous rocks and mélanges containing igneous olitholiths similar to those expose in the Osa Peninsula.

Observations from Costa Rica show that, along intra-oceanic convergent margins, igneous accretionary complexes may remain unobservable in numerous cases and their occurrence widely underestimated. This raises an important interrogation about the validity of the consensual model of erosive margin (e.g. von Huene & Scholl 1991; Clift & Vannucchi 2004), because unidentified (seamount) accretion may have occurred along many intra-oceanic convergent margins devoid of accreted sediments. Hence, some intra-oceanic subduction systems presently regarded as erosive may actually be accretionary over long periods of time, even if accreted sediment lacks and local and/or temporary erosion presently occurs.

Accretion v. erosion: the case of the southern Central American margin

Development of the Osa Igneous Complex in cross-section view (Figs 5 & 15) illustrates the southern Costa Rican margin experienced at least two events of episodic growth (emplacement of the Inner Osa Igneous Complex and Riyito–Ganado Units) separated by periods of tectonic erosion or low-rate accretion (erosion of the arc basement and formation of the Güerra and Vaquedano Units). On map views of southern Central America (Fig. 2) and southern Costa Rica (Fig. 2b) it appears igneous accretionary complexes characterized by distinct geochemical affinities (Hauff *et al.* 2000; this study) are randomly distributed along the forearc. They have a fairly limited along-strike extension at the scale of the entire margin that relates to smaller size of accreted seamount fragments. Old events of tectonic erosion are well illustrated on a map view by arc-related and arc basement rocks occurring close to the present-day trench on Coiba Island or to the Osa Igneous Complex in southern Costa Rica (Buchs *et al.* in prep.; this study). They may also be marked by limited along-strike extension of the Güerra and Vaquedano units in Osa. Hence, without implying any sediment accretion, it seems the southern Central American margin experienced variations in accretion/erosion rates in both time (i.e. seaward-landward variations seen on cross-section view) and space (along-strike variations seen on map view).

Exposed igneous complexes along the forearc of Costa Rica and Panama record subduction from the Maastrichitian to the Eocene. Over this *c.* 25 Ma-long period, the regime was globally characterized by no sediment accretion, seamount subduction and subduction erosion. This is a situation very similar to the present-day margin. However, it is clear that in some parts of the old margin, seamount accretion led to the construction of several-kilometres thick accretionary complexes (Fig. 2). Therefore, understanding the erosive/accretionary nature of the margin between the Maastrichtian and Eocene is only achievable by estimating the amount of material added to and removed from the margin over the period of interest. This seems, however, to be extremely difficult to carry out in the case of the southern Central American margin

for four main reasons: (1) occurrences of on-land arc basement and accretionary complexes are scarce relative to margin length; (2) current geophysical techniques cannot make a clear, precise distinction between composite, igneous accretionary complexes and the arc basement; (3) precise estimation of the total amount of eroded material is not possible because the shape of the margin prior to erosive events is unknown; and (4) the eroded material subducted into the mantle (?) cannot be observed with current geophysical techniques, prohibiting estimation of the exact volume of this material.

Since the Late Eocene, evolution of the margin is poorly constrained by the geological record. Sole on-land occurrence of Late Eocene and younger accretionary complex comes from the Osa Peninsula where the Osa Mélange indicates an overall growth of the margin by sediment accretion. Other areas close to the trench have been imaged trough seismic profile and apparently lack large, sedimentary accretionary wedges and are composed by basaltic rocks (Ranero & von Huene 2000). Accreted fragments of seamounts may actually compose most of the younger part of the margin. Hence, difficulties similar to older South Central American accretionary complexes exist in quantifying net accretion or erosion along the margin since the Eocene.

In conclusion to preceding observations, it seems that even if the present-day margin apparently lacks active accretionary prisms composed of sediments, seamount accretion and tectonic erosion may coexist and/or quickly alternate through time. This is in very good accordance with older geological record present in exposed accretionary complexes along southern Central America. Because of possible local growths of the margin by seamount accretion and difficulties in estimating volumes of eroded material, it is not clear if seamount subduction is associated with net growth or erosion of the margin over long periods of time. As a consequence, there is currently no way to unequivocally define the southern Central American margin as erosive or accretionary.

Summary and conclusions

The Osa and Burica Peninsulas are parts of the outer forearc of southern Central America that were recently uplifted in response to the subduction of the Cocos Ridge. Rocks exposed in these peninsulas are of significant interest for understanding key processes of convergent margins such as tectonic accretion, tectonic erosion and seismogenesis along the subduction zone. The basement rocks of the Osa and Burica Peninsula were studied through a multidisciplinary approach combining a geological survey, sedimentology, palaeontology, geochemistry, structural observations and remote tectonic analyses.

The basement comprises the Osa Igneous Complex that extends from the inner Osa Peninsula to the Burica Peninsula and, possibly, further into the Montijo Gulf (western Panama) and NW off Osa and the Osa Mélange, exposed in the seaward side of the Osa Peninsula and Caño Island (Figs 2 & 5). The Osa Igneous Complex is an accretionary complex made up of Late Cretaceous to Middle Eocene fragments of an oceanic plateau and seamounts. The Inner Osa Igneous Complex comprises fragments of a Coniacian–Santonian oceanic plateau that accreted in the Early–Middle Paleocene and was followed by emplacement of the Outer Osa Igneous Complex. The Outer Osa Igneous Complex comprises Coniacian–Santonian to Middle Eocene seamount fragments that accreted during (1) a period of low rate accretion/tectonic erosion between the Middle–Late Paleocene and the Middle Eocene (Güerra Unit), (2) a period of net growth in the Middle Eocene (Ganado and Riyito Units) and (3) another period of low rate accretion/tectonic erosion in the Middle–Late Eocene, prior to the emplacement of the Osa Mélange.

The Osa Mélange is an accretionary complex comprising (1) the San Pedrillo Unit, (2) the Cabo Matapalo Unit and (3) the Punta Salsipuedes Unit (Figs 2 & 5). The San Pedrillo Unit is a polygenetic mélange predominantly made up of deformed fine-grain detrital sediments and olistostromic deposits including Late Cretaceous to Late Eocene material. This material was sedimentarily removed from the forearc area in the Late Eocene and essentially comprises rocks derived from the Palaeo–South Central American volcanic arc and the Palaeo–Osa Igneous Complex. Landslide formation along the margin was presumably triggered by seamount subduction. Underplating at shallow level of the San Pedrillo Unit resulted from a combination of high, catastrophic sediment supply into the trench and quick dewatering of the subducted sediments, possibly driven to depth on the trailing edge of subducting seamounts. The Cabo Matapalo Unit may have formed in a similar way during the Miocene, whereas the Punta Salsipuedes Unit was formed by off-scraping and frontal accretion of pelagic sediments in the Miocene (Di Marco *et al.* 1995). Further work is required to better constrain the origins of these two later units.

Detailed study of the southern Costa Rican margin shows that faulting induced by the subducting Cocos Ridge is not solely controlled by the topography of the descending plate (e.g. Sak *et al.* 2004, 2009), but equally by compression and possible reactivation of ancient suture zones between the units, leading to the formation of recent and/or active backthrusts. Faulting associated with the thrusts may greatly contribute to the morphology of the area but is not clearly seen on offshore seismic profiles. Suture zones were initially located along palaeo-décollements, which therefore represent fossil subduction zones. In Osa they provide an interesting on-land opportunity to study some remote processes of convergent margins such as fluid flows along the décollement or, possibly, through the outer crystalline forearc (Ranero *et al.* 2008).

The Osa Igneous Complex accreted onto the Caribbean Plate south western margin, which is composed of an arc built on an oceanic plateau (Buchs *et al.* in prep.). Hence, the Osa Igneous Complex is exotic with respect to the Caribbean Plate and should no longer be considered as being part of the CLIP. No clear genetic link exists between the oceanic plateau forming the arc basement and the igneous rocks exposed in the Osa Igneous Complex. It is unclear if these lavas were emplaced above the same hotspot or not. This raises questions about the association of accreted igneous complexes of northern South America with the CLIP.

Similarly to other well-studied convergent margins in the world, the southern Costa Rican margin shows the accretionary prisms develop by episodic growth. In the case of the Osa Igneous Complex, we identified pulses of accretion of large fragments of oceanic plateau and seamounts presumably related to episodic arrival of an oceanic plateau and seamounts and/or tectonic changes at a regional scale, separated by periods of low-rate accretion and tectonic erosion. Accretion of the San Pedrillo Unit was triggered by a catastrophic event along the margin, presumably a seamount subduction.

The geological record of the South Central American forearc between the Late Cretaceous to Middle Eocene is similar to and consistent with the current situation. No significant amount of sediment accretion occurred during the early stages of margin evolution when several seamounts arrived at the subduction zone. In the Osa Igneous Complex, northern Costa Rica and western Panama, growth of the margin is attested by the presence of accreted seamount fragments that certainly extend offshore and have not been recognized on seismic profiles. As a consequence, it is clear that net growth of the margin over time is not restricted to sediment accretion but may result from the presence of incoming seamounts along the margin. In absence of precise information concerning the amount of material removed from the margin by tectonic erosion, this indicates the present-day margin may be accretionary rather than erosive.

We are thankful to K. James and P. Mann for their accurate review and discussions that led to significant improvement of the quality manuscript. We thank R. Arculus and J. Hernandez for their constructive criticism during field work and geochemical interpretations. We greatly appreciated discussions and review on a former version of the manuscript by R. Arculus, O. Münthener and G. Stampfli. Many thanks are due to J. Pindell for discussions on Caribbean tectonics. Discussions with K. Hoernle, F. Hauff and P. van den Bogaard have been a source of inspiration for some ideas developed in this manuscript. We greatly appreciated the hospitality and enthusiasm at Marenco Beach and Rainforest Lodge. Geochemical analyses were performed at the Centre d'Analyse Minérale (University of Lausanne) by J.-C. Lavanchy and at the Institute of Mineralogy (University of Lausanne) with the help of F. Bussy and A. Ulianov. This study was carried out in the framework of two research projects of the Swiss National Science Foundation (no. 00021-105845 and 200021-105845). Earlier field studies were supported by the Herbette Foundation (University of Lausanne).

References

ABRATIS, M. 1998. *Geochemical variations in magmatic rocks from southern Costa Rica as a consequence of Cocos Ridge subduction and uplift of the Cordillera de Talamanca.* PhD thesis, University of Göttingen, Germany.

ABRATIS, M. & WÖRNER, G. 2001. Ridge collision, slab-window formation, and the flux of Pacific asthenosphere into the Caribbean realm. *Geology*, **29**, 127–130.

ALONSO, J. L., MARCOS, A. & SUAREZ, A. 2006. Structure and organization of the Porma Melange: progressive denudation of a submarine nappe toe by gravitational collapse. *American Journal of Science*, **306**, 32–65.

ARIAS, O. 2003. Redefinición de la Formación Tulín (Maastrichtiano–Eoceno inferior) del Pacífico Central del Costa Rica. *Revista Geológica de America Central*, **28**, 47–68.

AZÉMA, J., BUTTERLIN, J., TOURNON, J. & DE WEVER, P. 1983. Presencia de material volcano-sedimentario de edad Eoceno medio en la Península de Osa (provincia de Puntarenas, Costa Rica). *10a Conferencia Geológica del Caribe*, Cartagena, Colombia.

BANDINI, A. N., BAUMGARTNER, P. O. & CARON, M. 2006. Turonian radiolarians from Karnezeika, Argolis Peninsula, Peloponnesus (Greece). *Eclogae Geologicae Helvetiae*, **99**, 1–20.

BANDINI, A. N., FLORES, K., BAUMGARTNER, P. O., JACKETT, S.-J. & DENYER, P. 2008. Late Cretaceous and Paleogene Radiolaria from the Nicoya Peninsula, Costa Rica: a tectonostratigraphic application. *Stratigraphy*, **5**, 3–21.

BANDY, O. L. & CASEY, R. E. 1973. Reflector horizons and Paleobathymetric history, Eastern Panama. *Geological Society of America Bulletin*, **84**, 3081–3086.

BANGS, N. L. B., GULICK, S. P. S. & SHIPLEY, T. H. 2006. Seamount subduction erosion in the Nankai Trough and its potential impact on the seismogenic zone. *Geology*, **34**, 701–704.

BARCKHAUSEN, U., ROESER, H. A. & VON HUENE, R. 1998. Magnetic signature of upper plate structures and subducting seamounts at the convergent margin off Costa Rica. *Journal of Geophysical Research, B, Solid Earth and Planets*, **103**, 7079–7093.

BAUMGARTNER, P. O. 1986. Discovery of subduction-related melanges on Cano Island and Osa Peninsula (Pacific, Costa Rica, Central America). *Onzième réunion annuelle des sciences de la terre, Réunion Annuelle des Sciences de la Terre*, Clermont-Ferrand, France, 12.

BAUMGARTNER, P. O., MORA, C. R. *ET AL.* 1984. Sedimentación y paleogeografía del Cretácico y Cenozoico del litoral pacifico de Costa Rica. *Revista Geológica de América Central*, **1**, 57–136.

BAUMGARTNER, P. O., OBANDO, J. A., MORA, C. R., CHANNELL, J. E. T. & STECK, A. 1989. Paleogene accretion and suspect terranes in southern Costa Rica (Osa, Burica, Central America). *Transaction of the 12th Caribbean Geological Conference*, St Croix, Virgin Islands, 529.

BAUMGARTNER, P. O., FLORES, K., BANDINI, A. N., BAUMGARTNER-MORA, C. & BUCHS, D. M. 2008. Terranes of NW-Costa Rica and the Hess Escarpment: a pre-Campanian paleo-plate boundary. *18th Caribbean Geological Conference*, March 2008, Santo Domingo, Dominican Republic.

BERRANGÉ, J. P. 1989. The Osa group: an auriferous Pliocene sedimentary unit from the Osa Peninsula, southern Costa Rica. *Revista Geológica de America Central*, **10**, 67–93.

BERRANGÉ, J. P. & THORPE, R. S. 1988. The geology, geochemistry and emplacement of the Cretaceous Tertiary Ophiolitic Nicoya Complex of the Osa Peninsula, Southern Costa-Rica. *Tectonophysics*, **147**, 193–220.

BERRANGÉ, J. P., BRADLEY, D. R. & SNELLING, N. J. 1989. K/Ar age dating of the ophiolitic Nicoya Complex of the Osa Peninsula, southern Costa Rica. *Journal of South American Earth Sciences*, **2**, 49–59.

BUCHS, D. M. 2003. *Etude géologique et géochimique de la région du Golfo Dulce (Costa Rica): Genèse et évolution d'édifices océaniques accrétés à la marge de la plaque caraïbe.* DEA thesis, Université de Lausanne, Switzerland.

BUCHS, D. M. 2008. *Late Cretaceous to Eocene geology of the South Central American forearc area (southern Costa Rica and western Panama): Initiation and evolution of an intra-oceanic convergent margin.* PhD thesis, Université de Lausanne, Switzerland.

BUCHS, D. M. & STUCKI, J. 2001. *Etude géologique, géochimique et structurale du prisme d'accrétion de la péninsule d'Osa, Cosa Rica.* Diploma thesis, Université de Lausanne, Switzerland.

BUCHS, D. M. & BAUMGARTNER, P. O. 2003. The mélange of Osa–Caño (Costa Rica): an access to the sedimentary processes recorded in an emerged Middle Eocene to Middle Miocene accretionary prism. *10th Meeting of Swiss Sedimentologists (SWISS SED)*, Fribourg, Switzerland.

BUCHS, D. M. & BAUMGARTNER, P. O. 2007. Comment on 'From seamount accretion to tectonic erosion: Formation of Osa Mélange and the effects of Cocos Ridge subduction in southern Costa Rica' by VANNUCCHI, P. *ET AL. Tectonics*, **26**, TC3009; doi: 3010.1029/2006TC002032.

BUCHS, D. M., GUEX, J., STUCKI, J. & BAUMGARTNER, P. O. 2009. Paleocene Thalassinidea colonization in deep-sea environment and the coprolite *Palaxius osaensis* n. ichnosp. in Southern Costa Rica. *Revue de Micropaléontologie*, **52**, 123–129.

BYRNE, T. & FISHER, D. 1987. Episodic growth of the Kodiak Convergent Margin. *Nature*, **325**, 338–341.

CARTER, R. M. 1979. Trench-slope channels from the New-Zealand Jurassic–Otekura Formation, Sandy Bay, South Otago. *Sedimentology*, **26**, 475–496.

CHANG, C. P., ANGELIER, J. & HUANG, C. Y. 2000. Origin and evolution of a melange: the active plate boundary and suture zone of the Longitudinal Valley, Taiwan. *Tectonophysics*, **325**, 43–62.

CLAGUE, D. A. & DALRYMPLE, G. B. 1987. The Hawaiian–Emperor volcanic chain, Part I, Geologic evolution. *In*: DECKER, R. W., WRIGHT, T. L. & STAUFFER, P. H. (eds) *Volcanism in Hawaii*. US Government Printing Office, US Geological Survey, Professional Paper, **1350**, 5–54.

CLIFT, P. & VANNUCCHI, P. 2001. Controls on tectonic accretion versus erosion in subduction zones: Implications for the origin and recycling of the continental crust. *Reviews of Geophysics*, **42**, RG2001.

CLOOS, M. 1984. Flow melanges and the structural evolution of accretionary wedges. *In*: RAYMOND, L. A. (ed.) *Melanges: their Nature, Origin and Significance*. Geological Society of America, Boulder, CO, Special Papers, **198**, 71–79.

CLOOS, M. & SHREVE, R. L. 1988. Subduction-channel model of prism accretion, melange formation, sediment subduction, and subduction erosion at convergent plate margins, Part 1. Background and description. *Pure and Applied Geophysics*, **128**, 455–500.

COATES, A. G., JACKSON, J. B. C. *ET AL.* 1992. Closure of the Isthmus of Panama, the near-shore marine record of Costa Rica and western Panama. *Geological Society of America Bulletin*, **104**, 814–828.

COLLINS, L. S., COATES, A. G., JACKSON, J. B. C. & OBANDO, J. A. 1995. Timing and rates of emergence of the Limon and Bocas del Torro basins: Caribbean effects of Cocos Ridge subduction? *In*: MANN, P. (ed.) *Geologic and Tectonic Development of the Caribbean Plate Boundary in Southern Central America*. Geological Society of America, Boulder, CO, Special Papers, **295**, 263–289.

COLLOT, J. Y. & FISHER, M. A. 1989. Formation of forearc basins by collision between seamounts and accretionary wedges: an example from the New Hebrides subduction zone. *Geology*, **17**, 930–933.

COLLOT, J. Y., LEWIS, K., LAMARCHE, G. & LALLEMAND, S. 2001. The giant Ruatoria debris avalanche on the northern Hikurangi margin, New Zealand: results of oblique seamount subduction. *Journal of Geophysical Research, B, Solid Earth and Planets*, **106**, 19271–19297.

CONEY, P. J. 1989. Structural aspects of suspect terranes and accretionary tectonics in Western North-America. *Journal of Structural Geology*, **11**, 107–125.

CORRIGAN, J., MANN, P. & INGLE, J. C. JR. 1990. Forearc response to subduction of the Cocos Ridge, Panama–Costa Rica. *Geological Society of America Bulletin*, **102**, 628–652.

DEMETS, C. 2001. A new estimate for present-day Cocos–Caribbean Plate motion; implications for slip along the Central American volcanic arc. *Geophysical Research Letters*, **28**, 4043–4046.

DENGO, G. 1962. Tectonic-igneous sequence in Costa Rica. *In*: ENGEL, A. E. J., JAMES, H. J. & LEONARD, B. F. (eds) *Petrologic Studies (A Volume in Honor of A. F. Buddington)*. Geological Society of America Boulder, Co, 133–161.

DENYER, P. & BAUMGARTNER, P. O. 2006. Emplacement of Jurassic–Lower Cretaceous radiolarites of the Nicoya Complex (Costa Rica). *Geologica Acta*, **4**, 203–218.

DENYER, P., BAUMGARTNER, P. O. & GAZEL, E. 2006. Characterization and tectonic implications of Mesozoic–Cenozoic oceanic assemblages of Costa Rica and Western Panama. *Geologica Acta*, **4**, 219–235.

DICKINSON, W. R. 2008. Accretionary Mesozoic–Cenozoic expansion of the Cordilleran continental margin in California and adjacent Oregon. *Geosphere*, **4**, 329–353.

DI MARCO, G. 1994. *Les terrains accrétés du Costa Rica: évolution teconostratigraphique de la marge occidentale de la Plaque Caraïbe*. Mémoires de Géologie, Lausanne, **20**.

DI MARCO, G., BAUMGARTNER, P. O. & CHANNELL, J. E. T. 1995. Late Cretaceous–Early Tertiary paleomagnetic data and a revised tectonostratigraphic subdivision of Costa Rica and western Panama. *In*: MANN, P. (ed.) *Geologic and Tectonic Development of the Caribbean Plate Boundary in southern Central America*. Geological Society of America, Boulder, CO, Special Papers, **295**, 1–27.

DISERENS, M.-O. 2002. *Upper Cretaceous and Paleogene radiolarian biostratigraphy of Southern Costa Rica; radiolarian faunas from the Rincon Block, Golfito and Burica Terranes, Osa-Cano Accretionary Complex and Herradura Block*. DEA thesis, Université de Lausanne, Switzerland.

DOMINGUEZ, S., MALAVIEILLE, J. & LALLEMAND, S. E. 2000. Deformation of accretionary wedges in response to seamount subduction: Insights from sandbox experiments. *Tectonics*, **19**, 182–196.

DOUBLEDAY, P. A., LEAT, P. T., ALABASTER, T., NELL, P. A. R. & TRANTER, T. H. 1994. Allochthonous oceanic basalts within the Mesozoic Accretionary Complex of Alexander Island, Antarctica – remnants of Proto-Pacific Oceanic-Crust. *Journal of the Geological Society*, **151**, 65–78.

DUMITRICA, P. 1975. Cenomanian radiolaria at Podul Dimbovitei. Micropaleontological guide to the Mesozoic and Tertiary of the Romanian Carpathians. Paper presented at *14th European Micropaleontological Colloquium*, Romania, Bucharest, Institute of Geology and Geophysics.

DUPUIS, C., HÉBERT, R. *ET AL.* 2005. The Yarlung Zangbo Suture Zone ophiolitic mélange (southern Tibet): new insights from geochemistry of ultramafic rocks. *Journal of Asian Earth Sciences*, **25**, 937–960.

FISHER, D. M., GARDNER, T. W., MARSHALL, J. S., SAK, P. B. & PROTTI, M. 1998. Effect of subducting seafloor roughness on fore-arc kinematics, Pacific Coast, Costa Rica. *Geology*, **26**, 467–470.

FISHER, D. M., GARDNER, T. W., SAK, P. B., SANCHEZ, J. D., MURPHY, K. & VANNUCCHI, P. 2004. Active thrusting in the inner forearc of an erosive convergent margin, Pacific Coast, Costa Rica. *Tectonics*, **23**, TC2007.

FLORES, K. 2003. *Propuesta tectonoestratigráfica de la región septentrional del golfo de Nicoya, Costa Rica.* Licenciatura thesis, Escuela Centroamericana de Geología, Universidad de Costa Rica.

FOREMAN, H. P. 1975. Radiolaria from the North Pacific, Deep Sea Drilling Project, Leg 32. *Initial Reports of the Deep Sea Drilling Project*, **32**, 579–676.

FOREMAN, H. P. 1977. Mesozoic Radiolaria from the Atlantic Basin and its borderlands. *In*: SWAIN, F. M. (ed.) *Stratigraphic Micropaleontology of Atlantic Basin and borderlands.* Developments in Paleontology and Stratigraphy, **6**. Elsevier, Amsterdam, 305–320.

FOREMAN, H. P., HEEZEN, B. C. ET AL. 1973. Radiolaria from DSDP Leg 20, Initial reports of the Deep Sea Drilling Project, covering Leg 20 of the cruises of the drilling vessel Glomar Challenger, Yokohama, Japan to Suva, Fiji September-November 1971. *Initial Reports of the Deep Sea Drilling Project*, **20**, 249–303.

GRÄFE, K., FRISCH, W., VILLA, I. M. & MESCHEDE, M. 2002. Geodynamic evolution of southern Costa Rica related to low-angle subduction of the Cocos Ridge: constraints from thermochronology. *Tectonophysics*, **348**, 187–204.

HAMPTON, M. A., LEE, H. J. & LOCAT, J. 1996. Submarine landslides. *Reviews of Geophysics*, **34**, 33–59.

HAUFF, F., HOERNLE, K. A., VAN DEN BOGAARD, P., ALVARADO, G. E. & GARBE-SCHONBERG, D. 2000. Age and geochemistry of basaltic complexes in Western Costa Rica: Contributions to the geotectonic evolution of Central America. *Geochemistry Geophysics Geosystems*, **1**; doi: 10.1029/1999GC000020.

HENNINGSEN, D. 1963. Notes on stratigraphy and paleontology of upper Cretaceous and Tertiary sediments in southern Costa Rica. *American Association of Petroleum Geologists Bulletin*, **50**, 562–566.

HOERNLE, K. & HAUFF, F. 2007. Oceanic igneous complexes. *In*: BUNDSCHUH, J. & ALVARADO, G. (eds) *Central America, Geology, Resources, Hazards.* Taylor & Francis, Balkema, **1**, 523–548.

HOERNLE, K., VAN DEN BOGAARD, P. ET AL. 2002. Missing history (16–71 Ma) of the Galpapagos hotspot: implications for the tectonic and biological evolution of the Americas. *Geology*, **30**, 795–798.

HOERNLE, K., HAUFF, F. & VAN DEN BOGAARD, P. 2004. 70 m.y. history (139–69 Ma) for the Caribbean large igneous province. *Geology*, **32**, 700.

HOLLIS, C. J. & KIMURA, K. 2001. A unified radiolarian zonation for the Late Cretaceous and Paleocene of Japan. *Micropaleontology*, **47**, 235–255.

HÜHNERBACH, V., MASSON, D. G., BOHRMANN, G., BULL, J. M. & WEINREBE, W. 2005. Deformation and submarine landsliding caused by seamount subduction beneath the Costa Rica continental margin: new insights from high-resolution sidescan sonar. *In*: HODGSON, D. M. & FLINT, S. S. (eds) *Submarine Slope Systems: Processes and Products.* Geological Society, London, Special Publications, **244**, 195–205.

ISOZAKI, Y., MARUYAMA, S. & FURUOKA, F. 1990. Accreted Oceanic Materials in Japan. *Tectonophysics*, **181**, 179–205.

KERR, A. C., WHITE, R. V., THOMPSON, P. M. E., TARNEY, J. & SAUNDERS, A. S. 2003. No oceanic plateau; no Caribbean Plate? The seminal role of an oceanic plateau in Caribbean Plate evolution. *In*: ARTOLINI, C., BUFFLER, R. T. & BLICKWEDE, J. (eds) *The circum-Gulf of Mexico and the Caribbean, Hydrocarbon Habitats, Basin Formation, and Plate Tectonics.* AAPG Memoirs, **79**, 126–168.

KOBAYASHI, K., CADET, J. P. ET AL. 1987. Normal faulting of the Daiichi–Kashima seamount in the Japan Trench revealed by the Kaiko-I Cruise, Leg-3. *Earth and Planetary Science Letters*, **83**, 257–266.

KODAIRA, S., TAKAHASHI, N., NAKANISHI, A., MIURA, S. & KANEDA, Y. 2000. Subducted seamount imaged in the rupture zone of the 1946 Nankaido earthquake. *Science*, **289**, 104–106.

KOLARSKY, R. A. & MANN, P. 1995. Structure and neotectonics of an oblique-subduction margin, southwestern Panama. *In*: MANN, P. (ed.) *Geologic and Tectonic Development of the Caribbean Plate Boundary in Southern Central America.* Geological Society of America, Boulder, CO, Special Papers, **295**, 131–157.

KOLARSKY, R. A., MANN, P. & MONTERO, W. 1995a. Island arc response to shallow subduction of the Cocos Ridge, Costa Rica. *In*: MANN, P. (ed.) *Geologic and Tectonic Development of the Caribbean Plate Boundary in Southern Central America.* Geological Society of America, Boulder, CO, Special Papers, **295**, 235–262.

KOLARSKY, R. A., MANN, P., MONECHI, S., MEYERHOFF, H. D. & PESSAGNO, E. A. JR. 1995b. Stratigraphic development of southwestern Panama as determined from integration of marine seismic data and onshore geology. *In*: MANN, P. (ed.) *Geologic and Tectonic Development of the Caribbean Plate Boundary in southern Central America.* Geological Society of America, Boulder, CO, Special Papers, **295**, 159–200.

KUKOWSKI, N. & ONCKEN, O. 2006. Subduction erosion – the 'normal' mode of fore-arc material transfer along the Chilean Margin? *In*: ONCKEN, O., CHONG, G. ET AL. (eds) *The Andes – Active Subduction Orogeny.* Springer, Berlin, 217–236.

LALLEMAND, S. E., SCHNUERLE, P. & MALAVIEILLE, J. 1994. Coulomb theory applied to accretionary and non-accretionary wedges: possible causes for tectonic erosion and/or frontal accretion. *Journal of Geophysical Research, B, Solid Earth and Planets*, **99**, 12033–12055.

LE PICHON, X., HENRY, P. & LALLEMANT, S. J. 1993. Accretion and erosion in subduction zones: the role of fluids. *Annual Review of Earth and Planetary Sciences*, **21**, 307–331.

MACMILLAN, I., GANS, P. B. & ALVARADO, G. 2004. Middle Miocene to present plate tectonic history of the southern Central American volcanic arc. *In*: DILEK, Y. & HARRIS, R. (eds) *Continental Margins of the Pacific Rim. Tectonophysics*, **392**, 325–348.

MANN, P. & KOLARSKY, R. A. 1995. East Panama deformed belt: structure, age, and neotectonic

454 D. M. BUCHS *ET AL.*

significance *In*: MANN, P. (ed.) *Geologic and Tectonic Development of the Caribbean Plate Boundary in Southern Central America*. Geological Society of America, Boulder, CO, Special Papers, **295**, 111–130.

MANN, P. & TAIRA, A. 2004. Global tectonic significance of the Solomon Islands and Ontong Java Plateau convergent zone. *Tectonophysics*, **389**, 137–190.

MARUYAMA, S. & LIOU, J. G. 1989. Possible depth limit for underplating by a seamount. *Tectonophysics*, **160**, 327–337.

MCDONOUGH, W. F. & SUN, S. S. 1995. The composition of the Earth. *Chemical Geology*, **120**, 223–253.

MEDLEY, E. W. 1994. *The engineering characterization of melanges and similar block-in-matrix rocks (Bimrocks)*. PhD thesis, University of California.

MEDLEY, E. W. 2001. Orderly Characterization of Chaotic Franciscan Melanges. *Engineering Geology*, **19**, 20–33.

MENDE, A. 2001. Sedimente und Architektur der Forearc- und Backarc-Becken von Südost-Costa Rica und Nordwest-Panamá. *Profil*, **19**, 1–130.

MENDE, A. & ASTORGA, A. 2007. Incorporating geology and geomorphology in land management decisions in developing countries: a case study in Southern Costa Rica. *Geomorphology*, **87**, 68–89.

MESCHEDE, M. 2003. The Costa Rica convergent margin: a textbook example for the process of subduction erosion. *Neues Jahrbuch Fur Geologie Und Palaontologie-Abhandlungen*, **230**, 409–428.

MESCHEDE, M. & FRISCH, W. 1994. Geochemical characteristics of basaltic rocks from the Central American ophiolites. *Profil*, **7**, 71–85.

MESCHEDE, M. & BARCKHAUSEN, U. 2001. The relationship of the Cocos and Carnegie ridges: age constraints from palaeogeographic reconstructions. *International Journal of Earth Sciences*, **90**, 386–392.

MESCHEDE, M., ZWEIGEL, P., FRISCH, W. & VOELKER, D. 1999. Melange formation by subduction erosion; the case of the Osa melange in southern Costa Rica. *Terra Nova*, **11**, 141–148.

MOORE, J. C. 1989. Tectonics and hydrogeology of accretionary prisms: Role of the decollement zone. *Journal of Structural Geology*, **11**.

MORA, C., BAUMGARTNER, P. O. & HOTTINGER, L. 1989. Eocene shallow water carbonate facies with larger foraminifera in the Caño Accretionary Complex, Caño Island and Osa Peninsula (Costa Rica, Central America). *12th Caribbean Geological Conference*, St Croix, Virgin Islands, 122.

MORELL, K. D., FISHER, D. M. & GARDNER, T. W. 2008. Inner forearc response to subduction of the Panama Fracture Zone, southern Central America. *Earth and Planetary Science Letters*, **265**, 82–95.

NELSON, B. K. 1995. Fluid-flow in subduction zones – evidence from Nd and Sr-isotope variations in metabasalts of the Franciscan Complex, California. *Contributions to Mineralogy and Petrology*, **119**, 247–262.

NISHIMURA, A. 1992. Paleocene radiolarian biostratigraphy in the Northwest Atlantic at Site-384, Leg-43, of the Deep-Sea Drilling Project. *Micropaleontology*, **38**, 317–362.

NISHIMURA, K. & ISHIGA, H. 1987. Radiolarian biostratigraphy of the Maizuru Group in Yanahara area,

Southwest Japan. *Memoirs of the Faculty of Science, Shimane University*, **21**, 169–188.

OBANDO, J. A. 1986. *Sedimentología y tectónica del Cretácico y Paleógeno de la region de Golfito, Península de Burica y Península de Osa, Provincia de Puntarenas, Costa Rica*. Licenciatura thesis, Escuela Centroamericana de Geología, Universidad de Costa Rica.

O'DOGHERTY, L. 1994. *Biochronology and Paleontology of Mid-Cretaceous Radiolarians from Northern Apennines (Italy) and Betic Cordillera (Spain)*.Mémoires de Geologie, Lausanne.

OKAMURA, Y. 1991. Large-scale melange formation due to seamount subduction; an example from the Mesozoic accretionary complex in central Japan. *Journal of Geology*, **99**, 661–674.

PAL, T., CHAKRABORTY, P. P., GUPTA, T. D. & SINGH, C. D. 2003. Geodynamic evolution of the outer-arc-forearc belt in the Andaman Islands, the central part of the Burma–Java subduction complex. *Geological Magazine*, **140**, 289–307.

PARK, J. O., TSURU, T. *ET AL.* 1999. A subducting seamount beneath the Nankai accretionary prism off Shikoku, southwestern Japan. *Geophysical Research Letters*, **26**, 931–934.

PESSAGNO, E. A. JR. 1976. Radiolarian zonation and stratigraphy of the upper Cretaceous portion of the Great Valley Sequence, California Coast Ranges. *Micropaleontology, Special issue*, **2**, 1–95.

PETTERSON, M. G., NEAL, C. R. *ET AL.* 1997. Structure and deformation of north and central Malaita, Solomon Islands: tectonic implications for the Ontong Java Plateau Solomon arc collision, and for the fate of oceanic plateaus. *Tectonophysics*, **283**, 1–33.

PETTERSON, M. G., BABBS, T. *ET AL.* 1999. Geological-tectonic framework of Solomon Islands, SW Pacific: crustal accretion and growth within an intra-oceanic setting. *Tectonophysics*, **301**, 35–60.

PHINNEY, E. J., MANN, P., COFFIN, M. F. & SHIPLEY, T. H. 2004. Sequence stratigraphy, structural style, and age of deformation of the Malaita accretionary prism (Solomon Arc–Ontong Java Plateau convergent zone). *Tectonophysics*, **389**, 221–246.

PINDELL, J., KENNAN, L., MARESCH, W. V., STANEK, K. P., DRAPER, G. & HIGGS, R. 2005. Plate kinematics and crustal dynamics of circum-Caribbean arc-continent interactions; tectonic controls on basin development in proto-Caribbean margins. *In*: *Caribbean–South American Plate Interactions, Venezuela*. Geological Society of America, Boulder, CO, Special Papers **394**, 7–52.

RANERO, C. R. & VON HUENE, R. 2000. Subduction erosion along the Middle America convergent margin. *Nature*, **404**, 748–755.

RANERO, C. R., VON HUENE, R., WEINREBE, W. & BARCKHAUSEN, U. 2007. Convergent margin tectonics: a marine perspective. *In*: BUNDSCHUH, J. & ALVARADO, G. E. (eds) *Central America – Geology, Resources and Hazards*. Taylor & Francis, Balkema, 239–276.

RANERO, C. R., GREVEMEYER, I. *ET AL.* 2008. Hydrogeological system of erosional convergent margins and its influence on tectonics and interplate seismogenesis. *Geochemistry Geophysics Geosystems*, **9**, Q03S04; doi: 10.1029/2007GC001679.

RAYMOND, L. A. 1984. Classification of melanges. *In*: RAYMOND, L. A. (ed.) *Melanges: their Nature, Origin and Significance*. Geological Society of America, Boulder, CO, Special Papers, **198**, 7–20.

RAYMOND, L. A. & TERRANOVA, T. 1984. Prologue: the melange problem: a review. *In*: RAYMOND, L. A. (ed.), *Melanges: their Nature, Origin and Significance*. Geological Society of America, Boulder, CO, Special Papers, **198**, 1–5.

REIDEL, W. R. & SANFILIPPO, A. 1974. Radiolaria from the southern Indian Ocean, DSDP Leg 26. *Initial Reports of the Deep Sea Drilling Project*, **XXVI**, 771–813.

SAK, P. B., FISHER, D. M. & GARDNER, T. W. 2004. Effects of subducting seafloor roughness on upper plate vertical tectonism; Osa Peninsula, Costa Rica. *Tectonics*, **23**, TC1017; doi: 1010.1029/2002TC001474.

SAK, P. B., FISHER, D. M., GARDNER, T. W., MARSHALL, J. S. & LaFEMINA, P. C. 2009. Rough crust, forearc kinematics, and Quaternary uplift rates, Costa Rican segment of the middle American Trench. *Geological Society of America Bulletin*, **121**, 992–1012.

SALLARÈS, V., DANOBEITIA, J. J., FLUEH, E. R. & LEANDRO, G. 1999. Seismic velocity structure across the middle American landbridge in northern Costa Rica. *Journal of Geodynamics*, **27**, 327–344.

SALLARÈS, V., DANOBEITIA, J. J. & FLUEH, E. R. 2001. Lithospheric structure of the Costa Rican Isthmus: Effects of subduction zone magmatism on an oceanic plateau. *Journal of Geophysical Research, B, Solid Earth*, **106**, 621–643.

SANFILIPPO, A. & RIEDEL, W. R. 1973. Cenozoic Radiolaria (exclusive of theoperids, artostrobiids and amphipyndacids) from the Gulf of Mexico, DSDP Leg 10. *In*: WORZEL, J. L., BRYANT, W. *ET AL.* (eds) *Initial Reports of the Deep Sea Drilling Project*. US Government Printing Office, **10**, 475–611.

SANFILIPPO, A. & RIEDEL, W. R. 1985. Cretaceous Radiolaria. *In*: BOLLI, H. M., SAUNDERS, J. B. & PERCH-NIELSEN, K. (eds) *Plankton Stratigraphy*, 573–630.

SANFILIPPO, A. & NIGRINI, C. 1998. Upper Paleocene– lower Eocene deep-sea radiolarian stratigraphy and the Paleocene/Eocene series boundary. *In*: AUBRY, M.-P., LUCAS, S. G. & BERGGREN, W. A. (eds) *Late Paleocene–Early Eocene Biotic and Climatic Events in the Marine and Terrestrial Records*. Colombia University Press, New York, 244–276.

SCHAAF, A. 1985. Un nouveau canevas biochronologique du Crétacé inférieur et moyen: les biozones à radiolaires. *Sciences Géologiques (Strasbourg)*, **38**, 227–269.

SENO, T. 2007. Collision vs. subduction: from a viewpoint of slab dehydration. *In*: DIXON, T. H. & MOORE, C. (eds) *The Seismogenic Zone of Subduction Thrust Faults. MARGINS Theoretical and Experimental Earth Science Series*. Columbia University Press, 601–623.

SENO, T. 2008. Conditions for a crustal block to be sheared off from the subducted continental lithosphere: What is an essential factor to cause features associated with collision? *Journal of Geophysical Research, B, Solid Earth*, **113**, B04414.

SHERVAIS, J. W., SCHUMAN, M. M. Z. & HANAN, B. B. 2005. The Stonyford volcanic complex: a forearc seamount in the northern California Coast Ranges. *Journal of Petrology*, **46**, 2091–2128.

SINTON, C. W., DUNCAN, R. A. & DENYER, P. 1997. Nicoya Peninsula, Costa Rica: a single suite of Caribbean oceanic plateau magmas. *Journal of Geophysical Research, B, Solid Earth*, **102**, 15507–15520.

SINTON, C. W., DUNCAN, R. A., STOREY, M., LEWIS, J. & ESTRADA, J. J. 1998. An oceanic flood basalt province within the Caribbean Plate. *Earth and Planetary Science Letters*, **155**, 221–235.

SMITH, W. H. F. & SANDWELL, D. T. 1997. Global sea floor topography from satellite altimetry and ship depth soundings. *Science*, **277**, 1956–1962.

TAIRA, A., MANN, P. & RAHARDIAWAN, R. 2004. Incipient subduction of the Ontong Java Plateau along the North Solomon trench. *Tectonophysics*, **389**, 247–266.

TAKETANI, Y. 1982. Cretaceous radiolarian biostratigraphy of the Urakawa and Obira areas, Hokkaido. *Science Reports of the Tohoku University*, **52**, 1–76.

THUROW, J., MOULLADE, M. *ET AL.* 1988. The Cenomanian/ Turonian boundary event (CTBE) at Hole 641A, ODP Leg 103 (compared with the CTBE interval at Site 398). *Proceedings of the Ocean Drilling Program, Scientific Results*, **103**, 587–634.

TOURNON, J. 1984. Magmatisme du Mésozoïque à l'actuel en Amérique Centrale: l'exemple du Costa Rica, des ophiolites aux andésites. *Mémoire des sciences de la terre, Université Pierre et Marie Curie*, **84**, 335.

TRENKAMP, R., KELLOGG, J. N., FREYMUELLER, J. T. & MORA, H. P. 2002. Wide plate margin deformation, southern Central America and northwestern South America, CASA GPS observations. *Journal of South American Earth Sciences*, **15**, 157–171.

UNDERWOOD, M. B. & BACHMAN, S. B. 1982. *Sedimentary Facies Associations within Subduction Complexes*. Geological Society, London, Special Publications, **10**, 537–550.

VANNUCCHI, P., SCHOLL, D. W., MESCHEDE, M. & KRISTIN, M. R. 2001. Tectonic erosion and consequent collapse of the Pacific margin of Costa Rica; combined implications from ODP Leg 170, seismic offshore data, and regional geology of the Nicoya Peninsula. *Tectonics*, **20**, 649–668.

VANNUCCHI, P., RANERO, C. R., GALEOTTI, S., STRAUB, S. M., SCHOLL, D. W. & McDOUGALL-RIED, K. 2003. Fast rates of subduction erosion along the Costa Rica Pacific margin: implications for non-steady rates of crustal recycling at subduction zones. *Journal of Geophysical Research, B, Solid Earth*, **108**.

VANNUCCHI, P., FISHER, D. M., BIER, S. & GARDNER, T. W. 2006. From seamount accretion to tectonic erosion; formation of Osa melange and the effects of Cocos Ridge subduction in southern Costa Rica. *Tectonics*, **25**, TC2004; doi: 2010.1029/2005TC001855.

VANNUCCHI, P., FISCHER, D. M. & GARDNER, T. W. 2007. Reply to comment by David M. Buchs and

Peter O. Baumgartner on 'From seamount accretion to tectonic erosion: Formation of Osa Mélange and the effects of the Cocos Ridge subduction in southern Costa Rica'. *Tectonics*, **26**, TC3010; doi: 3010.1029/2007TC002129.

VANNUCCHI, P., REMITTI, F. & BETTELLI, G. 2008. Geological record of fluid flow and seismogenesis along an erosive subducting plate boundary. *Nature*; doi: 10.1038/nature06486.

VERMA, S. P. 1992. Seawater alteration effects on Ree, K, Rb, Cs, U, Th, Pb and Sr-Nd-Pb isotope systematics of midocean ridge basalt. *Geochemical Journal*, **26**, 159–177.

VISHNEVSKAYA, V. S. 2001. *Jurassic to Cretaceous radiolarian biostratigraphy of Russia*. GEOS, Moscow, Thesis.

VISHNEVSKAYA, V. S. 2007. New radiolarian species of the family pseudoaulophacidae riedel from the upper Cretaceous of the Volga region. *Paleontological Journal*, **41**, 489–500.

VON HUENE, R. & LALLEMAND, S. 1990. Tectonic erosion along the Japan and Peru convergent margins. *Geological Society of America Bulletin*, **102**, 704–720.

VON HUENE, R. & SCHOLL, D. W. 1991. Observations at convergent margins concerning sediment subduction, subduction erosion, and the growth of continental-crust. *Reviews of Geophysics*, **29**, 279–316.

VON HUENE, R., BIALAS, J. *ET AL.* 1995. Morphotectonics of the Pacific convergent margin of Costa Rica.; Geologic and tectonic development of the Caribbean Plate boundary in southern Central America. *In*: MANN, P. (ed.) *Geologic and Tectonic Development of the Caribbean Plate Boundary in Southern Central America*. Geological Society of America, Boulder, Co, Special Papers, **295**, 291–307.

VON HUENE, R., RANERO, C. R. & VANNUCCHI, P. 2004. Generic model of subduction erosion. *Geology*, **32**, 913–916.

WALTHER, C. H. E. 2003. The crustal structure of the Cocos Ridge off Costa Rica. *Journal of Geophysical Research, B, Solid Earth and Planets*, **108**, 2136; doi: 2110.1029/2001JB000888.

WERNER, R., HOERNLE, K., VAN DEN BOGAARD, P., RANERO, C., VON HUENE, R. & KORICH, D. 1999. Drowned 14-m.y.-old Galapagos Archipelago off the coast of Costa Rica: implications for tectonic and evolutionary models. *Geology*, **27**, 499–502.

Rethinking the origins of the red chert at La Désirade, French West Indies

HOMER MONTGOMERY[1]* & ANDREW C. KERR[2]

[1]*Science and Mathematics Education, University of Texas at Dallas, Richardson, TX 75075, USA*

[2]*School of Earth and Ocean Sciences, Cardiff University, Cardiff CF10 3YE, UK*

*Corresponding author (e-mail: mont@utdallas.edu)

Abstract: La Désirade in the Lesser Antilles contains one of the rare fragments of Jurassic oceanic crust known on Caribbean islands. Others in the northeastern Caribbean occur on Puerto Rico and Hispaniola. These fragments each include radiolarian-bearing chert that has been linked to an origin in the Pacific Ocean. Of these, a fragment in Sierra Bermeja, Puerto Rico is clearly of Pacific origin as it contains Lower Jurassic radiolarians that predate the opening between North and South America. Red ribbon chert at El Aguacate, Dominican Republic is essentially identical to widespread radiolarites found in accreted material of the Pacific basin and from Pacific Ocean ODP Site 801. Extensive sampling in the Atlantic basin has produced no Jurassic radiolarites. Thus, based on age (the older of the Sierra Bermeja outcrops) and lithology (El Aguacate), two of these fragments are definitely of Pacific origin. Re-evaluation of the chert/pillow lava sequence on La Désirade in light of recent discoveries at spreading ridges has resulted in a revised interpretation of their probable origin. A wide range of features of these cherts indicate pelagic and hydrothermal sedimentation at an Upper Jurassic spreading ridge, one that almost assuredly existed in the eastern Pacific realm. These features include: the types of chert found on the island, lack of argillaceous partings, small outcrop size, discontinuous chert bodies, presence of limestone squeeze-ups into pillow lavas and indications of hydrothermal activity, including epidotization of basalt migrating outward from pillow margins with chert rinds that record pelagic and hydrothermal sedimentation at an Upper Jurassic spreading ridge, one that almost assuredly existed in the extreme eastern Pacific realm.

Mesozoic red cherts containing Jurassic radiolarian microfossils occur on La Désirade, Puerto Rico and Hispaniola (Fig. 1). These plate fragments have been interpreted in terms of a Pacific origin (Schellekens *et al.* 1990; Montgomery *et al.* 1992, 1994*a*, *b*; Montgomery & Pessagno 1999). Red chert of Puerto Rico and Hispaniola occurs in tectonic mélange as two types, intra-lava jasper and chert and ribbon chert above basalt. The rather extensive Upper Jurassic fragment at El Aguacate in the Dominican Republic is a genuine red ribbon chert strikingly similar to those in California (Montgomery *et al.* 1994*a*). A small upper Jurassic red chert body in Sierra Bermeja, Puerto Rico is not of the red ribbon variety. A second small outcrop in Sierra Bermeja presents some of the oldest rocks known from islands on the Caribbean Plate as it contains Lower Jurassic radiolarians (Montgomery *et al.* 1994*b*). This few square metre outcrop has been dismembered and metamorphosed so thoroughly that the tests for a true red ribbon chert, discussed below, cannot be applied with confidence to these rocks.

Well-studied Pacific Ocean floor strata provide comparative models for the poorly understood cherts of the Caribbean. Franciscan mélanges commonly consist of large tectonic blocks of pillow lava with red interpillow jasper and radiolarian ribbon chert capping the lava. Jasper deposition was intimately associated with active spreading, whereas the ribbon chert formed after the basalt had moved away from the spreading centre. Thus, it is vital to precisely define which types of red chert are present in the Caribbean. If all the chert on La Désirade is interpillow jasper, why, then, should there be any difference between Pacific and other oceanic crust with regard to red chert?

Montgomery *et al.* (1992) described the red chert of La Désirade as the Franciscan-type red ribbon variety because initial observations showed that the chert wedges are red, some display bedding, and a few intervals contain radiolarians. These initial conclusions were, however, somewhat misplaced since the assumption was made that the chert was simply dismembered red ribbon radiolarite and so an *in situ* hydrothermal origin was not considered. La Désirade cherts are markedly different from Franciscan red ribbon cherts in several critical ways. The observations and ideas contained in this paper stem from several subsequent visits to La Désirade and from new work on typical sediment

From: JAMES, K. H., LORENTE, M. A. & PINDELL, J. L. (eds) *The Origin and Evolution of the Caribbean Plate.*
Geological Society, London, Special Publications, **328**, 457–467.
DOI: 10.1144/SP328.18   0305-8719/09/$15.00 © The Geological Society of London 2009.

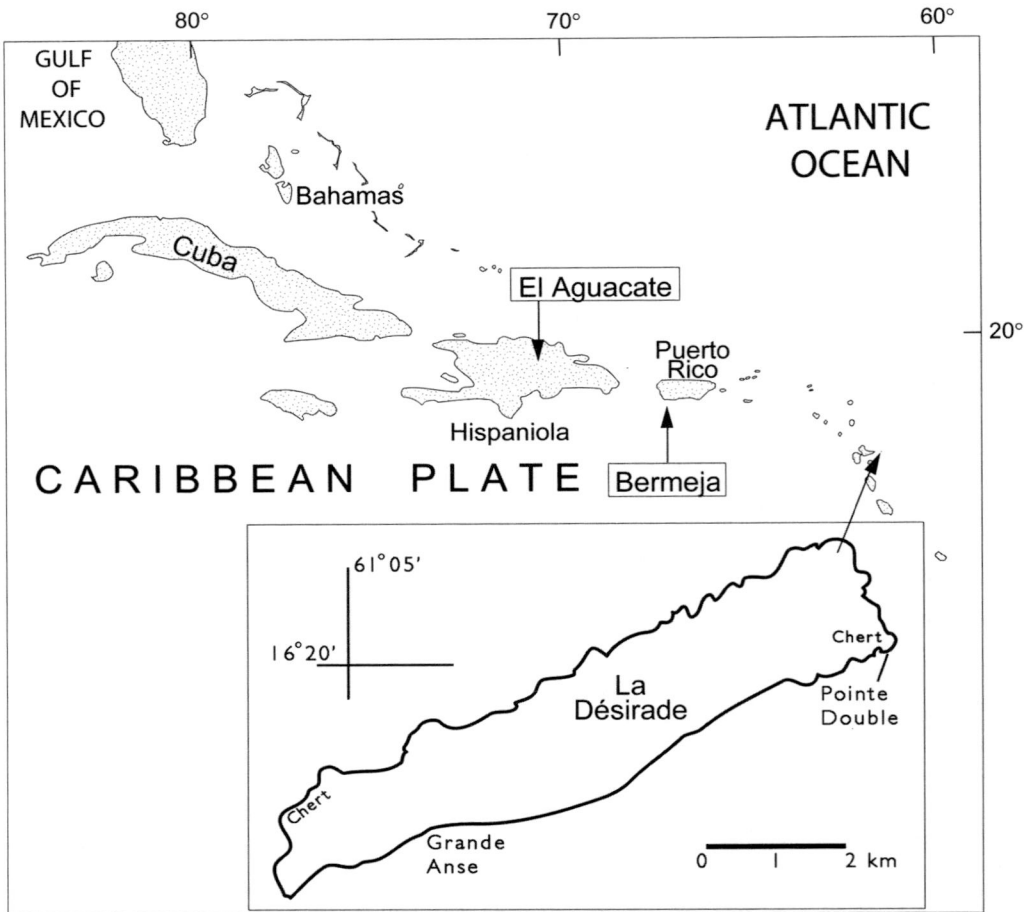

Fig. 1. Location map of La Désirade, French West Indies.

behaviour at a Mid-Ocean Ridge (MOR). The latter includes the actions of strong the bottom currents and the extreme mobility of new 'fluffy' pelagic sediment (Gooday 2003).

Reassessing the origin of the La Désirade basement rocks with a focus on chert also includes revisiting the tectonic implications. The initial study by Montgomery *et al.* (1992) noted that the red chert of La Désirade appeared to include miniature slivers of red ribbon chert (RRC) similar in character to classic localities such as the Marin Headlands of California. Accordingly, the cherts were described as being of red ribbon-type and so characteristic of an ophiolitic association deposited on oceanic crust distant from the spreading ridge (Montgomery *et al.* 1992). As we will show in this paper, further evaluation the La Désirade chert reveals that they appear more like chert deposits

formed at a MOR. We will also discuss if the red chert (that includes radiolarians) at La Désirade really provides good evidence that La Désirade is a Pacific fragment or if it could be proto-Caribbean in origin.

Geological setting

The present day Lesser Antilles arc is composed of a chain of active and dormant volcanoes most of which are capped with limestone. The arc is bounded on the east by a trench that marks the eastern boundary of the Caribbean Plate. La Désirade is a small, elongate island (2 × 9 km) located off the eastern tip of Guadeloupe in the French West Indies (Fig. 1). La Désirade represents the easternmost exposure of Jurassic igneous rocks on the Caribbean Plate. The nearest Jurassic outcrop is in the Greater Antilles, in Puerto Rico, 675 km to the NW.

The geological setting of La Désirade is discussed in detail by Mattinson *et al.* (2008). The complex is composed of volcanic, sedimentary and plutonic rocks. The oldest rocks include pillow basalt, red chert and pinkish tan to brick red limestone and are best exposed along the coast at the eastern end of the island. The red chert on both ends of the island contains essentially identical Upper Jurassic radiolarian faunas. Subsequent to chert and basalt deposition, crosscutting diabasic and microdioritic dykes were emplaced. Localized rhyolitic/trondhjemitic magmatism occurred during this period of dyke emplacement. Both in ophiolites and in modern oceanic crust, where they have been sampled on the seafloor, the basaltic lavas/dykes and gabbroic plutonics are earliest, while more siliceous magma develops locally at a later stage in the magmatic history when open-system fractionation (i.e., with continuous replenishment by basaltic magma) changes to closed-system fractionation (i.e., fractional crystallization without replenishment), allowing more siliceous magma to form. Field observations confirm that this appears to have been the case for La Désirade. The modifier 'meta' is utilized in the description of the rock units of the igneous complex to reflect hydrothermal and deuteric alteration of the rocks that was coeval with their formation. Resting unconformably on the igneous complex, a slab of Neogene nummulitid limestone caps much of the island rising to an elevation of approximately 275 m.

Chert is present as isolated bodies in the lavas; mostly in what appears to have been localized bathymetric flows a few metres across. The best-bedded exposures are also those with unambiguous stratigraphic position as they clearly lie beneath pillow lavas (Fig. 2). These cherts occur low in the section although some are present as wedges in tilted fault blocks of basalt. The blocks may be slightly higher in the section although this has not been structurally confirmed. The chert is always associated with basalt, and the basalt pile is clearly not capped by bedded chert. Isolated pockets of chert occur in the lava. Limestone is present as interpillow deposits having been squeezed up between extruding pillows. We interpret these squeeze-ups as reliable geopetals. As will be discussed, the stratigraphic relationship between chert and basalt on La Désirade has important tectonic implications.

The distinctive radiolarian *Vallupus hopsoni* is present in chert at both ends of the island, indicating that the associated lavas are coeval. The radiolarian fauna belongs to latest Jurassic Zone 4, Subzone 4 beta – lower Upper Tithonian (Montgomery *et al.* 1992). Mattinson *et al.* (2008) obtained a U–Pb zircon age of 143.74 ± 0.33 Ma (Berriasian) for the overlying trondhjemite so the underlying pillow basalt/chert/limestone sequence is older than this.

Palaeoceanography

Red radiolarian chert is common in the Circum-Pacific because this belt is characterized by subducting plate boundaries. Thus, older (Mesozoic)

Fig. 2. Red chert wedge deposited in a bathymetric low within the pillow lava section at La Désirade. Bedding is well defined, but there are no argillaceous partings. Radiolarians are rare. Some limestone is present and it squirted up into voids between basalt pillows. Locality at southeast end of the island.

oceanic crust and its overlying sedimentary strata, which include radiolarian ribbon cherts, are commonly exposed in accretionary wedges above subduction zones (e.g. the Franciscan Complex of California) and in collided island arcs (western Pacific). Similarly, the abundance of red radiolarian chert in the Alpine–Himalayan orogenic belt is due to the closure of the Neo-Tethys ocean basin and continental collision that expose Mesozoic oceanic rocks with radiolarian ribbon cherts. The reason that red radiolarian chert is not exposed above sea level in the Atlantic region is because Atlantic oceanic crust is the product of lithospheric extension (plate divergence) spreading outward from the Mid Atlantic Ridge (MAR). The earliest-formed basement and capping sedimentary strata remain unexposed beneath the deepest parts of the ocean, farthest from the MAR. Only drilling into Jurassic oceanic crust, which occurs beneath very thick sediment cover at the margins of the southern and equatorial Atlantic (not in the younger north Atlantic) might reveal such red cherts, if they exist. However, no DSDP-ODP drill sites have penetrated Atlantic oceanic crust with red radiolarian chert. In the central Atlantic at one of the more important locations (Site 534) that might contain Jurassic radiolarians, the comparison rocks are claystone (Bartolini & Larson 2001). This Middle Jurassic interval at Site 534 is only slightly younger than that at Site 801 in the Pacific that does contain red radiolarite. Thus, the presence of red cherts in the Circum-Pacific and Alpine–Himalayan belts, and non-exposure around the Atlantic, is determined by their respective tectonic histories.

Red siliceous rocks and associated biogenic components

Red siliceous rocks may be divided into distinct categories based on the mode of formation. The various types and modes of formation are reasonably well understood.

Primary ferruginous hydrothermal sediment

Primary ferruginous hydrothermal sediment is a product of low-temperature hot springs that vent at spreading ocean ridges (Goulding *et al.* 1998; Dias & Barriga 2006) along both the spreading axis and on the upper flanks of the ridge. Ferrous oxides (including hydroxide), nontronite and other substances precipitate around the vents, gradually building low mounds. These have a pelagic component, especially of settling radiolarians, that becomes incorporated during mound growth. The radiolarian component is enhanced where ooze is washed and redeposited from surrounding bathymetric highs.

Some hydrothermal mounds are preserved today in pristine condition, mainly on the ridge flanks where they were not buried by lava flows (as were those along the neovolcanic ridge axis). They are preserved in ophiolites as ferruginous umbers above the youngest lava but beneath sedimentary strata. A detailed discussion of the characteristics of umbers is presented by Richards & Boyle (1986). Both the mounds and the umbers are low-density material with abundant open pore space. The original uncompacted sediment is thought to have been water-saturated and highly incompetent ('fluffy'), similar to newly deposited calcareous ooze observed by submersibles at modern ocean ridges. Such ooze is highly mobile in ocean floor currents (Sarnthein & Faugeres 1993) due to the low density of the delicate, hollow structures of radiolarians. Spines and other ornamentation inhibit adhesion and further facilitate mobility.

Intralava sedimentary rocks

Intralava red jasper is deposited as ferruginous hydrothermal mounds containing a pelagic component, especially radiolarians, within the axial zone of volcanism along the spreading center. Here, after a period of accumulation, pelagic sediments are buried beneath new lava flows. An immediate result of loading is that the loose 'fluffy' sediment squirts up into fractures and fills interpillow open spaces (Fig. 3). If the original 'mound' was thin, only a discontinuous layer of ferruginous sediment (now jasper) marks the base of the flow. If the original mound was several metres thick, then the buried mound is preserved as lenses of intralava sediment. Burial beneath a new lava flow also results in: (1) recrystallization from 'cooking'; (2) oxidation to submicroscopic red hematite, red colouration of the metasediment; and (3) silicification by hydrothermal fluids in open pore space, converting ferruginous sediment to dense ferruginous chert or jasper. The bright red coloration of the jasper is conspicuous in the dark-coloured lava. Note that pelagic limestone (white, light grey or pink) commonly accompanies red jasper as intralava sediment, even in the same lava flow, since most ocean spreading ridges have a crestal depth between 2500 and 3000 m above the calcium carbonate compensation depth (CCD). The co-occurrence of intralava jasper and pelagic limestone is due to hydrothermal sediment from internal sources mixing with pelagic sediment that is deposited from higher in the water column.

The sampled Jurassic limestone on La Désirade (Table 1) is low in Fe_2O_3 and is fairly siliceous. This observation coupled with the presence of radiolarians and planktonic foraminifera indicates a pelagic rather than a hydrothermal source for the silica. As for the siliceous nature, Ogg *et al.* (1992),

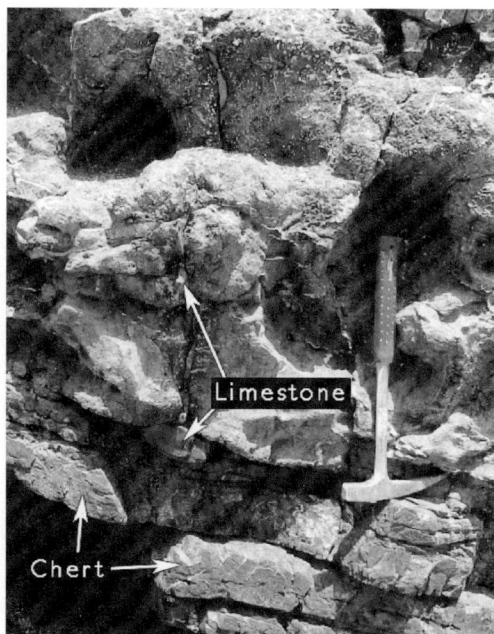

Fig. 3. The red rock parallel to the hammer handle is limestone. The bedded rock below the hammer is chert. 'Fluffy' calcareous ooze squirted up into interpillow voids. Both the limestone and the chert contain the same radiolarians.

discussing Jurassic carbonate preservation in the central equatorial Pacific, postulated that the CCD remained above 2700 m at that time. Below the CCD only siliceous organisms remain. Siliceous limestone formed above the CCD. Thus, the presence of calcareous plankton indicates that La Désirade chert/pillow lava sequence was deposited in waters slightly above the CCD and in a pelagic realm.

Intralava red radiolarian chert

Because of its slow deposition rate radiolarian ooze forms only a thin film blanketing lavas in the axial eruptive zone. Axial separation and spreading carries opposing oceanic crustal segments away from the eruption zone before significant thicknesses of radiolarian ooze accumulate. Thus, intralava radiolarian cherts are uncommon whereas radiolarian ribbon chert can build up to tens of metres thick on top of the lavas during subsequent seafloor spreading (Ogg et al. 1992).

There are exceptions to this general situation. Strong bottom currents scouring the axial neovolcanic zone of spreading ocean ridges wash the light 'fluffy' radiolarian ooze from the seafloor, especially from bathymetric highs, redepositing it in depressions. Thus, continuous slow pelagic sedimentation, and continuous stripping of ooze from high-standing seafloor and redeposition in adjacent depressions, causes the buildup of pelagic sediment ponds, some a few metres deep. Some ponded sediment develops thin bedding, from repeated depositional pulses. Later, when buried beneath a new lava flow, the locally ponded radiolarite becomes an intralava lens of radiolarian chert. Here again, burial beneath new lava was accompanied by loading (with intrusion into the overlying lava flow), recrystallization from heating, and oxidation to red hematitic chert.

Thus, intralava chert derived from radiolarian ooze and intralava jasper produced as hydrothermal sediment can resemble each other. Differences include a higher radiolarian content of the former and a higher Fe-oxide content of the latter. However, since gradations between the two end member sediments are common, so are gradations between the end-member radiolarian cherts and jaspers. Distinguishing the two in the field can be challenging; however, on La Désirade, cherts apparently derived from ooze have high radiolarian content, whereas no radiolarians occur in the assumed hydrothermal jaspers.

Intralava jaspers and radiolarian chert lenses are clearly independent of geographic position, since hot springs, and thus hydrothermal sediments, can exist wherever new oceanic crust forms at spreading centres. Furthermore, recrystallization, oxidation, and silicification accompanying burial beneath new lava flows takes place everywhere along oceanic spreading centres.

Table 1. *Chemical analyses*

		SiO_2	Fe_2O_3	MnO	MgO	CaO
LD92/15	Mostly carbonate	12.65	1.71	0.14	1.78	45.43
LD92/14	Mostly carbonate	13.26	2.36	0.10	2.59	43.71
LD92/09	Mostly carbonate	27.73	1.18	0.16	1.91	38.67
LD92/06	Mostly carbonate	4.40	1.74	0.12	0.85	51.98
LD06/03	Chert	69.07	18.13	0.07	0.96	0.49
LD06/02	Chert	74.81	13.28	0.07	1.85	0.95

Analysis of chert and carbonate of La Désirade.

Radiolarian ribbon chert on oceanic crustal lava

During the Jurassic, when the open-ocean CCD was around 2700 m (Ogg *et al.* 1992), pelagic sediment reaching the deeper Pacific ocean floor was pelagic radiolarian ooze (siliceous), not pelagic limey ooze (calcareous) (Ogg *et al.* 1992). As illustrated by studies at ODP Site 801 in the Pacific basin, Tithonian radiolarian abundance is positively correlated with a high-productivity zone close to the palaeo-equator (Ogg *et al.* 1992). Limey ooze formed on higher parts of the ridge axis but not at deeper locations such as axial discontinuities (e.g. overlapping spreading centres) or where fracture zones offset the ridge axis. Only radiolarian ooze was deposited on the ridge crest neovolcanic zone as well as on its deeper flanks. These ooze deposits became the red radiolarites described from ODP Site 801.

Outside the axial zone of spreading ridges, where burial beneath new lava flows does not occur, radiolarian oozes, including minute clay components, accumulated below the CCD, slowly form deposits up to tens of metres thick. Ogg *et al.* (1992) described a mechanism for the generation of red ribbon chert by deposition of alternating radiolarian-rich and clay-rich layers, in which the clay fraction originates as aeolian or distal turbiditic sediment. The ooze does not lithify directly to radiolarian chert; it initially compacts to porous, low-density radiolarite exhibiting 'proto-stratification' and then undergoes diagenesis to radiolarian ribbon chert. Diagenesis involving silica migration results in segregation of the clay and siliceous components (Ogg *et al.* 1992). Thus, siliceous, radiolarian-rich layers alternate with thin argillaceous partings. The ribbon structure should not be confused with the thin bedding present in intralava red cherts, which is depositional and not diagenetic. Diagenetic segregation is similar to metamorphism at higher temperatures that results in laminated phyllites and schists. Despite being produced by a seemingly simple diagenetic process, red ribbon deposits are only known from sites that include Jurassic submarine lavas. Volcanic eruptions generated the necessary silica for radiolarian test formation.

Significance of red coloration

The reddish coloration of radiolarian chert and jasper is caused by submicroscopic hematite, the product of oxidation due to vaporization of entrapped seawater within growing submarine lava piles at spreading centres. The Fe-bearing sediments are reddened during formation of intralava jaspers and cherts, which include varieties such as the brick red and reddish brown cherts that overlie

Fig. 4. Chert lens with a darker (Mn-rich) base and lighter-colored red (Fe-rich) upper portion. Southeast end of the island.

ophiolitic basalts and modern oceanic crust, as at ODP Site 801. Oxidation in this latter case is caused by well-oxygenated bottom waters.

Blackish or dark grey coloration is imparted by manganese oxide, since Mn, like Fe, is derived from low-temperature MOR hot springs. However, Mn reaches saturation at somewhat higher temperatures and tends to precipitate earlier at hot springs. For example, umbers that overlie pillow basalts in the Samail ophiolite of Oman, reflect progressive cooling of the depositing solutions, and grade upward from black (Mn-rich) to red (Fe-rich) coloration, indicating a decreasing Mn:Fe ratio of those off-axis hydrothermal sediments. Manganese ore deposits in Franciscan pillow lavas of California (Crerar *et al.* 1982) are products of hydrothermal deposition when temperatures were too high to produce iron ferric-hydroxides. Circulating bottom waters along spreading ocean ridges sometimes precipitate minute quantities of Mn oxide before ferric oxide, perhaps explaining upward gradation from dark grey to reddish radiolarian cherts on submarine lavas at some Franciscan localities (personal observation). At La Désirade hot hydrothermal waters mixing with cold waters precipitated a great deal of iron (throughout the basalt section) and visible manganous oxides (see Fig. 4), especially low in the section.

Hydrothermal alteration

Of interest, particularly because the base of the section has not been much discussed in the literature, are the epidotized basalts below the lowest

Fig. 5. Fractures in basalt at the base of the section along the beach. Coloration banding is present with the fracture itself containing a few centimetres thick rind of brick red jasper.

chert and pillow lava layer on the SE end of the island. Epidotized basalt is also common at the Point Sal remnant of the Coast Range Ophiolite near the base of the basaltic layer (personal observation). Such secondary epidote is deposited by circulating hydrothermal fluids filling voids and replacing some of the basalt. Metasomatic epidotization is a common feature at deep levels of oceanic basalts, where hydrothermal solutions circulate. Red coloration of the epidotized basalt on La Désirade has two different origins. (1) Hydrothermal sediment, reddened by heating and oxidation, was buried by new lava flows and squirted up into the flow. The contact between the hydrothermal sediment and the basalt is sharp. The fractures are filled with jasper and are variable in width, generally from less than 1 to approximately 3 cm. (2) Fluids present during the accumulation of the basaltic layer locally oxidized the lava along fractures that are now brick-red in colour. A prominent colour banding is present extending outwards from the fractures (Fig. 5).

Palaeogeographic significance of red coloration in intralava jasper and chert v. supralava ribbon chert

Intralava sediments, reddened by oxidation caused by vaporization of entrapped seawater, probably occur wherever oceanic crust is formed at spreading centres, regardless of age or geographic location. Supralava ribbon cherts are reddened by oxidation from bottom waters in regions of good oceanic circulation. They would not be expected where oceanic circulation was poor. We are speculating, but such low oxygen conditions may have existed

in the proto-Atlantic where the still narrow, newly opening rift basin impeded circulation, possibly explaining the apparent absence of Jurassic red ribbon chert in the Atlantic realm.

The chert protolith on La Désirade accumulated in bathymetric lows during periods when seawater was actively pumping through a hydrothermal system and pillow lavas were not erupting locally. As noted, the chert on La Désirade is enriched in Fe_2O_3 (Table 1), an indication that the parent was ferruginous hydrothermal sediment (Richards & Boyle 1986). At fast-spreading ridges lava flows form within hours or days (Massoth et al. 1995) with tens to hundreds of years between eruptions. In the case of slow-spreading ridges, thousands to tens of thousands of years may separate eruptions. Quiescent periods provide sufficient opportunity for hydrothermal sediment accumulation between lava flows.

Sedimentology

The dearth of radiolarians in the sedimentary rocks of La Désirade suggests exceedingly slow accumulation in shallow bathymetric depressions without much disturbance. Alternatively, if the cherts are of hydrothermal origin, they may have accumulated rapidly since work by Hayman (1989) revealed that many ridge-crest hydrothermal vents are short-lived, on the order of tens of years.

Accumulation of carbonate deposits would also follow this pulsed and rapid timetable. The pelagic sediment is deposited at a steady rate, which is much greater in the highly productive equatorial biological realm than farther north or south (Ogg et al. 1992). This pelagic sediment contains both calcareous and siliceous micro-skeletons, with a much higher proportion of the former. Following deposition, the siliceous skeletons dissolve during diagenesis leaving only the majority calcareous organisms (e.g., foraminifera). The preferential occurrence of limestone as intrapillow sedimentary rock was probably the result of infilling by the exceedingly light, fluffy and remobilized pelagic calcareous ooze. When covered by a new lava flow the sediments were too mobile to remain as interbeds. They mostly intruded (squirted up into) the over-riding flow. In short, the ferruginous hydrothermal sediments are not as fluffy and mobile as the calcareous ooze; therefore they commonly remain as bathymetric low fillings between pillowed flows, whereas the overridden calcareous ooze mostly squirts up into new cracks and between pillows.

Fe_2O_3 concentration would have been much higher in the original hydrothermal, ferruginous umber sediment, but subsequent and variable silicification during burial beneath new hot lava flows converted ferruginous umber to jasper (low Si:Fe ratio) or even chert (higher Si:Fe ratio), resulting in variable Fe_2O_3 concentrations (cf. Richards & Boyle 1986).

Biogenic components and their behaviours

Radiolarians are far less common in La Désirade red cherts than in typical red ribbon chert such as in the Franciscan or even at El Aguacate, where radiolarians are abundant in numerous intervals. Ogg et al. (1992) defined radiolarite as 'any siliceous sediment that has more than 25% radiolarians. The degree of silicification and type of bedding are not part of this definition.' Other than in a few rich micro-intervals, radiolarians are mostly absent or rare in the La Désirade red cherts.

The jaspers and red cherts are gradational with respect to Si:Fe ratio and are basically hydrothermal, not pelagic. The source of their silica is secondary. They are silicified by pore-filling deposition from the circulation of silica-saturated solutions during the heating and oxidation that accompanies burial beneath new lava flows (Richards & Boyle 1986). There is commonly a minor radiolarian component, the radiolarians that accompany the pelagic calcareous rain forming ooze. However, this 'sprinkling' is variable. Although radiolarians cannot be called plentiful in the chert at La Désirade, in a few locations radiolarians are relatively abundant where they were washed off sea bottom bathymetric highs and redeposited in lows. If those resedimented radiolarians come to rest amidst hydrothermal sediment, they are chemically stable with their host. They are not dissolved and become a permanent component. In contrast, the accompanying calcareous plankton is dissolved.

Discussion

Outcropping chert and limestone associated with pillow lavas present complicated relationships. Classic RRC (circum-Pacific material) is a thick and aerially extensive, rhythmically bedded rock with thin argillaceous partings (e.g. Bailey et al. 1964; Murchey 1984; Murchey & Jones, 1984). La Désirade chert is composed of several small blocks of only a few to a few tens of square metres in exposure. There are no argillaceous partings. Bedding is occasionally good but cannot be generally described as such. The RRC at El Aguacate is extensive with exposure over approximately 3 km^2. Well-defined shale partings are present (Fig. 6).

La Désirade chert is present almost exclusively below pillow lavas. Only a few small patches of chert crop out on top of the hills. None of these outcrops can be confirmed to be other than exposures within near vertical beds. Nothing on La Désirade

Fig. 6. Red ribbon chert at El Aguacate, Dominican Republic illustrating well defined argillaceous partings.

is remotely similar to the characteristic chert above basalt present in RRC (perfect keels on the pillows confirm right-side-up). Calcareous sediment was squeezed up between the pillows suggesting submarine eruption of pillows, onto pockets of unlithified pelagic material (Fig. 3).

The presence of limestone suggests deposition above the CCD, a condition that is not characteristic of Mesozoic RRC but is characteristic of a spreading ridge at approximately 2500–3000 m depth (Ogg et al. 1992). Limey ooze (>30% carbonate content) composed of calcareous pelagic plankton (and radiolarians) settles onto the submarine ridge crest (active neotectonic zone of lava eruption) when it is above the CCD, whereas only siliceous ooze (>30% siliceous content) from pelagic radiolarians accumulates below the CCD. The two situations are distinguishable in the rock record.

Because Ca-saturation is achieved only in the ocean's upper zone and Ca-undersaturation increases with depth, the open-ocean CCD was shallower in the Jurassic than in the Cretaceous and Cenozoic, permitting radiolarian ooze to settle directly onto spreading ridge crests, below the elevated CCD, at least along deeper portions of the ridge axis (i.e. near ridge-axis discontinuities and transform-fault intersections). Thus, intralava radiolarian chert could form between Jurassic oceanic crustal lavas, but not those of Cretaceous or younger age. In contrast, intralava hydrothermal jaspers should occur in all oceanic crustal lavas, irrespective of age.

On La Désirade, radiolarians are in overall low abundance or, most commonly, nonexistent in the siliceous rocks, while they are rather common in some of the limestone. Bedded radiolarian chert as exemplified in the Franciscan Complex (Murchey 1984) varies in radiolarian richness from zero to more than 50% in numerous intervals. Consequently, much of the La Désirade chert is likely to represent jasper of hydrothermal origin. The jasper contains radiolarians in only a few thin intervals periodically sprinkled with remobilized radiolarian ooze. Richards & Boyle (1986) noted this prominent dearth of radiolarians in jasper whose origin they ascribed to an original sedimentary origin mostly from hydrothermal exhalation. Radiolarian ooze was probably reworked into small depressions from adjacent bathymetric highs by strong bottom currents. Most of the chert is barren, as would be expected of mostly hydrothermal deposits. Where present, even though relatively rare overall, radiolarian abundance can be great (Fig. 7). Radiolarians in the chert and the limestone are identical so it is clear that radiolarian deposits first underwent reworking into jasper and were then mobilized into the overlying limestone before the eruption of pillow lavas onto these deposits. Much of this

Fig. 7. Thin sections of the densest radiolarian beds discovered in the Désirade chert. Radiolarians were Probably remobilized by being swept into bathymetric depressions by bottom currents.

scenario is plausible only at a site of active volcanism, either at a mid-ocean ridge or back-arc basin – Pacific or otherwise.

Bioturbated siliceous deposits drilled during ODP Leg 129 in the Pacific formed in areas conducive to the existence of burrowing fauna and as a result lumpy chert predominates. Since the chert at La Désirade does not share this feature, it is reasonable to suppose that conditions were unsuitable for burrowing fauna. Such biologically hostile conditions are likely to have existed in material that was rapidly resedimented into bathymetric lows in an active field of basalt extrusion characteristic of an active spreading centre.

Does the red radiolarian chert answer our fundamental tectonic question? Could the La Désirade basement rocks represent oceanic crust derived from the Pacific? Perhaps, if it is clearly associated with the cherts of Puerto Rico and Hispaniola. Could it represent proto-Caribbean oceanic crust? It is essentially assured that the La Désirade chert is a piece of Caribbean forearc material that was once the backarc of the westfacing pre-subduction polarity reversal arc that existed from Chortis to Ecuador. Once the polarity reversal occurred, La Désirade was located in the new forearc position and has remained there ever since.

Aside from this reinterpretation of the La Désirade chert, the authentic Oxfordian to Tithonian-aged red ribbon chert at El Aguacate in the Dominican Republic and the dismembered and metamorphosed Lower Jurassic material at Bermeja must both be of Pacific origin.

Conclusions

The origins of various chert bodies of La Désirade are more comprehensible in light of findings at Pacific Site 801 and at MOR environments. We now better understand the sedimentary history of La Désirade's Jurassic ocean basement. Simply assuming the chert is highly dismembered red ribbon radiolarite (Montgomery et al. 1992) was a flawed concept. Important observations leading to this conclusion for La Désirade are as follows.

With or without radiolarians, small bodies of red chert and jasper were deposited within intralava bathymetric lows, with no shaly partings. All known examples of red radiolarian ribbon chert have argillaceous partings and are found overlying submarine lava. There are no rocks at La Désirade approaching either the dimensions of typical red ribbon chert, nor is there any outcropping slab of chert on top of the basalt flows.

Accumulating siliceous and carbonate sediments were hostile environments to infauna since no nodular sedimentary deposits are present. No evidence has been observed in any of the chert at La Désirade of trace fossil activity. An active interpillow environment would be hostile compared with slow deposition atop basalt flows farther from rifting. Accumulations of sediment were buried rapidly and certainly before lithification, as indicated by the presence of carbonate squirting from below into overlying lavas.

We remain uncertain how much of the red chert at La Désirade was derived from hydrothermal sediment with a radiolarian component and how much was resedimented, ponded radiolarian ooze. Given the relative rarity of radiolarians and the exceedingly few observed intervals where they are present, we suspect the hydrothermal component is greater. However, intralava limestone, deposited above CCD, seems to represent the majority of genuine pelagic sediment. Thus, the chert, radiolarian-bearing or not, is probably mostly jasper, a type of hydrothermal sediment with a minor radiolarian component.

The primary sediments, hydrothermal and siliceous pelagic ooze, were reddened and lithified to jasper and chert by heating, oxidation, and recrystallization during burial beneath lava flows. Oceanic circulation in the overlying water column was not a factor.

La Désirade provides a compelling example of the dynamic nature of Late Jurassic sedimentation in the bathymetric depressions associated with ocean floor rifting. Deposition was rapid, given our understanding of vent history, and the sediment had both local and pelagic sources. Siliceous sediment was remobilized into bathymetric lows and then deformed by overflowing pillows. Fluffy accumulations of pelagic carbonate sediment squirted up into interpillow spaces of newly extruded overlying lavas.

The ideas about ridge origins presented herein evolved from long conversations with C. Hopson. His understanding of the intricacies of the Coast Range Ophiolites is extraordinary.

References

BAILEY, E. H., IRWIN, W. P. & JONES, D. L. 1964. Franciscan and related rocks, and their significance in the geology of western California. *California Division of Mines and Geology Bulletin*, **183**, 177.

BARTOLINI, A. & LARSON, R. 2001. The Pacific microplate and Pangea supercontinent in the Early–Middle Jurassic. *Geology*, **29**, 735–738.

CRERAR, D. A., NAMSON, J., CHYI, M. S., WILLIAMS, L. & FEIGENSON, M. D. 1982. Manganiferous cherts of the Franciscan assemblage: I. General geology, ancient and modern analogues, and implications for hydrothermal convection at oceanic spreading centres. *Economic Geology*, **77**, 519–540.

DIAS, A. S. & BARRIGA, F. J. A. S. 2006. Mineralogy and geochemistry of hydrothermal sediments from the serpentinite-hosted Saldanha hydrothermal field (36°34′N; 33°26′W) at MAR. *Marine Geology*, **225**, 157–175.

GOODAY, A. J. 2003. Introduction to meiobenthos studies. *Berichte zur Polar- und Meeresforschung*, **470**, 1–50. World Wide Web Address: http://eprints.soton.ac.uk/1294/

GOULDING, H. C., MILLS, R. A. & NESBITT, R. W. 1998. Precipitation of hydrothermal sediments on the active TAG Mound; implications for ochre formation. *In*: MILLS, R. A. & HARRISON, K. (eds) *Modern Ocean Floor Processes and the Geological Record*. Geological Society, London, Special Publications, **148**, 201–216.

HAYMAN, R. M. 1989. Hydrothermal processes and products on the Galapagos Rift and East Pacific Rise. *In*: WINTERER, E. L, HUSSONG, D. M. & DECKER, R. W. (eds) *The Eastern Pacific Ocean and Hawaii*. The Geology of North America. Geological Society of America, Boulder, CO, 125–144.

MASSOTH, G. J., BAKER, E. T. *ET AL*. 1995. Observations of manganese and iron at the co-axial seafloor eruption site, Juan de Fuca Ridge. *Geophysical Research Letters*, **22**, 151–154.

MATTINSON, J., PESSAGNO, E., MONTGOMERY, H. & HOPSON, C. 2008. Late Jurassic age of oceanic basement at La Désirade Island, Lesser Antilles arc. *In*: WRIGHT, J. & SHERVAIS, J. (eds) *Ophiolites, Arcs, and Batholiths: A Tribute to Cliff Hopson*. Geological Society of America Special Paper, **438**, 175–190.

MONTGOMERY, H., PESSAGNO, E. A. JR & MUNOZ, I. 1992. Jurassic (Tithonian) radiolaria from La Désirade (Lesser Antilles): preliminary paleontological and tectonic implications. *Tectonics*, **11**, 1426–1432.

MONTGOMERY, H., PESSAGNO, E. A. JR, LEWIS, J. A. & SCHELLEKENS, J. H. 1994a. Paleogeography of the Jurassic fragments in the Caribbean. *Tectonics*, **13**, 725–732.

MONTGOMERY, H., PESSAGNO, E. A. JR & PINDELL, J. L. 1994b. A 195 Ma terrane in a 165 Ma ocean: Pacific origin of the Caribbean Plate. *GSA Today*, **4**, 1–6.

MONTGOMERY, H. & PESSAGNO, E. A. JR. 1999. Cretaceous microfaunas of the Blue Mountains, Jamaica, and of the Northern and Central Basement Complexes of Hispaniola. *In*: MANN, P. (ed.) *Caribbean Basins*. Sedimentary Basins of the World, **4**, Elsevier Science, Amsterdam, **10**, 237–246.

MURCHEY, B. M. 1984. Stratigraphy and lithostratigraphy of chert in the Franciscan Complex, Marin Headlands, California. *In*: BLAKE, M. C. JR (ed.) *Franciscan Geology of Northern California*. Pacific Section S.E.P.M., **43**, 51–70.

MURCHEY, B. M. & JONES, D. L. 1984. Age and significance of chert in the Franciscan Complex in the San Francisco Bay Region. *In*: BLAKE, M. C. JR (ed.) *Franciscan Geology of Northern California*. Pacific Section S.E.P.M., **43**, 23–30.

OGG, J. G., KARL, S. M. & BEHL, R. J. 1992. Jurassic through Early Cretaceous sedimentation history of the central equatorial Pacific and of ODP Sites 800 and 801. *Proceedings. Ocean Drilling Program, Scientific Results*, **129**, 571–613.

RICHARDS, H. G. & BOYLE, J. F. 1986. Origin, alteration and mineralization of inter-lava metalliferous sediments of the Troodos Ophiolite, Cyprus. *In*: GALLAGHER, M. J., IXER, R. A., NEARY, C. R. & PRICHARD, H. M. (eds) *Metallogeny of Basic and Ultrabasic Rocks*. Institute of Mineralogy and Metallurgy, London, 21–31.

SARNTHEIN, M. & FAUGERES, J. C. 1993, Radiolarian contourites record Eocene AABW circulation in the equatorial East Atlantic. *Sedimentary Geology*, **82**, 145–155.

SCHELLEKENS, J., MONTGOMERY, H., JOYCE, J. & SMITH, A. 1990. Late Jurassic to Late Cretaceous development of island arc crust in southwestern Puerto Rico. *Transactions of the Caribbean Geological Conference*, **12**, 268–281.

Accreted oceanic terranes in Ecuador: southern edge of the Caribbean Plate?

ETIENNE JAILLARD[1,2]*, HENRIETTE LAPIERRE[†], MARTHA ORDOÑEZ[3], JORGE TORO ÁLAVA[4], ANDREA AMÓRTEGUI[2] & JÉRÉMIE VANMELLE[2]

[1]*IRD-LMTG, Observatoire Midi-Pyrénées, 14 av. Edouard Belin, 31400 Toulouse, France*

[2]*LGCA, Maison des Géosciences, BP 53, 38041 Grenoble Cedex 09, France*

[3]*Petroproducción, CIG-Guayaquil, km 6,5 vía a la Costa, Guayaquil, Ecuador*

[4]*Petroproducción, CIG-Quito, av. 6 de Diciembre y G. Cañero, PO Box, 17-01-1006, Quito, Ecuador*

[†]Deceased January 2006

Corresponding author (e-mail: Etienne.Jaillard@ujf-grenoble.fr)

Abstract: The western part of Ecuador is made from several oceanic terranes, which comprise two oceanic plateaus, of Early ($c.$ 120 Ma) and Late Cretaceous age ($c.$ 90 Ma), respectively. The older oceanic plateau was accreted to the Andean margin in the Late Campanian ($c.$ 75 Ma). Fragments of the Turonian–Coniacian plateau were accreted to the Ecuadorian margin in the Late Maastrichtian ($c.$ 68 Ma, Guaranda terrane) and Late Paleocene ($c.$ 58 Ma, Piñón–Naranjal terrane). The Guaranda terrane received either fine-grained oceanic sediments of Coniacian–Maastrichtian age, or island arc/back-arc volcanic suites of Middle Campanian–Middle Maastrichtian age. The Piñón–Naranjal terrane recorded a comparable history, completed in the Maastrichtian–Paleocene, either by pelagic cherts, or by island arc products (Macuchi arc). The Late Cretaceous plateau of Ecuador is interpreted as part of the Caribbean oceanic plateau (COP), because their evolutions are comparable. If so, the COP was not formed by the Galápagos hotspot, but on the Farallón oceanic plate, south of Ecuador and close to the South American margin. The COP belonged to the Farallón plate, until a subduction zone separated both plates in the Middle Campanian, giving way to a well-developed Mid Campanian–Mid Maastrichtian island arc. Accretion in the Late Maastrichtian triggered a change in the subduction system, and the development of a new arc system of Late Maastrichtian–Late Paleocene age, which crosscut the South America–COP plate boundary. The last accretion occurred in the Late Paleocene.

Although the Andes of South America form a continuous mountain belt and result from processes related to oceanic subduction, the geological evolution varies latitudinally along the chain. The northern Andes differ from the central Andes, by the lack of manifestations of oceanic subduction between latest Jurassic and Eocene times ($c.$ 140–40 Ma), and by the presence of magmatic material of oceanic origin in their western part (Gansser 1973; Goossens & Rose 1973). These oceanic units presently constitute the Coastal and Western Cordillera of Ecuador and Colombia and are separated from the continental margin by depressions (Inter Andean valley of Ecuador), usually filled with Cenozoic to Recent deposits, which obscure their geometric relations with the Andean margin (Fig. 1).

This oceanic material is widely accepted as representing Cretaceous oceanic terranes, accreted to the Andean continental margin between Late Cretaceous and Eocene times (e.g. Feininger & Bristow 1980; Lebrat et al. 1987; Reynaud et al. 1999; Kerr et al. 2002; Jaillard et al. 2004; Luzieux et al. 2006). Early petrographic and geochemical works on the magmatic basement of western Ecuador allowed the recognition of oceanic floor basalts, and island arc basalts and andesites (Kehrer & Van Der Kaaden 1979; Egüez 1986; Lebrat et al. 1987; Wallrabe-Adams 1990). In the same way, early workers proposed that accretions occurred in the Campanian, Paleocene and Eocene (Feininger & Bristow 1980; Lebrat et al. 1987; Daly 1989; Bourgois et al. 1990).

More recently, detailed geological survey (McCourt et al. 1998; Hughes et al. 1998; Boland et al. 2000; Fig. 1), systematic sampling and analysis of magmatic rocks (Reynaud et al. 1999; Lapierre et al. 2000; Kerr et al. 2002; Mamberti et al. 2003, 2004), associated with radiometric

From: JAMES, K. H., LORENTE, M. A. & PINDELL, J. L. (eds) *The Origin and Evolution of the Caribbean Plate.* Geological Society, London, Special Publications, **328**, 469–485.
DOI: 10.1144/SP328.19   0305-8719/09/$15.00 © The Geological Society of London 2009.

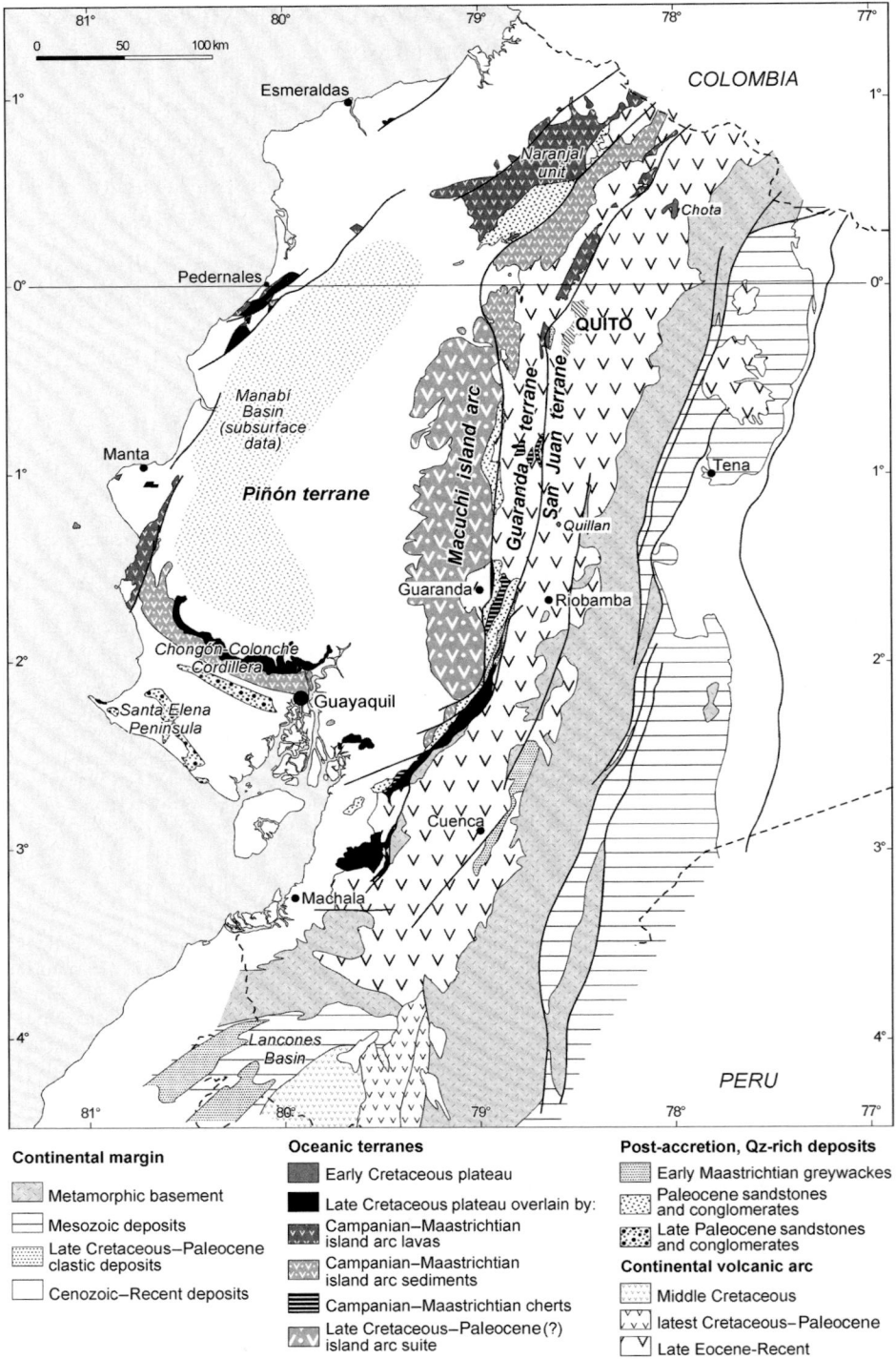

Fig. 1. Geological map of western Ecuador (simplified from Litherland *et al.* 1994; McCourt *et al.* 1998; Hughes *et al.* 1998; Boland *et al.* 2000; Kerr *et al.* 2002).

dating (Spikings *et al.* 2001, 2005; Vallejo *et al.* 2006) and stratigraphic and sedimentological studies (Jaillard *et al.* 2004, 2005, 2008; Toro & Jaillard 2005) allowed the recognition of several tectonic units and refinement of their Late Cretaceous and Palaeogene, pre- to post-accretion, evolution. Two oceanic plateaus have been identified, each one overlain by more differentiated lavas and volcaniclastic deposits of island arc affinity (Fig. 1). Stratigraphic studies enabled us to reconstruct stratigraphic series for each terrane, and to determine accretionary events in the Late Campanian, Late Maastrichtian and latest Paleocene.

The aim of this paper is to present an overview of the nature, age, stratigraphy and tectonic setting of the oceanic tectonic units accreted in Ecuador, to describe their Late Cretaceous and Palaeogene accretion and tectonic evolution, and to examine the implications of these data on the geodynamic evolution of the eastern Pacific area and the Caribbean Plate.

Oceanic terranes

San Juan terrane

The San Juan terrane is the easternmost oceanic unit (eastern part of the Pallatanga terrane of McCourt *et al.* 1998; Kerr *et al.* 2002), which crops out along a narrow belt in the western part of the Inter Andean valley, and at the eastern border of the Western Cordillera (Fig. 1). It yielded an Sm/Nd isochron of 123 ± 13 Ma (Lapierre *et al.* 2000) and a poor Ar/Ar integrated age of 105 Ma (Mamberti *et al.* 2004). These dates may reflect either the age of the oceanic crust, through which the plateau was poured out, or the age of the oceanic plateau itself. The latter interpretation has been preferred until now (Mamberti *et al.* 2004).

Oceanic rocks. Detailed petrographic and geochemical analysis of the San Juan magmatic rocks has been presented by Mamberti *et al.* (2004). Along the San Juan section (20 km SW of Quito), it comprises mainly ultramafic rocks: peridotites, layered cumulates and layered and isotropic gabbros. The gabbros and peridotites of the San Juan suite exhibit nearly flat rare earth elements (REE) patterns, but light rare earth elements (LREE) are slightly enriched in the gabbros, and slightly depleted in the peridotites. Moreover, the latter present very low concentrations (0.07–3 chondrites). On the basis of the REE pattern, mineralogy, and lithological assemblage of the San Juan section, these rocks have been interpreted as formed in deep magma chambers of an oceanic plateau (Mamberti *et al.* 2004). Isolated outcrops along the Inter Andean Valley of Central (Quillan) and Northern Ecuador (Chota, Fig. 1) suggest that the San Juan terrane extends eastward below the Inter Andean valley (Samper & Mollex 2001; fig. 2).

Syn-accretion deposits. The San Juan terrane is tectonically associated with a fine- to medium-grained turbiditic series of a lithic and feldspathic arenites of Early Maastrichtian age (Yunguilla Fm, Bristow & Hoffstetter 1977; Jaillard *et al.* 2004; Fig. 2, right), sourced both by volcanic and crystalline areas (Toro & Jaillard 2005). Although their geometric relations with the San Juan terrane are not visible, these deposits are interpreted as postdating the accretion of the San Juan terrane to the Andean margin, which occurred, therefore, in Late Campanian times (Hughes & Pilatasig 2002; Kerr *et al.* 2002; Jaillard *et al.* 2004, 2008). This tectonic event is further documented on the continental margin of southern Ecuador and northern Peru, by the widespread deposition of a coarsening and shallowing upward sequence of conglomerates of

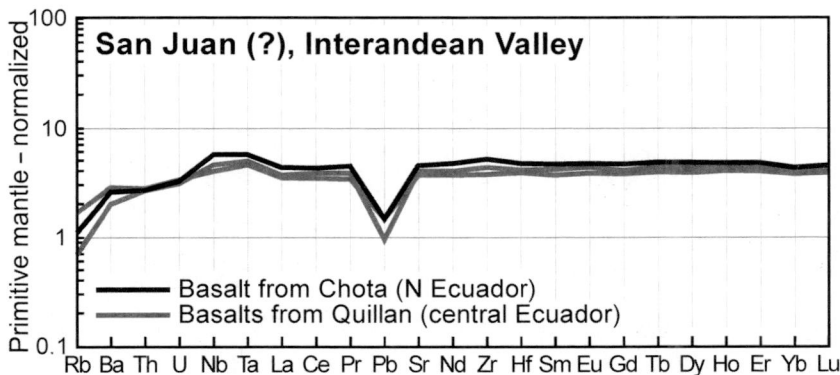

Fig. 2. Multi-element diagram (Sun & McDonough 1989) for the San Juan (?) oceanic plateau of the Inter Andean Valley of central (Quillan) and northern (Chota) Ecuador (from Samper & Mollex 2001). Location shown in Figure 1.

Late Campanian age, unconformably overlain by transgressive marine shales of Early Maastrichtian age (Taipe *et al.* 2004; Jaillard *et al.* 1999, 2005; Fig. 5).

The Yunguilla Fm is in turn unconformably overlain by a succession of shales and quartz-rich micaceous sandstones of Early to Middle Paleocene age (Saquisilí Fm, Hughes *et al.* 1998; Toro & Jaillard 2005; Fig. 3, right).

Guaranda terrane

The Guaranda terrane (Mamberti *et al.* 2003; western part of the Pallatanga terrane of McCourt *et al.* 1998; Kerr *et al.* 2002) is separated from the San Juan terrane by a major fault (Pujilí Mélange of Hughes & Pilatasig 2002), and forms the eastern part of the Western Cordillera (Fig. 1). Although not directly dated, it is overlain by pelagic cherts bearing Santonian to Maastrichtian radiolaria (Boland *et al.* 2000), thus supporting a Late Cretaceous, pre-Campanian age.

Oceanic evolution. Detailed description of the petrography and geochemistry of the magmatic basement can be found in Mamberti *et al.* (2003). Besides the hyaloclastites, pillow basalts, dolerites and gabbros that are common to all magmatic basement units, the Guaranda terrane is marked by

the frequent occurrence of high-Mg basalts, ankaramites (8–17% MgO) with large clinopyroxene phenocrysts, and picrites (21–27% MgO) (Mamberti *et al.* 2003). Basalts exhibit nearly flat REE plots, while ankaramites are slightly enriched in LREE, and picrites are highly depleted in LREE. All these rocks are interpreted as oceanic plateau basalts originated by a mantle plume (Kerr *et al.* 2002; Mamberti *et al.* 2003). Picrites would have been extracted from a depleted mantle source located in the hot plume tail, whereas ankaramites and Mg-rich basalts would have ascended from the less hot edges of the plume (Mamberti *et al.* 2003).

Since petrographic and geochemical signatures are similar to those of the Caribbean Plate (Kerr *et al.* 2002; Mamberti *et al.* 2003), the Guaranda terrane is interpreted as belonging to the Caribbean Plate and therefore to have been formed in the Turonian–Coniacian (*c.* 92–86 Ma, Sinton *et al.* 1998), although magmatism in the Caribbean Plate lasted at least until the Campanian (*c.* 75 Ma, Mauffret *et al.* 2001). The overlying rocks differ in northern and central Ecuador (Fig. 1).

In the Western Cordillera of northern Ecuador, lavas and volcaniclastic products of the Río Cala island arc (Boland *et al.* 2000; Fig. 4c) are thought to grade laterally into Campanian–Maastrichtian volcaniclastic deposits (Natividad unit, Hughes *et al.* 1998; Boland *et al.* 2000; Kerr *et al.* 2002).

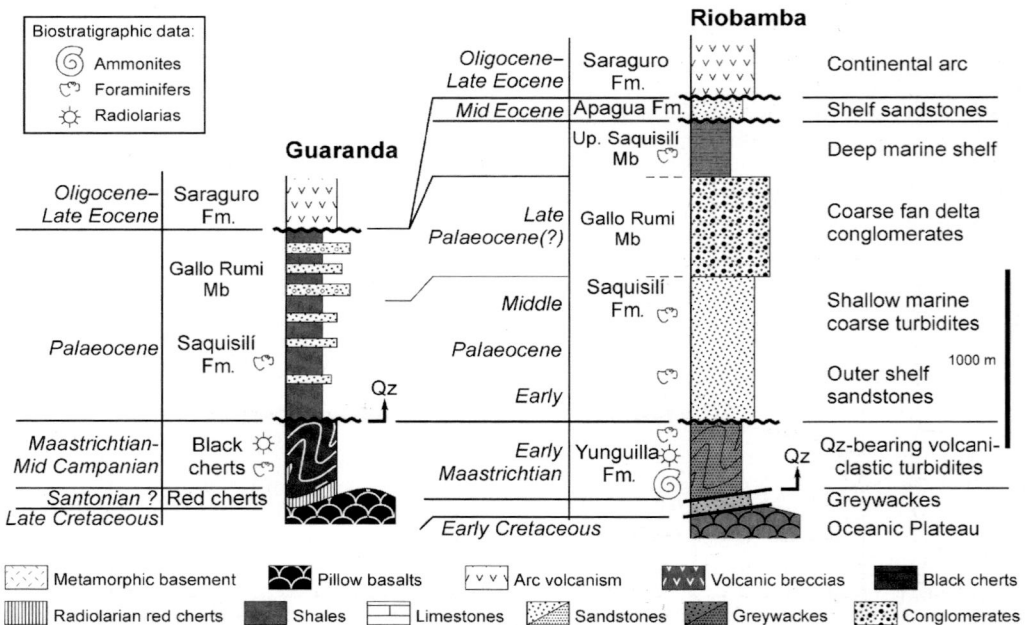

Fig. 3. Stratigraphic successions of the San Juan (right) and Guaranda (left) terranes along the Riobamba–Guaranda section (from Jaillard *et al.* 2004). Location shown in Figure 1. Qz, appearance of detrital quartz.

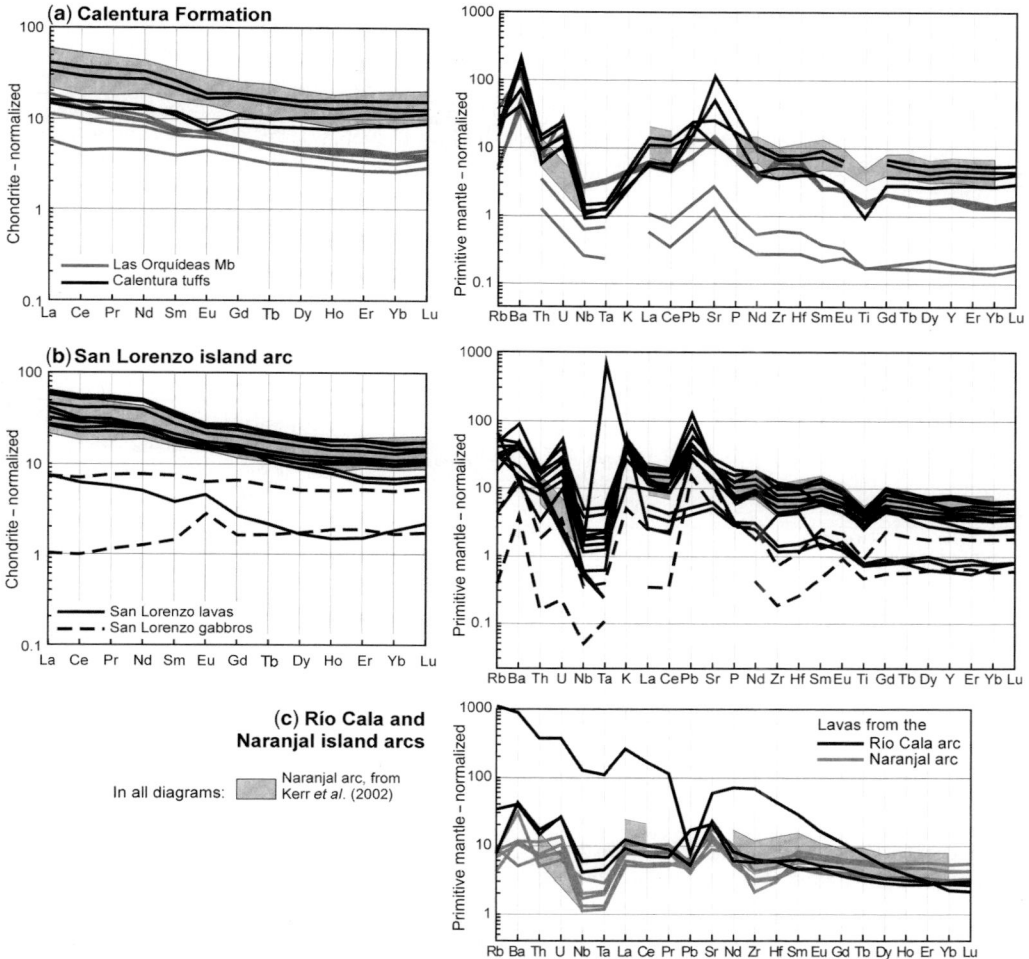

Fig. 4. Rare earth element (left) and multi-element (right) diagrams (Sun & McDonough 1989) for the Calentura Formation of the Guayaquil area (**a**), the San Lorenzo island arc of the Manta and Pedernales areas (**b**), and the Río Cala and Naranjal formations of northwestern Ecuador (**c**) (from Reynaud *et al.* 1999; Pourtier 2001; Mamberti 2001; Samper & Mollex 2001; Vanmelle 2004).

Some pillow basalts, associated with red cherts dated by Santonian to Campanian radiolarians, suggest a back-arc basin setting (La Portada unit, Kerr *et al.* 2002).

In central Ecuador, the oceanic plateau basement is overlain first by red cherts associated with subordinate black bituminous limestones, and then by fine-grained, radiolarian-rich black cherts of Mid Campanian to Maastrichtian age (Jaillard *et al.* 2004; Fig. 3, left). Therefore, the outpouring of the Guaranda oceanic plateau is followed either by island arc development (Río Cala and Naranjal formations, Boland *et al.* 2000; Fig. 4c) and local back-arc extension (northern Ecuador), or by deposition of autochthonous, siliceous pelagic mudstones without significant volcanic input (central and southern Ecuador).

Syn-accretion deposits. In central Ecuador, the latest Cretaceous oceanic cherts are unconformably overlain by micaceous quartz-sandstones of Early and Middle Paleocene age (Saquisilí Fm, Hughes *et al.* 1998). The abrupt and massive arrival of detrital quartz on an oceanic succession is interpreted as due to the accretion of the Guaranda terrane, which occurred in the Middle to Late Maastrichtian (Jaillard *et al.* 2004). The Paleocene succession follows with coarsening-upward conglomerates ascribed to the Late Paleocene (Gallo Rumi Mb), deposited in shallow fan deltas in the

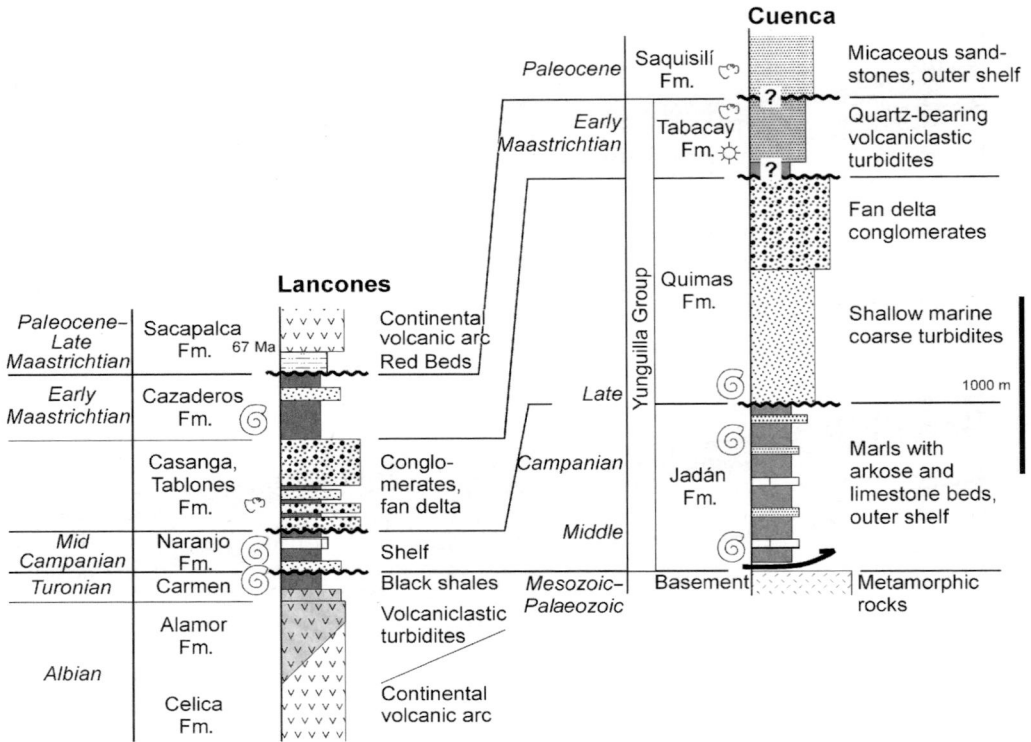

Fig. 5. Stratigraphic successions of the Cuenca and Lancones areas (from Jaillard *et al.* 1999, 2008). Location shown in Figure 1; caption as in Figure 3.

east or deeper turbidite fans in the west (Toro 2006). It ends up with fine-grained shales and sandstones of latest Paleocene or Early Eocene age (Fig. 3, left). This series is unconformably overlain by transgressive litharenites, of shallow shelf to turbiditic environment, dated as Middle Eocene (Apagua Fm, Egüez 1986; Santos & Ramírez 1986).

In the Western Cordillera of northern Ecuador, the accretion of the Cretaceous igneous units is postdated by a sharp unconformity at the base of coarse conglomerates of Eocene age, or below transgressive limestones and sandstones of probable Middle Eocene age (Boland *et al.* 2000). The age of accretion is assumed to be the same as in central Ecuador.

Farther south, on the continental margin of northern Peru and southern Ecuador, this tectonic event is recorded by a disconformity between Early to Middle Maastrichtian marine shales and sandstones, and Paleocene sandstones (Fig. 3), associated with a well marked hiatus (Taipe *et al.* 2004; Jaillard *et al.* 1999, 2005, 2008).

Piñón terrane

In this paper, we will consider that all magmatic outcrops located in the coastal area (Guayaquil, Manta, Pedernales and Esmeraldas areas, Fig. 1) that are not island-arc lavas, belong to the Piñón Formation. We ascribe also the western part of the Western Cordillera of northern Ecuador (part of Naranjal unit of Boland *et al.* 2000; Naranjal plateau of Kerr *et al.* 2002) to the Piñón Formation.

Oceanic evolution. The Piñón basement was classically ascribed to the Early Cretaceous (Goossens & Rose 1973; Jaillard *et al.* 1995a). It has been recently dated west of Guayaquil, where interpillow sediments yielded Coniacian–Campanian radiolarians and foraminifers (Vanmelle *et al.* 2008; Fig. 7, right).

The Piñón Formation consists mainly of basalts, pillow basalts, dolerites and small gabbroic intrusions. Basalts are usually marked by low Si ($<50\%$) and Al_2O_3 (13.5–15.8%) contents, moderate MgO (5–9%) and high CaO (9–13%) and TiO_2 (1–2%) values. Rare earth elements plots exhibit a flat pattern, with slight depletion in light REE, typical of Oceanic Plateaux basalts (Reynaud *et al.* 1999; Pourtier 2001; Fig. 6). Multi-element diagrams exhibit depletion either in high field strength (HFS) elements (Guayaquil area), or in large ionic lithophile elements (LILE) (Manta and Pedernales

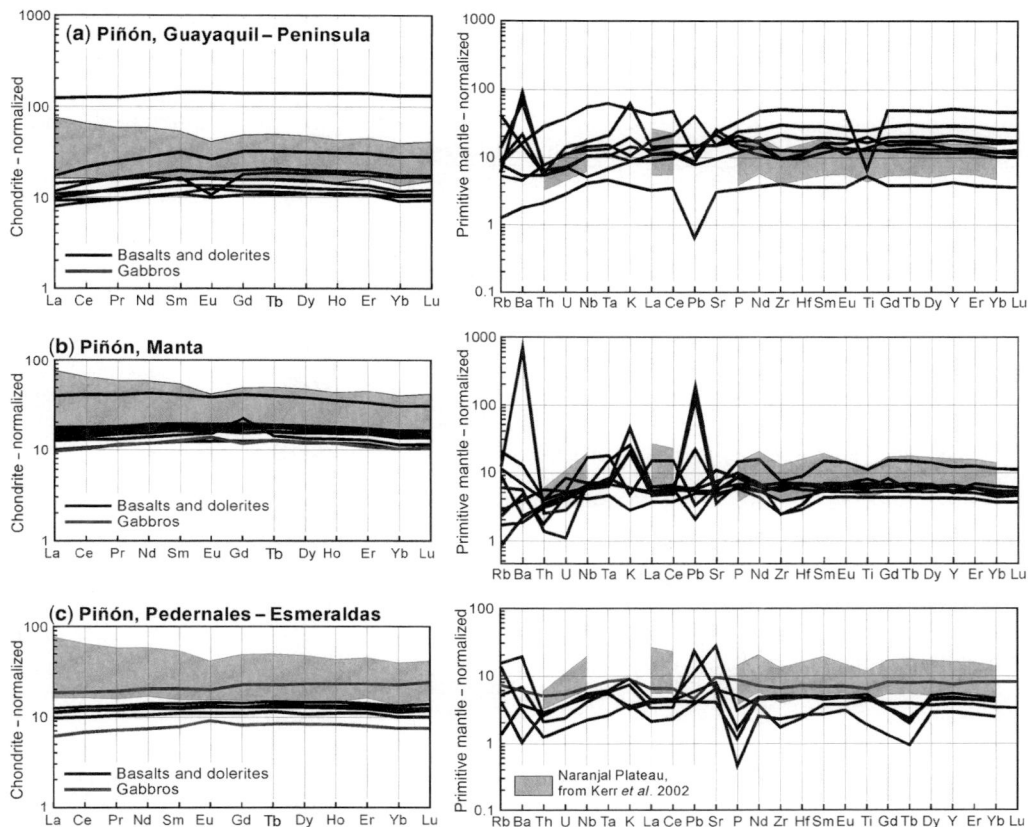

Fig. 6. Rare earth element (left) and multi-element (right) diagrams (Sun & McDonough 1989) for the Piñón oceanic plateau of southern (**a**), central (**b**) and northern (**c**) coastal Ecuador (from Ethien 1999; Pourtier 2001; Vanmelle 2004). Location shown in Figure 1.

areas, Fig. 6b, c). From the Pedernales area, Mamberti (2001) described a Mg-rich (21.2%) picrite, depleted in Al_2O_3 (10.6%) and TiO_2 (0.25%), exhibiting spectacular quenched olivine textures. This rock is geochemically very similar to Gorgona picrites (strong depletion in LREE, very low Nb and Ta content), thus supporting the idea that some of the Colombian and Ecuadorian terranes belong to the same oceanic plateau.

Near Guayaquil, the Piñón basement is overlain by a 30–200 m thick series of andesitic breccias, tuffs, and basaltic lavas (Las Orquídeas Mb, Reynaud et al. 1999; Fig. 4a). Although its geochemical signature is that of a primitive island arc, such a geodynamic setting is unlikely for two main reasons. On one hand, it is difficult to imagine that subduction-related magmatism immediately followed the formation of the Piñón oceanic plateau, since 1 or 2 Ma are necessary before the downgoing slab reaches the magma

generation depth. On the other hand, since the overlying deposits do not bear any evidence of arc activity, subduction-related magmatism would have ceased for about 10 Ma, before to resume with the development of the Mid Campanian–Maastrichtian island arc (c. 80–68 Ma, see below). We propose that the Las Orquídeas lavas and breccias might have an origin comparable to that of the arc-like lava quoted by Haase et al. (2005) on the East-Pacific ridge (Vanmelle et al. 2008).

This volcanic layer is overlain (Calentura Fm; Fig. 7) first by black siliceous limestones of Coniacian age, and then by undated radiolarian-rich, red siliceous mudstones, probably equivalent to the Santonian–Campanian red mudstones of the Guaranda terrane of northern Ecuador. These are in turn overlain by welded tuffs and marls of Middle Campanian age (top of Calentura Fm), which grade upward into a thick, coarsening-upward

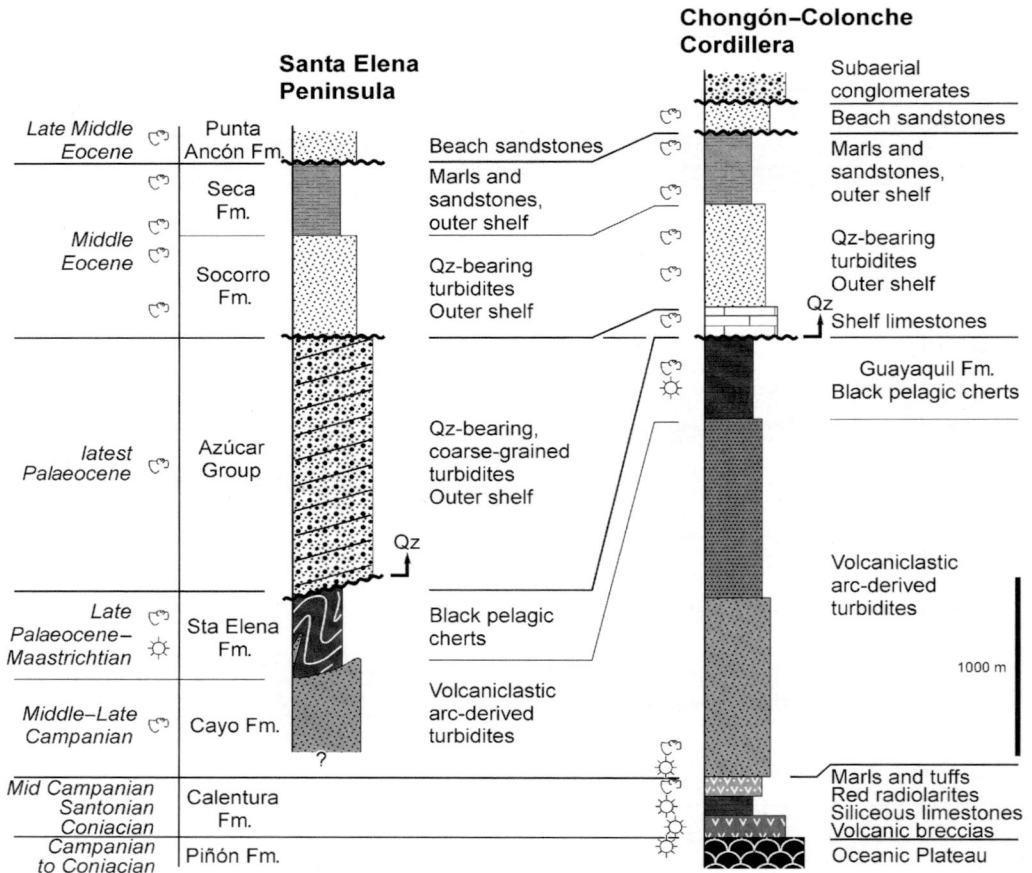

Fig. 7. Stratigraphic successions of the Piñón terrane in southwestern Ecuador (from Jaillard *et al.* 1995*a*, Vanmelle *et al.* 2008). Location shown in Figure 1; caption as in Figure 3. Qz, appearance of detrital quartz.

series of volcaniclastic turbidites (Cayo Fm, Fig. 7), the base of which yielded Middle Campanian microfauna (Vanmelle *et al.* 2008). These Campanian tuffs and volcaniclastic products exhibit a typical island arc geochemical affinity, more evolved than those of the Las Orquídeas Mb (Fig. 4a). Farther west, the thick volcaniclastic series grades laterally into pillow basalts, andesites and volcanic breccias of island arc signature (San Lorenzo Fm, Lebrat *et al.* 1987; Fig. 4b), associated with limestones of Middle Campanian to Middle Maastrichtian age (Jaillard *et al.* 1995*a*; Reynaud *et al.* 1999).

In northern Ecuador, the western part of the Western Cordillera exhibits a magmatic basement with oceanic plateau affinity (Naranjal plateau, Kerr *et al.* 2002), overlain by lavas and volcaniclastic products of island arc affinity (Naranjal arc, Fig. 4c), which are stratigraphically associated with pelagic oceanic purple-grey siliceous mudstones of

Late Campanian–Maastrichtian age (Boland *et al.* 2000). The Naranjal arc is correlatable with the San Lorenzo Fm of southwestern Ecuador, and with the Ricaurte island arc of southern Colombia (Spadea & Espinoza 1996).

In the Guayaquil area and west of it, the oceanic sedimentation ends up with fine-grained black siliceous cherts of Middle Maastrichtian to Late Paleocene age (Guayaquil Fm, Jaillard *et al.* 1995*a*; Keller *et al.* 1997; Fig. 7, right).

Syn-accretion deposits. Southwest of Guayaquil, the Paleocene pelagic cherts are highly deformed (Santa Elena Fm), and unconformably overlain by a thick succession of coarse-grained, quartz-rich high density turbidites of latest Paleocene age (Azúcar Gp, Jaillard *et al.* 1995*a*; Fig. 7, left). This unconformity that emphasizes the arrival of continent-deriving detritism, is interpreted as

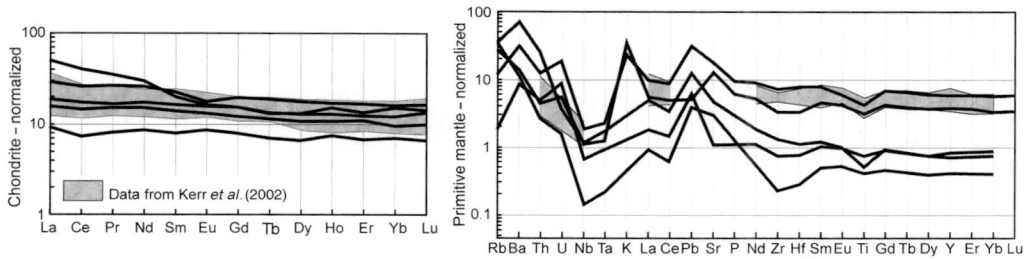

Fig. 8. Rare earth element (left) and multi-element (right) diagrams form the Macuchi island arc (from Cosma *et al.* 1998; Pourtier 2001).

sealing the accretion of the Piñón terrane to the continental margin. Farther north (Manabí Basin; Fig. 1), Paleocene quartz-rich deposits identified in subsurface data, and resting on Cretaceous magmatic basement (Deniaud 2000), are interpreted in the same way. In the Western Cordillera of northern Ecuador, cherts grading upward into massive sandstones of Late Paleocene age (La Cubera Fm, Boland *et al.* 2000) are regarded as announcing the Eocene accretion (Kerr *et al.* 2002). We propose that this succession reflects the accretion of the Piñón terrane in the Late Paleocene, as for the whole coastal terrane. On the Guaranda terrane, the Late Paleocene accretionary event is marked by a coarsening upward sequence of high density turbidites (Gallo Rumi Mb, Jaillard *et al.* 2004).

West of Guayaquil, the Azúcar Gp is in turn folded and unconformably overlain by a classical Middle Eocene forearc sedimentary sequence. The latter comprises from base to top (Fig. 7): diachronic transgressive shelf limestones of latest Early Eocene to Early Middle Eocene age, sandstones and turbidites, outer shelf marls of Middle Eocene age, and coarse lithic sandstones of late Middle Eocene (Bartonian) to Late Eocene age (Whittaker 1988; Jaillard *et al.* 1995a).

Macuchi island arc

The Macuchi terrane constitutes the western side of the Western Cordillera of central Ecuador (McCourt *et al.* 1998; Kerr *et al.* 2002; Fig. 1). Although classically ascribed to the Eocene (Egüez 1986; Henderson 1981), it might be older (McCourt *et al.* 1998; Kerr *et al.* 2002). The Macuchi terrane remains a poorly understood unit of the Western Cordillera of Ecuador.

Oceanic evolution. This unit consists mainly of volcaniclastic deposits (greywackes, lithic siltstones, volcanic clast-bearing breccias, cherts etc.) with intercalations of pillow basalts, andesites, hyaloclastites and tuffs, and intrusions of dolerites and

andesites (Egüez 1986; Kerr *et al.* 2002). Most samples exhibit geochemical signatures of an island arc, with typical Nb, Ta and Ti negative anomalies (Kehrer & Van Der Kaaden 1979; Kerr *et al.* 2002; Chiaradia & Fontboté 2002; Fig. 8). Some Mg-rich samples may be interpreted either as primitive arc products (Kerr *et al.* 2002), or as back-arc basin lavas (Hughes *et al.* 1998). Samples with 'MORB affinities' mentioned by Lebrat *et al.* (1987) from the western part of the Macuchi arc probably correspond to rocks from the Guaranda and/or San Juan Oceanic Plateaus.

Syn-accretion deposits. The Macuchi magmatic suite is associated with a forearc sedimentary sequence of Middle to Late Eocene age, which comprises from base to top: transgressive shelf limestones (Unacota unit), turbiditic litharenites (Apagua Fm) and fan delta conglomerates (Rumi Cruz Mb; Egüez 1986; Hughes *et al.* 1998; Toro 2006). Although the contact between these sediments and the Macuchi arc are often tectonized, we consider that the former are the stratigraphic cover of the latter. A comparable sequence is known in the central and northern parts of the Western Cordillera and throughout the coastal area (Fig. 9). Therefore, the Piñón, Naranjal, Macuchi and Guaranda terranes underwent a similar evolution in the Middle Eocene (Santos *et al.* 1986), indicating that accretions in Ecuador were achieved by Middle Eocene times.

An Eocene age for the Macuchi island arc is therefore unlikely. The Macuchi volcanic arc rocks are overlain by a forearc sedimentary sequence of Middle Eocene age, and yielded a latest Cretaceous FT age (68 ± 11 Ma, Spikings *et al.* 2005). They are, therefore, of pre-Middle Eocene age, probably of Late Cretaceous to Paleocene age. Furthermore, the Macuchi lavas contain xenoliths of red siliceous cherts similar to those deposited in the Santonian–Campanian on the Guaranda and Piñón terranes. This indicates on one hand, that the Macuchi island arc is younger

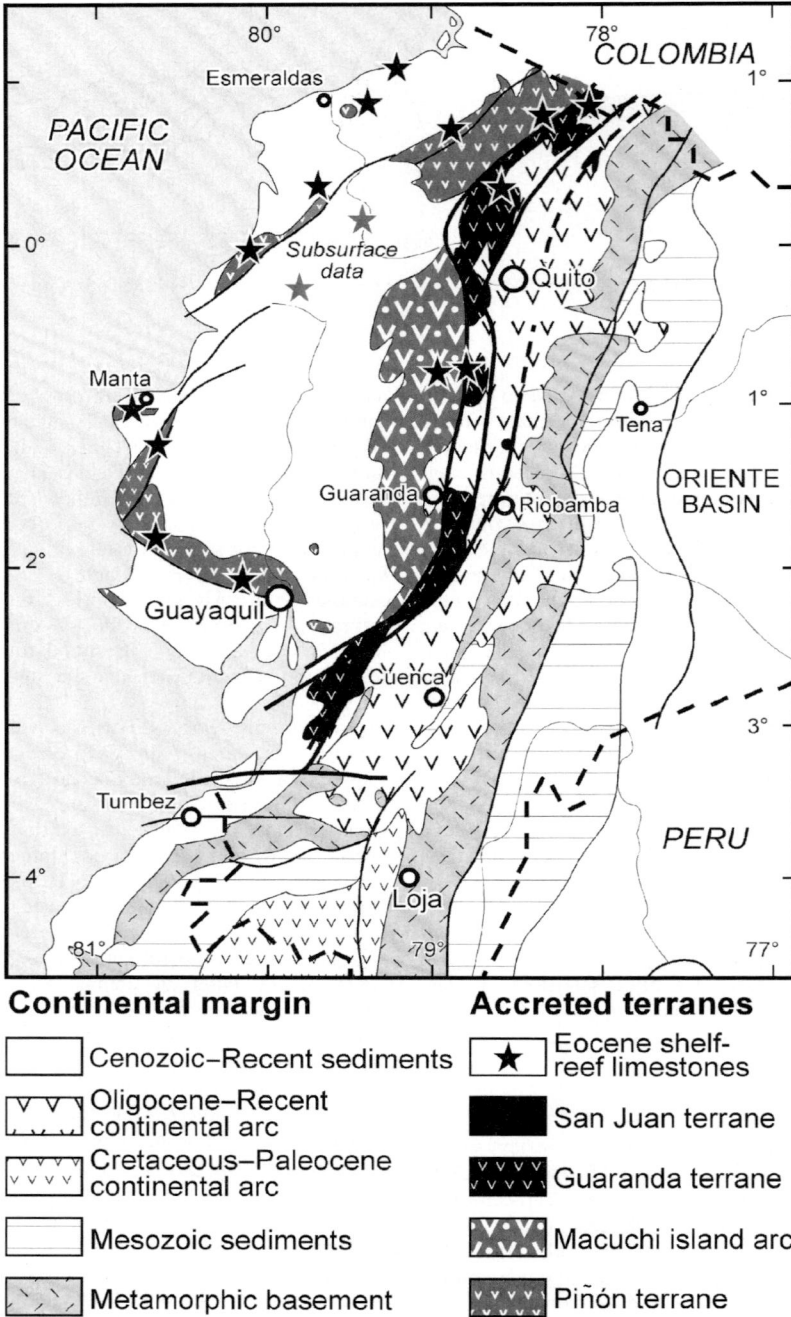

Fig. 9. Distribution of Middle Eocene, transgressive shelf limestones on the accreted terranes (adapted from Santos *et al.* 1986).

than Early Campanian, and on the other hand, that it overlies a series comparable to that of the Late Cretaceous oceanic plateau of Ecuador. The latter interpretation is supported by the lead isotopic compositions of the Macuchi island arc lavas, which suggest contamination by a Pb-rich basement, possibly the Caribbean oceanic plateau (Chiaradia & Fontboté 2002; Chiaradia *et al.* 2004).

Because major changes in the sedimentation (regional disconformities, abrupt appearance of coarse clastics and of lithic component in the detrital input) occurred at the Middle to Late Eocene transition, the Macuchi terrane has been interpreted as accreted at that time (Egüez 1986; Bourgois *et al.* 1990; Hughes & Pilatasig 2002). However, the fact that the Middle Eocene transgressive facies are common to all Late Cretaceous accreted oceanic plateau fragments (Santos *et al.* 1986; Fig. 9), indicates that the accretions were achieved by Middle Eocene times. Moreover, the Macuchi terrane is located between terranes accreted in the Late Maastrichtian to the east, and in the Late Paleocene to the west. As a consequence, unless the western Piñón terrane migrated hundreds of kilometres northward since the Eocene, the Macuchi terrane was necessarily accreted before the latest Paleocene. Note that this would imply that its oceanic evolution had ceased by that times, supporting a pre-Late Paleocene age for the Macuchi island arc.

Therefore, we propose that the Macuchi island arc is of Late Cretaceous to Paleocene age, overlies the Late Cretaceous oceanic plateau, and has been accreted in the Late Paleocene. In this interpretation, the Macuchi island arc is not a distinct, individualized terrane. It would be coeval with, and would represent a northward extension of, the Maastrichtian–Paleocene Sacapalca arc of southern Ecuador (Jaillard *et al.* 1996; Hungerbühler *et al.* 2002; Figs 1 & 5).

Geodynamic implications

Preliminary remarks

The Late Cretaceous oceanic plateau of Ecuador is usually considered as belonging to the Caribbean Oceanic Plateau (COP) (Kerr *et al.* 1996, 2002; Lapierre *et al.* 2000). This interpretation is consistent with petrographic, geochemical and radiometric data, as well as with the new stratigraphic data (Luzieux *et al.* 2006; Vanmelle *et al.* 2008). However, the Caribbean Plate is not restricted to the Caribbean Plateau, since the former comprises different magmatic suites of distinct ages (e.g. Donnelly *et al.* 1990a; Mauffret *et al.* 2001) and exhibits highly variable crustal structure (Mauffret & Leroy 1997). Moreover, the nature, age and

structure of the Caribbean Plate is mainly known by means of field studies of its highly tectonized margins, and little is known on the stratigraphy and palaeogeography of the plate as a whole. Therefore, the evolution of the Late Cretaceous oceanic plateau accreted in Ecuador does not necessarily reflect the evolution of the whole Caribbean Plate, and cannot be generalized to the latter.

No magmatic arc is known on the Ecuadorian–Colombian continental margin between the Late Jurassic and the Late Eocene (*c.* 140–40 Ma; Aspden *et al.* 1987; Jaillard *et al.* 1995b). This suggests that no oceanic plate was subducting beneath the NNE-trending margin of the northern Andes during the Cretaceous and Paleocene. Two consequences must be emphasized. First, this implies that the accretion of oceanic terranes was not preceded by subduction, and therefore, that the oceanic terranes were not separated from the Andean margin by wide oceanic basins. Second, since subduction did occur beneath the Chilean and Peruvian margins, it can be inferred that the Farallón oceanic plate shifted in a north- to NNE-ward direction during most of the Cretaceous, and that the Ecuadorian–Colombian margin acted mainly as a dextral transform margin. Although speculative, most plate tectonic reconstructions of the Farallón plate motion (Pilger 1984; Pardo-Casas & Molnar 1987) support the latter intrepretation, since they assume that the Farallón plate shifted northward during latest Cretaceous times at a velocity of between 5 and 10 cm/year, and progressively turned clockwise, until shifting in an easterly direction.

Finally, most of the accreted terranes of Ecuador underwent clockwise rotations after the Late Cretaceous, the amount of which varies from 20 to 120° (Roperch *et al.* 1987; Luzieux *et al.* 2006). Therefore, the Cretaceous orientation of the island arcs must be restored through a significant counterclockwise rotation.

Birth place of the COP

The COP is of Turonian–Coniacian age (*c.* 90 Ma) and is overlain by a Mid Campanian–Mid Maastrichtian island arc (*c.* 80–70 Ma), well developed in the coastal area (Cayo–San Lorenzo Fms), and in the northern Western Cordillera (Río Cala, Naranjal arcs). If we accept that the Las Orquídeas volcanics do not represent subduction-related magmatism, this means that, between 90 and 80 Ma, the COP was erupted through, and belonged to, the Farallón Oceanic Plate. The onset of subduction related magmatism in the Middle Campanian (*c.* 80 Ma) implies that, from then on, the COP was separated from the Farallón Plate by a subduction zone, and constituted an individualized tectonic plate.

This, together with the fact that the COP first accreted to the Ecuadorian–Colombian margin in Late Maastrichtian times, implies that the COP was born in the Pacific realm. If we accept that the Farallón plate moved northward before 80 Ma, the COP must have formed 500–1000 km south of the latitude of Ecuador as suggested by palaeomagnetic data (Luzieux *et al.* 2006). Moreover, since no subduction related magmatism was recorded on the Ecuadorian margin before the Late Cretaceous and Paleocene accretions, the COP was located close to the continental margin (Fig. 10).

As a consequence, the COP is likely not to have been originated by the Galápagos hot-spot, since the latter was located 2500–3000 km west of Ecuador in the Late Cretaceous (e.g. Scotese *et al.* 1988; Pindell & Kennan 2001).

Late Cretaceous–Paleocene evolution of the COP and Ecuadorian margin

Between 90 and 80 Ma, the COP underwent a 500–1000 km shift in a NNW to NE direction, before being stabilized by the creation of a subduction zone at its southern tip (Fig. 10). At that time, the Farallón plate (and COP) was bounded to the north by a SW-ward subduction zone evidenced by the Late Cretaceous Great Antilles Arc (Pindell *et al.* 2005). The north- to NE-trending latest Cretaceous San Lorenzo volcanic arc grades eastward into the volcaniclastic turbidites of the Cayo Fm, interpreted as backarc deposits. Taking into account the subsequent clockwise rotations, the island arc and associated subduction zone were originally orientated NW–SE to east–west, and most likely located at the southern edge of the Piñón terrane (Fig. 11). The creation of this southern subduction zone may be related to the collision of the northern edge of the COP against the Central American continental blocks (Donnelly *et al.* 1990b; Iturralde-Vinente 1998; Pindell & Kennan 2001), and appears grossly coeval with the accretion of the San Juan terrane in Ecuador (Late Campanian, *c.* 75 Ma, Jaillard *et al.* 2008). It was associated with a significant clockwise rotation (Roperch *et al.* 1987; Luzieux *et al.* 2006).

The accretion of the Guaranda terrane occurred in the Late Maastrichtian (*c.* 68 Ma; Fig. 11). As

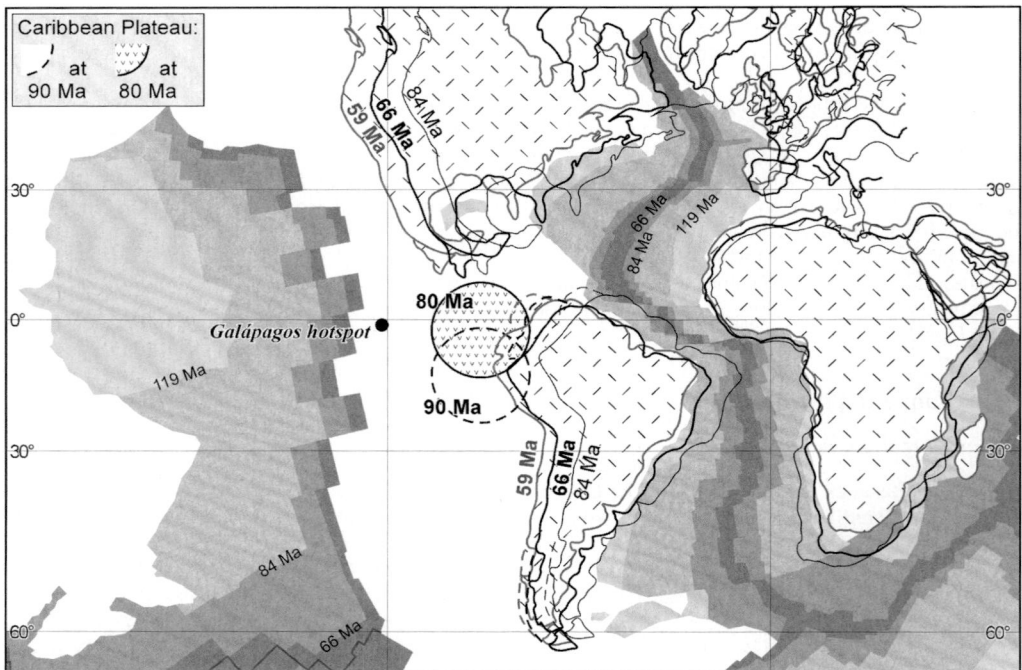

Fig. 10. Proposed evolution of the Late Cretaceous oceanic plateau of Ecuador during Late Cretaceous–Paleocene times. Reconstructed maps are simplified from Scotese *et al.* (1988). Geometry and age of the oceanic floor correspond to the 59 Ma map. 90 Ma: formation of the COP SW of Ecuador; 80 Ma: creation of a subduction zone south of the COP, probably due to collision with Central America, allowing development of the Campanian–Maastrichtian island arcs; 66 and 59 Ma: ongoing collision/accretion of the COP to northern South America.

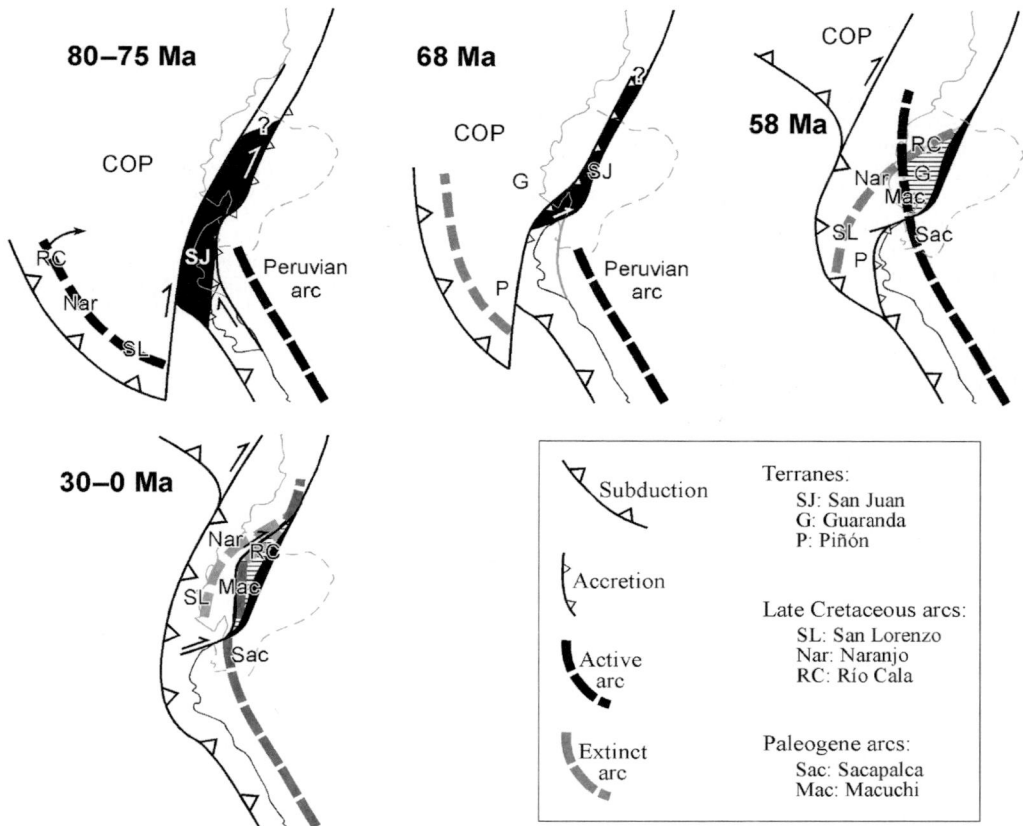

Fig. 11. A possible scenario for successive accretions in Ecuador. Late Campanian (80–75 Ma): collision of the COP with Central America creates a subduction zone and the related arc south of the COP; accretion of San Juan terrane against the Andean margin. Late Maastrichtian (68 Ma): accretion of Guaranda terrane, development of a Palaeogene arc. Late Paleocene (58 Ma): accretion of Piñón terrane, dextral movements along the sutures. Neogene (c. 30–0 Ma): Ongoing dextral movements, development of a Neogene arc (not drawn) superimposed on former ones.

for the Late Campanian accretionary event, the Late Maastrichtian tectonic event is known all around the Caribbean area, and is coeval with the collision of the COP in Central America (Maya Block; Donnelly *et al.* 1990*b*) and Cuba (Iturralde-Vinente 1998). As for the Late Campanian tectonic event, this accretionary event seems to have been followed by a re-organization of the subduction zones. We propose that the resumption of arc magmatism on the continental margin of southern Ecuador (Sacapalca Arc, Hungerbühler *et al.* 2002; Figs 1 & 5) was coeval with the development of the Macuchi island arc, on the not yet accreted Piñón–Naranjal oceanic terrane, and reflects a new subduction geometry. Since accretion was associated with clockwise rotation (Roperch *et al.* 1987; Luzieux *et al.* 2006), the new, Palaeogene arc is oblique to the Late

Cretaceous island arc (Fig. 11). This suggests that subduction, which was already occurring in northern Peru, became possible farther north, possibly because of the NE-ward shift of the COP (Pindell *et al.* 2005).

The Piñón terrane, last fragment of the COP, was accreted to the Ecuadorian margin in the Late Paleocene (c. 58 Ma; Fig. 11). Two interpretations must be discussed. First, part of the COP was still located at 3° lat. S at that time, or even 5° lat. S (Pécora *et al.* 1999), and accreted to the margin. As the COP is assumed to have collided with the North American plate in Cuba (c. 23° lat. N.) in the Late Maastrichtian; this would imply that the COP was about 3000 km long. However, the presence of an arc as early as the Late Maastrichtian in southern and central Ecuador implies that subduction had already occurred beneath the palaeo-Ecuadorian margin, and therefore, that the COP shifted farther north (Fig. 11).

In a second interpretation, the formerly accreted COP fragment has been split into an eastern (Guaranda) terrane, and a western (Pinón) terrane, the latter being underthrust beneath the former; this implies that the COP had been torn into several tectonic slices, probably bounded by NNE- to NE-trending dextral faults (Fig. 11). The Guayaquil, Manta–Naranjal, and Pedernales–Esmeraldas areas may represent such tectonic slices, since they are separated from each other by important NE-trending faults (Fig. 1), their magmatic basement present distinct petrographic features, their deformation styles differ and their stratigraphic successions are comparable but slightly different.

In our interpretation, no accretion occurred in Ecuador during the Middle to Late Eocene timespan. The compressional deformations recorded at that time in Ecuador are more likely to be related to the Incaic phase, known at this time in the whole Andean domain (Steinmann 1929; Mégard 1984; Jaillard *et al.* 2000).

Summary and conclusions

Three main oceanic terranes can be identified in Ecuador, which were accreted between Late Campanian and Late Paleocene times. (1) The San Juan terrane is made of an Early Cretaceous oceanic plateau, accreted in the Late Campanian (*c.* 75 Ma). (2) The Guaranda terrane is made from a Coniacian oceanic plateau (*c.* 90 Ma), overlain by either Campanian–Maastrichtian island arc products or Santonian–Maastrichtian pelagic cherts, which was accreted in the Late Maastrichtian (*c.* 68 Ma). (3) The Pinón–Naranjal terrane also comprises a Late Cretaceous oceanic plateau overlain by Late Cretaceous island arc suites, and accreted in the Late Paleocene (*c.* 58 Ma). We propose that the Macuchi island arc rests on the Pinón–Naranjal oceanic plateau, is of Late Maastrichtian–Late Paleocene age, and was accreted to the composite Andean margin in the Late Paleocene.

The Late Cretaceous oceanic plateau of Ecuador was formed on the Farallón Plate, south of Ecuador and close to the South American plate. It shifted 500–1000 km northward before being separated from the Farallón Plate by a north-dipping subduction zone in the Middle Campanian. Then, a fragment of this newly formed oceanic plate collided with the Ecuadorian margin in the Late Maastrichtian, inducing a re-organization of the subduction zones and the development of a new arc system (Sacapalca–Macuchi). New fragments of the COP were accreted to the Ecuadorian margin in the Late Paleocene. This scenario is compatible with

the idea according to which the Late Cretaceous oceanic plateau of Ecuador belonged to the Caribbean Plate, provided that this plate was rather large (*c.* 3000 km), and split into several fragments during its accretionary history.

This paper is dedicated to Professor Henriette Lapierre, who died abruptly in January 2006, after having worked on magmatic rocks of Ecuador for more than 10 years. We are indebted to A. Dhondt, P. Bengtson, N. Jiménez and J. Suárez for numerous palaeontological determinations. Most of the results presented here were obtained thanks to the financial support of the Institut de Recherche pour le Développement and the Dyeti funding programme. Thanks are due to Jorge Ferrer M. and Berend van Hoorn for their helpful reviews.

References

ASPDEN, J. A., MCCOURT, W. J. & BROOK, M. 1987. Geometrical control of subduction-related magmatism: the Mesozoic and Cenozoic plutonic history of Western Colombia. *Journal of the Geological Society, London*, **144**, 893–905.

BOLAND, M. P., MCCOURT, W. J. & BEATE, B. 2000. *Mapa geológico de la Cordillera Occidental del Ecuador entre 0°–1°N, escala 1/200.000*. Ministerio de Energía y Minas-BGS publs, Quito.

BOURGOIS, J., EGÜEZ, A., BUTTERLIN, J. & DE WEVER, P. 1990. Evolution géodynamique de la Cordillère Occidentale des Andes d'Equateur: la découverte de la formation éocène d'Apagua. *Comptes Rendus à l'Académie des Sciences, Paris II*, **311**, 173–180.

BRISTOW, C. R. & HOFFSTETTER, R. 1977. *Ecuador. Lexique Stratigraphique International*, V. CNRS, Paris, 5a2.

CHIARADIA, M. & FONTBOTÉ, L. 2002. Lead isotope systematics of Late Cretaceous–Cenozoic Andean arc magmas and associated ores between 8°N and 40°S: evidence for latitudinal mantle heterogeneity beneath the Andes. *Terra Nova*, **14**, 337–342.

CHIARADIA, M., FONTBOTE, L. & BEATE, B. 2004. Cenozoic continental arc magmatism and associated mineralization in Ecuador. *Mineralium Deposita*, **39**, 204–222.

COSMA, L., LAPIERRE, H. *ET AL.* 1998. Pétrographie et géochimie de la Cordillère Occidentale du Nord de l'Équateur (0°30′S): implications tectoniques. *Bulletin de la Société géologique de France*, **169**, 739–751.

DALY, M. C. 1989. Correlations between Nazca/Farallón plate kinematics and Forearc basin evolution in Ecuador. *Tectonics*, **8**, 769–790.

DENIAUD, Y. 2000. *Enregistrement sédimentaire et structural de l'évolution géodynamique des Andes équatoriennes au cours du Néogène: étude des bassins d'avant-arc et bilans de masse*. Géologie Alpine, Mémoire, **32**.

DONNELLY, T. W., BEETS, D. *ET AL.* 1990a. History and tectonic setting of Caribbean magmatism. *In:* DENGO, G. & CASE, J. E. (eds) *The Caribbean Region*. The Geology of Northern America, **H**. Geological Society of America, Boulder, CO, 339–374.

DONNELLY, T. W., HORNE, G. S., FINCH, R. C. & LÓPEZ-RAMOS, E. 1990*b*. Northern Central America: the Maya and Chortis blocks. *In*: DENGO, G. & CASE, J. E. (eds) *The Caribbean Region*. The Geology of Northern America, **H**. Geological Society of America, Boulder, CO, 37–76.

EGÜEZ, A. 1986. *Evolution Cénozoïque de la Cordillère Occidentale septentrionale d'Equateur (0°15'S-1°10'S): les minéralisations associées*. Unpublished thesis University Paris VI, 116 p., Paris.

ETHIEN, R. 1999. *Pétrologie, minéralogie et géochimie de basaltes et dolérites crétacés à affinités de Plateau Océanique d'Équateur Occidental*. Unpublished maîtrise memoir, University of Grenoble I.

FEININGER, T. & BRISTOW, C. R. 1980. Cretaceous and Paleogene history of coastal Ecuador. *Geologische Rundschau*, **69**, 849–874.

GANSSER, A. 1973. Facts and theories on the Andes. *Journal of the Geological Society, London*, **129**, 93–131.

GOOSSENS, P. J. & ROSE, W. I. 1973. Chemical composition and age determination of tholeitic rocks in the basic Cretaceous Complex, Ecuador. *Geological Society of America Bulletin*, **84**, 1043–1052.

HAASE, K. M., STRONCIK, N. A., HÉKINIAN, R. & STOFFERS, P. 2005. Nb-depleted andesites from the Pacific–Antarctic Rise as analogs for early continental crust. *Geology*, **33**, 921–924.

HENDERSON, W. G. 1981. The volcanic Macuchi Formation, Andes of Northern Ecuador. *Newsletters on Stratigraphy*, **9**, 157–168.

HUGHES, R. A. & PILATASIG, L. F. 2002. Cretaceous and Cenozoic terrane accretion in the Cordillera Occidental of the Andes of Ecuador. *Tectonophysics*, **345**, 29–48.

HUGHES, R. A., BERMUDEZ, R. & ESPINEL, G. 1998. *Mapa geológico de la Cordillera Occidental del Ecuador entre 0°–1°S, escala 1:200.000*. CODIGEM-Ministerio de Energía y Minas-BGS publs, Quito, Nottingham.

HUNGERBÜHLER, D., STEINMANN, M. ET AL. 2002. Neogene stratigraphy and Andean geodynamics of southern Ecuador. *Earth Science Reviews*, **57**, 75–124.

ITURRALDE-VINENTE, M. A. 1998. Sinopsis de la constitución geológica de Cuba. *Acta Geologica Hispanica*, **33**, 9–56.

JAILLARD, E., ORDOÑEZ, M. ET AL. 1995*a*. Basin development in an accretionary, oceanic-floored forearc setting: southern coastal Ecuador during Late Cretaceous to Late Eocene times. *In*: TANKARD, A. J., SUÁREZ, R. & WELSINK, H. J. (eds) *Petroleum Basins of South America*. American Association of Petroleum Geologists, Memoirs, **62**, 615–631.

JAILLARD, É., SEMPÉRÉ, T., SOLER, P., CARLIER, G. & MAROCCO, R. 1995*b*. The role of Tethys in the evolution of the Northern Andes between Late Permian and Late Eocene times. *In*: NAIRN, A. E. M., RICOU, L.-E., VRIELYNK, B. & DERCOURT, J. (eds) *The Tethys Ocean*. Ocean Basins and Margins, **8**. Plenum Press, New York, 463–492.

JAILLARD, É., LAUBACHER, G., BENGTSON, P., DHONDT, A. & BULOT, L. 1999. Stratigraphy and evolution of the Cretaceous forearc 'Celica-Lancones Basin' of Southwestern Ecuador. *Journal of South American Earth Sciences*, **12**, 51–68.

JAILLARD, E., HÉRAIL, G., MONFRET, T., DÍAZ-MARTÍNEZ, E., BABY, P., LAVENU, A. & DUMONT, J.-F. 2000. Tectonic evolution of the Andes of Ecuador, Peru, Bolivia and northernmost Chile. *In*: CORDANI, U. G., MILANI, E. J., THOMAZ, F. & CAMPOS, D. A. (eds) *Tectonic Evolution of South America*. Proceedings of the 31st International Geological Congress, Rio de Janeiro, 481–559.

JAILLARD, E., ORDOÑEZ, M., SUÁREZ, J., TORO, J., IZA, D. & LUGO, W. 2004. Stratigraphy of the Late Cretaceous–Paleogene deposits of the Western Cordillera of Central Ecuador: Geodynamic implications. *Journal of South American Earth Sciences*, **17**, 49–58.

JAILLARD, É., BENGTSON, P. & DHONDT, A. 2005. Late Cretaceous marine transgressions in Ecuador and northern Peru: a refined stratigraphic framework. *Journal of South American Earth Sciences*, **19**, 307–323.

JAILLARD, É., BENGTSON, P. ET AL. 2008. Sedimentary record of terminal Cretaceous accretions in Ecuador: the Yunguilla Group in the Cuenca area. *Journal of South American Earth Sciences*, **25**, 133–144.

KEHRER, W. & VAN DER KAADEN, G. 1979. Notes on the Geology of Ecuador with special reference to the Western Cordillera. *Geologische Jahrbuch*, **B35**, 5–57.

KELLER, G., ADATTE, T. ET AL. 1997. The Cretaceous–Cenozoic boundary event in Ecuador: reduced biotic effects due to eastern boundary current setting. *Marine Micropaleontology*, **31**, 97–133.

KERR, A. C., TARNEY, J., MARRINER, G. F., NIVIA, A. & SAUNDERS, A. D. 1996. The geochemistry and tectonic setting of Late Cretaceous Caribbean and Colombian volcanism. *Journal of South American Earth Sciences*, **9**, 111–120.

KERR, A. C., ASPDEN, J. A., TARNEY, J. & PILATASIG, L. F. 2002. The nature and provenance of accreted terranes in Western Ecuador: geochemical and tectonic constraints. *Journal of the Geological Society, London*, **159**, 577–594.

LAPIERRE, H., BOSCH, D. ET AL. 2000. Multiple plume events in the genesis of the peri-Caribbean Cretaceous Oceanic Plateau Province. *Journal of Geophysical Research*, **105**, 8403–8421.

LEBRAT, M., MEGARD, F., DUPUY, C. & DOSTAL, J. 1987. Geochemistry and tectonic setting of pre-collision Cretaceous and Paleogene volcanic rocks of Ecuador. *Geological Society of America Bulletin*, **99**, 569–578.

LITHERLAND, M., ZAMORA, A. ET AL. 1993. *Mapa geológico de la República del Ecuador, escala 1:1,000,000*. Geological Survey Publications, Keyworth, UK.

LUZIEUX, L., HELLER, F., SPIKINGS, R., VALLEJO, C. & WINKLER, W. 2006. Origin and Cretaceous history of the coastal Ecuadorian forearc between 1°N and 3°S: paleomagnetic, radiometric and fossil evidence. *Earth and Planetary Science Letters*, **249**, 400–414.

MAMBERTI, M. 2001. *Origin and evolution of two Cretaceous oceanic plateaus accreted in Western Ecuador (South America), evidenced by petrology, geochemistry and isotopic chemistry*. PhD thesis, Universities of Lausanne–Grenoble.

MAMBERTI, M., LAPIERRE, H. *ET AL.* 2003. Accreted fragments of the Late Cretaceous Caribbean–Colombian Plateau in Ecuador. *Lithos*, **66**, 173–199.

MAMBERTI, M., LAPIERRE, H., BOSCH, D., JAILLARD, É., HERNANDEZ, J. & POLVE, M. 2004. The Early Cretaceous San Juan plutonic suite, Ecuador: a magma chamber in an Oceanic Plateau. *Canadian Journal of Earth Sciences*, **41**, 1237–1258.

MAUFFRET, A. & LEROY, S. 1997. Seismic stratigraphy and structure of the Caribbean igneous province. *Tectonophysics*, **283**, 61–104.

MAUFFRET, A., LEROY, S. *ET AL.* 2001. Prolonged magmatic and tectonic development of the Caribbean Igneous Provience revealed by a diving submersible survey. *Marine Geophysical Researches*, **22**, 17–45.

MCCOURT, W. J., DUQUE, P., PILATASIG, L. F. & VILLAGÓMEZ, R. 1998. *Mapa geológico de la Cordillera Occidental del Ecuador entre 1°–2° S., escala 1/200.000.* CODIGEM-Min. Energ. Min.-BGS publs, Quito.

MÉGARD, F. 1984. The Andean orogenic period and its major structures in Central and Northern Peru. *Journal of the Geological Society of London*, **141**, 893–900.

PARDO-CASAS, F. & MOLNAR, P. 1987. Relative motion of the Nazca (Farallón) and South America plate since Late Cretaceous times. *Tectonics*, **6**, 233–248.

PÉCORA, L., JAILLARD, E. & LAPIERRE, H. 1999. Accrétion paléogène et décrochement dextre d'un terrain océanique dans le Nord du Pérou. *Comptes Rendus de l'Académie des Sciences, Paris, Earth Planetary Sciences*, **329**, 389–396.

PILGER, R. H., JR. 1984. Cenozoic plate kinematics, subduction and magmatism. *Journal of the Geological Society of London*, **141**, 793–802.

PINDELL, J. L. & KENNAN, L. 2001. Kinematic evolution of the Gulf of Mexico and Caribbean. *In: Petroleum Systems of Deep-water Basins: Global and Gulf of Mexico Experience, SEPM Gulf Coast Section, Proceedings of the 21st Annual Research Conference.* Society for Sedimentary Geology (SEPM), 193–220.

PINDELL, J. L., KENNAN, L., MARESCH, W. V., STANEK, K. P., DRAPER, G. & HIGGS, R. 2005. Plate kinematics and crustal dynamics of circum-Caribbean arc-continent interactions: tectonic controls on basin development in Proto-Caribbean margins. *In:* AVÉ LALLEMANT, H. G. & SISSON, V. B. (eds) *Caribbean–South American Plate Interactions, Venezuela.* Geological Society of America, Boulder, CO, Special Papers, **394**, 7–52.

POURTIER, E. 2001. *Pétrologie et géochimie des unités magmatiques de la côte équatorienne: implications géodynamiques.* Unpublished DEA thesis, University of Aix-Marseille.

REYNAUD, C., JAILLARD, E., LAPIERRE, H., MAMBERTI, M. & MASCLE, G. H. 1999. Oceanic plateau and island arcs of Southwestern Ecuador: their place in the geodynamic evolution of northwestern South America. *Tectonophysics*, **307**, 235–254.

ROPERCH, P., MÉGARD, F., LAJ, C., MOURIER, T., CLUBE, T. & NOBLET, C. 1987. Rotated oceanic blocks in Western Ecuador. *Geophysical Research Letters*, **14**, 558–561.

SAMPER, A. & MOLLEX, D. 2001. *Pétrologie, minéralogie et géochimie des laves de la Cordillère occidentale équatorienne.* Unpublished Maîtrise Memoir, University of Grenoble I.

SANTOS, M. & RAMIREZ, F. 1986. La Formación Apagua, una nueva unidad eocénica en la cordillera occidental ecuatoriana. *Actas IV Congreso Ecuatoriano de Ingenieros en Geología, Minería y Petróleo, Quito,* **I**, 179–190.

SANTOS, M., RAMIREZ, F., ALVARADO, G. & SALGADO, S. 1986. Las calizas del Eoceno medio del occidente ecuatoriano y su paleogeografía. *Actas IV Congreso Ecuatoriano de Ingenieros en Geología, Minería y Petróleo, Quito,* **I**, 79–90.

SCOTESE, C. R., GAHAGAN, L. M. & LARSON, R. L. 1988 Plate tectonic reconstructions of the Cretaceous and Cenozoic ocean basins. *In:* SCOTESE, C. R. & SAGER, W. W. (eds) *Mesozoic and Cenozoic Plate Reconstructions. Tectonophysics*, **155**, 27–48.

SINTON, C. W., DUNCAN, R. A., STOREY, M., LEWIS, J. & ESTRADA, J. J. 1998. An oceanic flood basalt province within the Caribbean Plate. *Earth and Planetary Science Letters*, **155**, 221–235.

SPADEA, P. & ESPINOSA, A. 1996. Petrology of Late Cretaceous volcanic rocks from the southernmost segment of the Western Cordillera of Colombia (South America). *Journal of South American Earth Sciences*, **9**, 79–90.

SPIKINGS, R., WINKLER, W., SEWARD, D. & HANDLER, R. 2001. Along-strike variations in the thermal and tectonic response of the continental Ecuadorian Andes to the collision with heterogeneous oceanic crust. *Earth and Planetary Science Letters*, **186**, 57–73.

SPIKINGS, R., WINKLER, W., HUGHES, R. A. & HANDLER, R. 2005. Thermochronology of allochtonous terranes in Ecuador: unraveling the accretionary and post-accretionary history of the Northern Andes. *Tectonophysics*, **399**, 195–220.

STEINMANN, G. 1929. *Geologie von Peru.* Karl Winter, Heidelberg.

SUN, S. S. & MCDONOUGH, W. F. 1989. *Chemical and Isotopic Systematics of Oceanic Basalts: Implications for Mantle Composition and Processes.* Geological Society, Special Papers, **42**, 313–345.

TAIPE, E., JAILLARD, E. & JACAY, J. 2004. Estratigrafia y evolución sedimentológica de la serie del Cretáceo superior de la Península de Paita. *Boletin de la Sociedad Geológica del Perú*, **97**, 7–27.

TORO, J. 2006. *Enregistrement des surrections liées aux accrétions de terrains océaniques: Les sédiments crétacé-paléogènes des Andes d'Equateur.* PhD thesis, University of Grenoble 1.

TORO, J. & JAILLARD, E. 2005. Provenance of the Upper Cretaceous to Upper Eocene clastic sediments of the Western Cordillera of Ecuador: tectonic and geodynamic implications. *Tectonophysics*, **399**, 279–292.

VALLEJO, C., SPIKINGS, R. A., LUZIEUX, L., WINKLER, W., CHEW, D. & PAGE, L. 2006. The early interaction between the Caribbean Plateau and the NW South American Plate. *Terra Nova*, **18**, 264–269.

VANMELLE, J. 2004. *Arcs insulaires crétacés installés sur un plateau océanique crétacé inférieur à l'Ouest de l'Équateur: lithostratigraphie et caractérisation pétro-géochimique*. Unpublished maîtrise memoir, University of Grenoble I.

VANMELLE, J., VILEMA, W. *ET AL*. 2008. Pre-collision evolution of the Piñón oceanic terrane of SW Ecuador: stratigraphy and geochemistry of the 'Calentura Formation'. *Bulletin de la Société Géologique de France*, **179**, 433–444.

WALLRABE-ADAMS, H.-J. 1990. Petrology and geotectonic development of the Western Ecuadorian Andes: the Basic Igneous Complex. *Tectonophysics*, **185**, 163–182.

WHITTAKER, J. E., 1988. *Benthic Cenozoic Foraminifera from Ecuador. Taxonomy and Distribution of Smaller Benthic Foraminifera from Coastal Ecuador (Late Oligocene–Late Pliocene)*. British Museum (Natural History) Publications, London.

Dextral shear, terrane accretion and basin formation in the Northern Andes: best explained by interaction with a Pacific-derived Caribbean Plate?

LORCAN KENNAN[1]* & JAMES L. PINDELL[1,2]

[1]*Tectonic Analysis Ltd, Chestnut House, Duncton, West Sussex GU28 0LH, UK*

[2]*Department of Earth Science, Rice University, Houston, TX 77002, USA*

Corresponding author (e-mail: lorcan@lorcankennan.com)

Abstract: The structure, stratigraphy and magmatic history of northern Peru, Ecuador and Colombia are only adequately explained by Pacific-origin models for the Caribbean Plate. Inter-American models for the origin of the Caribbean Plate cannot explain the contrasts between the Northern Andes and the Central Andes. Persistent large magnitude subduction, arc magmatism and compressional deformation typify the Central Andes, while the Northern Andes shows back-arc basin and passive margin formation followed by dextral oblique accretion of oceanic plateau basalt and island arc terranes with Caribbean affinity. Cretaceous separation between the Americas resulted in the development of a NNE-trending dextral–transpressive boundary between the Caribbean and northwestern South America, becoming more compressional when spreading in the Proto-Caribbean Seaway slowed towards the end of the Cretaceous. Dextral transpression started at 120–100 Ma, when the Caribbean Arc formed at the leading edge of the Caribbean Plate as a result of subduction zone polarity reversal at the site of the pre-existing Trans-American Arc, which had linked to Central America to South America in the vicinity of the present-day Peru–Ecuador border. Subsequent closure of the Andean Back-Arc Basin resulted in accretion of Caribbean terranes to western Colombia. Initiation of flat-slab subduction of the Caribbean Plate beneath Colombia at about 100 Ma is associated with limited magmatism, with no subsequent development of a magmatic arc. This was followed by northward-younging Maastrichtian to Eocene collision of the trailing edge Panama Arc. The triple junction where the Panama Arc joined the Peru–Chile trench was located west of present-day Ecuador as late as Eocene time, and the Talara, Tumbes and Manabi pull-apart basins directly relate to its northward migration. Features associated with the subduction of the Nazca Plate, such as active calc-alkaline volcanic arcs built on South American crust, only became established in Ecuador, and then Colombia, as the triple junction migrated to the north. Our model provides a comprehensive, regional and testable framework for analysing the as yet poorly understood collage of arc remnants, basement blocks and basins in the Northern Andes.

Supplementary material: A detailed geological map is available at http://www.geolsoc.org.uk/SUP18364

The geology of the Northern Andes, from their southern end in northernmost Peru to their northern end in northern Colombia and westernmost Venezuela, provides numerous tests of whether the Caribbean Plate was formed more or less *in situ* and has migrated only a short distance to its present position (e.g. James 2006) or originated in the eastern Pacific and is relatively far-travelled (e.g. Pindell 1993; Pindell *et al.* 1988, 2005, 2006; Pindell & Kennan 2001, 2009).

These two classes of model for the origin of the Caribbean have very different implications for the geology of northern South America, the subject of this paper, and for southern Mexico and the Chortís Block of Guatemala and Honduras. In particular, Pacific-origin and Inter-American models for Caribbean models have different implications for relationships between active and fossil plate boundaries, predicted 'stacking order' of terranes and arcs, spatial relations between terrane boundaries, and expected magmatic history and geochemistry. Geological data strongly support an eastern Pacific origin for the Caribbean Plate; although there are variations in detail between the predictions of different Pacific-origin models, these are of second-order significance compared with the differences with *in situ*-origin models for the Caribbean oceanic lithosphere. Below, we interpret the geology of the Northern Andes showing how it supports the case for a Pacific-origin for the Caribbean, and also highlight the role of some faults and shear zones which extend south of the 'traditional' view of the Northern Andes into northern Peru, and have not been incorporated into models

From: JAMES, K. H., LORENTE, M. A. & PINDELL, J. L. (eds) *The Origin and Evolution of the Caribbean Plate.*
Geological Society, London, Special Publications, **328**, 487–531.
DOI: 10.1144/SP328.20   0305-8719/09/$15.00 © The Geological Society of London 2009.

published to date. The analysis presented here is based on some tectonic first principles, dissection of geological maps and integration of geochemical and geochronological data with palinspastic restorations of Andean deformation (e.g. Pindell *et al.* 1998). We attempt to clarify and resolve some of the problems raised by our previous models and derivatives (e.g. the synthesis of Moreno & Pardo 2003) and anchor the geology of the Northern Andes in the context of the entire circum-Caribbean region, including Mexico and the Central Andes.

We focus on the Aptian to Middle Eocene in this paper. The pre-Aptian history of the region is reviewed briefly below since it provides the starting template for subsequent deformation. Significant new geochronological data have become available for this interval but there are as yet, to our knowledge, few if any regional quantitative structural studies of Cretaceous and older deformation. The new data have not previously been integrated into regional-scale tectonic models of the Caribbean region. The Maastrichtian and Cenozoic has been the subject of numerous recently published quantitative structural and stratigraphic studies (e.g. Montes *et al.* 2003, 2005; Gómez *et al.* 2003, 2005; Restrepo-Pace *et al.* 2004) and interactions between the Caribbean, Farallon and South American Plates for this period are relatively well-understood and there is little significant disagreement between models for Eocene and younger time. Many of the structures active since the Paleocene were also active during the Cretaceous and this paper aims to tie together structures mapped in Peru, Ecuador and Colombia, show how they accommodated Caribbean–South America relative motion. The model presented here provides a comprehensive, regional and testable framework for analysing the collage of arc remnants and associated basement fragments and basins in the Northern Andes which can be tested with future geological observations.

Overview of regional context

Pacific-origin models for the Caribbean Plate imply strong Cretaceous interaction with the Northern Andes, and this is reflected in the structure, stratigraphy, uplift and magmatic history of northern Peru, Ecuador and Colombia. In contrast, inter-American models for the origin of the Caribbean Plate do not imply this interaction and cannot adequately explain the dramatic contrasts in Cretaceous orogenesis and magmatism between the Northern Andes and the Central Andes (central Peru, Bolivia, northern Chile and northern Argentina). The Central Andes show evidence of persistent large magnitude east-directed subduction of the Farallon Plate or its precursors, associated more or less continuous

arc magmatism and dominantly compressional or extensional deformation, without significant strike–slip offsets in the arc or forearc. In contrast, the Northern Andes has a protracted history of back-arc basin and passive margin formation followed by accretion of oceanic plateau basalt and island arc terranes, combined with large magnitude dextral shear. Regional plate reconstructions (see Pindell & Kennan 2009) show that the Caribbean Plate originated in the easternmost Pacific and in the Indo-Atlantic hot spot reference frame has moved slowly to the NNW since the Middle Cretaceous. Relative motion between the Caribbean Plate and southern Mexico was ENE-directed. Ongoing separation between the Americas, however, resulted in the NNE-trending boundary between the Caribbean and northwestern South America being dominated by almost pure dextral strike–slip until spreading in the Proto-Caribbean Seaway slowed at about 84 Ma and stopped at about 71 Ma. Dextral shearing between the Caribbean and northwestern South America started at about 120 Ma, when the Caribbean Arc (sometimes referred to as the 'Great Arc of the Caribbean') formed at the leading edge of the Caribbean Plate as a result of subduction zone polarity reversal at the site of the pre-existing Trans-American Arc (see Pindell & Kennan 2009), which had linked to Central America to South America in the vicinity of the present-day Peru–Ecuador border. This was followed by oblique closure of the Andean Back-Arc Basin and accretion of Caribbean terranes to western Colombia. Remnants of the Caribbean Arc are found immediately west of the Central Cordillera in Colombia and appear to be of pre-Albian age, as also seen in Cuba, Hispaniola and Margarita. The oldest $^{40}Ar/^{39}Ar$ plateau ages in the Northern Andes suggest that cooling associated with dextral shear initiated no later than Middle Albian time, in agreement with cooling ages in Caribbean Arc fragments throughout the Caribbean region. However, most cooling ages in the accreted Caribbean Arc terranes in the Western Cordillera, and in the Cordillera Real, Central Cordillera shear zone to the east, are Santonian or younger and probably reflect enhanced uplift as Caribbean–South American motion became more compressional following the end of spreading in the Proto-Caribbean Seaway. This resulted in South America over-riding the Caribbean Plate above a low angle subduction zone, driving accretion of the Western Cordillera. Limited magmatism is associated only with the onset of subduction; subsequent magmatism in the region was driven by subduction of the Farallon Plate or Nazca Plate (after *c.* 23 Ma, Meschede & Barckhausen 2000).

Dextral shearing continued during the diachronous collision of the Panama Arc with Ecuador and Colombia between Maastrichtian and Eocene time, and has continued at a slower rate since then as a result of oblique subduction of the Farallon Plate and Miocene and younger Nazca Plate. Regional plate reconstructions suggest that the Caribbean Arc at the leading edge of the Caribbean spanned the gap between southern Yucatán and northwest Colombia by Maastrichtian time, and thus we propose (see below) that all the Late Cretaceous arc fragments accreted in Ecuador during and after Maastrichtian time pertain to the trailing edge (Costa Rica–Panama Arc) of the Caribbean Plate rather than to its leading edge (Caribbean Arc). As late as Eocene time, the triple junction between South America, the Caribbean, and the Farallon Plate, where the Greater Panama Arc joined western South America, was located west of present-day Ecuador, and strike–slip pull-apart basins such as the Talara, Tumbes and Manabi Basins directly relate to the northward migration of the triple junction. Features associated with the subduction of the Nazca Plate, such as active calc-alkaline volcanic arcs built on South American crust, only became established in Ecuador, and then Colombia, as the triple junction migrated to the north.

Plate boundaries and the importance of terrane stacking order in the Northern Andes

The relationships between the major plates and active plate boundaries in the Northern Andes (Fig. 1a) are key to assessing whether the Northern Andes were deformed by a Caribbean Plate that arrived in its present position from the SW during the Cretaceous and Palaeogene, or were driven by oblique subduction of the Farallon Plate or Nazca Plate, as they have been since at least Neogene time. The Lesser Antilles Arc forms the eastern boundary of the Caribbean Plate, where it overrides Atlantic lithosphere, and the Panama Arc forms its western boundary, where the Cocos and Nazca Plates (which formed from the Farallon Plate at about 23 Ma, Meschede & Barckhausen 2000) are subducting under the Americas roughly toward the NE and ENE, respectively. The southern end of the Panama Arc is a particular focus of this paper, and we propose below that related terranes extend south of westernmost Colombia into Ecuador and possibly offshore northernmost Peru. The southern edge of the Caribbean Plate is complex, defined by both anastomosing dextral shear zones and subduction beneath northern South America (e.g. Pindell et al. 1998). The Western Cordillera terranes in the Northern Andes comprise slivers of oceanic plateau basalt and island arc volcanic rocks (e.g. Kerr et al. 2003) and associated sedimentary

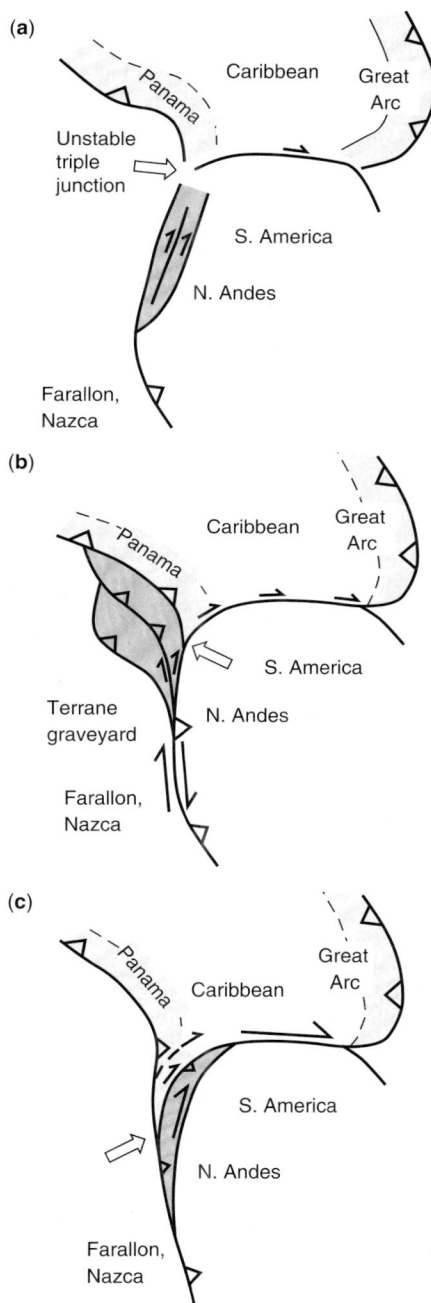

Fig. 1. (a) Sketch map of the northern Andes and Caribbean showing the major plates and plate boundaries. (b) Sketch map of terrane 'stacking order' predicted by Inter-American models for the origin of the Caribbean. (c) Sketch map of terrane 'stacking order' predicted by Pacific-origin models, and which best fits the geological maps.

rocks, accreted to South American basement and then subjected to large magnitude dextral shear. Active dextral shear (e.g. Trenkamp *et al.* 2002) is being driven by oblique ENE-directed subduction of the Nazca Plate, and many papers (e.g. Moberly *et al.* 1982) explicitly assume that oblique subduction of the Farallon Plate also explains dextral shear and terrane accretion as far back as the Cretaceous. The detailed relationships between these Western Cordillera terranes, the Panama Arc, and other terranes in the area are, however, more consistent with a Caribbean origin.

The terrane stacking order predicted by Inter-American models for the origin of the Caribbean, in which the Caribbean Plate and Panama Arc are restored only *c.* 300–400 km to the west relative to South America, implies that the Colombian Western Cordillera terranes should lie outboard of the meeting point of Panama and South America (Fig. 1b). There should be a Farallon-related calc-alkaline volcanic arc of Early Cretaceous and younger age along the Northern Andes as far north as Panama. Furthermore, because the Northern Andes in this view would have been subject to a protracted history (>100 Ma) of oblique Farallon Plate subduction beneath South America, there should be an Alaska-style terrane graveyard outboard of Panama.

In contrast, Pacific-origin models, in which the Caribbean and Panama have both moved ≫1500 km from west to east with respect to the Americas, predict that the dextral shear in the Northern Andes is the result of relative northeastward migration of the Caribbean Plate and the Caribbean–Andes–Farallon triple junction (trailing edge) relative to South America. Thus, the Panama Arc, and any slivers derived from Panama which were stranded farther south, should lie outboard of accreted Northern Andes terranes (Fig. 1c). Farallon Plate influence (for instance, establishing a calc-alkaline volcanic arc which persists to the present) would only be established as the triple junction migrates to the north and thus should be diachronous from south to north and much younger than predicted by Inter-American models.

Interpretative outline of key geological elements of the Northern Andes

The purpose of this brief outline of key terranes and faults is to introduce geological elements (Fig. 2) and the terrane-origin classification and terrane boundary nomenclature (Fig. 3 and Table 1) used in the plate reconstructions presented below, to outline their relationships to each other, and to justify some of our novel interpretations of those elements. Features younger than Eocene obscure

many of the inferred relationships and are not shown on these maps. The lateral and cross-strike relationships between key geological elements are somewhat clearer on an expanded, dissected geological map (Fig. 4).

Boundary B1: Sub-Andean Fault, Cimarrona Fault

The Sub-Andean and Cosanga Faults in Ecuador (Litherland *et al.* 1994) and the Cimarrona Fault in Colombia separate the para-autochthonous and allochthonous Cordillera Real (Ecuador) and Central Cordillera (Colombia) to the west from the Magdalena Basin, Eastern Cordillera and Subandean terranes to the east. The latter have not undergone significant northward lateral displacement with respect to *in situ* South American basement. Palinspastic reconstructions (see below) suggest that the non-Caribbean portions of the Santa Marta and Guajira peninsulas should also be considered as more or less *in situ* South America, in addition to the Perijá Range, Santander Massif and Eastern Cordillera. Shortening within the Perijá Range (Kellogg 1984), the Eastern Cordillera, Cordillera Real and Subandes as far south as Peru is essentially east-directed, involving thin-skinned shortening (e.g. Dengo & Covey 1993; Roeder & Chamberlain 1995) and inversion of basement-cored pre-existing Mesozoic rifts (e.g. Cooper *et al.* 1995; Baby *et al.* 2004). Cenozoic uplift in the Santander Massif is due to sinistral transpressive, and links the Perijá and Mérida Andes to the north with the Eastern Cordillera to the south across the Santa Marta–Bucaramanga Fault. Granitoid plutons in these terranes range in age from Triassic to Middle Jurassic (e.g. Tschanz *et al.* 1974; Dörr *et al.* 1995).

Terranes T1a and T1b: para-autochthonous terranes

The eastern part of the Colombian Central Cordillera (Maya-Sánchez 2001; Maya-Sánchez & Vásquez-Arroyave 2001) and most of the Ecuadorian Cordillera Real (Litherland *et al.* 1994) comprise para-autochthonous terranes with affinity to the basement of the Magdalena Basin. They include Neoproterozoic, Grenvillian gneisses and schists, unmetamorphosed to low-grade metamorphic Palaeozoic sedimentary rocks (Restrepo-Pace 1992; Restrepo-Pace *et al.* 1997) with a thin Cretaceous cover section comparable to Colombian Cordillera Oriental and to the foreland east of the Andes, intruded by plutons ranging in age from *c.* 235 Ma to 160 Ma, latest Triassic to Middle Jurassic. In Colombia, these include the Segovia, San Lucas, Sonsón and Ibagué batholiths of Colombia

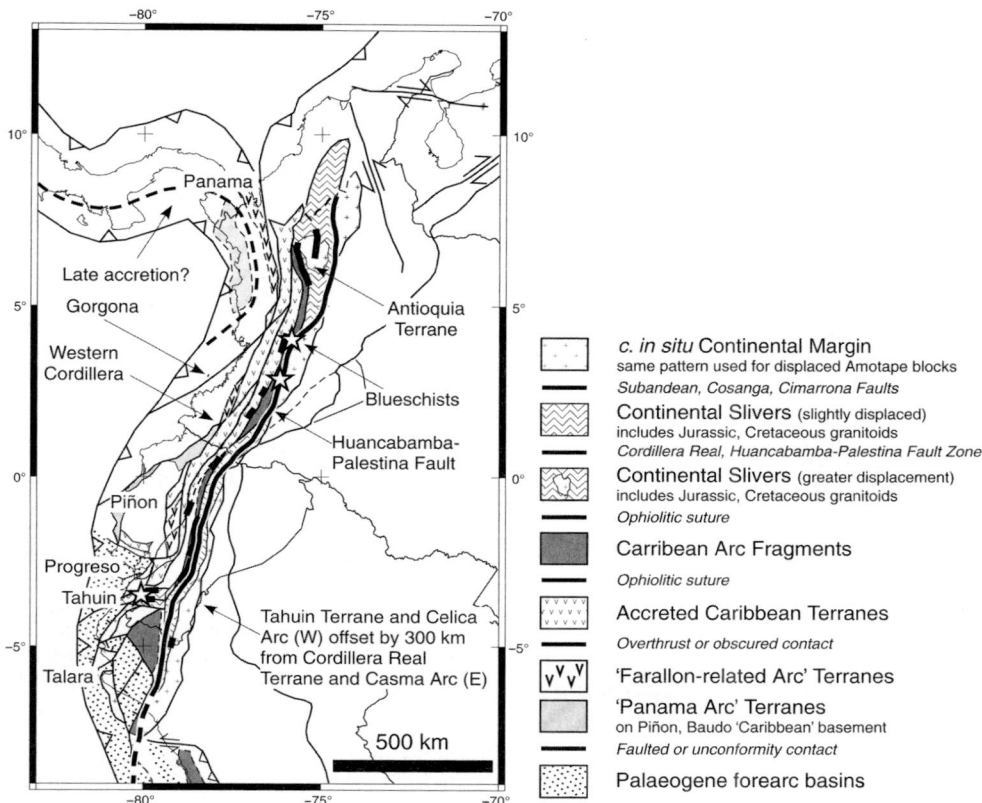

Fig. 2. Simplified geological map of the Northern Andes. Major ophiolitic sutures are commonly associated with blueschists. Note that the Cenozoic Huancabamba–Palestina Fault Zone in part reworks the line of a previously closed Andean Back-arc Basin, and separates para-autochthonous South American rocks to the east from allochthonous South American fragments (such as the Antioquia Terrane and much of the Cordillera Central of Colombia) and accreted Caribbean Arc and Caribbean plateau basalt terranes to the west. The terranes to the west of the fault originated at least 500 km to the SW of their present positions or in the eastern Pacific. A more detailed map (modified from Zamora and Litherland 1993; Schibbenhaus & Bellizzia 2001; Gómez *et al.* 2007*a*, *b*) showing geological and morphotectonic features referred to in the text is available in the online supplementary material.

(e.g. González 2001; Villagómez *et al.* 2008) and the Abitagua and Zamora plutons of Ecuador (e.g. Litherland *et al.* 1994). In Ecuador, the western part of this belt (Terrane T1b) comprises laterally discontinuous belts of moderately to highly sheared rocks including the southern Loja Terrane Palaeozoic sedimentary rocks, Jurassic Salado volcanic rocks and metasedimentary rocks, intruded by the 143 Ma Azafrán Granitoids (Noble *et al.* 1997), the youngest pre-Caribbean pluton. Palaeozoic components of this belt were formed during the assembly of Pangaea, and the younger granitoids, volcanic rocks and associated red beds were formed in two distinct but gradational tectonic contexts: (1) during rifting within and behind a Triassic–Jurassic volcanic arc formed on the west side of the Americas and (2) during the Middle

Jurassic opening of the western end of Tethys between the Americas, in which the Colombian margin formed as the conjugate to southeastern Chortís. The composite arc was broad, extending into the present-day Subandes and foreland, and included the volcanic rocks of La Leche, Oyotún, Colán and Sarayaquillo Formations of northern Peru (e.g. Rosas *et al.* 2007) and the Santiago, Misahualli, Chapiza and Yaupi Formations of Ecuador (e.g. Jaillard *et al.* 1990; Gaibor *et al.* 2008). Middle Jurassic to Cretaceous stratigraphy east of the Central Cordillera and Cordillera Real shows no indication of any adjacent arc from Latest Jurassic until Maastrichtian time (although thin distal tuff bands characterize the Turonian–Campanian Upper Villeta Formation), and is inferred to have been deposited east of a wide

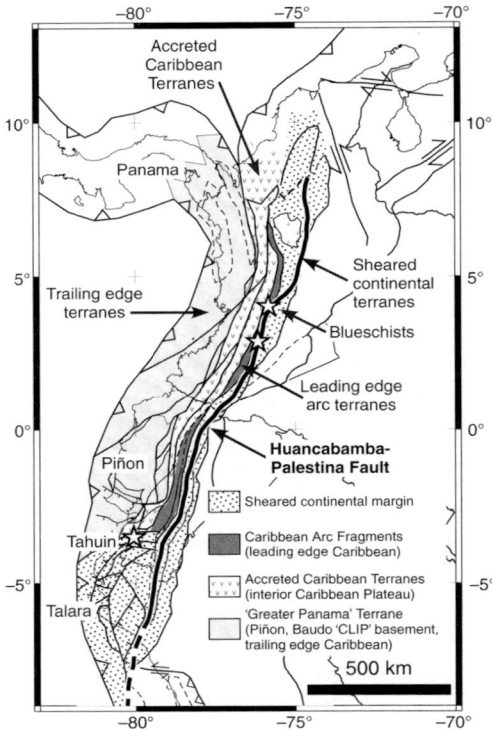

Fig. 3. Simplified terrane map of the Northern Andes. Analysis of terrane affinity suggests four major groupings from east to west: (1) para-autochthonous to allochthonous continental margin fragments; (2) 'Caribbean Arc' fragments derived from the leading edge of the Caribbean Plate, associated back-arc basin volcanic basement and sedimentary fill, and HP/LT rocks; (3) plateau basalts episodically accreted from the interior of the Caribbean Plate; and (4) 'Greater Panama' island arc fragments derived from the trailing edge of the Caribbean Plate.

back-arc basin referred to herein as the Colombian Marginal Seaway (as defined by Pindell 1993 and discussed further below), which linked to the Proto-Caribbean Seaway and western Tethys.

Boundary B2: Huancabamba–Palestina Fault Zone

The boundary between moderately and intensely sheared portions of the Central Cordillera and Cordillera Real is marked by an anastomosing zone of brittle faulting which can be traced more or less continuously from northern Peru to the Lower Magdalena Valley, northeast of Medellín (Fig. 5). This zone is well exposed in much of Colombia, where a narrow zone of large magnitude dextral brittle faults, the Palestina Fault Zone (e.g.

Feininger 1970), can be traced from east of Medellín, where it separates the Antioquia Terrane and the para-autochthonous Serranía San Lucas. North of $6°30'N$, there are four significant strands which all merge to the south: the Otú–Pericos, Nus, Bagre and Palestina Faults. These are probably kinematically related. The Otú–Pericos Fault defines the eastern edge of a ductile shear zone which is the precursor to the brittle fault zone. There is a pronounced jump in metamorphic grade across the Otú–Pericos Fault from greenschist (with lenses of much older granulites) on the east side to amphibolite on the west side (Maya-Sánchez 2001). Feininger (1970) mapped a $c.$ 27 km displacement on one major strand of the Palestina Fault Zone, but we suspect the total offset across the fault zone is much larger. Towards Ibagué farther south, it merges with the San Jeronimo Fault and can be traced south towards Pasto before disappearing beneath an extensive Palaeogene and younger ignimbrite cover (Gómez et al. 2007a, b). In Ecuador it is more difficult to track the fault precisely because the young cover is more widespread than in Colombia and there are multiple candidate fault strands with brittle fault breccias and mylonites in a narrow belt of the Cordillera Real some 25 km wide. In Ecuador, we suggest the equivalent fault trend in the north includes the La Sofia Fault and associated mylonite zones within and on the east side of the Azafrán Granite, the Baños Fault (although this is cross-cut by $c.$ 60 Ma granitoids) and the lineaments which separate the Alao–Paute Terrane from the Guamote Terrane to the west. Dextral brittle faulting is still active (e.g. Machette et al. 2003) along the western margin of the Cordillera Real and strands of the Huancabamba–Palestina Fault Zone within the Cordillera, defining the eastern boundary of a zone of pull-apart and ramp basin formation of which the Inter-Andean Depression near Quito and the Loja, Nabon and Cuenca Basins are parts (e.g. Winkler et al. 2005). Here the fault zone is loosely synonymous with the eastern limit of the 'Dolores–Guayaquil Megashear' (e.g. Moberly et al. 1982), which is usually inferred to link to the Peru–Chile Trench in the vicinity of the Gulf of Guayaquil (e.g. Jaillard et al. 1995).

However, faults which were active during the Late Cretaceous to Palaeogene extend farther south. Recently published geological maps suggest that close to the Peru border the Huancabamba–Palestina Fault Zone includes the southern end of the Las Aradas Fault, between the southern Loja and Alao–Paute terranes (Litherland et al. 1994). The sharply defined Palanda Fault, between the southern Loja Terrane and the Zumba Basin, may also accommodate some Palaeogene brittle faulting. There is often a significant mismatch between older and newer maps and between

Table 1. *Nomenclature and geological summary for key terranes and their boundaries*

Feature	Geological summary
B1	Sub-Andean Fault, Cimarrona Fault
T1a	Para-autochthonous South America fragments. Middle Jurassic (*c.* 160 Ma) San Lucas, Ibague, Abitagua, and Zamora granitoids, Palaeozoic and Neoproterozoic metasedimentary rocks, unconformable Cretaceous cover
T1b	Moderately sheared terranes including South Loja Terrane (Palaeozoic sedimentary rocks), Salado Terrane (Jurassic volcanic rocks) and the 144 Ma Azafrán granite
B2	Huancabamba–Palestina Fault, which can be traced offshore south of the Talara Basin in Peru (Cenozoic, brittle, anastomosing with trace of B3 San Jeronimo, Baños Faults)
T2a	Displaced South American terranes: Antioquia (Palaeozoic sedimentary rocks, *c.* 95–85 Ma Cretaceous granitoids), North Loja Terrane (Palaeozoic sedimentary rocks), Triassic Tres Lagunas granite
T2b	Olmos Massif (Precambrian?), Tahuin (Metamorphic, Palaeozoic, ophiolite protoliths), Amotape (Palaeozoic sedimentary rocks), Triassic Marcabeli and Limon Playa granitoids, Raspas Blueschists (and eclogite), Celica (Cretaceous volcanic arc), Palaeogene forearc basins (e.g. Talara, Progreso)
B3	San Jeronimo Fault, Baños Fault, associated with slivers of ultramafic rock
T3	Blueschists (including Jambalo), Arquia-Chaucha (Metasedimentary rocks with Palaeozoic–Cretaceous protoliths), Quebradagrande (mixed metasedimentary rocks and Cretaceous Caribbean Arc volcanic rocks), Guamote (Jurassic back-arc basin fill sediment), Alao (Jurassic volcanic rocks, possible back-arc basin basement), Chaucha (possible older continental basement to Cretaceous arc)
B4a	Cauca–Almaguer (Romeral) Fault Zone (east side of Cauca–Patia Basin), Pujili melange (?)
T4a	Amaime high-pressure metabasic rocks, 91 Ma Buga Batholith, Bolivar Complex, San Juan terrane. Caribbean Large Igneous Province oceanic plateau basalts extruded at 88–95 Ma.
B4b	Cauca–Patia Fault Zone (east side of Western Cordillera, west side of Cauca-Patia Basin)
T4b	Western Cordillera, Volcanic and Barroso Formations, Guaranda terrane (Late Cretaceous volcaniclastic rocks and lavas), San Jacinto and Sinu accretionary prisms
B5a	Atrato, Urumita Suture, parts of Pallatanga Fault Zone
T5a	Baudo–Choco (72–78 Ma Caribbean LIP?), Dabeiba Arc (48 Ma), Rio Cala Arc (<84 Ma) in Ecuador
B5b	Mulaute, Toachi and Chimbo Fault Zones
T5b	Macuchi (Palaeogene) arc and underlying Piñon basement (88 Ma Caribbean Plateau)
B5c	Puerto Ventura, Canande, Buenaventura Faults (parallel faults breaking up forearc into sub-terranes)
T5c	Naranjal, San Lorenzo Arcs (Island-arc lavas 86–65 Ma) and overlying volcaniclastic sedimentary rocks. Timbiqui Arc (Paleocene, previously accreted Caribbean LIP basement). Eocene and younger forearc basins
T5d	Gorgona (?88 Ma far-travelled plateau, originated *c.* 26°S)

Peruvian and Ecuadorian maps. However, the most recent maps (e.g. León *et al.* 1999) show that north–south-trending brittle faults also extend as far south as the city of Trujillo in northern Peru. The fault zone bounds north–south-trending slivers of Palaeozoic, Cretaceous and Paleocene rocks exposed where erosional windows cut through younger Late Palaeogene ignimbrites (possibly also reflecting Andean reactivation of the north–south-trending faults). The easternmost faults, near Huancabamba, are south of the Palanda Fault, and the westernmost faults, which define the eastern truncation of the Celica–Chignia volcanic arc near Morropón, are south of the Las Aradas Fault.

South of Olmos, the brittle fault zone swings towards the SW and merges with the Peru Trench south of the Sechura Basin. Basement depth maps (Wine *et al.* 2001) show that the Peruvian forearc

is markedly different either side of this fault zone; to the north, the Talara, Tumbes and Progreso Basins have a distinctive dextral pull-apart structural character, while the forearc basins to the south are bounded by simple thrust-related anticlines parallel to the Peru Trench, with little or no indication of strike–slip. On its SE side, the brittle fault zone abruptly truncates the Jurassic La Leche volcanic arc and the Cretaceous Casma arc. On its NW side, the fault zone truncates the Palaeozoic rocks of the Amotape Block, and isolates the Cretaceous Celica–Chignia volcanic rocks close to the Peru–Ecuador border. Analysis of relations between fault strands, Palaeogene arc volcanic rocks, and older rocks beneath indicate that dextral strike–slip offset of about 300 km, perhaps 350 km, dismembered the NW end of the central Andean Cretaceous arc and forearc prior to

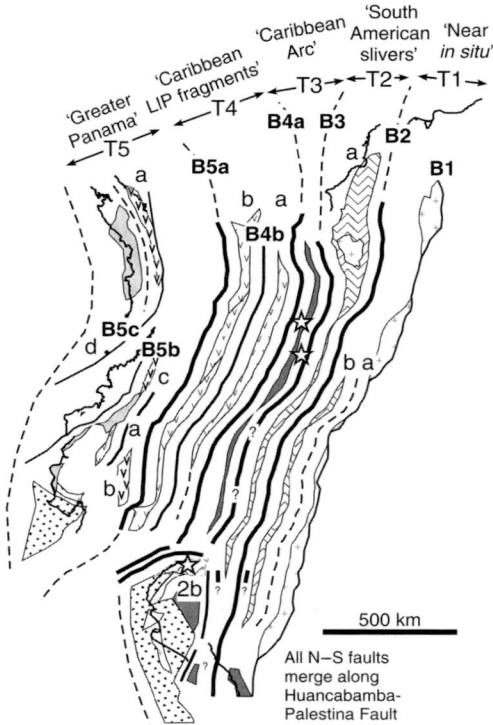

Fig. 4. Dissected present day geological map showing the major sutures and stacking order of the major terranes in the Northern Andes. The legend is the same as for Figure 2. Numbered features are listed in Table 1 and described in the text.

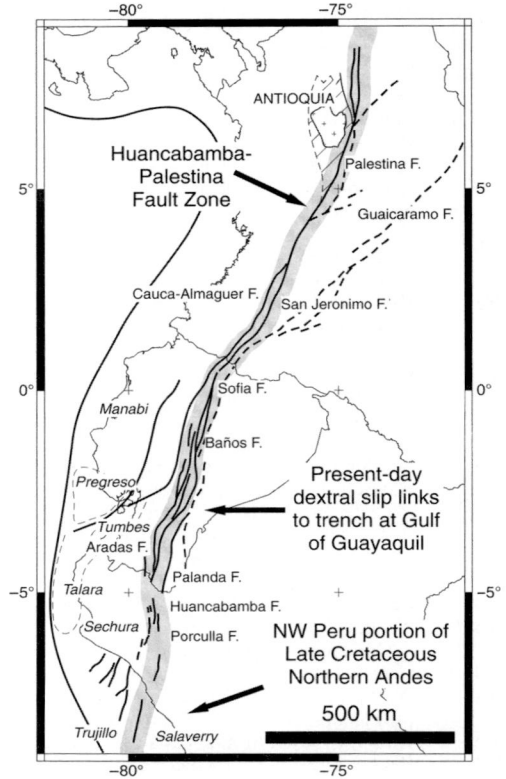

Fig. 5. Major brittle fault traces which define the Huancabamba–Palestina Fault Zone. Selected fault traces and basins (in italics) are named. This narrow brittle fault zone accommodated about 250–300 km dextral slip from Maastrichtian to Middle Eocene time, following a period of ductile shearing at the eastern margin of the Caribbean Plate. Much of its trace is obscured by Late Eocene and younger ignimbrites. From Late Eocene time, most brittle faulting through Ecuador linked to the trench through the Gulf of Guayaquil and faults in Peru became inactive. In Colombia, Late Eocene and younger faulting cut NE through the Upper Magdalena Valley and Eastern Cordillera, as far as the Guaicaramo Fault.

the Late Eocene. Restoration of these inferred offsets allows a new and relatively simple interpretation of the geology and tectonic role of metamorphic belts in Ecuador during the Cretaceous. Only since the Eocene has the eastern limit of significant Northern Andes shearing been linked to the Peru–Chile trench at the Gulf of Guayaquil; prior to then the 'Northern Andes' geological province is considered to have extended into northwestern Peru.

Terranes T2a and 2b: Displaced, sheared terranes west of the Huancabamba–Palestina Fault

The western parts of the Central Cordillera and Cordillera Real are commonly more intensely sheared than their eastern parts, and individual rock units are more laterally discontinuous and much narrower (a few kilometres wide). T2a rocks include slivers of sheared Triassic granitoid (Tres Lagunas, 227 Ma, Noble *et al.* 1997), sheared Palaeozoic sedimentary

rocks and schists (northern Loja terrane, which we infer to have been separated from similar rocks to the south) and the Middle Jurassic and possibly younger Alao–Paute metavolcanic rocks. The latter may be slightly younger (*c.* 160 Ma) than the granites of Terrane 1 (*c.* 170–190 Ma) and show no indication of granitoid intrusion. Geochemical data indicate a supra-subduction zone character (see Fig. 14 below), and there are some associated volcaniclastic turbidite sediments, which led Moreno & Pardo (2003) to explicitly tie them to the previously proposed Andean Back-Arc

Basin (e.g. Pindell 1993; Pindell & Erikson 1994; Pindell & Tabbutt 1995). K–Ar dates (Litherland *et al.* 1994) for Terrane 1 and Terrane 2 rocks are typically bimodal, being either Late Jurassic or older (probably protolith ages) or Albian or younger (probably exhumation and cooling ages).

The largest of the T2 terranes is the *c.* 100 km wide teardrop-shaped Antioquia Terrane in the northern Central Cordillera of Colombia, cored by the Antioquia Batholith, and bounded to the NE by the Otú–Pericos Fault. It is probably continuous with 'Tahami Terrane' basement of the floor of the Lower Magdalena Basin to the north, in which Jurassic thermal and magmatic events appear to be absent (Cardona *et al.* 2006), in contrast to the eastern parts of the Central Cordillera, east of the Otú–Pericos Fault and southern part of the Palestina Fault. The basement to the terrane comprises Neoproterozoic to Palaeozoic schists identical to the eastern part of the Central Cordillera and the floor of the Middle Magdalena Basin and Eastern Cordillera, and an unconformable Cretaceous clastic cover sequence is similar to those in Colombia and Ecuador. The terrane is clearly a fragment of the outer South American margin, but originated 500–700 km south of its present position. This Antioquia Batholith is relatively undeformed, unlike older Terrane 2 granitoids, and has recently been dated at 88–95 Ma (U–Pb zircon, Villagómez *et al.* 2008). Similar ages have also been obtained from the Altavista Stock and San Diego Gabbro (Correa *et al.* 2006), possibly the Sabanalarga Batholith ($^{40}Ar/^{39}Ar$ data only, Vinasco-Vallejo *et al.* 2003), the Aruba Batholith (White *et al.* 1999) and the Pujili granite in Ecuador (Vallejo *et al.* 2006). Only the last two of these have a close association with oceanic plateau basalts, but granites of this age appear to be restricted to close to the ductile boundary between the Caribbean and South American Plates, suggesting a Caribbean connection and separate origin from the older Jurassic plutons to the east. Furthermore, this plutonic episode appears in isolation. No further plutonism is recorded in the area until latest Cretaceous or Paleocene time. Detailed geological maps (e.g. González 2001; Maya-Sánchez 2001; Maya-Sánchez & Vásquez-Arroyave 2001) reveal a complex internal structure of north–south-trending, steeply dipping, sheared lozenges of interleaved Cretaceous sediments and schists with Palaeozoic and Mesozoic protoliths and ages of initial metamorphism. There is a central zone of sheared Mesozoic ultramafic rocks and gabbros, suggesting that the terrane was assembled during the Cretaceous and then intruded by the Antioquia Batholith after which shearing was concentrated around the terrane margins. At the western margin of the Antioquia Terrane, the Late Triassic Medellín Dunites were emplaced onto Neoproterozoic–Palaeozoic rocks above a low-angle thrust, possibly associated with the closure of the Andean Back-Arc Basin during Aptian–Albian time, prior to being intruded by the Antioquia Batholith.

The Antioquia Terrane is the only T2 terrane with associated internal slivers of ultramafic rocks, which elsewhere always lie west of T2 rocks. This may be a result of strike–slip duplexing of T2 and T3 rocks prior to intrusion of the Antioquia Batholith. Otherwise, there are no ultramafic rocks stranded between various components of T2 or between T2 and T1 rocks and no indication of associated subduction–accretion complexes. We think it unlikely that discrete subduction zones (as shown by Litherland *et al.* 1994; Chiaradia & Paladines 2004) separated, for instance, the Late Jurassic Alao and earlier Jurassic Salado metavolcanic rocks of Ecuador.

The southwesternmost portion of Terrane 2 comprises a basement of possibly Precambrian (Olmos Massif) to Palaeozoic metasedimentary rocks which appear to underlie a fragmentary Aptian–Albian volcanic arc and Palaeogene forearc basins (from south to north, Sechura, Talara and Tumbes). Close to the coast at the Peru–Ecuador border, Carboniferous rocks show a gradation from relatively undeformed in a landward or southern position (Amotape Massif of Peru, parts of the Tahuín Group of Ecuador) to highly sheared and exhumed garnet–biotite–Al-silicate schists of the Tahuín Group in the north, adjacent to Terrane 3 (Aspden *et al.* 1995). Metamorphic conditions in the Tahuín schists were of low-pressure Abukuma type. The age of initial metamorphism is pre-Aptian. The Aptian–Albian Celica volcanic arc overlies these Palaeozoic rocks and is sharply truncated by the Huancabamba–Palestina Fault Zone to the east and the ductile shear zones of the El Oro Metamorphic Belt of Ecuador to the north. To the SW it is transitional with forearc sediments deposited on low-grade Palaeozoic strata in the Amotape Hills. All these southwestern Terrane 2 rocks share a clear affinity with those of the more or less *in situ* Peruvian continental margin to the south. We classify T1 and T2 rocks together as 'sheared continental margin' (Fig. 3).

B3 San Jeronimo Fault, Baños Fault Zone, Zanjon–Naranjo Fault Zone

West of Medellín, the San Jeronimo Fault separates the Antioquia Terrane from Terrane 3 Quebradagrande Complex and Arquia Complex rocks. To the north the San Jeronimo Fault merges with the Romeral Fault, and south of Ibagué, it merges with the Huancabamba–Palestina Fault Zone.

In Ecuador, we tentatively place this boundary along the Baños Fault, east of the 'Peltetec Ophiolite' and Alao Terrane. In northernmost Ecuador, the boundary may follow the neotectonically active El Angél Fault and pass east of Ibarra, where a narrow 'Peltetec Ophiolite' inlier lies west of Loja Terrane para-autochthonous rocks. We caution that a range of whole rock $^{40}Ar/^{39}Ar$ ages of 53–1300 Ma suggest that unrelated rock suites along this fault zone may have been unjustifiably mapped together (Richard Spikings, pers. comm. 2008).

In southwestern Ecuador, we trace this terrane boundary as the Zanjon–Naranjo Fault Zone (Aspden et al. 1995) which separates a mixed terrane of Cretaceous metavolcanic rocks and high-pressure, low-temperature (HP/LT) schists to the north from the Las Piedras amphibolites (with Late Triassic volcanic protoliths) and the Tahuín schists with Palaeozoic protoliths to the south.

The key to identifying this terrane boundary is the sharp separation between variably deformed rocks to the east which show no indication of volcanic arc influence later than Middle Jurassic, and very heterogeneous, highly sheared rocks to the west which include Cretaceous arc volcanic rocks, back-arc basin metavolcanic rocks and HP/LT schists. We suggest (see maps presented below) that this boundary and associated slivers of ultramafic rock mark the site of the former Colombian Marginal Seaway back-arc basin, which did not start to close until c. 120 Ma.

Terrane T3: mixed blueschists, back-arc basin floor and Cretaceous volcanic arc fragments

Terrane 3 comprises rocks found on the western side of the Colombian Central Cordillera, east of the Cauca–Almaguer Fault, and as scattered outcrops within the floor of the Inter-Andean Depression of Ecuador between the Pallatanga Fault (Ecuadorian equivalent to the Cauca–Almaguer Fault) and Peltetec Fault and may include the Alao Terrane rocks of the westernmost Cordillera Real. In Colombia, Terrane 3 is well defined on recent maps (González 2001; Maya-Sánchez 2001; Maya-Sánchez & Vásquez-Arroyave 2001; Gómez et al. 2007a, b) and can be subdivided into the Arquia Complex in the west (graphitic schists, garnet schists and amphibolites, with Palaeozoic and Mesozoic mixed protoliths) and the Quebradagrande Complex in the east (metatuffs, pillow basalts and volcaniclastic sediments, all of Early Cretaceous age), separated by the Silvia–Pijao Fault. These rocks can be traced to Pasto, within 75 km of the Ecuadorian border, where they are buried beneath the Cenozoic volcanic cover.

Geochemical data on volcanic components of the Quebradagrande Complex suggest a tholeitic to andesitic island arc to back-arc origin with some influence of underlying continental crust (e.g. Nivia et al. 2006), consistent with the interpretation of Moreno & Pardo (2003). Depositional ages from fauna range from Berriasian to Aptian. The protoliths of the Arquia Complex schists include continental and ultramafic rocks which may be the basement to the Quebradagrande arc with K–Ar whole rocks ages suggesting a range of protolith ages from Neoproterozoic to Late Palaeozoic. Both these packages of rock under went Barrovian metamorphism no earlier than Aptian time (McCourt et al. 1984). K–Ar ages and $^{40}Ar/^{39}Ar$ ages both suggest the onset of subsequent cooling between 120 and 100 Ma (Maya-Sánchez 2001; Maya-Sánchez & Vásquez-Arroyave 2001; Vinasco-Vallejo et al. 2003). The oldest $^{40}Ar/^{39}Ar$ plateau ages are 102–115 Ma in the Arquia Complex and 90–100 Ma in the Quebradagrande Complex. Most reported $^{40}Ar/^{39}Ar$ ages are somewhat younger, typically less than 85–90 Ma and often as young as Late Cretaceous or Cenozoic, suggesting a protracted deformation and exhumation event starting at about Albian time.

Also present in T3 are numerous, separate and laterally discontinuous packages of HP/LT rocks, including blueschists and eclogites (see Pindell et al. 2005 for a more detailed review) with both volcanic and continental (Palaeozoic schist) protoliths. These are tectonically interleaved with the Arquia Complex (e.g. at Pijao and Barragán) or form larger massifs isolated between the Quebradagrande in the west and para-autochthonous T2 terranes (e.g. at Jambaló). Mafic protoliths of possible Early Cretaceous age appear to have reached peak metamorphism not later than Aptian time, based on reported K–Ar ages (McCourt et al. 1984; Maya-Sánchez 2001; Maya-Sánchez & Vásquez-Arroyave 2001) from separate localities of 125 ± 15, 110 ± 10, 120 ± 5 (whole rock) and 104 ± 14 Ma (hornblende). As elsewhere in the Central Cordillera, associated structures appear dominantly vertical to steeply east-dipping. We consider it unlikely that the Jambaló HP/LT rocks were once a nappe thrust over the Quebradagrande and suggest that the protolith for the HP/LT rocks may have formed in a back-arc basin to the east of (relative, present-day coordinates) the Quebradagrande arc and west of the Cordillera Central continental margin. There are no arc or back-arc rocks mapped between the Jambaló HP/LT rocks and the Proterozoic Quintero Gneiss in the Central Cordillera to the east.

The continuation of T3 rocks into Ecuador may comprise the Alao Terrane metavolcanic rocks and the 'Peltetec Ophiolite'. Geochemical data from some samples indicate an MORB-origin (Litherland *et al.* 1994; Richard Spikings, pers. comm. 2008), while others have supra-subduction characteristics (Litherland *et al.* 1994; Etienne Jaillard, pers. comm. 2006). The poorly constrained Middle Jurassic age of the Alao rocks (Litherland *et al.* 1994) suggests that the older Jurassic arc, founded on attenuated continental crust, was split by an intra-arc basin which evolved into the younger Andean Back-arc Basin, of which the Quebrada-grande Complex may be a part.

In central Ecuador, there are few outcrops beneath the younger cover of the Inter-Andean Depression. Poorly dated Early Jurassic to Early Cretaceous sediments of the Guamote Terrane (Litherland *et al.* 1994) may represent fragments of a back-arc basin fill, with sediment derived largely from the continental margin to the east. To the SW of the Guamote Terrane, there is a 50 km long slice of continental metamorphic rock mapped northeast of Chaucha (Dunkley & Gaibor 1997). The Bulubulu and Pallatanga Faults define its northwestern boundary and its eastern margin is faulted against Eocene volcaniclastic rocks or buried by younger volcanic rocks. It is mapped as 'La Delicia schist' by Zamora & Litherland (1993) and interpreted as forming the basement to a 'Chaucha Terrane' by Litherland *et al.* (1994). It appears to comprise schists with Late Palaeozoic or Jurassic protoliths similar to those in the Cordillera and in the El Oro Metamorphic Belt to the south. Litherland *et al.* (1994) interpreted the Chaucha Terrane as a fragment of continent which collided with the Ecuadorian margin in a Late Jurassic to Early Cretaceous closure of the Colombian Marginal Seaway. In contrast, Pindell (1993), suggested it may comprise the basement to the southeasternmost end of the Trans-American Arc, and delayed closure of this portion of the Colombian Marginal Seaway until Middle Cretaceous or later.

We assign the northern part of the El Oro Metamorphic Complex in southern Ecuador (Aspden *et al.* 1995) to Terrane 3. Highly deformed schists (Palenque Melange) and sheared Triassic granitoids are separated from the Terrane 2 Tahuín schists and granitoids by the Chilca blueschists and eclogites (e.g. Bosch *et al.* 2002) and the Raspas 'ophiolite', found immediately north of the Zanjon–Naranjo Fault Zone. Contextual interpretation is made difficult by widespread Cenozoic volcanic cover and by significant northward dextral translation along the Huancabamba–Palestina Fault Zone (see below) associated with *c.* 60° clockwise rotation during the Late Cretaceous (e.g. Mourier *et al.* 1988). The Palenque Melange is commonly assumed to comprise slices of low-grade metamorphic rock within a forearc subduction complex trenchward of the HP/LT belt (e.g. Bosch *et al.* 2002) driven by east-directed subduction beneath South America, but comparison with the relationships observed in Colombia suggests that the protoliths of the HP/LT rocks might have formed in a basin now sandwiched between two blocks founded on Palaeozoic continental crust: an arc to the west and a continental passive margin to the east. There are no constraints on thermal history other than a single 76 Ma K–Ar age on hornblende (Feininger & Silberman 1982) and a 68 Ma zircon fission track age (Spikings *et al.* 2005). There is a single reported age for possibly syn-metamorphic phengite in the HP/LT rocks of 132 ± 5 Ma, but it is unclear if too much weight should be attached to this single data point, given that phengite is prone to problems with excess ^{40}Ar. The resolution of the origin of these HP/LT rocks will require better constraint on both protolith and peak metamorphic ages.

We classify all these Terrane 3 rocks as sheared and accreted fragments of the Trans-American Arc and its successor Caribbean Arc (Fig. 3), the SE end of which was separated from continental South America by the Colombian Marginal Seaway until this was destroyed during the Late Cretaceous (e.g. as shown by Pindell *et al.* 2005, 2006). Note that in contrast to previously published models (e.g. Moreno & Pardo 2003, based on Pindell & Barrett 1990; Pindell 1993) we do not continue an oceanic-floored back-arc south across the Peru–Ecuador border, because the restoration of offset on the Huancabamba–Palestina Fault Zone puts the Celica and Casma Arc adjacent to each other, without any evidence of a back-arc basin axis between them. Caribbean Arc remnants show substantial evidence of continental crust contamination in Aptian and younger intrusive and extrusive rocks, and there is a close spatial association throughout the circum-Caribbean region with HP/LT and Barrovian metamorphic rocks with both continental and oceanic protoliths (summarized in Pindell *et al.* 2005). Exhumation of metamorphic terranes and volcanism evolving towards a more calc-alkaline character has been continuous since then. The relatively simple geologic history points to the existence of a west-dipping subduction zone which has persisted since *c.* 120 Ma.

Boundary B4 Cauca–Almaguer Fault Zone, Pujili mélange–Pallatanga Fault Zone, Jubones Fault Zone

The Cauca–Almaguer Fault Zone in Colombia (broadly synonymous with 'Romeral Fault', a term deprecated on current Colombian geological maps) is a fundamental boundary separating rocks with some continental affinities in the east, albeit separated from the Central Cordillera by an ophiolite or blueschist suture, from far-travelled oceanic plateau remnants in the west. In southern Colombia and Ecuador the fault remains the site of significant dextral strike–slip (Trenkamp et al. 2002) and field relationships have probably been substantially modified since the Cretaceous. In Colombia, the fault can be followed continuously west of the Arquia Complex, but is difficult to trace unambiguously into Ecuador. South of Pasto in southern Colombia, the Cauca–Almaguer, Silvia–Pijao and San Jeronimo Faults all pass beneath Cenozoic cover rocks. Recently published maps of Colombia (Gómez et al. 2007a, b) suggest that the Cauca–Almaguer Fault trends SW–NE and crosses the Ecuador border between Volcán Chiles and Tulcán, passing north of Ibarra and through or just west of the Chota Basin (Winkler et al. 2005). As defined in this paper, the fault must pass to the west of the Guamote Terrane, which has continental margin or back-arc basin affinities and thus the Cauca–Almaguer Fault is not equivalent to the Peltetec Fault. In the absence of any information on the basement of most of the Inter-Andean Depression, we map this terrane boundary as the eastern edge of the Pujili Mélange and the Pallatanga–Bulubulu Fault Zone, which define the western side of the Inter-Andean depression (e.g. Hughes & Pilatasig 2002) and the western edge of slices of continental-affinity schist of the 'Chaucha Terrane'. This is consistent with the presence of metamorphic xenoliths at Guagua Pichincha volcano near Quito (Richard Spikings, pers. comm. 2008). By our definition, this boundary does not exactly coincide with the locus of neotectonic dextral faulting, which steps east across the Inter-Andean Depression towards the Chingual Fault and Subandean Fault close to the Ecuador–Colombia border. Farther south, an equivalent boundary in the El Oro Metamorphic Belt is the Jubones Fault Zone (Aspden et al. 1995).

Field observations (e.g. Alfonzo et al. 1994; Kerr et al. 1998) and seismic lines (Ecopetrol, unpublished data) from Colombia suggest that Late Cretaceous through Cenozoic structuring involved underthrusting of Western Cordillera rocks to the east beneath the Central Cordillera and a similar model has been proposed for Ecuador (Jaillard et al. 2005). Although the surface trace of the terrane boundary lies at the western edge of the Inter-Andean Depression, the contact may be relatively low-angle and east-dipping, albeit dismembered and modified by subsequent higher-angle dextral strike–slip.

Terrane 4: parts of the Western Cordilleras of Colombia and Ecuador, basement of Cauca–Patia Valley and western foot of Central Cordillera

Stratigraphic terminology and geological maps for Terrane 4 are in a state of flux, and recently published high precision isotopic ages are leading to substantial revisions. Among recent reviews and syntheses we note Kerr et al. (1997, 2002a, 2003), Hughes & Pilatasig (2002), Mamberti et al. (2003, 2004), Jaillard et al. (2004), Vallejo et al. (2006) and Vallejo (2007). In Colombia, this terrane comprises all but the southwesternmost parts of the Western Cordillera in addition to parts of western foot of the Central Cordillera. It corresponds to the eastern parts of the Dagua–Piñon Terrane (Cediel et al. 2003) or Calima Terrane (Toussaint & Restrepo 1994). The terrane comprises fault-bounded slivers of oceanic plateau basalts and associated ultramafic rocks and sediments which originated as part of the Caribbean Large Igneous Province (or CLIP). In Colombia, the Amaime basalts in the westernmost part of the Central Cordillera (Terrane T4a) are separated from the Bolivar Complex and the Western Cordillera (Terrane T4b) by the Cauca–Patia Fault Zone, but this appears not to correspond to a fundamental Mesozoic feature. In Ecuador, this terrane is equivalent to the 'Pallatanga Terrane' and 'San Juan Terrane' (sensu Vallejo 2007) and possibly the 'Guaranda Terrane' (sensu Jaillard et al. 2004). Terrane 4 rocks are much less areally extensive in Ecuador than in Colombia, and are limited to discontinuous outcrops immediately west of the Pallatanga Fault.

Most crystallization ages from the plateau basalts are close to 90 Ma as elsewhere in the Caribbean. Geochemistry indicates an intra-oceanic plateau origin without subduction zone influence (Kerr et al. 1997). Small granitoids intruding the plateau basalts which pre-date terrane accretion include the c. 90 Ma Buga tonalite (Richard Spikings, pers. comm. 2008) and Buritica Tonalite (González & Londoño 2000) in Colombia, and the c. 86 Ma Pujili granite of Ecuador (Vallejo et al. 2006). Older ages include a poor 99 Ma $^{40}Ar/^{39}Ar$ plateau age (Mamberti et al. 2004) and a 123 ± 13 Ma Sm–Nd internal isochron age, but recent U–Pb zircon crystallization ages from these sites confirm the c. 90 Ma age of eruption (Vallejo 2007).

Within the Western Cordillera of Colombia, dominant structure is top to the west thrusting, interleaving volcanic rocks and sediments, with an overprint of dextral transpressive shear. Structures are cross-cut by numerous, but volumetrically minor, latest Maastrichtian to Early Eocene plutons which post-date accretion of Caribbean basalts west of the Central Cordillera (e.g. González 2002). Both Caribbean rocks (including fragments of the Caribbean Arc) and South American rocks are intruded by suites of identical plutons as far north as Santa Marta and Guajira, indicating that the leading edge of the Caribbean Plate was located close to the Venezuela border by that time (e.g. Cardona *et al.* 2008). The thrust fabric is cut by ENE-trending dextral strike–slip faults which truncate individual thrust sheets and Palaeogene intrusive and extrusive rocks, and appear to allow north–south lengthening of the Cordillera during and since Palaeogene time. In Ecuador and Colombia, there are numerous sub-terranes bounded by en-echelon strike–slip faults which trend slightly clockwise of the overall north–south-trend of the major terrane bounding faults, also consistent with an overall dextral transpressional structural context. The accretion ages summarized above are somewhat younger than the oldest $^{40}Ar/^{39}Ar$ plateau and K–Ar ages from schists and deformed igneous rocks in Terranes 2 and 3, which range from c. 105–89 Ma.

Boundary B5: Atrato and Urumita Sutures, parts of Pallatanga Fault Zone

There is a sharp morphologic break between the Western Cordillera and present-day forearc terranes. The boundary is clearest in the north as the Atrato and Urumita sutures east of the Panama and Baudó blocks, where terranes to the west comprise plateau basalts (some possibly as young as 78 Ma) but with overlying intra-oceanic island arc volcanic rocks, unlike Terrane 4 where plateau basalts and sediments show no arc influence. A traverse across the Western Cordillera near Cali (Kerr *et al.* 1997) shows that the boundary lies west of the Cordillera. In the Rio Timbiqui, the boundary coincides with a belt of Cretaceous ultramafic rocks which separate the Palaeogene Timbiqui Arc to the west from the plateau basalts to the east. South of 2°S, the boundary between the Cretaceous Ricaurte arc (Spadea & Espinosa 1986) and plateau basalts is defined by the north–south-trending Falla de Cuercuel (Gómez *et al.* 2007a, b), which approximately coincides with the northern limit of the Mulaute, Toachi and Chimbo Fault Zones (e.g. Hughes & Pilatasig 2002; Kerr *et al.* 2002a, b) to the east of the Campanian–Maastrichtian Naranjal Arc (Kerr *et al.* 2002a, b) and the Palaeogene Macuchi Arc. However, recent mapping in Ecuador (Boland *et al.* 1998; Vallejo 2007) shows that intra-oceanic island arc rocks, albeit with Caribbean Plateau basement extend as far east as the Pallatanga Fault Zone in northernmost

Ecuador, raising the possibility that island arc rocks may be more widespread in the southern Colombian portion of the Western Cordillera than currently mapped.

Several north–south- to ENE-trending strike–slip faults also cut the forearc and Western Cordillera terranes, such as the Buenaventura, Garrapata and Ibagué Faults in Colombia and the Puerto Cayo and Canande Faults in Ecuador, which appear to accommodate north–south lengthening, east–west thinning, and rotation of these terranes since Late Cretaceous time. The Mulaute–Toachi–Chimbo Fault Zone is a steeply dipping reverse fault separating the Eocene Macuchi Arc from Terrane 5 arc rocks in the Western Cordillera. The presence of sheared Eocene granitoids and fission track data (Richard Spikings, pers. comm. 2008) indicates that the feature may be a Miocene or younger fault zone rather than a significant terrane suture.

Terrane 5: 'Greater Panama Arc' and forearc trailing edge Caribbean terranes

Since Pindell & Dewey (1982), the majority of evolutionary models for Panama have considered the Buenaventura Fault at about 4°N as the southern limit of the Caribbean lithosphere from which the far-travelled Panama Arc terrane (trailing edge of Caribbean Plate) has been accreted to western Colombia, with the suture lying on the east side of Dabeiba Arch. Panama Arc crust has been indenting Colombia since the Eocene at the latitude of the Upper Magdalena Basin, driving the Oligocene development of the Chusma thrustbelt along the western Upper Magdalena Valley (Butler & Schamel 1988; Pindell *et al.* 1998). Seismic tomography (Van der Hilst & Mann 1994) supports the view that Caribbean lithosphere now reaches the Buenaventura Fault but not farther south. Here, we explore and develop the idea that additional buoyant parts of an originally longer Panama Arc were clipped off from their underlying lithosphere and accreted to the Northern Andes well south of Buenaventura, and we refer to the collective group of such additional terranes as the 'Greater Panama Arc' (Fig. 3).

The western parts of the Western Cordillera and the forearc of the present-day Andes of Ecuador and central and southern Colombia makes up Terrane 5. Here, boninites, tholeites and calc-alkaline basalts of intra-oceanic island arc origin overlie oceanic plateau basalts. Outcrops are relatively limited, comprising fault slices in the Western Cordillera and forearc. Most of this terrane is buried by Palaeogene and younger forearc basin sediments, such as those of the Progreso, Santa Elena and Manabi forearc basins in Ecuador (e.g. Jaillard *et al.* 1995) and the Pacifico forearc basin in Colombia.

The oceanic plateau basement is particularly well developed in the Piñon Formation of Ecuador and in the Serranía de Baudó in northwestern Colombia. The latter appears conspicuously younger than other plateau fragments, dated at 73–78 Ma, in contrast to typical CLIP ages of 88–93 Ma elsewhere (e.g. $^{40}Ar/^{39}Ar$ ages reported in Kerr *et al.* 1997), indicating a protracted history of plateau basalt eruption in the Caribbean. The young age, however, may be coincident with the passage of the rear of the Caribbean Plate over the Galapagos Hotspot (Pindell & Kennan 2009). The age of the Piñon basement has previously been considered as Aptian, but is now considered to be earliest Coniacian on the basis of identical *c.* 88 Ma crystallization ages from the forearc (Luzieux 2007) and Western Cordillera near Quito (Vallejo 2007).

Numerous fragments of oceanic island arc volcanic rocks and associated volcaniclastic sediments are associated with the Piñon basement. However, outcrop is poor and often only exposed briefly, and field relationships are thus not clear. Recent fieldwork, palaeomagnetism and biostratigraphy have led to substantial revisions and simplifications (Vallejo *et al.* 2006; Luzieux 2007; Vallejo 2007). The oldest island arc elements in Terrane 5 are found in the La Portada Formation (Vallejo 2007), in which primitive island arc boninites are associated with sediments from which Santonian to Campanian foraminifera have been recovered. These volcanic rocks are broadly coeval with the volcaniclastic Calentura Formation, which overlies the Piñon basement of the forearc. Younger arc elements include the San Lorenzo and Naranjal Formations in Ecuador and the Ricaurte Formation of southernmost Colombia. All yield low-precision Maastrichtian radiometric ages and all are associated with volcaniclastic sediments with well-constrained Campanian–Maastrichtian faunal ages, including the Mulaute and Natividad Formations in the Western Cordillera and the Cayo Formation in the forearc.

The geochemistry of these rocks is consistent with formation above an intra-oceanic subduction zone with no continental influence. Proximity to the South American margin is indicated by the first appearance of terrane-linking quartz-rich sediments which derive from the unroofing of the Cordillera Real in latest Campanian or Maastrichtian time in the Western Cordillera (e.g. Jaillard *et al.* 2004). In Ecuador, accretion of Terrane 5 rocks appears to be in the form of large thrust east-dipping sheets subcreted to the South American continental margin during discrete Late Campanian and Late Maastrichtian events, predating the Yunguilla and Saquisili Formations, respectively (Jaillard *et al.* 2008). However, field relationships are not always clear, and 'basement' and 'cover' rocks are usually found in adjacent fault slivers without clear unconformable contacts. In our view they do indicate palaeogeographic proximity, but not necessarily final terrane accretion, nor do they indicate the end of dextral strike–slip motion between now adjacent blocks. Eocene tuff-bearing quartzose strata coeval with the Saraguro Ignimbrites do overlap and post-date plateau basalt accretion, but not the end of dextral strike–slip.

Paleocene–Middle Eocene arc fragments are also found in Terrane 5 at the foot of the Western Cordillera, including the Macuchi arc (Terrane 5a in Fig. 4), the Timbiqui arc of southwestern Colombia, and the Dabeiba arc of northwestern Colombia and Panama. The structure of the forearc portion of Terrane 5 appears to be much simpler than the subduction complex in the Western Cordillera. There is relatively little internal shortening, and the dominant features on geological maps are an array of north–south-trending to SW–NE-trending dextral strike–slip faults which may link with those which cut across the Western Cordillera (the largest of these are distinguished on Fig. 4). Forearc narrowing as a result of strike–slip on these faults brings Cretaceous arc fragments to within 75 km to the present-day trench, and Palaeogene arc fragments step east to the foot of the Western Cordillera about 200 km east of the trench.

We introduce and clarify the concept of a 'Greater Panama' terrane (Fig. 3). This comprises the Panama Arc (defined as reaching the Buenaventura Fault by Pindell & Dewey 1982), and the Santonian to Maastrichtian Rio Cala, Ricaurte, Naranjal, and San Lorenzo arc remnants in Colombia and Ecuador. Together with the associated Piñon basement we propose that these rocks as having formed above a single northeast-dipping subduction zone at the trailing edge of the Caribbean Plate. The presently fragmented outcrop of these arcs reflects post-accretion faulting and block rotation after accretion of this terrane to the South American margin from latest Cretaceous to Palaeogene time. We suggest that mainly the buoyant upper crustal parts of the lithosphere were clipped off and accreted, while the deeper upper mantle lithosphere continued to be obliquely subducted, such that the accreted terranes now lie far south of their parental lithospheric root, which now lies north of the Buenaventura Fault. Previous models, which assumed that a triple junction at the end of the Panama Arc migrated northwards outside the Piñon forearc block, would not explain the presence of Cretaceous arc rocks on the Piñon block unless those arcs had formed at the leading edge of the Caribbean Plate. However, the Caribbean Arc appears to have initiated at about 120 Ma, about 35 Ma before the arcs on 'Greater Panama', following a regional change in subduction polarity which allowed oceanic crust of the Colombian Marginal Seaway and Proto-Caribbean Seaway to subduct west beneath the Caribbean.

In contrast, volcanism in the Western Cordillera arc terranes started only at about 85 Ma, following eruption of the CLIP oceanic plateau basalts at *c*. 90 Ma. There is no close association with fragments of continental rock, nor any geochemical evidence of continental contamination of arc volcanic rocks. The assemblage of Late Cretaceous plateau basalt basement and overlying arc terranes appears essentially the same from Guayaquil to Panama and to Costa Rica. Internal structural character and expression on gravity and magnetic field maps is similar, suggesting that they share a common history related to a single plate boundary at the trailing edge of the Caribbean Plate. The terrane was episodically and diachronously accreted as fragments were detached from the migrating Caribbean lithospheric root and stranded in northwestern Ecuador and southwestern Colombia. The map view of these accretion events is discussed further below.

Following accretion of these terranes to the South American margin, eastward or northeastward subduction of the Farallon Plate continued beneath them, as subduction continued beneath the intra-oceanic Costa Rica–Panama arc farther north, resulting in the formation of the Macuchi and Timbiqui volcanic arcs above oceanic basement that had already accreted to the South American margin. These arcs have tholeiitic to calc-alkaline chemistry, superficially suggesting an intra-oceanic island arc origin, but this is inconsistent with their position inboard of forearc terranes that were already adjacent to South America and receiving continent-derived sediment. New field and laboratory data (Vallejo 2007; Richard Spikings, pers. comm. 2008) shows that locally the Macuchi arc is overlain by, rather than faulted against, the mostly continent-derived Angamarca Formation clastic sediments, and that the Angamarca contains some detritus of Macuchi origin. Thus these rocks were adjacent during the Eocene. The *c*. 40 Ma exhumation event identified by Spikings *et al*. (2005) thus does not indicate the accretion of an far-travelled terrane but is probably a reflection of increased coupling with the subducting Farallon Plate. The chemistry of the Macuchi and Timbiqui arcs contrasts with the Paleocene and Eocene ignimbrites of northernmost Peru and southernmost Ecuador, which erupted during and after accretion of Terrane 5, but which show a strong influence of underlying South American continental crust. The oldest of these ignimbrites cross the major terrane boundaries but also show breaks and offsets that indicate that substantial strike–slip was yet to occur. The youngest of these formations overlap the terrane boundaries south of the Gulf of Guayaquil, indicating that Late Eocene and younger strike–slip was confined to Colombia and Ecuador and did not affect Peru.

A note on the pre-Cretaceous history of the Northern Andes and its influence on subsequent deformation

Granulite-grade metamorphic rocks as old as at least Neoproterozoic (*c*. 1 Ga), and greenschist to amphibolite facies Early to Late Palaeozoic metamorphic rocks and arc volcanic rocks are common elements of the Central and Eastern Cordilleras of Colombia (e.g. Restrepo-Pace 1992; Restrepo-Pace *et al*. 1997). However, the history of this part of South America is not one of a long-lived proto-Andean subducting margin which extended as far north as the Guajira Peninsula. During the Late Precambrian, following Grenvillian orogeny, Colombia lay adjacent to eastern North America and Greenland in the core of the Rodinia super-continent (e.g. Li *et al*. 2008). Subsequently, Rodinia broke up, leaving western South America facing the Iapetus Ocean during the Early Palaeozoic. Terranes now in Mexico and Central America, such as Chortís, Yucatán and Oaxaquia rimmed northern South America. The evolution of the Iapetus Ocean and younger Rheic Ocean is complex and subject to ongoing controversy (e.g. contrast Dalziel 1997; MacNiocaill *et al*. 1997; Van Staal *et al*. 1998; Keppie & Ramos 1999; Cocks & Torsvik 2006, among many). Laterally discontinuous, sometimes short-lived, subduction zones were established along both Gondwanan and Laurentian margins of Iapetus at various times, and some peri-Gondwanan terranes (the best understood is Avalonia) were detached from Gondwana during the opening of the Rheic Ocean, and accreted to Laurentia prior to the onset of Rheic Ocean closure. Chortís, Yucatán and Oaxaquia are thought to have remained on the South American side of the Rheic Ocean and the Palaeozoic volcanic arcs and metamorphic rocks in the Central Cordillera of Colombia (Restrepo-Pace 1992) probably reflect poorly understood arc accretion or back-arc basin opening and closure events between the southern end of Chortís and the proto-Peruvian trench which faced Iapetus and/or Rheic Oceans. Final closure of the Rheic Ocean during the Permian resulted in the assembly of Pangaea, with the Acatlán and Ouachita deformed belts representing the suture between Laurentia and Chortís/Oaxaquia and Yucatán, respectively.

Prior to the breakup of Pangaea during the Triassic to Early Jurassic, a single Pacific-facing arc became established along the west side of the Americas (Fig. 6). Breakup of Pangaea occurred between Colombia–Venezuela and Chortís–Yucatán, along a line south of the Rheic suture and possibly following an older back-arc basin trend. Motion of South America away from North America was towards the SE, resulting in the opening of an oceanic tract between the Americas by Late Jurassic time. This was separated from the Pacific domain by a lengthening Trans-American arc (see additional maps in Pindell & Kennan 2009), whose position is approximately defined by a flowline of South America away from southernmost Chortís (Fig. 6). Critically, it is kinematically impossible for the spreading centres in the Proto-Caribbean Seaway and Colombian Marginal Seaway (see Fig. 7) to cross the Inter-American trench (in contrast to the model of Jaillard et al. 1990) and for subduction of Pacific Plates beneath South America to have occurred beneath central or northern Colombia after about 160 Ma, which is

the age of the youngest Colombian supra-subduction rift-related granitoids. Only in Ecuador are younger granitoids and arc volcanic rocks found on undisputed para-autochthonous South American crust. The Colombian magmatic record after 160 Ma is very sparse and does not indicate the persistence of subduction immediately west of the Colombian margin.

The future Northern Andean margin of Ecuador and Colombia is defined by the continent–ocean boundary which formed as a result of Pangaea break up and is quite distinct from the trend of slightly older proto-Andean and Cordilleran Permian to Early Jurassic margins of the Americas. Subsequently, the western end of the Proto-Caribbean Seaway was modified by the Late Jurassic and Early Cretaceous opening of oceanic back-arc basins which extended the areas of oceanic crust into Mexico and towards the Peru–Ecuador border (Fig. 7). The ages of these back-arc basin seaways are poorly constrained. The oldest reasonably well-dated arc or back-arc volcanic rocks (K–Ar ages only) and sediments

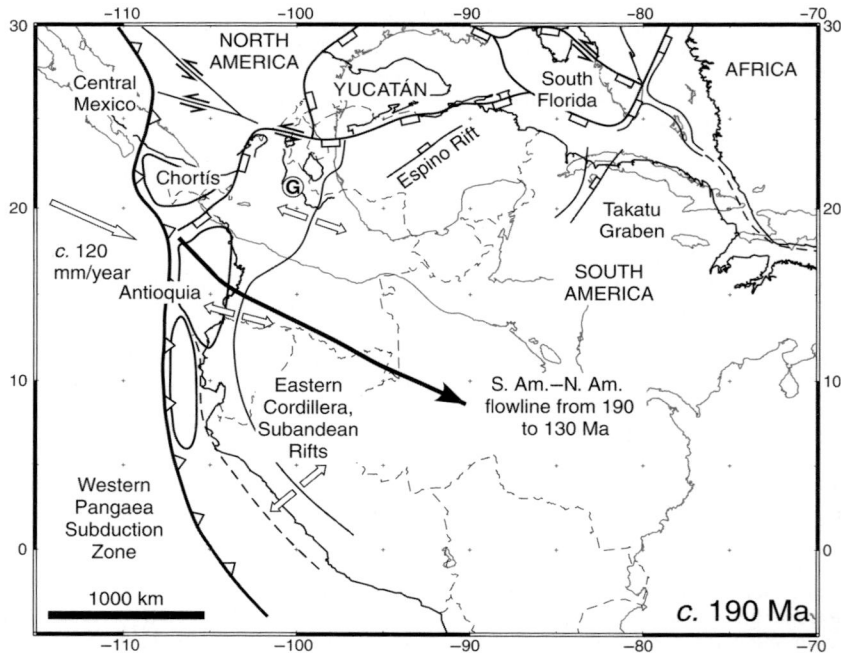

Fig. 6. A 190 Ma reconstruction of South America (after Pindell & Kennan 2009) against a fixed North America, closing the Equatorial and Central Atlantic oceans. Present-day coastlines are shown in grey. Pangaea started to break up shortly before this time, causing the lengthening of the subduction zone to the west. The 190–130 Ma flowline for South America away from North America defines the approximate boundary between a Proto-Caribbean realm of spreading between the Americas, and a zone of subduction or strike–slip between Pacific plates and both North and South America. It is kinematically impossible for the Proto-Caribbean spreading centre to extend into the Pacific across this subduction zone or for long-lived subduction to occur on the Colombian margin northeast of the palaeoposition of Antioquia shown.

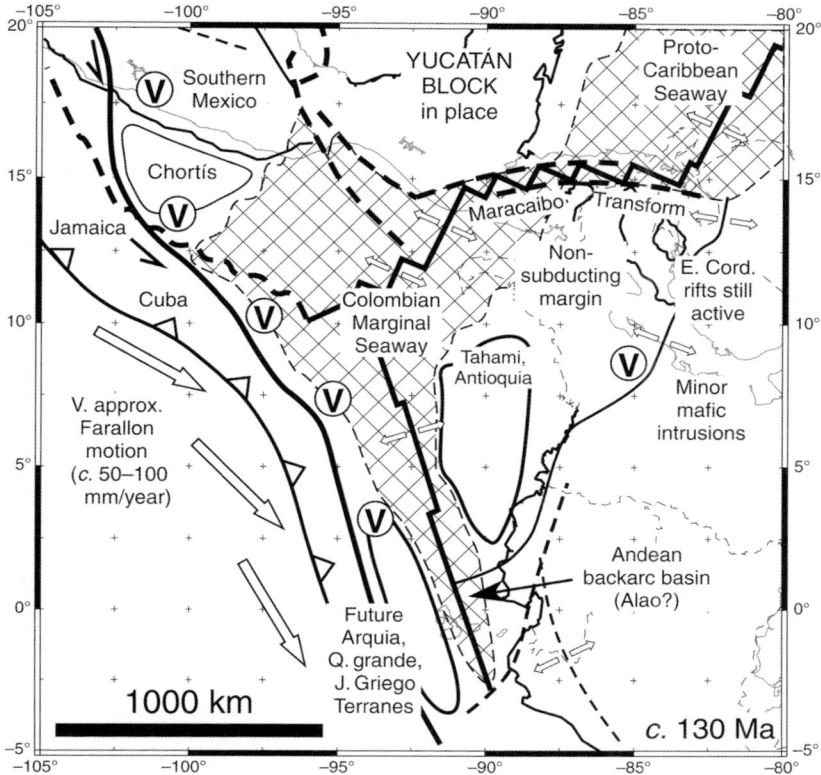

Fig. 7. A 130 Ma reconstruction, relative to a fixed North America (modifed from Pindell & Kennan 2009). Present-day coastlines are shown in grey. The Proto-Caribbean Seaway is interpreted to be part open, and linked to the Colombian Marginal Seaway by this time. Early Cretaceous back-arc basins parallel to the trench to the west and oblique to the Proto-Caribbean Seaway had formed in southern Mexico, southern Colombia and Ecuador. Limited plate motion data from the Pacific indicates that, if the plate to the west was the Farallon Plate, subduction was highly oblique to pure sinistral strike–slip opposite the Northern Andes (a mirror of present-day California). The basement to the future Caribbean Plate probably lay just to the west of the area shown in this map. This setting typically results in slab break-off and foundering into the mantle. As a result there may have been little or no subducted slab present beneath the Andean Back-arc Basin by Late Early Cretaceous time.

are earliest Cretaceous (Nivia 1996). The Aburra Ophiolite (which includes the Medellín Dunites) has also been proposed as part of the floor of the Andean Back-arc Basin (Correa 2007). However, the well-constrained 217–228 Ma age (U–Pb on zircons extracted from associated plagiogranites) of these rocks is c. 90 Ma older than any other Andean Back-Arc Basin rocks, and some 30 Ma older than basalt associated with ocean floor formation between the Americas (Sabalos Basalt, Cuba: Pszczolkowski 1999; Pszczolkowski & Myczynski 2003). If a back-arc basin rock, it must have formed in a basin parallel to the Trans-American arc. However, the lack of arc rocks of this age to the west suggests to us an alternative origin as an accreted ocean crust fragment in the Trans-American subduction zone. Only much later

was it thrust west onto the Antioquia Terrane, which we suggest lay close to the trench in Triassic–Early Jurassic time, seaward of a broad zone of subduction influenced granitoid intrusion and arc magmatism that stretched from northern Mexico to Ecuador. The geochemistry of the Aburra Ophiolite is consistent with MORB subsequently mixed with subduction related magmas, and its age is strikingly similar to accreted ophiolites and fossiliferous sediments in the forearc Cochimi Terrane (Viscaino Peninsula) of Baja California (Rangin et al. 1983; Sedlock 2003). Baja California lay not more than c. 1000–1500 km to the north at that time (Fig. 6), and may be a good analogue for the Aburra Ophiolite.

In previous papers (e.g. Pindell & Erikson 1994; Pindell & Tabbutt 1995), the Colombian margin east of the Palestina Fault has been characterized as essentially a passive or Atlantic-type margin (all rocks to the west are to greater or lesser degree allochthonous and 'suspect'). However, it is clear that some rifting did remain active in the Eastern Cordillera rifts of Colombia until about Albian time (Sarmiento-Rojas *et al.* 2006), possibly kinematically linked to the nearby Proto-Caribbean spreading centre which lay to the north of the Maracaibo Transform (Fig. 7), and there is low volume, rift-related (no supra-subduction zone geochemical signature) associated mafic magmatism (e.g. Vásquez & Altenberger 2005). Thus, we consider the 'passive margin' approximation still more or less valid, in the sense that the Colombia and Ecuador margins were not associated with adjacent subduction and have no volcanic arc built on them in the interval 160–50 Ma. The Colombian Marginal Seaway and Andean Back-arc Basin appears to have been wide enough to isolate the margin from the influence of the Trans-American Arc and the younger Caribbean Arc. Some thin foreland tuffs could be derived from hot spot volcanism within the coeval Napo Formation of Ecuador (Baby *et al.* 2004). A 130 Ma reconstruction (Fig. 7) is the starting point for subsequent Caribbean evolution involving subduction polarity reversal and closure of the Andean Back-arc Basin followed by eastward advance of the Caribbean Plate relative to the Americas.

Summary of critical contrasts between the Northern Andes and the Central Andes

In the northern Andean autochthonous continental basement, Jurassic magmatism can be plausibly related to eastward subduction of the Farallon Plate or its precursors beneath South America. However, Cretaceous magmatism is absent and a magmatic arc only became well established during Eocene time, after a *c.* 100 Ma hiatus. The only exceptions are the *c.* 95–85 Ma intrusions along the boundary between Western Cordillera and Central Cordillera rocks. All these rocks lie west of major dextral strike–slip faults and ductile shear zones and were not in their current positions when intruded. All are associated with remnants of the Caribbean Arc or the Andean Back-Arc Basin and are more plausibly related to the Aptian and younger Caribbean subduction zone which consumed the Proto-Caribbean Seaway or to the slightly younger onset of low-angle subduction of Caribbean crust beneath the Colombian margin (see maps and discussion below). There are no arc-associated intrusive or proximal extrusive rocks on the continental margin inboard of the Huancabamba–Palestina Fault Zone.

Only a single dismembered 'Greater Panama' arc terrane, built on Caribbean-like crust, is present outside the Western Cordilleras. Eastward subduction associated with Late Cretaceous to Palaeogene accretion of this terrane from the Caribbean crust did not form an associated volcanic arc on the continental crust to the east; the subduction was probably of low-angle geometry as it remains today in the north (Van der Hilst & Mann 1994; Pindell & Kennan 2009). The terrane has been dismembered by block faulting and rotation but there is no evidence of sutures, which would indicate that it comprises separate subterranes formed at multiple plate boundaries.

In contrast to the Northern Andes, in Peru (south of 7°S), Cretaceous intrusive and extrusive magmatism appears to be more or less spatially continuous and persisted into and through the Cenozoic (e.g. Pitcher & Cobbing 1985; Mukasa 1986; Jaillard & Soler 1996). Lava flows and tuffs are found to the east of the arc in the Cretaceous section, in contrast to the Northern Andes. Granite emplacement and andesite extrusion during the Cretaceous gave way to intrusion of smaller stocks and widespread ignimbrite volcanism during the Early Cenozoic. There are no known accreted Caribbean-type plateau or volcanic arc terranes in the forearc, which comprises mostly Palaeozoic rocks, accreted trench fill and cover sediments (e.g. von Huene *et al.* 1988; von Huene & Lallemand 1990; Wine *et al.* 2001). South of the boundary between the Talara and Salaverry forearc basins there are no major strike–slip fault zones within the forearc or arc.

Regional structural styles are very different. The Northern Andes shows both subduction complex formation and dextral strike–slip deformation of Late Cretaceous to Palaeogene age, evolving from ductile to brittle, and there is evidence of significant north–south lengthening of accreted terranes and northward terrane migration. In contrast, in the Central Andes a back-arc basin started to close at about 100–120 Ma (e.g. Cobbing *et al.* 1977) and east-directed thrusting progressing from west to east from Late Cretaceous to the present-day, which no indication of regional-scale strike–slip deformation and terrane migration south of Trujillo.

The boundary between autochthonous South America and allochthonous terranes

Central Cordillera–Cordillera Real Cretaceous dextral shear zone

The Central Cordillera of Colombia and the Cordillera Real of Ecuador are interpreted here to form a continuous major ductile shear zone, about 50–100 km wide, which separates mixed mafic, volcanic and high-grade metamorphic terranes with Caribbean affinity from the former South American passive margin. Protolith ages range from neo-Proterozoic through Palaeozoic to Jurassic, but these rocks were all strongly deformed (or redeformed) and exhumed starting at about 120–100 Ma. Structure, regardless of protolith, is dominated by steep to vertical cleavages and more or less horizontal stretching lineations (e.g. Litherland *et al.* 1994), oriented parallel to the more or less north–south trend of the shear zone. There is also evidence of significant cross-strike shortening in some areas (e.g. Pratt *et al.* 2005). The offset on the shear zone during Middle to Late Cretaceous is not directly known but can be inferred from internally consistent sequential palaeotectonic maps (see below). The Antioquia Batholith lies on the west side of this shear zone and has a similar age and host rock to a batholith trend on autochthonous South American crust which terminates against the east side of this ductile shear zone close to the Peru–Ecuador border, suggesting a total offset of several hundred kilometres is possible.

The age of onset of shearing is difficult to constrain precisely, in part because of the dating methods used, and because it is not clear whether the ages record partial resetting due to Ar loss, synmetamorphic (peak or retrogressive) mineral growth, or cooling of older mineral grains below closure temperature. In Colombia, K–Ar cooling ages of schists with a variety of protolith and types range from *c.* 120 to 90 Ma (e.g. McCourt *et al.* 1984; Maya-Sánchez 2001; Maya-Sánchez & Vásquez-Arroyave 2001). HP/LT rocks at Jambaló and Pijao have yielded ages of 125–104 Ma. ^{40}Ar/^{39}Ar ages from Central Cordillera schists are generally no older than *c.* 85 Ma (e.g. Maya-Sánchez 2001; Maya-Sánchez & Vásquez-Arroyave 2001), but some samples give plateau ages as old as 106 Ma (e.g. Sabanalarga Batholith, Vinasco-Vallejo *et al.* 2003). Together with the Aptian–Albian ages, the youngest faunal ages from associated sediments, the ages suggest that metamorphism and shearing starting at about 120–100 Ma. In Ecuador, most protolith ages are recorded by Rb–Sr (Litherland *et al.* 1994) and appear to be no younger than *c.* 145–150 Ma. K–Ar ages show a spread from *c.* 145 Ma into the Cenozoic but are strikingly bimodal, with most older than 145 Ma or falling between 100 and 60 Ma. There are no reported ages from 100 to 115 Ma, and only a single hornblende age in the interval 115–120 Ma. A few hornblende and biotite ages fall in the range 125–145 Ma and have been used to define an earliest Cretaceous 'Peltetec Event' for which there is little other evidence. These ages may be partially reset. To date there are few ^{40}Ar/^{39}Ar plateau ages in Ecuador older than

c. 80 Ma (Richard Spikings, pers. comm. 2008). We know of no U–Pb ages which could constrain the age of growth of higher-grade metamorphic minerals, and which might more precisely date the Aptian–Albian event.

We suggest that unroofing during dextral shearing resulted in the *c.* 120–60 Ma spread of ages in Colombia, and 100–60 Ma ages in Ecuador. The paucity of ages between cooling ages of 100–120 Ma and protolith ages of >145 Ma suggests that the Aptian–Abian ages are not simply partial resets (the interpretation of Vallejo *et al.* 2006), particularly since both K–Ar and ^{40}Ar/^{39}Ar ages from the accreted arc (Quebradagrande) and blueschist (Jambaló) fragments to the west also show evidence for an exhumation event starting at *c.* 120 Ma, as do terranes throughout the circum-Caribbean region (e.g. Pindell *et al.* 2005). Recently published zircon and apatite fission track dates (Spikings *et al.* 2001, 2005) show that unroofing became more clearly established in Ecuador in the interval 85–60 Ma, but regional palaeotectonic modelling suggests that it would be difficult to justify delaying the onset of cooling or shearing in Ecuador until 20 Ma after it apparently started in Colombia. The dominance of younger age in Ecuador may reflect the interaction between regional scale plate kinematics and subtle differences in palaeogeographic configuration between Colombia and Ecuador prior to 85 Ma (see below).

To our knowledge, this dextral shear zone has not previously been explicitly traced into Peru, where many of the rocks adjacent to those mapped as Mesozoic metamorphic rocks in Ecuador are commonly mapped as Precambrian. South of the border, rocks of the Ecuadorian Cordillera Real can be followed to about 6°S near Olmos, where they terminate abruptly against the north–south-trending brittle faults of the Huancabamba–Palestina Fault Zone and are juxtaposed with Cretaceous carbonate and clastic back-arc basin strata in Coastal Batholith intrusions. Cordillera Real rocks appear to have near-identical origin and geological history, including dextral ductile shearing and juxtaposition with HP/LT rocks, to those in the east–west-trending El Oro Metamorphic Belt of Ecuador. The El Oro Belt must have formed at a significant plate boundary but today comprises a plate boundary fragment only *c.* 300 km long (Fig. 2), truncated against the Huancabamba–Palestina Fault Zone at 3.5°S.

Brittle faulting on the Huancabamba–Palestina Fault Zone

During the latest Cretaceous and Early Cenozoic, shearing along the Central Cordillera–Cordillera

Real Cretaceous dextral shear zone became brittle, as recorded in the youngest K–Ar ages and the oldest zircon and apatite fission track ages. The brittle Huancabamba–Palestina Fault Zone (Fig. 5) follows the older ductile shear zone along most of its length in Colombia and Ecuador, except in the far north, where brittle faulting stepped about 25 km to the east (from the Otú–Pericos Fault onto the northern Palestina Fault). In Peru, brittle faulting cuts south and east of the El Oro Metamorphic Belt, into the formerly more or less *in situ* Coastal Batholith rocks of the northernmost Central Andes and adjacent forearc and back-arc. Rocks now on the west side of the brittle fault zone include the Amotape basement of the Talara Basin, the Celica Cretaceous volcanic arc, and the basement of the Sechura forearc basins. We propose that this block was translated north and rotated clockwise to bring the El Oro Metamorphic Belt into its present position relative to the Cordillera Real. If the El Oro Belt is restored to the southern end of the Cordillera Real shear zone, all the major ductile shear zones plausibly link to a single major plate boundary.

Paleocene–Eocene dextral brittle faulting on the Huancabamba–Palestina Fault Zone resulted in the opening of the Talara pull-apart basin in northernmost Peru, creating accommodation space for several kilometres of delta and slope sediments (e.g. Jaillard & Soler 1996) which unconformably overlie Albian to Maastrichtian forearc carbonates and turbidites (Jaillard *et al.* 1999) with mixed volcaniclastic (from uplifted Celica Arc) and continental (from erosion of Cordillera Real ductile shear zone) provenance. Paleocene and Early Eocene volcanic rocks (Llamas and Sacapalca Formations) show offsets smaller than those of Cretaceous rocks, and the faults are covered by widespread late Middle Eocene ignimbrites which show essentially no fault offset (Porculla and Saraguro Formations). The point where this dextral brittle shear has linked to the Peru–Ecuador trench has migrated north and west of the Talara area since the Eocene, through the Tumbes Basin in the Oligocene, into the Gulf of Guayaquil and Buenaventura Bay areas at present, leaving formerly migrating terranes south of Guayaquil docked to western South America. Since the Eocene, most of the dextral shear in Colombia seems to have stepped east of the Palestina Fault, into the Upper Magdalena Valley and across the Eastern Cordillera and Llanos Foothills. Neotectonic fault offsets are relatively small on both Cauca–Almaguer and Palestina Faults but significantly larger within the Eastern Cordillera (Paris *et al.* 2000; Trenkamp *et al.* 2002).

Summary of offset estimates and rates of Eocene and older dextral shear

There are several direct independent estimates for the amount of offset during latest Cretaceous to Eocene phase of brittle faulting on the Huancabamba–Palestina Fault Zone:

- Restoring the El Oro Metamorphic Belt to the southern end of the Cordillera Real, to define a single continuous ductile dextral shear zone which links to the Peru Trench, indicates an offset of 250–300 km.
- The Celica volcanic arc, straddling the Peru–Ecuador border and lying immediately south of the El Oro Metamorphic Belt, is similarly offset by approximately 300 km from the similar Casma arc of north-central Peru.
- Cretaceous carbonates and clastic rocks derived from a volcanic arc founded on continental crust cover both of these arc fragments, with a 300 km gap between them where Palaeozoic strata are found on either side of the brittle fault zone.
- An elongate sliver of Palaeozoic strata south of Talara (Paita area) may match similar Palaeozoic basement in the Salaverry forearc basin. These are separated by approximately 250–300 km across a prominent oblique tear bounding the southern edge of the Talara and Sechura pull-apart basins.

These estimates collectively suggest a dextral shear rate of *c.* 7–10 km/Ma between forearc terranes and the interior of South American during the earlier latest Cretaceous to Eocene brittle phase. Fault patterns in Colombia suggest that much of this brittle shear passed east of the Antioquia Terrane. The Cauca–Almaguer Fault between Cali and Medellín is dominated by a subduction accretion structural style reflecting east-directed underthrusting of Western Cordillera rocks beneath the Central Cordillera. In contrast, anastomosing patterns of brittle, high-angle faults characterize the Silvia–Pijao and San Jeromino Fault Zones in southern Colombia. North of Armenia (*c.* 4.5°N), this brittle faulting and associated pull-apart basins (the largest is near Manizales) and restraining bend pop-ups swing to the NE and follow the Palestina and Otú–Pericos faults and other north–south-trending fault strands between Antioquia and the Serranía San Lucas.

Eocene to present Nazca–South America relative plate motion history is well known (e.g. Pardo-Casas & Molnar 1987). The strike–slip component of this plate motion matches the rates of post-glacial (*c.* 10 000 years) moraine offsets, seismicity and present-day GPS fault slip measurements (Trenkamp *et al.* 2002), indicating about

250–350 km dextral offset at a rate of c. 5–7 km/Ma since 50 Ma. In northern Peru, this displacement probably occurred entirely seaward of Talara Basin, and is consistent with the observation of fragments of Piñon Block ultramafic fragments in conglomerates at Paita (Pecora et al. 1999), about 300 km south of the Piñon Block today. In Ecuador, this offset occurred on either of both of the Mulaute–Toachi–Chimbo Fault Zone, between the Macuchi Terrane and Pallatanga Terrane, or within the Central Andean Depression. Maastrichtian and younger total brittle offset of the Piñon forearc of Ecuador relative to South America is thus estimated to be up to 650 km and Antioquia must have moved north relative to the Llanos Basin by a similar amount. The slightly higher estimated offset rate from Maastrichtian to Eocene time is consistent with faster and more oblique subduction of the Farallon Plate beneath South America prior to Early Eocene time (Pardo-Casas & Molnar 1987; our own calculations based on the results of Doubrovine & Tarduno 2008).

Total offset must be significantly greater than the post-Maastrichtian brittle offset, but cannot be easily constrained. Relative plate motion data suggests that the dextral component of Farallon–South America motion was even higher during Aptian–Campanian time than since the Maastrichtian (Engebretson et al. 1985), and thus, the rate of ductile shearing could have been higher than 7–10 km/Ma. This is qualitatively consistent with the intense internal dextral ductile shear fabric. Most of the dextral offset since c. 120 Ma has occurred east of the Antioquia Terrane, or within it prior to the intrusion of the Antioquia Batholith.

Palinspastic palaeotectonic reconstructions of northwestern South America

Eocene palinspastic map

Reconstruction of the relationships between the Caribbean Plate and northwestern South America, and the testing the interpretations made above, requires the construction of palinspastic palaeotectonic maps which are based as far as possible on reliable estimates of strike–slip fault offset and shortening or extension measurements derived from the analysis of geological maps, balanced or semi-balanced cross-sections and other data sources which are independent of the Pacific-origin Caribbean model. We have constructed palinspastic maps of the Colombian region, progressively restoring the effects of Neogene and later Palaeogene deformation to produce a palinspastic map of Colombia during the Middle Eocene (Fig. 8,

modified from Pindell et al. 1998, 2000). The map accounts for post-Eocene strike–slip fault offsets, shortening in the Eastern Cordillera, Perijá and Mérida Andes, and northward expulsion of the Maracaibo Block towards the Caribbean. Terranes that were accreted after the Eocene are also removed. The map-view process is analogous to balancing a structural cross-section. No input to this map was derived from larger-scale plate models. The same basic process (and same offset estimates) can be applied semi-schematically to the dissected geological map (Fig. 4). The resulting 42 Ma map (Fig. 9) restores post-Eocene dextral motion of the Ecuador and Colombia forearc terranes and the Central Cordillera. This shear linked the trench north of the Talara Basin, was distributed from the Gulf of Guayaquil through the Cauca–Almaguer and San Jeronimo Fault Zones and stepped across the Eastern Cordillera towards the Llanos Basin. The map shows the Macuchi Arc (Eocene) and the Piñon Block (Cretaceous) adjacent to the Western Cordillera in Ecuador, south of their present positions. The Panama Arc has not yet accreted to the Colombian Margin, but trends more or less east–west and lies west of the Chusma thrust belt of the Upper Magdalena Valley. The Urumita Suture has not yet formed and there is a v-shaped tract of Caribbean Plate still visible between the Panama Arc and the Sinú accretionary prism in NW Colombia where Caribbean crust is subducting beneath South American continental crust and already-accreted terranes (San Jacinto Belt and northernmost Western Cordillera). Subsidence is coming to an end in the Talara Basin (e.g. Jaillard & Soler 1996) and brittle dextral faulting is stepping out to the west and initiating opening of the Tumbes and Progreso forearc basins.

Late Cretaceous palinspastic map

This in turn provides the basis for a semi-schematic restoration of the region to its 71 Ma configuration (Fig. 10). The proposed 250–300 km Late Cretaceous to Eocene dextral strike–slip offset on the Huancabamba–Palestina Fault Zone has been restored, restoring a continuous Aptian Celica–Casma Arc. We also restore a continuous 'Greater Panama' arc at the rear of the Caribbean Plate, which meets the Peru Trench near Paita in the Sechura Basin. The Tahuín–El Oro Metamorphic Belt is shown at the southern end of the Cordillera Real ductile shear zone, which until this time defined the eastern edge of the Caribbean Plate. The map also restores some of the clockwise rotations and northward terrane migration indicated by palaeomagnetic data (e.g. Roperch et al. 1987).

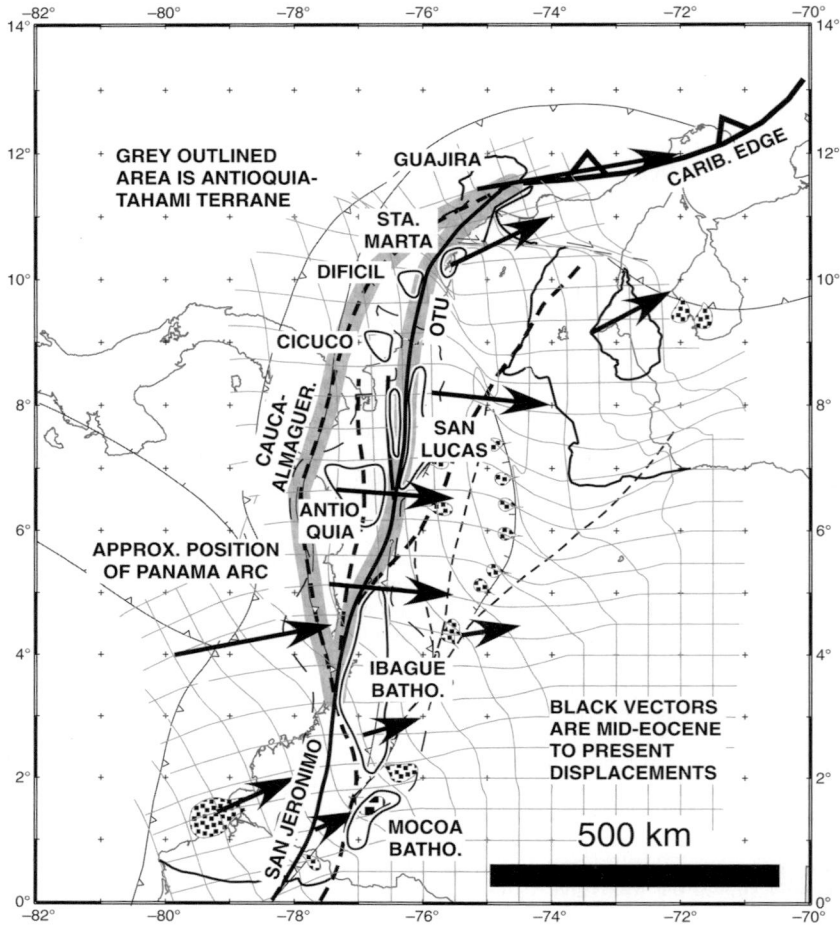

Fig. 8. Early Middle Eocene (c. 46 Ma) reconstruction of the Northern Andes, modified from Pindell *et al.* 1998), showing a palinspastic latitude-longitude grid which undoes the effect of subsequent deformation. Also shown are major blocks noted in the text. Black arrows summarize total 46–0 Ma displacement. The reconstruction differs from that of Montes *et al.* (2005) in applying lower magnitude post-46 Ma dextral shear, and interpreting palaeomagnetic rotations of blocks in northern Colombia and Venezuela as local, fault-related, rather than regional effects.

We infer the existence of a broader tract of Caribbean crust west of southern Colombia and Ecuador, between the Piñon–Panama Arc and the Cordillera Real shear zone. We do not draw the 'Greater Panama' arc and associated back-arc basin parallel to the present-day Peru–Colombia trench. At this time, the leading edge of the Caribbean (the Lesser Antilles Arc) lay to the north, and slices cut from the eastern end this arc had already been incorporated into the Cordillera Real–Central Cordillera shear zone (T2 'slivers' and T3 'Caribbean Arc' on Fig. 10). The only truly far-travelled terrane in this view is the Gorgona Island Terrane, which may originate far to the south of northern Peru (MacDonald *et al.*

1997; Kerr & Tarney 2005). This is perhaps the earliest reconstruction that we can confidently generate from local geological data, without constraint coming in part from the Pacific-origin Caribbean model presented below (which links together data drawn from far beyond the Northern Andes). This reconstruction is broadly similar to that of Montes *et al.* (2005); although we accounted for some block rotations in different ways, they also place the Antioquia Terrane opposite the Ecuador–Colombia border west of the present-day coastline. However, our maps do differ (e.g. contrast our Fig. 14 with Montes *et al.* 2005, fig. 16a) in the Late Cretaceous placement of the northern Maracaibo and Santa Marta areas.

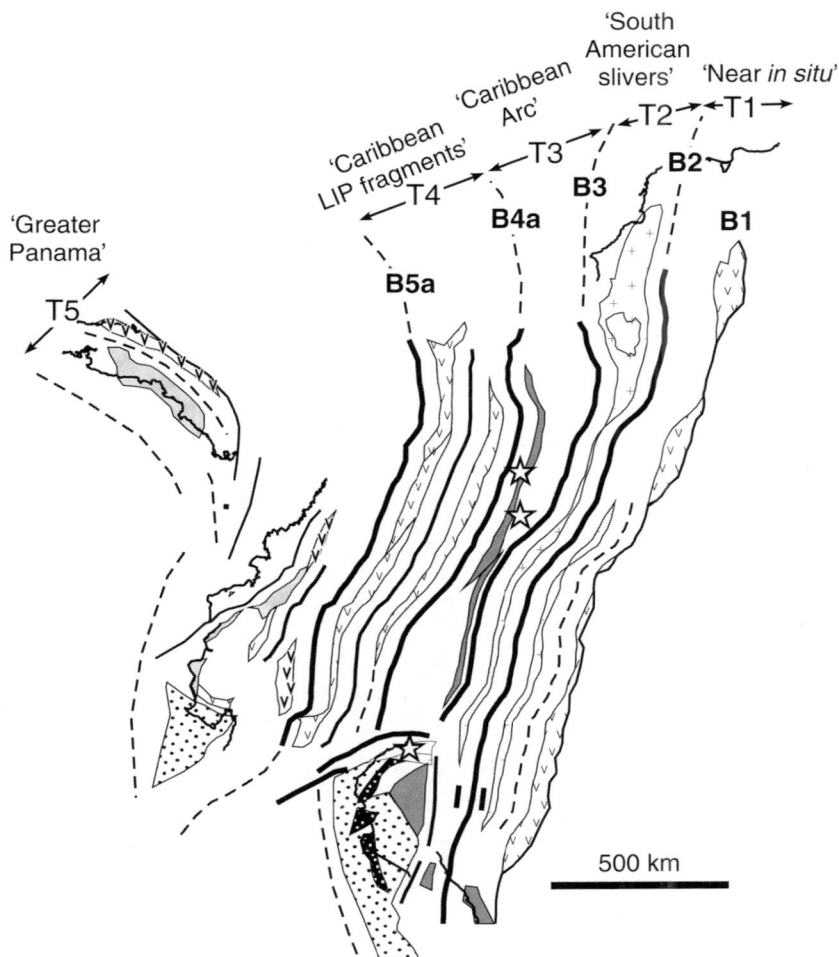

Fig. 9. Dissected geological map at 42 Ma based on detailed palinspastic maps of Colombia (Pindell *et al.* 1998). The Panama arc has been pulled back to the SW and lies opposite the Upper Magdalena Valley, and dextral pull-aparts in the Tumbes and Progreso Basins have been closed.

Reconstructions of the relationship between the Caribbean Plate and northwestern South America

Basic constraints on large-scale plate reconstructions

In building regional palaeotectonic maps, parts of which are presented below, we honour circum-Caribbean geological control such as ages, stratigraphy, geochemistry, incorporating data from Peru to Trinidad, Mexico to the Virgin Islands, the Lesser Antilles, Costa Rica to Panama and ODP/DSDP boreholes. We account for geophysical data such

as tomography, palaeomagnetic rotations, palaeo-magnetic latitudinal drift and seismicity patterns. This breadth of data helps us to avoid local *ad hoc* or non-unique explanations for geological features when data of adjacent geographic areas may lead us to larger-scale, geometrically simpler, explanations of the observations which have greater predictive value and testability.

We assume little or no Late Cretaceous shape change within the interior of the Caribbean oceanic lithosphere. Turonian–Coniacian oceanic plateau basalts were erupted at a sites of half graben formation in the interior of the plate, at its margins, onto possibly Early Aptian oceanic

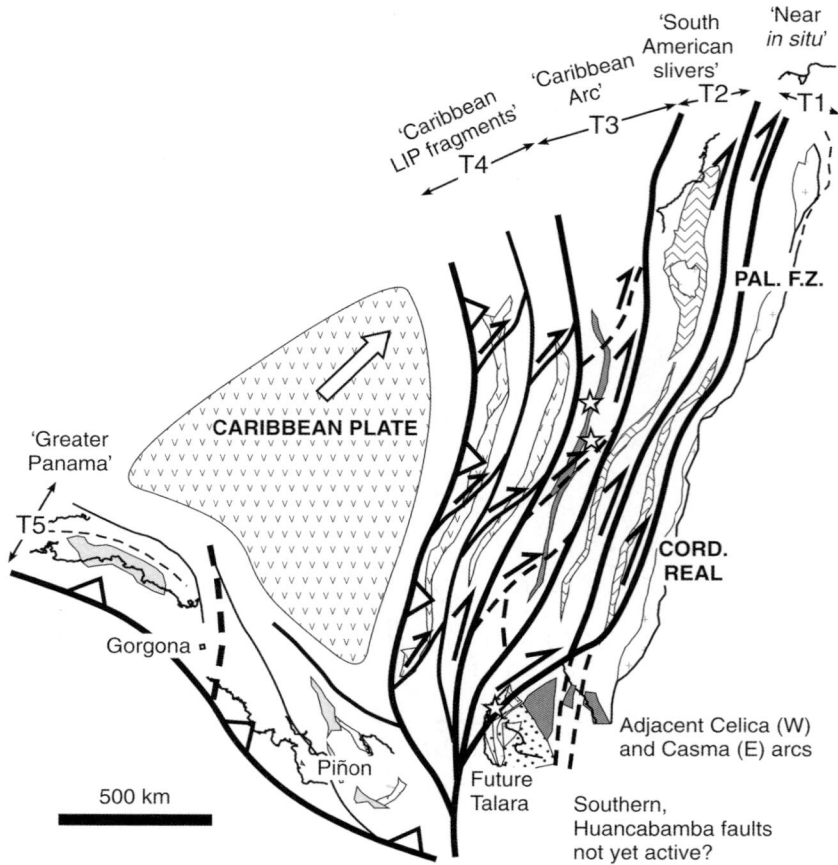

Fig. 10. Dissected geological map at 71 Ma. The Panama arc has been pulled back to opposite western Ecuador, and offsets between distinctive lithologic units on the Huancabamba–Palestina Fault Zone to the east of the Talara Basin have been restored. The Talara Basin has not yet started to open (it will form as the Panama triple junction passes by on its northward migration). Note that the Amotape–Tahuín metamorphic belt is restored south by the same distance as the Celica Arc and is interpreted as the southern end of a formerly continuous ductile shear zone between Caribbean and South American crust.

basement (Driscoll & Diebold 1999; Kerr *et al.* 2003). Since 90 Ma at least the assumption of no significant shape change is justified and prior to then it is as yet unquantifiable. The models presented also account for the entire area of the Caribbean Plate (Fig. 11), including areas subducted beneath Colombia and, possibly, the Nicaragua Rise, which are only visible with seismic tomography (van der Hilst 1990; van der Hilst & Mann 1994). This is a particularly important constraint on the origin of the Caribbean Plate because it shows that the plate is simply too large to have formed in the Early Cretaceous within the narrow gap between North America and South America that existed at that time (Pindell 1993).

Caribbean Plate motion in the hot spot reference frame

Pindell & Kennan (2009) analyse and update the Aptian–Eocene Caribbean evolutionary model of Pindell (1993) in the Müller *et al.* (1993) Indo-Atlantic hot spot reference frame, incorporating more recently published plate rotation parameters. Since 84 Ma, which is the period when the Caribbean lithosphere has had eastern and western inward-dipping subduction boundaries, the Caribbean oceanic lithosphere has moved less than 200 km east or west in the hot spot reference frame. Pindell & Kennan (2009) conclude that this is because of the anchoring effect of the Caribbean's

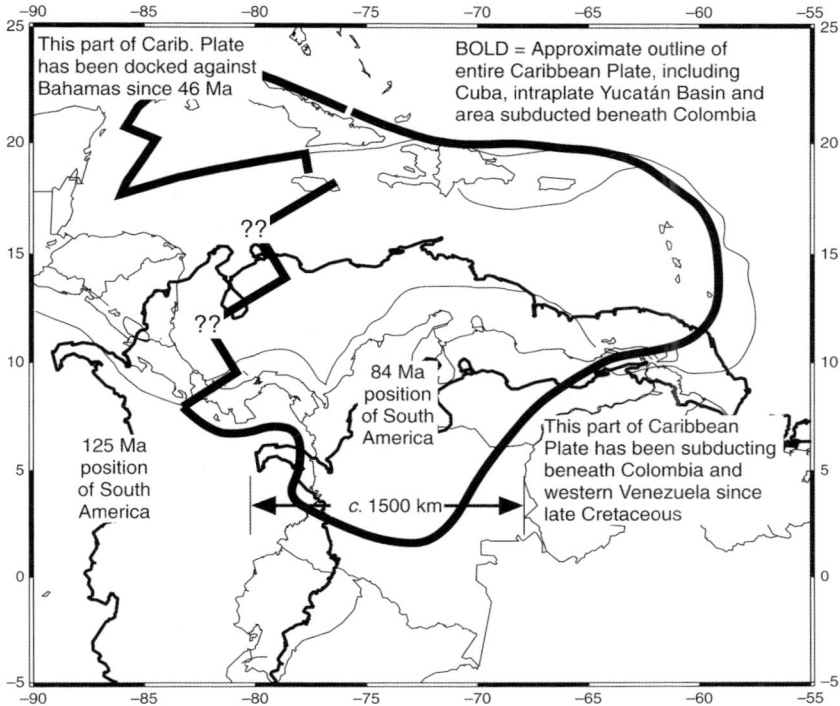

Fig. 11. The entire area of the Caribbean Plate (heavy line), including portions subducted beneath the Nicaragua Rise and Colombia, is much larger than the area at the surface today. A minimum 1500 km dextral offset between the southern Caribbean and northern South America is needed to restore the Caribbean crust which has been subducted beneath Colombia to the surface. Note that at c. 125 Ma, before formation of the basement to the Caribbean Plateau basalts, the inter-American gap was a fraction of the total width of known Caribbean crust (South America position calculated with respect to a fixed North America). In the Late Cretaceous, South America was at its farthest from North America; even at that time there was insufficient space to accommodate known Caribbean crust within the inter-American gap.

opposing Lesser Antilles/Aves Ridge and Middle American subduction zones in the mantle, which has been especially effective since the Eocene. However, according to the reconstructions, the Caribbean lithosphere moved northwards some 1100 km in the hot spot reference frame between 84 and 46 Ma. This motion is in good agreement with palaeomagnetic data and projection of the maps in other hot spot reference frames (e.g. Torsvik *et al.* 2008). Prior to about 90 Ma, there are significant differences between published reference frames, and the northward motion from 120 to 84 Ma shown in our maps is much reduced in palaeomagnetic-controlled reference frames. A key assumption in our new maps is that the total number of major plates and plate boundaries is similar to today, that major plate boundaries are 'relatively long' (i.e. 1000 km or more) and that subduction zones and substantial oceanic island arcs do not appear and disappear abruptly in space

and time; the scale of this region is somewhat smaller than the western Pacific.

Excerpts of the new maps, covering only the Northern Andes area, are presented below (Figs 12–14, 16–18). We present the maps working backwards in time from the relatively well-constrained Eocene reconstruction through increasingly uncertain, but internally consistent, Cretaceous reconstructions.

A 46 Ma Northern Andes reconstruction

This map (Fig. 12) shows the Macuchi arc (Eocene) and Piñon Block (plateau basalt and overlying Late Cretaceous oceanic island arc) c. 300 km south of their present positions and about to be covered by Eocene continent-derived, volcaniclastic terrane-linking sequences (e.g. Jaillard *et al.* 2004). By this time, the southern part of the Panama Arc had been accreted to the Ecuadorian margin as a result of a

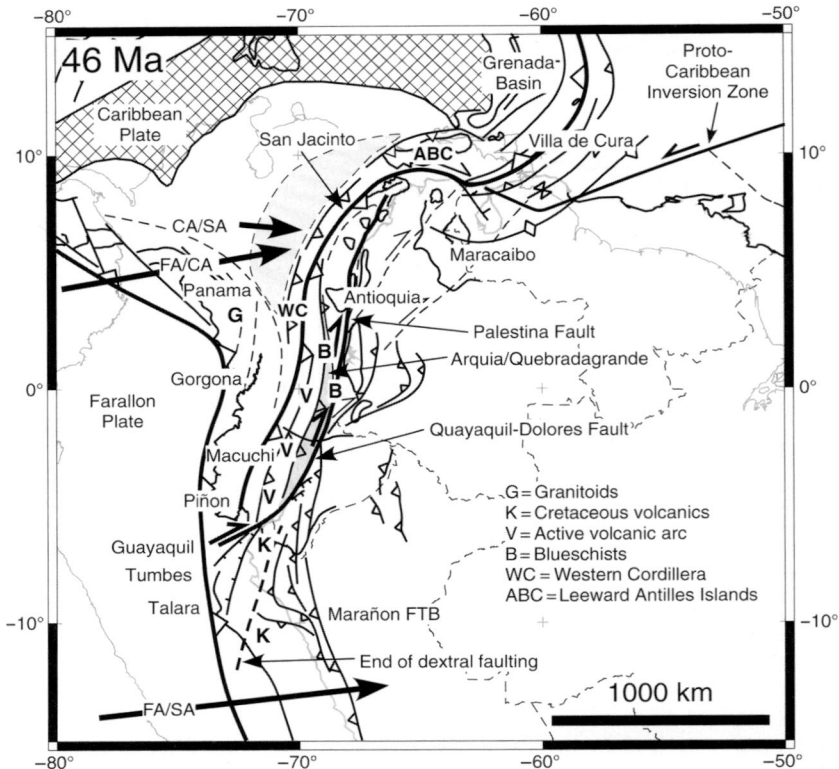

Fig. 12. A 46 Ma Caribbean reconstruction in the Müller *et al.* (1993) hotspot reference frame (as are all subsequent reconstructions). Present day coastlines are shown in light grey in the background. The hatched area is present day surface Caribbean Plate in its position at the time of this map. The grey area NW of Colombia is Caribbean Plate subducted between 46 and 33 Ma. Black arrows show relative plate motions for this interval (Abbreviations: NA, North America; SA, South America; CA, Caribbean; FA, Farallon; HS, Indo-Atlantic Hotspots). Note that Caribbean–South America motion is *c*. 23 km/Ma, more or less orthogonal to the Northern Andes, resulting in little dextral shear on the Huancabamba–Palestina Fault Zone. Oligocene and Early Miocene relative motions become more dextral oblique, resulting in renewed dextral shearing at that time. South of the Panama triple junction, the Farallon Plate was subducting at a high angle to the South American margin at about 110 km/Ma.

250% higher rate of subduction of the Caribbean beneath Ecuador and then Colombia compared to the Late Cretaceous (up from 11 to 23–30 km/ Ma). The Piñon Block and overlying Late Cretaceous volcanic arc was clipped off from its underlying Caribbean lithosphere, which continued to move towards the north, and became part of the hanging wall of the Peru trench. Following the accretion of the Piñon Block, the Ecuadorian margin was exposed to rapid, *c*. 100 km/Ma, east-directed subduction of the Farallon Plate (Pardo-Casas & Molnar 1987; our own calculations based on the results of Doubrovine & Tarduno 2008), resulting in stronger inter-plate coupling at the subduction interface. This resulted in a widespread *c*. 40 Ma cooling and exhumation event in Ecuador revealed by fission track studies

(Spikings *et al.* 2005). This event has previously been interpreted as the accretion of an intra-oceanic Macuchi arc (Hughes & Pilatasig 2002). However, here we show the Macuchi arc as built on the edge of South America, on a basement of Caribbean Plateau basalt slices (consistent with field observations, Richard Spikings, pers. comm. 2008; following the interpretation of Jaillard *et al.* 2008) that had already been accreted by Eocene time.

Rapid Palaeogene subduction of the Farallon Plate also resulted in the development of a bimodal volcanic arc on the continental margin in the latest Eocene, but only to the south of the triple junction. North of the triple junction, the Dabeiba calc-alkaline oceanic island arc of Panama had not yet accreted to the Colombian margin. No Farallon-related arc had yet developed on the Colombian

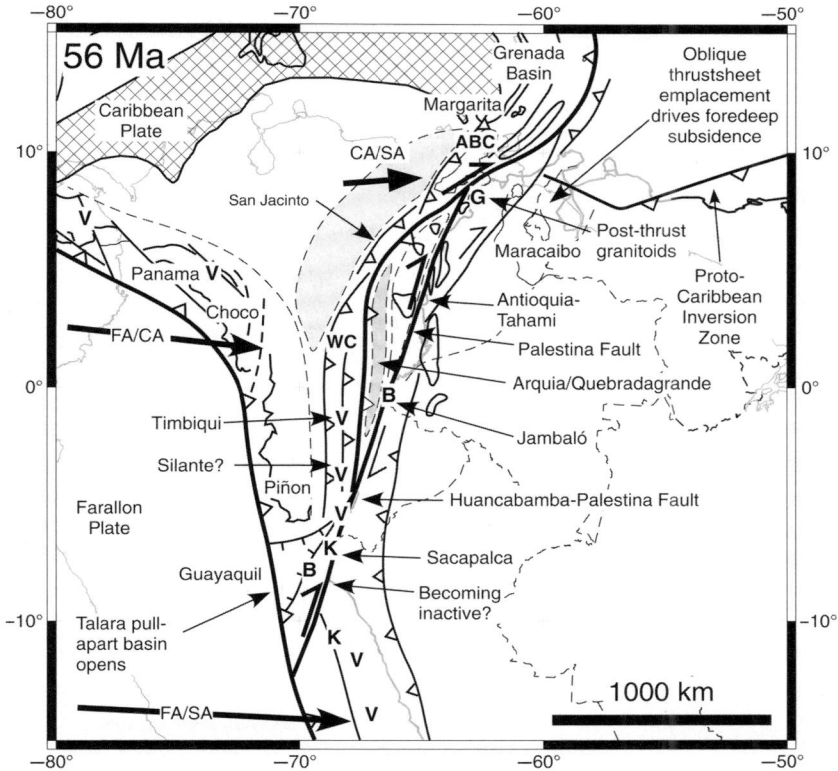

Fig. 13. A 56 Ma Caribbean reconstruction. The grey area NW of Colombia is Caribbean Plate subducted between 56 and 46 Ma. Note that post-56 Ma Farallon motion relative to the Americas is becoming oriented more east–west, and we expect a significant drop in the rate of dextral faulting on the Huancabamba–Palestina during the Eocene, compared with the Late Cretaceous, and an increase in the rate of subduction beneath Colombia. Caribbean–South America motion in the Northern Andes is partitioned into c. 30 km/Ma subduction and up to c. 15 km/Ma dextral shear. South of the Panama triple junction, the Farallon Plate was subducting at a high angle to the South American margin at about 130 km/Ma.

margin north of the Panama triple junction; subduction of the Caribbean Plate since the Maastrichtian was too slow and too low-angle to drive a volcanic arc but was sufficient to build the San Jacinto accretionary prism (Pindell & Tabbutt 1995). This reconstruction shows up to 1500 km of Caribbean crust yet to subducted beneath the South American Margin (see also Fig. 11); south of the Caribbean– South America–Farallon triple junction, subduction of Farallon lithosphere beneath South America was far faster, with correspondingly high volumes of arc volcanism (Pindell & Tabbutt 1995).

The direction of subduction beneath Ecuador and Colombia, first of the Caribbean Plate and then of the Farallon Plate, was almost orthogonal to the margin; with little dextral component to subduction, dextral strike–slip on the Huancabamba–Palestina Fault appears to have slowed and

stopped at about this time. East-directed underthrusting of Colombia at this time culminated in the regional development of a Late Eocene unconformity, the most spectacular manifestation of which is the consistent eastward tilting of the Central Cordillera and basement to the Middle Magdalena Valley east of the Palestina Fault. The same event seals the Eocene wrench structures of the Cimitarra–Opón area of the Middle Magdalena Valley. This tilting event may have led to the erosion of much of any earlier Paleocene–Eocene proximal foredeep section adjacent to the Palestina Fault. However, overall there is little evidence of significant east-directed shortening and accommodation space formation associated with Palaeogene underthrusting of the Caribbean Plate, and Central Cordillera detritus may have been transported farther to the east (for instance, as low-gradient Pepino Formation conglomerate

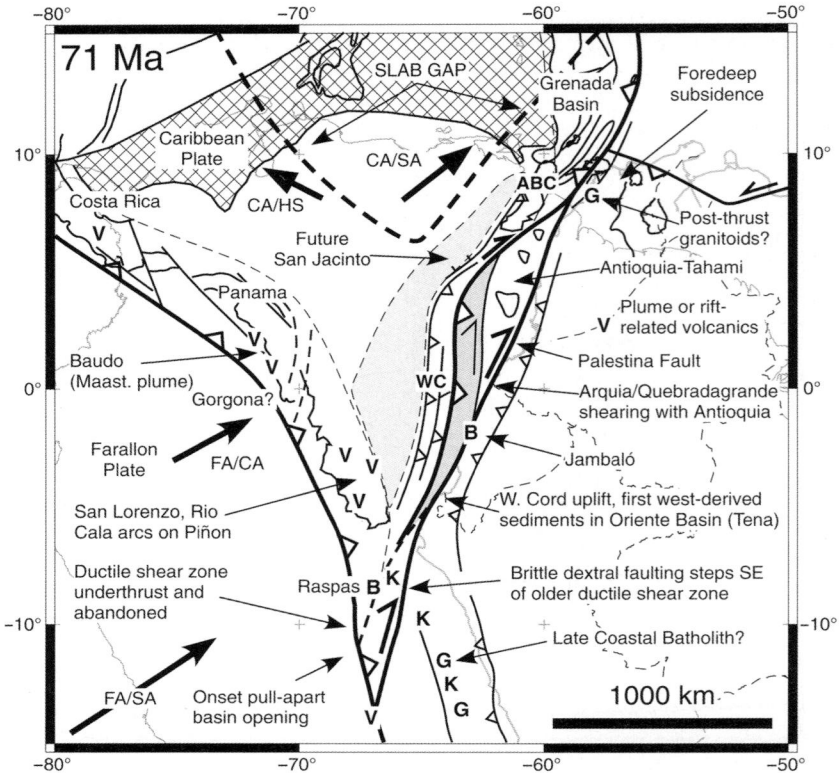

Fig. 14. A 71 Ma Caribbean reconstruction. The light grey area NW of Colombia is Caribbean Plate subducted between 71 and 56 Ma and the dark grey area is the accreted Arquia-Quebradagrande Complex former Caribbean Arc. Caribbean–South America motion in the Northern Andes is partitioned into *c.* 11 km/Ma subduction and up to *c.* 22 km/Ma dextral shear. South of the Panama triple junction, the Farallon Plate was subducting at very oblique angle to the South American margin. The trench-normal component of this motion was about 40 km/Ma and the dextral component was about 38 km/Ma.

fans in the Putumayo Basin) and ultimately fed the sandy Misoa shelf to the north (Pindell *et al.* 1998).

At the same time, ongoing opening of the Grenada intra-arc basin was allowing the Villa de Cura portion of the Caribbean forearc to impinge towards the SE onto the central Venezuelan margin, driving foredeep subsidence first in the Maracaibo area and then in the Guárico Basin.

A 56 Ma Northern Andes reconstruction

An earliest Eocene reconstruction (Fig. 13) restores about 40% of the total dextral slip on the brittle Huancabamba–Palestina Fault, still active and accommodating at least 50% of the dextral trench-parallel component of Caribbean oblique sub-duction beneath Colombia. North of Ibagué we also show an eastern splay from the Palestina Fault along the Cimitarra–Opón–Perijá trend linking to the leading edge of the Caribbean Plate

north of Maracaibo. The southern end of the dextral strike–slip fault system is defined by the active pull-aparts of the Talara–Tumbes area on the Peru–Ecuador border. The dextral component of Farallon–South America subduction in this area was lower than that of Caribbean–South America subduction farther north (about 5–10 km/Ma compared with about 15 km/Ma) and this resulted in north–south stretching of the Ecuador–Colombia forearc, and creation of accommodation space in the Manabi, Manglares and other forearc basins.

Following emplacement of Caribbean thrust sheets in the Guajira area, post-thrust granitoids were intruded which probably represent the south-ernmost magmatism associated with the Caribbean (Lesser Antilles) subduction zone. The map also shows the Piñon Block at the former trailing edge of the Caribbean essentially accreted to the Ecuadorian margin. Forearc basins on the Piñon Block were, for the first time, adjacent to areas of eroding quartz-rich rocks of the Cordillera Real on the South American margin and became the site of continental clastic deposition (such as the Angamarca Group and Azúcar Formation, Jaillard *et al.* 1995, 2004). Inboard of the forearc basins, the Silante, Timbiqui (and later Macuchi) volcanic arcs were established, driven by subduction of the Farallon Plate. The apparently intra-oceanic island arc geochemical character of these rocks reflects, in our opinion, the nature of the underlying basement (accreted Caribbean oceanic plateau basalt) through which subduction zone melts rose and the contribution of melted subducted oceanic sediment. The arc was not allochthonous to South America (in contrast to the model of Hughes & Pilatasig 2002). On adjacent South American continental crust to the east and south of the Huancabamba–Palestina Fault Zone volcanic rocks of this age are typically laterally extensive ignimbrite shields (Sacapalca, Saraguro and Llamas Formations). The Western Cordillera of Colombia was essentially accreted by this time and is overlain by Eocene clastic sediments (Chimborazo Formation, Nivia 2001) and intruded by Eocene and younger Farallon-related stocks.

A 71 Ma Northern Andes reconstruction

A Late Campanian reconstruction (Fig. 14) shows the southern part of the 'Greater Panama' Terrane, including the Rio Cala Arc and its Caribbean oceanic plateau basement, yet to be accreted to the Western Cordillera of Ecuador. Volcaniclastic flysch was being deposited in the narrow trough between the arc and the trench at the foot of the western Cordillera. To the east, in the Western Cordillera, the earliest accreted Caribbean Plateau basalt terranes were overlapped by quartz-bearing continent-derived Yunguilla Formation flysch at about this time (Jaillard *et al.* 2008). Jaillard *et al.* (1999) distinguished this basin as the 'Paita–Yunguilla' basin because its fill unconformably overlies the fill of the earlier Lancones Basin (see below), although its trend is broadly similar. Field relationships in the western Cordillera suggest that there may have been several discrete events during the Campanian and Maastrichtian in which separate sheets of Caribbean Plateau basalt underthrust and uplifted the South American margin. Fission track data suggest that terrane accretion and exhumation was diachronous, and younged from *c.* 75 Ma in the south to *c.* 65 Ma in the north of Ecuador (Spikings *et al.* 2005; Richard Spikings, pers. comm. 2008). However, in distinction, we interpret this diachroneity as being a result of northward-younging oblique accretion of the trailing edge of the Caribbean to South America, not its leading edge.

The San Lorenzo volcanic arc of westernmost Ecuador is interpreted here as a minor fault-bounded remnant of the Greater Panama Arc. Dextral slip on the Canande Fault Zone during the Palaeogene stripped the forearc from the San Lorenzo area and carried it to the NE, bringing the present-day trench much closer to the San Lorenzo outcrop, and causing the Palaeogene Timbiqui arc to form inboard of the older San Lorenzo and Naranjal Arc fragments and causing the Macuchi Arc to form just west of the accreted Rio Cala Arc. As the Piñon Block migrated north during the latest Cretaceous and Palaeogene, the Peru forearc was exposed to the Peru Trench above the subducting Farallon Plate, leading to transtensional collapse of the margin and to initial subsidence of the Talara pull-apart basin and subsequent development of the Tumbes and Progreso Basins to the north. The dextral component of both Farallon–South America and Caribbean–South America relative motion was significantly higher during Late Campanian–Palaeocene time than during the Eocene.

We have suggested that the Cordillera Real was continuous with the Tahuín–El Oro belt prior to the onset of brittle faulting on the Huancabamba–Palestina Fault Zone. At this scale, its position at the eastern edge of the Caribbean Plate is clear, as is its role in linking the leading edge of the Caribbean, which was overthrusting the Santa Marta Massif in Northern Colombia, to its trailing edge in the Talara area. We show the Antioquia Terrane about 300 km south of its present position, an absolute minimum for the brittle dextral shear that passes east of it. The role of the ductile Cordillera Real Shear Zone and younger brittle Huancabamba–Palestina Fault Zone in partitioning oblique Caribbean Plate subduction is shown schematically in Figure 15.

The overall position and orientation of the Caribbean Plate at this time is constrained by the need to initiate overthrusting of ophiolites onto the Yucatán Block in Guatemala (Rosenfeld 1993), to drive

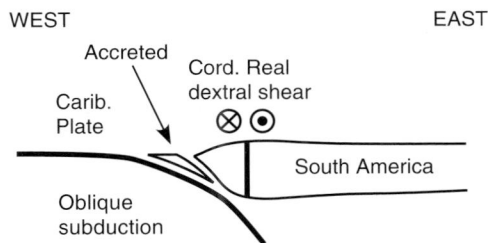

Fig. 15. Schematic cross section through Ecuador, highlighting the partitioning of oblique subduction of the Caribbean Plate into more or less orthogonal subduction and terrane accretion in the Western Cordillera, and inboard dextral shear. The latter was becoming brittle by the end of the Cretaceous and becoming more focused towards the western edge of the Cordillera Real.

accretion of Caribbean arc and forearc rocks and uplift of metamorphic rocks in the northwestern Sierra de Santa Marta and Guajira areas of northernmost Colombia (Case *et al.* 1984; Stéphan 1985), to initiate Caribbean-driven subsidence in northernmost Colombia and in the Maracaibo Basin of westernmost Venezuela (Pindell & Kennan 2009), and intrude Palaeogene post-collisional granitoids (e.g. Cardona *et al.* 2008). The Cimarrona Formation clastic sediments in the Middle Magdalena Basin of Colombia (Villamil 1999) were derived from the Central Cordillera to the west, which was unroofing rapidly by 80 Ma (Villagómez *et al.* 2008). Thus, the accreting plateau basalt terranes west of Cauca–Almaguer Fault must have been derived from the interior of the Caribbean Plate and not from the floor of the Proto-Caribbean Seaway, and the accreted volcanic arc terranes outboard of the plateau basalts in Ecuador must have derived from the trailing edge of the Caribbean Plate. This stands in contrast to the model of Vallejo *et al.* (2006, fig. 4) which shows the leading edge of the Caribbean Plate interacting with Ecuador, and too far to the SW relative to the Americas to interact with Yucatán or Guajira.

Regional plate motions suggest why peak exhumation may be some 30–40 Ma later than our preferred onset of shearing between the Caribbean Plate and South America and why accretion of Caribbean oceanic plateau basalt terranes to the Ecuadorian margin occurred during the Maastrichtian. At *c.* 84 Ma, spreading between the Americas slowed dramatically and at *c.* 71 Ma it ceased. Until this time, South America had been moving southeast relative to North America, opening the Proto-Caribbean Seaway (e.g. Pindell *et al.* 2006). While Caribbean terranes were migrating east relative to North America, inter-American spreading more or less matched the eastward component of Caribbean–South America motion and Caribbean–South America motion was extremely oblique or essentially dextral strike–slip (Figs 14, 16–18). However, once spreading stopped, there must have been a marked increase in the coupling of the Caribbean and South American Plates, resulting in dextral oblique subduction with a very significant component of inter-plate compression. The area of Caribbean crust subducted from 71–46 Ma is significantly larger than that shown on earlier reconstructions (Figs 16, 17 & 18).

A 84 Ma Northern Andes reconstruction

Recent work (Bandini *et al.* 2008; P. Baumgartner & D. Buchs, pers. comm. 2008) has shown that arc volcanism in Costa Rica and Panama did not start before this time. Thus, this is the oldest map on which we show an active arc at the trailing edge of the Caribbean Plate. New dates on the Orquideas, Rio Cala and Naranjal volcanic rocks in Ecuador (Vallejo 2007) suggest that none are older than about 85 Ma on this intra-oceanic plate boundary.

This map (Fig. 16) fully restores the Celica and Casma arcs as a single continuous volcanic arc which was inactive and degrading at this time and shedding volcaniclastic detritus into the Lancones Basin. This basin is typically described as 'forearc' because of its position between the Celica arc to its east and the Amotape Hills to the west; the latter appears to have been a source for quartz-rich sediment. However, there is no indication of an active volcanic arc to the east of the basin at this time. Rather, it may have occupied a back-arc position relative to the northeast-migrating newly active Greater Panama Arc, which lay outboard of the inactive Celica arc. Volcaniclastic flysch on the Piñon Block is derived directly from the active Naranjal and San Lorenzo volcanic arcs (fragments of the eastern end of the Panama Arc), and there is no indication of continent-derived sediment input. We infer the existence of a complexly deforming back-arc basin between Piñon, autochthonous Ecuador and NW Peru, producing the fragmentary arc remnants we see today, and which was deep enough to isolate the Piñon Block from South American margin sediments until Palaeogene time.

Northwestward migration of the Caribbean Plate is partitioned into minor subduction and more or less vertical dextral ductile shear through the Cordillera Real–Central Cordillera shear zone. $^{40}Ar/^{39}Ar$ ages indicate that unroofing had definitely begun by the time of this reconstruction. The position of the Caribbean Plate at this time is constrained by incipient interaction with Mexico and southern Yucatán. The positioning of the Caribbean Plate shown here (Fig. 16) exposes a part of the former Colombian margin which faced NW toward the Colombian Marginal Seaway. This is also consistent with apparent persistence of more or less passive margin conditions in the Middle Magdalena Basin until this time, with no west-derived sediment.

We restore the Antioquia Terrane to a position close to the Colombia–Ecuador border at the time of intrusion of the Antioquia Batholith (and Altavista Stock, Buga and Pujili Granites, see above), and speculate that it was intruded between 95 and 85 Ma as a result of the onset of subduction of the Caribbean Plate beneath South America (see Figs 16 & 17). The origin of the magma may be melting of the leading edge of the Caribbean slab early during subduction (e.g. Nikolaeva *et al.* 2008). The slow rate and low angle of subduction inhibit subsequent normal hydration melting of a

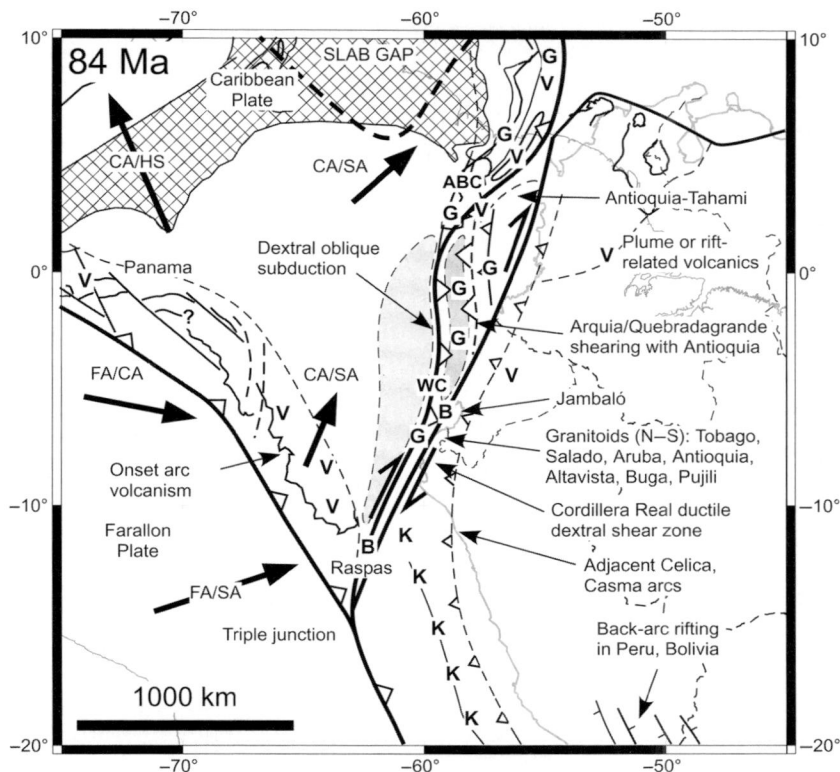

Fig. 16. An 84 Ma Caribbean reconstruction. The light grey area NW of Colombia is Caribbean Plate subducted between 84 and 71 Ma. Caribbean–South America motion in the Northern Andes is partitioned into *c.* 13 km/Ma subduction and up to *c.* 27 km/Ma dextral shear. South of the Panama triple junction, the Farallon Plate was subducting at very oblique angle to the South American margin. The trench-normal component of this motion was about 50 km/Ma and the dextral component was about 25 km/Ma.

mantle wedge beneath Colombia, so there is no significant younger Caribbean-related arc in Colombia. Both deformed continental margin (Antioquia Terrane) and already-accreted Caribbean plateau basalts (Amaime Formation, Buga area) were intruded. The age and geochemistry of the Colombian batholiths and stocks are similar to the Aruba Batholith in the Leeward Antilles Arc (White *et al.* 1999) and the Salado granite of Margarita (Stöckhert *et al.* 1995; Maresch *et al.* 2009), both of which are shown not far west of the inferred palaeoposition of Antioquia. The Aruba Batholith is in a position where it may derive from either the same mechanism as the Antioquia Batholith, or may be the southernmost melt related to subduction of Proto-Caribbean Seaway oceanic crust and sediment beneath the leading edge of the Caribbean. The Salado Granite and the granitoids of Tobago are probably of the latter origin.

Note that this reconstruction puts the Leeward Antilles (or ABC Islands) adjacent to the Lower Magdalena Basin, in the position of the present-day San Jacinto and Sinú accretionary prisms. This portion of the margin was approximately parallel to Caribbean–South America motion at this time and may have been the site of a major lateral ramp that cut across the Western Cordillera, allowing a northern fragment (ABC Islands) to move northeast with the Caribbean, bringing the site of Caribbean subduction much closer to the edge of continental crust than it was farther south. Note that we show the leading edge of the Caribbean as not yet having reached the Guajira Peninsula.

In northwestern Peru, ongoing Caribbean or Farallon Plate motion relative to South America drove deformation of the fill of the Albian–Coniacian Lancones 'forearc basin' prior to the deposition of the Cazaderos Formation, which is broadly correlative to the Yunguilla Formation farther north (Jaillard *et al.* 1999).

The Caribbean Plateau basalts were erupted onto pre-existing Caribbean oceanic crust between

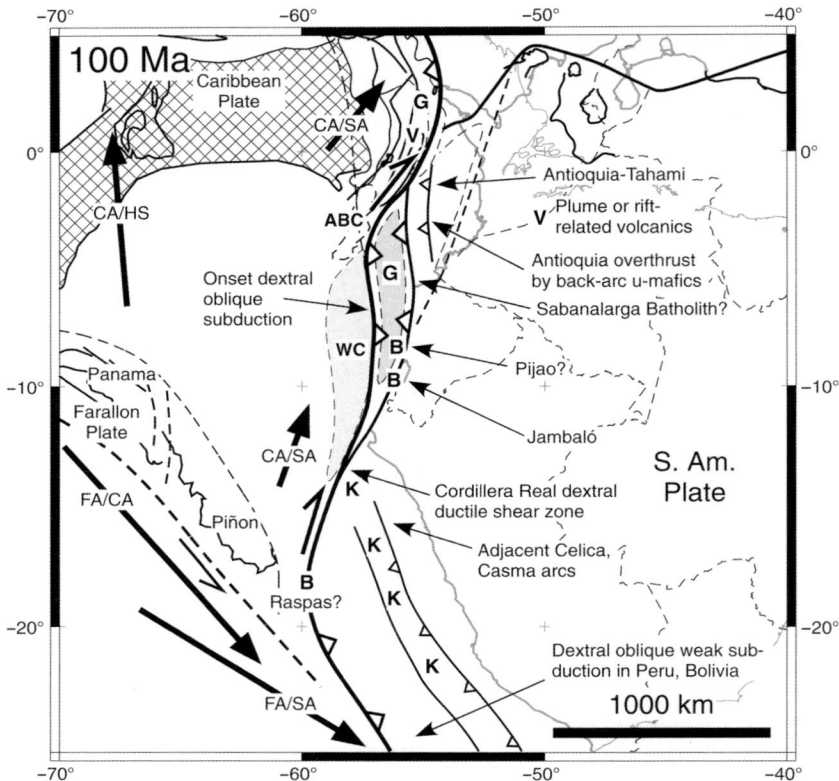

Fig. 17. A 100 Ma Caribbean reconstruction. The light grey area is Caribbean Plate subducted between 100 and 84 Ma. Caribbean–South America motion is *c.* 25 km/Ma pure strike–slip in Ecuador, driving transpressive obduction of back-arc rocks onto Antioquia in Colombia. Motion of the Farallon Plate relative to South America is very poorly constrained, but is estimated to be about 100–120 km/Ma directed to the ESE.

the Caribbean Arc (Antilles) to the east and the Panama Arc to the west about 6 Ma before the time shown on this map. The origin of these basalts is discussed in more detail in a companion paper (Pindell & Kennan 2009). Our calculations of relative motions between Pacific and Indo-Atlantic hot spots place the Galapagos hot spot in the vicinity of the central or western Caribbean at 84 Ma. Thus, if it existed at this time, it may have been the source of the Caribbean Large Igneous Province.

A 100 Ma Northern Andes reconstruction

A 100 Ma reconstruction (Fig. 17) pre-dates the onset of Middle American subduction in Panama and Costa Rica, and restores the Caribbean farther south with respect to northwestern South America. The approximate position of the Caribbean Plate is also controlled by early interaction of the north end of Caribbean Arc in Cuba and Jamaica with

the Chortís Block (Siuna Terrane of Rogers *et al.* 2007), generally believed to have lain adjacent to southern Mexico during the Late Albian. This, and the estimated width of the Inter-American gap (interpolated between 124 and 84 Ma reconstructions assuming a constant rate of spreading in the Proto-Caribbean Seaway) places the eastern end of the Caribbean Arc at the then NE end of the Central Cordillera ductile shear zone close to Pasto in southern Colombia. The lithosphere behind (SW of) the Caribbean Arc was probably that of the Farallon Plate, and the oldest parts of the Caribbean layered igneous province or its under-lying basement (95–100 Ma; Mamberti *et al.* 2004; Villagómez *et al.* 2008) pre-date the formation of the Middle American plate boundary (see also Pindell & Kennan 2009).

Geological maps of the Medellín area suggest that the Palaeozoic core of the Antioquia Terrane may comprise eastern and western parts, separated by one or more fault-bounded slivers of gabbro

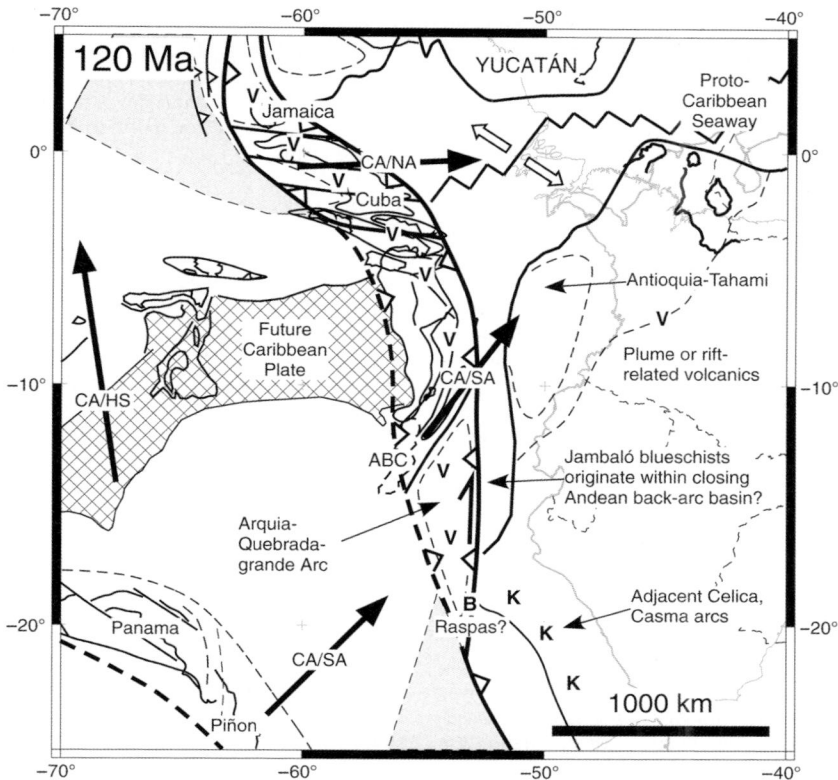

Fig. 18. A *c.* 120 Ma Caribbean reconstruction. The light grey areas SW of Chortís and Jamaica, and west of Peru, is oceanic crust subducted between 120 and 100 Ma. Caribbean–South America motion is estimated at *c.* 40 km/Ma parallel to the northern Andean margin of Colombia.

and ultramafic rock. Thus, we speculate that eastern Antioquia may have lain close to the continent–ocean boundary on the Colombian passive margin, and that western Antioquia (including the Palaeozoic Cajamarca Complex and Sabanalarga Batholith) may have originated farther south in the vicinity of central Ecuador, and that they were juxtaposed by dextral shearing prior to the intrusion of the Antioquia Batholith.

The back-arc basin which once separated the Caribbean Arc and the passive margin in southern Colombia and Ecuador (see Fig. 7 above) is inferred to have been closed by this time, and its site is marked by a newly formed ductile shear zone of gabbroic, ultramafic (including parts of the Peltetec 'ophiolite') and blueschist (e.g. Jambaló) slivers which are all that remains of subducted and sheared ocean crust that formerly floored the back-arc. In the Medellín area, at least, Triassic mafic and ultra-mafic rocks (Aburra Ophiolite) found within the back-arc or arc to the west seem to have been thrust east over older metamorphic

basement, consistent with the subduction polarity shown here (Fig. 18). We suggest that the short duration of subduction beneath the Caribbean Arc of the floor of the Andean Back-Arc Basin (*c.* 20–25 Ma) before its closure may be sufficient to produce the Jambaló and Pijao blueschists. Maresch & Gerya (2005) show that less than 5 Ma and 100 km of subduction can lead to production of both lawsonite and epidote blueschists. Exhumation in Colombia is almost certainly related to the dextral strike–slip superimposed on the closed back-arc.

On the west side of the back-arc suture, the Quebradagrande and Arquia Terranes of Colombia are inferred to be the sheared remains of the eastern end of the Caribbean Arc or the precursor Trans-American Arc, accreted to western Colombia as the Caribbean Arc passed to the NE. The youngest arc remains are slightly older than the age of onset of dextral shearing. This reconstruction also shows that following closure of the back-arc and thrusting of associated rocks onto the Antioquia Terrane, the Caribbean Plate started to subduct beneath

Antioquia, reusing the pre-existing east-dipping sub-duction zone beneath the Quebradagrande and Arquia Terranes, leading to intrusion of the Antioquia Batholith and the onset of exhumation.

In order for large magnitude subduction of Caribbean lithosphere to be possible beneath Colombia, any subducted oceanic crust of the former back-arc basin must have torn away from the continental crust to the east. Such a tear is probably diachronous and northward-younging, driven by northward migration of the leading edge of the Caribbean along the Colombian margin (the STEP model of Govers & Wortel 2005). Such a tear greatly enhances the possibility for migration of melts related to subduction of the Proto-Caribbean Seaway, into the Cordillera Real, Cordillera Central ductile shear zone, in addition to facilitating the onset of subduction of Caribbean lithosphere beneath Colombia.

Outboard of northern Peru and Ecuador, the Costa Rica–Panama trench may have been a transform margin separating the Caribbean and Farallon Plates at this time. Relative plate motions are very uncertain prior to the Campanian. Our own calculations (based on the work of Doubrovine & Tarduno 2008) indicate that the Farallon Plate may have been moving to the SE (at a low angle to the Peru margin) while the Caribbean Plate was moving to the NE, and that there may in consequence have been a subduction zone bounding the northwest Caribbean (see Pindell & Kennan 2009).

By the time of this reconstruction, the contiguous Celica and Casma Arcs appear to have become inactive and starting to erode, resulting in deposition of the Alamor and Copa Sombrero Formation volcaniclastic rocks in the Lancones Basin. We suggest that the Piñon Block was close to the Peru Trench at this time, and that the future triple junction between the trench outboard of the Greater Panama Arc and the Peru Trench lay close to Chimbote. The trench normal component of Farallon Plate sinistral-subduction may have increased between 110 and 100 Ma, driving the Albian metamorphism and deformation in the Peruvian back-arc basin north of Lima (Bussell 1983), and subsequently, the angle of subduction may have been too low to drive active arc volcanism during Cenomanian–Santonian time. The possible low rate of trench-normal subduction (about 50 km/Ma compared with 80–100 km/Ma during the Campanian–Maastrichtian) may also have been a factor.

There is little evidence for strong coupling of the plates at this time from the subsidence history of the Marañon or Oriente Basins of northern Peru and Ecuador. Tectonic subsidence curves from northern Peru (Jaillard 1993) show no enhanced subsidence in the Late Cretaceous, while curves from the northern Oriente in Ecuador (Thomas

et al. 1995) show no enhanced subsidence until about 72 Ma. It is possible that these areas were east of the flexural influence of the Caribbean deformation. The Late Cretaceous stratigraphic record in western areas is too poor to reconstruct subsidence history.

A 120 Ma Northern Andes reconstruction

So far our discussion has addressed the northward migration of Caribbean lithosphere once south-dipping subduction beneath the Great Arc had begun. However, this time is about when the Caribbean Arc and its SW-dipping subduction zone became active (Fig. 18). The evidence for this is four-fold: (1) arc plutonism and volcanism of the 'Antillean Cycle' began in the Albian, and needed some finite period (perhaps 10–15 Ma at our estimated subduction rate) to become well established; (2) nearly all circum-Caribbean HP/LT metamorphic complexes appear to have formed at about 110–125 Ma, suggesting that the associated west-dipping subduction zone itself began then but is not older; (3) there appears to have been an Aptian change in the direction of Farallon Plate motion with respect to North America, from SE to the east and the northeast (Engebretson *et al.* 1985; our own calculations based on the results of Doubrovine & Tarduno 2008), which could have made the early Caribbean boundary more compressional; and (4) the initial opening of the Equatorial Atlantic and westward drift of northern South America occurred during the Early Aptian, as well as a westward acceleration of North America (Pindell & Dewey 1982; Pindell 1993), which also could be expected to trigger plate boundary changes along the Cordilleran arc.

Two popular models for the advent of the Caribbean Arc are: (1) polarity reversal from SW-facing to NE-facing of an Trans-American Arc linking Chortís to Ecuador or northern Peru (Pindell 1993); and (2) onset of SW-dipping subduction at a transform/fracture zone plate boundary in the same position (Pindell *et al.* 2005, their figs 7c, d). In the transform model, slivers of arc, forearc and subduction complex from the west-facing, Late Jurassic–Early Cretaceous Mexico Arc (Baldwin & Harrison 1989, 1992; Sedlock *et al.* 1993; Sedlock 2003) may have been translated along the sinistral transform to the SE such that they were amalgamated into the Caribbean Arc when southwestward subduction began (Pindell & Kennan 2009). It is not the goal of this paper to resolve these ongoing uncertainties, but it may be helpful for the Northern Andes discussion if we can constrain where tectonic elements in these models may once have been located. A third model delays polarity reversal until about 85 Ma, calling on the

Caribbean Plateau to choke the SW-facing subduction zone (Burke 1988; Kerr *et al.* 2003). However, this model fails to explain the Aptian onset of Caribbean HP/LT metamorphism noted above, and we do not consider it further. If we can conceive of plausible models for the advent of the Caribbean Arc, then we can also generate a template for Early Aptian Caribbean palaeogeography.

We consider three ways of estimating where the Trans-American Arc or transform intersected South America. First, we can determine the northward limit of Early Cretaceous Central Andean arc magmatism, and the transition into a more or less passive Colombian margin. The Celica Arc is the northernmost known Early Cretaceous Central Andean magmatism, and hence the Early Cretaceous palinspastic position we estimate for that terrane should be about where the Trans-American Arc or transform intersected South America (e.g. Pindell & Tabbutt 1995). Second, we consider which blueschist complexes might have been formed by eastward dipping subduction on the outside of the Central Andean Early Cretaceous arc. The Raspas complex of southern Ecuador is the only HP/LT terrane that may lie outside the Early Cretaceous arc, and it restores very near to the Celica Arc. If this is the case, however, the quantity of older continental crust and intrusive rock, as opposed to deformed Cretaceous and older sediment, in the associated 'subduction complex' is atypically large. Third, somewhat indirect and model-dependent, is to determine how far down the Andes the southwestern tip of North American continental crust lay across from South America in Triassic reconstructions of western Pangaea. Most reconstructions place Chortís south of Mexico and outside of Colombia and northern Ecuador (e.g. Pindell & Dewey 1982), such that an arc or transform spanning Chortís to South America would remain in that position during and after rifting and seafloor spreading in the Proto-Caribbean. Everywhere north of this initial position on both Proto-Caribbean margins would remain facing the Proto-Caribbean Seaway until the time of Caribbean interaction, after initiation of the Caribbean Arc. The likely initial position for the southern tip of Chortís is, again, very near to the estimated positions of the Celica Arc. We conclude that the Early Cretaceous plate boundary spanning the North and South American cordillera intersected South America in NW Peru as shown here (Fig. 18).

If the Early Cretaceous plate boundary intersected South America as shown, then the Caribbean oceanic lithosphere must have originated from west of the Trans-American Arc or transform. In the Colombian and Venezuelan basins and the Beata Ridge, true oceanic basement (Diebold 2009)

beneath the areas of Turonian–Coniacian basaltic extrusions has not been dated. Thus it is possible that Aptian–Albian constructional plate boundaries between unnamed plates/platelets were situated in the area to the SW of the Trans-American Arc, but it is just as likely that this region comprised a single swath of Farallon crust.

Prior to the onset of southwest-dipping subduction (Fig. 7) and after (Fig. 18), we show the inter-American plate boundary linking Chortís and NW Peru. Prior to about 120 Ma, this boundary could have been dominated by sinistral transcurrent motion, or by oblique subduction. This trend incorporates various continental protoliths in its roots: at the northwestern portion of the arc (Cuba) Grenvillian age rocks (Renne *et al.* 1989) presumably derive and were carried south from Grenvillian terranes of Mexico/Chortís. At the southeastern end of the arc, we show the Celica–Casma Arc close to the Peru–Ecuador border, at the SE end of a thumb-like promontory of continental rock and associated arc volcanic rocks rifted away from the Ecuador–Colombia margin by probable Neocomian back arc spreading. This trend, possessing the western (continental) flank of the intra-arc rift, may include the precursors of the Arquia and Quebradagrande Complexes in Colombia, the Juan Griego Group of Margarita, and possibly the Tinaco Complex of Central Venezuela (Bellizzia 1985). In Ecuador, it is possible that the opening of a narrow oceanic back-arc was younger and post-dated Middle to Late Jurassic deformation of Cordillera Real rocks and the formation of angular unconformities at the base of Hauterivian and younger foreland strata ('Jurua', 'Nevadan' and possibly 'Peltetec' events, all poorly dated). We speculate that the high-pressure metavolcanic Rinconada Terrane of Margarita originated within the southern part of the Andean Back-Arc Basin and was then overthrust by the Caribbean Arc shortly after the subduction polarity reversal.

We are not aware of evidence in the Northern Andes (or elsewhere) which supports the collision of separate intra-oceanic island arc at the leading edge of the a very far-travelled Caribbean Plate (Mann 2007, modified from Dickinson & Lawton 2001) with the older Trans-American Arc. We know of no examples of thrust slices of possible Trans-American Arc origin lying to the east of primitive island arc rocks which lie at the eastern edge of the Caribbean Plate.

Figure 17 implies that the Andean shear zone at the east end of the Great Arc should have become active in the Aptian and propagated and lengthened north along the Ecuadorian and Colombian margins through Middle and Late Cretaceous time. If the intersection of the nascent Caribbean Arc with the South American margin is as far south as shown, the rocks now in Colombia which do show evidence for Aptian–Albian events (including the Arquia and Quebradagrande Complexes and the Sabanalarga Batholith) probably originated close to the Peru–Ecuador border. Rocks in a more inboard position or more northerly starting position within the Cordillera Real–Cordillera Central Shear Zone would only start to cool as they were entrained in the lengthening shear zone. Thus, the area from which we might expect to find cooling ages as old as 120 Ma will be much more limited than that for younger ages because the shear zone was relatively short at that time. Fragments of the 120 Ma shear zone may, however, now be preserved as areally limited tectonic lenses along the entire length of the Northern Andes from Amotape to Medellín and beyond.

There is independent evidence that a Proto-Cordillera Real was growing during the Early Cretaceous. First, palaeontological data indicates that the Napo Formation was isolated from the Pacific at this time (Vallejo et al. 2002) and, second, petrographic and detrital zircon U–Pb and fission track age data from the Hollín and Napo Formations support a partial western sediment source at that time (Ruiz et al. 2004; Winkler et al. 2008; Martin-Gombojav & Winkler 2008).

Discussion

The model presented here provides a comprehensive, regional and testable framework for analysing the collage of arc remnants and basins in the Northern Andes and explains many of the geological features of the Northern Andes. This paper places much of the progress that has been made in the understanding of the geology of the region into the general Pacific-origin Caribbean model of Pindell (1993), Pindell & Tabbutt (1995) and Pindell & Kennan (2009), improving in particular on the palinspastic restoration and subsequent motions, prior to Caribbean tectonism, of various blocks and terranes.

How many arcs are needed to explain the observations in northwestern South America?

Various regional syntheses have depicted the origin and evolution of Northern Andes arc fragments in two-dimensional cross-sections, rather than in map view, without addressing the issues of lateral continuity (e.g. Kerr et al. 2002a, b; Jaillard et al. 2005), subduction geometry (e.g. Hughes & Pilatasig 2002), or the effect of large strike–slip offsets on the cross-sections. In some models, every arc fragment is associated with a distinct subduction zone (e.g. Cediel et al. 2003), yet the map scale of the

fragments is such that they could simply represent fault-bounded blocks of slightly different ages and geochemistry within a single larger arc. Furthermore, the cross-section depictions result in maps in which arcs and back-arc basins are shown parallel to their present orientations, which is not consistent with palaeomagnetic data or with present-day analogues. Thus, models often miss the possibility of arc migration from out of the plane of cross-section. Arc migration along a margin can easily result in, for example, contrasts in subduction polarity between discrete peri-continental arcs and dramatic temporal changes in subduction geometry in any given cross-section.

The western Pacific-like multiple arc scenario (Fig. 19) is implausible; the scale of individual subducting slabs is too small and short-lived to produce subduction zone magmatism and the westward drive of the Americas during the Late Cretaceous would tend to hinder the trench retreat

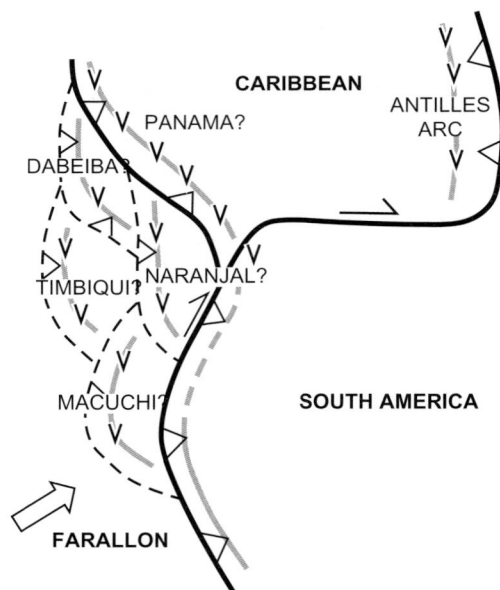

Fig. 19. Cartoon shown what northern South America might have looked like if there were numerous intra-oceanic plate boundaries, such as in the present-day western Pacific. Each arc fragment would be relatively short (<1000 km) and have its own associated subducting slab and independent magmatic and metamorphic history. In our view, this is unlikely, not only because the Americas were driving west relative to Pacific mantle, but because of the similar timing of events in all arc remnants in western Colombia and Ecuador.

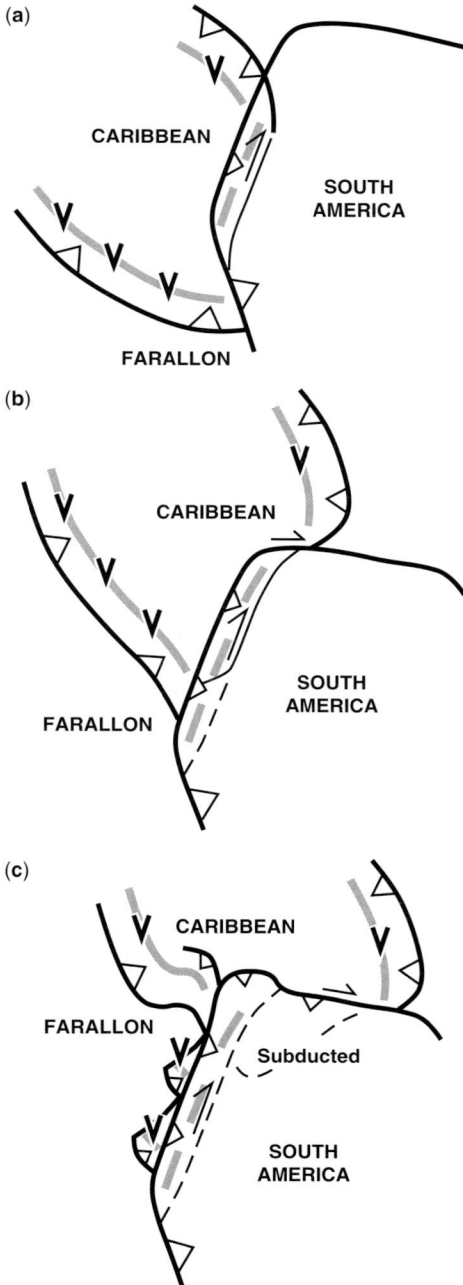

necessary to form discrete back-arc basins and subduction zones. Is there a simpler model that can explain all the relevant observations? In contrast (Fig. 20), we suggest there may have only been three significant plates near the Northern Andes (South America, Caribbean and Farallon) and only two arcs (at the leading and trailing edges of the Caribbean Plate) during the Late Cretaceous to Middle Cenozoic. This simple model requires no more major plates or plate boundaries than exist today. The eastern side of the Caribbean was defined by a zone of slow subduction and accretion (Western Cordillera during and after Late Cretaceous time) without an associated volcanic arc. The 'forearc-like' subduction complex sliver was separated from South America by broad dextral shear zone (Cordillera Real and Central Cordillera). All the fragments of volcanic arc (Quebradagrande Complex) and continental basement (Arquia Complex) with sedimentary cover along the west side of the dextral shear zone could have been derived from the east end of the Caribbean Arc, at the leading edge of Caribbean. Blueschists and mafic rocks between the Caribbean Arc remnants and the shear zone could have derived from the back-arc basin (Colombian Marginal Seaway) that was being overridden by the Caribbean Arc as it migrated north.

Caribbean Plateau basalts in the Western Cordillera of Ecuador and Colombia were accreted during slow subduction of the Caribbean Plate beneath the Cordillera Real shear zone once the leading edge of the Caribbean had passed to the north. The rate of trench-orthogonal subduction became much higher during the Maastrichtian as the Americas stopped separating, leading to accretion of Caribbean terranes in both Ecuador and Colombia, and to enhanced exhumation in the Cordillera Real dextral shear zone. The subduction accretionary prism structural style is most clearly seen in the San Jacinto and Sinú belts of Colombia because these areas have not yet been overprinted by a further accretion of Caribbean trailing edge terranes. Cretaceous arc slivers in the westernmost part of the Western Cordilleras of Ecuador and Colombia were derived from a single 'Greater Panama Arc' as it migrated with the Caribbean

Fig. 20. Major plate boundaries for Cretaceous, Early Cenozoic as proposed in this paper. We suggest that the number of significant plate boundaries has remained the same as present, but that they have migrated with respect to northern South America, stranding remnant arc fragments as they passed. (**a**) Late Cretaceous, showing a north-lengthening dextral shear zone at the east edge of the Caribbean Plate; (**b**) Early Cenozoic, showing northward migration of the trailing edge of the Caribbean past northern Peru, northward lengthening of the dextral shear zone, incorporating former Colombian passive margin rocks. (**c**) Middle Cenozoic showing that strike–slip at the east side of the Caribbean Plate is sufficient to dissect a once-continuous trench and trailing edge arc into a series of short arc fragments and bounding ophiolite or accretionary prism zone remnants.

Plate towards the north. Accretion started earlier in southern Ecuador (*c.* 75 Ma) than in northern Ecuador (*c.* 65 Ma) and Colombia (Palaeogene in the south, younging to Early Miocene in the north). Accretion of 'Greater Panama' was associated with fragmentation, strike–slip faulting and block rotation and resulted in the fragmented arc remains seen today beneath an extensive forearc cover. Accretion of the 'Greater Panama' terrane was also accompanied by northward younging of a volcanic arc on South America driven by subduction of the Farallon Plate (Pindell & Tabbutt 1995; Pindell *et al.* 1998).

Fundamental problems with inter-American models for the origin of the Caribbean Plate

'*In situ*' or 'Inter-American' models for the origin of the Caribbean Plate cannot account for:

- The *c.* 70 Ma hiatus in continental arc magmatism in Colombia and Ecuador, because in these models this margin lies south of the Trans-American Arc and is thus always exposed to subduction of 'Pacific' Plates from the west; unlike the thick, buoyant Caribbean Plate, the Farallon and Nazca Plates comprise normal oceanic crust and flat slab subduction zones and associated volcanic hiatuses are transient (<20 Ma) and on a shorter length scale than the Northern Andes.
- Terrane stacking order, with accreted Western Cordillera terranes always lying inboard of Panama terranes which are known to have formed at the trailing edge of the Caribbean Plate. A Farallon Plate origin for the accreted terranes would be expected were the inter-American Caribbean model correct.
- The accretion of plateau basalt fragments identical to those in the Caribbean. These fragments would have to originate west of the Costa Rica–Panama plate boundary, and therefore would not be of Caribbean origin.
- The accretion of multiple arc fragments to western Ecuador and Colombia. To do so requires the invention of multiple and arbitrary new 'Pacific' volcanic arcs and plate boundaries for which there is no independent evidence.
- The timing and magnitude of ductile and brittle shearing and associated blueschist metamorphism and accretion in Colombia and Ecuador, when nothing similar is seen in Peru and farther south.
- Palaeomagnetic data which places Caribbean-derived terranes near or south of the magnetic equator, when the gap between the Americas

lay at least 10° north of the palaeo-equator at the time of Caribbean Plateau basalt eruption.

Only a model in which the Caribbean Plate originated in the easternmost Pacific, immediately west of the Trans-American volcanic arc, satisfactorily explains all these observations.

Uncertainties in this Pacific-origin Caribbean model and possible future research directions

There remain many questions and uncertainties within this iteration of the Pacific-origin Caribbean model. Detailed integration of metamorphic petrology and geochronology from the wider circum-Caribbean region supports the *c.* 120 Ma differentiation of the Caribbean Plate from the Farallon Plate and generation of associated HP/LT rocks from protoliths derived from the Proto-Caribbean Seaway rather than from Pacific oceanic crust subducting towards the east. However:

- The *c.* 120 Ma age for the onset of shearing at the eastern edge of the Caribbean in the northern Andes is dependent mostly on relatively old and poorly documented K–Ar ages, with few modern ^{40}Ar/^{39}Ar (some are as old as Middle Albian) or fission track ages in critical areas (especially Colombia). The almost total lack of K-ages in the range 150–120 Ma suggests that the widespread post-120 Ma K–Ar ages may not simply be partial resets in Jurassic schists and plutons. Also suggestive is the lack of ages from the Quebradagrande sediments and volcanic rocks younger than about 120 Ma.
- It is not clear whether many of the ages in the literature reflect cooling of older minerals or growth of new metamorphic minerals. Better integration of updated geochronology with metamorphic petrography and microstructure is clearly required.
- Our model implies diachronous, northward-younging subduction of the oceanic crust of the Colombian Marginal Seaway beneath the leading edge arc. If this crust was coupled to the continental crust of the passive margin, there may be a northward migrating inflection in tectonic subsidence curves starting in northern Peru at *c.* 120 Ma (although areas with well-known stratigraphy may be beyond the influence of Caribbean-driven flexural subsidence). Robust flexural backstripping may allow better determination of the timing and rate of tectonic subsidence in adjacent foreland areas and may reveal something of the age and nature of Late Cretaceous tectonic events to the west.

Testable predictions of the Pacific origin Caribbean model

Our plate reconstructions make the following predictions, supported by preliminary observations of the geology of the northern Andes:

- We expect to find no pre-Caribbean subduction-related magmatic rocks within or east of the Cordillera Real–Cordillera Central shear zone that post-date westward rifting of the south end of the Trans-American Arc from interior South America during the Jurassic.

- We predict that the Peru Bank (offshore Tumbes Basin) lies at the tail of the 'Greater Panama Terrane' and comprises Caribbean plateau basalt and overlying arc volcanic rocks rather than continental crust similar to the onshore Amotape Block (e.g. Martinez *et al.* 2005).

- We predict that all volcanic arc remnants in Ecuador west of the Pallatanga Fault formed at the trailing edge of the Caribbean Plate as part of 'Greater Panama', and that none predate *c.* 85 Ma or formed at the leading edge of the Caribbean Plate as part of the older Caribbean Arc. We expect a belt of accreted interior Caribbean oceanic plateau basalts, reflecting subduction of the Caribbean Plate under northern South America, will always separate leading edge 'Caribbean Arc' (such as the Quebrada-grande Complex) and trailing edge 'Greater Panama' fragments, and that evidence for pre-existing Early Cretaceous Proto-Caribbean oceanic crust (in the form of ultramafic or basaltic rocks) will be found between Caribbean Arc fragments and the Cordillera Real–Cordillera Central Shear Zone.

- Panama triple junction migration north along the Andes allowed the Farallon Plate to be subducted directly beneath parts of the Northern Andes that were previously shielded by the northward migrating Caribbean Plate. We expect all ages on associated volcanic rocks to show the same pattern of slow northward migration of the onset of volcanism. *Ad hoc* multi-platelet models (Fig. 19) would not produce smooth, predictable variations in features such as the age of terrane accretion or the onset of volcanism in the post-accretion arc.

Conclusions

We subtly redefine the 'Northern Andes' to include much of northwestern Peru at least as far south as Trujillo, bounded by the southern continuation of the Huancabamba–Palestina Fault Zone and of the ductile Cretaceous shear zone (Cordillera Central/Real–Tahuín–El Oro belt). Pre-Eocene shear zones and orogenic events north of Trujillo are related to the interaction of the Caribbean Plate and South America, and were not driven by interaction of the Farallon and South American Plates and the Cretaceous and Early Cenozoic history of the Marañon, Oriente and Putumayo Basins is probably a Caribbean-driven history. Only Pacific-origin models for the Caribbean can adequately explain the geology of northwestern South America for this interval. Far from being more complex than *in situ* models, Pacific-origin models are geometrically simpler and therefore provide simpler, more testable predictions. Pacific-origin models introduce no arbitrary tectonic elements or mechanisms to explain tectonic, metamorphic and magmatic history, but rely on plates and plate boundaries still visible today and on well understood geological processes.

We dedicate this paper to our academic supervisor, Professor J. F. Dewey FRS, whose love for geology and its relationship to plate kinematics inspired and defined a standard for the kind of work presented in this paper. We are grateful for exposure to data while collaborating with Ecopetrol and PDVSA on research programs that are far more detailed than the regional story told here. We are also grateful to: E. Jaillard, IRD and University of Grenoble, who introduced L. Kennan to problems in northern Peru and southern Ecuador while on sabbatical at Oxford University; R. George who helped J. Pindell to define the fundamental significance of the Otú-Pericos and Palestina faults and the allochthonous nature of an Antioquia Terrane while collaborating with Ecopetrol in 1997; P. Baumgartner, D. Buchs and K. Flores at the University of Lausanne for discussions and conducting a workshop on Central American and western Colombian geology with J. Pindell in 2008; K. Stanek at the University of Freiberg for U–Pb dating of zircons from the Antioquia Batholith; R. Spikings of the University of Geneva, who critically read early drafts and pointed out numerous recently published data sources; and K. James and M. A. Lorente, who organized the Sigüenza conference in 2006 where an early version of this paper was presented. J. Rosenfeld, P. Restrepo-Pace and D. Truempy kindly reviewed this paper and their comments have greatly improved its quality.

References

ALFONZO, C. A., SACKS, P. E., SECOR, D. T., RINE, J. & PEREZ, V. 1994. A Tertiary fold and thrust belt in the Valle del Cauca Basin, Colombian Andes. *Journal of South American Earth Sciences*, **7**, 387–402.

ASPDEN, J. A., BONILLA, W. & DUQUE, P. 1995. *The El Oro Metamorphic Complex, Ecuador: Geology and Economic Mineral Deposits*. British Geological Survey, Overseas Geology and Mineral Resources, **67**.

BABY, P., RIVADENEIRA, M. & BARRAGÁN, R. (eds) 2004. *La cuenca Oriente: geología y petróleo*. Travaux de l'Institut Français d'Etudes Andines, **144**.

BALDWIN, S. L. & HARRISON, T. M. 1989. Geochronology of blueschists from west-central Baja California and the timing of uplift in subduction complexes. *Journal of Geology*, **97**, 149–163.

BALDWIN, S. L. & HARRISON, T. M. 1992. The $P-T-t$ history of blocks in serpentinite–matrix mélange, west-central Baja California. *Geological Society of America Bulletin*, **104**, 18–31.

BANDINI, A. N., FLORES, K., BAUMGARTNER, P. O., JACKETT, S.-J. & DENYER, P. 2008. Late Cretaceous and Paleogene Radiolaria from the Nicoya Peninsula, Costa Rica: a tectonostratigraphic application. *Stratigraphy*, **5**, 3–21.

BELLIZZIA, A. 1985. Sistema montanosa del Caribe – una Cordillera aloctona en la parte norte de America del Sur. *Memorias del VI Congreso Geologico Venezolano*. Sociedad Venezolana de Geologia, Caracas, **10**, 6657–6835.

BOLAND, M. P., McCOURT, W. J. & BEATE, B. 1998. *Mapa geologico de la Cordillera Occidental del Ecuador entre 0°–1°N (Scale 1:200 000)*. Corporación de Desarrollo e Investigación Geológico, Minero, Metalúrgica (CODIGEM) – British Geological Survey.

BOSCH, D., GABRIELE, P., LAPIERRE, H., MALFERE, J. L. & JAILLARD, E. 2002. Geodynamic significance of the Raspas metamorphic complex (SW Ecuador): geochemical and isotopic constraints. *Tectonophysics*, **345**, 83–102.

BURKE, K. 1988. Tectonic evolution of the Caribbean. *Annual Review of Earth and Planetary Sciences*, **16**, 210–230.

BUSSELL, M. A. 1983. Timing of tectonic and magmatic events in the Central Andes of Peru. *Journal of the Geological Society, London*, **140**, 279–286.

BUTLER, K. & SCHAMEL, S. 1988. Structure along the eastern margin of the Central Cordillera, Upper Magdalena Valley, Colombia. *Journal of South American Earth Sciences*, **1**, 109–120.

CARDONA, A., CORDANI, U. & MACDONALD, W. 2006. Tectonic correlations of pre-Mesozoic crust from the northern termination of the Colombian Andes, Caribbean region. *Journal of South American Earth Sciences*, **21**, 337–354.

CARDONA, A., DUQUE, J. F. *ET AL.* 2008. Geochronology and tectonic implications of granitoids rocks from the northwestern Sierra Nevada de Santa Marta and surrounding basins, northeastern Colombia: Late Cretaceous to Paleogene convergence, accretion and subduction interactions between the Caribbean and South American plates. *Abstract Volume of the 18th Caribbean Geological Conference*, 24–28 March 2008, Santo Domingo, Dominican Republic. World Wide Web Address: http://www.ugr.es/~agcasco/igcp546/DomRep08/Abstracts_CaribConf_DR_2008.pdf.

CASE, J. E., HOLCOMBE, T. L. & MARTIN, R. G. 1984. Map of geologic provinces in the Caribbean region. *In*: BONINI, W., HARGRAVES, R. & SHAGAM, R. (eds) *The Caribbean – South American Plate Boundary and Regional Tectonics*. Geological Society of America, Boulder, CO, Memoirs, **162**, 1–30.

CEDIEL, F., SHAW, R. P. & CÁCERES, C. 2003. Tectonic assembly of the northern Andean Block. *In*: BARTOLINI, C., BUFFLER, R. T. & BLICKWEDE, J. F. (eds) *The Circum-Gulf of Mexico and the Caribbean; Hydrocarbon Habitats, Basin Formation, and Plate Tectonics*. American Association of Petroleum Geologists, Memoirs, **79**, 815–848.

CHIARADIA, M. & PALADINES, A. 2004. Metal sources in mineral deposits and crustal rocks of Ecuador (1°N–4°S): a lead isotope synthesis. *Economic Geology*, **99**, 1085–1106.

COBBING, E. J., OZARD, J. M. & SNELLING, N. J. 1977. Reconnaissance geochronology of the crystalline basement rocks of the Coastal Cordillera of southern Peru. *Geological Society of America Bulletin*, **88**, 241–246.

COCKS, L. R. M. & TORSVIK, T. H. 2006. European geography in a global context from the Vendian to the end of the Palaeozoic. *In*: GEE, D. G. & STEPHENSON, R. A. (eds) *European Lithosphere Dynamics*. Geological Society, London, Memoirs, **32**, 83–95.

COOPER, M. A., ADDISON, F. T. *ET AL.* 1995. Basin development and tectonic history of the Llanos Basin, Colombia. *In*: TANKARD, A. J., SUAREZ-SORUCO, R. & WELSINK, H. J. (eds) *Petroleum Basins of South America*. American Association of Petroleum Geologists, Memoirs, **62**, 659–665.

CORREA, A. M. 2007. *Petrogênese e evolução do Ofiolito de Aburrá, Cordilheira central dos Andes Colombianos*. PhD Thesis, Universidade de Brazilia.

CORREA, A. M., PIMENTEL, M. *ET AL.* 2006. U–Pb zircon ages and Nd–Sr isotopes of the Altavista stock and the San Diego gabbro: new insights on Cretaceous arc magmatism in the Colombian Andes. *V South American Symposium on Isotope Geology*, 24–27 April 2006, Punta del Este, Uruguay. World Wide Web Address: http://www.vssagi.com/igcp478/AbstractsVSSAGI/212.pdf.

DALZIEL, I. W. D. 1997. Neoproterozoic–Paleozoic geography and tectonics: review, hypothesis, environmental speculation. *Geological Society of America Bulletin*, **109**, 16–42.

DENGO, C. A. & COVEY, M. C. 1993. Structure of the Eastern Cordillera of Colombia: implications for trap styles and regional tectonics. *American Association of Petroleum Geologists Bulletin*, **77**, 1315–1337.

DICKINSON, W. R. & LAWTON, T. F. 2001. Carboniferous to Cretaceous assembly and fragmentation of Mexico. *Geological Society of America Bulletin*, **113**, 1142–1160.

DIEBOLD, J. 2009. Submarine volcanic stratigraphy and the Caribbean LIP's formational environment. *In*: JAMES, K. H., LORENTE, M. A. & PINDELL, J. L. (eds) *The Origin and Evolution of the Caribbean Plate*. Geological Society, London, Special Publications, **328**, 797–806.

DÖRR, W., GRÖSSER, J. R., RODRIGUEZ, G. I. & KRAMM, U. 1995. Zircon U–Pb age of the Paramo Rico tonalite–granodiorite, Santander Massif (Cordillera Oriental, Colombia) and its geotectonic significance. *Journal of South American Earth Sciences*, **8**, 187–194.

DOUBROVINE, P. V. & TARDUNO, J. A. 2008. A revised kinematic model for the relative motion between

Pacific Oceanic Plates and North America since the Late Cretaceous. *Journal of Geophysical Research*, **113**, B12101.

DRISCOLL, N. W. & DIEBOLD, J. B. 1999. Tectonic and stratigraphic development of the eastern Caribbean: new constraints from multichannel seismic data. *In*: MANN, P. (ed.) *Caribbean Basins*. Sedimentary Basins of the World, **4**. Elsevier Science, Amsterdam, 591–626.

DUNKLEY, P. N. & GAIBOR, A. 1997. *Mapa geologico de la Cordillera Occidental del Ecuador entre 2°–3°S (Scale 1:200000)*. Corporación de Desarrollo e Investigación Geológico, Minero, Metalúrgica (CODIGEM) – British Geological Survey.

ENGEBRETSON, D. C., COX, A. & GORDON, R. G. 1985. *Relative Motions between Oceanic and Continental Plates in the Pacific Basin*. Geological Society of America, Boulder, CO, Special Papers, **206**.

FEININGER, T. 1970. The Palestina Fault, Colombia. *Geological Society of America Bulletin*, **81**, 1201–1216.

FEININGER, T. & SILBERMAN, M. L. 1982. *K–Ar Geochronology of Basement Rocks on the Northern Flanks of the Huancabamba Deflection, Ecuador*. United States Geological Survey, Open File Reports, **82-206**.

GAIBOR, J., HOCHULI, J. P. A., WINKLER, W. & TORO, J. 2008. Hydrocarbon source potential of the Santiago Oriente Basin, SE of Ecuador. *Journal of South American Earth Sciences*, **25**, 145–156.

GÓMEZ, E., JORDAN, T. E., ALLMENDINGER, R. W., HEGARTY, K., KELLEY, S. & HEIZLER, M. 2003. Controls on architecture of the Late Cretaceous to Cenozoic southern Middle Magdalena Valley Basin, Colombia. *Geological Society of America Bulletin*, **115**, 131–147.

GÓMEZ, E., JORDAN, T. E., ALLMENDINGER, R. W. & CARDOZO, N. 2005. Development of the Colombian foreland-basin system as a consequence of diachronous exhumation of the northern Andes. *Geological Society America Bulletin*, **117**, 1272–1292.

GÓMEZ, J., NIVIA, A. ET AL. 2007a. *Atlas Geologico de Colombia (Scale 1:500 000)*. Instituto Colombiano de Minería y Geología (INGEOMINAS), Bogotá. World Wide Web Address: http://www.ingeominas.gov.co/content/view/761/316/lang,es/.

GÓMEZ, J., NIVIA, A. ET AL. 2007b. *Geological Map of Colombia (Scale 1:1 000 000)*. Instituto Colombiano de Minería y Geología (INGEOMINAS), Bogotá. World Wide Web Address: http://www.ingeominas.gov.co/component/option,com_docman/task,doc_download/gid,5284/.

GONZÁLEZ, H. 2001. *Mapa geologico del departamento de Antioquia y memoria explicitiva (Scale 1:400 000)*. Instituto Colombiano de Minería y Geología (INGEOMINAS), Bogotá. World Wide Web Address: http://productos.ingeominas.gov.co/productos/.

GONZÁLEZ, H. 2002. *Cuarzodiorita de Mistrató, Catálogo de las unidades litoestratigráficas de Colombia*. Instituto Colombiano de Minería y Geología (INGEOMINAS), Bogotá. World Wide Web Address: http://productos.ingeominas.gov.co/productos/.

GONZÁLEZ, H. & LONDOÑO, A. C. 2002. *Tonalita de Buriticá, Catálogo de las unidades litoestratigráficas*

de Colombia. Bogotá, Instituto Colombiano de Minería y Geología (INGEOMINAS), Bogotá. World Wide Web Address: http://productos.ingeominas.gov.co/productos/.

GOVERS, R. & WORTEL, M. J. R. 2005. Lithosphere tearing at STEP faults: response to edges of subduction zones. *Earth and Planetary Science Letters*, **236**, 505–523.

HUGHES, R. A. & PILATASIG, L. F. 2002. Cretaceous and Tertiary terrane accretion in the Cordillera Occidental of the Andes of Ecuador. *Tectonophysics*, **345**, 29–48.

JAILLARD, E. 1993. L'évolution tectonique de la marge péruvienne au Sénonien et Paléocène et ses relations avec la géodynamique. *Bulletin de la Societe Geologique de France*, **164**, 819–830.

JAILLARD, E. & SOLER, P. 1996. Cretaceous to Early Paleogene tectonic evolution of the northern Central Andes (0–18°S) and its relations to geodynamics. *Tectonophysics*, **259**, 41–53.

JAILLARD, E., SOLER, P., CARLIER, G. & MOURIER, T. 1990. Geodynamic evolution of the northern and central Andes during Early to Middle Mesozoic time: a Tethyan model. *Journal of the Geological Society, London*, **147**, 1009–1022.

JAILLARD, E., ORDOÑEZ, M., BENITEZ, S., BERRONES, G., JIMENEZ, N., MONTENEGRO, G. & ZAMBRANO, I. 1995. Basin development in an accretionary in an accretionary, oceanic-floored fore-arc setting: Southern coastal Ecuador during Late Cretaceous–Late Eocene time. *In*: TANKARD, A. J., SUAREZ-SORUCO, R. & WELSINK, H. J. (eds) *Petroleum Basins of South America*. American Association of Petroleum Geologists, Memoirs, **62**, 615–631.

JAILLARD, E., LAUBACHER, G., BENGTSON, P., DHONDT, A. V. & BULOT, L. G. 1999. Stratigraphy and evolution of the Cretaceous forearc Celica–Lancones basin of southwestern Ecuador. *Journal of South American Earth Sciences*, **12**, 51–68.

JAILLARD, E., ORDOÑEZ, M., SUAREZ, J., TORO, J., IZA, D. & LUGO, W. 2004. Stratigraphy of the Late Cretaceous–Paleogene deposits of the cordillera occidental of central Ecuador: geodynamic implications. *Journal of South American Earth Sciences*, **17**, 49–58.

JAILLARD, E., GUILLIER, B., BONNARDOT, M.-A., HASSANI, R., LAPIERRE, H. & TORO, J. 2005. Orogenic buildup of the Ecuadorian Andes. *6th International Symposium on Andean Geodynamics*, Barcelona, Spain, 12–14 September 2005, 404–407. World Wide Web Address: http://irdal.ird.fr/PDF/ISAG_2005/isag05_404-407.pdf.

JAILLARD, E., BENGTSON, P., ORDOÑEZ, M., VACA, W., DHONDT, A., SUÁREZ, J. & TORO, J. 2008. Sedimentary record of terminal Cretaceous accretions in Ecuador: the Yunguilla Group in the Cuenca area. *Journal of South American Earth Sciences*, **25**, 133–144.

JAMES, K. 2006. Arguments for and against the Pacific origin of the Caribbean Plate: discussion, finding for an inter-American origin. *Geologica Acta*, **4**, 279–302.

KELLOGG, J. N. 1984. Cenozoic tectonic history of the Sierra de Perijá, Venezuela–Colombia, and adjacent basins. *In*: BONINI, W. E., HARGREAVES, R. B. &

SHAGAM, R. (eds) *The Caribbean–South America Plate Boundary and Regional Tectonics*. Geological Society of America, Boulder, CO, Memoir, **162**, 239–261.

KEPPIE, J. D. & RAMOS, V. A. 1999. Odyssey of terranes in the Iapetus and Rheic oceans during the Paleozoic. *In*: RAMOS, V. A. & KEPPIE, J. D. (eds) *Laurentia–Gondwana Connections before Pangea*. Geological Society of America, Boulder, CO, Special Papers, **336**, 267–276.

KERR, A. C. & TARNEY, J. 2005. Tectonic evolution of the Caribbean and northwestern South America: the case for accretion of two Late Cretaceous oceanic plateaus. *Geology*, **33**, 269–272.

KERR, A. C., MARRINER, G. F. ET AL. 1997. Cretaceous basaltic terranes in western Colombia: elemental, chronological and Sr–Nd isotopic constraints on petrogenesis. *Journal of Petrology*, **38**, 677–702.

KERR, A. C., TARNEY, J., NIVIA, A., MARRINER, G. F. & SAUNDERS, A. D. 1998. The internal structure of oceanic plateaus: inferences from obducted Cretaceous terranes in western Colombia and the Caribbean. *Tectonophysics*, **292**, 173–188.

KERR, A. C., ASPDEN, J. A., TARNEY, J. & PILATASIG, L. F. 2002*a*. The nature and provenance of accreted oceanic terranes in western Ecuador: geochemical and tectonic constraints. *Journal of the Geological Society*, **159**, 577–594.

KERR, A. C., TARNEY, J. ET AL. 2002*b*. Pervasive mantle plume head heterogeneity: evidence from the Late Cretaceous Caribbean–Colombian oceanic plateau. *Journal of Geophysical Research*, **107**, B7.

KERR, A. C., WHITE, R. V., THOMPSON, P. M. E., TARNEY, J. & SAUNDERS, A. D. 2003. No oceanic plateau – no Caribbean Plate? The seminal role of an oceanic plateau in Caribbean Plate evolution. *In*: BARTOLINI, C., BUFFLER, R. T. & BLICKWEDE, J. F. (eds) *The Circum-Gulf of Mexico and the Caribbean; Hydrocarbon Habitats, Basin Formation, and Plate Tectonics*. American Association of Petroleum Geologists, Memoirs, **79**, 126–168.

LEÓN, W., PALACIOS, O., SÁNCHEZ, A. & VARGAS, L. 1999. *Memoria explicativa del mapa geologico del Peru (Scale 1:1 000 000)*. Instituto Geologico, Minero y Metalurgico (INGEMMET), Lima.

LI, Z. X., BOGDANOVA, S. V. ET AL. 2008. Assembly, configuration, and break-up history of Rodinia: a synthesis. *Precambrian Research*, **160**, 179–210.

LITHERLAND, M., ASPDEN, J. A. & JAMIELITA, R. A. 1994. *The Metamorphic Belts of Ecuador*. British Geological Survey Overseas, Memoirs, **11**.

LUZIEUX, L. 2007. *Origin and Late Cretaceous-Tertiary evolution of the Ecuadorian forearc*. PhD thesis, ETH, Zurich.

MacDONALD, W. D., ESTRADA, J. J. & GONZALEZ, H. 1997. *Paleoplate Affiliations of Volcanic Accretionary Terranes of the Northern Andes*. Geological Society of America, Boulder, CO, Abstracts with Programs, **29**, 245.

MACHETTE, M. N., EGÜEZ, A., ALVARADO, A. & YEPES, H. 2003. *Map of Quaternary Faults and Folds of Ecuador and its Offshore Regions (Scale 1:1 250 000)*. United States Geological Survey, Open File Reports, **03-289**.

MacNIOCAILL, C., VAN DER PLUIJM, B. A. & VAN DER VOO, R. 1997. Ordovician paleogeography and the evolution of the Iapetus ocean. *Geology*, **25**, 159–162.

MAMBERTI, M., LAPIERRE, H., BOSCH, D., JAILLARD, E., ETHIEN, R., HERNANDEZ, J. & POLVE, M. 2003. Accreted fragments of the Late Cretaceous Caribbean–Colombian Plateau in Ecuador. *Lithos*, **66**, 173–199.

MAMBERTI, M., LAPIERRE, H., BOSCH, D., JAILLARD, E., HERNANDEZ, J. & POLVE, M. 2004. The Early Cretaceous San Juan Plutonic Suite, Ecuador: a magma chamber in an oceanic plateau? *Canadian Journal of Earth Sciences*, **41**, 1237–1258.

MANN, P. 2007. Overview of the tectonic history of northern Central America. *In*: MANN, P. (ed.) *Geologic and Tectonic Development of the Caribbean Plate Boundary in Northern Central America*. Geological Society of America, Boulder, CO, Special Papers, **428**, 1–19.

MARESCH, W. V. & GERYA, T. V. 2005. Blueschists and blue amphiboles: how much subduction do they need? *International Geology Review*, **47**, 688–702.

MARESCH, W. V., KLUGE, R., BAUMANN, A., PINDELL, J. L., KRÜCKHANS-LUEDER, G. & STANEK, K. 2009. The occurrence and timing of high-pressure metamorphism on Margarita Island, Venezuela: a constraint on Caribbean–South America interaction. *In*: JAMES, K. H., LORENTE, M. A. & PINDELL, J. L. (eds) *The Origin and Evolution of the Caribbean Plate*. Geological Society, London, Special Publications, **328**, 703–739.

MARTIN-GOMBOJAV, N. & WINKLER, W. 2008. Recycling of Proterozoic crust in the Andean Amazon foreland of Ecuador: implications for orogenic development of the Northern Andes. *Terra Nova*, **20**, 22–31.

MARTINEZ, E., FERNÁNDEZ, J., CALDERON, Y., HERMOZA, W. & GALDOS, C. 2005. *Tumbes and Talara Basins Hydrocarbon Evaluation*. Perupetro, Lima, Peru. World Wide Web Address: http://www.perupetro.com.pe/home-e.asp.

MAYA-SÁNCHEZ, M. 2001. *Distribución, Facies y Edad de las Rocas Metamórficas en Colombia, Memoria Explicitiva: Mapa metamorfico de Colombia*. Instituto Colombiano de Minería y Geología (INGEOMINAS), Bogotá. World Wide Web Address: http://productos.ingeominas.gov.co/productos/MEMORIA/Memoria%20MMC.pdf.

MAYA-SÁNCHEZ, M. & VÁSQUEZ-ARROYAVE, E. 2001. *Mapa metamorfico de Colombia (Scale 1:2 000 000)*. Instituto Colombiano de Minería y Geología (INGEOMINAS), Bogotá. World Wide Web Address: http://productos.ingeominas.gov.co/productos/OFICIAL/georecon/geologia/escmilln/pdf/Metamorfico.pdf.

McCOURT, W. J., ASPDEN, J. A. & BROOK, M. 1984. New geological and geochronological data from the Colombian Andes: continental growth by multiple accretion. *Journal of the Geological Society, London*, **141**, 831–845.

MESCHEDE, M. & BARCKHAUSEN, U. 2000. Plate tectonic evolution of the Cocos–Nazca spreading center. *In*: SILVER, E. A., KIMURA, G., BLUM, P. & SHIPLEY, T. H. (eds) *Proceedings of the Ocean Drilling Program, Scientific Results*, **170**, 1–10. World Wide Web Address: http://www-odp.tamu.edu/publications/170_SR/chap_07/chap_07.htm.

MOBERLY, R., SHEPHERD, G. L. & COULBOURN, W. T. 1982. Forearc and other basins, continental margin of northern and southern Peru and adjacent Ecuador and Chile. In: LEGGETT, J. K. (ed.) Trench–Forearc Geology; Sedimentation and Tectonics in Modern and Ancient Active Margins. Geological Society, London, Special Publications, 10, 171–189.

MONTES, C., RESTREPO-PACE, P. A. & HATCHER, R. D. 2003. Three dimensional structure and kinematics of the Piedras–Girardot fold belt: surface expression of transpressional deformation in the northern Andes. In: BARTOLINI, C., BUFFLER, R. T. & BLICKWEDE, J. F. (eds) The Circum-Gulf of Mexico and the Caribbean; Hydrocarbon Habitats, Basin Formation, and Plate Tectonics. American Association of Petroleum Geologists, Memoirs, 82, 849–873.

MONTES, C., HATCHER, R. D. & RESTREPO-PACE, P. A. 2005. Tectonic reconstruction of the Northern Andean blocks; oblique convergence and rotations derived from the kinematics of the Piedras–Girardot area, Colombia. Tectonophysics, 399, 221–250.

MORENO, M. & PARDO, A. 2003. Stratigraphical and sedimentological constraints on western Colombia: Implications on the evolution of the Caribbean Plate. In: BARTOLINI, C., BUFFLER, R. T. & BLICKWEDE, J. F. (eds) The Circum-Gulf of Mexico and the Caribbean; Hydrocarbon Habitats, Basin Formation, and Plate Tectonics. American Association of Petroleum Geologists, Memoirs, 79, 891–924.

MOURIER, T., LAJ, C., MÉGARD, F., ROPERCH, P., MITOUARD, P. & FARFAN, A. 1988. An accreted continental terrane in northwestern Peru. Earth and Planetary Science Letters, 88, 182–192.

MUKASA, S. B. 1986. Zircon U–Pb ages of super-units in the Coastal batholith, Peru: implications for magmatic and tectonic processes. Geological Society of America Bulletin, 97, 241–254.

MÜLLER, R. D., ROYER, J.-Y. & LAWVER, L. A. 1993. Revised plate motions relative to the hotspots from combined Atlantic and Indian Ocean hotspot tracks. Geology, 21, 275–278.

NIKOLAEVA, K, GERYA, T. V. & CONNOLLY, J. A. D. 2008. Numerical modelling of crustal growth in intraoceanic volcanic arcs. Physics of the Earth and Planetary Interiors, 171, 336–356.

NIVIA, A. 1996. The Bolívar mafic–ultramafic complex, SW Colombia: the base of an obducted oceanic plateau. Journal of South American Earth Sciences, 9, 59–68.

NIVIA, A. 2001. Mapa geologico del departamento del Valle de Cauca y memoria explicitiva (Scale 1:250 000). Instituto Colombiano de Minería y Geología (INGEOMINAS), Bogotá. World Wide Web Address: http://productos.ingeominas.gov.co/productos/.

NIVIA, A., MARRINER, G. F., KERR, A. C. & TARNEY, J. 2006. The Quebradagrande Complex: a lower Cretaceous ensialic marginal basin in the Central Cordillera of the Colombian Andes. Journal of South American Earth Sciences, 21, 423–436.

NOBLE, S. R., ASPDEN, J. A. & JEMIELITA, R. 1997. Northern Andean crustal evolution: new U–Pb geochronological constraints from Ecuador. Geological Society of America Bulletin, 109, 789–798.

PARDO-CASAS, F. & MOLNAR, P. 1987. Relative motion of the Nazca (Farallon) and South American plates since Late Cretaceous time. Tectonics, 6, 233–248.

PARIS, G., MACHETTE, M. N., DART, R. L. & HALLER, K. M. 2000. Map and Database of Quaternary Faults and Folds in Colombia and its Offshore Regions (Scale 1:1 250 000). United States Geoogical Survey, Open File Reports, 00-0284.

PECORA, L., JAILLARD, E. & LAPIERRE, H. 1999. Accretion paleogene et decrochement dextre d'un terrain oceanique dans le Nord du Perou. Comptes Rendus de l'Academie des Sciences, Serie IIa, Sciences de la Terre et des Planetes, 329, 389–396.

PINDELL, J. L. 1993. Regional synopsis of Gulf of Mexico and Caribbean evolution. In: PINDELL, J. L. & PERKINS, R. F. (eds) Transactions of the 13th Annual GCSSEPM Research Conference: Mesozoic and Early Cenozoic Development of the Gulf of Mexico and Caribbean Region, 251–274.

PINDELL, J. & DEWEY, J. F. 1982. Permo–Triassic reconstruction of western Pangea and the evolution of the Gulf of Mexico/Caribbean region. Tectonics, 1, 179–211.

PINDELL, J. L. & BARRETT, S. F. 1990. Geological evolution of the Caribbean region; a plate tectonic perspective. In: DENGO, G. & CASE, J. E. (eds) The Caribbean Region. Decade of North American Geology, H. Geological Society of America, Boulder, CO, 405–432.

PINDELL, J. L. & ERIKSON, J. P. 1994. Mesozoic passive margin of northern South America. In: SALFITY, J. A. (ed.) Cretaceous Tectonics of the Andes. Vieweg, Earth Evolution Sciences International Monograph Series, 1–60.

PINDELL, J. L. & TABBUTT, K. D. 1995. Mesozoic–Cenozoic Andean paleogeography and regional controls on hydrocarbon systems. In: TANKARD, A. J. R., SUAREZ-SORUCO, R. & WELSINK, H. J. (eds) Petroleum Basins of South America. American Association of Petroleum Geologists, Memoirs, 62, 101–128.

PINDELL, J. L. & KENNAN, L. 2001. Kinematic evolution of the Gulf of Mexico and Caribbean. In: FILLON, R. H., ROSEN, N. C. ET AL. (eds) Transactions of the 21st GCSSEPM Annual Bob F. Perkins Research Conference: Petroleum Systems of Deep-Water Basins, 193–220.

PINDELL, J. L. & KENNAN, L. 2009. Tectonic evolution of the Gulf of Mexico, Caribbean and northern South America in the mantle reference frame: an update. In: JAMES, K. H., LORENTE, M. A. & PINDELL, J. L. (eds) The Origin and Evolution of the Caribbean Plate. Geological Society, London, Special Publications, 328, 1–55.

PINDELL, J. L., HIGGS, R. & DEWEY, J. F. 1998. Cenozoic palinspastic reconstruction, paleogeographic evolution, and hydrocarbon setting of the northern margin of South America. In: PINDELL, J. L. & DRAKE, C. L. (eds) Paleogeographic Evolution and Non-glacial Eustasy, northern South America. SEPM (Society for Sedimentary Geology), Special Publication, 58, 45–86.

PINDELL, J. L., KENNAN, L. & BARRETT, S. F. 2000. Kinematics: a key to unlocking plays. Part 2 of a series: 'Regional Plate Kinematics: Arm Waving, or

Underutilized Exploration Tool'. *American Association of Petroleum Geologists Explorer*, July 2000. World Wide Web Address: http://www.aapg.org/explorer/geophysical_corner/2000/gpc07.cfm.

PINDELL, J. L., KENNAN, L., MARESCH, W. V., STANEK, K. P., DRAPER, G. & HIGGS, R. 2005. Plate-kinematics and crustal dynamics of circum-Caribbean arc-continent interactions, and tectonic controls on basin development in Proto-Caribbean margins. *In*: AVÉ-LALLEMANT, H. G. & SISSON, V. B. (eds) *Caribbean–South American Plate Interactions, Venezuela*. Geological Society of America, Boulder, CO, Special Papers, **394**, 7–52.

PINDELL, J. L., KENNAN, L., STANEK, K. P., MARESCH, W. V. & DRAPER, G. 2006. Foundations of Gulf of Mexico and Caribbean evolution: eight controversies resolved. *Geologica Acta*, **4**, 89–128.

PITCHER, W. S. & COBBING, E. J. 1985. Phanerozoic plutonism in the Peruvian Andes. *In*: PITCHER, W. S., ATHERTON, M. P., COBBING, E. J. & BECKINSALE, R. D. (eds) *Magmatism at a Plate Edge*. Blackie, Glasgow, 19–25.

PRATT, W. T., DUQUE, P. & PONCE, M. 2005. An autochthonous geological model for the eastern Andes of Ecuador. *Tectonophysics*, **399**, 251–278.

PSZCZOLKOWSKI, A. 1999. The exposed passive margin of north America in western Cuba. *In*: MANN, P. (ed.) *Caribbean Basins*. Sedimentary Basins of the World, **4**, Elsevier Science, Amsterdam, 93–122.

PSZCZOLKOWSKI, A. & MYCZYNSKI, R. 2003. Stratigraphic constraints on the Late Jurassic–Cretaceous paleotectonic interpretations of the Placetas Belt in Cuba. *In*: BARTOLINI, C., BUFFLER, R. T. & BLICKWEDE, J. F. (eds) *The Circum-Gulf of Mexico and the Caribbean; Hydrocarbon Habitats, Basin Formation, and Plate Tectonics*. American Association of Petroleum Geologists, Memoirs, **79**, 545–581.

RANGIN, C., GIRARD, D. & MAURY, R. 1983. Geodynamic significance of Late Triassic to Early Cretaceous volcanic sequences of Vizcaino Peninsula and Cedros Island, Baja California, Mexico. *Geology*, **11**, 552–556.

RENNE, P. R., MATTINSON, J. M., HATTEN, C. W., SOMIN, M. L., ONSTOTT, T. C., MILLAN, G. & LINARES, E. 1989. $^{40}Ar/^{39}Ar$ and U–Pb evidence for Late Proterozoic (Grenville-age) continental crust in north-central Cuba and regional tectonic implications. *In*: ONSTOTT, T. C. (ed.) *Recent Advances on the Precambrian Geology of South and Central America and the Caribbean. Precambrian Research*, **42**, 325–341.

RESTREPO-PACE, P. A. 1992. Petrotectonic characterization of the Central Andean Terrane, Colombia. *Journal of South American Earth Sciences*, **5**, 97–116.

RESTREPO-PACE, P. A., RUIZ, J., GEHRELS, G. E. & COSCA, M. 1997. Geochronology and Nd isotopic data of Grenville-age rocks in the Colombian Andes: New constraints for Late Proterozoic–Early Paleozoic paleocontinental reconstructions of the Americas. *Earth and Planetary Science Letters*, **150**, 427–441.

RESTREPO-PACE, P. A., COLMENARES, F., HIGUERA, C. & MAYORGA, M. 2004. A fold-and-thrust belt along the western flank of the Eastern Cordillera of Colombia; style, kinematics, and timing constraints derived from seismic data and detailed surface mapping. *In*: MCCLAY, K. R. (ed.) *Thrust Tectonics*

and Hydrocarbon Systems. American Association of Petroleum Geologists, Memoirs, **82**, 598–613.

ROEDER, D. & CHAMBERLAIN, R. L. 1995. Eastern Cordillera of Colombia: Jurassic–Neogene crustal evolution. *In*: TANKARD, A. J., SUAREZ-SORUCO, R. & WELSINK, H. J. (eds) *Petroleum Basins of South America*. American Association of Petroleum Geologists, Memoirs, **62**, 633–645.

ROGERS, R. D., MANN, P. & EMMET, P. A. 2007. Tectonic terranes of the Chortís block based on integration of regional aeromagnetic and geologic data. *In*: MANN, P. (ed.) *Geologic and Tectonic Development of the Caribbean Plate Boundary in Northern Central America*. Geological Society of America, Special Papers, **428**, 65–88.

ROPERCH, P., MEGARD, F., LAJ, C., MOURIER, T., CLUBE, T. & NOBLET, C. 1987. Rotated oceanic blocks in Western Ecuador. *Geophysical Research Letters*, **14**, 558–561.

ROSAS, S., FONTBOTE, L. & TANKARD, A. J. 2007. Tectonic evolution and paleogeography of the Mesozoic Pucará Basin, central Peru. *Journal of South American Earth Sciences*, **24**, 1–24.

ROSENFELD, J. H. 1993. Sedimentary rocks of the Santa Cruz Ophiolite, Guatemala – a proto-Caribbean history. *In*: PINDELL, J. L. & PERKINS, R. F. (eds) *Transactions of the 13th Annual GCSSEPM Research Conference: Mesozoic and Early Cenozoic Development of the Gulf of Mexico and Caribbean Region*, 173–180.

RUIZ, G. M. H., SEWARD, D. & WINKLER, W. 2004. Detrital thermochronology – a new perspective on hinterland tectonics, an example from the Andean Amazon Basin, Ecuador. *Basin Research*, **16**, 413–430.

SARMIENTO-ROJAS, L. F., VAN WESS, J. D. & CLOETINGH, S. 2006. Mesozoic transtensional basin history of the Eastern Cordillera, Colombian Andes: Inferences from tectonic models. *Journal of South American Earth Sciences*, **21**, 383–411.

SCHIBBENHAUS, C. & BELLIZZIA, A. (coordinators) 2001. *Geological Map of South America (Scale 1:5 000 000)*. CGMW-CPRM-DPNM-UNESCO, Brazilia. World Wide Web Address: http://ccgm.free.fr/.

SEDLOCK, R. L. 2003. Geology and tectonics of the Baja California peninsula and adjacent areas. *In*: JOHNSON, S. E., PATERSON, S. R., FLETCHER, J. M., GIRTY, G. H., KIMBROUGH, D. L. & MARTÍN-BARAJAS, A. (eds) *Tectonic Evolution of Northwestern México and the Southwestern USA*. Geological Society of America, Boulder, CO, Special Papers, **374**, 1–42.

SEDLOCK, R. L., ORTEGA, G. F. & SPEED, R. C. 1993. *Tectonostratigraphic Terranes and Tectonic Evolution of Mexico*. Geological Society of America, Boulder, CO, Special Papers, **278**, 1–153.

SPADEA, P. & ESPINOSA, A. 1996. Petrology and chemistry of Late Cretaceous volcanic rocks from the southernmost segment of the Western Cordillera of Colombia (south America). *Journal of South American Earth Sciences*, **9**, 79–90.

SPIKINGS, R. A., WINKLER, W., SEWARD, D. & HANDLER, R. 2001. Along-strike variations in the thermal and tectonic response of the continental Ecuadorian Andes to the collision with heterogeneous oceanic crust. *Earth and Planetary Science Letters*, **186**, 57–73.

SPIKINGS, R. A., WINKLER, W., HUGHES, R. A. & HANDLER, R. 2005. Thermochronology of allochthonous terranes in Ecuador: Unravelling the accretionary and post-accretionary history of the Northern Andes. *Tectonophysics*, **399** 195–220.

STÉPHAN, J. F. 1985. Andes et Chaine Caribe sur la transversale de Barquisimeto (Venezuela): Evolution Géodynamique. *In*: MASCLE, A. (ed.) *Géodynamique des Caribes*. Editions Technip, Paris, 505–529.

STÖCKHERT, B., MARESCH, W. V. ET AL. 1995. Crustal history of Margarita Island (Venezuela) in detail: constraint on the Caribbean Plate tectonic scenario. *Geology*, **23**, 787–790.

THOMAS, G., LAVENU, A. & BERRONES, G. 1995. Évolution de la subsidence dans le Nord du bassin de l'Oriente équatorien (Critacé supérieur à Actuel). *Comptes Rendus de l'Académie des Sciences, Serie IIa, Sciences de la Terre et des Planetes*, **320**, 617–624.

TORSVIK, T. H., MÜLLER, R. D., VAN DER VOO, R., STEINBERGER, B. & GAINA, C. 2008. Global plate motion frames: toward a unified model. *Reviews of Geophysics*, **46**, RG3004.

TOUSSAINT, J.-F. & RESTREPO, J. J. 1994. The Colombian Andes during Cretaceous times. *In*: SALFITY, J. A. (ed.) *Cretaceous Tectonics of the Andes*. Vieweg, Earth Evolution Sciences, International Monograph Series, 61–100.

TRENKAMP, R., KELLOGG, J. N., FREYMUELLER, J. T. & MORA, H. P. 2002. Wide plate margin deformation, southern Central America and northwestern South America, CASA GPS observations. *Journal of South American Earth Sciences*, **15**, 157–171.

TSCHANZ, C. M., MARVIN, R. F., CRUZ, B. J., MEHNERT, H. H. & CEBULA, G. T. 1974. Geologic evolution of the Sierra Nevada de Santa Marta, northeastern Colombia. *Geological Society of America Bulletin*, **85**, 273–284.

VALLEJO, C. 2007. *Evolution of the Western Cordillera in the Andes of Ecuador (Late Cretaceous–Paleogene.* PhD thesis, ETH, Zurich.

VALLEJO, C., HOCHULI, P. A., WINKLER, W. & VON SALIS, K. 2002. Palynological and sequence stratigraphic analysis of the Napo Group in the Pungarayacu 30 well, Sub-Andean Zone, Ecuador. *Cretaceous Research*, **23**, 845–859.

VALLEJO, C., SPIKINGS, R., LUZIEUX, L., WINKLER, W., CHEW, D. & PAGE, L. 2006. The early interaction between the Caribbean Plateau and the NW South American Plate. *Terra Nova*, **18**, 264–269.

VAN DER HILST, R. 1990. *Tomography with P, PP, and pP delay-time data and the three-dimensional mantle structure below the Caribbean region.* PhD Thesis, University of Utrecht.

VAN DER HILST, R. & MANN, P. 1994. Tectonic implications of tomographic images of subducted lithosphere beneath northwestern South America. *Geology*, **22**, 451–454.

VAN STAAL, C. R., DEWEY, J. F., MACNIOCAILL, C. & MCKERROW, W. S. 1998. The Cambrian–Silurian tectonic evolution of the northern Appalachians and British Caledonides: history of a complex, west and southwest Pacific-type segment of Iapetus. *In*: BLUNDELL, D. J. & SCOTT, A. C. (eds) *Lyell: The Past is the Key to the Present*. Geological Society, London, Special Publications, **143** 199–242.

VÁSQUEZ, M. & ALTENBERGER, U. 2005. Mid-Cretaceous extension-related magmatism in the eastern Colombian Andes. *Journal of South American Earth Sciences*, **20**, 193–210.

VILLAGÓMEZ, D., SPIKINGS, R., SEWARD, D., MAGNA, T. & WINKLER, W. 2008. Thermotectonic history of the Northern Andes. *7th International Symposium on Andean Geodynamics*, 2–4 September 2008, Nice, 573–576. World Wide Web Address: http://www-geoazur.unice.fr/ISAG08/Soumissions/PDF/573-576_Villagomez_et_al.pdf.

VILLAMIL, T. 1999. Campanian–Miocene tectonostratigraphy, depocenter evolution and basin development of Colombia and western Venezuela. *Palaeogeography, Palaeoclimatology, Palaeoecology*, **153**, 239–275.

VINASCO-VALLEJO, C. J., CORDANI, U. & VASCONCELOS, P. 2003. Application of the $^{40}Ar/^{39}Ar$ methodology in the study of tectonic reactivations of shear zones: Romeral fault system in the Central cordillera of Colombia. *IV South American Symposium on Isotope Geology*, 24–27 August 2003, Salvador, Bahia, Brazil, 138–144. World Wide Web Address: http://www.brasil.ird.fr/sympIsotope/Papers/ST1/ST1-29-Vinasco.pdf.

VON HUENE, R. & LALLEMAND, S. 1990. Tectonic erosion along the Japan and Peru convergent margins. *Geological Society of America Bulletin*, **102**, 704–720.

VON HUENE, R., SUESS, E. ET AL. 1988. Ocean Drilling Program Leg 112, Peru continental margin; Part 1, Tectonic history. *Geology*, **16**, 934–938.

WHITE, R. V., TARNEY, J. ET AL. 1999. Modification of an oceanic plateau, Aruba, Dutch Caribbean: Implications for the generation of continental crust. *Lithos*, **46**, 43–68.

WINE, G., ARCURI, J., MARTINEZ, E., MONGES, C., CALDERON, Y. & GALDOS, C. 2001. *A study on the remaining undiscovered hydrocarbon potential of the Trujillo offshore basin, Peru*. Proyecto de asistencia para la reglamentacion del sector energetico del Peru (PARSEP), distributed by Perupetro, Lima, Peru. World Wide Web Address: http://www.perupetro.com.pe/home-e.asp.

WINKLER, W., VILLAGÓMEZ, D., SPIKINGS, R., ABEGGLEND, P., TOBLERE, S. & EGÜEZ, A. 2005. The Chota basin and its significance for the inception and tectonic setting of the inter-Andean depression in Ecuador. *Journal of South American Earth Sciences*, **19**, 5–19.

WINKLER, W., VALLEJO, C., LUZIEUX, L., SPIKINGS, R. & MARTIN-GOMBOJAV, N. 2008. Timing and causes of the growth of the Ecuadorian cordilleras, as inferred from their detrital record. *7th International Symposium on Andean Geodynamics*, 2–4 September 2008, Nice, France, 587–591. World Wide Web Address: http://www-geoazur.unice.fr/ISAG08/Soumissions/PDF/587-591_Winkler_et_al.pdf.

ZAMORA, A. & LITHERLAND, M. 1993. *Mapa geologico de la Republica del Ecuador (Scale 1:1 000 000)*. Corporación de Desarrollo e Investigación Geológico, Minero, Metalúrgica (CODIGEM), British Geological Survey.

Presence of high-grade rocks in NW Venezuela of possible Grenvillian affinity

SEBASTIÁN GRANDE* & FRANCO URBANI

Universidad Central de Venezuela, Escuela de Geología, Minas y Geofísica, Ciudad Universitaria, Caracas 1053, Venezuela

Corresponding author (e-mail: sgrande52@gmail.com)

Abstract: High-grade metamorphic rocks – marble, charnockite, meta-anorthosite, metapelite, clinopyroxenite and garnet amphibolite – have been found in northwestern Venezuela. They occur as: (a) xenoliths in the Oligo-Miocene lavas of Cerro Atravesado, Central Falcón; (b) possibly olistoliths in Nuezalito Formation, NW Portuguesa; (c) in Cerro El Guayabo, an elongated east–west oriented hill in the Nirgua Complex, Yaracuy; (d) rounded clasts of marble in the basal conglomerate of Casupal Formation, Falcón; (e) rounded clasts of anorthosite and sillimanite gneiss in a conglomerate of Matatere Formation, Lara; and (f) basement cores extracted from La Vela Gulf, Falcón. These high-grade rocks probably suffered a retrograde metamorphism to amphibolite facies of Palaeozoic age, and an even more retrograde event to the greenschist facies during Early Cenozoic, together with strong shearing and hydrothermal alteration. They indicate the possible existence of an extensive high-grade basement, or a mosaic of such blocks, under northwestern Venezuela, especially below the Falcón petroleum basin. Similar rocks crop out extensively in northern and central Colombia, Mexico, Ecuador and Peru. This is the first time they are described from Venezuela. Their high-grade lithology, pre-Mesozoic positions, and the tectonic evolution of Northern South America allow interpretation of a possible Grenvillian affinity, related to the super-continents of Rodinia and Pangaea.

Geologists who worked in Venezuela in the second half of the twentieth century should have suspected that no Middle–Late Proterozoic rocks were reported from the north of the country. Apart from recently discovered kimberlite and lamprophyre dykes and sills in the Guaniamo region, dated at 830–710 Ma (Channer *et al.* 2001), there is no record of the billion year long interval from ±1600 Ma (Roraima diabase, Parguaza rapakivi granite and Peña de Mora augengneiss) to ±560 Ma (Caparo and Guaremal granites). This work shows that high- to medium-grade metamorphic rocks probably of Middle–Late Proterozoic age are present.

Location of the high- to medium-grade rocks

High- to medium-grade metamorphic rocks in NW Venezuela occur in the following diverse situations:

(1) As xenoliths of dolomitic marble, mafic granulite and anorthosite in the Oligocene–Miocene basanitic lavas of Cerro Atravesado, central Falcón Basin (Fig. 1a), where two subvolcanic intrusive bodies or diatremes contain numerous crustal xenoliths and ultramafic mantle nodules of spinel peridotite (Grande 2009).

(2) As boulders in the Nuezalito Formation, which crops out in Bocoy and Riecito riverbeds, NW Portuguesa, described by Skerlek (1976) as a graphitic phyllite that contains olistoliths a variety of rocks ranging from very low-grade volcanics to high-grade metamorphics. Field inspection by the authors failed to recognize olistoliths in the phyllite. The high-grade rocks occur only as waterworn boulders in river beds (Fig. 1b) (Grande *et al.* 2007a).

(3) As outcrops in the 4 km elongated hill of Cerro El Guayabo, in NE Yaracuy state (Fig. 1c), which contains a metamorphic suite of mafic charnockite with boudins of felsic charnockite and dolomitic marble, intruded by clinopyroxenite dykes (this block is incorrectly mapped as a peridotite–serpentinite (Mo, Mesozoic ophiolite) body in the Nirgua Complex (Kn, see Fig. 1c) (Grande *et al.* 2007b). The hill is a transpressive tectonic block bounded by two right-lateral east–west faults.

(4) As decimetre-sized rounded boulders of phlogopite marbles in the basal conglomerates of the Miocene Casupal Formation, Yaracuy (Fig. 1c) (Lozano & Mussari 2008).

(5) As centimetre-sized rounded clasts of anorthosite, leucogabbro and sillimanite gneiss in the flysch units of Paleocene Matatere Formation, in Lara state (Fig. 1d) (Valletta & Martínez 2008).

From: JAMES, K. H., LORENTE, M. A. & PINDELL, J. L. (eds) *The Origin and Evolution of the Caribbean Plate.*
Geological Society, London, Special Publications, **328**, 533–548.
DOI: 10.1144/SP328.21   0305-8719/09/$15.00 © The Geological Society of London 2009.

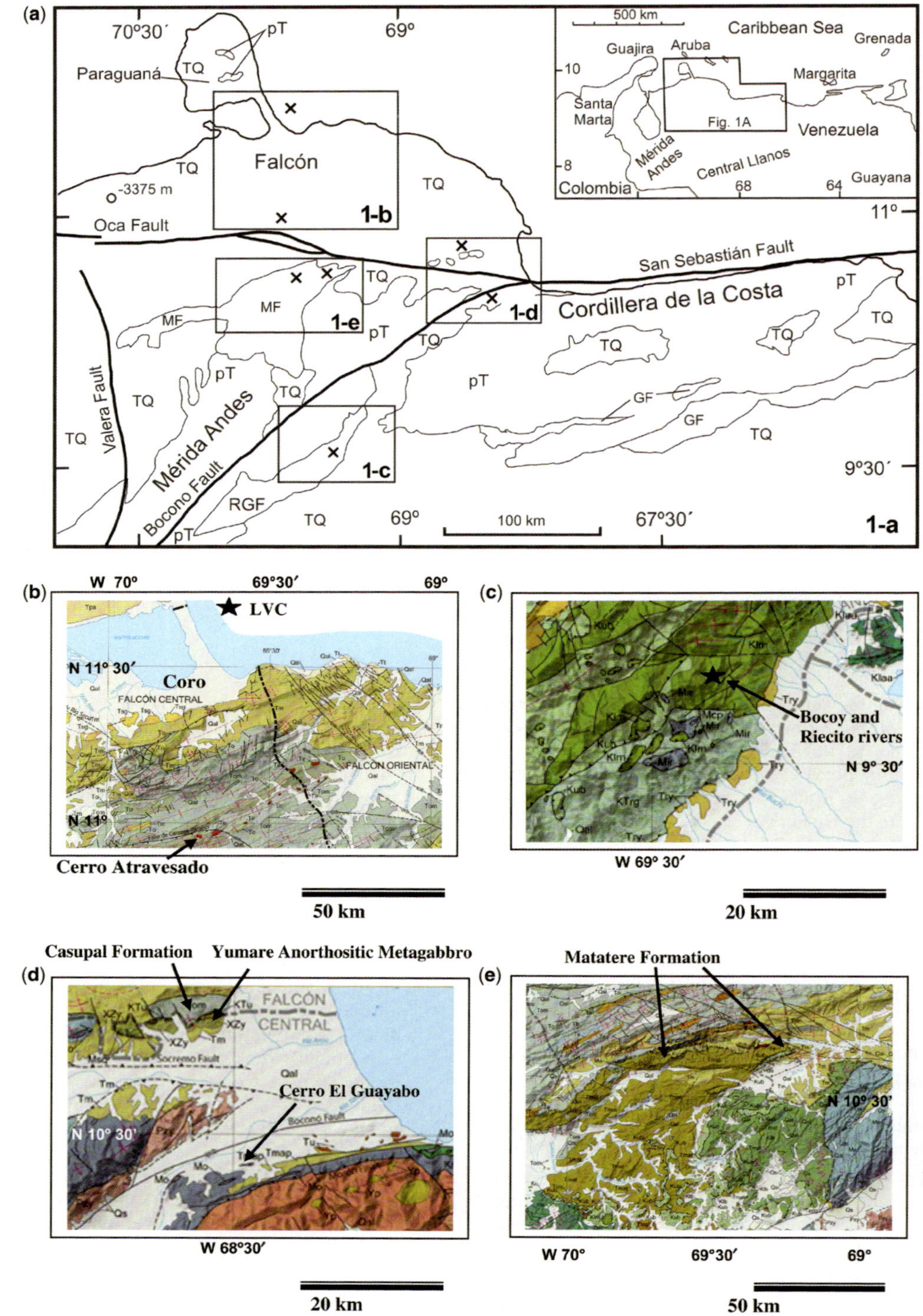

Fig. 1. Location maps in NW Venezuela: (**a**) General geotectonic map of NW Venezuela. TQ: Cenozoic-Quaternary sedimentary rocks (non-flysch formations) and sediments; MF: Palaeogene Matatere flysch; GF: Palaeogene Guárico flysch; RGF: Río Guache flysch and Nuezalito Formation; pT: pre-Cenozoic sedimentary, igneous and

(6) As cores extracted in the early 1970s by the Corporación Venezolana del Petróleo in the basement of Golfo de la Vela, Falcón state, at 8730 + 900 feet depth (Fig. 1a), which contain similar dolomitic marble, associated with high-grade pelitic gneiss, felsic and mafic granulite (or charnockite), amphibolite, garnet amphibolite and clinopyroxenite (Vazquez 1975; Mendi et al. 2005).

These widespread high- to medium-grade metamorphic rocks probably indicate the presence of an extended basement, or a mosaic of such blocks in northwestern Venezuela. Lithologically this basement is not related to the Guiana Shield; instead it is similar to the Grenville Province of southeastern Canada and northeastern USA, where it crops out as a gigantic orogenic belt, profoundly eroded as to show its deepest roots, as granulite facies rocks. The rock suite of the Grenvillian Orogen is like that found in NW Venezuela, with dolomitic marble (with serpentinized forsterite, diopside and phlogopite) and various types of charnockitic and anorthositic rocks being abundant. In Venezuela these rocks are dismembered and tectonized, while in Canada they are part of the Canadian Shield. In the USA they occur as tectonized basement inliers in the Appalachian Orogen. Also they crop out in a unit called Los Mangos Granulite in the western flank of the Sierra Nevada of Santa Marta, in Northern Colombia, and in the Oaxacan block, in Southern Mexico.

Lithology of the high- to medium-grade rocks

The following lithologies have been identified to date. Mineralogical compositions appear in Table 1.

Dolomitic marble

This is widespread and conspicuous. It occurs as loose boulders in Bocoy and Riecito riverbeds, as xenoliths in the lava of Cerro Atravesado, in cores retrieved from La Vela Gulf, as boulders of the basal conglomerates of the Miocene Casupal Formation and in outcrops and boudins in Cerro El Guayabo (Figs 2 & 3).

Texture ranges from very coarse to medium grain; some are slightly foliated to ultramylonite,

but most are granoblastic. Almost all contain well developed phlogopite crystals and small pyrite cubes. Some contain abundant pseudomorphs of serpentinized forsterite with idioblastic diopside, somewhat chloritized. These could be considered as silica undersaturated marbles. Others contain some quartz and only diopside, with little or no phlogopite, such as those of Cerro El Guayabo. These contain porphyroclasts of diopside and of orthoclase inverted to twinned microcline. The carbonate matrix is mylonitic to ultramylonitic (Fig. 3b).

Felsic charnockite or mangerite

This has been found so far in boudins that crop out in Cerro El Guayabo and in cores from La Vela Gulf. These are rocks formed of large tabular crystals of salmon-pink alkali feldspar, with some sodic oligoclase (An10–20), quartz and quite oxidized clino- and orthopyroxene (Fig. 4a). In Cerro El Guayabo these quartz–feldspatic rocks contain no hydrated primary minerals, such as amphibole or mica, but contain pseudomorphs of pyroxene almost totally altered to stilpnomelane, chlorite, limonite and leucoxene. Some xenoliths in the lavas of Cerro Atravesado, and some boulders in the Río Bocoy area, have the same lithology, and under the naked eye could be easily confused with normal granitic rocks.

Mafic charnockite or enderbite, and anorthosite

These occur in Cerro El Guayabo, as xenoliths in Cerro Atravesado lavas and as rounded clasts in turbiditic conglomerate of Matatere Formation (Valletta & Martínez 2008), along with all the igneous, metamorphic and other lithic content of Matatere of probable Northern Andes origin. These are gabbroic-looking rocks, with granoblastic texture. Metamorphic nature is revealed by the antiperthitic plagioclase and other complex substitution textures in this same mineral, such as inclusions of fibrous quartz and myrmekites, presence of some residual pyroxene and scarce retrograde biotite (Fig. 4b). Anorthosite shows abundant plagioclase with typical adcumulus to mesocumulus textures, with this mineral as the cumulus phase, with straight borders, polygonalized grains and no zoning.

Fig. 1. (*Continued*) metamorphic rocks; 'x', crop out areas of high grade rocks; 'o', Jurassic redbeds assigned to La Quinta Formation, and depth in metres in an exploratory well in the Dabajuro Platform, western Falcon. (**b**) Cerro Atravesado: the westernmost of a chain of subvolcanic Cenozoic intrusions in Central Falcón basin, Tir; also shown as an asterisk * the position of wells at Golfo de La Vela (LVC). (**c**) Río Bocoy–Río Riecito basin, NW Portuguesa: Klm (dark green) includes Nuezalito Formation (at the *); (**d**) Cerro El Guayabo, NE Yaracuy: a granulitic tectonic block incorrectly identified as a 'Mesozoic ophiolite', in dark violet colour and with symbol Mo, included in Kn, the Cretaceous Nirgua Suite; also shown the outcrop areas of Casupal Formation Tom and the Yumare Anorthositic Metagabbro XZy; (**e**) Extensive crop out area of Matatere Formation Tmat (dark green), in northern Lara. From Hackley et al. (2005).

Table 1. *Petrography of medium- to high-grade rocks from NW Venezuela*

Sample	Q	Kfs	Pl	Cpx	Bt	Sil	Grt	Hb	Chl	Car	Phl	Srp	Ttn	Ap+Ep	Op	Lithology
YA-250B	6	–	–	40	–	–	–	–	–	54	–	–	–	–	–	Dolomitic marble
YA-253E	3	–	–	18	–	–	–	–	–	75	–	–	–	–	4	
YA-254A	1	–	–	15	–	–	–	–	–	76	–	–	–	–	8	
YA-254B	1	4	–	26	–	–	–	–	–	69	–	–	–	–	–	
YA-254C	–	2	–	28	–	–	–	–	–	72	–	–	Tz	–	–	
PO-37	–	–	–	9	–	–	–	–	–	70	7	15	–	–	Tz	
FA-38	2	–	–	13	–	–	–	–	–	81	–	–	–	Tz	Tz	
FA-50 X1	–	–	–	32	–	–	–	–	–	50	17	–	1	–	–	
FA-51B X1	–	–	–	36	–	–	–	–	–	49	15	–	–	–	–	
YA-01	3	–	–	20	–	–	–	–	–	77	–	–	–	–	–	
YA-02	–	–	–	40	–	–	–	–	–	60	7	39	–	–	–	
LVC-12	–	–	–	–	–	–	–	–	–	54	–	–	–	–	–	
LVC-18	30	28	30	–	1	–	7	–	–	–	–	–	Tz	Tz	Tz	Metapelite
LVC-22c	45	20	–	–	–	–	17	–	8	–	–	–	–	–	8	
LVC-22a	40	22	20	–	–	3	10	–	4	–	–	–	–	–	1	Metapsammite
LVC-22b	50	30	–	–	–	–	10	–	10	–	–	–	Tz	Tz	Tz	
YA-250A	1	91	2	2	–	–	–	–	–	–	–	–	–	–	4	Felsic granulite
YA-250E	–	60	38	–	–	–	–	–	–	–	–	–	–	–	–	
YA-251A	–	53	5	28	1	–	–	–	–	–	–	–	4	–	1	
YA-251B	15	79	3	Tz	–	–	–	2	Tz	–	–	–	Tz	–	3	
YA-252B	4	83	10	–	–	–	–	–	–	–	–	–	–	–	1	
YA-253B	12	81	7	–	–	–	–	–	–	–	–	–	–	–	3	
LVC-22f	35	17	20	10	–	–	–	–	10	–	–	–	–	Tz	8	
LVC-22 g	–	40	17	20	–	–	3	–	10	Tz	–	10	–	Tz	Tz	
LVC-17	–	–	30	–	20	–	–	45	5	–	–	–	–	Tz	Tz	Amphibolite
LVC-22d	15	10	25	–	–	–	–	50	–	–	–	–	–	–	Tz	
FA-52	2	–	92	–	–	–	–	–	5	1	–	–	–	–	–	Metaanorthosite
FA-39	5	–	50	45	–	–	–	–	–	–	–	–	1	–	Tz	Mafic granulite
YA-250-D1	–	–	62	37	–	–	–	–	–	–	–	–	2	1	–	
YA-253C	–	–	40	42	–	–	–	15	–	–	–	–	–	–	–	
YA-250E	–	–	2	85	–	–	–	12	3	–	–	–	–	1	–	Hb clinopyroxenite
YA-250F	–	–	–	67	8	–	–	24	3	–	–	–	–	–	–	
YA-250-D2	3	–	1	95	1	–	–	–	–	4	–	–	–	–	2	Clinopyroxenite
YA-252A	3	–	1	91	–	–	–	–	–	3	–	–	–	–	8	
LVC-22e	–	5	Tz	85	1	–	–	–	–	10	–	Tz	Tz	Tz	–	

Provenance of rock samples: LVC-12 to LVC-22, cores drilled from La Vela platform, northern Falcon State; FA-38, 39, FA-50 X1, FA-51BX1, FA-52, xenoliths in the Cerro Atravesado lavas, Central Falcon State; PO-37, PO-53 and PO-61, boulders in Río Bocoy area, NW Portuguesa State; YA-01 and YA-02, loose blocks, Cerro El Guayabo, Yaracuy State; YA-250 to YA-254, outcrops, Cerro El Guayabo, Yaracuy State.

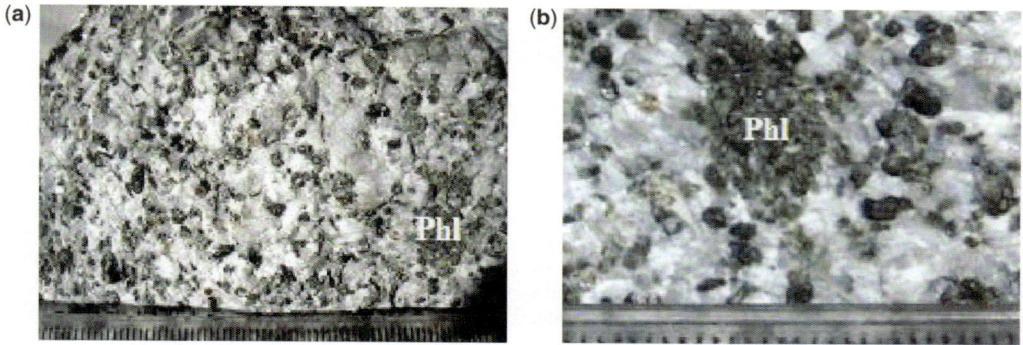

Fig. 2. (**a**) PO-37. Forsterite (serpentinized)–diopside–phlogopite marble, boulder in Río Bocoy. (**b**) Phlogopite concentrations in the preceding marble (detail) that could represent metamorphosed K-rich clay nodules in a colony of Proterozoic stromatolitic algae (see arrow). Scale in mm.

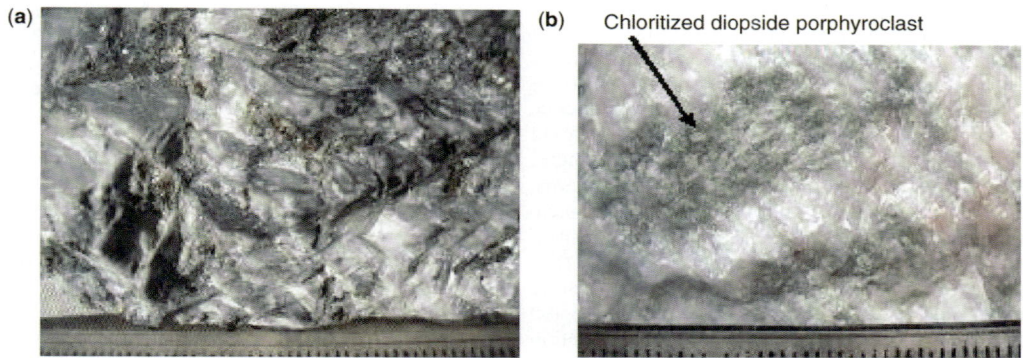

Fig. 3. (**a**) YA-249. Coarse phlogopite–forsterite–diopside marble, boulder in the basal conglomerate of Casupal Formation, of Miocene age. (**b**) YA-254B. Mylonitic marble in Cerro El Guayabo. The faint greenish areas are chloritized diopside porphyroclast (see arrow) in a matrix of finely ground carbonate (see also Fig. 9c for a photomicrography of this same rock). Scale in millimetres.

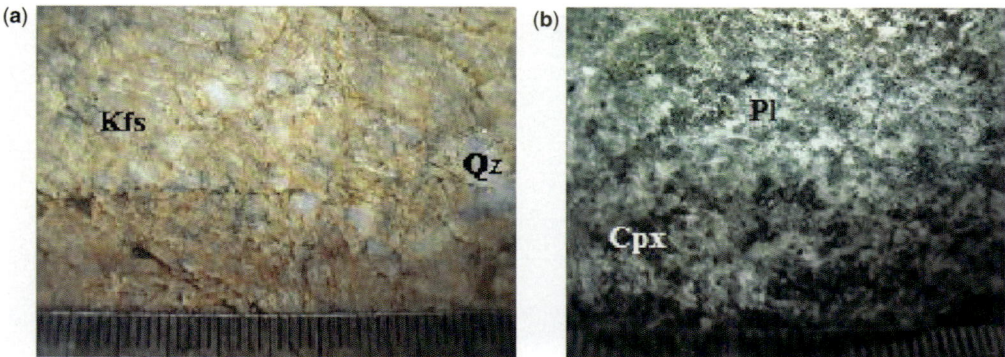

Fig. 4. (**a**) YA-253B. Felsic granulite or charnockite (mangerite), Cerro El Guayabo. Note the abscence of micas in this quartz–feldspatic rock, and the probable presence of altered pyroxenes. (**b**) YA-250D. Mafic granulite or charnockitic gabbronorite (enderbite), Cerro El Guayabo. Note the 'gabbroic' look of this rock. Scale in millimetres.

Metapelite and metapsammite

These rocks are known from the cores of La Vela Gulf and as clasts in conglomerate of Matatere Formation (Valletta & Martínez 2008). These are garnet–sillimanite–quartz–feldspathic gneiss (kinzigites?) and garnet–quartz–feldspathic gneiss. Some are granitic looking rocks, but sample LVC-22d (Mendi 2005) shows big and fractured garnet crystals and large prismatic laths of sillimanite, thus being high-grade metapelites or metapsammites formed from the same sedimentary suite as the marble found in these cores and elsewhere.

Clinopyroxenite

This rock occurs in Cerro El Guayabo and in the cores of La Vela Gulf. They are decimetre-thick highly sheared bodies cross-cutting the charnockite of Cerro El Guayabo, where they show reaction zones with the enclosing mafic and felsic charnock-ites (Fig. 5). These zones are of probable metaso-matic origin and consist of large hornblende prisms, granular epidote and feathery biotite in contact with metric marble boudins or blocks, there-fore could be interpreted as pyroxene-rich skarns. These bodies could also represent thin ultramafic in rusions probably intruded into an anorthosite–n angerite–charnockite assemblage, and they produced contact metamorphism in the marble units. Metasomatism extends to the pyroxenite and the enclosing charnockite as pervasive epidotization and amphibolitization. It is not possible to estimate the real thickness or length of these bodies, because intense tectonism has deformed them into boudins and sheared fragments, like the related marble and felsic charnockite.

Petrography

Dolomitic marble (Fig. 6)

This is a spectacular rock, of medium to very coarse grain, granoblastic to slightly foliated to ultramylo-mitic. Its very simple mineralogy consists of no more than three or four essential minerals and scarce accesories. Protoliths were not pure limestones since abundant mafic silicates, such as diopside and forsterite (Fig. 6a), and ubiquitous phlogopite indicate at least marly or possibly clayey dolostone protoliths. Most marbles of Cerro El Guayabo contain some quartz, therefore they were silica super-saturated rock (Fig. 6b). Accordingly they contain mostly diopside or K-feldspar but neither forst-erite nor phlogopite. Its terrigenous nature might indi-cate sedimentation on a margin not far from shore.

XRD analysis shows the presence of 2–10% dolomite. Forsterite has been totally serpentinized to orthochrysotile, 60_{C-1} polytype, formed at rela-tively low-temperature. Clinopyroxene is low-Fe diopside, somewhat chloritized. The following are concise petrographic descriptions of this marble in thin section.

- *Carbonate*: calcite and minor amounts of dolo-mite (recognized by XRD). Granular, frequently with mortar texture and strongly ground, with intense deformations and folding of the polysyn-thetic twinning. Carbonate sometimes substi-tutes diopside, leaving fragments of this mineral in optical continuity.
- *Diopside*: subhedral to euhedral, quite well pre-served, with very thin parallel exsolution lamel-lae of orthopyroxene. Some calcite-substituted crystals have skeletal textures, with fragments in optical continuity. Other examples show mortar texture. Diopside also forms conspicuous

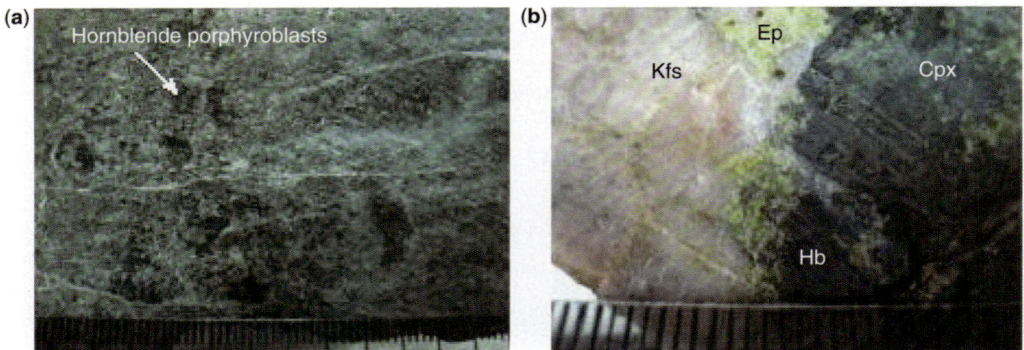

Fig. 5. (**a**) YA-250F. Clinopyroxenite, Cerro El Guayabo. Note the large blasts of hornblende associated with biotite (see arrow). (**b**) YA-250D. Clinopyroxenite dyke in mangerite, Cerro l Guayabo, showing a metasomatic reaction zone of hornblende and epidote. Scale in millimetres.

Serpentinized forsterite   1 mm

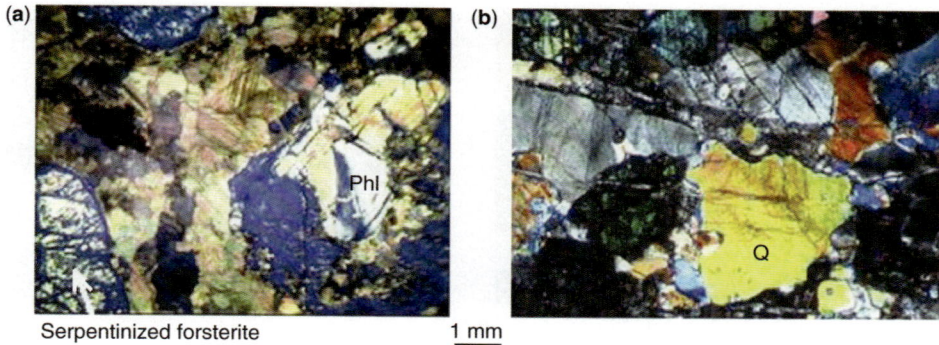

Fig. 6. (**a**) PO-37, with crossed polars (XP). Forsterite–phlogopite–diopside marble, loose boulder in Río Bocoy. Note undulatory extinction in phlogopite, total serpentinization of the nodular forsterite (see arrow), high birefringence of diopside and mortar texture of the carbonate. (**b**) YA-250B, XP. Quartz-diopside marble, Cerro El Guayabo; quartz (Q), in yellow, shows strong undulatory extinction.

chloritized porphyroclasts in strongly mylonitic marbles, as in Cerro El Guayabo (Fig. 3b).
- *Forsterite* (serpentinized): in nodular grains, totally altered to antigorite. These large sub-rounded nodules show a complex reaction corona of interbedded rings of orthochrysotile and carbonate. Its presence indicates a silica-undersatured protolith.
- *Quartz*: appears as xenoblastic grains in two samples from Cerro El Guayabo. Marbles with quartz lack forsterite and only contain K-feldspar, diopside and some phlogopite.
- *Microcline*: present as xenoblastic porphyroclasts in two quite tectonized samples from Cerro El Guayabo. It shows tartan twinning, possibly an inversion from orthoclase which is stable at higher temperatures. It is also perthitic, being originally a somewhat Na-rich orthoclase generated from potassic clays.
- *Phlogopite*: moderate pleochroism from yellowish to almost colourless, micaceous habit, with well-defined leaflets, little altered to almost isotropic chlorite in its borders and internally, parallel to cleavage planes. Tectonic deformation lends common undulatory extinction to the leaflets (Fig. 6a). Its presence is probably related to continental potassic clays.
- *Titanite*: found in El Guayabo marbles, as xenoblastic grains associated with diopside, and in a xenolith of Cerro Atravesado as inclusions inside phlogopite.
- *Opaques*: generally pyrite, in small cubic crystals or anhedral grains.

Felsic granulite or syenitic charnockite (mangerite)

This rock has a superficial 'granitic' aspect and a characteristic salmon-pink colour, due to the abundant alkali feldspar stained by Feoxides, which are also crosscut by numerous calcite veinlets. It is quite widespread in northern Venezuela, present in La Vela Gulf cores (Mendi *et al.* 2005, samples LCV-22f–g) and as boudins in Cerro El Guayabo, where it has been intruded by clino-pyroxenite dykes. It is a felsic rock with quartz and sodic plagioclase containing no primary hydrated minerals, but instead highly altered pyroxene grains, almost unrecognizable. It was quite surprising to find in a low-grade metamorphic belt such as Cordillera de la Costa, typically formed by micaceous gneiss, metagranite, amphibolite and schistose rock, a lithology so rich in feldspars without any amphibole or mica. These rocks might be confused with pyroxene syenites but complex internal textures show them to be felsic granulites or charnockites (according to the probable presence of orthopyroxene).

- *Alkali feldspar*: hypidioblastic to xenoblastic, with abundant flame-shaped perthitic exsolution lamellae of albite, which practically equal in volume the host microcline crystal, being therefore mesoperthite (Fig. 7a). The microcline shows local sparse tartan twinning. The albite lamellae have various shapes; most form long isolated droplets arranged in parallel bands. Locally the droplets get thicker and the plagioclase shows polysynthetic twinning. This alkali feldspar was originally Na-rich orthoclase that inverted to microcline and developed mesoperthitic texture on lowering the temperature and increasing deformation. Also trainlets of microgranular quartz are observed cutting through the large mesoperthite crystals, a typical cataclastic effect shown by many metagranitic rocks of Cordillera de la Costa.

Fig. 7. (**a**) YA-251B, XP. Mangerite or charnockitic quartz-syenite, Cerro El Guayabo. Note the mesoperthitic feldspar with a great proportion of albitic exsolution lamellae. (**b**) PO-53, uncrossed polars (UP). Enderbite or charnockitic gabbronorite, Río Riecito. Note a possible association of clino and orthopyroxene characteristic of these anhydrous metaigneous rocks.

- *Quartz*: xenoblastic, scarce to abundant, so that the rocks vary from granulitic to charnockitic syenites or quartz-syenites, according to the presence of recognizable orthopyroxene.
- *Plagioclase*: calcic–oligoclase (An32), scarce, with undulatory extinction. It occurs in grains surrounding the mesoperthite and it is not an exsolution product. Altered to calcite and sericite.
- *Clinopyroxene*: neutral, prismatic and xenoblastic ameboidal, altered to opaques and titanite.
- *Orthopyroxene*: quite scarce, alters readily to limonite and leucoxene. In some samples can be identified by its brownish colour, lower birefringence compared with clinopyroxene, and parallel extinction.
- *Titanite*: slightly pleochroic in brownish hues, idio- to hypidioblastic grains with polysynthetic twinning. Product of alteration of probable clinopyroxene.
- *Zircon–xenotime*: brownish, idioblastic with prismatic habit. The zircon crystals show epitaxial growth of yellowish xenotime.
- *Opaques*: limonite as fracture-filling veinlets, and leucoxene as alteration of mafic minerals.

Mafic granulite or grabbronoritic charnockite (enderbite)

This is a dark mafic rock, formed by (Na, Ca)-plagioclase and pyroxenes, sometimes altered to hornblende and biotite (Fig. 7b). Found first by Brueren (1949), Muessig (1978), McMahon (2001) and Escorihuela & Rondon (2002) as decimetric xenoliths in the lava of Cerro Atravesado then considered as 'gabbroic' rocks. On close analysis they reveal complex metamorphic textures due to their polycyclic nature. Some samples have also been found in the Bocoy River area; however they form the bulk in Cerro El Guayabo. Its origin is igneous,

but rocks were metamorphosed to high grade in the granulite facies, and possibly to medium grade.

- *Plagioclase*: andesine–oligoclase (An32), anhedral, with complex twinning laws. It shows subophitic texture, being partially surrounded by the pyroxenes. Often saussuritic or altered to carbonates. Also antiperthitic and with inclusions of fibrous quartz.
- *Quartz*: anhedral, has a fibrous habit when included in plagioclase, often amoeboidal in isolated crystals.
- *Clinopyroxene*: hypidioblastic, altered to Feoxides and chlorite. It partially surrounds the plagioclase laths, and also it is included in the poikiloblastic hornblende crystals.
- *Orthopyroxene*: recognizable, though quite altered, in sample FA-53 form Riecito River (Fig. 7b). Brownish, sometimes altered to stilpnomelane or chlorite. In other samples its presence is quite doubtful.
- *Hornblende*: pleochroic from light to dark green. Forms large poikiloblasts that enclose grains of pyroxene and plagioclase. It only appears conspicuously in the mafic granulites of Cerro El Guayabo.
- *Titanite*: xenoblastic, associated with clinopyroxene and embedded in hornblende.
- *Apatite*: xenoblastic, associated with clinopyroxene and plagioclase; also embedded in hornblende.
- *Opaques*: magnetite; leucoxene, and limonite as alteration products of titanite and mafic minerals.

Clinopyroxenite

Numerous small bodies and boudins (possibly disrupted intrusive dykes or skarns) of clinopyroxenite cut the mangerite and enderbite of Cerro El Guayabo (Fig. 8a), generating millimetric reaction

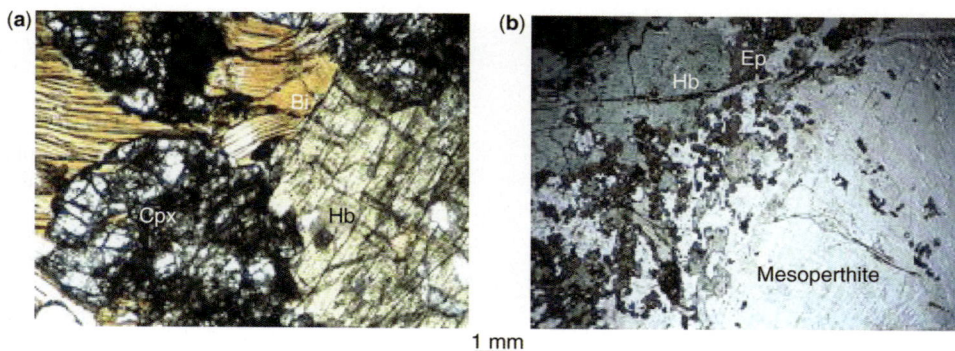

Fig. 8. (a) YA-250F, XP. Biotite–hornblende clinopyroxenite 'dyke' rock, Cerro El Guayabo, showing retrograde metamorphism to amphibolite facies, or perhaps a pervasive hydrothermal alteration. (b) YA-250-D2, UP. Metasomatic reaction zone between clinopyroxenite dyke and felsic granulite or charnockite, Cerro El Guayabo. Note the development of hornblende and epidote in the contact of both lithologies (clinopyroxene is located outside of the photomicrograph field, towards the upper left).

zones rich in hornblende and epidote probably due to the different chemistry of the ultramafic intrusives and the mafic or felsic host rock (Fig. 8b). Some show alteration to hornblende and/or biotite. Only one sample, YA-252A, has quartz associated with xenoblastic magnetite. Others contain abundant carbonate and epidote as products of alteration of calcic–plagioclase and clinopyroxene, but being in the contact between mafic granulite and marble they could also be metaskarns.

- *Clinopyroxene*: neutral, xenoblastic, with thin parallel exsolution lamellae of orthopyroxene. Cut by carbonate veinlets produced by alteration of the plagioclase embedded in it.
- *Hornblende*: light to moderate pleochroism in green hues, hypidioblastic, altered in its borders and cleavage planes to fibrous–acicular actinolite. Being itself a product of alteration of clinopyroxene occurs conspicuously in reaction zones with the enclosing granulites (charnockites), associated with epidote.
- *Biotite:* strongly pleochroic from orange-tan to colourless (Fe-poor, Ti rich) (Fig. 8a), micaceous to feathery habit. Dispersed and without orientation, which could mean a post-tectonic, or perhaps hydrothermal–metasomatic origin. Associated with mafic minerals and included in them.
- *Epidote*: light pleochroism in greenish hues, granular, associated with hornblende as an alteration product of clinopyroxene (Fig. 8b).
- *Quartz*: xenoblastic, scarce. Included in large crystals of clinopyroxene associated with magnetite.
- *Opaques*: leucoxene as product of alteration of pyroxene; magnetite associated with quartz in anhedral grains.

Metamorphic petrogenesis

Medium- to high-grade assemblages in these rocks have suffered low-grade retrograde metamorphism and/or low-temperature hydrothermal alteration. The lowest grade mineralogy was formed during retrograde metamorphism to the chlorite zone of the greenschist facies, or perhaps as low-temperature hydrothermal metasomatic reactions during tectonic emplacement in main fault zones probably during the Mid Cenozoic. The primary minerals, when they persist, are quite altered, or appear as pseudomorphs. This happens to forsterite, which has been totally serpentinized, to diopside and phlogopite, which are slightly chloritized, possibly to orthopyroxene which is also chloritized and altered to stilpnomelane, and to the saussuritized and carbonated (Na, Ca)-plagioclase. Alkali feldspar is tartan twinned microcline mesoperthite, but originally it could have been Na-orthoclase, later inverted and affected by exsolution at lower temperatures. Clinopyroxene of the presumed clinopyroxenite dykes (now boudins) in Cerro El Guayabo has suffered strong retrograde alteration to hornblende and biotite; the scarce plagioclase present is totally carbonated. Notwithstanding it is possible to recognize in these various rock types the following primary metamorphic mineral associations, which allow recognition of a specific metamorphic facies:

- Granulite facies

  (a) In carbonate rocks:

      Calcite + dolomite + diopside + phlogopite + quartz

      Calcite + dolomite + forsterite + diopside + phlogopite

      Calcite + quartz + diopside + K-feldspar

(b)   In pelitic-psammitic rocks (only in La Vela platform):

Orthoclase + quartz + garnet + sillimanite

Orthoclase + quartz + garnet

(c)   In quartz-feldspatic rocks:

Mesoperthite + plagioclase + quartz + clinopyroxene

Mesoperthite + quartz + clinopyroxene ± orthopyroxene

(d)   In mafic rocks:

Plagioclase + clinopyroxene + garnet + quartz

Plagioclase + clinopyroxene + hornblende ± biotite

(e)   In ultramafic rocks:

Clinopyroxene ± hornblende ± biotite

Clinopyroxene + plagioclase ± quartz ± magnetite

Cataclastic and hydrothermal effects

By Mid Cenozoic, during the Caribbean Orogeny (Pindell & Kennan 2007), these rocks suffered pervasive cataclastic deformation, probably accompanied by hydrothermal alteration or replacement reactions. The following cataclastic or hydrothermal effects are noted:

- diopside crystals in marbles with mortar, skeletal or porphyroclastic textures, sometimes substituted by carbonate, leaving residual in optical continuity (Fig. 9a);
- folded and deformed polysynthetic twins in carbonate of marble (Fig. 9b);
- calcite with mortar texture (Fig. 9b);

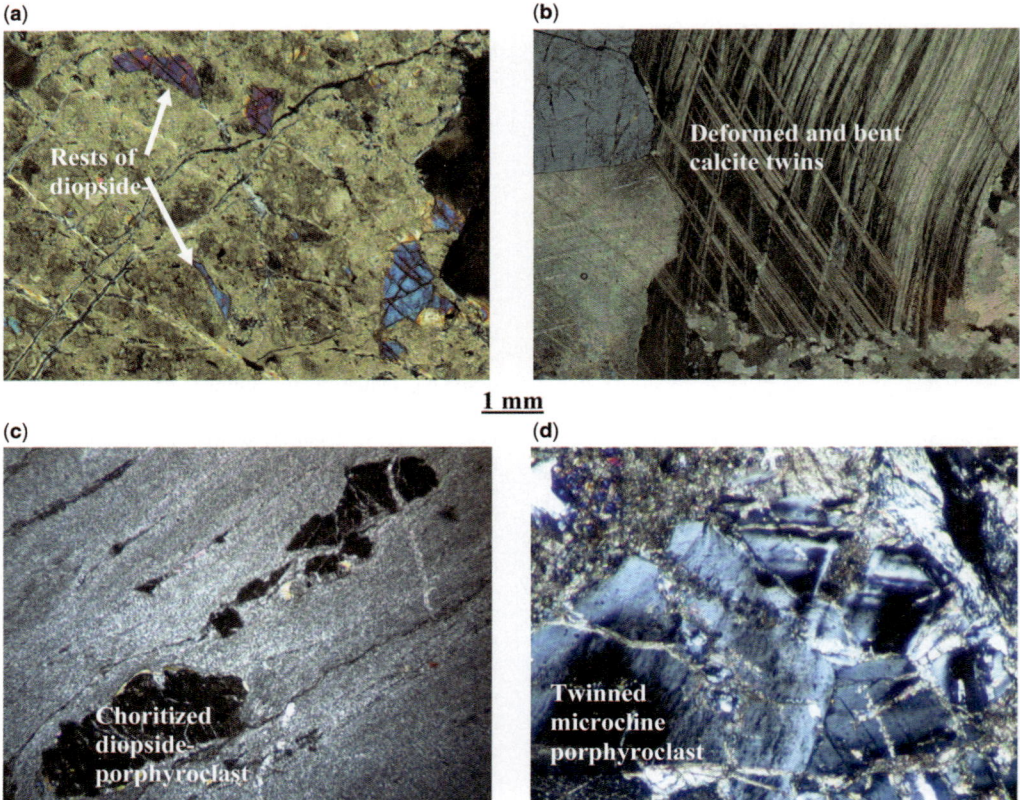

Fig. 9. Post-metamorphic effects. (**a**) PO-37, XP. Diopside replaced by carbonate leaving rests in optical continuity, Río Bocoy. (**b**) PO-61, XP. Carbonate with mortar texture and deformed polysynthetic twins, Río Bocoy. (**c**) YA-254B, XP. Mylonitic marble showing large porphyroclasts of chloritized diopside in a finely ground carbonate matriz. Some of these rocks can be considered as ultramylonites, Cerro El Guayabo. (**d**) YA-253B, XP. Mylonitic marble. Detail showing a fractured porphyroclast of microcline with tartan twinning (possibly inverted orthoclase).

- mylonitic rocks, almost ultramylonitic marbles, with diopside and K-feldspar 'augen' or porphyroclasts in a matrix of finely ground carbonate (Fig. 9c);
- bent phlogopite leaflets in marbles, with undulatory extinction (Fig. 6a);
- trainlets of quartz cutting mesoperthite crystals in mangerites and microcline porphyroclasts in marbles (Fig. 9d).

Presence of a medium- to high-grade basement in NW Venezuela

With these new findings we propose the existence in NW Venezuela of a previously unrecognized basement, or a mosaic of tectonic blocks, of high-grade rocks not related to the Imataca Supersuite, or any other unit of the Guiana Shield. These rocks can be correlated with similar lithologies and metamorphic assemblages of the Grenville Province. Such rocks also are recognized in the Oaxaca Massif, in Mexico, and units from Colombia to Argentina. The vicissitudes of plate displacements during the last 3 Ga have been described in the work of Rogers (1996), where he concludes that the present configuration of continental blocks results from a succession of supercontinents that underwent Wilson cycles every 500 Ma or so. The last of these continents was the well-known Jurassic Pangaea, but previously a large Late Proterozoic supercontinent, Rodinia, formed 750 Ma ago, about 0.5 Ga earlier (Fig. 10a).

Just as the formation of Pangaea involved the collision and terminal suturing of all continental masses present in the Palaeozoic, the assembly of Rodinia may have involved similar collision and suturing of all or most the continental blocks existing during the Early–Middle Proterozoic. Assembly of Rodinia concluded with the Grenville Orogeny, that of Pangaea concluded with the Caledonian–Appalachian Orogeny. All these processes of opening and closing of ocean basins are complete Wilson cycles, which generated rifts and passive margins, microcontinents, migratory volcanic arcs, active margins and collisional orogens (Hildebrand & Easton 1995). As these events happened on both sides of the present Atlantic Ocean, they can be referred as Circum-Atlantic tectonic processes.

Grenvillian Orogenic Belt

Rocks from the Grenvillian Orogen are widely distributed on the globe, from eastern Canada and USA, to southwestern USA (Texas–New Mexico), Oaxaca (Mexico), as inliers in the Andean cordillera from Argentina to Colombia (Cardona et al. 2005), and from dozens of other localities. It was a giant and global orogenic belt. The most recent reconstructions involve a length of more than 15 000 km, a mean width of 1000 km and a possible elevation above sealevel approaching 11 km (Fig. 10b).

The Grenville Orogen in eastern North America is a 2000 km-long, 500 km-wide metamorphic belt, built of three tectonic units separated by main thrust faults (Fig. 11). These are the Parautochthonous and the Allochthonous Polyclyclic belts and the Allochthonous Monoclyclic Belt (Carr et al. 2000).

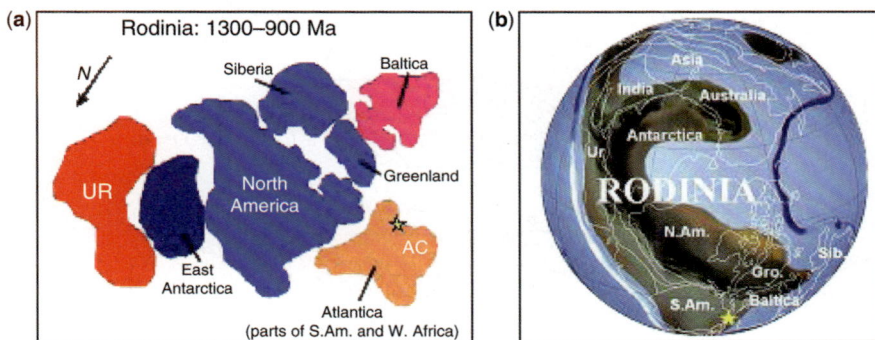

Fig. 10. (**a**) Reconstruction of the Late Proterozoic supercontinent of Rodinia. It encompassed all the continental blocks that existed at the time. Note the position of Atlantica (actual Amazonia–NW Africa, part of western Gondwanaland) and Laurentia (actual North America plus Baltic and Siberian cratons). The eastern margin of Atlantica (today's western South America) would contain parts or blocks of the Grenville Orogen, which is very well preserved in the Canadian Shield, in Ontario and Québec. (**b**) Visualization of the gigantic Grenvillian Orogenic Belt during the Late Proterozoic, just before it started rifting. The relation between the actual northwestern margin of South America and the Grenville Province in the actual eastern margin of North America is now obvious. The yellow star marks the relative position of the Guiana Shield, AC is the Amazon Craton. From Rogers (1996).

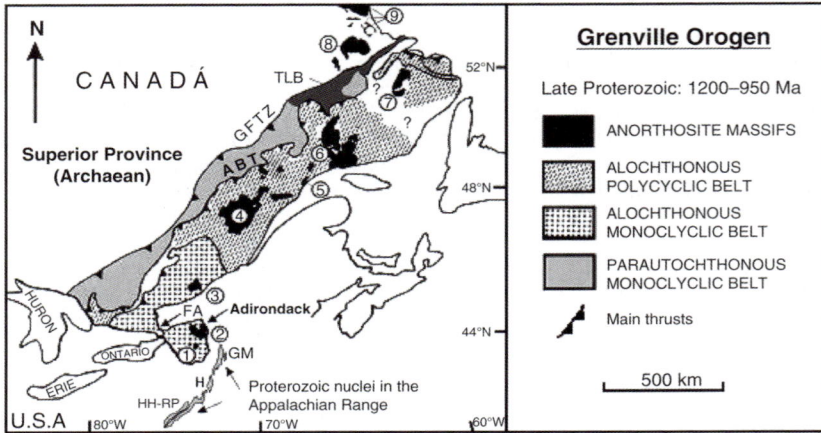

Fig. 11. Geological map of the Grenville Province in eastern North America, showing the three metamorphic belts that compose it. The high-grade rocks form El Guayabo and elsewhere in northwestern Venezuela could be related to the two easternmost belts of this orogen, including the anorthositic massifs. From Carr *et al.* (2000).

A proposed evolutionary model of this belt (Wasteneys *et al.* 1999) is shown in Figure 12. It involves multiple collisions between microcontinents and continental margins, where the extensive development of passive margins allowed the deposition of widespread carbonate platforms with stromatolithic reefs and colonies, subsequently metamorphosed to high-grade dolomitic marbles, which are the most defining lithology of this belt worldwide, together with other high-grade rocks.

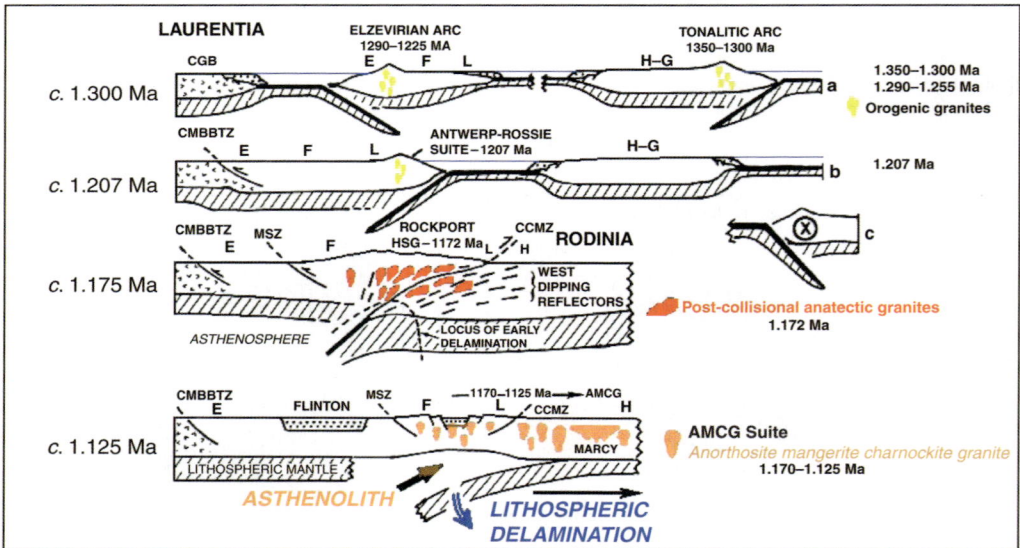

Fig. 12. Tectonic evolution model of the Grenvillian Orogen showing succesive stages. First the Elzevirian microcontinent (EFL) collides with Laurentia (CGB); a new subduction develops in the east in another microcontinent (L) generating a tonalitic arc in its easten margin (HG). The ocean basin is destroyed and this block collides with the previous composite suture (EFL). This last collision caused the total closure of the intervening ocean basins and formed the main Grenvillian Orogen and the Rockport postcollisonal anatectic granites. Delamination started in the subdued passive margin root (under FL), and proceeded to the east to L and H, generating the extensive Anorthosite–Mangerite–Charnockite–Granite Suite (AMCG Suite). Modified from Wasteneys *et al.* (1999).

The basement rocks of North and South America, including some pre-Caribbean microcontinental blocks such as Chortís and Maya, contain lithologies from this prodigious metamorphic belt. The common feature of worldwide Grenvillian rocks is their medium- to high-grade metamorphism, including rock types such as dolomitic marble, charnockite and other quite anhydrous rocks. To explain this profusion of charnockitic rocks Wasteneys *et al.* (1999) developed a model involving what is known as post-collisional lithospheric delamination.

This delamination or foundering of a continental lithospheric mantle root can occur in response to a major collisional event, that is, a macrocontinental suture, such as the Grenvillian, Caledonian, Uralian or Himalayan sutures. In these orogens the crust can have a root over 70 km thick, underlain by a lithospheric mantle root that reaches a depth of some 200 km. This cold, dense lithospheric root below the continental crust acts like an anchor in the continental lithosphere and eventually it can detach due to its higher density compared with underlying asthenosphere. Detachment is thought to start in the roots of the former passive margin, partially subducted, and to proceed in a foreland direction (Fig. 12). This slowly creates a space into which buoyant and hot asthenosphere flows, entering into contact with the base of the continental crust, causing rifting of the continental crust and a dramatic rise in isotherms.

This results in high-grade metamorphism and copious anhydrous magma generation. In this stage, along many parts of the Grenvillian Orogen in all continents, the AMCG Suite (anorthosite–mangerite–charnockite–granite/granophyre) formed some 1.15 Ga ago. This suite has been recognized in such separated places as Labrador (Canada), NE and SW USA, Ukraine, India, Siberia and various localities in western South America. Although supporting age data remain to be gathered, now, for the first time, this lithological suite has been identified in northwestern Venezuela. Although the rocks of the Yumare Anorthositic Metagabbro have been known for decades, and have been studied recently in detail, the complete lithological

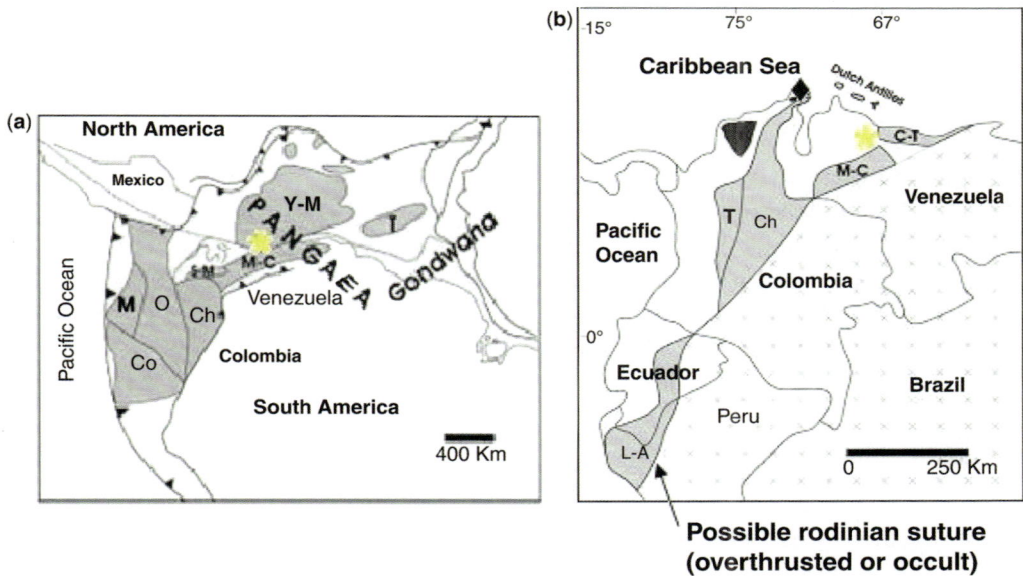

Fig. 13. Pre-Mesozoic terrains and continental basement exposures in northwestern South America and Caribbean region. (**a**) Reconstruction of Pangaea including microcontinental blocks. (**b**) Actual position of these blocks in the cordilleras in northwestern South America. Terrains: Santa Marta and Goajira (in black); L.A, Laja-Amotape (Southern Ecuador); T, Tahalí, Ch, Chibcha and S-M, Santa Marta (in Colombia); M-C, Mérida–Caparo; C-T, Caucagua–El Tinaco (Venezuela); Co, Chortís; Y-M, Yucatán–Maya; M, Mixteca; O, Oaxaquia (México). The Amazon Craton is shown cross hatched and it is not related with any of the mentioned terrains, which are Grenvillian or Appalachian. The Guiana–Brazil Shield should be sutured to a mosaic of Grenvillian blocks, which comprise the terrains (in grey) in the western margin of South America (in white and black), but the suture is covered or overthrusted by younger Andean and Caribbean foreland or back-arc belts. The star in (a) marks an inferred former position of the Yumare and El Guayabo blocks, and (b) its actual position. From Pindell & Kennan (2007).

association of dolomitic marble, high-grade meta-pelite, garnet granulite and clinopyroxenite has never been recognized before. An intimate relation must exist between the outcrops of Cerro El Guayabo and the nearby Yumare Anorthositic Metagabbro, since both tectonically emplaced blocks carry the complete AMCG suite, although quite dismembered and deformed.

The ubiquitous marbles are metasedimentary carbonate-rich rocks (marls?) that probably represent stromatolitic reefs or platforms, colonial carbonate algae as are known today in Australia (Papineau et al. 2005). They are known from Proterozoic South African and Australian greenstone belts where low deformation and very low-grade metamorphism preserve original organic structures. Colonies formed big, mushroom-shaped laminated bodies 1–2 m high, and 0.5–1 m in diametre. Inclusions of clay nodules in the insterstices between the algal colonies are frequent. Evidently in these rocks organic structures in high-grade metamorphism (granulite facies) became totally destroyed by recrystallization at very high temperatures (750–830 °C). In the Río Bocoy area the marble shows nodules or concentrations of pure phlogopite with a diametre of 1–2 cm (Fig. 2b). These could be the product of metamorphism of nodules rich in continental clays, such as K-illite, in an Fe-poor environment such as a carbonate platform.

Relation between the Grenville Orogen and northern South America

In his reconstruction of Rodinia (Fig. 10), Rogers (1996) shows that during the Late Proterozoic (750 Ma) South America was in a quite different position with respect to North America than today. 'South America' as a structured continental block formed during the Pan-African Orogeny, which concluded with the assembly of Gondwanaland, some 600 Ma ago. Rodinian South America was a fragment of Gondwana, made up of the Amazon Craton or the Guiana–Brazilian Shield. The reconstruction suggests that the western margin of present day South America (deprived of young orogens such as the Andes and the Caribbean ranges) was sutured to the Grenville Province, possibly forming part of the Allochthonous Monocyclic Belt (mainly metaigneous rocks) and the Allochthonous Polycyclic Belt (mainly metasedimentary rocks) (Fig. 11).

Cardona et al. (2005) and Pindell & Kennan (2007) (Fig. 13) proposed the existence of a suture, possibly ophiolitic, under the western margin of this continent. It lies, perhaps, buried below the young Andean Orogen and the imbricated

foreland nappes of that range, and the Caribbean Range nappes, in northwestern Venezuela. This occluded suture crops out in Rondonia State of SW Brazil, in the Sunsas Orogen. It is possible that the innermost unit of the Grenville Province (the Autochthonous Polycyclic Belt) is not represented in northern South America, because it is mostly recycled Archaean crust of the Superior Province. No Archaean rocks have been found so far in these regions. This discussion indicates an intimate relationship between the Grenville Orogen and the NW and western margins of South America.

A few hundred Ma after its assembly, Rodinia began to separate along a wide and branched system of rifts and fractures, generating the continental blocks that later would be sutured again, colliding in quite different directions, provoking at least three successive orogenies during the Palaeozoic: Taconic, Acadian and Appalachian (Alleghenian) of North America (Hatcher 1987; Fig. 14), and Caledonian, Variscan and Hercynian of Europe, respectively.

The Caledonian–Appalachian Orogen was not a minor mountain range, on a par with the Grenvillian Orogen, being just a little less global, that is Circum Atlantic. Notwithstanding, Caledonian–Appalachian age rocks have been found from northern South America (Mérida Andes), to Greenland, Scotland, Scandinavia, Southern Europe and NW

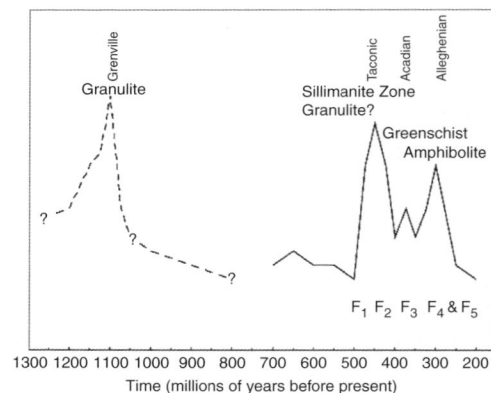

Fig. 14. Metamorphic events and facies in the Appalachian Mountains from the Middle Proterozoic to Mesozoic. Grenvillian rocks, included as basement blocks in this orogen, show characteristic granulite facies rocks; of the three Palaeozoic progenies shown, only the Taconic is suspected to have reached this high grade facies, but only locally, being more typical medium to low grade facies rocks in the Acadian and Alleghanian (Appalachian) orogens. Also shown are the five main deformation stages of the Appalachian Orogenic Belt. From Hatcher (1987).

Africa (Atlas Mountains). Early to Late Palaeozoic granitic rocks spanning all the three described orogenies are common in the Mérida Andes and Perijá Range (Martín Bellizzia, 1968; Cardona et al. 2005). They also crop out in the Venezuelan Cordillera de la Costa and Goajira and Paraguaná peninsulas. However, the Caledonian metamorphic belts, which locally could reach high-grade and even migmatization, are more characteristically recognized for their medium- to low-grade associations, that is, they are typically Barrovian, as described by John Barrow in the late nineteenth century in the Scottish Highlands. These are the famous and well-known Barrovian metamorphic zones, chlorite, biotite, almandine, staurolite, kyanite and sillimanite, that crop out extensively in the Iglesias Suite (upper amphibolite facies) and Tostós Suite (greenschist facies) in the Mérida Andes (Grauch 1972) (Mérida-Caparo Block in Fig. 13), which are also starting to be recognized in northern Venezuela.

Conclusions

The rocks described in this work, although lacking geochronological data, can be ascribed almost certainly to the Grenville Orogen, rather than to the Caledonian–Appalachian orogens, due to the pervasive presence of granulitic and charnockitic rocks and high-grade dolomitic marble, among other very high temperature, very low P_{H_2O} lithologies such as the AMCG Suite. Caledonian–Appalachian rocks are characterized by the almandine amphibolite and greenschist facies, formed in quite hydrated conditions with $P_{lit} \approx P_{H_2O}$ (Hatcher 1987). The observed alteration of clinopyroxenites and mafic granulites to hornblende and biotite could have possibly been an effect caused by a retrogression during the Caledonian–Appalachian orogeny on the high-grade assemblages of Grenvillian affinity. Other retrograde effects formed epidote, chorite and other low-grade minerals, but as yet it cannot be ascertained if they are of metamorphic, cataclastic or metasomatic origin. Some pyroxenites in contact with marbles could be pyroxene-rich metaskarns.

The high-grade rocks here described are not related in any sense to the high-grade rocks of the Archaean Imataca Supersuite, in the northern Guiana Shield of southern Venezuela. There is a dolomitic marble unit in that area, the Guacuripia Marble, but its lithological association is quite different. It includes Mn-rich metasediments and Banded Iron Formation, among other typically Late Archaean rocks. These rocks are some 2 Ga older than Grenvillian.

The presence of these proposed Grenvillian basement rocks, cropping out in Cerro El Guayabo, drilled in the basement of La Vela Gulf (Mendi et al. 2005), as a source of olistolithic blocks or conglomerate boulders in the Nuezalito, Matatere and Casupal formations, and revealed by xenoliths in lavas intruded into the basement of the Central Falcón Basin (Grande et al. 2007a, b; Grande 2008), suggests the existence of a Late Proterozoic continental block, or a mosaic of blocks, inside and to the north of the outstandingly triangular block delimited by the Oca, Boconó and Santa Marta faults (James 2006). Large outcrops of grenvillian rocks exist in central Colombia, besides the well-known outcrops of granulite and anorthosite in Los Mangos, in the western side of Sierra Nevada de Santa Marta (Ordoñez et al. 2002), dated at 1.2–0.9 Ga.

A boulder of felsic granulite was found in a conglomerate of the Soebi Blanco Formation in the island of Bonaire was dated by the U–Pb method giving an age of 1.15 Ga (Priem et al. 1986), typically Grenvillian, representing the granulite facies metamorphism. About the basement of the Falcón basin at that time the observations available were those of Muessig (1978) and Feo-Codecido et al. (1984) who said that only Mesozoic low-grade rocks were present; therefore they propose a 300 km eastward displacement of Bonaire with respect to South America.

However the recent findings of granulite facies in La Vela basement rocks and as xenoliths in the Cenozoic lavas of Falcon make this hypothesis not the only possibility. All this encourages us to propose the existence of a Proterozoic high-grade basement under most of Falcón basin, probably as far SE as northeastern Yaracuy. This basement could be of Grenvillian affinity, therefore with a Late Proterozoic age.

In this study we have described the petrology, petrography and some textural and field aspects of recently found rocks. Geochemical, isotopic and geochronological studies are necessary to improve knowledge about them. The mangerites are rich in zircon, which makes them ideal for U–Pb dating.

This work was possible thanks to a partial contribution of the GEODINOS G-2002000478 (UCV-FUNVISIS-FONACIT) project. Its appearance in this Special Publication of the Geological Society, London, was possible with the help of K. H. James. This paper benefited from critical reviews by K. H. James, J. L. Pindell and A. Kerr.

References

BRUEREN, J. 1949. Geological Report CPMS-310. Paraiso-Manaure (Center Falcón). Compañía Shell de Venezuela, unpublished report.

CARDONA, A., CORDANI, U. & MACDONALD, W. 2005. Tectonic correlations of pre-Mesozoic crust from the northern termination of the Colombia Andes, Caribbean region. *Journal of South American Earth Sciences*, **21**, 357–354.

CARR, S. D., EASTON, R. M., JAMIESON, R. A. & CULSHAW, N. G. 2000. Geologic transect across the Grenville orogen of Ontario and New York. *Canadian Journal of Earth Sciences*, **37**, 193–216.

CHANNER, D. M., EGOROV, A. & KAMINSKY, F. V. 2001. Geology and structure of the Guaniamo diamondiferous kimberlite sheets, south-west Venezuela. *Revista Brasileira de Geosciencias*, **31**, 615–630.

ESCORIHUELA, N. & RONDÓN, J. 2002. Estudio de las rocas ígneas presentes en el centro de la Cuenca de Falcón. *Geos, UCV, Caracas*, **37**, 58–59 + 247 pp. on CD.

FEO-CODECIDO, G., SMITH, F. D., ABOUD, N. & DI GIACOMO, E. 1984. Basement and Paleozoic rocks of the Venezuelan llanos basins. *In*: BONINI, W. E., HARGRAVES, R. B. & SHAGAM, R. (eds) *The Caribbean-South American Plate Boundary and Regional Tectonics*. Geological Society of America, **162**, 175–187.

GRANDE, S. J. 2009. Estudio petrográfico de los xenolitos presentes en las lavas del cerro Atravesado, Falcón central. *Revista de la Facultad de Ingeniería, UCV, Caracas*, in press.

GRANDE, S. J., URBANI, F. & MENDI, D. 2007a. Mármoles de alto grado probablemente grenvillianos en el noroeste de Venezuela. *Geos, UCV, Caracas*, **39**, 128–129.

GRANDE, S. J., URBANI, F. & MENDI, D. 2007b. Presencia de un basamento Grenvilliano de alto grado en el noroeste de Venezuela. *Memorias IX Congreso Geológico Venezolano, UCV, Caracas*, in *Geos, UCV, Caracas*, **39**, 90 + 59 pp. on DVD.

GRAUCH, R. 1972. A Late Palaeozoic metamorphism in Central Venezuelan Andes (short version). *Memorias VI Conferencia Geológica del Caribe, Margarita, Venezuela*, 304–306.

HACKLEY, P., URBANI, F. & GARRITY, C. 2005. *Geological shaded relief map of Venezuela 1:750.000*. United States Geological Survey, Open File Report 2005–1038. World Wide Web Address: http://pubs. usgs.gov/of/2005/1038.

HATCHER, R. D. JR. 1987. Tectonics of the Southern and Central Appalachian Internides. *Annual Reviews in Earth and Planetary Sciences*, 337–362.

HILDEBRAND, R. S. & EASTON, R. M. 1995. An 1161 Ma suture in the Frontenac terrane, Ontario segment of the Grenville orogen. *Geology*, **23**, 917–920.

JAMES, K. H. 2006. Arguments for and against the Pacific origin of the Caribbean Plate: discussion, finding for an inter-American origin. *In*: ITURRALDE-VINENT, M. A. & LIDIAK, E. G. (eds) *Caribbean Plate Tectonics. Geological Acta*, **4**, 279–302.

LOZANO, F. & MUSSARI, A. 2008. Geología de los macizos ígneo-metamórficos del norte de Yumare, estados Yaracuy y Falcón. *Geos, UCV, Caracas*, **40**, 286.

MARTÍN-BELLIZZIA, C. 1968. Edades isotópicas de rocas Venezolanas. *Bol. Geol. Minis. Minas. e Hidrocarb.*, **X**, 356–379.

MCMAHON, C. E. 2001. *Evaluation of the effects of oblique collision between the Caribbean and South American plates using geochemistry from igneous and metamorphic bodies of Northern Venezuela*. PhD thesis, Notre Dame University, Indiana, USA. [Reprinted in *Geos, UCV, Caracas*, **39**, 195–196 + 227 pp. on DVD, 2007.]

MENDI, D., CAMPOSANO, L. & BAQUERO, M. 2005. Petrografía de rocas del basamento de la ensenada de la Vela. Notas de avance. *Geos, UCV, Caracas*, **38**, 32–33 + presentation of 42 slides on CD.

MUESSIG, K. W. 1978. The central Falcon igneous suite, Venezuela: alkaline basaltic intrusions of Oligocene–Miocene age. *Geologie en Mijnbouw*, **57**, 261–266.

ORDOÑEZ, O., PIMENTEL, M. & MORAES, R. (2002). Granulitas de Los Mangos, un fragmento grenvilliano en la parte oriental de la Sierra Nevada de Santa Marta. *Revista de la Academia Colombiana de Ciencias Exactas, Físicas y Naturales*, **26**, 169–179.

PAPINEAU, D., WALKER, J. J., MOJZSIS, S. & PACE, N. R. 2005. Composition and Structure of Microbial Communities from Stromatolites of Hamelin Pool in Shark Bay, Western Australia. *Applied Environmental Microbiololgy*, **71**, 4822–4832.

PINDELL, J. L. & KENNAN, L. 2007. Cenozoic kinematics and dynamics of oblique collision between two convergent margins: the Caribbean–South America collision in Eastern Venezuela and Trinidad and Barbados. *In*: KENNAN, L., PINDELL, J. L. & ROSEN, N. (eds) *The Paleogene of the Gulf of Mexico and Caribbean Basins; Processes, Events, and Petroleum Systems. Proceedings, 27th Bob F. Perkins Research Conference*, Gulf Coast Section of the Society of Economic Paleontologists and Mineralogists, Houston, TX, 458–553.

PRIEM, H. N. A., BEETS, D. J. & VERDURMEN, E. A. TH. 1986. Precambrian rocks in an Early Tertiary conglomerate in Bonaire, Netherlands Antilles (southern Caribbean borderland): evidence for a 300 km eastward displacement relative to South American mainland? *Geologie en Mijnbouw*, **65**, 35–40.

ROGERS, J. 1996. A History of the continents in the past three billion years. *Journal of Geology* **104**, 91–107.

SKERLEK, G. M. 1976. *Geology of the Acarigua area*. PhD thesis, Princeton University. [Reproduced in *Geos, UCV, Caracas*, **39**, 199–200 + 315 pp. on DVD.]

VALLETTA, G. & MARTÍNEZ, J. 2008. Petrografía de las facies gruesas de la Formación Matatere y otras unidades del centro-occidente de Venezuela. *Geos, UCV, Caracas*, **40**, 277.

VÁSQUEZ, E. 1975. Results of explorations in La Vela bay. *Proceedings IX World Petroleum Congress*, Chichester, UK, **3**, 195–197.

WASTENEYS, H., MCLLELLAND, J. & LUMBERS, S. 1999. Precise zircon geochronology of the Adirondack Lowlands and implications for revising plate tectonic models of the Central Metasedimentary Belt and Adirondack Mountains, Grenville Province, Ontario and New York. *Canadian Journal of Earth Sciences*, **36**, 967–984.

The Cabo de la Vela Mafic–Ultramafic Complex, Northeastern Colombian Caribbean region: a record of multistage evolution of a Late Cretaceous intra-oceanic arc

M. B. I. WEBER[1]*, A. CARDONA[2], F. PANIAGUA[1], U. CORDANI[3],
L. SEPÚLVEDA[1] & R. WILSON[4]

[1]*Universidad Nacional de Colombia, Calle 65 No. 78-28, Facultad de Minas,
M1-324, Medellín, Colombia*

[2]*Smithsonian Tropical Research Institute, Balboa, Ancón, Panama – ECOPETROL,
Piedecuesta, Colombia*

[3]*Institute of Geoscience, USP, Rua do Lago 562, Cidade Universitária,
05508-080 São Paulo, Brazil*

[4]*Department of Geology, University of Leicester, University Road, Leicester LE1 7RH, UK*

*Corresponding author (e-mail: mweber@unalmed.edu.co)

Abstract: Ophiolite-related rocks accreted to Caribbean Plate margins provide insights into the understanding of the intra-oceanic evolution of the Caribbean Plate and its interaction with the continental margins of the Americas. Petrological, geochemical and isotope (K–Ar, Sr and Nd) data were obtained in serpentinites, gabbros and andesite dykes from the Cabo de la Vela Mafic–Ultramafic Complex from the Guajira Peninsula, in the northernmost Colombian Caribbean region. Field relations, metasomatic alteration patterns and whole rock–mineral geochemistry combined with juvenile isotope signatures of the different units suggest that gabbros and serpentinites formed in a slow-spreading supra-subduction zone that was brought to shallower depths and subsequently evolved to an arc setting where andesitic rocks formed with little sediment input. The tectonomagmatic evolution of the Cabo de la Vela Mafic–Ultramafic Complex involved an intra-oceanic arc that evolved from pre-Campanian time to 74 Ma. Relationships with other units from the Guajira Peninsula show either the existence of a mature arc basement or a series of coalesced allochthonous arcs, juxtaposed before accretion onto the passive continental margin of South American in pre-Eocene times.

Ophiolitic rocks are fundamental to understanding of the complex dynamics and evolution of oceanic crust and its interaction with continental margins to form so-called Cordilleran-type orogens (Shervais 2001; Beccaluva *et al.* 2004).

In the Caribbean area, studies of ophiolitic complexes have identified remnants of rift margins, ocean plateaus and arc systems formed in oceanic domains that evolved since the Mesozoic, following the break-up of Pangaea (Giunta *et al.* 2002, 2006). However, different models show lack of concensus. Testing of regional models through local relationships has often not been successful, possibly due to lack of detailed information (Iturralde-Vinent & Lidiak 2006 for discusions).

A few ophiolite-related rocks have been described from the Colombian Caribbean (MacDonald 1964; Lockwood 1965; Alvarez 1967); however, their formation setting and their implications for the geotectonic models of the Caribbean have not been discussed.

We present integrated petrological, mineral chemistry, whole rock geochemistry, Nd and Sr isotopes and K–Ar geochronology data from the Cabo de la Vela Mafic–Ultramafic Complex of the southwestern Caribbean. They allow reconstruction of a Late Cretaceous, arc-related magmatic system that was accreted to the continental margin of South America in pre-Eocene times. The implication of this and other available regional data is discussed within the framework of the Caribbean Plate tectonic evolution.

Geological setting

The Meso-Cenozoic history of the Colombian Andes in the Caribbean region is characterized by interaction between the northwestern border of

From: JAMES, K. H., LORENTE, M. A. & PINDELL, J. L. (eds) *The Origin and Evolution of the Caribbean Plate*.
Geological Society, London, Special Publications, **328**, 549–568.
DOI: 10.1144/SP328.22   0305-8719/09/$15.00 © The Geological Society of London 2009.

South America and the Caribbean and Nazca plates (Fig. 1). Multiple plate boundaries and Meso-Cenozoic transpressive tectonics controlled accretion of oceanic terranes to the continental margin. Discrete tectonic blocks record this complex evolution (Pindell 1993; Toussaint 1996; Montes *et al.* 2005).

The Guajira Peninsula, in northeasternmost Colombia (Fig. 2), is characterized by several isolated massifs with correlatable geology, surrounded by broader flat lands and Cenozoic basins (Mac-Donald 1964; Lockwood 1965; Alvarez 1967). Within these massifs at least three main lithotectonic belts can be identified. From oldest to most recent they include the following elements (Fig. 2):

1. A composite, Late Mesoproterozoic and Palaeozoic metamorphic domain, which includes medium and high-grade units. This is intruded by Jurassic magmatism, similar to the parautochthonous basement of the Andes (Cordani *et al.* 2005; Cardona-Molina *et al.* 2006).

2. A weakly deformed belt of Mesozoic sedimentary rocks with the same depositional characteristics and ages as the autochthonous South American margin (Villamil 1999).

3. A sequence of two different, Cretaceous, low-grade metavolcano-sedimentary metamorphic units. The northernmost unit has intercalated mafic and ultramafic plutonic rocks, intruded by Eocene magmatism (MacDonald 1964; Lockwood 1965; Alvarez 1967; Pindell 1993).

Fragments of high-pressure metamafic and meta-sedimetary rocks in Miocene conglomerates on the northwestern fringe of the central massif (Fig. 2) may represent a probably Mid-Cretaceous exhumed subduction-accretion complex (Green *et al.* 1968; Zapata *et al.* 2005).

An association of mafic and ultramafic rocks occurs in the Cabo de la Vela region, isolated from the main massifs and close to the zone of Caribbean–South American plate interaction. The following sections describe the geological and geochemical characteristics of this mafic–ultramafic

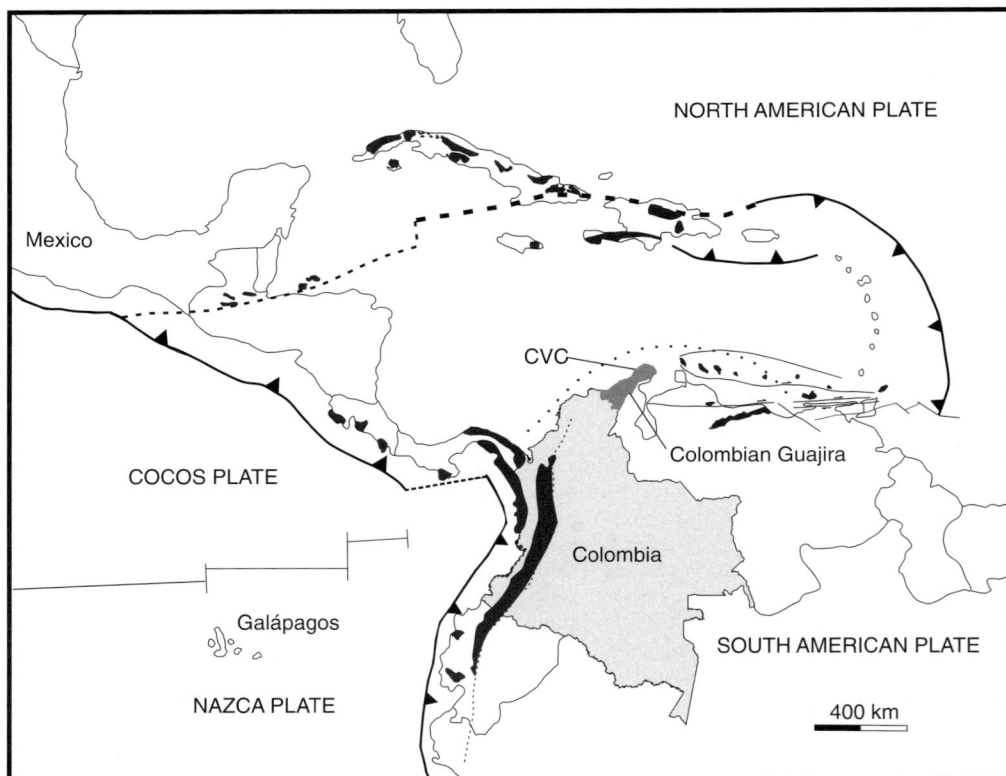

Fig. 1. Tectonic framework and location of the Caribbean region. The Guajira Peninsula and the CVC are shown. The distribuiton of ophiolite rocks within the Caribbean region is included (data from: Guinta *et al.* 2002; Lewis *et al.* 2006).

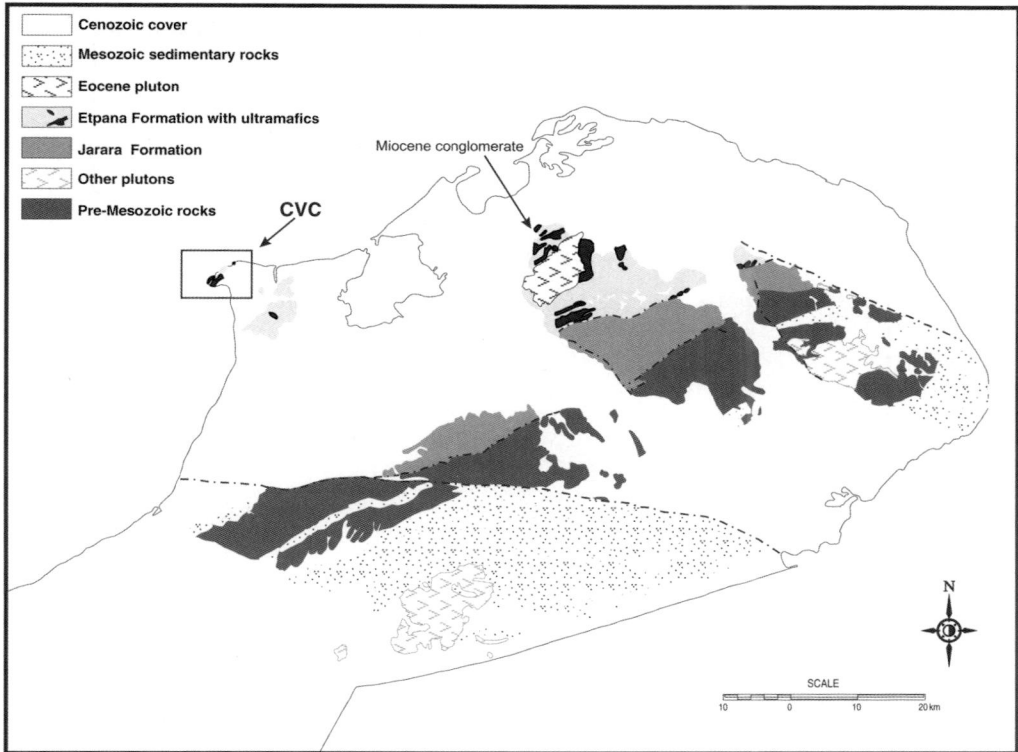

Fig. 2. Simplified geological map of the northern Guajira Peninsula, showing major lithostratigraphic units.

complex and discuss their significance to interpretations of tectonomagmatic evolution.

Cabo de la Vela Mafic–Ultramafic Complex

The Cabo de la Vela is a cape, built of a series of isolated small hills (150 m) where different mafic and ultramafic units are exposed (Alvarez 1967). In general they include a sequence of serpentinites, gabbros and mafic volcanic units. The serpentinite ultramafic unit contains gabbro and leucograbbros as lenses and pods, cross-cut by basalt dykes (Fig. 3). Miocene sediments locally overlie this sequence (Alvarez 1967).

To the SE, a lower and flat-land region of Cenozoic sediments extends for 5 km, separating the Cabo de la Vela from the massifs of the Guajira region (MacDonald 1964; Lockwood 1965; Alvarez 1967; Fig. 2). Similarities between the plutonic mafic and ultramafic rocks of the Cabo de la Vela region and the Cretaceous units of the massifs suggest a geological link between them (review in Alvarez 1967).

The Bouguer gravity map of northern Colombia (Fig. 4) shows a positive anomaly centred on the Cabo de la Vela, extending 100 km offshore to the northwest and 30 km onshore to the southeast (Kellogg et al. 1991). This anomaly is consistent with a NW–SE elongated dense body, with a narrow and gently sloping northwestern face and a broad and steep southeastern flank, and suggests wide extension of this mafic–ultramafic rock complex (Nieto & Ojeda, pers. comm., 2006).

The following section introduces the informal stratigraphic term Cabo de la Vela Mafic–Ultramafic Complex (CVC) and describes the field and petrographic elements of the constituent units.

Ultramafic rocks

Ultramafic rocks of the Cabo de la Vela area are mainly serpentinites. The only exposure of a partially serpentinized rock (sample FP-34C) shows meso-scale foliation, with original grains of olivine, spinel and clinopyroxene. Bastite minerals suggest the former presence of another pyroxene

M. B. I. WEBER *ET AL.*

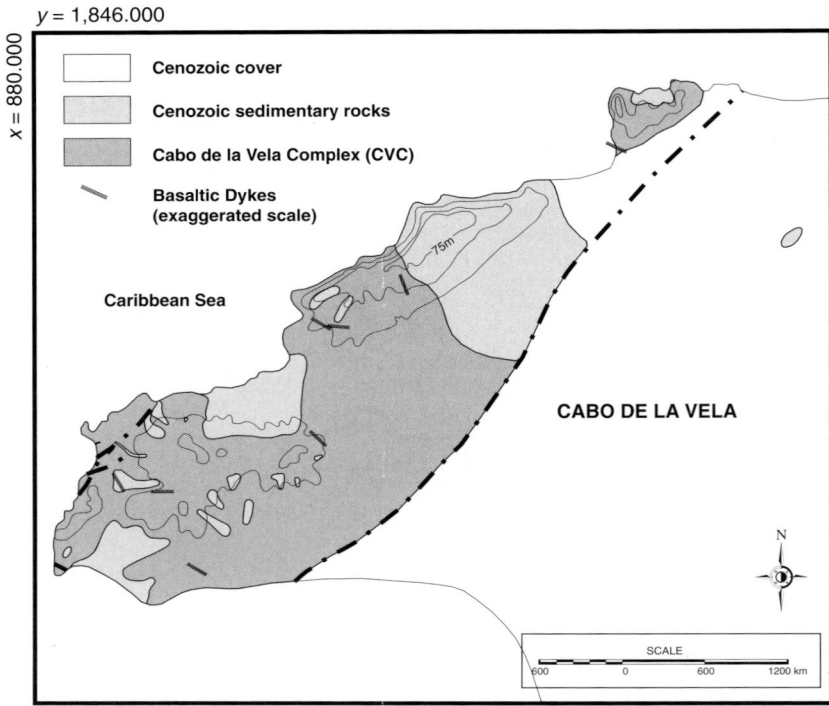

Fig. 3. Geology of the Cabo the Vela area.

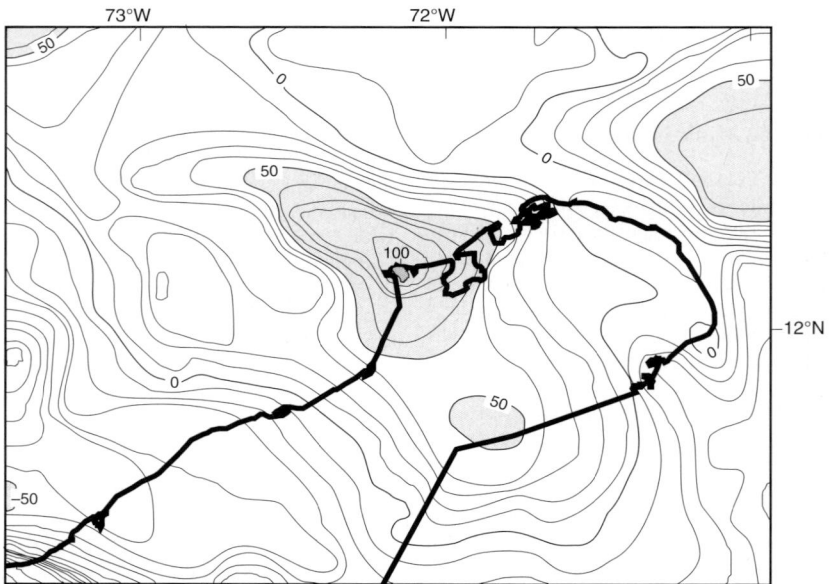

Fig. 4. Bouguer gravity map of the Guajira area (Kellogg *et al.* 1991).

(orthopyroxene?), which indicates that the original ultramafic rock was a wherlite. Relict textures in this rock are characterized by larger grains surrounded by fine-grained material and by deformed cleavages, indicating that the ultramafic protolith was a mantle tectonite (Fig. 5) (Nicolas & Rabinowicz 1984).

The serpentinites are mainly greenish to brownish rocks, in many cases mottled by the silky-looking bastite minerals. Microscopic analysis shows pseudomorph textures, such as antigorite mesh and lizardite–antigorite hourglass. Non-pseudomorphic textures include interpenetrating and intergrowth textures, both in the same sample

Fig. 5. (a) Gabbro intrusions in serpentinite. (b) Continuous basaltic andesite dyke within serpentinite. The arrow indicates the direction of the dyke. (c) Serpentinized wherlite (sample FP-34C) with relict olivine (Ol). Note the presence of bastite (Bst), as replacement of pyroxenes. (d) Gabbro tectonite. Note the amphibole inclusions in the central pyroxene (Cpx) crystal, and secondary amphibole on the recrystallized pyroxene grain margins (Hbl). The two upper pyroxenes and plagioclase (plg) have been completely recrystallized and have a granoblastic polygonal texture. (e) Basaltic Andesite with twinned pyroxene (Cpx) phenocryst in a hornblende (Anf) and saussuritized plagioclase (Sauss) matrix.

and both defined by antigorite. Other local minerals formed during serpentinzation are bastite, magnetite, magnesite and brucite. Relict minerals are brown spinel and occasional amphibole. The presence of amphibole indicates a hydrated mantle protolith.

Bastite may replace clinopyroxene, with opaques defining relict bend cleavages, and perhaps orthopyroxene without opaques. The presence of two pyroxenes and the foliated texture suggests that the protolith was also a wehrlitic tectonite.

In addition, there are at least four generations of veining, formed by either perpendicular or parallel to the wall opening precipitation of crysotile.

Gabbros and hornblendites

Gabbroic rocks occur in the ultramafites as small, discontinuous, irregular bodies, pods and lenses with maximum dimensions of 1 m × 100 m. They are coarse-grained to pegmatitic, and contain dark green pyroxene (diopside) and plagioclase (labradorite) crystals up to 15 cm in size.

These rocks generally show well-defined mineral banding and lineation. Most show evidence of deformation and in thin section the overall texture is granoblastic polygonal due to deformation–recrystallization of large crystals to smaller polygonal crystals at the edges (Fig. 5). The various degrees of recrystallization always involve more plagioclase than pyroxene, which is sometimes preserved in large crystals. These relict grains have numerous oriented inclusions of small brown amphibole blebs that impart a characteristic lustre to hand-specimen. The euhedral shape of the blebs, their colour and the fact that they formed before high temperature deformation all indicate magmatic origin (Coogan *et al.* 2001). Evidence of deformation of the larger grains before recrystallization is conspicuous, marked by bent and kinked cleavages and by alignment of the small amphibole blebs that disappear with recrystallization.

The fact that pyroxene and plagioclase ductile deformation and recrystallization annealed with granoblastic textures, without changes in the original mineralogy assemblage, is indicative of deformation at granulite–amphbolite facies conditions (Seyler *et al.* 1998).

Hornblendites are found generally on the edges of the gabbro dykes and sometimes as small metre-sized patches within the serpentinites. They are made up exclusively of coarse-grained dark brown and less commonly bright green amphibole. Field observations show that these rocks formed by complete replacement of pyroxene and plagioclase of the gabbros and possibly the ultramafics. Several stages of replacement are present. Early stages are seen as coronas around pyroxenes but at more advanced stages the reaction front replaces all of the minerals (Fig. 5). Petrographically these rocks show decussate textures and the nature of the replacement of these hornblendites indicates that they formed through metasomatism during annealing and therefore after deformation.

Gabbros and hornblendites have been rodingitized throughout the whole complex, and transitional rocks are common. Massive rodingites comprise Ca-rich minerals such as chlorite ± hydrogrossular ± vesuvianite ± epidote ± albite ± tremolite-actinolite ± prehnite. Some rodingites are zoned, with a chlorite margin, in which the relicts of the original pyroxene can still be identified. The mineral assemblages suggest that these rocks formed at greenschist facies and prehnite–pumpellyite facies conditions (Frost 1975; Dubińska 1995; Früh-Green *et al.* 1996).

The transition in facies, as well as different styles of complex superimposed deformation and fracturing during fluid penetration, indicate that rodingitization occurred from ductile to brittle conditions.

Mafic volcanic dykes

Dykes are continuous and undeformed and reach up to 2 m in width, cutting the previously described lithologies. They constitute fine-grained, phaneritic to aphanitic, green to grey-green rocks, generally with porphyritic texture.

Plagioclase (andesine) and clinopyroxene (diopside) with intergranular to subophitic texture are seen in thin section (Fig. 5). Subhedral crystals of clinopyroxene and plagioclase comprise the phenocrystal phase. Several of these rocks also contain brown amphibole as a matrix mineral and as uralitic replacement of pyroxene. The more felsic rocks contain quartz in the matrix. All minerals are generally altered to saussurite and chlorite and pumpellyite is common.

The suggested crystallization order would be pyroxene and plagioclase–opaques–amphibole–quartz, and petrographically may resemble basalts or andesites.

Analytical techniques

Whole rock geochemistry

Eleven samples were analysed for major and trace elements by X-ray fluorescence (XRF) and inductively coupled mass spectrometry (ICP-MS) at the Mineralogy and Petrology Department of the Institute of Geosciences of the University of Sao Paulo and at ACME analytical laboratories in Canada.

Samples were crushed with an iron steel crusher and pulverized in an agate mill. Sample preparation for XRF included microreduction to obtain pressed powder pellets, and fused glass discs for major and trace element determination. Major and selected trace element analysis were carried out in a wavelength dispersive Philips PW 2400 XRF spectrometer with detection limits generally of the order of 1–10 ppm for trace elements, following the methodology described by Mori *et al.* (1999). For other trace elements including rare earths (REE), the analyses were carried out by ICP-MS at ACME analytical laboratories (Canada), after lithium metraborate/tetraborate fusion and nitric acid digestion of a 0.2 g sample.

Mineral chemistry

Mineral analyses were obtained from carbon-coated polished thin sections using a Jeol 8600S electron microprobe at the Department of Geology, University of Leicester, UK, using an accelerating voltage of 15 kV and a probe current of 30 nA with a beam diameter of 5–10 μm. Quantitative background-corrected results were standardized against a combination of synthetic materials and well-characterized natural minerals and corrected for matrix effects using a ZAF correction procedure. Minimum detection limits under the analytical conditions used range from 0.01 wt% for Na_2O to 0.04 wt% for FeO.

K–Ar geochronology

Three whole rock samples were analysed by the K–Ar method at the Centre of Geochronological Research of the University of São Paulo (CPGeo-USP). Two aliquots from the same sample were separated for the K and Ar analysis. Potassium analyses of each pulverized sample were carried out in duplicate, coupled to an ultra-vacuum system. A spike of ^{38}Ar was added and the gas was purified in titanium and copper ovens. Final argon determinations were carried out in a Reynold-type gas spectrometer. Analytical precision for K is of 5% whereas for Ar it is around 0.5%. Decay constants for calculation are after Steiger & Jager (1977).

Nd–Sr isotopes

Six whole rock samples were analysed by Sm–Nd and Rb–Sr methods at the Centre for Geochronological Research of the University of São Paulo (CPGeo-USP). For the Sm–Nd method the analytical procedures followed Sato *et al.* (1995). Isotopic ratios $^{143}Nd/^{144}Nd$ were obtained in a multi-collector mass spectrometer Finnegan Mat, with analytical precision of 0.0014% (2σ). Experimental error for the $^{147}Sm-^{144}Nd$ ratios is of the order of 0.1%. La Jolla and BCR-1 standards yielded $^{143}Nd/^{144}Nd = 0.511849 \pm 0.000025$ (1σ) and 0.512662 ± 0.000027 (1σ) respectively during the period in which the analyses were performed. The ε_{Nd} were calculated following De Paolo (1988), and the constants used include $^{143}Nd/^{144}Nd$ (CHUR) $= 0.512638$ and $^{147}Sm-^{144}Nd$ $(CHUR)_0 = 0.1967$.

Rb–Sr analyses followed procedures presented by Tassinari *et al.* (1996). Rb and Sr values were obtained by X-ray fluorescence, and $^{87}Sr/^{86}Sr$ ratios were done with mass spectrometer VG-sector mass spectrometer and corrected for isotopic fractionation during thermal ionization with a $^{87}Sr/^{86}Sr = 01194$.

Geochemistry

Whole rock geochemistry and mineral chemistry data were obtained from selected samples in order to understand its tectonic setting of the CVC. Data analyses are presented in Table 1.

Mineral chemistry

Four samples, a partially serpentinized a peridotite, a serpentinite, a gabbro and a basalt, were selected for mineral chemical analysis. Analysed minerals were spinel, olivine, pyroxene, plagioclase and amphibole.

Spinel

Dark brown spinel in the partially serpentinized sample (FP-34C) and in serpentinites is armoured by magnetite grains, possibly formed during the serpentinization process. Original spinel has an allotriomorphic, interstitial texture. Preserved spinel cores have Cr/Cr + Al composition ranges from 0.5 to 0.6 and Fe/Fe + Mg around 0.5. TiO_2 compositions vary from 0.07 to 0.14 wt% and Al_2O_3 cluster around 25 wt%. In the Cr/Cr + Al v. Mg/Mg + Fe diagram (Fig. 6a) the CVC data do not fall within the abyssal spinel peridotite data of Dick & Bullen (1984) but do overlap the spinel data from the Mariana fore-arc of hole 780, ODP-Leg 125 from Parkinson & Pearce (1998).

Pyroxene

Pyroxene is seldom preserved in the ultramafic rocks. Only relict pyroxenes were found in sample FP-34C, where composition is Fe-rich diopside (Fig. 6b) and TiO_2 was found to be very low.

Clinopyroxene from an analysed gabbroic sample is mainly diopside, falling within the same compositional area as the relict pyroxenes from the serpentinites (Morimoto *et al.* 1988) with high

Table 1. *Representative mineral analyses*

Spinel

Sample Type	LS-32 SER	FP-34C PER
SiO$_2$	0.00	0.03
TiO$_2$	0.11	0.05
Al$_2$O$_3$	25.52	26.28
Cr$_2$O$_3$	39.05	41.75
FeO	23.36	22.39
MnO	0.32	0.13
MgO	10.86	11.13
CaO	0.00	0.00
NiO	0.11	
K$_2$O	0.00	
NiO	0.00	
Total	99.32	101.80
Si	0.00	0.01
Ti	0.02	0.01
Al	7.44	7.47
Cr	7.64	7.96
V	0.00	0.00
Fe(iii)	0.88	0.53
Fe(ii)	3.95	3.99
Mn	0.07	0.03
Mg	4.00	4.00
Ca	0.00	0.00
Zn	0.00	0.00
Total	24.00	24.00
Cr/Cr + Al	0.51	0.52
Fe/Fe + Mg	0.55	0.53
Mg/Mg + Fe	0.45	0.47
Cr/Cr + Al	0.51	0.52

Olivine

Sample	FP-34C PER border	FP-34C PER centre
SiO$_2$	40.16	40.48
TiO$_2$	0.00	0.00
Al$_2$O$_3$	0.01	0.00
Cr$_2$O$_3$	0.00	0.00
FeO	8.69	9.37
MnO	0.11	0.11
MgO	49.88	50.03
CaO	0.02	0.03
NiO	0.05	0.01
K$_2$O	0.00	0.02
NiO	0.42	0.43
Total	99.33	100.48
Si	0.99	0.99
Ti	0.00	0.00
Al	0.00	0.00
Cr	0.00	0.00
Fe(ii)	0.18	0.19
Mn	0.00	0.00
Mg	1.83	1.82
Ni	0.01	0.01
Ca	0.00	0.00
Total	3.01	3.01
Fo	91.00	90.39
Fa	8.89	9.50
Tp	0.11	0.12

Pyroxene

Sample	FP-34C PER	LS22 GBR	LS11 AND
SiO$_2$	53.26	53.51	51.74
TiO$_2$	0.00	0.08	0.08
Al$_2$O$_3$	1.50	1.59	1.47
Cr$_2$O$_3$	0.36	3.23	12.59
FeO	1.88	0.00	0.00
MnO	0.06	0.10	0.40
MgO	17.95	16.70	9.31
CaO	24.34	24.07	23.45
NiO	0.19	0.44	0.89
K$_2$O	0.00	0.00	0.00
Total	99.56	99.75	99.92
Si	1.94	1.96	1.98
Al	0.06	0.04	0.02
Al	0.00	0.03	0.04
Fe(iii)	0.03	0.06	0.06
Cr	0.05	0.00	0.00
Ti	0.00	0.00	0.00
Fe(ii)	−0.01	0.04	0.34
Mn	0.00	0.00	0.01
Mg	0.97	0.91	0.53
Ca	0.95	0.94	0.96
Na	0.01	0.03	0.07
K	0.00	0.00	0.00
Al tot	0.06	0.00	0.00
Total	4.09	4.02	4.02
Wo	48.68	47.48	48.74
En	49.96	45.84	26.93
Fs	0.66	5.10	20.98
Ac	0.70	1.57	3.35

Amphibole

Sample	LS18 HBL	LS22 GBR	LS11 AND
SiO$_2$	47.46	49.40	51.51
TiO$_2$	1.04	0.70	0.42
Al$_2$O$_3$	11.55	9.63	5.63
Cr$_2$O$_3$	0.02	0.04	0.00
FeO	5.15	5.08	12.18
MnO	0.11	0.06	0.40
MgO	18.46	18.78	7.11
CaO	12.39	12.85	18.74
NiO	2.06	1.75	2.58
K$_2$O	0.06	0.05	0.01
NiO	0.08	0.11	0.00
Total	98.37	98.44	98.58
Si	6.65	6.90	7.53
Ti	0.11	0.07	0.05
Al	1.91	1.58	0.97
Cr	0.00	0.00	0.00
Fe$_2$	0.60	0.59	1.49
Mn	0.01	0.01	0.05
Mg	3.85	3.91	1.55
Ca	1.86	1.92	2.94
Na	0.56	0.47	0.73
K	0.01	0.01	0.00
Ni	0.01	0.01	0.00
Total	15.57	15.48	15.30

SER, Serpentinite; PER, serpentinized peridotite; HBL, hornblendite; GBR, gabbro; AND, andesite.

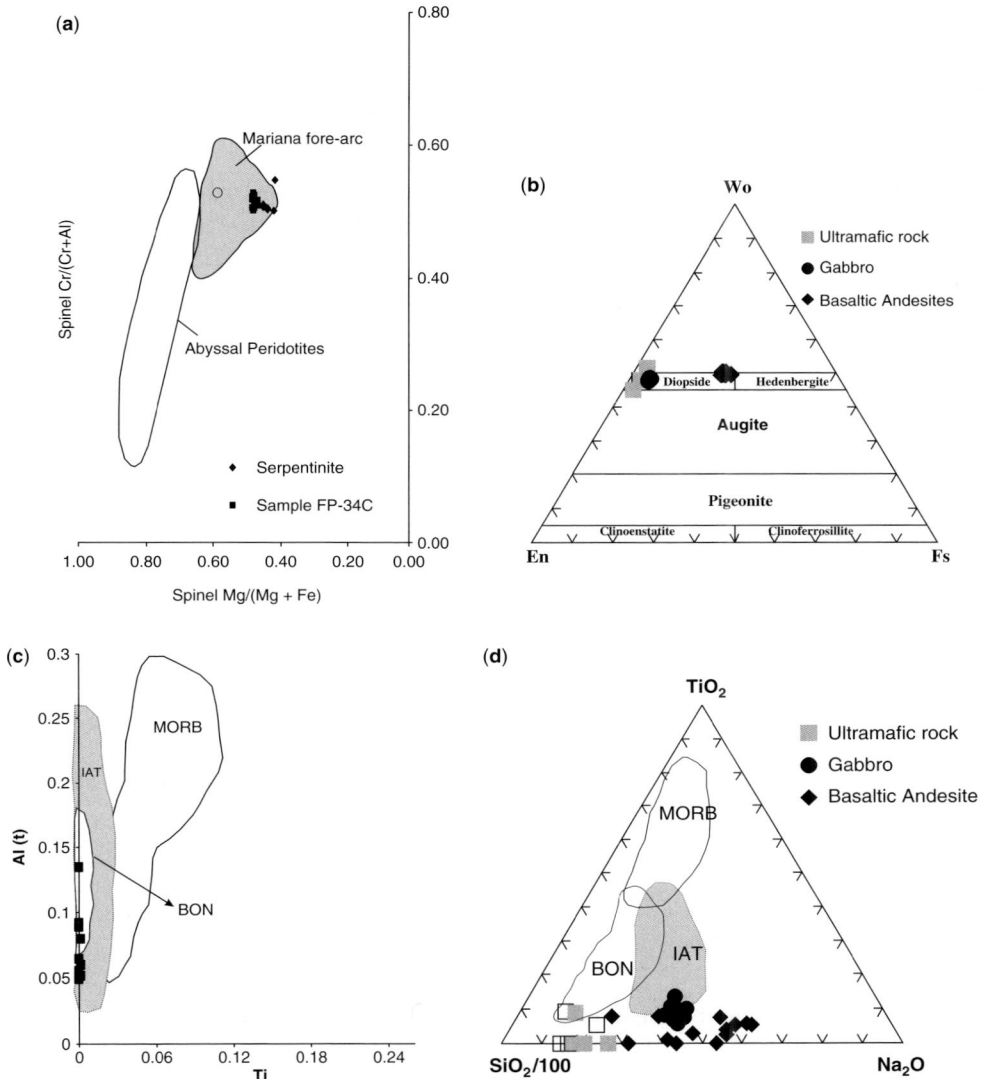

Fig. 6. (**a**) Cr# [Cr/(Cr + Al)] v Mg# [Mg/(Mg + Fe)] relation in spinels from a completely serpentinized and one partially serpentinized rock (sample FP-34C) of the CVC. For comparison, the spinel composition from peridotites with different refractory character are also plotted. Field for spinels from abyssal spinel peridotites (from Dick & Bullen 1984) and Izu–Bonin–Mariana forearc (hole 780, ODP-Leg 125, Parkinson & Pearce 1998) are shown; open circle subduction-related Oman ophiolite (Lippard *et al.* 1986). (**b**) Clinopyroxene classification for the CVC ultramafics, gabbros and basaltic dykes. (**c**) Co-variation diagram of Al(t) v Ti (atomic ratios) of studied pyroxenes in ultramafic rocks, indicating their tectonic settings. (**d**) TiO$_2$-Na$_2$O-SiO$_2$/100 in clinopyroxene (wt%) (IAT, Island Arc Tholeiite; BON, boninite; MORB, mid-ocean ridge basalt; after Beccaluva 1989).

Mg no. between 0.90–0.91, high SiO$_2$ and extremely low TiO$_2$.

Compositional patterns of magmatic clinopyroxene are consistently used to infer possible tectonic settings of ophiolitic mafic and ultramafic rocks

(e.g. Nisbet & Pearce 1977; Capedri & Venturelli 1979; Hebert & Laurent 1987; Beccaluva *et al.* 1989).

In bivariant diagrams that use TiO$_2$ in clinopyroxene as a marker of the degree of fusion and different tectonic settings, the Cabo de la Vela gabbroic rocks plot within island arc fields (Fig. 6b, c). Together with the low TiO$_2$ content of pyroxene from ultramafic rocks this suggests a supra-subduction setting.

Other minerals

Relict olivine is preserved within mesh cells from sample FC-34C. The composition is Fo90–91. Plagioclase compositions for the gabbros were determined petrographically, and comprise mainly labradorite–andesine. Chemical analyses for feldspar in some volcanic samples indicate spilitization for these rocks, as plagioclase is replaced by albite.

Brown amphibole of metasomatic origin, which replaces minerals in the gabbros, and amphiboles from one of the mafic dykes were analysed, both being magnesiohornblendes (Hawthorne 1981).

Major and trace element geochemistry

Five volcanic rocks, together with three gabbros and three hornblendites representative of the main rock types of the CVC, were selected for major and trace element analyses.

Element data are presented in Table 2. Element analyses for LILES in the gabbros are in general below detection limits.

SiO$_2$ values range from 45.87 to 52.67 wt% for gabbros, from 47.61 to 53.64 wt% for hornblendites, and cluster around 52.5 wt% for the mafic dykes. MgO wt% values are variable but are highest in the hornblendite samples (18.79–22.12%) and lowest for the mafic rocks (4.88–5.37%).

According to the immobile elements (Nb, Y, Zr and Ti) classification of Winchester & Floyd (1977) (not shown), all these rocks are of mafic intermediate composition and plot in the andesite field.

Since most of the samples show signs of alteration, chemical interpretation is based on high field strength elements and transitional metals considered to be immobile (Pearce & Cann 1973; Winchester & Floyd 1977).

In the V v. Ti/1000 of Shervais (1982), all samples plot towards the arc-related rocks field, whereas in the Ti v. Zr plot after Pearce (1982), the basaltic andesite dyke samples fall within the volcanic arc field (figures not shown). Gabbros and hornblendites have particularly low Zr.

Additional constraints on the geotectonic setting can be obtained from REE and multi-element diagrams presented in Figure 7b, c. Gabbros show a strongly depleted LREE pattern compared with HREE when plotted on the chondrite-normalized REE diagram (Fig. 7b), with Nd–Lu values ranging from 6.67 to 11.00. They have a notable positive Eu anomaly, indicating that plagioclase

fractionation was an important factor in the genesis of these rocks. Gabbro REE patterns roughly resemble modern MORB-type plutonic rocks. The multi-element diagram shows negative Zr and Ti anomalies as well as positive Sr and Ba spikes, indicative of a subduction-related component in the source.

Volcanic dykes show flat patterns, with a small negative Eu anomaly, when plotted on the chondrite-normalized REE diagram. Apparent slight variations in the LREE, with La–Lu ranging from 9.06 to 21.94, indicate small degrees of differentiation. This is in agreement with the petrographic observations, where the LREE enriched samples contain amphibole and quartz in addition to pyroxene and plagioclase.

The multi-element diagram for basaltic andesites shows a decrease from LILE-enriched to HFSE-depleted and contrasts with the gabbros. The LILE enrichment is generally attributed to element mobility during alteration; however, the increased contents of relatively immobile elements like Th and La suggest that this pattern is inherited from the original magmatic source. The basaltic rocks also show Nb, Zr and a weak Ti negative anomaly, which together with positive Sr and Ba anomalies and the already mentioned pattern, characterize subduction-related magmas.

Hawkesworth *et al.* (1993*a*, *b*), subdivided island arc basalts into two groups on the basis of LREE/HREE, using La–Yb ratios to discriminate between predominantly intra-oceanic arcs (La–Yb < 5) and arcs developed near continental margins (La–Yb > 5) (Fig. 8). The CVC basaltic rocks fall within the low La–Yb island arc group and overlap the data from the Mariana arc.

K–Ar geochronology

Previous age constraints from this region were restricted to stratigraphical relationships with Miocene sediments. Three andesitic dykes were dated by the K–Ar whole rock method. Obtained ages (Table 3) overlap within error. Their differences may be related to minor hydrothermal alteration shown by saussuritization of plagioclase. We consider *c.*74 Ma to be the age of dyke intrusion into the mafic–ultramafic unit based on the more precise analytical quality. Low K contents in the hornblendites preclude reliable K–Ar data to constrain the pre-dyke intrusion history. However, cross-cutting relations and depth of formation indicate that the gabbroic and ultramafic rocks must have been exhumed before intrusion of the dykes, and therefore a significant lapse of time must have passed between the formation of these two different units.

Table 2. Major and trace element analyses

Sample	LS-6	LS-18	LS-22	LS-26	LS-34	LS-44	LS-45	LS-51	LS-53	LS-59A	LS-62
SiO_2	49.03	47.61	45.87	52.91	52.49			52.03		52.67	53.64
TiO_2	1.011	1.092	0.302	0.732	0.892			0.384		0.521	0.372
Al_2O_3	7.31	10.09	14.75	14.95	16.15			16.6		13.23	4.29
Fe_2O_3	7.65	5.83	3.1	11.21	10.89			6.5		9.22	3.7
MnO	0.116	0.071	0.053	0.181	0.19			0.11		0.183	0.07
MgO	18.95	18.79	14.61	5.37	4.88			7.83		11.5	22.12
CaO	11.69	12.04	14.58	8.17	8.34			10.39		11.11	12.42
Na_2O	1.03	1.44	1.19	4.33	4.74			4.41		2.19	0.7
K_2O	0.06	0.02	0.02	0.79	0.58			0.08		0.03	0.01
P_2O_5	0.016	0.007	0.006	0.095	0.144			0.01		0.012	0.007
LOI	2.5	2.21	4.37	1.89	1.39			1.98		0	2.17
Total	99.36	99.2	98.85	100.63	100.69			100.32		100.67	99.5
Type	HBL	HBL	GBR	BA	BA	BA	BA	GBR	BA	GBR	HBL
K	498	166	166	6558	4815	–	–	664	–	249	83
Ba	3.9	2.1	9.1	272.6	553.8	264.7	506.9	18.6	539.7	7	0
Rb	0.7	0	1.9	11	8.4	3.4	8.9	1	7.8	0	0.6
Sr	14.6	29.8	269.6	219.5	293.9	188.7	283.7	240.6	272.2	111.7	5.1
Cs	0	0	0	0.7	0.4	0.2	0.3	0.3	0.2	0	0
Ga	12.7	13	12.3	17.7	19.2	15.9	18.9	17.7	16.6	13.3	4.4
Ta	0.4	0.1	0	0.6	0.3	0.3	0.3	0.5	1.1	0.9	0.5
Nb	2.2	0	0	0.5	1.3	13.8	1.3	0	1.4	0	0
Hf	1.4	0	0	1.1	1.8	1.7	1.4	0	1.7	0	0.5
Zr	36.6	9.4	3.1	36.9	51.2	45.4	51.7	5.9	48.2	10.4	9.3
Ti	6061	6547	1810	4388	5348	–	–	2302	–	3123	2230
Y	24.9	18.8	6.3	17.5	19.3	23.7	20	9.4	22	13.8	10.9
Th	0.7	0	0	0.4	1.2	0.4	1.7	0	1.2	0	0
U	0	0	0	0.2	0.5	0.3	0.6	0	0.3	0	0
Cr	733	390	132	38	40	0	0	74	0	502	5054
Ni	129.1	91.9	68.6	16.9	17.1	7.9	16.9	73.4	13.2	53.2	26.5
Co	71.1	69	32.3	51.1	42.8	88.2	43.5	55.1	81.1	83.4	52.5
Sc	41	60	38	37	31	–	–	35	–	40	37
V	277	422	149	324	259	308	262	200	366	233	261
Cu	0.6	3.4	2	117.3	91.3	108	90.7	51.1	114.9	30.7	0.4
Pb	0	0	0	0.8	1.3	2.6	1.1	0	0.5	0	0
Zn	7	4	3	65	57	54	55	10	63	4	3
W	118.1	154.1	66.6	511.4	199.7	3671.9	212.1	392.6	333.4	901.3	651.8
Mo	0.3	0.2	0.1	0.9	0.4	5	0.4	0.4	0.6	0.7	0.5

(Continued)

Table 2. *Continued*

Sample	LS-6	LS-18	LS-22	LS-26	LS-34	LS-44	LS-45	LS-51	LS-53	LS-59A	LS-62
Cl	1425	548	351	160	23	0	0	732	0	279	342
La	2.5	0	0	2.9	6.8	3.8	6.4	0.5	4	0	0.5
Ce	9.3	1.2	1	7.6	14.2	8.6	13.5	1.4	10.1	1.4	1.7
Pr	1.55	0.34	0.15	1.24	1.95	1.44	2.03	0.26	1.49	0.3	0.31
Nd	7.8	1.6	0.8	6.6	9.3	7.3	10.3	1.3	7.2	2.2	1.9
Sm	2.7	1.3	0.4	2.1	2.9	2.5	2.5	0.8	2.6	1	0.9
Eu	0.81	0.39	0.28	0.66	0.86	0.75	0.91	0.55	0.77	0.58	0.42
Gd	3.67	2.21	0.69	2.81	3.14	3.63	3.28	1.1	3.53	1.65	1.43
Tb	0.73	0.41	0.15	0.48	0.55	0.6	0.53	0.25	0.59	0.34	0.28
Dy	4.43	3.06	1.09	2.86	3.32	3.59	3.09	1.4	3.09	2.18	1.6
Ho	0.92	0.69	0.23	0.69	0.73	0.88	0.72	0.39	0.84	0.56	0.38
Er	2.25	1.84	0.64	1.7	1.78	2.08	1.92	1.07	2.18	1.41	1.1
Tm	0.41	0.25	0.1	0.27	0.26	0.36	0.29	0.15	0.33	0.24	0.19
Yb	2.95	1.96	0.64	1.96	2.16	2.53	2.12	1.09	2.38	1.55	1.16
Lu	0.43	0.24	0.11	0.32	0.31	0.36	0.29	0.14	0.35	0.2	0.15

HBL, hornblendite; GBR, gabbro; BA, basaltic andesite.

Sr–Nd isotopes

Andesitic rocks and gabbros were analysed for Rb–Sr and Sm–Nd isotopic ratios. Initial ratios were calculated for the 74 Ma K–Ar age. Results are presented in Table 3 and Figure 9. The ε_{Nd} and initial $^{87}Sr/^{86}Sr$ values for the basaltic andesite dykes range between 4.1 and 7.3 and 0.7038 and 0.7041 respectively. Gabbros have ε_{Nd} of 9.9 and initial $^{87}Sr/^{86}Sr$ of 0.7029 and 0.7031. When compared with the basaltic dykes, the gabbros overlap the MORB field (Fig. 9), indicating derivation from a depleted source, whereas the basaltic dykes plot similar to primitive island arc rocks found in other suites in the circum-Caribbean realm. Isotopic variation in the basalt dykes could be explained by different input of sedimentary components into the subduction-zone magmas. Interestingly, the Mariana back-arc samples plot between the gabbros and the basaltic dykes from the CVC, which could be interpreted as a higher input of the sedimentary component for the basaltic dykes in the CVC. This is further seen by slight off-set towards higher Sr ratios of the CVC and the Caribbean plutons when compared with the Mariana back-arc basalts.

When compared with the highly negative ε_{Nd} data from pre-Mesozoic basement and Jurassic plutons from the Guajira region (Cordani *et al.* 2005; Cardona-Molina *et al.* 2006), the data fall away from older crustal signatures. Therefore assimilation of older crust is precluded. Unpublished isotopic data from the Eocene continental-arc magmatic rocks intruding Cretaceous metamorphic complexes of the Guajira region, possibly related to the CVC, show mixing with older basement rocks or sediments, but this is not seen in the CVC rocks.

These isotopic characteristics clearly indicate a primitive mantle source and a minor sediment input in the subduction-related magma source of the basaltic rocks. The differences between the isotopic ratios of gabbros and basaltic dykes may be explained by different input at source. The gabbros formed by partial fusion of mantle in a suprasubduction-zone environment, with limited input from the subducted slab. Input from the subducting plate was greater in basaltic andesites and shifts the isotopic ratios towards lower ε_{Nd} and higher Sr, probably reflecting a changing tectonic configuration.

Tectonomagmatic setting of the Cabo de la Vela Complex

Based on the lack of units typical of complete ophiolite sequences and the unclear relation with the margin where it was accreted, the Cabo de la Vela

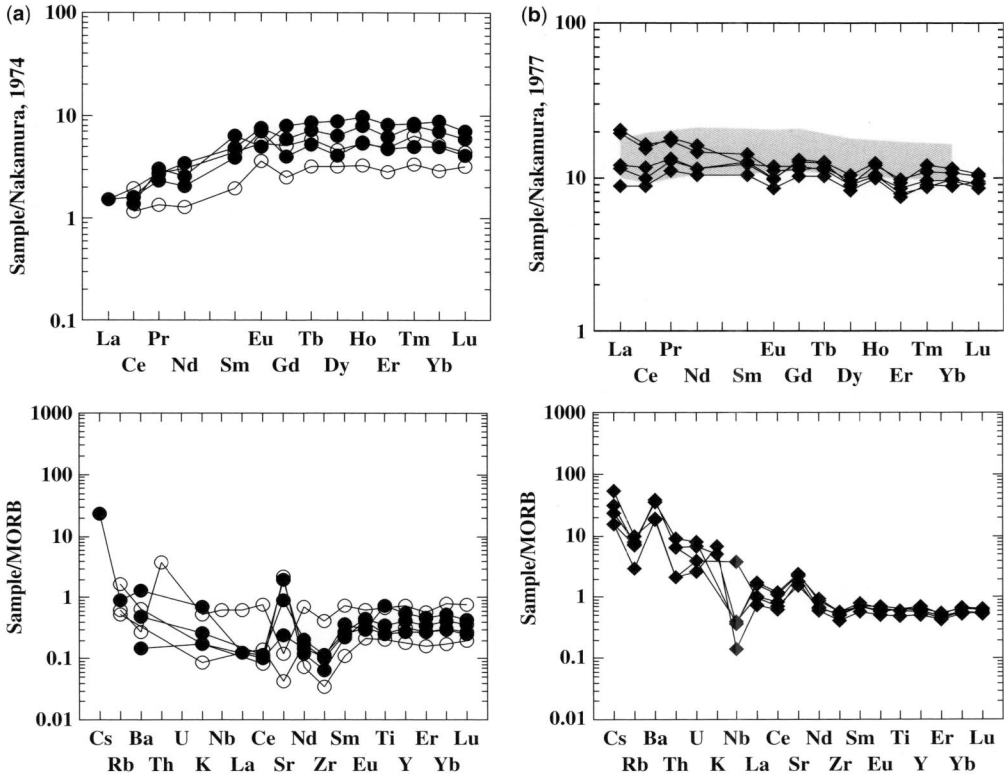

Fig. 7. Geochemistry of gabbros from the CVC. Normalizing values are from Nakamura (1974) and Pearce (1983). (**a**) REE patterns normalized to chondrite and multi-element patterns normalized to MORB for CVC gabbros. Open circles represent hornblendites, closed circles represent gabbros. (**b**) REE patterns normalized to chondrite and multi-element patterns normalized to MORB for CVC basaltic andesite dykes. The grey fields outline the compositional range of the southern Mariana Trough (Gribble *et al.* 1996).

complex represents a Cordilleran-type ophiolite, as do many other Caribbean ophiolites (Beccaluva *et al.* 1996, 2004).

Field and petrological evidence indicate that the Cabo the la Vela rocks followed a complex succession of events, recording the dynamic oceanic tectonic cycles common of ophiolite rocks (Shervais 2001).

The oldest unit comprises mantle rock of wherlitic composition. Coarse-grained gabbros and troctolites intrude this peridotite, recording continuous uplift in a slow spreading setting. Both units show evidence of correlatable high temperature deformation as well as hydration. Mineral chemistry of spinel and pyroxene (Figs 6a, b) from peridotites and gabbros indicates that these units correspond to a tectonite formed in a supra-subduction zone environment (Kenemetsky *et al.* 2001; Okamura *et al.* 2006). The MORB geochemical and more juvenile isotopic signature from the gabbros

combined with the supra-subduction signature from the mineral chemistry are more akin to a back-arc tectonic setting where both MORB or subduction zone signatures are common (Saunders & Tarney 1984).

The presence of hornblendites and different generations of pervasive serpentinization and rodingitization events are indicative of several hydrothermal overprints ranging from deep to shallow crustal depths and recording a continuous tectonic exhumation history in an oceanic setting. This, together with lateral heterogeneity of hydrothermal alteration, may be taken as evidence of a slow spreading ridge environment (Cannat *et al.* 1992; Cannat 1996). Absence of the other typical ophiolitic units, like cumulate gabbros and intrusive basaltic sheeted dykes, may be explained by erosion or tectonic removal after emplacement onto the margin, but nevertheless fits nicely into the evidence of a slow spreading ridge environment for these rocks.

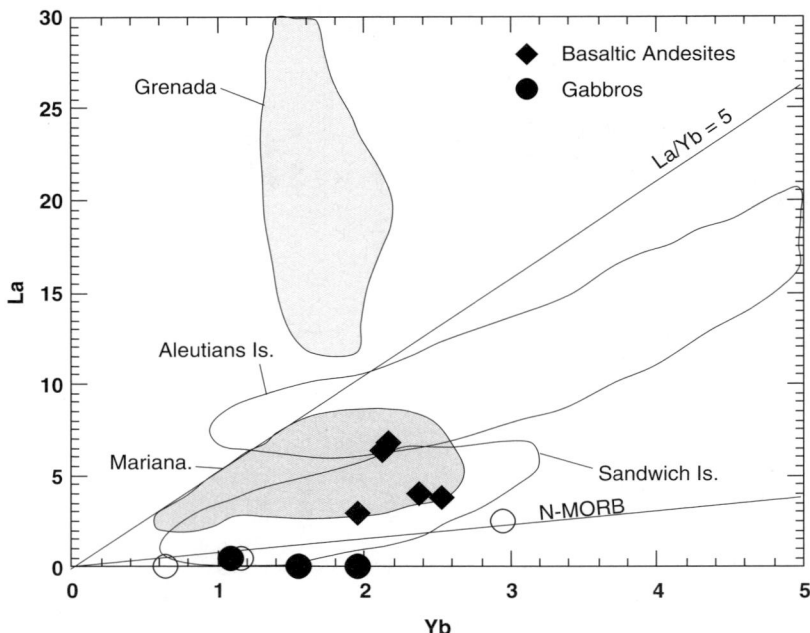

Fig. 8. La v. Yb in basalts (SiO$_2$ < 55%) in intra-oceanic and continental margin island arc basalts, after Hawkesworth *et al.* (1993*a*, *b*). Other data fields from Jolly *et al.* (2006) and Taylor & Martinez (2003).

Table 3. *Geochronological and isotopical data*

K–Ar Sample	Rock	% K	K error (%)	Ar^{40} Rad ccSTP/g ($\times 10^{-6}$)	Ar^{40}Atm (%)	Age (Ma)	± (Ma)
LS-34	Basalt	0.5013	0.5000	1.46	71.94	73.5	4.2
LS-44	Basalt	0.1711	0.5000	0.47	68.79	69.5	4.9
LS-53	Basalt	0.5162	1.7499	1.58	77.84	77.3	5.4

Sm–Nd Sample	Rock	Sm (ppm)	Nd (ppm)	^{147}Sm/^{144}Nd	Error	^{143}Nd/^{144}Nd	Error	f$_{Sm/Nd}$	ε (74Ma)
LS-51	Gabro	0.661	1.459	0.2739	9	0.513183	17	0.39	9.90
LS-22	Gabro		0.761			0.513228	12	− 1.00	
LS-34	Basalt	0.680	8.892	0.0462	2	0.512774	41	− 0.76	4.05
LS-44	Basalt	2.462	7.340	0.2028	7	0.512950	8	0.03	6.02
LS-26	Basalt	1.986	6.057	0.1983	7	0.513011	13	0.01	7.25

Rb–Sr Sample	Rock	Rb (ppm)	Sr (ppm)	Rb^{87}/Sr^{86} (X)	Error	Sr^{87}/Sr^{86} (Y)	Error	^{87}Sr/^{86}Sr 74 Ma
LS-51	Gabro	0.76	254.29	0.0087	1	0.70291	6	0.7029
LS-22	Gabro	15.28	228.79	0.1932	125	0.70338	7	0.7032
LS-34	Basalt	8.12	272.99	0.0861	7	0.70386	10	0.7038
LS-44	Basalt	3.00	166.68	0.0520	4	0.70396	7	0.7039
LS-26	Basalt	10.58	211.38	0.1448	12	0.70406	1	0.7041

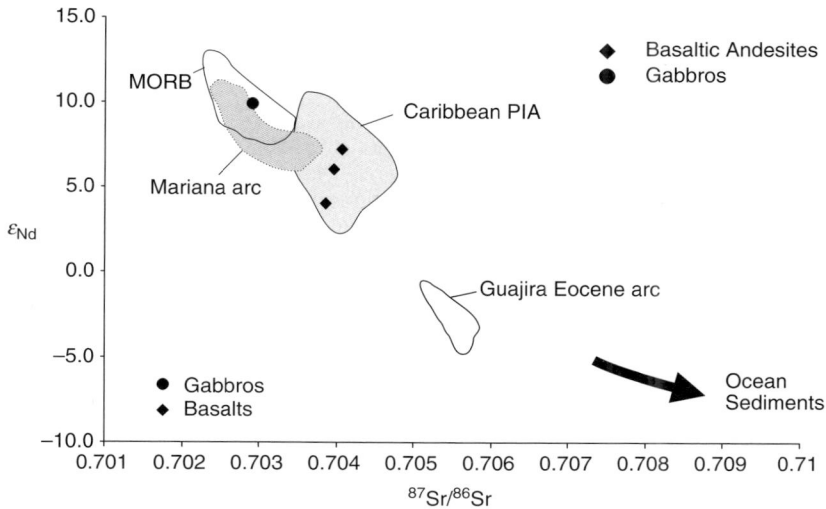

Fig. 9. ε_{Nd} v. $^{87}Sr/^{86}Sr$ for gabbros and basalts of the CVC. Other fields depicted are from the Mariana arc (Gribble *et al.* 1996 and sources listed therein), MORB (White & Hofmann 1982) and Caribbean PIA (Jolly *et al.* 2006). Also shown is the unpublished data field for the Eocene arc magmatism.

The history of the andesitic dykes differs from the previously discussed ultramafics–gabbro unit. The dykes show neither deformation nor pervasive hydrothermal alteration, and must therefore correspond to another stage of the evolution history. Geochemical data show that these rocks may represent an intraoceanic island arc with poor sediment input or older crust contamination. Sr and Nd isotopic comparison between the andesites and the gabbros neatly shows the differences between magma sources and confirms that the CVC plutonic and volcanic rocks are two separate units, formed at different stages within an intraoceanic environment. A possible explanation for such magmatic variations could include the presence of different mantle sources (MORB and supra-subduction) related to different phases of migration of the arc in a long term subduction environment (Stern 2002).

These events are constrained by the 74 Ma K–Ar crystallization age obtained for the volcanic dykes and imply that the ultramafic rocks and the gabbros of the CVC were emplaced before the Campanian.

Modern-day analogues for slow-spreading supra-subduction zone environments are the Mariana Trough and the Lau back-arc (Gribble *et al.* 1996; Taylor & Martinez 2003). Basalts from both back-arc basins include arc-like components and MORB-like end-members. Mantle flow and convection induce mixing of previously depleted mantle sources and produce a range of compositions that can vary through time between end-members (Taylor & Martinez 2003). This could well be the case in the CVC, whereby the ultramafic and gabbroic units represent of a more MORB-like end-member source and the andesitic dykes unit a later, more subduction-related end-member source of the same arc.

As described in the geological setting, two different Cretaceous low-grade volcano-sedimentary metamorphic units are exposed in the Guajira Peninsula, the Jarara Formation to the SE and the Etpana Formation to the NW, described in great detail by Alvarez (1967) and Lockwood (1965). These are temporally and geologically linked to the Cabo de la Vela Mafic–Ultramafic Complex of this paper (Fig. 2).

The lithostratigraphic characteristics of the metamorphic protoliths from these two units include a sequence of mainly siliclastic sediments (pelites to rudites) with intercalations of mafic tuffs and lavas. The presence of thick quartzite sequences attest to the mature nature of the metasedimentary protoliths. Differences between the units are that the Jarara Fm includes marbles and more abundant volcanic rocks, whereas the Etpana Fm has intermixed serpentinites and gabbros. These metasediments are part of the same basin, with the Etpana Fm representing the deeper sedimentary environment (Lockwood 1965).

Fossil ages range from Turonian to Maastrichtian. Nearby Cretaceous units of the autochthonous margin of South America lack the volcanic and predominant siliclastic components of the Etpana and Jarara Fms (McDonald 1964; Villamil 1999). It is therefore also possible to assume that the

Jarara–Etpana sequence formed in an allochthonous arc position.

The contact between these Cretaceous units and basement rocks has been described as a shear-zone (MacDonald 1964; Lockwood 1965; Alvarez 1967) that formed during the collision of the arc with South America. The presence of undeformed Eocene continental arc plutonism in this region and the regionally correlatable plutons of the Santa Marta Massif, that clearly intrude the South American margin (Tschanz *et al.* 1974; Cardona *et al.* 2008), also suggest that accretion occurred before this magmatic event.

The relation between the Jarara–Etpana sequence and the CVC is hidden. However, as Alvarez (1967) pointed out, the ultramafic and gabbroic rocks found in the CVC resemble the intercalated mafic and ultramafic rocks of the Etpana Fm Reconnaissance field and petrographic observations

of the Etpana Formation indicate a strong similarity to the CVC in the nature and distribution of serpentinization and rodingization (Arredondo *et al.* 2005).

The available information suggests that these units are tectonically mixed and metamorphosed during the late collisional event. Nevertheless, additional data is needed to test this model.

A tentative model for the Early Cretaceous to Eocene evolution of the CVC derived from the above data is presented in Figure 10. It includes three main stages:

1. Initiation of an ocean–ocean subduction zone, forming a slow-spreading back-arc basin in pre-Campanian times. This back-arc, represented by the CVC ultramafic and gabbroic units, was progressively exhumed as a consequence of the slow spreading dynamics.

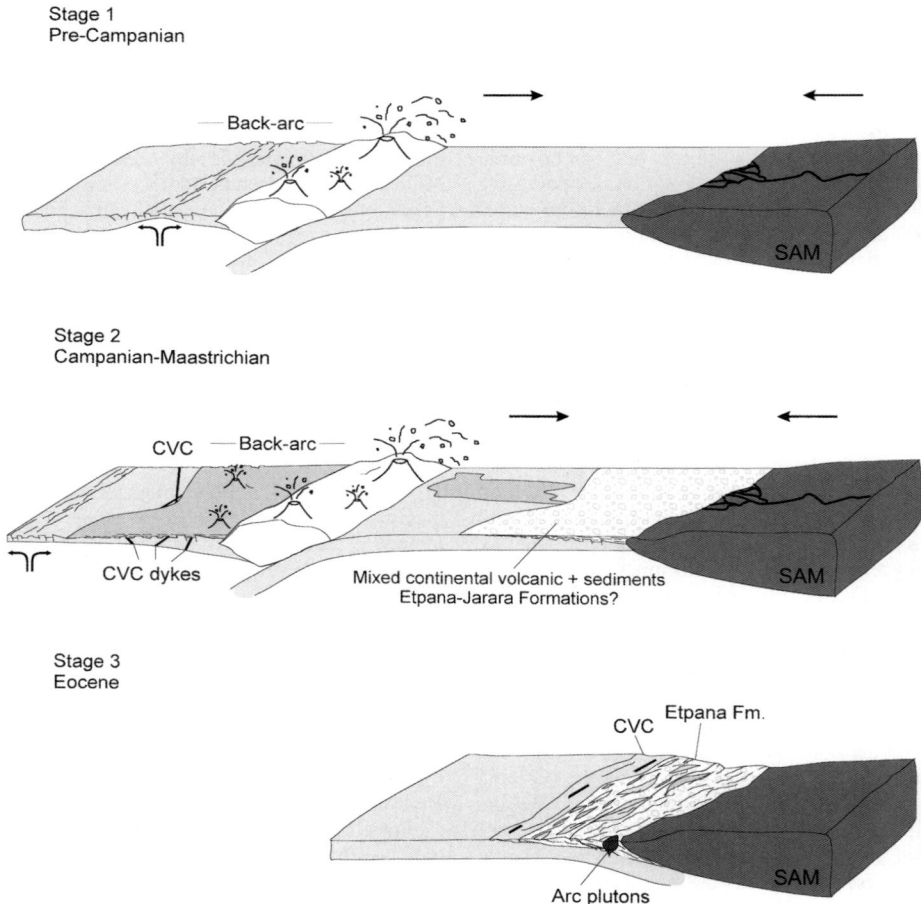

Fig. 10. Three-stage model for the formation of the Cabo de la Vela Mafic-Ultramafic Complex. SAM represents the South American margin (modified from Giunta *et al.* 2006).

2. Subduction zone configuration changed and arc-like magmatism developed upon a migrating spreading centre, represented by the andesitic dykes of the CVC. The Eptana and Jara Fms were deposited on the continental platform between this arc and the CVC. Continuous ocean subduction carried the arc towards the passive continental margin of South America.
3. The units were accreted onto the South American margin. Following the accretion the Caribbean Plate began to subduct under South America. This event is constrained by the age of Eocene magmatism that intruded the already deformed Etpana and Jarara Fms at 47 Ma (Lockwood 1965). Cardona-Molina *et al.* (2006) deduced that this could have occurred between the Late Cretaceous and the Paleocene, based on Ar/Ar spectra of basement rocks.

Caribbean realm

The tectonic evolution of the Caribbean is related to Jurassic–Early Cretaceous formation of oceanic Proto-Caribbean crust following the separation of North and South America. Subsequent development of a multistage intraoceanic-arc or several intra-oceanic arcs in either a near mid-American or a Pacific position happened from Late Cretaceous to Recent times, linked to the migration of thickened oceanic plateau crust from the west between the Americas, the passage of which left behind fragments within the continental margin (Pindell 1993; Pindell & Kennan 2001; reviews in Giunta *et al.* 2002, 2006; James 2006; Pindell *et al.* 2006).

The Cabo de la Vela Mafic–Ultramafic Complex and the associated Etpana–Jarara Fms of the Guajira region record the tectonic evolution of an island arc of Campanian and older age (whole rock K–Ar age of 74 Ma for the basaltic andesites, and the older ultramafic and gabbroic units). Other remnants of magmatic arcs and subduction complexes recording ocean–ocean convergence in the southern Caribbean are the Early Cretaceous Villa de Cura and Dos Hermanas units from Venezuela, the Santa Marta Schists of Colombia and the Washikemba Formation in Bonaire (MacDonald *et al.* 1971; Beccaluva *et al.* 1996; Giunta *et al.* 2002; Thompson *et al.* 2004). Petrographical and geochemical comparison, as well as age constrains, show similarities which indicate that the Dos Hermanas unit or the Washikemba Formation (Giunta *et al.* 2002; Thompson *et al.* 2004) could be correlated to the CVC.

Ages similar to those obtained for the CVC have been recorded throughout the Lesser Antilles. In Curaçao, Sinton *et al.* (1998) report a *c.* 76 Ma Ar/Ar age for a dolerite sill that intrudes the Albian–Turonian Curaçao lava succession. Late Cretaceous turbidites overlying this sequence contain a significant population of euhedral zircons in the range of 70–87 Ma (Wright 2004). In Aruba, Priem *et al.* (1986), based on various Rb–Sr and K–Ar age determinations, suggest that a *c.* 72 Ma thermal event was responsible for the isotopic resetting of the Aruba Batholith. These data suggest an important *c.* 70–76 Ma magmatic event in the Dutch Antilles that correlates with the CVC.

Both the segmented distribution of Cretaceous arc-related and other intra-oceanic units (including the CVC) that can be correlated along the continental margin of northern South America and the Caribbean, and the systematic age variation from SW to east whereby the collision record becomes younger from Ecuador to Colombia and Venezuela, are compatible with the north-easterly migration of the Caribbean Plate and the arc fronts between the Americas, and the accretion and subsequent disruption within an oblique continental margin (Pindell, 1993; Toussaint, 1996; Avé-Lallement & Sisson, 2005; Vallejo *et al.*, 2006).

This project was partially funded by the National University of Colombia, Medellin.

The authors wish to acknowledge the contributions of Jorge Gomez from the Colombian Geological Survey (INGEOMINAS) and the staff of the Geochronological Research Centre (CP-Geo) and geochemical laboratories of the University of Sao Paulo. We also thank G. Y. Ojeda and M. Nieto from the ICP for their help with gravimetric data and C. Jaramillo for constructive comments on the initial draft of the paper.

We are also grateful to K. H. James for inviting and encouraging us to present this paper and G. Giunta for the constructive review and suggestions to improve the manuscript.

References

ALVAREZ, W. 1967. *Geology of the Simarua and Carpintero area, Guajira Peninsula, Colombia.* PhD thesis, Princeton University.
ARREDONDO, L. F., WEBER, M. *ET AL.* 2005. Petrografía de las ultramafitas y rodingitas de la Serranía de Jarara, Península de la Guajira-Colombia. *X Congreso Colombiano de Geología – Simposio de Geología Regional.*
AVÉ-LALLEMENT, H. G. & SISSON, V. B. 2005. Exhumation of eclogites and blueschist in northern Venezuela: constraints from kinematic analysis of deformation structures. *In:* AVÉ LALLEMENT, J. G. & SISSON, V. B. (eds) *Caribbean–South American Plate Interactions, Venezuela.* Geological Society of America, Boulder, CO, Special Papers, **394**, 193–306.
BECCALUVA, L., MACCCIOTTA, G., PICCARDO, G. B. & ZEDA, O. 1989. Clinopyroxene composition of

ophiolite basalts as petrogenetic indicator. *Chemical Geology*, **77**, 165–182.

BECCALUVA, L., COLTORTI, M. *ET AL.* 1996. Cross sections through the ophiolitic units of th Southern and Northern Margins of the Caribbean Plate, in Venezuela (Northern Cordilleras) and Central Cuba. *Ofioliti*, **21**, 85–103.

BECCALUVA, L., COLTORTI, M., GIUNTA, G. & SIENA, F. 2004. Tethyan vs. Cordilleran ophiolites: a reappraisal of distinctive tectono-magmatic features of supra-subduction complexes in relation to the subduction mode. *Tectonophysics*, **393**, 163–174.

CANNAT, M. 1996. How thick is the magmatic crust at show spreading oceanic ridges? *Journal of Geophysical Research*, **101**, 2847–2857.

CANNAT, M., BIDEAU, D. & BOGAULT, H. 1992. Serpentnized peridotites and gabbros in the Mid-Atlantic Ridge axial valley at 15°37′N and 16°52′N. *Earth and Planetary Science Letters*, **109**, 87–106.

CAPEDRI, S. & VENTURELLI, G. 1979. Clinopyroxene composition of ophiolitic metabasalts in the Mediterranean area. *Earth and Planetary Science Letters*, **43**, 61–73.

CARDONA, A., DUQUE, J. *ET AL.* 2008. Geochronology and tectonic implications of granitoids rocks from the northwestern Sierra Nevada de Santa Maria and surrounding basins, northeastern Colombia: Late Cretaceous to paleogene convergence, accretion and subduction interactions between the Caribbean and South American plates. *18th Caribbean Geological Conference*, Dominican Republic, 24–29th March.

CARDONA-MOLINA, A., CORDANI, U. & MACDONALD, W. 2006. Tectonic correlations of pre-Mesozoic crust from the northern termination of the Colombian Andes, Caribbean region. *Journal of South American Earth Sciences*, **21**, 337–354.

COOGAN, L. A., MACLEOD, C. J. *ET AL.* 2001. Whole-rock geochemistry of gabbros from the Southwest Indian Ridge: constraints on geochemical fractionations between the upper and lower oceanic crust and magma chamber processes at (very) slow-spreading ridges. *Chemical Geology*, **178**, 1–22.

CORDANI, U. G., CARDONA, A., JIMENEZ, D., LIU, D. & NUTMAN, A. P. 2005. Geochronology of Proterozoic basement inliers from the Colombian Andes: tectonic history of remmants from a fragmented grenville belt. *In*: VAUGHAN, A. P. M., LEAT, P. T. & PANKHURST, R. J. (eds) *Terrane Processes at the Margins of Gondwana*. Geological Society, London, Special Publications, **246**, 329–346.

DE PAOLO, D. J. 1988. Age dependence of the composition of continental crust: evidence from Nd isotopic variations in granitic rocks. *Earth and Planetary Science Letters*, **90**, 263–271.

DICK, H. J. B. & BULLEN, T. 1984. Chromian spinel as a petrogenetic indicator in abyssal and alpine-type peridotites and spatially associated lavas. *Contributions to Mineralogy and Petrology*, **86**, 54–76.

DUBIŃSKA, E. 1995. Rodingites of the Eastern Part of the Jordanów–Gogolów Serpentinite Massif, Lower Silesia, Poland. *The Canadian Mineralogist*, **33**, 585–608.

FROST, B. R. 1975. Contact metamorphism of serpentinites: chloritic black-wall and rodingite at Paddy-go-easy, Central Cascades, Washington. *Journal of Petrology*, **16**, 272–313.

FRÜH-GREEN, G. L., PLASS, A. & LÉCUYER, C. 1996. Petrologic and stable isotope constraints on hydrothermal alteration and serpentinization of the EPR shallow mantle at Hess Depp (Site 895). *In*: MÉVEL, C. *ET AL.* (eds) *Proceedings of the Ocean Drilling Program, Scientific Results*, **147**. Ocean Drilling Program, Texas A&M University, 255–291.

GIUNTA, G., BECCALUVA, L., COLTORTI, M., SIENA, F. & VACCARO, C. 2002. The southern margin of the Caribbean Plate in Venezuela: tectono-magmatic setting of the ophiolite units and kinematic evolution. *Lithos*, **63**, 19–40.

GIUNTA, G., BECCALUVA, L. & SIENA, F. 2006. Caribbean Plate margin evolution: constraints and current problems. *Geologica Acta*, **4**, 265–277.

GREEN, D. H., LOCKWOOD, J. P. & KISS, E. 1968. Eclogite and almandine–jadeite–quartz rock from the Guajira Peninsula, Colombia, South America. *American Mineralogist*, **53**, 1320–1335.

GRIBBLE, R. F., STERN, R. J., BLOOMER, S. H., STÜBEN, D., O'HEARN, T. & NEWMAN, S. 1996. MORB mantle subduction components interact to generate basalts in the southern Mariana Trough back-arc basin. *Geochimica et Cosmchimica Acta*, **60**, 2153–2166.

HAWKESWORTH, C. J., GALLAGHER, K., HERGT, J. M. & MCDERMOTT, F. 1993*a*. Mantle slab contributions in arc magmas. *Annual Reiews in Earth and Planetary Science*, **21**, 175–204.

HAWKESWORTH, C. J., GALLAGHER, K., HERGT, J. M. & MCDERMOTT, F. 1993*b*. Trace element fractionation processes in the generation of island arc basalts. *In*: TARNEY, J., PICKERING, K. T., KNIPE, R. J. & DEWEY, J. F. (eds) *Melting and Melt Movement in the Earth*. Clarendon Press/The Royal Society, London, 393–405.

HAWTHORNE, F. C. 1981. Crystal chemistry of the amphiboles. *Reviews in Mineralogy and Geochemistry*, **9A**, 1–102.

HEBERT, R. & LAURENT, R. 1987. Mineral chemistry of the plutonic section of the Troodos ophiolite: new constraints for genesis of arc-related ophiolites. *In*: MALPAS, G. J. & MOORES, E. M. (eds). *Troodos 1987 Ophiolites and Oceanic Lithosphere*, 149–163.

ITURRALDE-VINENT, M. A. & LIDIAK, E. G. (eds) 2006. Caribbean Plate tectonics: stratigraphic, magmatic, metamorphic and tectonic events (UNESCO/IUGS IGCP Project 433). *Geologica Acta*, **4**, 1–341.

JAMES, K. H. 2006. Arguments for and against the Pacific origin of the Caribbean Plate: discussion, finding for an inter-American origin. *Geologica Acta*, **4**, 279–302.

JOLLY, W. T., LIDIAK, E. G. & DICKIN, A. P. 2006. Cretaceous to Mid-Eocene pelagic sediment budget in Puerto Rico and the Virgin Islands (northeast Antilles Island arc). *Geologica Acta*, **4**, 35–62.

KELLOGG, J. N., GODLEY, V. M., ROPAIN, C., BERMÚDEZ, A. & AIKEN, C. L. V. 1991. *Gravity Field of Colombia, Eastern Panama and Adjacent Marine*

Areas. The Geological Society of America, Boulder, CO, Map and Chart Series **MCH 070**.

KENEMETSKY, V. S., CRAWFORD, A. J. & MEFFRE, S. 2001. Factors controlling chemistry of magmatic spinel: an empirical stud of associated olivine, Cr-spinel and melt inclusions from primitive rocks. *Journal of Petrology*, **42**, 345–352.

LEWIS, J. F., DRAPER, G., PROENZA, J. A., ESPAILLAT, J. & JIMÉNEZ 2006. Ophiolite-related ultramafic rocks (serpentinites) in the Caribbean region: a review of their occurrence, composition, origin, emplacement and Ni-laterite soil formation. *Geologica Acta*, **4**, 237–263.

LIPPARD, S. J., SHELTON, A. W. & GASS, I. G. 1986. *The Ophiolite of Northern Oman*. Geological Society Memoirs, **11**. Blackwell, Oxford, 178.

LOCKWOOD, J. P. 1965. *Geology of the Serranía de Jarara area, Guajira Península, Colombia*. PhD thesis, Princeton University.

MACDONALD, W. D. 1964. *Geology of the Serranía de Macuira area, Guajira Península, Colombia*. PhD thesis, Princeton University.

MACDONALD, W., DOOLAN, B. & CORDANI, U. 1971. Cretaceous–Early Tertiary metamorphic age values from the South Caribbean. *Geological Society of America Bulletin*, **82**, 1381–1388.

MONTES, C., HATCHER, R. D. & RESTREPO-PACE, P. 2005. Tectonic reconstruction of the northern Andean blocks: oblique convergence and rotations derived from the kinematics of the Piedras–Girardot area, Colombia. *Tectonophysics*, **399**, 221–250.

MORI, P. E., REEVES, S., TEIXEIRA, C. & HAUKKA, M. 1999. Development of a fused glass disc XRF facility and comparison with the pressed powder pellet technique at Instituto de Geociencias, São Paulo University. *Revista Brasileira de Geociências*, **29**, 441–446.

MORIMOTO, N., FABRIES, J. ET AL. 1988. Nomenclature of pyroxenes. *American Mineralogist*, **62**, 53–62.

NAKAMURA, N. 1974. Determination of REE, Ba, Fe, Mg, Na and K in carbonaceous and ordinary chondrites. *Geochimica et Cosmochimica Acta*, **38**, 757–775.

NICOLAS, A. & RABINOWICZ, M. 1984. Mantle flow pattern at oceanic spreading centres: relation with ophiolitic and oceanic structures. *In*: GASS, I. G., LIPPARD, S. J. & SHELTON, A. W. (eds) *Ophiolites and Oceanic Lithosphere*. Geological Society, London, Special Publications, **13**, 147–151.

NISBET, E. G. & PEARCE, J. A. 1977. Clinopyroxene composition in mafic lavas from different tectonic settings. *Contributions to Mineralogy and Petrology*, **63**, 149–160.

OKAMURA, H., ARAI, S. & KIM, Y.-U. 2006. Petrology of forearc peridotite from the Hahajima Seamount, the Izu-Bonin arc, with special reference to chemical characteristics of chromian spinel. *Mineralogical Magazine*, **70**, 15–26.

PARKINSON, I. J. & PEARCE, J. A. 1998. Peridotites from the Izu–Bonin–Mariana forearc (ODP Leg 125): evidence for mantle melting and melt–mantle interaction in a supra-subduction zone setting. *Journal of Petrology*, **39**, 1577–1618.

PEARCE, J. A. 1982. Trace element characteristics of lavas from destructive plate boundaries. *In*: THORPE, R. S. (ed.) *Andesites*. Wiley, Chichester, 525–548.

PEARCE, J. A. & CANN, J. R. 1973. Tectonic setting of basic volcanic rocks determined using trace element analyses. *Earth and Planetary Science Letters*, **19**, 290–300.

PINDELL, J. L. 1993. Evolution of the Gulf of Mexico and the Caribbean. *In*: DONOVAN, S. K. & JACKSON, T. A. (eds) *Caribbean Geology: An Introduction*. University of the West Indies Publisher's Association, Kingston, 13–39.

PINDELL, J. L. & KENNAN, L. 2001. Kinematic evolution of the Gulf of Mexico and Caribbean. *Transactions, Petroleum Systems of Deep-water Basins: Global and Gulf of Mexico Experience. GCSSEPM 1st Annual Research Conference*, Houston, TX. Gulf Coast Section Society of Economic Paleontologists and Mineralogists Foundation, 159–192.

PINDELL, J. L., KENNAN, L., STANEK, K. P., MARESCH, W. V. & DRAPER, G. 2006. Foundations of Gulf of Mexico and Caribbean evolution: eight controversies resolved. *Geologica Acta*, **4**, 303–341.

PRIEM, H. N. A., BEETS, D. J., BOELRIJK, N. A. I. M. & HEBEDA, E. H., 1986. On the age of the Late Cretaceous tonalitic/gabbroic batholith on Aruba, Netherlands Antilles (southern Caribbean borderland). *Geologie en Mijnbouw*, **65**, 247–256.

SATO, K., TASSINARI, C. C. G., KAWASHITA, K. & PETRONILHO, L. 1995. O método geocronológico Sm–Nd no IG/USP e suas implicações. *Anais da Academia brasileira de Ciencias*, **67**, 315–336.

SAUNDERS, A. D. & TARNEY, J. 1984. Geochemical characteristics of basaltic volcanism within back-arc basin. *In*: KOKELAAR, B. P. & HOWELLS, M. F. (eds) *Marginal Basin Geology: Volcanic Associated Sedimentary and Tectonic Processes in Modern and Ancient Marginal Basins*. Geological Society, London, **16**, 59–76.

SEYLER, M., PAQUETTE, J.-L., CEULENEER, G., KIENAST, J.-K. & LOUBET, M. 1998. Magmatic underplating, metamorphic evolution, and ductile shearing in a Mesozoic lower crustal–upper mantle unit (Tinaquillo, Venezuela) of the Caribbean Belt. *Journal of Geology*, **106**, 35–58.

SHERVAIS, J. W. 1982. Ti–V plots and the petrogenesis of modern and ophiolitic lavas. *Earth and Planetary Science Letters*, **59**, 101–118.

SHERVAIS, J. W. 2001. Birth, death, and resurrection: the life cycle of suprasubduction zone ophiolites. *Geochemistry, Geophysics, Geosystems*, **2** (paper number 2000GC000080).

SINTON, C. W., DUNCAN, R. A., STOREY, M., LEWIS, J. & ESTARADA, J. J. 1998. An oceanic flood basalt province within the Caribbean Plate. *Earth and Planetary Science Letters*, **155**, 221–235.

SISSON, V. B., ERTAN, I. F. & AVÉ LALLEMANT, H. G. 1997. High pressure (~2000 MPa) glaucophane-bearing pelitic schist and eclogite from Cordillera de la Costa belt, Venezuela. *Journal of Petrology*, **38**, 65–83.

STEIGER, R. H. & JAGER, E. 1977. Subcommission on geochronology: convention on the use of decay constants in geo- and cosmochronology. *Earth and Planetary Science Letters*, **36**, 359–362.

STERN, R. J. 2002. Subduction zones. *Reviews of Geophysics*, **10**; doi: 1029/2001RG000108.

STÖCKHERT, B., MARESCH, W. V. ET AL. 1995. Crustal History of Margarita Island (Venezuela) in detail: constraint on the Caribbean Plate tectonic scenario. *Geology*, **23**, 787–790.

TASSINARI, C. C. G., MEDINA, J. G. C. & PINTO, M. C. S. 1996. Rb–Sr and Sm–Nd geochronology and isotope geochemistry of central Iberian metasedimentary rocks (Portugal). *Geologie en Mijnbouw*, **75**, 69–79.

TAYLOR, B. & MARTINEZ, F. 2003. Back-arc basin basalt systematics. *Earth and Planetary Science Letters*, **210**, 280–297.

THOMPSON, P. M. E., KEMPTON, P. D. ET AL. 2004. Elemental, Hf–Nd isotopic and Geochronological constraints on an island arc sequence associated with the Cretaceous Caribbean plateau: Bonaire, Dutch Antilles. *Lithos*, **74**, 91–116.

TOUSSAINT, J. F. 1996. *Evolución geológica de Colombia. Cretácico*. Universidad Nacional de Colombia, Medellín.

TSCHANZ, C., MARVIN, R., CRUZ, J., MENNERT, H. & CEBULA, E. 1974. Geologic evolution of the Sierra Nevada de Santa Marta. *Geological Society of America Bulletin*, **85**, 269–276.

VALLEJO, C., SPIKINGS, R. A., LUZIEUX, L., WINKLER, W., CHEW, D. & PAGE, L. 2006. The early interaction between the Caribbean Plateau and the NW South American Plate. *Terra Nova*, **18**, 264–269.

VILLAMIL, T. 1999. Campanian–Miocene tectono-stratigraphy, depocenter evolution and basin development of Colombia and Western Venezuela. *Paloegeography, Palaeoclimatology, Paleoecology*, **153**, 239–275.

WINCHESTER, J. A. & FLOYD, P. A. 1977. Geochemical discrimination of different magma series and their differentiation products using immobile elements. *Chemical Geology*, **20**, 325–343.

WRIGHT, J. E. 2004. Aruba and Curaçao: remnants of a collided Pacific oceanic plateau? Initial geologic results from the BOLIVAR Project. *American Geophysical Union*, Fall Meeting 2004, abstract no. T33B-1367.

ZAPATA, G., WEBER, M. ET AL. 2005. Análisis petrográfico de las rocas de alta presión de la Serranía de Jarara, La Guajira y sus implicaciones tectónicas. *X Congreso Colombiano de Geología – Simposio de Geología Regional*, 68–69.

Key issues on the post-Mesozoic Southern Caribbean Plate boundary

FRANCK A. AUDEMARD

FUNVISIS, Apartado postal 76880, Caracas 1070-A, Venezuela
(e-mail: faudemard@funvisis.gob.ve)

Abstract: A lithospheric-scale, geodynamic model of the Caribbean southern boundary takes into account tectonic and stratigraphic elements previously unconsidered or not well constrained. The Falcón, Bonaire, Blanquilla and Grenada basins are parts of a former single back-arc basin associated with the migrating Mesozoic Caribbean Arc in Late Eocene–Oligocene times. Spreading of the crescent-shaped basin was triggered by declining Caribbean eastward motion in response to increased convergence between the two Americas at around 38–33 Ma. Dextral wrenching along the southern boundary started in western Venezuela at 17–15 Ma and progressed eastward to the El Pilar Fault by 12 Ma. Eastward motion of the Caribbean relative to South America was earlier accommodated by east-progressing, non-partitioned oblique subduction and collision, as illustrated today in eastern Venezuela around the Los Bajos–El Soldado fault system. Extrusion of the Maracaibo and Bonaire blocks towards the NNE started around 5 Ma, thus the dextral Boconó fault is a young feature. Total plate boundary Cenozoic dextral wrenching recorded onshore Venezuela does not exceed 100 km and is more likely to be in the order of 55 km.

This paper discusses the Cenozoic–Quaternary evolution of the southern Caribbean Plate boundary zone using data from onshore and offshore Venezuela. The subject has been controversial for decades. The Caribbean region (Caribbean Plate) is complex. It is bounded by the very large Atlantic plate, oceanic in the east and continental in the north and south (North and South America respectively), and by the Cocos oceanic plate in the west. The large number and variety of models proposed over 30 years illustrates this complexity (see summary by Morris *et al.* 1990 and papers in this volume). A few of these considered the Cenozoic geodynamic evolution of the southern Caribbean Plate boundary (Stephan *et al.* 1990; Audemard 1993, 1998; Pindell 1994; Pindell *et al.* 1998; James 2002, 2005*a, b*).

Most models show northern Venezuela as part of the Atlantic passive margin before deformation during eastward-younging collision of an island arc through the Cenozoic and Quaternary. Major Cenozoic tectonic phases were Paleocene–Early Eocene collision, Middle Eocene compression, Late Eocene–Oligocene extension/transtension, Middle-to-Late Miocene compression/transpression and onset of ongoing Plio-Quaternary compression and/or wrenching.

The paper begins with a review and a definition of the most important geological elements and later attempts lithospheric-scale geodynamic reconstruction of Cenozoic and Quaternary events in Venezuela.

Definition of the South American–Caribbean Plate boundary zone

Present-day South Caribbean boundary

Hess & Maxwell (1953) proposed that the boundary was a simple dextral wrench system. Over 20 years of neotectonic analyses, integrating surface geology, geomorphology, microtectonics, seismotectonics and palaeoseismology, combined with data from conventional geological studies and on and offshore seismic reflection data, allow a more precise view. Following maps of Soulas (1986) and Beltrán (1993, 1994), Audemard *et al.* (2000) presented a Quaternary fault map of Venezuela and information about each fault and/or fault section (length, attitude, age, sense of slip, slip rate, geomorphic expression, latest activity from geological data). The report is accessible at the USGS web page (http://pubs.usgs.gov/of/2000/ofr-00-0018/ofr-00-0018.pdf). The more we learn about the boundary zone, the more complex it seems.

There is wide consensus that the Caribbean Plate plate is moving east relative to South America and recent GPS results strongly support this (Freymueller *et al.* 1993; Pérez *et al.* 2001*a, b*; Weber *et al.* 2001). However, northern South America reflects interaction of the Caribbean, South American and Nazca plates and the Panamá microplate. Deformation along the boundary is driven by oblique convergence (Silver *et al.* 1975; Pérez & Aggarwal 1981; Speed 1985; Lugo & Mann

From: JAMES, K. H., LORENTE, M. A. & PINDELL, J. L. (eds) *The Origin and Evolution of the Caribbean Plate.*
Geological Society, London, Special Publications, **328**, 569–586.
DOI: 10.1144/SP328.23   0305-8719/09/$15.00 © The Geological Society of London 2009.

1992; Russo & Speed 1992). Today this is more intense in the west than in the east where wrenching dominates.

Northern Venezuela lies in the interaction zone between the South America and Caribbean Plate, while western Venezuela and northern Colombia comprise a number of interplaying tectonic blocks or microplates (Fig. 1). The Caribbean–South America Plate boundary from Colombia to Trinidad is over a 100 km wide active transpression zone on and offshore northern Venezuela (Audemard 1993; Singer & Audemard 1997; Pindell et al. 1998; Audemard et al. 2000, 2005; Ysaccis et al. 2000). Significant positive relief is present along the Coastal and Interior ranges of north-central and northeastern Venezuela. Further west, the southern Caribbean boundary broadens to as much as 600 km and includes several small tectonic blocks or microplates (Fig. 1). The triangular Maracaibo block is bounded by the left-lateral Santa Marta–Bucaramanga fault in Colombia and the right-lateral

Boconó fault in Venezuela. The dextral Oca fault separates it in the north from the Bonaire Block. Both the Maracaibo and Bonaire blocks are being extruded northward. Extrusion is driven by the collision and suturing of the Chocó Block against the Pacific side of northern South America (Duque-Caro 1978, 1990; Audemard 1993, 1998), confirmed by GPS plate motion studies (Freymueller et al. 1993; Kellogg & Vega 1995; Kaniuth et al. 1999; Trenkamp et al. 2002). The Maracaibo and Bonaire blocks override the Caribbean Plate. North of the Netherlands Leeward Antilles south-dipping, amagmatic, flat subduction has developed in the last 5 Ma. The onset of this escape tectonics is a key issue and is discussed later.

Much of present day dextral slip along the southern Caribbean boundary seems to be focused along the major Boconó–San Sebastián–El Pilar–Los Bajos–Warm Spring fault system where geological and geodetic data indicate slip rate of 8–10 mm/annum. Most authors see this system as

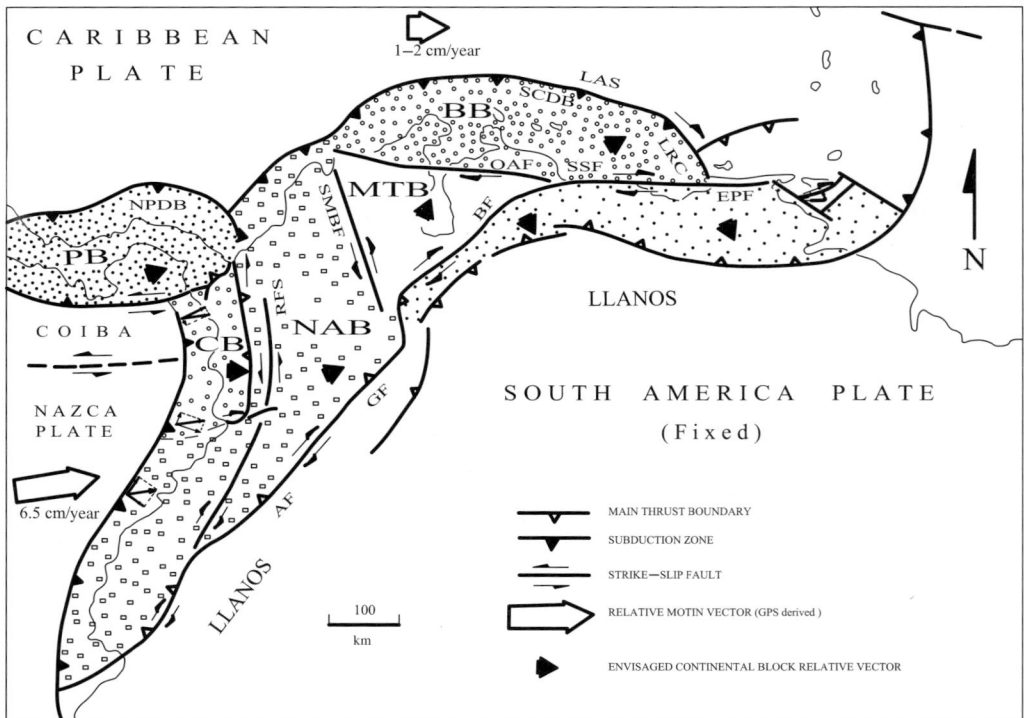

Fig. 1. Major geodynamic features along the present-day southern Caribbean Plate boundary zone. Width of the plate boundary zone is very variable along-strike, but never confined to a single feature. Major tectonic blocks within the plate boundary zone are: Bonaire (BB), Chocó (CB), Maracaibo (MTB), North Andean (NAB) and Panamá (PB). Some major faults are also reported: Algeciras (AF), Boconó (BF), El Pilar (EPF), Guaicaramo (GF), Romeral (RFS), Santa Marta–Bucaramanga (SMB), San Sebastián (SSF) and Oca–Ancón (OAF). Other features are: Leeward Antilles subduction (LAS), Los Roques Canyon (LRC), North Panamá deformation belt (NPDB) and Southern Caribbean deformation belt (SCDB).

the plate boundary (Hess & Maxwell 1953; Schubert 1979; Stephan 1985; Pérez *et al.* 2001*a*, *b*). Others note that deformation is distributed over a 100 km or more wide zone and interpret orogenic float in the Andes (Audemard 1991; Jácome 1994; Audemard & Audemard 2002), across the Falcón Basin (Porras 2000) and eastern Venezuela (Ysaccis 1997; Ysaccis *et al.* 2000). According to this understanding the zone is flanked by both A- and B-type subduction. Yet others recognize southeast-directed A-subduction or underthrusting below the Mérida Andes (Kellogg & Bonini 1982; De Toni & Kellogg 1993; Colletta *et al.* 1996, 1997).

The boundary zone exhibits strain partitioning from the Mérida Andes in the west to the Interior range in the east. In the Andes, partitioning occurs between the dextral Boconó Fault and thrust faults on both mountain flanks (Audemard & Audemard 2002). The north-central Coastal Range also exhibits strain partitioning. Dextral slip occurs in the core along the San Sebastián, La Victoria faults and minor synthetic Riedel shears while transverse shortening is accommodated by frontal thrust faults in the south (Guarumen basin; Audemard 1999). A sub-sea, mirror thrust fault system may exist to the north and the easternmost portion of the Leeward Antilles Subduction zone must account for some shortening as well.

Farther east, strain partitioning involves NNW–SSE-trending shortening over a 250 km wide zone from north of La Blanquilla to the active thrust front of the Interior Range. Slab detachment (Russo & Speed 1992; Russo 1999) associated with an incipient A-subduction, results in the largest onshore negative Bouguer anomaly in the world, south of the Interior Range. Dextral slip concurrently occurs along the main east–west striking El Pilar Fault and along the NW–SE striking Los Bajos–El Soldado faults and minor parallel and/or synthetic Riedel shear faults. Partitioning might also occur on Trinidad.

Although transpression is the dominant process at plate boundary scale, transtension also occurs in two relay, pull-apart basins, the Cariaco Trough (Schubert 1979, 1984) and the Gulf of Paria (Babb 1997; Ysaccis 1997) at the ends of the El Pilar Fault. The age of opening of these basins constrains the timing of wrenching onset.

Most of El Pilar Fault strike–slip motion transfers via the Los Bajos and El Soldado synthetic Riedel shears to the Warm Springs Fault of the Central Range of Trinidad (Weber *et al.* 2001). However, this shallow crustal deformation does not exclude a deeper plate boundary, as indicated by instrumental seismicity and as suggested by Pérez & Aggarwal (1981) and Sobiesiak *et al.* (2002, 2005).

Cenozoic South Caribbean boundary

The southern Caribbean Plate boundary was located along the east–west-striking Oca–Ancón system of northwestern Venezuela from around 17 to 15 Ma, when transpression started, to around 5 to 3 Ma (Audemard 1993, 1998). The Boconó Fault became involved when escape tectonics began and slip slowed along the Oca–Ancón system (Audemard 1993, 1998). Age constraints are discussed later.

The Cenozoic–Quaternary evolution of the southern Caribbean Plate boundary began with the initiation of oblique, type-B subduction (NW-dipping, South America-attached oceanic lithosphere beneath the eastward migrating Caribbean Plate). It evolved into east-younging oblique-collision that emplaced SSE-vergent Caribbean nappes above passive-margin nappes on undeformed South America passive margin. As collision became unsustainable, partitioned transpression took over (Audemard 1993, 1998). Subduction–transpression migrated like a wave breaking along a beach from west to east and is still active in Eastern Venezuela and Trinidad. Consequently, the boundary became increasingly strike–slip and less compressional with time.

Quantitative motion across the Southern Caribbean boundary

Present-day motion

Several authors have predicted that the Caribbean Plate moves east at about 20 mm/annum relative to South America (Jordan 1975; Rosencrantz *et al.* 1988; Stein *et al.* 1988; Calais *et al.* 1992) and GPS studies have confirmed this (Pérez *et al.* 2001*a*, *b*; Weber *et al.* 2001; Trenkamp *et al.* 2002). In the east, dextral slip along the plate boundary transfers to subduction in the Lesser Antilles.

GPS slip magnitudes remain rather constant from west to east but orientations change. ESE oblique convergence occurs between San Andrés Island, east of Honduras, and stable South America, confirming predicted N075°W convergence (Jordan 1975; Minster & Jordan 1978) and seismotectonic models (Pennington 1981; Van der Hilst & Mann 1994). Data from eastern Venezuela indicate almost pure wrenching between the Caribbean Plate and South America [086 ± 2° with respect to the Central range of Trinidad (Weber *et al.* 2001) and 084 ± 2°E with respect to South America (Pérez *et al.* 2001*a*)].

Vector orientations indicate transtension, as postulated by Robertson & Burke (1989), Algar & Pindell (1993) and Pindell *et al.* (1998), and inferred by Weber *et al.* (2001). Stress tensors of

Colmenares & Zoback (2003) indicate the same. However, Choy et al. (1998) determined stress tensors indicating NW–SE compressive wrenching. Several authors came to this same result using geological data (Beltrán & Giraldo 1989; Audemard 1993, 2000; Audemard et al. 2000, 2005).

In order to resolve this inconsistency Audemard et al. (2005) proposed that extrusion of small triangular blocks defined by ENE–WSW sinistral faults (Laguna Grande in Araya Peninsula; Punta Charagato on Cubagua Island) and east–west dextral faults (El Pilar, Coche, North Coast) occurs in the direction indicated by GPS data (N084 ± 2°E). This harmonizes Quaternary left-lateral slip on ENE–WSW trending faults, movement in the direction N084 ± 2°E and transpression at plate-boundary scale instead of transtension.

GPS results published by Pérez et al. (2001a, p. 70, fig. 1) indicate compression across the boundary south of the main dextral system (Audemard et al. 2005) where GPS rates are as high as 20–25% of the main dextral rate (18–20 mm/annum), the same order of magnitude as those calculated from long-term geological criteria (Audemard et al. 2000). In this area strain is partitioned between the major east–west dextral faults and ENE–WSW trending thrusts and folds of the Interior Range and the Margarita–Blanquilla platform.

Slip of 8–10 mm/annum along the dextral Boconó–San Sebastián–El Pilar–Warm Spring faults, from the southern Mérida Andes to Trinidad, accounts for up to 50% of the 20 mm/annum dextral relative motion between the Caribbean and South America plates. Greatest slip along the Boconó Fault from Late Quaternary geological markers is about 9 mm/annum (Soulas 1986; Audemard 1997a; Audemard et al. 1999). The Warm Spring Fault accounts for half of the dextral motion across Trinidad.

Pérez et al. (2001a) indicate that strain is distributed across a zone at least 110 km wide, with 68% (almost 14 mm/annum) of this occurring in a 30 km-wide fault zone involving the El Pilar Fault and sub-parallel faults to the north.

Weber et al. (2001) claim that the Warm Springs Fault is the southern plate boundary in Trinidad. However, there is still 8 mm/annum to be accommodated between southwestern most Trinidad and continental South America. This implies the presence of another major fault south of Trinidad, as proposed by many authors (Soulas 1986; Beltrán 1993, 1994; Audemard et al. 2000).

Former motion

Before escape tectonics began at around 5–3 Ma, relative motion between the Americas was convergent. This is discussed by Sclater et al. (1977),

Pindell & Dewey (1982) and Pindell et al. (1988, 1998). Convergence, similar to the present-day convergence both in magnitude and direction, has been in effect since 49 Ma according to Pindell et al. 1988, 1998) or 38 Ma according to Ladd (1976) and Sclater et al. (1977) (36 Ma originally proposed by Pindell & Dewey 1982). While Pindell & Dewey (1982) and Pindell et al. (1988, 1998) described constant rate and direction of convergence, Sclater et al. (1977) noted a direction shift at 21 Ma. It implied a change from mainly wrenching to transpression along the plate boundary. Data from northern Venezuela indicate compression or transpression since the Eocene, with intervals of localized transtension or extension.

Onset of tectonic escape in the west can be estimated at some time between the beginning of Chocó accretion (Middle Miocene, Duque-Caro 1990), the effects of this in the Eastern Cordillera of Colombia (10.5 Ma, Cooper et al. 1995), similar to the Mérida Andes (Audemard & Audemard 2002) and development of the Middle America land bridge, showing coupling of Chocó with South America. First Caribbean–Pacific planktonic faunal divergence occurred by 6.2 Ma (Keller et al. 1989), mammal exchange occurred by 3.3 Ma (Gingerich 1985), planktonic faunas show Caribbean–Pacific separation by 3.1 Ma (Keigwin 1978) and final gateway closure occurred by 1.8 Ma (Keller et al. 1989). These data indicate that the dextral Boconó Fault became active not earlier than 10 Ma and more likely closer to 3–5 Ma.

Further constraint on the dextral-slip onset along the Boconó Fault comes from the regional structure of northwestern South America. A long, dextral fault system, involving the Pallatanga, Algeciras, Guaicaramo and Boconó faults separates the Ecuadorian, Colombian and Venezuelan Andes from cratonal South America (Case et al. 1971; Dewey 1972; Pennington 1981; Stephan 1982; Audemard 1993, 1998; Freymueller et al. 1993; Ego et al. 1996). Opening of the pull-apart Jambelí Graben in the Gulf of Guayaquil, Ecuador is related to northward escape of northwestern South America (Audemard 1993, 1998; Fig. 1). Basin fill began in the Late Miocene (Benítez 1986).

Total slip along the South America–Caribbean boundary

Different models (in-place or Pacific) for the origin of the Caribbean Plate imply drift of a few hundreds or few thousands of kilometres, respectively. The plate travelled along northern Venezuela during its post-Jurassic eastward progression, leaving accreted nappes or terrains abandoned along the boundary (Falcón to Paria Peninsula, Northern Range of

Trinidad, Tobago). Therefore, many authors have claimed a migration in the order of at least a 1000 km. This dextral relative motion could only have happened along an oblique subduction zone (Audemard 1993, 1998) below the accreted terrains or nappes of northern Venezuela. Striations of 45° pitch on the south-vergent Cantagallo thrust (Audemard *et al.* 1988), bounding the Villa de Cura Nappe in the south, attest to dextral-reverse slip on the almost east–west thrust fault, paralleling the present-day plate boundary. Beck (1986) reported similar faults farther east.

Audemard & Giraldo (1997) discussed the amount of wrenching along the three major dextral onshore fault systems and concluded that it could not sum to more than 150 km and probably was less than 100 km.

Total dextral slip on the Oca–Ancón Fault

Whether total dextral slip on the Oca–Ancón system derived from onshore geology, but not from geodynamic data, is still controversial. Rod (1956) suggested tens of kilometres of offset from observations on Toas Island at the outlet of Lake Maracaibo. Doolan & MacDonald (1971) noted 20 km of Neogene offset of Mesozoic schists on the Guajira Peninsula. Feo-Codecido (1972) derived a post-Oligocene offset ranging between 15 and 20 km. Vásquez & Dickey (1972) estimated 14.1 km of dextral slip during the last 5 Ma, based on the amount of shortening in the Falcón anticlinorium. Case & MacDonald (1973) estimated 40 km offset of metamorphic units on the Guajira Peninsula since the Middle Miocene. Fault scarps and offset drainage indicate continued activity. Using the same criteria, Tschanz *et al.* (1974) estimated displacement of 65 km. For Janssen (1979) isopachs at Cretaceous Cogollo level in the Maracaibo Basin indicated displacement of more than 50 km. Soulas *et al.* (1987), Audemard (1993) and Audemard *et al.* (1994) measured post-Oligocene dextral slip in the order of 30 ± 3 km, from apparent displacement of Oligocene rocks in the core of the Falcón anticlinorium (geological map of Bellizzia *et al.* 1976). Pindell *et al.* (1998) estimated at least 90 km of offset, summing 65 km displacement along the fault with 25 km of lateral transfer or shortening within the Sierra de Perijá. However, the latter may be the consequence of slip along the Oca Fault.

Total dextral slip on the Boconó Fault

Pindell *et al.* (1998) estimated at least at 130 km of post-Middle Eocene dextral slip along the Boconó Fault from the apparent position of Palaeogene nappe fronts. Stephan (1982), using the same data,

proposed 80 km of offset. These values are unreliable because they use sub-horizontal markers to estimate horizontal offset. Two lines of evidence call for the careful consideration of these offset estimates. Striae pitch of up to 10–15° shows that not all Boconó Fault slip is horizontal (Giraldo 1985; Audemard 1997*a*). Moreover, differential uplift and erosion at the northern end of the Mérida Andes imply erosional retreat of the Palaeogene nappe fronts.

Giraldo (1989) and Audemard & Giraldo (1997) used steeply dipping, pre-fault markers to determine offset similar to the determination of Rod (1956). These estimates range between 20 and 30 km (Fig. 2). James (1990) estimated 300 km of dextral slip from palaeogeographic reconstructions but also noted that faults other than the Boconó might have been active in the past. More recently, Audemard & Audemard (2002), using the relative location of flexural depocentres on both sides of the Mérida Andes, proposed a dextral slip in the order of 30 km (Audemard *et al.* 2007, fig. I-15).

Total dextral slip on the El Pilar Fault

Dextral slip estimated along this fault is very variable. Rod (1956) estimated that dextral motion could be larger than a 100 km, but did not provide details. Metz (1968) calculated less than 15 km of offset. For Vierbuchen (1977), it ranged between 25 and 70 km. The maximum slip ever proposed is about 150 km (Giraldo 1993, 1996). Using similar data, but taking into account transverse crustal shortening, Audemard & Giraldo (1997) obtained an estimate of 55 km (Fig. 3). This figure seems to fit well with the vectorial addition of the independent motions of the Oca–Ancón Fault and Boconó Fault.

Total strike–slip along the South America–Caribbean Plate boundary zone

From the above, there seems to be no onshore fault or fault combination that can account for the eastward migration of the Caribbean Plate relative to South America of one to few thousands of kilometres. This supports the idea that the boundary functioned as an oblique subduction zone, as proposed by Audemard (1993, 1998, 2000) and illustrated today in Eastern Venezuela and Trinidad.

Audemard & Giraldo (1997) used two different approaches to determine total dextral slip. A simple tectonic reconstruction of northern Venezuela (Fig. 3) combined movement along major dextral faults and shortening across the boundary. Shortening is estimated at 40–50 km in the Andes (Audemard 1991; Pindell *et al.* 1998;

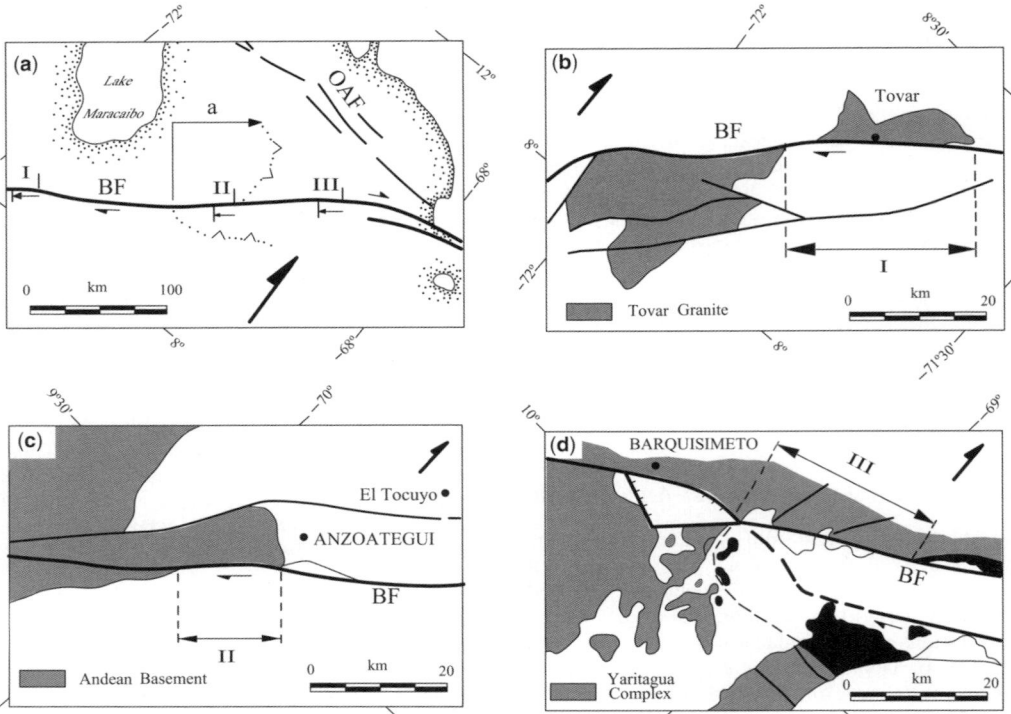

Fig. 2. Measurement of dextral offset along the Boconó fault (BF) on very different geological objects (from Audemard & Giraldo 1997), which ranges between 20 and 30 km: (**a**) relative location of the measured offsets along the Boconó Fault; 'a' identifies the offset derived by Stephan (1982) from nappe fronts. (**b**) Dextral offset (I) determined from the Tovar granite, in the southern Andes. (**c**) Dextral offset (II) measured on the Andean basement outcrops. (**d**) Dextral offset (III) measured on the Yaritagua Complex outcrops.

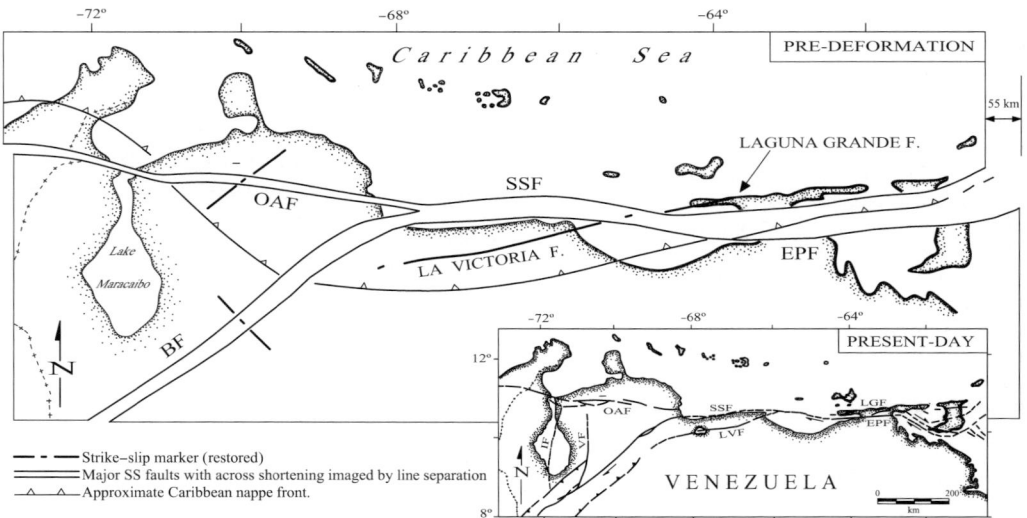

Fig. 3. Estimation of dextral wrenching accommodated by major onshore transcurrent faults in Cenozoic and Quaternary times, which seems to be in the order of some 55 km. Pre-deformation configuration for Trinidad is not realistic.

Audemard & Audemard 2002), 70 km in the Interior Range of eastern Venezuela (Passalacqua *et al.* 1996) and 10–15 km for the Falcón basin of western Venezuela. Dextral offset is 30 ± 3 km on the Oca–Ancón Fault, and 30 km on the Boconó Fault. Total displacement is in the order of 55 km, much less than the value proposed by Giraldo (1993, 1996), who did not take into account across-boundary shortening. This retro-deformation analysis indicates that La Victoria and Laguna Grande faults appeared to be a single feature before wrenching, which happens to be the hypothesis proposed by Giraldo (1996). A separate estimate of 51 ± 3 km is obtained by vector addition in the east–west direction of slip along the Oca–Ancón and Boconó faults of 30 ± 3 and 21 km (sine 45° × 30 km), respectively. Agreement between the two different approximations supports their reliability. Furthermore, restoration to the pre-offset configuration results in an almost straight system (pre-deformation stage in Fig. 3), as proposed by Albarracín (1988) and Audemard (1993).

Northward escape of the Maracaibo Block appears to be only in the order of 21 km, similar to the dextral component of the Boconó Fault in the east–west direction. This is a very minor portion of the 400 km of convergence postulated by van der Hilst & Mann (1994) and Pindell *et al.* (1998).

Lateral-slip onset

The date of onset of wrenching along the plate boundary is also controversial. Since two different wrench systems have operated, they are discussed separately.

Along the proto-boundary

The South America–Caribbean boundary used to be linear, west to east, along the Oca-Ancón, San Sebastían and El Pilar faults. The following data show that movement commenced, over some 8 Ma, from Early Miocene in Colombia to Late Miocene in Eastern Venezuela.

According to Audemard (1993, 2001) movement on the Oca-Ancón Fault began around the Early-to-Middle Miocene boundary (17–15 Ma), when compression started to invert the Late Eocene–Early Miocene Falcón Basin. Similar ages have been proposed by Tschanz *et al.* (1974), based on the age of the associated basins flanking the Santa Marta block (Early Miocene), and by Zambrano *et al.* (1971), Mann & Burke (1984) and Mathieu (1989). Igneous intrusions in the Falcón Basin young to the NE from 23 to 15 Ma ($^{40}Ar/^{39}Ar$ dating; McMahon 2000) indicate

progressive closure of magma conduits along with migrating compression.

There is no age control on the San Sebastián Fault.

Activation age of the El Pilar Fault seems well constrained. Earliest sedimentation in the Paria pull-apart basin shows that its eastern tip was active in the Late Miocene (12 Ma; Babb 1997). Giraldo (1996) proposed a similar onset age in the Barcelona region (Middle to Late Miocene; 15–10 Ma). Similarly, the Blanquilla flow path of Pindell *et al.* (1998) shows a change in eastern Venezuela from transpression to dominant wrenching between 12 and 10 Ma.

Along the current boundary

The main wrench movement within the western plate boundary zone shifted to the Boconó Fault around 5–3 Ma (Audemard 1993; Audemard & Audemard 2002). This resulted from collision and effective suturing of the Chocó Block with northwestern South America.

Major tectonic phases of the South America–Caribbean boundary

Although the Southern Caribbean boundary has been subject to compression throughout its Cenozoic and Quaternary evolution, important changes occurred, from oblique subduction to oblique collision and later to (partitioned or not) transpression.

Emplacement of the Caribbean nappes and their rotation

The southern Caribbean boundary evolved from oblique subduction to oblique collision diachronously along northern Venezuela from west to east. During oblique collision, Caribbean Plate nappes were overthrust onto the Mesozoic South America passive margin earlier on the west than on the east (e.g. Stephan 1982; Blanco *et al.* 1988; Audemard 1991; Pindell *et al.* 1998). This is shown by the age of the oldest sediments deposited in the foreland basins ahead of the nappes. Several detailed palinspatic reconstructions agree fairly well on this (Stephan *et al.* 1990; Pindell *et al.* 1998; Pindell & Kennan 2001). From west to east, the sedimentation starts in the Middle Eocene with the Gobernador and Pagüey formations in the Guanarito region, and with the Guárico Formation in the Guárico region (Blanco *et al.* 1988). Further to the east, the first deposits are Early Oligocene (La Pascua and Roblecito formations). In eastern Venezuela, the first foreland units are Middle to

Late Oligocene in age (Naricual and Quebradón, or Merecure formations; Blanco *et al.* 1988).

During emplacement, the nappes rotated 90° clockwise (Skerlec & Hardgraves 1980; Gose 1988). In contrast, Cretaceous rocks of the Perijá range only show 45° clockwise rotation (Perarnau *et al.* 1988; Gose *et al.* 1989). The Caribbean nappes rotated an additional 45° with respect to northwestern South America. The Cretaceous sedimentary nappes of Interior Range (northeastern Venezuela) have rotated clockwise by only 12° (Gose *et al.* 1989). Molasse deposits of the Quiamare formation show that the Interior Range began to rise in Late Miocene. Rotation occurred late in the eastward Caribbean Plate progression. The Mérida Andes show no rotation (Perarnau *et al.* 1988; Gose *et al.* 1989; Castillo *et al.* 1991). Neither does the Early–Middle Miocene igneous suite of central Falcón, implying that nappe rotation was mostly accomplished before Falcón Basin opening and intrusion.

Consequently, there is a regional pattern of rotation diminishing and younging toward the east along northern Venezuela. Recent studies on the Netherlands Leeward Antilles support this pattern (Beasrdsley & Avé Lallemant 2007). Two different mechanisms have been proposed to explain the eastward progression. Avé Lallemant (1997) has proposed that arc-parallel stretching occurred, evidenced by NW–SE trending horsts and grabens of the Leeward Antilles islands. Audemard (1993, 1998) suggested that oblique collision occurred along a fragmented arc bounded by NW–SE major dextral shears. Large dextral shears (Río Guárico, Tácata, Urica, Los Bajos–El Soldado faults) became Riedel shears to the major dextral system when transpression started, as imaged in the active tectonics map of Venezuela (Soulas 1986; Beltrán 1993, 1994; Audemard *et al.* 2000). Probably, both mechanisms played a major role during eastward drift of the Caribbean Plate. Arc-parallel stretching occurred in the Bonaire Block, north of the system, while block rotations occurred within the boundary zone. Both mechanisms helped accommodate north–south shortening across the zone.

Leeward Antilles subduction

Subduction on the northern boundary of the Bonaire Block occurs from the Uraba Gulf in the west to the Los Roques Canyon in the east. It might be partly inherited from Cretaceous South America type-B subduction. Leeward Antilles Subduction is rather young and is different from the subduction that led to collision of the Panamá Arc against western South America along the Chocó Block and San Jacinto terranes (Fig. 1). Original Caribbean

subduction, prior to the Panamá Arc–South America collision, used to trend roughly north–south along the western coast of South America. This major plate boundary was rather straight and was as old as Cretaceous in certain models (Pindell & Dewey 1982; Beck 1985; Pindell *et al.* 1988; Stephan *et al.* 1990; Taboada *et al.* 2000).

Taboada *et al.* (2000) show two subduction slabs in this region, both composed of abnormally thick Caribbean oceanic lithosphere, at different depths in asthenosphere. This is also imaged by tomography (van der Hilst 1990; van der Hilst & Mann 1994). This could suggest that the slabs are of different age, Leeward Antilles subduction being much younger. Duque-Caro (1978) proposed that Leeward Antilles subduction began around 10 Ma ago, whereas Audemard (1993, 1998) suggests 5–3 Ma, directly related to the expulsion of the Maracaibo Block. Subduction along the Leeward Antilles is incipient. A 30° S-dipping slab has descended 400–450 km below northwestern South America (Kellogg & Bonini 1982) and is just beginning to enter the asthenosphere. This implies convergence not larger than 100 km (using 30° dip from Kellogg & Bonini 1982) or 170 km (using 17° dip from van der Hilst & Mann 1994). Subtracting 21 km of northward escape of the Maracaibo Block, some 80 km of convergence has to be attributed to Americas convergence. If Leeward Antilles subduction is younger than 10 Ma (Duque-Caro 1978) or 5 Ma (Audemard 1993, 1998), the maximum convergence rate between the Americas is in the range of 16–30 mm/annum (80–150 km in 5 Ma). This is much faster than the average convergence (9 mm/annum for the last 20 Ma) implied by the Pindell *et al.* (1998) Americas-convergence path. However, poor constraints on upper continental (or intermediate-crust) lithosphere thickness, slab dip, hypocentre locations, and tomographic resolution could account for this difference. Indications that Leeward Antilles subduction is relatively young come from: (1) seismic profiles and bathymetry (e.g. Silver *et al.* 1975; Talwani *et al.* 1977; Mascle *et al.* 1979; Kellogg & Bonini 1982; Ruiz *et al.* 2000); (2) Pliocene–Pleistocene deformation of the accretionary prism west of Santa Marta (Ruiz *et al.* 2000); (3) lack of significant sedimentation in Los Roques canyon in the last few tens of millions years; (4) paucity of intermediate thrust earthquakes, up to 200 km deep, under the Maracaibo Basin (Orihuela & Cuevas 1992; Malavé & Suárez 1995).

The Proto-Caribbean Plate arc

The Aves Ridge, the Lesser Antilles arc and the extinct arc of the Leeward Antilles used to belong to a single large arc (Mesozoic Caribbean Arc of

Bouysse 1988). Audemard (1993, 1998) and Pindell *et al.* (1998) include the Villa de Cura Nappe of north-central Venezuela in this original arc. It was split into two by Paleocene (Bouysse 1988) to Eocene (Pinet *et al.* 1985) rifting and spreading in the Grenada back-arc Basin. Bird *et al.* (1993) discussed various models of Grenada Basin opening and concluded that east–west spreading occurred. Pinet *et al.* (1985) suggested that the basin opening north of Grenada progressed from south to north, from Eocene to Oligocene–Early Miocene. Bouysse (1988) noted that most subsidence predates the Miocene, implying an Oligocene–Early Miocene age.

The Falcón–Bonaire and other Oligocene basins

The Falcón Basin, today inverted into an ENE–WSW trending anticlinorium, extends eastward into the Bonaire basin (e.g. Hunter & Ferrell 1972; Díaz de Gamero 1977; Mascle *et al.* 1979; Audemard 1993, 2001). This basin used to be an east–west to ENE–WSW trending graben (e.g. González de Juana *et al.* 1980; Audemard 1993). It subsided rapidly during the Oligocene–Early Miocene. Upper Eocene sedimentary rocks crop out in eastern Falcón (González de Juana *et al.* 1980; Audemard 1995, 2001). The Falcón Basin was surrounded by land on three sides and opened to the east (or ENE). Sedimentation in the Bonaire part of the basin started in Late Eocene (Beck 1983). This suggests that the Falcón–Bonaire basin opened from east to west (Audemard 1993, 1995, 1998, 2001). The Blanquilla Basin was also of the same age (Ysaccis 1997). Before inversion it lay between the Bonaire and Grenada basins, north of Margarita Island.

Since these basins used to constitute a single, crescent-shaped back-arc basin, Audemard (1995, 1997b) proposed that it formed as the consequence of an orogenic collapse (see also Porras 2000). Opening of the basin subdivided the Caribbean Plate, with the Leeward Antilles and the Aves swell outboard and the Villa de Cura nappe to the south (Audemard 1993, 1998; Pindell *et al.* 1998). Audemard (1993, 1998) related the basin opening to resumption of convergence between the Americas at around 33–38 Ma. This slowed eastward motion of the Caribbean Plate, triggered roll-back of the Atlantic slab at the leading edge and back-arc spreading from the Grenada Basin to the Falcón Basin.

This hypothesis differs from the pull-apart model of Pindell & Dewey (1982) Muessig (1984), Stephan (1985) and many others. Geometry necessary to produce local transtension has not yet been convincingly demonstrated for the Falcón

Basin. There is no satisfactory evidence of an active Late Eocene–Oligocene strike–slip fault. Isolated igneous bodies in central Falcón trend WSW–ENE (Brueren 1949; Muessig 1984; McMahon 2000). These instrusions were dated at 22.9 ± 0.9 Ma (K–Ar, Muessig 1978) or between 23 and 15 Ma ($^{40}Ar/^{39}Ar$; McMahon 2000). They show no rotation. If intruded during transtension, the extensional faults, used as conduits, of an east–west dextral system, such as the Oca–Ancón fault system, should strike NW–SE, but not WSW–ENE. Finally, this trend of igneous bodies must delineate the axis of rifted crust.

Middle and Late Miocene tectonism responsible for inversion of the Falcón Basin also induced an unconformity in the Bonaire (Campos 1981; Biju-Duval *et al.* 1982; Beck 1983), Blanquilla (Ysaccis *et al.* 2000) and Grenada basins (Pinet *et al.* 1985; Bouysse 1988) basins. The inversion marks the beginning of partitioned transpression (Audemard 1993, 2001). This is evidenced by: (1) closure of igneous conduits in central Falcón; (2) inversion of the Falcón rift; (3) major unconformities throughout the basins (Bonaire, Blanquilla, Grenada) along the plate boundary; (4) opening of still active large pull-apart basins (Cariaco, Gulf of Paria); (5) diachronous wrenching (Oca–Ancón, San Sebastián, El Pilar, Warm Springs); and (6) a second, Late Miocene–Pliocene phase of Cenozoic orogenic collapse of the Caribbean nappe pile (Audemard 2002). It produced isolated intracontinental (occasionally early marine) basins, larger in size from west to east (Bejuma, Nirgua, Lake Valencia, Santa Lucía–Ocumare del Tuy, Barlovento–Ensenada de Barcelona, and eventually Gulf of Cariaco). Audemard (2002) suggested that wrenching slightly reduces coupling between the two plates, conforming with flowlines of Pindell *et al.* (1998, p. 60, fig. 12), diminishing compression across the boundary and inducing collapse of the higher portion of the orogen. The two major phases of Cenozoic orogenic collapse could also result from fast uplift and exhumation in response to detachment of a subducting slab.

Geodymic evolution at lithospheric scale

Following the above, Cenozoic evolution of the South America–Caribbean Plate boundary evolved in the following seven stages presented.

Late Cretaceous–Paleocene (65 Ma)

By this time, the Caribbean Great Arc had entered between the two Americas and was colliding with the Yucatán and South America passive margins (Fig. 4). In the south, the Caribbean Plate slid around the NW corner of South America, turning

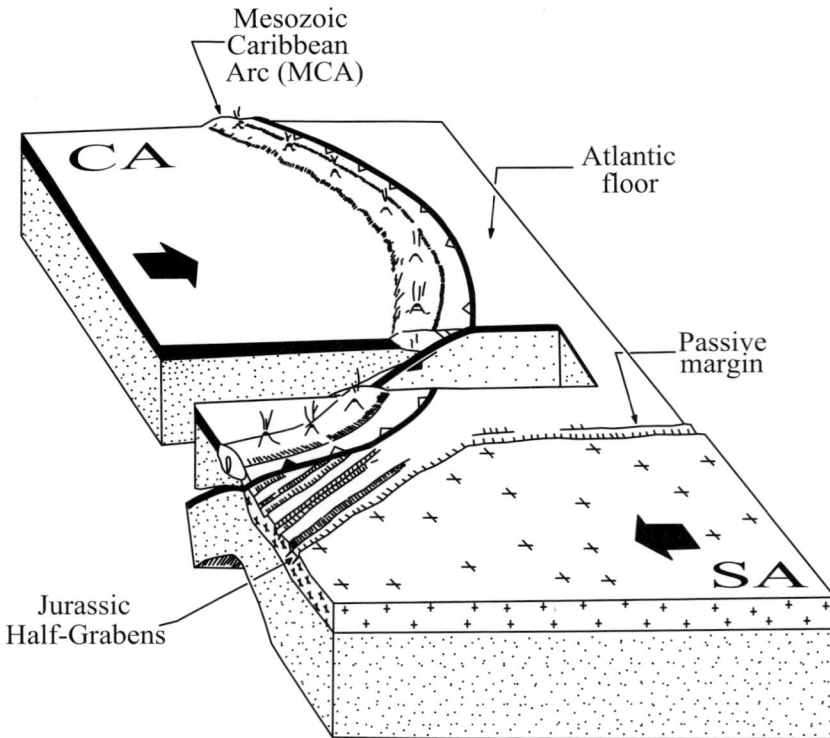

Fig. 4. Block diagram at lithospheric scale of the geodynamic reconstruction of the southern Caribbean Plate boundary zone at the Late Cretaceous–Paleocene boundary (from Audemard 1993).

progressively towards an east–west direction and diachronously docking against the northern South America passive margin. In western Venezuela, docking continued along a NW–SE direction until the Late Eocene.

Migration slowed almost to a complete stop, due to convergence of the Americas, and induced roll-back of Atlantic subduction below the arc in the Paleocene–Eocene, when back-arc spreading in the Grenada Basin started. Opening of this basin started near Grenada and progressed north and south. Volcanism ceased along the (abandoned) Aves Ridge.

Paleocene–Early Eocene (60–50 Ma)

After slight clockwise rotation in response to collision along the NW margin of South America, the Caribbean Plate progressed eastward between the Americas, colliding obliquely with northern South America. Overthrusting and foreland basin subsidence migrated eastward in western Venezuela (Fig. 5). Clockwise rotation of the Caribbean nappes against South America progressed eastward.

Middle Eocene (50–38 Ma)

Oblique collision continued with progressive clockwise docking of the Caribbean Plate against the South America (Fig. 6). This east-younging docking is a step-by-step process, during which discrete pieces of the tectonic pile are overthrusted. In particular, a fragment of the Caribbean Great Arc, the Villa de Cura Nappe was thrusted onto north-central Venezuela. Foredeep flysch (Guárico Formation) along the leading edge of the nappes was folded and thrusted southeastward. Spreading continued in the Grenada Basin towards the north and south ends.

Late Eocene (38–35 Ma)

Overthrusting emplaced the Caribbean nappes in north-central Venezuela (Fig. 7). Southward propagation of the Grenada Basin reached the Bonaire basin and the eastern Falcón region (Fig. 7).

Oligocene–Early Miocene (35–17 Ma)

The Grenada, Blanquilla, Bonaire and Falcón basins subsided strongly. Nappe emplacement deformed

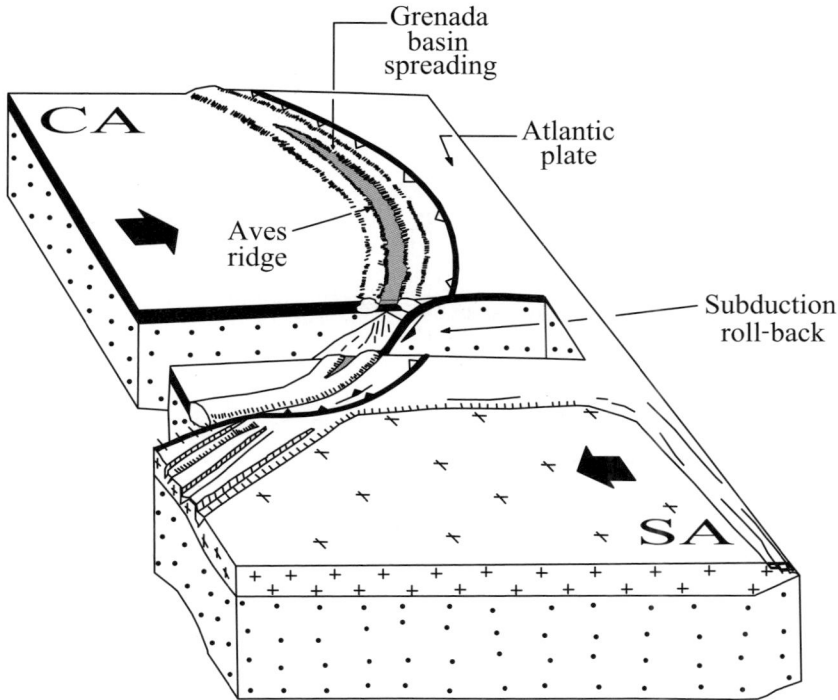

Fig. 5. Block diagram at lithospheric scale of the geodynamic reconstruction of the Southern Caribbean Plate boundary zone during the Paleocene–Early Eocene (60–50 Ma; from Audemard 1993).

their foreland basins and the Guárico basin. Basalts intruded the WSW–ENE axis of the Falcón Basin (23–15 Ma).

Middle–Late Miocene boundary (17–15 Ma)

Caribbean nappe emplacement continued along the southern Caribbean boundary in the Guárico basin. However, a drastic change in the dynamics of the boundary occurred around this time. All the subsiding basins of Grenada–Falcón show a regionally significant unconformity, coeval with inversion onset of the Falcón basin. The discrete block mechanism that enabled emplacement of the Caribbean nappes no longer allowed displacement of the Caribbean Plate to the east. Caribbean Plate–South America coupling had become too high. Instead, wrenching reactivated the weakened axis of the former Caribbean Plate and the Caribbean nappes by the Oligocene–Early Miocene back-arc spreading. Dextral slip occurred along 500 km of the former southern margin of Falcón–Bonaire basin (Fig. 8). The Oca–Ancón fault system and San Sebastián Fault located along the graben southern margin, whereas the El Pilar Fault reactivated an old high-angle thrust fault. This tectonic

inheritance was enabled by parallelism of the collision zone to Caribbean Plate motion.

During the Late Miocene Caribbean nappe emplacement occurred in the Maturín Basin. The allochthonous province of north-central Venezuela was reactivated (Middle–Late Miocene). The El Pilar Fault began dextral movement at around 12 Ma, with consequent opening of the Gulf of Paria and deepening of the Cariaco Trough. A second phase of orogenic collapse began along the Caribbean nappe, generating small basins between the Coastal and Interior ranges. They are mostly filled with continental deposits but early marine incursions occurred. The Gulf of Cariaco could be part of this event. A modest orogenic pulse occurred in the Mérida Andes, coeval with events in the Eastern Cordillera of Colombia.

Pliocene (5–3 Ma)

Caribbean Plate boundaries assumed their present configuration. The dextral southern boundary jumped from the Oca–San Sebastián–El Pilar fault system to the Boconó–San Sebastián–El Pilar system (Fig. 9). This shift reflects the last stages of Panamá arc collision and suturing against

F. A. AUDEMARD

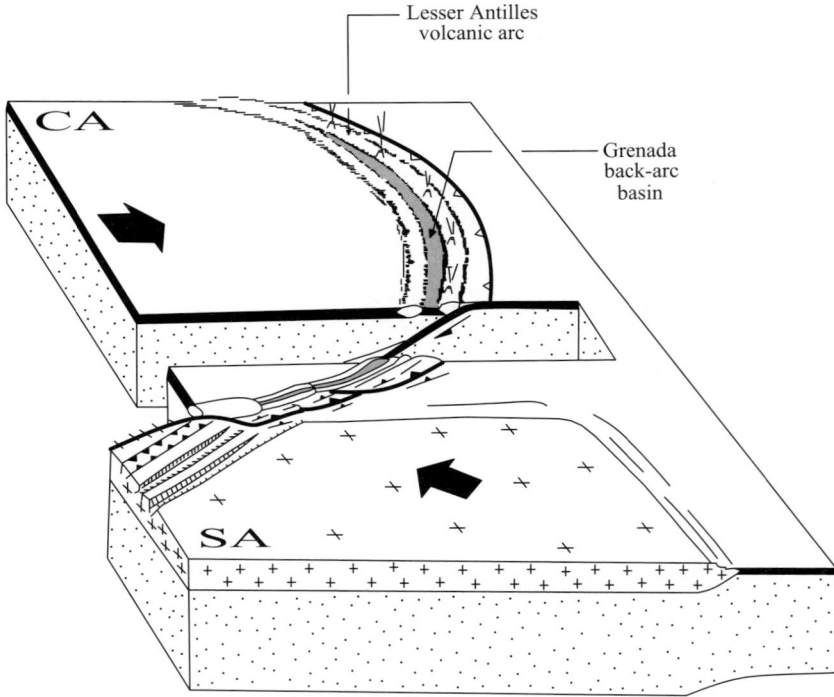

Fig. 6. Block diagram at lithospheric scale of the geodynamic reconstruction of the Southern Caribbean Plate boundary zone, during the Middle Eocene (50–38 Ma; from Audemard 1993).

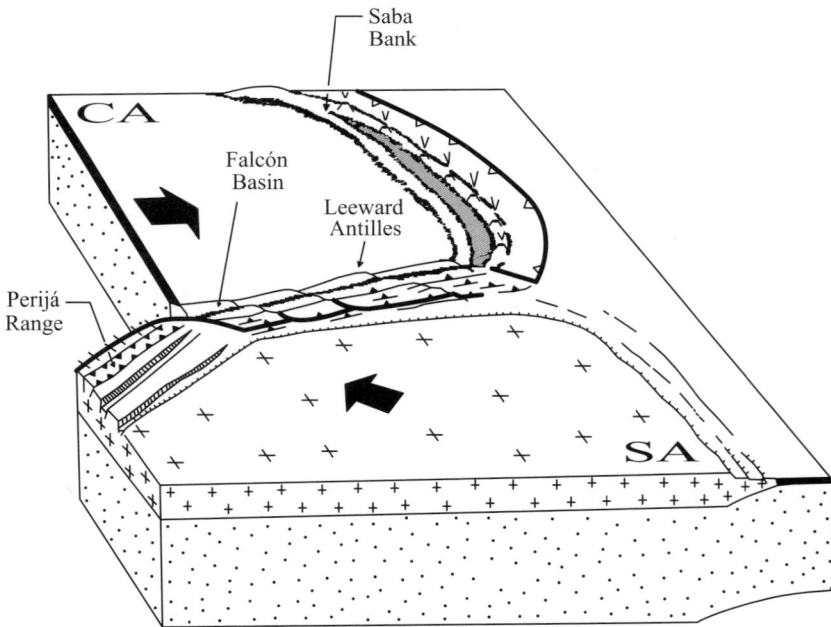

Fig. 7. Block diagram at lithospheric scale of the geodynamic reconstruction of the Southern Caribbean Plate boundary zone, during the Late Eocene (38–35 Ma; from Audemard 1993).

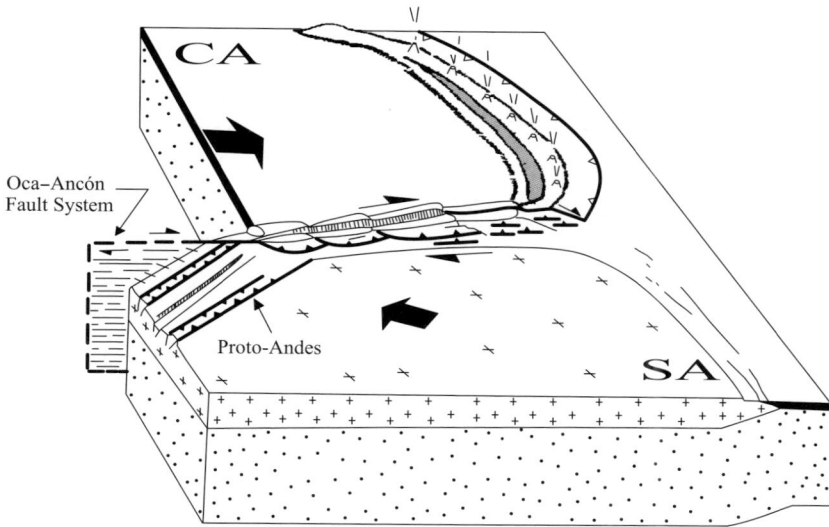

Fig. 8. Block diagram at lithospheric scale of the geodynamic reconstruction of the Southern Caribbean Plate boundary zone at the Early–Middle Miocene boundary (17–15 Ma; from Audemard 1993).

northwestern South America. It resulted in (1) NNE escape of the Maracaibo and Bonaire blocks bounded by the SW–NE Pallatanga–Dolores–Algeciras–Guaicaramo–Boconó fault system; (2) Flat-slab, amagmatic, B subduction of the Venezuela Basin floor under the Bonaire Block in northwestern Venezuela and northern Colombia, as well as the formation of the South Caribbean Deformed Belt; (3) fast uplift and lateral-shortening of the Mérida Andes; and (4) 30 km dextral movement along the Boconó Fault.

Conclusions

Constraints on the Cenozoic geodynamic evolution of the Southern Caribbean Plate boundary zone are revealed by active tectonics in Venezuela and major tectonic and stratigraphic evolution observed in basins and ranges. The boundary was wide and complex throughout the Cenozoic and Quaternary. It is characterized by partitioned tranpression and evolved from oblique subduction to oblique collision. Although transpression has dominated the

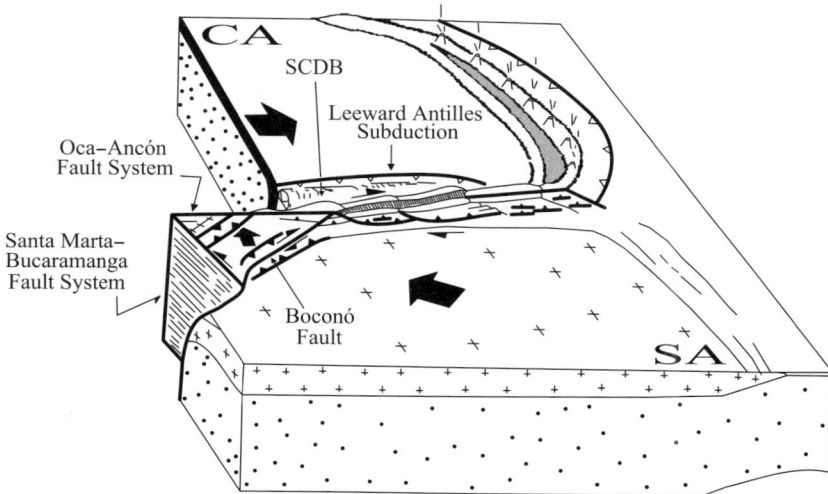

Fig. 9. Block diagram at lithospheric scale of the geodynamic reconstruction of the Southern Caribbean Plate boundary zone, during the Pliocene (5–3 Ma; from Audemard 1993).

boundary, (1) two major phases (Late Eocene–Early Miocene, Late Miocene–Early Pleistocene) of orogenic collapse occurred within the former Caribbean Plate or the Caribbean nappe pile and major transtensional basins formed (Cariaco Trough, Gulf of Paria). (2) The early phase was triggered by Atlantic subduction roll-back. The younger stage resulted from reduction of coupling due to shift from oblique collision to transpression. (3) The reported total amount of dextral slip along major fault systems in onshore Venezuela cannot account for the 1000+ km of Caribbean Plate migration along northern South America. The plate boundary formerly had to function in oblique subduction to account for large displacements. (4) Onset of uplift of the Mérida Andes, dextral wrenching of the Boconó Fault, activation of the Leeward Antilles subduction along northern Colombia and northwestern Venezuela and tectonic escape of the Maracaibo and Bonaire blocks is much younger (last 5 Ma) than usually proposed, and resulted from suturing of the Panamá arc against South America.

For a contribution such as this, it is impossible to thank individuals who may have contributed in some way, because it gathers many years of research and discussion on the southern margin of this very complex geological object. Thanks to you all! Nonetheless, I can acknowledge, of course, the unrestricted funding provided by FUNVISIS over 20 years, for active tectonics research and congress attendance. I sincerely thank M. Peña for the illustrations and K. H. James for greatly improving a former version of this contribution. This paper is a contribution to projects IGCP-433 and FONACIT 2001002492, G-2002000478 and PI-2003000090.

References

ALBARRACÍN, J. 1988. La Falla de Oca, su origen y desplazamiento. *Proceedings IV Congreso Venezolano de Geofísica*, Caracas, 443–450.

ALGAR, S. & PINDELL, J. L. 1993. Structure and deformation history of the northern range of Trinidad and adjacent areas. *Tectonics*, **12**, 814–829.

AUDEMARD, F. E. 1991. *Tectonics of Western Venezuela*. PhD thesis, Rice University, Houston, TX.

AUDEMARD, F. A. 1993. *Néotectonique, Sismotectonique et Aléa Sismique du Nord-ouest du Vénézuéla (Système de failles de Oca-Ancón)*. PhD thesis, Université Montpellier II, 369 pp.

AUDEMARD, F. A. 1995. La Cuenca Terciaria de Falcón, Venezuela Noroccidental: Síntesis Estratigráfica, Génesis e Inversión Tectónica. *Proceedings IX Congreso Latinoamericano de Geología, Caracas, Venezuela* (on Diskette).

AUDEMARD, F. A. 1997a. Holocene and Historical Earthquakes on the Boconó Fault System, Southern Venezuelan Andes: Trench Confirmation. *Journal of Geodynamics*, **24**, 155–167.

AUDEMARD, F. A. 1997b. Tectónica activa de la región septentrional de la cuenca invertida de Falcón, Venezuela occidental. *Proceedings VIII Congreso Geológico Venezolano*, Porlamar, **1**, 93–100.

AUDEMARD, F. A. 1998. Evolution géodynamique de la façade nord sud-americaine: Nouveaux apports de l'histoire géologique du bassin de Falcón, Vénézuéla. *Proceedings of 3rd Geological Conference of the Geological Society of Trinidad and Tobago and 14th Caribbean Geological Conference*, Port of Spain, Trinidad, 1995, **2**, 327–340.

AUDEMARD, F. A. 1999. Morpho-structural expression of active thrust fault systems in humid tropical foothills of Colombia and Venezuela. *Zeitschrift für Geomorphologie*, **118**, 1–18.

AUDEMARD, F. A. 2000. Major Active Faults of Venezuela. *Proceedings of 31st International Geological Congress*, Rio de Janeiro, Brazil (extended abstract; on CD-RoM).

AUDEMARD, F. A. 2001. Quaternary tectonics and stress tensor of the northern inverted Falcón Basin, northwestern Venezuela. *Journal of Structural Geology, Special Memorial Issue to Paul Hancock*, **23**, 431–453.

AUDEMARD, F. A. 2002. Syn-sedimentary extensional tectonics in the River Tuy basin, northern Venezuela: implications on basin genesis and southern Caribbean geodynamics. *Proceedings of XI Congreso Venezolano de Geofísica*, Caracas, Venezuela, (extended abstract; on CD-RoM).

AUDEMARD, F. A. & GIRALDO, C. 1997. Desplazamientos dextrales a lo largo de la frontera meridional de la placa Caribe, Venezuela septentrional. *Proceedings of VIII Congreso Geológico Venezolano*, Porlamar, **1**, 101–108.

AUDEMARD, F. E. & AUDEMARD, F. A. 2002. Structure of the Mérida Andes, Venezuela: relations with the South America–Caribbean geodynamic interaction. *Tectonophysics*, **345**, 299–327.

AUDEMARD, F. A., COSTA, C. & DE SANTIS, F. 1988. Observaciones geológicas sobre el 'Corrimiento' de Cantagallo, entre El Pao y San Juan de los Morros, Venezuela. Abstract volume of *V Congreso Venezolano de Sismología e Ingeniería Sísmica*.

AUDEMARD, F. A., SINGER, A., RODRÍGUEZ, J. A. & BELTRAN, C. 1994. Definición de la traza activa del sistema de fallas de Oca-Ancón, Noroccidente de Venezuela. *Proceedings of VII Congreso Venezolano Geofísica*, Caracas, Venezuela, 43–51.

AUDEMARD, F. A., MACHETTE, M., COX, J., DART, R. & HALLER, K. 2000. *Map and Database of Quaternary Faults and Folds in Venezuela and its Offshore Regions*. USGS Open-File Reports, **00-0018** (accessible from USGS webpage; open file reports ofr-00-0018).

AUDEMARD, F. A., PANTOSTI, D. ET AL. 1999. Trench investigation along the Merida sectión of the Boconó fault (central Venezuelan Andes), Venezuela. *Tectonophysics*, **308**, 1–21.

AUDEMARD, F. A., ROMERO, G., RENDÓN, H. & CANO, V. 2005. Quaternary fault kinematics and stress tensors along the southern Caribbean from fault-slip data and focal mechanism solutions. *Earth Science Reviews*, **69**, 181–233.

AUDEMARD, F. A., CARRILLO, E. & BECK, C. 2007. *Fieldtrip Guidebook for International Workshop on*

'Blind Dip-slip Faulting and Strain Partitioning in an Active Orogen: The Mérida Andes Case, Venezuela', Santo Domingo, State of Mérida, Venezuela, 5–9 March 2007.

AVÉ LALLEMANT, H. 1997. Transpression, displacement partitioning, and exhumation in the eastern Carribean/South America plate boundary zone. *Tectonics*, **16**, 272–289.

BABB, S. 1997. *Tectonics and sedimentation of the Gulf of Paria and Northern Basin, Trinidad*. PhD thesis, The University of Texas at Austin, TX.

BEARDSLEY, A. & AVÉ LALLEMANT, H. 2007. Oblique collision and accretion of the Netherlands Leeward Antilles to South America. *Tectonics*, **26**, TC2009.

BECK, C. 1983. Essai sur l'évolution géodynamique des Caraïbes sud-orientales. *Bulletin Societé Géologique de France*, **25**, 169–183.

BECK, C. 1985. Caribbean colliding, Andean drifting and the Mesozoic–Cenozoic geodynamic evolution of the Caribbean. *Proceedings of VI Congreso Geológico Venezolano*, Caracas, **10**, 6575–6614.

BECK, C. 1986. *Géologie de la Chaîne Caraïbe au méridien de Caracas (Venezuela)*. Societé Géologique du Nord Publications, **14**.

BELLIZZIA, A., PIMENTEL, N. & BAJO, R. 1976. *Mapa geológico-estructural de Venezuela*. Scale 1 : 500 000. Ministerio de Energia y Minas, Ed. Foninves, Caracas.

BELTRÁN, C. 1993. *Mapa Neotectónico de Venezuela*. Scale 1 : 2 000 000. Funvisis.

BELTRÁN, C. 1994. Trazas activas y síntesis neotectónica de Venezuela a escala 1 : 72 000 000. *Proceedings of VII Congreso Venezolano de Geofísica*, Caracas, 541–547.

BELTRÁN, C. & GIRALDO, C. 1989. Aspectos neotectónicos de la región nororiental de Venezuela. *Proceedings of VII Congreso Geológico Venezolano*, Barquisimeto, **3**, 1000–1021.

BENÍTEZ, S. 1986. Síntesis geológica del graben de Jambeli. *IV Congreso Ecuatoriano de Geología, Minería y Petróleo*, **1**, 137–160.

BIJU-DUVAL, B., MASCLE, A., ROSALES, H. & YOUNG, G. 1982. *Episutural Oligo–Miocene Basins Along the North Venezuelan Margin*. American Association of Petroleum Geologists, Memoirs, **34**, 347–358.

BIRD, D., HALL, S., CASEY, J. & MILLEGAN, P. 1993. Interpretation of magnetic anomalies over the Grenada Basin. *Tectonics*, **12**, 1267–1279.

BLANCO, B., GÓMEZ, E. & SÁNCHEZ, J. 1988. Evolución tectono-sedimentaria del Norte de los Estados Anzoategui, Guárico y Portuguesa. *Proceedings IV Congreso Venezolano Geofísica*, Caracas, Venezuela, 151–159.

BOUYSSE, P. 1988. Opening of the Grenada back-arc basin and evolution of the Caribbean Plate during the Mesozoic and Early Paleogene. *Tectonophysics*, **149**, 121–143.

BRUEREN, J. 1949. *Geological report CPMS-310. Paraiso-Maraure (central Falcon)*. Unpublished Maraven S.A. Report.

CALAIS, E., MERCIER DE LEPINAY, B., SAINT-MARC, P., BUTTERLIN, J. & SCHAAF, A. 1992. La limite de plaques décrochante nord caraïbe en Hispaniola: évolution paléogéographique et structurale cénozoïque. *Bulletin Societé Géologique de France*, **163**, 309–324.

CAMPOS, V. 1981. *Une transversale de la Chaîne Caraïbe et de la marge vénézuélienne dans le secteur de Carúpano (Vénézuéla oriental): Structure géologique et évolution géodynamique*. Thèse Doctorat 3ème Cycle, Université de Bretagne Occidentale, Brest.

CASE, J. & MACDONALD, W. 1973. Regional gravity anomalies and crustal structure in northern Colombia. *Bulletin of the Geological Society of America*, **84**, 2905–2916.

CASE, J. E., DURÁN, L., LÓPEZ, A. & MOORE, W. 1971. Tectonic investigations in western Colombia and eastern Panamá. *Bulletin Geological Society of America*, **82**, 2685–2712.

CASTILLO, J., GOSE, W. & PERARNAU, A. 1991. Paleomagnetic result from Mesozoic strata in the Mérida Andes, Venezuela. *Journal of Geophysical Research*, **96**, 6011–6022.

CHOY, J., MORANDI, M. T. & PALME DE OSECHAS, C. 1998. Determinación de patrones de esfuerzos tectónicos para el Oriente de Venezuela–Sureste del Caribe a partir de mecanismos focales. *Proceedings of IX Congreso Venezolano de Geofísica*, Caracas (CD-RoM; paper no. 17)

COLLETTA, B., ROURE, F., DE TONI, B., LOUREIRO, D., PASSALACQUA, H. & GOU, Y. 1996. Tectonic inheritance and structural styles in the Merida Andes (western Venezuela). *3rd International Symposium on Andean Geodynamics*, Saint-Malo, France, 323–326.

COLLETTA, B., ROURE, F., DE TONI, B., LOUREIRO, D., PASSALACQUA, H. & GOU, Y. 1997. Tectonic inheritance, crustal architecture, and contrasting structural styles in the Venezuelan Andes. *Tectonics*, **16**, 777–794.

COLMENARES, L. & ZOBACK, M. D. 2003. Stress field and seismotectonics of northern South America. *Geology*, **31**, 721–724.

COOPER, M. A., ADDISON, F. T. ET AL. 1995. Basin development and tectonic history of the Llanos Basin, Eastern Cordillera, and Middle Magdalena Valley, Colombia. *Bulletin of the American Association of Petroleum Geologists*, **79**, 1421–1443.

DE TONI, B. & KELLOGG, J. 1993. Seismic evidence for blind thrusting of the northwestern flank of the Venezuelan Andes. *Tectonics*, **12**, 1393–1409.

DEWEY, J. 1972. Seismicity and tectonics of western Venezuela. *Bulletin Seismological Society of America*, **62**, 1711–1751.

DÍAZ DE GAMERO, M. L. 1977. Estratigrafía y micropaleontología del Oligoceno y Mioceno inferior del centro de la Cuenca de Falcón, Venezuela. *GEOS*, **22**, 3–60.

DOOLAN, B. & MACDONALD, W. 1971. Structure and metamorphism of schists of the Santa Marta area, Colombia. *Proceedings of I Congreso Colombiano de Geología*, Bogotá, 187–206.

DUQUE-CARO, H. 1978. *Major Structural Elements of Northern Colombia*. American Association of Petroleum Geologists Memoirs, **29**, 329–351.

DUQUE-CARO, H. 1990. The Chocó block in the northwestern corner of South America; structural, tectonostratigraphic and paleogeographic implications. *Journal of South American Earth Sciences*, **3**, 1–14.

EGO, F., SÉBRIER, M., LAVENU, A., YEPES, H. &
EGUES, A. 1996. Quaternary state of stress in the
northern Andes and the restraining bend model
for the Ecuadorian Andes. *Tectonophysics*, **259**,
101–116.

FEO-CODECIDO, G. 1972. Breves ideas sobre la estructura
de la Falla de Oca, Venezuela. *Proceedings of VI Caribbean Geological Conference*, Margarita, Venezuela,
191–202.

FREYMUELLER, J. T., KELLOGG, J. N. & VEGA, V. 1993.
Plate motions in the north Andean region. *Journal of
Geophysical Research*, **98**, 21853–21863.

GINGERICH, P. D. 1985. South American mammals in the
Palaeocene of North America. *In*: STEHLI, F. G. &
WEBB, S. D. (eds) *The Great American Biotic Inter-
change*. Plenum Press, New York, 123–137.

GIRALDO, C. 1985. *Néotectonique et sismotectonique de
la région d'El Tocuyo-San Felipe (Vénézuéla centro-
occidental)*. PhD thesis, Université de Montpellier II.

GIRALDO, C. 1989. Valor del desplazamiento dextral acu-
mulado a lo largo de la falla de Boconó, Andes Venezolanos. *GEOS*, **29**, 186–194.

GIRALDO, C. 1993. New ideas about displacement and
history of El Pilar fault, Eastern Venezuela. *AAPG/
SVG International Congress and Exhibition* (Poster).

GIRALDO, C. 1996. Hipótesis acerca del desplazamiento
de la falla de El Pilar, Venezuela nororiental. *Proceedings of 8th Congreso Venezolano de Geofísica*,
Maracaibo, 387–392.

GONZÁLEZ DE JUANA, C., PICARD, X. & ITURRALDE, J.
1980. *Geología de Venezuela y de sus cuencas petrolíferas*. Ediciones Foninves, Caracas.

GOSE, W. 1988. Caribbean paleomagnetism: where do
we stand and where should we go? *Proceedings IV
Congreso Venezolano de Geofísica*. Caracas,
Venezvela, 451–456.

GOSE, W., CASTILLO, J. & PERARNAU, A. 1989. Resultados paleomagnéticos de rocas sedimentarias cretácicas
de Venezuela. *Proceedings of VII Congreso Geológico
Venezolano*, Barquisimeto, Venezuela, **2**, 726–730.

HESS, H. & MAXWELL, J. 1953. Caribbean Research
Project. *Bulletin Geological Society of America*, **64**, 1–6.

HUNTER, V. & FERRELL, A. 1972. Redefinición de
algunas unidades estratigráficas del Oligoceno de
Falcón Central. *Proceedings of III Congreso Geológico Venezolano*, Caracas, **2**, 807–816.

JÁCOME, M. 1994. *Interpretación geológica, sísmica y
gravimétrica de un perfil transandino*. Undergraduate
Thesis, Universidad Simón Bolívar, Caracas,
Venezuela.

JAMES, K. H. 1990. The Venezuelan hydrocarbon habitat.
In: BROOKS, J. (ed.) *Classic Petroleum Provinces*.
Geological Society, London, Special Publications,
50, 9–35.

JAMES, K. H. 2002. *A Simple Synthesis of Caribbean
Geology*. World Wide Web Address: http://www.ig.
utexas.edu/CaribPlate/CaribPlate.html.

JAMES, K. H. 2005*a*. Arguments for and against the Pacific
origin of the Caribbean Plate and arguments for an
in situ origin. Transactions of the 16th Caribbean Geological Conference, Barbados. *Caribbean Journal of
Earth Science*, **39**, 47–67.

JAMES, K. H. 2005*b*. A simple synthesis of Caribbean
geology. Transactions of the 16th Caribbean

Geological Conference, Barbados. *Caribbean
Journal of Earth Science*, **39**, 69–82.

JANSSEN, F. 1979. *Structural style of north-western
Venezuela*. Unpublished Company Report, EPC.
6270. Maraven S.A., Caracas.

JORDAN, T. 1975. The present-day motion of the Caribbean Plate. *Journal of Geophysical Research*, **80**,
4433–4439.

KANIUTH, K., DREWES, H. *ET AL.* 1999. *Position changes
due to recent crustal deformations along the
Caribbean–South American Plate boundary derived
from CASA GPS project*. General Assembly of the
International Union of Geodesy and Geophysics
(IUGG), Birmingham, UK. Poster at Symposium G1
of International Association of Geodesy.

KEIGWIN, L. 1978. Pliocene closing of the Isthmus of
Panamá, based on biostratigraphic evidence from
nearby Pacific Ocean and Caribbean Sea cores.
Geology, **6**, 630–634.

KELLER, G, ZENKER, C. E. & STONE, S. M. 1989.
Late Neogene history of the Pacific–Caribbean
gateway. *Journal of South American Earth Sciences*,
2, 73–108.

KELLOGG, J. & BONINI, W. 1982. Subduction of the
Caribbean Plate and basement uplifts in the overriding
South-American plate. *Tectonics*, **1**, 251–276.

KELLOGG, J. & VEGA, V. 1995. *Tectonic Development of
Panama, Costa Rica, and the Colombian Andes: Con-
straints from Global Positioning System Geodetic
Studies and Gravity*. Geological Society of America,
Boulder, CO, Special Papers, **295**, 75–90.

LADD, J. 1976. Relative motion of South America with
respect to North America and Caribbean tectonics.
Bulletin of the Geological Society of America, **87**,
969–976.

LUGO, J. & MANN, P. 1992. Colisión oblicua y formación
de una cuenca foreland durante el Paleoceno tardío al
Eoceno medio; Cuenca de Maracaibo, Venezuela.
*Proceedings of III Congreso Geológico de España
and VIII Congreso Latinoamericano de Geología*,
Salamanca, **4**, 60–64.

MALAVÉ, G. & SUÁREZ, G. 1995. Intermediate-depth
seismicity in northern Colombia and western Venezuela and its relationship to Caribbean Plate subduction. *Tectonics*, **14**, 617–628.

MANN, P. & BURKE, K. 1984. Cenozoic rift formation in
the northern Caribbean. *Geology*, **12**, 732–736.

MASCLE, A., BIJU-DUVAL, B. *ET AL.* 1979. Estructura y
evolución de los márgenes este y sur del Caribe
(análisis de los problemas del Caribe). *Bulletin du
Bureau de Recherches Géologiques et Minières*, **3/4**,
171–184.

MATHIEU, X. 1989. *La Serrania de Trujillo-Ziruma aux
confins du Bassin de Maracaibo, de la Sierra de
Falcon et de la Chaine Caraïbe*. PhD thesis, Université
de Brest.

MCMAHON, C. 2000. *Evaluation of the effects of oblique
collision between the Caribbean and South American
plates using geochemistry from igneous and metamorphic bodies of northern Venezuela*. PhD thesis,
University of Notre Dame, Notre Dame, IN.

METZ, H. 1968. Geology of the El Pilar fault zone, State of
Sucre, Venezuela. *Proceedings of 4th Caribbean Geological Conference*, Trinidad, 193–198.

MINSTER, J. & JORDAN, T. 1978. Present-day plate motions. *Journal of Geophysical Research*, **83**, 5331–5354.

MORRIS, A., TANER, I., MEYERHOFF, H. A. & MEYERHOFF, A. A. 1990. Tectonic evolution of the Caribbean region; alternative hypothesis. *In*: DENGO, G. & CASE, J. E. (eds) *The Geology of North America*. Geological Society of America, Boulder, CO, **H**, 433–457.

MUESSIG, K. 1978. The central Falcon igneous suite, Venezuela: alkaline basalts intrusions of Oligocene–Miocene age. *Geologie Mijnbow*, **52**, 261–262.

MUESSIG, K. 1984. *Paleomagnetic Data on the Basic Igneous of the Central Falcon Basin, Venezuela*. Geological Society of America, Boulder, CO, Memoirs, **162**, 231–237.

ORIHUELA, N. & CUEVAS, J. 1992. Modelaje sismo-gravimétrico de perfiles regionales del Caribe Central. *XIII Caribbean Geological Conference*, Cuba (Abstract).

PASSALACQUA, H., FERNÁNDEZ, F., GOU, Y. & ROURE, F. 1995. *Crustal Architecture and Strain Partitioning in the Eastern Venezuela Ranges*. American Association of Petroleum Geologists, Memoirs, **62**, 667–679.

PENNINGTON, W. 1981. Subduction of the Eastern Panama Basin and Seismotectonics of Northwestern South America. *Journal of Geophysical Research*, **86**, 10753–10770.

PERARNAU, A., CASTILLO, J. & GOSE, W. 1988. Paleo-magnetismo de unidades del Cretáceo en Los Andes y la Serranía de Perijá: implicaciones tectónicas. *Proceedings of IV Congreso Venezolano Geofísica*, Caracas, Venezuela, 399–404.

PÉREZ, O. & AGGARWAL, Y. 1981. Present-day tectonics of southeastern Caribbean and northeastern Venezuela. *Journal of Geophysical Research*, **86**, 10791–10805.

PÉREZ, O., BILHAM, R. *ET AL*. 2001*a*. Velocidad relativa entre las placas del Caribe y Sudamérica a partir de observaciones dentro del sistema de posicionamiento global (GPS) en el norte de Venezuela. *Interciencia*, **26**, 69–74.

PÉREZ, O., BILHAM, R. *ET AL*. 2001*b*. Velocity field across the southern Caribbean Plate boundary and estimates of Caribbean/South American Plate motion using GPS geodesy 1994–2000. *Geophysical Research Letters*, **28**, 2987–2990.

PINDELL, J. L. 1994. Evolution of the Gulf of Mexico and the Caribbean. *In*: DONOVAN, S. K. & JACKSON, T. A. (eds) *Caribbean Geology: An Introduction*. University of the West Indies Publishers Association/University of the West Indies Press, Kingston, Jamaica, 13–39.

PINDELL, J. L. & DEWEY, J. 1982. Permo-Triassic reconstruction of western Pangea and the evolution of the Gulf of Mexico/Caribbean region. *Tectonics*, **1**, 179–211.

PINDELL, J. L. & KENNAN, L. 2001. Kinematic evolution of the Gulf of Mexico and Caribbean. World Wide Web Address: http://www.ig.utexas.edu/CaribPlate/CaribPlate.html.

PINDELL, J. L., CANDE, S. *ET AL*. 1988. A plate-kinematic framework for models of Caribbean evolution. *Tectonophysics*, **155**, 121–138.

PINDELL, J. L., HIGGS, R. & DEWEY, J. 1998. Cenozoic palinspatic reconstruction, paleogeographic evolution and hydrocarbon setting of the northern margin of South America. *In*: PINDELL, J. & DRAKE, C. (eds) *Paleogeographic Evolution and Non-glacial Eustasy, Northern South America*. Society for Sedimentary Geology, Special Publications, **58**, 45–85.

PINET, B., LAJAT, D., LE QUELLEC, P. & BOUYSSE, P. 1985. Structure of Aves ridge and Grenada basin from multichannel seismic data. *Symposium Géodynamique des Caraïbes*, Paris. Ed. Technip, 53–64.

PORRAS, L. 2000. Evolución tectónica y estilos estructurales de la región costa afuera de las cuencas de Falcón y Bonaire. *Proceedings of VII Congreso Bolivariano Exploración Petrolera en las Cuencas Subandinas*, Caracas, Venezuela, 279–292.

ROBERTSON, P. & BURKE, K. 1989. Evolution of the southern Caribbean Plate boundary in the vicinity of Trinidad and Tobago. *Bulletin of the American Association of Petroleum Geologists*, **73**, 490–509.

ROD, E. 1956. Strike-slip faults of northern Venezuela. *Bulletin of the American Association of Petroleum Geologists*, **40**, 457–476.

ROSENCRANTZ, E., ROSS, M. & SCLATER, J. 1988. Age and spreading history of the Cayman Trough as determined from depth, heat flow, and magnetics anomalies. *Journal of Geophysical Research*, **93**, 2141–2157.

RUIZ, C., DAVIS, N., BENTHAM, P., PRICE, A. & CARVAJAL, D. 2000. Structure and tectonic evolution of the South Caribbean basin, southern offshore Colombia: a progressive accretionary system. *Proceedings of VII Simposio Bolivariano Exploración Petrolera en las Cuencas subandinas*, Caracas, Venezuela, 334–355.

RUSSO, R. 1999. Dynamics and deep structure of the southeastern Caribbean–South America plate boundary zone: relationship to shallow seismicity. *AGU Spring Meeting*, Boston, MA, S228 (abstract).

RUSSO, R. & SPEED, R. C. 1992. Oblique collision and tectonic wedging of the South American continent and Caribbean terranes. *Geology*, **20**, 447–450.

SCHUBERT, C. 1979. El Pilar fault zone, northeastern Venezuela: brief review. *Tectonophysics*, **52**, 447–455.

SCHUBERT, C. 1984. Basin formation along Boconó–Morón–El Pilar fault system, Venezuela. *Journal of Geophysical Research*, **89**, 5711–5718.

SCLATER, J. B., HELLINGER, S. & TAPSCOTT, C. 1977. The paleobathymetry of the Atlantic Ocean from the Jurassic to the Present. *Journal of Geology*, **85**, 509–552.

SILVER, E., CASE, J. & MACGILLARY, H. 1975. Geophysical study of the Venezuela borderland. *Bulletin of Geological Society of America*, **86**, 213–226.

SINGER, A. & AUDEMARD, F. A. 1997. *Aportes de Funvisis al desarrollo de la geología de fallas activas y de la paleosismología para los estudios de amenaza y riesgo sísmico*. Academia de las Ciencias Naturales, Matemáticas y Físicas, Publicación Especial, **33**, 25–38.

SKERLEC, G. M. & HARGRAVES, R. B. 1980. Tectonic significance of paleomagnetic data from northern Venezuela. *Journal of Geophysical Research*, **85**, 5303–5315.

SOBIESIAK, M., ALVARADO, L. & VÁSQUEZ, R. 2002. Sismicidad reciente del Oriente de Venezuela. *Proceedings XI Congreso Venezolano de Geofísica* (CD-RoM format).

SOBIESIAK, M., ALVARADO, L. & VÁSQUEZ, R. 2005. Recent seismicity in northeastern Venezuela and tectonic implications. *Revista de la Facultad de Ingeniería de la Universidad Central de Venezuela*, **20**(4), 43–52.

SOULAS, J. P. 1986. Neotectónica y tectónica activa en Venezuela y regiones vecinas. *Proceedings of VI Congreso Geológico Venezolano*, **10**, 6639–6656.

SOULAS, J. P., GIRALDO, C., BONNOT, D. & LUGO, M. 1987. *Actividad cuaternaria y características sismogénicas del sistema de fallas Oca-Ancón y de las fallas de Lagarto, Urumaco, Río Seco y Pedregal. Afinamiento de las características sismogénicas de las fallas de Mene Grande y Valera. (Proyecto COLM)*. Unpublished Report, Intevep, Funvisis.

SPEED, R. 1985. Cenozoic collision of the lesser Antilles arc and continental South America and the origin of the El Pilar Fault. *Tectonics*, **4**, 41–69.

STEIN, S., DEMETS, C. ET AL. 1988. A test of alternative Caribbean Plate relative motion models. *Journal of Geophysical Research*, **93**, 3041–3050.

STEPHAN, J.-F. 1982. *Evolution géodynamique du domaine Caraïbe, Andes et chaîne Caraïbe sur la transversale de Barquisimeto (Vénézuéla)*. Thése d'etat, Paris.

STEPHAN, J.-F. 1985. Andes et chaîne caraïbe sur la transversale de Barquisimeto (Vénézuela). Evolution géodynamique. *Proceedings of Symposium Géodynamique des Caraïbes*, Paris, 505–529.

STEPHAN, J.-F., MERCIER DE LEPINAY, B. ET AL. 1990. Paleogeodynamics maps of the Caribbean: 14 steps from Lias to Present. *Bulletin Societé Géologique de France*, **6**, 915–919.

TABOADA, A., RIVERA, L. A. ET AL. 2000. Geodynamic of the northern Andes: subductions and intracontinental deformation (Colombia). *Tectonics*, **19**, 787–813.

TALWANI, M., WINDISCH, C., STOFFA, P., BUHL, P. & HOUTZ, R. 1977. Multichannel seismic study in the Venezuelan Basin and the Curaçao Ridge. *In*: TALWANI, & PITMAN, (eds) *Island Arcs, Deep Sea Trenches and Back-arc Basins*. American Geophysical Union, Washington, DC, 83–98.

TRENKAMP, R., KELLOGG, J., FREYMUELLER, J. & MORA, H. 2002. Wide plate margin deformation,

southern Central America and northwestern South America, CASA GPS observations. *Journal of South American Earth Sciences*, **15**, 157–171.

TSCHANZ, C., MARTIN, R., CRUZ, J., MEHNERT, H. & CEBULA, G. 1974. Geologic evolution of the Sierra Nevada de Santa Marta, northeastern Colombia. *Bulletin of the Geological Society of America*, **85**, 273–284.

VAN DER HILST, R. 1990. *Tomography with P, PP and pP delay-time data and the three dimensional mantle structure below the Caribbean region*. Geologica Ultraiectina, **67**. University of Utrecht, Netherlands.

VAN DER HILST, R. & MANN, P. 1994. Tectonic implication of tomographic images of subducted lithosphere beneath northwestern South América. *Geology*, **22**, 451–454.

VÁSQUEZ, E. & DICKEY, P. 1972. Major faulting in northwestern Venezuela and its relation to global tectonics. *Proceedings of VI Conferencia Geológica del Caribe*, Margarita, Venezuela, 191–202.

VIERBUCHEN, R. 1977. New data relevant to the tectonic history of the El Pilar Fault. *Proceedings of 8th Caribbean Geological Conference*, Curaçao, 213–214.

WEBER, J., DIXON, T. ET AL. 2001. GPS estimate of relative motion between the Caribbean and South American plates, and geologic implications for Trinidad and Venezuela. *Geology*, **29**, 75–78.

YSACCIS, R. 1997. *Tertiary evolution of the northeastern Venezuela offshore*. PhD thesis, Rice University, Houston, TX.

YSACCIS, R., CABRERA, E. & DEL CASTILLO, H. 2000. El sistema petrolífero de la Blanquilla, costa afuera Venezuela. *Proceedings VII Congreso Bolivariano Exploración Petrolera en las Cuencas Subandinas*, Caracas, 411–425.

ZAMBRANO, E., VÁSQUEZ, E., DUVAL, B., CATREILLE, M. & COFFINIERES, B. 1971. Síntesis paleogeográfica y petrolera del Occidente de Venezuela. *Proceedings IV Congreso Geológico Venezolano. Boletín de Geología*, Publicación Especial, **5**, 483–552.

Polyphase development of the Falcón Basin in northwestern Venezuela: implications for oil generation

MARVIN BAQUERO[1]*, JORGE ACOSTA[2], ELÍAS KASSABJI[1], JOSÉ ZAMORA[3], JUAN CARLOS SOUSA[1], JOSMAT RODRÍGUEZ[1], JACQUELINE GROBAS[1], LUÍS MELO[1] & FREDERIC SCHNEIDER[4]

[1]*PDVSA-Exploración, Gerencia de Evaluación del Sistema Petrolífero, Guaraguao, Puerto La Cruz 6023, Anzoátegui, Venezuela*

[2]*EXGEO-CGG-VERITAS, Guaraguao, Puerto La Cruz 6023, Anzoátegui, Venezuela*

[3]*PDVSA-INTEVEP, Exploración y Caracterización de Yacimientos, Los Teques 1201, Miranda, Venezuela*

[4]*Beicip-Franlab, PDVSA, La Tahona 1083, Miranda, Venezuela*

*Corresponding author (e-mail: baqueroms@pdvsa.com; mbaquero75@gmail.com)

Abstract: A multi-event tectonic episode that affected the Caribbean and South American Plate boundaries as well as Cenozoic oil generation is based on new structural and geochemical data from the western Falcón Basin, Venezuela. It involves Late Cretaceous to Middle Eocene emplacement of the Lara Nappes followed by Late Eocene to Early Miocene tectonic collapse and graben formation, Middle Miocene inversion and out of sequence thrusting. Oil-source rock correlation of seeps in the northern part of the basin suggests a Cenozoic siliciclastic source rock deposited under suboxic to anoxic conditions. Potential Cenozoic source rocks and Late Cretaceous La Luna Formation were used to evaluate the generation conditions using one- and two-dimensional thermal modelling. A heat flow of *c.* 190 mW m^{-2} was reached during the Oligocene–Early Miocene in the central part of the basin. As a result the Cretaceous source rock is overmature, while the primary Cenozoic source rocks are in the oil window. The thermal modelling also suggests that hydrocarbon accumulations are mainly located on the flanks of the graben, with small amounts possible in the centre, due to erosion during basin inversion. This modelling is highly consistent with the proposed polyphase tectonic model.

The Falcón Basin (Audemard 1993) is situated within the Caribbean–South American Plates boundary area (Fig. 1). The framework of the basin is a result of the polyphase tectonics and it is rather complex. The Falcón Basin also referred to as the 'Falcón Anticlinorium' (González de Juana *et al.* 1980) is an inverted graben which consists of WSW–ENE trending fold and thrust belts. The Falcón Basin is separated from the prolific oil-producing Maracaibo Basin by a NNW–SSE trending range known as the Serranía de Trujillo.

A coastal plain bounds the basin from the north and the Lara Nappes form the south. The Lara Nappes display southeastward tectonic transport, evidenced by folded and faulted structures that involve synorogenic Palaeogene turbidite deposits and low-grade metamorphic Cretaceous rocks. Additionally, 'klippe structures' associated with the metamorphic rocks indicate that they were emplaced during the Paleocene. SW–NE trending strike–slip faults are also present (in the south) and are interpreted as structures associated with the Andean Boconó Fault system.

Although, the petroleum system in the Maracaibo Basin is well established with La Luna Fm as the main source rock, the Misoa Fm as the principal reservoirs and the Paují Fm representing the seal rock, no petroleum system has yet been established for the Falcón Basin, primarily because of the tectonic and stratigraphic complexity. A lack of regional seismic and well data has hampered efforts to unravel this frontier basin. During 2006–2007 Petróleos de Venezuela (PDVSA) conducted an aggressive field programme to address key issues such as the tectono-stratigraphic evolution of the western side of the basin and its relationship to, and evidence for, a potential hydrocarbon system. This paper presents a new structural model of the Falcón Basin and discusses its integration with the geochemical parameters of the petroleum system.

From: JAMES, K. H., LORENTE, M. A. & PINDELL, J. L. (eds) *The Origin and Evolution of the Caribbean Plate.* Geological Society, London, Special Publications, **328**, 587–612.
DOI: 10.1144/SP328.24   0305-8719/09/$15.00 © The Geological Society of London 2009.

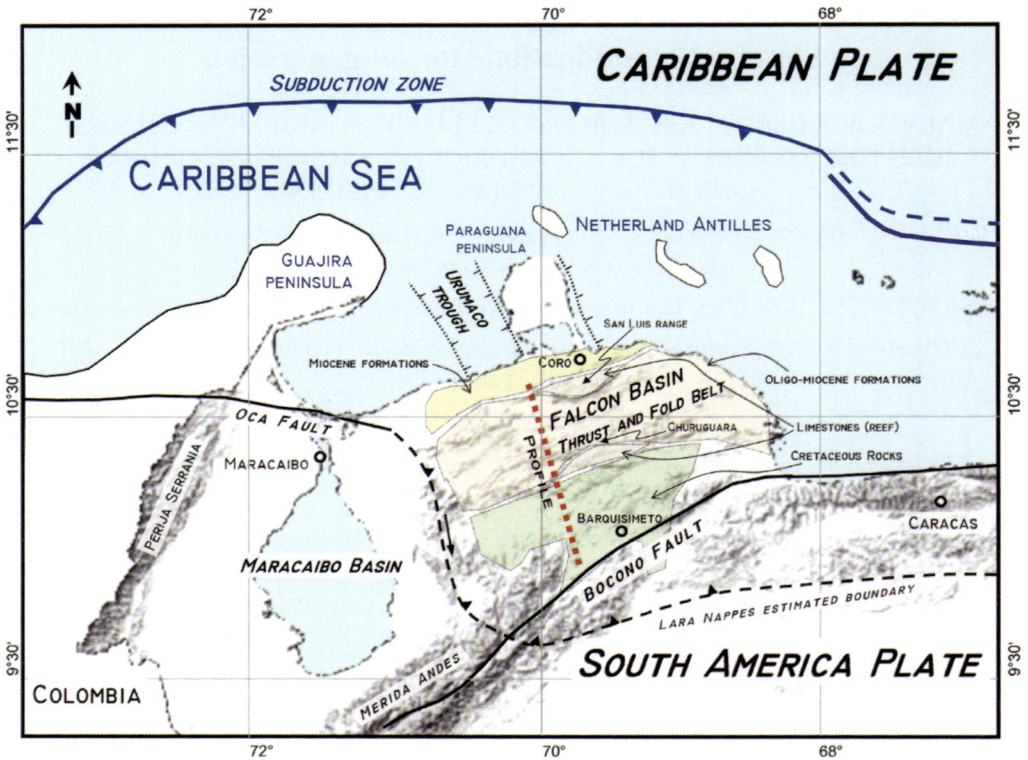

Fig. 1. Tectonic framework of the Falcón Basin.

Tectono-stratigraphy

The tectonic evolution of the Falcón Basin involves Late Cretaceous to Middle Eocene southeastward emplacement of intensively deformed, low-grade metamorphic rocks, such as slates, phyllite and metasandstones, of Cretaceous and Palaeogene age. This event marks the first interaction of the Caribbean and South American Plates in the region.

Emplacement of the nappes created a deep foreland basin filled with thick turbidate of the Paleocene–Middle Eocene Matatere Fm and equivalent units. Large Cretaceous olistoliths strongly suggest continuous tectonic activity (Fig. 2). The second tectonic event entailed Late Eocene to Early Miocene tectonic collapse and graben formation that accounts for the initial configuration of the Falcón Basin. This event was recorded by shallow marine and non-marine deposition stacked in multiple regressive pulses punctuated by marine transgressions. The section begins with conglomerates and very fine grained sandstones of the Late Eocene Santa Rita (SR) and shales of the Jarillal (Ja) Fms, deposited in continental to shallow marine environments (less than 200 m water

depth). A local transgression at the top of the syn-extensional Jarillal Fm marks regression during the Late Eocene (Pitelli & Molina 1989). During the Oligocene a complex stack of sequences accumulated in fluvial to deep-water environments. Fine-grained sandstones interbedded with black shales of the lower part of the El Paraíso Fm (EP) documents regression over a deltaic plain that gave way northward to marine conditions. Increase in shale content toward the top of the El Paraíso Fm (EP) records commencement of marine transgression (González de Juana *et al.* 1980). The overlying, deep-water, thick shaly sequence of the Pecaya Fm (Pe) formed in water as deep as 1500 m and was associated with pervasive extension and basin-wide subsidence (Fig. 3). The sagging phase of this extensional event is documented by deposition along the flanks of the basin. Shallow marine to continental sands of the Castillo Fm are capped by the carbonate reefal sequence of the Churuguara Fm (Ca).

Early–Middle Miocene basaltic sills and dykes intrude to the east of the basin (23–15.4 Ma, Muessig 1978, 1984; McMahon 2000), marking renewed extension and probably maximum

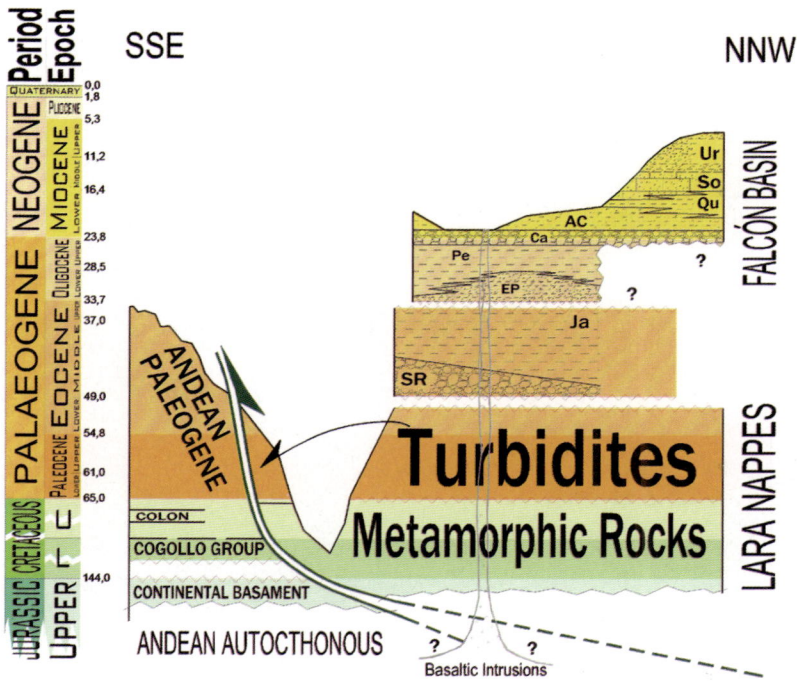

Fig. 2. Stratigraphic scheme for the northwestern of the Falcón Basin, see Figure 1 for location (profile).

stretching of the crust (Fig. 4). They are overlain by shallow marine sandstones of the Agua Clara Fm.

Middle Miocene tectonic inversion resulted in northward migration of the depocentre, which filled with interbedded conglomerate, sandstones and mudstones that represent shallow marine to fluviatile syn-orogenic sequences. The fluvial Cerro Pelado Fm, mainly formed by sandstones and fine-grained conglomerates, changes upward to claystones of the transgressive Querales Fm

Fig. 3. Outcrop that shows synsedimentary structures in the Pecaya Formation.

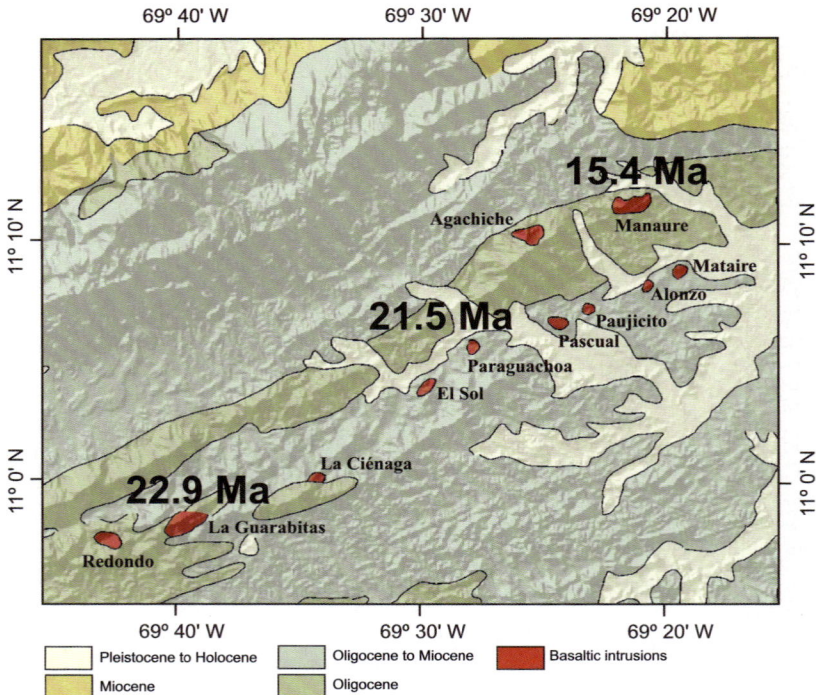

Fig. 4. Digital geological map of the Central Falcón modified from Garrity *et al.* (2006), with the age of intrusive bodies (modified from McMahon 2000).

(Qu), which in turn is overlain by sandstones of the delta-front deposits sequence of the Socorro Fm (So) and the shallow marine sequence of the Caujarao and La Vela Fms. The cycle is capped by alluvial fan conglomerates of the Coro Fm.

Structure

Finite strain analysis making use of strike and dip data from Applegate *et al.* (1956), Gaenslen *et al.* (1960), McDaniel (1960), Wheeler *et al.* (1960), McDaniel *et al.* (1961) and Gorman *et al.* (1966) defines a NW–SE tectonic transport direction (Fig. 5). This tectonic transport direction coincides with the previously proposed emplacement direction of the Lara Nappe during the Paleocene–Eocene (Stephan 1985; Lugo & Mann 1997).

Synsedimentary normal faults were plotted in a stereonet and then dip corrected to obtain their direction of generation (Fig. 6). Eigenvectors of: 279/66 (0.06), 77/21 (0.38) and 172/17 (0.76), indicate NNW–SSE extension, assuming that no rotation has affected the region.

This analysis also suggests that the tectonic transport direction has been relatively constant since the Paleocene and therefore plain strain

deformation is expected. This fact makes unlikely the hypothesis of the Falcón Basin as a pull-apart structure as proposed by Muessig (1984), Mann & Burke (1984), Boesi & Goddard (1991), Macellari (1995) and Ostos *et al.* (2005).

Accordingly, a 118 km long NW–SE oriented cross-section was constructed in order to understand the evolution of the region. New field data collected during this study combined with the Creole data were used in section construction. We assumed the reactivation of pre-existent normal faults and formation of a new thrust system, based on field observation of thrust systems, normal faulting (e.g. Fig. 3) and changes in stratigraphic thickness and facies found across some regional faults (Fig. 7).

The surface geological data (bedding attitude, faults and stratigraphic relationships) enable confident reconstruction of the structural configuration to a depth of about 4 km. Below this the reconstruction is more speculative because of the lack of seismic data. Several scenarios were considered. However, the most plausible is a detachment level at 17 km depth from which reverse faults propagate (Fig. 7). This detachment level might correspond to the Conrad discontinuity, which ramps up to the south and merges with an imbricate thrust system

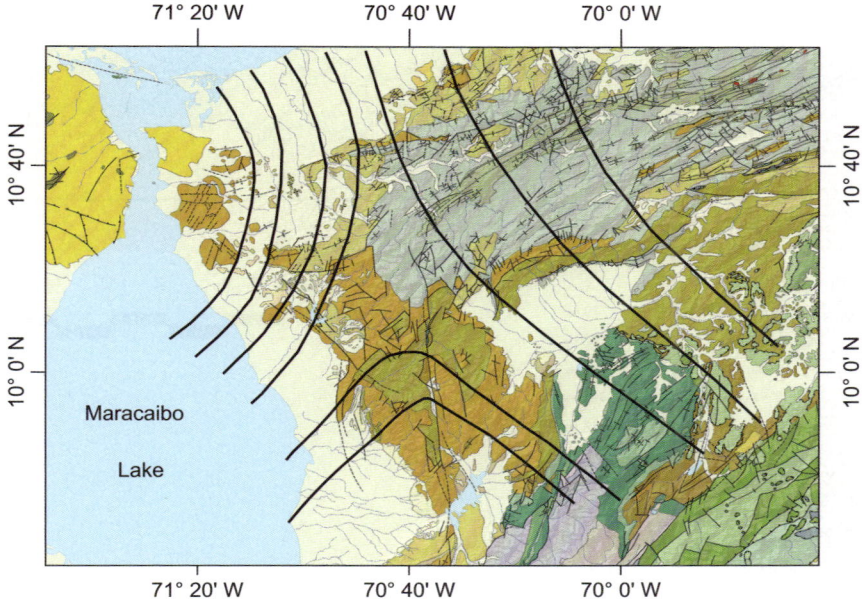

Fig. 5. Digital geological map of the region modified from Garrity *et al.* (2006), with finite strain lines (modified from Benguigui *et al.* 2006).

affecting low-grade metamorphic Cretaceous and Palaeogene rocks (Matatere Fm). Since these units show no metamorphism in the central and northern parts of the basin, metamorphism in the south appears to result from tectonic loading that ceased with the emplacement of the imbricate thrust system. Absence of metamorphism further north suggests that the imbricate system developed prior to the normal and reverse faulting observed in the central and northern part of the region.

Different stratigraphic units overlie turbidate Palaeogene rocks of the Matatere Fm across the

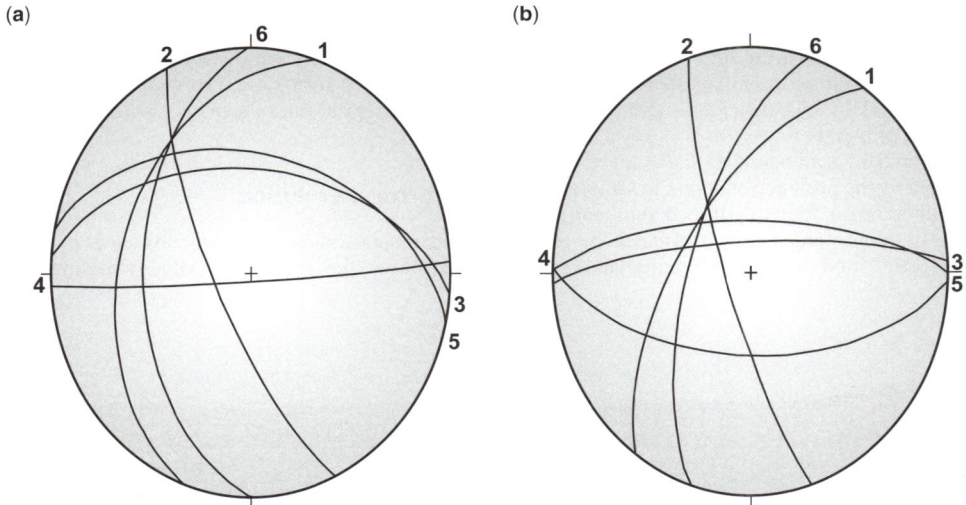

Fig. 6. (**a**) Stereonet plot of normal synsedimentary normal faults in the Pecaya Formation. (**b**) Same faults dip corrected.

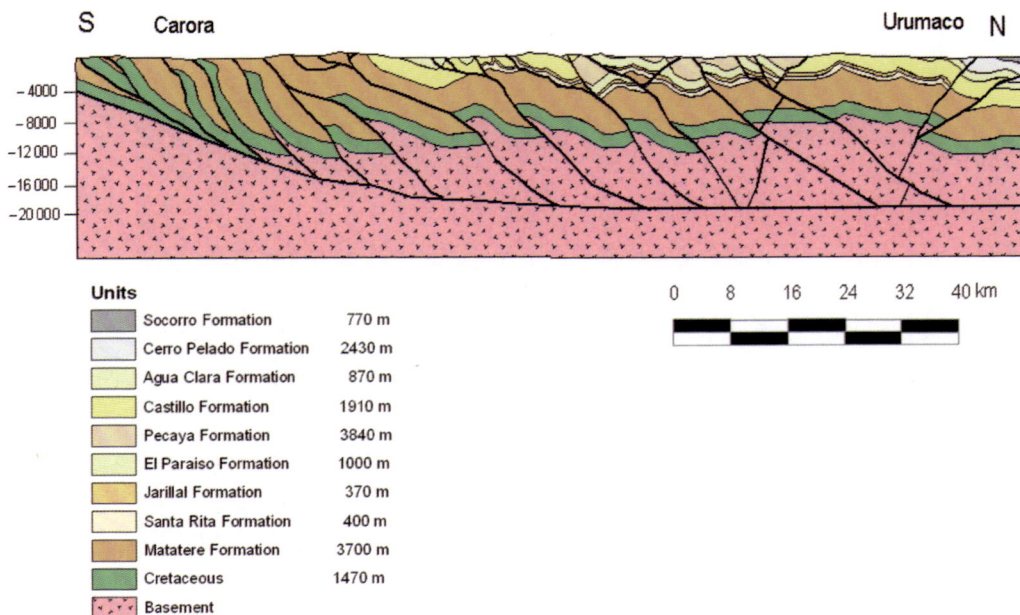

Fig. 7. Cross section; see Figure 1 for location (profile).

section (Fig. 6). In the south and north, the unit is followed unconformably by the Castillo Fm, whereas in the basin centre the unit is followed unconformably by the conglomerate and shale of the Santa Rita and Jarillal Fms. In turn, the Santa Rita and Jarillal Fms are overlain by the sandstone and shale of the Castillo Fm to the north and south; however, in the central part of the basin these Eocene units are overlain by the sandstones of the El Paraíso and shale of the Pecaya Fms, followed by the Castillo Fm. These stratigraphic relationships and profound unconformities document changes in subsidence and uplift during the evolution of a graben.

The graben south boundary faults reactivated, as indicated by the presence of short cut faults (Fig. 7), and minor thrust faults disturbed the central part of the basin. A folded region affected by normal faults reactivated as reverse faults and reverse

faults that affect pre-existent structures are interpreted for the north part of the section (Fig. 7). This interpretation is supported by seismic profiles north and west of the study cross-section. The structural relationship observed in the north part of the region and the presence of folded folds affecting the sandstone and conglomerate of the Cerro Pelado Fm suggest the development of out-of-sequence faults after the inversion of the basin. Therefore, four different structural events are recognized: generation of an imbricate thrust system to the south, graben formation, inversion of the basin and development of out of sequence structures.

Structural evolution

The understanding of the four structural events was useful for the restoration and validation of the

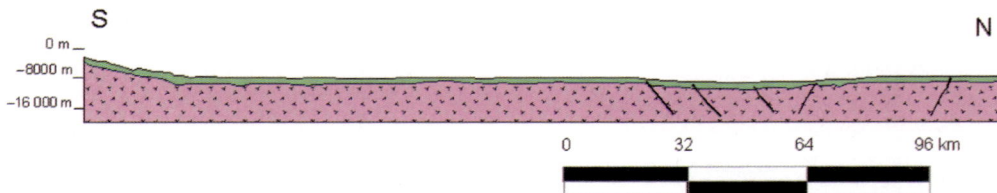

Fig. 8. Restored section at the Late Cretaceous.

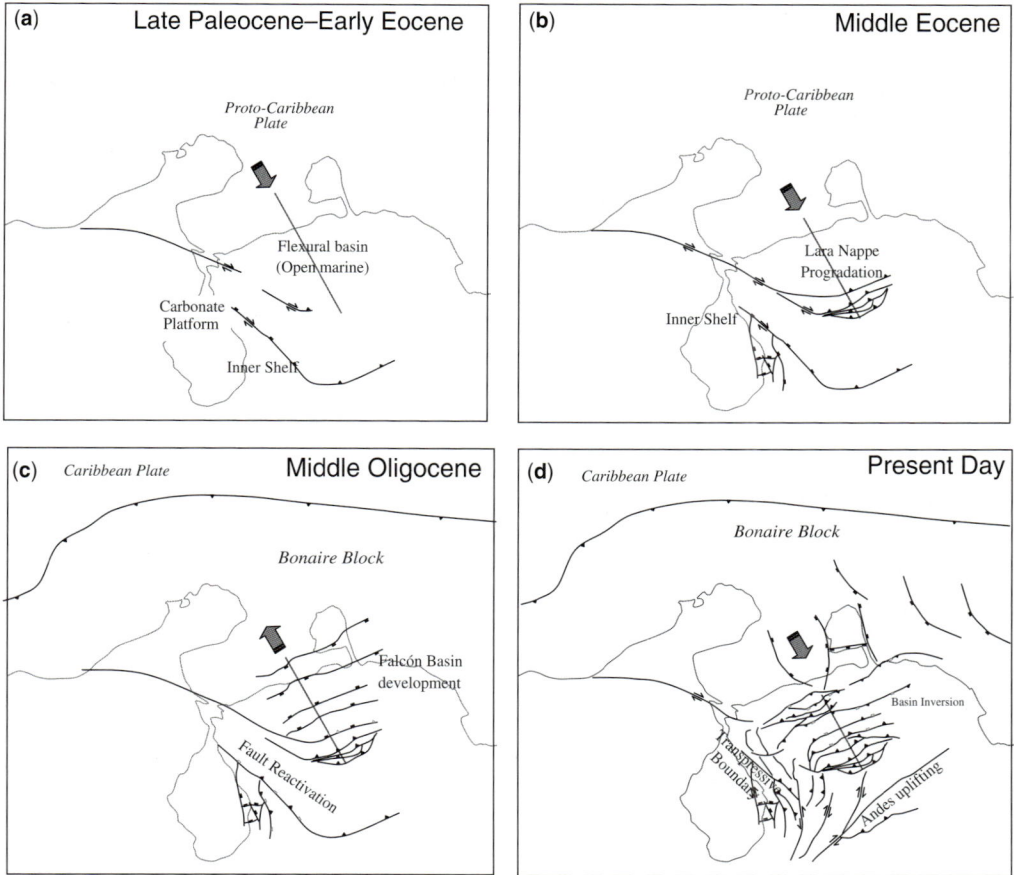

Fig. 9. Schematic structural evolution of the Falcón Basin.

section, which was undertaken with 2D Move, a back-stripping Midland Valley software. Several scenarios were considered during the process of the cross section balancing; however, the best constrained model implied the following evolution for the region since the Paleocene–Early Eocene: the presence of at least 236 km-long, relatively flat Cretaceous rocks (Fig. 8).

During the Late Cretaceous? to Middle Eocene the proto-Caribbean Plate moved southward, the

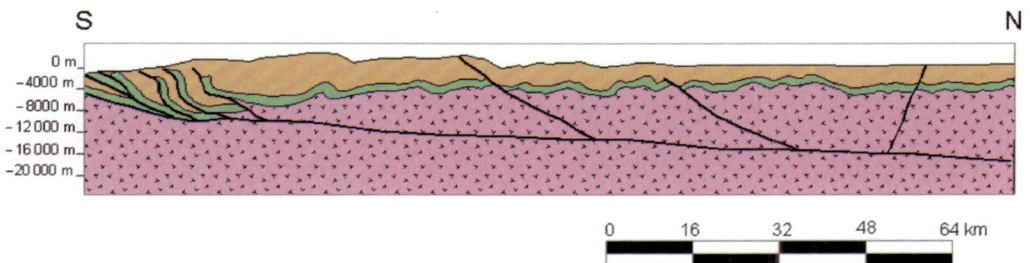

Fig. 10. Restored section at the Late Eocene.

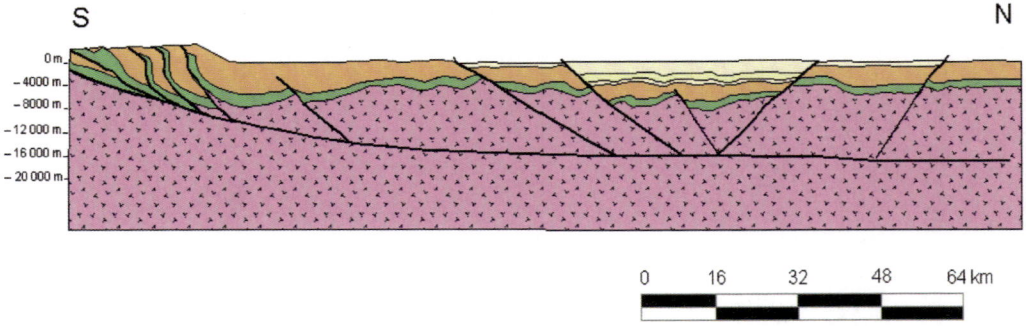

Fig. 11. Restored section at the Middle Oligocene.

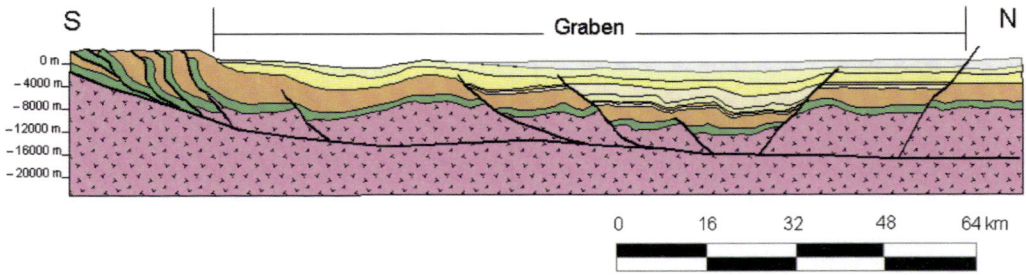

Fig. 12. Restored section at the late Early Miocene.

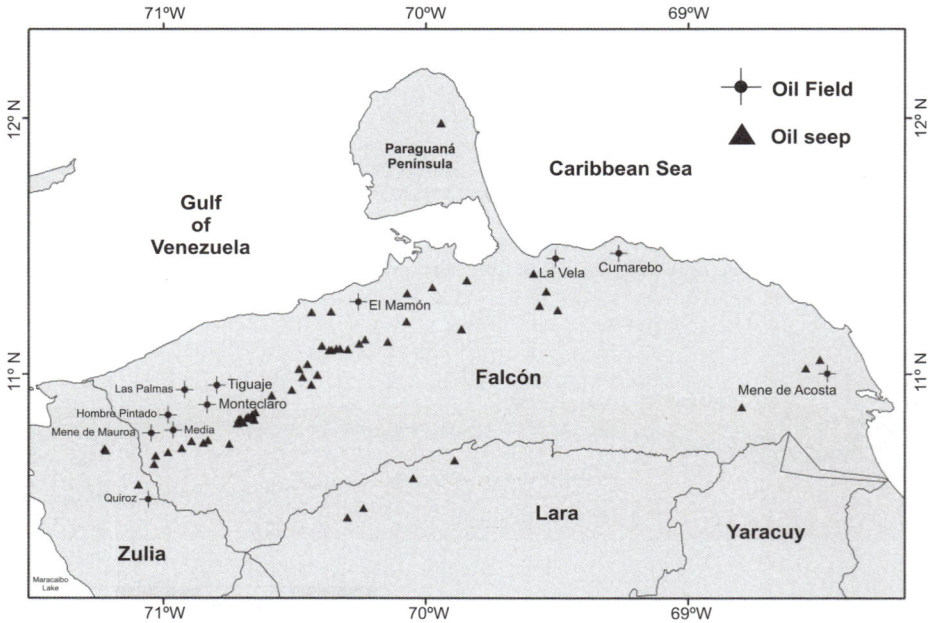

Fig. 13. Main oil fields and oil seeps in the region.

Cretaceous units were folded and the front broke up into a series of imbricated thrusts known as the Lara Nappes (Fig. 9). This displacement occurred along a flat detachment surface located within pre-Cretaceous rocks as it is observed in seismic profiles located to the north of the Falcón Basin. The imbricate system consists of a series of thrust sheets of different dimensions that develop in a normal sequence. The region shortened at least 40 km, being 25 km of the first sheet and 15 km of the

second (Fig. 10). This minimum shortening does not include in the proposed model minor scale thrusting generated during contraction. The third slice shortened the region by 8 km, while the fourth shortened it by 14 km. The fourth pulse of contraction resulted mainly in folding and uplifting of the central part of the section rather than thrusting. This behaviour is explained as a result of the flat detachment level and the greater friction in the front of the deformation. As contraction continued,

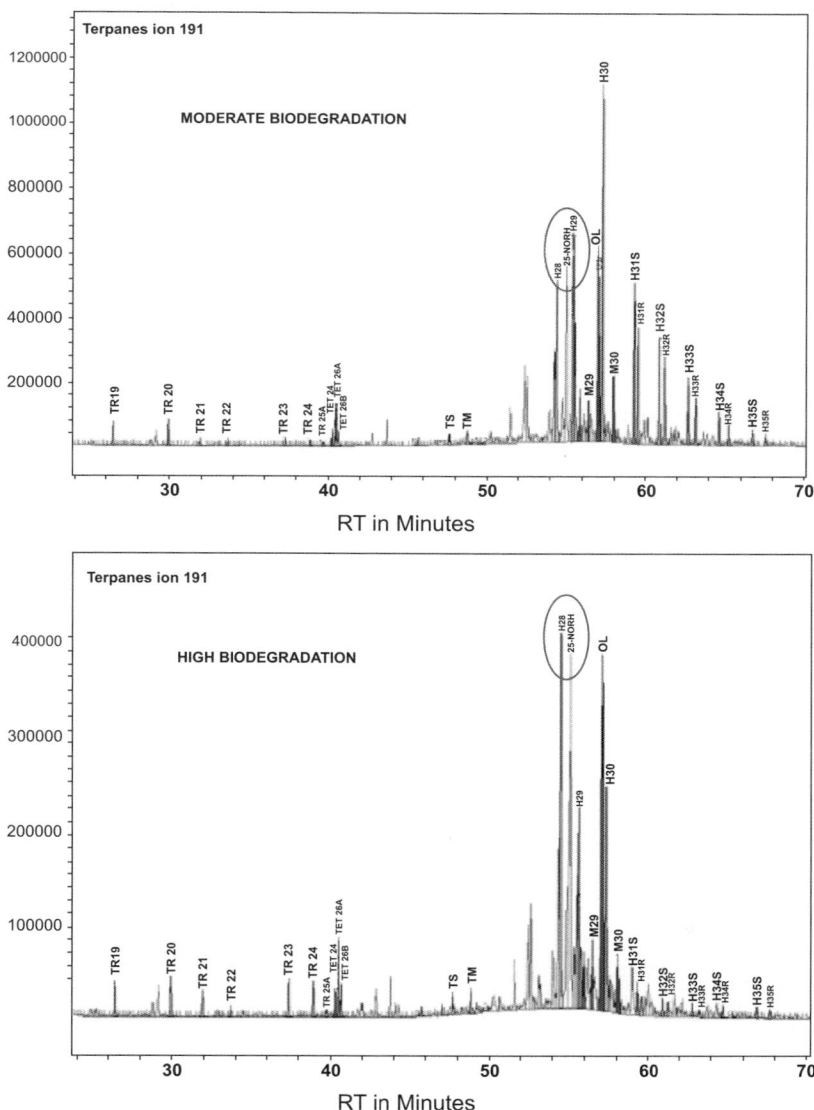

Fig. 14. Fragmentogram of the ion m/z 191 from the Mene de Mauroa, showing different degree of biodegradation (see the demethylated hopanes in the red circle).

a fifth slice formed with another 14 km of shortening. Consequently, the total shortening of the region during the Paleocene?–Early Eocene is about 74 km, representing 31% of the original section (Fig. 9a, b).

Several events occurred in the region during the Late Eocene–Oligocene? First, as a result of the shortening thickness, the increment of the crust made it unstable in the path of continuous movement of the proto-Caribbean Plate. Hence incipient low-angle subduction began to the north of the region, forming the current Caribbean Plate to the north and the Bonaire Block to the south (Fig. 9c). Then, local stress release occurred due to cessation of the action of the proto-Caribbean Plate in the region of the present day Falcón State. This is accompanied by dramatic erosion of the Lara Nappes that led to unroofing and generation of a zone of decompression sub-parallel to the new subducted slab of the Caribbean Plate (analogous to the front of slab roll-back). This event was accompanied by reactivation of pre-existing normal faults and tectonic collapse. Other normal faults developed in the central and northern parts of the region during the Late Eocene (Figs 9c & 10). To simplify the model, these faults are assumed to be sigmoidal and non-rotational. Furthermore, restoration suggests that they developed before Late Middle Eocene deposition of the Santa Rita Fm. A graben structure (Falcón Basin) developed where the Santa Rita and Jarillal Fms lie unconformably over the Matatere Fm (Fig. 11).

Alluvial fan deposits of the Santa Rita Fm developed near to basin fault boundaries suggest significant footwall uplift. During the Late Middle Eocene to the Early Late Eocene almost 6 km of extension occurred and a widespread marine transgression began, documented by deposits of the Jarillal Fm. It is assumed that the north and south part of the region remained emergent due to isostatic flexure as sediments loaded the central part of the basin. Thus the Jarillal is absent from the north and south ends of the section.

Extension continued throughout the Early to Middle Oligocene with a crestal collapse graben in the central part of the basin accommodating thick, bathyal shales of the Pecaya Fm. Uplift toward the south occurred in isostatic response. Crestal collapse continued until the Early Miocene (Fig. 11). Low extension (12 km, 7.5%) and high subsidence are explained as a result of the reactivation of basement structures.

Widespread, transitional shallow marine to continental sands and mudstones and shales of the Castillo Fm formed over the shaly sequence during the Late Oligocene–Early Miocene? (Fig. 12), marking the beginning of post-extension sedimentation. Sands and shales (?) of the shallow marine Agua Clara Fm followed. A 2.5 km shortening of the section suggests minor basin flexuring due to sediment loading. The post-extension event ended during the Middle Miocene with a delta front deposition of the Cerro Pelado Fm conglomerates and sands (Fig. 12).

At the end of the Middle Miocene, as a result of convergence of the Caribbean Plate from the north (Fig. 9d), compression began, characterized by reactivation of pre-existing faults and inversion of

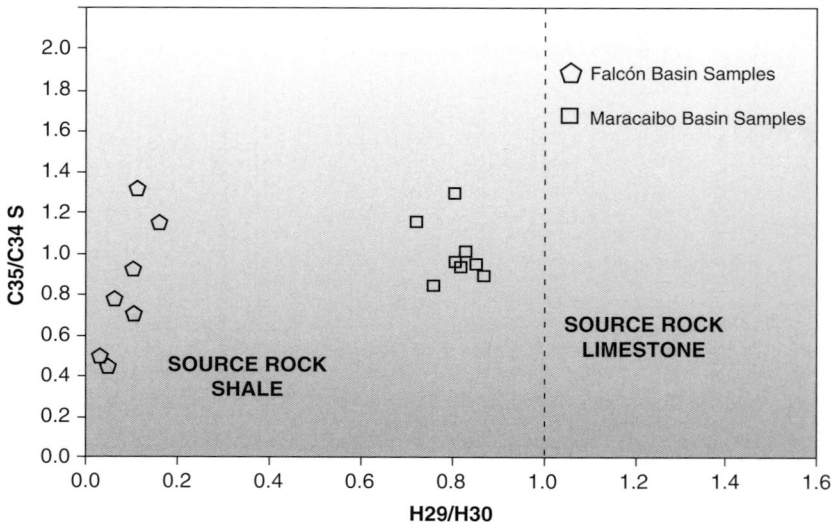

Fig. 15. Source rock lithology from the ratio regular hopans v. norapane.

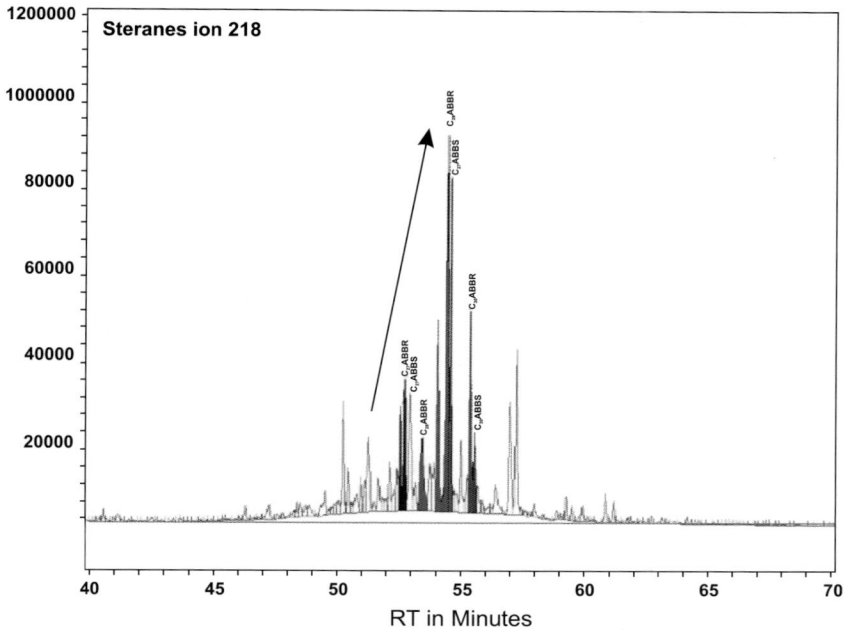

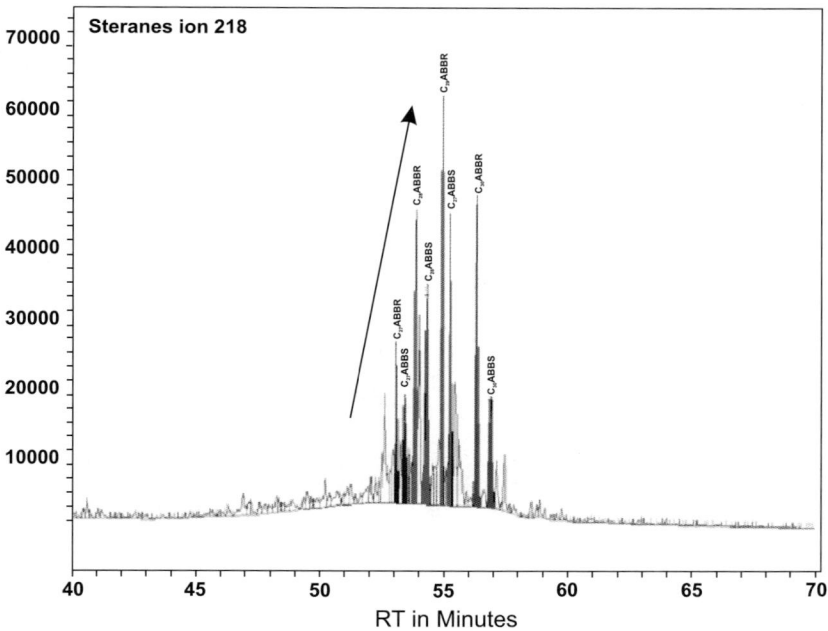

Fig. 16. Fragmentogram of ion m/z 218 from oilseeps taken near to the Tiguaje field.

the basin. Initially, this folded the northern part of the region, with pre-existent north-verging faults reactivating and generating a small basin to the north where deposition of shales of the Querales Fm commenced. With increased deformation during the Middle Miocene, south-verging pre-existing faults started to back-rotate, reactivated in normal sequence. Minor short cuts developed in the graben boundary faults and new reverse faults propagated from the detachment level. Two new

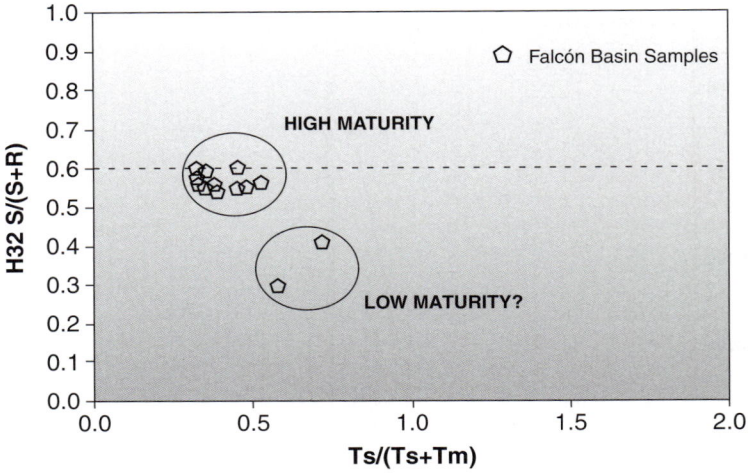

Fig. 17. Thermal maturity of the Falcón basin crudes as inferred from the ratio proposed by Peters *et al.* (2005).

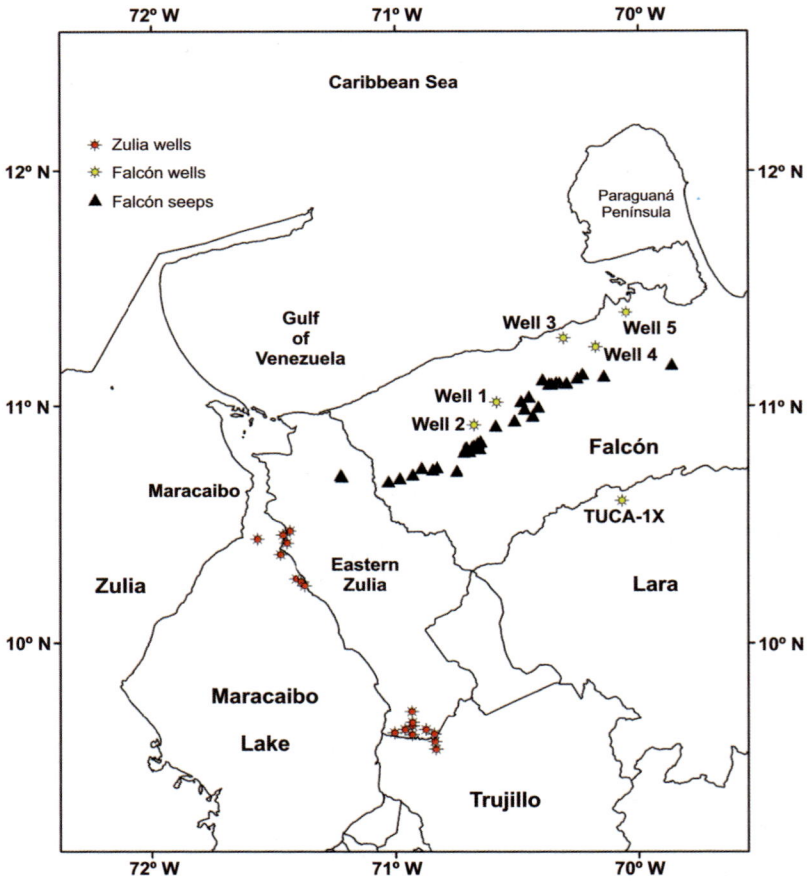

Fig. 18. Relative location of the wells in Falcón and Maracaibo Basin.

basins developed, to the north the Urumaco Trough and to the south a deformation front foredeep ancestor of the present day Carora intermountain basin. The exact detachment level of these faults is difficult to determine because there are no subsurface data in the region. However, seismic profiles in the Urumaco Trough suggest a pre-Cretaceous detachment that might correspond in the section with the Conrad discontinuity.

During the Late Miocene–Pleistocene?, the reactivated, south-verging, pre-existing faults reached the Paleocene?–Eocene imbricate sequence and inversion ceased. However, because of the continued convergence of the Caribbean Plate, out-of-sequence faults generated to the north of the region cross-cut pre-existing faults. These new faults also propagated from the proposed detachment level, affecting the central part of the former crestal collapse graben. During this time the Mérida Andes were uplifted along another detachment level that propagated from the south (De Toni & Kellogg 1993). The two detachment levels merged beneath the northern part of the region, forming a triangular zone. A lack of data has made it difficult to constrain the order of Miocene–Pleistocene? events. However, it is possible that the out-of-sequence faults in the north

and the Mérida Andes uplift occurred simultaneously. A 31% shortening (54 km) is estimated during this contractional event.

Source rock and thermal maturity

The most important oil fields in western Falcón are Mene de Mauroa, Media, Hombre Pintado, Las Palmas, Tiguaje, Monte Claro and Mamón (Fig. 13). Additionally, several oil seeps occur in the region (Moore *et al.* 1951; McDaniel *et al.* 1961; Gorman *et al.* 1966). Oils vary significantly in geochemical characteristics, suggesting different generation conditions.

Biodegradation

Mene de Mauroa oils are moderately to highly biodegraded. Biodegradation is inferred from the absence of *n*-paraffins and isoprenoids in the chromatogram of the saturate fraction GC-MS and corroborated by the presence of demethylated hopanes in the fragmentogram at m/z 191 (Fig. 14) and 177. Demethylated hopanes occur also in the Mene de Quiroz field, where they are even more abundant. Chromatography and mass

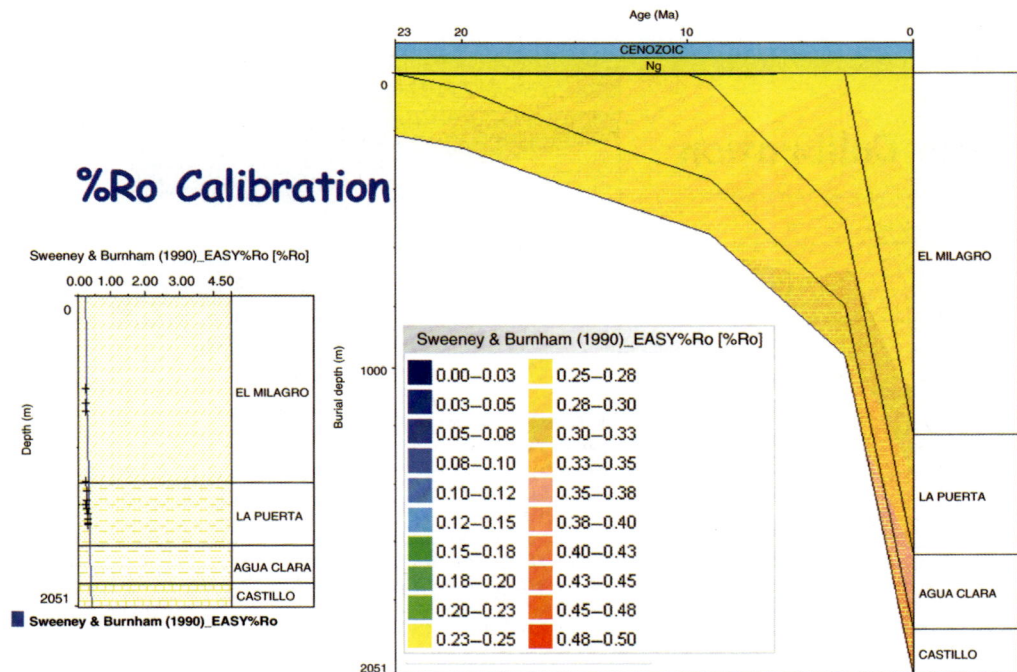

Fig. 19. Burial history of well 1.

%Ro Calibration

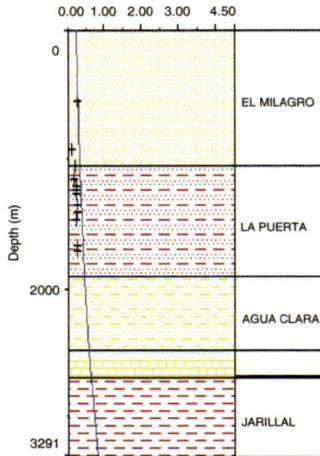

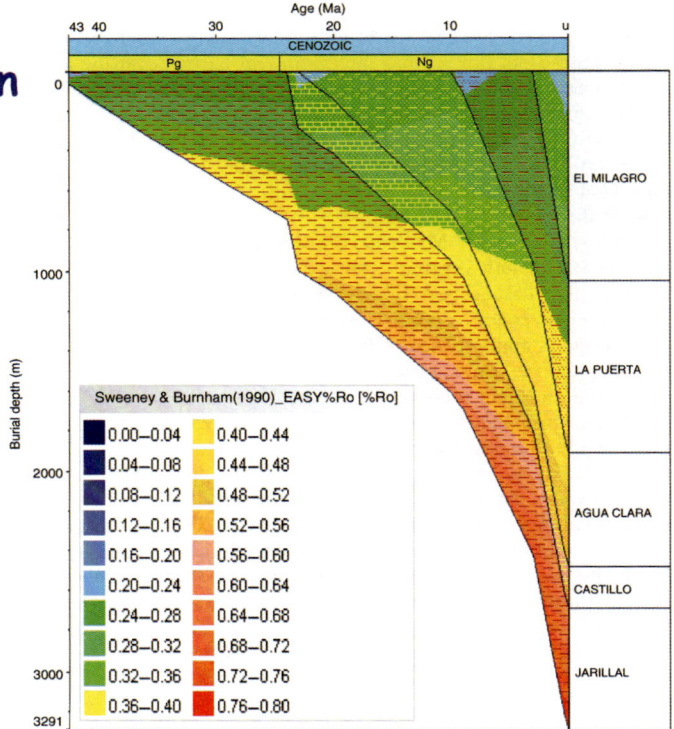

Fig. 20. Burial history of well 2.

%Ro Calibration

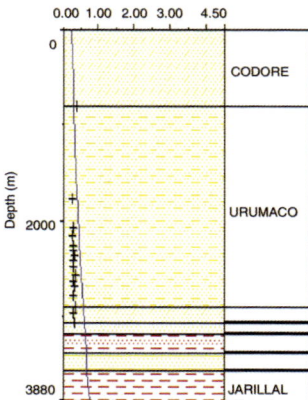

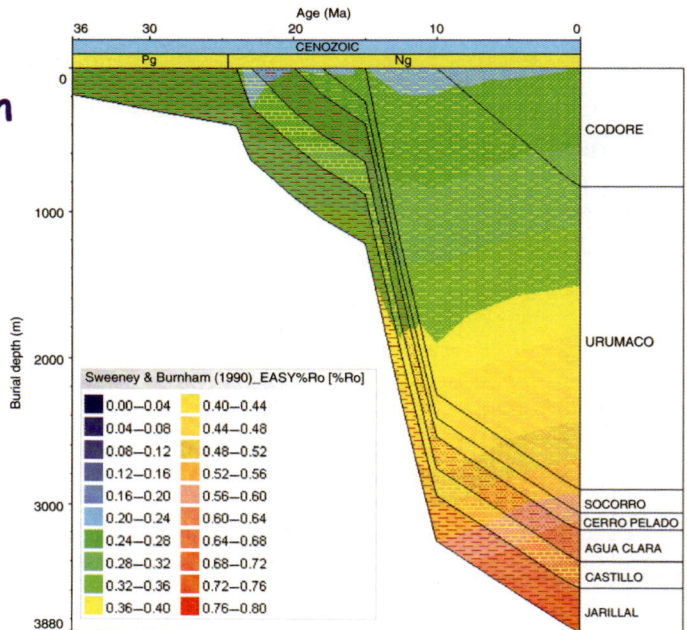

Fig. 21. Burial history of well 3.

%Ro Calibration

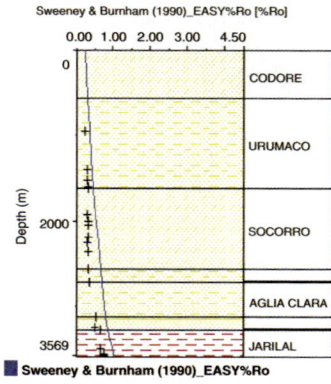

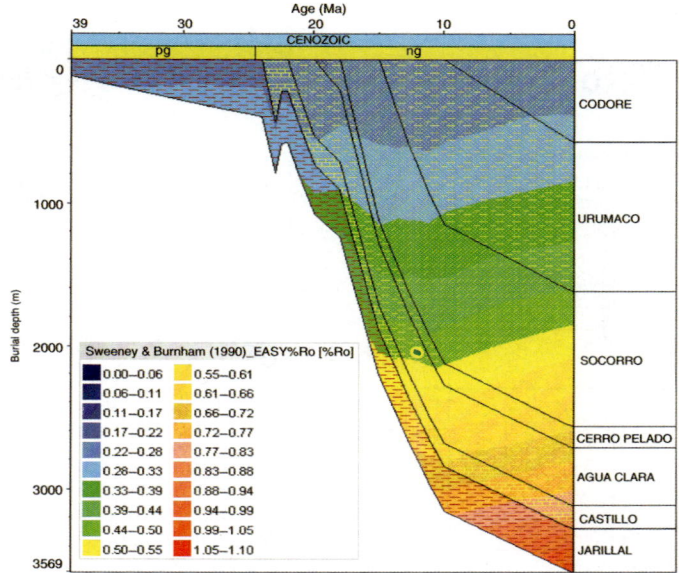

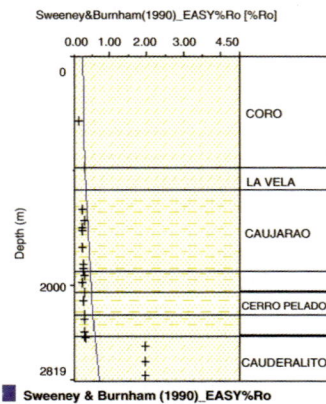

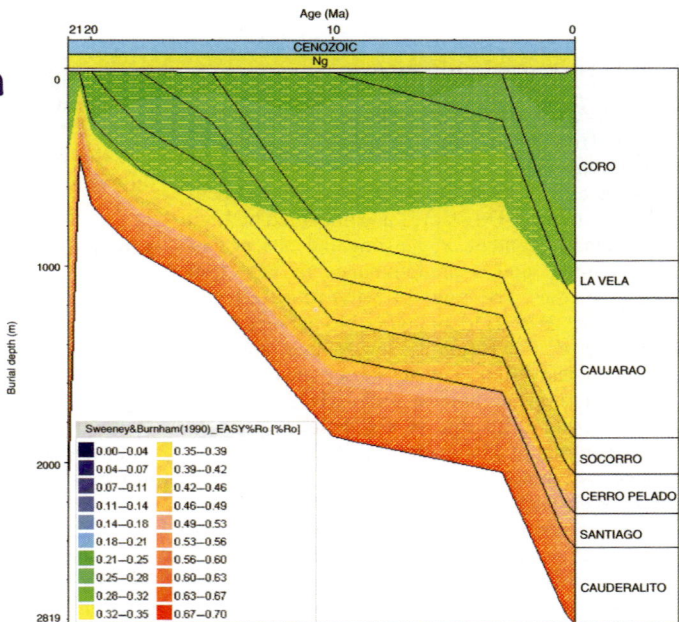

Fig. 22. Burial history of well 4.

spectrometry (m/z 191 and 177) of the saturate ratio of Hombre Pintado field oils also suggest high levels of biodegradation.

Chromatography of Tiguaje oils shows a moderate biodegradation, in contrast with fragmentograms that show a strong biodegradation. This suggests mixing of biodegraded and unaltered oils and perhaps active charging from a nearby kitchen (Toro & Escandón 1990). Additionally, variation in the 25-norhopane and bisnorhopane concentrations

%Ro Calibration

Fig. 23. Burial history of well 5.

%Ro Calibration

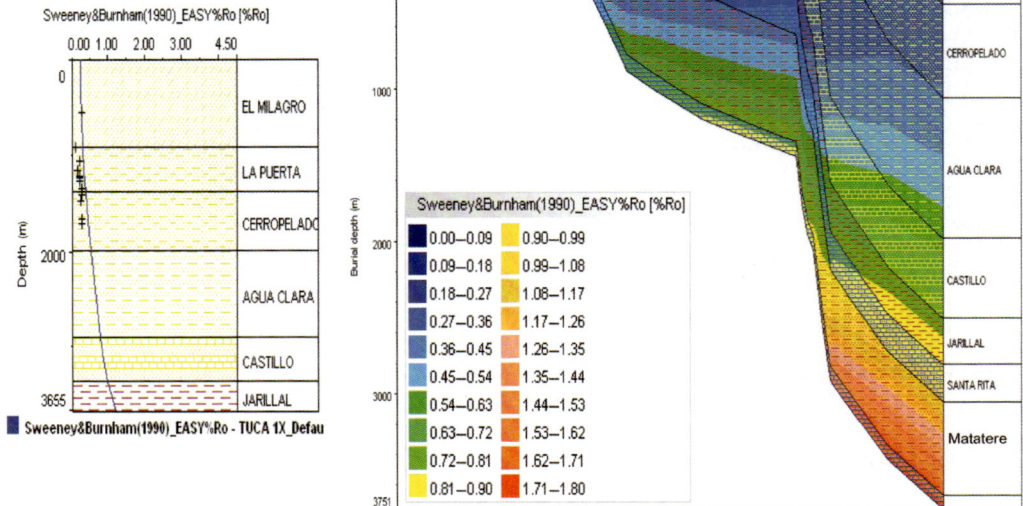

Fig. 24. Burial history of pseudowell TUCA-1X.

in samples of oil-seeps collected near the Hombre Pintado and Tijuage fields shows moderate to high biodegradation.

Source rock predominance

Analysis of hydrocarbon saturate fraction biomarkers (fragmentogram m/z 191) in two oil-seep samples near Hombre Pintado and Tiguaje fields shows that tricyclic terpane decrease relative to pentacyclic terpanes. This fact is accompanied by high 18α-oleanane content, suggesting the absence of algal material and a predominance of terrestrial organic matter. The ratio of tetracyclic terpanes and hopanes H29/H30 and other biomarkers in the oil seep near the Hombre Pintado field suggests organic facies variation of the source rock.

The norhopane:hopane ratio (H29:H30) of field oils and oil seeps of the Falcón Basin is lower than that of Maracaibo Basin oils, indicating a clear difference in lithology and depositional environment of the source rock (Fig. 15). A lacustrine environment with siliciclastic material input is inferred for the source rock of the Falcón Basin, radically different from the marine carbonate La Luna source of the Maracaibo Basin.

Another parameter indicative of the organic matter type is the intensity of the regular steranes

(%C27, %C28 and %C29). The GC-MS analysis (the m/z 218 fragmentogram; Fig. 16) shows a predominance of C29 over C27 sterane, suggesting terrestrial organic matter content.

These results are in agreement with Ghosh *et al.* (1997) and Toro & Escandón (1990), who postulated a fluvial deltaic to coastal plain environment. Toro & Escandón (1990) also considered that the source rock was at the beginning of the oil window.

Maturity

Maturity of crudes established by Boesi & Goddard (1991) suggested Eocene and possibly Oligocene source rocks for northwestern Falcón. They noted that the oil of the Tiguaje field is located at the liquid hydrocarbon generation window, whereas Hombre Pintado oils show a lower degree of maturity. Samples analysed in this study suggest that most of them are near to thermal equilibrium, as evidenced by the ratio $S:S+R$ of the isomer of C32 hopane (from the ion m/z 191 distribution), indicating high maturity. However, the ratio $Ts:(Ts+Tm)$ shows the opposite results. This apparent inconsistency is due to differences in biodegradation, making the determination of maturity ambiguous (Fig. 17).

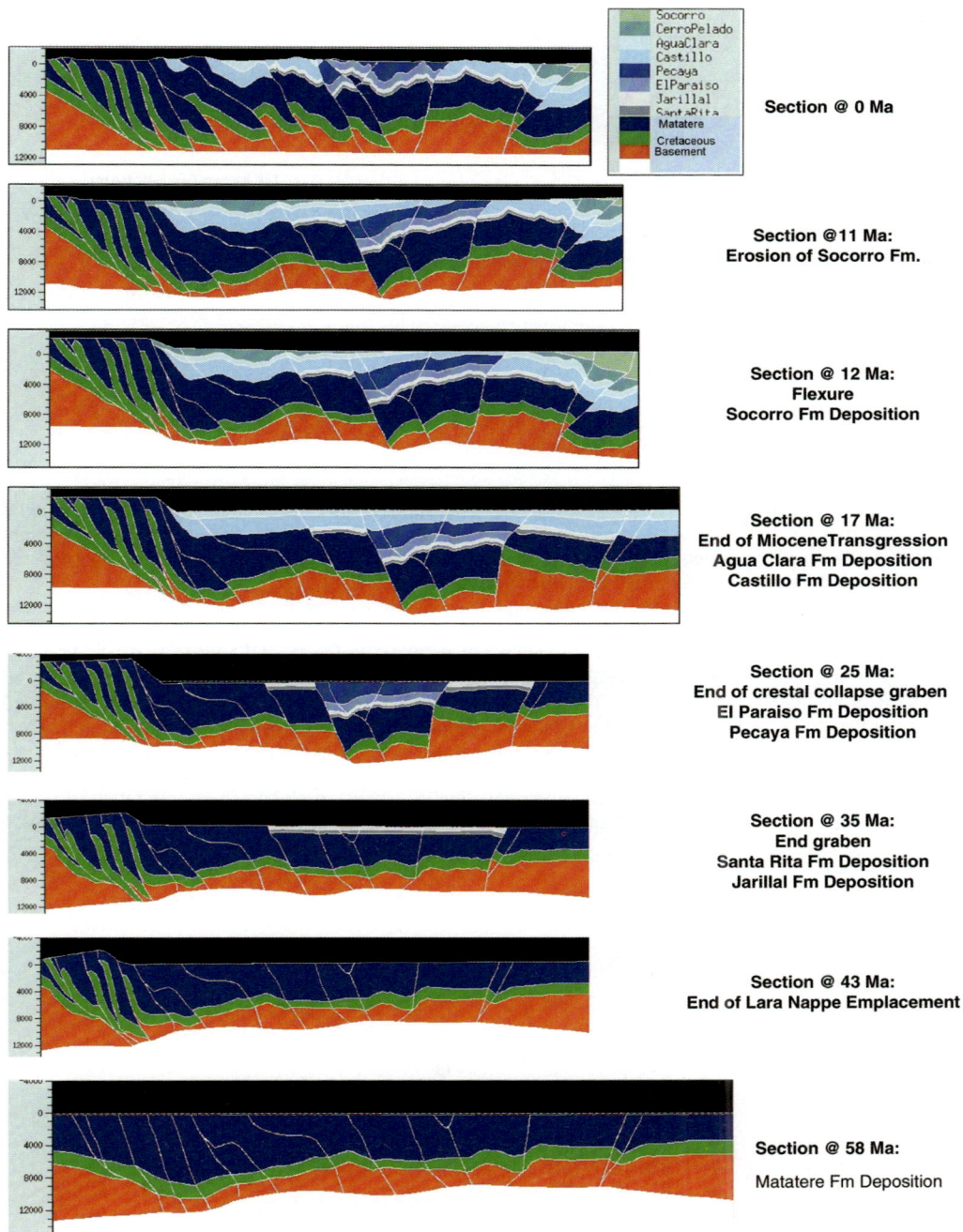

Fig. 25. Kinematic restoration with Ceres Software. From 25 Ma the scale is modified.

$$y = 25.8x + 28.014$$

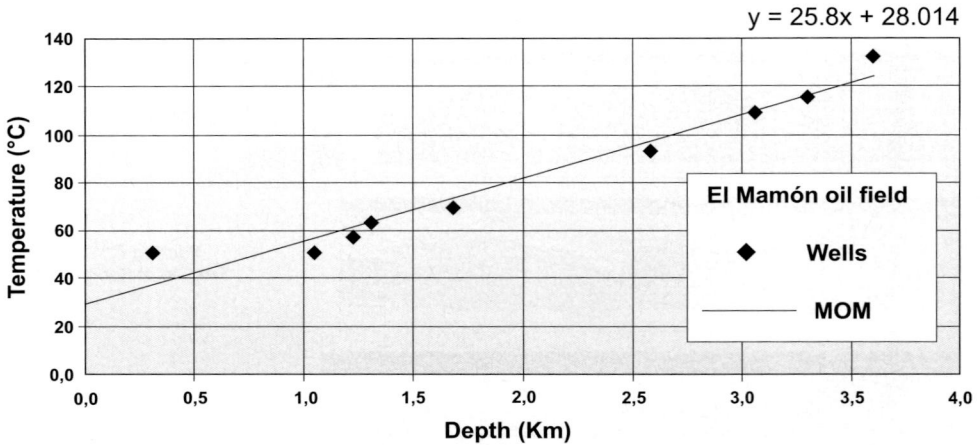

Fig. 26. Geothermal gradient used for the calibration.

Thermal calibration

Once the structural evolution of the region was established and Cenozoic rocks were documented as the source, oil burial history diagrams and thermal calibration of six wells (five drilled and one pseudowell) were modelled using %Ro values and T_{max} from Rock Eval pyrolysis (Figs 18–24). The structural model indicates that the Falcón Basin formed as a result of a tectonic collapse since the Late Eocene, coeval with an important crustal thinning as proposed by Sousa *et al.* (2005). This tectonic event was accompanied by an anomalous increase in the heat flow of the region due to mantle upwelling (incipient delamination process?) that played an important role in the source rock maturation as will be explained later.

The modelling shows that, in order to match the vitrinite reflectance values as closely as possible, the heat flow during the Oligocene to Early Miocene would have to be increased from 140 to 200 mW m^{-2} in the five wells and then decreased progressively to 40 mW m^{-2} at present. This heat flow anomaly is responsible for the high maturity of the Cenozoic units and explains the presence of hydrocarbons derived from clastic–terrigenous organic matter observed to the north of the basin.

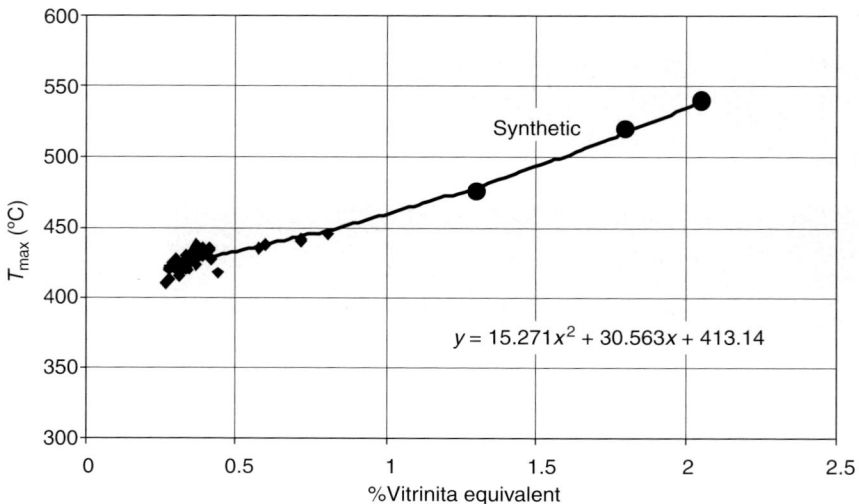

$$y = 15.271x^2 + 30.563x + 413.14$$

Fig. 27. Correlation of the calculated T_{max} and %Ro.

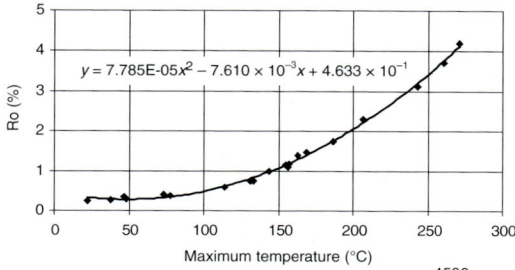

(a) Best correlation Ro (%) $-T_{max}$ through Ceres

$$y = 7.785\text{E-}05x^2 - 7.610 \times 10^{-3}x + 4.633 \times 10^{-1}$$

(b) Relationship between observed and calculated depth out of Ro (%) with a gradient of 25.8°C/km. Presence of a non significant erosion (<400 m)

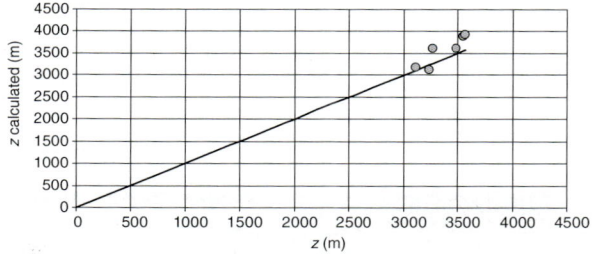

Fig. 28. Evaluation of the maximum erosion from temperature data.

Thus wells with high %Ro data are located in the northern part of the basin. High maturity in wells 2, 4 and 5, about 0.7% vitrinite values, occur at an average depth of 3000 feet. However, in well 3 this vitrinite value is reached at 3500 feet and in well 1, which was drilled to only 2061 feet, the maximum vitrinite values only reached 0.4%.

In order to model the heat flow in the central part of the basin, a pseudowell (TUCA-1X) was generated using T_{max} from Rock Eval pyrolysis of surface samples and calibrated using data from well 2. As a result, vitrinite values of 0.7% were found at a depth of 2000 feet, suggesting higher heat flows in the centre of the basin where the

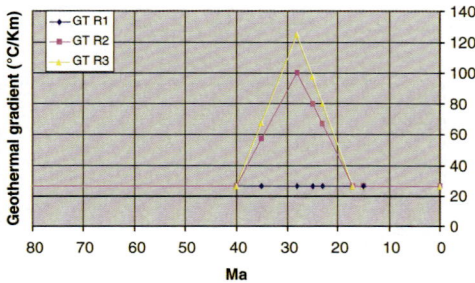

(a) Scenario: Increase of the geothermal gradient and dramatic deacrease after 28 Ma

(b) Scenario: Increase of the geothermal gradient and dramatic decrease with greater amplitude

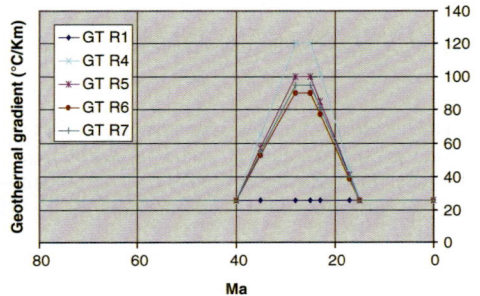

Formation	GT R1	GT R2	GT R3	GT R4	GT R5	GT R6	GT R7
Santa Rita	0.7	1.4	2.1	3.5	2.6	2	2.2
Jarillal	0.6	1	1.4	2.4	1.8	1.4	1.5

Maximum gradient = 95°C km^{-1} – 190 kW m^{-2}

Fig. 29. Geothermal gradient through time.

Formations	Number of samples	Average T_{max} (ºC)	Easy %Ro	T (°C)	Depth (gradient) (m)	Depth [% Ro (z)] (m)
Cretaceous	5	501	1.5	173	5500	5140
Matatere	5	495	1.4	168	5310	4890
Santa Rita	2	547	2.2	205	6730	5950
Jarillal	2	509	1.6	179	5730	5350
Pecaya	2	523	1.8	189	6120	5540
Castillo	4	413	0.2			

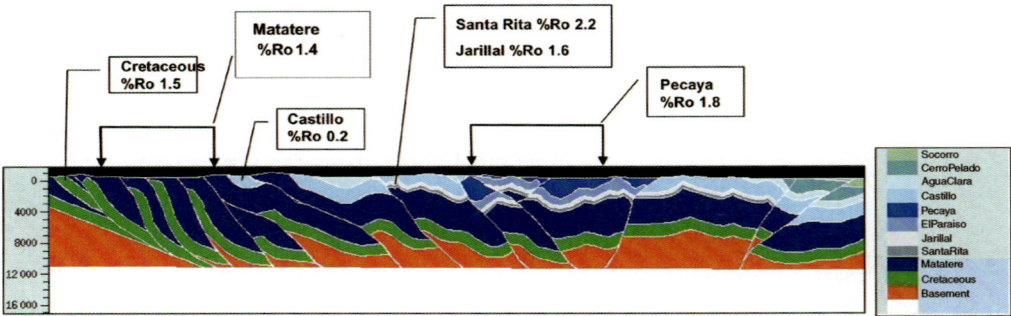

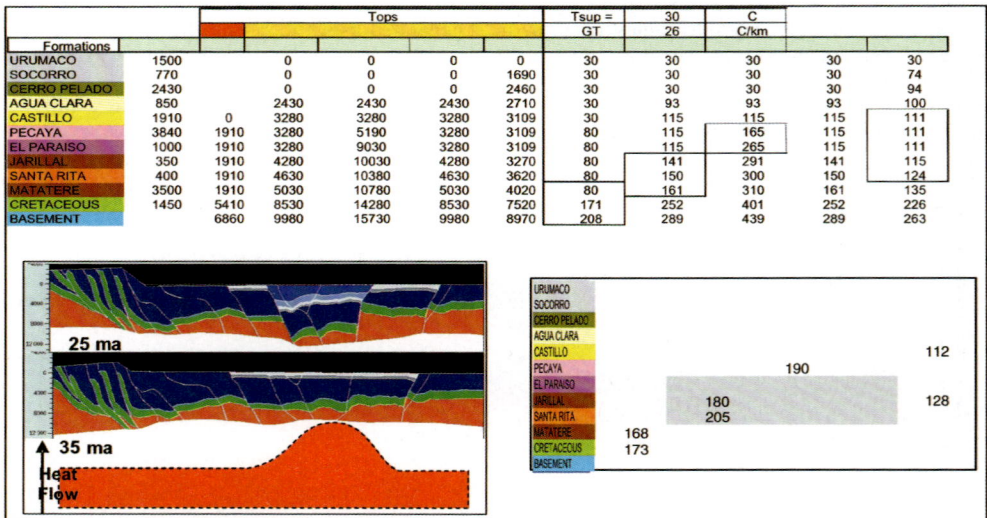

Cretaceous %Ro 1.5

Matatere %Ro 1.4

Santa Rita %Ro 2.2
Jarillal %Ro 1.6

Castillo %Ro 0.2

Pecaya %Ro 1.8

Legend: Socorro, CerroPelado, AguaClara, Castillo, Pecaya, ElParaiso, Jarillal, SantaRita, Matatere, Cretaceous, Basement

Formations			Tops				Tsup = 30	GT 26	C C/km		
URUMACO	1500		0	0	0	0	30	30	30	30	30
SOCORRO	770		0	0	0	1690	30	30	30	30	74
CERRO PELADO	2430		0	0	0	2460	30	30	30	30	94
AGUA CLARA	850		2430	2430	2430	2710	30	93	93	93	100
CASTILLO	1910	0	3280	3280	3280	3109	30	115	115	115	111
PECAYA	3840	1910	3280	5190	3280	3109	80	115	165	115	111
EL PARAISO	1000	1910	3280	9030	3280	3109	80	115	265	115	111
JARILLAL	350	1910	4280	10030	4280	3270	80	141	291	141	115
SANTA RITA	400	1910	4630	10380	4630	3620	80	150	300	150	124
MATATERE	3500	1910	5030	10780	5030	4020	80	161	310	161	135
CRETACEOUS	1450	5410	8530	14280	8530	7520	171	252	401	252	226
BASEMENT		6860	9980	15730	9980	8970	208	289	439	289	263

25 ma

35 ma

Heat Flow

URUMACO, SOCORRO, CERRO PELADO, AGUA CLARA, CASTILLO — 112, PECAYA, EL PARAISO, JARILLAL — 190, SANTA RITA — 180 / 205 — 128, MATATERE — 168, CRETACEOUS — 173, BASEMENT

Fig. 30. Synthesis of the temperature and maximum depth of each unit along the section.

Castillo and Jarillal Fms source rocks are in the oil window. An analogue is found in Taiwan, where the heat flow reached 240 mW m^{-2} as a result of mantle upwelling and delamination (Lin 1998, 2000).

Following these results a two-dimensional thermal calibration was undertaken using CERES, a numerical prototype software for hydrocarbon potential evaluation in complex areas developed by Beicip Franlab. The three main phases of deformation of the Falcón Basin were taken into account. However, some parts of the structural evolution, such as the out-of-sequence thrusts, could not be modelled. This is because at this stage the software does not have an algorithm to model the offset of the normal faults generated by these out-of-sequence structures. In addition, a flexural basin model was considered to smooth the structures and the

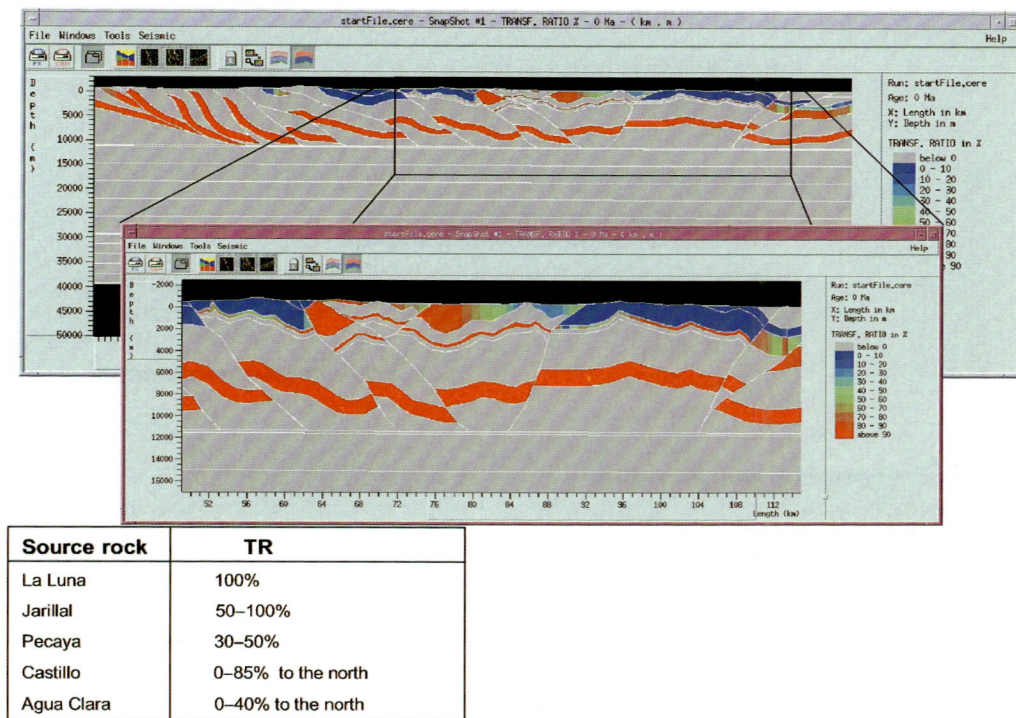

Source rock	TR
La Luna	100%
Jarillal	50–100%
Pecaya	30–50%
Castillo	0–85% to the north
Agua Clara	0–40% to the north

Fig. 31. Transformation ratio of the putative source rocks.

variation of the angles of the detachment faults and keep the section balanced (Fig. 25). Because the El Mamón oil field (MOM) was closer to the north part of the section, the bottom borehole temperature (BHT), vitrinite reflectance (%Ro) and T_{max} from three wells were used.

A geothermal gradient of $25.8\,°C\,km^{-1}$ was obtained for three wells of the El Mamón oil field and this value was extrapolated along the section (Fig. 26). Several calculations were required to determine the variation of the geothermal gradient through time. First the Vitrinite equivalent (easy %Ro) was obtained from three pseudowells similar to TUCA-1X (Fig. 27), calibrated from the T_{max} from Rock Eval pyrolysis of outcrop samples. Then the maximum burial of the north part of the section was estimated (Fig. 28a). Less than 400 m erosion was found from the correlation of the vitrinite equivalent (%Ro) with the geothermal gradient of the El Mamón oil field (Fig. 28b). Finally, two different scenarios were modelled to test the variation of the thermal gradient through time and particularly between 40 and 18 Ma, when the tectonic collapse that generated the Falcón Basin took place. Within the first scenario a dramatic decrease in the geothermal gradient occurred after

reaching a temperature of $125\,°C\,km^{-1}$ (Fig. 29a). Within the second scenario the maximum gradient of temperature considered was $95\,°C\,km^{-1}$, but with longer amplitude in time. The last scenario fits better the surface data and is equivalent to a heat flow of $190\,mW\,m^{-2}$ (Fig. 29b). This result shows an excellent correlation with the values of geothermal gradient through time and confirm the anomalous increment of the heat flow during the Oligocene–Early Miocene.

Once the temporal and spatial geothermal gradient was established, they could be plotted along the transect, as seen in Figure 30. The Vitrinite equivalent values of the La Luna and Matatere Fms indicate that this region was buried to a depth of up to 2000 m towards the south part of the section, assuming a geothermal gradient of $26\,°C\,km^{-1}$ and considering that the thermal anomaly had no influence in this part of the region. Hence if the southern part of the region was buried under this amount of sediment (that corresponds with the thickness of the Castillo Fm, as suggested by some outcrops near the town of Carora), it must have been eroded during the Andean deformation. Additionally, the average %Ro of the Santa Rita and Jarillal Fms to the south part of the graben

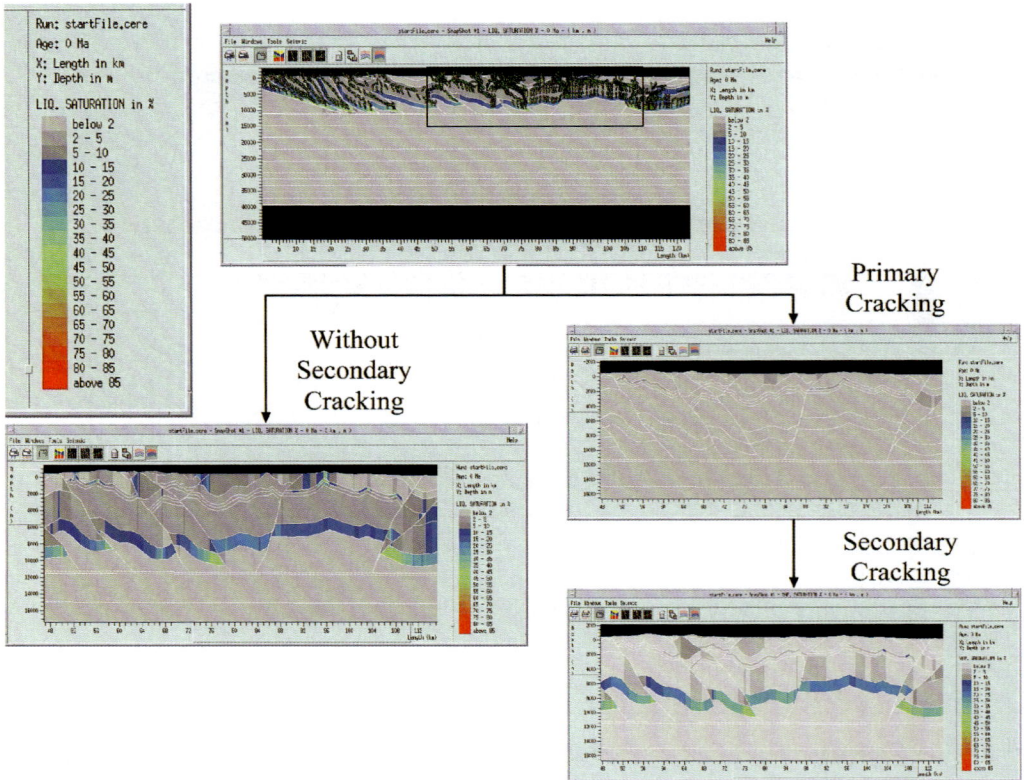

Fig. 32. Hydrocarbon saturation with and without secondary cracking.

are 2.2 and 1.6%, respectively (Fig. 30). This means that the geothermal gradient for these units increased around 28 Ma, because these temperatures do not correspond with the depth of the units at that time.

Assuming a gradient of 26 °C km^{-1}, the modelling also shows that the Pecaya Fm in the central part of the graben reached an average %Ro of 1.8%, a normal value for the depth that this unit was buried. This implies either the basin was hotter before this unit was deposited, and that the gradient was normalized during and after its deposition, or that the thermal anomaly did not reach the central part of the graben. To the south, the Castillo Fm is immature, showing that the unit was deposited after the formation of the graben and therefore was not affected by the increase in the temperature.

Discussion and petroleum system modelling

Simulation of oil generation is done using the kinetic parameters of the potential source rocks and the total organic carbon (%TOC) values. Then the petroleum system efficiency is tested (source, reservoir, seal rocks and timing). The transformation ratio (TR) for the La Luna Fm is 100% (Fig. 31), which indicates the potential of this unit to generate hydrocarbon. Likewise, the high temperature which affected the region caused Cenozoic rocks to have enough potential to generate hydrocarbon and saturate Miocene traps.

Several scenarios were considered to evaluate the evolution of the petroleum system, including thermal and structural variations. The first case considers the processes of primary and secondary cracking and their relationship with respect to the kinetics of the transformation ratio (Fig. 32): some accumulation is still taking place at the flanks of the central graben when a simulation of primary cracking is considered. Very little hydrocarbon accumulation takes place when secondary cracking is taken into account.

Hydrocarbon accumulation is only possible within the source rock itself and no hydrocarbon expulsion occurred if only one source rock (La Luna Fm) is considered (Fig. 33). Because of the high temperature that affected the region, only

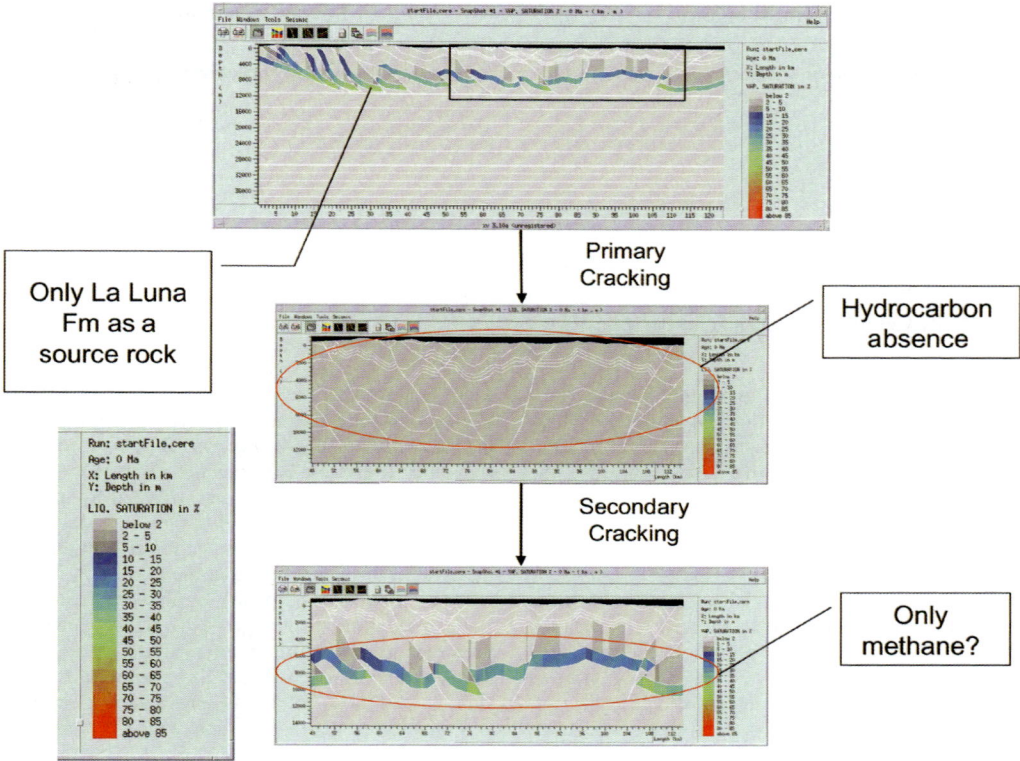

Fig. 33. Hydrocarbon saturation with only La Luna Formation as source rock.

methane gas is expected, assuming the most optimistic scenario.

The second case compares differing structural scenarios where a lower detachment angle for the Lara Nappes and smaller sedimentation ratio for the Socorro Fm are considered (compare Fig. 34a, b). An acceptable thermal calibration and a lower temperature gradient, which reaches 70 °C km^{-1} in the central part of the graben, is obtained from the modelling. However, no significant differences in the predicted accumulations of hydrocarbons based on these two structural scenarios were

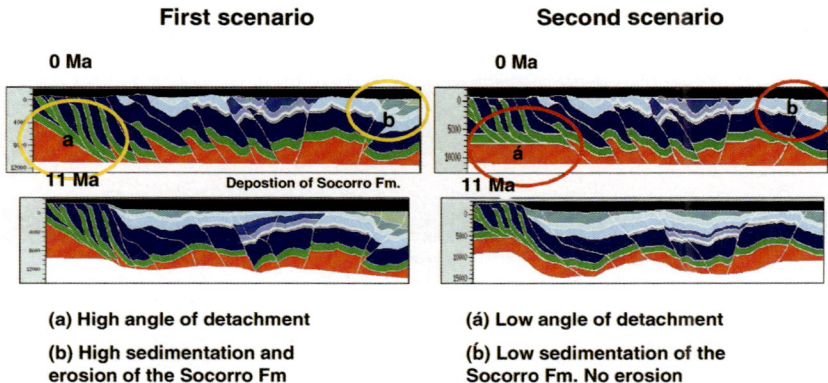

Fig. 34. Geological scenarios.

First scenario   **Second scenario**

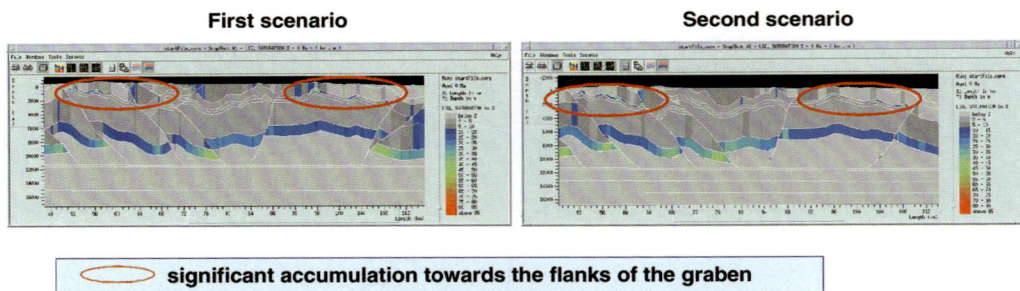

significant accumulation towards the flanks of the graben

Fig. 35. Hydrocarbon accumulation in both scenarios.

found (Fig. 35), and it is apparent that the most important element is the thermal anomaly which had predominance over the structural and sedimentological variations that could have occurred in the region (Fig. 35).

Tectonic implications

Recognition of an Oligocene–Early Miocene thermal anomaly seems to be inconsistent with the amount of extension in the region during that time. However, the presence of pre-existing normal faults in the lower crust would favour tectonic collapse. It is widely known that these structures

were generated along northern Venezuela during Cretaceous times, when a passive margin formed in the region. These normal faults were subsequently reactivated during the Oligocene–Early Miocene, when the region was affected by unroofing, allowing mantle upwelling that caused an incipient delamination process and an anomalous increase in the temperature in the region.

The idea of incipient Caribbean Plate flat slab subduction (dip angle <20 °S) is consistent with recent gravimetric results (Sousa *et al.* 2005). These authors proposed that the frontal part of the slab is currently below the Paraguaná Península at about 15 km depth (Fig. 36), which is where the

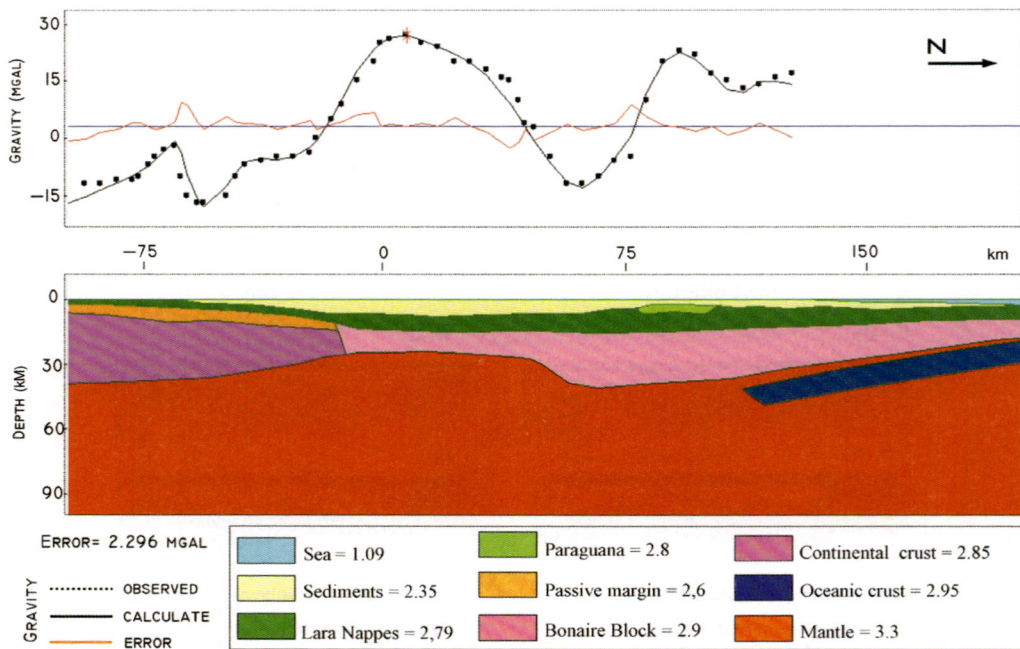

Fig. 36. Gravimetric model for the subduction in northwestern Venezuela (After Sousa *et al.* 2005).

bulk of the seismicity occurs. Hence, the occurrence of high temperatures and the action of the subducted flat slab enables the basin inversion that fashioned the region.

Conclusion

The tectono-stratigraphic model developed for this region reveals generation of a graben as a result of a tectonic collapse, creating the Falcón Basin. Significant increase in the thermal gradient occurred during the Oligocene–Early Miocene, with heat flow reaching 190 mW m^{-2} (due to incipient delamination?). This brought Cenozoic source rocks into the oil window and produced the basin's crude oil. Near-shore and platform clastic sediments with terrestrial organic plant matter are the sources of the crudes. The oils are different from those of the Maracaibo Basin, formed from organic matter in marine carbonate rocks deposited in an anoxic environment.

We thank PDVSA for logistic support and permission to publish the results, D. Hamilton, J. Rafalska, T. Jerzykiewicz, H. Krause, F. Audemard, J. Díaz, F. Urbani and K. H. James for improving the quality of the manuscript, and A. Aleman for exchange of constructive comments and discussions.

References

APPLEGATE, A. V., LESNIAK, S. W., AGUERREVERE, S. E., MCDANIEL, E. L. & PIERCE, A. C. 1956. *Mapa D-4-A Geología de superficie.* Creole Petroleum Corporation, Caracas. Mapa escala 1 : 100 000. [*In*: Universidad Central de Venezuela, Escuela de Geología, Minas y Geofísica, Laboratorio de Cartografía Geológica Digital. *Colección de mapas geológicos de Venezuela*, **1**, CD, 2007.]

AUDEMARD, F. A. 1993. *Néotectonique, sismotectonique et aléa sismique du nord-ouest du Vénézuéla (système de failles d'Oca–Ancón).* PhD thesis, Université Montpellier II, Montpellier, France.

BENGUIGUI, A., PAEZ, A., NUÑEZ, M. & ACOSTA, J. 2006. Diferenciación de dominios estructurales al noreste de la Cuenca de Maracaibo. *Memorias IX Simposio Bolivariano de Exploración Petrolera en las cuencas Subandinas* (CD).

BOESI, T. & GODDARD, D. 1991. A new geology model related to the distribution of hydrocarbon source rocks in the Falcón Basin, Northwestern, Venezuela. *In*: BIDDLE, K. T. (ed.) *Active Margin Basins.* American Association of Petroleum Geologists, Memoirs, **52**, 303–319.

DE TONI, B. & KELLOGG, J. 1993. Seismic evidence for blind thrusting of the Northwestern flank of the Venezuelan Andes. *Tectonics*, **12**, 1393–1409.

GAENSLEN, G. J., MCDANIEL, E. L. & APPLEGATE, A. V. 1960. *Mapa D-4-C Geología de superficie.* Creole Petroleum Corporation, Caracas. Mapa escala 1:100 000. [*In*: Universidad Central de Venezuela,

Escuela de Geología, Minas y Geofísica, Laboratorio de Cartografía Geológica Digital. *Colección de mapas geológicos de Venezuela*, **1**, CD, 2007.]

GARRITY, C. P., HACKLEY, P. C. & URBANI, F. 2006. *Digital geologic map and GIS database of Venezuela.* US Geological Survey Data Series Reports, **199**, edited version 1.0. World Wide Web Address: http://pubs.usgs.gov/ds/2006/199.

GHOSH, S., PESTMAN, P., MELENDEZ, L., TRUSKOWSKI, I. & ZAMBRANO, E. 1997. Evaluación Tectonoestratigráfica y Sistemas Petrolíferos de la Cuenca de Falcón, Venezuela Noroccidental. *Memorias del 8vo Congreso Venezolano.* Sociedad Venezolana de Geología, **I**, 317–329.

GONZÁLEZ DE JUANA, C., ITURRALDE DE AROZENA, M. & PICARD CADILLAT, X. 1980. *Geología de Venezuela y de sus cuencas petrolíferas.* Foninves, Caracas, 407.

GORMAN, J. M., REEDER, L. R. *ET AL.* 1966. *Mapa C-4 Geología de superficie.* Creole Petroleum Corporation, Caracas. Mapa escala 1 : 100 000. [*In*: Universidad Central de Venezuela, Escuela de Geología, Minas y Geofísica, Laboratorio de Cartografía Geológica Digital. *Colección de mapas geológicos de Venezuela*, **1**, CD, 2007.]

LIN, C. H. 1998. Tectonic implications of an aseismic belt beneath the Eastern Central Range of Taiwan: crustal subduction and exhumation. *Journal of the Geological Society of China*, **41**, 441–460.

LIN, C. H. 2000. Thermal modelling of continental subduction and exhumation constrained by heat flow and seismicity in Taiwan. *Tectonophysics*, **324**, 189–201.

LUGO, J. & MANN, P. 1995. Jurassic–Eocene tectonic evolution of Maracaibo Basin, Venezuela. *In*: TANKARD, A. J., SUÁREZ, R. & WELSINK, H. J. (eds) *Petroleum Basins of South America.* American Association of Petroleum Geologists, Memoirs, **62**, 699–725.

MACELLARI, C. E. 1995. Cenozoic sedimentation and tectonics of the southwestern Caribbean pull-apart basin, Venezuela and Colombia. *In*: TANKARD, A. J., SUÁREZ, R. S. & WELSINK, H. J. (eds) *Petroleum Basins of South America.* American Association of Petroleum Geologists, Memoirs, **62**, 757–780.

MANN, P. & BURKE, K. 1984. Neotectonics of the Caribbean. *Reviews of Geophysics and Space Physics*, **22**, 309–362.

MCDANIEL, E. L. 1960. *Mapa D-4-D Geología de superficie.* Creole Petroleum Corporation, Caracas. Mapa escala 1:100 000.

MCDANIEL, E. L., APPLEGATE, A. V., GAENSLEN, G. J., WHEELER, C. B. & PIERCE, A. C., 1961. *Mapa D-4 Geología de superficie.* Creole Petroleum Corporation, Caracas. Mapa escala 1 : 100 000. [*In*: Universidad Central de Venezuela, Escuela de Geología, Minas y Geofísica, Laboratorio de Cartografía Geológica Digital. *Colección de mapas geológicos de Venezuela*, **1**, CD, 2007.]

MCMAHON, C. E. 2000. *Evaluation of the effects of oblique collision between the Caribbean and South American plates using geochemistry from igneous and metamorphic bodies of Northern Venezuela.* PhD thesis, Notre Dame University, IN.

MOORE, B. C., HENRY, L. W. *ET AL.* 1951. *National Petroleum Convention.* The Technical Office of Hydrocarbons of the Ministry of Mines and Hydrocarbons, United States of Venezuela, Caracas.

MUESSIG, K. W. 1978. The central Falcón igneous suite, Venezuela: alkaline basaltic intrusions of Oligocene–Miocene age. *Geologie en Mijnbouw*, **57**, 261–266.

MUESSIG, K. W. 1984. Structure and Cenozoic tectonics of the Falcón Basin, Venezuela, and adjacent areas. *In*: BONINI, W. E., HARGRAVES, R. B. & SHAGAM, R. (eds) *The Caribbean–South American Plate Boundary and Regional Tectonics.* Geological Society of America, Boulder, CO, Memoirs, **162**, 217–230.

OSTOS, M., YORIS, F. & AVÉ LALLEMANT, H. G. 2005. Overview ot the southeast Caribbean-South American plate boundary zone. *In*: AVÉ LALLEMANT, H. G. & SISSON, V. B. (eds) *Caribbean–South American Plate Interactions, Venezuela.* Geological Society of America, Boulder, CO, Special Papers, **394**, 53–89.

PETERS, K., WALTERS, C. & MOLDOWAN, M. 2005. *The Biomarker Guide. Biomarkers and Isotopes in Petroleum Exploration and Earth History.* Cambridge University Press, Cambridge, **2**, 1132.

PITELLI, R. & MOLINA, A. 1989. *El Eoceno Medio Tardío y Eoceno Tardío de la parte occidental de la cuenca de Falcón.* Sociedad Venezolana de Geología, Boletín, **36**, 5–12.

SOUSA, J. C., RODRIGUEZ, J., GIRALDO, C., RODRÍGUEZ, I., AUDEMARD, F. A. & ALEZONES, R. 2005. An integrated geological-geophysical profile across northwestern Venezuela. Presented at the *Sixth International Symposium on Andean Geodynamics*, Barcelona, Spain.

STEPHAN, J. 1985. Andes et chaine Caraibe sur la transversale de Barquisimeto (Venezuela) Evolution Geodynamique. *In*: MASCLE, A. (ed.) *Geodynamique des Caraibes.* Editions Technip, Paris, 505–529.

SWEENEY, J. J. & BURNHAM, A. K. 1990. Evaluation of a simple model of vitrinite reflectance based on chemical kinetics. *AAPG Bulletin*, **74**, 1559–1570.

TORO, C. & ESCANDÓN, M. 1990. *Caracterización geoquímica de menes y crudos de pozos abandonados ubicados en la subcuenca oriental del Estado Falcón, Venezuela.* World Wide Web Address: http://www.pdvsa.com/lexico/posters/te.htm.

WHEELER, C. B., MCDANIEL, E. L. & PIERCE, A. C. 1960. *Mapa D-4-B Geología de superficie.* Creole Petroleum Corporation, Caracas. Mapa escala 1:100 000. [*In*: Universidad Central de Venezuela, Escuela de Geología, Minas y Geofísica, Laboratorio de Cartografía Geológica Digital. *Colección de mapas geológicos de Venezuela*, **1**, CD, 2007.]

Caribbean–South America oblique collision model revised

ROGER HIGGS

Geoclastica Ltd, 6 Ockham Court, 22 Bardwell Road, Oxford OX2 6SR, UK
(e-mail: rogerhiggs@geoclastica.com)

Abstract: The 'Caribbean oblique collision model' invokes a three-stage evolution for Ecuador–Colombia–Venezuela–Trinidad: (1) Jurassic rifting; (2) Oxfordian–Neogene passive margin subsidence beside the spreading (until Campanian) Proto-Caribbean Ocean; (3) Campanian to Miocene oblique collision of the Caribbean Arc from the west, producing a craton-verging thrust belt and foreland basin. The timing of these stages is incorrect. Rifting ended 70 Ma later (Coniacian), with Cuba breaking away from Venezuela–Trinidad. Brief Proto-Caribbean spreading (Santonian–Campanian) was followed by slow (amagmatic) subduction below Venezuela and Trinidad, driven by inter-Americas convergence, so the passive margin lasted just 10–15 Ma in eastern Venezuela, not 140 Ma (*sic*). Convergence changed from WSW to SSE in Paleocene time, causing Proto-Caribbean subduction under Colombia too. Subduction in Venezuela–Trinidad drove upper crustal nappes cratonward, metamorphosing overridden rift deposits of the Inner Nappe (Cordillera de la Costa) and feeding Campanian–Miocene olistostromes to a *Proto-Caribbean* 'pre-arc' foreland basin. The Caribbean Arc collided with Ecuador to Guajira from Campanian time and passed 'Guajira corner' in Early Oligocene time, not Paleocene, then migrated SE forming the Gulf of Venezuela–Falcón, diachronous (Oligo–Miocene), transform-related extensional basin, followed by oblique collision (obduction) in central Venezuela to Trinidad driving a Mio–Pliocene *Caribbean* foreland basin. Caribbean relative motion switched to eastward near 2.5 Ma, not 12 Ma as is widely believed. The new plate boundary follows the Eastern Cordillera–Mérida Andes–San Sebastián–El Pilar–Trinidad Central Range fault system. Pull-apart basins date from 2.5 Ma at the Cariaco and intra-Gulf of Paria stepovers. Elsewhere the boundary is characterized by transpressional uplift, overwhelmed in some areas (e.g. Gulf of Barcelona; greater Gulf of Paria) by subsidence due to dissolution of inferred, buried, Neocomian rift halite since the Middle Miocene (climate change). The revised timings of events, and the revival of the geosyncline-era concept of Late Cretaceous–Palaeogene orogeny in northern South America, will affect petroleum exploration.

In the popular 'Caribbean–South America dextral oblique collision model' (Pindell *et al.* 2006, p. 329), hereafter abbreviated to 'Caribbean model', the region comprising NW Colombia, northern Venezuela and Trinidad (Fig. 1) is interpreted as a passive margin (Jurassic rifting, then Jurassic–Cretaceous Proto-Caribbean spreading), evolving diachronously into a foreland basin from western Venezuela to Trinidad by oblique Paleocene–Miocene collision with the relatively SE-migrating Caribbean Plate frontal arc (Dewey & Pindell 1986; Pindell *et al.* 1988). The model builds on the idea that the Caribbean Plate migrated eastward from the Pacific (Wilson 1966).

The Caribbean model has strongly influenced exploration in this prolific oil and gas province, having met with wide acceptance (but see James 2006 for an *in situ* interpretation of the Caribbean Plate). Many recent publications use the Caribbean model as a foundation, reiterating the idea of a passive margin converting diachronously to a Caribbean foreland basin, in Paleocene time in western Venezuela (e.g. Mann *et al.* 2006; Escalona & Mann 2006; Gorney *et al.* 2007), and Oligo–Miocene in Trinidad and eastern Venezuela

(e.g. Bartok 2003; Boettcher *et al.* 2003; Jacome *et al.* 2003; Roure *et al.* 2003; Sageman & Speed 2003; Hung 2005; Locke & Garver 2005). However, the Caribbean model requires several fundamental corrections, as summarized by Higgs (2006) and discussed in detail here, based on exhaustive synthesis of the literature.

Accurate interpretation of the geology of northern South America is important for two reasons. First, the region is literally and figuratively central in tectonic reconstructions of the western part of the Pangaea supercontinent (Pindell & Dewey 1982; Pindell 1985); geological interpretations of the area affect those of formerly adjacent regions. Second, the region has large oil reserves, of great importance in the current world climate of political instability, surging demand, record high prices and concerns over security of supply. Sound tectono-sedimentary understanding is essential for prediction of new oilfields and extensions to old ones. The new model presented here will affect exploration in Ecuador, Colombia, Venezuela and Trinidad, changing interpretations and predictions of subsidence history, sand fairways, structure and palaeoheatflow.

From: JAMES, K. H., LORENTE, M. A. & PINDELL, J. L. (eds) *The Origin and Evolution of the Caribbean Plate*.
Geological Society, London, Special Publications, **328**, 613–657.
DOI: 10.1144/SP328.25   0305-8719/09/$15.00 © The Geological Society of London 2009.

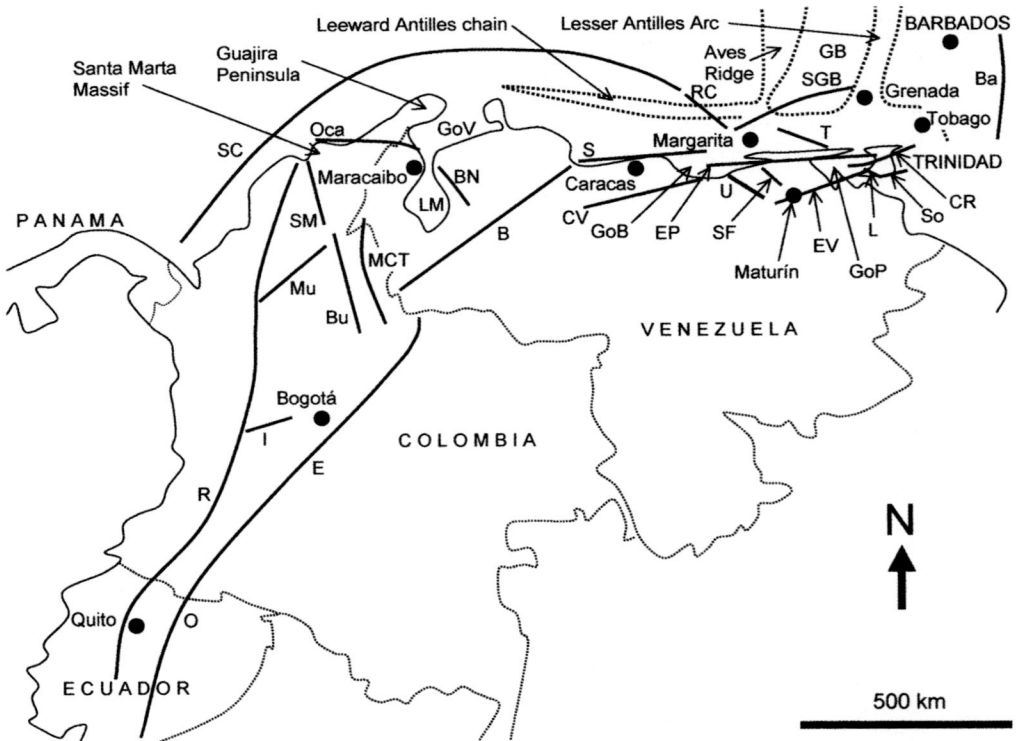

Fig. 1. Location map and major faults of northwestern South America. GB, Grenada Basin; GoB, Gulf of Barcelona; GoP, Gulf of Paria; GoV, Gulf of Venezuela; LM, Lake Maracaibo. Faults: B, Boconó; Ba, Barbados accretionary prism frontal (thrust); BN, Burro Negro; Bu, Bucaramanga; CR, Central Range–Warm Springs; CV, central Venezuela frontal (thrust); E, East Andean frontal (thrust); EP, El Pilar; EV, eastern Venezuela frontal (thrust); I, Ibague; L, Los Bajos; MCT, Mercedes–Caño Tomas; Mu, Murrucucú; O, Oriente; R, Romeral; RC, Roques Canyon; S, San Sebastián; SC, South Caribbean; SF, San Francisco; SGB, South Grenada Basin; So, south Trinidad frontal (thrust); SM, Santa Marta; T, Testigos; U, Urica. Dotted outline of Aves Ridge, Leeward Antilles extinct arc and Lesser Antilles Arc is the approximate 1500 m isobath. Tobago and Barbados mark the leading (eastern) edge of Caribbean crystalline crust (see text). Main sources: Parnaud *et al.* (1995*a*, *b*); Babb & Mann (1999); Villamil (1999); Audemard *et al.* (2000); Paris *et al.* (2000); Escalona & Mann (2006).

Longer rifting (younger Proto-Caribbean Ocean)

Pangaean rifting in Ecuador, Colombia, Venezuela and Trinidad supposedly ended in Late Jurassic time (Pindell & Dewey 1982; Pindell 1985; Stéphan 1985), in the Oxfordian according to Pindell & Kennan (2001*b*). However, the following 13 lines of evidence, among others, collectively prove that rifting lasted much longer, until the end of Coniacian time.

(1)   There is diverse, indirect evidence for a thick (kilometres), Neocomian, 'Carib Halite Formation', deposited in a 'Carib Graben' reaching from Colombia to Trinidad, but largely dissolved underground in Neogene

time following mountain uplift (Carib Graben inversion; Higgs 2009).

(2)   In Colombia, subsidence analysis of outcrops, boreholes and seismic profiles in the Eastern Cordillera and Upper Magdalena Basin indicates that the entire pre-Santonian Cretaceous succession accumulated by syndepositional normal faulting, i.e. in a rift setting (Fabre 1987; Cooper *et al.* 1995, fig. 18; Branquet *et al.* 2002, fig. 12; Duarte 2003; Sarmiento-Rojas *et al.* 2006). Rifting of this age here is further supported by heat-flow modelling (Gómez *et al.* 2002), magmatism of extensional composition (Moreno *et al.* 2002; Vásquez & Altenberger 2005), and evidence for dual, eastern and western provenance (Guerrero *et al.* 2000).

(3) Farther south, Cretaceous rifting in the Putumayo Basin and in Ecuador (Oriente Basin) is supported by irregular isopachs (Caballos and Villeta T and U sands; Bejarano *et al.* 1991), intraplate intrusive magmatism (Moreno *et al.* 2002; Barragan *et al.* 2005) and dual provenance (Nevers *et al.* 1991).

(4) In the Central Cordillera of Colombia, granitoid plutons of Cretaceous age (124–168 Ma, K–Ar; Aspden *et al.* 1987) are interpretable as rift related, dominated by the Antioquia Batholith. This plutonic episode, and a Jurassic one, are generally attributed to subduction (Aspden *et al.* 1987; Cooper *et al.* 1995; Kennan & Pindell 2007), despite their generally inconclusive geochemical signature (Aspden *et al.* 1987; González 2001) and the failure to produce an emergent continental-margin arc (Cooper *et al.* 1995). The Cretaceous plutons are coeval with overlying shallow marine deposits (Cooper *et al.* 1995; González 2001), indicating that the plutonism occurred in a subsiding basin, interpreted here as a rift, as suggested by Aspden *et al.* (1987) for *some* of the Jurassic plutons, based on geochemistry. Notwithstanding regional evidence for autochthoneity (Aspden *et al.* 1987), the Antioquia Batholith has been interpreted as strike–slipped far (hundreds of kilometres) from the south (Pindell & Kennan 2001*b*; Kennan & Pindell 2007) to conform with the previously inferred passive-margin setting of this region (Pindell & Dewey 1982; Pindell 1985; Pindell & Tabbutt 1995). According to Jaimes & de Freitas (2006, p. 486), the passive-margin 'paradigm' has caused a 'bias in prevailing models in which tectonism and unconformity are unreasonable or unexpected', hence 'a variety of evidence may have been overlooked, treated as local or anomalous, or entirely disregarded' (see also Cooney & Lorente 1997).

(5) In the Venezuelan Andes and productive basins (Maracaibo, Barinas, Eastern Venezuela), irregular Lower and Upper Cretaceous isopachs likewise support a syn-rift setting, up to and including the classical source-rock formations (La Luna, etc.; see below) and their cratonward sandier equivalents. Notable 'thicks' and 'thins' include the Mérida Arch, El Baúl-Machete Arch and Espino Graben (Patterson & Wilson 1953; Renz 1959; Feo-Codecido *et al.* 1984; Salvador 1986; Kiser 1989; Savian & Scherer 1993; Audemard & Serrano 2001; Erlich *et al.* 2003). The cratonward limit is interpreted as a fault by some authors (Canoa-Tigre

Fms in Pérez de Mejía *et al.* 1980, fig. 1–13a & b). El Baúl Arch Palaeozoic basement is locally covered directly by Paleocene strata, suggesting that the Arch was positive during Cretaceous deposition (Belotti *et al.* 2003).

(6) Tigre Fm sands contain staurolite and glaucophane (Escalona 1985). The passive margin model requires sediment derivation from the Guyana Shield in the south (e.g. Villamil & Pindell 1998). Staurolite does occur today far to the SE, both *in situ* and in placer deposits in Surinam and French Guiana (de Roever *et al.* 1981; Prost 1989). In Venezuela, staurolite *facies* metavolcanic rocks are widespread on the Shield (e.g. Pimentel 1984), but staurolite *per se* has only been mentioned as a trace component of modern river sands (Grosz 1989 pers. comm. in Wynn 1993). Glaucophane is unknown on the Shield, consistent with its absence in Archaean and Early Proterozoic rocks worldwide (Liou *et al.* 1990). In any case, in Cretaceous time the Shield was blanketed, as far north as the Orinoco (Briceño & Schubert 1990) and probably beyond (Feo-Codecido *et al.* 1984), by the Proterozoic Roraima Fm (e.g. LEV 1997), in which neither mineral has been reported. On the other hand, in *northern* Venezuela, staurolite occurs in the Juan Griego Group and Manicuare Fm, and glaucophane in the Tacagua and El Copey Fms (González de Juana *et al.* 1980; Avé Lallement *et al.* 1993; Stöckhert *et al.* 1995; Avé Lallement 1997; LEV 1997; Sisson *et al.* 1997). These metamorphic units, generally viewed as Jurassic–Cretaceous on insecure radiometric evidence, are here regarded as Palaeozoic (cf. Benjamini *et al.* 1987; Stöckhert *et al.* 1995), belonging to a Hercynian collisional–orogenic fringe in Trinidad and northern Venezuela (Feo-Codecido *et al.* 1984; Bartok 1993; Speed *et al.* 1997). The Tacagua and El Copey were included in a 'Margarita coastal ophiolite nappe' (Bellizzia & Dengo 1990) and assigned a Proto-Caribbean or Caribbean-related oceanic origin (Sisson *et al.* 1997; Speed & Smith-Horowitz 1998). Instead these units are probably Palaeozoic, as suggested for the Tacagua by Benjamini *et al.* (1987), and represent oceanic crust metamorphosed by subduction then obducted as a thrust sliver during Hercynian continental collision, i.e. a Phoibic Ocean remnant (Scotese & McKerrow 1990). The Bocas Complex, in a Carúpano Basin borehole about 10 km north of

Paria Peninsula, was correlated with the El Copey (Castro & Mederos 1985) and thus misinterpreted as Caribbean related (Pereira 1985; Speed & Smith-Horowitz 1998). The Bocas contains mineralogical indications of retrogradation from blueschist (Speed & Smith-Horowitz 1998). In conclusion, the Tigre glaucophane and staurolite can be interpreted as derived from Hercynian basement exposed on local half-graben shoulders.

(7)   Lateral facies changes are consistent with Early Cretaceous fault control. Isolated carbonate banks in Colombia, Venezuela and Trinidad are surrounded at least partially by shales (LEV 1997; Saunders 1997a; Villamil & Pindell 1998; Babb & Mann 1999; Erlich et al. 2003). Trinidad shales of this age (Cuche, Lopinot, Toco formations) contain limestone blocks interpreted as olistoliths derived from a neighbouring carbonate bank (Hutchison 1938; Barr 1952), suggesting steep bank margins and active normal faults.

(8)   Late Cretaceous deposition of anoxic source-rock facies (Ecuador to Trinidad; Napo–Villeta–La Luna–Querecual–Naparima Hill formations) on an ocean-facing middle shelf to upper slope (Pindell & Dewey 1982, fig. 18; Villamil & Pindell 1998) over a 10–15 Ma time span (Cenomanian through Coniacian; Erlich et al. 2003) is impossible because oceanic wind- and wave-driven circulation prevents long-term thermohaline stratification. This long time span included Oceanic Anoxic Event 2, but this event was too brief (<2 Ma; Tsikos et al. 2004) to explain the longevity of source-rock deposition in northern South America. Sixty years ago, Hedberg (1950) suggested that the Querecual Fm reflects restricted circulation behind a northern barrier, interpreted by Maresch (1974) as a continental-margin arc despite absence of tuffs. Active faulting during La Luna deposition has been postulated previously (Erlich et al. 1999; Macsotay et al. 2003).

(9)   Thin (centimetres to decimetres) debrites, conglomerates and slump beds in these source-rock formations (Macellari & De Vries 1987; Erlich et al. 2003; Macsotay et al. 2003) are consistent with tectonism, though masked by overall shaliness due to high eustatic sea levels (Haq et al. 1988). Reworked intraformational clasts in Luna and Querecual debrites suggest sea-floor fault scarps up to 50 m high (Macsotay et al. 2003).

(10) Load casts and 'pseudoconcretions' in the Luna and Querecual formations suggest syn-depositional seismicity (Macsotay et al. 2003), in agreement with Pratt (2001).

(11) The 700 m thickness of the Querecual Fm (LEV 1997), decompacted by a realistic shale factor of at least 3 (based on lamination curvature around early-diagenetic concretions; author's observations), gives 2 km deposited in about 13 Ma (approximately entire Cenomanian–Coniacian; LEV 1997), implying a subsidence rate of c. 150 m/Ma, an order of magnitude faster than a typical passive margin. Erikson & Pindell (1993) assumed a 14 Ma span and only a 1.4 decompaction factor (1000 m), giving 70 m/Ma, but reiterated the passive-margin interpretation.

(12) Across the supposed palaeo-shelf, the NW–SE extent of the La Luna Fm exceeds 300 km (Zambrano et al. 1970, 1971), excluding shortening in the Mérida and Perijá mountains. This is wider than any normal modern passive-margin middle shelf to upper slope and far wider than most.

(13) In the allochthon of central and eastern Venezuela, Cretaceous formations (some metamorphosed) contain conglomerates and volcanics, consistent with rift deposition, followed by metamorphism under an outer-margin nappe emplaced in Palaeogene time (see below).

The conclusion is that rifting ended in Late Cretaceous time, not Late Jurassic, a difference of nearly 70 Ma, with dramatic implications for oil exploration, as discussed below. The final rift deposits were the La Luna and equivalent anoxic facies (Cenomanian–Coniacian). Succeeding Santonian–Campanian shallow marine deposits contain bedded chert (upper Villeta; upper La Luna; lower Río Chavez-San Antonio; upper Naparima Hill; Macellari 1988; LEV 1997; Erlich et al. 2003). The chert, previously attributed to upwelling (Villamil et al. 1999), may instead reflect the proximity of Proto-Caribbean ridge volcanism. These upper strata are bioturbated and organically leaner, consistent with bottom oxygenation on the new passive margin. In Colombia the base of the cherty interval has locally been interpreted as unconformable (Maughan et al. 1979; Macellari & De Vries 1987, fig. 4), possibly indicative of a breakup unconformity (Falvey 1974). The late start of Proto-Caribbean spreading (Santonian) means that if La Luna-equivalent source rocks are present in the Barbados accretionary prism (scraped off subducting Proto-Caribbean), as inferred from analysis of Barbados oils (Hill & Schenk 2005), only upper La Luna equivalents are likely, but possibly richer than their South American counterparts as they were deposited in a young ocean basin (relatively shallow, within oxygen-minimum zone?).

The total duration of rifting was about 150 Ma, assuming it began in the Late Triassic (*c.* 229 Ma U–U, La Quinta Fm volcanics; LEV 1997) and ended in Coniacian time, around 85 Ma. This is much longer than typical rift episodes worldwide, whose duration seldom exceeds 50 Ma (Allen & Allen 1990). A possible explanation is that rifting spanned (and bridged) two hemiglobal superplume events: (1) a Late Triassic–Early Jurassic event responsible for initial fragmentation of Pangaea (227–183 Ma; Vaughan & Storey 2007); and (2) a Mid-Cretaceous event (120–80 Ma; Vaughan & Storey 2007) responsible for the Caribbean Plateau overthickened oceanic crust (90–75 Ma; Larson 1991), which moreover formed close to northern South America (e.g. Meschede & Frisch 1998). A speculative 'Colombia Mantle Plume' may have been centred in the Central Cordillera plutonic belt (Fig. 2b).

Seismic profiles and wireline correlation panels in the Oriente Basin (Ecuador) and Eastern Cordillera indicate that Early and Middle Cretaceous sedimentation was interrupted by compressional interludes (Balkwill *et al.* 1995; Higgs 2002; Jaimes & de Freitas 2006). This may reflect development of a south-dipping subduction zone at the northern edge of the microcontinental block (including Cuba) rifting away from South America (see below; cf. Fig. 2c), causing intermittent compression within the rift. Thin (centimetres) Albian–Coniacian tuffs (Villamil 1998) in the Colombian rift succession testify to the remoteness of both this arc and the Inter-American Arc (see below).

Missing 'Inter-Americas Ocean' and detachment of Cuba from South America

The later start (Santonian) of Proto-Caribbean spreading requires an earlier ocean to accommodate more than 2000 km of NW–SE divergence between North and South America (Middle Jurassic–Campanian), interpreted from Atlantic magnetic anomalies and fracture zones (Pindell *et al.* 1988). A vanished, Jurassic–Cretaceous 'Inter-Americas Ocean' is proposed here, separating the Yucatán–Bahamas margin from a 'Chortís–Greater Antilles Superterrane' (CGAS), then still joined to South America (Fig. 2a, b). The superterrane comprised two components: (a) the later Chortís Terrane, consisting of Nicaragua, Honduras and the Nicaragua Rise (summaries in Mann 1999; James 2006) and considered here to include Jamaica (see discussion in Pindell *et al.* 2005), and (b) the metacontinental terranes of Cuba and Hispaniola (Pindell *et al.* 2005), and probably below Puerto Rico. In Santonian time, CGAS

became detached from the Ecuador–Colombia–Venezuela–Trinidad margin by Proto-Caribbean spreading. In this model the Proto-Caribbean Ocean fronted all four countries, similar to the *Early* Cretaceous reconstructions of Ross & Scotese (1988, figs 6 & 7). In contrast, Colombia and Ecuador have been viewed as facing an Andean back-arc ocean that connected northward to the Proto-Caribbean and tapered southward. This 'Colombian Marginal Seaway' (Pindell 1993) is interpreted to have opened from Late Jurassic until Aptian time above an east-dipping subduction zone and closed diachronously northward in Aptian to Maastrichtian time following subduction-polarity reversal (see below; Pindell 1993; Pindell & Tabbutt 1995; Pindell & Kennan 2001*b*). In this model, the Colombian source rocks (Villeta–La Luna) were seen as passive-margin deposits of the closing back-arc ocean.

Northwestward CGAS migration and Proto-Caribbean opening (Fig. 2c) were partly accommodated by southward subduction of the Inter-Americas Ocean at a 'Cuba–Hispaniola Arc' built on CGAS, active since at least Aptian–Albian time based on the age of (1) the Colombia–Ecuador compressive episodes mentioned above, and (2) the oldest arc volcanics in Cuba–Hispaniola (Pindell & Barrett 1990; Kerr *et al.* 1999). Subduction continued beyond the change from inter-Americas divergence to convergence (Campanian; see below), resulting in Paleocene–Eocene arc magmatism in Cuba and Hispaniola (Pindell & Barrett 1990). Intra-arc spreading split Cuba–Hispaniola longitudinally to form the Paleocene–Eocene (Rosencrantz 1990) Yucatán interarc basin. Volcanism and Yucatán Basin spreading ended during collision of Cuba against the Yucatán Platform–Bahama Platform reentrant (Fig. 2d & e) in Early Eocene time (Pindell & Kennan 2001*b*). The CGAS superterrane was later disrupted by strike–slip opening of Cayman Trough (Fig. 2f).

Thus, Cuba is a piece of South America detached from Venezuela–Trinidad (Klitgord & Schouten 1986, fig. 10) that migrated relatively NW to collide with North America (Yucatán–Bahamas). In a different model, Cuba is viewed as a fragment of southwestern *North* America (Chortís, *after* accretion; see below) carried NE in the Caribbean arc accretionary wedge and reaccreted to North America as the arc expanded into the Yucatán–Bahamas reentrant (Pindell & Kennan 2001*b*; Pindell *et al.* 2005, 2006). The blueschist Escambray Terrane (Cuba), viewed in this model as a sliver clipped off North America and metamorphosed in the Caribbean subduction zone (Pindell *et al.* 2005, 2006), is reinterpreted here as a slice of CGAS continental margin metamorphosed in the (north-verging) *Cuba–Hispaniola* subduction

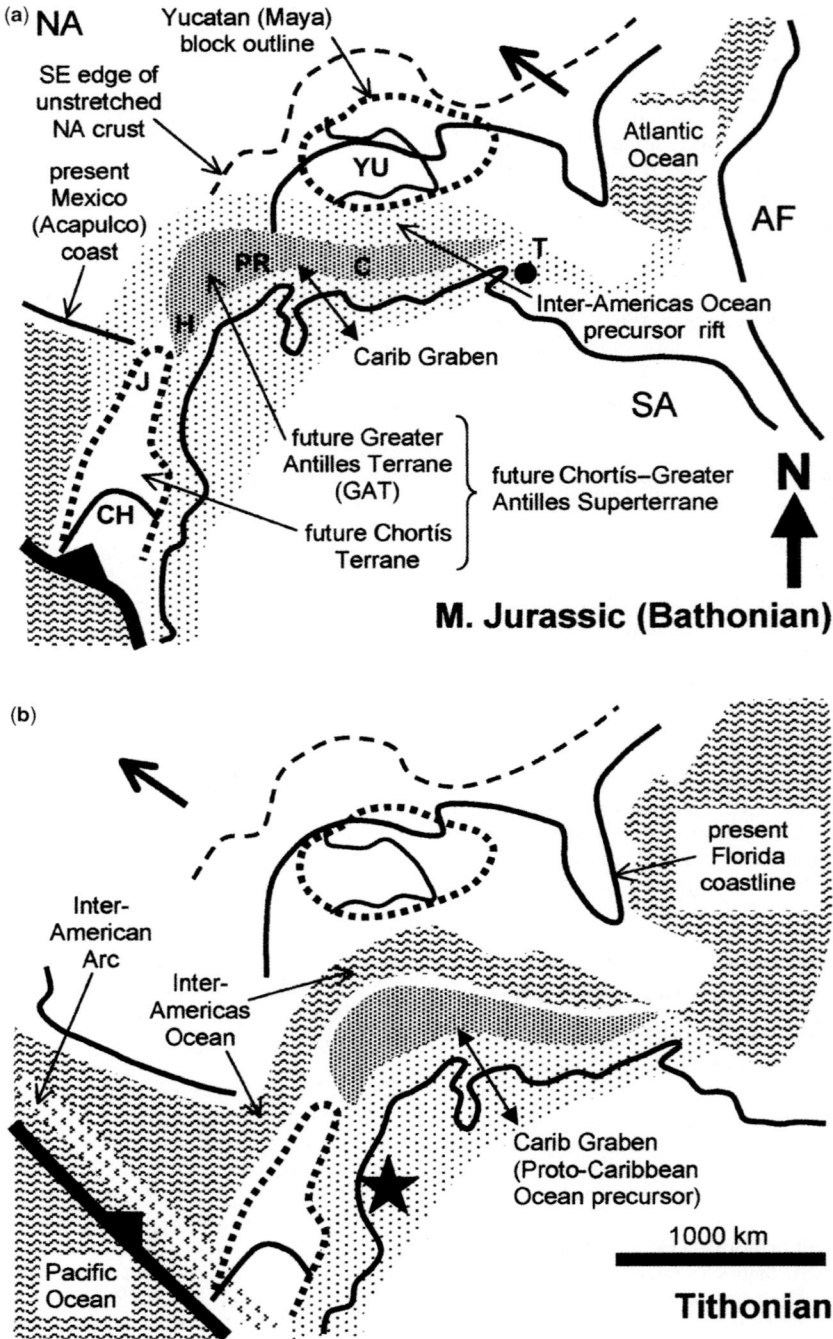

Fig. 2. Sequential plate reconstructions, Caribbean-Gulf of Mexico region, from Jurassic to Miocene time. Wavy pattern is oceanic crust. Relative positions of North and South America plates (NA, SA) through time from Pindell & Kennan (2001), figs 1–18 (which also show stretched v. unstretched crust). Open-headed arrows indicate NA relative motion direction. AF, Africa Plate. In the vector triangles, the Caribbean (CA)–NA vector was obtained by summing the two vectors: (1) CA–SA (eastward, then SE, then east again; see text); and (2) NA–SA

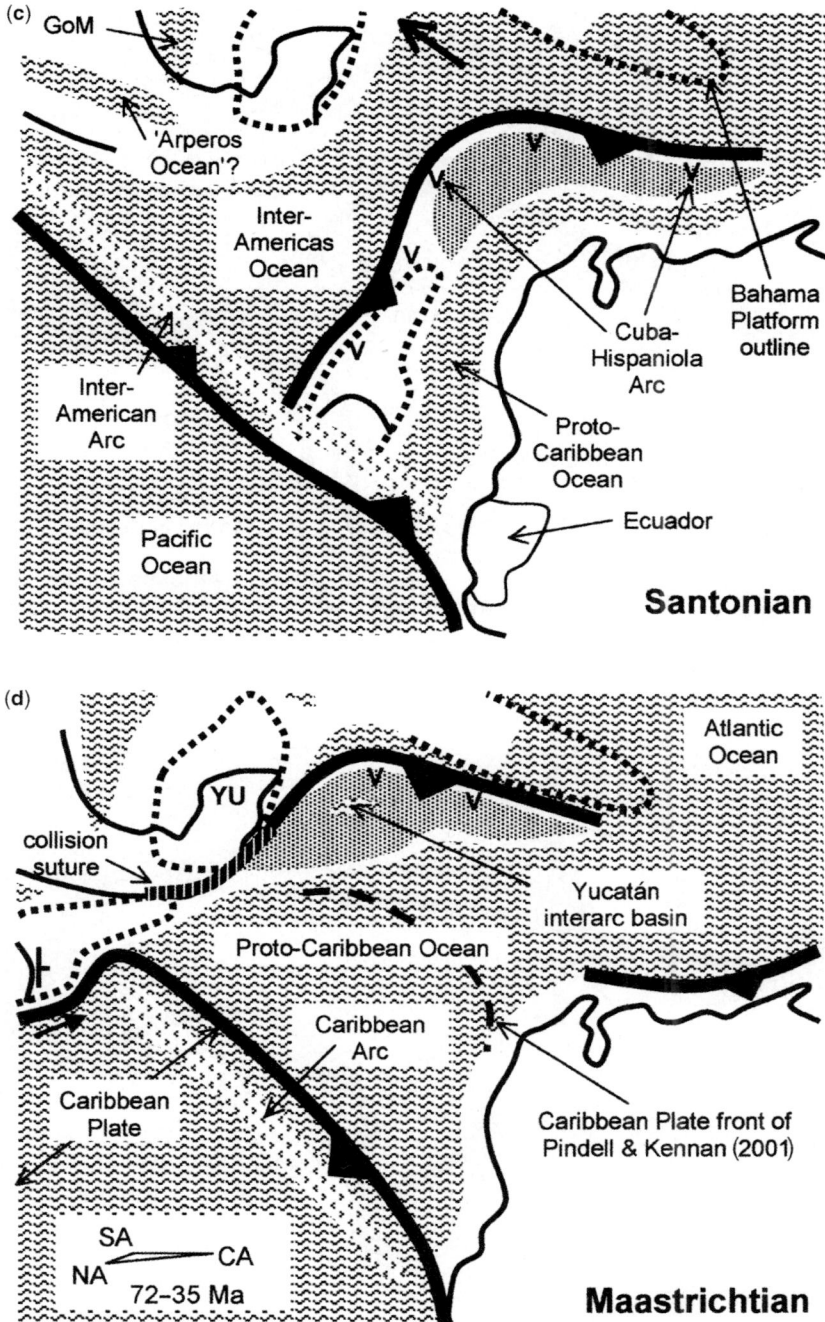

Fig. 2. (*Continued*) (Müller *et al.* 1999, fig. 9). C, Cuba continental terranes; CH, Chortís Block; GoM, Gulf of Mexico; H, Hispaniola; J, Jamaica; PR, Puerto Rico; T, Trinidad; YU, Yucatan Block. The black star in (b) represents a speculative 'Colombia mantle plume', interpreted here as responsible for Jurassic–Cretaceous granitoid magmatism of the Central Cordillera–Santa Marta belt, conventionally attributed to a continental-margin arc (see text).

(e)

Yucatán reentrant

C

H   PR

Proto-Caribbean Ocean

Caribbean Arc

Falcon reentrant

Caribbean Plate front of Pindell & Kennan (2001)

Farallon Plate

Caribbean Plate

Amaime suture

Costa Rica-Panama Arc

Middle Eocene

(f)

J

Lesser Antilles Arc

C

H   PR

Cayman Trough

Caribbean Plate

Honduras-Nicaragua

Nicaragua Rise

South Caribbean Fault

Caribbean Plate front of Pindell & Kennan (2001)

Costa Rica-Panama Arc

Cocos-Nazca Plates

SA

NA

35–2.5 Ma   CA

Late Miocene

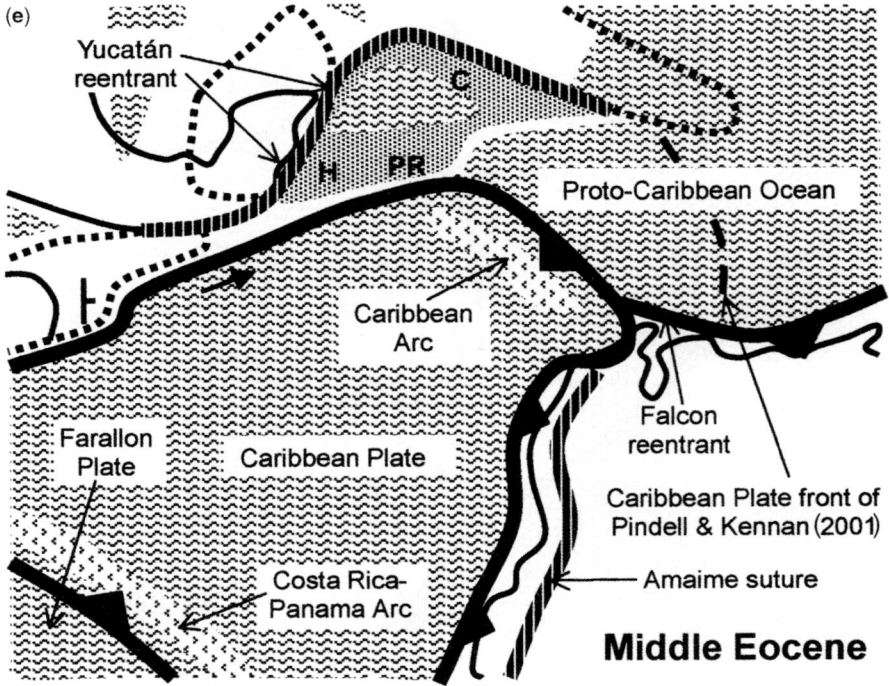

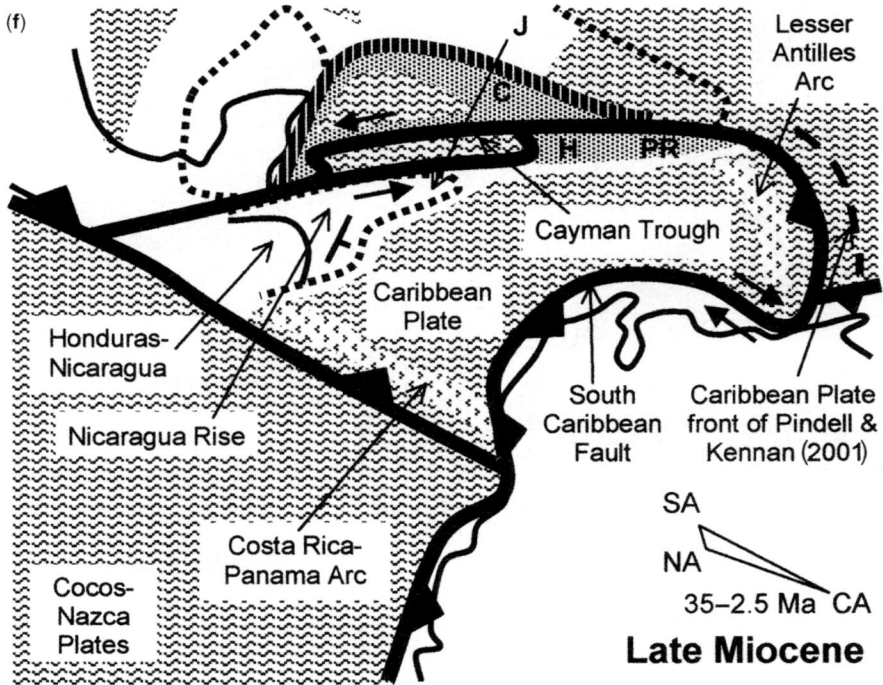

Fig. 2. *Continued.*

zone and ultimately exhumed by CGAS–Bahamas collision.

Campanian birth of Caribbean Plate

The growing Jurassic–Campanian gap between North and South America was bridged by a NW-trending Inter-American Arc (IAA; Pindell & Kennan 2003), separating the NE-subducting Farallon Plate (Pacific Ocean) from (A) the shrinking Inter-Americas Ocean, (B) the end-on CGAS migrating NW and (C) the expanding Proto-Caribbean Ocean (Fig. 2c). The IAA connected at each end to the continental-margin arcs of western North and South America, merging into the latter at the Ecuador–Peru coastal bend. The IAA later switched polarity, forming the Caribbean Arc that migrated eastward relative to the Americas, subducting Proto-Caribbean lithosphere. Polarity reversal is widely supposed to have occurred in Aptian time, following Pindell & Dewey (1982), but this is impossible if there was then (Aptian) no Proto-Caribbean Ocean (Santonian birth) to subduct. Campanian reversal is preferred here (Duncan & Hargraves 1984; Burke 1988; discussion in Pindell et al. 2006), during a 72 Ma plate reorganization which terminated inter-Americas divergence, as described below. The Chortís sector of CGAS collided against southern Mexico at this time (Fig. 2d), welding onto North America, leaving the inter-Americas gap occupied only by the Proto-Caribbean Ocean, making polarity reversal easier (no need to subduct CGAS).

Eastward relative migration of the Caribbean Arc caused diachronous collision with South America, obducting the Amaime–Ruma and Villa de Cura–Margarita (La Rinconada Gp)–Tobago forearc nappes (see below). These nappes consist of oceanic (MORB) and arc rocks, largely metamorphosed, some in blueschist facies (Stöckhert et al. 1995; Kerr et al. 1997; Snoke et al. 2001; Pindell et al. 2005). The protolith age of these rocks in Margarita and Tobago islands (Fig. 1) is Early and Mid-Cretaceous (Snoke et al. 2001; Pindell et al. 2005), pre-dating both Proto-Caribbean spreading (Santonian–Campanian) and Caribbean Arc onset (Campanian), therefore contrary to previous interpretations (Stöckhert et al. 1995; Pindell et al. 2005), the Margarita MORB rocks (Rinconada) cannot be of Proto-Caribbean origin and their metamorphism is unrelated to subduction at the Caribbean Arc. Instead the Margarita history is reconstructed here as follows (radiometric dates from Stöckhert et al. 1995 and Pindell et al. 2005):

(1)   Rinconada Lower Cretaceous MORB formed by Inter-Americas Ocean spreading (cf. Fig. 2b & c);

(2)   114–105 Ma Guayacán–Matasiete intrusions into Rinconada rocks at the Inter-American Arc (IAA);

(3)   Late Cretaceous metamorphism and mylonitization in the NE-dipping IAA subduction zone, implying prior transference to the forearc, presumably due to arc retreat by tectonic erosion;

(4)   86 Ma arc-related, non-metamorphosed El Salado Granite intrusion, implying westward trench jump, or an increase in slab dip;

(5)   inheritance into the Caribbean forearc at the 72 Ma polarity reversal, requiring that the Caribbean Arc nucleated west of the IAA, not upon it, consistent with the young age of the Proto-Caribbean (shallower dip of warm slab);

(6)   Miocene obduction of the Caribbean forearc nappe onto the Venezuela margin Hercynian basement (Juan Griego Gp) and its rift/passive margin cover (Los Robles etc.; see below).

Several million years elapsed after polarity reversal before the newly subducting Proto-Caribbean slab reached melting depth and Caribbean Arc magmatism began. Hence the arc can be no older than Maastrichtian, so the early-obducted, southern portion of the Amaime forearc nappe (Ecuador–SW Colombia), emplaced in Campanian–Early Maastrichtian time as shown below, had no Caribbean Arc in its rear. Northward, a progressively maturing arc is predicted, accreted diachronously to NW Colombia, probably underlying the inner Sinú or outer San Jacinto belt (cf. Flinch 2003; Flinch et al. 2003).

No 'Great Arc'

Conventionally, Cuba and Hispaniola volcanics are assigned to a 'Great Arc of the Caribbean' (Burke 1988), interpreted to have been smeared along the southern and northern margins (respectively) of North and South America by eastward oblique obduction, driving diachronous, Cenozoic, hydrocarbon-producing foreland basins (Pindell & Kennan 2001b, summarizing interpretations of Pindell and co-authors since 1982). This model requires dynamically complex sideways (north, south) expansion of the Caribbean Plate into the Yucatán and Falcón reentrants as it migrated east. Instead, a separate Cuba–Hispaniola arc is invoked here, riding on the CGAS. After this arc accreted to North America, as already described, a second arc, for which the formal name Caribbean Arc is proposed, simply migrated past (Fig. 2). The southern end of the Caribbean Arc was progressively obducted onto South America (see below), leaving only a small sector active today (Lesser Antilles Arc).

Proto-Caribbean interaction with South America, Campanian–Pliocene

In latest Campanian time (chron 32, 72 Ma), North America motion relative to South America changed from NW to between WNW and WSW (Müller et al. 1999, fig. 9; see also Pindell et al. 1988). WSW motion agrees well with geological developments in Venezuela and Trinidad, as shown below. In Ecuador–Colombia, the NE-trending youthful passive margin converted to a transtensional basin complex bordering the Proto-Caribbean (Fig. 3). Westerly sourced Maastrichtian sands (Fig. 3; Cimarrona, Monserrate Fms; Villamil 1999; Gómez et al. 2003) are interpreted here as derived from reactivated outer-margin rift blocks.

Venezuela–Trinidad northern nappe mountains

In Venezuela and Trinidad, the WSW motion caused (1) Proto-Caribbean oblique (sinistral) subduction at the SE-trending Falcón continental margin sector reaching from 'Guajira corner' to Golfo Triste (Higgs 2009, figs 1, 2), and (2) highly oblique sinistral subduction at the Caracas–Trinidad sector, trending slightly north of east (c. 080°, Fig. 3). Proto-Caribbean subduction under northern South America was tentatively suggested by Sykes et al. (1982) and Wielchowsky et al. (1991) and treated in more depth by Pindell et al. (1991, 1998, 2006). Subduction in both sectors pushed an outer-continental margin block cratonward as an 'Outer Nappe', comprising former passive-margin basement and its Jurassic–Cretaceous rift- and post-rift cover (uplifted as coastal mountains), whose inboard counterparts were metamorphosed under the nappe (Fig. 4b). Many authors have similarly proposed Late Cretaceous or Paleocene northern uplift in Trinidad and central Venezuela, to explain metamorphism (Maresch 1974) and north-derived sediments (Hedberg 1937; Kugler 1953; Suter 1960; Bell 1971, 1972; Kugler & Saunders 1967; Salvador & Stainforth 1968; Persad 1985; Tyson & Ali 1991; Wielchowsky et al. 1991; Higgs 2000), although allowance for palinspastic restoration must be made in Trinidad, where, as shown below, about 50 km of dextral slip has occurred on the Central Range Fault since 2.5 Ma (cf. 100 km since 12 Ma, Pindell & Kennan 2001a). This is the 'Cretaceous–Eocene boundary' orogenesis of Senn (1940, p. 1563), the 'Cretaceous-to-Miocene geosynclinal orogeny' of Salvador & Stainforth (1968, p. 36), and the 'Orogénesis del Cretácico Superior'

of González de Juana et al. (1980, p. 180). This end-Cretaceous orogeny in Venezuela and Trinidad has been disputed by other authors, who viewed the metamorphic rocks as allochthonous, transported from the west in the Caribbean accretionary prism and obducted onto South America in Eocene to Neogene time (Dewey & Pindell 1986; Pindell et al. 1988). In Trinidad, Algar (1998) interpreted the metamorphism as occurring in situ in Miocene time under a Caribbean nappe.

In central and eastern Venezuela, certain metasedimentary units (Juan Griego Gp, Manicuare Fm, El Tinaco Complex; LEV 1997) are interpreted here as Outer Nappe basement, of Hercynian metamorphic age, containing meta-ophiolite slices (Tacagua, El Copey formations). The Outer Nappe overrode a later 'Inner Nappe', whose basement includes the Yumare, Sebastopol and Dragon metaplutonics (LEV 1997), again probably Hercynian as inferred for the Dragon Gneiss (Speed et al. 1997).

The Outer Nappe load metamorphosed the Inner Nappe cover (e.g. Caracas Gp of north-central Venezuela; Caribbean Gp of Trinidad's Northern Range), whose Jurassic and Cretaceous ages are based largely on megafossils and are thus imprecise (Kugler 1953; Bellizzia 1986; LEV 1997). This cover, previously viewed as post-rift, passive-margin shelf and slope deposits (Maresch 1974; Algar 1998; Pindell & Kennan 2001a), is reinterpreted here as mainly of rift origin, consistent with many formations containing conglomerates and/or volcanics, e.g. Aragüita, Aroa, Carorita, Carúpano, Chuspita, Conoropa, Grande Riviere, Güinimita, Las Brisas, Las Mercedes, Las Placitas, Los Naranjos, Mamey, Maracas, Muruguata, Pueblo Nuevo, San Quintín, Sans Souci, Toco, Tucutunemo, Tunapui, Urape (Kugler 1953; Barr 1962; Wadge & Macdonald 1985; Bellizzia 1986; Jackson et al. 1991; Potter 1997; LEV 1997). Similarly, the non-eroded remnants of the Outer Nappe cover (non-metamorphosed) include conglomerates and volcanics (e.g. Mejillones Complex, Barremian–Santonian; Pilancones Fm, Albian–Cenomanian; LEV 1997). Some of the formations with volcanics are reasonably constrained to Mid or Late Cretaceous age (Aragüita, Las Placitas, Mejillones), yet volcanics are unknown in strata of this age in the adjacent autochthon to the south (i.e. mountains and basins of the rest of Venezuela and Trinidad), and are apparently limited to the Jurassic (Ipire, La Quinta formations; LEV 1997). Confinement of Cretaceous volcanism to the north suggests that extension was greater there, nearer the centre of the graben (cf. Higgs 2009, fig. 1). The lack of (wind-blown) tuffs in the south suggests that the northern volcanism was largely submarine, in agreement with the facies of the three listed formations (dark shale/phyllite, chert, pillow lava, etc.; LEV

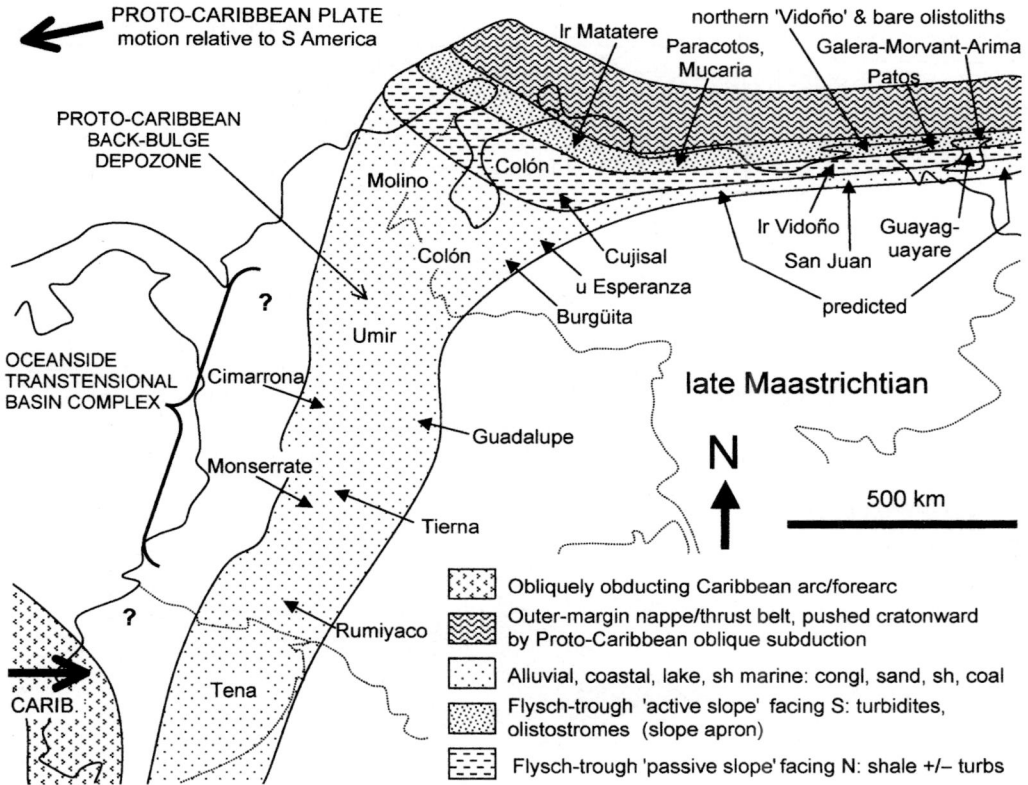

Fig. 3. Late Maastrichtian palaeogeographic map of northwestern South America. The Proto-Caribbean Foreland Basin, then largely confined to Venezuela–Trinidad, comprised a northern flysch trough flanked in the south by a north-facing slope and shelf. Main sources of formation names in this and subsequent maps: Ghosh & Odreman (1987); Kiser (1989, 1994); Dashwood & Abbotts (1990); Cooper *et al.* (1995); Higgs *et al.* (1995); Macellari (1995); Macsotay *et al.* (1995); LEV (1997); Saunders (1997*a, b*); Pindell *et al.* (1998); Villamil (1999); Kugler (2001). Solid arrows showing positions of formations also reflect approximate sediment-supply directions. Heavy arrows indicate plate-motion directions (not rates) relative to South America. Insignificant at this scale is palinspastic restoration for (see text): (i) shortening of the foreland basin by thrust-front advance; (ii) distributed NW–SE shortening in Colombia and western Venezuela; (iii) *c.* 50 km eastward relative movement of the 'Northern Andes Block', including the 'Maracaibo Block', since 2.5 Ma; and (iv) about 50 km of strike–slip along the Central Range Fault of Trinidad. 'Predicted' means non-exposed, either buried in thrust sheets, or deeply buried under (or immediately ahead of) the present-day frontal thrust. sh, shallow marine; congl, conglomerate; sh, shale; turbs, turbidites.

1997). Further south, Cretaceous extensional magmatism is known in the Eastern Cordillera of Colombia, as mentioned above.

The maximum metamorphic grade of the rift cover is lower greenschist (Frey *et al.* 1988; LEV 1997), as in other cases of intracontinental, thrust-related metamorphism (Warr *et al.* 1991, 1996). The estimated maximum burial depth of 13 km in Trinidad (Weber *et al.* 2001b) is compatible with the likely 15–20 km thickness of the overrunning nappe; i.e. the thickness of the seismogenic upper crust (Fig. 4; Jackson 2002).

The metamorphosed inferred rift interval is interpreted here to include the Siquisique volcanics

(Bartok *et al.* 1985), the volcanic Sans Souci Fm (Wadge & Macdonald 1985), the Tinaquillo rift peridotite (Ostos *et al.* 2005*a*), and dacite porphyry intrusions in the presumed Lower Cretaceous Tunapui Fm metasediments (see Carúpano, Intrusivas Graníticas Jóvenes de, in LEV 1997). While one of the dacite intrusions gave a K–Ar age of 5 Ma (Sifontes & Santamaría 1972; Santamaría & Schubert 1974), only one field sample was dated and thin sections suggest decomposition and hydrothermal metasomatism (Sifontes & Santamaría 1972). Therefore this date is suspect and a Jurassic or Cretaceous age (rift magmatism) is more likely.

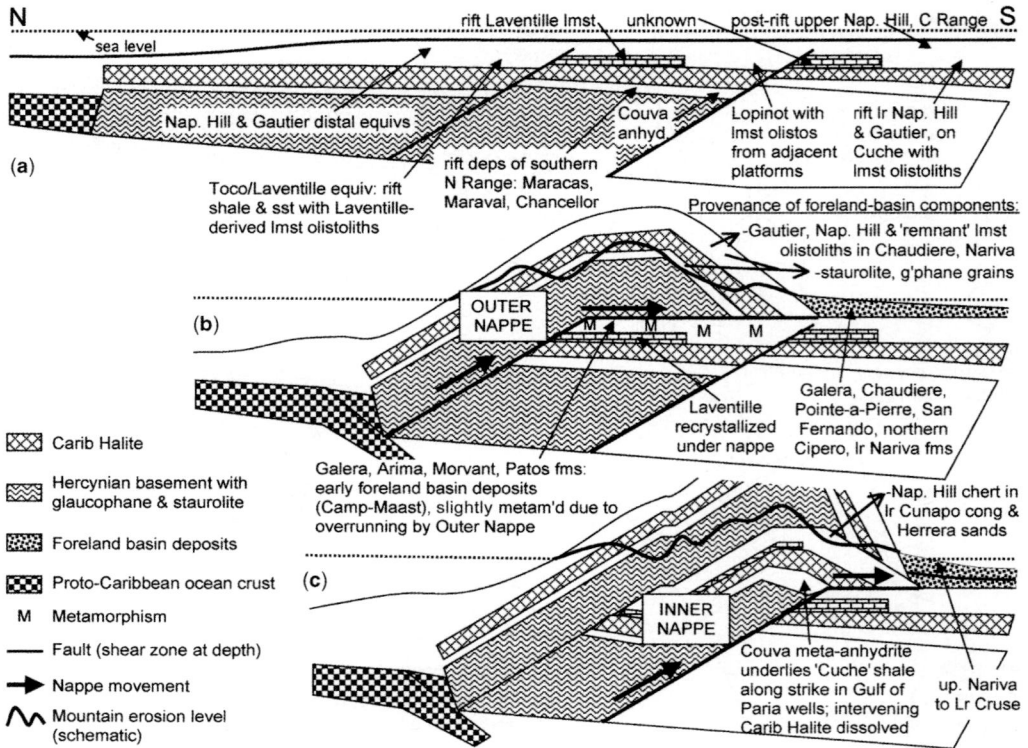

Fig. 4. Schematic north–south palaeo-section across northern South America continental margin at longitude of western Trinidad, at three different times, showing development of Outer Nappe and Inner Nappe. This sketch is equally applicable to western, central and eastern Venezuela, except Inner Nappe was uplifted earlier (Paleocene) in western and central Venezuela (see text) and formation names are different. Assumed detachment at base of seismogenic upper crust. For simplicity, this figure neglects (i) thrusting on younger detachments, (ii) possible halokinesis (Higgs 2009), (iii) cannibalization and nappe-overriding of early foreland-basin deposits and (iv) advance of thrust belt beyond nappe front. Basement is inferred to be Hercynian in the north and Precambrian in the south.

In Margarita the Juan Griego Gp can be interpreted as Hercynian basement of the Outer Nappe tail, predicted to underlie the entire northern shelf of eastern Venezuela and Trinidad, overrun by the Caribbean forearc nappe in Miocene time (see below). The Caribbean nappe is represented in Margarita by the La Rinconada Gp, thrust onto (Bellizzia & Dengo 1990) and metamorphosing Jurassic–Cretaceous sediments and volcanics (El Piache, Los Robles Fms; Manzanillo Mbr; LEV 1997) interpretable as rift cover deposited on Juan Griego basement rocks. Both the Rinconada and Juan Griego show high *P–T* metamorphism, previously taken as indicating that they were metamorphosed together, in Cretaceous time, in the IAA subduction zone (Stöckhert *et al.* 1995), rather than separate Cretaceous (IAA) and Hercynian subduction zones, respectively. The poorly exposed Rinconada–Griego contact is associated with mylonite at one locality, interpreted as a ductile thrust by

Chevalier *et al.* (1988), consistent with the Stöckhert *et al.* (1995) model. Instead, the mylonite is interpreted here as an intra-Rinconada, IAA product *truncated* by the Caribbean thrust (brittle), juxtaposing it against the Juan Griego.

Many authors have interpreted the Cordillera de la Costa belt (*sensu* Bellizzia & Dengo 1990) as 'terranes' removed from northwestern South America and metamorphosed in the Caribbean accretionary prism which transported them far (hundreds of kilometres) to the east before obduction back onto South America (Algar & Pindell 1993; Stöckhert *et al.* 1995; Avé Lallement 1997; Speed & Smith-Horowitz 1998; Pindell & Kennan 2001*a*; Avé Lallement & Sisson 2005). These and the other metamorphic rocks discussed above are instead interpreted here as transported less than 100 km in the Outer and Inner Nappes. Margin-parallel (ENE) stretching lineations in metarift strata of this belt (Carúpano, Tunapui Fms;

Caribbean Gp), generally ascribed to Caribbean dextral transpression (Algar & Pindell 1993; Avé Lallement 1997; Weber et al. 2001b; Cruz et al. 2003; Avé Lallement & Sisson 2005), may in fact reflect Proto-Caribbean-driven *sinistral* transpression (Fig. 3) under the Outer Nappe.

In summary, Outer and Inner Nappe emplacement was driven by Proto-Caribbean subduction. Thus, tectonically speaking, the 'Sistema Montañoso del Caribe' (Bellizzia 1986) is more accurately termed the 'Sistema Montañoso del *Proto*-Caribe'. Emplacement of the younger, Inner Nappe ended when the Proto-Caribbean tectonic drive was terminated, diachronously eastward, by arrival of the Caribbean arc (see below).

Siquisique

In the Siquisique ophiolite complex, slightly metamorphosed rift (aborted ocean) volcanics and shales with Middle Jurassic ammonites are associated with (meta-) conglomerates, sandstones and shales (Bartok et al. 1985) yielding Barremian and probably Hauterivian ammonites (Stéphan 1985). The long-controversial history of Siquisique (allochthonous v. autochthonous; Bartok et al. 1985) can be unravelled as follows.

(1) Jurassic and Cretaceous rift volcanics, conglomerates and shales were deposited.
(2) These deposits were overrun by the Outer Nappe (later completely eroded), causing slight metamorphism.
(3) Moderate southward transportation (tens of kilometres) in the Inner Nappe followed, onto turbidites and olistostromes that now crop out nearby (lower Matatere Fm; Fig. 3). The Matatere was derived in part from the same nappe, based on 'La Luna' olistoliths (Renz et al. 1955) that are slighly metamorphosed (Barquisimeto Fm; see below). Olistoliths of granite and crystalline metamorphics (Renz et al. 1955) may have come from remnant Outer Nappe basement klippen on the Inner Nappe.
(4) Burial under the Oligo–Miocene Falcón extensional basin took place.
(5) Pliocene northwestward thrusting in the Falcón fold-thrust belt occurred, leading to exhumation. More details of Stages 3–5 are given below.

Thus the Siquisique complex is relatively autochthonous (Bartok et al. 1985), rather than part of a west-derived Caribbean nappe (Stéphan 1985; Dewey & Pindell 1986; Macellari 1995; Pindell et al. 1998; Gorney et al. 2007). This conclusion is important for Pangaean reconstructions and palaeobiogeographic studies since the Siquisique

Middle Jurassic ammonites have been invoked (Bartok et al. 1985) as evidence of the pre-oceanic 'Hispanic Corridor' connecting the Pacific Ocean and Tethys Sea (Smith 1983).

Proto-Caribbean foreland basin

Emplacement of the Outer Nappe and subsequently the Inner Nappe drove an inboard thrust belt and Proto-Caribbean foreland basin, both migrating south, reaching along strike from western Venezuela (Maracaibo-Lara-Falcón) to Trinidad (Fig. 3). The basin contains olistostromes ('wildflysch' of Kugler 1953) and turbidites ('flysch'), deposited on a south-dipping slope apron flanking a deep-sea trough. Some of the turbidites are shallow-water types (Myrow et al. 2002) with hummocky cross stratification (author's observations, e.g. Galera, Caratas, Los Jabillos formations), presumably relatively proximal. This northern 'active slope' faced a 'passive slope' and shelf on the opposite side of the trough. The earliest known olistostromal units assignable to this trough are the Campanian and/or Maastrichtian Mucaria, Paracotos, Galera, Arima and Morvant formations (Fig. 3; Saunders 1972, 1997a, b; Potter 1973, 1997; Kugler 1974; Macsotay et al. 1995; LEV 1997). All of these formations except the most southerly (Mucaria) are slightly metamorphic (phyllitic), having been overrun by the Outer Nappe that supplied them. Also assignable to this age and tectonic setting is the Patos Conglomerate Mbr (Fig. 3), considered 'possibly Neocomian to Aptian' by Kugler (1974, p. 477), but this estimate was based on fossils in conglomerate clasts and in thin limestone beds in phyllitic shale (Kugler 1974), probably turbidites or tempestites; therefore the fossils were probably reworked.

Supporting the proposed Campanian or Maastrichtian start of nappe uplift and foreland-basin subsidence, Sisson et al. (2005) found Late Cretaceous inherited zircon fission-track ages in Miocene sandstones of the central Venezuela thrust belt, and attributed these ages to Maastrichtian uplift of a continental-margin source area by Proto-Caribbean subduction. Estimated unroofing was as much as 10 km, but this assumed a 25 °C/km geothermal gradient, unrealistically low if rifting ended shortly before (Coniacian).

Nappe emplacement also produced a forebulge that migrated across the former passive margin ahead of the foreland basin. Initial forebulge uplift was likewise Campanian, as shown by the following three phenomena, described from west to east.

(1) In western Venezuela, the Tres Esquinas Mbr, a glauconitic, phosphatic super-condensed section (Ghosh 1984) interpretable as forebulge submerged-arch deposits, separates

(upper) La Luna Fm passive-margin mid-shelf cherts and concretionary shales (Campanian and older; LEV 1997) from overlying Colón Fm shales (Maastrichtian) interpreted here as foreland-basin marine 'passive' shelf and slope deposits (Fig. 3). Minor reverse faults cut the Luna–Colón contact and appear to die out upward in the lower Colón (Cooney & Lorente 1997), attributable to syn-Colón fault growth in a regionally compressive setting after the forebulge had migrated past. Back-bulge subsidence (Decelles & Giles 1996) may have accommodated late Campanian and Maastrichtian deposition in Colombia (e.g. Umir Fm, Fig. 3).

(2) In eastern Venezuela, an intra-Campanian unconformity (Villamil & Pindell 1998) separates San Antonio Fm cherts, shales and sandstones from overlying San Juan Fm sandstones dominated by hummocky cross stratification (commonly faint suggesting seismic liquefaction; author's observations). The unconformity was tentatively attributed (Villamil & Pindell 1998) to shallowing caused by initial compressive stress on the margin prior to the onset of Proto-Caribbean subduction. The unconformity is interpreted here as due to continentward migration of the Proto-Caribbean forebulge, leaving a subaerial unconformity (ravinement surface) between passive-margin mid-shelf deposits and foreland basin 'passive shelf' strata (Fig. 3).

(3) In Trinidad, a missing Late Campanian faunal zone (Saunders & Bolli 1985, fig. 9) possibly reflects a thin (centimetres to decimetres), as yet unsampled (not exposed or cored), forebulge condensed section (cf. Tres Esquinas Mbr), separating Naparima Hill Fm passive-margin outer-shelf cherts and shales from foreland-basin 'passive slope' marine shales (Guayaguayare Fm; Fig. 3).

Paleocene relative motion change: effect in Colombia, Venezuela, Trinidad

In Late Paleocene time (chron 25), North America motion relative to South America changed from WSW to SSE (Müller et al. 1999). SSE motion continued until the Quaternary Period, except for a SW episode during Middle Eocene time (chrons 21–18; Müller et al. 1999). Along the Colombia margin, the SSE motion caused the Proto-Caribbean to subduct sinistrally, inducing outer-margin-nappe obduction and Proto-Caribbean foreland basin subsidence (Fig. 5). By then the Caribbean nappe had already advanced part-way northward, obliquely, along the Colombian margin (Fig. 5; see below). Development of the Middle Magdalena Valley

Unconformity (MMVU) by eastward tilting, due to 'eastward propagation of Central Cordillera uplift' during Paleocene to Middle Eocene time (Gómez et al. 2003, fig. 16), is attributed here to Proto-Caribbean nappe push.

Along the Guajira–Gulf of Venezuela–Falcón and Caracas–Trinidad continental-margin sectors, subduction changed to dextral- and near-orthogonal respectively, interrupted during chrons 21–18 (SW motion) by near-orthogonal and sinistral subduction (Figs 5–11). Deposition of turbidites and olistostromes continued in the Proto-Caribbean foreland basin. The next olistostromes, exposed south of the Campanian–Maastrichtian meta-olistostromes, are of Paleocene–Eocene age: Matatere (continued), Río Guache, Los Cajones–Garrapata (sensu Macsotay et al. 1995) and Chaudiere formations (Kugler 1953; LEV 1997). Following Bell (1971, 1972), James (2000, p. 32) interpreted this near-continuous belt of turbidites and olistostromes from western Venezuela to Trinidad as indicating a 'roughly east–west trending foreland basin along the entire Venezuela-Trinidad margin … bounded to the north by a zone of inversion'.

Paleocene–Eocene Maracaibo foredeep

Proto-Caribbean nappe loading drove kilometric Paleocene–Eocene subsidence in the Maracaibo–Lara foredeep, accommodating the Matatere–Trujillo–Misoa suite of formations (Figs 5 & 6). The Misoa Fm, deposited on the 'passive shelf', is possibly the western hemisphere's largest conventional-oil reservoir (Higgs 1996). Subsidence was attributed to Caribbean nappe loading by Dewey & Pindell (1986) and Pindell et al. (1988). However, the Misoa thickens and palaeobathymetrically deepens NE (Zambrano et al. 1970, 1971), toward Matatere wildflysch thrusted SW (Stéphan 1977); loading was therefore from the NE. Thus, for the Caribbean nappe to be responsible would imply loading on a SE-trending lateral ramp connecting two sectors of the nappe front (Dewey & Pindell 1986, fig. 2B), begging the questions (a) why did these two frontal sectors not produce even thicker Paleocene–Eocene foredeeps, and (b) can a lateral load induce kilometric subsidence? Moreover, this 'problematic' model (Lugo & Mann 1995, p. 717) requires a much faster Paleocene–Eocene Caribbean Arc eastward migration rate relative to North America than relative to South America (e.g. Lugo & Mann 1995, fig. 23), an impossibility given that inter-American relative motion at that time was essentially north–south (Müller et al. 1999). Nevertheless, the model is widely accepted (e.g. Lugo & Mann 1995; Stéphan et al. 1990; Parnaud et al. 1995a; Ostos et al. 2005b; Escalona & Mann 2006).

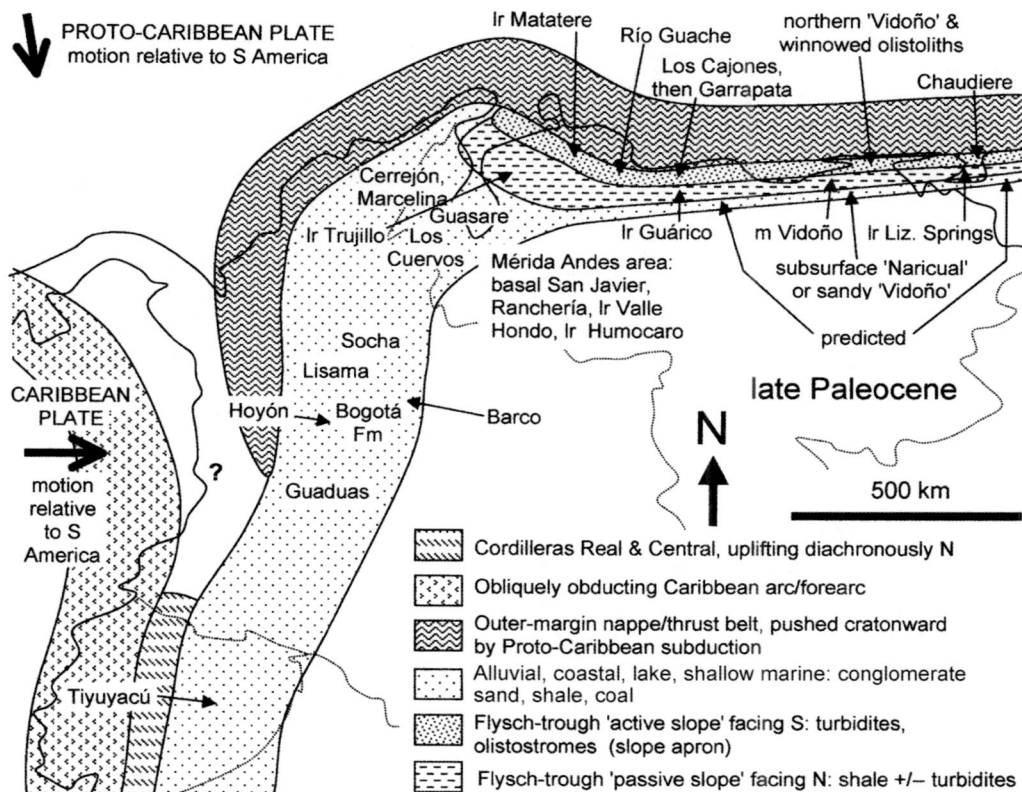

Fig. 5. Late Paleocene palaeogeographic map of northwestern South America. See Figure 3 caption for further details.

A supposed lateral ramp, suitably positioned for the problematic Dewey & Pindell (1986) model but not mentioned by them, was proposed earlier by Stéphan (1977). The 'Transversal de Barquisimeto' was thought to connect two sectors of the supposed Caribbean nappe, namely Guajira ('Ruma metamorphic zone' of MacDonald *et al.* 1971) and the 'Napa de Lara', the latter comprising the entire igneous-metamorphic complex of central Venezuela (Stéphan 1977). [The Lara Nappe is not to be confused with the 'complexe tectono-sédimentaire de Lara' (Stéphan 1985) lying further west, also known as the 'Surco de Barquisimeto' (Stéphan 1977), essentially comprising the Matatere Fm,] The Transversal de Barquisimeto is reinterpreted here as the *Proto*-Caribbean nappe frontal thrust trace. Thus, Stéphan (1977, 1985) mis-tied the Caribbean and Proto-Caribbean nappes. The true Caribbean 'transversal' is a palaeo-transform fault separating the Falcón and Bonaire Basins, formed when the Caribbean Arc passed here in Oligocene time (see below).

Eastern Venezuela olistostromes

In eastern Venezuela there is a perceived gap in the olistostrome belt (Bell 1971, 1972; James 1997, 2000). Olistostromes are in fact present, but much reduced by erosion. In the far north of the Eastern Serranía mountains, Metz (1968, pp. 289–290) described a 'problematic ?post-Eocene sequence' with Lower Cretaceous and Paleocene microfaunas interpreted by him as reworked. Vierbuchen (1984) described this as an 'unnamed Eocene? formation' with 'olistoliths' (his fig. 2) and 'boulder beds' (p. 194), associated with shales. In the same area are fields of limestone–breccia and sandstone boulders, variably rounded (author's observations), interpretable as olistoliths exhumed from shale (cf. central Venezuela 'exolistolitos' of Vivas & Macsotay 1995*a*, p. 98). These perhaps include the sandstone blocks mentioned as possible olistoliths by Vivas & Macsotay (1995*b*), which they assigned to the Maastrichtian to Upper Paleocene Río Chávez Fm. The limestone–breccia olistoliths are pale yet

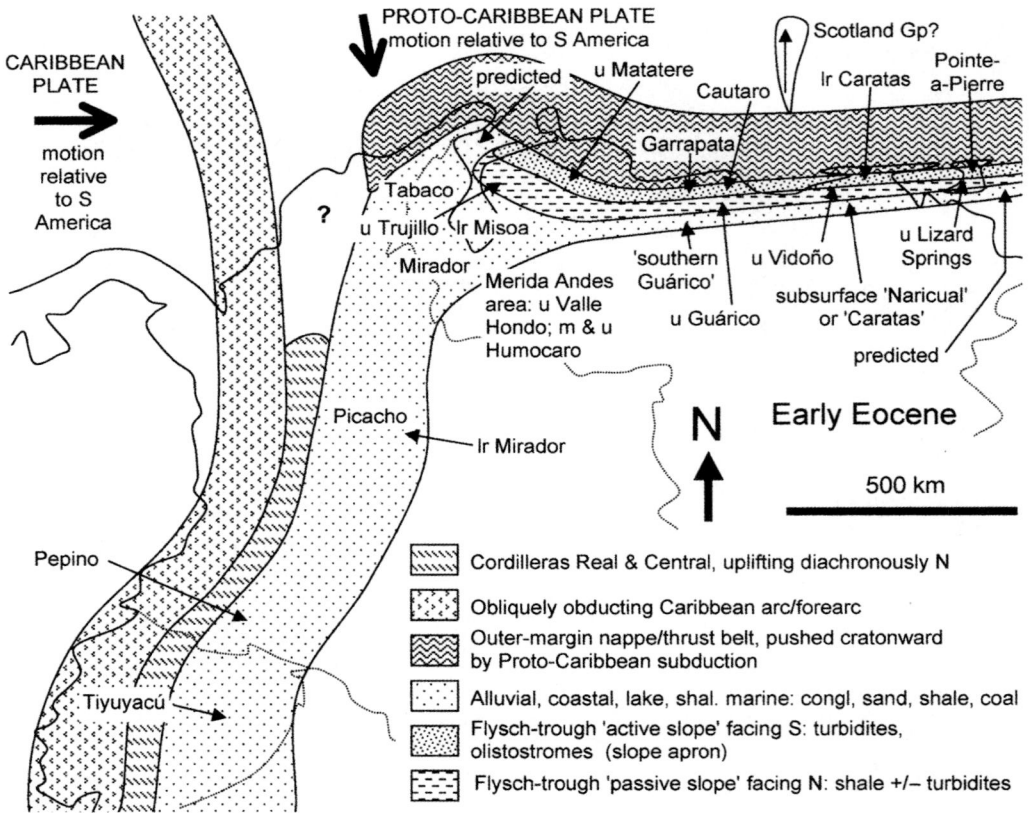

Fig. 6. Early Eocene palaeogeographic map of northwestern South America. See Figure 3 caption for further details.

lack obvious fossils, possibly due to recrystallization, suggesting northern derivation from the Tunapui Fm (Araya Peninsula), interpreted by the author as the Inner Nappe (lightly metamorphosed) equivalent of the Eastern Serranía Morro Blanco Mbr (Barranquín Fm; LEV 1997). A shale exposure with decimetric floating clasts (author's observations) occurs near outcrops mapped by Vierbuchen (1984) as Vidoño Fm, of Campanian–Eocene regional age (LEV 1997).

In the southern part of the Eastern Serranía, the Eocene–Lower Miocene interval (Caratas, Los Jabillos, Areo, Naricual, Capiricual formations) comprises mostly shale with shallow- and deep-sea turbidites, variably amalgamated and burrowed (author's observations). Two or more discontinuous limestones occur at or near the top of the Caratas (González de Juana *et al.* 1980, p. 485 and figs VI-14 & 79), including the Tinajitas Mbr, interpreted below as calciturbidites and calcidebrites. The Areo Fm includes olistostromes described below. Thus the entire Caratas–Capiricual interval

can be interpreted as north-derived Proto-Caribbean active-slope deposits (Figs 3, 5–9 & 12), rather than passive margin Caratas–Areo overlain by Caribbean foreland basin deposits (Pindell 1991), the generally accepted view (e.g. Hung 2005). An intra-Eocene unconformity (Hedberg 1950), pivotal in oil exploration, is widely assumed at outcrop, separating the Caratas and Los Jabillos formations and expanding southward in the subsurface, where it supposedly separates southward-onlapping Oligo–Miocene strata (Los Jabillos–Capiricual equivalents) from north-dipping, erosionally truncated Upper Cretaceous (Tigre Fm) through Vidoño–Caratas equivalents (e.g. Parnaud *et al.* 1995*b*, fig. 4, after Pérez de Mejía *et al.* 1980, fig. 1-13a). However, the outcrop unconformity is doubtful: no angularity has ever been proven and the supposedly missing Upper Eocene fossils are now known in the Caratas Fm (González de Juana *et al.* 1980, fig. VI-79; LEV 1997). The subsurface 'unconformity', difficult to locate because the entire section is sandy with few fossils

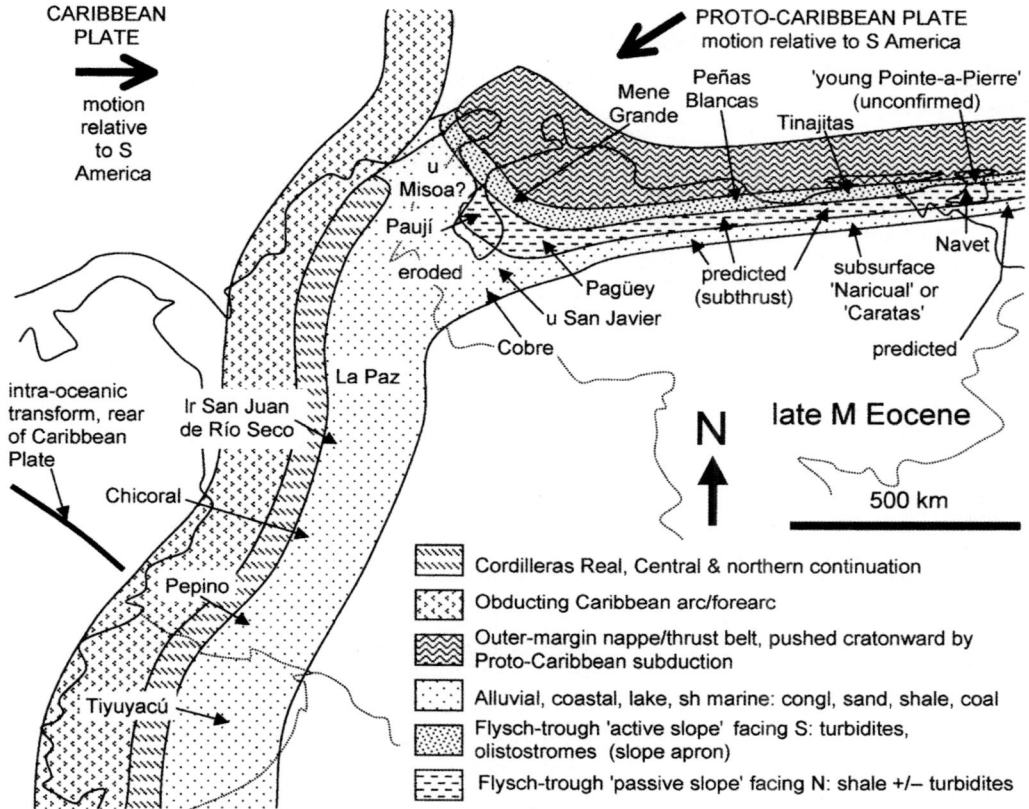

Fig. 7. Late Middle Eocene palaeogeographic map of northwestern South America. See Figure 3 caption for further details.

(Parnaud *et al.* 1995*b*, fig. 4), is interpretable instead as a Campanian–Miocene diachronous basal sand onlapping the (sandy) Tigre Fm, reflecting the south-advancing Proto-Caribbean forebulge (Fig. 12). The first *Caribbean* deposits belong to the Quiamare Fm (Fig. 10).

Regarding olistostromes in this region, the so-called 'Boulder Bed' at 'Kilometer 12' (Hedberg 1950, plate 8, fig. 3) consists of sandy mudstone containing variably rounded clasts of algal-orbitoidal limestone. This has been interpreted as a right-way-up, structural repetition of the Kilometer 11 Tinajitas Mbr overturned type section, hence the isoclinal 'Tinajitas Syncline' concept (Creole 1955; Rosales 1967). However, the type section is different, consisting of graded limestone beds (centimetres to metres) and lacking boulders (author's observations; Sageman & Speed 2003, fig. 7). Both exposures were reinterpreted by Vivas (in Macsotay *et al.* 1985) as a (repeated) basal Los Jabillos olistostrome, of Early Oligocene age

(bivalve), but this contradicts Middle Eocene micro-fossils at Kilometer 11 (Galea 1985; Sageman & Speed 2003). Kilometer 12 is reinterpreted here as an *overturned* olistostrome and assigned to the *Areo* Fm, based on (a) way-up indicators in adjacent strata, negating the syncline, and (b) compatibility of the enclosing sandstones and shales with the Areo (LEV 1997), mapped nearby (Creole 1955). The Tinajitas boulders show tension gashes whose orientation differs from one clast to another, indicating cannibalization from lithified, strained, *in situ* Tinajitas strata, uplifted behind the south-advancing thrust front. Other bouldery intervals at Kilometer 12, also interpretable as olistostromes, are dominated by clasts resembling typical Los Jabillos and (finer, glauconitic) Caratas sandstones. The entire boulder-bearing package was interpreted differently, as a right-way-up transgressive lag, incised-valley fill, or fault-scarp talus, by Pindell & Erikson (2001, stop 8). At the nearby Los Montones Quarry, Caratas and Tinajitas boulders float in sandy

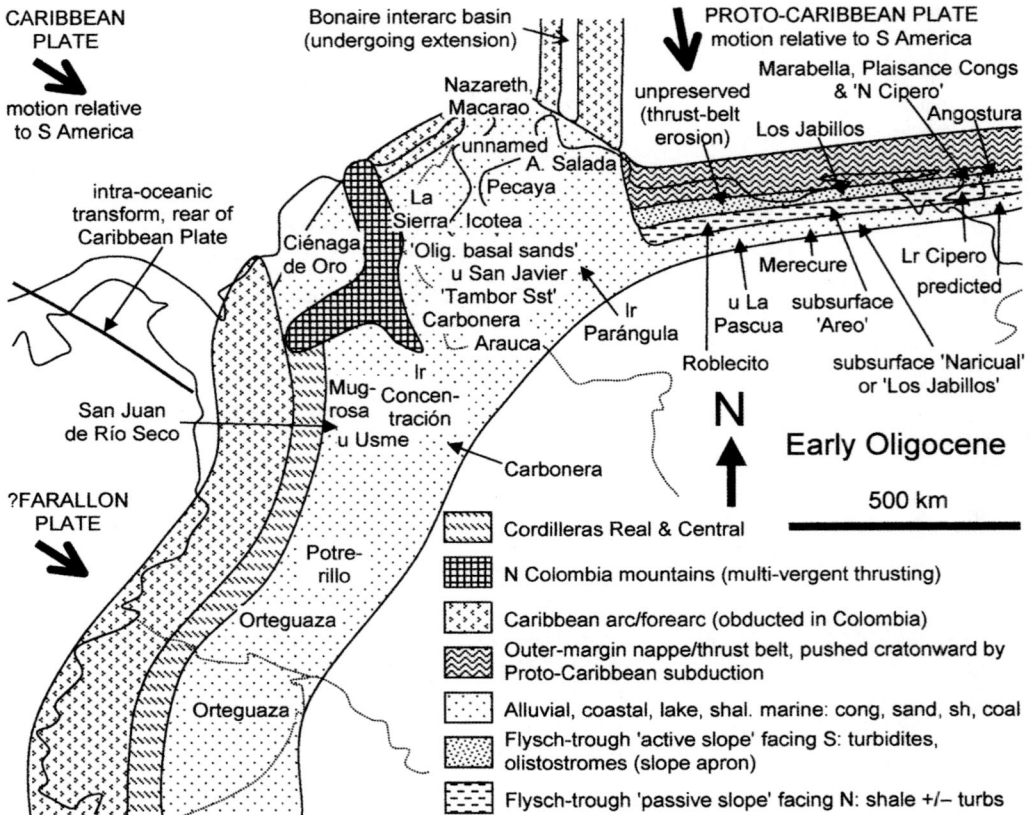

Fig. 8. Early Oligocene palaeogeographic map of northwestern South America. See Figure 3 caption for further details.

mudstone (author's observations; Galea 1985; Vivas in Macsotay *et al.* 1985). Creole (1955) again viewed this as a distinct facies of the Tinajitas Mbr, a facies interpreted by Galea (1985) as slope debrites derived from coeval shelf-edge carbonates. Los Montones is interpreted here as another Areo olistostrome of the Proto-Caribbean foreland basin.

Eocene SW motion and Maracaibo 'Eocene Unconformity'

During the chron 21–18 interval in western Venezuela, the Proto-Caribbean thrust front advanced SW to within the area of modern Lake Maracaibo. The youngest known strata deposited ahead of, then overrun by, the thrust front belong to the Mene Grande Fm (Upper Middle Eocene; LEV 1997). When the thrust belt began to subside again in Early Oligocene time, driven by Santander loading from the west (see below), the first deposits were the Icotea Fm (Fig. 8). These strata overlie the renowned 'discordancia post-eocena' (Zambrano

et al. 1971), commonly known as the 'Eocene Unconformity', whose origin has been unclear (Pindell *et al.* 1998).

Age span of Proto-Caribbean foreland basin

Olistostromes in central Venezuela range from Campanian or Maastrichtian to at least Eocene (Paracotos–Mucaria to Garrapata; Macsotay *et al.* 1995; LEV 1997), and possibly Lower Oligocene if the 'Tememure Fm' is genuinely olistostromal (Vivas & Macsotay 1995*a*) rather than a structural complex (Peirson 1965; LEV 1997) or a rauhwacke (Higgs 2009). In Trinidad, the oldest olistostromes are again Campanian or Maastrichtian (Galera–Arima–Morvant), but the youngest are Upper Miocene (Lower Cruse; Saunders 1997*a*; Kugler 2001). This apparently later termination of olistostrome deposition eastward reflects the progressively later arrival of the Caribbean nappe (Figs 5–11), causing a change from an underfilled foreland basin to an overfilled (no flysch trough) *Caribbean* foreland basin.

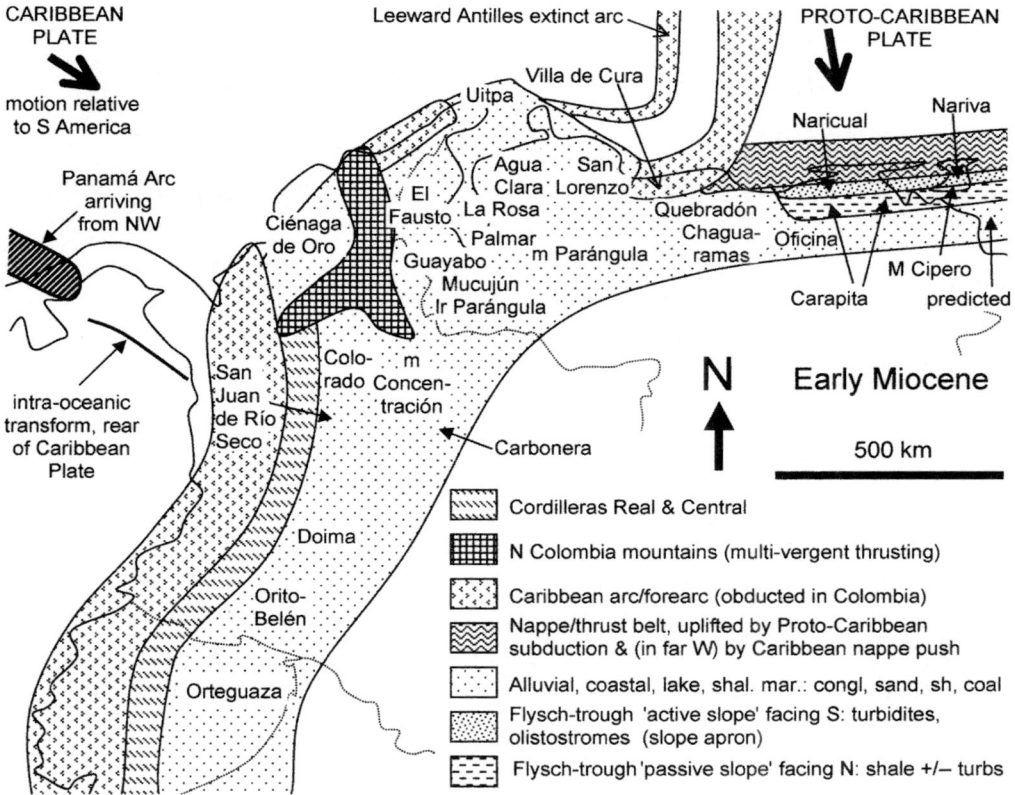

Fig. 9. Early Miocene palaeogeographic map of northwestern South America. See Figure 3 caption for further details.

Wildflysch northern source

Northern derivation of the Venezuela–Trinidad wildflysch (Kugler 1953; Suter 1960; Bell 1967, 1971, 1972) is confirmed by three lines of evidence: (a) southward fining; (b) olistolith lithologies; and (c) heavy minerals. Southward fining was recognized long ago, in the form of coeval shales to the south, seen to interdigitate in Trinidad (Kugler 1953). The shales contain turbidites in some cases but not olistostromes (Kugler 1953; LEV 1997), e.g. the Paleocene–Eocene Trujillo, Guárico (*sensu* Macsotay *et al.* 1995) and Lizard Springs formations (Figs 5 & 6).

Olistolith lithologies are described in many publications (Kugler 1953, 1956, 1959, 1961, 1996, 2001; Renz *et al.* 1955; Bell 1967; González de Juana *et al.* 1980; Macsotay *et al.* 1995). In Trinidad and eastern Venezuela, olistoliths mostly resemble local, non-metamorphosed Jurassic–Cretaceous formations (i.e. autochthon) and can be interpreted as eroded off the Outer Nappe (Fig. 4). In central and western Venezuela, most olistoliths

closely match Outer Nappe basement (e.g. El Tinaco Complex) schist, gneiss and granitoid rocks, and Inner Nappe Jurassic–Cretaceous sedimentary–volcanic cover (rift/post-rift) slightly metamorphosed under the Outer Nappe. This confinement of igneous and metamorphic clasts to the west suggests deeper nappe unroofing, consistent with the westward increase in inter-Americas, north–south convergence (Campanian–Recent Pindell *et al.* 1988; Müller *et al.* 1999). Notably, foliated metavolcanics typical of the (Caribbean) Villa de Cura nappe (e.g. Smith *et al.* 1999) have *not* been reported among the clasts of the adjacent Los Cajones and Garrapata wildflysch (Bell 1967; author's observations), reflecting arrival of Villa de Cura long after deposition of these Paleocene–Eocene formations (Miocene; Fig. 9).

Turning to heavy minerals, glaucophane and/or staurolite occur in Paleocene to Miocene sandstones in Trinidad and in the Caratas and Los Jabillos formations of Venezuela (Hedberg 1937; Feo-Codecido 1956; Suter 1960; Kugler 2001, citing unpublished oil company reports by Griffiths), all interpreted

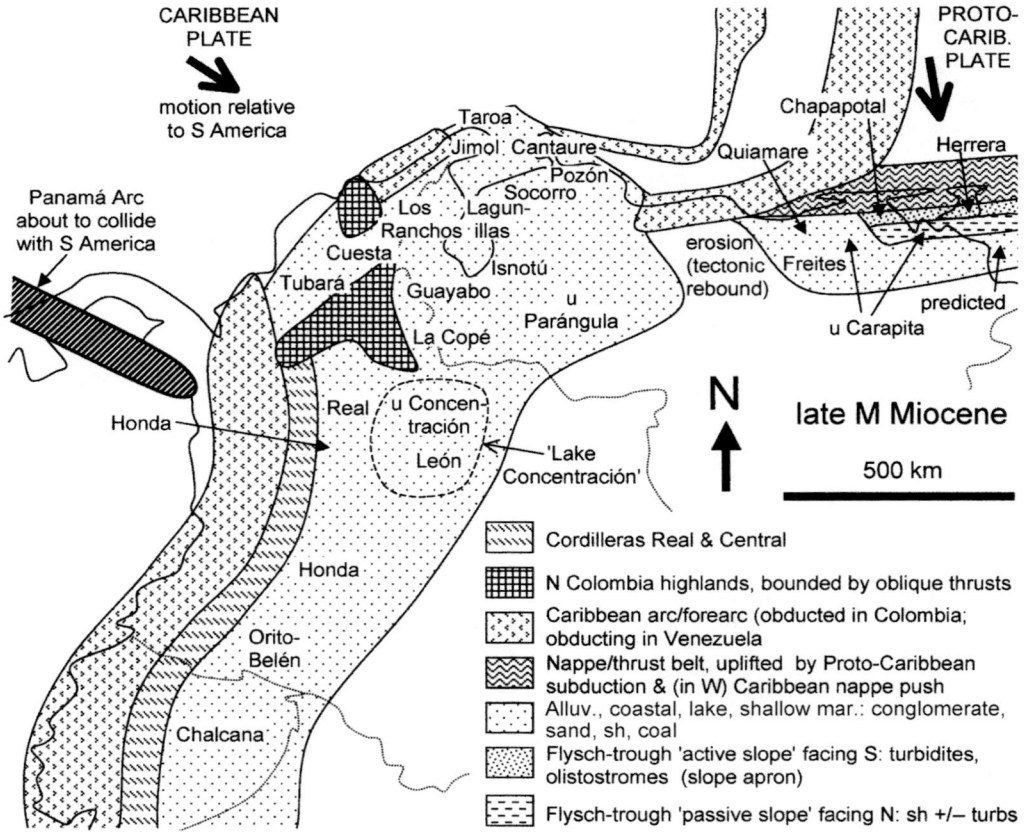

Fig. 10. Late Middle Miocene (12 Ma; *Gt. fohsi robusta* zone) palaeogeographic map of northwestern South America. See Figure 3 caption for further details.

here as north-derived (Figs 5–10 & 12). The Villa de Cura klippe contains glaucophane (Smith *et al.* 1999), but was not emplaced until Miocene time and cannot have supplied glaucophane to the Oligocene (oldest reported occurrence) sands in Trinidad. As discussed above, these minerals are thought to come from Hercynian basement in the north, though in Trinidad they were probably recycled via Jurassic–Cretaceous rift cover of the Outer Nappe (Fig. 4), because the basement was not exposed in Paleocene time, as shown by the lack of basement olistoliths. [Similarly, Precambrian zircon in the Oligocene Angostura Fm of Trinidad (Fig. 8; K. Meyer 2007, oral comm.) does not negate a northern source, as the zircon could have been supplied northward *initially*, from rift-shoulder Precambrian outcrops near the northern limit of shield-type basement (cf. Fig. 4), into rift deposits that were later uplifted and eroded off the Outer Nappe.] In Barbados, Senn (1940) interpreted Eocene Scotland Gp sandstones as derived from Venezuela–Trinidad

coastal mountains. Glaucophane and staurolite in these sandstones (Senn 1940, citing Hedberg private report) support this interpretation. Eastward relative displacement of the Scotland Gp, in the Caribbean accretionary prism, means that the southern source lay to the west in Venezuela, (Fig. 6; Dickey 1980; Pudsey & Reading 1982).

Mega-olistoliths

Olistoliths in the Los Cajones Fm include phyllite and recrystallized limestone, not known *in situ*, interpreted here as northern representatives of Cretaceous formations well-known in the autochthon (Périja, Mérida and Eastern Serranía mountains), metamorphosed under the Outer Nappe, then released as olistoliths upon uplift of the Inner Nappe. Some of these olistoliths are so numerous and large (e.g. 30 m thick, >1 km long; Renz *et al.* 1955) that they have their own formation or member names (González de Juana *et al.* 1980), like the Cojedes

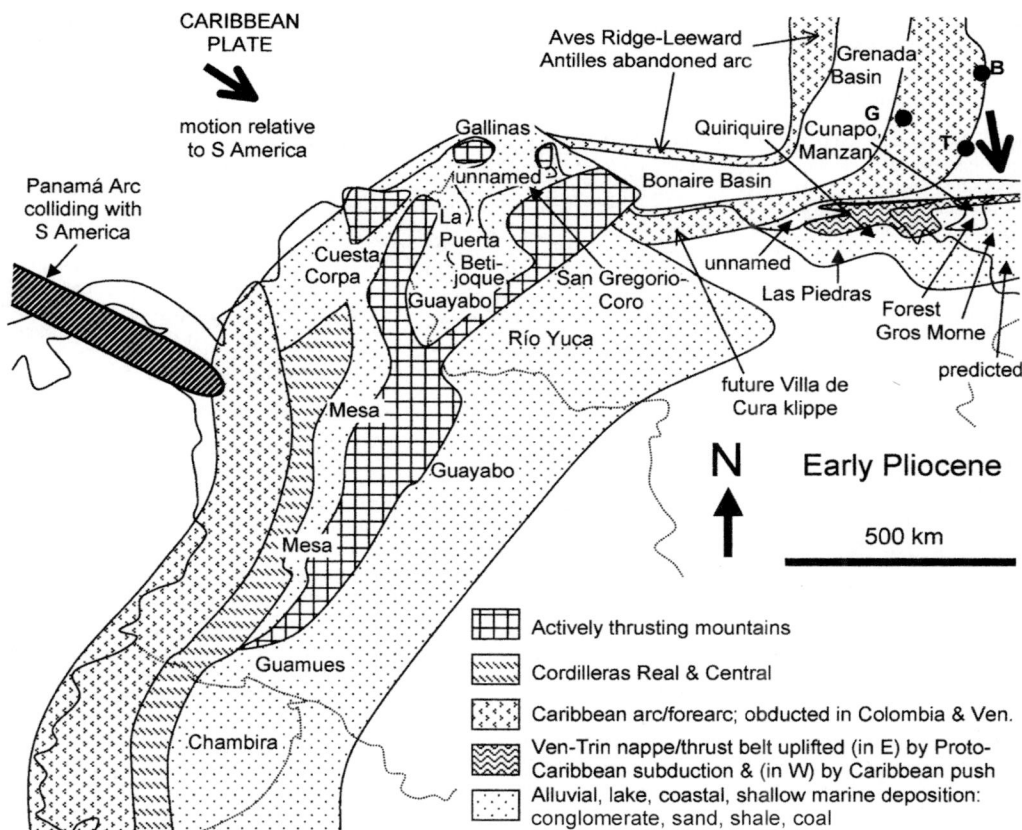

Fig. 11. Early Pliocene palaeogeographic map of northwestern South America. B, Barbados; G, Grenada; T, Tobago. See Figure 3 caption for further details. Not shown on the Caribbean nappe are post-obduction basins in Colombia (e.g. Cauca-Patía, Pacífico) or Venezuela (e.g. Tuy, Carúpano).

(interpretable as meta-Barranquín Fm), Morro del Faro (El Cantil) and Mapuey (Querecual, LEV 1997). The large size, similar to large olistoliths in other basins (Stow *et al.* 1996), implies that these are 'slip masses' (Kugler 1953; i.e. slide blocks) rather than clasts carried in a mudflow. Examples in Trinidad can exceed 900 m in length (Kugler 1953). The Barquisimeto Fm of western Venezuela, a low-grade-metamorphic version of the La Luna Fm (same age and depo-facies), occurs both as mega-olistoliths in the Matatere Fm and *in situ* (González de Juana *et al.* 1980; LEV 1997).

In eastern Venezuela, the Paso de las Peñas Quarry (Socas 1991; Pindell & Erikson 2001, stop 10) exposes a sandstone body at least 200 m long and 20 m high (base and sides not exposed), interpreted here as a mega-olistolith for three reasons: (a) it is anomalously hard, with stylolites and tight quartz cement (author's observations) suggesting burial to much greater depth than the soft overlying

shale (Areo Fm; Creole 1955); (b) the upper surface has metre-scale irregular relief, onlapped by sandy intervals in the overlying shale; (c) the upper surface is draped by a thin (<1 m) glauconite-rich bed, under the onlapping cover, attributable to perched (shallower), condensed deposition on top of the olistolith, possibly within reach of storm-wave-induced winnowing currents, while the sides were onlapped by mud and sand deposited from bottom-hugging turbidity currents. The mega-olistolith may be derived from the San Juan or Los Jabillos Fm, both characterized by thick (metres to tens of metres) sandstones lacking obvious lamination (author's observations). Socas (1991) interpreted the sandstone as *in situ* Los Jabillos, unconformably overlain by Naricual Fm shales, a previously unknown relationship, the Los Jabillos always conformable under Areo (LEV 1997). Pindell & Erikson (2001) viewed the sandstone as a new, western Areo sandy facies overlain by Naricual shales.

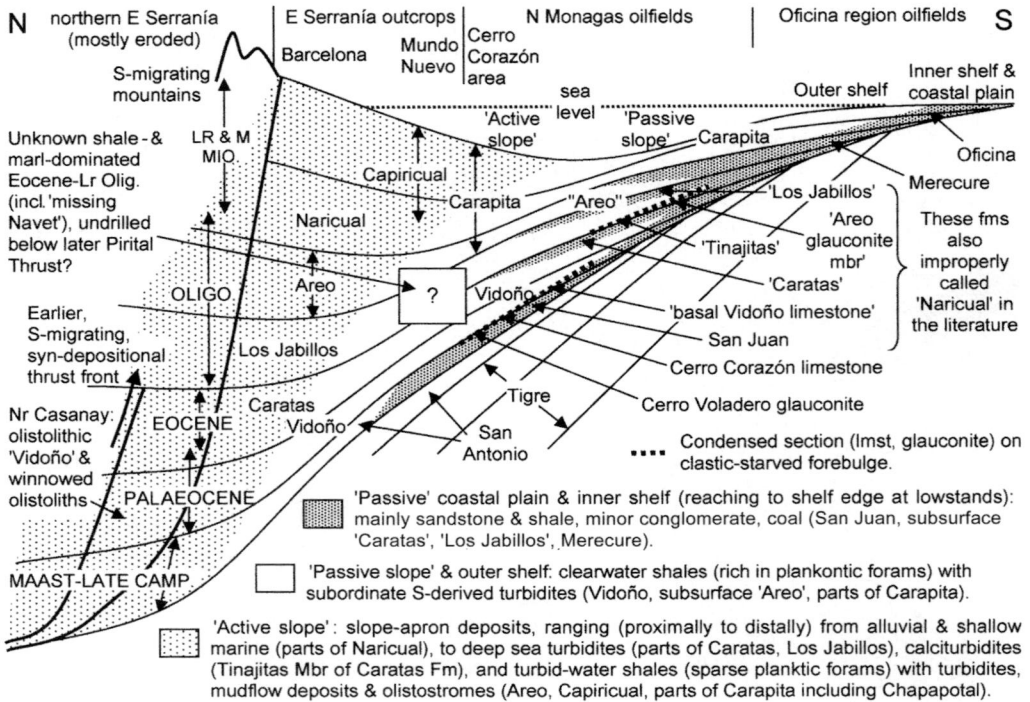

Fig. 12. Interpretive north–south section through Maturín city (Fig. 1) showing Late Campanian–Middle Miocene sequence stratigraphy and tectono-sedimentary environments of the eastern Venezuela sector of the Proto-Caribbean Foreland Basin. The profile comprises a northern deep-water trough and a southern shelf. For simplicity, each successive depositional profile (time line) is identical to the last, but is simply rotated and displaced to the right of the previous one, to simulate a foreland basin migrating south toward a (migrating) point of zero subsidence on the flank of a (migrating) forebulge. The right-hand side thus depicts long-term onlap onto eroded Cretaceous strata of the south-migrating forebulge. Names of formations are shown, and of informal units in general industry usage (e.g. Sams 1995; Colmenares et al. 1997; Rodríguez et al. 2004); note how Palaeogene informal terminology in northern Monagas oilfields (El Furrial etc.) is inappropriately taken from the Eastern Serranía outcrop type sections, based on time equivalence established by microfossils. Two forebulge excursions northward (orogenward), possibly due to orogen thickening, produced two condensed sections on the displaced forebulge (submerged arch): (1) basal Vidoño limestone and correlative Cerro Corazón limestone and Cerro Voladero glauconite; and (2) subsurface 'Tinajitas' limestone and associated 'Areo glauconite member'.

Naparima Hill in Trinidad, the type locality of the Naparima Hill Fm (Kugler 1956), is a possible mega-olistolith (Dr T. England, oral comm. 1996), consistent with five observations: (a) its dimensions (at least hundreds of metres cubed; Kugler 1996), similar to other large olistoliths worldwide (Stow et al. 1996), though perhaps abnormally thick; (b) the geologically small and isolated nature of this inlier (Kugler 1961; Saunders 1997a); (c) its upper contact with Chaudiere shales rich in floating Naparima Hill pebbles and cobbles (St Joseph Conglomerate), supposedly an unconformity (missing microfossils; Saunders 1997a; Kugler 2001), redundant in the mega-olistolith model; (d) lack of evidence for subaerial exposure at the contact (e.g. palaeosoil or transgressive lag; author's observations); and (e) presence of smaller Naparima Hill olistoliths in many Trinidad formations, from the Chaudiere (Paleocene) to Karamat (Miocene; Kugler 2001). The inferred Naparima Hill Thrust bounding the base and one side of the Naparima Hill Fm at Naparima Hill (Kugler 1996) does not preclude a mega-olistolith. Chert Hill, an areally smaller and equally isolated Naparima Hill Fm outcrop (Kugler 1961; Saunders 1997a), again 'unconformable' with the Chaudiere (Kugler 1996), may be another mega-olistolith.

Orphan olistoliths

Paleocene to Miocene olistostromes in Trinidad include olistoliths unmatchable to any Trinidad formation *in situ*. These orphan olistoliths, called 'remnant formations' by Suter (1951, p. 192), are of four main kinds (Salvador & Stainforth 1968), as follows.

(1) Olistoliths of Lower Cretaceous shelly limestone, lithologically and faunally indistinguishable from the El Cantil Fm of eastern Venezuela (Salvador & Stainforth 1968), are interpreted here as recycled from a Lower Cretaceous rift olistostrome uplifted in the Outer Nappe (Fig. 4a & b). This olistostrome was south-derived; the olistoliths came from a penecontemporaneous carbonate bank on the adjacent rift block (future Inner Nappe; Laventille Fm in Fig. 4a). Supporting this idea, the Bocas-1 borehole along strike to the west found, in the Mejillones sedimentary-volcanic complex (see below), an El Cantil-like limestone interval *c.* 100 m thick, possibly out of stratigraphic order (Speed & Smith-Horowitz 1998) and absent from seven other Carúpano Basin boreholes reaching the Mejillones (Castro & Mederos 1985; Speed & Smith-Horowitz 1998). Subsequently the Outer Nappe was pushed south and uplifted (Campanian onward), completely eroding the donor olistostrome, while the nappe overran and metamorphosed the Laventille, recrystallizing it beyond recognition as the original olistolith source, and making it hard to date (fossils recrystallized; Kugler 1974; Potter 1974). Olistoliths in the Laventille-age Toco and Lopinot formations (Potter 1974, 1997; Saunders 1997*a*, *b*) were also derived from an adjacent carbonate platform (Barr 1962; Potter 1974) but are recrystallized, indicating that these are Inner Nappe formations, metamorphosed under the Outer Nappe (Lopinot shales, Fig. 4a, are phyllitic).

(2) Upper Cretaceous dark siliceous limestones, probably Gautier and Naparima Hill equivalents (Salvador & Stainforth 1968), can be interpreted as eroded from the Outer Nappe (Fig. 4a).

(3) Olistoliths of Paleocene highly glauconitic sandstones (e.g. Bontour Sandstone; Saunders 1997*a*), formerly considered Maastrichtian (Salvador & Stainforth 1968), are interpreted here as cannibalized from earlier foreland-basin deposits caught up in the advancing thrust belt and entirely eroded.

(4) Paleocene fossiliferous limestones (e.g. remnant Soldado Fm; Kugler & Caudri 1975; Kugler 2001) can be interpreted the same way.

All lithologies except type (3) occur *in situ* in eastern Venezuela formations (Salvador &

Stainforth 1968). Type 1 is typical of the El Cantil Fm, whose equivalent in the Trinidad autochthon is the shale-dominated Cuche Fm, containing type (1) olistoliths derived from an adjacent, penecontemporaneous carbonate platform (Hutchison 1938; Barr 1952). Type (2) lithology is typical of the San Antonio Fm (Salvador & Stainforth 1968). Type (4) Paleocene limestone occurs in the subsurface south of the Eastern Serranía, separating the San Juan and Vidoño Fms (Fig. 12), and crops out as molluscan coquina at Cerro Corazón (Hedberg 1950; Salvador & Stainforth 1968; author's observations). This basal Vidoño limestone is interpreted here as shallow-marine arch deposits (condensed section) reflecting a northward 'pulling-in' of the Proto-Caribbean forebulge onto the 'passive' shelf, due to tectonic rejuvenation in the nappe-thrust belt (Sinclair *et al.* 1991), associated with the Paleocene change of subduction obliquity. The limestone is correlated here with the Cerro Voladero glauconite, a sand-grade, nearly pure glauconite unit 15 m thick (author's observations), interpretable as a different manifestation of the same condensed section, deposited on the forebulge flank, where the water was too deep (sub-photic zone) for benthonic molluscs to survive (cf. forebulge facies model for Tinajitas Mbr by Sageman & Speed 2003). The same coquinoid–glauconite forebulge association can be inferred in northern Trinidad but was entirely eroded by thrust-belt advance, explaining Bontour and Soldado olistoliths (i.e. types 3 and 4) documented by Kugler (2001) in Eocene through Miocene formations of central and southern Trinidad (e.g. Marac Quarry; Kugler 1953).

A similar but younger (Eocene) paired limestone–glauconite condensed section occurs in eastern Venezuela, comprising (a) the subsurface 'Tinajitas' limestone (Sams 1995), overlain by and passing laterally into (b) the 'Areo glauconite member', up to 18 m thick (Aguado 1993; Sams 1995). These deposits are attributed to another 'pulling in' of the forebulge, due to the Middle Eocene (chrons 21–18) plate-motion adjustment.

Three less common orphan lithologies in Trinidad are: (a) indurated, pink, quartzose sandstone clasts in the Patos Conglomerate (Kugler 1974) and in the Plaisance Conglomerate Mbr (Late Eocene; Saunders 1997*a*; author's observations), possibly derived from reddened Jurassic strata (cf. Ipire, La Quinta formations) eroded off the Outer Nappe (not metamorphosed); (b) dacitic pebbles in the Tamana Fm (Kugler 1953), possibly from Jurassic or Lower Cretaceous rift volcanics of the Outer Nappe, entirely eroded; (c) clasts of Lower Turonian ammonite coquina in the Plaisance Conglomerate (Reyment 1972). The age of this coquina partially corresponds to a hiatus reported between the Gautier and Naparima Hill formations by Saunders & Bolli (1985), later thought to be shorter or absent (Saunders 1997b). The hiatus may instead be the same ammonitic condensed section, too thin (centimetres to decimetres) to detect readily in drill cuttings. The Early Turonian age coincides precisely with the 91.5 Ma 'major' eustatic condensed section, representing the highest sea level attained in post-Palaeozoic time (Haq et al. 1988). The coquina is again interpreted as derived from a disappeared (eroded), northern, Outer Nappe source.

Olistolithic olistoliths

Olistoliths inside olistoliths are known in Trinidad. One example is a Gautier Fm olistolith within a Chaudiere shale olistolith, enclosed by Nariva shales (Kugler 1953). Another is the celebrated Soldado Fm olistolith contained in a San Fernando olistolith forming Soldado Rock, encased in Miocene ?Karamat shales (Kugler & Caudri 1975). The first example involves recycling of older foreland-basin deposits (Chaudiere) into younger ones (Nariva), i.e. cannibalization, reflecting thrust-front advance. The second involves 'double cannibalization'.

Unusual turbidites/debrites

Cherry conglomerate. The Plaisance Conglomerate contains beds of 'cherry-cake conglomerate' (author's observations). The name was coined by Barr (1962) for Toco Fm examples; other such beds occur in eastern Venezuela in the coeval Güinimita Fm (González de Juana et al. 1972), and in the Cenozoic 'Lecherías Beds' (Higgs in Pindell & Erikson 2001). The cherry-pebble beds are interpreted as high-density turbidites and debrites, the pebbles being shallow-diagenetic, siderite-cemented concretions winnowed from contemporaneous sea-floor muds (Higgs in Pindell & Erikson 2001), possibly by turbidity flows eroding channel banks. The pebbles were subsequently haematized (reddened) by subsurface diagenesis or tropical weathering. This interpretation is supported by the presence of variably reddened sideritic concretions in many shales in the region, e.g. Areo, Chaudiere, Cuche, Toco, Vidoño formations (Barr 1952; author's observations).

Redeposited carbonates. The Proto-Caribbean foreland basin includes Eocene and Miocene algal-orbitoidal limestone members and formations, including the Peñas Blancas and Tinajitas in Venezuela (Fig. 7) and in Trinidad the Vistabella, Tamana, Concord and several intra-Brasso limestones (Kugler 1959, 1961, 2001; González de Juana et al. 1980; LEV 1997; Saunders 1997a, b).

These limestones are interpreted here as north-derived debrites and high-density turbidites, supplied by nearshore carbonate factories during arid episodes of low siliciclastic supply. Previous authors interpreted essentially *in situ* shallow-marine deposition, at or near the factory (Galea 1985; Erlich et al. 1993), on a forebulge related either to Caribbean nappe approach (Pindell et al. 1998) or subduction of South America northward at an intra-oceanic plate boundary (Sageman & Speed 2003).

Guárico Fm

The Guárico Fm (Maastrichtian–Eocene) dominates the central Venezuela foothills (González de Juana et al. 1980; LEV 1997). Redefined by Macsotay et al. (1995) to exclude the Los Cajones and other olistostromal members, the turbiditic Guárico Fm is interpreted here as south-derived, following Bell (1967), deposited on the southern, passive slope of the Proto-Caribbean flysch trough (Fig. 6), consistent with lateral transition into a southern thick-bedded sandstone facies (Bell 1967) and with a lack of reported staurolite or glaucophane (Menéndez 1965; Bell 1967). Palaeocurrents flowed mainly NE (Bell 1967; Albertos et al. 1989), obliquely down the ENE-striking slope, consistent with Coriolis deflection to the right in the northern hemisphere. Palaeomagnetic studies indicate negligible vertical-axis rotation of the formation as a whole (Blin et al. 1990), thus palaeocurrent indicators are reliable. Despite the evidence for derivation of the Guárico turbidites from the south (Bell 1967), the opposite is generally assumed (Peirson et al. 1966; Albertos et al. 1989; LEV 1997).

Age of Inner Nappe mobilization

In western and central Venezuela, Inner Nappe uplift was in progress in Paleocene or Eocene time, as shown by olistoliths of Inner Nappe metarift cover in the Matatere and Los Cajones Fms, indicating erosional breaching of the overlying Outer Nappe already. This conclusion agrees with radiometric (Ar/Ar) and fission-track dates from central Venezuela interpreted as indicating post-metamorphic cooling (uplift) of the Caracas Gp and its basement from Middle Eocene to Middle Miocene time (Sisson et al. 2005 review and new dates).

Inner Nappe uplift began much later in Trinidad. Basement-derived crystalline olistoliths are absent in the Chaudiere Fm (e.g. Kugler 2001), implying that the Outer Nappe was not yet deeply unroofed (Paleocene), suggesting that the Inner Nappe had not yet moved. This inference is consistent with shelly limestone 'remnant' olistoliths in the Chaudiere (Barr & Saunders 1968), derived from the Outer Nappe rift cover as explained above. Inner Nappe initial uplift is dated by an influx of chert in Lower Miocene conglomerates exposed in the Central Range (Cunapo Fm, type area; Kugler 1961, 2001). This 'Guaico Conglomerate Mbr' (Saunders 1997a) shows sedimentological evidence for deposition by high-density gravity flows (author's observations), consistent with the south-migrating slope apron model described above. The chert clasts are interpreted here as north-derived, from upper Naparima Hill Fm chert-rich strata (cf. Erlich *et al.* 2003) that cropped out at the front of the uplifting Inner Nappe (Fig. 4c), but were later completely(?) eroded. The Naparima Hill Fm is indeed missing at the sub-Miocene unconformity under northern Caroni Basin and northern Gulf of Paria (Kugler 1961 cross sections; Flinch *et al.* 1999), suggesting that the Inner Nappe frontal thrust lies between here and the Central Range conglomerates. If the Fishing Pond Thrust (Kugler 1961) under mid-Caroni Basin is the fault concerned (Higgs 2009, fig. 3 inset), then the 'Guaico' conglomerate in the subsurface must reach no further north than here, not to the El Pilar Fault (Kugler 1961; Saunders 1997a). Moreover, the 'Guaico' name is inappropriate, as the conglomerates in the Guaico-1 borehole in northern Caroni Basin contain much less chert (Kugler 2001) and are probably a separate, younger formation (Upper Miocene), fringing the Manzanilla and Springvale formations (Kugler 1961) of the Caroni halite-dissolution basin (Higgs 2009).

Early and Middle Miocene fission track ages from the Caribbean Gp (Weber *et al.* 2001b) support the Early Miocene start of Inner Nappe uplift in Trinidad inferred from the chert influx. Also consistent are Early Miocene Ar/Ar ages obtained by Foland *et al.* (1992) and Speed *et al.* (1997) from the Caribbean Gp and from the basement (Dragon Gneiss) nearby in Paria Peninsula, although they were interpreted by these authors as metamorphism rather than cooling ages.

Later uplift of the Inner Nappe in Trinidad than Venezuela suggests less north–south shortening in the east, consistent with eastwardly decreasing inter-Americas convergence. Thrust-belt shortening ahead of the nappe would probably also have been diachronous, uplifting earlier-deposited foreland basin strata, no earlier than Eocene time in the west (post-Garrapata) and Miocene in Trinidad (cf. Beck *et al.* 1990, fig. 7). Indeed, Pérez de Armas (2005) suggested Eocene uplift of the central Venezuela thrust belt, based on fission-track ages in Guárico Fm sandstones, moreover attributing the uplift to Proto-Caribbean subduction. The same model was applied to Oligocene fission-track ages from Cretaceous (Barranquín) sandstones in the eastern Venezuela thrust belt (Locke & Garver 2005).

Age of metamorphism, Caracas–Caribbean Gps

The age of metamorphism of the Inner Nappe rift- and passive-margin cover (Caracas, Caribbean Gps and related formations), by burial under the Outer Nappe, is constrained by the time of initial uplift of each nappe, dated above from foreland-basin stratigraphy, clast compositions and radio-metric/fission-track data. The possible time span is thus intra-Campanian until Paleocene or Eocene in central Venezuela and Campanian to Early Miocene in Trinidad. Many authors attributed the metamorphism to burial under the *Caribbean* accretionary prism, either nearly *in situ* or far to the west (references above). The metamorphism is attributed here to *Proto*-Caribbean parautochthonous nappe stacking, prior to the Caribbean leading edge arrival in central Venezuela and Trinidad during Early Miocene and Pliocene time respectively (Figs 9 & 11).

Amount of Proto-Caribbean subduction and nappe overlap

Proto-Caribbean subduction was too slow ($<1\ cm\ a^{-1}$) for the downgoing slab ever to reach sufficient depth for arc magmatism (Pindell *et al.* 1991). This reflects the limited inter-Americas north–south convergence, totalling only 300 km (Campanian to Recent) at the longitude of western Venezuela and 100 km at Trinidad near the rotation pole (Müller *et al.* 1999, fig. 11). Thus the Proto-Caribbean foreland basin is of a little-known 'pre-arc' type, seldom preserved, usually obliterated by arc magmatism.

The (Inner, Outer) nappe travel distance cannot exceed the amount of inter-Americas shortening. A minimum measure of this travel is the outcrop width of metamorphosed Jurassic–Cretaceous sedimentary–volcanic strata (Inner Nappe cover), measured perpendicular to the continental margin (trend NW–SE in the west; near east–west in the east). This width reflects the minimum travel of the Outer Nappe (formerly suprajacent, hence metamorphism). The width is *c.* 200 km in the west, measured beside and parallel to the Boconó Fault (e.g. Pimentel 1984, map units 14, 16 & 17), consistent with (a) net inter-Americas SW–NE convergence of *c.* 250 km at this longitude up until Recent time (Müller *et al.* 1999, fig. 11), and (b) termination of Proto-Caribbean nappe motion before this due to the shielding effect of the 'intruding' Caribbean Plate, progressively later eastward (Oligocene in Falcón; Miocene in central Venezuela).

Caribbean interaction, Ecuador to Guajira, Campanian–Eocene

Amaime Nappe

At its 72 Ma birth, the Caribbean Plate was fronted by a NW-trending subduction zone (future Caribbean Arc) at the site of the former IAA (Fig. 2c & d), and was trailed by the nascent Panamá–Costa Rica Arc, connected to South America by a transform fault (Pindell & Barrett 1990, plate 12D). The Caribbean front migrated east relative to South America (Pindell & Kennan 2001b), colliding obliquely (dextrally) along the NE–trending Ecuador–Colombia margin as far as Guajira. The oblique collision obducted a forearc nappe diachronously along this margin (Pindell 1993), driving an overfilled Caribbean foreland basin that superseded the Proto-Caribbean foreland basin progressively later northward (Figs 3, 5–7).

Caribbean nappe remnants include (south to north) the Pallatanga (Kerr et al. 2002), Amaime (Aspden & McCourt 1986) and Ruma (MacDonald et al. 1971) terranes. The Romeral Fault (Fig. 1) is generally regarded as the Amaime suture but in places is a west-verging backthrust or a strike–slip fault, offsetting the nappe sole thrust (e.g. Restrepo & Toussaint 1988; Flinch 2003), hence earlier perceptions of the Amaime terrane underthrusting South America (e.g. Pindell & Barrett 1990). As the Amaime Nappe sutured progressively along the margin, Caribbean ocean lithosphere continued to converge on South America, and was thus forced to subduct at a backthrust (Fig. 2e; Pindell & Erikson 1994). This backthrust is the South Caribbean Marginal Fault (accretionary wedge; Mann 1999) and its southern continuation, the latter masked by later Panamá collision. The NE tip of this fault advanced along strike, scissor style (cf. Pindell & Tabbutt 1995; Pindell et al. 1998), shadowing the migrating nappe suture point and connected to it by an east–west transform fault (parallel to relative plate motion) that necessarily jumped north intermittently.

Amaime suturing was considered Campanian and Maastrichtian by Pindell (1993), reflecting the idea that the Caribbean Arc was well past Guajira corner by Paleocene–Eocene time, driving Maracaibo–Lara foredeep subsidence. However, this model is untenable (see above). Five observations suggest that the Caribbean Arc transit along Colombia lasted longer, from Campanian time until c. 35 Ma (latest Eocene; Figs 2d, e, 3 & 5–7).

(1)   The Ruma complex in Santa Marta and Guajira contains diorite intrusions with 50 and 48 Ma K–Ar ages respectively (Green et al. 1968; MacDonald et al. 1971). If the diorites were formed at the Caribbean Arc, they cannot have been obducted onto South America much earlier than 50 Ma.

(2)   Renewed sedimentation on the MMVU began in Middle Eocene time, by westward onlap, ascribed by Gómez et al. (2003) to loading by early west-verging thrust uplift of the Eastern Cordillera, based on unconformities of that age there. Instead the MMVU subsidence is attributed here to Caribbean nappe loading, and the unconformities to Caribbean forebulge uplift.

(3)   In the South Caribbean Fault accretionary wedge, northernmost Sinú sector, west of Guajira, the detachment is interpreted to lie near the base of the Oligocene succession (Flinch et al. 2003, fig. 2B & C), consistent with subduction not starting here (i.e. Guajira nappe obduction not ending) until Late Eocene time (cf. Fig. 7).

(4)   SE-younging Oligocene–Miocene subsidence occurred along the NW–SE-trending Guajira–Falcón palaeo-margin and can be interpreted as transform-related block faulting as the Caribbean Arc passed by (see below).

(5)   Enigmatic, isolated Upper Eocene marine shales in SE Falcón (Cerro Misión Fm; LEV 1997), unconformable under Miocene strata, can be interpreted as distal Caribbean foreland basin deposits, with subsidence driven by the Guajira sector of the Caribbean nappe.

Caribbean Plate rear boundary

By the time the Caribbean leading edge reached Guajira corner (35 Ma), the rear boundary was at its present (Panamá) position in mid-Colombia (cf. Miocene arrival in Pindell 1993), having travelled NE along the Ecuador–Colombia margin. During transit this rear boundary was a NW-trending transform fault and (further offshore) the co-linear Panamá Arc (cf. Pindell & Barrett 1990, plate 12e & f). Southwest of the transform, Pacific Ocean lithosphere was subducting under Ecuador and Colombia, continuing today. Northeastward migration of the transform along South America, passing the Ecuador–Colombia border in Middle Eocene time (Fig. 7), had three effects:

(1)   Birth of the modern Colombian continental-margin arc (Pacific slab dip steeper than Caribbean; Van der Hilst & Mann 1994), starting in Middle Miocene time in the far SW (Van Houten 1976), once the slab had reached melting depth, and predicted to young NE along-strike.

(2)   Corresponding retro-arc foreland basin subsidence in the joint Putumayo–Upper

Magdalena Basin (Fig. 8), replacing the Caribbean foreland basin, presumably diachronously;

(3) Start of (*in situ*) Eocene–Quaternary subsidence in the little known Chocó–Pacífico–Tumaco and Cauca–Patía Basins (Alfonso *et al.* 1994; ANH 2005*a*, *b*; Barrero *et al.* 2007). The latter, of uncertain origin (Alfonso *et al.* 1994), is interpreted here as an inner-forearc basin of the Colombia arc, consistent with basin onset younging NE, along strike (ANH 2005*a*). The former is also a forearc basin (ANH 2005*b*; Barrero *et al.* 2007). Diachronous Oligo–Miocene Panamá Arc collision along southern Colombia (Pindell & Tabbutt 1995, figs 7–9) has difficulty explaining these basins and conflicts with palaeobiogeographic evidence that Panamá–Colombia collision did not start until Middle Miocene time (Coates *et al.* 2004; see also Higgs 2009).

Circa 35 Ma birth of Gulf of Venezuela–Falcón Basin

At or before the time the Caribbean Arc (or more precisely, the forearc frontal edge) reached Guajira corner, in latest Eocene or earliest Oligocene time (*c.* 35 Ma), Caribbean relative motion must have changed from east to SE, otherwise the Villa de Cura sector of the Caribbean forearc nappe could not have arrived, after obduction (later), at such a westerly position (Fig. 9). After passing the corner, the arc migrated SE along the SE-trending Guajira–Gulf of Venezuela–Falcón continental margin sector (Outer Nappe tail-end), reactivating the former continent-ocean boundary palaeofault as a SE-lengthening transform, here named the Falcón–Bonaire Fault (FBF), forming one arm of a migrating triple junction (transform/Caribbean subduction/Proto-Caribbean subduction). This model contrasts with Caribbean/Proto-Caribbean 'trench-trench' collision proposed for both the Guajira–Falcón and the Caracas–Trinidad sector (Pindell *et al.* 2006; Pindell & Kennan 2007*b*, *c*). Masking the FBF are complex younger tectonic effects (Gorney *et al.* 2007).

As the arc migrated along Falcón, a longitudinal interarc basin opened, separating the Leeward Antilles–Aves Ridge abandoned arc and the Lesser Antilles Arc precursor (Figs 8 & 9). Interarc sea-floor spreading occurred in the north (Grenada Basin; Bird *et al.* 1999), but only block faulting in the south (Bonaire Basin; Gorney *et al.* 2007). Extension in both basins was inferred by Pindell *et al.* (1998) to have begun in Early Eocene time.

The Gulf of Venezuela and Falcón continental borderland region is characterized by Oligo–Miocene block faulting (Macellari 1995). Subsidence is interpreted here to have begun progressively later southeastward, based on published biostratigraphic ages of the first strata overlying a prominent unconformity: Early Oligocene in Guajira (including its NE shelf) to central Falcón; Late Oligocene or Early Miocene in SE Falcón (Díaz de Gamero 1977 stratigraphic tables at http://www.pdvsa.com/lexico/correlac/falcon.htm; Macellari 1995, fig. 2). Extensional magmatism occurred in central Falcón during Early and earliest Middle Miocene time (Muessig 1978; McMahon thesis cited by Gorney *et al.* 2007). The diachronous block faulting is attributed here to transform-related extension caused by the Caribbean arc migrating past. The Burro Negro Fault to the SW (Fig. 1; Escalona & Mann 2006) may have controlled the inboard limit of extension. The Cenozoic Queen Charlotte Basin in Canada is similar in size and interpreted transform-related origin, except that no arc is involved (Dietrich *et al.* 1993). No arc tuffs have been reported in the Gulf of Venezuela–Falcón Basin, consistent with the passing arc being characterized by interarc extension (submarine volcanism) rather than arc volcanism. Thermal subsidence may have diachronously followed block faulting. Previous authors interpreted the Gulf of Venezuela–Falcón Basin as a pull-apart formed by *east–west* Caribbean relative motion (Muessig 1978, 1984; Boesi & Goddard 1991), on Caribbean nappe basement, i.e. *after* the Caribbean Arc had passed (Pindell 1993; Macellari 1995; Pindell *et al.* 1998). According to Gorney *et al.* (2007), the Falcón Basin is an intra-arc extensional basin formed during Caribbean Arc obduction.

Circa 35 Ma onset of 'Andean' (Caribbean) orogeny, NW Colombia

Also near the start of Oligocene time (*c.* 35 Ma), Caribbean subduction at the South Caribbean Fault is interpreted here to have slowed by choking, SW of the Santa Marta Fault (Fig. 1), when Caribbean Plateau thickened ocean crust first arrived (cf. present distribution of Plateau, Meschede & Frisch 1998, fig. 2); this choking may have caused the Caribbean relative-motion change. Some of the subsequent Caribbean–South America convergence (NW–SE) was thence accommodated by shortening in the overriding plate, with two important effects.

(1) The Murrucucú and Santa Marta faults (Fig. 1) were activated, as near-orthogonal and sinistral thrusts respectively, forming the reentrant Lower Magdalena Basin by combined loading (Figs 8 & 9). These faults were interpreted as

normal by Villamil (1999, fig. 12). The Santa Marta is a sinistral reverse fault according to Paris *et al.* (2000). Complex Oligocene block faulting in this basin, previously ascribed to regional transtension and block rotation (Reyes *et al.* 2000), is attributed here to forebulge extension (Decelles & Giles 1996) on the two merging forebulges, succeeded by Mio-Quaternary flexural subsidence. Outboard, in the San Jacinto belt (Flinch 2003), the Oligocene and younger section belongs to the same foreland basin (distal), as shown by near-identical stratigraphy (Reyes *et al.* 2000). Northwest-vergent thrusting in San Jacinto is related to development of the adjacent Sinu accretionary wedge (Flinch 2003).

(2) Eastward thrusting began on the Caño Tomas–Tarra–Mercedes–Labateca Fault (Paris *et al.* 2000; Corredor 2003), earlier and reaching further south than envisaged by Pindell *et al.* (1998), uplifting the Santander Massif (hanging wall) and driving a 'Catatumbo–Apure Foreland Basin' (Figs 8–10), as shown by marked WSW thickening of the Catatumbo fill (Audemard 1991, figs 14 & 15). Parallel to the Santander Massif is the Palmar–Avispa basement uplift under southwestern Lake Maracaibo (Brondijk 1967), interpreted here as the initial Santander forebulge, overlain by Catatumbo–Apure Basin strata that merge NE into Falcón–Gulf of Venezuela Basin deposits (Figs 8–10).

Caribbean oblique collision with central Venezuela–Trinidad

After traversing the Falcón margin, the Caribbean Arc collided obliquely along the Caracas–Trinidad margin (Fig. 2f), progressively obducting the Villa de Cura–Margarita (Rinconada)–Tobago forearc nappe. On the nappe, forearc-basin strata of Eocene age in Margarita (Speed & Smith-Horowitz 1998), and Eocene–Oligocene under the Carúpano Basin (see below), escaped erosion during uplift; they merge NE into the present Tobago Trough forearc basin (Speed & Smith-Horowitz 1998). Drag due to oblique collision rotated the arc approximately 90° into near-parallelism with the continental margin (Skerlec & Hargraves 1980). Rotation affected all four components of the arc complex: forearc; Lesser Antilles Arc precursor; Bonaire interarc basin; and Leeward Antilles abandoned-arc sector (Higgs 2009, fig. 2). The latter and its northward prolongation (Aves Ridge) yield Maastrichtian–Paleocene granitoid K–Ar ages (Santamaría & Schubert 1974), consistent

with Maastrichtian birth of the Caribbean Arc (see above). The granitoids intrude metamorphics in the Leeward Antilles (Santamaría & Schubert 1974), probably IAA forearc rocks.

In Middle Miocene time, after the forearc had already begun obducting onto central Venezuela (Fig. 10), Lesser Antilles Arc volcanism started (*c.* 12 Ma; Speed *et al.* 1993), signalling the end of inter-arc extension. The new arc grew on the rifted eastern half of the previous arc, whose obducted southern continuation, shown by gravity anomalies and bathymetric highs, curves westward as the Testigos–Frailes–Tortuga island chain (Mann 1999).

The present southern limit of the Caribbean forearc nappe runs as follows (starting in the W): (a) ENE along the south front of Villa de Cura (Pimentel 1984; Smith *et al.* 1999); (b) NE offshore across Barcelona Bay, south of boreholes that reached volcanic basement (Bellizzia 1986; Goddard 1988); (c) between Araya and Margarita (La Rinconada basal thrust); (d) ENE along the Carúpano Basin; and (e) ESE along the North Coast Basin, passing between Tobago and Trinidad, possibly south of the Alice-1 well which bottomed in metaigneous rock (Speed & Smith-Horowitz 1998; Robertson & Burke 1989). In the fourth sector (Carúpano), the nappe front separates two groups of offshore boreholes: (a) a NW group that bottomed in arc volcanics giving Eocene K–Ar ages (Los Testigos Complex) or in Eocene to Middle Miocene clastics with rare limestones and pyroclastics, interpreted as forearc-basin deposits (Castro & Mederos 1985; Speed & Smith-Horowitz 1998); and (b) a SE group that reached the Mejillones Complex, unmetamorphosed Lower and Upper Cretaceous volcanics, limestone, chert and clastics, including dark shale of euxinic aspect (Castro & Mederos 1985) of exploration interest. The Mejillones Complex was assigned to a Caribbean-related terrane by Speed & Smith-Horowitz (1998), but is reinterpreted here as Outer Nappe rift cover, underlain in one well by MORB metabasalts (Bocas Complex) correlated with the El Copey Fm of the nearby Araya–Paria onshore region (Castro & Mederos 1985; Speed & Smith-Horowitz 1998), which is interpreted here as Outer Nappe Hercynian basement. The Mejillones Complex is equivalent in age and tectonic setting to the Los Robles metastrata of Margarita, but unmetamorphosed as it lies ahead of the Caribbean nappe front.

Caribbean nappe loading again produced a Caribbean foreland basin, diachronously (Middle Miocene–Pliocene) superseding the Proto-Caribbean one (Figs 10 & 11). The central Venezuela sector of the new basin was later entirely eroded by east-younging tectonic rebound, hence the overall east-younging (Oligocene–Recent) map pattern along the central and eastern Venezuela foredeep (e.g. Pimentel 1984). Rebound reflects the Caribbean load becoming diachronously severed by the South Caribbean Fault (northern Venezuela sector). The fault, which accommodated (by subduction) continued Caribbean convergence as in Colombia, is oriented more easterly in Venezuela, outboard of the accreted arc complex (Fig. 1). The migrating fault tip shadowed the east-migrating Caribbean nappe suture point, connected to it by a SE-trending transform fault that jumped progressively eastward. The resulting set of abandoned NW–SE transforms dissects the accreted arc complex. These were interpreted as Oligocene to Holocene normal faults by Gorney et al. (2007). The Charallave Fault (Pimentel 1984) cutting the Villa de Cura klippe is the onshore projection of one such fault, offset by dextral slip on the east–west San Sebastián Fault (Fig. 1) since the 2.5 Ma change from oblique to near-transcurrent plate motion (see below). Each transform jump eastward signified the transference, to the South America Plate, of a further sector of the arc complex, severing the Caribbean load. Transforms connected southeastward to the frontal thrust.

East of Bonaire Basin, the South Caribbean subduction zone was offset southward, via the Roques Canyon transform fault (Pindell et al. 1998), to another subduction zone and accretionary prism, the 'southern Grenada Basin deformation belt' (SGB in Fig. 1; Speed & Smith-Horowitz 1998). This jump reflects the relative subductability of Grenada Basin oceanic lithosphere under South America, unlike Bonaire Basin stretched-arc lithosphere along-strike. The continentward projection of the Roques Canyon Fault is the Urica Fault, offset about 50 km dextrally on the El Pilar Fault since 2.5 Ma (Fig. 1). The Urica Fault ceased activity in Late Miocene or Pliocene time according to Salvador & Stainforth (1968). Late Miocene deactivation due to SGB initiation is invoked here.

Seismic and biostratigraphic data indicate that significant thrust-belt shortening ended in Late Miocene time in the western sector of eastern Venezuela (NW Monagas state), but continued in the Plio-Quaternary in the eastern sector and Trinidad (Duerto & McClay 2002; Pindell & Kennan 2007a, c). A Caribbean 'bow-wave' model (Pindell & Kennan 2007a, c) was proposed to explain this diachronism, involving a Caribbean relative-motion change from ESE to eastward (transcurrent) near 10 Ma, and a plate boundary jump inboard from the SGB to the El Pilar Fault. However, the bow-wave model has three difficulties: (a) abundant evidence for the plate shift being much later (2.5 Ma; see below); (b) improbability that highly oblique loading (east of the east-migrating El Pilar Fault tip, where the 'boat's' bow and south side met) could have produced the kilometric Pliocene

foredeep subsidence south of Trinidad (Di Croce et al. 1999, fig. 15c); and (c) thrusts in the South Grenada Basin accretionary prism interpreted by Ysaccis et al. (2000, fig. 5) as terminating upward in the Lower Pliocene interval, negating the proposed 10 Ma abandonment of subduction here. Anticlines bounded by these thrusts 'developed mainly during the Middle Miocene to Early Pliocene' (translated from Ysaccis et al. 2000, p. 414). An alternative to the bow-wave model is as follows.

By the time the eastwardly obducting Caribbean forearc first impinged on eastern Venezuela (c. 14 Ma), south-vergent Proto-Caribbean thrusting had already reached as far south as the eastern Venezuela frontal thrust (Fig. 1) and the Trinidad Central Range Fault, forming a continuous highland belt (Eastern Serranía–northern Gulf of Paria–northern half of Trinidad). Near 11 Ma, after the frontal thrust reached southernmost Trinidad, highland collapse began by underground dissolution (wetter climate since 13 Ma start of Gulf Stream) of the Carib Halite, forming supraorogen basins (Gulf of Paria and Trinidad except Northern Range) that shortened syn-depositionally but too slowly for uplift to outpace solution subsidence (Higgs 2009). As the Caribbean nappe 'impingement point' progressed eastward, thrust-belt north–south shortening accelerated correspondingly (diachronously), inverting these basins. In Trinidad, Caribbean-driven shortening began to replace Proto-Caribbean shortening around 3.5 Ma, inverting the solution basin, whose youngest strata are Lower Pliocene (Kugler 2001, p. 90, Talparo–Erin Fms). As the trailing suture point passed, accommodation of further inter-plate convergence was transferred diachronously (starting c. 11 Ma in the far west) to the SGB. East-jumping dextral transforms can again be inferred, cutting (a) the accreted forearc (though masked in Carúpano Basin by post-2.5 Ma deposits; but see unnamed ESE-trending fault east of Margarita in Pindell & Kennan 2007c, fig. 1b, here named Testigos Fault in Fig. 1) and (b) the thrust belt (e.g. San Francisco, Los Bajos faults). Los Bajos transform displacement was 3–10 km dextrally, as shown by offset of a syncline cored by Erin Fm strata (Kugler 1961; Wilson 1968); this displacement spanned about 1 Ma, from say 3.5 Ma (post-Erin) until the 2.5 Ma plate-boundary jump. Each transform segment in the thrust belt was active shortly before as a transfer fault (sinistral, transpressive, east-jumping), joined to the frontal thrust, separating the thrust belt into a western sector (Caribbean) and an eastern sector (Proto-Caribbean). Seismic interpretation in southern Trinidad confirms significant shortening starting in Late Pliocene time, but continuing in Quaternary time (Pindell & Kennan 2007a, c), past the 2.5 Ma plate shift, because the Trinidad sector of the new plate boundary is transpressively oblique (Central Range Fault, c. 070°) to the eastward plate motion. In summary, peak (Caribbean-driven) shortening youngs eastward (NW Monagas c. 14–11 Ma; southern Trinidad c. 3.5–0 Ma) and can be explained without recourse to an ad hoc bow-wave model.

After obduction, the Villa de Cura sector of the Caribbean nappe rebounded along with its foreland. Rebound still affects the present-day mountains of central Venezuela (including the Villa de Cura klippe), probably added to an element of uplift by (highly oblique) plate-boundary transpression since 2.5 Ma (see below). In contrast, the eastern sector of the nappe (Barcelona Bay to Tobago) is largely below sealevel, except in Margarita and Tobago islands. This subsidence reflects dissolution of halite in the Caribbean Nappe substrate (i.e. in the tail of the Proto-Caribbean Outer Nappe), forming other supraorogen basins (e.g. Gulf of Barcelona, Carúpano, North Coast; Higgs 2009). An anomalous normal-fault shallow earthquake near Tobago in 1997 (Weber 2005) might reflect active halite dissolution.

Late Miocene uplift of Eastern Cordillera

The Panamá Arc began colliding against Colombia, end-on, in Middle Miocene time (Fig. 10; Coates et al. 2004). Late Miocene uplift of the Eastern Cordillera (Gómez et al. 2003) is attributed here to this collision. Near-orthogonal bivergent thrusting drove foreland-basin subsidence on both sides (Middle and Upper Magdalena; Llanos). Panamá convergence with South America was additionally accommodated by oroclinal buckling of the arc, causing it to overthrust the Caribbean Plate (Mann & Corrigan 1990). The collision also terminated deep-water circulation between the Pacific and Caribbean, producing a wetter climate in Venezuela and Trinidad, conducive to underground halite dissolution (Higgs 2009).

Mérida Andes–Falcón Pliocene orogeny

Along-strike from the partially choked (since c. 35 Ma) Colombian sector of the South Caribbean Fault, choking is inferred to have begun in the adjacent Santa Marta-Guajira sector at about 5 Ma, based on evidence reviewed below. Again some of the Caribbean convergence (NW–SE) was transferred to the overriding plate, with the following observed or inferred effects.

(1) The sinistral slip rate on the Santa Marta Fault was reduced.
(2) The Bucaramanga Fault (Fig. 1) was activated as a west-verging sinistral thrust, driving

subsidence in the northernmost Middle Magdalena Valley.
(3) The Mérida Andes were uplifted by near-orthogonal bivergent thrusting (Boconó Fault Zone), bifurcating the preceding Catatumbo–Apure Foreland Basin. The Mérida and Barinas flanking foreland basins (Zambrano et al. 1970, 1971) were initiated, recorded by influx of coarser, proximally thickening deposits (Betijoque, Río Yuca Fms). In most previous interpretations, the age of initial Mérida Andes uplift was Miocene, based on diverse techniques (review in Audemard 2003). Using additional, indirect evidence and assumptions, Audemard (2003) inferred Late Miocene onset, but with the bulk of uplift occurring in the last 3–5 Ma. Uplift starting near 5 Ma is advocated here, based on three criteria: (a) pre-uplift strata of probable Middle Miocene age are preserved as steeply dipping intramontane erosional remnants of the Catatumbo–Apure Basin (La Copé Fm, compare Figs 10 & 11; Macellari 1984; Higgs et al. 1995); (b) 17 out of 22 Mérida Andes apatite fission-track ages obtained by Kohn et al. (1984) are 4.9 Ma and younger; and (c) the Betijoque Fm may be entirely Pliocene (e.g. González de Juana et al. 1980, fig. VI-41, p. 541; Audemard 1991, fig. A11), and thus the Río Yuca too (presumed coeval; LEV 1997), though these alluvial strata, difficult to date, are commonly considered Upper Miocene–Pliocene (LEV 1997). At the SW end of the Mérida Andes, the Tachira structural and topographic depression (Meier et al. 1987) may reflect continued loading by the Santander Massif, retarding uplift, until Santander thrusting was terminated by the 2.5 Ma plate-boundary jump. Santander loading (5–2.5 Ma) along strike contributed to subsidence of the Mérida and Barinas foreland basins.
(4) Uplift of the Perijá and Lara-Falcón ranges began, and uplift in Santa Marta and Guajira was rejuvenated. All but Guajira are known to verge mainly NW (Kellogg 1984; Boesi & Goddard 1991; ANH 2005c; Mora & García 2006). In the Sierra de Perijá, Kellogg (1984, p. 247) inferred a 'Pliocene age for the major uplift' from stratigraphic relationships and fission-track ages. Falcón–Lara uplift was Pliocene according to Macellari (1995, fig. 20). Here the thrust front advanced rapidly, possibly detached at the Carib Halite level (cf. Higgs 2009, fig. 4), and perhaps reached as far NW as Guajira before the plate-boundary jump at 2.5 Ma. The southwestern lateral ramp may have been the Burro Negro

Fault. Despite the regional shortening, syn-orogenic solution subsidence occurred in the Gulf of Venezuela, Lower Guajira and Cesar–Ranchería Basins, outweighing probable simultaneous uplift of the sub-halite substrate (Higgs 2009). Perijá backthrusting (Duerto et al. 2006) was insufficient to assist Mérida foreland basin subsidence, as shown by NW thinning of Mio-Pliocene isopachs (Audemard 1991, fig. 15).

Caribbean motion shift, 2.5 Ma

Some time after the 5 Ma start of choking in western Venezuela a plate reorganization occurred. The Caribbean Plate assumed its current eastward relative motion (c. 085°), measured by GPS studies (Pérez et al. 2001; Weber et al. 2001a; Trenkamp et al. 2002). The plate boundary jumped from the South Caribbean–Roques–South Grenada Basin–Testigos–Los Bajos–South Trinidad fault system to its present position, the Eastern Cordillera Frontal–Chitaga–Boconó–Sebastián–Pilar–Central Range fault linkage (Fig. 1; Molnar & Sykes 1969; Dewey 1972; Audemard et al. 2000; Weber et al. 2001a). Modern and historical large earthquakes (Paige 1930; Dewey 1972; Pennington 1981; Pérez & Aggarwal 1981; Russo et al. 1993; Audemard et al. 2000; Paris et al. 2000) have occurred on all these faults except the Central Range Fault, whose apparent aseismicity suggests either locked or creeping behaviour (Weber et al. 2001a; Weber 2005, 2007), possibly lubricated by buried halite. Other major faults, including those of the previous plate boundary, are inactive but remain conspicuous as they are not long abandoned (e.g. Bucaramanga, Santa Marta, South Caribbean, Roques, Urica).

By virtue of the plate-boundary jump, a region named here the Northern Andes Block (NAB) was annexed and now moves essentially east with the Caribbean Plate, as shown by GPS (Pérez et al. 2001; Trenkamp et al. 2002). The NAB is bordered in the south by an uncertain plate-boundary sector (Molnar & Sykes 1969; Paris et al. 2000), probably the ENE-trending oblique-dextral Ibagué Fault (Fig. 1; Paris et al. 2000), interpreted here as a transform that may connect WSW to the Buenaventura Fault (Ingeominas 1988). A minimum dextral offset of 30 km on the Ibagué Fault (Montes et al. 2005) is consistent with 2 cm a^{-1} of Caribbean eastward relative motion since 2.5 Ma (see below). The NAB is the northern part of the 'North Andean Block' (Kellogg 1984) and the synonymous 'Cordilleran terrane' (Dewey & Pindell 1985). The NAB embraces the 'Maracaibo Block' (sensu Mann et al. 2006), transected by the Oca Fault, whose east–west extent is contentious. The only obvious

sector of the Oca exists in the west, sharply separating the Perijá–Santa Marta mountains from the Lower Guajira Basin (Rod 1956; Paris et al. 2000). This may be a Cretaceous intragraben fault, reactivated in a north-down sense to form the basin by differential halite solution during regional post-5 Ma uplift. To the east, no continuous linear projection of the Oca Fault crossing Falcón is evident on topographic or geological maps (e.g. Bellizzia 1976; dashed line in Pimentel 1984).

The Eastern Cordillera and Boconó faults are now dextral thrusts (NE trend; east–west motion). A kink at Mérida trends ENE and is thus not a releasing bend; supposed pull apart there (Schubert 1980) is more likely a halite-solution basin (Higgs 2009). The trend of the San Sebastián–El Pilar faults (c. 080°) causes near-transcurrent dextral transpression (Audemard et al. 2000), as shown by raised Quaternary beach- and shallow-marine deposits in central Venezuela, Araya, Margarita, Coche island and the Northern Range (Méndez 1997; Sisson et al. 2005; Weber 2005). However, transpression is widely masked by Neogene solution basins (Gulfs of Barcelona and Paria, Carúpano, North Coast). The Gulfs rupture the central and eastern Venezuela mountains, whose ongoing collapse is also indicated by other supramontane solution basins (e.g. Valencia, Caracas, Santa Lucía, Tuy, San Juan Graben; Higgs 2009). Superimposed on the Gulf of Paria solution basin is a young (2.5–0 Ma) pull apart in the rightward stepover between the El Pilar Fault and the Central Range/Warm Springs Fault (i.e. northern Gulf-western Caroni). An alternative, Late Miocene onset of pull apart (11 or 12 Ma; Algar & Pindell 1993; Pindell et al. 1998; Pindell & Kennan 2001a; Pindell et al. 2005), invoked moreover to explain the entire Gulf of Paria Basin, requires plate transcurrence from that time, contrary to the evidence listed below that transcurrence began near 2.5 Ma. Similar stepover between the San Sebastían and El Pilar Faults resulted in the Cariaco Trough pull-apart basin (Schubert 1982).

The following twelve observations and inferences from northern South America and elsewhere (Atlantic; northern Caribbean) collectively indicate that the change to c. 085° Caribbean relative motion occurred in Late Pliocene time (c. 2.5 Ma).

(1)   Uplift accelerated about 2.5 Ma ago in the Eastern Cordillera and Mérida Andes, due to focusing of the plate boundary (previously a 500 km-wide belt of regional distributed shortening) along this bivergent thrust belt. Intensified uplift of the Eastern Cordillera from Late Pliocene time is indicated by tilting of the Middle Magdalena Basin and

by palynological studies in the Bogotá Basin (Van der Hammen *et al.* 1973; Gómez *et al.* 2003; Torres *et al.* 2005). In the Mérida Andes, the start of faster uplift is dated by an influx of (?Plio-) Quaternary conglomerates on both flanks (Carvajal, Guanapa Fms; LEV 1997). A relatively recent start of rapid uplift is also consistent with (a) survival of erosional remnants, at high altitude near Mérida, of a palaeosoil formed at much lower elevations (Giegengack 1984), and (b) insufficient altitude for glaciation until Late Pleistocene time (Schubert & Vivas 1993).

(2) The Cariaco Trough was considered probably Quaternary by Schubert (1982). Plio-Quaternary deposits here are much thicker than underlying Upper Miocene deposits (Goddard 1988), consistent with post-2.5 Ma pull-apart superimposed on post-11 Ma halite-solution subsidence in this and the encompassing Barcelona Bay–Tortuga Platform area.

(3) The calculated east–west pull-apart extension in the Gulf of Paria is 50 km (Weber 2005), equating to the current relative plate velocity of 2 cm/year (Weber *et al.* 2001*a*) for 2.5 Ma.

(4) The kinked shelf edge east of Trinidad (e.g. Case & Holcombe 1980) can be restored to near alignment (NW–SE) by removing an assumed 50 km of dextral offset along trend with the Central Range Fault.

(5) The Roques Canyon and Urica faults come into alignment if 50 km of dextral slip on the El Pilar Fault is retracted (Fig. 1). This agrees well with the view that El Pilar displacement 'has been estimated at as much as 1000 km, although a new reconstruction of the South Caribbean boundary amounts to only 55 km of strike–slip' (Audemard *et al.* 2000, p. 62).

(6) Restored westward by the same 50 km, the Maracaibo Block is out of the way of Villa de Cura nappe southeastward emplacement (Figs 9–11; cf. Pimentel 1984). This restoration implies that *apparent* dextral offset along the bounding Boconó Fault Zone is only 35 km (sine 045° fault trend × 50 km), compared with previous estimates of 290 and 100 km of dextral slip (Dewey & Pindell 1985, 1986), supporting the objection of Salvador (1986, p. 699) that the Mérida Arch pre-Cretaceous basement high crosses the Andes nearly orthogonally 'with no major horizontal displacement'. The 35 km value is close to the 0–40 km estimates of most earlier authors (summary in Salvador

1986). Glacial moraines about 10 000 years old are offset 66 m dextrally by the main strand of the Boconó Fault Zone (Schubert & Sifontes 1970). Extrapolating this rate gives 17 km since 2.5 Ma, consistent with the calculated 35 km for the entire fault zone. Thus, the popular concept of northward 'escape' of the Maracaibo Block (Mann & Burke 1984), incorporated in Pangaea reconstructions (Pindell & Dewey 1982; Pindell 1985), is untenable. The block moved SE from 5 to 2.5 Ma, as described above, orthogonal to the Boconó Fault Zone. There is no evidence, and no evident reason, for significant northward movement during Mesozoic–Cenozoic time.

(7) In the Gulf of Paria, the uppermost seismically defined unit ('Sequence B' or 'Sequence 5'), dated as Mid or Late Pliocene through Quaternary, overlies an unconformity (Payne 1991; Babb & Mann 1999). The unconformity is interpreted here to reflect Caribbean-driven uplift superseded by subsidence (by halite solution and pull apart) when shortening was dramatically reduced at 2.5 Ma.

(8) In eastern central Trinidad, unconformably based Quaternary (and Late Pliocene?) deposition in Nariva Swamp (e.g. Kugler 1961) is attributed here to post-2.5 Ma subsidence by transpression on the adjacent Central Range Fault (NW dip; Kugler 1961), loading the footwall.

(9) In the South Caribbean accretionary prism, even the frontal thrusts do not appear to affect the Quaternary section (Ruiz *et al.* 2000; Flinch *et al.* 2003), consistent with accretion ending at 2.5 Ma.

(10) At the Caribbean–*North* America plate boundary, the Puerto Rico–Virgin Islands Arch is of post-Early Pliocene age (Mann *et al.* 2005). This regional-scale fold and uplift reflect the predicted 2.5 Ma change from nearly pure strike slip here (Fig. 2f) to the present-day transpression. Along-strike to the west, transpressive uplift of Hispaniola was already in progress before 2.5 Ma, due to the restraining bend there (e.g. Dolan *et al.* 1998).

(11) A reorganization of spreading direction and rate occurred along the entire Atlantic spreading ridge at 2.5 Ma (Klitgord & Schouten 1986). An anomalous, Pliocene surge in subsidence rates around the north Atlantic passive margins (Cloetingh *et al.* 1990) may be related to this change. The reorganization would necessarily have affected Caribbean relative motion, given

that the North and South America plates border the ridge and sandwich the Caribbean.
(12) Sumner & Westbrook (2001) interpreted other marine geological data as indicating a change in North America–South America motion, imprecisely dated (5–3 Ma).

The last two of these factors suggest that the inferred change in Caribbean–South America motion at 2.5 Ma has a larger-scale, possibly global cause (Cloetingh *et al.* 1990). The change to eastward Caribbean relative motion terminated the Caribbean foreland basin loading force in eastern Venezuela and Trinidad, apart from Nariva Swamp. Nevertheless, over the former basin area, subsidence continues in the Deltana–Columbus Channel–southern Columbus basins, presumably by compaction, combined with eastward gravitational extension toward the Atlantic Ocean in Columbus Basin (Bevan 2007). In contrast, eastern Venezuela is rebounding west of Maturín city (Fig. 1), eroding into what remains of the Caribbean foreland basin fill (westward-aging Pleistocene to Miocene outcrop; Pimentel 1984).

Caribbean Plate absolute velocity: control on arc type

The Caribbean Plate moves east relative to South America at about the same rate (*c.* 2.0 cm a^{-1} Weber *et al.* 2001*a*) that South America drifts west relative to the mantle; therefore the Caribbean Plate is nearly stationary in the mantle reference frame (Pindell *et al.* 2006). These conditions can be assumed to have applied since the 2.5 Ma plate-motion change.

Barbados lies on the Caribbean leading edge, 1150 km east of the longitude of Guajira corner, which it passed *c.* 35 Ma ago, assuming that the Caribbean Arc has maintained an approximately north–south orientation. Thus the average eastward component of Caribbean velocity relative to South America since 35 Ma is 3 cm a^{-1}, after subtracting the 150 km width of Bonaire–Grenada interarc basin, assuming this is entirely post-35 Ma and that Grenada spreading was arc-normal (Bird *et al.* 1999) rather than arc-parallel (Pindell & Barrett 1990). Simultaneously, the Americas drifted west relative to the mantle at 2–3 cm a^{-1} throughout Cenozoic time (Pindell *et al.* 2006). Thus, Caribbean average eastward absolute velocity, if any, was less than typical trench rollback rates (1–2 cm a^{-1}; Conrad & Lithgow-Bertelloni 2006), so the Caribbean Arc would have been 'extensional' or 'neutral' in the Dewey (1980) classification. The Bonaire and Grenada basins confirm extensional behaviour.

From the 72 Ma plate reorganization until 35 Ma, the Caribbean leading edge migrated from Ecuador to Guajira corner, amounting to *c.* 1100 km of eastward travel component, again giving an average eastward relative velocity of 3 cm a^{-1}. At no stage, then, could a 'compressional' arc with Andean relief form.

Pangaean reconstruction

The classic reconstruction of western Pangaea (Pindell & Dewey 1982; Pindell 1985) needs modifying to accommodate: (a) north–south shortening in Venezuela and Trinidad by Proto-Caribbean nappe advance; (b) relative motion of the Maracaibo Block only 50 km eastward, rather than 100–300 km NE; and (c) placement of Cuba between Yucatán and South America.

Concluding remarks and exploration implications

The 'Caribbean model' substantially misinterprets the post-Palaeozoic geology of northern South America, misjudging (1) the duration of Pangaea rifting and Proto-Caribbean seafloor spreading, (2) the timing and extent of pre-Caribbean orogeny by Proto-Caribbean subduction and (3) the timing of Caribbean Arc transit. Crucially for exploration, these criteria seriously affect predictions and models of burial and thermal history (organic maturation; sandstone diagenesis), and of palaeogeography in each tectonic stage (reservoir fairways). Highlighting the severity of the problem, the old model misinterprets the classic La Luna source rock as passive-margin rather than rift deposits, and the prolific conventional-oil reservoirs of western and eastern Venezuela (Misoa, Merecure; Parnaud *et al.* 1995*a*, *b*) as Caribbean foreland basin deposits rather than Proto-Caribbean, with far-reaching consequences in exploring for look-alikes. Appreciation that there are multiple generations of foreland basin in Ecuador, Colombia, Venezuela and Trinidad will help explorers develop more sophisticated and reliable models of oil and gas generation, migration and entrapment.

In Venezuela and Trinidad, most petroleum explorationists know (and care) little about the coastal metamorphic basement and its lightly metamorphosed Jurassic–Cretaceous cover, partly because of the general (erroneous) belief that these rocks are highly allochthonous and thus have little bearing on hydrocarbon exploration in the adjacent basins. These northern regions now require systematic re-examination from a petroleum standpoint, with the resulting data being integrated into future exploration models.

I gratefully acknowledge careful reviews by S. Burley, K. James, M. Keeley, L. Kennan and H. Krause. Many colleagues of the former Maraven (Petróleos de Venezuela) generously shared their knowledge during my initial years in Venezuelan geology (1990–1993). I thank M. Keeley and T. England for introducing me to Colombia and Trinidad respectively. J. Pindell generously included me in many subsequent projects in Colombia, Venezuela and Trinidad; our countless heated, usually friendly, geological discussions benefited me immeasurably. F. Hosein, K. James, H. Krause and M. A. Lorente kindly invited me to present these ideas at conferences in Spain, Trinidad and Venezuela in 2006–2007. This and the accompanying paper owe much to the support of Ashley Ellison.

References

AGUADO, B. 1993. Variations in sedimentology and mineralogy of condense sequences: an example from Venezuela. *American Association of Petroleum Geologists, Annual Convention, Program and Abstracts*, 66.

ALBERTOS, M., YORIS, F. & URBANI, F. 1989. Estudio geológico y análisis petrográfico-estadístico de la Formación Guárico y sus equivalentes en las secciones Altagracia de Orituco–Agua Blanca–Gamelotal–San Francisco de Macaira, Estados Guárico y Miranda. *VII Congreso Geológico Venezolano, Memorias*, 290–314.

ALFONSO, C. A., SACKS, P. E., SECOR, D. T., RINE, J. & PÉREZ, V. 1994. A Tertiary fold and thrust belt in the Valle del Cauca Basin, Colombian Andes. *Journal of South American Earth Sciences*, 7, 387–402.

ALGAR, S. 1998. Tectonostratigraphic development of the Trinidad region. *In*: PINDELL, J. L. & DRAKE, C. (eds) *Paleogeographic Evolution and Non-glacial Eustasy, Northern South America*. SEPM, Special Publications, 58, 87–109.

ALGAR, S. T. & PINDELL, J. L. 1993. Structure and deformation history of the Northern Range of Trinidad and adjacent areas. *Tectonics*, 12, 814–829.

ALLEN, P. A. & ALLEN, J. R. 1990. *Basin Analysis: Principles and Applications*. Blackwell, Oxford.

ANH 2005a. *Cauca–Patia Basin*. Agencia Nacional de Hidrocarburos, Bogotá, Colombia, promotional brochure.

ANH 2005b. *Pacifico Basin*. Agencia Nacional de Hidrocarburos, Bogotá, Colombia, promotional brochure.

ANH 2005c. *Cesar-Rancheria Basin*. Agencia Nacional de Hidrocarburos, Bogotá, Colombia, promotional brochure.

ASPDEN, J. A. & MCCOURT, W. J. 1986. Mesozoic oceanic terrane in the Central Andes of Colombia. *Geology*, 14, 415–418.

ASPDEN, J. A., MCCOURT, W. J. & BROOK, M. 1987. Geometrical control of subduction-related magmatism: the Mesozoic and Cenozoic plutonic history of Western Colombia. *Journal of the Geological Society, London*, 144, 893–905.

AUDEMARD, F. E. 1991. *Tectonics of western Venezuela*. Doctoral thesis, Rice University.

AUDEMARD, F. A. 2003. Geomorphic and geologic evidence of ongoing uplift and deformation in the Mérida Andes, Venezuela. *Quaternary International*, 101–102, 43–65.

AUDEMARD, F. E. & SERRANO, I. C. 2001. Future petroliferous provinces of Venezuela. *In*: DOWNEY, M. W., THREET, J. C. & MORGAN, W. A. (eds) *Petroleum Provinces of the Twenty-First Century*. American Association of Petroleum Geologists, Memoirs, 74, 353–372.

AUDEMARD, F. A., MACHETTE, M. N., COX, J. W., DART, R. L. & HALLER, K. M. 2000. *Map and Database of Quaternary Faults in Venezuela and its Offshore Regions*. US Geological Survey Open-File Reports, 00-018.

AVÉ LALLEMENT, H. G. 1997. Transpression, displacement partitioning, and exhumation in the eastern Caribbean/South America plate boundary zone. *Tectonics*, 16, 272–289.

AVÉ LALLEMENT, H. G. & SISSON, B. 2005. Exhumation of eclogites and blueschists in northern Venezuela: constraints from kinematic analysis of deformation structures. *In*: AVÉ LALLEMANT, H. G. & SISSON, V. B. (eds) *Caribbean–South American Plate Interactions, Venezuela*. Geological Society of America, Special Papers, 394, 193–206.

AVÉ LALLEMENT, H. G., SISSON, V. B & WRIGHT, J. E. 1993. Structure of the Cordillera de la Costa belt, north-central Venezuela: implications for plate tectonic models. American Association of Petroleum Geologists–Sociedad Venezolana de Geólogos Conference, Caracas. *American Association of Petroleum Geologists Bulletin*, 77, 304 (abstract).

BABB, S. & MANN, P. 1999. Structural and sedimentary development of a Neogene transpressional plate boundary between the Caribbean and South America plates in Trinidad and the Gulf of Paria. *In*: MANN, P. (ed.) *Caribbean Basins*. Elsevier, Amsterdam, 495–557.

BALKWILL, H. R., RODRIGUE, G., PAREDES, F. I. & ALMEIDA, J. P. 1995. Northern part of Oriente Basin, Ecuador: reflection seismic expression of structures. *In*: TANKARD, A. J., SUÁREZ, R. & WELSINK, H. J. (eds) *Petroleum Basins of South America*. American Association of Petroleum Geologists, Memoirs, 62, 559–571.

BARR, K. W. 1952. Limestone blocks in the Lower Cretaceous Cuche Formation of the Central Range, Trinidad, B. W. I. *Geological Magazine*, 89, 417–425.

BARR, K. W. 1962. The geology of the Toco district, Trinidad, West Indies. Part 1. *The Quarterly Bulletin of the Overseas Geological Surveys*, 8, 379–415. [Reprinted 1963 as: *The Geology of the Toco District, Trinidad, West Indies*. HMSO, London.]

BARR, K. W. & SAUNDERS, J. B. 1968. An outline of the geology of Trinidad. *Fourth Caribbean Geological Conference, Trinidad, 1965, Transactions*, 1–10.

BARRAGAN, R., BABY, P. & DUNCAN, R. 2005. Cretaceous alkaline intra-plate magmatism in the Ecuadorian Oriente Basin: geochemical, geochronological and tectonic evidence. *Earth and Planetary Science Letters*, 236, 670–690.

BARRERO, D., PARDO, A., VARGAS, C. A. & MARTINEZ, J. F. 2007. *Colombian Sedimentary Basins: Nomenclature, Boundaries and Petroleum Geology, a New Proposal*. Agencia Nacional de Hidrocarburos, Bogotá.

BARTOK, P. 1993. Prebreakup geology of the Gulf of Mexico–Caribbean: its relation to Triassic and Jurassic rift systems of the region. *Tectonics*, **12**, 441–459.

BARTOK, P. 2003. The peripheral bulge of the Interior Range of the Eastern Venezuela Basin and its impact on oil accumulations. *In*: BARTOLINI, C., BUFFLER, R. T. & BLICKWEDE, J. F. (eds) *The Circum-Gulf of Mexico and the Caribbean: Hydrocarbon Habitats, Basin Formation, and Plate Tectonics*. American Association of Petroleum Geologists, Memoirs, **79**, 925–936.

BARTOK, P. E., RENZ, O. & WESTERMANN, G. E. G. 1985. The Siquisique ophiolites, northern Lara State, Venezuela: a discussion on their Middle Jurassic ammonites and tectonic implications. *Geological Society of America Bulletin*, **96**, 1050–1055.

BECK, C., OGAWA, Y. & DOLAN, J. 1990. Eocene paleogeography of the southeastern Caribbean: relations between sedimentation on the Atlantic abyssal plain at Site 672 and evolution of the South America basin. *In*: MOORE, J. C., MASCLE, A. ET AL. *Proceedings of the Ocean Drilling Program, Scientific Results*, **110**, 7–15.

BEJARANO, A., REYES, R. & VILLEGAS, E. 1991. Caracterización y evaluación de parámetros de registros de pozos en la Cuenca del Putumayo. *IV Simposio Bolivariano de la Exploración Petrolera en las Cuencas Subandinas, Memorias*, **1**, Trabajo 11.

BELL, J. S. 1967. *Geology of the Camatagua area, Estado Aragua, Venezuela*. Doctoral thesis, Princeton University.

BELL, J. S. 1971. The tectonic evolution of the central part of the Venezuelan coast ranges. *In*: DONNELLY, T. W. (ed.) *Caribbean Geologic, Tectonic, and Petrologic Studies*. Geological Society of America, Boulder, CO, Memoirs, **130**, 117–180.

BELL, J. S. 1972. Geotectonic evolution of the southern Caribbean area. *In*: SHAGAM, R. ET AL. (eds) *Studies in Earth and Space Sciences*. Geological Society of America, Boulder, CO, Memoirs, **132**, 369–386.

BELLIZZIA, A. 1976. *Mapa Geológico y Estructural de Venezuela*. Venezuela Ministerio de Energía y Minas, 1 : 500 000.

BELLIZZIA, A. 1986. Sistema Montañoso del Caribe – una cordillera alóctona en la parte norte de America del Sur. *Sociedad Venezolana de Geólogos, VI Congreso Geológico Venezolano, Memorias*, **10**, 6657–6836.

BELLIZZIA, A. & DENGO, G. 1990. The Caribbean mountain system, northern South America; a summary. *In*: DENGO, G. & CASE, J. E. (eds) *The Caribbean Region*. Geology of North America, **H**. Geological Society of America, Boulder, CO, 167–175.

BELOTTI, H., CONFORTO, G., SILVESTRO, J., RODRIGUEZ, J. & KRAEMER, P. 2003. Sistema petrolero Terciario Pagüey–Pagüey Inferior (!) en la Subcuenca de Guarumen, Venezuela. *VIII Simposio Bolivariano de la Exploración Petrolera en las Cuencas Subandinas, Memorias*, **1**, 214–226.

BENJAMINI, C., SHAGAM, R. & MENENDEZ, A. 1987. (Late?) Paleozoic age for the 'Cretaceous' Tucutenemo Formation, northern Venezuela: stratigraphic and tectonic implications. *Geology*, **15**, 922–926.

BEVAN, T. 2007. Structural cross sections of the Columbus Basin, Trinidad & Tobago: an integrated geological model based on a comprehensive 3D seismic database (extended abstract). *Fourth Geological Conference of the Geological Society of Trinidad and Tobago, abstracts CD*.

BIRD, D. E., HALL, S. A., CASEY, J. F. & MILLEGAN, P. S. 1999. Tectonic evolution of the Grenada Basin. *In*: MANN, P. (ed.) *Caribbean Basins*. Elsevier, Amsterdam, 389–416.

BLIN, B., SICHLER, B. & STEPHAN, J. F. 1990. Contribution of paleomagnetism in the relations between the Piemontin and internal nappes of the Venezuelan Caribbean chain: synthesis and new data. *In*: LARUE, D. K. & DRAPER, G. (eds) *12th Caribbean Geological Conference, St Croix, Transactions*, 453–460.

BOESI, T. & GODDARD, D. 1991. A new geologic model related to the distribution of hydrocarbon source rocks in the Falcón Basin, northwestern Venezuela. *In*: BIDDLE, K. T. (ed.) *Active Margin Basins*. American Association of Petroleum Geologists, Memoirs, **52**, 303–319.

BOETTCHER, S. S., JACKSON, J. L., QUINN, M. J. & NEAL, J. E. 2003. Lithospheric structure and supracrustal hydrocarbon systems, offshore eastern Trinidad. *In*: BARTOLINI, C., BUFFLER, R. T. & BLICKWEDE, J. F. (eds) *The Circum-Gulf of Mexico and the Caribbean: Hydrocarbon Habitats, Basin Formation, and Plate Tectonics*. American Association of Petroleum Geologists, Memoirs, **79**, 529–544.

BRANQUET, Y., CHEILLETZ, A., COBBOLD, P. R., BABY, P., LAUMONIER, B. & GIULIANI, G. 2002. Andean deformation and rift inversion, eastern edge of Cordillera Oriental (Guateque–Medina area), Colombia. *Journal of South American Earth Sciences*, **15**, 391–407.

BRICEÑO, H. O. & SCHUBERT, C. 1990. Geomorphology of the Gran Sabana, Guayana Shield, southeastern Venezuela. *Geomorphology*, **3**, 125–141.

BRONDIJK, J. F. 1967. 'Eocene' formations in the southwestern part of the Maracaibo Basin. *Asociación Venezolana de Geología, Minería y Petróleo, Boletín Informativo*, **10**, 35–50.

BURKE, K. 1988. Tectonic evolution of the Caribbean. *Annual Review of Earth and Planetary Sciences*, **16**, 210–230.

CASE, J. E. & HOLCOMBE, T. L. 1980. *Geologic–Tectonic Map of the Caribbean Region*. US Geological Survey Miscellaneous Investigations Series, Map I-1100, 1 : 2 500 000.

CASTRO, M. & MEDEROS, A. 1985. Litoestratigrafía de la Cuenca de Carúpano. *Sociedad Venezolana de Geólogos, VI Congreso Geológico Venezolano, Memorias*, **1**, 201–225.

CHEVALIER, Y., STEPHAN, J.-F. ET AL. 1988. Obduction et collision pré-Tertiaire dans les zones internes de la Chaîne Caraïbe vénézuélienne, sur le transect Ile de Margarita–Péninsule d'Araya. *Comptes Rendus de la Académie des Sciences, Paris*, **307**(II), 1925–1932.

CLOETINGH, S., GRADSTEIN, F. M., KOOI, H., GRANT, A. C. & KAMINSKI, M. 1990. Plate reorganization: a cause of rapid Late Neogene subsidence and sedimentation around the North Atlantic? *Journal of the Geological Society, London*, **147**, 495–506.

COATES, A. G., COLLINS, L. S., AUBRY, M.-P. & BERGGREN, W. A. 2004. The geology of the Darien,

Panama, and the Late Miocene–Pliocene collision of the Panama arc with northwestern South America. *Geological Society of America Bulletin*, **116**, 1327–1344.

COLMENARES, O., CHATELLIER, J.-Y., MARQUEZ, P. & LAING, J. F. 1997. Synsedimentary tectonic events revealed by palynology and core studies, the vital proof for a previously tentative model, El Furrial Field, eastern Venezuela. *In*: CHATELLIER, J.-Y. & AQUINO, R. (eds) *First Core Workshop, Hydrocarbon Basins of Venezuela*. VIII Congreso Geológico Venezolano, I Congreso Latinoamericano de Sedimentología, trabajo VIII.

CONRAD, C. P. & LITHGOW-BERTELLONI, C. 2006. How mantle slabs drive plate tectonics. *Science*, **298**, 207–209.

COONEY, P. M. & LORENTE, M. A. 1997. Implicaciones tectónicas de un evento estructural en el Cretácico Superior (Santoniense–Campaniense) de Venezuela occidental. *Sociedad Venezolana de Geólogos, VIII Congreso Geológico Venezolano, Memorias*, **1**, 195–204.

COOPER, M. A., ADDISON, F. T. *ET AL.* 1995. Basin development and tectonic history of the Llanos Basin, Eastern Cordillera, and Middle Magdalena Valley, Colombia. *American Association of Petroleum Geologists Bulletin*, **79**, 1421–1443.

CORREDOR, F. 2003. Eastward extent of the Late Eocene–Early Oligocene onset of deformation across the northern Andes: constraints from the northern portion of the Eastern Cordillera fold belt, Colombia, and implications for regional oil exploration. *VIII Simposio Bolivariano de la Exploración Petrolera en las Cuencas Subandinas, Memorias*, **1**, 34–45.

CREOLE 1955. *Geología del superficie*. Creole Petroleum Corporation, unpublished maps D-10a-d, D-11a-d, 1 : 50 000.

CRUZ, L., TEYSSIER, C. & FAYON, A. 2003. Exhumation and deformation in the Venezuelan Paria Peninsula, SE Caribbean–South American plate boundary. *Geological Society of America, Abstracts with Programs*, **35**, A548.

DASHWOOD, M. F. & ABBOTTS, I. L. 1990. Aspects of the petroleum geology of the Oriente Basin, Ecuador. *In*: BROOKS, J. (ed.) *Classic Petroleum Provinces*. Geological Society, London, Special Publications, **50**, 89–117.

DECELLES, P. G. & GILES, K. A. 1996. Foreland basins. *Basin Research*, **8**, 105–123.

DE ROEVER, E. M. F., LATTARD, D. & SCHREYER, W. 1981. Surinamite: a beryllium-bearing mineral. *Contributions to Mineralogy and Petrology*, **76**, 472–473.

DEWEY, J. W. 1972. Seismicity and tectonics of western Venezuela. *Bulletin of the Seismological Society of America*, **62**, 1711–1751.

DEWEY, J. F. 1980. Episodicity, sequence, and style at convergent plate boundaries. *In*: STRANGWAY, D. W. (ed.) *The Continental Crust and its Mineral Deposits*. Geological Association of Canada, Special Papers, **20**, 553–573.

DEWEY, J. F. & PINDELL, J. L. 1985. Neogene block tectonics of eastern Turkey and northern South America: continental applications of the finite difference method. *Tectonics*, **4**, 71–83.

DEWEY, J. F. & PINDELL, J. L. 1986. Neogene block tectonics of Turkey and northern South America: continental applications of the finite difference method: Reply. *Tectonics*, **5**, 703–705.

DIAZ DE GAMERO, M. L. 1977. Estratigrafía y micropaleontología del Oligoceno y Mioceno inferior del centro de la cuenca de Falcón. *GEOS, Universidad Central de Venezuela*, **22**, 2–50.

DI CROCE, J., BALLY, A. W. & VAIL, P. 1999. Sequence stratigraphy in the Eastern Venezuelan Basin. *In*: MANN, P. (ed.) *Caribbean Basins*. Elsevier, Amsterdam, 419–476.

DICKEY, P. A. 1980. Barbados as a fragment of South America ripped off by continental drift. *9th Caribbean Geological Conference, Transactions*, 51–52.

DIETRICH, J. R., HIGGS, R., ROHR, K. M. & WHITE, J. M. 1993. The Tertiary Queen Charlotte Basin: a strike–slip basin on the western Canadian continental margin. *In*: FROSTICK, L. E. & STEEL, R. J. (eds) *Tectonic Controls and Signatures in Sedimentary Successions*. International Association of Sedimentologists, Special Publications, **20**, 161–169.

DOLAN, J. F., MULLINS, H. T. & WALD, D. J. 1998. Active tectonics of the north-central Caribbean: oblique collision, strain partitioning, and opposing subducted slabs. *In*: DOLAN, J. F. & MANN, P. (eds) *Active Strike–Slip and Collisional Tectonics of the Northern Caribbean Plate Boundary Zone*. Geological Society of America, Boulder, Co, Special Papers, **326**, 1–61.

DUARTE, L. M. 2003. Historia de la Cuenca Cretacea del VSM en un marco cronoestratigráfico. Implicaciones ambientales. *VIII Simposio Bolivariano de la Exploración Petrolera en las Cuencas Subandinas, Memorias*, **2**, 223–232.

DUERTO, L. & MCCLAY, K. 2002. 3D geometry and evolution of shale diapirs in the Eastern Venezuelan Basin. *American Association of Petroleum Geologists, Search and Discovery Articles*, **10026**, http://www.search anddiscovery.net/documents/duerto/index.htm

DUERTO, L., ESCALONA, A. & MANN, P. 2006. Deep structure of the Mérida Andes and Sierra de Perijá mountain fronts, Maracaibo Basin, Venezuela. *American Association of Petroleum Geologists Bulletin*, **90**, 505–528.

DUNCAN, R. A. & HARGRAVES, R. B. 1984. Plate tectonic evolution of the Caribbean in the mantle reference frame. *In*: BONINI, W. E., HARGRAVES, R. B. & SHAGAM, R. (eds) *The Caribbean–South American Plate Boundary and Regional Tectonics*. Geological Society of America, Boulder, CO, Memoirs, **162**, 81–94.

ERIKSON, J. P. & PINDELL, J. L. 1993. Analysis of subsidence in northeastern Venezuela as a discriminator of tectonic models for northern South America. *Geology*, **21**, 945–948.

ERLICH, R. N., FARFAN, P. F. & HALLOCK, P. 1993. Biostratigraphy, depositional environments, and diagenesis of the Tamana Formation, Trinidad: a tectonic marker horizon. *Sedimentology*, **40**, 743–768.

ERLICH, R. N., MACSOTAY, O., NEDERBRAGT, A. J. & LORENTE, M. A. 1999. Palaeoecology, palaeogeography and depositional environments of Upper Cretaceous rocks of western Venezuela. *Palaeogeography, Palaeoclimatology, Palaeoecology*, **153**, 203–238.

ERLICH, R. N., VILLAMIL, T. & KEENS-DUMAS, J. 2003. Controls on the deposition of Upper Cretaceous organic carbon-rich rocks from Costa Rica to Suriname. *In*: BARTOLINI, C., BUFFLER, R. T. & BLICKWEDE, J. F. (eds) *The Circum-Gulf of Mexico and the Caribbean: Hydrocarbon Habitats, Basin Formation, and Plate Tectonics*. American Association of Petroleum Geologists, Memoirs, **79**, 1–45.

ESCALONA, N. 1985. Relaciones estratigráficas con el método de minerales pesados, area Machete, Faja Petrolífera del Orinoco. *Sociedad Venezolana de Geólogos, VI Congreso Geológico Venezolano, Memorias*, **1**, 536–587.

ESCALONA, A. & MANN, P. 2006. Tectonic controls of the right-lateral Burro Negro tear fault on Paleogene structure and stratigraphy, northeastern Maracaibo Basin. *American Association of Petroleum Geologists Bulletin*, **90**, 479–504.

FABRE, A. 1987. Tectonique et génération d'hydrocarbures: un modèle de l'évolution de la Cordillère Orientale de Colombie et du Bassin des Llanos pendant le Crétacé et le Tertiaire. *Archives des Sciences*, **40**, 145–190.

FALVEY, D. A. 1974. The development of continental margins in plate tectonic theory. *Australian Petroleum Exploration Association Journal*, **14**, 95–106.

FEO-CODECIDO, G. 1956. Heavy-mineral techniques and their application to Venezuelan stratigraphy. *American Association of Petroleum Geologists Bulletin*, **40**, 984–1000.

FEO-CODECIDO, G., SMITH, F. D., JR., ABOUD, N. & DE DI GIACOMO, E. 1984. Basement and Paleozoic rocks of the Venezuelan Llanos basins. *In*: BONINI, W. E., HARGRAVES, R. B. & SHAGAM, R. (eds) *The Caribbean–South American Plate Boundary and Regional Tectonics*. Geological Society of America, Boulder, CO, Memoirs, **162**, 175–187.

FLINCH, J. F. 2003. Structural evolution of the Sinu–Lower Magdalena area (northern Colombia). *In*: BARTOLINI, C., BUFFLER, R. T. & BLICKWEDE, J. F. (eds) *The Circum-Gulf of Mexico and the Caribbean: Hydrocarbon Habitats, Basin Formation, and Plate Tectonics*. American Association of Petroleum Geologists, Memoirs, **79**, 776–796.

FLINCH, J. F., RAMBARAN, V. *ET AL*. 1999. Structure of the Gulf of Paria pull-apart basin (Eastern Venezuela-Trinidad). *In*: MANN, P. (ed.) *Caribbean Basins*. Elsevier, Amsterdam, 477–494.

FLINCH, J. F., AMARAL, J., DOULCET, A., MOULY, B., OSORIO, C. & PINCE, J. M. 2003. Onshore–offshore structure of the northern Colombia accretionary complex. *American Association of Petroleum Geologists International Conference, Barcelona*, extended abstract.

FOLAND, K. A., SPEED, R. & WEBER, J. 1992. Geochronologic studies of the Caribbean mountains orogen of Venezuela and Trinidad. *Geological Society of America, Abstracts with Programs*, **24**, A148.

FREY, M., SAUNDERS, J. & SCHWANDER, H. 1988. The mineralogy and metamorphic geology of low-grade metasediments, Northern Range, Trinidad. *Journal of the Geological Society, London*, **145**, 563–575.

GALEA, F. A. 1985. *Biostratigraphy and Depositional Environment of the Upper Cretaceous–Eocene Santa Anita Group (North Eastern Venezuela)*. Free University Press, Amsterdam.

GHOSH, S. K. 1984. Late Cretaceous condensed sequence, Venezuelan Andes. *In*: BONINI, W. E., HARGRAVES, R. B. & SHAGAM, R. (eds) *The Caribbean–South American Plate Boundary and Regional Tectonics*. Geological Society of America, Boulder, CO, Memoirs, **162**, 317–324.

GHOSH, S. K. & ODREMAN, O. 1987. Estudio sedimentológico-paleoambiental del Terciario en la zona del Valle de San Javier, Estado Mérida. *Sociedad Venezolana de Geólogos, Boletínes*, **31**, 36–46.

GIEGENGACK, R. 1984. Late Cenozoic tectonic environments of the central Venezuelan Andes. *In*: BONINI, W. E., HARGRAVES, R. B. & SHAGAM, R. (eds) *The Caribbean–South American Plate Boundary and Regional Tectonics*. Geological Society of America, Boulder, CO, Memoirs, **162**, 343–364.

GODDARD, D. 1988. Seismic stratigraphy and sedimentation of the Cariaco Basin and surrounding continental shelf, northeastern Venezuela. *11th Caribbean Geological Conference, Barbados, Transactions*, **34**, 1–21.

GOMEZ, E., JORDAN, T. E. & HEGARTY, K. 2002. New exploration and production opportunities in Colombia: lessons from basin-analyses studies and a look forward. *American Association of Petroleum Geologists Annual Meeting, Program and Abstracts*.

GOMEZ, E., JORDAN, T. E., ALLMENDINGER, R. W., HEGARTY, K., KELLEY, S. & HEIZLER, M. 2003. Controls on architecture of the Late Cretaceous to Cenozoic southern Middle Magdalena Valley Basin, Colombia. *Geological Society of America Bulletin*, **115**, 131–147.

GONZALEZ, H. 2001. *Mapa Geológico del Departamento de Antioquia, Memoria Explicativa*. Ministerio de Minas y Energía, Colombia.

GONZÁLEZ DE JUANA, C., MUÑOZ, N. G. & VIGNALI, M. 1972. Reconocimiento geológico de la península de Paria, Venezuela. *IV Congreso Geológico Venezolano, Memorias, Boletín de Geología, Publicaciónes Especiales*, **5**, 1549–1588.

GONZÁLEZ DE JUANA, C., ITURRALDE, J. M. & PICARD, X. 1980. *Geología de Venezuela y de sus Cuencas Petroliferas*. Ediciones Foninves, Caracas.

GORNEY, D., ESCALONA, A., MANN, P., MAGNANI, M. B. & BOLIVAR STUDY GROUP 2007. Chronology of Cenozoic tectonic events in western Venezuela and the Leeward Antilles based on integration of offshore seismic reflection data and on-land geology. *American Association of Petroleum Geologists Bulletin*, **91**, 653–684.

GREEN, D. H., LOCKWOOD, J. P. & KISS, E. 1968. Eclogite and almandine–jadeite–quartz rock from the Guajira Peninsula, Colombia, South America. *The American Mineralogist*, **53**, 1320–1335.

GUERRERO, J., SARMIENTO, G. & NAVARRETE, R. 2000. The stratigraphy of the west side of the Cretaceous Colombian basin in the Upper Magdalena Valley. *Geología Colombiana*, **25**, 45–110.

HAQ, B. U., HARDENBOL, J. & VAIL, P. R. 1988. Mesozoic and Cenozoic chronostratigraphy and cycles of sea-level change. *In*: WILGUS, C. K., HASTINGS, B. S., KENDALL, C. G.ST.C., POSAMENTIER, H. W.,

ROSS, C. A. & VAN WAGONER, J. C. (eds) *Sea-Level Changes: an Integrated Approach*. SEPM, Special Publications, **42**, 71–108.

HEDBERG, H. D. 1937. Stratigraphy of the Rio Querecual section of northeastern Venezuela. *Geological Society of America Bulletin*, **48**, 1971–2024.

HEDBERG, H. D. 1950. Geology of the Eastern Venezuela Basin (Anzoátegui–Monagas–Sucre–eastern Guárico portion). *American Association of Petroleum Geologists Bulletin*, **61**, 1173–1216.

HIGGS, R. 1996. A new facies model for the Misoa Formation (Eocene), Venezuela's main oil reservoir. *Journal of Petroleum Geology*, **19**, 249–269.

HIGGS, R. 2000. The Chaudiere and Nariva 'wildflysch' of central Trinidad: a modern sedimentological perspective. *Geological Society of Trinidad and Tobago/ Society of Petroleum Engineers, Conference, Programme and Abstracts*, 17.

HIGGS, R. 2002. Tide-dominated estuarine facies in the Hollin and Napo ("T" and "U") formations (Cretaceous), Sacha field, Oriente Basin, Ecuador: Discussion. *American Association of Petroleum Geologists Bulletin*, **86**, 329–334.

HIGGS, R. 2006. Colombia-Venezuela-Trinidad "Caribbean Oblique Collision Model" revised. *Sociedad Venezolana de Geofísicos, XIII Congreso Venezolano de Geofísica, Memorias*. CD and World Wide Web Address: http://www.congresogeofisica-sovg.org/dyncat.cfm?catid=4378, (accessed February 2008).

HIGGS, R. 2009. The vanishing Carib Halite Formation (Neocomian), Colombia–Venezuela–Trinidad prolific petroleum province. *In*: JAMES, K. H., LORENTE, M. A. & PINDELL, J. L. (eds) *The Origin and Evolution of the Caribbean Plate*. Geological Society, London, Special Publications, **328**, 659–686.

HIGGS, R., PINDELL, J. & ODREMAN, O. 1995. Mesozoic–Cenozoic tectonics and sedimentation in the Venezuelan Andes region, and implications for petroleum exploration. *IX Congreso Latinoamericano de Geología, Caracas, Venezuela, Field Trip Guidebook*, **6**.

HILL, R. J. & SCHENK, C. J. 2005. Petroleum geochemistry of oil and gas from Barbados: implications for distribution of Cretaceous source rocks and regional petroleum prospectivity. *Marine and Petroleum Geology*, **22**, 917–943.

HUNG, E. 2005. Thrust belt interpretation of the Serranía del Interior and Maturín subbasin, eastern Venezuela. *In*: AVÉ LALLEMENT, H. G. & SISSON, V. B. (eds) *Caribbean–South American Plate Interactions, Venezuela*. Geological Society of America, Boulder, CO, Special Papers, **394**, 251–270.

HUTCHISON, A. G. 1938. Una nota sobre el Cretáceo de Trinidad. *Boletín de Geología y Minería, Caracas, Venezuela*, **2**, 226–238.

INGEOMINAS 1988. *Mapa Geológico de Colombia*. Instituto Nacional de Investigaciones Geológico-Mineras, 1:1 500 000.

JACKSON, J. 2002. Strength of the continental lithosphere: time to abandon the jelly sandwich? *GSA Today*, **12**, 4–10.

JACKSON, T. A., SMITH, T. E. & DUKE, M. J. M. 1991. The geochemistry of a metavolcanic horizon in the

Maracas Formation, Northern Range, Trinidad: evidence of ocean floor basalt activity. *Second Geological Conference of the Geological Society of Trinidad and Tobago, 1990, Transactions*, 42–47.

JACOME, M. I., KUSZNIR, N., AUDEMARD, F. & FLINT, S. 2003. Tectono-stratigraphic evolution of the Maturin Foreland Basin, Eastern Venezuela. *In*: BARTOLINI, C., BUFFLER, R. T. & BLICKWEDE, J. F. (eds) *The Circum-Gulf of Mexico and the Caribbean: Hydrocarbon Habitats, Basin Formation, and Plate Tectonics*. American Association of Petroleum Geologists, Memoirs, **79**, 735–749.

JAIMES, E. & DE FREITAS, M. 2006. An Albian–Cenomanian unconformity in the northern Andes: evidence and tectonic significance. *Journal of South American Earth Sciences*, **21**, 466–492.

JAMES, K. H. 1997. Distribution and tectonic significance of Cretaceous–Eocene flysch–wildflysch deposits of Venezuela and Trinidad. *Sociedad Venezolana de Geólogos, VIII Congreso Geológico Venezolano, Margarita, Memorias*, **1**, 415–421.

JAMES, K. H. 2000. The Venezuelan hydrocarbon habitat, part 1: tectonics, structure, palaeogeography and source rocks. *Journal of Petroleum Geology*, **23**, 5–53.

JAMES, K. H. 2006. Arguments for and against the Pacific origin of the Caribbean Plate: discussion, finding for an inter-American origin. *Geologica Acta*, **4**, 279–302.

KELLOGG, J. N. 1984. Cenozoic tectonic history of the Sierra de Perijá, Venezuela-Colombia, and adjacent basins. *In*: BONINI, W. E., HARGRAVES, R. B. & SHAGAM, R. (eds) *The Caribbean–South American Plate Boundary and Regional Tectonics*. Geological Society of America, Boulder, CO, Memoirs, **162**, 239–261.

KENNAN, L. & PINDELL, J. 2007. Dextral shear, terrane accretion and basin formation in the Northern Andes: explained only by interaction with a Pacific-derived Caribbean Plate. *In*: KENNAN, L., PINDELL, J. & ROSEN, N. (eds) *The Paleogene of the Gulf of Mexico and Caribbean Basins: Processes, Events, and Petroleum Systems*. GCSSEPM Foundation, 27th Annual Research Conference, Transactions, CD.

KERR, A. C., MARRINER, G. F. *ET AL*. 1997. Cretaceous basaltic terranes in western Colombia: elemental, chronological and Sr–Nd isotopic constraints on petrogenesis. *Journal of Petrology*, **38**, 677–702.

KERR, A. C., ITURRALDE-VINENT, M. A., SAUNDERS, A. D., BABBS, T. L. & TARNEY, J. 1999. A new plate tectonic model of the Caribbean: Implications from a geochemical reconnaissance of Cuban Mesozoic volcanic rocks. *Geological Society of America Bulletin*, **111**, 1581–1599.

KERR, A. C., ASPDEN, J. A., TARNEY, J. & PILATASIG, L. F. 2002. The nature and provenance of accreted terranes in western Ecuador: geochemical and tectonic constraints. *Journal of the Geological Society, London*, **159**, 577–594.

KISER, G. D. 1989. *Relaciones estratigráficas de la ceunca Apure/Llanos con areas adjacentes, Venezuela suroeste y Colombia oriental*. Sociedad Venezolana de Geólogos, Monografías, **1**.

KISER, G. D. 1994. Santa Marta Massif: a major element in the tectonic-sedimentary evolution of northern South America. *V Simposio Bolivariano de la Exploración*

Petrolera en las Cuencas Subandinas, Memoria, 317–335.

KLITGORD, K. M. & SCHOUTEN, H. 1986. Plate kinematics of the central Atlantic. In: VOGT, P. R. & TUCHOLKE, B. E. (eds) *The Western North Atlantic Region*. The Geology of North America, **M**. Geological Society of America, Boulder, CO, 351–378.

KOHN, B. P., SHAGAM, R., BANKS, P. O. & BURKLEY, L. A. 1984. Mesozoic–Pleistocene fission-track ages on rocks of the Venezuelan Andes and their tectonic implications. In: BONINI, W. E., HARGRAVES, R. B. & SHAGAM, R. (eds) *The Caribbean–South American Plate Boundary and Regional Tectonics*. Geological Society of America, Boulder, CO, Memoirs, **162**, 365–384.

KUGLER, H. G. 1953. Jurassic to recent sedimentary environments in Trinidad. *Bulletin Association Suisse des Géologie et Ingeneur du Pétrole*, **20**, 27–60.

KUGLER, H. G. 1956. *Trinidad*. Centre National de la Recherche Scientifique, Paris, Lexique Stratigraphique International, **5**.

KUGLER, H. G. 1959. *Geological Maps of Trinidad*. Ministry of Energy, Trinidad, 1 : 50 000.

KUGLER, H. G. 1961. *Geological Map and Cross-Sections of Trinidad*. Petroleum Association of Trinidad, 1 : 100 000. Republished as part of Kugler 1996.

KUGLER, H. G. 1974. The geology of Patos Island (East Venezuela). *Eclogae Geologicae Helvetiae*, **67**, 469–478.

KUGLER, H. G. 1996. *Treatise on the Geology of Trinidad, Part 3. Detailed Geological Maps and Sections*. Natural History Museum, Basel, Switzerland.

KUGLER, H. G. 2001. *Treatise on the Geology of Trinidad, Part 4. The Palaeocene to Holocene Formations of Trinidad*. Ed. by BOLLI, H. M. & KNAPPERTSBUSCH, M., Natural History Museum, Basel, Switzerland.

KUGLER, H. G. & SAUNDERS, J. G. 1967. On Tertiary turbidity-flow sediments in Trinidad. *Asociación Venezolana de Geología, Minería y Petroleo, Boletín Informativo*, **10**, 243–259.

KUGLER, H. G. & CAUDRI, C. M. B. 1975. Geology and paleontology of Soldado Rock, Trinidad (West indies). Part 1: Geology and biostratigraphy. *Eclogae Geologicae Helvetiae*, **68**, 365–430.

LARSON, R. L. 1991. Latest pulse of Earth: evidence for a Mid-Cretaceous superplume. *Geology*, **19**, 547–550.

LEV 1997. *Léxico Estratigráfico de Venezuela, 3ra edición*. Ministerio de Energía y Minas, Boletín de Geología, Publicaciónes Expeciales, **12**. World Wide Web Address: http://www.pdvsa.com/lexico/lexicoh.htm, (accessed November 2008).

LIOU, J. G., MARUYAMA, S., WANG, X. & GRAHAM, S. 1990. Precambrian blueschist terranes of the world. *Tectonophysics*, **181**, 97–111.

LOCKE, B. D. & GARVER, J. I. 2005. Thermal evolution of the eastern Serranía del Interior foreland fold and thrust belt, northeastern Venezuela, based on apatite fission-track analyses. In: AVÉ LALLEMENT, H. G. & SISSON, V. B. (eds) *Caribbean–South American Plate Interactions, Venezuela*. Geological Society of America, Boulder, CO, Special Papers, **394**, 315–328.

LUGO, J. & MANN, P. 1995. Jurassic–Eocene tectonic evolution of Maracaibo Basin, Venezuela.

In: TANKARD, A. J., SUÁREZ, R. & WELSINK, H. J. (eds) *Petroleum Basins of South America*. American Association of Petroleum Geologists Memoirs, **62**, 699–725.

MACDONALD, W. D., DOOLAN, B. L. & CORDANI, U. G. 1971. Cretaceous–Early Tertiary metamorphic K–Ar age values from the south Caribbean. *Geological Society of America Bulletin*, **82**, 1381–1388.

MACELLARI, C. 1984. Late Tertiary tectonic history of the Táchira Depression, southwestern Venezuelan Andes. In: BONINI, W. E., HARGRAVES, R. B. & SHAGAM, R. (eds) *The Caribbean–South American Plate Boundary and Regional Tectonics*. Geological Society of America, Boulder, CO, Memoirs, **162**, 333–341.

MACELLARI, C. E. 1988. Cretaceous paleogeography and depositional cycles of western South America. *Journal of South American Earth Sciences*, **1**, 373–418.

MACELLARI, C. E. 1995. Cenozoic sedimentation and tectonics of the southwestern Caribbean pull-apart basin, Venezuela and Colombia. In: TANKARD, A. J., SUÁREZ, R. & WELSINK, H. J. (eds) *Petroleum Basins of South America*. American Association of Petroleum Geologists, Memoirs, **62**, 757–780.

MACELLARI, C. E. & DE VRIES, T. J. 1987. Late Cretaceous upwelling and anoxic sedimentation in northwestern South America. *Palaeogeography, Palaeoclimatology, Palaeoecology*, **59**, 279–292.

MACSOTAY, O., VIVAS, V., PIMENTEL, N. & BELLIZZIA, A. 1985. Estratigrafía y tectónica del Cretáceo–Paleógeno de las islas al norte de Puerto la Cruz-Santa Fé y regiones adyacentes. Excursión. *Sociedad Venezolana de Geólogos, VI Congreso Geológico Venezolano*, **10**, 7125–7174.

MACSOTAY, O., VIVAS, V. & MOTICSKA, P. 1995. Biostratigraphy of the Piemontine Nappe of north-central Venezuela: Senonian to Eocene gravitational sedimentation. *Venezuela Ministerio de Energía y Minas, Boletín de Geología, Publicaciónes Especiales*, **10**, 114–123.

MACSOTAY, O., ERLICH, R. N. & PERAZA, T. 2003. Sedimentary structures of the La Luna, Navay and Querecual Formations, Upper Cretaceous of Venezuela. *Palaios*, **18**, 334–348.

MANN, P. 1999. Caribbean sedimentary basins: classification and tectonic setting from Jurassic to present. In: MANN, P. (ed.) *Caribbean Basins*. Elsevier, Amsterdam, 3–31.

MANN, P. & BURKE, K. 1984. Neotectonics of the Caribbean. *Reviews of Geophysics and Space Physics*, **22**, 309–362.

MANN, P. & CORRIGAN, J. 1990. Model for Late Neogene deformation in Panama. *Geology*, **18**, 558–562.

MANN, P., HIPPOLYTE, J.-C., GRINDLAY, N. R. & ABRAMS, L. J. 2005. Neotectonics of southern Puerto Rico and its offshore margin. In: MANN, P. (ed.) *Active Tectonics and Seismic Hazards of Puerto Rico, the Virgin Islands and Offshore Areas*. Geological Society of America and Boulder, CO, Special Papers, **385**, 173–214.

MANN, P., ESCALONA, A. & CASTILLO, M. V. 2006. Regional geologic and tectonic setting of the Maracaibo supergiant basin, western Venezuela. *American Association of Petroleum Geologists Bulletin*, **90**, 445–477.

MARESCH, W. V. 1974. Plate tectonics origin of the Caribbean mountain system of northern South America: discussion and proposal. *Geological Society of America Bulletin*, **85**, 669–682.

MAUGHAN, E. K., ZAMBRANO, F. O., MOJICA, J., ABOZAGLO, J., PACHON, F. P. & DURAN, R. R. 1979. *Paleontologic and Stratigraphic Relations of Phosphate Beds in Upper Cretaceous Rocks of the Cordillera Oriental, Colombia*. US Geological Survey Open-File Report, 79-1525.

MEIER, B., SCHWANDER, M. & LAUBSCHER, H. P. 1987. The tectonics of Táchira: a sample of north Andean tectonics. *In*: SHAER, J. P. & RODGERS, J. (eds) *The Anatomy of Mountain Ranges*. Princeton University Press, Princeton, NJ, 229–237.

MENDEZ, J. 1997. *El Cuaternario en Venezuela*. World Wide Web Address: http://www.pdvsa.com/lexico/q00w.htm (accessed August 2008).

MESCHEDE, M. & FRISCH, W. 1998. A plate-tectonic model for the Mesozoic and Early Cenozoic history of the Caribbean Plate. *Tectonophysics*, **296**, 269–291.

METZ, H. L. 1968. Stratigraphic and geologic history of extreme northeastern Serrania del Interior, State of Sucre, Venezuela. *Fourth Caribbean Geological Conference, Port of Spain, 1965, Transactions*, 275–292.

MOLNAR, P. & SYKES, L. R. 1969. Tectonics of the Caribbean and Middle America regions from focal mechanisms and seismicity. *Geological Society of America Bulletin*, **80**, 1639–1684.

MONTES, C., HATCHER, R. D., JR. & RESTREPO-PACE, P. A. 2005. Tectonic reconstruction of the northern Andean blocks: oblique convergence and rotations derived from the kinematics of the Piedras-Girardot area, Colombia. *Tectonophysics*, **399**, 221–250.

MORA, A. & GARCIA, A. 2006. *Cenozoic Tectonostratigraphic Relationships between the Cesar Subbasin and the Southeastern Lower Magdalena Valley Basin of northern Colombia*. American Association of Petroleum Geologists, Search and Discovery Article, **30046**. World Wide Web Address: http://www.searchanddiscovery.com/documents/2006/06134mora/index.htm

MORENO, M., CONCHA, A., PATARROYO, P. & NAVARRETE, A. 2002. Magmatic basic events during the Early Cretaceous in the Colombian Eastern Mountain Range. *5th International Symposium on Andean Geodynamics*, Toulouse, France, extended abstracts.

MUESSIG, K. W. 1978. The central Falcon igneous suite, Venezuela: alkaline basaltic intrusions of Oligocene–Miocene age. *Geologie en Mijnbouw*, **57**, 261–266.

MUESSIG, K. W. 1984. Structure and Cenozoic tectonics of the Falcón Basin, Venezuela, and adjacent areas. *In*: BONINI, W. E., HARGRAVES, R. B. & SHAGAM, R. (eds) *The Caribbean–South American Plate Boundary and Regional Tectonics*. Geological Society of America, Boulder, CO, Memoirs, **162**, 217–230.

MÜLLER, R. D., ROYER, J.-Y., CANDE, S. C., ROEST, W. R. & MASCHENKOV, S. 1999. New constraints on the Late Cretaceous/Tertiary plate tectonic evolution of the Caribbean. *In*: MANN, P. (ed.) *Caribbean Basins*. Elsevier, Amsterdam, 33–59.

MYROW, P. M., FISCHER, W. & GOODGE, J. W. 2002. Wave-modified turbidites: combined-flow shoreline and shelf deposits, Cambrian, Antarctica. *Journal of Sedimentary Research*, **72**, 641–656.

NEVERS, G. M., DORMAN, J. H., HARRISON, P. J. & ROJAS, O. 1991. Recent exploration results. *IV Simposio Bolivariano de la Exploración Petrolera en las Cuencas Subandinas, Memorias*, **1**, Trabajo 8.

OSTOS, M., AVÉ LALLEMENT, H. G. & SISSON, V. B. 2005a. The alpine-type Tinaquillo peridotite complex, Venezuela: fragment of a Jurassic rift zone? *In*: AVÉ LALLEMENT, H. G. & SISSON, V. B. (eds) *Caribbean–South American Plate Interactions, Venezuela*. Geological Society of America, Boulder, CO, Special Papers, **394**, 207–222.

OSTOS, M., YORIS, F. & AVÉ LALLEMENT, H. G. 2005b. Overview of the southeast Caribbean-South American plate boundary zone. *In*: AVÉ LALLEMENT, H. G. & SISSON, V. B. (eds) *Caribbean–South American Plate Interactions, Venezuela*. Geological Society of America, Boulder, CO, Special Papers, **394**, 53–89.

PAIGE, S. 1930. The earthquake at Cumana, Venezuela, January 17, 1929. *Bulletin of the Seismological Society of America*, **20**, 1–10.

PARIS, G., MACHETTE, M. N., DART, R. L. & HALLER, K. M. 2000. *Map and Database of Quaternary Faults and Folds in Colombia and its Offshore Regions*. US Geological Survey Open-File Report, **00-0284**.

PARNAUD, F., GOU, Y., PASCUAL, J.-C., CAPELLO, M. A., TRUSKOWSKI, I. & PASSALACQUA, H. 1995a. Stratigraphic synthesis of western Venezuela. *In*: TANKARD, A. J., SUÁREZ, R. & WELSINK, H. J. (eds) *Petroleum Basins of South America*. American Association of Petroleum Geologists Memoirs, **62**, 681–698.

PARNAUD, F., GOU, Y., PASCUAL, J.-C., TRUSKOWSKI, I., GALLANGO, O. & PASSALACQUA, H. 1995b. Petroleum geology of the central part of the Eastern Venezuela basin. *In*: TANKARD, A. J., SUÁREZ, R. & WELSINK, H. J. (eds) *Petroleum Basins of South America*. American Association of Petroleum Geologists, Memoirs, **62**, 741–756.

PATTERSON, J. M. & WILSON, J. G. 1953. Oil fields of the Mercedes region, Venezuela. *American Association of Petroleum Geologists Bulletin*, **37**, 2705–2733.

PAYNE, N. 1991. An evaluation of post-Middle Miocene geological sequences, offshore Trinidad. *Second Geological Conference of the Geological Society of Trinidad and Tobago, Transactions*, 70–87.

PEIRSON, A. L. 1965. Geology of the Guárico mountain front. *Asociación Venezolana de Geología, Minería y Petróleo, Boletín Informativo*, **8**, 183–212.

PEIRSON, A. L., SALVADOR, A. & STAINFORTH, R. M. 1966. The Guárico Formation of north-central Venezuela. *Asociación Venezolana de Geología, Minería y Petróleo, Boletín Informativo*, **9**, 181–224.

PENNINGTON, W. D. 1981. Subduction of the Eastern Panama Basin and seismotectonics of northwestern South America. *Journal of Geophysical Research*, **86**, 10753–10770.

PEREIRA, J. G. 1985. Evolución tectónica de la cuenca de Carúpano durante el Terciario. *Sociedad Venezolana de Geólogos, VI Congreso Geológico Venezolano, Memorias*, **4**, 2618–2648.

PEREZ, O. J. & AGGARWAL, Y. P. 1981. Present-day tectonics of the southeastern Caribbean and northeastern Venezuela. *Journal of Geophysical Research*, **86**, 10791–10804.

PEREZ, O. J., BILHAM, R. *ET AL*. 2001. Velocity field across the southern Caribbean Plate boundary and estimates of Caribbean/South-American Plate motion using GPS geodesy 1994–2000. *Geophysical Research Letters*, **28**, 2987–2990.

PEREZ DE ARMAS, J. 2005. Tectonic and thermal history of the western Serranía del Interior foreland fold and thrust belt and Guárico basin, north-central Venezuela: implications of new apatite fission-track analysis and seismic interpretation. *In*: AVÉ LALLEMENT, H. G. & SISSON, V. B. (eds) *Caribbean–South American Plate Interactions, Venezuela*. Geological Society of America, Boulder, CO, Special Papers, **394**, 271–314.

PEREZ DE MEJIA, D., KISER, G. D., MAXIMOWITSCH, B. & YOUNG, G. A. 1980. Geología de Venezuela. In: FELDER, B. (ed.) *Evaluación de Formaciones en Venezuela*. Schlumberger, Caracas, 1 : 1–1 : 23.

PERSAD, K. M. 1985. Outline of the geology of the Trinidad area. *Fourth Latin American Geological Congress, Trinidad, 1979, Transactions*, 738–758.

PIMENTEL, N. 1984. *Mapa Geológico Estructural de Venezuela*. Ministerio de Energía y Minas, 1 : 2 500 000.

PINDELL, J. L. 1985. Alleghenian reconstruction and subsequent evolution of the Gulf of Mexico, Bahamas, and Proto-Caribbean. *Tectonics*, **4**, 1–39.

PINDELL, J. L. 1991. Geologic rationale for hydrocarbon exploration in the Caribbean and adjacent regions. *Journal of Petroleum Geology*, **14**, 237–257.

PINDELL, J. L. 1993. Regional synopsis of Gulf of Mexico and Caribbean evolution. *In*: PINDELL, J. L. & PERKINS, R. (eds) *Mesozoic and Early Cenozoic Development of the Gulf of Mexico and Caribbean Region*. GCSSEPM Foundation, 13th Annual Research Conference, Proceedings, 251–274.

PINDELL, J. & DEWEY, J. F. 1982. Permo-Triassic reconstruction of western Pangea and the evolution of the Gulf of Mexico/Caribbean region. *Tectonics*, **1**, 179–211.

PINDELL, J. L. & BARRETT, S. F. 1990. Geological evolution of the Caribbean region; a plate-tectonic perspective. *In*: DENGO, G. & CASE, J. E. (eds) *The Caribbean Region*. The Geology of North America, **H**, Geological Society of America, Boulder, CO, 405–432.

PINDELL, J. L. & ERIKSON, J. P. 1994. The Mesozoic passive margin of northern South America. *In*: SALFITY, J. A. (ed.) *Cretaceous Tectonics of the Andes*. Vieweg, Wiesbaden, 1–60.

PINDELL, J. & TABBUTT, K. D. 1995. Mesozoic–Cenozoic Andean paleogeography and regional controls on hydrocarbon systems. *In*: TANKARD, A. J., SUÁREZ, R. & WELSINK, H. J. (eds) *Petroleum Basins of South America*. American Association of Petroleum Geologists, Memoirs, **62**, 101–128.

PINDELL, J. & ERIKSON, J. 2001. *Stratigraphy and Structure of the Northwestern Serranía del Interior Oriental, Venezuela: Constraints on Dynamic Evolution,*

Eastern Venezuela/Trinidad. Geological Society of Trinidad and Tobago, Field Guide.

PINDELL, J. & KENNAN, L. 2001*a*. Processes and events in the terrane assembly of Trinidad and Eastern Venezuela. *In*: FILLON, R. H., ROSEN, N. C. & WEIMER, P. (eds) *Petroleum Systems of Deep-Water Basins: Global and Gulf of Mexico Experience*. GCSSEPM Foundation, 21st Annual Research Conference, Transactions, 159–192.

PINDELL, J. & KENNAN, L. 2001*b*. Kinematic evolution of the Gulf of Mexico and Caribbean. *In*: FILLON, R. H., ROSEN, N. C. & WEIMER, P. (eds) *Petroleum Systems of Deep-Water Basins: Global and Gulf of Mexico Experience*. GCSSEPM Foundation, 21st Annual Research Conference, Transactions, 193–220.

PINDELL, J. & KENNAN, L. 2003. Synthesis of Gulf of Mexico and Caribbean tectonic evolution: Pacific origin model for Caribbean lithosphere (abstract). *American Association of Petroleum Geologists International Conference*, Barcelona.

PINDELL, J. & KENNAN, L. 2007*a*. Bow-wave model for deformation and foredeep development since 10 Ma, eastern Venezuela and Trinidad. *Fourth Geological Conference of the Geological Society of Trinidad and Tobago*. extended abstracts CD.

PINDELL, J. & KENNAN, L. 2007*b*. Cenozoic Caribbean–South America tectonic interaction: a case for prism–prism collision in Venezuela, Trinidad, and Barbados Ridge. *Fourth Geological Conference of the Geological Society of Trinidad and Tobago*. extended abstracts CD.

PINDELL, J. & KENNAN, L. 2007*c*. Cenozoic kinematics and dynamics of oblique collision between two convergent plate margins: the Caribbean–South America collision in eastern Venezuela, Trinidad and Barbados. *In*: KENNAN, L., PINDELL, J. & ROSEN, N. (eds) *The Paleogene of the Gulf of Mexico and Caribbean Basins: Processes, Events, and Petroleum Systems*. GCSSEPM Foundation, 27th Annual Research Conference, Transactions, CD.

PINDELL, J. L., CANDE, S. C. *ET AL*. 1988. A plate-kinematic framework for models of Caribbean evolution. *Tectonophysics*, **155**, 121–138.

PINDELL, J. L., ERIKSON, J. P. & ALGAR, S. 1991. The relationship between plate motions and sedimentary basin development in northern South America: from a Mesozoic passive margin to a Cenozoic eastwardly progressive transpressional orogen. *Second Geological Conference of the Geological Society of Trinidad and Tobago, 1990, Transactions*, 191–202.

PINDELL, J. L., HIGGS, R. & DEWEY, J. F. 1998. Cenozoic palinspastic reconstruction, paleogeographic evolution and hydrocarbon setting of the northern margin of South America. *In*: PINDELL, J. L. & DRAKE, C. (eds) *Paleogeographic Evolution and Non-glacial Eustasy, Northern South America*. SEPM, Special Publications, **58**, 45–85.

PINDELL, J., KENNAN, L., MARESCH, W. V., STANEK, K.-P., DRAPER, G. & HIGGS, R. 2005. Plate-kinematics and crustal dynamics of circum-Caribbean arc–continent interactions: tectonic controls on basin development in Proto-Caribbean margins. *In*: AVÉ LALLEMENT, H. G. & SISSON, V. B. (eds)

Caribbean–South American Plate Interactions, Venezuela. Geological Society of America, Boulder, CO, Special Papers, **394**, 7–52.

PINDELL, J., KENNAN, L., STANEK, K.-P., MARESCH, W. V. & DRAPER, G. 2006. Foundations of Gulf of Mexico and Caribbean evolution: eight controversies resolved. *Geologica Acta*, **4**, 303–341.

POTTER, H. C. 1973. The overturned anticline of the Northern Range of Trinidad near Port of Spain. *Journal of the Geological Society, London*, **129**, 133–138.

POTTER, H. C. 1974. Observations on the Laventille Formation, Trinidad. *Verhandlungen der Naturforschenden Gesellschaft in Basel*, **84**, 202–208.

POTTER, H. C. 1997. Notes on the Northern Range and changes from H. G. Kugler's map. *Trinidad and Tobago Ministry of Energy, Explanatory Notes for 1997 Geological Map of Trinidad*, 13–22.

PRATT, B. R. 2001. Septarian concretions: internal cracking caused by synsedimentary earthquakes. *Sedimentology*, **48**, 189–213.

PROST, M. T. 1989. Coastal dynamics and chenier sands in French Guiana. *Marine Geology*, **90**, 239–267.

PUDSEY, C. J. & READING, H. G. 1982. Sedimentology and structure of the Scotland Group, Barbados. *In*: LEGGETT, J. K. (ed.) *Trench-Forearc Geology: Sedimentation and Tectonics on Modern and Ancient Active Plate Margins*. Geological Society, London, Special Publications, **10**, 291–308.

RENZ, O. 1959. Estratigrafía del Cretáceo en Venezuela occidental. *Venezuela Ministerio de Minas y Hidrocarburos, Boletín de Geología*, **5**, 3–38.

RENZ, O., LAKEMAN, R. & VAN DER MUELEN, E. 1955. Submarine sliding in western Venezuela. *American Association of Petroleum Geologists Bulletin*, **39**, 2053–2067.

RESTREPO, J. J. & TOUSSAINT, J. F. 1988. Terranes and continental accretion in the Colombian Andes. *Episodes*, **11**, 189–193.

REYES, A., MONTENEGRO, G. & GÓMEZ, P. D. 2000. Evolución tectonoestratigráfica del Valle Inferior del Magdalena, Colombia. *VII Simposio Bolivariano de la Exploración Petrolera en las Cuencas Subandinas, Memoria*, 293–309.

REYMENT, R. 1972. Some Lower Turonian ammonites from Trinidad and Colombia. *Geologiska Föreningens I Stockholm Förhandlingar*, **94**, 357–368.

ROBERTSON, P. & BURKE, K. 1989. Evolution of southern Caribbean Plate boundary, vicinity of Trinidad and Tobago. *American Association of Petroleum Geologists Bulletin*, **73**, 490–509.

ROD, E. 1956. Strike–slip faults of northern Venezuela. *American Association of Petroleum Geologists Bulletin*, **40**, 457–476.

RODRIGUEZ, J., ANGULO, S. & DELGADO, M. 2004. Sedimentological model of the Palaeocene to Lower Miocene Naricual Formation, Carito Norte and Carito Oeste oilfields, Monagas State, eastern Venezuela. *American Association of Petroleum Geologists Annual Convention, Program and Abstracts.*

ROSALES, H. 1967. *Geología del area Barcelona–Río Querecual.* Asociación Venezolana de Geología, Minería y Petróleo, Guia de la Excursión (Field Guide).

ROSENCRANTZ, E. 1990. Structure and tectonics of the Yucatan Basin, Caribbean Sea, as determined from seismic reflection studies. *Tectonics*, **9**, 1037–1059.

ROSS, M. I. & SCOTESE, C. R. 1988. A hierarchical tectonic model of the Gulf of Mexico and Caribbean region. *Tectonophysics*, **155**, 139–168.

ROURE, F., BORDAS-LEFLOCH, N. *ET AL.* 2003. Petroleum systems and reservoir appraisal in the sub-Andean basins (E Venezuela and E Colombian foothills). *In*: BARTOLINI, C., BUFFLER, R. T. & BLICKWEDE, J. F. (eds) *The Circum-Gulf of Mexico and the Caribbean: Hydrocarbon Habitats, Basin Formation, and Plate Tectonics*. American Association of Petroleum Geologists, Memoirs, **79**, 750–775.

RUIZ, C., DAVIS, N., BENTHAM, P., PRICE, A. & CARVAJAL, D. 2000. Structure and tectonic evolution of the south Caribbean basin, southern offshore Colombia: a progressive accretionary system. *VII Simposio Bolivariano de la Exploración Petrolera en las Cuencas Subandinas, Memoria*, 334–355.

RUSSO, R. M., SPEED, R. C., OKAL, E. A., SHEPHERD, J. B. & ROWLEY, K. C. 1993. Seismicity and tectonics of the southeastern Caribbean. *Journal of Geophysical Research*, **98**, 14299–14319.

SAGEMAN, B. B. & SPEED, R. C. 2003. Upper Eocene limestones, associated sequence boundary, and proposed Eocene tectonics in eastern Venezuela. *In*: BARTOLINI, C., BUFFLER, R. T. & BLICKWEDE, J. F. (eds) *The Circum-Gulf of Mexico and the Caribbean: Hydrocarbon Habitats, Basin Formation, and Plate Tectonics*. American Association of Petroleum Geologists, Memoirs, **79**, 874–890.

SALVADOR, A. 1986. Comments on 'Neogene block tectonics of eastern Turkey and northern South America: continental applications of the finite difference method' by DEWEY, J. F. & PINDELL, J. L. *Tectonics*, **5**, 697–701.

SALVADOR, A. & STAINFORTH, R. M. 1968. Clues in Venezuela to the geology of Trinidad, and vice versa. *Fourth Caribbean Geological Conference, Port of Spain, 1965, Transactions*, 31–40.

SAMS, R. H. 1995. Interpreted sequence stratigraphy of the Los Jabillos, Areo, and (subsurface) Naricual formations, northern Monagas area, Eastern Venezuela Basin. *Sociedad Venezolana de Geología, Boletín*, **20**, 30–40.

SANTAMARÍA, F. & SCHUBERT, C. 1974. Geochemistry and geochronology of the southern Caribbean–northern Venezuela plate boundary. *Geological Society of America Bulletin*, **7**, 1085–1098.

SARMIENTO-ROJAS, L. F., VAN WESS, J. D. & CLOETINGH, S. 2006. Mesozoic transtensional basin history of the Eastern Cordillera: inferences from tectonic models. *Journal of South American Earth Sciences*, **21**, 383–411.

SAUNDERS, J. B. 1972. Recent paleontological results from the Northern Range of Trinidad. *VI Conferencia Geológica del Caribe, Margarita, Venezuela, Memorias*, 455–459.

SAUNDERS, J. B. 1997a. *Trinidad Stratigraphic Chart and Geological Map.* Trinidad and Tobago Ministry of Energy, 1 : 100 000.

SAUNDERS, J. B. 1997*b*. General explanatory notes & changes from H. G. Kugler's 1959 map. *Trinidad and Tobago Ministry of Energy, Explanatory Notes for 1997 Geological Map of Trinidad*, 2–12.

SAUNDERS, J. B. & BOLLI, H. M. 1985. Trinidad's contribution to world biostratigraphy. *Fourth Latin American Geological Congress, Trinidad, 1979, Transactions*, 781–795.

SAVIAN, V. & SCHERER, W. 1993. Mapa isópaco de la Formación La Luna. World Wide Web Address: http://www.pdvsa.com/lexico/image/lunaisop.gif (accessed August 2008).

SCHLANGER, S. O. & JENKYNS, H. C. 1976. Cretaceous oceanic anoxic events: causes and consequences. *Geologie en Mijnbouw*, **55**, 179–184.

SCHUBERT, C. 1980. Late Cenozoic pull-apart basins, Boconó fault zone, Venezuelan Andes. *Journal of Structural Geology*, **2**, 463–468.

SCHUBERT, C. 1982. Origin of Cariaco Basin, southern Caribbean Sea. *Marine Geology*, **47**, 345–360.

SCHUBERT, C. & SIFONTES, R. S. 1970. Boconó Fault, Venezuelan Andes: evidence of postglacial movement. *Science*, **170**, 66–69.

SCHUBERT, C. & VIVAS, L. 1993. *El Cuaternario de la Cordillera de Mérida, Andes Venezolanos*. Universidad de Los Andes/Fundación Polar, Mérida.

SCOTESE, C. R. & MCKERROW, W. S. 1990. Revised world maps and introduction. *In*: MCKERROW, W. S. & SCOTESE, C. R. (eds) *Palaeozoic Palaeogeography and Biogeography*. Geological Society, London, Memoirs, **12**, 1–21.

SENN, A. 1940. Paleogene of Barbados and its bearing on history and structure of Antillean-Caribbean region. *American Association of Petroleum Geologists Bulletin*, **24**, 1548–1610.

SIFONTES, R. S. & SANTAMARÍA, F. 1972. Rocas intrusivas jovenes en la región de Carúpano. *VI Conferencia Geológica del Caribe, Margarita, Venezuela, Memorias*, 122–125.

SINCLAIR, H. D., COAKLEY, B. J., ALLEN, P. A. & WATTS, A. B. 1991. Simulation of foreland basin stratigraphy using a diffusion model of mountain belt uplift and erosion: an example from the central Alps, Switzerland. *Tectonics*, **10**, 599–620.

SISSON, V. B., ERTAN, I. E. & AVÉ LALLEMENT, H. G. 1997. High-pressure (∼2000 MPa) kyanite- and glaucophane-bearing pelitic schist and eclogite from Cordillera de la Costa belt, Venezuela. *Journal of Petrology*, **38**, 65–83.

SISSON, V. B., AVÉ LALLEMENT, H. G. ET AL. 2005. Overview of radiometric ages in three allochthonous belts of northern Venezuela: old ones, new ones, and their impact on regional geology. *In*: AVÉ LALLEMENT, H. G. & SISSON, V. B. (eds) *Caribbean–South American Plate Interactions, Venezuela*. Geological Society of America, Boulder, CO, Special Papers, **394**, 91–117.

SKERLEC, G. M. & HARGRAVES, R. B. 1980. Tectonic significance of paleomagnetic data from northern Venezuela. *Journal of Geophysical Research*, **85**, 5303–5315.

SMITH, P. L. 1983. The Pliensbachian ammonite *Dayiceras dayiceroides* and Early Jurassic paleogeography. *Canadian Journal of Earth Sciences*, **20**, 86–91.

SMITH, C. A., SISSON, V. B., AVÉ LALLEMENT, H. G. & COPELAND, P. 1999. Two contrasting pressure–temperature–time paths in the Villa de Cura blueschist belt, Venezuela: possible evidence for Late Cretaceous initiation of subduction in the Caribbean. *Geological Society of America Bulletin*, **111**, 831–848.

SNOKE, A. W., ROWE, D. W., YULE, J. D. & WADGE, G. 2001. *Petrologic and Structural History of the Southern Caribbean*. Geological Society of America, Special Papers, **354**.

SOCAS, M. 1991. *Estudio sedimentológico de la Formación Naricual, Estado Anzoátegui*. BS thesis, Universidad Central de Venezuela.

SPEED, R. C. & SMITH-HOROWITZ, P. L. 1998. The Tobago terrane. *International Geology Review*, **40**, 805–830.

SPEED, R. C., SMITH-HOROWITZ, P. L., PERCH-NIELSEN, K. V. S., SAUNDERS, J. B. & SANFILIPPO, A. B. 1993. *Southern Lesser Antilles Arc Platform: Pre-Late Miocene Stratigraphy, Structure and Tectonic Evolution*. Geological Society of America, Special Papers, **277**.

SPEED, R. C., SHARP, W. D. & FOLAND, K. A. 1997. Late Paleozoic granitoid gneisses of northeastern Venezuela and the North America–Gondwana collision zone. *Journal of Geology*, **105**, 457–470.

STÉPHAN, J. F. 1977. El contacto Cadena Caribe–Andes Merideños entre Carora y El Tocuyo (Edo. Lara): observaciones sobre el estilo y la edad de las deformaciones cenozoicas en el occidente venezolano. *V Congreso Geológico Venezolano, Memorias*, 789–816.

STÉPHAN, J. F. 1985. Andes et chaîne caraïbe sur la transversale de Barquisimeto (Venezuela): évolution géodynamique. *In*: MASCLE, A. (ed.) *Géodynamique des Caraïbes*. Editions Technip, Paris, 505–529.

STÉPHAN, J. F., MERCIER DE LEPINAY, B. ET AL. 1990. Paleogeodynamic maps of the Caribbean: 14 steps from Lias to present. *Bulletin de la Societé Géologique de France*, **8**, 915–919.

STÖCKHERT, B., MARESCH, W. V. ET AL. 1995. Crustal history of Margarita Island (Venezuela) in detail: constraint on the Caribbean Plate tectonic scenario. *Geology*, **23**, 787–790.

STOW, D. A. V., READING, H. G. & COLLINSON, J. D. 1996. Deep seas. *In*: READING, H. G. (ed.) *Sedimentary Environments: Processes, Facies and Stratigraphy*. 3rd edn. Blackwell Science, Oxford, 395–453.

SUMNER, R. H. & WESTBROOK, G. K. 2001. Mud diapirism in front of the Barbados accretionary wedge: the influence of fracture zones and North America–South America plate motions. *Marine and Petroleum Geology*, **18**, 591–613.

SUTER, H. H. 1951. *The General and Economic Geology of Trinidad, B. W. I*. Colonial Geology and Mineral Resources, **2**(3), 177–217, **2**(4), 271–307.

SUTER, H. H. 1960. *The General and Economic Geology of Trinidad, B. W. I.*, 2nd edn. HMSO, London.

SYKES, L. R., MCCANN, W. R. & KAFKA, A. L. 1982. Motion of Caribbean Plate during last 7 million years and implications for earlier Cenozoic movements. *Journal of Geophysical Research*, **87**, 10656–10676.

TORRES, V., VANDENBERGHE, J. & HOOGHIEMSTRA, H. 2005. An environmental reconstruction of the sediment infill of the Bogotá basin (Colombia) during the last 3 million years from abiotic and biotic proxies. *Palaeogeography, Palaeoclimatology, Palaeoecology*, **226**, 127–148.

TRENKAMP, R., KELLOGG, J. N., FREYMUELLER, J. T. & MORA, H. P. 2002. Wide plate margin deformation, southern Central America and northwestern South America, CASA GPS observations. *Journal of South American Earth Sciences*, **15**, 157–171.

TSIKOS, H., JENKYNS, H. C. ET AL. 2004. Carbon-isotope stratigraphy recorded by the Cenomanian–Turonian Oceanic Anoxic Event: correlation and implications based on three key localities. *Journal of the Geological Society, London*, **161**, 711–719.

TYSON, L. & ALI, W. 1991. Cretaceous to Middle Miocene sediments in Trinidad. Field Trip Guide. *Second Geological Conference of the Geological Society of Trinidad and Tobago, 1990, Transactions*, 266–277.

VAN DER HAMMEN, T., WERNER, J. H. & VAN DOMMELEN, H. 1973. Palynological record of the upheaval of the northern Andes: a study of the Pliocene and Lower Quaternary of the Colombian Eastern Cordillera and the early evolution of its high-Andean biota. *Review of Palaeobotany and Palynology*, **16**, 1–122.

VAN DER HILST, R. & MANN, P. 1994. Tectonic implications of tomographic images of subducted lithosphere beneath northwestern South America. *Geology*, **22**, 451–454.

VAN HOUTEN, F. B. 1976. Late Cenozoic volcaniclastic deposits, Andean foredeep, Colombia. *Geological Society of America Bulletin*, **87**, 481–495.

VASQUEZ, M. & ALTENBERGER, U. 2005. Mid-Cretaceous extension-related magmatism in the eastern Colombian Andes. *Journal of South American Earth Sciences*, **20**, 193–210.

VAUGHAN, A. P. M. & STOREY, B. C. 2007. A new supercontinent self-destruct mechanism: evidence from the Late Triassic–Early Jurassic. *Journal of the Geological Society, London*, **164**, 383–392.

VIERBUCHEN, R. C. 1984. The geology of the El Pilar fault zone and adjacent areas in northeastern Venezuela. *In*: BONINI, W. E., HARGRAVES, R. B. & SHAGAM, R. (eds) *The Caribbean–South American Plate Boundary and Regional Tectonics*. Geological Society of America, Memoirs, **162**, 189–212.

VILLAMIL, T. 1998. Chronology, relative sea-level history and a new sequence stratigraphic model for basinal Cretaceous facies of Colombia. *In*: PINDELL, J. L. & DRAKE, C. (eds) *Paleogeographic Evolution and Non-glacial Eustasy, Northern South America*. SEPM, Special Publications, **58**, 161–216.

VILLAMIL, T. 1999. Campanian–Miocene tectonostratigraphy, depocenter evolution and basin development of Colombia and western Venezuela. *Palaeogeography, Palaeoclimatology, Palaeoecology*, **153**, 239–275.

VILLAMIL, T. & PINDELL, J. L. 1998. Mesozoic paleogeographic evolution of northern South America: foundations for sequence stratigraphic studies in passive margin strata deposited during nonglacial times. *In*: PINDELL, J. L. & DRAKE, C. (eds) *Paleogeographic Evolution and Non-glacial Eustasy,*

Northern South America. SEPM, Special Publications, **58**, 283–318.

VILLAMIL, T., ARANGO, C. & HAY, W. W. 1999. Plate tectonic paleooceanographic hypothesis for Cretaceous source rocks and cherts of northern South America. *In*: BARRERA, E. & JOHNSON, C. C. (eds) *Evolution of the Cretaceous Ocean–Climate System*. Geological Society of America, Boulder, CO, Special Papers, **332**, 191–202.

VIVAS, V. & MACSOTAY, O. 1995a. Formación Tememure: unidad olistostrómica Eocene superior–Oligoceno inferior en el frente meridional de la Napa Piemontina Venezuela nor-central. *Venezuela Ministerio de Energía y Minas, Boletín de Geología, Publicaciónes Especiales*, **10**, 95–113.

VIVAS, V. & MACSOTAY, O. 1995b. Dominios tectono-estratigráficos del Cretácico–Neógeno en Venezuela nororiental. *Venezuela Ministerio de Energía y Minas, Boletín de Geología, Publicaciónes Especiales*, **10**, 124–152.

WADGE, G. & MACDONALD, R. 1985. Cretaceous tholeiites of the northern continental margin of South America: the Sans Souci Formation of Trinidad. *Journal of the Geological Society, London*, **142**, 297–308.

WARR, L. N., GREILING, R. O. & ZACHRISSON, E. 1996. Thrust-related very low grade metamorphism in the marginal part of an orogenic wedge, Scandinavian Caledonides. *Tectonics*, **15**, 1213–1229.

WARR, L. N., PRIMMER, T. J. & ROBINSON, D. 1991. Variscan very low-grade metamorphism in southwest England: a diastathermal and thrust-related origin. *Journal of Metamorphic Geology*, **9**, 751–764.

WEBER, J. C. 2005. *Neotectonics in the Trinidad and Tobago, West Indies Segment of the Caribbean–South American Plate Boundary*. Geological Institute of Hungary, Occasional Papers, **204**, 21–29.

WEBER, J. C. 2007. Neotectonics in Trinidad and Tobago. *Fourth Geological Conference of the Geological Society of Trinidad and Tobago*, extended abstracts CD.

WEBER, J. C., DIXON, T. H. ET AL. 2001a. GPS estimate of relative motion between the Caribbean and South American plates, and geologic implications for Trinidad and Venezuela. *Geology*, **29**, 75–78.

WEBER, J. C., FERRILL, D. A. & RODEN-TICE, M. K. 2001b. Calcite and quartz microstructural geothermometry of low-grade metasedimentary rocks, Northern Range, Trinidad. *Journal of Structural Geology*, **23**, 93–112.

WIELCHOWSKY, C. C., RAHMANIAN, V. D. & HARDENBOL, J. 1991. A preliminary tectonostratigraphic framework for onshore Trinidad (abstract). *Second Geological Conference of the Geological Society of Trinidad & Tobago, Transactions*, **41**.

WILSON, J. T. 1966. Are the structures of the Caribbean and Scotia Arc regions analagous to ice rafting? *Earth and Planetary Science Letters*, **1**, 335–338.

WILSON, C. C. 1968. The Los Bajos Fault. *Fourth Caribbean Geological Conference, Port of Spain, 1965, Transactions*, 87–90.

WYNN, J. C. 1993. Placer titanium and other heavy minerals. *In: Geology and Mineral Resource Assessment* of the Venezuelan Guayana Shield. US Geological Survey, Bulletin **2062**, 89–90.

YSACCIS, R., CABRERA, E. & DEL CASTILLO, H. 2000. El sistéma petrolífero de la Cuenca de la Blanquilla, costa afuera Venezuela. *VII Simposio Bolivariano de la Exploración Petrolera en las Cuencas Subandinas, Memoria*, 411–425.

ZAMBRANO, E., VASQUEZ, E., DUVAL, B., LATREILLE, M. & COFFINIERES, B. 1970. Synthèse paléogéographique et pétrolière du Venezuela occidental. *Revue du l'Institut Français de Pétrole*, **25**, 1449–1492. [Republished by Editions Technip.]

ZAMBRANO, E., VASQUEZ, E., DUVAL, B., LATREILLE, M. & COFFINIERES, B. 1971. Síntesis paleogeográfica y petrolera del occidente de Venezuela. *IV Congreso Geológico Venezolano, Memorias, Boletín de Geología, Publicaciónes Especiales*, **5**, 483–552.

The vanishing Carib Halite Formation (Neocomian), Colombia–Venezuela–Trinidad prolific petroleum province

ROGER HIGGS

Geoclastica Ltd, 6 Ockham Court, 22 Bardwell Road, Oxford OX2 6SR, UK
(e-mail: rogerhiggs@geoclastica.com)

Abstract: Literature survey reveals evidence of a thick (*c.* 3 km), largely dissolved, Berriasian–Valanginian 'Carib Halite Formation', deposited in a Jurassic to Coniacian 'Carib Graben' from Colombia through Venezuela to Trinidad. From Campanian time the graben inverted into the (partly metamorphosed) south-verging nappe/thrust-belt of northern South America (Guajira to Trinidad), and later the bivergent Eastern Cordillera and Mérida Andes. Outcropping halite is confined to Colombia. Numerous lines of evidence suggest *buried* halite in Colombia, Venezuela and Trinidad: drowning coastal geomorphology (halite-solution subsidence); subaerial and submarine closed depressions (solution pits); saline springs; mud-volcano-fluid analysis; heat-flow anomalies; gravity anomalies; and thrust-belt structural style. Other data suggest thick (kilometres) *vanished* halite: a Berriasian regional faunal gap; highly organic shales/phyllites (solution residues); intense fracturing (solution collapse); metamorphic-grade discontinuities (solution weld); and Neogene supraorogen basins attributable to buried-halite solution, for example Gulf of Venezuela, Monagas thrust belt, Gulf of Paria and onshore Trinidad. Halite solution by circulating meteoric water is inferred to have begun near 11 Ma, reflecting climate change (wetter) accompanying the onset of the Gulf Stream, induced by collision of the Panamá Arc against Colombia, interrupting deep-sea Caribbean–Pacific interchange. The Carib Halite concept has important implications for exploration in the oil-rich thrust belts of Colombia, Venezuela and Trinidad. There is also potential for new finds of emeralds and other evaporite-associated minerals.

Based on a plate tectonic model of the Caribbean–Gulf of Mexico region and on the presence of drilled anhydrite in the Gulf of Paria separating Trinidad from Venezuela, Pindell & Kennan (2001*b*, fig. 4) postulated a 'thick salt' basin throughout Trinidad and NE Venezuela, coeval with the Callovian–Oxfordian Louann Salt of the Gulf of Mexico. This idea is expanded here by the suggestion that a 'Carib Halite Formation' accumulated in an intra-Pangaean rift complex ('Carib Graben') from Trinidad to Colombia (Figs 1 & 2). The idea is based on literature survey (Higgs 2006) indicating subsurface halite hidden by deep burial and/or solution, except for outcropping diapirs near Bogotá in the Eastern Cordillera, Colombia. The South American, southern 'half' of the Carib Graben separated from its counterpart in the Chortís–Greater Antilles region during Proto-Caribbean ocean spreading (Higgs 2009, fig. 2). The halite is post-Louann, earliest Cretaceous in age, coinciding with a profound eustatic low (Haq *et al.* 1988), confining halite deposition to the topographically lowest subgrabens of the Carib Graben. As shown below, an Eastern Cordillera subgraben contained halite, interfingering with marine rift deposits toward the palaeo-Pacific ocean. Northward the halite-depositing area widened, embracing much of NW, north-central and NE Venezuela, and all of Trinidad. Much of this area was inverted from Late Campanian time by Proto-Caribbean

amagmatic subduction, forming a south-verging continental nappe- and thrust-belt (Higgs 2009). An Outer Nappe moved southward over an Inner Nappe, metamorphosing rift cover now exposed as the Caracas and Caribbean Gps (among other units), containing evidence for thick dissolved halite discussed below. An inboard fold and thrust belt corresponds to the halite southern fringe, where no metamorphism occurred. The Eastern Cordillera–Mérida Andes trough was inverted bivergently from the Miocene to the present day. In the Maracaibo Basin, the deposited halite was probably thin in general, based on the sharp NW limit of the Mérida Andes (inverted subgraben) and the lack of diapirs on seismic profiles in the tectonically little-disturbed SW region (Cooney & Lorente 1997; Castillo & Mann 2006).

Halite is very prone to partial or complete subsurface solution, almost from the time of deposition (Warren 1999). Solution can occur by lateral inward withdrawal or by vertical retreat of the base or top (Gustavson *et al.* 1980; Warren 1999). Solution is enhanced by tectonic uplift, allowing meteoric (rain) water to circulate deep (kilometres) and far (tens to hundreds of kilometres) into a basin (Warren 1999; Zielinski *et al.* 2007). Most ancient evaporite deposits show thinning, especially near the edges where flushing by meteoric and basinal waters is most likely to occur (Warren 1999). Evaporite intervals

From: JAMES, K. H., LORENTE, M. A. & PINDELL, J. L. (eds) *The Origin and Evolution of the Caribbean Plate.*
Geological Society, London, Special Publications, **328**, 659–686.
DOI: 10.1144/SP328.26   0305-8719/09/$15.00 © The Geological Society of London 2009.

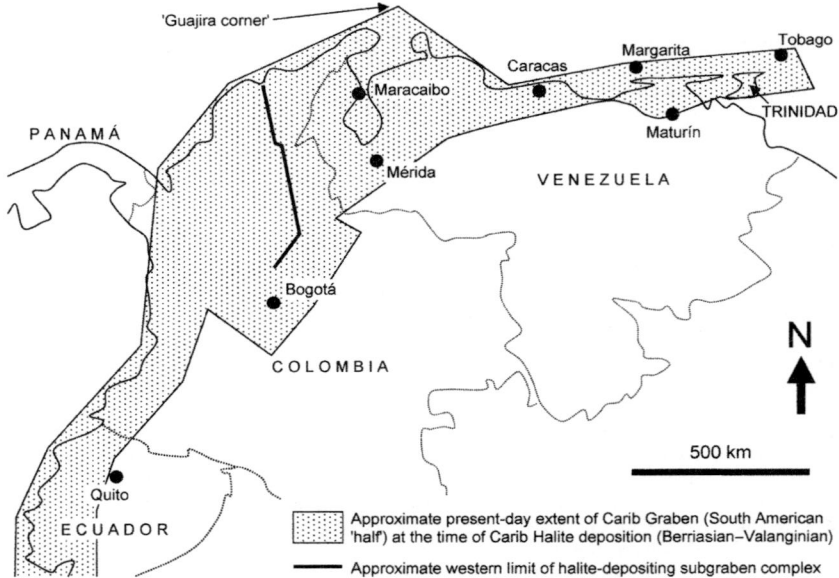

Fig. 1. Map showing interpreted extent of the Carib Graben (South American portion) at the time of Carib Halite deposition (Berriasian–Valanginian). Also shown is the inferred western limit of the main halite-depositing subgraben complex. The 'missing half' of the graben (not shown) was separated from South America by Proto-Caribbean spreading (Santonian–Campanian), and is now predicted to underlie the southern fringe of the detached continental pieces, namely eastern Nicaragua Rise, Cuba, Hispaniola and Puerto Rico (Higgs 2009, fig. 2). No halite has yet been reported from these regions. The zigzag shape of the modern Colombia–Venezuela–Trinidad continental margin (note 'Guajira corner') is inherited from Proto-Caribbean opening. Halite in Trinidad and Venezuela interfingered SW with marine deposits in Colombia (see text). The inner boundary faults of the graben at that time, reactivated later as thrusts and lateral ramps, were the following: southern Romeral; Upper Magdalena 'Falla 2'; Eastern Cordillera frontal; Chitaga–Pamplona; Mérida–central Venezuela frontal; Urica; eastern Venezuela frontal; Soldado; Trinidad south coast (for locations see: Babb & Mann 1999; F. A. Audemard *et al.* 2000; Paris *et al.* 2000; Duarte 2003). The graben expanded cratonward during Late Cretaceous time, by backstepping of the boundary (Higgs 2009).

can dissolve completely, leaving thin solution residues. These remnants, and consequently an important episode of basin history, are easily overlooked (Kyle 1991; Warren 1999).

Halite indicators discussed below show that the depositional area included the Eastern Cordillera of Colombia, much of western Venezuela and the entire nappe-thrust belt of Venezuela and Trinidad (Higgs 2009). Nevertheless, despite a century of oil exploration in northern South America, the Carib Halite remains undetected except near Bogotá. There are three reasons. First, solution has removed the halite at outcrop and over large areas underground. Second, there is a scarcity of drilling in the mountain interiors, particularly where metamorphic rocks crop out, and of deep boreholes (reaching Neocomian) in the foothills thrust belts (Mérida, Barinas, central and eastern Venezuela) and below the Neogene supraorogen basins (e.g. Gulf of Paria and onshore Trinidad). A few Gulf of Paria wells have possibly penetrated the Carib Halite solution weld. Third, detection by seismic

reflection is frustrated not only by poor data quality due to structural complexity and/or great depth (kilometres), but also by acoustic impedance similarity where the enclosing strata are indurated shale or sandstone. Halite diapirs, readily visible on seismic profiles in undeformed basins like the Gulf of Mexico, in orogenic belts tend to be distorted, squeezed shut or covered by thrust sheets (Hudec & Jackson 2006). Seismic *refraction* is ineffective, because halite velocity is not exceptionally high and overlaps that of sandstone and limestone. Gravity, on the other hand, does suggest halite locally in Trinidad and Venezuela (see below).

Age of the Carib Halite

A Berriasian and Lower Valanginian age of the Carib Halite is based on the following three lines of evidence:

(1)   There is a near-complete or complete lack of Berriasian fossils in Venezuela and Trinidad,

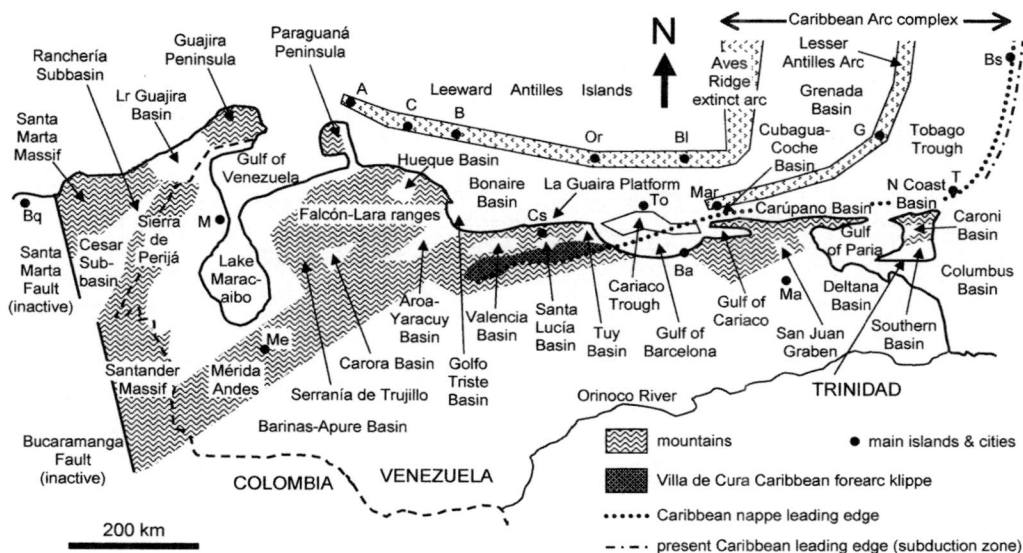

Fig. 2. Location map showing mountains and basins of Venezuela and Trinidad. Cities: Ba, Barcelona; Bq, Barranquilla; Cs, Caracas; M, Maracaibo; Ma, Maturín; Me, Mérida. Islands: A, Aruba; Bl, Blanquilla; B, Bonaire; Bs, Barbados; C, Curaçao; G, Grenada; Mar, Margarita; Or, Orchila; T, Tobago; To, Tortuga.

consistent with strata of this age being almost all halite that has disappeared.

(2) Berriasian and Valanginian marine fossils occur near the Bogotá halite diapirs but in beds of uncertain structural relationship (McLaughlin 1972). Shale xenoliths and/or ruptured interbeds in the halite have younger, Hauterivian and/or Barremian palynological ages (López *et al.* 1991), favouring the xenolith interpretation.

(3) The Late Berriasian–Early Valanginian extreme eustatic lowstand (Hardenbol *et al.* 1998 after Haq *et al.* 1988) is likely to have facilitated halite deposition in the Carib Graben by restricting the connection to the world ocean. López *et al.* (1991) noted the probable correlation between the eustatic low and the Bogotá halite.

Known evaporites in NW Colombia, Venezuela and Trinidad

Halite

Lower Cretaceous halite has been mined at several localities near Bogotá. McLaughlin (1972) argued, contrary to most contemporaneous literature, that the halite masses were not diapirs but were in their correct stratigraphic position. This is difficult to

reconcile with his cross sections that show the halite as local bodies confined to thrusted anticlinal cores and pinching out laterally. López *et al.* (1991) reiterated the traditional diapir interpretation, in which case the stratigraphic base of the halite is unknown. The Bogotá halite is interpreted here as coeval and originally contiguous with the Carib Halite. There are no published reports of halite in Venezuela or Trinidad.

Anhydrite/gypsum

In Trinidadian waters of the Gulf of Paria, three boreholes penetrated the anhydrite-dominated Couva Marine Fm (Saunders 1997*b*), loosely dated as Upper Jurassic or Lower Cretaceous from limited palaeontological and radiometric data (Bray & Eva 1987; Eva *et al.* 1989). One well proved a thickness exceeding 2 km (Bray & Eva 1987), although seismic data suggest that thrusting is involved (Flinch *et al.* 1999). The anhydrite was interpreted as a possible diapir by Algar & Pindell (1993), but anhydrite is too dense for diapirism. Anhydrite may reach westward to the Venezuelan side of the Gulf, where a thrust-duplicated thickness of about 4 km is interpreted seismically (Flinch *et al.* 1999). The Couva anhydrite is inferred here to be of Tithonian and/or Berriasian age, immediately predating the Carib Halite (Fig. 3). Rather

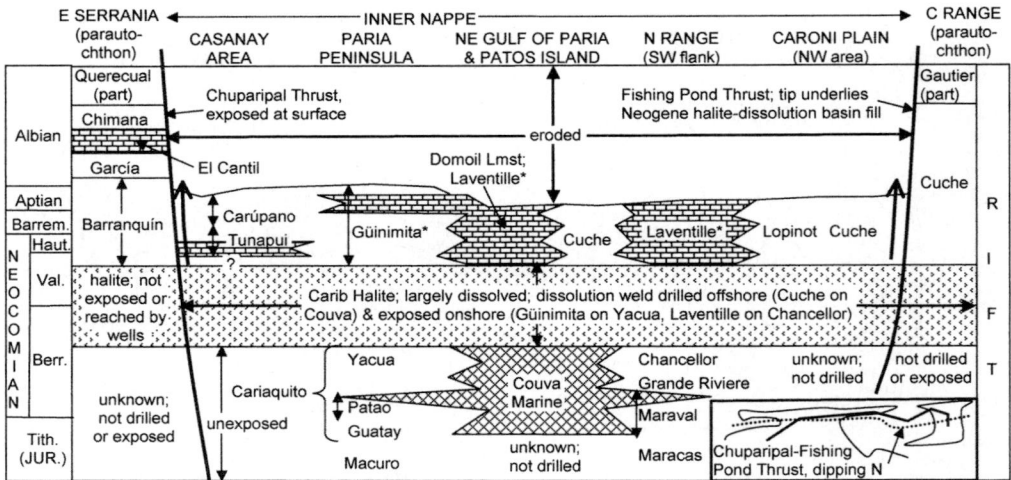

Fig. 3. Interpreted Uppermost Jurassic–Lower Cretaceous stratigraphy of northern fringe of eastern Venezuela and Trinidad, showing inferred stratigraphic position of the Carib Halite Fm. The entire interval shown is interpreted as rift phase (Higgs 2009). In the Northern Range, Maracas under Maraval rather than vice versa is after Kugler (1961*a*; contrast Potter 1973; Saunders 1997*a*). Casanay and Paria successions from González de Juana *et al.* (1968, 1972, 1980), except age ranges adjusted (see text). Gulf of Paria data in part from Persad (1983), Babb & Mann (1999) and Flinch *et al.* (1999). Chuparipal and Fishing Pond thrusts from Metz (1968*a*) and Kugler (1961*a, b*). Nappe interpretation from Higgs (2009). Inner Nappe successions show low-grade (downward increasing) metamorphism, due to former burial under Outer Nappe, hence overmature Cuche and recrystallization of Güinimita and Laventille limestones (impedes dating; see text). The only firm age in the metasediments below the Carib Halite is a Tithonian ammonite in the Maraval Fm (Potter 1997). Couva anhydrite correlation with Patao anhydrite from Saunders (1997*b*). Patao-equivalent anhydrite tongue near top of Maraval based on Potter (1973) report of collapse folds.

than recording 'a succession of spills of Atlantic Ocean water into a fault bounded basin' (Bray & Eva 1987), the Couva anhydrite may reflect evaporite deposition confined to the extreme end of a long, *Pacific*-connected, marine gulf of otherwise normal salinity occupying the Carib Graben. An eastern connection to the Atlantic may have been blocked by the Bahama Platform, sliding past northwestward along a transform fault (cf. Pindell & Kennan 2001*b*, figs 4–6; Higgs 2009, fig. 2). Subsequent eustatic lowering, impairing the marine connection via Colombia, and possibly combined with increasing aridity, led to Carib Halite deposition along much of the graben, as discussed below.

In Trinidad, the Laventille Fm (Fig. 3) outcrop contains gypsum (hydrated anhydrite) lenses up to 10 m thick (Kugler 1959; Potter 1974; Rodrigues & Scott 1985), interpreted below as part of the Carib Halite solution residue.

Gypsum beds 2–120 m thick, of kilometric (min.) lateral extent, crop out in the Río Negro Fm of the Mérida Andes and in Nirgua Fm ('Fase') and Cariaquito Fm (Patao Mbr; Fig. 3) metasediments of the Cordillera de la Costa and Paria Peninsula, all three of confirmed or probable Early Cretaceous age (Bellizzia & Rodríguez

1976; González de Juana *et al.* 1968, 1972, 1980; LEV 1997). In the central Venezuela thrust belt, anhydrite–gypsum lenses and masses of uncertain stratigraphic affinity occur in the Tememure Fm or structural complex (Evanoff 1951; Rodriguez 1985; Rosales & Macsotay 1985; Vivas & Macsotay 1995*a*; LEV 1997). In western Venezuela, anhydrite cement in Gobernador Fm sandstones (Toro & Mahmoudi 1997) may reflect reprecipitation after solution of deeper anhydrite.

Evidence for buried halite

'Drowning coast' geomorphology

The Gulf of Barcelona and adjoining Gulf of Cariaco (Fig. 2) interrupt the Coastal and Interior ranges of Venezuela, suggesting mountain collapse despite the current Caribbean–South America transpression, albeit highly oblique (Higgs 2009). Paige (1930, p. 9) noted that 'The eastern portion of the north coast of Venezuela including the north coast of Trinidad is physiographically "drowned". The Gulf of Cariaco is but a flooded valley, in part at least of structural origin'. In the adjacent eastern end of Barcelona Bay, 'one finds the drowning-coast physiography of the Gulf of Santa Fé' (González de Juana *et al.* 1980, p. 738, author's translation). Folds can be correlated from islands in the gulf to the mountains to the east (González de Juana *et al.* 1980), testifying to collapse of the ground between. The Gulfs of Barcelona and Cariaco are interpreted as halite-solution basins reflecting large-scale solution of buried halite since about 11 Ma (see below).

Raised beaches in the eastern Northern Range of Trinidad indicate uplift (Weber 2005), consistent with transpression. In contrast, alluviating valleys and deeply indented bays of the western Northern Range Peninsula and Paria Peninsula indicate drowning (Paige 1930; González de Juana *et al.* 1972; Kugler 1961*a*; Saunders 1997*a*; Ritter & Weber 2005; Weber 2005). Weber (2005) related drowning to pull-apart in the Gulf of Paria, in progress since Caribbean Plate relative motion changed from SE to nearly due east, *c.* 2.5 Ma (Higgs 2009). However, the El Pilar northern master fault of the pull-apart runs south of the peninsulas. Instead, drowning is attributed here to subsurface halite solution, outweighing transpressive uplift. Continuing uplift in the east implies that large-scale solution has not begun there yet.

In Colombia, the northwestern coast of Guajira Peninsula (Fig. 2) is highly indented, again suggesting coastal drowning, possibly due to buried-halite solution.

Closed depressions

In eastern Venezuela, surficial closed depressions on the Mesa Fm (Hedberg 1950) in the Morichito Basin (supraorogen, at Eastern Serranía mountain front) resemble those formed by solution of buried halite (Gutierrez *et al.* 2001) or limestone (karst), supporting the suggestion below that the Morichito is a Neogene halite-withdrawal and -solution basin.

The Dragon's Mouth marine strait interrupts the Paria Peninsula–Northern Range highland chain (Fig. 2), again suggesting mountain collapse. Horst-and-graben bathymetry (Wall & Sawkins 1860; Van Andel & Postma 1954; Bassinger *et al.* 1971) indicates recent faulting. Kugler (2001, pp. 273, 283) referred to 'graben-like troughs' between islands in the Dragon's Mouth and to an 'extraordinary deep depression . . . where depths of as much as 1200 feet were observed'. This and other closed bathymetric depressions (Kugler 1961*a*) are attributed here to active solution of buried halite. Similarly, 'peculiar' closed troughs in the adjacent Gulf of Paria (Van Andel & Postma 1954, p. 25) are interpretable as solution depressions. North of Dragon's Mouth, surficial pockmarks on the Trinidad shelf (Rietman &

Pallister 2007) possibly have the same origin. Neogene basins under the shelf and the Gulf of Paria are attributed below to solution subsidence.

Intramontane plains

In western and central Venezuela, the intramontane La González (Mérida), Carora, Hueque, Yaracuy, Lake Valencia and Caracas plains (González de Juana *et al.* 1980) are interpreted as the surface expression of halite-solution basins.

Saline springs

Saline springs commonly indicate active subsurface halite solution (Gustavson *et al.* 1994; Warren 1999). In Colombia, saline springs are associated with outcropping halite and also extend for hundreds of kilometres beyond (McLaughlin 1972; López *et al.* 1991; Branquet *et al.* 2002). In Venezuela, thermal springs are common in all highland regions, from the Perijá, Mérida and Falcón mountains in the west to the Eastern Serranía (Urbani 1977, 1989, 1991). The few springs for which chemical analyses have been published include examples with Na–Cl composition (Urbani 1989), compatible with halite solution. A few springs are saline enough to deposit gypsum and ephemeral halite (Urbani 1989, 1991).

Reported saline springs are less common in Trinidad. The Warm Springs are located 250 m from where the Warm Springs Fault intersects the Gulf of Paria coast (Kugler 1959, 1996), consistent with the idea that Gulf subsidence is partly due to halite solution. The spring water is slightly saline, attributed to a deep source mixing with near-surface fresh water (Suter 1960). In SE Trinidad, Salt Spring on the Salt Spring Fault feeds the Salt River (Kugler 1959, 1961*a*, 1996). There appear to be no published water analyses. The salt spring water is responsible for almost an acre of land bare of trees (Kugler 2001).

Heat-flow anomalies

Abnormal temperatures occur in a few 'hot wells' in Venezuela and Trinidad. The anomalies could indicate the chimney effect (Mello *et al.* 1995) of a halite or anhydrite (high thermal conductivity) local thickening, such as a diapir, a subgraben fill or a dissolving residual body.

In western Venezuela, a borehole on the SE Falcón shelf encountered a temperature of 177 °C at 3584 m (PDV 1997*a*). The expected temperature at this depth is only 115 °C (assuming 25 °C average surface temperature and normal global geothermal gradient of 25 °C km^{-1}). The Centro and Lago fields of Lake Maracaibo have anomalously high thermal gradients (González de Juana *et al.* 1980); they lie near the edge of a Jurassic intragraben low (Lugo & Mann 1995), where halite could be present.

In central Venezuela, high heat flow of unknown cause characterizes the northernmost Guárico Subbasin south of the thrust front (F. E. Audemard & Serrano 2001). Here, the Yucal–Placer Gasfield has the highest (56 °C km^{-1}) geothermal gradient known in the Eastern Venezuela Basin (Daal & Lander 1993). Seismic profiles (Aymard *et al.* 1985) show no indication of a chimney-like evaporite body below the deepest drilled rocks. The southern depositional limit of the Carib Halite is thought to coincide with, and control the position of, the Central Serranía thrust front, nearby (10–30 km) to the north. A possible explanation of the heat-flow anomaly is a mega-chimney effect due to thick (kilometres) halite under the thrust belt. At the eastern extreme of the same subbasin, subsurface temperatures in the Anaco area are higher than those to the north and south (Funkhouser *et al.* 1948). Again, a mega-chimney effect of halite under the adjacent Eastern Serranía mountains may be responsible. High heat flow due to upward flow of groundwater recharged from the mountains (Zielinski *et al.* 2007) is not likely, since no hot springs are reported in the Guárico Subbasin (Urbani 1991).

Borehole temperature logs suggest that 'hydrothermal brines' ascend the Pirital Thrust under Morichito Basin (Roure *et al.* 2003, fig. 13d), possibly dissolving halite. Fluid-inclusion analysis of a Morichito Fm (Neogene) sandstone sample from about 2 km depth indicated that quartz cementation took place at a temperature of 140 °C (Roure *et al.* 2003), much higher than the value of about 75 °C expected at that depth and consistent with high *palaeo*-heat flow.

Anomalously hot wells in the Gulf of Paria and southern Trinidad (Rodrigues 1990) include those on the Avocado High (Babb & Mann 1999), underlain by the Couva anhydrite, explaining the high heat flow there.

Gravity anomalies

In western Venezuela, a closed, undrilled gravity low in the Gulf of Venezuela (Fig. 2; Pimentel 1984) coincides with the Urumaco Trough (Boesi & Goddard 1991; Macellari 1995), but might also reflect a buried halite body (e.g. a subgraben fill). The Gulf is interpreted below as a halite-solution basin.

At the north extreme of the Eastern Serranía, the Carúpano gravity low was attributed by Graterol & Wong (1978) to a buried basin. Some of Venezuela's most Na–Cl-rich springs, capable of precipitating halite, lie along the El Pilar Fault (Urbani 1989) within the outline of the gravity anomaly. The

anomaly is attributed here to a halite diapir or solution remnant.

In eastern Venezuela and southern Trinidad, a gravity super-low trends ENE (Bonini 1978), enclosing a belt of mud volcanoes (Higgins & Saunders 1974; Duerto & McClay 2002). The oil-rich Monagas–southern Trinidad thrust belt (Ablewhite & Higgins 1968; Aymard *et al.* 1990) coincides with the anomaly. The axis of the anomaly lies slightly (<10 km) north of the buried deformation front (e.g. Pimentel 1984) of the combined Proto-Caribbean/Caribbean orogen (Higgs 2009). Thick Oligo-Miocene shales (Carapita Fm; Parnaud *et al.* 1995b) contribute to the anomaly but cannot alone explain it (Bonini 1978; Russo & Speed 1992; Passalacqua *et al.* 1995). A subgraben containing thick (kilometres) halite is invoked here, deep beneath a Neogene halite-solution basin atop the thrust belt (see below). The southern limit of the halite coincides with (controls) the position of the deformation front, by thrust reactivation of the graben-bounding normal fault. Supporting this model, an undrilled Jurassic–Cretaceous graben, here named Maturín Graben, has been interpreted seismically under the Monagas basal thrust (Duerto & McClay 2002, figs 1–3; Roure *et al.* 2003, fig. 3). This feature is interpretable as the southernmost subgraben of the Carib Graben complex. Thick halite here can explain: (a) the Neogene basin (solution subsidence); (b) the gravity anomaly; (c) possible velocity pull-up (Roure *et al.* 1994); and (d) the enigmatic Santa Barbara–Masacua Fault, a steeply south-dipping, Neogene, normal fault offsetting Cretaceous formations by at least 1 km (De Sisto 1964; Lamb & Sulek 1968). This can be interpreted as a Jurassic–Cretaceous fault delimiting the north side of the subgraben, reactivated from 11 Ma by differential halite solution.

In Trinidad a local 'puzzling gravity minimum' (Nettleton 1949, p. 1159) occurs inside the superlow. This anomaly is possibly also halite related.

Mud-volcano fluid composition

Trinidad mud-volcano fluids are contributed in part by a deep (kilometres) source of water with an Na:Cl ratio of 1.0 (Dia *et al.* 1999), strongly suggestive of dissolving halite (Haese *et al.* 2006; Heeschen *et al.* 2008) and compatible with halite solution under Trinidad (A. Dia 2008, pers. comm.).

Fold- and thrust-belt structural style

Fold–thrust belts detaching on evaporites typically show long, symmetrical, upright folds with wide synclines and sharper anticlines (Davis & Engelder 1985). These features broadly characterize the

Falcón, Eastern Serranía and central and southern Trinidad fold-thrust belts (Bellizzia 1976; Kugler 1961a). The Eastern Serranía structural style was likened by Vivas (1986, thesis cited by Chevalier 1993) to that of the classic, evaporite-based Jura mountains.

At the front of the Eastern Serranía, seismic and well data suggest that a major décollement may exist 'in hypothetical Late Jurassic-Lower Cretaceous salt layers' (Roure et al. 2003, p. 753). Except where missing by solution, the inferred Carib Halite would underlie the Barranquín, the oldest formation drilled or exposed in eastern Venezuela (Fig. 3). Thick halite below the Eastern Serranía would solve the space problem identified by Hung (2005), whereby if basement is not involved in the thrusting, as appears likely, a thick (kilometres) interval of pre-Barranquín rocks must occupy the space between the lowest thrust and the undeformed top of the basement, dipping gently north under the Serranía. Indeed, Morichito Basin seismic profiles show evidence for about 4 km of Carib Halite at depth (see below).

Detachment in the Gulf of Paria and Trinidad has been interpreted as the top (Roure et al. 1994) or base (Flinch et al. 1999) of the Couva anhydrite.

Evidence for vanished (dissolved) halite

Enigmatic southern anticline

On seismic profiles, the deformation front from NW Monagas (El Furrial Oilfield region; Aymard et al.

1990) and southern Trinidad coincides with a curious anticline of uncertain origin. Unlike fault-bend folds, the anticline is symmetrical and relatively sharp. The anticline is fault-cored according to some authors (Kugler 1961b, cross section 7; Di Croce et al. 1999), while others attribute it to intrusion of a wall-like mud diapir (Pindell & Kennan 2001a; Duerto & McClay 2002). The interpretation preferred here is thrust-fault reversal by underground solution of the halite held responsible for the gravity super-low (Figs 4 & 5). Strata in the northern (hanging) wall dip north due to normal-fault drag, having subsided into lateral juxtaposition (across the fault) with older strata dipping south due to earlier thrust drag, defining an 'inverted-thrust anticline'. Thinning occurs on both flanks of the anticline, toward the axial fault (Duerto & McClay 2002, 2007). Similar, smaller-scale thinning and uptilting of Pleistocene beds toward a normal fault in Italy was interpreted as drag by Massari et al. (2002). Locally a second, parallel anticline is present (Duerto & McClay 2007), indicating a second (intra-graben) inverted thrust. Mud volcanoes are fed by injection of overpressured Oligo-Miocene mud up the fault plane in the anticline core (Dia et al. 1999; Kugler 2001, cross section 7; Duerto & McClay 2007), rather than directly by mud diapirs (Higgins & Saunders 1974). Mud-volcano lineaments (Higgins & Saunders 1974; Duerto & McClay 2002) confirm more than one contributing fault.

During solution subsidence, uplift by *slow* north–south shortening due to Proto-Caribbean

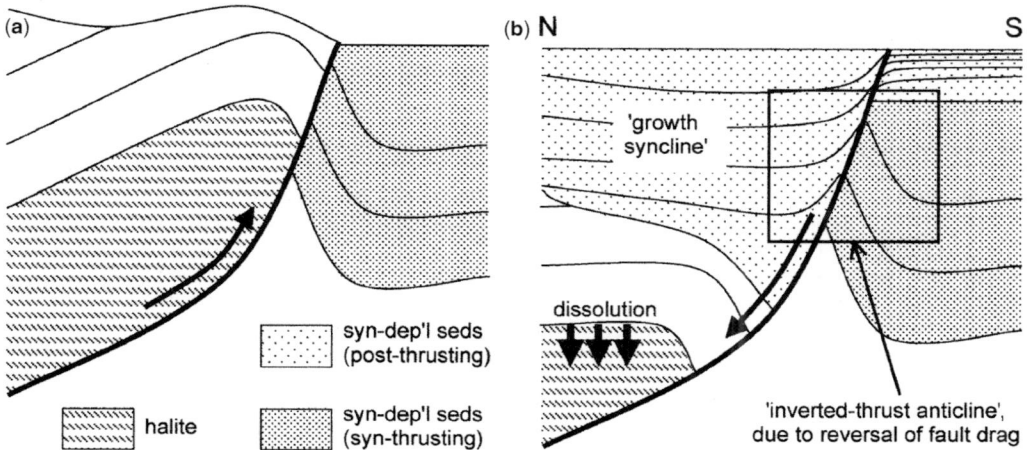

Fig. 4. Sketch sections showing inferred origin of enigmatic structure at the deformation front of Monagas thrust belt (Venezuela), interpreted as an 'inverted-thrust anticline'. (**a**) Thrusting on a sub-halite décollement. Foreland-basin strata beside the thrust dip south due to fault drag. (**b**) Following a change to a wetter climate, underground halite solution causes reversal of fault movement above the halite, producing a solution basin with opposed, northward fault-drag dip. During solution, thrusting can continue below the halite (see text). dep'l, depositional; seds, sediments.

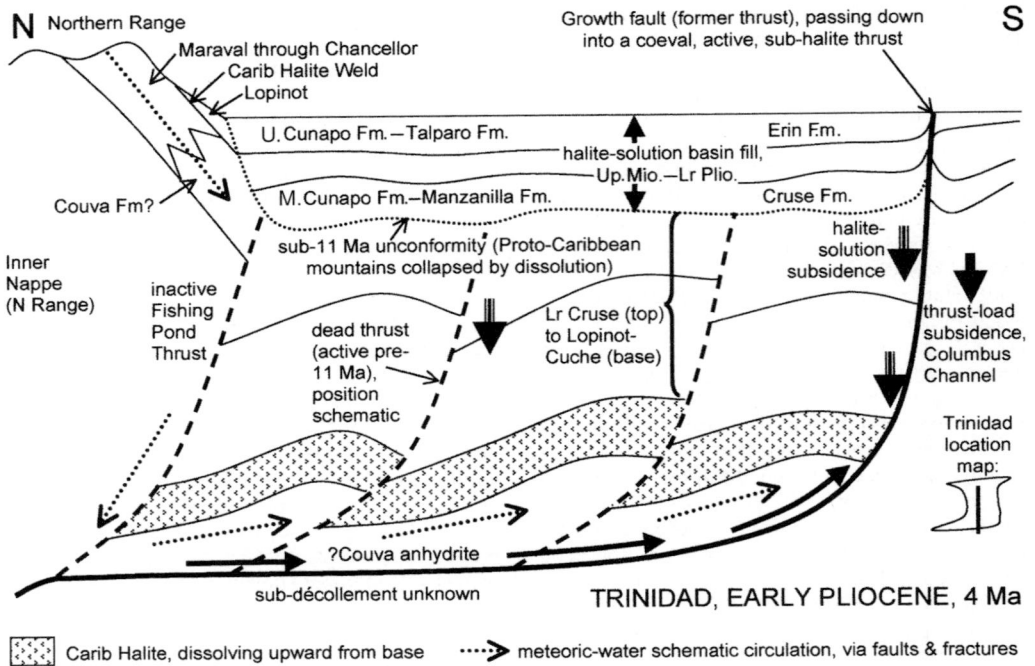

Fig. 5. Interpreted sketch section (north–south) through Trinidad at 4 Ma (Early Pliocene). From Campanian time until the Caribbean Arc arrived from the west (c. 3.5 Ma), Trinidad was a shortening thrust belt driven by slow Proto-Caribbean subduction southward. Uplift was overtaken by halite-dissolution subsidence from about 11 Ma, following a change to a rainier climate shortly before (see text). Nevertheless, Proto-Caribbean convergence continued, and is inferred to have caused syn-depositional shortening in the solution-basin fill, shown here schematically as growth folds with amplitude increasing downward. At 3.5 Ma, post-dating this diagram, Caribbean arrival transpressively shortened the solution-basin fill; uplift outpaced solution subsidence and the basin was split by erosion (bare Central Range) into the present northern (Caroni) and southern portions. The Carib Halite, shown dissolving upward here in Trinidad, can potentially dissolve downward too further west (Gulf of Paria, Eastern Serranía), where it is overlain by Cuche-shale-equivalent fractured limestone or sandstone aquifers (Laventille–Domoil–Barranquín; Fig. 3). Note the 'inverted-thrust anticline' at upper right (see Fig. 4). For simplicity, diagram omits (**a**) thrusting on post-halite décollements, and (**b**) intra-solution-basin growth faults.

subduction (Higgs 2009) continued below the sub-halite decollement level, but above was outpaced by solution subsidence. Arrival of the transpressively colliding Caribbean Arc produced a wave of *fast* shortening, migrating east, diachronously accelerating uplift and inverting the solution basin. At any longitude, shortening died after the wave had passed, that is following passage of the Caribbean nappe suture point (Higgs 2009). Because of this diachronism, the details of the frontal anticline and solution-basin age vary from west to east as follows. In the far west (NW Monagas), the Caribbean wave occurred c. 14–10 Ma, hence solution subsidence (11–0 Ma; see below) took place atop a dead thrust; simultaneous subsidence south of the fault was by compaction (thrust loading over; Fig. 4). Slightly to the east, in the Maturín region (Fig. 2), the Caribbean wave was later

(c. 12–8 Ma) and simply delayed the start of solution subsidence. Further east the wave *interrupted* solution subsidence and uplifted the basin, therefore an unconformity is predicted within the Upper Miocene–Quaternary section in NE Monagas, younging eastward across southern Gulf of Paria. During pre-uplift solution subsidence, the frontal fault was hybrid: a south-verging thrust passing upward, along the fault plane, into a down-to-the-north growth fault (Fig. 5); meanwhile foreland subsidence (south of the fault) was by loading plus compaction. In contrast, when solution subsidence resumed after uplift ended, the thrust was dead, so foreland subsidence was by compaction alone (Fig. 4). Compaction subsidence is eventually overtaken diachronously by tectonic rebound (Higgs 2009), explaining Pleistocene outcrop in the west v. Recent subsidence in the east (e.g.

Pimentel 1984). Further east, in Trinidad, the Caribbean wave arrived c. 3.5 Ma (Higgs 2009), inverting this sector of the solution basin diachronously (younger eastward across Trinidad). The frontal anticline was modified by renewed thrusting and offset by the Los Bajos transform fault (Higgs 2009). At the 2.5 Ma Caribbean relative motion change, the Trinidad sector of the plate boundary jumped from the south coast to the Central Range Fault, whose ENE trend is sufficiently oblique to Caribbean eastward relative motion for transpressive shortening and uplift to continue in southern Trinidad (Higgs 2009).

Caribbean-driven overpressuring probably initiated the Venezuela–Trinidad mud volcanoes, progressively later eastward. Mud volcanism in eastern Venezuela may have begun in Late Miocene or Early Pliocene time (Duerto & McClay 2007, p. 17). Mud volcanism is still active in both countries, indicating that overpressure persists.

Berriasian faunal gap in Venezuela–Trinidad

As mentioned above, the Carib Halite is interpreted to be Berriasian and Valanginian in age, partly because Berriasian fossils are almost entirely unknown in Venezuela and Trinidad. In contrast, in Colombia, Berriasian and Valanginian marine strata are well known in the Eastern Cordillera (Campbell 1962; Etayo-Serna et al. 1976; Branquet et al. 2002). The Carib Halite is inferred to have interfingered with these deposits in the Colombia–Venezuela border region, implying increasingly marine conditions toward the SW end of the Carib Graben, as discussed below. In the extreme north of Colombia, Berriasian and/or Valanginian fossils in Guajira (Renz 1960; Rollins 1965 cited by MacDonald & Opdyke 1972; MacDonald 1968) are interpreted here to reflect a northward marine incursion into the Carib Graben.

In the Perijá and Mérida mountains, crustaceans, palynomorphs, plants and radiometrically dated volcanics in the La Quinta Fm indicate a Late Triassic and Jurassic age (LEV 1997). The oldest confirmed Cretaceous fossils are Aptian (Renz 1982), although the Río Negro Fm contains non-marine fossils that are probably as old as Barremian (LEV 1997) or possibly Hauterivian (Useche et al. 1985, citing Macsotay 1975). To the north, Middle and Upper Jurassic ammonites occur in Paraguaná Peninsula (Fig. 2), Siquisique (Lara state) and Guajira (Renz 1960; MacDonald 1968; Bartok et al. 1985), in metasediments assignable to the Inner Nappe (Higgs 2009). The oldest reported Cretaceous fossils are Hauterivian and Barremian (MacDonald 1968; Stéphan 1985). Thus Berriasian and Valanginian fossils are unknown in western Venezuela. In

the Cesar Subbasin of Colombia (Fig. 2), an inferred faunal gap spans the Berriasian and most of the Valanginian, separating the La Quinta and Río Negro Fms (Mora & García 2006). In the Ranchería Subbasin, the Río Negro is thought to be Barremian or Early Aptian in age, implying a Neocomian faunal gap (Cáceres et al. 1980).

In central Venezuela, at the western end of the Coastal Range, tintinnids in metasediments are loosely dated in the range Tithonian–Hauterivian (Nirgua 'Fase' and Aroa Fm; LEV 1997). In the Caracas Gp, no faunal ages are known between Kimmeridgian bivalves and Albian ammonites (Bellizzia 1986; LEV 1997). In the thrust belt to the south, two microfaunas in the range Berriasian–Hauterivian and Berriasian–Valanginian were discovered in limestone olistoliths or fault slices (Furrer 1972; Beck & Furrer 1977). Based on these age ranges, the limestones could be as young as Hauterivian and Valanginian, and thus do not necessarily negate the Berriasian faunal gap, or they might represent a genuine Berriasian thin (decimetres to metres?) former interbed in the Carib Halite, reflecting a marine incursion coeval with the one in Guajira. Another limestone olistolith yielded a Tithonian gastropod (Vivas & Macsotay 1995a).

In eastern Venezuela, the oldest reported fossil is Valanginian, in the Barranquín Fm (Macsotay et al. 1985; Vivas & Macsotay 1995b; LEV 1997). The Cariaquito Fm (Patao Mbr) has yielded fauna suggesting an age within the range Neocomian–Barremian (González de Juana et al. 1968). In the overlying Güinimita Fm (Fig. 3), the oldest known fossils are Barremian (González de Juana et al. 1968, 1972).

The only known Jurassic fossil in Trinidad is a Tithonian ammonite from the Maraval Fm (Fig. 3; Hutchison 1939; Spath 1939). The Couva anhydrite contains unspecified (micro-?) fossils or palynomorphs of Upper Jurassic or Lower Cretaceous age (Bray & Eva 1987; Eva et al. 1989). The oldest firmly dated Cretaceous fossils in Trinidad are Valanginian (Saunders 1997a).

Collapse fracturing

The Chama Valley Breccia of the Mérida Andes is a mosaic pack-breccia (Giegengack 1984) like many evaporite-solution collapse breccias (Warren 1999). The breccia underlies the Quaternary La González Basin, of probable halite-solution origin (see below). The breccia, made from La Quinta Fm in one area and basement metamorphics elsewhere, was tentatively attributed to thrusting by Giegengack (1984), although he knew of no analogue. The collapse-breccia interpretation preferred here requires that the La Quinta and basement were first thrust over the Carib Halite, during uplift of the Mérida Andes.

In the Maracaibo Basin, intense fracturing responsible for exceptional oil productivity from Aptian–Albian Cogollo Gp carbonates in La Paz Field was attributed to tectonism by Nelson *et al.* (2000). However, the Río Negro and La Quinta rift-fill formations lie directly below (Nelson *et al.* 2000), therefore Cogollo fracturing may partly reflect solution collapse. Residual halite may account for anomalously saline formation water in La Paz (see below). Méndez (1989) described breccias at the base of the Cogollo Gp (Apón Fm) in three Lake Maracaibo wells and in Mérida Andes outcrops, interpreting them as karst collapse breccias, but halite-solution collapse is an alternative possibility.

In eastern Venezuela, El Cantil Fm limestones (Fig. 3) below the Morichito Basin are described as highly tectonized (PDV 1997*b*), suggesting intense fracturing. A collapse-breccia interpretation is supported by seismic evidence for basin-fill 'roll-over', attributed below to lateral halite withdrawal. Further north, near El Pilar, megabreccias and a structurally chaotic zone (Metz 1968*b*) may reflect solution collapse. Nearby strewn boulders of brecciated limestone (Higgs 2009) may come from a collapse breccia. Güinimita Fm rocks on the Paria Peninsula show considerable distortion (González de Juana *et al.* 1968), possibly due in part to halite solution.

In Trinidad, the Laventille Fm is highly faulted (González de Juana *et al.* 1968; Potter 1974; Rodrigues & Scott 1985). Thick cataclastic zones along the southern foot of the Northern Range were related by Weber *et al.* (2001) to faulting. Those involving the Laventille Fm are reinterpreted here as collapse phenomena, consistent with a jump in metamorphic grade across the contact with the underlying Chancellor Fm and a possible solution residue (see below). In the Northern Range there are 'Collapse folds ... (possibly) ... due to ... the solution of an evaporite that may have been present near the top of the Maraval Formation' (Potter 1973, p. 133). These folds could reflect a dissolved tongue of Couva-age halite or anhydrite in the Maraval Fm (Fig. 3).

No other potential collapse breccias are known to the author in Venezuela and Trinidad. This apparent scarcity may reflect modest fracturing by a gradual letting-down of the overburden as dissolution proceeds, implying top-down or base-up dissolution enabled by an aquifer above or below, rather than focused dissolution (e.g. edge-inward, or along cross-cutting faults), which can form breccias by collapse of solution cavities (Warren 1999). Indeed, the Carib Halite was widely under- and/or overlain by aquifers in the form of orogenically

fractured sandstone, limestone or schist/gneiss (e.g. Las Brisas, Las Mercedes, Chancellor, Laventille, La Quinta, Río Negro, Barranquín fms; Fig. 3).

Solution residues

Large-scale halite solution produces a non-laminated residue of formerly dispersed quartzose clay–silt, calcite–dolomite rhombs and graphite (former phytoplankton); plus deformed sandstone, shale, limestone or anhydrite former interbeds, if any (Warren 1999). The residue, too thin (tens of centimetres to tens of metres) to resolve seismically, corresponds to the solution 'weld' potentially recognizable on seismic profiles (cf. Jackson & Cramez 1989).

In Colombia (Eastern Cordillera), a Berriasian–Valanginian interval with gypsum and black shales (Branquet *et al.* 2002) could be partly halite-solution residue. Emeralds geochemically related to halite solution (see below) are locally associated.

In western Venezuela, the contact between the La Quinta and Río Negro sand-rich formations is conventionally interpreted as an unconformity (González de Juana *et al.* 1980; LEV 1997), based on probable missing time (faunal gap) and slight angular discordance on seismic profiles (e.g. Parnaud *et al.* 1995*a*, fig. 4B). However, the basal 200 m of the Río Negro Fm is mainly calcareous shale, locally carbonaceous, with beds of sandstone, limestone and gypsum (Useche & Fierro 1972; Useche *et al.* 1985). Stunted bivalves suggest hypersalinity (García Jarpa *et al.* 1980). The gypsum occurs as lenses (García Jarpa *et al.* 1980), suggesting that it may have partially dissolved. Based on these characteristics, the basal Río Negro interval could be (or include) the Carib Halite solution residue, representing a thick (kilometres) missing halite interval. The angular discordance may be an artefact of greater rotational subsidence before (and during?) halite deposition than after.

In eastern Venezuela, highly graphitic phyllites in the Upper Cariaquito Fm (Yacua Mbr; González de Juana *et al.* 1972) could represent the Carib solution residue, consistent with a discontinuity in metamorphic grade (see below) and marked disturbance of the overlying Güinimita Fm.

In Trinidad, the Lower Laventille Fm includes black graphitic shales associated with gypsum (Kugler 1959; Potter 1974, 1997; Rodrigues & Scott 1985), interpretable as the Carib solution residue.

Metamorphic-grade discontinuities

All of the following cases involve metamorphosed rift strata of the Inner Nappe (Higgs 2009). In central Venezuela, the Nirgua–Aroa formational contact has been interpreted as gradational (Aroa Fm webpage in LEV 1997) or faulted ('Chivacoa' map link on 'Nirgua, Fase' webpage), but examination is hindered by tropical vegetation. The Nirgua metamorphic grade is higher overall (schist, gneiss, amphibolite, quartzite, marble; v. phyllite, limestone, schist), raising the possibility that the Carib Weld separates these formations, consistent with their available age control (within the range Tithonian–Hauterivian, see above). Gypsum beds in the Nirgua, mentioned above, may represent solution residue. In the Caracas Gp, the Carib Weld might separate the Las Brisas and Las Mercedes fms, consistent with age constraints, the perceived concordant contact (sedimentary or tectonic; LEV 1997) and contrasting metamorphic grade (gneiss and schist, v. schist and phyllite; Bellizzia 1986; LEV 1997).

In Paria Peninsula, the Cariaquito–Güinimita formational contact is interpreted as the Carib Weld (Fig. 3), based on a probable solution residue and two separate criteria suggestive of a metamorphic-grade break compatible with solution of an intervening halite unit about 3 km thick (see below): (a) schist interbeds occur throughout the Cariaquito, whereas phyllite is the highest metamorphic grade in the Güinimita (González de Juana et al. 1968); and (b) limestones in the Middle Cariaquito (Patao Mbr) are dolomitic and lack obvious fossils (González de Juana et al. 1968), probably due to metamorphic recrystallization, whereas Güinimita limestones only 1 km concordantly above have a prolific fauna (González de Juana et al. 1968). The contact was initially interpreted by González de Juana et al. (1968) as transitional but locally faulted. Later the same authors concluded that the relationship was discordant (González de Juana et al. 1972).

In the Northern Range, Weber et al. (2001, p. 93) interpreted a 'sharp, east–west-trending major thermal discontinuity' as a south-dipping normal fault, which they correlated with the putative Arima Fault (Kugler 1961a, b; Saunders 1997a), supposed to separate lower greenschist Chancellor Fm (Frey et al. 1988) from Laventille Fm 'phyllitic shales' and recrystallized limestone (Potter 1974). Instead of a fault, this strike-parallel contact is interpreted here as the Carib Weld; the thermal discontinuity reflects c. 3 km of missing halite.

In the Gulf of Paria, the drilled Cuche-on-Couva contact (Fig. 3; Flinch et al. 1999) is interpreted as the Carib Weld, consistent with the available age control. Metamorphism of the Cuche is unknown in Trinidad, even here in the Inner Nappe, possibly because the overburden (Outer Nappe) in this frontal position was comparatively thin, tapering south (Higgs 2009, fig. 4b). A reference to the

Cuche being 'metamorphic in Caroni Basin' (Pindell & Kennan 2001a, fig. 3b) is erroneous; it is only overmature (Persad et al. 1993; Pindell & Kennan 2001a, p. 165).

Possible diapirs and diapiric solution breccias

In the SW part of the Gulf of Venezuela, halite diapirs rooted in Jurassic strata were interpreted on seismic profiles by an oil company in the 1960s (Dr Wolfgang Scherer, pers. comm., 2008). These features have not been drilled.

Interpreted mud diapirs on northern Maracaibo Basin seismic profiles (Escalona & Mann 2006) are reinterpreted here as diapiric solution breccias (Warren 1999), based on two lines of evidence: (a) two of the illustrated diapirs are shown as rooting between the 'Cretaceous' and 'Pre-Cretaceous' (Escalona & Mann 2006, fig. 10), a stratigraphic level unknown for thick shales anywhere in Venezuela but consistent with lowest Cretaceous halite; and (b) the two diapirs coincide with the Burro Negro Fault (Escalona & Mann 2006), possibly a former intragraben fault controlling thicker halite deposition on its NE side (Higgs 2009). The lack of a strong seismic reflection from the diapirs suggests that the halite has dissolved, leaving a diapir-shaped chaotic breccia of fragmented intrahalite beds, entrained wallrock xenoliths and collapsed roofrock (Warren 1999). Nearby in eastern Maracaibo Basin, 'turtles, weldings and isolated subbasins' attributed to shale tectonics (Nunez et al. 2005) may instead reflect flow or solution of halite.

On nearby Toas Island, straddling another possible intragraben fault (Oca), a 'mélange' attributed by Pimentel (1975) to fault drag may instead be a diapiric solution breccia. The Toas Granite, at least 2 km long and enigmatically 3 km above regional basement, has been interpreted as a fault sliver in a flower structure (Rod 1956). Alternatively, the granite mass(es) could be a solution-breccia 'gigaclast' (>1 km; Jackson et al. 2003) detached from a graben wall by the ascending diapir. This implies 3 km of vertical transport, comparable to 5 km for interpreted diapiric solution gigabreccias in Africa (Jackson et al. 2003), and far less than the 10–15 km height of many Gulf of Mexico diapirs (illustrations in Worrall & Snelson 1989; Jackson 1995).

A similar feature is the El Baño Granite (Habicht 1960), about 50 km east of the Burro Negro diapirs. This small (hundreds of metres) granite outcrop was interpreted as a strike–slip fault sliver raised even higher (5 km) above regional basement. Instead the granite might belong to a diapiric solution breccia, along with adjacent Cretaceous strata, which are highly fractured (Habicht 1960; author's observations). The El Baño hot springs occur here, of unstated salinity (Urbani 1991).

Rauhwacke

Rauhwackes are fault breccias along halite-lubricated thrusts, after solution of the halite (Warren 1999). In Colombia, emeralds occur along thrusts in an 'unusual polygenetic breccia' (Branquet *et al.* 1999, p. 183), possibly a rauhwacke.

In the central Venezuela frontal thrust belt, the chaotic Tememure Fm or structural complex (LEV 1997) is possibly a rauhwacke sheet. The contained blocks, interpreted as olistoliths by Vivas & Macsotay (1995*a*), range from Uppermost Jurassic to Eocene in age and are largely correlatable to formations of the Eastern Serranía (Beck 1978; Vivas & Macsotay 1995*a*). The shale matrix is Eocene and Oligocene, at least locally. These matrix and block ages are compatible with a thrust-fault breccia, whether or not halite was involved. Berriasian or Valanginian limestone bodies (Beck & Furrer 1977) and lenticular anhydrite–gypsum bodies of latest Jurassic or earliest Cretaceous age (Beck 1978; Rosales & Macsotay 1985; Vivas & Macsotay 1995*a*) may represent ruptured (limestone) and partly dissolved (anhydrite) former interbeds within the Carib Halite.

In Paria Peninsula, the Cariaquito Fm contains 'massive banded gypsum of about 25 m thickness with an intercalation of 3 m of Rauwacke' (Kugler 1953, p. 30).

Rollover structure

The Morichito was interpreted as a piggyback basin on the Pirital thrust sheet by Roure *et al.* (1994). Seismic profiles show Miocene clinoforms apparently downlapping northward onto an unconformity that truncates progressively younger Cretaceous and Palaeogene strata southward (Roure *et al.* 1994). Miocene microfaunal palaeobathymetry shallows northward (Cobos 2005), therefore the apparent downlaps are northward onlaps rotated structurally. The clinoforms have been attributed to tilting of the basin and depocentre migration during thrusting (Roure *et al.* 1994; Cobos 2005). An alternative interpretation, namely rollover caused by northward retreat (by flow or solution) of buried halite, is supported by the following four factors.

(1)   Halite-withdrawal (flow) rollovers in the Gulf of Mexico and the Santos Basin of Brazil (Ge *et al.* 1997) resemble the Morichito geometry. Flow rather than solution is inferred for the Morichito Basin, as the downlapping interval comprises *Middle* Miocene strata (Cobos 2005), predating the climatically induced

Late Miocene onset of halite solution regionally in Venezuela and Trinidad (see below).

(2)   A seismically transparent interval directly above the thrust under northern Morichito Basin (Roure *et al.* 1994, fig. 6) is interpretable as halite. The interval thickens northward from zero to about 2 s two-way-time, implying a thickness of *c.* 4 km, possibly thickened by flow. The hypothetical infra-Barranquín evaporite detachment of Roure *et al.* (1994, 2003) would thus be at the base of the Carib Halite.

(3)   The El Cantil Fm beneath the Morichito Basin is strongly tectonized (PDV 1997*b*), possibly reflecting solution-related collapse.

(4)   There is evidence for anomalously high palaeo-heat flow, as described above.

Two published seismic profiles from the Gulf of Paria show apparent clinoforms in the Mio–Pliocene section (i.e. predating 2.5–0 Ma pull apart; Higgs 2009), dipping toward a basement high (Scott 1985, fig. 12a; Babb & Mann 1999, fig. 29). This configuration is consistent with halite withdrawal by flow or solution.

Gypsum veins

Gypsum veins are common in roof strata overlying dissolved or dissolving halite (Gustavson *et al.* 1994). Solution produces fractures in collapsing overburden in which gypsum derived by hydration of subsurface anhydrite can precipitate (Gustavson *et al.* 1994). Many Trinidad and Venezuela outcrops, despite lacking bedded gypsum or anhydrite, show thin (millimetres, rarely centimetres) gypsum veins and, in soils, dispersed gypsum crystals interpreted here as disaggregated veins (author's observations; Wall & Sawkins 1860; Suter 1960; González de Juana *et al.* 1980; LEV 1997; Kugler 2001). Wall & Sawkins (1860, p. 90) noted that in Trinidad gypsum is 'extensively diffused . . . in the joints of shales and clays in almost every portion of the country'.

While gypsum can form in soils due to oxidation of pyrite (Logan & Nicholson 1997), five lines of evidence suggest that this is not the case here: (a) the gypsum occurs as discrete crystals or crystal clusters (Logan & Nicholson 1997, fig. 11), rather than veins; (b) pyrite oxidation is typical of arid and semi-arid climates (Logan & Nicholson 1997), unlike the seasonally wet–dry tropical climate of Trinidad and much of Venezuela; (c) gypsum occurs in many Trinidad outcrops in which no pyrite is reported (Kugler 2001); (d) gypsum veins in Venezuela commonly occur in outcrops lacking soil cover (author's observations); and (e) exceptionally, veins can reach 4 cm width (author's observations), suggesting a volume of gypsum larger than pyrite oxidation can supply.

Oils with evaporitic signatures

Evaporite-related shales are increasingly recognized for their source potential (Warren 2005). Halite-solution residual shales can have exceptionally high (concentrated) organic content. The Bogotá halite contains black laminated shale 'very rich in organic material' (López et al. 1991, p. 24, author's translation). In the central Venezuela thrust belt, organic-rich black shales (Rodriguez 1985) occur in contact with Tememure gypsum bodies, one of which contains a 'band of organic material' (Rosales & Macsotay 1985, fig. 4, author's translation).

In eastern Venezuela, the Querecual Fm cannot account for all the oil in the Orinoco heavy-oil belt (F. E. Audemard & Serrano 2001). The source potential of older sections is unproven (James 2000a) but Goodman et al. (2002) postulated probable Jurassic hypersaline-lacustrine source rocks deposited in rifts. Geomark (1993, cited by James 2000a) related oils with a hypersaline signature to a possible Jurassic–Lower Cretaceous source. A 'saline to hyper saline source' is recognized at Pedernales Oilfield (Duerto & McClay 2007, p. 26). The hypothetical sub-Querecual source could be shales associated with the Carib Halite, including solution residue.

In the Carúpano Basin, unexplained condensate occurs at Río Caribe Field in a biogenic gas trend (James 2000b). The source may be halite-related shale in the tail of the Outer Nappe, below the deepest drilled level of the Mejillones Complex rift deposits (Higgs 2009).

In the Gulf of Paria, a possible Jurassic source has been postulated (Bevan 2007). The unknown source rock of Couva Marine and South Domoil oilfields, considered to be Cenozoic based on abundant oleanane (Rodrigues 1995), may instead be a shale related to the Carib Halite (dissolved at Couva Marine; Fig. 3) or Couva anhydrite. Although oleanane is generally viewed as confined to Upper Cretaceous and Cenozoic strata (Moldowan et al. 1994; Peters et al. 2004), recent research suggests that angiosperms first appeared earlier than previously thought, in the Jurassic or even Triassic (Hochuli & Feist-Burkhardt 2004).

In the Northern Range, graphitic schists and shales occur in the Upper Chancellor and Lower Laventille Fm respectively (Rodrigues & Scott 1985). These formations are inferred to have enclosed the now-dissolved Carib Halite in the frontal part of the Inner Nappe (Fig. 3; Higgs 2009, fig. 4). The graphite indicates a high original content of organic matter, metamorphosed by burial under the Outer Nappe.

Anomalous oilfield brine salinity

Scarcity of reports of hypersaline formation water in Venezuela and Trinidad may reflect dilution by meteoric water circulating underground. Water in the Cogollo Gp fractured-limestone reservoir of La Paz Field is up to twice as saline as modern seawater and contains much higher Na and Cl concentrations than fractured-basement water in the same field (Nelson et al. 2000), suggesting contact with dissolving halite. Cogollo water at Centro Field is equally saline (Bockmuelen et al. 1983). Both fields lie on or beside Jurassic intra-graben lows (Lugo & Mann 1995) that could contain residual halite.

Saline fluid inclusions

In the Coastal Range, enigmatic hypersaline (up to 224 ppt) fluid inclusions of unknown source occur in a quartz-calcite vein cutting metamorphic rock (Sisson et al. 2005). The vein-hosting fracture formed in Miocene time (c. 15 Ma; Sisson et al. 2005). The brine could be formation water that acquired its high salinity by dissolving buried (meta-) halite after the c. 13 Ma start of deep circulation of meteoric water (see below).

Barite cement in sandstone

Barite cement in sandstones can indicate solution of a deeper evaporite (Warren 1999). Barite was reported by Sutton (1946) among heavy minerals from the Guasare, Misoa and ?Mene Grande Fms of western Venezuela. The barite must represent either cement (freed by crushing), supporting the Carib Halite concept, or detrital grains derived from a donor rock related or unrelated to evaporites. For barite found in Paleocene through Pliocene sandstones in Trinidad (Suter 1960) the same applies.

Mineralization

Emeralds in Colombia were precipitated from evaporite-dissolving groundwater (Ottaway et al. 1994). A genetic link with the (largely dissolved?) Carib Halite is clear, because the emeralds: (a) contain NaCl-rich fluid inclusions; (b) occur in veins in Berriasian, Valanginian and Hauterivian shales; (c) are associated with gypsum bodies; (d) occur on the thrusted flanks of the Eastern Cordillera, at the original borders of the Cretaceous rift trough (cf. Fabre 1987); and (e) were formed by high-temperature hydrothermal circulation involving a geothermal gradient elevated by halokinesis (Giuliani et al. 1995). Venezuela and Trinidad may thus also have potential for emerald discoveries.

Copper, fluorite, iron, lead, silver, uranium and zinc mineralization can be associated with evaporite solution (Sanford 1990; Kyle 1991; Warren 1997, 1999). In the Mérida Andes, Cu impregnations occur in the La Quinta Fm and at its contact with the Río Negro (González de Juana *et al.* 1980; LEV 1997). Although the La Quinta lies below the inferred stratigraphic level of the Carib Halite, thrusting over the halite (see above) or downward groundwater flow could explain these shows. The La Quinta also has possible U anomalies (Tarache 1980, thesis cited by LEV 1997). In central Venezuela, concentrations of Ag, Cu, Fe, Pb and Zn occur in the Nirgua 'Fase' (LEV 1997). In eastern Venezuela there are Ag enrichments within the Carúpano gravity low (Vierbuchen 1984). In Trinidad, minor ore deposits and shows of fluorite, hematite and magnetite occur in the Maracas and Laventille fms (Barr 1952, 1985; Kugler 1959; Potter 1997).

Neogene supraorogen basins

The mountain ranges of Colombia, Venezuela and Trinidad carry Neogene basins. Some are relatively small and surrounded by mountains, including the Bogotá, Cesar–Ranchería, La González, Carora,

Hueque, Aroa–Yaracuy, Valencia, Santa Lucía, Caracas, Tuy, Gulf of Cariaco, Morichito and San Juan basins (Fig. 2). Other basins are large and rupture or terminate a mountain range (Gulf of Barcelona, Gulf of Paria). In still other cases, the basin is broader than the adjacent mountains, such as the Gulf of Venezuela Basin beside the Paraguaná and Guajira Peninsulas; and the eastern Venezuela and Trinidad northern shelf basins (Tortuga–Margarita, Cubagua–Coche, Carúpano–North Coast) encircling Margarita and Tobago islands (Fig. 2). Finally, the whole of Trinidad, apart from the Northern Range, is a Neogene basin formed on collapsed mountains, subsequently inverted.

These basins are ascribed here to halite solution, based on two attributes: (a) they all show at least one of the features listed above suggestive of halite solution; and (b) seismic profiles show that the fill is characterized by syn-depositional listric normal faults, anomalous in the Neogene convergent plate-tectonic context (Higgs 2009). Similar, artificial basins were produced by Ge & Jackson (1998) in models of sediment deposition above dissolving halite (Fig. 6). In these experiments, listric growth faults developed in both 'synkinematic' sediments deposited during halite removal and in 'prekinematic' roof strata. The experimental faults all

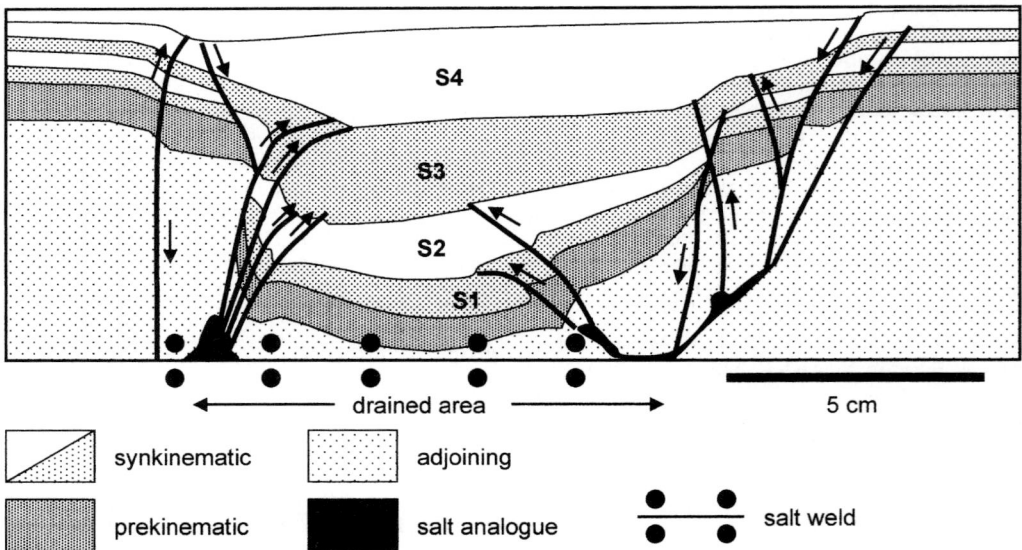

Fig. 6. Physical model simulating a halite body with an initial sub-rectangular cross section (e.g. graben fill or diapiric salt wall), largely removed by solution, leaving minor remnants only (black). Redrawn from Ge & Jackson (1998); see their fig. 16 for five preceding stages of this experiment. Subsidence caused by simulated solution (drainage of silicone polymer) produced a basin containing 'synkinematic' sediments; layers S1–S4 were added sequentially. The underlying 'prekinematic' layer was originally tabular and horizontal. Note: (1) steep, syn-depositional normal and reverse faults, terminating downward at the halite analogue or at its solution 'weld'; and (2) a pseudo-flower structure at both of the former walls of the modelled halite body.

terminated downward at the level of the withdrawing pseudo-halite. Growth faults in the Neogene basins are inferred to terminate at the base of the Carib Halite or at the solution weld. Thus, solution basins should partly mimic the original block-faulted 'thins' and 'thicks' of the halite depositional basin (complex of subgrabens), modified by halokinesis. The experiments simulated a linear diapiric intrusion ('salt wall'; Ge & Jackson 1998) adjacent to halite-free zones evacuated by flow to the diapir. The model also serves to simulate a halite-filled graben or subgraben. Experiments simulating continuous tabular halite failed to produce faults (Ge & Jackson 1998). Thus, the presence of faults in the South American inferred halite-solution basins suggests that the halite had steep sides, probably of two kinds: intragraben faults; and sides of diapiric walls ascending from an irregular regional blanket.

The earliest-formed basins embrace all of the Late Miocene foram zones (González de Juana *et al.* 1980; Beck 1985; Castro & Mederos 1985; Pérez de Mejía & Tarache 1985; Goddard 1988; Robertson & Burke 1989; LEV 1997; Saunders 1997a; Babb & Mann 1999; Kugler 2001). This means that subsidence began around 11.4 Ma (base of N15 zone; Berggren *et al.* 1995), implying, in the solution-basin model, a change to a rainier climate shortly before. The probable cause of climate change was collision of the Panamá Arc against Colombia, which interrupted deep-water Caribbean-Pacific interchange by 12.8 Ma (Middle Miocene; Coates *et al.* 2004), thereby initiating the modern Gulf Stream (Mullins *et al.* 1987) alongside Venezuela and Trinidad. In Venezuela a remarkable Upper Miocene–Lower Pliocene upwelling fish fauna in the Cubagua–Coche Basin (Aguilera & Rodrigues 2001) may reflect this oceanographic change. No Middle or Late Miocene climate change has been reported before in Venezuela or Trinidad, despite continental formations that probably straddle the 13–11 Ma change (Palmar–Isnotú, Parángula Fms; LEV 1997). Rainfall may have been highest during Late Miocene to Quaternary post-glacial 'pluvial' episodes, the last of which ended about 5000 years ago (González de Juana *et al.* 1980). This may better explain solution basins in currently semi-arid areas (Gulf of Venezuela; Carora Basin).

The solution-basin model explains the dilemma of *apparent* extensional basins (growth-faulted) of Late Miocene to Quaternary age on a nappe-thrust belt that has been shortening since Campanian time in central Venezuela to Trinidad (Higgs 2009). Solution subsidence generally outpaced the rate of uplift induced by shortening. Four types of supraorogen basin can be recognized.

Basins unconformable on collapsed highlands, now re-uplifting. In Trinidad, Upper Miocene–Pliocene

formations (Manzanilla to Talparo in the north; 'Upper Cruse' to Erin in the south) are unconformable on an earlier thrust belt and have, in turn, been folded, thrusted and uplifted (e.g. Kugler 1961a, b). The youngest strata are Lower Pliocene (Kugler 2001), therefore the second deformation began in Pliocene or Quaternary time. As explained above, the Mio–Pliocene basin is attributable to halite-solution subsidence, beginning around 11 Ma, of a Proto-Caribbean south-vergent thrust belt that reached to the Trinidad present south coast (Fig. 5). Arrival of the Caribbean Arc at *c.* 3.5 Ma caused an acceleration of shortening, inverting the solution basin.

Basins on collapsed highlands, still subsiding (excluding Caribbean nappe). Basins of this type include the Bogotá, Cesar–Ranchería, Lower Guajira, Golfo Triste, Caracas, southern Gulf of Barcelona, southern Cubagua–Coche, southern Carúpano–North Coast, Monagas thrust belt (supra), Gulf of Paria and Trinidad eastern shelf. Seismic profiles (Bassinger *et al.* 1971; Pereira 1985; Scott 1985; Goddard 1988; Robertson & Burke 1989; Payne 1991; Foley & Ballard 1997; Babb & Mann 1999; Flinch *et al.* 1999; Porras 2000; Pindell & Kennan 2001a; Duerto & McClay 2002, 2007; Mora & García 2006) tend to show listric syn-depositional growth faults.

Bogotá Basin has a Plio–Quaternary fill (Van der Hammen 1985; Torres *et al.* 2005). Presence of saline springs and halite (McLaughlin 1972) suggests subsidence by halite solution. A Late Pleistocene erosional unconformity between fluvial and underlying lacustrine deposits (Torres *et al.* 2005) suggests that uplift may now be underway. This could indicate halite exhaustion already, consistent with the Carib Halite here being only a thin (hundreds of metres?) distal tongue, compatible with the thinness of the basin fill (*c.* 500 m; Torres *et al.* 2005).

The youngest distinct unconformity in the Cesar Subbasin is sub-uppermost Oligocene or Lower Miocene (Mora & García 2006), interpreted here to indicate subsidence by halite solution or lateral flow beginning earlier than in most other basins, locally collapsing the western Santa Marta mountains which had been uplifting by westward thrusting from latest Eocene or earliest Oligocene time (*c.* 35 Ma; Higgs 2009). Solution would imply that the mountains were already high enough, and the climate wet enough in this western region (unlike the rest of the study area) to induce deep underground penetration of meteoric water. In the adjacent Ranchería Subbasin, Quaternary deposits lie unconformably on Eocene strata (Cáceres *et al.* 1980), suggesting that solution subsidence began in the Quaternary, locally collapsing the Perijá and eastern Santa Marta highlands uplifting from 5 Ma (Higgs 2009).

The Lower Guajira Basin (ANH 2005; Barrero et al. 2007), interpreted as flexural by F. A. Audemard et al. (2000), is attributed here to halite solution. The basin is sharply bounded in the south by the Oca Fault, interpreted as a syn-Carib Halite intragraben fault, reactivated by differential solution during post-5 Ma Perijá–Santa Marta uplift.

In the Mérida Andes, the La González Basin or Chama Graben contains Quaternary continental deposits (Schubert 1980; Kohn et al. 1984). Solution subsidence is consistent with thermal springs (Urbani 1991) and a possible collapse breccia mentioned above. A fault bounding the SE flank of the basin has a normal displacement exceeding 3 km (Giegengack 1984), enigmatic in an active orogen. A pull-apart origin (Schubert 1980) is incompatible with eastward plate convergence (Higgs 2009). The late onset of subsidence in this and the Bogotá Basin reflects the late start (c. 5 Ma) of uplift in the Mérida Andes and of strong uplift in the Eastern Cordillera (Higgs 2009).

The Quaternary Caracas graben (González de Juana et al. 1980) lacks reported saline springs or gypsum veins (Singer & Feliziani 1986; Urbani 1991), but is interpreted here as a halite-solution basin based on its irregular outline (Bellizzia 1976), anomalous isopach thickenings (Kantak et al. 2005) and proximity to the Santa Lucía and Tuy Basins with gypsum shows (González de Juana et al. 1980; Beck 1985).

The northern part of the Gulf of Barcelona overlies the Caribbean nappe and is dealt with separately below. Part of the southern basin fill may be exposed in Barcelona city. Here the almost unfossiliferous 'Lecherías Beds', interpreted as low- and high-density turbidites and debrites (Higgs, in Pindell & Erikson 2001), were mapped by Creole (1955) as the Naricual Fm, implying an Oligo–Miocene age (LEV 1997). An Oligocene marine-slope foram(s) was reported from one locality (T. King pers. comm. in Pindell & Erikson 2001). The Lecherías Beds were viewed as Oligocene oceanic trench deposits obducted in the Caribbean accretionary prism by Pindell & Erikson (2001). Instead these beds are interpreted here as Upper Miocene or Pliocene solution-basin deposits, folded and uplifted by movement on the Urica Fault, which passes through Barcelona (Munro & Smith 1984) and ceased activity in Late Miocene time (Higgs 2009). The Oligocene foram is considered redeposited. The near-absence of trace fossils (author's observations) and scarcity of marine microfossils negates an ocean-trench origin and may indicate a closed depression poorly connected to the sea, with sub- or supra-normal salinity (arid

episode?). Gypsum veins in the Lecherías Beds (author's observations) support the solution-basin model.

In front of the Eastern Serranía, south of the Pirital Thrust and east of the Urica Fault, Upper Miocene to Quaternary deposits unconformably overlie the Monagas thrust belt. The supra-thrust cover is interpreted as a halite-solution basin, as described above. Shale withdrawal (Duerto & McClay 2002, 2007) may have contributed to the subsidence.

The Monagas solution basin merges east into the Gulf of Paria solution basin. Here, solution subsidence from c. 11 Ma was interrupted, as in Monagas, by an east-younging wave of Caribbean transpressive uplift (see above). In the eastern Gulf, uplift began at c. 5 Ma and was terminated by the 2.5 Ma change to near-transcurrency, allowing Gulf-wide solution subsidence to reassert itself, combined with pull-apart in the northern Gulf (Higgs 2009). A regional near-base-Quaternary unconformity in the Gulf (Payne 1991; Babb & Mann 1999; Flinch et al. 1999) is attributable to the 2.5 Ma change.

Basins near-conformable on an earlier basin, and still subsiding. Gulf of Venezuela subsidence by halite solution is supported by, or consistent with: (a) the Urumaco Trough gravity low; (b) Quaternary fault activity in the Trough and along-strike to the SE (F. A. Audemard et al. 2000); (c) gypsum veins/crystals in nearby Falcón onshore outcrops (González de Juana et al. 1980); and (d) nearby saline springs (Urbani 1991). A sub-Upper Miocene unconformity in the Gulf (González de Juana et al. 1980) can be interpreted as separating the extensional and post-extensional (thermal) phases of the preceding, transform-related Falcón–Gulf of Venezuela Basin (Higgs 2009). A second, less marked unconformity near the base of the Pliocene (above the G. acostaensis zone; González de Juana et al. 1980) is inferred here to mark the start of regional, Caribbean-driven shortening responsible for the Falcón–Lara– Paraguaná–Guajira highlands (5–2.5 Ma; Higgs 2009). One or more of these highlands provided catchments for deeply penetrating rainwater, such that shortening-driven upwarping (non-angular unconformity) in the Gulf was eventually outpaced, above the halite, by solution subsidence. At 2.5 Ma regional shortening ended but solution subsidence has continued in the Gulf and is also causing collapse of the Falcón–Lara highlands, as shown by Quaternary minibasins (Carora, Hueque; Fig. 2), gypsum veins/crystals and saline springs.

Basins on the Caribbean nappe. The Caribbean forearc-nappe front runs ENE along the south edge of the Villa de Cura Klippe (Smith *et al.* 1999), crosses the Gulf of Barcelona, then between Araya and Margarita, along the Carúpano–North Coast Basin, and south of Tobago (Fig. 2; Higgs 2009). Basins now overlying the nappe are the Tuy, northern Gulf of Barcelona, Cariaco Trough, Tortuga–Margarita Shelf, northern Cubagua–Coche and northern Carúpano–North Coast. Several seismic profiles and line drawings have been published (Lidz *et al.* 1968; Lattimore *et al.* 1971; Peter 1972; Schubert 1982; Pereira 1985; Bellizzia 1986; Goddard 1988; Robertson & Burke 1989; Tyson *et al.* 1991).

The Tuy Basin rests unconformably on Villa de Cura metavolcanic basement (González de Juana *et al.* 1980; Pimentel 1984). Any Villa de Cura forearc-basin cover was completely eroded during and after obduction, prior to nappe subsidence by halite solution in the substrate (Outer and Inner *Proto*-Caribbean nappe complex). Tuy Basin subsidence began in Late Miocene time (González de Juana *et al.* 1980). To the west, the Santa Lucía and Valencia Basins lie north of the Villa de Cura Klippe, on Inner and Outer Nappe basement, indicating that by the time subsidence began here, in the Late Miocene (Santa Lucía; Beck 1985) and Quaternary (Valencia; González de Juana *et al.* 1980), the klippe had already been isolated by erosion. Such rapid erosional breaching of the Villa de Cura nappe since its Middle Miocene emplacement (Higgs 2009) suggests that it was relatively thin (kilometres), consistent with the present remaining thickness (*c.* 1 km; Smith *et al.* 1999).

The other, more easterly supra-Caribbean-nappe basins overlie Eocene to Miocene strata (Castro & Mederos 1985; Pereira 1985; Goddard 1988) interpretable as forearc-basin deposits, as suggested for one of these basins previously (Carúpano; Speed & Smith-Horowitz 1998). Above is an interval that is Upper Miocene to Recent in the west (Barcelona Basin; Goddard 1988) and nearly all Pliocene to Recent in the east (North Coast Basin; Robertson & Burke 1989). These are interpreted here as solution-basin deposits. The eastward younging reflects diachronous emplacement of the Caribbean nappe onto the tail of the Outer-Inner nappe complex, subsiding by halite solution. The basal age at any longitude post-dates passage of the Caribbean nappe suture point (Higgs 2009), indicating that, during obduction, uplift by internal shortening of the nappe outweighed solution subsidence of the substrate.

Alongside the Cubagua–Coche Basin, Upper Miocene to Quaternary marine strata have been uplifted above sea level (Cubagua and Castillo de Araya Fms of Araya Peninsula and Margarita; LEV 1997). Similarly, the Pliocene Rockly Bay Fm is exposed on Tobago (unconformable on

crystalline basement; Saunders & Muller-Merz 1985). These occurrences may reflect local halite exhaustion, allowing uplift by acute transpression or (remote from plate boundary) rebound to overtake.

Subsidence in the Cariaco Trough, separating the Gulf of Barcelona from the Tortuga–Margarita shelf (Fig. 2), is by combined halite solution and (since 2.5 Ma) pull-apart (Higgs 2009).

Morichito Basin. The Morichito is a hybrid halite-related basin. Middle Miocene 'rollover' subsidence was by northward *flow* of buried halite (see above), collapsing this sector of the Eastern Serranía thrust belt, which had previously been uplifted by Proto-Caribbean convergence. The onset of Caribbean-induced compression here near 14 Ma induced Pirital out-of-sequence thrusting (Roure *et al.* 1994) and accelerated north–south shortening, causing uplift and erosional truncation of the Morichito Basin early fill, assumed here to be *early* Middle Miocene. Subsidence resumed in Late Miocene time (intra-Morichito unconformity, Cobos 2005) at *c.* 10 Ma, after the Caribbean nappe suture point had passed, resulting in rebound that was easily outweighed by halite solution under the Pirital Thrust. This Morichito upper interval (Upper Miocene–Quaternary) onlaps both northward and southward (Cobos 2005). A sub-Pliocene unconformity involving a downward shift in onlap (Cobos 2005) is interpreted here as eustatic. Thus, halite solution is causing the Eastern Serrania mountains to collapse centripetally from the south (Morichito), west (Gulf of Barcelona), north (Gulf of Cariaco) and east (Gulf of Paria, San Juan Graben; Fig. 2).

Alternative interpretations. Other subsidence mechanisms have been suggested for some of the halite-solution basins. Subsidence between diverging strike–slip faults was proposed for the Santa Lucía and Tuy basins (Schubert 1988), and pull-apart for the La González, Yaracuy and Valencia Basins (Schubert 1980, 1988; F. A. Audemard *et al.* 2000). However, pull-apart is negated by a lack of suitably positioned faults and/or by inappropriate plate-motion vectors, before and after the 2.5 Ma relative-motion change (Higgs 2009). Moreover these basins have irregular, non-rhomboidal outlines.

The Carúpano–North Coast Basin, whose origin was considered 'not yet clear' by Speed & Smith-Horowitz (1998, p. 809), was attributed by them to gravitational extension, reducing the slope of a Late Neogene high-mountain belt above a crustal detachment. Similarly, Pindell & Kennan (2001*a*, p. 12) envisaged that the eastern Venezuela-Trinidad orogenic belt since 12 Ma 'has failed and collapsed eastwards, probably largely under the influence of gravity toward the Atlantic Ocean, but certainly prodded by Caribbean relative plate motion'. These gravitational models imply crustal overthickening like that responsible for high elevation (5 km) and extension in the Tibetan Plateau (Dewey 1988; Burchfiel *et al.* 1992). However, the Trinidad–NE Venezuela crust is not abnormally thick (Pindell & Kennan 2001*a*; Boettcher *et al.* 2003; VanDecar *et al.* 2003).

Two other mechanisms proposed for extensional collapse of mountains are reduction in plate convergence (e.g. Western Cordillera of North America, Constenius 1996) and 'corner flow' above a retrograding subducting slab (Neogene Italian Appenines, Cavinato & Decelles 1999). However, unlike these examples the Venezuela-Trinidad basins lack volcanism.

Lastly, subsidence by thrust loading can be ruled out. Besides all of the listed evidence for halite solution, none of the basins shows evidence of the fill thickening into a bordering thrust.

Carib Halite environment, climate, thickness

Depositional environment

Prior to the Berriasian–Valanginian lowstand, the Carib Graben was a Jurassic seaway connecting Colombia to Trinidad (González de Juana *et al.* 1980), as shown by Middle and Upper Jurassic ammonites, bivalves and gastropods in the Eastern Cordillera, Guajira, NW Venezuela, central Venezuela and Trinidad (Renz 1960; Campbell 1962; MacDonald 1968; Díaz de Gamero 1969; Urbani 1969; Bartok *et al.* 1985; Vivas & Macsotay 1995*a*). The marine strata containing these fossils were deposited on, and seaward of, Girón and La Quinta red beds.

The lowstand would have restricted the connection between the Carib Graben and the ocean, favouring halite deposition. At the Colombia end of the graben, there is evidence that the halite interfingered with marine sediments and with anhydrite. The Guavio, Macanal and Rosablanca Fms of the Eastern Cordillera and Middle Magdalena Basins contain Berriasian–Valanginian marine fossils, anhydrite/gypsum and dark shales (Cruz & Vargas 1972; Branquet *et el.* 2002; Maya *et al.* 2004; López 2005). Some of the shale and anhydrite/gypsum might be dissolution residue. A salt spring may indicate buried halite in the Guavio Fm (Branquet *et al.* 2002), a suggestion supported by the Bogotá halite diapirs nearby. The marine fossils imply a Colombian opening to the palaeo-Pacific Ocean, probably across the Central

Cordillera region NW of Bogotá, where Berriasian–Valanginian marine strata are known (López 2005; cf. palaeogeographic maps of Etayo-Serna *et al.* 1976, fig. 11 and Cooper *et al.* 1995, fig. 8). To the SW the Upper Magdalena–southern Eastern Cordillera region was non-depositional (Duarte 2003; López 2005). In contrast, the Trinidad end of the graben lacks evidence for interfingering marine sediments. This suggests that the graben during the Berriasian–Valanginian lowstand was a long (*c.* 2000 km) hypersaline gulf, connected to the Pacific at one end but blocked off from the Atlantic at the other by a basement horst, or by the Bahama Platform. No modern analogue is known. A possible Miocene analogue is the coincidentally named Gharib Halite of the Red Sea rift (Griffin 1999).

Climatic implications

Deposition of extensive evaporites may seem incongruous, given that northern Venezuela was, like today, near 10°N latitude in earliest Cretaceous time (e.g. Pindell & Kennan 2001*b*, fig. 4), where a humid sub-equatorial climate would normally be expected. However, equatorial winds in western Pangaea blew from the west according to Jurassic global palaeoclimate models (Chandler *et al.* 1992) and confirmed by cross-bedding orientations in Lower Jurassic aeolian deposits of the southwestern USA (Loope *et al.* 2004). Cordilleran arc mountains along the western Pangaea margin (e.g. Scotese 2002 palaeogeographic map) produced a rain shadow (Loope *et al.* 2004), enabling halite precipitation in the Carib Graben.

Depositional thickness and subsidence rate

Seismic profiles in subsiding halite-solution basins across the region, from the Gulf of Venezuela to the North Coast Basin, excluding the Cariaco Trough and northern Gulf of Paria which have additional pull-apart subsidence, show that the average thickness of the Upper Miocene–Quaternary section is around 2 km (González de Juana *et al.* 1980; Pereira 1985; Goddard 1988; Robertson & Burke 1989), approximating the thickness of dissolved halite. The thickest interval in Monagas solution basin is about 4 s two-way time (Duerto & McClay 2002, 2007), equating to roughly 6 km, but this may partly reflect shale-withdrawal subsidence. The average *depositional* thickness of halite exceeded the calculated 2 km, because continuing subsidence suggests the halite is not yet exhausted. Intrabasinal thickness variations across growth faults suggest differential solution, possibly across the flanks of diapirs or intragraben deeps. Onshore Trinidad the solution package also averages about 2 km (Manzanilla–Talparo, 'Upper Cruse'–Erin; Kugler 1961*b*), but this is truncated by erosion. Based on these values, an average regional halite depositional thickness of 3 km is assumed, enough to explain the Carib Weld metamorphism discontinuity. For comparison, reconstructions of the Louann halite to its pre-halokinesis configuration suggest an average depositional thickness of 4–6 km (Humphris 1978, in Jackson 1995, fig. 15; Worrall & Snelson 1989, fig. 20b; Peel *et al.* 1995, figs 4d & 5e).

The 3 km Carib Halite estimate, and the Berriasian–Valanginian ultra-lowstand duration of about 3 Ma (Hardenbol *et al.* 1998), imply a subsidence rate of 1 km/Ma, consistent with rifting but yielding an impossible total rift-fill thickness if extrapolated over the *c.* 150 Ma span of rifting (Higgs 2009). Thus, subsidence over the life of the Carib Graben must have been pulsed, reflecting variable rates of extension; also the average subsidence rate may have decreased with time. In addition, Carib Halite deposition may have been longer, possibly corresponding to the long-term eustatic low from Late Tithonian until Hauterivian time (146–129 Ma; Hardenbol *et al.* 1998).

Halite-dissolving fluid circulation

Halite-dissolving meteoric water enters the subsurface along aquifers and faults (Warren 1999). The details of the circulation system in the case of the Carib Halite are unknown. The system may have resembled that of the Banff Hot Springs in the Canadian Rocky Mountains, where rainwater is interpreted to flow down permeable beds in a thrust ramp to a depth of *c.* 3 km, resurfacing along the next thrust behind (Grasby & Lepitski 2002). Alternatively the circulation could be confined to a single fault plane, with recharge and discharge at different sites along the surface trace (López & Smith 1995). Numerous hot saline springs along the El Pilar Fault in eastern Venezuela are consistent with either model. Based on geothermometry and models of fluid heat-loss during ascent, Urbani (1989) suggested that the water need only have come from depths of 300–1000 m to account for the high (surface boiling) temperature of some of these springs.

Comparison with other halite-solution basins

Thickness of dissolved halite

Interpreted halite-solution basins of Devonian to Quaternary age occur on other continents in various tectonic settings, although the total number described is small (Gustavson *et al.* 1980; Gustavson & Finley 1985; Gustavson 1986; Hopkins 1987; Heward 1990; Higgs 1990; Anderson & Knapp 1993; Gutierrez 1996, 2004; Warren 1999; Gutierrez *et al.* 2001; Kirkham *et al.* 2002). In no case does the thickness of the dissolved or dissolving halite (max. 1–2 km, generally far less) approach the 3 km inferred for the Carib Halite. An interpreted halite-related Neogene basin in Algeria resembles the Carib solution basins in thickness and half-graben style, but halite withdrawal by flow is envisaged, rather than solution (Vially *et al.* 1994, fig. 26; Letouzey *et al.* 1995, fig. 17b). The withdrawn halite is inferred to have extruded diapirically at the surface where it dissolved (Letouzey *et al.* 1995).

Depth of halite solution

In the few published examples, the interpreted depth of halite solution is less than 600 m. In contrast, in some of the Carib basins, solution is occurring under kilometres of overburden, comprising the solution-basin fill and the pre-kinematic strata.

Rate of dissolution

Halite solution rates can be remarkably high. Calculations based on river solute loads in west Texas indicate that subsurface Permian halite intervals totalling hundreds of metres are potentially capable of receding laterally by tens of kilometres per million years, and vertically by hundreds of metres (Gustavson *et al.* 1980). These rates are all the more impressive given that the present (nonglacial) climate in west Texas is much drier than that of Trinidad and most of Venezuela. Based on these figures, the average rate of vertical dissolution implied for the Carib Halite is modest, on the order of 2 km in 11 Ma (180 m/Ma; 0.18 mm/a^{-1}).

Unique scale and extent of Carib Halite solution

The thickness and lateral extent of dissolved or dissolving Carib Halite and the inferred depth of solution exceed those of any other known examples. Four fortuitous factors are: (a) kilometric halite thickness; (b) high (palaeo-) rainfall; (c) location within a nappe/thrust belt, providing a large hydraulic head and extensive fracturing, in turn promoting deep circulation of meteoric water; and (d) fractured sandstone, limestone or schist aquifers immediately below and/or above the halite, connecting it to highland catchments. Other cases of buried halite worldwide lack some of these favourable factors. Of 33 examples of evaporites involved in compressive deformation, mostly including halite (compilation by Letouzey *et al.* 1995, fig. 1), many are in arid regions, to the extent of halite occurring undissolved at outcrop (e.g. Zagros; Salt Range). Only one example is within 10° of the equator, in eastern Peru where, despite a wet climate and halite thick enough to form diapirs (Alemán & Marksteiner 1993), no solution subsidence has been invoked, possibly reflecting comparatively recent or rapid uplift (exceeding halite-solution subsidence), or non-aquifer beds below and above.

In Thailand the Khorat Basin has multiple Cretaceous halite intervals at depths ranging from 0 to 1.5 km (Warren 1999). Solution subsidence is limited to thin (tens of metres?) Quaternary basins near the periphery. In this case, three of the four favourable factors are lacking: the halite is much thinner (cumulative <200 m); there is no active thrusting; and the enveloping strata are mudstone and 'pedogenically cemented . . . sands and shales' (Warren 1999, p. 125).

Buried halite thicker than 1.5 km in one borehole occurs in Guatemala–southern Mexico (Viniegra 1971; Bishop 1980), in a high-rainfall, high-relief setting. Saline springs and collapse breccias have been attributed to evaporite solution (Blount & Moore 1969; Viniegra 1971; Bishop 1980). No large-scale solution subsidence has been reported, but two lakes overlie a salt dome partly collapsed by meteoric-water ingress (Bishop 1980).

Thick (kilometres?) diapiric halite occurs in Cuba (Meyerhoff & Hatten 1968). No solution subsidence has been reported. The youngest strata at outcrop in the diapir regions are Miocene (Meyerhoff & Hatten 1968), indicating that no halite-solution subsidence is in progress. Cuba is mainly of low relief. The most recent orogeny ended in Eocene time (Cuba–Bahamas collision; Pindell & Kennan 2001*b*), unlike northern South America, where orogeny lasted until Mio–Pliocene time in Falcón, central/eastern Venezuela and Trinidad, and continues in the Eastern Cordillera and Mérida Andes.

Impact on oil and mineral exploration

Halite is well known for its association with oil and gas reserves (e.g. Jackson *et al.* 1995). There is also growing appreciation of the importance of recognizing *vanished* halite in petroleum and mineral exploration (Warren 1997, 1999). The Carib Halite concept is important for oil and gas exploration in Colombia, Venezuela and Trinidad for two reasons: (a) it supports the idea that Jurassic rifting continued through most of Cretaceous time (Higgs 2009), with implications for heat flow (maturation) and subsidence history modelling; and (b) it predicts evaporite-related source rocks, anomalous heat

flow, and structures related to décollement and halokinesis. Mining companies should also reevaluate the region in the light of the Carib Halite concept.

Reviews by P. Bartok, C. Giraldo and L. Kennan, and meticulous editing by K. H. James are gratefully acknowledged. Others who have generously shared their thoughts, positive and negative, on the Carib Halite are T. Bevan, M. Burrus, P. DeCelles, J. Frampton, S. Ghosh, M. Keeley, A. King, P. Pestman and J. Pindell. I especially thank A. Ellison for her unfailing help and support.

References

ABLEWHITE, K. & HIGGINS, G. E. 1968. A review of Trinidad, West Indies, oil development and the accumulations at Soldado, Brighton Marine, Grande Ravine, Barrackpore–Penal and Guayaguayare. *In*: SAUNDERS, J. B. (ed.) *Fourth Caribbean Geological Conference, Port of Spain, 1965, Transactions*, 41–74.

AGUILERA, O. & RODRIGUES, D. 2001. An exceptional coastal upwelling fish assemblage in the Caribbean Neogene. *Journal of Palaeontology*, **75**, 732–742.

ALEMAN, A. M. & MARKSTEINER, R. 1993. Structural styles in the Santiago fold and thrust belt, Peru: a salt related orogenic belt. *Second International Symposium on Andean Geodynamics, Oxford, Memoirs*, 147–153.

ALGAR, S. T. & PINDELL, J. L. 1993. Structure and deformation history of the Northern Range of Trinidad and adjacent areas. *Tectonics*, **12**, 814–829.

ANDERSON, N. L. & KNAPP, R. 1993. An overview of some of the large scale mechanisms of salt solution in western Canada. *Geophysics*, **58**, 1375–1387.

ANH 2005. *Onshore Guajira Basin*. Agencia Nacional de Hidrocarburos, Bogotá, Colombia, promotional brochure.

AUDEMARD, F. E. & SERRANO, I. C. 2001. Future petroliferous provinces of Venezuela. *In*: DOWNEY, M. W., THREET, J. C. & MORGAN, W. A. (eds) *Petroleum Provinces of the Twenty-First Century*. American Association of Petroleum Geologists, Memoirs, **74**, 353–372.

AUDEMARD, F. A., MACHETTE, M. N., COX, J. W., DART, R. L. & HALLER, K. M. 2000. *Map and Database of Quaternary Faults in Venezuela and its Offshore Regions*. US Geological Survey, Open-File Reports, **00–018**.

AYMARD, R., QUIJADA, J. & CORIAT, M. 1985. Campo Yucal–Placer, trampa estratigráfica gigante de gas en la cuenca oriental de Venezuela. *Sociedad Venezolana de Geólogos, VI Congreso Geológico Venezolano, Memorias*, **4**, 2779–2803.

AYMARD, R., PIMENTEL, L. *ET AL.* 1990. Geological integration and evaluation of northern Monagas, Eastern Venezuelan Basin. *In*: BROOKS, J. (ed.) *Classic Petroleum Provinces*. Geological Society, London, Special Publications, **50**, 37–53.

BABB, S. & MANN, P. 1999. Structural and sedimentary development of a Neogene transpressional plate boundary between the Caribbean and South America plates in Trinidad and the Gulf of Paria.

In: MANN, P. (ed.) *Caribbean Basins*. Elsevier, Amsterdam, 495–557.

BARR, K. W. 1952. Limestone blocks in the Lower Cretaceous Cuche Formation of the Central Range, Trinidad, B.W.I. *Geological Magazine*, **89**, 417–425.

BARR, K. W. 1985. Graded bedding and associated phenomena in the Northern Range of Trinidad. *Fourth Latin American Geological Congress, Trinidad, 1979, Transactions*, 117–134.

BARRERO, D., PARDO, A., VARGAS, C. A. & MARTINEZ, J. F. 2007. *Colombian Sedimentary Basins: Nomenclature, Boundaries and Petroleum Geology, a New Proposal*. Agencia Nacional de Hidrocarburos, Bogotá.

BARTOK, P. E., RENZ, O. & WESTERMANN, G. E. G. 1985. The Siquisique ophiolites, northern Lara State, Venezuela: a discussion on their Middle Jurassic ammonites and tectonic implications. *Geological Society of America Bulletin*, **96**, 1050–1055.

BASSINGER, B. G., HARBISON, R. N. & WEEKS, L. A. 1971. Marine geophysical study northeast of Trinidad–Tobago. *American Association of Petroleum Geologists Bulletin*, **55**, 1730–1740.

BECK, C. M. 1978. Polyphasic Tertiary tectonics of the Interior Range in the central part of the Western Caribbean Chain, Guárico State, northern Venezuela. *Geologie en Mijnbouw*, **57**, 99–104.

BECK, C. M. 1985. New data about recent tectonics in the central part of the Caribbean chain: the Santa Lucia–Ocumare del Tuy graben, Miranda State, Venezuela. *Fourth Latin American Geological Congress, Trinidad, 1979, Transactions*, 59–68.

BECK, C. M. & FURRER, M. A. 1977. Sobre la existencia de sedimentos marinos no metamorfizados del Neocomiense en el noreste del Estado Guárico, Venezuela Septentrional. *V Congreso Geológico Venezolano, Memorias*, **1**, 135–147.

BELLIZZIA, A. 1976. *Mapa Geológico y Estructural de Venezuela*. Venezuela Ministerio de Energía y Minas, 1:500 000.

BELLIZZIA, A. 1986. Sistema Montañoso del Caribe – una cordillera alóctona en la parte norte de America del Sur. *Sociedad Venezolana de Geólogos, VI Congreso Geológico Venezolano, Memorias*, **10**, 6657–6836.

BELLIZZIA, A. & RODRÍGUEZ, D. 1968. Consideraciones sobre la estratigrafía de los Estados Lara, Yaracuy, Cojedes y Carabobo. *Venezuela Ministerio de Minas, Boletín de Geología*, **9**, 516–563.

BERGGREN, W. A., KENT, D. V., SWISHER, C. C. III & AUBRY, M.-P. 1995. A revised Cenozoic geochronology and chronostratigraphy. *In*: BERGGREN, W. A., KENT, D. V., AUBRY, M.-P. & HARDENBOL, J. (eds) *Geochronology Time Scales and Global Stratigraphic Correlation*. SEPM, Special Publications, **54**, 129–212.

BEVAN, T. 2007. Structural cross sections of the Columbus Basin, Trinidad & Tobago: an integrated geological model based on a comprehensive 3D seismic database. *Fourth Geological Conference of the Geological Society of Trinidad and Tobago, Postconference Extended Abstracts*, CD.

BISHOP, W. F. 1980. Petroleum geology of northern Central America. *Journal of Petroleum Geology*, **3**, 3–59.

BLOUNT, D. N. & MOORE, C. H. JR. 1969. Depositional and non-depositional carbonate breccias, Chiantla Quadrangle, Guatemala. *Geological Society of America Bulletin*, **80**, 429–442.

BOCKMEULEN, H., BARKER, C. & DICKEY, P. A. 1983. Geology and geochemistry of crude oils, Bolivar coastal fields, Venezuela. *American Association of Petroleum Geologists Bulletin*, **67**, 242–270.

BOESI, T. & GODDARD, D. 1991. A new geologic model related to the distribution of hydrocarbon source rocks in the Falcón Basin, northwestern Venezuela. *In*: BIDDLE, K. T. (ed.) *Active Margin Basins*. American Association of Petroleum Geologists, Memoirs, **52**, 303–319.

BOETTCHER, S. S., JACKSON, J. L., QUINN, M. J. & NEAL, J. E. 2003. Lithospheric structure and supracrustal hydrocarbon systems, offshore eastern Trinidad. *In*: BARTOLINI, C., BUFFLER, R. T. & BLICKWEDE, J. F. (eds) *The Circum-Gulf of Mexico and the Caribbean: Hydrocarbon Habitats, Basin Formation, and Plate Tectonics*. American Association of Petroleum Geologists, Memoirs, **79**, 529–544.

BONINI, W. E. 1978. Anomalous crust in the Eastern Venezuela Basin and the Bouguer gravity anomaly field of northern Venezuela and the Caribbean borderland. *Geologie en Mijnbouw*, **57**, 117–122.

BRANQUET, Y., CHEILLETZ, A., GIULIANI, G., LAUMONIER, B. & BLANCO, O. 1999. Fluidized hydrothermal breccia in dilatent faults during thrusting: the Colombian emerald deposits. *In*: MCCAFFREY, K. J. W., LONERGAN, L. & WILKINSON, J. J. (eds) *Fractures, Fluid Flow and Mineralization*. Geological Society, London, Special Publications, **155**, 183–195.

BRANQUET, Y., CHEILLETZ, A., COBBOLD, P. R., BABY, P., LAUMONIER, B. & GIULIANI, G. 2002. Andean deformation and rift inversion, eastern edge of Cordillera Oriental (Guateque–Medina area), Colombia. *Journal of South American Earth Sciences*, **15**, 391–407.

BRAY, R. & EVA, A. 1987. Age, depositional environment, and tectonic significance of the Couva Marine evaporite, offshore Trinidad. *Tenth Caribbean Geological Conference, Cartagena, Colombia, Transactions*, 372 (abstract).

BURCHFIEL, B. C., CHEN, Z. ET AL. 1992. *The South Tibetan Detachment System, Himalayan Orogen: Extension Contemporaneous with & Parallel to Shortening in a Collisional Mountain Belt*. Geological Society of America, Boulder, CO, Special Papers, **269**.

CÁCERES, H., CAMACHO, R. & REYES, J. 1980. The geology of the Ranchería Basin. *In: Geological Field Trips Colombia 1980–1989*. Colombian Society of Petroleum Geologists and Geophysicists, Geotec, Bogotá, 1–31.

CAMPBELL, C. J. 1962. A section through the Cordillera Oriental of Colombia between Bogotá and Villavicencio. *In: Geological Field Trips, Colombia, 1959–1978*. Colombian Society of Petroleum Geologists and Geophysicists, Geotec, Bogotá, 89–118.

CASTILLO, M. V. & MANN, P. 2006. Cretaceous to Holocene structural and stratigraphic development in south Lake Maracaibo, Venezuela, inferred from well and three-dimensional seismic data. *American Association of Petroleum Geologists Bulletin*, **90**, 529–565.

CASTRO, M. & MEDEROS, A. 1985. Litoestratigrafía de la Cuenca de Carúpano. *Sociedad Venezolana de Geólogos, VI Congreso Geológico Venezolano, Memorias*, **1**, 201–225.

CAVINATO, G. P. & DECELLES, P. G. 1999. Extensional basins in the tectonically bimodal central Appenines fold-thrust belt, Italy: response to corner flow above a subducting slab in retrograde motion. *Geology*, **27**, 955–958.

CHANDLER, M. A., RIND, D. & RUEDY, R. 1992. Pangaean climate during the Early Jurassic: GCM simulations and the sedimentary record of paleoclimate. *Geological Society of America Bulletin*, **104**, 543–559.

CHEVALIER, Y. 1993. A cross section from the oil-rich Maturin sub-basin to Margarita Island: the geodynamic relations between South American and Caribbean plates. *American Association of Petroleum Geologists-Sociedad Venezolana de Geólogos, Conference, Caracas, Field Trip Guidebook*, **1**.

COATES, A. G., COLLINS, L. S., AUBRY, M.-P. & BERGGREN, W. A. 2004. The geology of the Darien, Panama, and the Late Miocene–Pliocene collision of the Panama arc with northwestern South America. *Geological Society of America Bulletin*, **116**, 1327–1344.

COBOS, L. S. 2005. *Structural interpretation of the Monagas foreland thrust belt, Eastern Venezuela.* American Association of Petroleum Geologists, Search and Discovery Articles, **30031**, World Wide Web Address: http://www.searchanddiscovery.net/documents/2005/cobos/index.htm.

CONSTENIUS, K. N. 1996. Late Palaeogene extensional collapse of the Cordilleran foreland fold and thrust belt. *Geological Society of America Bulletin*, **108**, 20–39.

COONEY, P. M. & LORENTE, M. A. 1997. Implicaciones tectónicas de un evento estructural en el Cretácico Superior (Santoniense–Campaniense) de Venezuela occidentale. *Sociedad Venezolana de Geólogos, VIII Congreso Geológico Venezolano, Memorias*, **1**, 195–204.

COOPER, M. A., ADDISON, F. T. *ET AL.* 1995. Basin development and tectonic history of the Llanos Basin, Eastern Cordillera, and Middle Magdalena Valley, Colombia. *American Association of Petroleum Geologists Bulletin*, **79**, 1421–1443.

CREOLE 1955. *Geología del superficie.* Creole Petroleum Corporation, unpublished maps D-10a-d, D-11a-d, 1:50 000.

CRUZ, J. A. & VARGAS, R. 1972. Informe sobre los yesos en la Formación Rosa Blanca en la Mesa de Los Santos. *Ingeominas, Colombia, Boletín Geológico*, **20**, 105–129.

DAAL, J. Q. & LANDER, R. 1993. Yucal-Placer Field, Venezuela. *In*: BEAUMONT, E. A. & FOSTER, N. H. (compilers). *Treatise of Petroleum Geology, Atlas of Oil and Gas Fields, Structural Traps VIII,* American Association of Petroleum Geologists, 307–328.

DAVIS, D. M. & ENGELDER, T. 1985. The role of salt in fold-and-thrust belts. *Tectonophysics*, **119**, 67–88.

DE SISTO, J. 1964. The Santa Barbara Fault of northern Monagas. *Asociación Venezolana de Geología, Minería y Petroleo*, **7**, 98–110.

DEWEY, J. F. 1988. Extensional collapse of orogens. *Tectonics*, **7**, 1123–1139.

DIA, A. N., CASTREC-ROUELLE, M., BOULÈGUE, J. & COMEAU, P. 1999. Trinidad mud volcanoes: where do the expelled fluids come from? *Geochimica et Cosmochimica Acta*, **63**, 1023–1038.

DIAZ DE GAMERO, L. 1969. Identificación y significación cronoestratigráfica de los pelecípidos de la Formación Las Brisas. *Asociación Venezolana de Geología, Minería y Petroleo, Boletín Informativo*, **12**, 455–464.

DI CROCE, J., BALLY, A. W. & VAIL, P. 1999. Sequence stratigraphy in the Eastern Venezuelan Basin. *In*: MANN, P. (ed.) *Caribbean Basins.* Elsevier, Amsterdam, 419–476.

DUARTE, L. M. 2003. Historia de la Cuenca Cretacea del VSM en un marco cronoestratigráfico. Implicaciones ambientales. *VIII Simposio Bolivariano de la Exploración Petrolera en las Cuencas Subandinas, Memorias*, **2**, 223–232.

DUERTO, L. & MCCLAY, K. 2002. *3D geometry and evolution of shale diapirs in the Eastern Venezuelan Basin.* American Association of Petroleum Geologists, Search and Discovery Articles, **10026**, World Wide Web Address: http://www.searchanddiscovery.net/documents/duerto/index.htm.

DUERTO, L. & MCCLAY, K. 2007. Strain partitioning in the shale fold and thrust belt, Eastern Venezuelan Basin. *Fourth Geological Conference of the Geological Society of Trinidad and Tobago, Post-Conference Extended Abstracts*, CD.

ESCALONA, A. & MANN, P. 2006. Tectonic controls of the right-lateral Burro Negro tear fault on Paleogene structure and stratigraphy, northeastern Maracaibo Basin. *American Association of Petroleum Geologists Bulletin*, **90**, 479–504.

ETAYO-SERNA, F., RENZONI, G. & BARRERO, D. 1976. Contornos sucesivos del mar Cretáceo en Colombia. *Primer Congreso Colombiano de Geología, 1969, Memorias*, 217–252.

EVA, A. A., BURKE, K., MANN, P. & WADGE, G. 1989. Four-phase tectonostratigraphic development of the southern Caribbean. *Marine and Petroleum Geology*, **6**, 9–19.

EVANOFF, J. 1951. Geología de la región de Altagracia de Orituco. *Venezuela Ministerio de Energía y Minas, Boletín de Geología*, **1**, 237–264.

FABRE, A. 1987. Tectonique et génération d'hydrocarbures: un modèle de l'évolution de la Cordillère Orientale de Colombie et du Bassin des Llanos pendant le Crétacé et le Tertiaire. *Archives des Sciences*, **40**, 145–190.

FLINCH, J. F., RAMBARAN, V. *ET AL.* 1999. Structure of the Gulf of Paria pull-apart basin (Eastern Venezuela–Trinidad). *In*: MANN, P. (ed.) *Caribbean Basins.* Elsevier, Amsterdam, 477–494.

FOLEY, D. C. & BALLARD, J. H. 1997. Cretaceous through Miocene tectono-stratigraphic sequences in the Gulf of Paria. *Sociedad Venezolana de Geólogos, I Congreso Latinamericano de Sedimentología, Memorias*, **1**, 261–267.

FREY, M., SAUNDERS, J. & SCHWANDER, H. 1988. The mineralogy and metamorphic geology of low-grade metasediments, Northern Range, Trinidad. *Journal of the Geological Society, London*, **145**, 563–575.

FUNKHOUSER, H. J., SASS, L. C. & HEDBERG, H. D. 1948. Santa Ana, San Joaquín, Guario, and Santa Rosa oil fields (Anaco fields), central Anzoátegui, Venezuela. *American Association of Petroleum Geologists Bulletin*, **32**, 1851–1908.

FURRER, M. A. 1972. Fossil tintinnids in Venezuela. *VI Conferencia Geológica del Caribe, Margarita, Memorias*, 451–454.

GARCIA JARPA, R., GHOSH, S. *ET AL.* 1980. Correlación estratigráfica y síntesis paleoambiental del Cretáceo de los Andes Venezolanos. *Venezuela Ministerio de Energía y Minas, Boletín de Geología*, **14**, 3–88.

GE, H. & JACKSON, M. P. A. 1998. Physical modeling of structures formed by salt withdrawal: implications for deformation caused by salt solution. *American Association of Petroleum Geologists Bulletin*, **82**, 228–250.

GE, H., JACKSON, M. P. A. & VENDEVILLE, B. C. 1997. Kinematics and dynamics of salt tectonics driven by progradation. *American Association of Petroleum Geologists Bulletin*, **81**, 398–423.

GIEGENGACK, R. 1984. Late Cenozoic tectonic environments of the central Venezuelan Andes. *In*: BONINI, W. E., HARGRAVES, R. B. & SHAGAM, R. (eds) *The Caribbean–South American Plate Boundary and Regional Tectonics*. Geological Society of America, Memoirs, **162**, 343–364.

GIULIANI, G., CHEILLETZ, A., ARBOLEDA, C., CARRILLO, V., RUEDA, F. & BAKER, J. H. 1995. An evaporitic origin of the parent brines of Colombian emeralds: fluid inclusion and sulphur isotope evidence. *European Journal of Mineralogy*, **7**, 151–165.

GODDARD, D. 1988. Seismic stratigraphy and sedimentation of the Cariaco Basin and surrounding continental shelf, northeastern Venezuela. *11th Caribbean Geological Conference, Barbados, Transactions*, **34**, 1–21.

GONZALEZ DE JUANA, C., MUÑOZ, N. G. & VIGNALI, M. 1968. On the geology of eastern Paria (Venezuela). *In*: SAUNDERS, J. B. (ed.) *Fourth Caribbean Geological Conference, Port of Spain, 1965, Transactions*, 25–29.

GONZALEZ DE JUANA, C., MUÑOZ, N. G. & VIGNALI, M. 1972. Reconocimiento geológico de la península de Paria, Venezuela. *IV Congreso Geológico Venezolano, Memorias, Boletín de Geología, Publicaciónes Especiales*, **5**, 1549–1588.

GONZALEZ DE JUANA, C., ITURRALDE, J. M. & PICARD, X. 1980. *Geología de Venezuela y de sus Cuencas Petroliferas*. Ediciones Foninves, Caracas.

GOODMAN, E. D. G., SUMMA, L. L., RICHARDSON, M., NORTON, I. O. & GREEN, A. R. 2002. Genetic approach to understanding a complex hydrocarbon system: Caribbean colossus of East Venezuela. *American Association of Petroleum Geologists Bulletin*, **86**, 1848 (abstract).

GRASBY, S. E. & LEPITZKI, D. A. W. 2002. Physical and chemical properties of the Sulphur Mountain thermal springs, Banff National Park, and implications for endangered snails. *Canadian Journal of Earth Sciences*, **39**, 1349–1361.

GRATEROL, V. & WONG, J. 1978. Anomalía de Bouger residual en la región nor oriental de Venezuela. *Geologie en Mijnbouw*, **57**, 377 (abstract).

GRIFFIN, D. L. 1999. The Late Miocene climate of northeastern Africa: unravelling the signals in the sedimentary succession. *Journal of the Geological Society, London*, **156**, 817–826.

GUSTAVSON, T. C. 1986. Geomorphic development of the Canadian River Valley, Texas Panhandle: an example of regional salt solution and subsidence. *Geological Society of America Bulletin*, **97**, 459–472.

GUSTAVSON, T. C. & FINLEY, R. J. 1985. Late Cenozoic geomorphic evolution of the Texas Panhandle and northeastern New Mexico – case studies of structural controls on regional drainage development. *The University of Texas at Austin, Bureau of Economic Geology, Reports of Investigations*, **148**.

GUSTAVSON, T. C., FINLEY, R. J. & McGILLIS, K. A. 1980. *Regional solution of Permian salt in the Anadarko, Dalhart, and Palo Duro Basins of the Texas Panhandle*. The University of Texas at Austin, Bureau of Economic Geology, Reports of Investigations, **106**.

GUSTAVSON, T. C., HOVORKA, S. D. & DUTTON, A. R. 1994. Origin of satin spar veins in evaporite basins. *Journal of Sedimentary Research*, **A64**, 88–94.

GUTIERREZ, F. 1996. Gypsum karstification induced subsidence: effects on alluvial systems and derived geohazards (Catalayud Graben, Iberian Range, Spain). *Geomorphology*, **16**, 277–293.

GUTIERREZ, F. 2004. Origin of the salt valleys in the Canyonlands section of the Colorado Plateau: evaporite-solution collapse versus tectonic subsidence. *Geomorphology*, **57**, 423–435.

GUTIERREZ, F., ORTÍ, F. *ET AL.* 2001. The stratigraphical record and activity of evaporite solution subsidence in Spain. *Carbonates and Evaporites*, **16**, 46–70.

HABICHT, K. 1960. La sección de el Baño, Serranía de Trujillo, Estado Lara. *III Congreso Geológico Venezolano, Memorias, Boletín de Geología, Publicaciónes Especiales*, **3**, 192–213.

HAESE, R. R., HENSEN, C. & DE LANGE, G. J. 2006. Pore water geochemistry of eastern Mediterranean mud volcanoes: implications for fluid transport and fluid origin. *Marine Geology*, **225**, 191–208.

HAQ, B. U., HARDENBOL, J. & VAIL, P. R. 1988. Mesozoic and Cenozoic chronostratigraphy and cycles of sea-level change. *In*: WILGUS, C. K., HASTINGS, B. S. *ET AL.* (eds) *Sea-Level Changes: an Integrated Approach*. SEPM, Special Publications, **42**, 71–108.

HARDENBOL, J., THIERRY, J., FARLEY, M. B., JACQUIN, T., DE GRACIANSKY, P.-C. & VAIL, P. R. 1998. Mesozoic and Cenozoic sequence chronostratigraphic framework of European basins. *In*: DE GRACIANSKY, P.-C., HARDENBOL, J., JACQUIN, T. & VAIL, P. R. (eds) *Mesozoic and Cenozoic Sequence Stratigraphy of European Basins*. SEPM, Special Publications, **60**.

HEDBERG, H. D. 1950. Geology of the Eastern Venezuela Basin (Anzoátegui–Monagas–Sucre-eastern Guárico portion). *American Association of Petroleum Geologists Bulletin*, **61**, 1173–1216.

HEESCHEN, K., HAECKEL, M. *ET AL.* 2008. Origin of fluids and salts at Mercator mud volcano, Gulf of Cadiz. *Geophysical Research Abstracts*, **10**.

HEWARD, A. P. 1990. Salt removal and sedimentation in Southern Oman. *In*: ROBERTSON, A. H. F., SEARLE, M. P. & RIES, A. C. (eds) *The Geology and Tectonics of the Oman Region*. Geological Society, London, Special Publications, **49**, 637–652.

HIGGS, R. 1990. *Sedimentology and Petroleum Geology of the Artex Member (Charlie Lake Formation), Northeastern British Columbia.* Canada, Province of British Columbia, Ministry of Energy, Mines and Petroleum Resources, Petroleum Geology Special Papers, **1990-1**.

HIGGS, R. 2006. Colombia–Venezuela–Trinidad 'Caribbean Oblique Collision Model' revised. *Sociedad Venezolana de Geofísicos, XIII Congreso Venezolano de Geofísica, Memorias*, CD. World Wide Web Address: http://www.congresogeofisica-sovg.org/dyncat.cfm?catid=4378 (accessed February 2008).

HIGGS, R. 2009. Caribbean–South America oblique collision model revised. *In*: JAMES, K. H., LORENTE, M. A. & PINDELL, J. L. (eds) *The Origin and Evolution of the Caribbean Plate.* Geological Society, London, Special Publications, **328**, 613–657.

HIGGINS, G. E. & SAUNDERS, J. B. 1974. Mud volcanoes – their nature and origin. *Verhandlungen der Naturforschenden Gesellschaft in Basel*, **84**, 101–152.

HOCHULI, P. A. & FEIST-BURKHARDT, S. 2004. A boreal early cradle of angiosperms? Angiosperm-like pollen from the Middle Triassic of the Barents Sea (Norway). *Journal of Micropalaeontology*, **23**, 97–104.

HOPKINS, J. C. 1987. Contemporaneous subsidence and fluvial channel sedimentation: Upper Mannville C Pool, Berry Field, Lower Cretaceous of Alberta. *American Association of Petroleum Geologists Bulletin*, **71**, 334–345.

HUDEC, M. R. & JACKSON, M. P. A. 2006. Advance of allochthonous salt sheets in passive margins and orogens. *American Association of Petroleum Geologists Bulletin*, **90**, 1535–1564.

HUNG, E. 2005. Thrust belt interpretation of the Serranía del Interior and Maturín subbasin, eastern Venezuela. *In*: AVÉ LALLEMENT, H. G. & SISSON, V. B. (eds) *Caribbean–South American Plate Interactions, Venezuela.* Geological Society of America, Boulder, CO, Special Papers, **394**, 251–270.

HUTCHISON, A. G. 1939. A note upon the Jurassic in Trinidad, B.W.I. *American Association of Petroleum Geologists Bulletin*, **23**, 1243.

JACKSON, M. P. A. 1995. Retrospective salt tectonics. *In*: JACKSON, M. P. A., ROBERTS, D. G. & SNELSON, S. (eds) *Salt Tectonics: A Global Perspective.* American Association of Petroleum Geologists, Memoirs, **65**, 1–28.

JACKSON, M. P. A. & CRAMEZ, C. 1989. Seismic recognition of salt welds in salt tectonic regimes. *GCSSEPM Gulf Coast Section, 10th Annual Research Conference, Program and Extended Abstracts*, 66–71.

JACKSON, M. P. A., ROBERTS, D. G. & SNELSON, S. (eds) 1995. *Salt Tectonics: A Global Perspective.* American Association of Petroleum Geologists, Memoirs, **65**.

JACKSON, M. P. A., WARIN, O. N., WOAD, G. M. & HUDEC, M. R. 2003. Neoproterozoic allochthonous salt tectonics during the Lufilian orogeny in the Katangan Copperbelt, central Africa. *Geological Society of America Bulletin*, **115**, 314–330.

JAMES, K. H. 2000a. The Venezuelan hydrocarbon habitat, part 1: tectonics, structure, palaeogeography and source rocks. *Journal of Petroleum Geology*, **23**, 5–53.

JAMES, K. H. 2000b. The Venezuelan hydrocarbon habitat, part 2: hydrocarbon occurrences and generated-accumulated volumes. *Journal of Petroleum Geology*, **23**, 133–164.

KANTAK, P., SCHMITZ, M. & AUDEMARD, F. 2005. Sediment thickness and a west-east geologic cross section in the Caracas valley. *Revista de la Facultad de Ingeniería de la U.C.V.*, **20**, 85–98.

KIRKHAM, R. M., STREUFERT, R. K., KUNK, M. J., BUDAHN, J. R., HUDSON, M. R. & PERRY, W. J. JR. 2002. Evaporite tectonism in the lower Roaring Fork River Valley, west-central Colorado. *In*: KIRKHAM, R. M., SCOTT, R. B. & JUDKINS, T. W. (eds) *Late Cenozoic Evaporite Tectonism and Volcanism in West-Central Colorado.* Geological Society of America, Boulder, CO, Special Papers, **366**, 73–99.

KOHN, B. P., SHAGAM, R., BANKS, P. O. & BURKLEY, L. A. 1984. Mesozoic–Pleistocene fission-track ages on rocks of the Venezuelan Andes and their tectonic implications. *In*: BONINI, W. E., HARGRAVES, R. B. & SHAGAM, R. (eds) *The Caribbean–South American Plate Boundary and Regional Tectonics.* Geological Society of America, Boulder, CO, Memoirs, **162**, 365–384.

KUGLER, H. G. 1953. Jurassic to recent sedimentary environments in Trinidad. *Bulletin Association Suisse des Géologie et Ingeneur du Pétrole*, **20**, 27–60.

KUGLER, H. G. 1959. *Geological Maps of Trinidad.* Ministry of Energy, Trinidad, 1 : 50 000.

KUGLER, H. G. 1961a. *Geological Map of Trinidad.* Petroleum Association of Trinidad, 1 : 100 000. [Republished as part of Kugler 1996.]

KUGLER, H. G. 1961b. *Geological Cross-Sections of Trinidad.* Petroleum Association of Trinidad, 1 : 100 000. [Republished as part of Kugler 1996.]

KUGLER, H. G. 1996. *Treatise on the Geology of Trinidad, Part 3. Detailed Geological Maps and Sections.* Natural History Museum, Basel.

KUGLER, H. G. 2001. *Treatise on the Geology of Trinidad, Part 4. The Palaeocene to Holocene Formations of Trinidad.* Ed. by BOLLI, H. M. & KNAPPERTSBUSCH, M. Natural History Museum, Basel.

KYLE, J. R. 1991. Evaporites, evaporitic processes and mineral resources. *In*: MELVIN, J. L. (ed.) *Evaporites, Petroleum and Mineral Resources.* Developments in Sedimentology, **50**. Elsevier, Amsterdam, 477–533.

LAMB, J. L. & SULEK, J. A. 1968. Miocene turbidites in the Carapita Formation of eastern Venezuela. *Fourth Caribbean Geological Conference, Port of Spain, 1965, Transactions*, 111–119.

LATTIMORE, R. K., WEEKS, L. A. & MORDOCK, L. W. 1971. Marine geophysical reconnaissance of continental margin north of Paria peninsula, Venezuela. *American Association of Petroleum Geologists Bulletin*, **55**, 1719–1729.

LETOUZEY, J., COLLETTA, B., VIALLY, R. & CHERMETTE, J. C. 1995. Evolution of salt-related structures in compressional settings. *In*: JACKSON, M. P. A., ROBERTS, D. G. & SNELSON, S. (eds) *Salt Tectonics: A Global Perspective.* American Association of Petroleum Geologists, Memoirs, **65**, 41–60.

LEV 1997. *Léxico Estratigráfico de Venezuela*, 3rd edn. Ministerio de Energía y Minas, Boletín de Geología,

Publicaciónes Expeciales, **12**. World Wide Web Address: http://www.pdvsa.com/lexico/lexicoh.htm (accessed November 2008).

LIDZ, L., BALL, M. & CHARM, W. 1968. Geophysical measurements bearing on the problem of the El Pilar Fault in the northern Venezuelan offshore. *Bulletin of Marine Science*, **18**, 545–560.

LOGAN, W. S. & NICHOLSON, R. V. 1997. Origin of dissolved groundwater sulphate in coastal plain sediments of the Rio de la Plata, eastern Argentina. *Aquatic Geochemistry*, **3**, 305–328.

LOOPE, D. B., STEINER, M. B., ROWE, C. M. & LANCASTER, N. 2004. Tropical westerlies over Pangaean sand seas. *Sedimentology*, **51**, 315–322.

LOPEZ, E. 2005. *Chronostratigraphic correlation charts of Colombia*. Ingeominas, Colombia, Informes.

LOPEZ, D. L. & SMITH, L. 1995. Fluid flow in fault zones: analysis of the interplay of convective circulation and topographically driven groundwater flow. *Water Resources Research*, **31**, 1489–1503.

LOPEZ, C., BRICENO, A. & BUITRAGO, J. 1991. Edad y origen de los diapiros de sal de la Sabana de Bogotá. *IV Simposio Bolivariano de la Exploración Petrolera en las Cuencas Subandinas, Memorias*, **1**, 19:1–40.

LUGO, J. & MANN, P. 1995. Jurassic-Eocene tectonic evolution of Maracaibo Basin, Venezuela. *In*: TANKARD, A. J., SUAREZ, R. & WELSINK, H. J. (eds) *Petroleum Basins of South America*. American Association of Petroleum Geologists, Memoirs, **62**, 699–725.

MACDONALD, W. M. 1968. Estratigrafía, estructura y metamorfismo de las rocas del Jurasico Superior, peninsula de Paraguaná, Venezuela. *Venezuela Ministerio de Minas e Hidrocarburos, Boletín de Geología*, **9**, 441–457.

MACDONALD, W. D. & OPDYKE, N. D. 1972. Tectonic rotations suggested by paleomagnetic results from northern Colombia, South America. *Journal of Geophysical Research*, **77**, 5720–5730.

MACELLARI, C. E. 1995. Cenozoic sedimentation and tectonics of the southwestern Caribbean pull-apart basin, Venezuela and Colombia. *In*: TANKARD, A. J., SUÁREZ, R. & WELSINK, H. J. (eds) *Petroleum Basins of South America*. American Association of Petroleum Geologists, Memoirs, **62**, 757–780.

MACSOTAY, O., ALVAREZ, E., RIVAS, D. & VIVAS, V. 1985. Geotérmia tectónica en la región El Pilar–Casanay, Venezuela nor-oriental. *Sociedad Venezolana de Geólogos, VI Congreso Geológico Venezolano, Memorias*, **2**, 881–917.

MASSARI, F., RIO, D. ET AL. 2002. Interplay between tectonics and glacio-eustasy: Pleistocene succession of the Crotone basin, Calabria (southern Italy). *Geological Society of America Bulletin*, **114**, 1183–1209.

MAYA, M., BUENAVENTURA, J. & SALINAS, R. 2004. *Estado del conocimiento de la exploración de esmeraldas en Colombia*. Ingeominas, Colombia, Informes.

MCLAUGHLIN, D. H. JR. 1972. Evaporite deposits of Bogotá area, Cordillera Oriental, Colombia. *American Association of Petroleum Geologists Bulletin*, **56**, 2240–2259.

MELLO, U. T., KARNER, G. D. & ANDERSON, R. N. 1995. Role of salt in restraining the maturation of subsalt source rocks. *Marine and Petroleum Geology*, **12**, 697–716.

MENDEZ, J. 1989. Porosidades en el Grupo Cogollo y su relación con los ambientes depositacionales. *VII Congreso Geológico Venezolano, Memoria*, 867–889.

METZ, H. L. 1968*a*. Stratigraphic and geologic history of extreme northeastern Serrania del Interior, State of Sucre, Venezuela. *Fourth Caribbean Geological Conference, Port of Spain, 1965, Transactions*, 275–292.

METZ, H. L. 1968*b*. Geology of the El Pilar Fault Zone, State of Sucre, Venezuela. *Fourth Caribbean Geological Conference, Port of Spain, 1965, Transactions*, 293–298.

MEYERHOFF, A. A. & HATTEN, C. W. 1968. Diapiric structures in central Cuba. *In*: BRAUNSTEIN, J. & O'BRIEN, G. D. (eds) *Diapirism and Diapirs*. American Association of Petroleum Geologists, Memoirs, **8**, 315–357.

MOLDOWAN, J. M., HUIZINGA, B. J., DAHL, J. E., FAGO, F. J., TAYLOR, D. W. & HICKEY, L. J. 1994. The molecular fossil record of oleanane and its relationship to angiosperms. *Science*, **265**, 768–771.

MORA, A. & GARCÍA, A. 2006. Cenozoic tectono-stratigraphic relationships between the Cesar Sub-basin and the southeastern Lower Magdalena Valley Basin of northern Colombia. American Association of Petroleum Geologists, Search and Discovery Articles, **30046**. World Wide Web Address: http://www.searchanddiscovery.com/documents/2006/06134mora/index.htm.

MULLINS, H. T., GARDULSKI, A. F., WISE, S. W. JR. & APPLEGATE, J. 1987. Middle Miocene oceanographic event in the eastern Gulf of Mexico: implications for seismic stratigraphic succession and Loop Current/Gulf Stream circulation. *Geological Society of America Bulletin*, **98**, 702–713.

MUNRO, S. E. & SMITH, F. D. JR. 1984. The Urica fault zone, northeastern Venezuela. *In*: BONINI, W. E., HARGRAVES, R. B. & SHAGAM, R. (eds) *The Caribbean–South American Plate Boundary and Regional Tectonics*. Geological Society of America, Boulder, CO, Memoirs, **162**, 213–215.

NELSON, R. A., MOLDOVANYI, E. P., MATCEK, C. C., AZPIRITXAGA, I. & BUENO, E. 2000. Production characteristics of the fractured reservoirs of the La Paz field, Maracaibo basin, Venezuela. *American Association of Petroleum Geologists Bulletin*, **84**, 1791–1809.

NETTLETON, L. L. 1949. Geophysics, geology, and oil finding. *American Association of Petroleum Geologists Bulletin*, **33**, 1154–1160.

NUNEZ, M. ET AL. 2005. Deformation due to shale tectonic in northwestern Venezuela. *American Association of Petroleum Geologists, Annual Convention, Program and Abstracts*.

OTTAWAY, T. L., WICKS, F. J., BRYNDZIA, L. T., KYSER, T. K. & SPOONER, E. T. C. 1994. Formation of the Muzo hydrothermal emerald deposit in Colombia. *Nature*, **369**, 552–554.

PAIGE, S. 1930. The earthquake at Cumana, Venezuela, January 17, 1929. *Bulletin of the Seismological Society of America*, **20**, 1–10.

PARIS, G., MACHETTE, M. N., DART, R. L. & HALLER, K. M. 2000. Map and database of Quaternary faults and folds in Colombia and its offshore regions. *US Geological Survey Open-File Report*, **00-0284**.

PARNAUD, F., GOU, Y., PASCUAL, J.-C., CAPELLO, M. A., TRUSKOWSKI, I. & PASSALACQUA, H. 1995a. Stratigraphic synthesis of western Venezuela. *In*: TANKARD, A. J., SUÁREZ, R. & WELSINK, H. J. (eds) *Petroleum Basins of South America*. American Association of Petroleum Geologists, Memoirs, **62**, 681–698.

PARNAUD, F., GOU, Y., PASCUAL, J.-C., TRUSKOWSKI, I., GALLANGO, O. & PASSALACQUA, H. 1995b. Petroleum geology of the central part of the Eastern Venezuela basin. *In*: TANKARD, A. J., SUÁREZ, R. & WELSINK, H. J. (eds) *Petroleum Basins of South America*. American Association of Petroleum Geologists, Memoirs, **62**, 741–756.

PASSALACQUA, H., FERNANDEZ, F., GOU, Y. & ROURE, F. 1995. Crustal architecture and strain partitioning in the eastern Venezuelan ranges. *In*: TANKARD, A. J., SUÁREZ, R. & WELSINK, H. J. (eds) *Petroleum Basins of South America*. American Association of Petroleum Geologists, Memoirs, **62**, 667–679.

PAYNE, N. 1991. An evaluation of post-Middle Miocene geological sequences, offshore Trinidad. *Second Geological Conference of the Geological Society of Trinidad and Tobago, Transactions*, 70–87.

PDV 1997a. *Golfo Triste*. Petróleos de Venezuela-Intevep website. World Wide Web Address: http://www.pdv.com/lexico/camposp/cp059.htm (accessed August 2008).

PDV 1997b. *Pirital–Orocual–Manresa*. Petróleos de Venezuela-Intevep website, World Wide Web Address: http://www.pdv.com/lexico/camposp/cp053.htm (accessed August 2008).

PEEL, F. J., TRAVIS, C. J. & HOSSACK, J. R. 1995. Genetic structural provinces and salt tectonics of the Cenozoic offshore U.S. Gulf of Mexico: a preliminary analysis. *In*: JACKSON, M. P. A., ROBERTS, D. G. & SNELSON, S. (eds) *Salt Tectonics: A Global Perspective*. American Association of Petroleum Geologists, Memoirs, **65**, 153–175.

PEREIRA, J. G. 1985. Evolución tectónica de la cuenca de Carúpano durante el Terciario. *Sociedad Venezolana de Geólogos, VI Congreso Geológico Venezolano, Memorias*, **4**, 2618–2648.

PEREZ DE MEJIA, D. & TARACHE, C. 1985. Sintesis geológica del Golfo de Paria. *Sociedad Venezolana de Geólogos, VI Congreso Geológico Venezolano, Memorias*, **5**, 3243–3277.

PERSAD, K. 1983. Petroleum potential of the Mesozoic of Trinidad and Tobago. *Geological Society of Trinidad and Tobago, Newsletter*, **5**. World Wide Web Address: http://www.gstt.org/publications/news/news5/mesozoic%20oil.htm.

PERSAD, K. M., TALUKDAR, S. C. & DOW, W. G. 1993. Tectonic control in source rock maturation and oil migration in Trinidad and implications for petroleum exploration. *In*: PINDELL, J. L. & PERKINS, R. (eds) *Mesozoic and Early Cenozoic Development of the Gulf of Mexico and Caribbean Region*. GCSSEPM Foundation, 13th Annual Research Conference, Proceedings, 237–249.

PETER, G. 1972. Geologic structure offshore north-central Venezuela. *VI Conferencia Geológica del Caribe (6th Caribbean Geological Conference), Margarita, Venezuela, Memorias*, 283–294.

PETERS, K. E., WALTERS, C. C. & MOLDOWAN, J. M. 2004. *The Biomarker Guide*, 2nd edn. Cambridge University Press, Cambridge.

PIMENTEL, N. R. 1975. Excursión No. 3 – Falla de Oca: Isla de Toas y San Carlos. *II Congreso Latinoamericano de Geología, Memorias, Boletín de Geología, Publicaciónes Especiales*, **7**, 326–338.

PIMENTEL, N. 1984. *Mapa Geológico Estructural de Venezuela*. Ministerio de Energía y Minas, 1 : 2 500 000.

PINDELL, J. L. & ERIKSON, J. 2001. *Stratigraphy and Structure of the Northwestern Serranía del Interior Oriental, Venezuela: Constraints on Dynamic Evolution, Eastern Venezuela/Trinidad*. Geological Society of Trinidad and Tobago, Field Guide.

PINDELL, J. L. & KENNAN, L. 2001a. Processes and events in the terrane assembly of Trinidad and Eastern Venezuela. *In*: FILLON, R. H., ROSEN, N. C. & WEIMER, P. (eds) *Petroleum Systems of Deep-Water Basins: Global and Gulf of Mexico Experience*. GCSSEPM Foundation, 21st Annual Research Conference, Transactions, 159–192.

PINDELL, J. L. & KENNAN, L. 2001b. Kinematic evolution of the Gulf of Mexico and Caribbean. *In*: FILLON, R. H., ROSEN, N. C. & WEIMER, P. (eds) *Petroleum Systems of Deep-Water Basins: Global and Gulf of Mexico Experience*. GCSSEPM Foundation, 21st Annual Research Conference, Transactions, 193–220.

PORRAS, L. R. 2000. Evolución tectónica y estilos estructurales de la región costa afuera de las cuencas de Falcón y Bonaire. *VII Simposio Bolivariano de la Exploración Petrolera en las Cuencas Subandinas, Caracas, Memoria*, 279–292.

POTTER, H. C. 1973. The overturned anticline of the Northern Range of Trinidad near Port of Spain. *Journal of the Geological Society, London*, **129**, 133–138.

POTTER, H. C. 1974. Observations on the Laventille Formation, Trinidad. *Verhandlungen der Naturforschenden Gesellschaft in Basel*, **84**, 202–208.

POTTER, H. C. 1997. Notes on the Northern Range & changes from H.G. Kugler's map. *Trinidad and Tobago Ministry of Energy, Explanatory Notes for 1997 Geological Map of Trinidad*, 13–22.

RENZ, O. 1960. Geología de la parte sureste de la Peninsula de la Goajira (República de Venezuela). *III Congreso Geológico Venezolano, Memoria, Boletín de Geología, Publicaciónes Especiales*, **3**, 317–347.

RENZ, O. 1982. *The Cretaceous Ammonites of Venezuela*. Maraven, Petróleos de Venezuela S.A.

RIETMAN, J. L. & PALLISTER, B. J. 2007. Characteristics and significance of a large pockmark field, northern offshore Trinidad. *Fourth Geological Conference of the Geological Society of Trinidad and Tobago, Abstracts*, CD.

RITTER, J. B. & WEBER, J. C. 2005. Geomorphology and Quaternary geology of the Northern Range, Trinidad: recording Quaternary subsidence and uplift associated with a pull-apart basin. *Geological Society of America, Abstracts with Program*, **37**, 427.

ROBERTSON, P. & BURKE, K. 1989. Evolution of southern Caribbean Plate boundary, vicinity of Trinidad and Tobago. *American Association of Petroleum Geologists Bulletin*, **73**, 490–509.

ROD, E. 1956. Strike–slip faults of northern Venezuela. *American Association of Petroleum Geologists Bulletin*, **40**, 457–476.

RODRIGUEZ, S. E. 1985. Secuencias de evaporitas en zonas tectonizadas del Guárico nororiental, Cordillera de la Costa, Venezuela. *Fourth Latin American Geological Congress, Trinidad, 1979, Transactions*, 250–257.

RODRIGUES, K. 1990. Significance of geothermal gradients in petroleum exploration in Trinidad. *In:* LARUE, D. K. & DRAPER, G. (eds) *12th Caribbean Geological Conference, St Croix, Transactions*, 444–452.

RODRIGUES, K. 1995. The Couva Marine oil: a unique, terrestrially-sourced, Tertiary oil in Trinidad. *14th Caribbean Geological Conference, Port of Spain, Transactions*, 521–531.

RODRIGUES, K. & SCOTT, J. P. 1985. The structural geology of the Laventille area east of Port of Spain. *Fourth Latin American Geological Congress, Trinidad, 1979, Transactions*, 106–115.

ROSALES, T. & MACSOTAY, O. 1985. Yeso-anhidrita como evidencia de tectónica horizontal en Altagracia de Orituco, Estado Guárico. *Sociedad Venezolana de Geólogos, VI Congreso Geológico Venezolano, Memorias*, **2**, 1091–1119.

ROURE, F., CARNEVALI, J. O., GOU, Y. & SUBIETA, T. 1994. Geometry and kinematics of the North Monagas thrust belt (Venezuela). *Marine and Petroleum Geology*, **11**, 347–362.

ROURE, F., BORDAS-LEFLOCH, N. *ET AL.* 2003. Petroleum systems and reservoir appraisal in the sub-Andean basins (E Venezuela and E Colombian foothills). *In:* BARTOLINI, C., BUFFLER, R. T. & BLICKWEDE, J. F. (eds) *The Circum-Gulf of Mexico and the Caribbean: Hydrocarbon Habitats, Basin Formation, and Plate Tectonics*. American Association of Petroleum Geologists, Memoirs, **79**, 750–775.

RUSSO, R. M. & SPEED, R. C. 1992. Oblique collision and tectonic wedging of the South American continent and Caribbean terranes. *Geology*, **20**, 447–450.

SANFORD, R. F. 1990. Hydrogeology of an ancient arid closed basin: implications for tabular sandstone-hosted uranium deposits. *Geology*, **18**, 1099–1102.

SAUNDERS, J. B. 1997a. *Trinidad Stratigraphic Chart and Geological Map*. Trinidad and Tobago Ministry of Energy, 1:100 000.

SAUNDERS, J. B. 1997b. General explanatory notes & changes from H. G. Kugler's 1959 map. *Trinidad and Tobago Ministry of Energy, Explanatory Notes for 1997 Geological Map of Trinidad*, 2–12.

SAUNDERS, J. B. & MULLER-MERZ, 1985. The age of the Rockly Bay Formation, Tobago. *Fourth Caribbean Geological Conference, Port of Spain, 1965, Transactions*, 339–344.

SCHUBERT, C. 1980. Late Cenozoic pull-apart basins, Boconó fault zone, Venezuelan Andes. *Journal of Structural Geology*, **2**, 463–468.

SCHUBERT, C. 1982. Origin of Cariaco Basin, southern Caribbean Sea. *Marine Geology*, **47**, 345–360.

SCHUBERT, C. 1988. Neotectonics of La Victoria Fault Zone, north-central Venezuela. *Annales Tectonicae*, **2**, 58–66.

SCOTESE, C. R. 2002. *Late Jurassic*. World Wide Web Address: http://www.scotese.com/late1.htm (accessed August 2008).

SCOTT, J. P. 1985. The continental margin around Trinidad and Tobago – its exploration possibilities. *Fourth Latin American Geological Congress, Trinidad, 1979, Transactions*, 1031–1047.

SINGER, A. & FELIZIANI, P. 1986. Excursión 6, Geología urbana de Caracas. *Sociedad Venezolana de Geólogos, VI Congreso Geológico Venezolano*, **10**, 7043–7124.

SISSON, V. B., KESSLER, R., CHAIKA, C., HUANG, S. & UNGER, L. M. 2005. Exhumation history of two high-pressure belts, northern Venezuela, based on fluid inclusions in quartz and calcite veins. *In:* AVÉ LALLEMANT, H. G. & SISSON, V. B. (eds) *Caribbean–South American Plate Interactions, Venezuela*. Geological Society of America, Boulder, CO, Special Papers, **394**, 157–171.

SMITH, C. A., SISSON, V. B., AVÉ LALLEMENT, H. G. & COPELAND, P. 1999. Two contrasting pressure-temperature-time paths in the Villa de Cura blueschist belt, Venezuela: possible evidence for Late Cretaceous initiation of subduction in the Caribbean. *Geological Society of America Bulletin*, **111**, 831–848.

SPATH, L. F. 1939. On some Tithonian ammonites from the Northern Range of Trinidad, B.W.I. *Geological Magazine*, **76**, 187–189.

SPEED, R. C. & SMITH-HOROWITZ, P. L. 1998. The Tobago terrane. *International Geology Review*, **40**, 805–830.

STEPHAN, J. F. 1985. Andes et chaîne caraïbe sur la transversale de Barquisimeto (Venezuela): évolution géodynamique. *In:* MASCLE, A. (ed) *Géodynamique des Caraïbes*. Editions Technip, Paris 505–529.

SUTER, H. H. 1960. *The General and Economic Geology of Trinidad, B.W.I.*, 2nd edn. HMSO, London.

SUTTON, F. A. 1946. Geology of Maracaibo Basin, Venezuela. *American Association of Petroleum Geologists Bulletin*, **30**, 1621–1741.

TORO, M. & MAHMOUDI, M. 1997. Evaluación sedimentológica y petrográfica de la Formación Gobernador, area de Barinas norte, Venezuela. *In:* CHATELLIER, J.-Y. & AQUINO, R. (eds) *Hydrocarbon Basins of Venezuela. VIII Congreso Geológico Venezolano, I Congreso Latinoamericano de Sedimentología, Asociación Venezolana de Sedimentólogos, First Core Workshop*, **III**, 1–9.

TORRES, V., VANDENBERGHE, J. & HOOGHIEMSTRA, H. 2005. An environmental reconstruction of the sediment infill of the Bogotá basin (Colombia) during the last 3 million years from abiotic and biotic proxies. *Palaeogeography, Palaeoclimatology, Palaeoecology*, **226**, 127–148.

TYSON, L., BABB, S. & DYER, B. 1991. Middle Miocene tectonics and its effects on Late Miocene sedimentation in Trinidad. *Second Geological Conference of the Geological Society of Trinidad and Tobago, Transactions*, 26–40.

URBANI, F. 1969. Primera localidad fosilífera del Miembro Zenda de la Formación Las Brisas: Cueva El Indio, La Guairita, Estado Miranda. *Asociación Venezolana de Geología, Minería y Petroleo, Boletín Informativo*, **12**, 447–454.

URBANI, F. 1977. Geochímica de las aguas termales del area de El Pilar-San Antonio del Golfo, Edo. Sucre. *V Congreso Geológico Venezolano, Memorias*, **3**, 1061–1065.

URBANI, F. 1989. Geothermal reconnaissance of northeastern Venezuela. *Geothermics*, **18**, 403–427.

URBANI, F. 1991. *Fuentes de aguas termales en Venezuela*. World Wide Web Address: http://www.pdv.com/lexico/menes/aguter.htm (accessed August 2007).

USECHE, A. & FIERRO, I. 1972. Geología de la región de Pregonero, Estados Táchira y Mérida. *IV Congreso Geológico Venezolano, Memorias, Boletín de Geología, Publicaciónes Especiales*, **5**, 963–998.

USECHE, A., FIERRO, I., ODREMAN, O. & KISER, G. D. 1985. Excursión 2, Andes meridionales: San Antonio–Rubio–San Cristóbal–San Joaquín de Navay. *VI Congreso Geológico Venezolano, Guia de la Excursión*, **10**, 6880–6916.

VAN ANDEL, TJ. & POSTMA, H. 1954. *Recent Sediments of the Gulf of Paria*. Verhanderlingen der Koninklijke Nederlandse Akademie van Wetenschappen, Afd. Naturkunde, **20**, 1–245.

VAN DECAR, J. C., RUSSO, R. M., JAMES, D. E., AMBEH, W. B. & FRANKE, M. 2003. Aseismic continuation of the Lesser Antilles slab beneath continental South America. *Journal of Geophysical Research*, **108**, 2043; ESE18:1–12.

VAN DER HAMMEN, T. 1985. The Plio-Pleistocene climatic record of the tropical Andes. *Journal of the Geological Society, London*, **142**, 483–489.

VIALLY, R., LETOUZEY, J., BÉNARD, F., HADDADI, N., DESFORGES, G., ASKRI, H. & BOUDJEMA, A. 1994. Basin inversion along the North African Margin, the Saharan Atlas (Algeria). *In*: ROURE, F. (ed.) *Peri-Tethyan Platforms*. Éditions Technip, Paris, 79–118.

VIERBUCHEN, R. C. 1984. The geology of the El Pilar fault zone and adjacent areas in northeastern Venezuela. *In*: BONINI, W. E., HARGRAVES, R. B. & SHAGAM, R. (eds) *The Caribbean–South American Plate Boundary and Regional Tectonics*. Geological Society of America, Boulder, CO, Memoirs, **162**, 189–212.

VINIEGRA, F. 1971. Age and evolution of salt basins of southeastern Mexico. *American Association of Petroleum Geologists Bulletin*, **55**, 478–494.

VIVAS, V. & MACSOTAY, O. 1995a. Formación Tememure: unidad olistostrómica Eocene superior-Oligoceno inferior en el frente meridional de la Napa Piemontina Venezuela nor-central. *Venezuela Ministerio de Energía y Minas, Boletín de Geología, Publicaciónes Especiales*, **10**, 95–113.

VIVAS, V. & MACSOTAY, O. 1995b. Dominios tectono-estratigráficos del Cretácico–Neógeno en Venezuela nororiental. *Venezuela Ministerio de Energía y Minas, Boletín de Geología, Publicaciónes Especiales*, **10**, 124–152.

WALL, G. P. & SAWKINS, J. G. 1860. *Report on the Geology of Trinidad; Part I of the West Indian Survey*. Geological Survey, Memoirs.

WARREN, J. K. 1997. Evaporites, brines and base metals: fluids, flow and 'the evaporite that was'. *Australian Journal of Earth Sciences*, **44**, 149–183.

WARREN, J. 1999. *Evaporites: Their Evolution and Economics*. Blackwell Science, Oxford.

WARREN, J. 2005. Evaporitic source rocks: a geological response to biological cycles of 'feast or famine' in layered brines. American Association of Petroleum Geologists, Annual Convention, Program and Abstracts.

WEBER, J. C. 2005. Neotectonics in the Trinidad and Tobago, West Indies segment of the Caribbean-South American plate boundary. *Geological Institute of Hungary, Occasional Papers*, **204**, 21–29.

WEBER, J. C., FERRILL, D. A. & RODEN-TICE, M. K. 2001. Calcite and quartz microstructural geothermometry of low-grade metasedimentary rocks, Northern Range, Trinidad. *Journal of Structural Geology*, **23**, 93–112.

WORRALL, D. M. & SNELSON, S. 1989. Evolution of the northern Gulf of Mexico, with emphasis on Cenozoic growth faulting and the role of salt. *In*: BALLY, A. W. & PALMER, A. R. (eds) *The Geology of North America – An Overview*. The Geology of North America, **A**, Geological Society of America, Boulder, CO, 97–137.

ZIELINSKI, G. W., BJORØY, M., ZIELINSKI, R. B. L. & FERRIDAY, I. L. 2007. Heat flow and surface hydrocarbons on the Brunei continental margin. *American Association of Petroleum Geologists Bulletin*, **91**, 1053–1080.

A structuring event of Campanian age in western Venezuela, interpreted from seismic and palaeontological data

P. M. COONEY[1]* & M. A. LORENTE[2]

[1]Red Sky Energy Pty Ltd Level 25, 2 Chifley Square, Sydney, NSW 2000, Australia

[2]Apdo 281, Jumilla 30520, Spain

*Corresponding author (e-mail: pcooney@romtech.com.au)

Abstract: A period of structuring, uplift, non-deposition and/or erosion in the Campanian in Western Venezuela, different from the generally known Late Cretaceous event is proposed to explain: (1) a varying time gap (10 to 1 million years) from east to west across the Maracaibo basin between La Luna and Colón formations; (2) a correlating time gap of 11 million years between the Santonian and Upper Campanian sediments in the Barinas basin; (3) structuring at the Top La Luna seismic horizon in the southwestern, west and central parts of the Maracaibo basin which is not reflected in the overlying section; (4) different thickness patterns in the isopach maps for the units underlying the Top La Luna seismic level and the immediately overlying section at least in area of the Colón Unit; (5) an abrupt change in vitrinite reflectance values in the SW of the basin from 0.47–0.60% above to 1.09–1.80% below the top of La Luna Formation; and (6) fission track ages in the range 70–80 Ma in the Circum-Caribbean. The predominant north–south trend of this structuring suggests that it is related to changes on the dynamics of the South American plate boundary during Campanian that may have involved a major igneous and volcanic event registered 70–80 Ma.

This paper is an extended and updated of work first presented at the VIII Congreso Geologico Venezolan (1997). This was the first documentation of a Campanian tectonic event that ended the La Luna Formation Cycle in Western Venezuela. This occurred earlier than and is different from the transition from passive to compressive tectonics of the northern South America Plate margin (Parnaud et al. 1995). The original abstract, published in Spanish, is difficult to find and has been generally overlooked. This version is intended to promote discussion of the precise timing and origin of events that influenced the evolution of northern South America.

Traditionally, the Late Cretaceous in western Venezuela is seen as a time of continuous, fine-grained marine sedimentation on an extensive passive margin, marked only by change from the anoxic–dysoxic conditions of the Cenomanian–Santonian La Luna sea to the normal marine, Campanian to Maastrichtian Colón sea.

Seismic data show numerous unconformities in the upper part of the Colón–Mito Juan Formation. These are dated as Late Campanian–Maastrichtian (Pindell & Tabbutt 1995; Parnaud et al. 1995 – Cretaceous Super sequence B, sequences K4 and K5 of the passive margin), younger than the 'Top La Luna' event discussed here.

Biostratigraphy indicates four major Late Cretaceous palaeogeographic provinces in the Maracaibo Basin: Trujillo–Lara to the east, Lake Maracaibo in the centre, the Perijá Range to the west and the Táchira–Mérida part of the North Andean Foothills in the south (Fig. 1). This paper concentrates on seismic data from the southwestern part of the Maracaibo Basin on the Venezuela–Colombia border (Fig. 1). We also refer to recently published seismic and palaeontological data that support points raised by our work.

Geological setting

The Colón Unit of the southwestern Maracaibo basin covers an area of some 3000 km^2 (Fig. 1) between the north–south trending Perijá Range to the west and the NE-trending Venezuelan Andes to the SE. The sedimentary section, overlying Permo-Carboniferous metamorphic rocks or Jurassic red beds, comprises about 600–800 m of Lower–Middle Cretaceous, mixed clastic–carbonate rocks followed by around 750 m of Upper Cretaceous Colón shales, overlain by 4000 m (basin centre) to 0 m (eroded basin margin) of almost entirely clastic Cenozoic rocks. The Upper Cretaceous shales, frequently overpressured, act as a major décollement between zones of very different structural styles. The Cenozoic is characterized by low-angle thrusts, which sole out in the Cretaceous shales, and the Lower Cretaceous by generally high-angle, basement-involved reverse faults that cross

From: JAMES, K. H., LORENTE, M. A. & PINDELL, J. L. (eds) The Origin and Evolution of the Caribbean Plate.
Geological Society, London, Special Publications, **328**, 687–703.
DOI: 10.1144/SP328.27   0305-8719/09/$15.00 © The Geological Society of London 2009.

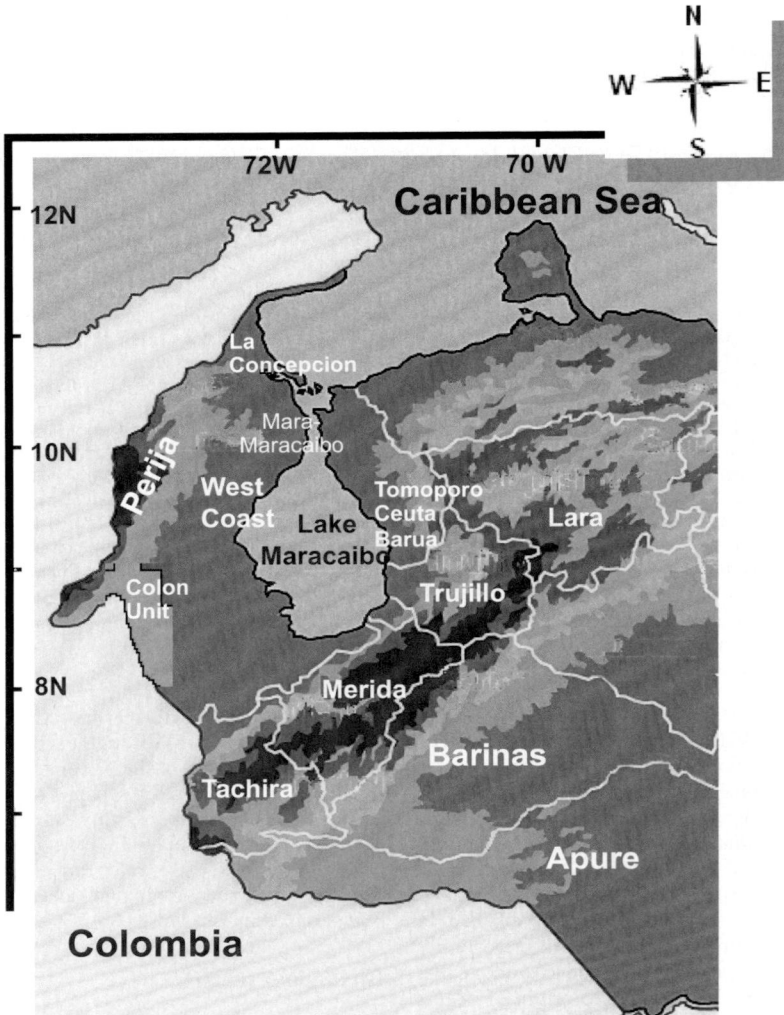

Fig. 1. Map showing location of the Colón unit.

the Lower Cretaceous and die upwards into the same shales.

Proved reserves of the SW Maracaibo basin total more than 750 million barrels of oil. These reserves occur mainly in fields located along two major structural trends, the north–south Los Manueles–Las Cruces–Petrolea high and the NW-trending Tibú–Socuavo–Sardinata high. Eighty-five per cent of the oil is reservoired in the Cenozoic section, mainly the Paleocene Barco and the Eocene Mirador formations; more than 98% is believed to be derived from Cretaceous source rocks of the Cenomanian–Campanian La Luna Formation (Talukdar & Marcano 1994).

Most of the observed structuring is post Middle Miocene to Recent in age, but there also is evidence for earlier structuring episodes (Cooney & Lorente 1997).

La Luna Formation and the Tres Esquinas Member

A unit called the Tres Esquinas Member immediately overlies the La Luna Formation along the west and south flanks of the basin. Many recent authors, consider this to be part of the La Luna (Erlich *et al.* 1997, 1999*a*, *b*, 2000; De Romero *et al.* 2003; Rey *et al.* 2004).

The Tres Esquinas crops out as a transitional, 3–5 m thick unit of dark coloured shales and fine to very fine grained sandstones with abundant glauconite, pyrite and phosphate layers. It is considered in most Venezuelan literature to be a classic condensed section but its origin and palaeoenvironmental significance are not completely understood (Erlich *et al.* 2000; Parra *et al.* 2000, 2003).

Parnaud *et al.* (1995) placed the last maximum flooding surface of the Maracaibo Basin upper Cretaceous within the Tres Esquinas Member. Erlich *et al.* (2000) identified different types of phosphate beds (following Föllmi *et al.* 1992) and suggested that phosphorites of the central Merida Andes are condensed, formed by winnowing over long periods of time, while allochtonous phosphorites, transported as gravity flows, prevail in the northern and southern Merida Andes. Some coarse-grained phosphorites are poorly sorted, often graded and show grains with micritic rims and frequent microboring, suggesting 'periods of slow or no deposition, interrupted by rapid and or catastrophic reworking and redistribution' (Erlich *et al.* 2000). Reworking of phosphate and glauconite grains and lateral lithological variations observed in the type section of the Perijá are attributed to variable bottom-water oxygenation, uneven seafloor topography, bottom currents and turbidity currents triggered by seismicity (Parra *et al.* 2003).

According to Rey *et al.* (2004), this member marks the culmination of highstand deposition that ended the La Luna Fm sedimentary cycle and an upper sequence boundary is present at the base of the overlying Colón Fm. These authors interpreted an environment of stratified/weakly stratified water column, with common oxygenation episodes and intermittent upwelling, in a shallowing basin and a cooling climate, based on increases in benthic foramifera, with infaunal forms, vanadium–nickel and phosphorous values and high Ba–Al ratios.

According to this paper the Tres Esquinas Member is not a classic 'condensed section'. Instead it is a complex unit, with important lateral variations, marking tectonic activity that terminated the La Luna Formation cycle.

Seismic evidence, southwestern Maracaibo Basin

In this work the Tres Esquinas Member is included in the so-called Top La Luna seismic horizon. Its 3–5 m thickness and its lithology do not allow seismic differentiation from the La Luna Fm.

Seismic data in the southwestern part of the Maracaibo Basin are of fair to good and occasionally very good quality and reveal attributes not previously seen in the Upper Cretaceous Colón and Mito Juan formations. The Mito Juan formation represents the final shallower-water, coarser-grained phase of the deposition of the Colón Formation shales deposition. It is almost impossible to distinguish between these two units in wells or on seismic. Accordingly, they are treated here as a single unit.

Seismic data show onlap of the lowermost Mito Juan and Colón formations section onto an already structured 'Top La Luna' reflection (Figs 3, 4, 5 & 9, seismic levelled at top Mito Juan/Colón to remove later structuration), while sub-horizontal apparently undisturbed reflections are seen infilling structural lows or grabens. The bounding, apparently high-angle, reverse faults do not appear to be, as originally interpreted, inverted normal faults. The seismic data show that much of the structuring seen at the 'Top La Luna' seismic horizon is not reflected at all or dies out very rapidly in the overlying Mito Juan/Colón section.

Reflection definition and continuity are generally poor in the Colón and Mito Juan formations due to the lack of lithology contrasts. Nonetheless angular onlap by the Colón and Mito Juan formations onto the 'Top La Luna' reflections is common. It has been observed by some seismic interpreters (Parra *et al.* 2000, 2003; Gallango *et al.* 2002) but not many others (e.g. Mann *et al.* 2006). We believe that the displacement seen on the 'Top La Luna' surface on the illustrated sections is largely pre-cenozoic. There is no indication of compaction such as cycle splitting off structure or cycle thinning over the structures. The Colón and Mito Juan formations act as a décollement zone between two distinct structural styles. The section is frequently overpressured and might be expected to behave plastically. However, at outcrop and in well cuttings these shales are brittle and splintery. It is doubtful that such a rock would accommodate much plastic deformation. Our experience is that structures in the Colón area reflect imbricate faulting rather than shale diapirism and we suggest that complete absence of structuring higher in the Colón and Mito Juan formations unit implies early structuration, seen in the 'Top La Luna' seismic level. Syntectonic sedimentation is a more reasonable solution to the structural space problems than the plastic flow generally invoked up to now.

The high angle of the numerous reverse faults that break up the 'Top La Luna' seismic reflection suggests basement involvement (Fig. 9). Here they must sole out well below the deepest mapable reflections. Their predominantly north–south alignment indicates east–west compression. However, the overall fault pattern suggests some lateral movement (Fig. 5). Occasionally substantial rollover at 'Top La Luna' seismic reflection level gives rise to fault bend type folds, but fault propagation type folds appear more common. All these structures may be related to the subduction of the Caribbean Plate beneath South America, as proposed by Burke *et al.* (1984) and Pindell & Tabbutt (1995). They could also be related to uplift of the Central Cordillera of Colombia and/or to an igneous–volcanic episode related to thickening of the Caribbean (James 2005*a*, *b*). Interpretation depends on the model used to explain the origin of the Caribbean Plate.

Further evidence for structuring in the Late Cretaceous comes from the Colón and Mito Juan formations isopach map (Fig. 7). This was derived by multiplying the time thickness by an average interval velocity calculated from well data. The map shows good agreement with the more regional maps of Vásquez & Dickey (1972) based on well and outcrop data, and of Lugo & Mann (1995). The isopach shows long, linear 'thicks' and 'thins' superimposed on regional thickening to the west. At some locations, such as the Tarra anticline in the SE, thickening reflects repeated section from faulting and/or steep dips. However, most features result from thinning or thickening over pre-existing highs or lows. This picture is very different from northward thinning of the underlying La Luna and Capacho formations (Fig. 6) and from the WNW thickening of the overlying Orocue Group (Fig. 8).

Seismic evidence from other areas of the Maracaibo Basin

Generally speaking, an angular relationship between seismic reflections (onlap, downlap, truncation) on the surface of contact represents a time gap or unconformity (Mitchum *et al.* 1977).

Such a relationship exists where reflections within and at the base of the Colón Formation onlap 'Top La Luna' seismic horizon reflections. The same relationship is illustrated in published sections from other areas of the Maracaibo basin (Gallango *et al.* 2002, fig. 9; Parra *et al.* 2003, fig. 11). Other published figures also show this relationship is present, although it is not interpreted as such (e.g. Castillo & Mann 2006*a*, fig. 6) the same happens with the thinning of the Colón Formation ('shale') by onlap onto a structure in the southern Maracaibo Basin (Castillo & Mann,

2006*a*, fig 22). Gallango *et al.* (2002) show such structures in Perijá area.

The 'Top La Luna' seismic marker thus appears to be an unconformity surface or a surface of non-deposition structured prior to deposition of the Colón Formation. On the other hand there is little if any evidence of significant erosion and no evidence of the karst topography as seen at the deeper Apon Formation level (Castillo & Mann 2006*b*). This may indicate that structuring occurred while the sediments were below erosion base/sealevel. There is no reason why this structuring could not have been incurred at the same time as a major transgression such as that at the base of the Colón Formation. Certainly the work of Parra *et al.* (2003) strongly suggests that the Tres Esquinas Member at the very top of the La Luna Formation was deposited in widely varying water depths, while the variations in the thickness of the overlying Colón Formation (Cooney & Lorente 1997) suggests deposition onto a structured if not eroded surface.

Seismic expression in the Barinas Basin

Seismic data from the Barinas Basin suggest minor low angle truncation at the surface between Navay and Burguita formations (De Guerra *et al.* 1996) equivalents of the Colón and La Luna formations.

Palaeontological evidence

We begin this section by highlighting the change in the last years of the age of the Campanian–Maastrichtian boundary and the stage equivalence of the biostratigraphic zonations used for that period of time and in this review.

The Campanian–Maastrichtian boundary

The Campanian–Maastrichtian boundary, important in the Maracaibo basin, has gone through revision and changes by the ICS since 2001, due to designation of a new Global Standard Section and Point (GSSP) for the base of the Maastrichtian (Odin & Lamaurelle 2001; Gradstein *et al.* 2004), as shown in Table 1 (simplified and modified after Ogg 2004).

Table 1. *Base of the Maastrichtian Stage GSSP age and GSSP location*

Stages	Age (Ma) GTS 2004	Derivation of age	GSSP and location	Publication
Base Maastrichtian stage	70.6 ± 0.6	Estimated placement relative to Ar–Ar calibrated Sr-curve	115.2 m level in Grande Carrière quarry, Tercis-les-Bains, Landes province, SW France	Odin (2001), Odin & Lamaurelle (2001)

Designation of this new GSSP also produced a change in the Campanian, since the age of the top shifts accordingly. This is relevant to our work, since the age of the boundary is 3.4 million years younger than previous calibrations. Figure 2 shows the new stage equivalence of the relevant foraminifera zones. The boundary now lies within the *Gansserina gansseri* zone while previously it occurred at the boundary between the *Globotruncanita calcarata* and *Globotruncanella havanaensis* zones. This is an important shift of two entire foraminifera zones.

In terms of magnetic polarity zones a shift of one entire zone is involved. Previously the Campanian–Maastrichtian boundary was close to the top of zone C33. Now it is close to the base of zone C31.

Calibration between the biostratigraphic zones and the magnetic polarity zones has not changed, so the absolute age of the foraminifer zones

remains as in Bralower *et al.* (1995). For this reason we refer to biostratigraphic zones (unchanged) and stage assignation is an interpretation based on the last accepted ICP equivalence (changed recently). Unfortunately many authors of past publications refer mainly to the stage rather than to the biostratigraphic or magnetic polarity zone. This can be confusing when comparing data from publications prior to the ICP change with those from later publications.

Age of La Luna Formation

The La Luna Formation has been considered one of the most typical representatives of the widespread Cretaceous 'anoxic' event (OAE). Its Cenomanian to Santonian age was originally based upon ammonites (Liddle 1928; Sutton 1946; Renz 1959, 1977, 1982) and the unit was traditionally considered to

(a)

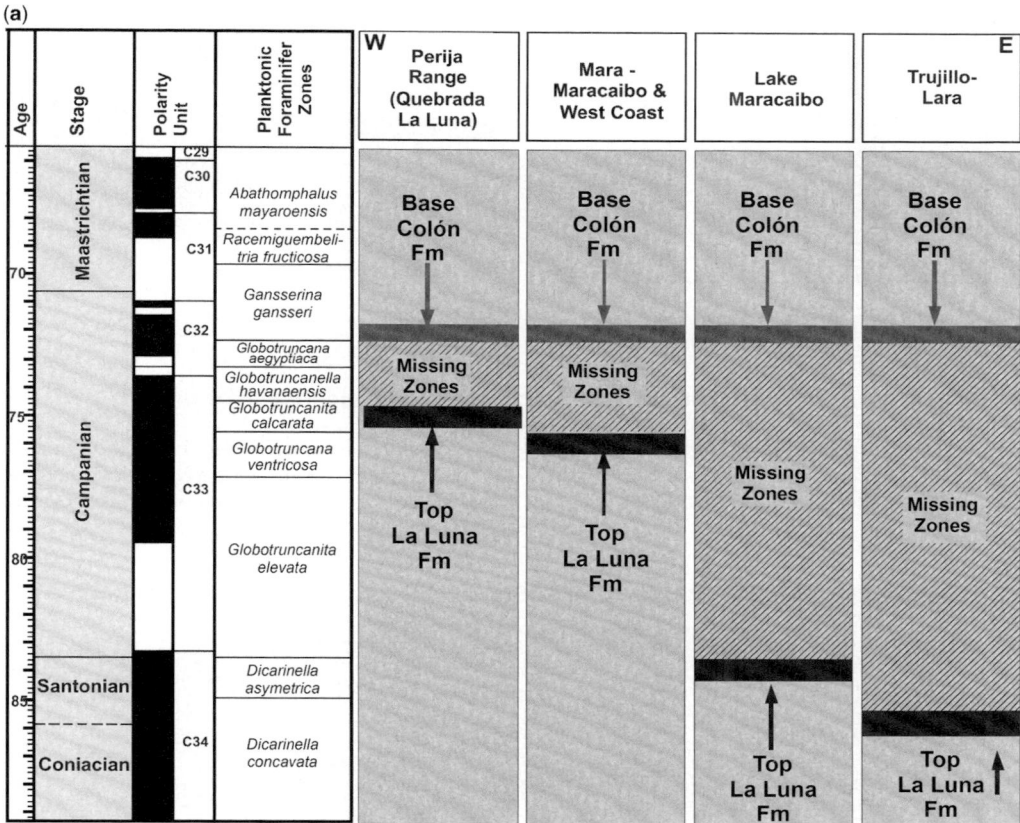

Fig. 2. Schematic chronostratigraphic section showing age variation of the La Luna Formation top as well as the Colón Formation base. (**a**) West–east section approximately across the centre of the Maracaibo Basin, clearly shows the younging trend of the top of La Luna Formation. (**b**) NW–SE section across the Maracaibo Basin; the older ages of the top La Luna in the centre of Lake Maracaibo and centre of North Andes Foothills are apparent, where the formation is thinner than in the adjacent areas. Time scales created with TS-Creator (Ogg & Lugowski 2006).

(b)

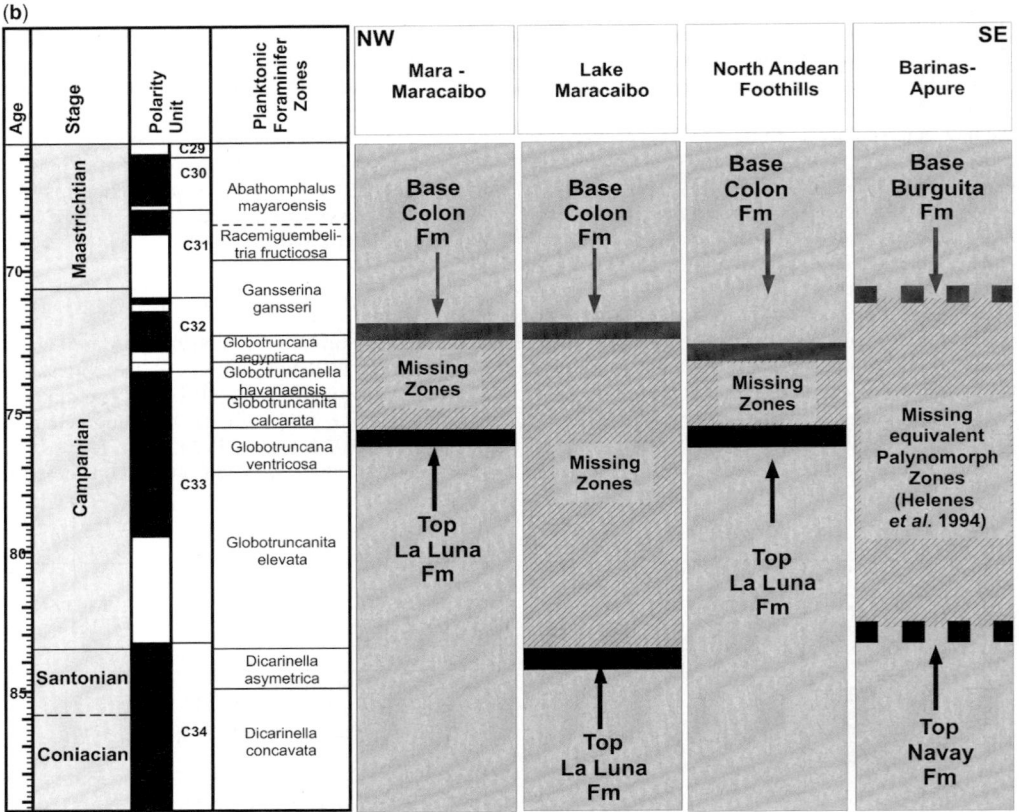

Fig. 2. *Continued.*

be isochronous. However, over the years diachroneity has been recognized (Boesi *et al.* 1988, 1993; Galea Alvarez 1989; Truskowski *et al.* 1995; Lorente *et al.* 1997; Davis *et al.*1999; De Romero *et al.* 2003; Zapata *et al.* 2003).

Age of La Luna Formation top

The diachronous nature of the La Luna Formation top has given rise to controversy. Some authors associated this with an unconformity between the

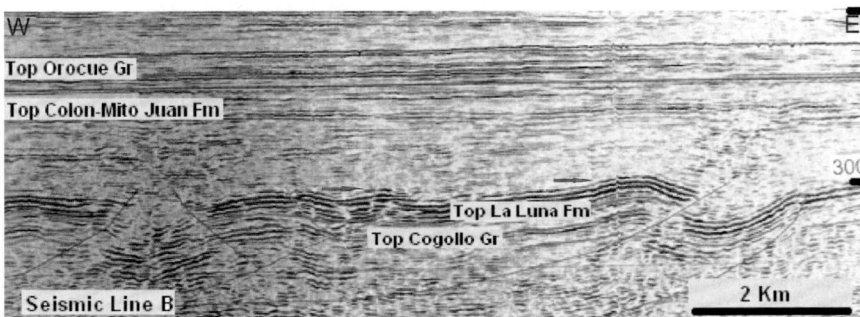

Fig. 3. Seismic line segment (line B on Fig. 8) datumned on the horizon 'Top of Colón–Mito Juan' showing onlap onto the central Top La Luna high and infill of the graben to the west by undeformed horizontal beds.

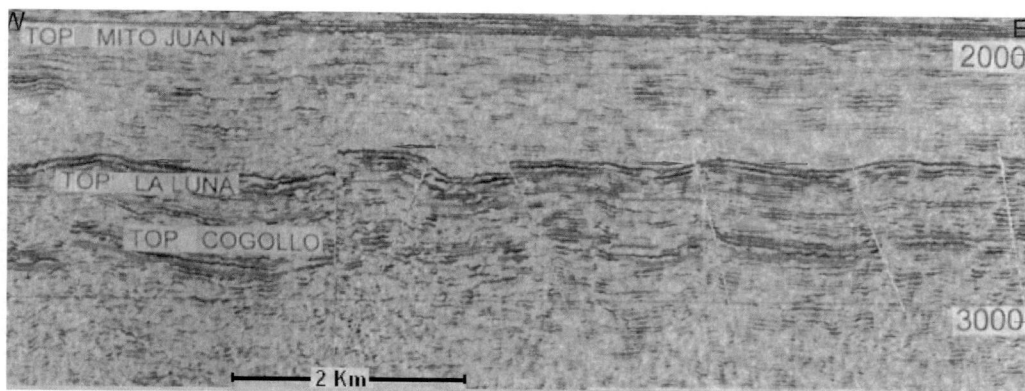

Seismic Line C

Fig. 4. Seismic line segment (line C on Fig. 8) datumned on the horizon 'Top of Colón–Mito Juan' showing onlap by originally horizontal reflections onto Top La Luna uplifted fault blocks.

La Luna and the overlying Colón Formation (De Romero & Odreman 1996; Lorente *et al.* 1997; Cooney & Lorente 1997; Gallango *et al.* 2002; Parra *et al.* 2003).

In this paper we discuss the diachronous nature of the formation top, which in our opinion shows a general westward younging trend, and the isochronous nature of the hiatus top (Fig. 2a & b). Palaeontological, age-related evidence, ranging from ammonites to foraminifera (Galea Alvarez 1989) to calcareous nannoplankton (De Romero & Galea Alvarez 1995; Farias *et al.* 2000), can be geographically summarized as follows.

To the east, in the Trujillo/Lara area (Figs 1 & 2), the presence of the ammonites species *Peroniceras* aff. *canaense* and *Hauericeras* aff. *gardeni* (Renz 1982) of Coniacian and Early Santonian age, and the presence of the *Dicarinella concavata* zone in the Chejendé section (Mérida Andes, Truskowski, pers. comm. 1997) date the formation top as not younger than Santonian. Studies by De Romero *et al.* (2003) in La Peña–San Felipe Section, Trujillo, showed that the top of the Timbetes Member (top of La Luna) has been eroded. However, the presence of calcareous nanofossils *Micula staurophora* and *Lithastrinus moratus* and the absence of *Micula concavata* in the preserved uppermost part of that unit suggest a Middle to Late Coniacian age. According to these authors, the presence of the planktic foraminifera *Dicarinella concavata* and the absence of *D. asymetrica* are consistent with that age, since the *D. concavata* zone crosses the Coniacian–Santonian boundary.

In the central area of Lake Maracaibo (Figs 1 & 2), the presence of the *D. asymetrica* zone (Fuenmayor 1989; Hambalek & Pittelli 1994*a, b*) dates the uppermost La Luna Formation as not younger than Late Santonian (zone ages after Bralower *et al.* 1995; stage age after Gradstein *et al.* 2004).

In the Mara La Paz and La Concepcion oil fields to the north west of Lake Maracaibo Fuenmayor (1989) reported the presence of the Late Santonian *D. asymetrica* zone. The same author identified in that area the operational *Eponides*-17 zone (equivalent to the *G. calcarata–G. elevata* planktonic zones according to him) in the Socuy Member, a calcareous interval that some authors also include in the La Luna Formation (Lorente *et al.* 1997; Erlich *et al.* 2000). However, in his generalized Range Chart for the area, Fuenmayor does not show the species *G. calcarata* but shows instead species of more extensive ranges such as *Guembelitria cretacea, Pseudoguembelina punctulata, P. costulata, P. plummerae* and *Hedgerbella* cf. *holmedelensis.*

Further to the west, in the Perijá Range (Figs 1 & 2), the Santonian Calcareous Nannoplankton zones NK15 to NK16 (in the sense of Sissingh 1997) have been reported. At the same time the (scarce) presence of the key species *Globotruncanita calcarata* (*Globotruncanita calcarata* is also referred by some authors as *Radotruncana calcarata*, a synonymy situation beyond the scope of this work) in Quebrada La Luna (type section, in the Perijá) dates the upper levels as Campanian (Lorente *et al.* 1997). The overlying Tres Esquinas Member in this area was dated based on the presence of the calcareous nannoplankton zone NK22 as Late Campanian (Lorente *et al.* 1997).

In the Los Manueles, Tarra and Rosario Oil fields of the Colón District Fuenmayor (1989) reports in La Luna Formation an operational *Cibicides*-28

COLÓN UNIT

VENEZUELA

COLOMBIA

===== REVERSE FAULT
DOTTED LINE DOWNDIP

FAULT TRACES
TOP LA LUNA
REFECTOR

0   10   20
                km

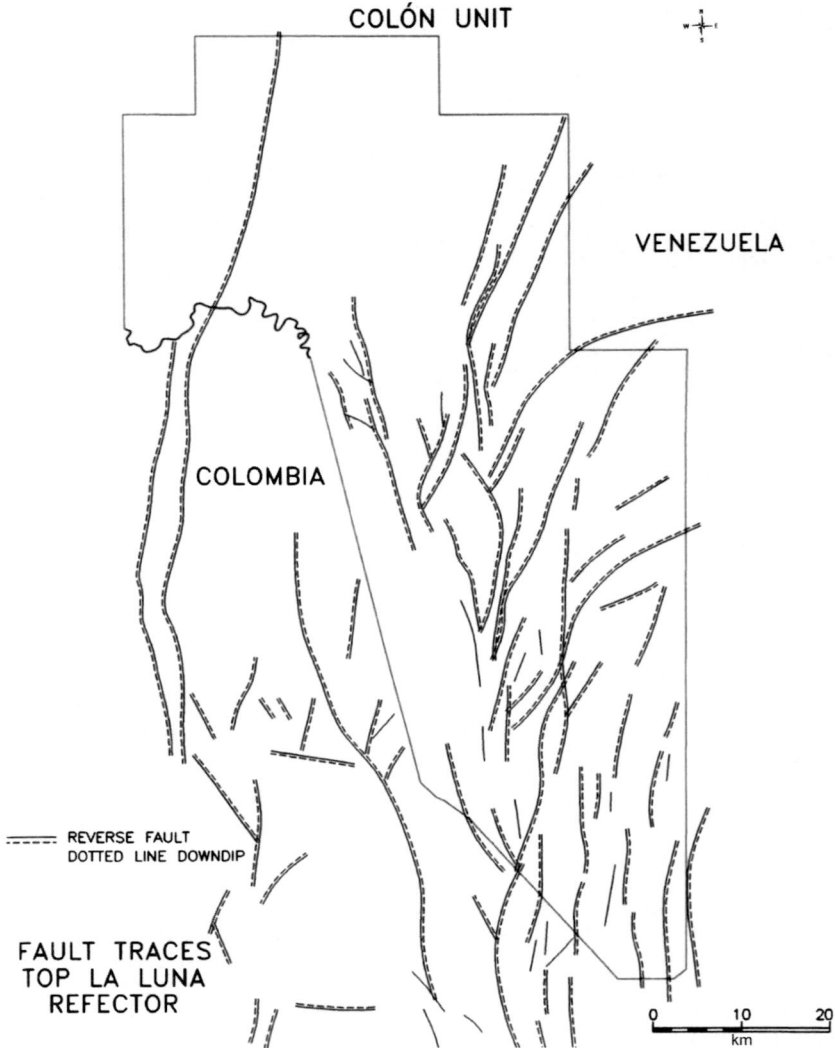

Fig. 5. Fault trace pattern at the Top La Luna reflector. Note the general north–south trend and the bifurcation to the north, suggesting strike–slip.

zone, equivalent according to him to the *Dicarinella asymetrica–D. concavata* zones of Coniacian–Santonian age. For the Tres Esquinas Member he reports *Gyroidina*-12 (equivalent to *G. ventricosa–G. elevata* plankton zones). If the equivalence with the plankton zones proposed by that author is correct, then the age is Campanian.

To the south, in the Táchira Andes, the ammonite *Texanites texanus* of Lower to Middle Santonian age (Renz 1982; Gradstein *et al.* 1995) and *Paralenticeras sieversi* (Sutton 1946, in Macellari & De Vries 1987) indicates a Santonian age for the upper part of the La Luna Formation. This is older than in the Quebrada La Luna section but similar to ages reported from the subsurface below Lake Maracaibo and younger than those reported to the east from Chejendé.

In the same Táchira Trough, but further to the south, in Quebrada Bellaca, Ford & Houbolt (1963) identified *Globotruncanita calcarata* in thin sections from La Luna Formation. This species dates the Formation as Campanian in this area, the same as the youngest age reported from the Perija Range.

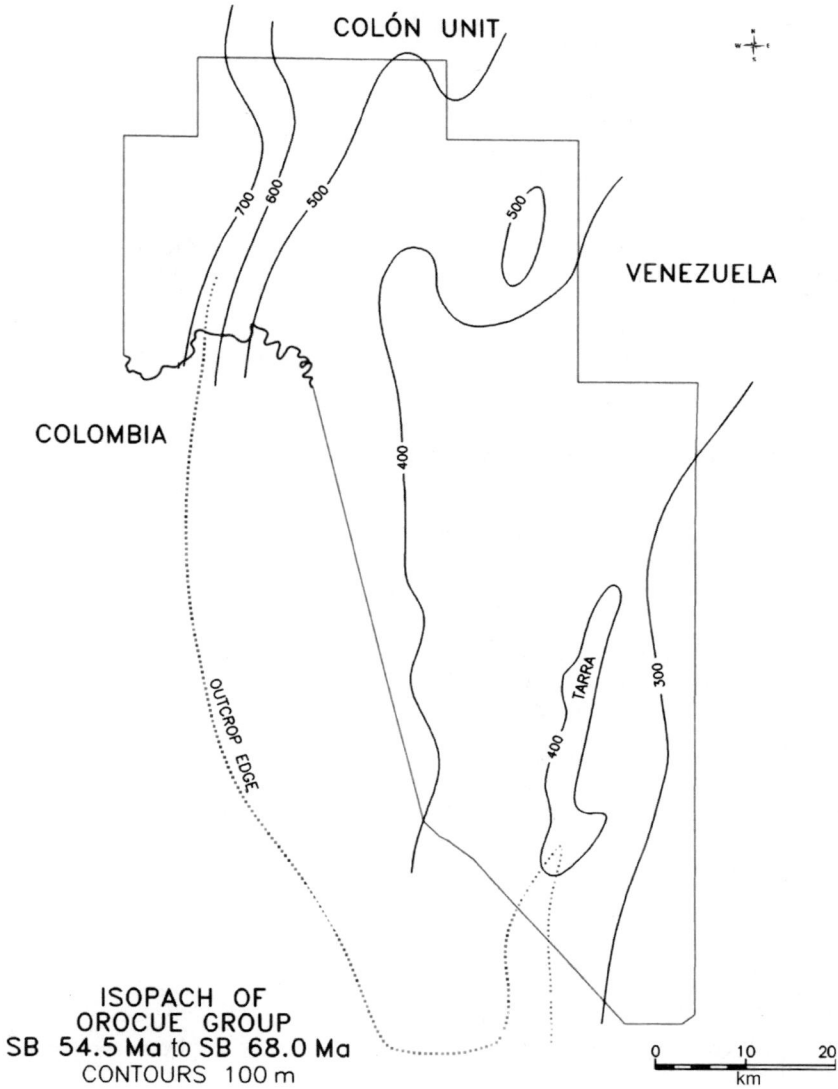

Fig. 6. Isopach of the Orocue group (Paleocene) showing general thickening to the west.

For the rest of Northern Andean Foothills the Tres Esquinas Member has been reported to include the *G. ventricosa* zone (Lorente *et al.* 1997) of Campanian age.

Colón Formation

The overlying Colón Formation is dated using planktonic foraminifera everywhere in the subsurface of Lake Maracaibo and in outcrops of the Perija Range and the Venezuelan Andes. In these areas the formation contains only the *Gansseri gansseri* zone of Late Campanian–Maastrichtian age.

The oldest section of the Colón Formation occurs in wells from the northern Andes foothills, where the *Globotruncana aegyptiaca* zone has been confirmed (Lorente *et al.* 1997).

The ubiquitous absence of the *Globotruncanella havanaensis* zone and the partial or total absence of the *Dicarinella asymetrica, Globotrucanita elevata, Globotruncana ventricosa, Globotruncanita calcarata* and *Globotruncana aegyptiaca* zones

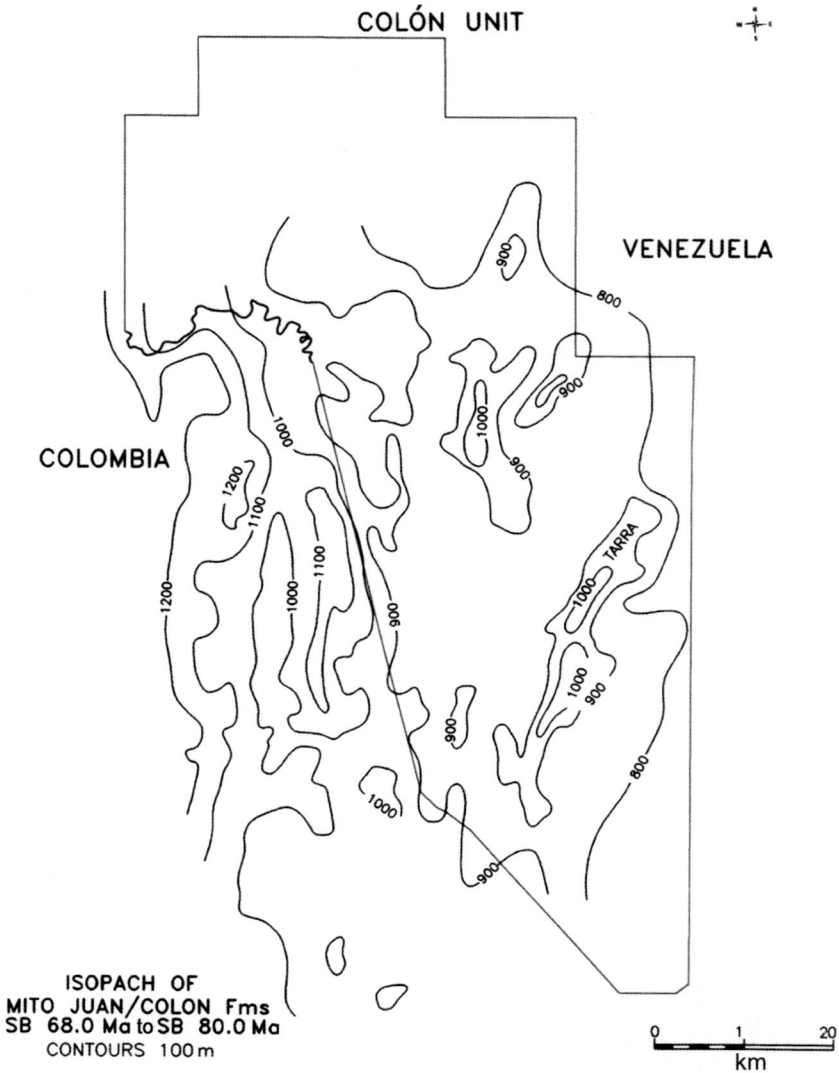

Fig. 7. Isopach of the combined Mito Juan–Colón formation (Upper Campanian–Maastrichtian) showing an overall thickening to the west upon which have been superimposed linear north–south trending thickness variations.

from parts of the Maracaibo Basin (Figs 1 & 2) allows us to infer the existence of a hiatus between the top of La Luna Formation and the base of Colón Formation. This ranges from ±13 Ma in the area of Chejendé in Trujillo state and ±2 Ma in the Perija Range (Quebrada Maraca, Quebrada La Luna) and the Tachira Trough (times estimated by equivalence between the planktonic zones of Caron 1985; Bralower *et al.* 1995 and the new ICP calibration of Odin & Lamaurelle 2001; Odin 2001; Gradstein *et al.* 2004).

This Campanian regional hiatus (Lorente *et al.* 1995, 1996, 1997; Cooney & Lorente 1997; Erlich *et al.* 1999*a*) was recognized by an assembly of micropalaeontologists convened as a contribution to the IGCP 381 in Puerto La Cruz, Venezuela (Galea-Alvarez *et al.* 2000).

A Campanian hiatus has also been recognized in the Barinas–Apure Basin, SE of the Mérida Andes, by Helenes *et al.* (1994, 1998). Based on palyno-morph associations, it occurs between the Navay and the Burguita formations. The presence of

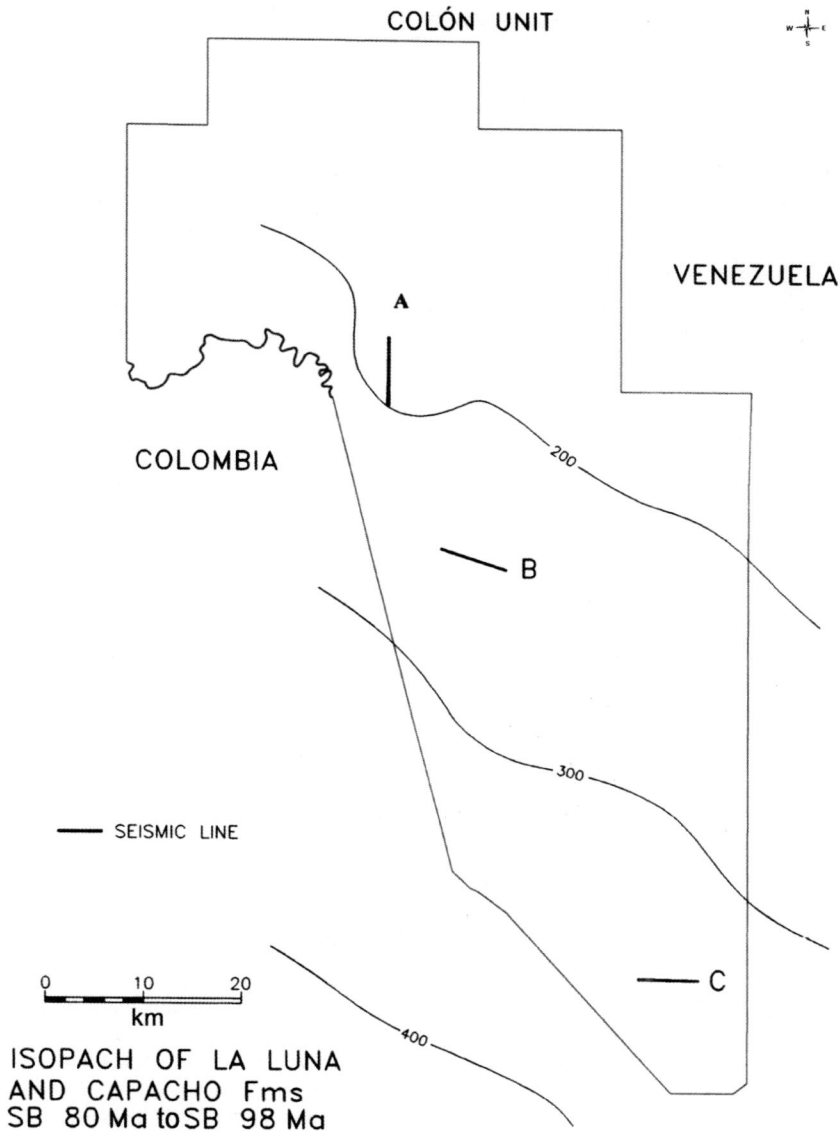

Fig. 8. Isopach of the combined La Luna and Capacho formations (Cenomanian–Campanian) showing a general thickening to the SW. This diagram also shows the location of the illustrated seismic line segments.

Campanian dinoflagellates *Senegalinum bicavatum, Palaeoperidinium cretaceum, Impagidinium grande, Odontochitina porifera, Canningia senonica* and *Andalusiella gabonensis* in the Navay Formation assemblage and the presence of the sporomorph *Foveotriletes margaritae* and absence of Early Maastrichtian markers in the overlying Burguita Formation suggest a 12 Ma hiatus. The hiatus has also been recognized in the basin by Erlich *et al.* (1999*a*) in the Rio Santo Domingo outcrop, Maporal and Paez Fields and in the Conso-3 well.

Remarks about the biostratigraphic evidence

The authors consider that available biostratigraphic information points to a general westward younging, from Coniacian to Campanian, in the age of the top

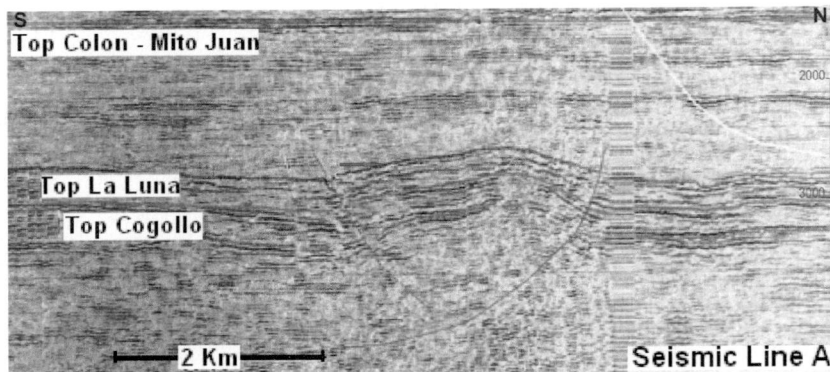

Fig. 9. Seismic line segment (line A on map in Fig. 8) showing horizontal reflections onlapping Top La Luna high. Note that the bounding faults of the Top La Luna structure do not extend into the overlying Mito Juan–Colón Formation, which appears to be gently draped over the pre-existing high.

of La Luna Formation (Fig. 2a). Although this trend is not uniform in a NW–SE section (Fig. 2b), the oldest ages occur in Lake Maracaibo and in the Lara–Trujillo areas, where the thickness of the formation is less than 100 m (maps of Savian 1993; Savian & Scherer 1997).

The hiatus ranges from about 13 Ma in the east (*Dicarinella asymetrica* zone in contact with *Gansserina gansseri* zone) to about 2 Ma in the west and south of the basin (*Globotruncanita calcarata* zone in contact with zones *Globotruncana aegyptiaca/ Gansserina gansseri*) (Fig. 2b). Further evidence of the regional extension of the hiatus comes from the correlative 12 Ma hiatus between Campanian and Maastrichtian strata in the Barinas Basin (combining palynological data) where seismic data suggest minor, low-angle truncation (De Guerra *et al.* 1996).

In the east of the Maracaibo Basin the hiatus extends up to 72 Ma. In the west it ranges to about 75–73 Ma. In the Barinas Basin it extends to 71 Ma. These ages suggest a probable age of 75–73 Ma for the event that produced the uplift, erosion and/or non-sedimentation affecting the Maracaibo and Barinas basins.

Vitrinite and fission track data

A marked change in vitrinite reflectance maturity levels occurs between the Leon to Mito Juan–Colón formations section and the underlying La Luna–Tibu formations in outcrop sections immediately to the west of the Colón Block in Colombia. Reflectance values increase steadily downwards from 0.47 to 0.60% and then abruptly change to 1.20–1.80% (Cooney & Lorente 1997). The

change may be, at least in part, structurally related major uplift and to erosion (Shagam *et al.* 1984), which has not been recognized before.

Apatitite and zircon fission-track ages (AFT and ZFT) from granite samples from the eastern part of the Central Cordillera of Colombia indicate rapid cooling between 80 and 70 Ma (Villagomez *et al.* 2006). K–Ar ages of 70–77 ± 5 Ma were obtained from igneous rocks of Cordillera de la Costa (Granito de Guaremal), Curaçao and Isla Los Hermanos (Santamaria & Schubert 1972). We suggest that these indicate Cenomanian igneous activity related to the tectonic event we describe.

Discussion

In this article, as in our previous work (Cooney & Lorente 1997), we propose an episode of uplift, structuring and some erosion of Campanian age in Western Venezuela to explain the following observations.

(1) A systematic variation from east to west, across the Maracaibo basin, in the age of the top of the La Luna Formation.

(2) A correlative 12 Ma time gap between the Campanian and Maastrichtian sections accompanied by some erosion in the Barinas Basin to the south.

(3) In the SW and central parts of the Maracaibo Basin, structuring at the Top La Luna level which is not reflected shallower in the section. The Top La Luna seismic reflection is onlapped by the lowest beds of the Mito Juan–Colón Formation.

(4) Very different thickness patterns in the unit underlying the Top La Luna (La Luna/

Capacho Fm), the immediately overlying (Mito Juan–Colón Fm) and the next higher element (Orocue Group) in the Colón Unit.

(5) A Caribbean Plate-wide event of Late Campanian–Early Maastrichtian (74–68 Ma) age (Iturralde Vinent 2007).

The roughly north–south faulting direction seen in the Colón area indicates east–west compression. This may be explained in different ways. It could be related to tectonic uplift in northern and eastern Colombia during the Campanian and Maastrichtian, possibly a consequence of, if we use the Pacific Origin Model for the Caribbean Plate:

(1) subduction of the original Caribbean Plate beneath South America as illustrated by Burke *et al.* (1984) and Pindell & Tabbutt (1995);
(2) collision of the Pacific island arcs with western Colombia;
(3) changes in plate geodynamics in the Caribbean, possibly associated with the Romeral suture and the development of the Cordillera Occidental in Colombia (Case *et al.* 1984; Chicangana 2005);
(4) a combination of the above.

If we use the '*In situ*' Origin Model (James 2005*a*, *b*), the possible causes are:

(1) a movement during the Late Campanian of the Bolivar and Bonaire blocks;
(2) a Cretaceous Caribbean flood basalt event (Donnelly 1973, 1989; Donnelly *et al.* 1990); one of the extrusion events reported from the ODP Hess Escarpment Site 1001 (Sigurdsson *et al.* 1996) is dated at 77 Ma and is covered by Maastrichtian sediments, which seems to be coeval our structuring event and the associated hiatus;
(3) an early pulse of dextral movement occurring along the NE-trending Sierra Perijá and Mérida Andes, perhaps triggered by Farallon convergence, which would be a mechanism to generate north–south trending compression in the Colón district (James, pers. comm.);
(4) a combination of the above.

According to the Pacific Origin Model, Parnaud *et al.* (1995) and Pindell & Tabbutt (1995) consider that the transition from a passive to an active margin took place in Late Cretaceous–Early Cenozoic time, coinciding in our area with the deposition of the Mito Juan–Colón Formation and the Orocue Group. The 'Top La Luna' occurs at the base of their depositional unit K6, in our area equivalent to the Mito Juan–Colón Formation and is onlapped by it. Therefore their proposed transition is younger than the structural event we are proposing.

Surprisingly there is, with one notable exception, little lithological or electrical log expression of these surfaces and they apparently represent channels cut into silts and shales and in turn filled by silts and shales. The exception is the Catatumbo Formation sandstone, which in the West Tarra field can be up to 30 m thick and is the main producer in that field. This unit is a classic incised valley fill and the valley itself can be seismically mapped. The channels and unconformities suggest a very shallow marine to non-marine environment of deposition. With this exception the boundary between the Colón–Mito Juan unit and the overlying Orocue Group is very gradational, being expressed by a very slow and gradual upwards increase in sand and silt content. The indicated shallow water depths suggest that deposition was able to keep up with the high rates of subsidence, of the order of 65–150 m/Ma (Lugo & Mann 1995) associated with the development of the Perija Foredeep. The rapid increase in sedimentation rates from 8 to 30 m/Ma in the La Luna Formation probably corresponds to the change from passive to active margin described by Parnaud *et al.* (1995). These authors considered the change to be of Maastrichtian age but we believe it started during the Campanian.

Hydrocarbon implications

At least two periods of hydrocarbon generation, migration and accumulation are evidenced by two main types of oil in the Colón Unit, both derived from the La Luna source rock. Oil reservoired in the Cenozoic has an American Petroleum Institute (API) gravity of 18–32° and is generally undersaturated. Oil in Cretaceous reservoirs has an API of 40–45° and is generally oversaturated, with very high gas/oil ratios (GORs). The first could be immature to mature and the latter late mature to early overmature (biodegradation is not involved).

The early structuring episodes are important. Basin modelling suggests that oil generation could have started as early as the beginning of the Cenozoic in the Perijá Foredeep. Oil generated at this time migrated into early formed structures associated with the Late Cretaceous structuring event. Such locations are defined by onlap in the lowest part of the Colón Formation. Oil then remigrated vertically and horizontally into Cenozoic reservoir rocks during severe Late Miocene to Recent structuring episodes. Reburial or the continuing generation of hydrocarbons from an already mature La Luna Formation source in recent deeps such as the Andean Trough has resulted in the generation of the present day overmature Cretaceous reservoired oils. Overpressured Colón Formation shales prevented this oil from migrating vertically into the Cenozoic reservoirs. Migration through the Colón Formation only occurred when major structural movements broke down the seal.

Conclusions

The Late Campanian appears to be a significant period of structuring, uplift, non-deposition and at least local erosion in the NW border of the South American Plate passive margin. This structuring may be related to a period of obduction of the original Caribbean Plate onto South America, which took place about 80 Ma (Coletta *et al.* 1990, Pindell & Tabbutt 1995) or to a period of stress re-accommodation due to an increase in the extension of the passive margin borders of the Tethys Sea as a result of changes in seafloor spreading rates.

Absence of significant erosion evidence on seismic data in SW Zulia state may indicate that the structuring here occurred while the sediments were below erosion base or sealevel (Pindell & Tabbutt 1995; Parra *et al.* 2003).

This could have occurred at the same time as major transgression. The work of Parra *et al.* (2003) strongly suggests that the Tres Esquinas Member at the very top of the La Luna Formation was deposited in widely varying water depths, while thickness variations in the overlying Colón Fm (Cooney & Lorente 1997) suggest deposition onto a structured if not eroded surface.

Hydrocarbons may have accumulated in these older structures whose locations can be defined by onlap patterns in the lowest part of the overlying Colón–Mito Juan Formation. These older structures were drastically modified or obliterated by younger tectonism.

The original work was published as an extended abstract at the VIII Congreso Geologico Venezolano, Margarita Island 1997. We thank C. Macellari, J. Kellogg, H. Krause, K. H. James and an anonymous reviewer for their comments and suggestions that helped to improve this and an earlier version of this paper.

References

BOESI, T., ROJAS, G., DURAN, I., GALEA, F., LORENTE, M. A. & VELASQUEZ, M. 1988. Estudio estratigrafico del Flanco Norandino en el sector Lobatera–El Vigia. *Memoria III Simposio Bolivariano Exploracion Petrolera de las Cuencas Subandinas*, Caracas, 2–41.

BOESI, T., LORENTE, M. A., MOMPART, L., TESTAMARK, S. & FALCON, J. 1993. Facies and sedimentary environments of the La Luna Formation in San Pedro del Rio, Appendix 2. *In*: BOESI, T., TESTAMARK, J. S. & ODREMAN, O. (eds) *Creatceous and Paleogene Sedimentation in the Southwestern Venezuela Andes. American Association of Petroleum Geologist–Sociedad Venezolana de Geologos International Congress and Exhibition*, Caracas, *Excursion Guidebook*, Field Trip, **4**.

BRALOWER, T. J., LECKIE, R. M., SLITER, W. V. & THEIRSTEIN, H. R. 1995. An integrated Cretaceous microfossil biostratigraphy. *In*: BERGGREN, W. A.,

KENT, D. V., AUBRY, M. P. & HARDENBOL, J. (eds) *Geochronology, Time Scales and Global Stratigraphic Correlation*. Society of Economic Paleontologists and Mineralogists, Special Publications, **54**, 65–79.

BURKE, K., COOPER, C., DEWEY, J. F., MANN, P. & PINDEL, J. L. 1984. *Caribbean Tectonics and Relative Plate Motions*. Geological Society of America, Boulder, CO, Memoirs, **162**, 31–60.

CARON, M. 1985. Cretaceous planktic foraminifera. *In*: BOLLI, H. M., SAUNDERS, J. B. & PERCH-NIELSEN, K. (eds) *Plankton Stratigraphy*, **1**, *Planktic Foraminifera, Calcareous Nannoplakton and Calpionellids*. Cambridge University Press, Cambridge, 17–86.

CASE, J. E., HOLCOMBE, T. L. & MARTIN, R. G. 1984, *Map of Geologic Provinces in the Caribbean region. Caribbean Tectonics and Relative Plate Motions*. Geological Society of America, Boulder, CO, Memoirs, **162**, 1–30.

CASTILLO, M. V. & MANN, P. 2006a. Cretaceous to Holocene structural and stratigraphic development in south Lake Maracaibo, Venezuela, inferred from well and three-dimensional seismic data. *American Association of Petroleum Geologists Bulletin*, **90**, 529–565.

CASTILLO, M. V. & MANN, P. 2006b. Deeply buried, Early Cretaceous paleokarst terrane, southern Maracaibo Basin, Venezuela. *American Association of Petroleum Geologists Bulletin*, **90**, 567–580.

CHICANGANA, G. 2005. The Romeral Fault System: a shear and deformed extinct subduction zone between oceanic and continental lithospheres in northwestern South America. *Earth Sciences Research Journal*, **9**, 51–66.

COLETTA, B., HEBRARD, F., LETOUZEY, J., WERNER, P. & RUDKIEWICZ, J. L. 1990. Tectonic style and crustal structure of the Eastern Cordillera (Colombia) from a balanced cross section. *In*: LETOUZEY, J. (ed.) *Petroleum and Tectonics in Mobile Belts*. Editions Technip, Paris, 81–100.

COONEY, PH. & LORENTE, M. A. 1997. Implicaciones tectonicas de un evento estructural en el Cretacico Superior (Santoniense–Campaniense) de Venezuela Occidental. *Memorias del VIII Congreso Geologico Venezolano, Isla de Margarita*. Sociedad Venezolana de Geologos, **I**, 195–204.

DAVIS, C., PRATT, L., SLITER, W., MOMPART, L. & MURAT, B. 1999. Factors influencing organic carbon and trace metal accumulation in the Upper Cretaceous La Luna Formation of western Maracaibo Basin, Venezuela. *In*: BARRERA, E. & JOHNSON, C. C. (eds) *Evolution of the Cretaceous Ocean Climate System*. Geological Society of America, Boulder, CO, Special Papers, **332**, 203–230.

DE GUERRA, C., AQUINO, R. & FIGUERA, L. 1996. Chronostratigraphy and regional distribution of the Cretaceous–Tertiary and Upper Cretaceous (Late Campanian–Santonian) unconformities, Barinas Basin, Southwestern Venezuela. *XVI Coloquio Interfilial de Bioestratigrafía*. Maraven S.A., Caracas.

DE ROMERO, L. M. 1991. *Estudio Bioestratigrafico del Miembro Tres Esquinas. Edad y Ambiente de Sedimentacion*. Thesis, Universidad de Los Andes, Merida, Venezuela.

DE ROMERO, L. M. & GALEA-ALVAREZ, F. A. 1995. Campanian Bolivinoides and microfacies from the La Luna Formation, western Venezuela. *Marine Micropaleontology*, **26**, 385–404.

DE ROMERO, L. M. & ODREMAN, O. E. 1996. Correlation of the Upper La Luna Formation, western Venezuela (abstract). *Abstract Volume, Fifth International Cretaceous Symposium and Second Workshop of Inoceramids*, Freiberg, 20.

DE ROMERO, L., TRUSKOWSKI, I. *ET AL.* 2003. An integrated Calcareous microfossil biostratigraphic framework for the La Luna Formation, western Venezuela. *Palaios*, **18**, 349–366.

DONNELLY, T. W. 1973. *Late Cretaceous Basalts from the Caribbean, a Possible Flood-basalt Province of Vast Size*. EOS, **54**.

DONNELLY, T. W. 1989. Geologic history of the Caribbean and Central America. *In*: BALLY, A. W. & PALMER, A. R. (eds) *An Overview*. The Geology of North America, **A**. Geological Society of America, Boulder, CO, 299–321.

DONNELLY, T. W., BEETS, D. *ET AL.* 1990. History and tectonic setting of the Caribbean magmatism. *In*: DENGO, G. (ed.) *The Caribbean Region*. The Geology of North America, **H**. Geological Society of America, Boulder, CO, 339–374.

ERLICH, R. N., NEDERBRAGT, A. J. & LORENTE, M. A. 1997. Origin and depositional environments of Turonian–Maastrichtian organic-rich and phosphatic sediments of western Venezuela. *VI Simposio Bolivariano Exploracion Petrolera en las Cuencas Subandinas, Caratgena de Indias, Colombia*. Asociación Colombiana de Geologos y Geofísicos del petroleo. Memorias, **I**, 478–524.

ERLICH, R. N., MACSOTAY, O., NEDERBRAGT, A. J. & LORENTE, M. A. 1999a. Palaeoceanography, palaeoecology,and depositional environments of Upper Cretaceous rocks of western Venezuela. *Palaeogeography, Palaeoclimatology, Palaeoecology*, **153**, 203–238.

ERLICH, R. N., MACSOTAY, O., NEDERBRAGT, A. J. & LORENTE, M. A. 1999b. Geochemical characterization of oceanographic and climatic changes recorded in upper Albian to lower Maastrichtian strata, western Venezuela. *Cretaceous Research*, **20**, 547–581.

ERLICH, R. N., MACSOTAY, O., NEDERBRAGT, A. J. & LORENTE, M. A. 2000. Birth and death of the Late Cretaceous 'La Luna Sea' and the origin of the Tres Esquinas phosphorites. *Journal of South American Earth Sciences*, **13**, 21–45.

FARIAS, A., PILLOUD, A., CRUX, J., CANACHE, M. & TRUSKOWSKI, I. 2000. Biostratigraphic study of calcareous nannofossils of the La Luna Formation and its lateral equivalents in western Venezuela (abstract). *Extended Abstracts, Society of Economic Paleontologists and Mineralogists Research Conference: Paleogeography and Hydrocarbon Potential of the La Luna Formation and Related Anoxic Systems*, CD, Caracas.

FÖLLMI, K. B., GARRISON, R. E., RAMIREZ, P. C., ZAMBRANO-ORTIZ, F., KENNEDY, W. J. & LEHNER, B. L. 1992. Cyclic phosphate-rich sucessions in the Upper Cretaceous of Colombia. *Paleogeography, Paleoclimatology, Paleoecology*, **93**, 151–182.

FORD, A. & HOUBOLT, J. J. H. C. 1963. *Las Microfacies del Cretaceo de Venezuela Occidental*. E. J. Brill, Leiden.

FUENMAYOR, A. N. 1989. *Manual de Foraminiferos de la Cuenca de Maracaibo*. Maraven, S. A. Filial de Petroleos de Venezuela, Maracaibo.

GALEA ALVAREZ, F. A. 1989. Microfacies, edad y ambiente de sedimentación de la Formación La Luna, Flanco Norandino, Venezuela. *Contribuciones de los Simposios Sobre el Cretácico de America Latina*. Parte A, Eventos y Registro Sedimentario, Buenos Aires, A57–A73.

GALEA-ALVAREZ, F. A., TRUSKOWSKI, I. *ET AL.* 2000. *Cretaceous Planktonic Foraminifers Biostratigraphy from Venezuela*. Research Contribution to *SAMC News*, **17**. World Wide Web Address: http://www.rzuser.uni-heidelberg.de/~dc8/samc/News17.htm#10.

GALLANGO, O., NOVOA, E. & BERNAL, A. 2002. The petroleum system of the central Perija fold belt, western Venezuela. *American Association of Petroleum Geologists Bulletin*, **86**, 1263–1284.

GRADSTEIN, F. M., AGTERBERG, F. P. *ET AL.* 1995. A Triassic, Jurassic and Cretaceous timescale. *In*: BERGGREN, W. A., KENT, D. V., AUBRY, M.-P. & HARDENBOL, J. (eds) *Geochronology, Timescales and Global Stratigraphic Correlation*. Society of Economic Palaeontologists and Mineralogists, Special Publication, **54**, 95–126.

GRADSTEIN, F. M., OGG, J. G. *ET AL.* 2004. *A Geologic Time Scale 2004*. Cambridge University Press, Cambridge.

HAMBALEK, N. & PITTELLI, R. 1994a. *Analisis micropaleontologico de alta resolucion de la Fm La Luna, del pozo VLE-738, Lago de Maracaibo (nucleo 1 al 6, 16033' a 16300')*. Maraven S.A., Internal Report, Caracas, Venezuela.

HAMBALEK, N. & PITTELLI, R. 1994b. *Analisis micropaleontologico de alta resolucion de la Fm La Luna, del pozo VLB-704, Lago de Maracaibo (nucleo 1 al 3, 134863' a 13707')*. Maraven S.A., Internal Report, Caracas, Venezuela.

HELENES, J., DE GUERRA, C. & VASQUEZ, J. 1994. Estratigrafia por secuencias del Cretacico Superior en el subsuelo del area de Barinas. *V Simposio Bolivariano Exploracion Petrolera en las Cuencas Subandinas*. Memorias, 29–39.

HELENES, J., DE GUERRA, C. & VASQUEZ, J. 1998. Palynology and chronostratigraphy of the Upper Cretaceous in the Barinas Area, western Venezuela. *American Association of Petroleum Geologists Bulletin*, **82**, 1308–1328.

ITURRALDE-VINENT, M. A. 2007. Pricipales eventos tectónicos de la Placa del Caribe y su entorno. *Field Workshop of Caribbean Geology, 2da Convención Cubana de Ciencias de la Tierra*, Havana, Cuba. UNESCO/IUGS-IGCP Project 546, abstracts GEO10-01.

JAMES, K. H. 2005a. Arguments for and against the Pacific origin of the Caribbean Plate and arguments for an *in situ* origin. Transactions of the 16th Caribbean Geological Conference, Barbados. *Caribbean Journal of Earth Science*, **39**, 47–67.

JAMES, K. H. 2005b. A simple synthesis of Caribbean Geology. *Transactions of the 16th Caribbean*

Geological Conference, Barbados. World Wide Web Address: http://www.searchanddiscovery.net/documents/2004/james/index.htm.

LIDDLE, R. A. 1928. *The Geology of Venezuela and Trinidad.* J. P. MacGowan, Forth Worth.

LORENTE, M. A., DURAN, I. & RUIZ, M. 1995. Late Cretaceous in Western Venezuela: a new biostratigraphical approach. *Extended Abstracts, Program Book, International Geological Correlation Program 362 Annual Meeting*, Maastricht, The Netherlands.

LORENTE, M. A., RULL, V., RUIZ, M., DURAN, I., TRUSKOWSKI, I. & DI GIACOMO, E. 1997. Nuevos aportes para la datacion de los principales eventos tectonicos y unidades litoestratigraficas de la Cuenca de Maracaibo, Venezuela Occidental. *Boletin Ministerio de Energia y Minas. Direccion General Sectorial de Minas y Geologia*, **XVIII**, 33–50.

LUGO, J. & MANN, P. 1995. Jurassic–Eocene Tectonic Evolution of the Maracaibo Basin, Venezuela *In*: TANKHURST, A. J., SUÁREZ, R. & WELSINK, H. J. (eds) *Petroleum Basins of South America.* American Association of Petroleum Geologists, Memoirs, **62**, 699–725.

MACELLARI, C. E. & DE VRIES, T. 1987. Late Cretaceous upwelling and anoxic sedimentation in Northwestern South America. *Paleogeology, Paleoclimatology, Paleoecology*, **59**, 279–292.

MANN, P., ESCALONA, A. & CASTILLO, V. 2006. Regional geologic and tectonic setting of the Maracaibo supergiant basin, western Venezuela. *American Association of Petroleum Geologists Bulletin*, **90**, 445–477.

MITCHUM, R. M., JR, VAIL, P. R. & THOMPSON, S. 1977. Seismic stratigraphy – applications to hydrocarbon exploration. *In*: *Seismic Stratigraphy and Global Changes of Sea Level. Part 2. The Depositional Sequence as a Basic Unit of Stratigraphic Analysis.* American Association Petroleoum Geologists, Memoirs, **26**.

ODIN, G. S. (ed.) 2001. *The Campanian–Maastrichtian Stage Boundary: characterisation at Tercis les Bains (France): Correlation with Europe and other continents.* International Union Geological Sciences Special Publications (Monographs), **36**. Developments in Palaeontology and Stratigraphy Series **19**. Elsevier Sciences, Amsterdam.

ODIN, G. S. & LAMAURELLE, M. A. 2001. The global Campanian–Maastrichtian stage boundary. *Episodes*, **24**, 229–238.

OGG, J. 2004. *Overview of Global Boundary Stratotype Sections and Points (GSSPs), Status on June 2004.* ICP. World Wide Web Address: http://www.stratigraphy.org/.

OGG, J. & LUGOWSKI, A. 2006. *TS-Creator visualization of enhanced Geologic Time Scale 2004 database (version 2.1; 2006).* World Wide Web Address: http://www.stratigraphy.org and/or http://www.chronos.org.

PARNAUD, F., GOU, Y., PASCUAL, J. C., CAPELLO, M. A., TRUSKOWSKI, I. & PASSALACQUA, H. 1995. Stratigraphic synthesis of Western Venezuela. *In*: TANKARD, J., SUÁREZ, R. & WELSINK, H. J. (eds) *Petroleum Basins of South America.* American Association Petroleoum Geologists, Memoirs, **62**, 681–698.

PARRA, M., MOSCARDELLI, L. & LORENTE, M. A. 2000. Lateral microfacies changes and their

paleoenvironment implications in the Tres Esquinas Member, La Luna Formation, western Venezuela (Abstract). *Society of Economic Paleontologists and Mineralogists Research Conference, 'Paleogeography and Hydrocarbon Potential of the La Luna Formation and Related Anoxic Systems'*, CD, Caracas.

PARRA, M., MOSCARDELLI, L. & LORENTE, M. A. 2003. Late Cretaceous anoxia and lateral microfacies changes in the Tres Esquinas member, La Luna Formation, western Venezuela. *Palaios*, **18**, 321–333.

PINDELL, J. L. & TABBUT, K. D. 1995. Mesozoic–Cenozoic Andean paleogeography and regional controls on hydrocarbon systems. *In*: TANKARD, A. J., SUÁREZ, R. & WELSIN, H. J. (eds) *Petroleum Basins of South America.* American Association Petroleoum Geologists, Memoirs, **62**, 101–128.

RENZ, O. 1959. Estratigrafia del Cretaceo en Venezuela occidental. *Boletín Geologico del Ministerio de Minas e Hidrocarburos* Venezuela, **5**, 3–48.

RENZ, O. 1977. The lithologic units of the Cretaceous in western Venezuela. *Memorias V Congreso Geologico Venezolano*, Caracas, **I**, 45–58.

RENZ, O. 1982. *The Cretaceous Amonites of Venezuela.* Birkhäuser, Basel.

REY, O., SIMO, J. A. & LORENTE, M. A. 2004. A record of long- and short-term environmental and climatic change during OAE3: La Luna Formation, Late Cretaceous (Santonian–Early Campanian), Venezuela. *Sedimentary Geology*, **170**, 85–105.

SANTAMARIA, F. & SCHUBERT, C. 1972. Geochemistry and geochronology of the Southern Caribbean–Northern Venezuela plate boundary. *Geological Society of America Bulletin*, **85**, 1085–1098.

SAVIAN, V. 1993. *Geología del Cretácico de la Cuenca de Maracaibo.* Trabajo Especial de Grado, Universidad Central de Venezuela.

SAVIAN, V. & SCHERER, W. 1997. Geometría y Litofacies del Cretácico de la Cuenca de Maracaibo. *Codigo Geológico de Venezuela. Afiches/Posters.* World Wide Web Address: www.pdvsa.com/lexico/bibgeol/bg3428.htm.

SHAGAM, R., KOHN, B. P. ET AL. 1984. Tectonic implications of Cretaceous–Pliocene fission-track ages from rocks of the circum-Maracaibo Basin region of western Venezuela and eastern Colombia. *In*: *Caribbean Tectonics and Relative Plate Motions.* Geological Society of America, Boulder, CO, Memoirs, **162**, 385–412.

SIGURDSSON, H., LECKIE, R. M. & ACTON, G. D. 1996. Caribbean ocean history and the Cretaceous/Tertiary boundary event. *Ocean Drilling Program, Leg 165.* Preliminary Report.

SISSINGH, W. 1977. Biostratigraphy of Cretaceous calcareous nannoplankton. *Geologie en Mijnbouw*, **56**, 37–65.

SUTTON, F. A. 1946. Geology of Maracaibo Basin, Venezuela. *American Association of Petroleum Geologists Bulletin*, **30**, 1621–1741.

TALUDKAR, S. C. & MARCANAO, F. 1994. Petroleum systems of the Maracaibo Basin, Venezuela. *In*: MAGOON, I. B. & DOW, W. G. (eds) *The Petroleum System: From Source to Trap.* American Association of Petroleum Geologists, Memoirs, **60**, 463–481.

TRUSKOWSKI, I., GALEA, F. & SLITER, W. 1995. Cenomanian hiatus in Venezuela. *Geological Society of*

America Annual Meeting. Abstracts with Programs, New Orleans, 303.

VASQUEZ, E. & DICKEY, P. 1972. Major faulting in north-western Venezuela and its relation to global tectonics. *Conferencia Geologia del Caribe VI, Porlamar, Julio 1971,* C. Petzall, Editora, Cromotip, Caracas, 191–203, figs 2 & 7.

VILLAGOMEZ, D., SEWARD, D. & SPIKINGS, R. 2006. Accretionary and post-accretionary cooling, exhumation and tectonic history of the central and western Andes of Colombia. *4th Swiss Geoscience Meeting,* Bern. Abstracts. World Wide Web Address: geoscience-meeting.scnatweb.ch/sgm2006/SGM06_abstracts/08_OS_Min_Pet/Villagomez_Diego_Poster.pdf.

ZAPATA, E., PADRON, V., MADRID, I., KERTZNUS, V., TRUSKOWSKI, I. & LORENTE, M. A. 2003. Bioestratigraphic, sedimentologic and chronostratigraphic study of the La Luna Formation (Late Turonian–Campanian) in the San Miguel and Las Hernandez Sections, western Venezuela. *Palaios,* **18,** 367–377.

The occurrence and timing of high-pressure metamorphism on Margarita Island, Venezuela: a constraint on Caribbean–South America interaction

WALTER V. MARESCH[1*], ROLF KLUGE[2,3], ALBRECHT BAUMANN[2],
JAMES L. PINDELL[4,5], GABRIELA KRÜCKHANS-LUEDER[2,6] & KLAUS STANEK[7]

[1]*Institute of Geology, Mineralogy & Geophysics, Ruhr-University Bochum,
44780 Bochum, Germany*

[2]*Institute of Mineralogy, Münster University, Corrensstrasse 24, 48149 Münster, Germany*

[3]*Present address: AQUANTA Hydrogeologie GmbH & Co. KG, Kirchplatz 1, 48301
Nottuln, Germany*

[4]*Tectonic Analysis Ltd. Chestnut House, Duncton, Sussex GU28 0LH, UK*

[5]*Department of Earth Science, Rice University, Houston, TX 77002, USA*

[6]*Present address: Tornescher Weg 150, 25436 Uetersen, Germany*

[7]*Institute of Geology, TU Bergakademie Freiberg, 09596 Freiberg, Germany*

*Corresponding author (e-mail: walter.maresch@rub.de)

Abstract: The metamorphic rock sequences exposed on the Island of Margarita, Venezuela, located in the southeastern corner of the Caribbean Plate margin, are composed of a high-pressure/low-temperature (HP/LT) nucleus subducted to at least 50 km depth, now structurally overlain by lower-grade greenschist-facies units lacking any sign of high-pressure subduction-zone metamorphism. The HP/LT nucleus involves protoliths of both oceanic (metabasalts and intimately associated carbonaceous schists of the La Rinconada unit; peridotite massifs) and continental affinity (metapelites, marbles and gneisses of the Juan Griego unit). All HP/LT units were joined together prior to the peak of high-pressure metamorphism, as shown by their matching metamorphic pressure–temperature evolution. The metamorphic grade attained produced barroisite as the regional amphibole. Glaucophane is not known from Margarita. Contrary to a widely propagated assumption, there are no *major* nappe structures *post-dating* HP/LT metamorphism anywhere *within* the high-pressure nucleus of Margarita Island. U–Pb zircon dating of key tonalitic to granitic intrusive rocks provides the following constraints: (1) the Juan Griego unit is heterogeneous and contains Palaeozoic as well as probable Mesozoic protolith; (2) the peak of HP/LT metamorphism, that is maximum subduction, is younger than 116–106 Ma and older than 85 Ma, most probably *c.* 100–90 Ma, a time span during which the southeastern Caribbean/South American border was clearly a passive margin. The assembly of Margaritan protoliths and their HP/LT overprint occurred far to the west in northwestern South America, a scenario completely in accord with the details of the Pacific-origin model outlined by Pindell & Kennan. Juxtaposition of the greenschist-facies units occurred after exhumation into mid-crustal levels after *c.* 80 Ma.

During a workshop on the origin and development of the Caribbean area in Sigüenza, Spain, in 2006, two major trains of thought were followed. As recapitulated in more detail in other papers in this volume, one can be described as the 'Pacific-origin' model. As outlined in a number of publications (Pindell & Dewey 1982; Pindell 1993; Pindell *et al.* 2005 and numerous references therein), kinematic models and reconstructions can be used to develop in considerable detail how the Mesozoic separation of the Americas produced passive margins that were then overridden from west to east by allochthonous arc and oceanic complexes related to the 'bow' of the progressing Caribbean Plate. The 'intra-American' or '*in situ*' model on the other hand also calls for intra-American rifting to produce the space needed for the Caribbean Plate, but most subsequent tectonic interaction between the plate and the margins of North and South America are ascribed to predominantly north–south relative movement (e.g. Meschede & Frisch 1998; Giunta *et al.* 2002; James 2006). A recent proposal by Higgs (2009) combines

From: JAMES, K. H., LORENTE, M. A. & PINDELL, J. L. (eds) *The Origin and Evolution of the Caribbean Plate.*
Geological Society, London, Special Publications, **328**, 705–741.
DOI: 10.1144/SP328.28 0305-8719/09/$15.00 © The Geological Society of London 2009.

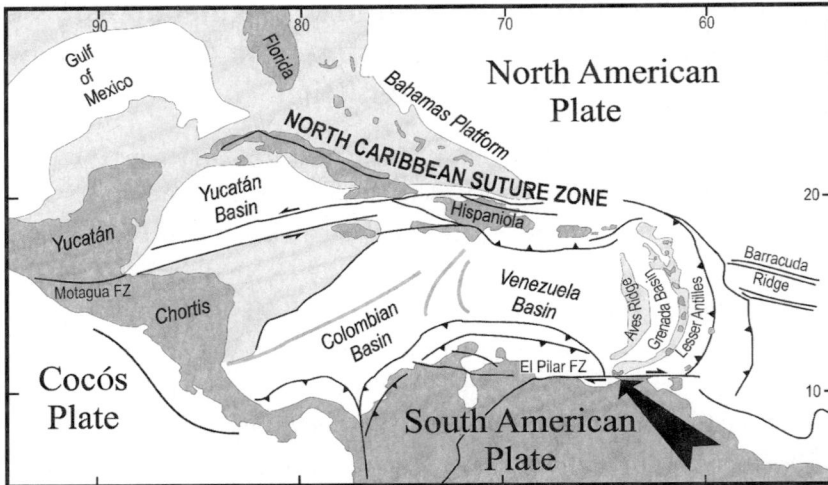

Fig. 1. Location of Margarita Island (arrow) with respect to the Caribbean area.

initial elements of the '*in situ*' model with later overriding of a 'Pacific-origin' Caribbean Plate.

The collision zones festooning the margins of the Caribbean Plate with North and South America commonly entrain high-pressure (HP) metamorphic rocks, or, more accurately, rocks that have experienced high-pressure/low-temperature (HP/LT) metamorphism. It is important to recall this more precise designation, because it is not the absolute pressure that is important here, but the relatively high ratio of pressure to temperature, which is higher than in the normal geotherm found in Earth. Consequently, only large-scale rapid subduction of cold crustal material into Earth's mantle can produce such pressure–temperature conditions, and, vice versa, the presence of such rocks must point to a former process of subduction. Subduction can be a long-lived process in an intra-oceanic setting between two oceanic plates (e.g. Lesser Antilles island arc), or along a continental margin between a continental and an oceanic plate (e.g. Andean system). HP/LT rocks may be brought to the surface during on-going subduction by return-flow processes, or by arc–continent or continent–continent collision, leading to a cataclysmic end of subduction and rapid uplift/exhumation of deeply buried HP/LT rocks. The individual history of each scenario will be recorded in the pressure–temperature–time–deformation ($P–T–t–d$) development of the HP/LT rocks available for study. Thus, the study of exposed metamorphic rocks in such suture zones allows the 'depth dimension' to be incorporated into terrane analysis. Understanding the timing, origin and geodynamic setting of the high-pressure metamorphism recorded by these rocks is a key element in understanding Caribbean Plate interaction with the Americas as well as for testing and refining the above regional models. The concomitant study of

associated magmatic rocks from the subduction zone yields information on input, output, conveyor belts and plant facilities of the 'subduction-zone factory'.

The Island of Margarita (Fig. 1), the largest of the Caribbean islands belonging to the Republic of Venezuela, is located in the SE corner of the present Caribbean Plate. Margarita offers excellent exposure of a variety of HP/LT metamorphic rocks that are perforated by datable igneous intrusions. Depending on an author's viewpoint, the exposed rocks may be representative of as much as 70 000 km^2 of coastal Venezuela, Trinidad and Tobago. Because Margarita is located so far to the east along the southern Caribbean margin, the HP/LT rocks exposed here represent an excellent test case for discriminating between the two competing models outlined above.

Here we briefly review the complex geology of Margarita Island, concentrating on the early, subduction-related history. We will point out those observations critical for an understanding of Caribbean regional development and reiterate some important observations that appear to have been lost in the literature during the last few decades. Magmatic activity provides an excellent geochronological reference system for Margarita tectonic history. We present petrographical and analytical details for ages briefly mentioned in earlier publications (Stöckhert *et al.* 1995) that allow the timing of HP/LT metamorphism to be constrained.

Geological setting, metamorphic development and structural relationships

The metamorphic rocks of Margarita Island have been studied for more than 60 years (early summary by Kugler 1948; Hess & Maxwell 1949; compilation by González de Juana 1968; Taylor 1960*a, b*; Maresch 1971, 1973, 1975; Vignali, 1979; Bellizzia, 1985; Chevalier 1987, 1993; Guth & Avé Lallemant, 1991; Macsotay *et al.* 1997, and many references therein). Rekowski & Rivas (2005) have recently presented an exhaustive overview of previous work that also incorporates observations and data from a number of unpublished Venezuelan theses. The above studies have often been guided by drastically changing ideas on geodynamic processes and physico-chemical controls of metamorphism. Early systematic mapping (Taylor 1960*a, b*; González de Juana 1968; Maresch 1971, 1973) was characterized by the constraints imposed by the call for strict adherence to stratigraphical principles, even though the pre-Cenozoic rocks were strongly deformed and clearly metamorphosed. In contrast, Chevalier (1987, 1993) interpreted Margarita in terms of extreme nappe development and postulated at least eight nappes that are considered to dissect the metamorphic section on Margarita Island. Chevalier (1987, 1993) considered that the serpentinite lenses found in the metamorphic sequences can be interpreted to define nappe boundaries and to be the lubricants at the base of allochthonous units. A lower nappe system was interpreted as a stack or duplex of imbricated units metamorphosed under HP/LT conditions, with an upper nappe system characterized by lower grades of metamorphism originating from the most external domains of the orogen. This complex structural architecture, composed mainly of a meta-ophiolitic complex and its presumed metasedimentary cover, was interpreted to be allochthonous above para-autochthonous metamorphic units of continental affinity. However, a number of postulated nappes possess an equivalent lithological inventory, and, above all, comparable metamorphic grades and similar pressure–temperature–time–deformation paths. Their juxtaposition, whether as nappes or ductile imbrications in an accretionary complex must have occurred prior to peak high-pressure metamorphism. In terms of the *timing* of the actual phase of high-pressure metamorphism it is not critical whether the stratified rock complex involved represents an original stratigraphical sequence or a superposition of nappes. In addition, recent detailed seismic investigation of the present-day Venezuelan margin during the course of the Bolivar project (e.g. Clark *et al.* 2008) shows Margarita Island at high structural level within the Caribbean crust. There is no present direct link to South American basement, so that *all* of the rock sequences exposed on Margarita Island must be considered to be allochthonous with respect to South American basement (cf. Chevalier 1987). The issue concerns the degree of allochthoneity, and for the above reasons we believe that a reappraisal of the metamorphic section on Margarita Island is called for, in order to place the new geochronological data presented here

into proper tectonic and palaeogeographic perspective. In the following we propose to separate and describe the metamorphic section in two parts (Fig. 2): a *nucleus* composed of rocks that have experienced similar high-pressure metamorphism and similar pressure–temperature–time–deformation paths, and a *periphery* where the metamorphic grade never exceeded the greenschist facies. The latter corresponds to the 'El Piache', 'Los Robles' and 'Matasiete-Guayamuri' nappes of Chevalier (1987, 1993). Our own work suggests that the lower-grade Los Robles nappe is much less extensive on Macanao (Fig. 2) than shown by Chevalier (1987, 1993), but this is the only significant difference in terms of map pattern and not of particular import for the present discussion. Urbani (2007) has recently proposed a systematic nomenclature for the rocks of Margarita Island based on stratigraphical code and 'lithodemic unit' terminology. We see some problems with this terminology that still require clarification, inasmuch as in our opinion (outlined below) the hierarchy of units and subunits should be modified in several cases. In addition, direct translation of the Spanish suggestion into English causes some linguistic problems. Thus, the direct translation of 'Metamáficas de La Rinconada', that is 'La Rinconada Metamafics', in which an adjective is used as a noun, would not be accepted by a number of publications (such as this one). In this paper, the petrological *content* of the metamorphic rock units and not their nomenclature will be the primary objective, as well as the record and the history these rocks are telling us. In the following we will therefore simply use the neutral designation 'unit', in order not to preempt future formal nomenclature schemes.

The high-pressure metamorphic nucleus

The high-pressure metamorphic nucleus forms the cores and central highlands of the two parts of Margarita Island, the western Peninsula called Macanao, and the eastern main part that is generally referred to as Paraguachoa (Figs 1 & 2). Hess & Maxwell (1949) first recognized a metamorphic nucleus they called the 'Juan Griego Group' composed of two principal rock sequences they termed the 'quartzose division', a series of mainly metasedimentary schists and gneisses, and the 'greenstone division', a series of metabasaltic rocks. In later studies (see summaries in Maresch 1971, 1973, and Chevalier 1987), the name 'Juan Griego Group' was restricted to the 'quartzose division', and Maresch (1971, 1973, 1975) introduced the term 'La Rinconada Group' for the 'greenstone division' to avoid connotations of lithology and/or metamorphic grade. According to Urbani (2007), the 'Juan Griego Group' should now be formally referred to as the 'Asociación Metamórfica Juan Griego', and the 'La Rinconada Group' as the

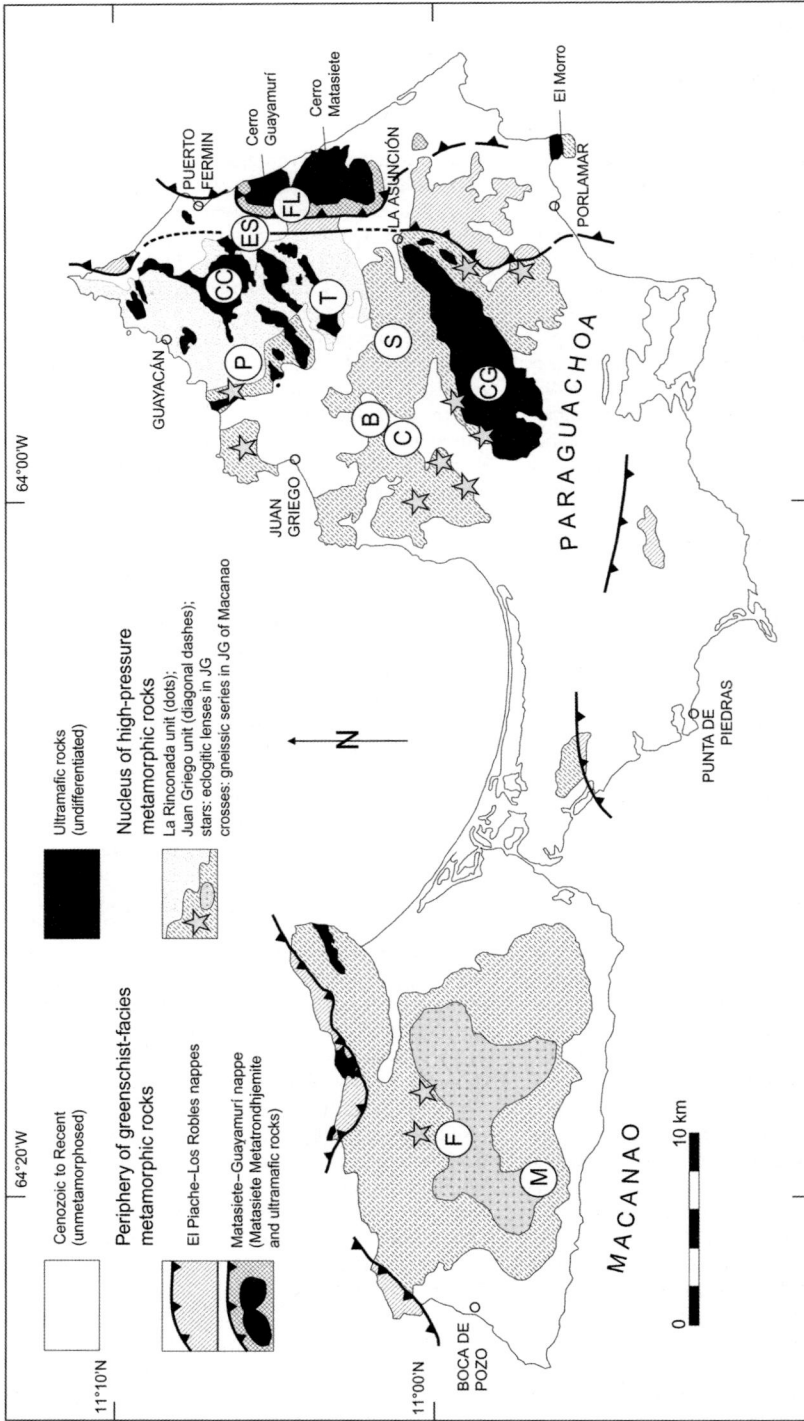

Fig. 2. Geological sketch map of Margarita Island indicating the main units of the high-pressure metamorphic nucleus and the greenschist-facies periphery discussed in text (based on maps of Taylor 1960*a*, *b*; González de Juana & Vignali 1972; Maresch 1971, 1973, 1975; Chevalier 1987; and own observations). Only the most prominent of the many localities of pods and lenses of eclogite and amphibole eclogite within the Juan Griego unit (stars) are shown. Localities mentioned in the text: B, Bolívar and El Maco (includes eclogite body); C, Calle Nueva and Boquerón (includes amphibole–eclogite locality); CC, Cerro Chico; CG, Cerro Grande or Cerro El Copey; ES, El Salado; F, San Francisco; FL, Flandes; M, El Manglillo; P, Pedro González; S, San Sebastián (includes numerous small eclogitic bodies); T, Tacarigua Valley.

'Metamáficas de La Rinconada' within a 'Complejo Metaofiolítico Paraguachí'. Since the study of Hess & Maxwell (1949) and the mapping of Taylor (1960*a*, *b*), a broad antiformal structure involving both Juan Griego and La Rinconada type rocks is recognized on Paraguachoa that generally plunges to the SW.

In the following discussion, emphasis will be placed on the fact that the entire high-pressure metamorphic nucleus of Margarita Island represents a relatively coherent sequence that has experienced similar conditions of high-pressure metamorphism and a similar pressure–temperature– path with time. In recent years, the terms 'mélange' and 'knockers' of blueschist or eclogite have been widely used in discussing the eclogite- bearing rock series of the Venezuelan Coast Ranges (e.g. Guth & Avé Lallemant 1991; Sisson *et al.* 1997; Avé Lallemant & Sisson 2005). To many readers this may raise connotations of higher-grade blocks swimming in a lower-grade matrix, as suggested by classical usage in the Fran- ciscan of California. However, as Sisson *et al.* (1997) state in their description of the Puerto Cabello area of the Cordillera de la Costa: 'Because all the lithologies in this outcrop record high-P conditions, this metamorphic mélange formed before or during peak metamorphism in a Mid-Cretaceous subduction zone'. This is an important point to emphasize, because in the context of this paper, it must be stressed that the HP/LT metamorphic nucleus of Margarita Island was amalgamated and its units juxtaposed *prior* to the peak of high-pressure metamorphism. In a similar vein, we will avoid the term 'knocker' in this paper. Although the pods and lenses of eclogi- tic rocks found in the mica schists of the Juan Griego unit at least differ significantly in bulk-rock chemistry and mineralogy from their host rocks, the lenses of eclogitic rocks found in the La Rinco- nada unit are an integral part of that metabasic series and owe their presence to only subtle differ- ences in rock chemistry and water activity (e.g. Maresch 1977; Maresch & Abraham 1981). It is also important to stress that true two-phase eclogites (i.e. consisting only of omphacite + garnet) are restricted to a few occurrences of eclogite lenses within the Juan Griego unit. All others contain at least amphibole as a major additional phase of the eclogite assemblage, irrespective of any further amphiboles that may have formed during later depressurization reactions. Classical blueschists are not found on Margarita Island. No glaucophane or true alkali amphibole has so far been convin- cingly documented. Winchite, a composition between glaucophane and actinolite, has been analysed from one eclogitic pod within the Juan Griego unit (Maresch *et al.* 1982). The

'Ca-glaucophane' mentioned by Chevalier (1987), named on the basis of a superseded old classification scheme, is either barroisite or actinolite. In fact, the amphibole barroisite, a sodic–calcic composition between glaucophane and hornblende *sensu lato* (e.g. Leake *et al.* 1997), is the typical amphibole of all high-pressure rocks of the metamorphic nucleus of Margarita Island (e.g. Maresch & Abraham 1981; Beets *et al.* 1984; Maresch *et al.* 1985; Chevalier, 1987). As suggested by Ernst (1979) and corroborated in numerous field-based studies as well as in experiments on MORB composition (e.g. Ernst & Liu 1998), this amphibole is indicative of HP/LT metamorphism, but at temperatures some- what higher than those at which glaucophane is stable in metabasic rocks.

There has been considerable discussion in the past on whether metamorphism of the high-pressure nucleus and its later lower-pressure overprinting occurred during a single metamorphic cycle or was due to polycyclic processes (e.g. Navarro 1981; Blackburn & Navarro 1977). An early, poorly defined, distinct, higher-pressure/lower- temperature blueschist [*sic*] event is thought to have been followed by a later lower-pressure/ higher-temperature overprint in which all metabasic rocks then went through eclogite, amphibole– eclogite and garnet–amphibolite stages. Vignali (1979), Talukdar & Loureiro (1982) and Stephan *et al.* (1980), for instance, consider eclogite blocks like those in the Juan Griego unit to be 'tectonic injections' or synsedimentary 'olistoliths' that were emplaced into the original sedimentary proto- lith in already metamorphosed form before final metamorphism of the aggregate to schists and gneisses. Maresch & Abraham (1981), Beets *et al.* (1984) and Chevalier (1987) have discussed and weighed these arguments on the basis of mineral zonations and reaction textures in considerable detail. Although Chevalier (1987) did not specifi- cally recognize a high-pressure metamorphic imprint in the Juan Griego-type rocks, he noted a general similarity in the post-peak metamorphic evolution and the concomitant structural and fabric development in both La Rinconada and Juan Griego-type rocks (see also summary by Rekowski & Rivas, 2005). Irrespective of any cir- cumstantial evidence postulated on the basis of regional interpretations, it can be concluded, as already discussed by Stöckhert *et al.* (1994, 1995) and developed further below, that the only rocks providing evidence for more than one distinct cycle of metamorphism are the gneissic series of Macanao. The high-pressure metamorphic imprint on the Margaritan metamorphic nucleus developed in a single collisional cycle followed by later, increasingly heterogeneous overprinting at decreas- ing depths during exhumation.

Juan Griego unit and Macanao orthogneiss. The
Juan Griego unit is characterized by a series of
schists and gneisses of predominantly sedimentary
origin. Attempts have been made to recognize a
systematic stratigraphy in terms of marble horizons,
'chloritic' (i.e. Al-rich), 'graphitic', '(quartzo)-
feldspathic' v. 'non-feldspathic' sequences (Taylor
1960*a*, *b*; González de Juana & Vignali 1972;
Vignali 1979). Chevalier (1987) suggests a basal
series of quartzofeldspathic schists and gneisses
followed upward by metaconglomerates, lenticular
marbles, garnet–mica schists, graphitic schists and
massive marbles (the El Piache marbles), although
these units have not everywhere been mapped out
in a systematic way. Urbani (2007) also includes
the 'Mármol de El Piache' within the 'Asociación
Metamórfica Juan Griego'. However, we agree
with other workers who have preferred to associate
the El Piache marbles with the lower-grade series of
the metamorphic periphery. Macsotay *et al.* (1997)
have postulated that on Macanao the feldspathic
sequences represent a klippe that has been thrust
over the rest of the sequence. The age of the Juan
Griego unit is not constrained by fossils.

In a detailed and systematic petrological study of
Juan Griego metasediments, Krückhans-Lueder
(1996) found a number of localities throughout the
outcrop area with relatively Al-rich metapelitic
schists with diagnostic assemblages containing
garnet, chloritoid, staurolite and kyanite. Because
of the very heterogeneous degree of overprinting
along the *P–T*–time–deformation path experienced by these rocks, the study of domain equilibria
allows different segments of this path to be reconstructed. Logically, relics of the earliest part of
the high-pressure stage are the rarest and most difficult to find. Quantitative *P–T* determinations based
on detailed microanalytical study are shown in
Figure 3 for three key areas in Macanao (Manglillo
locality, Fig. 2) and central Paraguachoa (San
Sebastián and Boquerón localities, see Fig. 2).
Together with qualitative analyses of changing
mineral parageneses, keyed to calculated petrogenetic grids (e.g. Spear & Cheney 1989), the diagnostic *P–T* path of Figure 3 was constructed by
Krückhans-Lueder (1996). In agreement with the
results on eclogite pods found there (see below),
the mica schists in central Paraguachoa in the
general area of Boquerón-El Maco (see Fig. 2)
may have reached temperatures ≥650 °C. Although
no signs of anatexis are evident there, the incipient
breakdown of staurolite to garnet can be observed
in thin section. All other localities studied agree
well with the path drawn for the Manglillo locality.
The results in Figure 3 can be considered to be the
first decisive evidence for the fact that the *mica
schists* of the Juan Griego unit record a high-
pressure metamorphic history.

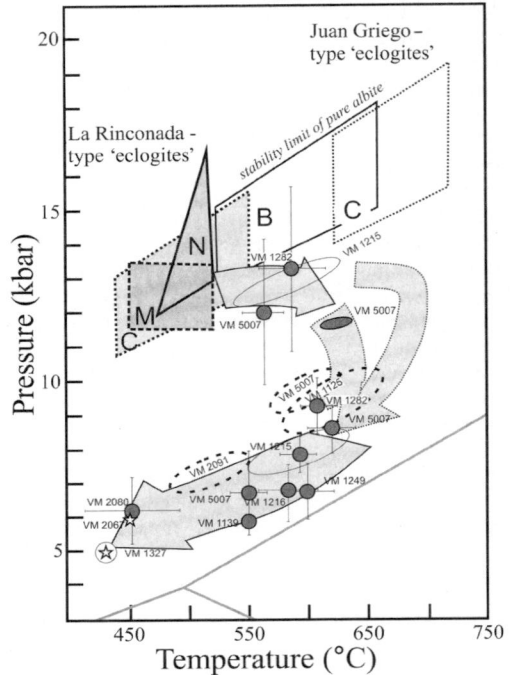

Fig. 3. Summary of available information on the *P–T*
conditions for rocks of the metamorphic nucleus.
Ellipses: *P–T* conditions estimated by conventional
thermobarometry; filled circles/ellipses: *P–T* conditions
estimated with the method of multivariant equilibria (see
Krückhans-Lueder 1996, for details). La Rinconada unit:
estimates for two coarse-grained amphibole–paragonite
eclogites sampled near Guayacán (Fig. 2). M, Maresch &
Abraham (1981); N, Newton (1986; recalculation of
Maresch & Abraham data); C, Chevalier (1987). Juan
Griego unit metabasic bodies: B, range covered by nine
samples of eclogite/amphibole–eclogite (Bocchio *et al.*
1996); C, Chevalier (1987) sample of eclogite from El
Maco. Juan Griego unit gneisses (Krückhans-Lueder
1996): Manglillo locality (VM 2091,VM 2080),
San Francisco locality (VM 1327; VM 2067). Juan
Griego Group Al-rich schists (Krückhans-Lueder 1996):
Manglillo locality (VM 1215, VM 1216, VM 1282, VM
5007), San Sebastian locality (VM 1125, VM 1139),
Boquerón locality (VM 1249). For localities see
Figure 2. For further details see text. Albite stability limit
after Holland (1980).

Rounded pods and lenses of basic rocks, generally from centimetres to metres in size, now metamorphosed to eclogite, amphibole–eclogite or
amphibolite, are common in the Juan Griego unit
outcrop area. Chevalier (1987, 1993) suggests that
these are usually localized within his basal quartzofeldspathic horizon. However, in a stratigraphically
higher sequence of garnet-mica schists, Chevalier

(1987, 1993) also describes a second type of amphibole-rich bodies with the shapes of 'olistoliths' or 'intrusions' into the sedimentary protolith (the 'metadiorites' of Taylor, 1960a, b). The largest body found on Macanao exceeds 1 km in size. Bocchio et al. (1990) also noted that typical sills and cross-cutting dykes 15–200 cm thick can be recognized locally. Major and minor element chemical compositions were reported by Bocchio et al. (1990). The compositions indicate a tholeiitic source and ocean-floor affinity, much like the more voluminous La Rinconada unit (see below). Nevertheless, there are subtle differences that preclude these basic bodies simply being tectonic inclusions derived from the La Rinconada unit.

In a later study, Bocchio et al. (1996) analysed a suite of 20 samples of such basic bodies of both types and found that most of them contain omphacite or relicts thereof. Many garnets show complex zoning patterns with helicitic whorls indicative of strong deformation during growth, followed by static tempering. The results of $P–T$ determinations on nine of these samples are shown in Figure 3, as well as analogous data for an eclogite pod near El Maco from Chevalier (1987). Accurate $P–T$ estimates on eclogitic assemblages are generally less diagnostic and more difficult than on Al-rich metapelites. The ranges shown in Figure 3 are based principally on exchange equilibria between garnet and omphacite to yield temperatures and on jadeite contents of omphacite to yield pressures. However, in the general absence of plagioclase, these are minimum pressures only. The corresponding upper pressures are suggested to be the stability limit of pure albite, present as stable relict inclusions in some eclogite garnets (Bocchio et al. 1996), and the general occurrence of albite in the surrounding schists. As a result, the pressure ranges in Figure 3 are relatively broad. The temperatures indicated in Figure 3 also mirror large ranges in temperatures indicated by the compositionally strongly zoned garnets (Blackburn & Navarro 1977; Chevalier 1987; Bocchio et al. 1996), indicating a complex, long-lived growth history. In general terms, the higher temperatures of 650–700 °C are found in eclogite pods of central Paraguachoa, in agreement with a corresponding 'thermal high' observed in the surrounding metapelites. Within error, the pressures appear compatible between the eclogitic pods and the enclosing mica schists.

Gneissic rocks included within the sequence traditionally mapped as the Juan Griego unit predominate in the high hills of central Macanao, where they form the backbone of the central massif (e.g. González de Juana & Vignali 1972; Vignali 1979; Chevalier 1987, 1993). Macsotay et al. (1997) have suggested that on Macanao these gneisses represent a klippe thrust over the mica schists. On the basis of our own work, we cannot substantiate this as a general conclusion. In the near-sealevel El Manglillo area (Fig. 2), gneiss, granitic orthogneiss and Al-rich mica schists are closely associated.

The gneissic rocks have not been studied very systematically so far. Stöckhert et al. (1994, 1995) have pointed out that some of these gneissic rock types display features of older, high-grade crystalline basement, such as deformed, in part very coarse-grained quartz–feldspar–muscovite-rich schlieren. Macsotay et al. (1997) give petrographic descriptions of a number of feldspar-rich samples, all of which they consider to be metasedimentary, despite the common occurrence of perthitic K-feldspar. Microanalytical data are not reported. González de Juana & Vignali (1972) also pointed to numerous bodies of pegmatitic character in the south-central part of Macanao. Macsotay et al. (1997) suggest that these are metatrondhjemite dykes. All these leucocratic rocks thus appear to be interesting targets for future zircon dating via the U–Pb method.

On the basis of the brief petrological overview of Kluge (1996) and Krückhans-Lueder (1996), who also present microanalytical data, the gneissic rocks of the Juan Griego unit can be broadly divided into two types. One is a typical coarse-grained augengneiss with porphyroclasts of perthitic potassium feldspar. Zircon from this rock type has been dated, and more detailed mineralogical information on the studied samples is given in a following section. Kluge (1996) informally referred to all the gneissic rocks in the Juan Griego unit as Macanao orthogneiss; however, at the present level of knowledge, this designation should probably be restricted to the augengneisses sensu stricto. The second gneiss type lacks the typical porphyroclasts of perthitic potassium feldspar and is richer in quartz-rich schlieren and layers. Carbonaceous material is common. Depending on the amount of mica, the rocks can appear either compact or well-foliated with a distinct planar fabric. Even in rock-types poor in mica, the fabric is characterized by a distinct alternation of quartz-feldspar layers. Plagioclase in most of these rocks is recrystallized albite with less than 4 mol% An-component. However, in some specimens porphyroclasts of an older generation are found that may reach oligoclase composition with 26 mol% An-content. Some contain helicitic whorls of graphite stringers with no relationship to the external foliation and conspicuous overgrowths of later albite. Gneisses with such porphyroclasts are often distinctly greyish in colour. Such rocks appear difficult to reconcile with a purely magmatic origin (see also Macsotay et al. 1997). Some of these gneisses also contain larger plagioclase crystals choked with nests of epidote needles, suggesting partial reequilibration of an originally more basic plagioclase indicative of a higher-grade origin of the original rock. A common feature of many of these rocks is the conspicuous growth of late porphyroblasts of albite that overgrow the fabric of the rock. Accessory minerals are almandine-rich garnet, biotite, chlorite, stilpnomelane, titanite and zoisite.

Pressure and temperature estimates for the metamorphism of the Juan Griego unit gneisses are difficult, because of the simple mineralogy involved, and the question of defining equilibrium assemblages. The $P-T$ conditions estimated (Fig. 3) by Krückhans-Lueder (1996) describe the shear-dominated greenschist overprint also seen in the Juan Griego schists (Fig. 3, Krückhans-Lueder 1996; Stöckhert *et al.* 1994, 1995; Chevalier 1987, 1993). Figure 4 shows phengite compositions for white mica from three samples of gneiss. The Si-contents are conspicuous indications of metamorphic pressures (Massonne & Schreyer 1987; as updated by Massonne & Szpurka 1997), although these are usually minimum pressures unless the stable coexistence of both biotite and K-feldspar in the critical assemblage can be proven. The latest phengites show Si $<$ 3.3 a.p.f.u. (atoms per formula unit), consistent with conditions shown in Figure 3. However, the compositions of the cores of large phengite crystals oriented in the foliation and especially of phengites protected by albite overgrowths from later shearing reach $\geq$3.45 Si a.p.f.u. Such high Si-contents are essentially the same as in the Juan Griego mica schists, where a complete $P-T$ path can be constructed (Fig. 3). Assuming temperatures analogous to those of the Juan Griego schists, minimum pressures of 11–13 kbar are indicated for earlier stages of the $P-T$ development of the gneisses. Although this metamorphic record is much less complete than in the Juan Griego aluminous schists, it is entirely compatible with the $P-T$ path shown in Figure 3.

Figures 5 and 6 summarize bulk chemical data obtained for various samples of Macanao orthogneiss by Kluge (1996). These are the only data at present available. Based on the assumption that in fact all gneisses originally derived from magmatic protolith, the normative Streckeisen Q–A–P proportions calculated from X-ray fluorescence analysis indicate compositions that vary from alkali-feldspar granite to granodiorite, with most samples lying in the granite field.

La Rinconada unit. The La Rinconada is a relatively coherent unit extending approximately 15 km in north–south direction in northeastern Paraguachoa (Fig. 2). It consists predominantly of amphibole gneiss *sensu lato*, with omphacite occurring in the amphibole assemblage in the northern part of this outcrop area (Maresch 1971,

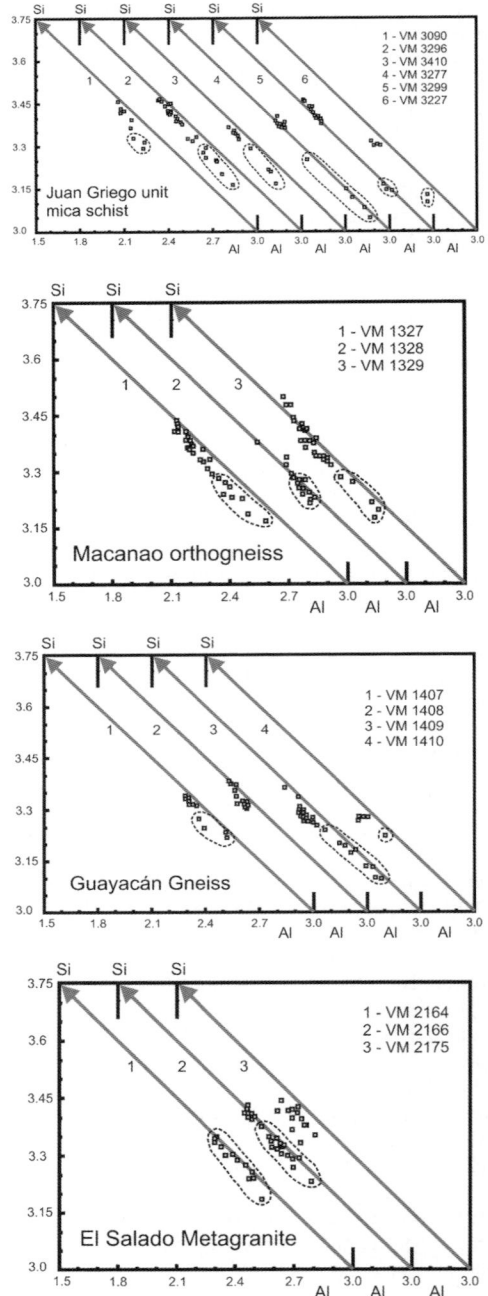

Fig. 4. Al v. Si contents ('atoms per formula unit') for phengites. The arrow indicates the trend of the ideal inverse Tschermaks substitution $Si(Mg,Fe^{2+})Al^{IV}_{-1}Al^{VI}_{-1}$. Higher Si contents correlate with increasing pressures of metamorphism. Encircled data points indicate secondary mica crystals (small, i.e. $<$100 μm; recrystallized rims; crystals growing across foliation).

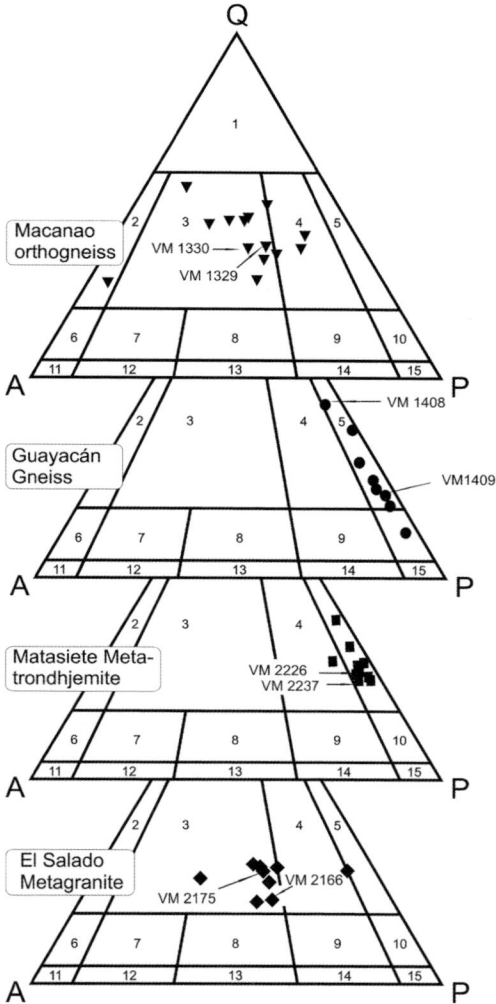

Fig. 5. Streckeisen Q–A–P plots of magmatic rocks discussed in text. Occupied fields: 2, alkali-feldspar granite; 3, granite; 4, granodiorite; 5, tonalite; 10, quartz diorite. U–Pb data have been obtained in this study for the numbered samples.

1973, 1975; Chevalier 1987, 1993). In coarse-grained layers in this northern part, distinctive amphibole–paragonite eclogites are found. Barroisite, a sodic–calcic amphibole typical of high-pressure metamorphism (see above), is the dominant amphibole type of the entire sequence. Interlayers of carbonaceous mica schist and quartzite are present. Towards the contact with the Juan Griego unit, these interlayers become more common until they predominate as carbonaceous mica schists and carbonaceous micaceous quartzite

of the Juan Griego unit. These contact relations are critical and are discussed further below.

The relative internal coherence of the La Rinconada unit has been documented on the basis of systematic north–south variations in mineral assemblages, systematic reaction relationships shown by characteristic reaction textures, and subtle changes in amphibole chemistry, both regional and with time during a homogeneous exhumation path (Maresch 1977; Maresch et al. 1985; Maresch & Abraham 1981; Beets et al. 1984). There are no indications of major breaks in the sequence that would confirm the existence of major thrusts dissecting the La Rinconada unit *after* peak metamorphic conditions were reached (cf. Chevalier 1987, 1993).

$P-T$ estimates for two samples of amphibole–paragonite eclogite from northern Paraguachoa are shown in Figure 3. The calculations from three different studies agree well and fit the summary $P-T$ path obtained for the metapelites of the Juan Griego unit. Nevertheless, it is obvious that these eclogites last equilibrated at lower temperatures than the Juan Griego schists and also the small eclogite pods found within them. This is not surprising, considering that the La Rinconada eclogitic rocks are only slightly foliated and appear to have evolved from coarse-grained, H_2O-poor gabbros (Maresch & Abraham 1981). As noted above (see Fig. 3), the mica schists of the Juan Griego unit and the eclogite pods within them show evidence of reequilibration over a much wider temperature range. The strongly isothermal exhumation paths originally shown by Maresch & Abraham (1981), Beets et al. (1984) and Chevalier (1987, 1993) for these eclogites are no longer proposed here. These were based on the supposed necessity of staying within the stability limit of antigorite, in keeping with the associated serpentinized ultramafic rocks. Recent study of the latter (Koller, pers. comm. 2006) has shown that in fact much of the olivine in these massifs is metamorphic rather than of mantle origin, so that the upper stability of antigorite was exceeded.

The geochemistry of the metabasic series of the La Rinconada unit has been investigated in a number of studies (e.g. Navarro 1974; Bocchio et al. 1990; Beets et al. 1984; Giunta et al. 2002; Ostos & Sisson 2005, and references therein). The geochemical signature is typically that of ocean-floor basalts and low-K tholeiites, that is MORB, but there are clear tendencies towards island-arc basalts. Chondrite-normalized REE patterns are typically flat and have been interpreted as typical of N-MORB and E-MORB (Ostos & Sisson 2005). Samples from the La Rinconada unit show conspicuous variations from low TiO_2/high $Cr + Ni$ (MORB signature) to high TiO_2/low $Cr + Ni$ contents (IAT signature). Ostos & Sisson

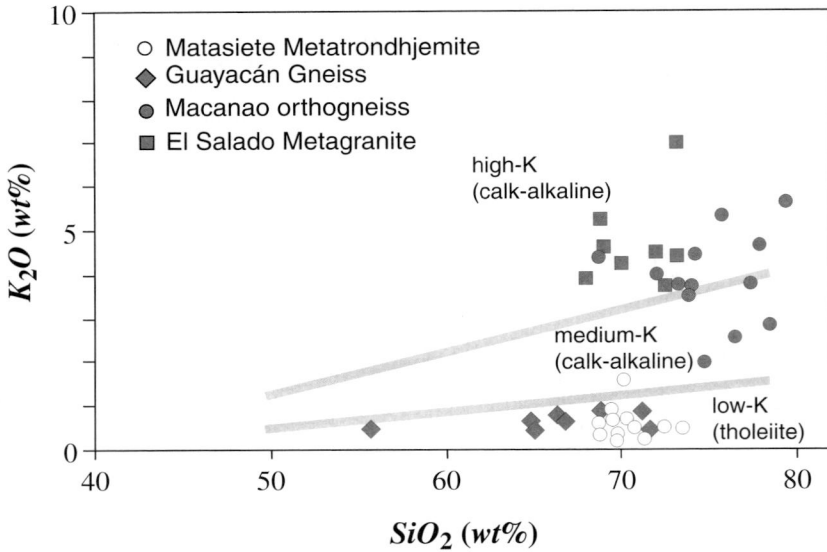

Fig. 6. K_2O-SiO_2 classification diagram (Le Maitre 1989) for magmatic rocks discussed in text.

(2005) relate the IAT signature to an origin in an island-arc or backarc basin close to a volcanic arc. On the other hand, Bocchio *et al.* (1990) suggest that this trend could be indicative of strong variations in cumulate content and differentiation processes on a local scale (gabbroic sills and small intrusions interspersed with flows and pillows). Giunta et al (2002) stress that the combination of MORB chemistry with flat REE patterns suggests an analogy with tholeiitic magmatism at oceanic plateaus. In gauging this apparent ambiguity it should not be forgotten that the La Rinconada unit has experienced a strong metamorphic imprint in a subduction zone and has been subjected to strong ductile deformation with unclear consequences as to relative element mobility. On the basis of the finely laminated nature of many of the non-eclogitic La Rinconada rocks, Maresch (1971, 1973, 1975) originally called for the presence of abundant pyroclastic rocks in the section. The obvious lack of a pervasive pelitic or calcareous sedimentary contamination in the analyses suggests instead that this fabric is due to extensive deformation of original basic basaltic/gabbroic rocks.

Guayacán Gneiss. A widespread leucocratic rock type of tonalitic composition was described and called the Guayacán Orthogneiss by Maresch (1971, 1975), redefined as Guayacán Gneiss by Urbani (2007). Masses of Guayacán Gneiss occur throughout the outcrop area of the La Rinconada unit and also the contact area with the Juan Griego

unit as mapped by Maresch (1971, 1973, 1975), where they intrude *both* amphibole gneisses and mica schists. Thus, both rock types must have already been juxtaposed at the time of intrusion, which thus stitches together in time the La Rinconada and Juan Griego units. Exposures as large as 3.5×1.5 km^2 are known from northern Paraguachoa, but in many cases the intrusive bodies are thin lit-par-lit intercalations in the host rocks and zones of such mixed rocks can extend for tens to hundreds of metres (Maresch 1971, 1973, 1975). The rocks may vary from equigranular in some of the larger bodies to predominantly gneissic. They have been deformed and metamorphosed together with the host rocks, producing a relatively simple assemblage of untwinned albite, intergrown phengite and paragonite, clinozoisite and amphibole. In many samples this amphibole has been partially to totally replaced by chlorite. Although a primary igneous intrusive origin for the Guayacán Gneiss is not evident from the present texture and mineralogy, Maresch (1975, pp. 865–868) has documented the contact relationships with the enclosing rocks in detail and suggested that the Guayacán protoliths were highly mobile intrusions into host rocks of reduced strength, that is at relatively elevated temperatures.

The analytical data presented by Kluge (1996) allow some important conclusions on pressures and temperatures of metamorphism to be drawn. Both the amphibole and phengite compositions corroborate that the Guayacán Gneiss was metamorphosed at high pressures analogous to the metabasic rocks of the La Rinconada unit, as field relationships (Maresch 1975) would suggest. Although most of the amphiboles show strong secondary alteration to greenschist-type assemblages, analysed remnants are of barroisite composition (Fig. 7), as also documented by Chevalier (1987). This sodic-calcic, high-pressure amphibole is also the amphibole found in the La Rinconada amphibole eclogites and high-pressure metabasic assemblages (Maresch 1975; Maresch & Abraham 1981; Beets et al. 1984; Chevalier 1987). The Si-contents of phengites enclosed and protected within albite porphyroblasts are as expected higher than in the interstitial micas. Assuming a temperature of at least 500 °C (analogous to the La Rinconada and Juan

Griego units), the minimum pressure for the stability of the enclosed phengites is >9 kbar. The composition of coexisting phengite with paragonite (e.g. Guidotti et al. 1994) suggests temperatures of mica re-equilibration for the interstitial aggregates down to 400–500 °C (Kluge 1996), in accord with similar estimates for phengite–paragonite pairs from Guayacán Gneiss by Maresch (1971, 1973).

Figures 5 and 6 summarize bulk chemical data obtained for various samples of Guayacán Gneiss by Kluge (1996). The samples plot mainly in the field of tonalite, grading into quartz diorite. They are of tholeiitic affinity. Recalculated plagioclase compositions correspond to oligoclase. This, together with the leucocratic nature of the rock suggests that

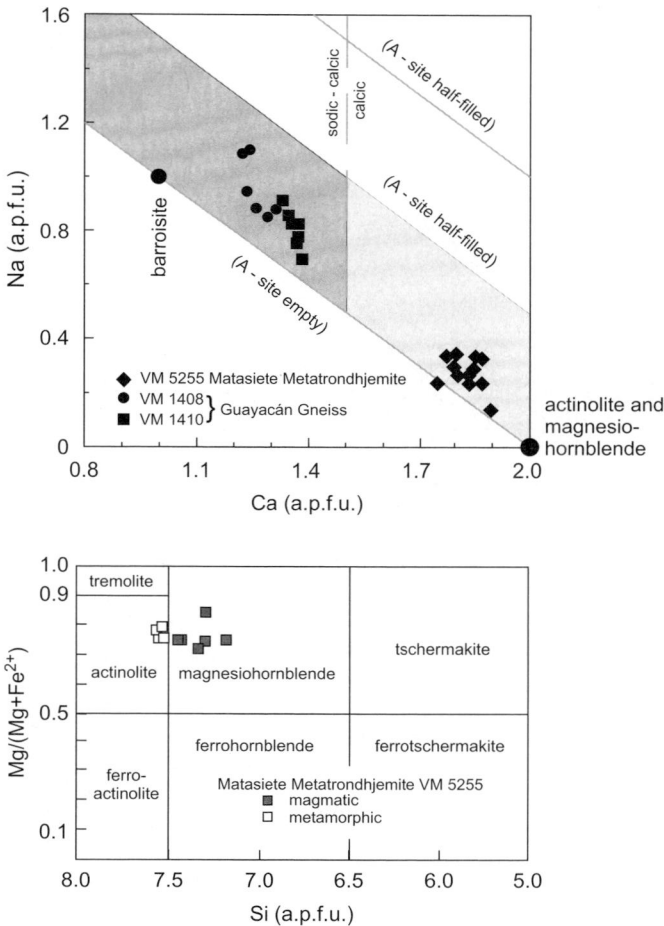

Fig. 7. Compositions of amphiboles in Guayacán Gneiss and Matasiete Metatrondhjemite, according to Leake et al. (1997).

the original rock could be termed a trondhjemite or plagiogranite. Further analyses of Guayacán Gneiss are available from Chachati & Macsotay (1985), Ostos (1990) and Ostos & Sisson (2005). Chachati & Macsotay (1985) have suggested that the Guayacán Gneiss represents metamorphosed 'quartz-dioritic apophyses' of a 'plagiogranitic' magma. Use of the term plagiogranite tends to imply an origin by processes near a mid-ocean ridge. Nevertheless, because of complete recrystallization during high-pressure metamorphism and the obvious late-stage growth of albite porphyroblasts indicating considerable major element mobility, major element criteria must be used with caution. Giunta *et al.* (2002) present both major and minor element data for two samples. One sample is described as a diorite with within-plate-tholeiitic affinity, the other as a quartz diorite related to tonalitic arc magmatism. They are considered to be the differentiates of the basaltic volcanism that led to the protoliths of the La Rinconada unit. Ostos (1990) used Nb–Y and Rb–Y–Nb discrimination plots (Pearce *et al.* 1984) to suggest a trend toward a volcanic arc origin. On the other hand, the low

$^{87}Sr/^{86}Sr$ ratio of 0.703, comparable to normal abyssal tholeiites, led Ostos (1990) and Ostos & Sisson (2005) to postulate a partial melting product of oceanic crust in a subduction zone. As discussed further below, the observations of García-Casco (2007), Lázaro & García-Casco (2008) and García-Casco *et al.* (2008) on the formation of trondhjemitic magmas in subducted metabasic rocks of eastern Cuba may provide a realistic model for such an origin of the Guayacán Gneiss. All authors stress the distinct chemical similarity to the Matasiete Metatrondhjmite (see below).

El Salado Metagranite. Maresch (1971, 1973, 1975) originally introduced the designation 'El Salado Granite' for a sheared, potassium-rich intrusive exposed primarily in the hills east of the village of Santa Ana (see Fig. 8). The rock varies from granular, where it can be described as a leucogranite, to highly sheared, where it takes on the aspect of an augengneiss. It does not show evidence for a high-pressure metamorphic overprint. The type locality, near the village of El Salado, is the northeastern mass shown in Figure 8. Intrusive contacts have

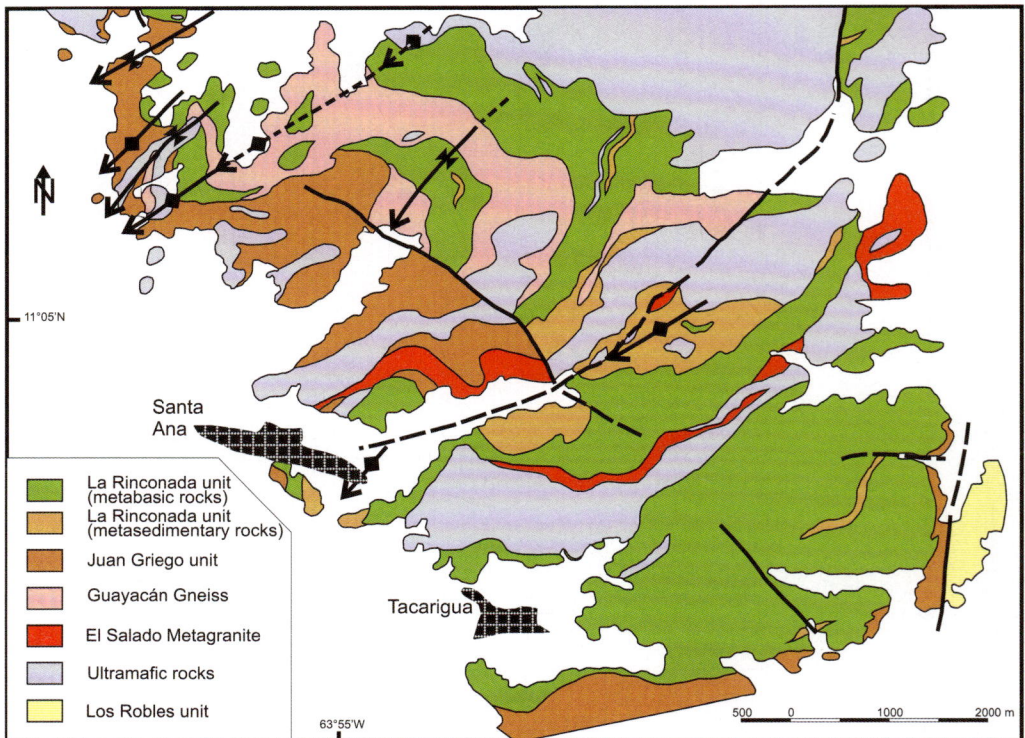

Fig. 8. Detailed map of lithologies and large-scale structures of the contact between the La Rinconada unit and the Juan Griego unit (excerpt from Maresch 1975, Fig. 2).

been observed by us with ultramafic rocks, La Rinconada metabasic sequences and with Juan Griego unit metasediments (see Fig. 8). In contrast, Chevalier (1987) considers all bodies of El Salado orthogneiss to be allochthonous bodies uprooted and separated from their original host rocks. The age of the El Salado intrusion is thus a key element in pinpointing the age of high-pressure metamorphism on the one hand, because according to our observations it intrudes high-pressure metamorphic rocks, and also because it again (analogous to the Guayacán Gneiss) 'stitches' together in time the La Rinconada and Juan Griego units on the other.

The composition of the phengite in these rocks yields valuable insight into both the conditions of magmatic crystallization as well as later ductile shearing. The large phengite crystals of sample VM 2175 (Fig. 4), which are interpreted as magmatic in origin, also deviate most strongly from the ideal trend of the Tschermaks substitution, indicating a relatively high proportion of the ferromuscovite component, as might be expected in a product from a late-stage granitic magma. This introduces some uncertainty into the phengite barometer. Nonetheless, at magmatic temperatures, pressures of $5-10$ kbar are indicated. The large range of phengite compositions down to $Si = 3.2$ a.p.f.u. mirrors continuous reequilibration enhanced by ductile shearing during exhumation to levels of $2-3$ kbar.

Figures 5 and 6 summarize bulk chemical data obtained for various samples of El Salado Metagranite by Kluge (1996). The samples plot mainly in the field of granite, grading into K-rich granodiorite. They are clearly calc-alkaline. One outlier in the tonalite field is from a highly sheared and foliated rock in which the bulk analysis may have been biased. Four additional analyses are cited in Ostos (1990) and Ostos & Sisson (2005). These analyses are enigmatic, because the K_2O contents are very low, varying between only 0.64 and 1.80 wt%, thus actually placing them in the low-K tholeiitic field of Figure 6 and in the same range as the tondhjemitic intrusions of Guayacán and Matasiete (see below). In fact, there is little significant difference between the analyses listed as Guayacán, Matasiete and El Salado in Ostos (1990) and Ostos & Sisson (2005), and a sampling problem may be involved there.

Contact relationships between the Juan Griego unit, the La Rinconada unit and the El Salado Metagranite. This critical question has become a key element in a number of discussions. Whereas earlier authors (e.g. Hess & Maxwell 1949; Taylor 1960*a*, *b*; Maresch 1971, 1973, 1975) originally suggested stratigraphical continuity between the Juan Griego unit and the La Rinconada unit, others (e.g. Bellizzia 1985; Chevalier 1987, 1993; Chevalier *et al.* 1988; Speed & Smith-Horowitz

1998; Higgs 2009, and references therein) have since then accepted the postulated existence of a major nappe thrust between the two units, with La Rinconada type rocks thrust over Juan Griego-type 'parautochthonous basement'. This difference has far-reaching consequences for tectonic scenarios in the southern Caribbean area and is also central to the tectonic scenario proposed by Higgs (2009) for the Caribbean as a whole.

The starting-point for a discussion on such a problem should be a fundamental description of the contact as observed in the field. The critical areas were mapped at a scale of $1:25\,000$ at a traverse interval of <250 m, and this map is readily available (Maresch 1973, 1975). Figure 8 shows an excerpt of the map published in Maresch (1975) of the key area NE of Santa Ana and Tacarigua. Nevertheless, Higgs (2009) dismisses the La Rinconada–Juan Griego contact as 'poorly exposed' and postulates a nappe structure of major proportions involving rocks of very different age. Similarly, as already indicated above, intrusive El Salado Metagranite contacts with La Rinconada and Juan Griego type-rocks were observed and mapped by Maresch (1971, 1973, 1975). However, Chevalier (1987) concludes that the El Salado bodies represent fault-bounded fragments (Chevalier 1987, p. 153), basing this conclusion on his tectonic model of the structural architecture of Margarita.

The constraint involved in mapping for the Ministerio de Minas e Hidrocarburos in 1968–1970 was that a stratigraphic system should be developed. Consequently, Maresch (1975) wrote that 'Detailed mapping during the present study has indicated that the contact between the La Rinconada unit and the overlying Juan Griego unit (as defined by Jam & Méndez Arocha, 1962, and used in this study), is in fact transitional (cf. Taylor, 1960*a*, *b*), and that the transition is effected as rare layers of carbonaceous mica schist become more and more prominent going up in section until they constitute 100% of the exposures. The contact is here defined as the upper contact of the highest interlayer of amphibole gneiss of at least 100 m stratigraphic thickness in the stratigraphic section below the feldspathic unit of the Juan Griego unit. The contact is easily recognized, as the transition from predominantly amphibole gneiss to predominantly carbonaceous mica schist is relatively abrupt. Only in the low hills northeast of Santa Ana . . . does extensive interlayering occur.' On the basis of Figure 8, the following conclusions can be drawn:

(1) The contact between the La Rinconada and Juan Griego units was arbitrarily defined by Maresch (1971, 1973, 1975) for classical formal reasons within a stratigraphic

framework. In the light of modern paradigms, tectonic juxtaposition is a viable alternative. In terms of the rock types involved, conspicuous Juan Griego-type brown carbonaceous mica schist is found within greenish to bluish La-Rinconada-type metabasic rocks and vice versa (Fig. 8; see also Fig. 2 of Maresch 1975). There is no sharp lithological boundary attributable to a thrust plane, whether hidden or not, nor is there a recognizable sole thrust between the two rock types.

(2) The structural development of the metamorphic nucleus (e.g. Chevalier 1987; Guth & Avé Lallemant 1991; Stöckhert *et al.* 1994, 1995; Rekowski & Rivas 2005) calls for an initial deformational phase producing a foliation that is subsequently refolded. High-pressure metamorphic conditions outlasted this refolding. A subsequent major ductile phase during greenschist-facies conditions produced a pervasive, now ENE–WSW subhorizontal stretching lineation with parallel fold axes. This latest ductile deformation is responsible for the present ENE–WSW grain of the island and the broad antiform found on Paraguachoa. In detail there are subsidiary smaller antiforms and synforms that are evident from the map of Figure 8. These structures are common to and conformable in both the Juan Griego type schists and the La Rinconada type metabasic series. Although the early deformational history is better preserved in the metabasic rocks as compared with the metasediments, there is no discernible break in the common deformational history of Juan Griego and La Rinconada-type lithologies (Guth & Avé Lallemant 1991; Chevalier 1987, 1993; Stöckhert *et al.* 1994, 1995).

(3) The *structurally* lower strata exposed in antiforms are La Rinconada-type rocks, whereas the *structurally* higher strata exposed in synforms are Juan Griego-type rocks. This contradicts the metabasic-rock-over-micaschist nappe structure postulated by Bellizzia (1985), Chevalier (1987, 1993), Chevalier *et al.* (1988), Speed & Smith-Horowitz (1998) and others, as well as the overthrusting of La Rinconada over Juan Griego required in the model of Higgs (2009).

(4) El Salado Metagranite is preferentially found associated with the antiformal structure east of Santa Ana. The granite intrudes both La Rinconada-type metabasic rocks, Juan Griego-type mica schists and ultramafic rocks. It has been deformed and sheared together with the enclosing rocks under the late greenschist-facies conditions experienced by the entire package (Stöckhert *et al.* 1994, 1995; Chevalier 1987).

(5) Ultramafic rocks show a general concordance with the metamorphic grain of the enclosing rocks. Even the large mass of dunite making up Cerro Chico (NE part of map area in Fig. 8) is marginally entrained into the regional fabric. Ultramafic rocks are enclosed both within La Rinconada-type metabasic and Juan Griego-type metasedimentary rocks. As described below, these rocks have also been metamorphosed together with the enclosing metabasic and metasedimentary rocks.

(6) Guayacán Gneiss is found in intimate intrusive relationship with La Rinconada-type amphibole gneiss, La Rinconada-type mica schist and Juan Griego type metasediments (Fig. 8: low hills north of Santa Ana).

The above observations are a clear indication that the Juan Griego unit, the La Rinconada unit and at least the ultramafic rocks now associated with them were metamorphosed together under high-pressure conditions. The 'sole thrust' described by Chevalier (1987, 1993) and Chevalier *et al.* (1988) near Bahia de Plata (NW of Pedro González in Fig. 2) separates a serpentinite body from Juan Griego-type mica schists. The actual present contact between the Juan Griego-type schists and the La Rinconada-type metabasic rocks in this area is the nearly vertical NNW-trending Punta Corey fault 1 km to the east (Maresch 1975, Fig. 2, sectors B8–B9). The Juan Griego schists between the 'sole thrust' and the Punta Corey fault contain beautiful examples of eclogitized basic dykes and/or sills.

A system of ductile shear zones 50–250 m in width is exposed along the entire northern coast of Paraguachoa east of Bahia de Plata (e.g. the Manzanillo Shear Zone of Maresch 1971, 1973, 1975; Chevalier 1987). The matrix is serpentinite with many examples of typical metasomatic blackwall alteration zones (talc, chlorite and actinolite schist). Blocks of high-pressure rocks are common, suggesting that the shear zones were active during peak high-pressure metamorphism. However, these shear zones do not separate rocks of distinctly different metamorphic grade nor (east of Guayacán) rocks of different lithology (all La Rinconada type). The topographical trend indicates that for most of its length it must be a mainly vertical structure. Although a prominent feature, it cannot be a thrust of major proportions post-dating the peak of high-pressure metamorphism.

Of particular interest are black garnet and epidote amphibolites (Maresch 1971, 1973, 1975; Chevalier 1987) that are associated both with this shear zone system and are also found in the margins of the Cerro Chico peridotite massif. These enigmatic rocks (see also Taylor, 1960*a, b*) appear to resemble the restites described by Lázaro & García-Casco (2008) formed after trondhjemitic melt was extracted from basic crust (see below).

The age of the Guayacán Gneiss, also metamorphosed under high-pressure conditions, constrains the minimum age at which Juan Griego- and La Rinconada-type rocks must have come together. Conversely, the El Salado Metagranite, which also intrudes all these units but has not been subjected to high-pressure metamorphism, provides a minimum age for this high-pressure event.

Ultramafic rocks associated with the high-pressure metamorphic nucleus. With the exception of the ultramafic massifs of Cerros Guayamuri and Matasiete (Fig. 2) and scattered smaller bodies in the greenschist periphery (El Morro on Paraguachoa, northern Macanao), all ultramafic bodies are associated with the high-pressure metamorphic nucleus of Margarita Island. Petrographic descriptions are given in Maresch (1971, 1973, 1975) and Chevalier (1987, 1993). Extensive zones of black-wall alteration accompany almost all occurrences of ultramafic rocks (Maresch 1971, 1973, 1975). Examples of bulk rock and/or mineral compositions are given in Chevalier (1987) and Giunta *et al.* (2002). In general terms, they can be described as variously serpentinized, olivine-rich spinel peridotites. So far unpublished author data indicate that the spinel compositions are typical of a supra-subduction-zone setting. Wehrlite and clinopyroxenite are relatively common. Harzburgite has been described only from a locality between Cerro Guayamuri and Cerro Matasiete (see below) outside of the high-pressure nucleus of Margarita. The ultramafic rock bodies of the high-pressure nucleus are conformable with the regional deformational fabric (e.g. Fig. 8) and have been deformed together with the enclosing rocks of the La Rinconada and Juan Griego units.

Chevalier (1987) already noted that the olivines in some dunites are slightly more iron-rich than expected for a mantle peridotite. In our on-going study we have also found that metamorphic olivine is common and that samples of slightly serpentinized 'dunite' can actually be *metamorphosed* former antigorite-rich rocks and not pristine mantle rocks (F. Koller, pers. comm. 2007). We have also found porphyroblastic titanian clinohumite, a mineral that, depending on its specific composition, can be an indicator of metamorphic pressures (e.g. López Sánchez-Vizcaino *et al.* 2005). Certainly, such rocks can have experienced temperatures near the upper stability of antigorite at $\geq 600\,°C$ (e.g. Wunder & Schreyer 1997) and provide the potential for analysing more detailed $P-T$ paths. These observations corroborate that the ultramafic rocks associated with the La Rinconada

and Juan Griego units share a common metamorphic and deformational history with them. Further study is needed. However, it is clear that one single lithodemic designation for *all* the ultramafic rocks of Margarita Island (the 'Metaultramáficas de Cerro El Copey' of Urbani 2007) does not do justice to these rocks.

Greenschist-facies periphery and Matasiete Metatrondhjemite

One purpose of this paper is to underline the fact that there is a first-order break in the metamorphic grade and metamorphic history of rock units on Margarita Island that allows the grouping of rock units with a conspicuous higher-pressure (and also higher-temperature) history on the one hand, and a low-pressure and only greenschist-facies temperature realm on the other. This difference must mirror a first-order difference in terms of geodynamic record. On this specific point we are in accord with the nappe concept of Chevalier (1987, 1993). As summarized above, the high-pressure nucleus of Margarita Island involves the units Juan Griego, La Rinconada and Guayacán Gneiss. The El Salado Metagranite intrudes this package, without itself showing a high-pressure metamorphic overprint. The thrusts postulated *within* these units (e.g. Chevalier 1987, 1993; Chevalier *et al.* 1988), irrespective of whether there is agreement on all of them, are of secondary importance with respect to the first-order discontinuity seen between high-pressure nucleus and greenschist-facies periphery.

Following the nomenclature of Chevalier (1987, 1993) and Urbani (2007), we group together as the greenschist-facies periphery the rocks of the Los Robles nappe of Chevalier (1987, 1993; the 'Asociación Metamórfica de Los Robles of Urbani, 2007) and the Matasiete–Guayamuri nappe of Chevalier (1987, 1993). We also include the rocks of the El Piache–La Asunción nappe of Chevalier (1987, 1993), because only greenschist-facies rocks without any documented prior high-pressure imprint appear to be involved. There is therefore at present no realistic basis for lumping together these rocks with the 'Asociación Metamórfica Juan Griego' (cf. Chevalier 1987, 1997; Urbani 2007). In addition, we include a series of low-grade metavolcanic rocks from northeastern Paraguachoa, the Manzanillo unit (Taylor 1960*a, b*; Maresch 1971, 1973, 1975). Lithologically, the rocks of the greenschist-facies periphery are a series of heterogeneous metasediments, including thick sequences of marbles, as well as intermediate to basic metapyroclastic/metavolcanic deposits (e.g. Hess & Maxwell 1949; Taylor 1960*a, b*; Jam & Méndez Arocha 1962; González de Juana 1968; Vignali 1979; Chevalier 1987, 1993; Rekowski & Rivas

2005). Biotite has been observed; garnet is absent. Many metavolcanic rocks still show relicts of volcaniclastic fragments. No evidence anywhere of an earlier high-pressure overprint is observed. Although no detailed pressure–temperature estimates are available, the mineral assemblages observed indicate maximum temperatures distinctly below 500 °C and pressures not exceeding 5–6 kbar (e.g. Bucher & Frey 2002).

The term Matasiete Trondhjemite was used by Turner & Verhoogen (1960) for the intrusive body introduced to the literature by Hess & Maxwell (1949) as 'Matasiete soda granite porphyry'. Urbani (2007) suggests the official designation Matasiete Metatrondhjemite. The main outcrop area is a narrow 8 × 0.5 km north–south strip at the base of the ultramafic massifs of Cerro Matasiete and Cerro Paraguachi on the east coast of Paraguachi (Fig. 2). Satellite bodies exposed in Quaternary sediments of the eastern coastal plain extend approximately 3 km both to the north to Puerto Fermin and to the south to Agua de Vaca. Chachati & Macsotay (1985) suggest that the small satellite body SE of Cerro Matasiete (Fig. 2) should be an independent intrusive body ('Metagranodiorita de Agua de Vaca' of Urbani 2007). Small, metre-sized bodies of similar rocks are tectonically incorporated into serpentinites exposed near the north slope of Cerro Guayamuri and also in the fishing harbour of El Morro 10 km to the south. The present western limit of the main exposure is an active fault scarp that has beheaded the eastward-draining valley between the two above ultramafic massifs. Basic dykes and xenoliths from centimetres to metres in size are found in many parts of the intrusive. As described below, equigranular intrusive rocks composed primarily of magmatic minerals are rare, and composed primarily of sodic plagioclase, quartz and up to 5% hornblende. However, much of the massif actually represents the well-exposed basal zone of a major nappe structure (Guillet & Cannat 1984). Progressive deformation and recrystallization leads to mylonites and a greenschist-facies mineralogy dominated by albite, actinolite, chlorite and epidote. No evidence of high-pressure minerals or a high-pressure metamorphic history has been found in these rocks.

Figures 5 and 6 summarize bulk chemical data obtained for various samples of Matasiete Metatrondhjemite (Kluge 1996). Almost all samples plot in the field of tonalite, with great similarity to Guayacán Gneiss. Chachati & Macsotay (1985) classify the Matasiete Metatrondhjemite as a plagiogranite geochemically correlative with the Guayacán Gneiss. Both rock types show low-K tholeiitic affinity. Ostos (1990) and Ostos & Sisson (2005) interpret Nb–Y and Rb–Y + Nb discrimination plots (Pearce *et al.* 1984) as well as low

$^{87}Sr/^{86}Sr$ ratios to postulate origin by partial melting of oceanic crust in a subduction zone.

Contact relationships between periphery and nucleus. The tectonic boundary shown in Figure 2 between the El Piache–Los Robles nappe and the high-pressure metamorphic nucleus is based primarily on a recognizable sudden change in metamorphic grade. Neither a thrust nor a normal detachment fault is directly exposed. North of La Asunción the tectonic boundary is an active recent fault that is part of the recent graben structure developed west of Cerro Matasiete. When mica schists on both sides of the tectonic boundary are carbonaceous and weathered, it can be locally difficult (e.g. northern Macanao, north of La Asuncion) to distinguish rocks of the Juan Griego unit from rocks of the greenschist-facies periphery in the field. However, the presence of garnet is usually a sufficient indicator.

Guillet & Cannat (1984) and Chevalier (1987, 1993) have pointed out that the Massif underlain by the ultramafic rocks of Cerros Guayamuri and Matasiete (Fig. 2) represents a well-exposed nappe structure, with the Matasiete Metatrondhjemite representing the sole of the structure that has been thrust SSW over the low-grade schists and phyllites of the El Piache–Los Robles nappe. Although Los Robles rocks are exposed to the west of Cerro Matasiete, the original thrust contact there is obscured by the younger fault noted above. Nevertheless, as the contact is approached, the extent of shearing in the trondhjemite increases dramatically, and mylonites/ultramylonites are found. The relationship is more clearly displayed in outcrops along the coast on the northeast flank of Cerro Guayamuri (the 'demi-fenêtre' of Chevalier 1987), where slivers of Matasiete Metatrondhjemite thinned to only a few metres in thickness can be observed between serpentinite and Los Robles phyllites and metavolcanic rocks. Strongly sheared Matasiete Metatrondhjemite is ubiquitous, and mylonites are found along the entire north–south extent of Matasiete occurrences. Equigranular rocks with magmatic fabric are restricted to a few localities such as in satellite bodies within Quaternary sediments southeast of Cerro Matasiete (Guillet & Cannat 1984) or as loose blocks near the cisterns of Flandes in the valley between the two ultramafic hills.

Guillet & Cannat (1984) suggest that nappe thrusting took place at temperatures of quartz plasticity below 500 °C, that is in the greenschist facies. The temperatures of >1000 °C quoted by Chevalier *et al.* (1988) for thrusting quite clearly must refer to earlier deformation of the ultramafic rocks within the upper mantle (Guillet & Cannat 1984). This conclusion is borne out by the greenschist-facies overprint shown by the trondhjemite itself, as well as the epidote–chlorite–actinolite assemblage developed in basic dykes and xenoliths from centimetres to metres in size found in many parts of the intrusive.

Summary of field relationships

On the basis of the above analysis two main features stand out:

(1) There is no major thrust structure between La Rinconada-type and Juan Griego-type rocks.

(2) The La Rinconada-type and Juan Griego-type rocks and the ultramafic rocks associated with them have all experienced the same high-pressure metamorphic event and have all followed a similar pressure (depth)–temperature–time–deformation path.

The logical conclusion is, as stated by Stöckhert et al. (1995), that these two rock sequences came into contact before the maximum of the high-pressure event, and have formed a cohesive unit since then, as borne out by the structural studies of Chevalier (1987, p. 413).

The original melts of the Guayacán Gneiss intruded both La Rinconada-type and Juan Griego-type rocks. After crystallization, these intrusive bodies then experienced the same high-pressure metamorphic conditions as the enclosing La Rinconada- and Juan Griego-type rocks. The age of this intrusive activity therefore determines the minimum age at which the two units came together, and at which subsequent high-pressure metamorphism took place. Sparse geochemical evidence suggests that these melts are arc-related rather than being typical MOR plagiogranites (see above). Thus, the origin of the Guayacán melts must be seen in relation to subduction processes. Juxtaposition of the La Rinconada and Juan Griego units, Guayacán magmatism and high-pressure metamorphism could be a relatively short-lived and closely connected event.

The end of active subduction and the beginning of exhumation of the metamorphosed composite is constrained by the age of the calc-alkaline and clearly arc-related intrusions of the El Salado Metagranite melts.

Although the Guayacán and Matasiete magmas are clearly similar in their chemistry and plausibly of similar origin, the Matasiete intrusion was never subducted to depths sufficient for high-pressure metamorphism. As an integral part of the rocks of the greenschist-facies periphery, it evolved independently until the greenschist-facies nappes were faulted into contact with the high-pressure metamorphic nucleus when this nucleus reached a regional greenschist-facies level during exhumation.

U–Pb geochronology

Description of samples

Macanao orthogneiss. Two samples of augengneiss corresponding compositionally to granite, VM 1329 and VM 1330 (Fig. 5), were selected for multi-grain U–Pb dating of zircon populations (Kluge 1996) from the San Francisco locality (see Fig. 2) in central Macanao. Zircons from sample VM 2065, taken approximately 250 m to the west of the above, were later analysed with the SHRIMP technique for corroboration.

Oriented porphyroclasts of partly perthitic, twinned microcline form lenticular augen from millimetres to centimetres in size. The perthitic exsolution bodies are unusually rounded, indicating tempering and reequilibration subsequent to their formation. The porphyroclasts are set in a granoblastic to lepidoblastic matrix (150–600 μm grainsize) of largely recrystallized quartz, potassium feldspar, albite and white mica that flows around the porphyroclasts and penetrates them along fractures. The matrix K-feldspar is untwinned and free of perthitic exsolution. Locally, a quartz foam structure with 120° grain-boundary triple junctions suggests late-stage stress-free annealing. Late porphyroblasts of albite also overgrow the foliation. Green-brown biotite is observed as a late product in the interstices between quartz and feldspar. Chlorite is observed. Titanite and zircon are accessories. Sample VM1329 contains small 0.3 mm, zoned garnets rich in grossular (50–54 mol%) and almandine (35–38 mol%), with spessartine varying from 14.5 mol% in the core to 7 mol% in the rim, indicative of growth zoning (Krückhans-Lueder 1996).

Guayacán Gneiss. Two samples were selected from relatively large bodies of gneiss in northern Paraguachoa, one near the fishing village of Guayacán (VM 1409), the other from the locality Pedro González (Fig. 2). The samples are light-coloured and equigranular, with oriented laths of greenish amphibole. Under the microscope, the most distinctive feature is a 'pavement' of albite porphyroblasts ($<An_{03}$) 1–2 cm in diameter, enclosing rounded blebs of quartz as well as clinozoisite. The arrangement of inclusions mimics an older foliation. Twinning in albite is rare. Unoriented large crystals of phengite (150–600 μm) are enclosed in albite porphyroblasts, whereas aggregates of phengite and paragonite (20–100 μm) form interstitial lepidoblastic nests. The contact between such mica nests and albite is very irregular, and the plagioclase develops a very thin rim in which anorthite contents reach An_{16}. Amphibole with blue-green pleochroism is partially to totally replaced by chlorite, clinozoisite and albite. Idiomorphic to subidiomorphic

aggregates of clinozoisite up 1000 μm in length are found enclosed in albite porphyroblasts and mica aggregates. Titanite and zircon are accessory minerals.

Matasiete Metatrondhjemite. Two samples of Meta-trondhjemite from the village of Flandes between Cerros Guayamuri and Matasiete (Fig. 2) were selected. Sample VM 2226 (Fig. 5) represents one of the rare examples of equigranular rocks with recognizable magmatic fabric. These are dominated by millimetre-sized, zoned plagioclase crystals set in a matrix of mosaic-textured quartz exhibiting deformation bands and grain boundary migration. Plagioclase is rhythmically zoned, with An-richer cores, and compositions varying between An_{30} and almost pure albite. Fine polysynthetic albite twinning is common. An-rich zones are strongly saussuritized. Magmatic amphibole, constituting up to 5 vol% of the rock, is commonly twinned, with compositions corresponding to Si-rich, Al-poor magnesiohornblende (Fig. 7), which can be taken as a possible indication of a relatively shallow level of magma crystallization (e.g. Johnson & Rutherford 1989).

As deformation and overprinting increases, these amphiboles are topotactically replaced by a secondary generation of actinolite, even though the change in composition is small (Fig. 7). This is a key observation, because the often-quoted ages of 71 ± 5 Ma [Martín Bellizzia (1968) or 72–74 Ma (Santamaria & Schubert 1974; ages recalculated by Guth & Avé Lallemant 1991] obtained from the Matasiete Trondhjemite are based on K–Ar dating of amphibole concentrates. Rather than dating the age of intrusion, these results mirror the age of the greenschist-facies overprint, and indirectly therefore provide a constraint on the timing of nappe overthrusting. As the greenschist overprint proceeds, clear albite rims with blebs of quartz form as overgrowths on the magmatic plagioclase. Quartz, white mica and chlorite crystallize in the pressure shadows of increasingly comminuted plagioclase porphyroclasts. Primary rutile is rimmed by titanite. Sample VM 2237 (Fig. 5) is an example of strongly deformed Matasiete Metatrondhjemite.

With increasing degrees of shearing, amphibole is replaced by aggregates of chlorite, clinozoisite and titanite, which become progressively dragged out into lenticular greenish layers in which white mica also appears from the breakdown of primary plagioclase. With the production of subparallel quartz- and quartz-feldspar layers, a mylonitic fabric is finally achieved.

El Salado Metagranite. Two samples of El Salado Metagranite were chosen for U–Pb dating (Fig. 5). Sample VM 2175 shows a magmatic texture and was taken from the type locality near El Salado (Fig. 2). Sample VM 2166 is an augengneiss sampled from the end of the valley east of Tacarigua.

According to the descriptions and analyses of Maresch (1971, 1973, 1975; Kluge 1996), the least sheared examples of El Salado are typically heterogranular in aspect and composed mainly of microperthitic microcline (typically $Or_{92-94}Ab_{06-08}An_{00}$) and polysynthetically twinned, saussuritized plagioclase. The latter is typically $Ab_{96-98}Or_{1-3}An_{1-2}$, but crystals with anorthite contents up to sodic oligoclase have been analysed. These feldspar grains occur as anhedral crystals up to 1 cm in diameter and as smaller interstitial fragments. Even the granular examples exhibit a matrix of recrystallized untwinned albite, potassium feldspar and deformed quartz showing grain-boundary migration. Large deformed crystals of white mica and brown biotite appear magmatic in origin, as are apatite, allanite and zircon.

When highly sheared, El Salado Granite is distinctly foliated, and is given a 'knotty' appearance by augen composed of crushed and broken fragments of feldspar distorting the foliation. The latter is characterized by a compositional layering, where lenticular layers (0.5–2 mm thick) of polycrystalline quartz alternate with layers of recrystallized feldspar + white mica + epidote (drawn-out former augen) and layers of newly formed white mica + green-brown biotite + epidote + titanite ± chlorite. Very rare garnet can show inclusion-free cores with poikiloblastic rims enclosing matrix minerals or even atoll structures. At least the inclusion-rich rims and atoll garnets should be a product of the metamorphic overprint. The composition is highly variable ($Sps_{28-59}Alm_{10-35}Grs_{12-28}Adr_{9-22}Prp_{0-3}$), with no discernible zoning trends.

Sample preparation and analytical procedures

Samples of *c.* 30 kg were processed by hammer, hydraulic press, jaw crusher, roller mill, sieves and Wilfley-table to obtain zircon concentrates. The crystals were separated with a magnetic separator and with bromoform and methylene iodide and subsequently sieved into several size fractions. Some fractions were subdivided by hand-picking according to morphological criteria, degree of metamictization and occurrence of inclusions. These fractions were ultrasonically purified for 1 min with diluted HNO_3 at room temperature.

Chemical extraction of U and Pb from the zircon fractions was carried out according to the standard method of Krogh (1973). A $^{235}U-^{208}Pb$ mixed spike was used for the determination of the U and Pb concentrations by the isotope dilution method. U and Pb were loaded on Re single filaments by

Table 1. *U–Pb concentration and isotope data of selected metaigneous rocks from Isla Margarita*

Sample	Number in figures	Sieve fraction (μm)	Type*	Zircon weight (mg)	Concentrations U (ppm)	Pb (tot) (ppm)	Pb (rad) (ppm)	^{206}Pb (rad) (μmol g^{-1})	Measured atomic ratios $^{208}Pb/^{206}Pb$	$^{207}Pb/^{206}Pb$	$^{206}Pb/^{204}Pb$	Calculated atomic ratios $^{206}Pb/^{238}U$	$^{207}Pb/^{235}U$	$^{207}Pb/^{206}Pb$	Apparent ages (Ma) $^{206}Pb/^{238}U$	$^{207}Pb/^{235}U$	$^{207}Pb/^{206}Pb$
Macanao orthogneiss																	
VM 1330	1	100–125	E, G	0.97	1450	60.02	44.72	169.8	0.41043	0.13356	182	0.0280	0.2063	0.0533	178.3	190.4	343
VM 1330	2	80–90	I, M	4.03	1178	37.50	32.66	126.2	0.28751	0.08799	405	0.0257	0.1835	0.0518	163.3	171.1	277
VM 1330	3	90–100	E, I	0.91	517	23.22	22.36	87.1	0.22630	0.06673	1015	0.0404	0.2913	0.0523	255.3	259.2	299
VM 1330	4	100–125	E	1.19	542	25.18	23.08	90.0	0.24846	0.07730	596	0.0398	0.2895	0.0527	251.7	258.2	317
VM 1330	5	90–100	E, I, M	1.16	828	43.53	31.51	120.2	0.42434	0.14411	162	0.0349	0.2596	0.0540	221.1	234.3	369
VM 1330	6	90–100	E, G	0.51	1309	51.98	43.55	166.3	0.33281	0.09935	299	0.0304	0.2110	0.0503	193.3	194.4	208
Guayacán Gneiss																	
VM 1408	1	62–80	E	4.3	420	6.86	6.75	29.0	0.09653	0.05414	2253	0.0165	0.1104	0.0484	105.8	106.3	118
VM 1408	2	80–90	E	3.5	420	6.77	6.73	29.0	0.08659	0.05108	4843	0.0165	0.1094	0.0480	105.7	105.5	101
VM 1408	3	100–125	E	1.3	430	7.01	6.92	29.7	0.10055	0.05566	1905	0.0166	0.1096	0.0480	106.0	105.6	97
VM 1408	4	125–160	E	3.3	250	4.18	4.11	17.7	0.09174	0.05423	2376	0.0170	0.1228	0.0480	108.8	108.5	101
VM 1409	5	50–60	E	1.5	130	2.47	2.41	9.83	0.15765	0.05657	1764	0.0182	0.1209	0.4820	116.1	115.9	111
VM 1409	6	62–80	E	5.4	126	2.38	2.25	9.17	0.19004	0.06815	1742	0.0175	0.1166	0.4830	111.8	112.0	116
VM 1409	7	80–90	E	4.1	112	2.01	1.99	8.15	0.15852	0.05500	2138	0.0174	0.1157	0.0481	111.5	111.2	105
VM 1409	8	90–100	E	3.6	123	2.29	2.22	9.06	0.16272	0.05695	1590	0.0177	0.1164	0.0477	113.1	111.8	84
Matasiete Metatrondhjemite																	
VM 2237	1	80–90	M, E	2.23	193	3.48	3.44	14.1	0.15764	0.06032	1306	0.0176	0.1190	0.0491	112.4	114.2	152
VM 2237	2	80–90	E	2.14	338	6.22	6.03	25.0	0.15322	0.05997	1212	0.0178	0.1170	0.0479	113.4	112.4	91
VM 2237	3	100–125	E	2.40	166	3.14	3.04	12.4	0.63055	0.25249	71	0.0179	0.1242	0.5020	114.6	118.9	205
VM 2237	4	80–90	E, B	1.78	178	3.22	3.20	13.2	0.16589	0.06568	847	0.0178	0.1187	0.0484	113.8	113.9	117
VM 2226	5	90–100	E	1.37	152	2.79	2.71	11.3	0.15936	0.06722	770	0.0178	0.1184	0.0481	114.0	113.6	106
VM 2236	6	50–62	E	1.55	189	3.58	3.35	14.0	0.19680	0.07923	475	0.0176	0.1175	0.0483	112.8	112.8	112
El Salado Metagranite																	
VM 2175	1	90–100	E	0.74	1984	48.62	25.46	111.2	0.47106	0.20976	91	0.0134	0.0898	0.0484	86.1	87.3	121
VM 2175	2	90–100	E, B	6.79	2168	58.27	27.61	120.3	0.53930	0.23592	78	0.0134	0.0884	0.0482	85.3	86.0	107
VM 2175	3	50–62	E	2.23	2218	44.45	28.91	124.2	0.35096	0.15571	137	0.0134	0.0892	0.0482	85.9	86.8	109
VM 2166	4	100–125	E, B	0.19	3092	89.74	39.14	170.5	0.61290	0.26331	68	0.0132	0.0852	0.0467	84.7	83.0	34

* E, euthedral crystals; M, metamict crystals; I, inclusions (unidentified); G, graphitic inclusion; B, zircon fragments.

means of the Ta_2O_5 suspension and silica gel–H_3PO_4 techniques.

Isotope ratios were measured on thermal ionization mass spectrometers (TIMS), either on a Teledyne SS 12″, 90° with a single Faraday cup, or on a VG Sector 54. A correction of 0.12% per mass unit was applied to the Pb isotope ratios according to measurements of the NBS SRM 982 lead standard.

Error ellipses in the concordia diagrams were calculated according to Ludwig (1980), taking into account the internal precision ($2\sigma_M$) of the isotope ratio measurements, the estimated uncertainty of the U–Pb ratio in the spike (c. 0.15%), the individual error magnification from the spike–sample ratio, and the estimated uncertainties in the isotope composition of the initial Pb (c. 1%) and blank Pb (c. 1%). The Pb blank was <1 ng; the blank composition ($^{208}Pb/^{204}Pb = 37.7$, $^{207}Pb/^{204}Pb = 15.52$, $^{206}Pb/^{204}Pb = 17.72$) was measured in mixtures of aliquots of the reagents used for analysis. Discordia regression lines were calculated according to York (1969) on a 95% confidence level. Element concentrations and ages were computed with the constants recommended by Steiger & Jäger (1977).

U–Pb analyses of the zircons separated from sample VM 2065 were obtained with a SHRIMP-II ion microprobe (Center of Isotopic Research, VSEGEI, St Petersburg, Russia). The zircons were mounted together with chips of the TEMORA (Middledale Gabbroic Diorite, New South Wales, Australia) and 91500 (Geostandard zircon) reference zircons in epoxy resin, polished, and observed both in BSE and CL mode on a scanning electron microscope. Each analysis consisted of five scans through the mass range with a spot diameter of about 25 μm; the primary beam intensity was about 10 nA.

The data have been reduced in a manner similar to that described by Williams (1998, and references therein), using the SQUID Excel Macro of Ludwig (2000). The Pb–U ratios have been normalized relative to a value of 0.0668 for the $^{206}Pb–^{238}U$ ratio of the TEMORA reference zircons, equivalent to an age of 416.75 Ma (Black et al. 2003). Uncertainties given for individual analyses (ratios and ages) are at the 1σ level; however the uncertainties in calculated concordia ages are reported at the 2σ level. The Ahrens–Wetherill (Wetherill 1956) concordia plot has been prepared using ISOPLOT/EX (Ludwig 1999).

Geochronological results

Analytical data are given in Tables 1 and 2. The results are summarized in Table 3.

Macanao orthogneiss. Six zircon fractions of sample VM 1330 were analysed. The crystals are

Table 2. *U–Th–Pb concentrations and isotope data for zircons from Macanao orthogneiss (sample VM 2065)*

Shot number	Concentrations					Measured atomic ratios				Calculated atomic ratios (^{204}Pb corrected)						Apparent ages (Ma)					
	U (ppm)	Th (ppm)	$^{232}Th/^{238}U$	Radiogenic ^{206}Pb (ppm)	Common ^{206}Pb (%)	Total $^{238}U/^{206}Pb$	Percentage error	Total $^{207}Pb/^{206}Pb$	Percentage error	$^{207}Pb/^{235}U$	Percentage error	$^{206}Pb/^{238}U$	Percentage error	ρ	$^{204}corr$ $^{206}Pb/^{238}U$	1σ error	$^{204}corr$ $^{207}Pb/^{206}Pb$	1σ error			
1.1	476	357	0.78	17.2	0.42	23.79	0.7	0.0518	2.2	0.28	4.2	0.0419	0.7	0.175	264.3	1.9	125	97			
1.2	357	145	0.42	13.0	0.00	23.64	0.9	0.0539	2.6	0.31	2.7	0.0423	0.9	0.339	267.1	2.4	367	58			
2.1	1065	586	0.57	40.0	0.42	22.88	0.6	0.0540	1.8	0.30	3.5	0.0435	0.6	0.180	274.6	1.7	229	81			
2.2	448	198	0.46	17.3	0.29	22.25	0.9	0.0499	2.9	0.29	4.0	0.0448	0.9	0.219	282.6	2.4	81	93			
3.1	2600	1290	0.51	90.5	0.21	24.67	0.8	0.0526	1.2	0.28	1.9	0.0405	0.8	0.419	255.6	2.0	239	40			
5.1	114	47	0.43	4.1	-1.27	24.01	2.5	0.0517	4.8	0.36	9.2	0.0422	2.6	0.279	266.3	6.7	658	190			
5.2	472	174	0.38	18.0	0.32	22.57	0.8	0.0518	2.1	0.30	3.5	0.0442	0.8	0.232	278.6	2.2	163	80			
7.1	572	234	0.42	20.7	0.46	23.76	0.7	0.0525	2.1	0.28	5.4	0.0419	0.8	0.144	264.6	2.0	145	125			
8.2	579	326	0.58	21.1	-0.13	23.56	0.8	0.0529	2.5	0.32	3.2	0.0425	0.8	0.254	268.3	2.1	368	70			
9.1	289	112	0.40	11.3	0.29	22.06	1.1	0.0525	3.4	0.31	4.8	0.0452	1.1	0.224	285.1	3.0	205	108			

SHRIMP shots performed on the basis of CL images. In the case of two shots the first refers to the core, the second to the rim of the crystal (cf. Fig. 11). Error in standard calibration is 0.47%. Analyst: Nikolay Rodionov, Center of Isotopic Research, VSEGEI, St Petersburg, Russia.

Table 3. *Summary of U–Pb zircon ages of selected metaigneous rocks from Isla Margarita*

	Sample	Location*	Zircon fraction[†]	Age (Ma)	Comment[‡]
Macanao orthogneiss	VM 1330	359 775/1 217 725	1–6	55	LCIA
			1–6	315	UCIA
	VM 2065	359 040/1 217 800		271.0 ± 6.6	SHRIMP
Guayacán Gneiss	VM 1408	400 300/1 227 440	1–3	106	ccd
			4	107	ccd
	VM 1409	400 720/1 231 400	6–8	112	ccd
			5	116	ccd
Matasiete Metatrondhjemite	VM 2237	407 260/1 225 080	1–4	114	ccd
	VM 2226	407 400/1 225 040	5–6	114	ccd
El Salado Metagranite	VM 2175	405 600/1 226 550	1–3	86	ccd
	VM 2166	402 685/1 223 040	4	85	ccd

*Locations in coordinate system UTM, zone 20, NADIR 27 (Caribbean). [†]See Table 1.
[‡]LCIA, UCIA: lower and upper concordia intercept ages, respectively; ccd: concordant.

characterized by euhedral prismatic crystals with small pyramid facies and variable length to width ratios from 2 : 1 to 5 : 1 (Fig. 9a). Translucent zircons, in part with very small inclusions, show light pink colours. Other crystals with rough surfaces are yellowish-brown in colour due to metamict alteration, or greyish to opaque due to disseminated inclusions of probable graphite. Some zircons show

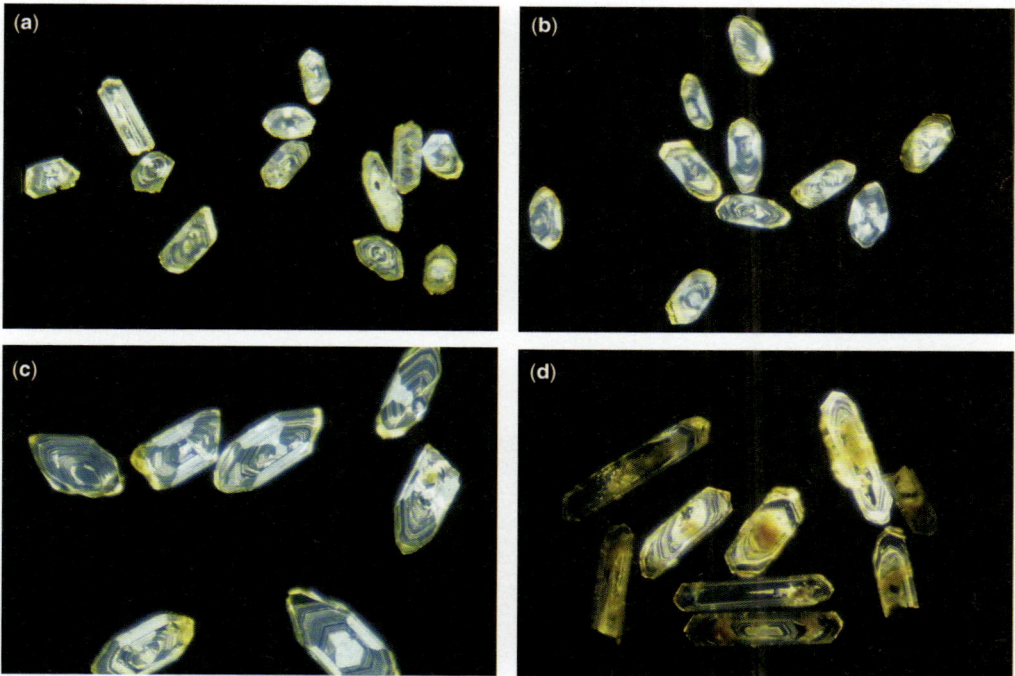

Fig. 9. Cathodoluminescence images of separated zircons. (**a**) Macanao orthogneiss, sample VM 1330, size fractions 62–80 μm. Note distinct oscillatory growth zoning, as well as the hollow channel parallel c-axis in long-prismatic crystal in upper left corner of grain mount. (**b**) Guayacán Gneiss, sample VM 1408, size fraction 62–80 μm. Note the oscillatory growth zoning and the distinctly stronger luminescence of prism faces as compared to pyramid faces. (**c**) Matasiete Metatrondhjemite, sample VM 2237, size fraction 100–125. Growth zoning and luminescence as in Guayacán Gneiss zircons. (**d**) El Salado Metagranite, sample VM 2175. Size fraction 90–100 μm. Note oscillatory growth zoning and crystal regions with strong metamictization. For further descriptions see text.

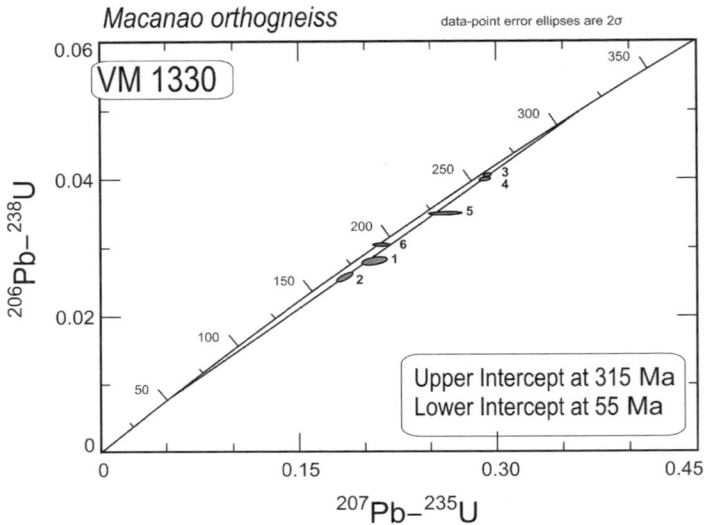

Fig. 10. Concordia diagram for zircons of the Macanao orthogneiss. Six fractions (see text) were analysed.

cracks, or occur as relict fragments with pyramid faces broken off. The cathodoluminescence images show distinct growth zonation patterns, but no relics of older cores can be observed.

The zircons with metamict alterations or graphitic inclusions (fractions 1, 2, 5, and 6 in Fig. 10) have distinctly higher U concentrations (828–1450 ppm) and significantly lower $^{206}Pb/^{204}Pb$ ratios (162–405) than the translucent light pink ones (fractions 3 and 4: U = 517–545 ppm, $^{206}Pb/^{204}Pb$ = 596–1015).

Because of the relatively low $^{206}Pb/^{204}Pb$ ratios, the U–Pb isotope data of zircon fractions 1, 2, 5 and 6 are characterized by relatively high analytical error. Therefore the data points plot on a poorly defined discordia (Fig. 10). The upper concordia

intercept age of 315 Ma indicates crystallization of the zircons in a granitic magma during the Permo-Carboniferous and may be interpreted as the intrusion age of at least the sampled part of the pluton. The lower concordia intercept age of 55 Ma is coincident with the final stages of greenschist-facies overprinting discussed by Stöckhert *et al.* (1995), and can be explained by partial lead loss during greenschist metamorphism and the ductile deformation of the high-pressure metamorphic nucleus of Margarita Island subsequent to high-pressure metamorphism.

To corroborate the Palaeozoic age indicated by these results, a suite of zircon crystals from sample VM 2065, from a locality approximately 250 m west of the above samples, was analysed with the SHRIMP technique (Figs 11 & 12). Ten

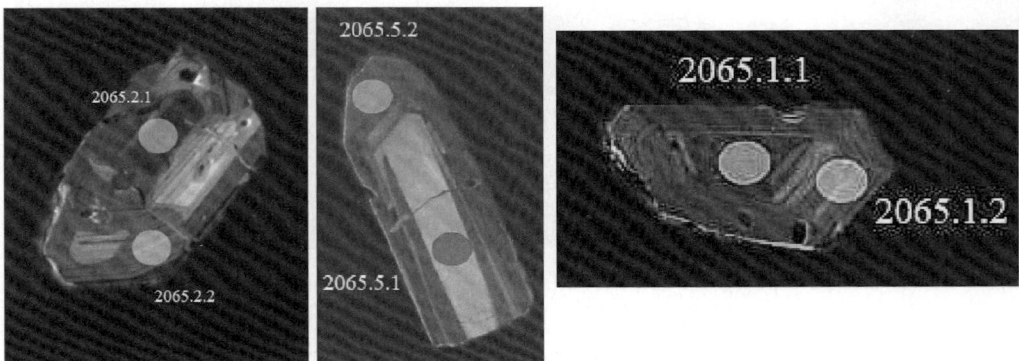

Fig. 11. Example of cathodoluminescence zircon images from sample VM 2065 (Macanao orthogneiss) showing SHRIMP analysis 'shots'. The shot numbers refer to those in Table 2. The diameter of the shots is about 25 μm.

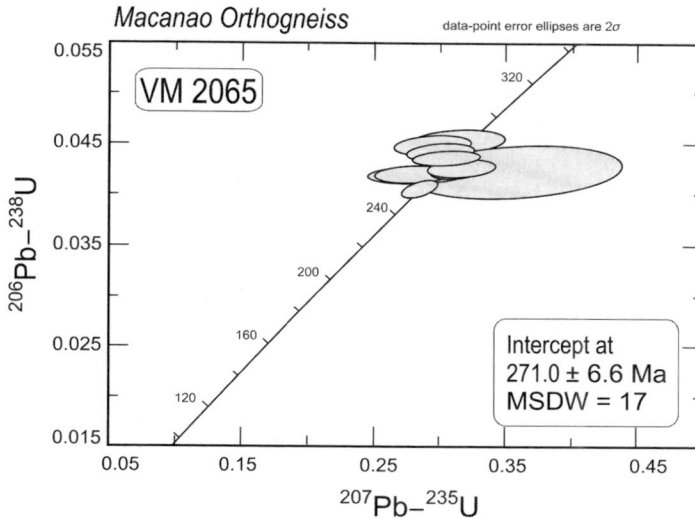

Fig. 12. Concordia diagram for zircons of the Macanao orthogneiss analysed by SHRIMP (sample VM 2065). Isotopic data of the SHRIMP measurements are given in Table 2. For further discussion see text.

'shots' in seven crystals yielded a relatively consistent picture. The radiogenic Pb content of these crystals was low, but the corresponding Tera–Wasserburg plot (not shown) for the uncorrected data yielded a well-defined regression line with a consistent intercept age at 271.0 ± 6.6 Ma (MSWD = 17). No significant core–rim variations are evident in this sample. This method thus yields somewhat younger ages than the multi-grain results reported above. Although the difference is difficult to interpret at this stage (methodology? inhomogeneous magmatic rock suite?), and further work is warranted, there seems little doubt that the augengneisses of the Juan Griego unit represent Palaeozoic continental crust.

Guayacán Gneiss. Four zircon fractions each of the two samples VM 1408 and VM 1409, from two separate bodies approximately 5 km apart, were analysed. The zircon populations in both samples are uniform in their appearance (Fig. 9b), and show strong similarities with the zircons of the Matasiete Metatrondhjemite (see below). The crystals are clear, short-prismatic with rounded pyramid tips, and in some cases show irregular, rough surfaces. Cathodoluminescence patterns exhibit a distinct oscillatory growth zonation; no inherited cores can be observed.

U and Pb concentrations of the zircon fractions 62–125 μm of sample VM 1408 (fractions 1–3 in Fig. 13) are restricted to narrow ranges of 420–430 and 6.9–7.0 ppm, respectively, in contrast to fraction 125–160 μm (no. 4 in Fig. 13) with 250 ppm U and 4.2 ppm Pb. However, U–Pb

ratios are similar in all fractions (60–62). The data points of the zircon fractions 1–3 are concordant within error limits at 106 Ma (Fig. 13), whereas fraction 4 plots slightly above the concordia at 109 Ma. This may be an analytical artefact. Another possibility is that weak lead loss of fractions 1–3 could have shifted these data points slightly down along the concordia to younger ages.

The U and Pb concentrations of zircons from sample VM 1409 (112–130 and 2.0–2.5 pmm, respectively) are much lower than those from sample VM 1408, but again the U–Pb ratios (53–56) are similar. The data points are concordant within error limits, but reveal a certain spread from 116 to 112 Ma, for which there is no plausible explanation. The position of data point number 8 above the concordia in Figure 13 may be an analytical artefact due to incomplete mixing of spike and sample solution.

Nevertheless, the U–Pb data from zircons of the Guayacán Gneiss samples point to a crystallization age between 116 and 106 Ma, suggesting that intrusion of Guayacán-type magmas was active during this time.

Matasiete Metatrondhjemite. Four zircon fractions from sample VM 2237 (deformed) and two from sample VM 2226 (equigranular) were analysed. The fractions are uniform in both samples. The translucent crystals exhibit short-prismatic shapes (Fig. 9c). Occasionally, metamict zones or rims and some dark inclusions (ore minerals?) can be observed. The surfaces of the zircons appear to be rough. In general, the zircons of the Matasiete

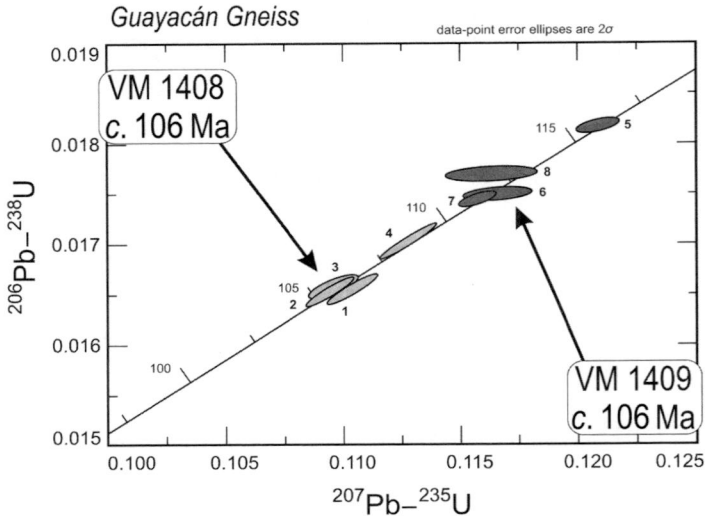

Fig. 13. Concordia diagram for zircons of Guayacán Gneiss. Zircon fractions 1–4 are from sample VM 1408 (Pedro González), and zircon fractions 5–8 from sample VM 1409 (Guayacán). For further discussion see text.

Metatrondhjemite are quite similar to those of the Guayacán Gneiss. Another common feature of both the Matasiete zircons and the zircons of sample VM 1409 of the Guayacán Gneiss are the low U (152–338 ppm) and corresponding Pb (2.8–6.2 ppm) concentrations and similar U–Pb ratios (53–56).

All U–Pb isotope data are concordant at 114 Ma within error limits (Fig. 14). Concordance of zircons is common in intermediate magmatic rocks. The age of 114 Ma is the crystallization age of the zircons, and since the Matasiete Meta-trondhjemite is considered to be a near-surface intrusion (Kluge 1996), it is also the intrusion age. These results show that the magmatism leading to the intrusion of the Guayacán Gneiss and the Matasiete Metatrondhjemite was essentially contemporaneous.

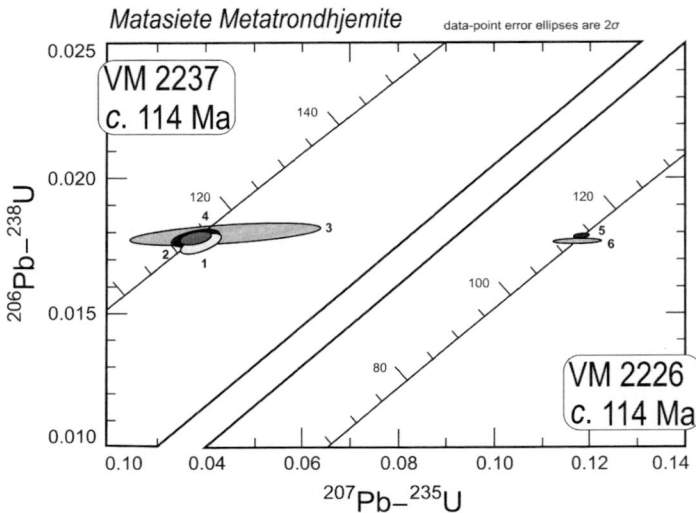

Fig. 14. Concordia diagrams for zircons of Matasiete Metatrondhjemite. Zircon fractions 1–4 are from sample VM 2237 (strongly deformed), and zircon fractions 5–6 are from sample VM 2226 (equigranular).

El Salado Metagranite. Three zircon fractions from sample VM 2175 (magmatic texture) and one from sample 2166 (augengneiss) were separated for analysis. All zircon populations are the same in shape and colour (Fig. 9d). Pyramids are developed at both ends of the crystals, and the length to width ratios are from 3:1 to 5:1. The zircons are yellowish-brown due to metamict alterations. The surfaces of the crystals show leaching. Very fine cracks occur on the prism faces. Cathodoluminescence images show growth zonation, and inherited cores cannot be detected.

The very high U concentration in these zircons (1984–3092 ppm) is probably the cause for the metamict alteration. The high concentration of common Pb is documented in very low $^{206}Pb/^{204}Pb$ ratios, and by the fact that only about half of the total Pb concentration in the zircons is radiogenic (Table 1). This may be explained by the metamictization, which caused damage in the crystal lattice, providing pathways for fluids so that common Pb could penetrate the zircons.

All zircon fractions are concordant within error limits at 85–86 Ma (Fig. 15). Because of the high amount of common Pb, this age bears some uncertainty. It is, however, coincident with a Rb–Sr isochron age obtained for a feldspar-dominated thin-slab section from the El Salado Metagranite. This age of 82.6 ± 6.1 was obtained from sample VM 2164, located mid-way between the two samples analysed above and documents the age of ductile shearing in the metagranite. This datum will be presented and discussed in a forthcoming paper in which the post-El Salado exhumation

history of the high-pressure metamorphic nucleus will be documented in detail by Rb–Sr, K–Ar, Ar–Ar and fission-track methods. The age of 85–86 Ma obtained by U–Pb on zircon is therefore considered here to be the intrusion age of the El Salado granite.

Discussion

Constraints provided by age dating

On the basis of the geochronological results reported here, the field relationships discussed above can now be put into an explicit time frame and the following conclusions drawn:

(1) The Juan Griego unit contains granitic rocks of Palaeozoic age. Further work is needed to ascertain whether other gneissic rocks of the Juan Griego unit are also of Palaeozoic age. These gneissic series are the only rocks that show evidence of more than one cycle of metamorphism.

(2) The age of the mica schists of the Juan Griego unit is as yet not known. They could be Palaeozoic as well, or Mesozoic sediments deposited on the above Palaeozoic 'basement', or later tectonically juxtaposed with these gneisses. In any case, all these rocks were together during subduction and high-pressure metamorphism.

(3) Amalgamation of La Rinconada-type rocks and Juan Griego-type ('composite') rocks occurred *before* or at the *latest* during the interval of Guayacán intrusive activity between 116 and 106 Ma.

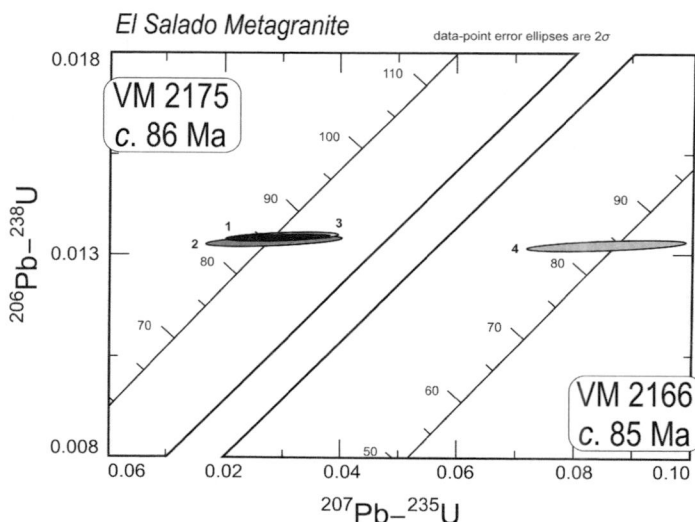

Fig. 15. Concordia diagrams for zircons of the El Salado Metagranite. Zircon fractions 1–3 are from sample VM 2175 (magmatic texture, type locality), and zircon fraction 4 from sample VM 2166 (augengneiss).

(4) An origin of Guayacán-type magmas in a subduction zone setting (see below) implies that
subduction of La Rinconada- and Juan Griego-
type rocks was in progress at 116–106 Ma.

(5) Before or during peak high-pressure metamorphism, the ultramafic rocks were added
to the ensemble.

(6) Subduction of the high-pressure Margaritan
metamorphic nucleus was relatively short
lived, the *P–T* path of Figure 3 suggesting a
'collision-type' exhumation path (Ernst 1988)
for this rather large rock unit. In other words,
exhumation of these rocks was essentially
coeval with the termination of this subduction
zone.

(7) Exhumation was proceeding during the intrusion
of the El Salado magmas at 86–85 Ma, which
'stitched' the rocks of the high-pressure
metamorphic nucleus together. El Salado
Metagranites intruded at levels progressively
shallower than 35 km (see above), and no
high-pressure metamorphic overprint is
observed in these rocks. Pervasive shearing
and recrystallization occurred under greenschist-facies conditions.

(8) Therefore, the high-pressure metamorphic
subduction event during which the gneisses
and schists of the Juan Griego unit, the metabasic series of the La Rinconada unit and the
now associated ultramafic rocks were juxtaposed and metamorphosed together is constrained to the time interval between 116–106
and 85 Ma. By allowing for the kinetics of subduction and exhumation, the metamorphic peak
of the high-pressure nucleus of Margarita Island
can be realistically placed at *c.* 100–90 Ma.

(9) The subsequent complex exhumation history
of the high-pressure Margaritan metamorphic
nucleus has been outlined in several stages
by Stöckhert *et al.* (1995) and will be documented on the basis of structural studies and
available Rb–Sr, K–Ar, Ar–Ar and fission-
track dating in a forthcoming paper. The
main feature of this exhumation path is its discontinuous nature. The high-pressure metamorphic rocks were exhumed rapidly until a
greenschist facies level was reached. They
remained in the intermediate crust, continuously being subjected to penetrative shearing,
for more than 30 Ma. Final rapid uplift
into the brittle upper crust was initiated at
55–50 Ma.

(10) The age of overthrusting (or juxtaposition by
extensional detachment) of the greenschist
periphery is not known, but the regional
ENE–WSW grain of the high-pressure
metamorphic nucleus is also found in the
greenschist periphery (Chevalier 1987, 1993;

Rekowski & Rivas 2005), so that juxtaposition
between *c.* 80 and 50 Ma appears logical. The
K–Ar age of 70–74 Ma obtained on the
greenschist-facies amphiboles in the Matasiete
mylonites (Martín Bellizzia 1968; Santamaria
& Schubert 1974; Guth & Avé Lallemant
1991) may well date the timing of juxtaposition of the Guayamuri–Matasiete over the
El Piache–Los Robles nappe (Fig. 2).

Although the Matasiete Metatrondhjemite and the
Guayacán Gneiss share similar zircons, a similar
age and a similar chemistry, the Matasiete Metatrondhjemite did not share the same high-pressure
P–T–t path (i.e. subduction evolution) as the
Guayacán Gneiss and its country rocks. It was
added to the high-pressure ensemble at a later
time. An unequivocal interpretation of the origin
of these trondhjemitic magmas is not possible at
present, but this open question does not significantly
affect the general conclusions drawn above. If the
magmas were to represent typical MOR plagiogranites, as assumed by Maresch *et al.* (2000), then
their age would date the age of the La Rinconada
metabasic series, and subduction and high-pressure
metamorphism would have to have commenced
after 106 Ma. However, field relations (intrusive
into carbonaceous mica schists) and the now available
geochemical data do argue for an island-arc-related
origin. The limited trace element data (e.g. Y, La,
Yb, Nb, Sr) provided by Giunta *et al.* (2002) and
Ostos & Sisson (2005) are quite compatible with
the trondhjemite magmas described as anatectic products of wet melting of oceanic crust during early subduction in eastern Cuba (Lázaro & García-Casco
2008). Such an origin would also elegantly explain
the enigmatic black garnet and epidote amphibolites
found locally in northern Paraguachoa. On the
other hand, such magmas would then have to be extremely mobile to reach shallow levels and produce an
intrusion such as the Matasiete Metatrondhjemite.

Implications for the regional tectonic development of northern South America

With the chronological and geochemical criteria on
Margarita's litho-units and their metamorphism presented and reviewed herein, we may now constrain
the origin and tectonic history of Margarita as
follows (see Figs 16 & 17). Given that the protolith
of the Macanao orthogneiss was intruded during
Permo-Carboniferous times, it may be inferred
that at least this component of Margarita lay
originally within the zone of Permo-Carboniferous
orogenesis. Unfortunately, although predominantly
buried by Mesozoic strata today, Permo-
Carboniferous granitic and metamorphic (generally
Barrovian, not HP/LT) terrane is understood to

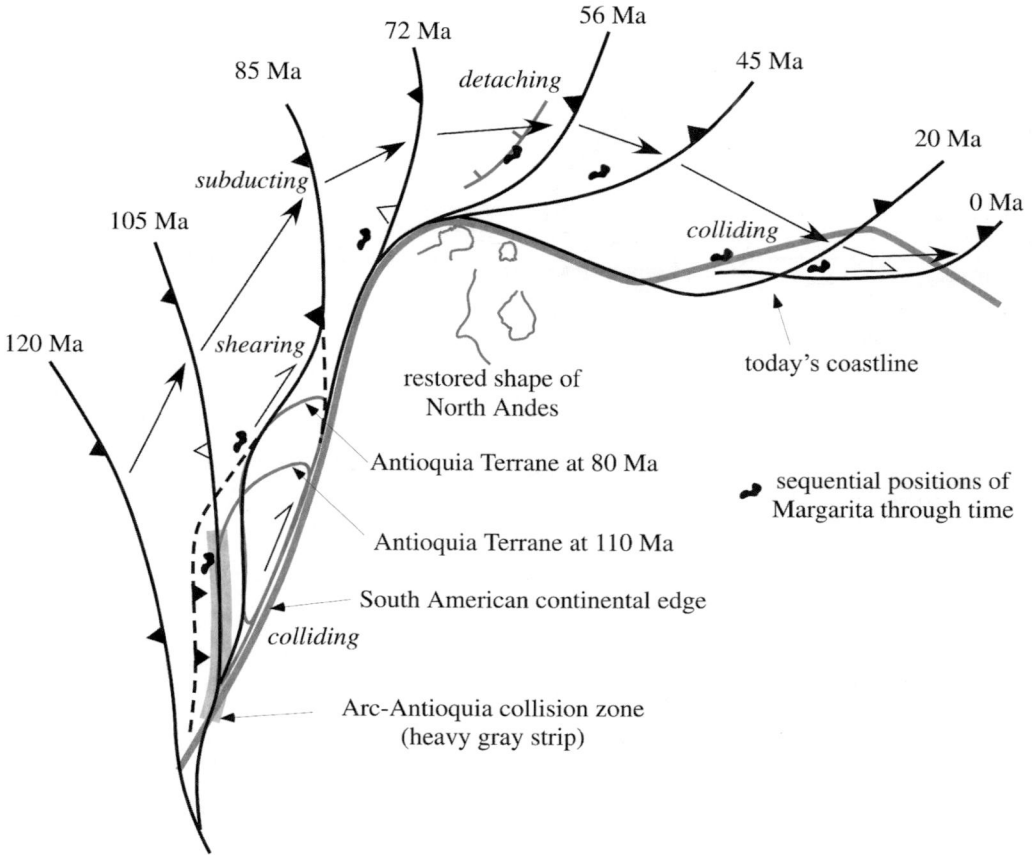

Fig. 16. Tectonic model for Margarita (see also Fig. 17): by 120 Ma, the Caribbean Arc is converging with the western flank of the Antioquia Terrane, which harbours the Juan Griego continental protoliths (continental crust and sediments, including the Macanao orthogneiss). The backarc basin that closes is floored by La Rinconada oceanic crust and sediments. Ultramafic rocks underlie La Rinconada and may also be exposed at rifted continental margin. Anatexis of subducting oceanic crust leads to the production of Guayacán/Matasiete-type trondhjemitic magmas between 116 and 106 Ma. All units undergo HP/LT metamorphism together until collision and subduction culminate at 100–90 Ma. Concurrently, eastward-dipping subduction of Farallon lithosphere beneath the newly accreted arc and Antioquia terrane renews (dashed line) and allows continued relative plate motion. The units of the 'Margarita high-pressure metamorphic nucleus' lie just north of this new subduction zone such that they migrate north-northeastward with the Caribbean Arc along the Antioquia–Caribbean shear zone. Other HP/LT complexes remain attached to western Antioquia. In the transform setting, Margarita undergoes intense dextral shear and progressive exhumation throughout Late Cretaceous. By 86 Ma, the HP/LT complex is intruded by El Salado island arc magmas due to the subduction of Proto-Caribbean crust beneath the leading edge of the Caribbean Plate. The HP/LT rocks continue their exhumation to mid-crustal levels, perhaps by upward wedging of tectonic slivers but more likely by axis-parallel extension during transpression, as a strong stretching lineation occurs on Margarita (Chevalier 1987, 1993; Chevalier *et al.* 1988; Guth & Avé Lallemant 1991; Stöckhert *et al.* 1994, 1995; Rekowski & Rivas 2005). Dextral shearing continues. By 56 Ma, Margarita is rounding the Guajira salient, and asymmetric intra-arc rifting in the Grenada Basin begins (see normal fault), which allows the forearc to maintain progressive transpressional collision with the South American margin, even though the arc does not hug the margin so closely. Margarita lies on the footwall of this low angle detachment, and is extruded from beneath the eastern flank of the Aves Ridge Arc, now extinct, further exhuming the HP/LT rocks and setting zircon fission track clocks between 53 and 50 Ma (Stöckhert *et al.* 1995). Margarita then rides passively on the eastern flank of the Caribbean Plate, just north of Villa de Cura, until collision with central and eastern Venezuela occurs in the Early and Middle Miocene. Apatite fission track ages of about 20–25 Ma (Maresch *et al.* 2000) may record this interaction with South America. Since 10 Ma, Margarita has moved due east with respect to South America.

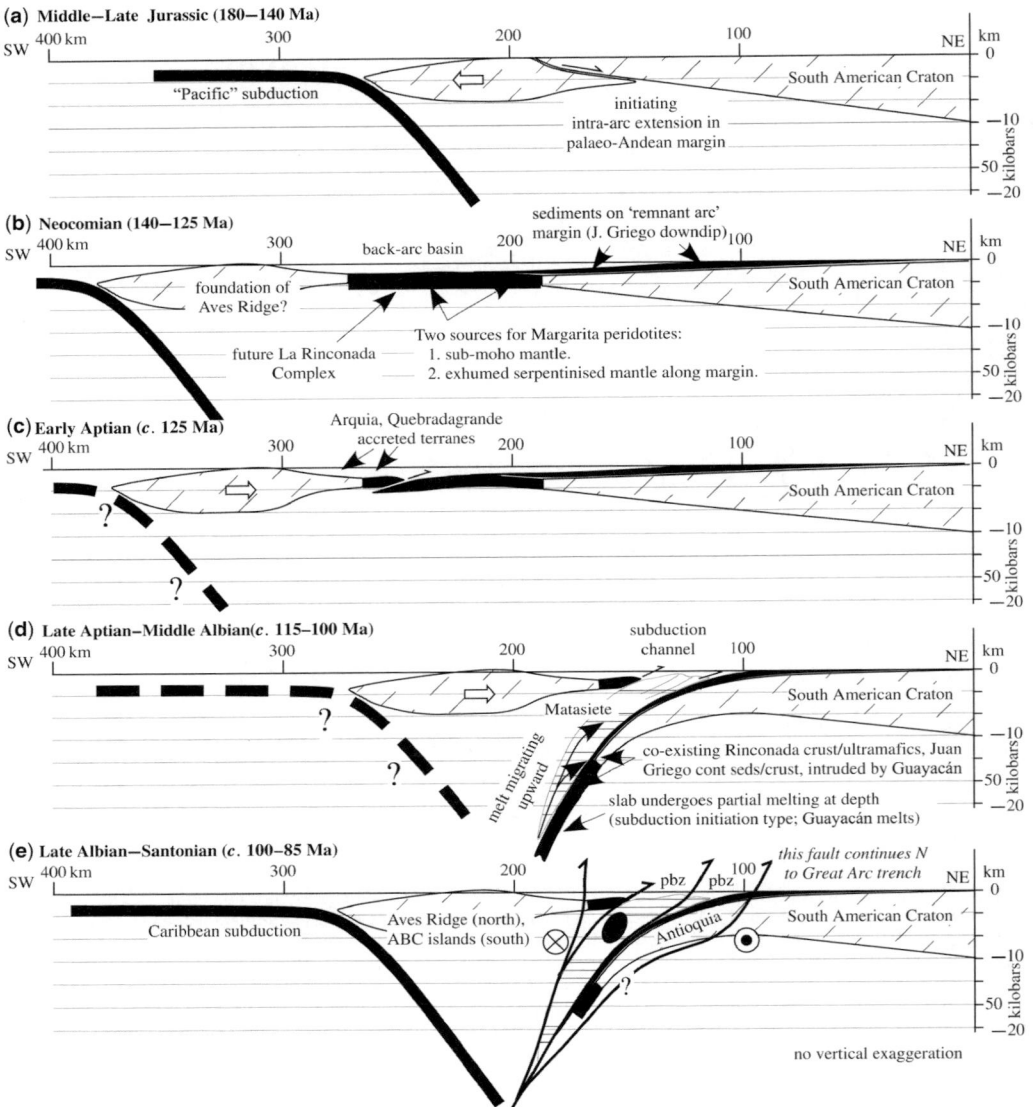

Fig. 17. Cross sections of five stages of Cretaceous development on Margarita Island (see Fig. 16 for palaeogeographic positioning). (**a**, not yet represented on Margarita) Late Jurassic rifting that led to the Andean intra-arc basin, employing a low-angle detachment extensional style. (**b**) Neocomian opening of the Andean backarc basin, and deposition of marginal sediments. (**c**) Aptian initial backarc basin closure, with active arc overthrusting backarc, leading to arc-continent collision. It is uncertain if Pacific subduction still continued at this stage (dashed). (**d**) Albian subduction of backarc basin floor and generation of HP/LT conditions. Slab melts at depth, providing rising magma for Guayacán and Matasiete 'trondhjemites'. Guayacán chilled at depth with intrusive contact with the rest of Margarita's HP nucleus, whereas the Matasiete is chilled much shallower. (**e**) dextral transpressive phase in which this flank of the Caribbean Plate, the offshore arc, and the Antioquia Terrane migrated northward along the Andean flank. A wide Caribbean–North Andean transpressive plate boundary (pbz) developed from the Aves Ridge/Aruba–Blanquilla (ABC) trend arc's forearc to east flank of the Antioquia Terrane. Mélange and adjacent crustal slivers are forced upwards to crustal levels by 85 Ma, where Margarita's HP nucleus (grey oval) was intruded by El Salado arc granites, derived from subduction of Proto-Caribbean crust from the north, or by renewed subduction of Pacific (now Caribbean) crust from the west. After this time, as the Aves Ridge/ABC migrated out of the cross section, Caribbean lithosphere was overridden by South America.

underlie much of the northern South American margin (Feo-Codecido *et al.* 1984), the Yucatán Platform (Lopez-Ramos 1975), and the western Pangaean arc system of the Central Cordillera of Colombia, Ecuador, and northern Peru (McCourt *et al.* 1984), all of which were geographically contiguous in the Permo-Triassic reconstruction of western Pangaea (Rowley & Pindell, 1989). Hence, this potential constraint is not of great value on its own. However, the facts that HP/LT metamorphism of the Macanao orthogneiss and other units of the Margaritan metamorphic nucleus can be constrained to the interval 106–86 Ma (and not partly of Palaeozoic age as speculated by Higgs 2009), and also that arc activity was occurring in Margarita (El Salado Granite) by 86 Ma, steer us toward the interpretation of a Pacific origin for Margarita, which in turn points us to the northern Andes option for the source of the Permo-Carboniferous protolith of the Macanao orthogneiss.

The reasoning for this conclusion is as follows. First, concerning the 86 Ma El Salado arc magmatism, which we expect is a small representation of a larger volume of such magmatism and presumably volcanism along a greater arc, there is no record of coeval tuff deposition in the nearby passive margin section of the Serranía del Interior of Venezuela (e.g. Hedberg 1950; Erikson & Pindell 1998), which today lies within only 50–100 km of Margarita. Spatial separation between the passive margin and the arc/forearc complex, of which Margarita and probably the allochthonous nappes of NW Araya were part, seems required. We note here that we consider all the nappes NW of the Laguna Grande–Punta Los Carneros Fault on Araya (Chevalier 1987) to be allochthonous with respect to the parautochthonous, South American slope sediments, that is the Paria–Northern Terrane, which they are thrust over. Only the allochthonous units show Cretaceous metamorphic cooling ages (Sisson *et al.* 2005). All units to the east are characterized by Oligocene–Miocene metamorphic cooling ages pertaining to the collision between the leading edge of the Caribbean Plate and the South American margin (Foland *et al.* 1992; Pindell *et al.* 2009). This requirement of spatial separation between the Caribbean arcs and the Proto-Caribbean passive margins is a Caribbean-wide phenomenon including the Bahamas and the Yucatán Block (Pindell 1990, 1993). Second, concerning the Mid-Cretaceous HP/LT metamorphism, the relative motion history between North and South America remained divergent until the Maastrichtian (Pindell *et al.* 1998; Müller *et al.* 1999), with little chance of HP/LT metamorphism occurring within the inter-American region in Cretaceous times. If a subduction zone were to have formed within the inter-American region at that time, then it would need to have co-existed with a seafloor spreading ridge where spreading outpaced subduction. In such a small area, this seems highly unlikely, and again is countered by the lack of arc-related tuffs around the Proto-Caribbean passive margins, not to mention the lack of tectonic deformation (apart from gravity slumping) in those margins as well. Further, HP/LT metamorphic rocks of medial Cretaceous age also occur along the northern Caribbean; thus, an *ad hoc* scenario of a co-existing ridge and southerly Caribbean trench would need to be mirrored in the northern Caribbean as well, requiring the co-existence of two subduction zones with a spreading ridge between them. We flatly do not accept such a complicated scenario in such a small area as the Proto-Caribbean Basin, especially in the light of the passive character and lack of arc-related tuffs of the adjacent Proto-Caribbean margins. Thus, of the options available, the only sound interpretation is that the Permo-Carboniferous protolith of the Macanao orthogneiss and the development of the HP/LT metamorphism and island arc activity derive from farther west than the Proto-Caribbean Basin of Pindell (1985), along the Pacific flank of the northern Andes (Ecuador–Colombia).

By all accounts, for example the seismic and crustal velocity work of Clark *et al.* (2008), Margarita now lies within the Caribbean Plate, well within and associated with the regional terrane including Tobago and the Carupano Platform that is considered to be an integral part of the Caribbean Plate back to Early Cretaceous times (Speed & Smith-Horowitz 1998; Snoke *et al.* 2001). If we accept that the origin of the Macanao protolith was along the western flank of the northern Andes, and that the HP/LT metamorphism (106–86 Ma) and island arc activity (86 Ma) on Margarita (as well as in the onshore Villa de Cura Group; Smith *et al.* 1999) occurred to the west of Colombia, then it is safe to accept both that Margarita's geology must fit within Caribbean history, as well as help to define Caribbean history, back to Early Cretaceous times. By integrating the field relations, metamorphic petrology and *P–T–t* path data of the various litho-units on Margarita Island outlined herein with the updated Pacific origin model for Caribbean development by Pindell & Kennan (2009), the following history for the rocks on Margarita, as well as for much of the southeastern Caribbean Plate, is proposed.

In the Neocomian–Barremian interval (141–125 Ma), intra-arc rifting occurred within the west-facing Andean arc system from southwestern Colombia to northern Peru. This development was probably driven by rollback of the subducting oceanic plate (Farallon?), given that northern South America was not yet moving west in the mantle reference frame and hence that the arc could have been 'extensional' in the sense of Dewey (1980). We consider that the oceanic floor of the back-arc basin was the site of seafloor spreading responsible for the La Rinconada basaltic protoliths and much of Margarita's serpentinites and other ultramafic rocks. The latter could be the normal sub-Moho peridotites subjacent to the La Rinconada basic crust or perhaps continental mantle exhumed during early rifting in the backarc basin (e.g. Whitmarsh *et al.* 2001). Because this was an intra-arc basin, parts or all of these basalts as well as the ultramafic rocks could carry a supra-subduction, arc-like geochemical character. The intra-arc basin separated an offshore active arc from the western South American margin of the intra-arc basin (remnant arc). Because the rifting had developed within the hanging wall of the Andean continental arc, which in this area had been affected by Permo-Carboniferous orogenesis (Ramos 1988; Dalmayrac *et al.* 1980), the offshore active arc probably possessed fault blocks, rafts or large swaths of Permo-Carboniferous crust within its core. Given this setting, two possible origins for the Permo-Carboniferous protolith of the Juan Griego continental crust on Margarita (Macanao orthogneiss) are: (1) the roots of this offshore active arc; and (2) the rifted eastern flank of the intra-arc basin (i.e. the remnant arc). The protoliths of the metasediments on Margarita that form both an integral part of the Juan Griego unit and are also intimately associated with the La Rinconada unit may simply be clastic input into the back-arc basin. Given that this basin will be narrow (300 km?) and probably restricted in terms of circulation, the deposition of large amounts of carbonaceous clastic strata and limestone is reasonable. A simple and intimate depositional relationship of such sediments with the other units might be expected on the rifted margins and centre of the intra-arc basin. For reasons discussed later, we prefer that the Juan Griego represents the eastern flank of the basin (remnant arc).

By 120–125 Ma (Early Aptian, just after marine magnetic anomaly M0), Equatorial South America began to rift from Africa (Pindell & Dewey 1982) and northern South America began to migrate west in the mantle reference frame. This would have caused the arc system along the western northern Andes to become convergent in the sense of Dewey (1980). Intra-arc spreading presumably ceased as the intra-arc basin began to close (Pindell 1993). A concurrent northwestward acceleration of North America can be argued from Atlantic spreading data as well (Klitgord & Schouten 1986; Pindell *et al.* 1988). This hemispheric scale westward acceleration of the Americas (i.e. both Atlantic basins) may have been a primary cause for the onset of SW-dipping subduction beneath

the Caribbean Arc spanning the greater Colombian Marginal Seaway between Colombia and Central America at that time, as well as of closure of the Andean intra-arc basin noted above (Bourgois *et al.* 1982; Aspden & McCourt 1986; Pindell 1993; Pindell & Tabbutt 1995). Another possible cause was a predicted change in azimuth of the Farallon Plate with respect to North America from southeast to northeast at about the same time (Engebretson *et al.* 1985; Pindell 1993), but more recent evaluations of Pacific–North America kinematics no longer necessarily support such a change (Steinberger 2000; Tarduno & Gee 1995; Pindell & Kennan 2009). Whatever the cause, with the onset of SW-dipping subduction beneath the Caribbean Arc, the western South American margin (remnant arc margin) entered the newly established Caribbean subduction zone as the former drifted west, and the intra-arc basin was closed. If the basin had been 300 km or wider, then, as discussed by Maresch & Gerya (2005), a true phase of west-dipping subduction would have been established prior to this collision, with the corresponding establishment of an isothermal structure conducive to HP/LT metamorphism. The combination of young oceanic crust subducting behind a recently active arc could lead to a relatively 'warm' subduction zone in which blueschist conditions are skirted and barroisite-bearing amphibolites and eclogites predominate.

Three reasons suggest that the crustal sliver containing the Macanao orthogneiss (the unequivocal Permo-Carboniferous crustal component of the Juan Griego unit) represents the rifted flank of the remnant arc margin that choked the west-dipping subduction zone upon backarc closure, as opposed to the roots of the active arc. First, existing data on the Juan Griego high-pressure metasediments do not appear to show affinity to volcanogenic protoliths, which would be expected had the Macanao sliver represented the active arc flank of the basin. However, more work is needed on this issue to resolve this better. Second, the *P–T* path data (Fig. 3) are most consistent with subduction and HP/LT metamorphism being arrested by collision (for instance by continental choking), rather than tracking the *onset* of subduction of continental crust, which might have been the case had the Macanao sliver originated in the active arc. Third, there is no record of island-arc magmatism on Margarita at the time of intra-arc basin closure (see next paragraph), which might be expected if the Macanao sliver had represented the active arc. What we seem to have on Margarita is an intimate association of ?thinned continental basement (Macanao orthogneiss slivers), adjacent MOR-type basalts with possible ?backarc geochemical affinity and associated ultramafic rocks (La Rinconada-type rocks), as well as clastic/carbonate type sedimentation (Juan Griego-type rocks). Such an association could well have been situated at the thinned continent-ocean boundary (Maresch 1974) of the eastern margin of the intra-arc basin, a setting well known to cause choking of subduction zones upon backarc basin closure (Dewey & Bird 1970). Concerning which part of the Andean margin was involved in this backarc closure, we follow the Caribbean model of Pindell & Kennan (2009) and Kennan & Pindell (2009) and accept that the origin of the Macanao Permo-Carboniferous protolith sliver was the western flank of the Antioquia Terrane, in which Permo-Carboniferous intrusive and metamorphic rocks are known (González 1980; Restrepo et al. 1991), before the Antioquia Terrane was displaced transpressively northwards by several hundred kilometres to its present position with respect to interior South America.

In an attempt to refine our picture of backarc closure further, in addition to the Macanao orthogneiss, the La Rinconada metabasalts, the Juan Griego metasediments, and the ultramafic rocks, the Guayacán Gneiss (crystallization at 116–106 Ma) is the fifth unit to be intimately interrelated in the high-pressure metamorphic nucleus on Margarita. These rocks show primary intrusive contacts with Juan Griego-type metasediments and La Rinconada-type metabasalts. For reasons outlined earlier, there is much to be said for their origin as anatectic melts of subducted hydrous metagabbros/metabasalts, possibly equivalents of the La Rinconada unit. Many features described by García-Casco (2007), García-Casco et al. (2008) and Lázaro & García-Casco (2008) for the migmatites and amphibolite restites in eastern Cuba are observed in the La Rinconada type rocks. However, the temperatures required to melt the metabasalts are higher than those experienced by the Juan Griego and La Rinconada units; thus, we must assume that the Guayacán melts originated from subducted metabasalts deeper in the subduction zone system, and that these melts migrated essentially unmodified up the subduction zone system to a position where they could also intrude subducted sediments undergoing metamorphism. As discussed at length earlier, the metabasic and metasedimentary host rocks, the trondhjemitic intrusives, the Macanao orthogneiss and probably the associated ultramafic rocks of the Margaritan high-pressure nucleus show similar HP/LT overprints and analogous $P-T-t$ paths. Indeed, if the presumably shallow-level Matasiete Metatrondhjemite intrusion is of similar origin as the Guayacán Gneiss, then considerable mobility of these trondhjemitic melts must be invoked.

The next phase in the history of Margarita achieved three developments (Figs 16 & 17): (1) the composite HP/LT assemblage began the process of rapid exhumation by 85 Ma, either by counterflow in the subduction channel or by other exhumation mechanisms of crustal zones along the plate edge; (2) all the rocks of Margarita were geographically situated in a position where they could continue to migrate northwards into the Caribbean region rather than stay behind with other HP/LT complexes such as the Pijao Complex (eclogites and glaucophane schists) on the SW flank of the Antioquia Terrane (Núñez & Morillo 1978), which may be correlative; and (3) a strong high-angle dextral shear was imparted in all the units of the HP/LT complex (Chevalier 1987, 1993; Chevalier et al. 1988; Guth & Avé Lallemant 1991; Stöckhert et al. 1994, 1995; Rekowski & Rivas 2005). We suggest that all three developments were effected by 'the Margarita high-pressure metamorphic nucleus assemblage' migrating transpressionally along and past the northern end of the Antioquia Terrane with which collision had occurred, remaining primarily on the NW flank of the plate boundary such that it continued to move more or less with the Caribbean lithosphere. A possible compressive component of this shear may have driven wedges of forearc rock upwards where pressures were less; alternatively, the axis-parallel extension exhumation mechanism of Avé Lallemant (1997) may have operated along this western Andean margin to effect a significant part of the exhumation.

By the time the El Salado Granite was intruded at 85 Ma (Santonian), the HP̂/LT complex had reached crustal levels (i.e. <10 kbar), and thus the El Salado does not show a high-pressure overprint. However, the dextral shearing continued beyond the time of its emplacement. Because the Margarita rock assemblage was situated on the NW flank of the Caribbean–Andean shear zone at this time, it was in a position, like the entire Caribbean Arc, to receive arc magmas from the subduction of Proto-Caribbean crust. The El Salado Metagranite has clear arc affinity.

Juxtaposition of the high-pressure metamorphic nucleus with the greenschist-facies periphery (i.e. El Piache–Los Robles nappe, Matasiete–Guayamuri nappe) is conceivable at any time after greenschist-facies levels had been reached by the metamorphic nucleus. As discussed earlier, the age of the greenschist-facies mylonitization (74–70 Ma) of the Matasiete Metatrondhemite may date overthrusting of the Matasiete–Guayamuri nappe over the El Piache–Los Robles nappe. Two features are of importance. The Matasiete–Guayamuri nappe contains rocks similar to the La Rinconada unit (metatrondhjemite, spinel peridotites). These could well have originated in a similar setting, without being subducted, however. The El Piache–Los Robles nappes, although in part composed of sedimentary protolith similar to the Juan Griego unit, on the other hand contain a variety of metavolcanic – especially metapyroclastic rocks – from basic to intermediate chemistry. We can only speculate that these may pertain to a volcanogenic portion of the original intra-arc basin fill.

By 55 Ma, Margarita was rounding the Guajíra salient (Fig. 16), and asymmetric intra-arc rifting in the Grenada Basin began to separate the Caribbean forearc, comprising the Tobago, Margarita and Villa de Cura components, from the Caribbean magmatic arc axis, comprising the Aves Ridge and the Leeward Antilles islands. The Grenada and Bonaire basins were sites of this oblique dextral extension, which allowed the forearc terranes to converge with the South American margin more than the magmatic arc did. Pindell *et al.* (2005) and Pindell & Kennan (2009) interpret the Margarita–Testigos Ridge to have formed the footwall of a large-throw, low-angle detachment during this stage, such that the ridge was extruded from beneath the eastern flank of the Aves Ridge. The effect of this low angle detachment was a rapid tectonic unroofing, albeit in a submarine setting where Late Eocene turbiditic sedimentation ensued (Punta Mosquito Fm and other flysches). We believe this unroofing led to cooling as dated by the numerous zircon fission track ages of 50–55 Ma in our data set (Stöckhert *et al.* 1995). Likewise, the Tobago Terrane is in turn envisioned by Pindell & Kennan (2009) as having been extruded from beneath the Margarita–Testigos Ridge. Both 'extrusion zones' are today highly inverted, but the geometry of Eocene sediment within the Blanquilla and Caracolita sub-basins (Ysaccis 1997) do appear to fit an Eocene low-angle extrusion model. Rifting in this intra-arc basin was probably completed by the late Middle or Late Eocene (Pindell & Kennan 2009), after which time Margarita and the other forearc terranes, situated on the southeastern flank of the Caribbean Plate, converged with dextral obliquity upon the central and eastern Venezuelan margin. The collision culminated in the Middle Miocene (Pindell *et al.* 1998). Apatite fission track ages of about 20–25 Ma on Margarita (Maresch *et al.* 2000) may record initial interaction with the South American margin, which underthrusted and uplifted the Caribbean forearc as the collision progressed, leading to deep erosion and probable removal of several kilometres of Eocene–Oligocene turbidites. Since the 10 Ma change in Caribbean–South American azimuth of motion (Algar & Pindell 1993; Pindell *et al.* 1998), Margarita has moved due east with respect to South America as part of the Caribbean Plate, and continues so today (Perez *et al.* 2001).

Stepping back now, there are a number of other continental rock assemblages, in addition to the Macanao orthogneiss, within the mobile Caribbean suture belt along northern South America. These include the Dragon Gneiss of the Paria Peninsula with an intrusive protolith age of 321 $+13/-29$ Ma (Speed *et al.* 1997), the multi-component Tinaco–Caucagua terrane of the Caribbean Mountains (Bellizzia 1985; Bellizzia & Dengo 1990; Sisson *et al.* 2005), some 'Grenvillian' and Lower Palaeozoic occurrences in the Cordillera de la Costa Belt of the Caribbean Mountains (Sisson *et al.* 2005), and several granulitic and marble-bearing blocks of the Lara region of western Venezuela that are inferred to be Grenvillian in age (Grande & Urbani 2009). It is beyond the scope of this paper to conceive of specific origins for these units, other than to say that all of them potentially are allochthonous along with Margarita and the Caribbean Plate, and derive from west of northern South America, either representing the roots of the active arc outside the Andean intra-arc basin, or the western margin of South America, where Grenville-aged rocks are common. The protolith age of the Dragon Gneiss, in particular, appears to be close to that of the protolith of the Macanao orthogneiss, and the two lie fairly near one another today. If a more definite correlation can be made for these two units with further work, then new constraints may be placed on the primary architecture of the Eastern Venezuelan thrust belt; namely, that the Dragon Gneiss might represent an erosional klippe of a once-larger Margarita (Caribbean) hanging wall thrust (eroded since the Miocene) that was obducted onto the former passive margin sedimentary section of the Paria Terrane in the Oligocene, thereby driving the Mid-Cenozoic greenschist metamorphism of the latter, as well as shortening the Serranía del Interior Oriental to the south.

We thank B. Stöckhert for collaboration. S. Sergeyev and N. Rodionov (St Petersburg) supported us with the SHRIMP measurements. L. Kennan and G. Draper provided excellent reviews that greatly improved presentation, highlighted gaps in the flow of logic, and painstakingly pin-pointed the niceties of journal style. Y. Chevalier provided an aggressive defence of various alternative views on the geology of Margarita Island that we hope we have answered. We look forward to future discussions. We are also grateful to K. James and M. A. Lorente for organizing the June 2006 Sigüenza Caribbean meeting, which prompted us to summarize and review existing data on Margarita Island in view of contrasting ideas on the origin of the Caribbean area. This is a contribution to IGCP-546 'Subduction zones of the Caribbean'. The German Science Foundation provided generous support at various stages in this research. J. Pindell's contribution to this paper was supported by NSF BOLIVAR Program Grant EAR-0003572 to Rice University.

References

ALGAR, S. T. & PINDELL, J. L. 1993. Structure and deformation history of the northern range of Trinidad and adjacent areas. *Tectonics*, **12**, 814–829.

ASPDEN, J. A. & MCCOURT, W. J. 1986. Mesozoic oceanic terrane in the Central Andes of Colombia. *Geology*, **14**, 415–418.

AVÉ LALLEMANT, H. G. 1997. Transpression, displacement partitioning, and exhumation in the eastern Caribbean/South American plate boundary zone. *Tectonics*, **16**, 272–289.

AVÉ LALLEMANT, H. G. & SISSON, V. B. 2005. Prologue. *In*: AVÉ LALLEMANT, H. G. & SISSON, V. B. (eds) *Caribbean–South American Plate Interactions, Venezuela*. Geological Society of America, Boulder, CO, Special Papers, **394**, 1–5.

BEETS, D. J., MARESCH, W. V. *ET AL.* 1984. Magmatic rock series and high pressure metamorphism as constraints on the tectonic history of the southern Caribbean, *In*: BONINI, W. E., HARGRAVES, R. B. & SHAGAM, R. (eds) *The Caribbean–South American Plate Boundary and Regional Tectonics*. Geological Society of America, Boulder, CO, Memoirs, **162**, 95–130.

BELLIZZIA, A. 1985. Sistema montañoso del Caribe – una cordillera alóctona en la parte norte de América del Sur. *Sociedad Venezolana de Geólogos, VI Congreso Geológico Venezolano*, Memorias, **10**, 6657–6836.

BELLIZZIA, A. & DENGO, G. 1990. Caribbean mountain system, northern South America. *In*: DENGO, G. & CASE, J. E. (eds) *The Caribbean Region*. Decade of North American Geology, **H**. Geological Society of America, Boulder, CO, 167–176.

BLACK, L. P., KAMO, S. L. *ET AL.* 2003. TEMORA 1: a new zircon standard for U–Pb geochronology. *Chemical Geology*, **200**, 155–170.

BLACKBURN, W. H. & NAVARRO, E. 1977. Garnet zoning and polymetamorphism in the eclogitic rocks of Isla de Margarita, Venezuela. *Canadian Mineralogist*, **15**, 257–266.

BOCCHIO, R., DE CAPITANI, L., LIBORIO, G., MARESCH, W. V. & MOTTANA, A. 1990. The eclogite-bearing series of Isla Margarita, Venezuela: geochemistry of metabasic lithologies in the La Rinconada and Juan Griego Groups. *Lithos*, **25**, 55–69.

BOCCHIO, R., DE CAPITANI, L., LIBORIO, G., MARESCH, W. V. & MOTTANA, A. 1996. Equilibration conditions of eclogite lenses from Isla Margarita, Venezuela: implications for the tectonic evolution of the metasedimentary Juan Griego Group. *Lithos*, **37**, 39–59.

BOURGOIS, J., CALLE, B., TOURNON, J. & TOUSSAINT, J. F. 1982. The ophiolitic Andes megastructures on the Buga–Buenaventura transverse (Western Cordillera–Valle Colombia). *Tectonophysics*, **82**, 207–229.

BUCHER, K. & FREY, M. 2002. *Petrogenesis of Metamorphic Rocks*. Springer, Berlin.

CHACHATI, B. & MACSOTAY, O. 1985. Estudio geodinámico y geoquímico de rocas meta-acidas de Paraguaychoa, Venezuela Nororiental. *VI Congreso Geológico Venezolano*, **III**, Tema II, 1586–1622.

CHEVALIER, Y. 1987. *Les zones internes de la chaîne sud-caraibe sur la transect: Ile de Margarita–Pénsula d'Araya (Venezuela)*. PhD thesis, Université de Bretagne Occidentale, Brest.

CHEVALIER, Y. 1993. A cross section from the oil-rich Maturin sub-basin to Margarita Island. *Código Geológico de Venezuela*. World Wide Web Address: http://www.pdvsa.com/lexico//excursio/exc-93.htm.

CHEVALIER, Y., STEPHAN, J.-F. *ET AL.* 1988. Obduction et collision pré-Tertiaire dans les zones internes de la Chaîne Caraïbe vénézuélienne, sur le transect Ile de Margarita–Péninsule de Araya. *Comptes rendues de l' Academie des Sciences Paris*, **307**, 1925–1932.

CLARK, S. A., ZELT, C. A., MAGNANI, M. B. & LEVANDER, A. 2008. Characterizing the Caribbean–South American plate boundary at 64°W using wide-angle seismic data. *Journal of Geophysical Research*, **113**, B07401.

DALMAYRAC, B., LAUBACHER, G. & MAROCCO, R. 1980. Caractères généraux de l'évolution géologique des Andes péruviennes. *Travaux et documents de l'ORSTOM* **124**.

DEWEY, J. F. 1980. Episodicity, sequence, style at convergent plate boundaries. *In*: STRANGWAY, D. W. (ed.) *The Continental Crust and its Mineral Deposits*. Geological Association of Canada, Waterloo, Special Papers, **20**, 553–573.

DEWEY, J. F. & BIRD, J. M. 1970. Mountain belts and the new global tectonics. *Journal of Geophysical Research*, **75**, 2625–2647.

ENGEBRETSON, D. C., COX, A. & GORDON, R. G. 1985. *Relative Motions Between Oceanic and Continental Plates in the Pacific Basin*. Geological Society of America, Boulder, CO, Special Papers, **206**.

ERIKSON, J. P. & PINDELL, J. L. 1998. Cretaceous through Eocene sedimentation and paleogeography of a passive margin in northeastern Venezuela. *In*: PINDELL, J. L. & DRAKE, C. (eds) *Paleogeographic Evolution and Non-glacial Eustacy: North America*. SEPM Special Publications. Society for Sedimentary Geology, Tulsa, OK, 217–259.

ERNST, W. G. 1979. Coexisting sodic and calcic amphiboles from high-pressure metamorphic belts and the stability of barroisitic amphibole. *Mineralogical Magazine*, **43**, 269–278.

ERNST, W. G. 1988. Tectonic history of subduction zones inferred from retrograde blueschist $P-T$ paths. *Geology*, **16**, 1081–1084.

ERNST, W. G. & LIU, J. 1998. Experimental phase-equilibrium study of Al- and Ti-contents of calcic amphibole in MORB, a semiquantitative thermobarometer. *American Mineralogist*, **83**, 952–969.

FEO-CODECIDO, G., SMITH, F. D. J., ABOUD, N. & DI GIACOMO, E. 1984. Basement and Paleozoic rocks of the Venezuelan Llanos basins. *In*: HARGRAVES, R. B., SHAGAM, R. & BONINI, W. E. (eds) *The Caribbean–South American Plate Boundary and Regional Tectonics*. Geological Society of America, Boulder, CO, Memoirs, **264**, 175–187.

FOLAND, K. A., SPEED, R. & WEBER, J. 1992. Geochronologic studies of the hinterland of the Caribbean mountains orogen of Venezuela and Trinidad. *Geological Society of America Abstract with Programs*, **24**, 149.

GARCÍA-CASCO, A. 2007. Magmatic paragonite in trondhjemites from the Sierra del Convento mélange, Cuba. *American Mineralogist*, **92**, 1232–1237.

738   W. V. MARESCH *ET AL.*

GARCÍA-CASCO, A., LÁZARO, C. *ET AL*. 2008. Partial melting and counterclockwise *P–T* path of subducted oceanic crust (Sierra del Concento Mélnage, Cuba). *Journal of Petrology*, **49**, 129–161.

GIUNTA, G., BECCALUVA, L., COLTORTI, M., SIENA, F. & VACCARO, C. 2002. The southern margin of the Caribbean Plate in Venezuela: tectono-magmatic setting of the ophiolitic units and kinematic evolution. *Lithos*, **63**, 19–40.

GONZÁLEZ, H. 1980. Geología de las planchas 167 (Sonsón) y 187 (Salamina). *Boletín Geológico INGEO-MINAS*, **23**, 1–174.

GONZÁLEZ DE JUANA, C. 1968. *Guía de la excursion geológica a la parte oriental de la Isla de Margarita*. Asociación Venezolana de Geología, Mineralogía y Petrología, Caracas.

GONZÁLEZ DE JUANA, C. & VIGNALI, M. 1972. Rocas metmamórficas e ígneas en la Peninsula de Macanao, Margarita, Venezuela. *Transactions Caribbean Geological Conference*, **VI**, Margarita, Venezuela, 1971, 63–68.

GRANDE, S. & URBANI, F. 2009. Presence of high-grade rocks in NW Venezuela of possible Grenvillian affinity. *In*: JAMES, K. H., LORENTE, M. A. & PINDELL, J. L. (eds) *The Origin and Evolution of the Caribbean Plate*. Geological Society, London, Special Publications, **328**, 533–548.

GUIDOTTI, C. V., SASSI, F. P., BLENCOE, J. G. & SELVERSTONE, J. 1994. The paragonite–muscovite solvus: I. *P–T–X* limits derived from the Na–K compositions of natural, quasibinary paragonite–muscovite pairs. *Geochimica Cosmochimica Acta*, **58**, 2269–2275.

GUILLET, P. & CANNAT, M. 1984. Cinematique de mise en place de l'unité ultrabasique du Cerro Matasiete, Ile de Margarita, Venezuela. *Comptes rendues de la Academie des Sciences Paris*, **299**, Série II, 133–138.

GUTH, L. R. & AVÉ LALLEMANT, H. G. 1991. A kinematic history for eastern Margarita Island, Venezuela. *In*: LARUE, D. K. & DRAPER, G. (eds) *Transactions 12th Caribbean Geological Conference, 1989, St Croix*. Miami Geological Society, 472–480.

HEDBERG, H. D. 1950. Geology of the Eastern Venezuela Basin (Anzoategui–Monagas–Sucre–Eastern Guarico portion). *Bulletin of the Geological Society of America*, **61**, 1173–1216.

HESS, H. H. & MAXWELL, J. C. 1949. Geological reconnaissance of the Island of Margarita, Pt. I. *Geological Society of America Bulletin*, **60**, 1857–1868.

HIGGS, R. 2009. Caribbean–South America oblique collision model revised. *In*: JAMES, K. H., LORENTE, M. A. & PINDELL, J. L. (eds) *The Origin and Evolution of the Caribbean Plate*. Geological Society, London, Special Publications, **328**, 613–657.

HOLLAND, T. J. B. 1980. The reaction albite = jadeite+ quartz determined experimentally in the range 600–1200 °C. *American Mineralogist*, **65**, 129–134.

JAM, L. P. & MÉNDEZ AROCHA, M. 1962. Geología de las Islas Margarita, Coche, y Cubagua. *Memórias Sociedad Ciencias Natural de La Salle*, **22**, 51–93.

JAMES, K. H. 2006. Arguments for and against the Pacific origin of the Caribbean Plate: discussion, finding for an inter-American origin. *Geológica Acta*, **4**, 279–302.

JOHNSON, M. C. & RUTHERFORD, M. J. 1989. Experimental calibration of the aluminium-in-hornblende geobarometer with application to Long Valley Caldera (California) volcanic rocks. *Geology*, **17**, 837–841.

KENNAN, L. & PINDELL, J. 2009. Dextral shear, terrane accretion and basin formation in the Northern Andes: best explained by interaction with a Pacific-derived Caribbean Plate. *In*: JAMES, K. H., LORENTE, M. A. & PINDELL, J. L. (eds) *The Origin and Evolution of the Caribbean Plate*. Geological Society, London, Special Publications, **328**, 487–531.

KLITGORD, K. & SCHOUTEN, H. 1986. Plate kinematics of the central Atlantic. *In*: VOGT, P. R. & TUCHOLKE, B. E. (eds) *The Western North Atlantic Region*. Decade of North American Geology, **M**. Geological Society of America, Boulder, CO, 351–378.

KLUGE, R. 1996. *Geochronologische Entwicklung des Margarita-Krustenblocks, NE Venezuela*. PhD Thesis, University of Münster, Germany.

KROGH, T. E. 1973. A low-contamination method for hydrothermal decomposition of zircon and extraction of U and Pb for isotopic age determinations. *Geochimica Cosmochimica Acta*, **37**, 485–494.

KRÜCKHANS-LUEDER, G. E. 1996. *Petrologie und Geochemie der Juan–Griego–Einheit, Insel Margarita (Venezuela)*. PhD Thesis, University of Münster, Germany.

KUGLER, H. G. 1948. *Report on our present knowledge of the geology of Margarita, Cubagua and Coche*. North Venezuelan Petroleum Company, Geological Report **103**, Pointe-à-Pierre, Trinidad.

LÁZARO, C. & GARCÍA-CASCO, A. 2008. Geochemical and Sr–Nd isotope signatures of pristine slab melts and their residues (Sierra del Convento mélange, eastern Cuba). *Chemical Geology*, **255**, 120–133.

LEAKE, B. E., WOOLLEY, A. R. *ET AL*. 1997. Nomenclature of amphiboles: report of the subcommittee on amphiboles of the International Mineralogical Association Commission on New Minerals and Mineral Names. *European Journal of Mineralogy*, **9**, 623–651.

LE MAITRE, R. W. 1989. *A Classification of Igneous Rocks and a Glossary of Terms*. Blackwell Scientific, Oxford.

LOPEZ-RAMOS, E. 1975. Geological summary of the Yucatan Peninsula. *In*: NAIRN, A. E. M. & STEHLI, F. G. (eds) *Ocean Basins and Margins*, **3**. *The Gulf of Mexico and Caribbean*. Ocean Basins and Margins. Plenum Press, New York, 257–282.

LÓPEZ SÁNCHEZ-VIZCAINO, V., TROMMSDORFF, V., GÓMEZ-PUGNAIRE, M. T., GARRIDO, C. J., MÜNTENER, O. & CONNOLLY, J. A. D. 2005. Petrology of titanian clinohumite and olivine at the high-pressure breakdown of antigorite serpentinites to chlorite harzburgite (Almirez Massif, S. Spain). *Contributions to Mineralogy and Petrology*, **149**, 627–646.

LUDWIG, K. R. 1980. Calculation of uncertainties for U–Pb-isotope data. *Earth Planetary Science Letters*, **46**, 212–220.

LUDWIG, K. R. 1999. *User's Manual for Isoplot/Ex, Version 2.10, A Geochronological Toolkit for Microsoft Excel*. Berkeley Geochronology Center, Berkeley, CA, Special Publications, **1a**.

LUDWIG, K. R. 2000. *SQUID 1.00, A User's Manual.* Berkeley Geochronology Center, Berkeley, CA, Special Publications, **2**.

MACSOTAY, O., CHACHATI, B. & AVAREZ, E. 1997. Eventos de sedimentación, intrusión y sobrecorrimiento en Macanao, Estado Nueva Esparta, Venezuela nor-oriental. *Memorias del VIII Congreso Geológico Venezolano*, **11**, 17–24.

MARESCH, W. V. 1971. *The metamorphism of northeastern Margarita Island, Venezuela.* PhD thesis, Princeton University, Princeton, NJ.

MARESCH, W. V. 1973. Metamorfísmo y estructura de Margarita Oriental, Venezuela. *Boletín Geológico del Ministerio de Minas e Hidrocarburos*, **XII**, 3–172.

MARESCH, W. V. 1974. Plate tectonics origin of the Caribbean Mountain System of northern South America: discussion and proposal. *Geological Society of America Bulletin*, **85**, 669–682.

MARESCH, W. V. 1975. The geology of northeastern Margarita Island, Venezuela: a contribution to the study of Caribbean Plate margins. *Geologische Rundschau*, **64**, 846–883.

MARESCH, W. V. 1977. Similarity of metamorphic gradients in time and space during metamorphism of the La Rinconada Group, Margarita Island, Venezuela. *GUA Papers in Geology, Amsterdam*, **1**, 110–111.

MARESCH, W. V. & ABRAHAM, K. 1981. Petrography, mineralogy and metamorphic evolution of an eclogite from the Island of Margarita, Venezuela. *Journal of Petrology*, **22**, 337–362.

MARESCH, W. V. & GERYA, T. V. 2005. Blueschists and blue amphiboles: How much subduction do they need? *International Geology Review*, **47**, 688–702.

MARESCH, W. V., MEDENBACH, O. & RUDOLPH, A. 1982. Winchite and the actinolite–glaucophane miscibility gap. *Nature*, **296**, 731–732.

MARESCH, W. V., ABRAHAM, K., BOCCHIO, R. & MOTTANA, A 1985. Systematic compositional variations in amphiboles from the La Rinconada Group metabasalts, Margarita Island, and their paleotectonic implications. *Transactions IV Latin American Geological Conference, Trinidad & Tobago*, 1979, 398.

MARESCH, W. V., STÖCKHERT, B. ET AL. 2000. Crustal history and plate tectonic development in the southern Caribbean. *Zeitschrift für Angewandte Geologie*, **SH1**, 283–289.

MARTÍN BELLIZZIA, C. 1968, Edades isotópicas de rocas Venezolanas. *Boletín de Geología, Caracas*, **X**, 356–379.

MASSONNE, H.-J. & SCHREYER, W. 1987. Phengite geobarometry based on the limiting assemblage with K-feldspar, phlogopite, and quartz. *Contributions to Mineralogy and Petrology*, **96**, 212–224.

MASSONNE, H.-J. & SZPURKA, Z. 1997. Thermodynamic properties of white micas on the basis of high-pressure experiments in the system $K_2O-MgO-Al_2O_3-SiO_2-H_2O$ and $K_2O-FeO-Al_2O_3-SiO_2-H_2O$. *Lithos*, **41**, 229–250.

MCCOURT, W. J., ASPDEN, J. A. & BROOK, M. 1984. New geological and geochronological data from the Colombian Andes continental growth by multiple accretion. *Journal of the Geological Society, London*, **141**, 831–845.

MESCHEDE, M. & FRISCH, W. 1998. A plate-tectonic model for the Mesozoic and Early Cenozoic history of the Caribbean Plate. *Tectonophysics*, **296**, 269–291.

MOTTANA, A., BOCCHIO, R., LIBORIO, G., MORTEN, L. & MARESCH, W. V. 1985. The eclogite-bearing metabasaltic sequence of Isla Margarita, Venezuela: a geochemical study. *Chemical Geology*, **50**, 351–368.

MÜLLER, R. D., ROYER, J.-Y., CANDE, S. C., ROEST, W. R. & MASCHENKOV, S. 1999. New constraints on the Late Cretaceous/Tertiary plate tectonic evolution of the Caribbean. *In*: MANN, P. (ed.) *Caribbean Basins. Sedimentary Basins of the World*, **4**. Elsevier Science, Amsterdam, 33–59.

NAVARRO, E. 1974. *Petrogenesis of the eclogitic rocks of Isla de Margarita, Venezuela.* PhD thesis, University of Kentucky, Lexington, KY.

NAVARRO, E. 1981. Relaciónes mineralógicas en las rocas eclogíticas de la Isla de Margarita, Estado Nueva Esparta. *GEOS (Caracas)*, **26**, 3–44.

NEWTON, R. C. 1986. Metamorphic temperatures and pressures of Group B and Group C eclogites. *In*: EVANS, B. W. & BROWN, B. H. (eds) *Blueschists and Eclogites*. Geological Society of America Memoir, **164**, 17–30.

NÚÑEZ, A. & MURILLO, A. 1978. Esquistos de glaucofano en el municipio de Pijao (Quindío). *Resúmenes II Congreso Colombiano de Geología*, Medellín.

OSTOS, R. M. 1990. Evolución tectónica del margen de Sur-Central del Caribe basado en datos geoquimicas. *GEOS*, **30**, 1–294.

OSTOS, R. M. & SISSON, V. B. 2005. Geochemistry and tectonic setting of igneous and metaigneous rocks of northern Venezuela. *In*: AVÉ LALLEMANT, H. G. & SISSON, V. B. (eds) *Caribbean–South American Plate Interactions, Venezuela*. Geological Society of America, Boulder, CO, Special Paper, **394**, 119–156.

PEARCE, J. A., HARRIS, N. B. W. & TINDLE, A. G. 1984. Trace element discrimination diagrams for the tectonic interpretation of granitic rocks. *Journal of Petrology*, **25**, 956–983.

PEREZ, O. J., BILHAM, R. ET AL. 2001. Velocity field across the southern Caribbean Plate boundary and estimates of Caribbean/South-American Plate motion using GPS geodesy 1994–2000. *Geophysical Research Letters*, **28**, 2987–2990.

PINDELL, J. L. 1985. Alleghanian reconstruction and subsequent evolution of the Gulf of Mexico, Bahamas, and Proto-Caribbean. *Tectonics*, **4**, 1–39.

PINDELL, J. L. 1990. Geological arguments suggesting a Pacific origin for the Caribbean Plate. *In*: LARUE, D. K. & DRAPER, G. (eds) *Transactions of the 12th Caribbean Geologic Conference, St Croix, 1989*. Miami Geological Society, Miami, FL, 1–4.

PINDELL, J. L. 1993. Regional synopsis of Gulf of Mexico and Caribbean evolution. *In*: PINDELL, J. L. & PERKINS, R. F. (eds) *13th Annual Research Conference: Mesozoic and Early Cenozoic Development of the Gulf of Mexico and Caribbean Region*. Gulf Coast Section Society of Economic Paleontologists and Mineralogists Foundation, Houston, TX, 251–274.

PINDELL, J. L. & DEWEY, J. F. 1982. Permo-Triassic reconstruction of western Pangea and the evolution of the Gulf of Mexico/Caribbean region. *Tectonics*, **1**, 179–211.

PINDELL, J. L. & TABBUTT, K. D. 1995. Mesozoic–Cenozoic Andean paleogeography and regional controls on hydrocarbon systems. *In*: TANKARD, A. J., SUÁREZ, S. R. & WELSINK, H. J. (eds) *Petroleum Basins of South America*. American Association of Petroleum Geologists, Memoirs, **62**, 101–128.

PINDELL, J. L. & KENNAN, L. 2009. Tectonic evolution of the Gulf of Mexico, Caribbean and northern South America in the mantle reference frame: an update. *In*: JAMES, K. H., LORENTE, M. A. & PINDELL, J. L. (eds) *The Origin and Evolution of the Caribbean Plate*. Geological Society, London, Special Publications, **328**, 1–55.

PINDELL, J. L., CANDE, S. C. *ET AL.* 1988. A plate-kinematic framework for models of Caribbean evolution. *Tectonophysics*, **155**, 121–138.

PINDELL, J. L., HIGGS, R. & DEWEY, J. F. 1998. Cenozoic palinspastic reconstruction, paleogeographic evolution, and hydrocarbon setting of the northern margin of South America. *In*: PINDELL, J. L. & DRAKE, C. L. (eds) *Paleogeographic Evolution and Non-glacial Eustasy, Northern South America*. SEPM (Society for Sedimentary Geology), Tulsa, OK, Special Publications, **58**, 45–86.

PINDELL, J. L., KENNAN, L., MARESCH, W. V., STANEK, K. P., DRAPER, G. & HIGGS, R. 2005. Plate-kinematics and crustal dynamics of circum-Caribbean arc-continent interactions, and tectonic controls on basin development in Proto-Caribbean margins. *In*: AVÉ-LALLEMANT, H. G. & SISSON, V. B. (eds) *Caribbean–South American Plate Interactions, Venezuela*. Geological Society of America, Boulder, CO, Special Papers, **394**, 7–52.

PINDELL, J. L., KENNAN, L., WRIGHT, D. & ERIKSON, J. 2009. Clastic domains of sandstones in central/eastern Venezuela, Trinidad, and Barbados: heavy mineral and tectonic constraints on provenance and paleogeography. *In*: JAMES, K. H., LORENTE, M. A. & PINDELL, J. L. (eds) *The Origin and Evolution of the Caribbean Plate*. Geological Society, London, Special Publications, **328**, 743–797.

RAMOS, V. A. 1988. Plate tectonic setting of the Andean Cordillera. *Episodes*, **22**, 183–190.

REKOWSKI, F. & RIVAS, L. 2005. Integración geológica de la isla de Margarita, estado Nueva Esparta. *GEOS*, Universidad Central de Venezuela, Caracas, **38**, 97–98.

RESTREPO, J., TOUSSAINT, J. F. *ET AL.* 1991. Precisiones geocronologicas sobre magmatismo antina y su marco tectónico. *Memorias*, **I**, Manizalez, 1–22.

ROWLEY, D. B. & PINDELL, J. L. 1989. End Paleozoic–Early Mesozoic western Pangean reconstruction and its implications for the distribution of Precambrian and Paleozoic rocks around Meso-America. *Precambrian Research*, **42**, 411–444.

SANTAMARÍA, F. & SCHUBERT, C. 1974. Geochemistry and geochronology of the Southern Caribbean–Northern Venezuela plate boundary. *Geological Society of America Bulletin*, **85**, 1085–1098.

SISSON, V. B., AVÉ LALLEMANT, H. G. *ET AL.* 2005. Overview of radiometric ages in three allochthonous belts of northern Venezuela; old ones, new ones, and their impact on regional geology.

In: AVÉ LALLEMANT, H. G. & SISSON, V. B. (eds) *Caribbean–South American Plate Interactions, Venezuela*. Geological Society of America, Boulder, CO, Special Papers, **394**, 91–118.

SISSON, V. B., ERTAN, I. E. & AVÉ LALLEMANT, H. G. 1997. High-pressure (~2000 MPa) kyanite- and glaucophane-bearing pelitic schist and eclogite from Cordillera de la Costa belt, Venezuela. *Journal of Petrology*, **38**, 65–83.

SMITH, C. A., SISSON, V. B., AVÉ LALLEMANT, H. G. & COPELAND, P. 1999. Two contrasting pressure–temperature–time paths in the Villa de Cura blueschist belt, Venezuela; possible evidence for Late Cretaceous initiation of subduction in the Caribbean. *Geological Society of America Bulletin*, **111**, 831–848.

SNOKE, A. W., ROWE, D. W., YULE, J. D. & WADGE, G. 2001. *Petrologic and Structural History of Tobago, West Indies: A Fragment of the accreted Mesozoic Oceanic-arc of the Southern Caribbean*. Geological Society of America, Boulder, CO, Special Papers, **354**.

SPEAR, F. S. & CHENEY, J. T. 1989. A petrogenetic grid for pelitic schists in the system $SiO_2–Al_2O_3–FeO–MgO–K_2O–H_2O$. *Contributions to Mineralogy and Petrology*, **101**, 149–164.

SPEED, R. C. & SMITH-HOROWITZ, P. L. 1998. The Tobago terrane. *International Geology Review*, **40**, 805–830.

SPEED, R. C., SHARP, W. D. & FOLAND, K. A. 1997. Late Paleozoic Granitoid Gneisses of Northeastern Venezuela and the North America–Gondwana Collision Zone. *Journal of Geology*, **105**, 457–470.

STEIGER, R. E. & JÄGER, E. 1977. Subcommission on geochronology: convention on the use of decay constants in geo- and cosmochronology. *Earth Planetary Science Letters*, **36**, 359–362.

STEINBERGER, B. 2000. Plumes in a convecting mantle: models and observations for individual hotspots. *Journal of Geophysical Research*, **105**, 11127–11152.

STEPHAN, J. F., BECK, C. M., BELLIZZIA, A. & BLANCHET, R. 1980. La chaîne Caraïbe du Pacifique à l'Atlantique. *Mémoire du BRGM, Paris*, **115**, 38–59.

STÖCKHERT, B., MARESCH, W. V. *ET AL.* 1994. Tectonic history of Isla Margarita, Venezuela – a record of a piece of crust close to an active plate margin. *Zentralblatt für Geologie und Paläontologie*, **Heft 1993**, 485–498.

STÖCKHERT, B., MARESCH, W. V. *ET AL.* 1995. Crustal history of Margarita Island (Venezuela) in detail: constraint on the Caribbean Plate tectonic scenario. *Geology*, **23**, 787–790.

TALUKDAR, S. & LOUREIRO, D. 1982. Geología de una zona ubicada en el segmento norcentral de la cordillera de la Costa, Venezuela. Metamorfísmo y deformación. Evolución del margen septentrional de Suramerica en el marco de la tectonica de placas. *GEOS*, **27**, 15–77.

TARDUNO, J. A. & GEE, J. 1995. Large-scale motion between Pacific and Atlantic hotspots. *Nature*, **378**, 477–480.

TAYLOR, G. C. 1960*a*. *Geology of the Island of Margarita, Venezuela*, PhD dissertation, Princeton University, Princeton, NJ.

TAYLOR, G. C. 1960b. Geología de la isla de Margarita, Venezuela. *Boletín Geología, Publicación Especial*, **3**, 838–893.

TURNER, F. J. & VERHOOGEN, J. 1960. *Igneous and Metamorphic Petrology*, McGraw-Hill, New York.

URBANI, F. 2007. Las regions de rocas igneas y metamórficas del norte de Venezuela. *Memorias IX Congreso Venezolano de Geología, Caracas*, **39**, 93.

VIGNALI, M. 1979. Estratigrafía y estructura de las cordilleras metamórficas de Venezuela oriental (peninsula de Araya-Paria e isla de Margarita). *GEOS, Caracas*, **25**, 19–66.

WETHERILL, G. W. 1956. Discordant uranium–lead ages. *Transactions American Geophysical Union*, **37**, 320–326.

WHITMARSH, R. B., MANATSCHAL, G. & MINSHULL, T. A. 2001. Evolution of magma-poor continental margins from rifting to seafloor spreading. *Nature*, **413**, 150–154.

WILLIAMS, I. S. 1998. U–Th–Pb Geochronology by Ion Microprobe. *In*: MCKIBBEN, M. A., SHANKS III, W. C. & RIDLEY, W. I. (eds) *Applications of Microanalytical Techniques to Understanding Mineralizing Processes*. Reviews in Economic Geology, **7**, 1–35.

WUNDER, B. & SCHREYER, W. 1997. Antigorite: high pressure stability in the system $MgO–SiO_2–H_2O$ (MSH). *Lithos*, **41**, 213–227.

YORK, D. 1969. Least squares fitting of a straight line with correlated errors. *Earth and Planetary Science Letters*, **5**, 320–324.

YSACCIS, B. 1997. *Tertiary evolution of the northeastern Venezuela offshore*. PhD thesis, Rice University, Houston, TX.

Clastic domains of sandstones in central/eastern Venezuela, Trinidad, and Barbados: heavy mineral and tectonic constraints on provenance and palaeogeography

JAMES L. PINDELL[1,2]*, LORCAN KENNAN[1], DAVID WRIGHT[3] & JOHAN ERIKSON[4]

[1]*Tectonic Analysis Ltd, Chestnut House, Duncton, West Sussex GU28 0LH, UK*

[2]*Department of Earth Science, Rice University, Houston, TX 77002, USA*

[3]*Department of Geology, University of Leicester, Leicester LE1 7RH, UK*

[4]*Department of Natural Sciences, St. Joseph's College, Standish, ME 04084, USA*

*Corresponding author (e-mail: jim@tectonicanalysis.com)

Abstract: Current models for the tectonic evolution of northeastern South America invoke a Palaeogene phase of inter-American convergence, followed by diachronous dextral oblique collision with the Caribbean Plate, becoming strongly transcurrent in the Late Miocene. Heavy mineral analysis of Cretaceous to Pleistocene rocks from eastern Venezuela, Barbados and Trinidad allow us to define six primary clastic domains, refine our palaeogeographic maps, and relate them to distinct stages of tectonic development: (1) Cretaceous passive margin of northern South America; (2) Palaeogene clastics related to the dynamics of the Proto-Caribbean Inversion Zone before collision with the Caribbean Plate; (3) Late Eocene–Oligocene southward-transgressive clastic sediments fringing the Caribbean foredeep during initial collision; (4) Oligocene–Middle Miocene axial fill of the Caribbean foredeep; (5) Late Eocene–Middle Miocene northern proximal sedimentary fringe of the Caribbean thrustfront; and (6) Late Miocene–Recent deltaic sediments flowing parallel to the orogen during its post-collisional, mainly transcurrent stage. Domain 1–3 sediments are highly mature, comprising primary Guayana Shield-derived sediment or recycled sediment of shield origin eroded from regional Palaeogene unconformities. In Trinidad, palinspastic restoration of Neogene deformation indicates that facies changes once interpreted as north to south are in fact west to east, reflecting progradation from the Maturín Basin into central Trinidad across the NW–SE trending Bohordal marginal offset, distorted by about 70 km of dextral shear through Trinidad. There is no mineralogical indication of a northern or northwestern erosional sediment source until Oligocene onset of Domain 4 sedimentation. Paleocene–Middle Eocene rocks of the Scotland Formation sandstones in Barbados do show an immature orogenic signature, in contrast to Venezuela–Trinidad Domain 2 sediments, this requires: (1) at least a bathymetric difference, if not a tectonic barrier, between them; and (2) that the Barbados deep-water depocentre was within turbidite transport distance of the Early Palaeogene orogenic source areas of western Venezuela and/or Colombia. Domains 4–6 (from Late Oligocene) show a strong direct or recycled influence of Caribbean Orogen igneous and metamorphic terranes in addition to substantial input from the shield areas to the south. The delay in the appearance of common Caribbean detritus in the east, relative to the Paleocene and Eocene appearance of Caribbean-influenced sands in the west, reflects the diachronous, eastward migration of Caribbean foredeep subsidence and sedimentation as a response to eastward-younging collision of the Caribbean Plate and the South American margin.

Supplementary material: Location maps and detailed heavy mineral data tables are available at http://www.geolsoc.org.uk/SUP18365.

Current kinematically rigorous Cenozoic models for the evolution of northeastern South America invoke a Palaeogene tectonic phase due to inter-American convergence followed by diachronous dextral oblique collision with the Caribbean Plate (Pindell *et al.* 1991, 2006; Perez de Armas 2005), which became strongly transcurrent in the Late Miocene (Pindell *et al.* 1998, 2005). In eastern Venezuela and Trinidad, the collision and subsequent shear between the Caribbean Plate and South America (Fig. 1) juxtaposed, imbricated and offset former palaeogeographic elements, hindering reconstruction of former depositional systems. Correct reconstruction of these is important because certain palaeogeographic aspects pertain directly to petroleum exploration. For example, fluvial–depositional

From: JAMES, K. H., LORENTE, M. A. & PINDELL, J. L. (eds) *The Origin and Evolution of the Caribbean Plate.*
Geological Society, London, Special Publications, **328**, 743–797.
DOI: 10.1144/SP328.29   0305-8719/09/$15.00 © The Geological Society of London 2009.

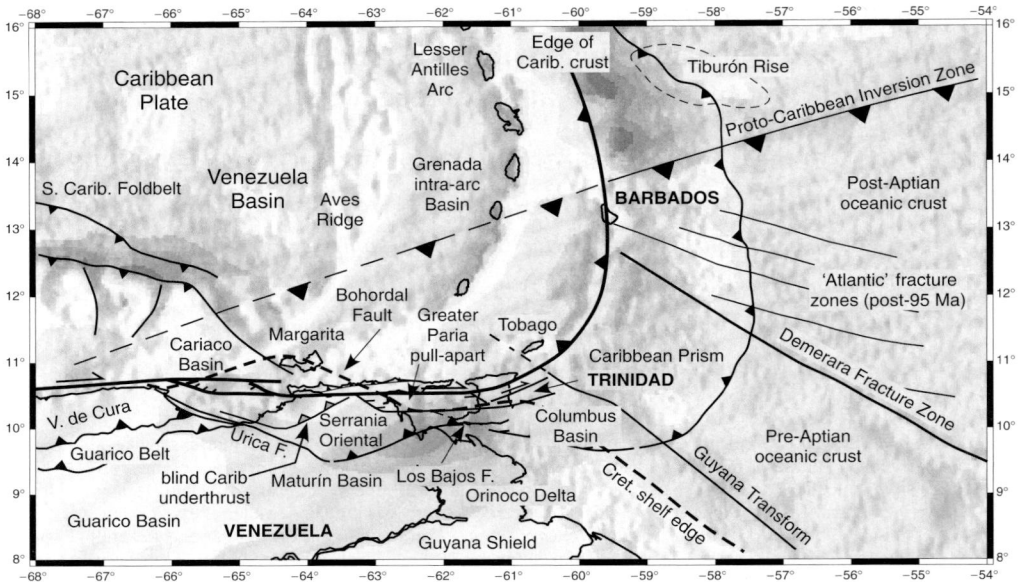

Fig. 1. Key tectonic features of the Eastern Caribbean region, including central and eastern Venezuela and Trinidad, and key features referred to in the text. The background for the map is the satellite gravity of Smith & Sandwell (1997). Today, Caribbean crust lies east of Tobago and under Barbados. This crust and the Caribbean accretionary prism have over-ridden much of the proposed Proto-Caribbean Inversion Zone, which is only exposed today in the area south of Tiburón Rise. The positions shown on this map are based on three-dimensional restoration of features mapped on seismic tomography (after Pindell & Kennan 2007*a*).

systems and turbidite fairways may contain good quality reservoir sandstone packages that can drive hydrocarbon exploration efforts.

Previous work has demonstrated three general principles regarding the provenance of sandstones in northeastern South America (see Figs 2–4 for stratigraphic summaries): (1) South America was the predominant source for most clastic sediments, which are typically quartz prone and mineralogically mature; (2) a less mature Caribbean mineral association began in the Caribbean foredeep basins at the following times: Paleocene in western Venezuela, Eocene in central Venezuela, and Oligocene in eastern Venezuela–Trinidad, indicating the diachronous advance of the Caribbean Plate from the west; and (3) the Palaeogene clastic units of Barbados were derived from relatively high-grade metamorphic (e.g. sillimanite-bearing) rocks, probably from the South American craton or Andes, but also include minerals of probable Caribbean origin such as glaucophane (Senn 1940; Gonzales de Juana *et al.* 1980; Kasper & Larue 1986; Socas 1991; Kugler 2001). These observations allow the regional stratigraphic units to be understood in the context of the basic Caribbean–South America collisional model (e.g. Dewey & Pindell 1986; Kasper & Larue 1986; Pindell *et al.* 1988, 1998). However, there has

been no systematic attempt to use heavy minerals to test such concepts as the diachroneity of collision, or the possible existence of a Proto-Caribbean prism/thrustbelt (proposed by Pindell *et al.* 1991, 2006). This study attempts to fill this void, and to define the cause-and-effect relationship between the regional stratigraphic units and tectonic evolution.

We have collected and analysed the heavy mineral content of 118 sandstone and siltstone samples from the Cretaceous and Cenozoic of eastern Venezuela, Trinidad and Barbados, supported by petrographic examinations of sand grain composition of these and many other samples, to determine sandstone composition variations through the stratigraphic column. In addition, we have studied thin sections and obtained X-ray diffraction data (T. Rieneck & W. Maresch, Universität Bochum, Germany pers. comm. 2007) on field and core samples from the three countries on distinctive red, rounded pebbles and less rounded rip-up clasts, informally referred to as 'cherries'. In Trinidad, these are particularly characteristic of the Cretaceous (Barremian–Albian) Cuche Formation in the Central Range and probably the similarly aged Toco Formation on the north coast, and are also found in (albeit less-oxidized) coarse intervals of the younger Gautier Formation. They are also found in

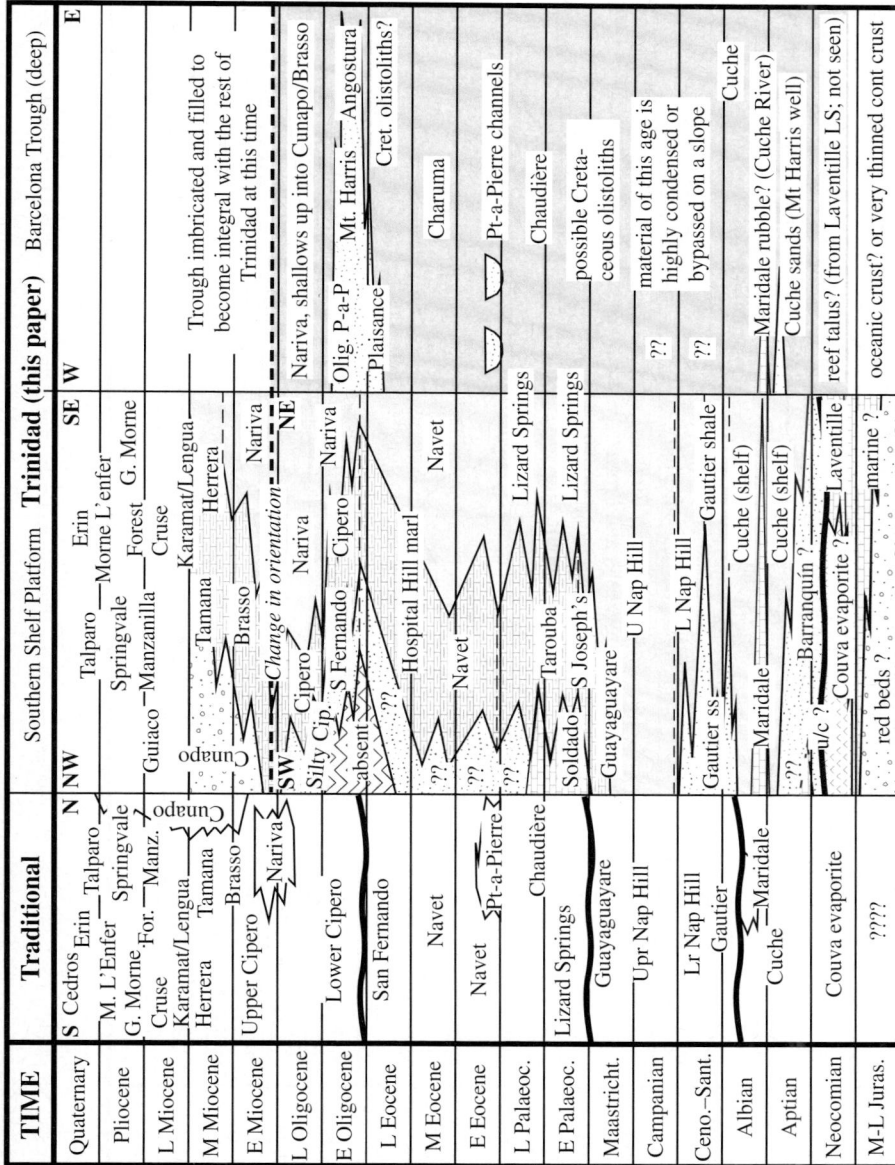

Fig. 2. Stratigraphic chart for Trinidad showing relationships between key formations both across- and along-strike, and contrasting our view with more traditional stratigraphic schemes (e.g. Carr-Brown & Frampton, 1979; Saunders *et al.* 1998). Ages for some of the formations are shown as different from the published literature. The revisions are based on our own unpublished faunal ages, our observations of field relationships and our interpretations of unpublished, proprietary, seismic lines. Ages assigned to most of our samples are based on this framework. Some samples were assigned to units different from those in published maps based on distinctive heavy mineral or petrographic characteristics. The chart shows the contrasts between the southern Trinidad Platform and the eastern end of the Barcelona Trough (see Figs 11 & 12 for location) to the NW, which formed during the Palaeogene above the former passive margin. The platform overlies less stretched continental basement and lies to the north of the Early Cretaceous shelf edge, probably marked by a reef trend at the palinspastically restored position of the footwall of the Central Range. The trough lies to the north of the platform, over highly stretched continent, transitional crust, or possibly oceanic crust of Late Jurassic age and from its inception was significantly deeper. The stratigraphy of the trough is now found in the highest thrust sheets of the Central Range. Locations for key stratigraphic sections, sample sites and major morphotectonic divisions in Trinidad are included in the supplementary material.

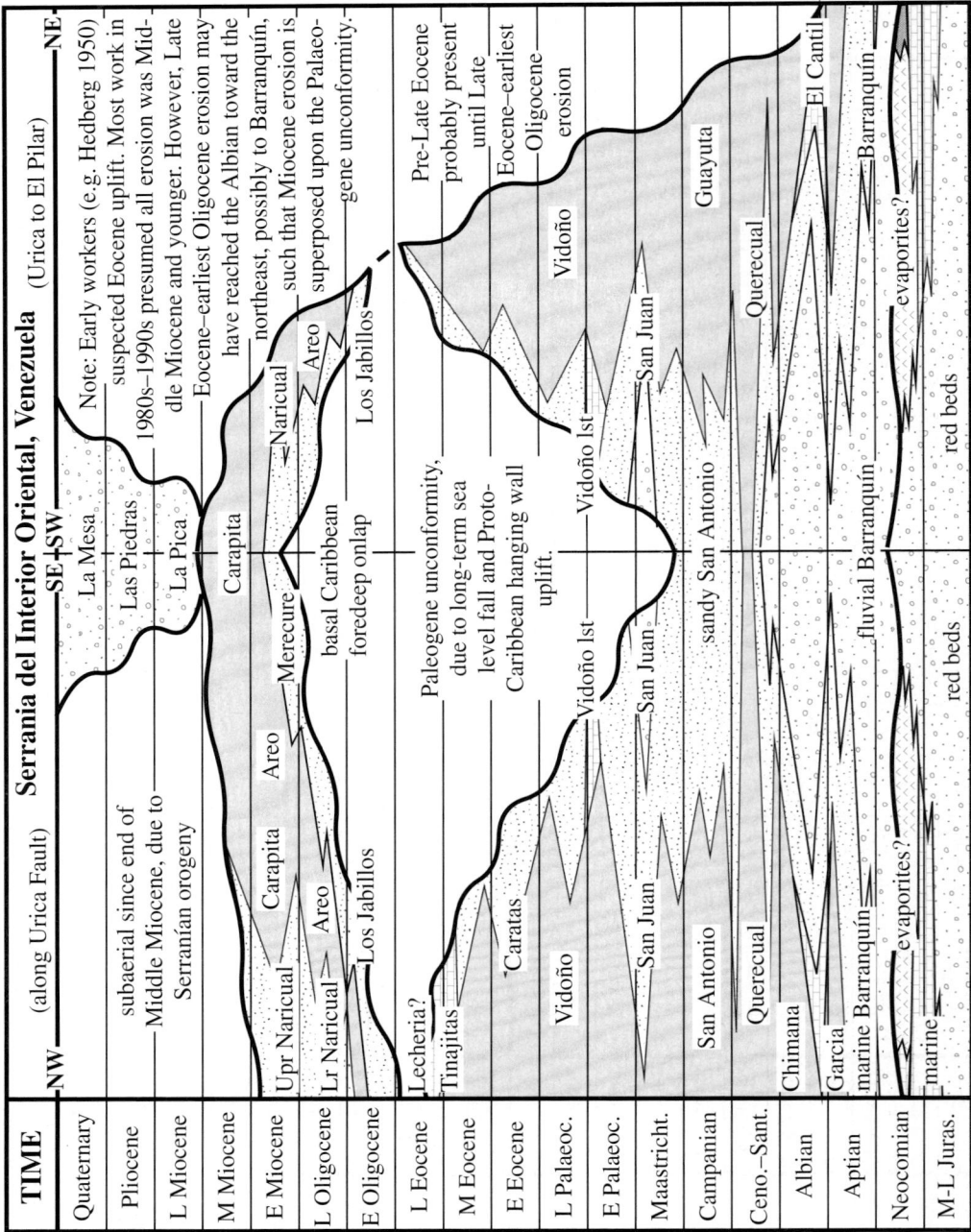

Fig. 3. Stratigraphic chart for Venezuela showing relationships between key formations both across- and along-strike for a NW–SE profile from the Barcelona area to the foreland near Urica, and for a SW–NE profile from Urica towards the northeastern Serranía. The approximate location of the profile is included in the supplementary material.

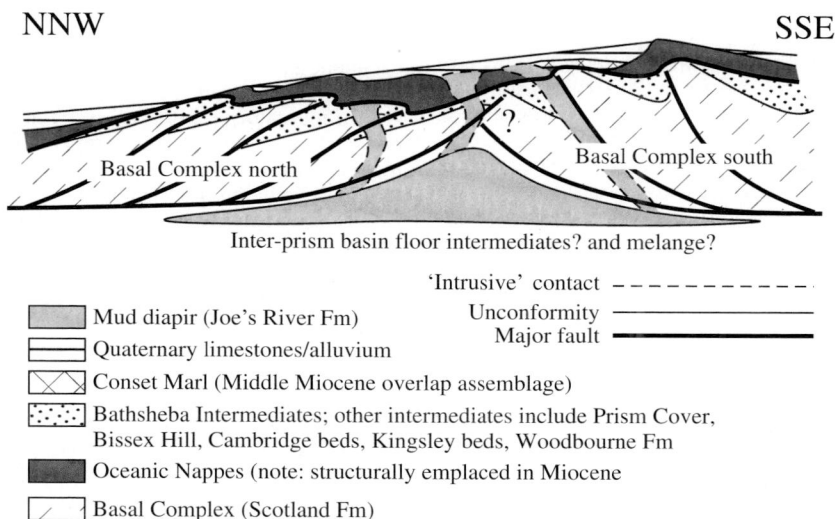

Fig. 4. Schematic structural cross-section of Barbados, summarizing possible stratigraphic relationships. Few reliable ages are available for pre-Miocene strata. According to Senn (1940), the Scotland Group section from top to bottom included the members: Belle Hill/Mount All; Windy Hill/Chalky Mount; Walkers; Murphy's and Morgan Lewis. Speed (2002) aggregated all these into an undifferentiated 'Basal Complex'. Further, we suggest that there may be a juxtaposition of the once-separate Caribbean and Proto-Caribbean accretionary prisms (Basal Complex north and south, verging south and north, respectively). Locations for key stratigraphic sections and sample sites included in the supplementary material.

the (possibly) Late Eocene to earliest Oligocene Plaisance Conglomerate and the basal part of the Mount Harris section of the eastern Central Range in Trinidad, and we have found them in one outcrop of the Oligocene–Early Miocene Nariva Formation sandstone. In Venezuela, they occur in the (possibly) Early Oligocene Lechería beds north of Barcelona. Additionally, we have examined (XRD, thin sections) the dark 'clasts' of similar size and shape in the Galera Formation shales that have been previously considered possible precursors for the 'cherries' of younger formations (Higgs 2006, 2009).

The new heavy mineral and petrographic analyses augment those previously published, and along with the XRD results and previously published modal point count data are integrated with plate kinematic data and regional structural relationships to constrain Mesozoic–Cenozoic clastic distribution patterns and, in turn, palaeogeographic evolutionary models for northern South America.

Heavy mineral study rationale and laboratory methods

Background to heavy mineral studies

Heavy mineral analysis allows the efficient reconstruction of source area lithology and provides information on sand provenance and the direction of sand supply, all key to palaeogeographic reconstruction. Diagnostic minerals provide clues to the correlation of sequences linked by a common provenance, and the differentiation of those that were derived from different source lithologies. During the sedimentary cycle, original heavy mineral assemblages may undergo changes controlled by various modifying factors in the sedimentary environment, such as: (1) hydraulic processes during transport, producing preferential sorting according to size, shape and density related to the differing densities of the individual species; and (2) post-depositional dissolution, due to the low resistance of the majority of heavy minerals to either prolonged acidic or alkaline geochemical conditions either preceding or following burial. Mineral persistence during diagenesis is directly related to their chemical stability and there is typically a progressive decline in the abundance and diversity of heavy mineral species with increasing depth of burial and increasing age. Ultimately, sediments reach a stage of high mineralogical maturity, where the heavy mineral fraction comprises only ultrastable detrital minerals such as zircon, tourmaline, rutile and apatite. These modifying factors need to be evaluated in any heavy mineral study.

Sample preparation

Analytical work was performed in the geochemical laboratory of the Department of Geology in Leicester University, using the methods described in Mange & Maurer (1992). Sample preparation involved: (1) disaggregation of the consolidated sandstones by rock crushing, using a mortar and pestle; (2) removing drilling mud from the cuttings by wet sieving, using detergent, followed by drying; (3) acid digestion to dissolve carbonates both in cores and cuttings by means of 10% acetic acid which leaves acid-sensitive apatite intact; (4) wet sieving using a sieve of 0.063 mm mesh to remove remaining clay and silt particles; (5) drying, followed by standard sieving, retaining the 0.063–0.210 mm size fractions; (6) oil stained samples and cuttings rich in organic particles were cleaned using chloroform; and (7) heavy mineral separation was performed in bromoform (specific gravity of 2.89) using the centrifuge and partial freezing method.

Microscopy

A split of each heavy mineral sample was mounted in liquid Canada balsam on a glass slide for the microscopic investigation. The viscous consistency of liquid Canada balsam facilitates rolling the grains to obtain the required orientation that helps identification and accurate observation of grain morphology. Grain counting was made along parallel bands on the slide, described as the 'ribbon counting' method by Galehouse (1971). Components of the non-opaque heavy mineral suite were counted, excluding micas. However, the presence and abundance of various micas, opaque grains and authigenic phases, together with associated lithic fragments, organic particles etc., were recorded.

The proportion of garnet largely depends on the grain size of the particular deposit, the effects of dissolution processes during diagenesis, mechanical fracture along cleavage planes producing several 'grains' from one original, and, to a certain extent, accidental fracturing of the often large grains during rock crushing. It is therefore advantageous to count garnet separately, thereby avoiding the masking effect of varying garnet quantities on the associated minerals. Colourless and orange to pink varieties were distinguished and their frequencies recorded as a percentage of the total number of grains counted per sample. Anatase, a predominantly authigenic phase, was treated similarly.

During the first stage of grain point-counting the number of individual heavy mineral species was recorded (conventional, species-level analysis) in parallel with the point counting of selected varietal types of zircon, tourmaline and apatite, allocated to

relevant categories (high-resolution heavy mineral analysis). When the total of the individual species, excluding garnet, reached 200, counting of the varieties continued until 75–100 grains each of zircon, tourmaline and apatite varietal types were recorded, respectively. This permits a reliable estimate of heavy mineral abundance. For recording the grain counting, a specially designed HYPER-CARD program was used. Two datasheets were completed for each sample, one for the species-level (overall) mineralogy and one for the varietal study. The raw data were transferred to spreadsheets and recalculated to number percentages.

Heavy mineral data

Trinidadian heavy mineral assemblages are summarized in Tables 1 and 2, Venezuelan heavy mineral assemblages are summarized in Tables 3 and 4 and Barbadian heavy mineral assemblages in Table 5. Detailed heavy mineral data and location maps are included in the online supplementary materials. We have augmented our data with published sources and the limited older industry data that have been released. Both sources are generally non-quantitative. A limited number of modern quantitative studies remain proprietary. Where sediment ages are well-constrained, the results are discussed by age, from oldest to youngest. Figures 2–4 summarize the stratigraphy of Trinidad, Venezuela and Barbados, respectively.

Trinidad

Trinidadian samples were collected from numerous outcrops and from several well cores. Although exposure is generally poor, there is good palaeontological age control at many outcrops, and field data from past industry mapping and augering (down to 20 m) was made available to us. Where possible (usually), samples were collected from active stream cuts, or from recent quarries, in order to get below the worst of the tropical weathering. The Trinidad data (Tables 1 and 2) appear to be immediately separable into two age-dependent assemblages, with only three possible exceptions as noted below. The large number of samples and the availability of absolute and relative age control allow the data to be displayed as a synthetic stratigraphic column (Fig. 5) that accentuates the contrasts between the two assemblages. Early Rupelian (Early Oligocene) and older rocks are characterized by mature zircon, tourmaline and rutile (ZTR) assemblages. Some garnet and apatite was also found in core samples from the Late Albian–Cenomanian Gautier Formation and in one sample of the similar-aged Toco Formation on

Table 1. *Summary of heavy minerals from Trinidadian formations – this study*

Formation	Age	Heavy minerals
Cuche sandstone	Cretaceous	ZTR only
Gautier sandstone	Cretaceous	ZTR, garnet, apatite, possible spinel
Toco	Cretaceous	ZTR, garnet, apatite
Naparima Hill sandstone	Cretaceous	ZTR only
Chaudière	Paleocene–Eocene	ZTR, possible rare apatite
Pointe-a-Pierre	Paleocene–Eocene	ZTR only
Charuma phacoids	Paleocene–Eocene (?)	ZTR, apatite, garnet, chloritoid
Plaisance Conglomerate	Early Oligocene	ZTR, rare epidote
Mt Harris sandstone	Early Oligocene	ZTR, rare epidote
Nariva	Late Oligocene	ZTR, garnet, chloritoid, staurolite, kyanite, glaucophane, apatite, epidote, corundum, monazite
Cunapo	Miocene	ZT only (extremely low recovery of heavy minerals)
Brasso	Miocene	ZTR and chloritoid. Characteristic blue tourmaline
Herrera	Miocene	ZTR, staurolite, epidote, garnet, apatite, sphene, monazite, anatase. Rare hornblende, kyanite
Basal Manzanilla*	Miocene	ZTR only
Lower Manzanilla	Miocene	ZTR only
Upper Manzanilla	Miocene	ZTR, apatite, epidote, clinozoisite, kyanite, chloritoid, chlorite
Cruse	Miocene–Pliocene	ZTR, apatite, staurolite, garnet, kyanite, chloritoid
Talparo	Pleistocene	ZTR, staurolite, sphene, kyanite, glaucophane, epidote, sillimanite, xenotime

*Appears to have been mismapped by Kugler (1996) as Cunapo Conglomerate.

Table 2. *Summary of heavy minerals from Trinidadian formations – other sources*

Formation*	Age	Heavy minerals	Reference
Naparima Hill Argilite	Cretaceous	ZTR, trace garnet, staurolite, kyanite, epidote	Edelman & Doeglas (1934)
Lr Lizards Springs	Paleocene	ZTR, trace garnet, staurolite, kyanite, epidote	Edelman & Doeglas (1934)
San Fernando	Early Oligocene	ZTR, garnet in some samples, trace staurolite, kyanite, epidote	Edelman & Doeglas (1934)
Bamboo or Flat Rock Silt	Early Oligocene	ZTR, trace garnet, staurolite, epidote, glaucophane in one sample	Edelman & Doeglas (1934)
Moruga	Miocene	ZTR, apatite, staurolite, garnet, blue topaz	Kugler (2001)
Forest	Pliocene	ZTR, apatite, garnet, chloritoid, epidote, staurolite, kyanite, andalusite, glaucophane, spinel.	Kugler (2001)
Morne L'Enfer	Pliocene	ZTR, epidote, garnet, chloritoid, staurolite, kyanite, andalusite, topaz, anatase, glaucophane	Kugler 2001
Erin	Pleistocene	ZTR, epidote, staurolite, kyanite, andalusite, amphibole, topaz, anatase. Rare glaucophane	Kugler (2001)

*There appears to be no public-domain heavy mineral data from the Mayaro or Springvale Formations or from the Angostura sandstones.

Table 3. *Summary of heavy minerals from Venezuelan formations – this study*

Formation	Age	Heavy minerals
Barranquín	Cretaceous	Zircon, tourmaline, rutile (hereafter ZTR). Minor staurolite, epidote
San Juan	Cretaceous	ZT only
Caratas	Eocene	ZT only
Lechería	Early Oligocene	ZT only
Los Jabillos	Early Oligocene	ZT only
Lr Naricual	Early Oligocene	ZT, rare kyanite
Areo	Late Oligocene	ZT, minor garnet
Quebradón	Miocene	ZTR, apatite, staurolite, garnet

the north coast. In contrast, probably Late Rupelian and definitely Late Oligocene, and younger rocks are consistently dominated by an immature assemblage of labile minerals, including staurolite, aluminium silicates and glaucophane, in addition to some apatite. Abundant garnet and chloritoid are particularly characteristic. Tropical weathering of outcrops cannot explain this apparent abrupt maturity contrast. Within a given formation, similarities between borehole and field samples, notwithstanding weathering and/or diagenesis, indicate there is a primary compositional difference, and therefore a difference in provenance, between Early and Late Oligocene sediment source rocks. In addition, we sampled both younger and older formations with identical weathering effects, some from immediately adjacent outcrops, and always found dramatic contrasts in heavy mineral assemblages. In some formations for which we have many samples from both outcrop and wells (e.g. Late Oligocene–Early Miocene Nariva Formation), it is clear that there is a geographical grouping of samples in which there seems to be a correlation between increasing tropical weathering and decreased content of certain unstable heavy minerals (Fig. 6). There may also be an element of subtle variation in original heavy mineral content within the Nariva. However, the primary contrast between mature and immature sands is still clear.

Higgs (2006, 2009), reported that previous authors had identified staurolite and other non-ZTR minerals throughout the Palaeogene strata in Trinidad (relying heavily on the synthesis of Suter 1960) and proposed a northern, rather than Guayana Shield, source for both Palaeogene and Neogene strata. However, our examination of unpublished industry heavy mineral studies (Petrotrin archive files mostly from the 1930s–1950s) show the same mature–immature contrast that we have found, as also reported by Kugler (1996, but based on a 1950s synthesis) and Illing (1928). Trace to low-abundance staurolite, kyanite, chloritoid and very rare glaucophane have been found in

a few samples (and many were exotic blocks associated with mud volcanoes; their origins and age were thus poorly known), and there is little justification for considering these minerals as characteristic of the Palaeogene as a whole.

There are three potential exceptions in our data to the apparent Oligocene character change for Trinidad. First, the samples from the Middle Eocene 'Charuma Silt Member' of the Pointe-a-Pierre Formation at its type locality yield an immature heavy mineral signature (Table 1). Although this is the type locality of this stratigraphic unit, exceptional recently cleared exposure on the day of our collection showed that the Charuma section comprised sheared, sandstone phacoids within a scaly clay gouge zone, with sections and rafts of silty clay. There is no undisrupted bedding in the outcrop and the silty clay carries a Middle Eocene 'Gaudryina' fauna. There are no fresh outcrops of this formation and attempts to separate heavy minerals from other sites where stratigraphy and field relationships are clear (e.g. from Piparo Gorge, where adjacent beds are mapped as typical Pointe-a-Pierre sandstone and show zircon and tourmaline only) failed to yield any, much less immature, heavy minerals. Although this unit has been drilled in several wells, there are no longer any cuttings available and the drilled section rarely if ever included sandstones (John Keens-Dumas pers. comm.). Faults juxtaposed the sampled outcrops with the Navet marls and Nariva sandstones, the latter with the same complex mineral signature and similar textural characteristics in thin section. We suspect that contamination is possible in our Charuma sample due to the shearing; the sample could be Oligocene Nariva contaminated by Eocene Charuma or Navet fauna, or it could be true Charuma (Eocene) contaminated by Nariva minerals. If this were not the type section of the Charuma beds, we would consider the outcrop as structurally disrupted enough to pay it little attention. Alternatively, the chloritoid, kyanite and garnet present in the samples could represent a

Table 4. *Summary of heavy minerals from Venezuelan formations – other sources*

Formation*	Age	Heavy minerals	Reference
Pre-Cretaceous	Palaeozoic, Jurassic	ZTR, hornblende, tremolite, glaucophane (or non-HP/LT blue-grn amph?). One report of chloritoid	Escalona (1985)
Canoa	Cretaceous	ZT, epidote, zoisite, magnetite, ilmenite. Minor kyanite, staurolite, other amphiboles	Escalona (1985)
Tigre	Cretaceous	ZT, epidote, kyanite, staurolite, glaucophane (or non-HP/LT blue-grn amph?). Minor magnetite, ilmenite, other amphiboles	Escalona (1985)
Mito Juan	Cretaceous	ZTR, garnet, chloritoid, ilmenite, leucoxene	PDVSA (2005)
Garrapata	Eocene (?)	Amphiboles, pyroxenes in association with volcanic rock fragments	PDVSA (2005)
Los Cajones	Paleocene–Eocene (?)	ZTR, epidote, apatite, magnetite, leucoxene, in association with volcanic and schist fragments	PDVSA (2005)
Guárico	Paleocene–Eocene	ZTR, garnet, trace chloritoid, anatase. (Guarumen-Ortiz sandstone)	Kamen-Kaye (1942)
Misoa (El Mene)	Eocene	ZTR, staurolite (upper member only), rare garnet	Feo-Codecido (1955)
Cobre	Eocene	ZTR, garnet, staurolite	PDVSA (2005)
La Pascua	Early Oligocene	ZTR. Minor or trace staurolite, kyanite, hornblende, other amphiboles	Escalona (1985)
Lr Roblecito	Late Oligocene	ZTR. Minor staurolite, kyanite	Escalona (1985)
Upr Roblecito	Late Oligocene	ZTR, staurolite, kyanite, glaucophane	Escalona (1985)
Merecure	Late Oligocene–Miocene	ZTR, minor garnet, minor chloritoid	PDVSA (2005); Feo-Codecido (1955)
Carapita	Late Oligocene–Miocene	ZTR, epidote, glaucophane	Feo-Codecido (1955)
Capaya	Miocene	ZTR, staurolite, glaucophane	PDVSA (2005)
Chaguaramas	Miocene	ZTR, staurolite, kyanite, andalusite, sillimanite, glaucophane, chloritoid	PDVSA (2005)
Oficina	Miocene	ZTR, staurolite, kyanite, garnet, chloritoid	PDVSA (2005); Hedberg *et al.* (1947)
Freites	Miocene	ZTR, some garnet, chloritoid, staurolite, kyanite, glaucophane	Feo-Codecido (1955)
La Pica	Miocene	ZTR, epidote, some garnet, chloritoid, staurolite, kyanite, glaucophane	Feo-Codecido (1955)
Las Piedras	Pliocene	ZTR, kyanite, chloritoid, corundum, hornblende. Minor sillimanite, staurolite, epidote, garnet, kyanite, glaucophane	PDVSA (2005); Hedberg *et al.* (1947)
La Mesa	Pleistocene	ZTR, staurolite, kyanite, andalusite, sillimanite, magnetite	Hedberg *et al.* (1947)
Rio Caroni	Holocene	ZTR, ilmenite, staurolite	Wynn (1993)

*We recovered no heavy minerals, nor are there reports in the literature of heavy minerals from the Paleocene–Eocene Vidoño Formation.

Table 5. *Summary of heavy minerals from Scotland Group members – this study*

Member	Heavy minerals
Morgan Lewis	ZTR, epidote, clinozoisite, apatite, garnet (abundant), chloritoid, kyanite, staurolite
Walkers	ZTR, clinozoisite (very abundant), epidote, apatite, garnet, chloritoid, kyanite (abundant), staurolite, sphene; rare tremolite, chrome spinel, lawsonite
Chalky Mount	ZTR, epidote (minor), clinozoisite, apatite, garnet (abundant), chloritoid (abundant), kyanite (abundant), staurolite
Windy Hill	ZTR, epidote, clinozosite (abundant), garnet, chloritoid, staurolite
Mount All	ZTR, epidote (rare), clinozosite (abundant), garnet, chloritoid, kyanite, staurolite
Belle Hill	ZTR, epidote (rare), clinozosite, garnet, kyanite, staurolite
Bathsheba	ZTR, epidote, clinozoisite

rare flush of first cycle labile minerals from the shield into the Trinidad area. Trace kyanite and garnet are both reported from Palaeogene sections close to top Cretaceous and Late Eocene unconformities in Trinidad (Edelman & Doeglas 1934) and all these minerals are present on the shield to the south (see below).

The second exception is a lignite-bearing sandstone collected from an isolated fault-bounded outcrop in the Chaudière River (on the north flank of Mount Harris), which also has a characteristic Nariva heavy mineral signature. Although mapped as Paleocene Chaudière Formation by Kugler (1996), bedding in the outcrop dips in the opposite direction to proven Paleocene beds nearby, and is closely associated with poorly exposed breccias, suggesting it may be fault-bounded. Abundant lignite is not found in formations older than Nariva. 'Chaff-like plant remains and wisps of carbonaceous matter' (Kugler 1996) are described from the well-cemented sandstones at the Pointe-a-Pierre type locality, but heavy minerals from this site were extremely mature (other than one unpublished report of a single kyanite grain in one of several samples). In hand specimens, these samples are friable compared with adjacent Chaudière sandstones, and in thin section, these samples are texturally identical to Nariva sandstones collected elsewhere, being fine-grained and well-sorted, with poorly rounded grains. Nariva is mapped along-strike several hundred metres west of this outcrop and we suspect it continues unmapped along the north flank of Mount Harris.

The third exception is a sample from some 200 m south of the Mount Harris picnic site of the eastern Central Range, also mapped as Chaudière or Pointe-a-Pierre Formation (Algar 1993; Kugler 1996). We have identified chloritoid and blue tourmaline here, both characteristic of the younger Brasso Formation. Thin sections are texturally indistinguishable from Nariva or Brasso Formation sandstones collected elsewhere, and quite unlike nearby sandstones mapped as Pointe-a-Pierre Formation. Furthermore, a sample of interbedded claystone yielded a single foraminifer no older than earliest Miocene (J. Frampton pers. comm. 2006). We provisionally interpret this sample as Brasso sandstone in the footwall of a thrust carrying Early Oligocene Mount Harris sandstone in its hanging wall. There are mapped Brasso outcrops within a few hundred metres of this site.

From the above, the three apparent exceptions to the Oligocene heavy mineral character change appear to be explicable by faulting and mis-mapping of fault-bounded Neogene rocks as Palaeogene. This is not surprising, as up to 150 km of dextral shear must have passed through Trinidad since 10 Ma (Pindell & Kennan 2007*a*).

Venezuela

Signature minerals such as kyanite, staurolite and glaucophane appear, at first glance, more common in Venezuelan samples (Tables 3 and 4). However, our samples show a fundamental difference between Eastern Venezuela v. Central and

Fig. 5. Heavy mineral varieties from Trinidad samples, sorted by relative age. Note the very mature assemblages in Cretaceous through Early Oligocene rocks; trace staurolite and kyanite has also previously been reported from a few samples of this age. Regional facies patterns and provenance data both indicate a broadly 'southern provenance'. There is a very pronounced provenance break in late Early Oligocene and younger samples, marked by the appearance of immature mineral assemblages derived from in part from the Caribbean Orogen. The orogen was to the west at the time

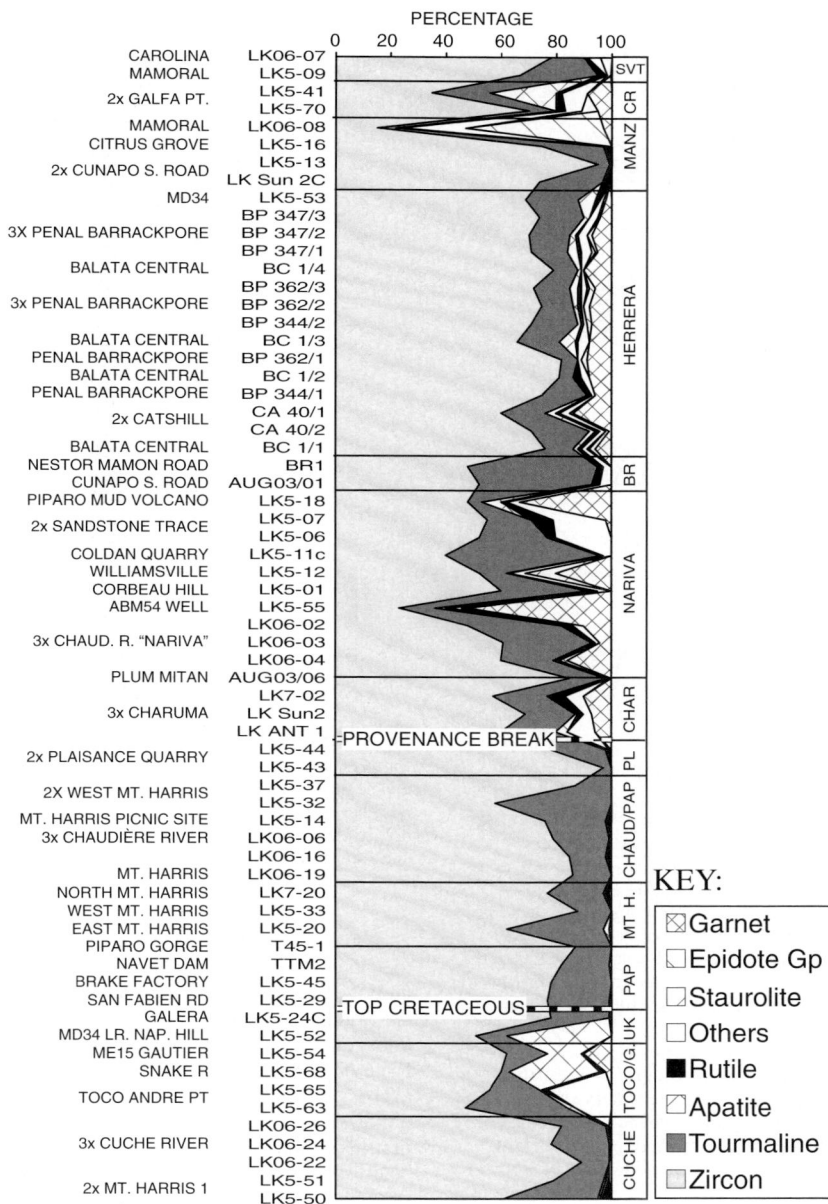

Fig. 5. (*Continued*) but sediment of this age was transported down the axis of an orogen-parallel drainage system, much like the present-day Orinoco. In both wells and outcrops it is possible to find adjacent examples of these two assemblages in rocks which appear to have similar weathering or diagenetic characteristics, indicating that they are primary differences. Late Middle Miocene and younger strata show a more mixed provenance, with Caribbean detritus diluted either by new south-derived input into the basin or recycled Palaeogene rock which had been incorporated into the edge of the Caribbean orogen. Sediment recycling may also contribute to the cleaner signature. Our few Late Pliocene and Pleistocene samples (Mamoral sample from Springvale Formation and Carolina sample from Talparo Formation), are strikingly more immature than slightly older Herrera and Cruse samples. Abbreviations: UK, Upper Cretaceous; PAP, type Pointe-a-Pierre; MT H., Mount Harris sandstone; CHAUD/PAP, mapped as Chaudière/ Pointe-a-Pierre but probably equivalent to or younger than MT H.; PL, Plaisance Conglomerate (possibly equivalent to basal MT H.); CHAR, Charuma 'silt'; BR, Brasso; MANZ, Manzanilla Formation in Caroni Basin; CR, Cruse Formation in Southern Basin; SVT, undifferentiated Springvale–Talparo.

Nariva Formation
'Striped' sands

Outside in:
LK06/4
LK06/3
LK06/2
All Chaud. River

Nariva Formation
'Sealed' sands

Outside in:
LK05/55, ABM54
LK05/12, Williamsville
LK05/18, Piparo MV

Nariva Formation
Sandstone Trace

Outside in:
LK05/1, Corbeau Hill
LK05/11, Guaracara

Nariva Formation
Other outcrops

Outside in:
LK05/6
LK05/7

KEY:
⊠ Garnet
⊠ Epidote Gp
⊡ Staurolite
☐ Others
■ Rutile
⊠ Apatite
■ Tourmaline
⊡ Zircon

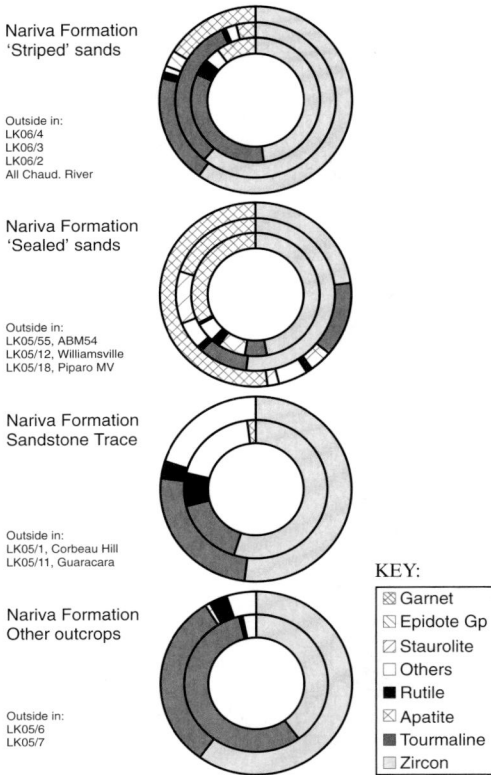

Fig. 6. Variations in heavy mineral abundance in the
Nariva Formation. Significant changes in relative
abundance of key mineral in Nariva samples probably
reflect post-depositional leaching by acidic fluids
(basinal or tropical soil origin) superimposed on some
original, unquantifiable, but probably limited
compositional variation. Fairly fresh 'striped' sandstones
from the Chaudière River show intermediate amounts of
garnet and some chloritoid and kyanite, but 'sealed'
samples from outcrops, core or mud volcanoes where
intense surface weathering has not happened show very
high levels of garnet, with some apatite and staurolite
also preserved. The Williamsville Quarry sample is the
only one with preserved glaucophane (very unstable).
Weathered outcrop samples in the Piparo area show
moderate preservation of chloritoid but no garnet and
only trace apatite. The Sandstone Trace samples show
intense feldspar breakdown and pervasive clay cement
but otherwise do not show deep weathering. Chloritoid is
particularly abundant but garnet is rare and apatite and
staurolite are absent. The relative ages of these samples
are not clear. All are associated with earliest Miocene
shales in adjacent outcrops.

Western Venezuela: in the east, as in Trinidad, Late
Maastrichtian through Early Oligocene sandstones
(i.e. San Juan, Caratas, and Los Jabillos Formations)
are highly mature (ZTR-dominated, Fig. 7). There is

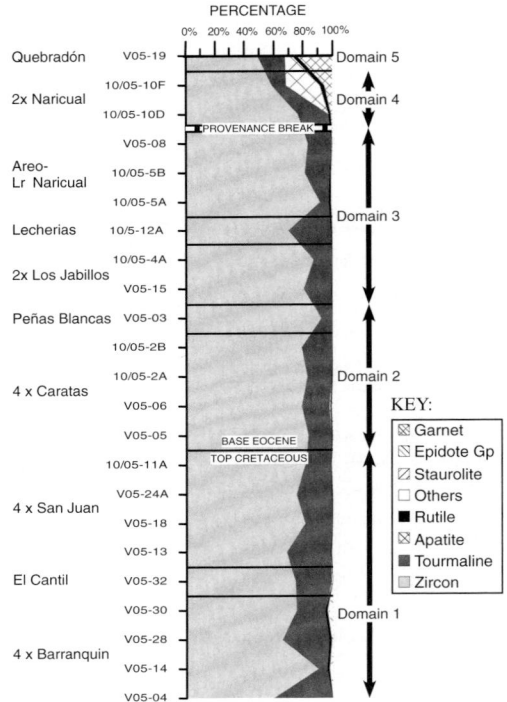

Fig. 7. Heavy mineral varieties from East Venezuela
samples, sorted by apparent relative age. Trace kyanite
and staurolite are seen in the Barranquín Formation, but
all other Cretaceous to Early Oligocene samples are
highly mature. Kyanite reappears in the Areo, and garnet
and staurolite in Naricual and younger samples.

an apparent break in sandstone provenance along
the Palaeogene South American margin at the
Gulf of Barcelona/Urica Fault. East of this break,
the South American shelf and slope section did not
know about the impending arrival of the Caribbean
Plate until at least the latest Early Oligocene. This
accords with the concept of eastward younging
oblique collision between the two plates that was
first identified by the eastward migration of Caribbean foredeep subsidence along the margin
(Pindell 1985; Pindell & Barrett 1990).

Only trace glaucophane and staurolite have been
reported from pre-Cretaceous and Late Cretaceous
samples from wells in the foreland south of
Caracas. Even in Central Venezuela, where orogenesis has commonly been thought to have begun in
the Cretaceous, signature minerals are in much
higher abundance in Late Oligocene and younger
strata, such as the Upper Roblecito and Quebradón
Formations, than they are in earlier autochthonous
and para-autochthonous Caribbean foredeep strata
such as the Guárico Formation. The Early Oligocene

La Pascua and Lower Roblecito Formations, which onlap an earlier Palaeogene hiatus, are characterized by a mature ZTR assemblage with only trace kyanite and staurolite suggesting, as in Trinidad, an important provenance break of intra-Oligocene age.

In Central Venezuela, the volcanic, serpentinitic, and metamorphic content of the Garrapata and Los Cajones Formations (Early Eocene; Macsotay et al. 1995) indicate that Caribbean uplift, erosion and redeposition of clastic materials in a trough between the two plates was underway at that time, but these units are entirely allochthonous by an uncertain distance as there is uplift, rather than subsidence, of that age in the foreland. Table 4 appears to suggest that the Guárico Formation was receiving northern/Caribbean-derived detritus as well. However, Peirson et al.'s (1966) original mapping of the Los Cajones and Garrapata as members of the Guárico is invalid, based on more recent field studies. Vivas & Macsotay (1997) propose the Los Cajones and Garrapata units as distinct formations, with no syn-sedimentary intercalation. Further, Perez de Armas (2005) states that the contact between the internal (Los Cajones/Garrapata) and external (Mucaria/Guárico) zones of the Guárico Fold-Thrust Belt is always a fault. Our own field studies support this view entirely, and suggest that the three isolated mapped occurrences of Los Cajones strata within the Guárico Belt south of the Don Alonso and Guárico faults (as shown by Bellizzia & González 1971) are not Los Cajones Formation, but rather sections of Guárico that have been intensely deformed with two intersecting cleavages, giving the false appearance of sedimentary rubble in shaly matrix. Thus, the Guárico and the Los Cajones/Garrapata Formations appear to have been initially deposited in different depocentres that merged over time as the Caribbean thrustbelt advanced upon South America. To our knowledge, it is not permissible with current data to consider the Garrapata and Los Cajones strata as members of the Guárico Formation, nor to claim that the volcanic, serpentinite and metamorphic clasts and minerals of the Los Cajones/Garrapata units have anything to do with the Guárico Formation south of the Don Alonso–Guárico Fault. Our thin sections from Guárico samples show no such contamination, and we find nothing in the literature to counter this mature characterization of the Guárico. However, with the approach of the Leeward Antilles Arc in the Palaeogene, it would not be surprising if the Guárico were eventually shown to contain crystals of airfall tuff, as is the case with the Oligocene in Trinidad (Algar et al. 1998).

We also highlight the occurrence of chloritoid in the Maastrichtian–(possibly) Paleocene Mito Juan Formation of the SW Maracaibo area, which we believe is the only known occurrence of this mineral in Cretaceous through Early Oligocene strata in Venezuela (PDVSA 2005). However, it is characteristic of Late Oligocene and younger sandstones in eastern Venezuela–Trinidad, with rare exceptions.

Barbados

In Barbados, Palaeogene terrigenous turbidites of South American affinity (bearing mainly quartz with high-grade metamorphic clasts and minerals) occur in the Scotland District (stratigraphic and structural relations are summarized in Fig. 4). Senn (1940) and Poole & Barker (1983) subdivided the 'Scotland Beds' into Lower and Upper Scotland units. Senn further considered the Lower Scotland to comprise the Morgan Lewis and the Walkers sections, and the Upper Scotland to comprise the Murphy's, Chalky Mount and Mount All sections. Speed (2002) considered all these Scotland beds as packets of accretionary prism material in his 'Basal Complex'. The age range of the Scotland beds is Late Paleocene through Middle Eocene, based on foraminifera and radiolaria (Speed 2002) and pollen (Pindell & Frampton 2007, citing D. Shaw 2007).

Senn (1940) listed the following heavy minerals (citing an unpublished report by Hollis Hedberg) from 113 samples of the Scotland Formation: black opaque minerals, leucoxene, zircon, tourmaline, garnet, staurolite, sillimanite, kyanite, andalusite, topaz, glaucophane (in five of the samples only), epidote, zoisite-clinozoisite, rutile, anatase, brookite, chloritoid, hypersthene, augite, titanite and corundum. Unfortunately, the locations of the samples were not specified and thus the data, as reported by Senn at least, cannot be used to attribute certain heavy mineral signatures to specific Scotland sections.

In our data (Table 5 and Fig. 8), characteristic metamorphic minerals are found in all but the Bathsheba sample, which was considered as the 'Intermediate Unit' (possible Early Miocene) by Barker et al. (1987) rather than true Scotland. Apatite is absent in Mount All, Belle Hill and Bathsheba samples, and chloritoid is only absent in the Belle Hill and Bathsheba samples. The various Barbados lithostratigraphic units do not appear to be characterized by very distinct heavy mineral signatures. However, there is tremendous diversity in relative proportions within and between units that suggests the mixing, during transportation, of two end member sediment sources, one mineralogically and texturally mature and one immature. This would be expected if the Scotland Formation (Basal Complex) were fed by a foreland trunk river system flowing between the developing Caribbean Orogen and the Guayana Shield (Kasper & Larue

PERCENTAGE

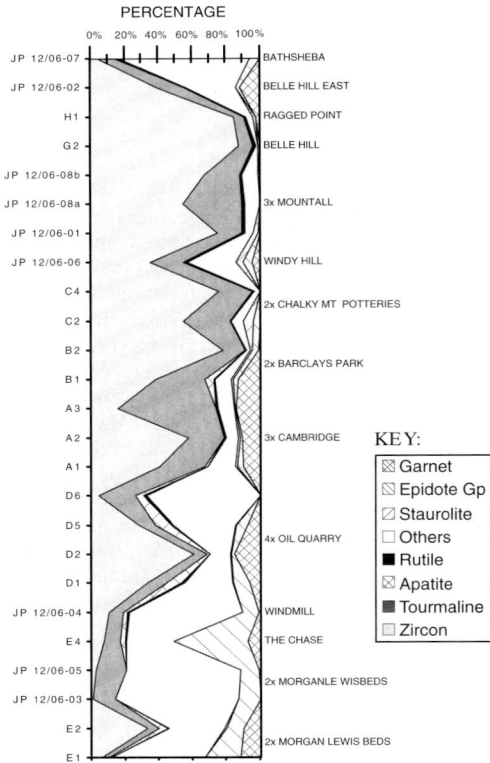

Fig. 8. Heavy mineral varieties from Barbados samples, sorted by apparent relative age. All but the youngest sample, from Bathsheba, show mixed South America and Caribbean provenance. The Bathsheba sample is swamped by epidote group minerals but is otherwise mineralogically mature, with only zircon, tourmaline and rutile.

1986). As will be seen, given the Late Paleocene–Middle Eocene age of the Scotland beds, the source area was probably situated in Colombia and/or western and possibly central Venezuela. The almost ubiquitous chloritoid, common staurolite, kyanite and glaucophane are in striking contrast to strata of the same age in east Venezuela and Trinidad, which show none of the mixed provenance seen in Barbados. If the Basal Complex of Barbados is not very far travelled (i.e. <300 km), then its depocentre must have been deeper than, if not isolated from, the deep-water deposits of eastern Venezuela and Trinidad, and also been depositionally downstream of western or central Venezuela, where Caribbean allochthons were being unroofed. However, the Basal Complex may be farther travelled than 300 km and the mature south-derived component of the sands may derive from Colombia or western Venezuela, where, for example,

sillimanite-bearing basement was exposed between the times of Guaduas (Early Paleocene) and La Paz (Late Eocene) Formation deposition (Pindell *et al.* 1998), coeval with Scotland deposition.

Potential source areas for characteristic heavy minerals

Our approach to the heavy mineral data has been to identify the signature minerals that characterize particular formations or groups of formations. Our sampling is too sparse to look in detail at the ratios of particular minerals, or to more than qualitatively interpret the relative abundances of particular mineral varieties or grain shapes (typical of detailed studies on a particular well or formation). Instead, we attempt to tie these minerals to well-described possible sediment source regions in Colombia, Venezuela and Trinidad (Fig. 9 and Table 6). Identification of sediment source regions would be greatly aided by modern fission track studies of both potential source areas and of detrital grain populations in foreland sediments (an approach used in, for example, Ecuador by Ruiz *et al.* 2004, 2007). Modelled unroofing history of detrital grains could then be compared with that of sediment source areas and combined with heavy mineral data could lead to less ambiguous identification of sediment sources, where there is more than one possible source for particular minerals, but where those source regions have different unroofing histories.

The heavy minerals present in the Late Cretaceous to Palaeogene Venezuelan, Trinidadian and Barbadian sediments can derive from old metamorphic or igneous source areas, or may be eroded and recycled from Cretaceous or Cenozoic strata. Sufficient information is available to constrain potential sediment sources, especially when combined with additional information such as sedimentary, structural or radiometric constraints on age of uplift, cooling and development of associated unconformities. Potential source areas for selected key heavy minerals are readily identified (Table 6). Some minerals, such as sphene, epidote and apatite, are so ubiquitous in greenschist or amphibolite facies terranes or widespread plutons and associated aureoles as to be non-diagnostic of sediment source. Others, such as garnet, chloritoid, staurolite and kyanite, are geographically widespread but absent in a few critical localities of significance for palaeogeographic reconstruction. They may be present in small quantities in sediments of Palaeogene and older age, but dramatic changes in their abundance also correlate with the appearance of distinctive larger clasts, such as circum-Caribbean volcanic rocks, which suggest

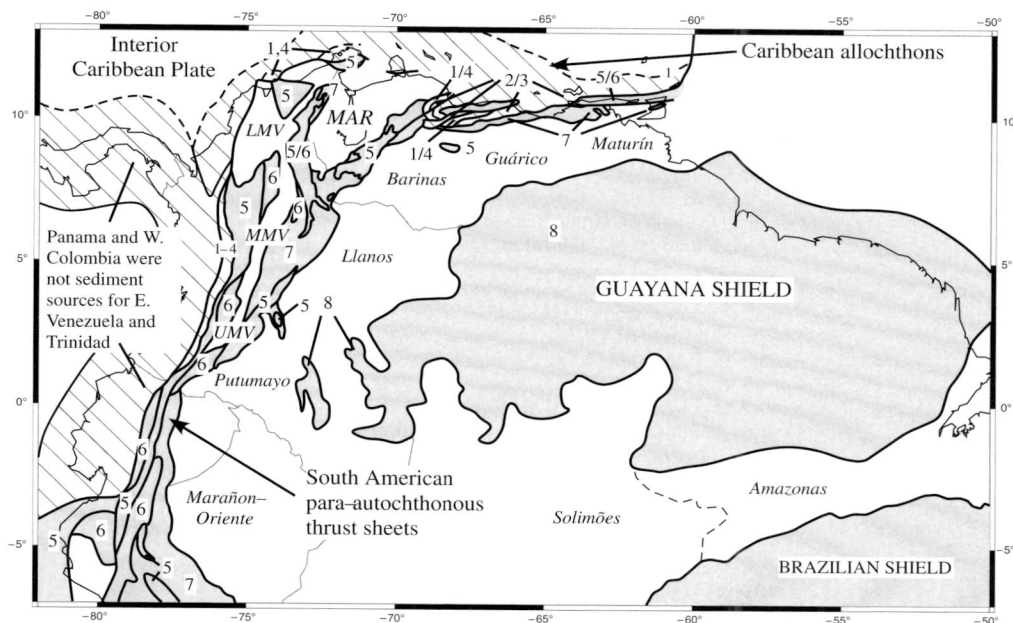

Fig. 9. Maps showing location of sediment source areas classified in the text. Caribbean allochthons (cross-hatch pattern) comprise: (1) Palaeozoic–Mesozoic igneous and sedimentary protoliths with greenschist to amphibolite facies metamorphism; (2) Precambrian–Palaeozoic protoliths, HP/LT metamorphism; (3) Mesozoic mixed protoliths, HP/LT metamorphism; and (4) Mesozoic mixed protoliths, greenschist-facies or lower grade metamorphism. Andean Para-autochthons, shield areas (grey) comprise: (5) Precambrian–Palaeozoic protoliths, greenschist–amphibolite facies, locally granulite; (6) Mesozoic mixed protoliths, greenschist-facies or lower-grade metamorphism; (7) Mesozoic sediments, unmetamorphosed, Cenozoic uplift and unroofing; and (8) Precambrian metaigneous and metasedimentary rocks.

they are palaeogeographically diagnostic. Some minerals, such as glaucophane, appear to be very strongly associated with Caribbean provenance. Other useful indicators include the presence of rare but distinctive species (e.g. chrome spinel and lawsonite) and mineral associations indicative of a particular paragenesis (e.g. clinozoisite with glaucophane, epidote and chrome spinel is indicative of a metabasic blueschist source lithology).

The major possible sediment source areas can be classified based on distinctive mineralogies and discrete tectonic origins. Four of these classes are unambiguously associated with terranes at the leading edge of the allochthonous Caribbean Plate and four appear to belong to autochthonous or parautochthonous South America (Fig. 9). There are important mineralogical contrasts (the reader is referred to the references in Table 6 for mineral-specific data sources) between these two groups of potential sediment source areas that support the palaeogeographic models we discuss later. The regional plate tectonic context and detailed local geological evolution of many of the Andean source areas discussed below are discussed in

more detail in companion papers (Kennan & Pindell 2009; Pindell *et al.* 2009).

Class 1: Far-travelled Caribbean greenschist and amphibolite-grade metamorphic terranes with mostly Palaeozoic or older sedimentary (primary) and plutonic or volcanic (secondary) protoliths

These terranes all lie outboard, north or west, of accreted Caribbean oceanic or arc terranes. Where dated, they typically have a Palaeozoic initial age of metamorphism related to the final assembly of Pangaea. Initial cooling from a subduction zone-related orogenic event is Albian to early Late Cretaceous (Stöckhert *et al.* 1994; Sisson *et al.* 2005), indicating an origin to the west of Colombia because only passive margin conditions existed at that time along northern South America (Villamil & Pindell 1998). Class 1 rocks were rifted from South America and then incorporated into the leading edge of the Caribbean Plate west of

Table 6. *Possible source areas for selected characteristic heavy minerals*

Mineral	Sediment source area	Reference
Glaucophane	Yaritagua complex (Lara)	PDVSA (2005)
	Cordillera de la Costa HP/LT belt	Sisson *et al.* (1997)
	Antimano Formation	PDVSA (2005)
	Villa de Cura Group	Smith *et al.* (1999)
	El Copey Formation, Araya Peninsula	PDVSA (2005)
	Guayana Shield (very doubtful, location not specified)	Kamen-Kaye (1937)
Chloritoid	Pastora, Botanamo Proterozoic greenschists and metavolcanics, Guayana Shield	Sidder & Mendoza (1995)
	Margarita Island HP/LT belt	Stöckhert *et al.* (1995)
	Manicuare Formation, Araya Peninsula	Schubert (1971)
	Cordillera de la Costa HP/LT belt	Sisson *et al.* (1997)
	Santa Marta Metamorphic Belt (Colombia)	Maya-Sánchez (2001)*
	Cordillera Central, Colombia (west of Palestina Fault)	Maya-Sánchez (2001)
	Amaime–Cauca blueschist belt (Colombia)	Maya-Sánchez (2001)
Lawsonite	Villa de Cura Group	Smith *et al.* (1999)
	Bocas Complex (offshore Venezuela)	PDVSA (2005)
Staurolite	Margarita Island HP/LT belt (Juan Griego)	Stöckhert *et al.* (1995)
	Cordillera de la Costa HP/LT belt	Sisson *et al.* (1997)
	Los Torres Association, Merida Andes	PDVSA (2005)
	El Aguila Formation, Merida Andes	PDVSA (2005)
	Iglesias Complex, Merida Andes	PDVSA (2005)
	Macuira Formation gneisses, southeast Guajira (Colombia)	Maya-Sánchez (2001)
	Santa Marta Metamorphic Belt (Colombia)	Maya-Sánchez (2001)
	Silgara Formation, Santander Massif (Colombia)	Maya-Sánchez (2001)
	Cordillera Central Colombia (west of Palestina Fault)	Maya-Sánchez (2001)
	Bakhuis granulite belt, Suriname	Delor *et al.* (2003*a*)
	Greenschist belts, French Guiana	Delor *et al.* (2003*b*)
	Rio Caroní sediment (Guayana Shield)	Wynn (1993)
Kyanite	Eastern Guayana Shield	PDVSA (2005)
	Manicuare Formation, Araya Peninsula	Schubert (1971)
	Margarita Island	Stöckhert *et al.* (1995)
	Cordillera de la Costa (Central Venezuela)	Sisson *et al.* (1997)
	Iglesias Complex, Merida Andes	PDVSA (2005)
	Silgara Formation, Santander Massif (Colombia)	Maya-Sánchez (2001)
	Caldas Gneiss (Central Cordillera, Colombia)	Maya-Sánchez (2001)
Sillimanite	Imataca Archaean granulite gneisses, Pastora, Botanamo Proterozoic greenschists, overlying Roraima sediments, Guayana Shield	Sidder & Mendoza (1995)
	Cabriales Gneiss (Central Venezuela)	PDVSA (2005)
	Santa Marta Metamorphic Belt (Colombia)	Maya-Sánchez (2001)
	Silgara Formation, Santander Massif (Colombia)	Maya-Sánchez (2001)
	Guaicaramo, Garzón Massifs (Eastern Cordillera, Colombia)	Maya-Sánchez (2001)
	Cordillera Central Colombia (west of Palestina Fault)	Maya-Sánchez (2001)
	Guiana Shield (SE Colombia)	Maya-Sánchez (2001)
Andalusite	Cerrajón Schist, El Baúl Uplift	PDVSA (2005)
	Cerro Azul, El Aguila Formations, Mérida Andes	PDVSA (2005)
	Iglesias Complex, Mérida Andes	PDVSA (2005)
	Pastora, Botanamo Proterozoic greenschists and metavolcanics, Guayana Shield	Sidder & Mendoza (1995)
	Roraima and Mapare Guayana Shield quartzites	PDVSA (2005); Sidder & Mendoza (1995)

(*Continued*)

Table 6. *Continued*

Mineral	Sediment source area	Reference
	Northwest Guajira schists (Colombia)	Maya-Sánchez (2001)
	Santa Marta Metamorphic Belt (Colombia)	Maya-Sánchez (2001)
	Cordillera Central, Colombia (west of Palestina Fault)	Maya-Sánchez (2001)
	Guayana (or Guiana) Shield (SE Colombia)	Maya-Sánchez (2001)
Garnet	Shield amphibolite facies terranes	PDVSA (2005)
	Hato Viejo Formation (Cambrian, Guárico Wells)	PDVSA (2005)
	Mérida Andes amphibolite facies terranes	PDVSA (2005)
	Cordillera de la Costa HP/LT belt	Sisson *et al.* (1997)
	El Tinaco Allochthon	Oxburgh (1966)[†]
	Las Brisas, Formation, Caracas Group	PDVSA (2005)
Titanite	Ubiquitous in metagranites and schists from Araya–Paria and west, not source specific	PDVSA (2005); Bellizzia (1985)
Apatite	Ubiquitous, not source-specific	PDVSA (2005)
Blue tourmaline	El Baúl Arch 'cornbunianites' (or tourmalinites)	PDVSA (2005)
	Guayana Shield 'cornubianites' (or tourmalinites)	PDVSA (2005)

*Text to accompany Maya-Sánchez and Vásquez-Arroyave (2001).
[†]This area originally interpreted as Las Mercedes Formation, Caracas Group but now mapped as part of the El Tinaco 'basement' complex (e.g. Bellizzia, 1985).

Colombia in the Early Cretaceous and comprise the basement to the 'Great Arc of the Caribbean'. Examples include the schists and gneisses of the northwest Guajíra Peninsula, parts of the Santa Marta Massif, and the Arquia Complex of western Colombia. We also interpret the El Tinaco Complex as a far-travelled Palaeozoic basement fragment (an idea proposed by Bellizzia 1985), because its geological history is typical of circum-Caribbean terranes and unlike that of the South American autochthon in central Venezuela. The associated Tucutunemo Formation includes Permian limestones that are not local to central Venezuela (Benjamini *et al.* 1986, 1987). These rocks are the westernmost potential sediment source area for sillimanite and andalusite, prior to eastward migration with other Caribbean terranes.

Class 2: Far-travelled Caribbean high-pressure, low-temperature (HP/LT) metamorphic terranes with mostly Palaeozoic or older major sedimentary (primary) and plutonic or igneous (secondary) protoliths

Class 2 rocks are situated in the same areas as Class 1. Well-known examples include the Juan Griego high-pressure rocks of Margarita, some Cordillera de la Costa metasedimentary protoliths and possibly much of the Manicuare Formation (Schubert 1971) of the Araya Peninsula. The garnet and glaucophane schists of the Yaritagua Complex of Lara, Venezuela (protolith age unknown) may also belong in this grouping (since pre-Caribbean

HP/LT rocks are unknown in northern South America). All are intimately associated with HP/LT metaigneous rocks and are often near ophiolite remnants inferred to be part of the Caribbean suture zone. By 100–120 Ma they were juxtaposed with younger igneous rocks (Class 3, below) and share the same metamorphic, structural and exhumation history (Stöckhert *et al.* 1995; Pindell *et al.* 2005). On the north side of the Venezuela and Trinidad foredeep basin, Margarita and the Cordillera de la Costa are the easternmost identified potential sources for staurolite and the Manicuare Formation is the easternmost identified source for garnet and kyanite. The limited drainage from these areas into the Orinoco suggests that they are secondary sources for foredeep staurolite, kyanite and garnet, at least for the last few million years, compared with the Mérida Andes (Class 5 below), which drain directly into the headwaters of the Orinoco.

Class 3: Far-travelled Caribbean HP/LT metamorphic terranes with Cretaceous volcanic and volcano-sedimentary protoliths and initial age of metamorphism

Class 3 rocks include the Rinconada metabasic rocks (Margarita), the Villa de Cura blueschist belt, and the Jambaló, Barragán and Pijao blueschists of Colombia. The Cretaceous protolith age is constrained by the oldest U–Pb zircon ages from Margarita (for a very detailed review of the geological evolution of Margarita, see Maresch *et al.* 2009). Detailed metamorphic and geochronological studies indicate protolith eruption followed rapidly by juxtaposition with continental-affinity rocks, high-pressure blueschist–eclogite metamorphism at 100–120 Ma, onset of exhumation and cooling in the Mid-Cretaceous, and intrusion at mid-crustal levels by granitoids by 85–90 Ma. These two HP/LT rock types are the only identified sources of the glaucophane that characterizes Late Oligocene and younger sediments in the study area and are also likely sources of kyanite and chloritoid. The easternmost glaucophane source is found at Tres Puntas on the Araya Peninsula, mapped as part of the El Copey Formation. The Villa de Cura, Cordillera de la Costa and Manicuare Formation are the easternmost identified chloritoid sources. There is a report of possible lawsonite in the Bocas-1 well (Escalona 1985), just north of the Paria Peninsula.

Class 4: Far-travelled Caribbean lower-grade metamorphic terranes with Cretaceous volcanic and volcano-sedimentary protoliths and initial age of metamorphism

Class 4 rocks include the schists of the Quebradagrande Complex and NW Guajíra Peninsula, both in Colombia, and the schists and metaigneous rocks of Tobago (Snoke *et al.* 2001). The Cretaceous rocks which unconformably overlie the allochthonous El Tinaco 'basement' of Central Venezuela also belong in this grouping. The Cojedes (with a basal conglomerate on El Tinaco Complex basement), Pilancones and Araguita Formations comprise Aptian–Albian clastic sediments and carbonates, with intercalated Albian and younger volcaniclastic and extrusive rocks (Bellizzia 1985). In the offshore Bocas-1 well along the northern Paria Peninsula, limestones thought to be Albian overlie metavolcanic schists (Ysaccis 1997), and the well may also contain a thin section of Mejillones Formation arc volcanic rocks. The association of low-grade metasedimentary rocks unconformably overlain by limestones and pillow basalts is typical of many Aptian–Albian circum-Caribbean arc terranes (e.g. Lebron & Perfit 1993; Pindell *et al.* 2005, 2006) and indicates a far-travelled Caribbean origin for the El Tinaco and the Bocas rocks. Low-grade metavolcanic rocks mapped as El Copey Formation on the Araya Peninsula may also belong in this class. They are spatially associated with fragments of Mid-Cretaceous basalt that structurally overlie low-grade continental-affinity Mesozoic metasedimentary rocks (Class 6 below) and may be the remnants of the sole thrust of the Caribbean accretionary prism. The Sans Souci basalts of northern Trinidad may also be of this origin (Algar & Pindell 1993). Metamorphism is typically prehnite–pumpellyite,

greenschist or lower amphibolite facies and fossils are sometimes well preserved. The terranes are potential sources of pyroxene, hornblende, apatite, chlorite and apatite. There are few metapelites in these terranes, and no reports of chloritoid, garnet, staurolite or aluminum-silicates.

Class 5: Parautochthonous and authochthonous greenschist to amphibolite facies metamorphic terranes with Precambrian to Palaeozoic protoliths and typically Late Palaeozoic (Carboniferous–Permian) initial metamorphism

Rocks sharing the same Pangaea-assembly origin as Class 1 are common inboard of the Caribbean allochthons. They are found within thrust sheets driven ahead of accreted Caribbean rocks or uplifted during Late Oligocene and younger 'Andean orogeny', as 'basement' slices beneath Mesozoic metasedimentary rocks, and as basement arches in the foreland. As such, they typically have significantly younger cooling ages than similar rocks associated more closely with the Caribbean Arc, and those ages become younger from west to east. Examples include:

- Metamorphic belts intimately associated with the accretion of Caribbean terranes, such as parts of the Central Cordillera, Santa Marta Massif and Guajíra Peninsula of Colombia in the west. Garnet, sillimanite and kyanite-bearing schists are common in these belts. In the west, these terranes started to exhume following Late Cretaceous onset of subduction of Caribbean lithosphere beneath Colombia and accretion of Caribbean allochthons, while in the east Caribbean-associated unroofing is of Miocene or younger age.
- Greenschist and amphibolite facies belts which started to unroof at the Late Oligocene onset of regional Andean orogeny in western South America, which include parts of the central Cordillera, the eastern Cordillera of Colombia, the Santander Massif and the Mérida Andes. With the exception of the Colombian central Cordillera, which was unroofed in the Middle Eocene and then mostly re-buried in Late Eocene–Oligocene time (Pindell *et al.* 1998), there is no evidence for earlier Palaeogene unroofing of these belts. Andalusite is commonly associated with low-pressure contact metamorphism adjacent to Late Palaeozoic plutons. A large area of the Mérida Andes drains directly into the upper Orinoco and this may have been the primary source for Late Oligocene and younger staurolite and Al-silicates. In contrast, these minerals

are restricted to the Cordillera de la Costa of Central Venezuela and, at least at the present day, this area does not drain into the Orinoco. In the past, the Caribbean Mountains source areas may have been larger and may have drained south into a palaeo-Orinoco.

- Sheared granitoids and metasedimentary rocks such as the Sebastopol Gneiss (Central Venezuela) and the Dragon Gneiss of the Paria Peninsula (Eastern Venezuela) are inferred to be the basement of Caracas Group and to structurally underlie far-travelled Caribbean rocks. Megafeldspar augengneisses are associated with greenschist facies (chlorite, biotite, muscovite bearing) schists and phyllites, some granites and associated hornfels contact metamorphic rocks. Reported protolith ages (mostly Rb–Sr) range c. 450–166 Ma and K–Ar and fission track cooling ages are as young as Miocene. There are no reported occurrences of garnet or staurolite within these 'basement' gneisses and sillimanite is only reported in one place, adjacent to the Cordillera de la Costa. There is no evidence that Palaeogene unroofing was sufficient to fully exhume these rocks and their mineralogy indicates they are not a source for higher grade metamorphic minerals (in contrast to the model of Higgs 2006, 2009).
- Poorly dated Palaeozoic rocks are also known from the foreland El Baúl Arch, where low-pressure metavolcanic and metasedimentary rocks, intruded by Carboniferous or Permian granitoids, crop out within the foreland basin. Andalusite is typical of contact metamorphic aureoles. The age of uplift of this arch is uncertain and it seems likely to be only a secondary source of andalusite, possibly only in the Late Neogene.
- On the north side of the Guayana Shield there is an abrupt transition from Precambrian and Palaeozoic rock. Within the Palaeozoic terranes there may be one or more sutures related to Pangaea assembly which could be the source of some blue amphibole (identified in probable error as glaucophane) and chloritoid, found in small quantities in both pre-Cretaceous (possibly Jurassic) and Late Cretaceous passive margin strata. Such rocks have not been identified in the relatatively few deep wells and, thus, this idea remains unproven. Low-grade or non-metamorphic Palaeozoic clastic rocks are also reported from the Guárico Basin subsurface (Cambrian Hato Viejo and Carrizal Formations, PDVSA 2005). Heavy mineral assemblages in these rocks are mature, with some garnet. The strata are notably micaceous. These Palaeozoic rocks appear to be absent from the basement of the Maturín Basin, east of 65°W.

Class 6: Parautochthonous greenschist-facies or lower-grade metamorphic terranes with mostly Mesozoic protoliths

In the Caribbean Mountains of Central Venezuela (66–68° W, Fig. 1), the Caracas Group comprises greenschist facies schists in an anticlinal structural window beneath the Villa de Cura HP/LT allochthons. Limited K–Ar and fission track data indicate that peak metamorphism pre-dates the arrival of Caribbean allochthons dated by nearby foredeep subsidence (Pindell et al. 1991). Similar rocks are found in the Araya and Paria Peninsulas north of the Serranía Oriental and in the Northern Range of Trinidad, where cooling appears to be Late Oligocene and younger. Quartzites and marbles are common. Greenschist facies phyllites and schists are characterized by muscovite, epidote, chlorite and graphite. There are no reports of staurolite and chloritoid or Al-silicates. Garnet is not known east of the westernmost Araya Peninsula. Thus, as with the 'basement' gneisses above, they are not a viable northern source region for sediments with higher-grade metamorphic minerals.

Class 7: Mesozoic sediments, unmetamorphosed, Cenozoic uplift and unroofing ages

Unmetamorphosed Cretaceous strata of the Colombian eastern Cordillera and the Subandean fold-thrust belts of Ecuador and Colombia unconformably overlie metamorphic and plutonic rocks of Jurassic and older age. They uplifted significantly for the first time during the Neogene (e.g. Villamil 1999) and may have contributed significantly to Early Miocene and younger rocks in the study area. There are no published heavy mineral data on these rocks, but we expect they are broadly similar to rocks of the same age (i.e. mineralogically mature) examined in the course of this study.

Class 8: Heterogeneous Precambrian metamorphosed sedimentary and igneous rocks in the Guayana Shield

A wide variety of minerals are reported from the amphibolite and granulite facies rocks of the Guayana Shield in Venezuela (alternatively spelled Guyana or Guiana in neighbouring countries), and from associated granitoids and their metamorphic aureoles, including all three Al-silicates, garnet, cordierite and chloritoid (Sidder & Mendoza 1995; Schruben et al. 1997). Glaucophane is reported, but no locality given, by Kamen-Kaye (1937), but is otherwise unknown anywhere in the shield (Salomon Kroonenberg, pers. comm. 2008). Riebeckite and other blue-green amphiboles, not characteristic of HP/LT metamorphism, are present (Schruben *et al.* 1997) and we suspect that these may have been mistaken for glaucophane. Tropical weathering and laterite formation result in a low-diversity heavy mineral assemblage in river sediment regardless of rock type in the drainage basin. We expect garnet to be more stable than staurolite, chloritoid (in acid water conditions), Al-silicates and, especially, glaucophane. None of these minerals are likely to survive more than two cycles of sedimentation. There appears to be only one published report on recent river sediment on the shield. The Río Caroní, which flows north from the Guayana Shield at *c.* 63° W, crosses a wide variety of metasedimentary and metavolcanic terranes. Heavy mineral residues comprise ZTR, ilmenite and, notably, staurolite (Wynn 1993). Staurolite is not reported *in situ* from the Venezuelan portion of the shield, but is present to the SE in contact aureoles around granitoids within the greenstone belts of French Guiana (Delor *et al.* 2003*a*), continuous with those of Guyana (Cole & Heesterman 2002) and southeastern Venezuela, and in the high-temperature rocks of the Bakhuis Metamorphic Complex of Suriname (Delor *et al.* 2003*b*). Chloritoid is widespread, and appears to be a product of staurolite breakdown during near isobaric cooling. Although the shield must have been an important sediment source on the Cretaceous passive margin (below), we suspect that it only made a relatively small contribution to the characteristic unstable heavy minerals found in the Neogene basins because of the effects of tropical weathering. Although areas of low elevation have much lower denudation rates than areas of higher elevation (e.g. Wilkinson & McElroy 2007), the enormous area of the shield resulted in a large volume of sand, deposited in southwest to NE-flowing fluvial–deltaic systems fringing the foreland basin in Venezuela and Trinidad up to the present day, and these sediments constantly dilute the heavy minerals derived from more exotic peri-Caribbean terranes.

Palinspastic reconstruction, tectonic elements, and plate boundary/thrustbelt evolution

In order to understand original facies distributions and the relationships of sediments to possible sediment source areas, it is essential to palinspastically restore plate movements and structural deformations back in time, and to portray former sedimentation patterns on appropriate palinspastic basemaps. Post-depositional deformation can juxtapose sediments and facies of very different origins, rotate palaeoflow indicators, and change the orientation of sandstone fairways. In NE South America, there have been three superimposed deformation phases since the Jurassic creation of the passive margin, from youngest to oldest: (1) eastwest-oriented Caribbean–South American dextral strike–slip since about 10 Ma; (2) Early and Middle Miocene southeastward dextral oblique collision between the Caribbean and South American crusts; and (3) the Palaeogene development of the Proto-Caribbean Inversion Zone (Pindell *et al.* 1991, 1998). A map restoring phase 1 deformation should be used to show late Middle Miocene facies belts, and a map restoring phases 1 and 2 deformations should be used to show Oligocene facies belts, and so on.

From 2001 to 2007, Petrotrin provided the opportunity for us to work with much of the relevant seismic and well data in and around Trinidad and Tobago in conjunction with our field studies in eastern Venezuela, Trinidad and Barbados. We have critically examined most structures in the region to assign them to the correct phase of deformation, and have tested and refined estimates of shortening and strike–slip offset. This is important because most workers have previously combined the structures of phases 1 and 2 into a single model of ongoing transpressive collision between the Caribbean and South America, thus blurring the superposition of distinct structural styles.

Our palinspastic maps address the stratal level at which deposition was occurring for the indicated age of the map. We account for the depth to fault detachment and only apply the restoration to strata within fault hanging walls or to terranes that are allochthonous relative to South America. This principle has some important consequences for understanding geological development, particularly in southern Trinidad where Late Miocene through Pliocene eastward extension soled into a detachment above previously deformed Middle Miocene and older strata, while a similar magnitude of dextral shear soling into an intra-Cretaceous or base-Cretaceous detachment was occurring on the Point Radix–Darien Ridge fault zone through central Trinidad. Thus, palinspastic grids appropriate for Cretaceous levels show large magnitude offset across the Point Radix Fault, while grids for Late Neogene strata show much smaller offset, although the deformation is of Late Neogene age.

The methods of palinspastic map construction in northern South America have been described elsewhere (e.g. Pindell *et al.* 1998, 2000), and the detailed strain estimates used to construct the maps shown here are discussed in Pindell & Kennan (2007*a*). We have constructed palinspastic latitude–longitude grids for *c.* 12 Ma (Fig. 10), at the transition to phase 1 strike–slip-dominated plate boundary movements and for *c.* 25 Ma (Fig. 11), restoring the effects of transcurrent motions and oblique collision of the Caribbean Plate in eastern Venezuela and Trinidad. The estimated position of the crystalline leading edge of the Caribbean Plate is shown in both maps. These reconstructions are based on careful assessment of shortening, extension and strike–slip offsets, but reconstructions for times before 25 Ma are subject to greater uncertainty, as are strain estimates in central and western Venezuela. However, south of the deformation front shown for 25 Ma (Fig. 11), the basemap for older reconstructions remains the same.

A general tectonic elements map of the Caribbean–South America collision zone (Fig. 12), using the 25 Ma palinspastic restoration, includes the migrating Caribbean trench and arc, the migrating Caribbean foredeep on the South American margin, and the Proto-Caribbean Inversion Zone of northern South America. The concepts of arc-passive margin collision (Speed 1985) and of a migrating Cenozoic arc-passive margin collision (Dewey & Pindell 1986; Pindell *et al.* 1988) have been well accepted, but the existence of a Proto-Caribbean Inversion Zone ahead of the Caribbean Plate remains more speculative, and is one of the features on which this paper may shed some light. The Proto-Caribbean Inversion Zone was probably initiated in the latest Maastrichtian, and certainly by the Paleocene, by which time motion between North and South America had become convergent (Pindell *et al.* 1988; Müller *et al.* 1999). Based on seismic tomographic images (Van der Hilst 1990) of subducted Atlantic/Proto-Caribbean lithosphere below the eastern and southern Caribbean Plate, we have previously suggested (Pindell *et al.* 1991, 2006; Pindell & Kennan 2001, 2007*a*) that the convergence was accommodated, prior to the arrival of the Caribbean Plate from the west, at a newly formed south-dipping 'Proto-Caribbean Inversion Zone' beneath northern South America. Today, only the eastern end of this inversion zone has not yet been subducted beneath the Caribbean Plate and remains visible at the Earth's surface, projecting east from the Lesser Antilles trench. There, an ENE-trending ridge (south) and trough (north) pair with some 3 km of buried basement relief is situated between the Caribbean crystalline limit and the Late Maastrichtian western Atlantic magnetic anomaly 30, and cuts across the previously formed regional pattern of Atlantic fracture zones (Speed *et al.* 1984). The linearity, basement relief/and cross cutting relationship of this structure are suggestive of a north-vergent thrust/inversion zone plate boundary. Beneath the Caribbean Plate, the

seismic tomography suggests that this plate boundary once continued WSW to at least the Golfo de Triste along Venezuela, at the NE end of the Mérida Andes. From there, the plate boundary may have continued along the limit of the continental crust to the north of Maracaibo and northern Colombia. It is not yet clear if deformation in the Mérida Andes of possibly Palaeogene age tied into this structure as well. Accepting our projection of the Proto-Caribbean Inversion Zone below the present day Caribbean Plate (Fig. 12), we suggest that the Proto-Caribbean hanging wall would have had positive but submarine bathymetric expression created by basement uplift about 2–3 km, as basement does today in the not-yet-subducted area south of Tiburón Rise (see basement structure map in Speed *et al.* 1984; Pindell & Kennan 2007*a*).

We propose the existence of the 'Barcelona Trough', situated north of the South American margin and south of the Proto-Caribbean Inversion Zone and accretionary prism, that formed during the Paleocene (Fig. 12). The trough was the site of significant Palaeogene clastic deposition, initially deep water but shallowing upward by Late Oligocene time. It was also the site of initial Oligocene south-vergent structural shortening in Trinidad, as the originally north-vergent Proto-Caribbean Ridge (hanging wall of Proto-Caribbean Inversion Zone) was incorporated into the greater migrating Caribbean orogen, and then backthrust, with respect to the Proto-Caribbean Inversion Zone, southeastward toward the southern Trinidadian platform. The Central Range of Trinidad restores to a former position in the vicinity of the present day Paria Peninsula. The northern part of the Early Cretaceous carbonate platform (subsequently drowned) in the Serranía Oriental restores some 100 km farther NW of the Early Cretaceous southern Trinidad platform (Cuche Formation); there must have been a significant marginal offset and escarpment to the east of the northern part of the Serranía Oriental. This NW-trending feature has been referred to as the Bohordal Escarpment (Pindell *et al.* 1991; Pindell & Kennan 2001). However, because the strata that defined it have been incorporated into the Serranía–Nariva fold-thrust belt, its original position can only be estimated from structural restoration of facies changes (e.g. El Cantil carbonate in Venezuela to Cuche shale in Trinidad).

Any model for the tectonic history of northern South America (Fig. 13) must accommodate Cenozoic convergence between the Americas, which increased westward from zero in the area of magnetic anomaly 30 (*c.* 66 Ma), ENE of Barbados, to some 450 km at Guajíra, Colombia. At crustal levels, as the Caribbean lithosphere arrived in any north–south cross section during its eastward

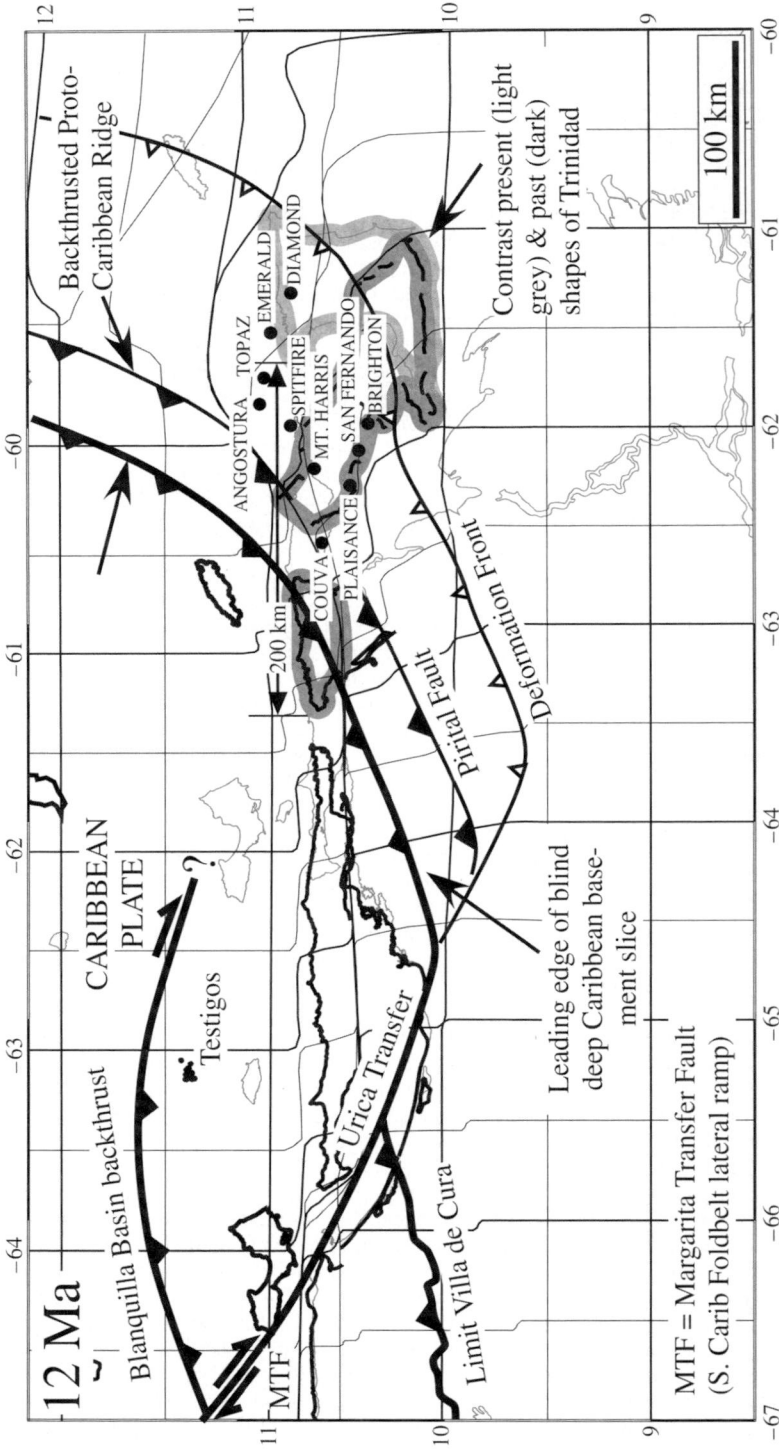

Fig. 10. Palinspastic reconstruction for 12 Ma, close to the end of Middle Miocene orogeny showing the geometry of the orogen before Late Miocene and younger deformation? Distorted latitude–longitude grids account for cumulative deformation and the resulting map is analogous to a restoration of a balanced structural cross-section. Restoring even Late Miocene and younger deformation dramatically distorts the shape of Trinidad (palaeoshape shown with bold lines). To make this map we used relatively conservative shortening and shear estimates south of El Pilar, Caroni Fault. The position of the Northern Range is harder to constrain because of uncertainties in the amount and age of internal strain. The map provides a geographic framework for plotting and reconstructing Middle Miocene deformation.

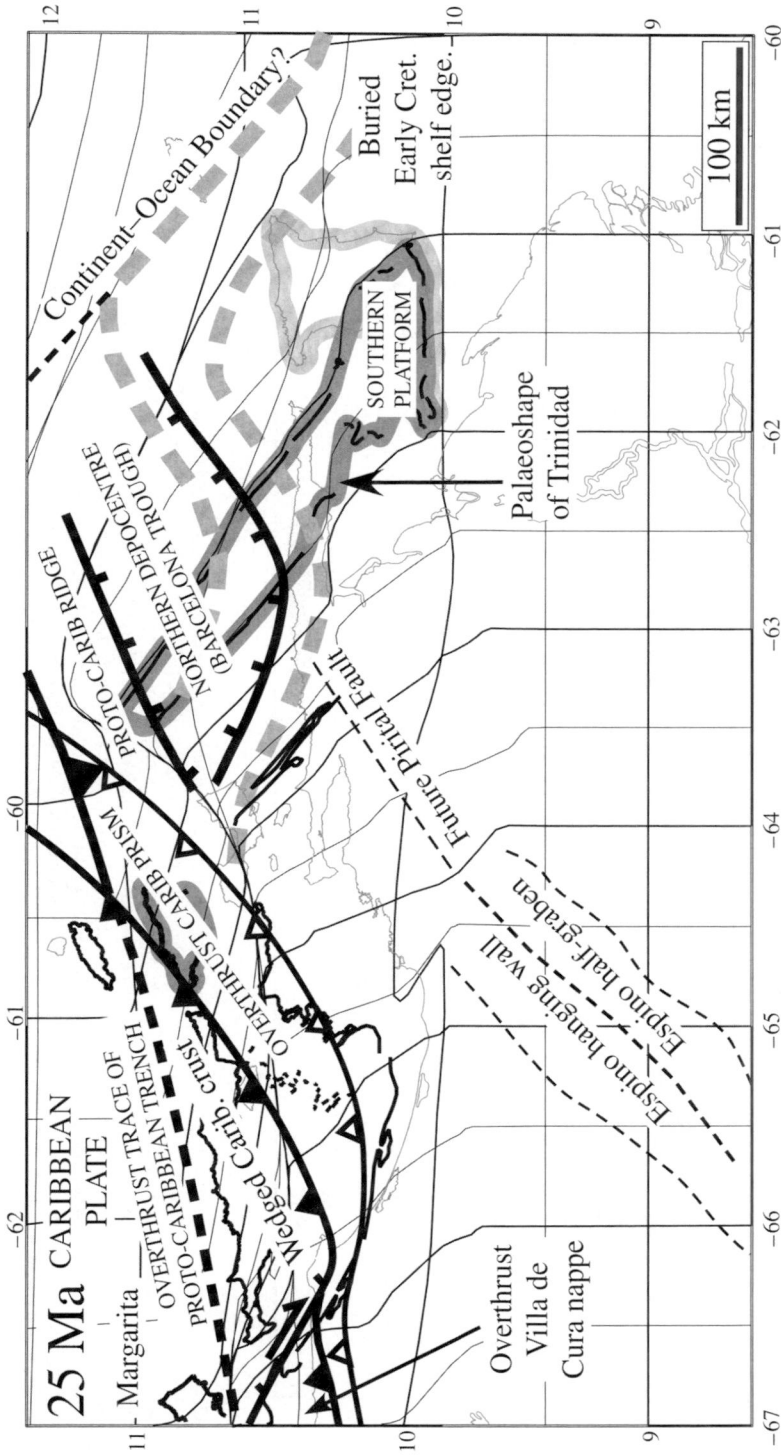

Fig. 11. A 25 Ma palinspastic reconstruction, showing the geometry of the orogen before Early and Middle Miocene deformation. Note that the distortion in the shape of Trinidad indicates that apparent north–south facies changes between the Central Range and Southern Basin in present-day geographic coordinates are in fact NW–SE facies changes parallel to the Bohordal Escarpment. One important effect of the retro-deformation is to place the northern depocentre or Barcelona Trough adjacent to the Serranía Oriental, such that Palaeogene sands can be derived from the west or SW without crossing the southern Trinidad Platform pelagic shelf, where carbonates prevailed. Large Early Cretaceous clasts (such as in the Plaisance conglomerate) can derive from either/both the NE facing Bohordal slope or the north-facing Central Range slope through incision of bypass surfaces with little Late Cretaceous or Palaeogene cover. The map supports the hypothesis of a point source for sediment from what we refer to as the 'Espino–Maturín River'. The proposed 'Espino–Maturín River' was almost certainly the source of the San Juan Sand Lobe of Maastrichtian age (see Fig. 15 below).

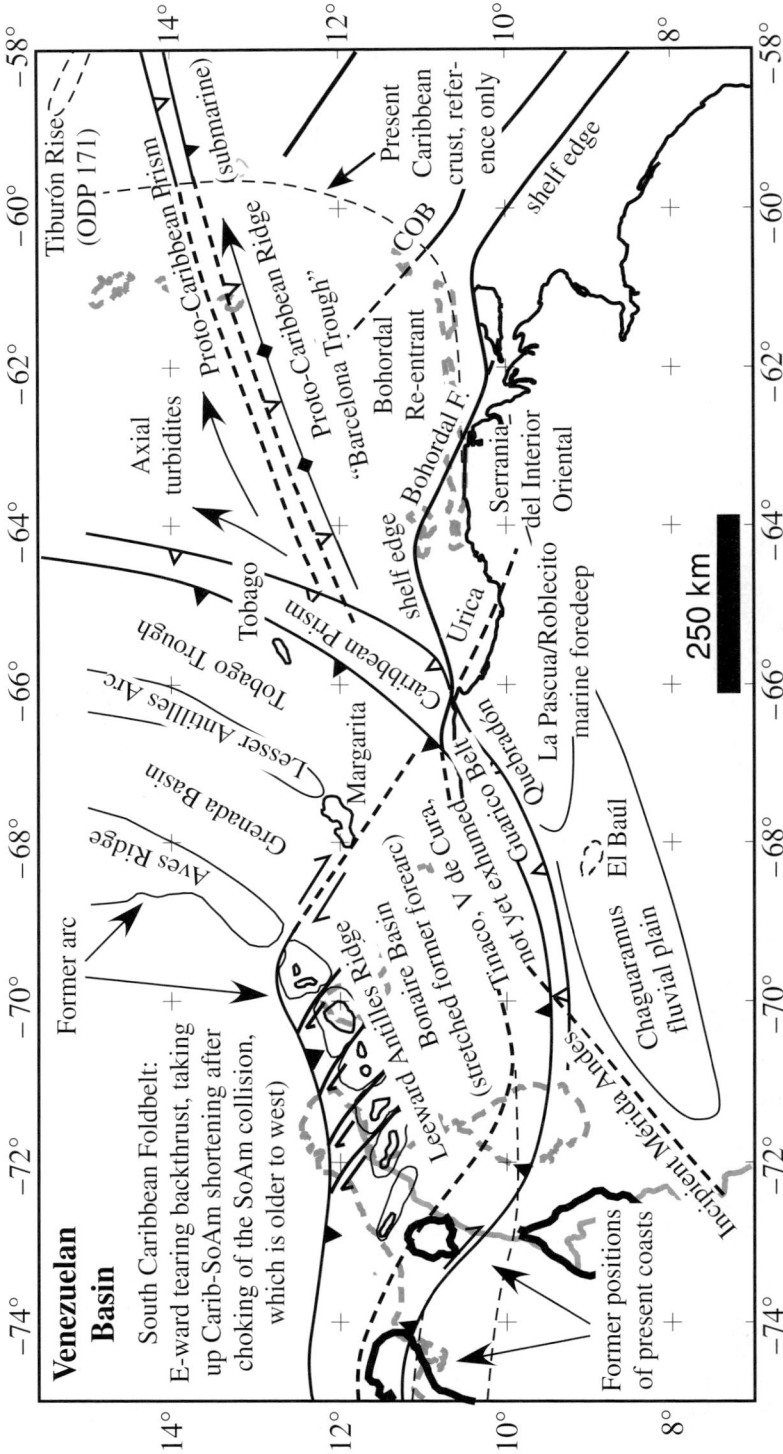

Fig. 12. Mid-Cenozoic tectonic elements of the Cenozoic Caribbean–Proto-Caribbean trench–trench collision, compiled from features identified and defined in Pindell *et al.* (1998, 2006). This reconstruction shows a v-shaped remnant of the Proto-Caribbean Seaway in the NE, bounded by the Caribbean and Proto-Caribbean accretionary prisms. The Proto-Caribbean Ridge is thought to have developed as the South American Plate ramped up over underthrust Proto-Caribbean crust and initiated prism formation. It acted as a bathymetric barrier to sediment flow between the Barcelona and Proto-Caribbean basins, still visible on basement structure maps to the east of the present day Caribbean Prism and south of the Tiburón Rise. From Guajira to the Serranía Oriental (overridden by the Caribbean by the time of this reconstruction), the hanging wall of the trench was continental, and from Serranía Oriental eastwards the trench was intra-oceanic. It is not clear whether a narrow oceanic or a thinned continental forearc existed north of the Serranía or in the Bohordal re-entrant. Caribbean collision with the Proto-Caribbean Inversion Zone was diachronous from west to east, reaching the Guajira area in the Maastrichtian–Paleocene, Maracaibo in the Eocene, Central Venezuela in the Oligocene, and Eastern Venezuela in the Miocene.

migration relative to the Americas, inter-American convergence was partly taken up at the southern Caribbean Plate boundary, giving the southern Caribbean its predominantly compressive character. However, at depth beneath the Caribbean Plate, the original Proto-Caribbean lithosphere must also be shown to balance in north-south cross sections from North to South America (Pindell & Kennan 2007a).

About 100–150 km of shortening had already accumulated at the Proto-Caribbean Inversion Zone at any point by the time it was overridden by the Caribbean Plate and about 70 km of underthrusting had been achieved near eastern Venezuela between latest Maastrichtian and latest Eocene time. Such significant magnitudes of convergence must be manifested in the geology of northern South America, including the heavy mineral data.

If the Mesozoic passive margin slope sedimentary section at the position of the Proto-Caribbean Inversion Zone at the time of its end-Cretaceous inception was a typical 10 km thick, then 100–150 km of north–south convergence by Late Eocene time would have produced a significant accretionary prism at least 15 km thick (possibly partially subaerial) with incipient metamorphism at deep levels (Fig. 13a). The slope and rise strata include the Late Jurassic–Cretaceous Caracas Group of the Caribbean Mountains and the Paria–Northern Range terrane (Speed 1985, 2002; Algar & Pindell 1993). Note that the rocks of the Araya Peninsula differ from those of Paria in that they were metamorphosed in the Cretaceous (Sisson et al. 2005) rather than in the Oligocene (see below), and thus are allochthonous of Caribbean origin, thrust onto the western Paria terrane. The position of this accretionary belt, with an isostatically estimated relief of perhaps 3 km above the Proto-Caribbean seafloor to the north (i.e. 1–2 km subsea), can be crudely reconstructed from seismic tomography data (see Pindell & Kennan 2007a).

As the Caribbean and South America hanging walls converged, the peripheral bulges ahead of them would become yoked (Fig. 13b) and no longer roll or migrate ahead of either. In that case, further convergence would lead to hanging wall uplift in both margins, generating low stand fans along each (Fig. 13c). Further convergence would progressively load the Proto-Caribbean until it sank into the mantle, allowing collision between the Caribbean and South American crust (Fig. 13d). In northern South America, the Caribbean forearc initially overrode South America, but later, after accretion, thrust polarity has reversed, especially in the west (Fig. 13e).

$^{40}Ar/^{39}Ar$ cooling ages on F1 metamorphic micas of 35–40 Ma from the Caracas Group (Sisson et al. 2005), 14–30 Ma from the Paria terrane (Speed et al. 1997) and 23–26 Ma from the Northern Range (Foland et al. 1992) are interpreted by those authors to record the peak of metamorphism coincident with the development of F1 foliation, from which the rocks have progressively cooled by unroofing due to erosion judging from zircon and apatite fission track studies (Algar et al. 1998; Sisson et al. 2005; Cruz et al. 2007). In at least the Paria Peninsula and the Northern Range, this F1 foliation dips south and carries an east–west stretching lineation (Algar & Pindell 1993; Cruz et al. 2006), and in Paria, at least, the early deformation involved top-to-the-west shear (Cruz et al. 2006). The F1 foliation in the Northern Range has been related to north-vergent deformation (Potter 1973; Algar & Pindell 1993), consistent with a north-facing accretionary prism. This may also be the case for the Paria Peninsula, but Cruz et al. (2007) have derived a different (and much younger) explanation for the south-dipping F1 foliation there, the timing of which we question.

The peak metamorphic ages noted above closely match the time of Caribbean oblique arrival at our estimated position of the Proto-Caribbean prism, slightly older to the west (i.e. 100–300 km to the NW of the present positions of these terranes based on palinspastic reconstructions and our estimated position of the Proto-Caribbean Inversion Zone). The Caribbean forearc basement (Villa de Cura complex and its eastward continuation in the western Gulf of Barcelona subsurface; Ysaccis 1997) and the Caribbean accretionary pile (e.g. the allochthonous Caramacate, Los Cajones, Garrapata Formations of the Caribbean Mountains; the allochthonous Manicuare and Copey Formations of the Araya Peninsula, and the Sans Souci allochthon of the Northern Range) were thrust onto the preexisting Proto-Caribbean prism. In central Venezuela and the western Gulf of Barcelona, the Villa de Cura forearc basement wedge itself was thrust over the Proto-Caribbean prism by some 50–100 km, whereas to the east in the Paria–Northern Range, where prism–prism collision began later and in a more oceanic setting, only the Caribbean prism, and not the forearc basement, appears to have been thrust onto the Proto-Caribbean prism.

In central Venezuela, the forearc Villa de Cura/Tinaco Nappe overthrust the Caracas Group and accreted it to its base. With continued shortening, the Villa de Cura/Tinaco/Caracas composite nappe terrane was thrust southward onto, or wedged into, the Guárico Belt of the outer South American shelf margin in the Middle Eocene, which in turn shortened during Middle Eocene time, when peak metamorphism was reached in the Caracas Group (35–40 Ma). In the Late Eocene–Early Oligocene (30–35 Ma), the imbricated Guárico Belt was entirely detached from its basement and thrust to

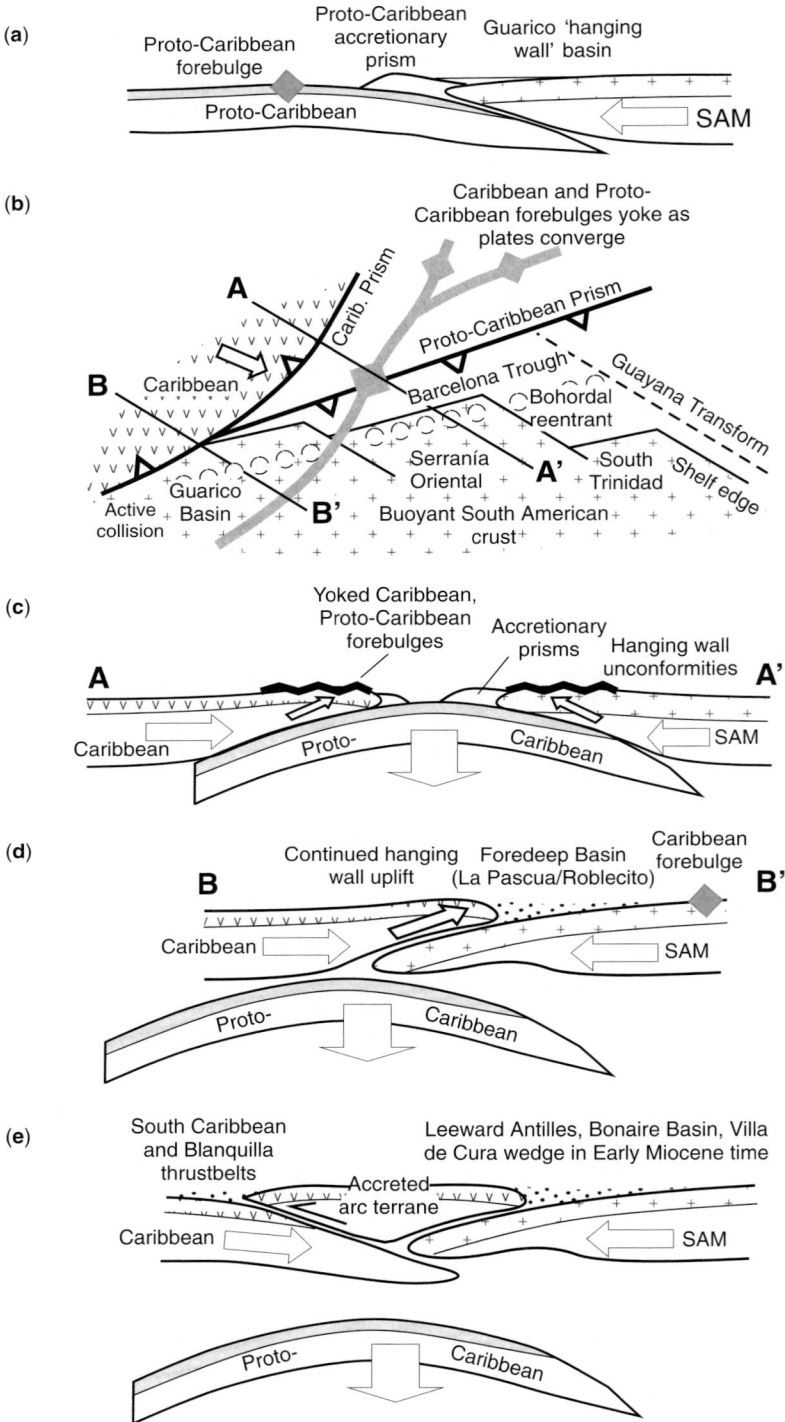

Fig. 13. Schematic history of Caribbean–South American interactions. (**a**) Onset of Proto-Caribbean underthrusting prior to Caribbean arrival, producing Proto-Caribbean accretionary prism. Guárico Basin lies in hanging wall of South

the south, driving subsidence in the foredeep Guárico Basin (La Pascua–Roblecito Formations). In the Early Oligocene, the overthrust South American crust began to choke the collision zone and further advance of the Caribbean nappe pile slowed considerably. The bulk of continued shortening between the Caribbean and South America transferred to new structures in the north, including the north-vergent South Caribbean Foldbelt, and the Margarita Fault or Transfer Zone (Ysaccis 1997) served as the lateral ramp along which that shortening was transferred northwest (on Fig. 12 the Margarita Fault is shown linking to the eastern end of the Villa de Cura Nappe at this time). Both the South Caribbean Foldbelt detachment and the Margarita Fault sole into an intra-crustal detachment that deepens south beneath the Leeward Antilles arc. In Central Venezuela, this south-dipping detachment cut the entire Caribbean lithosphere, allowing the lower footwall of the Caribbean to begin underthrusting the crust of the South American hanging wall (Fig. 13e). About 100 km of further shortening has occurred on this structure since the Oligocene, and a slab of Caribbean lithospheric mantle (not crust) is visible on Central Venezuelan seismic tomographic images beneath the Caribbean Mountains (Van der Hilst 1990; Pindell & Kennan 2007a).

To the east, rather than the Caribbean forearc basement overthrusting the Proto-Caribbean prism, it appears that the Tobago forearc terrane was wedged beneath the Paria–Northern Range terrane. This Early Oligocene event imparted or enhanced the south-dipping F1 foliation in the Paria-Northern Range prism, along with its east-west stretching lineation and top-to-west sense of shear (dextral, north-vergent backthrusting). As a result of this wedge insertion and backthrusting, cooling in Paria–Northern Range was probably initiated by a combination of cooling from below, related to the cold Caribbean forearc slab beneath Paria and the Northern Range, and the surface uplift and erosion that ensued above the wedge. Concurrently, south-directed thrusting of the older Proto-Caribbean prism was initiated, resulting in Late Oligocene syn-tectonic deposition of Upper Naricual Formation of the NW Serranía, and initial imbrication of Early Oligocene strata in the Bohordal re-entrant (possibly including the Angostura Formation, N. Evans, oral presentation to 4th Geological Conference of the Geological Society of Trinidad and Tobago 2007). On the east side of the Margarita Fault, backthrusting within the Caribbean crust also developed in the Early and Middle Miocene at the same time as Serranía thrusting; the Blanquilla foldbelt between Margarita and La Blanquilla shortened some tens of kilometres prior to 10 Ma (Ysaccis 1997), constricting the original southern end of the Grenada Basin. At 10 Ma, the Blanquilla backthrust was positioned entirely west of the Bohordal transfer zone, and backthrusting farther east (NW flank of the Tobago Ridge) is only hinted at locally; it is apparent that, to the east of the Bohordal transfer zone (Fig. 12), the Bohordal re-entrant with its thinner continental or oceanic crust and thick overlying shaly Palaeogene clastic section offered less resistance to south-directed thrusting than the Serranía to the west.

In the light of this dynamic history and our heavy mineral data, we propose the existence of six clastic domains associated with distinct tectonic settings in northern South America. Four domains are present on a Mid-Cenozoic reconstruction (Fig. 12), but two are younger. Three of the domains are not geographically fixed but sweep diachronously from west to east with the migrating Caribbean–South American collision, causing strata of different domains to be superimposed. The six clastic

Fig. 13. (*Continued*) America and thus may be tectonically active but not receiving Caribbean erosional detritus, except possibly airfall tuffs). (**b**) Map of Caribbean–South American oblique convergence above Proto-Caribbean lithosphere for late Middle Eocene time. The obliquity causes both the hanging wall uplift (position shown by fine circles) and the Caribbean forebulge (heavy grey line) to migrate eastwards. Because the trends of the hanging wall uplift and Caribbean forebulge are different, complex interference patterns will result. Here, the Guárico Basin hanging wall uplift is already transgressed by the Caribbean foredeep, but the Serranía Oriental is undergoing hanging wall uplift. Due to the Bohordal re-entrant, the southern Trinidad platform is too far from the Proto-Caribbean Inversion Zone to experience hanging wall uplift. With further convergence, hanging wall uplift will affect the Bohordal re-entrant (shown in dashed circles), and the Caribbean forebulge will pass through Trinidad in the Earliest Oligocene. (**c**) Mechanism of diachronous hanging wall uplift (Caratas, Cautauro, and Peñas Blancas/Tinajitas Formations record shallowing up, but Middle Eocene in Guárico Basin is removed by erosion) and tectonic production of lowstand wedges above an intermediary plate (Proto-Caribbean lithosphere). The two forearcs load the Proto-Caribbean and cause it to sink into the mantle, but not without concurrent uplift of the forearcs. Line of section is A–A' in part (b). (**d**) Result of continuing convergence, with consumption of Proto-Caribbean lithosphere, and the Caribbean overthrusting South America. Migrating Caribbean foredeep develops on South America with Caribbean forebulge beyond the foredeep. Line of section is B–B' in part (b). (**e**) Future (Late Oligocene) tectonics of section B–B'; South America chokes the north-dipping trench, and backthrusting develops outboard of the Leeward Antilles Arc, which takes up most subsequent convergence between the two plates.

domains and their respective proposed source areas are:

- Cretaceous passive margin clastics, derived from the Guayana Shield to the south;
- Palaeogene clastics associated with development and dynamics of the Proto-Caribbean Inversion Zone and the Barcelona Trough, derived from the pre-existing passive margin section and/or shield;
- cratonic side of the migrating Caribbean fore-deep fill, derived from reworked existing marginal strata above the craton, with some first cycle input from the craton;
- orogenic side of Caribbean foredeep fill, derived from the subaerial parts of the migrating Caribbean–South America collision belt and deposited down depositional dip in the foredeep basin axis;
- proximal, syn-orogenic detritus, derived from the uplifting Caribbean–South America collision zone, and deposited adjacent to the mountain front, in wedge top and piggyback basins;
- post-orogenic foreland fill, derived from the east-flowing Orinoco River.

Clastic domains

Domain 1: Cretaceous passive margin

Northern South America persisted as a passive margin until the Maastrichtian, following Jurassic rifting from eastern Yucatán and Neocomian dextral motion along the Guyana Transform (Pindell & Kennan 2007*a*). Sequence- and litho-stratigraphic patterns indicate a northward-deepening palaeo-bathymetry with marginal offsets: the passive margin (Fig. 14) faced the Proto-Caribbean Seaway, an arm of the Central Atlantic (Pindell & Barrett 1990; Erikson & Pindell 1998). Overall, as expected in a thermally subsiding passive margin, successive transgressions step farther into the continent through time, and along the Guyana–Suriname margin former prograding shelf edges have been drowned to form two or three packages of strata back-stepping landward from the continent–ocean boundary.

Seismic, well and sequence stratigraphic data (e.g. Erlich *et al.* 2003; Erlich & Keens-Dumas 2007) indicate that the Early Cretaceous shelf edges were close to the present-day Central Range in Trinidad, and the Guyana Transform offshore from Trinidad to Suriname. Subsequently, the shelf edges in younger Cretaceous strata stepped south (or landward), being located in Maastrichtian time near the south coast of Trinidad and just offshore from Venezuela to Suriname. The heavy mineral signature of Trinidadian Cretaceous rocks

is dominated by zircon, tourmaline and rutile. The paucity of labile minerals (common in the rocks of the eastern Guayana Shield) indicates intense weathering in the sediment source region.

Of particular interest are the red 'cherries' which characterize the Albian Cuche and Toco Formations in Central and Northern Ranges in Trinidad, respectively, and similar red mudstone clasts in the Cenomanian Gautier and Santonian Lower Naparima Hill Formations in the subsurface of the southern basin. In thin section these contain angular silt to fine quartz sand with some muscovite embedded in an iron oxide matrix, and the most likely source is from laterites or intensely oxidized coastal plain mudstones and siltstones, with access to the Shield for mica and fine-grained angular quartz clasts, eroded by streams or coastal incision. Limited XRD data indicates that calcareous nodules within the shales of the Galera Formation shales are neither the same nor sufficiently iron-rich to be precursors to 'cherries' in various younger Cenozoic formations and sections.

The presence of coarse sandstones and cherries in the Cuche Formation of the Central Range has been used by Higgs (2006, 2009) to propose a northern sediment source as old as Aptian. However, the Cuche strata of the Central Range are somewhat older (Barremian–Early Aptian) than shaly 'Cuche Formation' reached in the few deep wells in southern Trinidad, which may be correlative with the Late Aptian to Early Albian Garcia shale in Venezuela. The sediments of the Central Range Cuche could be derived from a southern sediment source if the shelf edge was at or near their site of deposition and subsequently back-stepped to the south. The Cenomanian Gautier and Santonian Lower Naparima Hill Formations have been extensively cored in southern Trinidad and include wedges of sandstone and conglomerate very similar to the coarser Cuche facies (Erlich & Keens-Dumas 2007). These were deposited during lowstands or periods of wetter climate in their sediment source areas. Facies variations and the geometry of slump folds and slump faults, fold and fault vergence, ripples, and pebble orientation and imbrication in oriented cores all indicate southward derivation and that palaeobathymetry deepened to the north/northwest (Larue 1995; Sprague *et al.* 1996; unpublished industry data). To the south of Trinidad, these sands and intervening shale-dominated formations give way to a sand and conglomerate updip fringe (Di Croce *et al.* 1999). We interpret the coarse Cuche facies as south-derived flushes of material encased in shale.

Barranquín and younger Cretaceous clastic sections were sourced from the Precambrian rocks of the Guayana Shield, and from the now-buried Palaeozoic rocks (Hercynian, related to the

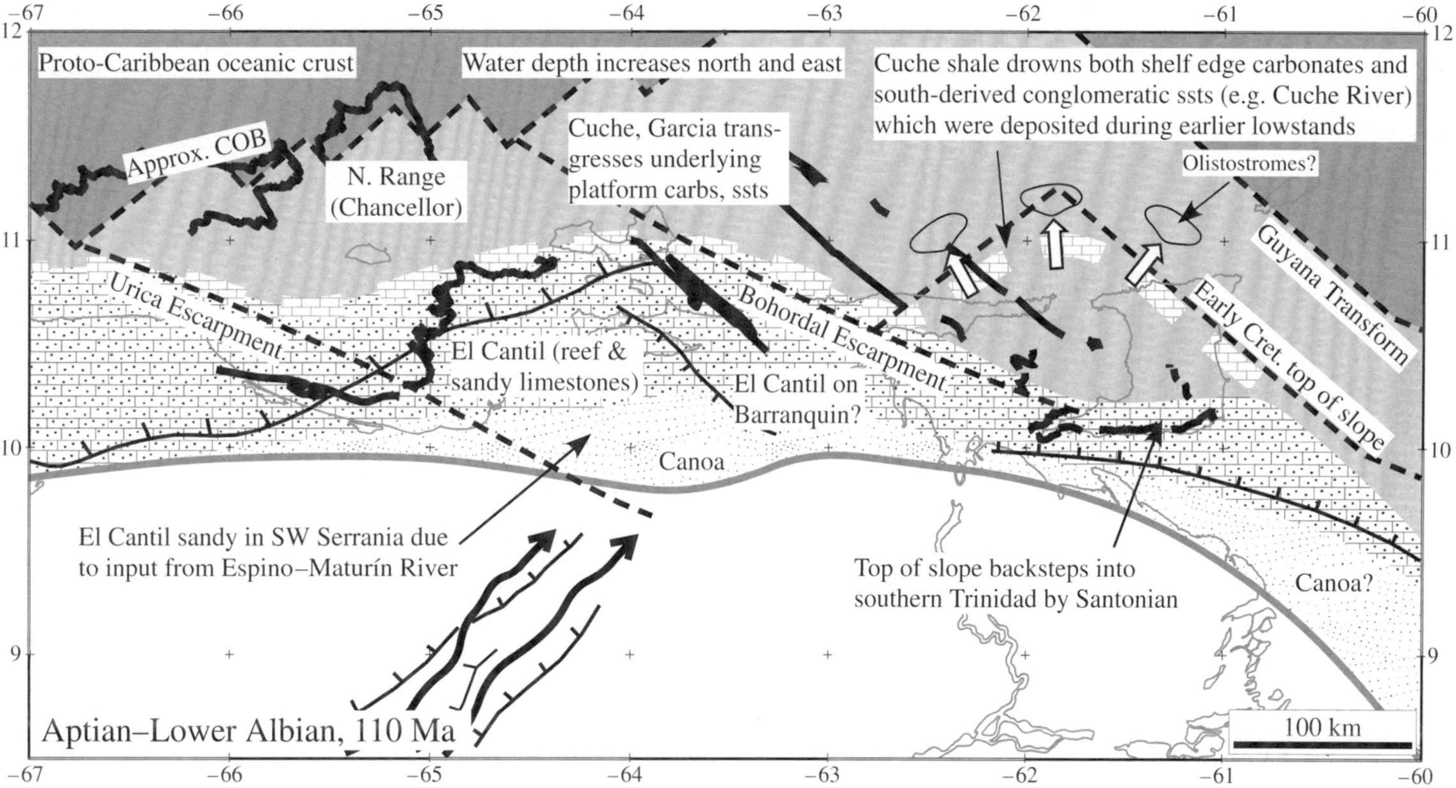

Fig. 14. Palaeogeographic reconstruction for Aptian to Early Albian time. Northern Venezuela and Trinidad were part of the Domain 1 passive margin at this time, facing the Proto-Caribbean Seaway, with continent–ocean boundary (COB) probably located near the palinspastically restored position of the Araya–Paria Peninsulas and the Northern Range. A former shelf (represented by the Barranquin and Morro Blanco Formations in Venezuela) was drowned by the Garcia Shale transgression and only in Venezuela did a widespread carbonate shelf re-establish itself (El Cantil Formation). In Trinidad, Early Cuche sandstones, limestones and conglomerates were buried by the Cuche Shale and succeeding Gautier Shale in the Central Range, also near a drowned shelf edge. We interpret the outcrops in the Cuche River as south-derived, part of a pre-transgression reef or reef-fringing trend, analogous to similar south-derived facies of Cenomanian–Santonian age in southern Trinidad. They are overlain conformably by the Late Aptian Maridale Marl (Garcia equivalent). Strata older than the Maridale Marl have not been drilled in southern Trinidad, but to the south in Venezuela there is a thick sand and conglomerate clastic fringe of Early Aptian and older age overlying the Guayana Shield.

assembly of Pangaea) and Early Mesozoic rift fill (Espino Graben and equivalents). The trace stauro-lite in the Barranquín can be tied to the Shield where it is found today in Río Caroní sands (Wynn 1993). Only to the west, in the Guárico fore-land basin, do Cretaceous formations show small amounts of minerals such as kyanite, chloritoid and glaucophane. The kyanite and chloritoid are both proven in Guayana Shield rocks to the south. The origin of the glaucophane is unclear, but may be from suture zones within the Precambrian, or between the shield and heterogeneous Palaeozoic terranes that underlie the foreland basins of western and central Venezuela. Cretaceous sandy sediments of Late Albian–Cenomanian age in Domain 1 occasionally include garnet and apatite (e.g. Gautier and Toco Formations). This coincides with a well-developed regional lowstand (Late Cen-omanian, Villamil & Pindell 1998) and may indicate faster transport of minerals from source to basin, preserving the intermediate-stability garnet and apatite, but not less stable species.

Domain 2: Palaeogene clastics associated with development and dynamics of the Proto-Caribbean Inversion Zone and the Barcelona Trough

Inter-American convergence began at the proposed Proto-Caribbean Inversion Zone in the latest Cretac-eous (Fig. 15) and this inversion zone and its associ-ated accretionary prism probably led to the development of a submarine ridge that separated the original Proto-Caribbean oceanic basin and the Barcelona Trough. By Paleocene–Early Eocene time (Fig. 16), the Barcelona Trough narrowed westward and tapered out in the Guárico Trough of onshore Venezuela, and was open to the deep Atlantic Ocean toward the east. Unlike the younger Caribbean foredeep there is no diachronous shift in age of its sedimentary fill with respect to South America. The Proto-Caribbean Inversion Zone continued westward, along the northern flank of the structurally growing Northern Range, Paria, and Caracas Group accretionary belt, until the point, which migrated east through the Cenozoic, where it was progressively overthrust by the Carib-bean trench. The Proto-Caribbean Seaway to the north of the prism, but not the Barcelona Trough, received Early Palaeogene to orogen-derived heavy minerals and lithic clasts from the Carib-bean–South America collision zone updip to the southwest.

The two trenches bounding the Proto-Caribbean basin (Figs 16 & 17) were logical sites for long dis-tance down-dip axial transport of Caribbean oro-genic clastic sediments, but much of that detritus may have spread out across the Proto-Caribbean interior as well. The Late Paleocene–Middle Eocene Scotland Formation or Basal Complex of the Barbados accretionary prism, with its glauco-phane, sillimanite and other minerals typical of arc-continent collision, undoubtedly relates to the Proto-Caribbean clastic province somewhere far to the west of Barbados today, as coeval sandstones in eastern Venezuela and Trinidad bear no such min-erals. A question that is not yet resolved is whether, or how much of, the Basal Complex was accreted at the Proto-Caribbean prism v. the Caribbean accre-tionary prism (Fig. 17) prior to their collision and merger in the Middle Miocene (Pindell & Frampton 2007). The occurrence of glaucophane in the Eocene–Oligocene section in DSDP wells on the Tiburón Rise (Mascle *et al.* 1988*a*, *b*) suggests that some orogenic heavy minerals were able to reach even farther NE within the Proto-Caribbean/ Atlantic Basin than the original depositional site of the Barbados Scotland beds. The Paleocene– Early Eocene Matatere Formation of the accreted Lara Nappes of western Venezuela also carries an orogenic signature (PDVSA 2005), and probably represents a more proximal portion of the Caribbean orogenic detrital fairways.

In contrast to the trench-bounded Proto-Caribbean province, the Barcelona Trough was, beginning in the Maastrichtian, shallower than the Proto-Caribbean basin floor and a logical site for deposition of South America-derived progradation of mature clastics as a result of long-term and/or episodic relative sealevel fall. Our data suggest that the Palaeogene sands of eastern Venezuela and Trinidad derive from the Shield or are reworked from Cretaceous passive margin sediments such as the Barranquín, Canoa and Temblador Formations. Recycled Cretaceous sediment is likely to be cleaner than first-cycle sediment sourced directly from the shield, explaining the typical lack of staur-olite, garnet or apatite in the Paleocene through Early Oligocene strata. Suter (1960) reported staur-olite in some Palaeogene samples, drawing on unpublished oil industry work by J. C. Griffiths (the original reports can no longer be found). Our own examination of the few available unpublished industry reports from that era show that trace staur-olite and kyanite has occasionally been found in southern Trinidadian Late Cretaceous (Naparima Hill) to Eocene or Early Oligocene (San Fernando) strata. We found only one report on a single kyanite grain from Pointe-a-Pierre grits at the type section.

Paleocene–Early Eocene strata of the Barcelona Trough (Fig. 16) include the shale dominated fine distal turbidites of the Guárico Formation of Central Venezuela, the northern facies (Pindell & Kennan 2001) of the Vidoño Formation, and the

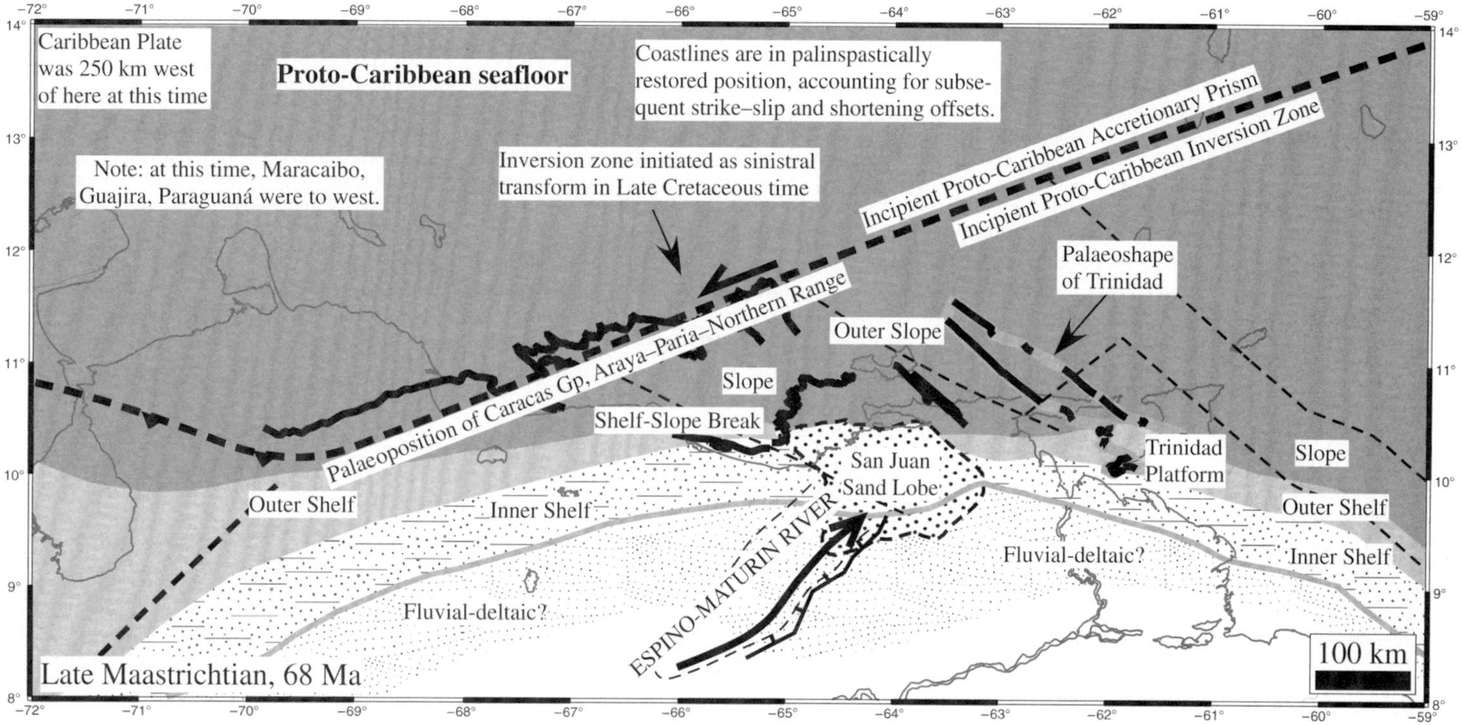

Fig. 15. Palaeogeographic reconstruction for the Maastrichtian (68 Ma) of northeastern South American continental passive margin (Domain 1). Present day coastlines are shown with fine grey lines and restored positions of coastlines are shown with heavy black lines. The palaeogeographic map is drawn for the period immediately prior to the onset of underthrusting on our proposed 'Proto-Caribbean Inversion Zone'. The map shows the buried positions of the Early Cretaceous shelf-slope break beneath the Central Range of Trinidad, which continued to support a relatively shallow 'Trinidad Platform'. Sedimentary facies indicate the existence an inner slope close to the south coast, but the ultimate drop to abyssal depths lay north of the buried older shelf edge. The buried 'Bohordal Fault' is inferred to bound the Serranía Platform on the east, and the ancestral Urica Fault is thought to separate the Serranía from Central Venezuela. Both were active as sinistral transfer faults during Early Cretaceous and older rifting. The continent–ocean boundary to the east of Trinidad is well constrained from present-day Columbus Channel to Suriname. Prior to Jurassic opening of the Proto-Caribbean Seaway, the deep basement of Cuba and the Bahamas Platform lay NE of this line. Sedimentary facies are shown as fluvial fringe, shallow marine inner shelf, outer shelf or upper slope (light grey) and mid-slope and deeper (dark grey). The interpretation is based on well data and our own field work. The southern fluvial fringe is mostly eroded and onlapped by La Pascua and younger formations (Domain 3, southern onlap edge of Caribbean foredeep, see Fig. 17). Of particular note is the San Juan Sand Lobe (isopachs based on Di Croce 1989), which appears to originate from the axis of the Jurassic Espino Graben. The southern boundary faults of this graben (inactive by this time) may have controlled topography and the course of an 'Espino–Maturín River'. The faults were subsequently inverted during the Late Cenozoic (as the Anaco Thrust).

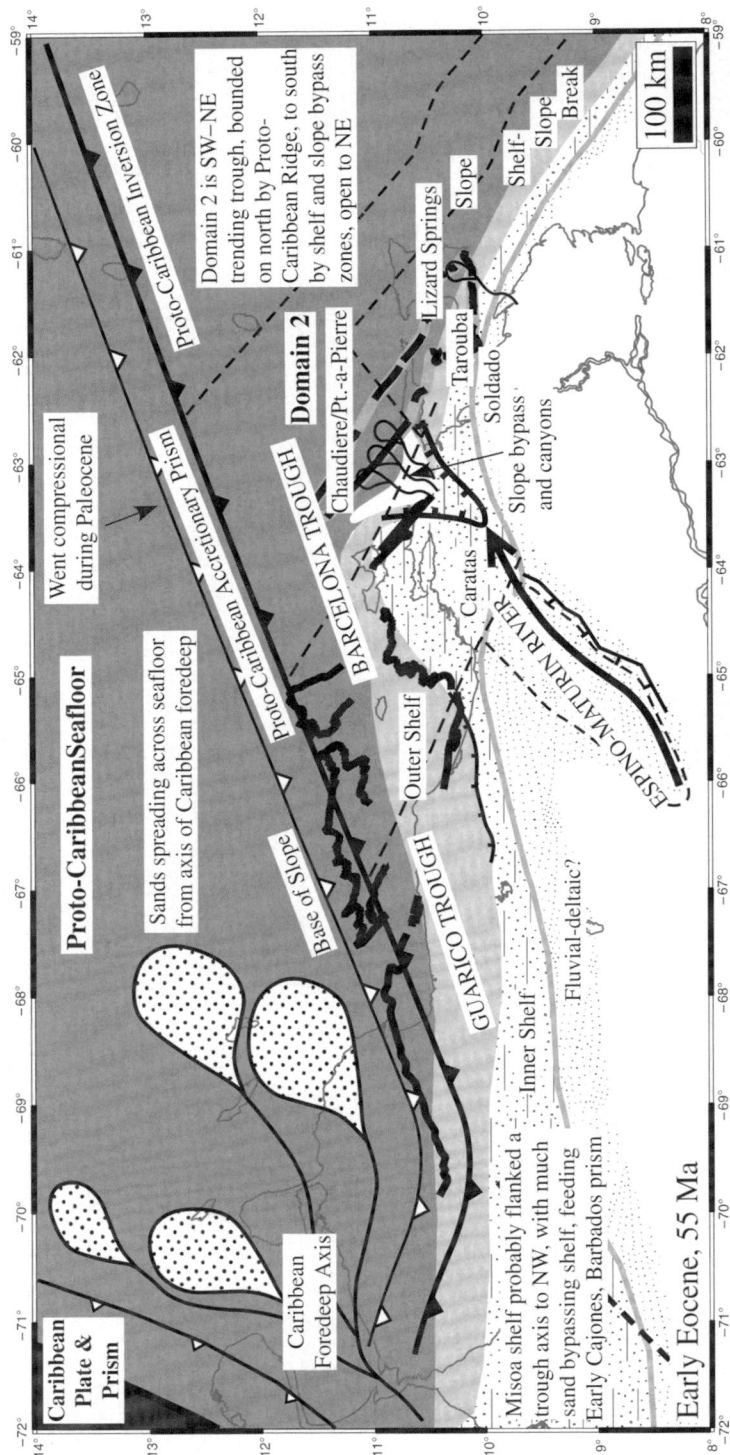

Fig. 16. A reconstruction for Earliest Eocene time (*c.* 55 Ma) shows the effect of the initiation of Proto-Caribbean underthrusting on the north side of the former passive margin. Clastic Domain 2 developed to the south of the uplifted Proto-Caribbean Ridge and associated accretionary prism, to the north of the former Cretaceous slope in the Central Range, and east of the Bohordal Fault, with sedimentary facies and thickness indicating substantially greater waters within the Barcelona Trough than in the Serranía to the SW. At this time, the Caribbean Plate lay about 800 km to the west and was not influencing subsidence and sedimentation in eastern Venezuela and Trinidad. There is no indication of a northern source for any of the sediment in the Barcelona Trough. Instead, reconstructions indicate a point source in the embayment between the Bohordal Fault and Central Range slopes, which may have been fed by the 'Espino–Maturín River' (which previously sourced the San Juan Sand Lobe). The area of the future northeast Serranía unconformity is shown dashed. At this time basement was probably uplifted immediately south of the Proto-Caribbean Inversion Zone, resulting in deposition of a thin and shallow Caratas section, with exposure and erosion only occurring during the latest Eocene to earliest Oligocene.

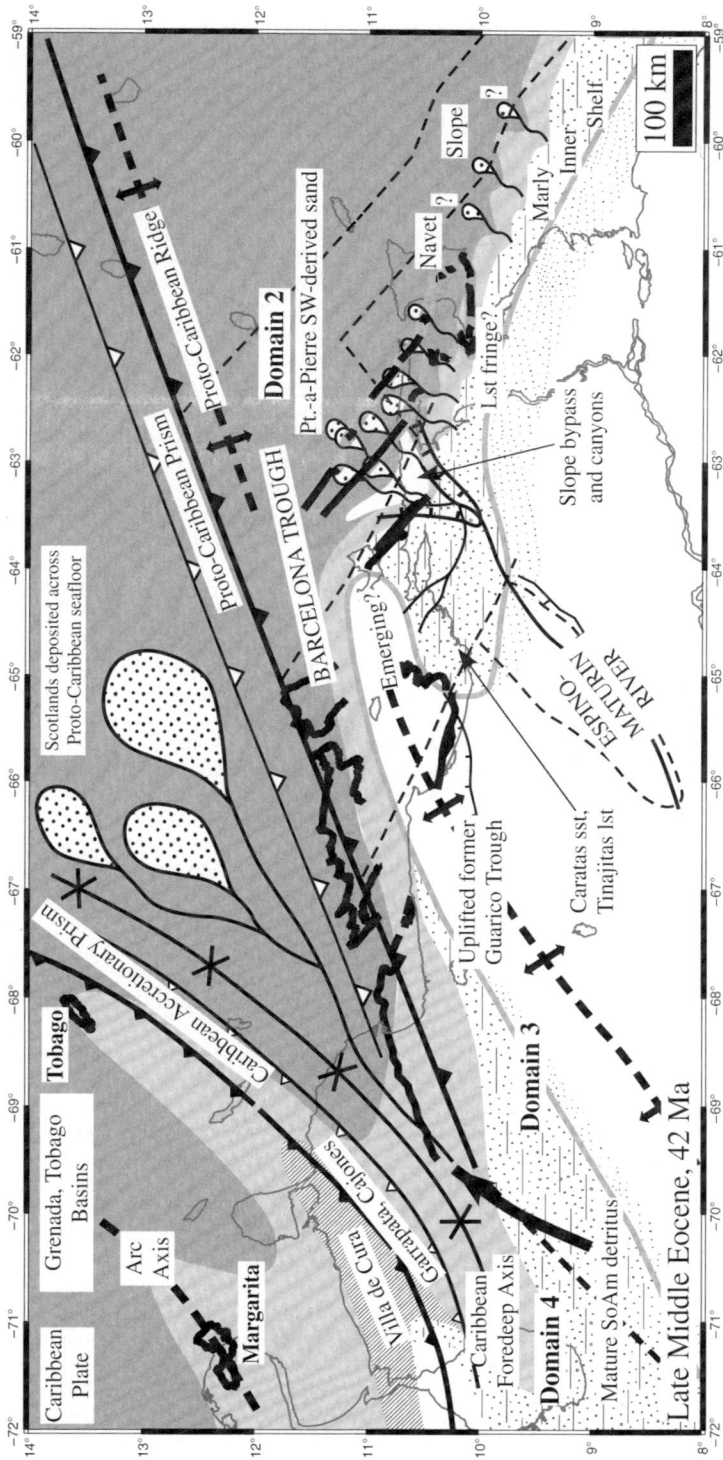

Fig. 17. A late Middle Eocene (c. 42 Ma) reconstruction shows a possible interpretation of the depositional context for the Scotlands of Barbados, if we accept Middle Eocene foraminifera and pollen in the Scotlands as non-reworked and honour the apparent absence of any Late Eocene fauna. The majority if not all of the Scotland sands were probably deposited about 600–700 km downstream of the onshore end of the Caribbean foredeep axis ahead of the leading edge of the Caribbean Plate. It is also possible that some of the Scotland strata were deposited to the north of the Proto-Caribbean Inversion Zone, particularly if the Proto-Caribbean Inversion Zone axis was overfilled, such that accommodation space was only available to the east along the Proto-Caribbean Trench. The presence of trace glaucophane as far as Tiburón Rise to the east of the present-day Barbados Prism shows that Caribbean derived debris may have been spreading over the entire Proto-Caribbean seafloor. A trench associated with a Proto-Caribbean Inversion Zone seems a logical place to expect accumulation of Caribbean-derived turbidites. In the southern Scotlands there is a coarser fraction (conglomerates containing what appear to be Rio Chavez shales and carbonates) that is clearly of South American provenance, and there is also abundant mature quartz sand together with Caribbean immature orogenic heavy minerals. Some of this South American component could come from the proposed northeastern Serranía unconformity. Alternatively, south-derived sediment from the western Guayana Shield may have been transported onto Proto-Caribbean Seafloor along the foredeep axis of the Caribbean. Carbonate and shale clasts may be derived in this case from South American margin rocks already incorporated into the leading edge of the Caribbean orogen in western Venezuela. The map also shows the Caribbean forebulge rolling from west to east. Associated uplift drives a diachronous tectonic lowstand, in this case uplifting the former Guárico Trough. Detritus uplifted off the forebulge may have been deposited to the NW, in the northeastern Barcelona Trough, or to the south and transported to the Trinidad region by the proposed Espino–Maturín River.

Chaudière Formation of central Trinidad, long noted to have lithological and faunal similarities (e.g. Salvador & Stainforth 1968). The Guárico, northern Vidoño and Chaudière Formations carry detrital muscovite, but apparently no heavy minerals other than ZTR. Muscovite is the only mica capable of surviving lateritic weathering intense enough to break down staurolite, garnet and other metamorphic minerals (Anand & Gilkes 1987), so a South American source for the Barcelona Trough sediment is entirely reasonable, the most likely being the Guayana Shield or perhaps reworked Cretaceous Canoa Formation incised inner margin strata, both to the south of the Barcelona Trough. In the northern Serranía Oriental, the northern Vidoño facies beds carry 1–5 m blocks of chert/hemipelagite of probable olistostromal origin. The blocks resemble the underlying Campanian–Maastrichtian San Antonio/Río Chavez Formation of the Serranía platform, and thus are probably south derived. On the eastern flank of Mount Harris, eastern Central Range of Trinidad, either the Chaudière or the Plaisance (field relations do not allow determination of which) also carries 1–5 m probably allochthonous blocks, identified as Maastrichtian Guayaguayare and Upper Cretaceous Naparima Hill and Gautier Formations (Kugler 2001), again well known to the south; transport direction was presumably northwards into the deeper parts of the basin, possibly via shelf-edge canyons or even shelf edge slump complexes. Nothing in these formations suggests or requires a northward derivation of material for these northward deepening units.

Over much of the Serranía Oriental, the glauconitic Vidoño Formation (upper slope/outer shelf) coarsens upward into the glauconite-bearing shelf sands and shales of the Early to Middle Eocene Caratas Formation, which, in turn, is capped over much of the western and southern Serranía Oriental by the Tinajitas Limestone. Because this shallowing upward trend appears to occur over only about 100 m of section, we suspect the shallowing was tectonically driven (i.e. due to hanging wall uplift and/or the Caribbean forebulge). Unfortunately the age of this event is poorly constrained. Mixed assemblages Middle and Late Eocene larger foraminifera are common (Sageman & Speed 2003; O. Macsotay, pers. comm. 1995; T. King, pers. comm. 2008) and the oldest age of overlying strata permits a latest Eocene or Early Oligocene age.

To the west of the Urica Fault trend in Central Venezuela, the Guárico Formation (south of Don Alonso–Guárico Fault) appears not to extend above the Early Eocene level, yet the foredeep subsidence south of the Guárico Formation does not begin until the latest Eocene or Early Oligocene (La Pascua Formation onlap, younging

southeastward). Middle Eocene conditions here are thus not clear. This may be due to extreme telescoping at the frontal Guárico thrust such that a more distal Middle Eocene section is entirely overridden. However, by analogy with the Serranía Oriental, it may be that the Middle Eocene was a time of uplift and erosion in Central Venezuela, such that the imbrication of the Guárico Formation by Caribbean collision post-dated the Proto-Caribbean hanging wall/Caribbean forebulge unconformity. Further, deformations during this pre-Caribbean uplift may have produced a reported angular unconformity beneath the late Middle Eocene Peñas Blancas limestones at the easternmost end of the Guárico Belt, especially near the Urica fault trend, a Jurassic transfer fault repeatedly rejuvenated in the Cenozoic. Our own observations of this reported unconformity along the Piritú ridge, SW of the Gulf of Barcelona suggest Late Miocene shale diapirism and differential deformation of the mechanically rigid limestone and the underlying and more deformable terrigenous beds. Thus it remains unclear to us how much significance should be placed on this reported 'Eocene angular unconformity'. The late Middle or early Late Eocene may then have been the time of initial tectonic emplacement of the overlying Los Cajones/Garrapata allochthonous nappe onto the Guárico strata, the subsequent shortening of which: (1) imbricated the Querecual, Mucaria, and Guárico Formations by perhaps 100 km (Perez de Armas 2005) with no Middle Eocene sections known at any of the Guárico imbricates; and (2) produced the autochthonous latest Eocene–Oligocene La Pascua/Roblecito flexural foredeep basin. The Late Eocene Orupé Formation conglomerates of numerous lithologies in red sandy matrix from the Caribbean Mountains is a heavily eroded molassic orogenic fringe, now known in local positions, small outliers and remnant erosional blocks (Macsotay *et al.* 1995).

In central Trinidad, Early–early Middle Eocene deep-water turbidite fans and channels of the glauconite-free Pointe-a-Pierre Formation are spatially associated with the shales and silts of the Chaudière Formation, which probably served as the basin floor on which the Pointe-a-Pierre channels and fans developed. This setting was situated north of the coeval pelagic/shaly southern shelf/upper slope margin of Trinidad (Lizard Springs and Navet Formations), and was deeper than that of the northern Vidoño Formation. Unlike the exposed Caratas Formation of the Serranía, the Caratas interval of the Maturín subsurface is gritty and lacks glauconite (unpublished oil industry data). Since sand could not have derived from southern Trinidad where coeval pelagic sediments and shale prevailed, the Maturín subsurface is the logical equivalent and updip source for the Pointe-a-Pierre sands of central Trinidad. These relationships support the existence of a Bohordal re-entrant as a palaeogeographic feature.

One final note concerning the Caratas Formation (and equivalents in Trinidad) is that the complete lack of a Caribbean influence in the sandstone mineralogy argues against the proximity of the Caribbean Plate during the Eocene. The Caratas is not 'flysch' in the orogenic sense as portrayed by James (2006), but rather records the reworking of sands along a north facing continental margin.

Diachronous Caribbean basal foredeep unconformity in the northeastern South American margin. In the Guárico (foreland) Basin south of Caracas and west of Urica Fault, a hiatus separates eroded Upper Cretaceous passive margin section from southwardly transgressive ?latest Eocene–Oligocene La Pascua sandstones, which shale upward into the Oligocene Roblecito Formation. In the Serranía Oriental and Maturín Basin to the east, all but two known sections have a regional hiatus separating the Maastrichtian San Juan, the Paleocene Vidoño and the Early to Middle Eocene Caratas Formations from the overlying southwardly transgressive Oligocene Los Jabillos/Merecure sandstones, which deepen up into the Areo and/or Carapita Formations (Caribbean foredeep). In the south flank of the Maturín subsurface, this basal foredeep onlap may be as young as Early Miocene (Di Croce *et al.* 1999). In southern Trinidad, a hiatus separates variably eroded/incised Campanian–early Late Eocene Naparima Hill, Guayaguayare, Lizard Springs and Navet Formations from the Early Oligocene San Fernando and the late Early Oligocene Basal Cipero Formations. Because of their overlap in age (Late Eocene is common to each), these hiatuses clearly correlate with each other, the main difference being the eastward-younging time of southward marine transgression. Rarely do these hiatuses display any discernable angularity.

The foredeep onlap is clearly diachronous eastward and records the eastward relative migration of the Caribbean foredeep basin and plate along the South American margin (Dewey & Pindell 1986; Pindell & Kennan 2009; Pindell *et al.* 1988). These authors interpreted the underlying hiatus as being due to peripheral bulge uplift within a passive margin, which is typically about 5% of foredeep subsidence in continental crust (i.e. about 250 m for a 5 km foredeep deflection). However, the recognition that South America was probably an active rather than a passive margin before Caribbean arrival (Pindell *et al.* 1991, 1998, 2006) suggests a modification to this explanation for the unconformity. If South America were an active margin before the collision (i.e. the

foot of the margin was either a site of inversion or incipient subduction), homoclinal hanging wall uplift of both the South American and Caribbean margins, as the Caribbean and South American plates converged above and loaded the intervening Proto-Caribbean lithosphere, could have generated greater uplift and fluvial incision than a peripheral bulge within a passive margin (Pindell *et al.* 1991, 2006; Pindell & Kennan 2007*a*). Further, although hanging wall uplift would also be diachronous, it would be of greatest magnitude toward the trench, although an initially deep-water setting may still have remained submarine, and it would predate peripheral bulge uplift because South America could not be flexed by the Caribbean Plate to produce a forebulge until the Caribbean lithosphere had actually begun to cross the position of the Proto-Caribbean Inversion Zone, by which time the hanging wall uplift would necessarily have been erased by sinking of the Proto-Caribbean slab into the mantle beneath it (Fig. 13). Thus, in cross section, the forebulge can be expected to emerge some distance landward from the zone of earlier hanging wall uplift. When the hanging wall uplift produces emergent conditions that are continuous with the interior of the continent, the forebulge will simply be superposed on and enhance a pre-existing unconformity, making incision deeper and non-marine conditions last longer. Where the hanging wall uplift has not exposed the depositional surface to erosion, such as when a marine continental shelf lies too far from the trench to be uplifted sufficiently, then the Caribbean forebulge may be the first and only mechanism of uplift at that shelf. This may have been the case in southern Trinidad, where the Caribbean basal foredeep unconformity spans the least time and incises the least section (San Fernando Formation on the Navet Hospital Hill member), and which lay some 300 km SE of the probable trace of the Proto-Caribbean Inversion Zone, which is too distant for hanging wall uplift to have operated there. Because of the tectonic uplift involved with these mechanisms, a Type-1 unconformity is expected in most parts of the shelf margin (eastern Guárico Basin, Serranía Oriental and Trinidad), where the shoreline retreated to near or past the original shelf edge.

Lowstand wedges correlative with the Caribbean basal foredeep unconformity. With the development of the Caribbean basal foredeep unconformity, coeval lowstand channels and fans carrying South American detritus from the shield and marginal section can be expected outboard of the unconformity, and this detritus should carry material ranging in age from Precambrian to the depositional age. Obtaining the actual depositional age is therefore challenging, especially as so much of it is coarse, but can be achieved by (1) identifying the youngest zircons in sandstones (Algar *et al.* 1998); (2) bracketing the age of the detritus between adjacent datable shales or carbonates; and (3) dating intra-fan or intra-channel shales that record a true depositional age. Thus far, dating efforts point strongly to an earliest Oligocene depositional age for the bulk of the low-stand detritus.

Along the Central Range trend of Trinidad, low-stand fan candidates include the Plaisance Conglomerate (mostly Cretaceous–Palaeogene South American margin clasts with coarse, quartz sand matrix; the 300 m thick Mount Harris section of the eastern Central Range; and the 250–300 m thick Angostura Sandstone (Early Oligocene according to BHP-Billiton 2003) in the eastern offshore projection of the Central Range. The Plaisance Conglomerate is traditionally considered a member of the San Fernando Formation (Kugler 1996, but see below). All three units lie entirely north of the Central Range Fault, and carry the distinctive red mudstone 'cherries' discussed earlier for Cretaceous formations dating back to the Albian, and thus are South American derived, probably denoting direct fluvial communication with lateritic weathering sites on a coastal plain. In addition, many, but not all, of the ZTR quartz arenitic sandstone outcrops once mapped in the Central Range as Early Eocene Pointe-a-Pierre Formation (Saunders *et al.* 1998) are probably earliest Oligocene equivalents of the Mount Harris, Plaisance and Angostura beds, if we accept the 31–33 Ma detrital zircon fission track ages from fresh euhedral crystals of probable volcanic ash origin as the depositional age (Algar *et al.* 1998). Of particular interest is the report of entirely Proterozoic U–Pb ages from detrital zircons (Kevin Meyer, oral presentation to 4th Geological Conference of the Geological Society of Trinidad and Tobago 2007), which points strongly to a southerly South American derivation for this and all the noted similar units, whereas a provenance from hypothetical uplifted northern terranes (such as the model of Higgs 2006, 2009) should be dominated by Palaeozoic–Mesozoic ages.

South of the Central Range Fault in Trinidad's Central Range, the Plaisance Conglomerate does not occur, but the traditional definition of the San Fernando Formation would demand that the other three members (Glauconitic Sandstone, Glauconitic Siltstone, and the Vistabella Limestone members) be considered as possible low-stand candidates like the Plaisance. Additional low-stand candidates south of the Central Range Fault are the Marabella Conglomerate and the ?Late Eocene 'limestone debris' beds of Soldado Rock. In places these units can be shown to occupy incised valley positions; at Marabella, the Glauconitic Sandstone

Member fills a trough that cuts through the Navet and into the Paleocene Lizard Springs. Elsewhere in southern Trinidad the San Fernando, if present, may lie on rocks extending down to the Santonian. The Glauconitic Sandstone member of the San Fernando Formation contains *Cassigerinella chipolensis* (Jenkins 1964), indicating an Early Oligocene age (P18 biozone). This is the youngest determined age known to us for the San Fernando, but it could be a maximum age as some of the beds appear to be reworked. In addition, at least one locality shows that the San Fernando can be truncated by incision; at Marabella Village near San Fernando City, the Glauconitic Sandstone member is incised and overlain by the Marabella Conglomerate consisting of 95% rounded clasts of Naparima Hill argillites and 5% Maastrichtian sandstone, most less than 8 cm across, the latter of which is only known to the south and west (San Juan Formation, Eastern Venezuela). Where San Fernando members are absent, mainly in southern onshore Trinidad, the medial Oligocene Lower Cipero Formation is the first unit to be deposited above the Eocene–Late Cretaceous formations, such as on the Naparima Hill Formation in southeasternmost Trinidad. The Lower Cipero clearly belongs to the Caribbean foredeep basin, which continues up-section through the Middle and Upper Cipero, until about 10 Ma (end Middle Miocene).

Given the above, erosional incision is deepest and the duration of missing time is greatest (potentially the whole of the Late Eocene and some of the Oligocene, some 7 or 8 Ma) at the basal San Fernando unconformity. We consider this unconformity to be correlative with that of the Serranía Oriental, indicating regional incision of the Late Eocene shelf and upper slope. We view the three southern members of the San Fernando as post-incision transgressive valley fill and onlap (thus defining the basal Caribbean foredeep basin) following low-stand basinal deposition associated with the unconformity, as is suggested by the high glauconite content of the southern San Fernando (i.e. condensed section). This in turn suggests that the low-stand wedge of the Plaisance Conglomerate and its correlatives to the north of the Central Range Fault (Mount Harris Formation) are older than the southern San Fernando members and that the Plaisance (and Mount Harris) should not be considered members of the San Fernando Formation. This should not be surprising given that there is virtually no lithological similarity between these formations. In this interpretation, Caribbean foredeep onlap is marked by the southern members of the San Fernando Formation, which is transitional to the basal Cipero. The local occurrence of the Marabella Conglomerate, which cuts into the San Fernando, is probably a local submarine channel where scour and fill occurred in an otherwise onlapping and deepening setting, perhaps the last vestige of high-energy fluvial discharge as the foredeep migrated south. The proximity of the Marabella Conglomerate to the Saint Joseph Conglomerate (Early Paleocene basal Lizard Springs deposited on incised Naparima Hill argillites, presumably from the nearby shoreline to the south) suggests the area was the site of repeated channel incision and filling. The following points highlight the reasons for removing the Plaisance Conglomerate as a member of the San Fernando Formation:

(1) The Plaisance Conglomerate occurs only in the Central Range trend, north of the Central Range Fault, and there is no stratigraphic contact with the other members to the south. Our palinspastic reconstructions restore the Plaisance some 70–100 km to the WNW of its occurrences today, putting its westerly extent adjacent to the eastern flank of the reconstructed Serranía Oriental, and its entire occurrence within the deep water Bohordal re-entrant. Highly quartzose formations with ZTR heavy mineral assemblages (southern non-glauconitic Caratas, the San Juan, and possibly the Barranquín Formations) were exposed in the Serranía as sources for the quartz matrix in the Plaisance, and these are not known in the area of the Trinidadian portion of the basal foredeep unconformity. Further, a river may have bypassed the Maturín Basin when it was subaerial and delivered its load directly to the Bohordal re-entrant.

(2) The Plaisance carries well-rounded limestone and sandstone boulders up to 50 cm across of Serranía-type passive margin strata dating down to the Albian; such incision in the Late Eocene is possible in the Serranía, but there is no indication of such incision in southern Trinidad. Employing the concept of the hanging wall uplift during Caribbean convergence, uplift and subaerial exposure in the Serranía may have been greatest in the northern and eastern Serranía (possibly eroding as deep as the Albian level), homoclinally decreasing southward (Pindell & Kennan 2001). It is possible, given today's geological map pattern and the cooling history of the Barranquín Formation in the Serranía, which suggests that uplift was greatest in the north and began in the Eocene (Locke & Garver 2005), that southward flowing streams carried such boulders to a Maturín trunk river, which then carried them to the Bohordal re-entrant.

(3) The southern San Fernando members do not contain red 'cherries' or significant quartz as

the Plaisance does, suggesting different source areas. Besides reworked limestone debris, the main clastic component of the southern members is granular glauconite, suggesting a 'Vidoño-like' source area where the Caratas was never deposited, possibly along the Columbus Channel. In the absence of palinspastic reconstructions of shortening and strike–slip, this lithological difference has caused some workers (e.g. Senn 1940; Kugler 1953; Higgs 2009) to infer the existence of a Palaeogene northern subaerial clastic source area. In terms of heavy minerals, the San Fernando contains rare zircon and tourmaline and trace kyanite and staurolite, but no other signature minerals (Edelman & Doeglas 1934, unpublished oil industry report).

Higgs (2006) proposed a northern source area (e.g. his Margarita Coastal Nappe) for Palaeogene Central Range clastics based on the presence of staurolite, but the chlorite, biotite and glaucophane also found in this 'Margarita Coastal Nappe' are entirely absent in Trinidadian Palaeogene strata. Data from the Caroní River on the Guayana Shield (Wynn 1993) clearly shows that trace staurolite in some Palaeogene samples can be derived from the Guayana Shield. Furthermore, structural and geochronological data unequivocally demonstrate that the igneous and Palaeozoic basement elements of the HP/LT metamorphic terranes of Margarita had been juxtaposed by about Albian time, and were both intruded by c. 85 Ma plutons of the Caribbean Arc (Maresch et al. 2009); their HP/LT metamorphism and P–T–t paths are similar to other Caribbean HP/LT complexes associated with the Caribbean Arc. Thus we conclude that they lay far to the west in Palaeogene time, and that the Juan Griego rocks are not a viable source for staurolite in the Barcelona Trough. In addition, there are no tuffs in the South American Cretaceous that would be expected if Margarita and nearby island arc rocks (the Patao–Mejillones–Tobago High, to the north of Araya–Paria) were not completely allochthonous (Pindell 1993; Stöckhert et al. 1995). The red 'cherries' in the Plaisance unit are indistinguishable (XRD results) from those in Aptian to Cenomanian Formations from southern Trinidad and thus it is most reasonable to accept a southern derivation for all post-Maastrichtian 'cherries'. The reappearance of cherries in Early Oligocene sections is not evidence of a new orogen to the north of Trinidad. Our tests on dark clasts within the Maastrichtian Galera Formation, which we entertained as possible precursors to 'cherries' developed on a northern Trinidad high, indicate that they do not contain sufficient iron to produce the cherries (W. Maresch, pers. comm. 2007). Higgs also suggested that the

Espino–Maturín River hypothesis was non-viable because the coeval Caratas sands in the Serranía Oriental is highly glauconitic; however, this is not true of the Caratas in the Maturín subsurface (unpublished industry data), which is coarse-grained, gritty and mineralogically mature, similar in character to the Pointe-a-Pierre sands of Trinidad.

To summarize the low-stand wedge for Trinidad, we propose a west to east progradation of fans in latest Eocene to Early Oligocene (Mount Harris Formation, with Plaisance, 'Oligocene Pointe-a-Pierre', Mount Harris sandstone, and Angostura sandstone members). Olistostromes of former outer passive margin (slope) material from the northern flank of the southern Trinidad shelf (drowned by Aptian time, Fig. 14) and slope may be expected to be interbedded with the primary west-to-east clastic trough fill, although from a different (southern) origin. The intensification of quartz-sand clastic input in the earliest Oligocene may relate to a coeval subaerial exposure of the Serranía Oriental surface, such that shield-derived sands that had been coursed onto a marine shelf now bypassed the Serranía platform and Maturín area to a delta setting (point source) at the SW corner of the Bohordal re-entrant.

In Eastern Venezuela, low-stand wedge candidates are limited to three depositional units; the Lechería beds (informal), the Lower Narical Formation (Socas 1991), and the 'reworked Tinajitas' seen in the Tinajitas Syncline at Via Alterna. All of these are in the vicinity of Barcelona, suggesting that this area may be the only part of the Serranía that remained below sea level during the low-stand.

The Lechería unit (informal usage) is a structurally isolated occurrence of micaceous, thin to thick-bedded, fine to coarse-grained quartzose turbidites with large intra-formational rip up clasts/bedsets, interbedded with shales, and characterized by distinctive tabular to rounded red mudstone clasts (or 'cherries'). It is remarkably similar to the Plaisance unit in Trinidad, but was mapped by Creole petroleum industry geologists in the 1950s as Oligocene–Early Miocene Narical Formation, despite great differences with that formation, because there was no other formation in the Serranía Oriental to which correlation could be made. The Lechería beds have a mature heavy mineral assemblage (ZTR with minor garnet, unlike the Narical which is orogenic with basalt fragments), and an Oligocene fauna (T. King, pers. comm. 2001). We infer an Early Oligocene age because the Lechería beds lack the lignite/coal and mineralogical immaturity typical of the Late Oligocene Narical and younger formations nearby.

Socas (1991) studied the Narical Formation near Barcelona and proposed a new sub-unit,

the Lower Narical member of the Narical Formation, comprising undated sandstones with a mature heavy mineral assemblage in the immediate area of the well-established 'Upper Narical'. The two sub-units are not in apparent stratigraphic contact. Lignite and coal are absent, diagenesis is better developed and bedding plane dips are often greater in the Lower Narical. In these respects, it is more tempting to correlate the Lower Narical and Lechería, but we did not see the haematitic 'cherries' in the outcrops of Lower Narical we visited.

At Via Alterna in the Tinajitas Syncline, the position of the Tinajitas limestone is occupied by limestone of a redeposited nature, comprising mass flow bed sets mainly of mobile rhodoliths (R. Higgs & J. Pindell, unpublished field observations). This occurrence of the Tinajitas differs from the more typically platformal and more probably *in situ* occurrences of the Peñas Blancas/Tinajitas limestone. Thus, although the 'Tinajitas' here is in normal stratigraphic position, its age is not necessarily late Middle Eocene (top of Caratas Formation) as is the *in situ* Tinajitas. The Via Alterna Tinajitas beds may thus be (1) an unique version (carbonate only) of the lowstand wedge (latest Eocene or earliest Oligocene), (2) a basal bed of the overlying Los Jabillos Formation (base of onlapping Caribbean foredeep section), or (3) an end Middle Eocene debris flow derived directly from a Tinajitas or Peñas Blancas platform carbonate. We favour the last option; the lack of 'cherries' and of coarse quartz hints against the lowstand under discussion herein, especially when such lithologies (Lechería) occur very nearby, and the lack of quartz argues against the Los Jabillos option.

To the west, along the foothills of the eastern parts of the Caribbean Mountains, descriptions of the Early Oligocene Tememure Formation (Vivas & Macsotay 1995) are very similar to our observations of the Lechería beds. Although we have not studied the Tememure ourselves, we consider the two units to be equivalent and deposited along strike from each other, probably as low-stand fans, a proposal that O. Macsotay considered entirely feasible (O. Macsotay, pers. comm. 2008).

In summary for Eastern Venezuela, the Lechería beds north of Barcelona probably represent a low-stand wedge of Early Oligocene age. These beds are not seen farther inland, because the rest of the Serranía and Guárico Basin were subaerial at the time of deposition. Palinspastically, the Lechería beds lie along with the northern fringe of the Serranía WNW to the NW part of the Gulf of Barcelona. Given that position, the source area for the Lechería beds could be the pre-La Pascua unconformity of the northeastern Guárico Basin just as well as the Serranía Oriental, although the apparent absence of staurolite or kyanite distinguishes it from the La Pascua

Formation (see discussion of Domain 3, the Caribbean foredeep onlap fringe, below). The Lechería beds may give us a control point for the occurrence of latest Eocene–Early Oligocene lowstand wedges well west of Trinidad, hinting that similar wedges once flanked the northern fringe of the Serranía Oriental. Further, because such beds are absent at the Late Eocene hiatus in all the Serranía Oriental river sections, they also provide further evidence that the Serranía Oriental was subaerially exposed for at least some if not all of the Late Eocene. Sandstones containing ferruginous mudstone clasts (which we interpret as 'cherries') are also reported from a well east of the Serranía in the Venezuelan part of the Gulf of Paria (unpublished oil company data), probably providing another control point for this low-stand trend.

Domain 3: Southern onlap assemblage of the Caribbean foredeep

The diachronous collision of the Caribbean Plate with northern South American involved first the accretion of the Caracas Group, Paria and Northern Range slope and rise strata followed by further convergence that led to the emplacement of allochthonous terranes and the accreted strata onto the South American margin (Fig. 18). In this paper, we will not explore the thrust history of this orogen in detail (see Pindell et al. 2009), treating it simply as 'the allochthon'.

In central Venezuela, west of the Urica Fault, the Caribbean forearc basement (Villa de Cura Nappe) and its accretionary prism (e.g. Paleocene–Middle Eocene Caramacate, Escorzonera, Los Cajones, Morro del Faro, and Garrapata units south of Villa de Cura) approached the Proto-Caribbean Prism (Caracas Group) in an offshore position that was outboard from the Guárico Trough behind the Proto-Caribbean Prism. Hanging wall uplift elevated the Proto-Caribbean Prism and Guárico Trough in the late Middle Eocene, eroding any Middle Eocene section. Caribbean advance then overthrust the Proto-Caribbean prism with peak metamorphism occurring at 40–35 Ma. The composite Caribbean forearc, accretionary prism and the subcreted Proto-Caribbean Prism, together advanced southeastward onto the Guárico Belt over South American basement in the Late Eocene, initiating foredeep subsidence in the Guárico Basin (La Pascua Formation). Further motion continued to shorten the Guárico Belt, below which the detachment was probably at the Albian level thus imbricating the Querecual, Mucaria, and Guárico Formations, as well as the Oligocene through Middle Miocene foredeep basin further south (Perez de Armas 2005). Domain 3 is represented

in the Guárico Basin by the latest Eocene/Early Oligocene onlapping sands of the La Pascua Formation and by the overlying earliest Oligocene Lower Roblecito shales. The heavy mineral assemblage in the La Pascua is dominated by ZTR, with minor staurolite and kyanite, which are readily sourced from Palaeozoic or shield rocks of South America, or more likely recycled from the Cretaceous Canoa or Tigre Formations, over which it transgressed. There is no sign of the more labile heavy minerals seen in the underlying Cretaceous, such as glaucophane (reported from the Tigre Formation) or garnet (reported from the Gautier Formation).

In the Serranía Oriental, the Domain 3 basal foredeep onlap is marked by the Early Oligocene Los Jabillos Formation inner shelf sandstones, which unconformably overlie the Caratas Formation and deepen upward into the Upper Oligocene Areo Formation shale. Our heavy mineral data for the Areo Formation (Fig. 7) show ZTR and trace kyanite, but the Los Jabillos Formation shows only ZT, all south-derived, as this landward flank of the Caribbean foredeep was situated south of the foredeep axis. The transgression was diachronous to the south, reaching the Maturín subsurface in the earliest Miocene (Merecure Formation: Di Croce et al. 1999; PDVSA 2005), advancing east at about the same 15–20 km/Ma rate as the Caribbean Plate.

Foredeep subsidence following the passage of the forebulge may have been accompanied by instability on the southern flank of the basin. Thus, although we are proposing the Lechería beds comprise a low stand wedge deposited before the Los Jabillos foredeep transgression, they could instead post-date the Los Jabillos Formation, and be derived from the craton as the foredeep developed but before the Naricual orogenic fringe sediment had arrived from the north. The depositional instability may reflect thrust belt advance, rather than gravity-driven slumping during a lowstand. 'Cherries', which are found in strata as old as Albian (Gautiér Formation), and as young as Early Miocene (a single outcrop of Nariva Formation), certainly seem to be an indicator of southern provenance. However, the overall sedimentary character, clast content and heavy minerals of the Lechería are so like the Plaisance Conglomerate and other units of the Mount Harris Formation, as well as the Tememure Formation, that we prefer to equate them and assign a pre-Los Jabillos age.

In southern Trinidad, the Late Oligocene 'Lower Cipero', 'Arenaceous Cipero' (referring to Foraminifera) or 'Silty Cipero' overlies the San Fernando Formation and rocks as old as the Naparima Hill argillites without angular unconformity. These strata appear to lie at the base of the Caribbean

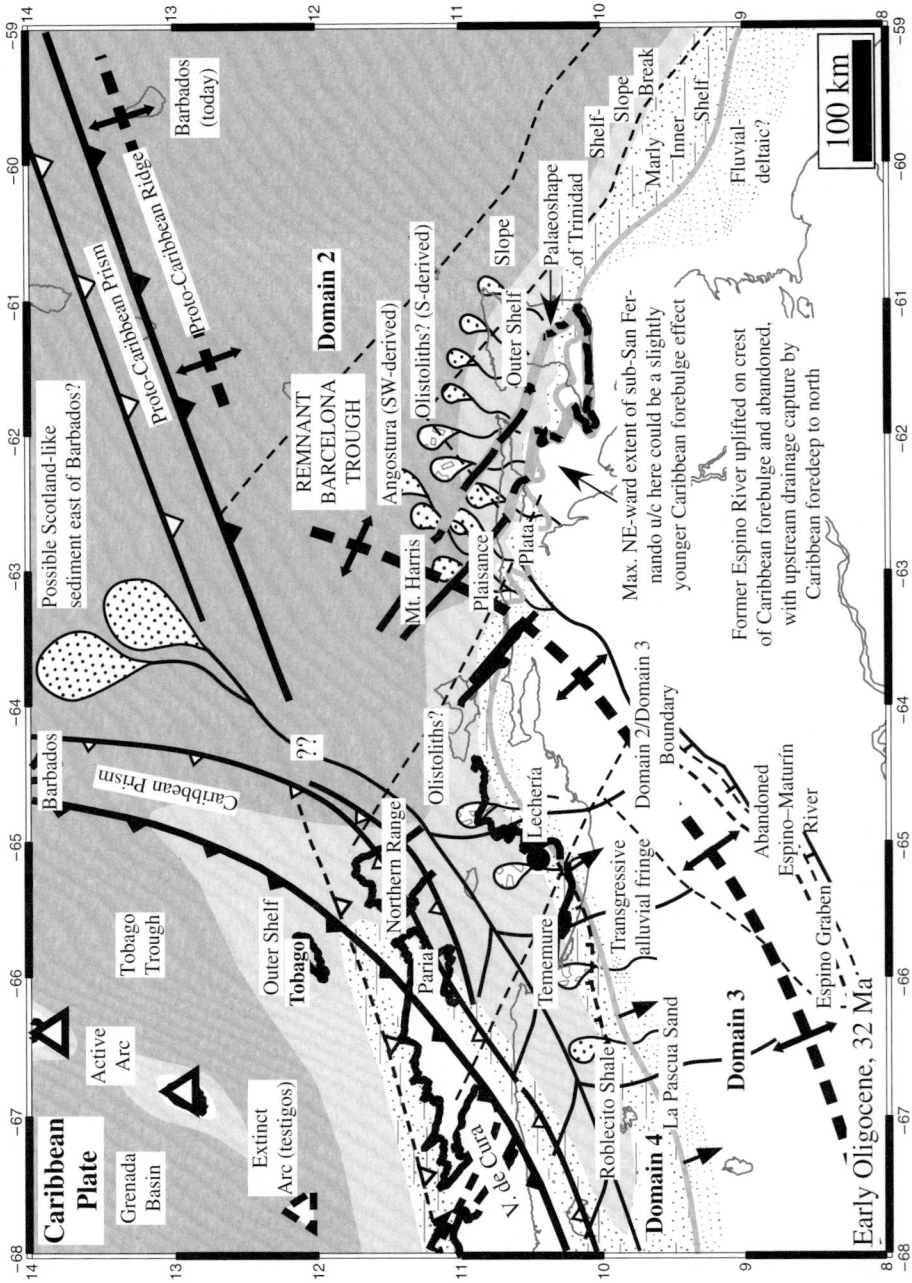

Fig. 18. *Continued.*

foredeep section, and locally overlie the San Fernando Formation incised channel fill, which we thus include in the San Fernando at the base of the Domain 3 Caribbean foredeep section. The limited sand content of both the San Fernando and the Lower Cipero in most places reflects the lithology of the underlying Navet, Lizard Springs and Upper Cretaceous shales and marls across which erosion, incision and subsequent foredeep transgression took place. Cipero facies include several shelly limestone 'reefs'. As elsewhere, we would expect Domain 3 sediments in Trinidad to be mineralogically mature. We were unable to sample any sands from proven basal Cipero Formation, and one sand sample from the San Fernando which was either San Fernando Formation (most probable), or possibly overlying Lower Cipero, yielded only trace zircon and tourmaline. However, Edelman & Doeglas (1934) collected two samples from the Bamboo Silt member of the Lower Cipero from a location assigned to the P19 biozone (ex. *G. ampliapertura* biozone, latest Early Oligocene) by Kugler (2001). Their heavy mineral residues were dominated by ZTR, as expected, but the heavy mineral residues also contained about 5% chloritoid, 3% garnet, trace kyanite and a single glaucophane. This signature is more like the Nariva of Domain 4 (see below). The sampled section is highly faulted, and observed bedding often contradicts the mapped faunal biozones and its stratigraphic age was the subject of debate by early workers (e.g. Stainforth 1948). Given the proven presence of chloritoid, kyanite and garnet on the Guayana Shield to the south, it is possible, though we consider it unlikely, that this is a unique occurrence of unweathered labile minerals in the Domain 3

transgressive fringe. Alternatively it is possible that the P19 fauna has been reworked into sediment of Nariva age, because the Bamboo Silt can be interpreted as stratigraphically overlying marls of the P21 (ex-*G. opima opima* biozone, Early Nariva time). Given the local complexity of bedding and poor exposure we also cannot rule out the possibility that Nariva sand has been mixed with Cipero marl during faulting. Sand is not characteristic of the Bamboo Silt, which mostly comprises foetid-smelling clay and marl with abundant large foraminifera (Stainforth 1948).

In summary, the loading of the South American crust by the advancing Caribbean forearc and growing orogen caused diachronous west-to-east migrating drowning and southward marine onlap of the pre-existing hanging wall and/or forebulge unconformity (Fig. 18). Clastic Domain 3 comprises the strata associated with this onlap, and its detritus is south-derived.

Domain 4: Distal syn-orogenic detritus, axial feed along the Caribbean Foredeep

Domain 4 comprises the axial foredeep tract between the Domain 3 transgressive belt (distal foreland) to the south and the advancing Caribbean thrust belt and its associated proximal clastic fringe to the north (Fig. 18). Sandstone petrography and heavy mineral data support the view that the Guayana Shield and its sedimentary cover continued to contribute detritus to the migrating foredeep, probably from a migrating forebulge ahead of the basin. However, Domain 4 also shows a significant contribution from arc volcanic, metamorphic and ophiolitic terranes of the Caribbean Plate and/or

Fig. 18. A reconstruction for Early Oligocene time (c. 32 Ma) captures the end of sedimentation in Domain 2 (Barcelona Trough) and the onset of sedimentation at the leading edge of the SE-migrating Caribbean foredeep in the Serranía. By earliest Oligocene time a significant area of subaerial exposure had developed in the NE Serranía and sediment eroded from this surface was deposited directly over the Bohordal shelf edge into the Barcelona Trough, or transported via tributaries of the proposed 'Espino–Maturín River' and discharged into the Trough at a point source upstream of the restored position of the Plaisance Conglomerate. The onset of Caribbean foredeep subsidence drowned this non-angular unconformity. In central Venezuela the onlap strata are assigned to the La Pascua Formation, younging south and east into the Los Jabillos Formation of the Serranía (latest Eocene in the NW to Late Oligocene in the El Furrial area). South of its onlap edge there must have been a subtle flexural forebulge developing, which probably resulted in capture of much of the drainage of the Espino River into the foredeep. The Plaisance conglomerate and Angostura sandstones, which were abruptly buried by shales, represent the culmination of the Espino drainage and the uplift of its drainage basin on the crest of the forebulge and the ensuing shaly sedimentation is the eastern distal and entirely submarine equivalent to the onlap seen in the Serranía. As the forebulge crossed the trace of the 'Espino–Maturín River' its drainage would have been captured by the Caribbean foredeep axis and mature south-derived sediment deflected to the northeast. We tentatively show the Lechería as comprising detritus from this captured drainage system. This map shows a tentative interpretation of the youngest of the Scotlands of Barbados, honouring a tentative Early Oligocene faunal age for the coarsest facies. In this case the Caribbean foredeep may have been channelling sediment comprising both Caribbean-derived and shield derived sediments, including mixed heavy minerals, distinctive red 'cherries' and limestone and shale clasts towards Barbados at this time shortly before being caught in the vice between the waning Proto-Caribbean prism and the advancing Caribbean accretionary prism. Alternatively, if we accept only older Scotland ages, Caribbean foredeep sediment may have been accumulating only south of the Proto-Caribbean Accretionary Prism by this time.

of the collision zone with South America. The first consistent appearance of characteristic Caribbean orogenic detritus in autochthonous positions along the margin youngs from west to east, beginning in the Maastrichtian in the western parts of the Colón Formation in northern Colombia and in the Paleocene–Early Eocene in the western parts of the Misoa Formation in the Maracaibo Basin (Van Andel 1958), continuing in the latest Early Oligocene Upper Roblecito of central Venezuela (Escalona 1985), in the Late Oligocene Naricual Formation of the Serranía Oriental and Late Oligocene–Early Miocene Nariva Formation of Trinidad, a diachroneity spanning some 40 million years and reflecting the diachronous emplacement of Caribbean allochthons onto the South American autochthon. In contrast to the typically mineralogically mature, shield-derived or reworked sediments of Domains 1–3, those in Domain 4 include lithic fragments (chert, basalt, metavolcanic and metamorphic rocks) and diverse heavy minerals. Glaucophane, in particular, appears to be diagnostic of a Caribbean provenance from unroofed HP/LT terranes in the collision zone. It first appears in our study area in the Upper Roblecito and Chaguaramas Formations (Venezuela) and in the Nariva Formation (Trinidad). It is unlikely to survive recycling from the trace glaucophane reported from some, but not all, Late Cretaceous sediments by Escalona (1985). The reappearance of minerals such as garnet, staurolite, kyanite, chloritoid and apatite in Late Oligocene and younger sediments strongly suggests it derived from Class 1 and Class 5 terranes of the Caribbean Orogen (see also Bellizzia & Dengo 1990) and the associated parautochthonous rocks in the flanking thrust belts during collision. The lack of these minerals in Domain 3 strata also argues against a southern provenance for Domain 4 minerals.

Barbados as the downstream eastern continuation of Domain 4. The texturally mature, fine to coarse quartz sandstones and minor conglomerates of the Scotland beds of Barbados carry the signature glaucophane and high-grade continental metamorphic minerals of Domain 4. The age of the Scotland turbidites has been debated. Favouring a Paleocene–Middle Eocene age are numerous arenaceous foraminifera (some reworked, some probably *in situ*, T. King, pers. comm. 2008) in shales and pelagic radiolaria from hemipelagic beds (Speed 2002) and pollen analyses on eight dispersed samples (Pindell & Frampton 2007). Favouring an Oligocene depositional age are Oligocene zircon fission track cooling ages (Baldwin *et al.* 1986) and two tentative Oligocene faunal ages (Pindell & Frampton 2007, citing T. King, pers. comm. 2008) from conglomerate clasts collected by us below Chalky

Mount on the east coast. Further, these same conglomerates carry red 'cherries' that are similar to, but less oxidized, than the 'cherries' in Late Eocene–earliest Oligocene Plaisance and Lechería beds of Venezuela and Trinidad. Probably the strongest data are the radiolaria and pollen ages (Late Paleocene–Middle Eocene), but the question remains open and it is possible that the Scotland unit has scattered Oligocene packets of sand/conglomerate within dominantly older material, either tectonically interleaved or deposited within incised channels. Indeed, our palaeogeographic maps show no specific reason to expect clastic supply to be shut off from the Scotlands, other than possible bathymetric building up of prism morphology, in Late Eocene time.

Concerning depositional position, the Scotland material clearly did not bypass Eastern Venezuela or Trinidad, nor does it bear any similarity with the shale dominated, south-derived Guárico Formation of Central Venezuela other than age. Neither could the huge volumes of highly mature quartz derive from the Caribbean arc, where quartz is practically lacking. Thus, the Scotland turbidites were derived from western Venezuela and/or Colombia, which is the nearest position from which quartz sands with an orogenic influence could derive.

The structural style of the Scotlands is that of an accretionary prism ahead of a trench (Speed 2002). In single-trench models for the Caribbean (e.g. Dewey & Pindell 1986; Speed 2002), and retracting the present day plate configuration back in time, the Scotlands represents the Caribbean Prism only. However, in double-trench models with Caribbean and Proto-Caribbean prisms (e.g. this paper), it becomes a question to which prism, or both, the Scotlands were accreted. Pindell & Frampton (2007) took this issue up, and we will not deal with it here because heavy minerals do not provide an answer; both prisms should have the orogenic influence, as they both flank the Proto-Caribbean Basin. Suffice it to say that Caribbean orogenic material was carried into the Proto-Caribbean Basin from NW South America, where Domain 4 was first initiated in the Maastrichtian, by a significant river draining probably the Palaeogene Andean foredeep (Pindell *et al.* 1998), and was accreted at one or both of the accretionary prisms flanking the Proto-Caribbean Basin. In the two-prism model (this paper), a suture between the Caribbean and Proto-Caribbean prisms should exist somewhere in the greater Barbados Ridge today (Pindell & Frampton 2007).

Oligocene jump in Domain 4 deposition from Barbados towards Trinidad. Upon Early Oligocene prism–prism collision north of the Serranía Oriental (contrast Figs 18 & 19), the Proto-Caribbean

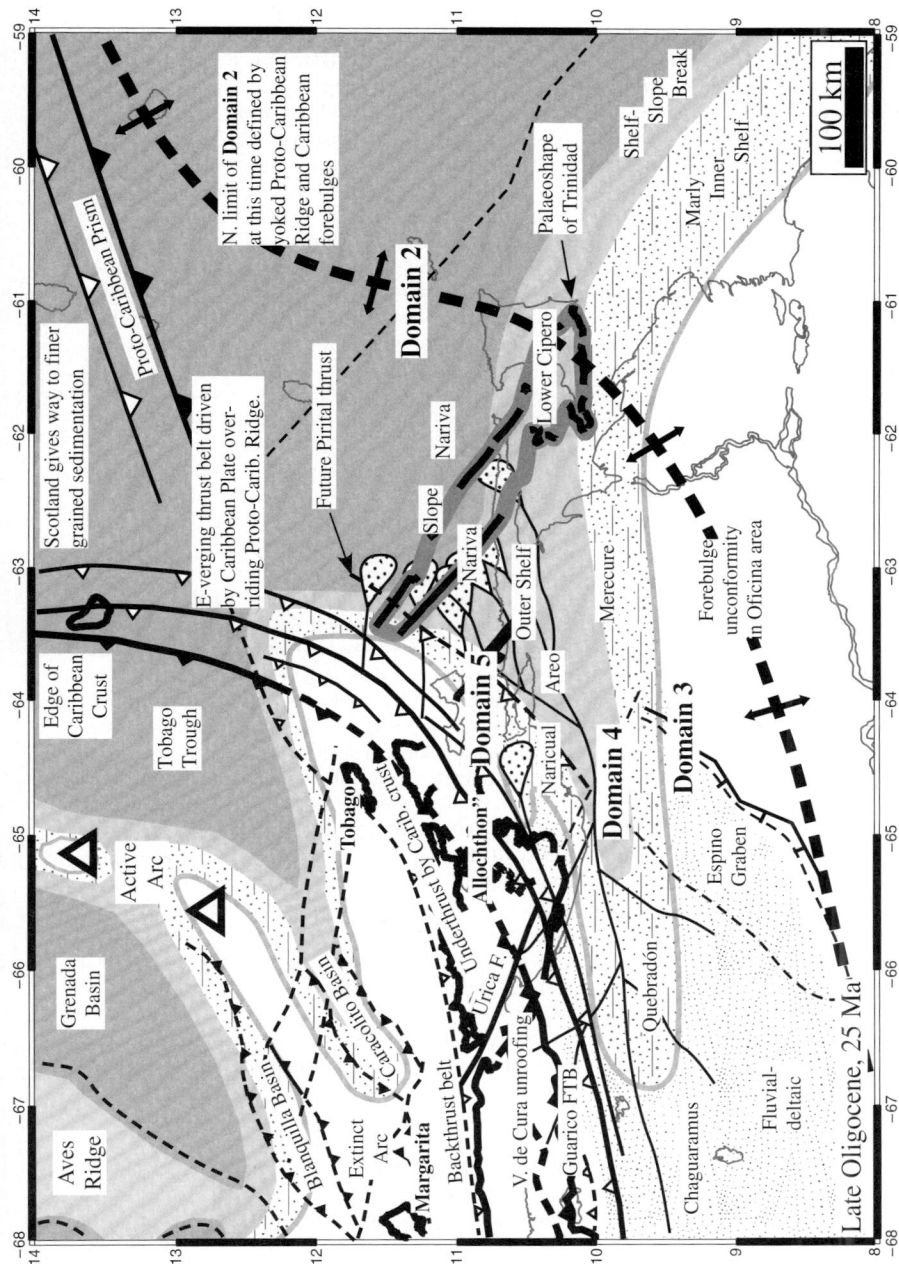

Fig. 19. Reconstruction for Late Oligocene time (c. 25 Ma). The leading edge of the Caribbean crust had migrated sufficiently far to the SE to interact strongly with eastern Venezuela. East of the Urica Fault, rather than overthrusting the margin, the leading edge of Caribbean crust seems to have wedged between crystalline basement (below) and passive margin sediment (above) and started to drive shortening to the SE in the passive margin section. The onset of shortening in the Serranía pushed the foredeep axis to the south far enough that the downstream end of the foredeep now drained into the former Domain 2 Barcelona Trough. This in turn led to the abandonment of coarse-grained sedimentation in the Scotland of Barbados shortly before the accretion of the former Proto-Caribbean and advancing Caribbean prisms. We see important along-strike and across-strike facies variations in the Caribbean foredeep. Domain 3 south-derived mature sandstones are initially buried by axial foredeep (Domain 4) shale-dominated sections, with minor sandstone content. The section becomes sandier approaching the active thrust front (Domain 5). To the west, in shallower marine to alluvial settings foredeep axis sediment tends to be sandier, and the transition from coarse to fine facies in the foredeep axis migrates from west to east as the trough fills in. In Trinidad, sandstone facies in the Nariva formation become more abundant in northern or western outcrops, and younger in the formation, relative to shale and marly facies developed towards the east end of the foredeep axis. These sandstones carry a distinctive heavy mineral signature indicating input from high-pressure metamorphic terranes that is absent from Domain 2 and Domain 3 sediments.

inter-prism trough was closed. Proto-Caribbean Prism rocks and the Serranía Oriental Cretaceous passive margin were incorporated into an enlarged composite Caribbean Prism and thrust towards the SE. As a result, the river that had formerly fed detritus down the Caribbean foredeep axis and into the Proto-Caribbean basin was deflected abruptly eastward into the Serranía Oriental (Naricual Formation of the northwestern and western Serranía) and Trinidad downstream (Nariva Formation), defining a new position for the palaeo-Orinoco River. Thus, further deposition near Barbados comprised reworking of the older prism materials ('Intermediate' or prism cover strata; Barker *et al.* 1986; Speed 2002) as the two prisms collided. Syndepositional structural growth in the northwestern Serranía resulted in slight angular unconformity between the Upper Naricual Formation deltaic sediments and the older strata of Domains 1–3. More regionally, the deflection of Caribbean foredeep axial drainage into the Serranía Oriental and Trinidad (Fig. 19) superimposed clastic Domain 4 sands of the Upper Naricual, Upper Areo, Carapita and Nariva Formations) onto the earlier south-derived Domain 2 and 3 sediments, probably by downlap, or sidelap (as the Guayana Shield was an important sediment source), during west to east initial progradation (Pindell *et al.* 1998). Heavy mineral assemblages for all Domain 4 sediments in the Serranía and Trinidad show an unambiguous Caribbean orogen signature, including characteristic glaucophane, chloritoid, staurolite, garnet and Al–silicates.

A reinterpretation of the Nariva Formation of Trinidad. Most of the Nariva Formation in Trinidad comprises unctuous mud. Where exposed at the surface by tectonism, it is usually highly sheared and tightly folded, but in subsurface sections such deformation is far less pervasive. The general lack of planktonic foraminifera indicates turbid water conditions consistent with the muddy character, but the depth of deposition remains subject to considerable debate. Conventional literature (e.g. Kugler 2001) implies a thickness up to 1500 m and a 1000–2000 m deep basinal setting. However, the Nariva is intensely imbricated and our studies of seismic and well data suggest that depositional thickness is closer to 500 m. The turbid water conditions implied by the foraminifera could pertain to fairly shallow shelf or upper slope environments if sedimentation occurred in front of a strong fluvial influence, such as the Upper Naricual delta. Turbid waters appear to have been confined to the axis of the Nariva depocentre. To the SE in Trinidad, turbidity waxed and waned, resulting in turbid Nariva facies interfingering with the open marine marly Middle Cipero facies of coeval age. If coastal currents along the Guayana margin were NW-directed,

as they are today, this may also have helped to maintain clearer water conditions in southern Trinidad, deflecting Nariva sands and muds towards the area SE of present-day Barbados. Nariva-aged mudstones are of structural importance, serving as a major décollement surface in onshore central Trinidad, and in parts of the Barbados Prism, including the Trinidadian ultra-deep exploration blocks.

The onset of Nariva Formation mudstone and sandstone deposition appears to be gradual rather than related to sudden, regionally synchronous, relative sealevel change. Thus, we consider that the best definition for the base of the Nariva Formation is the first consistent appearance of signature heavy minerals such as chloritoid, glaucophane, garnet, kyanite, and staurolite in medial Oligocene time. 'Trinidadian' Nariva deposition probably began in the Bohordal re-entrant of the Barcelona Trough in the Middle Oligocene downstream from the Upper Naricual delta and Areo Formation shelf. By the Late Oligocene, sediment deposition and imbrication of Nariva and older strata at the leading edge of the advancing Caribbean orogen resulted in the Barcelona Trough beginning to shallow upward. The approach of the tectonic load and flexure of the crust to the southeast allowed Nariva facies to encroach diachronously southward into the clear-water Cipero marls of the Southern Basin. The growing orogen built a northern subaerial margin to the Naricual–Nariva depocentre, but no subaerially eroded conglomerates are known from the Nariva in the basin axis, suggesting that Nariva sands may largely be axially derived from the Naricual Delta. The first indications of north-derived detritus are in the Cunapo and Brasso Formation conglomerates of the Central Range which downlap onto and bury the Nariva (Kugler 2001; our own data). Claystones within the Brasso have foraminiferal assemblages characteristic of clear water (Wilson 2003, 2004), suggesting that the Brasso depocentre was a perched piggyback basin and that the turbid water mudstones of the axial Nariva derive from sediment plumes downstream of the Naricual delta.

Nariva Formation sandstones form only a small proportion of the whole formation, and sandy sections are only locally more than a few tens of metres thick, such as in the Brighton Marine and Nariva Hill areas. The distinctive immature Caribbean orogen heavy mineral assemblage of the Nariva Formation is seen in: (1) white sands which in outcrop appear to be very clean, with little mica and no plant material; (2) in outcrop and well samples of 'dirty' sands with lignite and some disseminated mica; (3) sheared and broken blocks of sandstone within sheared Nariva muds which have often been interpreted as olistostromes; and (4) highly sheared phacoids in the type section of the Charuma Silt member of the Pointe-a-Pierre Formation. Intense deformation and sediment remobilization mean it is often difficult to determine if multiple Nariva sands are individual beds or structural repetitions of the same bed. Outcrops of this formation are generally poor quality. Our field observations show some features not easily reconciled with the conventional deepwater view, such as trough-cross-bedding (Corbeau Hill), thin red claystone layers and perfectly preserved leaves in extremely leaf-rich beds (Guaracara Road Quarry), perhaps suggesting a shelf-depth setting above storm wave base. Lignite beds up to a metre thick also occur, but no associated rootlets or palaeosols have been found, ruling out an *in situ* shoreface or lagoonal origin. Disseminated lignite and plant fragments are a feature of almost all Nariva sandstones and may have been transported as suspended load from the coaly Naricual Formation to the west. We have also found 'cherries' in the Nariva at the Guaracara Road quarry and in the Naricual at Río Aragua Este near San Francisco, probably the most easterly sandy occurrence in the Naricual Formation *sensu stricto* in the Serranía Oriental.

The Nariva Formation is conventionally considered to be olistostromal (e.g. Kugler 2001), with exotic blocks as old as Aptian occurring within Nariva muds. No pre-Nariva sandstones carry the 'orogenic' signature, and there are no associated Nariva conglomerates (other than the aforementioned 'cherries'), and none of the exotic blocks in the Nariva show evidence of rounding during transport. Thus, the 'orogenic' sandstone blocks at the Williamsville 'building site' and elsewhere are almost certainly of intraformational origin, and we favour a tectonically sheared origin for the sand bed disaggregation, which is not seen in the subsurface away from fault zones. The shaly matrix at Williamsville is highly sheared and transposed to the point where original sedimentary laminations are only rarely seen. We have found outcropping 'floating blocks' of pre-Nariva lithologies embedded in Nariva mudstone only in areas that are clearly intensely tectonically sheared. Exotic blocks are seen neither in unsheared Nariva outcrops, nor in well core. 'Slip masses' of pre-Nariva lithologies (such as the limestones along Plum Road or at Stack Rock, Cuche sandstones, Naparima Hill argillites and Chaudière Formation shales) are widely reported in the literature and in older oil company reports and were used to infer an adjacent, uplifting and eroding northern highland. Our studies of well data, published maps (Kugler 1996) and the original field sheets on which these maps were based, and our own fieldwork show that many or all of the 'olistostromes' were interpreted from the existence of exotic blocks at the surface only. Blocks are commonly embedded in Nariva-derived regolith and

are much smaller than drawn on the maps. The blocks are often found downstream of *in situ* outcrops of older formations that could be the source of some of the reported 'olistoliths'. Large, deep-rooted, mud volcanoes in central Trinidad associated with major thrusts and shear zones also carry blocks which were subsequently winnowed at the surface of the mud volcano and transported downstream of their site of 'eruption'.

True olistostromes may exist in the Nariva Formation but we have yet to find a convincing example either in the field or in a well where a sedimentary origin is visible through tectonic shearing overprint, or where shearing is absent. We are confident that we have identified several isolated Nariva 'orogenic' sandstone blocks encased in a sheared, silty matrix from which Eocene benthonic foraminifera are reported (Charuma Silt type section along the Central Range Fault Zone, Kugler 1996, 2001). We have also recovered chloritoid grains from the matrix at this site, but have not been able to recover fauna. In this area, clastic facies of Eocene through Early Miocene age are fault-juxtaposed with multiple slivers of Eocene Navet Marl and Middle Miocene Biche limestone. Significant dextral strike–slip fault displacement has occurred on this fault zone in the Pliocene–Quaternary.

Domain 5: Proximal, syn-orogenic wedgetop and piggyback deposits on the evolving orogen

Domain 5 comprises syn-orogenic detritus subaerially eroded from the rising orogen and deposited on the northern, proximal fringe of the Caribbean foredeep basin and in piggyback basins within the growing orogen (Figs 20 & 21). As with Domains 3 and 4, Domain 5 is diachronous, younging from west to east and north to south as turbidite and alluvial channels and fans propagated. Domain 5 strata were deposited in bathymetrically shallower conditions due to thrust imbrication and foredeep infilling beneath them. Clastic delivery was probably orthogonally downslope into the foredeep basin at first, but those that reached the basin axis became merged with Domain 4 strata, where basin floor currents may then have carried them along axis to the east for some unclear distance.

In the western Guárico Basin north of El Baúl, Domain 5 is represented by the now deeply eroded Orupé Formation (Macsotay *et al.* 1995), a possibly Late Eocene (but also possibly younger) multi-component conglomerate in sandy matrix with abundant rounded chert, volcanic rocks, and metamorphic clasts. In the eastern Guárico Basin, Domain 5 is marked by the Upper Oligocene–Middle Miocene marine Quebradón Formation, but

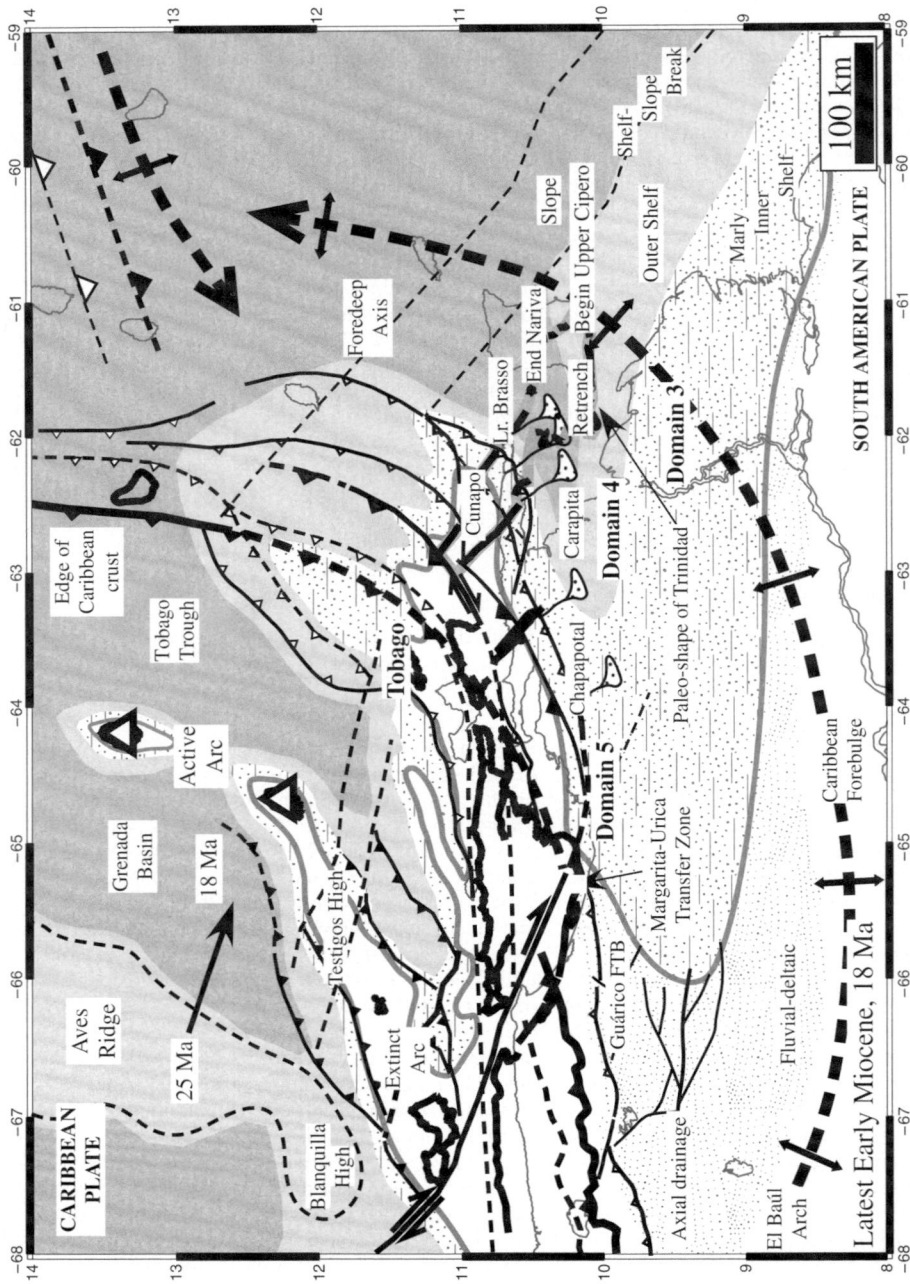

Fig. 20. Reconstruction for earliest Middle Miocene time (c. 18 Ma) shows the Caribbean prism as having over-ridden the Proto-Caribbean prism north of Tobago. This was followed by wedging of the crystalline Caribbean basement under the merged prisms, forming the characteristic 'oceanics' blind roofthrust over backthrust structural style seen from Tobago to Barbados. To the south, the Caribbean crust had accreted parts of the Proto-Caribbean Ridge in front of Tobago and was probably driving thin-skinned basement thrust sheets within the Serranía and Gulf of Paria resulting in a dramatic shoaling of the Serranía and Trinidad thrust belts, where unconformities developed in the north and former deep marine thrust belts (Nariva) were uplifted above imbricated thrust sheets of Cretaceous strata. The shallow marine Brasso Formation was deposited above the thrust belt, and was dominantly carbonate on structural highs and conglomeratic in structural lows, defined by transtensional lateral ramps. By this time, the Caribbean forebulge had swept through Trinidad sediments of Late Nariva age were deposited in central and southern Trinidad, while Brasso facies were deposited north of the thrust front in what is now the Central Range.

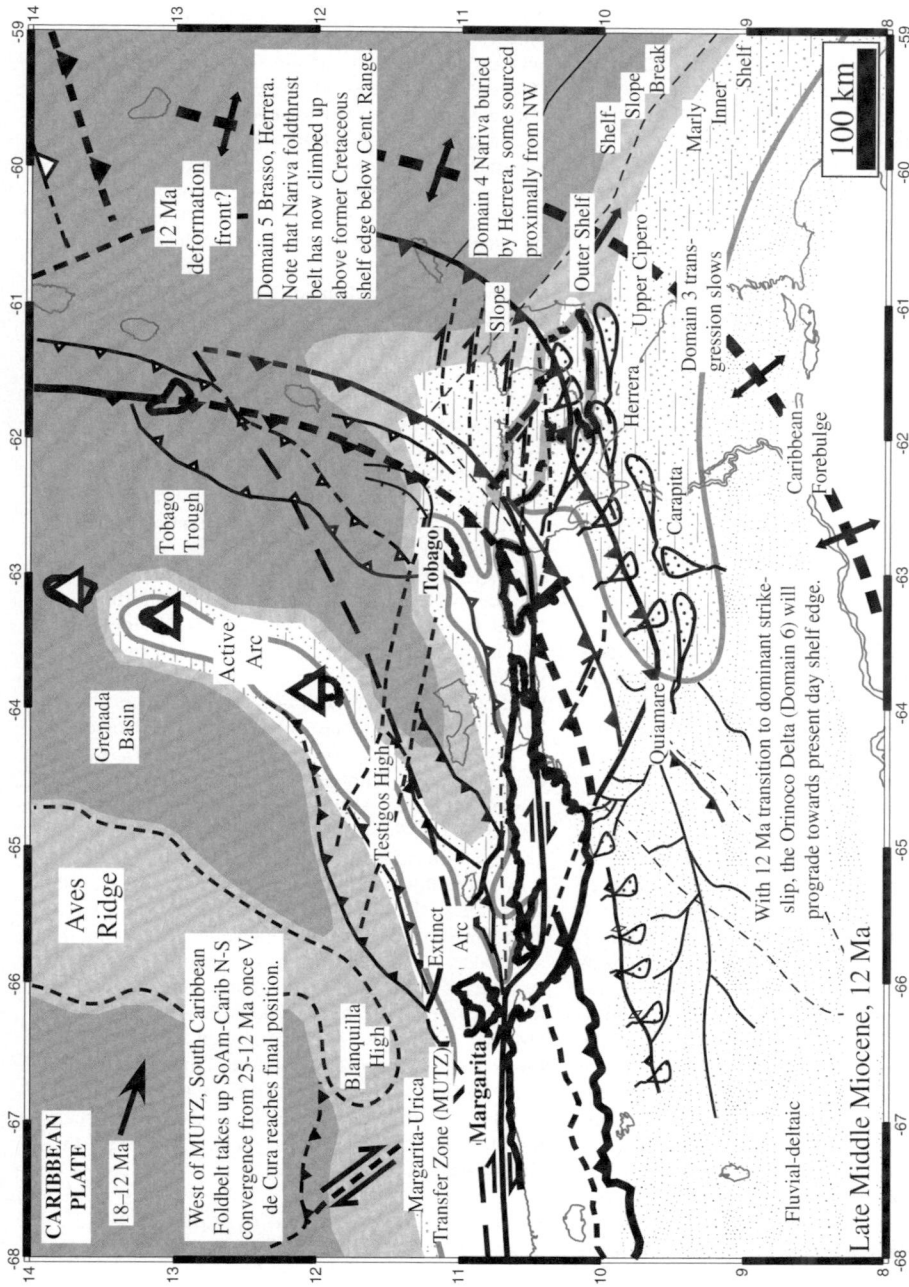

Fig. 21. Reconstruction at the culmination of Middle Miocene orogeny in Venezuela and Trinidad at about 10–12 Ma shows substantial areas of the former foreland basin incorporated into the orogen. Proximal fine through coarse-grained clastic and carbonate facies accumulated on the north side of the foreland basin, deposited in shallower water conditions than in the axial trough to the south. Sands in Domain 5 were sourced from nearby pre-Miocene outcrops, including reworked foreland basin sediment. Significant east–west extension in the orogen is indicated by thick carbonate and clastic sediments deposited in waterdepths far less than the sediment thickness. True basement subsidence is indicated by continued foredeep subsidence beyond the deformation front, and additional accommodation space resulted from thinning of the allochthons. With the end of SE-directed relative plate motion we see continued subsidence driven by distant loads of the Caribbean Plate but not in most areas by active continued foreland shortening. As a result, Domain 6 Orinoco sediments, although of the same origin as older foredeep axis sediments, are able to overstep and bury the thrust front active during deposition of Domain 5 sediments.

as the foredeep became filled it was taken over by the non-marine, eastwardly prograding and shallowing Chaguaramus Formation. Nearer to the Urica Fault, Domain 5 is marked by the Middle Miocene Quiamare Formation, especially its El Pilar Member in the north (Vivas & Macsotay 1989).

In the southern flank of the Serranía Oriental, Domain 5 strata include the middle and upper parts (Middle Miocene) of the Carapita Formation, and the Chapopotal turbiditic member of the Carapita in the Maturín subsurface which forms a coarser reservoir than the surrounding Carapita shale (Lamb & Sulek 1968). The Morichito Basin above the Pirital thrust was once thought to be Middle Miocene, but now is considered Late Miocene (Roure *et al.* 2003) and is now assigned to Domain 6 (below).

In central Trinidad and in the southern Caroni Basin subsurface, Domain 5 strata include the Early and Middle Miocene Cunapo Conglomerate (alluvial and submarine fans), the late Early through Middle Miocene Brasso (shelf carbonates with clastic channels), and middle and late Middle Miocene Tamana (reefal and redeposited limestones) Formations, and the Retrench and Herrera sandstone fairways of the Cipero Formation. Black chert conglomerates and sandstones are characteristic of the Cunapo, Brasso and Tamana Formations. Downstream to the SE, the Chapopotal, Retrench and Herrera sandstones and minor conglomerates reached the Middle Miocene foredeep axis and are interbedded with the younger elements of Domain 4. None of these units contain metamorphic clasts to our knowledge, as the Paria and Northern Range were not yet unroofed.

The Middle Miocene Serranía–Naparima thrust belt of eastern Venezuela and Trinidad was a continuous thrustbelt prior to the low-angle detachment faulting that has formed the Gulf of Paria basin since the Late Miocene. The black chert pebbles/grains of the Cunapo Conglomerate, Chapopotal turbidites, Brasso and Tamana channels, and Herrera turbidites derive from the San Antonio Formation in the Serranía Oriental. It may also have derived from a possible former Upper Cretaceous section above the Northern Range/northern Caroni Basin, but as the Northern Range/northern Caroni Basin restores palinspastically back to a position north of the Serranía Oriental in the Middle Miocene (Fig. 10), this point is moot. We concur with Erlich *et al.* (1993) that subaerial highlands to the NW fed conglomerates and finer detritus to the Brasso/Tamana shelf, with bypass channels carrying some of this detritus farther south into the Upper Cipero foredeep basin, where they fine and become the characteristic 'salt and pepper' Herrera turbidites, the 'pepper' being largely chert, the 'salt' largely quartz, plus other lithic grains.

At least the northern Serranía Oriental (and hypothetical cover of Northern Range) had become subaerial in the Early Miocene, supplying the rounded Cretaceous chert pebbles in Early Miocene portions of the Cunapo Conglomerate. Further, thrust imbrication had caused littoral to neritic depositional depths by the late Early Miocene in Trinidad, as shown by the Brasso and Tamana Formations (Erlich *et al.* 1993). Accommodation space for these strata was locally provided by axis-parallel extensional structures (half grabens) cutting obliquely (ESE-trending) across the orogen (Pindell & Kennan 2007*b*). Tamana limestone thicknesses reach over 100 m yet much of its depositional environment is probably in less than 20 m water depth. Thus, Tamana depocentres must have subsided during the southeastward thrusting which otherwise produced uplift. The Tamana may thus be used to estimate positions where thinning of the thrust belt was achieved by low-angle detachments; this process is a function of the oblique collision and must happen to some degree in all highly oblique thrust belts. These low angle detachment basins crossing the thrust belt were also probably the means by which conglomeratic channels reached the southern foredeep basin. At the downstream ends of some of these extensional structures (e.g. Middle Miocene stage of extension on the Los Bajos Fault: Pindell & Kennan 2007*b*), there are unusually coarse Herrera facies (e.g. Herrera conglomerates elevated by diapiric mud in the core of the Southern Range anticline at Galfa Point). The heavy mineral signature of the Herrera sandstones is broadly similar to the slightly older Nariva Formation, but with a lower proportion of exotic minerals to ZTR. Although there was probably continued first-cycle sediment input from far-travelled Caribbean terranes at the back of the thrustbelt, it is possible that weathering and recycling of Nariva sandstones which were incorporated into the advancing thrust wedge resulted in some loss of labile minerals. A further cause for dilution may be capture of drainage on the Guayana Shield that had previously been directed south or east. The absence of slivers of Herrera-aged sandstones in the Nariva outcrop belt of central Trinidad strongly suggests that the Nariva had already been incorporated into the orogenic wedge by Herrera time, such that the Herrera turbidites mostly bypassed the Nariva Belt.

Domain 6: Orinoco Delta overlap assemblage

Oblique collision between the Caribbean Plate and South America culminated by about 10 Ma (Fig. 21) when the azimuth of Caribbean Plate motion relative to South America changed from

ESE-directed to east-directed, after which east–west transcurrent (simple shear) tectonics took over (Pindell *et al.* 1998; Pindell & Kennan 2007*a, b*). However, foredeep loading continued in the Maturín and Columbus basins after 10 Ma, even in the absence of significant south-directed thrusting, because progressively thicker parts of the Caribbean lithosphere continued to move east, further loading the NW-dipping, down-flexed South American lithosphere beneath the Maturín Basin and Trinidad (Pindell & Kennan 2007*b*). This subsiding basin has been fed primarily from west to east by the Orinoco River or precursors (which may have been located closer to the Caribbean Mountains and had a higher portion of sediment input from there) since about 10 Ma, with lesser contributions from north or south. Primary Late Miocene–Pliocene west to east shifts in palaeoenvironment were controlled by changes in tectonic loading in the north, eastward gravitational collapse of the orogen and eustasy (Wood 2000). The sands of the Orinoco trunk drainage system are highly variable, carrying detritus from multiple first-cycle and second-cycle sources including the Serranía Oriental and Caribbean Mountains to the north, El Baúl Arch and Mérida Andes in the west, the Guayana Shield to the south, and the cratonic cover of the interior plains. Domain 6 sediments post-date the primary accretion of the Caribbean Orogen to South America but remain strongly controlled by subsequent (post-10 Ma) dextral strike–slip, transtensional basin growth, and dextral transpression. Domain 6 sediments are particularly thick in the Gulf of Paria transtensional pull-apart basin, and in the Eastern Columbus Channel, where eastward-migrating gravity driven normal faulting provides sediment accommodation space. The influence of our proposed Proto-Caribbean Ridge is also clear up to the present. Thick distal Domain 6 sediments of Plio–Pleistocene age from the Orinoco are largely confined to the south of the Proto-Caribbean basement ridge ENE of Barbados (Speed *et al.* 1984; Dolan *et al.* 2004). The incorporation of these into the leading edge of the Barbados accretionary ridge has resulted in the prism being dramatically wider and thicker to the south of the Proto-Caribbean basement ridge than to the north.

Our samples of Domain 6 strata are few and widespread but suggest some issues that should be the subject of future work. First, the Lower Manzanilla in Eastern Trinidad appears to lack an orogenic signature of chloritoid and other minerals seen in the Brasso. A nearby conglomerate, currently mapped as Brasso or Cunapo has a similar signature and is probably part of the Lower Manzanilla section. Second, sand beds at Plum Mitan Road citrus grove, currently mapped as Eocene Pointe-a-Pierre,

have a clean ZTR assemblage but sedimentary features in common with younger strata and a possible Late Miocene foraminifer; we have assigned these sands to the Lower Manzanilla Formation. Third, the Upper Manzanilla shows minor staurolite input, and we have provisionally assigned undated sandstone at Mapapire Road to this or to the slightly younger Springvale Formation. The Middle Cruse Formation sandstones at Galfa Point have a hybrid signature consistent with their age. The Pleistocene Talparo Formation is distinct, with a stronger orogenic signature and reappearance of glaucophane, possibly signifying erosional downcutting to older levels in the Guárico/Maturín foreland to the west. We suspect that the cleaner samples represent the deposits of delta lobes which were sourced more directly from the southwest in the Guayana Shield, while more orogenic signatures represent the deposits of delta lobes which had a significant input from streams directly draining the Caribbean Allochthons.

Summary

Heavy mineral and sandstone petrographic data have been applied to existing tectonic concepts and palaeogeographic models to refine our understanding of basinal and palaeogeographic evolution in Eastern Venezuela, Trinidad, Barbados and the SE Caribbean. The result is an internally consistent and stratigraphically dynamic regional evolutionary framework involving plate motions that can be used for understanding other aspects of evolution, as well as for extending or predicting elements of hydrocarbon plays in petroleum exploration. Six clastic domains of different genetic origin are identified, each of which pertains to a primary stage of palaeogeographic evolution.

Domain 1, the Cretaceous passive margin, comprises clastic sediments, including distinctive red mudstone clasts, derived entirely from the Guayana Shield or its sedimentary cover (Precambrian and Palaeozoic 'basement', Mesozoic cover) on its south side. Intense tropical weathering results in mature ZTR assemblages, with local traces of staurolite, garnet and apatite, all of which are found in the shield to the south, or in present-day rivers crossing the shield. Red ironstone clasts ('cherries') have been found in a number of Cretaceous formations and are thought to derive from laterites or oxidized palaeosols developed over older rocks or Cretaceous alluvial plain sediments to the south.

Domain 2 comprises Palaeogene, largely redeposited sands, conglomerates and shales of continental origin associated with basin formation behind the Proto-Caribbean Inversion Zone along northern South America. We propose the 'Barcelona Trough' paralleling the margin from Central

Venezuela to NE of Trinidad as a clastic basin along northern South America that was distinct from the rest of the Proto-Caribbean basin, the two separated by the accretionary and basement hanging wall ridge of the north facing Proto-Caribbean Inversion Zone. Continental erosional detritus (reworking, and lowstand fans) produced by initial Caribbean interactions also fit within this domain, the interactions being (1) hanging wall uplift as the Caribbean and Proto-Caribbean prisms converged; and (2) forebulge uplift ahead of the migrating Caribbean foredeep basin.

Domain 3 comprises the cratonic side or distal foreland of the migrating Caribbean foredeep basin. Sediments are reworked from existing cratonic cover and share the mature heavy mineral signature of Domains 1 and 2. There is little sign of exotic shield derived minerals, despite Caribbean forebulge rejuvenation of relief, pointing to the continuing efficiency of tropical weathering.

Domain 4 occupies the axis of the migrating Caribbean foredeep basin, the sands being of dual provenance from both the Shield/cratonic cover to the south of the basin and also the developing orogenic topography to the north or NW. These sands include the first sediments for which an explicit northern source, both mineralogically and geographically, can be justified, albeit reworked and mixed with south-derived sands along the basin axis. They carry a distinctive, immature heavy mineral assemblage including minerals derived from high-pressure blueschists and accreted Palaeozoic greenschist and amphibolite facies rocks. Sediments of Paleocene to Middle Eocene (and possibly Oligocene) in Barbados fit into this domain, lying down-dip (although on the ocean floor) of the Palaeogene Caribbean foredeep basin axis. During Late Oligocene time, accretion of the Serranía Oriental into the greater Caribbean accretionary belt closed off the Caribbean foredeep–Proto-Caribbean Basin connection, deflecting the axis of the Caribbean foredeep southwards into the Serranía Oriental and Trinidad. These Caribbean foredeep sediments are superimposed (by downlap or sidelap) on the mature sandstones of Domain 3 in the Serranía Oriental and Trinidad.

Domain 5 consists of proximal, syn-orogenic detritus on the advancing orogenic wedge and in piggyback basins, derived from the uplifting and diachronous Caribbean–South America collision zone. These strata are generally coarse grained immediately adjacent to the subaerial part of the orogen, but tend to fine to the SE. They carry first-cycle detritus from the Caribbean Orogen and also probably reworked Domain 4 sediments caught up in the thrust belt, but these are diluted with reworked debris from mature sediments within the orogen. Clasts of organic rich chert from uplifted Cretaceous

strata are common. As with Domains 3 and 4, these sediments are diachronous from west to east, but are not younger than *c.* 10 Ma. The olistostromal Early Eocene turbidites of the Los Cajones and Garrapata Formations of Central Venezuela can be considered in Domain 5 but at a time when the Caribbean forearc had not yet collided with the central Venezuelan autochthon.

Domain 6 comprises post-collisional, but syn-transcurrent, foreland fill of the Orinoco fluvial, deltaic and pro-deltaic settings, prograding from west to east. To the south, these sediments merge with sediments from minor rivers draining the Guayana Shield, and towards the north they are confined by and contributed to by the Caribbean Mountains, the Serranía Oriental, the Paria Peninsula and the Northern Range of Trinidad. Domain 6 strata show a very mixed provenance, with variable but continued input from the Caribbean Orogen or recycled Domain 4 and 5 sediments upstream, together with possible first cycle material coming from the Shield.

Domains 1–3 are mature and derived from South America, while Domains 4–6 show affinity to lithologies of the allochthonous Caribbean Plate and its accreted terranes. The transition in Eastern Venezuela and Trinidad occurred in the Oligocene, marking the onset of Caribbean collision with South America in that position. This transition in northern Colombia and western Venezuela occurs in Paleocene, attesting to the diachronous nature of the Caribbean–South America collision, which is itself a consequence of the Pacific origin of the Caribbean Plate. However, in the mantle reference frame, the relative motion is due to the westward drift of the Americas relative to a Caribbean Plate that has undergone little west to east motion.

The correct interpretation of the palaeogeography of all these domains depends critically on robust palinspastic restorations, with well-constrained estimates of shortening and strike–slip offsets derived from interpretations of regional seismic, well, and field datasets. Palaeogeographies built from facies distribution in present-day coordinates are misleading at best, and may lead to highly erroneous predictions of sand fairway orientation and fining direction. In some cases, the misinterpretation of structural mixing as olistostromal deposition may lead to the expectation of non-existent proximal facies (and land masses). The palaeogeographic development outlined in this paper could be greatly improved with additional dating of sediments using fauna or direct dating of detrital grains. Future combined studies of the thermal history of grain populations in the sediments and the multiple source areas for characteristic heavy minerals may reduce the ambiguity in current interpretations.

We cheerfully thank J. Frampton and B. Carr-Brown (Biostratigraphic Associates, Trinidad), J. Keens-Dumas and C. Lakhan (Petrotrin), T. King and M. Bolivar (Biostratigraphic Associates, Venezuela) for providing palaeontological age data on field and core samples over many years; W. Maresch and T. Reinecke (Universität Bochum, Germany) for XRD lab work on siltstone 'cherries'; and M. Mange for second opinions on certain heavy mineral identifications. We also are indebted to M. A. Navas, R. Maraj, K. Latter, L. Barker, R. Higgs, F. Urbani and S. Grande for field assistance in Venezuela, Trinidad or Barbados during this work, and to A. Ramlackhansingh for discussion of the structural context of many of the samples. S. Kroonenberg provided valuable insights into the geology of the Suriname portion of the Guayana Shield. This paper derives from an industry funded research program by Tectonic Analysis Ltd that was supported in stages from 2000 to 2007 by BP, BHP-Billiton, Chevron, Total, Anadarko/Kerr, Primera, Marathon, Repsol, Petro-Canada, TED, Phillips, Hess, Petrotrin, ENI, Shell, Venture, Talisman and Tullow, for which we are grateful, as well as from funds provided by the BOLIVAR study program (NSF grant EAR-0003572) at Rice University. We are particularly indebted to Petrotrin for providing a comprehensive seismic and well database and other files from which we built the palinspastic reconstructions so important for the palaeogeographic reconstructions presented in this paper. K. James and M. A. Lorente organized the Sigüenza conference in 2006 which inspired us to write this paper.

References

ALGAR, S. 1993. *Structure, stratigraphy, and thermochronologic evolution of Trinidad.* PhD thesis, Dartmouth College.
ALGAR, S. T. & PINDELL, J. L. 1993. Structure and deformation history of the northern range of Trinidad and adjacent areas. *Tectonics*, **12**, 814–829.
ALGAR, S., HEADY, E. C. & PINDELL, J. L. 1998. Fission-track dating in Trinidad: implications for provenance, depositional timing and tectonic uplift. *In*: PINDELL, J. L. & DRAKE, C. (eds) *Palaeogeographic Evolution and Non-glacial Eustacy: North America.* SEPM (Society for Sedimentary Geology), Special Publication, **58**, 111–128.
ANAND, R. R. & GILKES, R. J. 1987. Muscovite in Darling Range bauxitic laterite. *Australian Journal of Soil Research*, **25**, 445–450.
BALDWIN, S. L., HARRISON, T. M. & BURKE, K. 1986. Fission track evidence for the source of accreted sandstones, Barbados. *Tectonics*, **5**, 457–468.
BARKER, L., GORDON, J. & SPEED, R. C. 1987. A study of the Barbados Intermediate Unit, surface and subsurface. *11th Caribbean Geological Conference*, Transactions, Bridgetown, Barbados, 36:1–36:28.
BELLIZZIA, A. 1985. Sistema montanosa del Caribe – una Cordillera aloctona en la parte norte de America del Sur. *Memorias del VI Congreso Geologico Venezolano.* Sociedad Venezolana de Geologia, Caracas, **10**, 6657–6835.
BELLIZZIA, A. & GONZÁLEZ, L. A. 1971. *Geological Map of San Juan de los Morros* (Sheet 6745, Scale 1:100,000, 1st edn). Ministerio de Energia y Minas, Caracas, Venezuela.
BELLIZZIA, A. & DENGO, G. 1990. Caribbean mountain system, northern South America. *In*: DENGO, G. & CASE, J. E. (eds) *The Caribbean Region.* The Geology of North America, **H**. Geological Society of America, Boulder, CO, 167–176.
BENJAMINI, C., SHAGAM, R. & MENÉNDEZ, A. 1986. Formación Tucutunemo. *Memorias del VI Congreso Geologico Venezolano.* Sociedad Venezolana de Geologia, Caracas, **10**, 6551–6574.
BENJAMINI, C., SHAGAM, R. & MENÉNDEZ, V. A. 1987. (Late?) Palaeozoic age for the 'Cretaceous' Tucutunemo Formation, northern Venezuela; stratigraphic and tectonic implications. *Geology*, **15**, 922–926.
BHP-BILLITON 2003. *Briefing Paper – Angostura Development, Trinidad and Tobago.* World Wide Web Address: http://www.bhpbilliton.com/bbContent Repository/News/RelatedContent/Angostura120303. pdf.
CARR-BROWN, B. & FRAMPTON, J. 1979. An outline of the stratigraphy of Trinidad. *Transactions of the 4th Latin American Geological Congress*, 7–15 July 1979, Port of Spain, Trinidad, 7–19.
COLE, E. & HEESTERMAN, L. J. 2002. *Geological Map of Guyana* (Scale: 1:3 500 000). Guyana Geology and Mines Commission, Georgetown, Guyana.
CRUZ, L., FAYON, A., TEYSSIER, C. & WEBER, J. 2007. Exhumation and deformation processes in transpressional orogens: the Venezuelan Paria Peninsula, SE Caribbean–South American plate boundary. *In*: TILL, A. B., ROESKE, S. M., SAMPLE, J. C. & FOSTER, D. A. (eds). *Exhumation Associated with Continental Strike-Slip Fault Systems.* Geological Society of America, Special Paper, **434**, 149–165.
DELOR, C., LAHONDÈRE, D. ET AL. 2003a. Transamazonian crustal growth and reworking as revealed by the 1:500 000 scale geological map of French Guiana (2nd edn). *Géologie de la France/Geology of France and Surrounding Areas*, **2–4**, 5–57.
DELOR, C., DE ROEVER, E. W. F. ET AL. 2003b. The Bakhuis ultrahigh-temperature granulite belt: II. Implications for Late Transamazonian crustal stretching in a revised Guiana Shield framework. *Géologie de la France/Geology of France and Surrounding Areas*, **2–4**, 207–230.
DEWEY, J. F. & PINDELL, J. L. 2006. Tectonic significance of the Caribbean Plate as a reference frame since 100 Ma: the 'Backbone' explained, 'Backbone of the Americas', 3–7 April 2006, Mendoza, Argentina. *Geological Society of America Abstracts with Programs*, Speciality Meetings, **2**, 118. World Wide Web Address: http://gsa.confex.com/gsa/06boa/finalprogram/abstract_100635.htm.
DOLAN, P., BURGGRAF, D. ET AL. 2004. Challenges to Exploration in Frontier Basins – the Barbados Accretionary Prism. *American Association of Petroleum Geologists International Conference*, 24–27 October 2004, Cancun, Mexico. World Wide Web Address: http://aapg.confex.com/aapg/can2004/techprogram/A90080.htm.
DI CROCE, J. 1989. *Analisis Sedimentologico de la Formacion San Juan en la Cuenca Oriental de Venezuela*

(*Estados Anzoategui y Monagas*). M.Sc. thesis, Universidad Central de Venezuela.

DI CROCE, J., BALLY, A. W. & VAIL, P. 1999. Sequence stratigraphy of the Eastern Venezuelan Basin. *In*: MANN, P. (ed.) *Caribbean Basins*. Sedimentary Basins of the World, **4**. Elsevier Science, Amsterdam, 419–476.

EDELMAN, C. H. & DOEGLAS, D. J. 1934. *Report on the heavy minerals of some Cretaceous and Eocene sediments from Trinidad*. Unpublished oil industry report held in Petrotrin library, Santa Flora.

ERIKSON, J. P. & PINDELL, J. L. 1998. Sequence stratigraphy and relative sea-level history of the Cretaceous to Eocene passive margin of northeastern Venezuela and the possible tectonic and eustatic causes of stratigraphic development. *In*: PINDELL, J. L. & DRAKE, C. (eds) *Paleogeographic Evolution and Non-glacial Eustacy: North America*. SEPM (Society for Sedimentary Geology), Special Publications, **58**, 261–281.

ERLICH, R. N. & KEENS-DUMAS, J. 2007. Late Cretaceous Palaeogeography of northeastern South America: implications for source and reservoir development. *In*: *Proceedings of the 4th Geological Society of Trinidad and Tobago Geological Conference, 'Caribbean Exploration – Planning for the Next Century*, 17–22 June 2007, Port of Spain, Trinidad. World Wide Web Address: http://www.gstt.org.

ERLICH, R. N., FARFAN, P. F. & HALLOCK, P. 1993. Biostratigraphy, depositional environments, and diagenesis of the Tamana Formation, Trinidad: a tectonic marker horizon. *Sedimentology*, **40**, 743–768.

ERLICH, R. N., VILLAMIL, T. & KEENS-DUMAS, J. 2003. Controls on the deposition of Upper Cretaceous organic carbon-rich rocks from Costa Rica to Suriname. *In*: BARTOLINI, C., BUFFLER, R. T. & BLICKWEDE, J. (eds) *The Circum-Gulf of Mexico and the Caribbean: Hydrocarbon Habitats, Basin Formation, and Plate Tectonics*. American Association of Petroleum Geologists, Tulsa, OK, Memoirs, **79**, 1–45.

ESCALONA, N. 1985. Relaciones estratigráficas con el método de minerales pesados, Área Machete, Faja Petrolífera del Orinoco. *Memorias del VI Congreso Geologico Venezolano*. Sociedad Venezolana de Geologia, Caracas, **1**, 536–587.

FEO-CODECIDO, G. 1955. Heavy mineral techniques and their application to Venezuelan stratigraphy. *American Association of Petroleum Geologists Bulletin*, **40**, 984–1000.

FOLAND, K. A., SPEED, R. & WEBER, J. 1992. Geochronologic studies of the hinterland of the Caribbean Mountains Orogen of Venezuela and Trinidad. *Geological Society of America Abstracts with Programs*, **24**, A148.

GALEHOUSE, J. S. 1971. Point counting. *In*: CARVER, R. E. (ed.) *Procedures in Sedimentary Petrology*. Wiley Interscience, New York, 385–407.

GONZALEZ DE JUANA, C., ITURRALDE DE AROZENA, J. M. & PICARD CADILLAT, X. 1980. *Geologia de Venezuela y de sus Cuencas Petroliferas*. Ediciones FONINVES, Caracas.

HEDBERG, H. D. 1950. Geology of the eastern Venezuela basin (Anzoátegui, Monagas, Sucre, eastern Guárico portion). *Geological Society of America Bulletin*, **61**, 1173–1216.

HEDBERG, H. D., SASS, L. C. & FUNKHOUSER, H. J. 1947. Oil fields of Greater Oficina area, central Anzoategui, Venezuela. *American Association of Petroleum Geologists Bulletin*, **31**, 2089–2169.

HIGGS, R. 2006. Venezuela–Trinidad–NW Colombia geohistory revised: long-lived (Campanian–Quaternary) nappe-thrust belt and stacked foreland basins (Proto-Caribbean, Caribbean), capped by supraorogen salt-dissolution basins (11–0 Ma). Extended abstract. *XIII Congreso Venezolano de Geofisica*, 22–25 October 2006, Caracas, Venezuela. World Wide Web Address: http://www.congresogeofisica-sovg.org/dyncat.cFormation?catid=4378.

HIGGS, R. 2009. Caribbean–South America oblique collision model revised (Campanian–Recent) on a Jurassic–Cretaceous passive margin. *In*: JAMES, K. H., LORENTE, M. A. & PINDELL, J. L. (eds) *The Origin and Evolution of the Caribbean Plate*. Geological Society, London, Special Publications, **328**, 613–657.

ILLING, V. C. 1928. Geology of the Naparima region of Trinidad. *Geological Society of London Quarterly Journal*, **84**, 1–56.

JAMES, K. 2006. Arguments for and against the Pacific origin of the Caribbean Plate: discussion, finding for an inter-American origin. *Geologica Acta*, **4**, 279–302.

JENKINS, D. G. 1964. Panama and Trinidad Oligocene rocks. *Journal of Palaeontology*, **38**, 606.

KAMEN-KAYE, M. 1937. Reconnaissance Geology in State of Anzoategui, Venezuela, South America. *American Association of Petroleum Geologists Bulletin*, **21**, 233–245.

KAMEN-KAYE, M. 1942. 'Ortiz Sandstone' and 'Guarumen Sandstone Group' of North-Central Venezuela. *American Association of Petroleum Geologists Bulletin*, **26**, 126–133.

KASPER, D. A. & LARUE, D. K. 1986. Palaeogene quartzose sandstones of Barbados and palaeogeography of northern South America. *Tectonics*, **5**, 837–854.

KENNAN, L. & PINDELL, J. L. 2009. Dextral shear, terrane accretion and basin formation in the Northern Andes: best explained by interaction with a Pacific-derived Caribbean Plate. *In*: JAMES, K. H., LORENTE, M. A. & PINDELL, J. L. (eds) *The Origin and Evolution of the Caribbean Plate*. Geological Society, London, Special Publications, **328**, 487–531.

KUGLER, H. G. 1953. Jurassic to recent sedimentary environments in Trinidad. *Bulletin de l'Association Suisse des Géologues et l'Ingénieurs du Petrole*, **20**, 27–60.

KUGLER, H. G. 1996. *Treatise on the Geology of Trinidad. Detailed Geological Maps and Sections*. Natural History Museum, Basel.

KUGLER, H. G. 2001. *Treatise on the Geology of Trinidad, part 4: The Paleocene to Holocene Formations of Trinidad*. KNAPPERTSBUSCH, M. & BOLLI, H. M. (eds) Natural History Museum, Basel.

LAMB, J. L. & SULEK, J. A. 1968. Miocene turbidites in the Carapita Formation of eastern Venezuela. *In*: SAUNDERS, J. B. (ed.) *Transactions of the Fourth Caribbean Geological Conference*, 28 March to 12 April 1965, Port-of-Spain, Trinidad and Tobago. Caribbean Printers, Arima, Trinidad, 111–119.

LARUE, D. K. 1995. Non-conventional methods of determining clastic reservoir orientation using paleocurrent/paleoslope studies in oriented core–application to subsurface of Trinidad, *In: Proceedings of the 3rd Geological Conference of the Geological Society of Trinidad and Tobago*, 777–813.

LEBRON, M. C. & PERFIT, M. R. 1993. Stratigraphic and petrochemical data support subduction polarity reversal of the Cretaceous Caribbean island arc. *Journal of Geology*, **101**, 389–396.

LOCKE, B. & GARVER, J. 2005. Thermal evolution of the eastern Serranía del Interior foreland fold and thrust belt, northeastern Venezuela, based on apatite fission track analyses. *In*: AVÉ LALLEMANT, H. G. & SISSON, V. B. (eds) *Caribbean–South American Plate Interactions, Venezuela*. Geological Society of America, Special Papers, **394**, 315–328.

MACSOTAY, O., VIVAS, V. & MOTICSKA, P. 1995. *Lithostratigraphy of the Piemontine Nappe of North Central Venezuela: Senonian to Eocene Gravitational Sedimentation*. Boletin de Geológia del Minsterio de Energia y Minas, Publicación Especial, **10**, 114–123.

MANGE, M. A. & MAURER, H. F. W. 1992. *Heavy Minerals in Colour*. Chapman and Hall, London.

MARESCH, W. V., KLUGE, R. ET AL. 2009. The occurrence and timing of high-pressure metamorphism on Margarita Island, Venezuela: A constraint on Caribbean–South America interaction. *In*: JAMES, K. H., LORENTE, M. A. & PINDELL, J. L. (eds) *The Origin and Evolution of the Caribbean Plate*. Geological Society, London, Special Publications, **328**, 705–741.

MASCLE, A., MOORE, J. C. ET AL. 1988a. Site 671 – Barbados Ridge. *Proceedings of the Ocean Drilling Program, Initial Report*, **110**, 67–204.

MASCLE, A., MOORE, J. C. ET AL. 1988b. Site 672 – Barbados Ridge. *Proceedings of the Ocean Drilling Program, Initial Report*, **110**, 205–310.

MAYA-SÁNCHEZ, M. 2001. *Distribución, Facies y Edad de las Rocas Metamórficas en Colombia, Memoria Explicitiva: Mapa metamorfico de Colombia*. Instituto Colombiano de Minería y Geológia (INGEOMINAS), Bogotá. World Wide Web Address: http://productos.ingeominas.gov.co/productos/MEMORIA/Memoria%20MMC.pdf.

MAYA-SÁNCHEZ, M. & VÁSQUEZ-ARROYAVE, E. 2001. *Mapa metamorfico de Colombia* (Scale: 1:2 000 000). Instituto Colombiano de Minería y Geológia (INGEOMINAS), Bogotá. World Wide Web Address: http://productos.ingeominas.gov.co/productos/OFICIAL/georecon/geologia/escmilln/pdf/Metamorfico.pdf.

MÜLLER, R. D., ROYER, J.-Y., CANDE, S. C., ROEST, W. R. & MASCHENKOV, S. 1999. New constraints on the Late Cretaceous/Tertiary plate tectonic evolution of the Caribbean. *In*: MANN, P. (ed.) *Caribbean Basins*. Sedimentary Basins of the World, **4**, Elsevier Science, Amsterdam, 33–59.

OXBURGH, E. R. 1966. Geology and metamorphism of Cretaceous rocks in eastern Carabobo state, Venezuelan coast ranges. *Caribbean Geological Investigations*. Geological Society of America, Boulder, CO, Memoirs, **98**, 241–310.

PDVSA 2005. *Codigo Estratigráfico de las Cuencas Petroleras de Venezuela*. World Wide Web Address: http://www.pdvsa.com/lexico/index.html.

PEIRSON, A. L., SALVADOR, A. & STAINFORTH, R. M. 1966. The Guárico formation of north-central Venezuela. *Boletin Informativo – Asociacion Venezolana de Geologia, Mineria y Petroleo*, **9**, 181–224.

PEREZ DE ARMAS, J. 2005. Tectonic and thermal history of the western Serranía del Interior foreland fold and thrust belt and Guárico Basin, north-central Venezuela: implications of new apatite fission track analysis and seismic interpretation. *In*: AVÉ LALLEMANT, H. G. & SISSON, V. B. (eds) *Caribbean–South American Plate Interactions, Venezuela*. Geological Society of America, Boulder, CO, Special Paper, **394**, 271–314.

PINDELL, J. L. 1985. Alleghanian reconstruction and subsequent evolution of the Gulf of Mexico, Bahamas, and Proto-Caribbean. *Tectonics*, **4**, 1–39.

PINDELL, J. L. 1993. Regional synopsis of Gulf of Mexico and Caribbean evolution. *In*: PINDELL, J. L. & PERKINS, R. F. (eds) *Transactions of the 13th Annual GCSSEPM Research Conference: Mesozoic and Early Cenozoic Development of the Gulf of Mexico and Caribbean Region*, 251–274.

PINDELL, J. L. & BARRETT, S. F. 1990. Geological evolution of the Caribbean region; a plate tectonic perspective. *In*: DENGO, G. & CASE, J. E. (eds) *The Caribbean Region. The Geology of North America*, **H**. Geological Society of America, Boulder, CO, 405–432.

PINDELL, J. L. & KENNAN, L. 2001. Processes and events in the terrane assembly of Trinidad and eastern Venezuela. *In*: FILLON, R. H., ROSEN, N. C. ET AL. (eds) *Transactions of the 21st GCSSEPM Annual Bob F. Perkins Research Conference: Petroleum Systems of Deep-Water Basins*, 159–192.

PINDELL, J. L. & FRAMPTON, J. 2007. Geology and Tectonic Evolution, Scotland District, Barbados, field trip guide. *Proceedings of the 4th Geological Society of Trinidad and Tobago Geological Conference – 'Caribbean Exploration – Planning for the Next Century*, 17–22 June 2007, Port of Spain.

PINDELL, J. & KENNAN, L. 2007a. Cenozoic kinematics and dynamics of oblique collision between two convergent plate margins: the Caribbean–South America collision in eastern Venezuela, Trinidad, and Barbados. *In*: KENNAN, L., PINDELL, J. L. & ROSEN, N. C. (eds) *Transactions of the 27th GCSSEPM Annual Bob F. Perkins Research Conference: The Paleogene of the Gulf of Mexico and Caribbean Basins: Processes, Events and Petroleum Systems*, 458–553.

PINDELL, J. L. & KENNAN, L. 2007b. Bow-wave model for deformation and foredeep development since 10 Ma, eastern Venezuela and Trinidad. *Proceedings of the 4th Geological Society of Trinidad and Tobago Geological Conference – Caribbean Exploration – Planning for the Next Century*, 17–22 June 2007, Port of Spain.

PINDELL, J. L. & KENNAN, L. 2009. Tectonic evolution of the Gulf of Mexico, Caribbean and northern South America in the mantle reference frame: an update. *In*: JAMES, K. H., LORENTE, M. A. & PINDELL, J. (eds) *The Origin and Evolution of the Caribbean Plate*. Geological Society of London, Special Publications, **328**, 1–55.

PINDELL, J. L., CANDE, S. C. ET AL. 1988. A plate-kinematics framework for models of Caribbean evolution. *Tectonophysics*, **155**, 121–138.

PINDELL, J. L., ERIKSON, J. P. & ALGAR, S. T. 1991. The relationship between plate motions and the sedimentary basin development in northern South America: from a Mesozoic passive margin to a Cenozoic eastwardly-progressive transpressional orogen. *In*: *Transactions of the Second Geological Conference of the Geological Society of Trinidad and Tobago*, 191–202.

PINDELL, J. L., HIGGS, R. & DEWEY, J. F. 1998. Cenozoic palinspastic reconstruction, paleogeographic evolution, and hydrocarbon setting of the northern margin of South America. *In*: PINDELL, J. L. & DRAKE, C. L. (eds) *Palaeogeographic Evolution and Non-glacial Eustasy, northern South America*. SEPM (Society for Sedimentary Geology), Special Publications, **58**, 45–86.

PINDELL, J. L., KENNAN, L., MARESCH, W. V., STANEK, K. P., DRAPER, G. & HIGGS, R. 2005. Plate-kinematics and crustal dynamics of circum-Caribbean arc-continent interactions, and tectonic controls on basin development in Proto-Caribbean margins. *In*: AVÉ-LALLEMANT, H. G. & SISSON, V. B. (eds) *Caribbean–South American Plate Interactions, Venezuela*. Geological Society of America, Boulder, CO, Special Papers, **394**, 7–52.

PINDELL, J. L., KENNAN, L., STANEK, K. P., MARESCH, W. V. & DRAPER, G. 2006. Foundations of Gulf of Mexico and Caribbean evolution: eight controversies resolved. *Geologica Acta*, **4**, 89–128.

PINDELL, J. L., KENNAN, L. & DEWEY, J. F. 2009. Tectonic evolution of the Gulf of Mexico, Caribbean and northern South America in the mantle reference frame: an update. *In*: JAMES, K. H., LORENTE, M. A. & PINDELL, J. L. (eds) *The Origin and Evolution of the Caribbean Plate*. Geological Society of London, Special Publications, **328**, 1–55.

POOLE, E. G. & BARKER, L. H. 1983. Map: Geology of Barbados, Scale 1:50,000, Series D.O.S. 1229, Edition 1-D.O.S. Published by the Government of the United Kingdom (Directorate of Overseas Surveys) for the Government of Barbados, © Barbados Government.

POTTER, H. C. 1973. The overturned anticline of the Northern Range of Trinidad near Port of Spain. *Journal of the Geological Society of London*, **129**, 133–138.

ROURE, F., BORDAS-LEFLOCH, N. *ET AL.* 2003. Petroleum systems and reservoir appraisal in the sub-Andean basins (eastern Venezuela and eastern Colombian foothills). *In*: BARTOLINI, C., BUFFLER, R. T. & BLICKWEDE, J. (eds) *The Circum-Gulf of Mexico and the Caribbean: Hydrocarbon Habitats, Basin Formation, and Plate Tectonics*. American Association of Petroleum Geologists, Tulsa, OK, Memoirs, **79**, 750–775.

RUIZ, G. M. H., SEWARD, D. & WINKLER, W. 2004. Detrital thermochronology – a new perspective on hinterland tectonics, an example from the Andean Amazon Basin, Ecuador. *Basin Research*, **16**, 413–430.

RUIZ, G. M. H., SEWARD, D. & WINKLER, W. 2007. Evolution of the Amazon Basin in Ecuador with special reference to hinterland tectonics: data from zircon fission-track and heavy mineral analysis. *Developments in Sedimentology*, **58**, 907–934.

SAGEMAN, B. B. & SPEED, R. C. 2003. Upper Eocene limestones, associated sequence boundary, and proposed Eocene tectonics in eastern Venezuela. *In*: BARTOLINI, C., BUFFLER, R. T. & BLICKWEDE, J. F. (eds) *The circum-Gulf of Mexico and the Caribbean; Hydrocarbon Habitats, Basin Formation, and Plate Tectonics*. American Association of Petroleum Geologists, Tulsa, OK, Memoirs, **79**, 874–890.

SALVADOR, A. & STAINFORTH, R. M. 1968. Clues in Venezuela to the geology of Trinidad, and vice versa. *In*: SAUNDERS, J. B. (ed.) *Transactions of the Fourth Caribbean Geological Conference*, 28 March to 12 April 1965, Port-of-Spain, Trinidad and Tobago. Caribbean Printers, Arima, Trinidad, 31–40.

SAUNDERS, J. B., ROBERTS, C. L., ALI, W. M. & EGGERTSON, E. B. (eds) 1998. *Geological Map of Trinidad and Tobago*. Ministry of Energy and Energy Industries, Trinidad.

SCHRUBEN, P. G., WYNN, J. C., GRAY, F. & COX, D. P. 1997. *Geology and Resource Assessment of the Venezuelan Guayana Shield at 1 : 500,000 scale – a Digital Representation of Maps published by the U.S. Geological Survey*. USGS Digital Data Series, **DDS-46**.

SCHUBERT, C. 1971. Metamorphic rocks of the Araya Peninsula. *Geologische Rundschau*, **60**, 1571–1600.

SENN, A. 1940. Palaeogene of Barbados and its bearing on history and structure of Antillean–Caribbean region. *American Association of Petroleum Geologists Bulletin*, **24**, 1548–1610.

SIDDER, G. B. & MENDOZA, V. 1995. Geology of the Venezuelan Guayana Shield and its relation to the entire Guayana Shield. *US Geological Survey Bulletin*, **2124**, B1–B41.

SISSON, V. B., ERTAN, I. E. & AVÉ LALLEMANT, H. G. 1997. High-pressure (~2000 MPa) kyanite- and glaucophane-bearing pelitic schist and eclogite from the Cordillera de la Costa belt, Venezuela. *Journal of Petrology*, **38**, 65–83.

SISSON, V. B., AVÉ-LALLEMANT, H. G. *ET AL.* 2005. Overview of radiometric ages in three allochthonous belts of northern Venezuela: Old ones, new ones, and their impact on regional geology. *In*: AVÉ-LALLEMANT, H. G. & SISSON, V. B. (eds) *Caribbean–South American Plate Interactions, Venezuela*. Geological Society of America, Boulder, CO, Special Papers, **394**, 91–117.

SMITH, W. H. & SANDWELL, D. T. 1997. Global sea floor topography from satellite altimetry and ship depth soundings. *Science*, **277**, 1956–1962.

SMITH, C. A., SISSON, V. B., AVÉ LALLEMANT, H. G. A. & COPELAND, P. 1999. Two contrasting pressure–temperature–time paths in the Villa de Cura blueschist belt, Venezuela: possible evidence for Late Cretaceous initiation of subduction in the Caribbean. *Geological Society of America Bulletin*, **111**, 831–848.

SNOKE, A. W., ROWE, D. W., YULE, J. D. & WADGE, G. 2001. *Petrologic and Structural History of Tobago, West Indies: A Fragment of the Accreted Mesozoic Oceanic-arc of the Southern Caribbean*. Geological Society of America, Boulder, CO, Special Papers, **354**.

SOCAS, M. 1991. *Estudio Sedimentologico de la Formacion Naricual, Estado Anzoategui*. Geological Engineering Thesis, Universidad Central de Venezuela.

SPEED, R. C. 1985. Cenozoic collision of the Lesser Antilles arc and continental South America and the origin of the El Pilar Fault. *Tectonics*, **4**, 41–69.

SPEED, R. C. 2002. Field trip to the Scotland District: Exposed example of an accretionary prism. *In: Transactions of the 16th Caribbean Geological Conference*, Barbados, 16–23 June, 2002.

SPEED, R. C., WESTBROOK, G. K. *ET AL.* 1984. *Atlas 10, Ocean Margin Drilling Program, Lesser Antilles and Adjacent Ocean Floor*. Marine Science Institute Publications, Woods Hole, MA.

SPEED, R. C., SHARP, W. D. & FOLAND, K. A. 1997. Late Paleozoic granitoid gneisses of northeastern Venezuela and the North America–Gondwana collision zone. *Journal of Geology*, **105**, 457–470.

SPRAGUE, A. R., LARUE, D. K., FAULKNER, B. L., SYKES, M. & ADEN, L. 1996. Sequence stratigraphy, facies architecture and reservoir distribution, Cretaceous lowstand fan reservoirs, Southern Basin, Onshore Trinidad, Association Round Table. *American Association of Petroleum Geologists Bulletin*, **80**, 1338.

STAINFORTH, R. M. 1948. Description, correlation and palaeoecology of Cenozoic Cipero Marl Formation, Trinidad. *American Association of Petroleum Geologists Bulletin*, **32**, 1292–1330.

STÖCKHERT, B., MARESCH, W. V. *ET AL.* 1995. Crustal history of Margarita Island (Venezuela) in detail: Constraint on the Caribbean Plate tectonic scenario. *Geology*, **23**, 787–790.

SUTER, H. H. 1960. *The General and Economic Geology of Trinidad*. British West Indies, Overseas Geological Surveys, Mineral Resources Division, London.

VAN ANDEL, T. H. 1958. Origin and classification of Cretaceous, Paleocene, and Eocene sandstones of western Venezuela. *American Association of Petroleum Geologists Bulletin*, **42**, 734–763.

VAN DER HILST, R. D. 1990. *Tomography with P, PP and pP delay-time data and the three-dimensional mantle structure below the Caribbean region*. PhD thesis, University of Utrecht.

VILLAMIL, T. 1999. Campanian–Miocene tectonostratigraphy, depocenter evolution and basin development of Colombia and western Venezuela.

Palaeogeography, Palaeoclimatology, Palaeoecology, **153**, 239–275.

VILLAMIL, T. & PINDELL, J. L. 1998. Mesozoic paleogeographic evolution of northern South America: foundations for sequence stratigraphic studies in passive margin strata deposited during non-glacial times. *In:* PINDELL, J. L. & DRAKE, C. (eds) *Paleogeographic Evolution and Non-glacial Eustacy: Northern South America*. SEPM (Society for Sedimentary Geology), Special Publications, **58**, 283–318.

VIVAS, V. & MACSOTAY, O. 1989. Miembro El Pilar de la Formación Quiamare: Ejemplo de molasa orogénica Neogena de Venezuela nororiental. *GEOS*, **29**, 108–125.

VIVAS, V. & MACSOTAY, O. 1995. *Formacion Tememure: Unidad olistostromica Eoceno Superior–Oligoceno Inferior en el frente meridional de la Napa Piemontina*. Boletin de Geológia del Minsterio de Energia y Minas, Publicación Especial, **10**, 95–113.

VIVAS, V. & MACSOTAY, O. 1997. Reinterpretación de la cobertura sedimentaria Cretácico–Palaeocena de la napa de Villa de Cura, Venezuela nor-central. *Memorias del VIII Congreso Geológico Venezolano*. Sociedad Venezolana de Geologia, Caracas, **2**, 516–525.

WILKINSON, B. H. & MCELROY, B. J. 2007. The impact of humans on continental erosion and sedimentation. *Geological Society of America Bulletin*, **119**, 140–156.

WILSON, B. 2003. Foraminifera and paleodepths in a section of the Early to Middle Miocene Brasso Formation, central Trinidad. *Caribbean Journal of Science*, **39**, 209–214.

WILSON, B. 2004. Benthonic foraminiferal palaeoecology across a transgressive–regressive cycle in the Brasso Formation (Early–Middle Miocene) of central Trinidad. *Caribbean Journal of Science*, **40**, 126–138.

WOOD, L. J. 2000. Chronostratigraphy and tectonostratigraphy of the Columbus Basin, eastern offshore Trinidad. *American Association of Petroleum Geologists Bulletin*, **84**, 1905–1928.

WYNN, J. C. 1993. Placer titanium and other heavy minerals. *US Geological Survey Bulletin*, **2062**, 89–90.

YSACCIS, R. 1997. *Cenozoic evolution of the northeastern Venezuela offshore*. PhD thesis, Rice University, Houston, TX.

Submarine volcanic stratigraphy and the Caribbean LIP's formational environment

JOHN DIEBOLD

Lamont-Doherty Earth Observatory, 61 Rte 9W, Palisades, NY 10964-8000, USA
(email: johnd@ldeo.columbia.edu)

Abstract: In 1995, R/V *Maurice Ewing* cruise EW9501 collected 5200 km of high-quality multichannel reflection profiles and 104 wide angle sonobuoy profiles in the Venezuelan Basin and over the Beata Ridge. These data provide images of the entire Caribbean crust, from seafloor to Moho, and verify earlier models of a two-'layer' crust. The EW9501 data also image the internal structure of these layers, revealing previously unseen features. The upper crustal 'layer' is actually a complex sequence, whose stratigraphy is analysed to show that the thickest part of the Caribbean volcanic plateau was experiencing east–west compressional deformation during the last stages of its emplacement.

The origins of the Caribbean volcanic plateau, which has been characterized as a Large Igneous Province, or LIP (Coffin & Eldholm 1994) remain shrouded and controversial, despite decades of study. Drilling has only penetrated the uppermost crust of the Caribbean Sea in a few places, so that nearly all of the geological and petrological studies of the Caribbean LIP (CLIP) have been based on samples and locales around its periphery. Many of these studies are reminiscent of the tale of six blind men examining an elephant. Each palpates some appendage, while none is able to look at the elephant itself. As a result, six completely different interpretations are made. Here, I present results from one set of seismic reflection data, acquired in 1995, which sees beneath the elephant's skin and reveals some skeletal structure.

In early 1995, R/V *Maurice Ewing* collected 5200 km of multichannel seismic reflection data (Fig. 1) over the Venezuelan Basin, Beata Ridge, and the basins immediately to the west of the ridge scarp (Driscoll & Diebold 1998, 1999; Diebold *et al.* 1999). The purpose of this paper is to revisit a part of that data set which clearly shows the effects of east–west compression during the emplacement of the thickest part of the CLIP– the flanks of Beata Ridge. How and why this compressional episode took place is an enigma whose solution contains as-yet unknown details of the province's tectonic history.

The Caribbean has been the locus of active seismic surveys for many decades. Single channel analogue reflection data acquired in the mid-1960s showed that, unlike the top surface of oceanic crust seen elsewhere, the upper surface of acoustic 'basement' is unusually smooth (Ewing & Ewing 1962). Two-ship explosive refraction profiles showed that the underlying crust is unusually thick, consisting of an upper layer, having variable acoustic velocity in general slower than that of 'normal' oceanic crust, and a lower, more homogeneous layer, with higher velocities (Officer *et al.* 1957, 1959; Edgar *et al.* 1971). The conclusions of the early workers were summarized by Edgar *et al.* (1971): 'the Caribbean crust, although *definitely not continental*, is significantly different from the crust found in an average ocean basin' (emphasis added). Following earlier conventions, the smooth Caribbean 'basement' horizon was named B″ ('B double-prime' following 'B prime' in the Pacific, and simply 'B' in the Atlantic). The advent of multichannel seismic (MCS) profiling in 1974 began to show that the most deeply sedimented part of the Venezuelan Basin is floored by rough basement, necessitating the terminology 'rough B″' and 'smooth B″'. These early MCS data also showed layering beneath smooth B″ (Talwani *et al.* 1977). The first clear images of the Moho, and therefore of the vertical entirety of the Caribbean crust, were published by Diebold *et al.* (1981), who also identified a triangular zone of thin but apparently oceanic crust, devoid of overlying basalt flows, in the SE Venezuelan Basin. The identification as oceanic crust was supported by interpretation of two-ship MCS refraction profiles.

Subsequently, results from scientific drilling and petrological/stratigraphic studies on land have shown that most of the Caribbean Plate is a large igneous province (LIP) (Donnelly *et al.* 1973). Radiometric dating of DSDP and ODP drill cores and of samples from accreted terranes on Haiti, Curaçao and Western Colombia indicate that the LIP formed by massive basalt flooding between 91 and 88 Ma (Sinton *et al.* 1998). Those authors, and others, also identified a second episode of magmatism at 76 Ma. The flood basalts, with their

From: JAMES, K. H., LORENTE, M. A. & PINDELL, J. L. (eds) *The Origin and Evolution of the Caribbean Plate.*
Geological Society, London, Special Publications, **328**, 799–808.
DOI: 10.1144/SP328.30   0305-8719/09/$15.00 © The Geological Society of London 2009.

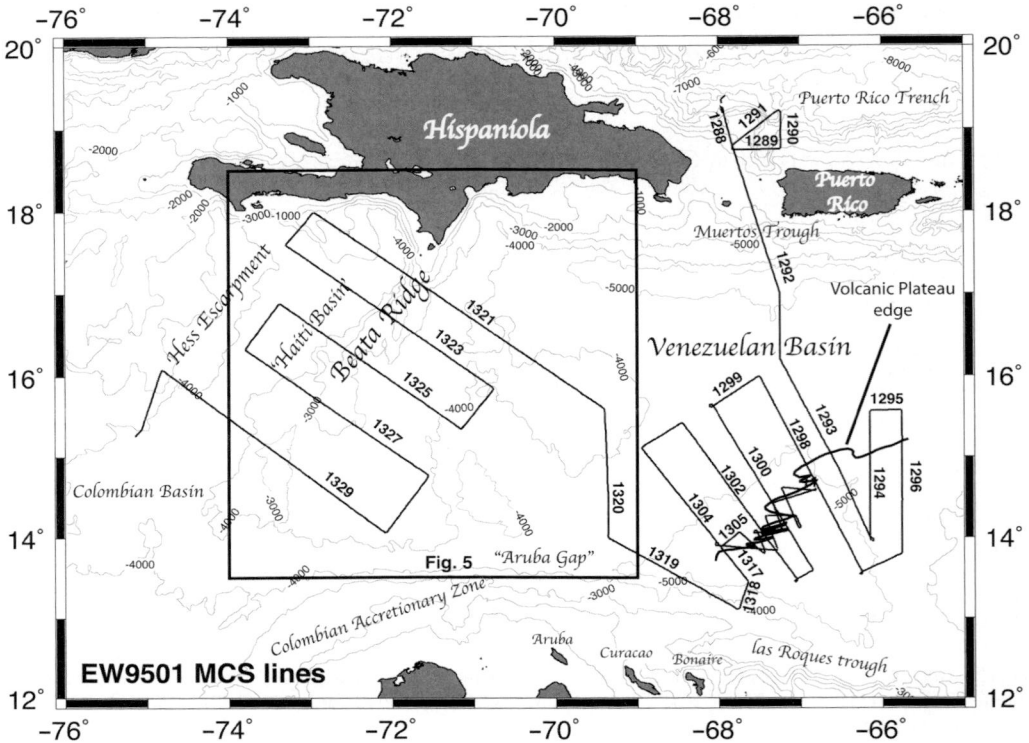

Fig. 1. Location map showing all *Maurice Ewing* EW9501 multichannel reflection profiles. This study concentrates on parts of three profiles on the eastern flank of Beata Ridge.

smooth tops and layered structure, form the upper seismic 'layer' discovered in the early two-ship refraction studies. The identification of the entire smooth-topped upper crustal sequence throughout the Venezuelan Basin as basalts has been extrapolated from its sampled locations by extensive velocity measurements and MCS imaging (Diebold *et al.* 1981, 1999).

Since its identification, two large-scale questions have dominated research on the origin of the CLIP. The first discusses the source of the basalts; were they sourced from a mantle plume or not? The second concerns the physical location of the plateau at the time of formation and its subsequent movement and evolution in a plate tectonic sense. Active seismic reflection and refraction data cannot by themselves answer these questions fully, but they can contribute several important constraints to the process. The R/V *Maurice Ewing* survey, EW9501 (Fig. 1), and resulting publications (Driscoll & Diebold 1998, 1999; Diebold *et al.* 1999) addressed three questions: the nature of the CLIP's SE boundary in the Venezuelan Basin; the origin of linear magnetic anomalies identified just

NW of that boundary; and the possibility that the shape and altitude of Beata Ridge was due to post-rift lithospheric flexure.

Detailed seismic stratigraphy shows that Beata Ridge has undergone three major phases of deformation. Most recently, a north–south belt of transpressive buckling formed the 'Taino Ridge' (Leroy 1995; Leroy & Mauffret 1996). This deformation plays an important part in the debate about the present-day tectonics of the Caribbean Plate (Driscoll & Diebold 1999).

Earlier, a phase of roughly east–west extension caused rifting, creating what Mauffret *et al.* (2001) called the 'Haiti Basin' and uplift of the Beata ridge as the result of lithospheric flexure (Diebold *et al.* 1999; Driscoll & Diebold 1998). This is, doubtlessly, the most significant deformational episode visited upon the Beata Ridge. Mauffret *et al.* (2001) dated this uplift as Campanian (*c.* 80 Ma) and attributed it to underplating.

The principal subject of this paper is a yet earlier, previously unreported, compressional episode, which apparently coincided with emplacement of the uppermost and most voluminous volcanic extrusive elements of the CLIP and its immediate aftermath. It has been previously demonstrated that, at the SE boundary of the CLIP, the uppermost and presumably last observable volcanic flows formed within a mildly extensional regime (Diebold *et al.* 1981, 1999). For these extrusive events to be contemporaneous requires a marked heterogeneity in the tectonic stress experienced by the plateau during its emplacement. It seems likely that interpreting this sequence of 'push-me–pull-you' tectonics is crucial to the understanding of the origin and development of the CLIP, and may help answer the 'Pacific' v. 'In-place' question.

Here we show that east–west compression (a) was coincident with emplacement of the uppermost, final and most voluminous volcanic sequences of the CLIP; (b) was concentrated in the thickest part of Beata Ridge, and (c) formed the north–south striking fabric that was reactivated during recent compression (Leroy & Mauffret 1996).

Data

The 5200 km of EW9501 MCS lines were acquired with a 20-element source array with total volume of 8500 cubic inches and a 160-channel, 4 km digital hydrophone array, with record lengths of 12 s (Diebold *et al.* 1999). Shot spacing was 50 m and the data were binned into 12.5 m CMP gathers, for a nominal fold of 40. During the survey, 103 sonobuoys were deployed at regular intervals; 81 of these produced useful records from which velocity–depth functions were determined by interactive ray tracing with one-dimensional models. Trial, or 'brute' stacking velocities were determined by applying sonobuoy-derived interval velocities to a layered model whose horizons were picked from plots of the near-channel recordings. This process was repeated to refine velocities as necessary. All data were time-migrated but in some cases, particularly towards the southern end of the Beata Ridge, where the lines (oriented so as to cross the main escarpment orthogonally) cross local structure obliquely, unmigrated stacked sections have a better appearance. The Moho reflection is seen in all of the lines over the Venezuelan Basin but less often beneath the flanks of Beata Ridge, due to interference from the seafloor multiple. Future surveys, with longer hydrophone arrays, will do a better job of defeating this interference.

Interpretation

Nature and source of the plateau

The Caribbean volcanic plateau, as illuminated by EW9501 (Diebold *et al.* 1999) comprises two distinct seismic sequences (Fig. 2). These sequences correlate with the double-layering determined by two-ship refraction profiles (Officer *et al.* 1957, 1959; Edgar *et al.* 1971). The upper sequence appears layered, as first observed by Stoffa *et al.* (1981), while the lower sequence is generally devoid of reflecting horizons. The frequency content of the EW9501 reflection sections is limited to perhaps 40 Hz, allowing a resolution limit of about 20 m at velocities of $c.$ 5 k ms^{-1}. At this resolution, the upper volcanic sequence is dominated by sub-horizontal reflections, many of which are coherent and traceable for tens of kilometres. In contrast, the lower sequence appears homogeneous with a strong, low-frequency reflection at its top, and, where the seafloor multiple does not interfere, the Moho at its base. Illustrating all this, EW9501 sonobuoy 68 (Fig. 2) returned a nearly complete crustal section. The interpreted two-way time of the top of the upper volcanic sequence is a bit early, due to a small rise in the B″ surface at that position. Velocities vary considerably within the upper sequence, typically increasing with depth of burial, though in one or two places low velocity zones are evident in the sonobuoy records (fig. 8 of Diebold *et al.* 1999). The transition between the upper and lower sequences often appears as a zone with relatively high velocities on the EW9501 sonobuoy records – perhaps due to the presence of picrites, as observed in Curaçao by Klaver (1987) and others elsewhere. If we take sonobuoy 68's 5.80/6.44 km s^{-1} horizon as the upper/lower boundary, then the upper sequence is 3.45 km thick and the lower 9.35 km thick at this position. Should one wish instead to include the 5.80 km s^{-1} layer in the lower sequence, the sequence thicknesses would be 2.83 and 9.97 km.

In general, the thickness and volume of the upper volcanic sequence increases from zero in the SE to 10 km or more near the crest of the Beata Ridge. In the Venezuelan Basin, EW9501 and earlier MCS data show the upper sequence extending beyond the limits of the lower sequence, appearing to flow onto the pre-existing thin oceanic crust (Fig. 1; Diebold *et al.* 1981). Despite the relatively high-quality of the EW9501 reflection data, it is not easy to typify the overall distribution of the lower sequence. Its thickness is quite variable, forming ridges and domes, but is never as thin as the apparently oceanic crust of the SE Venezuelan Basin. This suggests that the upper volcanic sequence was not simply superimposed upon pre-existing oceanic crust. Rather, original crust must have been thickened by, and likely entrained within, the material forming the lower sequence. Diebold *et al.* (1999) made observations that enabled them to conclude the lower sequence predates the upper, but this relationship may not be

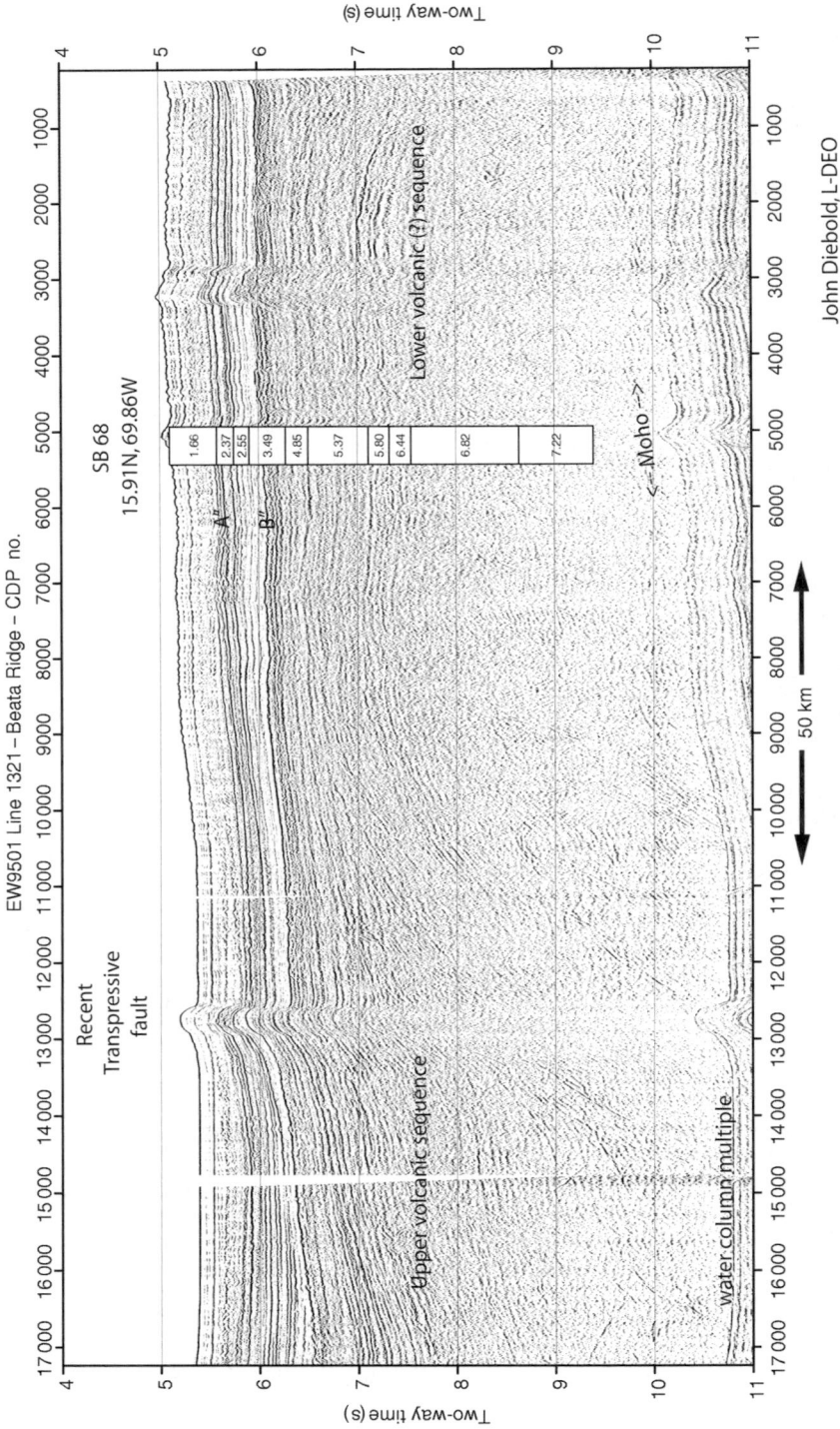

Fig. 2. The easternmost portion of EW9501 MCS line 1321, identifying the major volcanic sequences that are discussed. Superimposed is the velocity column for one of the 86 sonobuoy records taken during the survey and subsequently interpreted. This result shows typical velocities for the upper and lower volcanic sequences, and that the lower sequence, while homogeneous to reflection, is layered and of high velocity from a refraction point of view. Location is shown in Figure 5.

universally so, especially since it cannot be determined whether the entire lower sequence is itself contemporaneous.

These observations, and velocities from expanding spread profiles (Diebold *et al.* 1981) agree well with models for a plume-generated oceanic plateau (cf. Farnetani *et al.* 1996; Kerr *et al.* 1997*a*) or what has been observed in some places on land. For example, Klaver (1987) reported a 5 km-thick sequence of Cretaceous basalt flows on Curaçao, dominated by pillow basalts, but also included hyaloclastics, and a few non-vesicular sills and flows. From Figures 2–4, it is very easy to imagine an analogue in the upper volcanic sequence of the Beata Ridge and the rest of the CLIP; however the continuity of reflectors suggests a sequence more dominated by volcanic flows, with subordinate intercalations of hyaloclastics, pyroclastics and ash. By analogy with results from ODP site 1001 samples (Sigurdsson *et al.* 1997), much of the

massive basalts is likely to have formed the inner sections of 'inflated flows' that may reach thicknesses of 74 m, well above the resolution of the EW9501 data. These flows may reach tens of kilometres in length as a result of lava injection into preexisting, insulating, extrusives (Umino *et al.* 2006). On the other hand, samples taken nearby (Mauffret *et al.* 2001; Révillon *et al.* 2000) are dominated by dolerite sills and gabbros, with only occasional pillow basalts. One such suite of samples was collected during Mauffret *et al.*'s dive NB11, which was located less than 10 km from the NW edge of Figure 3. This dichotomy suggests either a sampling scheme favouring a few competent sills, or a high degree of lateral heterogeneity in the CLIP, one in which transitions between flow-dominated emplacement and sill-dominated emplacement (such as that seen in the vicinity of CDP 10 000 of line 1321; Fig. 2) can take place within a short distance of much thicker sequences of flows.

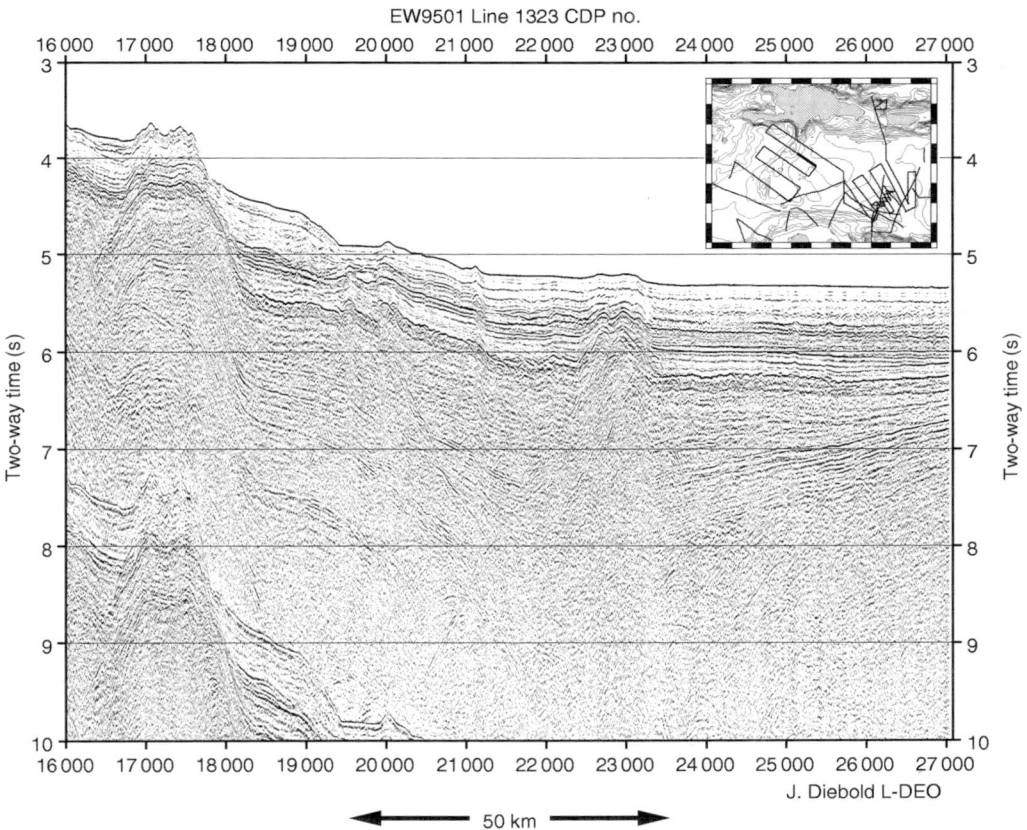

Fig. 3. The easternmost portion of EW9501 MCS line 1323, next SSW of 1321 (Fig. 2). Unlike 1321, the upper volcanic sequence here is consistently thick, and is distorted in a number of ways described in the text. This is an unmigrated stacked section.

(a)

(b)

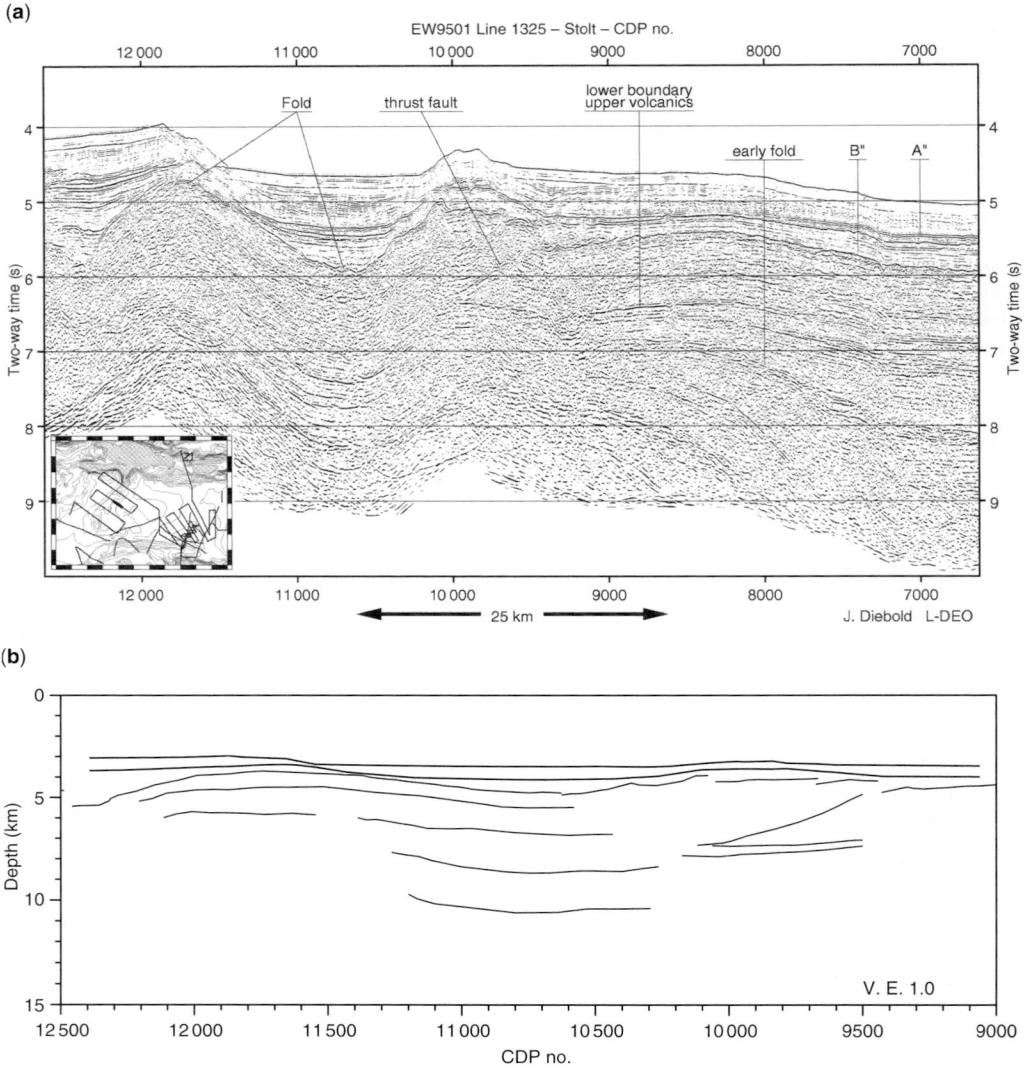

Fig. 4. (**a**) EW9501 line 1325, Post-stack time migration. (**b**) Line drawing of the migrated horizons in (a), converted to depth at true scale. Fold amplitude averages 1200 m, wavelength is 22 500 m. (**c**) EW9501 MCS line 1325, horizon B″ has been flattened to clarify the onlapping of the uppermost volcanic flows against previously deformed volcanics around CDP 8000.

A volcanic structure can be seen between CDPs 22 500 and 23 250 on EW9501 line 1323 (Fig. 3). Although its vertical extent is at least 5 km, this volcano has little surface expression. The seamount is surrounded by a flexural moat into which the upper sequence thickens and onto whose flanks it onlaps. These relationships show that the moat deepened during extrusion of the upper sequence, suggesting the seamount was only recently emplaced at that time and, hence, could have contributed some of the extrusive material. The Venezuelan basin is dotted by dozens of similar volcanic edifices to that imaged in Figure 3, which project above horizon B″ leading Donnelly (pers. comm.) to consider them to represent a late, perhaps final, phase of volcanism. The fact that these seamounts are the only obviously volcanic features identified in MCS profiles suggests that they might well, however, be the conduits for all of the main upper volcanic sequence.

(c)

EW9501 Line 1325 — Horizon B" flattened

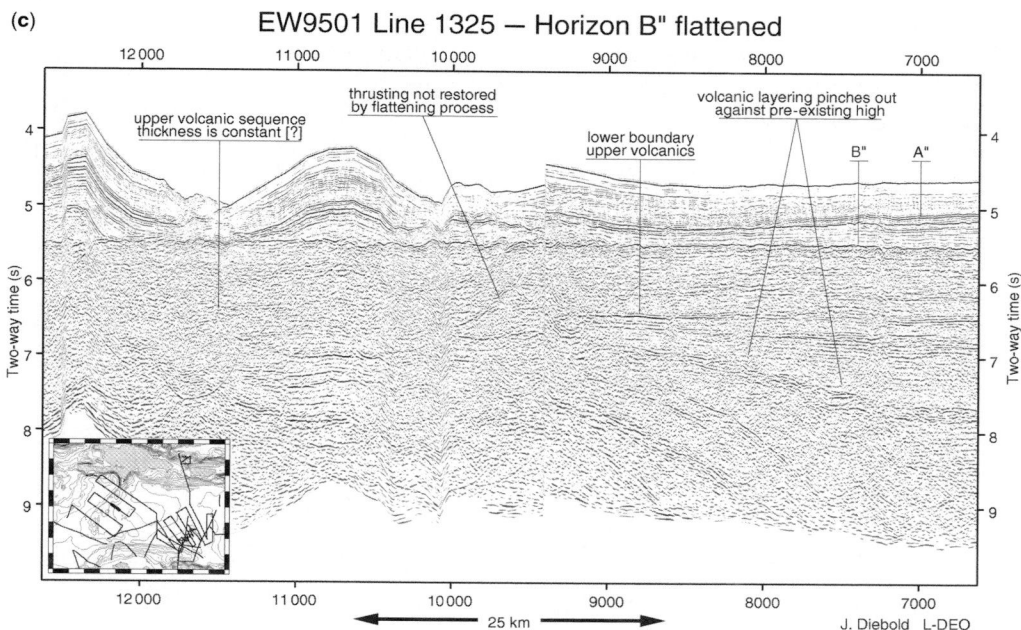

Fig. 4. (*Continued*).

Compression at the eastern Beata Ridge

Figures 3–5 focus on some remarkable, compressional features of the eastern Beata ridge flank. Figure 3 shows the easternmost portion of EW9501 MCS line 1323 parallel to and south of line 1321 (Fig. 2). Upslope of the seamount volcano on line 1323, two thrust faults and a basement ridge, formed by compressive buckling, are clearly seen at CDPs 19 500–20 000. These features are superimposed upon a regional flexure presumably resulting from extension that uplifted Beata ridge and thinned the Haiti Basin (Mauffret & Leroy 1999a). So there is evidence for compression and extension; which came first, and why? Kerr *et al.* (1997b) interpreted evidence on land for deformation of the volcanic plateau and attributed it to plateau/arc collision, but in most plate tectonic reconstructions this occurs after plateau construction. A different scenario is revealed by the volcanic stratigraphy of EW9501 MCS line 1325 (Fig. 4a). Here, a well-developed fold of the upper volcanic sequence (CDPs 10 000–12 500) is flanked by a thrust fault which faces in the opposite direction to those on adjacent line 1323 (Fig. 3). None of the overlying sediment appears to be folded, indicating that folding took place soon after emplacement of the uppermost volcanic sequence.

In Figure 4c, a simple vertical trace shifting operation has been applied to flatten horizon B". Assuming that horizon B" was once flat, or at least planar, this operation removes the distortion due to the B" fold. Between CDPs 7500 and 9000, the underlying sequence onlaps what was a previously folded monocline. This indicates that compression and folding were taking place during the emplacement of the upper 5 km of volcanics. The onlapping nature of the sequence layering also indicates that it is predominated by aerially large-scale flows, accompanied, perhaps, by turbidites.

Discussion

The Beata Ridge is roughly fan-shaped, narrow and tall towards the north, where it shallows to Beata Island, but deep and broad towards the south, where its surface expression is eventually lost beneath the sediments of the Aruba Gap (Figs 1 & 5). Similarly, as compressional features like those seen on EW9501 MCS lines 1323 and 1325 are confined to a narrow zone in the northernmost line, 1321 (Driscoll & Diebold 1999), the zone of compression also broadens southwards. The gentler, southeastern, flank of the Beata Ridge escarpment, which strikes NE–SW, exhibits a strong, roughly north–south grain (Fig. 5). Although the seismic lines are too far apart to allow firm correlation of the folds and faults discussed here, it seems reasonable to propose that

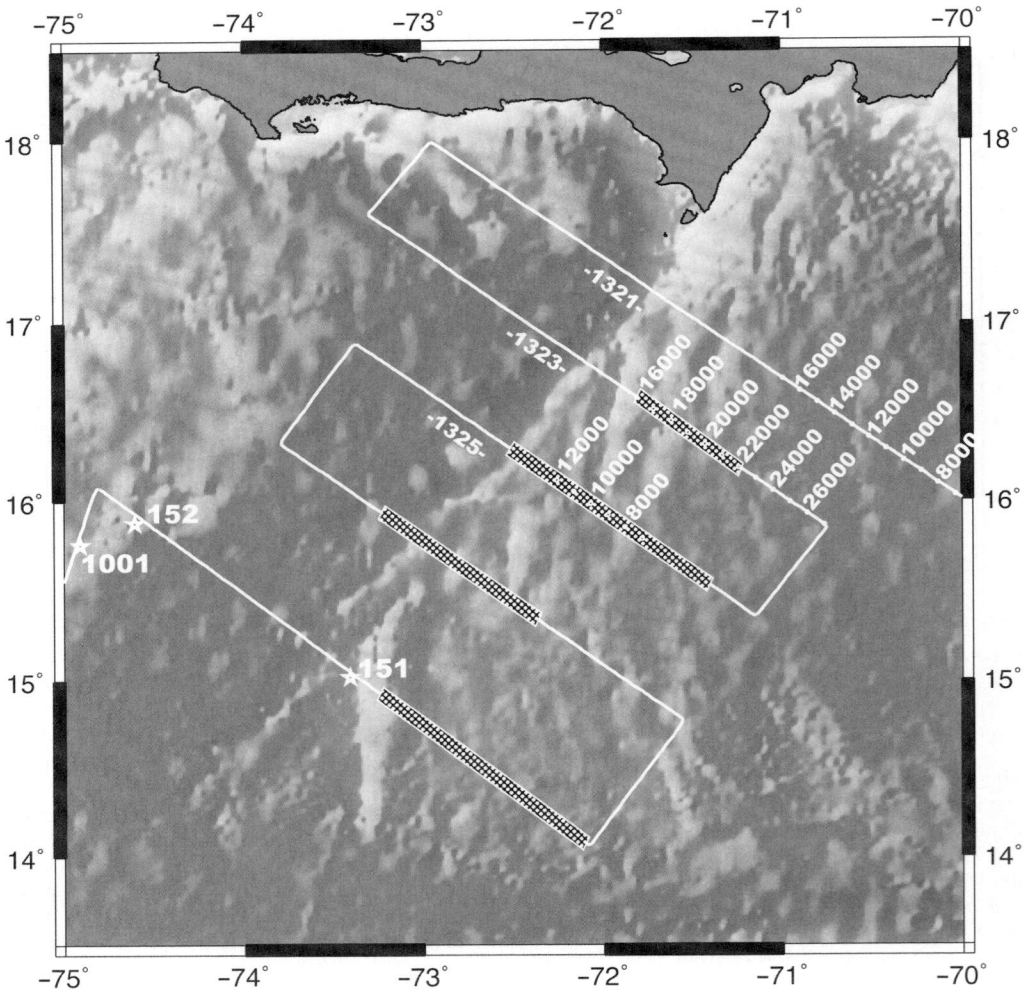

Fig. 5. Detailed map of EW9501 MCS lines over Beata Ridge. CDP numbers for Figures 2–5 are annotated. Areas distorted and compressed by folding are stippled.

they strike parallel to this grain. I also propose that extension, contemporaneous with the formation of the Haiti Basin, exploited this grain, forming grabens like the prominent example at 73.2°W, between 14 and 15.5°N. Later still, a compressional or transpressive regime was imposed, and structures such as 'Taino Ridge' exploited the pre-existing north–south linear compressive features described here.

The most puzzling aspect of these observations is the cause of the deformation and its timing. The lack of folded sediments, particularly apparent in Figure 4, may be explained by uplift of the ridge, as proposed by Mauffret *et al.* (2001). This could have resulted in non-deposition and/or erosion.

Brittle deformation, which accompanies folding of the uppermost volcanic sequence, suggests that cooling was underway during the latter stages of compression. This may not have been the case for the pre-existing deeper fold imaged in Figure 4c, suggesting that the compressional episode may have spanned the emplacement of the upper volcanic sequence and the period immediately afterwards. If this deformation is to be related to insertion of a Pacific-origin plateau between North and South America, its timing does not fit well with early models that show the plateau west of any tectonic influences at the time of its emplacement (e.g. Burke *et al.* 1978; Duncan & Hargraves 1984). Those models suggest the plateau collided with and choked an east-dipping subduction zone, causing the reversal of the subduction direction, but have difficulty explaining how the thin leading edge of CLIP and even thinner, probably Jurassic, oceanic crust in the SE Venezuelan Basin can have induced and survived the subduction reversal (Diebold *et al.* 1981; Mauffret & Leroy 1999*a*; Diebold *et al.* 1999). More recent models (Meschede & Frisch 1998; Pindell *et al.* 2006) envisage the plateau as being emplaced between an Albian, east-dipping subduction zone and a west-dipping one – the Great Caribbean Arc, closing the timing gap and side-stepping the problem of a subduction reversal. Nonetheless, with these models it remains unclear how the thinned leading edge can have formed *in situ*. Mauffret & Leroy (1999*a*) offer the explanation that Aves ridge predates the CLIP, and was the locus of a subduction reversal, protecting the trailing Venezuelan Basin crust from deformation.

Conclusions

Multichannel seismic reflection profiles from R/V *Maurice Ewing* cruise EW9501 have been presented here and in several previous publications. The profiles reveal structural and compositional details of the Caribbean Plateau in the Venezuelan Basin and in the area of its greatest thickness, the Beata Ridge. Everywhere in the EW9501 profiles, the plateau is seen to comprise two separate sequences, distinguished by physical appearance, stratigraphic position and seismic velocity. The two sequences could be contemporaneous, the lower sequence formed mainly by intrusion, the upper by extrusion. The fact that the entire body appears to have folded in response to compression and vertical loading suggests this, but the reflection profiles otherwise offer little evidence either way. Only the upper sequence has been sampled by drilling. In agreement with those samples, and with outcrops in the Netherlands Antilles, the upper sequence is interpreted as a series of flood-like basalts, comprising massive flows, hyaloclastics, pillows and ash. It includes sub-sequences, whose boundary horizons can be tracked for hundreds of kilometres, and whose fluidity is readily apparent. The thickness of the upper sequence varies from near zero, in the SE Venezuelan Basin (Diebold *et al.* 1999) to as much as 10 km.

Since the fluid and voluminous basalts of the upper sequence have the demonstrated ability to fill stratigraphic lows, resulting in onlapping patterns such as those seen in Figures 3 and 4, many of the intra-sequence horizons must have been horizontal or subhorizontal at the time of emplacement. Tilting, folding, and folding of these horizons betrays their subsequent deformation. In the east–west direction, the Beata Ridge exhibits evidence

for both extensional and compressional deformation. The primary extensional features are the uplift of the ridge, related to rift-induced lithospheric rebound, and tilted blocks in the rifted ridge flank. Compressional features include the early phase of ductile buckling and small-scale brittle faulting illustrated in Figures 3 and 4 and the later formation of transpressional horsts described by Leroy & Mauffret (1996) and Mauffret & Leroy (1999*b*).

Arguments presented here are intended to demonstrate that early compression coincided with the emplacement of the upper, extrusive volcanic sequence in the area of the eastern flank of the Beata Ridge, well before formation of any of the extensional features, which include the uplift of the ridge. Radiometric dates of *c.* 91 Ma (Sinton *et al.* 1998) for flood basalt emplacement coupled with the Mauffret *et al.* (2001) early date (Campanian) for the onset of Beata–Lower Nicaragua Rise rifting support this. Production of further basalts, including some sampled at DSDP Site 152 and ODP Site 1001 (Sinton *et al.* 2000) has been dated at roughly 76 Ma and attributed to partial melting in response to lithospheric extension (Sinton *et al.* 1998), and it is possible that some of these are imaged in the profiles presented here. Many plate tectonic reconstructions (Pindell *et al.* 1988; Pindell 1994) show the Caribbean plateau just beginning to move between the North and South American plates at this time. It is possible that the forces causing the early compression were related to the collision of the eastern edge of the plateau with the edges of North and South America, and that subsequent extension was the result of back-arc-basin forming forces after the initiation of the current, west-dipping zone.

Useful comments and editorial improvements were made by B. Bally, K. H. James and G. Eagles. The field work and data processing was supported by NSF grant no. OCE-93-02578. L-DEO contribution no. 7198.

References

BURKE, K., FOX, P. J. & SENGOR, A. M. C. 1978. Buoyant ocean floor and the evolution of the Caribbean. *Journal of Geophysical Research*, **83**, 3949–3954.
COFFIN, M. F. & ELDHOLM, O. 1994. Large igneous provinces: crustal structure, dimensions and external consequences. *Reveiws in Geophysics*, **32**, 1–36.
DIEBOLD, J. B., STOFFA, P. L., BUHL, P. & TRUCHAN, M. 1981. Venezuelan Basin crustal structure. *Journal of Geophysical Research*, **86**, 7901–7923.
DIEBOLD, J., DRISCOLL, N. & EW-9501-Science Team. 1999. New insights on the formation of the Caribbean basalt province revealed by multichannel seismic images of volcanic structures in the Venezuelan Basin. *In*: MANN, P. (ed.) *Caribbean Basins*. Sedimentary Basins of the World. Elsevier Science, Amsterdam, 561–589.

DONNELLY, T. W., MELSON, W., KAY, R. & ROGERS, J. W. 1973. Basalts and dolerites of Late Cretaceous age from the central Caribbean. *Initial Reports of the Deep Sea Drilling Project.* US Government Printing Office, Washington, DC, 1137.

DRISCOLL, N. W. & DIEBOLD, J. B. 1998. Deformation of the Caribbean region: one plate or two? *Geology,* **26**, 1043–1046.

DRISCOLL, N. W. & DIEBOLD, J. B. 1999. Tectonic and stratigraphic development of the eastern Caribbean: new constraints from multichannel seismic data. *In*: MANN, P. (ed.) *Caribbean Basins. Sedimentary Basins of the World.* Elsevier Science, Amsterdam, 591–626.

DUNCAN, R. A. & HARGRAVES, R. B. 1984. Plate-tectonic evolution of the Caribbean region in the mantle reference frame. *In*: BONINI, W., HARGRAVES, R. B. & SHAGAM, R. (eds) *The Caribbean–South American Plate Boundary and Regional Tectonics.* Geological Society of America, Boulder, CO, Memoirs, **162**, 81–93.

EDGAR, N. T., EWING, J. I. & HENNION, J. 1971. Seismic refraction and reflection in Caribbean Sea. *America Association of Petroleum Geologists Bulletin,* **55**, 833–870.

EWING, J. & EWING, M. 1962. Reflection profiling in and around the Puerto Rico Trench. *Journal of Geophysical Research,* **67**, 4729–4739.

FARNETANI, C. G., RICHARDS, M. A. & GHIORSO, M. S. 1996. Petrological models of magma evolution and deep crustal structure beneath hotspots and flood basalt provinces. *Earth and Planetary Science Letters,* **143**, 81–94.

KERR, A. C., TARNEY, J., MARRINER, G. F., NIVIA, A. & SAUNDERS, A. D. 1997a. The Caribbean–Colombian Cretaceous Igneous Province: the internal anatomy of an oceanic plateau. *In*: MAHONEY, J. J. & COFFIN, M. F. (eds) *Large Igneous Provinces: Continental, Oceanic, and Planetary Flood Volcanism.* American Geophysical Union, Washington, DC, Geophysical Monographs, 123–144.

KERR, A. C., MARRINER, G. F. ET AL. 1997b. Cretaceous basaltic terranes in western Colombia: elemental, chronological and Sr–Nd isotopic constraints on petrogenesis. *Journal of Petrology,* **38**, 677–702.

KLAVER, G. T. 1987. *The Curaçao Lava Formation: an ophiolitic analogue of the anomalously thick layer 2B of the Mid-Cretaceous oceanic plateaus in the western Pacific and central Caribbean.* PhD thesis, University of Amsterdam.

LEROY, S. 1995. *Structure et origine de la plaque Caribe. Implications geodynamiques.* These de l'Universite Paris 6.

LEROY, S. & MAUFFRET, A. 1996. Intraplate deformation in the Caribbean region. *Journal of Geodynamics,* **21**, 113–122.

MAUFFRET, A. & LEROY, S. 1999a. Seismic stratigraphy and structure of the Caribbean igneous province. *Tectonophysics,* **283**, 61–104.

MAUFFRET, A. & LEROY, S. 1999b. Neogene intraplate deformation of the Caribbean Plate at the Beatta ridge, in Caribbean sedimentary basins. *In*: MANN, P. (ed.) *Caribbean Basins. Sedimentary Basins of the World.* Elsevier Science, Amsterdam, 627–669.

MAUFFRET, A., LEROY, S., VILA, J.-M., HALLOT, E., DELEPINAY, B. M. & DUNCAN, R. A. 2001. Prolonged magmatic and tectonic development of the Caribbean Igneous Province revealed by a diving submersible survey. *Marine Geophysical Researches,* **22**, 17–45.

MESCHEDE, M. & FRISCH, W. 1998. A plate-tectonic model for the Mesozoic and Early Cenozoic history of the Caribbean Plate. *Tectonophysics,* **296**, 269–291.

OFFICER, C. B. J., EWING, J. I., EDWARDS, R. S. & JOHNSON, H. R. 1957. Geophysical investigations in the eastern Caribbean: Venezuelan Basin, Antilles island arc, and Puerto Rico Trench. *Geological Society of America Bulletin,* **68**, 359–378.

OFFICER, C., EWING, J., HENNION, J., HARKINDER, D. & MILLER, D. 1959. Geophysical investigations in the eastern Caribbean – summary of the 1955 and 1956 cruises. *In*: AHRENS, L. M. ET AL. (eds) *Physics and Chemistry of the Earth,* **3**. Pergamon, London, 17–109.

PINDELL, J. L. 1994. Evolution of the Gulf of Mexico and the Caribbean. *In*: DONOVAN, S. K. & JACKSON, T. A. (eds) *Caribbean Geology: An Introduction.* University of the West Indies Publishers Association/University of the West Indies Press, Kingston, Jamaica, 13–39.

PINDELL, J. L., CANDE, S. C. ET AL. 1988, A plate-kinematic framework for models of Caribbean evolution. *Tectonophysics,* **155**, 121–138.

PINDELL, J. L., KENNAN, L., STANEK, K. P., MARESCH, W. V. & DRAPER, G. 2006. Foundations of Gulf of Mexico and Caribbean Evolution: eight controversies resolved. *Geologica Acta,* **4**, 303–341.

RÉVILLON, S., HALLOT, E., ARNDT, N. T., CHAUVEL, C. & DUNCAN, R. A. 2000. A complex history for the Caribbean plateau: petrology, geochemistry, and geochronology of the Beatta Ridge, south Hispaniola. *Journal of Geology,* **108**, 641–661.

SIGURDSSON, H., LECKIE, R. M. ET AL. 1997. *Proceedings of ODP, Initial Reports of the Deep Sea Drilling Project,* **165**. Ocean Drilling Program, College Station, TX.

SINTON, C. W., DUNCAN, R. A., STOREY, M., LEWIS, J. & ESTRADA, J. J. 1998. An oceanic flood basalt province within the Caribbean Plate. *Earth and Planetary Science Letters,* **155**, 221–235.

SINTON, C. W., SIGURDSSON, H. & DUNCAN, R. A. 2000. Geochronology and petrology of the igneous basement at the lower Nicaragua Rise, Site 1001. *In*: LECKIE, R. M., SIGURDSSON, H., ACTON, G. D. & DRAPER, G. (eds) *Proceedings of Ocean Drilling Program, Scientific Results,* **165**. Ocean Drilling Program, College Station, TX.

STOFFA, P. L., MAUFFRET, A., TRUCHAN, M. & BUHL, P. 1981. Sub-B″ layering in the southern Caribbean: the Aruba Gap and Venezuela Basin. *Earth and Planetary Science Letters,* **53**, 131–146.

TALWANI, M., WINDISCH, C., STOFFA, P. L., BUHL, P. & HOUTZ, R. E. 1977. Multichannel seismic study in the Venezuelan Basin and the Curacao Ridge. *In: Island Arcs, Deep Sea Trenches and Back-Arc Basins.* Maurice Ewing Series, I. American Geophysical Union, 83–98.

UMINO, S. L., NONAKA, M. & KAUAHIKAUA, J. 2006. Emplacement of subaerial pahoehoe lava sheet flows into water: 1990 Kupaianaha flow of Kilauea volcano at Kaimu Bay, Hawaii. *Bulletin of Volcanology,* **69**, 125–139.

Magma source evolution beneath the Caribbean oceanic plateau: new insights from elemental and Sr–Nd–Pb–Hf isotopic studies of ODP Leg 165 Site 1001 basalts

ANDREW C. KERR[1]*, D. GRAHAM PEARSON[2] & GEOFF M. NOWELL[2]

[1]*School of Earth and Ocean Sciences, Cardiff University, Park Place, Cardiff CF10 3YE, UK*

[2]*Department of Earth Sciences, Durham University, South Road, Durham DH1 3LE, UK*

*Corresponding author (e-mail: kerrA@cardiff.ac.uk)

Abstract: Ocean Drilling Project Leg 165 sampled 38 m of the basaltic basement of the Caribbean Plate at Site 1001 on the Hess Escarpment. The recovered section consists of 12 basaltic flow units which yield a weighted mean Ar/Ar age of 80.9 ± 0.9 Ma. The basalts (6.4–8.5 wt% MgO) are remarkably homogeneous in composition and are more depleted in incompatible trace elements than N-MORB. Depleted initial radiogenic isotope ratios (ε_{Nd} +11.1 to +11.9; ε_{Hf} +15.2 to +16.9; $^{87}Sr/^{86}Sr$ 0.7025–0.7028; $^{206}Pb/^{204}Pb$ 18.34–18.50; $^{207}Pb/^{204}Pb$ 15.42–15.51; $^{208}Pb/^{204}Pb$ 37.64–37.90) reveal a long-term history of depletion. Although the Site 1001 basalts are superficially similar to N-MORB, radiogenic isotopes in conjunction with incompatible trace element ratios show that the basalts have more similarity to the depleted basalts and komatiites of Gorgona Island. This chemical composition strongly implies that the Site 1001 basalts are derived from a mantle plume-depleted component and not from depleted ambient upper mantle. Therefore the Site 1001 basalts are, both compositionally and tectonically, a constituent part of the Caribbean oceanic plateau. Mantle melt modelling suggests that the Site 1001 lavas have a composition which is consistent with second-stage melting of compositionally heterogeneous mantle plume source material which had already been melted, most likely to form the 90 Ma basalts of the plateau. The prolonged residence (>10 Ma) of residual mantle plume source material below the region confirms computational model predictions and places significant constraints on tectonic models of Caribbean evolution in the Late Cretaceous.

Although most oceanic crust away from oceanic islands and their hotspot tracks averages 6–7 km thick, substantial areas of the oceanic basins preserve crustal thicknesses substantially in excess of these values. These sometimes vast features (e.g. Ontong Java and Kerguelen Plateaus) have become known as oceanic plateaus because, due to their over-thickened basaltic crust, they are more buoyant and are elevated *c.* 4 km above oceanic crust of average thickness. Significantly, the first extensive area of over-thickened crust to be recognized by seismic refraction and reflection surveys was in the Caribbean basin (Officer *et al.* 1957; Edgar *et al.* 1971). Following DSDP drilling in the Caribbean during Leg 15, Donnelly (1973) proposed that the Caribbean represented a widespread 'oceanic flood basalt province'. This term subsequently fell largely into abeyance when Kroenke (1974) recognized that the Ontong Java plateau represented an area of elevated thickened oceanic crust which he termed 'oceanic plateau'.

Caribbean–Colombian oceanic plateau

Of all the extant plateaus in the present ocean basins, the Caribbean–Colombian oceanic plateau is by far the best exposed. The reasons for this are abundantly clear and stem from the ultimate derivation of the plateau from the Pacific realm, its subsequent collision with the northwestern continental margin of South America and its movement between the diverging Americas to form the core of the Caribbean Plate (Duncan & Hargraves, 1984; Burke, 1988; Pindell & Barrett 1990; Pindell *et al.* 2006). The accretion of the plateau around the margins of the Caribbean and in northwestern South America during this movement has resulted in the uplift and exposure of extensive plateau sequences in Curaçao, Aruba, Jamaica, Hispanola, Costa Rica, Colombia and Ecuador (see Kerr *et al.* 2003 for a review). The intimate association of accreted island arc rocks with these plateau sequences testifies to a Caribbean Plate which has been highly mobile, throughout its tectonic history (Kerr *et al.* 2003) and is completely at variance with the somewhat esoteric *in situ* models of Caribbean evolution (e.g. Meschede & Frisch, 1998). While there are differences between workers over precise details (e.g. Duncan & Hargraves 1984; Burke 1988; Pindell & Barrett 1990; Mauffret & Leroy 1997; Sinton *et al.* 1998; Lapierre *et al.* 1999; Kerr *et al.* 2003; Hoernle *et al.* 2004; Giunta *et al.* 2006;

From: JAMES, K. H., LORENTE, M. A. & PINDELL, J. L. (eds) *The Origin and Evolution of the Caribbean Plate.*
Geological Society, London, Special Publications, **328**, 809–827.
DOI: 10.1144/SP328.31   0305-8719/09/$15.00 © The Geological Society of London 2009.

Pindell et al. 2006), the overwhelming weight of scientific evidence supports a Pacific-derived origin for the Caribbean Plate. In contrast, models which propose that the Caribbean Plate formed in situ are based on little substantive data and important lines of evidence, in particular geochemistry, are at best misunderstood and at worst deliberately misrepresented (e.g. James, 2006). Accordingly, this paper adopts the widely accepted, near-universal view that the Caribbean Plate formed in the Pacific in the Cretaceous, and has had a highly mobile tectonic history.

The thickness (up to 20 km; Case et al. 1990; Mauffret & Leroy 1997) and areal extent of the Caribbean oceanic crust (c. 800 000 km^2) is entirely consistent with its formation by extensive melting of a hot mantle plume, as is the occurrence of high-MgO lavas in accreted sections in Colombia and Curaçao (e.g. Kerr et al. 1996b; 2002; Hauff et al. 2000; Révillon et al. 2002; Kerr 2005). Furthermore, although lavas from the plateau span an age range from 73 to 94 Ma, a significant peak in volcanic activity occurs c. 91 Ma with a lesser peak from 78 to 80 Ma (see figure 12 in Kerr et al. 2004). This voluminous outpouring of melt at c. 91 Ma also supports a hot mantle plume origin for a significant proportion of the Caribbean Plate. As reviewed by Kerr et al. (2003), the geochemistry of lavas and sills from the c. 91 Ma phase of the Caribbean plateau shows little evidence for derivation from a shallow upper mantle source region at either mid-ocean ridge or a back-arc setting. Furthermore, the absence of any continental crustal signature in Caribbean oceanic plateau lavas rules out formation close to a rifted continental margin.

Structure of the Caribbean Plate and DSDP Leg 15 drill sites

The internal tectonic structure of the Caribbean Plate is considerably complex and is entirely consistent with derivation of the plate from the Pacific region and its subsequent insertion between the two Americas. The reader is referred to the excellent summary of internal Caribbean Plate structure given by Mauffret & Leroy (1997), who proposed a revised nomenclature of basins and rises in the central Caribbean (Fig. 1). In general terms, the central Caribbean is divided into the Colombian and Venezuelan basins. These are separated by the Beata Ridge, a tectonically complex north–south trending topographic high (Fig. 1) with crustal thicknesses up to 23 km (Mauffret & Leroy, 1997). The Venezuelan Basin (including the Puerto Rico and Dominican sub-basins) is bounded to the east by the Aves Ridge (widely regarded to be the

remnants of an extinct Paleocene–Eocene island arc; Fox et al. 1971) and to the north and south by two deformed belts, the Muertos Trench and the Venezuela deformed belt (Fig. 1). The Colombian Basin is bounded to the south by the Colombia–Panama deformed belt and to the NW by the Hess Escarpment. The southern Nicaragua Rise and the Cayman Trough lie NW of the Hess Escarpment (Fig. 1).

Seismic studies have shown that the of the top of the igneous basement (a reflector horizon known as B′) is relatively smooth over much of the plate (Officer et al. 1959), and these areas (which are generally the thicker parts of the plate) have been interpreted to represent areas flooded by oceanic plateau lavas (Donnelly et al. 1973). However, in some places the B′ reflector is much rougher and several explanations have been advanced to explain this: either the rough basement represents older extended (possibly Jurassic) oceanic lithosphere with no oceanic plateau cover (e.g. the southern part of the of the Venezuelan and Colombian basins; Bowland & Rosencrantz 1988; Mauffret & Leroy 1997) or alternatively this basement may have been formed by extension after formation of the plateau (e.g. between the Beata Ridge and the Hess Escarpment; Diebold & Driscoll 1999).

The Deep Sea Drilling Project (DSDP) drilled the Caribbean Plate at Sites 146–154, and five of these sites penetrated the basement of the plateau (Donnelly et al. 1973; fig. 1). Sites 146 and 150 were drilled at the western edge of the Venezuela Basin close to the Taïno Ridge. At Site 146 two dolerite sills were encountered near the base of the hole (the upper sill, 1.3 m thick and a lower sill which was penetrated to a depth of 16 m). At Site 150 the drilling also terminated 11 m into a dolerite sill. Ar/Ar dating of the sills from these two sites yielded a weighted mean age of 93.7 ± 1.9 Ma (The dates from Sinton et al. (1998) have been normalized to an age of 28.34 ± 0.28 Ma for Taylor Creek rhyolite sanidine 85G003 (Renne et al. 1998), to enable comparison with the Ar/Ar dates from ODP Leg 165 and the Beata Ridge) (Sinton et al. 1998). At Site 153 near the Aruba Gap, at the southern end of the Beata Ridge, c. 10 m of basalt were cored, but only a few metres were actually recovered. Although not able to be radiometrically dated, the basalts at Site 153 are overlain by Coniacian (89.3–85.8 Ma) limestone with ash layers. Site 151, near the southern end of the Beata Ridge, only penetrated 4 m of basalt which have not yielded a reliable Ar/Ar age. However, gabbros, dolerites and basalts sampled by submersible from the Beata Ridge (Mauffret et al. 2001) have yielded Ar/Ar ages which fall into two groups: nine samples range

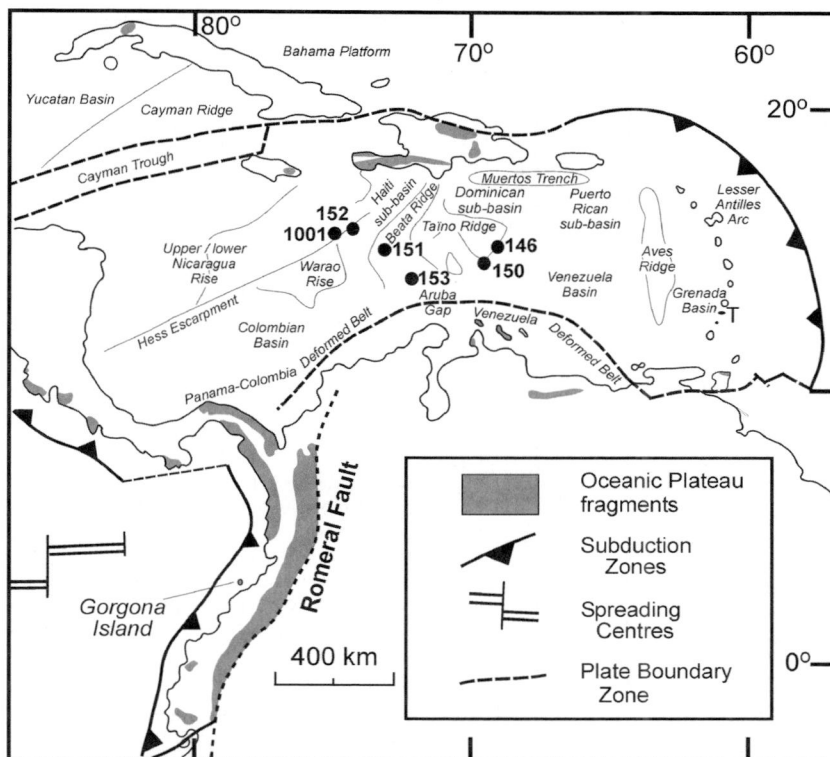

Fig. 1. Map of the Caribbean showing the location of Leg 15 and Leg 165 drill sites that penetrated basement.

from 81.0 to 74.2 Ma and two samples range from 56.2 to 55.3 Ma (Révillon *et al.* 2000*b*). Site 152 is located on thinned crust at the base of the Hess Escarpment and penetrated 7 m into basaltic basement, with good core recovery. Although unable to be dated by Ar/Ar, the Site 152 basalts contain some marble inclusions which contain recognizable Early Campanian (83.5–79 Ma) foraminifera (Donnelly *et al.* 1973).

ODP Leg 165

Although the primary aims of ODP Leg 165 were to sample the sedimentary succession in the Caribbean, in particular around the K–T boundary proximal ejecta blanket, the Leg was also intended to shed new light on the formation of the Caribbean basement. To this end drilling at Site 1001A penetrated 37.7 m into the basaltic basement with 20.5 m of core recovered (Sigurdsson *et al.* 1997). A second hole, 1001B (30 m to the south of 1001A) drilled into the basement to a depth of just 3 m. Site 1001 lies 35 km to the WSW of Site 152 on the rim of the Hess Escarpment, at the extreme

southeastern edge of the Lower Nicaraguan Rise (Fig. 1). Consequently, at a water depth of 3260 m, Site 1001 is *c.* 640 m shallower than Site 152. The basalts at Site 1001 are overlain by Campanian clayey limestones containing nanofossils belonging to the CC21 biozone indicating a minimum age of 77 Ma (Sigurdsson *et al.* 1997). This date is consistent with three Ar/Ar ages of 81.3 ± 5.4, 80.8 ± 1.3 and 81.0 ± 1.2 Ma (weighted mean age 80.9 ± 0.9 Ma) reported by Sinton *et al.* (2000) from the basalts sampled at Site 1001.

The basaltic sequence in Hole 1001A comprises 12 flows of variable thickness (0.15–6 m), with associated hyaloclastite layers. The lava flows are both pillowed and more massive and frequently possess highly vesicular and glassy margins. Most of the basalts are aphyric; however, some flows contain small amounts of plagioclase and clinopyroxene phenocrysts (see Fig. 2 and Sigurdsson *et al.* 1997 for more details). Only a limited amount of shipboard geochemical data has ever been published on these basalts (Sigurdsson *et al.* 1997; Sinton *et al.* 2000) and this paper reports the first comprehensive elemental and isotopic data from the basalts drilled at Site 1001. We will also

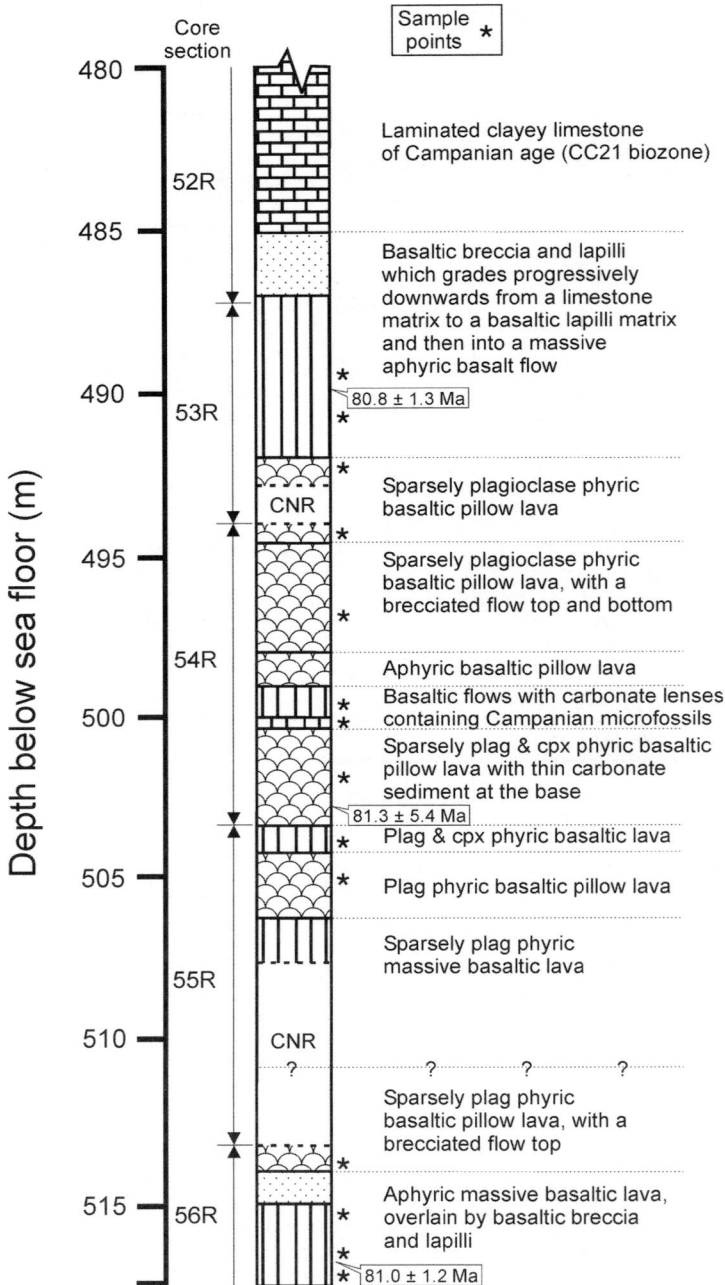

Fig. 2. Stratigraphic section through the 12 basaltic units penetrated at Site 1001 showing summary petrology, core locations sampled for the present study and locations (with dates) of units dated by ^{40}Ar/^{39}Ar (from Sinton *et al.* 2000). CNR indicates that core was not recovered from this interval. Plag, plagioclase feldspar; cpx, clinopyroxene. (Diagram modified from Sigurdsson *et al.* 1997.)

explore the implications of these new data for our understanding of the formation of the thickened Caribbean oceanic plate and the tectonic evolution of the region.

Analytical methods

Major elements were analysed by shipboard XRF on an ARL8420 and full analytical details can be found in Sigurdsson *et al.* (1997). Trace elements were prepared and run at the Northern Centre for Isotopic and Elemental Tracing, Department of Earth Sciences, Durham University via ICPMS using the procedures described in Ottley *et al.* (2003). Replicate dissolutions of in-house and international rock standards are better than 8% relative for concentrations of most trace element of greater atomic mass than Ni. Replicate dissolutions ($n = 9$) of the BE-N international basalt standard give better than 0.6% RSD for Sm–Nd, 1.3% RSD for Lu–Hf, 2.7% for Rb–Sr and 1.2% for Zr–Hf ratios. Powders for trace element determinations were leached for 20 min in 0.5 M HCl at 30 °C in an ultrasonic bath prior to dissolution. Powders were successively centrifuged and washed before analysis. Because of potential mass loss during leaching, concentrations measured are minima. Comparison between unleached powders analysed by shipboard XRF and leached powders analysed by ICPMS indicates that Sr abundances agree to within 5%. Rb abundances are too close to the XRF detection limit for meaningful comparison.

Sr–Nd–Hf isotopic compositions were determined at the Northern Centre for Isotopic and Elemental Tracing, Durham University using a Thermo Electron Neptune multi-collector plasma mass spectrometer. Procedures for the analysis of basaltic rocks in this laboratory are detailed in Thompson *et al.* (2007) and are summarized below. The long-term performance of the Durham Neptune for the above elements is summarized in Nowell *et al.* (2003). Prior to HF–HNO$_3$ dissolution all powders for isotopic analysis were leached in 6M HCl for 30 min in an ultrasonic bath at 40 °C. Substantial radiogenic Sr was removed during this leaching procedure, as is common with basalts that may contain carbonate surficial and vug-infill contamination.

The basic measurement protocol used for each element on the Neptune comprises a static multi-collection routine of one block of 50 cycles with an integration time of 4 s per cycle, total analysis time 3.5 min. After chemistry, Sr samples were taken up in 1 ml of 3% HNO$_3$ and introduced into the Neptune using an ESI PFA50 nebulizer and a dual cyclonic Scott Double Pass spraychamber. With this sample introduction set-up, and the normal H skimmer cone, the sensitivity for Sr on the Neptune is typically *c.* 60 V/total Sr ppm at an uptake rate of 90 µl/min. Prior to analysis a small aliquot was first tested to establish the Sr concentration of each sample by monitoring the size of the ^{84}Sr beam (^{88}Sr was too high in non-diluted aliquot to measure directly) from which a dilution factor was calculated to yield a beam of approximately 20 V ^{88}Sr. Instrumental mass bias was corrected for using a ^{88}Sr/^{86}Sr ratio of 8.375209 (the reciprocal of the ^{86}Sr/^{88}Sr ratio of 0.1194) and an exponential law. The samples analysed here were run in a single session during which the average ^{87}Sr/^{86}Sr value for NBS987 was 0.710262 ± 0.000012 (16.3 ppm 2SD; $n = 9$).

Following chemistry the REE cuts containing the Nd fraction were taken up in 1 ml of 3% HNO$_3$ and introduced into the Neptune using an ESI PFA50 nebulizer and a dual cyclonic Scott Double Pass spraychamber. With this sample introduction set-up, and the normal H skimmer cone, the sensitivity for Nd on the Neptune is 60–80 V total Nd ppm at an uptake rate of 90 µl/min. Instrumental mass bias was corrected for using a ^{146}Nd/^{145}Nd ratio of 2.079143 (equivalent to the more commonly used ^{146}Nd/^{144}Nd ratio of 0.7219) and an exponential law. The ^{146}Nd/^{145}Nd ratio is used for correcting mass bias since at Durham Nd isotopes are measured on a total REE-cut from the first-stage cation columns and this is the only Ce- and Sm-free stable Nd isotope ratio. This approach requires a correction for isobaric interferences from Sm on ^{144}Nd, ^{148}Nd and ^{150}Nd. The correction used is based on the method of Nowell & Parrish (2001). The accuracy of the Sm correction method during analysis of a total REE fraction is demonstrated by repeat analyses of BHVO-1, which gave an average ^{143}Nd/^{144}Nd ratio of 0.512982 ± 0.000007 (13.5 ppm, 2 SD, $n = 13$) after Sm correction, identical to the TIMS ratio of 0.512986 ± 0.000009 (17.5 ppm, 2 SD; $n = 19$) obtained by Weis *et al.* (2005). Leg 165 samples were analysed in a single session during which the average ^{143}Nd/^{144}Nd value for pure and Sm-doped J&M standard was 0.511112 ± 0.000012 (16.3 ppm, 2 SD; $n = 17$).

For analysis, Hf samples were taken up in 0.5 ml 3% HNO$_3$–1M HF and introduced using an ESI PFA50 nebulizer together with a Cetac Aridus desolvator. With this sample introduction set-up, and the high sensitivity X skimmer cone, the sensitivity for Hf on the Neptune was 400–450 V/total Hf ppm at an uptake rate of 90 µl/min. Instrumental mass bias was corrected for using a ^{179}Hf/^{177}Hf ratio of 0.7325 and an exponential law. Corrections for isobaric interferences from Yb and Lu on ^{176}Hf were made by monitoring $^{172-173}$Yb and ^{175}Lu and using the approach of Nowell & Parrish

(2001), although in practice the corrections are negligible. Leg 165 samples were analysed in a single session during which the JMC 475 standard gave an average value of 0.282149 ± 0.000005 (16.1 ppm, 2 SD; $n = 9$).

Pb isotope ratios were measured at the Department of Terrestrial Magnetism, Carnegie Institution, Washington, DC using a VG 354 TIMS instrument. Samples were leached with 6M HCl in an ultrasonic bath at 20 °C for 30 min, then washed with water prior to dissolution. Pb was separated using 50 ml of AG1-X8 (200–400 mesh). Samples were loaded on columns in 0.5 M HBr and Pb eluted with 0.5 M HNO_3. Two column passes were made. Total blanks were between 50 and 80 pg and represented $\ll 1\%$ of all Pb analysed. All samples were blank corrected. Corrections for fractionation during mass spectrometry were made relative to SRM981.

Geochemical data

These tholeiitic flows from Site 1001 display a relatively restricted range in composition with MgO contents varying from 6.4 to 8.5 wt% (Table 1, Fig. 3). However, as noted by Sigurdsson *et al.* (1997), the highest MgO content is found in the most-altered topmost flow, and since this elevated MgO is accompanied by a decrease in CaO, the most likely cause is the secondary formation of Mg-rich smectite and albitization of plagioclase causing Ca loss. If this MgO value is discounted then the range in MgO reduces even further (6.4–7.9 wt%). Within this slight variation it is noticeable that the flows nearer the bottom of the hole are slightly more evolved. This is particularly evident when Cr (and to a lesser extent Ni) contents are considered (Fig. 3). Although the extent of alteration in these lavas is relatively minor, Rb, Ba, Sr and K do not display good linear trends when plotted against a relatively immobile index of fractionation such as Zr and so cannot be used to assess petrogenesis. In terms of their more immobile incompatible trace elements, site 1001 lavas display very coherent liquid lines of descent when plotted against Zr (Fig. 3). This coherency is obvious from chondrite normalized rare earth element (REE) plots and N-MORB-normalized ('Normal' Mid-Ocean Ridge Basalt, erupted well away from a transform fault) multi-element diagrams (Fig. 4). Figure 4 also highlights the markedly depleted nature of the 1001 samples with respect to the most incompatible trace elements, particularly in Figure 4a, which shows that the samples are more depleted than average N-MORB. The N-MORB-normalized plot for the 1001 samples is otherwise relatively unremarkable in that it only shows a slight enrichment

in Nb and depletion in Zr (Fig. 4a). Perhaps not surprisingly the samples from Leg 165 are similar in composition, in their immobile trace element systematics, to those from the nearby DSDP Site 152 (Figs 3–5).

The similarly of Site 1001 samples to those from Site 152 is further illustrated by overlapping incompatible element ratios (Fig. 5). Furthermore, on the Nb/Y v. Zr/Y diagram (Fitton *et al.* 1997) it can be seen that both sets of samples plot below the Iceland 'tramline' field and at the more depleted end of the present-day East Pacific Rise (EPR) N-MORB field (Fig. 5a). Interestingly, on this diagram the 1001/152 samples compositionally overlap with those from the Reykjanes Ridge south of Iceland. However, significantly there is little overlap with the similarly aged Beata Ridge, or most of the *c.* 90 Ma lavas from the Caribbean Plateau (Fig. 5). An exception to this is the depleted *c.* 90 Ma Gorgona komatiite–basalt suite that has a comparable range in trace element ratios to Site 1001/152 samples (Fig. 5).

The very depleted nature of the 1001 samples is further borne out by their radiogenic isotope ratios, particularly initial (i = 81 Ma) ε_{Nd} values which range from $+11.1$ to $+11.9$ (Fig. 6a & b). In contrast to many samples (including Site 152) from the Caribbean plateau which have variably elevated $(^{87}Sr/^{86}Sr)i$ values due to alteration (and the apparent inability of acid leaching to remove this effect; cf. Kerr *et al.* 1996*b*; Révillon *et al.* 2002), the Site 1001 samples have a narrow range of $^{87}Sr/^{86}Sr$ (0.7025–0.7028), which suggests that these ratios represent magmatic values. ε_{Hf} for Site 1001 samples span a slightly wider range $(+15.2$ to $+16.9)$ than ε_{Nd} (Fig. 6b) and on an Nd–Hf diagram, like the Sr–Nd diagram, partially overlap with the field for Gorgona komatiites.

Although the 1001 samples show a similar extent of depletion to EPR basalts, they form a sub-parallel array to the EPR field that extends to higher initial $^{206}Pb/^{204}Pb$ (18.34–18.50) values for a given $^{207}Pb/^{204}Pb$ or $^{208}Pb/^{204}Pb$ ratio (Fig. 6c & d). Thus, although Site 1001 samples with the highest $^{207}Pb/^{204}Pb$ and $^{208}Pb/^{204}Pb$ overlap with the field for EPR N-MORB and Gorgona komatiites and basalts, the compositional divergence at lower $^{207}Pb/^{204}Pb$ and $^{208}Pb/^{204}Pb$ values marks the Site 1001 samples out as distinctive. The one existing Pb isotope analysis of basalt from Site 152 (Hauff *et al.* 2000) has significantly more radiogenic Pb isotope systematics than the samples from Site 1001 (Fig. 6). The nature of this somewhat anomalous Pb isotope signature for the Site 1001 samples can be investigated further using Pb–Nd isotope systematics. Figure 7 shows the three Pb isotope ratios plotted against initial ε_{Nd} and reveals that the divergence from the EPR N-MORB array in Pb isotope

Table 1. *Major element, trace element and radiogenic isotope data for Site 1001 basalts*

	53R-2, 75–80	53R-3, 68–71	53R-4, 93–96	54R-1, 44–47	54R-3, 37–40	54R-4, 93–98	54R-5, 81–84	54R-6, 126–130	55R-1, 72–75	55R-2, 81–83	56R-1, 132–134	56R-2, 77–80	56R-3, 74–78	56R-3, 131–134
SiO_2 (wt%)	50.23	49.66	49.68	49.90	49.32	49.10	49.40	49.30	49.90	49.00	49.80	49.80	49.30	49.10
TiO_2	1.41	1.29	1.29	1.30	1.31	1.23	1.25	1.19	1.37	1.58	1.57	1.48	1.47	1.39
Al_2O_3	16.11	14.89	15.70	15.73	14.60	15.30	15.00	15.63	15.70	14.30	14.45	14.50	14.70	14.30
Fe_2O_3	10.59	12.10	11.75	11.62	11.70	11.20	11.38	11.20	10.40	12.30	11.92	11.50	12.60	13.48
MnO	0.18	0.20	0.19	0.20	0.20	0.20	0.19	0.20	0.20	0.23	0.23	0.22	0.23	0.21
MgO	8.53	7.20	6.36	6.40	6.86	6.55	7.20	7.38	7.86	7.42	6.80	6.98	7.41	6.99
CaO	11.28	12.98	13.30	13.22	13.73	14.67	13.03	13.13	12.56	12.37	12.47	13.02	12.40	12.34
Na_2O	1.29	2.08	2.08	1.98	2.11	2.07	2.14	2.05	2.28	2.22	2.29	2.34	2.28	2.11
K_2O	0.10	0.11	0.27	0.29	0.59	0.51	0.09	0.10	0.08	0.20	0.53	0.20	0.10	0.08
P_2O_5	0.09	0.11	0.08	0.08	0.09	0.09	0.09	0.07	0.09	0.11	0.11	0.10	0.10	0.09
Total	99.81	100.62	100.70	100.72	100.51	100.92	99.77	100.25	100.44	99.73	100.17	100.14	100.57	100.09
L.O.I.	1.75	1.78	1.65	2.19	3.14	3.31	1.94	1.29	2.36	1.36	1.35	1.45	1.32	0.98
Sc (ppm)	52	48	50	48	50	48	48	47	54	54	56	50	52	47
V	368	355	360	353	334	321	343	346	378	396	368	346	392	375
Cr	289	296	285	285	271	251	299	269	272	195	208	163	173	168
Co	56	63	57	65	63	60	57	74	59	64	57	65	64	66
Ni	108	116	158	146	125	100	108	98	94	104	67	88	84	81
Cu	137	124	50	35	21	49	120	125	139	182	53	36	147	141
Zn	100	100	106	98	86	81	79	85	96	100	96	93	98	106
Ga	17.1	16.0	16.6	16.9	16.4	15.6	15.6	16.2	16.9	16.8	17.7	16.3	18.0	16.7
Rb	0.2	0.2	5.0	4.7	8.8	8.6	0.3	0.1	0.1	4.2	11.8	3.9	0.3	0.3
Sr	99	99	104	101	101	92	98	94	98	92	99	98	98	94
Y	31.0	32.9	32.5	32.5	33.4	30.8	31.5	32.4	32.5	37.7	40.1	36.4	35.0	35.5
Zr	66	65	65	65	66	58	62	61	69	75	81	71	74	71
Nb	1.73	1.60	1.32	1.31	1.24	1.61	1.51	1.64	1.69	1.66	2.05	2.06	1.89	1.91
Cs	0.01	0.02	0.08	0.10	0.11	0.15	0.02	0.01	0.01	0.04	0.24	0.04	0.01	0.02
Ba	7.8	5.7	8.0	14.7	64.5	12.9	6.0	4.8	6.9	5.8	10.2	8.2	8.6	8.3
La	1.58	1.57	1.52	1.49	1.60	1.47	1.48	1.50	1.62	1.75	1.88	1.93	1.86	1.86
Ce	5.44	5.51	5.29	5.31	5.60	4.99	5.20	5.12	5.61	6.34	6.68	6.57	6.38	6.30
Pr	1.09	1.10	1.10	1.09	1.13	1.02	1.06	1.04	1.11	1.28	1.34	1.29	1.28	1.27

(*Continued*)

Table 1. *Continued*

	53R-2, 75–80	53R-3, 68–71	53R-4, 93–96	54R-1, 44–47	54R-3, 37–40	54R-4, 93–98	54R-5, 81–84	54R-6, 126–130	55R-1, 72–75	55R-2, 81–83	56R-1, 132–134	56R-2, 77–80	56R-3, 74–78	56R-3, 131–134
Nd	7.15	7.08	7.12	7.06	7.23	6.67	6.74	6.61	7.29	8.46	8.78	8.30	8.17	7.99
Sm	2.72	2.80	2.72	2.79	2.81	2.60	2.64	2.67	2.87	3.30	3.50	3.19	3.13	3.08
Eu	1.08	1.08	1.07	1.07	1.10	1.02	1.04	1.05	1.12	1.27	1.34	1.21	1.25	1.19
Gd	4.20	4.31	4.27	4.23	4.39	4.12	4.18	4.27	4.38	5.11	5.32	4.77	4.80	4.73
Tb	0.80	0.81	0.81	0.79	0.83	0.77	0.77	0.80	0.81	0.95	0.99	0.89	0.93	0.88
Dy	5.07	5.25	5.19	5.16	5.32	4.93	5.07	5.22	5.25	6.17	6.52	5.79	5.81	5.66
Ho	1.14	1.15	1.13	1.14	1.20	1.11	1.14	1.12	1.15	1.38	1.42	1.26	1.27	1.25
Er	3.19	3.24	3.22	3.25	3.34	3.10	3.16	3.20	3.23	3.84	4.06	3.63	3.56	3.52
Tm	0.52	0.52	0.51	0.51	0.52	0.50	0.51	0.50	0.51	0.61	0.64	0.57	0.57	0.55
Yb	3.42	3.35	3.33	3.30	3.40	3.22	3.24	3.24	3.31	3.92	4.11	3.64	3.75	3.60
Lu	0.54	0.55	0.54	0.55	0.55	0.52	0.53	0.53	0.54	0.63	0.66	0.60	0.60	0.59
Hf	1.95	1.93	1.94	1.89	1.93	1.76	1.82	1.80	2.02	2.24	2.26	2.10	2.17	2.08
Ta	0.56	1.25	0.10	0.26	0.08	0.95	1.15	0.78	0.13	0.30	0.14	1.55	0.51	0.79
Pb	1.14	1.59	1.24	0.79	0.75	0.52	0.44	0.48	0.43	0.56	0.51	0.55	0.67	1.16
Th	0.09	0.08	0.07	0.07	0.07	0.07	0.07	0.08	0.09	0.09	0.09	0.09	0.10	0.09
U	0.03	0.03	0.07	0.04	0.30	0.18	0.02	0.03	0.05	0.08	0.57	0.26	0.04	0.04
Leached measured values														
$^{143}Nd/^{144}Nd$	0.513266	0.513232	0.513248	0.513250	0.513232	0.513237	0.513235	0.513240	0.513237	0.513251	0.513231	0.513225	0.513231	0.513234
$^{176}Hf/^{177}Hf$	0.283259	0.283253	0.283241	0.283246	0.283249	0.283248	0.283231	0.283245	0.283233	0.283234	0.283235	0.283218	0.283237	0.283211
$^{87}Sr/^{86}Sr$	0.702673	0.702661	0.702742	0.702761	0.702813	0.702926	0.702665	0.702506	0.702758	0.702831	0.702884	0.702877	0.702810	0.702804
$^{206}Pb/^{204}Pb$	18.381	18.512	–	–	–	–	18.387	–	18.504	18.542	–	–	–	–
$^{207}Pb/^{204}Pb$	15.428	15.471	–	–	–	–	15.428	–	15.518	15.465	–	–	–	–
$^{208}Pb/^{204}Pb$	37.650	37.850	–	–	–	–	37.647	–	37.919	37.789	–	–	–	–

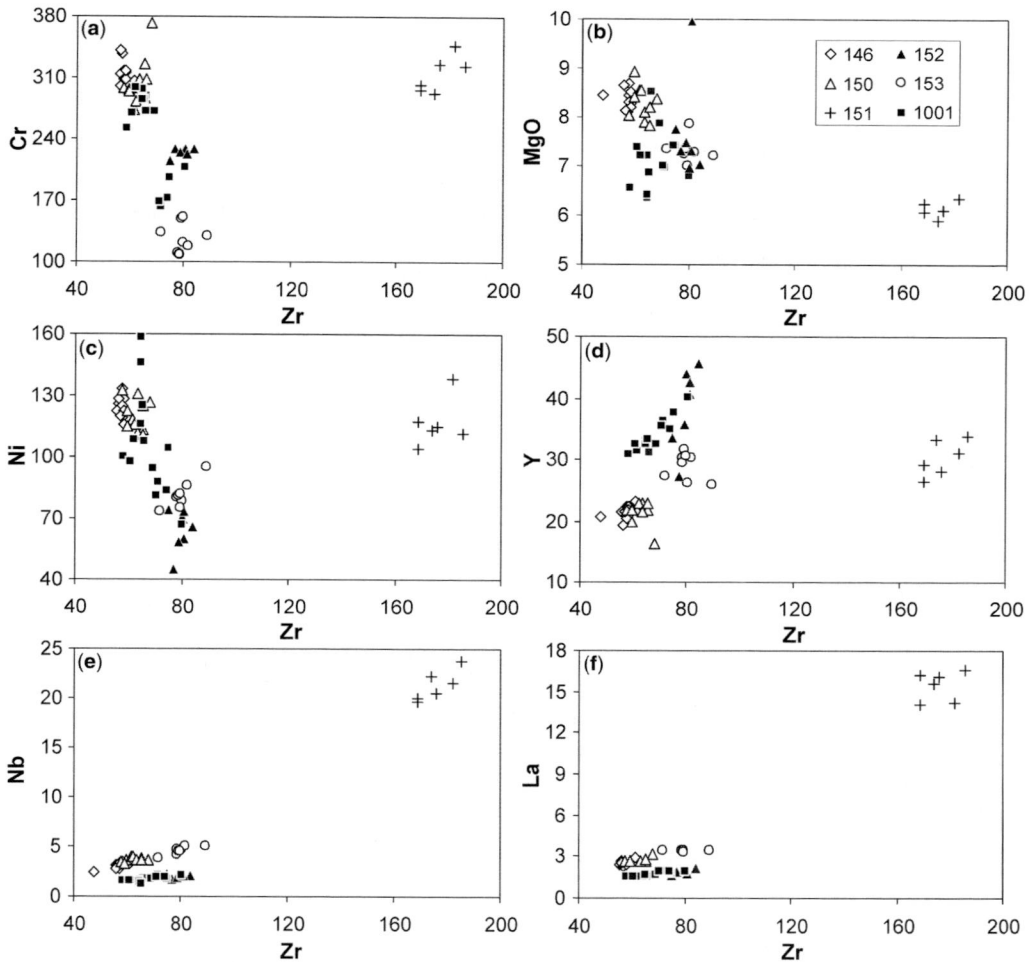

Fig. 3. Plots of MgO and representative trace elements v. Zr, for ODP Site 1001 and DSDP Sites 146, 150, 151, 152 and 153. DSDP data from Sinton *et al.* (1998), Hauff *et al.* (2000) & Kerr *et al.* (2002).

space is primarily due to elevated $^{206}Pb/^{204}Pb$ ratios and (to a lesser extent) higher $^{208}Pb/^{204}Pb$ ratios. In contrast, Site 1001 $^{207}Pb/^{204}Pb$ values are more comparable to MORB as the partial overlap between the EPR and Site 1001 data on the ε_{Nd} v. $^{207}Pb/^{204}Pb$ plot reveals.

Finally, the vast majority of samples from the Caribbean Plateau (including the Beata Ridge) are more enriched than those from Site 1001, which are much more MORB-like in composition and quite unlike a typical oceanic plateau (Fig. 6). Of all the rocks thus far analysed from the Caribbean plateau and its accreted sections, the Site 1001/ 152 samples bear most similarity to Gorgona komatiites and basalts in terms of trace element ratios and radiogenic isotopes signatures.

Discussion

In attempting to assess the Cretaceous tectonic evolution of the Caribbean and its accreted margins, the temporal and spatial evolution of the mantle source regions of the oceanic plateau and associated island arc rocks has been of key importance (cf. White *et al.* 1999; Lapierre *et al.* 2000; Kerr *et al.* 2003; Escuder-Viruete *et al.* 2006; Jolly *et al.* 2007). In terms of the Caribbean oceanic plateau, the marked heterogeneity of the mantle (plume) source region supplying the *c.* 90 Ma lavas and intrusions has been well established (e.g. Hauff *et al.* 2000; Kerr *et al.* 2002). However, what is less clear is if this (or a similar) heterogeneity is also found in the younger Campanian

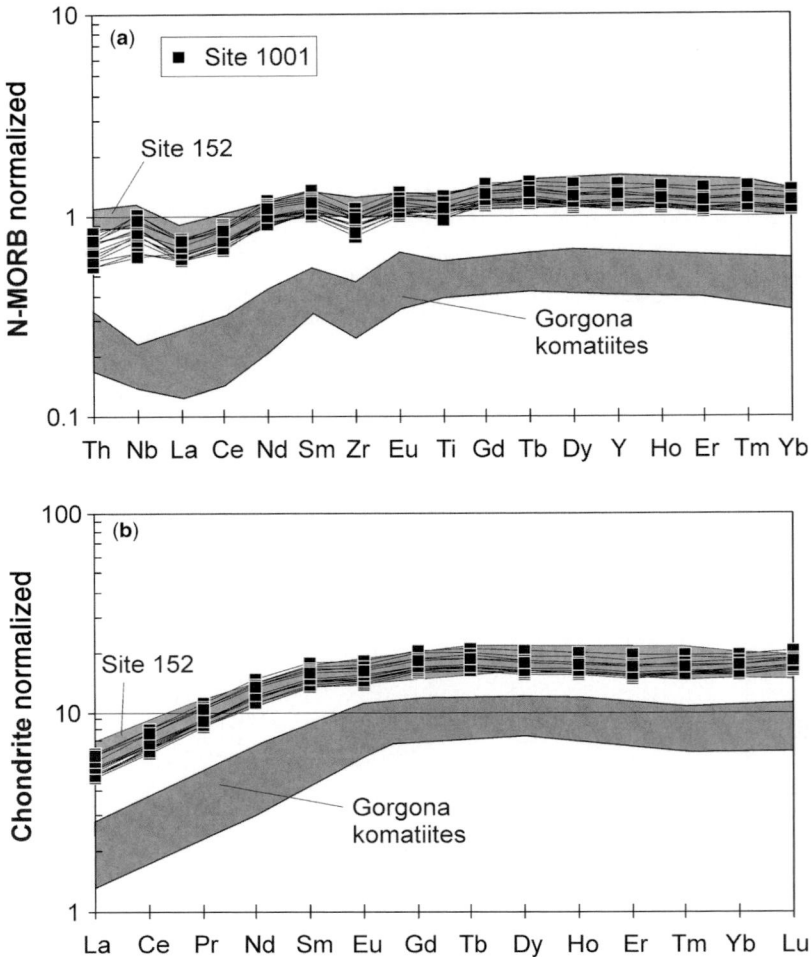

Fig. 4. (**a**) N-MORB normalized multi-element plot of Site 1001 and Site 152 samples; (**b**) chondrite-normalized REE plot of Site 1001 samples, with fields for Site 152 (for data sources see Fig. 3) and Gorgona komatiites (Kerr *et al.* 1996*a*; Revillon *et al.* 2000*a*; Kerr 2005). Normalizing values from Sun & McDonough (1989).

age plateau-like volcanic rocks in the Caribbean region. This is potentially highly significant because if lavas of Campanian age are found with a similar compositional range to the *c.* 90 Ma Caribbean plateau, then it is likely that both events were derived from a compositionally similar mantle plume source region. Moreover, it suggests that this mantle plume-source region may have existed below the region for >10 Ma.

A key question which we will address in the following discussion is the nature of the mantle source region of the Site 1001/152 lavas and, given the highly depleted composition of the lavas, this basically distils down to two options: either the lavas are derived from depleted N-MORB source

asthenospheric mantle material (DMM) or they are derived from melting of highly depleted plume mantle (DPM).

Deep-mantle derived depleted components in mantle plumes have been recognized for over 10 years (Kerr *et al.* 1995; Thirlwall, 1995; Nowell *et al.* 1998; Kempton *et al.* 2000; Hauff *et al.* 2000; Fitton *et al.* 2003) and, although the model is not without its critics (e.g. Hanan *et al.* 2000), there is an increasing volume of persuasive evidence to show that depleted signatures in mantle plumes are chemically distinctive from the depleted ambient upper mantle, i.e. the source of N-MORB. Although depleted lavas are relatively rare in the main accreted plateau terranes in Colombia

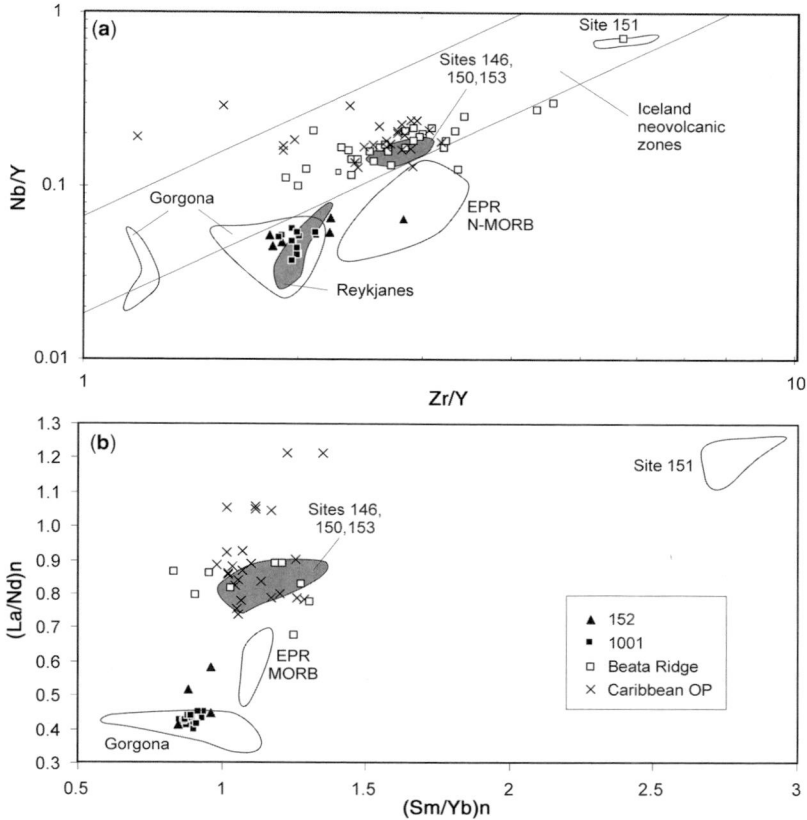

Fig. 5. (a) Zr/Y v. Nb/Y and (b) (Sm/Yb)n v. (La/Nd)n variation in Site 1001 basalts in comparison with EPR N-MORB, Gorgona komatiites and Caribbean basalts. Data sources: DSDP Leg 15, see Figure 3; Gorgona komatiites, see Figure 4; Beata Ridge, Revillon *et al.* (2000*b*); *c.* 90 Ma Caribbean oceanic plateau lavas, Hauff *et al.* (2000); EPR N-MORB, Mahoney *et al.* (1994); Chauvel & Blichert-Toft (2001) & Sims *et al.* (2002). Reykjanes Ridge data and the field for Iceland neovolcanic zones are from Fitton *et al.* (1997).

and around the Caribbean margins, Kerr *et al.* (1997, 2002) have shown that these depleted lavas were derived from significant depleted components within their mantle plume source regions. The most depleted lavas in the accreted oceanic plateau material found around the Caribbean and in NW South America are the high MgO (>18 wt%) komatiites and picrites of Gorgona Island (Kerr 2005). All recent studies have clearly demonstrated that these depleted lavas are compositionally distinct from MORB (Kerr *et al.* 1996*a*; Arndt *et al.* 1997; Révillon *et al.* 2000*a*; 2002). Furthermore, using parameterized experimental data on mantle melting phase relations, Herzberg & O'Hara (2002) calculated that the primary magmas of the Gorgona komatiites contained 18–20 wt% MgO and issued from a source with a potential temperature of 1520–1570 °C, i.e. *c.* 250 °C hotter than ambient upper mantle

(generally taken to be *c.* 1280 °C; McKenzie & Tickle, 1988). This elevated mantle source temperature is entirely consistent with an origin in a hot mantle plume. Accordingly, in our assessment of the nature of the mantle source region of Site 1001/152 lavas, it is vital that we make comparisons with both depleted, and clearly mantle plume-derived, Gorgona lavas and N-MORB from the East Pacific Rise.

Although the Site 1001/152 samples have MORB-like REE patterns (Fig. 4b) an N-MORB-normalized plot of the data clearly shows that the samples are more depleted in the most incompatible trace elements than average N-MORB (Fig. 4a). Furthermore, it can also be seen on this diagram that, unlike average N-MORB, the Site 1001/152 samples possess a small positive Nb anomaly, a feature which is often associated with plume-derived lavas (cf. Weaver 1991). Although

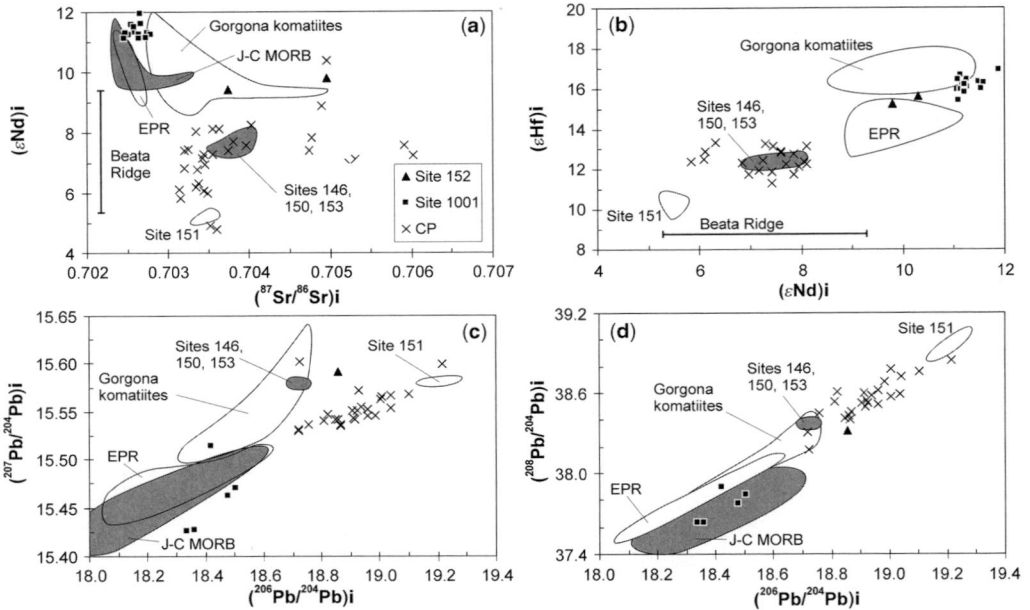

Fig. 6. Plots of (**a**) ($^{87}Sr/^{86}Sr$)i v. (ε_{Nd})i; (**b**) (ε_{Nd})i v. (ε_{Hf})i; (**c**) ($^{206}Pb/^{204}Pb$)i v. ($^{207}Pb/^{204}Pb$)i; (**d**) ($^{206}Pb/^{204}Pb$)i v. ($^{208}Pb/^{204}Pb$)i for Site 1001, EPR N-MORB, Gorgona komatiites and Caribbean Plateau (CP) basalts. CP data from Sinton *et al.* (1998); Hauff *et al.* (2000); Kerr *et al.* (2002); Geldmacher *et al.* (2003); Thompson *et al.* (2004). EPR N-MORB data from Mahoney *et al.* (1994); Chauvel & Blichert-Toft (2001); Sims *et al.* (2002). Jurassic–Cretaceous (J–C) Pacific MORB data from Castillo *et al.* (1992); Janney & Castillo (1997); Mahoney *et al.* (2005). Caribbean oceanic plateau, Sites 146, 150 151 and 153 along with Gorgona lavas are age-corrected to 90 Ma, Site 1001/152 lavas are age-corrected to 81 Ma.

Gorgona komatiites have lower concentrations of incompatible trace elements than Site 1001/152 due to the dilution effect of their higher MgO, there is a striking similarity in the shape of the REE patterns for Gorgona komatiites and depleted basalts and Site 1001/152 lavas. As has been shown, this compositional similarity is also clearly evident on incompatible trace element ratio plots (Fig. 5), where the divergence from EPR N-MORB compositions is also very apparent. It is, however, noticeable that Gorgona komatiites do not have a positive Nb anomaly (Fig. 4b) and so their source region, although similar, is not identical to that of Site 1001/152 basalts. This is unsurprising, given the evidence for marked hetero-geneity of Cretaceous mantle plume sources from basalts and high-MgO lavas in the Caribbean–NW South American region (e.g. Hauff *et al.* 2000; Kerr *et al.* 2002).

The Zr/Y v. Nb/Y diagram has been used effec-tively elsewhere (e.g. Fitton *et al.* 1997; Weis & Frey 2002) and in the Caribbean (e.g. Kerr *et al.* 1997, 2002; Thompson *et al.* 2004) to distinguish mantle plume-derived lavas from N-MORB. However, Gorgona komatiites and depleted basalts

(which we have seen are clearly derived from a deep-derived mantle plume) plot both above *and* below the lower tramline (Fig. 5a), and overlap with the EPR N-MORB field. This apparent paradox most likely stems from the fact that the diagram was originally devised by Fitton *et al.* (1997) to resolve source components in Icelandic magmas. Consequently, the plume compositional field as represented by the tramlines on Figure 5a is solely defined on the basis of basalts and picrites from Iceland. Thus, although the diagram has been used elsewhere in the world to successfully assess plume v. DMM components, the fact that some clearly plume-derived Gorgona komatiites plot outside the 'plume' field provides a salutary lesson that such discrimination diagrams should not be used blindly in isolation but that other geochemical and petrological evidence must also be taken into account.

As reviewed in the previous section, radiogenic isotope ratios are equivocal in resolving the nature of the depleted source component which melted to form the basalts at Site 1001. Figures 6 and 7 show that while initial ε_{Nd}, $^{87}Sr/^{86}Sr$ and $^{207}Pb/$$^{204}Pb$ for Site 1001 display considerable overlap

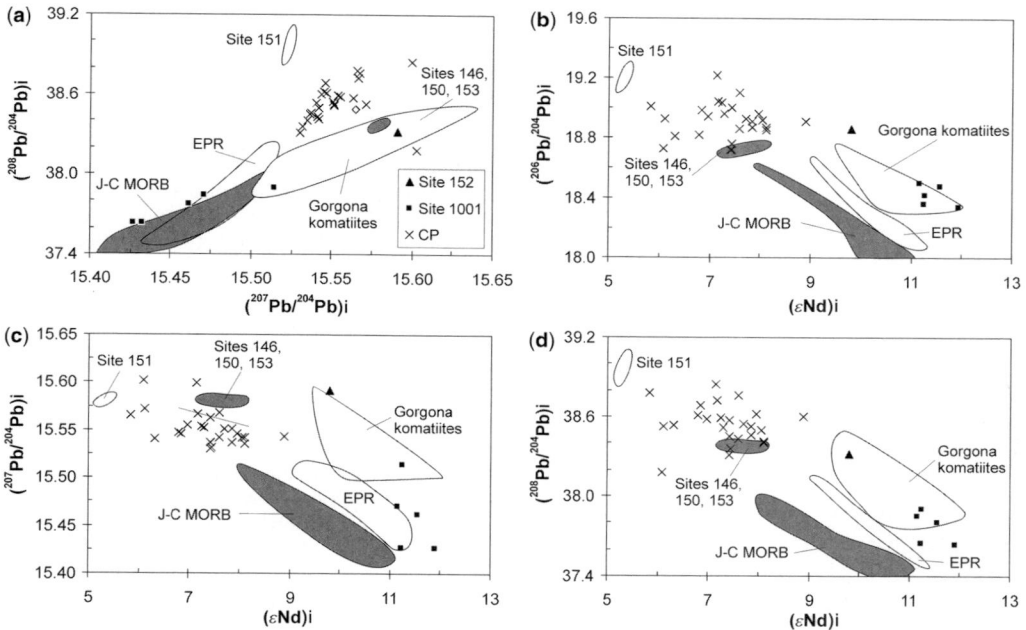

Fig. 7. Plots of (**a**) $(^{207}Pb/^{204}Pb)i$ v. $(^{208}Pb/^{204}Pb)i$; (**b**) $(\varepsilon_{Nd})i$ v. $(^{206}Pb/^{204}Pb)i$; (**c**) $(\varepsilon_{Nd})i$ v. $(^{207}Pb/^{204}Pb)i$; (**d**) $(\varepsilon_{Nd})i$ v. $(^{208}Pb/^{204}Pb)i$ for Site 1001, EPR N-MORB, Gorgona komatiites and Caribbean Plateau (CP) basalts. Data sources are as in Figure 6.

with present-day EPR N-MORB, Jurassic–Cretaceous Pacific MORB and Gorgona komatiites, initial $^{206}Pb/^{204}Pb$ and $^{208}Pb/^{204}Pb$ ratios combined with initial ε_{Nd} values, are more compositionally similar to Gorgona lavas. This inability of Sr, Nd and Pb radiogenic isotopes to resolve depleted source components is one of the reasons why depleted mantle plume signatures went undetected until the mid-1990s. One of the difficulties in using Sr and Pb isotopes to identify such signatures in lavas that have experienced slight alteration is the possibility of a strong influence of alteration effects, even after leaching. This is well demonstrated by the diffuse Nd–Sr isotope compositions of many of the Gorgona and Caribbean samples compared with those from Site 1001. In recent years, however, the analysis of Hf isotopes has become more routine. The Lu–Hf isotope system is more robust to the effects of seawater alteration and this property has led to ε_{Hf} being shown to be a particularly effective discriminant of depleted mantle components, particularly when used in conjunction with incompatible trace element ratios (e.g. Kempton *et al.* 2000; Thompson *et al.* 2004). Plots of ε_{Hf} against Zr/Nb and Zr/Y (Fig. 8) clearly distinguish between Gorgona komatiites and EPR N-MORB and show that Site 1001/152 basalts are most similar in terms of their Zr/Nb,

Zr/Y and ε_{Hf} values to depleted Gorgona rocks, and have only limited compositional overlap with the majority of EPR N-MORB.

Thus, the balance of evidence suggests that the mantle source region of the 81 Ma basalts from Site 1001/152 on the Hess Escarpment was dominated by depleted plume material, probably mixed with a some depleted upper MORB-source mantle, in order to account for the spread of data points on Figure 8, between Gorgona and EPR N-MORB.

It has been proposed that the Gorgona komatiites formed by *c.* 15% melting of a moderately depleted plume source, with any enriched components either being swamped by later melting or being swept outwards to the margins of the plume (Arndt *et al.* 1997; Révillon *et al.* 2000*a*). In contrast, the basalts comprising most of the accreted sections of the Caribbean plateau (including those sampled from the Beata Ridge and drilled at Leg 15 Sites 146, 150 and 153) with their flat-to-moderately LREE depleted REE patterns, are the fractionated products of high-MgO melts, into which the early enriched melt component has been incorporated (e.g. Kerr *et al.* 1996*b*, 1997; Sinton *et al.* 1998; Hauff *et al.* 2000). The 90 Ma Caribbean plateau picrites found on Curaçao have essentially flat REE patterns and are representative of the

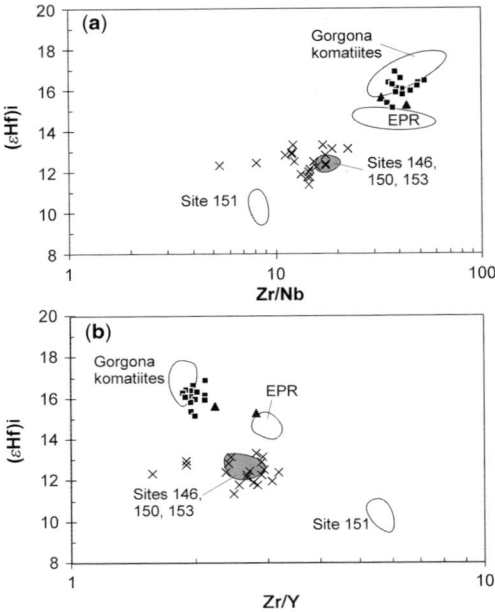

Fig. 8. (a) Zr/Nb-(ε_{Hf})i and (b) Zr/Y-(ε_{Hf})i variation in Site 1001 basalts in comparison with EPR N-MORB, Gorgona komatiites and Caribbean basalts. Data sources are as in Figure 6.

parental magma of the vast majority of the preserved basaltic lavas in the Caribbean plateau (Kerr *et al.* 1996*b*). Modelling suggests that these parental picritic melts can be generated by *c.* 10% pooled fractional melting of a moderately depleted, spinel-bearing, mantle source region (composition taken from Salters & Stracke 2004). The widespread formation of lavas with near-flat REE patterns over a

large proportion of the Caribbean plateau means that there will be a considerable volume of residual mantle from this melting. Calculations, assuming 10% melting and an estimated original melt volume of 4 million cubic km (Kerr 1998), show that this residual mantle would have occupied a volume of the order 36 million cubic km, equivalent to a sphere with a diameter of over 400 km.

Even >10 Ma after the formation of the main part of the plateau, it is likely that a substantial proportion of this residual mantle still remained below the region. There are two main reasons for this: (1) mantle residual from mantle melting is more buoyant than the original un-melted mantle material. This is known as 'depletion buoyancy' and modelling has shown that this buoyancy prevents advection of the residual mantle away from the top of the plume (e.g. Manglik & Christensen 1997). In the Caribbean case additional buoyancy will also be provided by residual heat from the plume. (2) Advection of residual mantle from formation of the 90 Ma phase of the Caribbean plateau is also likely to be restricted by the close proximity of thick continental lithospheric roots of North and South America at that time, as well as by the developing peripheral subduction zones around much of the plateau's margin (cf. Figure 7a in Pindell *et al.* 2006). Thus, it is possible that the depleted lavas from the Hess Escarpment are derived from re-melting of the same mantle plume that had already produced the main phase of oceanic plateau volcanism in the region *c.* 10 Ma previously.

In order to test this model we have calculated the composition of residual mantle left after the extraction of *c.* 10% pooled fractional melt of a moderately depleted mantle source region (Table 2, taken from Salters & Stracke 2004).

Table 2. *Results from pooled fractional melt modelling of two-stage mantle melting*

	Initial mantle composition	Residue after 10% melting + 3% melt	4% melting	6% melting	8% melting
La (ppm)	0.234	0.089	2.22	1.49	1.12
Ce	0.772	0.336	8.03	5.56	4.20
Pr	0.131	0.065	1.40	1.03	0.79
Nd	0.713	0.389	7.50	5.79	4.63
Sm	0.270	0.158	2.72	2.18	1.79
Eu	0.107	0.066	1.03	0.85	0.71
Gd	0.395	0.242	3.80	3.13	2.62
Tb	0.075	0.046	0.71	0.59	0.50
Dy	0.531	0.335	4.89	4.11	3.49
Ho	0.122	0.076	1.13	0.95	0.80
Er	0.371	0.231	3.44	2.87	2.43
Tm	0.060	0.037	0.55	0.46	0.39
Yb	0.401	0.251	3.60	3.03	2.59
Lu	0.063	0.040	0.55	0.47	0.40

To make the model more physically realistic, we assumed that 3% of this melt was not extracted and remained with the residual mantle (Table 2). A similar approach was adopted by Révillon et al. (2000a) in modelling the composition of Gorgona lavas. This residual mantle plus non-extracted melt was then used as a source composition to model the formation of the lavas from Site 1001/152. With 4–8% pooled fractional melting, this modelling produces REE patterns which are very similar in shape to those of the Hess Escarpment (Fig. 9). Although the overall modelled concentrations of elements are generally too low to explain the composition of the Hess Escarpment lavas, it is likely that these lavas have undergone a considerable degree of fractional crystallization, which would have elevated their REE concentrations. Figure 9 shows that 20% fractional crystallization of a 6% mantle melt composition can produce REE concentrations not dissimilar to the Site 1001/152 lavas. Although this model is non-unique, it does nonetheless demonstrate that rocks resembling those found on the Hess Escarpment can be generated under reasonable geological conditions from a previously melted, residual mantle plume source region. Although somewhat beyond the scope of the present study, the ultimate origin of the depleted source component responsible for the Site 1001/152 basalts, and depleted plume related lavas elsewhere in the world, is still quite uncertain. Recent combined Os–Hf work on xenoliths from Hawaii (Bizimis et al. 2007) has shed new light on the depleted nature of the Hawaiian plume, identifying depleted residual material as an intrinsic component of the Hawaiian plume. In the Hawaiian

case, the melting age of these components appears to be considerably older than the age of melting beneath Hawaii, but the presence of such material indicates clearly that old depleted components can be integral to the upwelling flux of plumes. Future analysis of combined Os and Hf isotopes on the depleted Caribbean oceanic plateau lavas may well yield significant information on the origin of their depleted source.

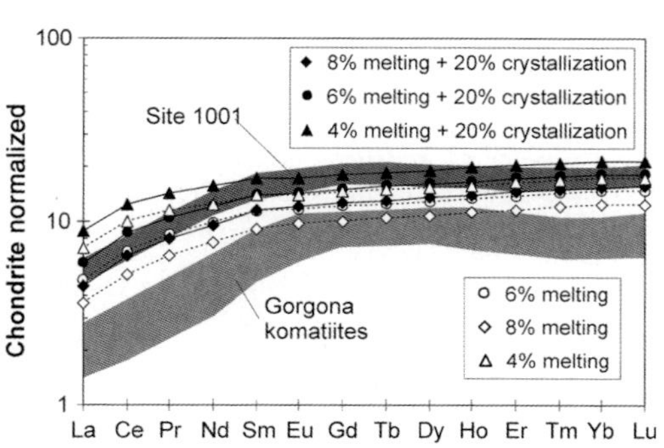

It is interesting that oceanic plateau derived lavas and intrusions of a similar age have been found on the Beata Ridge, yet these are not depleted in the most incompatible trace elements (although they are derived from a source which has a long-term history of depletion – ε_{Nd} +9.0; Révillon et al. 2000b). The generation of magmas in both these locations c. 81 Ma was probably not initiated by asthenospheric processes, e.g. the arrival of a new surge of hot plume asthenosphere below the region, since this would be unlikely to produce magmas as depleted as those at Site 1001/152. Rather, the formation of magmas at the Beata Ridge and the Hess Escarpment c. 81 Ma is more likely to be related to plate-scale tectonic processes, e.g. lithospheric extension and renewed decompression melting of previously melted plume mantle which was still hotter than ambient upper mantle (cf. Sinton et al. 1998; Hauff et al. 2000). The less depleted rocks found at the Beata Ridge may simply be a reflection of either smaller degrees of partial melting than those on the Hess Escarpment, stemming from thicker lithosphere below the former, or may indicate that the source of the Beata Ridge lavas had undergone less previous melt extraction.

The recognition that the mantle underlying the Caribbean Plate c. 10 Ma after the main phase of plateau formation is in large part residual from this previous melting event has profound implications for plate tectonic models of the Caribbean. The near absence of a depleted MORB source mantle signature in the basalts of the Caribbean plateau in conjunction with the apparent long-term (>10 Ma) availability of a deep mantle-plume-derived source region below the Caribbean severely calls into question most aspects of the slab window model (Pindell et al. 2006, 2009) for the origin of the Caribbean Plateau. For the slab-window model of Pindell et al. to be correct, given the mantle source constraints outlined in this paper, it would require the 'highly fortuitous' arrival of a mantle plume directly below the slab c. 91 Ma. The 'Atlantic convection cell' invoked by Pindell et al. (2006, 2009) in their slab window model is presumably of depleted MORB-source upper mantle composition (although they do not explicitly state this). They do however state 'we would expect the resulting lavas [from their slab window model] to have the

Fig. 9. Chondrite normalized plot showing the results of pooled fractional melt modelling along with compositional fields for Site 1001 basalts and Gorgona komatiites. Hollow symbols represent 4–8% melting of previously melted (10%) depleted source mantle containing 3% residual melt. See Table 2 for data. Solid symbols show the composition of these mantle melts after 20% fractional crystallization. See text for more details.

"oceanic plateau" geochemical signature'. We note, however, that if the source region invoked in their model is ambient upper mantle and not a deep mantle plume, then the magmas which are formed will not have an oceanic plateau geochemical signature. This represents a fundamental flaw in the slab window model of Pindell *et al.* (2006, 2009).

Another major problem with the slab window model of Caribbean Plateau formation is that magmas formed in a slab window are likely to possess a clear subduction signature. Generally, any input from a subduction zone will be much easier to detect when it is added to a magma depleted in incompatible trace elements. Thus, if any subduction-derived component is present in the source region of the Caribbean plateau it should be most readily observed in the most depleted lavas (i.e. those from Site 1001/152), in the form of elevated LREE contents, resulting in a negative Nb anomaly on normalized multi-element diagrams. However, as noted above, the lavas from the present study do not have elevated Th and LREE contents and instead possess a positive Nb anomaly (Fig. 4a). This trace element evidence in combination with their depleted radiogenic isotope signatures rules out any subduction component whatsoever in the source region of these lavas and again poses a fundamental problem for the slab window model of Pindell *et al.* (2006, 2009). It could be argued that, because the lavas from Site 1001/152 were erupted in the centre of the Caribbean Plate, they are derived from a mantle source region which is less likely to be contaminated by subduction-influenced mantle than CCOP lavas nearer the edge of the plate. However we would contend that this is not the case, as no subduction signature has been found in any of the Caribbean oceanic plateau lavas (Kerr *et al.* 2003). This view is corroborated by a recent study of similarly depleted late-stage basalts dredged from the Manihiki Plateau, which has shown that the mantle source region of these oceanic plateau basalts was probaly contaminated by *c.* 2% subducted sediment from a mantle wedge (Ingle *et al.* 2007). This work on the Manihiki Plateau basalts unequivocally demonstrates just how easily oceanic plateau basalts can acquire a subduction related signature and poses grave difficulties for the 'slab window model' of Caribbean plateau petrogenesis.

Conclusions

(1) The *c.* 81 Ma depleted basalts recovered during drilling at ODP Site 1001 (and DSDP Site 152), although superficially similar to N-MORB, have trace element signatures and radiogenic isotope ratios which are inconsistent with

their derivation from ambient depleted upper mantle.

(2) The Site 1001 basalts are most similar to the mantle plume-derived *c.* 90 Ma Gorgona komatiite-basalt suite, indicating that the source region of the Site 1001 basalts is more likely to have been a depleted mantle plume, probably mixed with some depleted MORB source upper mantle, and did not comprise solely depleted MORB source upper mantle.

(3) Melt modelling reveals that mantle material, residual from *c.* 10% melting of an initially moderately depleted plume source region, can produce melts with similar trace element ratios to the Site 1001 basalts.

(4) This implies the long-term residence (>10 Ma) of residual (deep mantle-derived) plume source material below the region, and renders completely improbable Caribbean tectonic models which invoke melting of depleted upper mantle.

We would like to thank J. Pindell, J. Tarney, L. Kennan, A. Hastie, I. Neill and G. Draper for numerous discussions on Caribbean geology. Constructive comments by two anonymous reviewers helped improve the manuscript. We are grateful to C. Ottley for help with trace element determinations and to R. Carlson for access to the DTM Pb isotope facilities during a visit by D. G. P. Pearson.

References

ARNDT, N. T., KERR, A. C. & TARNEY, J. 1997. Dynamic melting in plume heads: the formation of Gorgona komatiites and basalts. *Earth and Planetary Science Letters*, **146**, 289–301.

BIZIMIS, M., GRISELIN, M., LASSITER, J. C., SALTERS, V. J. M. & SEN, S. 2007. Ancient recycled mantle lithosphere in the Hawaiian plume: osmium–hafnium isotopic evidence from peridotite mantle xenoliths. *Earth and Planetary Science Letters*, **257**, 259–273.

BOWLAND, C. L. & ROSENCRANTZ, E. 1988. Upper crustal structure of the western Colombian Basin, Caribbean Sea. *Geological Society of America Bulletin*, **100**, 534–546.

BURKE, K. 1988. Tectonic evolution of the Caribbean. *Annual Review of Earth and Planetary Sciences*, **16**, 201–230.

CASE, J. E., MACDONALD, W. D. & FOX, P. J. 1990. Caribbean crustal provinces; Seismic and gravity evidence. *In*: DENGO, G. & CASE, J. E. (eds) *The Caribbean Region*. The Geology of North America, **H**. Geological Society of America, Boulder, CO, 15–36.

CASTILLO, P. R., FLOYD, P. A. & FRANCE-LANORD, C. 1992. Isotope geochemistry of Leg 129 basalts: implications for the origin of the widespread Cretaceous volcanic event in the Pacific. *In*: LARSON, R. L., LANCELOT, Y., FISHER, A. & WINTERER, E. L. (eds) *Proceedings of the Ocean Drilling Program*,

Scientific Results. Ocean Drilling Program, Texas A&M University, College Station, TX, 405–414.

CHAUVEL, C. & BLICHERT-TOFT, J. 2001. A hafnium isotope and trace element perspective on melting of the depleted mantle. *Earth and Planetary Science Letters*, **190**, 137–151.

DIEBOLD, J. & DRISCOLL, N. 1999. New insights on the formation of the Caribbean basalt province revealed by multichannel seismic images of volcanic structures in the Venezuelan Basin. *In*: MANN, P. (ed.) *Caribbean Basins*. Sedimentary Basins of the World. Elsevier Science, Amsterdam, 561–589.

DONNELLY, T. W. 1973. Late Cretaceous basalts from the Caribbean, a possible flood basalt province of vast size. *EOS*, **54**, 1004.

DONNELLY, T. W., MELSON, W., KAY, R. & ROGERS, J. W. 1973. *Basalts and Dolerites of Late Cretaceous Age from the Central Caribbean*. Initial Reports of the Deep Sea Drilling Project **15**. US Government Printing Office, Washington, DC, 989–1004.

DUNCAN, R. A. & HARGRAVES, R. B. 1984. Plate tectonic evolution of the Caribbean region in the mantle reference frame. *In*: BONINI, W. E., HARGRAVES, R. B. & SHAGAM, R. (eds) *The Caribbean–South America Plate Boundary and Regional Tectonics*. Geological Society of America, Boulder, CO, Memoirs, 81–93.

EDGAR, N. T., EWING, J. I. & HENNION, J. 1971. *Seismic Refraction and Reflection in the Caribbean Sea*. American Association of Petroleum Geology, Tulsa, OK, **55**, 833–870.

ESCUDER-VIRUETE, J. E., DE NEIRA, A. D. *ET AL.* 2006. Magmatic relationships and ages of Caribbean Island arc tholeiites, boninites and related felsic rocks, Dominican Republic. *Lithos*, **90**, 161–186.

FITTON, J. G., SAUNDERS, A. D., KEMPTON, P. D. & HARDARSON, B. S. 2003. Does depleted mantle form an intrinsic part of the Iceland plume? *Geochemistry Geophysics Geosystems* **4**, article no. 1032; doi: 10.1029/2002GC000424.

FITTON, J. G., SAUNDERS, A. D., NORRY, M. J., HARDARSON, B. S. & TAYLOR, R. N. 1997. Thermal and chemical structure of the Iceland plume. *Earth and Planetary Science Letters*, **153**, 197–208.

FOX, P. J., SCHREIBER, E. S. & HEEZEN, B. C. 1971. The geology of the Caribbean crust: Tertiary sediments, granitic and basic rocks from the Aves Ridge. *Tectonophysics*, **12**, 89–109.

GELDMACHER, J., HANAN, B. B. *ET AL.* 2003. Hafnium isotopic variations in volcanic rocks from the Caribbean Large Igneous Province and Galapagos hot spot tracks. *Geochemistry Geophysics Geosystems* **4**, article no. 1062; doi:10.1029/2002GC000477.

GIUNTA, G., BECCALUVA, L. & SIENA, F. 2006. Caribbean Plate margin evolution: constraints and current problems. *Geologica Acta*, **4**, 265–277.

HANAN, B. B., BLICHERT-TOFT, J., KINGLSEY, R. & SCHILLING, J.-G. 2000. Depleted Iceland mantle plume geochemical signature: artefact or multicomponent mixing? *Geochemistry Geophysics Geosystems* **1**; doi: 10.1029/1999GC000009.

HAUFF, F., HOERNLE, K., TILTON, G., GRAHAM, D. W. & KERR, A. C. 2000. Large volume recycling of oceanic lithosphere over short time scales: geochemical constraints from the Caribbean Large Igneous Province. *Earth and Planetary Science Letters*, **174**, 247–263.

HERZBERG, C. & O'HARA, M. J. 2002. Plume-associated ultramafic magmas of Phanerozoic age. *Journal of Petrology*, **43**, 1857–1883.

HOERNLE, K., HAUFF, F. & VAN DEN BOGAARD, P. 2004. 70 m.y. history (139–69 Ma) for the Caribbean large igneous province. *Geology*, **32**, 697–700.

INGLE, S., MAHONEY, J. J. *ET AL.* 2007. Depleted mantle wedge and sediment fingerprint in unusual basalts from the Manihiki Plateau, central Pacific Ocean. *Geology*, **35**, 595–598.

JAMES, K. H. 2006. Illogical arguments for and against the Pacific origin of the Caribbean Plate: discussion, finding for an inter-American origin. *Geologica Acta*, **4**, 279–302.

JANNEY, P. E. & CASTILLO, P. R. 1997. Geochemistry of Mesozoic Pacific mid-ocean ridge basalt: constraints on melt generation and the evolution of the Pacific upper mantle. *Journal of Geophysical Research: Solid Earth*, **102**, 5207–5229.

JOLLY, W. T., SCHELLEKENS, H. & DICKIN, A. P. 2007. High-Mg andesites and related lavas from southwest Puerto Rico (Greater Antilles Island Arc): Petrogenetic links with emplacement of the Late Cretaceous Caribbean mantle plume. *Lithos*, **98**, 1–26.

KEMPTON, P. D., FITTON, J. G. *ET AL.* 2000. The Iceland plume in space and time: a Sr-Nd-Pb-Hf study of the North Atlantic rifted margin. *Earth and Planetary Science Letters*, **177**, 255–271.

KERR, A. C. 1998. Oceanic plateau formation: a cause of mass extinction and black shale deposition around the Cenomanian–Turonian boundary. *Journal of the Geological Society, London*, **155**, 619–626.

KERR, A. C. 2005. La Isla de Gorgona, Colombia: a petrological enigma? *Lithos*, **84**, 77–101.

KERR, A. C., SAUNDERS, A. D., TARNEY, J., BERRY, N. H. & HARDS, V. L. 1995. Depleted mantle plume geochemical signatures: no paradox for plume theories. *Geology*, **23**, 843–846.

KERR, A. C., MARRINER, G. F. *ET AL.* 1996a. The petrogenesis of Gorgona komatiites, picrites and basalts: new field, petrographic and geochemical constraints. *Lithos*, **37**, 245–260.

KERR, A. C., TARNEY, J., MARRINER, G. F., KLAVER, G. T., SAUNDERS, A. D. & THIRLWALL, M. F. 1996b. The geochemistry and petrogenesis of the Late Cretaceous picrites and basalts of Curaçao, Netherlands Antilles: a remnant of an oceanic plateau. *Contributions to Mineralogy and Petrology*, **124**, 29–43.

KERR, A. C., MARRINER, G. F. *ET AL.* 1997. Cretaceous basaltic terranes in western Colombia: elemental, chronological and Sr–Nd constraints on petrogenesis. *Journal of Petrology*, **38**, 677–702.

KERR, A. C., TARNEY, J. *ET AL.* 2002. Pervasive mantle plume head heterogeneity: evidence from the Late Cretaceous Caribbean–Colombian oceanic plateau. *Journal of Geophysical Research – Solid Earth*, **107**, article no. 2140; doi 10.1029, 2001JB000790.

KERR, A. C., WHITE, R. V., THOMPSON, P. M. E., TARNEY, J. & SAUNDERS, A. D. 2003. No Oceanic Plateau – no Caribbean Plate? The seminal role of an oceanic plateau in Caribbean Plate evolution. *In*:

BARTOLINI, C., BUFFLER, R. T. & BLICKWEDE, J.
(eds) *The Gulf of Mexico and Caribbean Region:
Hydrocarbon Habitats, Basin Formation and Plate
Tectonics*. American Association of Petroleum Geologists, Tulsa, OK, Memoirs, **79**, 126–168.

KERR, A. C., TARNEY, J., KEMPTON, P. D., PRINGLE, M.
& NIVIA, A. 2004. Mafic pegmatites intruding
oceanic plateau gabbros and ultramafic cumulates
from Bolívar, Colombia: evidence for a 'wet' mantle
plume? *Journal of Petrology*, **45**, 1877–1906.

KROENKE, L. W. 1974. Origin of continents through
development and coalescence of oceanic flood basalt
plateaus. *EOS*, **55**, 443.

LAPIERRE, H., DUPUIS, V. ET AL. 1999. Late Jurassic
oceanic crust and Upper Cretaceous Caribbean
plateau picritic basalts exposed in the Duarte igneous
complex, Hispaniola. *Journal of Geology*, **107**,
193–207.

LAPIERRE, H., BOSCH, D. ET AL. 2000. Multiple plume
events in the genesis of the peri-Caribbean Cretaceous
oceanic plateau province. *Journal of Geophysical
Research – Solid Earth*, **105**, 8403–8421.

MAHONEY, J. J., SINTON, J. M., MACDOUGALL, J. D.,
SPENCER, K. J. & LUGMAIR, G. W. 1994. Isotope
and trace element characteristics of a super-fast spreading ridge: East Pacific rise, 13–23°S. *Earth and Planetary Science Letters*, **121**, 173–193.

MAHONEY, J. J., DUNCAN, R. A., TEJADA, M. L. G.,
SAGER, W. W. & BRALOWER, T. J. 2005. A Jurassic–
Cretaceous boundary age and mid-ocean-ridge-type
mantle source for Shatsky Rise. *Geology*, **33**,
185–188.

MANGLIK, A. & CHRISTENSEN, U. R. 1997. Effect of
mantle depletion buoyancy on plume flow and
melting beneath a stationary plate. *Journal of Geophysical Research – Solid Earth*, **102**, 5019–5028.

MAUFFRET, A. & LEROY, S. 1997. Seismic stratigraphy
and structure of the Caribbean igneous province.
Tectonophysics, **283**, 61–104.

MAUFFRET, A., LEROY, S., VILA, J. M., HALLOT, E., DE
LEPINAY, B. M. & DUNCAN, R. A. 2001. Prolonged
magmatic and tectonic development of the Caribbean
Igneous Province revealed by a diving submersible
survey. *Marine Geophysical Researches*, **22**, 17–45.

MCKENZIE, D. P. & TICKLE, M. J. 1988. The volume and
composition of melt generated by extension of the
lithosphere. *Journal of Petrology*, **29**, 625–679.

MESCHEDE, M. & FRISCH, W. 1998. A plate-tectonic
model for the Mesozoic and Early Cenozoic history
of the Caribbean Plate. *Tectonophysics*, **296**, 269–291.

NOWELL, G. M. & PARRISH, R. 2001. Simultaneous
acquisition of isotope compositions and parent/daugh-
ter ratios by non-isotope dilution solution-mode
plasma ionisation multi-collector mass spectrometry
(PIMMS). *In*: HOLLAND, J. G. & TANNER, S. D.
(eds) *Plasma Source Mass Spectrometry: the New Millennium*. The Royal Society of Chemistry, Cambridge,
298–310.

NOWELL, G. M., KEMPTON, P. D. & NOBLE, S. R. 1998.
High precision Hf isotope measurements of MORB
and OIB by thermal ionisation mass spectrometry:
insights into the depleted mantle. *Chemical Geology*,
149, 211–233.

NOWELL, G. M., PEARSON, D. G., OTTLEY, C. J.,
SCHWEITERS, J. & DOWALL, D. 2003. Long-term
performance characteristics of a plasma ionisation
multi-collector mass spectrometer (PIMMS): the
ThernoFinnigan Neptune. *In*: HOLLAND, J. G. &
TANNER, S. D. (eds) *Plasma Source Mass Spectrometry: Applications and Emerging Technologies*,
The Royal Society of Chemistry, Cambridge,
307–320.

OFFICER, C. B. J., EWING, J. I., EDWARDS, R. S. &
JOHNSON, H. R. 1957. Geophysical investigations
in the eastern Caribbean: Venezuelan Basin, Antilles
island arc, and Puerto Rico Trench. *Geological
Society of America Bulletin*, **68**, 359–378.

OFFICER, C., EWING, J., HENNION, J., HARKINDER, D. &
MILLER, D. 1959. Geophysical investigations in
the eastern Caribbean – summary of the 1955 and
1956 curises. *In*: AHRENS, L. M. ET AL. (eds)
Physics and Chemistry of the Earth. Pergamon,
London, 17–109.

OTTLEY, C. J., PEARSON, D. G. & IRVINE, G. J. 2003.
A routine method for the dissolution of geological
samples for the analysis of REE and trace elements
via ICP-MS. I *In*: HOLLAND, J. G. & TANNER, S. D.
(eds) *Plasma Source Mass Spectrometry: Applications
and Emerging Technologies*. The Royal Society of
Chemistry, Cambridge, 221–230.

PINDELL, J. L. & BARRETT, S. F. 1990. Geological evolution of the Caribbean region: a plate tectonic perspective. *In*: DENGO, G. & CASE, J. E. (eds) *The Geology of
North America*. Geological Society of America,
Boulder, CO, 405–432.

PINDELL, J., KENNAN, L., STANEK, K. P., MARESCH,
W. V. & DRAPER, G. 2006. Foundations of Gulf of
Mexico and Caribbean evolution: eight controversies
resolved. *Geologica Acta*, **4**, 303–341.

PINDELL, J. L., KENNAN, L., WRIGHT, D. & ERIKSON, J.
2009. Clastic domains of sandstones in central/eastern
Venezuela, Trinidad, and Barbados: heavy mineral and
tectonic constraints on provenance and palaeography.
In: JAMES, K. H., LORENTE, M. A. & PINDELL,
J. L. (eds) *The Origin and Evolution of the Caribbean
Plate*. Geological Society, Special Publications,
London, **328**, 743–797.

RENNE, P. R., SWISHER, C. C., DEINO, A. L., KARNER,
D. B., OWENS, T. L. & DEPAOLO, D. J. 1998. Interca-
libration of standards, absolute ages and uncertainties
in Ar-40/Ar-39 dating. *Chemical Geology*, **145**,
117–152.

RÉVILLON, S., ARNDT, N. T., CHAUVEL, C. & HALLOT,
E. 2000a. Geochemical study of ultramafic volcanic
and plutonic rocks from Gorgona Island, Colombia:
the plumbing system of an oceanic plateau. *Journal
of Petrology*, **41**, 1127–1153.

RÉVILLON, S., HALLOT, E., ARNDT, N. T., CHAUVEL, C.
& DUNCAN, R. A. 2000b. A complex history for the
Caribbean plateau: Petrology, geochemistry, and
geochronology of the Beata Ridge, south Hispaniola.
Journal of Geology, **108**, 641–661.

RÉVILLON, S., CHAUVEL, C. ET AL. 2002. Heterogeneity
of the Caribbean plateau mantle source: Heterogeneity
of the Caribbean plateau mantle source: Sr, O and He
isotopic compositions of olivine and clinopyroxene

from Gorgona Island. *Earth and Planetary Science Letters*, **205**, 91–106.

SALTERS, V. J. M. & STRACKE, A. 2004. Composition of the depleted mantle. *Geochemistry Geophysics Geosystems*, **5**, article no. Q05B07; doi: 10.1029/2003GC000597.

SIGURDSSON, H., LECKIE, R. M. *ET AL.* 1997. Site 1001. *In*: SIGURDSSON, H., LECKIE, R. M. & ACTON, G. D. (eds) *Proceedings of the Ocean-Drilling Program; Initial Reports; Caribbean Ocean History and the Cretaceous/Tertiary Boundary Event; Covering Leg 165 of the Cruises of the Drilling Vessel JOIDES Resolution, Miami, Florida, to San Juan, Puerto Rico, sites 998–1002, 19 December 1995–17 February 1996.* Texas A&M University, Ocean Drilling Program. College Station, TX, 291–357.

SIMS, K. W., GOLDSTEIN, S. J. *ET AL.* 2002. Chemical and isotopic constraints on the generation and transport of magma beneath the East Pacific Rise. *Geochimica et Cosmochimica Acta*, **66**, 3481–3504.

SINTON, C. W., DUNCAN, R. A., STOREY, M., LEWIS, J. & ESTRADA, J. J. 1998. An oceanic flood basalt province within the Caribbean Plate. *Earth and Planetary Science Letters*, **155**, 221–235.

SINTON, C. W., SIGURDSSON, H. & DUNCAN, R. A. 2000. Geochronology and petrology of the igneous basement at the lower Nicaraguan Rise, Site 1001. *In*: GARMAN, P. (ed.) *Proceedings of the Ocean Drilling Program, Scientific Results. Leg 165*. Texas A&M University, Ocean Drilling Program. College Station, TX, 233–236.

SUN, S.-S. & MCDONOUGH, W. F. 1989. Chemical and isotope systematics of oceanic basalts: implications for mantle composition and processes. *In*: SAUNDERS,

A. D. & NORRY, M. J. (eds) *Magmatism in the Ocean Basins*. Geological Society, London, Special Publications, 313–345.

THIRLWALL, M. F. 1995. Generation of the Pb isotopic characteristics of the Iceland plume. *Journal of the Geological Society*, **152**, 991–996.

THOMPSON, P. M. E., KEMPTON, P. D. *ET AL.* 2004. Hf–Nd isotope constraints on the origin of the Cretaceous Caribbean plateau and its relationship to the Galapagos plume. *Earth and Planetary Science Letters*, **217**, 59–75.

THOMPSON, R. N., RICHES, A. J. V. *ET AL.* 2007. Origin of CFB magmatism: multi-tiered intracrustal picrite-rhyolite magmatic plumbing at Spitzkoppe, Western Namibia, during Early Cretaceous Etendeka magmatism. *Journal of Petrology*, **48**, 1119–1154.

WEAVER, B. L. 1991. The origin of ocean island basalt end-member compositions: trace element and isotopic constraints. *Earth and Planetary Science Letters*, **104**, 381–397.

WEIS, D. & FREY, F. A. 2002. Submarine basalts of the northern Kerguelen Plateau: interaction between the Kerguelen Plume and the southeast Indian Ridge revealed at ODP site 1140. *Journal of Petrology*, **43**, 1287–1309.

WEIS, D., KIEFFER, B. *ET AL.* 2005. High-precision isotopic characterization of USGS reference materials by TIMS and MC-ICP-MS. *Geochemistry Geophysics Geosystems*, **7**, Q08006; doi:10.1029/2006GC001283.

WHITE, R. V., TARNEY, J. *ET AL.* 1999. Modification of an oceanic plateau, Aruba, Dutch Caribbean: implications for the generation of continental crust. *Lithos*, **46**, 43–68.

Index

Note: Page numbers in *italic* denote figures. Page numbers in **bold** denote tables.